PHOTOMETRIC
AND
FLUOROMETRIC
METHODS
OF
ANALYSIS

METALS

FOSTER DEE SNELL

PHOTOMETRIC AND FLUOROMETRIC METHODS OF ANALYSIS

METALS

PART 1

FOSTER DEE SNELL, Ph.D., Sc.D.
Chairman of the Board, Emeritus
Foster D. Snell, Inc.

A WILEY-INTERSCIENCE PUBLICATION

JOHN WILEY & SONS, New York • Chichester • Brisbane • Toronto

Library of Congress Cataloging in Publication Data:

Snell, Foster Dee, 1898–
  Photometric and fluorometric methods of analysis.

"A Wiley-Interscience publication."
  Includes bibliographical references.
  1. Metals—Analysis. 2. Photometry. 3. Fluorimetry. I. Title.

QD132.S53   546′.31   77-25039
ISBN 0-471-81014-2

Printed in the United States of America

10 9 8 7 6 5 4 3 2 1

PREFACE

Not very many years ago Dr. Herbert A. Laitinen, as editor of *Analytical Chemistry*, editorialized on the seven ages of an analytical method. The methods represented by the title of this book are in their fifth age.

Color methods passed through (1) comparison with a series of standards, (2) balancing methods typified by the DuBoscq Colorimeter, (3) filter photometry, (4) spectrophotometry as typified by the Beckman DU instrument, and (5) automatic methods as typified by the AutoAnalyzer. Each development has been a step forward in ease of determination and in attainable accuracy.

Present terminology refers to methods as colorimetric, photometric, spectrophotometric, absorptiometric, and fluorometric; and a few other variants are used, as well.

Each chapter in this book is a quasi-monograph on determination of one element, with citation of substantially all of the literature for more than 20 years. Material from earlier references is not cited, since it is likely that all significant methods will have been updated in the more than 8700 references during that period. For simplicity, there is a minimum of cross-referencing between chapters and that not by page number. On the assumption that the user has a basic background, the book does not tell how to prepare a calibration curve: that follows from the method of development of the sample. It does not specify the size of cuvette, which will vary according to the intensity of color developed because of variation in size and quality of sample. It does not usually describe the equipment because apparatus will vary from laboratory to laboratory, and each operator knows what he has available and the limitations imposed.

The preparation of this book has been made possible by the accumulation of abstracts and tear sheets over many years, currently at a rate approaching 1000 a year. That figure is indicative of how dynamic the subject still is, in spite of development of other sophisticated methods of analysis. Thus it appears that senescence, the seventh stage as foretold by Dr. Laitinen, is still beyond the horizon.

The accumulation of references has been performed or supervised by Dr. Cornelia T. Snell, an author in this field. Editorial assistance has been provided by Mrs. Helen Schmidt and Miss Helen Nettleton, both of whom have assisted in earlier volumes. The index has been prepared by Mrs. Joan Battle. Cooperation by the library of the Chemists' Club and the library of Foster D. Snell, Inc., and particularly by the latter in the reproduction of references prior to the Xerox era, is acknowledged. Finally I appreciate the assistance of many people at Booz Allen and Hamilton, Inc., which in many ways has helped to bring these volumes to completion.

FOSTER DEE SNELL

New York, New York
June 1978

CONTENTS, PART 1

TABLES

FIGURES

CHAPTER ONE
LEAD

The dominant method for reading lead is by dithizone. And since the colors of lead dithizonate and excess dithizone are so far separated, there is no real need to use a monocolor technic. Even the classical lead sulfide is read photometrically, and aside from a host of other reagents, lead ion is read in the ultraviolet range and by the flame photometer.

The great difficulty in accurate determination of small concentrations of lead is the almost universal distribution of this element; thus reagents often require special purification, and special precautions are necessary to remove adsorbed lead from glassware.

Since organic lead compounds are often determined as lead ion, it appears appropriate to include them here.

For collection of lead by precipitation of manganese dioxide, the optimum acidity is 0.008 *M*.[1] Recovery is in the 80–90% range.

Lead in industrial water supplies is determined by dithizone,[2] as is that recovered from atmospheric air.[3a]

A technic for determination of copper by its catalytic effect on the iodine azide reaction is also applicable to lead. For details, see Copper.

Procedure. *Isolation with Caproic Acid.* To the sample solution in hydrochloric acid, add 40 ml of methanol and dilute to 80 ml.[3b] Adjust to pH 7.5 with ammonium hydroxide or hydrochloric acid. Shake with 20 ml, 10 ml, and 10 ml of a 1.5% solution of caproic acid in chloroform. Evaporate the chloroform and heat

[1]G. F. Reynolds and F. S. Tyler, *Analyst* **89**, 579–86 (1964).
[2]British Standards Institution, *BS 2690*, Pt. **14**, 15 pp (1972).
[3a]Artur Strusinski and Halina Wyszynska, *Rocz. panstw. Zakl. Hig.* **22**, 649–656 (1971).
[3b]R. Pietsch, *Anal. Chim. Acta* **53**, 287–294 (1971).

to drive off excess caproic acid. Dissolve in an appropriate acid or ash according to the reagent to be used for development.

DITHIZONE

Dithizone, (diphenylthiocarbazone, phenylazothionoformic acid) forms intensely colored dithizonates with many metals concentratable by extraction into chloroform or carbon tetrachloride or occasionally into mixtures of other solvents. Therefore it is not surprising that it is the major reagent for determination of lead, and it is important for several other elements. The complexes are largely keto in structure rather than enol because of the greater solubility in that form.

The methods fall into three general categories.[3c]

Monocolor Method. By this technic the metal dithizonate is extracted into the organic phase with all or part of the excess of the green dithizone. By shaking with an alkaline solution, the free dithizone is extracted, leaving only the metal dithizonate to be read photometrically. The excess dithizone is more easily extracted from carbon tetrachloride than from chloroform. By extraction at pH 11.5, the excess dithizone is concentrated in the aqueous phase,[3d] approaching a monocolor rather than a mixed color method.

Mixed Color Method. The organic extract is read with the excess dithizone remaining. This is by far the most common method, since the narrow wave band of the spectrophotometer often excludes the green of the free dithizone. An alternative is to read the excess dithizone. For either version of this technic, the metal dithizonate must absorb in a distinctly different range from dithizone. Thus it is applicable to the red of lead dithizonate but not to greenish platinum and palladium dithizonates.

Reversion Method. By this technic, the metal dithizonate is extracted by organic solvent, and separated from the aqueous layer; then the metallic ion is extracted with very dilute acid, leaving the equivalent amount of dithizone in the solvent layer. Thus it is an indirect method in that the amount of dithizone equivalent to the metal is read.

Since the metal ion reacts with a hydrogen of dithizone, it follows that the extraction is related to the pH of the aqueous solution and that the metal dithizonate as a salt of a relatively weak acid can be decomposed and the metallic ion extracted by an aqueous solution of a strong acid as in the reversion method. Often metal thizonates can be separated by appropriate pH adjustment of the extraction medium.[4] This is illustrated by the equilibrium curve at various pH levels in chloroform (Figure 1-1).

[3c]C. W. Zuelke, *Org. Chem. Bull.* (Eastman Kodak Co.) **21** (4) (1949).
[3d]L. J. Snyder, *Anal. Chem.* **19**, 684–687 (1947).
[4]H. J. Wichman, *Ind. Eng. Chem., Anal. Ed.* **11**, 66 (1939).

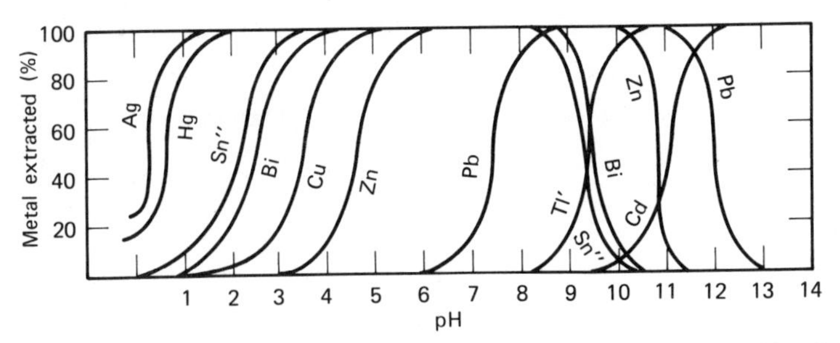

Figure 1-1 Provisional equilibrium curves of some metal dithizonates in chloroform.

The following tabulation shows the metals extractable with dithizone, although that reagent is not important for determination of all of them.[5]

Element	Degree of Oxidation	Optimum pH for Extraction
Manganese	2	11
Iron	2	> 7
Cobalt	2	8–9
Nickel	2	8–9
Copper	1, 2	1
Zinc	2	8.3
Palladium	2	0
Silver	1	1
Cadmium	2	13
Indium	3	3–6
Tin	2	6–9
Tellurium	4	1
Platinum	2	0
Gold	3	1
Mercury	1, 2	− 1
Thallium	1, 3	9–12, 3–4
Lead	2	8.5–11
Bismuth	3	> 2
Polonium	4	0

Carbon tetrachloride and chloroform are the usual extractants,[6] but benzene and toluene are often used, and there are others. Carbon tetrachloride is a better solvent for the dithizonate than chloroform or benzene. Increase of pH or concentration of solvent increases the velocity of extraction, which is much greater in carbon tetrachloride than in chloroform.[7] The partition coefficient is greatly influenced by the type of buffer. The reagent itself is insoluble in water but soluble in ammonium hydroxide.

The corresponding di-β-naphthylthiocarbazone offers greater sensitivity with

[5]Henry Freiser, *Chemist-Analyst* **50**, 62 (1961).
[6]Shigeharu Ichikawa, Minora Nanjo, and Shizuo Kano, *Eisei Shikenjo Hokoku* **77**, 427–428 (1959).
[7]Antonio Catino and Alda Addone, *Rass. Chim.* **10**, 13–21 (1958).

Table 1-1 **Comparative Properties of Dithizone and the Dinaphthyl Analogue**

Reagent	Dithizone Maximum (nm)	Dithizone Sensitivity (liters/cm mole)	Dinaphthyl Compound Maximum (nm)	Dinaphthyl Compound Sensitivity (liters/cm mole)
Dithizone	620	32,000		
Dinaphthyl compound			650	67,000
Bismuth	490	80,000	520	114,000
Mercuric	485	70,000	510	86,000
Zinc	535	92,600	560	121,000
Silver	462	27,200	500	48,000
Lead	520	68,600	550	80,000
Cadmium	520	88,000	545	107,000
Thallous	510	33,200	540	38,000
Cupric	550	45,200	570	78,000
Cobalt	542	59,200	560	66,000
Nickel	480	30,400	510	59,500
Indium	510	87,000	540	123,000
Stannous	520	54,000	56/	60,000
Auric	450	31,600	475	83,500
Palladous	450	34,400	52/	52,800
Praseodymium	490	31,600	510	26,000

some metals.[8] Table 1-1 lists the wavelengths at which each absorbs, including some metals for whose determination dithizone is not commonly used. The reagent is somewhat unstable in chloroform solution, but the impurities do not react with the metal ions.

Selenazone, the selenium analogue of dithizone, is 3-selena-1,5-diphenylformazan. Like dithizone, it extracts many cations from aqueous solution.[9] The diselenide is more selective than dithizone, and it extracts only silver, mercuric, nickel, and palladous ions.

By formation of more stable complexes, the extraction of a metal by thizone may be inhibited. Thus in alkaline cyanide solution, silver, mercury, cadmium, copper, nickel, and zinc are not extractable by thizone, whereas lead, thallium, bismuth, and tin are extractable. Similarly, only mercury, gold, and copper are extractable from weakly acid thiocyanate solution. Masking agents also include ethylene diamine tetraacetate (EDTA), thiosulfate, and even halides.

The extreme sensitivity of reading metal dithizonates carries with it the complementary problem of requiring extreme purity of reagents to avoid large blanks. Likewise, oxidizing agents must be absent, since they convert dithizone to intensely yellow diphenylthiocarbodiazone, which is soluble in organic solvents. Hydroxylamine or sulfites reduce this to dithizone, and in fact it is often an impurity in dithizone. A purified form is offered by Eastman Kodak Co., but it must be protected from oxidation. Small details of variation in laboratory technic can cause significant errors.

[8]R. J. DuBors and Samuel B. Knight, *Anal. Chem.* **36**, 1313–1320 (1964); cf O. W. Lombardi, *ibid.* **36**, 415–418 (1964).
[9]R. S. Ramakrishna and H. M. N. H. Irving, *Anal. Chim. Acta* **48**, 251–266 (1969).

The brilliantly colored dithizonates include those of gold, platinum, palladium, silver, mercury, stannous, bismuth, copper, zinc, cobalt, nickel, lead, thallous, and cadmium ions. It reacts with many ions, and this property is both a virtue and a defect, the latter because of the necessity of isolation of the test substance.

The useful range is usually 0.001–0.2 mg/ml with accuracy within 1–5%.

Since dithizone solutions are both heat and light sensitive, it follows that such reagents should be stored in brown bottles in the dark and refrigerated. Chloroform should contain 0.5% of ethanol as a stabilizer. Such a solution will keep for a month in the refrigerator.

Dithizone itself is read by the inversion method; thus the absorption in carbon tetrachloride or chloroform is at a maximum at 620 nm with a smaller value at 446 nm.[10] The fully oxidized material absorbs at 390 nm.

Bismuth and thallium interfere by being extracted with the lead. Phosphates of titanium and calcium or magnesium prevent complete extraction of the lead by dithizone. Tartrate or citrate prevents extraction of indium at normal concentrations. Diethylammonium diethyldithiocarbamate or sodium diethyldithiocarbamate are often used to separate lead prior to its determination with dithizone.[11a] Bismuth can be extracted from lead with dimethylammonium diethyldithiocarbanate in solutions 3 *N* with sulfuric acid and at least 2 *N* with hydrochloric acid. An alternative is 4–6 *N* hydrochloric acid. Then at 1.5–2 *N* lead is extracted but thallium is not. In the dithiocarbamate extraction of lead under these conditions there is no interference by titanium, ferrous ion, thorium, magnesium, manganese, vanadium, cerium, hexavalent uranium, aluminum, calcium, pentavalent antimony, trivalent chromium, or zirconium. Some zinc, cadmium, tin, platinum, and indium accompany the lead. Large amounts of tin could prevent complete extraction of the lead. The bismuth extract at the high acidity contains copper, ferric ion, molybdenum, and tungsten with part of the germanium, hexavalent chromium and platinum, and a trace of cadmium.

Dibutyl–lead diacetate in urine is extracted as the dithizonate with toluene after buffering to pH 4.75.[11b] The reading at 485 nm may be corrected for the absorption of dithizone at that wavelength. Alternatively the excess dithizone may be extracted at pH 12 with an equal volume of 0.8% sodium hydroxide solution before reading.

The dithizone method has been applied to fruit, vegetables, and soil in demonstration that the produce did not take up additional lead from soil with a high lead content.

Special attention to the glassware involves cleaning with nitric and sulfuric acid, rinsing with tap water, then rinsing with distilled water,

In the extraction by dithizone, lead is adsorbed on glassware from an alkaline solution, necessitating a second step to recover it with 1 : 99 nitric acid.

Beer's law applies up to at least 1 mg of lead per ml of dithizone extract. Reading is against a sample reverted with sulfuric acid in addition to subtraction of a blank.

[10]H. G. C. King and G. Pruden, *Analyst* **96**, 146–148 (1971).
[11a]J. C. Gage, *Ibid* **80**, 789 (1955); H. V. Hart, *ibid.* **76**, 692 (1951); H. G. Lockwood, *ibid.* **79**, 143 (1954); N. Strafford, P. F. Wyatt, and F. G. Kershaw, *ibid.* **70**, 232 (1945), *ibid.* **78**, 624 (1953); S. L. Tompsett, *ibid.* **81**, 330 (1956).
[11b]G. Gras and F. Fauran, *Ann. Pharm. Fr.* **30**, 545–554 (1972).
[11c]D. R. Crudgington, J. Markland, and J. Vallance, *J. Assoc. Pub. Anal.* **11** (4), 120–126 (1973).

Procedures

Iron and Steel.[12] Digest a 1-gram sample with 10 ml of 1:1 nitric acid and 10 ml of perchloric acid. Evaporate almost to dryness and cool. Take up the residue in 20 ml of water and dilute to 250 ml. Dilute a 25-ml aliquot to 40 ml and add in succession 2 ml of 20% tartaric acid, 10 ml of saturated sodium sulfite solution, and 10 ml of 20% potassium cyanide solution. Heat to 80° to reduce the iron to the ferrous state and cool. Add 10 ml of ammonium hydroxide and shake for 1 minute with 10 ml of benzene containing 0.1 mg of dithizone per ml. Discard the aqueous layer and wash the benzene layer with 50 ml of 1 : 100 ammonium hydroxide and 2 ml of saturated solution of sodium sulfite to remove excess dithizone. Read at 520 nm and correct for a blank.

Ascorbic acid is an alternative to sulfite for reducing iron to the ferrous state.[13] Lead can then be extracted as the dithizonate in the presence of large amounts of copper, zinc, and cadmium. Hydroxylamine hydrochloride also serves that function.[14] If necessary to avoid interference by bismuth, the lead can be back-extracted from the dithizone extract by a phthalate–hydrochloric acid buffer solution for pH 3.4, leaving the bismuth in the carbon tetrachloride. The aqueous extract adjusted to pH 9.6 is then re-extracted by dithizone in carbon tetrachloride.

For cast iron the lead can be separated from iron and almost all interfering elements by column chromatography on Wofatite L-150 resin in chloride form. The sample in 1 : 1 hydrochloric acid is followed by a strong acid wash to remove iron. Thereafter the lead is eluted by 0.005 *M* hydrochloric acid for determination by dithizone using the inversion technic.[15]

To determine lead in iron-bearing material, dissolve a sample containing about 0.015 mg of lead in hydrofluoric acid and aqua regia.[16] Evaporate to dryness and convert to chlorides by repeated evaporation to dryness with hydrochloric acid. Take up in 1 : 15 hydrochloric acid and remove bismuth, ferric, thallic, and cupric ions by repeated extractions with cupferron in chloroform. Adjust the aqueous phase to 1.5 *N* in hydrochloric acid and extract lead and thallous ions with 10 ml and 10 ml of 1% diethylammonium diethyldithiocarbamate in chloroform. Ash the extract with nitric and perchloric acids, and apply the dithizone method to a solution of the ash.

An alternative is to extract the iron from 9 *N* hydrochloric acid with amyl acetate.[17] The lead is then precipitated by ammonium hydroxide, with sodium phosphate and calcium chloride reacting to produce calcium phosphate as a collector in isolating lead from copper, nickel, and cobalt present as their ammonia complexes. Determination is then as lead dithizonate. In 0.5–2 *N* hydrochloric acid lead forms a complex anion that can be separated on a cation exchange resin from

[12]Shigeo Wakamatsu, *Jap. Anal.* **5**, 509–512 (1956); cf British Iron and Steel Research Association Metallurgy Division, *Open Rep.*, **MG/D/554/67**, 1967, 3 pp; **MG/D/555/67**, 1967, 5 pp; cf Theo Kurt Willmer, *Arch. Eisenhüttenw.* **29**, 159–164 (1968).

[13]D. Filipov, I. Nachev, and Kh. Nachev, *C. R. Acad. Bulg. Sci.* **18**, 813–816 (1965).

[14]J. A. Stobart, *Analyst* **90**, 278–282 (1965).

[15]Zofia Waclawik, *Pr. Inst. Odlew.* **20**, 136–148 (1970).

[16]Jesse J. Warr and Frank Cuttitta, *U.S. Geol. Survey, Prof. Papers* **400B**, B483–489 (1960).

[17]N. P. Strel'nikova and V. N. Pavlova, *Zavod. Lab.* **26**, 63 (1960).

iron, copper, nickel, and cobalt. The iron content may also be extracted with ethyl or isopropyl ether.[18]

Heat-Resisting Alloys.[19] This technic is applicable to both iron-based and nickel-based types. Dissolve 1 gram of sample in 5 ml of nitric acid, add 10 ml of hydrochloric acid, and evaporate to dryness. Add 3 ml of hydrochloric acid, evaporate to dryness, and repeat that step twice more. Dissolve the residue in 10 ml of 1:5 hydrochloric acid and pass through a column of AN-31 resin that has been washed with 1:5 hydrochloric acid. Elute lead and zinc with 100 ml of 0.02 *N* hydrochloric acid at 1 ml/minute. Evaporate to 50 ml and adjust to pH 9.5 with ammonium hydroxide. Extract with 10 ml, 5 ml, and 5 ml of 0.0001% dithizone solution in carbon tetrachloride. Wash the combined extracts with 10-ml portions of 0.5% potassium cyanide solution until no more color is extracted, then with 5 ml of water. Dilute to 50 ml and read at 520 nm against a blank.

Copper and Brass.[20] Dissolve a 1-gram sample in a minimun of nitric acid and boil off the oxides of nitrogen. Add 35 ml of 10% ammonium citrate solution, and add ammonium hydroxide until nearly neutral. Cool and complex copper and zinc with 2 ml of 20% potassium cyanide solution. Add phenol red, free from chelating agents, as indicator and, if necessary, more ammonium hydroxide to make the solution red.[21] Extract bismuth and lead with 5 ml, 5 ml, and 5 ml of chloroform containing 0.03 mg of dithizone per ml. Back-extract the metal ions from the combined chloroform extracts by shaking with 10 ml of 1:15 nitric acid. Adjust the extract to pH 2 with ammonium hydroxide. Extract the bismuth with 3 ml and 3 ml of chloroform containing 0.01 mg of dithizone per ml and discard these extracts. Neutralize the aqueous layer with ammonium hydroxide. Add 5 ml of 10% ammonium citrate solution and 2 ml of 20% potassium cyanide solution. Extract the lead with 5 ml and 5 ml of chloroform containing 0.03 mg of dithizone per ml. Wash the combined extracts with 5 ml and 5 ml of 1% ammonium citrate solution containing 0.1% of potassium cyanide to remove excess dithizone. Read at 520 nm.

Aluminum.[22] As an acid reagent, mix 5 parts of sulfuric acid, 1 part of nitric acid, and 5 parts of water. As ammonium citrate solution, react 500 grams of citric acid with 570 ml of ammonium hydroxide and dilute to 1 liter.

Digest a 1-gram sample with 20 ml of 20% sodium hydroxide solution until reaction ceases. Add 20 ml of the acid reagent and heat at 100° until the solution is clear. Dilute to 50 ml. Mix a 10-ml aliquot with 10 ml of the ammonium citrate reagent, 10 ml of ammonium hydroxide, and 2 ml of 10% potassium cyanide solution. Extract with 20 ml of 0.1% dithizone solution in carbon tetrachloride. Then extract with 10 ml and 10 ml of carbon tetrachloride. Read the combined extracts at 520 nm against carbon tetrachloride.

[18]Siegfried Meyer and Othmar G. Koch, *Arch. Eisenhüttenw.* **31**, 711–715 (1960).

[19]V. I. Kurbatova, V. V. Stepin, V. I. Ponosov, E. V. Novikova, G. N. Emasheva, and L. Ya. Kalashnikova, *Zavod. Lab.* **37**, 413–414 (1971).

[20]R. Socolovschi, *Rev. Chim.* (Bucharest) **11**, 348–349 (1960); cf R. P. Hair and E. J. Newman, *Analyst* **89**, 42–48 (1964).

[21]Kenneth M. Hallam, *Anal. Chem.* **34**, 1339 (1962).

[22]M. C. Steele and L. J. England, *Metallurgia* **59**, 153–156 (1959).

Aluminum-Bronze Alloys.[23] Dissolve a 2-gram sample with 1:1:3 sulfuric acid–nitric acid–water. Heat until oxides of nitrogen have been expelled, then add 10 ml of hydrochloric acid. Evaporate to sulfur trioxide fumes and cool. Boil with 100 ml of water until soluble salts are dissolved. Allow to stand for an hour; filter, and wash the lead sulfate on the filter with 1:19 sulfuric acid.

Dissolve the lead sulfate from the filter with 10 ml of hot 30% ammonium acetate solution and wash with hot water. Cool and dilute to 250 ml. Mix a 5-ml aliquot of the sample solution with 20 ml of 30% ammonium acetate solution and 10 ml of ammonium hydroxide. Dilute to about 75 ml and extract with 25 ml of a 0.01% solution of dithizone in carbon tetrachloride. Read the extract at 535 nm against a blank.

Bismuth.[24] Dissolve a 2-gram sample in 15 ml of 1:1 nitric acid and evaporate to dryness. Cool, add 15 ml of hydrobromic acid and evaporate to dryness. Heat to $280° \pm 50°$ and cool. Again evaporate with 15 ml of hydrobromic acid and bake until the yellow color of bismuth bromide disappears. Take up the residue in 10 ml of 1:3 hydrobromic acid and add 10 ml of water. Add 1.5 ml of a solution containing 1.5 ppm of zinc ion to assist in the extraction of interfering ions. Add 1:1 ammonium hydroxide to pH 3.5. Add 1 ml of a phthalate buffer for pH 3.5 containing 1% of hydroxylamine hydrochloride. Shake for 5 minutes with 20 ml of a solution of 60 mg/liter of dithizone in carbon tetrachloride. Discard the dithizone extract, repeat the extraction, and discard.

Add to the aqueous layer 2 ml of 25% diammonium hydrogen citrate solution, 5 ml of ammonium hydroxide, and 2 ml of 10% potassium cyanide solution. Shake for 2 minutes with 10 ml of carbon tetrachloride containing 20 mg of dithizone per ml. Wash the carbon tetrachloride layer four times with 10-ml portions of 0.5% ammonium chloride solution adjusted to pH 9.5 and containing 0.1% of potassium cyanide. Read at 530 nm. The method will determine 1 ppm of lead.

Tin.[25] Decompose a 1-gram sample by heating with 10 ml of sulfuric acid. Then at 200–230° bubble air through the solution and add 10 ml of hydrobromic acid to volatilize tin, antimony, and arsenic. Evaporate to dryness and cool. Take up the residue by heating with 10 ml of 1:1 nitric acid and dilute to a known volume.

Mix a 10-ml aliquot with 10 ml of water, 1 ml of 10% hydroxylamine hydrochloride solution, 1 ml of 10% sodium potassium tartrate solution, and a drop of phenolphthalein indicator. Make just alkaline with ammonium hydroxide and add 10 ml of 5% potassium cyanide solution. Extract with 10 ml of benzene containing 0.1 mg of dithizone per ml and read at 520 nm. This method is also applicable to tin-base babbitt metal.

For determination of lead in the tin coating of cans, the cans were treated with bromine in hydrobromic acid, the iron reduced to the ferrous form, and the lead extracted at pH 9.5 by dithizone in chloroform for reading at 518 nm.[26]

[23] M. Freegarde and Mrs. B. Allen, *Analyst* **85**, 731–735 (1960).
[24] Koichi Nishimura, Akira Tsuchibuchi, and Tomoko Aoyama, *Jap. Anal.* **13**, 220–225 (1964).
[25] Kazuo Ota and Shigeru Mori, *ibid.* **5**, 442–445 (1956).
[26] K. G. Bergner and H. Miethke, *Deut. Lebensm. Rundsch.* **63**, 49–51 (1967).

Tin-Base Babbitt Metal. Dissolve a 0.5-gram sample containing 0.04–0.2% of lead in 20 ml of hydrochloric acid to which 2 ml of nitric acid has been added.[27] Evaporate nearly to dryness to get rid of the nitric acid, dissolve in water and dilute to 100 ml. Mix a 5-ml aliquot with 5 ml of hydrochloric acid and extract the tin and antimony with 25 ml and 25 ml of methyl isobutyl ketone. Partially evaporate the aqueous layer to drive off organic solvent, and cool. Neutralize with ammonium hydroxide. Add 5 ml of 5% hydrazine hydrochloride solution, 5 ml of 5% sodium potassium tartrate solution, 10 ml of ammonium hydroxide, and 5 ml of 5% potassium cyanide solution. Extract with 20 ml of benzene containing 2 mg of dithizone. Wash the extract with 5 ml, 5 ml, and 5 ml of 1:10 ammonium hydroxide containing 1% of potassium cyanide. Filter the benzene extract to remove entrained moisture and read at 520 nm against a blank.

Silver Alloys.[28] BISMUTH AND LEAD. Dissolve a sample expected to contain 0.05–0.2 mg of bismuth and associated amounts of lead in 5 ml of 1:1 nitric acid. Add 75 ml of water and 2 ml of 2% aluminum nitrate solution, and heat nearly to boiling. Add ammonium hydroxide dropwise until a dark blue color is obtained. Filter after 20 minutes and wash with 1:10 ammonium hydroxide.

Dissolve the precipitate in 20 ml of 1:1 nitric acid and dilute to 50 ml with water. Adjust a 10-ml aliquot to pH 2.5 by addition of 1:1 ammonium hydroxide and extract the bismuth with 5 ml and 5 ml of 0.001% dithizone solution in chloroform. At this point the lead remains in the aqueous layer. Back-extract the bismuth from the combined chloroform extracts by shaking for 5 minutes with 20 ml of 1:1 nitric acid. Neutralize the acid extract with 1:1 ammonium hydroxide. Add 5 ml of 10% potassium cyanide solution and extract with 5 ml and 5 ml of 0.001% dithizone solution in chloroform. Wash the combined extracts with 10 ml of 1:7 ammonium hydroxide and read the bismuth at 500 nm.

Neutralize the sample solution from which bismuth has been removed to pH 9.7 and extract with 5 ml and 5 ml of the dithizone solution. Wash the combined extracts with 10 ml and 10 ml of 0.5% potassium cyanide solution and read at 520 nm, correcting for a blank.

Gold.[29] Lead in gold is determined photometrically after isolation of iron. Dissolve a 1-gram sample in 6 ml of hydrochloric acid and 2 ml of nitric acid. Evaporate to dryness and take up the residue in 10 ml of 1:5 hydrochloric acid. Extract the gold with 10 ml and 10 ml of isopentyl acetate, and discard. Evaporate the solution to about 1 ml and dilute to 10 ml with 1:1 hydrochloric acid. Extract the iron with 10 ml and 10 ml of isopentyl acetate as a sample for determination with bathophenanthroline.

Add 1 ml of 1:1 sulfuric acid to the aqueous layer and evaporate to sulfur trioxide fumes. Take up the residue in water and dilute to 10 ml. Shake for 1 minute with 10 ml of benzene containing 0.1 mg of dithizone. Wash the extract

[27]Nobuo Tajima and Moriji Kurobe, *Jap. Anal.* **9**, 884–886 (1960).

[28]Z. Skorko-Trybula and J. Chwastowska, *Chem. Anal. (Warsaw)* **8**, 859–864 (1963).

[29]G. Ackermann and J. Köthe, *Z. Anal. Chem.* **231**, 252–261 (1967); Zygmunt Marczenko, Krzysztof Kasiura, and Maria Krasiejko, *Chem. Anal.* (Warsaw) **14**, 1277–1287 (1969); cf. Krzysztof Kasiura, *ibid.* **20**, 809–816 (1975).

with 10 ml and 10 ml of 1:100 ammonium hydroxide to remove excess dithizone and read at 520 nm.

Electrolytic Manganese.[30] Dissolve a 0.1-gram sample in 5 ml of 1:1 nitric acid. Add 10 ml of 50% ammonium citrate solution adjusted to pH 9 with ammonium hydroxide as a masking agent for manganese, and make alkaline with ammonium hydroxide. Add 10 ml of 5% potassium cyanide solution as a masking agent for heavy metals, even more if much zinc and copper are present. To reduce ferric ion, add 3 ml of 20% hydroxylamine hydrochloride solution adjusted to pH 2 with hydrochloric acid. Heat to boiling and cool. Adjust to pH 9.5–11 with ammonium hydroxide. Extract with 5 ml, 5 ml, and 5 ml of a solution of 0.1 mg of dithizone per ml in carbon tetrachloride. The last extract should remain green. Shake the combined extracts with 10-ml portions of 0.5% potassium cyanide solution until no further color is extracted, usually three or four extractions. Wash the lead dithizonate in carbon tetrachloride with 5 ml of water. Dilute with carbon tetrachloride to 50 ml, read at 520 nm, and correct for a blank.

Ferromanganese. Dissolve a 0.1-gram sample in 10 ml of 1:1 nitric acid and boil off the oxides of nitrogen. Add 10 ml of 50% ammonium citrate solution adjusted to pH 9 with ammonium hydroxide as a masking agent for manganese. Add 10 ml of 20% hydroxylamine hydrochloride solution adjusted to pH 2 with hydrochloric acid. Boil for 2–3 minutes and cool. Adjust to pH 9.5 with ammonium hydroxide. Extract with 5 ml, 5 ml, and 5 ml of a solution of 0.1 mg of dithizone per ml in carbon tetrachloride. The last extract should remain green. Back-extract the combined extracts with 10 ml of 1:100 nitric acid and repeat the extraction with dithizone in carbon tetrachloride. Lead and zinc are now in the aqueous phase. Discard the solvent layer, which contains copper and bismuth. Extract the aqueous phase with a few ml of carbon tetrachloride to collect any residual dithizone and discard. Adjust the aqueous phase with 1:1 ammonium hydroxide to pH 9.5. Add 5–10 ml of 5% potassium cyanide solution and 3 ml of 20% hydroxylamine hydrochloride adjusted to pH 2 with hydrochloric acid. Set aside for 15 minutes and complete as for electrolytic manganese from "Extract with 5 ml, 5 ml, and 5 ml of a solution of 0.1 mg...."

Manganese, Manganese Ores, and Ferromanganese.[31] Dissolve 1–5 grams of sample in 10–15 ml of hydrochloric acid. Add 10 ml of 8 *N* hydrochloric acid and cool. Prepare a chromatographic column of 10–15 grams of Anionite AV-17. Wash the column with 50 ml of 8 *N* hydrochloric acid at the rate of 0.5 ml/minute. Filter the sample solution into the column and follow it with 60 ml of 8 *N* hydrochloric acid. Iron, copper, cobalt, zinc, and lead are retained on the column; manganese, nickel, chromium, and aluminum are in the effluent. Elute cobalt, iron, and copper from the column with 150 ml of 2.5 *N* hydrochloric acid at the rate of 1 ml/minute. Evaporate this solution to 50 ml and dilute to 100 ml. Use aliquots for

[30]Hidehiro Goto and Kichinosuke Hironawa, *J. Jap. Chem. Soc.* **21**, 65–68 (1957); *Sci. Rep. Res. Inst., Tohoku Univ.* **10**, 10–19 (1958).
[31]Hidehiro Goto and Kichinosuke Hirohawa, *J. Jap. Chem. Soc.* **21**, 65–68 (1957); V. V. Stepin, T. L. Barbash, E. V. Silaeva, L. G. Rakhmatulina, and O. V. Morozova, *Tr. Vses. Nauchn.-Issled. Inst. Stand. Obraztsov.* **3**, 124–132 (1967).

determination of copper by diethyldithiocarbamate, cobalt by nitroso-R salt, and iron by sulfosalicylic acid or 1:10 phenanthroline.

Elute the lead and zinc from the column with 100 ml of 0.02 *N* hydrochloric acid at 1 ml/minute. Evaporate this to 50 ml, dilute to 100 ml, and use as a sample for determination of zinc or of lead by dithizone. As an appropriate technic for lead, see Electrolytic Manganese, starting at "Adjust to pH 9.5–11 with ammonium hydroxide."

Nickel-Chromium Alloys.[32] Dissolve a 1-gram sample containing 0.005–0.05 mg of lead in 50 ml of 1:1 hydrochloric-nitric acids and evaporate to dryness. Take up the residue with 3 ml of 1:1 nitric acid and dilute to 200 ml. The acidity should approximate 0.1 *N*.

Impregnate 6–7 grams of pumice with 5% cadmium acetate solution, expose to hydrogen sulfide, and wash with water. Use this to fill a chromatographic column.

Pass the sample solution through this cadmium sulfide column, which retains lead while nickel, chromium, aluminum, iron, titanium, cobalt, zirconium, zinc, and manganese pass through. After washing the column with water, dissolve the lead and cadmium sulfides with 5 ml of nitric acid. Dilute to 20 ml and add 5 ml of 5% potassium cyanide solution to mask residual heavy metals. Complete as for Electrolytic Manganese, starting at "Adjust to pH 9.5–11 with ammonium hydroxide."

Ferrochrome. LEAD, ZINC, AND BISMUTH.[33] Dissolve a 2-gram sample in 50 ml of 1:1 hydrochloric acid and 2 ml of nitric acid. Evaporate nearly to dryness. Add 5 ml of hydrochloric acid and again evaporate nearly to dryness. Take up in 30 ml of 1:5 hydrochloric acid.

Pass through a column of AN-31 Anionite in the chloride form and follow with 80 ml of 1:5 hydrochloric acid. Iron, chromium, nickel, aluminum, and copper are not adsorbed. Elute zinc with 200 ml of 0.65 *N* hydrochloric acid, then lead with 200 ml of 0.02 *N* hydrochloric acid, and finally bismuth with 300 ml of *N* nitric acid. Concentrate each eluate for determination by dithizone.

For lead, concentrate to 20 ml. Add 5 ml of 5% potassium cyanide solution to mask residual heavy metals. Complete as for Electrolytic Manganese starting at "Adjust to pH 9.5–11 with ammonium hydroxide."

Ferromolybdenum. Dissolve a 1-gram sample in 30 ml of 1:1 nitric acid. Add 5 ml of hydrochloric acid and evaporate nearly to dryness. Complete as for ferrochrome from "Add 5 ml of hydrochloric acid...."

Uranium Metal.[34] Dissolve a 2-gram sample with 10 ml of 1:1 hydrochloric acid and 2 ml of 30% hydrogen peroxide solution. Evaporate to dryness and take up the residue in 10 ml of 1:100 nitric acid. To this add 4 ml of 50% ammonium citrate solution, 1 ml of 10% sodium sulfite solution, and 3 ml of ammonium

[32] W. Stolper, *Neue Hütte* **8**, 424–427 (1963).

[33] V. I. Kurbatova, V. V. Stepin, V. I. Ponosov, N. A. Zobnina, E. V. Novikova, G. N. Emasheva, and L. Ya. Kalashnikova, *Trudy Vses. Nauchn.-Issled. Inst. Standt. Obraztsov. Spektr. Etalonov*, **7**, 98–102(1971).

[34] A. B. Crowther, UK AEA, Ind. Group **SCS-R-3**, 3 pp (1948); Ken Saito and Tsugio Takeuchi, *Jap. Anal.* **10**, 152–156 (1961).

hydroxide. Extract the lead with 5 ml of benzene containing 0.5 mg of dithizone per ml. Wash the extract with 10 ml of 0.5% sodium sulfite solution containing 0.1% of potassium cyanide. This removes bismuth and thallium from the benzene layer. Then wash with 10 ml of 0.5% sodium sulfite solution and read at 520 nm against a blank.

Indium.[35] Dissolve a 1-gram sample in nitric acid and evaporate to dryness. Take up the residue in 5*N* hydrobromic acid to give 3–20 ppm of lead. Extract the bulk of the indium from this solution with several successive portions of isopropyl ether.Complete as for Electrolytic Manganese from "Add 10 ml of 50% ammonium citrate...." Results tend to be slightly low.

Selenium.[36] Dissolve a 0.5-gram sample in 10 ml of hydrochloric acid and 2 ml of nitric acid. Evaporate to dryness and dissolve the residue in 2 ml of hydrochloric acid. Add 5 ml of water and 30 ml of 20% sodium citrate solution. Neutralize to phenolphthalein with 10% sodium hydroxide solution and add 3 ml of 0.5 *M* EDTA. Add 2 ml of 1% sodium diethyldithiocarbamate solution and 6 drops of 10% potassium cyanide solution. After 10 minutes extract the lead with 25 ml and 5 ml of chloroform. Reextract the lead from the chloroform with 10 ml and 10 ml of 1:100 nitric acid. Add 5 ml of an ammonium cyanide solution containing 2% of potassium cyanide and 15% of ammonium hydroxide. Shake for 30 seconds with 2 ml of chloroform containing 5 mg of dithizone per liter. If the extract is not greenish, extract with further dithizone solution and combine the extracts. Read the lead dithizonate at 595 nm.

Tellurium.[37] This technic provides for determination of both bismuth and lead. Dissolve a 2-gram sample of tellurium containing about 5 ppm of bismuth and lead in 25 ml of 1:1 hydrochloric acid and 5 ml of nitric acid. Evaporate nearly to dryness and add 5 ml of 1:1 hydrochloric acid. Repeat this operation to reduce to tetravalent tellurium, and finally take up in 10 ml of 1:1 hydrochloric acid. Make alkaline with 25 ml of ammonium hydroxide diluted with 70 ml of water. Add 10 ml of 10% potassium cyanide solution and adjust to pH 10 with ammonium hydroxide. Extract with 10 ml and 10 ml of chloroform containing 0.05 mg of dithizone per ml.

Reextract the bismuth and lead from the combined dithizonate solutions with 10 ml and 10 ml of 1:50 sulfuric acid. Add 10 ml of phthalate buffer for pH 4 and 1 ml of 5% hydrazine hydrochloride. Adjust to pH 4 with 1:1 ammonium hydroxide and dilute with water to 50 ml. Extract with 10 ml of chloroform containing 0.2 mg of dithizone by shaking for 1 minute and read the bismuth at 490 nm.

Use 10 ml of the dithizone solution and 5 ml of chloroform to wash the aqueous phase; discard the washings. Add 5 ml of ammonium hydroxide, 1 ml of 10% potassium cyanide solution, and 10 ml of the dithizone solution, and shake for 1 minute. Read the lead in the chloroform layer at 520 nm.

[35]Koichi Nishimura and Teruo Imai, *Jap. Anal.* **13**, 423–429 (1964).
[36]N. P. Strel'nikova, G. G. Lystsova, and G. S. Dolgorukova, *Zavod. Lab.* **28**, 1319–1321 (1962).
[37]Yoshihiro Ishihara, Hajime Kishi, and Hideo Komuro, *Jap. Anal.* **11**, 932–936 (1962); cf. C. R. Veale and R. G. Wood, *Analyst* **85**, 371–374 (1960); Barbara Kasterka and Krystyna Kwiatkowska-Sienkiewicz, *Chem. Anal.* (Warsaw) **18**, 1153–1161 (1973).

Niobium-Tantalum Compounds.[38a] Heat a 0.1-gram sample with 5 ml of 85% phosphoric acid at 250°. Cool, take up in water, and dilute to 50 ml. There should have been no significant loss of phosphoric acid in heating. Pass through an 11×150 mm column containing 7–8 grams of KU-3 resin at 1 ml/minute. Wash the column with water until neutral, then elute the lead with 30 ml of 1:3 hydrochloric acid. Add 5 ml of nitric acid and evaporate to dryness. Add 2 ml more of nitric acid and take to dryness. Take up the residue in 1:100 nitric acid and dilute to 50 ml. Mix a 1-ml aliquot with 1 ml of 1:5:1 5% potassium cyanide solution–30% ammonium citrate solution–25% hydrazine hydrochloride solution. Shake with 1 ml of chloroform containing 0.01 mg of dithizone and read the extract at 520 nm.

Triuranium Octaoxide.[38b] Digest 1 gram of sample with 2 ml of perchloric acid and 2 ml of hydrobromic acid. Evaporate to dryness and take up in 50 ml of 2 *N* hydrobromic acid. Pass through a column of 100-200 mesh Dowex 1-X8 in chloride form at 1 ml/minute. Wash out uranium and most other ions with 2 *N* hydrobromic acid. Then elute lead with 1:1 hydrochloric acid. Only the noninterfering indium ion will also be eluted. Evaporate the eluate to dryness. Take up with hydrochloric acid and develop with dithizone.

Yellow Cake. Dissolve 1 gram of sample in 5 ml of hydrobromic acid. Proceed as above for triuranium octaoxide from "Evaporate to dryness and ...."

Monazite.[39a] Transfer a 0.05-gram sample containing about 0.15 mg of lead to a 100-ml round-bottom flask and add 3 ml of sulfuric acid. Place a short-stemmed funnel in the neck of the flask to minimize loss of acid vapors and heat to 150–200° for an hour. Let cool and boil with 30 ml of 1:2 nitric acid to drive off oxides of nitrogen. Add 20 ml of hot water and digest at 100° for 15 minutes. If the residue is inappreciable, cool and dilute to 100 ml.

If appreciable residue remains, filter through paper and wash with hot 1:99 nitric acid. Place the paper in a platinum crucible, wet the paper and residue with sulfuric acid, and ash at 500°. Add hydrofluoric and nitric acid and evaporate to dryness. Add small portions of nitric acid and evaporate to dryness several times to ensure complete removal of hydrofluoric acid. Take up the residue in a small volume of hot 1:9 nitric acid and add to the filtrate. Dilute to 100 ml.

To a 10-ml aliquot add 5 ml of 6% sulfurous acid, 5 ml of 50% sodium citrate solution adjusted to pH 9.2 with ammonium hydroxide, and a few drops of *m*-cresol indicator solution. Add ammonium hydroxide dropwise until the solution is purple. Then add 4.5 ml of ammonium hydroxide and 2.5 ml of 10% potassium cyanide solution to give approximately pH 9.2.

Shake for 1 minute with 10 ml and 10 ml of a solution containing 18 mg of dithizone per liter of chloroform. Extract the aqueous phase with 10 ml of chloroform and add this to the dithizone extracts.

If a significant amount of bismuth is present, strip the lead from the chloroform

[38a]Yu-wei Ch'en and Kuei-ying Ch'i, *K'o Hsueh Tung Pao* **1962** (3), 38–39.
[38b]J. Korkisch and H. Gross, *Mikrochim. Acta* **11**, 413–424 (1975).
[39a]R. A. Powell and C. A. Kinser, *Anal. Chem.* **30**, 1139–1141 (1958); cf. K. F. Chung and J. P. Riley, *Anal. Chem. Acta* **28**, 1–29 (1963).

extract by shaking for 30 seconds with 25 ml of a buffer solution for pH 3.4. If bismuth is substantially absent, use 25 ml of 1:99 nitric acid to strip the lead.[39b] Wash the aqueous phase by shaking for 2 minutes with 25 ml of chloroform. Add 5 ml of 1% solution of potassium cyanide in 1:6 ammonium hydroxide. Extract the lead by shaking for 2 minutes with 15 ml of a solution containing 8 mg of dithizone per liter of chloroform. Read at 520 nm against water and subtract a blank.

Additional Procedures for Dithizone

Igneous Minerals. LYNDOCHITE.[40a] As a diethylammonium diethyldithiocarbamate reagent, dilute 3 ml of redistilled diethylamine to 10 ml with chloroform and add slowly with stirring 9 ml of chloroform containing 1 ml of redistilled carbon bisulfide. Dilute 1:19 with chloroform for use as the dithiocarbamate reagent.[40b]

Heat 0.1 gram of finely ground sample dried at 110° with 20 ml of sulfuric acid near the boiling point. Cool and dilute to 100 ml. Take 5 ml of the well-shaken suspension as a 0.005-gram sample. Add 10 ml of hydrochloric acid and digest at 100° for 20 minutes. Decant through a medium-porosity sintered-glass filtering tube. Repeat the digestion three more times. Dilute the combined digestates to 100 ml.

Mix a 5-ml aliquot with 15 ml of 1:1 hydrochloric acid to give a solution approximately 5.3 *N* with hydrochloric acid. Extract by shaking for 30 seconds with 5 ml, 5 ml, and 5 ml of the dithiocarbamate reagent. The bismuth is in this extract.

Add 50 ml of water to the aqueous residue to reduce the acidity to approximately 1.5 *N* and extract with 5 ml, 5 ml, and 5 ml of the dithiocarbamate reagent. Combine these extracts and evaporate to dryness under a heat lamp. The lead is in this fraction.

Add 0.5 ml of sulfuric acid, heat, and add 1.5 ml of nitric acid dropwise to destroy the organic matter. An alternative is to heat with 0.4 ml of perchloric acid and 1 ml of nitric acid. Dilute with 10 ml of water and boil off oxides of nitrogen. Cool, dilute to 25 ml, and filter through cotton. Add an amount of solution containing 5 mg of dithizone per liter in chloroform corresponding to the expected lead content. Add 2% solution of potassium cyanide in 15:85 ammonium hydroxide to give pH 9–10 and shake for 30 seconds. Withdraw the dithizone extract and extract successively with 1-ml portions of the dithizone reagent until the extracts remain green. Dilute the extracts to 25 ml. Revert a portion of the extract by mixing with an equal volume of 1:6 sulfuric acid. Read the sample against the reverted portion at 595 nm.

Apatite. Heat a 0.5-gram sample in a covered platinum crucible with 0.5 ml of perchloric acid and 10 drops of nitric acid for about an hour. Remove the cover and heat to fumes of perchloric acid. Add hydrofluoric acid to remove silica, and

[39b]K. Bambach and R. E. Burkey, *Ind. Eng. Chem., Anal. Ed.* **14**, 904–907 (1942).
[40a]A. D. Maynes and W. A. E. McBryde, *Anal. Chem.* **29**, 1259–1263 (1957); Heinrich M. Koester, *Ber. Deut. Keram. Ges.* **46**, 247–253 (1969).
[40b]N. Strafford, P. F. Wyatt, and F. G. Kershaw, *Analyst* **70**, 232–246 (1945).

evaporate to dryness. Rinse down the cover and the sides of the crucible with water, add 0.5 ml of perchloric acid, and again take to dryness. Take up the residue in hydrochloric acid and filter through glass wool. Make up the volume of acid to 27 ml, add 5 ml of water, and proceed as described for lyndochite from "Extract by shaking for 30 seconds with...."

Feldspar. Heat a 0.5-gram sample in a platinum crucible with 0.5 ml of perchloric acid and 2.8 ml of hydrofluoric acid. After effervescence has ceased, heat to fumes of perchloric acid and cool. Add 2.1 ml of hydrofluoric acid and evaporate to dryness. Dissolve the residue with hydrochloric acid, filter through glass wool, make up the volume to 34 ml of hydrochloric acid. Add 6 ml of water and proceed as described for lyndochite from "Extract by shaking for 30 seconds with...."

Sphene. Mix a 0.06-gram sample with 0.5 gram of sodium carbonate in a platinum crucible. Cover and heat gradually to the full heat of a Meker burner for 8 minutes. Cool and add 10 ml of 1:19 sulfuric acid, covering at once. After effervescence has ceased, add 1.4 ml of hydrofluoric acid and evaporate nearly to dryness. Rinse down the cover and sides of the crucible with water and heat to near dryness. Transfer to a beaker with water, completing the transfer with hydrochloric acid. Evaporate to near dryness and take up with hydrochloric acid. Filter through glass wool, make the volume of acid up to 27ml, and add 5 ml of water. Complete as described for lyndochite from "Extract by shaking for 30 seconds with 5 ml, 5 ml,..." 1:9 dithiocarbamate solution.

Marine Sediments.[41] Grind to under 120 mesh. Digest a l-gram sample with 4 ml of perchloric acid and 15 ml of hydrofluoric acid in a covered platinum crucible overnight at 100°. Evaporate gently until fumes disappear, then almost to dryness under an infrared heater. Add 2 ml of perchloric acid and repeat the evaporation. Dissolve the residue in 1 ml of perchloric acid and 15 ml of water, then dilute to 100 ml. Complete as for lindochite from "Mix a 5-ml aliquot with 15 ml of...."

Airborne Lead. Draw 15 liters of air containing 0.1–0.8 mg of lead per cubic meter through a Millipore AA filter at 3–5 liters/minute. Extract the filter with 2.5 ml of 1:19 nitric acid containing 0.06% of hydrogen peroxide.[42] Add to 15 ml of solution containing 0.3% of potassium cyanide, 0.6% of sodium metabisulfite, 0.5% of ammonium citrate, and 32.5% of ammonium hydroxide. Extract with 5 ml of 0.004% solution of dithizone in 1,1,1-trichloroethane, and read at 520 nm.

Biological Material.[43] For this technic bismuth and thallium interfere seriously; 1 mg of tin or 10 mg of chromium causes a small negative error. Prepare an ammonium citrate solution by dissolving 167 grams of citric acid in 100 ml of water and 150 ml of concentrated ammonium hydroxide; neutralize to bromothymol blue.

Digest a 5-gram sample with 5 ml of 1:4 sulfuric acid, 2 ml of nitric acid, and 3

[41] R. Chester and M. J. Hughes, *Trans. Inst. Min. Met.* **B77**, B37–B41(1968).
[42] D. M. Groffman and R. Wood, *Analyst* **96**, 140–145 (1971).
[43] A. Dyfverman, *Ark. Kem.* **27**, 79–85 (1967).

ml of perchloric acid. Add 1-ml portions of nitric acid from time to time until the digestate is colorless. Cool and add 10 ml of water. Add 5 ml of the ammonium citrate solution and boil if turbid. Add 10 ml of water, 10 ml of ammonium hydroxide, 2 ml of 10% potassium cyanide solution, and 2 ml of 5.2% hydroxylamine sulfate solution. Shake for 1 minute with 3 ml of 0.2% solution of dithizone in chloroform. Wash the extract twice with 1:5:94 ammonium hydroxide – 10% potassium cyanide solution – water. Read at 515 nm against a blank.

URINE.[44] As a buffer for pH 9 mix 50 ml of a solution containing 12.369 grams of boric acid and 14.911 grams of potassium chloride per liter with 21.4 ml of 0.8% sodium hydroxide solution; dilute to 200 ml. As citrate-cyanide solution, dissolve 350 grams of citric acid monohydrate in 200 ml of water and add 2 ml of thymol blue indicator solution. Add 1:3 ammonium hydroxide until the indicator turns green. Dilute to about 1400 ml and add 20 grams of hydroxylamine hydrochloride. Dilute to 2 liters.

Evaporate a 50-ml sample containing not more than 0.01 mg of lead to dryness with 5 ml of nitric acid, and ash at 500°. If charring occurs, cool and repeat the treatment with nitric acid and ashing. Take up the ash in 30 ml of 1:16 nitric acid by boiling. Cool and dilute to about 50 ml.

Add 20 ml of citrate-cyanide solution to the sample and at once add ammonium hydroxide or nitric acid until the color matches that of the pH 9.0 buffer solution with thymol blue indicator. Add 10 ml of chloroform containing 5 mg of dithizone per liter and shake for 60 seconds. Read the extract at 510 nm against chloroform. There is no interference from calcium, magnesium, orthophosphate, or reasonable amounts of bismuth.

BLOOD.[45] As a citrate-cyanide reagent, combine 30 ml of 10% potassium cyanide solution, 3 ml of 50% ammonium citrate solution, and 10 ml of ammonium hydroxide. Dilute to 1 liter. As a wash solution, dilute 10 ml of 10% potassium cyanide and 10 ml of ammonium hydroxide to 1 liter.

Mix a 3-ml sample with 5 ml of nitric acid, 3 ml of perchloric acid, and 0.5 ml of sulfuric acid. Set aside for 30 minutes to minimize later frothing. Heat at $200 \pm 5°$ until the evolution of sulfur trioxide fumes ceases. Let cool, and add 3 ml of 50% ammonium citrate solution, 5 ml of ammonium hydroxide, and 10 ml of the citrate-cyanide reagent. Shake for 30 seconds with 0.1 ml of 0.1% solution of dithizone in chloroform and 10 ml of chloroform. Wash the chloroform layer with four successive 20-ml portions of the wash solution. Filter through paper and read at 525 nm against chloroform. Subract a blank.

Alternatively,[46] to prepare a citrate-cyanide reagent, dissolve 100 grams of sodium carbonate, 150 grams of sodium citrate, 20 grams of potassium cyanide, and 20 grams of sodium hydroxide in water. Dilute to 1 liter. Mix 1 ml of blood

[44] E. V. Browett and R. Moss, *Analyst* **90**, 715–726 (1965); cf P. Vintner, *J. Med. Lab. Technol.* **21**, 281–286 (1964); M. L. Santi, *Accad. Med. Genova* **71**, 43–46 (1956); G. Machata and H. Neuninger, *Wien. Med. Wochschr.* **110**, 39–41 (1960); O. M. Gulina, *Gigiera Tr. Prof. Zabol.* **4** (11), 58–60 (1960); J. M. Dick, R. W. Ellis, and J. Steel, *Brit. J. Ind. Med.* **18**, 283–286 (1961); Morris B. Jacobs and Jeanne Herndon, *Am. Ind. Hyg. Assoc. J.* **22**, 372–376 (1961).

[45] P. Vinter, *J. Med. Lab. Technol.* **21**, 281–286 (1969).

[46] Eleanor Berman, *Am. J. Clin. Pathol.* **36**, 549–551 (1961).

with 10 ml of 5% trichloroacetic acid solution. Centrifuge and separate the supernatant layer. Add 5 ml of the citrate-cyanide reagent. Extract with 10 ml of chloroform containing 8 mg of dithizone per liter and read at 520 nm.

As another technic[47a] the blood is digested with sulfuric and nitric acid; iron is extracted with amyl acetate before formation of the lead dithizonate and extraction with carbon tetrachloride.

Beer.[47b] Mix 100 ml of nondecarbonated beer with 10 ml of nitric acid and evaporate at 100° to a syrup, adding more nitric acid as may be necessary to complete the oxidation. Adjust to pH 9 with ammonium hydroxide. Add a solution of 5 mg of dithizone per liter in chloroform, dropwise with shaking, until the color changes to purple. Withdraw the chloroform layer. Wash the aqueous layer with 2 ml of chloroform and add this to the extract. Wash the extract with 10 ml of water adjusted to pH 9. Dry the extract with sodium sulfate, dilute to 10 ml with chloroform, and read at 530 nm.

Fish.[47c] Oxidize a 50-gram sample with sulfuric, nitric, and perchloric acids. Dilute to a concentration N in acidity and extract the interfering mercury with dithizone; for details see the chapter on mercury. A minor amount of lead is lost with the mercury.

After extraction of the mercury, mix an aliquot of the solution with 25% solution of sodium citrate and adjust to pH 9.25 ± 0.25 with ammonium hydroxide. Add 10% solution of potassium cyanide and extract with portions of 0.001% solution of dithizone in chloroform until violet lead dithizonate is no longer extracted. Extract free dithizone from the combined extracts with a 0.1% solution of potassium cyanide in 1% solution of ammonia. Dilute to a known volume and read at 510 nm.

Organic Polymers.[48] Although the dithizonates are usually extracted into a solvent, the lead dithizonate is also soluble in 70% acetone for reading. Ethanol and methanol are alternative solvents. As a buffer and complexing mixture, dissolve 3 grams of potassium cyanide, 3 grams of hydroxylamine hydrochloride, and 3 grams of sodium barbital in 300 ml of water and dilute to 1 liter with acetone.

Heat a 0.5-gram sample with 50 ml of nitric acid nearly to dryness over a period of 30–45 minutes. Cool and add 5 ml of 1:1 nitric acid and 2 ml of perchloric acid. Boil gently to white fumes; then simmer until the solution is colorless. Cool, add water, and neutralize with 1:1 ammonium hydroxide. Add acetone to make it 70% in that solvent and dilute to a known volume with 70% acetone. Mix 25 ml of the buffer and complexing mixture, 10 ml of 0.01% dithizone in 70% acetone, and an aliquot of the sample solution containing not more than 10 mg of lead. Dilute to 25 ml with 70% acetone and read at 500 or 600 nm against a blank.

Plant Material.[49] Prepare a citrate-cyanide reagent by mixing 100 ml of 10%

[47a]Kai-ki Han, *Ann. Biol. Clin.* (Paris) **17**, 168–172 (1950); cf. Kata Voloder, Nikola Ivicic, and Bozena Svigir, *Mikrochim. Acta* **1971**, 341–347.
[47b]Harold E. Weissler and Kamal P. Yadev, *Proc. Am. Soc. Brew. Chem.* **1969**, 51–53.
[47c]Regina Gajewska, Michal Nabrzyski, and Eullia Lipka, *Bromatol. Chem. Toksykol.* **9**, 1–5 (1976).
[48]D. G. M. Diaper and A. Kuksis, *Can. J. Chem.* **35**, 1278–1284.
[49]Z. Kerin, *Mikrochim. Acta* **1968**, 927–933; cf K. Riebartsch and G. Gottschalk, *Z. Anal. Chem.* **241**, 179–185 (1965).

ammonium citrate solution with 200 ml of 10% potassium cyanide solution, adjust to pH 9.5 with ammonium hydroxide, and dilute to 1 liter.

Mix 1–2 grams of dried sample with 5 ml of water and 15 drops of 0.01 *N* sulfuric acid and dry at 80°. Add 5 ml of nitric acid and boil until colorless. Cool and dilute to 10 ml. Neutralize the sample solution to pH 9.5 as indicated by thymol blue by adding ammonium hydroxide. Add 20 ml of the citrate-cyanide reagent. Extract with 5 ml and 2 ml of 0.02% dithizone solution in carbon tetrachloride. Dilute the combined extracts with carbon tetrachloride to either 10 or 25 ml. Extract excess dithizone from the carbon tetrachloride layer with 5-ml portions of 1% potassium cyanide–0.1% ammonium chloride solution until the extracts are no longer green. Filter the carbon tetrachloride extract to remove entrained moisture and read at 516 nm against carbon tetrachloride.

Dialdehyde Starch.[50] Char a 5-gram sample with 2 ml of sulfuric acid and ash until carbon-free. Take the ash up in 1 ml of hydrochloric acid and dilute to about 50 ml. Add sodium peroxide to pH 10 and concentrate to about 10 ml. Acidify to pH 4 and dilute to 10 ml as a sample for further treatment of aliquots for chromium by diphenylcarbizide and for lead by dithizone.

Periodic Acid. Evaporate a 10-ml sample and ignite at 600° for 16 hours. Further treat as for the ash of dialdehyde starch.

Organic Lead in Air.[51] After absorption of organic lead in activated carbon, the specified technic extracts 98–100% of it. Lead dithizonate is read, then decomposed by diethyldithiocarbamate. An alternative agent for the decomposition is disodium EDTA which requires longer shaking. The method is applicable in the range of 10 parts of organic lead per trillion parts of air; and it is much more sensitive than an earlier one that relied on a scrubber of iodine crystals.[52]

As cyanide-citrate reagent, dissolve 20 grams of potassium cyanide, 40 grams of dibasic ammonium citrate, and 200 grams of anhydrous sodium sulfite in water. Dilute to 1 liter and add 600 ml of ammonium hydroxide.

Add 10 grams of 30-50 mesh activated carbon and a small wad of glass wool to the scrubber illustrated in Figure 1-2. Seal the dry glass joint without stopcock grease by surrounding it with electrical tape to avoid leaks. Connect the top of the carbon scrubber to a Millipore membrane filter, type H-2, to remove particulate matter. Connect the bottom of the scrubber to a vacuum pump and collect approximately 200 cubic meters of air at 0.7 cubic feet/minute.

Remove the scrubber from the system, disassemble it, and invert it to pour the carbon into a 500-ml Erlenmeyer flask. Add the glass wool to the flask. Use a freshly prepared mixture of 75 ml of hydrochloric acid and 25 ml of nitric acid to wash out the scrubber and to wash down the sides of the flask. Digest overnight, which should take it nearly to dryness. Add 30 ml of nitric acid and heat at 100° for an hour. Mix with 100 ml of water and let stand for 2–3 hours. Decant the supernatant layer into a 500-ml Erlenmeyer flask and filter the residue through paper into the flask. Wash the residue on the filter three times with water.

[50] L. T. Black, E. B. Lancaster, and H. G. Maister, *Cereal Chem.* **40**, 66–71 (1963).
[51] Louis J. Synder, *Anal. Chem.* **39**, 591–594 (1967).
[52] Louis J. Snyder and S. R. Henderson, *ibid.* **33**, 1175–1180 (1961); U.S. Patent 3,371,446 (1963).

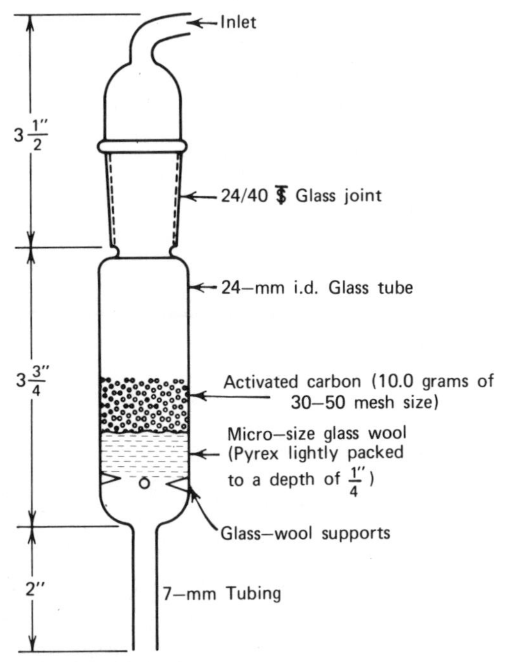

Figure 1-2 Organic lead-in-air scrubber for composite sampling.

Add 20 ml of 2:3 nitric-perchloric acid to the flask and heat at 150° to perchloric acid fumes. If organic matter remains unoxidized, repeat this step with 10 ml of the mixed acid. Cool and add 20 ml of 1:4 nitric acid and about 25 ml of water. After 30 minutes at 30–40° for complete solution, transfer to a 250-ml absorption cell (Figure 1-3) 50 ml of buffer solution and 20 ml of cyanide-citrate reagent. Set aside for 20 minutes for reduction of the sample. Add 5 ml of dithizone solution containing 60 mg/liter in chloroform. Shake for 30 seconds, and read the chloroform layer at 510 nm against air. If the absorbance reading exceeds 2, add an additional 5 ml of dithizone solution, shake, and reread. Add 5 ml of 0.2% solution of sodium diethyldithiocarbamate, shake for 15 seconds, and read the chloroform layer within 30–45 seconds. The sodium diethyldithiocarbamate has quantitatively decomposed the lead dithizonate, and the difference between the two readings measures lead. Subtract a blank. On longer standing some dithizone is decomposed.

Tetraethyl, Triethyl, Diethyl, and Ionic Lead[53]

Since tetraethyl lead does not react with dithizone, the reagent may be used for the three ionic forms and by difference for tetraethyl lead. Other alkylated lead dithizonates absorb at the same wavelength as the ethyl compounds.

All conditions must be rigidly standardized. Unless protected from light, the triorganodithiozonates are converted to lead dithizonates at a rate depending on the organic radical and the amount of exposure. Organic compounds of zinc,

[53]S. R. Henderson and L. J. Snyder, *Anal. Chem.* **33**, 1172–1175 (1961).

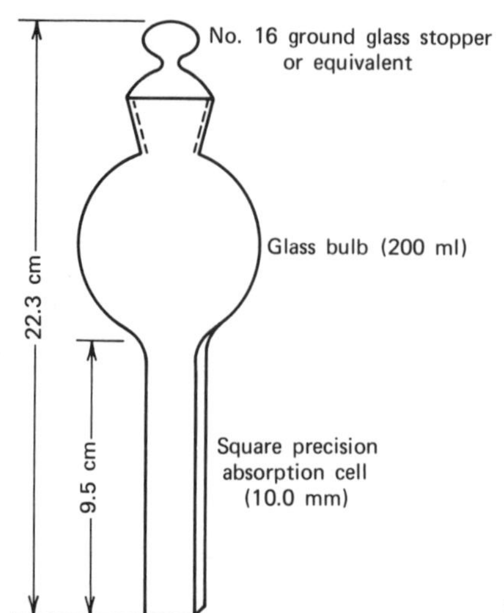

Figure 1-3 Modified absorption cell for lead.

mercury, thallium, tin, and bismuth may interfere. Ethylmercury ion is not removed by complexing with cyanide. Organomanganese interferes. Decomposed dithizone increases the blank at 424 nm.

Dyes and colored additives may interfere in the application to leaded fuels but can be corrected in the blank. As distinguished from the red color of lead dithizonate, triethyl lead dithizonate is canary yellow and diethyl lead dithizonate is orange. Gas chromatography is appropriate for separation of lead alkyls for determination by dithizone.[54]

A technic has been developed for determining 1–20 ppm alkyl-lead contamination of nonleaded gasoline outside the laboratory.[55] A sample diluted with carbon tetrachloride is shaken with potassium bromide, potassium bromate solution, and hydrochloric acid to convert tetraalkyl-lead compounds to dialkyl-lead salts. To mask other heavy metals, tetraethylenepentamine is added in sodium sulfite–ammonium hydroxide to buffer the solution to pH 11–11.5. The lead is then extracted with dithizone in chloroform–carbon tetrachloride and read at 485 nm.

Triethyl, Diethyl, and Ionic Lead. As an ammoniacal buffer solution, dissolve 20 grams of ammonium citrate to solubilize cations, 8 grams of potassium cyanide to complex interfering cations, and 98 grams of sodium sulfite to retain a reducing system in 320 ml of water. Dilute to 2 liters with ammonium hydroxide.

Add 50 ml of water, 20 ml of the ammoniacal buffer solution, and 25 ml of dithizone solution containing 60 mg/liter in chloroform to a special absorption cell (Figure 1-3). Mix vigorously until clear and read as a blank at 424, 500, and 540 nm.

Add an aliquot of sample containing about 0.1 mg of lead and acid not equivalent to more than 2 ml of 1 : 9 nitric acid. Wrap the cell in opaque cloth to protect the light-sensitive thizonates, and shake vigorously until the three forms of

[54]W. W. Parker, C. Z. Smith, and R. L. Hudson, *Anal. Chem.* **33**, 1170–1171 (1961).
[55]S. T. Holding and G. Williams, *J. Inst. Petrol.* **58**, 230–236 (1972).

ionic lead have been extracted into the chloroform phase as dithizonates. Read at 425, 500, and 540 nm and subtract the blank from each.

$$a = \text{corrected absorption at 424 nm}$$

$$b = \text{corrected absorption at 500 nm}$$

$$c = \text{corrected absorption at 540 nm}$$

$$x = \text{triethyl lead} = 245a - 77b + 41c \qquad (1)$$

$$y = \text{diethyl lead} = -55a + 192b - 203c \qquad (2)$$

$$z = \text{inorganic lead} = 26a - 90b + 194c \qquad (3)$$

Tetraethyl Lead. By oxidation of tetraethyl lead with iodine in chloroform, it is converted almost instantaneously to predominantly triethyl lead. Add to an absorption cell 50 ml of water, 5 ml of chloroform, 5 drops of 0.1 *N* iodine solution, and 20 ml of the ammoniacal buffer solution. This is the blank. To a similar cell add 5 ml of water, 5 ml of chloroform, and 5 drops of 0.1% iodine solution.

To each cell add a sample volume containing approximately 0.2 mg of lead as tetraethyl lead. Shake the second cell for 5 seconds, and immediately add and shake with 20 ml of ammoniacal buffer to reduce the excess iodine. Add 20 ml of dithizone solution to each and shake until equilibrium is attained. Read each at 424, 500, and 540 nm, and subtract the readings of the first cell from those of the second cell.

$$\text{tetraethyl lead} = 216a + 25b + 32c \qquad (4)$$

A reduction in the volume of dithizone solution gives greater sensitivity but requires an alteration of the calculations.

Fractionated Lead Alkyls. Gasoline. With a Perkin-Elmer Fractometer, Model 154, use a 1 foot × $\frac{1}{4}$ inch column packed with Nujol on 35-80 mesh Chromosorb with a helium flow of 100 ml/minute, a temperature of 70°, and a chart speed of $\frac{1}{2}$ inch/minute. The absorption cells are as described in the previous technic. The sample is 100 ml.

Add 5 ml of a 0.2 *N* solution of iodine in methanol, 6.25 grams of iodine per 250 ml of solvent, to a cell. Deliver the appropriate fraction from the gas chromatograph into the iodine solution for 3.75 minutes as tetramethyl lead.

Add 5 ml of iodine solution to two additional cells. To one add 0.1 ml of sample for determination of total methyl- and tetraethyl lead. The other is a reagent blank. Add 3 ml of 25% potassium iodide solution to each. Heat each at 100° for 3 minutes and cool.

As a reagent, dissolve 12.5 grams of ammonium citrate, 5 grams of potassium cyanide, and 30 grams of sodium sulfite in 200 ml of water and dilute to 1 liter with ammonium hydroxide.

Add to each cell 50 ml of water and 35 ml of reagent and shake. Add 25 ml of a dithizone solution containing 60 mg/liter of chloroform. After shaking vigorously for 1 minute, read each at 520 nm. The three absorbers are, respectively, tetramethyl lead, tetramethyl lead plus tetraethyl lead, and a blank. Determine tetraethyl lead by difference.

Tetraethyl Lead. Gasoline and Naphtha.[56] Tetraethyl lead in gasoline or naphtha is decomposed by treatment with bromine. The reaction proceeds rapidly to diethyl lead bromide. Further treatment with bromine in aqueous solution completes the decomposition to lead ion for reading as lead dithizonate.

As a buffer solution for pH 10, dissolve separately in water 20 grams of potassium cyanide, 6 grams of ammonium citrate, and 40 grams of anhydrous sodium sulfite. Mix, add 200 ml of ammonium hydroxide and dilute to one liter. A second buffer solution for pH 10.5 differs only in the use of 600 ml of ammonium hydroxide.

Select a sample size from the following table. Add 30% by volume bromine solution in carbon tetrachloride until the color persists for at least 5 minutes and let stand for 10 minutes. Heat on a hot plate until bromine is evaporated. Cool and add 25 ml of dilute nitric acid prepared by dilution of 8 ml of concentrated acid to 1 liter with water. Shake for 2 minutes. Drain the aqueous extract. Repeat the extraction with 25 ml of acid and combine the extracts.

Expected Lead (ppb)	Sample (ml)	Thizone Solution (ml)
10–100	100	5
50–150	100	10
100–1000	100	25
100–2000	50	25
1000–20,000	5	25

Add 1 ml of bromine water and evaporate to approximately 10 ml. Dilute to 50 ml and add 10 ml of the buffer solution for pH 10. If the solution is colored at this point, extract with chloroform until a colorless solution is obtained. Add the volume of dithizone solution containing 60 mg/liter of chloroform as called for in the table and shake for 2 minutes.

As a reading blank take 50 ml of the nitric acid and if the sample was decolorized by extracting with chloroform, extract it with the same amount. Add the same amounts of buffer solution and dithizone solutions as were added to the sample and shake for the same length of time. Read the extract of the sample at 510 nm against the reading blank.

Miscellaneous Dithizone Methods

For lead in tap water,[57] additions are a chelating agent, potassium cyanide, sodium citrate, ammonium hydroxide, and dithizone. Chloroform extracts the lead di-

[56] Margaret E. Griffing and Adele Rozek, L. J. Snyder and S. R. Henderson, *Anal. Chem.* **29**, 190–195 (1957).

[57] R. L. Waroquier, *Allg. Prakt. Chem.* **18**, 68–69 (1967).

thizonate with some zinc and copper dithizonate. Washing with a dilute chelating solution removes the zinc and copper and excess dithizone for a monocolor reading at 520 nm. If the water is soft, added trisodium phosphate optimizes extraction.

For inorganic lead in air,[58] filter an amount containing 0.5–50 μg through paper. Wet-ash and take up in 1:1 nitric acid. Adjust to pH 8.5–11.5 with ammonium hydroxide and add cyanide. Extract with dithizone in carbon tetrachloride.

For lead in chromium[59] dissolve the sample in hydrochloric acid and take to fumes to form chromic acid. Cool, take up in water, and add acetic acid. By adding an aqueous solution of a barium salt, coprecipitate barium chromate and lead chromate. Filter and treat with perchloric acid and hydrochloric acid to volatilize the chromic acid as chromyl chloride. This leaves a residue for determination of lead by dithizone.

For the sum of lead, copper, and zinc in water add to 100 ml of sample 1 ml of 2.3% solution of sodium tartrate, 1 ml of 1.25% solution of sodium sulfite, 2 ml of 5% solution of hydroxylamine hydrochloride, and 1 ml of 0.65% solution of potassium cyanide.[60] Add sufficient 2.5 M sodium carbonate–M sodium bicarbonate to adjust to pH 10.5, extract with 10 ml of 0.16 mM dithizone in carbon tetrachloride, and read at 532 nm.

For collection of lead by precipitation of manganese dioxide, the optimum acidity is 0.008 M.[61] Recovery is in the 80–90% range. Lead in industrial water supplies[62] is determined by dithizone. A technic for determination of copper by its catalytic effect on the iodine azide reaction is also applicable to lead. For details, see Copper.

For analysis of silver,[63] dissolve 10 grams in nitric acid. Add lanthanum nitrate solution and coprecipitate gold, lead, bismuth, iron, and aluminum along with the lanthanum by adding ammonium hydroxide. Filter and wash. Dissolve the precipitated hydroxides in 10 ml of 1:1 hydrochloric acid and dilute to 50 ml. Determine lead in an aliquot by dithizone.

For analysis of tantalum, dissolve the sample with nitric and hydrofluoric acids.[64] Extract lead, cadmium, iron, and copper as diethyldithiocarbamates with chloroform. Determine lead by dithizone.

For determination of lead in vanadium and niobium,[65] the sample solution contains tartrate and cyanide at pH 11–12. Extract lead, bismuth, and cadmium as diethyldithiocarbamates in chloroform. Extract the chloroform phase with 1:5 hydrochloric acid to remove lead and cadmium, leaving bismuth in the chloroform. Then by splitting the acid extract determine lead by the dithizone method in the presence of cyanide and hydrazine. Determine cadmium by dithizone in alkaline cyanide solution in the other fraction.

[58]T. J. Kneip, R. S. Ajemian, J. R. Carlberg, J. Driscoll, H. Freiser, L. Kornreich, K. Kumler, and R. J. Thompson, *Health Lab. Sci.* **9**, 79–83 (1972); cf. Artur Strusinski and Halina Wyszynska, *Rocz. Panstw. Zal. Hig.* **22**, 649–656 (1971).

[59]Nobuhisa Matano and Akeira Kawase, *Trans. Nat. Inst. Metals* (Tokyo) **4**, 111–116 (1962).

[60]Takashi Ashizawa, Shigenori Sugizaki, Hizao Asahi, Hiroko Sakai, and Keiichi Shibata, *Jap. Anal.* **19**, 1333–1340 (1970).

[61]G. F. Reynolds and F. S. Tyler, *Analyst* **89**, 579–586 (1964).

[62]British Standards Institution, *BS 2690*, Pt. **14**, 15 pp, 1972.

[63]Z. Marczenko and K. Kasiura, *Chem. Anal.* (Warsaw) **9**, 87–95 (1964).

[64]V. A. Nazarenko, E. A. Biryuk, M. B. Shustova, G. G. Shifareva, S. Ya. Vinkovetskaya, and G. V. Flyantikova, *Zavod. Lab.* **32**, 267–269 (1966).

[65]V. A. Nazarenko and E. I. Biryuk, *ibid.*, **25**, 28–30 (1959).

In analysis of tungsten ores and concentrates, lead is read as the dithizonate.[66]

For lead in pyrites, digest a sample containing 0.005–0.015 mg of lead in hydrofluoric acid and aqua regia.[67] Convert to chlorides by repeated evaporations with concentrated hydrochloric acid. Add 20 mg of copper as the chloride in 75 ml of 1:25 hydrochloric acid. Add hydrazine hydrochloride solution to reduce the iron, and precipitate sulfides with hydrogen sulfide. Filter the sulfides of copper, bismuth, and tin; redissolve and reprecipitate. Boil off hydrogen sulfide from the combined filtrates and cool. Extract lead by repeated shaking with 1% diethylammonium diethyldithiocarbamate in chloroform. Wet-ash the combined extracts with nitric and perchloric acids and complete by isolation of lead dithizonate for reading at 520 nm.

For lead in calcium fluoride,[68] remove fluoride ion by treatment with boric acid in perchloric acid. Then extract lead as the dithizonate in the presence of ammonium citrate and potassium cyanide, and read. For 0.1 mg of lead there is no interference by an equal amount of bismuth or thallium or by 1 mg of iron or copper.

For lead in gelatin, pass the sample solution through a column of cation exchange resin, Dowex 50-W, to extract the metal ions. Elute the metal ions with 1:3 hydrochloric acid. Extract the lead at pH 9.5–10 in the presence of citrate and cyanide with 0.002% dithizone solution in carbon tetrachloride, and read at 520 nm.

The dithizone method has been applied to fruit, vegetables, and soil in demonstrating that the produce did not take up additional lead from soil with a high lead content.[69a]

For determination of lead in fruit juice, acidify 10 ml with hydrochloric acid and heat to the boiling point.[69b] Cool and adjust to pH 9.5 with ammonium hydroxide. Extract the lead with a solution of dithizone in chloroform and read at 510 nm. For orange juice, a blank is read because of natural pigments extracted with chloroform.

Dibutyl–lead diacetate in urine is extracted as the dithizonate with toluene after buffering to pH 4.75.[70] The reading at 485 nm may be corrected for the absorption of dithizone at that wavelength. Alternatively, the excess dithizone may be extracted at pH 12 with an equal volume of 0.8% sodium hydroxide solution before reading.

Wet-ash 2 ml of blood or 5 ml of urine with nitric acid and sulfur dioxide.[71] Evaporate to dryness, take up in water, and add hydroxylamine hydrochloride. Neutralize with ammonium citrate and complex interfering metal ions with potassium cyanide. Extract lead with dithizone in chloroform. Bismuth interferes.

[66]H. E. Peterson, W. L. Anderson, and M. R. Howcroft, *Rep. Invest. U.S. Bur. Mines* No. **6148**, 30 pp (1963).

[67]Frank Cuttitta and Jesse J. Warr, *U.S. Geol. Survevey, Prof. Papers* **400**, (B185–196) (1960; cf. T. S. Harrison and W. W. Foster, *Metallurgia* **73**, 141–146, 191–195 (1966).

[68]A. B. Crowther, F. Warburton, and D. J. Grimshaw, *UKAEA, Ind. Group Hdqr.* **SCS-R-9**, 6 pp (1959).

[69a]D. R. Crudgington, J. Markland, and J. Vallance, *J. Assoc. Pub. Anal.* **11** (4), 120–126 (1973).

[69b]Teresa Zawadzka, *Roczn. Panst. Zakl. Hig.* **27**, 411–418 (1976).

[70]G. Gras and F. Fauran, *Ann. Pharm. Fr.* **30**, 545–554 (1972).

[71]R. Truhaut and C. Boudene, *Pharm. Biol.* **2** (16), 167–170 (1960).

For lead in drugs,[72] decompose a 1-gram sample with 10 ml of sulfuric acid, adding hydrogen peroxide and/or perchloric acid as required. Evaporate to dryness and take up in 20% acetic acid. Adjust to pH 2.7 and extract copper with dithizone in chloroform. Add ammonium acetate buffer solution for pH 4.9 and extract the zinc with dithizone in chloroform. Finally, adjust to pH 8.5 in the presence of ammonium citrate for extraction and further determination of lead as the dithizonate.

With respect to wet-ashing, a collaborative study[73] of lead in various forms of milk by that method with nitric and perchloric acids indicates that previously published values may have been too high by a magnitude of an order or more.

Dithizone is applicable to calcarous compounds with the AutoAnalyzer at up to 20 samples per hour.[74] Dissolve the sample in hydrochloric acid and pass it through Amberlite resin LA-1 in heptane. Extract the lead with a buffer solution containing 10% of ammonium citrate, 0.2% of hydroxylamine hydrochloride, 16% of ammonium hydroxide, and 0.25% of potassium cyanide. Extract the lead from this buffer solution with 0.001% dithizone in chloroform and read at 510 nm.

Lead in canned food[75] is coprecipitated with strontium, and the precipitate is dissolved in sodium hexametaphosphate solution. The lead is extracted and read as the dithizonate.

A study of lead in rib bone revealed good agreement between emission spectroscopy, atomic absorption after isolation of the lead by dithizone, and atomic absorption after isolation of the lead from acid or alkaline solution by diethyldithiocarbamate, reading as lead dithizonate or reading as lead chloride.[76] Reading as lead dithizonate or determination by emission spectroscopy is preferred.

For analysis of rayon spinning baths, the large amount of zinc present can be precipitated with oxine before determining lead in the filtrate by dithizone.[77]

For zirconium and zircaloy, adjust a citrate-cyanide solution to pH 9–9.5. Extract the lead with dithizone in chloroform and reextract the lead into an aqueous alkaline medium containing ammonia and cyanide. From this again extract the lead dithizonate for reading 50–500 ppm per 1-gram sample.[78]

In the presence of cyanide and citrate, the optimum pH for extraction of lead by dithizone in carbon tetrachloride is 10.8–10.9, but a variation of 0.2 unit from that has little effect.[79]

For lead in zircon,[80] fuse a portion of powdered sample containing 0.008–0.015 mg of lead in platinum with 1 : 1 sodium carbonate–borax. Dissolve the melt in 3 : 7 nitric acid for determination as lead dithizonate.

[72]J. Hollos, *Gyogyszerszet* **7**, 204–210 (1963); cf. J. Hollos and I. Horvath-Gabai, *Pharmazie* **20**, 207–210 (1965).
[73]Manuel Brandt and Jerome M. Bentz, *Microchem. J.* **16**, 113–120 (1971).
[74]F. J. Bano and R. J. Crossland, *Analyst* **97**, 823–827 (1972).
[75]V. P. Sagakova and A. I. Lyubivaya, *Tr. Ukr. Nauch.-Issled. Inst. Konserv: Prom.* **1962** (4), 62–67.
[76]J. S. Hislop, A. Parker, G. S. Spicer, and W. S. W. Webb. *Rep. UKAEA, AERE – R7321*, 14 pp. (1973).
[77]Hiroyuki Yoshikawa and Masami Morinaka, *Toyo Rayon Shuko* **10**, 145–148 (1955).
[78]G. W. Goward and B. B. Wilson, *USAEC*, **WAPD-CTA (GLA)-155** (Rev. 1) 6 pp (1957).
[79]O. B. Mathre and E. B. Sandell, *Talenta* **11**, 295–314 (1964).
[80]Frank Cuttitta and Jesse J. Warr, *U.S. Geol. Survey, Prof. Papers* **400B**, B486–B487 (1960).

ARSAZAN

The complex of lead with arsazan can be extracted with butanol for reading after suitable masking of interferences.[81]

Procedures

Water. As a buffer solution, mix 40 ml of 10% ammonium hydroxide, 130 ml of 10% ammonium chloride solution, and 80 ml of water. Acidify 20 ml of water to Congo red with 1:1 hydrochloric acid. Add 0.2 ml of 10% sulfosalicylic acid solution, 0.1 ml of 10% thiourea solution, and 2 ml of the buffer solution. Add 4 ml of butanol, 0.5 ml of 0.01% solution of arsazan in butanol, and 0.2 ml of 1% potassium ferrocyanide solution. Shake and read the organic layer for 0.5–10 μg of lead.

Bronze. Dissolve a 0.5-gram sample in 25 ml of sulfuric acid and 5 ml of water. Add 200 ml of 1:5 sulfuric acid and dilute to 500 ml. Mix a 2-ml aliquot with 6 ml of water and neutralize with ammonium hydroxide to a clear blue. Make slightly acid with hydrochloric acid and add 0.2 ml of 10% sulfosalicylic acid solution, 3 ml of 10% thiourea solution, 2 ml of butanol, 1 ml of 0.01% solution of arsazan in butanol, 0.2 ml of 1% solution of potassium ferrocyanide, and 2 ml of buffer solution. Read. The standard must contain an appropriate amount of copper.

AZO DYES OF ANTHRANILIC ACID

The dyes so developed obey Beer's law for 0.01–0.2 mg of lead per ml of the final solution and are stable for more than an hour.[82]

Procedure. To a sample solution containing 0.5–10 mg of lead, add 15 ml of neutral 3% sodium anthranilate solution. Filter the precipitated lead anthranilate and wash with water saturated with lead anthranilate. Dissolve the precipitate in 10 ml of 1:5 hydrochloric acid. Diazotize the liberated anthranilic with 5 ml of 4% sodium nitrite solution. Add 10 ml of a 3% ethanolic solution of 3-hydroxy-2-naphthoic acid and 2 ml of 1:6 ammonium hydroxide. Dilute to 50 ml, and after 30 minutes read the orange-red color at 500 nm.

An alternative reagent to couple with the diazo compound is a 3% solution of 7-hydroxynaphthalene-1,3-disulfonic acid in 0.56% potassium hydroxide solution. This gives a lemon-yellow color, read at 430 nm.

---

[81] A. M. Lukin, L. S. Chernaya, G. S. Petrova, and A. Y. Sosnina, *Zavod. Lab.* **28**, 398–401 (1962).
[82] V. Armeanu, P. Costinescu, and C. G. Calin, *Rev. Chim. (Bucharest)* **19**, 290 (1968).

CHLOROANILATE

When lead is reacted with sodium chloroanilate[83a] by the technic below, the absorption of 0.05–0.25 mg of lead is linear.

For determination of lead in solution add excess of a solution of 2-mercaptobenzothiazole in dimethylformamide containing 1 mg per ml.[83b] Centrifuge to separate the lead complex. Determine the excess of reagent in dilute acid by adding excess of mercuric chloroanilate and read the liberated chloranilic acid at 530 nm.

Procedure. Slurry 30 grams of a strongly basic anion exchange resin with 1:12 hydrochloric acid and transfer to a chromatographic column. The sample should be in 1:12 hydrochloric acid. Pass it through the column at 1 drop/second and wash with 40 ml of 1:12 hydrochloric acid. Discard the eluate and washings. To recover the lead, pass 140 ml of water through the column. Adjust the eluate to pH 4.0 and dilute to 150 ml.

To a 10-ml aliquot add 1 ml of acetate buffer for pH 4.63 and 2 ml of methyl Cellosolve. Mix and add 1 ml of fresh 1% sodium chloroanilate solution. Mix, set aside for 30 minutes, and centrifuge. Decant to waste, add 5 ml of 50% isopropyl alcohol to the precipitate, and shake. Centrifuge, decant, and remove the residual supernatant layer with filter paper.

If the lead is more than 50 μg, add 1 ml of acetate buffer solution for pH 4.63 and 5 ml of 5% solution of sodium EDTA. Stopper, shake, and let stand for 10 minutes. Read at 530 nm against a blank.

If the lead is less than 50 μg, add 0.3 gram of ammonium chloride and shake. Centrifuge, decant, and remove residual supernatant layer with filter paper. Complete as for more than 50 μg from "...add 1 ml of acetate buffer solution...."

DIETHYLDITHIOCARBAMATE

The complex of lead with diethyldithiocarbamate is extracted into carbon tetrachloride at a high pH. Then the carbon tetrachloride is shaken with aqueous copper sulfate solution to convert the complex to the copper salt for reading at 435 nm.[84] Bismuth, cadmium, and thallium interfere. As little as 2.5 μg per ml of solvent is readable.[85a]

For analysis of platinum-rhodium alloy chromatograph the nitric acid sample solution on a column of Catex S. Elute platinum and rhodium with 1:30 nitric acid, bismuth with 1:24 hydrochloric acid, and thereafter the lead with 1:8

[83a]Edwin A. Wynne, Russel D. Burdick, and Leonard H. Fine, *Anal. Chem.* **33**, 807–808 (1961).
[83b]H. Raber, E. Gagliari, and H. Meisterhofer, *Mikrochim. Acta.* **1976, II** (1–2), 71–73.
[84]J. F. Tertoolen, D. A. Detmar, and C. Buijee, *Z. Anal. Chem.* **167**, 401–408 (1959); R. Keil, *ibid.* **229**, 117–118 (1969); L. B. Sokolovich, Yu. L. Lel'chuk, and G. A. Detkova, *Izv. Tomsk. Politekh. Inst.* **163**, 130–133 (1970); Jan Inglot and Maria Kawecka, *Pr. Inst. Hutn.* **20**, 381–385 (1968).
[85a]Emiko Sudo, *Sci. Rep. Tohoku Univ.* **6** (2), 137–141 (1959).

hydrochloric acid. React the lead with sodium diethyldithiocarbamate for chloroform extraction and read at 435 nm.[85b] The reagent is applicable to solutions of aluminum alloys.[85c]

The formation of copper diethyldithiocarbamate is applicable to determination of lead, copper, bismuth, cadmium, zinc, manganese, and iron.[85d] The metals are prepared as the diethyldithiocarbamates and extracted into chloroform. Washing the chloroform solution with 0.01% sodium hydroxide solution removes excess reagent. On shaking with 5% copper sulfate solution, the ion exchange reactions convert all the diethyldithiocarbamates to the copper complex for reading at 436 nm.

Procedure. Mix 25 ml of sample containing up to 0.2 mg of lead with 5 ml of 50% solution of ammonium citrate. Add 20% sodium hydroxide solution to raise the pH well above 10.5, and add 5 ml of 10% potassium cyanide solution. Mix with 5 ml of 0.2% solution of sodium diethyldithiocarbamate and set aside for 10 minutes. Add 10 ml of carbon tetrachloride and shake for 1 minute. Filter the carbon tetrachloride extract into 5 ml of 5% solution of copper sulfate. Shake for 1 minute and filter the carbon tetrachloride phase. Read at 435 nm against a blank.

Cast Iron.[86] Dissolve a 1-gram sample in 10 ml of 3 : 1 hydrochloric-nitric acids. Evaporate to dryness and heat to about 130°. Cool, moisten with hydrochloric acid, and again take to 130°. Take up the residue in 30 ml of 1 : 1 hydrochloric acid; cool and filter. Wash the filter with five 5-ml portions of 1 : 1 hydrochloric acid.

Extract the bulk of the iron from the combined filtrate and washings with 50 ml of 1 : 1 methyl isobutyl ketone–isopentyl acetate. Evaporate the aqueous layer almost to dryness. Dissolve the residue by heating with 5 drops of hydrochloric acid and 25 ml of water.

To an aliquot of the sample solution, add 0.2 gram of ascorbic acid, 10 ml of 20% sodium potassium tartrate solution, and 2 drops of 0.1% phenolphthalein solution. Neutralize with ammonium hydroxide and add 7 ml more of ammonium hydroxide. Add 1 gram of potassium cyanide and 5 ml of 0.2% solution of sodium diethyldithiocarbamate.

Shake for 3 minutes with 15 ml and 10 ml of carbon tetrachloride. Filter the combined extracts and shake for 3 minutes with 10 ml of 5% copper sulfate solution. Filter the carbon tetrachloride layer and read at 435 nm against carbon tetrachloride. Correct for a blank.

Drugs.[87] HEAVY METALS. Heat 0.5 gram of solid sample or 1 gram of an injection solution with 2 ml of sulfuric acid until fumes of sulfur trioxide are

[85b] Vladimir Widtmann, *Hutn. Listy* **25**, 733–735 (1970).
[85c] C. M. Dozinel, *Metallurgia* **20**, 242–245 (1966).
[85d] A. B. Blank, N. T. Sizonenko, and A. M. Bulgakova, *Zh. Anal. Khim.* **18**, 1046–1050 (1963); V. B. Sokolovich, Yu. L. Lel'chuk, and G. A. Detkova, *Izv. Tomsk. Politekh. Inst.* **163**, 130–133 (1970).
[86] Z. Vecera and B. Bieber, *Hutn. Listy* **20**, 888–889 (1965).
[87] G. Peinhardt, F. Scholz, and J. Wattengel, *Pharmazie* **29**, 482–483 (1974).

evolved. Cool, add 2 ml of 30% hydrogen peroxide, and again heat to sulfur trioxide fumes. Repeat this step until the material is substantially colorless. Cool, take up in 5 ml of water, add 10 ml of Tris buffer solution for pH 8.5 and 2 ml of 0.025% solution of sodium diethyldithiocarbamate. Extract with 5 ml, 5 ml, and 5 ml of chloroform. Shake the combined chloroform extracts for 2 minutes with 2 ml of 0.5% solution of cupric sulfate pentahydrate and 5 ml of the Tris buffer solution. Extract the aqueous layer with 5 ml more of chloroform and combine with the previous extract. Filter and read at 430 nm.

1,5–DIPHENYLCARBOHYDRAZIDE

As an indirect method, lead is precipitated as the chromate and the anion is determined by 1,5–diphenylcarbohydrazide. The method will determine 0.01 mg of lead in the sample taken.

Procedure. *Air.*[88a] Pump through filter paper for 30 minutes at 7–10 liters/minute or for 5–8 hours at one liter/minute. Ash the paper at 500° and dissolve the ash in 5 ml of 1:25 nitric acid. Filter and evaporate to dryness. Dissolve the residue in 5 ml of a 1% solution of sodium acetate in 0.5% acetic acid. Heat and precipitate the lead by adding 1 ml of 1% potassium chromate solution. After 30 minutes, filter the precipitated lead chromate and wash with water. Dissolve the precipitate with 3 ml of hydrochloric acid, and dilute the solution and washings to 40 ml. Add 1 ml of a 1% solution of 1,5–diphenylcarbohydrazide in glacial acetic acid. Dilute to 50 ml, set aside for 30 minutes, and read at 540 nm.

EOSIN AND 1,10-PHENANTHROLINE

As a ternary complex with 1,10-phenanthroline and lead, eosin is preferable to erythrosine or rose bengal.

Procedure. *Air.* TETRAETHYL LEAD.[88b] Prepare three absorption vessels each containing 2 grams of activated carbon. Pass air at 3 liters per minute for 2 hours. Shake the carbon with 10 ml of ethanol for 30 minutes and filter. Wash with 2 ml of ethanol. Add a few crystals of iodine to the filtrate and set aside for 20 minutes. Evaporate to dryness. Add 3 ml of 6 *N* nitric acid and again evaporate to dryness. Add 2 ml of 0.01 *N* nitric acid and 15 ml of water to the residue. Set aside for 10 minutes, then adjust to pH 6.5 with 0.01 *N* potassium hydroxide and dilute to 30 ml. Add 1 ml of 0.05 *M* 1,10-phenanthroline in chloroform and 1 ml of 5 m*M*

[88a] V. Smolcic, *Arh. Hig. Rada Toksikol.* **17**, 309–316 (1966).
[88b] M. M. Tananaiko and N. S. Bilenko, *Zavod. Lab.* **42**, 761–763 (1976).

eosin. Set aside for 10 minutes. Extract the lead complex by adding 4 ml of chloroform and after 5 minutes read at 550 nm.

FLAME PHOTOMETRY

This technic[89] by reading at 405.8 nm in comparison with a standard was developed primarily for determination of lead in gasoline. The method is also applicable to manganese. A modified technic adds lithium naphthenate to the sample to read at 670.8 nm as an internal standard.[90] For routine determination without a standard, compare the intensity at 405.8 nm with that at 402.0 nm as a base value.[91] The method is also applicable to lead alloys dissolved in sulfuric acid, complexed with ammonium citrate, and neutralized with ammonium hydroxide.[92] By use of a mean of three readings, accuracy is $\pm 5\%$.

The flame photometric technic varies in lead emission with the type of gasoline but is rapid and highly accurate when unleaded base stock is available for comparison. Thus standards prepared from one base stock cannot be used with another base stock.[93] The best precision is obtained with an oxyhydrogen flame and a small-bore capillary. Isooctane is preferable to kerosene as a diluent because it gives less capillary fouling. Such fouling is avoided by frequent use of acetone. Emission at 405.8 nm is linear for as much as 25 ml of tetraethyl lead per gallon. Manganese interferes only at tetraethyl lead concentrations under 0.35 ml per gallon. For lower concentrations, the lead emission is read at 368.3 nm and corrected for background emission at 370 nm. Because of the hazard in handling concentrated solutions of tetraethyl lead to prepare standards, lead naphthenate is used with equally satisfactory results.

The scope of flame photometry can be extended and the ratio of line to background increased by selecting a limited area of the flame to be viewed when using organic solvents and an oxyhydrogen flame. In the case of lead, the height of maximum emission depends on the particular compound present.[94] Thus it is higher in the flame for the series tetraethyl lead, tetramethyl lead, lead naphthenate.

For extraction of lead as the 1,10 phenanthroline complex for flame photometry, a perchlorate solution buffered to pH 7 with 0.1 *M* ammonium acetate is desirable.[95] Nitrobenzene is the most efficient solvent and is also suitable as the aspiration solvent.

[89]Garland W. Smith and Alton K. Palmby, *Anal. Chem.* **31**, 1798–1802 (1959).
[90]Tomio Okada, Tadao Ueda, and Tsunehide Kohzuma, *Bunko Kenkyu* **4**, 30–33 (1956).
[91]Wilhelm Linne and Hans Dieter Wülfken, *Erdol Kohle, Erdgas, Petrochem.* **10**, 757–758 (1957).
[92]C. L. Chakrabarti, R. J. Magee, W. F. Pickering, and C. L. Wilson, *Talenta* **9**, 145–152 (1962).
[93]P. T. Gilbert, Jr., *ASTM Spec. Tech.* Publ. No. **116**, 77 (1952); J. H. Jordan, Jr. *Petrol. Refiner* **32**, 139 (1953); cf. V. W. Linne and H. D. Wülfken, *Erdol Kohle Erdgas Petrol*, **10**, 757 (1957).
[94]Bruce E. Buell, *Anal. Chem.* **34**, 635–640 (1962).
[95]Alfred A. Schilt, Rose L. Abraham, and John E. Martin, *ibid.* **45**, 1808–1811 (1973).

Procedures

Gasoline. The equipment requires a Beckman 2400 DU equipped with a Beckman 4200 flame photometer and a Beckman 4300 photomultiplier. Either a 6-volt storage battery or a Beckman DU power supply 23700 is suitable. The Beckman aspirator-type oxyhydrogen burner 4050 with small-bore capillaries is used. Rotameters in the oxygen and fuel gas lines establish constant flame conditions. Stainless steel tubing is preferable to plated tubing, and a Neoprene disk surrounding the burner capillary prevents ignition of fumes from the sample cup.

Prepare a standard containing the equivalent of 3 ml of tetraethyl lead per gallon equivalent to 3.171 grams of lead per gallon by dissolving chemically analyzed lead naphthenate in isooctane.

Warm the instrument until galvanometer readings become constant. Light the oxyhydrogen burner and allow 5 minutes for the flame to come to equilibrium. Set the selector switch at 0.1 and the photomultiplier to give adequate amplification. Use a slit width approximating 0.03 nm.

Aspirate a portion of the reference standard into the flame and adjust the wavelength standard to the point of maximum lead emission, 405.8 nm. Adjust the concave mirror in the burner housing to give the maximum lead emission. Remove the unburned reference standard, and clean the capillary of the burner by aspirating acetone into the flame.

Dilute 5 ml of sample to 10 ml with the reference standard and another 5 ml of sample to 10 ml with isooctane. Aspirate the sample diluted with reference standard into the flame; set the wavelength scale at 400 nm and the transmittance scale at zero. With the shutter open, adjust the dark current control until the galvanometer needle coincides with the last fiducial mark to the left of zero on the galvanometer scale, an arbitrary selection. Record the closed-shutter dark current galvanometer reading, which must not change during the subsequent steps. While the sample is still being aspirated, open the shutter and set the transmittance scale at 100%. Adjust the wavelength scale for maximum emission at 405.8 nm. Adjust the sensitivity scale until the galvanometer needle coincides with the fiducial mark. If the closed-shutter dark current galvanometer reading changes after the first adjustment of the sensitivity control, make a second approximation of the dark current and sensitivity control.

Aspirate the sample diluted with isooctane into the flame and adjust the transmittance scale until the galvanometer needle returns to the fiducial mark and record this reading.

Calculation. Tetraethyl lead in ml per gallon $= AB/(100 - B)$, where $A =$ tetraethyl lead concentration in ml per gallon of the reference standard and $B =$ transmittance scale reading.

Alloys.[96] The flame emission technic is applicable to lead, bismuth, zinc, and cadmium, in particular for 1–2 ppm in steel.

As ammonium citrate solution, dissolve 700 grams of citric acid in a minimal volume of water, add 800 ml of ammonium hydroxide, and dilute to 1500 ml.

Dissolve a 1-gram sample in hydrochloric acid, adding nitric acid if necessary. Filter and dilute to 50 ml. Add 50 ml of the ammonium citrate solution and 50 ml of 10% hydroxylamine hydrochloride solution. Adjust the pH to 6.8–7.2 by addition of ammonium hydroxide. Extract with 10 ml, 10 ml, and 10 ml of 0.1% solution of dithizone in chloroform. Combine the chloroform extracts and extract the metal ions with 10 ml of 1 : 10 hydrochloric acid. Use this extract for flame photometry with an acetylene-air flame. More than 2% of cobalt or 0.2% of copper interferes.

LEAD CHLORIDE

By reading lead chloride at selective wavelengths in the ultraviolet, agreement is obtained with electrolytic and gravimetric methods with no danger of loss of lead sulfate as in the gravimetric procedure. Wet-ashing gives in general higher values than dry-ashing. Since the absorption is affected by the acid concentration, accurate duplication in preparation of the 1 : 1 hydrochloric acid is necessary.[97]

Other heavy metals, particularly mercury and bismuth, also absorb in the ultraviolet, but their absorption in the 250–280 nm range is low and nonselective. Therefore, insofar as these elements might be present as minor components, they are corrected by the calculation. This has as a primary function correcting for iron interference. Likewise vanadium, chromium, and tin show no selective absorption. The chlorides of calcium, magnesium, aluminum, and zinc have negligible absorption above 240 nm. Of copper, manganese, cobalt, and nickel, only copper absorbs appreciably above 250 nm.

For the best results, the iron content should be lower than lead, but it can be tolerated up to twice the lead content. Absence of more than a light yellow coloration shows that it is not excessive. If excessive it can be extracted from the solution in 1 : 1 hydrochloric acid with ether without affecting the lead content. It can also be reduced to ferrous ion by titanous chloride. Copper would normally be too low to interfere.

Interferences such as copper and ferric iron can be extracted with triisooctylamine.[98] This solvent also removes less common interferences such as molybdenum, mercury, tin, cadmium, thallium, and tellurium. Copper is a special case, requiring a relatively large extractant–sample ratio and long shaking. The other metals are easily extracted. Titanium is not extracted but seldom presents a serious problem. Bismuth and ferrous iron are not extracted but do not interfere. Beer's law is followed up to 16 mg of lead per ml.

The chloride concentration must be rigidly standarized because both the maximum absorption and the wavelength for maximum absorption are affected.

[96]Ulrich Bohnstedt, *DEW Tech. Ber.* **11**, 101–105 (1971).
[97]C. Merritt, H. M. Hershenson, and L. B. Rogers, *Anal. Chem.* **25**, 572 (1953).
[98]Frank H. Ilcewicz, Richard B. Holtzman, and Henry F. Lucas, Jr., *ibid.* **36**, 1132–1135 (1964).

Procedures

Rubber Products and Compound Materials.[99] Homogenize the sample on a rubber mill and weigh a sample of 1–10 mg if more than 1% of lead is expected, more for less than that lead content. Dry-ash at 550° for about 10 minutes and at 900° for 2 minutes. Alternatively ash with 3–5 drops of perchloric acid and 1 ml of nitric acid.[100a] Cool, dissolve the ash in 2 ml of 1 : 1 hydrochloric acid, and dilute to 10 ml with the same acid.

Read at 250, 270, and 289 nm against 1 : 1 hydrochloric acid.

$$\Delta A_{270}^{S} = (A_{270} - A_{289}) - ([A_{250} - A_{289}]/2)$$

$$\%\ \text{lead} = (\Delta A_{270}^{S} \times 100)/\Delta A_{270}^{\text{Pb}} \times \text{sample conc. mg/ml.}$$

Zinc Oxide. For white rubber products, lead in the zinc oxide must be low, to avoid a brown discoloration by reaction with the sulfur in vulcanization. In contrast with the electrolytic procedure which is not accurate below 0.01% of lead, the photometric technic is accurate as low as 0.001%.

Select a sample of the maximum size allowable in 10 ml of 1 : 1 hydrochloric acid according to the following tabulation.

Normal Lead Sulfate Content (%)	Sample Weight (grams of zinc oxide)
less than 0.004	1–2
less than 0.03	0.4
less than 0.1	0.1
less than 1	0.01

Dissolve in the minimum possible amount of hydrochloric acid, completing the solution dropwise. Dilute to 5 ml with water, then to 10 ml with 1 : 1 hydrochloric acid. Complete as for rubber products from "Read at 250, 270,. . . ."

Bone. Ash in platinum at 620°. Dissolve a 10-gram sample in 11 *N* hydrochloric acid and dilute with that acid to 50 ml. Extract a 10-ml aliquot of the sample solution with 5 ml, 5 ml, and 5 ml of 20% solution of triisooctylamine in chloroform by shaking for 10 minutes. Wash the sample solution with 5 ml, 5 ml, and 5 ml of chloroform. Discard the extracts and washings and read at 271 nm. Subtract a blank.

Polyvinyl Chloride.[100b] Extract with hydrochloric acid and read at 270 nm.

[99] K. E. Kress, *ibid.* **29**, 803–807 (1957).
[100a] K. E. Kress, *ibid.* **27**, 1618 (1955).
[100b] S. Grossman and J. Haslam, *J. Appl. Chem. (London)* **7**, 636–644 (1957).

LEAD IODIDE

The formation of colloidal lead iodide is an appropriate measure of lead in the presence of less than 3% of chromium, molybdenum, and nickel and less than 1% of vanadium.[101a] The colloidal lead iodide in 50% phosphoric acid stabilized by gum arabic is read at 410 nm.[101b]

Procedure. *Plain Carbon and Low-Alloy Steels.* As a sample, take 2 grams if 0.025–0.25% of lead is present or 1 gram if 0.25–0.5% is present. Dissolve in 8.5 ml of perchloric acid and 11.5 ml of water. If tellurium is present, add 2 ml of 10% solution of stannous chloride dihydrate in 1:9 hydrochloric acid. Boil for 3 minutes and cool. Reduce the copper by addition of 0.5 gram of reduced iron. Maintain at 20° for 5 minutes; dilute to 50 ml and filter.

Prepare a mixture of 25 ml of phosphoric acid, a 5-ml aliquot of the sample, 1 ml of 1% gelatin solution, and 10 ml of water. Cool to about 15° and add 5 ml of 2.5% solution of sodium iodide and 0.5 ml of the stannous chloride reagent defined in the previous paragraph. Dilute to 50 ml and after 10 minutes read at 410 nm against a blank from which sodium iodide was omitted.

4-(2-PYRIDYLAZO)RESORCINOL

When lead ion reacts with 4-(2-pyridylazo)resorcinol, commonly referred to as PAR, 0–5 mg per ml can be read at 520 nm.[102] The corresponding complex with cobalt absorbs at 510 nm, that with uranium at 530 nm. Beer's law applies for 0.01–0.2 mg of lead per ml. The color is stable for 24 hours, provided not less than 8 equivalents of the reagent is present.

For lead in food additives, ash the dried sample at 475°. Extract the ash with hot 1:1 hydrochloric acid.[103] Next form the copper diethyldithiocarbamate complex and extract with chloroform. Read at 436 nm for copper. Destroy the complex and add potassium iodide. Extract the lead with 4-(2-pyridylazo)resorcinol in methyl isobutyl ketone and read at 530 nm.

Procedures

Iron and Steel.[104] Dissolve a 1-gram sample with 10 ml of 1:1 nitric acid and 10 ml of perchloric acid. Evaporate nearly to dryness. Take up the residue in 20 ml of water and 1 ml of hydrochloric acid. Extract the ferric salt with 5 ml and 5 ml of isopentyl acetate. Add 1.5 gram of potassium iodide and extract the lead iodide with 5 ml and 5 ml of methyl isobutyl ketone.

[101a] O. G. Koch, *Anal. Chim. Acta* **62**, 462–463 (1972).
[101b] Imre Bozsai and Endre Kopocsy, *Magy. Kem. Foly.* **62**, 12–15 (1956).
[102] F. H. Pollard, P. Hanson, and W. J. Geary, *Anal. Chim. Acta* **20**, 26–31 (1950).
[103] E. Kroeller, *Deut. Lebensm. Rundsch.* **66**, 190–192 (1970).
[104] R. M. Dagnall, T. S. West, and P. Young, *Talenta* **12**, 583–592 (1965).

Reextract the lead with 10 ml of an ammonia–ammonium chloride buffer solution containing 10% of potassium cyanide, for pH 10. Add 1 ml of 0.01% solution of 4-(2-pyridylazo)resorcinol to form a 1 : 1 complex with lead, and read at 520 nm.

Lead Sulfide Film.[105] Decompose a 0.3-gram sample by heating with 10 ml of 1 : 120 hydrochloric acid and boil off the hydrogen sulfide. Pass the solution through a column of KU-2 cation active resin in the hydrogen form and wash the column with water. Elute the lead from the column with 3 ml of 1 : 3 hydrochloric acid and wash through with water. Neutralize the eluate with ammonium hydroxide and dilute to 10 ml. Add 5 ml of an ammonia–ammonium chloride buffer solution containing 10% of potassium cyanide, for pH 10. Add 1 ml of 0.01% solution of PAR, dilute to 25 ml, and read at 520 nm.

SULFARSAZAN

Sulfarsazan is 5-nitro-2-[3-(4-*p*-sulfophenylazophenyl)-1-triazeno] benzene. The complex with lead in sodium tetraborate solution absorbs at 420 nm.[106] Copper is masked by urea, zinc by ferrocyanide, and iron by ammonium oxalate. Large amounts of iron are precipitated as the hydroxide. If necessary for separation from interfering ions, the lead may have been extracted by dithizone and reextracted with very dilute hydrochloric acid.[107]

Procedures

Acidify 5 ml of neutral sample soultion with 2 drops of 0.1 *N* hydrochloric acid. Add 2 ml of 1% potassium ferrocyanide solution and 3 ml of 5% solution of sodium tetraborate. At this point the pH should be about 8. Add 2 ml of a 0.05% solution of sulfarsazan and dilute to 20 ml. Read at 420 nm against a blank.

Titanium Dioxide.[108] Dissolve a 1-gram sample in 20 ml of sulfuric acid and 15 grams of ammonium sulfate by heating. Cool and dilute to about 50 ml. To mask titanium, aluminum, iron, and chromium, add 20 ml of 50% potassium citrate solution. Add a couple of drops of phenolphthalein indicator solution and neutralize with 1 : 3 ammonium hydroxide. Add 10 ml of 1% sodium diethyldithiocarbamate solution and 10 ml of chloroform. Shake for 3 minutes and remove the chloroform extract. Repeat the extraction with 5 ml portions of chloroform as long as color is extracted. Evaporate the combined extracts to about 5 ml. Add 20 ml of perchloric acid and 2 ml of 30% hydrogen peroxide. Evaporate to dryness and

[105] A. N. Stepanova, M. I. Bulatov, and V. B. Aleskovskii, *Izv. Vyssh. Ucheb, Zaved, Khim. Khim. Tekhnol.* **15**, 33–37 (1972).
[106] E. A. Kyuregyan, *Izv. Akad. Nauk Arm. SSR, Geol. Georgr. Nauk* **16**, 163–166 (1963); cf. A. M. Lukin, and G. S. Petrova, *Zh. Anal. Khim.* **15**, 295–298 (1960).
[107] A. I. Markova, *Zh. Anal. Khim.* **17**, 952–954 (1962).
[108] V. A. Malevannyi and A. V. Zholnin, *Khim. Volokna* **1967**, 72–73.

ignite. Take up the residue in 10 ml of 1:5 nitric acid and evaporate to about 5 ml. Add 2 ml of 10% thiourea solution and to mask copper, cadmium, and zinc add 1 ml of potassium ferrocyanide solution. Dilute to 25 ml with 10% solution of sodium tetraborate.

Filter part of the contents of the flask and take a 5-ml aliquot. Raise to pH 9–9.6 by dropwise addition of 1:3 ammonium hydroxide and add 0.3 ml of 0.05% solution of sulfarsazan. Dilute to 7 ml with 10% sodium tetraborate solution and read at 546 nm against a blank.

Water.[109] LEAD OR ZINC. Mix 200 ml of sample with 5 ml of 30% hydrogen peroxide and heat for 10 minutes. Add hydrochloric acid until acid to Congo red paper and boil for 3 minutes. Cool and add 10-ml increments of a saturated solution of sodium acetate until it turns Congo red paper blue. Add an additional 5 ml. Add 20 ml of 3% solution of sodium diethyldithiocarbamate. Extract with 20 ml and 20 ml of 2:1 toluene–isoamyl alcohol, shaking for 30 seconds with each portion. Combine the organic layers and extract lead and zinc with 20 ml and 20 ml of 2 *N* hydrochloric acid. Evaporate the acid extract to about 1 ml and dilute to 100 ml. Take a 10-ml aliquot and add 4 drops of saturated solution of ammonium molybdate to mask manganese. Add saturated ammonium chloride solution until Congo red paper is turned from blue to red and 0.5 ml in addition. Add 2 ml of 0.05% sulfarsazan solution in 0.05 *M* sodium tetraborate. For reading lead, add 0.4 ml of 1% solution of potassium cyanide. Read at 500 nm for lead or zinc.

SULFIDE

The go–no go use of colloidal lead sulfide as a check on maximum permissible amounts of lead is greatly refined to become a photometric method for lead.

In analysis of platinum metals lead is separated from other base metals and in ammonium acetate solution determined by the color of lead sulfide.[110]

For lead in free-cutting steel,[111] dissolve a sample in tartaric acid solution. Add excess sodium hydroxide solution, followed by potassium cyanide solution, to mask interfering ions. Then add sodium sulfide in aqueous glycerol to form colloidal lead sulfide for reading at 546 nm.

Concentrate lead and copper in natural waters by passing through a column of cation exchange resin at a rate of 1 liter/hour.[112] Desorb the ions by washing with 1:2 hydrochloric acid. Concentrate the effluent for reading lead as the sulfide and copper as the dithizonate.

[109]M. A. Yagodnitsyn, *Gig. Sanit.* **1970** (11), 62–63.
[110]V. G. Levian and T. P. Yufa, *Anal. Blagorod. Metal., Akad. Nauk SSSR, Inst. Obshch. Neorg. Khim. U.S. Kurnakova* **1959**, 65–69.
[111]K. H. Koch, K. Ohls, and G. Riemer, *Z. Anal. Chem.* **237**, 167–172 (1968).
[112]V. B. Aleskovskii, R. I. Libina, and A. D. Miller, *Tr. Leningr. Tekhnol. Inst. Lensoveta* **48**, 5–11 (1958).

Procedures

Paint.[113] Ash a sample expected to contain 0.1–0.3 mg of lead. Extract the ash with 1:6 nitric acid and neutralize the extract with 20% sodium hydroxide to pH 8.3. Add 1 ml of 2% sodium sulfide solution, dilute to a known volume, and read at 546 nm.

Edible Oils and Fats.[114] If the lead, copper, iron, and manganese do not exceed 50 ppm, dissolve a 100-gram sample in 150 ml of petroleum ether. Add 100 ml of 1:4 nitric acid and shake at 120 cycles/minute for 1 hour. Filter the aqueous layer, evaporate to dryness, and ash the residue. Moisten it several times with 1:4 nitric acid and ignite each time. Take up the residue in 5 ml of 1:4 nitric acid. Add 20% potassium hydroxide solution to distinct alkalinity. Add 1 ml of 2% sodium sulfide solution, dilute to a known volume, and read at 546 nm.

For higher concentrations of heavy metals, reflux 5–10 grams of sample with 10 ml of 1:4 nitric acid for 40–45 minutes, using an air condenser. Cool and shake with 10 ml of petroleum ether. Filter the aqueous layer and wash the petroleum ether with water. Dilute the filtrate and washings to a known volume to take an aliquot. Evaporate to dryness and ash the residue. Proceed as in the previous paragraph from "Moisten it several times with...."

Canned Foods.[115] Ash a 25-gram sample. Take up the ash in 5 ml of 1:1 hydrochloric acid and evaporate to dryness. Dissolve in 2 ml of 1:9 hydrochloric acid and add 3 ml of water. Filter and wash the filter with 20 ml of water. Add 8% sodium hydroxide solution until a white precipitate begins to appear and clarify with 2 ml of 1:9 sulfuric acid. Add 2 ml of a solution containing 0.5% of calcium chloride and 0.5% of barium chloride. Centrifuge to separate lead, calcium, and barium sulfates, and decant the supernatant layer. Wash the sulfates with 3 ml, 3 ml, and 3 ml of 5% hydrogen peroxide. Wash the solids free from sulfuric acid with 50% ethanol.

Take up the solids with 2 ml of saturated potassium hydroxide solution by heating at 100° for 5 minutes. Dilute with water to 10 ml. Centrifuge and decant the supernatant layer. Repeat the extraction of the solids with 1 ml of the potassium hydroxide solution. Add 1 ml of 2% sodium sulfide solution to the combined extracts. Dilute to 25 ml and read at 546 nm. Subtract a blank.

Urine.[116a] Mix 100 ml of urine with 30 ml of 30% hydrogen peroxide and 3 ml of nitric acid. Evaporate to dryness and ash. Take up the residue in 5 ml of 25% acetic acid and 5 ml of 25% ammonium acetate solution, grinding any insoluble matter. Add 5 drops of nitric acid and 4 ml of a hot 2% solution of gelatin in 25%

[113]E. Hoffmann, *Z. Anal. Chem.* **208**, 423–425 (1965).

[114]C. G. Macarovici, V. Farcasan, G. Schmidt, V. Bota, M. Macarovici, A. Dorutiu, I. Pirvu, and E. Tesler, *Stud. Univ. Babes-Bolyai. Cluj. Chim.* **1961**, 103–108.

[115]B. Ya. Krasnaya, *Konserv. Ovoshchesush. Prom.* **1961**, 35–36.

[116a]B. I. Leonov, *Tr. Kishinev. Med. Inst.* **6**, 267–270 (1957).

acetic acid. Mix with 1 ml of 20% formaldehyde solution and 3 drops of 10% sodium sulfide solution. Dilute to 2.5 ml with water and read at 546 nm.

TETRAPHENYLPORPHINETRISULFONIC ACID

In alkaline solution this reagent complexes with lead to form an absorption band well separated from that of the reagent. At pH 9.8–10.5 with 10-fold reagent concentration, reaction is complete in 1 minute at 70°. The complexes with cupric, cadmium, ferrous, manganous, mercuric, and indium ions interfere. Indium and manganous ions can be masked by cyanide ion. Beer's law is followed for 0.05–5 μg of lead per liter.

Procedure.[116b] To 20 ml of sample solution containing 1–10 μg of lead, add 0.5 ml of 20% potassium cyanide solution and 1 ml of borate buffer solution for pH 10.2. Heat to 70°, add 2.5 ml of 0.1 m*M* tetraphenylporphinetrisulfonic acid, and maintain at 70° for 5 minutes. Cool, dilute to 25 ml, and read at 464 nm.

THIOMALIC ACID

When lead ion is reacted with thiomalic acid, which is mercaptosuccinic acid, Beer's law applies over the range of 0.5–40 ppm. Cations that precipitate in alkaline solution interfere. The technic is more rapid but less sensitive than that with dithizone.

Procedure.[117] Mix a sample containing 0.1 mg of lead with 20 ml of 0.1% thiomalic acid solution and 20 ml of 0.1 *M* borate buffer solution for pH 10.5. Dilute to 100 ml and read at 305 nm.

To a sample solution containing about 0.1 mg of lead, add 5 ml of 10% sodium cyanide solution, 5 ml of 5% sodium tartrate solution, and 1 ml of 1% solution of **ammonium phenazo-4-carbodithioate**. Extract with 10 ml of chloroform, and convert the lead complex to a nickel compound by shaking the organic layer with a solution containing 0.29% of nickel nitrate hexahydrate and 0.282% of sodium potassium tartrate. Read at 575 nm. Thallium and bismuth give the same reaction,

[116b]Jun-ichi Itoh, Makoto Yamahira, Takao Yotsuyanagi, and Kazuo Aomura, *Jap. Anal.* **25**, 781–784 (1976).
[117]Yasuhisa Hayashi, Shigeo Hara, and Isao Muraki, *Bull. Chem. Soc. Jap.* **38**, 1214–1216 (1965).

but there is no interference from nickel, silver, copper, cobalt, mercury, zinc, or cadmium.

For lead in salts of zinc, cadmium, or bismuth dissolve 0.5 gram in water, add 0.25 ml of 0.2 N nitric acid, and dilute to 50 ml.[118] Mix a 2.5-ml aliquot with 10 ml of 0.1% solution of 9,9′-**bixanthene** solution in 1,4-dioxan and dilute to 25 ml with the same solvent. Irradiate with an ultraviolet lamp and read the fluorescence at 510 nm. Beer's law is followed for 5–80 μg/ml and readings at 0.5–100 μg/ml are reliable.

The 1:1 complex of lead with **2-(5-bromo-2-pyridylazo)-5-diethylaminophenol** has a maximum absorption at 575 nm.[119] It conforms to Beer's law for 1–4.5 μg/ml. The complex may be extracted into chloroform. At much lesser sensitivity, the antipyridylazo complex can be read at 520 nm or the 2-(2-thioazolylazo)- analogue at 575 nm.

The blue 1:1 complex of **bromopyrogallol red** with lead at pH 5.5 is read at 630 nm.[120] It is stable for 12 hours and obeys Beer's law for 0.8–14.6 ppm. There is interference by uranyl ion, copper, lanthanum, cerium, and thorium.

Lead precipitated as the **chromate** can be dissolved in 1:2 hydrochloric acid and read at 267.5 nm.[121] As an alternative, the chromate is determined by *s*-diphenylcarbizide reading at 540 nm as shown earlier in this chapter.

The complexes of **O, O′-dichlorodithizone**, which is 1,5-bis-(2-chlorophenyl)-3-mercaptoformazan, with lead, zinc, cadmium, and mercury are extractable into carbon tetrachloride for reading.[122]

For lead in mM samples add **diethylenetriaminepentaacetic acid**[123] in a solution buffered with acetate at pH 2–5; read at 244 nm.

As a kinetic method, the rate of oxidation of **3,4-dihydroxyazobenzene** by persulfate in a borate buffer for pH 8.15 will determine 10–30 ng of lead per ml with accuracy to better than 15%.[124] Only cupric, silver, ferric, and chromic ions interfere.

Lead gives an intense blue to purple with **6,13-dihydroxy-1,4,8,11-pentacene-quinone-2,9-disulfonic acid.**[125] For determination, mix 1 ml of sample solution containing 2–20 μg of lead with 1 ml of 5% sodium acetate solution, 1 ml of water, and 3 drops of 0.05% solution of the reagent. Read at 630 nm against a blank. Manganese, ferric ion, barium, calcium, and magnesium interfere.

For determination of lead with **diphenylcarbazone**, extract into a 0.05% solution of dithizone in benzene.[126] Filter the benzene extract through cotton. Take an aliquot equivalent to not more than 8 mg of lead. Shake with 4 ml of 1:99 nitric acid and centrifuge. To 3 ml of the aqueous layer, add 0.35 ml of 20% potassium cyanide solution. Then add 3 ml of 0.02% solution of diphenylcarbazone in xylene. Shake vigorously and centrifuge. If bismuth is absent, read at 525 nm after a

[118] P. A. Shiryaev and V. I. Rigin, *Zavod. Lab.* **41**, 917–918 (1975).
[119] S. I. Gusev and E. M. Nikolaeva, *Zh. Anal. Khim.* **24**, 1674–1678 (1969).
[120] S. C. Dhupar, K. C. Srivastava, and Samir K. Banerji, *J. Chin. Chem. Soc. (Taipei)* **20**, 145–150 (1973).
[121] Tonu Nozaka and Nuoru Veno, *J. Jap. Chem. Soc.* **70**, 484–487 (1958); A. I. Krupkin, *Tr. Kom. Akad. Nauk SSSR, Inst. Geokhim. Anal. Khim.* **8**, 204–209 (1958).
[122] R. S. Ramakrishna and M. Fernandopulle, *Anal. Chim. Acta* **41**, 35–41 (1968).
[123] P. S. Edgaonkar, M. Atchayya, and P. R. Subbaraman, *Indian J. Chem.* **13**, 400–402 (1975).
[124] E. Vasinskiene and S. Kalesnikaite, *Zh. Anal. Khim.* **25**, 87–90 (1970).
[125] H. Junek and H. Wittmann, *Mikrochim. Acta* **1962**, 114–118.
[126] N. Trinder, *Analyst* **91**, 587–590 (1966).

standardized interval of time. If bismuth is present, read 2 hours after adding the color-producing reagent.

For reading of lead by **fluorescein**, an appropriate concentration is 0.1–0.6 ppm in 3 : 10 hydrochloric acid 0.8 *M* with potassium chloride.[127] When activated at 270 nm, the fluorescence at 480 nm persists for at least 15 minutes. Interference is caused by a fiftyfold concentration of bismuth, tetravalent chromium, cupric ion, ferric ion, hexavalent molybdenum, thallous ion, pentavalent vanadium, ascorbic acid, or sulfite. Chromium, iron, and vanadium can be masked as well as a fivefold concentration of molybdenum, a tenfold concentration of copper, and a thirtyfold concentration of bismuth.

Both diethyl lead and dimethyl lead give violet complexes with **glyoxal bis-(2-hydroxyanil)** when extracted into chloroform.[128] Lead and trialkyl lead ions in moderation do not interfere. For diethyl lead, treat 50 ml of sample with 0.5 ml of *M* ammonium citrate, 0.2 gram of ascorbic acid, and 2 drops of 0.1% solution of thymolphthalein in ethanol. Adjust to pH 9 with ammonium hydroxide or nitric acid and add 10 ml of 0.1 *M* ammonium chloride buffer solution for pH 8.8. Shake for 1 minute with 10 ml of chloroform and 1.5 ml of 0.01 *M* glyoxal bis-(2-hydroxyanil) in methanol. Read the organic layer at 680 nm. For dimethyl lead the technic is identical.

Triethyl lead ion complexes with **1-hydroxy-4-(4-nitrophenylazo)-2-naphthoate**[129a] to give a product extractable with chloroform. The optimum pH for extraction from 1% sodium chloride solution is 8.1–8.3; the maximum absorption is at 440 nm. For about 0.06 mg of triethyl lead ions, 2 ml of 0.01 *M* ethylenediamine-*N N'*-bis-2-hydroxyphenylacetic acid masks 0.1 mg of dimethyl lead and diethyl lead ions and 1.8 mg of lead ion. Cupric and ferric ions are masked by 2 ml of 0.01 *M* 1,2-diaminocyclohexane-*N N N' N'*-tetraacetate. The method will determine 0.2 ppm of triethyl lead.

Lead forms a 1 : 1 complex with **2-(2-hydroxyphenyl)benzothiazoline**[129b] which at pH 10.8 is extracted into toluene for reading in 0.03–0.3 m*M* at 410 nm.

Lead is determinable by **methyl *N*-[α-(8-hydroxyquinoline-7-yl)benzyl]anthranilate**, which is 8-hydroxy-7-[D-(2-methoxycarbonylanilino)benzyl]quinoline.[129c] It is extracted from a solution at about pH 12 by the reagent in chloroform and read at 410 nm. The method will determine 0.2 micromole of lead in 25 ml. Alkali metals and calcium do not interfere. Citrate masks aluminum and indium. Cyanide masks copper, zinc, cadmium, cobalt, and nickel. Formaldoxime masks manganese and ferric ions. The presence of magnesium or thallium necessitates reextraction procedures. A high concentration of tartrate is required to prevent extraction of bismuth, and this greatly reduces the sensitivity.

Methylthymol blue forms a 1 : 1 complex with lead which at pH 5.8 in a hexamine–nitric acid buffer absorbs at 600 nm.[130] The most suitable concentration is 0.8–9 ppm. Potassium fluoride prevents interference by aluminum, thallium, and

[127] G. F. Kirkbright and C. G. Saw, *Talenta* **15**, 570–574 (1968).
[128] Shin-Ichi Imura, Kiyoshi Fututaka, and Toshihiko Kawaguchi, *Jap. Anal.* **18**, 1008–1013 (1969).
[129a] Shin-Ichi Imura, Hazime Aoki, and Toshi-nari Sakai, *Jap. Anal.* **20**, 704–708 (1971).
[129b] E. Uhlemann and V. Pohl, *Anal. Chim. Acta* **65**, 319–328 (1973).
[129c] Gerhard Roebisch, *Anal. Chim. Acta* **47**, 539–546 (1969).
[130] K. C. Srivastava and S. K. Banerji, *Chim. Anal. (Paris)* **51**, 28–31 (1969).

zirconium. Ascorbic acid avoids interference by ferric, thallic, and palladous ions. Traces of bismuth, nitrilotriacetic acid, or EDTA interfere.

As a modified **ninhydrin**, mix 1 ml of 0.1 *M* ascorbic acid, 50 ml of a citrate-phosphate buffer for pH 5.0, 1 ml of 1 : 19 ammonium hydroxide, and 10 ml of 0.1 *M* ninhydrin.[131] Heat for 10 minutes at 100° and cool. Filter the precipitate and air dry. Recrystallize from 50% propanol. A complex formed by lead with a solution of this reagent containing a fivefold excess is readily extracted by chloroform for reading at 550–560 nm in the range of 1–5 μg/ml.

Lead **perchlorate** in *N* perchloric acid has a maximum absorption at 208 nm which can be read with accuracy to about 5%.[132] Ferric, bismuth, antimony, and tin ions cause a positive error. The absorption is attributed to a hydrated lead ion. In the same solution bismuth can be read at 222 nm and ferric ion at 240 nm.

When read at 302 nm, lead perchlorate in 4.3 *N* hydrobromic acid conforms to Beer's law up to 15 ppm and remains constant for 5 hours.[133] There is no interference by cobalt, manganese, nickel, aluminum, or arsenic. Ferric ion, bismuth, stannic ion, and pentavalent antimony interfere.

When a lead solution at pH 6–8 is passed through a column of powdered **iron**, the effluent contains an equivalent of ferrous ion.[134] This can be read at 512 nm with 1 : 10 phenanthroline or oxidized and determined as ferric ion. The same reaction is given by ions of cobalt, nickel, copper, silver, or mercury. Sodium, magnesium, and manganese do not interfere, but carbon dioxide and strong oxidizing agents do.

The complex of lead with **1,10-phenanthroline**[135] adds tetrabromo-, tetraiodo-, or dichlorotetraiodofluorescein, extractable at pH 9 with chloroform. The maximum absorption is at 540 nm or in 1 : 1 chloroform-acetone fluorescence at 580 nm. Fluorescence will determine 0.05–10 μg in 10 ml, one order of magnitude more sensitive than spectrophotometry.

Lead produces a stable reproducible color with **1-(2-pyridylazo)-2-naphthol** in methanol at pH under 6.[136] The reagent is added at 0.004 *M* in methanol and should not be more than 50 times the concentration of lead. Reading is at 530 nm and Beer's law applies at 0.8–6.6 μg of lead per ml.

Lead and **quinalizarin** form a blue-black at pH 6.2–6.4 which can be read at 533 nm.[137] At pH 8.2–8.8 the color is blue and is read at 655 nm.

Ruhemann's purple, [2-(3-hydroxy-1-oxoinden-2-ylimino)indan-1,3-dione] can be used for determination of lead, mercury, and bismuth.[138] Mix the sample solution with an aqueous solution of the sodium salt of the reagent. Add a buffer solution for lead at pH 4.5, for bismuth at 4.3, for mercury at 5.5. Extract with chloroform and read lead or bismuth at 560 nm or mercury at 530 nm.

Lead in zinc **sulfate**[139] is determined in the ultraviolet range. Precipitate the lead

[131] A. Ya. Burkina and P. L. Senov, *Farm. Mosk.* **20** (6), 57–62 (1971).
[132] Masayoshi Ishibashi, Yuroku Yamamoto, and Kazuo Hiiro, *Jap. Anal.* **7**, 582–585 (1958); *Bull. Inst. Chem. Res. Kyoto Univ.* **36**, 24–29 (1958).
[133] Yuroku Yamamoto, *J. Chem. Soc. Jap.* **80**, 1426–1428 (1959).
[134] Eugeniusz Klaczko, *Chem. Anal. (Warsaw)* **16**, 1291–1298 (1971).
[135] D. N. Lisitsyna and D. P. Shcherbov, *Zh. Anal. Khim.* **28**, 1203–1205 (1973).
[136] D. Degoiu, A. Kriza and L. Baloiu, *Anal. Univ. Buc., Ser. Stiint. Nat. Chim.* **13**, 165–174 (1964).
[137] A. F. Nemirovskaya, *Tr. Novocherk. Politekh. Inst.* **143**, 45–53 (1963).
[138] A. Ya. Burkina and P. L. Senov, *Zh. Anal. Khim.* **27**, 473–478 (1972).
[139] M. J. Maurice and S. M. Ploeger, *Z. Anal. Chem.* **179**, 246–258 (1961).

as the sulfate from the solution of a 50-gram sample. Centrifuge and decant from the lead sulfate. Dissolve the precipitate in glacial acetic acid, make up as 1 : 1 hydrochloric acid, and read at 270 nm.

Lead ion forms a brown-yellow with **thioacetamide** with a maximum at pH 10.[140] The full color develops in 5 minutes and can be read at 490 nm above 1 ppm. Iron interference is avoided by addition of sodium potassium tartrate. High concentrations of sodium and potassium ion have little effect, but ammonium salts can modify the color to yellow.

Lead ion is extracted as the complex with **thio-2-thenoyltrifluoroacetone**, which is 1, 1, 1-trifluoro-4-mercapto-4-(2-thienyl)but-3-ene-2-one. Mix a sample solution with 10 ml of a buffer solution for pH 6.5 and dilute to 25 ml with water.[141] Shake for 10 minutes with 10 ml of 2 m*M* thio-2-thenoyltrifluoroacetone in carbon tetrachloride. Read the extract at 480 nm against a reagent blank. The extract may contain 1–40 ppm lead and is stable for 60 hours. Trivalent bismuth, stannous ion, titanic ion, telluride ion, EDTA, and oxalate must be absent. Colored complexes are formed with mercury, cadmium, palladium, cobalt, nickel, cerium, and the noble metals; therefore they may interfere.

Lead at 0.1–2.5 μg/ml is read in the **ultraviolet** range at 270 nm.[142] For a water sample, 2.25% of sodium chloride is added to reduce background absorption. The absorption is linear for 0.1–2.5 μg/ml.

The complex of lead with **xylenol orange**[143a] is 1 : 1 at pH 2.5–4.5. The maximum absorption is at 580 nm. With ammonium fluoride, potassium ferrocyanide, and ascorbic acid present, it is best to use pH 4.5 to avoid interference of small amounts of aluminum, bismuth, copper, zinc, cadmium, mercury, tin, beryllium, iron, cobalt, and nickel.

To determine lead in water supplies by xylenol orange, the technic requires isolation of copper and zinc.[143b] First adjust the sample to 0.1 *N* with hydrochloric acid and extract copper with dithizone in carbon tetrachloride. Reextract the copper from the dithizonate with 1 : 1 hydrochloric acid. Adjust the acid extract to pH 5.6, add xylenol orange solution, and read at 580 nm.

After extraction of the copper, adjust the sample solution to pH 8.5 and extract lead and zinc with dithizone in carbon tetrachloride. Reextract the lead and zinc from their dithizonates with 0.05 *N* hydrochloric acid. Divide the extract into two aliquots. To one for lead, add potassium ferrocyanide to mask the zinc; add xylenol orange solution, and read at 572 nm. In the other aliquot, mask lead with sodium thiosulfate, add xylenol orange solution and read at 570 nm. The error is less than 25% for each.

[140]G. G. Karanovich, *Tr. Vses. Nauch.-Issled. Inst. Khim. React.* **1959** (23); *Sb. Statei Vses. Nauchn.-Issled. Inst. Khim. Reak.* **1959** (23), 96–101.

[141]S. B. Akki and S. M. Khopkar, *Bull. Chem. Soc. Jap.* **45**, 167–170 (1972).

[142]Sera Skurnik-Sarg, M. Zidon, I. Zak, and Y. Cohen, *Isr. J. Chem.* **8**, 545–559 (1970).

[143a]P. V. Marchenko, *Ukr. Khim. Zh.* **30**, 224–227 (1964); cf. A. Cabrera Martin, M. H. Fernandez, J. L. Perel Fernandez, and F. Burriel-Marti, *Quim. Anal.* **28**, 38–42 (1974).

[143b]T. V. Gurkina and A. M. Igoshin, *Zh. Anal. Khim.* **20**, 778–781 (1965).

CHAPTER TWO
THALLIUM

Thallium resembles gallium and indium in many properties. It is commonly associated with lead, zinc, iron, tellurium, and alkalies. It forms mono- and trivalent ions, and like many other ions, the one of higher valence liberates iodine from hydriodic acid. The methods of determination are numerous, but the trend in recent years has been toward formation of a complex of the trivalent ion with a dyestuff that is extractable into an organic solvent for reading. Rosaniline derivatives represented by methyl violet and crystal violet are typical. Thallic ion can be read in the ultraviolet and the characteristic green of thallium determined by the flame photometer.

Illustrative of its activity amounts under 20 mg of indium, beryllium, hexavalent uranium, iron, zinc, gallium, cadmium, copper, nickel, lead, aluminum, chromium, titanium, manganese, molybdenum, tungsten, and vanadium can be separated from 25 ml of solution at pH 2.5 containing 0.1–1 mg of thallous ion per ml by passage through a 150 × 16 mm column of anion exchange resin in carbonate form. After the above-mentioned ions have passed through the column, the thallous ion is eluted with 50 ml of water.[1a]

For preconcentration of 0.1–4 μg of thallic ion in 500 ml of solution, add 25 mg of ferric ion.[1b] Adjust to pH 8.5 with ammonium hydroxide, filter and dissolve in 30 ml of 2 N sulfuric acid.

Procedure. *Isolation by Dithizone.*[2] Mix 50 ml of neutral inorganic solution with a few drops of 10% hydroxylamine hydrochloride solution to ensure the thallous condition. Add 0.5 gram of ammonium citrate and 0.5 gram of potassium cyanide.

[1a] V. D. Eristavi and M. G. Mgaloblishvii, *Zh. Anal. Khim.* **28**, 375–377 (1973).
[1b] Emiko Kaneko, *Jap. Anal.* **25**, 299–301 (1976).
[2] L. A. Haddock, *Analyst* **60**, 394–399 (1935).

Add 1:1 ammonium hydroxide to raise to pH higher than 9. Extract with 15-ml portions of 0.1% solution of dithizone in chloroform as long as successive extracts are not pure green. Wash the combined extracts, which contain the thallium, with an equal volume of water. Evaporate the chloroform extracts to dryness. Add 1 ml of concentrated sulfuric acid and take to dryness to destroy the dithizone. During this process add 30% hydrogen peroxide dropwise. Cool and take up in 20 ml of water as a thallium sample freed from heavy metals.

ACRIDINE YELLOW

After oxidation of thallous ion to thallic with chlorine, the 1:1 complex with acridine yellow, which is 3,6-diamino-2,7-dimethylacridine hydrochloride, is extracted with butyl acetate or with dichloroethane. The absorption at 0.1–5 μg of thallium per ml of extract conforms to Beer's law. Nitrate, auric, and antimonic ions interfere.

Procedure. *Aluminosilicate Minerals.*[3] Decompose 1 gram of sample with excess hydrofluoric acid and evaporate to dryness. Add small portions of hydrochloric acid and evaporate to dryness several times to remove fluorides. Dissolve in 25 ml of 1:3 hydrochloric acid. Add 1 ml of 10% solution of sodium nitrite and set aside for 5 minutes. Add 0.5 ml of saturated solution of urea to destroy excess nitrite ion. Shake for 2 minutes with 25 ml of butyl acetate. Wash the organic layer with 5 ml of 1:3 hydrochloric acid; then evaporate the solvent. Take up the residue in 0.25 *N* hydrochloric acid and dilute to 25 ml with that acid. Mix a 5-ml aliquot with 1 ml of 0.1 m*M* acridine yellow and dilute to 10 ml with 0.25 *N* hydrochloric acid. The pH should be 0.4–1. Extract for 2 minutes with 6 ml of dichloroethane. Read the extract at 495 nm.

ACRIFLAVINE

The complex of acriflavine and thallium chloride is extracted for reading by fluorescence at 2.5–500 μg/liter. The intensity is constant for 5 hours. Interference by mercuric, auric, and antimonic ions must be avoided by a preliminary extraction of thallium from 3 *N* hydrochloric acid with butyl acetate. There is also interference by cupric, ferric, aluminum, and gallium ions.

Procedure. To 5 ml of solution[4] containing 0.1–5 μg of thallium, add 1 ml of 0.025% solution of acriflavine and dilute to 10 ml with 1:5 hydrochloric acid. Adjust to pH 1.5 and shake for 1 minute with 6 ml of dichloroethane. Dry the

[3] L. A. Grigoryan, V. Zh. Artaruni, and V. M. Tarayan, *Arm. Khim. Zh.* **27**, 188–192 (1974).
[4] V. M. Tarayan, L. A. Grigoryan, F. V. Mirzoyan, and *Zh.* V. Sarkisyan, *ibid.* **26**, 996–1000 (1973).

organic phase with sodium sulfate and read the fluorescence at 490 nm as activated by a mercury vapor lamp.

BRILLIANT GREEN

The reaction with brilliant green (basic green 1) is more sensitive than that with methyl violet.[5] More than 1 ppm of antimony or 50 ppm of iron interferes, as do phenacetin, acetylsalicylic acid, and phenazone.

Procedures

Indium.[6] Dissolve 2 grams of sample in 25 ml of 1:1 hydrochloric acid. Add 0.2 gram of copper wire, evaporate to about 5 ml, and remove the copper wire. Add 5 ml of 1:1 hydrochloric acid, 5 drops of bromine water, and 5 drops of a 10% solution of phenol in acetic acid. Extract the thallic chloride with 5 ml and 5 ml of isopropyl ether. Wash the combined organic extracts with 5 ml and 5 ml of 1:3 hydrochloric acid, each containing a drop of bromine water. This wash removes auric, antimonic, ferric, gallic, and indic ions. Then shake with 5 ml of 0.3 *M* hydrochloric acid and 5 ml of 0.01% solution of brilliant green. Discard the aqueous phase, dilute the isopropyl ether to a known volume to contain less than 2 ng of thallium per ml, and read at 630 nm.

Rocks.[7] Add 5 ml of hydrofluoric acid to 2 grams of sample in platinum and evaporate to dryness. Repeat with 5 ml of hydrofluoric acid and 1 ml of sulfuric acid until no more sulfur trioxide is evolved. Dissolve in 10 ml of hot water, add 5 ml of hydrobromic acid, and evaporate to dryness. Moisten with hydrobromic acid saturated with bromine and again take to dryness. Repeat. Dissolve the residue in 25 ml of *N* hydrobromic acid saturated with bromine. Extract with 25 ml and 25 ml of ether saturated with *N* hydrobromic acid. Wash the combined ether extracts with 3 ml of *N* hydrobromic acid. Evaporate the ether extract. Add 2 ml of hydrochloric acid with 2 ml of bromine water and evaporate to dryness. Repeat three times. Dissolve the residue in 3 ml of *N* hydrochloric acid and 2 ml of bromine water. Heat to drive off the bromine and cool. Add 1 ml of 0.01% solution of brilliant green and dilute to 25 ml. Extract with 10 ml of ethyl acetate and read at 630 nm.

Ores. Dissolve 1 gram with aqua regia and evaporate excess acid. Add 5 ml of sulfuric acid and evaporate to a wet paste. Boil with 50 ml of water; filter and wash with hot 1:50 sulfuric acid. Evaporate the filtrate to dryness. Proceed as for rocks from "Dissolve in 10 ml of hot water, add 5 ml of hydrobromic acid...."

[5] N. T. Voskresenskaya, *Zh. Anal. Khim.* **11**, 585–589 (1956).
[6] Z. Marczenko, H. Kalowska, and M. Jojski, *Talenta* **21**, 93–97 (1974).
[7] N. T. Voskresenskaya, *Zavod. Lab.* **24**, 395–398 (1958).

Urine.[8] Add 2 ml of concentrated sulfuric acid to a 50-ml sample and boil for a few minutes. Add 5 ml of concentrated nitric acid and evaporate. If it appears that this amount of acid is insufficient to decolorize, add more. Evaporate to a colorless residue evolving fumes of sulfur trioxide. Cool, add 5 ml of water, and again evaporate to sulfur trioxide fumes. Repeat this step.

Cool and take up in 25 ml of water. Add 0.2 ml of saturated bromine water. After 1 minute add 0.1 ml of a saturated aqueous solution of sulfosalicylic acid to destroy excess bromine. Mix with 1 ml of 2% solution of brilliant green. Shake for 60 seconds with 5 ml and 5 ml of toluene. Adjust the combined extracts to a known volume and read at 640 nm against toluene.

Alternatively,[9] take a 10-ml sample containing not more than 0.016 mg of thallium. Add 5 ml of water and 12 ml of a mixture containing 10 ml of saturated bromine water and 2 ml of 25% sodium hydroxide solution. Mix and add 10 ml of hydrochloric acid. Mix, and add 2 ml of 0.1% solution of brilliant green in 75% ethanol. Extract with 2.5 ml, 2.5 ml, and 2.5 ml of toluene for 2 minutes. Centrifuge the combined extracts to separate water and read the reddish-brown complex at 640 nm against toluene. If a green solution is formed instead of the complex, repeat with a sample to which urea has been added.

Blood.[10] Dilute 0.1 or 0.2 ml of sample with water to 1 ml and add 4 ml of 1:2 hydrochloric acid. Heat at 100° for 5 minutes. Cool and add 2 ml of 6.9% sodium nitrite solution. After 5 minutes add 0.3 ml of 1% solution of brilliant green in 75% ethanol. Mix and add 2.5 ml of toluene at once. Shake for 15–30 seconds and centrifuge to separate the layers. Read at 640 nm against toluene.

Feces. Mix a 0.2–0.3 gram sample with water to make 2 grams. Add 8 ml of 1:1 hydrochloric acid, mix well, and filter on a fluted filter. Treat 5 ml of filtrate as for blood from "Heat at 100° for 5 minutes...."

CRYSTAL VIOLET

"Crystal violet" groups several closely related derivatives of pararosaniline. Crystal violet is pure hexamethylpararosaniline chloride, color index 42555. Methyl violet 2B is similar but is mostly the pentamethyl derivative, color index 42535. Methyl violet 6B is a pentamethyl benzyl derivative, color index 42536. Gentian violet is a U.S. Pharmacopeia (USP) name for the hexamethyl product with usually pentamethyl and tetramethylene compounds. Their metal compounds would be 1:1 in all cases, differing only to a minor extent in the wavelength of maximum absorption. In each case the specific member of the family recommended is specified.

All these dyestuffs form complexes with thallium which are extractable with organic solvent. The acidity of the solution extracted should not exceed 0.5 *N*. The

[8] J. N. M. deWolf and J. B. Lenstra, *Pharm. Weekbl.* **99**, 377–382 (1964).
[9] M. Ariel and D. Bach, *Analyst* **88**, 30–35 (1963).
[10] N. V. Reis, *Lab. Delo* **3** (6), 12–16 (1957).

color with methyl violet is stable for at least 20 hours.[11] Cadmium and zinc do not interfere, but antimony, tin, and bismuth should be separated. Antimony, mercury, and gold can be removed with a copper coil.[12]

Toluene is the usual solvent for extraction of the dye complex, but it is more soluble in benzene with identical absorbance.[13] The preferred method of isolation of thallium from antimony is extraction of thallic bromide with ether. Antimony and mercury are precipitated as thiocyanates.[14] In the presence of large amounts of cadmium, chlorine is preferable to bromine to oxidize the thallium. Chromium is complexed with tartaric acid. Other oxidizing agents for converting thallous to thallic ion are persulfate, hydrogen peroxide, and ferric salts.[15] The maximum absorption with methyl violet is at 540–620 nm.[16]

By an alternative technic,[17] the thallium is oxidized in 1 : 1 hydrochloric acid with ceric sulfate and hydroxylamine hydrochloride is added. At a known volume, ethyl violet, crystal violet, or malachite green is added, and the color is extracted with benzene for reading at an appropriate wavelength.

If iron is present, error can be introduced by oxidation-reduction reactions; these are best avoided by not heating the sample in which the valence of thallium has been adjusted, usually to the thallic state.[18]

Crystal violet and malachite green are more sensitive than methyl violet or brilliant green.[19] Basic fuchsin reacts similarly to crystal violet for extraction with butyl acetate for reading at 555 nm.[20]

For thallium, mercury, and gold, an organic sample is wet-ashed, taken up in acid, and determined by crystal violet.[21]

When thallic ion reacts with potassium iodide in an acid medium, the liberated iodine can react at pH 3–5 with crystal violet or brilliant green for reading to 0.01 μg/ml.[22] The same reaction is given by chlorate, chlorite, bromate, iodate, periodate, nitrite, permanganate, persulfate, dichromate, ferricyanide, ferric, cupric, vanadate, ceric, and auric ions, as well as by hydrogen peroxide.

As a diethyldithiocarbamate, thallium is extractable from tungsten with chloroform. In complexing with methyl violet, there is interference by trivalent gold, boron, iron, bismuth, and antimony; by bivalent tin, cadmium, and mercury; and by chromate.[23] Extracted with toluene, one reads 0.08–4 μg/ml.

[11]G. G. Shemeleva and V. I. Petrasheu, *Tr. Novocherk. Politekh. Inst.* **41**, 35–40 (1956); G. G. Shemeleva, *Tr. Kom. Anal. Khim., Akad. Nauk SSSR, Inst. Geokhim. Anal.* **8**, 135–140 (1958).

[12]I. A. Blyum and I. A. Ul'yanova, *Zavod. Lab.* **23**, 283–284 (1957); I. A. Blyum, S. T. Solv'yan, and G. N. Shebalkova, *ibid.* **27**, 950–956 (1961).

[13]S. M. Milaev, *Sb. Nauch. Tr. Vsesz. Nauch.-Issled. Gornomet. Inst. Tsvet. Met.* **1958** (3), 258–265; cf. G. G. Shemeleva and V. I. Petrashen, *Tr. Novocherk. Politekh. Inst.* **41**, 35–40 (1956).

[14]G. V. Efremov and A. M. Blokhin, *Vestn. Lening. Univ.* **14**, (22), *Ser. Fiz. Khim.* (4) 148–151 (1959).

[15]N. I. Bashilova and T. V. Khomutova, *Zh. Anal. Khim.* **24**, 999–1004 (1969).

[16]S. D. Gur'ev, *Sb. Nauch. Tr. Gos. Nauch.-Issled. Inst. Tsvet. Met.* **1955** (10), 371–377.

[17]Tsutomo Matsuo and Shunsuke Funada, *Jap. Anal.* **12**, 521–526 (1963).

[18]N. I. Bashilova and N. I. Nelyapina, *Zh. Anal. Khim.* **19**, 1516–1519 (1964).

[19]Kh. Ya. Levitman and L. S. Korel', *Izv. Akad. Nauk SSR, Ser. Khim. Nauk* **1972** (5), 63–69.

[20]V. M. Tarayan, E. N. Ovsepyan, and V. Zh. Artsruni, *Uch. Zap. Eravansk. Gos. Univ.* **1969,** 65–69.

[21]E. L. Kothny, *Analyst* **94**, 198–203 (1969).

[22]E. I. Ramanauskas, L. V. Bunikene, M. S. Sapragonene, A. K. Shulyunene, and M. V. Zhilenaite, *Zh. Anal. Khim.* **24**, 244–246 (1969).

[23]S. D. Gur'ev, *Sbo. Nauch. Tr. Gos. Nauch.-Issled. Inst. Tsvet. Met.* **1955** (10), 371–377.

Procedure. Adjust a 10-ml sample to pH 2–3 with sulfuric acid, add 2 ml in excess, and evaporate to sulfur trioxide fumes.[24] Take up in 5 ml of 1 : 1 hydrochloric acid and oxidize thallous to thallic ion with 3 drops of 10% solution of sodium nitrite. Dilute to 25 ml and add 0.5 ml of 0.2% solution of methyl violet. Shake for 2 minutes with 10 ml of benzene and read the extract at 570 or 620 nm.

Tungsten [25]

THALLIUM 1–10 PPM. Add 2 ml of hydrofluoric acid to 1 gram of powdered sample. Add nitric acid dropwise until solution is complete. Then add 4 ml of 30% solution of tartaric acid and adjust to pH 8–11 with 20% potassium hydroxide solution. Add 1 ml of 2% solution of sodium diethyldithiocarbamate. At this point the volume will be 25–30 ml. Extract with 20 ml and 20 ml of chloroform for 2 minutes each. Wash the combined extracts with 3 ml and 3 ml of glycine buffer solution at pH 10. Evaporate the chloroform extract to dryness. Add 1.8 ml of sulfuric acid and 20 ml of 30% hydrogen peroxide. Heat to sulfur trioxide fumes to destroy residual organic matter. Add 2 ml of water and again evaporate to sulfur trioxide fumes. Add 20 ml of water, 2 ml of 1 : 9 hydrobromic acid, and 0.5 ml of bromine water. Boil off excess bromine; cool and dilute to 49 ml. Add 1 ml of 0.05% methyl violet solution in 0.05 *N* sulfuric acid, freshly prepared by dilution of a 0.1% solution of the dye. Extract for 1 minute with 5 ml of benzene and read the organic layer at 605 nm against benzene.

THALLIUM 0.1–1 PPM. Add 25 ml of hydrofluoric acid to 10 grams of powdered sample, then add nitric acid dropwise until solution is complete. Add 40 ml of 30% tartaric acid solution and 2 ml of 0.05 *N* EDTA. Adjust to pH 8–11 with 60% potassium hydroxide solution. When cool add 10 ml of 2% solution of sodium diethyldithiocarbamate. The volume will be 200–300 ml. Shake for 2 minutes each with four 20-ml portions of chloroform. Wash the combined extracts with 5 ml and 5 ml of glycine buffer solution at pH 10 and complete as for thallium, 1–10 ppm, from "Evaporate the chloroform extract to dryness...."

Zinc and Zinc-Base Alloys.[26] Dissolve 1 gram of sample in 5 ml of 1 : 9 bromine-hydrobromic acid and 5 ml of water. Evaporate until the color of excess bromine has disappeared; cool; and dilute to about 25 ml. Extract thallic bromide with 25 ml of isopropyl ether. Wash the extract with 25 ml and 25 ml of 1.8 *M* hydrobromic acid containing 0.1 ml of bromine water. Shake the organic phase with 10 ml of 0.05 *M* hydrobromic acid freshly mixed with 10 ml of 0.002% crystal violet solution. Adjust the organic phase to 25 ml with isopropyl ether and read at 595 nm against a reagent blank. The organic phase may contain up to 0.05 mg of thallium. The method will determine as low as 5 ppm of thallium in the sample.

[24] I. G. Raikova, L. S. Tsvetkova, V. A. Manikhvatova, and A. I. Ul'yanov, *Nauch. Tr. Nauch.-Issled. Proekt. Inst. Redk. Prom.* **1973**, (47), 233–234.
[25] Karoly Vadasdi, Piroska Buxbaum, and Andras Salamon, *Anal. Chem.* **43**, 318–322 (1971).
[26] P. A. Chainani, P. Murugaiyan, and C. Venkateswarlu, *Anal. Chim. Acta* **57**, 67–72 (1972).

Zinc Electrolyte.[27] Heat a 5–10 ml sample with 0.5 ml of 10% ferric chloride solution and a few drops of 30% hydrogen peroxide. Add hot 10% sodium hydroxide solution to precipitate iron, manganese, and thallium as the hydroxides. Filter the precipitate and wash three times with hot 3% sodium hydroxide solution containing a few drops of hydrogen peroxide; then wash twice with hot water. Dissolve the precipitate in 1:3 hydrochloric acid and evaporate to dryness. Add 1 ml of 1:1 hydrochloric acid and again evaporate to dryness. Take up in 5 ml of 1:1 hydrochloric acid and add 2–3 drops of 10% sodium nitrite solution. Set aside for 5 minutes, then add 5 ml of water. Dilute to about 25 ml with water adding 0.5 gram of tetrasodium pyrophosphate and 1 ml of 0.2% crystal violet solution. Extract for 30 seconds with 10 ml of toluene and read at 570 nm.

Tin Products.[28] Mix 1 gram of sample with 15 ml of hydrofluoric acid and 5 ml of nitric acid. Evaporate almost to dryness and repeat the treatment. Take up the residue in 15 ml of hot *N* hydrobromic acid and filter. Wash the filter three or four times with hot *N* hydrobromic acid. Evaporate the filtrate to dryness and take up the residue in 5 ml of hydrobromic acid. Add a few drops of 30% hydrogen peroxide and evaporate to dryness. Add 5 ml of hydrobromic acid and again take to dryness. Take up the residue in *N* hydrobromic acid containing a slight excess of bromine. Extract the thallium with 10 ml of butyl acetate by shaking for 1 minute, and discard the aqueous layer. Wash the organic layer with 5 ml and 5 ml of *N* hydrobromic acid.

Extract the thallium from the organic phase with 10 ml of ammonium thiocyanate solution and wash the organic phase with 10 ml of water. To the aqueous extract and washings, add a few drops of hydrochloric acid and nitric acid to destroy thiocyanates and evaporate to dryness. Add 0.2 ml of 30% hydrogen peroxide and evaporate. Take up the residue by heating with 20 ml of 1:15 hydrochloric acid. Cool and add 2.5 ml of 10% solution of ferric chloride hexahydrate. Add 0.5 ml of 30% hydrogen peroxide. Dilute to about 40 ml with 1:15 hydrochloric acid and add 0.5 ml of 0.2% methyl violet solution. Extract with 10 ml and 10 ml of toluene and read the combined extracts at 610 nm.

Lead and Lead Oxide.[29] THALLIUM AND ANTIMONY. Dissolve a 2-gram sample in nitric acid; dilute and precipitate lead by addition of saturated solution of sodium chloride. Filter the lead chloride. Evaporate the filtrate to dryness and take up the residue in 1/ ml of 1:30 nitric acid. Add bromine water in slight excess and extract the excess bromine with 10 ml of carbon tetrachloride. Add 1 ml of 0.1% methyl violet solution and shake for 1 minute with 10 ml and 10 ml of benzene. Dilute the combined extracts with benzene to 25 ml, filter, and read at 600 nm against benzene. If more than 1 μg of gold is present, increase the nitric acid to 1:2. If more than 10 μg of mercury is present, increase the nitric acid to 1:5.

For antimony, add 1 ml of sulfuric acid and 1 ml of nitric acid to the aqueous phase from which thallium has been extracted. Evaporate to dryness and take up

[27] S. D. Gur'ev, and E. P. Shkrobot, *Anal. Rud Tsvet. Met. Prod. Pererab.* **1956** (12), 79–88.
[28] L. S. Kiparisova, *Uch. Zap. Tsent. Nauch.-Issled. Inst. Olovyan. Prom.* **1965**, 24–25.
[29] M. Cyrankowska and J. Downarowicz, *Chem. Anal.* (Warsaw) **12**, 137–142 (1967).

the residue in 10 ml of 8 *N* hydrochloric acid. Add 3 drops of 0.5% solution of ferric chloride and discharge the color with a drop or two of 10% solution of stannous chloride in 8 *N* hydrochloric acid. Oxidize with 0.5 ml of 10% sodium nitrite solution. After 5 minutes add 1 ml of 25% urea solution. Complete as for thallium from "Add 1 ml of 0.1% methyl...."

Lead Alloys.[30] Dissolve the sample in nitric acid, precipitate lead by addition of sodium chloride solution, and filter. To an aliquot expected to contain 0.005–0.65 mg of thallium, add 1 ml of 0.25% solution of triphenyltetrazolium chloride and 5 ml of benzene. Add 0.2 ml of 0.1% solution of crystal violet and 5 ml of 0.16 *N* hydrochloric acid. Shake for 2 minutes to replace the triphenyltetrazolium ion with the dye cation and read at 610 nm. Auric ion interferes.

Antimony Alloys. Dissolve a 2-gram sample in 5 ml of 1:1 hydrochloric acid with periodic additions of 30% hydrogen peroxide. Add 6 grams of sodium citrate and dilute to 15 ml. Heat at 100° for 10 minutes and cool. Add 1 ml of 0.2% solution of brilliant green and cool. Extract with 5 ml and 5 ml of chloroform. Repeat the addition of brilliant green solution and the extractions. Add 10 ml of 1:5 hydrochloric acid to the combined extracts and heat at 100° to permit the chloroform vapor to pass through the aqueous phase. Add 30% hydrogen peroxide to bleach the brilliant green in the thallium solution. Add 1 ml of 0.2% crystal violet solution and extract with 10 ml and 10 ml of toluene. Read at 610 nm against toluene. The method is sensitive to 1 ppm of thallium.

Cadmium.[31] If less than 0.002% of thallium is present, use a 4-gram sample, otherwise use 2 grams. Dissolve in 20 ml of 1:9 sulfuric acid, adding a minimal amount of nitric acid. Evaporate to sulfur trioxide fumes, add 10 ml of water, and cool. Neutralize with 15% solution of sodium hydroxide and add an equal volume of water. Filter the precipitate of basic salts of antimony, tin, and bismuth. Add 1.5 ml of *N* hydrochloric acid and 1 ml of freshly prepared chlorine water. Boil off excess chlorine, cool, and dilute to 50 ml. Add to a separatory funnel 10 ml of toluene, 1 ml of *N* hydrochloric acid, and 2.8 ml of 0.1% solution of methyl violet. Add an appropriate aliquot of sample and shake for 30 seconds. After 10 minutes read the toluene layer at 610 nm.

Cadmium Alloys.[32] Dissolve a 2-gram sample in 5 ml of 1:1 hydrochloric with periodic additions of 30% hydrogen peroxide. Dilute to 15 ml with water and heat for 10 minutes at 100°. Cool and dilute to 150 ml. Add 1.5 ml of 0.2% crystal violet solution and extract with 10 ml and 10 ml of toluene. Read at 610 nm against toluene.

Cadmium Oxide.[33] Dissolve a 1-gram sample in 25 ml of 1:5 hydrochloric acid. Mix with 1 ml of 20% sodium nitrite solution and heat at 100° for 10 minutes. Destroy the excess nitrite by adding 1 ml of saturated solution of urea. Dilute to

[30]A. Aleksandrov and A. Dimitrov, *Mikrochim. Acta* **1972**, 680–686 (1972).
[31]G. G. Shemeleva, *Fiz-Khim. Methody Anal. Kontr. Proizvod,* **1961**, 151–154.
[32]S. A. Lomonosov and F. Ya. Mil'shtein, *Zavod. Lab.* **33**, 14–16 (1967).
[33]S. A. Lomonosov, V. N. Podchainova, and V. E. Rybina, *ibid.* **31**, 420 (1965).

200 ml in which the acid concentration should approximate 0.2 *N*. Add 1.5 ml of 0.2% crystal violet solution and extract with 10 ml and 10 ml of toluene. Read at 610 nm against toluene.

Thallium-Activated Sodium Iodide Crystals.[34] Dissolve a 2-gram sample in water, add 2 ml of nitric acid, and evaporate to dryness. Add 1 ml of sulfuric acid and take to dryness. Take up the residue in 10 ml of 1 : 1 hydrochloric acid and add sodium nitrite to convert the thallium to the trivalent form. Add 0.5 ml of 0.2% methyl violet solution and extract with 10 ml and 10 ml of toluene. Read the combined extracts at 610 nm.

Ores.[35] Treat 1–3 grams of finely ground sample with excess hydrochloric acid and nitric acid until reaction is complete, and evaporate to dryness. Add 5 ml hydrochloric acid and take to dryness. Repeat. Heat with 30 ml of water to extract the soluble matter. Add excess of bromine water and filter. Wash the residue on the filter with hot water.

If the sample is high in thallium or contains gold or antimony, evaporate to dryness, take up in 10 ml of 1 : 5 hydrochloric acid, and pass twice through a copper reductor. Omit this paragraph in the absence of gold and antimony if the thallium content is low. If considerable iron is present, add a few drops of phosphoric acid.

Add 1 gram of potassium bromide per 50 ml of sample solution. Extract the thallium with 10 ml and 10 ml of ether. Evaporate the combined ether extracts to dryness and take up the residue in 10 ml of 1 : 5 hydrochloric acid. Add bromine water in slight excess and after 30 seconds add a few crystals of sulfosalicylic acid to decolorize or, in the presence of ferric ion, to give a violet color. Add about 0.5 gram of potassium bromide, 1 ml of 0.1% solution of crystal violet, and 10 ml of water. Extract with 15 ml of benzene and read the extract at 610 nm.

Slag.[36] Dissolve 0.1–1 gram of finely ground sample in a mixture of hydrofluoric, nitric, and sulfuric acids. Evaporate to dryness and add 1 ml of 1 : 1 hydrochloric acid to the residue. Warm, and add 10–30 ml of hot water. Boil to extract the soluble salts, filter insoluble matter, and wash well. Complete as for ores from "Add 1 gram of potassium bromide...."

Silicate Rocks.[37] The ion exchange resin recommended, Wofatit P, adsorbs thallium quantitatively from a pH below 1. The presence of 5% of citric acid converts antimony into a complex anion. Mercuric ion is not adsorbed. Trivalent gold is partially adsorbed, and the presence of more than 0.005% interferes. The iron, copper, tin, and zinc normally found in such materials as pyrites, galena,

[34]A. M. Bulgakova, A. M. Volkova, and G. S. Plotnikova, *Tr. Vses. Nauch.-Issled. Inst. Khim. Reak.* **1959**, (23), 102–105.

[35]V. Patrovsky, *Chem. Listy* **57**, 961–964 (1963); cf. G. V. Efremov and Chzhi-Gu Syui, *Vestn. Leningr. Univ.* **13**, (16), *Ser. Fiz e Khim* (3), 156–159 (1958); Pin-chin Hyang, *Hua Hsueh Shih Chich* **1959,** 144–145.

[36]V. A. Oshman, *Tr. Ural. Nauch.-Issled. Proekt. Inst. Med. Prom.* **1963**, 417–422.

[37]B. N. Panchev, *Izv. Geol. Inst. "Strashimir Dimitrov"* **12**, 237–243 (1963).

chalcopyrite, and sphalerite are partially adsorbed by the resin and eluted with the thallium but do not interfere.

Heat a 1-gram sample of silicate rock with 10 ml of nitric acid and 20 ml of hydrofluoric acid. If necessary to remove all the silica add more hydrofluoric acid. Evaporate to dryness. Add 5 ml of nitric acid, evaporate to dryness, and repeat that step. Take up with dilute hydrochloric acid, add 5 ml of hydrochloric acid, and evaporate to dryness. Repeat this evaporation with 5 ml of hydrochloric acid twice more.

Take up the residue in 20 ml of 0.05 *N* hydrochloric acid containing 5% of citric acid, heat to boiling, and let the residue settle. Filter and wash the residue several times with the hydrochloric-citric acid solution.

Prepare a 16×1 cm column of Wofatit P resin, well washed with hydrochloric-citric acid. Pass the filtrate at 120 ml/hour through the column, and wash the column several times with 5% citric acid solution. Then wash the column thoroughly with water.

Elute the thallium with 100 ml of 1:1 hydrochloric acid at 40 ml per hour and take an appropriate aliquot. To the thallium solution, add 1–4 ml of *N* hydrobromic acid and 2 ml of bromine water. Dilute to 20 ml and boil off excess bromine. Add 2 drops of 0.01% methyl violet solution and dilute to 25 ml. Extract with 5 ml of amyl acetate and read at 610 nm.[38]

Complex Mixtures.[39] Thallium at 10–40 μg can be determined in the presence of 5 mg of antimony or bismuth, 20 mg of cadmium or copper, and 30 mg of lead, zinc, ferrous ion, or aluminum.

Make the sample solution 1:4 with sulfuric acid and evaporate to fumes of sulfur trioxide. Dilute to *N* with sulfuric acid and boil with a copper spiral for 20–30 minutes, to deposit bismuth and antimony. After removing the spiral add 8–10 ml of 30% hydrogen peroxide and 4–5 drops of sulfuric acid. Evaporate to fumes of sulfur trioxide and cool. Add 10–15 ml of water and make faintly pink with 0.05 *N* potassium permanganate. Dilute to a 0.1–0.15 *N* concentration of sulfuric acid. Add 2 ml of 0.1% potassium bromide solution, 1.5 ml of 0.2% methyl violet solution, and 20 ml of toluene. Shake for 2 minutes, let separate for 15 minutes, and read the toluene extract at 595 nm.

Urine.[40] To a 500-ml sample, add 10 ml of 1:9 sulfuric-nitric acid and a couple of drops of *N*-octanol as an antifoaming agent. Heat to sulfur trioxide fumes and cool. Add more nitric acid until a final residue in sulfuric acid is white. Cool, take up in water, and dilute to about 20 ml. Add 1 ml of hydrobromic acid and heat at 180° until a light brown color appears. Cool, make up to 20 ml with water, and heat to dissolve the solids. Add 1 ml of 4.4% hydrobromic acid and a few drops of bromine water. Heat at 180° until colorless, partially cool, and dilute to 50–75 ml. When cool, add 10 ml of amyl acetate. Add 1 ml of 0.02% methyl violet solution in 0.1 *N* sulfuric acid and shake for 60 seconds. Centrifuge the solvent layer to separate water and read at 585 nm against amyl acetate.

[38]N. T. Voskresenskaya, *Zh. Anal. Khim.* **11**, 585–589 (1956).
[39]G. Popa, Ri Zin Su, and C. Patroescu, *Rev. Roum. Chim.* **12**, 969–974 (1967).
[40]E. Campbell, Morris F. Milligan, and Jean A. Lindsey, *Am. Ind. Hyg. Assoc. J.* **20**, 23–25 (1959); cf. Morris B. Jacobs, *ibid.* **23**, 411–413 (1962).

Alternatively,[41] acidify the sample with sulfuric acid and filter through activated carbon to remove mercury. Mix 10 ml of the filtered sample with 3 ml of bromine water, 2 ml of 20% solution of sulfosalicylic acid, and 10 ml of 1 : 1 sulfuric acid. Extract with 5 ml and 3 ml of 0.05% solution of crystal violet in chloroform. Dilute the combined extracts to 10 ml with ethanol and read at 590 nm.

Tissue. The technics for urine are applicable to an appropriate sample of tissue homogenate.

Air. Collect the suspended thallium by drawing an appropriate volume of air through filter papers. Digest the paper in a centrifuge tube with 5 ml of 1:9 sulfuric-nitric acid at 180° until disintegrated. Cool and add 2 ml of 1:1:8 nitric-sulfuric-perchloric acid at 350°. If necessary add more perchloric acid until the solution is colorless. Evaporate to a final white residue in sulfuric acid. Complete as for urine from "Cool, take up in water...."

DIBENZYLDITHIOCARBAMATE

Thallium dibenzyldithiocarbamate can be read in carbon tetrachloride at 0.01–0.05 mg per ml. A preliminary extraction eliminates interference from copper and bismuth.

Procedure. Take a sample containing 0.1–1.5 mg of thallium in 0.5 *N* sulfuric acid.[42] If more than 0.5 mg of copper or 0.2 mg of bismuth is present, add 0.5 ml of 0.5% sodium sulfite solution to reduce the thallium to the monovalent state. Extract the copper and/or bismuth with four successive 5-ml portions of 0.1% zinc dibenzyldithiocarbamate in carbon tetrachloride and discard the extracts. Add excess saturated bromine water to the aqueous layer to oxidize thallous ion to thallic. Destroy the excess bromine with a few drops of 1% phenol solution, and immediately extract the thallium with 10 ml of 0.1% zinc dibenzyldithiocarbamate solution in carbon tetrachloride. Read at 438 nm.

DIETHYLDITHIOCARBAMATE

As with several other ions, thallic ion can be isolated as the diethyldithiocarbamate. The sensitivity is increased more than twentyfold by converting to the copper complex.[43] Dithizone in strongly alkaline solution is appropriate for separation of thallium from bismuth.[44]

[41] O. Wawschinek, W. Beyer, and B. Paletta, *Mikrochim. Acta* **1968**, 201–204.
[42] Kazuyoshi Hagiwara, Hiroshi Suzuki, and Isao Muraki, *Jap. Anal.* **10**, 607–612 (1961).
[43] R. Keil, *Z. Anal. Chem.* **258**, 97–99 (1972).
[44] G. Ackermann and W. Angermann, *ibid.* **250**, 353–357 (1970).

Procedure. Adjust a 25-ml sample solution of thallic ion in inorganic acid to pH 0.3–0.6. Add 0.25 gram of potassium bromate and evaporate to about half the volume, cool, and adjust to a known volume. To an aliquot containing up to 0.1 mg of thallium, add 10 ml of a masking solution containing 5% of tetrasodium EDTA, 5% of triethanolamine, and 7% of ammonium citrate. Adjust to pH 10.6–11.1 with 20% sodium hydroxide solution. Add 2 ml of 25% potassium cyanide solution and 5 ml of 0.3% sodium diethyldithiocarbamate solution. After 10 minutes extract with 10 ml of carbon tetrachloride, avoiding exposure to strong light. Filter the carbon tetrachloride layer into 5 ml of an ammoniacal solution of cupric sulfate. Shake and read the organic phase at 436 nm against carbon tetrachloride. Subtract a blank.

DITHIOOXAMIDE

Dithiooxamide, also known as rubeanic acid, forms in ammoniacal solution a 2:3 complex with thallic ion.[45] It also reacts with thallous ion.

Procedures

Reducing Medium. To the sample solution, add 10 ml of 1:15 ammonium hydroxide, 10 ml of 0.5% gelatin solution and 4 ml of 2% solution of hydrazine sulfate. After 5 minutes add 1 ml of 0.01 *M* dithiooxamide. Dilute to 100 ml and set aside for 30 minutes before reading with a blue filter.

Oxidizing Medium. Treat as for reducing medium, replacing the hydrazine sulfate solution with 30% hydrogen peroxide.

DITHIZONE

After separation by column chromatography, thallous ion can be read as the dithizone complex at 505 nm.[46] The alternative complex with di-β-naphthylthiocarbazone is read at 540 nm and is only moderately more sensitive.[47] A typical technic is that for manganese ores.[48]

For analysis of autopsy tissue, dry a 10-gram sample and digest with nitric and

[45] L. S. Mikhalevich and Kh. Ya. Levitman, *Izs. Akad. Nauk. Beloruss. SSR, Ser. Khim. Nauk* **1966**, (3) 55–58.
[46] K. Kasiura, *Chem. Anal.* (Warsaw) **13**, 849–855 (1968); Arne Dyfverman, *Anal. Chim. Acta* **21**, 357–365 (1959); D. Jamrog and J. Piotrowski, *Med. Pr.* **9**, 299–330 (1958).
[47] Richard J. Dubois and Samuel B. Knight, *Anal. Chem.* **36**, 1313–1320 (1964).
[48] Roy S. Clarke, Jr., and Frank Cuttitta, *Anal. Chim. Acta* **19**, 555–562 (1958); Frank Cuttitta, *U.S. Geol. Surv. Profess. Papers No.* **424-c**, 384–385 (1961).

sulfuric acid.[49] Dilute with water and add an excess of chlorine water. Extract thallic chloride with ether, then back-extract with aqueous sulfurous acid. Interfering metals such as lead, arsenic, mercury, copper, bismuth, zinc, and barium must be absent before extraction with dithizone for reading at 510 nm.

Procedure. *Manganese Ores.* Digest a sample of manganese ore expected to contain 0.1–0.5 mg of thallium with 5 ml of hydrobromic acid, 5 ml of water, and 2 ml of hydrofluoric acid for 90 minutes. Evaporate to dryness and ensure that only bromides are present by evaporating to dryness several times with hydrobromic acid. Dissolve the residue in 16 ml of hydrobromic acid and dilute with water to 50 ml, giving a sample solution in 1.5 *N* acid.

Mix a 5-ml aliquot with 5 ml of water and chill. Extract the thallium with four successive 10-ml portions of prechilled ether. Wash the combined ether extracts with 10 ml of *N* hydrobromic acid. Evaporate the ether extracts and ash with 10 ml and 10 ml of 1 : 1 nitric acid to destroy the last traces of organic matter. Take up the residue in 1% hydrochloric acid and dilute to 30 ml. Extract with 5 ml and 5 ml of 0.001% dithizone in chloroform. Combine the extracts and extract the thallium with 5 ml and 5 ml of 1 : 100 nitric acid. Read the liberated dithizone in the chloroform at 505 nm. Lead, bismuth, and tin interfere seriously.

FLAME PHOTOMETRY

Applying thallium in 80% acetone containing 0.5 *N* hydrochloric acid gives a fiftyfold increase in flame emission.[50] The appropriate wavelength is 377.6 nm, the background, 376 nm or 380 nm. Ether extraction from 5.5 *N* hydrochloric acid eliminates all interference. Appropriate media are air-acetylene or hydrogen.[51] The limit of detection in ether is 0.13 ppm.[52]

When silver is dissolved in nitric acid, the silver is reduced by formic acid and removed as an amalgam before reading thallium by flame photometry.[53]

Butyl acetate will extract the product of thallic ion, bromine, and hydrobromic acid as thallobromic acid from aqueous solution for flame photometry at 535 nm.[54]

With a 5 cm $\times$ 0.5 mm burner viewed lengthwise, the optimum height of the red zone was 4 mm for reading at 535 nm. Addition of 0.1% of potassium chloride enhances the signal by suppressing ionization in the flame.[55]

For biological material, wet-ash with nitric acid and extract as thallic bromide with methyl isobutyl ketone.[56] Read the flame emission at 535 nm.

[49] S. N. Tewari, S. P. Harpalani, and S. S. Triparthi, *Mikrochim. Acta* **1**, 13–18 (1975).
[50] Helmut Bode and Hurst Fabian, *Z. Anal. Chem.* **170**, 387–399 (1959).
[51] J. Malinowski, K. Kancewicz, and S. Szymczak, *Pol. Acad. Sci. Inst. Nuclear Res. Rept.* No. 113/VIII, 7 pp. (1959).
[52] H. Brandenberger and H. Bader, *Helv. Chim. Acta* **47**, 353–358 (1964).
[53] J. Meyer, *Z. Anal. Chem.* **231**, 241–252 (1967).
[54] T. V. Gurkina and E. Ya. Litvinova, *Zh. Anal. Khim.* **24**, 374–378 (1969).
[55] E. E. Pickett and S. R. Koirtyohann, *Spectrochim. Acta* **B 24**, 325–333 (1969).
[56] A. S. Curry, J. F. Read, and A. R. Knott, *Analyst* **94**, 744–753 (1969).

Procedures

Silicates.[57] Decompose a 1-gram sample in sulfuric and hydrofluoric acids. Evaporate to dryness and take up the residue in 10 ml of hydrobromic acid. Add 10 ml of *N* hydrobromic acid saturated with bromine and evaporate to dryness. Repeat this treatment three times. Take up the residue in 5 ml of *N* hydrobromic acid, add 25 ml of *N* hydrobromic acid saturated with bromine, and boil off the bromine. Dilute with water to 30 ml and extract thallium bromide with 10 ml of pentyl acetate saturated with *N* hydrobromic acid. Separate the organic layer, wash it with 10 ml of *N* hydrobromic acid, and dry with calcium chloride. Read by flame photometry at 377.6 nm using oxygen at 10 psi and hydrogen pressure at 0.75 psi. The technic will detect 0.2 ppm of thallium.

Mercury.[58] To a 100-gram sample, add 2 ml of water and dissolve by slowly adding 75 ml of nitric acid. Then add 110 ml of formic acid and heat gently until the precipitated mercury is agglomerated into a single mass. Decant the liquid through filter paper and evaporate to dryness under a moderate vacuum. Dissolve in 5 ml of 1 : 1 hydrochloric acid and read by flame photometry at 377.6 nm.

Organs and Body Fluids.[59] Wet-ash a 2-gram sample of tissue (use somewhat more of body fluids) with 20 : 5 : 1 nitric perchloric-sulfuric acid. Avoid charring by periodic addition of 30% hydrogen peroxide and nitric acid as necessary. Then boil the clear digest to evolution of sulfur trioxide. Cool, dilute with water, and again evaporate to the evolution of sulfur trioxide. Repeat this step twice more. Chill in ice, dilute somewhat with water, and neutralize with ammonium hydroxide. Adjust with water to about 5 ml. Add a drop of bromine and 0.5 ml of hydrobromic acid. Boil off the excess of bromine and cool. Extract the thallium bromide with 5 ml of methyl hexyl ketone saturated with hydrobromic acid. Wash the solvent phase with 5 ml of *N* hydrobromic acid, centrifuge, and read at 377.6 nm.

3–HYDROXY–7, 3–DIPHENYLTRIAZINE

This reagent forms an orange-red 3–1 complex with thallic ion which is insoluble in water but is soluble in 70% ethanol for reading at 422 nm.[60] The full color is developed at pH 5–8 and obeys Beer's law at 0.02–0.16 mg of thallium per ml. The reagent is also applicable to photometric determination of copper, iron, cobalt, nickel, palladium, and molybdenum. Fluoride complexes excess iron, titanium, thorium, and zirconium to avoid interference. Indium, copper, and nickel are filtered out as insoluble precipitates. The color begins to fade after 5–6 hours. Unless complexed, 3 μg of ferric ion interferes.

[57] A. Fornaseri and A. Penta, *Metg. Ital.* **55**, 437–441 (1963).
[58] J. Meyer, *Z. Anal. Chem.* **219**, 147–160 (1966).
[59] William B. Stavinoha and Joe B. Nash, *Anal. Chem.* **32**, 1695–1697 (1960); W. J. Wilson and R. Hausman, *J. Lab. Clin, Med.* **64**, 154–159 (1964).
[60] S. C. Shome, H. R. Das, and B. Das, *Anal. Chem.* **38**, 1522–1524 (1966).

Procedure. Adjust the sample solution to pH 5–7 by addition of 10% solution of sodium acetate or sodium potassium tartrate and dilute to 25 ml. Chill in ice and add a few ml of ethanol. Add 2–4 ml of 0.02 *M* solution of the reagent. Add ethanol and water to give a final 70% concentration of ethanol. After 30 minutes read at 422 nm against a blank.

IODINE

Thallic ion is reduced to thallous by potassium iodide liberating iodine. Various conventional methods of determination of iodine are applicable. Thus one can develop the color with starch for reading at 575 nm.[61] Extraction for direct reading is highly satisfactory.[62]

As a variant of this general method, thallium is precipitated as the orthoperiodate from nitric acid. The filtered precipitate is dissolved in hydrochloric acid for treatment with potassium iodide to release an equivalent of iodine.[63]

For reading by luminescence, a sample containing more than 0.5 mg of thallium per ml mixed with 9 volumes of saturated salt solution is activated at 240 nm and read at 420 nm.[64a]

Procedure. To a neutral sample free from trivalent ions that liberate iodine and containing about 0.5 mg of thallium, add 1 gram of ammonium chloride and dilute to about 20 ml. Add 10 ml of hydrochloric acid and 25 ml of a 10% solution of disodium phosphate in saturated bromine water. The phosphate eliminates the effect of any iron present, and the bromine oxidizes thallium to the trivalent form. Boil vigorously for a few minutes to drive off most of the excess bromine; but note that prolonged boiling will cause reduction of thallic ion.

Cool and add 0.25 ml of 25% solution of phenol in acetic acid, to destroy a last trace of bromine. Dilute to about 60 ml. Add 5 ml of 0.2% potassium iodide solution and 20 ml of carbon bisulfide. Shake for 30 seconds and read the organic layer at 600 nm.

Lead.[64b] As an iodide reagent dissolve 1.2 gram of potassium iodide in 120 ml of water. Add 30 ml of 0.1% solution of sodium sulfite and 50 ml of 6.8% solution of sodium acetate adjusted to pH 6 ± 0.1 with acetic acid.

Dissolve 5 grams of sample containing more than 5 μg of thallium in 15 ml of 1 : 1 nitric acid. Add 1 ml of 3% hydrogen peroxide and simmer for 30 minutes.

[61] Harry H. Ackeman, *J. Ind. Hyg. Toxocol.* **30**, 300–302 (1948); V. S. Fikhengol'ts and N. P. Kozlova, *Zavod. Lab.* **21**, 407–408 (1955).

[62] L. A. Haddock, *Analyst* **60**, 394–399 (1935).

[63] N. A. Verdizade and M. G. Iskenderov, *Uch. Zap. Azerb. Univ., Ser. Khim. Nauk* **1969**, 19–22.

[64a] R. Bock and E. Zimmer, *Z. Anal. Chem.* **198**, 170–173 (1963).

[64b] Sakakibara Iwao, Masanao Sakakibara, and Toshimi Yamamoto, *Anal. Chim. Acta* **83**, 251–258 (1976).

Add 1% potassium permanganate solution dropwise to a pale pink, then decolorize by dropwise addition of 0.1% sodium nitrite solution. Dilute to 50 ml adding 2.5 ml of *M* hydrobromic acid. Shake for 5 minutes with 5 ml of freshly prepared 6% solution of trioctylamine in benzene. Shake the organic phase sequentially with 20 ml and 20 ml of *N* nitric acid, 20 ml of *M* perchloric acid and finally with 20 ml of the iodide reagent. Read the organic phase at 400 nm against the trioctylamine solution and subtract a blank.

MALACHITE GREEN

When the 1 : 1 complexes of methylene blue or methylene green with chlorthallate ion in 0.1–1.5 *N* hydrochloric acid are extracted, they absorb at the same wavelength and will determine 6 ppm of thallium in a sample.[67] They are more sensitive than the crystal violet family or brilliant green; they operate over a wider range of acidity, and the extracts are more stable. Antimony interferes.

Procedures

Cadmium Metal. Dissolve a 1-gram sample with 5 ml of nitric acid and 5 ml of hydrochloric acid. Evaporate to less than 2 ml and cool. Add 5 ml of hydrochloric acid, 1 ml of hydrobromic acid, and 5 ml of bromine water. Evaporate to dryness. Take up in 20 ml of 0.5 *M* hydrobromic acid and extract the thallium with 20 ml and 20 ml of ether. Wash the combined ether extracts with 20 ml of 0.5 *M* hydrobromic acid. Add 2 ml of nitric acid, 2 ml of perchloric acid, and a few drops of sulfuric acid to the ether extract and evaporate to dryness. Take up the residue in 15 ml of 1 : 18 sulfuric acid and 11 ml of water. Mix vigorously with 1 ml of 0.1% ceric sulfate solution and 1 ml of 10% potassium bromide solution for 5 minutes, to convert to the thallic form. Add 1 ml of 1% solution of hydroxylamine hydrochloride and 1 ml of 0.2% solution of malachite green. Extract with 5 ml of benzene and read at 625 nm against a blank.

Urine.[66] Evaporate 20 ml of urine and 20 ml of nitric acid. Take up the residue in 10 ml of *N* hydrobromic acid. Add saturated bromine water to slight excess and remove the unreacted portion by dropwise addition of 1% phenol solution. Extract the thallic bromide with 5 ml and 5 ml of isopropyl ether. Evaporate the combined extracts to dryness and take up in 10 ml of 1 : 7 hydrochloric acid. Add 1 ml of 0.2% malachite green solution and extract with 10 ml of benzene. Read at 635 nm against benzene and subtract a blank.

---

[65]Bunshiro Kominami and Hutaba Ono, *Jap. Anal.* **18**, 578–582 (1969).
[66]Toshio Suzuki, *ibid.* **14**, 130–134 (1965).

METHYLENE BLUE OR METHYLENE GREEN

When the 1 : 1 complexes of methylene blue or methylene green with chlorthallate ion in 0.1–1.5 *N* hydrochloric acid are extracted, they absorb at the same wavelength and will determine 6 ppm of thallium in a sample.[67] They are more sensitive than the crystal violet family or brilliant green; they operate over a wider range of acidity, and the extracts are more stable. Antimony interferes.

Procedures

Ores. Dissolve a 0.5-gram sample in 10 ml of 3 : 1 hydrochloric-nitric acid and evaporate to dryness. Add 3 ml, 3 ml, and 3 ml of hydrochloric acid and evaporate to dryness after each addition, to drive off nitrates. Take up the residue in 10 ml of 1 : 3 hydrochloric acid. If antimony is present, displace it with a copper coil. Filter if necessary.

To 5 ml of the sample solution, add 1 ml of 10% sodium nitrite solution, let stand for 5 minutes, and add 0.5 ml of a saturated solution of urea. Dilute to 10 ml and add 1 ml of 0.02% solution of methylene blue (basic blue 9) or methylene green. (basic green 5). Adjust the acidity to the range of 0.1–0.5 *N* hydrochloric acid and extract with 10 ml of 1 : 1 dichloroethane-trichloroethylene. Read at 655 nm against a reagent blank. The thallium content of the extract should be 0.2–3.5 μg/ml.

Indium Metal. Dissolve a 1-gram sample in 20 ml of hydrochloric acid and evaporate to dryness. Take up the residue in 10 ml of 1 : 3 hydrochloric acid. Mix 5 ml with 1 ml of 10% sodium nitrite solution. After 5 minutes add 0.5 ml of saturated urea solution and 3 ml of 0.5 *M* EDTA. Proceed as for ores from "Dilute to 10 ml and add...."

MELDOLA BLUE

The complex of thallic ion with meldola blue (basic blue 6) is extracted with benzene-acetone for reading.[68] Turquoise blue is also suitable for the determination.[69]

Procedure. Adjust 8 ml of sample solution containing 2–16 mg of thallium in *N* hydrobromic acid with 1 : 1 ammonium hydroxide to pH 3–5. Add 0.5 ml of 0.2% solution of meldola blue and extract with 5 ml and 5 ml of 4 : 1 benzene-acetone.

[67] V. M. Tarayan, E. N. Ovsepyan, and V. Zh. Artsruni, *Dokl. Akad. Nauk Arm. SSR* **47**, 27–30 (1968); *Zavod. Lab.* **35**, 1435–1437 (1969); *Uch. Zap. Erevan. Gos. Univ.* **1970**, 35–41.
[68] A. T. Pilpenko and Nguyen Dyk Tu, *Ukr. Khim. Zh.* **35**, 303–305 (1969).
[69] G. V. Efremov and V. A. Galibin, *Uch. Zap. Leningr. Gos. Univ. A. A. Zhdanova* **211**, *Ser. Khim. Nauk* **15**, 83–86 (1957).

Read the combined extracts at 570 nm. The method is sensitive to 2 μg of thallium in the 10 ml of extract.

1–NAPHTHOL AND *N*–PHENYL–*p*–PHENYLENEDIAMINE

Thallic ion oxidizes a mixture of 1–naphthol and *N*–phenyl–*p*–phenylenediamine to give a blue color with a maximum absorption at 590 nm, extractable by organic solvents.[70] The optimum pH for the reaction is 5. The color body contains two atoms of thallium to one molecule of each of the other reactants.

Procedure. *Zinc or Titanium Products.* Dissolve a 1-gram sample in 1 : 1 sulfuric acid and evaporate to dryness. Take up the residue in water and filter any insoluble matter. Wash the filter with 1 : 100 sulfuric acid and evaporate the filtrate and washings to about 50 ml. Add excess of bromine water. Boil off the excess of bromine, and complete its removal by dropwise addition of 25% solution of phenol in acetic acid. Add sufficient phosphoric acid to mask ferric ion. Dilute to 100 ml with water.

To a 2-ml aliquot of the sample solution add 1 ml each of m*M* 1-naphthol and *N*-phenyl-*p*-phenylenediamine in isopentyl alcohol. Add 0.5 ml of acetate buffer solution for pH 5 and shake for 1 minute. After 10 minutes remove the organic layer, and after a further 10 minutes read it at 590 nm against isopentyl alcohol.

NILE BLUE

An organic extract of the complex of thallium with Nile blue (basic blue 12) is read at 650 nm.[71] It is applicable to 0.006–0.5% in the sample.

Procedure. *Ores and Concentrates.* Treat a 0.5-gram sample with 10 ml of aqua regia and evaporate to dryness. Add 3 ml, 3 ml, and 3 ml of hydrochloric acid and evaporate to dryness after each addition to eliminate nitrates. Take up the residue in 1 : 3 hydrochloric acid and dilute to 50 ml with that acid. Add a coil of copper wire to displace antimony, gold, and mercury. Filter if necessary.

To a 5-ml aliquot, add 1 ml of 10% sodium nitrite solution and let stand for 5 minutes. Add 0.5 ml of a saturated solution of urea and adjust to approximately pH 3 by dropwise addition of 20% sodium hydroxide solution. Add 1 ml of 0.01% solution of Nile blue and dilute to 10 ml. Extract with 10 ml of 1 : 1 dichloroethane-trichloroethylene and read the extract at 650 nm against a reagent blank.

---

[70] A. P. Kreshkov, L. P. Senetskaya, and A. M. Karagodina *Zh. Anal. Khim.* **21**, 413–419 (1966).
[71] V. M. Tarayan, E. V. Ovsepyan, and V. Zh. Artsruni, *Arm. Khim. Zh.* **22**, 992–997 (1969).

8-HYDROXYQUINOLINE

The reagent 8-hydroxyquinoline is also known as 8-quinolinol and as oxine. Trivalent thallium as well as gallium is precipitated above pH 3.8 and 3.1, respectively, by oxine;[72] a chloroform solution is read in the region of 400–402 nm. Other appropriate extractants are benzene and methyl isobutyl ketone, to give maxima at 410 and 400 nm, respectively.[73] Differences in molar absorption in the three solvents are not significant. The absorption is constant above pH 4.5, but above pH 6 a tendency to deposit oxine develops. Acetate buffering at pH 5.5 is desirable. The absorption of the extract is constant for 2–7 minutes, then gradually decreases. Beer's law is followed up to 15 μg of thallium per ml of extract.

In the extraction of thallic oxinate there is interference by zinc, aluminum, chromium, copper, nickel, cobalt, titanium, uranium, thorium, niobium, iron, molybdenum, gallium, and indium. Since thallous oxinate is not extracted by benzene or methyl isobutyl ketone, the interferences are extracted from the thallous form, which is later oxidized for determination as thallic oxinate.

Procedures

Separation from Interfering Ions. Adjust the sample solution to 2 *N* with hydrochloric acid and take 10 ml containing 0.01–0.15 mg of thallium. Extract with 10 ml of methyl isobutyl ketone by shaking for 30 seconds, and evaporate the organic layer to dryness. This contains thallium and the interfering elements. Heat with a few ml of nitric acid and perchloric acid to destroy organic matter. Cool and dissolve the residue in 5 ml of 1 : 15 nitric acid, warming if necessary. Add 5 ml of 1.64% sodium acetate solution and 2 ml of 2% oxine solution. Adjust to pH 5.5 with 24% sodium hydroxide solution or 1 : 1 nitric acid. Extract with 10 ml of methyl isobutyl ketone. Interfering members of group 2 are in the solvent and are discarded.

Evaporate the aqueous layer, which contains the thallium in thallous form, almost to dryness. Take up with 5 ml of 1 : 15 nitric acid, heat to about 90°, and add saturated bromine water until the solution is yellowish red to oxidize thallous ion to thallic. Heat for about 50 seconds to remove excess bromine. If excess is present when oxine is added, it causes a reddish-brown precipitate. Cool at once. Excess heating above 70° will pyroreduce thallic ion to thallous ion.

Development. To 5 ml of thallic sample solution in 0.5–1 *N* nitric or hydrochloric acid, add 5 ml of 1.64% sodium acetate solution and 2 ml of 2% oxine solution. Add 24% sodium hydroxide solution or 1 : 1 nitric acid to adjust to pH 5.5. Within 30 minutes add 10 ml of benzene or methyl isobutyl ketone and shake for 30 seconds. Discard the aqueous layer and filter the organic layer. Read at 410 nm for benzene or 400 nm for methyl isobutyl ketone against the solvent, and subtract a blank.

---

[72]Therold Moeller and Alvin J. Cohen, *Anal. Chem.* **22**, 686–690 (1950).

[73]Hidehiro Goto, Yachiyo Kakita, and Norio Ichinose, *J. Chem. Soc. Jap., Pure Chem. Sect.* **88**, 638–643 (1967).

o-PHENYLENEDIAMINE

The oxidation of *o*-phenylenediamine by thallic ion in acid solution instantly develops color that is stable for more than 2 hours.[74] The pH should be 1.9–5.6. Beer's law applies over the range 0.58–36.7 μg/ml.

Procedure. To 25 ml of neutral sample solution, add 1 ml of 1 : 1 sulfuric acid and sufficient bromine water to give a permanent color. Decolorize the solution of thallic bromide by bubbling air through it for 30 minutes. Dilute to a known volume expected to have about 0.8 mg of thallium per ml at about pH 1. Mix 10 ml with 1 ml of 0.5% *o*-phenylenediamine solution and dilute to 20 ml. Read at 460 nm against a reagent blank.

4-(2-PYRIDYLAZO)RESORCINOL

Thallium reacts with 4-(2-pyridylazo)resorcinol, often abbreviated as PAR, to give a red, water-soluble 1:2 complex absorbing at 520 nm.[75] The reagent absorbs at 390 nm. Germanium and indium form complexes absorbing at 510 nm. The complexes are stable for 48 hours at up to 90°. The optimum pH values are 3.5–4.5 for thallium or indium and 3.7–4.5 for gallium. There is interference by thorium, zirconium, hexavalent uranium, pentavalent vanadium or niobium, ferric ion, bivalent cobalt, nickel, palladium, copper, zinc, cadmium, and aluminum.

An alternative reagent is 4-(2-thiazolylazo)resorcinol at pH 2.8–3 read at 540–550 nm.[76]

Procedure. To a solution containing 1–2 ppm of thallium, add sufficient solution of PAR so that there are more than 5 moles to 1 of the test substance. Adjust to pH 4 ± 0.2 and let stand for 30 minutes. Read at 520 nm against water.

QUERCETIN

Quercetin, which is tetrahydroxyflavanol, is applicable to 2–16 μg of thallium per ml of aqueous ethanol or acetone in zinc and cadmium solutions.[77] Chloride, bromide, iodate, acetate, and tartrate interfere.

Procedure. Mix a 10-ml sample with 1 ml of 1 : 8 sulfuric acid. Add 0.1 gram of

[74] M. Papafil, M. Furnica, and D. Furnica, *Anal. Stiint. Univ. Al. I. Cuza*, **11**, 133–140 (1965).

[75] C. D. Dwivedi, K. N. Munshi, and A. K. Dey, *Chemist-Analyst* **55**, 13 (1966).

[76] M. Hnilickova and L. Sommer, *Talenta* **16**, 83–94 (1969).

[77] A. P. Golovina and V. G. Tiptsova, *Zh. Anal. Khim.* **17**, 525–566 (1962); cf. R. Manczyk, J. Terpilowski, and M. Kopacz, *Diss. Pharm. Warsaw* **18**, 643–649 (1966).

ammonium persulfate and 2 drops of 0.1 *N* silver nitrate. Heat for a couple of minutes to oxidize thallium to the trivalent form, then boil to decompose excess persulfate. Add 10 ml of phthalate buffer solution for pH 5 and 0.5 ml of 0.5% quercetin solution in acetone. Shake for 2 minutes with 10 ml of amyl, butyl, or ethyl acetate. Read at 510 nm 5 minutes after shaking was begun.

RHODAMINE B

When thallium is oxidized by bromine to the thallic state, it combines with rhodamine B (basic violet 10) to give a complex extractable with benzene.[78] Dithizone in chloroform serves to extract thallous ion from a cyanide solution at pH 10, separating it from antimony, gold, iron, and tungsten. The thallium is then extracted from the chloroform phase with 1 : 100 nitric and fumed with sulfuric acid. After destruction of organic matter with persulfate and oxidation to the thallic form with bromine, the rhodamine B complex is extracted. The complex can also be extracted with diisopropyl ether.[79] For complete extraction of thallium, the solution must be at least *N* in hydrobromic acid.[80]

Alternatively,[81] thallium can be adsorbed as the chlorthallate ion on De-Acidite FF, an anion exchange resin. An appropriate medium is 0.1*N* hydrochloric acid. Other elements are eluted with 0.5 *N* nitric and hydrochloric acids, after which the thallium can be eluted with sulfurous acid.

The reagent has been applied for determination of thallium in the residue from acid decomposition of selenium.[82] Extraction was with diisopropyl ether. The rhodamine B chlorothallate can also be read by fluorescence at 580 nm.[83]

For determination of very minor amounts of thallium, the principle is applied by extraction of a halide complex of the element with a triphenylmethane dye for separation from interfering ions followed by displacement of the triphenylmethane dye by a rhodamine dye. Thus the technic combines the selectivity of the triphenylmethane dye with the sensitivity of the rhodamine dye.[84] As an example, the complex of crystal violet with thallic chloride is extracted with benzene from 2 *M* phosphoric acid, and the dye is substituted in a reducing medium with butylrhodamine B. A 1-gram sample will permit detection of 2×10^{-6} % of thallium with accuracy to 20%. Silver, tungsten, and vanadium interfere.

For determination of thallium in silicate rocks by fluorometry, digest 1 gram of sample with 1 : 1 hydrofluoric acid-nitric acid.[85] Add 5 ml of hydrochloric acid and evaporate to dryness. Take up the residue in ammoniacal solution at pH 10.5 ± 0.5. Add ascorbate, citrate, and cyanide as masking agents, then extract the thallium with dithizone in chloroform. Back-extract with dilute nitric acid and evaporate to

[78] Hiroshi Onishi, *Bull. Chem. Soc. Jap.* **29**, 945 (1956); *ibid.* **30**, 567–571 (1957).
[79] J. F. Woolley, *Analyst* **83**, 477–479 (1958).
[80] R. E. Van Aman and J. H. Kanzelmeyer, *Anal. Chem.* **33**, 1128–1129 (1961).
[81] A. D. Matthews and J. P. Riley, *Anal. Chim. Acta* **48**, 25–34 (1969).
[82] Masuo Miyamoto, *Jap. Anal.* **10**, 98–102 (1961).
[83] B. Hiroshi Onishi, *Bull. Chem. Soc. Jap.* **30**, 827–828 (1957).
[84] I. A. Blyum and A. I. Chuvileva, *Zh. Anal. Khim* **25**, 18–25 (1970).
[85] Marian M. Schneppe, *Anal. Chim. Acta* **79**, 101–108 (1975).

dryness. Treat the residue with sulfuric acid and 30% hydrogen peroxide to destroy organic matter. Take up in dilute hydrochloric acid and oxidize to thallic ion with bromine water. Add aluminum chloride to enhance the fluorescence and add rhodamine B. Activate at 560 nm and read the fluorescence of the chlorothallate complex at 580 nm. Beer's law is followed for up to 0.2 μg of thallium per ml.

Procedure. Mix 5 ml of neutral sample with 25 ml of 1:11 hydrochloric acid.[86] Add 1 ml of 0.1% ceric sulfate solution and 1 ml of 10% solution of potassium bromide. Mix and let stand for 10 minutes. Add 0.01% solution of hydroxylamine sulfate until the yellow color of residual ceric sulfate disappears and 1 ml in excess. Add 1 ml of 0.1% rhodamine B solution. After 5 minutes extract with 5 ml of benzene and read at 560 nm.

Lead.[87] Dissolve 1 gram in 10 ml of 1:3 nitric acid and boil off brown fumes. Add 50 ml of water and 1 ml of 1% potassium permanganate solution. Boil gently in the presence of silicon carbide granules and add 1 ml of 1% manganese nitrate solution. Boil for 2 minutes, add another 1 ml of 1% manganese nitrate solution, and boil 2 minutes longer. Cool to 60–70° and filter on paper. Wash well and discard the filtrate. Heat the paper and contents with 4 ml of sulfuric acid, 10 ml of nitric acid, and 0.5 ml of perchloric acid. When the paper has been destroyed and the nitric acid expelled, the solution will begin to darken with permanganate ion. Heat to copious fumes of sulfur trioxide fumes to expel all the nitric and perchloric acids. Add about 0.1 gram of hydrazine sulfate and evaporate to 1 ml over a flame. Cool and add 10 ml of 1:9 hydrochloric acid and 1 ml of saturated bromine water. Heat at 80–90° until the bromine is expelled. Cool to $25 \pm 0.5°$ and mix with 10 ml of 1:9 hydrochloric acid. Add 2 ml of 0.1% rhodamine B solution, 1 ml of 33% butyl Cellosolve solution, and 15 ml of benzene. Shake for 1 minute, and centrifuge the benzene layer to remove water droplets. Read at 565 nm against benzene.

Zinc or Cadmium.[88] Add a sample containing up to 0.025 mg of thallium to 1 ml of sulfuric acid and 4 ml of hydrobromic acid per gram of zinc or 2 ml per gram of cadmium. When reaction abates, add 5 ml of hydrobromic acid and heat gently to minimize loss of hydrobromic acid. Add 1 ml of a ceric sulfate solution containing 2 mg of the tetrahydrate to oxidize thallous ion to thallic ion. Dilute to about 30 ml, and shake for 30 seconds with 15 ml of isopropyl ether. Discard the aqueous layer. Add 20 ml of 0.01% rhodamine B solution in 1:30 hydrochloric acid and shake for 30 seconds. Read the organic layer at 555 nm against isopropyl ether. The hydrobromic acid must be at least M to obtain complete extraction of thallic bromide into isopropyl ether. Antimony begins to interfere above 3 M hydrobromic acid.[89]

[86] V. Miketukova and J. Kohlicek, *Z. Anal. Chem.* **208**, 7–15 (1965).
[87] C. L. Luke, *Anal. Chem.* **31**, 1680–1682 (1959).
[88] R. L. Van Aman and J. H. Kanzelmeyer, *Anal. Chem.* **33**, 1128–1129 (1961); cf. British Standards Institution, *BS 3630*, Pt. **10** (1967).
[89] R. E. Van Aman, F. D. Hollibaugh, and J. L. Kanzelmeyer, *Anal. Chem.* **31**, 1783–1785 (1959).

Ores.[90] Mix 0.1–0.2 gram of sample with 7 ml of aqua regia and heat for 15 minutes. Evaporate to dryness, add 2 ml of hydrochloric acid, and evaporate to dryness. Repeat twice more to eliminate nitric acid. Take up the residue in 5 ml of hydrobromic acid and evaporate nearly to dryness. Repeat three times and take to dryness. Take up the residue in 5 ml of *N* hydrobromic acid and boil until the bromine is removed. Filter and dilute to 25 ml with *N* hydrobromic acid.

To separate from antimony and mercury, extract a 1-ml aliquot for 90 seconds with 2 ml of ether saturated with *N* hydrobromic acid. Repeat the extraction. Extract the combined ether extracts with 1 ml of *N* hydrobromic acid saturated with ether. Repeat the extraction of the ether extracts. Discard the ether and combine the hydrobromic acid solutions. If antimony and mercury are absent, omit this paragraph.

Mix the extracted hydrobromic acid solution or a 1-ml aliquot from the 25-ml dilution with 0.1 ml of 0.1% rhodamine B in 5 *N* hydrobromic acid, and extract with 1 ml of benzene. Read at 560 nm after 5 minutes. The thallium in 1 ml of benzene should not be less than 5 μg.

Silicates.[91] Roast 1 gram of finely powdered sample for 30 minutes. Cool and moisten with water. Add 1 ml of perchloric acid, let stand for 15 minutes, and add 15 ml of hydrofluoric acid. Heat to white fumes. Add 0.5 ml of perchloric acid and 10 ml of hydrofluoric acid and heat almost to dryness. Repeat this step three more times. Dissolve the residue in 15 ml of boiling water and cool. Add 5 ml of 25% sodium citrate solution and neutralize by dropwise addition of 10% sodium hydroxide solution. Add 2 ml of 20% potassium cyanide solution, 1 ml of fresh 1% solution of ascorbic acid, and 0.2 ml of 10% sodium hydroxide solution. Dilute to 30 ml.

To extract the thallium add 0.5 ml of 0.01% solution of dithizone in chloroform and shake for 3 minutes. The pink color should disappear from the chloroform layer. Wash the aqueous layer with 5 ml of chloroform. Shake the chloroform layer and washings with 10 ml and 10 ml of 1 : 100 nitric acid to back-extract the thallium. Wash the chloroform layer with 10 ml of water. Add 2 ml of 1 : 4 sulfuric acid to the aqueous extracts and evaporate to white fumes. If the residue is brown, decolorize by heating with 20 mg of ammonium persulfate and cool. Add 10 ml of 1 : 5 hydrochloric acid and 2 ml of bromine water. Heat nearly to boiling to drive off excess bromine and cool. Add 10 ml of 1 : 5 hydrochloric acid and 2.5 ml of 0.2% solution of rhodamine B. Extract with 20 ml, 10 ml, and 10 ml of benzene. Filter the combined benzene extracts to remove entrained water, dilute to 50 ml, and read at 560 nm.

Urine or Tissue.[92] Treat 200 ml of urine or 25 grams of tissue with 20 ml of nitric acid. Ash to a white residue adding hydrogen peroxide as necessary. Carbonization must not occur or thallium will be lost. Take up the residue in 20 ml of 1 : 10 hydrochloric acid. Add 0.2 gram of sodium sulfite to the clear solution and let stand for 3 hours. Boil off sulfur dioxide, cool, and dilute to 50 ml with water. Mix

[90]E. P. Stolyarov, *Vest. Leningr. Univ.* **14** (10); *Ser. Fiz. Khim.* **2**, 149–152 (1959).
[91]J. Minczewski, E. Wieteska, and Z. Marczenko, *Chem. Anal.* (Warsaw) **6**, 515–522 (1961).
[92]J. Duvivier, R. Versie, J. Bonnard, and A. Noirfalise, *Clin. Chim. Acta* **9**, 454–460 (1964).

a 10-ml aliquot with 2 ml of 2.5% cadmium sulfate solution and 10 ml of 4% sodium hydroxide solution containing 3.2% of sodium sulfide. Centrifuge after 2 hours and discard the supernatant liquid. Wash the precipitate with 5 ml of 4% sodium hydroxide solution. Dissolve the precipitate in 25 ml of 1 : 1 hydrochloric acid and boil off the hydrogen sulfide. Cool to 40° and add 0.3 ml of bromine. Let stand for 10 minutes and boil off the excess bromine. Evaporate to about 5 ml and chill with ice. Add 1 ml of 0.05% solution of rhodamine B in hydrochloric acid. Extract with 25 ml, 15 ml, and 15 ml of benzene. Dilute the benzene extracts to 50 ml and read at 540 nm against benzene.

Alternatively,[93] boil 200 ml of urine, 20 grams of homogenized tissue or 20 ml of blood with 5 ml of sulfuric acid and 5 ml of nitric acid, with addition of hydrogen peroxide and more nitric acid until the solution is no more than a faint yellow. Take to fumes of sulfur trioxide and cool. Add 10 ml of water and 5 ml of saturated ammonium oxalate solution. Boil to drive off oxides of nitrogen, cool, and dilute to 50 ml. At this point the concentration of sulfuric acid should approximate 2.5 *N*.

Mix an aliquot containing 1–9 μg thallium, usually 25 ml, with 1 ml of 0.1% ceric sulfate solution in 1 : 12 sulfuric acid and 1 ml of 10% potassium bromide solution. Mix and let stand for 10 minutes. Add 0.01% solution of hydroxylamine hydrochloride until the yellow of ceric ion disappears, and add 1 ml in excess. Add 3 ml of 0.05% rhodamine B solution. At this point the concentration of sulfuric acid should be 2–3 *N*. Extract with 5 ml of benzene for 3 minutes and read at 560 nm against a blank.

RHODAMINE 6 ZH

This reagent, which is basic red 1, complexes with thallous ion.

Procedure. *Thallium-Tellurium Mixture.*[94] Evaporate a sample containing thallium and tellurium in 1 : 1 hydrochloric acid, which may also contain a little nitric acid, to a moist residue and take it up in *N* hydrobromic acid. Dilute an aliquot of the solution with 2 *N* hydrobromic acid to contain 0.5–1 μg of thallium per ml. Mix 1 ml of this solution with 5 ml of 1 : 1 sulfuric acid, 1 ml of 2 *N* hydrobromic acid, and 0.5 ml of 40% potassium bromide solution. Add 0.5 ml of titanium trichloride to reduce the thallium to the monovalent state. After 10 minutes add 1 ml of 0.1% solution of rhodamine 6 ZH. Shake for 30 seconds with 5 ml of benzene. After 4 minutes read the fluorescence of the extract at 570 nm.

SAFRANINE

Up to 8 μg of thallium can be determined in the presence of 5 mg of gallium or indium with safranine (basic red 2) with less than ±5% error.

[93] V. Miketukova and K. Kacl, *Cesk. Farm.* **14**, 470–474 (1965).

[94] T. Vesiene, *Tr. Akad. Nauk Lit. SSR, Ser. B* [**2** (65)] 101–103 (1971).

Procedure. To a nonreducing solution containing 1–30 μg of thallium, add 1 ml of bromine water to oxidize the thallium and evaporate to dryness. Take up the residue in water acidified with hydrobromic acid. Add 2 ml of 6.8 *N* hydrobromic acid, 2 ml of *N* potassium bromide solution, and 0.5 ml of 0.2% solution of safranine. Dilute with water to 20 ml and extract with 3 ml, 3 ml, and 3 ml of benzene. Dilute the combined extracts to 10 ml with benzene and read at 490 or 510 nm.

2-THENOYLTRIFLUOROACETONE AND DIANTIPYRINYLTOLUENE

The reagents above react with thallic ion to give a color for extraction and reading.[95] The absorption conforms to Beer's law for 0.7–10 μg of thallic ion per ml of extract. Excess of the ferric complex can be extracted with benzene without affecting the thallic complex.

Procedure. Adjust a sample 0.03–0.6 μ*M* with thallium to pH 1–1.5. Make the solution 0.02 *M* with 2-thenoyltrifluoroacetone and 0.1 *M* with diantipyrinyltoluene. Dilute to 10 ml and set aside for 30 minutes. Extract with 10 ml of 1 : 1 isoamyl alcohol–chloroform and read at 492 nm.

o-TOLIDINE

Liberated iodine from decomposition of thallic iodide to thallous iodide is determined as iodine by *o*-tolidine, a normal reagent for halogens.[96]

Procedure. *Sodium Iodide.* Treat a 0.5-gram sample with 4–5 drops of sulfuric acid and evaporate the acid to remove iodine compounds. Take up the residue in 50 ml of water and add 10 ml of bromine water. Then add 1.5 ml of 10% sodium hydroxide solution and 1.5 ml of 10% sulfuric acid. Decolorize the solution with a few drops of 5% phenol solution and dilute to 250 ml. Let stand for 3 hours and dilute to 400 ml. Mix a 10-ml aliquot with 0.12 ml of 10% solution of *o*-tolidine in acetic acid at pH 2.5–3.1. Read the blue-green absorption at not more than 5 μg of thallium per ml.

[95] A. I. Busev and V. Z. Filip, *Vest. Mosk. Gos. Univ., Ser. Khim.* **1969**, 99–101.
[96] A. P. Kilimov, Yu. V. Naboĭkin, and A. M. Volvova, *Tr. Vses. Nauch-Issled. Inst. Khim. Reak.* **1958** (22), 124–127.

XYLENOL ORANGE

The 1 : 1 chelate of thallic ion with xylenol orange is formed at pH 4 ± 0.2.[97] Beer's law applies for 0.02–0.15 mg in 25 ml. The color develops instantaneously, but it fades slowly, requiring standardization of the operation. There is serious interference by chloride, bromide, iodide, and EDTA. Colored solutions are formed with the trivalent ions of bismuth, gallium, indium, and iron, and by the tetravalent ions of thorium, vanadium, and zirconium. Addition of fluoride ion avoids interference from thorium and zirconium.

Procedure. To a sample solution containing 0.02–0.15 mg of thallium add 5 ml of 1 : 8 acetic acid and 4 ml of 0.001 *M* xylenol orange. Dilute to 25 ml, and after 15 minutes read at 520 nm against a reagent blank.

Chlorthallic ion complexes with **acridine orange** (solvent orange 15) at pH 0.8–2.5.[98] The product is extractable with butyl acetate for reading the absorption at 495 nm or the fluorescence at 520 nm. The latter is readable from 3 ng to 6 μg of thallium per ml of extract. Interference by mercury, auric ion, and stibnic ion is avoided by plating out in a copper wire. Indium can be preextracted from 1 : 3 hydrochloric acid with butyl acetate.

Bis(amine)tetrakis(thiocyanato)chromate complexes with thallium.[99] For this determination, make a 50-ml sample containing 10–100 mg of thallium nearly to *N* with nitric acid. Add an excess of 2% solution of the reagent in 50% ethanol. Filter off the precipitate on a porous crucible and wash with 25% ethanol. Dry the precipitate, and dissolve in an appropriate volume of dimethylformamide, usually 50 ml. Read with a green filter.

Thallium in trivalent form reacts with ***p*-aminophenol** in acid solution to produce a violet color.[100] This develops in 20 minutes, lasts 2–4 hours, and is read at 570 nm.

When thallium has been oxidized to the trivalent state by bromine and the excess has been removed by addition of phenol, **aminopyrine** will give a red color for reading at 510 nm.[101]

To a sample solution containing about 0.1 mg of thallium, add 5 ml of 10% sodium cyanide solution, 5 ml of 5% sodium tartrate solution, and 1 ml of 1% solution of **ammonium phenazo-4-carbodithioate**. Extract with 10 ml of chloroform, and convert the lead complex to a nickel compound by shaking the organic layer with a solution containing 0.29% of nickel nitrate hexahydrate and 0.282% of

[97] Makoto Otomo, *Bull. Chem. Soc. Jap.* **38**, 1044–1046 (1965); Chandra D. Dwivedi and Arun K. Dey, *Mikrochim. Acta* **4**, 708–711 (1968).

[98] L. A. Grigoryan, F. V. Mirzoyan, and V. M. Tarayan, *Zh. Anal. Khim.* **28**, 1962–1965 (1973).

[99] Didina Oprescu, Csaba Varhelyi, and Ion Ganescu, *Stud. Univ. Babes-Bolyai, Ser. Chem.* **17**, 23–27 (1972).

[100] V. P. Gladyshev and G. A. Tolstikov, *Zavod. Lab.* **22**, 1166–1168 (1956).

[101] Yuichiro Kamemoto, *J. Chem. Soc. Jap.* **78**, 604–607 (1957).

sodium potassium tartrate. Read at 575 nm. Lead and bismuth give the same reaction.

For thallic ion by **astrazon blue** (basic blue 5), make the sample solution 2 *M* with hydrobromic acid and add excess of 0.15 m*M* dye solution.[102] Extract with 2:1 benzene-methyl ethyl ketone, and read the extract at 620 nm over a range of 1–6.7 μg per ml of solvent. There is interference by ferric, indium, gallium, and antimony ions. The complex of the dye and thallic bromide absorbs at 600 nm in aqueous solution and at 630 nm in benzene.[103]

Thallic ion forms a 1:2 complex with **6-(2-benzothiazolylazo)-5-bromo-*m*-cresol** in *N* nitric acid containing 80% of ethanol.[104] The color is stable for 30 minutes and is read at 630 nm. Beer's law is followed for 2–24 μg of thallium per ml. There is interference by mercuric ion, antimonous ion, chloride, and bromide.

Thallic ion reacts with **1,5-bis(1-benzylbenzamidazol-2-yl)-3-methylformazan** in chloroform.[105] The technic is to extract the thallium from an aqueous solution at pH 3. Gold must be absent. The 1:1 complex formed has a maximum absorption at 656 nm.

Thallic bromide buffered at pH 3–5 reacts with 3 μ*M* 4:1 **bipyridyl-ferrous chelate.**[106a] An extract at pH 5 with 1,2-dichloroethane is read at 324 nm. Mercuric, auric, and iodide ions must be absent.

At pH 2.5–5, thallic ion forms a blue precipitate with **5-(1-benzylbenzimidazol-2-yl)-3-methyl-1-(4-nitrophenyl)formazan.**[106b] This is extractable with nonpolar solvents such as carbon tetrachloride and benzene for reading at 670 nm. Beer's law is followed for 0.3–5 μg of thallium per ml.

To a sample solution containing 2–23 μg of thallium, add 4 ml of hydrobromic acid and 2 ml of 0.4% solution of **brilliant cresyl blue.**[107] Dilute to 10 ml and extract with 10 ml of 6:1 benzene–methyl ethyl ketone. Read the 1:1 complex at 635 nm. Beer's law is followed for 0.2–2.3 μg of thallium per ml. Aluminum interferes; iron must not exceed the thallium content.

Thallous ion is determinable by the violet fluorescence at 430 nm of the **chlorothallate ion.**[108] Excitation is at 250 nm. To minimize interference, oxidation to the thallic form by hydrogen peroxide is followed by extraction from 1:5 hydrochloric acid by ethyl ether. This separates from all interfering ions except bismuth, gold, pentavalent antimony, and tetravalent platinum. After evaporation of the ether, the solution in 3.3 *N* hydrochloric acid plus 0.8 *N* potassium chloride is reduced with sulfur dioxide. A large excess of sulfate ion is tolerated but more than 100-fold excess of nitrate ion interferes. When the thallous solution is mixed with 9 volumes of saturated sodium chloride solution, the fluorescence is a linear function below 0.05 μg/ml.[109]

[102]Cecelia Constantinescu, *Rev. Chim.* (Bucharest) **23**, 495–497 (1972).
[103]Ana Serbanescu, *ibid.* **24** (6), 475–478 (1973).
[104]S. I. Gusev, I. M. Shevaldina, and G. A. Kurepa, *Zh. Anal. Khim.* **30**, 279–283 (1975).
[105]S. A. Lomonosov, Yu. A. Rybakova, V. N. Podchainova, and N. P. Bednyagina, *Zh. Anal. Khim.* **19**, 1062–1066 (1964).
[106a]K. Kotsuji, Y. Yoshimura, and S. Eeda, *Anal. Chim. Acta* **42**, 225–231 (1968).
[106b]M. I. Dement'eva and V. N. Podchainova, *Zavod. Lab.* **41**, 1319–1321 (1975).
[107]Ana Serbanescu, *Rev. Chim.* (Bucharest) **26**, 771–775 (1975).
[108]G. F. Kirkbright, T. S. West, and C. Woodward, *Talenta* **12**, 517–524 (1965).
[109]R. Bock and E. Zimmer, *Z. Anal. Chem.* **198**, 170–173 (1963).

An indirect method for thallium is to precipitate as the **chromate**, dissolve the precipitate in hydrochloric acid, and determine the chromium by diphenyl carbizide.[110] If aluminum, beryllium, or chromium is present, the element is masked by sodium sulfosalicylate before precipitation.

As another indirect method, thallium is precipitated as a double **cobalt ammonium chloride.**[111] The precipitate is converted to cobalt sulfide. The latter, dissolved in nitric acid, is determined by nitroso-R salt.

Thallic ion forms a 1:1 complex with **α,α-diantipyrinyl-4-dimethylaminobenzyl alcohol.**[112] The complex can be extracted into 4:1 carbon tetrachloride–nitrobenzene from an acid medium containing excess bromide ion. Reading at 550 nm will then determine 2 ppm of thallium.

Thallic halide forms a 1:1 complex with **3-(4-diethylaminophenylazo)-1,4-dimethyl-1,2,4-triazolium.**[113] Desirable conditions are 0.1–1 *N* sulfuric acid, 0.1–1 *M* potassium bromide, and 0.1 m*M* reagent. The complex is extracted into benzene for reading at 548 nm and obeys Beer's law up to 8 μg of thallium per ml. Pentavalent antimony and auric ion interfere.

Thallic ion forms a 1:2 complex with **5-diethylamino-2-(2-pyridylazo)phenol** and its **5-bromo** and **3,5-dibromo** analogues.[114] A benzene solution of the dibromo compound absorbs at a maximum at 480 nm, and the thallic complex absorbs at 580 nm. It conforms to Beer's law over the range of 0.2–1.6 μg/ml with accuracy to ±5%. There is no interference by large amounts of aluminum, lead, cadmium, and silver, nor by small amounts of zinc and antimony.

Other dyes with complexes with thallic ion in hydrochloric acid and their maxima are **dimethylthionine** (azure A), 645 nm; **trimethylthionine** (azure C), 660 nm; and **toluidine blue** (basic blue 17), 660 nm.[115a] They are extracted into dichloroethane-trichloroethylene, the first from 0.1–1 *N*, the others from 0.01–1 *N* hydrochloric acid. The color with toluidine blue is read at 660 nm over the range 0.05–1 μg per ml of extract. Gold, mercury, and antimony interfere. It is necessary to have 0.1–2 μg of thallium per ml of extract. The colors are stable for 10 hours. Similarly, with a substituted phenoxazinium sulfate the reading is at 650 nm.

Fuchsine (basic violet 14) gives a complex for extraction with dichloroethane-trichloroethane or butyl acetate for reading at 545 nm. Only mercury, antimony, and nitrate interfere.

The optimum conditions for determination of thallous ion with **ethylrhodamine B** are 10 *N* sulfuric acid, 10 *N* bromide ion, and 0.05 mg of reagent per ml. Extraction with benzene from 8 *N* sulfuric acid containing 20 mg of bromide per ml and 0.4 mg of reagent provides preconcentration.[115b]

Thallic ion forms complexes with **flavonol-2-sulfonic acid**, which is 2-(3-hydroxy-4-oxo-4H-1-benzopyran-2-yl)benzenesulfonic acid.[116] At pH 2.2–2.6, the

[110] H. Zimmer, *ibid.* **165**, 268–271 (1959).
[111] Toru Nozaki, *J. Chem. Soc. Jap.* **77**, 493–496 (1956).
[112] S. I. Gusev, A. S. Pesis, and E. V. Sokolova, *Zh. Anal. Khim.* **20**, 67–71 (1965).
[113] P. P. Kish, S. G. Kremeneva, and E. E. Monich, *ibid.* **29**, 1741–1747 (1974).
[114] S. I. Gusev and G. A. Kurepa, *Zh. Anal. Khim.* **24**, 1148–1151 (1969).
[115a] V. M. Tarayan, E. N. Ovsepyan, and V. Zh. Artsruni, *Arm. Khim. Zh.* **21**, 819–820 (1968); *Zh. Anal. Khim.* **25**, 691–694 (1970).
[115b] I. A. Bochkareva and I. A. Blyum, *Zh. Anal. Khim.* **30**, 874–882 (1975).
[116] Katsumi Yamomoto and Kiyoshi Takamizawa, *J. Chem. Soc. Jap., Pure Chem. Sect.* **88**, 345–348 (1967).

complex is 1 : 1 with an absorbancy maximum at 368 nm. At pH 6–7.6 a 2 : 1 complex absorbs at 372 nm. At pH 7 the method is applicable to 0.3–7 μg of thallium per ml.

Thallic bromide forms a 1 : 1 complex with **furfurol green**, which is the hydrochloride of α,α-bis-(4-dimethylaminophenyl)furfuryl alcohol.[117] The complex is almost insoluble in water but dissolves in organic solvents. Desirable conditions are 1.38 *N* hydrobromic acid, 0.125% solution of the reagent, extraction with 2 : 1 benzene–methyl ethyl ketone, and reading at 640 nm. The maximum absorption by the reagent is at 627 nm. Beer's law is followed for 0.2–2.5 mg per ml of extract. There is interference by auric, indium, ferric, gallium, and germanium ions.

When thallium is precipitated as the **hexanitrodiphenylaminate** by the magnesium salt of the reagent, the solution of the precipitate in acetone is read at 400 nm.[118a] The reagent also precipitates potassium, rubidium, and cesium.

Thallium is extracted from lead, zinc, and cadmium by an **isopropylcarbaminate** reagent.[118b] The solution in ether is read at 315 nm.

As an indirect method mix a solution containing 0.3–1.7 mg of thallous ion with an excess of m*M* sodium tetraphenylborate.[118c] Add 1 ml of 0.01 *N* **luciginin**, which is 10,10′-dimethyl-9,9′biacridinium dinitrate, and extract the thallium complex with methyl isobutyl ketone. Read the excess luciginin at 410 nm. Silver and mercury interfere.

Thallous ion in a borate buffer solution for pH 10.5 reacts with **mercaptosuccinic acid** for absorption at 300 nm.[119] Lead, bismuth, mercury, and indium interfere seriously.

The 1 : 3 complex formed by thallic ion with **N-methylanabasine-(α′-azo-7)-8-hydroxyquinoline** in acid solution has a maximum absorption at 570 nm.[120] A fortyfold amount of most common ions does not interfere.

Thallium oxidized to the tetravalent state reacts with **methyl pyridinium iodide**.[121a] This is buffered at pH 2 in 70% ethanol and read for 10–20 ppm of thallium at 400 nm. The interferences are numerous. Those by ferric and cupric ions are suppressed by EDTA and zinc ion.

Trivalent thallium is extracted from 0.5 *N* sodium bromide in 10 *N* sulfuric acid with 0.3 *N* bis(2-ethylhexyl) hydrogen phosphate in heptane.[121b] Treat a portion of the extract containing 3–25 μg of thallium with 0.01% solution of **methyl red** in 100 : 1 benzene-ethanol. A 1 : 1 : 1 complex is formed by thallic ion, the extracting reagent, and the dye. Read at 555 nm. Beer's law is followed for 0.5–4 μg of thallium per ml.

Thallium can be precipitated as the **molybdophosphate** which, reduced at 100° in *N* perchloric acid with hydrazine sulfate, forms soluble heteropoly-blue.[122] This is read at 808 nm over a range of 2–16 ppm against a reagent blank. Alternatively, the

[117]Cecelia Constantinescu, *Rev. Chim.* (Bucharest) **24**, 919–922 (1973).

[118a]D. S. Gorbenko-Germanov and R. A. Zenkova, *Zh. Anal. Khim.* **20**, 1020–1022 (1965).

[118b]Hans Pohl, *Z. Erzbergb. Metalhüttenw.* **9**, 530–531 (1956).

[118c]I. Sarudi and J. Inczedy, *Z. Anal. Chem.* **282**, 48 (1976).

[119]Yasuhisa Hayashi and Isao Muraki, *Jap. Anal.* **14**, 516–519 (1965).

[120]M. Yusupov, R. Kh. Dzhiyanbaeva, and Sh. T. Talipov, *Nauch. Tr. Tashk. Univ.* **1967**, 61–69.

[121a]D. Betteridge and J. H. Yoe, *Anal. Chim. Acta*, **27**, 1–8 (1962).

[121b]L. G. Anokhina, T. F. Rodina, T. P. Borovkova, and I. S. Levin, *Zh. Anal. Khim.* **29**, 2290–2292 (1974).

[122]L. G. Hargis and D. F. Boltz, *Anal. Chem.* **37**, 240–247 (1965).

precipitate is dissolved in a borate buffer solution for pH 9 and read at 227 nm for 1–5 ppm or 209 nm for 0.5–3 ppm against the buffer solution.

Mix 5 ml of 10 m*M* solution of thallic ion with 1 ml of hydrobromic acid and 2 ml of 0.05% solution of *o*- or *p*-**nitral green.**[123] Dilute to 10 ml and shake for 2 minutes with 10 ml of benzene. Read at 645 nm for the *ortho* compound or 655 nm for the *para* compound. Beer's law is followed for 0.05–6.1 μg of thallic ion per ml of extract. Indium, auric, ferric, germanium, gallium, and antimonous ions interfere.

In a survey of some 14 reagents, **phenylfluorone** with gelatin present as a stabilizer or **methylthymol blue** was recommended for thallium.[124]

1-(2-Pyridylazo)2-naphthol forms a 2:1 complex with thallic ion.[125] Extracted from 0.1–0.5 *N* solution of a strong acid by 0.3 *N*-bis(2-ethylhexyl) hydrogen phosphate in heptane, the thallium is reacted with m*M* reagent in benzene. By extraction with a tenfold excess of PAN in solution in chloroform at pH 2.4–5, the crimson thallic complex is read at 570 nm.[126] Beer's law is followed for 0.5–24 μg of thallic ion per ml. The optimum pH is 1.5–4.[127] There is interference by cupric ion, bismuth, mercury, chloride, bromide, and iodide.

For thallium, mix a sample solution containing 5–15 μg of thallium with 6 ml of hydrobromic acid. Add 0.5 ml of a 0.4% solution of **pyronine G** and dilute to 10 ml. Extract the 1:1 complex with 5 ml of benzene and read at 520 nm.[128]

Thallic ion forms a 2:1 complex with **6-(2-quinolylazo)-3,4-xylenol** at pH 0.8–5.[129] This is extracted with chloroform for reading at 615 nm. Beer's law is followed for 1–30 μg of thallium per ml. Bromides and iodides interfere.

Thallium forms a 1:1 complex in 60% ethanol with **rezarson** which is 5-chloro-3-(2,4-dihydroxyphenylazo)-2-hydroxybenzenearsonic acid.[130] The reagent is applied in 50% ethanol at pH 1–2.2 for reading at 495 nm.

A violet complex of thallic ion with **solochrome cyanine R** at a pH round 3.5 is read in the range of 1–4 ppm.[131]

Sulfanilic acid, *p*-toluidine, and ***p*-anisidine** serve as photometric reagents for thallic ion.[132] The thallium is precipitated by ammonium hydroxide with aluminum hydroxide as a carrier; then it is filtered and washed. After heating with the reagent solution at 70°, the precipitate is dissolved in 1:18 sulfuric acid and read at 420 nm. The absorbance is linear up to 0.16 mg/ml.

[123]Cecelia Constantinescu and Simon Fisel, *Rev. Roum. Chim.* **19**, 329–335, 509–515 (1974).

[124]E. A. Bashirov, M. K. Akhmedli, E. L. Glushchenko, A. M. Sadykova, and L. I. Zykova, *Uch. Zap. Azerb. Gos. Univ. Ser. Khim. Nauk*, **1970**, 27–34.

[125]T. F. Rodina, V. S. Kolomiichuk, and I. S. Levin, *Zh. Anal. Khim.* **28**, 1090–1092 (1973).

[126]K. Rakhmatullaev and Kh. Tashmamatov, *Manuscript No.* **748-74**, deposited at Vsesoyuznyi Institut Nauchnoi i Tekhnicheskoi Informatsii, Moscow, 6 pp (1974).

[127]S. I. Gusev, L. G. Dazhina, G. A. Kurepa, and L. V. Poplevina, *Uch. Zap. Perm. Gos. Univ.* **1973** (289), 246–254.

[128]A. Serbanescu, *Ann. Chim. Anal.* **2**, 34–37 (1972); cf. S. Fisel, A. Craciun-Ciobanu, I. Popescu, and M. Poni, *Rev. Roum. Chim.* **9**, 559–563 (1964).

[129]K. Rakhmatullaev, Kh. Tashmamatov, and M. A. Rakhmatullaeva, *Zh. Anal. Khim.* **29**, 1020–1022 (1974).

[130]A. M. Lukin, N. A. Kaslina, V. I. Fadeeva, and G. S. Petrova, *Vest. Mosk. Gos. Univ., Ser. Khim.* **1972** (2), 247–249.

[131]Arun P. Joshi and Kailash N. Munshi, *Microchem. J.* **12**, 447–453 (1967).

[132]G. Frepper, *Z. Anal. Chem.* **193**, 179–186 (1963).

A 1 : 1 complex of **tetradecyldimethylbenzylammonium chloride** with thallic iodide in neutral solution extracted into dichloroethane is read at 395 nm.[133]

To a solution containing 12.5–175 μg of thallic ion, add 4 ml of m*M* **4-(2-thiazolylazo)-3-hydroxynaphthalene-2,7-disulfonic acid.**[134] Adjust to pH 2.3–2.8, dilute to 25 ml, and read at 580–590 nm. Oxalate interferes seriously. Interference by gallium can be prevented by adding a tenfold amount of tartrate. Indium interferes.

At pH 1.–2.8 the complex of thallic ion with **2-(2-thiazolylazo)-1-naphthol** is stable for 30 minutes for reading at 610 nm.[135]

To a sample containing 0.1–1 millimole of thallium, add 3 ml of *M* hydroxylamine hydrochloride, 10 ml of 16.8% potassium hydroxide solution containing 6.5% potassium cyanide, and 10 ml of 3 m*M* benzene solution of **thiodibenzoylmethane,** which is 2-(thiobenzoyl) acetophenone.[136] Shake for 5 minutes and separate the organic layer. Wash with 20 ml and 20 ml of 16.8% potassium hydroxide solution. Read the benzene layer at 415 nm against a reagent blank.

For the determination of 20–300 ppm of trivalent thallium in the presence of monovalent thallium, add the sample solution to 0.015 *M* potassium **thiocyanate** solution containing 0.025 *M* pyridine.[137] Adjust to pH 5.2–5.5 and read at 405 nm. There is serious interference by gallium, indium, aluminum, cupric, silver, ferric, uranyl, molybdate, vanadate, and bromide ions.

As indirect method, thallium is precipitated from cyanide solution by **thionalide,** which is thioglycollic β-aminonaphthalide.[138] A solution of the precipitate reduces phosphotungstomolybdic acid to give a molybdenum blue.

Thiosalicylamide added as 0.01 *M* solution in ethanol to thallic ion in 2–3.5 *N* sulfuric acid reacts in a 1 : 2 ratio to form an orange-red precipitate.[139] A chloroform extract of the complex is read at 460 nm.

Tris-(2,2′-bipyridyl)ferrous tetraphenylborate reacts at pH 3–10 with thallous ion to liberate tris-(2,2′-bipyridyl)ferrous ion, which is read at 522 nm.[140a] Silver and mercuric ions react similarly. The reaction will determine 5–50 ppm of thallium. Chloride interferes.

In 10 m*M* **uranyl sulfate** containing *M* phosphoric acid at 25° the quenching of the fluorescence due to the uranyl ion by 0.5–400 μM thallous ion conforms to Beer's law.[140b] There is interference by halides, silver, ferrous ion, mercurous ion, and stannous ion.

Victoria blue 4R (basic blue 8) reacts in hydrochloric acid with thallic chloride to give a complex readily extractable from sulfuric acid.[141] Suitable solvents are

[133]Tsutomo Matsuo, Junichi Shida, and Toshitsugu Sasaki, *Jap. Anal.* **16**, 546–551 (1967).

[134]A. I. Busev, T. N. Zholondkovskaya, L. S. Krysina, and N. A. Colubkova, *Zh. Anal. Khim.* **27**, 2165–2169 (1972).

[135]S. I. Gusev, G. A. Kurepa, and I. M. Shevaldina, *ibid.* **29**, 1535–1538 (1974).

[136]E. Uhlemann and B. Schuknecht, *Anal. Chim. Acta* **69**, 78–84 (1974).

[137]R. S. Ramakrishna and M. E. Fernandopulle, *ibid.* **60**, 87–92 (1972).

[138]Richard Berg and W. Roebling, *Angew. Chem.* **48**, 430–432, 597–601 (1935); Richard Berg, E. S. Fahrenkamp, and W. Roebling, *Mikrochem., Festchr. von Hans Molish* **1936**, 42–51.

[139]S. C. Shome and M. Mazumdar, *Anal. Chim. Acta* **46**, 155–158 (1969).

[140a]M. C. Mehra and P. Obrien, *Microchem. J.* **19**, 384–389 (1974).

[140b]Masataka Moriyasu, Yu Yokoyama and Shigero Ikeda, *Jap. Anal.* **24**, 257–261 (1975).

[141]P. P. Kish and E. E. Monich, *Zh. Anal. Khim.* **25**, 272–276 (1970).

benzene and its homologues, chloroform, bromobenzene, anisole, or propyl benzoate. Absorption maxima at 556 and 608 nm conform to Beer's law over the range of 0.1–10 μg/ml.

Victoria pure blue BO (basic blue 7) is also applicable to 1–7ml of sample 3.5–5.5 *M* with hydrobromic acid.[142] Add 0.4 ml of 0.05% solution of the dye and dilute to 10 ml. Extract with 10 ml of 6 : 1 benzene–ethyl methyl ketone and read the extract at 625 nm for 0.1–3 μg of thallic ion per ml. Gold, iron, and indium interfere at 0.1 m*M*.

[142]Cecelia Constantinescu, *Rev. Chim.* (Bucharest) **24**, 565–567 (1973); Cecelia Constantinescu and G. Constantinescu, *Revue Roum. Chim.* **21**, 571–583 (1976).

CHAPTER THREE
SILVER

The various methods for determination of silver by binary complexes are subject to criticisms.[1a] The most common method, using *p*-dimethylaminobenzylidenerhodanine, is very sensitive to change of acidity. Dithizone requires careful standardizing of laboratory technic, and the complex is photosensitive. Pyrogallol red has only a low sensitivity, like 2-amino-6-methylthio-4-pyrimidine carboxylic acid, which has also many interferences. Most of the drawbacks of binary complexes are eliminated by ternary complexes, which are generally more sensitive.

For various mineralogical forms of silver in galena extract successively cerargyrite with a 2.5% solution of ammonia, isomorphous silver sulfide with a 2% solution of EDTA and 0.3% hydrogen peroxide in a 1.25% solution of ammonia, perargyrite and argentite with hydrochloric acid, metallic silver with a 10% solution of ferric sulfate in 2% nitric acid and hessite with 1 : 1 hydrochloric acid-nitric acid.[1b] Fluorometry with eosin is appropriate for the first and the last two. Spectrophotometry with eosin is appropriate for the second and third and last.

2-AMINO-6-METHYLTHIO-4-PYRIMIDINECARBOXYLIC ACID

The silver chelate of this reagent[2] is suitable for reading at 375 nm and pH 10. The interferences are numerous. The color is stable for a month. The maximum sensitivity at pH 11 is to be avoided because of slow formation of silver oxide, giving a bleaching effect.

[1a]Mohamed T. El-Ghamry and Roland W. Frei, *Anal. Chem.* **40**, 1986–1990 (1968).
[1b]A. A. Antipina, M. I. Timerbulatova, D. N. Lisitsyna, and D. P. Shcherbov, *Zh. Anal. Khim.* **29**, 2181–2185 (1974).
[2]Okkung K. Chung and Clifton E. Meloan, *ibid.* **39**, 383–385 (1967).

Procedure. As reagent, dissolve 0.463 gram of 2-amino-6-methylthio-4-pyrimidinecarboxylic acid in 110 ml of dimethyl formamide and dilute to 250 ml with water. Add five pellets of sodium hydroxide to avoid precipitation at under 20°. The sample should contain no more than 50 mg of silver, with chloride and periodate absent. Add an approximately fivefold excess of the reagent and adjust the pH to 10 ±0.1 by addition of 20% sodium hydroxide solution. Dilute to a known volume and read at 375 nm against a blank.

BRILLIANT GREEN

The complex of silver with brilliant green (basic green 1) can be extracted from alkaline cyanide solution. Beer's law applies at 0.05–0.9 μg of silver per ml of the extract.[3] Brilliant green combines with the anion tetraphenylborate as a solid reagent.[4] Shaken with a solution containing 0.25–4 ppm of silver or mercuric ion, it releases the colored cation.

Procedure. *Sulfide Minerals.* Decompose a 1-gram sample with nitric and perchloric acids. Cool, dilute to 25 ml, and filter. Evaporate the filtrate to dryness. Take up the residue in 5 ml of 1 : 5 nitric acid and add 5 mg of tellurium. Dilute to 10 ml, heat to 70–80°, and add 5% stannous chloride solution dropwise until precipitation is complete. Filter and wash with hot 1 : 5 hydrochloric acid followed by hot water. Dissolve the precipitate with 5 ml of hot 1 : 1 nitric acid and evaporate to dryness. Take up the residue in 0.5 ml of 0.4% sodium hydroxide solution and 1 ml of 0.1% potassium cyanide solution. Dilute to a known volume. Dilute an aliquot to 30 ml. Add 10 ml of benzene and 2.5 ml of 0.2% brilliant green solution. Shake for 15 seconds and let the layers separate. Read the benzene layer at 640 nm against benzene.

BROMOPYROGALLOL RED

A compound of pyrogallol red with silver is read in aqueous solution at 635 nm. Ferrous ion is masked by 1 : 10-phenanthroline, uranyl and thorium ions by fluoride, pentavalent niobium by hydrogen peroxide, and many other ions by EDTA. In 100-fold excess over silver there is interference by auric, cyanide, and thiosulfate ions.[5]

[3] N. V. Markova, N. S. Sumakova, T. V. Yakubtseva, and A. K. Poltorykhina, *Tr. Tsent. Nauchn.-Issled. Gornorazved. Inst. Tsvet. Redk. Blagorodn. Met.* **1969**, 244–250.
[4] M. C. Mehra and C. Bourque, *Analysis* **3**, 299–302 (1975).
[5] R. M. Dagnall and T. S. West, *Talenta* **11**, 1533–1541 (1964).

Procedure. The sample should not exceed 40 ml containing 1–10 μg of silver. Add to a mixture of 1 ml of 0.1 *M* EDTA, 1 ml of 0.001 *M* 1,10-phenanthroline, 1 ml of 20% ammonium acetate solution, and 2 ml of 0.0001 *M* bromopyrogallol red. Dilute to 50 ml with water and read at 635 nm within 30 minutes against a blank.

COPPER DIETHYLDITHIOCARBAMATE

When a solution of silver is brought into contact with a solution of copper diethyldithiocarbamate in organic solvent, the greater stability of the silver derivative causes copper to be displaced. The result is a decrease in the color of the reagent proportional to the silver content.[6] The silver must not be present as bromide. Palladium and gold interfere.

When an aqueous mercuric solution is added to a known amount of cupric diethyldithiocarbamate complex at pH 8.5–5, the mercury will displace copper from the complex, and the reduction of color of the complex is a measure of the mercury.[7] Silver reacts similarly but can be masked by bromide. Therefore the technic is applicable by difference to silver.

Procedure. As reagent, make a solution containing 10 mg of copper as copper sulfate alkaline with 1 : 1 ammonium hydroxide. Add 4.2 ml of 1% solution of sodium diethyldithiocarbamate and dilute to 50 ml. Extract with 50 ml of chloroform and dilute the extract with chloroform to an absorbance of 0.5.

The sample solution should contain 50–100 mg of silver per ml.[8] Adjust to pH 3.72 by adding 50% sodium acetate solution and dilute to 100 ml with a buffer solution for that pH. Extract with 10 ml and 5 ml of the reagent and dilute with chloroform to 25 ml. Read the extract and the reagent at 435 nm against chloroform; the difference is a measure of the silver.

Copper Alloys. As reagent add to 20 ml of copper sulfate solution containing 10 mg of copper per ml, 1 gram of sodium nitrate and 1 gram of disodium EDTA dihydrate. Add sufficient 40% sodium hydroxide solution to raise to pH greater than 10, followed by 10 ml of 0.1% solution of sodium diethyldithiocarbamate. Shake with 100 ml of carbon tetrachloride and filter the organic layer through cotton into an actinic flask. Mix with a further 100 ml of carbon tetrachloride. Prepare this reagent fresh daily.

Take a 2.5-gram sample if it contains 0.001–0.01% of mercury, a 1-gram sample if it contains 0.1% of mercury. Transfer to a 100-ml flask fitted with a condenser and a water-filled trap. Dissolve the sample in 20 ml of 1 : 1 nitric acid, applying

[6] L. N. Krasil'nikova, M. G. Efimova, and L. S. Ponomareva, *Sb. Nauch. Tr. Vses. Nauchn.-Issled. Gornomet. Inst. Tsvet. Met.* **1965**, 9–16.
[7] T. W. F. Tertoolen, C. Buijze, and G. J. Van Kolmeschate, *Chemist-Analyst* **52**, 100–101 (1963).
[8] S. E. Kreimer, A. S. Lomekhov, and V. Stovoga, *Zh. Anal. Khim.* **17**, 674–677 (1962).

the trap when violent reaction ceases. Heat to complete solution, cool, and dilute to 50 ml.

To an aliquot containing 0.01–0.1 mg of mercury, add a pinch of sulfamic acid. If gas is evolved, add more sulfamic acid until evolution ceases. Add 1 gram of sodium citrate for every 100 mg of copper present. Add 4 grams of sodium nitrate and 10 ml of an acetate buffer solution containing 28 grams of sodium acetate and 12 ml of acetic acid per 140 ml. Adjust to pH 4.5–5 by dropwise addition of 40% sodium hydroxide solution. Add 5 ml of 0.2% potassium bromide solution and shake. Add 6 ml of 30% hydrogen peroxide and shake.

Add 50 ml of the reagent and shake for 4 minutes. Filter part of the organic phase through cotton and read at 435 nm against carbon tetrachloride, at the same time reading the reagent. The difference is a measure of the mercury present, the silver having been masked by the addition of potassium bromide.

Transfer a 20-ml aliquot of sample, which should contain 0.01–0.1 mg of mercury plus silver. Follow the technic for mercury, omitting addition of potassium bromide solution. The reading is mercury plus silver. Determine silver by difference.

Lead and Lead Alloys.[9] As reagent, use ammonium hydroxide to make alkaline a solution containing 5 mg of copper as nitrate or sulfate, and add 5 ml of 0.1% sodium diethyldithiocarbamate solution. Extract with 50 ml of benzene and dilute the separated extract to an absorption of 0.5.

Dissolve a 10-gram sample with 20 ml of perchloric acid and 5 ml of nitric acid. Evaporate to white fumes. To drive off tin and antimony, add 5 ml of hydrobromic acid, and again evaporate to white fumes. Cool and add 20 ml of water. Warm, and precipitate lead and silver by dropwise addition of 1 ml of 1:5 hydrochloric acid. Filter, and wash well with water to remove the bulk of the lead chloride. Dissolve the precipitate in 10 ml of 25% citric acid solution and add 15 ml of 1:1 ammonium hydroxide. Adjust to pH 8 ± 0.2 by addition of ammonium hydroxide or nitric acid.

Shake with 10 ml of reagent for 10 minutes. Read the reagent and the extract at 435 nm against benzene. The technic will determine 0.5 ppm of silver. Mercury would interfere but is removed in washing the chloride precipitate. Up to 0.2–0.3 gram of lead in the solution for extraction will not interfere.

Gold and Silver. Dissolve 5 grams of the gold and silver button from lead assay in 65 ml of 1:3 nitric acid. Evaporate to about 5 ml and cool. Add 50% sodium acetate solution until neutral to methyl orange. Add acetate buffer solution for pH 4 to make a volume of 40 ml. Shake the solution for 2 minutes with 20 ml of copper diethyldithiocarbamate solution of known concentration in carbon tetrachloride. After 20 minutes, read the extract at 453 nm against the solvent. Correct the reduction in absorption by a blank.

Cyanide Solutions. As the stock reagent solution, dissolve 0.04 gram of copper sulfate in 40 ml of water. Add 1:3 ammonium hydroxide to pH 9 and 4.5 ml of 10% sodium diethyldithiocarbamate solution. Extract with 20 ml and 20 ml of

[9]Tadao Hattori and Toshiaki Kuroha, *Jap. Anal.* **11**, 723–726 (1962).

toluene. Wash the combined extracts twice with water and dilute to 500 ml with toluene. As a working reagent, dilute 15 ml of the stock solution to 200 ml with toluene.

Mix a volume of sample solution containing 0.01–0.05 mg of silver with an equal volume of nitric acid.[10] Evaporate to about 2 ml and add 8 ml of 1 : 1 sulfuric acid. Evaporate to a little under 1 ml and cool. Add 5 ml of water and a drop of methyl orange indicator. Add 50% sodium acetate solution to the orange-yellow of pH 4. Dilute to 20 ml with 1% acetate solution, add 10 ml of the working reagent and shake for 2 minutes. Let stand for 30 minutes and read at 435 nm against a blank.

COPPER TETRAETHYLTHIURAM DISULFIDE

Silver displaces cupric ion in the yellowish-brown cupric tetraethylthiuram disulfide, known as mercupral.[11] Addition of EDTA and tartaric or citric acid prevents interference by large amounts of lead, antimony, tin, and bismuth. Mercury gives the reaction. Cobalt and nitric acid and large amounts of nickel interfere.[12] Moderately concentrated solutions of ceric ion decolorize the reagent.[13] The method is applicable to lead and to cable-sheathing alloys.[14]

Procedures

Ores. To prepare the reagent, mix equal volumes of saturated solutions of tetraethylthiuram disulfide in 50% ethanol and cupric sulfate pentahydrate in ethanol. After 48 hours separate the brown crystals of mercupral formed.

Add 5 ml of 1 : 1 nitric acid to 1 gram of finely ground sample. Let stand for 15 minutes, add 5 ml of hydrofluoric acid, and evaporate to dryness. Moisten the residue with a few drops of sulfuric acid and fuse with about 6 parts of potassium bisulfate. Take up in water and add 2 grams of EDTA. If antimony is present also, add a small amount of tartaric acid. Stir and make alkaline with 1 : 1 ammonium hydroxide. A brown color will be due to an iron complex. Filter, wash the precipitate, acidify the filtrate with acetic acid, and dilute to 200 ml.

Shake an aliquot expected to contain 10–30 μg of silver with an equal volume of 0.001% solution of mercupral in benzene for 30 seconds. The color of the benzene layer is decreased by silver ion. Read at 440 nm against benzene and correct for a blank.

Thorium Nitrate.[15] Dissolve 5 grams of thorium nitrate in 50 ml of water. Add 2 ml of ethanolic *p*-dimethylaminobenzylidenerhodanine. Precipitation of silver is quantitative, and less than 2 mg of thorium is coprecipitated. Filter and wash with

[10] I. F. Goryunova, *Nauch. Tr. Irkutsk. Gos. Nauchn.-Issled. Inst. Redk. Tsvet. Met.* **1968**, 63–70.
[11] V. Patrovsky, *Chem. Listy* **57**, 268–270 (1963).
[12] A. L. Gershuns and L. Z. Kalmykova, *Zavod. Lab.* **26**, 152–153 (1960).
[13] Jan Michal and Jaroslav Zyka, *Chem. Listy* **51**, 56–62 (1957).
[14] A. Schottak and H. Schweiger, *Z. Erzbergb. Metallhüttenw.* **19**, 180–185 (1966).
[15] Shizo Hirano, Atsushi Mizuike, and Yusuke Ujihira, *Jap. Anal.* **12**, 160–163 (1963).

water. Ash with 1 ml of 1:5 sulfuric acid and 1 ml of perchloric acid. Take up the residue in 5 ml of 1:10 sulfuric acid and add 1 gram of EDTA. Complete as for ores from "If antimony is present...."

---

p-DIMETHYLAMINOBENZYLIDENERHODANINE

Silver is precipitated by this reagent, but under appropriate conditions it can be read as a colloid.[16] Addition of a protective colloid such as starch, gelatin, or gum arabic broadens the range that conforms to Beer's law.[17] In reading the red suspension of silver with *p*-dimethyl- or *p*-diethylaminobenzylidenerhodanine, precautions must be taken against loss of silver by adsorption on lead sulfate and the walls of the equipment.[18] The latter can be by pretreatment with a water repellent. After separation of interfering elements on a tellurium carrier, the reagent is preferred to dithizone.[19] This reagent is suitable for determination of silver in water.[20]

Procedure.[21a] To a neutral solution containing 2–10 μg of silver, add 5 ml of 0.25 *N* nitric acid. Dilute to 17 ml, add 5 ml of 0.1 m*M* ethanolic *p*-dimethylaminobenzylidenerhodanine, and dilute to 25 ml. Read against water at 580 nm within the next 30 minutes.

Uranium. Dissolve a 5-gram sample in a minimal amount of nitric acid. Dilute with water and neutralize with 1:1 ammonium hydroxide to pH 2.8–3. Dilute to about 90 ml and add 5 ml of a 0.05% ethanolic solution of the reagent. Dilute to 100 ml and set aside in the dark for 15 minutes. Filter on a sintered-glass crucible and wash with the reagent diluted twentyfold with water. Then wash on the filter with hot ethyl acetate.

Dissolve from the filter with 5 ml, 5 ml, and 5 ml of hot nitric acid. Evaporate to dryness and take up in 1 ml of 1:11 nitric acid and 15 ml of water. Add 0.5 ml of 0.025% ethanolic reagent. Dilute to 25 ml and read at 495 nm.

As an alternative dissolve 5 grams of sample in nitric and sulfuric acids.[21b] Dilute to 5 ml. Add thallic sulfate solution and potassium iodide solution to coprecipitate the silver with thallic iodide. Filter, wash, and dissolve in sulfuric acid and hydrogen peroxide. Evaporate to dryness and dissolve the residue in 60% acetic acid. Add 0.004% solution of *p*-dimethylaminobenzylidenerhodanine and read at 580 nm.

[16] E. B. Sandell and J. J. Neumayer, *Anal. Chem.* **23**, 1863–1865 (1951); R. D. Shukla, R. V. Gokhale, L. M. Mahajan, and S. M. Jogdeo, *Indian J. Chem.* **11**, 199–200 (1973).
[17] M. Z. Yampol'skii, *Uch. Zap., Kursk. Gos. Ped. Inst.* **1958** (7), 73–82.
[18] Nabi Bukksh and Akram Khattak, *Pak. J. Sci. Ind. Res.* **5**, 86–90 (1962).
[19] Bogna Wichrowska, *Rocz. Panstw. Zakl. Hig.* **22**, 561–570 (1971).
[20] S. Fiala, *Vodn. Hospod.*, **14**, 372–373 (1959).
[21a] Rakhila Borisova, Maria Koeva, and Elena Topalova, *Talenta* **22**, 791–796 (1975).
[21b] Daido Ishii and Tsugio Takeuchi, *Jap. Anal.* **11**, 171–175 (1962).

Lead and Lead-Antimony Alloy.[22] Dissolve a 1-gram sample by heating with 5 ml of 1 : 2 nitric acid and 0.5 gram of tartaric acid. Filter if necessary—for example, if metastannic acid is formed. Dilute to about 20 ml and neutralize with 20% sodium hydroxide solution. Add 10 ml of 1 : 6 nitric acid, 2 ml of 5% ascorbic acid solution, 2 ml of 5% gum arabic solution, and 5 ml of 0.05% *p*-dimethylaminobenzylidenerhodanine solution. Dilute to 50 ml, set aside for 10 minutes, and read at 475 nm. Tartaric acid masks antimony and gum arabic stabilizes the colloid.

Electrolytic Copper.[23] Dissolve a 0.5-gram sample in 25 ml of 1 : 50 nitric acid. Impregnate filter paper with a 1% solution of dithizone in chloroform and dry. Pass the sample solution through the paper three times to fix the silver as dithizonate. Wash the paper with 5 ml of 1 : 150 nitric acid. Then decompose the paper and dithizonate with 1 ml of sulfuric acid and 5 ml of nitric acid. Take up the residue in water and complete as for lead and lead-antimony alloys from "Dilute to about 20 ml...." The technic will determine 2 ppm in the sample.

Silver may also be extracted as the dithizonate in carbon tetrachloride; the dithizonate may be decomposed and the silver determined with *p*-aminobenzylidenerhodanine.[24]

By an inversion technic, the silver is precipitated by the reagent, then decomposed by thiocyanate. The liberated dimethylaminobenzylidenerhodanine in carbon tetrachloride is read at 450 nm.[25] A suitable range is 0.05–6 μg of silver per ml.

Ores, Minerals, and Soil. Digest 1 gram of finely divided sample with 15 ml of nitric acid at 100° for 30 minutes. Cool, add 10 ml of water, and filter. Dilute the filtrate and washings to 50 ml.

Make a 5-ml aliquot alkaline with 1 : 1 ammonium hydroxide. Add 5 ml of 25% ammonium citrate solution, 3 ml of glacial acetic acid, 5 ml of water, and 0.2 ml of a 0.1% solution of *p*-dimethylaminobenzylidenerhodanine in acetic acid. Mix and let stand for 15 minutes. Add 5 ml of carbon tetrachloride and shake for 30 seconds. The silver complex forms a film at the interface between the phases. Without disturbing the film, draw off the carbon tetrachloride phase and discard it. Again shake the aqueous phase with 5 ml of carbon tetrachloride to extract excess of the reagent and discard the carbon tetrachloride. Transfer the film to a 10-ml flask, avoiding more than a minimum of the aqueous phase. Use a small portion of carbon tetrachloride to complete the transfer. Add 2 drops of 10% ammonium thiocyanate solution and shake to transfer the free *p*-dimethylaminobenzylidenerhodanine to the carbon tetrachloride layer. Remove the aqueous layer as by absorption in filter paper and dilute the carbon tetrachloride to 10 ml for reading at 450 nm against carbon tetrachloride.

Galena, Sphalerite, or Chalcopyrite.[26] Adsorb 10–100 mg of silver from a weakly

[22]Masami Murano, *Jap. Anal.* **11**, 735–739 (1962).
[23]Katsuaki Fukuda and Atsushi Mizuike, *ibid.* **18**, 1130–1131 (1969).
[24]Shizo Hirano, Atsushi Mizuike, Yoshio Iida, and Yoichi Hasegawa, *ibid.* **12**, 61–63 (1963).
[25]K. M. S. Bhattathiripad and R. G. Joshi, *Z. Anal. Chem.* **242**, 247–248 (1968).
[26]N. V. Markova and T. V. Yakubtseva, *Tr. Tsent. Nauchn.-Issled. Gornorazved. Inst. Tsvet. Redk. Blagorod. Met.* **1971**, 175–178.

acidic solution on a column of EDE-10P anionic resin in chloride form. Wash with 1:5 hydrochloric acid. Elute the silver with successively 1:3, 1:2, and 1:1 ammonium hydroxide. Evaporate to dryness. Successively add 5 ml, 5 ml, and 5 ml of nitric acid and evaporate to dryness with each portion. Take up the residue in 18 ml of 0.2 *N* nitric acid, and add 1 ml of 0.5% gelatin solution to avoid coagulation of the precipitate. Add 1 ml of 0.01% methanolic reagent and read at 470 nm.

Zinc Sulfide Phosphors.[27] Treat a 1-gram sample with 10 ml of 30% hydrogen peroxide and evaporate to dryness. Add 5 ml of nitric acid and evaporate to dryness. Take up the residue in 1 ml of nitric acid and dilute to 100 ml. Mix a 10-ml aliquot with 0.2 ml of a 0.05% methanolic solution of 4-dimethylaminobenzylidenerhodanine. Set aside for 10 minutes and read at 470 nm.

Fixing Baths.[28] Dilute 1 ml of sample containing 0.5–20 μg of silver to 100 ml. To a 5-ml aliquot add 0.5 ml of 1:1 ammonium hydroxide, 0.1 ml of 0.5% copper sulfate solution, and 5 ml of 3% solution of hydrogen peroxide. Mix well and add 10 ml of 1:1 ammonium hydroxide. Add 1 ml of 0.5% solution of polyvinyl alcohol and 10 ml of 0.01% methanolic reagent. Dilute to 50 ml and read at 530 nm. The original sample may contain up to 40% of sodium thiosulfate pentahydrate.

DITHIZONE

Dithizone is a major reagent for determination of silver. Silver dithizonate is extractable over the pH range from about 1 to more than 11. Dithizone is also useful in some cases for isolation of silver for determination by other reagents. The reagent is normally sensitive to 1 μg, but this can be modified by the volume of extracting solution or the sensitivity of the spectrophotometer.[29] Di-β-naphthylcarbazone is nearly twice as sensitive as dithizone in reaction with silver.[30] The maximum absorption is shifted from 500 to 462 nm.

For mercury and silver in a solution of a purified uranium compound, adjust to pH 3 and add EDTA. Extract with 0.025 *M* dithizone in chloroform. Read the mercury at 485 nm, the silver at 620 nm.[31]

For analysis of trade waste for silver, first extract the silver contaminated by other metals with dithizone in carbon tetrachloride.[32] Extract the silver from that solution with an aqueous thiocyanate solution. Dry, ash, and dissolve in dilute nitric acid. Finally, above pH 1 extract silver as the dithizonate for reading at 620 nm.

Gold can be extracted by ethyl acetate in the presence of hydroxylamine without

[27] Yoshihide Kotera, Tadao Sekine, and Masao Takahashi, *Bull. Chem. Soc. Jap.* **39**, 2523–2526 (1966).
[28] Jaromir Vrbsky and Jaroslav Fogl, *Chem. Prum.* **20**, 323–325 (1970).
[29] R. Carson and E. G. Walliczek, *Proceedings of the SAC Conference, Nottingham* **1965**, pp. 305–309.
[30] R. J. DuBois and Samuel B. Knight, *Anal. Chem.* **36**, 1313–1320 (1964).
[31] J. Marecek and E. Singer, *Z. Anal. Chem.* **203**, 336–339 (1964).
[32] T. B. Pierce, *Analyst* **85**, 166–177 (1960).

loss of silver from the solution.[33] The dithizone method is appropriate for determination of traces of silver in cesium, potassium, and sodium hydroxides.[34]

Extract silver from 0.1–4 *N* sulfuric acid, hydrochloric acid, hydrobromic acid, or perchloric acid with 15 m*M* bismuth bis(2-ethylhexyl)phosphorodithioate in octane or carbon tetrachloride.[35] Wash the organic layer with 4 *N* sulfuric acid, then back-extract the silver with 2.5 *M* ammonium thiocyanate. Add 4 : 5 nitric acid–1 : 1 sulfuric acid and heat to decompose the thiocyanate, giving a solution for determination of silver by dithizone.

For thermoelectric materials, dissolve tellurides and selenides of lead, bismuth, and antimony in nitric acid.[36] For those of germanium, add sodium oxalate, and for those of mixed lead and tin, use sulfuric acid and hydrogen peroxide. Mask interfering ions with ammonium citrate and EDTA. Adjust to pH 4–5 and extract silver with dithizone in carbon tetrachloride. Read silver dithizonate at 462 nm and unreacted dithizone at 620 nm. Mercury interferes.

As an inversion technic, silver in 1 : 3 sulfuric acid is extracted by dithizone in benzene.[37] Shaking with aqueous sodium chloride decomposes the silver dithizonate. A strong dithizone solution in benzene will reextract the silver, after which the excess dithizone can be washed out with dilute ammonium hydroxide for reading as a monocolor technic. As an alternative, an indirect inversion method is to extract with dilute dithizone in benzene, wash out the excess of dithizone with dilute ammonium hydroxide, remove the silver by extraction with sodium chloride solution, and read the liberated dithizone.

The use of EDTA to mask a large content of copper is unsatisfactory because this acid interferes with the extraction of silver.

Procedures

Separation of Silver, Mercury, and Copper. These three metals can be extracted at pH 1–2 by dithizone with modification of the procedure according to the relative proportions of the elements.[38]

Adjust the sample solution to pH 2 if chloride is absent. Silver cannot be extracted at pH 2 in the presence of chloride. With not more than 1% chloride, silver can be completely extracted at pH 3.5.

SILVER. Extract the sample with successive portions of 13-ppm dithizone in carbon tetrachloride, using a stronger solution if more than 0.05 mg of the metals is expected. When an extract remains green, discard the aqueous layer. To separate the silver, extract the mixed dithizonates in carbon tetrachloride with 3 ml and 3 ml of 10% sodium chloride solution in 0.015 *N* hydrochloric acid. Dilute the extract

[33] Masuo Miyamoto, *Jap. Anal.* **9**, 869–873 (1960).
[34] M. M. Godneva and R. D. Vodyannikova, *Zh. Anal. Khim.* **20**, 831–835 (1965).
[35] Yu. M. Yukhin, I. S. Levin, N. M. Meshkova, and N. E. Kozlova, *Zh. Anal. Khim.* **30**, 261–264 (1975).
[36] V. Fano and L. Zanotti, *Anal. Chim. Acta* **72**, 419–422 (1974).
[37] Masao Kawahata, Keiichi Mochizuki, and Takeshi Misaki, *Jap. Anal.* **11**, 1017–1020 (1962).
[38] Harald Friedeberg, *Anal. Chem.* **27**, 305–306 (1955).

containing the silver to 60 ml and extract the silver with 13-ppm dithizone in carbon tetrachloride. Read at 620 nm.

Extract the mercury and copper from the carbon tetrachloride phase with 3 ml and 3 ml of 1 : 1 hydrochloric acid. Discard the carbon tetrachloride and neutralize the acid phase with ammonium hydroxide. Acidify to pH 1.5–2 and add 1 ml of 0.01 *N* EDTA solution, more if more than 0.05 mg of copper is expected. This masks the copper.

MERCURY. Extract mercury with 13-ppm dithizone in carbon tetrachloride and read at 485 nm.

COPPER. To the extract from which mercury was removed, add 1 ml of 11.1% solution of anhydrous calcium chloride, 3 ml of 25% ammonium citrate solution that has been adjusted to pH 9 with ammonium hydroxide, and sufficient ammonium hydroxide to raise to pH 9. If more than 0.02 mg of copper is present, take an aliquot. Extract with 13-ppm dithizone in carbon tetrachloride and wash the combined extracts with water to remove traces of EDTA. Extract the copper with 3 ml and 3 ml of 1 : 1 hydrochloric acid. Neutralize the combined acid extracts with 1 : 1 ammonium hydroxide. Acidify to pH 2–3, extract with the dithizone solution, and read at 550 nm.

LARGE AMOUNTS OF COPPER PRESENT. For silver, add 2 equivalents of EDTA per unit of copper. Extract the silver with five portions of dithizone solution. Some copper is coextracted. Complete as above from "To separate the silver...." Results will be low.

For mercury, add 2 equivalents of EDTA per unit of copper. Extract the mercury with 5 portions of dithizone solution. Some copper is coextracted. Extract the mercury from the dithizonate with 3 ml and 3 ml of 1 : 1 hydrochloric acid. Neutralize with ammonium hydroxide and acidify to pH 1.5–2. Add1 ml of 0.01 *N* EDTA, extract the mercury as dithizonate, and read at 485 nm.

LARGE AMOUNTS OF SILVER PRESENT. Extract copper and mercury as for lesser amounts of silver, adding sufficient sodium chloride to precipitate the silver and ignoring the precipitate in making the extractions.

LARGE AMOUNTS OF MERCURY PRESENT. For silver, extract the mercury in 10% sodium chloride solution and 0.015 *N* hydrochloric acid. Extract with 100 ppm dithizone solution. Then dilute and determine silver as previously.

Without Extraction.[39] The sample should be about 0.02 *N* with nitric acid. Add 15 ml of acetone. Add dropwise a 0.05% solution of dithizone in acetone until the color is altered to a dirty yellow. Dilute to 25 ml with acetone and read at 465 nm against one drop of dithizone reagent diluted to 25 ml with acetone. Beer's law applies up to 75 μg of silver with a sensitivity to 0.5 μg of silver per ml in the sample. Alternatively, read at 603 nm, with Beer's law applying up to 55 μg of silver.

[39]Mikhail E. Macovschi, *Zh. Anal. Khim* **25**, 1226 (1970).

As a modified technic,[40] mix an aqueous sample containing 0.01–1 mg of silver with 10 ml of 0.005% ethanolic dithizone solution, 5 ml of 1 : 1 ammonium hydroxide, and 5 ml of 0.5% solution of disodium EDTA. Color develops with lead, magnesium, zinc, cadmium, and cobalt ions, but it is unstable and fades within 5 minutes. Nickel is masked by EDTA. Read at 562 nm.

With Extraction

Copper.[41] This technic assumes high purity copper with silver a trace. As a collector, dissolve 1 gram of thallium in 10 ml of 1 : 18 sulfuric acid in the presence of a platinum wire and dilute to 100 ml with water. As a dithizone reagent, dissolve 20 mg in 500 ml of carbon tetrachloride. Purify by shaking with 100 ml of 1 : 70 sulfuric acid followed by 100 ml of 6% sodium sulfite solution.

Dissolve an appropriate sample in nitric acid and dilute to 10 ml with water. Add 5 ml of the thallium collector solution and 5 ml of 0.5% solution of potassium bromide. Centrifuge and decant. Add 10 ml of 1 : 1 sulfuric acid to the precipitate and evaporate to 2–3 ml. Cool and dilute to 10 ml. Add 5 ml of 5% EDTA solution and 5 ml of 10% sodium acetate solution. Adjust to pH 5 with 1 : 10 ammonium hydroxide or 1 : 10 sulfuric acid.

Extract the silver with 3 ml and 3 ml of the dithizone reagent. Treat the combined extracts with 10 ml of 1 : 18 sulfuric acid and 2 ml of 0.1% potassium permanganate solution to back-extract the silver. Decolorize the aqueous phase by addition of 5% sodium sulfite solution. Extract the silver from this aqueous phase by 5 ml of a 1 : 5 dilution of the dithizone reagent and read at 620 nm. This will detect 7×10^{-6}% of silver per gram of sample.

An alternative technic provides for separation by column chromatography.[42] Dissolve 1 gram of sample with 1 ml of nitric acid and 5 ml of hydrochloric acid. Dilute to 100 ml. Treat 80-100 mesh Dowex 1-X8 with 1 : 20 hydrochloric acid and wash thoroughly with water. Charge a 20×0.6 cm column with the resin, and pass the solution through the column at 2 ml/minute to adsorb copper and silver. Wash the column with 50 ml of 1 : 25 hydrochloric acid to remove the copper. Then elute the silver with 70 ml of 1 : 9 nitric acid. Add 2 ml of 1 : 2 sulfuric acid and evaporate to white fumes. Cool and take up in 30 ml of water. Add 2 ml of 3% EDTA solution and adjust to pH 2 with 1 : 1 ammonium hydroxide. Extract the silver with 3 ml and 3 ml of carbon tetrachloride containing 2 mg of dithizone per liter, and read the combined extracts at 620 nm.

Gold.[43] Dissolve a 1-gram sample in 6 ml of hydrochloric acid and 2 ml of nitric acid. Evaporate to dryness and take up the residue in 10 ml of 1 : 5 hydrochloric acid. Extract the gold with 10 ml and 10 ml of isopentyl acetate and discard. Evaporate the solution to about 2 ml, and dilute to 10 ml with 1 : 1

[40]Shigeo Hara, *Jap. Anal.* **7**, 142–147 (1958).

[41]W. Angermann and H. Bastius, *Neue Hütte* **9**, 36–39 (1964); cf. A. I. Makar'yants, T. V. Zaglodina, and E. D. Shuvaloya, *Anal. Rud. Tsvet. Met. Prod. Pererab.* **1956** (12), 130–137.

[42]Masuo Miyamoto, *Jap. Anal.* **10**, 321–326 (1961).

[43]G. Ackermann and J. Köthe, *Z. Anal. Chem.* **231**, 252–261 (1967); Zygmunt Marczenko, Krzysztof Kasiura, and Maria Krasiejko, *Chem. Anal. (Warsaw)* **14**, 1277–1287 (1969); Krzysztof Kasiura, *ibid.* **20**, 809–816 (1975).

hydrochloric acid. Extract the iron with 10 ml and 10 ml of isopentyl acetate as a sample for determination with bathophenanthroline. Acidify the aqueous layer with sulfuric acid, add 1 ml in excess, and evaporate to sulfur trioxide fumes. Complete as for the first technic for copper from "Cool and dilute to 10 ml."

Lead.[44] Dissolve a 10-gram sample in 50 ml of 1:3 nitric acid and add 5 ml of sulfuric acid. Dilute to 100 ml and let the lead sulfate settle. Take a 10-ml aliquot of the supernatant liquid, which should contain 1–10 μg of silver, and add 10% urea solution to destroy oxides of nitrogen. Neutralize to methyl orange with 8% sodium hydroxide solution. Add as a buffer 5 ml of 1:1 0.1 *N* acetic acid–0.1 *N* sodium acetate. Add 2 ml of 0.1 *M* EDTA to sequester the residual lead. Extract with 5-ml portions of 0.001% dithizone in chloroform until the last extract shows no yellow. Wash the combined extracts with 2 ml of 0.1 *M* EDTA I in 0.01 *N* ammonium hydroxide. Dilute the chloroform layer to a known volume, let stand in the dark for 1 hour, and read at 470 nm.

For silver present as a trace in high purity lead, the lead is complexed with EDTA and the other impurities by ammonium citrate. The silver is then extracted at pH 4.6 with dithizone and read by the mixed color method.[45]

Palladium.[46] Dissolve a 0.5-gram sample in a minimal volume of aqua regia and evaporate to dryness. Add four 5-ml portions of hydrochloric acid, and evaporate to dryness each time. Take up the residue in 25 ml of hydrochloric acid and 17 ml of water. Add 8 ml of 20% ammonium thiocyanate solution. Extract palladium by vigorous shaking with 50 ml and 20 ml of methyl isobutyl ketone and discard the extracts.

Add 2 drops of sulfuric acid and 20 ml of nitric acid to the aqueous layer and evaporate it to dryness. Sequentially add 2 ml of sulfuric acid, 2 ml of nitric acid, and 2 ml of perchloric acid, evaporating to dryness after each addition. Take up the residue in 35 ml of water and add 5 ml of 1% ethanolic dimethylglyoxime. Extract with 20 ml of chloroform to remove a trace of palladium and discard. Extract the aqueous phase with 10 ml of benzene containing 0.5 mg of dithizone by shaking for 3 minutes. Extract the silver from the benzene with 7 ml and 7 ml of 10% sodium chloride solution in 1:80 hydrochloric acid. Mix the combined silver extracts with 20 ml of a buffer solution for pH 9 consisting of a 6% solution of ammonium chloride in a 1.5% solution of EDTA. Extract with 10 ml of the dithizone in benzene. Wash this extract by shaking for 2 minutes with 2% sodium hydroxide solution. Read at 470 nm against a blank. Gold and mercury interfere.

Lead, Copper, and Gold Concentrates.[47] Treat 2 grams of sample with 20 ml of nitric acid and set aside for 30 minutes. Heat to drive off oxides of nitrogen; then add 8 ml of 1:1 sulfuric acid and evaporate almost to dryness. Add 5 ml of water and evaporate again. Take up in 1:3 sulfuric acid and dilute to 25 ml with that acid. Dilute a 5-ml aliquot to 40 ml with 1:3 sulfuric acid. Extract with 10 ml and

[44]R. Socolovschi, *Rev. Chim.* (Bucharest) **10**, 712–713 (1959); cf. H. Jednzejewska and M. Malusecka, *Chim. Anal.* (Paris) **12**, 579–584 (1967).
[45]B. L. Jangida, *J. Sci. Ind. Res. (India)* **20 B** (2), 80 (1961).
[46]Teruo Imai, *Jap. Anal.* **15**, 109–113 (1966).
[47]S. Beleva and R. Dancheva, *Khim. Ind., Sofia* **36**, 64–66 (1964).

10 ml of 0.001% solution of dithizone in benzene. Wash the combined extracts with 25 ml, 25 ml, and 25 ml of 0.2% ammonium hydroxide. Dilute to 25 ml with benzene, centrifuge, and read at 435 nm. If copper is present also, read at 584 nm and subtract as a correction from the reading at 435 nm.

Alloy Steel.[48] Dissolve a 1-gram sample by heating with 20 ml of 1:1 sulfuric acid, adding 30% hydrogen peroxide to expedite the reaction. Cool and dilute to 100 ml. Mix a 10-ml aliquot with 10 ml of 3:5 sulfuric acid. Add 0.2% solution of EDTA, using at least 2 equivalents for each equivalent of copper. Extract the silver by shaking with five successive 5-ml portions of 0.01% solution of dithizone in benzene. Some copper is also extracted. Reextract the silver from the combined solution of dithizonates with 3 ml of 10% sodium chloride solution in 0.015 *N* hydrochloric acid. Dilute the aqueous extract to 60 ml and reextract the silver with 5 ml and 5 ml of 0.01% dithizone in benzene. Wash the combined extracts with 10 ml and 10 ml of 1:70 sulfuric acid and read at 620 nm against benzene. Correct for a blank.

Minerals.[49] Dissolve or disintegrate a 0.5-gram sample in 5 ml of nitric acid. Add 5 ml of water and evaporate to dryness. Take up with 5 ml of 1:1 nitric acid, add 1 ml of 10% EDTA solution, and dilute to 10 ml. Neutralize with ammonium hydroxide, filter, and ash. Treat the residue with a few drops of sulfuric acid and 2 ml of hydrofluoric acid. Take to dryness and fuse with 1 gram of potassium bisulfate. Take up the melt in water and combine with the previous filtrate. Adjust to pH 2–3 by addition of 10% sodium acetate solution. Shake with 15 ml of 0.001% solution of dithizone in chloroform for 1 minute. Wash the extract with 15 ml and 15 ml of 1:9 ammonium hydroxide containing a little EDTA; read at 490 nm.

Galena Ores.[50a] Dissolve 0.02–0.06 gram of ore in 2 ml of nitric acid. Evaporate nearly to dryness and take up the residue in 50 ml of 1:99 nitric acid. Extract with 15 ml of 0.001% solution of dithizone in chloroform. Extract silver from the copper and mercury dithizonates with a 5% solution of sodium chloride in 0.015 *N* hydrochloric acid. Adjust the aqueous phase to pH 4.6 by addition of 0.16% sodium acetate solution. Reextract the silver with dithizone. Revert by extracting the silver with 2 *N* hydrochloric acid and read the dithizone in the organic layer at 610 nm.

Cadmium-Mercury-Tellurium Semiconductors.[50b] Dissolve the layer with nitric acid from a mica support. Dilute 10-fold with water and extract the mercury and the silver impurity with 0.005% solution of dithizone in carbon tetrachloride. Read the extract at 462 nm as the sum of silver and mercury. Add sulfuric acid, nitric acid, and potassium sulfate. Evaporate to dryness to eliminate up to 1 mg of mercury. Dissolve the residue in *N* sulfuric acid and extract the silver with 0.001% solution of dithizone in carbon tetrachloride. Back-extract the silver with *N*

[48]Masao Kawahata, Heiichi Mochizuki, and Takeshi Misaki, *Jap. Anal.* **11**, 1017–1020 (1962).
[49]V. Patrovsky, *Rudy* **12**, 207–208 (1964).
[50a]E. J. Bounsall and W. A. E. McBryde, *Can. J. Chem.* **38**, 1488–1494 (1960).
[50b]Z. Marczenko and E. Podsiadlo, *Mikrochim. Acta* **II**, 3–4 (1976).

hydrochloric acid. Add EDTA to mask residual mercury and adjust to pH 4 with ammonium hydroxide. Reextract the silver with 0.001% solution of dithizone in carbon tetrachloride and read at 462 nm. Mercury can be read by difference. Results tend to be low by 10%.

FLAME PHOTOMETRY

Silver in the flame photometer emits in the ultraviolet at 328 and 338.3 nm.[51] With the oxyacetylene flame, an optimum intensity is approximated at 5 cubic feet/hour of acetylene and 5.5 cubic feet/hour of oxygen. This corresponds to combustion of the carbon to the monoxide rather than the dioxide. The two lines are of similar intensity. The following are typical settings.

Selector switch position	0.1
Sensitivity control ERA (% adjust)	0
Phototube resistor (megohms)	22
Phototube RCA1 P 28 (volts/dynode)	60
Slit (nm)	0.30
Spectral bandwidth (nm)	0.17

Mirror blocked except for hydrogen flames
Fuel flow (cubic feet/hr)
Hydrogen 13.5, Oxygen 4.7
Acetylene 5.0, Oxygen 5.5

There is considerable background emission with the oxyacetylene flame but essentially none with an oxyhydrogen flame. The use of an organic solvent for the silver multiplies the emission with the oxyacetylene flame by factors ranging around 5 times and with the oxyhydrogen flame around 9 times. The only common anions having a significant effect are carbonate and oxalate. With cations, the effects are shown in Table 3-1.

At 2660° the emission at 338.3 nm increases with increasing ethanol content, but at 1850° the increase is very slight.[52]

The presence of perchloric acid in the feedstock raises the temperature of an oxyhydrogen flame and the emission of silver.[53] A similar effect of acetic acid is due to its modification of the surface tension of the solution. Cyanide and ammonia have no effect.

With gold dissolved in aqua regia, neutralized with ammonia, the detection limit in the oxyhydrogen flame is 0.2 μg/ml for silver.[54a]

[51] John A. Dean and Charles B. Stubblefield, *Anal. Chem.* **33**, 382–386 (1961).
[52] E. Pungor and I. Konkoly-Thege, *Ann. Univ. Sci. Budap., Rolando Eotvos Nominatae, Sect. Chim.* **2**, 477–484 (1960).
[53] Erno Pungor and Ilona Konkoly-Thege, *Magy. Kem. Foly.* **62**, 225–228 (1956).
[54a] Lyn Jarman, E. Manolitsis, and M. Matic, *J. S. Afr. Inst. Mining Met.* **62**, Pt. 2, 773–779 (1962).

Table 3-1 Cation Interference Effects in Flame Photometry of Silver
Present, 50 ppm of Silver

Cation Tested	Concentration (ppm)	Percent Error Measured at 328.0 nm	Percent Error Measured at 338.3 nm
Aluminum	5,000	−1	−3
Ammonium (or NH_3)	5,000	−8	−7
	2,000	−2	−5
Cadmium	5,000	−5	−2
Calcium	5,000	−4	0
Cerium	5,000	3	5
	2,000	−1	3
Chromium	1,000	−3	−3
Cobalt	5,000	1	[a]
Copper	5,000	0	−2
Iron	5,000	1	3
Lead	1,000	−2	−2
Magnesium	5,000	−7	−8
	2,000	−5	−5
Manganese	5,000	−2	−6
Nickel	5,000	−2	[a]
Potassium	5,000	−5	−6
Sodium	10,000	1	−2
Tin	5,000	−14	−16
Zinc	5,000	−4	−5
	2,000	−1	0

[a] Direct spectral interference.

Procedures

Blister Copper.[54b] Dissolve a known weight of sample in concentrated nitric acid. Dilute with 10 volumes of water and precipitate the silver with sodium chloride solution. Filter the silver chloride and wash with water. Dissolve the silver chloride in 1 : 1 ammonium hydroxide and read by flame photometry at 338.3 nm. No interferences are known.

Thorium-Uranium Slurry. Dissolve a 0.5-gram sample in 10 ml of 1 : 1 nitric acid and a few drops of hydrofluoric acid. Dilute to 50 ml to give about 0.1 mg of silver, copper, or palladium per ml.[55]

[54b] N. McN. Galloway, *Analyst* **83**, 373–374 (1958).
[55] C. A. Meinz and P. D. LaMont, *US AEC*, **TID-7568**, Pt. 1, 150–156 (1958).

MICHELER'S THIOKETONE

Micheler's thioketone, 4,4′-bis(dimethylamino)thiobenzophenone, is read at 520 nm under conditions, including extraction, that vary with the silver content.[56] The complex is not stable to light, necessitating work with actinic glassware. Halogens in amounts equivalent to the silver do not interfere. There is interference by thiosulfate, thiocyanate, sulfide, phosphate, and reducing agents such as ascorbic acid. Tartaric acid and EDTA may be used to mask other cations. Other noble metals form the complex, but their complexes are stable to light. Therefore in the presence of other noble metals, read the absorption, expose to light to destroy the silver complex and make the determination by difference.

Procedures

5–30 μg. To the sample solution, add 1 ml of acetate buffer solution for pH 3 and dilute nearly to 10 ml. Adjust to pH 3 ±0.2 with 1 : 1 ammonium hydroxide or with acetic acid. Add 1 ml of m*M* Micheler's thioketone in ethanol and 15 ml of ethanol. Mix and make up to 25 ml with ethanol. After 20 minutes read at 520 nm against a blank.

1–10 μg. Add 1 ml of acetate buffer solution for pH 3 and dilute to 20 ml with water. Adjust to pH 3 ±0.2. Extract with 10 ml of 0.2 m*M* Micheler's ketone in isoamyl alcohol by shaking for 1 minute. Filter the organic layer and read at 520 nm.

Less than 1 μg. Add a couple of drops of the acetate buffer solution for pH 3 to the sample. Dilute to 8 ml, adjust to pH 3 ±0.2, and extract with 2 ml of the more dilute reagent. Filter the extract and read at 520 nm.

NAPHTHO[2,3-D]TRIAZOLE

This reagent is used to read silver at 0.1–3 ppm photometrically or to measure its quenching effect on fluorescence.[57] The reagent is prepared by diazotizing naphthalene-2,3-diamine in acetic acid and recrystallizing the product.

Procedures

Photometric. As masking solution, make 0.05 *M* sodium tartrate also 0.5*M* in EDTA. As a buffer, add 20% sodium hydroxide solution to 0.6% hexamine solution until the pH reaches 10.5. As the reagent solution, dissolve 1 mg of naphtho [2,3-D]-triazole per ml in 0.025% sodium hydroxide solution.

[56] K. L. Cheng, *Mikrochim. Acta* **1967**, 820–827.

[57] Garry L. Wheeler, John Andrejack, James H. Wiersma, and Peter F. Lott, *Anal. Chim. Acta* **46**, 239–245 (1969).

To 25 ml of sample solution containing 1–3 ppm of silver, add 5 ml of reagent, 10 ml of masking solution, and 10 ml of buffer. After waiting 3–5 minutes for the yellow color to develop, read at 436 nm against a reagent blank.

Fluorimetric. Prepare as for the photometric determination, substituting 10 ml of water for the masking solution. Activate at 362 nm and read at 406 nm, 5 minutes after adding the reagent. In the absence of the masking reagent this is subject to interference.

NILE BLUE

The complex of silver with nile blue (basic blue 12) can be extracted with chloroform for reading.[58] Many of the possible interferences are eliminated by evaporation with sulfuric acid. The range called for below conforms to Beer's law.

Procedure. Evaporate the sample nearly to dryness with sulfuric acid; cool. Take up with water and again evaporate nearly to dryness. Take up in water and adjust to pH 2–3 by adding 10% sodium hydroxide solution. Dilute to a known volume; and take an aliquot expected to contain 5–35 μg of silver.

Add 10 ml of potassium iodide solution containing about 60 μg of iodide ion and 10 ml of a 0.005% solution of Nile blue chloride or sulfate in 1 : 140 sulfuric acid. Dilute to 50 ml and extract with 5 ml, 5 ml, and 5 ml of chloroform. Dilute the combined extracts to 25 ml and read at 626 nm.

8-HYDROXYQUINOLINE-5-SULFONIC ACID

Potassium persulfate oxidizes silver to a trivalent state that reacts with 8-hydroxyquinoline-5-sulfonic acid, also known as oxine-5-sulfonic acid, to give a fluorescent product.[59] Ions tolerated in ppm are lithium (100), aluminum (75), lanthanum (50), cobalt (20), nickel (10), and ferric (10). More than 1 ppm of cupric, mercuric, or palladous ion causes quenching, and more than 1 ppm of tetravalent zinconium or hafnium increases the intensity.

Procedure. Mix a sample containing 0.1–10 μg of silver with 10–25 *M* excess of the reagent. Adjust to pH 1.5–3.5 and add 5 ml of 0.1 *M* potassium persulfate. Dilute to 10 ml, and after 60–90 minutes excite at 375 nm for reading at 485 nm against a reagent blank. The intensity is linear at 0.0125–5 ppm.

---

[58] W. Likussar and H. Raber, *ibid.* **50**, 173–175 (1970).
[59] D. E. Ryan and B. K. Pal, *ibid.* **44**, 385–389 (1969).

PERMANGANATE

A dynamic method determines silver by its catalytic effect in conversion of manganous ion to permanganate by persulfate. The method will determine 0.01 ppm of silver in a mineral.

Procedures

Natural Materials.[60] Ignite 1 gram of sample in platinum at 700° for 30 minutes. Add 2 ml of perchloric acid and 10 ml of hydrofluoric acid, and evaporate nearly to dryness. Dissolve in 40 ml of 0.5 *M* perchloric acid and cool. Add 1 ml of 10% ascorbic acid to separate silver from platinum, palladium, gold, and copper. Extract with 10 ml of 0.005% solution of dithizone in 1 : 19 carbon tetrachloride–benzene. Wash the extract with 10 ml of 0.5*N* sulfuric acid. Extract the silver from the solvent with 15 ml of 0.6 *N* hydrochloric acid. Add 1 ml of 1 : 1 sulfuric acid and 1 ml of 2% potassium sulfate solution. Add 30% hydrogen peroxide dropwise and evaporate to sulfur trioxide fumes. Take up in 5 ml of water adding 1 ml of 1 : 2 phosphoric acid, and 1 ml of 0.073% manganese sulfate solution. Dilute to 8 ml. Add 2 grams of potassium persulfate, heat at 100° for 5 minutes and chill in ice water. After 10 minutes, filter and read the permanganate at 540 nm.

Sulfides. Ignite 0.1 gram and dissolve in 4 ml of aqua regia. Add 5 ml of 1 : 4 perchloric acid and evaporate to dryness. Repeat that step. Complete as for natural materials from "Dissolve in 40 ml of 0.5 *M* perchloric acid...."

1,10-PHENANTHROLINE AND BROMOPYROGALLOL RED

Silver reacts with these compounds to form a ternary complex.[61] Only thiosulfate and excessive amount of gold interfere.

Procedure. To a sample solution containing 0.01–0.05 mg of silver, add an excess of EDTA. If more than 0.25 mg of gold is present, add an excess of bromide ion to form the trivalent bromide compound. Add sufficient mercuric ion to mask cyanide, thiocyanate, and iodide. Add 1 ml of 20% ammonium acetate solution, 5 ml of 0.001 *M* 1,10-phenanthroline, 1 ml of 0.1 *M* EDTA, and 1 ml of *M* sodium nitrite. Dilute to 25 ml and extract with 20 ml of nitrobenzene. Shake the organic extract with 25 ml of 0.0001 *M* bromopyrogallol red in water, separate the layers, and read the organic layer within 30 minutes at 590 nm against a reagent blank.

---

[60] Yu. I. Grosse and A. D. Miller, *Zavod. Lab.* **40**, 262–263 (1974); cf. L. I. Pets, *Tr. Tallin. Politekh. Inst., Ser. A* **1972**, 79–86.
[61] R. M. Dagnall and T. S. West, *Talenta* **11**, 1627–1631 (1964).

1,10-PHENANTHROLINE AND EOSINE

Eosine is 2,4,5,7-tetrabromofluorescein. With 1,10-phenanthroline it forms a ternary complex with silver at pH 4–8 in water. At pH 8 the complex is extractable with nitrobenzene. Each is read at 550 nm.[62] In organic solvent the complex is about 50% more sensitive. Only tetravalent iridium and cyanide interfere seriously. The reaction is instantaneous. For extraction with chloroform at pH 6, it is desirable to have acetone present to prevent formation of a phase-boundary film. The acetone also increases the fluorescence of the extract.[63] Thus it is read by absorption at 540 nm or activated at 300 nm for reading by fluorescence at 580 nm.

A similar reaction is carried out with 3′, 6′-dichloro-2,4,5,7-tetraiodofluorescein for reading at 570 nm and with 3′, 4′, 5′, 6′-tetrachloro-2,4,5,7-tetraiodofluorescein for reading at 585 nm.[64] In those cases palladium must be absent.

Procedures

In Water. Take up to 20 ml of sample solution, an amount expected to contain 0.005–0.1 mg of silver. Add 1 ml of *M* EDTA, 1 ml of 0.01 *M* 1 : 10-phenanthroline, 1 ml of 9.6% solution of sodium acetate trihydrate as a buffer solution for pH 7, and 15 ml of 0.0001 *M* eosine. Dilute to 50 ml and read at 550 nm. The color is stable for 15 days.

In Solvent. Take up to 20 ml of the sample solution expected to contain 0.005–0.1 mg of silver. Add 1 ml of *M* EDTA, 1 ml of 0.01 *M* 1 : 10-phenanthroline, 1 ml of an 18% solution of disodium phosphate dodecahydrate in *M* EDTA, as a buffer solution for pH 8, and 15 ml of 0.0001 *M* eosine. Dilute to 25 ml and add 25 ml of nitrobenzene. Mix by inversion shaking for about a minute and let separate over a 15-minute period. Read at 550 nm against a reagent blank. The color is stable for 12 hours.

1, 10-PHENANTHROLINE AND PYROGALLOL RED

After separation from interferences and concentration by dithizone extraction, silver is determined by the reagents above.[65] Small amounts of copper and tin coextracted do not interfere. Beer's law is obeyed for 1–10 μg of silver in 50 ml, but if there is less than 3 μg of silver present, the loss in extraction may reach 8%.

Procedure. *Tellurium.* Dissolve a 0.5-gram sample in 50 ml of 1 : 3 nitric acid.

[62]Mohamed T. El-Ghamry and Roland W. Frei, *Anal. Chem.* **40**, 1986–1990 (1968); Merrie N. White and Donald J. Lisk, *J. Assoc. Off. Anal. Chem.* **53**, 1055 (1970).
[63]D. N. Lisitsyna and D. P. Sheherbov, *Zh. Anal. Khim.* **25**, 2310–2314 (1970).
[64]K. P. Stolyarov and V. V. Firyulina, *ibid.* **26**, 1731–1735 (1971).
[65]Jan Dobrowolski and Stanislaw Szwabski, *Chem. Anal. Warsaw* **15**, 1033–1035 (1970).

Raise to pH 2 by dropwise addition of 20% sodium hydroxide solution. Extract the silver by shaking with successive 1-ml portions of 0.0001% dithizone solution in chloroform as long as the original green color is modified. Combine the chloroform extracts and extract the silver from them with 5 ml of 1 : 10 hydrochloric acid. Neutralize this acid extract with 2% sodium hydroxide and dilute to 250 ml. Mix a 25-ml aliquot with 1 ml of 0.01 *M* EDTA, 1 ml of 0.1 m*M* 1,10-phenanthroline, and 2 ml of 0.1 m*M* pyrogallol red in 1% ammonium nitrate solution. Dilute to 50 ml and set aside for 30 minutes. Read at 595 nm against a blank.

---

m- AND *p*-PHENYLENEDIAMINE

The coupling of *m*-phenylenediamine and *p*-phenylenediamine under the influence of silver ion is a measure of the latter.[66] The technic given is also applicable to toluene-2,4-diamine and *N*,*N*-dimethyl-*p*-phenylenediamine. It is a four-electron oxidation similar to the coupling that occurs in some color photographic processes and is applicable not only in water or in methanol, as in the procedure that follows, but also in other solvents with various periods of development; isopropanol, 5 minutes at the boiling point; acetone, 1 minute at 25°; water containing a pH 7 buffer solution, 5 minutes at 25°. In ethyl acetate the reagents are introduced in methanol leading to a final concentration of 1 : 1 ethyl acetate–methanol. Development then takes 10 minutes. In toluene the reagents are introduced in isopropanol ending with 50% toluene, 49.5% isopropanol, 0.5% water, which develops in 1 minute.

The dyestuff produced is probably structurally similar to indamine. Most of the colors developed were stable up to an hour, but that in acetone begins to fade immediately and must be read within 1 minute. In water the color varies with pH but has a maximum absorption at pH 6.5–7.

Oxidizing materials that interfere because they produce the same color rapidly are manganese dioxide, iodate, chlorplatinate, chloraurate, and ferric and ceric ions. Oxygen and mercury react too slowly to interfere. Although bromide and iodide precipitate the silver, a 200-fold excess of chloride is tolerated. There is no interference by chlorate, acetate, ferrous, or nitrate ions.

Procedure. Equip a 10-ml volumetric flask with a small reflux condenser. Add 5 ml of sample in methanol, 1 ml of 0.1 *M* sodium acetate in methanol, 1 ml of 0.1% solution of *M*-phenylenediamine in methanol, and 1 ml of 0.1% solution of *p*-phenylenediamine in methanol. Boil for 10 minutes, cool, and dilute to volume. Filter out the silver and read at 605 nm.

With other solvents the respective wavelengths are as follows: isopropanol, 615 nm; acetone, 605 nm; water, 550 nm; 1 : 1 methanol-water, 587 nm; 1 : 1 ethyl acetate–methanol, 605 nm; and 100 : 99 : 1 toluene-isopropanol-water, 610 nm.

---

[66] Robert L. Rebentus and Vaughn Levin, *Anal. Chem.* **40**, 2053–2054 (1968).

PYROGALLOL RED

The reaction of silver with pyrogallol red will tolerate tenfold molar excess of common ions if anthranilic acid–*N*,*N*-diacetic acid is added.[67] Mercury is an exception. If present, it must be titrated with 0.001 *M* EDTA in a separate sample. This is buffered to pH 7 with pyridine acetate and uses xylenol orange as the indicator. Then an exact equivalent of EDTA is added to the test specimen. Copper and aluminum interfere. The system conforms to Beer's law up to 10 ppm.

Procedures

By Extraction. As the first extraction reagent, dissolve 3.2 grams of salicylic acid and 30 ml of dibutylamine in methyl isobutyl ketone and dilute to 1 liter with that solvent. The second extraction reagent differs in containing 70 ml of dibutylamine per liter.

SILVER, 1–10 PPM. Mix a 10-ml sample at approximately pH 7 with 0.4 ml of 0.1 *M* anthranilic acid-*N*,*N*-diacetic acid as the sodium salt. Add 1 ml of 5 *M* sodium nitrite and 5 ml of the first extraction reagent. Shake for 1 minute. Separate the organic layer and wash with 10 ml of water. Then add 5 ml of methyl isobutyl ketone and 5 ml of 0.0042% solution of pyrogallol red in absolute ethanol. After 20 minutes dilute to 50 ml with methyl isobutyl ketone and read at 390 nm against a reagent blank.

SILVER, 0.01–0.1 PPM. Treat a 2-liter separatory funnel with water repellent to avoid loss of silver by adsorption. To 1 liter of sample, add 5 ml of 5 *M* sodium nitrite solution, 0.4 ml of the 0.1 *M* anthranilic acid–*N*,*N*-diacetic acid sodium salt, and 50 ml of the second extraction reagent. Shake for 1 minute and cool in ice water until clear. Discard the aqueous layer. To the organic layer, add 5 ml of methyl isobutyl ketone and 5 ml of 0.0042% ethanolic pyrogallol red. Dilute to 50 ml with methyl isobutyl ketone and set aside for 90 minutes. Read at 390 nm against a reagent blank.

Without Extraction.[68] To an aliquot of sample, add 1 ml of 35% sodium nitrite solution and dilute to about 75 ml. Add 10% sodium acetate until the pH is 7–7.5. Add 5 ml of 0.0001 *M* pyrogallol red and mix. After 90 minutes dilute to 100 ml and read at 390 nm against a blank. This technic is applicable for up to 0.085 mg of silver. The only anions that interfere are halides. Most cations interfere but can be masked by an appropriate amount of EDTA.

The same technic can be applied with bromopyrogallol red, with reading at 440 nm.

[67] R. M. Dagnall and T. S. West, *Anal. Chim. Acta* **27**, 9–14 (1962).
[68] R. M. Dagnall and T. S. West, *Talenta* **8**, 711–719 (1961).

RHODANINE

Photometric determination of silver in photographic film with rhodanine is accurate to better than 2.5%.[69a]

Procedure. Treat 0.5 sq dm of positive or negative film at around 25° with 25 ml of 1.185% sodium thiosulfate solution in subdued light, stirring occasionally. Wash the film three times with water and dilute the extract to 100 ml. Check that extraction is complete by treating the washed film with hydrogen sulfide. Mix a 5-ml aliquot of the sample with 10 ml of 1 : 1 ammonium hydroxide and 10 ml of 0.01% rhodanine solution. Dilute to 100 ml and read at 550 nm.

X-Ray Film. Proceed as above but dilute the extract to 1 liter and use a 2-ml aliquot.

THIOTHENOYLTRIFLUOROACETONE

The captioned reagent is 1,1,1-trifluoro-4-mercapto-4-(2-thienyl)but-3-en-2-one. Beer's law is followed for 5–75 μg of silver per ml in carbon tetrachloride. The complex is stable for 72 hours. There is serious interference by indium, mercury, bismuth, palladium, and zinc. Copper must be preextracted at pH 4.5 with acetylacetone.

Procedure.[69b] Adjust a sample solution containing 0.05–0.7 mg of silver to pH 1 with nitric acid or ammonium hydroxide and dilute to 25 ml. Add 5 ml of butanol and extract with 5 ml of m*M* solution of the reagent in carbon tetrachloride. Read the organic phase at 460 nm.

TOLUENE-3,4-DITHIOL

Toluene-3,4-dithiol gives an insoluble complex with silver that although colloidal, has particles so fine that it appears transparent.[70] The yellow color does not reach a maximum for more than 24 hours, but a practical technic calls for reading after 1 hour when the color is 65% developed. Lead, mercury, tin, and arsenic must be absent. A dispersing agent serves to stabilize the colloid. The presence of thioglycollic acid in the reagent expedites the development of color. The absorbance does not conform to Beer's law. At 4–40 μg per ml of the final solution, the precision is ±5–6%.

[69a] I. Hincak, *Kem. Ind.* **17**, 827–829 (1968).
[69b] K. R. Solamka and Shripad M. Khopkar, *Mikrochim. Acta* **II**, (1-2), 41–48 (1976).
[70] J. P. Dux and W. R. Fairheller, *Anal. Chem.* **33**, 445–447 (1961).

Procedures

Solutions. Mix 4 ml of 1 : 1 sulfuric acid with a volume of neutral sample solution containing 0.12–2.5 mg of silver. Add water to about 40 ml and 2 drops of 30% dodecyl sodium sulfate. Put in a shaker at gentle agitation, and if swirling is not obtained, add a little more water. This should give mixing without excessive foaming. Add 1 ml of a 0.3% solution of the dithiol in 2% sodium hydroxide solution to which has been added 8 drops of thioglycollic acid per 50 ml. When mixing is complete, add 8 drops of the dodecyl sodium sulfate solution. After swirling for about 1 minute, remove from the shaker and dilute to 50 ml. After 1 hour, read at 416 nm against water. No blank is necessary.

Organic Matter. Mix a sample containing 0.12–2.5 mg of silver with 25 ml of nitric acid and 2 ml of sulfuric acid. Digest to destroy the organic matter, and drive off the nitric acid by heating to fumes of sulfur trioxide. Complete as for solutions from "Add water to about 40 ml and...."

1-Amidino-2-thiourea forms a yellow 1 : 1 complex with silver.[71] At not more than 0.1 *N* ammonium or sodium hydroxide, the method will determine as little as 0.4 μg of silver per ml. Read at 400 nm, it conforms to Beer's law for 1–12 μg of silver per ml. Mercuric and auric ions interfere.

Silver in argyrol or protargol is determined by wet-ashing the sample and reacting the neutral solution in ethanol with **benzoyl-4-(4-ethoxyphenyl) thiosemicarbazide.**[72] Adjust a sample containing up to 1 mg of silver to about pH 5 and dilute to 20 ml with ethanol. Add 10 ml of 5% solution of the reagent in ethanol; dilute to 50 ml and read.

Heating a silver solution at pH 3–5 and 50° for 10 minutes with potassium persulfate and **2,2′-bipyridyl** or **2,2′,6′,2″-terpyridyl** gives bivalent silver complexes.[73] The reading at 460 nm conforms to Beer's law up to 0.03 mg of silver per ml.

In spite of the instability of **crystal violet** to cyanides, by careful standardization it can be used to extract silver cyanide into benzene.[74] Make a sample that is 10^{-6}–10^{-4} *M* with silver 10^{-3} *M* with sodium cyanide and 0.4% with sodium hydroxide. Shake 10 ml of this solution with 10 ml of benzene and 1 ml of 10^{-3} *M* crystal violet for exactly 15 seconds. Centrifuge for 60 seconds. Read the benzene layer at 600 nm 30–40 minutes later.

For determination of silver in galena it is precipitated by ***p*-diethylaminobenzylidinerhodanine** in the presence of EDTA. When collected and washed with ethanol, the precipitate is dissolved with potassium cyanide and read at 470 nm.[75]

[71] R. A. Nadkayni and B. C. Haldar, *J. Indian Chem. Soc.* **42**, 473–478 (1965).
[72] R. Craciuneanu and L. Chiorean, *Pharmazie* **27**, 108–109 (1972).
[73] E. Gagliardi and P. Presinger, *Mikrochim. Ichnoanal. Acta* **1964**, 1175–1180.
[74] James J. Markham, *Anal. Chem.* **39**, 241–242 (1967).
[75] N. Bukhshand and A. Khattak, *Pak. J. Sci. Ind. Res.* **5**, 86–90 (1962).

To a neutral sample containing 0.1–1 mg of silver as the nitrate, add 2 ml of 0.5% solution of **N-ethyl-N-hydroxyethyl-p-phenylenediamine sulfate.**[76] Mix for 10 minutes and dilute to 100 ml. Read at 530 nm. There is interference by ferric, chromic, nitrous, chloride, and bromide ions.

Silver forms a 1:1 complex with **1,5-di-2-naphthylthiocarbazone.**[77] This is extractable by chloroform from *N* sulfuric acid, which is pH 0, up to pH 10 for reading at 505 nm. If the sample solution contains selenium or tellurium, it is masked with EDTA. If the sample solution contains mercuric ion, it must be 0.5 *N* with hydrochloric acid.

Silver at 0.0002–0.1 mg causes development of a pink color in the following reagent.[78] To the sample, add 1 ml of 1:150 nitric acid, 0.5 ml of 3% mercuric nitrate solution, 1 ml of 20% ammonium persulfate solution, 0.1 mg of copper sulfate, 0.5 ml of 1% **dimethylglyoxime** solution, and 1 ml of 10% pyridine solution. Dilute to 20 ml before reading.

The optimum conditions for determination of silver with **ethylrhodamine B** are 6 *N* sulfuric acid, *N* bromide ion, and 0.1 mg of reagent per ml.[79a] Extraction with benzene from 8 *N* sulfuric acid containing 20 mg of bromide per ml and 0.4 mg of reagent provides preconcentration.

In 0.25 *M* potassium bromide at pH 2 in dilute sulfuric acid, silver forms a 1:1:1 complex of silver ion, **ethyl violet**, and bromide ion when extracted with a fivefold excess of the dye in toluene.[79b] Read at 615 nm, the color obeys Beer's law for 0.1–1 μg of silver per ml and is stable for 2 hours. Up to twofold bismuth is tolerated.

If a silver solution is passed through a column of powdered iron, it displaces its equivalent of **ferrous ion**, which may be read with 1,10-phenanthroline at 512 nm or oxidized to the ferric state for determination.[80] The same reaction is given by lead, cobalt, nickel, copper, and mercuric ions. Sodium, magnesium, and manganese do not interfere, but carbon dioxide and strong oxidizing agents do.

Silver forms a 1:1 complex with **glyoxal dithiosemicarbazone** in perchloric acid solution at pH 1.1 containing EDTA.[81] Beer's law is followed for 0.05–0.4 m*M*. There is interference by bromide, iodide, thiocyanate, and thiosulfate.

After precipitation of silver with *p*-dimethylaminorhodanine, it is complexed with **1,10-phenanthrolein and eosin** or a related fluorescein derivative.[82] The complex is extractable with chloroform at pH 9. The absorption is read at 540 nm. The fluoresence in 1:1 acetone-chloroform at 580 nm is 10 times as sensitive and permits determination of 0.005–1 μg of silver per ml.

Silver forms a stable orange-red 1:2:1 complex with **1,10-phenanthroline** and **3,4,5,6-tetrachlorofluorescein.**[83] The maximum absorption at 540 nm is constant at

[76]I. Antonescu and M. Caramlau, *Bul. Inst. Politeh. Iasi* **14**, 195–199 (1968).
[77]V. G. Tipsova, A. M. Andreichuk, and L. A. Bazhanova, *Zh. Anal. Khim.* **21**, 1179–1182 (1966).
[78]B. A. Gerasimov and E. P. Gokieli, *Tr. Gruz. Sel'sk.-Kloz. Inst.* **1964**, 369–371.
[79a]I. A. Bochkareva and I. A. Blyum, *Zh. Anal. Khim.* **30**, 874–882 (1975).
[79b]N. L. Shestidesyatnaya, L. I. Kotelyanskaya, and I. A. Chuchulina, *ibid.* **30**, 1303–1309 (1975).
[80]Eugeniusz Kloczko, *Chem. Anal. Warsaw*, **16**, 1291–1298 (1971).
[81]B. W. Budsinsky and J. Svec, *Anal. Chim. Acta* **55**, 115–124 (1971).
[82]D. P. Scherbov and D. N. Lisitsyna, *Zavod. Lab.* **39**, 656–658 (1973); D. N. Lisitsyna and D. P. Shcherbov, *Zh. Anal. Khim.* **28**, 1203–1205 (1973).
[83]Itsuo Mori, Takehisa Enoki, and Toyoko Mano, *Jap. Anal.* **22**, 1202–1209 (1973).

pH 9.8–11.5. The maximum fluorescence is at 535 nm, constant for pH 10.5–11.8. Beer's law is followed photometrically for 5.4–11.6 μg of silver per 10 ml and fluorimetrically for 0.5–6 μg of silver per 10 ml. EDTA masks interference by bismuth, thorium, zinc, ferric ion, and zirconium, but iodide, cyanide, and thiosulfate interfere seriously. For spectrophotometry, stabilize with polyvinylpyrrolidone; for fluorometry, with gelatin.

Ternary complexes are also formed by silver and 1 : 10-phenanthroline with pyrogallol red or bromopyrogallol red as shown elsewhere.[84]

Silver serves to catalyze persulfate oxidation of **1, 10-phenanthroline.**[85] All metals forming stable complexes with 1, 10-phenanthroline interfere. Mix 10 ml of a solution containing up to 15 μg of silver per ml as the nitrate with 15 ml of glacial acetic acid and 5 ml of 0.1 *M* 1, 10-phenanthroline. Add a small excess of solid ammonium persulfate and dilute with water to 50 ml. After 10 minutes, read at 420 nm. Beer's law applies for that range.

Silver is extracted as a 1 : 1 : 1 complex by shaking a solution at pH 4.3–6 containing **1, 10-phenanthroline** with **thiothenoyltrifluoroacetone** in xylene.[86] Wash out excess reagent from the organic layer with a borate buffer solution for pH 11.5 and read at 360 nm against a reagent blank.

Picolinaldehyde thiosemicarbazone forms a stable 1 : 1 complex with silver ion at pH 5–6.[87] At pH 5.5 in the presence of EDTA, it is read at 316 nm. There is interference by an equimolar amount of mercuric ion, double the amount of titanium or uranium, 5 times the molar amount of copper or hafnium, and 8 equivalents of scandium. There is interference by equimolar bromide, iodide, or cyanide and 3 equivalents of chloride ion.

In the presence of disodium **2, 6-pyridinedicarboxylate**, the silver ion at pH 2.5–4 causes an oxidation by ammonium persulfate to produce bis(2, 6-pyridinedicarboxylate) divalent silver anion.[88a] This, in the presence of an acetate buffer solution, for pH 3 can be read at 366, 578, or 920 nm for 0.002–0.25 mg of silver per ml. Nickel, cobalt, and halides must be absent.

A 1 : 1 complex is formed by silver ion in a 0.004% solution of **1-(2-pyridylazo-2-naphthol** dissolved in 4 m*M* sodium hydroxide containing 0.04% of sodium tartrate.[88b] Extracted with 7 : 3 benzene-isobutanol the complex is read at 540 nm and obeys Beer's law for 0.1–2.8 μg of silver per ml. The color is stable for 15 hours. Zirconium and ferric ion are masked by fluoride; acetate, oxalate, or citrate masks phosphate. There is interference by copper, cobalt, nickel, zinc, cadmium, mercuric ion, thiocyanate, cyanide, and EDTA.

To a solution containing less than 40 μg of silver, add 5 ml of 0.1% solution of tartaric acid. Adjust to pH 10.5 with 0.2% solution of sodium hydroxide and 1 ml of 0.1% solution of **4-(2-pyridylazo)resorcinol**, often referred to as PAR. Dilute to 25 ml and read at 510 nm.[89] Beer's law is followed for 0.16–1.6 μg of silver per ml.

[84]N. V. Lukashenkova, N. S. Talmacheva, and E. P. Shkrobot, *Zavod. Lab.* **39**, 341–343 (1973).
[85]F. Vydra and V. Markova, *Chem. Listy* **57**, 958–961 (1963).
[86]Masakazu Deguchi and Tsutomu Inamori, *J. Pharm. Soc. Jap.* **95**, 1010–1012 (1975).
[87]D. J. Leggett and B. W. Budesinsky, *Microchem. J.* **16**, 87–93 (1971).
[88a]Heinrich Hartkamp, *Z. Anal. Chem.* **184**, 98–107 (1961).
[88b]M. C. Eshwar and B. Subrahmanyam, *Zh. Anal. Khim.* **31**, 2319–2322 (1976).
[89]M. C. Eshwar and B. Subrahmanyam, *Z. Anal. Chem.* **272**, 44 (1974).

Bismuth, aluminum, zirconium, thorium, and vanadium can be masked with fluoride ion; lead can be masked with phosphate. Interfering ions that must be absent are copper, nickel, cobalt, cadmium, zinc, iron, and cyanide.

Adjust a sample solution of not more than 20 ml, containing 0.1–12 μg of silver to 0.1 *N* sulfuric acid and 6 m*M* potassium iodide.[90] Add 0.5 ml of m*M* **malachite green** and extract with 5 ml of benzene. Read the organic phase at 640 nm. Beer's law is obeyed. Cadmium can be masked with EDTA, cupric ion with thiourea.

The reaction of silver with **thioacetamide** at pH 9–14 gives a brown-yellow color suitable for reading at 490 nm.[91] Ferric ion must be masked with sodium potassium tartrate.

Silver reacts with **rubeanic acid**, which is **dithiooxamide** in neutral solution.[92] On addition of ammonia, the silver becomes a sulfide. In the presence of gelatin, either form is read at 470 nm. The diethyl, methyl, and benzyl derivatives are equally satisfactory.[93]

Mix a sample solution containing 5–100 μg of silver with 10 ml of acetic acid–ammonium acetate buffer solution for pH 6 and dilute to 25 ml. Shake for 10 minutes with 10 ml of m*M* **thiodibenzoylmethane**, which is 2-thiobenzoylacetophenone, in benzene, and read the organic layer at 420 nm against a reagent blank.[94] Beer's law is obeyed for 0.5–10 μg of silver per ml. There is serious interference by uranium, copper, cobalt, cadmium, nickel, and mercury. Titanium, beryllium, and thallium are masked with fluoride; cadmium with EDTA.

Oxidation of **sulfanilic acid** by persulfate can serve as a measure of the catalytic effect of silver suitably complexed.[95] At pH 4.35, 2,2′-bipyridyl acts as the activator. Reading is at 535 nm. **Silver sulfide** produced in the presence of collodian and methanol is read at 350 nm.[96]

The 1:1 complex of silver with **thiothenoyltrifluoroacetone** is extracted at pH 6.8±0.1 with a solution of the reagent in xylene containing some pyridine.[97] Because excess reagent reduces the sensitivity, wash the organic phase with 1:9 pyridine–sodium tetraborate–sodium hydroxide buffer solution for pH 11.5. Then read at 360 nm against a reagent blank. Beer's law is followed for 0.3–3 μg of silver in the organic phase.

Tris-(2,2′-bipyridyl)ferrous tetraphenylborate reacts at pH 3–10 with silver ion to liberate tris-(2,2′-bipyridyl)ferrous ion, which is read at 522 nm.[98] Thallous and mercuric ions react similarly. The reaction will determine 5–50 ppm of silver. Chloride interferes.

By use of the silver as a catalyst for oxidation of **tropaeolin O** by persulfate with reading at 314 nm, it is possible to determine silver down to 65 pg/ml.[99]

[90] N. L. Shestidesyatnaya and L. I. Kotelyanskaya, *Ukr. Khim. Zh.* **41**, 883–887 (1975).
[91] G. G. Karanovich, *Sb. Statei Vses. Nauchn.-Issled. Inst. Khim. Reakt.* **1959** (23), 96–101.
[92] Kh. Ya. Levitman and E. V. Gorskaya, *Belorus. Politekh. Inst. I. V. Stalina, Sb. Nauch. Tr.* **1959**, No. 87, 55–65.
[93] J. Xavier and Priyadaranjan Ray, *J. Indian Chem. Soc.* **35**, 432–444 (1958); *Sci. and Culture* (Calcutta) **21**, 694–695 (1956).
[94] R. R. Mulye and S. M. Khopkar, *Anal. Chim. Acta* **76**, 204–207 (1975).
[95] P. R. Bontschev, A. Alexiev, and B. Dimitrova, *Talenta* **16**, 597–602 (1969).
[96] M. Ziegler and H. Sbrzesny, *Z. Anal. Chem.* **175**, 321–324 (1960).
[97] Masakazu Deguchi, Michiyuki Harada, and Mikio Yashiki, *J. Pharm. Soc. Jap.* **94**, 1025–1027 (1974).
[98] M. C. Mehra and P. Orien, *Microchem. J.* **19**, 384–389 (1974).
[99] E. Jankauskiene and E. Jasinskiene, *Zh. Anal. Khim.* **24**, 527–530 (1969).

CHAPTER FOUR
MERCURY

Methods for determination of mercury still lean very heavily on dithizone. Formation of other binary complexes with dyestuffs are less significant than with some other elements. And as is the case with other elements, a host of other reagents have been applied—a few of them old and familiar, but many of them new in recent decades.

Since some inorganic compounds of mercury are volatile, precautions must be taken in preparation of samples that are not required with the majority of elements. And since mercury is poisonous, this property to some extent dictates the nature of the samples examined.

Mercury in urine, blood, feces, tissue, and air is commonly determined on the pretreated sample by diphenylcarbazone, dithizone, or di-2-naphthylthiocarbazone.[1] For suspected mercury poisoning, urine is analyzed, and this analysis is supplemented by the more difficult determination in blood and tissue. Some organic mercury derivatives are important and are included here.

BINDSCHEDLER'S GREEN

The bromo complex of mercury oxidizes the leuco base 4,4′-bis(dimethylamino) diphenylamine to form the dyestuff known as Bindschedler's green.[2] This is extractable with 1,5-dichloroethane at pH 1.6–2.6. A citrate buffer solution for the pH also serves as a salting-out agent. The extracted color is stable for 2 hours. Stannous and stannic ions interfere. Antimony precipitates the reagent.

[1] L. Vignoli, R. Badre, M. C. Morel, and J. Ardorino, *Chim. Anal.* (Paris) **45**, 53–59 (1963).
[2] Masihiro Tsubouchi, *Anal. Chem.* **42**, 1087–1088 (1970).

Procedure. *Water Containing Ethyl Mercuric Phosphate.* As 10^{-3} M color reagent, dissolve the leuco base in 0.01 N sulfuric acid. As a buffer solution for pH 2, add a few drops of 1 : 1 sulfuric acid to 0.05 M sodium citrate.

Mix 20 ml of sample, 10 ml of sulfuric acid, and 10 ml of 30% hydrogen peroxide. Heat under a condenser until foaming ceases, and let cool. Adjust to about pH 2 with 20% sodium hydroxide solution and dilute to 500 ml.

Mix 10 ml of the oxidized sample solution containing less than 10^{-5} M of mercuric ion, 1 ml of 0.05 M potassium bromide solution, 5 ml of the citrate buffer solution, 1 ml of the color reagent, and 2 ml of 0.01 M ammonium persulfate solution. Dilute to about 25 ml, extract with 10 ml of dichloroethane for 1 minute, and let stand for 5 minutes. Dry the organic layer with sodium sulfate, and read at 740 nm against a reagent blank.

BRILLIANT GREEN

Brilliant green is a dyestuff that forms a 1 : 2 ion-association complex with mercuric iodide, which is extractable with benzene. A 330-fold excess of the dye is desirable. A suitable hydrochloric acid medium for extraction is under pH 2.5, preferably pH 1. Beer's law holds for 0.1–3 μg of mercury per ml of extract. Thallic, auric, ferric, antimonic, perchlorate, thiocyanate, and nitrate ions interfere.

As a related technic, brilliant green combines with the anion tetraphenylborate as a solid reagent.[3] Shaken with a solution containing 0.25–4 ppm of silver or mercuric ion, it releases the colored cation for reading at 623 nm.

Procedure.[4] Dilute a sample containing up to 20 μg of mercuric ion to about 21 ml. Add 3 ml of 1.66% solution of potassium iodide and 3 ml of mM brilliant green. Adjust to pH 0.7–1.5 with 0.9 M sulfuric acid and dilute to 30 ml. Extract with 10 ml of benzene and dry the extract with sodium sulfate. Read at 640 nm against a reagent blank.

CAPROLACTAM

The turbidity that mercuric ion forms with caprolactam in the presence of potassium iodide can be read photometrically.[5] Beer's law holds for 0.03–0.1 mg of mercuric ion in the final suspension. The density decreases with temperature to an optimum at 5°.

[3] M. C. Mehra and C. Bourque, *Analysis* **3**, 299–302 (1975).
[4] Tsuguo Sawaya, Hajime Ishii, and Tsugikatsu Odashima, *Jap. Anal.* **22**, 318–322 (1973); cf. V. M. Tarayan, E. N. Ovsepyan, and N. S. Karimyan, *Dokl. Akad. Nauk Armyan. SSR* **49**, 242–245 (1969).
[5] Halina Sikorska-Tomicka, *Chem. Anal.* (Warsaw) **5**, 269–275 (1960).

Procedure. Chill 1 ml of sample solution containing 0.3–1 mg of mercuric ion to 0° in an ice bath. With vigorous stirring, add dropwise 0.5 ml of sulfuric acid and leave in the ice bath for 5 minutes. Add 0.5 ml of 2% potassium iodide solution, 1 ml of 2.5% gum acacia solution, and 2 ml of a 30% solution of caprolactam. Dilute to 10 ml with 2.5 *N* sulfuric acid. Transfer to a bath at 20° and read.

COPPER DIETHYLDITHIOCARBAMATE

When an aqueous mercuric solution is added to aqueous cupric diethyldithiocarbamate, the reduction in absorption by displacement of the copper is a measure of the mercuric ion.[6] The same reaction is given by silver. The combined value for mercury and silver less that for mercury with the silver masked gives silver by difference. For mercury by that technic, see Silver. The technic is also applicable to organomercurials.

Copper bis(2-hydroxyethyl)dithiocarbamate is applicable much like copper diethydithiocarbamate.[7] Reading at 432 nm determines 0.01–0.2 mg of mercury in 20 ml. Ferric, cupric, iodide, and cyanide ions interfere.

Procedure. *Phenylmercury Acetate or Phenylmercury Chloride.* The technic separates the organomercury compound from inorganic mercury ion.[8] It is not applicable to iodides but can be applied to ethylmercury acetate. Take a sample solution that contains 0.04–0.36 mg of phenylmercury acetate or chloride. Adjust it to 10 ml containing 4% of acetic acid and 4% of methanol. Extract the organomercurial into 20 ml of chloroform. Shake the chloroform extract with 10 ml of a *M* acetate buffer for pH 4.5 and 5 ml of 0.0004 *M* solution of copper diethyldithiocarbamate for 2 minutes. Read the organic layer at 436 nm.

COPPER THIURAMATE

Copper thiuramate, also known as mercupral, is the reaction product of thiuram, which is tetramethylthiuram disulfide, and cupric acetate. In benzene solution it is bleached by displacement of the copper by mercuric ion.[9] Organomercury ion is not strong enough to do this. Therefore mercupral is a reagent for measuring

[6] J. W. F. Tertoolen, C. Buijze, and G. J. Van Kolmeschate, *Chemist-Analyst* **52**, 100 (1963); cf. Seiichiro Hikime, Hitoshi Yoshida, and Masahiro Yamomoto, *Jap. Anal.* **10**, 508–513 (1961); Sumio Komatsu and Sadao Kuwano, *J. Chem. Soc. Jap., Pure Chem. Sect.* **83**, 1262–1264 (1962); Yoshimasa Tanaka and Noriko Shido, *Kumamoto Pharm. Bull.* **1962**, 292–295.

[7] Seiichiro Hikime, Hitoshi Yoshida, and Masahiro Yamamoto, *Jap. Anal.* **10**, 508–513 (1961).

[8] Hiroshi Takehara, Takuya Takeshita, and Itsuo Hara, *Jap. Anal.* **15**, 332–338 (1966).

[9] Jan Michal and Jaroslav Zyka, *Chem. Listy* **51**, 56–62 (1957); E. I. Vail, A. P. Mirnaya, I. A. Rastrepina, and L. V. Sigalova, *Zavod. Lab.* **27**, 1465–1467 (1961).

mercuric contamination of organomercury products or for total mercury after suitable treatment. Cerium decolorizes the reagent. For preparation of the reagent, see Silver.

Procedures

Liquors. FREE MERCURIC ION. Dilute a sample containing 1 mg of mercury (e.g., 2 ml of a liquor containing 0.5 gram of mercury per liter) to 100 ml. Mix a 5-ml aliquot with 10 ml of a reagent containing 40 mg of copper thiuramate per liter in benzene. Shake for 2 minutes and read the benzene layer at 435 nm.

Total Mercury. Reflux 100 ml of sample with 0.5 gram of potassium nitrate for 5 minutes to decompose the organomercury compound. Cool and proceed as for liquors.

Ethylmercuric Chloride. Heat 0.5 gram of sample with 2 ml of nitric acid for 10 minutes. Dilute with water and destroy the residual nitric acid by additions of hydrogen peroxide. Filter, dilute to 100 ml, and proceed as for liquors.

CRYSTAL VIOLET

A 1:1 complex is formed by crystal violet with either chloromercurate ion or bromomercurate ion.[10] Beer's law holds for the ranges given below.

Procedures

Chloro Complex. To a sample containing mercuric ion in hydrochloric acid at pH 1.5, add 1.5 ml of 1.2% sodium chloride solution and 2.5 ml of 0.1% solution of crystal violet. Extract with 5 ml and 5 ml of benzene. Adjust the extract to contain 0.19–4.68 μg of mercury per ml and read at 605 nm 15–20 minutes after the phases have separated.

Bromo Complex. To a sample containing mercuric ion in hydrobromic acid at pH 1.14, add 1 ml of 1.2% potassium bromide solution and 2 ml of 0.1% solution of crystal violet. Extract with 5 ml and 5 ml of 4:1 ether-benzene. Adjust the combined extracts to 0.19–1.8 μg of mercury per ml and read at 595 nm after 15–20 minutes.

---

[10] V. M. Tarayan, E. N. Ovespyan, and S. P. Lebedeva, *Arm. Khim. Zh.* **23**, 1085–1890 (1970); cf. E. L. Kothny, *Analyst* **94**, 198–203 (1969).

DIETHYLDITHIOCARBAMATE

Although mercuric ion is determined by displacement of cupric ion from copper diethyldithiocarbamate, mercuric diethyldithiocarbamate is also read in the ultraviolet.[11] The latter can be extracted with carbon tetrachloride from pH 4–11, but the reagent is also extracted below pH 8.5. If iron is present, it coprecipitates mercury above pH 10. By the technic specified, there is no interference by 1 mg of the majority of common elements. It tolerates 0.5 mg of silver, 0.2 mg of ruthenium, 0.1 mg of lead, 0.01 mg of thallous ion, and 2 μg of copper. Permanganate is reduced.

Procedure. As reagent, add 2 drops of 20% sodium hydroxide solution to a 0.2% solution of diethyldithiocarbamate. As 5% buffered EDTA solution, dissolve 50 grams with 25 grams of sodium carbonate in 800 ml of water. Add sufficient 20% sodium hydroxide to adjust to pH 9–9.5, and dilute to a liter.

Heat a sample solution containing 10–60 μg of mercury with 1 ml of nitric acid at 100° for 10 minutes, to ensure that all is in the mercuric state. Add 5 ml of buffered EDTA solution and adjust to pH 9.3–10 by adding 20% sodium hydroxide solution. Add 1 ml of 1% potassium cyanide solution, and 1 ml of 0.2% reagent solution. Extract with 10 ml of carbon tetrachloride. Read the carbon tetrachloride layer at 278 nm against carbon tetrachloride within the next 20 minutes.

If the sample solution contains more than 0.2 mg of ruthenium, adjust the sample volume to 5 ml. Using hydrochloric acid or 20% sodium hydroxide solution, adjust to pH 2–5. Warm and add 3–4 drops of 6% sodium hypochlorite solution. Heat for 10 minutes to volatilize the ruthenium. If the solution remains brown, showing that ruthenium is still present, add a few more drops of sodium hypochlorite solution and continue heating, but avoid taking to dryness. Use absorbant tissue to remove any ruthenium oxide that deposits on the walls of the beaker. When the solution is clear, add 2 drops of hydrochloric acid, warm for 2 minutes, add 2–3 drops of 6% sulfurous acid, and continue to heat for 5 minutes before proceeding as above.

5-(4-DIMETHYLAMINOBENZYLIDINE) RHODANINE

This reagent is sensitive enough to determine mercuric ion in a solution in which dithizone shows negative results.

Procedure. *Sodium Chloride Electrolysis Solutions.*[12] Water condensed from the hydrogen liberated by electrolysis is a typical sample. Appropriately dilute with water, such as 10–200 ml. Add 10 ml of 1% gelatin solution and with slow stirring,

[11] E. A. Kakkila and G. R. Waterbury, *Anal. Chem.* **32**, 1340–1342 (1960).

[12] A. Tanu, G. Teodorescu, A. Dinescu, and L. Cocargeanu, *Bul. Inst. Politeh. G. Gheorghin-Dej Buc.* **27**, 61–63 (1965).

add 5 ml of 0.02% solution of 5-(4-dimethylaminobenzylidine)rhodanine in acetone by delivery by a pipet below the surface of the solution. Add 1 ml of 1:9 nitric acid per 25 ml of the final volume and dilute to that volume. Read within 2 hours.

DI-2-NAPHTHYLTHIOCARBAZONE

This analogue is used much as is diphenylthiocarbazone, usually referred to as dithizone.

Procedures[13]

Air. Pass the air at 2 liters/minute through a mixture of 70 ml of water, 10 ml of sulfuric acid, and 20 ml of 3.2% potassium permanganate solution. Set aside for 24 hours to complete the reaction. To the contents of the absorber, add a saturated solution of hydroxylamine hydrochloride dropwise, until the solution is decolorized plus 0.5 ml excess. With shaking, add 5 ml of 10% urea solution to remove oxides of nitrogen. Add 1 ml of 2% EDTA to mask interfering metals. Add 50% sodium acetate solution dropwise until the pH is 1.5. Add dropwise fresh 0.005% solution of di-2-naphthylthiocarbazone in chloroform until the color is green to green-violet, indicating excess of the reagent; shake vigorously for 3 minutes. Dilute the solvent layer to 10 ml with chloroform and filter, set aside for 15 minutes and read at 515 nm.

Cereals.[14] This is a field test. Digest 0.5 gram of sample with 2 ml of water, 3 ml of sulfuric acid, and 10 ml of saturated solution of potassium permanganate. Heat to 100° and complete the digestion at room temperature for 8 hours. Add 10% solution of hydroxylamine hydrochloride to destroy the residual permanganate, and adjust to pH 1.5 with 10% sodium hydroxide solution. Add 1 ml of 10% solution of disodium EDTA to mask interfering heavy metals. Add 1 ml of saturated solution of urea and 5 ml of buffer solution for pH 1.5. Extract mercury with 10 ml of 0.005% solution of di-2-naphthylcarbazone in chloroform; read by visual comparison.

Medicaments.[15] This technic was developed for determining phenylmercuric borate and nitrate. Wet-ash the sample as necessary. To a sample containing 0.1–1.5μg of mercury, add 0.2 ml of saturated solution of hydroxylamine sulfate and 2 ml of a glycine–citrate–hydrochloric acid buffer solution for pH 1.5. Dilute to 10 ml and shake with 5 ml of 0.005% solution of the captioned reagent in chloroform. Set aside for 15 minutes and read at 515 nm.

---

[13]R. J. DuBois and S. B. Knight, *Anal. Chem.* **36**, 1316–1320 (1964); cf. V. G. Tiptsova, A. M. Andreichuk, and L. A. Bazhanova, *Zh. Anal. Khim.* **20**, 1200–1203 (1965).
[14]W. Raffke and Barbara Wirthgen, *Nahrung* **20**, 41–45 (1976).
[15]D. Melle-Robert, M-F. Fonteret, and S. Fleury, *An. Pharm. Fr.* **28**, 465–476 (1970).

4,4–DINITRODIAZOAMINOBENZENE

The red alkaline solution of the captioned reagent, 1,3-bis(4-nitrophenyl)triazen, is decolorized by mercuric ion.[16] The mercuric complex has a ratio of 1:3 and is stable for 4 hours. Beer's law applies up to 40 μg of mercury per ml. Cadmium and silver interfere by also reducing the color, but pentavalent antimony and trivalent arsenic intensify the color.

Procedure. Adjust the sample to be 2 *N* with sulfuric acid. Extract with 5 ml and 5 ml of 0.001% solution of dithizone in carbon tetrachloride. Combine the organic phases and back extract the mercury with 8 ml of *N* hydrobromic acid. Heat this extract until any carbon tetrachloride has been driven off and cool. Neutralize to nitrophenol with 1.2% potassium hydroxide solution. Add 7 ml of a solution of 28.7 mg per liter of the captioned reagent in 0.6% potassium hydroxide solution. Read at 510 nm against a reagent blank.

DITHIZONE

For a general discussion of dithizone, see Lead. In the dithizone technic, large amounts of chloride ion do not interfere.[17] An equivalent amount of EDTA effectively masks zinc, lead, and cadmium, as well as up to 1 mg of copper. 1,2-Diaminocyclohexane-*N,N,N′,N′*-tetraacetic acid is more effective in masking copper. Sodium bisulfite removes oxidative agents.

Passage of a chloroform solution of mercuric dithizonate through an aluminum oxide column removes excess dithizone, at the same time ensuring that all the dithizonate is in the enol form.[18] Silver must be absent. Interference by copper is avoided by extracting the mercury dithizonate from 1–2 *N* sulfuric acid. Other common interfering ions are retained. A recommended column is 0.7×5 cm of 100-325 mesh alumina containing 10% moisture.[19]

For isolation of mercury dithizonate, a sulfuric acid digestate is applied to a column of poly(chlorotrifluoroethylene) having on it a 0.001% solution of dithizone in chloroform as the stationary phase. Wash the column with 1:5 sulfuric acid equlibrated with 0.01% dithizone solution, then with 0.4% sodium hydroxide solution similarly equilibrated. Wash with 0.2% sodium hydroxide solution and with 10% acetic acid. Finally, strip the orange-yellow mercury dithizonate from the column with chloroform and read at 483 nm.[20]

An alternative technic for removal of interfering ions is to pass the digestate in sulfuric acid through a column of anionic resin and wash well with water. Mercuric ion is retained and as much as 0.3–0.5 mg of copper, lead, cadmium, thorium, zinc,

[16]Shih-ti Ts'ao, Ch'an Ma, and Hsu-ch'in Yen, *Acta Chim. Sin.* **32**, 82–88 (1966).
[17]Takio Kato, Shinsuke Takei, and Akio Okagami, *Technol. Rep. Tohoku Univ.* **21**, 291–305 (1957).
[18]M. D. J. Isaacs, P. Morries, and R. E. Stuckey, *Analyst* **82**, 203–206 (1957).
[19]Takashi Ashizawa, *Jap. Anal.* **10**, 443–448 (1961).
[20]H. Woidich and W. Pfannhauser, *Z. Anal. Chem.* **261**, 31 (1972).

and nickel pass through. Elute the mercury with 0.002 *M* thiourea in 0.01 *M* hydrochloric acid for dithizone determination.[21a]

For selective adsorption of mercury from a solution such as sea water use polystyrene beads crosslinked with 2% of divinylbenzene.[21b] Swell them in a solution of dithizone in chlorobenzene. Adjust the sample with sulfuric acid or nitric acid to contain 1–10 ppm of mercury at pH 1. Pass through a column of the beads at 100–300 ml per hour to adsorb mercury as an orange dithizonate. Elute the dithizonate with 50 ml of 8 *N* hydrochloric acid and read at 490 nm. The technic is effective in separation of mercury from zinc, cadmium, ferrous ion, and lead. Cupric ion requires masking with EDTA.

Mercury in urine can be deproteinized with trichloroacetic acid. At this lower pH, addition of potassium bromide forms potassium bromomercurate.[22] On raising the pH to 5 with a formate buffer, the mercury is extracted as the dithizonate.

Reproducible results may not result if part of the mercuric ion is reduced by free hydroxylamine before the pH is reduced to 1.5.[23] To use the monocolor method, excess dithizone can be washed out of the organic layer with 0.4% sodium hydroxide. The layer is then washed with 25% acetic acid to improve stability of the mercury dithizonate.

Both mercury and copper dithizonates can be read in the ash of biological material such as preserved fish or shellfish.[24] Wet-ash the sample with nitric and sulfuric acids and hydrogen peroxide. After dilution, remove oxidizing substances with hydroxylamine hydrochloride. Extract oily residues with trichloroethylene, then with chloroform. Add sodium hydroxide solution and citrate buffer solution to adjust to pH 2.5. Extract the copper and mercury with 0.001% dithizone in chloroform. Remove excess dithizone by shaking with 0.8% sodium hydroxide solution. Read copper dithizonate at 571 nm against a reagent blank. Decompose the copper dithizonate by shaking the chloroform extract with 9 *N* ammonium hydroxide and read the mercury at 494 nm. The copper can be reextracted from the ammoniacal extract by adjusting to pH 4 and using dithizone. It is then read at 530 nm.

For mercury at 1–10 ppm in paper, burn the sample in an oxygen bomb and analyze the acid bomb washings by dithizone.[25]

Mercury in ointments as the element or its acid-soluble salts is extracted with either nitric or hydrochloric acid to give a solution for determination by dithizone.[26a]

For 4–12 ppm of mercury in the spent liquor from a cell for generating chlorine and sodium hydroxide extract with dithizone in chloroform.[26b] Read the mercury complex at 500 nm and free dithizone at 600 nm.

For mercury in a solution of selenium adjust the pH of the solution to 4.5–5 before dithizone extraction.[27]

[21a]J. F. Kopp and R. G. Keenan, *Am. Ind. Hyg. Assoc. J.* **24**, 1–10 (1963).
[21b]Shunsuke Ide, Tairoku Yano, and Keihei Ueno, *Jap. Anal.* **25**, 820–823 (1976).
[22]E. J. Cafruny, *J. Lab. Clin. Med.* **57**, 468–472 (1961).
[23]Ana Loffler and Margareta Putinaru, *Rev. Chim.* (Bucharest) **13** (6), 374–375 (1962).
[24]H. Woidich and W. Pfannhauser, *Z. Lebensm.-Unters. m. Forsch.* **149**, 1–7 (1972).
[25]L. G. Borchardt and B. L. Browning, *Tappi* **41**, 669–671 (1958).
[26a]G. Bussman, *Pharm. Acta Helv.* **38**, 690–701 (1963).
[26b]P. B. Janardhan and S. Rajeswari, *Curr. Sci.* **45**, 373–374 (1976).
[27]E. N. Pollock, *Talenta* **11**, 1548–1550 (1964).

For mercury and silver as dithizonates in the solution of a purified uranium compound, adjust to pH 3 and add EDTA.[28] Extract with 0.025 *M* dithizone in chloroform. Read mercury at 484 nm, and silver at 620 nm.

Mercury dithizonate can be separated from those of other metals such as copper, lead and zinc by thin-layer chromatography on Kieselgel G.[29] The developer is benzene. Scrape off the red zone of mercury dithizonate, dissolve in methanol, centrifuge, and read at 483 nm.

The dithizone method is appropriate for determining mercury in beer.[30]

A solution of dithizone in carbon tetrachloride is appropriate for isolation of copper, mercury, cadmium, and thallium from strongly alkaline solution.[31] Then the mercury must be separated for determination by dithizone.

For mercury in selenium[32] dissolve 5 grams of sample in nitric acid and dilute to 100 ml. Decompose nitrite with ammonium sulfamate and adjust to pH 5–5.5. Extract mercury, silver, and copper with dithizone in chloroform. Divide the organic layer into halves. Back-extract the mercury from one portion with a solution of sodium chloride in dilute hydrochloric acid. Read both portions at 600 nm and mercury by the difference.

For mercury in canned tuna, it is recommended that the instability of the mercury dithizonate be bypassed by reading the unconsumed dithizone in the aqueous phase.[33]

For determination of mercuric chloride in activated carbon catalysts in the presence of mercurous ion and elementary mercury, extract the mercuric chloride with 0.1 *N* cadmium nitrate or aluminum nitrate in 3 : 7 ethanol-methanol.[34] Then determine it in the filtered extract at pH 4 with dithizone.

A detailed procedure is given elsewhere for cadmium by its bleaching of cadion. The procedure is applicable to mercury with this variation.[35] The reaction takes place slowly and is temperature dependent. Therefore it is carried out at 20° and read after 30 minutes. It follows Beer's law up to 1.4 ppm of mercury in the final solution. Cyanide completely inhibits the reaction, and iodide, arsenite, stannous ion, and antimony interfere.

For isolation of mercuric ion from many other ions, make the solution 0.1–3 *M* perchloric acid or 0.05–5 *N* sulfuric acid.[36] Extract with 6–10 m*M* bi-bis(2-ethylhexyl)phosphorodithioate in octane. Wash the extract successively with 4*N* sulfuric acid, 2.5 *M* ammonium thiocyanate in *N* sulfuric acid, and 1% ammonium fluoride solution. Then back-extract the mercuric ion with 5 *N* hydrobromic acid. Make the extract 10 m*M* with EDTA to mask bismuth, adjust to pH 4, extract mercuric ion with dithizone in chloroform, and read at 620 nm.

For isolation of mercuric ion from organic mercurials, convert it to the ethylenediamine complex at pH 7.1 of a phosphate buffer solution.[37] Shake with 0.1 mm

[28] J. Marecek and E. Singer, *Z. Anal. Chem.* **203**, 336–339 (1964).
[29] J. Baümler and S. Rippstein, *Mitt. Lebensm. Hyg.* (Bern.) **54**, 472–478 (1963).
[30] Bronislawa Legatowa, *Rocz. Panstw. Zakl. Hig.* **23**, 429–431 (1972).
[31] G. Ackermann and W. Angermann, *Z. Anal. Chem.* **250**, 353–357 (1970).
[32] Satoru Uehara and Shuichi Hamada, *Kogyo Kagaku Zasshi* **63**, 1580–1583 (1960).
[33] A. Henrioul, F. Henrioul, and F. Henrioul, *Ann. Falsif. Expert. Chim.* **65**, 274–278 (1972).
[34] G. S. Lisetskaya, M. N. Yakumova, and V. I. Golub, *Ukr. Khim. Zh.* **39**, 1173–1176 (1973).
[35] P. Chavanne and Cl. Geronimi, *Anal. Chim. Acta* **19**, 442–447 (1958).
[36] Yu. M. Yukhin, I. S. Levin, and B. Sh. Litvinova, *Zh. Anal. Khim.* **30**, 1091–1094 (1975).
[37] A. Gorgia and D. Monier, *Anal. Chim. Acta* **55**, 247–251 (1971).

microbeads of sodium-calcium glass for selective adsorption of the mercuric complex. Adsorption is about 85% complete but is reproducible for 1–20 m*M*. Desorb by shaking with dilute sulfuric-acetic acid and a solution of dithizone in carbon tetrachloride. Desorption is about 70% complete but reproducible. The dithizone reading will determine down to 1 μg of mercuric ion in 100-fold excess of an organic mercurial such as chlormerodrin.

Pink wheat is so stained to indicate that it carries an organic mercurial fungicide.[38] Ash a 0.25-gram sample in oxygen. Take up the residue in 1 : 35 sulfuric acid containing hydroxylamine hydrochloride. Extract with dithizone in carbon tetrachloride. Extract the excess dithizone with ammonium hydroxide and read the dithizonate at 490 nm. Recoveries average 88%.

In determination by dithizone after treatment with hypochlorite and hydroxylamine hydrochloride, discrepancies have been attributed to use of acetone to rinse the glassware and the conversion of traces of it to 2-chloro-2-nitrosopropane.[39]

For determination by reversion, extract mercuric ion from 0.25 *N* sulfuric acid with dithizone in chloroform.[40] Read a portion of the extract at 496 nm against chloroform. Shake another portion with an equal volume of a solution containing 2.04% of potassium acid phthalate and 6% of potassium iodide to revert to dithizone, and read again. The change in absorption measures the mercury.

For separation of mercury from silver and copper by dithizone based on differences in pH, see Silver. For separation of silver, mercury, and copper and their individual determination by dithizone, see Silver.

Procedure.[41] As reagent I, dissolve 18.61 grams of disodium EDTA, 47.25 grams of monochloroacetic acid and 45.24 grams of diammonium citrate in about 1500 ml of water. Adjust to pH 3.8–4 with ammonium hydroxide, filter, and dilute to 2 liters. As reagent II, dissolve 18.6 grams of disodium EDTA and 94.5 grams of monochloroacetic acid in about 600 ml of water. Adjust to pH 7 ± 0.5, filter, and dilute to 1 liter.

Reflux a sample containing 0.05–0.3 mg of mercury with 8 ml of nitric acid for 5 minutes to ensure that all is in the bivalent form. Cool and dilute to 50 ml. Take an aliquot containing 5–30 μg of mercury and not more than 2.25 m*M* with diverse metal ions. Add 20 ml of reagent I, and adjust to pH 2.75 ± 0.1 with ammonium hydroxide or hydrochloric acid. Add 10 ml of water and shake for 30 seconds with 10 ml of 0.0011% solution of dithizone in xylene. Wash the organic phase with water.

Back-extract the mercury from the organic phase by shaking with 5 ml of 1 : 1 hydrochloric acid followed by 5 ml of 1 : 3 hydrochloric acid. Wash with not more than 10 ml of water and combine the extracts and washings. Add 5 ml of reagent II, followed by ammonium hydroxide, to adjust to pH 2.75 ± 0.1. Add 10 ml of water and extract with 10 ml of 0.0011% solution of dithizone in carbon tetrachloride. Filter the extract through cotton and read at 510 nm against a blank.

[38] L. Jones and G. Schwartzman, *J. Assoc. Off. Agr. Chem.* **46**, 879–881 (1963).
[39] D. F. Lee and J. A. Roughan, *Analyst* **94**, 306–307 (1969).
[40] H. Irving, G. Andrew, and E. J. Risdon, *J. Chem. Soc.* **1949**, 541–547.
[41] Stanley S. Yamamura, *Anal. Chem.* **32**, 1896–1897 (1960).

If copper and other cations are known to be low, the steps from "Add 20 ml of reagent I..." to "Add 5 ml of reagent II..." can be omitted.

Indium.[42] Dissolve a 1-gram sample in hydrobromic acid and dilute to about 5 *N*. Extract the indium with isopropyl ether and discard. Acidify the aqueous phase with perchloric acid, and evaporate to dense fumes to remove hydrobromic acid. Cool, take up in water, and adjust to pH 1 with hydrochloric acid. Dilute to a known volume.

Complete as for the second method for ores, starting at "To an aliquot...."

Copper Alloys.[43] By this technic up to 0.02 mg of mercury is recovered in the presence of 100 mg of copper in *N* sulfuric acid. Changes in the amount of copper, especially below 30 mg, affect the result. The presence of up to 8 mg of stannic ion or 10 mg of bismuth and lead does not affect the result. Oxidizing agents and silver must be absent.

Treat a 0.1-gram sample with 11 ml of 1:1 sulfuric acid and 4 ml of 30% hydrogen peroxide. Heat until the sample is dissolved and the hydrogen peroxide decomposed. Cool and dilute to 100 ml. Take a 50-ml aliquot and add enough saturated copper sulfate solution to bring the copper content to about 100 mg. Dilute to 100 ml and shake with 5 ml of 0.001% dithizone in chloroform for 2 minutes. Separate the extract and shake the aqueous layer with 5 ml of chloroform for 2 minutes. Combine the extracts and read at 495 nm.

Mercury Indium Telluride.[44] Dissolve 0.05 gram of sample in 3 ml of hydrochloric acid and 1 ml of nitric acid by warming. Evaporate to incipient dryness at 180°. This material must not go to dryness in these evaporations or there will be loss of mercury. Cool, add 1 ml of hydrochloric acid, and evaporate nearly to dryness. Add 3 ml of water and evaporate nearly to dryness to drive off the excess hydrochloric acid. Take up in water and dilute to 25 ml.

Dilute a 1-ml aliquot of the sample solution to 25 ml with water and take a 1-ml aliquot. Add 2 ml of 1:12 hydrochloric acid and 1 ml of 1:16 acetic acid. Add 10 ml of 0.003% dithizone solution in carbon tetrachloride and shake for 30 seconds. Read the solvent layer at 500 nm against a reagent blank. The mercury dithizonate in carbon tetrachloride is stable for 3 hours and should be at 0–4 ppm. There is no interference by the indium or tellurium.

Ores.[45] Mix 1 gram of finely ground sample with 3 grams of iron filings. To a procelain crucible, add a 5-mm layer of barium carbonate. Place the sample mixed with iron filings on it. Cover with a 3–4 mm layer of copper filings and over that an 8–10 mm layer of ignited magnesium oxide. Apply a gold cover and heat for 10–15 minutes. The mercury is collected in elementary form on the cover.

Cool and rinse the cover with carbon tetrachloride. Dissolve the mercury from

[42]K. Kasiura, *Chem. Anal.* (Warsaw) **11**, 141–149 (1966).
[43]Akinari Ichiryu and Toshio Sawada, *Jap. Anal.* **12**, 429–435 (1963).
[44]E. J. Workman, *Analyst* **97**, 703–707 (1972).
[45]K. F. Gladysheva, *Sb. Tr. Vses. Nauch-Issled. Gorn-met. Inst. Tsvet. Met.* **1962**, 325–330; cf. Z. Hasek, *Hutn. Listy* **20**, 420–427 (1965).

the cover with 2 ml of nitric acid. Dilute to an appropriate volume such as 25, 100, or 250 ml according to the amount of mercury expected.

Take an aliquot expected to contain 3–4 μg of mercury. Add about 1 gram of urea and 10 ml of 0.2 *M* EDTA. Add 1:10 ammonium hydroxide to a methyl orange end point and 10 ml of acetate buffer for pH 4.7. Extract with 1-ml portions of 0.003% dithizone solution in carbon tetrachloride until the green of extract is no longer discolored. Read the combined extracts at 490 nm.

Alternatively,[46] treat a 1-gram sample with 5 ml of nitric acid. Warm and add 5 ml of 1:1 sulfuric acid. Evaporate to a small volume and cool. Add 50 ml of water, filter, and dilute to a known volume. To an aliquot containing about 15 μg of mercury, add 10 ml of 0.1 *M* EDTA and 5 ml of 20% ammonium thiocyanate solution. Add 10% sodium acetate solution until the pH is 2–3. Shake for 1 minute with 15 ml of 0.001% dithizone solution in benzene. Separate the dithizone extract and remove the excess dithizone by shaking with 15 ml and 15 ml of 1:9 ammonium hydroxide containing a little EDTA. Filter and read at 490 nm against a reagent blank.

For determination of mercury in minerals at a concentration of about 0.1%, the dithizone method is more consistent than spectrography but slower.[47]

Soil.[48] Reflux a l-gram sample with 5 ml of sulfuric acid containing excess potassium permanganate. Heat without reflux until substantially dry, and cool. Take up in 25 ml of water and add 30% hydrogen peroxide until all the manganese dioxide and excess potassium permanganate have been destroyed. Dilute to 100 ml.

To a 10-aliquot, add 1 ml of 1% EDTA solution, 1 ml of 20% sodium sulfite solution, and 0.1 ml of 1% potassium thiocyanate solution. Extract by shaking for a minute with 4 ml of 0.0005% dithizone solution in chloroform, and read at 496 nm against chloroform.

As an alternative, for mercury in soil, pelletize 1 gram of dry sample with 0.4 gram of cellulose acetate.[49] Burn three such pellets in an oxygen flask, absorbing the result in 0.1 *N* hydrochloric acid. Then burn three more pellets, absorbing in the same portion of acid. Thereafter determine by dithizone.

As another alternative[50] digest 40 grams of sample of particle diameter less than 2 mm under reflux with 50 ml of sulfuric acid, 0.1 gram of selenium, and 5 ml of nitric acid. Distill the mercury from the reaction mixture with a stream of dry hydrogen chloride gas, collecting the product by water traps. Extract with dithizone and read by the monocolor method.

Coal[51]

MERCURIC CHLORIDE. Shake a 1-gram sample for 30 minutes with 5 ml of water

[46]V. Patrovsky, *Rudy* **12**, 207–208 (1964). Florica Popea and Margareta Jemaneanu, *Acad. Rep. Pop. Rom., Stud. Cercet. Chim.* **8**, 607–616 (1960).
[47]Eva Komarkova, Jan Michal, and Petr Novotny, *Z. Anal. Chem.* **258**, 342–346 (1972).
[48]A. E. Vasilevskaya and V. P. Shcherbakov, *Pochvovedenie* **1963**, 96–97.
[49]Merrie N. White and D. J. Lisk, *J. Assoc. Off. Anal. Chem.* **53**, 530 (1970).
[50]J. A. Pickard and J. T. Martin, *J. Sci. Food Agr.* **14**, 706–709 (1963).
[51]A. E. Vasilevskaya and V. P. Shcherbakov, *Dopov. Akad. Nauk Ukr.SSR* **1963**, 1494–1496.

and 1 ml of ethanol. Centrifuge the coal, make another extraction and combine the extracts. Add 1 ml of 1% EDTA solution, 1 ml of 20% sodium sulfite solution, and 0.1 ml of 1% potassium thiocyanate solution. Shake for 1 minute with 4 ml of 0.0005% dithizone solution in chloroform. Read at 496 nm against chloroform.

MERCURIC OXIDE. Shake the coal from which mercuric chloride has been extracted with 5 ml of 1:72 sulfuric acid and 1 ml of ethanol. Centrifuge the coal and make another extraction. Wash with water and centrifuge. Complete as for mercuric chloride from "...combine the extracts."

MERCURIC SULFIDE. Add 5 ml of saturated sodium sulfide solution to the extracted and washed coal and shake for 1 hour. Add 1 ml of ethanol and centrifuge. Repeat the extraction with sodium sulfide and ethanol. After centrifuging, wash the sample with water. Combine the extracts and washings and evaporate to dryness. Take up the resudue in 1 ml of sulfuric acid containing enough potassium permanganate to end with a slight pink color. Dilute to about 10 ml and neutralize by dropwise addition of 20% sodium hydroxide solution. Complete as for mercuric chloride from "Add 1 ml of 1% EDTA...."

ELEMENTARY MERCURY. Shake the extracted coal for 30 minutes with 5 ml of a 0.2% solution of iodine in 2% potassium iodide solution. Centrifuge and repeat twice more. Dilute the combined extracts to 25 ml and use a 10-ml aliquot. Continue as for mercuric chloride from "Add 1 ml of 1% EDTA...."

TOTAL MERCURY. The sum of the foregoing is correct for total mercury. The figures for mercuric chloride and mercuric oxide are low, compensated by mercuric sulfide and elementary mercury being high.

Alternatively,[52] for total mercury, reflux with saturated ethanolic sodium sulfide a 1-gram sample ground to less than 1 mm. Evaporate to dryness. Take up the residue in 1 ml of sulfuric acid containing sufficient potassium permanganate to give a slight pink color. Dilute to about 20 ml, neutralize by dropwise addition of 20% sodium hydroxide solution, and dilute to 25 ml. Treat a 10-ml aliquot with 1 ml of 10% hydroxylamine hydrochloride solution; use more if the permanganate color is not fully discharged. Complete as for mercuric chloride from "Add 1 ml of 1% EDTA...."

Coal Tar.[53] Heat a 1-gram sample with 5 ml of sulfuric acid under a reflux condenser. Add potassium permanganate in small amounts until a pink color persists. Cool, dilute to about 10 ml, and add 1 gram of ammonium persulfate. Continue to digest until the liquid is colorless. Add 10% hydroxylamine hydrochloride solution until residual oxidizing agent and manganese dioxide are destroyed. Dilute to 100 ml and take a 10-ml aliquot. Add 1 ml of 1% EDTA solution and 1 ml of 1% potassium thiocyanate solution. Extract by shaking for 1 minute with 5 ml of 0.0005% dithizone solution in chloroform and read at 490 nm.

[52]A. E. Vasilevskaya, V. P. Shcherbakov, and E. V. Karakozova, *Zh. Anal. Khim.* **19**, 1200–1203 (1964); cf. M. D. Schlesinger and Hyman Schultz, *Tech. Prog. Rep. U.S. Bur. Mines*, **TPR 43**, 4 pp (1971).
[53]V. P. Shcherbakov and A. E. Vasilevskaya, *Zh. Anal. Khim.* **19**, 308–311 (1964).

Sulfuric Acid.[54] Dilute a weighed 1-ml sample to 100 ml with water. Mix a 1-ml aliquot with 10 ml of 0.1 *M* EDTA and neutralize to methyl orange with 5% sodium hydroxide solution. Add 10 ml of 1 : 1 *N* sodium acetate–*N* acetic acid as buffer. Add 0.5 ml of 30% ammonium thiocyanate solution and cool. Extract with 5 ml and 5 ml of 0.00005% solution of dithizone in carbon tetrachloride. Read at 490 nm against a reagent blank.

Water.[55] Preserve a 5-liter sample by adding 20 ml of nitric acid. Add 11 mg of cadmium nitrate as carrier and precipitate with hydrogen sulfide. Filter, and wash with water containing hydrogen sulfide. Dissolve the precipitate in 1 ml of nitric acid and 2 ml of hydrochloric acid and boil off the hydrogen sulfide. Add 2 ml of 20% solution of hydroxylamine hydrochloride and adjust to pH 4.5. Extract copper with 5 ml of 0.1 *M* trifluoroacetylacetone in chloroform. Make the solution 0.25 *N* with sulfuric acid and extract the mercury with 5 ml of 0.001% dithizone solution in chloroform. Read at 480 nm.

As another technic[56] for water analysis, treat the gross sample with an excess of sodium sulfide and sodium hydroxide. This forms a stable mercuric disulfide ion. Evaporate to dryness, and take up the residue in acid for adding appropriate masking agents and extraction with dithizone.

Latex Paint and Paint Films. Mix 0.25 gram of paint or 0.125 mg of paint film with 15 ml of 1 : 35 sulfuric acid. Stir vigorously and heat nearly to boiling. Add 25 ml of water, filter, and dilute to 100 ml.

Mix a 10-ml aliquot with 20 ml of chloroform and add 0.2% solution of dithizone in chloroform dropwise with shaking until the yellow-orange color of mercury dithizonate in the chloroform layer shows the greenish tinge of excess dithizone. Shake for 1 minute and separate the chloroform layer. Dilute to 25 ml with chloroform and read at 475 and 605 nm. Use the value at 605 nm to correct the absorption of mercuric dithizonate at 475 nm for an overlap by a peak at 455 nm. Use phenylmercury acetate to prepare a standard curve. Organomercurials may amount to about 0.5% of mercury in polyvinyl acetate, acrylic, and styrene–butadiene latex paints.

Organic Matter.[57] The wet decomposition of organic matter calls for the use of the special equipment (Figure 4-1). The flask has a 250-ml capacity, the reservoir 150–200 ml. A 200° thermometer is required.

Add a weighed sample to the oxidation flask. Allowing for that in a wet sample, add 20 ml of water. Add 5 ml of sulfuric acid and 50 ml of nitric acid. If the sample contains more than 10 grams of dry solids, increase the nitric acid by 5 ml for each additional gram of dry solids. Add glass beads to prevent bumping. After any

[54] I. A. Solferman, *Zavod. Lab.* **31**, 164 (1965).

[55] Naoichi Ota, Minoru Terai, and Masanori Isokawa, *J. Chem. Soc. Jap., Pure Chem. Sect.* **91**, 351–354 (1970); cf. Charles T. Elly, *J. Water Pollut. Control Fed.* **45**, 940–945 (1973).

[56] A. E. Vasilevskaya, V. P. Shcherbakov, and A. V. Levchenko, *Zh. Anal. Khim.* **18**, 811–815 (1963).

[57] Report of the Subcommittee on Metallic Impurities in Organic Matter of the Society for Analytical Chemistry, *Analyst* **90**, 515–530 (1965); cf. IUPAC Commission for the Determination of Trace Elements in Food, *Pure Appl. Chem.* **10**, 77–81 (1965); cf. Julio Cesar Meridio, *Rev. Fac. Cienc. Quim., Univ. Nac. La Plata* **33**, 111–120 (1960–1961); N. A. Smart and A. R. C. Hill, *Analyst* **94**, 143–147 (1969).

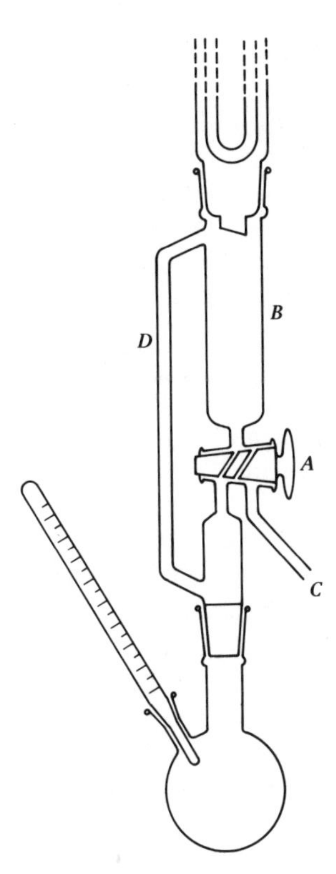

Figure 4-1 Apparatus for wet decomposition of organic matter.

initial reaction has ceased, heat cautiously collecting the distillate in *B* with *A* closed. When the thermometer reaches 116°, which is close to the boiling point of nitric acid, withdraw the contents of *B* through the drain tube *C* into a measuring cylinder.

Continue heating and collecting distillate in *B*. When the contents of the flask start to darken, return some of the contents of *B* to the flask. Continue this procedure, maintaining excess nitric acid in the flask until the contents ceases to darken. Heat until sulfur trioxide fumes are evolved; let cool. Add the contents of *B* to the first distillate, which was drawn off and retained. This will total 80–90 ml.

Titrate 1 ml of the distillate as a guide and dilute to approximately *N* acid, a volume of about 400 ml. Heat to boiling and remove from the heat. Add one-tenth of the volume of 10% hydroxylamine hydrochloride solution, let stand for 15 minutes, and cool to room temperature. If fat is present, extract with carbon tetrachloride and discard this extract.

As dithizone solution, dilute 2 ml of 0.05% stock solution to 100 ml with carbon tetrachloride unless copper is present. If there is copper, replace the carbon tetrachloride by chloroform and alter the wavelength for later reading to 492 nm. Shake for 1 minute with 10 ml of the dithizone solution. Draw off the dithizone extract and continue to extract with successive 1-ml portions until two successive ones remain green. To the combined extracts add 10 ml of 0.1 *N* hydrochloric acid and 1 ml of 5% sodium nitrite solution. Shake for 1 minute and discard the lower layer. The mercury is now in the aqueous layer. Add 1 ml of 10% hydroxylamine hydrochloride solution and set aside for 15 minutes with occasional shaking. Add 1 ml of 10% urea solution and 1 ml of 2.5% EDTA solution.

Add 0.5 ml of the dithizone solution, shake for 10 seconds, and run the lower layer into 5 ml of 1:3 acetic acid. This acid avoids the light sensitivity of mercury dithizonate.[58] Repeat the extractions with 0.5-ml portions of dithizone solution until the separated layer is greenish orange. Then use 0.2-ml portions, shaking for 30 seconds until the layer is a mixed greyish, indicating that mercury is fully extracted. Dilute the dithizone extracts to 4 ml with solvent. Read at 485 nm against a blank.

Food Colors.[59] The method provides for isolation of mercury at under 0.5 μg per gram of sample by distillation into the extraction chamber of a Sohxlet extractor surrounded by a condenser. A dropping funnel provides for additions to the digestion flask.

Treat a 5-gram sample with 10 ml of nitric acid and 20 ml of 1:1 sulfuric acid. Referring to the apparatus shown in Figure 4-1, distill with the condensate collected in *B* until white fumes are formed. Withdraw the contents of *B* through *C* and reserve. With stopcock *A* closed, add to chamber *B* 25 ml of nitric acid, or 40 ml if the sample is a triphenylmethane dye. By suitable adjustment of stopcock *A*, add this to the flask at one drop per second. Adjust the rate of heating so that brown vapors are formed and no carbonization occurs. If white fumes form or evolution of brown vapor ceases, increase the rate of addition of nitric acid. When the nitric acid is exhausted from *B*, close stopcock *A* and increase the rate of heating so that distillation into chamber *B* occurs. Let cool and return the distillate to the flask, adding the distillate earlier reserved. Again distill until white fumes are evolved. Let the flask cool, and add 5 ml of nitric acid and 5 ml of perchloric acid. Boil for 10 minutes, then increase the heat, distilling until white fumes form. Continue to heat until no residual organic matter remains. If on cooling the material is not colorless, reheat.

Return the contents of the extractor chamber to the flask, rinsing the condenser and the adapter with water. Reflux the combined liquids. Add 5 ml of 40% urea through the condenser and boil for 15 minutes. Cool and dilute to 500 ml.

Adjust the acidity of a 200-ml aliquot to approximately *N* and heat to boiling. Remove from the heat, add 20 ml of 10% hydroxylamine hydrochloride solution, and let stand for 15 minutes to cool.

Complete as for organic matter from "As dithizone solution dilute...."

Foods.[60] Typical samples for wet oxidation are 50 grams of fish muscle, 15 grams of milk powder, or rice. If the sample contains less than 1 μg of mercury, additional samples must be subjected to wet oxidation and combined.

Digest the sample with sulfuric, nitric, and perchloric acids using usual precautions with respect to rate of addition of acids and temperature of heating. Filter and dilute to 250 ml. Check a 1-ml portion with standard sodium hydroxide solution to see that the acidity is 1–1.4 *N* and return to the flask.

[58] J. F. Reith and K. W. Gerritsma, *Rec. Trav. Chim. Pays-Bas* **64**, 41 (1945).
[59] H. Onrust, *Chem. Weekbl.* **53**, 383–386 (1957).
[60] Michal Nabrzyski, *Anal. Chem.* **45**, 2438–2440 (1973).

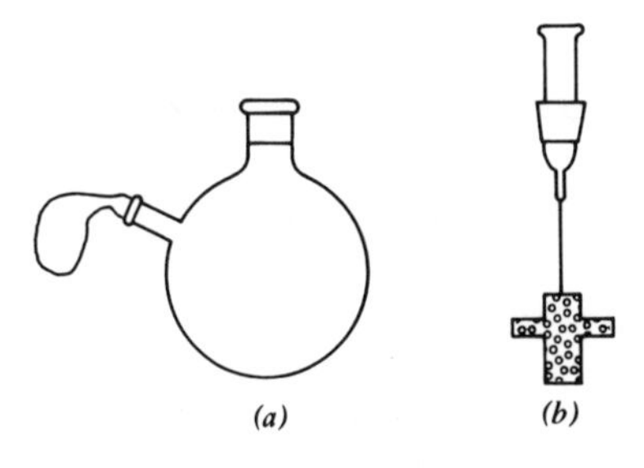

Figure 4-2 Combustion equipment for organic samples. (*a*) Flask. (*b*) Platinum holder open. (*c*) Platinum holder closed.

Divide the digestate into two approximately equal portions and evaporate each to approximately 50 ml in the range of 80–90°. Add 100 ml of water to each and evaporate at 80–90° to about 100 ml. Add 1 gram of hydroxylamine hydrochloride to each and cool.

To each portion of the sample, add 2 ml of 1:2 acetic acid and 2 ml of chloroform. Shake for 30 seconds and discard the chloroform extract. Shake with 2 ml and 2 ml of 0.001% solution of dithizone in chloroform. Combine the dithizone extracts and dilute to 10 ml with chloroform. Filter and read at 490 nm against a reagent blank. To avoid interference by copper, the dithizone extraction should be conducted at pH 0.

APPLES.[61] The modified 5-liter Schöniger flask used (Figure 4-2) is round bottomed and is made of borosilicate glass. A 40/35 standard taper, female ground joint is sealed on the neck. A 1×4 cm side arm is sealed to the flask as illustrated. A rubber balloon, about 8 cm long, is secured to the side arm with a rubber band.

The platinum holder is constructed by sealing a 12-cm length of No. 16 B. and S. gauge platinum wire onto a 40/35 standard taper, male ground joint drawn out 8 cm below the ground portion as shown in (b). A piece of perforated platinum sheet 2.5 cm wide and 7.5 cm long, bent as shown, is electrically welded to the last 3 cm of this wire. Two platinum tabs (1.25×2.5 cm) are welded on, to form the two remaining sides of the holder and to keep the sample compressed as shown in (c). The bottom of the resulting holder is 3 cm deep.

Blend apples in their own juice until a homogeneous mixture is obtained. Spread a 10-gram portion of the mixture on a 3×3 inch square of cellophane supported on a watch glass. Place the watch glass and sample in a desiccator containing sulfuric acid. Several samples may be dried at once. Evacuate the desiccator and allow the samples to remain under vacuum overnight. Remove the sample from the desicca-

[61]Walter H. Guttenmann and Donald J. Lisk, *Agr. Food Chem.* **8**, 306–308 (1960).

tor and wrap the cellophane around it by folding the opposite sides up. Place the package in the partially open platinum holder and compress by bending the tabs around it. Insert a 0.5 × 10 cm fuse cut from filter paper into the sample. Place 200 ml of 0.1 *N* hydrochloric acid and a magnetic stirrer in the combustion flask. Thoroughly purge the flask with oxygen. Light the fuse and place the holder in the flask as soon as about 1 inch of the fuse has burned. Hold the top on until combustion is complete and the balloon has collapsed. Start the magnetic stirrer, and operate it fast enough to wash the inside surface of the flask with the splashing solution. Stirring from 7 to 10 minutes is usually sufficient. Remove the flask from the stirrer, and tilt it to rinse the upper portion of the flask. Remove the platinum holder. Pour the absorbing solution into a 500-ml separatory funnel while rinsing the platinum holder. Rinse the flask, holder, and balloon with 50 ml of 0.1 *N* hydrochloric acid and add the rinse solution to that in the funnel. Repeat with a second 50-ml portion of the acid.

To the combined acid solution, add 10 ml of 20% hydroxylamine hydrochloride solution and 5 ml of 30% acetic acid and mix. Add 25 ml of a solution of 4 mg/liter of dithizone in chloroform and shake for 1 minute. Drain the lower layer through a cotton pledget and read at 490 nm against the dithizone solution.

OYSTERS.[62] Homogenize 10 oysters. Mix 20 grams of homogenate expected to contain 5–50 μg of mercury and a few drops of antifoaming compound with 15 ml of water. Add 20 ml of sulfuric acid and 0.5 gram of potassium permanganate. After the reaction subsides, add further 0.5-gram portions of permanganate until a purple color persists. Add 50 ml of water and 10% hydroxylamine hydrochloride solution until a clear solution is obtained. Filter through glass wool and dilute to 1 liter. Extract a 50-ml aliquot with 25 ml of a solution containing 10 μg of dithizone in chloroform and read at 490 nm.

FISH.[63] Place 1 gram of freshly ground sample on top of two 1.5-inch squares of filter paper. After air drying for about 3 hours, dry overnight in a vacuum desiccator over sulfuric acid. Transfer the lower square of paper to the top of the sample and fold into about a 0.75-inch square packet. Add 100 ml of 0.2 *N* hydrochloric acid and 10 ml of 0.02 *M* potassium permanganate to the combustion flask described for determination of mercury in apples. Burn as described there.

Magnetically stir the absorbing solution for 20 minutes and rinse the sample holder with it. Filter and wash with 50 ml of 0.2 *N* hydrochloric acid. Add 5 ml of 20% hydroxylamine hydrochloride solution and let stand for 10 minutes. Extract with 10 ml of a solution of 4 mg/liter of dithizone in chloroform. Filter the extract through a pledget of glass wool and read at 490 nm against the dithizone solution.

As an alternative[64] in the analysis of canned fish, reflux with sulfuric acid, nitric acid, and hydrogen peroxide. Then add to the clear solution hydroxylamine hydrochloride to reduce oxidizing materials, 1 : 3 ammonium hydroxide with cool-

[62] Jack Mayer, *Bull. Environ. Contam. Toxicol.* **5**, 383–388 (1970).
[63] Carl A. Bache, Colin McKone, and Donald J. Lisk, *J. Assoc. Off. Anal. Chem.* **54**, 741–743 (1971).
[64] M. Nagy, *Deut. Lebensm. Rundsch.* **67**, 297–300 (1971).

ing to raise to pH 3, and acetate buffer solution to further raise to pH 4.6. Filter, add EDTA solution to mask interferences, and extract as mercury dithizonate.

RICE.[65] Supply the reaction flask with a 300×18 mm borosilicate condenser packed with 80 mm of 6-mm Raschig rings to avoid loss of mercury compounds by volatilization. Add a 5-gram sample and a few beads to the flask. Add 35 ml of nitric acid through the condenser. Warm gently to start the reaction. After frothing ceases, heat until nearly dry. If taken to dryness, ignition may occur. Add 10 ml of perchloric acid and 10 ml of nitric acid and evaporate to fumes. Add 10 ml of water through the condenser and again take to fumes to drive off oxides of nitrogen. Add 25 ml and 25 ml of water through the condenser. Filter and wash.

Mix 5 ml of 10% hydroxylamine hydrochloride solution and 3 ml of 30% acetic acid. Dilute to 200 ml, extract with 5 ml of chloroform, and discard the extract.

For 0–20 μg of mercury, use 15 ml of dithizone solution at 5 mg/liter of chloroform. For 20–50 μg of mercury, use 25 ml of dithizone solution at 10 mg/liter of chloroform. Shake for 15 seconds with the dithizone solution and filter through cotton. Read at 490 nm against the dithizone solution.

Urine.[66] To 50 ml of urine chilled to 0°, add 10 ml of sulfuric acid dropwise. Chill and add 50 ml of 6% potassium permanganate solution dropwise with cooling. After digesting for 12–16 hours, cautiously add 10 ml of 20% hydroxylamine hydrochloride solution to clarify the solution and dissolve precipitated manganese dioxide. Let stand for 2 hours. Then shake for 1 minute with 20 ml of 0.001% solution of dithizone in carbon tetrachloride. Wash the organic layer with 25 ml and 25 ml of 1:3 ammonium hydroxide. Filter and read at 490 nm 45 minutes after the first washing. The curve does not conform to Beer's law.

Alternatively,[67] digest a 25-ml sample under reflux with sulfuric acid and an excess of potassium permanganate. Cool and dilute to a known volume. To an aliquot containing less than 3 μg of mercury, add 2 grams of citric acid. Add 20 ml of 5% EDTA solution to chelate copper and bismuth. Add 10 mg of potassium iodide to prevent extraction of silver, and a drop of 0.5% ethanolic bromocresol green. Add 10% sodium hydroxide solution to pH 4–4.5 to avoid interference by iodide ion and add 0.2 gram of sodium sulfite to inhibit extraction of gold and palladium.

Add 1 ml of 20% hydrooxylamine hydrochloride solution. Extract with successive 2-ml portions of 0.00015% dithizone solution in chloroform until the green is

[65]Ernest A. Epps, Jr., *J. Assoc. Off. Anal. Chem.* **49**, 793–795 (1966); cf. Sabina Hordynska, Bronislawa Legatowa, Kazimiera Kobylecka, Danuta Rozycka, and Maria Strycharska, *Rocz. Panstw. Zakl. Hig.* **20**, 391–401 (1969).

[66]Morris B. Jacobs and Ana Singerman, *J. Lab. Clin. Med.* **59**, 871–877 (1962).

[67]F. N. Kudsk, *Scand. J. Clin. Lab. Invest.* **16**, 575–583, 670–676 (1964); Chu-liang Chen, *Shang Hai Hsuch Pao*, **1958**, 71–73; Juan F. Saredo and Nelida Saredo Arrechea, *An. Fac. Quim. Farm., Univ. Rep. Orient. Urug.* **6**, 81–98 (1960); Donald L. Sorby and Elmer M. Plein, *J. Am. Pharm. Assoc.* **99**, 160–162 (1960); A. Seguro Serra, *Rev. Port. Farm.* **17**, 448–449 (1967).

no longer discolored. Dilute the extracts to 15 ml with dithizone solution and wash with 50 ml of 1:99 sulfuric acid containing 0.2 gram of sodium sulfite. Read the chloroform layer at 490 nm against chloroform. Shake the chloroform layer with 10 ml of 40% potassium bromide solution and read again. The increase in absorbtion measures the mercury. This reversion technic will tolerate 10 μg of gold, 5 μg of palladium, and 5 mg of copper in the aliquot, but excessive iodide will inhibit extraction of mercury.

One may also read at 610 nm the dithizone consumed rather than the yellow-orange mercury dithizonate at 490 nm.[68]

Tissue. Reflux 25 ml of homogenized sample with 10 ml of sulfuric acid and 5 ml of nitric acid until liquefied. Filter to remove fat. Add an excess of potassium permanganate and heat at 100° until all organic matter is destroyed. Remove excess permanganate and any manganese dioxide by dropwise addition of 30% hydrogen peroxide. Cool and dilute to a known volume.

Extract as for urine from "To an aliquot containing less...."

Aromatic Mercury Compounds by Dithizone

The use of organic derivatives of mercury dictates their inclusion in this inorganic volume. The organic mercury ester is saponified with alkali to give the organic mercury derivative, which then reacts with dithizone much as mercuric ion does.[69] The unreacted green dithizone is determined to give the organic mercurial by difference rather than reading the yellow phenylmercury dithizonate. There are no interferences. The technic as stated will not give quantitative results with ethyl mercury.

Another technic oxidizes the mercury compounds with hydrogen peroxide to mercuric ion. This is then reacted with ditolylmercury to form 2 moles of the monorganic mercury compound, which is reacted with dithizone.[70] Only silver and bismuth interfere.

By hydrogenation in alkaline solution with zinc, the mercury can be isolated as an amalgam.[71] This can be dissolved in acid and give a sample solution substantially free from interference. As an example, take a sample of food or a pharmaceutical product containing about 0.025 mg of mercury. Add 50 ml of water, 5 grams of sodium hydroxide pellets, 1 gram of zinc, and a few drops of octyl alcohol to control foaming. Reflux for 5 hours and cool. Separate the amalgam and wash with water. Dissolve the amalgam by refluxing with 20 ml of nitric acid to produce the sample solution.

[68]V. Miketukova and K. Kacl, *Arch. Toxikol.* **20**, 242–252 (1964).
[69]V. L. Miller, Donna Lillis, and Elizabeth Csonka, *Anal. Chem.* **30**, 1705–1706 (1958).
[70]Dorothy Polley and V. L. Miller, *ibid.* **27**, 1162–1164 (1955); V. L. Miller and Frank Swanberg, Jr., *ibid.* **2.**, 391–392 (1957).
[71]J. A. Gautier and F. Pellerin, *Prod. Pharm.* **13**, 149–152 (1958).

Procedures

Water. As concentrated dithizone reagent dissolve 0.1 gram in 100 ml of chloroform. For daily use, dilute with chloroform at 1:30 to give a 0.0033% solution. The technics that follow provide for use of this dilute reagent for up to 30 μg of mercury.

Neutralize a 200-ml sample[72] to *p*-nitrophenol and make 0.2 *N* with hydrochloric acid. Make the sample pink by addition of a 6% solution of potassium permanganate and add 0.2 ml in excess. If cyanide or thiocyanate is present, add 0.1 gram of mercurous chloride at this point as a masking agent. It will not alter the results reported as organic mercury compounds if the later chloroform extract is thoroughly washed with 1:120 hydrochloric acid.

Shake with 60 ml of chloroform for 3 minutes and discard the chloroform phase. Decolorize the aqueous phase with 10% solution of hydroxylamine hydrochloride and extract the aqueous phase with 60 ml, 60 ml, and 60 ml of chloroform. Wash the combined chloroform extracts with 50 ml, 50 ml, and 50 ml of 0.1 *N* hydrochloric acid. Extract the mercury compound with 20 ml and 20 ml of 0.5 *N* ammonium hydroxide and combine the extracts. Add sufficient hydrochloric acid under the surface of the liquid to lower to pH not more than 1. Cool if necessary for not more than 5 minutes.

Back-extract with 11 ml of chloroform and shake for 1 minute. Wash the chloroform layer with 20 ml of 1:10 hydrochloric acid and shake for 45 seconds. Shake the chloroform layer with 25 ml of 1:60 acetic acid. Add 1 ml of the dilute dithizone reagent and shake for 30 seconds. Dilute the chloroform layer to 11 ml and read the unreacted dithizone at 620 nm.

Alternatively,[73] add 5 ml of 1:1 sulfuric acid to a 100-ml sample. Add 2 ml of saturated solution of potassium permanganate and set aside for 1 hour. Add 2 ml of 10% hydroxylamine hydrochloride solution, 3 ml of 2% solution of EDTA, and 1 ml of 20% solution of ammonium thiocyanate. Shake for 1 minute with 10 ml of 0.002% solution of dithizone in carbon tetrachloride. Wash the extract with 10 ml and 10 ml of 1:1 ammonium hydroxide to remove excess dithizone and read at 485 nm.

Urine. Mix a 10-ml sample with 20 ml of 4% sodium hydroxide solution in a 250-ml standard taper flask. Attach a reflux condenser and heat at 100° for about 30 minutes. Let cool and add 25 ml of 5% potassium permanganate solution. Mix 12.5 ml of filtered 30% solution of hydroxylamine sulfate with 12.5 ml of ammonium hydroxide and add to the flask. After mixing add 5 ml of 30% ammonium sulfamate solution and cool the reaction with running water. Complete as for water from "Add sufficient hydrochloric acid under the surface...."

Kidney, Liver, Muscle, or Spleen. This determination requires an absorption column. Seal 75 mm of 0.5 mm i.d. capillary to 85 mm of 10 mm i.d. tubing with a

[72]Tetsuro Murakami and Toshiaki Yoshinaga, *Jap. Anal.* **20**, 878–884 (1971).
[73]Ni-wen Kuan and Nan-chiang Li, *Chem. Bull.* (Chin.) **1974** (1), 15–19.

reservoir of 75 mm of 18 mm i.d. tubing at the top. Treat Hyflo Supercel with 1:10 hydrochloric acid three or four times to remove iron. The final treatment should give no more than a faint pink with thiocyanate. Wash free of acid with water and dry at 100°. Place glass-fiber filter paper in the bottom of the column and pack with purified Supercal to a depth of 55 mm. Pack 5 mm of calcium carbonate on top.

Treat a 1-gram sample in a 250-ml standard taper flask with 20 ml of 4% sodium hydroxide solution. Attach a reflux condenser and heat at 100° until the tissue is dissolved and for 10 minutes longer. Treat as for urine from "Let cool and add 25 ml...."

Peroxide Oxidation. URINE. As oxidation catalyst, add 1 gram of ferric chloride hexahydrate and 2 grams of chromium potassium sulfate to 100 ml of 1:1 hydrochloric acid. As peroxide test reagent, shake 10 grams of tetanyl sulfate for several hours with 50 ml of water and 20 grams of sulfuric acid. Centrifuge to obtain a clear solution. As ditolylmercury reagent, dissolve 20 mg by refluxing with 200 ml of redistilled absolute ethanol. Stored in a dark bottle, it is stable for 6 weeks.

Transfer a 100-ml sample to a standard taper flask and add 10 ml of catalyst mixture. Add 11–12 ml of 50% hydrogen peroxide and connect with a condenser. Heat at 100° until no further bubbles are evolved from the peroxide. The solution becomes green. Continue to heat for 30 minutes and let cool.

To be sure that all the peroxide has decomposed, take a few drops of the peroxide test reagent on a spot plate and add a drop of the digestate. No yellow or orange color should develop. If bismuth or silver is present filter the digest through glass-fiber filter paper or a medium porosity sintered-glass filter; otherwise filtration is unnecessary.

Transfer the digest or an aliquot to a separatory funnel and add 1 ml of glacial acetic acid. If emulsions may develop later because of separation of slightly soluble iron compounds, add 1 gram of citric acid at this point. Add 1:1 ammonium hydroxide to adjust to pH 3–3.5 as shown by indicator paper. Add 1 ml of ditolylmercury reagent and shake 10–15 times. One minute later add 10 ml of chloroform and shake vigorously for 1 minute. After separation occurs, drain the chloroform layer into a separatory funnel containing 25 ml of 1:60 acetic acid. Add 1 ml of 0.03% dithizone in chloroform and shake for 30 seconds. Separate the chloroform layer, dilute to 10 ml with chloroform, and read at 620 nm.

When it is necessary to determine both mercury and an organic mercurial in the sample, first wet-ash the sample with perchloric acid and determine mercury by dithizone.[74a] Then prepare a standard containing equimolar amounts of mercuric ion and the organomercury compound known to be present, with variations of each component in each direction retaining a constant total concentration of mercury. Determine a pH at which there is complete extraction of both mercuric ion and the organic mercurial. Extract with dithizone in carbon tetrachloride. Then from the total mercury determination and that of the mixture as read at 620 nm, each component can be determined.

[74a] A. M. Kiwan and M. F. Fouda, *Anal. Chim. Acta*, **40**, 517–520 (1968).

EDTA

Mercury can be extracted as the 1:1:2 complex of mercuric ion, EDTA, and Capriquat, which is methyltrioctylammonium chloride. By the technic given, the tolerance levels are 0.1 mg for silver and stannous ion; 0.5 mg for bismuth, calcium, chromic ion, cupric ion, nickel, and chloride; 1 mg or more for other common ions.

Procedure.[74b] To a sample solution containing less than 80 μg of mercuric ion, add sequentially 0.4 ml of 0.01 *M* disodium EDTA and 10 ml of a phosphate buffer solution for pH 7. Dilute to 30 ml and add 20 ml of 0.01 *M* Capriquat in 1,2-dichloroethane. Shake for 10 minutes; then dry 10 ml of the organic layer with sodium sulfate. Shake with 10 ml of 0.03 *M* potassium iodide, dry again, and read the 1:1:2 mercuric-EDTA-Capriquat complex at 333 nm.

FERRIC TRIS-(4-METHYLPYRIDINE-2-THIONE-1-OXIDE)

Mercuric ion reacts with the ferric complex of tris-2-mercaptopyridine 1-oxide to give a determination by difference.[75] There is interference by mercurous, cupric, bismuth, stannic, silver, palladium, hexavalent molybdenum, nitrate, and thiocyanate ions.

Procedure. To prepare the reagent, add 20% excess of a 1% solution of 1-hydroxy-4-methylpyridine-2-thione in 0.4% sodium hydroxide solution to a ferric solution at pH 1.5–3.5. Heat at 100° for 10 minutes. Filter, wash, and dry.[76]

To a sample solution containing 0.05–0.2 mg of mercuric ion at about pH 4, add 5 ml of 0.05 *M* potassium acid phthalate. Extract with 10 ml of a 0.08 *M* solution of the reagent in chloroform, filter the extract through glass wool, and read at 519 or 588 nm.

FERROCYANIDE

Mercuric ion and weakly complexed mercuric compounds react with potassium ferrocyanide to liberate an equivalent of ferrous ion.[77] This is then determined as an indirect measure of mercury. Colored compounds are formed with the reagent by iodides, nitroso compounds, manganese, cobalt, and nickel compounds, and of course ferrous compounds. Silver and palladium give the same reaction as mercuric

[74b] Akira Shimazaki, Masao Sugawara, and Tomihito Kambara, *Bull. Chem. Soc. Japan* **49**, 3327–3328 (1976).
[75] M. Edrissi, A. Massoumi, and J. A. W. Dalziel, *Microchem. J.* **15**, 579–584 (1970).
[76] M. Edrissi and A. Massoumi, *ibid.* **16**, 353–358 (1971).
[77] Zbigniew Prasal, *Chem. Anal.* (Warsaw) **7**, 617–624 (1962).

ion and zinc; copper and bismuth precipitate the reagent. If ferric ion is present in the sample, complex it with sodium fluoride.

Procedure. As a preliminary, titrate a 1-ml sample to pH 3.6–4 with 0.2 *M* sodium acetate solution, using bromophenol blue indicator. Treat a 1-ml sample with the required amount of sodium acetate solution without indicator. Add 1 ml of 0.0018 *M* potassium ferrocyanide. After 3 minutes add 0.4 ml of 0.5% ethanolic 2,2′-bipyridine. Dilute to 10 ml, set aside for 15 minutes, and read at 495 nm.

FERROUS 2,2′-BIPYRIDYL

The bromomercurate ion is extractable with 1,2-dichloroethane as a complex with ferrous 2,2′-bipyridyl at pH 4–5.[78] The background is less than with the iodomercurate. There is no interference by zinc, cobalt, ferric ion, lead, or bismuth. Up to 0.01 mg per ml of copper and cadmium can be masked with 0.004 *M* nitrilotriacetic acid.

Procedure. To 5 ml of aqueous sample containing 0.005–0.05 mg of mercuric ion, add 5 ml of 1.2% solution of potassium bromide, 5 ml of 0.005*M* ferrous-2,2′-bipyridyl complex, and 1 ml of *M* acetate buffer solution for pH 4.5. Shake with 10 ml of 1,2-dichloroethane for 3 minutes and allow 30 minutes for complete separation. Dry the organic layer with sodium sulfate. Read at 526 nm against the solvent.

6–METHOXY–3–METHYL–2–[4–(*N*–METHYLANILINO)PHENYLAZO]BENZOTHIAZOLIUM CHLORIDE

The captioned reagent combines with 0.33–50 m*M* mercuric bromide to give an extractable blue compound.[79] The solution should be 0.1–9 *N* in sulfuric acid, 0.1–7 *M* in phosphoric acid, or 0.1–3 *N* in acetic acid. The preferred solvent is 3:1 toluene-cyclohexanone. The maximum absorption at 642 nm is more sensitive than dithizone for determining mercury. Beer's law holds for 0.1–8 μg of mercury per ml of extract. It will determine less than 0.01 ppm of mercury in potassium hydroxide, sulfuric acid, or phosphoric acid, and less than 0.1 ppm in acetic or hydrochloric acid. Equal amounts of auric or thallic ion interfere, and iodide or thiocyanate must be absent.

[78] Yoroku Yamomoto, Sumiyoshi Kikuchi, Yasuhisa Hayashi, and Takahiro Kumamaru, *Jap. Anal.* **16**, 931–936 (1967); cf. Yuroku Yamamoto and Keiya Kotsuji, *Bull. Chem. Soc. Jap.* **37**, 594–595 (1964).
[79] P. P. Kish and G. M. Vitenko, *Zavod. Lab.* **38**, 5–8 (1972).

Procedures

General. Adjust the sample to 2 *N* in sulfuric acid and make it m*M* with potassium bromide. Add 2 ml of 0.04% solution of the captioned reagent and extract with 6 ml of 3 : 1 toluene-cyclohexanone. Centrifuge the extract and read at 640 nm against a reagent blank.

Potassium Hydroxide. Dissolve 15 grams of sample in 37 ml of water. Neutralize by gradual addition in increments of 40.6 ml of 1 : 1 sulfuric acid. Cool and add 0.5 ml of *M* potassium bromide and 3 ml of 0.04% solution of the reagent. Extract as for the general technic.

Acids. Prepare the sample according to the following table.

Acid (ml)		Water (ml)	*M* KBr	0.04% Reagent Solution (ml)
Sulfuric	20	53	0.5	2
Phosphoric	25	30.4	0.6	4
Acetic	7	30	0.4	2.6
Hydrochloric	2	7.8	0.12	1

Extract as for the general technic.

4-METHOXY-2-(2-THIAZOLYLAZO)PHENOL

This reagent forms 1 : 1 and 2 : 1 complexes with mercuric ion.[80] By the technic below, it follows Beer's law for 10–70 m*M*. Bromide, iodide, cyanide, thiocyanate, sulfide, cupric, nickel, and zinc ions interfere.

Procedure. Mix an appropriate aliquot of sample with 20 ml of a monopotassium phosphate–disodium phosphate buffer solution for pH 8.4 and dilute to 50 ml. Shake with 10 ml of 0.1 m*M* reagent in chloroform for 2 minutes and read at 638 nm.

METHYLENE BLUE

With mercury in the form of mercuric iodide, the cation of methylene blue forms a complex extractable with chloroform.[81] Beer's law applies to ±5% for 0.2–2 μg of

[80]Fumiaki Kai, *Anal. Chim. Acta* **44**, 242–246 (1969).
[81]N. Ganchev and B. V. Atanasova, *C. R. Acad. Bulg. Sci.* **21**, 359–361 (1968).

mercury per ml in the chloroform extract. Low concentrations of mercury can be concentrated by precipitation with cadium sulfide as a collector, since cadmium does not interfere.

Procedure. Mix 4 ml of sample solution containing 0.002–0.02 mg of mercury with 1 ml of 0.4 m*M* solution of methylene blue containing 7.5 μg of iodine as potassium iodide. Add 5 ml of 1:35 sulfuric acid and extract with 10 ml of chloroform. Wash the chloroform layer with a mixture containing 5 ml of 0.025 *M* methylene blue, 0.6 mg of iodine as potassium iodide, and 0.5 ml of 1:35 sulfuric acid per 100 ml. Read the chloroform extract at 600 nm against a blank. An alternative[82a] is to have the mercury as the bromide, extract with 3:2 dichloroethane-trichloroethylene, and read at 660 nm.

Air.[82b] Pass the sample at 2 l per min for 30–40 minutes through two absorbers in series, each containing a solution of bromine in *N* sulfuric acid. Heat the combined extracts at 100° to remove bromine completely. Cool and extract with 10 ml of 3:2 1,2-dichloroethane-trichloroethylene containing 0.01% of methylene blue. Read the extract at 660 nm. Alternatively the solvent may contain 0.0025% solution of methylene green.

METHYL VIOLET

Whereas crystal violet is used by extraction, the homologue methyl violet is read directly in aqueous solution.[83a] The method is applicable for up to 0.03 mg of mercuric ion in the sample. Cadmium, cupric, and ferric ions interfere. Replacement of the potassium iodide by potassium bromide avoids this by up to 0.06 mg of the interfering ion per ml.

Procedure. To 3 ml of sample at pH 1.4, add 1.3 ml of 0.01% solution of methyl violet and 0.5 ml of 2.5% solution of potassium iodide. Dilute to 5 ml and read.

MICHLER'S THIOKETONE

This reagent is 4,4′-bis(dimethylamino)thiobenzophenone. Mercuric ion forms a red photolabile complex with it. Beer's law is followed for 7 ng to 1.5 μg per ml. There is interference by gold, palladium, silver, and platinum.

[82a]V. M. Tarayan and N. S. Karimyan, *Arm. Khim. Zh.* **26**, 643–648 (1973).
[82b]V. M. Tarayan and N. S. Karimyan, *Zavod. Lab.* **41**, 405–406 (1975).
[83a]M. P. Anan'evskaya and V. I. Petrashen, *Nauch. Tr. Novocherk. Politekh. Inst. S. Ordzhonikidze* **25**, 246–251 (1955).

Procedure.[83b] Mix 10 ml of sample solution with 20 ml of 0.2 *M* acetate buffer for pH 5.8 in an actinic flask. Add 15 ml of 0.25 m*M* Michler's thioketone in propanol and dilute to 50 ml. Set aside for 10 minutes and read, avoiding exposure to sunlight.

NEUTRAL RED

The complex of neutral red with mercuric ion is extractable into nitrobenzene.[84] Silver, stannous and stannic ions, cyanide, iodide, iodate, thiocyanate, bichromate, and nitrate interfere.

Procedure. To a sample containing less than 0.03 m*M* mercury, add 2 ml of 0.005 *M* potassium bromide, 2 ml of 0.4 *M* acetate buffer solution, and 5 ml of 0.03% solution of neutral red. Dilute to about 25 ml and shake for 2 minutes with 10 ml of nitrobenzene. Read the extract at 552 nm.

NICKEL ANTIPYRINYLTHIOFORMIC ACID

Mercuric ion displaces nickel from a chloroform solution of the reagent. The reduction in absorption is then a measure of the mercury.[85] Silver, cupric, auric, cyanide, thiosulfate, and sulfite ions interfere, as does EDTA.

Procedure. As reagent, shake 100 ml of m*M* antipyrinylthioformic acid in chloroform with 150 ml of 0.1 *M* nickel acetate adjusted to pH 7.5 with 1:1 ammonium hydroxide. Read the chloroform solution at 575 nm, and repeat the shaking with nickel acetate solution until the extinction of the reagent is constant. Wash with water; then dilute with chloroform to an extinction of approximately 0.4.

Add a sample containing 0.01–0.06 mg of mercury to a separatory funnel. Add 2 ml of an ammonium acetate buffer solution for pH 2.6 and dilute to 10 ml. Shake with 5 ml, 2.5 ml, and 2.5 ml of the reagent, and combine these chloroform extracts. Read at 575 nm against the unreacted reagent.

[83b] Gerhard Ackermann and Heins Roeder, *Talenta* **24**, 99–103 (1977).
[84] Masahiro Tsubouchi, Taeko Nakamura, and Masaya Tanaka, *J. Chem. Soc. Jap., Pure Chem. Sect.* **91**, 1095 (1970); *Anal. Chim. Acta* **54**, 143–148 (1971).
[85] B. Sawicki, *Mikrochim. Acta* **1967**, 176–179.

5-NITRO-2-FURALDEHYDE SEMICARBAZONE

This reagent, which is furacine, gives an orange complex with mercuric ion at pH 3.5–7.5.[86] Beer's law applies up to 0.2 m*M*. The technic is applicable to samples containing up to 0.5% copper, 2% zinc, 0.4% lead, and 5.6% iron. The color is destroyed by chloride, iodide, bromide, sulfide, sulfite, and thiosulfate ions.

Procedure. *Ores.* Decompose a 5-gram sample by heating with 10 ml of sulfuric acid. Add 1 ml of nitric acid and evaporate to sulfur trioxide fumes. Cool and take up in about 25 ml of water. Boil off oxides of nitrogen; cool, filter, and dilute to 100 ml.

To a 20-ml aliquot, add 2 ml of 0.05% gelatin solution. Add 5 ml of 1% sodium fluoride solution to mask ferric ion. Add 10 ml of m*M* furacine and adjust to pH 3.5–4.5 with 10% sodium hydroxide solution. Dilute to 50 ml and read at a selected wavelength in the 420–480 nm range against a reagent blank.

NITROFURAZONE

In an acid medium, mercuric ion forms a 1 : 2 complex with nitrofurazone, which is 5-nitro-2-furaldehyde semicarbazone.[87] Oxalic, ascorbic, and citric acids, hydroxylamine sulfate, and stannous ion interfere.

Procedure. *Mercuric Oxide.* Mix a 0.2-gram sample with 25 ml of water and 1 ml of 1 : 1 nitric acid. Dilute to about 90 ml, adjust to pH 2 by addition of 10% sodium hydroxide, and dilute to 100 ml. Mix a 0.5-ml aliquot with 2 ml of 0.05% gelatin solution and add 10 ml of m*M* nitrofurazone. Read at 420 nm against a blank.

N-PHENYLBENZOHYDROXAMIC ACID

The precipitate that is formed by the captioned reagent with mercuric ion is soluble in chloroform to give a yellow solution.[88] Mercurous ion does not react in this manner. Chloride, cyanide, and EDTA must be absent.

Procedure. Dilute a sample solution containing 0.375–1.3 mg of mercuric ion to 30 ml. Add sodium acetate solution to adjust to pH 6.2. Extract with four

[86] K. N. Bagdasarov, L. G. Anisimova, and O. A. Tataev, *Zavod. Lab.* **34**, 390–392 (1968).
[87] K. N. Bagdasarov, L. G. Anisimova, and O. A. Tataev, *Zh. Anal. Khim.* **23**, 1002–1007 (1968).
[88] B. Das and S. C. Shome, *Anal. Chim. Acta* **35**, 345–350 (1966).

successive 3-ml portions of 0.8% reagent in chloroform, shaking each for 20 minutes. Dilute the combined extracts to 25 ml with chloroform and read at 340 nm against a reagent blank.

4-(2-PYRIDYLAZO)RESORCINOL

Mercuric ion forms a soluble orange complex under the conditions described below.[89] Provision is made for many extraneous metals, but bismuth, copper, zinc, cadmium, yttrium, cobalt, and nickel interfere.

Procedure. Take a sample containing 0.025–0.225 mg of mercuric ion. Add 5 ml of 30% sodium citrate solution to mask magnesium, calcium, aluminum, gallium, lead, and scandium. Add 2.5 ml of 2% sodium fluoride solution to mask beryllium, indium, stannic, and zirconium ions. Add 2.5 ml of sodium carbonate–sodium bicarbonate buffer solution for pH 10 and 5 ml of 0.025% solution of 4-(2-pyridylazo)resorcinol. Dilute to 25 ml and read at 500 nm against a reagent blank.

RHODAMINE B

In the complex of mercuric ion and rhodamine B (basic violet 10), addition of potassium bromide stabilizes the color of the complex and improves sensitivity.[90] Beer's law applies up to 0.04 mg of mercuric ion in the final 50 ml of reaction mixture. In that volume less than 15 μg of antimony does not interfere. Bismuth does not interfere, but most heavy metal ions cause a positive error.

Quenching of the fluorescence of rhodamine B is a measure of mercuric ion but not of organically bound mercury.[91a] The solution is 0.4 m*M* potassium iodide, pH less than 4, excitation at 486 nm for reading at 586 nm, reading for up to 10 m*M* mercuric ion. There is interference by thallic, palladous, platinic, bismuth, ferric, cadmium, and antimony ions, as well as by cysteine and albumin.

Procedure. The sample should be 25 ml in 1–2 *N* hydrochloric acid. Add 10 ml of 0.2% rhodamine B solution and 3 ml of *N* potassium bromide solution. Dilute to 50 ml. Extract 25 ml with 10 ml of 1:1 benzene-dioxan. After 30 minutes read the extract at either 550 or 562 nm.

---

[89] Joichi Ueda, *J. Chem. Soc. Jap., Pure Chem. Sect.* **92**, 418–421 (1971); cf. M. C. Eshwar and S. G. Nagarkar, *Z. Anal. Chem.* **260**, 289 (1972).

[90] Hiromu Imai, *J. Chem. Soc. Jap., Pure Chem. Sect.* **90**, 275–279 (1969); cf. A. I. Ivankova' and D. P. Shcherbov, *Tr. Kazk. Nauchn.-Issled. Inst. Mineral. Syr'ya* **1962**, 227–231.

[91a] Gen-Ichiro Oshima and Kinzo Nagasawa, *Chem. Pharm. Bull.* (Tokyo) **18**, 687–692 (1970).

RHODAMINE 6G

This reagent (basic red 1) determines mercury at 200–1000 ppm. EDTA masks cupric, lead, bismuth, and ferric ions. Chromate is reduced with hydrazine sulfate. Stannous and arsenite ions are oxidized with bromine. Platinic ion is masked with sulfite and palladous ion by ammonia. Silver is precipitated as the iodide.

Procedure.[91b] As an iodide reagent dissolve 5 grams of potassium iodide, 5 grams of potassium acid phthalate, and 0.1 gram of sodium thiosulfate pentahydrate in 250 ml of water. Mix a sample solution containing 5–25 μg of mercuric ion with 2 ml of the iodide reagent, 5 ml of 0.005% solution of rhodamine 6G, and 1 ml of 1% gelatine solution. Dilute to 25 ml and read at 575 nm against a reagent blank.

SELENOCYANATE

The reaction of mercury with selenocyanate ion is analogous to that with thiocyanate. The reagent precipitates copper or nickel in the presence of pyridine. Beer's law applies for 5–40 ppm.

Procedure.[92] To prepare the reagent, heat 10 grams of potassium cyanide with 20 grams of selenium in 200 ml of water at 100° for about 15 minutes. Cool, filter, and dilute to 1 liter. Mix 5 ml of sample in less than *N* nitric acid with 2 ml of 24% sodium hydroxide solution. Add 2 ml of reagent, heat at 100° for 6 minutes, and cool. Read at 400 nm against a reagent blank.

THIAMINE

The thiochrome produced by mercury is read by fluorescence.[93] For mercuric ion, Beer's law is followed for 0.05–1.6 μg/ml; for mercurous ion, 0.2–1.6 μg/ml. There is interference by silver, stannous, bismuth, ferric, iodide, thiosulfate, sulfite, sulfide, ferrocyanide, ferricyanide, manganate, chromate, and dichromate ions. In the absence of interferences, the sensitivity is about 10 times that of the dithizone method, the limit of determination being 0.05 μg/ml for mercuric ion and 0.2 μg/ml for mercurous ion.

Procedure. Mix 2 ml of a hydrochloric acid–borax buffer solution for pH 7.7

[91b] T. V. Ramakrishna, G. Aravamudan and M. Vijayakumar, *Anal. Chim. Acta* **84**, 369–375 (1976).

[92] F. Bosch Serrat, D. Rodriguez Polo, and J. L. Guardiola Saenz, *Inf. Quim. Anal. Pura apl. Ind.* **25**, 67–72 (1971).

[93] Yasuhiro Yamane, Motoichi Miyazaki, Takahiro Kasamatsu, Noriko Murakami, Sumiko Kito, and Yayoi-cho Chiba, *Jap. Anal.* **22**, 192–196 (1973).

with a sample solution containing 0.1–5 μg of mercuric ion.[94] Add 1 ml of 0.03 *M* thiamine hydrochloride. Dilute to 10 ml, activate at 375 nm, and read at 440 nm after an hour. The solution must be less than 0.02 *M* in foreign salts. The fluorescence is quenched to an equivalent degree by cyanide, iodide, sulfide, or EDTA.

By Extraction. Shake 5 ml of sample containing 0.25–8 μg of mercury with 1 ml of a thiamine reagent containing either 0.5 or 5 μg/ml, according to whether the mercury content is high or low. Then add 4 ml of borate buffer solution for pH 9.5 and again shake for 1 minute. Heat at 100° for 5 minutes and cool in an ice bath. Add 3.5 grams of sodium chloride and extract the thiochrome by shaking with 6 ml of isobutyl alcohol. Activate the organic layer at 375 nm and read the fluorescence at 430 nm.

THIOSALICYLAMIDE

Mercuric ion reacts with salicylamide in dilute ethanol for photometric determination.[95]

Procedure. By addition of hydrochloric acid and sodium acetate solution, adjust the sample to pH 2.5. Add 3 ml of 0.02 *M* ethanolic thiosalicylamide. Add water and ethanol to adjust to 25 ml at 50% ethanol. After 30 minutes read at 355 nm.

THIOTHENOYLTRIFLUOROACETONE

The reagent, which is 1,1,1-trifluoro-4-mercapto-4-(2-thienyl)but-3-en-2-one, forms an extractable complex with mercury in an acid medium.[96a] Conformity to Beer's law extends from 0.0095–0.14 mg of mercury/ml. Silver, cyanide, and ascorbic acid must be absent, and only small amounts of lead, stannous, bismuth, and titanic ions can be tolerated. After the complex is extracted, the excess reagent is extracted with a borate buffer solution for pH 10.5–12. Mercurous ion is extracted along with mercuric if the reagent is stronger than 2 m*M*. Chloride ions suppress the extraction from an acid medium unleess a citrate buffer solution for pH 3–4 is present. Diaminocyclohexanetetraacetate, along with hydroxylamine hydrochloride, masks most interfering ions.

Procedure.[96b] Adjust a 25-ml sample of solution containing about 0.76 mg of

[94] J. Holzbecher and D. E. Ryan, *Anal. Chim. Acta* **64**, 333–336 (1973).
[95] M. Mazumdar and S. C. Shome, *Anal. Chim. Acta* **56**, 149–153 (1971).
[96a] Hiroshi Hashitani and Kazuo Katsuyama, *Jap. Anal.* **19**, 355–361 (1970); K. R. Solanke and S. M. Khopkar, *Indian J. Chem.* **11**, 485–487 (1973).
[96b] Kamihiko Itsuki and Hideo Komuro, *Jap. Anal.* **19**, 1214–1218 (1970).

mercury to pH 2–3 with 1 : 1 hydrochloric acid or 1 : 1 ammonium hydroxide. Add 10 ml of a m*M* solution of the reagent in carbon tetrachloride and shake for 10 minutes. Read the organic layer at 370 nm against a reagent blank.

Waste Water. TOTAL MERCURY. Digest 50 ml with 7 ml of nitric acid, 3 ml of hydrochloric acid, and 2 ml of 30% hydrogen peroxide at 90° for an hour to decompose organic mercury compounds. Add 5 ml of 50% ammonium citrate solution, 5 ml of 5% solution of diaminocyclohexanetetraacetate in 2% sodium hydroxide solution, and 1 ml of 10% solution of hydroxylamine hydrochloride. Add ammonium hydroxide to adjust to pH 3–4, and shake with 10 ml of a solution containing 5 mg of the captioned reagent in 100 ml of benzene. Wash the organic layer with 50 ml of 0.1 *M* borate solution at pH 11. Read at 365 nm against benzene.

TRIBENZYLAMINE

Mercury, cadmium, and zinc are extracted stepwise with tribenzylamine. The sample solution may contain 1–30 μg of mercury, 1–25 μg of cadmium, and 1–25 μg of zinc. Copper, silver, lead, and bismuth interfere.

Procedures

Mercury.[96c] Adjust a sample solution to pH 4 with potassium hydroxide solution and add potassium bromide to make the solution 0.05 *M*. Equilibrate 0.088 *M* tribenzylamine in chloroform with an equal volume of 0.5 *M* hydrobromic acid. Shake the sample for 5 minutes with 8 ml of this solution. Separate the organic phase and add 2 ml of m*M* dithizone in chloroform. Shake with 10 ml of 5 m*M* EDTA in *N* potassium hydroxide. Read the organic phase at 490 nm.

Cadmium. Add to the mercury-free solution 1 ml of 10% sodium bisulfite solution, 1 ml of 6 *M* potassium iodide, and 1 ml of 3 *N* sulfuric acid. The sample should now be 0.5 *M* with iodide ion at pH 1.5. Add thiourea to mask copper, lead, and bismuth. Silver interferes.

Extract for 3 minutes with the equilibrated solution of tribenzylamine. If the ratio of zinc to cadmium exceeds 100 : 1, wash the organic layer containing cadmium with 20 ml of a solution containing 8.4% of potassium iodide and 0.5% of sodium bisulfite in 0.3 *N* sulfuric acid.

Add 2 ml of m*M* dithizone in chloroform to the organic phase and wash it with 10 ml of *N*-potassium hydroxide. Read cadmium in the organic phase at 510 nm.

Zinc can be determined in the mercury- and cadmium-free solution with dithizone.

[96c] Kayoko Nakamura and T. Ozawa, *Anal. Chim. Acta* **86**, 147–156 (1976).

TRIMETHYLTHIONINE

The 1 : 1 bromo complex of mercuric ion with trimethylthionine is extractable for reading.[97] Beer's law holds for 0.1–3.3 μg of mercury per ml of the aqueous solution.

Procedures

From Hydrobromic Acid. The sample should be in hydrobromic acid at pH 2–4. Mix 8 ml with 1 ml of 0.01% trimethylthionine solution. Extract with 10 ml of 4 : 3 dichloroethane-trichloroethylene. Read at 655 nm.

From Sulfuric Acid. The sample should be in 0.5–2.5 *N* sulfuric acid. To 8 ml, add 0.5 ml of 0.1 *M* potassium bromide solution and 1 ml of 0.01% reagent solution. Extract with 10 ml of the mixed solvent and read at 655 nm.

VICTORIA BLUE B

Victoria blue B (basic Blue 26) forms complexes as HgX_3R, where X is chloride, bromide, iodide, or thiocyante.[98] The complexes are extracted into a modified benzene for reading at 632–634 nm. There is interference by silver, auric, thallic, and antimonic ions.

Procedure. To 2 ml of sample solution, add 0.2 ml of 0.01 *M* potassium bromide, 0.8 ml of 1 : 1 sulfuric acid, and 0.5 ml of m*M* Victoria blue B. Dilute to 10 ml and extract with 6 ml of 5 : 1 benzene–methyl ethyl ketone. Read at 634 nm.

Antipyrine in the presence of ammonium thiocyanate forms a white precipitate with mercuric ion. This is extracted by 3 : 1 amyl alcohol–ethyl acetate at pH 1.8–2.3 for reading at 325 nm.[99] There is interference by copper, cobalt, tin, iron, nickel, and lead.

Mercuric ion forms a ternary complex in nitric acid with iodide ion and **2-antipyrinylazo-5-diethylaminophenol.**[100] In a chloroform extract, the maximum absorption at 0.5–10 μg/ml is at 600 nm. Lead, bismuth, cadmium, and nickel interfere.

[97] V. M. Taravan, E. N. Ovsepyan, and N. S. Karimyan, *Arm. Khim. Zh.* **24**, 121–127 (1971).

[98] A. T. Pilipenko, P. P. Kish, and G. M. Vitenko, *Ukr. Khim. Zh.* **37**, 1149–1154 (1971); *ibid* **38**, 479–482 (1972).

[99] Emiko Sudo, *J. Chem. Soc. Jap., Pure Chem. Sect.* **74**, 919–920 (1953).

[100] I. V. Kolosova, *Izv. Vyssh. Ucheb. Zaved., Khim. Khim. Tekhnol.* **12**, 1329–1332 (1969).

The compounds of **antipyrinyl-bis[4-(benzylmethylamino)phenyl]-4-dimethyl-aminophenyl-menthanol** with mercuric chloro or bromomercury complex are extractable into benzene or toluene for reading at 580 nm.[101] An optimum concentration for the extraction is 0.4–0.6 *N* sulfuric acid, and the range of mercury concentrations is 0.005–0.05 mg for 10 ml of extract. Cadmium and trivalent iron or antimony interfere. A corresponding reaction occurs when the compound is not bis. Then the maximum absorption is at 585 nm, the optimum sulfuric acid concentration 0.6–1.5 *N*, and the range 0.001–0.06 mg per 10 ml. Then stannic and thiocyanate ions interfere.

Complexes are formed between mercuric ion and **azoxine H,** which is 5-hydroxy-4-(8-hydroxy-7-quinolylazo)naphthalene-2,7-sulfonic acid, or **azoxine C,** which is 3-(8-hydroxy-7-quinolylazo)naphthalene-1,5-disulfonic acid at pH 1.6. The 1:2 complexes are read at 540 nm.[102] Iron, copper, and nickel can be masked by potassium oxalate.

Mercury is determined indirectly through the **bichromate.**[103] To a sample containing 0.1–1 mg of mercury per ml, add powdered ammonium dichromate and a couple of drops of pyridine. After 5 minutes of vigorous agitation, filter the precipitate of $[Hgpy_2]Cr_2O_7$ on fritted glass. Wash with a solution containing 0.5 gram of potassium dichromate and 0.5 ml of pyridine per liter, followed by acetone containing traces of pyridine. Dissolve the precipitate with 2 ml of 8% sodium hydroxide solution. Acidify with acetic acid, dilute to a known volume, and read at 440 nm.

The effect of mercuric ion in catalyzing the reaction of **4,4′-bipyridyl with ferrocyanide** ion is used for its determination in potassium sulfoterephthalate.[104] To carry out the reaction, dissolve 0.3 gram of sample in 40 ml of acetate buffer solution for pH 3.2. Add 1 ml of 2% solution of bipyridyl in the acetate buffer and follow with 0.5 ml of 1% solution of potassium ferrocyanide. Dilute to 50 ml with the buffer solution and incubate at $65° \pm 1°$ for 3 minutes. Cool rapidly by immersion in ice water, and read at 527 nm against a blank.

Mercuric iodide in potassium iodide solution forms a complex with the **2,2′-bipyridyl-ferrous ion chelate.**[105] At pH 6.3 this is extractable with dichloroethane for reading at 526 nm. Beer's law applies to 0.4–2.4 ppm, and interference by other ions is avoided by adding EDTA.

To 10 ml of sample solution containing 0.1–3.4 μg of mercuric ion per ml, add 1 ml of 0.1 *M* potassium bromide.[106] Add 2 ml of m*M* **brilliant yellow,** and 1 ml of 5 *M* sulfuric acid. Extract with 10 ml of benzene, toluene, or 1:2 carbon tetrachloride–benzene. Read at 640 nm against a salt of the dye.

After separation from some interfering ions by crystal violet, mercuric ion can be determined by extraction as a **butylrhodamine B** complex.[107] The bromomercur-

[101] A. I. Busev and L. S. Khintibidze, *Zh. Anal. Khim.* **22**, 857–862 (1967).

[102] A. I. Chenkesov, V. S. Tonkoshkurov, A. I. Postoronko, and V. N. Ryzhov, *ibid.* **25**, 466–473 (1970).

[103] Florin Modreanu, *Acad. Rep. Pop. Rom., Fil. Iasi, Stud. Cercet. Stiint. Ser. I*, **6**, No. 3/4, 231–235.

[104] Alicja Milosz and Henryk Krzystek, *Chem. Anal.* (Warsaw) **16**, 1085–1089 (1971).

[105] Keiya Kotsuji, *Bull. Chem. Soc. Jap.* **38**, 402–406 (1965); cf. Vinka Karas and Tomislav Pinter, *Croat. Chem. Acta* **30**, 141–147 (1958).

[106] V. M. Tarayan, E. N. Ovsepyan, and N. S. Karimyan, *Izv. Vyssh. Ucheb. Zaved., Khim. Khim. Tekhnol.* **16**, 358–361 (1973).

[107] I. A. Blyum, N. A. Brushtein, and L. A. Oparina, *Zh. Anal. Khim.* **26**, 48–54 (1971).

ate–crystal violet complex is first extracted with benzene from a nitric or hydrochloric acid solution. Then 0.6 *N* sulfuric acid back-extracts the mercury and leaves most of the interference behind. The acid solution is treated with ascorbic acid to mask thallic, rhenium, and antimony ions. Then potassium bromide and crystal violet are again added, and mercury is extracted with benzene. Finally the benzene layer is shaken with butylrhodamine B, potassium bromide, and sufuric acid to displace the crystal violet. The butylrhodamine B–mercury complex is read photometrically at 560 nm for 2–10 μg of mercury or fluorimetrically at 590 nm for less than 2 μg. The double extraction decreases the interference of gold fivefold.

Mercuric ion forms a complex with **α-(4-chloro-1,3-diphenyl-5-pyrazolone-4-yl)-4,4′ bis(dimethylamino)benzhydrol.**[108] This is best extracted with benzene from 4*N* sulfuric acid containing about 0.1% potassium bromide for reading at 644 nm. Beer's law applies to 0.1–2 μg of mercury per ml. Nitrate, chloride, and elements that complex with bromides interfere.

Mercury vapor is measured by the color developed when the air is drawn through filter paper sprayed or brushed with ethanolic **cuprous iodide.**[109] A creamy white to yellow-orange develops for reading by reflectance at 510 nm against a barium sulfate standard. A 2-hour exposure can measure about 15 μg of mercury per liter of air.

The complexes of ***o,o′*-dichlorodithizone,** which is 1,5-bis(2-chlorophenyl)-3-mercaptoformazan, with lead, zinc, cadmium, and mercury are extractable into carbon tetrachloride for reading.[110]

The complex of mercuric chloride with **diphenylcarbazone** is read at 570 nm.[111] The color is stable for 2 hours at pH 6–8. The method is more satisfactory if the color is extracted with benzene and read at 562 nm.[112] For reading organic mercury compounds such as phenylmercury borate, add to the neutral sample solution 1 ml of 5% sodium bicarbonate solution and 0.5 ml of 1% ethanolic **diphenylcarbazone** or **diphenylcarbazide.**[113]

The optimum conditions for determination of mercuric ion with **ethylrhodamine B** are 7 *N* sulfuric acid, 2.5 *N* bromide ion, and 0.05 mg of reagent per ml.[114] Extraction with benzene from 8 *N* sulfuric acid containing 20 mg of bromide per ml and 0.4 mg of reagent provides preconcentration.

Mercuric ion in the range of 2–16 ppm is determined at 460 nm by its decolorization of the complex of **ferric thiocyanate.**[115] There is interference by cerium, chromium, silver, and halogens.

Passing a solution containing mercuric ions at pH 6–8 through a column of powdered iron delivers an equivalent of **ferrous iron** in the effluent.[116] Cobalt,

[108]V. P. Zhivopistsev and A. P. Lipchina, *Uch. Zap. Perm. Gos. Univ.* **1968**, 174–182.

[109]L. Palalau, *Rev. Chim.* **19**, 54–58 (1968); cf. D. P. Shcherbov and K. M. Konovalova, *Zavod. Lab.* **23**, 663–665 (1957).

[110]R. S. Ramakrishna and M. Fernandopulle, *Anal. Chim. Acta* **41**, 35–41 (1968).

[111]Nasim Mufti and Naheed Fatima, *Pak. J. Sci. Ind. Res.* **2**, 1920 (1959).

[112]Hideo Seno and Yachio Kakita, *J. Chem. Soc. Jap.* **82**, 452–455 (1961); cf. N. A. Ugol'nikov and Z. P. Ikonnikova, *Dokl. 7-oi Nauch. Konf. Posvyaschch. 40-letiyu Velikolut. Oktyaler. Sots. Revolyutsii, Tomsk Univ.* **1957** (2), 180.

[113]A. Ichim and Eugenia Radovici, *Igiena* **21**, 683–686 (1972).

[114]I. A. Bochkareva and I. A. Blyum, *Zh. Anal. Khim.* **30**, 874–882 (1975).

[115]Toshio Shibazaki and Masanobo Koibuchi, *J. Pharm. Soc. Jap.* **88**, 140–144 (1968).

[116]Eugeniusz Kloczko, *Chem. Anal.* (Warsaw) **16**, 1291–1298 (1971).

nickel, lead, copper, and silver give the same reaction. Carbon dioxide and oxidants interfere.

Mercuric ion forms a 1 : 1 complex with **glyoxal dithiosemicarbazone** in perchloric acid solution at pH 1.1 containing EDTA.[117a] The reduction in absorption by the reagent is read. Beer's law is followed for 0.05–0.4 m*M*. There is interference by bromide, iodide, thiocyanate, and thiosulfate. Silver can be masked with chloride.

As reagent, mix 2.5 ml of 10 m*M* potassium **hexacyanochromate** with either 2.5 ml of 10 m*M* *p*-anisidine or 5 ml of 10 m*M* *p*-phenylenediamine hydrochloride.[117b] Add 1 ml of sample solution containing mercuric ion up to 10 m*M*. Dilute to 50 ml with a buffer solution for pH 3. For reading, dilute an aliquot to 10 ml with the buffer solution and read at 400 nm. There is interference by mercuric, chromic, vanadyl, and molybdate ions.

Mercuric iodide as the **iodide complex** can be read in 2.5 m*M* potassium iodide in 96% methanol over the range of 0.02–0.1 m*M* mercury content.[118] The equilibrium in water is not favorable. The maximum absorption is at 301.5 nm. There is no interference by chloride, bromide, or acetate ions, nor by ethylmercury cations up to 50% of the mercuric ion.

Mercuric iodide can be extracted from potassium iodide solution in *N* hydrochloric acid with isoamyl alcohol or 3 : 1 isoamyl alcohol–ethyl acetate and read at 305 nm.[119] The mixed solvent gives a somewhat higher absorption.

Mercury in air is absorbed in a solution containing cupric chloride, sodium sulfite, potassium iodide, and sodium bicarbonate. If the air contains sulfur dioxide, however, that must first be removed by passing the air through a column of AV-17 in carbonate or hydroxide form containing 15–20% water.[120]

Mix a sample solution containing 1–200 μg of mercuric ion with an equal volume of 1 : 3 sulfuric acid and add 0.5 ml of 0.1 *M* potassium iodide.[121a] Shake for 1 minute with 10 ml of octanol and discard the aqueous layer. Wash with 3 ml, 3 ml, and 3 ml of water. Back-extract the mercury with 3 ml, 3 ml, and 3 ml of 3 *M* potassium iodide. Add 0.5 ml of 0.1 *M* sodium thiosulfate and dilute to 10 ml or 25 ml according to the mercury content. Read at 322 nm.

Janus green forms a 2 : 1 complex with mercuric chloride. At pH 2–7 the complex is extracted with 3 : 1 benzene–ethylene dichloride for reading at 625 nm.[121b] Beer's law is obeyed for 0.8–20 μg of mercury per ml. Auric ion, stannic ion, and thiourea interfere. Silver is masked with thiocyanate. Ascorbic acid masks bismuth, indium, and ferric ion.

As a reagent prepare 0.1 m*M* **lead xanthate** in chloroform by the reaction of lead acetate and potassium ethylxanthate.[122] To the sample solution containing 40–120 μg of mercuric ion, add 100 μg of cupric ion. Dilute to 30 ml, adjust to pH 2.6, and shake with 10 ml of the reagent solution. Mercuric ion displaces copper from the

[117a]B. W. Budsinsky and J. Svec, *Anal. Chim. Acta* **55**, 115–124 (1971).

[117b]W. U. Malik and K. D. Sharma, *Z. Anal. Chem.* **276**, 379 (1975).

[118]M. H. Abraham, G. F. Johnston, and T. R. Spaulding, *J. Inorg. Nucl. Chem.* **30**, 2167–2171 (1968); cf. A. J. Pappas and H. B. Powell, *Anal. Chem.* **39**, 579–581 (1967).

[119]Hidehiro Goto and Setsuko Suzuki, *Sci. Rep. Res. Inst., Tohoku Univ.* **6**, 130–136 (1954).

[120]M. K. Zagorskaya, O. A. Karlova, and R. P. Varlamov, *Zavod. Lab.* **37**, 539 (1971).

[121a]Ivan Kressin, *Talenta* **19**, 197–202 (1972).

[121b]I. L. Bagbanly, U. Kr. Rustamov, and K. D. Rashidov, *Azerb. Khim. Zh.* **1974**, 122–125.

[122]A. L. J. Rao and Chander Shekhar, *Indian J. Chem.* **13**, 628–629 (1975).

cupric complex, to permit reading the result at 420 nm and interpretation from a copper calibration curve by difference. Silver interferes.

Mercuric ion forms a 1:4 complex with **2-mercaptobenzimidazole.**[123] In *M* perchloric acid containing 10 % of ethanol and 10% of dimethylformamide, this is extracted with chloroform as a 1:4:2 mercuric-reagent-perchlorate complex. This ternary complex is read at 340 nm and obeys Beer's law for 1–10 μg of mercury per ml. Palladium interferes. Bismuth can be read in the extract at 490 nm.

Metalphthalein, which is **o**-cresolphthalein, forms a purple 1:1 complex with mercuric ion in a weakly alkaline solution, pH 9.6–10.2.[124] As a typical example, mix 50 ml of sample with 10 ml of 0.5 m*M* metalphthalein and 20 ml of a buffer solution of 0.1 *M* sodium carbonate–0.2 *M* sodium bicarbonate. Read at 583 nm for 0.005–0.2 mg of mercury in the sample. Silver must be absent. Interference by nickel, zinc, copper, calcium, magnesium, manganese, and cadmium can be avoided by reading against a similar sample containing 1 ml of 0.1 *M* potassium iodide. There is interference by cyanide, iodide, sulfide, aluminum, lead, and cobalt ions.

Mercury as a halide or thiocyanate is extracted at as low as 0.01 ppm from acids or alkalies with **4-(6-methoxy-3-methylbenzothiazolin-2-ylazo)-*N*-methyldiphenylamine** in 3:1 toluene-cyclohexane and read at 642 nm.[125]

Methyl green forms a ternary complex with mercuric chloride, best extractable with benzene, or with mercuric bromide, best extractable with toluene.[126] There is interference with the chloro complex by auric ion, thallium, antimony, and lead. Only cadmium interferes with the bromo complex.

Methylthmol blue forms a 1:1 chelate with mercuric ion at pH 6.[127] To the sample containing 0.5–7.5 mg of mercuric ion, add sufficient hexamethylenediamine nitrate to adjust to pH 6, then 5 ml of 0.001 *M* methylthymol blue. Dilute to 25 ml and read at 630 mn. To correct for interference by less than 25% zinc, copper, iron, or lead based on the mercury content, mask the latter with thiosemicarbazide and read the interferences as a correction. For less than 6 *M* chloride, mask it with 1.5 ml of 0.01 *M* silver nitrate.

Mercuric ion forms a 1:1 complex with reduced **nicotinamide–adenine-dinucleotide** (NAD).[128] In doing so, the color of reduced NAD is decreased and can be read at 340 nm. As an alternative, by activation at 340 nm the suppression of fluorescence at 470 nm is accurate to ±15%. Iodide, bromide, and chloride interfere. The method is not applicable in the presence 50–100 times the molar concentration of organomercury compounds.

At pH 4–4.5, mercuric ion forms a crimson 1:1 complex with **6-(4-nitrophenylazo)pyridoxol.**[129] Beer's law is followed for 0.5–5 μg/ml. For ores containing copper, iron, lead, and antimony, dissolve 1 gram with 2 ml of sulfuric acid

[123]K. A. Uvarova, A. A. Kovalenko, T. I. Zubtsova, and Yu. I. Usatenko, *Zh. Anal. Khim.* **30**, 274–278 (1975).
[124]Sumio Komatsu and Toshiaki Nomamura, *J. Chem. Soc. Jap., Pure Chem. Sect.* **88**, 542–545 (1967).
[125]P. P. Kish, I. I. Zimomrya, I. I. Pogoida, Yu. K. Onishchenko, and G. M. Vitenko, *Tr. Khim. Khim. Tekhnol.* (Gor'kii) **1973** [4 (35)], 71–73.
[126]V. M. Tarayan, E. N. Ovsepyan, and S. P. Lebedeva, *Zh. Anal. Khim.* **26**, 1745–1751 (1971).
[127]Nobuhiko Iritani and Taketsuni Miyahara, *Jap. Anal.* **12**, 1183–1188 (1963).
[128]A. Gorgia and D. Monnier, *Anal. Chim. Acta* **54**, 505–510 (1971).
[129]K. N. Bagdasarov, Yu. M. Gavrilko, and L. P. Kolesnikova, *Zavod. Lab.* **39**, 929–930 (1973).

followed by 0.5 ml of nitric acid. Evaporate to fumes, take up in 20 ml of water, and boil off oxides of nitrogen. Filter and dilute to 100 ml with an acetate buffer solution for pH 4–4.5. Add a l-ml aliquot to 20 ml of the buffer solution, 1 ml of N potassium oxalate to mask the copper, and 5 ml of mM reagent. Dilute to 50 ml with the buffer solution and read with a green filter.

A. 1:2:1 complex formed of mercuric ion–**1,10-phenanthroline and bromophenol blue** is extracted with chloroform at pH 6–8 for reading at 610 nm.[130] A 2.5–3 fold excess of dye and 500-fold excess of 1,10-phenanthroline is required. Beer's law is followed for 0.2–2 mg of mercuric ion per ml. There is interference by ferrous ion, nickel, cobalt, copper, cadmium, zinc, manganous ion, and silver.

A complex of mercuric ion is formed with **6-phenyl-2,3-dihydro-as-triazine-3-thione.**[131] Mix 1 ml of sample, 1 ml of 4 N tartaric acid, 1 ml of 0.001 M reagent in 16% sodium hydroxide solution, and 1 ml of 16% sodium hydroxide solution. Extract with 5 ml, 3 ml, and 2 ml of chloroform. Adjust the combined extracts to 10 ml and read at 430 nm against chloroform.

1-(2-Pyridylazo)-2-naphthol forms a 1:1 complex with mercuric ion.[132a] Buffered at pH 8.5 with sodium carbonate–sodium bicarbonate solution, this is extracted with chloroform. The maximum absorption at 555 nm is stable for 3 hours and obeys Beer's law up to 0.05 mg of mercury per 10 ml of chloroform. Copper, zinc, nickel, cobalt, ferric ion, pentavalent arsenic, and trivalent antimony interfere.

Mercuric, cadmium, and zinc ions form deep red 1:2 complexes with **1-(2-quinolylazo)acenaphthylen-2-ol.**[132b] They are extractable with organic solvents such as carbon tetrachloride.

Mercury precipitated by **Reinecke salt** can be dissolved in anhydrous pyridine for reading.[133] An alternative is to dissolve in aqueous potassium iodide stabilized with ascorbic acid.[134] Beer's law applies for 0.08–2 mg of mercury per liter, and the color is stable for several hours.

Ruhemann's purple, which is 2-(3-hydroxy-1-oxoinden-2-ylimino)indan-1,3-dione, forms a violet 2:1 complex with mercuric ion.[135] This complex is extractable with benzene at pH 4–7, but the reagent is not extractable below pH 6. For 10 ml of sample containing less than 0.03 mg of mercury, add 5 ml of 0.1 mM reagent and 2 ml of 0.5 M acetate buffer solution for pH 6. Dilute to 10 ml and shake with 10 ml of benzene for 5 minutes. Read at 530 nm. There is interference by transition metal ions, zinc, and cadmium; also by anions forming stable mercury complexes (e.g., iodide, bromide, and EDTA.) For more details of determination of mercury by Ruhemann's purple, see Lead.

The 1:2 complex of mercury with **2-(3′-sulfobenzoyl)pyridine-2-pyridylhydrazone,** which is α-2-pyridyl-α-(2-pyridylhydrazono)toluene-*m*-sulfonic acid, is water soluble and has a sensitivity approching that of dithizone, 1-(2-pyridylazo)-2-naphthol,

[130] M. M. Tananaiko and L. I. Gorenshtein, *Izv. Vyssh. Ucheb. Zaved., Khim. Khim. Tekhnol.* **19** (1), 12–14 (1976).

[131] Rostam H. Maghssoudi and Fadhil A. Shamsa, *Anal. Chem.* **47**, 550–552 (1975).

[132a] S. Shibata, *Anal. Chim. Acta* **25**, 348–359 (1961); Li-shu Ho, Ching-nan Kuo, Chih-sheng Shih, and Wu Chiang, *Chem. Bull.* (Peking), **1965**, 250–253.

[132b] Ishwar Singh, Y. L. Mehta, B. S. Garg, and R. P. Singh, *Talenta* **23**, 617–618 (1976).

[133] Jerzy Wayers, *Chem. Anal.* (Warsaw) **5**, 95–99 (1960); *Diss. Pharm.* **12**, 29–34 (1960).

[134] Tadeusz Pelczar and Jerzy Weyere, *Diss. Pharm.* **13**, 243–246 (1961).

[135] Makoto Kanke, Yoshio Inoue, Hirotaka Watanabe, and Setsuko Nakamura, *Jap. Anal.* **21**, 622–626 (1972); cf. A. Ya. Burkina and P. L. Senov, *Zh. Anal. Khim.* **27**, 473–478 (1972).

or 4-(2-pyridylazo)resorcinol.[136] The spectrum of the ligand does not overlap that of the complex.

The complex of mercury with 1,10-phenanthroline adds **tetrabromo-**,tetraiodo- or dichlorotetraiodo-**fluorescein**, extractable at pH 9 with chloroform.[137] The maximum absorption is 540 nm or in 1 : 1 chloroform-acetone, fluorescence at 580 nm. Fluorescence will determine 0.05–10 μg in 10 ml, which is one order of magnitude more sensitive than spectrophotometry.

To a 5-ml sample containing 0.05–0.5 millimole of mercury, add 5 ml of 0.1 *M* EDTA, 10 ml of 1 : 1 nitric acid, and 10 ml of 3 m*M* benzene solution of **thiodibenzoylmethane,** which is 2-(thiobenzoyl)acetophenone.[138] Shake for 5 minutes and separate the organic layer. Wash it with 20 ml and 20 ml of 16.8% potassium hydroxide solution. Read the benzene layer at 360 nm against a reagent blank.

To determine mercuric ion extract at pH 6.4 with 0.01 *M* **thiomaltol**, which is 3-hydroxy-2-methylpyran-4-one, in benzene,[139] wash unused reagent from the organic phase with a glycine buffer solution at pH 12.6 and read at 420 nm against a reagent blank. Copper must be masked with tartrate.

Mercuric ion produces an intense yellow color with **titan yellow** (direct yellow 9).[140] The desirable pH of the solution is 6.8–9.5. For 0.1 mg of mercuric ion, add 4 ml of 1% titan yellow solution. Read at 456 nm. Beer's law applies for 1–28 μg of mercury per ml.

Tris-(2,2′-bipyridyl)ferrous tetraphenylborate reacts at pH 3–10 with mercuric ion to liberate tris-(2,2′-bipyridyl)ferrous ion, which is read at 522 nm.[141] Silver and thallous ions react similarly. The reaction will determine 5–50 ppm of mercuric ion. Chloride interferes.

Mercury at 0.022–3.2 m*M* can be determined by **variamine blue B.**[142] There is an ion-pair complex between $HgBr_3^-$ and the cation which results from oxidation of the dyestuff in a sodium citrate buffer solution for pH 2.9. The resulting blue complex is extracted with nitrobenzene and read immediately at 605 nm. For maximum color development, the dye must be at least in threefold *M* excess over the mercuric ion. Serious interference occurs with stannous, stannic, antimonous, iodide, cyanide, and thiocyanate ions.

Xylenol orange at pH 7 in the presence of hexamine forms a ternary complex with mercuric ion which can be read at 590 nm for the range 5–20 m*M*.[143] Replacement of the amine by a citrate buffer solution gives a maximum at 580 nm for 2–9 ppm.

[136] J. E. Going and C. Sykora, *Anal. Chim. Acta* **70**, 127–132 (1974).
[137] D. N. Lisitsyna and D. P. Shcherbov, *Zh. Anal. Khim.* **28**, 1203–1205 (1973).
[138] E. Uhlemann and B. Schuknecht, *Anal. Chim. Acta* **69**, 79–84 (1974).
[139] Bernd Schuknecht, Erhard Uhlemann, and Gisela Wilke, *Z. Chem.* **15**, 285–286 (1975).
[140] D. Negoiu and A. Kriza, *An. Univ. Bucur. Ser. Stiint. Nat., Chim.* **11**, 115–120 (1962).
[141] M. C. Mehra and P. O'Brien, *Microchem. J.* **19**, 384–389 (1974).
[142] Masahiro Tsubouchi, *Bull. Chem. Soc. Jap.* **43**, 2812–2815 (1970).
[143] H. H. Walker and J. A. Poole, *Talenta* **16**, 739–743 (1969); A. Cabrera-Martin, J. L. Peral-Fernandez, S. Vincente-Perez, and F. Burriel-Marti, *ibid.* 1023–1036.

CHAPTER FIVE
COPPER

If lead is the most universally distributed element, as has been claimed by some, copper could be described as one of the most universally analyzed elements, a statement based on the volume of literature related to its determination in trace amounts. The old standby method for the determination of copper by sodium diethyldithiocarbamate has developed numerous variants in the use of derivatives from which the metal is displaceable by copper. Because of their family relationship, they have been grouped along with the closely related dibenzyldithiocarbamates as a subchapter. Likewise 1,10-phenanthroline has spawned a group of widely used methods.

Traces of copper in food and beverages can promote rancidity and off-flavors. In blood, urine, and tissue they can be significant in medical diagnosis and biochemical research. Control of quality of many products such as paper, petroleum, and alloys necessitates analysis for copper. In general, photometric determinations require less expensive instrumentation than atomic absorption or atomic emission methods and give greater sensitivity.

However numerous the reagents, they are necessarily incomplete. Thus we omit 40 new anils of quinaldehyde and its 4-phenyl analogue, which form 1:2 colored complexes with cuprous ion in the range of pH 5.5–6.5.[1] 2,2-Biquinoline and 2,9-dimethyl-1,10-phenanthroline (neocuproine) appear to be the most specific, the most sensitive, and the easiest to use.

The formation of copper diethyldithiocarbamate is applicable to determination of the sum of lead, copper, bismuth, cadmium, zinc, manganese, and iron.[2] The metals are prepared as the diethyl dithiocarbamates and extracted into chloroform. Washing the chloroform solution with 0.01% sodium hydroxide solution removes

[1] A. L. Gerchuns and I. A. Rastrepina, *Tr. Kom. Anal. Khim.* **17**, 242–250 (1969).
[2] A. B. Blank, N. T. Sizonenko, and A. M. Bulgakova, *Zh. Anal. Khim.* **18**, 1046–1050 (1963); V. B. Sokolovich, Yu. L. Lel'chuk, and G. A. Detkova, *Izv. Tomsk. Politekh. Inst.* **163**, 130–133 (1970).

excess reagent. On shaking with 5% copper sulfate solution, the ion exchange reaction converts all the diethyldithiocarbamates to the copper complex for reading at 436 nm.

For concentration of copper and nickel from sodium chloride, dissolve a 100-gram sample in a liter of water.[3] Add 0.12 mM of ferric ion followed by 0.4% sodium hydroxide to pH 8.5–9 for copper or 9.5–10 for nickel. Stir for 10–15 minutes and set aside for 24 hours. Filter the precipitate and dissolve in 1 : 5 hydrochloric acid.

For recovery of copper,[4] zinc, and iron at about 5 μg/liter, cobalt at 0.025 μg, and chromium at 0.5 μg, heat the sample to 80° and slowly, with vigorous agitation, add 10 ml of a saturated acetone solution of 5,7-dibromo-8-hydroxyquinoline per liter. Boil for 10 minutes to drive off acetone, and let stand overnight. Filter with suction.

Iron, copper, manganese, and aluminum can be determined in the same solution in that sequence.[5] Iron is extracted with chloroform as the 1,10-phenanthroline–ferrous complex and read at 490 nm. Addition of sodium diethyldithiocarbamate forms the copper complex, extracted into butyl acetate and read at 440 nm. The aqueous phase is made alkaline with ammonium hydroxide extracted with chloroform, and the chloroform is discarded. Addition of sodium diethyldithiocarbamate forms the manganese complex in the aqueous phase, extracted with isobutyl acetate for reading at 345 nm. Finally the aluminum is extracted by carbon tetrachloride as the hydroxyquinolate for reading at 400 nm. The iron may be up to 0.02 mg/ml, copper or manganese 0.03 mg/ml, and aluminum 0.08 mg/ml.

For copper in urban air, filter at 75 cubic meters/hour through a glass-fiber filter.[6] Extract the filter with benzene; discard the extract, dry the filter, and grind it. Extract a sample of the filter with 1 : 1 hydrochloric acid–7 N nitric acid and develop with diethyldithiocarbamate or with dithizone.

For copper and iron in blood serum and water by 2,4-BDTPS, which is 2,4-bis[5,6-bis(4-phenylsulfonic acid)-1,2,4-triazin-3-yl]pyridine, see Iron.

ACETYLACETONE

This reagent, which is 2,4-pentanedione, forms a 1 : 1 complex with cupric ion in acid solution.[7] It is read in either the visible or the ultraviolet region. Cobalt interferes in a molar ratio greater than 1 : 1. Ferric ion at no more than a 1 : 2 molar ratio can be masked by fluoride. Tin precipitates and adsorbs some copper. In the ultraviolet, ferric ion over 10% of the copper interferes as a 1 : 3 ferric ion–acetylacetone chelate.

[3] A. L. Gavrilynk and V. T. Chuiko, *Vish. L'viv. Univ., Ser. Khim.* **1971**, 42–44, 83.
[4] J. P. Riley and G. Topping, *Anal. Chim. Acta* **44**, 234–246 (1969).
[5] J. Paul and Sharad M. Paul, *Microchem. J.* **19**, 204–209 (1972).
[6] Halina Wyszynska, Artur Strusinski, and Maria Borkowska, *Rocz. Panstw. Zakl. Hig.* **21**, 483–491 (1970).
[7] A. H. I. Ben-Bassat and Gila Frydman-Kupler, *Chemist-Analyst* **51**, 44–45 (1962); *ibid.* **52**, 8–9 (1963).

Procedures

Visible Range. Dissolve the sample in nitric acid and dilute to contain 0.5–2 mg of cupric ion per ml. Mix 1 ml of this sample solution, 2 ml of acetylacetone, 2 ml of ethanol, and 5 ml of 9:1 *M* acetic acid–*M* sodium acetate as a buffer solution. Shake and filter. Wash the filter with 5 ml of the buffer, then discard the filtrate and washings. Moisten the precipitate with ethanol and dissolve with 5 ml of chloroform. Dilute to 25 ml with ethanol and read at either 660 or 570 nm.

Ultraviolet Range. Dilute 10 ml of sample containing 0.005–0.01 mg of copper to 50 ml with 2 m*M* acetylacetone. Read at 300 or 305 nm against a reagent blank.

ACETYLACETONE DIOXIME

This reagent, which is the dioxime of 2,4-pentadione, forms a 1:1 chelate with copper at pH 5.5 in the presence of excess reagent.[8] Nickel and cobalt absorb at 600 nm but not at 550 nm; thus moderate amounts are tolerated at the latter wavelength. Even then, nickel interferes if more than 3 times the copper, and cobalt if more than 9 times. Tin interferes by precipitating. Chromic ion must not exceed 0.01 *M*. Ferric ion can be masked by fluoride, but this decreases the sensitivity.

Procedure. Adjust a solution of the sample in dilute nitric acid to 0.5–1 mg of copper per ml and pH 5–6. Mix 5 ml of sample solution, 5 ml of an acetate buffer solution for pH 5.5, 10 ml of 0.1 *M* reagent, and 5 ml of water. If nickel and cobalt are absent, read at 600 nm against a reagent blank. If either of them is present, read at 550 nm.

N-ACETYLANABASINE THIOCYANATE

This reagent is applicable to copper in steel, cast iron, aluminum, and so on.[9] The 1:2:2 complex of cuprous ion, thiocyanate, and *N*-acetylanabasine is an amorphous precipitate extractable with chloroform. Beer's law applies at 0.8–40 μg of copper per ml. EDTA decomposes the complex, and tartrate and citrate weaken the color.

[8] A. H. I. Ben-Bassat, Y. Sa'at, and S. Sarel, *Bull. Soc. Chim. Fr.* **1960**, 948; A. H. I. Ben-Bassat and Y. Sa'at, *Chemist-Analyst* **49**, 108–109 (1960).
[9] Sh. T. Talipov, K. G. Nigai, and E. L. Abramova, *Zavod. Lab.* **29**, 804 (1963); *Nauch. Tr. Tashk. Univ.* **1964**, 58–62.

Procedure. To an aliquot of the sample solution at pH 4, add 8 ml of 10% ammonium thiocyanate solution. Add 7.4% solution of ammonium fluoride dropwise to mask aluminum and iron until the solution is colorless. Add 10 ml of an acetate buffer solution for pH 4.6 and 0.2 gram of *N*-acetylanabasine. Dilute to 50 ml and extract with 8 ml, 8 ml, and 8 ml of chloroform. Adjust the combined extracts to 25 ml and read at 410 nm.

ALIZARIN BLACK SN

Alizarin black SN (mordant black 25) forms a purple 1:1 complex with cupric ion.[10] Excess EDTA prevents formation of a 2:1 complex. Low concentrations of chromic ion, lanthanum, nickel, stannous ion, and zirconium are tolerated. Bismuth, ferric ion, and thorium must be removed; this is conveniently done by coprecipitation with aluminum hydroxide in the presence of glycine and sodium chloride. Cyanide and cobalt must be absent. Addition of methanol, ethanol, or isopropanol to the solution of the chelate increases the extinction.[11]

Procedure. As a buffer solution, dissolve 7.5 grams of glycine and 5 grams of sodium chloride in 500 ml of water and add 15 ml of ammonium hydroxide. Add 4% sodium hydroxide solution to raise to pH 11.2, and dilute to 1 liter. To an aliquot of sample solution containing no more than 0.035 mg of copper, add 2 ml of 0.01 *M* disodium EDTA, 10 ml of buffer solution, and 50 ml of 0.0001 *M* solution of alizarin black SN. Dilute to 100 ml, set aside for 1 hour, and read at 650 nm against a reagent blank.

AMMONIA

When a tin-base alloy is dissolved in hydrochloric and nitric acids, phosphoric acid prevents precipitation of tin when the solution is made ammoniacal.[12] The method is designed for alloys containing 1–10% copper. In a tightly stoppered flask, the color is unchanged for 24 hours. Nickel, cobalt, and chromium interfere but are rarely present in tin-base alloys. The technic is applicable to ferric chloride etching baths, with the iron masked by tartrate.[13]

For reading copper in Schweitzer's reagent, the sample is diluted with 1:5 ammonium hydroxide and read at 625 nm with accuracy to 0.09 gram of copper per liter.[14] Although the cuprammonium absorption is ordinarily read around 620

[10]M. Hosain and T. S. West, *Anal. Chim. Acta* **33**, 164–172 (1965).
[11]E. Gagliardi and M. Khadem-Awal, *Mikrochim. Acta* **1969**, 882–887.
[12]George Norwitz, *Anal. Chem.* **20**, 469–470 (1948).
[13]F. J. Conrad and B. T. Kenna, *Plating* **53**, 763–764 (1966).
[14]Konrad Macher and Henryk Nadziakiewicz, *Chem. Anal.* (Warsaw) **7**, 599–604 (1962); cf. T. G. Malkina and V. N. Podchainova, *Tr. Ural. Lesotekh. Inst.* **1969**, 391–393 (1970).

nm, it is 10 times more sensitive in the ultraviolet at 220 nm.[15] For copper naphthenate, dissolve the sample in benzene and extract the copper by shaking with 1:4 hydrochloric acid.[16] The extract is then made ammoniacal and read. Copper in a solution of platinum metals is read as the cuprammonium complex.[17]

Procedures

Alloy Steels.[18] Dissolve 0.1 gram of sample in 10 ml of a mixture of 3 parts of hydrochloric acid and 2 parts of 1:1 nitric acid. Add 35 ml of ammonium hydroxide and dilute to 50 ml. Filter and read at 615 nm.

Tin-Base Alloys. To a 1-gram sample, add 10 ml of hydrochloric acid followed by 10 ml of nitric acid. Warm to dissolve, and add 30 ml of phosphoric acid. Heat at 100° for 15 minutes, in which time the solution should be a clear green. Let cool somewhat and add 200 ml of water. While swirling, add 100 ml of ammonium hydroxide. Cool to room temperature and dilute to 500 ml. Read at 580 nm against water and subtract a reagent blank.

Etching Baths. Mix 5 ml of sample, 25 ml of water, 1 ml of hydrochloric acid, and excess bromine. Boil off the unreacted bromine, cool, and dilute to 200 ml. Dissolve 50 grams of sodium tartrate in the sample and add 200 ml of ammonium hydroxide. Dilute to 1 liter and read at 625 nm against a blank.

Bronze.[19] Dissolve a 1-gram sample in 10 ml of 1:1 nitric acid and evaporate nearly to dryness. Take up in 50 ml of water and make alkaline with ammonium hydroxide. Dilute to a known volume that has an absorption corresponding to 0.2–0.4 mg/ml. Read at 615 nm against a solution containing 10 mg of copper as the ammonia complex per 50 ml.

Aluminum.[20] The following methods apply to pure and remelted aluminum and specified alloys. For pure aluminum, Osmagal, and Zieral, dissolve a 10-gram sample in 50 ml of 25% sodium hydroxide solution. Dilute to 200 ml, filter, and wash with hot water. The copper is in the residue. Dissolve it in 20 ml of 1:1 nitric acid. Boil off nitrogen oxides and add 80 ml of 10% ammonium chloride solution. Add 20 ml of ammonium hydroxide, dilute to 250 ml, and read at 625 nm.

For Okadur, dissolve and proceed as above, but after solution in sodium hydroxide, add 1 ml of 3% hydrogen peroxide.

Aluminum Alloys.[21] Treat a 0.4-gram sample with 15 ml of 1:1 hydrochloric

[15] N. M. Paunovic and M. M. Paunoric, *Bull. Sci., Cons. Acad. RPF Yougosl.* **5**, 99–100 (1960).
[16] A. G. Bhattacharya, A. Bose, and J. Dutta, *Sci. Cult.* **34**, 31–32 (1968).
[17] V. G. Levian and T. P. Yufa, *Anal. Blagorod. Metal., Akad. Nauk SSSR, Inst. Obshche. Neorg. Khim. U.S. Kurnakova* **1959**, 65–69.
[18] V. F. Mal'tsev, *Proizvod. Trub.* **1962** (6), 160–163.
[19] T. G. Malkiad and V. N. Podchainova, *Zh. Anal. Khim.* **19**, 668–670 (1964).
[20] M. Bednara, *Z. Erzberg. Mettalhüttenw.* **5**, 149–152 (1952).
[21] M. M. Menkina and L. A. Fridman, *Mashinostroit. Beloruss.* (Minsk) *Sb.* **1956** (1), 119–120.

acid, add 3 ml of nitric acid, and boil until solution is complete. Add 100 ml of water and 40 ml of ammonium hydroxide. Dilute to 200 ml. Filter, discarding the first 25 ml, and read at 615 nm.

Alternatively,[22] dissolve a 1-gram sample in 100 ml of 5% sodium hydroxide solution. Add 10 ml of 10% sodium sulfide solution and heat to 100°. Dissolve the precipitate in 20 ml of 1:1 nitric acid and dilute to 100 ml. Mix a 10-ml aliquot with 10 ml of 20% ammonium chloride solution and add 10 ml of 40% citric acid solution. Neutralize with ammonium hydroxide and add 20 ml excess. Dilute to 100 ml and read at 625 nm. The solution from dissolving the sulfides is also suitable for determination of iron, nickel, and manganese.

Plating Bath.[23] As a reagent, dissolve 200 grams of ammonium persulfate in 800 ml of ammonium hydroxide. Dilute 100 ml of this to 250 ml and add 0.5 ml of the plating bath. After 2 minutes read at 700 nm.

Copper Cyanide Bath.[24] Add 1 ml of the bath and 2 ml of hydrochloric acid to 60 ml of water and boil until white crystals appear. Cool and add about 5 ml of ammonium hydroxide and 10 ml of 5% EDTA solution. Boil for 5 minutes. Cool, add 5 ml of ammonium hydroxide, dilute to 100 ml, and read at 740 nm.

Wine.[25] Pass a 200-ml sample through a column 1.5 cm in diameter containing 10 ml of KU-2V cationite in hydrogen form, and wash the column with water. Elute with 100 ml of 1:9 hydrochloric acid and wash with 100 ml of water. Evaporate the eluate and washings to dryness and take up in 5 ml of nitric acid. Evaporate to dryness and add 0.5 ml of sulfuric acid to the residue. While warming, add 10 ml of water dropwise. Add 5 ml of 1:3 ammonium hydroxide and filter the precipitated ferric hydroxide. Wash the filter with 1 ml of 1:30 ammonium hydroxide and dilute the filtrate to 25 ml. Read at 615 nm.

ASCORBIC ACID REDUCTION OF ISOPOLYMOLYBDATE

The catalytic effect of cupric ion on the titled reaction is extremely sensitive.[26] The pH of the reduced molybdate solution should be 1.85. The reaction is highly temperature dependent. Against a standard of reproducibility to 2% permissible foreign ions in ppm are as follows: vanadic, 3, cobalt, 5, nickel, 5, zinc, 8, manganese, 8, ferric, 2, magnesium, 5, chloride or bromide, 100, iodide, 40, nitrate, 45, oxalate, 3, acetate, 30, chlorate, 3. Lead causes precipitation, and there is complete interference by aluminum, chromic, fluoride, and silicate ions.

[22]A. Bradvarov and M. Dimitrova, *Mashinostroene* (Sofia) **9** (9), 32–33 (1960).
[23]G. Kopczyk, *Galvanotechnik* **53**, 25–26 (1962).
[24]Kazwo Hasegawa, *J. Metal Finish. Soc. Jap.* **13**, 403–405 (1962).
[25]G. I. Beridze and G. D. Gudzhedzhiani, *Vinodel. Vinograd. SSSR* **1962** (7), 5–7.
[26]R. W. Heller, Jr., and J. C. Guyon, *Anal. Chem.* **40**, 773–776 (1968).

Procedure. As a reagent, mix 20 ml of 10% solution of sodium molybdate dihydrate with 2.8 ml of hydrochloric acid and dilute to 1 liter. Store in polyethylene and let stand at least 36 hours before use.

Take a neutral 25-ml sample containing 5–30 μg of copper. Start a timer, and add 50 ml of the reagent. Two minutes later add 2 ml of 5% ascorbic acid solution. Dilute to 100 ml and place in a 25 ± 0.1° bath. Read at 755 nm against water after 60 minutes for development.

AZIDE ION

Azide ion combines with cupric ion at pH 2.5–6.4 to absorb at 385 nm.[27] Ammonium, nickel, cobalt, chromium, aluminum, bismuth, tin, antimony, molybdenum, silver, titanium, lead, uranium, cyanide, phosphate, oxalate, tartrate, and succinates interfere. Azide ion is also used in combination with pyridine, a parallel to the thiocyanate-pyridine technic.[28] Gold and cobalt must be absent. Nickel and palladium give color with the reagent.

Procedures

Azide Alone. Adjust a 50-ml sample containing 0.1–0.5 mg of cupric ion to pH 5–6. If iron is present, add 10 ml of saturated solution of sodium fluoride. Add 5 ml of 10% solution of sodium azide and dilute to 100 ml. Read at 385 nm.

Azide and Pyridine. Adjust a sample containing 0.03–0.1 mg of copper to pH1–2. Add 1 ml of 1.12 *M* pyridine and 2 ml of 6.5% solution of sodium azide. Add ammonium hydroxide to raise the pH to 5–7 and dilute to 10 ml. Shake with 10 ml of chloroform for 30 seconds, filter through cotton, and read the extract at 325 nm.

AZOAZOXY BN

Azoazoxy BN, which is 2-hydroxy-2′-(2-hydroxy-1-naphthylazo)-5-methylazoxybenzene, forms a red-violet complex with cupric ion in alkaline solution.[29] There is interference by EDTA, calcium, cobalt, and by a tenfold excess of cyanide. The absorbance is a straight line up to 25 m*M*.

[27] Francesco Maggio and Francesco Paolo Cavasino, *Ann. Chim.* (Rome) **51**, 1392–1398 (1961); Teruyuki Kanie, *Jap. Anal.* **7**, 510–512 (1958); Kazimierz Kapitanczyk, Zbigniew Kurzawa, and Zygmunt Pryminski, *Chem. Anal.* (Warsaw) **6**, 23–27 (1971).
[28] Ray G. Clem and E. H. Huffman, *Anal. Chem.* **38**, 926–928 (1966).
[29] E. Wietska, *Chem. Anal.* (Warsaw) **13**, 413–419 (1968).

Procedures

Acetic Acid. Add 3 drops of sulfuric acid to a 100-gram sample and evaporate to sulfur trioxide fumes. Take up the residue in 6 ml of 1 : 1 hydrochloric acid and add 2 ml of 5 m*M* oxalic acid. Neutralize with 16% sodium hydroxide solution and add 1 ml in excess. Extract the copper with 3 ml of 0.2 m*M* azoazoxy BN in carbon tetrachloride. Further extract the aqueous phase with 1 ml and 1 ml of carbon tetrachloride. Combine the extracts and read at 580 nm.

Oxalic Acid. Add 0.5 ml of 0.25 *M* sulfuric acid to a 10-gram sample and sublime the oxalic acid. Take up the residue in 6 ml of 1 : 1 hydrochloric acid and carry through as for acetic acid until the copper is in the carbon tetrachloride phase, using in this case 5 ml, 2 ml, and 1 ml of carbon tetrachloride. Then add 2 ml of 0.5 m*M* EDTA to the aqueous layer, extract with 2 ml of carbon tetrachloride, and add to the previous extracts. Read at 580 nm.

BATHOCUPROINE

The reaction product of cuprous ion and 1 : 10-phenanthroline in acid solution is colorless, but adding methyl groups present in bathocuproine, which is 2,9-dimethyl-4,7-diphenyl-1,10-phenanthroline, inhibits chelation with iron and many other metals. Color development is promoted by the added phenyl groups while the specificity for cuprous ion is retained.[30a] The chelate is readily extractable into hexanol, and pH control is not critical; with the technic specified, other metals do not form colored complexes, and stability of the chelate is not a problem. As applied to pulp, paper, and pulping liquors, dry-ashing is unsatisfactory but wet-ashing gives good results. Because of the sensitivity of the method, water redistilled from glass, glassware rinsed with 1 : 1 nitric acid, and reagents preextracted with bathocuproine in 1-hexanol are essential. The absorption at 479 nm conforms to Beer's law up to 30 μg of copper in 16 ml of hexanol.[30b] Cyanide, sulfide, cobalt, zinc, and ferrous ion interfere.

For copper in butter fat, digest the sample at 80° with nitric acid.[31] Cool and remove the fatty layer with petroleum ether. Add perchloric acid to the aqueous layer and digest until colorless. Dilute the acid layer and add ammonium hydroxide to pH 4.5. Add hydroxylamine hydrochloride to reduce the iron and develop the copper with bathocuproine. Extract the complex with hexanol and read at 479 nm.

Bathocuproine can be sulfonated with chlorosulfonic acid to give water solubility and to obviate the need for extraction.[32]

[30a]Leroy G. Borchardt and John P. Butler, *Anal. Chem.* **29**, 414–419 (1957).
[30b]W. E. Dunbar and A. A. Schilt, *Talenta* **19**, 1025–1031 (1972).
[31]Timmen and Jutta Bluethgen, *Z. Lebensm.-Unters. Forsch.* **153**, 283–288 (1973).
[32]James W. Landers and Bennie Zak, *Tech. Bull. Regist. Med. Technol.* **28**, 98–100 (1958); *Am. J. Clin. Pathol.* **29**, 590–592 (1958).

Procedures

Pulp, Paper, and Pulping Liquors. Weigh a 1-gram sample of pulp and paper or take a 10-ml sample of pulping liquors. Add 20 ml of nitric acid, 5 ml of perchloric acid, and a boiling chip. Warm gently until most of the carbonaceous matter is decomposed, and heat further to fumes of perchloric acid. Cool, dilute to 25 ml, and boil off oxides of chlorine. Neutralize with ammonium hydroxide as indicated by Congo red paper, add 10 drops in excess, and cool. Add in order, 2 ml of 10% hydroxylamine hydrochloride solution, 1 ml of 0.01 *M* bathocuproine in 1-hexanol, and 5 ml of 1-hexanol. Shake for 2 minutes and read the hexanol layer at 470 nm against a blank.

Tellurium.[33] Dissolve 0.5 gram of a sample containing more than 0.5 ppm of copper in 3 ml of hydrochloric acid and 2 ml of nitric acid. Evaporate to dryness and take up in 5 ml of 1:1 hydrochloric acid. Add 20 ml of 20% ammonium citrate solution followed by ammonium hydroxide to adjust to about pH 6. Add 2 ml of 10% hydroxylamine hydrochloride solution. Mix with 1 ml of 0.1% ethanolic bathocuproine solution and let stand for 5 minutes. Shake with 5 ml and 5 ml of chloroform and read at 477 nm. Tellurium does not interfere.

For tellurium containing 0.05–0.5 ppm of copper, dissolve a 5-gram sample in 50 ml of 1:1 nitric acid. Evaporate nearly to dryness, take up in 20 ml of water, and filter off the bulk of the tellurous acid. Proceed as for pulp from "Add 20 ml of 20% ammonium citrate...."

Edible Oils and Shortenings.[34] Reflux a sample containing 1–10 μg of copper for 30 minutes with 50 ml of 0.01% solution of EDTA in 1:3 hydrochloric acid. Cool, filter, and add 10 ml of nitric acid. Evaporate to dryness at 100° and take up the residue in 10 ml of water. Add 2 ml of 10% solution of hydroxylamine hydrochloride and raise the pH to 7±0.5 with ammonium hydroxide. Add 10 ml of 10% solution of ammonium acetate and 1 ml of 0.36% solution of bathocuproine in hexanol. Add 5 ml of hexanol and shake for 2 minutes. Read the organic layer at 479 nm against a reagent blank.

Protein.[35] To 1 ml of sample solution containing 2–5 mg of copper, add 1 ml of 0.1% bathocuproine in glacial acetic acid. Mix and add 0.05 ml of 5% solution of hydroxylamine hydrochloride and 0.95 ml of ethanol. Centrifuge and read at 479 nm. This will detect 2 nanomoles per mg of protein.

---

[33]Nobuhisa Matano and Akira Kawase, *Jap. Anal.* **11**, 346–351 (1962).

[34]Rudolph E. Deck and Kay K. Kaiser, *J. Am. Oil Chem. Soc.* **47**, 126–128 (1970).

[35]D. E. Griffiths and David C. Wharton, *J. Biol. Chem.* **236**, 1850 (1961); David C. Wharton and Mildred Rader, *Analyt. Biochem.* **33**, 226–229 (1970).

α-BENZOINOXIME

For a solution containing copper and iron in 1:1 hydrochloric acid, extract ferric ion along with auric ion with 4-methylpentan-2-ol.[36] Adjust to pH 3.5 and make the solution 2 *M* with pyridine. Add 1 ml of 1% α-benzoinoxime in 10% sodium hydroxide solution. Extract with chloroform and read at 440 nm for 2–10 μg of copper per ml in the extract. There is interference by ruthenium, stannous ion, and complexing anions.

Procedure. *Edible Oils.*[37a] Reflux 10 grams of sample with 10 ml of 1:6 nitric acid for 45 minutes, using an air condenser. Cool, shake with 10 ml of petroleum ether, and discard the extract. Filter the aqueous layer and dilute to a known volume. To an aliquot containing up to 1 mg of copper, add 25 ml of a mixture of 50 ml of 60% sodium potassium tartrate solution and 3 ml of 10% sodium hydroxide solution. Raise to pH 11.5–12 with 10% sodium hydroxide solution; add 1 ml of 1% solution of σ-benzoinoxime in 1% sodium hydroxide solution. Extract with 20 ml, and 10 ml of chloroform. Dilute the filtered extracts to 50 ml with chloroform and read at 440 nm.

For lesser copper content, dissolve 50 grams of sample in 150 ml of petroleum ether. Shake at 120 cycles/minute with 100 ml of 1:2.5 nitric acid for 1 hour. Filter the aqueous layer, evaporate to dryness, and calcine. Take up in 5 ml of 1:2.5 nitric acid and determine as above from "To an aliquot containing up to 1 mg....."

1-BENZOYL-3-(2-PYRIDYL)-2-THIOUREA

The 1:2 complex of cupric ion with this reagent obeys Beer's law at 0.5–3.5 μg of copper per ml. Up to 30 μg of ferric ion can be masked with triethanolamine. There is interference by chromate, manganous ion, vanadate, tetravalent thorium, and cerium ions.

Procedure.[37b] Mix 5 ml of sample solution containing about 25 μg of copper with 4 ml of ethanolic 0.1 m*M* 1-benzoyl-3-(2-pyridyl)-2-thiourea. Adjust to pH 5 and dilute to 25 ml. Shake for 2 minutes with 5 ml and 5 ml of chloroform. Adjust the combined extracts to 10 ml and read the 1:2 copper-reagent complex at 340 nm against a reagent blank. The complex can also be read in ethanol at 430 nm.

---

[36] S. B. Gawali and V. M. Shinde, *Span. Sci.* **9**, 451–458 (1974).

[37a] C. G. Macarovici, V. Farcasan, G. Schmidt, V. Bota, M. Macarovici, A. Dorutiu, I. Pirvu, and E. Tesler, *Stud. Univ. Babes-Bolyai, Cluj, Chim.* **1961** (2), 103–108.

[37b] D. N. Wandalkar and A. P. Joshi, *Z. Anal. Chem.* **280**, 220 (1976); cf. Mrinal K. Das and A. K. Majumdar, *Microchem. J.* **15**, 540–544 (1970).

2,2′-BICINCHONINIC ACID

With a tenfold excess of this reagent, Beer's law applies for 1–9 μg of cuprous ion per ml.[38] Antimony, nickel, iron, and cobalt interfere. The reagent has been applied to copper at 560 nm in cesium iodide in the presence of iron and ascorbic acid.[39]

Procedures

Variamine Blue Added. This addition substantially intensifies the color of the complex.[40] To a sample solution containing 15–90 μg of cupric ion, add 5 ml of acetate buffer solution for pH 4.5 and 4 ml of 0.02% solution of reduced variamine blue. Add 2 ml of 0.1% solution of sodium 2,2′-bicinchoninate and set aside for 10 minutes. Read at 555 nm against a reagent blank.

Aluminum Alloys.[41] Dissolve a 1-gram sample in 20 ml of 1:1 hydrochloric acid, adding a few drops of 30% hydrogen peroxide. Boil off excess hydrogen peroxide, filter, and dilute to 100 ml.

To a 5-ml aliquot, add 2 ml of 10% hydroxylamine hydrochloride solution and 2 ml of 0.1% solution of 2,2′-bicinchoninic acid in 2% potassium hydroxide solution. Add a saturated solution of tetrasodium pyrophosphate until the initial precipitate is redissolved. Check with an external indicator that the pH is 6. Dilute to 25 ml and read at 560 nm against a developed standard containing 0.8 μg of copper per ml.

Nickel.[42] Dissolve a 2-gram sample in 40 ml of aqua regia and evaporate almost to dryness. Add 20 ml of hydrochloric acid and again evaporate almost to dryness. Take up the residue in 20 ml of 8 *N* hydrochloric acid.

Prepare a column of anionite TM and wash it with 25 ml of 8 *N* hydrochloric acid. Pass the sample solution through the column at 1 ml/minute and wash the column with 25 ml of the acid. Elute the copper with 120 ml of 0.5 *N* hydrochloric acid. Evaporate the eluate to about 30 ml and dilute to 50 ml.

To a 3-ml aliquot, add 2 ml of 10% hydroxylamine hydrochloride solution and 2 ml of a 0.1% solution of 2,2′-bicinchoninic acid in 2% potassium hydroxide solution. Add 1:20 ammonium hydroxide to adjust the pH to 6 and dilute to 25

[38] A. L. Gershuns, *USSR Patent* 126, 295 (1960); A. L. Gershuns, A. A. Verezubova, and Zh. A. Tolstykh, *Izv. Vyss. Ucheb. Zaved., Khim. Khim. Tekhnol.* **4** (1), 25–27 (1961). N. N. Lapin and A. I. Vovk, *Izv. Vyssh. Ucheb. Zaved. Khim. Khim. Tekhnol.* **9**, 27–30 (1966); A. L. Gershuns and V. L. Koval', *Vest. Khar'k. Univ., Ser. Khim.* **1970**, 69–71.

[39] A. B. Blank, V. G. Chepurnaya, L. P. Eksperiandova, and V. Ya. Vakulenko, *Zh. Anal. Khim.* **29**, 1705–1709 (1974).

[40] Franciszek Buhl, Malgorzata Palka, and Barbara Piwowarska, *Chem. Anal.* (Warsaw) **19**, 361–366 (1974).

[41] T. G. Malkina, V. N. Podchainova, and V. N. Stashkova, *Tr. Vses. Nauch.-Issled. Inst. Stand. Obraztsov.* **3**, 98–100 (1968); *Tr. Ural. Lesotekh. Inst.* **1969**, 387–390.

[42] V. N. Podchainova and T. G. Malkina, *Tr. Vses. Nauchn.-Issled. Inst. Stand. Obraztov.* **3**, 100–104 (1967).

ml. Read at 560 nm against a developed solution containing 0.8 μg of copper per ml.

Products of Lead Metallurgy. Dissolve a 0.2-gram sample in 40 ml of 1 : 1 nitric acid. Add 10 ml of 1 : 2 sulfuric acid and evaporate to fumes. Take up in 25 ml of water and filter. Develop as for nickel from "To a 3-ml aliquot...."

Copper Alloys.[43] Dissolve a 0.2-gram sample in 4 ml of nitric acid, adding water if desired. Filter any metastannic acid and dilute to 250 ml. Dilute a 10-ml aliquot to 250 ml. Mix a 25-ml aliquot of this dilution with 1 ml of 10% hydroxylamine hydrochloride solution and 1 ml of 50% solution of sodium potassium tartrate tetrahydrate. Add 5 ml of a buffer solution for pH 6 containing 33 ml of 2 *M* acetic acid and 467 ml of 2 *M* sodium acetate. Add 5 ml of 0.2% solution of the reagent in 0.7% potassium hydroxide solution and dilute to 50 ml. Read at 570 nm against a copper reference solution.

Antimony.[44] Moisten a 5-gram sample with water and dissolve with 30 ml of aqua regia. Add 5 ml of hydrobromic acid and evaporate to dryness. Take up the residue with 10 ml of 1 : 1 hydrochloric acid and dilute. Take a 5-ml aliquot containing 10–30 μg of copper and add 2 ml of 10% hydroxylamine hydrochloride solution. Add 2 ml of a 0.1% solution of 2,2′-bicinchoninic acid in 2% potassium hydroxide solution. Add ammonium hydroxide to adjust to pH 6 and dilute to 25 ml. Read at 560 nm.

Magnesium.[45] Dissolve a 0.5-gram sample in 7.5 ml of 1 : 1 hydrochloric acid. Add 1 ml of 10% hydroxylamine hydrochloride solution, 5 ml of 50% sodium potassium tartrate solution, and 5 ml of acetate buffer solution for pH 6. Then add 2 ml of 0.1% solution of 2,2′-bicinchoninic acid in 2% potassium hydroxide solution and dilute to 25 ml. Read at 570 nm. The sample can be expected to contain 0.005–0.04 mg of copper.

Zinc or Cadmium.[46] Dissolve a 1-gram sample in 15 ml of 1 : 1 nitric acid. Add 1.5 ml of 15% solution of hydroxylamine hydrochloride and 2 ml of 0.1% 2,2′-bicinchoninic acid solution. Add 1 : 20 ammonium hydroxide until a violet color develops. Dilute to 50 ml with a buffer for pH 6 containing 55% sodium acetate trihydrate solution and 10% 6% acetic acid. Read at 560 nm.

Tin-Lead Solder. Disintegrate 0.1 gram of sample in 15 ml of 1 : 1 nitric acid. Boil off the oxides of nitrogen and filter the metastannic acid. Wash well with water and dilute to 50 ml. Mix a 5-ml aliquot with 10 ml of 15% solution of hydroxylamine hydrochloride and 2 ml of 0.1% solution of 2,2′-bicinchoninic acid. Add 1 : 20 ammonium hydroxide to develop the violet color and dilute to 50 ml with the acetate buffer solution described for zinc or cadmium. Read at 560 nm.

[43] V. N. Tikhonov, *Izv. Vyssh. Ucheb. Zaved., Khim. Khim. Tekhnol.* **15**, 307–308 (1972).
[44] V. N. Podchainova, T. G. Malkina, and L. S. Studenskaya, *Tr. Vses. Nauchn.-Issled. Inst. Stand. Obraztsov Spektr. Etalonov*, **2**, 18–21 (1965).
[45] V. N. Tikhonov and I. S. Mustafin, *Zh. Anal. Khim.* **20**, 390–392 (1965).
[46] K. I. Fridman and O. L. Turchina, *Khim. Prom., Inf. Nauk-Tekh. Zb.* **1965** (2), 74–75.

Soldering Alloys.[47] Dissolve a 0.1-gram sample in 5 ml of nitric acid. Dilute, neutralize to pH 6 with 10% sodium hydroxide solution, and dilute to 100 ml. To a 5-ml aliquot, add 2 ml of 10% hydroxylamine hydrochloride solution and 2 ml of 5% tartaric acid solution. Add 2 ml of 0.1% solution of 2,2′-bicinchoninic acid in 2% potassium hydroxide, and again adjust to pH 6. Dilute to 25 ml and read at 540 nm against a standard containing 0.1 mg of copper per 25 ml.

Minerals.[48] To a 2-gram sample in platinum, add 2 ml of sulfuric acid and 5 ml of hydrofluoric acid, and evaporate to sulfur trioxide fumes. Add 10 ml of hydrofluoric acid and repeat. Then add 5 ml of hydrofluoric acid and repeat. Add 50 ml of 1:3 hydrochloric acid, heat, filter, and dilute to 100 ml.

To a 10-ml aliquot add 5 ml of 20% hydroxylamine hydrochloride solution and heat until the yellow color of ferric ion disappears. Cool, and add 2 ml of 0.1% solution of 2,2′-bicinchoninic acid in 2% sodium hydroxide solution. Add 1 ml of 15% citric acid solution, then ammonium hydroxide, to raise to pH 6. Dilute to 25 ml with acetate buffer solution for pH 6 and read at 560 nm.

Ore. Heat 20 grams of sample with 50 ml of hydrochloric acid and add a few ml of hydrofluoric acid from time to time, as long as silica is present. Filter, fuse the insoluble residue with potassium pyrosulfate, and dissolve the melt in 1:18 sulfuric acid. Combine the solutions and add freshly precipitated zinc sulfide in excess. Raise the pH to about 5 by adding ammonium hydroxide. Filter and heat the precipitate with excess of 1:18 sulfuric acid. Filter and dissolve the residue in nitric acid. Evaporate the combined solutions to sulfur trioxide fumes and cool. Take up in water and dilute to 50 ml.

Mix a 10-ml aliquot with 5 ml of 2% ascorbic acid solution to reduce the copper. Add 4% sodium hydroxide solution to adjust to pH 4.6. Add 10 ml of 2 *N* acetate buffer for that pH and 5 ml of 0.1% solution of 2,2′-bicinchoninic acid in 2% sodium hydroxide solution. Dilute to 50 ml and read at 357 nm after 5 minutes.

Copper-Vanadium Filter Cake.[49a] This technic will determine 20% of copper within ±0.5%. Dissolve a 1-gram sample with 5 ml of nitric acid and 30 ml of 1:1 sulfuric acid. Evaporate to sulfur trioxide fumes, rinse down the wall of the container, and evaporate to about 10 ml. Dilute to 100 ml, filter, and ignite the filter. Fuse the residue with 2 grams of sodium carbonate and 1 gram of sodium tetraborate decahydrate. Take up the melt in water, add 15 ml of 1:1 sulfuric acid, and combine with the original filtrate. Dilute to 250 ml and dilute a 10-ml aliquot to 200 ml.

To 10 ml of the final dilution add 1 ml of 10% hydroxylamine hydrochloride solution and 1 ml of 10% sodium potassium tartrate solution. Add ammonium hydroxide to about pH 5, followed by 5 ml of an acetate buffer solution for pH 6. Add 5 ml of 0.2% solution of 2,2′-bicinchoninic acid in 0.7% potassium hydroxide solution. Dilute to 50 ml and read at 560 nm against a developed solution containing 0.5 mg of copper.

[47]A. S. Arkhangel'skaya and G. P. Batalina, *Zavod. Lab.* **34**, 408–409 (1968).
[48]G. S. Plyusnin and V. V. Serebrennikov, *Tr. Tomsk. Gos. Univ.* **192**, 132–133 (1968).
[49a]V. N. Tikhonov and M. Ya. Grankina, *Zh. Anal. Khim.* **21**, 1016–1018 (1966).

Water.[49b] To a sample containing 1–100 μg of copper add 2 drops of 0.1% solution of 2,2′-bicinchoninic acid in 2% solution of potassium hydroxide. Dilute to 25 ml and read the 1:2 cuprous-reagent complex at 540 nm.

Waste Water. To a sample containing 1–100 μg of copper add 2 ml of 10% hydroxylamine hydrochloride solution, 2 ml of 0.1% solution of the reagent, and sufficient saturated solution of tetrasodium pyrophosphate to give pH 6. Dilute to 25 ml and read.

Serum.[50] Mix 5 ml of sample with 2 ml of nitric acid and 3 ml of sulfuric acid. Heat to sulfur trioxide fumes and add dropwise 1 ml of 10% hydrogen peroxide. Evaporate almost to dryness. Add 5 ml of acetate buffer solution for pH 5.7 and adjust to about pH 6 with 30% sodium hydroxide solution. Add 1 ml of 15% solution of hydroxylamine hydrochloride and 2 ml of 0.1% solution of the disodium salt of the reagent. Dilute to 25 ml with the buffer solution, and after 15 minutes read at 570 nm. The final solution can be expected to contain 0.1–1 μg of copper per ml.

2,2′-BIQUINOLYL-4,4′-DICARBOXYLIC ACID

This reagent which is cuproine dicarboxylate, reacts with cuprous ion in an acetate buffer solution for pH 4.3–4.6 to absorb at 560 nm.[51] Ascorbic acid reduces the cupric and ferric ions. With tartaric acid as a masking agent, it tolerates a 1000-fold excess of manganese, bismuth, nickel, or titanium, a 100-fold excess of cobalt, or a tenfold excess of vanadium.

Procedures

Steel.[52] Dissolve a 0.1-gram sample in 20 ml of 1:1 hydrochloric acid. Add 5 ml of nitric acid, evaporate to half-volume, and filter. Dilute the filtrate to 50 ml. Mix a 3-ml aliquot with 10 ml of 15% hydroxylamine hydrochloride solution and heat to ensure complete reduction of ferric ion. Cool, and add 2 ml of a 0.1% solution of the reagent in 2% potassium hydroxide solution. Add 1:24 ammonium hydroxide to adjust to pH 6. Dilute to 25 ml with an ammonium acetate buffer solution for pH 6 and read at 560 nm.

Molybdenum Wire. The sample has a binary coating with nickel, tin, or aluminum. Heat a piece of the wire 2–6 cm long with 2 ml of 1:1 hydrochloric acid at 100° for 30 minutes. Complete as for steel from "Cool and add 2 ml ...."

[49b] T. G. Malkina, *Trudy Inst. Khim. ural.' nauch. Tsentr. Akad. Nauk SSSR* **1974** (27), 33–40.
[50] Z. Gregorowicz, G. Kwapulinska, and B. Piwowarska, *Chem. Anal.* (Warsaw) **13**, 887–890 (1968).
[51] N. N. Lapin, A. T. Slyusarev, and A. G. Efimenko, *Zavod. Lab.* **29**, 807 (1963).
[52] I. S. Mustafin, N. S. Frumina, and V. S. Kovaleva, *ibid.* **29**, 782–785 (1963).

BIS[2,6-DIMETHYLMORPHOLINO(THIOFORMYL)]DISULFIDE

This reagent is suitable for determination of copper in digestates of biological materials as well as in the solution of high purity aluminum.[53] A 1:1 complex is formed. Beer's law is followed by 0.5–5 μg of copper per ml of the diluted extract. Ferric and nitrate ions must be absent.

Procedure. The sample should be 10 ml in 1:1 hydrochloric acid. Extract with 2 ml, 2 ml, and 2 ml of 0.001 *M* solution of the reagent in tributyl phosphate. Dilute the combined extracts to 10 ml with methanol and read at 430 nm.

BIURET

Copper, nickel, and cobalt can be developed with biuret.[54] Chromium as the chromate is produced in the same solution. All four substances are read on the same sample and solved by simultaneous equations. The biuret shows no absorption over the range used. Beer's law applies for 2–56 ppm of copper, 0.5–50 ppm of cobalt, 2–66 ppm of nickel, and 0.01–1 ppm of chromium. A large concentration of biuret in strongly alkaline solution avoids precipitation of cobalt and nickel hydroxides. The pH must exceed 12.4. The complexes form immediately and are stable for months. In steel analysis the iron is extracted from strong hydrochloric acid solution with ether. When necessary because of a very high percentage of one metal, a corresponding amount is added to the blank; that is; differential spectrophotometry is employed.

Ions that interfere by precipitating are ferrous, ferric, lead, magnesium, cadmium, manganese, mercuric, and bismuth. Highly colored ions such as permanganate interfere. When ammonia is present in amounts exceeding 1000 ppm it forms competing complexes.

Procedure. As a reagent, dissolve 30 grams of purified biuret and 80 grams of sodium hydroxide in water and dilute to one liter. Store in polyethylene. Because of decomposition to ammonia and urea, do not use if more than 3 days old.

Steel. COPPER, COBALT, NICKEL, AND CHROMIUM. Select a sample so that the weights of the four metals fall in the following approximate ranges: copper, 0.3–6 mg, cobalt, 0.2–5 mg, nickel, 0.2–7 mg, chromium, 0.015–1 mg. If the concentration of one or more of the metals is above the recommended range, employ the technic of differential spectrophotometry.

Dissolve the sample in 5 ml of hydrochloric acid. Add 3 drops of nitric acid to oxidize iron to the ferric state. Dilute with 1:1 hydrochloric acid and extract the iron with 30 ml and 30 ml of ether. Boil the solution to drive off dissolved ether,

[53] O. Wawschinek and H. H. Tagger, *Anal. Chim. Acta* **35**, 109–110 (1966).
[54] Vaughn K. Gustin and Thomas R. Sweet, *Anal. Chem.* **33**, 1942–1944 (1961).

then add sodium peroxide to oxidize chromium to the hexavalent state. Again boil to decompose excess peroxide, and cool. Add solid sodium carbonate as long as carbon dioxide is evolved. Then add hydrochloric acid dropwise to dissolve any precipitate. Add to 50 ml of the biuret reagent and dilute to 100 ml. Read at 333, 380, 412, and 570 nm against a 1:1 dilution of the biuret reagent.

Calculations. The simultaneous equations required are as follows:

$$C_{Co} = 53.93\,A_{333} + 5.62\,A_{412} - 13.65\,A_{380} - 30.10\,A_{570}$$

$$C_{Cu} = 288.9\,A_{570} + 18.31\,A_{380} - 55.39\,A_{333} - 27.92\,A_{412}$$

$$C_{Ni} = 139.9\,A_{412} - 16.69\,A_{333} - 25.15\,A_{380} - 10.27\,A_{570}$$

$$C_{Cr} = 1.568\,A_{380} + 0.732\,A_{570} - 1.141\,A_{333} - 0.882\,A_{412}$$

BRILLIANT GREEN

Brilliant green (basic green 1) is a suitable reagent for analysis of metals and alloys for copper.[55]

Procedures

Nickel, Chromium, Ferrochrome. As a basic solution, dissolve 113 grams of potassium sulfate and 38 grams of potassium chloride in hot water, cool, add 100 ml of 1:35 sulfuric acid, and dilute to 1 liter.

Dissolve a 1-gram sample in aqua regia. Add 5 ml of sulfuric acid and evaporate to fumes. Take up in water and dilute to 100 ml. Mix a 5-ml aliquot of the sample with 13 ml of the basic solution. Add 20 ml of benzene, 0.1 ml of 15% titanous chloride solution, and 2 ml of 0.2% ethanolic solution of brilliant green. Shake for 30 seconds, let stand in the dark for 15 minutes, and read the extract at 640 nm against a blank.

Ferromolybdenum. Dissolve 1 gram in 1:3 nitric acid. Proceed as above from "Add 5 ml of sulfuric acid and ...."

---

[55] L. N. Makogonova, V. I. Kurbatova, and G. N. Emasheva, *Tr. Vses. Nauchn.-Issled. Inst. Stand. Obraztsov. Specktr. Etalonov.* **5**, 98–104 (1969).

CUPRIZONE

This reagent is also known as bicyclohexanone oxalyldihydrazone, and it has diverse applications ranging from alloys to vegetation.[56] It has been applied to pulp and paper[57] and to nonferrous metals and alloys.[58] The complex with inorganic or organic bivalent copper is effective for extraction with methanol from petroleum distillates.[59] Appropriate samples contain 0.1–2 μg of copper. By use of silver bromide as a collector, copper cuprizonate is isolated from lead or nickel solutions.[60] When copper is determined in leaded bronze by electrolytic deposition a trace left in the electrolyte can be determined by this reagent.[61]

The reagent is applicable to alloys having tin, lead, and antimony as major components.[62] There is no interference by 20 mg of antimony, 10 mg of tin, or 100 mg of lead. Ammonium tartrate prevents precipitation of hydroxides of tin and lead.

The reagent is applicable to copper down to 2.5 μg per gram of cadmium sulfide, as well as to titanium,[63] zirconium, and their alloys.[64] Borate buffer solutions provide better pH control than ammonium hydroxide and ammonium citrate.[65] Excess ammonia can prevent formation of the complex.

The method is applicable to the products of decomposition of soil, sediment, and rocks.[66] It has been applied to determination of 0.0005–0.125% of copper in high purity niobium, tantalum, molybdenum, and tungsten.[67]

In general, adjust a solution containing 5–10 grams of lead to pH 5 or one containing 1–5 grams of nickel to pH 7 and add excess of cuprizone in methanol.[68] Add silver nitrate solution followed by sodium bromide to precipitate 0.2 gram of silver bromide as a collector for the copper complex. Filter and dissolve the precipitate in sodium iodide solution. Adjust to pH 4.5, extract the copper with cuprizone in chloroform, and read with a limitation of 0.03 ppm. This will determine down to 0.3 ppm of copper in lead or 1.4 ppm in nickel.

Procedure. As a buffer solution for pH 9.1, mix 400 ml of 0.5 *M* boric acid with 60 ml of 0.5 *M* sodium hydroxide. To an acid sample containing up to 0.05 mg of

[56] G. Nilsson, *Acta Chem. Scand.* **4**, 205 (1950); R. E. Peterson and M. E. Bollier, *Anal. Chem.* **27**, 7 (1955).
[57] C. V. Wetlesen and G. Gran, *Sven. Papperstidrn.* **55**, 212 (1952).
[58] R. K. Rohde, *Anal. Chem.* **38**, 7 (1966).
[59] C. E. Lambdin and W. V. Taylor, *Anal. Chem.* **14**, 2196–2197 (1968).
[60] E. Jackwerth and E. Doering, *Z. Anal. Chem.* **255**, 194–201 (1971).
[61] British Standards Institution, *BS 1748*, Pt. **11** (1964).
[62] Saichiro Onuki, Kunihiko Watanuki, and Yukichi Yoshino, *Jap. Anal.* **14**, 339–442 (1965); A. Schottak and H. Schweiger, *Z. Erzbergb. Metallhüttw.* **19**, 180–185 (1966).
[63] F. Cabane-Brouty, *Anal. Chim. Acta* **47**, 511–516 (1969).
[64] D. F. Wood and R. T. Clark, *Analyst* **83**, 509–516 (1958); cf. William K. Murray, *U.S. Dept. Comm., Office Tech. Serv. PB Rept.* **138, 591**, 12 pp (1958).
[65] E. Somers and J. L. Garraway, *Chem. Ind.* (London) **1957**, 395.
[66] G. A. Nawlan, *U.S. Geol. Survey Prof. Paper* **525 D**, 189–191 (1965).
[67] E. M. Penner and W. R. Inman, *Talenta* **10**, 407–412 (1963).
[68] E. Jackwerth and E. Doering, *Z. Anal. Chem.* **255**, 194–201 (1971).

copper, add 5 ml of 10% ammonium citrate. Add a drop of neutral red indicator solution and neutralize with 12% sodium hydroxide solution. Add 5 ml of the buffer solution and 0.5 ml of 0.5% solution of the reagent in 50% ethanol. Dilute to 50 ml and read at 595 nm after 10 minutes and within 3 hours.

Steel.[69] For this technic the sample must contain no cobalt, less than 20% nickel, and less than 0.5% copper. Dissolve 0.5 gram of sample in 5 ml of hydrochloric acid, 5 ml of nitric acid, and 5 ml of perchloric acid. Evaporate to copius fumes of perchloric acid and cool. Take up in 50 ml of water and filter. Dilute the filtrate and washings to 250 ml.

To a 5-ml aliquot, add 2.5 ml of 20% citric acid solution, 11 ml of 1 : 10 ammonium hydroxide, and 2 ml of 0.5% ethanolic reagent. Dilute to 100 ml and after 30 minutes read at 600 nm against a blank. The final pH should be 8.8–8.9.

For broader applicability, the copper is extracted from the solution of high alloy steel with dithizone and later read with the titled reagent.[70] For copper and nickel in buffered solutions of low alloy steel, cuprizone is added for reading the copper at 600 nm, disodium ethyl bis(5-tetrazolylazo) acetate is added to the same solution for reading nickel at 510 nm.[71]

Bismuth.[72] Dissolve 2 grams in 10 ml of nitric acid and 10 ml of perchloric acid. Heat to the point that fumes just begin, cool, and dilute to 100 ml. Mix a 10-ml aliquot with 2 ml of 25% citric acid solution and add ammonium hydroxide to pH 8.8–9.1. Add 10 ml of 40% solution of acetaldehyde and 10 ml of 0.25% solution of cuprizone. Dilute to 50 ml, let stand for 30 minutes, and read at 540 nm against a blank.

Solder Alloys.[73] Treat a 1-gram sample with 30 ml of hydrochloric acid and a few drops of 30% hydrogen peroxide. Evaporate almost to dryness. Add 80 ml of 2.5 *N* hydrochloric acid in 65% ethanol and filter. Dilute to 100 ml.

Since aluminum must first be isolated, the determination is included at this point. Prepare a 100×11 mm column of 100-200 mesh Diaion SA-100 resin in chloride form. Pass an aliquot containing 0.01–0.02 mg of aluminum through the column, and wash the column with 80 ml of the ethanolic hydrochloric acid. Concentrate the percolate and add 10% sodium hydroxide solution to bring it to pH 6. Add 2 ml of 0.1% solution of chrome azurol S and dilute to 100 ml with an acetate buffer solution for pH 6. Read at 550 nm.

Elute the copper from the column with 50 ml of 1 : 5 hydrochloric acid. Add 10 ml of 10% ammonium tartrate solution and adjust to pH 8.5–9.7 with 30% sodium hydroxide solution. Add 10 ml of a borate buffer solution for pH 9 and 20 ml of 0.04% solution of cuprizone. Dilute to 100 ml and read at 590 nm.

[69] I. R. Capelle, *Chim. Anal.* (Paris) **42**, 69–77, 122–135, 181–190 (1960); cf. C. U. Wetlesen, *Anal. Chim. Acta* **16**, 268–270 (1957); cf. Siegfried Meyer and Othmar G. Koch, *Arch. Eisenhüttenw.* **32** (2), 67–70 (1961).
[70] C. R. Elliott, P. F. Preston, and J. H. Thompson, *Analyst* **84**, 237–239 (1959).
[71] M. Freegarde and Miss B. Jones, *Analyst* **84**, 716–719 (1959).
[72] P. Niviere, *Ann. Falsif. Expert. Chim.* **58**, 311–318 (1965).
[73] Saichiro Onuki, Kunihiko Watanuki, and Yukichi Yoshino, *Jap. Anal.* **15**, 924–928 (1966).

Type Metal or White Metal.[74] Dissolve a sample not heavier than 0.2 gram in 50 ml of 1 : 1 hydrochloric acid. Add a few ml of 30% hydrogen peroxide to expedite solution, and boil to decompose excess peroxide. Dilute to 250 ml and take an aliquot containing about 1 mg of copper. Add 10 ml of 10% ammonium tartrate solution and 10% sodium hydroxide solution to raise to pH 9. Add 10 ml of borate buffer solution for pH 9 and 20 ml of 0.04% solution of the reagent in 10% ethanol. Dilute to 100 ml and read at 590 nm.

Fume and Dust.[75] Collect samples on Millipore AA filter paper and dissolve in 1 ml of 1 : 1 hydrochloric acid. Add 5 ml of buffer solution containing 5% of citric acid in 1 : 4 ammonium hydroxide. Add 2.5 ml of 0.25% solution of cuprizone in 50% ethanol, dilute to 25 ml, and read at 590 nm.

Middle Distillate. As reagent, dissolve 0.2% of cuprizone in anhydrous methanol by warming. As a buffer solution, dissolve 77 grams of ammonium acetate in about 800 ml of water, add 1 : 1 ammonium hydroxide to adjust to pH 9, and dilute to 1 liter.

Transfer a 100-gram sample to a 250-ml glass-stoppered separatory flask. Add 8 ml of the reagent and 2 ml of the buffer. Shake vigorously for 1 minute, venting occasionally to release the pressure developed. Let settle for 15 minutes or centrifuge. Transfer the methanol phase to a 25-ml volumetric flask. Repeat the extraction with the same amounts of reagents as before. Transfer to the volumetric flask and dilute to volume with methanol. Read at 606 nm against reagent blank. Beer's law is followed up to 2 μg of copper per ml. The color is stable for 4 hours.

Because the copper compounds are often adsorbed on the surface of the container, it is recommended that the sample be taken in a suitable unit container, such as a 4-ounce bottle. Then it is recommended that the sample container be rinsed with the reagent–buffer solution mixture before adding it to the sample.

Lubricating Oils. Replace the methanol with *N*-methyl-2-pyrrolidone and proceed as for middle distillate.

Animal and Poultry Feed.[76] Dry 5 grams of a premix or 50 grams of a complete feed for 2 hours at 130°. Stir for 5 minutes with 50 ml of carbon tetrachloride. Centrifuge and decant the clear upper layer. Add another 50 ml of carbon tetrachloride, agitate, and filter. After evaporation of the solvent, boil the solid residue with 50 ml of 1 : 1 hydrochloric acid for 15 minutes. Add 0.5 gram of activated carbon, dilute to 100 ml, and filter. Treat an aliquot as for plants, starting at "Add neutral red indicator and neutralize...."

Plants.[77] Char a sample expected to contain 0.02–0.06 mg of copper and ignite to a white ash at 550–600°. Add 2 ml of nitric acid and 3 ml of hydrochloric acid, and evaporate to dryness at 100°. Add 10 ml of 1 : 5 nitric acid and digest for 30 minutes. Filter, and do not exceed 30 ml in washing. Add neutral red indicator and

[74]British Standards Institution, *BS 3908*, Pt. **4**, 6 pp (1967).
[75]Her Majesty's Factory Inspectorate, *H. M. Stationery Office* 10 pp (1971).
[76]A. Amati and A. Minguzzi, *Ind. Agr.* **8**, 249–257 (1970).
[77]K. R. Middleton, *Analyst* **90**, 234–240 (1965).

neutralize with an alkaline citrate solution containing 20% of sodium citrate dihydrate in 8% sodium hydroxide solution. Add 5 ml of ammonium acetate buffer solution for pH 9. Cool and add 1 ml of a 0.5% solution of the reagent in 50% ethanol. Dilute to 50 ml and read at 600 nm against a reagent blank.

Leaves.[78] Add 5 grams of ammonium citrate and 60 ml of ammonium hydroxide to water and dilute to 500 ml. As a borate buffer solution, dissolve 12.4 grams of boric acid in 400 ml of water, and add 1.2 grams of sodium hydroxide in 60 ml of water. As the reagent, prepare a 0.1% solution of cuprizone in 50% ethanol and dilute it 2:5 with water before use.

For up to 0.05 mg of copper, immerse the leaf for 15 minutes in 10 ml of 1:35 sulfuric acid containing Lissapol DNB as a wetting agent, swirling at 5-minute intervals. To the extract, add 5 ml of the ammoniacal ammonium citrate solution, 5 ml of the borate buffer solution, and 10 ml of the diluted reagent solution. Swirl and read at 595 nm.

CUPROINE

Cuproine is also called 2,2′-biquinolyl and 2,2′-biquinoline. The resulting solution with the copper reduced to the cuprous form can be read directly, or the color can be concentrated by extraction with pentanol.[79] The complex of cuprous ion in spirits with this reagent can be developed without ashing. In the analysis of multivitamin preparations, the complex is extracted with amyl alcohol.[80] The reagent forms a ternary complex with cuprous ion and bromophenol blue.

For analysis of gold by this reagent after solution in aqua regia, the chloraurate complex may be removed by extraction with tributyl phosphate from 3.6 *N* hydrochloric acid, or the gold may be adsorbed on an anion exchange resin from 0.6 *N* hydrochloric acid. Either method determines 0.5 ppm or more in the solution freed from gold.[81]

The reagent is standard in the United Kingdom for analysis of nickel[82] for electronic tubes and valves and is also applicable to beryllium.[83] It is applicable to a solution of silicates in sulfuric and hydrofluoric acid, as well. In that case the aliquot of the sample solution should contain less than 0.08 mg of copper.[84]

For copper in milk with this reagent, the proteins are precipitated with trichloroacetic acid, carrying down the fatty matter, and copper is determined in the resulting serum.[85]

For preparation of a sample of marine sediments for determination of copper by 2,2′-biquinolyl, see Lead.

[78]R. B. Sharp, *ibid.* **91**, 212–213 (1966).
[79]N. A. Marchenko, Z. S. Raiber, and S. K. Lipko, *Zavod. Lab.* **28**, 1192 (1962).
[80]Lilo E. Guerello, *Proanalisia* **1970** (6), 29–33.
[81]T. W. Steele and A. Hyde, *J. S. Afri. Inst. Mining Met.* **62**, Pt. 2, 764–772 (1962).
[82]British Standards Institution BS **3727**: Pt. 6, 7 pp (1964).
[83]*UKAERE* **AM 21**, 2 pp (1959).
[84]Heinrich M. Koester, *Ber. Deut. Keram, Ges.* **46**, 247–253 (1969).
[85]A. de Vleeschauer, A. van den Baere, and M. Naudts, *Meded. Landbouwhogesch. Opzoekingssta. Staat Gent* **23** (1), 55–63 (1958).

Procedures

Iron and Steel.[86] Treat a 0.5-gram sample with 10 ml of hydrochloric acid and 5 ml of nitric acid. If tungsten is present filter off the tungsten oxide. If the sample is high in chromium, first dissolve it in hydrochloric acid and later add the nitric acid as an oxidizing agent. In any case add 10 ml of perchloric acid, evaporate to fumes, and continue heating for 3 minutes. Add 10 ml of hydrochloric acid and boil off liberated chlorine. If the sample contains up to 0.6% of copper, dilute to 100 ml, for 0.6–1.5% of copper to 250 ml, or for 1.5–4% of copper to 500 ml. Filter off graphite or silica.

Take as an aliquot 10 ml for a copper content under 0.3%; otherwise, use 5 ml. Add 5 ml of 20% ascorbic acid solution and 25 ml of 0.06% solution of cuproine in dimethylformamide. Dilute to 50 ml and read at 545 nm.

Aluminum, Magnesium, Zinc, etc.[87] Dissolve a 0.5-gram sample in 10 ml of hydrochloric acid and 5 ml of nitric acid. Evaporate to dryness and take up in 5 ml of 1:5 hydrochloric acid. Add 5 ml of 10% tartaric acid solution and 5 ml of 10% hydroxylamine hydrochloride solution. Dilute to about 60 ml and neutralize to pH 5–6 with 30% sodium hydroxide solution. Dilute to 100 ml. Shake an appropriate aliquot, usually 50 ml, with 6 ml and 6 ml of 0.02% solution of cuproine in isoamyl alcohol for 10 minutes. Read the combined layers at 545 nm, subtracting a blank.

Plutonium Solution.[88] Adjust the pH of a solution containing 5–12 μg of copper by adding sufficient 10% sodium hydroxide solution to produce a precipitate; then add sufficient nitric acid to just dissolve the precipitate. Add 2 ml of 10% hydroxylamine hydrochloride solution and 5 ml of saturated solution of sodium acetate. Extract with 5 ml of 0.1% solution of cuproine in pentanol. Dry the extract by filtering through paper and read at 550 nm. The technic will detect 2 μg of copper.

Ammoniacal Plating Solution. Dilute the sample to about 50 mg of copper per liter. Mix 5 ml of the diluted sample with 5 ml of 40% hydroxylamine hydrochloride solution and 5 ml of 50% tartaric acid solution. Add 5 ml of an acetate buffer solution for pH 5–6 and extract with 20 ml of 0.02% solution of cuproine in isopentyl alcohol. Read the extract at 545 nm.

Sodium Chloride.[89] Dissolve a sample of about 60 grams in 200 ml of water.

[86]British Standards Institution BS **1121**, Pt. **36**, 7 pp (1956); C. E. A. Shanahan and R. H. Jenkins, *Analyst* **86**, 166–167 (1961); H. J. Graebner, *Z. Anal. Chem.* **182**, 401–408 (1961); T. S. Harrison and W. W. Foster, *Metallurgia* **73**, 141–146 (1966); Heinrich M. Koester, *Ber. Deut. Keram. Ges.* **46**, 247–253 (1969); A. Funke and H. J. Laukner, *Z. Anal. Chem.* **249**, 26–30 (1970); Detlef Thierig, *Arch. Eisenhüttw.* **43**, 475–477 (1972).
[87]A. L. Gershuns and Yu. V. Bashkevich, *Zavod. Lab.* **23**, 787–788 (1957); cf. Friedrich Oehlmann, *Chem. Tech.* (Berlin) **9**, 599–600 (1957); A. J. Leeband F. Hecht, *Z. Anal. Chem.* **168**, 101–106 (1959).
[88]*UKAEA* Rep. PG 651 (W), 6 pp (1965).
[89]H. Bahr and I. Lipinska, *Chem. Anal.* (Warsaw) **13**, 399–403 (1968).

Add 2 ml of 10% tartaric acid solution and 2 ml of 10% hydroxylamine hydrochloride solution. Add 10 ml of saturated solution of sodium acetate, 40 ml of ethanol, and 10 ml of 0.1% solution of cuproine in freshly distilled chloroform. Shake for 3 minutes, filter the chloroform through paper, and read at 550 nm.

Cement.[90] Heat a 1-gram sample with 1 gram of ammonium chloride and 10 ml of hydrochloric acid at 100° for 30 minutes. Add 20 ml of hot water and filter. Dilute to a known volume and take an aliquot. Add 5 ml of 10% solution of hydroxylamine hydrochloride and 6 ml of 10% tartaric acid solution. Add ammonium hydroxide to pH 5.5 and dilute to 50 ml. Shake for 2 minutes with 10 ml of isoamyl alcohol containing 0.8 mg of cuproine and read the extract at 550 nm.

Seawater.[91] To 900 ml of filtered sample, add 5 ml of 25% solution of hydroxylamine hydrochloride and 10 ml of 13.6% solution of sodium acetate trihydrate. Extract with 6 ml of 0.03% solution of cuproine in hexanol. Add 2 ml of the 25% solution of hydroxylamine hydrochloride, and extract, with an additional 3 ml of the cuproine solution. To the combined extracts, add 0.5 ml of 1% solution of hydroquinone in ethanol and dilute to 10 ml with hexanol. Read at 540 nm.

Sodium Hydroxide and Sodium Hypochlorite Solutions.[92] Neutralize a sample containing about 5 μg of copper with hydrochloric acid and evaporate to dryness. Take up in water and dilute to 50 ml. To a 20-ml aliquot add 10 ml of 10% solution of sodium potassium tartrate made alkaline to litmus with ammonium hydroxide. Add 10 ml of 10% hydroxylamine hydrochloride solution and neutralize to litmus with ammonium hydroxide. Add 40 ml of 0.025% solution of cuproine in butanol and shake. Read the butanol layer at 450 nm against a blank.

Spirits.[93] Shake 10 ml of sample with either 1 ml of 2.5% cysteine hydrochloride solution in 0.5 *N* hydrochloric acid or with 1 ml of 10% solution of hydroxylamine hydrochloride. After allowing to stand for 20 minutes, add 10 ml of 0.02% ethanolic cuproine, set aside for 10 minutes and read at 546 nm against a blank.

Brandy.[94] Evaporate a 10-ml sample to about 3 ml and dilute to 10 ml with water. Add 1 ml of 2.5% solution of cysteine in 0.5 *N* hydrochloric acid and 10 ml of a solution of 2 mg of cuproine in 10 ml of pentanol. Shake for 1 minute, centrifuge, filter the pentanol layer, and read at 550 nm, against a blank.

Fat.[95] Mix a 3-gram sample with 10 ml of 1:6 sulfuric acid and simmer for 5 minutes. For margarine containing emulsifying agents, double the concentration of the sulfuric acid. Add 5 ml of water, extract with 20 ml of chloroform and discard the chloroform. Extract the aqueous acid phase with 5 ml and 5 ml of a 0.1% solution of diethylammonium diethyldithiocarbamate. Combine these extracts and

[90] Hajime Ishii, Hisahiko Einaga, and Eiichi Aoki, *Jap. Anal.* **15**, 825–829 (1966).
[91] J. P. Riley and P. Sinhaseni, *Analyst* **83**, 299–300 (1958).
[92] F. Mains and R. E. Ragett, *Chemist-Analyst* **50**, 4–6 (1961).
[93] E. Szobolotzky, *J. Inst. Brew.* **76**, 245–249 (1970).
[94] V. Chioffi and G. Osti, *Boll. Lab. Chim. Prov.* **16**, 633–639 (1965).
[95] R. Gaigl, *Z. Lebensm. Unters.* **119**, 506–509 (1963).

back-extract the copper with 5 ml of 5:1:4 nitric acid–30% hydrogen peroxide–water. Evaporate this acid extract to dryness and take up the residue in 10 ml of a buffer solution containing 10% of sodium potassium tartrate tetrahydrate, 5% of sodium acetate trihydrate, and 3% of hydroxylamine hydrochloride. Extract the copper with 5 ml and 5 ml of 0.02% solution of cuproine in pentanol. Read the combined extracts at 545 nm against pentanol. The graph is straight for 0.5–25 mg of copper, is specific for copper, and will detect 0.3 μg.

With Bromophenol Blue.[96] Mix a sample solution containing 4–16 μg of copper with 5 ml of 10% tartaric acid solution and 5 ml of 10% hydroxylamine hydrochloride solution. Adjust to pH 5.7–5.9 with 1:1 ammonium hydroxide and dilute to 50 ml. Shake with 10 ml of 0.05% solution of cuproine in isoamyl alcohol. Separate the organic phase and wash it with 5 ml of ammoniacal ammonium chloride buffer solution for pH 8.8. Shake with 10 ml of 50 m*M* bromophenol blue; then wash with water and centrifuge. Read the ternary complex in the clear organic layer at 605 nm against isoamyl alcohol.

DIBUTYLAMMONIUM CARBONATOCUPRATE

After copper forms polyoxyethylene glycol saccharate, it is converted to the titled compound.[97] Beer's law is obeyed for 1–10 mg of copper per 100 ml. Silver and mercury must be absent.

Procedure. Prepare dibutyl ammonium carbonate solution by passing carbon dioxide through a cooled mixture of 98:48:4 by volume dibutylamine-dichloromethane-methanol until saturated.

To 50 ml of slightly acid sample, add ammonium acetate to adjust to pH 3.5 Add 5 ml of 20% solution of polyoxyethylene glycol, 15 ml of 5% solution of sodium saccharate, and 15 ml of 10% solution of hydroxylamine hydrochloride. The latter may need to be increased in the presence of a large amount of iron. After 5 minutes extract with 7 ml and 7 ml of dichloromethane. Add an excess of 1% solution of bromine in dichloromethane. Add 2 ml of the dibutylammonium carbonate solution, adjust to a known volume and read at 575 nm.

DICUPRAL

Dicupral is the trivial name for **tetraethylthiuram disulfide,** which though not structurally a dithiocarbamate in reacting with cupric ion, forms the corresponding diethyldithiocarbamate.[98] The maximum extinction of the copper complex is developed in 7:3 ethanol-water adjusted to 0.1 *M* with perchloric acid and maintained at 50° for 1 hour.

[96] Keiichi Sekine and Hiroshi Onishi, *Anal. Lett.* **7**, 187–194 (1974).
[97] M. Ziegler and J. Holland, *Z. Anal. Chem.* **194**, 249–255 (1963).
[98] K. Lesz and T. Lipiec, *Chem. Anal.* (Warsaw) **11**, 523–529 (1966).

As applied to water analysis,[99] the sensitivity of this reagent is equal to that obtained by oxalyl dihydrazide and better than that by neocuproin, bathocuproin, or sodium diethyldithiocarbamate.

The maximum intensity is reached in 15 minutes and is stable for 3 hours. Beer's law is followed for 0.04–2.5 ppm.[100]

For analysis of blood serum, mix 2 ml of sample with 2 ml of 20% trichloroacetic acid solution and 1 ml of 10% ascorbic acid solution.[101] Centrifuge and develop by addition of dicupral in glacial acetic acid, thereby avoiding development of turbidity.

The reagent is applicable to the digestate of serum, urine, and brain sections.[102] After decomposition with perchloric acid, dilute to 2.2–6.7 *N* with that acid. Add 0.04% solution of the reagent in 80% acetone. Read at 422 nm after 15 minutes and within 3 hours.

The reagent is applicable by ether extraction to wine, syrups, juices, ciders, and so on, without wet-ashing.[103] It is suitable for steels and iron ore.[104]

For isolation of copper from water by Dowex 50 W-X2 for determination by dicupral, see Lead Diethyldithiocarbamate in this chapter.

Procedures

Zinc.[105] Dissolve a 1-gram sample containing 0.005–0.05% of copper in 5 ml of hydrochloric acid, adding 30% hydrogen peroxide dropwise during the operation. Boil for 2 minutes to decompose excess hydrogen peroxide. When cool, add 2 ml of 1% ascorbic acid solution and dilute to 100 ml. Add 20 ml of ethanol to a 10-ml aliquot. Add 5 ml of 0.3% solution of dicupral in ethanol and dilute to 50 ml with ethanol. After 5 minutes read at 420 nm against a blank.

Carbonate Minerals.[106] Dissolve a 5-gram sample in a minimal amount of nitric acid. The excess must not exceed 2 ml in 50 ml of solution. Boil off carbon dioxide and dilute to either 50 or 100 ml. To a 10-ml aliquot add 3 ml of ethanol, and 3 ml of 0.01 *M* ethanolic dicupral. Dilute to 50 ml with ethanol and read at 435 nm.

Water. To 100 ml of sample add 5 ml of hydrochloric acid and evaporate to about 20 ml. Add 3 ml of 5% potassium fluoride solution to mask iron and 1 ml of 5% solution of disodium EDTA. Dilute to 50 ml and add 40 ml of ethanol. Add 3 ml of 0.3% ethanolic dicupral and dilute to 100 ml with ethanol. After 15 minutes read at 435 nm.

Liquid Foods. Mix 10 ml of sample solution containing not more than 0.02 mg

[99]A. Bilikova and J. Zyka, *Chem. Listy* **59**, 91–94 (1965).
[100]Yasuo Matsuba, *J. Jap. Biochem. Soc.* **37**, 351–359 (1965).
[101]Ruth Watkins, Lawrence M. Wiener, and Bennie Zak, *Microchem. J.* **16**, 14–23 (1971); Ruth Watkins and Bennie Zak, *ibid.* 300–310 (1971).
[102]Y. Matsuba and Y. Takahashi, *Anal. Biochem.* **36**, 182–191 (1970).
[103]H. Novozamska and J. Zyka, *Prum. Potravin.* **15**, 520–521 (1964).
[104]Sergej Gomiscek, *Rud.-Met. Zb.* (1), 51–62 (1959).
[105]J. Balcarek, *Hutn. Listy* **23**, 208–209 (1968).
[106]Miroslav Vesely and Zdenek Sulcek, *Chem. Listy* **52**, 2010–2012 (1958).

of copper with 1 ml of 0.01 *M* ethanolic dicupral. After 20 minutes, shake with 20 ml of ether. Read the ether layer at 445 nm.

4,4′-DIHYDROXY-2,2′-BIQUINOLINE

This reagent[107] forms a 2 : 1 complex with copper in strongly alkaline solution. The complex can be read directly, but greater sensitivity can be obtained by extraction with isoamyl alcohol. Many ions precipitate from the strongly alkaline solution. Even 1 ppm of cyanide interferes seriously. Tartrate can be used to mask iron and some other ions.

Procedures

Sodium. As a 0.004 *M* solution of the reagent, add 3 drops of 24% sodium hydroxide solution to 0.115 gram of the reagent, take up in ethanol, and dilute to 100 ml with that solvent.

Dissolve a clean 9-gram sample in methanol. Evaporate to a thick paste and slowly add 25 ml of water. Boil off the residual methanol, cool, and dilute to 50 ml. Mix a sample expected to contain 0.005–0.1 mg of copper with 2 ml of 10% solution of hydroxylamine hydrochloride and 2 ml of 0.004 *M* solution of the reagent. Extract with 5 ml and 1 ml of isoamyl alcohol. To the combined extracts, add 1 ml of 10% hydroxylamine hydrochloride solution in 50% ethanol and dilute to 10 ml with ethanol. Read against a blank at 525 nm.

As an alternative, the extraction may be bypassed. To the sample containing 0.005–0.1 mg of copper, add 1 ml of 10% hydroxylamine hydrochloride solution and 2 ml of the reagent solution. Dilute to 10 ml with 24% sodium hydroxide solution and read at 528 nm against a blank.

Potassium. This requires additional precautions in preparation of the sample. Chill 100 ml of methanol to 0° in an ice bath. Add 16 grams of potassium as pea-sized chunks, one at a time. Continue as for sodium from "Evaporate to a thick paste."

6,7-DIMETHYL-2,3-DI-(2-PYRIDYL)QUINOXALINE

Copper forms a reddish-brown 1 : 2 complex with this reagent.[108] By extracting the complex into nitrobenzene, less than 2% error in determining 0.3 mg of copper is caused by 10 mg of vanadate ion, 20 mg of ceric ion, 3 mg of iodide, or 40 mg of aluminum.

The binary complex adds methyl orange to give a ternary complex in the ratio 1 : 2 : 1.[109a] Perchlorate must be absent.

[107] Alfred A. Schilt and William C. Hoyle, *Anal. Chem.* **41**, 344–347 (1969).
[108] F. C. Trusell and W. F. McKenzie, *Anal. Chim. Acta* **40**, 350–351 (1968).
[109a] David Foster and Fred Trusell, *Anal. Chim. Acta* **47**, 154–155 (1969).

Procedures

Binary Complex. Mix an acid solution of sample containing 0.05–0.5 mg of copper with 20 ml of 10% solution of hydroxylamine hydrochloride and 5 ml of a solution containing 15% of tartaric acid and 15% of sodium chlorate. Add 2 ml of a saturated solution of the reagent in ethanol and add 10% sodium hydroxide solution until the pH is raised to 4. Extract with 15 ml of nitrobenzene and dilute the extract with ethanol to 50 ml. Read at 514 nm.

Ternary Complex. Dissolve a sample containing 0.01–0.1 mg of copper in an appropriate acid. If nitric or perchloric acid is used, add 3 ml of sulfuric acid and heat to sulfur trioxide fumes. Cool, dilute to about 40 ml, and add 40% sodium hydroxide to about pH 3. Cool, and add 20 ml of 10% hydroxylamine hydrochloride solution, 5 ml of a 15% solution of tartaric acid, and 5 ml of 0.001 *M* methyl orange solution. Adjust to pH 5 and add 2 ml of 1.25% solution of the reagent in 1,2-dichloroethane. Extract with 15 ml and 15 ml of 1,2-dichloroethane, dilute the combined extracts to 50 ml with ethanol, and read at 418 nm against a reagent blank.

4-(4, 5-DIMETHYL-2-THIAZOLYLAZO)-2-METHYL-RESORCINOL

The complex of cupric ion with this reagent obeys Beer's law for 0.02–0.25 ppm of copper. Aluminum is masked with fluoride ion and ferric ion with tartrate.

Procedure.[109b] Mix the neutral sample solution with 10 ml of 0.1 m*M* methanolic reagent, 3 ml of methanol, and 2 ml of acetate buffer solution for pH 5.2. Dilute to 25 ml and read at 550 nm.

1,4-DIMETHYL-1,2,4-TRIAZOLIUM-(3-AZO-4)-*N,N*-DIETHYLANILINE

This reagent, which is 3-(4-diethylaminophenylazo)-1,4-dimethyl-1,2,4-triazolium, complexes with cupric iodide at pH 1–7.[110] As extracted by chlorobenzene, Beer's law is followed for 0.1–4 μg of copper per ml. There is interference by an equal amount of auric ion, stannous ion, cadmium, or bismuth and by tenfold amounts of selenic or mercuric ion.

Procedures

Ferrous Ammonium Sulfate. Dissolve 1 gram of sample in the minimum volume of water. Add 0.4 ml of 16.6% solution of potassium iodide and 0.4 ml of 5 m*M* solution of the captioned reagent. Dilute to 10 ml with acetate buffer solution for

[109b] C. Sanchez-Pedreno, V. Gonzalez Diaz, and J. J. Arias, *An. Quim.* **73**, 71–78 (1977).
[110] P. P. Kish, A. I. Busev, and I. I. Pogoida, *Zavod. Lab.* **39**, 1302–1304 (1973).

pH 3 and extract with 6 ml of chlorobenzene. Read the organic layer at 554 nm against a reagent blank.

Aluminum Alloys. Dissolve 0.1 gram of sample in 5 ml of 1:1 sulfuric acid and 0.1 ml of nitric acid. Evaporate nearly to dryness, add 5 ml of water, and again evaporate nearly to dryness. Dissolve in water and dilute to 100 ml. Dilute 1 ml to 5 ml. Proceed as for ferrous aluminum sulfate from "Add 0.4 ml of 16.6% solution...."

---

sym–DIPHENYLCARBAZIDE

This reagent, which is 1,5-diphenylcarbohydrazide, in basic solution forms a 2:1 combination with cupric ion, which is extremely sensitive.[111] The technic is applicable up to 12 μg per ml of extract. It conforms to Beer's law for 0.01–0.26 ppm. Precise buffering is essential. The agent in ethanol is only stable for an hour due to air oxidation. The rate of color development increases with temperature. Excess reagent increases the rate of reaction, but large excesses can lead to erratic results. Chromium interferes at the same concentration as the copper, ceric ion at 1 ppm. Addition of 500 ppm of citrate prevents interference by precipitation of ferric, manganese, and cadmium ions. Mercuric ion at 10 ppm is masked by iodide. Molybdenum can be masked with oxalate. Such organic materials as purines and pyrimidines inhibit the reaction. The color is extractable by benzene or isoamyl alcohol.

Procedure. Mix 1 ml of sample solution containing 0.01–0.26 μg of copper with 1 ml of a buffer solution for pH 9. Dilute to 3 ml and add 0.2 ml of a 0.1% ethanolic solution of diphenylcarbazide. Read at 495 nm from 4–7 minutes later, and correct for a blank.

Boiler Water. As a buffer for pH 9.1, mix 50 ml of 0.2*M* boric acid in 0.2 *M* potassium chloride with 24 ml of 0.2 *M* sodium hydroxide and dilute to 200 ml. To 25 ml of sample, add 1.5 ml of 1:2 nitric acid. Add 20% sodium hydroxide solution to approximately pH 9.1. Add 5 ml of buffer solution and 6 ml of 0.2% solution of *sym*-diphenylcarbazide in ethanol containing 0.02% of ascorbic acid. Dilute to 50 ml, heat at 30° for 10 minutes, cool, and read at 500 nm. The presence of ascorbic acid stabilizes the reagent for 2 days.

Whole Blood, Serum, or Tissue. To 1 ml of blood or serum, or tissue equivalent to about 0.4 gram of dry solids, add 1 ml of nitric acid. Heat and later add 1 ml of 1:1 nitric-perchloric acid. Evaporate to dryness. Take up the residue in 0.5 ml of 8% solution of sodium citrate dihydrate and neutralize to phenolphthalein with 4% sodium hydroxide solution. Add 10 ml of 25% sodium carbonate solution and 6 ml of a solution of 20 mg of *sym*-diphenylcarbazide per 100 ml of benzene. Stopper the

[111]Roger W. Turkington and Frederick M. Tracy, *Anal. Chem.* **30**, 1699–1701 (1958); Miroslav Makva and Marta Janockova, *Chem. Prum.* **11**, 139–140 (1961); Shinji Sato and Kaoru Sakai, *J. Chem. Soc. Jap., Ind. Chem. Sect.* **64**, 597–600 (1961); D. Mikac-Devic, *Clin. Chim. Acta* **7**, 788–793 (1962); R. E. Stoner and W. Dasler, *Clin. Chem.* **10**, 845–852 (1964).

tube and invert 200 times in 2 minutes. Read the benzene extract at 540 nm against a reagent blank.

Serum.[112a] As a buffer, mix 88.5 ml of 0.1 *N* hydrochloric acid with 10.3 grams of barbitone sodium in 500 ml of water and dilute to 1 liter.

Mix 0.2 ml of sample with 0.3 ml of 3 : 1 : 1 nitric-perchloric-sulfuric acid. Heat at a low temperature, then raise to 260° until white fumes disappear. Take up in 1 ml of water and neutralize to thymol blue with 1 : 10 ammonium hydroxide. Mix with 5 ml of the buffer solution and 0.2 ml of 1.5% solution of *sym*-diphenylcarbazide in ethanol. After 15 minutes shake for 2 minutes with 5 ml of chloroform and read the extract at 550 nm against a reagent blank.

Urine. Mix 6 ml with 0.4 ml of the 3 : 1 : 1 acid mixture and complete as for serum.

N, N'-DIPHENYLTHIOCARBAMOYLHYDROXAMIC ACID

This reagent is 1-hydroxy-1, 3-diphenylthiourea. Both copper and molybdenum react but copper can be masked with thiourea.

Procedure. *Steel.*[112b] Dissolve the sample by heating with 25 ml of 1 : 4 sulfuric acid. When reaction ceases add 2 ml of nitric acid and continue heating to decompose carbides. Add 1 ml more of nitric acid and evaporate to sulfur trioxide fumes. Dissolve the salts in 30 ml of water at 100°. Cool and adjust the pH to 2–3 with 20% sodium hydroxide solution.

To an aliquot containing less than 0.1 mg of copper and molybdenum, add 0.5 gram of ascorbic acid. Dilute to 100 ml. Extract molybdenum and copper from a 50-ml aliquot with the reagent in chloroform. To the other 50-ml aliquot add thiourea to mask copper and extract molybdenum with the reagent in chloroform. Determine copper by difference.

1,5-DIPHENYLCARBAZONE

Diphenylcarbazone is as sensitive as dithizone and less subject to interferences, at least as applied to soil extracts.[113] The red complex with cupric ion is extractable with benzene.[114] Iron is masked as an ammonium or phosphate complex. The sensitivity to 0.2–2 μg of copper makes the reagent applicable to samples of 50–300 mg of food products.[115]

[112a] D. Mikac-Devic, *Clin. Chim. Acta* **26**, 127–130 (1969).
[112b] V. P. Maklakova, *Trudy Inst. Khim. ural.' nauch. Tsentr Akad. Nauk SSSR* **1974** (27), 113–115.
[113] H. R. Geering and J. F. Hodgson, *J. Assoc. Off. Anal. Chem.* **49**, 1057–1060 (1966).
[114] L. N. Lapin and N. V. Reis, *Zh. Anal. Khim.* **13**, 426–429 (1958); L. N. Lapin and M. A. Rish, *Tr. Tadzh. Uchit. Inst.* **4**, 71–78 (1957).
[115] L. N. Lapin and I. G. Priev, *Vopr. Pitan.* **17**, No. 1, 68–72 (1958).

For analysis of cheese or milk, digest with sulfuric acid and develop the solution at pH 6.5–8.5 with 1,5-diphenylcarbazone.[116] Extract the violet complex with benzene for reading at 0.4–2.5 μg of copper per ml in the extract at 550 nm.

Procedure. *Soil Solution.* To 50 ml of sample solution, add 1 ml of 1:1 hydrochloric acid and 10 ml of 0.025% solution of dithizone in carbon tetrachloride. Shake for 10 minutes and drain the extract. Wash the aqueous phase with 5-ml portions of carbon tetrachloride until no more green color is extracted. Evaporate the combined extracts at 100°. Add 2 ml of perchloric acid to the residue and take to dryness. Take up the residue in 5 ml of 0.01 *N* hydrochloric acid. Dilute to 25 ml and take a 20-ml aliquot. Add a drop of phenolphthalein indicator solution and adjust to a faint pink with 1:75 ammonium hydroxide. Add 5 ml of 13.6% solution of monopotassium phosphate and 5 ml of 0.0036% solution of 1,5-diphenylcarbazone in benzene. Shake for 10 minutes, centrifuge, and discard the aqueous phase. Read the benzene layer at 550 nm against benzene.

O,O-DIPHENYLPHOSPHORODITHIOATE

Copper forms a 1:2 complex with the reagent above, extractable with carbon tetrachloride or dichloroethane.[117]

Procedure. To a neutral sample solution containing 0.002–0.025 mg of copper, add 5 ml of 5 *N* sulfuric acid and dilute to 25 ml. Add 10 ml of dichloroethane and a threefold excess of 0.3 m*M* reagent. Shake for 2 minutes, and separate the organic layer. Extract with 6 ml and 6 ml of the solvent. Filter the combined extracts, dilute to 25 ml with solvent, and read at 420 nm against the solvent.

DIPYROPHOSPHATOCUPRATE

The dipyrophosphatocuprate at 3–10 ppm is read in the ultraviolet range at 241 nm.[118] The di- compound predominates at pH 7–10. The solution conforms to Beer's law at 5–10 ppm. At least 0.005 *M* pyrophosphate is essential, and 0.02 *M* is desirable. No change in absorption occurs in 2 weeks. There is serious interference by ferrous or ferric ion, auric and stannous ions, lead, chromic ion, nitrate, and EDTA. Ammonia must be absent because mixed ammine complexes are formed.

Procedure. Adjust a sample containing not more than 0.6 mg of cupric ion to pH

[116] Friedrich Kiermeier and Guenther Weiss, *Z. Lebensm.-Unters. Forsch.* **141**, 150–158 (1969).
[117] A. I. Busev and A. N. Shishkov, *Zh. Anal. Khim.* **23**, 181–185 (1968).
[118] Mary Louise Nebel and D. F. Boltz, *Anal. Chem.* **36**, 144–146 (1964).

8–10 and dilute to 25 ml. Add 10 ml of 0.1 *M* sodium pyrophosphate solution and dilute to 50 ml. Read at 241 nm against a reagent blank.

8,8′-DIQUINOLYLDISULFIDE

This reagent reacts with cuprous ion to give the cuprous complex of 8-mercaptoquinoline.[119] The procedure is applicable to 0.2–8 μg of copper per ml. There is interference by gold, stannous, titanous, thiocyanate, and cyanide ions. Ascorbic acid slowly reduces the reagent.

Procedures

Aluminum.[120] This provides for sequential determination of copper, iron and manganese in the same sample.

Dissolve 1 gram of sample in 30 ml of 1 : 1 hydrochloric acid. Add a few drops of nitric acid, boil for 2 minutes, and filter. Add 8 grams of tartaric acid to the filtrate, raise to pH 5 with ammonium hydroxide, and dilute to 100 ml. Take a 25-ml aliquot and add 5 ml of 20% ascorbic acid. Raise the pH to 5 with ammonium hydroxide and add 10 ml of 0.2% solution of 8,8′-diquinolyldisulfide in chloroform. Shake for 2 minutes and filter the extract through cotton. Read copper at 432 nm.

To the aqueous phase, add 2 ml of 0.3% solution of 8-mercaptoquinoline and adjust to pH 4. Extract the resulting complex with 10 ml and 10 ml of chloroform. Filter the combined extracts through cotton and read iron with a blue filter.

To the remaining aqueous phase, add 2 ml of 0.3% solution of 8-mercaptoquinoline and adjust to pH 9 with ammonium hydroxide. Extract the manganese complex by shaking with 10 ml of chloroform for 1 minute. Read the extract with a blue filter.

Zinc. Dissolve a 1-gram sample in 20 ml of 1 : 1 hydrochloric acid. Add 2 ml of 30% hydrogen peroxide and boil to destroy excess peroxide. Evaporate to incipient crystallization. Take up in hot water containing a few drops of 1 : 1 hydrochloric acid and dilute to 100 ml. To a 10-ml aliquot add ammonium hydroxide to adjust to pH 5. Add 5 ml of 20% ascorbic acid solution to reduce to cuprous ion followed by 10 ml of a chloroform solution of 8,8′–diquinolyldisulfide containing 0.4 gram per 130 ml. Shake for 2 minutes, filter through cotton, and read at 432 nm against a blank.

Blood, Tissue. As a reagent mixture, combine 10 ml of 20% ascorbic acid

[119] J. Bankovskis, A. Ievins, P. Bockaus, and G. Dambite, *Latv. PSR Zinat. Akad. Vest., Kim. Ser.* **1961** (1), 47–53; J. Bankovskis, A. Ievins, E. Luksa, and P. Bockaus, *Zh. Anal. Khim.* **16**, 150–157 (1961); M. Z. Kharkover, L. S. Studenskaya, and L. P. Korshunova, *Tr. Vses. Nauchn.-Issled. Inst. Stand. Obraztsov.* **3**, 71–75 (1967).

[120] M. Z. Kharkover and V. F. Barkovskii, *Tr. Vses. Nauchn.-Issled. Inst. Stand. Obraztsov.* **3**, 77–81 (1967).

solution, 30 ml of 50% solution of sodium acetate 5 ml of 25% solution of sodium tartrate, 7 ml of 50% solution of urea, and 50 ml of water.

To 1 ml of blood or 0.5 gram of tissue, add 1 ml of an 80% solution of magnesium nitrate, 5 drops of sulfuric acid, 0.5 ml of fuming nitric acid, and 1 ml of 30% hydrogen peroxide solution. Evaporate almost to dryness, cool, and add 5 drops of fuming nitric acid and 5 drops of 30% hydrogen peroxide. Repeat the previous step and evaporate almost to dryness. Take up the residue in 2 ml of water and transfer to a separatory funnel. Wash in with four successive 2.5-ml portions of the mixed reagent. If necessary add more 50% solution of sodium acetate to adjust to pH 4–5. Extract with 3 ml of a 0.2% solution of 8,8′-diquinolyldisulfide in chloroform, filter, and read at 432 nm.

DISODIUM ETHYL BIS(5-TETRAZOLYLAZO)ACETATE TRIHYDRATE

The reagent above forms a colored 1 : 1 complex with cupric ion and also a colored 2 : 1 complex with nickel.[121] Copper has peaks at 268, 300, and 535 nm; nickel at 335 and 505 nm. Colored complexes are also formed with silver, ferrous and ferric ions, manganese, zinc, cadmium, chromium, and aluminum. The reagent has peaks at 270 and 410 nm. Only ferrous ion and cobalt interfere with the copper peak at 535 nm. That of ferrous ion is eliminated by oxidation to ferric ion. Cobalt interference can be eliminated by a method of successive substitution. The pH must be 5–8. The reagent will detect 0.2 ppm of copper.

For nickel reading at 550 nm is preferable because the absorption by the reagent at that wavelength is so small that it can be neglected. Only copper, zinc, and cadmium interfere. Common anions do not interfere. The pH must be 6–9 for nickel.

Procedure. For preparation of the reagent, add 28.4 ml of 20% sodium nitrite solution to a chilled solution of 9.7 grams of 5-aminotetrazole monohydrate in 40 ml of 20% sodium hydroxide solution. Add crushed ice. Prepare a mixture of 28.4 ml of hydrochloric acid, 120 ml of water, and 200 grams of crushed ice. Dump the first mixture into the second all at once with vigorous stirring. After 20 minutes add a chilled solution of 50 grams of sodium acetate trihydrate in 100 ml of water. Then at once add 13 ml of ethyl acetoacetate. Maintain the intensely red reaction mixture at 0–5° while stirring for 2 hours. Near the end of the 2 hours, dissolve 100 grams of sodium chloride in the mixture. In a refrigerator at 5° the product crystallizes slowly over a period of 2 days. Filter the orange-red product by suction. The product can be crystallized slowly at 5° from a large volume of ethanol.

Dilute an acid solution of 0.1 gram of sample to 50 ml and filter. Adjust to pH 6 and dilute to 100 ml. Mix a 1-ml aliquot with 5 ml of 0.0001 *M* reagent and read at 535 nm for copper. For nickel, read at 550 nm.

---

[121] Hans B. Jonassen, Virginia C. Chamblin, Vernon L. Wagner, Jr., and Ronald A. Henry, *Anal. Chem.* **30**, 1660–1663 (1958).

DITHIOCARBAMATES

Because several dithiocarbamates are significant reagents for copper, they are grouped into a subchapter. Thus these follow successively as reagents, followed by some as minor reactants.

Sodium diethyldithiocarbamate
Diethylammonium diethyldithiocarbamate
Bis(carboxymethyl)dithiocarbamate
Tetramethylenedithiocarbamate
Lead diethyldithiocarbamate
Zinc diethyldithiocarbamate
Antimony dibenzyldithiocarbamate
Zinc dibenzyldithiocarbamate
Cuprethol[bis-(2-hydroxyethyl)dithiocarbamic acid]

Sodium Diethyldithiocarbamate

Copper diethyldithiocarbamate can be read in ammonium citrate solution.[122] Nickel gives an interfering greenish yellow, and lead causes a precipitate.[123] After complexing nickel and lead with EDTA, the carbamate complex can be extracted with chloroform, carbon tetrachloride,[124] butyl acetate,[125] pentyl acetate,[126] or toluene.[127]

Phosphorus acids do not interfere. At pH 9 in the presence of a large amount of citrate, nickel, cobalt, and iron are masked by a small amount of EDTA. The method will determine less than 0.2 ppm of copper in high purity thorium oxide or nitrate. The yellow complex develops almost immediately and can be read at 450 nm without extraction.[128] It can be stabilized as a suspension by starch.

The reagent is applicable to bearing metals,[129] ferroalloys of manganese, silicon, and molybdenum,[130] titanium, zirconium and their alloys;[131] beryllium compounds,[132] and sodium chloride[133] as well as to copper separated from atmospheric

[122] J. A. Brabson, P. A. Schaeffer, A. Truchan, and L. Deal, *Ind. Eng. Chem., Anal. Ed.* **18**, 554–556 (1946).
[123] George Horwitz, Joseph Cohen, and Martin A. Everett, *Anal. Chem.* **36**, 142–144 (1964); Jan Inglot and Maria Kewecka, *Pr. Inst. Hutn.* **20**, 381–385 (1968).
[124] A. Chira, *Rev. Chim.* (Bucharest) **13**, 494 (1962).
[125] H. Gielczewska and M. Kleczynska, *Pr. Inst. Mech. Precyz.* **15**, 41–46 (1967).
[126] H. van Diun and C. Brons, *Neth. Milk Dairy J.* **17**, 323–333 (1963).
[127] S. N. Mladenovic and B. Filipovic, *Glas. Hem. Drus. Beogr.* **29**, 45–49 (1964).
[128] Giuseppe Signorelli, *Ann. Chim.* (Rome) **51**, 952–958 (1961).
[129] M. E. Garate and M. T. Garate, *Chim. Anal.* (Paris) **40**, 77–79 (1958).
[130] Takuya Imai and Shinko Nagumo, *Kogyo Kagaku Zasshi* **59**, 886–888 (1956).
[131] D. F. Wood and R. T. Clark, *Analyst* **83**, 509–516 (1958).
[132] J. Walkden, *UKAERE* **AM53**, 3 pp (1959).
[133] Jose Maria Rafols Rovira, *Info. Quim. Anal.* (Madrid) **11**, 99–102 (1957); B. Riva, *Ann. Chim.* (Rome) **53**, 1898–1908 (1963).

air.[134] After slow electrodeposition of a 0.1–0.25 gram sample of very pure copper, the residual amount remaining in the electrolyte is determined by sodium diethyldithiocarbamate.[135]

For copper in an electrolyte containing the following substances in grams per liter: fluoroboric acid, 60–100; lead, 40–60; tin, 6–12; resorcinol, 5–7; gelatin, 0.5–6; antimony, 0.1–0.25; and copper, 0.1–5, the electrolyte is concentrated by heating with nitric and sulfuric acid and the copper diethyldithiocarbamate is formed in an aliquot and extracted with toluene.[136] For separation from lead and bismuth, copper is extracted from a sample solution containing excess hypochlorous acid before determination as the diethyldithiocarbamate.[137]

For foodstuffs after digestion with nitric and sulfuric acids, addition of ammonium citrate and EDTA left only bismuth as interfering at pH 9.[138] This is solved by reading before adding cyanide, then after it has broken down the copper complex, the bismuth is read alone.

The method is appropriate for analysis of the solution of the product of dry-ashing of bread[139] and applied to the product of wet-ashing of wheat flour and bran[140] as well as to a solution of the ash of rubber accelerators[141] and for water analysis.[142]

To determine the total heavy metals, copper, lead, bismuth, cadmium, zinc, manganese, and iron, the materials are extracted as diethyldithiocarbamates in chloroform. Excess reagent is removed by washing with 0.01% sodium hydroxide solution. Then ion exchange treatment of the chloroform solution with 5% copper sulfate solution converts all to copper diethyldithiocarbamate, which is read at 436 nm as the equivalent of the group.[143]

Deposits in an internal combustion engine occur from fuel and organometallic additives in the lubricant.[144] Digest a 0.5-gram sample of material insoluble in petroleum ether and in 1 : 1 benzene-ethanol with sulfuric acid and nitric aicid. Dilute with water, filter, and dilute to 250 ml. Use an aliquot for determination of copper by sodium diethyldithiocarbamate.

For 5–50 μg of copper per gram in reactor graphite, burn in air or oxygen at 800–900°.[145a] For 0.5–5 μg/gram add a known amount of copper. Dissolve the ash in acid and determine as the diethyldithiocarbamate.

For copper at 0–20 ppm in lithium metal, hydride or hydroxide complex with

[134] Halina Wyszynska, Artur Strusinski, and Maroa Borkowska, *Rocz. Panstw. Zakl. Hig.* **21**, 483–491 (1970); Artur Strusinski and Halina Wyszynska, *ibid.* **22**, 649–656 (1971).

[135] T. J. Murphy and J. K. Taylor, *Anal. Chem.* **37**, 929–931 (1965).

[136] A. F. Oparina and E. S. Bruile, *Zh. Prikl. Khim. Leningr.* **45**, 1368–1369 (1972).

[137] J. A. Corbett, *Metallurgia* **65**, 43–47 (1962).

[138] Antonino Poretta, *Ind. Conserve* (Parma) **32**, 81–82 (1957); cf. B. Ticha and M. Friml, *Listy Cukrov.* **89** (3), 69–72 (1973).

[139] C. P. Czerniejewski, C. W. Shank, W. G. Bechtel, and W. B. Bradley, *Cereal Chem.* **41**, 65–72 (1964).

[140] M. Jasinska and A. Zechalko, *Acta Pol. Pharm.* **24**, 533–538 (1967).

[141] M. Uchman and W. Stachowski, *Chem. Anal.* (Warsaw) **13**, 655–658 (1968).

[142] B. Tuck and E. M. Osborne, *Analyst* **85**, 105–110 (1960).

[143] A. B. Blank, N. T. Sizonenko, and A. M. Bulgakova, *Zh. Anal. Khim.* **18**, 1046–1050 (1963).

[144] T. Hammerich and H. Gondermann, *Erdöl Kohle Erdgas, Petrochem.* **16**, 303–308 (1963).

[145a] Krystna Gorcynska, Danuta Ciecierska, and Halina Waledziak, *Chem. Anal.* (Warsaw) **2**, 52–60 (1957).

diethyldithiocarbamate.[145b] Add ammonium hydroxide to precipitate iron and thus avoid rapid fading of the color. Then extract with isoamyl alcohol. Interfering amounts of nickel can be pre-extracted with chloroform as nickel dimethylgloxime.

A solution of dithizone in carbon tetrachloride is appropriate for isolation of copper, mercury, cadmium, and thallium from strongly alkaline solution. The copper must then be isolated for determination by sodium diethyldithiocarbamate.

For determination of copper, zinc, and lead in the same sample dry, ash at 475° and extract with hydrochloric acid.[146] Develop copper with diethyldithiocarbamate, extract with chloroform, and read at 436 nm. Destroy the complex in the chloroform extract with hydrochloric acid, and develop zinc with 1-(2-pyridylazo)-2-naphthol and read at 450 nm. Then lead is extracted in the presence of potassium iodide into methyl isobutyl ketone and is developed with 4-(2-pyridylazo)resorcinol for reading at 530 nm.

When copper and manganese are present, adjust to pH 0.7 with perchloric acid and add sodium diethyldithiocarbamate. Set aside for 5 minutes and extract the copper complex with butyl acetate for reading at 440 nm.[147] Then make the aqueous phase just alkaline with ammonium hydroxide. Add more sodium diethyldithiocarbamate and set aside for 10 minutes. Extract the manganese with isobutyl acetate and read at 345 nm. Either element can be read for up to 30 μg per ml.

For copper in arsenic solution, add ammonia and sodium potassium tartrate and develop with sodium diethyldithiocarbamate.[148]

Traces of copper diethyldithiocarbamate can be estimated by the catalytic effect on sodium azide–potassium iodide–iodine in 5 : 1 methanol-chloroform.[149] After a 5-minute reaction period, the reaction mixture is read at 420 nm against the original mixture. Lead or cadmium can be determined in the same way, and 0.06 ppm of copper in 1 gram of sample has been determined.

For isolation of copper from niobium with diethyldithiocarbamate followed by column chromatography, see Zinc.

Procedures

By Extraction.[150] To a sample solution containing 0.0005–0.32 mg of copper in not more than 15 ml of reagent, add a drop of 0.1% thymol blue solution and neutralize with 1 : 6 ammonium hydroxide. Add 5 ml of *M* tartaric acid and 5 ml of 0.25 *M* EDTA. Add ammonium hydroxide to pH 9.2 ± 0.4. Add 2 ml of 0.1 *M* sodium diethyldithiocarbamate and dilute to 30–35 ml. Extract with successive 5-ml portions of trichloroethylene as long as the extracts are colored, and dilute to 25 or 50 ml according to the intensity of color of the combined extracts. Adjust to $20 \pm 2°$ and read at 440 nm within 30 minutes.

[145b] H. A. Friedman, *U. S. At. Energy Comm.* **Y-962**, 11 pp (1957).
[146] E. Kroeller, *Deut. Lebensm. Rundschau* **66**, 190–192 (1970).
[147] Sharad M. Shah and J. Paul, *Microchem. J.* **17**, 119–124 (1972).
[148] L. B. Kristaleva, *Zavod. Lab.* **25**, 1294–1295 (1959).
[149] A. K. Babko, L. V. Markova, and M. U. Prikhod'ko, *Zh. Anal. Khim.* **21**, 935–939 (1966).
[150] G. Gottschalk, *Z. Anal. Chem.* **194**, 321–323 (1963).

Without Extraction.[151] To a sample solution containing 0.005–0.15 mg of copper, add an equal volume of butyl Carbitol. Adjust to pH 5–6 with hydrochloric acid or ammonium hydroxide. Add 2 ml of 0.1% sodium diethyldithiocarbamate solution and read at 436 nm. With a molar ratio of reagent to copper of 30 : 1, the color is stable for 5–15 minutes, but at a ratio of 300 : 1 it is stable for 24 hours. Beer's law is followed for 0.1–9 μg/ml.

Metals, Ores, and Alloys

ALUMINUM.[152] Digest a 1-gram sample with 20 ml of 20% sodium hydroxide solution. Add 20 ml of 5 : 1 : 5 sulfuric acid–nitric acid–water. Boil until clear and dilute to 50 ml. To a 10-ml aliquot, add 2 ml of ammonium citrate solution containing 500 grams of citric acid and 500 grams of ammonium hydroxide per liter. Add 20 ml of 1 : 3 ammonium hydroxide, 10 ml of 5% solution of gum acacia, and 10 ml of 0.2% solution of sodium diethyldithiocarbamate. Dilute to 100 ml and read at 440 nm against a blank. The method is applicable to high purity aluminum.[153] The system can be automated for up to 3% copper in aluminum.[154]

Alternatively,[155] dissolve the sample in hydrochloric acid and evaporate to dryness. Take up the residue in 7.6% ammonium thiocyanate solution. Add to a 10×0.5 cm column packed with 1 μg polytetrafluoroethylene having tributyl phosphate as the stationary phase. Copper, gallium, cobalt, and ferric ion are retained, but aluminum passes through. Wash the column with 50 ml of 7.6% ammonium thiocyanate solution at 0.5–1 ml/minute.

Elute the copper and iron with 1 : 24 hydrochloric acid and determine the copper by dithizone, the iron by 1-nitroso-2-naphthol.

ALUMINUM ALLOYS.[156a] The method contemplates copper, 1–5%, bismuth, 0.1%, and traces of arsenic, antimony, and cadmium. To a 0.4-gram sample, slowly add 12 ml of 1 : 5 hydrochloric acid; after the initial reaction subsides, add 6 ml of 1 : 3 nitric acid. Warm to complete solution. Dilute to a known volume and complete as for aluminum from "To a 10-ml aliquot...."

COPPER ALLOYS.[156b] Dissolve a sample by boiling with 5 ml of 1 : 4 sulfuric acid and 1 ml of nitric acid. Dilute to a known volume. To a 15-ml aliquot add 4 ml of *M* tartaric acid and adjust to pH 5–5.5. Add 10 ml of 0.5 *M* benzoic acid in chloroform containing 0.25 *M* pyridine. Shake, and separate the organic phase.

[151] J. Artigas and M. Ciutat, *An. Real Soc. Espan. Fis. Quim.* (Madrid) **57B**, 445–452 (1961).
[152] M. C. Steele and L. J. England, *Metallurgia* **59**, 153–156 (1959).
[153] T. B. Gorodentseva, G. S. Dolgorukova, K. F. Vorozhbitskaya, V. A. Verbitskaya, L. S. Studenskaya, and V. S. Shvarev, *Tr. Vses. Nauchn.-Issled. Inst. Stand. Obraztsov. Spektr. Etalonov* **2**, 34–39 (1965); G. Matelli and V. Vicentini, *Allumin. Nuova Metall.* **38**, 559–566 (1969).
[154] H. Schilling, *Z. Anal. Chem.* **261**, 394 (1972); cf. J. M. Carter and G. Nickless, *Analyst* **95**, 148–152 (1970).
[155] I. P. Alimarin, N. I. Ershova, and T. A. Bol'shova, *Vest. Mosk. Gos, Univ., Ser. Khim.* **1969**, 79–84.
[156a] Giorgio DeAngelis and Maria Gerardi, *Ric. Sci. Rend. Sez.* ⊆2], **1** (1), 67–86 (1961); cf. V. N. Podchainova, *Tr. Ural. Politekh. Inst., Sb.* **1956** (57), 32–37; Friedrich Oehlmann, *Chem. Tech.* (Berlin) **8**, 544–595 (1956).
[156b] I. V. Pyatnitskii and T. A. Slobodenyuk, *Ukr. Khim. Zh.* **40**, 1333–1335 (1974).

Add solid sodium diethyldithiocarbamate, dilute to 25 ml with chloroform and read at 400 nm.

IRON AND STEEL.[157] Dissolve a 0.1-gram sample in 10 ml of 2:1:10 sulfuric acid–nitric acid–water and dilute to 100 ml. To a 10-ml aliquot, add 10 ml of 50% citric acid and 10 ml of ammonium hydroxide. Cool and add 10 ml of 0.1% solution of sodium diethyldithiocarbamate. Shake with 40 ml of carbon tetrachloride for 1 minute and read the extract at 430 nm. As applied to iron and steel, the diethyldithiocarbamate method is as accurate as gravimetric determination by the German referee method and requires less time.[158]

ALLOY STEEL, NICKEL, AND COBALT PRESENT.[159] Dissolve a 0.5-gram sample in 5 ml of nitric acid and evaporate nearly to dryness. Take up in water and dilute to 50 ml. To a 10-ml aliquot, add 2 ml of 3% solution of tetrasodium pyrophosphate, 5 ml of 20% citric acid solution, and 5 ml of 3% ammonium persulfate solution. Precipitate nickel with 1 ml of ammoniacal solution of dimethylglyoxime and filter. To the filtrate add 3 ml of 0.1% solution of sodium diethyldithiocarbamate and extract with 3 ml, 3 ml, and 3 ml of chloroform. Dilute the combined extracts to 10 ml and read at 440 nm.

ELECTROLYTIC NICKEL.[160a] COPPER, IRON, AND COBALT. These three metals are determined by extraction from a solution of a single sample. Dissolve a 10-gram sample in 25 ml of nitric acid and evaporate to a syrup. Dilute to 250 ml with water.

Dilute a 25-ml aliquot to 50 ml, add 5 ml of 0.2% solution of sodium diethyldithiocarbamate in chloroform, and shake for 2 minutes. Extract with 5-ml portions of chloroform as long as the extracts are colored, and dilute to 25 ml. Read at 435 nm.

To the aliquot from which copper was extracted, add 50% solution of sodium acetate until the pH reaches 5. Mix with 30 ml of an acetate buffer solution for pH 5.57 and 2 ml of ethyl acetate. Extract the iron by shaking for 5 minutes with 5 ml, 5 ml, and 5 ml of chloroform. Dilute the extracts to 25 ml and read iron.

After extraction of the iron, add hydrochloric acid dropwise to adjust to pH 2. Add 5 ml of 40% ammonium thiocyanate and 10 ml of 2% solution of diantipyrylmethane in 1:10 hydrochloric acid. Extract the cobalt precipitate with 10 ml of chloroform. Repeat the extraction as long as color is obtained. Combine the chloroform extracts and add 15 ml of 0.2 *N* sodium acetate and 5 ml of 0.5% solution of nitroso-R salt. Shake for 2 minutes to transfer the cobalt to the aqueous layer. Again extract the chloroform layer with 15 ml of 0.2 *N* sodium acetate and 5 ml of 0.5% solution of nitroso-R salt. Combine the aqueous extracts

[157]Ohiko Kammori and Kimio Mukaewaki, *Nippon Kinsoku Gakkaishi* **18**, 420–424 (1954); cf. C. Vandael, *ATB Met.* **1** (8), 221–224 (1959).
[158]H. Pohl, *Materialprüfüng* **1**, 177–178 (1959).
[159]L. B. Kristaleva and R. D. Taesnitskaya, *Uch. Zap. Tomsk. Gos. Univ. V. V. Kuibysheva* **1959** (29), 126–129; W. B. Sobers, *Foundry* **88**, 110–113 (1960); V. I. Ptukh, G. I. Chadova, A. S. Aksel'rod and K. N. Mamina, *Zavod. Lab.* **43**, 30–31 (1977).
[160a]S. E. Kreimer, A. V. Stogova, and A. S. Lomekhov, *Zavod. Lab.* **27**, 386–387 (1961).

and add 5 ml of nitric acid. After 20 minutes, dilute to 200 ml and read cobalt at 525 nm.

For down to 30 ppm of copper in a solution of a nickel-based alloy, add EDTA to mask the nickel and develop copper with diethyldithiocarbamate.[160b] Alternatively, down to 20 ppm of copper form the cuprous-2,2′-biquinolyl complex.

COPPER, NICKEL, AND COBALT.[161] Make alkaline to phenol red a sample solution, typically containing 0.002–0.36 mg of copper, 0.0001–0.01 mg of nickel, and 0.002–0.2 mg of cobalt, by addition of sodium carbonate solution. If iron or manganese is present, add 4 ml of 4% solution of sodium pyrophosphate decahydrate. Acidify with citric acid, then add sodium carbonate solution to a faint red. Add 3 ml of 0.1% sodium diethyldithiocarbamate solution and extract with 10 ml of carbon tetrachloride by shaking for 3 minutes. Filter the organic phase and read copper at 436 nm, nickel at 328 nm, and cobalt 367 nm. A correction for cobalt is required in determining nickel. If a yellow precipitate is formed by uranium, redissolve it with aluminum nitrate solution.

ANTIMONY.[162] This technic will determine 0.00005–0.002% of copper in antimony. As a tartrate reagent, dissolve 40 grams of tartaric acid in 100 ml of water and neutralize with 50 ml of ammonium hydroxide. As a sequestrant, dissolve 25 grams of EDTA in water, add 6.5 grams of ammonium hydroxide, and dilute to 250 ml.

Dissolve a 0.5-gram sample in 1 ml of nitric acid, 1.5 ml of hydrochloric acid, and 6.5 ml of the tartrate reagent. Evaporate to half volume and add 10 ml of the sequestrant solution. Add 5 ml of 50% solution of ammonium citrate and neutralize to the cresol red end point. Add 5 ml of 0.1% solution of sodium diethyldithiocarbamate and extract with 5 ml, 5 ml, and 5 ml of chloroform. Read the combined extracts at 436 nm against chloroform.

CADMIUM AND ZINC.[163] Dissolve a sample containing 2–15 μg of copper in 2:3 nitric acid and boil off oxides of nitrogen. Dilute until the cadmium or zinc is less than 10% of the content. Neutralize to pH 4.5–5 with 1:4 ammonium hydroxide. If the sample was zinc, add 2 ml of 20% ammonium citrate solution. Add 1 ml of 0.3% solution of the reagent and extract with 10 ml of carbon tetrachloride by shaking for 2 minutes. Read the extract at 435 nm.

TIN.[164] Dissolve a 5-gram sample in 20 ml of hydrochloric acid and dilute to 130 ml. Mask the copper with 15 ml of 0.25 *M* EDTA, precipitate stannous tin with

[160b]Y. Toita, H. Nagai, K. Sekine, and H. Onishi, *Rep. Japan Atom. Energy Res. Inst.* **JAERI-M-5094**, 17 pp (1973).

[161]Antonia Borges Pimenta, *Rev. Port. Quim.* **14**, 46–53 (1972); cf. D. I. Ryabchikov and A. I. Lazarev, *Renii, Tr. Vses. Soveshch. Probl. Reniya, Akad. Nauk SSSR., Inst. Met.* **1958**, 267–274; V. N. Podchainova, *Tr. Kom. Anal. Khim., Akad. Nauk SSSR, Inst. Geokhim. Anal. Khim.* **11**, 146–164 (1960).

[162]Jan Provaznik and Miroslav Knizek, *Chem. Listy* **55**, 79–82 (1961).

[163]K. B. Kladnitskaya and A. N. Grissevich, *Ukr. Khim. Zh.* **27**, 803–807 (1961); *Zavod. Lab.* **27**, 1343–1345 (1961); cf. A. Chira, *Rev. Chim.* (Bucharest) **13**, 494 (1962).

[164]Sashichi Ikeda and Hinoshi Nagai, *Jap. Anal.* **7**, 76–79 (1958).

40 ml of saturated solution of oxalic acid, and filter. Add 3 ml of 1% ferric ammonium sulfate solution to the filtrate to demask the cupric ion. Add 25 ml of acetate buffer solution for pH 4.5 and 2 ml of 1% solution of sodium diethyldithiocarbamate. Extract with 10 ml and 10 ml of carbon tetrachloride and read at 440 nm.

THORIUM.[165] Dissolve a sample of the metal or its compounds equivalent to 2.5 grams of thorium oxide in platinum with 10 ml of 1 : 1 nitric acid and 3 drops of hydrofluoric acid. Evaporate to dryness at 100°. Take up the residue by heating with 5 ml of 1 : 5 nitric acid. Add 5 ml of 50% citric acid solution and stir in 10 ml of 1 : 1 ammonium hydroxide. Add 4 ml of 1% EDTA solution and dilute to about 40 ml. Check that the pH is around 9 to thymol blue. Add 1 ml of 0.2% solution of sodium diethyldithiocarbamate and shake with 5 ml of isoamyl acetate for 2 minutes. Filter the extract through a dry paper and read at 430 nm against a blank.

URANIUM.[166] Dissolve a sample containing up to 10 μg of copper in nitric acid and evaporate nearly to dryness. Take up the residue in 5 ml of 1 : 4 nitric acid. Add 5 ml of 50% citric acid and follow with ammonium hydroxide to pH 9. Add 3 ml of 1% solution of EDTA and dilute to about 25 ml. When cool, add 1 ml of 1% solution of sodium diethyldithiocarbamate and shake with 5 ml of butyl acetate for 3 minutes. This solvent gives a higher absorption than chloroform or carbon tetrachloride. Wash the extract with 5 ml of 2 *N* sulfuric acid and read at 440 nm.

Selenium.[167] Dissolve a 0.5-gram sample in 10 ml of hydrochloric acid and 2 ml of nitric acid. Evaporate to dryness and take up the residue in 2 ml of hydrochloric acid. Add 5 ml of water and 30 ml of 20% sodium citrate solution. Neutralize to phenolphthalein with 20% sodium hydroxide solution and add 3 ml of 0.5 *M* EDTA solution. Add 2 ml of 1% sodium diethyldithiocarbamate solution. After 10 minutes extract with 25 ml and 5 ml of chloroform and read at 440 nm.

Tellurium.[168] Dissolve a 1-gram sample in 3 ml of hydrochloric acid and 2 ml of nitric acid. Evaporate to dryness and take up in 5 ml of 1 : 10 hydrochloric acid. Add ammonium hydroxide to raise to pH 10. Add 5 ml of ammonium citrate buffer solution for pH 10 and 1 ml of 0.5 *M* EDTA solution. Add 0.5 ml of 0.5% sodium diethydithiocarbamate solution. Extract with 10 ml of carbon tetrachloride and read at 430 nm.

Triuranium Octaoxide.[169] Digest 1 gram with 2 ml of perchloric acid and 2 ml of hydrobromic acid. Evaporate to dryness and take up in 50 ml of 9 : 1 methanol–1.5 *N* hydrobromic acid. Pass at 1 ml/minute through a column of 100-200 mesh Dowex 1-X8 in chloride form, which has been equilibrated with the solvent mixture. Wash with the solvent mixture; then elute copper with 3 *N*

[165]Tomoyuki Kukoyama and Norio Ichinose, *ibid.* **10**, 107–111 (1961).
[166]Masami Suzuki, *ibid.* **8**, 395–397 (1959).
[167]N. P. Strel'nikova, G. G. Lystsova, and G. S. Dolgorukova, *Zavod. Lab.* **28**, 1319–1321 (1962).
[168]Teru Yuasa, *Jap. Anal.* **11**, 359–360 (1962).
[169]J. Korkisch and H. Gross, *Mikrochim. Acta* **11**, 413–424 (1975).

hydrochloric acid and develop with diethyldithiocarbamate. Lead and indium accompany the copper but do not interfere.

Yellow Cake. Dissolve 1 gram of sample in 5 ml of hydrobromic acid. Proceed as for triuranium octaoxide from "Evaporate to dryness and ...."

Thallium Nitrate.[170] Dissolve 2 grams of thallium nitrate in water. Add 1 ml of 0.01 *M* EDTA and dilute to 55 ml. Add 15 ml of 19.6% solution of sodium iodide, shake for 3 minutes, and centrifuge. Evaporate an aliquot of the clear layer to dryness with 1 ml of sulfuric acid. Dissolve the residue in 2 ml of hydrochloric acid, dilute to 20 ml, and neutralize with 1 : 1 ammonium hydroxide. Dilute to 40 ml and add 50 ml of acetate buffer solution for pH 4.5. Add 2 ml of 1% solution of sodium diethyldithiocarbamate and extract with 5 ml, 5 ml, and 5 ml of chloroform. Read the extract at 430 nm.

Thallium Sulfate. Proceed as for the nitrate but dilute to 70 ml before precipitation of thallous iodide.

Cadmium Nitrate.[171] Dissolve 5 grams of sample in water and dilute to 50 ml. Mix a 5-ml aliquot with 4 ml of a 25% solution of EDTA in 1 : 4 ammonium hydroxide. If not alkaline to phenolphthalein, add more ammonium hydroxide. Add 2 ml of a fresh 0.2% solution of sodium diethyldithiocarbamate and extract with 3 ml of heptane. Read at 271 nm. Large amounts of mercury, arsenic, and bismuth interfere.

Nickel–Zinc Ferrite.[172] These ferrites contain 0.001–0.1% copper and therefore call for 0.1–1 gram of sample. Add half the weight of sample as ammonium chloride and dissolve in 20 ml of hydrochloric acid. Dilute to 100 ml and take an aliquot to contain 0.02–0.05 mg of copper. Add 10 ml of 10% EDTA solution, then add ammonium hydroxide to pH 10–11. Add 5 ml of 1% solution of diethyldithiocarbamate to the red solution of sample. Extract with 10 ml, 10 ml, and 10 ml of ethyl acetate. Wash the combined extract with 10 ml of 1 : 5 ammonium hydroxide. Dilute to 50 ml and read at 440 nm.

Magnesium Chromium Ferrite. Fuse the sample with potassium pyrosulfate and dissolve the melt in water. Proceed as for nickel–zinc ferrite from "Dilute to 100 ml and ...."

Manganese Ores, Ferromanganese, and Manganese Metal. For isolation of copper for determination by diethyldithiocarbamate, see Lead.

Iron and Manganese Ores.[173] Dissolve 1 gram of finely powdered sample in 10

[170]E. Jackwerth, J. Lohmar, and G. Schwark, *Z. Anal. Chem.* **260**, 101–106 (1972).
[171]V. N. Podchainova, I. N. Liplavk, and L. N. Ushkova, *Zavod, Lab.* **38**, 411–412 (1972).
[172]V. T. Vasilenko, *ibid.* **31**, 1070–1071 (1965).
[173]J. Jankovsky and E. Pavlikova, *Sb. Pr. Ustavu Pro Vyzk. Rud* **IV**, 229–241 (1962).

ml of hydrochloric acid. Add 3 ml of nitric acid and evaporate to dryness. Add 5 ml of hydrochloric acid and again evaporate to dryness. Take up in 5 ml of hydrochloric acid, add 50 ml of hot water, and filter. Wash well with hot 1:9 hydrochloric acid and hot water. Ignite the filter in platinum. Add 1 ml of 1:1 sulfuric acid and 5 ml of hydrofluoric acid, and evaporate to sulfur trioxide fumes. Add 2 grams of potassium pyrosulfate and fuse. Dissolve the melt in 5 ml of 1:5 hydrochloric acid and add to the original filtrate. Filter if barium sulfate is present, and dilute to 250 ml.

To a 20-ml aliquot, add 15 ml of 0.25 *M* EDTA followed by ammonium hydroxide to give a red color and 2 drops in excess. Dilute to 200 ml and add 2 ml of 1% solution of sodium diethyldithiocarbamate. Shake for 3 minutes with 15 ml and 10 ml of carbon tetrachloride. Dilute the combined extracts to 50 ml with ethanol. Read at 430 nm against a blank.

Soil

COPPER.[174] Treat a 3-gram sample with 5 ml of sulfuric acid and 5 ml of hydrofluoric acid. Evaporate to sulfur trioxide fumes, cool, and take up in 10 ml of hydrochloric acid. Filter, add 10 ml of 25% sodium citrate solution, and adjust to pH 2 by adding 10% sodium hydroxide solution. Shake for 15 minutes with 10 ml of 0.05% dithizone solution. Repeat this extraction and combine the extracts. Add 1 ml of nitric acid, 1 ml of sulfuric acid, and 2 ml of 30% hydrogen peroxide. Evaporate to dryness. Take up the residue in 10 ml of 1:100 hydrochloric acid. Add 1 ml of 5% ammonium citrate solution and 3 ml of 10% EDTA solution. Add 10% sodium hydroxide to the phenolphthalein end point. Add 5 ml of fresh 0.2% solution of sodium diethyldithiocarbamate and 10 ml of carbon tetrachloride. Shake for 10 minutes and repeat the extraction with another 10 ml of solvent. Filter the combined extracts, dilute to 25 ml, and read at 436 nm.

ZINC. After extraction of the copper, add 5 ml of 25% sodium citrate solution to the aqueous phase followed by 10% sodium hydroxide solution to the phenolphthalein end point. Extract zinc and cobalt by shaking with 10 ml and 10 ml of 0.05% solution of dithizone in carbon tetrachloride. Shake the combined extracts with 50 ml and 50 ml of 0.02 *N* hydrochloric acid for 3 minutes to back-extract the zinc. Wash the combined aqueous extracts with 10 ml of carbon tetrachloride to remove traces of dithizone. Add 2 drops of methyl orange solution followed by 5% sodium acetate solution until the color changes to yellow. Add 0.2 ml more of 5% sodium acetate solution and 5 ml of a 50% solution of sodium thiosulfate pentahydrate. Extract the zinc with 2-ml portions of 0.05% solution of dithizone in carbon tetrachloride until the color of the extractant is no longer altered. Shake the combined extracts with 50 ml of 0.01% solution of ammonium hydroxide to remove free dithizone. Dilute the organic layer to 25 ml and read at 535 nm against carbon tetrachloride.

COBALT. For cobalt in the organic phase from which copper and zinc have been removed, add 1 ml of sulfuric acid, 1 ml of nitric acid, and 2 ml of 30% hydrogen peroxide. Evaporate to dryness. To the residue, add 1 ml of 0.01 *N* hydrochloric acid, 1 ml of 40% ammonium citrate solution, and 1 ml of 0.2% nitroso-R-salt

[174] K. V. Verigina, *Mikroelem. SSSR* **1963** (5), 50–549

solution. Heat at 100° for 5 minutes. Add 1 ml of nitric acid and 2 ml of hydrochloric acid and heat at 60° for 40 minutes to destroy excess of the reagent. Cool, dilute to 25 ml, and read at 425 nm.

Alternatively,[175] mix 5 grams of 70-mesh sample of soil with 5 grams of Zeo-Karb 215. Add 200 ml of water and shake for 2 hours at 25 ± 1°. Separate the soil from the resin by washing it through a 70-mesh sieve. Extract the resin with 25 ml of 2 *N* sulfuric acid as a sample for determination of copper, zinc, and cobalt.

NONCARBONATE SOIL.[176] EXTRACTABLE ELEMENTS. Shake 40 grams of soil for 1 hour with 400 ml of 10% potassium nitrate solution acidified to pH 3 with nitric acid. If colorless, evaporate to under 200 ml and dilute to 200 ml. If the extract is colored, add 5 ml of 30% hydrogen peroxide with 4 ml of nitric acid and evaporate to dryness. Take up in water and dilute to 200 ml for determination of copper with sodium diethyldithiocarbamate, cobalt with nitroso-R salt, zinc with dithizone, manganese with ammonium persulfate, boron with carmine, and molybdenum with thiocyanate.

CARBONATE SOIL. Extract as for noncarbonates, with sufficient added nitric acid to neutralize the carbonates.

Zinc Sulfide Phosphors.[177] This reagent is appropriate for determination of 0.5–1 ppm of copper. For larger amounts, a technic with zinc dibenzyldithiocarbamate is shown elsewhere.

Dissolve 0.1 gram of sample in 2 ml of hydrochloric acid and boil off hydrogen sulfide. Dilute to about 20 ml. Add 10 ml of 10% EDTA solution, 5 ml of *M* ammonium citrate, and 2 drops of 0.1% cresol red solution. Add ammonium hydroxide to the end point shown by the indicator and dilute to 50 ml. Add 5 ml of 0.1% solution of sodium diethyldithiocarbamate. Extract for 2 minutes each with 10 ml and 10 ml of chloroform. Dilute the extracts to 25 ml and read at 420 nm.

Red Phosphorus. Dissolve 200 grams of ammonium citrate in 400 ml of water and make just alkaline to litmus paper with 1 : 1 ammonium hydroxide. Shake for 1 minute with 1 ml of 0.1% solution of diethyldithiocarbamate and 25 ml of chloroform. Discard the chloroform layer and dilute to 500 ml as the 40% ammonium citrate solution.

Mix a 1-gram sample with 10 ml of water and 35 ml of nitric acid. Evaporate to about 20 ml and add 2 ml of perchloric acid. Heat to fumes of perchloric acid and fume for a couple of minutes. Add 25 ml of water, 5 ml of 40% ammonium citrate solution, and 1 ml of 5% EDTA solution. Add 2 drops of phenol red indicator and ammonium hydroxide until the first red appears. Add 1 ml of 0.1% sodium diethyldicarbamate solution. Extract with 10 ml, 5 ml, and 5 ml of chloroform. Dilute to 25 ml and read at 435 nm against a reagent blank.

EDTA.[178] Dissolve a sample equivalent to 3 grams of the free acid in 30 ml of

[175] D. K. Acquaye, A. B. Ankomah and I. Kanobo, *J. Sci. Food Agr.* **23**, 1035–1044 (1972).
[176] A. N. Gyul'akhmedov, *Izv. Akad. Nauk Azerb. SSR Ser. Biol. Med. Nauk* **1961**, (7), 57–63; *Mikroelem. SSSR* **1963** (5), 45–49.
[177] Yoshihide Kotera, Tadao Sekine, and Masao Takahashi, *Bull. Chem. Soc. Jap.* **39**, 2523–2526 (1966).
[178] F. Bydra and R. Pribil, *Chemist-Analyst* **51**, 76–77 (1962).

hot water. If the sample was tetrasodium EDTA, add 1 gram of ammonium chloride. Otherwise raise to pH 9 or above with 5 ml of ammonium hydroxide. Cool and add 5 ml of fresh 1% sodium diethyldithiocarbamate solution. Set aside for 5 minutes, then extract with 30 ml and 10 ml of chloroform. Dilute the combined extracts to 50 ml with ethanol and read at 420 nm against a blank in which 20 ml of 2.5% solution of potassium chloride serves as sample.

Copper Sodium Chlorophyllins.[179a] Mix 1 gram of sample with 5 ml of 10% ammonium carbonate solution and adjust to pH 5.2 with acetic acid. Let stand for 45 minutes and filter. Add 5 ml of 10% sodium acetate solution and raise to pH 8 by addition of ammonium hydroxide. Saturate with sodium chloride and add 1 ml of 0.1% solution of sodium diethyldithiocarbamate. Shake with 10 ml of isoamyl alcohol and read the organic layer at 436 nm.

Ores, Tailings, and Concentrates.[179b] Decompose the sample by fusion with sodium carbonate and zinc oxide. Dissolve the fusion in hydrochloric acid and dilute as necessary. To an aliquot add 5 ml of 25% solution of ammonium citrate, 1 ml of 2.5% solution of EDTA, 25 ml of 25% solution of ammonia, 5 ml of 0.2% solution of polyvinyl alcohol, and 10 ml of 0.2% solution of sodium diethyldithiocarbamate. Dilute to 100 ml and set aside for 10 minutes. Read at 520 nm. Ammonium citrate and EDTA mask iron, nickel, and cobalt.

Pulp.[180] Ash 10 grams of pulp, avoiding the use of burners containing copper. Take up the ash in 5 ml of 1 : 1 hydrochloric acid and evaporate to dryness at 100°. Dissolve in water. Treat any residue with 5 ml of 1 : 1 hydrochloric acid at 100°. To the combined extracts, add 10 ml of 5% EDTA solution and 5 drops of phenolphthalein indicator. Add ammonium hydroxide to the end point and cool. Add 5 ml of 0. 1% sodium diethyldithiocarbamate solution and 20 ml of carbon tetrachloride. Shake for 3 minutes, filter the organic phase through a cotton plug, and read at 440 nm against a reagent blank. Another reference reported variations due to dry ashing, but this is an official method.

Pulp, Paper, and Pulping Liquors. Prepare the sample by wet digestion as described under bathocuproine. To the cooled digestate, add 20 ml of 5% solution of EDTA. Dilute to about 60 ml, neutralize to Congo red paper with ammonium hydroxide, and add 5 ml in excess. Cool, and add 10 ml of 1% solution of ammonium diethyldithiocarbamate in chloroform, and shake for 2 minutes. Draw off the chloroform extract through a glass wool plug. Read at 435 nm against a blank.

Mineral Oils.[181] This technic will determine 0.002–0.2 mg of copper in the presence of 0.04–0.1 mg of iron. Extract 10 ml of oil with 10 ml, 10 ml, and 10 ml of acetic acid. Raise the combined extracts to above pH 9 by addition of 1 : 1 ammonium hydroxide. Mix with 1 gram of citric acid and 10 ml of 0. 02% sodium

[179a]M. Ney and P. Esch, *Ann. Falsif. Expert. Chim.* **63**, 92–96 (1970).
[179b]M. I. Kazartseva and T. K. Butorina, *Zavod. Lab.* **42**, 1275– (1976).
[180]Scandinavian Pulp, Paper, and Board Testing Committee. *SCANC* **12**, 62, *Norsk. Skogind.* **16**, 391–392 (1962); cf. British Standards Institution, *BS 4897,* 11 pp (1973).
[181]J. M. Howard and H. O. Spauschus, *Anal. Chem.* **35**, 1016–1017 (1963).

diethyldithiocarbamate solution. Extract the brown copper complex with 10 ml, 10 ml, and 10 ml of carbon tetrachloride. Shake the combined extracts with 10 ml of acetic acid to clarify them. Dilute the extract to 50 ml, read at 440 nm, and subtract a blank.

Liquid Fuels.[182] Shake a 500-ml sample for 10 minutes with 30 ml of 15% hydrochloric acid. Follow by extraction with 10 ml and 10 ml of water. Combine the extracts and wash with 30 ml of tetrachloroethylene. To the aqueous acid extract add 5 ml of 20% solution of ammonium citrate and 5 ml of 0.5% solution of EDTA. Neutralize to phenolphthalein with ammonium hydroxide and add 0.5 ml in excess. Wash with several portions of tetrachloroethylene and discard the washings. Add 2 ml of 0.1% solution of sodium diethyldithiocarbamate, and extract with 5 ml, 5 ml, and 5 ml of tetrachloroethylene. Dilute the combined extracts to 25 ml and read at 436 nm. The sample is expected to contain 0.003–0.05 mg of copper.

Wine.[183] Evaporate a 10-ml sample with 3 ml of 0.1 *N* sulfuric acid to dryness. Ash at 500–600° for 15 minutes. To the cooled residue, add 0.5 ml of nitric acid and 0.5 ml of sulfuric acid and heat at 100°. Add 5 ml of 20% ammonium acetate solution and raise to pH 8.3–8.4 with ammonium hydroxide. Add 1 ml of 0.1% solution of sodium diethyldithiocarbamate and 5 ml of methanol. Extract with 5 ml, 5 ml, 4.5 ml, and 4.5 ml of carbon tetrachloride. Dilute the combined extracts to 20 ml and read at 440 nm.

Saponifiable Oils.[184] As a reagent, dissolve 20 grams of citric acid in 80 ml of ethanol and dilute to 100 ml with carbon tetrachloride. Dissolve 2 grams of sample in 2 ml of carbon tetrachloride with 2 ml of the citric acid reagent and set aside for 20 minutes. Add 2 ml of 0.1% solution of sodium diethyldithiocarbamate in absolute ethanol and dilute to 10 ml with ethanol. Read at 436 nm.

Alternatively,[185] for wet-ashing, mix 20 ml in a silver dish with 5 ml of 2:1:1 perchloric-sulfuric-nitric acid. Ash below 150°. Take up the ash in 0. 5 ml of nitric acid, transfer to a platinum crucible, and take to dryness. Add 0. 5 ml of 10% sodium carbonate solution. Take to dryness and fuse. Take up the melt in water and dilute to 25 ml. Mix a 10-ml aliquot with 10 ml of acetate buffer solution for pH 5 and 2 ml of 1% solution of sodium diethyldithiocarbamate. Extract with 5 ml and 5 ml of carbon tetrachloride and read the combined extracts at 430 nm.

Milk, Butter, Cream.[186] To 50 grams of milk, 25 grams of cream plus 25 ml of water, or 25 grams of butter plus 25 ml of water, add 10 ml of hydrochloric acid. Heat at 60–70° for 20 minutes with frequent mixing and filter. Add 10 ml of acetate buffer solution for pH 4.5 and 2 ml of 1% solution of sodium diethyldithiocarbamate. Extract with 5 ml and 5 ml of chloroform and read the combined extracts at 430 nm.

[182]Deutscher Normenausschuss, Fachausschuss Mineralol -und Brenustoff-normung, *Erdol Kohle, Erdgas, Petrol,* **17**, 305–306 (1964).
[183]L. Laporta, *Boll. Lab. Chim. Prov.* **15**, 315–326 (1964).
[184]T. P. Labuza and M. Karel, *J. Food Sci.* **32**, 572–575 (1967).
[185]Tsugio Takeuchi and Tomatsu Tanaka, *J. Chem. Soc. Japan, Ind. Chem. Sect.* **64**, 305–307 (1961).
[186]Celina Barska, W. Bednarczyk, and Maria Luczak, *Int. Dairy Congr., Proc. 15th Congr., London* **3**, 1754–1760 (1959).

Milk Powder.[187] To avoid wet- or dry-ashing of milk powder, reconstitute it and precipitate fat and protein from a 50-ml sample by heating with an equal volume of 10% trichloroacetic acid solution at 100° for 15 minutes. Filter and complete as above from "Add 10 ml of acetate buffer. . . ."

Photographic Gelatin.[188] Ash a 2-gram sample at 450–500°. Take up the residue in 4 ml of 1:60 nitric acid and dilute to 50 ml. To a 10-ml aliquot, add 1 ml of 1:60 nitric acid, 1 ml of 1% EDTA solution, and 1 ml of 0.1% solution of sodium diethyldithiocarbamate. Extract with 5 ml of carbon tetrachloride and read the extract at 435 nm.

Plant Material.[189] COPPER AND ZINC. As ammoniacal nitrate reagent, mix 52.5 grams of nitric acid with water and add ammonium hydroxide to pH 9. Add 75 ml of 1 : 14 ammonium hydroxide and dilute to 1 liter. As a citrate buffer, dissolve 46.6 grams of citric acid in water. Add 50 ml of ammonium hydroxide and dilute to 1 liter, checking that the pH is 9 and adding more ammonium hydroxide if necessary. As EDTA-citrate solution, dissolve 17.28 grams of citric acid in water, neutralize with ammonium hydroxide, add 5 grams of disodium EDTA, and dilute to 1 liter.

Digest 1 gram of sample with 10 ml of nitric acid, 1 ml of sulfuric acid, and 2 ml of perchloric acid. After a clear digestate is obtained, fume for another 20 minutes. Cool and add 20 ml of water. Neutralize to methyl red with 1:3 ammonium hydroxide, add 2 ml of 1 : 10 hydrochloric acid, and dilute to 50 ml.

Mix an aliquot containing not more than 0. 02 mg of zinc or copper with 40 ml of the ammoniacal nitrate reagent. Shake for 2 minutes with 10 ml, 5 ml, and 5 ml of 0. 01% dithizone in carbon tetrachloride. Back-extract the zinc from the combined dithizone extracts by shaking with 50 ml of 0. 02 *N* hydrochloric acid for 2 minutes. Wash this zinc extract twice with 5-ml portions of carbon tetrachloride. Combine these washings with the organic solvent from which zinc was extracted. This contains the copper.

To the aqueous zinc solution, add 15 ml of the citrate buffer solution for pH 9. Extract with 10 ml and 5 ml of 0. 01% solution of dithizone in carbon tetrachloride. Dilute the combined extracts to 50 ml and read at 535 nm.

Evaporate the copper dithizonate in carbon tetrachloride to dryness. Add 2 drops of sulfuric acid and 5 ml of nitric acid. Heat to fumes and continue to heat for 10 minutes. Cool. Add 10 ml of EDTA-citrate solution and two drops of thymol blue indicator. Add 1 : 4 ammonium hydroxide until the color becomes green. Add 1 ml of 1% sodium diethyldithiocarbamate solution and 10 ml of carbon tetrachloride. Shake for 2 minutes and read the extract at 436 nm.

Serum.[190] Mix 6 ml of serum containing 1.8–16 μg of copper with 2 ml of 1 : 1 hydrochloric acid. After 5 minutes add 2 ml of 20% solution of trichloroacetic acid.

[187] C. C. J. Olling, *Neth. Milk Dairy J.* **17**, 295–305 (1963).

[188] A. D. Vinogradova, M. E. Kovalskaya, and V. I. Sheberstov, *Zh. Nauch. Prikl. Fotogr. Kinematogr.* **6**, 450–452 (1961).

[189] E. R. Page, *Analyst* **90**, 435–436 (1965).

[190] F. J. Garcia Canturri and M. Piedad Sanchez de Rivera, *An. Acad. Farm. (Madrid)* **30**, 51–76 (1964); cf. I. Dezso and T. Fulop, *Mikrochim. Acta* **1961**, 154–157.

Filter and dilute to 11 ml. Add 1 ml of 10% EDTA solution, 2 ml of ammonium hydroxide, and 1 ml of 0.4% sodium diethyldithiocarbamate solution. Extract with 10 ml of ethyl pentyl ether, and dry the extract with sodium sulfate. Read at 440 nm. The complex can also be extracted with isoamyl alcohol.[191]

Biological Samples. LIQUIDS.[192] For water or colorless liquids containing 1 ppm or more of copper, evaporate 50 ml to less than 10 ml and dilute to 15 ml. Mineral acid must be absent. Add 25 ml of acetic acid and mix. Add 50 mg of EDTA, mix, and set aside for 15 minutes. Add 10 mg of sodium diethyldithiocarbamate, mix, and set aside for 15 minutes. Filter, wash the residue with acetic acid, and dilute to 50 ml with acetic acid. Set aside for 15 minutes and read at 430 nm.

For colored liquids, evaporate 50 ml to dryness in platinum. Oxidize the residue with nitric acid, evaporate to dryness, and take up in 15 ml of water. Complete as above from "Add 25 ml of acetic acid and ...."

Solids. Ash 1 gram of sample containing at least 5 ppm of copper in platinum. Dissolve the ash in dilute hydrochloric acid and evaporate to dryness. If the residue is colored, add 2 ml of water and again evaporate. Take up in 15 ml of water and proceed as above from "Add 25 ml of acetic acid and...."

Diethylammonium Diethyldithiocarbamate

This reagent is more acid resistant than the disodium salt. It is applied to extraction of copper from a large excess of lead.[193] Bismuth must not exceed copper more than 100-fold or silver threefold. Iron must be absent. Beer's law is followed up to 0.04 mg of copper per ml.[194] As applied to a solution of indium, only a negligible amount of the major component is extracted if hydrochloric acid exceeds 5 *N*.[195]

The reagent is also used for isolation of copper from lead and lead cable-sheathing alloys, after which the ligand is destroyed for determination of the copper by cuprizone.[196]

Procedure. To 35 ml of sample solution containing a slight excess of perchloric acid, add 1 ml of 10% solution of hydroxylamine hydrochloride and 1 ml of 30% solution of sodium citrate. Add *m*-cresol purple indicator solution and 1:1 ammonium hydroxide until a yellow color indicates a pH of the order of 2. Add 10 ml of an acetate buffer solution for pH 4 and 2 ml of 0.1% ethanolic neocuproine. If zirconium or hafnium is present, bring to a boil and let cool. Extract with 10 ml of chloroform and filter the extract. Add 0.1 gram of crystalline form of the titled reagent and read at 435 nm.

[191] Shigeko Morimoto, *Nisshin Igaku* **45**, 216–230 (1958).
[192] M. E. M. S. de Silva, *Analyst* **99**, 408–412 (1974).
[193] Kaarina Lounamaa, *Z. Anal. Chem.* **150**, 7–13 (1956).
[194] C. L. Luke, *Anal. Chim. Acta* **32**, 286–287 (1965).
[195] Koichi Nishimura and Teruo Imai, *Jap. Anal.* **13**, 713–717 (1964).
[196] J. H. Thompson and M. J. Ravenscroft, *Analyst* **85**, 735–738 (1960).

Organic Samples.[197] According to the nature of the sample, destroy the organic matter, preferably by wet-ashing. Take up in water, remove insoluble matter by filtration, and dilute to an appropriate volume.

Add an aliquot to 10 ml of a solution of 20 grams of ammonium citrate and 5 grams of EDTA in 100 ml of water. Add bromothymol blue indicator followed by 1:1.5 ammonium hydroxide to a green or bluish green. Shake with 10 ml of 1% solution of diethylammonium diethyldithiocarbamate in carbon tetrachloride. Filter the extract through cotton and read immediately at 436 nm against a reagent blank.

Shake the extract with 5 ml of 5% potassium cyanide solution, which destroys the copper complex. Determine whether bismuth and tellurium are present. If so, repeat the determination, wash the carbon tetrachloride extract twice with 1% sodium hydroxide solution, and apply the reading thereafter as a correction of the copper absorption.

Bis(carboxymethyl)dithiocarbamate

The titled reagent is used as the tripotassium salt of (dithiocarboxy)iminodiacetic acid. Below pH 5 it tends to decompose. At higher pH levels it forms complexes with copper, silver, mercury, lead, cadmium, nickel, bismuth, and antimony.[198] The complex is applied to determination of copper.

Procedures

Aluminum Alloys. Boil a 0.2-gram sample with 10 ml of 1:1 hydrochloric acid and 2 ml of nitric acid. Cool, dilute, filter, and dilute to 250 ml. Mix a 10-ml aliquot with 10 ml of acetate buffer solution for pH 5. Mask the iron with 0.2 ml of saturated solution of sodium fluoride. Add 1 ml of 0.01 *M* reagent and dilute to 50 ml. Read at 453 nm.

Babbitt Metal. Dissolve a 0.2-gram sample in 10 ml of 1:1 nitric acid and evaporate to about 4 ml. Add 50 ml of water, filter, and wash the residue with hot 1:2 nitric acid followed by hot water. Neutralize with ammonium hydroxide and dilute to 250 ml. Complete as for aluminum alloys from "Mix a 10-ml aliquot...."

Tetramethylenedithiocarbamate

This is another member of the family of dithiocarbamate reagents.[199] It follows Beer's law for up to 0.05 mg of copper in 10 ml of extract.

[197] Metallic Impurities in Organic Matter Subcommittee, Analytical Methods Committee, Society for Analytical Chemistry, *Analyst* **88**, 253–258 (1963).
[198] F. M. Tulyupa, V. A. Pavlichenko, and Yu. I. Usatenko, *Ukr. Khim. Zh.* **36**, 204–213 (1970).
[199] E. Kovacs and H. Guyer, *Z. Anal. Chem.* **209**, 388–394 (1965).

Procedure. *Steel and Ferrous Alloys.* Dissolve 0.2 gram of sample containing 0.01–0.3% copper in 15 ml of 1:1 hydrochloric acid with 2 ml of nitric acid. Boil off nitrous fumes and dilute to 100 ml. To aliquot containing not more than 0.04 mg of copper, add 5 ml of 10% citric acid and 5 ml of 12% EDTA solution. Add ammonium hydroxide to pH 8.2 ± 0.2. Mix well with 2 ml of 2.5% solution of the reagent and extract with 10 ml of carbon tetrachloride. Read at 436 nm.

Lead Diethyldithiocarbamate

Lead diethyldithiocarbamate is a reagent for determination of copper by displacement.[200] Up to 10 mg of nickel or 20 mg of cobalt per ml does not interfere.[201]

A chloroform solution is used for extraction from 2 *N* hydrochloric acid. Ascorbic acid prevents interference by ferric ion. Complexes of bismuth, molybdenum, and tellurium are also extracted with the copper. They are removed by extracting the chloroform solution with a strongly alkaline solution of sodium tartrate presaturated with chloroform. Then filter the organic phase and read at 440 nm. A suitable copper content of the extract is 5–30 μg/ml. Oxidizing agents must be absent, and if bismuth is more than 20 times the copper content or tellurium more than 10 times, a second extraction with alkaline tartrate is essential.

For copper, nickel, and cobalt in aluminum, dissolve the sample in 1:1 hydrochloric acid and 1:1 nitric acid. Convert the solution to a citrate buffer at pH 8.5–9.[202] Add sodium diethyldithiocarbamate and extract the complexes of the three metals with benzene. Reextract the copper and nickel with 1.25 *N* nitric acid, leaving the cobalt to be read in the benzene phase at 325 nm. Adjust the acid phase to pH 8.5–9 and extract the copper selectively by lead diethyldithiocarbamate in chloroform for reading at 435 nm. Then extract the nickel with dimethylglyoxime in chloroform and read it at 375 nm. No other ingredient of aluminum interferes.

The reagent is applicable to analysis of sulfides and selenides of cadmium and zinc used as semiconductors.[203] Dissolve sulfides in 1:1 hydrochloric acid; dissolve selenides in sulfuric acid containing bromine. Evaporate to dryness and take up in an acetate buffer solution. Extract the copper with a solution of lead diethyldithiocarbamate in chloroform.

In analysis of textile materials containing 0.009–0.019% copper, the most satisfactory reagents were lead diethyldithiocarbamate or dithiooxamide.[204]

For analysis of plant material, ash 25 grams at 450°. Leach the ash with 10 ml and 10 ml of hydrochloric acid.[205] Filter and wash. Dilute to 100 ml and determine copper in a 25-ml aliquot with lead diethyldithiocarbamate.

[200] S. E. Kreimer and L. P. Butylkin, *Zavod. Lab.* **24**, 131–133 (1958); V. B. Aleskovskii, G. S. Semikozov, and I. P. Kalinkin, *Tr. Leningrd. Tekhnol. Inst. Lensoveta* **1960** (61), 144–149; I. Nachev and D. Filipov, *C. R. Acad. Bulg. Sci.* **24**, 757–760 (1971).
[201] I. P. Kalinkin and G. S. Smikozov, *Zavod. Lab.* **27** (1), 17–20 (1961).
[202] Toshiaki Kuroha, *Jap. Anal.* **20**, 1365–1369 (1971); cf. C. M. Dozinel, *Metall* **20**, 242–245 (1966).
[203] R. P. Pantaler, N. B. Lebed', and L. N. Semenova, *Tr. Kom. Anal. Khim.* **16**, 24–29 (1968).
[204] Pavel Bartusek, *Textil* **28**, 90–92 (1973).
[205] Milan Karvanek, *Sb. Vys. Sk. Chem.-Technol. Praze, E.* **23**, 13–21 (1969).

As the reagent, dissolve 0.2 gram of lead acetate in 50 ml of water and add 1 gram of sodium potassium tartrate. Add 4% sodium hydroxide solution until the reagent is clarified. Add 10 ml of 10% potassium cyanide and 0.25 gram of sodium diethyldithiocarbamate dissolved in 5 ml of water. Filter the precipitate, wash, and drain. Dissolve the precipitate in 200 ml of chloroform and wash this solution with water. Add to the chloroform solution 1 volume of amyl acetate and 8 volumes of additional chloroform.

Procedures

Lead.[206] Dissolve a sample expected to contain 0.01–0.05 mg of copper in 1:2 nitric acid and evaporate almost to dryness. Take up the residue in 50 ml of water and adjust to pH 1–4. Shake with 20 ml of 0.001% solution of lead diethyldithiocarbamate in toluene for 1 minute. Wash the extract with 25 ml and 25 ml of 1:1 hydrochloric acid and read at 436 nm.

Zinc Electrolyte.[207] As an ammonium citrate solution, dissolve 200 grams of citric acid in 150 ml of water and add 210 ml of ammonium hydroxide.

Add 1 ml of nitric acid to a 10-ml sample and boil for 5 minutes to decompose the dextrin content. Cool, add 1 ml of sulfuric acid, and evaporate until salts start to crystallize. Cool and add sufficient water to dissolve the salts. Add 5 ml of the ammonium citrate solution, 5 drops of a saturated solution of cresol red, and 1:1 ammonium hydroxide until the color is yellowish pink in reflected light. Dilute to 50 ml and cool. Shake for 2 minutes each with 5 ml, 5 ml, and 5 ml of 0.001% solution of lead diethyldithiocarbamate in chloroform. Dilute the combined extracts to 25 ml and read at 436 nm.

If the sample contains 0.15–0.3 mg of lead per liter, modify the technic by heating the extracts to about 35° to clarify them before reading. In that case double the concentration of the reagent.

Rhodium-Plating Electrolyte.[208] Neutralize 5 ml of sample containing about 1% rhodium with 8% sodium hydroxide solution. Add 5 ml of 10% sodium chloride solution and evaporate nearly to dryness. Add 5 ml of hydrochloric acid and evaporate nearly to dryness, repeating this treatment three more times. Take up the residue in 0.04 *N* hydrochloric acid and dilute to 100 ml with that acid.

Pack a 30×1 cm column with 0.25–0.5 mm KU-Z resin in hydrogen form. Pass the sample through the column at 30 drops/minute and wash the column with 50 ml of 0.04 *N* hydrochloric acid. Elute iron, nickel, and copper with 75 ml of the 0.04 *N* hydrochloric acid and wash the column with 50 ml of water. Evaporate the eluate and washings nearly to dryness, take up in water, and dilute to 25 ml.

Develop copper in an aliquot as for zinc electrolyte from "Add 5 ml of the ammonium citrate solution...." The iron can be developed with sulfosalicylic acid, the nickel with dimethylglyoxime.

[206] Karolev Asen, *Rudodob. Metal.* **24**, 37–39 (1969).

[207] G. S. Semikozov, E. G. Kruglova, and I. P. Kalinkin, *Izv. Vyssh. Ucheb. Zaved. Khim. Khim. Tekhnol.* **7**, 194–197 (1964).

[208] A. T. Pulpenko, L. N. Ivashchenko, and N. F. Falendysh, *Zavod. Lab.* **39**, 20–21 (1973).

Zinc and Cadmium Sulfides and Selenides.[209] Dissolve a 2-gram sample by heating with 5 ml of nitric acid. Add 10 ml of water and boil off the oxides of nitrogen. Add ammonium hydroxide until the precipitate initially formed has dissolved. Cool and extract the copper with 5 ml and 5 ml of 0.001% solution of lead diethyldithiocarbamate in chloroform. Combine the extracts and wash with 10 ml and 10 ml of 1:4 ammonium hydroxide followed by 10 ml of water. Dilute the chloroform extract to 25 ml and read at 436 nm.

Highly Pure Alkali–Metal Halide Crystals.[210] Copper, nickel, iron, and manganese are determined seriatim on the same solution.

As a sample of potassium chloride or potassium bromide, dissolve 5 grams in 5 ml of water. For sodium iodide, dissolve 5 grams in 5 ml of water, add 1 ml of 1:1 sulfuric acid, and evaporate to sulfur trioxide fumes. Take up in 5 ml of water. For ammonium fluoride dissolve 5 grams in 5 ml of water, add 1 ml of perchloric acid, and heat to dense fumes of perchloric acid. Take up in 5 ml of water. As a buffer use 4% sodium citrate solution containing 5 ml of 0.1% phenol red indicator per liter, neutralized with ammonium hydroxide. For copper, add 5 ml of the ammonium citrate buffer solution and dilute to 50 ml. Extract with 5 ml and 5 ml of lead diethyldithiocarbamate reagent in chloroform. Then extract with 5 ml of chloroform. Dilute the extracts to 25 ml and read at 436 nm.

For nickel, neutralize the aqueous phase from which copper was extracted to pH 8 with 1:20 ammonium hydroxide. Add 1 ml of 1% solution of dimethylglyoxime and set aside for 15 minutes. Extract the nickel complex with 5 ml, 5 ml, and 5 ml of chloroform. Back-extract the nickel from the combined chloroform extracts with 5 ml, 5 ml, and 5 ml of 0.5 *N* hydrochloric acid. To the combined acid extracts, add 0.2 ml of saturated bromine water, 1 ml of ammonium hydroxide, and 0.5 ml of 1% solution of dimethylglyoxime. Set aside for 15 minutes and read at 442 nm.

For iron, add 15 ml of 10% ascorbic acid solution and neutralize with ammonium hydroxide. Add 5 ml of 0.1% solution of 1-nitroso-2-naphthol in 0.008% potassium hydroxide solution and dilute to 150 ml. Set aside for 45 minutes; then extract the ferrous complex with 15 ml and 10 ml of isoamyl alcohol. Combine the extracts and read at 690 nm.

For manganese, extract with 10 ml of chloroform and discard the extract. Add 5 ml of 0.1% solution of sodium diethyldithiocarbamate and extract the manganese complex with 5 ml of chloroform. Read at 400 nm.

Gallium Arsenide.[211] As the reagent, dissolve 0.4 gram of lead acetate and 0.4 gram of sodium diethyldithiocarbamate in 250 ml of chloroform. Dissolve a 0.5 gram sample, which may contain as little as 0.1 ppm of copper, in 3 ml of hydrochloric acid and 2 ml of nitric acid. Evaporate to about 1 ml and add 10 ml of water. The solution should be approximately 0.5 *N* with residual hydrochloric

[209] I. P. Kalinkin and V. B. Aleskovskii, *Izv. Vyssh. Ucheb. Zaved. Khim. Khim. Tekhnol.* **6**, 553–556 (1963).

[210] A. B. Blank, A. M. Bulgakova, and N. T. Sizonenko, *Zh. Anal. Khim.* **16**, 715–719 (1961); cf. A. M. Bulgakova, A. B. Blank, A. K. Khurkryanskii, and G. S. Plotnikova, *Stsintill. Stsintilyats Mater. Vses. Nauchn.-Issled. Soveshch.* **1957**, 281–290; *Metody Anal. Veshchestv. Oboboi Chist. Monokrist. Gos. Kom. Sov. Min. SSSR Khim.* **1962** (1), 63–73.

[211] V. G. Goryushina, E. Ya. Biryukova, and L. S. Razumova, *Zh. Anal. Khim.* **23**, 1044–1046 (1968).

acid. If not, adjust at this point. Add about 0.5 gram of urea and heat to boiling. Cool and add 1 ml of the reagent solution. Shake with 5 ml of chloroform and read the organic phase at 425 nm.

Water.[212] Add 10 ml of 200-400 mesh Dowex 50W-X2 in hydrogen form to 1 liter of sample and stir for 15 minutes. Filter the resin on a sintered glass filter and wash with water. Elute the copper with 40 ml of 50% solution of calcium chloride. Extract the copper with 10 ml and 10 ml of 0.04% solution of the reagent. Filter the combined extracts, adjust to 20 ml, and 2 minutes later read at 425 nm.

The eluate from the resin can also be developed by cuprizone or dicupral.

Food.[213] COPPER, MERCURY PRESENT. Wet-ash and isolate the mercury with dithizone. Take an aliquot containing up to 15 μg of copper and adjust to pH 3 by dropwise addition of ammonium hydroxide. Extract copper with 1-ml portions of 0.01% solution of dithizone in chloroform until the last portion remains green. Add 50 ml of 0.1 N sulfuric acid to the combined extracts and add 25 mg of sodium nitrite. Heat at 75° to evaporate chloroform and decompose nitrous acid. Cool, and add 5 ml of 5% solution of sodium potassium tartrate. Adjust to pH 6 with 2 N ammonium hydroxide. Add 2 ml of chloroform, shake, and discard the chloroform extract. Extract the aqueous solution with 3 ml and 2 ml of 0.04% solution of lead diethyldithiocarbamate in chloroform. Filter the combined extract and read at 440 nm against a blank.

Fatty Acids.[214] COPPER, IRON, AND MANGANESE. Ash a 3-gram sample and increase the calcining temperature to 500°. Take up the residue in 3 ml of 1:3 hydrochloric acid. Reduce the iron by addition of 1 ml of 10% hydroxylamine hydrochloride solution. Add 1:5 ammonium hydroxide to pH 3.5. Shake with 10 ml and 5 ml of 0.03% solution of lead diethyldithiocarbamate in chloroform and wash with 5 ml of chloroform. Dilute the chloroform extracts to 25 ml, filter, and read copper at 436 nm.

To the aqueous solution from which copper was extracted, add 1 ml of 0.2% solution of 1,10-phenanthroline and set aside for 1 hour. Add ammonium hydroxide to raise to pH 6.5, then add 1 ml of 10% solution of potassium iodide. Extract the ferrous complex with 10 ml, 5 ml, and 5 ml of chloroform, dilute the extracts to 25 ml, and read iron at 490 nm.

To the iron-free aqueous solution, add 4 ml of 0.5% solution of sodium diethyldithiocarbamate. After 3 minutes, extract with 10 ml, 5 ml, and 5 ml of chloroform. Dilute these extracts to 25 ml and read manganese at 400 nm.

Rapeseed.[215] Ash a 5-gram sample at 500°. Treat the ash with 10 ml of 1:3 hydrochloric acid. Filter, wash well with the same acid, and dilute to 25 ml. Dilute a 5-ml aliquot to 20 ml with water and extract with 10 ml of a 0.001% solution of lead diethyldithiocarbamate in chloroform. Read the extract at 436 nm.

[212] M. Pavlik, *Chem. Prum.* **15**, 365–366 (1965).
[213] Michal Nabrzyski, *Anal. Chem.* **47**, 552–553 (1975).
[214] V. K. Chebotarev, F. M. Kozlova, and I. A. Kozlov, *Tr. Novocherek. Politekh. Inst.* **266**, 19–22 (1972); cf.M. Karvenak, *Sb. Vys. Sk. Chem-Technol., Odd Fak. Potrav. Technol.* **5**, 203–212 (1961).
[215] M. Karvanek, *Prum. Potravin.* **15**, 282–283 (1964).

Biological Material.[216] Mix 2 ml of serum, plasma, blood, or cerebrospinal fluid with 2 ml of 1 : 1 hydrochloric acid and let it stand for 10 minutes. Deproteinize by addition of 2 ml of 20% solution of trichloroacetic acid. Centrifuge and decant. Neutralize with ammonium hydroxide to the bromophenol blue end point. Extract with 5 ml of the prepared solution of lead diethyldithiocarbamate in chloroform and read at 440 nm against a blank.

Zinc Diethyldithiocarbamate

Copper displaces the zinc from the reagent above.[217] Bismuth at the twofold level or thallium at tenfold are not tolerated.

For copper in high purity chromium at around 5 ppm, the sample solution is buffered to about pH 3 with sodium acetate and extracted with zinc diethyldithiocarbamate in carbon tetrachloride.[218] This is read at 435 nm. The copper is then extracted with ammoniacal potassium cyanide solution, and the loss in absorption measures the copper.

Procedure.[219a] Adjust a 25-ml sample containing 0.25–1.75 μg of copper per ml to 0.1 *N* with hydrochloric acid. Extract with 5-ml portions of 0.01% dithizone solution until the extracts remain green. Wash the combined extracts with water. Back-extract copper from the dithizonate by shaking with 25 ml of 0.005% aqueous solution of zinc bis(hydroxyethyl)dithiocarbamate. Wash this aqueous layer with carbon tetrachloride. Extract the copper from it by shaking with 5-ml portions of 0.02% solution of zinc diethyldithiocarbamate in carbon tetrachloride until the extract remains colorless. Dilute the combined extracts to 25 ml with more extracting solution and read at 420 nm against a portion of the extracting solution that has been shaken with the bis(hydroxyethyl)dithiocarbamate solution.

Zinc. Dissolve 10 grams of metallic zinc in 110 ml of 2:3 nitric acid. Boil to drive off oxides of nitrogen and cool. Adjust to pH 4 with ammonium hydroxide and add 3 ml of 40% solution of ammonium citrate. Shake with 10 ml of 0.0001 *M* reagent in carbon tetrachloride for 5 minutes. Filter through paper and read at 436 nm.

Zinc Salts. Dissolve an appropriate sample in water and proceed as for zinc from "Adjust to pH 4. . . ."

Tantalum and Niobium.[219b] Dissolve a 1-gram sample containing 0.2–25 ppm of copper in nitric acid and hydrofluoric acid. Add 5 ml of 1 : 1 sulfuric acid and heat to sulfur trioxide fumes. Cool and take up in 10 ml of 25% solution of tartaric acid. Add 5 ml of 4% solution of boric acid and dilute to 50 ml. At this point the solvent

[216] J. Masopust, *Prakt. Probl. Lek. Chem. I. Sjezd Lek. Biochem. Pilsen* **1955**, 165–171.
[217] K. Ogiolda, *Chem. Anal.* (Warsaw) **10**, 611–618 (1965).
[218] Tadashi Yanagihara, Nobuhisa Matano, and Akira Kawase, *Jap. Anal.* **11**, 108–111 (1962).
[219a] Yoshimasa Tanaka and Kazuo Ito, *ibid.* **6**, 728–731 (1957).
[219b] Tsutomu Fukasawa and Takeshi Yamane, *Anal. Chim. Acta* **84**, 195–9 (1976).

may be 0.5–6 *N* sulfuric acid. Extract with 10 ml of 0.03% solution of zinc diethyldithiocarbamate in carbon tetrachloride and read the extract at 438 nm against a reagent blank for up to 25 μg of copper per ml. If bismuth is present wash it out of the extract with 7 *N* hydrochloric acid.

Drinking Water.[220] Acidify a 50-ml sample containing not more than 0.04 mg of copper with 5 ml of 1 : 1 sulfuric acid. Shake with 10 ml of 0.05% solution of the reagent in carbon tetrachloride. Filter the organic layer and read it at 430 nm against carbon tetrachloride.

Vegetable Oils.[221] Heat a 5-gram sample with 10 ml of sulfuric acid. Add 30% hydrogen peroxide in 0.5-ml portions from time to time until the solution is colorless. Dilute to 100 ml and follow the general procedure from "Adjust a 25-ml sample containing...."

Antimony Dibenzyldithiocarbamate

This is the most sensitive of the various dibenzyldithiocarbamates used for determining copper.[222] It is prepared by treating the carbamate with antimony chloride in chloroform and crystallized from chloroform and methanol. The solution in chloroform or carbon tetrachloride is stable to 5 *M* hydrochloric acid. It conforms to Beer's law for 0.25–2 μg of copper in the extract. Bismuth is extracted with the copper but can be removed by shaking the extract with 4.5 *N* hydrochloric acid.

Procedure. Mix the sample with a potassium acid phthalate–sodium hydroxide buffer solution for pH 4–5. Shake with four 5-ml portions of 0.05 m*M* solution of the reagent in carbon tetrachloride. Adjust the combined extracts to 25 ml and read at 440 nm.

Zinc Dibenzyldithiocarbamate

This reagent, also known as arazate, obeys Beer's law up to 2 μg of copper per ml of extract.[223] As applied to beer, the problem becomes complex. The interference by natural color increases exponentially with decrease in the copper content at less than 0.6 mg/liter but is negligible at higher levels.[224] Therefore the copper content is increased by 0.7 mg/liter to permit reliable results, almost unaffected by the natural color.

As applied to steel, bismuth interferes, but nickel and cobalt do not, and the

[220] D. C. Abott and J. R. Harris, *Analyst* **87**, 497–499 (1962).

[221] Toshihisa Manuta and Tomoyuki Mukoyama, *Jap. Anal.* **18**, 1312–1316 (1969).

[222] Yoshimasa Tanaka, Kei Okudaira, and Atsushi Sugii, *ibid.* **20**, 1240–1245 (1971).

[223] Leroy G. Borchardt and John P. Butler, *Anal. Chem.* **29**, 414–419 (1957).

[224] H. Trommsdorff, *Monatsschr. Brau.* **15**, 78–82 (1962), *ibid.* **16**, 101–104 (1963); cf. R. G. Ault, *J. Inst. Brew.* **76**, 347–348 (1970).

interference due to iron is negligible.[225] The color is stable for 1 hour. The reagent is applicable to feedstuffs,[226] plantlike material, and so on, and to soybean oil.[227]

Milk, deproteinized with trichloroacetic acid, can be used for determination of copper.[228] Carotene interferes but is extracted with toluene.[229] As applied to natural phosphate rock, the reagent will determine up to 10 μg in a 2-gram sample.[230] Iron is a major source of interference.

Procedures

Steel. Dissolve an appropriate sample in 10 ml of 3:3:14 sulfuric acid–phosphoric acid–water. Add a few drops of 10% hydrogen peroxide to destroy carbides. After partial cooling, add 10 ml of water and boil to destroy residual hydrogen peroxide. Dilute to a known volume and take an aliquot containing 0.01–0.05 mg of copper. Add 5 ml of 1 : 1 sulfuric acid and dilute to 50 ml. Extract with 10 ml and 10 ml of 0.05% zinc dibenzyldithiocarbamate solution in carbon tetrachloride. Dilute to 25 ml and read at 430 nm.

Zinc Sulfide Phosphors.[231] This technic is for 5–40 ppm of copper. One for lesser amounts by sodium diethyldithiocarbamate is given elsewhere. Dissolve 0.5 gram of sample in 5 ml of hydrochloric acid and boil off the hydrogen sulfide. By adding water and hydrochloric acid, adjust it to N in acid. Shake for 1 minute with 10 ml of 0.01% solution of zinc dibenzyldithiocarbamate in carbon tetrachloride. Read the extract at 420 nm.

Water.[232] Mix 50 ml of sample with 5 ml of 1 : 4 hydrochloric acid. Add 5 ml of 0.05% solution of arazate in carbon tetrachloride. Shake for 2 minutes, and filter the solvent layer through glass wool. Read at 435 nm against a blank. Results agree well with those by bathocuproine. More than 10 ppm of chlorine interferes.

Boiler Feed Water.[233] Adjust a 200-ml sample to 0.25 N by addition of 1 : 1 hydrochloric acid. Extract with 5 ml and 5 ml of 0.1% solution of the reagent in carbon tetrachloride. Filter the combined extracts and read against a blank at 438 nm 10–60 minutes after extraction. The reagent is preferred over neocuproine and sodium diethyldithiocarbamate.[234]

Alcoholic Beverages. This reagent is suitable for determination of copper in high

[225] Om P. Bhargava, *Talenta* **16**, 743–745 (1969).
[226] E. Schuurmans and A. A. Steiner, *Chem. Weekbl.* **54**, 198–202 (1958).
[227] G. R. List, R. L. Hoffman, W. F. Kwolek, and C. D. Evans, *J. Am. Oil Chem. Soc.* **45**, 872–875 (1968).
[228] A. C. Smith, *J. Dairy Sci.* **50**, 664–668 (1967).
[229] B. C. Armstrong and C. W. Dill, *ibid.* **51**, 1851–1853 (1968).
[230] Jeanine Kocher, *Chim. Anal.* (Paris) **44**, 161–165 (1962).
[231] Yoshihide Kotera, Tadao Sekine, and Masao Takahashi, *Bull. Chem. Soc. Jap.* **39**, 2523–2536 (1966).
[232] A. C. Smith, *J. Dairy Sci.* **55**, 39–41 (1972).
[233] A. L. Wisson, *Analyst* **87**, 884–894 (1962); UKAEA, *Report* **PG 434 (W)**, 4 pp (1963).
[234] J. Hissel and M. Cabot-Dethier, *Trib. Cebedeau* **15**, 549–554 (1962).

wines, spirits, gin, whiskey, brandy, rum, and wine.[235] To 10 ml having a proof no higher than 135, add 0.5 ml of 1:5 sulfuric acid. Shake briefly with 10 ml of 0.2% solution of arazate in carbon tetrachloride and release the pressure. Restopper and shake 100 times. Filter the extract through a plug of glass wool or cotton and read at 438 nm in 10–60 minutes.

Wood.[236] For preservative-treated wood, the sample must be microtomed to a thickness less than 300 μm. Soak an appropriate weight of sample for 5 minutes in 10 ml of hydrochloric acid. Wash with water and dilute to 100 ml. Shake with 10 ml of 0.1% solution of the reagent in carbon tetrachloride and read the extract at 435 nm.

Pulp, Paper, and Pulping Liquors. Wet-ash the sample as described under bathocuproine. To the cooled and diluted digestate add 5 ml of 5% sodium sulfite solution. Add 5 ml of 0.05% solution of zinc dibenzyldithiocarbamate in carbon tetrachloride and shake for 1 minute. Read the extract at 435 nm against a reagent blank.

For analysis of preserved softwoods, it is not necessary to ash.[237] Convert to sawdust and dry. Extract with 50 ml of 1:6 sulfuric acid and 10 ml of 30% hydrogen peroxide at 75° for 20 minutes. Mix with 150 ml of water and let cool and settle. Filter and determine copper with zinc dibenzyldithiocarbamate as previously.

Beer. Mix 25 ml of sample with 3 ml of 1:3 sulfuric acid and 1 ml of 30% hydrogen peroxide. Heat at 100° for 30 minutes and cool. Add copper sulfate solution containing 17.5 μg of copper and shake with 5 ml of 0.02% solution of the reagent in carbon tetrachloride for 5 minutes. Read the organic phase at 440 nm and subtract the correction for added copper.

Biological Material.[238] Wet-ash a sample with sulfuric and nitric acids. Evaporate to sulfur trioxide fumes, take up in water, and dilute to a known volume.

Dilute an aliquot containing less than 0.02 mg of copper to about 20 ml with water and add 1 ml of 1:1 sulfuric acid. Add 10 ml of 0.01% solution of the reagent in carbon tetrachloride and shake for 3 minutes. Filter the extract through cotton, rejecting the first ml of the filtrate, and read at 440 nm.

Urine.[239] Heat 100 ml with 20 ml of sulfuric acid until colorless. Dilute the digestate with 20 ml of water. Add 1 ml of 12% citric acid solution and 5 ml of 0.1% zinc dibenzyldithiocarbamate in carbon tetrachloride. Shake for 4 minutes and separate the solvent. Dry with sodium sulfate and read at 437 nm against a reagent blank.

Alternatively,[240] mix 25 ml with 1 ml of hydrochloric acid and set aside for 15

[235] Duane H. Strunk and A. A. Andreasen, *J. Assoc. Off. Anal. Chem.* **50**, 334–338 (1967).
[236] A. I. Williams, *Analyst* **93**, 611–617 (1968).
[237] A. I. Williams, *ibid.* **95**, 670–674 (1970).
[238] A. Klewska and M. Strycharska, *Chem. Anal.* (Warsaw) **12**, 1325–1328 (1967).
[239] D. Kilshaw, *J. Med. Lab. Technol.* **20**, 295–297 (1963).
[240] A. J. Giorgio, G. E. Cartwright, and M. M. Wintrobe, *Am. J. Clin. Pathol,* **41**, 22–26 (1964).

minutes. If bilirubin is present, heat at 94° for 15 minutes and cool. Shake with 10 ml of 0.015% solution of zinc dibenzyldithiocarbamate in carbon tetrachloride. Centrifuge the lower layer and read at 435 nm. Compare with a 2-μg standard reference and an extraction blank.

Blood.[241] Digest a 5-ml sample with 4 ml of nitric acid, 2 ml of perchloric acid, and 1 ml of sulfuric acid. Heat to fumes, cool, and take up with 10 ml of water. Boil, cool, and dilute to 30 ml. Shake for 2 minutes with 5 ml of 0.05% solution of zinc dibenzyldithiocarbamate in carbon tetrachloride. Read the extract at 435 nm.

Liver. Digest 3 grams of fresh liver or 1 gram of dried liver as for blood. Dilute the digestate to 100 ml and dilute 5 ml to about 30 ml for extraction with 10 ml of reagent.

Cuprethol

Cuprethol is bis-(2-hydroxyethyl)dithiocarbamic acid. As applied to beer, the reagent is simple, rapid, and reasonably accurate. It has been recommended for routine use, although it is not the best reagent below 0.1 ppm of copper.[242] The reagent simultaneously determines nickel and cobalt. Silver increases the value for cobalt. Manganese, aluminum, and tin decrease the value for nickel. At normal levels of less than 0.3 ppm, iron causes less than 0.05 ppm error in the value for copper and no significant effect on cobalt and nickel. The reagent is also applicable to distilled alcoholic beverages.[243]

Procedures

Copper, Nickel, and Cobalt in Beer. Prepare the reagent fresh daily by mixing 1 volume of 9.5% by volume carbon bisulfide in methanol with 3 volumes of 2% diethanolamine in methanol. As a buffer solution, dissolve 105 grams of sodium acetate trihydrate in water, add 65 ml of glacial acetic acid, and dilute to 1 liter.

Mix 50 ml of degassed beer with 25 ml of the buffer solution. Take a 20-ml aliquot as sample and another as blank. To the sample, add 2 ml of reagent; add 2 ml of 1% methanolic diethanolamine to the blank. After 10 minutes read copper at 445 nm. Read cobalt and nickel at 325 and 365 nm, respectively. The cobalt and nickel must be corrected for copper if the absorption at 445 nm exceeds 0.05.

Distilled Alcoholic Products. To 30 ml of sample, add 0.5 ml of 1 : 1 hydrochloric acid, 0.5 ml of 3% solution of sodium pyrophosphate decahydrate, and 7 ml of

[241] Takakata Matsubara, Tomoji Tanaka, and Toshiya Fushimizu, *Kumamoto Daigaku, Igaku-Bu, Kawakita Naikagaku-Kyoshitsu* **47**, 247–251 (1958); N. A. Brown and R. G. Hemingway, *Res. Vet. Sci.* **3**, 345–347 (1962).

[242] M. Pinkas, *Bull. Soc. Pharm. Lille* **1960**, No. 3, 93–97; P. Balatre and M. Pinkas, *Chim. Anal. (Paris)* **43**, 433–438 (1961); M. W. Brenner, M. J. Mayer, and S. R. Blick, *Proc. Am. Soc. Brew. Chem.* **1963**, 165–174; S. R. R. Blick and B. K. Blenkinship, *ibid.* **1965**, 187–193.

[243] D. H. Strunk and A. A. Andreasen, *J. Assoc. Off. Agr. Chem.* **48**, 478–482 (1965).

30% solution of sodium acetate. The pH should then be 5–6. After 5 minutes add 0.5 ml of a reagent prepared by mixing equal volumes of 2% solution of diethanolamine in methanol and a 1.5% by volume solution of carbon bisulfide in methanol. Dilute to 50 ml and after 10 minutes read at 433 nm against water.

Copper gives a yellow complex with **zinc di-(2-hydroxyethyl)dithiocarbamate**[244] at pH 5.5–9.5. The absorption with less than 0.07 mg in 25 ml is read at 470 nm and is extractable with tributyl phosphate. The reagent also gives complexes with bismuth, cobalt, nickel, and tellurium.

A variant on other salts of the reagent is the use of **cadmium diethyldithiocarbamate.**[245] As the reagent, dissolve 0.07 gram of cadmium acetate and 0.5 gram of sodium potassium tartrate in 150 ml of water. Add 10 ml of 10% solution of potassium cyanide and 0.125 gram of sodium diethyldithiocarbamate. Shake with 250 ml and 50 ml of chloroform, filter the extracts through paper, and dilute to 1 liter with chloroform.

For 1–10 parts of copper per billion, extract 400 ml of sample solution with 20 ml of the reagent at pH 2–8 for 10 minutes. There is no interference by 1 ppm of nickel, cobalt, iron, manganese, aluminum, zinc, calcium, and magnesium. If excess iron interferes, add citrate to the sample as a masking agent.

When a sample at pH 3 containing 0.1–3 μg of copper is shaken with 0.1% solution of **lead dibenzyldithiocarbamate**[246] in chloroform, the cupric ion displaces the lead for reading at 435 nm. Mercury, silver, and thallic ion interfere. Citrate and tartrate at 0.33 *M* mask ferric, nickel, cobalt, zinc, cadmium, manganous, stannous, and bismuth ions.

Lead bis(2-hydroxyethyl)dithiocarbamate[247] forms water-soluble complexes with many heavy metals. For cupric, nickel, and mercuric ions, the ratio of metal to the reagent is 1 : 2. The stability of the complexes decreases in the following order: auric, mercury, silver, copper, bismuth, lead, ferric, cadmium, cobalt, nickel, zinc, manganese. For determination of copper in solutions of iron, aluminum and lead alloys, or nonferrous metal ores, add sodium fluoride solution to prevent interference by ferric ion. Then add a buffer solution for pH 3. Add 0.01 *M* solution of the lead salt of the reagent in thirtyfold excess to prevent interference by nickel, manganese, cobalt, and ferrous ion and read at 413 nm.

Piperidinium pentamethylenedithiocarbamate[248] forms a 2 : 1 complex with copper. To a 10-ml sample containing about 0.02 mg of copper, add 10 ml of 10% sodium citrate solution and 2 ml of a 0.1% solution of the reagent. Dilute to 50 ml and adjust to pH 9–9.5. Shake for 3 minutes with 10 ml of carbon tetrachloride and read the extract at 434 nm against a blank. The complex obeys Beer's law up to 2.4

[244]Hitoshi Yoshida, Masahiro Yamamoto, and Seiichiro Hikime, *Jap. Anal.* **11**, 197–201 (1962).

[245]Kin'ya Sono, Yoshimichi Mitsukami, Tatsuo Nakashima, and Hiroto Watanabe, *ibid.* **14**, 1127–1133 (1965).

[246]F. Dittel, *Z. Anal. Chem.* **229**, 188–192 (1967).

[247]F. M. Tulyupa, G. E. Bekleshova, and M. A. Vitkina, *Zh. Anal. Khim.* **21**, 783–786 (1966).

[248]C. G. Ramachandran Nair, V. R. Satyanarayana Rao, and A. R. Vasudeva Murthy, *Mikrochim. Acta* **1961,** 741–748; cf. P. Dostal, J. Cernak, and J. Kartous, *Collct. Czech. Chem. Commun.* **33**, 1539–1548 (1968).

μg/ml and is photosensitive. Iron, cobalt, nickel, zinc, and manganese must be masked.

Dibutyldithiocarbamate[249] as 10 ml of 0.3% solution in carbon tetrachloride will extract 0.005–0.04 mg of cupric ion from 50 ml of 0.5–10 *N* sulfuric acid, 0.5–7 *N* hydrochloric acid, or 0.5–2 *N* nitric acid. Reading is at 432 nm. **Ethylphenyldithiocarbamate** performs similarly for reading at 438 nm. When 2 ml of 0.05% **dimethyldithiocarbamate** is added to a 40-ml solution at pH 9 containing 5 ml of 20% citric acid solution, the copper complex is extracted with chloroform and read at 430 nm. The complex of copper with **pyrrolidinedithiocarbamate**[250] is solvent extractable for reading.

Both cuprous and cupric ions form a yellow-brown chelate with **morpholinium 3-oxapentamethylenedithiocarbamate,**[251] extractable into chloroform for reading at 438 nm. Ammonium citrate and EDTA mask interfering elements.

When cupric ion is extracted from an acetate-buffered solution at pH 3.8 with 0.3 m*M* lead **diethyldiselenocarbamate**[252] in chloroform, followed by a chloroform wash, the combined extracts are read at 496 nm.

DITHIOOXAMIDE

This reagent, also known as rubeanic acid, combines with cuprous ion for reading at 425 nm over the range of 2–12 ppm.[253] Small amounts of cobalt and nickel are masked with EDTA. The reagent is applicable to alloys containing 0.05–6% copper.[254] The amounts of antimony, zinc, nickel, bismuth, or iron normally present in tin and tin-lead alloys do not interfere.

For determination of copper in such materials as serum, the organic matter is destroyed by wet digestion. Iron is eliminated as an interference by adding malonic acid. Then there is no need of extraction of the complex formed with the reagent.[255] The absorption conforms to Beer's law up to 8 μg/ml, and the blank is negligible. The reagent is applicable to green vegetables.[256a]

The complex follows Beer's law up to 20 ppm of cupric ion in 2 *N* sulfuric acid.[256b] The reagent must be added last, and heat is necessary for full color development. There is serious interference by aluminum, gold, cobalt, chromic, mercuric, tungstate, chromate, vanadate, chloride, and nitrate ions.

[249] Tomoyuki Mukoyama and Takuo Hasebe, *Jap. Anal.* **20**, 961–966 (1971).
[250] E. Kovacs and H. Guyer, *Z. Anal. Chem.* **186**, 267–288 (1962).
[251] W. Beyer and W. Likussar, *Mikrochim. Acta* **1967**, 721–724.
[252] A. I. Busev, Kh. Kirspuu, Kh. Kokk, and T. Tuisk, *Uch. Zap. Tartu. Gos. Univ.* **1969**, 153–161.
[253] Agnes Paul, *Anal. Chem.* **35**, 2119–2121 (1963).
[254] V. P. Zhivopistsev and E. A. Selezneva, *Uch. Zap. Perm. Univ.* **25**, 84–88 (1963).
[255] D. S. McCann, Patricia Burcar, and A. J. Boyle, *Anal. Chem.* **32**, 547–548 (1960); cf. Agnes Paul, *ibid.* **35**, 2119–2221 (1963).
[256a] Federico Minutilli and Piero Ruggieri, *Rass. Chim.* **11** (5), 14–16 (1959).
[256b] Dale A. Williams and D. F. Boltz, *Anal. Lett.,* **8**, 103–114 (1975).

Procedures

Steels.[257] As a buffer solution for pH 4–5, mix 400 grams of glacial acetic acid, 400 grams of sodium acetate trihydrate, and 200 ml of water. Dissolve a 1-gram sample in 15 ml of nitric acid and evaporate to dryness. Take up in 0.2 ml of nitric acid and 20 ml of water by heating. Cool and dilute to 100 ml.

To a 1-ml aliquot, add 2 ml of saturated solution of tetrasodium pyrophosphate and stir until clear. Add a drop of 2 *N* sulfuric acid and 2 ml of a buffer solution for pH 4–5. Mix and add 1 ml of 0.5% gelatin solution and 0.5 ml of 0.05% solution of dithiooxamide. Adjust to a known volume and read at 385 nm. The reagent is applicable to pig iron containing around 0.05% copper.[258]

Zinc.[259] As a buffer solution, mix 200 grams of ammonium acetate and 200 ml of 80% acetic and 100 ml of water. Dissolve a 5-gram sample in 40 ml of 1:1 hydrochloric acid and evaporate to half the volume. Add 1 ml of the buffer solution, at which point the pH should be 3.8–4.3. Add 2 ml of 1% gelatin solution containing a few drops of formaldehyde solution and 0.5 ml of 0.5% ethanolic dithiooxamide. Dilute to 50 ml and read at 690 nm.

Zinc Alloys. Dissolve a 5-gram sample in 2.5 ml of 1:1 hydrochloric acid, evaporate to half volume and dilute to 50 ml. Complete the analysis of a 20-ml aliquot as for zinc from "Add 1 ml of the buffer...."

Tin and Tin–Lead Alloys. Heat a 0.2-gram sample with 25 ml of 1:2 nitric acid and 4 grams of tartaric acid. When the sample starts to dissolve, add 0.5 ml of hydrochloric acid. When solution is complete, boil off oxides of nitrogen, cool, and dilute to 100 ml.

Adjust a 25-ml aliquot to pH 1–1.5 by addition of ammonium hydroxide. Then add 2.5 grams of sodium acetate trihydrate, which will raise the pH to 4–5. Add 10 ml of 0.5% gelatin solution and 2 ml of 0.5% ethanolic dithiooxamide. Dilute to 100 ml and read at 385 nm.

Lead.[260] Dissolve a 2-gram sample containing 0.024–0.110 gram of copper in 5 ml of nitric acid and dilute to 30 ml. Boil off oxides of nitrogen. Add ammonium hydroxide until turbidity appears. Stir in 2–3 ml of a suspension of cadmium sulfite; filter and wash the precipitate. Add 0.5 ml of hydrochloric acid to the precipitate in a porcelain crucible. Calcine and dissolve the residue in 1 ml of hydrochloric acid. Add 5 ml of water and neutralize with ammonium hydroxide. Acidify with acetic acid and dilute to 9 ml with acetate buffer solution for pH 4.2–4.3. Add 0.5 ml of 0.1% ethanolic dithiooxamide. Dilute to a known volume and read at 385 nm against a blank.

[257] V. N. Podchainova, *Tr. Ural. Politekh. Inst. im S. M. Kirova* **1960** (96), 117–123.
[258] Theo. Kurt Willmer, *Arch. Eishenhuttenw.* **29**, 159–164 (1958).
[259] K. N. Bagdasarov, *Uch. Zap. Rostov-na-Donu Gos. Univ.* **60** (10), 105–108 (1959).
[260] V. N. Podchainova and N. V. Krylova, *Tr. Ural. Politekh. Inst. S. M. Kirova* **1960** (96), 113–116.

Type Metal and Lead-Base Babbitt Metal.[261] Heat a 1-gram sample with 7 ml of sulfuric acid to a light grey precipitate. Cool, add 15 ml of 1:2 hydrochloric acid, and boil until the precipitate is white. Cool, add 30 ml of water, and ammonium hydroxide to a blue reaction to litmus. Add acetic acid until litmus is red. Add 5 ml of 0.5% gum arabic solution and 4 ml of 0.5% solution of thiooxamide in ethanol. Dilute to 100 ml, let settle, and read the supernatant layer at 650 nm.

Soap.[262] Reflux 100 grams of grated soap for 1 hour with 100 ml of water and 12 ml of sulfuric acid. Cool, filter, and wash the fatty acids with four 20-ml portions of 1:35 sulfuric acid. Dilute the filtrate and washings to 200 ml. Adjust a 30-ml aliquot to about pH 5 with ammonium hydroxide. Add 10 ml of acetate buffer solution for pH 4.5–4.8 and dilute to 49 ml. Add 0.5 ml of 0.1% ethanolic dithiooxamide and dilute to 100 ml. Read at 655 nm against a blank.

Serum. As a digestion reagent, mix 1 volume of perchloric acid with 9 volumes of nitric acid. As a buffered color reagent, dissolve 300 grams of sodium acetate trihydrate in 500 ml of water and filter. Add 280 ml of glacial acetic acid and 0.2 gram of gum arabic in 20 ml of water. Dissolve 0.1 gram of thiooxamide in 20 ml of ethanol, add to the mixture, and dilute to 1 liter. Dissolve 10 grams of malonic acid in 100 ml of water. Neutralize to a faint odor of ammonia and dilute to 500 ml.

To a 2-ml sample, add 4 ml of acid digestion mixture. Place in an aluminum heating block preheated to approximately 120° on a hot plate. Increase the temperature of the plate to incipient dryness in about 3 hours, with a final temperature of 260–290°. Remove the tubes from the block and cool. Wash down the side of the tube with 1 ml of ammonium malonate solution and warm briefly in hot water to ensure complete solution. Cool and add 3 ml of buffered color reagent and swirl. The final pH should be 4.2–4.3. After 30 minutes read at 385 nm against a reagent blank. A $3 \times 3 \times 12$ inch block carrying six pairs of duplicates, a reagent blank, and a $2\frac{1}{2}$ inch deep thermometer well has been suggested. This replaces Erlenmeyer flasks as a saving of time.

Vegetable Oil.[263] Extract a 130-gram sample in a liquid-liquid extractor with azeotropic (21%) hydrochloric acid for 12 hours. Neutralize the acid extract to Congo red with ammonium hydroxide. Develop an aliquot with dithiooxamide and read at 655 nm. Another aliquot may be used for determination of iron with sulfosalicylic acid.

DITHIZONE

Copper is one of the metals determined with dithizone. For more detail of dithizone and di-β-naphthylthiocarbazone as reagents, see Lead. As applied to

[261] James Brinn, *Metallurgia,* **63**, 52 (1961).
[262] A. Popov and B. Ivanova, *Zavod. Lab.* **35**, 161–162 (1969).
[263] A. D. Popov, B. S. Ivanova, and N. V. Yanishlieva, *Nahrung* **13**, 39–42 (1969).

titanium and titanium alloys, no chemical separation is required at pH 2, and the reagent will determine 0.0001–1% of copper.[264] A wavelength of 530 nm gives the greatest absorption by copper dithizonate, but at 515 nm there is the least interference by dithizone. Protected from light, the color is stable for 1 hour. Vanadium tends to oxidize the dithizone, but hydroxylamine avoids that. As applied to iron and steel, the solution is in hydrochloric acid with sodium sulfite present and extraction with 0.001% dithizone in carbon tetrachloride.[265]

In analysis of high purity lead, dissolve a 10-gram sample in nitric acid, and after appropriate dilution to pH 2, extract silver, bismuth, and copper with 0.005% dithizone solution in carbon tetrachloride.[266] Back-extract bismuth from the mixed dithizonates with *N* sulfuric acid. Strip the silver dithizonate with *N* hydrochloric acid. Evaporate the solvent, which contains the copper, to dryness, and take up in *N* hydrochloric acid for extraction of the copper as the dithizonate and reading at 550 nm.

For copper in molybdenum oxide, the solution of a pyrosulfate melt at pH 11–11.3 serves for coprecipitation with magnesium, and the solution of the precipitate at pH 0.9 is extracted with dithizone.[267]

For biological material, a digestate with sulfuric, nitric, and perchloric acids has citric acid and ammonia added.[268] With adjustment to pH 4–5 with sulfuric acid, the copper dithizonate is extracted by carbon tetrachloride. By shaking with permanganate in 0.1 *N* sulfuric acid, the copper is back-extracted. After adding an acetate-iodide-sulfite reagent, the copper is reextracted as the dithizonate for reading at 550 nm.

Ionic copper in commercial copper chlorophyllins can be separated by acidifying a solution of the sample to pH 3 with hydrochloric acid and filtering.[269] It can then be determined by dithizone.

As applied to gelatin[270] the solution is passed through a column of Dowex 50 W, a cation exchange resin. Copper is then eluted with 1 : 4 nitric acid. Adjusted to pH 2.1–7.2, the copper is extracted with 0.005% dithizone in carbon tetrachloride in the presence of sodium acetate and sodium thiosulfate.

For copper in liver, the dithizone method has been reported to be more accurate than those with neocuproine and cuprizone.[271]

For analysis of tantalum, dissolve the sample with nitric and hydrofluoric acids.[272] Extract lead, cadmium, iron, and copper as diethyldithiocarbamates with chloroform. Determine copper by dithizone.

For copper, nickel, cobalt, zinc, and cadmium as impurities in sodium chloride

[264]Howard W. Pender, *Anal. Chem.* **30**, 1915–1917 (1958).
[265]K. Nachev, D. Filipov, and I. Nachev, *C. R. Acad. Bulg. Sci.* **18**, 639–642 (1965).
[266]H. Jedrzejewska and M. Malusecka, *Chem. Anal.* (Warsaw) **12**, 579–584 (1967); cf. A. I. Makar'yants, T. V. Zaglodina, and E. D. Shuvaloya, *Anal. Rud Tsvet. Met. Prod. Pererab.* **1956** (12), 130–137; M. Cyrankowska and J. Downarowicz, *Chem. Anal.* (Warsaw) **10**, 1015–1027 (1965).
[267]Elizbieta Wieteska and Urszula Stolarczyk, *Chem. Anal.* (Warsaw) **15**, 183–189 (1970).
[268]E. J. Butler and G. E. Newman, *Clin. Chim. Acta* **11**, 452–460 (1965).
[269]Bernard M. Blank and Kenneth Morgareidge, *J. Am. Pharm. Assoc.* **46**, 73–74 (1957).
[270]J. Saulnier, *Sci. Ind. Photogr.* **35**, 1–7 (1964).
[271]E. J. Butler and D. H. S. Forbes, *Anal. Chim. Acta* **33**, 59–66 (1965).
[272]V. A. Nazarenko, E. A. Biryuk, M. B. Shustova, G. G. ShiFareva, S. Ya. Vinkovetskaya, and G. V. Flyantikova, *Zavod. Lab.* **32**, 267–269 (1966).

complex with EDTA, and extract from the solution with chloroform.[273] Evaporate to dryness, add sulfuric and nitric acids, and heat to sulfur trioxide fumes. Determine copper by dithizone.

On a 100×5 mm polytetrafluoroethylene column with dibutylphosphate as the stationary phase, gallium, cobalt, copper, and ferric ion are absorbed in *N* ammonium thiocyanate as the mobile phase.[274] Thus a 0.6-gram sample of aluminum is dissolved in hydrochloric acid, evaporated to dryness, and taken up in *N* ammonium thiocyanate. This is passed through the column at 0.5-1 ml/minute and washed with 50 ml of *N*-ammonium thiocyanate. Copper and iron are then eluted with 0.5 *N* hydrochloric acid, and the copper is determined with dithizone.

Concentrate copper and lead in natural waters by passing through a column of cation exchange resin at a rate of 1 liter/hour.[275] Desorb the ions by washing with 1:2 hydrochloric acid. Concentrate the effluent for reading lead as the sulfide and copper as the dithizonate.

For the sum of copper, lead, and zinc in water, add to 100 ml of sample 1 ml of 2.3% solution of sodium tartrate, 1 ml of 1.25% solution of sodium sulfite, 2 ml of 5% solution of hydroxylamine hydrochloride, and 1 ml of 0.65% solution of potassium cyanide.[276a] Add sufficient 2.5 *M* sodium carbonate–*M* sodium bicarbonate to adjust to pH 10.5, extract with 10 ml of 0.16 m*M* dithizone in carbon tetrachloride, and read at 532 nm.

Ash an organic sample containing 1–25 μg of copper.[276b] Separate copper chromatographically and determine as the dithizonate.

For separation of silver, mercury, and copper, and their individual determination by dithizone, see Silver.

For determination of copper and mercury by dithizone in the same sample of biological material, see Mercury.

Procedures

Without Extraction.[277] To 5 ml of slightly acid sample solution, add sequentially 15 ml of acetone and dropwise 0.05% dithizone in acetone until the red-violet color changes to grey-violet. Dilute to 25 ml with acetone and read against acetone containing 1 drop of the reagent per 25 ml. Beer's law is followed up to 0.015 mg of copper per ml.

Silver and Mercury Present.[278] Extract the sample solution at pH 2 with successive portions of a 13-ppm solution of dithizone in carbon tetrachloride until a portion remains green. If copper is high, it is permissible to use a higher

[273]Cezary Rozychi and Janusz Rogozinski, *Chem. Anal.* (Warsaw) **20**, 107–111 (1975).
[274]I. P. Alimarin, N. I. Ershova and T. A. Bol'shova, *Vest. Mosk. Gos. Univ., Ser. Khim.* **1969** (6), 79–84.
[275]V. B. Aleskovskii, R. I. Libina, and A. D. Miller, *Tr. Leningr. Tekhnol. Inst. Lensoveta* **48**, 5–11 (1958).
[276a]Takashi Ashizawa, Shigenori Sugizaki, Hisao Asahi, Hiroko Sakai, and Keiichi Shibata, *Jap. Anal.* **19**, 1333–1340 (1970).
[276b]Aleksandra Smoczkiewicz and Teresa Smoczkiewicz-Szczepanska, *Poznan. Towarozn. Przyj. Nauk, Wydz. Lek. Pr. Kom Farm.* **1**, 51–59 (1961).
[277]M. E. Macovschi and Viorica Nitu, *Rev. Roum. Chim.* **15**, 605–606 (1970).
[278]Harald Friedberg, *Anal. Chem.* **27**, 305–305 (1955).

concentration of dithizone. If chloride is present to not more than 1%, raise to pH 3.5 with ammonium hydroxide to be able to extract the silver. As a maximum, pH 5 may be required.

To separate the silver, extract the mixed dithizonates with 3 ml and 3 ml of 10% sodium chloride solution which is 0.015 *N* with hydrochloric acid. Then extract mercury and copper from the solution of dithizonates in carbon tetrachloride with 3 ml and 3 ml of 1 : 1 hydrochloric acid. Neutralize the acid extract to pH 1.5–2 with ammonium hydroxide. If more than 50 μg of copper is present, add 1 ml of 0.01 *N* EDTA solution.

To the solution free from silver and mercury, add 1 ml of 0.1 *N* calcium chloride solution. Add 3 ml of 25% ammonium citrate solution adjusted to pH 9 with ammonium hydroxide. Extract with portions of 13-ppm dithizone solution in carbon tetrachloride until the extracts no longer show a purple color. Wash the combined extracts with water to remove traces of EDTA. Extract the copper with 3 ml and 3 ml of 1 : 1 ammonium hydroxide. Reextract the copper with 3 ml and 3 ml of 13-ppm dithizone in carbon tetrachloride and read at 520 nm.

Gold.[279] Copper in gold is determined photometrically by dithizone after isolation of iron. Dissolve a 1-gram sample in 6 ml of hydrochloric acid and 2 ml of nitric acid. Extract the gold with 10 ml and 10 ml of isopentyl acetate and discard. Evaporate the solution to about 2 ml and dilute to 10 ml with 1 : 1 hydrochloric acid. Extract the iron with 20 ml and 10 ml of isopentyl acetate as a sample for determination with bathophenanthroline.

Add 1 ml of 1 : 1 sulfuric acid to the aqueous layer and evaporate to sulfur trioxide fumes. Take up the residue in water and adjust to pH 1. Shake for 1 minute with 10 ml of 0.002% solution of dithizone in carbon tetrachloride, and read at 520 nm.

Titanium and Titanium Alloys. Dissolve a 1-gram sample in 50 ml of 1 : 1 sulfuric acid. Oxidize by dropwise addition of 30% hydrogen peroxide and boil off the excess. Wash down the sides of the container, boil for a few minutes, and cool. Dilute according to the probable copper content and take an appropriate aliquot. Add 5 ml of 50% citric acid solution and 5 ml of 20% hydroxylamine hydrochloride solution. Adjust to pH 2 with ammonium hydroxide or sulfuric acid and cool if necessary. Add 25 ml of dithizone solution containing 8 mg per liter of carbon tetrachloride and shake for 10 minutes. Dilute the carbon tetrachloride layer to 25 ml with dithizone solution and let stand for 15 minutes. Read at 520 nm against a blank.

Uranium.[280] Adjust the sample solution to pH 1.5. Add hydrobromic acid until the solution is 5 *N* and extract the uranium with isopropyl ether. Silver and mercury are not extracted. Add perchloric acid and evaporate to fumes to drive off hydrobromic acid. Add 5 ml of water and again take to fumes. Take up in water

[279]G. Ackermann and J. Kothe, *Z. Anal. Chem.* **231**, 252–261 (1967); Zygmunt Marczenko, Krzysztof Kasiura, and Maria Krasiejko, *Chem. Anal.* (Warsaw) **14**, 1277–1287 (1969); cf. Krzysztof Kasiura, *ibid.* **20**, 809–816 (1975).

[280]J. Maracek and E. Singer, *Z. Anal. Chem.* **203**, 336–339 (1964); cf. Yoshio Morimoto, Takashi Ashizawa, and Kenji Miyahara, *Jap. Anal.* **11**, 61–65 (1962).

and dilute to a known volume. Take an aliquot containing 0.5–7.5 μg of copper. Complete as for titanium and titanium alloys from "Add 5 ml of 50% citric acid..." Palladium and gold interfere.

Indium.[281] Dissolve the sample in hydrobromic acid and adjust to 5 N. Remove the indium by extraction with isopropyl ether. Complete as for uranium from "Add perchloric acid...."

Ammonium Chloride.[282] Heat 10 grams of sample in a platinum crucible and take up the residue in 2 ml of 1:9 sulfuric acid. Heat at 100° for 10 minutes, cool, and add 8 ml of water. Add 5 ml of 0.001% solution of dithizone in carbon tetrachloride and shake for 5 minutes. Read the carbon tetrachloride layer at 520 nm.

Mercury Present in Substantial Amounts.[283] Adjust a sample containing 0.5–7.5 μg of copper to pH 2–2.5. Complex the copper with 1 ml of 0.1% solution of disodium nitrilotriacetic acid to avoid reduction of copper by iodide. Then add 2 ml of 16.6% solution of potassium iodide to mask the mercuric ion. Shake with 10 ml of 0.0001 M dithizone in carbon tetrachloride for 5 minutes. Filter the extract and read at 548 nm. The sample may contain 7 mg of mercury without interference. Large amounts of silver interfere.

Separation of Copper and Cobalt.[284] The technic was developed and applied to clover, as described below. The chromatographic column was 300×10 mm with a three-way stopcock at the bottom. Three grams of activated silicic acid added as a slurry in carbon tetrachloride produced a bed about 7.5 cm deep.

Boil a 1-gram sample of clover in 2 ml of water and 1 ml of nitric acid until reduced to about 1.5 ml. Cool, add 2 ml of sulfuric acid, and cool. Add 1 ml of nitric acid and evaporate to sulfur trioxide fumes. Repeat the last step until the liquid is colorless; then evaporate to about 0.5 ml. Cool and dilute with 2 volumes of water. Add a few drops of 50% sodium hydroxide solution until the indicator is yellow. Wash down the sides of the flask with water, add a few drops of 50% sodium hydroxide solution and 3 drops of 0.02% methyl red solution. Add 50% sodium hydroxide solution until the indicator is yellow. Wash down the sides of the flask with water, add a drop of sulfuric acid, and neutralize with N sodium hydroxide solution. Add 25 ml of ammonium hydroxide–ammonium chloride buffer solution for pH 9.6–9.8. Add 15 ml of 10% solution of sodium pyrophosphate and 2 ml of 0.1% solution of sodium diethyldithiocarbamate. Set aside for 1 minute, then shake with 3 ml, 3 ml, and 3 ml of carbon tetrachloride for 3 minutes each. Pass the combined extracts through a drying tube of anhydrous sodium sulfate, which is then washed with 1 ml of carbon tetrachloride. Add the carbon tetrachloride extract to the column to flow through at 0.5 ml/minute. Wash the column at least twice with carbon tetrachloride. Develop the chromatogram with 1:1 chloroform–carbon tetrachloride at 0.3–0.4 ml/minute. It requires 1–2 hours

[281] K. Kasiuaa, *Chem. Anal.* (Warsaw) **11**, 141–149 (1966).
[282] Shigehisa Nishio, *J. Chem. Soc. Japan* **77**, 1701–1704 (1956).
[283] Shinsuke Takei and Takio Kato, *Technol. Rep. Tohoku Univ.* **24**, 67–73 (1959).
[284] Dwight M. Smith and John R. Hayes, *Anal. Chem.* **31**, 898–902 (1959).

to develop the separate copper and cobalt chelate zones. Evaporate the separate eluates. Add 0.75 ml of sulfuric acid and 0.3 ml of perchloric acid to each. Heat at 200–250° until colorless. Cool, take up in an equal volume of water, and neutralize to methyl red indicator by dropwise addition of 50% sodium hydroxide solution. Make acid with 1 drop of sulfuric acid. Add *N* sodium hydroxide until the indicator is yellow again. Add 1 drop of hydrochloric acid to the copper solution. Dilute each to 25 ml.

For copper, mix 3 ml of 0.03 *N* hydrochloric acid with 2 drops of 10% hydroxylamine hydrochloride solution to minimize oxidation of the reagent. Add an aliquot of sample and shake for 10 minutes with 10 ml of 0.01% dithizone solution. Read the copper dithizonate at 510 nm against carbon tetrachloride.

For cobalt, add 25 ml of ammonium hydroxide–ammonium chloride buffer solution for pH 9.6–9.8 to an aliquot of the sample, shake for 5 minutes with 10 ml of 0.0025% dithizone solution. Read the extract at 545 nm against carbon tetrachloride.

Urine.[285] For 1–4 mg of copper per 100 ml, evaporate 100 ml of a 24-hour sample to dryness. Heat cautiously to 300° and cool. Add 5 ml of nitric acid. Evaporate and heat to 400°. Cool, add 5 ml of nitric acid, evaporate, and heat to 450°. Dissolve the ash in 2 ml of 1 : 1 hydrochloric acid and dilute to 50 ml. Add 3 drops of cresol red indicator and neutralize to a yellow color with ammonium hydroxide. Add 1 ml of a buffer solution containing 3.32% of disodium phosphate and 15.2% of citric acid. Shake for 10 minutes with 10 ml of 0.01% dithizone in carbon tetrachloride and read the extract at 515 nm.

Raw Gluten.[286] Ash a 5-gram sample and cool. Add a few ml of nitric acid, evaporate to dryness, and ignite at 450°. Cool and take up in 2 ml of hydrochloric acid. Dilute to 50 ml and take an aliquot. Dilute to 9 ml with 1 : 100 hydrochloric acid. Add 1 ml of a buffer solution containing 3.32% of disodium phosphate and 15.2% of citric acid. Extract with 10 ml of 0.01% dithizone in carbon tetrachloride. Clarify the extract by addition of 0.2 ml of absolute ethanol and read at 515 nm.

EDTA

The blue complex of cupric ion with EDTA is read in citrate buffer solution at pH 4.6.[287] Reducing agents, sulfide, iodide, thiocyanate, and cyanide interfere but are removed by heating to fumes with nitric acid. Weakly colored complexes of nickel, vanadium, and chromium are formed within 30 minutes in the cold. At pH 4.75–6.5 Beer's law is followed for 0.035–3.6 mg of copper per ml.

[285] Michal Misiak, Andrzej Gradzinski, and Stanislaw Sierawski, *Chem. Anal.* (Warsaw) **5**, 763–768 (1960).
[286] K. Roszek-Masiak, *ibid,* **9**, 1125–1128 (1964).
[287] G. Gottschalk, *Z. Anal. Chem.* **193**, 1–15 (1963); cf. N. A. Ramiah and Vishnu *Proc. Indian Acad. Sci.* **45A**, 112–116 (1957); cf. J. Areses Trapote, A. Concheiro Nine, and J. Rodriguez Vazquez, *Quim. Anal.* **30**, 277–280 (1976).

Procedure. To 50 ml of acid sample solution containing 0.3–32 mg of copper, add 8% sodium hydroxide solution until a slight turbidity appears. Add sequentially with mixing 10 ml of 2 *N* hydrochloric, nitric, phosphoric, or perchloric acid, 10 ml of 0.25 *M* EDTA solution, and 20 ml of *M* sodium citrate solution. Dilute to 100 ml and read within 60 minutes at 691 nm against water.

Oxidized Ores and Flotation Tailings.[288] Boil a 1-gram sample with 100 ml of 10% solution of EDTA for 40–45 minutes. Filter, wash the residue and dilute the filtrate to 100 ml. Read the copper–EDTA complex at 630 nm.

Silicate Rocks.[289] Transfer 0.5 gram of sample to platinum and mix with 2 ml of sulfuric acid and 10 ml of hydrofluoric acid. Evaporate to fumes and take up with 5 ml of water. Add 5 ml of hydrochloric acid and extract the iron with 5 ml of pentyl acetate. If the sample contains more than 3% iron, repeat this extraction. Evaporate the aqueous phase to fumes and nearly neutralize with 10% sodium hydroxide solution. Add 15 ml of 0.2 *M* EDTA. Adjust to pH 4 and boil for 5 minutes. Cool, dilute to 25 ml, and read copper at 735 nm.

ETHYLENEDIAMINE

Unlike many of the reagents for copper, ethylenediamine is selected to determine relatively large amounts, typically 0.2–8% of copper in ores.[290] It is preferable to ammonia for the purpose. If read at 540 nm, the result is not affected by variation in the ratio of copper to amine, provided the excess ethylenediamine does not exceed 500-fold. The pH must be at least 5, preferably 7–10. Temperatures of 10–40° do not affect the result, and the color is stable for at least 3 weeks. At 0.5 *M* chloride increases the absorption by 12%. With copper and nickel equimolar, the copper was in error by +10%. Cobalt interferes.

Procedures

Silicate Ore. Take a 200-mesh sample of 1–5 grams according to the probable copper content. Digest with 50 ml of 1 : 1 hydrochloric acid and 25 ml of nitric acid for 1 hour at 100°. Then evaporate to dryness. Take up the residue in 50 ml of 1 : 10 nitric acid. Heat to boiling and filter. Wash the residue with hot water slightly acidified with nitric acid and dilute to 250 ml.

To a 50-ml aliquot, add 10% sodium hydroxide solution dropwise to a slight turbidity, and clear it with a few drops of nitric acid. Add 5 ml of 0.5*M*

[288] S. G. Trofimova and E. V. Tyumentseva, *Tr. Tashk. Politekh. Inst.* **1966**, 147–151; A. N. Aizenberg, *Tr. Vses. Gos. Inst. Nauchn.-Issled. Proekt. Rabot. Ogneuporn, Prom.* **1971**, 125–135.
[289] K. P. Stolyarov and F. B. Agrest, *Zh. Anal. Khim.* **19**, 457–466 (1964).
[290] E. A. Tomic and J. L. Bernard, *ibid.* **34**, 632–635 (1962).

ethylenediamine and mix well. Check with indicator paper that the pH exceeds 8 and dilute to 100 ml. Read at 540 nm.

Lead-Base Alloy. Proceed as for silicate ore.

Wrought Aluminum. Proceed as for silicate ore.

Zinc-Base Alloy. Proceed as for silicate ore.

Solutions Containing Chloride. Mix with an equal volume of nitric acid and evaporate to dryness. Take up in 10 ml of 1:5 nitric acid and add 10% sodium hydroxide dropwise to a slight turbidity. Clear with a few drops of nitric acid. Add 1 ml of 0.5 *M* ethylenediamine and mix well. Dilute to 100 ml and read at 540 nm.

3, 3′–ETHYLENE-4, 4′–DIPHENYL–2, 2′–BIQUINOLYL

Cuprous ion forms a complex with the titled compound.[291] Cupric ion is reduced to cuprous by hydroxylammonium chloride unless silver is present, in which case reduction must be by sulfite. Citrate and thiosulfate interfere; common cations do not. Cyanide and amino-*N*-polyacetic acids inhibit the reaction.

Procedure. Adjust 10 ml of sample solution containing 0.005–0.11 mg of copper to pH 6 and add a few crystals of hydroxylamine hydrochloride. Add 5 ml of 2.5 m*M* solution of the reagent in isoamyl alcohol and 10 ml of the same solvent. Shake for 2 minutes, dry the organic layer with sodium sulfate, and read at 550 nm.

1-{4-(5-ETHYL-1, 3, 4-THIADIAZOL-2-YL)SULFAMOYL)PHENYL}-3-METHYL-5-(1-METHYLBENZIMIDAZOL-2-YL)FORMAZAN

The formazan above is a selective reagent. The complex with copper forms immediately and is read at 639 nm.[292] The color from nickel develops and is read as the additional absorption at the end of 9 minutes. Interference by ferric ion and by cobalt can be avoided by adding pyrophosphate and hydrogen peroxide, respectively.

Procedure. *Lake Sediments.*[293] Mix a 10-gram sample with 10 ml of sulfuric acid and heat to fumes. Let cool, add 5 ml of 30% hydrogen peroxide, and evaporate nearly to dryness. Take up the residue in 10 ml of 1:9 hydrochloric acid, filter, and

[291] E. Uhlemann and K. Waiblinger, *Anal. Chim. Acta* **41**, 161–164 (1968).
[292] L. F. Dubinina and V. N. Podchainova, *Zavod. Lab.* **38**, 1322 (1972).
[293] V. N. Podchainova and L. F. Dubinina, *Zh. Anal. Khim.* **27**, 242–245 (1972).

dilute to 250 ml. Evaporate a 20-ml aliquot nearly to dryness and add 10 ml of water. Add 1 ml of a 5% solution of the reagent in acetone followed by 3% potassium hydroxide solution to pH 5–6. Extract with 5 ml of chloroform and read the extract at 630 nm.

FERROZINE

Ferrozine is the disodium salt of 3-(2-pyridyl-5,6-bis(4-phenyl sulfonic acid)-1, 2,4-triazine. There is serious interference by iron, cobalt, and nickel with the determination of copper, but the maximum absorption by iron is sufficiently different to permit solution by simultaneous equations.

Procedures

Copper.[294] To 10 ml of sample solution containing 2.5–100 μg of copper, add 1 ml of 1% solution of hydroxylamine hydrochloride to reduce to cuprous ion. Boil for 10 minutes and cool. Add 1 ml of 0.514% solution of the reagent and 5 ml of phosphate buffer solution for pH 7. Dilute to 25 ml and read at 470 nm against a reagent blank.

Copper and Iron. As acid reagent, dissolve 5.14 grams of ferrozine and 100 grams of hydroxylamine hydrochloride in water. Mix with 500 ml of hydrochloric acid and dilute to 1 liter. Add 1 ml of reagent to 10 ml of sample solution containing 2.5–200 μg each of copper and iron. Heat at 100° for 10 minutes and cool. Add 5 ml of phosphate buffer solution for pH 7 and set aside for 5 minutes. Read copper at 470 nm and iron at 526 nm against a reagent blank and solve by simultaneous equations.

For determination of copper and iron by ferrozine, see Iron.

FAST GREY RA

Fast grey RA (mordant black 15), which is 1-azo-*m*-hydroxynaphthyl-2-hydroxy-3-nitrobenzene-5-sulfonic acid, forms with cuprous ion a 1:2 complex that absorbs at 555 nm.[295] Iron is reduced with ascorbic acid, and most common ions do not interfere.

Procedure. *Organs.* Digest 1 gram of dried sample with 5 ml of 1:1 sulfuric

[294] S. K. Kundra, Mohan Katyal, and R. P. Singh, *Anal. Chem.* **46**, 1605–1606 (1974).

[295] H. Khalifa, *Z. Anal. Chem.* **158**, 103–108 (1957); M. T. Fouad, Y. L. Awad, and M. E. Georgy, *Microchem. J.* **18**, 536–542 (1973); H. Khalifa, M. T. Fouad, Y. L. Awad, and M. E. Georgy, *ibid.* 617–621.

acid, 3 ml of nitric acid, and 3 ml of 30% hydrogen peroxide. Evaporate to dryness, add 5 ml of water, and again evaporate to dryness. Dissolve in hot water and dilute to 10 ml. Mix a 3-ml aliquot with 2 ml of 0.05 *N* nitric acid, 0.2 ml of *M* ascorbic acid, and 3 ml of 0.05% solution of the dye. Dilute to 10 ml and read at 555.5 nm against a reagent blank. The original work was on Egyptian camels.

FLAME PHOTOMETRY

For determination of copper in plant tissue, the sample material is wet-ashed. The copper is then extracted from a hydrochloric acid solution by dithizone in 1:1 chloroform-kerosene for flame photometric determination.[296] The strength of the hydrochloric acid is not significant. The kerosene in the solvent improves atomization. The conditions would also extract palladium, platinum, silver, and mercury. Only mercury might be present, as from a mercury-containing fungicide. Trial with 1:1 copper-mercury showed no interference. Agreement with the AOAC spectrophotometric method was good. An alternative is separation of copper by oxine.[297]

For urine, blood, or plasma, wet-ash with sulfuric and nitric acids and 30% hydrogen peroxide.[298] Dilute, neutralize to pH 4–5, and extract with dithizone in chloroform. Oxidize the copper dithizonate and excess dithizone with potassium permanganate, and follow with hydrogen peroxide. Read by flame photometry. A minimum of 5 μg of copper is required, which dictates 10 ml of normal blood or 250 ml of normal urine.

The complex of cupric ion with oxine can be extracted with methyl isobutyl ketone, amyl acetate, ethyl acetate, or chloroform for reading at 324.8 nm in an oxyhydrogen flame. The sensitivity is 6–8 times that for an aqueous solution.[299]

Extraction of the copper complex of 1,10-phenanthroline is most efficient with nitrobenzene from a solution containing chlorate ion buffered at pH 7 with 0.1 *M* ammonium acetate.[300] The extract is appropriate for flame photometry.

For a solution of a silica–potassium oxide–lead glaze in hydrofluoric and chloric acids, a 1:1 air-acetylene flame is required with 0.5% of disodium EDTA to suppress potassium oxide and lead oxide and perchloric acid in the calibrating solution.[301] In 1:5 nitric acid, copper is read at 20–200 ppm by flame photometry.[302]

In general, aliphatic esters and ketones, methyl phenyl esters, benzene, and toluene are satisfactory solvents for flame photometry of copper; they often increase the sensitivity as much as fortyfold.[303]

[296]H. F. Massey, *Anal. Chem.* **29**, 365–366 (1957); cf. H. Seiler, *Mitt. Geb. Lebensmittelunters Hyg.* **63**, 180–187 (1972).
[297]A. D. Berneking and W. G. Shrenk, *J. Agr. Food Chem.* **5**, 742–745 (1957).
[298]G. E. Newman and M. Ryan, *J. Clin. Pathol.* **15**, 181–184 (1962).
[299]Hidehiro Goto and Emiko Sudo, *Jap. Anal.* **10**, 175–181 (1961).
[300]Alfred A. Schilt, Rose L. Abraham, and John E. Martin, *Anal. Chem.* **45**, 1808–1811 (1973).
[301]H. Svejda, *Ber. Deut. Keram. Ges.* **48**, 116–120 (1971).
[302]C. A. Meinz and B. D. LaMont, *USAEC* **TID 7568**, Pt. 1, 150–156 (1958); cf. E. Pungor and I. Konkoly-Thege, *Ann. Univ. Sci. Budap., Rolando Eotvos Nominatae, Sect. Chim.* **2**, 477–484 (1960).
[303]Helmut Bode and Horst Fabian, *Z. Anal. Chem.* **162**, 328–336 (1958); cf. Hidehiro Goto and Emiko Sudo, *Jap. Anal.* **10**, 175–181 (1961).

Procedures

Copper, Manganese, and Iron. SEPARATION BY CHROMATOGRAPHY.[304] Use a column 25×1.1 cm. Suspend 100-200 mesh Dowex 1-X8 in hydrochloric acid and fill the column to 25 cm. Wash several times alternately with 25 ml of water and 25 ml of hydrochloric acid. Pass 25 ml of sample solution in 10 *N* hydrochloric acid through the column at 0.4 ml/minute. Elute the manganese with 25 ml of 1:1 hydrochloric acid. Then elute the copper with 70 ml of 2.5 *N* hydrochloric acid. Then elute the iron with 40 ml of 0.5 *N* hydrochloric acid.

Evaporate each of the eluates to dryness, making sure that the iron is not heated thereafter. Add 3 drops of hydrochloric acid to the copper residue, dilute to an appropriate volume, and determine by oxyacetylene flame at 324.7 nm. Similarly take up the manganese and read at 403.2 nm. Take up the iron in the same way, but add 1 ml of hydrochloric acid per 25 ml of the final solution and read at 386 nm.

Ferrous Alloys.[305] Dissolve a 0.5-gram sample in 30 ml of 1:3 hydrochloric acid, adding nitric acid in small amounts to oxidize the ferrous ion. If insolubles are present, filter and wash the residue with hot 1:99 hydrochloric acid. Make up to 100 ml and take an aliquot containing 0.015–0.15 mg of copper. Add 1 ml of 40% ammonium citrate solution for each 50 mg of sample represented by the aliquot. Add 1:1 ammonium hydroxide dropwise to raise to pH 3. Extract with four successive 5-ml portions of 1% salicylaldoxime in chloroform or *n*-amyl acetate, shaking each for at least 1 minute. Dilute the combined extracts to 25 ml and measure the flame background at 325 nm and the copper at 324.7 nm using an oxyacetylene flame. The organic solvent multiplies the spectral emission of copper by 10 as compared with an aqueous solution.

Aluminum Alloy.[306] COPPER, NICKEL, AND MANGANESE. As an acetate buffer solution for pH 6.75, dissolve 247 grams of ammonium acetate, 109 grams of sodium acetate, and 6 grams of acetic acid in water and dilute to 1 liter.

Dissolve the alloy sample with acid and evaporate to dryness. Take up the residue in 1:3 hydrochloric acid. Filter off insoluble material and dilute to an appropriate volume.

Take an aliquot of the sample solution containing 0.025–0.5 mg of the least abundant of the three elements and complex the metals with a 10% solution of sodium citrate dihydrate. Adjust to pH 6–6.5 with the acetate buffer solution. Add sufficient 10% sodium diethyldithiocarbamate solution to be in excess of the amount to react with the foregoing test substances plus any iron, lead, and zinc present. Of this reagent, 1 ml reacts with approximately 13 mg of divalent metals and 8 mg of trivalent metals. Shake with 10 ml of chloroform for 30 seconds and draw off the chloroform through cotton. Further extract with 5-ml portions of chloroform until a clear layer results. Dilute the combined extracts to 25 ml.

[304] W. G. Schrenk, Kenton Graber, and Russell Johnson, *Anal. Chem.* **33**, 106–108 (1961); W. G. Schrenk and Russell Johnson, *ibid.* **33**, 1799–1801 (1961).
[305] John A. Dean and J. Harold Lady, *ibid.* **28**, 1887–1889 (1956).
[306] John A. Dean and Carl Cain, Jr., *ibid.* **29**, 530–532 (1957).

Determine the elements by flame photometry as follows.

Element	Line Emission (nm)	Background (nm)
Copper	324.7	325.3
Manganese	403.2	401
Nickel	352.5	353.5

Bracket the unknown samples with standards in chloroform. When using a recording instrument, scan the regions as follows: copper, 323.5–325.5 nm; nickel, 351.5–353.5 nm; manganese, 401–404 nm.

The appropriate instrument settings are as follows:

Sensitivity control, turns from clockwise limit
Copper 5.0
Nickel 8.3
Manganese 8.0

Selector switch position	0.1
Phototube resistor (megohms)	22
Phototube (volts/dynode)	
Copper	60
Nickel	60
Manganese	50
Acetylene (cubic feet/hour)	2.15
Oxygen (cubic feet/hour)	4.55
Slit (mm)	0.030

If copper considerably exceeds manganese and nickel, the solution may be diluted with chloroform for the copper determination. If the copper exceeds 20 times that of the nickel or manganese, modify the procedure before the extraction as follows: after the initial evaporation to dryness, take up in *N* hydrochloric acid and add sufficient 5% solution of thioacetamide to precipitate about 90% of the copper present. Heat to 80° for a few minutes and filter. Boil off excess hydrogen sulfide and cool. Proceed as before for manganese and nickel from "Take an aliquot of the sample...." Determine copper on a smaller aliquot of an original sample.

Gold.[307] Dissolve the sample in aqua regia and evaporate to dryness. Add hydrochloric acid and again evaporate to dryness. Take up the residue in 1:1 hydrochloric acid and pass it through a column of strong base, ion exchange resin. Silver and gold pass through. Neutralize with ammonium hydroxide and make up to 1:9 ammonium hydroxide. Spray into an oxyhydrogen flame. The limit of detection is 0.4 μg/ml.

Nuclear-Grade Uranium.[308] Dissolve a 3-gram sample in 20 ml of 1:2 nitric

[307]Lyn Jarman, E. Manolitsis, and M. Matic, *J. S. Afr. Inst. Mining Met.* **62**, Pt. 2, 773–779 (1962).
[308]B. Podobnik, J. Korosin, and L. Kosta, *Z. Anal. Chem.* **218**, 184–192 (1966).

acid. Extract continuously with 200 ml of 10% solution of tributyl phosphate in kerosene and discard the extract. Evaporate the residual solution to dryness and take up in 10 ml of 1 : 100 hydrochloric acid for reading by flame photometry. The method will determine 2 ppm.

Cobalt Mattes and Concentrates.[309] IRON, COPPER, AND COBALT. The technic is applicable to 1–15% copper in the sample. Heat 1 gram with 10 ml of hydrochloric acid to liberate hydrogen sulfide. Add 2 ml of bromine to oxidize residual sulfur. Add 5 ml of nitric acid and evaporate to dryness. Take up the residue in 2 ml of hydrochloric acid and evaporate to dryness. Take up the residue in a minimal amount of 8 *N* hydrochloric acid. Filter through cotton. Rinse the beaker and wash the cotton with a minimal amount of 8 *N* hydrochloric acid. Extract the iron from the solution with 2 : 1 methyl ethyl ketone–chloroform and finally with 10 ml of chloroform, combining the extracts. Wash the combined extracts with 8 *N* hydrochloric acid.

Evaporate the iron extracts to dryness, using a current of air to complete the removal of solvent. As oxidizing reagent, mix 6 ml of 30% hydrogen peroxide with 3 ml of sulfuric acid and cool. Add this in small amounts to the residue in the beaker. After the carbon has been removed–if necessary, by adding more hydrogen peroxide–evaporate to sulfur trioxide fumes. Cool, take up in water, and boil to complete solution. Cool and dilute to 100 ml. Dilute 10 ml of this to 100 ml for reading the iron flame at 386 nm.

Heat the combined aqueous layers containing copper and cobalt to drive off residual solvent. Add 5 ml of nitric acid and evaporate to 5 ml. This is the solution for flame photometry of cobalt at 387.4 nm. Mix 10 ml of this solution with 9 ml of isopropanol and dilute to 100 ml for flame photometry of copper at 324.8 nm.

Plant Tissue. Treat a 5-gram sample containing less than 0.2 mg of copper with 40 ml of nitric acid. After foaming ceases, apply low heat until dry. If a large amount of charred material remains, repeat this step. Add 20 ml of nitric acid and 5 ml of perchloric acid and heat strongly until nearly dry. If charring occurs when the perchloric acid starts to boil, remove from the heat, add more nitric acid, and heat again. When almost all the acid has evaporated, remove from the heat and let cool. Add 20 ml of 1 : 10 hydrochloric acid, bring to a boil and remove from the heat. Dilute to 200 ml. Mix 5 ml of chloroform containing 1 mg of dithizone per ml with 5 ml of kerosene and shake vigorously for at least 3 minutes.

Filter some of the organic phase through paper and read at 324.8 nm by flame photometry.

FORMIC ACID HYDRAZIDE

The complex of copper with formic acid hydrazide is suitable for reading at 30 μg/ml.[310]

[309] N. McN. Galloway, *Analyst* **84**, 505–508 (1959).
[310] M. H. Hashmi, Abdur Rashid, Mohammad Umar, and Farooq Azam, *Anal. Chem.* **38**, 439–441 (1966).

Procedure. Mix a sample solution containing 0.3–1.5 mg of copper with 1 ml of 0.25 *N* sulfuric acid and dilute to 8 ml. Add 2 ml of 10% solution of formic acid hydrazide and read at 630 nm.

FURIL α-DIOXIME

This reagent is appropriate for solutions of minerals. As extracted into chloroform, it is sensitive to 0.6 μg of copper per ml.[311] Silver, gold, platinum, palladium, and cyanide interfere.

Procedure. To an aliquot of the sample solution, add 10 ml of 20% citric acid solution to mask elements such as iron, aluminum, and titanium that precipitate with ammonium hydroxide at pH 7–9.5. Add 3 ml of 1% ethanolic furil α-dioxime and raise nearly to pH 9.5 with ammonium hydroxide. Extract copper with 20 ml of chloroform. Back-extract the copper with 10 ml of 0.6 *N* sulfuric acid. This leaves cobalt and nickel in the discarded chloroform phase. To the copper phase, add 1 ml of 1% ethanolic α-dioxime and adjust to pH 7.2–9.5 by adding 1:20 ammonium hydroxide. Extract the copper complex with 10 ml and 2 ml of chloroform. Dilute the combined chloroform extracts to 15 ml and dry with sodium sulfate. Read against chloroform at 450 nm.

2-FUROYLTRIFLUOROACETONE

The copper complex of this reagent is extractable for reading.[312] Iron, nickel, cerium, palladium, and vanadium interfere.

Procedure. Mix a nitric acid solution of the sample containing 1–10 mg of copper with 10 ml of saturated solution of sodium acetate and dilute to 40 ml. Add 20% sodium hydroxide solution to adjust to pH 7. Add 2 ml of 5% ethanolic solution of the reagent and mix. After 10 minutes, extract with 5 ml and 3 ml of methyl isobutyl ketone. Dilute the combined extracts to 10 ml and read at 660 nm.

HYDROQUINONE

Copper can be measured by its catalytic effect on the oxidation of hydroquinone by hydrogen peroxide in the presence of pyridine at pH 7.61. Ferric ion interferes.

Procedure. *Medical Tinctures.*[313] Evaporate a 10-ml sample to dryness under an

[311] N. V. Benediktova-Lodochnikova, *Zh. Anal. Khim.* **18**, 1322–1325 (1963).
[312] Eugene W. Berg and M. C. Day, *Anal. Chim. Acta* **18**, 578–581 (1958).
[313] Tadeusz Pelczar, *Acta Pol. Pharm.* **26**, 549–554 (1969).

infrared lamp and ash at 700°. Cool, add 1 drop each of hydrochloric and nitric acids, and dry under infrared light. To the residue, add 1 ml of 0.02 *M* pyridine, 1 ml of 0.5 *M* hydrogen peroxide, 7 ml of a borate buffer solution for pH 7.61, and 1 ml of 0.02 *M* hydroquinone. Centrifuge, set aside for 10 minutes, and read the supernatant layer at 420 nm.

1-HYDROXY-1,3-DIPHENYLTHIOUREA

This reagent is *NN'*-diphenylthiocarbamoylhydroxamic acid. Cupric ion forms a brown 1:2 complex with it in 0.1–0.5 *N* sulfuric acid.

Procedure.[314a] To a sample solution containing 30–100 μg of copper, add 1 ml of 5 *N* sulfuric acid and dilute to 10 ml. Add 0.5 ml of 0.1 *M* ethanolic reagent. Extract with 5 ml of chloroform for 1 minute and read at 330, 364, or 400 nm.

6-(3-HYDROXY-3-METHYLTRIAZENO-2-BENZOTHIAZOLE

For determination in strongly alkaline solution the complex of the reagent with cupric ion remains stable but that with ferric ion decomposes.

Procedure. *Ferrotitanium.*[314b] Dissolve 0.1 gram of sample by heating with 20 ml of 1:4 sulfuric acid and 0.3 ml of nitric acid. Mix a 5-ml aliquot with 1 ml of 5 *M* glycerol, 2 ml of 8 *N* sodium hydroxide, and 0.2 ml of 0.8% ethanolic reagent. Dilute to 10 ml and heat at 100° for 30 minutes. Filter. Dissolve the precipitate in 13 ml of warm 1:1 hydrochloric acid. Add 2 ml of 0.1 *M* 2-naphthol. Heat at 100° for 30 minutes, cool, add 10 ml of acetone, and read at 490 nm against a reagent blank.

8-HYDROXYQUINOLINE

This reagent is also known as 8-quinolinol and as oxine. The complex of cupric ion with oxine is extractable with chloroform. The technic will determine 0.6 ppm of copper in high purity thorium compounds.[315]

As applied to a solution of uranium, any residual uranium and other interfering metals can be removed from the chloroform extract by washing it with 15% sodium

[314a]V. N. Podchainova and V. P. Maklakova, *Trudy Inst. Khim. Ural. Nauch. Tsentr. Akad. Nauk SSSR* **1974**, (27), 121–125.

[314b]E. D. Korotkaya, A. Kh. Klibus, V. Ya. Pochinok and S. K. Gaiduk, *Ukr. Khim. Zh.* **42**, 755–760 (1976).

[315]Kwok-yuen Chan and Ton-pol Tam, *Chemistry* (Taipei) **1969**, 55–61.

hydroxide solution.[316a] The narrow pH range of 3.1–3.3 for extraction of the cupric oxinate must not be disregarded.

Extraction with molten naphthalene of complexes with 8-hydroxyquinoline is applicable at varying pH levels to copper, zinc, magnesium, cadmium, bismuth, cobalt, nickel, molybdenum, uranium, aluminum, iron, palladium, and indium.[316b]

Procedure. *Thorium Compounds.* Dissolve a 10-gram sample in 20 ml of 1:1.5 nitric acid and extract the thorium with 30 ml and 30 ml of 1:1 tributyl phosphate–chloroform. Evaporate the aqueous solution to dryness and dissolve the residue in 25 ml of water. Mix a 10-ml aliquot with 2 ml of 2% solution of oxine, 5 ml of glacial acetic acid, and 100 ml of water. Adjust to pH 3.1–3.3 with 1:1 ammonium hydroxide, and extract the copper complex with 10 ml of chloroform. Wash aluminum from the extract by shaking with 50 ml of 15% sodium hydroxide solution. Dry with sodium sulfate and read at 500 nm.

LIX-64N

This reagent is a mixture of hydroxy-substituted oximes in a hydrocarbon solvent used in extractive metallurgy of copper. Beer's law is followed for 10–150 ppm of copper as the complex with the reagent dissolved in carbon tetrachloride.

Procedure.[317a] Mix a sample of filtered leach liquor containing 0.5–1 mg of copper with 5 ml of 10% solution of ascorbic acid and dilute to 20 ml. Adjust to pH 2 and extract with 10 ml of 10% solution of the reagent in carbon tetrachloride. Read the extract at 410 nm against a reagent blank.

LUMOCUPFERRON

The product of reaction of cupric ion in faintly alkaline solution with lumocupferron, which is α-(4-dimethylaminobenzylidene)hippuric acid, is read by fluorescence.[317b] The method will determine 0.001 μg of copper per ml.

Procedure. To 5 ml of sample containing up to 0.3 μg of copper, add 20 mM of

[316a] Kenji Motojima, Hiroshi Hashitani, and Hideyo Yoshida, *Jap. Anal.* **10**, 79–82 (1961).

[316b] T. Fujinaga, Masatada Satake, and Masaaki Shimizu, *Jap. Anal.* **25**, 313–318 (1976).

[317a] K. S. Koppiker and N. Maity, *Anal. Chim. Acta* **75**, 239–241 (1975).

[317b] S. U. Kreingol'd and E. A. Bozhevol' nov, *Zh. Anal. Khim.* **18**, 942–949 (1963); Yu. M. Martynov, E. A. Kreingol'd, and B. M. Maevskaya, *Zavod. Lab.* **31**, 1447 (1965); A. V. Konstantinov, L. M. Korobotchkin, and G. V. Anastas'ina, *Tr. Nov. Apparat. Metodikam. Perv. Mosk. Med. Inst.* **1967**, 167–173; cf. A. A. Obraztsov and V. G. Bocharova, *Tr. Voronezh. Gos. Univ.* **82**, 182–184 (1971).

reagent in 0.02 *M* ammonium acetate buffer solution for pH 10. After 10 minutes at 100° cool and read the fluorescence at 520 nm.

Human Skin. Heat the sample with 5 ml of sulfuric acid, 10 ml of perchloric acid, and 15 ml of nitric acid to a colorless solution, usually 1–1.5 hour. Evaporate to dryness and take up in 40 ml of hot water. Cool and adjust to pH 7–8. Dilute a 2.5-ml aliquot to 4.5 ml and add 0.5 ml of ammonium acetate buffer solution for pH 10. Add 0.2 ml of 0.015% solution of lumocupferron in acetone; after about 15 minutes, add 0.2 ml of m*M* EDTA. Read the bright green fluorescence at 400–680 nm.

Silicon Tetrachloride. Evaporate a 50-ml sample with carbon tetrachloride. Volatilize the small amount of silicon dioxide formed by addition of hydrofluoric acid. Evaporate to dryness with 0.5 ml of hydrochloric acid. As a reagent, mix 10 ml of a buffer solution for pH 10, 3 ml of 0.03% lumocupferron solution in acetone, and 87 ml of water. Add 5 ml of reagent to the residue and heat at 100° for 15 minutes. Cool and read the fluorescence when irradiated by a mercury lamp.

MAGNESON KhS

Magneson KhS is 5-chloro-2-hydroxy-3-(2-hydroxy-1-naphthylazo)-benzenesulfonic acid. It forms a 1:1 complex with cupric ion at pH 1.7–5.[318] The optimum is at pH 2. Beer's law holds for 0.25–10 μg of copper per ml. There is no interference by 2.5-fold amounts of titanium or bismuth or an equivalent amount of tantalum. Zirconium and molybdenum interfere. The color develops immediately and is stable for 24 hours.

Procedure. *Steel.* Dissolve a 0.1-gram sample in 5 ml of 1:1 hydrochloric acid and 0.5 ml of nitric acid. Evaporate to dryness with several successive additions of hydrochloric acid. Take up the residue in water, neutralize with ammonium hydroxide, filter, and dilute to 100 ml. Mix a 4-ml aliquot with 3 ml of 1:1 hydrochloric acid and evaporate to dryness. Take up the residue in 2 ml of 0.025 *N* hydrochloric acid. Add 2.4 ml of ethanol and 0.6 ml of 0.2% solution of the reagent in 50% ethanol. Read at 560 nm against a reagent blank.

MERCAPTOSUCCINIC ACID

Mercaptosuccinic acid forms a 1:1 complex with cuprous ion.[319] As read at 530 nm, it conforms to Beer's law for 1–30 ppm in the solution as developed. Hydrazine sulfate prevents autoxidation of the complex.

[318] A. M. Lukin and N. N. Vysokova, *Zavod. Lab.* **37**, 28–29 (1971).
[319] Yasuhisa Hayashi, Shigeo Hara, and Isao Muraki, *Jap. Anal.* **14**, 1040–1043 (1965).

Procedure. To a sample containing 0.1–3 mg of copper, add 10 ml of 0.05 *M* acetate buffer for pH 5.5, 2 ml of 1% hydrazine sulfate solution, and 10 ml of 0.6% solution of mercaptosuccinic acid. Dilute to 100 ml and read at 530 nm.

METHYL-2-PYRIDYL KETOXIME

In alkaline solution this reagent gives a yellowish-green color of a 2:1 complex with cuprous ion and a red-violet color of a 3:1 complex with iron. The reagent reduces cupric ion to cuprous, but conveniently both copper and iron are reduced by hydriodic acid. Color is developed in strongly alkaline solution and is stable for 24 hours. Beer's law is followed by copper at 0.5–37 ppm and by iron at 0.05–10 ppm. When both are present, each must be corrected for the other. For determination of copper, there is serious interference by cyanide, EDTA, iron, nickel, cobalt, manganese, and vanadium. For determination of iron, there is serious interference by cyanide, EDTA, copper, nickel, cobalt, manganese, and vanadium.

Procedures

Copper.[320] To 5 ml of acid sample solution containing 0.08–0.65 mg of copper, add 1 gram of sodium potassium tartrate and 10 ml of fresh 0.5% solution of methyl-2-pyridyl ketoxime. Raise to pH 10.5–11 by addition of 10% sodium hydroxide solution and dilute to 25 ml. Read at 410 nm.

Copper and Iron. Treat a 5-ml sample containing ferric and cupric ions with 1 ml of 2 *N* sulfuric acid. Add 5 ml of 5% potassium iodide solution and set aside in the dark for 10 minutes. Titrate out the free iodine by dropwise addition of 0.05 *N* sodium thiosulfate solution. Add 5 ml of 0.5% solution of methyl-2-pyridyl ketoxime and dilute to 20 ml. Add 10% sodium hydroxide to pH 11 and dilute to 25 ml. Read copper at 410 nm, iron at 525 nm. Calculate each by simultaneous equations.

MOLYBDENUM BLUE

The formation of molybdenum blue is a method of determining cupric ion. The maximum absorption is at pH 7, but the reaction is carried out at a lower pH.[321] The optimum temperature is 25°. The color is stable for 24 hours. Nickel, iron, and tin interfere. Lead, silver, and mercurous ions are precipitated as chlorides. No reducing agent is present when the color is developed.

Cupric ion catalyzes reduction of molybdate by ascorbic acid at 25° and pH 1.85.[322] After 60 minutes molybdenum blue is read at 755 nm.

[320] D. K. Banerjea and K. K. Tripathi, *Anal. Chem.* **32**, 1196–1200 (1960).
[321] S. K. Tobia, Y. A. Gawargious, and M. E. El-Shahat, *Anal. Chim. Acta* **39**, 392–393 (1967).
[322] R. L. Heller, Jr., and J. C. Guyon, *Anal. Chem.* **40**, 773–776 (1968).

Procedure. To a sample solution containing 0.05–0.3 mg of copper, add 2 ml of nitric acid and evaporate to dryness. Take up the residue in 2 ml of water and add 1 ml of 1 : 10 hydrochloric acid. If a precipitate forms, centrifuge and decant. Add 1 ml of 1 : 10 hydrochloric acid, 1 ml of 1% potassium cyanide solution, and 1 ml of 1% phosphomolybdic acid solution. Dilute to 10 ml and read at 725 nm.

NEOCUPROINE

The reaction product of cuprous ion and neocuproine, 2,9-dimethyl-1,10-phenanthroline, is extracted with chloroform for reading. The extraction of the cuprous-neocuproine complex into water-immiscible alcohols is specific and independent of pH over a wide range.[323] Mixed chloroform and ethanol as extracting solvent improves the sensitivity.[324] The range of copper content that can be measured is extended by reading the sample against a reference solution instead of a blank with increase in accuracy and precision. Sulfide, cyanide, and large amounts of phosphate prevent complete recovery of copper by complexing it. The procedure for titanium analysis provides for removal of chromium, since only 2 mg is tolerated in the aliquot. An alternative is complexing of the chromium with sulfite. The technic is also applicable to fuel oil and turbine oil. The application is primarily to diesel fuel.[325] The reagent is in isopropanol, with chloroform serving as a mutual solvent. Some mercaptans cause low results. Interference by mercaptobenzothiazole is overcome by a large excess of reagent. Butyl zymate causes the color to fade in 2–5 minutes. At above 0.1%, barium sulfonate or sodium sulfonate retard color development but it is complete in 4 hours.[326]

As applied to thorium nitrate in ammonium citrate solution at pH 6, after addition of hydroxylamine hydrochloride solution, the neocuproine complex is extracted with chloroform. At 457 nm Beer's law holds up to 5 μg per ml of extract.

The copper-neocuproine complex can be extracted with chloroform from plutonium solution buffered at pH 3–9 with citrate–hydrochloric acid.[327] Ethanol in the chloroform enhances the color.

For analysis of chromium, dissolve in hydrochloric acid and pass a stream of air through to oxidize to chromic ion. Add potassium chlorate to assist in solution.[328] An anion-exchange resin lets the chromium pass through with iron, copper, zinc, and cobalt retained on the column. Then copper is eluted with 2.5 *N* hydrochloric acid for determination with neocuproine. The reagent is applicable to bauxite and to silicate rocks.[329]

Results with neocuproine agree with those by zinc dibenzyldithiocarbamate.[330] If

[323] G. F. Smith and W. H. McCurdy, *ibid.* **24**, 371 (1952)
[324] A. R. Gahler, *ibid.* **26**, 577–579 (1954).
[325] David M. Zall, Ruth E. McMichael, and D. W. Fisher, *ibid.* **29**, 88–90 (1957).
[326] Nobuhisa Matano and Akira Kawasse, *Trans. Nat. Res. Inst. Met., Jap.* **4**, 117–119 (1962).
[327] J. W. Lindsay and C. E. Plock, *Talanta* **16**, 414–416 (1969).
[328] Tadashi Yanagihara, Nobuhisa Matano, and Akira Kawase, *Jap. Anal.* **9**, 439–443 (1960).
[329] M. F. Landi, *Met. Ital.* **52** (6), 336–340 (1960); D. E. Bodart, *Chem. Geol.* **6**, 133–142 (1970).
[330] P. D. Jones and E. J. Newman, *Analyst* **87**, 637–642 (1962).

much phosphate is present, interference is avoided by increasing pH and adding sodium citrate or by using of the reagent in 0.01 *N* hydrochloric acid.

In determination of copper in tellurium, selenium must be volatilized from bromide solution but not at temperatures exceeding 200° or copper is lost.[331a]

Propylene carbonate extracts cuprous ion as the yellow chelate with neocuproine and ferrous ion as the violet chelate with bis-2,4,6-tri(2-pyridyl)-1,3,5-triazine from a solution buffered to pH 4.75. The reading of the copper chelate at 596 nm requires correction for 0.123 times the reading for iron at 458 nm. The iron reading requires no correction. In determination of copper, there is interference by cobalt, manganous ion, nickel, stannous ion, EDTA, and nitrite. In determination of iron there is interference by manganous ion, citrate, and EDTA.

A 2:2:1 complex of cupric ion, neocuproine, and chrome azurol S forms a uniform dispersion in the presence of 0.05% of Triton X-100.[331b] In a buffer solution of sodium tetraborate and monopotassium phosphate for pH 6.4 with 0.04 m*M* neocuproine and 0.02 m*M* chrome azurol S, Beer's law is followed up to 0.035 m*M* copper. There is serious interference by ferric ion, tetravalent thorium, beryllium, aluminum, and citrate. Cobalt and nickel cause positive errors but they as well as iron can be masked with sodium fluoride. Beryllium is masked with sodium tartrate.

Procedures

Removal of Palladium.[332] Since more than 0.4 mg of palladium in a 0.5-gram sample interferes with determination by neocuproine, its removal is provided for as the palladium bromide complex by extraction with cyclohexanone.

Evaporate the sample in aqua regia to dryness. Add 5 ml of hydrochloric acid, evaporate to dryness, and repeat this step three more times. Add 1 ml of hydrobromic acid and 25 ml of 3:16 sulfuric acid. Extract with 25 ml and 25 ml of cyclohexanone, shaking 30 seconds with each. Combine the hexanone layers and wash with 50 ml, 50 ml, and 30 ml of 1:50 sulfuric acid. If the washings contain suspended cyclohexanone, combine them and shake with 15 ml of cyclohexanone. Wash this cyclohexanone extract with 20 ml of water and add it to the extracted sample. Now add the combined sulfuric acid washings to the extracted sample. Wash the combined aqueous layers with 25 ml of chloroform. Discard the chloroform and hexanone extracts. Add 3 ml of nitric acid and 3 ml of perchloric acid to the combined aqueous layers and evaporate to white fumes. Cool, take up in water, and use all or an aliquot as sample.

Copper and Iron.[333] To an acid solution of sample containing 40–200 μg of copper and 10–50 μg of iron in less than 100 ml, add sequentially 5 ml of *M* hydroxylamine hydrochloride, 5 ml of freshly prepared 0.5 *M* hydroquinone, and 5 ml of a buffer solution for pH 4.75, which is 2 *M* sodium acetate and 2 *M* acetic

[331a]Boris Nebesar, *Anal. Chem.* **36**, 1961–1964 (1964).

[331b]Ryoei Ishida and Koichi Tonosaki, *Jap. Anal.* **25**, 625–630 (1976).

[332]Teruo Imai, *Jap. Anal.* **15**, 321–328 (1966).

[333]B. G. Stephens, H. L. Felkel, Jr., and W. M. Spinelli, *Anal. Chem.* **46**, 692–694 (1974); cf. J. R. Hall, M. R. Litzow, and R. A. Plowman, *ibid.* **35**, 2124–2126 (1963).

acid. Adjust to pH 4.5–5 with sodium hydroxide solution. Add 2 ml of 0.02 *M* neocuproine in propylene carbonate and 3 ml of 0.02 *M* bis-2,4,6-tri(2-pyridyl)-1,3,5-triazine in propylene carbonate. Add 1 ml of *M* sodium perchlorate and 5 ml of saturated solution of sodium chloride. Add 15 ml of propylene carbonate for each 100 ml of solution and shake. Add additional 5-ml portions until a 5-ml nonaqueous phase separates. Shake, drain the organic phase, and dilute to 25 ml with propylene carbonate. Read at 458 and 596 nm. Subtract reagent blanks. Multiply the absorption at 596 nm by 0.123 and subtract from the absorption at 458 nm to give the corrected value for copper. The reading at 596 nm measures the ferrous ion content and requires no correction.

Electronic Nickel.[334] Dissolve a 0.25-gram sample in 10 ml of nitric acid and 5 ml of phosphoric acid by gentle heating. Wash down the sides of the container and add 10 ml of perchloric acid. Evaporate to fumes for 5 minutes and cool. Take up in 75 ml of water and dilute to 250 ml. Dilute a 25-ml aliquot to 50 ml. Add 1 ml of phosphoric acid, 10 ml of 30% sodium citrate solution, and 5 ml of 10% hydroxylamine hydrochloride solution. Stir well for 30 seconds and add 10 ml of 0.1% solution of neocuproine in ethanol. Add 1:1 ammonium hydroxide to adjust the pH to approximately 5.

Shake for 30 seconds with 10 ml and 5 ml of chloroform. Dilute the combined extracts to 25 ml with ethanol and read at 455 nm against a reagent blank. The color is stable for 4 days. Traces of cyanide interfere.

Steel, Cast Iron, Open-Hearth Iron, and Wrought Iron. The size of sample and aliquot as dictated by the copper content can be found from the following table.

Copper (%)	Sample Size (g)	Aliquot (ml)	Copper Content (mg)
0.01–0.15	1.0	20.0	0.01–0.150
0.10–0.25	1.0	10.0	0.05–0.125
0.20–0.50	0.5	10.0	0.05–0.125
0.40–1.00	0.5	5.0	0.05–0.125
0.80–2.00	0.25	5.0	0.05–0.125

For plain carbon steels and cast iron, dissolve in 15 ml of nitric acid. For low alloy steels and alloyed cast irons, dissolve in 10 ml of nitric acid and 5 ml of hydrochloric acid. For stainless steel dissolve in 10 ml of hydrochloric acid and 5 ml of nitric acid. In each case heat gently until dissolved. Wash down the sides of the container and add 0.5 ml of hydrofluoric acid if less than 1.5% silicon is present, 1 ml if more than 1.5%. Add 15 ml of perchloric acid and evaporate to fumes. Boil gently for 5 minutes and cool. Take up the residue in 75 ml of water. If silver is present, add 1 ml of hydrochloric acid. Filter if necessary. Wash the residue three times with 1% perchloric acid. Dilute to 200 ml and take the appropriate aliquot according to the table. Dilute to 50 ml. Add 10 ml of 30% sodium citrate solution and 5 ml of 10% hydroxylamine solution. Stir well and add

[334]ASTM **E107-60T**, cf. Thomas A. Downey, *Plating* **44**, 383–385 (1957).

10 ml of 0.1% ethanolic neocuproine. Mix and adjust to about pH 5 with 1:1 ammonium hydroxide.

Complete as for electronic nickel from "Shake for 30 seconds...."

Zirconium and Zirconium-Base Alloys.[335] Decompose a 1-gram sample in platinum or plastic with 1 ml of hydrofluoric acid in 15 ml of water. When solution is complete, add 0.5 ml of nitric acid, 25 ml of 5% boric acid solution, and 10 ml of 20% tartaric acid solution. Add ammonium hydroxide to pH 4–6 and dilute to 100 ml.

As an aliquot for less than 300 ppm of copper in the sample, take 50 ml; for 200–600 ppm, take 25 ml; for 400–1500 ppm, take 10 ml; and for 800–3000 ppm, take 5 ml. To the aliquot, add 5 ml of 10% hydroxylamine hydrochloride solution to reduce to cuprous form and 10 ml of 0.1% solution of neocuproine in ethanol. Extract for 30 seconds with 10 ml and 5 ml of chloroform. Add 3 ml of ethanol and dilute to 25 ml with ethanol. Read at 457 nm against a reagent blank.

Germanium.[336] Dissolve 2 grams of 150-mesh sample with 2 ml of perchloric acid, 10 ml of nitric acid, and 40 ml of hydrochloric acid. Evaporate to under 10 ml. When the solution begins to get cloudy add 15 ml of hydrochloric acid and evaporate to initiation of perchloric acid fumes. Add 15 ml more of hydrochloric acid and heat to perchloric acid fumes to expel the last of the germanium. Add 3 ml of hydrobromic acid and evaporate to 0.5 ml to expel any arsenic, antimony, and selenium present. Add 50 ml of water, followed by 2 ml of 10% hydroxylamine hydrochloride solution and 2 ml of 30% solution of sodium citrate dihydrate. Add ammonium hydroxide until Congo red paper turns red, pH 4. Add 2 ml of 0.1% solution of neocuproine in ethanol and extract with 15 ml of chloroform. Filter the extract and read at 457 nm against chloroform.

Germanium Oxide. Dissolve 3 grams of sample as described for germanium.

Beryllium or Beryllium Oxide.[337] Add 20 ml of 1:1 hydrochloric acid dropwise to a sample containing about 50 μg of copper. When the vigorous reaction has abated, heat to complete solution and evaporate to about 5 ml. If there is an insoluble residue, filter, wash, ash in the original beaker, and ignite at 500°. Fuse the ash with 1 gram of potassium bisulfate, dissolve in 10 ml of 1:9 sulfuric acid, and add to the filtrate. Adjust to contain about 5 ml of hydrochloric acid and 10 ml of sulfuric acid. Add 10 ml of 50% solution of citric acid and 5 ml of 10% solution of hydroxylamine hydrochloride. Dilute to 100 ml and cool. Adjust to pH 5 ± 0.1 by addition of 50% solution of sodium hydroxide while chilling in an ice bath. Add 10 ml of 0.1% solution of neocuproine in absolute ethanol and set aside for 2 hours. Shake for 1 minute with 10 ml of methyl isobutyl ketone. Centrifuge the organic layer and read at 457 nm against the solvent.

[335] ASTM Committee E3.

[336] C. L. Luke and Mary E. Campbell, *Anal. Chem.* **25**, 1588–1593 (1953).

[337] J. O. Hibbits, W. F. Davis, and M. R. Menke, *Talenta* **9**, 61–66, 101–103 (1960).

Tungsten and Tungsten Alloys.[338] For the determination of less than 0.03% copper, 0.09% iron, 0.15% nickel, and 0.04% cobalt, weigh a 2-gram sample into a platinum dish. Add 5 ml of hydrofluoric acid and 4 ml of nitric acid and warm gently. In case the heating has to be prolonged, add more hydrofluoric and nitric acids, but finish with 4–7 ml of solution. Place 50 ml of 20% ammonium tartrate solution in a flask and wash in the sample with 30 ml of water. Add 90 ml of 4.5% sodium borate solution and dilute to about 220 ml. Cool and dilute to 250 ml.

Mix a 50-ml aliquot with 5 ml of 10% hydroxylamine hydrochloride solution and 10 ml of 0.1% neocuproine solution in ethanol. Shake for 30 seconds with 10 ml and 5 ml of chloroform. Dilute to 25 ml with ethanol and read at 455 nm against a reagent blank.

Titanium.[339] Dissolve a 0.5-gram sample in 100 ml of 1:4 sulfuric acid and 5 ml of fluoboric acid by heating just below boiling for 30 minutes. Add nitric acid dropwise until the dark green color of the fluotitanate is discharged and a few drops in excess. Heat until the solution is clear and add 10 ml of perchloric acid. Evaporate rapidly to perchloric acid fumes, then add hydrochloric acid dropwise as long as orange fumes of chromyl chloride are evolved. Evaporate rapidly until the perchloric acid has been driven off and fumes of sulfur trioxide appear. Cool, add 50 ml of water, and warm to dissolve salts. Filter, and wash the residue with water. Cool to room temperature and dilute to 100 ml.

To a 25-ml aliquot add 10 ml of a solution containing 5% of hydroxylamine hydrochloride and 30% of sodium citrate pentahydrate. Add ammonium hydroxide dropwise to pH 4–6 as shown by indicator paper. Add 10 ml of 0.2% solution of neocuproine in ethanol. Shake for 30 seconds with 10 ml, 3 ml, and 2 ml of chloroform. Dilute the combined extracts to 25 ml with absolute ethanol. Read at 452 nm against a reagent blank or a known amount of copper carried through the procedure.

Bismuth.[340a] For high purity metals, follow the preceding method for titanium, but use a 2-gram sample and add 5 ml of 10% hydroxylamine hydrochloride solution, followed by 30 ml of 50% citric acid solution. The specified amount of citrate will mask antimony or it can be volatilized as the bromide.

Alternatively, dissolve a 1-gram sample in 10 ml of nitric acid.[340b] Pass the solution through a 15×2 cm column of 100-200 mesh Amberlite CG-400 in chloride form. Elute with 1:4 nitric acid, discarding the first 30 ml of eluate and retaining the next 20 ml as containing the copper. A typical method of development would be that given in the last paragraph for determination in titanium.

Thorium.[341] Convert to a solution of thorium nitrate and determine as under uranium compounds starting at "Adjust an aliquot...."

[338]George Norwitz and Herman Gordon, *Anal. Chem.* **37**, 417–419 (1965); cf. T. E. Green, *Rep. Invest. U.S. Bur. of Mines* **6277** (1963).
[339]Andrew J. Frank, Arthur B. Goulston, and Americo A. Deacutis, *Anal. Chem.* **29**, 750–754 (1957).
[340a]Yoshihiro Ishihara and Moritaka Koga, *Jap. Anal.* **11**, 246–247 (1962).
[340b]Takeshi Sakamoto, Toshio Nakashima, Katsuaki Fukuda, and Atsushi Mizuike, *ibid.* **19**, 1218–1223 (1970).
[341]Nobuhisa Matano and Ikira Kawase, *Trans. Nat. Res. Inst. Metals* (Tokyo) **4**, 117–119 (1962).

Zinc or Cadmium.[342] Dissolve a 3-gram sample with 20 ml of 1 : 1 nitric acid, boil off the oxides of nitrogen, and dilute to 100 ml. Dilute an aliquot containing not more than 0.02 mg of copper to 50 ml. Add 2 ml of 10% hydroxylamine hydrochloride solution and 2 ml of 30% sodium citrate solution. Then add 2 ml of 0.1% ethanolic neocuproine. Mix well and let stand for 15 minutes. Shake for 2 minutes with 10 ml of chloroform and read the extract at 457 nm against chloroform.

Lead. Proceed as for zinc, except use 5 ml of 30% ammonium acetate solution in place of the 2 ml of sodium citrate solution.

Tin. Digest a 1-gram sample with 10 ml of 48% hydrobromic acid and 2 ml of bromine water. Evaporate to dryness to volatilize the tin. Cool, add 10 ml of 1 : 1 nitric acid, and heat at 100°. Dilute to 100 ml and proceed as for zinc from "Dilute an aliquot containing...."

Tungsten Powder.[343] Add 10 ml of water to 10 grams of sample. Add 5 ml of 30% hydrogen peroxide, cooling the container in water if necessary to restrain the exothermic reaction. Add peroxide, gradually increasing the increments, until a clear, colorless solution is obtained when about 90 ml has been added. Near the end of the reaction it may be necessary to warm the container. Add 10 ml of 10% tartaric acid solution and neutralize with 20% sodium hydroxide solution. Adjust to pH 4 with 1 : 1 sulfuric acid. Boil for 30 minutes to decompose excess peroxide and cool. Add the following sequentially, mixing after each addition: 10 ml of 5% hydroxylamine hydrochloride solution, 10 ml of 30% sodium citrate solution, 10 ml of 0.1% ethanolic neocuproine, and 10 ml of chloroform. Shake for 1 minute and drain the chloroform into 5 ml of ethanol. Extract with 5 ml of chloroform and dilute the combined extracts to 25 ml with ethanol. Read at 457 nm against water. Subtract a reagent blank. One ppm of copper can be determined.

Zircaloy.[344] To an acid solution of the sample, add 10 ml of 20% solution of hydroxylamine hydrochloride, 10 ml of 30% sodium citrate solution, and 10 ml of 0.1% neocuproine solution. After allowing to stand for 10 minutes, adjust to pH 5, extract with chloroform, and read at 457 nm.

Gold.[345] Dissolve a 1-gram sample in 6 ml of hydrochloric acid and 4 ml of nitric acid. Boil off the acid until the remaining solution is dark brown. Cool and add 10 ml of 1 : 3 hydrochloric acid. Dilute with 20 ml of water. Extract the bulk of the gold with 5 ml and 5 ml of ethyl acetate, discarding the extracts. Add 5 ml of 10% solution of hydroxylamine hydrochloride. Boil for 20 minutes, and filter the remaining gold. Add 5 ml of ammonium acetate buffer solution for pH 4–5 and dilute to 50 ml. Add 2 ml of 0.1% ethanolic neocuproine and extract with 10 ml of

[342] Yoshihiro Ishihara and Yasuro Taguchi, *Jap. Anal.* **6**, 588–589 (1957).
[343] A. S. Prokopovitsh and T. E. Green, *U.S. Bur. Mines, Rep. Invest.* **5720**, 7 pp (1960).
[344] E. B. Read, P. R. Hicks, H. M. Lawler, E. Pollock, H. M. Read, and L. Zopatti, *USAEC* **NMI-1178**, 17 pp (1957); G. W. Goward and B. B. Wilson, *ibid.* **WAPD-CTNGLA-154** (Rev. 1), 6 pp (1957).
[345] Masuo Miyamoto, *Jap. Anal.* **9**, 748–753 (1960).

chloroform. Read at 460 nm. The method will determine as little as 1 μg of copper in the 1-gram sample.

Tellurium. As a compound buffer solution, dissolve 200 grams of citric acid monohydrate in about 600 ml of water. Dissolve 120 grams of hydroxylamine hydrochloride in 200 ml of water and filter. Mix the two solutions and dilute to 1600 ml. Add 200 ml of ammonium hydroxide, cool, and dilute to 2 liters. The pH is about 5.6. If purification becomes necessary, add 1 ml of 0.1% solution of the reagent in ethanol to 250 ml of buffer solution and extract with 10 ml, 10 ml, and 10 ml of chloroform. Thus purified, the buffer may become yellowish green, but that does not interfere.

Take a 0.5-gram finely ground sample of technical tellurium or a 1-gram sample of pure tellurium. Add 10 ml of 1 : 1 nitric acid and set aside to react. Heat gently and increase to expel brown fumes. After washing down the sides and cover of the container with water, evaporate at a temperature not exceeding 150° until the nitric acid is expelled, but do not bake. Add 7.5 ml of hydrochloric acid and swirl to give a perfectly clear yellowish-green solution.

While swirling gradually add 50 ml of the compound buffer solution. The temperature rises and some fumes appear. Cover and set aside for 10 minutes in a tray of running water. Add 6 ml of ammonium hydroxide with swirling and cool. Adjust to about 80 ml and shake for a minute with 5 ml of chloroform. Discard this extract. Add 5 ml of 0.1% solution of the reagent in ethanol and shake. After 2 minutes extract with chloroform as follows: 5 ml if the sample was pure tellurium, 10 ml if it was technical grade. Drain the extract through a glass wool plug. For pure tellurium this should be into 2 ml of ethanol, for technical tellurium into 5 ml of ethanol. Wash in with 1 ml of chloroform. Again extract, with 1.5 ml of chloroform for pure tellurium or 5 ml for the technical grade. Adjust to 10 ml for pure tellurium or 25 ml for the technical grade, and read at 457 nm against water. Correct for a blank.

Thermoelectric Compounds. These are usually of a bismuth telluride type. Dissolve a 0.1-gram sample in 4 ml of 1 : 1 nitric acid and evaporate to dryness. Add 2 ml of hydrobromic acid and evaporate to dryness for an expected 4 mg of selenium. If selenium is higher, use 2 ml of 1 : 10 bromine-hydrobromic acid. Take up in 2 ml of hydrochloric acid and add 25 ml of compound buffer solution. Add 2 ml of ammonium hydroxide and dilute to about 40 ml. Add 2 ml of 0.1% solution of the reagent in ethanol and shake. Complete as for tellurium from "After 2 minutes extract...."

Uranium Compounds.[346] Treat 2 grams of uranium tetrafluoride with aqua regia. Fuse any residue with potassium bisulfate to dissolve it. For uranium metal, ammonium diuranate, or uranium oxide, dissolve 5 grams in nitric acid.

Adjust an aliquot containing not more than 0.5 mg of copper to pH 0.5–1. Add 5 ml of 10% hydroxylamine hydrochloride solution and 5 ml of 30% sodium citrate solution. Add 5 ml of 0.1% solution of neocuproine in ethanol and adjust to pH 5

[346]UKAEA, *Report* **PG294(S)** 6 pp (1962).

±0.5. Shake for 30 seconds with 10 ml and 5 ml of chloroform and dilute the extracts to 25 ml with ethanol. Read at 457 nm against a reagent blank.

Germanium Dioxide.[347] Digest a 5-gram sample for 1 hour with 100 ml of 1 : 1 hydrochloric acid and evaporate to dryness. Take up the residue in 50 ml of water and add 2 ml of 10% hydroxylamine hydrochloride solution. Adjust the pH to 5.6–5.8 by adding 2 ml of 30% sodium acetate solution. Add 2 ml of 0.1% ethanolic neocuproine solution. After 10 minutes shake the solution for 5 minutes with 30 ml of isoamyl alcohol. Read the extract at 455 nm.

Products of the Alkali Industry.[348a] For sodium hydroxide, neutralize 100 grams with hydrochloric acid. For soda ash, similarly neutralize 50 grams. For crude sodium chloride or ammonium chloride, dissolve 50 grams in ammonium acetate buffer solution for pH 3–5. Evaporate 100 ml of hydrochloric acid and take up the residue in the ammonium acetate buffer solution.

For 1–25 μg of copper in the samples above, check that the pH is 3–5 and pass through a column of 40-50 mesh Dowex A-1 resin in sodium form. Elute the copper with 1 : 5 hydrochloric acid. Add 10 ml of 5% hydroxylamine hydrochloride solution containing 30% of sodium citrate. Adjust to pH 4–6 and add 5 ml of 0.1% ethanolic neocuproine solution. Extract with 10 ml of chloroform and read the extract at 460 nm.

Fertilizers. Zinc ion displaces cuprous ion from EDTA.[348b] As extraction solution, dissolve 14 grams of sodium hydroxide, 45 grams of tris(hydroxymethyl)aminomethane (TRAM) , and 50 grams of EDTA in water and dilute to 1 liter. As a buffer solution, mix 150 ml of water, 5 grams of zinc oxide, and 14 grams of nitrilotriacetic acid, with 18 grams of TRAM and dilute to 200 ml. Shortly before use, add 1 gram of hydroxylamine hydrochloride per 100 ml.

Boil 5 grams of sample with 400 ml of the extraction solution for 30 minutes. Cool, dilute to 500 ml, and filter. Take two aliquots containing 45–250 gmg of copper and no more than 12 ml of extraction solution. Add an amount of extraction solution to make the content to 12 ml. Add 10 ml of buffer solution to each. Then add 5 ml of 0.25% solution of neocuproine in 50% 2-propanol to one and 5 ml of 2-propanol to the other. Dilute each to 50 ml and read the one containing the color reagent against the other.

Fuel Oil or Turbine Oil. With chloroform, dilute a sample to 25 ml or, if the copper content is high, to 100 ml. To an aliquot, add 10 ml of a solution containing 0.1% of neocuproine and 0.1% of hydroquinone in isopropanol. Dilute to 25 ml with chloroform, let stand for 30 minutes, and read at 454 nm against a reagent blank.

Serum. COPPER AND IRON.[349] With a mixed reagent, both copper and iron are

[347]Hideo Baba, *Jap. Anal.* **5**, 631–634 (1956).
[348a]Hiroshi Imoto, *ibid.* **10**, 1354–1357 (1961).
[348b]E. R. Larsen, *Anal. Chem.* **46**, 1131–1132 (1974).
[349]J. V. Joossens and J. H. Claes, *Rev. Belg. Pathol. Med. Exp.* **31**, 254–264 (1965).

read in the same sample. Mix 2 ml of sample, 3 ml of 0.2 *N* hydrochloric acid, and 50 mg of ascorbic acid. Cover with Parafilm and let stand for 30 minutes. Add 1 ml of 30% trichloroacetic acid solution, let stand for 10 minutes, and centrifuge. Mix 4 ml of the supernatant layer with 0.5 ml of saturated solution of sodium acetate and 3 ml of a color reagent containing 40 mg of 4,7-diphenyl-1,10-phenanthroline and 0.1 gram of neocuproine per 100 ml of pentanol. Centrifuge, and read the upper layer at 436 nm for copper and at 546 nm for iron. Solve the content of copper and iron by simultaneous equations.

NEOCUPRIZONE

Neocuprizone is bis(ethyl acetoacetate)oxalyldihydrazone.[350] At pH 9 copper forms a water-soluble 1:2 blue complex with it. Platinum gives a blue complex, and the brown complexes with cobalt and nickel interfere. Beer's law is applicable to 0.02–0.75 mg of copper per 100 ml of solution.

Procedure. *Slag.* As a reagent, dissolve 0.5 gram of neocuprizone in 90 ml of 50% methanol, add 2 ml of acetone, and dilute to 100 ml with 50% methanol. Dissolve a sample expected to contain about 0.5 mg of copper in 30 ml of 1:6 nitric acid and 30 ml of 1:17 sulfuric acid. Evaporate to fumes and ignite to drive off the sulfuric acid. Take up the residue in 50 ml of 1:17 sulfuric acid, add water, boil briefly, and dilute to 100 ml. Filter and to 50 ml of filtrate add 10 ml of ammonia–ammonium citrate buffer solution for pH 9. Add 50 ml of reagent and set aside for 1 hour, to allow the color to develop. Dilute to 100 ml and read at 585 nm.

NICKEL DIETHYLPHOSPHORODITHIOATE

The reaction of cupric ion with nickel diethylphosphorodithioate is applicable to a residue from mercury.[351]

Procedures

Mercury. Dissolve a 10-gram sample in 20 ml of nitric acid and evaporate to dryness. Take up the residue in 30 ml of 1:3 hydrochloric acid, neutralize, and add excess ammonium hydroxide. Filter and dilute to 100 ml. Neutralize a 50-ml aliquot to litmus with 1:1 sulfuric acid. Add 2 ml of 1:17 sulfuric acid, boil for 20 minutes, and cool. Add 10 ml of 40% potassium bromide solution and 2 ml of

[350] G. Ackermann and W. Kaden, *Z. Anal. Chem.* **234**, 409–414 (1968).
[351] B. M. Lipshits, A. M. Andreichuk, and G. S. Agatonova, *Zavod. Lab.* **30**, 1075 (1964).

0.04% solution of the reagent. Extract with 5 ml and 5 ml of carbon tetrachloride and read at 420 nm.

Soil

TOTAL COPPER.[352] Calcine 2 grams of finely ground sample at 550° for 2 hours. Cool and add 30 ml of 3:1 hydrochloric acid–nitric acid. Take almost to dryness, add 2 ml of sulfuric acid, and evaporate to white fumes. Take up in 50 ml of water and boil for 15 minutes. Filter and wash the residue with 1:250 sulfuric acid. Dilute the filtrate and washings to 100 ml. Add 10 ml of carbon tetrachloride to a 50-ml aliquot; then add dropwise 2 ml of 0.125% solution of nickel diethylphosphorodithioate. After 1 minute filter the extract and read at 420 nm.

FREE COPPER. Shake 5 grams of soil for 30 minutes with 25 ml of *N* hydrochloric acid. Filter and add 1 ml of nitric acid to the filtrate. Evaporate to a moist residue, add 2 ml of sulfuric acid, and evaporate to white fumes. Complete as for total copper from "Take up in 50 ml...."

Water. To 1 liter add 2 ml of 0.005 *M* reagent. Extract with 10 ml and 10 ml of carbon tetrachloride. Filter the extracts through dry paper, dilute to 25 ml, and read at 420 nm.

Biological Samples. Ash a 5-gram dried sample at 500° after adding 0.5 ml of nitric acid. Take up in 5 ml of 1:10 nitric acid, add 20 ml of water, filter, and dilute to 100 ml. Complete as for total copper in soil from "Add 10 ml of carbon tetrachloride to...."

Food Products.[353] Ash 5–10 grams of air-dried sample in porcelain at a dull-red heat. Periodically accelerate the ashing by addition of a few drops of 3% hydrogen peroxide. Take up the ash in 5 ml of 1:1 hydrochloric acid and a drop of 3% hydrogen peroxide. Evaporate to dryness at 100°. Take up the residue in 5 ml of 2:5 hydrochloric acid and dilute to about 25 ml. Filter and dilute to 100 ml.

To a 50-ml aliquot add 10 ml of chloroform and 2 ml of 0.005 *M* nickel diethyl phosphorodithioate. Shake for 60 seconds. Filter the extract and read at 420 nm.

NITRITE

An optimum concentration of sodium nitrite for reading cupric ion is 4 *M* or in the region of 28%.[354] It conforms to Beer's law for 9.5–95 μg of copper per ml. The error for more than 0.03 mg of copper per ml is $\pm 2\%$, somewhat greater as the

[352] A. I. Busev and A. A. Nemodruk, *Pochvovedenie* **1967** 110–111; cf. A. I. Busev and M. I. Ivanyutin, *Vestn. Mosk., Univ., Ser. Mat., Mekh., Astron., Fiz. Khim.* **13** (2), 177–181 (1958); A. A. Nemodruk and V. V. Stasyuchenko, *Zh. Anal. Khim.* **16**, 407–411 (1961).
[353] V. S. Gryuner and L. A. Evmenova, *Vopr. Pitan.* **20**, 66–69 (1961).
[354] G. Burlacu, C. Ioan, and R. Smocot, *Iasi* **9**, 103–110 (1963).

content decreases. If ferric chloride is present, the reading must be made within 10 minutes because of decomposition of the nitrite.

Procedure. *Lead Concentrates.* Sample according to the expected copper content, 3.5 gram for 1% copper down to 0.5 gram for 10%. Dissolve the ground sample in 15 ml of 1 : 1 nitric acid. Evaporate to dryness and heat to drive off nitric acid. Digest the residue with water and filter off gangue and lead sulfate. Dilute the filtrate and washings to 250 ml.

Mix a 5-ml aliquot with 10 ml of 8 *M* sodium nitrite solution and dilute to 20 ml. Filter precipitated ferric hydroxide and read the filtrate. Low results with lead-rich ores may be due to adsorption of copper on lead sulfate.

OXALYLDIHYDRAZIDE

Oxalyldihydrazide forms a binary compound with cupric ion and a ternary compound on addition of acetaldehyde.[355] In the ternary compound, replacement of acetaldehyde by formaldehyde, propylaldehyde, benzaldehyde, or salicylaldehyde reduces the absorption. As applied to copper in water, cobalt must not exceed 1 ppm, chromic ion 5 ppm, and cyanide 30 ppm. The cyanide can be oxidized with hypochlorite.

The method will determine as little as 0.1 μg of copper in erythrocytes.[356] The color develops within 30 minutes and is stable for 24 hours. Manganous ion does not form a color with the reagent, but it does inhibit color development. Tetravalent manganese does not. Ammonium ion is necessary for color development. The optimum pH is 8.6–10.3. Ferric ion above 0.5 mg per 4 ml interferes strongly.

The reagent is also applied to copper-containing enzymes such as ascorbic acid oxidase and tyrosinase.[357] The copper is released by treatment with hydrochloric acid.

For determination in ingot zinc and zinc alloys, the solution contains ammonium citrate and excess ammonium hydroxide. Then acetaldehyde is used with the reagent.[358] That for copper in magnesium is similar.[359] The reagent with acetaldehyde is satisfactorily applied to ash of vegetables, meat, fish, and fruits.[360]

By use of the AutoAnalyzer, copper is developed with ammoniacal oxalyldihydrazide reagent and read at 546 nm.[361]

[355] R. Capelle, *Chim. Anal.* (Paris) **43**, 111–113 (1961); S. Jancovic, *Trib. CBEDEAU* **20**, 103–104 (1967).
[356] Harold Markowitz, G. S. Shields, W. H. Klassen, G. E. Cartwright, and M. M. Wintrobe, *Anal. Chem.* **33**, 1594–1598 (1961).
[357] George R. Stark and Charles R. Dawson, *ibid.* **30**, 191–194 (1958).
[358] British Standards Institution, *BS 3630*, Pt. **9**, 9 pp (1969).
[359] British Standards Institution, *BS 3907*, Pt. **3**, 8 pp (1966).
[360] V. P. Sagakova and A. I. Lyubivaya, *Konserv. Ovoshch. Prom.* **1961** (9), 35–37.
[361] H. Hirsch, *Z. Klin. Chem. Klin. Biochem.* **11**, 465–470 (1973); cf. A. Bernegger, H. Keller, and R. Wenger, *Z. Klin. Chem. Klin. Biochem.* **10**, 359–362 (1972).

Procedure. Prepare the reagent by mixing equimolecular amounts of oxalic acid and hydrazine monohydrate, each in 5 volumes of ethanol.[362] To a 10-ml sample containing 1–10 ppm of copper in 2 *N* hydrochloric acid, add 4 ml of *M* ammonium citrate at pH 9.[363] Add 2 ml of ammonium hydroxide, 2 ml of a saturated solution of oxaldihydrazide, and 5 ml of 40% acetic acid. Dilute to 25 ml, let stand for 30 minutes, and read at 542 nm.

Brass.[364] Dissolve a 1-gram sample in a mixture of hydrochloric, nitric, and tartaric acids. Adjust to 2 *N* when diluted to 50 ml. Complete as above from "To a 10-ml sample...."

High Purity Uranium.[365] For 2–30 ppm of copper in uranium, dissolve a 1-gram sample in 1 : 1 nitric acid and evaporate almost to dryness. Add 4 ml of water and 5 ml of 24.3% solution of ammonium citrate. This buffers to pH 9. Add 4 ml of ammonium hydroxide, 2 ml of saturated solution of oxalyldihydrazide, and 4 ml of 40% acetic acid. Dilute to 25 ml and read at 542 nm 35 ± 5 minutes later against a blank. Cobalt and EDTA interfere.

Water. Add 1 ml of nitric acid to a 100 ml sample containing less than 0.1 mg of copper and evaporate to about 25 ml. Cool, and add 10 ml of 25% ammonium citrate solution as a buffer for pH 9 and 2 ml of ammonium hydroxide. Add 2.5 ml of a 0.1% solution of oxalyldihydrazide and 5 ml of 45% acetaldehyde solution. Dilute to 100 ml and after 35 minutes read at 540 nm.

For isolation of copper from water by Dowex 50W-2, see Copper by Lead Diethyldithiocarbamate. As applied in water analysis at pH 9.2, the aldehyde is preferably a 40% solution of butyraldehyde in methanol.[366] The complex in chloroform is then read at 550 nm.

Beer.[367] Mix 35 ml of degassed beer with 1 ml of 20% trichloroacetic acid and heat at 100° for 30 minutes. Chill in ice, while in the bath, add sequentially with mixing, 1.5 ml of saturated solution of diammonium hydrogen citrate, 3.5 ml of saturated solution of oxalyldihydrazide, dropwise 3.5 ml of 40% acetaldehyde, and 2 ml of ammonium hydroxide. Dilute to 50 ml after 30 minutes and read at 542 nm against a reagent blank. The method is adequate but time-consuming when compared with the standard method of the American Society of Brewing Chemists.

Urine.[368] As a reagent, mix 10 ml of saturated solution of oxalyldihydrazide, 8 ml of 40% acetaldehyde, and 5 ml of ammonium hydroxide. Heat 50 ml of urine with 5 ml of nitric acid and 1 ml of sulfuric acid until reduced to a volume at which charring occurs. Add 5 ml of nitric acid and 1 ml of perchloric acid and heat until

[362] Gunnar Gran, *Anal. Chim. Acta* **14**, 150–152 (1956).
[363] M. N. Stojanovic and D. Stevancevic, *Acta Vet.* (Belgrade) **11** (1), 55–58 (1961).
[364] J. Le Polles, *Bull. Soc. Pharm. Bordeaux* **101** (3), 189–193.
[365] D. B. Stevancevic, *Z. Anal. Chem.* **165**, 348–354 (1959).
[366] R. Capelle, *Chim. Anal.* (Paris) **48**, 498–504 (1966).
[367] M. W. Brenner, M. J. Mayer, and S. R. Blick, *Proc. Am. Soc. Brew. Chem.* **1963**, 165–174.
[368] J. F. Wilson and W. H. Klassen, *Clin. Chim. Acta* **13**, 766–774 (1966).

water white, adding more acid if necessary. Take up in water, neutralize with ammonium hydroxide, and cool. Add 2 ml of the mixed reagent and dilute to 25 ml. Read at 542 nm after 45 minutes.

Blood. Transfer the freshly drawn sample to a tube containing 0.2 ml of heparin solution at 10 mg/ml. Centrifuge and remove the plasma by aspiration. Wash the cells twice with equal volumes of 0.85% sodium chloride solution, centrifuge, and determine the volume of packed blood cells. Mix a 3-ml aliquot with glass beads, 5 ml of nitric acid, and 1 ml of sulfuric acid. These steps should follow on the day the blood is drawn. Let the sample stand 4 hours or longer to avoid excessive foaming during digestion.

Heat until charring occurs and cool. Add 5 ml of nitric acid and 1 ml of perchloric acid, and heat for 30 minutes. If the solution is not water-clear, repeat the addition of acids.At room temperature add 8 ml of ammonium hydroxide with constant agitation. Boil the solution until a precipitate of ammonium sulfate begins to form, and let cool.

Add 2.5 ml of 1 : 20 ammonium hydroxide. After 1 hour, centrifuge the ferric hydroxide and remove the supernatant layer. Wash the ferric hydroxide with 1 ml of 1 : 20 ammonium hydroxide and add the washings to the reserved solution. Dissolve the ferric hydroxide in 0.2 ml of 1 : 1 hydrochloric acid and reprecipitate with 0.8 ml of ammonium hydroxide. Centrifuge and add the supernatant layer to the reserved solution.

Add sequentially the following: 0.1 ml of 40% phosphoric acid, 0.5 ml of ammonium hydroxide, 1 ml of saturated solution of oxalyldihydrazide, and 0.8 ml of 40% acetaldehyde. Dilute to 10 ml and read at 542 nm after 90 minutes and before 4 hours has elapsed. Correct for the reagent blank.

Serum.[369] To 0.5 ml of sample, add 2 ml of nitric acid and 2 ml of perchloric acid. Heat to dryness. Take up in 3 ml of *N* hydrochloric acid.

Alternatively, mix 1 ml of serum with 1.5 ml of 1 : 5 hydrochloric acid and 1.5 ml of 20% trichloroacetic acid solution. Centrifuge after 10 minutes and take 3 ml of the supernatant layer as the aliquot.

To the 3-ml portion add 0.3 ml of a saturated solution of oxalyldihydrazide and 2 ml of ammonium citrate buffer solution for pH 9. Then add sequentially 0.3 ml of a saturated solution of sodium pyrophosphate, 2 ml of 1 : 6 ammonium hydroxide, and 1 ml of 40% acetaldehyde solution. If the sample was dry-ashed, add one drop of 1 : 5 sodium silicate solution. Read at 542 nm.

Enzymes. To the sample, which has been treated with hydrochloric acid and contains 2–4 μg of copper, add successively 1.2 ml of ammonium hydroxide, 0.8 ml of saturated solution of oxalyldihydrazide, and 2 ml of cold 40% acetaldehyde solution. Dilute to 10 ml after cooling to room temperature. After 30 minutes read at 542 nm.

---

[369] J. Wiercinski, *Med. Wet.* **22**, 564–565 (1966); cf. S. G. Welshman, *Clin. Chim. Acta* **5**, 497–498 (1960); Eugene W. Rice and Betty S. Grogan, *J. Lab. Clin. Med.* **55**, 325–328 (1960); R. N. Beale and D. Croft, *J. Clin. Pathol.* **17**, 260–263 (1964).

OXINE BLUE

Oxine blue is the condensation product of 8-hydroxyquinoline and 2-(4-hydroxybenzoyl) benzoic acid; it complexes with copper, zinc, cadmium, and cobaltous ion. Although almost colorless, the complexes form the basis for determination of copper, zinc, and cobalt.[370] The zinc and copper complexes are decomposed by cyanide. The complex of zinc with cyanide is decomposed by formaldehyde to re-form the zinc complex with the reagent. The method will determine 0.1 μg of copper or cobalt and 0.2 μg of zinc in a volume of 5 ml.

Procedure. As a reagent, dissolve 46.4 mg of oxine blue in 500 ml of ethanol and dilute to 1 liter with water.

Cobalt and Zinc Absent. Mix a neutral sample solution containing 0.1–1.5 μg of cupric ion with 1 ml of *N* ammonium hydroxide and 1 ml of *M* sodium carbonate. Add excess of the reagent and dilute to 5 ml. Read at 590 nm.

Copper, Zinc, and Cobalt. Treat as in the absence of zinc and cobalt and read as value E_1. Add 0.03 ml of 0.5 *M* potassium cyanide and read as E_2. Add a drop of 3 *M* formaldehyde and read as E_3. The blank is E_0. Then $E_3 - E_1$ = copper, $E_3 - E_2$ = zinc,and $E_2 - E_0$ = cobalt. Added correction factors are required for copper and zinc.

1,10-PHENANTHROLINE AND ROSE BENGAL

The ternary complex of these two reagents with cuprous ion can be read by fluorescence for 1–6 μg of copper.[371] For copper in tellurium, it gives results down to 0.01 ppm in 5 grams, with a sensitivity to 42 μg in the final solution. The standard addition technic improves accuracy for very small amounts.

Procedure. Prepare a reagent 0.1 *M* in sodium nitrite, 0.01 *M* in EDTA,containing 30% sodium citrate and 1% hydroxylamine hydrochloride. To a sample containing 0.02–0.06 mg of copper, add 10 ml of the reagent. Add 1 ml of 0.1% solution of neocuproine and dilute to 25 ml. Shake for 1 minute with 25 ml of chloroform.

To 10 ml of the chloroform layer add 10 ml of m*M* 1,10-phenanthroline containing 0.1 m*M* of rose Bengal (acid red 94). Add 2 ml of a 20% solution of disodium phosphate and 10 ml of water. Shake and separate after 30 minutes. Read the chloroform layer at 570 nm against a blank or activate at 560 nm for reading the fluorescence at 570 nm.

---

[370]M. Wronski, *Chem. Anal.* (Warsaw) **12**, 1199–1204 (1967).

[371]B. W. Bailey, R. M. Dagnall, and T. S. West, *Talenta* **13**, 753–761, 1661–1665 (1966); Barbara Kasterka and Jan Dobrowolski, *Chem. Anal.* (Warsaw) **16**, 619–630 (1967).

6-PHENYL-2,3-DIHYDRO-1,2,4-TRIAZINE-3-THIONE

This reagent is specific in forming a 2:1 complex with cupric ion. The complex is extracted with chloroform at pH 4–14. Large amounts of ferrous, silver, mercurous, mercuric, and cobalt ions interfere. Beer's law is followed for 2.1–12.3 μg of copper per ml.

Procedure. To a sample solution containing 25–100 μg of cupric ion add 1 ml of 4 *N* tartaric acid and 3 ml of a fresh 0.001 *M* solution of m*M* 6-phenyl-2,3-dihydro-1,2,4-triazine-3-thione.[372] Shake with 1 ml of chloroform, add 1 ml of 16% sodium hydroxide solution, and shake again. Further extract with 3 ml and 2 ml of chloroform. Adjust the extracts to 10 ml and read at 500 nm against a reagent blank.

PICOLINALDEHYDE THIONAPHTHOYLHYDRAZONE 

The stable complex of this reagent with copper as extracted conforms to Beer's law up to 10 ppm.[373a] The pH may vary widely below 2. The molar ratio of reagent to metal should exceed 4:1.

Procedure. To a sample containing 0.02–0.2 mg of copper in less than 10 ml, add 12 ml of 0.1% solution of the lead salt of the reagent in 0.1 *N* hydrochloric acid. Adjust the acidity to *N* and dilute to 25 ml. Extract with 10 ml and 10 ml of benzene, dilute the extracts to 25 ml, and read at 480 nm

---

PICRAMINE EPSILON

This reagent is 8-hydroxy-7-(2-hydroxy-3,5-dinitrophenylazo)naphthalene-1,6-disulfonic acid. In a strongly acid medium it can form 1:1 and 1:2 cuprous complexes.[373b] Read at 508 nm, it is a mixture of the two. In 0.2 *N* hydrochloric or nitric acid, it will determine 0.25–40 μg of copper per 25 ml. The reagent is applicable to hot chromium-plating electrolytes and to effluents.[373c] Addition of ascorbic and phosphoric acid masks iron, molybdenum, and niobium.

Procedure.[373d] As a sample of an electrolyte containing up to 2 grams of copper

[372]Rostam H. Maghssoudi and Ahmed B. Fawzi, *Anal. Chem.* **47**, 1694–1696 (1975).
[373a]D. E. Ryan and Mohan Katyal, *Anal. Lett.* **2**, 515–522 (1969).
[373b]Yu. M. Dedkov, V. P. Koluzanova, and A. K. Vernadskii, *Zh. Anal. Khim.* **25**, 1482–1486 (1970).
[373c]F. I. Kotik, *Zavod. Lab.* **38**, 662–663 (1972).
[373d]V. I. Bogdanova and Yu. M. Dedkov, *Mikrochim. Acta* **1971**, 502–506.

per liter, dilute 10 ml to 200 ml; for copper at 2–5 grams/liter, dilute 5 ml to 200 ml. Take 0.6 ml as sample of the diluted electrolyte. As a sample of effluent take 15 ml for up to 2 mg of copper per liter, 5 ml for 2–10 mg/liter, or 1 ml for 10–20 mg/liter.

To the sample, add 2 ml of 1:3.5 hydrochloric acid, 0.04 gram of ascorbic acid, and 1 ml of phosphoric acid. Five minutes later add 2 ml of 0.1% picramine-epsilon solution and dilute to 25 ml. Read in the 480–500-nm range against a blank.

Iron-Based Alloys.[373e] Dissolve 0.2 gram of sample in 15 ml of aqua regia and evaporate to 5 ml. Dilute to a known volume and take an aliquot. Add 1 ml of 10% solution of ascorbic acid and set aside for 2 minutes. Add 1 ml of phosphoric acid and again set aside for 2 minutes. Add 1 ml of 0.1% solution of picramine-epsilon. Dilute to 50 ml with 0.1 *N* hydrochloric acid and read at 550 nm against a reagent blank.

PICRAMINE M

Picramine M is 3-(2-hydroxy-3,5-dinitrophenylazo)-6-(3-sulfophenylazo)chromotropic acid. It has a maximum absorption at 560 nm and forms a 1:1 complex with cupric ion absorbing at 640 nm.[374] At 0.04–0.65 μg of copper per ml it conforms to Beer's law. The optinum pH is 1.3–1.8. Iron, titanium, stannous ion, and EDTA interfere.

Procedure. Decompose a 1-gram sample with 10 ml of nitric acid, adding a few ml of hydrofluoric acid from time to time. Boil off the oxides of nitrogen; cool, dilute, and filter. Dilute to 50 ml. Evaporate an aliquot to dryness and take up the residue in 10 ml of 0.01 *N* hydrochloric acid. Extract the copper with 5 ml and 5 ml of 0.001% solution of dithizone in chloroform. Evaporate the combined extracts to dryness. Add 1 ml of sulfuric acid and a few drops of perchloric acid. Evaporate to the disappearance of sulfur trioxide fumes. Take up the residue in 1.5 ml of a 0.04% solution of picramine M. Dilute to 50 ml with a hydrochloric acid–potassium chloride buffer solution for pH 1.5 and read at 640 nm.

PICRAMINE R

Picramine R is 3-hydroxy-4-(2-hydroxy-3,5-dinitrophenylazo)naphthalene-2,7-disulfonic acid. This forms a 1:1 complex with cupric ion up to the acidity of 0.7 *M* hydrochloric acid.[375a] The maximum absorption is around 530 nm, but the complex

[373e] T. V. Matrosova, G. P. Rudkovskaya, and L. P. Lienkova, *Zavod. Lab.* **42**, 764–765 (1976).
[374] V. A. Mineeva, M. Yu. Yusupova, and D. N. Pachadzhanov, *Dokl. Akad. Nauk Tadz. SSR* **13**, 37–39 (1970).
[375a] V. I. Bogdanova and Yu. M. Dedkov, *Zavod. Lab.* **34**, 688–690 (1968); *Zh. Anal. Khim.* **23**, 1046–1048 (1968); cf. S. S. Goyal and J. P. Tandon, *Mikrochim. Acta* **1969**, 237–243.

is read at 555 nm to lessen absorption by excess reagent. More than 30 mg of barium in 25 ml precipitates as a sulfonate. Interference by niobium, zirconium, ferric ion, and hexavalent molybdenum can be masked with ascorbic acid.

Procedure. *Minerals.* Dissolve a 0.1-gram sample in 7 ml of nitric acid saturated with bromine and evaporate nearly to dryness. Add 10 ml of 1 : 2 hydrochloric acid and again evaporate nearly to dryness. Take up the residue in a minimal amount of hydrochloric acid.

If the sample contains no more than 7 μg of copper, neutralize with ammonium hydroxide and add 0.6 ml of 1 : 1 hydrochloric acid. Add 0.3 ml of 0.06% solution of the reagent, dilute to 5 ml, and read at 555 nm. For samples containing more copper, take an aliquot containing not more than 0.07 mg. Neutralize to Congo red with ammonium hydroxide and add 3 ml of 1 : 1 hydrochloric acid. Add 1 ml of 0.06% solution of the reagent, dilute to 25 ml, and read at 555 nm.

PIPERIDINE-1-CARBODITHIOATE

At 2-4 *M* trichloroacetic acid forms a 1 : 2 : 2 complex of cupric ion, the solvent acid and the reagent. Addition of lead ion stabilizes the complex. Hexahydroazepine can replace the above reagent.

Procedure. *Steel.*[375b] Dissolve 0.2 gram of sample in 20 ml of 6 *M* trichloroacetic acid by boiling with 0.3 ml of nitric acid. Filter and dilute to 100 ml. Mix an aliquot containing about 12 μg of copper with 2 ml of 10% ascorbic acid solution to mask iron. Add 2 ml of 20% lead acetate solution; 4 ml if the nickel content exceeds 10%. Add 10 ml of 6 *M* trichloroacetic acid and 0.5 ml of 0.5% solution of sodium piperidine-1-carbodithioate. Set aside for 30 minutes and read at 420 nm. A similar procedure can be carried out with hexahydroazepine-1-carbodithioate.

PRIVALOYLACETYLMETHANE

Privaloylacetylmethane is 5.5 dimethylhexane-2-4-dione. Copper is extracted as a complex with this reagent. Then the copper chelate is converted in the organic phase to stable copper diethyldithiocarbamate for reading. Beer's law holds for 2–55 μg per 10 ml of benzene. There is interference by ferric ion, palladous ion, EDTA, and citrate. Ferric ion can be corrected for by washing one developed sample with a solution containing sodium diethyldithiocarbamate and reading it against one from which that reagent was omitted from the wash water.[376]

[375b] K. A. Uvarova, Yu. I. Usatenko, T. S. Topalova and I. B. Frolova, *Zavod. Lab.* **42**, 653–655 (1976).
[376] Hideo Koshimura, *Jap. Anal.* **22**, 97–102 (1973).

Procedure. Adjust a sample solution containing more than 1 mg of cupric ion to about pH 5 with a *M* acetate buffer and dilute to 50 ml. Extract with 10 ml of 0.1 *M* privaloyacetylmethane in benzene by shaking for 10 minutes. Mix 5 ml of the organic layer with 5 ml of benzene and shake for 3 minutes with 20 ml of 0.025% solution of sodium diethyldithiocarbamate at pH 9.5. Read the organic phase at 436 nm against the other 5 ml, diluted with 5 ml of benzene and washed with water.

4-(2-PYRIDYLAZO)RESORCINOL

This reagent has the trivial designation of PAR. It forms a red 1:1 complex with cupric ion at pH 1.5–2.5.[377] The reagent absorbs at 395 nm, the complex at 540 nm. The solution conforms to Beer's law for 0.12–2.5 μg of copper per ml. Addition of 3% hydrogen peroxide solution masks large amounts of rare earths, zirconium, hafnium, thorium, indium, gallium, and aluminum. Iron is masked with fluoride.

The color is stabilized in borate buffer solution for pH 9.7 with tartrate and acetate present.[378] Then it will tolerate for each 0.03 mg of copper up to 1 mg of tetravalent vanadium, titanium, aluminum, octavalent osmium, hexavalent molybdenum, hafnium, beryllium, ceric ion, and thallium. The monomethyl derivatives of the reagent have no advantage over the parent structure.[379]

With the solution buffered at pH 4.5, the complex can be extracted with tributyl phosphate.[380a] Salting out makes the extraction complete. A similar color is produced by nickel, cobalt, iron, vanadium, indium, and gallium, Tartrate, citrate, and acetate affect the color. The graph is linear up to 0.04 mg of copper.

Procedure. Adjust sample to pH 4.5 with ammonium hydroxide and add 5 ml of phthalate buffer solution for pH 4.5. Add 3 grams of sodium chloride and 0.2 ml of 0.1% ethanolic solution of the reagent. Dilute to 50 ml and shake for 1 minute with 10 ml of tributyl phosphate. Dry the solvent layer with sodium sulfate and read at 550 nm.

Boiler Feed Water.[380b] As a reagent solution, dissolve 290 grams of sodium citrate in 500 ml of water, add 0.02 gram of PAR in 10 ml of ethanol diluted to 250 ml, and dilute to 1 liter. To 100 ml of sample, add 2 drops of 30% hydrogen peroxide and 5 ml of the reagent solution. Adjust to pH 7.1 ± 0.6 and set aside for 50 minutes. Read at 510 nm against a blank.

[377] O. A. Tataev, S. A. Akhmedov, and Kh. A. Akhmedova, *Zh. Anal. Khim.* **24**, 834–838 (1969).

[378] Kazumasa Ueda, Yoshikazu Yamamoto, and Shunzo Ueda, *J. Chem. Soc. Japan, Pure Chem. Sect.* **91**, 254–258 (1970).

[379] W. J. Geary and F. Bottomley, *Talenta* **14**, 537–542 (1967).

[380a] Huniko Terashima and Hideo Tomioka, *Jap. Anal.* **18**, 998–1003 (1969).

[380b] Zona Kowalska, *Chem. Anal.* (Warsaw) **16**, 197–201 (1971).

3-(2-PYRIDYL)-5,6-DIPHENYL-1,2,4-TRIAZINE

This reagent complexes both copper and iron, and the complexes can be extracted into amyl alcohol. The sum can be read photometrically and the copper then masked instantaneously by cyanide.[381] With the copper masked, the iron is read to give the copper by difference. The copper complex in isoamyl alcohol can fade because of air oxidation unless this is prevented by adding ascorbic acid. With a 200-ml sample, extracted into 10 ml of isoamyl alcohol and read in a 1-cm absorption cell, the method detects 1 ppb of iron and 4 ppb of copper. About 80 times those amounts is required for optimum relative accuracy. Cobalt interferes by forming a complex with the reagent, absorbing at 500 nm and bleached by sodium cyanide. The nickel complex is weakly colored, and since it is not bleached by cyanide, it interferes in determination of iron. Chromium and cobalt consume reagent to interfere with determination of iron. Nitrite, oxalate, and molybdate are tolerated at 500 ppm. In copper determination, nickel is tolerated at 50 ppm, thiosulfate at 20 ppm, and cobalt at less than 1 ppm. In iron determination chromium is tolerated at 200 ppm, cobalt at 50 ppm and nickel at 10 ppm.

The same procedure is applicable with 3-(4-phenyl-2-pyridyl)5,6-diphenyl-1,2,4-triazine reading at 480 and 561 nm but offers no advantage, and the reagent is more expensive and more difficult to prepare.

Procedures

Copper and Iron. As a 0.01 *M* solution of the reagent, add 2 drops of hydrochloric acid to 0.77 gram of the reagent and take up in 25 ml of ethanol.

Take a sample containing 4–80 μg of copper and/or 1–20 μg of iron. Add 5 ml of a *M* acetate buffer solution for pH 4.5, 2 ml of 10% hydroxylamine hydrochloride solution, 2 ml of 0.01 *M* solution of the reagent in ethanol and 2 ml of 50% solution of sodium perchlorate. Using indicator paper, check the pH as 4.5 ± 1 and adjust if necessary. Extract with 6 ml and 2 ml of isoamyl alcohol. Dilute the combined extracts to 10 ml with ethanol containing 0.5% of ascorbic acid. Read at 488 nm against a blank.

Add 5–10 mg of sodium cyanide to the isoamyl alcohol solution and read at 488 and 555 nm against a blank within 1 hour. The loss of absorption at 488 nm measures the copper, and the final absorption at 555 nm measures the iron.

Whole Milk. Add 25 ml of whole milk to a heated crucible at approximately 1 drop/second to evaporate without frothing. After moisture has evaporated, heat slowly to avoid loss by swelling, and ignite at 450–500° to a grey ash. Let cool, add 1 ml of nitric acid, evaporate to dryness, and again ignite at 450–500° for 1 hour. Let cool and take up the resulting white ash in 10 ml of 1 : 15 nitric acid. Complete as for copper and iron.

Wine. For dry-ashing, evaporate 2 ml of sample to dryness in a crucible and ash at 450–500°. Allow to cool and add 1 ml of nitric acid to this grey ash.

[381] Alfred A. Schilt and Paul J. Taylor, *Anal. Chem.* **42**, 220–224 (1970).

Evaporate and ignite at 450–500° for 1 hour. Let cool and take up this white ash in 10 ml of 1 : 15 nitric acid. Complete as for copper and iron.

For wet-ashing, add to 2 ml of wine, 5 ml of nitric acid and 5 ml of perchloric acid. Heat gently as long as brown fumes are evolved, then strongly until copious perchloric acid fumes are evolved. Let cool, and boil with 20 ml of water to drive off chlorine. Adjust to pH 4.5 ± 1 to indicator paper by adding about 4 ml of ammonium hydroxide. Complete as for copper and iron.

2-QUINOLINEALDEHYDE 2-QUINOLYLHYDRAZONE

This reagent forms a 1 : 1 water-insoluble purple complex with cupric ion.[382] This is extractable with nitrobenzene, chloroform, isoamyl alcohol or benzene. It forms over the pH range 2.95–9.45. Substantial excess of reagent must be present. Beer's law applies to 3.3–22.2 ppm of copper. There is interference by cadmium, cobalt, ferric ion, manganese, nickel, and zinc: all form insoluble colored products with the reagent. Thiosulfate, EDTA, cyanide, and thiocyanate interfere by forming copper complexes.

Procedure. Prepare the reagent by refluxing equimolar amounts of 2-quinolylhydrazine and quinoline-2-carboxaldehyde in ethanol. Recrystallize from pyridine. For use, dissolve 0.078 gram of the reagent in a minimum of hydrochloric acid and dilute to 500 ml with water.

To a sample containing 0.2–1.4 ppm of cupric ion, add at least a fourfold excess of reagent. Add 20% sodium hydroxide solution to raise to pH 10. Add 20 ml of 0.035 *M* sodium borate–sodium hydroxide buffer solution for pH 10.2. Dilute to 40 ml and extract with 10 ml of nitrobenzene. Read at 536 nm against a blank.

QUINOXALINES

The quinoxaline reagents are 2,3-di-2pyridyl-quinoxaline and 2,3-bis-(l-methyl-2-pyridyl) quinoxaline, together with 12 derivatives of each.[383] Their cuprous complexes are extractable with chloroform or isoamyl alcohol. They are cheaper and easier to make than the cuproine-type reagents. The first of the two cited above shows the greater sensitivity.

Procedure. To 1 ml of sample containing 1–100 ppm of cupric ion, add 10 ml of acetate buffer solution for pH 4.7. Add 1 ml of either 1% hydroxylamine hydrochloride solution or 1% ascorbic acid solution and 4 ml of 0.1% ethanolic solution

[382] Richard E. Jensen, Norman C. Bergman, and Richard J. Helvig, *Anal. Chem.* **40**, 624–626 (1968); G. G. Sims and D. E. Ryan, *Anal. Chim. Acta* **44**, 139–146 (1969).
[383] W. I. Stephen and P. C. Uden, *ibid.* **39**, 357–368 (1967).

of the reagent. Extract with 4 ml and 4 ml of isoamyl alcohol, dilute the combined extracts to 10 ml, and read at 525 nm against solvent.

SALICYLALDOXIME

The reaction product of salicyaldoxime and copper is extractable with chloroform for reading.[384]

Procedure. *Ferric Oxide.* Dissolve a 1-gram sample in 1:1 hydrochloric acid. Add 6 grams of tartaric acid to mask manganese, lead, and iron, and adjust to pH 5–8 by adding ammonium hydroxide. Dilute to 100 ml. Shake a 20-ml aliquot for 5 minutes with 5 ml of 0.025% solution of salicylaldoxime in chloroform and read the extract at 365 nm.

SOLOCHROMATE FAST GREY RA

This dye which is mordant black 15, has been applied to copper in animal serum.[385] The sample should contain 0.0025–0.02 mg of copper.

Procedure. To isolate from the globulins, extract a 5-ml sample with 2 ml, 1 ml, and 1 ml of 20% trichloroacetic acid solution at about 95°. Dilute the combined extracts to 100 ml with water. Mix a 5-ml aliquot with 2 ml of 0.2 *M* ascorbic acid and 3 ml of 0.005% solution of the dye. After 15 minutes read the complex at 555 nm.

TETRAETHYLENEPENTAMINE OR TRIETHYLENETETRAMINE

These reagents complex with cupric ion separating it from iron or aluminum, both of which precipitate. The method tolerates 2 mg of tetravalent vanadium, 20 mg of chromic ion, 80 mg of nickel and 180 mg of thiocyanate with the pentamine, and 1, 15, 65, and 20 mg respectively with the tetramine.

Procedure.[386] *Ores and Intermediates.* Decompose a sample by fusion in an iron crucible with sodium peroxide and sodium hydroxide; then dissolve in water. Alternatively, decompose by heating with sulfuric acid, nitric acid, and sodium sulfate. For copper oxides shake with a 15% solution of sodium sulfite in 7% sulfuric acid.

[384] K. N. Bagdasarov, O. D. Kashparova, and M. S. Chernov'yants, *Zavod. Lab.* **37**, 1046–1047 (1971).
[385] H. Khalifa, M. T. Foad, Y. L. Awad, and M. E. Georgy, *Microchem. J.* **17**, 266–272 (1972).
[386] B. W. Budesinsky, *Z. Anal. Chem.* **282**, 218 (1976).

Mix an aliquot of sample solution containing 1–35 mg of copper and not more than 70 milliequivalents of sulfuric or nitric acid with 25 ml of a 3% solution of either captioned reagent in 1 : 1 ammonium hydroxide. Dilute to 100 ml and filter. Read at 650 nm with the pentamine or at 590 nm with the tetramine.

2-THENOYLTRIFLUOROACETONE

This reagent is appropriate for determining copper in water.[387a] Tartrate, citrate, thiosulfate, cyanide, and thiocyanate interfere. The addition of fluoride masks aluminum, ceric and ferric ion and thallium.

A 1 : 2 : 1 cupric-diketone-triphenylphosphine complex is more readily extracted with benzene or chloroform than the binary metal-diketone complex.[387b] The triphenylphosphine should be 0.1 *M* and the copper 10 μM. The diketone component is 0.01 *M* 2-thenoyltrifluoroacetone, trifluoroacetylacetone, or hexafluoroacetylacetone.

A complex of copper is promptly formed with 0.02 *M* 2-thenoyltrifluoroacetone and 1.2 m*M* Capriquat, which is methyltrioctylammonium chloride, in a solution buffered at pH 4.75 with acetate.[387c] This can be extracted with benzene. Nickel and cobalt react similarly but slowly.

Procedure. Mix 50 ml of sample containing up to 0.01 mg of copper with 0.1 gram of fluoride as sodium fluoride and adjust to about pH 5.5. Extract with 10 ml of m*M* solution of the reagent in cyclohexane. Separate the extract and wash it with 10 ml of 0.004% solution of sodium hydroxide containing 1% of pyridine. Read at 340 nm against a reagent blank.

THIAMINE

The product of reaction of copper with thiamine can be isolated and used for fluorescent determination down to 0.1 μg/ml.[388a] Fluorescence is greatest at pH 6–8 and at low temperature. Both cuprous and cupric ions react, but cuprous is more sensitive. In an ice-cooled solution the absorption is linear up to 0.03 mg/ml.

Procedure. Add 10 mg of hydroxylamine sulfate to a 5-ml sample and adjust to pH 7 by dropwise addition of 20% sodium hydroxide solution. Chill in ice and add 2 ml of 0.005% thiamine solution, preadjusted to pH 11–11.5 and chilled in ice. Let stand for 10 minutes and shake with 5 ml of isoamyl alcohol containing 30 μg of

[387a] Hideo Akaiwa, *Jap. Anal.* **12**, 457–460 (1963); Hideo Akaiwa, Hiroshi Kawamoto, and Masanobu Abe, *Bull. Chem. Soc. Jap.* **44**, 117–120 (1971); cf. Shirpad M. Khopkar and Anil K. De, *Z. Anal. Chem.* **171**, 241–246 (1959).

[387b] V. G. Makarov, G. G. Goroshko and O. M. Petrukhin, *Zh. Anal. Khim.* **31**, 460–464 (1976).

[387c] Hiroshi Kawamoto and Hideo Akaiwa, *Jap. Anal.* **24**, 127–130 (1975).

[388a] Yashuhiro Yamane, Motoichi Miyazaki, and Masakatsu Ohtawa, *Jap. Anal.* **18**, 750–753 (1969).

2,6-di-tert-butyl-*p*-cresol for 1 minute. Dry the extract with sodium sulfate, activate at 290 nm, and read at 650 nm. Use 2.5 m*M* rhodamine B as a reference solution.

THIOCYANATE

Both cupric and ferric ions can be read as the thiocyanate in the same solution because of differences in their absorption maxima.[388b] To prevent reduction of copper to the cuprous form and its precipitation, an oxidizing agent is normally required. The use of 40–60% dioxane as solvent makes this unnecessary for several hours. There is interference by cobalt, nickel, mercuric ion, lead, bismuth, fluoride, chloride, iodide, nitrite, carbonate, sulfide, and oxalate.

For cobalt, copper, and iron as thiocyanates by simultaneous equations, see cobalt.

Procedure. Dissolve the sample in 1 : 1 nitric acid and boil off oxides of nitrogen. Dilute to a known volume so that a 25-ml aliquot will contain 0.2–2 mg of copper. To the aliquot, add 1 ml of nitric acid and 20 ml of dioxane. Add 1 ml of 15% solution of ammonium thiocyanate and dilute to 50 ml with dioxane. After 5 minutes read at 373 nm. Iron can be read at 488 nm.

THIOCYANATE AND PYRIDINE

With separation of the iron content, the color with thiocyanate and pyridine can be extracted with chloroform for reading.[389] The appropriate range for copper is 0.3–3 mg.[390a] A large excess of the reagents is needed, and extraction should be at pH 8. Small amounts of manganese, zinc, silver, and mercury ions are tolerated, but cobalt, cadmium, lead, ferric, and aluminum ions interfere. There is also interference by phosphate, bromide, iodide, cyanide, and cyanate.

Procedures

Iron Ore. For an ore containing 0.001–0.01% of copper, dissolve a 0.5-gram sample in 3 ml of hydrochloric acid and 2 ml of nitric acid. Add 4 ml of 1 : 1 sulfuric acid and evaporate to white fumes. Cool and take up in 10 ml of water. Filter, wash the precipitate with 1 : 100 sulfuric acid, and discard the paper. Neutralize with ammonium hydroxide and add 10 ml in excess. Heat to 100° to coagulate the precipitate and filter. Wash the precipitate with hot water. Take up the precipitate in 10 ml of 1 : 1 hydrochloric acid, reprecipitate with ammonium hydroxide, and filter through the same paper. Take a 20-ml aliquot containing up

[388b] Ivor Ilmet, *Chemist-Analyst* **54**, 71–72 (1965).
[389] Bela Simo, *Kohasz. Lapok* **91**, 341–343 (1958).
[390a] G. Spacu and J. Scherzer, *Acad. Rep. pop. Rom., Stud. cercet. Chim.* **4**, 210–225 (1957).

to 0.1 mg of copper. Add 1 drop of phenolphthalein indicator, 3 drops of *N* sulfuric acid, 6 ml of 10% ammonium thiocyanate solution, and 1 ml of 10% pyridine solution. Shake for 30 seconds, add 10 ml of chloroform, and again shake for 30 seconds. Read the yellowish-green chloroform extract.

To the sample solution, which may contain cobalt, nickel, and ferric ion in addition to cupric ion, add 5 ml of *M* pyridine.[390b] Adjust to pH 5.5 and dilute to 20 ml. Shake for 1 minute with 20 ml of 0.1 *M* salicylic acid in chloroform. Use the extract for determination of copper by pyridine and thiocyanate.

Steel.[391] Dissolve a 2-gram sample in 40 ml of 1:4 sulfuric acid. Add a few drops of nitric acid, and boil off the oxides of nitrogen. Filter if necessary, cool, and dilute to 100 ml.

Shake a 4-ml aliquot with 2 ml of 15% tartaric acid solution, 9 ml of 1:4 ammonium hydroxide, and 20 ml of 15% caproic acid solution in chloroform. Reextract the copper from the caproic acid with 10 ml and 10 ml of 1:4 nitric acid. Proceed as for iron ore from "Add 1 drop of phenolphthalein...."

Copper, Nickel, Cobalt, and Iron.[392] To the solution containing any or all of these elements, add either 25 ml of 10% tartaric acid solution or 25 ml of 1% citric acid solution. Add 6 ml of 50% magnesium nitrate solution, 8 ml of pyridine, and 8 ml of 10% potassium thiocyanate solution. Adjust to pH 5–7 with 10% sodium hydroxide solution. Extract with 10 ml, 10 ml, and 10 ml of chloroform. It may be necessary to add more pyridine. Read copper at 405 nm, nickel at 360 nm, cobalt at 335 nm, and iron at 375 nm. With multiple ions present, simultaneous equations are necessary.

THIO-2-THENOYLTRIFLUOROACETONE

This reagent is 1,1,1-trifluoro-4-mercapto-4-(2-thienyl)but-3-en-2-one. It chelates with cupric ion to give an olive-brown complex extracted in carbon tetrachloride.[393] Extraction is quantitative at pH 2–5. The absorption conforms to Beer's law at 1.23–12.35 μg of copper per ml at 490 nm. The complex is stable for 72 hours. A ratio of 1:1 is tolerated for cadmium, ruthenium, iron, nickel, and zirconium. Gold, mercury, oxalate, cyanide, and thiocyanate interfere seriously.

Procedure. Take an aliquot of sample in sulfate solution containing about 0.5 mg of copper. Adjust to pH 4 with 0.01 *N* sodium hydroxide or sulfuric acid and dilute

[390b] A. T. Pilipenko and N. A. D'Yachenko, *Izv. Vyssh. Ucheb. Zaved., Khim. Khim. Tekhnol.* **19**, 1007–1009 (1976).
[391] V. V. Sukhan and S. F. Skachkova, *Zavod. Lab.* **36**, 1029–1031 (1970).
[392] Gilbert H. Ayres and Stephen S. Baird, *Talenta* **7**, 237–247 (1961).
[393] V. M. Shinde and S. M. Khopkar, *Anal. Chem.* **41**, 342–344 (1969); Masakazu Duguchi and Mikio Yashiki, *Jap. Anal.* **20**, 317–321 (1971).

to 25 ml. Shake with 10 ml of 0.001 *M* reagent in carbon tetrachloride for 10 minutes. Read the organic layer at 490 nm against a reagent blank.

2-(TOLUENE-4-SULFONAMIDO)ANILINE

At pH 10–11, this reagent, which is 2-aminotoluene-*p*-sulfonanilid, forms a violet complex suitable for spectrophotometry.[394] Complexes are also formed with ferrous, cobalt, nickel, and zinc ions.

Procedure. *Alloys.* Dissolve a 1-gram sample in 25 ml of 1:1 hydrochloric acid–nitric acid according to the nature of the sample. Evaporate to about 10 ml and make the solution 6 *N* with hydrochloric acid. Extract the iron with 20 ml and 20 ml of ether. Filter and wash the precipitate if antimony or bismuth is present. Add 10% sodium hydroxide solution and dilute to a known volume.

To an aliquot expected to contain 0.02–0.1 mg of copper, add 10 ml of 1.57 m*M* reagent in 70% ethanol. Add 5 ml of borate buffer solution for pH 10 and dilute to 25 ml with 70% ethanol. Read within 2 minutes at 530 nm. As an alternative, extract immediately with 25 ml of benzene, adding sodium nitrite to break the emulsion, and read the benzene extract at 560 nm.

1,1,3-TRICYANO-2-AMINO-1-PROPENE

At pH 7.8–8.4, this reagent produces a fluorescence with cupric ion at 365 nm.[395] The reagent itself absorbs at 283 nm. Although the ratio of copper reacting with the reagent is 1:2, the excess of reagent must be rigidly controlled because it affects the rate of reaction.

Procedure. *Tissue.* Wet-ash and adjust the solution to pH 8.5 by addition of sodium hydroxide solution. Dilute to such a volume that the copper is 0.1–0.5 μg/ml. Mix 0.3 ml of m*M* reagent with 0.5 ml of 2 m*M* imidazole and add 1 ml of sample. Incubate for 15 minutes at 37–40°. Add 6 ml of water and a drop of 1:2 hydrochloric acid. Read the fluorescence at 365 nm.

THYRAM

Thyram is tetramethylthiuram disulfide, closely related to dicupral, which is the tetraethyl compound. It produces an extractable color with cupric ion which is

[394] D. Betteridge and R. Rangaswamy, *Anal. Chim. Acta* **42**, 293–310 (1968).
[395] Kim Ritchie and Joseph Harris, *Anal. Chem.* **41**, 163–166 (1969).

stable for 24 hours.[396] By analogy, the end product is presumably cupric dimethyldithiocarbamate.

Procedure. *Pyritic Residues.* Heat a 2-gram sample with 50 ml of 1:9 hydrochloric acid at 100° for 45 minutes. Cool, dilute to 500 ml, and filter. Heat a 10-ml aliquot to 100°, add 10 ml of freshly saturated solution of the reagent, and heat to 100°. Cool, and extract the resulting precipitate with 10 ml and 10 ml of chloroform. Dilute the combined extracts to 100 ml with ethanol and read.

TRIETHENETETRAMINE

Unlike several other reagents used for the photometric determination of copper, triethenetetramine has the virtue of relatively low sensitivity.[397] The maximum absorption with this reagent and cupric ion is at pH 5–10 in the absence of EDTA and at 9.5–11.5 in the presence of EDTA. The absorption conforms to Beer's law for 0.5–10 mg of copper in 25 ml. Color development is immediate and stability endures for weeks. Fluoride has little effect, but increasing absorption is shown by chloride-bromide-iodide without shifting the maximum. Tartrate and EDTA suppress most interferences. More than 0.01 m*M* of cyanide interferes. It is applicable to 0.5–10 mg of copper in 25 ml without interference by propylenediaminetetracetic acid in plating baths.[398a]

Procedure. To a solution containing up to 10 mg of copper, add 1 ml of *M* EDTA, 1 ml of *M* triethenetriamine, and 5 grams of potassium iodide. Dilute to about 20 ml and adjust to pH 10 ± 0.2 by addition of 10% sodium hydroxide solution. Dilute to 25 ml and read at 590 nm against a reagent blank.

Zinc-Base Alloy. Dissolve a 1-gram sample with 10 ml of 1:1 hydrochloric acid and 5 ml of 1:1 nitric acid. If solution is incomplete, add more nitric acid. Evaporate nearly to dryness and cool. Add an amount of *M* EDTA slightly in excess of that to complex all metals present. Add 1 ml of *M* tartaric acid and 2 ml of *M* triethenetriamine. Add 10 grams of potassium iodide and adjust to pH 10 ± 0.2 by adding 10% sodium hydroxide solution. Dilute to 50 ml, filter if necessary, and read at 590 against a reagent blank.

Aluminum-Base Alloy. Dissolve 0.25-gram sample, following instructions for a zinc-base alloy.

---

[396] B. V. Mikhal'chuk and Z. A. Sazonova, *Soobshch. Nauch.-Issled. Rabotakh Novai Tekh. Nauch. Inst. Udobren. Insektofungits.* **1958** (10), 76–79; cf. R. M. Agamirova, *Sb. Tr. Azerb. Med. Inst.* **1958** (4), 315–319.

[397] K. L. Cheng, *Anal. Chem.* **34**, 1392–1396 (1962).

[398a] B. L. Goydish, *Mikrochim. Acta* **1971**, 675–679.

1,1,1-TRIFLUORO-3-(2-THENOYL)-ACETONE

This reagent absorbs strongly at the same wavelength as its cupric complex. Beer's law is followed for up to 1 μg of cupric ion per ml of extract. The interference by 40-fold aluminum, ferric ion, or chromic ion is masked by fluoride.

Procedure.[398b] For masking purposes add 5 ml of 4% solution of potassium fluoride to the sample solution. Adjust to pH 5.4 with an acetate buffer solution and shake for 3 minutes with 10 ml of m*M* solution of the reagent in chloroform. Shake the organic phase for 4 minutes with 10 ml of m*M* sodium hydroxide to remove excess reagent. Read at 344 nm.

T-SULFONAMIDINE

The three-component system of T-sulfonamidine, which is *o*-(*p*-toluenesulfonamid)aniline, pyridine, and cupric ion, forms a golden yellow complex.[399] The maximum color is approximated in 20 minutes. The absorption conforms to Beer's law over a range of 2–16 ppm. No common ion interferes. Where precipitation of metal hydroxides may occur, as by ferric, cadmium, and manganese ions, it can be prevented by adding 1000 ppm of tartrate.

Procedure. To a sample containing 0.05–4 mg of copper, add 5 ml of 0.1% solution of the reagent in ethanol and 8 ml of pyridine. Mix and dilute to 25 ml with ethanol. After 20–30 minutes read at 455 mn against a reagent blank.

Copper Alloys. Dissolve 0.75 gram of sample in 25 ml of 1 : 1 nitric acid and dilute to 1 liter. Dilute a 50-ml aliquot to 500 ml and develop an aliquot by the procedure above.

VIOLURIC ACID

Violuric acid is 5-isonitrosobarbituric acid. It forms a yellow complex with cupric ion at pH 4.5–5.5.[400] Beer's law is not followed. Cobalt interferes. Ferric ion is masked with fluoride ion.

Procedure. *Aluminum Alloys.* As a buffer solution, mix 120 grams of ammonium acetate, 50 grams of sodium acetate, 10 ml of glacial acetic acid, and 350 ml of

[398b] Hideo Akaiwa, Hiroshi Kawamoto and Fujio Izumi, *Talenta* **23**, 403–404 (1976).
[399] John H. Billman and Robert Chernin, *Anal. Chem.* **36**, 552–554 (1964).
[400] Max Ziegler, *Z. Anal. Chem.* **164**, 387–390 (1958).

water. Dissolve a 0.2-gram sample containing 0.02–10% of copper in 10 ml of 1 : 1 hydrochloric acid. Add 0.5 ml of nitric acid to oxidize the iron and dilute to 250 ml. To an aliquot containing not more than 3.5 mg of copper, add 20 ml of 0.5% solution of violuric acid. Then add 15% sodium carbonate solution until the color is yellow. Add 10 ml of buffer solution and 2 ml of 10% potassium fluoride solution. Dilute to 100 ml and draw a vacuum to remove free carbon dioxide. Read at 415 nm.

ZINCON

Both copper and zinc react with Zincon, which is 2-carboxy-2′-hydroxy-5′-sulfoformazylbenzene.[401] Addition of EDTA blanks out the zinc for determination of copper. Since equal concentrations have equal absorption at 610 nm, only a single calibration curve is required for determination of both copper and zinc. Each follows Beer's law for 0.1–2.4 ppm. The maxima for copper and zinc are 595 and 620 nm, respectively. The absorption at 610 nm is stable for 1 hour. There is interference by cobalt, manganese, nickel, aluminum, molybdenum, beryllium, bismuth, cadmium, and iron. As applied to animal tissue, only iron would normally interfere. Under those conditions, calcium, magnesium, and phosphorus are present at levels that precipitate. These interferences can be removed.

Copper can be separated from nickel, aluminum, calcium, and magnesium by passing the solution in 11 *N* hydrochloric acid through a column of AV-17 resin.[402a] The copper is then eluted with 4 *N* hydrochloric acid.

A 0.1% solution of Zincon in 0.1 *N* ethanolic sodium hydroxide is more stable than an aqueous solution.[402b]

Procedures

Animal Tissue. As a reagent, dissolve 0.13 gram of Zincon in about 2 ml of 4% sodium hydroxide solution and dilute to 100 ml. As a Clark and Lubs buffer for pH 9, add 21.3 ml of 0.2 *N* sodium hydroxide to 50 ml of 0.2 *N* boric acid in 0.2 *N* potassium chloride and dilute to 200 ml.

Prepare a colorless solution of ash by digestion with nitric and perchloric acids. The aliquot for use should contain about 0.01 mg of copper and zinc. Interferences in it should not exceed 0.05 mg of iron, 0.002 mg of manganese, 0.02 mg of magnesium, 0.126 mg of potassium, and 0.1 mg of phosphorus. If the amounts present do not exceed those levels, or if there is no precipitate on adjustment to pH 9, the next paragraph for separation may be omitted and the buffer solution added to the sample.

If separation is necessary, make the sample alkaline with ammonium hydroxide and add 2 ml in excess. Add 1 ml of 24% solution of sodium hydroxide. Centrifuge

[401] J. T. McCall, G. K. Davis, and T. W. Stearns, *Anal. Chem.* **30**, 1345–1347 (1958); Robert H. Maier and J. Richard Kuykendall, *Chemist-Analyst* **47**, 4–5 (1958).
[402a] A. N. Aizenberg, *Tr. Vses. Gos. Inst. Nauchn.-Issled. Proekt. Rabot. Ogneuporn. Prom.* **1971**, 125–136.
[402b] A. G. Knight, *Proc. Soc. Water Treat. Exam.* **15**, 159–160 (1966).

for about 5 minutes, decant the supernatant layer, and save. Wash the precipitate with 5 ml of 1:4 ammonium hydroxide, centrifuge, and add to the first decantate. Heat the solution until the odor of ammonia is no longer detectable. Make just acid by dropwise addition of hydrochloric acid.

Add 10 ml of the buffer solution and, if necessary, 4% sodium hydroxide solution or 1:10 hydrochloric acid to adjust to pH 9. Add 3 ml of Zincon reagent and dilute to 50 ml. Read at 610 nm against a reagent blank. Add 3 drops of 4% disodium EDTA solution, which will complex 0.1 mg of zinc, and again read at 610 nm. The loss in absorption measures the zinc, and the absorption after adding EDTA measures the copper.

Water.[403] Mix 300 ml of sample containing up to 11.6 μg of cupric ion with 2.5 ml of 10 m*M* sodium citrate. This and EDTA added later mask lead, chromic ion, iron, and manganous ion. Adjust to pH 4–5 with 16% sodium hydroxide solution. Add 1 ml of 6.5 m*M* Zincon and 10 ml of 0.05 *M* borate buffer solution. Adjust to pH 8–8.5 with 16% sodium hydroxide solution. Add 5 ml of 0.1 *M* hydrogen peroxide and 5 ml of 0.1 *M* manganous ion. Set aside for 80 minutes for decomposition of any zinc complex as well as excess Zincon. Add 5 ml of 0.1 *M* EDTA in *M* potassium chloride, followed by 5 ml of 10 m*M* zephiramine. Dilute to 350 ml and extract with 10 ml of chloroform. Dry the organic layer with sodium sulfate and read at 625 nm against chloroform. Correct for a reagent blank.

Boiler Water.[404] Acidify 80 ml with 1 ml of 1:5 nitric acid. Use 10% sodium hydroxide solution or *M* tartaric acid to adjust to pH 3.5–9 and add 0.2 ml of 0.001 *M* solution of Zincon in 50% ethanol. Dilute to 100 ml and read at 600 nm. The method has been applied at pH 4 with an automatic analyzer plotting the absorption against time.

ZINC *O,O*-DIISOPROPYL PHOSPHORODITHIOATE

The copper complex of this reagent is extractable for reading.[405] More than 5 mg of silver or 100 mg of bismuth interferes.

Procedure. To prepare the reagent, react 10 grams of phosphorus pentasulfide with 100 ml of isopropanol by heating. Cool to 40° and pass a stream of nitrogen to expel hydrogen sulfide. Add 4 grams of zinc oxide and heat at 100° for 10 minutes. Filter while hot, and chill to crystallize. Recrystallize from isopropanol.

Adjust a sample containing 0.004–0.04 mg of copper to approximately 4 *N* with hydrochloric acid. Add 25 ml of 0.04% solution of the reagent in carbon tetrachloride and shake at once for 30 seconds. Read the extract at 404.7 nm.

---

[403] Masao Sugawara, Yukio Ozawa, and Tomihito Kambara, *Jap. Anal.* **23**, 1058–1062 (1974).
[404] Kazuo Tanno, *J. Chem. Soc. Jap., Ind. Chem. Sect.* **67**, 1200–1207 (1964).
[405] W. A. Forster, P. Brazenall, and J. Bridge, *Analyst* **86**, 407–410 (1961).

The reaction product of copper and **acetanilide** can be read photometrically in the ultraviolet.[406] To a neutral sample solution containing 0.15–0.9 mg of copper, add 1 ml of 1% acetanilide in ethanol. Dilute to 25 ml and adjust the pH with 1:4 ammonium hydroxide. Shake thoroughly with 10 ml of chloroform, dry the extract with sodium sulfate, and read at 340 nm.

The complex of **acetylpyridine oxime** with copper is read at 410 nm and pH 10–11.7.[407] Beer's law holds for 0.5–40 ppm. Iron, cobalt, nickel, uranium, vanadate, and chromate interfere strongly.

Cupric ion forms a 1:1 purple complex with **1-acetylthiosemicarbazide** at pH 4.5–8.7.[408] The calibration is linear up to 6 μg per 50 ml of solution containing 10 ml of buffer for pH 7.1, 10 ml of 0.5% starch solution, and m*M* reagent as read against a reagent blank. Cyanide and thiocyanate must be absent.

Alizarin black SN (mordant black 25) is a reagent for reading copper at 600 nm.[409] There is interference by uranium, bismuth, thorium, iron, molybdenum, and vanadium. The addition of a 1–3 carbon alcohol increases the absorption.

Alizarin red S, which is sodium 3-alizarinsulfonate, forms a 1:1 cupric complex having a maximum absorption at 500 nm.[410] The color is stable at pH 3.2–7.5. There is interference by many ions.

Adjust 25 ml of sample solution containing less than 40 μg of cupric ion to pH 4–5 with 2 ml of *M* acetate buffer.[411] Add 10 ml of 0.01 *M* solution of the **aluminum-cupferron** complex in chloroform and shake for 2 minutes. Shake the segregated organic phase with 10 ml of 0.04% sodium diethyldithiocarbamate and 10 ml of 0.1 *M* ammonium salt buffer solution for pH 9.5. Read the organic phase at 436 nm against a reagent blank. Beer's law is followed by 0.2–4 μg of copper per ml of extract. Ferric ion in the sample can be masked with fluoride, vanadate with hydrogen peroxide.

Dissolve 1 gram of aluminum in 30 ml of 2:1 hydrochloric acid. Then add 5 ml of sulfuric acid and dilute to 100 ml. Adjust an aliquot to pH 6 and add 2 ml of 10% ethanolic **aniline.**[412] Mix with an equal volume of 3% hydrogen peroxide and set aside for 5 minutes. Extract the copper complex with isoamyl alcohol and read.

Cupric ion is determined by adding 4 ml of 0.9% solution of **amidinophenylurea nitrate** and 10 ml of ammonium hydroxide and diluting to 50 ml.[413] Then 5–35 ppm is read at 317 nm within 2 hours.

5-Amino-3-(3-carboxy-1,2,4-triazol-5-ylazo)-4-hydroxynaphthalene-2,5-disulfonic acid, which is 5-(8-amino-1-hydroxy-3,6-disulfo-2-naphthylazo)-1,2,4-triazole-3-carboxylic acid, forms a 1:1 complex with copper having a maximum absorption at 580 nm.[414] The optimum pH is 6–9, and the absorption obeys Beer's law for

[406] Ajit Kumar Sarkar and Jyotirmoy Das, *J. Indian Chem. Soc.* **48**, 817–822 (1971).
[407] Keshav Kumar Tripathi and D. Banerjea, *Sci. Culture* (Calcutta) **23**, 611–612 (1958).
[408] Sumio Komatsu, Shigeki Matsuda, and Toshio Nakamura, *J. Chem. Soc. Jap., Pure Chem. Sect.* **91**, 731–734 (1970).
[409] E. Gagliardi and M. Khadem-Awal, *Mikrochim. Acta* **1969**, 882–887.
[410] Anil K. Mukherji and Arun K. Dey, *J. Indian Chem. Soc.* **34**, 461–466 (1957); *Bull. Chem. Soc. Jap.* **31**, 521–524 (1958).
[411] Yoshioni Sasaki, *Jap. Anal.* **25**, 108–112 (1976).
[412] M. I. Abramov, *Uch. Zap. Azerb. Univ. Ser. Khim. Nauk.* **1963**, 31–38.
[413] F. Capitan Garcia and F. Salinas, *Rev. Soc. Quim. Mex.* **13**, 13A–16A (1969).
[414] D. N. Pachadzhanov, M. A. Israilov, M. Yu. Yusupov, and A. Tabarov, *Izv. Akad. Nauk Tadzh. SSR, Otdel. Fiz.-Matem. Geol.-Khim. Nauk* **1972** (4), 42–45.

0.2–1 μg of copper per ml in the final solution. For determination, add to the sample 0.5 ml of 0.1% reagent solution and 10 ml of borate buffer solution for pH 7. Dilute to 25 ml and read against a blank. Nickel, cobalt, iron, zinc, zirconium, thorium, tin, and EDTA interfere.

Cupric ion forms a 1:1 complex with **2[(2-aminoethyl)amino]ethyl hydrochloride** above pH 6.9.[415] Typically, 2 ml of sample containing 0.63% copper is added to 10 ml of 20% sodium acetate solution and 2 ml of 2% solution of the reagent. On dilution to 25 ml, it is read at 619 nm over the range of 0.4–4.7 mg of copper per ml.

Cupric ion forms a yellow complex with ***N*-(2-aminoethyl)-*N*-[(2-pyridylmethyleneamino)ethyl]dithiocarbamic acid** in 50 ml of acetate buffer solution at pH 3.9–5.5 and containing 5 ml of 0.4% solution of Triton X-100.[416] Set aside for 45 minutes and read at 432 nm against a reagent blank. Beer's law is obeyed for 0.13-1 μg of copper per ml. There is serious interference by ferric ion, cobalt, nickel, and lead.

6-Amino-4-hydroxy-2-mercapto-5-nitrosopyrimidine forms a 1:1 complex with cupric ion at pH 6.6–8.5.[417] Because of instability, the extinction should be read after 1 hour but before 3 hours. There is interference by tartrate, oxalate, EDTA, cyanide, thiosulfate, thiocyanate, ferric, chromic, thorium, uranyl, manganese, zinc, cadmium, cobalt, and nickel ions. The solution conforms to Beer's law for 0.3–5.1 μg/ml and shows maximum absorption at 386 nm.

Aminomethylazo III, which is 3, 6-bis{2-[bis(carboxymethyl)]phenylazo} chromotropic acid, forms a 3:1 complex with cupric ion.[418] The absorption at 600 nm conforms to Beer's law for 5×10^{-7} to $5\times10^{-6}M$.

For determination of copper by sodium ***p*-aminosalicylate**, make the following additions to a sample containing 0.003–0.3 mg of copper:[419] 5 ml of 4 *M* reagent, 10 ml of 16% sodium hydroxide solution, 0.2 gram of sodium peroxide. Then 20 minutes later add 1.25 ml of *M* sodium sulfate. Dilute to 25 ml for reading. The brown color is stable for 30 minutes. Cobalt and arsenic interfere.

Ammonium pyrophosphate forms a complex with cupric ion that can be read at 650 nm.[420] The low absorption recommends it for major amounts of copper.

Copper forms an extractable complex with **ammonium pyrrolidine-1-carbodithioate.**[421] Mix a sample solution containing 0.001–0.08 mg of cupric ion with 10 ml of potassium phosphate buffer solution for pH 9. Add 5 ml of 0.15% solution of the reagent containing 0.2 ml of 1:1 ammonium hydroxide per liter. After 5 minutes extract with 10 ml of chloroform and read the extract at 269 nm. Alternatively, use an acetate buffer solution for pH 4.7 in place of the phosphate buffer and without extraction read at 435 nm.

Copper can be determined by ***p*-anisidine and potassium thiocyanate.**[422] The purple color in neutral or weakly acid solution is a ternary complex in which the copper is reduced to the cuprous form. Dissolve a 2-gram sample of aluminum,

[415] D. Negoiu, C. Vasilescu, and M. Marica, *Stud. Univ. Babes-Bolyal, Cluj. Chim.* **8**, 31–38 (1963).
[416] Yasoo Itoh, Masao Sugawara, and Tomihito Kambara, *Jap. Anal.* **24**, 571–574 (1975).
[417] S. Przeszlakowski and A. Waksmundzki, *Chem. Anal.* (Warsaw) **9**, 919–924 (1964).
[418] B. Budesinsky and K. Haas, *Z. Anal. Chem.* **214**, 325–331 (1965).
[419] Radu Ralea and Neculai Iorga, *An. Stiint. Univ. "Al. I. Cuza" Iasi, Sect.* **12**, 211–226 (1956).
[420] C. J. Keattch, *Talenta* **3**, 351–355 (1960).
[421] R. W. Looyenga and D. F. Boltz, *ibid.* **19**, 82–87 (1972).
[422] V. N. Podchainova, *Tr. Ural. Politekh. Inst. S. M. Kirova* **1957** (69), 119–125; **1959** (81), 192–211.

nickel, cadmium, or zinc in aqua regia and evaporate nearly to dryness. Take up in water, nearly neutralize with ammonium hydroxide, and reduce the iron with sodium thiosulfate. Add a cadmium salt and precipitate the mixed copper and cadmium sulfides. Filter and dissolve the cadmium sulfide on the filter by warm 1:9 hydrochloric acid. Dissolve the remaining copper sulfide with concentrated hydrochloric acid. Neutralize the copper solution with ammonium hydroxide and make faintly acid with acetic acid. Add a mixture of equal parts of 2 *N* potassium thiocyanate and 0.2 *N* *p*-anisidine. Read at 500 nm after 30 minutes. The color is extractable with chloroform.

The 1:1 complex of cupric ion with **7-antipyrinylazo-8-hydroxyquinoline** at pH 4–6 has a maximum absorption at 505 nm.[423] Beer's law is followed for 0.4–3 μg of copper per ml.

Copper and iron are determinable in a single sample of blood filtrate using **bathocuproine disulfonate** and tripyridyltriazine, respectively.[424] To the serum sample, add an equal volume of 10% trichloroacetic acid solution. To the centrifugate add 1 ml of saturated solution of sodium acetate and 10 mg of ascorbic acid. Read at 485 nm against a blank. As a color reagent, dissolve 0.1 gram of the reagent in 1 ml of acetic acid and dilute to 25 ml. To sample and blank add 0.1 ml of the color reagent and read at 485 nm.

If iron is to be determined as the blank for iron, read at 600 nm the solution already read for copper. Add 0.1 mg of 0.4% solution of tripyridyl-triazine in 4% acetic acid and read at 600 nm.

For copper in serum, [425a] treat with phosphate buffer solution for pH 5.5, then with Thesit, which is hydroxypolyethoxydodecane, and add hydrogen peroxide to release the copper. After reduction with ascorbic acid, the copper is developed with bathocuproindisulfonate and read at 492 nm.

Dissolve 1 gram of steel in 20 ml of 1:4 hydrochloric acid, adding 1 ml of nitric acid, and dilute to 250 ml. Adjust a 5-ml aliquot containing 8–60 μg of copper with ammonium hydroxide to pH 9–9.5. Add 1 ml of 0.02% solution of **benzimazole-2-carboxanilide oxime** in 4:1 chloroform-butanol and 4 ml of chloroform and shake.[425b] Read the 1:1 complex in the organic layer at 337 nm against a reagent blank. Ferric ion must be masked with fluoride.

A blue 1:1 complex is formed by cupric ion and **1-(benzimidazol-4-yl)-3-methyl-5-phenylformazan** which is read at 610 nm.[425c] Nickel, cobalt and cadmium also complex with this reagent. Beer's law is followed for 0.2–1.2 μg of cupric ion per ml.

Dissolve 0.1 gram of piston-rod material as above and dilute to 200 ml. Adjust a 5-ml aliquot to pH 10-11 with 5 ml of borate buffer solution. Add 1 ml of 0.02% solution of **1-benzyl-benzimidazole-2-aldoxime** in chloroform and 4 ml of chloroform. Extract the 1:1 complex and read at 337 nm against chloroform.

[423]Sh. T. Talipov, N. G. Smaglyuk, and A. Khodzhaev, Manuscript No. 6203-73, Vsesoyuznyi Institut Nachnoi i Tekhnicheskoi Informatsii, Moscow.

[424]Bennie Zak, Gerard A. Cavanaugh, and L. A. Williams, *Chemist-Analyst* **50**, 8–9 (1961).

[425a]Gerlinde Beyer and G. Hillman, *Z. Klin. Chem. Klin. Biochem.* **11** (3), 121–122 (1973).

[425b]K. N. Bagdasarov, M. S. Chernov'yants, T. M. Chernoivanova, and E. B. Tsupak, *Zavod. Lab.* **42**, 143–144 (1976).

[425c]V. N. Podchainova, N. P. Bednyazina, I. I. Shevelina and L. P. Sidorova, *Trudy Inst. Khim. Ural' Nauch. Tsentr Akad. Nauk SSSR* **1974**, (27), 30–33.

The 1 : 2 complex of copper with **2-(2-benzothiazolylazo)-1-naphthol** is extractable with chloroform above pH 6.[426] The maximum absorption is at 610 nm. A nickel complex absorbs at 634 nm, a cobalt complex at 620 nm.

The copper complex with **benzothiazole-2-yl-hydrazone of benzothiazole-2-aldehyde** absorbs at 476 nm.[427] The optimum pH is 5.42. Mix 2 ml of sample with 2 ml of 0.05% solution of the reagent in dioxan and dilute to 5 ml with dioxan. Read within 10 minutes. A corresponding complex with the **hydrazone of naphtho-[2,1-d] thiazole-2-aldehyde** absorbs at 462 nm, has an optimum pH of 4, and is similarly developed. Zinc and palladium interfere.

Benzothiazole-2-yl phenyl ketoxime as an orange complex with copper absorbs at 358 nm. Mix a 2-ml sample with 0.5 ml of an ammoniacal buffer solution for pH 9 and add 1 ml of a 0.2% solution of the reagent in ethanol. Dilute to 5 ml with dioxan and read after 30 minutes. Palladium and copper interfere.

Cupric ion is extracted at pH 6 by a solution of **1-benzoyl-3-methyl-2-thiourea** in benzene.[428] When read at 355 nm it obeys Beer's law for 0.1–0.5 mg of copper per ml. There is interference by vanadate, ferric ion, cobalt, nickel, and zinc. The pH is critical, copper is not extracted below pH 4.5, and the complex is hydrolyzed above pH 6.4.

Mix 10 ml of sample with 8 ml of an appropriate buffer solution to maintain a pH above 3. Add 10 ml of 0.1 *M* **benzoyltrifluoroacetone** in isoamyl alcohol.[429] Extract for 10 minutes, dilute the organic layer to 25 ml, and read at 375 nm. Ferric ion interferes.

Cupric ion forms a 1 : 1 complex with **biacetyl monothiosemicarbazone.** At 2–9 ppm in an acetic acid–sodium acetate buffer solution it is read at 410 nm.[430a] At pH 3.3, 1 part of copper can be read in the presence of 30 parts of nickel. Cobalt, tartrate, citrate, and oxalate interfere.

Cupric ion forms a 1 : 1 complex when treated with 0.25% ethanolic **biacetyl monoxime thiosemicarbazone.**[430b] In ammonium hydroxide–ammonium chloride buffer solution for pH 10, this is read after 1 hour at 345 nm. With acetic acid–sodium acetate buffer solution, also in 1 hour, the reading is at 340 nm. The alkaline solution is more sensitive. Beer's law is followed for 0.5–5 ppm of copper. Bismuth interferes in the acid medium.

Cuprous ion forms an intensely orange 1 : 2 complex with **3,3′-bi-(5,6-dimethyl-1,2,4-triazinyl).**[431a] To a 5-ml sample at about pH 4, containing 0.5–10 ppm of copper, add 2 ml of 10% hydroxylamine hydrochloride solution. Add 5 ml of buffer solution containing *M* sodium acetate–0.5 *M* acetic acid and a forty-fold excess of the reagent. Dilute to 50 ml and read at 444 nm. Ferrous ion, nickel, cobalt, and bivalent ruthenium also form colored complexes with the reagent.

A water-insoluble complex is formed by cuprous ion and **bis(di-isopropoxyphosphinothioyl)disulfide.**[431b] This is extractable with chloroform from pH 10 to

[426] Akira Kawase, *Jap. Anal.* **17**, 56–60 (1968).
[427] Toyozo Uno and Sumiyuki Akihama, *ibid.* **10**, 822–827 (1961).
[428] S. M. Kashikar and A. P. Joshi, *Curr. Sci.* **45**, 99–100 (1976).
[429] G. N. Rao and J. S. Thakur, *Z. Anal. Chem.* **271**, 286 (1974).
[430a] F. Sanchez Burgos, P. Martinez, and F. Pino Perez, *Inf. Quim. Anal. Pura Apl. Ind.* **23**, 17–27 (1969).
[430b] M. Valcarcel and D. Perez Bendito, *Inf. Quim. Anal. Pura Apl. Ind.* **24**, 49–62 (1970).
[431a] R. E. Jensen and R. T. Pflaum, *Proc. Iowa Acad. Sci.* **71**, 136–145 (1964).
[431b] A. N. Shishkov, N. K. Nikolov and A. I. Busev, *Zh. Anal. Khim.* **31**, 454–459 (1976).

0.8 N hydrochloric acid. The extract is read as 422 nm conforms to Beer's law for 0.4–6 μg per ml. Gold can be masked with thiosulfate. Palladium interferes. Stannous and titanous ions must be oxidized with bromine. Cupric ion can be reduced with ascorbic acid for reaction with this reagent.

For 115 ppm-0.17% of copper in steel or in aluminum alloys dissolve 0.4–4 grams of sample in aqua regia.[431c] Evaporate to dryness, take up in 10 ml of water and filter. Add 2 ml of 10% solution of hydroxylamine hydrochloride to reduce the copper to cuprous ion. Adjust the solution to pH 6.3. Extract the 1:2 copper-reagent complex with 10 ml of 0.04% solution of **4,4′-bis(4-ethoxycarbonyl)-2,2′-biquinoyl** in butanol and read at 556 nm. Beer's law is followed for 0.5–50 μg of cuprous ion per ml. Iron can be masked with fluoride ion.

Bipyridine complexes with copper at 685 nm[432a] as well as with iron and nickel at appropriate wavelengths. All three can be read in the same solution by simultaneous equations. The concentration of copper must be less than 10 μg/ml.

Cuprous ion forms a 1:2 complex with **2,2′-biquinolyl** which in turn forms a 1:1 complex with bromocresol purple.[432b] With the medium containing an acetate buffer for pH 5–6 the complex is extracted with 20:1 chloroform-isoamyl alcohol for reading at 406 nm. Beer's law is followed for 0.024–5 μg of copper per ml of extract. Thiosulfate and thiocyanate interfere.

A ternary complex of cupric ion–**bromophenol blue**–2,2′-biquinolyl is extracted with 20–1 chloroform–isopentyl alcohol from a medium buffered at pH 5 with acetate.[432c] The method is indirect in that the bromophenol blue is back-extracted with 0.01 N sodium hydroxide for reading at 593 nm. Beer's law is obeyed for 0.08–4 μg of copper per ml. There is interference by thiosulfate, perchlorate, thiocyanate, oxalate, and citrate.

Three ***N*-(bromophenyl derivatives of 2-thiocarbamoyldimedone**, which is 4.4-dimethyl-2,6-dioxycyclohexanethiocarboxamide, form chelates with cupric ion at pH 7–7.4.[433] Extracted into benzene, they are read at 405 nm for 0.1–6 μg/ml. Iron, nickel, manganese, silver, and gold interfere.

When a sample solution is mixed with 0.04 mM **calcichrome** at pH 5.2, either cuprous or cupric ion complexes.[434] Reading is at 535 or 310 nm. More than 1.25 moles of reagent per mole of copper is essential. Aluminum, iron, titanium, vanadium, zinc, nickel, oxalate, and thiosulfate interfere.

A 1:1 complex is formed by cupric ion with **2,2′-bis(carboxymethylmercapto)-diethyl ether**, which is 3-oxapentamethylenedithio)diacetic acid.[435] At 360 nm Beer's law is followed at 0.02–0.15 mM; at 660 nm at 0.41 mM.

1-(2-carboxy-4-sulfonatophenyl)-3-hydroxy-3-phenyltriazene, which is 2-(3-hydroxy-3-phenyltriazeno)-5-sulfobenzoic acid, forms a yellow 2:1 complex with cupric ion at pH 4.8–8.4.[436] It conforms to Beer's law for 0.3–3 ppm.

[431c]A. L. Gershuns and L. G. Grineva, *Zh. Anal. Khim.* **31**, 2048–2050 (1976).

[432a]Giordano Trabanelli, *Atti Accad. Sci. Ferrara* **35**, 133–142 (1957–1958).

[432b]A. L. Gershunes, L. P. Adamovich, and V. M. Skorobogatov, *Zh. Anal. Khim.* **29**, 1905–1911 (1974).

[432c]V. M. Skorobogatov, A. L. Gershuns, L. P. Adamovich, and I. M. Kogan, *Zh. Anal. Khim.* **30**, 1704–1706 (1975).

[433]R. Mocanu and S. Fisel, *An. Stiint. Univ. Al. I. Cuza* **17**, 9–11 (1971).

[434]Hajime Ishii and Hisahiko Einaga, *Bull. Chem. Soc. Jap.* **38**, 1416–1417 (1965); *ibid.* **39**, 1154–1160 (1966).

[435]E. Casassas, J. J. Arias, and A. Mederos, *An. Quim.* **69**, 1121–1131 (1973).

[436]A. K. Majundar and D. Chakraborti, *Anal. Chim. Acta* **53**, 393–400 (1971).

Cupric ion complexes with **2,2′-bis(carboxythio)diethyl ether.**[437] A sample is treated with 5 ml of 0.01 *M* reagent, diluted to 40 ml, adjusted to pH 4–6.75 with sodium hydroxide solution, and diluted to 50 ml. Beer's law is followed for 1.27–10.16 ppm when read at 360 nm or at 25.4–152.4 ppm when read at 660 nm. There is interference by nickel, hexavalent molybdenum, pentavalent vanadium, and more than tenfold excess of cyanide.

A 1:2:2 complex is formed as cupric ion–**catechol violet**–tridodecylammonium bromide.[438] This is produced by 4 ml of 0.5 *M* catechol violet with up to 6.5 μg of cupric ion in sodium acetate–phosphate buffer solution for pH 7.6 diluted to 50 ml. Shake with 10 ml of 0.6 m*M* solution of the quaternary compound in benzene; then separate the organic layer, dilute with ethanol, and read at 663 nm against a reagent blank.

Chrome azurol S (mordant blue 29) forms a red complex with copper at pH 8–12 which is more sensitive in the presence of zephiramine.[439] Read at 530 nm, the graph is linear for 1–30 μg of copper in 25 ml of solution containing 2 ml of 0.25% solution of the dye with a 5:2 mixture of 0.05 *M* sodium tetraborate and 0.8% sodium hydroxide to adjust to pH 10, followed by 4 ml of 1% solution of zephiramine. The absorption is constant for 4 hours. Less than 0.1 mg of ferric ion is masked by 1 mg of fluoride ion; chromic ion is preoxidized to chromate. Palladium interferes seriously. Citrate and EDTA prevent color formation.

Copper is also read without zephiramine.[440a] Then it gives a linear graph for 9–22 μg of copper in 10 ml of solution containing 0.5–1 ml of 0.001 *M* chrome azurol S and 2 ml of 0.5 *M* hexamine perchlorate buffer for pH 6.7. A similar color is given by cerous and beryllium ions, hexavalent uranium, scandium, and yttrium. A negative error is caused by ferric, ferrous, and stannous ions. Oxalate, tartrate, citrate, and EDTA affect the color.

A polar solvent causes a bathochromic shift between 590 and 610 nm in the absorption of the 1:2 copper–chrome azurol S complex in 4:1 solvent-water systems.[440b] The molar extinction tends to increase with decreasing dielectric constant of the solvent. With 0.34 m*M* chrome azurol S and 0.01 *M* hexamine, Beer's law is obeyed for 5–26 m*M* cupric ion.

For analysis for copper in copper-cobalt alloys, the **cupric ion** in solution in 10 *M* hydrochloric acid is read at 960 nm, provided the cobalt is only a fewfold greater than the copper.[441a]

For analysis of brass or bronze, dissolve a 2-gram sample in 20 ml of hot 30% nitric acid and dilute to 100 ml.[441b] Read at 825 nm. Cupric ion is read in 4:1 acetone–1:1 hydrochloric acid at 400 nm.[441c] If large amounts of iron, molybdenum, and vanadium are present, they must be preextracted.

[437]E. Casassas and J. J. Arias, *Inf. Quim. Anal. Pura Apl. Ind.* **27**, 151–157 (1973).
[438]Yoshio Shijo, *Bull. Chem. Soc. Jap.* **47**, 1642–1645 (1974).
[439]Yoshio Shijo and Tsugio Takeuchi, *Jap. Anal.* **15**, 1063–1067 (1966); Yoshizo Horiuchi and Hiroshi Nishida, *ibid.* **18**, 694–698 (1969).
[440a]Ryoei Ishida and Takeo Sawaguchi, *ibid.* **16**, 590–595 (1967).
[440b]Hiroshi Nishida and Taeko Nishida, *Jap. Anal.* **25**, 55–57 (1976).
[441a]G. Lindley, *R and D* **1963**, 23–26; cf. Kichinosuke Hirokawa, *Nippon Kinzoku Gakkaishi* **22**, 181–185 (1958).
[441b]A. A. Nemodruk, O. K. Primachek, and L. N. Trukhacheva, *Zh. Anal. Khim.* **31**, 23–26 (1976).
[441c]C. R. Walker and O. A. Vita, *Anal. Chim. Acta* **47**, 9–18 (1969).

For copper in zinc sulfate, the lead is precipitated from a solution of 50 grams as a sulfate. The solution is then made 1 : 1 with hydrochloric acid and read at 278 and 360 nm for determination of copper and iron by simultaneous equations.[442] Not more than 0.0001% antimony may be present. For 1–10 μg of copper per gram of sodium chloride, mix 8 ml of sample solution containing 2.5 grams of sodium chloride with 1 ml of 0.026% potassium cyanide solution. Dilute to 10 ml and read at 235 nm.[443] Auric ion interferes seriously; manganese, zinc, cadmium, nickel, and ferric ion slightly.

Copper, lead, and bismuth separated by paper chromatography are extracted as the chlorides and read in the ultraviolet. For details, see Lead.

Cuprotest, which is 6,7-dihydro-5,8-dimethyldibenzo-1,10-phenanthroline, is another reagent for cuprous ion.[444] The copper is reduced with ascorbic acid. Up to 0.05 mg/ml is extracted from a tartaric acid solution at pH 3–10 by a chloroform solution of Cuprotest and read at 554 nm. There is interference by EDTA, thiosulfate, cyanide, and sulfide.

Cyclohexanediaminetetraacetic acid, also known as Complexon IV, forms a 1 : 1 complex with copper which has a maximum absorption at 720 nm for reading 1–10 mg of copper per ml at pH 5–11.[445]

Mix a sample solution containing 0.1–0.4 mg of cupric ion with 10 ml of 0.1% solution of **cyclohexane-1,2-dione bis-thiosemicarbazone** in dimethylformamide.[446] Adjust to pH 4.5–6 with an acetate–acetic acid buffer solution and add 50 μg of ascorbic acid. Dilute to 50 ml and read at 467 nm. Addition of 10 ml of 0.1 *M* EDTA will mask ferrous, ferric, cobalt, nickel, cadmium, and zinc ions.

For determination of cupric ion by **diacetyldithiobenzhydrazone**, the sample is dissolved in 1 : 1 sulfuric acid and the reagent added in the same solvent.[447] The maximum absorption is at 530 and 375 nm. Antimony, arsenic, and gold interfere. The reagent is applicable to 0.006–0.488 mg/ml.

Both copper and ferrous ion are read at pH 12.5 with **2,5-diacetylpyridine dioxime** at 360 and 490 nm, respectively.[448] Beer's law applies to either one at 0.1–5 μg/ml. Hydroxylamine is used to reduce ferric ion.

1,2-Diaminoanthraquinone forms a 2 : 1 complex with cupric ion.[449] To the sample solution, add 15 ml of 2% solution of polyvinyl alcohol, 15 ml of ethanol, 5 ml of 0.085% solution of the color reagent in ethanol, and 4 ml of 16% sodium hydroxide solution. Dilute to 50 ml and read at 650 nm. It follows Beer's law for 1–7 ppm of copper in the developed solution.

Cupric ion forms a stable 1 : 1 complex in acid solution with **di-(2-aminoethyl)-ether-NNN′N′-tetraacetic acid.**[450] To a sample solution containing less than 1.6

[442] M. J. Maurice and S. M. Ploeger, *Z. Anal. Chem.* **179**, 246–258 (1961).

[443] A. Glasner, S. Sarig, D. Weiss, and M. Zidon, *Talenta* **19**, 51–57 (1972).

[444] G. Ackermann and W. Angermann, *ibid.* **15**, 79–85 (1968).

[445] Walter Nielsch, *Mikrochim. Acta* **1959**, 419–423; E. Jacobson and A. R. Selmer-Olsen, *Anal. Chim. Acta* **25**, 476–481 (1961).

[446] J. A. Munoz Leyva, J. M. Cano Pavon, and F. Pino Perez, *An. Quim.* **72**, 392–395 (1976).

[447] Gerhard Bähr and D. Thiele, *Chem. Tech.* (Berlin) **10**, 420 (1958).

[448] E. Gagliardi and P. Presinger, *Mikrochim. Ichnoan. Acta* **1965**, 1047–1052.

[449] F. Capitan Garcia, M. Roman Ceba, and F. Garcia-Sanchez, *Inf. Quim. Anal. Pura Apl. Ind.* **27**, 179–189 (1973).

[450] F. Bermejo-Martinez and A. G. Blas-Perez, *Anal. Chim. Acta* **27**, 459–464 (1962).

gram of copper, add 2 ml of 5% reagent as the sodium salt and adjust to pH 2 with 1:5 hydrochloric acid or 1:10 ammonium hydroxide. Add 10 ml of 0.2 *M* hydrochloric acid–potassium chloride buffer solution for pH 2 and dilute to 25 ml. Read at 720 nm. Tartrate, oxalate, benzoate, acetate, and thiosulfate interfere. Beer's law is followed up to 650 μg of copper per ml. For copper in silver solder, dissolve a 0.5-gram sample in 20 ml of 1:4 nitric acid, evaporate, take up in water, precipitate the silver by dropwise addition of dilute hydrochloric acid, and filter. Add 5 ml of sulfuric acid, evaporate to sulfur trioxide fumes, take up in water, dilute to 100 ml, and develop as previously.

To a slightly acid solution containing less than 20 mg of cupric ion, add 10 ml of 0.1 *M* **1,2-diaminopropane-NNN′N′-** tetraacetic acid,[451] dilute to 25 ml, and read the 1:1 complex at 730 nm. Chromic, nickel, and cobalt ions also complex with this reagent.

Copper forms a blue chelate with **diethylenetriaminepentaacetic acid**, which at 0.009–1.5 mg/ml is read at 650 nm and pH 6.5–13.[452] Ferric, cobalt, and nickel ions interfere at more than 0.04 mg/ml.

4,4′-Dihydroxy-2,2′-biquinolyl proved the most promising of 13 derivatives of 2,2′-biquinoyl in chelating cuprous ion in 24–48% sodium hydroxide solution.[453] As its 2:1 complex, it is desirably extracted into amyl alcohol for reading at 525 nm.

6,7-Dihydroxynaphthalene-2-sulfonic acid forms a 2:1 complex with cupric ion at pH 7.5–7.7 suitable for reading 0.3–0.7 μg per ml of copper at 345 nm.[454a]

A complex of copper with **2′,4′-dihydroxyvalerophenone oxime** is extracted with chloroform and read at 640 nm.[454b] Beer's law is obeyed for 45–80 ppm.

5-Dimethylamino-2-(2-thiazolylazo)phenol forms a 1:1 complex with copper at pH 1.5–4.8 but a 2:1 complex in alkaline solution.[455] The latter can be extracted with chloroform, butanol, or isopentyl alcohol for reading at 0.02–0.8 μg of copper per ml.

6,6′-Dimethyl-4,4′-diphenyl-2,2′-biquinolyl is suitable for determining cuprous ion.[456] To 25 ml of sample containing not over 0.02 mg of copper, add 0.1 gram of hydroxylamine hydrochloride, 0.2 gram of sodium acetate, and 5 ml of 0.00195 *M* reagent in isopentyl alcohol. Shake for 15 minutes and read the organic layer at 554 nm. Beer's law is followed for 0.02–0.8 μg of copper per ml.

1-(2,4-Dinitrophenyl)-2-acetylhydrazine, which is acetic acid-2,4-dinitrophenylhydrazide, forms a green complex with copper in faintly alkaline solution absorbing at 600 nm.[457] The complex is soluble in water or ethanol but not in chloroform.

The cupric complex with **2,4-dinitroresorcinol** is formed around pH 5 in the

[451] S. Vicente-Perez, L. Hernandez, and J. Rosas, *Quim. Anal.* **28**, 283–288 (1974).
[452] Bermejo Martinez and J. A. Rodriguez Campos, *Microchem. J.* **11**, 331–341 (1966); Masako Idemori and Toru Nozaki, *J. Chem. Soc. Jap., Pure Chem. Sect.* **86**, 77–82 (1965).
[453] A. A. Schilt and W. C. Hoyle, *Anal. Chem.* **41**, 344–347 (1969).
[454a] Yoshinaga Oka, Norimasa Nakazawa, and Hikaru Harada, *J. Chem. Soc. Japan, Pure Chem. Sect.* **86**, 1162–1166 (1965).
[454b] Jai Singh, S. P. Gupta, and O. P. Malik, *Indian J. Chem.* **13**, 1217–1220 (1975).
[455] J. Minczewski and K. Kasiura, *Chem. Anal.* (Warsaw) **10**, 719–727 (1965).
[456] Saburo Nakano, *J. Chem. Soc. Jap., Pure Chem. Sect.* **82**, 1256–1261 (1962); cf. T. G. Dmitrieva and A. L. Gershuns, *Tr. Kom. Anal. Khim.* **17**, 230–234 (1969).
[457] Laszlo Legradi, *Z. Anal. Chem.* **260**, 126–127 (1972).

presence of a fourfold excess of reagent. Addition of sodium perchlorate to make the solution 0.01 *M* intensifies the extinction, which is read at 370 nm.[458]

Ethanolic **diphenylcarbazone** added to a sample buffered with phosphate or acetate to pH 4–8 forms a copper complex.[459] Extracted with benzene, the complex is read at 558 nm. The reagent has a maximum at 470 nm. Beer's law holds up to 7.6 μg per 15 ml. Silver, nickel, cobalt, stannous ion, cyanide, and EDTA interfere.

The complex of cupric ion with sodium **diphenylphosphinodithioate** at pH 1–2.5 absorbs at 440 nm.[460] Stannous ion, antimony, and manganese interfere.

By adding potassium **dithiophosphate** to an acid solution of copper, the complex is extracted with carbon tetrachloride and read at 420 nm.[461]

4,4′-Bis-(4-ethoxycarbonylanalino)-2,2′-biquinolyl forms a 2:1 complex with cuprous ion.[462] A butanol extract conforms to Beer's law for 2–200 m*M*. The extinction is unchanged for days. For development, add hydroxylamine to the sample solution to reduce cupric ion, and shake with an equal volume of m*M* reagent in butanol. Read the extract at 556 nm.

Cupric ion forms a complex with **2,2′-(ethylene-diimino)diproprionic acid** at pH 3–9 which is read at 670 nm.[463a] The absorption conforms to Beer's law up to 400 ppm. The reagent must be in twofold excess. Nickel must not exceed copper. Permissible excess of halide, acetate, phosphate, cerium, or cobalt varies from 5- to 25-fold.

Cupric ion complexes with the **monoethyl ester of α-(ethylamino)-2-hydroxybenzylphosphonic acid** at pH 5.6.[463b] The stable green compound can be read at 410 nm or 760 nm. Beer's law is followed for 63–630 μg of cupric ion per ml.

In alkaline solution, **ethylidene oxalyldihydrazone** forms a stable complex with copper.[464] As applied to zinc and its alloys, dissolve a 10-gram sample in 1:1 hydrochloric acid, adding 30% hydrogen peroxide dropwise to complete the solution. Evaporate to a syrup and dilute to an appropriate volume: 50 ml for 0.002% of copper, 1 liter for 0.25% of copper. To a 10-ml aliquot, add 10 ml of ammonium hydroxide, 10 ml of 50% citric acid solution, and 10 ml of 0.15% solution of the reagent. Dilute to 50 ml, let stand for 5 minutes, and read at 605 nm.

The optimum conditions for determination of cuprous ion with **ethylrhodamine B** are 7 *N* sulfuric acid, 2.5 *N* bromide ion, and 0.1 mg of reagent per ml.[465] Extraction with benzene from 8 *N* sulfuric acid containing 20 mg of bromide per ml and 0.4 mg of reagent provides preconcentration.

In slightly alkaline solution the complex of copper with **fast sulfon black F** (acid black 32) is read at 630 nm.[466] It applies down to 0.08 ppm, but there is serious

[458]S. E. Zayan, R. M. Issa, and Jacqueline Maghrabi, *Microchem. J.* **18**, 662–669 (1973).

[459]Hideo Seno and Yachiyo Kakita, *Jap. Chem. Soc.* **82**, 1365–1367 (1961).

[460]L. K. Kabanova, S. V. Usova, and P. M. Solozhenkin, *Izv. Akad. Nauk Tadzh. SSR Otdel. Fiz.-Matem. Geol.-Khim. Nauk* **1971**, 47–56.

[461]A. I. Busev and M. I. Ivanyutin, *Vestn. Mosk. Univ. Ser. Mat. Mekh. Astron., Fiz. Khim.* **12** (5), 157–161.

[462]A. L. Gershuns and L. G. Grin'ova, *Visn. Khar'k. Univ. Khim.* 53–56 (1971).

[463a]J. J. R. Frausto da Silva, J. C. Goncalves Calado

[463b]Gheorghe Morait and Gheorghe Zuchi, *Revta Chim.* **27**, 233–235 (1976), and M. Legrand de Moura, *Talenta* **12**, 467–474 (1965).

[464]G. Lanfranco, *Metallurg. Ital.* **55**, 365–370 (1963).

[465]I. A. Bochkareva and I. A. Blyum, *Zh. Anal. Khim.* **30**, 874–882 (1975).

[466]R. Belcher, A. Cabrera-Martin, and T. S. West, *An. Real Soc. Esp. Fis. Quim. B* **59**, 281–284 (1963).

interference by nickel, manganese, and beryllium and slight interference by iron, chromium, cobalt, and trivalent arsenic.

For determination of copper in biological samples by **ferrocyanide**, wet-ash a 10-gram sample and evaporate to dryness. Take up in 20 ml of 1 : 20 hydrochloric acid and add 5 ml of 2% solution of stannous chloride. Add a 1-cm^2 piece of tin foil and let stand overnight.[467] Dissolve the foil, on which the copper will have plated out, in 4 ml of 1 : 5 nitric acid. Centrifuge to separate the metastannic acid, and wash the precipitate with 3 ml of 1 : 5 nitric acid. Evaporate the filtrate and washings to dryness and take up in 3 ml of hot water. Centrifuge and wash the precipitate with 3 ml of water. To the solution obtained, add 5 drops of 80% acetic acid containing 10% of sodium acetate and dilute to 9 ml. Add 1 ml of 5% solution of potassium ferrocyanide and read.

Copper retards the autoxidation of **2-furaldehyde** in alkaline solution.[468a] Thus a sample containing 0.04–0.8 μg of copper per ml is made 0.04 *M* with 2-furaldehyde and 0.4 *M* with sodium hydroxide. The oxidation products of the reagent are read at 400 nm after 4 minutes. Anions that complex with copper interfere.

Cupric ion forms a 2:1 complex with **2-furaldehyde benzothiazol-2-ylhydrazone**[468b] at pH 5.6–9.6. This is extractable with benzene for reading 0.1–1 μg per ml at 415 nm. There is interference by citrate, tartrate, mercuric ion, chromic ion, titanic ion, iodide, and thiocyanate. The reagent also complexes with silver, cobalt, mercuric ion, nickel, and zinc for photometric reading.

2-Furfurylidene-1-isonicotinoylhydrazine, formed by heating 4.5 grams of isoniazid and 4.1 grams of furfuraldehyde in 50 ml of ethanol, forms a 1 : 1 complex with copper.[469] At 0.5–6.3 ppm the complex formed at pH 0.5–6.3 is read at 385 nm.

In slightly acid solution **glycinecresol red** forms a red-violet complex with copper for reading at 550 nm.[470] Beer's law is followed for 1–16 μg of copper in 5 ml.

Copper serves as a catalyst to promote air oxidation of ammoniacal **guaicyl valerate**.[471] It will detect 0.1 μg in 1 gram of wet tissue.

A dichloroethane solution of ***O*-hexyl butylphosphonodithioate** extracts cupric ion from 0.1–5 *N* sulfuric acid solution.[472] Over the range of 0.2–1.2 μg of copper per ml of the extract, it follows Beer's law.

The 1 : 1 complex of copper in a large excess of **hydrobromic acid** absorbs at 273, 345, 515, and 590 nm.[473] Down to 1 μg/ml in 7 *N* hydrobromic acid, the latter two are appropriate for reading. When the copper content is high, 800 nm is preferable. The ultraviolet bands fade with time.

Copper forms a 1 : 1 complex with **1-(2-hydroxybenzylidene)-2-(2-picolinoyl)-hydrazine** at pH 4–5 having a maximum absorption at 420 nm.[474] The reaction is

[467]T. V. Marchenko and G. B. Dmitrieva, *Farm. Zh.* (Kiev) **16** (2), 58–60 (1961).
[468a]Enrique Cassas and Heraclio Torres, *Inf. Quim. An*
[468b]Tsugikatsu Odashima and Hajime Ishii, *Anal. Chim. Acta* **83**, 431–432 (1976).al. Pura Apl. Ind. **23**, 61–74 (1969).
[469]Nam Ho Paik and Yun Soo Choi, *Yakhak Hoeji* **9**, 18–22 (1965).
[470]*Rep. Sci. Police Res. Inst.* **16**, 196–199 (1963).
[471]M. Ya. Shapiro, *Vopr. Med. Khim.* **9**, 298–300 (1963).
[472]P. M. Solozhenkin, L. K. Kabanova, O. N. Grishina, and V. A. Bondarenko, *Izv. Akad. Nauk Tadz. SSR, Otdel. Fiz. Matem. Geol.-Khim. Nauk* **1970**, 48–54.
[473]Shigeki Matsuo, *J. Chem. Soc. Japan.* **82**, 32–35 (1961).
[474]Atsushi Sugii and Motoko Dan, *Jap. Anal.* **12**, 368–371 (1963).

carried out with 1 ml of 0.1% ethanolic reagent in 50% ethanol containing 0.1 *M* acetate buffer solution for pH 4.6–4.7. The graph is linear for 0.1–1.3 ppm of copper. There is interference by cobalt, iron, nickel, and silver.

2-Hydroxy-1,3-diphenyltriazene as a complex with copper is extracted with benzene at pH 4–12 for reading at 395 nm.[475] There is interference by palladium, iron, auric, and titanium ion.

The complex of copper with zinc **bis-(2-hydroxyethyl)-dithiocarbamate** is read at 432 nm.[476] Add 1 ml of 20% citric acid solution to the sample solution, adjust the pH to 5.5–9.5 with *N* hydrochloric acid or 1 : 3 ammonium hydroxide, add 1 ml of 0.5% solution of the reagent, and dilute to 100 ml. Read against a blank within 25 minutes.

2-Hydroxy-2′-(1-hydroxy-2-naphthazo)-5-methylazoxybenzene forms a 1 : 1 complex with copper at less than 3 μg/ml.[477] The solution should contain 1–4% sodium hydroxide for reading at 580 nm. The color is stable for 48 hours.

A yellow 1:2 complex of cupric ion with **4-hydroxyimino-3-methyl-1-phenyl-2-pyrazolin-5-one** is formed in sodium acetate solution at 30°.[478] Extracted into chloroform, the maximum absorption is at 395 nm. With twentyfold excess of reagent and shaking for 15 minutes, 0.062 mg of copper per ml is almost completely extracted. The technic tolerates up to 0.6 mg of nickel, 1.1 mg of manganese, and 0.53 mg of cobalt per ml.

Palladium is extracted from a solution containing 10–700 μg each of copper and palladium at pH 1 with twentyfold excess of **2′-hydroxy-4-methoxy-5′-methylchalcone oxime** in 10 ml of isobutyl alcohol for reading at 380 nm. Then the aqueous layer is adjusted to pH 5.8 and copper extracted with tenfold excess of reagent in benzene for reading at 375 nm.[479]

5-Hydroxy-3-methyl-5-(D-arabinotetrahydroxybutyl)thiazolidine-2-thione in 1% solution forms a pale yellow complex with silver but a deep yellow complex with cupric ion.[480] At pH 4.6–7 with an optimum at 4.7, copper is read at 432 nm. Beer's law is followed for up to 4 mg of copper per liter. Silver, citrate, tartrate, and EDTA interfere.

Cupric ion forms a 1:2 complex with **1-hydroxy-3-methyl-1-phenyl-2-thiourea**, which is *N*-methylaminothioformyl-*N*-phenylhydroxylamine,[481] in 50% ethanol at pH 5.4–8.9 for reading at 430 nm. The complex is stable for 18 hours and obeys Beer's law up to 1.3 ppm of copper.

The condensation product of **2-hydroxy-1-naphthaldehyde** and **4-chlorobenzyl dithiocarbamate** in 50% dimethylformamide, when activated at 470 nm, shows an intense fluorescence at 520 nm.[482] Added cupric ion complexes with the reagent to reduce the fluorescence permitting determination of 0.1–1.1 ppm. Cobalt and cadmium react similarly and must be absent. Silver, nickel, and zinc also interfere.

[475]Kazumasa Ueda, Yoshikazu Yamomoto, and Shunzo Ueda, *J. Chem. Soc. Jap., Pure Chem. Sect.* **91**, 1151–1155 (1970).

[476]Hitoshi Yoshida, Masahiro Yamamoto, and Seiichiro Hikime, *Jap. Anal.* **11**, 197–201 (1962).

[477]J. Minczewski and E. Wieteska, *Chem. Anal.* (Warsaw) **9**, 365–372 (1964).

[478]M. C. Patel, J. R. Shah, and R. P. Patel, *J. Prakt. Chem.* **314**, 181–183 (1972).

[479]B. K. Deshmukh and R. B. Kharat, *Z. Anal. Chem.* **276**, 299 (1975).

[480]J. A. Corbett, *Talenta* **13**, 1089–1096 (1966); B. K. Deshmukh, S. B. Gholse, and R. B. Kharat, *Z. Anal. Chem.* **279**, 363 (1976).

[481]S. P. Mathur and M. R. Bhandari, *Chem. Ind.* **1975**, 745–746.

[482]Imre Kasa and Gabor Bajnoczy, *Period. Polytech. Chem. Eng.* **18**, 289–294 (1974).

2-hydroxy-1-naphthaldehyde and **Girard's T reagent,** which is (carboxymethyl) trimethylammonium chloride hydrazide, combine to determine copper.[483] Make the sample solution 0.1 *M* with sodium perchlorate and 0.01 *M* with the combined reagent. Adjust to pH 3 with perchloric acid or sodium hydroxide solution. Dilute to volume and set aside for 30 minutes. Read at 408 nm against a blank. Beer's law applies up to 6 ppm of copper.

When a 1% solution of **2-hydroxy-1-naphthaldehyde-2-hydroxyanil** is added to a solution at pH 2–6 in 50% ethanol, copper is precipitated.[484] This will separate it from a tenfold concentration of cobalt and nickel. Filter and dissolve the precipitate in hot 1 : 18 sulfuric acid to liberate 2-aminophenol. Read at 270 nm. Beer's law applies to 0.1–1 mg of copper in the original solution. Iron interferes.

At pH below 3 cupric ion forms a 1 : 1 complex with **5-(2-hydroxy-1-naphthylazo)pyrazole-4-carboxylic acid.**[485] This is read at 525 nm and obeys Beer's law for 0.5–2.5 μg of copper per ml. There is interference by oxalate, thiosulfate, EDTA, and cyanide. At pH 10.4, cupric ion forms a 1 : 2 complex with the reagent which is read at 545 nm.

8-hydroxy-7-nitroquinoline-5-sulfonic acid is a reagent for copper.[486] At pH 3 the absorption of 0.24–16 m*M* solution is read at 625 nm. Oxalate interferes seriously, and the tolerance for fluoride is low.

For copper in the presence of iron, develop with **3-hydroxy-3-phenyl-1-(3-pyridyl)triazen,** extract with chloroform, and read at 347 nm.[487] Mask iron with sorbic acid. Cobalt must be absent because it is read in the same region as copper.

2-(3-hydroxy-3-phenyltriazeno)benzoic acid forms 1 : 1 complexes with several metals. That with copper at pH 2.2–3.8 absorbs at 400 nm for 0.5–4 ppm.[488] Nickel and palladium must be masked. Similarly, 0.25–2 ppm of nickel at pH 6.8–8.3 absorbs at 410 nm. Thiosulfate will mask copper and palladium for that determination. For reading 0.5–4 ppm of palladium in pH 2.4–3.5 at 410 nm, nickel and copper are masked with oxalic acid.

The product of **1-(2-hydroxypropyl)anabasine** and hydrogen peroxide irradiated in the ultraviolet shows a maximum fluorescence at 525 nm.[489] At pH 9–11 cupric ion displaces it to a shorter wavelength and decreases the intensity. For determining 0.05–0.5 μg of copper, appropriate conditions in a borate buffer solution for pH 10–11 are reagent, 3 m*M*, hydrogen peroxide, 2.5 m*M*, duration of reaction, 10 minutes. Many cations do not interfere up to 0.01 mg/ml, but uranate and palladous ions do.

Copper forms a 1 : 3 complex with **5-hydroxy-3-propyl-5-arabino-tetrahydroxylbutyl-3-thiazolidine-2-thione.**[490] Read 1 ml of 0.15% solution in 100 ml of water at pH 6–8 at 436 nm. Cyanide inhibits the reaction.

Nickel, copper, and palladium form 1 : 1 complexes with **2-(3-hydroxy-3-phenyl-**

[483] S. P. Rao and T. A. S. Reddy, *Z. Anal. Chem.* **277**, 127 (1975).
[484] Kiyoharu Isagai, *J. Chem. Soc. Jap., Pure Chem. Sect.* **87**, 566–569 (1966).
[485] Boleslaw Janik and Tadeusz Gancarczyk, *Chem. Anal.* (Warsaw) **15**, 397–404 (1970).
[486] S. K. Patel, D. P. Soni, and I. M. Bhatt, *Lab. Pract.* **20**, 651–652 (1971).
[487] E. D. Korotkaya, A. Kh. Klibus, V. E. Pochinok, and T. S. Shul'gach, *Izv. Vyssh. Ucheb. Zaved., Khim. Khim. Tekhnol.* **18**, 1695–1699 (1975).
[488] A. K. Majumdar and D. Chakraborti, *Z. Anal. Chem.* **257**, 33–36 (1971).
[489] L. E. Zel'tser, Z. T. Maksimycheva, and Sh. T. Talipov, *Akad. Nauk Uzbek. SSR* **1969**, 30–31; Z. T. Maksimycheva, Sh. T. Talipov, and L. E. Zel'tser, *Dokl. Akad. Nauk Uzbek. SSR* **1968** (10), 25–26.
[490] M. J. Stiff, *Analyst* **97**, 146–147 (1972).

triazeno)benzoic acid.[491] That of copper is read at 400 nm for 0.5–4 ppm at pH 2.2–3.8.

Copper in uranium can be determined with **8-hydroxyquinoline**. The complex at pH 3.1–3.3 is extracted into chloroform.[492] Shaking the extract with 15% sodium hydroxide solution removes some contaminants, including uranium. Read at 410 nm and subtract a background blank read at 580 nm. The residual uranium in the sample solution can be masked with acetic acid. The method determines copper down to 0.5 ppm, and the only interference in nuclear fuel is more than 0.2 mg of nickel.

With a sample containing 0.004–0.04 mg of cupric ion, 2 ml of 0.1 *M* acetate buffer solution for pH 4.8, which is m*M* with **8-hydroxyquinoline-5-sulfonic acid** and 10 ml of 5 m*M* **zephiramine** a 1:2:2 complex is formed.[493] After adjusting to pH 5.3–5.5 with a buffer solution, the complex is extracted by shaking for 5 minutes with 10 ml of chloroform and read at 410 nm. Beer's law is obeyed for 0.0044–0.044 mg of cupric ion in the extract.

Dissolve 0.2 gram of steel in 10 ml of 1:1 hydrochloric acid and 2 ml of nitric acid. Evaporate to dryness, take up in 5 ml of 4 *N* sulfuric acid, and dilute to 100 ml. To a 10-ml aliquot containing 10–100 μg of copper add 0.5 gram of ascorbic acid and 0.5 gram of citric acid. Add 5 ml of 0.01 *M* ***N*-hydroxythiocarbanilide**, which is *NN'*-diphenylthiocarbomoylhydroxamic acid, or 5 ml of ***N*-*p*-bromophenyl-*N'*-phenylcarbamoylhydroxamic acid** in chloroform, and shake for 2 minutes. Read the 1:2 complex of cupric ion and the reagent at 400–450 nm.[494] The reagents do not absorb appreciably. Beer's law is obeyed for 2–20 μg of copper in the extract. Ferric ion and vanadate are masked with ascorbic acid; tungstate and molybdate with citric acid.

Adjust an aqueous solution with 0.05 *M* chloroacetate buffer solution to pH 2.6. Develop 0.017–2.8 μg of cupric ion with 0.33 m*M* **3-hydroxy-4-(2-thiazolylazo)-naphthalene-2,7-disulfonic acid.**[495] Adjust the ionic strength to 0.5 with potassium nitrate and read at 528 nm. Many other cations interfere. Organic solvents enhance the absorption of the complex and decrease that of the reagent.

Copper forms a 1:2 complex with **β-ionone semicarbazone.**[496] β-Ionone is 4-(2,6,6-trimethyl-1-cyclohexene-1yl)but-3-en-2-one. Treat a sample solution containing 10–60 μg of cupric ion with 10 ml of acetate buffer for pH 4.7, 0.2 gram of sodium perchlorate monohydrate, and water to make 20 ml. Extract with 20 ml and 10 ml of 0.02% solution of the reagent in chloroform, dry the combined extracts with sodium sulfate, and read at 380 nm as the maximum absorption or at 400 nm. Oxalate, EDTA, cyanide, and several cations interfere.

When a sample solution containing copper is passed through a column of powered **iron**, the effluent contains an equivalent amount of ferrous ion.[497] This can be read at 510 nm by 1:10 phenanthroline or after oxidation to ferric ion at

[491] A. K. Majumdar and D. Chakraborti, *Z. Anal. Chem.* **257**, 33–36 (1971).
[492] Kenji Motojimo, Hiroshi Hashitani, and Hideyo Yoshida, *Jap. Anal.* **10**, 79–82 (1961).
[493] Tomihito Kambara, Michiyo Maeyama, and Kiyoshi Hasebe, *ibid.* **20**, 1249–1254 (1971).
[494] V. P. Maklakova, *Zh. Anal. Khim.* **25**, 257–259 (1970).
[495] M. Langova, V. Kuban, and D. Nonova, *Collect. Czech. Chem. Commun.* **40**, 1694–1710 (1975).
[496] M. Guzman, D. Perez Bendito, and F. Pino Perez, *Inf. Quim. Anal. Pura Apl. Ind.* **27**, 209–216 (1973).
[497] Eugeniusz Klozko, *Chem. Anal.* (Warsaw) **16**, 1291–1298 (1971).

458 nm by salicylhydroxamic acid. Cobalt, nickel, lead, silver, and mercuric ion react similarly.

Copper, nickel, zinc, and cobalt in a solution containing ammonium carbonate or sulfate are extracted with 0.5% solution of **kelex 100,** which is a β-alkalene-8-hydroxyquinoline in Solvesso 10 containing 10% isodecyl alcohol.[498] They are back-extracted with 15% sulfuric acid.

Mix 5 ml of sample solution containing up to 20 μg of cupric ion with 1 ml of m*M* **lead bis-(4-sulfobenzyl)dithiocarbamate** and read the 1:2 copper-ligand complex at 433 nm against a reagent blank.[499] The pH range is 2–10. There is interference by cobalt, ferric ion, bismuth, nickel, and high salt concentrations.

As a reagent, prepare 0.1 m*M* **lead xanthate** in chloroform by the reaction of lead acetate and potassium ethylxanthate.[500] Adjust the sample solution containing 10–50 μg of cupric ion to pH 2.6 and shake with 10 ml of the reagent solution. Read the cupric complex at 420 nm against a reagent blank. Silver interferes.

8-Mercaptoquinoline, also known as thiooxine, forms a 2:1 complex with cupric ion.[501] The compound is extractable from acid or alkaline solutions by chlorobenzene, bromobenzene, chloroform, amyl acetate, or isopropanol. Beer's law is followed up to 8 μg/ml, but palladium, ruthenium, and osmium must be absent.

To a sample containing 5–40 μg of copper, add 5 ml of acetate buffer solution for pH 4.5 and 1.5 ml of 0.1% ethanolic ***n*-methylanabasine-(α'azo-6)-*m*-aminophenol**, which is 5-amino-2-[3-(1-methyl-2-piperidyl)-2-pyridylazo]phenol.[502] Dilute to 25 ml and read at 520 nm. The curve is linear for 0.2–1.6 μg/ml. Tartaric acid masks 200-fold amounts of cobalt and nickel; sodium fluoride masks ferric ion.

***N*-Methylanabasine-α'-azo-1-naphthol-5-sulfonic acid** forms 2:1 complexes with copper, nickel, cobalt, zinc, cadmium, and manganese and a 1:1 complex with palladium.[503] For copper the optimum pH is 1–3; color develops immediately and is read at 590 nm, as compared to the reagent itself at 440 nm. Similar complexes are formed with *N*-methylanabasine-α'-azodiethylaminophenol.[504]

***N*-Methylanabasine-α'-diethylaminophenol**, known as **MAAF II**, which is diethylamino-2-[3-(1-methyl-2-piperidyl)-2-pyridylazo]phenol complexes with copper at an optimum pH of 5.[505a] The reagent also gives stable colored compounds with nickel, cobalt, bismuth, vanadium, indium, and gallium.

In aqueous ethanol or acetone at pH 1.2–3.2 with hydrochloric acid the cupric complex with **1-(1-methylbenzimidazol-2-ylazo)-2-naphthol** is read at 580–590

[498]G. M. Ritcey and B. H. Lucas, *C. I. M. Bull.* (Canada) **68** (754), 105–113 (1975).

[499]Yoshimasa Tanaka and Junichi Odo, *Jap. Anal.* **23**, 942–946 (1974).

[500]A. L. J. Rao and Chander Shekhar, *Indian J. Chem.* **13**, 628–629 (1975).

[501]J. Bankovskis and A. Ievins, *Zh. Anal. Khim.* **13**, 643–646 (1958).

[502]G. Kamaeva, Sh. T. Talipov, and R. Kh. Dzhiyanbaeva, *Nauch. Tr. Bukhar. Gos. Pedgog. Inst.* **1967**, 5–17.

[503]G. Kamaeva, Sb. T. Talipov, and R. Kh. Dzhiyanbaeva, *ibid.* **19**, 3–18 (1968); cf. G. Kamaeva, T. Talipov, R. Kh. Dzhiyanbaeva, and A. Tashkhodzhaev, *Uzbek. Khim. Zh.* **11**, 14–17 (1967).

[504]Sh. T. Shiripova, R. Kh. Dzhiyanbaeva, and Sh. T. Talipov, *Tr. Tashk. Gos. Univ.* **1968**, 39–45; cf. A. E. Martirosov, Sh. T. Talipov, and R. Rh. Dzhiyanbaeva, *ibid.* **1969**, 73.

[505a]Sh. T. Talipov, R. Kh. Dzhiyanbaeva, and A. E. Martirosov, *Izv. Akad. Nauk Kaz. SSR, Ser. Khim.* **1968** (6), 1–4.

nm.[505b] A 4-fold excess of reagent is necessary. Beer's law is followed for 0.05–2.5 μg per ml.

6-Methylpicolinaldehyde azine at pH 6–10 in a borate buffer solution first reduces cupric ion to cuprous and then forms a complex read at 480 nm.[506] As an extraction method, the pH is adjusted with phthalate–hydrochloric acid buffer solution, the copper reduced with ascorbic acid in the presence of sodium chlorate, and the copper extracted by a 0.05% solution of the reagent in nitrobenzene. There are many interferences with reading in an aqueous medium. By extraction, serious interference is by silver, mercurous, auric, palladous, selenic, platinic, EDTA, and oxalate ions.

6-Methylpicolinaldehyde hydrazone is suitable for determination of 1–7 ppm of copper.[507] The reagent reduces copper to the cuprous form. In a solution with a borate buffer solution for pH 8.6, it is read at 425 nm. Alternatively prepare a solution containing 0.05 gram of the reagent, 0.5 gram of ascorbic acid, and 7.5 grams of sodium chlorate monohydrate in 250 ml of the borate buffer solution. To a sample solution containing 0.01–0.075 mg of cupric ion, add an excess of the reagent solution and extract with nitrobenzene. Dry the extract with sodium sulfate and read at 435 nm. Only palladous, auric, and nickel ions interfere with the second technic, which is the preferable one.

Cuprous ion forms a 1:2 complex with **6-methylpyridine-2-aldoxime** at pH 4 to strong alkalinity.[508] This is read at 422 nm and pH 5, and in alkaline solution at 445 nm. Butanol extracts the complex from acid solution.

Cuprous ion forms an orange-yellow complex with **bis-(6-methyl-2-pyridyl)glyoxal dihydrazone.**[509] For alkalies or brine, add sodium chlorate and ascorbic acid, and extract the copper with a solution of the reagent in nitrobenzene. For milk, a preliminary deproteinizing with trichloroacetic acid is necessary. An alternative extracting solvent is chloroform. Reading is at 440 nm. The most serious interference is by palladium, auric ion, EDTA, and oxalate.

Cupric ion at pH 12 forms a 1:2 complex with **6-methyl-2-pyridylphenyl ketoxime**, which has a maximum absorption at 460 nm.[510] It can be extracted into amyl alcohol. For determination in the presence of 100-fold cobalt it must be strongly alkaline before the reagent is added. Ammonia or Tris buffer solution inhibits formation of the complex. Other complexes are formed at pH 7–11.5 and pH 3.5–6.5.

A 2:1 complex of cupric ion with **3-methyl-1-thiocarbamoyl-2-pyrazolin-5-one** is read at 410 nm.[511] It follows Beer's law for 0.005–0.08 mg of copper per ml. The complex is stable for a week in ethanol at pH 3.7–9.6. Ferric, cobalt, ferricyanide, and ferrocyanide ions interfere.

For determination of copper as the 1:2 complex with **6-(methyl)thiopicolinamide** ascorbic acid is first added to reduce to cuprous ion in 0.1 *M* hydrochloric acid.[512]

[505b]K. N. Bagdasarov, G. S. Vasil'eva, and Yu. M. Gavrilko, *Izv. Vyssh. Ucheb. Zaved., Khim. Khim. Tekhnol.* **19**, 688–690 (1976).cf. A. E. Martirosov, *Izv. Akad. Nauk Kaz. SSR, Ser. Khim.* **1968** (6), 1–4.
[506]M. Varcarcel, D. Perez Bendito, and F. Pino Perez, *Inf. Quim. Anal. Pura Apl. Ind.* **25**, 39–54 (1971).
[507]M. Valcarcel and F. Pino Perez, *ibid.* **26**, 116–125 (1972).
[508]Heinrich Hartkamp, *Z. Anal. Chem.* **126**, 185–194 (1960).
[509]M. Valcarcel and F. Pino Perez, *Analyst* **98**, 246–250 (1973).
[510]J. R. Pemberton and H. Diehl, *Talenta* **16**, 393–398 (1969).
[511]Boleslaw Janik and Halina Gorniak, *Chem. Anal.* (Warsaw) **16**, 1347–1354 (1971).
[512]O. Wawschinek, *Mikrochim. Ichnoanal. Acta* **1965**, 860–864.

The complex is then extracted with 0.001 *M* reagent in pentanol for reading at 495 nm. Beer's law applies up to 0.01 mg of copper per ml. Thiocyanate interferes.

The reaction of morpholine and carbon bisulfide produces **morpholinium morpholine-4-carbodithioate.**[513] In its application, hydrolyze serum at room temperature with hydrochloric acid for 20 minutes. Add trichloroacetic acid to deproteinize and filter. Add the reagent and read copper at 438 nm. Iron can be read in the same sample at 510 nm. Both complexes can be extracted at pH 2–6 with chloroform.

Murexide, which is ammonium 5,5′-nitrilobarbiturate, also known as ammonium purpurate, as a 1 : 1 complex with cupric ion has maximum absorption at 230 and 480 nm.[514]

Copper **myristate and palmitate** dissolved in pyridine can be read at 675 nm.[515] The result is linear but limited by the low solubility of the soaps.

For copper in high purity aluminum, dissolve 1 gram of sample in 10 ml of 10*N* sulfuric acid and dilute to 100 ml.[516] Add 5 ml of 0.0005 *M* **nickel diethyldithiophosphate** and extract for 3 minutes with 10 ml, 5 ml, and 5 ml of carbon tetrachloride. Dilute the combined extracts to 25 ml and read at 420 nm.

For copper in high purity indium, dissolve 1 gram in 20 ml of 1 : 1 sulfuric acid by heating. This normally requires 4 hours on a sand bath. Adjust to 100 ml in *N* sulfuric acid and proceed as above from"Add 5 ml of 0.0005 *M*...."

Nitro-neocuproine is more stable to acids than cuproine and more sensitive.[517]

Cupric ion forms a yellow-brown 1 : 2 complex instantly in 0.01 *M* **3-nitroso-4-hydroxy-5,6-benzocoumarin**, which is 1-hydroxy-2-nitrosonaphtho[2,1-*b*]pyran-3-one, at pH 7.[518] The solution in 60% 1,4-dioxan is read at 420 nm. Beer's law is followed for 0.63–2.41 ppm of copper. Interferences except nickel and cobalt can be masked.

A 1 : 1 complex of cupric ion with **methylthymol blue** has a maximum absorption at pH 6 and 580 nm.[519] This is applicable to 12.8–44.8 μg of copper in 25 ml. Aluminum at two-fold and ferric ion at five-fold are tolerated.

2-(5-nitro-2-pyridylazo)-1-naphthol is a reagent for cupric ion.[520] As an initial step, the copper is extracted at pH 3–4 as a complex with salicylaldoxime by chloroform. It is then reextracted into 0.1 *N* hydrochloric acid 0.05 *M* with ammonium chloride. After adjustment to pH 9.3, addition of an excess of the reagent in 1.4-dioxan precipitates the copper complex. This is extracted into chloroform for reading at 598 nm against a reagent blank. Beer's law applies for 0.5–10 m*M* of this complex in the final extract. Large amounts of complex-forming ions such as citrate may render extraction as the salicylaldoxime complex incomplete.

The complex of copper with **nitroso-R-salt**, when read at 480 nm, obeys Beer's law for 0.21–8.7 ppm.[521] The complex is stable at pH 2.5–6.5. Beryllium, thorium, ferric ion, cobalt, uranium, palladium, oxalate, citrate, and tartrate must be absent.

[513]W. Beyer, *Clin. Chim. Acta* **38**, 119–126 (1972); W. Beyer and W. Likusser, *Mikrochim. Acta* **1971**, 610–614.
[514]R. K. Chaturvedi, *Curr. Sci.* **29**, 128–129 (1960).
[515]Wahid U. Malik and Rizwanul Haque, *Z. Anal. Chem.* **189**, 179–182 (1962).
[516]A. I. Busev and N. P. Borgankova, *Zavod. Lab.* **27**, 13–15 (1961).
[517]E. Gagliardi and P. Höhn, *Microchim. Ichnoanal. Acta* **1964**, 1036–1042.
[518]Nitin Kohli and R. P. Singh, *Chemia* **28**, 661–663 (1974).
[519]E. A. Bashirov and T. E. Abdulaeva, *Uchen. Zap. Azerb. Gos. Univ., Ser. Khim. Nauk* **1973** (4), 21–24.
[520]Ingvar Dahl, *Anal. Chim. Acta* **62**, 145–152 (1972).
[521]Satendra P. Sangal, *J. Prakt. Chem.* **29**, 76–77 (1965).

Copper **octanoate, decanoate,** or **laurate** in organic solvent can be read at 680–690 nm for 22–63 mg of the soap per liter.[522]

To a sample containing 0.05–0.5 mg of cupric ion at pH 5–6, add 2 ml of acetate buffer solution for pH 5.5.[523] Extract with 3 ml and 3 ml of 0.01 *M* solution of **octyl-α-analinobenzylphosphonate** in chloroform. Dilute the combined extracts to 10 ml with chloroform and read at 371 nm.

When a neutral solution containing not more than 20 ppm of copper is treated with 0.08% **orotic acid** solution and buffered at pH 5.5, a 1 : 1 complex can be read at 330 nm.[524] Nickel, cobalt, zinc, and cadmium interfere, but lead can be tolerated.

For reading copper in the ultraviolet region as the **oxalate**, the sample solution should contain less than 0.25 mg of copper and 0.5–1.4 ml of perchloric acid.[525] Add 10 ml of 0.02 *M* ammonium oxalate and dilute to 25 ml. After 30 minutes, read at 255 nm. Ferric and vanadic ions interfere, as do also cyanides and ascorbic acid. If tin is present, it can be volatilized as stannic bromide by adding bromine and hydrobromic acid. If lead is present, it can be filtered off as the bromide. By using 10 *M* excess of oxalate, copper, nickel, and iron in a solution of a cupronickel alloy at pH 6 can be read at 730, 390, and 370 nm respectively and solved by simultaneous equations.

When copper and nickel are present in an alloy with up to 3% iron, dissolve and read as oxalate complexes at pH 6, copper at 730 nm, nickel at 390 nm, iron at 370 nm.[526] Simultaneous equations are necessary. Cyanide and ascorbic acid interfere.

Oximidobenzotetronic acid, which is 4-hydroxy-3-nitrosocoumarin, is a reagent for copper as the sulfate.[527] Mix 2.5 ml of sample containing 5–30 ppm of copper with 2.5 ml of m*M* solution of the reagent in dioxan. Adjust to pH 5.3–7.5 and dilute to 10 ml with dioxan. Dilute to 25 ml with water and read at 427 nm. Tartrate masks bismuth, antimony, stannous and stannic ions, and hexavalent tungsten. Moderate amounts of nickel, cobalt, ferrous and ferric ions, ceric, oxalate, and acetate are tolerated.

Periodic acid and copper form diperiododatocuprate in 0.025 *M* potassium hydroxide solution.[528a] At 415 nm the extinction follows a straight line for 0.5–5 ppm. With more than 0.001 *M* potassium periodate present, the color is stable for 2 hours.

The reaction of **phenol** in 20% sodium hydroxide solution with **chloramine T** is catalyzed by copper.[528b] After heating at 100° for about 2 hours, the mixture is diluted and read at 410 nm. It will show up to 0.6 μg of copper. EDTA must be absent.

The complex of copper with **1,10-phenanthroline** adds tetrabromo-, tetraiodo- or

[522]K. N. Mehotra, V. P. Mehta, and T. N. Nagar, *Z. Anal. Chem.* **245**, 323 (1969).

[523]B. Tamhina, M. J. Herak, and V. Jagodic, *Croat. Chem. Acta* **45**, 593–601 (1973).

[524]F. Capitan-Garcia and A. Arrebola, *Rev. Univ. Ind. Santander* **7**, 241–247 (1965).

[525]Toru Nozaki and Hirondo Kurihara, *J. Jap. Chem. Soc.* **82**, 710–712 (1961); cf. V. M. Bhuchar and Padma Narayan, *Indian J. Chem.* **6**, 534–538 (1968).

[526]V. M. Bhuchar and Padma Narayan, *Indian Chem. J.* **6**, 934–938 (1968); cf. V. M. Bhuchar and A. R. Sarkar, *Indian J. Technol.* **12**, 366 (1974).

[527]G. S. Manku, R. D. Gupta, A. N. Bhat, and B. D. Jain, *Mikrochim. Acta* **1970**, 836–840.

[528a]Eiichi Torikai and Youji Kawami, *Jap. Anal.* **10**, 908–910 (1961).

[528b]Yachiyo Kakita, Michiko Namiki, and Hidehiro Goto, *Talenta* **13**, 1561–1566 (1966).

dichlorotetraiodo-fluorescein, and is extractable at pH 9 with chloroform.[529] The maximum absorption is at 540 nm. In 1 : 1 chloroform-acetone read fluorescence at 580 nm. Fluorscence will determine 0.05–10 μg in 10 ml, which is one order of magnitude more sensitive than spectrophotometry.

Phenylacetic acid in chloroform will extract copper quantitatively from solution in aqueous hexamine.[530] The complex is read at 700 nm. Interference by iron, uranium, vanadium, cerium, nickel, and cobalt is masked by ammonium fluoride. Gold, palladium, and platinum are coextracted but do not interfere.

Cuprous ion forms a 1:2 complex with **4,4′-bis(4-phenylazoanalino)-2,2′-biquinolyl** in the presence of threefold excess of reagent and thirtyfold excess of hydroxylamine hydrochloride at pH 6.1–7.5.[531] Extracted with isoamyl alcohol, this is read at 552 nm. Beer's law is followed for 1–12 μg/ml.

A solution of *N*-phenylbenzohydroxamic acid in chloroform extracts copper and iron simultaneously from acid solution.[532] Reading at 400–470 nm permits calculation of each.

For less than 60 μg of copper, less than 100 μg of palladium, and less than 80 μg of gold, form the complex with ***syn*-phenyl-2-pyridyl ketoxime** at pH 10.[533] Extract with chloroform and read at 550, 410, and 455 nm, respectively. Calculate by simultaneous equations. There is serious interference by platinum. Prior evaporation with aqua regia destroys such ingredients of plating baths as pyridine, EDTA, ethylenediamine, cyanide, thiourea, sulfite, and tartrate.

Dissolve 10 mg of ferrite in 2 ml of boiling nitric acid. Add 2 ml of 1 : 1 sulfuric acid and heat to sulfur trioxide fumes. Dilute to 50 ml. Take an aliquot containing up to 0.04 mg of iron with 0.2 mg of copper and dilute to 10 ml. Add 1 ml of 1% ascorbic acid solution and 1 ml of 1% ethanolic ***syn*-phenyl-2-pyridyl ketoxime.**[534] Raise to pH 10 with sodium carbonate solution and extract with 3 ml and 3 ml of chloroform. Read at 475 and 550 nm, and determine copper and iron by simultaneous equations. Nickel and cobalt interfere seriously.

4-Phenyl-2-(pyridylmethyleneamino)phenol is synthesized from picolinaldehyde and 2-amino-4-phenylphenol. Although insoluble in water, it is soluble in ethanol and gives a red complex with copper.[535] The color intensity is constant at pH 3 and above. Complexes are formed with nickel and cobalt above pH 3.5 and with ferric ion above pH 6. For determination, mix 25 ml of sample solution containing 0.01–0.2 mg of copper with 10 ml of 0.5 *M* sodium acetate buffer for pH 3.5 and 5 ml of 0.05% ethanolic reagent. Dilute to 50 ml and read at 480 nm after 30 minutes.

The blue of copper at 0.11–0.25 mg per 100 ml on reaction with **1-phenylsemicarbazide** is proportional to the copper content.[536]

Picolinaldehyde-2′-hydroxyanil and its 5′-chloro- and 5′-methyl-derivatives form 1 : 1 water-soluble complexes with cupric ion.[537] At pH 3.5–4.4 they absorb at 454, 462, and 466 nm, respectively. Beer's law applies for up to 0.5 mg of copper for 50

[529]D. N. Lisitsyna and D. P. Shcherbov, *Zh. Anal. Khim.* **28**, 1203–1205 (1973).
[530]Jiri Adam and Rudolf Pribil, *Talenta* **19**, 1105–1111 (1972).
[531]A. L. Gershuns, P. Ya. Pustovar, and A. D. Gubin, *Ukr. Khim. Zh.* **42**, 378–381 (1976).
[532]I. G. Per'kov and Nguyen Van Nhi, *Zh. Anal. Khim.* **25**, 59–63 (1970).
[533]C. K. Bhaskare and S. G. Kawatkar, *Indian J. Chem.* **13**, 523–524 (1975).
[534]C. K. Bhaskare and S. G. Kawatkar, *Talenta* **22**, 189–193 (1975).
[535]Kyoharu Isagai and Kazuyo Isagai, *Jap. Anal.* **17**, 171–175 (1968).
[536]N. Lazarov and V. Mirchev, *Vopr. Med. Khim.* **8**, 305–306 (1962).
[537]Kiyoharu Isagai and Kazuyo Isagai, *J. Chem. Soc. Jap., Pure Chem. Sect.* **88**, 1292–1295 (1967).

ml containing two threefold M excess of the reagent in ethanol with an acetate buffer solution. The reaction is completed within 30 minutes, and nickel, cobalt, and iron do not interfere.

For cupric ion, **piconaldehyde semicarbazone**[538] has the following optima: buffer solution of boric acid–sodium hydroxide for pH 9.5; wave length, 350 nm; range for Beer's law, 0.3–1.5 ppm.

The absorption of the copper complex with **picolinaldehyde thiosemicarbazone** is constant for pH 8.9–10.7 and for nickel at pH 7–11.[539] Beer's law is followed up to 7 ppm of copper or 5 ppm of nickel. Interference by the reagent is minimal at 415 nm.

Copper can be determined by precipitating as the **picrate** and the excess precipitating agent determined.[540] As a reagent, neutralize saturated aqueous picric acid with 8% sodium hydroxide solution and mix 2 ml of it with 10 ml of 1:3 ammonium hydroxide and 28 ml of water. Then 90% saturate it with cuprammonium picrate.

Evaporate the sample solution containing 1–10 μg of copper to dryness. Mix with 1 ml of the reagent and filter after 30 minutes. Determine the residual picrate in the filtrate with methylene blue.

A 25 mM solution of **piperidine-1-carbothioate** in benzene extracts cupric ion at 0.5–4 μg/ml from up to 4 N hydrochloric acid for reading at 440 nm.[541] EDTA masks ferric, chromic, and molybdate ions. There is interference by antimonous, arsenous, mercuric, stannous, vanadate, thiocyanate, and nitrate ions.

For copper and iron in serum, deproteinize with hydrochloric and trichloroacetic acids. Centrifuge. For copper, warm at 37° for 10 minutes with a solution of **bis(piperidinothiocarbonyl)disulfide** in acetic acid and read copper at 420 nm.[542] Then add thiourea and mercaptoacetic acid. Warm at 37° for 10 minutes to reduce cupric and ferric ions and to mask the copper. Add ferrozine and ammonium chloride, warm at 37°, and read iron at 560 nm.

For copper in serum, treat with an equal volume of 2 M trichloroacetic acid and centrifuge after 30 minutes. To 1 ml of the supernatant layer, add 1 ml of 2 mM **bis[piperidino(thioformyl)]disulfide** in methanol, 1 ml of water, and 0.2 ml of 30% hydrogen peroxide. Dilute to 10 ml with methanol and read at 425 nm.[543] The developed solution should contain 80% methanol and should be 0.2 M with trichloroacetic acid. There is interference by silver, mercuric, chromate, and vanadate ions.

To 10 ml of a nearly neutral solution containing 0.1–1 mg of copper, add 5 ml of a solution containing 10% of **polyethyleneimine** adjusted to pH 1–2 with 1:5 hydrochloric acid.[544] Add a saturated solution of sodium acetate to adjust to pH 3–6. Add 5 ml of a buffer solution containing 200 grams of sodium chloride, 50

[538] M. P. Martinez Martinez, D. Perez Bendito, and F. Pino Perez, *An. Quim.* **69**, 747–756 (1973).
[539] J. M. Cano Pavon, J. Vazquez Allen, D. Perez Benedito, and F. Pino Perez, *Inf. Quim. Anal. Pura Apl. Ind.* **25**, No. 5, 149–158 (1971).
[540] N. Ganchev and A. Dimitrova, *ibid.* **1967**, 507–512.
[541] A. Varadarajulu and A. P. Rao, *Indian J. Chem.* **13**, 974–975 (1975).
[542] Hugh Y. Yee and Jesse F. Goodwin, *Clin. Chem.* **20**, 188–191 (1974).
[543] R. T. Lofberg, *Anal. Lett.* **2**, 439–448 (1969).
[544] M. Ziegler and L. Ziegler, *Talenta* **14**, 1121–1122 (1967).

grams of sodium acetate, and 160 ml of acetic acid per liter. Dilute to 25 ml and read at 645 nm.

To determine copper in the aqueous humor from the human eye, potassium iodide is reacted with the cupric ion.[545] The liberated iodine as a complex with **polyvinyl alcohol** is read at 490 nm. Beer's law is applicable at 0.03–0.8 mg of copper per 100 ml.

Copper can be read at 0.5–4 ppm, pH 3.5, 610 nm by **pontachrome azure B.**[546]

Potassium-1,2-dicarba-undecaborane forms a 2:1 complex with cupric ion in 4–12% sodium hydroxide solution.[547] This is extractable with tributyl phosphate or methyl ethyl ketone for reading at 490 nm. Beer's law is followed for 0.2–4 μg of copper per ml of extract. Silver interferes.

Copper is determined with **proline thymol blue.**[548]

Pyridine 2,6-dialdoxime forms a 1:1 green chelate with cupric ion.[549] At pH 3.8 it absorbs significantly at 600, 400, and 360 nm. The reagent also forms colored chelates with cobalt, manganese, nickel, and zinc.

Dissolve 1 gram of copper or brass in nitric and sulfuric acids. Neutralize to pH 8.5 and adjust to *M* tartaric acid at that pH. Extract the copper with *M* hexanoic acid in chloroform containing 0.5 *M* **pyridine** and read at 730 nm.[550] Then determine iron in the aqueous phase as the thiocyanate.

The copper compounds of **3-(2-pyridylazo)chromotropic acid** and the 3-3 derivative absorb at 580 and 570 nm, respectively.[551] The ranges are 0.5–2 and 1–4 ppm. Cerium, stannous and stannic ions, oxalate, phosphate, and fluorides interfere.

The anionic form of **4-(2-pyridylazo)-1-naphthol** in 70% ethanol complexes with cupric ion, as do also zinc, nickel, manganese, and cobalt.[552]

At pH 2.2–8, copper in 1:1 methanol–70% 1,4-dioxan complexes with **1-(2-pyridylazo)-2-phenanthrol.**[553] At 560 nm, this complex obeys Beer's law up to 3.2 ppm of copper. There is serious interference by thiosulfate, fluoride, nickel, cobalt, and zinc. Citrate at 50 ppm masks 5 ppm of cadmium, pentavalent vanadium, hexavalent uranium, palladium, and rhodium. At 50 ppm iodide masks 10 ppm of mercury. Fifty ppm of tartrate masks 5 ppm of iron.

Cuprous ion complexes with **2-pyridyl ketoxime.**[554] Dissolve the sample in acid and neutralize to pH 2.5–3 with 20% sodium hydroxide solution. Add 20 ml of 10% hydroxylamine hydrochloride solution and 5 ml of 0.01 *M* reagent. Extract with 15 ml of isopentyl alcohol. Dilute the extract to 25 ml, set aside for an hour, and read at 360 nm. Cobalt interferes.

Adjust 10 ml of a sample solution containing less than 0.2 mg of copper to pH 3–4 with 5 ml of acetate buffer solution.[555] Add 5 ml of 0.04% ethanolic solution of

[545]P. Calme, J. P. Gerhard, and E. Kraeminger, *Mikrochim. Acta* **1972**, 173–182.
[546]Yukiteru Katsube, Katsuya Uesugi, and John H. Yoe, *Bull. Chem. Soc. Jap.* **34**, 72–76 (1961).
[547]B. I. Nabivants, V. I. Stanko, N. A. Truba, and V. A. Brattsev, *Zh. Anal. Khim.* **28**, 897–901 (1973).
[548]T. S. West, *Ind. Chem.* **39**, 379–381 (1963).
[549]Saswati P. Bag, Quintus Fernando, and Henry Freiser, *Anal. Chem.* **35**, 719–722 (1963).
[550]I. V. Pyatnitskii, A. K. Boryak, and P. B. Mikhel'son, *Zh. Anal. Khim.* **30**, 900–905 (1975).
[551]A. K. Majumdar and A. B. Chatterjee, *J. Indian Chem. Soc.* **42**, 241–246 (1965).
[552]D. Betteridge, P. K. Todd, Q. Fernando, and H. Freiser, *Anal. Chem.* **35**, 729–733 (1963).
[553]A. K. Rishi, K. C. Trikha, and R. P. Singh, *Curr. Sci.* **44**, 122 (1975).
[554]F. Trusell and K. Lieberman, *Anal. Chim. Acta* **30**, 269–272 (1964).
[555]Kazuyo Isagai, Kiyoharu, Isagai and Sachiko Mori, *Jap. Anal.* **24**, 414–419 (1975).

1-(2-pyridylmethyleneamino)-2-naphthol, which is pyridine-2-aldehyde-2′-hydroxynaphthylimine. Set aside for 1 hour; then dilute to 25 ml. Read at 482 nm against a reagent blank. Beer's law is followed for 0.2–6 ppm of copper as diluted. Cobalt interferes. Nickel and ferric ion must not exceed 5 times the copper.

3, 3′-(Pyrrole-2, 5-diylazo)di(5-chloro-2-hydroxybenzenesulfonic) acid complexes with copper at pH 4–5 for reading at 660 nm.[556] The reagent also complexes with zinc, cadmium, and yttrium.

Quinizarin-2-sulfonic acid, which is 1,4-dihydroxy-9,10-dioxanthracene-2-sulfonic acid, at pH 4.58–5.37 forms a 1 : 1 complex with cupric ion absorbing at 480 nm.[557]

Cupric ion can be determined by its catalysis of the oxidation of **quinol** by hydrogen peroxide in the presence of pyridine.[558] The reaction is carried out at 25° in a phosphate-tartrate buffer solution for pH 7 and read at 470 nm. It will determine down to 4.2 ng of copper per ml or down to 0.9 ng if tartrate is absent.

The oxidation is also applied in the presence of 2,2′-bipyridyl.[559] For as low as 20 μg of copper per ml in salt solutions, or 0.1 ppb in seawater or hydrochloric acid, make the sample 0.5 *M* with hydrogen peroxide, 20 m*M* with **hydroquinone**, m*M* with 2,2′-bipyridyl, and adjust to pH 8.4–9 with sodium tetraborate. Read at 530 nm.

Cuprous ion complexes with **quinoline-2-aldoxime** at pH 5 for reading up to 8 μg of copper per ml at 467 nm.[560] There is interference by palladium, cyanide, thiocyanate, and nitrous ions.

Both cupric and zinc ion can be read as 1 : 2 complexes with **2-(2-quinolylazo)-1-naphthol** in 50% 1,4-dioxan and in carbon tetrachloride.[561]

Cupric ion at pH 5.5–8.7 mixed with methanolic **1-(2-quinolylazo)phenanthren-2-ol** and heated at 100° for 1 hour precipitates a bluish-green complex.[562] Extracted into chloroform, it is read at 590 nm. Beer's law is obeyed for 0.1–1.7 ppm, and the color is stable for 24 hours. Aluminum can be masked with fluoride, and ferric ion with tartrate. Cobalt and zinc interfere, and no more than a minimal amount of nickel, palladium, and osmium is tolerated.

Cupric ion forms a 1 : 2 complex with ***N*-8-quinolyl-5-bromosalicylideneimine**, which is 5-bromo-*N*-8-quinolylsalicylaldimine. At pH 1.5–6.5 this obeys Beer's law for 0.05–13 ppm of copper at a maximum absorption of 435 nm.[563] At pH 1.5 absorption by nickel and cobalt is negligible. Aluminum and cadmium interfere.

Cupric ion forms a 1 : 2 complex with ***N*-8-quinolyltoluene-*p*-sulfonamide** which absorbs strongly in chloroform at 370 nm.[564] The reagent does not absorb at that

[556] S. B. Savvin, Yu. G. Rozovskii, R. F. Propistsova, and E. A. Likhonina, *Izv. Akad. Nauk SSSR, Ser. Khim.* **1969**, 1364–1366.

[557] D. P. Joshi and D. V. Jain, *J. Indian Chem. Soc.* **47**, 1109–1110 (1970).

[558] F. Dittel, *Z. Anal. Chem.* **229**, 193–201 (1967); cf. I. F. Dolmanova and V. M. Peshkova, *Zh. Anal. Khim.* **19**, 297–302 (1964); V. M. Peshkova, E. K. Astakhova, I. F. Dolmanova, and V. M. Savostina, *Acta Chim. Hung.* **53**, 121–125 (1967).

[559] I. F. Dolmanova, V. P. Poddubienko, and V. M. Peshkova, *Zh. Anal. Khim.* **28**, 592–595 (1973).

[560] Nobuichi Oi., *J. Chem. Soc. Jap.* **79**, 1327–1330 (1958).

[561] Akira Kawase, *Anal. Chim. Acta* **58**, 311–322 (1972).

[562] R. N. Virmani, B. S. Garg, and R. P. Singh, *Indian J. Chem.* **10**, 225–226 (1972).

[563] Kazuko Katuki and Kiyoharu Isagai, *Jap. Anal.* **22**, 1470–1474 (1973).

[564] D. T. Haworth and J. H. Munroe, *Anal. Chem.* **41**, 529–531 (1969).

wavelength. An appropriate range is up to 0.1 m*M*. Iron, cadmium, zinc, lead, and manganous ion interfere.

Dissolve the ash of a sample containing 0.1–0.8 mg of copper in 3 ml of hydrochloric acid and dilute to 25 ml.[565] To remove interferences, add 25 ml of acetic acid followed by dropwise addition of a fresh solution of **Reinecke's salt** in excess. Filter, wash with water, and discard the precipitate. Add a small amount of solid hydroquinone, to reduce to cuprous ion, and reagent solution in slight excess. Heat to 90° and cool. Filter on fritted glass and wash with cold water. Dissolve with pyridine and dilute to 25 ml with that solvent. Read with a green filter. Beer's law is followed for 5–30 μg of copper per ml. Cupric reineckate is water soluble but cuprous reineckate is insoluble.

β-Resorcylidenethiosemicarbazide, which is 2,4-dihydroxylbenzaldehyde thiosemicarbazone forms a 1:1 complex with cupric ion having maximum absorption at 300 and 374 nm.[566] It is stable at pH 2.5–3.8 and determines 1 μg of copper per 5 ml of solution. A 100-fold excess of lead, manganous, hexavalent molybdenum, or tungsten ions, a 50-fold excess of uranous ion, a 25-fold excess of cadmium or zinc, or a 10-fold excess of nickel, ferric, or chromic ions can be masked with citric acid and sodium fluoride.

The 1:1 complex of copper with **rezarson**, which is 5-chloro-3-(2,4-dihydroxyphenylazo)-2-hydroxybenzenearsonic acid, requires 50% ethanol to prevent precipitation. Conditions are pH 1.2–3.2 and maximum absorption at 510 nm.[567]

At pH 4.5–6.5, **salicylaldehyde benzoylhydrazone** complexes with cupric ion in 50% acetone for reading at 385 nm.[568] Beer's law is followed for up to 5 ppm of copper.

Salicylaldehyde 2-hydroxyanil, which is 2-salicylideneaminophenol, precipitates copper above pH 4.[569] The precipitate so separated can be dissolved in 2 *N* sulfuric acid and the solution boiled to drive off salicylaldehyde; read as 2-aminophenol at 270 nm. The method is applicable up to 1 mg of copper per 50 ml. Iron in excess of the copper is masked with citrate; cobalt and nickel do not interfere.

Solid **salicylaldehyde semicarbazone** added as 2–3 moles compared to the copper at pH 3–5 forms a soluble green complex absorbing at 685 nm.[570] At room temperature, reaction is completed within an hour. Excess reagent is insoluble. Results for 1–20 mg of copper per 50 ml conform to Beer's law. There is interference by nickel and by more than 1.5 mg of cobalt in 50 ml. Iron must be precipitated before adding the reagent.

A complex of copper with **salicylfluorone** is formed at an optimum pH of 10 and read at 530 nm.[571]

A 1:1:1 complex is formed by cupric ion, pyridine, and **2-salicylideneamino-**

[565]J. Weyers, *Chem. Anal.* (Warsaw) **6**, 975–978 (1961).

[566]S. Stankoviansky, A. Beno, J. Carsky, and E. Kominakova, *Chem. Zvesti* **25**, 123–131 (1971).

[567]A. M. Lukin, N. A. Kaslina, V. I. Fadeeva, and G. S. Petrova, *Vest. Mosk. Gos. Univ., Ser. Khim.* **1972**, 247–249.

[568]D. K. Rastogi, S. K. Dua, Shri Prakash, and R. P. Singh, *Acta Chim. Hung.* **87**, 63–67 (1975).

[569]Kiyoharu Isagai, *J. Chem. Soc. Jap., Pure Chem. Sect.* **82**, 1172–1176 (1957).

[570]Kiyoharu Isagai, *ibid.* **87**, 570–573 (1966).

[571]O. A. Tataev, G. N. Bagdasarov, S. A. Akmedov, Kh. A. Mirzaeva, Kh. G. Buganov, and E. A. Yarysheva, *Izv. Sev-Kazk. Nauch. Tsent. Vyssh. Shk., Ser. Estestv. Nauk* **1973** (2), 30–32.

benzoic acid.[572] Dilute a sample containing up to 60 μg of cupric ion to 30 ml and add 1 ml of 5% solution of pyridine and 1 ml of acetate buffer solution for pH 5.6. Shake with 10 ml of 0.005% solution of the reagent in chloroform for 3 minutes. Dry the organic phase with sodium sulfate and read at 402 nm. There is interference by aluminum, chromic, titanic, vanadic, tartrate, and citrate ions.

Cupric ion forms an insoluble green 2:3 complex with **2-salicylideneaminobenzothiol.**[573] Precipitation is complete within 1 hour in a solution buffered with acetate to pH 3–4 in 60% ethanol at 40°. An appropriate amount of copper is 0.05–0.5 mg. The precipitate is filtered, washed, and boiled with 1:1 hydrochloric acid to decompose the complex and volatilize the salicylaldehyde. 2-Aminobenzenethiol is then read at 240 nm. Less than 0.5-fold amounts of nickel or cobalt do not interfere, but iron prevents precipitation of the complex.

Salicylideneamino-*o*-cresol precipitates copper in an acetate buffer solution at pH 4.5–5.3 within 1 hour.[574] Boiled with 2 *N* hydrochloric acid for 5 minutes, excess reagent is decomposed and salicylaldehyde is expelled from the complex. Cooled and diluted, the α-amino-*o*-cresol is read at 272 nm. Up to twentyfold cobalt and nickel do not interfere.

Salicylideneaminomethyl-2-naphthol precipitates copper at pH 5.8–6.8.[575] Boiled with 2 *N* hydrochloric acid, the 1-amino-2-naphthol is read at 276 nm.

1-Salicylideneamino-2-naphthol precipitates copper at pH 4–6.9.[576] Boiled with 2 *N* hydrochloric acid, the liberated 1-amino-2-naphthol is converted to 1,2-naphthoquinone for reading at 282 nm. Up to threefold amounts of nickel or cobalt or an equal amount of ferric ion are tolerated.

2-Salicylideneaminophenol forms a 1:1 complex with copper.[577] Shake 30 ml of sample solution containing 1 ml of *M* acetate buffer for pH 5.1 with 10 ml of 0.012% solution of the reagent in methyl isobutyl ketone for 10 minutes, dry the extract with sodium sulfate, and read at 423 nm. There is interference by ferric ion, reducable to ferrous ion, which is without effect. Tetra and pentavalent vanadium and hexavalent molybdenum interfere.

Copper can be precipitated with **salicylidene-2-hydroxybenzamine** and filtered. Heating the precipitate decomposes it and evaporates the salicylaldehyde. The resulting 2-hydroxybenzamine is read at 272 nm.[578]

1-Salicylidene-3-thiosemicarbazide will determine up to 7 μg of copper in 50 ml of solution at pH 2.5–3.5.[579] Ferric ion interferes. After addition of 0.1 gram of reagent to the sample at 60°, set aside for 1 hour, cool, dilute to 50 ml, and read at 630 nm.

Mix 50 ml of sample solution containing 10–120 μg of copper with 5 ml of m*M* **sarcosine cresol red** and 5 ml of 2 *M* acetate buffer solution for pH 4.8. Read the 1:1 complex at 570 nm.[580] Ferrous and gallium ions cause large positive errors.

[572]Hajime Ishii, Hisahiko Einaga, and Tsuguo Sawaya, *Jap. Anal.* **22**, 546–550 (1973).
[573]Kiyoharu Isagai and Katuko Koguchi, *J. Chem. Soc. Japan, Pure Chem. Sect.* **86**, 213–217 (1965).
[574]Kiyoharu Isagai, *ibid.* **82**, 1359–1361 (1957).
[575]*Ibid.* 447–450.
[576]*Ibid.* 450–455.
[577]Hajime Ishii and Hisahiko Einaga, *Jap. Anal.* **18**, 230–235 (1969).
[578]Kiyoharu Isagai, *J. Jap. Chem. Soc.* **82**, 1359–1361 (1961).
[579]Kiyoharu Isagai, *Jap. Anal.* **19**, 344–349 (1970).
[580]Tsutomo Matsuo, Junichi Shida, and Shin Sato, *ibid.* **20**, 693–697 (1971).

High results are caused by nickel, bismuth, titanium, and vanadium; low results by citrate and tartrate. Tin and antimony are volatilized as bromides; lead is precipitated as the sulfate.

Copper forms a 1:2 complex with **solochrome azurine BS** (mordant blue 1), which is read at 525 nm for 1–15 ppm.[581]

At pH 8.5, **solochrome fast navy 2R** (mordant blue 9) absorbs at 550 and 620 nm.[582] Copper alters the spectrum, as do manganese and zinc. For determination of any of these three metals, add 0.2 mg of the dye to the sample solution containing 0.5–4 μg of the ion at pH 8.5; dilute to 25 ml and read the decrease in absorption at 620 nm. EDTA masks interference by calcium, magnesium, manganese, lead, and zinc.

To a sample solution containing 0.0096–0.032 mg of copper, add 1 ml of 0.65% solution of **stilbazo** and dilute to 25 ml with an ammonium acetate buffer solution for pH 6.[583] After 10 minutes read the 1:2 greenish complex at 656 nm against water. There is interference by stannous, scandium, titanous, beryllium, gallium, germanium, manganous, aluminum, zinc, and arsenate ions, and by ascorbic acid or thiourea. Ammonium fluoride and sulfosalicylic acid will mask arsenate, gallium, scandium, calcium, stannous, and titanous ions. Lead, nickel, calcium, and ferric ions interfere.

At pH 4–10.5 a blue ternary complex is formed with cupric ion by a 3-fold excess of **succinimide** and a 4-fold excess of either isobutylamine or isopropylamine.[584] As read at 590 nm, Beer's law is followed for 14–178 ppm of copper with isobutylamine or 13–100 ppm with isopropylamine. Sulfate and phosphate interfere. The color is stable for more than 24 hours.

The product of cupric ion, **sucrose**, and alkali can be read at 246 nm.[585] Adding 2% sucrose and 0.4% sodium hydroxide will determine 1.5–13 mg of copper per ml. The addition of 2% of sodium sulfate increases the absorption by about 2.5%.

Sulfarsazan, which is 5-nitro-2-[3-(4-*p*-sulfophenylazophenyl)-1-triazenophenyl-1-triazeno]benzenearsonic acid, forms an anionic 1:1 cupric complex absorbing at 500 nm at pH 7.[586] The complex is stable for 22 hours, and on heating to 100° in the presence of threefold excess of reagent and 0.05% sodium tetraborate, it conforms to Beer's law for 0.0032–0.032 mg per 25 ml of final solution. An amount of zinc, nickel, or EDTA similar to the copper interferes. Interference is caused by ferric ion, aluminum, and oxalate.

In analysis of alloys, cupric ion is read at 820 nm, nickel ion at 395 nm, and cobalt ion at 515 nm in 0.5 *M* **sulfuric acid**.[587a] Simultaneous equations are then applied. Chloride ion interferes. Ferric ion amounting to less than 5% of the nickel can be masked with phosphoric acid in reading the latter. Higher concentrations of iron are extracted with isopropyl ether from 8 *M* hydrochloric acid solution before driving off the chloride by heating with sulfuric acid.

[581] C. L. Sharma, C. L. Tandon, and Surindar Kumar, *Z. Anal. Chem.* **255**, 368 (1971).
[582] A. A. Abd El Raheem, M. Z. El Sabban, and M. M. Dokhana, *ibid.* **188**, 96–109 (1962).
[583] E. A. Bashirov, M. K. Akhmedli, and T. E. Abdullaeva, *Uch. Zap. Azerb. Gos. Univ. Ser. Khim. Nauk* **1966**, 29–34.
[584] C. L. Sharma and Savita Sharma, *Z. Anal. Chem.* **280**, 219 (1976).
[585] S. S. M. A. Khorasni and A. M. Shafiqul, *Pak. J. Sci. Ind. Res.* **15**, 25–29 (1972).
[586] E. A. Bashirov and T. E. Abdullaeva, *Uch. Zap. Azerb. Gos. Univ., Ser. Khim. Nauk* **1970**, 53–57.
[587a] D. V. Jayawant and T. K. S. Murthy, *Indian J. Technol.* **9**, 396–400 (1971).

To a solution of cupric sample in hydrochloric acid, add methanolic 0.08% **tetraethylthiuram disulfide** and read at 422 nm against a blank.[587b] The color is stable for 15 minutes. Digest serum, urine, or brain tissue with perchloric acid and dilute to 2.2–2.7 *N* perchloric acid.[587c] Add 0.04% solution of the reagent in 20% acetone. Reaction develops in 15 minutes. Read at 422 nm within 3 hours.

Cupric ion forms a 1 : 2 complex with **1,2,3,4-tetrahydro-1-hydroxyiminophenazine** at pH above 4 with maximum absorption at 400–405 nm.[588] At less than pH 4 with ascorbic acid present, a red 1 : 2 complex is formed with cuprous ion having a maximum absorption at 505 nm. The latter will determine 2–100 μM copper with the reagent added in 20% ethanol.

Mix a sample containing 4–200 μg of copper with 1 ml of 5% solution of ***NNN'N'*-tetrakis-(2-hydroxypropyl)ethylenediamine.**[589] Add 4% sodium hydroxide solution to pH 10 and add 10 ml of borax buffer solution for pH 10. Dilute to 25 ml and read at 690 nm.

Copper at 0.42–43 μg per ml is read by the brown-red color with ***NN'N''N'''*-tetraphenyloxamidine.**[590] Mix the sample with 7 ml of 0.005 *M* reagent and add water and ethanol to make the solution 65% ethanol. Read after 10 minutes. The color is stable for 2 hours. Aluminum interferes if it is twice the copper content; iron greater than the copper produces large errors.

For up to 0.8 mg of copper, add 2% ethanolic **3-thianaphthenoyltrifluoroacetone**, which is 1-(benzo[b]thienyl)-4,4,4-trifluorobutane-1,3-dione, to the sample solution at pH 4 and heat at 100° for 15 minutes. Cool, extract with chloroform, and read the extract at 410 nm.[591a]

A blue 1 : 2 chelate of cupric ion with 0.1 m*M* **2-(2-thiazolylazo)-*p*-cresol** can be read for low concentrations of copper at 590 nm or 610 nm.[591b] Appropriate media are 50% ethanol at pH 8 or a mixture of benzene and 0.1 *M* tributyl phosphate buffered at pH 10 with 0.04 *M* sodium tetraborate. In the latter case, extraction is complete in 30 minutes. Hydrolysis of the complex in aqueous solution is prevented by addition of 0.05 *M* sodium acetate.

3-(2-Thiazolylazo)-*p*-cresol complexes with 5–60 μg of cupric ion at pH 6.5 for reading at 610 nm.[592]

A 1 : 1 complex of cupric ion with **2-(2-thiazolylazo)-1-naphthol** at pH 3 absorbs at 590 nm.[593] Zinc, cadmium, lead, cobalt, nickel, and antimony interfere. The 4-isomer forms a similar complex.

4-(2-Thiazolylazo)resorcinol complexes with cupric ion as well as with lead, zinc, cadmium, and bismuth.[594] To the sample, add 15 ml of 0.02% ethanolic solution of the reagent and 5 ml of *M* formate buffer for pH 3.1. Add 1 : 10 ammonium hydroxide to pH 3.4 and dilute to 50 ml. Read at 560 nm. **4-(4-Methyl-2-thiazo-**

[587b]S. Grys, *Mikrochim. Acta* **1976**, **I**, 147–152.

[587c]Y. Matsuba and Y. Takahashi, *Anal. Biochem.* **36**, 182–191 (1970).

[588]Carla Bertoglio Riolo, Teresa Fulle Soldi, and Giovanni Spini, *Talenta* **20**, 684–688 (1973).

[589]F. Bermejo Martinez, J. M. Grana, and J. A. Rodriguez Vazgez, *Quim. Anal.* **28**, 20–22 (1974).

[590]M. Papafil, D. Furnica, and N. Hurduc, *An. Stiint. Univ. "A. I. Cuza" Iasi Sect. I* **5**, 135–140 (1959).

[591a]J. R. Johnston and W. J. Holland, *Mikrochim. Acta* **1972**, 126–131.

[591b]L. Sommer, M. Langova, and V. Kuban, *Colln. Czech. Chem. Commun.* **41**, 1317–1333 (1976).

[592]S. I. Gusev, I. N. Glushkova, and L. A. Ketova, *Uch. Zap. Perm. Gos. Univ.* **1973** (289), 255–264.

[593]S. I. Gusev, L. A. Ketova, and I. N. Glushkova, *Zh. Anal. Khim.* **25**, 2099–2105 (1970).

[594]M. Hnilickova and L. Sommes, *Talenta* **13**, 667–687 (1966); S. I. Gusev, I. N. Glushkova, and L. A. Ketova, *Uch. Zap. Perm. Gos. Univ.* **1973** (289), 255–264.

lylazo)resorcinol complexes similarly at pH 3.1 for reading at 540 nm. There is interference by lead, ferric ion, cobalt, nickel, antimonous ion, and bismuth.

Copper is determined by **thioacetamide** by technics not unlike those for the sulfide.[595] The final reading is as a cupric-glycerine complex at 735 nm. Bismuth interferes.

For determination of copper by **2-thiobarbituric acid**, add to the neutral sample 10 ml of 0.3% solution of the reagent, 0.5 ml of 1 : 3 sulfuric acid, and 5 ml of 1% solution of gum acacia. Dilute to 25 ml and read the 1 : 4 copper-reagent complex at 400 nm against a reagent blank.[596] There is interference by the following ions: bismuth, ferric, antimonous, stannic, mercuric, silver, and lead. At less than 1 mg per 25 ml there is no interference by nitrate, sulfate, thiocyanate, fluoride, acetate, oxalate, or citrate. Beer's law is followed for 0.5–12 μg of copper per ml.

A 1 : 2 complex of cupric ion is extracted by 5 mM **thiobenzoic acid** in chloroform from solution at pH 2.5–5.[597] Beer's law is followed for 1–10 μg of copper per ml of extract when read at 365 nm.

To the sample solution, 0.05–0.5 mM with copper, add 5 ml of M sodium thiosulfate, 5 ml of 0.1 M EDTA, and 5 ml of M sodium hydroxide.[598] Shake for 3 minutes with 10 ml of mM **thiodibenzoylmethane** in hexane. Wash the organic phase with 20 ml and 20 ml of 0.1 N sodium hydroxide. Read at 410 nm. There is interference from cyanide and palladium in quantities equal to the copper.

To determine cupric ion, extract at pH 6 by 0.01 M **thiomaltol**, which is 3-hydroxy-2-methylpyran-4-one, in benzene.[599] Wash unused reagent from the organic phase with a glycine buffer solution at pH 12.6 and read at 410 nm against a reagent blank. Mercury must be masked with EDTA.

Treat a cupric solution with 5 ml of a 0.02% solution of **thiophen-2-aldehyde thiosemicarbazone** in 50% ethanol. Add 5 ml of acetic acid buffer solution for pH 4.7. After 45 minutes shake for 3 minutes with 10 ml of chloroform. Read at 372 nm. Beer's law applies for 0.2–1.2 ppm of copper.

For determination of copper as a 1 : 2 : 2 copper-pyridine-**thiosalicylic acid** complex a single extraction with chloroform at pH 5.5–7 gives 85–90% yield.[600]

Thiosemicarbazide complexes with copper.[601] As applied to malachite or chalcopyrite, dissolve a 0.75-gram sample in 5 ml of hydrochloric acid and 3 ml of nitric acid. Boil off oxides of nitrogen, dilute to 15 ml, and filter. Mix an aliquot with 10 ml of 0.1 M thiosemicarbazide, adjust to pH 3, and dilute to 25 ml. Read at 665 nm. There is also a peak absorption at 356 nm.[602] In 0.2 N hydrochloric acid containing 0.05 gram of reagent in 50 ml, Beer's law is followed for up to 10 ppm of copper. Ferric ion interferes.

The complex of copper with the **thiosemicarbazone of benzothiazol-2-yl phenyl**

[595]G. G. Karanovich, *Tr. Vses. Nauch. Issled. Inst. Khim. React.* **1959**, No. 23, 96–101; G. C. Krijn, C. J. J. Rouws, and G. den Boef, *Anal. Chim. Acta* **23**, 186–188 (1960).

[596]Halina Sikorska-Tomicka, *Mikrochim. Acta* **1969**, 718–719.

[597]R. A. Alekperov and M. M. Mamedov, *Zh. Anal. Khim.* **28**, 1001–1004 (1973).

[598]E. Uhlemann, B. Schuknecht, K. D. Busse, and V. Pohl, *Anal. Chim. Acta* **56**, 185–189 (1971).

[599]Bernd Schuknecht, Erhard Uhlemann, and Gisela Wilke, *Z. Chem.* **15**, 285–286 (1975).

[600]I. V. Pyatnitski and G. N. Trochinskaya, *Ukr. Khim. Zh.* **41**, 262–268 (1975).

[601]T. E. Abdullaeva, M. K. Akhmedli, and E. A. Bashirov, *Azerb. Khim. Zh.* **1968** (4), 106–110; cf. M. K. Akhmedli, E. A. Bashirov, and T. E. Abdullaeva, *Uch. Zap. Azerb. Gos. Univ., Ser. Khim. Nauk* **1971**, 17–23.

[602]Masahiro Morimoto and Akira Hirakoba, *Jap. Anal.* **14**, 1058–1061 (1965).

ketone is read at 446 nm in 50% ethanol.[603] The reaction is acetate buffered to pH 6.5. The reagent reacts with cobalt, nickel, cadmium, and zinc at pH 8. Silver, palladium, iron, and lead interfere.

The brown insoluble complex of copper with **thiotropolone**, which is 2-mercaptocyclohepta-2,4,6-trien-1-one, is extractable with nonpolar solvents.[604] Add 5 ml of 0.5 m*M* reagent in chloroform to a sample containing 8.7–31.6 μg of copper and adjust to pH 3–7.5 with acetate buffer solution. Add 5 ml of chloroform and shake. Centrifuge the organic phase and read at 450 nm. Beer's law is followed up to 4.55 ppm of copper. Iron is masked with fluoride, zinc with tartrate, and mercury with iodide.

The 1:2 chelate of cupric ion with **Tiron**, which is disodium-1,2-dihydroxybenzene-3,5-disulfonate, at pH 6.5–8.3 follows Beer's law at 0.05–0.2 mg/ml for reading at 380–420 nm.[605] At pH 4.3 the complex is 1:1 and has a maximum absorption at 435 nm.[606] In the presence of calcium, barium, or strontium, the complex is 1:1:2 copper–alkaline earth–Tiron and absorbs at 380 nm. There is interference by the following ions:[607] ferric, titanic, silver, auric, niobic, tantalic, uranic, ceric, molybdate, vanadate, and chromate, and by osmium tetroxide.

At pH 3–5.4, 10 *M* excess of **8-(toluene-*p*-sulfonamido)quinoline** forms a complex with cupric ion which is extracted with benzene and read at 450 nm.[608] The calibration is linear for 1–40 ppm. Borate, oxalate, citrate, thiosulfate, ferrocyanide, and ferricyanide interfere seriously.

To 1 ml of sample of a nonelectric plating bath containing propylenediaminetetraacetic acid, add 1 m*M* of **triethylenetetramine** and 5 ml of potassium iodide solution containing 1 gram/ml. Dilute to 10 ml and adjust to pH 10$\pm$0.2 with sodium hydroxide or hydrochloric acid solution. Dilute to 25 ml and read at 580 nm against a reagent blank.[609a] This is applicable to 0.5–10 mg of copper in 25 ml. Cyanide must not exceed 0.01 m*M*.

At pH 4–9 cupric ion complexes with **1,1,7-trimethyldiethylenetriamine.**[609b] As read at 640 nm it obeys Beer's law for 40–160 μg per ml. There is interference by aluminum and chromate.

Cupric ion can be determined by **variamine blue B** (azoic diazo component 35).[610] To a sample solution containing 0.01–0.1 mg of copper, add acetate buffer solution to adjust to pH 4–6. Add 1 ml of a solution containing 10 mg of the reagent per liter of 20% acetic acid. Dilute to 50 ml and read after 15 minutes. Strong oxidizing and reducing agents interfere.

For determination of copper by **xylenol orange**, see Lead. The absorption of the

[603] Toyozo Uno and Sumiyuki Akihama, *ibid.* **10**, 941–945 (1961).

[604] J. W. Srivastava and R. P. Singh, *J. Chin. Chem. Soc(taipei)* **21**, 275–279 (1974).

[605] Yoshinaga Oka, Norimasa Nakazawa and Hikaru Harada, *J. Chem. Soc. Jap., Pure Chem. Sect.* **86**, 1158–1162 (1965).

[606] S. Arribas Jimeno, M. L. Alvarez Bartalome, and M. Gonzalez Alvarez, *An. Quim.* **69**, 633–641 (1973).

[607] A. K. Majumdar and C. P. Savariar, *Anal. Chim. Acta* **21**, 53–57 (1959).

[608] Mohan Katyal, G. K. Sharma, and R. P. Singh, *J. Inst. Chem. India* **40**, 207–211 (1968).

[609a] B. L. Goydish, *Mikrochim. Acta* **1971**, 675–679.

[609b] A. Alvarez Devesa, Maria del Carmen Cisnaros Garcia, and J. M. Grana Molares, *Quim. Anal.* **30**, 263–266 (1976).

[610] Zbigniew Gregorowicz, *Z. Anal. Chem.* **171**, 246–250 (1959).

purple solution obeys Beer's law for 0.013–0.085 mg of copper in 25 ml containing 2.5–5.5 ml of 0.001 *M* reagent at pH 5.4–5.8.[611a]

Alternatively, when cupric ion oxidizes variamine blue at pH 2–4 in 60% ethanol the resulting cuprous ion complexes with 2,2′-biquinolyl for reading at 550 nm.[611b] Oxygen must have been removed from the sample solution by passing a stream of nitrogen. Beer's law is followed for 0.16–0.8 μg of copper per ml.

Cupric ion forms a 1:3 complex with **uramildiacetic acid**, which is (hexahydro-2,4,6-trioxopyrimidin-5-yl)iminodiacetic acid, at an optimum pH of 3. This is read at 775 nm.[611c] Beer's law is followed for 20–420 μg of copper per ml but preferably applied at 70–400 μg per ml. There should be less than 20 μg per ml of palladium, vanadate, thiocyanate, iodide, and oxalate.

Zephiramine, which is benzyldimethyl(tetradecyl)ammonium chloride, precipitates a copper-thiocyanate complex.[612] To a solution containing 0.02–0.2 mg of copper in 25 ml of neutral solution, add 10 ml of 1:25 hydrochloric acid, 5 ml of *M* potassium thiocyanate, and 10 ml of 0.01 *M* reagent. Shake with 10 ml of chloroform for 5 minutes and read at 440 nm. The absorption is unchanged for 3 hours. There is interference by ferric, cobalt, bismuth, and manganese ions.

Zincon, which is 2-carboxy-2′-hydroxy-5′-sulfoformazylbenzene, and cupric ion form a 1:1 complex at pH 5–10.5.[613] An appropriate buffer solution is perchloric acid and ammonia at pH 8 ± 1.5. Beer's law is followed up to 2.5 ppm of copper. A 1:1:2 complex is formed by cupric ion, Zincon, and zephiramine. To a sample containing 1.6–16 μg of cupric ion, add 1 ml of 0.065 m*M* Zincon and 3.5 ml of 5 m*M* zephiramine. Adjust to pH 8.3–8.5 with a borate–hydrochloric acid buffer solution, dilute to 50 ml, and shake with 10 ml of chloroform for 10 minutes. Allow 10 minutes for separation, dry the organic phase with sodium sulfate, and read at 625 nm against a reagent blank.

A complex of cupric ion is formed at pH 1–11 with **zinc tetramethylenediselenocarbamate**.[614] In chloroform at pH 1, it is read at 610 nm.

[611a]Makoto Otomo, *Jap. Anal.* **14**, 45–52 (1965).
[611b]Franciszek Buhl and Malgorzata Chwistek, *Chem. Anal.* (Warsaw) **21**, 421–426 (1976).
[611c]F. Bermejo Martinez and Mercedes Molina-Poch, *Microchem. J.* **20**, 7–16 (1975).
[612]Hiroshi Matsuo, Shokichi Chaki, and Shigeki Hara, *ibid.* **15**, 125–129 (1966).
[613]M. A. Dosal Gomez, J. A. Perez-Bustamente, and F. Burriel-Marti, *An. Quim.* **70**, 515–520 (1974).
[614]Janusz Terpilowski and Pawel Ladogorski, *Farm. Pol.* **30**, 1011–1015 (1974).

CHAPTER SIX
CADMIUM

For a method of removal of palladium from cadmium by extraction of the bromide complex with cyclohexanone, see Copper. A method for determination of copper by its catalytic effect on the iodine-azide reaction is also applicable to cadmium. For details, see Copper. For isolation of cadmium from samples of vanadium or niobium, see Lead.

Cadmium in excess of 0.02 ppm can be extracted as the iodide from a solution of uranium at pH 1–9.[1a] At 0.1–1 *M* iodide above pH 6, there may be present 1 mg of iron, 0.5 mg of nickel, copper, lead, manganese, cobalt, tin, and thorium, and up to 0.25 mg of zinc, bismuth, and chromium.

Extraction with molten naphthalene of complexes with 8-hydroxyquinoline is applicable at varying pH levels to cadmium, zinc, magnesium, copper, bismuth, cobalt, nickel, molybdenum, uranium, aluminum, iron, palladium, and indium.[1b]

Extraction of cadmium from 5 *N* sodium hydroxide as the diethyldithiocarbamate and back-extraction with 2 *N* hydrochloric acid avoids all interference other than by thallous ion.[1c]

ISOLATION OF CADMIUM, ZINC, COBALT, AND NICKEL

For isolation of cadmium, zinc, cobalt, and nickel from metals containing them as minor impurities, proceed as follows. Dissolve in an appropriate acid or mixture of acids.[2] Dilute with water and make strongly ammoniacal to complex these four metals, and filter. Evaporate the filtrate until excess ammonia has been driven off and acidify with hydrochloric acid. Then add 1 : 1 ammonium hydroxide to pH 8. Extract the four metals with a solution of dithizone in carbon tetrachloride.

[1a] Daido Ishii and Tsugio Takeuchi, *Jap. Anal.* **11**, 52–55 (1962).
[1b] T. Fujinaga, Masatada Sataka and Masaaki Shimizu, *Jap. Anal.* **25**, 313–318 (1976).
[1c] Sixto Bajo and Armin Wyttenbach, *Anal. Chem.* **49**, 158–161 (1977).
[2] Z. Marczenko, M. Mojski, and K. Kasiura, *Zh. Anal. Khim.* **22**, 1805–1837 (1967).

Separate zinc and cadmium from the other dithizonates by extraction with aqueous hydrochloric acid at pH 2.5. Concentrate this acid extract, make alkaline with 11% potassium hydroxide solution, and extract the cadmium as the dithizonate. Adjust to pH 8 and extract the zinc as the dithizonate.

Evaporate the carbon tetrachloride solution of dithizonates of cobalt and nickel to dryness. Add nitric acid and oxidize the organic matter. Then take up with water and separate the cobalt by extraction into chloroform as its complex with 2-nitroso-1-naphthol. Extract the nickel into carbon tetrachloride as its complex with furil α-dioxime.

3,5′-BIS[BIS(CARBOXYMETHYL)AMINOMETHYL]-4,4′-DIHYDROXY-*TRANS*-STILBENE

This reagent[3] is effective in determination of cadmium by fluorescence after isolation from most other cations. For 0.0005–0.025 mg of cadmium in the presence of 0.06 mg of zinc, the diethyldithiocarbamate extraction can be omitted.

Procedure. Mix a sample solution containing 0.0005–0.025 mg of cadmium with 5 ml of 20% solution of sodium potassium tartrate. Add ammonium hydroxide to bring to pH 11. Add 2 grams of potassium cyanide and 1 ml of 0.2% solution of sodium diethyldithiocarbamate. Extract the cadmium from the organic layer by shaking with 10 ml of 0.2 *N* hydrochloric acid. Neutralize the acid extract to pH 10 with ammonium hydroxide and add 5 ml of *M* hexamine. Add 5 ml of 0.1 m*M* reagent and dilute to 25 ml. Activate at 360 nm and read the fluorescence at 440 nm.

BROMOBENZOTHIAZO

Bromobenzothiazo is 1-(6-bromobenzothiazol-2-ylazo)-2-naphthol, which complexes with cadmium.[4] The reagent in xylene will extract 0.1 mg of cadmium from 100 ml of water. Silver and mercuric ion can be masked by thiosulfate, cobalt, and nickel, and copper by sulfonazo.[5] As applied to lead-zinc ores, the solution is passed through a cation active resin in chloride form, and the cadmium is eluted with water. When read with the reagent in slightly alkaline solution at 590 nm, the reagent shows no absorption.

Procedure. To 3.3 ml of sample containing about 0.01 mg of cadmium, add 0.3 ml of 40% solution of sodium potassium tartrate, 4 ml of xylene, 0.2 ml of 0.01 *M*

[3] B. Budesinsky and T. S. Wesr, *Analyst* **94**, 182–183 (1969).

[4] V. G. Brudz, D. A. Drapkina, K. A. Smirnova, N. I. Doroshina, Z. S. Sidenko, and G. S. Chishova, USSR Patent *146,088* (1962); D. A. Drapkina, V. G. Brudz, K. A. Smirnova, and M. A. Doroshina, *Zh. Anal. Khim.* **17**, 940–944 (1962).

[5] E. P. Shkrobot and L. M. Bakinovskaya, *Zavod. Lab.* **32**, 1452–1455 (1966).

bromobenzothiazo in xylene, and 1.5 ml of 10% sodium hydroxide solution. Shake for 2 minutes and read the xylene extract at 600 nm.

2-(6-BROMOBENZOTHIAZOL-2-YLAZO-*P*-CRESOL

The complex of cadmium with this reagent is extractable with chloroform, *o*-xylene, toluene, benzene, or carbon tetrachloride.[6]

Procedure. Mix the sample solution with 0.4 ml of 40% sodium potassium tartrate solution and 5 ml of 8% sodium hydroxide solution. Dilute to 10 ml and add 2 ml of m*M* reagent in chloroform. Add 8 ml of chloroform and shake for 1 minute. Read the chloroform phase at 610 nm against a blank.

CADION

Cadion reagent is 3-(*p*-nitrophenyl)-1-(*p*-phenylazophenyl)triazene. It forms a 3:1 complex with cadmium.[7] Beer's law is followed up to 0.6 ppm of cadmium in the final solution. Cyanide interferes. Silver and mercury must be prereduced to the elements in ammoniacal solution. The method depends on the bleaching of the reagent by reaction with cadmium. There is no effect of temperature over the range 5–30°, nor is the reaction time dependent.

Procedure. Prepare a 0.02% solution of cadion in 0.02 *N* ethanolic potassium hydroxide and filter after 24 hours. As a reagent, mix 10 ml of the 0.02% ethanolic solution of cadion, 25 ml of 11.2% solution of potassium hydroxide, 50 ml of ethanol, 5 ml of a 20% solution of sodium potassium tartrate, 1 ml of a 25% solution of poly(vinylpyrrolidinone), and 100 ml of water. Mix 10 ml of the reagent with the sample solution and dilute to 25 ml. Read the residual reagent at 560 nm as a measure of the cadmium in the sample. Reading at 505 nm has also been recommended.[8]

Alternatively,[9] mix the sample solution in dilute acetic acid with 2 ml of 0.25% gelatin solution and 0.75 ml of 0.02% solution of the reagent in 0.02 *N* ethanolic potassium hydroxide. Dilute to 25 ml. Of the bands at 225, 970, 760, and 460 nm of the cadmium complex, that at 460 nm is to be preferred.

---

[6]S. I. Gusev, M. V. Zhvakina, and I. A. Kozhevnikova, *Zh. Anal. Khim.* **26**, 1493–1498 (1971).

[7]P. Chavanne and C. Geronimi, *Anal. Chim. Acta* **19**, 377–3, (1958); E. Kroller, *Z. Leben. Unters. Forsch.* **125**, 401–405 (1964).

[8]V. P. Razumova, *Nauchn.-Tekhn. Inf. Byul. Lening. Politekhn. Inst.* **1960** (11), 90–94.

[9]V. P. Razumova, *Tr. Leningr. Politekh. Inst. M. I. Kalinina* **201**, 136–140 (1959).

CADION 2B

Cadion 2B is 1-(4-nitro-1-naphthyl)-3-(4-phenylazophenyl)triazene. It forms a blue colloidal complex in aqueous alkaline solution.[10]

Procedures

Biological Material. Digest 2 grams of organic matter, 10 ml of blood, or 100 ml of urine with 10 ml of sulfuric acid. When a clear colorless solution is obtained, often by adding small amounts of 30% hydrogen peroxide, evaporate to sulfur trioxide fumes. Add bromine water to oxidize iron, and again take to fumes. Take up with water and dilute to a known volume.

Dilute an aliquot containing not more than 4 mg of cadmium to 10 ml. Adjust to pH 10 by addition of ammonium hydroxide and extract the cadmium with 5 ml of 0.006% solution of dithizone in chloroform. Wash the extract with 10 ml of water.

Extract the cadmium from the chloroform layer with 10 ml of 1 : 100 hydrochloric acid. Evaporate to dryness and take up in 6 ml of water. Add 0.5 ml of 10% sodium tartrate solution, 1 ml of 0.02% solution of cadion 2B in 0.02 *N* ethanolic potassium hydroxide, and 1 ml of 11.2% potassium hydroxide solution. Dilute to 10 ml and after 30 minutes read at 530 nm.

Air. Isolate the cadmium from air in conventional absorbers using 3% nitric acid to not more than 9 μg/ml. Mix 2 ml of the nitric acid sample with 1 ml of 33% potassium hydroxide solution. Add 2 ml of 1% gelatin solution and 5 ml of 0.008% solution of cadion 2B in acetone. Dilute to 25 ml with water, shake for 30 seconds, and read at 605 nm.

1-(5-CHLORO-2-PYRIDYLAZO)-2-NAPHTHOL

This reagent complexes with cadmium.[11] Interference by cobalt, mercury, manganese, iron, zinc, and nickel is masked by cyanide. Formaldehyde prevents the cyanide from masking cadmium.

Procedure. To a 20-ml sample containing 0.001–0.01 mg of cadmium, add 2 ml of 10% potassium cyanide solution, 10 ml of 2% formaldehyde solution, 2 ml of m*M* reagent in methanol, and 5 ml of a borate buffer solution for pH 10. Let stand for 10 minutes and extract with 10 ml of chloroform. Dry the extract with sodium sulfate and read at 566 nm against a reagent blank.

---

[10] Hong-kang Dao, Shih-hsüan Jen and Nai-ching Wang, *Acta Chim. Sin.* **29**, 344–347 (1963); L. Truffert, M. Favert, and Y. LeGall, *Ann. Falsif. Expert. Chim.* **60**, 275–279 (1967).

[11] Shozo Shibata, Masamichi Furukawa, and Yoshio Ishiguro, *Mikrochim. Acta* **1972**, 721–727.

DIETHYLDITHIOCARBAMATE

For the use of diethyldithiocarbamate, EDTA, silver, auric, bismuth, cupric, cobalt, mercuric, manganous, nickel, lead, stannous, and zinc ions must be absent.[12] Limited amounts of chromic, ferric, ferrous, magnesium, stannic, and antimony ions are tolerated.

Procedure. To the sample solution containing 0.01–0.1 mg of cadmium, add 5 ml of 20% solution of potassium tartrate and 5 ml of 20% solution of potassium citrate. Add 10 ml of a buffer solution containing 0.3% ammonium chloride in 1:4 ammonium hydroxide, followed by 5 ml of 0.2% solution of sodium diethyldithiocarbamate. After a few minutes extract with 25 ml and 15 ml of chloroform. Dilute the combined extracts to 50 ml and read at 260 nm against a reagent blank.

DITHIZONE

In recovery of cadmium from the solution of a zinc sample, extract as the iodocadmate by a xylene solution of a water-insoluble secondary amine. To avoid complex formation by iron and precipitation of silver, chloride should be absent. Interfering amounts of copper can be extracted by neocuproine into isoamyl alcohol.

To separate cadmium in mg amounts from 2.5 gram of zinc in 2 *M* sodium acetate solution, extract with sodium diethyl diphosphorodithioate in carbon tetrachloride.[13] Then add dithizone to the organic phase and shake it with 2% solution of sodium hydroxide, to read the cadmium dithizonate at 520 nm.

For analysis of "yellow cake"(triuranium octaoxide), treat with hydrofluoric acid to remove silica, add hydrochloric and boric acids, and evaporate to dryness.[14] Dissolve the residue in 1.2 *N* hydrochloric acid and pass through a column of anion exchange resin, Dowex 1-X8, in chloride form. Elute zinc with 0.15 *M* hydrobromic acid, then cadmium with dithizone.

Cadmium in a sample of gold is determined by dithizone after sequential removal of gold and iron.[15]

In analysis of acid solutions of ores or of soil containing copper, zinc, lead, and iron for cadmium, neutralize to pH 7–8 with ammonium hydroxide.[16] Iron is complexed with sodium citrate. Hydroxylamine hydrochloride prevents manganese oxidation. Then extract with dithizone in carbon tetrachloride and wash with water, followed by acid extraction of the cadmium and its reextraction as the

[12]D. F. Boltz and E. A. Havlena, Jr., *Anal. Chim. Acta* **30**, 565–568 (1969); cf. Emiko Sudo, *Sci. Rep. Res. Inst., Tohoku Univ.* **6**, 142–146 (1954).

[13]H. Bode and K. Wulff, *Z. Anal. Chem.* **219**, 32–48 (1966).

[14]J. Korkisch and D. Dimitriadis, *Mikrochim. Acta* **1974**, 449–459.

[15]G. Ackermann and J. Köthe, *Z. Anal. Chem.* **231**, 252–261 (1967); Zygmunt Marczenko, Krzysztof Kasiura, and Maria Krasiejko, *Chem. Anal.* (Warsaw) **14**, 1277–1287 (1969).

[16]Florica Popea and Madeleine Gutman, *Acad. Rep. Po. Rom. Stud. Cercet. Chim.* **9**, 673–680 (1961).

dithizonate. The technic is applicable to the product of wet digestion of blood and urine.[17]

For analysis of tantalum, dissolve the sample with nitric and hydrofluoric acids.[18] Extract lead, cadmium, iron, and copper as diethyldithiocarbamates with chloroform. Determine cadmium by dithizone.

For copper, nickel, cobalt, zinc, and cadmium as impurities in sodium chloride, complex with EDTA and extract from the solution with chloroform.[19] Evaporate to dryness, add sulfuric and nitric acids, and heat to sulfur trioxide fumes. Determine cadmium by dithizone.

For cadmium at more than 0.005 ppm, extract at pH 3.3–3.7 with 0.01 *M* dithizone in chloroform.[20] Reextract from the chloroform with 0.1 *N* hydrochloric acid. Adjust the pH with sodium hydroxide, mask with potassium cyanide, and extract the cadmium as the dithizonate for reading at 518 nm.

For cadmium by extraction with tribenzylamine in a solution containing mercury, cadmium and zinc and determination with dithizone see mercury.

Procedures

Amine Extraction.[21] Make 10 ml of sample solution containing 0.001–0.03 mg of cadmium 0.3–0.5 *N* with hydrobromic acid. Extract with 10 ml of 2% tribenzylamine solution in dichloroethane. Wash the organic phase with 5 ml of 0.3 *N* hydrobromic acid. Mix the organic phase with 10 ml of 0.01% solution of dithizone in dichloroethane. Add 5 ml of 10% sodium hydroxide solution and 5 ml of 10% solution of sodium potassium tartrate. Shake well and separate the organic phase. Wash it with 5% sodium hydroxide solution until the washings are colorless. Filter through a dry filter, dilute to 25 ml with the solvent, and read at 520 nm.

Zinc.[22] As reagent, dissolve 50 mg of dithizone in 1 quart of carbon tetrachloride. As Amberlite reagent, prepare a 1% solution of the LA-2 grade in xylene. As wash solution, dilute 50 ml of 16.6% solution of potassium iodide to about 450 ml; add 1 ml of sulfuric acid, and dilute to 500 ml.

Dissolve 0.2 gram of zinc in 10 ml of 1 : 1 sulfuric acid and transfer to a separatory funnel. Add 5 ml of 16.6% potassium iodide solution and dilute to about 50 ml. Check the pH: if above 3, add sulfuric acid to drop to below that level. Add 10 ml of the Amberlite solution and shake for 20 seconds. Allow 15 minutes for separation of the layers and discard the aqueous phase. Shake for 20 seconds with the wash solution and discard the aqueous phase.

Shake the xylene solution for 20 seconds with 10 ml and 10 ml of 12.4% sodium carbonate solution to extract the cadmium, and discard the xylene. Mix the

[17]P. Sanz-Pedrero and M. D. Hermoso, *Med. Segur. Trab.* **11**, 28–36 (1963).
[18]V. A. Nazarenko, E. A. Biryuk, M. B. Shustova, G. G. ShiFareva, S. Ya. Vinkovetskaya, and G. V. Flyantikova, *Zavod. Lab.* **32**, 267–269 (1966).
[19]Cezary Rozychi and Januez Rogozinski, *Chem. Anal.* (Warsaw) **20**, 107–11 (1975).
[20]Yoshio Morimoto and Takashi Ashizawa, *Jap. Anal.* **10**, 1383–1386 (1961).
[21]A. I. Vasyutinskii, N. A. Kissl, and E. N. Matveeva, *Zh. Anal. Khim.* **23**, 1847–1848 (1968).
[22]J. R. Knapp, R. E. Van Aman, and J. H. Kanzelmeyer, *Anal. Chem.* **34**, 1374–1378 (1962); cf. P. V. Marchenko and A. I. Voronina, *Ukr. Khim. Zh.* **35**, 652–656 (1969).

combined extracts with 5 ml of 20% solution of sodium potassium tartrate tetrahydrate and 5 ml of 5% hydroxylamine hydrochloride solution. Let stand 2 minutes and add 15 ml of 24% sodium hydroxide solution. Shake for 30 seconds with 15 ml of the dithizone reagent. Filter through a cotton plug and read at 520 nm against a reagent blank.

Aluminum.[23] Dissolve 1 gram of sample in 28 ml of 20% sodium hydroxide solution. Add 10 ml of 20% solution of tartaric acid, 5 ml of 1% solution of potassium cyanide, and 2 ml of 10% solution of hydroxylamine hydrochloride. Dilute to 70 ml and extract with 15 ml and 10 ml of 0.008% solution of dithizone in chloroform.

Extract the cadmium from the combined chloroform extracts with 25 ml of 20% tartaric acid solution. Wash this extract with 5 ml of chloroform and discard all chloroform extracts.

To the acid aqueous extract, add 0.5 ml of 10% solution of hydroxylamine hydrochloride, 18 ml of 20% sodium hydroxide solution, and 1 ml of 0.5% potassium cyanide solution. Cool and extract with 15 ml of 0.003% solution of dithizone in chloroform. Read this extract at 517 nm.

Lead.[24] Dissolve a 10-gram sample in the minimum amount of nitric acid. Dilute with water, adjust to pH 2, and extract silver, bismuth, and copper with 0.005% dithizone solution in chloroform. Discard the extracts. Add sulfuric acid to the aqueous layer to precipitate lead. Make an aliquot of the aqueous layer alkaline and extract cadmium by dithizone.

Uranium.[25] Dissolve the sample in hydrochloric acid and hydrogen peroxide. Evaporate to dryness and take up in 0.01 *N* hydrochloric acid. Dilute to a known volume with that acid and take an aliquot.

Add 5 ml of 10% ammonium citrate solution and make alkaline with ammonium hydroxide. Add 5 ml of 10% potassium cyanide solution and 5 ml of 0.004% solution of dithizone. Extract with 5 ml, 5 ml, and 5 ml of carbon tetrachloride. Other elements are extracted.

Extract cadmium from the combined organic extracts with 10 ml of 0.01 *N* hydrochloric acid. To this acid extract, add 5 ml of 10% ammonium citrate solution, 5 ml of 10% sodium hydroxide solution, and 5 ml of 5% potassium cyanide solution. Add 5 ml of 0.004% solution of dithizone and extract with 5 ml and 5 ml of carbon tetrachloride. Wash the combined extracts with 10 ml of 1% sodium hydroxide solution and read at 528 nm.

Nuclear Grade Zincaloy-2.[26] Dissolve a 1-gram sample by heating with 15 ml of 1 : 1 hydrochloric acid. From time to time add four 0.5-ml portions of hydrofluoric acid and finally evaporate to dryness. Add 5 ml of hydrochloric acid and evaporate

[23]Shikao Hashimoto and Reizi Tanaka, *Jap. Anal.* **8**, 564–568 (1959).

[24]H. Jedrzejewska and M. Malusecka, *Chem. Anal.* (Warsaw) **12**, 574–584 (1967).

[25]Rafael H. Rodriguez Pasques and Julia F. Possidini de Albinati, *Anal. Asoc. Quim. Argent.* **44**, 90–103 (1956); Ken Saito, Daido Ishii, and Tsuguo Takeuchi, *Jap. Anal.* **9**, 299–305 (1960); Yoshio Morimoto and Takashi Azhizawa, *ibid.* **10**, 1383–1386 (1961).

[26]G. Ghersini and S. Mariottini, *Talenta* **18**, 492–496 (1971).

to dryness. Repeat this step twice. Take up the residue in 4 ml of hydrochloric acid and 30 ml of warm water. Dilute to 50 ml.

Equilibrate 0.1 *M* trioctylamine in cyclohexane by shaking for 5 minutes with an equal volume of 1 : 10 hydrochloric acid. Shake a 10-ml aliquot of sample with 10 ml of the trioctylamine solution. Separate the organic extract of cadmium and shake with 5 ml, 5 ml, and 5 ml of 10% tartaric acid solution for 3 minutes each. Extract the cadmium by shaking the washed organic layer with 10 ml and 10 ml of 1 : 8 sulfuric acid. Finally add 50 ml of 25% potassium hydroxide solution to the acid layer and extract the cadmium with 10 ml of 0.002% solution of dithizone in chloroform. To simplify separation of the chloroform extract, add 50 ml of water. Filter the chloroform layer and read at 520 nm against a reagent blank.

High Purity Zinc.[27] Dissolve a sample expected to contain about 0.01 mg of cadmium in 10 ml of 1 : 10 sulfuric acid, adding 5 drops of nitric acid. Dilute to 30 ml and add 3 ml of 50% solution of potassium iodide. Shake with 10 ml of a 2% solution of tribenzylamine in chloroform. Separate the organic layer and wash with 10 ml and 10 ml of 1 : 35 sulfuric acid containing 5% of potassium iodide. Reextract the cadmium from the chloroform layer with 10 ml of 4% sodium hydroxide solution containing 1 ml of 20% solution of sodium tartrate. To this alkaline cadmium solution, add 1 ml of 10% hydroxylamine hydrochloride solution and shake with 7 ml of 0.001% solution of dithizone in carbon tetrachloride. Dry the organic layer with sodium sulfate and read at 536 nm. The amounts of bismuth, lead, cobalt, nickel, copper, silver, mercury, gallium, thallium, and iron in such a material do not interfere.

Cast Iron.[28] Dissolve 0.5 gram in 5 ml of nitric acid and 12 ml of perchloric acid. Heat to white fumes, cool, add 50 ml of water, and filter. Add 10 ml of 50% citric acid solution, followed by ammonium hydroxide, to bring to pH 9 ± 0.2. Extract with 10 ml portions of 0.05% solution of dithizone in chloroform until extraction is complete. Wash the combined extracts with 5 ml of 10% ammonium citrate solution. Extract the chloroform solution of dithizonates with 20 ml of 0.1 *N* hydrochloric acid. This removes cadmium, zinc, and lead contaminated with some copper, nickel, and cobalt. To the acid extract add 3 ml of 5% solution of hydroxylamine hydrochloride, 5 ml of 20% solution of sodium potassium tartrate, and as a carrier, 5 ml of bismuth nitrate solution containing 0.1 mg of bismuth per ml. Neutralize with ammonium hydroxide, add 2 ml of 10% potassium cyanide solution, and add ammonium hydroxide to pH 10.4–10.6. Add 7 ml of 2% formaldehyde solution and 5 ml of 5% solution of sodium diethyldithiocarbamate. Shake for 3 minutes and extract the precipitate with 10 ml of carbon tetrachloride. Evaporate to about 2 ml and dilute to about 40 ml. Add 5 ml of 5% solution of hydroxylamine hydrochloride, 5 ml of 20% solution of sodium potassium tartrate, 6 ml of 20% sodium hydroxide solution, and 1 ml of 1% potassium cyanide solution. Shake with 10 ml of 0.005% solution of dithizone in chloroform for 10 minutes and read at 510 nm.

[27]P. V. Marchenko and A. I. Voronina, *Ukr. Khim. Zh.* **35**, 652–656 (1969); cf. D. Filipov, I. Nachev, and K. Nachev, *C. R. Acad. Bulg. Sci.* **18**, 537–540 (1965).
[28]Shizuya Maekawa and Yoshio Yoneyama, *Jap. Anal.* **10**, 732–740 (1961).

Beryllium.[29] Dissolve a sample containing about 10 μg of cadmium by suspending in water and gradually adding 10 ml of hydrochloric acid per gram of metal. If necessary when reaction ceases, add a few drops of 30% hydrogen peroxide to dissolve residual metal, and boil to destroy excess peroxide. Add 20 ml of 1:2 sulfuric acid and dilute to 150 ml. As a collector, add 5 ml of a nickel salt containing 5 mg of nickel per ml. Add 25 ml of 50% citric acid per gram of beryllium to complex the beryllium. Adjust to pH 8.25±0.1 with ammonium hydroxide. Dilute to 300 ml, add 25 ml of 2% benzotriazole solution, and digest at 100° for 2 hours. Let stand overnight and add filter pulp. Filter, and wash with a minimum amount of 1% ammonium citrate solution preadjusted to pH 8.25. Return the paper to the original beaker, add 25 ml of nitric acid and 25 ml of perchloric acid, and evaporate to dryness. Take up in 100 ml of 1:9 hydrochloric acid and cool.

Prepare a 1×12 cm column containing 8 cm of 200-400 mesh Dowex 1-X8. Wash the column with 100 ml of 1:4 nitric acid, then with 100 ml of water. Convert to the chloride form with 100 ml of 1:9 hydrochloric acid.

Pass the solution through the column and wash with 200 ml of 1:9 hydrochloric acid. Elute the cadmium with 125 ml of 1:4 nitric acid. Add 1 ml of sulfuric acid to the eluate and evaporate to dryness. Take up in 10 ml of water, adding successively 25 ml of 2% tartaric acid solution, 1 ml of 5% hydroxylamine hydrochloride solution, 5 ml of 40% sodium hydroxide solution containing 0.05% of potassium cyanide, and 20 ml of 0.015% solution of dithizone in chloroform. Shake for 1 minute, filter the extract through a cotton plug, and read at 518 nm against chloroform. Subtract a blank.

Beryllium Oxide. Dissolve a sample containing approximately 10 μg of cadmium in 20 ml of hot 1:2 sulfuric acid per gram of sample. Add 10 ml of hydrochloric acid and dilute to 150 ml. Complete as for beryllium from "As a collector, add 5 ml...."

Water.[30] Mix 500 ml of sample with 20 ml of 20% ammonium chloride solution, 5 ml of 50% sulfosalicylic acid solution, and 5 ml of 50% ammonium citrate solution. Neutralize to phenolphthalein with 50% sodium hydroxide solution. If the copper in the sample exceeds 0.01 mg, extract it with 2.5 ml of 0.2% solution of lead diethyldithiocarbamate in carbon tetrachloride. For preparation of that reagent, see its use for determination of copper. Wash the aqueous solution with 5 ml of carbon tetrachloride.

Add 40 ml of 10% sodium hydroxide solution to the aqueous sample and shake with 2 ml of 0.05% dithizone solution in carbon tetrachloride. Further extract with 5 ml and 5 ml of carbon tetrachloride.

Wash the combined organic extracts with 50 ml of water, 5 ml of 25% ammonium chloride solution, 20 ml of 10% sodium hydroxide solution, and 5 ml of 50% sulfosalicylic acid solution. Discard these washings.

[29] J. O. Hibbits, Silve Kallmann, H. Oberthin, and J. Oberthin, *Talenta* **B**, 104-108 (1961).

[30] E. M. Mal'kov, V. G. Kosyreva, and A. G. Fedoseeva, *Zavod. Lab.* **31**, 1327 (1965); cf. Takashi Ashizawa, *Jap. Anal.* **10**, 817–822 (1961); J. Ganotes, E. Larson, and R. Navone, *J. Am. Water Works Assoc.* **54**, 852–854 (1962); G. Ackermann and W. Angermann, *Z. Anal. Chem.* **250**, 353–357 (1970); Takashi Ashizawa and Koji Hosoya, *Jap. Anal.* **20**, 1416–1422 (1971).

Extract the cadmium from the organic phase by shaking with 60 ml of 0.06 *N* hydrochloric acid; discard the organic phase. To this aqueous acid extract, add 2.5 ml of 25% ammonium chloride solution and neutralize to phenolphthalein with 10% sodium hydroxide solution. Add 15 ml excess of the 10% sodium hydroxide solution and extract the cadmium with 2 ml of 0.05% solution of dithizone in carbon tetrachloride supplemented by extraction with 3 ml of carbon tetrachloride. Read the combined extracts at 510 nm.

Alternatively,[31] mix 500 ml of sample with 50 ml of hydrochloric acid and filter. Pass it through a column of Dowex 1-X8 resin, which will retain cadmium, zinc, and a few other elements in noninterfering amounts. Elute the zinc with 50 ml of 0.15 *N* hydrobromic acid. Then elute the cadmium with 50 ml of 1:7 nitric acid. Evaporate this eluate to dryness and dissolve in 10 ml of 1:14 nitric acid. Add 25 ml of a freshly prepared complexing solution containing 8% of sodium hydroxide, 0.2% of potassium cyanide, 1% of sodium potassium tartrate, and 0.8% of hydroxylamine hydrochloride. Extract the cadmium with 7 ml of 0.012% solution of dithizone in chloroform. Wash the organic extract with 5 ml of 6% sodium hydroxide solution, filter, and read at 490 nm.

Air.[32] As a complexing solution, mix 50 ml of 40% sodium hydroxide solution containing 1% of potassium cyanide, 10 ml of 25% sodium potassium tartrate solution, and 10 ml of 20% solution of hydroxylamine hydrochloride, and dilute to 250 ml.

Collect a sample containing 0.5–100 μg of cadmium by filtration on a paper disk. Ash the paper and dissolve the residue in 5 ml of 1:1 hydrochloric acid. Evaporate to dryness, dissolve in 1:20 nitric acid, and dilute to 10 ml with that acid. Add a 2-ml aliquot and 3 ml of 1:20 nitric acid to 25 ml of freshly prepared complexing solution. Extract by shaking for 30 seconds with 7 ml of 0.012% solution of dithizone in chloroform. Shake the organic layer for 30 seconds with 5 ml of 6% sodium hydroxide solution. If the wash solution turns yellow, repeat the washing. Read the dithizonate in chloroform at 490 nm.

Food.[33] Digest an appropriate sample with 3 ml of sulfuric acid and 10 ml of nitric acid, adding more nitric acid from time to time if necessary until a clear digestate is obtained. Perchloric acid may also be added. Evaporate to sulfur trioxide fumes, cool, and take up in 20 ml of water. Neutralize with 40% sodium hydroxide solution and dilute to 25 ml. Add 1 ml of 25% sodium potassium tartrate solution, 5 ml of 40% sodium hydroxide solution containing 1% of potassium cyanide to mask copper and nickel, and 1 ml of 20% solution of hydroxylamine hydrochloride. Extract for 1 minute each with 10 ml, 10 ml, 5 ml, 5 ml, and 5 ml of 0.015% dithizone solution in carbon tetrachloride. Collect the extracts in 25 ml of 2% tartaric acid solution. Shake for 2 minutes and discard the organic layer. Wash the tartaric acid extract with 5 ml of carbon tetrachloride. Add 0.25 ml of 20% hydroxylamine hydrochloride solution, 10 ml of 0.015% dithizone solution in

[31] J. Konkisch and D. Dimitriadis, *Talenta* **20**, 1295–1301 (1973).

[32] Ignacy Pines, *Chem. Anal.* (Warsaw) **15**, 103–110 (1970).

[33] The Metallic Impurities in Organic Matter Subcommittee of the Society for Analytical Chemistry, *Analyst* **94**, 1153–1158 (1969).

carbon tetrachloride, and 5 ml of 40% sodium hydroxide solution containing 1% of potassium cyanide. Shake for 1 minute and filter the extract. Make further extractions with 10 ml, 5 ml, 5 ml, and 5 ml of the 0.015% dithizone solution in carbon tetrachloride. Dilute the filtered extracts to 50 ml and read at 525 nm against a reagent blank.

ERIOCHROME GREY BL

This reagent forms a red complex with cadmium in an alkaline solution.

Procedure.[34] Take a sample solution containing 0.002–0.05 mg of cadmium. Add 0.4 ml of 0.1% solution of the reagent and dilute to 10 ml with a buffer solution for pH 11 containing 0.828% of ammonium chloride and 11.3% of ammonium hydroxide. Read after 30 minutes.

FERROUS–2,2′–BIPYRIDYL

Ferrous-2,2′-bipyridyl is a chelate that complexes with cadmium.[35] Beer's law is followed for 0.4–2.8 ppm. The color is stable at room temperature for many hours. EDTA interferes seriously, as does 0.4 ppm of lead, mercuric, or bismuth ion.

Procedure. As the reagent, mix 2 ml of 8 m*M* 2,2′-bipyridyl, 2 ml of 2m*M* ferrous ammonium sulfate, 2 ml of 3 *M* sodium acetate, and 7 ml of 0.3 *M* potassium iodide. Add the sample solution and adjust to pH 6.5 with dilute sulfuric acid or dilute sodium hydroxide solution. Extract the complex by shaking with 10 ml of dichloroethane. Shake the extract with 1 gram of sodium sulfate to clarify it. Read at 526 nm.

FLAME PHOTOMETRY

Using a Beckman DU spectrophotometer and photomultiplier, the minimum for cadmium by flame photometry at 326.1 nm is 0.5 ppm with an air-hydrogen flame.[36] Cadmium for flame photometry is efficiently extracted as the 1:10-phenanthroline complex from perchlorate solution buffered to pH 7 with 0.1 *M*

[34]Osvald Sokelle, *Chem. Listy* **56**, 1108–1110 (1962).
[35]Yuroku Yamamoto and Keiya Kotsuji, *Bull. Chem. Soc. Jap.* **37**, 594–595 (1964); Keiya Kótsuji, *ibid.* **38**, 988–992 (1965).
[36]Paul T. Gilbert, Jr., *Anal. Chem.* **31**, 110–114 (1959).

ammonium acetate.[37] The most effective solvent is nitrobenzene, which is also suitable to serve as the aspiration solvent.

Procedures

Alloys.[38] As an ammonium citrate buffer solution, disperse 700 grams of citric acid in water, add 800 ml of ammonium hydroxide, and dilute to 1500 ml.

Dissolve 1 gram of sample in 10 ml of hydrochloric acid, adding 5 ml of nitric acid if necessary. Evaporate to a small volume, dilute with water, and filter. Dilute to 50 ml and add 50 ml of ammonium citrate buffer solution and 10 ml of 10% hydroxylamine hydrochloride solution. Add ammonium hydroxide to pH 6.8–7.2. Extract with 10 ml, 10 ml, and 10 ml of 1% dithizone solution in chloroform. Extract the combined dithizone extracts with 10 ml of 1 : 10 hydrochloric acid and determine by flame photometry.

Mercury.[39] To 100 grams of sample, add 2 ml of water and dissolve by slowly adding 75 ml of nitric acid. Then add 110 ml of formic acid and heat gently until the precipitated mercury is agglomerated into a single mass. Decant the liquid through a filter paper and evaporate to dryness under a moderate vacuum. Dissolve in 5 ml of 1 : 1 hydrochloric acid and read by flame photometry at 326.1 nm.

8-HYDROXYQUINOLINE

8-Hydroxyquinoline is also known as 8-quinolinol and as oxine. At pH 5–13 cadmium forms an insoluble oxinate which is extractable with chloroform with a maximum absorption at 400 nm. Zinc forms a similar complex.

Procedure. *Anhydrite.*[40] Dissolve 0.1 gram of sample containing not more than 0.1 mg of cadmium in 10 ml of 7% hydrochloric acid. Add 2.2 ml of ammonium hydroxide to adjust to pH 7.6–8.6. Add 2 ml of 5% solution of oxine in 15% acetic acid. The yellow precipitate that forms includes cadmium oxinate, calcium oxinate, and ammonium oxinate. Immediately add 20 ml of chloroform and shake for 1 minute. Filter the organic layer to remove calcium oxinate and read at 400 nm.

---

[37] Alfred A. Schilt, Rose L. Abraham, and John E. Martin, *Anal. Chem.* **45**, 1808–1111 (1973).
[38] Ulrich Bohnstedt, *DEW Tech. Ber.* **11**, 101–105 (1971).
[39] J. Meyer, *Z. Anal. Chem.* **219**, 147–160 (1966).
[40] W. L. Medun, *Anal. Chem.* **32**, 632–634 (1960).

N-METHYLANABASINE-(α′-AZO-7)-8-HYDROXYQUINOLINE

This reagent, which is 8-hydroxy-7-[3-(1-methyl-2-piperidyl)-2-pyridylazo]quinoline, forms a red 2:1 complex with cadmium at pH 5–6.[41] It develops immediately and remains stable for 20 minutes. Although the maximum for the complex is at 545 nm, that for the reagent is at 445 nm. Beer's law is followed for 0.2–2 μg/ml. Zinc, nickel, lead, and iron interfere. Copper to double the amount of cadmium can be masked by 5 ml of 5% thiourea solution and up to 0.15 mg of aluminum can be masked by 2 ml of saturated sodium fluoride solution. Up to fortyfold magnesium does not interfere.

Procedure. Mix the sample solution with 2 ml of 0.2% solution of the reagent. Add 10 ml of buffer solution for pH 5.3 and dilute to 25 ml. Read at 545 nm against a blank.

1,10-PHENANTHROLINE

The 1:2 complex of cadmium with 1,10-phenanthroline associates with two iodide ions to give an insoluble complex that in turbidimetric form is read photometrically.[42] As applied to analysis of cadmium in indium metal, the latter does not interfere. The concentration of lead, thallous, nickel, and cobalt ions must not exceed that of cadmium.

Procedure. *Indium.* As sodium citrate solution, dissolve 250 grams of citric acid in water, neutralize with 10% sodium hydroxide solution, and dilute to 1 liter. As 0.1 *M* NTA solution, dissolve 19.11 grams of nitrilotriacetic acid in water, neutralize to pH 5 with 10% sodium hydroxide solution, and dilute to 1 liter. As 0.1 *M* 1,10-phenanthroline, dissolve 9.9114 grams of the monohydrate in hot water and adjust to pH 5 with acetic acid.

Dissolve a sample in 1:1 hydrochloric acid with the assistance of a few drops of nitric acid. Evaporate to dryness, take up in water acidified with hydrochloric acid, and dilute to a known volume. Mix 25 ml of 25% sodium citrate solution adjusted to pH 4.8 with 2.5 ml of 0.1 *M* nitrilotriacetic acid and 1 ml of 0.001 *M* sodium thiosulfate. Add an aliquot of the sample solution such that the contents will be 2.7×10^{-6} *M* in cadmium and up to 0.025 *M* in indium. Add 2 ml of 0.1 *M* 1,10-phenanthroline and 5 ml of *M* potassium iodide. Dilute to volume and after 50 minutes read at 533 nm against a blank.

---

[41]M. Yusupov, Sh. T. Talipov, R. Kh. Dzhiyanbaeva, and A. Z. Tatarskaya, *T. Tash. Gos. Univ.* **1967** 74–78.

[42]F. Vydra and K. Stulik, *Chemist-Analyst* **54**, 77–78 (1965).

1,10-PHENANTHROLINE AND BROMOPHENOL BLUE

Cadmium and zinc form a ternary complex of two atoms of the metal with one molecule of each of these reagents.[43] Copper must be absent. The complex is extractable with dichloroethane or chloroform.

Procedure. *Aluminum Alloy.* Dissolve a 0.1-gram sample in 15 ml of 1:1 nitric acid. Add 20 ml of water and filter. Add ammonium hydroxide to raise to pH 1 and dilute to 100 ml. Extract copper from a 10-ml aliquot with 0.01% solution of dithizone in chloroform and discard the extract. Add ammonium hydroxide to raise the aqueous phase to pH 7. Add 2 ml of a phosphate buffer solution for pH 7 and 2 ml of a 0.3% solution of 1,10 phenanthroline and dilute to 25 ml. Shake with 5 ml of chloroform for 2 minutes and read the extract at 610 nm. The result is the sum of zinc and cadmium.

1,10-PHENANTHROLINE AND EOSIN

The ternary complex of cadmium with 1,10-phenanthroline and eosin (acid red 87) is extractable for fluorescence or spectrophotometry.[44] The eosin may be replaced by iodoeosin or by erythrosin (acid red 51). The optimum pH is 8–8.5. When extracted with chloroform, addition of acetone increases the fluorescence. Palladium, silver, zinc, gallium, nickel, and ferric ion interfere.

Procedures

Photometric. Mix a neutral solution containing 1–6 μg of cadmium with 1 ml of a phosphate buffer solution for pH 9, 3 ml of 0.2% solution of 1,10-phenanthroline, and 0.2 ml of 0.1% solution of eosin. Dilute to 10 ml and shake for 1 minute with 6 ml of chloroform. Read the extract at 530 nm.

Fluorometric. For a sample containing 0.05–1 μg of cadmium, proceed as for photometry but reduce the 1,10-phenanthroline to 1 ml and read the fluorescence at 570 nm. Results are one order of magnitude more sensitive than those obtained with spectrophotometry.

---

[43] M. M. Tananaiko and N. S. Bilenko, *Izv. Vyssh. Ucheb. Zaved., Khim. Khim. Tekhnol.* **15**, 1693–1695 (1972).

[44] M. A. Matveets and D. P. Scherbov, *Zh. Anal. Khim.* **26**, 823–826 (1971); D. N. Lisitsyna and D. P. Shcherbov, *ibid.* **28**, 1203–1205 (1973).

1-(2-PYRIDYLAZO)-2-NAPHTHOL

Cadmium and zinc sulfides and selenides are converted to sulfates in ammoniacal buffer solution.[45] The complexes with 1-(2-pyridylazo)-2-naphthol, often referred to as PAN, are then extracted with chloroform at pH 8.7–10 for reading at 555 nm.[46] For determination in seawater, the cadmium is coprecipitated with copper as the sulfide and subsequently is separated by ion exchange.

Procedure.[47] Dilute an aliquot of slightly acid sample containing up to 25 μg of cadmium ion to 20 ml. Add 1 ml of 0.1% methanolic PAN and 5 ml of ammonium chloride–ammonium hydroxide buffer solution for pH 9.2. Set aside for 10 minutes and extract with 10 ml of chloroform. Read the red organic layer at 555 nm.

Nickel, Cadmium, and Zinc.[48] Dissolve 0.1 gram of sample in 2 ml of 2 : 1 nitric acid. Evaporate to dryness and take up in 20 ml of water. Add 5 ml of an ammonium hydroxide–ammonium chloride buffer solution for pH 10, 5 ml of 1% solution of potassium cyanide, and 1 ml of 0.1% methonolic solution of PAN. Add 2 ml of 40% formaldehyde solution to unmask cadmium. Extract the PAN complexes of cadmium and zinc with 10 ml of chloroform. Set aside for 5 minutes and read the organic layer at 540 nm as a measure of cadmium and zinc. Extract the chloroform solution with 10 ml of 0.1% solution of sodium diethyldithiocarbamate to remove the cadmium. Read at 540 nm for zinc and determine cadmium by difference.

PYRROLIDINEDITHIOCARBAMATE

The chelate of pyrrolidinedithiocarbamate is soluble in chloroform and several other organic solvents.[49] The same is true of chelates with cobalt, bismuth, and molybdenum. The extracts absorb in the ultraviolet region. A large excess of reagent is to be avoided because it causes a large blank. The complex is formed at pH 2–14. The following ions must be absent: silver, gold, bismuth, copper, cobalt, ferrous, ferric, mercuric, nickel, and lead. No more than 1 ppm of manganese or 5 ppm of stannous ion is permissible. This reagent will determine 1 ppm of cadmium in the presence of tenfold excess of zinc.

Procedure. As a buffer solution, dissolve 22.1 grams of sodium acetate trihydrate and 5 ml of glacial acetic acid in water and dilute to 1 liter.

[45] R. P. Pantaler, N. B. Lebed', and L. N. Semenova, *Tr. Kom. Anal. Khim.* **16**, 24–29 (1968).
[46] Masayoshi Ishibashi, Tsunenobu Shigematsu, Masayoshi Tabushi, Yasuharu Nishikawa, and Shiro Goda, *J. Chem. Soc. Jap., Pure Chem. Sect.* **83**, 295–298 (1962).
[47] S. Shibata, *Anal. Chim. Acta* **25**, 348–354 (1961).
[48] Wilhelm Berger and Heinz Elvers, *Z. Anal. Chem.* **171**, 255–261 (1959).
[49] Melvyn B. Kalt and D. F. Boltz, *Anal. Chem.* **40**, 1086–1091 (1968).

Mix a sample solution containing 0.01–0.15 mg of cadmium with 10 ml of buffer solution. If the sample contains less than 0.1 mg of cadmium, add 5 ml of 1% solution of the reagent and set aside for 5 minutes. If it contains more than 0.1 mg increase the amount of reagent to 10 ml. Extract with 25 ml and 15 ml of chloroform and add the extracts to 10 ml of 20% sodium hydroxide solution. Shake the extracts with the sodium hydroxide solution for at least 1 minute. Separate the chloroform phase, dilute to 50 ml, and read at 262 nm against a blank.

RHODAMINE B

In dilute sulfuric acid solution, cadmium iodide forms a 1:2 complex with rhodamine B (basic violet 10).[50] Beer's law is followed up to 0.6 μg of cadmium per ml. There is interference by citric, tartaric, and oxalic acids, thiourea, hydroxylamine, thiosulfate, and organic solvents. Diethyldithiocarbamate extractions separate cadmium from zinc, cobalt, and nickel.[51]

Procedure. *Magnesium or Aluminum.* Dissolve 0.5 gram of sample in hydrochloric acid assisted by later addition of 1 ml of nitric acid. Evaporate nearly to dryness and take up in 20 ml of water. Add 5 ml of 20% sodium citrate solution and 20 ml of water. Make neutral or slightly alkaline to Congo red paper with a few drops of ammonium hydroxide. Add 5 ml of 0.5% sodium diethyldithiocarbamate solution. Extract the complexes of cadmium, copper, bismuth, mercury, and antimony with 10 ml of chloroform. Add 2 ml of 0.082% solution of rhodamine B and extract with 5 ml of chloroform. Wash the combined extracts with water and back-extract the cadmium with 10 ml and 10 ml of *N* hydrochloric acid. Evaporate the acid extracts to dryness and take up in 2 ml of 3.6 *N* sulfuric acid. Add 2 ml of 0.08% solution of rhodamine B. If ferric ion is present, add 1 ml of 2% ascorbic acid. Then add 2.5 ml of *N* potassium iodide and dilute to 25 ml. Read the 1:2 cadmium iodide–rhodamine B complex at 550 nm.

SULFARSAZEN

Sulfarsazen is the product of replacement of the sulfonic acid group of cadion IREA by an arsonic group. Cadmium forms a red-orange 1:1 complex with the yellow sulfarsazen.[52a] Tolerances for determination of 2 mg of cadmium are lead, 2 mg; cobalt, nickel, copper, uranium, manganese, aluminum, and tin, 5 mg. The color is stable for several hours.

[50] A. I. Lazarev and V. I. Lazareva, *Zavod. Lab.* **25**, 783–786 (1959).
[51] D. I. Ryabahikov and A. I. Lazarev, *Renii, Tr. Vses. Sovshch. Probl. Reniya, Akad. Nauk SSSR, Inst. Mat.* **1958**, 267–274.
[52a] K. A. Smirnova, A. M. Lukin, G. S. Petrova, and N. N. Vysokova, *Tr. Vses. Nauchn.-Issled. Inst. Khim. Reakt.* **1969** (31), 7–13.

Procedure. *Less than 1 mg of Zirconium Present.* To 15 ml of sample containing not over 15 μg of cadmium, add 5 ml of 20% solution of sodium hydroxide and 10 ml of 0.01% solution of dithizone in carbon tetrachloride. Shake, separate the extract, and evaporate it to dryness.

Add 1 ml of 1 : 1 nitric acid and again evaporate to dryness. Take up the residue in 15 ml of water. Extract with 10 ml of 0.01% solution of dithizone in carbon tetrachloride. Evaporate to dryness and dissolve the residue in 2.75 ml of water. Add 0.5 ml of 4% solution of citric acid and 0.75 ml of 0.05% solution of sulfarsazen made faintly ammoniacal. Read at 520 nm against a blank.

$\alpha, \beta, \gamma, \delta$-TETRAPHENYLPORPHINE

This reagent is applied to determination of cadmium present as an impurity in zinc.[52b]

Procedure. *Zinc.* As reagent solution, reflux 0.1075 gram of the reagent with 1 liter of glacial acetic acid for 8 hours. Dilute 5 ml to 25 ml with acetic acid for use. Dissolve 0.05 gram of sample in 1 : 1 nitric acid and evaporate to dryness. Ignite at 500–600° for 15 minutes. Dissolve the residue in 5 ml of glacial acetic acid and dilute to 100 ml with acetic acid. To a 10-ml aliquot, add 5 ml of the reagent and dilute to 25 ml with acetic acid. Let stand for 1 hour and read at 551 nm.

THIOSEMICARBAZONE OF BENZOTHIAZOL-2-YL PHENYL KETONE

This reagent gives a visible absorption with copper, cobalt, nickel, cadmium, and zinc.[53] The result is a straight line for cadmium from 0.2–7.5 μg/ml. By reading nickel at pH 2.5 and 446 nm, cadmium at pH 4 and 418 nm, and zinc at pH 6 and 420 nm, the three metals can be determined by simultaneous equations. Silver, palladium, iron, and lead interfere.

Procedure. Mix 2 ml of sample with 0.5 ml of an ammoniacal buffer solution for pH 8. Add 1 ml of 0.1% ethanolic solution of the reagent and dilute to 5 ml with ethanol. Set aside for 30 minutes. Read at 418 nm.

[52b]Charles V. Banks and Ramon E. Bisque, *Anal. Chem.* **29**, 522–525 (1957).
[53]Toyozo Uno and Sumiyuki Akihama, *Jap. Anal.* **10**, 941–945 (1961).

XYLENOL ORANGE

The optimum pH for complexing cadmium with xylenol orange is 2–6.[54] The maximum absorption is at 580 nm. It is possible to determine cadmium, indium, and bismuth in aliquots of the same sample solution. Beer's law is followed at 0.02–0.12 mg per 25 ml. Many cations interfere.

A ternary complex[55] of cadmium in the presence of more than fifteenfold excess of xylenol orange and 750-fold excess of diphenylguanidine is extracted with butanol for reading at 480 nm.

Procedure. As a buffer solution, adjust a 20% hexamine solution to pH 6.3 with 1:1 nitric acid. Mix 10 ml of sample solution with 2.8 ml of m*M* xylenol orange solution. Add 5 ml of 15% solution of potassium nitrate and dilute to 25 ml with the buffer solution. After 15 minutes read at 575 nm.

Cadmium complexes with **acid chrome dark blue** at pH 10 in an ammoniacal buffer solution. As applied to cyanide cadmium plating baths, the cadmium is first separated as the sulfide.[56]

The blue 1:1 complex that **arsenazo III** forms with cadmium around pH 10 is read at 600 nm.[57] Beer's law is followed for 0.24–2.38 μg/ml. Interference by zinc is masked by pyridinium nitrate.

The yellow 1:1 complex of cadmium with **2-(benzothiazol-2-ylazo)-4-6-dichlorophenol** absorbs at 550 nm in 15% acetone buffered to pH 5 with citric acid and disodium phosphate.[58] There is a straight line absorption up to 0.045 mg of cadmium per ml, but the following ions interfere: ferric, mercuric, barium, calcium, cobaltous, cupric, lead, and zinc.

The 1:2 complex of cadmium with **5-benzoxazol-2-yl-1,3-diphenylformazan** is extracted at pH 7–10 with benzene for reading at 610 nm.[59a] Beer's law is followed for 1–18 μg of cadmium per ml. There is interference by an equivalent amount of copper, cobalt, nickel, silver, zinc, lead, and ferrous ion.

At 0.04–0.1 m*M* cadmium complexes with **benzylxanthate** ions at pH 7–8. A 1:2 cadmium-reagent complex is extracted with chloroform and read at 306 nm.[59b] Zinc forms a similar complex at pH 6–7.5.

Cadmium forms a red complex with **2,2′-bibenzothiazolyl.**[60a] Mix a 7-ml sample

[54]Makoto Otomo, *Bull. Chem. Soc. Jap.* **37**, 504–508 (1964); K. N. Bagdasarov, P. N. Kovalenko, and M. A. Shenyakina, *Zh. Anal. Khim.* **23**, 515–520 (1968).
[55]L. G. Anisimova, E. T. Beschetnova, and O. A. Tataev, *Zh. Anal. Khim.* **30**, 63–67 (1975).
[56]G. G. Lomakina and V. N. Tolmacheva, *Zavod. Lab.* **26**, 62 (1960).
[57]V. Mikhailova and L. Yurnkova, *Anal. Chim. Acta* **68**, 73–82 (1974).
[58]V. Armeanu and Elena Dragusin, *Rev. Roum. Chim.* **18**, 1475–1482 (1973).
[59a]L. V. Kholevinskaya and S. L. Mertsalov, *Zh. Anal.*
[59b]Kenjiro Hayashi, Yoshiaki Sasaki and Shinki Furusho, *Jap. Anal.* **24**, 151–155 (1975). *Khim.*, **30**, 265–268 (1975).
[60a]Masakazu Deguchi, *Jap. Anal.* **18**, 159–164 (1969).

containing 0.004–0.12 mg of cadmium with 2 ml of 0.2 *M* potassium chloride–sodium hydroxide buffer solution for pH 13.2. Add 1 ml of freshly prepared 0.1% solution of the reagent in dioxan and shake with 10 ml of 4:1 chloroform-pyridine. Read at 370 or 570 nm. Nickel, cobalt, zinc, manganese, copper, mercury, and silver can be masked with 1 ml of 5% potassium cyanide solution, but 0.4 ml of 10% formaldehyde must be added to avoid masking the cadmium.

Cadmium in 0.8 *N* sodium hydroxide complexes with **1-(6-bromobenzothiazol-2-ylazo)-2-naphthol.**[60b] This is extracted with toluene for reading.

To 20 ml of sample solution containing 0.5–10 μg of cadmium, add 0.7 ml of ethanolic 0.5% **2-(5-bromo-2-pyridylazo)-5-dimethylaminophenol,** and 5 ml of ammonium hydroxide–ammonium chloride buffer solution for pH 9.[60c] Extract the complex with 10 ml of 3-methyl butanol and read at 555 nm against a reagent blank. Beer's law is followed for 0.01–1 ppm of cadmium. Cupric ion, zinc, lead, nickel, and cobalt must be preextracted–for example, by dithizone.

Calcein, which is 3′,6′-dihydroxy-4′,5′-bis[*N,N*′(dicarboxymethyl)aminomethyl]fluoran, is effective for fluorescence of the cadmium complex by activation at 490 nm and reading at 520 nm.[61] Readings must be taken at pH 13.3 within 15 minutes after addition of the reagent. Anion exchange separation in 2 *N* hydrochloric acid removes numerous interfering ions before development.

The iodocomplex of cadmium with **crystal violet** can be collected at the phase interface between isopropyl ether and water at pH 1.1.[62] By draining off the water and adding acetone in the mixed solvent, the precipitate is dissolved and read at 580 nm. Bismuth and stannic ion must be absent.

The complexes of ***O,O*′-dichlorodithizone,** which is 1,5-bis-(2-chlorophenyl)-3-mercaptoformazan, with lead, zinc, cadmium, and mercury, is extractable into carbon tetrachloride for reading.[63]

Cadmium in a cyanide bath for cadmium plating can be precipitated as the sulfide, separated by centrifuging, dissolved in nitric acid, and read by **1,8-dihydroxy-2-(2-hydroxyphenylazo)-3,6-naphthalenedisulfonate.**[64] The aliquot must be equivalent to less than 0.08 mg of cadmium sulfate per ml. At pH 10–11 and tenfold excess of reagent, reading is at 600 nm.

The orange complex of cadmium with **2-(3,5-dimethylpyrazol-1-yl)-8-hydroxyquioline** can be extracted from an alkaline medium with chloroform.[65a] Activated at 425 nm, the reagent does not fluoresce, but the complex shows a maximum at 590 nm. The method will determine 0.05–0.5 ppm of cadmium in zinc salts.

To 1 ml of sample solution containing 1–5 m*M* of cadmium, add 1 ml of 0.005 *M* **EDTA** and 1 ml of 11.4% ammonium hydroxide–1.4% ammonium chloride buffer solution for pH 10. Dilute to 5 ml and read at 225 nm.[65b] Nitrate and acetate

[60b]V. G. Yakovleva, Yu. M. Ivanov, and A. M. Andreichuk, *Zh. Anal. Khim.* **31**, 884–887 (1976).
[60c]Shozo Shibata, Eijiro Kamata, and Ryozo Nakashima, *Anal. Khim. Acta* **82**, 169–174 (1976).
[61]A. J. Hefley and Bruno Jaselskis, *Anal. Chem.* **46**, 2036–2038 (1974).
[62]Jacqueline Courtot-Coupez and Pierre Guerder, *Bull. Soc. Chim. Fr.* **1961**, 1942–1944.
[63]R. S. Ramakrishna and M. Fernandopulle, *Anal. Chim. Acta* **41**, 35–41 (1968).
[64]G. G. Lomakina and V. N. Tolmachev, *Izv. Vyssh. Ucheb. Zaved., Khim. Khim. Tekhnol.* **3**, 819–822 (1960).
[65a]E. A. Bozhevol'nov, L. F. Fedorova, L. A. Krasavin, and V. M. Dziomko, *Zh. Anal. Khim.* **25**, 1722–1726 (1970).
[65b]J. R. Dunstone and E. Payne, *Analyst* **84**, 110–113 (1959).

must be absent. The same technic is applicable for calcium, magnesium, strontium, barium, and zinc. Since the measurement is of the decrease of absorption of EDTA due to complexing, each interferes.

Mix the sample solution containing 11–101 μg of cadmium with a solution of **glycinethymol blue** and add a hexamine buffer solution for pH 8.9.[65c] Dilute to 25 ml and read at 610 nm. The intensity of color is affected by pH, the nature of the buffer solution used, and the concentration of reagent.

The complex of cadmium with **glyoxal bis(2-hydroxyanil)** is extracted at pH 10.6–13.6 by 5:1 chloroform-pyridine for reading at 610 nm.[66a] The absorption of the reagent is negligible, and that with cadmium is stable for 60 minutes. Beer's law holds for 0.001–0.00.04 mg of cadmium in 10 ml of extract. Nickel and cobalt can be masked with potassium cyanide, after which, in the presence of up to 1 mg of cobalt or nickel, the cadmium is demasked with 1 ml of 10% formaldehyde.

To separate cadmium or zinc from iron, adjust the sample solution to pH 8–11 and extract with 15 ml of *M* hexanoic acid in chloroform that is 0.5 *M* with butylamine.[66b] Add to the separated organic phase 5 ml of 20% ethanolic butylamine and 3 ml of 1.5% ethanolic 4-(2-pyridylazo)resorcinol. Dilute to 50 ml with ethanol and read.

Alternatively, add 1.5 ml of *M* tartaric acid and adjust to pH 9–10. Extract with 15 ml of *M* tartaric acid and adjust to pH 9–10. Extract with 15 ml of *M* **hexanoic acid** in chloroform that is 0.5 *M* with butylamine. Back-extract the organic layer with 20 ml of 1:8 sulfuric acid and develop. Beer's law is followed for 1–15 m*M* zinc or 1–20 cadmium. 2-Bromohexanoic acid may also be used.

The **bis(4-hydroxybenzoylhydrazone) of glyoxal** as a chelate is a reagent for determining cadmium in the presence of citrate when other cations are present.[67] Beer's law is followed up to 50 m*M* by absorption or fluorescence.

The condensation product of **2-hydroxy-1-naphthaldehyde** and **4-chlorobenzyl dithiocarbamate** in 50% dimethylformamide, when activated at 470 nm, shows an intense fluorescence at 520 nm.[68] Added cadmium ion complexes with the reagent to reduce the fluorescence, permitting determination of 0.1–1.1 ppm. Cupric ion and cobalt react similarly, and these must be absent. Silver, nickel, and zinc also interfere.

8-Hydroxyquinoline-5-sulfonic acid is a reagent for determining cadmium by fluorescence.[69] Adjust a sample containing 0.01–2 μg of cadmium to pH 7.1–8.5 with ammonium hydroxide–ammonium chloride buffer solution. Add 4 ml of 0.003% solution of the reagent and dilute to 10 ml. The maximum fluorescence of cadmium is then at 520 nm. Zinc interferes because it has the same maximum.

Cadmium forms at 1:2 complex with ***N*-methylanabasine-α′-azo-1-naphthol-5-sulfonic acid.**[70] The optimum pH is 6–10, the color develops immediately, and reading is at 600 nm. Beer's law is followed for 1–25 μg/ml. The corresponding

[65c]M. K. Akhmedli, S. R. Alieva and A. M. Ayubova, *Azerb. Khim. Zh.* **1975**, (4), 106–110.
[66a]Nobuichi Oi, *Jap. Anal.* **9**, 770–773 (1960).
[66b]V. V. Sukhan, I. V. Pyatnitskii, and V. A. Frankovskii, *Udr. Khim. Zh.* **41**, 1308–1312 (1975).
[67]M. Lever, *Anal. Chim. Acta* **65**, 311–318 (1973).
[68]Imre Kasa and Gabor Bajnoczy, *Period. Polytech. Chem. Eng.* **18**, 289–294 (1974).
[69]D. E. Ryan, A. E. Pitts, and R. M. Cassidy, *Anal. Chim. Acta* **34**, 491–494 (1966).
[70]Sh. T. Sharipova, R. Kh. Dzhiyanbaeva, and Sh. T. Talipov, *Tr. Tashk. Gos. Univ.* 39–45.

-2-naphthol-6-sulfonic acid has an optimum pH at 9.2 and maximum absorption at 540 nm.[71] The reagent absorbs at 470 nm.

The complex of cadmium with **4-(2-pyridylazo)resorcinol** at pH 9.8–10.4 exhibits maximum absorption at 495 nm.[72] To a sample containing up to 0.04 mg of cadmium, add 10 ml of *M* sodium tartrate solution and 10 ml of 5 *M* ammoniacal buffer for pH 10. Add 2 ml of 0.5% solution of the reagent in 0.02% sodium hydroxide solution and dilute to 50 ml. At this dilution it conforms to Beer's law. There is interference by manganese, ferric, cobalt, copper, nickel, mercuric, gallium, indium, and lead ions. Stannic and yttrium ions can be masked by 4 ml of 5% sodium fluoride solution. A suitable method of separating cadmium from copper, manganese, cobalt, and nickel is by a strongly basic anion-exchange resin.

3,3′-(Pyrole-2,5-diylazo)di(5-chloro-2-hydroxybenzenesulfonic acid) complexes with cadmium at pH 6.5–7 for reading at 680 nm.[73a] The reagent also gives color with copper, zinc, and yttrium at appropriate wavelengths.

Mercuric, cadmium, and zinc ions form deep red 1 : 2 complexes with **1-(2-quinolylazo)acenaphthylen-2-ol.**[73b] They are extractable with organic solvents such as carbon tetrachloride.

To determine cadmium by its effect on the fluorescence of **8-quinolyl dihydrogen phosphate** or **carboxymethyl-8-hydroxyquinoline,** mix 0.5 ml of sample solution with 0.5 ml of m*M* reagent.[74] Add 1.5 ml of 50 m*M* Tris–hydrochloric acid buffer solution for pH 9.16 and 7.5 ml of methanol. Set aside for 30 minutes. Activate at 316 nm and read the fluorescence at 440 nm. Compare the fluorescence against quinine sulfate with and without added cadmium sample.

Cadmium can be isolated and read as the **reineckate.**[75] Add a solution of ammonium reineckate to the sample solution until an intense red color is present. After 30 minutes decant, and use ether to transfer the precipitate to a sintered-glass crucible. Dissolve the precipitate in acetone, dilute to 100 ml with water, and read.

Salicylaldehyde isonicotinoylhydrazone forms a 2 : 1 complex with cadmium as well as with manganese, nickel, cobalt, and zinc.[76] In 50% 1,4-dioxan as a medium, Beer's law is followed for 0.5–3 ppm at 390–400 nm.

As a reagent solution for cadmium and zinc, dissolve 50 mg of **3-salicylidenedithiocarbazoic acid methyl ester** in 5 ml of 0.1 *N* sodium hydroxide and dilute to 100 ml.[77] Alternatively, dissolve in 100 ml of dimethylformamide. Mix 1 ml of reagent solution with 15 ml of ammonium chloride–ammonium hydroxide buffer solution for pH 10, add the sample solution, and dilute to 20 ml. Yet another alternative is to mix 1 ml of reagent solution with 9 ml of dimethylformamide, add the sample solution, and dilute to 20 ml with the buffer solution. Activate at 335

[71]R. Kh. Dzhiyanbaeva, Sh. T. Talipov, U. Mansurkhozhdaev, and Sh. T. Talipov, *ibid.* **1968** (323), 21–27.

[72]Mitsugu Kitano and Joichi Ueda, *J. Chem. Soc. Jap., Pure Chem. Sect.* **91**, 760–762 (1970).

[73a]S. B. Savvin, Yu. G. Rozovskii, R. F. Propistsova, and E. A. Likhonina, *Izv. Akad. Nauk SSSR, Ser. Khim* **1969**, 1364–1366.

[73b]Ishwar Singh, Y. L. Mehta, B. S. Garg, and R. P. Singh, *Talenta* **23**, 617–618 (1976).

[74]Kinzo Nagasawa and Osamu Ishidaka, *Chem. Pharm. Bull.* (Tokyo) **22**, 375–384 (1974).

[75]G. Teodorescu and Rodica Catuneanu, *Rev. Chim.* (Bucharest) **21**, 12 (1970).

[76]G. S. Vasilikiotis and T. A. Kouimtzis, *Michrochem. J.* **18**, 85–94 (1973).

[77]Imre Kasa and Jeno Korosi, *Period. Polytech. Chem. Eng.* **17**, 241–255 (1973); *Magy. Kem. Foly.* **80**, 151–155 (1974).

nm and read the fluorescence at 465 nm. Beer's law is followed up to 7 μg of cadmium per ml and up to 5 μg of zinc per ml. There is interference by cobalt, nickel, copper, silver, and mercury.

The 1:2 complex of cadmium with **2-(3′-sulfobenzoyl)pyridine-2-pyridylhydrazone,** which is α-2-pyridyl-α-(2-pyridylhydrazono)toluene-*m*-sulfonic acid, is water soluble and has a sensitivity approaching that of dithizone, 1-(2-pyridylazo)-2-naphthol, or 4-(2-pyridylazo)resorcinol.[78] The spectrum of the ligand does not overlap that of the complex.

Cadmium is extracted[79a] from a sample solution at pH 6.5–7.5 by m*M* **thiothenoyltrifluoroacetone,** which is 1,1,1,-trifluoro-4-mercapto-4-(2-thienyl)but-3-en-2-one, in carbon tetrachloride with *M* ammonium nitrate added as a salting-out agent. Beer's law is followed at 0.9–19 μg of cadmium per ml of extract. Silver, iron, cerium, antimony, and zirconium must be absent or masked.

For cadmium as a ternary complex add 1,10-phenanthroline to the aqueous sample solution. Then extract at pH 5.6–10 with a solution of thiothenoyltrifluoroacetone in xylene as a 1 : 2 : 1 cadmium-acetone derivative—1,10-phenanthroline complex.[79b] Wash out the excess of the acetone derivative from the organic layer with sodium hydroxide-sodium tetraborate buffer solution for pH 11.5 and read the stable complex at 370 nm. Beer's law holds for 1–35 μg of cadmium per 10 ml of extract. There is interference by mercuric ion, cupric ion, zinc, and cobalt which can be masked with cyanide ion. There is also interference by manganous and by sulfide ions.

Cadmium and zinc form a chelate with **8-(toluene-*p*-sulfonamido) quinoline,** which on excitation at 265 or 370 nm fluoresces at 500 nm.[80] Beer's law is followed up to 0.1 *M*. Aluminum and copper interfere; a large excess of cyanide or chromate causes quenching.

Cadmium and zinc in a glycine buffer for pH 8.0–8.3 combine with **8-*p*-tolylsulfonylaminoquinoline** for reading by green fluorescence in ultraviolet light.[81] The presence of 0.01 mg of aluminum per ml extinguishes the fluorescence.

For cadmium by tribenzylamine see Mercury.

[78]J. E. Going and C. Sykora, *Anal. Chim. Acta* **70**, 127–132 (1974).
[79a]K. R. Solanke and S. M. Khopkar, *Sep. Sci.* **8**, 511–518 (1973).
[79b]Masakazu Deguchi and Noriyoshi Kiyskawa, *Eisei Kagaku* **22**, 308–311 (1976).
[80]D. T. Haworth and R. H. Boeckeler, *Mikrochem. J.* **13**, 158–164 (1968).
[81]V. M. Dziomko and G. V. Serebryakova, USSR Patent *120,029* (1959).

CHAPTER SEVEN
BISMUTH

The principal photometric determinations of bismuth are as the iodide and as the complex with thiourea. But it forms many binary complexes and some ternary ones of a reagent with the hexaiodobismuthate.

For isolation of bismuth in ferrochrome by column chromatography, see Lead. Bismuth, copper, and lead, separated by paper chromatography, are extracted as the chlorides and read in the ultraviolet range. For details, see Lead.

For analysis for bismuth in silver, dissolve 10 grams of sample in nitric acid.[1a] Add lanthanum nitrate solution and coprecipitate gold, lead, bismuth, iron, and aluminum, along with the lanthanum, by adding ammonium hydroxide. Filter and wash. Dissolve the precipitated hydroxides in 10 ml of 1 : 1 hydrochloric acid and dilute to 50 ml. Determine bismuth in an aliquot by dithizone.

Extraction with molten naphthalene of complexes with 8-hydroxyquinoline is applicable at varying pH levels to bismuth, zinc, magnesium, cadmium, copper, cobalt, nickel, molybdenum, uranium, aluminum, iron, palladium, and indium.[1b]

AMARANTH

A 3 : 1 complex of bismuth is formed with amaranth. This is insoluble and is filtered and treated with disodium phosphate solution to release an equivalent amount of amaranth. The dye is read as an indirect measure of the bismuth.

Procedure. *Alloys.*[2] Warm 100 mg of alloy with 0.1 gram of mercury until an amalgam is formed. Dissolve the cooled amalgam in 0.5 ml of saturated boric acid

[1a]Z. Marczenko and K. Kasiura, *Chem. Anal.* (Warsaw) **9**, 87-95 (1964).
[1b]T. Fujinaga, Masatada Satake, and Masaaki Shimizu, *Jap. Anal.* **25**, 313–318 (1976).
[2]Morris P. Grotheer and Jack L. Lambert, *Anal. Chem.* **30**, 1997–1999 (1958).

and a minimal amount of nitric acid, usually 0.5–1 ml. Keep the amalgam at 15–25° while dissolving. Then add to 500 ml of 2% boric acid solution and dilute to 1 liter. The pH must now be 2.3–3; if necessary, adjust by dropwise addition of nitric acid. Mix a 20-ml aliquot with 1 ml of 0.2% solution of amaranth.

The filtration apparatus consists of two 18/9 mm borosilicate glass socket joints, one fused very close to the bottom of a 25×100 mm test tube and the other having a porous borosilicate glass disk at its mouth. The lips of the joints and the porous disk are ground flat and smooth so that a tight seal is obtained when hardened filter paper is clamped between them.

Prepare a barium sulfate mat by reacting 5.0 ml of 0.02 *M* barium chloride with 5.0 ml of 0.02 *M* sodium sulfate and filtering the fine precipitate through Whatman No. 50 filter paper. Wash the mat several times.

Allow the amaranth-bismuth complex to stand for at least 5 minutes but not more than an hour. Filter through a freshly prepared calcium sulfate mat. Wash the bismuth-amaranth compound retained on the barium sulfate mat two to four times, depending on the amount of the compound, with 5-ml portions of 1% boric acid solution adjusted to pH 3–3.5. Pour 10 ml of 1% disodium hydrogen phosphate solution through the compound on the filter. The filtrate now contains released amaranth dye equivalent to the bismuth in the aliquot. Read this solution at 521 nm, the wavelength of maximum absorption for amaranth. Determine the concentration of bismuth when 10 ml of 1% dihydrogen phosphate solution was used as the eluent by

$$C = 1.45\,VA$$

where

C = mg of bismuth in the alloy sample (mg)

V = volume of the aliquot solution sample

A = absorption of the eluted dye solution at 520 nm

$$\%\,\mathrm{Bi} = 100\,C/\text{sample weight (mg)}$$

ARSENAZO III

Arsenazo III, which is trisodium 3-(2-arsonophenylazo)-4,5-dihydroxy-2,7-naphthalene disulfonate, forms a 1:2 complex with bismuth.[3] The medium should be at pH 0.1–3 with sulfuric acid or perchloric acid or at pH 2–3.5 with hydrochloric acid. It conforms to Beer's law with 0.2–2.5 μg of bismuth per ml. Ferric ion can be masked with ascorbic acid and stannous ion with fluoride. There is interference by substantial amounts of pentavalent arsenic or vanadium, hexavalent tungsten or molybdenum, chromic, chloride, perchlorate, oxalate, or EDTA ions, and by citric and tartaric acids.

[3] V. F. Barkovskii and Z. N. Povet'iva, *Zavod. Lab.* **35**, 555–556 (1969).

Procedure. *Copper Ores and Concentrates.* Dissolve 1 gram of sample in 20 ml of 1 : 1 nitric acid. Add 10 ml of 1 : 1 sulfuric acid and heat until fuming ceases. Take up with 10 ml of water and reevaporate. Take up in 12 ml of 1 : 5 sulfuric acid. Filter and dilute to 100 ml. Mix an aliquot containing 0.01–0.125 mg of bismuth with 5 ml of 4% solution of ascorbic acid, 3 ml of 0.1% solution of arsenazo III, and 1 ml of 1% solution of sodium fluoride. Raise to pH 1 with 10% sodium hydroxide solution, dilute to 50 ml, and read at 610 nm.

2,2′-BIPYRIDYL AND IODIDE

Bismuth complexes with iodide ion and 2,2′-bipyridyl.[4] Traces of cupric, ferrous, ferric, manganous, stannous, and chloride ion interfere.

Procedure. Mix a sample solution containing 0.1–1 mg of bismuth with 8 ml of 0.02 *M* 2,2′-bipyridyl, 0.1 gram of ascorbic acid, and 5 ml of 0.1 *M* potassium iodide containing 4% ascorbic acid. Add 30 ml of acetone and shake until solution is complete. Dilute with water to 50 ml and read at 455 nm.

BIS(THIOANTIPYRINYL)METHANE

This reagent is made by chlorinating diantipyrinylmethane and treating the product with sodium hydrosulfide. It forms a 2 : 1 complex with bismuth in 0.1–1 *M* sulfuric acid. The reagent has no absorption in the visible region.

Procedure. *Ores.*[5] Dissolve 1 gram in 8 ml of nitric acid and 22 ml of hydrochloric acid. Evaporate nearly to dryness, add 10 ml of hydrochloric acid, and evaporate to dryness. Take up in 2 ml of 1 : 1 hydrochloric acid, dilute to about 20 ml, and filter. Add ammonium hydroxide to neutralize and add 2 ml in excess. Filter and wash with 1 : 20 ammonium hydroxide. Dissolve the precipitate in 5 ml of 10 *N* sulfuric acid and dilute to 50 ml. To a 10-ml aliquot add the following: 2 ml of 5% ascorbic acid solution, 0.5 ml of 25% tartaric acid solution, and 2.5 ml of 0.25% solution of the reagent in 1 : 3 acetic acid. Dilute to 25 ml with 1 : 10 acetic acid and read at 535 nm.

CHLORPROMAZINE HYDROCHLORIDE

The complex of this reagent with hexaiodobismuthate is applicable to solutions of alloys and ores.[6] The technic described is applicable to 0.04–0.4 mg of bismuth in the 10-ml aliquot. Silver, mercuric, and ferric ions interfere.

[4]Francisek Buhl and Krystyna Kania, *Chem. Anal.* (Warsaw) **18**, 369–373 (1973).
[5]A. V. Dolgorev, Ya. G. Lysak, and A. P. Lukoyanov, *Zavod. Lab.* **40**, 247–249 (1974).
[6]H. Basinska and M. Tarasiewicz, *Chem. Anal.* (Warsaw) **13**, 1287–1294 (1968).

Procedure. *Wood's Metal.* Dissolve 0.03 gram in 10 ml of 1:1 nitric acid and dilute to 100 ml. Mix a 10-ml aliquot with 2.5 ml of 2% potassium iodide solution. Add 1 ml of 4% solution of the reagent and extract with 10 ml and 10 ml of chloroform. Dilute the combined extracts to 50 ml with chloroform and read at 485 nm.

m–CRESOLPHTHALEXON–S

This reagent, which is 3,3′-bis[bis(carboxymethyl)amino]methyl-*m*-cresolsulfonphthalein, forms a 1:1 complex with bismuth at pH 1–2.[7] There is interference by ferric ion, gallium, zirconium, thorium, orthophosphate, and chloride ion.

Procedure. *Copper and Copper Alloys.* Dissolve 5 grams of sample in 20 ml of 1:1 nitric acid and add 2 ml of 6% solution of ferric sulfate. Boil off oxides of nitrogen and dilute to about 500 ml. Precipitate ferric hydroxide as a collector by adding ammonium hydroxide to pH 8–9. Then dilute to about 800 ml and set in a warm place for 30 minutes. Filter the precipitate. If the sample contained aluminum, wash it out of the precipitate with 30% sodium hydroxide. Next wash with 1:3 ammonium hydroxide, then with water. Dissolve the precipitate in 10 ml of hot 1:1 nitric acid or, if the sample contained manganese, use hot concentrated nitric acid. Add ammonium hydroxide to raise to pH 7. Reduce the iron by adding 70 mg of ascorbic acid. Add 2.5 ml of m*M* solution of the reagent and dilute to 25 ml with acetate-nitrate buffer solution for pH 1. Read at 582 nm.

DIAZOTIZED 4–AMINONAPHTHALENE–1–SULFONIC ACID

This reagent will determine up to 0.007 mg of bismuth in conformity to Beer's law.[8] Osmium, ruthenium, ferric, and antimony ions interfere. The latter two can be masked.

Procedure. *Lead.* Dissolve 1 gram of sample in 10 ml of 1:4 nitric acid and evaporate to dryness. Take up the residue in 10 ml of water, add 1:1 hydrochloric acid as long as precipitation occurs, and filter. Neutralize the filtrate to phenolphthalein with ammonium hydroxide. Add 5 ml of 10% ammonium acetate buffer solution for pH 8.5 and 0.1% ethanolic 8-hydroxyquinoline until precipitation is complete. Filter and wash with 1:20 ammonium hydroxide. Dissolve the

[7] A. V. Grunin, *Zavod. Lab.* **39**, 1070 (1973).

[8] V. Armeanu, Polixenia Costinescu, and Camelia Georgeta Calin, *Chim. Anal.* **2** (3) 182–183 *Rev. Chim.* (Bucharest) (10) (1972).

precipitate from the filter with 3 ml of 1:5 hydrochloric acid and dilute to 50 ml. To a 5-ml aliquot, add 5 ml of 0.25% ethanolic reagent and 3 ml of 1:4 ammonium hydroxide. Read at 500 nm.

DIETHYLDITHIOCARBAMATE

Bismuth diethyldithiocarbamate is extractable from alkaline solution with isoamyl alcohol.[9] As applied to silicated limestones, the interfering ions are complexed with cyanide and EDTA.[10] The color of the complex in chloroform, carbon tetrachloride, ethyl acetate, and amyl acetate is not stable to light but is stable in *n*-pentanol and *n*-butanol.[11]

The reagent is applicable to nonferrous metals and their alloys containing 0.0001–0.5% bismuth by extracting with chloroform from a solution containing EDTA and cyanide.[12] The reading is at 405 nm.

In 50–65% acetone at pH 8.5–11 containing 1% of EDTA, the maximum is at 360 nm.[13] The curve is linear for 1–30 μg of bismuth per ml. Cupric, ferric, mercuric, antimonous, and stannic ions interfere under these conditions.

Platinum-rhodium alloy in nitric acid solution is chromatographed on Catex S.[14] Then platinum and rhodium are eluted with 0.5 *N* nitric acid, bismuth with 0.5 *N* hydrochloric acid, and lead with 1.5 *N* hydrochloric acid. From the eluate, bismuth is extracted as the diethyldithiocarbamate, which for sensitivity is converted to the copper complex.

For analysis of iron and steel, decompose the sample with nitric and perchloric acids.[15] The antimony is oxidized to pentavalency. Extract ferric ion with butyl acetate, which will be accompanied by telluric ion. Then adjust to pH 8.5 ± 0.2, extract bismuth as the diethyldithiocarbamate, and read at 372 nm.

For analysis of ferrotungsten, decompose the sample with nitric, hydrofluoric, and perchloric acids. Filter the precipitated tungstate and dissolve with sodium hydroxide solution to recover adsorbed bismuth. Then coprecipitate the bismuth along with ferric hydroxide. Dissolve this with acid and add to the prior acid filtrate. Proceed as above from "Extract ferric ion with butyl...."

Procedures

Cobalt Solutions. Adjust the sample solution to pH 3–6. Stir in small increments of 0.2 *M* disodium phosphate as long as the turbidity increases, and mix for 15 minutes. Centrifuge to separate the precipitate, which contains the bismuth, and

[9] V. T. Chuiko and N. I. Reva, *Zavod. Lab.* **33**, 1503 (1967).
[10] F. N. Ward and H. E. Crowe, *U.S. Geol. Survey Bull.* **1063-I**, 173–179 (1956).
[11] J. Kinnunen and B. Wennerstrand, *Chemist-Analyst* **45**, 109 (1956).
[12] H. Pohl, *Metall*, **28**, 113–115 (1964).
[13] Hitoshi Yoshida, *Jap. Anal.* **9**, 759–763 (1960).
[14] Vladimir Widtmann, *Hutn. Listy* **25**, 733–735 (1970).
[15] Shizuya Maekawa, Yoshio Yoneyama, and Eiichi Fujimori, *Jap. Anal.* **10**, 345–349 (1961); Zdenek Vecera and Boleslav Bieber, *Hutn. Listy* **16**, 607–609 (1961).

wash by decantation. Dissolve the precipitate in 5 ml of 1:1 hydrochloric acid. Add 1 ml of 10% solution of EDTA and 5 ml of saturated solution of tartaric acid. Add ammonium hydroxide to raise to pH 9–11 and then 1 ml of fresh 0.2% sodium diethyldithiocarbamate solution. Extract the bismuth complex with 3 ml, 3 ml, and 3 ml of amyl alcohol. Wash the combined extracts with water. Reextract the bismuth with 10 ml of 1:1 hydrochloric acid. Add ammonium hydroxide to raise to pH 9–11. Add 1 ml of fresh 0.2% sodium diethyldithiocarbamate solution and again extract the bismuth complex with 3 ml, 3 ml, and 3 ml of isoamyl alcohol. Combine these extracts, dilute to 10 ml, and read at 440 nm against a reagent blank.

Gold.[16] Dissolve 1 gram of sample in 1 ml of nitric acid and 3 ml of 1:1 hydrochloric acid. Prepare a mixture of 7 ml of ammonium hydroxide, 10 ml of 20% potassium cyanide solution, 5 ml of 3% EDTA solution, and 20 ml of water. Add the sample solution and cool. Dilute to 60 ml and add 3 ml of 2% solution of sodium diethyldithiocarbamate. Shake with 10 ml of carbon tetrachloride and read the extract at 366 nm.

2,6-DIMERCAPTO-3,5-DIPHENYLTHIOPYRAN-4-ONE

This reagent complexes with bismuth in 0.5 *N* hydrochloric acid.[17] The extinction is constant at pH 0.5–3; if the amount of bismuth is substantial, however, a protective colloid is required to prevent precipitation. The complexes formed are bismuth-to-reagent 1:3 and 2:3. To avoid interference by antimony, arsenic, silver, copper, lead, cadmium, palladium, mercury, tin, and ferric ion, cyanide and EDTA may be added to the sample solution from which bismuth is to be extracted as the diethyldithiocarbamate. If copper is less than 200 times the bismuth and lead, mercury, and silver are absent, the cyanide is not needed. The method is applicable to tin, ferrochrome, chromium, iron, and copper ores, and to steel.

Procedure. Dissolve 2 grams of sample in 20 ml of aqua regia. Evaporate to about 7 ml and neutralize to phenolphthalein with ammonium hydroxide. Add 20 ml excess of ammonium hydroxide and 3 ml of 1% solution of sodium diethyldithiocarbamate. Extract with 15 ml, 10 ml, and 10 ml of ethyl acetate. Wash the combined extracts with water and reextract the bismuth with 10 ml of 1:1 hydrochloric acid. Add 15 ml of 27.2% solution of sodium acetate trihydrate and as a protective colloid 2 ml of 0.5% solution of OP-10. Add 2 ml of 0.001 *M* reagent, dilute to 50 ml, and read at 360 nm.

---

[16] Masuo Miyamoto, *Jap. Anal.* **10**, 317–320 (1961).
[17] Yu. I. Usatenko, A. M. Arishkevich, and A. G. Akhmetshin, *Zavod. Lab.* **31**, 788–790 (1965).

DITHIZONE

Interference of other elements in determination of bismuth by dithizone, which is diphenylthiocarbazone, is avoided by first separating bismuth and lead as diethyldithiocarbamates and subsequent separation of the dithizonates.

In the presence of much lead at pH above 10, the use of magnesium EDTA is desirable to complex the lead.[18] For determination in cobalt, it is recommended that the bismuth in the sample in *N* hydrochloric acid be separated by electrolysis.[19]

Bismuth in gold is determined as the dithizonate after isolation of iron and gold.[20] For separation of bismuth from lead and zinc in refractory alloys and ferroalloys for determination by dithizone, see Lead. For determination of bismuth and lead in silver alloys by dithizone, see Lead.

Procedures

Separation by Iodide Extraction.[21] This technic is applicable to less than 5 μg of bismuth in the presence of much ferric, calcium, magnesium, and phosphate ion. Prepare a buffer solution containing 20 grams of diammonium citrate, 10 grams of sodium sulfite, 30 grams of potassium cyanide, 800 ml of water, and sufficient ammonium hydroxide to adjust to pH 9. Extract with 25 ml of 0.02% solution of dithizone in carbon tetrachloride followed by 25 ml of carbon tetrachloride. Then add ammonium hydroxide to raise to pH 10.5 and dilute to 1 liter. As the dithizone reagent, just before use dilute 5 ml of 0.01% solution to 100 ml with the buffer solution, thus producing a 0.0005% solution.

Dilute the sample to about 18 ml, adding perchloric acid to make the solution 2 *N* at 20 ml. Add 0.8 gram of sodium sulfite, 0.4 gram of ascorbic acid, and 1 ml of 0.14 *M* sodium iodide. Let stand until any free iodine is reduced to iodide and shake for 3 minutes each with 5 ml and 5 ml of isoamyl acetate. Shake the combined extracts with 10 ml of 2 *N* perchloric acid containing 0.4 gram of sodium sulfite, 0.2 gram of ascorbic acid, and 0.5 ml of 0.14 *M* sodium iodide. Discard this extract. Shake the isoamyl acetate phase with 15 ml of 0.0005% solution of dithizone for 1 minute. Filter the organic phase through glass wool and read at 480 nm against isoamyl acetate.

Water.[22] Filter 10 liters of sample through an acid-washed Whatman glass-fiber filter. If the sample is seawater, add 90 ml of hydrochloric acid; if it is nonsaline, add 550 ml. Wash De-Acidite FF resin, 100-200 mesh, 7–9% crosslinked, with 250 ml of 1 : 15 nitric acid, and with 50 ml of 1 : 25 hydrochloric acid. Prepare a 75×6 mm column. Pass the sample through the column at 120-170 ml/hour and discard

[18] Lajos Barcza, *Acta Chim. Acad. Sci. Hung.* **28**, 143–149 (1961).
[19] A. Lagrou and F. Verbeek, *J. Electroanal. Chem.* **10**, 68–75 (1965).
[20] G. Ackermann and J. Köthe, *Z. Anal. Chem.* **231**, 252–261 (1967); Zygmunt Marczenko, Krzysztof Kasiura, and Maria Krasiejko, *Chem. Anal.* (Warsaw) **14**, 1277–1287 (1969).
[21] H. A. Mottola and E. B. Sandell, *Anal. Chim. Acta* **25**, 520–524 (1961).
[22] J. E. Portmann and J. P. Riley, *ibid.* **34**, 201–210 (1966).

the effluent. Elute the bismuth with 285 ml of 1 : 15 nitric acid, discarding the first 35 ml. To the eluate, add 2 ml of 10% potassium cyanide solution, 2 ml of 10% solution of sodium acetate, and 20 ml of ammonium hydroxide. Extract the bismuth with 10 ml, 10 ml, and 10 ml of 0.006% solution of dithizone in chloroform. Wash the combined extracts with 5 ml and 5 ml of 5% solution of potassium acid phthalate. Shake each of these phthalate washings with 1 ml of 0.006% solution of dithizone in chloroform, and add these to the previous dithizone extracts.

Extract the bismuth from the dithizonates in chloroform with 5 ml, 5 ml, and 5 ml of 1 : 50 nitric acid. Add 0.5 ml of 10% solution of sodium citrate to these combined acid extracts and dilute to 30 ml with 1 : 50 nitric acid. Mix a 25-ml aliquot with 0.5 ml of 10% solution of potassium cyanide and 3.2 ml of 1 : 7 ammonium hydroxide. Extract the bismuth with 3 ml of 0.003% solution of dithizone in chloroform. Wash this extract with 5 ml of a 1% solution of potassium cyanide in 1 : 150 ammonium hydroxide and read at 495 nm. Determine a blank on water from which the bismuth has been removed by passage through the resin column.

Lead.[23] Dissolve a 5-gram sample in 40 ml of 1 : 4 nitric acid and evaporate to about 25 ml. Add ammonium hydroxide to pH 1–1.2. Add 1 ml of ferric nitrate solution containing 0.02 mg of iron. Add 2 ml of 1% solution of cupferron and extract with 5 ml of chloroform. Again add 1 ml of the ferric nitrate solution and cupferron solution, and extract with 5 ml of chloroform. Wash the combined chloroform extracts with 20 ml, 20 ml, and 20 ml of water.

Extract the bismuth from the washed chloroform extracts with 10 ml and 10 ml of 1 : 17 sulfuric acid. Repeat the separation above from "Add 1 ml of ferric..." to "...10 ml of 1 : 17 sulfuric acid." Wash the 20-ml combined extracts of bismuth in sulfuric acid with 3 ml and 3 ml of chloroform; discard the washings. To the sulfuric acid solution, add 5 ml of 10% solution of ammonium citrate and 1 ml of 20% solution of hydroxylamine hydrochloride. Add ammonium hydroxide to raise to pH 9.2; then add 2 ml of 5% potassium cyanide solution. Extract the bismuth with 10 ml of 0.002% solution of dithizone in chloroform and read at 495 nm.

Steel and Castings.[24] Dissolve a 2-gram sample in 10 ml of hydrochloric acid, adding small amounts of nitric acid if necessary to complete disintegration. Add 5 ml of sulfuric acid and heat to sulfur trioxide fumes. Dilute to 25 ml with water and filter off silica and, in some cases, part of the lead. Add 7 ml of sulfuric acid and dilute to 50 ml, at which point the solution should contain about 28% acid.

Take an aliquot expected to contain 0.5–10 μg of bismuth and add 10% ascorbic acid solution until the ferric ion has been reduced. Add 5 ml of 25% solution of sodium potassium tartrate, 20 ml of 20% solution of EDTA, and ammonium hydroxide to raise to pH 7.5–8. Add 30 ml of 20% potassium cyanide solution and dilute to 100 ml at pH 11–12, adding ammonium hydroxide as necessary. Add 10

[23]Yoshihiro Ishihara, Kiyoshi Shibata, Hazime Kishi, and Toru Hori, *Jap. Anal.* **11**, 91–95 (1962); cf. H. Jedrezejewska and M. Malusecka, *Chem. Anal.* (Warsaw) **12**, 574–584 (1967).

[24]D. C. Filipov and I. R. Nachev, *C. R. Acad. Bulg. Sci.* **20**, 109–112 (1967); cf. R. F. Statham, *Br. Steel Corp. Open Rep.* MG/CC/567/72, 4 pp.

ml of 10% solution of sodium diethyldithiocarbamate and extract the lead and bismuth complexes with 12 ml of carbon tetrachloride. Extract the organic phase with 5 ml of cuprammonium solution containing 5 drops of a 20% solution of sodium potassium tartrate. To the cuprammonium solution containing the lead and bismuth, add 3 ml of 0.5% solution of potassium cyanide followed by 1 : 1 hydrochloric acid to reduce to about pH 8.75. Extract the lead and bismuth with 10 ml of 0.18 m*M* dithizone in carbon tetrachloride. Extract lead from the dithizonates with 60 ml of m*M* hydrochloric acid. Then extract the bismuth from its dithizonate in carbon tetrachloride with 10 ml of 1 : 15 nitric acid. To this extract, add 4 ml of 20% sodium sulfite solution and sufficient ammonium hydroxide to raise to pH 7.5–8.5. Extract the bismuth as dithizonate with 7 ml of 0.042 *M* dithizone in carbon tetrachloride. Wash this extract with 50 ml of m*M* hydrochloric acid to separate any residual traces of lead dithizonate. Read the bismuth at 495 nm. The technic is also applicable to zinc, copper, and aluminum.

Tellurium.[25] Dissolve 1 gram of sample in 7 ml of hydrochloric acid and 3 ml of nitric acid. Evaporate nearly to dryness and take up with 10 ml of water. Add ammonium hydroxide to about pH 10 and add 10 ml of an ammoniacal buffer solution for pH 10. Add 1 ml of 1% potassium cyanide solution and 1 ml of 0.0025 *M* magnesium disodium EDTA to mask lead. Extract with 10 ml of 0.005% solution of dithizone in carbon tetrachloride and read at 495 nm.

FLAME PHOTOMETRY

For more details of the basic technic, see Lead.

Procedures

Steel.[26a] As ammonium citrate solution, disperse 700 grams of citric acid in a small volume of water, add 800 ml of ammonium hydroxide, and dilute to 1500 ml.

For 2–3 ppm of bismuth, dissolve a 1-gram sample in 10 ml of 1 : 1 hydrochloric acid, adding a minimal amount of nitric acid if necessary. Dilute somewhat, filter, and dilute to 50 ml. Add 50 ml of ammonium citrate solution and 50 ml of 10% solution of hydroxylamine hydrochloride. Add 1 : 1 ammonium hydroxide until the pH is 6.8–7.2. Extract with 10 ml, 10 ml, and 10 ml of 0.1% solution of dithizone in chloroform. Combine the chloroform extracts and extract the bismuth with 10 ml of 1 : 10 hydrochloric acid. Use this solution with an acetylene-air flame to read bismuth at 472.26 nm.

Bismuth-Antimony Alloys.[26b] Dissolve 0.4 gram of sample which may contain up to 30% of antimony in 5 ml of aqua regia. Dilute to 2 ml with 1 : 1 hydrochloric

[25]Teru Yuasa, *Jap. Anal.* **12**, 507–510 (1963).
[26a]Ulrich Bohnstedt, *DEW Tech. Ber.* **11**, 101–105 (1971).
[26b]Guenter Henrion, Dieter Marquardt and Hans-Otto Raguse, *Z. Chem.* **16**, 477–482 (1976).

acid. Dilute a 20-ml aliquot to 100 ml with propanol and read bismuth by flame photometry at 233.06 nm.

Using the bismuth line as an internal standard read antimony at 231.15 nm. For samples containing more than 30% of antimony add tin to the sample solution and read it at 235.48 nm as the internal standard.

IODIDE

Bismuth is conveniently separated from the lead, tin, and antimony of solders by column chromatography.[27] Then bismuth at 1–20 μg is extracted as the tetraiodobismuthate in isopentyl alcohol. The solution for extraction is customarily stabilized by one of several reducing agents: thiourea, although it is also a reagent for determination of bismuth, is frequently used for this purpose. The aqueous solution can also be read.

Bismuth is recovered from a solution of antimonial lead by electrolysis in 16% citric acid solution at pH 1–1.5.[28] Some antimony is codeposited. The solution of the deposit in 1 : 5 nitric acid is developed with potassium iodide and thiourea.

In 1 : 1 acetic acid–acetone as solvent, the maxima of the tetraiodobismuthate are at 345 and 490 nm with a minimum at 450 nm.[29]

For separation of bismuth in 0.1–6 *N* hydrochloric acid, complex copper with thiourea.[30] Then extract with 0.01 *N* *O*,*O*-bis-(2-ethylhexyl)phosphorodithioate in octane. Back-extract with 6 *N* nitric acid and develop with potassium iodide. Antimony threefold to the bismuth interferes.

For analysis of lead, the cupferron compound of bismuth is extracted with chloroform, reextracted with 1 : 18 sulfuric acid, and treated with sulfite as a reducing agent.[31] The tetraiodobismuthate formed on addition of potassium iodide is read at 460 nm. The technic is also applicable to aluminum alloys.[32]

In determination of bismuth in titanium dioxide, the color of the complex iodide is stabilized by the presence of hydrazine sulfate.[33] The reagent is also applied to analysis in rhenium.[34]

For analysis of heat-resisting alloys based on iron or nickel and containing about 2% titanium, dissolve 1 gram of sample in nitric acid, add hydrochloric acid, and evaporate to dryness.[35] Dissolve the residue in 2 *N* hydrochloric acid and apply to a column of AN-31 ion exchange resin prewashed with 2 *N* hydrochloric acid. Bismuth, zinc, and lead are adsorbed. Then elute zinc with 0.65 *N* hydrochloric

[27]Saichiro Onuki, *Jap. Anal.* **12**, 844–848 (1963).

[28]A. S. Andreev and N. P. Korets, *Zavod. Lab.* **22**, 538–540 (1956).

[29]E. Maggiorelli, *Farm.* (Pavia) *Ed. Prat.* **15**, 384–390 (1960).

[30]Yu. M. Yukhin, I. S. Levin, and N. M. Meshkova, *Z. Anal. Khim.* **30**, 99–102 (1975).

[31]J. A. Corbett, *Metallurgia* **65**, 43–47 (1962); cf. F. I. Makaryants, T. V. Zaglodina, and E. D. Shuvaloya, *Anal. Rud Tsvet. Metal. Prod. Pererab.* **1956**, (12), 130–137.

[32]Giorgio De Angelis and Maria Gerardi, *Ric. Sci. Rend. Sez. A* [2], **1** (1), 67–75 (1961).

[33]A. A. Tumanov and A. N. Sidorenko, *T. Khim. i Khim. Tekhnol.* (Cor'kii) **1962**, 378–383.

[34]D. I. Ryabchikov and A. I. Lazarev, *Renii, Tr. Vses. Soveshch. Probl. Rencya, Akad. Nauk SSSR, Inst. Met.* 1958, 267–274.

[35]V. I. Kurbatova, V. V. Stepin, V. I. Ponosov, E. V. Novikova, G. N. Emasheva, and L. Ya. Kalashnikova, *Zavod. Lab.* **37**, 413–414 (1971).

acid followed by lead with 0.02 *N* hydrochloric acid. Finally, elute bismuth with *N* nitric acid for determination as the iodo complex.

For determining bismuth in stainless steel and ferromolybdenum in the ppm range, a sulfuric acid solution is used.[36] Bismuth is collected with copper sulfide as a carrier. Silica in the ash of this precipitate is volatilized with hydrogen fluoride. The bismuth and lead in a solution of the ash are separated as dithizonates and ashed; the bismuth read as the tetraiodobismuthate. For this purpose, bismuth is precipitated from a solution of ferrotungsten by thioacetamide.[37]

Although the absorption is usually read around 450 nm, there is much greater sensitivity in the ultraviolet region, with a maximum around 330 nm.[38]

For analysis of tantalum, dissolve the sample with nitric and hydrofluoric acids.[39] Extract bismuth, zirconium, and zinc from the acid medium containing fluoride with cyclohexanone. Then determine bismuth as the tetraiodobismuthate.

Procedures

Solder Alloy. Dissolve a 0.1-gram sample in 2.5 ml of 1 : 1 nitric acid, 7.5 ml of 1 : 1 hydrochloric acid, and a few drops of bromine water. Evaporate to dryness and take up in 20 ml of 8 *N* hydrochloric acid containing a few drops of bromine water. Pass the solution through a 110×11 mm column of Diaion SA-100 in chloride form at 1 ml/minute. Wash the column with 40 ml of 8 *N* hydrochloric acid, to ensure that all lead has passed through, and with 50 ml of 2 *N* nitric acid to elute tin. Then elute the bismuth with 200 ml of 0.5 *N* nitric acid. Concentrate the eluate to about 15 ml and add 25 ml of 1 : 5 sulfuric acid. Add 2 ml of 10% potassium iodide solution and 10 ml of 25% solution of sodium hypophosphite. Extract with 10 ml of isopentyl alcohol, dry the extract with sodium sulfate, and read at 460 nm.

Iron and Steel.[40] Dissolve a 1-gram sample containing 0.0048–0.012% bismuth in 20 ml of 1 : 1 nitric acid and 15 ml of perchloric acid. Evaporate to white fumes, cool, dilute with hot water, and filter. Dilute to about 150 ml. Add 5 ml of 5% manganese sulfate solution and 3 ml of 3% potassium permanganate solution. Boil to precipitate manganese dioxide as a collector for bismuth. Filter. Dissolve the precipitate with 25 ml of 1 : 2 sulfuric acid and 1 ml of 30% hydrogen peroxide. Boil to decompose the excess peroxide, cool, and dilute to 50 ml.

To a 25-ml aliquot, add 2 ml of 10% potassium iodide solution, 10 ml of 25% sodium hypophosphite, and 10 ml of isoamyl alcohol. Shake for 1 minute and read the organic layer at 450 nm against a reagent blank.

An alternative is to pass the chloride solution of the sample through Dowex 2-X8

[36] C. G. Carlström and V. Pälvärinne, *Jerukontor. Ann.* **146**, 453–461 (1962).
[37] V. I. Kurbatova, L. Ya. Kalashnikova, G. N. Emasheva, I. N. Nikulina, and V. V. Feofanova, *Tr. Vses. Nauchn.-Issled. Inst. Stand. Obraztsov. Spektr. Etalonov,* **5**, 111–118 (1969).
[38] Hidehiro Goto and Setsuko Suzuki, *Sci. Rep. Res. Inst., Tohoku Univ.* **6**, 130 –136 (1954).
[39] V. A. Nazarenko, E. A. Biryuk, M. B. Shustova, G. G. ShiFareva, S. Ya. Vinkovetskaya, and G. V. Flyantikova, *Zavod. Lab.* **32**, 267–269 (1966).
[40] Tomio Chiba, *Jap. Anal.* **10**, 980–984 (1961).

and elute with thiourea in dilute sulfuric acid, after which the bismuth is isolated as the sulfide, dissolved, and read as the iodide.[41]

Cast Iron

ANTIMONY ABSENT.[42] Dissolve 1 gram of sample in 20 ml of 1:4 sulfuric acid. Add 2 ml of nitric acid, heat to oxidize the carbides, and boil off the oxides of nitrogen. Cool, filter, and dilute to 50 ml.

To a 20-ml aliquot add 1:1 ammonium hydroxide until a precipitate forms. Then add sufficient 1:1 sulfuric acid to redissolve the precipitate and 2 ml in addition. Add 10 ml of 20% solution of sodium citrate, 0.5 gram of ascorbic acid, 1 ml of 10% solution of thiourea, and 5 ml of 40% potassium iodide solution. Extract the bismuth as the iodide with 10 ml and 5 ml of pentyl acetate. Add 10 ml of 20% sodium citrate solution to the combined extracts and add 4% sodium hydroxide solution until just alkaline to phenolphthalein. Shake to reextract the bismuth. To the aqueous layer add 1 ml of 1:1 sulfuric acid, 0.1 gram of ascorbic acid, 0.5 ml of 10% thiourea solution, and 2 ml of 40% potassium iodide solution. Dilute to 25 ml and read against water at 435 nm. Separation of the bulk of the iron by extraction is a desirable step.[43]

ANTIMONY PRESENT.[44a] Dissolve 2 grams of sample in 30 ml of 2:1 hydrochloric acid, add 5 ml of nitric acid, and evaporate to dryness. Take up the residue in 20 ml of hydrochloric acid and 8.5 ml of water. Filter hot and wash the residue with 1:1 hydrochloric acid. Evaporate to dryness and take up the residue in 20 ml of 1:1 hydrochloric acid. Extract the iron and antimony by shaking with 80 ml and 30 ml of isobutyl acetate. Evaporate the aqueous solution to dryness and take up in 10 ml of 1:1 hydrochloric acid. To reduce the remaining iron to ferrous ion, add 20 ml of 10% stannous chloride solution. Add 20 ml of 20% potassium iodide solution and dilute to 100 ml with 1:1 hydrochloric acid. After 10 minutes read at 450 nm against a blank.

COPPER.[44b] Dissolve 25 grams of sample in 200 ml of 1:1 nitric acid by adding the acid in portions while heating. Dilute to 350 ml and add 10 ml of a 0.1% solution of iron in 1:1 nitric acid. Heat to 80°, neutralize with ammonium hydroxide, and add several ml after the blue amine is formed. Filter, washing with 1:19 ammonium hydroxide, then with water. Dissolve the precipitate by pouring 25 ml of hot 1:1 hydrochloric acid through the filter. Wash the paper with hot water. Add 2 ml of sulfuric acid to the filtrate and evaporate to 10 ml. Add 20 ml of hydrobromic acid and volatilize antimony and tin by evaporating to sulfur trioxide fumes. Let cool and again heat to sulfur trioxide fumes. Take up with 5 ml of water. If a precipitate of silica or lead sulfate appears dilute to 20 ml, evaporate to 10 ml, filter, wash, and evaporate to 5 ml. Cool to 15°. Add 10 ml of a solution containing 12.5% of potassium iodide and 5% of monosodium phosphate mono-

[41]N. Leontovitch, *Chim. Anal.* (Paris) **47**, 458–467 (1965).
[42]V. I. Lazareva and A. I. Lazarev, *Zavod. Lab.* **31**, 1437–1438 (1965).
[43]O. G. Koch, *Z. Anal. Chem.* **255**, 269–270 (1971).
[44a]J. Siekierska and A. Piotrowski, *Chem. Anal.* (Warsaw) **11**, 545–548 (1966).
[44b]George Norwitz and Michael Galan, *Anal. Chim. Acta* **83**, 289–295 (1976).

hydrate. Dilute to 25 ml and set aside for 10 minutes. Read at 463 nm within 1 hour.

BRASS. Dissolve 10 grams with 100 ml of 1:1 nitric acid and proceed as for copper.

Nonferrous Metals.[45] Dissolve 1 gram of sample in 5 ml of hydrochloric acid and 3 ml of nitric acid. Evaporate to dryness and add 3 ml of hydrochloric acid. Repeat that step three times and take up the residue in hot water. Dilute to about 50 ml and filter while hot. Wash the residue with hot 1:100 hydrochloric acid. Neutralize the filtrate with ammonium hydroxide until ferric hydroxide begins to form. Add 1 gram of sodium nitrite and further ammonium hydroxide until the odor of excess is perceptible. Filter and wash with 1% ammonium chloride solution. Dissolve the precipitate from the paper with 5 ml of 1:10 hydrochloric acid, reprecipitate with excess of ammonium hydroxide, and filter on the same paper. To the combined filtrates, add 20 ml of 10% sodium tartrate solution and neutralize with 8% sodium hydroxide solution. Add 2 ml excess of the sodium hydroxide solution, then 5 ml of 10% potassium cyanide solution, and 5 ml of 1% solution of sodium diethyldithiocarbamate. Extract with 10 ml and 10 ml of chloroform and evaporate the combined extracts to dryness. Destroy the residual organic matter by adding 2 ml of hydrochloric acid and 1 ml of nitric acid and evaporating to dryness. It may be necessary to repeat this step. Take up the residue in 10 ml of 1:10 hydrochloric acid. Add 0.5 ml of 5% ascorbic acid solution, 0.5 ml of 10% thiourea solution, and 2 ml of 40% potassium iodide solution. Dilute to 25 ml and read at 435 nm.

Metal Sludges.[46] The typical sample is of a copper or nickel sludge from electropurification. Dissolve 2 grams of air-dried sample in 6 ml of hydrochloric acid and 3 ml of nitric acid. Evaporate to dryness, add 5 ml of hydrochloric acid, and again evaporate to dryness. Again add 5 ml of hydrochloric acid and take to dryness. Take up the residue with 50 ml of 1:4 hydrochloric acid, filter, and wash on the filter with 1:50 hydrochloric acid. Add ammonium hydroxide until precipitation of ferric hydroxide begins and add 1 gram of sodium nitrite. Continue to add ammonium hydroxide until the odor of ammonia persists. Filter. The filtrate contains the platinum metals as nitrite complexes together with most of the copper and nickel. The precipitate contains the bismuth, the rest of the copper and nickel, along with iron, selenium, and tellurium. Wash the filter with 1% solution of ammonium chloride in 1:50 ammonium hydroxide.

Dissolve the precipitate in 5 ml of hydrochloric acid, dilute to about 25 ml, and reprecipitate with excess ammonium hydroxide. Filter on the same paper as before and wash. Dissolve the precipitate in 5 ml of hydrochloric acid and add 20 ml of 10% sodium tartrate solution to complex the iron. Add phenol red indicator and neutralize with 8% sodium hydroxide solution, adding 2 ml excess. Add 5 ml of 10% potassium cyanide solution to complex the copper and nickel, as well as any

[45]N. P. Strel'nikova and G. G. Lystsova, *Zavod. Lab.* **28**, 659 (1962); cf. M. Bednara, *Z. Erzberg. Metallhüttenw.* **5** 149–152 (1952).
[46]N. P. Strel'nikova and G. G. Lystsova, *Zavod. Lab.* **28**, 659 (1962).

residual platinum metals. Add 5 ml of 1% solution of sodium diethyldithiocarbamate and extract the bismuth with 10 ml and 10 ml of chloroform. Evaporate the combined extracts to dryness. Add 1 ml and 1 ml of nitric acid and evaporate to dryness after each addition. Add 1 ml and 1 ml of hydrochloric acid and evaporate to dryness after each addition. Take up the residue in 10 ml of 1 : 10 hydrochloric acid. Add 0.5 ml of 5% solution of ascorbic acid, 0.5 ml of 10% thiourea solution, and 5 ml of 20% potassium iodide solution. Dilute to 25 ml and read at 435 nm.

Ferroniobium.[47] Add 0.5 gram of sample to 2 ml of water in platinum; then add 5 ml of hydrofluoric acid. Add nitric acid dropwise until the sample is decomposed. Add 15 ml of sulfuric acid and evaporate to sulfur trioxide fumes. Transfer to a beaker and add 10 ml of 3.5% solution of sodium fluoride to mask niobium and tin. Add 10 ml of 10% potassium iodide followed by 10 ml of 10% thiourea solution to mask iron and copper. Dilute to 100 ml, filter, and read at 450 nm against water.

Niobium.[48] Dissolve 1 gram of sample in platinum by heating with 5 ml of hydrofluoric acid and 5 ml of nitric acid. Evaporate to a syrup and dilute with water to 150 ml. Add 10 ml of 1% solution of potash alum and make distinctly alkaline with ammonium hydroxide. Filter the hydroxides and wash well with 1 : 20 ammonium hydroxide. Dissolve the precipitate in 10 ml of 1 : 1 nitric acid and add 2 ml of sulfuric acid. Evaporate to sulfur trioxide fumes. Dilute to about 40 ml. Add 2.5 ml of 10% potassium iodide solution and 1 ml of 5% thiourea solution. Dilute to 50 ml and read at 450 nm.

Bismuth and Antimony in Iron Alloys. As a part of the procedure, the bismuth is developed with thiourea and the antimony by brilliant green (basic green 1).

Dissolve a 0.5-gram sample in 5 ml of 1 : 1 hydrochloric acid and 5 ml of 1 : 1 sulfuric acid and evaporate to sulfur trioxide fumes. Take up with 5 ml of water and reduce the iron and titanium with 2 ml of 10% solution of hydroxylamine sulfate. An alternative reducing agent is 3 ml of a solution containing 8% of thiourea and 1% of cupric sulfate. In either case boil the solution, cool, add 1.5 ml of 1 : 1 sulfuric acid, and dilute to 10 ml. Extract the bismuth and antimony with 10 ml of *N* 2-ethylhexyl pyrophosphate in hexane.

Bismuth. Reextract the bismuth from the organic phase with 5 ml, 5 ml, and 5 ml of 2% potassium iodide solution acidified with 3 drops of 1 : 1 sulfuric acid. Add 4 ml of 1 : 1 sulfuric acid and 2 ml of 5% solution of thiourea to the combined extracts. Dilute to 25 ml and read at 440 nm.

Antimony. Extract the antimony from the hexane phase with 10 ml of 1 : 10 hydrochloric acid. To oxidize, add 1 ml of 10% sodium nitrite solution and stir for 3 minutes. Add 1 ml of saturated solution of urea to destroy residual nitrite. Add

[47] N. I. Shishkina and Ya. G. Sakharova, *Tr. Ural' Nauchn.-Issled. Inst. Chern. Met.* **9**, 51–54 (1970).
[48] P. Ya. Yakovlev, G. P. Razumova, R. D. Malinova, and M. S. Dymova, *Zh. Anal. Khim.* **17** (1), 90–93 (1962).

0.5 ml of 0.2% solution of brilliant green and extract the antimony complex with 10 ml of benzene. Let the extract stand for 30 minutes and read at 640 nm.

Molybdenum, Copper, or Nickel Alloys. Dissolve 0.5 gram of sample in 5 ml of hydrochloric acid and 2 ml of nitric acid. Boil off oxides of nitrogen, add 4 ml of 1:1 sulfuric acid, and evaporate to sulfur trioxide fumes. Proceed as for bismuth and antimony in iron alloys from "Take up with 5 ml of water and...."

Tinning Baths.[49] The sample with bismuth as a minor ingredient and tin a major one also contains glue, phenol, and emulsifiers.

As 2-ethylhexylpyrophosphate, add 145 grams of phosphorus pentoxide in small increments to 300 ml of 2-ethylhexanol, controlling the temperature to below 80°. After 1 hour decant from unconsumed phosphorus pentoxide and dilute with heptane to approximate a normal solution.

Mix 2 ml of sample with 4 ml of 1:9 sulfuric acid and 0.5 ml of 5.5% solution of ammonium fluoride. Extract with 5 ml of the *N* ethylhexylpyrophosphate in heptane. Reextract bismuth from the organic layer with 10 ml and 10 ml of 2% solution of potassium iodide acidified with 0.3 ml of 1:4 sulfuric acid. To the combined extracts, add 4 ml of sulfuric acid, 4 ml of 25% tartaric acid solution, and 4 ml of 5% solution of thiourea. Dilute to 50 ml and read at 450 nm.

Tin-Bismuth Coatings. Weigh appropriate amounts of metal before and after coating to give 0.5 gram of coating. Dissolve in 10 ml of perchloric acid and 3 ml of nitric acid. Boil off the oxides of nitrogen and cool. Extract with 10 ml of the ethylhexylpyrophosphate reagent. Wash the organic extract with 5 ml and 5 ml of 1:11 sulfuric acid. Complete as for tinning baths from "Reextract bismuth...."

Arsenious Oxide and Arsenic.[50] Distill the arsenic from 0.5 gram of arsenious oxide or 1.5 gram of arsenic as arsenic trichloride in the presence of hydroxylamine. Take up the residue in 5 ml of nitric acid and dilute to 50 ml. Add 3 ml of 0.05 *M* thionalide in chloroform and extract with 5 ml and 5 ml of chloroform. Extract bismuth from the combined extracts with 10 ml of 1:5 hydrochloric acid. Add 5 ml of sulfuric acid, evaporate to sulfur trioxide fumes, and take up in 10 ml of water. Dilute to 25 ml, and add 2 ml of 10% potassium iodide solution and 10 ml of 25% solution of sodium hypophosphite. Extract with 10 ml of isoamyl alcohol and read at 460 nm.

Urine.[51] Digest 50 ml with 2 ml of sulfuric acid and 5 ml of nitric acid until colorless, adding more nitric acid if necessary. Concentrate to 5 ml and dilute to 10 ml. Add 0.6 ml of 1% solution of sodium bisulfite and 0.5 ml of 20% potassium iodide solution. Set aside for 15 minutes and read at 450 nm.

---

[49] A. A. Shatalova, I. S. Levin, I. I. Shmargolina, Z. S. Komashko, and M. V. Solov'eva, *Zavod. Lab.* **32**, 1320–1321 (1966); cf. G. N. Poryvaeva, *ibid.* **36**, 1051 (1970).
[50] L. B. Kristaleva and P. V. Kristalev, *Sb. Nauch. Tr. Perm. Politekhn. Inst.* **1963** (14) 68–70.
[51] Erich Flotow, *Pharm. Zentralhalle* **94**, 178–179 (1955).

IODIDE AND ANTIPYRINE

Bismuth iodide forms a complex with antipyrine which is extracted with chloroform.[52] By reading against calibration curves at 345 and 494 nm, interferences can be detected.

Procedure. *Organic Samples.* Decompose a sample containing 0.005–0.05 mg of bismuth with an appropriate acid or mixture and evaporate nearly to dryness. Take up with 5 ml of water and neutralize with 8% sodium hydroxide solution. Add 3 ml of 1:6 sulfuric acid and filter if necessary. Add 2 ml of *M* disodium citrate and dilute to 20 ml with 7% solution of sodium sulfate. Add 1 ml of 10% solution of ascorbic acid, 1 ml of 20% solution of potassium iodide, and 1 ml of 15% solution of antipyrine. After 30 minutes, extract with 10 ml of chloroform and read the extract at 494 nm. If the amount present proves to be less than 0.005 mg, read at 345 nm; at that wavelength, however, there is more chance of other ions interfering.

IODIDE AND 1,10-PHENANTHROLINE

The three-component system of bismuth, 1,10-phenanthroline, and iodide forms an extractable complex.[53] An equal amount of ferrous ion interferes; tenfold antimonous, cupric, or mercuric ion is tolerated.

Procedure. To a sample solution containing 0.01–0.08 mg of bismuth, add 5 ml of 0.01 *M* 1,10-phenanthroline and 5 ml of 2% ascorbic acid solution. Add 8 ml of 0.1 *M* potassium iodide solution, and extract with 5 ml and 5 ml of cyclohexanone. Dilute the combined extracts to 10 ml and read at 465 nm.

IODINE

After extraction of bismuth as the triiodide, the iodine in the compound can be displaced by bromine and the iodine read by the familiar starch–iodide blue.[54]

Procedure. Mix a sample solution containing less than 0.01 mg of bismuth in 0.1 *N* sulfuric acid with 2 ml of a solution containing 1 mg of iron per ml. Heat to 70° and add 5 ml of ammonium hydroxide. Filter after 5 minutes and wash with 1:50 ammonium hydroxide. Dissolve the precipitate with 10 ml of 2:5 sulfuric acid at

[52] P. Klantschnigg, *Farm.* (Pavia) *Ed. Prat.* **13**, 361–370 (1958).
[53] Franciszek Buhl and Hubert Skibe, *Chem. Anal.* (Warsaw) **17**, 258–288 (1972).
[54] Zygmunt Marczenko, Iwona Zoladek, and Andrzej Limbach, *ibid.* **14**, 741–747 (1969).

50°. Dilute the solution to 15 ml and add 0.5 ml of 2% ascorbic acid. After 1 minute add 1 ml of 0.3% potassium iodide solution and 5 ml of 1 : 1 sulfuric acid. Extract the bismuth triiodide with 20 ml and 10 ml of benzene. Reextract the bismuth from the benzene solution by shaking with 5 ml and 5 ml of 0.1 *N* sulfuric acid. Add to the combined aqueous phases 3 drops of saturated bromine water. After 1 minute, add 0.2 ml of 25% solution of phenol in acetic acid. After 3 minutes, add 0.5 ml of 1% solution of potassium iodide and 2 ml of 1% starch solution. Dilute to 25 ml and read at 590 nm against a blank.

METHYL GREEN

Bismuth forms a ternary complex with methyl green (basic green 5) and iodide ion which follows Beer's law for 0.1–40 μg of bismuth per 6 ml of extract.[55] Mercury, gold, and tellurium interfere.

Procedure. *Alloys.* Dissolve a sample containing 0.02–0.08 mg of bismuth in 3 ml of hydrochloric acid and 2 ml of nitric acid. After cooling, dilute to 100 ml with 1 : 17 sulfuric acid. Take a 25-ml aliquot and add 0.1 ml of 1.5% solution of sodium iodide, 0.4 ml of m*M* methyl green, and 6 ml of 1 : 17 sulfuric acid. Extract with 6 ml of 1 : 1 benzene-nitrobenzene and read the extract at 630 nm.

MOLYBDENUM BLUE

The development of molybdenum blue by ascorbic acid reduction of molybdate ion in sulfuric acid is catalyzed by cadmium ion; the reaction is so sensitive that it is in the ppb range.[56] As might be expected, it requires extreme control of conditions, and the blank is substantial. There is serious interference by cupric, ferric, antimony, arsenite, silicate, phosphate, vanadate, tungstate, ferrous, chromic, chromate, stannic, and fluoride ions.

In perchloric acid a complex mixed bismuth-phosphorus dimeric heteromolybdate is formed.[57] Although the wavelength for maximum absorption is the same as for the product discussed in the previous paragraph, they have little in common. The reaction is so rapid that it must be evaluated by a time versus absorption curve.[58] Tolerances in ppm are thallium, 950; cupric, 750; ferrous, 50; mercuric, 40; ferric, 40; arsenite, 850; acetate, 900; fluoride, 20; chloride, 100. All other tolerances investigated exceeded 1000 ppm.

[55] N. L. Shestidesyatna, P. P. Kish, and A. V. Merenich, *Zh. Anal. Khim.* **25**, 1547–1551 (1970); N. L. Shestidesyatna and P. P. Kish, *Ukr. Khim. Zh.* **38**, 489–490 (1972).
[56] John C. Guyon and Linda J. Cline, *Anal. Chem.* **37**, 1778–1779 (1965).
[57] H. D. Goldman and L. G. Hargis, *ibid.* **41**, 490–495 (1969).
[58] L. G. Hargis, *ibid.* **41**, 597–600 (1969); cf. Robert H. Campbell and M. G. Mellon, *ibid.* **32**, 52–57 (1960).

Procedures

In Sulfuric Acid. Prepare a reagent mixture of 200 ml of 5% sodium molybdate solution, 280 ml of 14.2% solution of sodium sulfate, and 18.6 ml of 1 : 1 sulfuric acid diluted to 2 liters. To a sample containing 5–100 ppb of bismuth, add 4 ml of 5% solution of ascorbic acid. Rapidly add 50 ml of the reagent mixture and dilute to 100 ml. Incubate at 25° and read at 725 nm exactly 30 minutes after mixing.

In Perchloric Acid. As a mixed reagent, dissolve in perchloric acid sufficient monopotassium phosphate to make it 6×10^{-4} *M* and sufficient sodium molybdate to make it 6×10^{-2} *M*. Neutralize a sample containing 0.02–0.3 mg of bismuth to pH 3.5–9; greater accuracy is unnecessary. Turbidity at this point will not be significant. Add 5 ml of the mixed reagent and let stand for 5 minutes to equilibrate. Dilute to 50 ml. To 2 ml of this solution in a 1-cm absorption cell, add rapidly 1 ml of 3.7% ascorbic acid. Record the values at 725 nm versus time until the change is no longer linear, which will be for less than 2 minutes.

PYROLIDINE DITHIOCARBAMATE

This reagent forms a chelate with bismuth as well as with cobalt, cadmium, and molybdenum.[59] All absorb in the ultraviolet range and are soluble in chloroform, methyl isobutyl ketone, and other organic solvents. The following ions must be absent: silver, auric, bismuth, copper, cobalt, ferrous, ferric, mercuric, nickel, lead. The extract conforms to Beer's law up to 4 ppm. If molybdenum is present in the sample, modification to above pH 6 avoids formation of its chelate.

Procedure. As reagent solution, dissolve 0.2 gram of pyrolidinedithiocarbamate in water containing 0.2 ml of 1 : 1 ammonium hydroxide and dilute to 100 ml. Prepare fresh daily. As a buffer solution, dissolve 22.1 grams of sodium acetate trihydrate in water, add 5 ml of glacial acetic acid, and dilute to 1 liter.

Add a sample solution containing 0.05–0.2 mg of bismuth to 10 ml of the acetate buffer solution. If a slight precipitate of the oxychloride forms, ignore it. Extract with 10 ml of reagent solution and set aside for 5 minutes. Extract with 25 ml and 15 ml of chloroform. Dilute the extracts to 50 ml and read at 357 nm against a reagent blank.

[59] Melvyn B. Kalt and D. F. Boltz, *ibid.* **40**, 1086–1091 (1968); cf. Herbert K. Y. Lau, Henry A. Droll, and Peter F. Lott, *Anal. Chim. Acta* **56**, 7–16 (1971); E. Kovaks and H. Guyer, *Z. Anal. Chem.* **186**, 267–288 (1962).

QUINOXALINE-2,3-DITHIOL

Bismuth forms a 1:5 complex with this reagent in 4–5 N sulfuric acid. This obeys Beer's law for up to 4 μg of bismuth per ml. There is interference by nickel, cobalt, ferric ion, copper, cadmium, silver, mercuric ion, and chloride. The reagent is not as sensitive toward bismuth as dithizone but is more selective.

Procedure. *Cast Iron or Steel.*[60] Dissolve 0.5 gram of sample in 20 ml of 1:1 hydrochloric acid with dropwise addition of nitric acid, and boil off oxides of nitrogen. Add 10 ml of ammoniacal EDTA solution and 10 grams of tartaric acid. Add 2 drops of phenolphthalein indicator solution, sufficient ammonium hydroxide to neutralize, and 20 ml of ammonium hydroxide in excess. Filter and wash. Add 3 ml of 1% solution of sodium diethyldithiocarbamate to the filtrate. Extract with 10 ml and 10 ml of ethyl acetate. Wash the combined bismuth extracts with 10 ml of water. Extract the bismuth from the ethyl acetate solution with 10 ml of 5 N hydrochloric acid. Add 2.8 ml of sulfuric acid and evaporate to sulfur trioxide fumes. Dissolve in 10 ml of water. Add 3 ml of 1% gelatin solution and 1 ml of 10 mM quinoxaline-2,3-dithiol. Dilute to 25 ml and read at 490 nm against a blank prepared from a bismuth-free sample.

RHODAMINE B

The complex of bismuth with potassium iodide is extractable, after which it is complexed with rhodamine B in organic solvent.[61a] It is applicable to 0.005–0.3 mg of bismuth. More than 2000-fold concentration of antimony, thallium, gallium, gold, mercury, or iron interferes.

Procedure. Adjust the sample to pH 5. Mix 1 ml with 0.5 ml of 10% potassium iodide solution and extract with 2 ml of 3:2 isobutyl methyl ketone–benzene. Centrifuge to separate the organic phase, and wash it with water acidified with hydrochloric acid to pH 5. Add 1 ml of 0.001% solution of rhodamine B in benzene and read the violet complex at 560 nm.

Semiconductors. For analysis of a semiconductor film approximating cesium bismuth disulfide, dissolve 5 mg from the glass base with 1:1 nitric acid and dilute to 100 ml.[61b] Add a 0.1-ml aliquot to 10 ml of 0.08 M potassium iodide. Add 0.2 mM rhodamine B in 10-fold excess. Extract the red complex of bismuth iodide and

[60] L. I. Chernomorchenko and G. A. Butenko, *Zavod. Lab.* **39**, 1448–1450 (1973).
[61a] R. Sivori and A. H. Guerrero, *An. Asoc. Quim. Argent.* **55**, 157–163 (1967).
[61b] N. L. Shestidesyatnaya and L. I. Kotelyanskaya, *Zh. Vses. Khim. Obshch.* **21**, 220–221 (1976).

rhodamine B at pH 4 with benzene containing 4% of tributyl phosphate. Read at 652 nm. Beer's law is followed for 1–12 μg per ml.

2-THIOBARBITURIC ACID

Bismuth forms a 1:3 complex with this reagent.[62] There is interference by ferric ion, copper, antimony, tin, mercury, silver, cadmium, EDTA, phosphate, and oxalate. Tartrate as a masking agent must not exceed 8 mg/ml.

Procedure. To a neutral sample solution containing 0.012–0.25 mg of bismuth, add 7 ml of 0.3% thiobarbituric acid solution followed by 2 ml of 1:15 nitric acid. Dilute to 25 ml, set aside for 10 minutes, and read at 390 nm against a reagent blank.

THIOCAPROLACTAM

Thiocaprolactam reacts with bismuth in an acid medium containing halide ions to form a complex of 3 moles of the lactam with 1 mole of bismuth and 6 moles of halide.[63] The complex is readily extractable by chloroform. Copper, EDTA, and about 10% of tartaric acid interfere, but 5 mg per ml of tartrate, oxalate, phosphate, and citrate is tolerated. When iodide ion is used, small amounts of lead, cupric, antimony, mercuric, ferric, arsenite, and stannous ions interfere. When bromide or chloride ion is used, the system tolerates 0.2 mg per ml of ferric, cupric, mercuric, lead, phosphate and acetate ions.

Corresponding complexes are obtained with pyrrolidine-2-thione or thio-oenantholactam, which is octahydroazocine-2-thione.[64]

Procedures

As the Iodide.[65] Make a sample solution containing 0.005–0.16 mg of bismuth *N* with hydrochloric acid. Add 1 ml of 1% ethanolic caprolactam and 1 ml of 1% potassium iodide solution. Dilute to 5 ml with water and extract with successive 0.5-ml portions of chloroform as long as the extract is colored. Dilute the combined extracts to 10 ml and read at 490 nm.

As the Bromide.[66] The 2:1:1 complex formed is thiolactam-bismuth bromide-

[62] Halina Sakorska-Tomicka, *Mikrochim. Acta* **1969**, 715–717.

[63] H. Sikorska-Tomicka, *Chem. Anal.* (Warsaw) **12**, 1291–1298 (1967); *Mikrochim. Acta* **1968**, 1106–1109.

[64] H. Sikorska-Tomicka, *Chem. Anal.* (Warsaw) **14**, 97–103 (1969).

[65] H. Sikorska-Tomicka, *Mikrochim. Ichnoanal. Acta* **1965**, 1160–1162.

[66] H. Sikorska-Tomicka, *Chem. Anal.* (Warsaw) **13**, 341–349 (1968).

hydrobromic acid. To a sample containing 0.01–0.2 mg of bismuth in 1:35 sulfuric acid, add 1 ml of 1% solution of potassium bromide and 2 ml of 1:3 sulfuric acid. Extract with 10 ml of 1% solution of caprolactam in chloroform. Read at 440 nm.

Tin and Tin Alloys. Dissolve 0.2–5 grams of sample according to bismuth content in 50 ml of 1:1 nitric acid by heating and dilute to 200 ml. Take a 5-ml aliquot and dilute to 100 ml with 1:15 nitric acid. Mix a 10-ml aliquot with 3 ml of 1% sodium chloride solution. Extract with 3 ml, 3 ml, and 3 ml of 3% solution of thiocaprolactam in chloroform. Dilute the combined extracts to 10 ml and read at 410 nm.

Tin. To the sample solution containing 1–30 mg of bismuth, add 1 ml of 1:6 sulfuric acid and 1 ml of 5% solution of sodium chloride, potassium bromide, or potassium iodide. Extract with 3 ml, 3 ml, and 3 ml of 0.9% solution of thiocaprolactam in chloroform. The last extract should not be colored. Dilute the extracts to 10 ml and read the chloride at 410 nm, the bromide at 440 nm, or the iodide at 490 nm.

THIOUREA

For reading bismuth complexed with thiourea in the presence of halide ion, the difference method is preferred.[67] The intensity of color of the complex of thiourea with bismuth is essentially the same over the range 0.5–1 N nitric acid.[68] If the acidity is more than 1.2 N, the color intensity is modified. Difference spectrophotometry will determine 1–9 mg of bismuth per 100 ml of solution containing 15 ml of 12% thiourea solution.[69] Large amounts of iron are masked with ascorbic acid, and 0.1% of lead or copper is permissible.

When bismuth, thiourea, and chloride or bromide are present, a 1:2:2 complex is formed, absorbing at 410 nm.[70] There is no interference by an equal amount of antimony or tellurium ion.

In the analysis of lead alloys, solution in nitric-hydrochloric-hydrobromic acid permits filtration of part of the lead as chloride.[71] The bismuth is then extracted as the diethyldithiocarbamate complex in chloroform. After isolation of the bismuth from that complex, it is read as the thiourea complex. The method is also applied to analysis of tin and agrees with the potassium iodide method.[72]

As applied to pharmaceuticals, only vanadium and antimony give a similar

[67] C. Mahr, *Z. Anal. Chem.* **241**, 133–135 (1968); cf. R. A. Karanov and A. N. Karolev, *Zavod. Lab.* **26**, 48–50 (1960).
[68] P. N. Kovalenko, *Fiz-Khim. Metody Anal. Kontr. Proizvod.* **1961**, 120–131.
[69] L. A. Chazova and G. V. Azadcheva, *Zavod. Lab.* **39**, 409 (1973).
[70] I. K. Guseinov and A. B. Abdullaeva, *Azerb. Khim. Zh.* **1970** (5–6), 114–119.
[71] J. H. Thompson and B. W. Peters, *Analyst* **84**, 180–182 (1959); cf. R. A. Karanov and A. N. Karolev, *Zavod. Lab.* **26**, 48–50 (1960).
[72] M. I. Shviger, V. P. Paklina, and A. S. Medvedva, *Zavod. Lab.* **24**, 16–17 (1958); K. N. Bagdasarov, S. K. Kasatkina, and S. V. Malygina, *Zh. Anal. Khim.* **23**, 1173–1178 (1968).

color, and the complex with thiourea can be read as low as 0.05 mg of bismuth.[73] Antimony can be masked by tartaric acid.[74]

Procedure. Dilute a sample solution containing 3–10 mg of bismuth and 1–4 grams of chloride or bromide ion to 100 ml with 1:20 nitric acid. Take two 10-ml aliquots and add a solution containing 1 mg of bismuth to one. To each, add 35 ml of 1:20 nitric acid saturated with thiourea. Read each against the nitric acid–thiourea reagent at 440 nm.

In Perchloric Acid.[75] When determined in perchloric acid, osmium, ruthenium, palladium, molybdenum, and antimony interfere. Mix a sample solution containing up to 0.4 mg of bismuth with 2.5 ml of perchloric acid. Add 25 ml of 12% solution of thiourea and dilute to 50 ml. Shake for 2 minutes with 10 ml of tributyl phosphate and read the organic phase at 470 nm.

When EDTA is present, mix 15 ml of sample containing about 15 mg of bismuth with 20 ml of perchloric acid and 5 ml of saturated solution of thiourea. Dilute to 50 ml and read at 265 nm.[76]

Collection by Titanium Phosphate.[77] Decompose a 1-gram sample containing less than 1 mg of bismuth with appropriate acids and evaporate to dryness. Take up the residue in 10 ml of hot 1:3 hydrochloric acid. Cool, filter, and wash well with cold 1:3 hydrochloric acid. Evaporate to a syrup, add 2 ml of nitric acid, and evaporate to dryness. Moisten the residue with nitric acid and take up with 50 ml of water. Add 1:1 ammonium hydroxide until hydroxides begin to precipitate. Add 1 gram of hydroxylamine sulfate and boil for 5 minutes to reduce ferric ion. Cool, and add 5 ml of 1:5 nitric acid containing 1 mg of titanium dioxide per ml. Add 2 ml of 5% solution of ammonium orthophosphate and 5 drops of 0.2% solution of methyl violet. Add 1:1 ammonium hydroxide to pH 2.5 as shown by a blue-violet color of the indicator. Dilute with hot water to 900 ml and boil for 5 minutes to coagulate the precipitate. Set aside for 1 hour and filter. Wash the filter with 0.2% ammonium chloride solution. Dissolve the precipitate on the filter with 10 ml of 5:4:11 nitric acid–30% hydrogen peroxide–water. Wash the filter thoroughly with 1:100 nitric acid. Evaporate the filtrate to about 25 ml and add 2 mg of copper sulfate, 2 mg of ferrous sulfate, and 5 ml of 20% tartaric acid solution. Boil until colorless, and cool. Add 10 ml of 10% thiourea solution, dilute to 50 ml, and read at 440 nm.

Separation from Copper and Nickel.[78] The sample solution should contain 0.002–0.016 mg of bismuth and about 3 grams of copper and nickel per liter. Mix 10 ml with 5 ml of perchloric acid. Extract with 15 ml of a mixture of 2-ethylhexyl-

[73]Theodore E. Byers, *J. Assoc. Offic. Agr. Chem.* **41**, 503–504 (1958); **42**, 470–471 (1959).
[74]E. P. Ozhigov, A. N. Dorokhina, and I. I. Mirkina, *Soobshch. Nauch.-Issled. Rabot. Chlenov Primorsk. Otdel. Vses. Khim. Obshch. D. I. Mendeeleva* **1957** (3), 79–87.
[75]Fumio Aoki and Hideo Tomioka, *Bull. Chem. Soc. Jap.* **38**, 1557 (1965).
[76]V. V. Rublev and N. I. Buzina, *Zh. Anal. Khim.* **27**, 303–306 (1972).
[77]V. A. Oshman and V. M. Volkov, *Tr. Ural'sk. Nauch.-Issled. Proekt. Inst. Med. Prom.* **1962**, 251–255.
[78]V. V. Rublev, *Zavod. Lab.* **34**, 1447 (1968).

dihydrogen phosphate and bis(2-ethylhexyl)hydrogen phosphate. Wash the extract with 15 ml of 0.1 *M* perchloric acid. Reextract the bismuth with 15 ml of 1:20 nitric acid. Saturate this extract with thiourea and dilute to 50 ml. Read at 440 nm.

Pure Lead and Galena.[79] Dissolve 0.5 gram of ore in excess hydrochloric acid or 0.5 gram of the metal or oxide in excess nitric acid. Add 3 ml of perchloric acid and evaporate to copious fumes. Dilute with water to 0.5 *M* acidity—which, depending on the loss in fuming, may be of the order of 50 ml. Add 0.5 gram of ascorbic acid to reduce ferric ion. If interference by antimony is suspected, add 2 ml of 10% solution of tartaric acid. Pass the solution through a column of 50-100 mesh Dowex 50W-X8 in hydrogen form, and wash the column with 0.5 *M* perchloric acid. The EDTA-bismuth chelate is very stable in acid solution. Elute the bismuth from the column with 0.5 *M* perchloric acid containing 0.008 *M* EDTA, and dilute to a known volume.

Take an aliquot of the eluate containing 0.01–0.3 mg of bismuth. Add successively 10 ml of 5 *M* perchloric acid, 0.2 gram of ascorbic acid, and 10 ml of 10% thiourea solution. Dilute to 50 ml, set aside for 1 hour, and read at 470 nm.

Permanent-Magnet Alloys.[80] These alloys contain iron, cobalt, nickel, aluminum, copper, and titanium. Dissolve 0.1 gram of sample in 10 ml of 1:1 hydrochloric acid and 5 ml of 1:1 nitric acid. Add 10 ml of 1:1 sulfuric acid and evaporate to sulfur trioxide fumes. Take up in 100 ml of hot water, add 20 ml of 30% sodium thiosulfate solution, and boil for 15 minutes to coagulate the resulting precipitate of copper and bismuth sulfides. Filter, wash with hot 1:200 sulfuric acid, and ash at 900°. Dissolve the ash in 20 ml of 1:9 nitric acid, filter the sulfur, and wash it with hot 1:9 nitric acid. Add 1 ml of 0.5% ferric chloride solution to the filtrate. Follow with 1:3 ammonium hydroxide until precipitation of ferric and bismuth hydroxides is complete. Filter and wash the precipitate 8–10 times with hot water. Dissolve the precipitate in 50 ml of hot 1:9 nitric acid. Add 2 grams of hydrazine sulfate to this solution and boil for 15 minutes to reduce the iron. Cool, add 20 ml of 10% thiourea solution, and dilute to 100 ml. Set aside for 10 minutes and read at 440 nm.

Iron and Copper Sulfides.[81] Dissolve a 1-gram sample in 10 ml of hydrochloric acid and 5 ml of nitric acid. Evaporate nearly to 1 ml, add 5 ml of nitric acid, and evaporate to about 1 ml. Repeat this addition and evaporation 3 times to remove all hydrochloric acid. If the sample was pyrites add 20 ml of 1:5 nitric acid to the moist residue, boil to dissolve the salts, and filter while hot. Wash the filter thoroughly with hot 1:100 nitric acid. If the sample was chalcopyrites, borites, enargites, or a copper ore boil with 150 ml of 1:30 nitric acid to dissolve.

Carefully add ammonium hydroxide until hydroxides start to precipitate, and 6 ml excess. Add 3 ml of a saturated solution of ammonium carbonate, heat to boiling, and filter. Redissolve from the filter with 5 ml of hot 1:1 nitric acid and

[79]Z. Sulcek, P. Povondra, and V. Kratochvil, *Collect. Czech. Chem. Commun.* **34**, 3711–3721 (1969); cf. L. N. Krasil'nikova, *Sb. Nauch. Tr. Vses. Nauch.-Issled. Gornomet. Inst. Tsvet. Met.* **1958** (3), 266–278; A. Schottak and H. Schweiger, *Z. Erzbergb. Metallhütt.* **19**, 180–185 (1966).
[80]E. Yu. Zel'tser and N. N. Kuznetsova, *Tr. Vses. Nauchn.-Issled. Inst. Elektromekh.* **35**, 157–160 (1971).
[81]S. A. Dekhtrikyan, *Izv. Akad. Nauk Arm. SSR, Nauki Zemle* **1964** (6), 53–57.

reprecipitate as before. Filter on the same paper and wash with hot 1:20 ammonium hydroxide. Dissolve the precipitate in hot nitric acid, evaporate almost to dryness, and take up with 20 ml of 1:5 nitric acid. Filter and wash thoroughly with hot 1:100 nitric acid. If copper is absent add 2 ml of 2% solution of cupric nitrate trihydrate. Add 0.3 ml of 2% solution of salicylic acid. Add sufficient solid ascorbic acid to discharge the color of the solution. Add 10 ml of 10% solution of thiourea, dilute to 50 or 100 ml, and read at 440 nm.

THORON I

Thoron I, which is also called thorin, is *o*-(2-hydroxy-3,6-disulfo-1-naphthylazo) benzenearsonic acid. It complexes with bismuth to give a color proportional to the latter at pH 1.75–2.15 and 0.04–0.64 mg of bismuth in 25 ml.[82] Iron interferes because of its color. In 25 ml, 1 mg of calcium, strontium, barium, copper, cobalt, mercury, lead, or silver is tolerated. Most multivalent cations interfere, and many anions have a bleaching effect.

Heating to fumes with equal parts of hydrochloric, nitric, and perchloric acids removes osmium, ruthenium, and germanium and reduces the amounts of selenium, mercury, and rhenium. Fuming with hydrochloric and hydrobromic acid completes removal of the last three elements and also removes arsenic, antimony, and tin. Thereafter the bismuth can be chromatographed on Dowex, further isolated by thizone extraction above pH 10, and extracted by acid for determination with thoron I.

Procedure. Make a neutral or slightly acid solution of the sample 0.02–0.04 *M* in perchloric acid and add 2 ml of 0.1% solution of thoron I. If 0.03–0.5 mg of bismuth is present, dilute to 25 ml; for 0.005–0.03 mg, dilute to 10 ml. Read at 535 nm against a reagent blank.

XYLENOL ORANGE

When coprecipitated with ferric hydroxide, 1–100 ppm of bismuth in copper or nickel is determined photometrically as a 1:1 complex with xylenol orange.[83] The maximum color development requires the rather narrow range of acidity of 0.08–0.15 *N*. Interference by small amounts of zirconium, thorium, and iron can be masked by fluoride ion and ascorbic acid.[84] Fluoride also prevents interference by aluminum.[85]

[82] K. N. Bagdasarov, P. N. Kovalenko, and A. A. Rabtsun, *Fiz.-Khim. Metody Anal. Kontr. Proizvod.*, **1961**, 115–119; H. A. Mottola, *Anal. Chim. Acta* **27**, 136–143 (1962); *Anal. Quim. Acta* **29**, 261–266 (1963).

[83] K. L. Cheng, *Talenta* **5**, 254–259 (1960).

[84] Hiroshi Onishi and Nasumi Ishiwatari, *Bull. Chem. Soc. Jap.* **33**, 1581–1584 (1960).

[85] K. N. Bagdasarov, P. N. Kovalenko, and M. A. Shemyakina, *Zh. Anal. Khim.* **23**, 515–520 (1968).

A minimum of 40 ppm of bismuth in 0.25 gram of uranium can be determined with this reagent at pH 1.1 reading at 560 nm.[86] The method is applicable to perchloric acid solutions of ferrous alloys from which the iron and antimony have been extracted with isobutyl acetate.[87]

Procedures

Copper and Nickel. Dissolve a 5-gram sample in 50 ml of 1 : 1 nitric acid. Cool, and add 10 ml of 0.06% ferric chloride solution. Warm to 70°, neutralize with ammonium hydroxide, and add 5 ml excess. Let stand for 30 minutes and filter. Dissolve the precipitate of ferric hydroxide in 12 ml of hot 1 : 5 hydrochloric acid and dilute to about 50 ml. Neutralize with ammonium hydroxide and add 5 ml excess. Filter, and this time dissolve the precipitate in 5 ml of 1 : 1 nitric acid. Evaporate to dryness and take up the residue in 1.2 ml of 1 : 15 nitric acid. Dilute to about 5 ml and boil with 1 ml of 10% ascorbic acid solution to reduce the iron. Cool and add 1 ml of 0.1% sodium fluoride solution to mask stannous ion. Add 0.4 ml of 0.1% solution of xylenol orange in 1 : 150 nitric acid. Dilute to 10 ml, set aside for 10 minutes, and read at 531 nm against a blank.

Copper, Brass, or Bronze.[88] Dissolve a 2-gram sample in 10 ml of 1 : 1 nitric acid, and boil until nitrogen oxides have been driven off. Dilute to 50 ml and add 5 ml of 2% solution of aluminum nitrate. Add ammonium hydroxide until precipitation of aluminum hydroxide is complete and copper is in solution as the ammonia complex. Filter, and wash the precipitate with hot 1 : 20 ammonium hydroxide. Dissolve the precipitate with 5 ml of 1 : 4 nitric acid followed by ammonium hydroxide to raise to pH 2–3. Add a few crystals of ascorbic acid, 0.2 ml of 1% sodium fluoride solution, and 0.2 ml of 0.08% solution of xylenol orange. Dilute to 25 ml with 0.1 *N* nitric acid and check that the pH is 1. Read at 536 nm.

Lead. Dissolve 4 grams of sample in 10 ml of 1 : 2 nitric acid and boil off the oxides of nitrogen. Add ammonium hydroxide and complete as for copper, brass, or bronze from "Add a few crystals of ascorbic...."

Steel or Cast Iron.[89] Dissolve a 0.5-gram sample containing less than 0.1% of bismuth in 15 ml of 1 : 5 sulfuric acid and filter while hot. Add 2 ml of 1 : 1 nitric acid to oxidize the ferrous ion, and boil off oxides of nitrogen. Dilute to 200 ml, add 3 ml of 5% manganese sulfate solution, and boil. Then, while boiling, add 3 ml of *N* potassium permanganate and boil for 5 minutes. Set aside, keep warm for 20 minutes, and filter off the manganese dioxide that has collected the bismuth. Wash several times with hot water. Dissolve the precipitate in 20 ml of 1 : 35 sulfuric acid containing a few drops of 30% hydrogen peroxide. Cool, and add 2 ml of 5% ascorbic acid solution. Add 10 ml of 0.05% solution of xylenol orange in 1 : 350

[86] B. Budesinsky, *Collect. Czech. Chem. Commun.* **28**, 1858–1866 (1963).
[87] Maria Lucco Borlera, *Atti Accad. Sci., Torino, Cl. Sci. Fis. Mat. Nat.* **105**, 439–452 (1971).
[88] V. N. Danilova and P. V. Marchenko, *Zavod. Lab.* **28**, 654–656 (1962).
[89] A. A. Amsheeva and D. V. Bezuglyi, *Zh. Anal. Khim.* **19**, 97–101 (1964).

sulfuric acid and dilute to 100 ml. Check that the pH is about 1.4, set aside for 15 minutes, and read at 531 nm against an iron solution containing no bismuth.

ZEPHIRAMINE AND IODIDE

Zephiramine, which is tetradecyldimethylbenzylammonium chloride, in neutral solution forms a 1 : 1 complex with the bismuth tetraiodide ion.[90] When extracted with 10 ml of dichloroethane, the absorption gives a straight line for 0.01–0.08 mg of bismuth. Since thallium gives the same type of complex, it interferes, as do copper, lead, stannic, ferric, ceric, arsenic, mercury, and antimony ions.

Procedure. Mix 2 ml of sample solution with 5 ml of 27% solution of sodium acetate trihydrate. Add 5 ml of 5% solution of potassium iodide and adjust to pH 5–6 with 1 : 8 acetic acid. Add 10 ml of 0.01 *M* zephiramine and dilute to 50 ml. Set aside for 10 minutes and extract with 10 ml of dichloroethane. Read at 490 nm.

ZINC DIBENZYLDITHIOCARBAMATE

The bismuth complex derived from the zinc dibenzyldithiocarbamate conforms to Beer's law up to 0.08 mg/ml in carbon tetrachloride.[91] There is serious interference by cupric and antimony ions and also by lead, silver, and mercury at the same level as the bismuth. Interference by the latter two is minimized by extraction from hydrochloric acid.

Procedure. Adjust the acidity of a solution containing up to 0.08 mg of bismuth to 25 ml at 0.2–2.5 *M* with nitric acid, 0.3–4 *M* with hydrochloric acid, or 0.3–3 *M* with sulfuric acid. Shake with 10 ml of 0.03% solution of the reagent in carbon tetrachloride for 90 seconds and read the organic layer at 370 nm against a reagent blank.

A complex of bismuth with **acid alizarin black SN** (mordant black 25) is extracted with butanol for reading at 600 nm.[92] Similar complexes are formed with copper, uranium, thorium, iron, molybdenum, and vanadium.

To a sample solution containing about 0.1 mg of bismuth, add 5 ml of 10%

[90] Tsutumo Matsuo, Junichi Shida, and Toshitsugu Sasaki, *Jap. Anal.* **16**, 546–551 (1967).
[91] Takeshi Yamane, Takeshi Suzuki, and Tomoyuki Mukoyama, *Anal. Chim. Acta* **62**, 137–143 (1972).
[92] E. Gagliardi and M. Khadem-Awal, *Mikrochim. Acta* **1969**, 882–887.

sodium cyanide solution, 5 ml of 5% sodium tartrate solution, and 1 ml of 1% solution of **ammonium phenazo-4-carbodithioate.** Extract with 10 ml of chloroform and convert the lead complex to a nickel compound by shaking the organic layer with a solution containing 0.29% of nickel nitrate hexahydrate and 0.282% of sodium potassium tartrate. Read at 575 nm. Lead and bismuth give the same reaction.

In nitric acid at pH 2, add 1% ethanolic **5-analino-3-mercapto-4-phenyl-1,2,4-triazole.**[93] Extract the orange-red complex with chloroform and read at 600°. Beer's law is followed for 30–400 μg of bismuth per ml. There is serious interference by EDTA and the following ions: silver, cupric, mercuric, stannous, antimony, phosphate, iodide, bromide, and fluoride. For bismuth in a solution of lead alloy, precipitate bismuth with cupferron, extract into chloroform, and back-extract with dilute nitric acid.

A tertiary complex of bismuth in perchloric acid solution is formed with **antipyrine** and 1-(2-pyridylazo)-2-naphthol.[94] At pH 2.7–3 it is extracted with 1:1 chloroform–isobutyl alcohol for reading at 570 nm. Similar ternary complexes of bismuth and antipyrine are formed and extracted as follows: with 4-(2-pyridylazo)resorcinol at pH 3–6 for extraction with chloroform and reading at 520 nm; with 2-ethylamino-5-(2-thiazolylazo)-*p*-cresol at pH 2.4–2.8 for extraction with 1,2-dichloroethane for reading at 590 nm, and with 5-(benzothiazol-2-ylazo)-2-methylamino-*p*-cresol at pH 2.4–2.8 for extraction with 1,2-dichloroethane and reading at 605 nm.

1,1-Antipyrinylethane and **1,1-antipyrinylbutane** complex with bismuth chloride and iodide.[95a] The iodide complexes can be extracted into chloroform and read at 495 nm. Beer's law is followed for 0.002–0.02 mg/ml. Cadmium and mercury interfere.

Bismuth ion forms a 1:2 complex when treated with 0.25% ethanolic **biacetyl monoxime thiosemicarbazone.**[95b] It is not stable in aqueous solution. Add sodium perchlorate as a salting-out agent. Extract with methyl isobutyl ketone from pH 2.5–3.5 and read at 335 nm. Beer's law is followed at 8–22 ppm of bismuth. Copper interferes.

At pH 0–0.9 bismuth forms a 1:1 complex with **1,2-bis(2-aminoethoxy)ethane-*NNN'N'*-tetraacetic acid** which is read at 265 nm.[95c]

To a sample containing 5–40 μg of bismuth, add 10 ml of 0.01 *M* **2,2'-bipyridyl** and 1 ml of 0.1 m*M* potassium iodide.[96] Dilute to 25 ml and adjust to pH 3. Extract with 5 ml and 5 ml of nitromethane. Add 1 ml of ethanol to the combined extracts and adjust to 10 ml. Read the 1:2 complex of bismuth iodide and bipyridyl at 460 nm against a blank. Stannous ion, copper, antimony, and ferrous ion interfere.

For determination of bismuth in lead alloys, the antimony is removed and bismuth read as the **bromide.**[97a] When bismuth perchlorate is dissolved in hydro-

[93] S. S. Chatterjee and H. S. Mahanti, *Indian J. Technol.* **13**, 45–46 (1975).
[94] E. M. Nevskaya, E. I. Shelikhina, and V. P. Antonovich, *Zh. Anal. Khim.* **30**, 1560–1565 (1975).
[95a] A. I. Busev, V. K. Akimov, and Said Alisha Saber, *ibid.* **25**, 918–923 (1970).
[95b] M. Valcarcel and D. Perez Bendito, *Inf. Quim. Anal. Pura Apl. Ind.* **24**, 49–62 (1970).
[95c] B. Karadakov, D. Venkova and A. Aleksieva, *Dokl. Bolg. Akad. Nauk* **29**, 1645–1647 (1976).
[96] Franciszek Buhl and Krystyna Kania, *Chem. Anal.* (Warsaw) **20**, 1055–1063 (1975).
[97a] British Standards Institution, *BS 3908*, Pt. **3**, 7 pp.

bromic acid, bismuth is read at 256 nm or, less satisfactorily, at 375 nm.[97b] Iron, lead, tin, and antimony interfere.

Dissolve 0.5 gram of aluminum alloy containing up to 1% of bismuth by heating with 20 ml of hydrobromic acid containing bromine.[98] Add 10 ml of perchloric acid and heat to crystallization, continue to heat for an additional 20 minutes. Dissolve in water and filter out silica. Discard if less than 1%; otherwise, decompose with hydrofluoric and nitric acids to recover adsorbed bismuth, dissolve in 0.2 ml of perchloric acid, and add to the main filtrate. Dilute to 100 ml. To an aliquot, add 25 ml of hydrobromic acid and 5 ml of 10% solution of ascorbic acid. Dilute to 50 ml and read at 375 nm against a reagent blank.

Brompyrogallol red and bismuth form a blue complex that stabilized in acid solution with gelatin, is read at 635 nm.[99]

Caprolactam complexes with the bismuth tetraiodide ion in nitric acid solution.[100] The complex extracted into chloroform has acetone added for reading at 420 nm. Mercuric, antimony, and chloride ions interfere in determining 0.02–0.2 mg of bismuth. Thiocaprolactam is more significant as a reagent.

A 1:1:2 complex is formed by basic bismuth ion–**catechol violet** and diphenylguanidine at pH 5–7.[101] This is read at 630 nm and obeys Beer's law for 0.5–5 μg of bismuth per ml. At pH 6 the complex can be extracted with butanol.

When bismuth is extracted with bis(2-ethylhexyl)phosphate from perchloric acid solution, an aliquot of the extract can be developed with 0.08 m*M* **catechol violet** or 0.07 m*M* **1-(2-pyridylazo)-2-naphthol** for reading in 80% ethanol at 560 and 565 nm, respectively.[102] Interference from ferric, thallic, indium, antimony, stannous, and stannic ions must be masked.

The complex of **diantipyrylmethane** with iodide and bismuth is extractable with chloroform.[103] To a sample solution containing 0.02–0.3 mg of bismuth, add 2 ml of 10% ascorbic acid solution, 5 ml of 10% potassium iodide solution, and 10 ml of 5% solution of the reagent in 0.5 *N* hydrochloric acid. Extract with 25 ml of chloroform and read. If 1–2% of copper is present, precipitate as the iodide and filter. Tolerances of common ions are substantial.

Bismuth forms a 1:1 complex with **2,5-dibenxoyl-3,4-dihydroxyselenophen** which is read at 490 nm in a pH 2.8 citrate buffer solution.[104] Beer's law is obeyed at 1–10 μg of bismuth per ml, and amounts of lead, mercuric, nickel, and aluminum ions not in excess of the bismuth do not interfere.

For bismuth in m*M* sample, add **diethylenetriaminepentaacetic acid** in a solution buffered with acetate at pH 2–5 and read at 277 nm.[105]

The complex of bismuth by reaction with sodium **diethylphosphinodithioate** can be extracted with carbon tetrachloride from 0.5–5 *N* sulfuric acid giving a maxi-

[97b]Yuroku Yamamoto, *J. Chem. Soc. Jap*. **80**, 875–878 (1959).
[98]C. Trabucchi and V. Vicentini, *Chim. Ind.* (Milan) **44**, 27–29 (1962).
[99]V. Suk and M. Smetanova, *Collect. Czech. Chem. Commun.* **30**, 2532–2537 (1965).
[100]H. Sikorska-Tomicka, *Z. Anal. Chem.* **187**, 258–262 (1962).
[101]N. L. Shestidesyatnaya, N. M. Milyaeva, and L. I. Kotelyanskaya, *Zh. Anal. Khim.* **30**, 522–527 (1975).
[102]Yu. M. Yukhin, I. S. Levin, and I. A. Bylkhovskaya, *ibid.* **28**, 532–535 (1973).
[103]V. P. Zhivopistsev and N. I. Zenkova, *Izv. Estestv. Inst. Perm. Univ.* **14**, 77–81 (1960).
[104]A. Balenovic-Solter, M. Tomaskovic, and Z. Stefanac, *Mikrochim. Acta* **1968**, 344–349.
[105]P. S. Edgaonkar, M. Atchayya, and P. R. Subbaraman, *Indian J. Chem.* **13**, 400–402 (1975).

mum absorption at 450 nm.[106] Copper interferes if present to more than 10% of the amount of bismuth. Tin, antimony, and manganese interfere. There is an analogous reaction with **diphenylphosphinodithioate.** The corresponding complex with ***O*-hexyl butylphosphonodithioate** is read at 430 nm.[107]

Dimercaptothiopyrone A, which is 2,6-dimercapto-3-amyl-1,4-thiopyrone, forms a stable complex for reading bismuth at 533 nm.[108] Other dimercaptothiopyrones complex with bismuth in 0.2 *M* to *M* hydrochloric acid for reading at 619 or 533 nm.[109] The 3,5-diethyl and diphenyl derivatives have been recommended for determination of bismuth.[110]

Mix a sample solution containing up to 90 μg of bismuth with 1.5 ml of 0.5 m*M* ethanolic **5-dimethylamino-2-(2-thiazolylazo)phenol** and 2 ml of 0.5 *M* 1,3-diphenylguanidine.[111] Adjust to pH 4.5 with 0.2 *M* acetate buffer solution and dilute to 50 ml. Set aside for 45 minutes and read the bluish-violet complex at 585 nm against a reagent blank. Beer's law holds up to 1.8 μg of bismuth per ml. It is necessary to mask lead, copper, tin, and antimony. There is serious interference by the following ions: titanic, ceric, ferrous, ferric, vanadic, zirconium, nickel, cobalt, tartrate, citrate, and oxalate.

1,4-Dimethyl-1,2,4-triazonium-(3-azo-4)-*NN*-diethylaniline, which is 3-(4-diethylaminophenylazo)-1,4-dimethyl-1,2,4-triazolium chloride, complexes with bismuth and potassium iodide.[112] The complex is extracted from 1.5 *M* sulfuric acid by 4:1 benzene-nitrobenzene for reading at 540 nm. In 5 ml of extract, Beer's law applies to 0.1–25 μg. There is interference by mercury, auric, silver, scandium, and tellurium ions.

The precipitate formed by **1,2-dimorpholinoethane** with bismuth may be dissolved in dimethylformamide and read at 436 nm.[113]

Like that with diphenylcarbazone, dithizone, the complex of bismuth with **di-2-naphthylthiocarbazone** is extractable with chloroform.[114] For more detail, see Lead.

Bismuth and **9-(2,4-dinitrophenyl)-2,4,6-trihydroxy-3*H*-xanthene-3-one** form a 1:1 complex in 10% ethanol at pH 3 for reading at 560 nm.[115] It should be stabilized with 0.04% of gelatin. There is interference by ferric, titanium, zirconium, niobium, tantalum, molybdenum, germanium, stannic, and antimonous ions. A 1:2 bismuth-ligand complex is also formed at pH 5.5. For separation from a lead-based solder, extract the bismuth as the diethyldithiocarbamate complex with chloroform; then back-extract with 6 *N* nitric acid.

***NN'*-Bis-(phenylamine thioformyl)hydrazine** is appropriate for development of

[106]L. K. Kabanova, S. V. Usova, and P. M. Solozhenkin, *Izv. Akad. Nauk Tadzh. SSR Otdel. Fiz.-Matem. Geol.-Khim. Nauk* **1971** (1), 47–56.

[107]P. M. Solozhenkin, L. P. Kabanova, O. N. Grishina, and V. N. Bondarenko, *ibid.* **1970** (4), 48–54.

[108]Yu. I. Usatenko and A. M. Arishkevich, USSR Patent *149,252* (1962).

[109]A. M. Arishkevich and Yu. I. Usatenko, *Tr. Dnepr. Khim.-Tekhnol. Inst.* **1963** (16) 27–34.

[110]Yu. I. Usatenko, A. M. Arishkevich, and A. G. Akhmetshin, *Zh. Anal. Khim.* **20**, 462–469 (1965).

[111]Chikao Tsurumi and Keiichi Furuya, *Jap. Anal.* **24**, 566–570 (1975).

[112]A. I. Busev, N. L. Shestidesyatnaya, and P. P. Kish, *Zh. Anal. Khim.* **27**, 298–302 (1972).

[113]E. Asmus, H. Hahne, and K. Ohls, *Z. Anal. Chem.* **196**, 161–166 (1963).

[114]R. J. DuBois and S. B. Knight, *Anal. Chem.* **36**, 1316–1320 (1964).

[115]V. P. Antonovich, E. I. Shelikhena, E. M. Nevskaya, and T. P. Yakovleva, *Zh. Anal. Khim.* **30**, 1566–1571 (1975).

bismuth extracted from paper-chromatographic spots.[116] To the sample in 5 ml of 1:5 hydrochloric acid, add 5 ml of 0.5% solution of the reagent in dimethylformamide and 1 ml of ammonium hydroxide. Dilute to 25 ml with ethanol and read at 400 nm.

In 0.8 *N* perchloric acid the complex of **EDTA** with bismuth is read at 265 nm.[117a] The similar complex with **1,2-diaminocyclohexane-*NNN'N'*-tetraacetic acid** is read at 267 nm. Maximum tolerable concentrations are ferric ion, 0.8 μg/ml, and copper, 0.2 mg/ml.

At pH 6 bismuth is extracted with m*M* **dithiodibenzoylmethane** in benzene for reading at 470 nm.[117b] This obeys Beer's law for 2.5–30 μg of bismuth per ml. There is interference by citrate, oxalate, EDTA, phosphate, and arsenate. Other interferences masked are cadmium, copper, and silver with cyanide; zirconium with fluoride; ruthenium and gold with thiocyanate. Iron can be pre-extracted with acetylacetone.

Bismuth of m*M* concentration can be extracted from 0.2–0.5 *N* hydrochloric acid by 0.1 *M* **bis-(2-ethylhexyl)amine** in carbon tetrachloride.[118] The solution may be 0.2 *M* with copper and/or iron. The extracted chloro complex is read at 328 nm.

Fast grey RA forms a violet-red complex with bismuth over the range 0.1–1 ppm, probably 3 moles of dye to 1 of the cation.[119] Reading is at 570 nm, pH 2–3, and fading begins after 20 minutes. There is interference by copper, nickel, cadmium, vanadium, zirconium, and iron. Reduction by ascorbic acid avoids that by iron.

Gallein forms a 1:1 complex with bismuth which can be read in the presence of large amounts of lead.[120] Dissolve the sample of lead in 1:2 nitric acid, evaporate to dryness, take up in 0.01 *N* nitric acid, and dilute to 200 ml with 1:17 acetic acid. Mix an aliquot with 3 drops of a solution of gallein in acetone and 10 ml of 1% gelatin solution. Dilute to 25 ml with 1:17 acetic acid and read at 574 nm. Beer's law is followed for 0.005–0.05 mg/ml. There is interference by stannic, stannous, antimony, iron, thallic, zirconium, chloride, and sulfate ions.

When **hematoxylin** is oxidized with hydrogen peroxide, it forms a 1:1 chelate with bismuth at pH 2.3.[121] The maximum absorption is at 550 nm. Beer's law is followed for 0.4–16.7 ppm of bismuth. A preferred range is 1.6–15 ppm. Aluminum, cupric, ferric, tetravalent vanadium, indium, and gallium ions interfere.

The complex formed by **bis(2-hydroxyethyl)dithiocarbamate** with bismuth at pH 5.5–9.5 in the presence of citrate ion and hydroxylamine has a maximum absorption at 363 nm.[122]

In m*M* ***N'*-(2-hydroxyethyl)ethylenediamine-*NNN'*-triacetic acid** adjusted to 0.8 *M* with perchloric acid, the complex with bismuth has a maximum absorption at 260 nm.[123a] Beer's law is followed for 2–36 μg of bismuth per ml. More than 0.8 μg

[116]E. Popper, E. Florean, and P. Marcu, *Rev. Roum. Chim.* **12**, 1229–1233 (1967).
[117a]B. P. Karadakov and Kh. R. Ivanova, *Zh. Anal. Khim.* **28**, 525–531 (1973).
[117b]B. N. Prabhu and S. M. Khopkar, *Indian J. Chem., A*, **14**, 198–199 (1976).
[118]N. A. Ivanov, *C. R. Acad. Bulg. Sci.* **25**, 779–782 (1972).
[119]H. Khalifa, *Anal. Chim. Acta* **17**, 318–321 (1957).
[120]A. I. Cherkesov, G. S. Akchurina, and A. S. Aleksandrovich-Mel'nikova, *Tr. Astrakh. Tekhn. Inst. Rybn. Prom. Khoz.* **1962** (8), 74–81; L. Kekedy and G. Balogh, *Stud. Univ. Babes-Bolyai, Cluj, Chim.* **7**, 131–138 (1962).
[121]O. Prakash and S. P. Mushran, *Indian J. Chem.* **6**, 270–271 (1968).
[122]Hitoshi Yoshida, Masahiro Yamamoto, and Seiichiro Hikime, *Jap. Anal.* **11**, 197–201 (1962).
[123a]B. P. Karadakov, P. P. Nenova and Kh. R. Ivanova, *Zh. Anal. Khim.* **31**, 2058–2061 (1976).

of ferric ion per ml and more than 1.2 mg of nitrate ion per ml interfere. To a 50-ml sample containing 0.6–13 ppm of bismuth, add 5 ml of 0.01 *M* reagent.[123b]

8-Hydroxy-7-[α-(2-methoxycarbonyl-analino)benzyl]quinoline, which is methyl *N*-[α-8-hydroxyquinolin-7-yl)benzyl]anthranilate, forms a 3:1 complex with bismuth in a solution containing sodium citrate, potassium cyanide, and sodium hydroxide at pH 12.2±0.4 as established by a glycine buffer solution.[124] The reagent in ethanol-free chloroform must be in tenfold excess over the bismuth for extraction. The extract is read at 400 nm. Lead interferes seriously.

Bismuth can be precipitated with **4-hydroxy-3-nitrophenylphosphonic acid** in 0.1–0.3 *N* nitric acid.[125] The filtered and washed precipitate is dissolved in *N* hydrochloric acid. Then 1 ml of 5% tartaric acid solution and 20 ml of 8% sodium hydroxide solution are added. Dilute to 250 ml and read at 430 nm.

Bismuth displaces the wavelength of maximum absorption of **mercaptoacetic acid** from 245 to 320 nm at pH 2–6 and to 340 nm at pH 8–12.[126] In the acid range a 1:1 complex is formed, in the alkaline range a complex of 2 bismuth to 1 mercaptoacetic acid. To a solution of bismuth in perchloric acid, add 0.1 *M* mercaptoacetic acid. Adjust to pH 3 and make the ionic strength 1. Read at 320 nm. Beer's law is followed up to 100 ppm of bismuth.

The complex of **2-mercapto-5-analino-1,3,4-thiadiazole** with bismuth is insoluble in water but soluble in ethanol, acetone, and alkalies.[127] The amorphous red-orange precipitate with bismuth is soluble in excess reagent and then proportional to the bismuth content. Addition of potassium iodide intensifies the color. As a typical reaction, mix 1 ml of a solution of bismuth chloride in 1:2 hydrochloric acid containing 1.265 mg of bismuth with 6 ml of freshly prepared 1% solution of the reagent in ethanol. Add 2 ml of freshly prepared 25% potassium iodide solution and dilute to 10 ml with ethanol. Nitrates and oxidizing agents interfere.

2-Mercaptobenzimidazole forms a ternary complex with bismuth and perchlorate ion.[128] This is extracted with chloroform from a solution containing 25% ethanol, 12 m*M* reagent, and 2 *M* perchloric acid. At 350 nm the extract conforms to Beer's law for 0.3–5.6 μg of bismuth per ml; at 490 nm it conforms for 0.6–12 μg/ml. Most ions do not interfere.

1,3-Bis-(8-mercapto-1,3-dimethyl-2,6-dioxypurpurin-7-yl)propane in 0.2*N* to *N* hydrochloric acid or 5–9 *N* sulfuric acid forms a 2:1 complex with bismuth which can be extracted by chloroform.[129] Aluminum, copper, and ferric ions precipitate with the reagent but are not extracted. The maximum absorption is at 405 nm. Mercury, stannous, pentavalent arsenic, tri-, and pentavalent antimony interfere.

2-Mercaptopyridine complexes with bismuth and with palladium.[130] The sample should contain 0.0425–1.125 mg of bismuth or 0.0125–0.875 mg of palladium. Adjust to pH 1.2–1.7 for bismuth or to pH 1.5–3 for palladium. For bismuth, add 3 ml of 1% ethanolic solution of the reagent; for palladium, add 2 ml of a 2%

[123b]Toru Nozaki and Kunzi Koshiba, *J. Chem. Soc. Jap., Pure Chem. Sect.* **88**, 1287–1291 (1967).
[124]Gerhard Roebisch, *Anal. Chim. Acta* **48**, 161–167 (1969).
[125]M. I. Mikhailenko, *Farm. Zh.* (Kiev) **15** (2), 22–27 (1960).
[126]B. P. Karadakov and S. A. Popova, *Khim. Ind.* **46**, 59–62 (1974).
[127]E. Popper, V. Junie, L. Popa, and L. Roman, *Rev. Chim.* (Bucharest) **11**, 341–343 (1960).
[128]K. A. Uvarova, Yu. I. Usatenko, and T. S. Chagir, *Zh. Anal. Khim.* **28**, 693–696 (1973).
[129]E. Asmus and G. Marsen, *Z. Anal. Chem.* **225**, 252–260 (1967).
[130]A. Izquirdo and E. Bosch, *Inf. Quim. Anal. Pura Apl. Ind.* **26**, 261, 271–276 (1972).

solution. Dilute to 25 ml and read at 420–430 nm. There is interference by platinic, octavalent osmic, cupric, and tetravalent iridium ions.

In a neutral medium, **mercaptosuccinic acid** forms a 1 : 1 complex with bismuth.[131] At 315 nm 0.5–70 ppm of bismuth in 20 ml of 0.1 *M* acetate buffer solution for pH 6.5, and 20 ml of 0.1% reagent solution per 100 ml, conform to Beer's law. The determination tolerates 100 ppm of cadmium or 75 ppm of arsenic and antimony.

Bismuth forms a 1 : 1 complex with ***N*-methylanabasine-α′-azodiethylaminophenol,** which is diethylamino-2-[3-(1-methyl-2-piperidyl)-2-pyridylazo]phenol.[132] The maximum absorption is at 580–590 nm. The optimum pH is 2–4; the color develops within 5 minutes and is very stable. It conforms to Beer's law for 0.2–3.2 μg/ml. Interference by copper is masked by 6% thiourea, that by tungsten by 1% tartaric acid. There is serious interference by iron, cobalt, vanadium, indium, gallium, hafnium, chloride, fluoride, and citrate.

A very similar complex is obtained with ***N*-methylanabasine-α′-azo-β-naphthol.**[133]

Adjust the sample to pH 2.5 with perchloric acid or sodium hydroxide solution. Extract as the yellow complex at a level of 2–12 μg of bismuth per ml with m*M* potassium ***O,O*-bis-[α-(3-methyl-5-oxo-1-phenyl-2-pyrazolin-4-yliden)-*p*-tolyl] phosphorodithioate** in chloroform.[134] Read at 410 nm. The complex is also soluble in benzene and acetone. Cupric ion interferes, and the concentration of many other cations must be controlled. By adding dithizone in chloroform to the extract, bismuth becomes the dithizonate and is read at 500 nm.

At pH 1.3–3 bismuth forms a red 1 : 3 complex with **4-[3-(1-methyl-2-piperidyl)-2-pyridylazo]resorcinol.**[135] The maximum absorption is at 520–530 nm, and Beer's law is obeyed for 0.8–10 μg/ml. To a sample solution containing 0.02–0.25 mg of bismuth, add 2.5 ml of 0.2% solution of the reagent and 10 ml of buffer solution for pH 1.44. Dilute to 25 ml and read against a blank. Interference by cupric ion is avoid by reducing it with thiourea before adding the reagent.

When bismuth is complexed with **methyl-4,5,6,7-tetrachloro-3′,4′,5′,6′-tetrahydroxyfluoran,** which is 3,4,5,6-tetrahydroxy-9-(2,3,4,5-tetrachloro-6-methoxycarbonylphenyl)xanthylium, on addition of persulfate the absorption of the reagent at 380 or 495 nm is proportionately decreased.[136] To the sample solution containing up to 0.017 mg of bismuth, add 1 ml of saturated solution of potassium persulfate and 1 ml of 1% solution of polyvinyl alcohol. Add 1.5% nitric acid to adjust to pH 1.2, and add 1.5 ml of m*M* reagent in methanol. Dilute to 10 ml with water. Maintain at 35° for 60 minutes, at which time reaction is complete. Compare with a blank. Only stannic ion interferes.

Methylthymol blue forms a blue chelate with bismuth.[137a] The absorption is a

[131] Isao Muraki, Shigeo Hara, and Yasuhisa Hayashi, *Jap. Anal.* **14**, 289–292 (1965).

[132] A. E. Martirosov, Sh. T. Talipov, R. Kh. Dzhiyanbaeva, and A. Z. Tatarskaya, *Tr. Tashk. Gos. Univ.* **1968** (323), 78–82; Sh. T. Talipov, R. Kb. Dzhiyanbaeva, and A. E. Martirosov, *Izv. Akad. Nauk Kaz. SSR, Ser. Khim.* **1968** (6), 1–4.

[133] Sh. T. Talipov, L. V. Chaprasova, R. Kh. Dzhiyanbaeva, and S. I. Sukhovskaya, *Izv. Akad. Nauk Kaz. SSR, Ser. Khim.* **1972** (2), 10–13.

[134] A. I. Busev and A. K. Panova, *C. R. Acad. Bulg. Sci.* **27**, 1243–1245 (1974).

[135] K. G. Nigai, Sh. T. Talipov, and I. Ya. Ivanova, *Nauch. Tr. Tashk. Univ.* **1964** (263), 63–68.

[136] Itsuo Mori, *Jap. Anal.* **20**, 1007–1010 (1971).

[137a] Takehisa Enoki, Itsuo Mori, and Yoko Izumi, *ibid.* **18**, 963–968 (1969).

straight line at 550 nm up to 0.02 mg of bismuth per 10 ml containing 0.05 mg of reagent at pH 1.5.

A 1:2:2 bismuth-methylthymol blue-diphenylguanidine adduct is extracted with butanol from a medium buffered with ammonium acetate at pH 5–7.[137b] The maximum absorption is at 560 nm but is read at 597 nm. Beer's law is followed at 1.4–12 μg of bismuth per ml in the presence of 5-fold molar excess of methylthymol blue and 800-fold excess of diphenylguanidine. Tartrate, fluoride, and aluminum interfere with the chelate but not with the adduct. Oxalate, EDTA, nitrilotriacetate, gallium, and ferric ion must be absent.

Methylxylenol blue and hexadecylpyridinium chloride complex with bismuth in an acid medium.[138] Fluoride ion prevents formation of a similar complex with thorium. Lead, copper, iron, and tin interfere. Adjust a sample solution containing up to 0.082 mg of bismuth to pH 1.2 with 1:33 nitric acid. Add 3 ml of 0.722% solution of methylxylenol blue, 1 ml of 0.01 *M* hexadecylpyridinium chloride, and 1.5 ml of 0.01 *M* sodium fluoride. Read at 585 nm.

After separation by thin-layer chromatography on cellulose, bismuth is extracted with 0.1 *N* hydrochloric acid, complexed with **morpholinium 3-oxapentamethylene-dithiocarbamate,** extracted at pH 2–7 into chloroform, and read at 365 nm.[139]

The reaction of bismuth with **nickel diethyldithiophosphate** to displace the nickel gives a complex absorbing at 330 and 400 nm:[140] the former for 0.05–0.3 mg, the latter for 0.15–1.1 mg. Saturate the sample with thiourea. Add 2 ml of 0.02 *M* reagent. Extract the reaction product with 5 ml and 5 ml of carbon tetrachloride and read at a wavelength according to concentration.

Bismuth in *N* **perchloric acid** is read at 222 nm up to 15 ppm by the usual method and up to 70 ppm by the differential method.[141] Ferric and aluminum ions interfere seriously. It is also possible to read lead at 208 nm and iron at 240 nm in the solution and determine the three by simultaneous equations.[142]

Bismuth, **phenylfluorone** or **salicylfluorone,** antipyrine, and an acid anion form a 1:2:2:1 complex for extraction with organic solvents and reading at 520 nm.[143] The desirable pH is 1.6–2.3 for the phenyl derivative and 2–2.7 for the salicyl form. Beer's law is followed for up to 5 μg/ml. The acid anion is perchlorate for either form or thiocyanate for the salicyl compound.

A ternary complex of bismuth is also formed with salicylfluorone and an acid anion at an optimum pH of 2.5. It is read at 490 nm.[144]

Phthalexon S, which is 3,3′-bis[bis(carboxymethyl)aminomethyl]phenolsulfonaphthalein, and the analogous ***m*-cresolphthalexon S** form 1:1 complexes with

[137b]N. L. Shestidesyatnaya, N. M. Milyaeva and L. I. Kotelyanskaya, *Zh. Anal. Khim.* **31**, 1176–1180 (1976).

[138]Takehisa Enoki, Itsuo Mori, Misako Yamazaki, and Michiko Inoue, *ibid.* **21**, 31–36 (1972).

[139]E. Gagliardi and W. Likussar, *Mikrochim. Acta* **1967**, 555–559.

[140]M. I. Ivanyutin and A. I. Busev, *Nauch. Dok. Vyssh. Shk., Khim. Khim. Tekhnol.* **1958** (1), 73–78.

[141]Yukoku Yamamoto, *Jap. Anal.* **8**, 513–518 (1959).

[142]Yoroku Yamamoto and Kazuo Hiiro, *Bull. Inst. Chem. Res., Kyoto Univ.* **30**, 24–29 (1958).

[143]O. A. Tataev and A. Sh. Shakhabudinov, *Izv. Vyssh. Ucheb. Zaved., Khim. Khim. Tekhnol.* **16**, 29–31 (1973).

[144]O. A. Tataev, G. N. Bagdasarov, S. A. Akmedov, Kh. A. Mirzaeva, Kh. G. Buganov, and E. A. Yarysheva, *Izv. Sev-Kavkaz. Nauch. Tsent. Vyssh. Shk., Ser. Estestv. Nauk*, **1973** (2), 30–32.

bismuth.[145] The respective maxima are at 540 and 570 nm, with optimum pH levels of 0.85 and 1.05. There is interference by ferric, zirconium, hafnium, thorium, gallium, and phosphate ions.

The yellow precipitate of bismuth with **potassium ethylxanthate** can be extracted into carbon tetrachloride for reading at 400 nm.[146a] The pH is desirably 4.

Bismuth forms a 1 : 1 complex with **1-(5-pyrazolylazo)-2-naphthol** in 0.5 *N* nitric acid containing 20% of 1,4-dioxan.[146b] This is read at 505 nm. Beer's law is followed for 1–3 μg of bismuth per ml. There is interference by cupric ion, vanadate, chloride, bromide, iodide, oxalate, tartrate, and citrate.

Mix a sample solution containing less than 40 μg of bismuth per ml with 5 ml of 0.1 *N* nitric acid, 3 ml of 5% solution of potassium thiocyanate, and 1 ml of 0.1% methanolic **1-(2-pyridylazo)-2-naphthol.**[146c] Extract with 10 ml of methyl isobutyl ketone and dry the extract with sodium sulfate. Read at 500 nm. Beer's law is followed for 0.5–4 μg of bismuth per ml.

The complex of bismuth with **4-(2-pyridylazo)resorcinol,** is extracted from 0.1–0.28 *N* nitric acid with tributylphosphate.[147] The reading at 530 nm is a straight line up to 0.025 mg of bismuth per ml of the extract containing 0.01% of reagent. More than 2 grams of lead, 1 mg of copper, and 0.3 mg of vanadium will be extracted with the bismuth but can be washed out with 1 : 100 nitric acid. More than 0.05 mg of antimony or 1 mg of hexavalent chromium must be absent.

Bismuth forms a 1 : 1 complex with **pyrocatochol violet** at pH 2.1–2.6 which can be read at 580 nm.[148] There is interference by tin, zirconium, titanium, gallium, indium, bivalent molybdenum, hexavalent tungsten, and complexing anions.

Mix 0.05–0.3 ml of sample solution with 0.5 ml of 1 : 1 sulfuric acid, 0.2 ml of 0.1 *M* sodium iodide, and 0.6 ml of m*M* **pyronine G.**[149] Dilute to 10 ml and extract the 1 : 1 complex of the reagent and iodobismuthate ion with 10 ml of 2 : 1 benzene–methyl ethyl ketone. Read at 530 nm. Beer's law is followed for 5–30 m*M* bismuth. Auric ion interferes.

The presence of thiourea promotes quantitative precipitation of bismuth by **Reinecke salt.**[150a] As applied to pharmaceuticals, dissolve a sample containing 20 mg of bismuth in 10 ml of 1 : 20 nitric acid and dilute to 100 ml. Dissolve 0.25 gram of thiourea in a 20-ml aliquot and add fresh Reinecke salt solution in excess. After 30 minutes, filter and wash the filter with ether. Dissolve in acetone or pyridine and read. Mercury, cadmium, silver, and auric ion interfere.

Bismuth in arsenic can be determined by flotation of a complex of bismuth bromide and **rhodamine 6G.**[150b] This is dissolved in ethanol and read at 530 nm. There is interference by auric, nitrate, and perchlorate ions.

Bismuth complexes with **Ruhemann's purple,** which is 2-(3-hydroxy-1-oxoin-

[145] A. V. Grunin and A. I. Cherkesov, *Zh. Anal. Khim.* **28**, 1346–1350 (1973).
[146a] T. Chakrabarty and A. K. De, *Z. Anal. Chem.* **242**, 152–159 (1968).
[146b] Boleslaw Janik and Tadeusz Gancarczyk, *Acta Pol. Pharm.* **33**, 485–488 (1976).
[146c] B. Subrahmaryam and M. C. Eshwar, *Chemia Analit.* **21**, 873–877 (1976).
[147] Hideo Tomioka and Kuniko Terashima, *Jap. Anal.* **16**, 698–702 (1967).
[148] Miroslav Malat, *Z. Anal. Chem.* **186**, 418–423 (1962).
[149] C. Patroescu and M. Matache, *Rev. Chim.* (Bucharest) **25**, 746–747 (1974).
[150a] J. Weyers and Z. Kahl, *Acta Pol. Pharm.* **23**, 229–233 (1966).
[150b] Jadwiga Chwastowska and Krzysztof, *Chem. Anal.* (Warsaw) **21**, 525–529 (1976).

den-2-yl-imino)indan-1,3-dione.[151] Buffered at pH 4.3, the extract with chloroform is read at 560 nm. Lead and mercury complex similarly.

At pH 2–2.2 **stilbazo** forms a 2:1 complex with bismuth for reading at 582 nm.[152] There is interference by tenfold ferric ion and by chloride.

Bismuth perchlorate in 1.8 *N* **sulfuric acid** is read at 227.5 nm; it conforms to Beer's law up to 90 ppm and has a stable absorption for 24 hours.[153a] Antimony interferes.

Bismuth in 2 *N* sulfuric acid containing iodide ion forms a complex with **tetraphenylarsonium chloride.**[153b] This is extractable with 1,2-dichloroethane for reading at 492 nm. Beer's law is obeyed for 3–30 μg of bismuth per ml. Cadmium, ferric, and antimonous ions interfere at 10-fold concentration; chromate, permanganate, and persulfate appreciably; thiocyanate and chloraurate ions slightly.

In dilute sulfuric acid, bismuth iodide forms an orange precipitate with **tetraphenylphosphonium bromide.**[154] This is extracted into chloroform for reading at 505 nm. There is interference by ferric, cupric, antimony, mercuric, cadmium, stannous, and stannic ions.

To 15 ml of 0.02% ethanolic **4-(2-thiazolylazo)resorcinol,** add dilute perchloric acid until a pure yellow is formed.[155] Add a sample solution containing up to 0.4 mg of bismuth and adjust to pH 2.8–4 with dilute ammonium hydroxide. Dilute to 50 ml and read bismuth at 530 nm. Beer's law is followed up to 8.3 mg of bismuth per ml. The color is stable for 24 hours. Copper forms a similar complex read at 560 nm.

Bismuth in the 2–25 ppm range is read as the **thiocyanate** at 470 nm.[156] For 0.01–0.1% of bismuth it is not necessary to separate from aluminum or lead. Below 0.01%, cation exchange is desirable. Optimum conditions include pH 1–2, thiocyanate more than 3.2 *M*, and 0.034–0.02 mg of bismuth in the developed solution.[157] Cobalt, nickel, antimony, and copper interfere. Separation of bismuth from interference may be by coprecipitation with cadmium sulfide, by extraction of interfering elements with methyl isobutyl ketone, or by coprecipitation with ferric hydroxide followed by extraction of the iron with methyl isobutyl ketone.

A complex of bismuth with thiocyanate and antipyrine is extracted at pH 1.8–2.3 by 1:4 amyl alcohol–ethyl acetate for reading at 325 nm.[158] Copper, cobalt, tin, iron, nickel, and lead interfere.

Thiosalicylic acid which is 2-mercaptobenzoic acid, forms a water-insoluble 1:1 complex with bismuth.[159] For determination, mix a sample containing 1–20 μg of bismuth with 5 ml of 1% solution of the reagent in 50% ethanol. Extract with chloroform and read at 385 nm. **Thiosalicylic acid amide** performs similarly. There is interference by the following ions: cupric, molybdate, cobaltous, vanadate,

[151] A. Ya. Burkina and P. L. Senov, *Zh. Anal. Khim.* **27**, 473–478 (1972).
[152] I. M. Korenman and T. E. Vorontsova, *Tr. Khim. Tekhnol.* (Gor'kii) **1968** [2(20)], 131–136.
[153a] Yuroku Yamomoto, *J. Chem. Soc. Jap.* **80**, 1256–1260 (1959).
[153b] N. K. Baishya and G. Baruah, *Curr. Sci.* **45**, (3), 94–95 (1976).
[154] Nobuhisa Matano and Akira Kawase, *Trans. Nat. Res. Inst. Metals* (Tokyo) **1**, 69–71 (1959).
[155] M. Huilickova and L. Sommer, *Talenta* **13**, 667–687 (1966).
[156] G. D'Amore and F. Corigliano, *Atti Soc. Pelorit. Sci. Fis. Mat. Nat.* **11**, 239–254 (1965).
[157] Cezary Rozycki and Jerzy Maksjan, *Chem. Anal.* (Warsaw) **15**, 391–396 (1970).
[158] Emiko Sudo, *J. Chem. Soc. Jap., Pure Chem. Sect.* **74**, 918–920 (1953).
[159] H. Sikorska-Tomicka and M. Lewicka, *Chem. Anal.* (Warsaw) **19**, 271–278 (1974).

hexavalent uranium, tungstate, antimonous, mercuric, palladous, trivalent rhodium, thallic, and trivalent ruthenium, as well as large excesses of phosphate, oxalate, and tartrate. Small amounts of ferric and cadmium ions are tolerated.

Bismuth may be extracted at pH 6 with **thiothenoyltrifluoroacetone,** which is 1,1,1-trifluoro-4-mercapto-4-(2-thienyl)but-3-en-2-one.[160] The reagent is used in carbon tetrachloride, and as read at 460 nm, it conforms to Beer's law for 2.5–30 μg/ml. Only cadmium, tin, iron, titamium, and nickel interfere seriously.

Bismuth complexes with 3 moles of perchloric acid and 5 moles of ***o*-tolylthiourea.**[161] A similar complex is formed with phenylthiourea. To the sample in perchloric acid, add the reagent in dichloroethane, separate the organic layer, and read at 455 nm.

Bismuth can be extracted with dichloromethane from 80 ml of 1:8 hydrochloric acid containing phosphoric acid to mask the iron and 5 ml of 5% **tributylammonium chloride.**[162] The extract is read at 330 nm.

To a solution containing 5–100 μg of bismuth, add 2 ml of 0.5% solution of gelatin, 10 ml of acetone, and 3 ml of color reagent.[163] The color reagent is mM **2,6,7-trihydroxy-9-phenylfluoren-3-one** in acetone, acidified to pH 1.5 with hydrochloric acid, or correspondingly, **salicylfluorone** at pH 2.5 or **9-(4-dimethylaminophenyl)-2,6,7-trihydroxyfluoren-3-one** at pH 2. For the first two, dilute to 50 ml and read at 540 nm. For the third, heat to boiling, cool, dilute to 50 ml, and read at 540 nm. All are stable for days. Beer's law is followed at 0.5–7.5, 0.18–9.3, and 0.1–6.2 μg of bismuth per ml, respectively. Common anions do not interfere.

Bismuth at 0.001–0.07 mg/ml can be measured by its degree of inhibition of the effect of alkaline phosphatase on **umbelliferone phosphate.**[164] The fluorescence of the umbelliferone released is read at 365 nm and pH 8. Decrease in the slope of the fluorescence-versus-time curve measures the concentration of inhibitor.

Zinc di-(2-hydroxyethyl)dithiocarbamate and bismuth in the presence of citrate form a yellow complex at pH 5.5–9.5.[165] The curve is linear at 363 nm up to 0.01 mg of bismuth per ml. The complex can be extracted by tributyl phosphate. Copper, nickel, and tellurium also complex with the reagent.

A complex of bismuth ion is formed at pH 1–11 with **zinc tetramethylenediselenocarbamate** and extracted with chloroform.[166]

[160]K. R. Solanke and S. N. Khopkar, *Anal. Lett.* **6**, 31–41 (1973).

[161]A. I. Busev, N. V. Shvedova, V. K. Akimov, and E. G. Fursova, *Zh. Anal. Khim.* **24**, 1833–1837 (1969).

[162]M. Ziegler and H. Schroeder, *Z. Anal. Chem.* **211**, 299–304 (1965).

[163]O. A. Tataev and A. Sh. Shakhabudinov, *Zh. Anal. Khim.* **28**, 147–148, 170–172 (1973).

[164]G. G. Guilbault, M. H. Sadar, and M. Zimmer, *Anal. Chim. Acta* **44**, 361–367 (1969).

[165]Hitoshi Yoshida, Masahiro Yamamoto, and Seiichiro Hikime, *Jap. Anal.* **11**, 197–201 (1962).

[166]Janusz Terpilowski and Pawel Ladogorski, *Farm. Pol.* **30**, 1011–1015 (1974).

CHAPTER EIGHT
ARSENIC

Unlike most of the metallic elements, arsenic forms some volatile compounds. That leads to the following methods of avoidance of interfering elements:

1. Evolution as arsine.
2. Distillation as a halide.
3. Extraction as a complex.
4. Masking.

The methods of determination in use until the 1950s were largely variants on the development of arsenomolybdate blue,[1a] thus differing from copper, for example, where the methods were and are multitudinous. More recently, however, the absorption of arsine by silver diethyldithiocarbamate has become popular because of its simplicity.

For arsenic in organometallic compounds such as tetraphenylarsonium chloride, for example, add benzoyl peroxide before combusting in an oxygen flask.[1b] This prevents alloying with the platinum holder. Absorb the combustion products in 0.01 *N* iodine to ensure complete oxidation to pentavalency. Then add ammonium molybdate and hydrazine sulfate for reading the reduced molybdoarsenate at 835 nm.

SEPARATION OF ARSENIC, ANTIMONY, AND TIN

Although a separation is often an integral part of the development of a sample, distillation serves as preliminary isolation. It provides for macro rather than micro samples, which can either be aliquoted or miniaturized.

[1a]A. Halasz and E. Pungor, *Talenta* **18**, 557–575 (1971); A. Halasz, E. Pungor, and K. Polyak, *ibid.* 577–586.
[1b]E. Celon, S. Degetto, G. Marangoni, and L. Sindellari, *Mikrochim. Acta* **I** (1), 113–120 (1976).

Arsenic, antimony, and tin can be separated from possibly interfering elements by precipitation as the sulfides in dilute sulfuric acid containing tartaric acid.[2a] The solution of the sulfides is then separated by the technic that follows.

The distillation equipment required is illustrated in Figure 8-1, where *A* is a 50-ml bulb for containing acid to be added in the course of the distillations. The tube from *A* extending into *B* is 25 cm long; this makes it possible to overcome the pressure of the carbon dioxide that aids in carrying over the vapors from the distilling flask *C*. The part of this tube extending into *B* should have a diameter not exceeding 3–4 mm, to prevent the acid from draining from it; *B* permits gauging the rate of flow of the acid. The side tube of *B* provides for introduction of a stream of carbon dioxide.

The distilling flask *C* has a capacity of 200 ml, a 2.5-cm diameter neck, and a thermometer well. A side tube provides for delivery of acid and carbon dioxide to the distilling flask. The other side tube connects the distilling flask to the condenser *D*.

The sample should not contain more than 100 mg of antimony or 200 mg of tin. When only antimony is to be determined, the tin can increase to 500 mg. Decompose the sample in a Kjeldahl flask with 20 ml of 1 : 1 nitric acid. Cool, and add 8 ml of sulfuric acid. Heat with constant shaking until all the nitric acid is distilled and sulfur trioxide is evolved. Cool, add 10 ml of water, and again heat to sulfur trioxide fumes, thus driving off the last of the nitric acid. Add 1 gram of flowers of sulfur and fume for 10 minutes. Cool, and add 25 ml of water and 10 ml of hydrochloric acid. Stir to dissolve the salts and filter off the sulfur. Wash with 30 ml of 1 : 1 hydrochloric acid and transfer the filtrate to the distilling flask *C* of Figure 8-1, using hydrochloric acid to transfer. Add hydrochloric acid to make the volume 100 ml.

For arsenic, provide 50 ml of water in the receiver with the condenser dipping into it. Pass carbon dioxide at 6–8 bubbles per second and boil gently until the volume in the distilling flask is reduced to about 50 ml. The temperature at this stage should be 110–111°, but if antimony is large some may be carried over and redistillation of the arsenic required. Elsewhere a method is given for distillation of arsenic from antimony at temperatures not exceeding 95°.

For antimony, add 7 ml of phosphoric acid to the distilling flask and provide a fresh receiver with 50 ml of water. Reduce the flow of carbon dioxide and increase the rate of heating until the temperature in the flask reaches 156°. Add hydrochloric acid from bulb *A* at 30–40 drops/minute and maintain the temperature in the distilling flask at 155–165°. Each mg of antimony requires distillation of 1.5–1.75 ml of hydrochloric acid.

For tin, let the flask cool to 140° and supply a new receiver with 50 ml of water. Add to bulb *A* 3 : 1 hydrochloric-hydrobromic acid. Supply this at 30–40 drops/minute while distilling at 140°. Each mg of tin requires distillation of about 1.25 ml of the mixed acid.

Alternatively, trivalent arsenic is extracted from 0.4–0.5 *N* hydrochloric acid with 0.05 *N* solution of dialkyl phosphorodithioates in decane.[2b] After washing the extract with hydrochloric acid, shake it with an equal volume of saturated bromine

[2a]Hidehiro Goto, Yachiyo Kakita, and Masahiko Sase, *Nippon Kakita Gakkaishi*, **21**, 385–387 (1957).
[2b]V. V. Sergeeva, I. S. Levin, L. I. Tishchenko, and V. S. Dan'kova, *Zh. Anal. Khim.* **28**, 2188–2191 (1973).

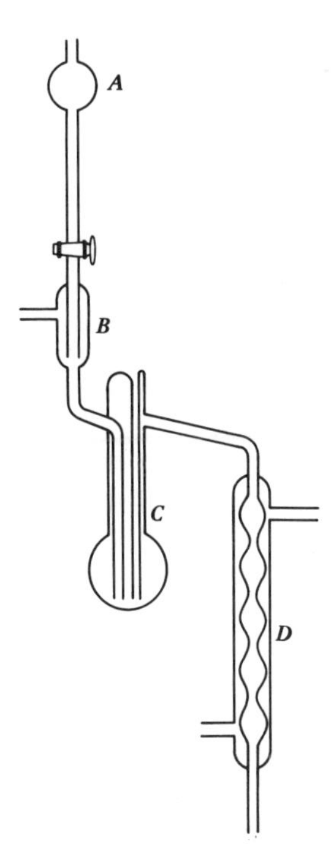

Figure 8-1 Distillation apparatus for separation of arsenic, antimony, and tin.

water. This decomposes the extractant and oxidizes the arsenic to pentavalent for determination as arsenomolybdate.

ARSENOMOLYBDATE BLUE

For arsenic and silicon,[3] mix 2 ml of sample with 2 ml of 1:5 hydrochloric acid and 2 ml of 5% solution of ammonium molybdate. Set aside for 15 minutes and add 2 ml of perchloric acid for pH 0.2. Shake for 30 seconds with 10 ml of 2-ethylhexanol to extract the silicon.

To determine the arsenic, add 1 ml of 1% hydrazine sulfate to the aqueous phase, set aside for 1 hour, and read at 690 nm.

To determine the silicon, shake the 2-ethylhexanol extract for 30 seconds with 6 ml of water. Separate the aqueous phase and add to it 0.5 ml of a solution containing 0.2% of 4-amino-3-hydroxynaphthalene-1-sulfonic acid, 2.4% of sodium sulfite heptahydrate, and 12% of sodium metabisulfite. After 40 minutes read the reduced molybdosilicic acid at 690 nm.

The arsenate, silicate, and phosphate can be separated by extraction and each determined as the heteropoly complex.[4] Mix 80 ml of neutral sample with 5 ml of 10% ammonium molybdate solution. Adjust to pH 4, add a slight excess of bromine water, and boil. Adjust to pH 1.7–1.9 and extract the phosphate with 35 ml of ether. Add 10 ml of hydrochloric acid and extract the silicate with 50 ml of

[3]J. Paul, *Anal. Chim. Acta* **35**, 200–205 (1966).
[4]W. Stanley Clabaugh and Audrey Jackson, *J. Res. Nat. Bur. Stand.* (U.S.) **62**, 201–205 (1959).

butanol. Add 10 ml of ammonium hydroxide and extract the arsenate with 25 ml of methyl isobutyl ketone. Reduce each in the solvent by shaking with 0.2 ml of 2% stannous chloride dihydrate solution in hydrochloric acid. Read each at 630 or 830 nm.

Both arsenate and phosphate in a sample can be determined by the following method.[5a] The sum of the two is determined as molybdenum blue. Then pentavalent arsenic is reduced to trivalent, which does not react and permits determination of the phosphate, which gives the arsenic by difference. The appropriate reducing agent is a 1:2:2 mixture of 3.5 *N* sulfuric acid–0.74 *M* sodium metabisulfite–0.56 *M* sodium thiosulfate. This reagent can be stored for 24 hours under refrigeration.

For determination, mix 50 ml of sample with 5 ml of the reagent and set aside for 15 minutes. Then add 2 ml of 1% ammonium molydenum solution and set aside for 90 minutes before reading at 690 nm. Similarly, develop another portion of sample without adding the reducing agent.

To determine arsenic at 1 ppm to 1% in concentrates of copper, nickel, molybdenum, lead, and zinc containing less than 150 mg of iron per sample, digest with mixed acids and bromine.[5b] Precipitate ferric hydroxide with ammonia as a carrier for the arsenic. Redissolve with hydrochloric acid and reprecipitate. Take up in 2 *N* hydrochloric acid and concentrate to a small volume. Add ferrous ion to reduce pentavalent arsenic to trivalent. Make the solution 11 *N* with hydrochloric acid, add potassium ethyl xanthate and extract the arsenic complex with chloroform. Only small amounts of copper, iron, and molybdenum are coextracted, not sufficient to interfere. Oxidize arsenic in the extract with bromine to pentavalency and back-extract with water. Develop as arsenomolybdate blue.

Evolution as Arsine

As an example of this method of separation for determination as arsenomolybdate, arsine is evolved from a biological digestate absorbed in solid mercuric iodide, eluted with iodine solution and the product used for color development.[6] Use of a condenser in the digestion avoids loss of arsenic trichloride. A stabilizer preserves the color for 1 hour. Beer's law is followed for up to 0.05 mg of arsenic by the technic below. The method is applied successfully to brain, liver, kidney, hair, muscle, and commercial cattle feed.

Arsenic is distilled from the solution of aluminum as arsine; it is absorbed in sodium carbonate and iodine solution and developed as arsenomolybdate blue.[7]

For arsenic in coal or coke, mix a 72-mesh sample with magnesium oxide and potassium permanganate and burn in oxygen.[8] Acid-extract the ash and evolve the arsenic as arsine. The latter, absorbed and oxidized to arsenate by iodine, goes to arsenomolybdenum blue with ammonium molybdate and hydrazine.

[5a]David L. Johnson, *Environ. Sci. Technol.* **5**, 411–414 (1971).

[5b]Elsie M. Donaldson, *Talenta* **24**, 105–110 (1977).

[6]W. T. Oliver and H. S. Funnell, *Anal. Chem.* **31**, 259–260 (1959); cf. Gustavo Ramirez Corredor, *An. Fac. Farm. Bioquim., Univ. Nac. Mayor San Marcos* (Lima) **8**, 550–554 (1957).

[7]T. L. Kurdenkova, *Teknol. Legk. Splavov. Nauchn.-Tekh. Byull. VILSA* **1972** (3), 122–125.

[8]S. R. Crooks and S. Wald, *Fuel* **39**, 313–322 (1960); Gary Eichorn, Walter Wolf, and Selman A. Berger, *Mikrochim. Acta* **1976**, **I**, 135–140.

Alternatively, simmer 1 gram of coal or coke for 30 minutes with 50 ml of 1:7 nitric acid. Filter and wash with 1:7 nitric acid. Add 14 ml of 1:7 sulfuric acid and evaporate to sulfur trioxide fumes. Take up in water and develop through arsine to arsenomolybdenum blue.[9a]

When water such as seawater, potable water, and effluents is treated with sodium borohydride the arsine evolved is absorbed in a solution of iodine and potassium iodide.[9b] This converts the arsenic to arsenate for development as arsenomolybdate blue. Silver, copper, nickel, cadmium, and bismuth must have been removed by Zerolit 225 in sodium form or preextracted with dithizone in chloroform. Organic arsenic compounds do not react with sodium borohydride.

It should be noted that although evolution of arsine is a method leading to arsenomolybdate blue, a very popular method that appears later uses the evolution of arsine for absorption by silver diethyldithiocarbamate.

Procedures

Biological Material. The required apparatus is depicted in Figure 8-2. As purified mercuric iodide, mix the commercial grade with excess 0.001 *N* iodine solution. Filter, and wash with cold water. Wash with ethanol, then with ether, and air dry.

As stock 0.02 *N* iodine solution, dissolve 2.54 grams of iodine and 8 grams of potassium iodide in water and dilute to 1 liter. Prepare fresh every 2 weeks. As stabilized 0.001 *N* iodine solution, dissolve 10 mg of alginate in 60 ml of water at

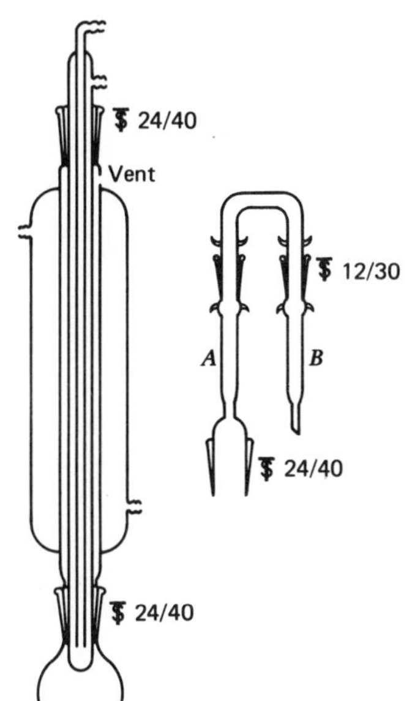

Figure 8–2 Flask fitted with cold finger condenser for evolution of arsine.

[9a]R. F. Abernethy and F. H. Gibson, *Rep. Invest. U.S. Bur. Mines* **RI7184**, 10 pp (1968).
[9b]M. G. Haywood and J. P. Riley, *Anal. Chim. Acta* **85**, 219–230 (1976).

70°. Cool to room temperature, add 5 ml of 0.02 *N* iodine solution, and dilute to 100 ml. Prepare fresh for use.

As absorption powder, dissolve 0.1 gram of purified mercuric iodide in 10 ml of acetone. Mix with 1 gram of cellulose powder and evaporate the acetone by stirring in a current of warm air.

As ammonium molybdate reagent, mix 14 ml of sulfuric acid with 60 ml of water. While still hot, stir in 1 gram of ammonium molybdate powder. When cool, dilute to 100 ml. Prepare fresh before use.

To 1 gram of macerated sample in a 125-ml extraction flask, add 15 ml of hydrochloric acid. Heat under the condenser at 100° until the sample is liquefied; continue heating for 1 hour longer. Rinse the condenser with five 10-ml portions of water and cool the flask and rinsings to 25°. Remove the condenser. Add 2 ml of freshly prepared 30% potassium iodide solution and 3 drops of 40% solution of stannous chloride in 1:2 hydrochloric acid to the flask. Plug the base of the scrubbing tube *A* with cotton and fill it with fine sand. Moisten the sand with 10% lead acetate solution and draw off any excess. Place a cotton plug in tube *B* and by tapping add 0.15 gram of absorption powder, followed by 0.15 gram of purified mercuric iodide. Lubricate the joints and fasten the joints of *A* and *B* with rubber bands.

After 15 minutes add 1 ml of 1:1000 Tween 80 solution and 7 or 8 drops of octyl alcohol to the flask. Add 5 grams of zinc powder, insert the absorption system, and leave in a 25° bath for 1 hour. Elute absorption tube *B* with four 2-ml portions of stabilized 0.001 *N* iodine solution into a 10-ml flask. Add 1 ml of molybdate reagent and mix. Add 0.4 ml of 0.15% solution of hydrazine sulfate and mix. Heat the flask in boiling water for 10 minutes and cool quickly in running water. Dilute to 10 ml and read at 720 nm against a reagent blank.

When the preceding technic is applied to urine, the urine must be used in development of the standard curve because an unidentified volatile compound lowers the color intensity developed.[10]

Alternatively, biological material is wet ashed and the arsine is absorbed in iodine solution.[11] Pentavalent arsenic can be reduced to the trivalent form by chromous sulfate before conversion to arsine (e.g., by electrolytic reduction) and absorption by silver nitrate or mercuric bromide.[12]

For evolution of arsine, the presence of a small amount of ferrous ion will expedite the evolution.[13]

Uranium.[14] As a reagent, mix 10 ml of 10% ammonium molybdate solution with 90 ml of 1:5 sulfuric acid and add 10 ml of 0.15% hydrazine sulfate solution. For use, dilute 10 ml of this to 100 ml and set aside for 24 hours before use.

Transfer 1 gram to a flask for evolution of arsine. To one absorber add 3 ml of 0.2% solution of iodine in 0.4% solution of potassium iodide diluted with 20 ml of water. To a second one add 20 ml of 0.25% solution of sodium bicarbonate. Add

[10] Eva Arato, *Egeszsegtudomany* **5**, 272–278 (1961).
[11] V. Del Vecchio, P. Valori, and A. M. Alasia, *Ig. Sanita Pubb.* (Rome) **18**, 3–17 (1962).
[12] A. K. Babko, A. T. Pilipenko, and A. L. Rozenfel'd, *Zavod. Lab.* **30**, 1060–1061 (1964).
[13] Sheng-chieh Hsü and Cheng-chi Wang, *Hua Hsueh, Shih Chieh* **13**, 468–470 (1958).
[14] Yoshio Morimoto and Takashi Ashizawa, *Jap. Anal.* **10**, 667–668 (1961).

1 : 1 hydrochloric acid to the sample until reaction ceases; then add 0.7 ml of 30% hydrogen peroxide and 20 ml of water to complete solution of the sample.

Add 5 grams of zinc and close the apparatus. After 1 hour transfer the contents of the absorbers to separate flasks. To each, add 10% solution of sodium bisulfite until the color of iodine disappears. Dilute each to 40 ml, add 7 ml of reagent, and dilute to 50 ml. Heat at 100° for 20 minutes, cool, and read at 665 nm.

Antimony.[15] As reagent, mix 10 ml of 0.25% solution of hydrazine sulfate with 1 gram of ammonium molybdate in 10 ml of water, 90 ml of 1 : 5 sulfuric acid, and 80 ml of water.

As magnesia mixture, dissolve 55 grams of magnesium chloride hexahydrate and 140 grams of ammonium chloride in 500 ml of water. Add 130.5 ml of ammonium hydroxide and dilute to 1 liter. Filter if not crystal clear.

Use 1 gram of sample or, if the arsenic content exceeds 0.0005%, reduce the size accordingly. Decompose with 5 ml of nitric acid. Add 5 grams of ammonium hydrogen tartrate and 80 ml of water. Heat to a clear solution, dilute to 150 ml, and cool. Add 10 ml of 1.92% solution of monopotassium phosphate and 10 ml of magnesia mixture. Add ammonium hydroxide until a precipitate begins to form; then add 7.5 ml in excess. Let stand overnight and filter. Wash on the filter with 1 : 100 ammonium hydroxide. Dissolve the precipitate from the paper into a distillation flask, using 50 ml of hot 1 : 10 sulfuric acid. Add 2 ml of 15% potassium iodide solution and 0.5 ml of 40% solution of stannous chloride in hydrochloric acid. After allowing to stand for 30 minutes to complete reduction of arsenic to the trivalent form, add 5 grams of granulated zinc to the cold solution. The distilling equipment is typified by Figure 8-1. The receiver is a centrifuge tube containing 1 ml of 1.5% mercuric chloride solution, 0.2 ml of 1 : 5 sulfuric acid, and 0.2 ml of 0.3% potassium permanganate solution. Pass hydrogen through the solution for 90 minutes. Add 5 ml of the molybdate reagent and heat at 100° for 15 minutes. Cool; extract with 3 ml, 3 ml, and 3 ml of isoamyl alcohol. Dilute the combined extracts to 10 ml and read at 840 nm against a blank.

Niobium. Dissolve 1 gram of sample in 10 ml of sulfuric acid and 4 grams of potassium sulfate by heating for 10–15 minutes. Cool, and gradually add 100 ml of 7% ammonium hydrogen tartrate solution. Add 2 ml of 30% hydrogen peroxide. Cool the yellow opalescent solution and treat as for antimony from "Add 10 ml of 1.92% solution...."

Vanadium. Dissolve a 1-gram sample in 10 ml of nitric acid. Add 10 ml of sulfuric acid and drive off the oxides of nitrogen. Cool, add 10 grams of oxalic acid, and evaporate to sulfur trioxide fumes. Add 100 ml of water and cool. Add 1 ml of 40% solution of stannous chloride in hydrochloric acid and 0.2 ml of 15% potassium iodide solution. Set aside for 30 minutes. Add 7 grams of granulated zinc and complete as for antimony from "The receiver is a centrifuge tube...."

Silicon. Dissolve 1 gram of sample in 5 ml of 10% sodium hydroxide solution in

[15] V. A. Nazarenko, G. V. Flyantikova, and N. V. Lebedeva, *Zavod. Lab.* **23**, 891–896 (1957).

a distilling flask. Add 30 ml of water and 15 ml of hydrochloric acid. Add 2 ml of 15% potassium iodide solution and 0.5 ml of 40% stannous chloride solution in hydrochloric acid. After 15 minutes add 5 grams of granulated zinc and proceed as for antimony from "The receiver is a centrifuge tube...."

Gallium and Indium. Dissolve 0.4 gram of sample in 4 ml of 1 : 1 nitric acid and 3 ml of sulfuric acid. Add 5 ml of water and evaporate to crystallization of salts. Repeat that step twice more. Add 5 ml of water and 0.25 ml of 25% solution of sodium metabisulfite; boil until sulfur dioxide has been expelled, adding more water if necessary. Dilute to 40 ml, warm to 40°, and add 1.5 ml of 20% potassium iodide solution. Add 1 ml of freshly prepared 5% solution of ascorbic acid and set aside for 15 minutes.

As an extraction agent, mix 20 ml of 1% solution of sodium diethyldithiocarbamate with 0.8 ml of 1 : 5 sulfuric acid at under 10°. Extract with 20 ml of chloroform at under 10° and store the chloroform solution of diethyldithiocarbamic acid at a temperature below 10° for 1 hour.

Extract with 5 ml and 5 ml of the carbamic acid reagent in chloroform. Combine the extracts and wash with 2.5 ml and 2.5 ml of *N* sulfuric acid. Evaporate the chloroform solution, add 1.5 ml of nitric acid, and evaporate to dryness. Add 2 ml of 7% magnesium nitrate solution to the residue and evaporate to a paste. Add 0.5 ml of nitric acid, evaporate to dryness, and ignite at 500°. Take up the residue in 0.2 ml of 1 : 5 sulfuric acid and dilute to 2 ml. Add 1 ml of 1.5% mercuric chloride solution and 0.2 ml of 0.3% potassium permanganate solution. Complete as for antimony from "Add 5 ml of the molybdate...."

Thallium. Dissolve 0.4 gram of sample in 1.5 ml of nitric acid. Add 6 ml of 1 : 5 sulfuric acid and evaporate to sulfur trioxide fumes. Add 5 ml of water and take to sulfur trioxide fumes four times. Complete as for gallium and indium from "Add 5 ml of water and 0.25 ml of 25% solution of sodium metabisulfite...."

Indium.[16] Treat a 1-gram, finely ground sample with 30 ml of 1 : 1 hydrochloric acid, 0.5 ml of 40% solution of stannous chloride in hydrochloric acid, and 2 ml of 20% potassium iodide solution. Absorb the evolved arsine in 1 ml of 1.5% mercuric chloride solution, 0.25 ml of 1 : 5 sulfuric acid, and 0.15 ml of 0.1 *N* potassium permanganate. When evolution is complete, add 1.5 ml of water, 0.3 ml of 1% solution of ammonium molybdate, and 0.3 ml of 0.3% solution of hydrazine sulfate. Heat at 100° for 10 minutes and cool. Shake with 0.4 ml of isopentyl alcohol and read at 840 nm.

Selenium.[17] Dissolve 2 grams of sample in a slight excess of nitric acid and evaporate to dryness. Dissolve the selenic acid in hydrochloric acid and an equal volume of saturated solution of sulfur dioxide. Set aside for at least 3 hours. Filter the selenium and wash with cold water. Pass an inert gas through the filtrate until sulfur dioxide is no longer detectable. Dilute to a known volume and take an aliquot expected to yield 2–40 μg of arsenic. Add an excess of nitric acid and

[16] G. V. Flyantikova, *ibid.* **32**, 529 (1966).
[17] James F. Reed, *Anal. Chem.* **30**, 1122–1124 (1958).

evaporate to light fumes of sulfur trioxide. Add 5 ml of water and again evaporate to sulfur trioxide fumes. Dissolve the residue in 2 ml of water and neutralize with ammonium hydroxide to pH 4, Congo red paper.

As a molybdate-hydrazine reagent, mix 10 ml of 1% ammonium molybdate solution in 5 *N* sulfuric acid with 1 ml of 0.15% solution of hydrazine sulfate and dilute to 100 ml. Add 5 ml to the neutralized sample solution and heat at 100° for 30 minutes. Cool, dilute to a known volume, and read at 808 nm against a reagent blank.

Polyphosphates.[18] As reagent, dissolve 8.15 grams of ammonium molybdate tetrahydrate in 60 ml of water. Mix 25 ml with 25 ml of 1 : 1 hydrochloric acid. Add 10 ml of mercury, shake vigorously for 5 minutes, and filter the reduced reddish-brown solution. Mix 30 ml of the original aqueous solution with 50 ml of hydrochloric acid, and 56 ml of sulfuric acid. Add 40 ml of the reduced molybdate solution with constant agitation and cooling. Finally dilute to 200 ml. This contains the molybdenum in the pentavalent and hexavalent states.

Place a sample expected to contain 0.3–10 μg of arsenic in an evolution flask.[19] Dilute to 35 ml and add 5 ml of 15% solution of potassium iodide. Add 0.5 ml of 40% solution of stannous chloride in hydrochloric acid and let stand for 15 minutes. Add 5 grams of granulated zinc. The absorbant is 1 ml of 1.5% mercuric chloride solution and 0.15 ml of 0.1% potassium permanganate solution. When evolution is complete, add 0.15 ml of the reagent to the contents of the absorber. Heat at 100° for 15 minutes and cool. Dilute to a known volume and read at 840 nm.

Coal and Coke.[20] Transfer a 1-gram sample to a Kjeldahl flask. Add 7 ml of sulfuric acid and 3.5 ml of nitric acid through a dropping funnel. After the initial reaction has subsided heat, with a free flame. When brown fumes are not being evolved, add further 0.3-ml portions of nitric acid dropwise through a dropping funnel until all visible carbonaceous matter has disappeared. After 90 minutes of reaction, the contents should be greenish yellow. Then heat more strongly to evolution of white fumes for 5 minutes. Add glass beads and 10 ml of water, and again heat to fumes for 10 minutes. Cool somewhat, add 0.2 ml of nitric acid, and fume for 10 minutes. Again add 10 ml of water and fume for 10 minutes. Again add 0.2 ml of nitric acid and fume for 10 minutes. Add 10 ml of water and fume for 20 minutes. Add 10 ml of water and fume for 10 minutes. These precautions ensure the absence of nitric acid.

Transfer with water to the evolution flask *A* of Figure 8-3, making a final volume of 35 ml. Add 2 ml of 15% potassium iodide solution and 0.5 ml of 40% solution of stannous chloride in hydrochloric acid. Let stand for 15 minutes. Meanwhile dilute 0.02 *N* iodine in 0.8% potassium iodide solution to 0.001 *N* and add to the absorber *D*. This contains helix *E* to ensure maximum absorption of arsine. Add cotton that has been wetted with saturated solution of lead acetate to the cones of delivery tube *B*. Moisten the joints of *B* and *C* with water. Insert the delivery tube into the

[18]Felipe Lucene-Conde and L. Prat, *An. Edafol. Fisiol. Veg.* (Madrid) **16**, 1–18 (1967).
[19]K. Bauer, *Kem. Ind.* (Zagreb) **9**, 235–238 (1960).
[20]A. Crawford, J. G. Palmer, and J. H. Wood, *Mikrochim. Acta* **1958**, 277–294.

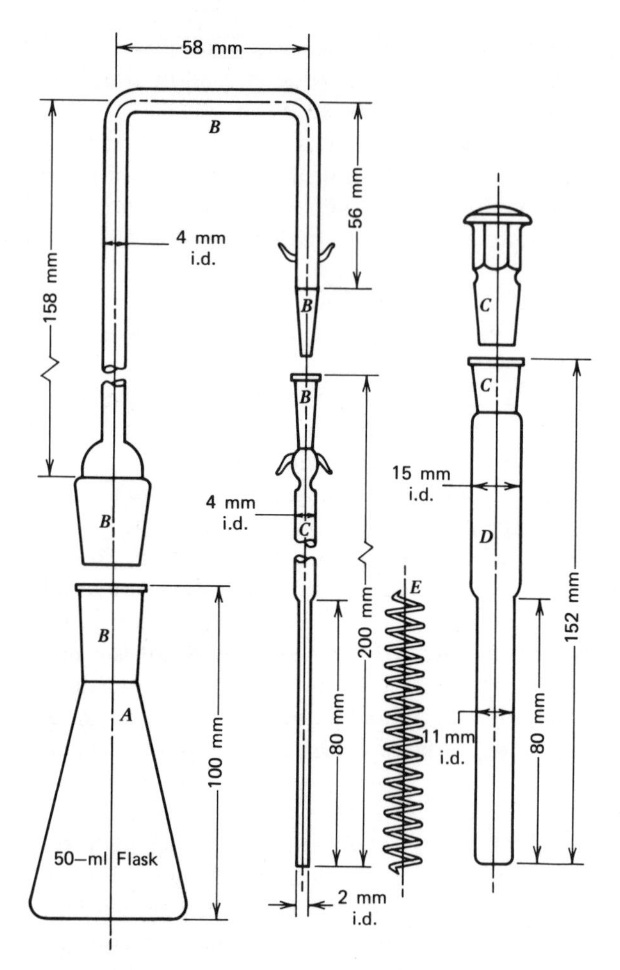

Figure 8-3 Arsine evolution and absorption apparatus.

absorption tube. Add 5 grams of zinc to evolution flask *A* and let evolution proceed for 1 hour.

Disconnect the absorption tube and add 0.5 ml of 1% ammonium molybdate solution in 5 *N* sulfuric acid. Insert the stopper and shake. Remove the helix and heat the absorber at 100° for 10 minutes. Cool, and read at 835 nm against water.

Soil.[21] Digest 1 gram of sample with 2 ml of sulfuric acid, increasing the acid if the clay content is high. After the organic matter has been oxidized and the residue is light grey, add 3 ml of perchloric acid and boil for 2 hours. Cool, and mix with 30 ml of 1:2 hydrochloric acid, 2 ml of 15% potassium iodide solution, and 0.5 ml of 40% solution of stannous chloride in hydrochloric acid. Let stand for 15 minutes and connect with the apparatus for distillation as arsine as typified by Figure 8-1. Add 5 grams of zinc and let the arsine distill for 1 hour into 30 ml of 0.001 *N*

[21]H. G. Small, Jr. and C. B. McCants, *Soil Sci. Soc. Am. Proc.* **25**, 346–348 (1961); Graham F. Collier, *J. Sci. Food Agr.* **24**, 1115–1117 (1973).

iodine solution. Add 2 ml of 1% ammonium molybdate solution in 1:6 sulfuric acid Add 0.8 ml of 0.15% solution of hydrazine sulfate and heat at 100° for 10 minutes. Cool and read at 840 nm.

Tobacco. Wet 1 gram of finely ground sample with 3 ml of water and digest with 10 ml of nitric acid at room temperature for 10 minutes. Heat to white fumes with 10 ml of nitric acid, 1 ml of sulfuric acid, and 1 ml of perchloric acid. Cool, add 5 ml of water, and again heat to white fumes. Cool, add 5 ml of water, and simmer for 2 hours. Complete as for soil from "Cool and mix with..."

Flue Gases.[22] Draw 100 liters at 10–15 liters/minute through a fine paper and then through an absorber containing 5 ml of 1% sodium hydroxide solution. This collects arsenates and oxides of arsenic. Put the paper in the sodium hydroxide solution, add 1 ml of 30% hydrogen peroxide, and heat at 100° for 15 minutes. Add 5 ml of nitric acid and heat until decomposition is complete, adding more acid if necessary. Complete as for antimony from "Add 5 grams of ammonium hydrogen tartrate..."

Meat and Poultry.[23] Transfer an appropriate weight of finely ground sample to a 50-ml Vycor crucible. Add 40% of the weight of sample as magnesium nitrate hexahydrate and mix until the magnesium nitrate is dissolved. An alternative is to blend and take an aliquot. Place in a cool furnace and gradually raise the temperature to 600°. Hold there until visible carbon is largely burned. Cool, moisten the ash with water, and add 3 ml of 1:4 nitric acid. Place in a cool furnace, gradually raise to 600°, and hold there about 1 hour. If necessary to get a white ash, repeat this treatment. Dissolve in 10 ml of 1:1.6 hydrochloric acid and transfer to a 125-ml flask with the assistance of 10-ml portions of the acid. Add 2 ml of 15% potassium iodide solution, then 1 ml of 40% stannous chloride solution in 1:1.6 hydrochloric acid. Let stand 15–30 minutes. Using Kingsley-Schaffert distilling apparatus (Corning Glass Works, No. 33680), place 7 ml of 0.001 *N* iodine solution in a cuvette. Place a small ball of absorbant cotton dampened with saturated lead acetate solution on top of the funnel. Lubricate the ground glass joint with water. Add 12.5 grams of zinc to the flask and distill for 1 hour into the cuvette chilled with crushed ice. Add 0.5 ml of 1.4% solution of ammonium molybdate in 14:86 sulfuric acid followed by 0.3 ml of 0.15% solution of hydrazine sulfate. Heat at 100° for 10 minutes, let cool in the dark for 1 hour, and read at 840 nm against water.

Distillation as Arsenic Halide

To avoid loss of arsenic in distillation as the halide, the presence of hydrogen peroxide in the absorber has been recommended.[24]

[22] N. V. Lebedeva, L. I. Vinarova, and M. F. Grigoa'eva, *Zavod. Lab.* **32**, 1208–1209 (1966).
[23] W. Hearon Buttrill, *J. Assoc. Off. Anal. Chem.* **56**, 1144–1148 (1973).
[24] D. N. Finkel'shtein and G. N. Krynchkova, *Zh. Anal. Khim.* **12**, 196–199 (1957).

When tungsten oxide is dissolved in sodium hydroxide for determination of arsenic phosphoric acid, hydrochloric acid, hydrazine sulfate, and potassium bromide are added. The arsenic is then distilled as a halide and developed as arsenomolybdenum blue.[25] Arsenic can be distilled as the bromide for determination as the arsenomolybdate.[26]

Procedure. Distill a 10-ml sample containing 0.005–0.03 mg of arsenic with 10 ml of sulfuric acid and 2 ml of 30% solution of potassium bromide. Cool, add 5 ml of water, and repeat the distillation. Add 5 ml of 2.5% solution of ammonium molybdate in 1 : 2.5 sulfuric acid and 2 ml of 0.5% solution of hydrazine sulfate. Dilute nearly to 50 ml, heat at 100° for 10 minutes, cool, dilute to 50 ml, and read at 800 nm.

An example of this variant of the arsenomolybdate blue technic is designed to determine arsenic in charge stocks and in catalysts poisoned by arsenic.[27] The arsenic is distilled as the trichloride. Hydrazine is used as the reducing agent. It is applicable to 0.02–40 μg of arsenic. Antimony, tin, tellurium, germanium, and selenium will distill with the arsenic; only germanium would interfere. If germanium were present, it is believed that it would be volatilized on evaporation of the arsenic with sulfuric acid. Phosphorus is not distilled, and silica would not interfere at the acidity used.

When arsenic trichloride is distilled in a stream of carbon dioxide, it may be absorbed in water or solutions of hydrogen peroxide, ammonium hydroxide, or sodium hydroxide.[28] The absorption of the arsenomolybdate complex is not affected by substantial concentrations of ammonium chloride, sodium sulfate, or potassium sulfate. Large amounts of sodium chloride or potassium chloride form addition compounds with arsenious chloride and reduce the color. The acidity must be balanced against the excess of ammonium molybdate.

Arsenic in lead and in cable-sheathing alloys is distilled as arsenic trichloride and used to produce arsenomolybdate blue.[29]

Petroleum Stocks. LOW OLEFINES. As a 0.08% ammonium molybdate solution, dissolve 1.67 grams of ammonium molybdate tetrahydrate in water containing 75 ml of hydrochloric acid, and dilute to 2 liters.

Mix 25 ml of 30% hydrogen peroxide with 25 ml of sulfuric acid in a 200-ml round-bottom flask. Cool to under 5° and add 40 ml of sample. Attach 2 Vigreux column and shake vigorously for 2 minutes. Heat under reflux, and after 15 minutes shake carefully with the column still attached. Then with the column still attached, shake as vigorously as possible while still hot for another 2 minutes. Reflux 15 minutes more and remove the column. Boil off unreacted sample and heat further to light fumes of sulfuric acid. If charring appears, add small increments of hydrogen peroxide. Finally, when colorless, cool and add 20 ml of water.

To the cool solution, add 1.5 grams of ferrous sulfate and 10 grams of sodium

[25]Zdenek Brezany, *Hutn. Listy* **26**, 669–671 (1971).
[26]G. Milazzo and E. Mezi, *R.C. 1st Suppl. Sanita* **25**, 542–555 (1962).
[27]David Liederman, J. E. Bowen, and O. I. Milner, *Anal. Chem.* **30**, 1543–1546 (1958).
[28]G. A. Butenko, V. P. Korzh, and E. M. Rodionova, *Zh. Anal. Khim.* **16**, 602–604 (1961).
[29]A. Schottak and H. Schweiger, *Z. Erzbergb. Metallhütt.* **19**, 180–185 (1966).

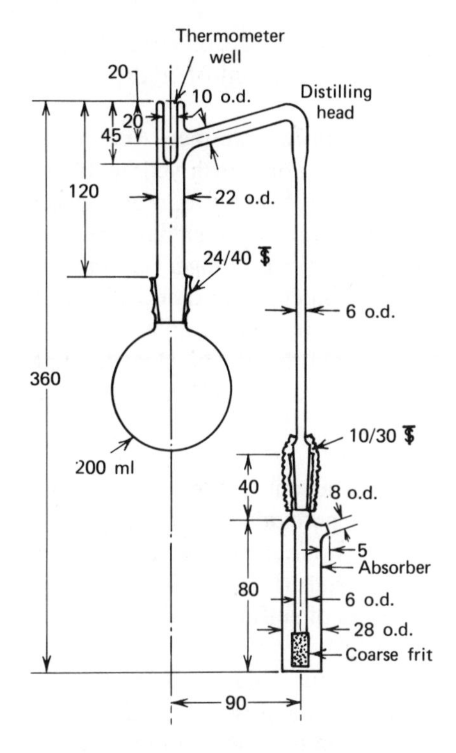

Figure 8–4 Wet oxidation apparatus; all dimensions in mm.

chloride. Without swirling, add 50 ml of hydrochloric acid. Quickly attach the distilling head of Figure 8-4 with 15 ml of nitric acid in the absorber. Transfer to a hood and put the absorber in an ice bath. Heat the flask gently, and gradually increase the heat during distillation reaching 50° in about 1 hour. Continue to heat to 100° when a few drops distill over. Cut off the heat, and at once remove the absorber. Transfer the contents of the absorber to a beaker, rinsing with 10 ml of water forced through the frit by the pressure from a rubber bulb. Evaporate the contents of the beaker at $135 \pm 10°$ and place in an oven at that temperature for 5 minutes. Cool and add 4.5 ml of 0.05% ammonium molybdate solution and 0.5 ml of 0.1% hydrazine sulfate solution. Swirl to mix and pour without rinsing into a 10-ml glass-stoppered cylinder. Heat at 100° for 10 minutes, pour into the 70-mm cell of Figure 8-5 and read at 840 nm against water.

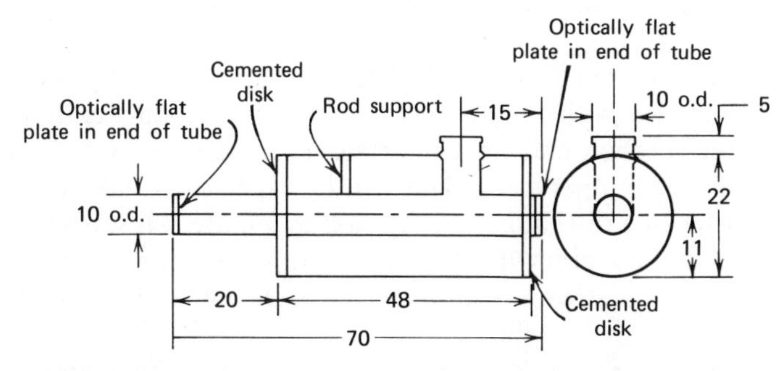

Figure 8–5 Spectrophotometer cell; all dimensions in mm. Paint with optically black paint all surfaces except 10-mm tube ends.

High Olefines. As a 0.05% ammonium molybdate solution, dissolve 0.555 gram of ammonium molybdate tetrahydrate in water containing 25 ml of hydrochloric acid and dilute to 1 liter.

Prepare the oxidizing reagent and add the sample as for low olefines. Attach the Vigreux column and shake until the aqueous layer becomes brownish yellow. If the reaction appears to become too vigorous, chill in ice water. When it darkens, cease shaking until the color lightens. Again shake until the aqueous layer darkens and stop until it lightens. Repeat this step until the aqueous layer no longer darkens on shaking but retains a reddish color. Cool, remove the condenser, and add 15 ml of 30% hydrogen peroxide. Heat strongly to drive off the unreacted sample. If severe carbonizing occurs, add nitric acid dropwise until the solution clears. Heat to light fumes of sulfurtrioxide and complete as for low olefines from "If charring appears add... ."

Catalysts. Take 0.6 gram in a platinum crucible for arsenic under 0.05% or 0.3 gram if over that level. Mix with 3 grams of sodium peroxide and cover with 1 additional gram of peroxide. Fuse at $500 \pm 25°$ for 1 hour. Leach with 25 ml of water and cool. Carefully acidify with 40 ml of 1:1 sulfuric acid. Transfer to a 200-ml round-bottom flask and evaporate to light fumes of sulfurtrioxide. Cool and add 20 ml of water.

Complete as for low olefines from "To the cool solution, add... ." If the arsenic is expected to exceed 0.04 mg, use 25 ml of nitric acid in the absorber and take an aliquot. To develop the color, use 15 ml of 0.08% acid molybdate solution and 2.5 ml of 0.1% hydrazine sulfate solution; dilute to 25 ml before heating and read in a 1-cm cell.

Stainless Steel.[30] The method is suitable for 10–300 ppm of arsenic in niobium-stabilized chrome-nickel stainless steel. Dissolve 0.25 gram of sample in perchloric acid and evaporate until the material begins to turn orange. Take up in 40 ml of 1:1 hydrochloric acid and transfer to a distilling flask. Add 0.5 gram of potassium bromide and 0.3 gram of hydrazine sulfate. Add 25 ml of water to the absorber. Pass a stream of carbon dioxide as a carrier and distill the arsenic as the trichloride.

Add 5 ml of nitric acid to the content of the absorber and evaporate to dryness. Take up in 5 ml of water and add 2 ml of 0.15% hydrazine sulfate solution. Add 5 ml of 1% solution of ammonium molybdate in 1:5 sulfuric acid. Dilute to 25 ml and heat at 100° for 15 minutes. Cool and read at 830 nm.

Antimony.[31] Add 3 ml of 1:1 hydrochloric acid to 1 gram of powdered sample. Add bromine dropwise with continuous stirring and cooling until the liquid is brown. Decant the solution and treat the residue in the same way, repeating until the sample is completely dissolved. This can be expected to take five or six such treatments. Transfer to a distilling flask typified by Figure 8-1, washing in with 1:1 hydrochloric acid. Add 1 gram of hydroxylamine hydrochloride and 0.5 gram of hydrazine sulfate. Pass a stream of nitrogen through the contents of the flask and

[30] UKAEA Report **PG473** (S), 7 pp (1963); cf. British Standards Institution, *BS 3908*, Pt. **3**, 7 pp.
[31] M. Kowalczyk, *Chem. Anal.* (Warsaw) **9**, 331–341 (1964).

distill the arsenious chloride into water until two-thirds of the volume has gone over. Transfer the distillate to a beaker, using 10 ml of nitric acid, and evaporate to dryness. Bake at 130° for 1 hour and cool.

As a reagent, mix 50 ml of 1% solution of ammonium molybdate tetrahydrate with 5 ml of 0.25% hydrazine sulfate solution. Add 8 ml of the reagent to the residue and heat at 100° for 15 minutes. Cool, dilute to 10 ml, and read at 840 nm against a 4:1 mixture of reagent and water. Subtract a blank.

Ores, Concentrates, and Sinter.[32] Fuse or sinter a 0.5-gram sample with 0.3 gram of sodium carbonate in platinum at 1000°. Cool and leach with 1:1 hydrochloric acid. Transfer to a distillation flask typified by Figure 8-1 and add 0.5 gram of potassium bromide and 0.3 gram of hydrazine sulfate. Dilute to about 40 ml with 1:1 hydrochloric acid. Distill the arsenic trichloride while passing a stream of carbon dioxide through the solution, absorbing it in water, until 65% of the volume has gone over. Neutralize the distillate with ammonium hydroxide and dilute to 100 ml.

Take a 20-ml aliquot and add 0.05 *N* potassium permanganate solution until a pink color persists. Add 2 ml of 0.15% hydrazine sulfate solution and 4 ml of 1.25% solution of ammonium molybdate in 1:5 sulfuric acid. Dilute to 50 ml and heat at 100° for 15 minutes. Cool and read at 700 nm against a reagent blank.

Iron Ore or Zinc Concentrate.[33] Dissolve a sample containing up to 0.25 mg of arsenic in 5 ml of nitric acid and 5 ml of 1:1 sulfuric acid. Evaporate to dryness and take up in 20 ml of water. Transfer to a distilling flask using 30 ml of hydrochloric acid. A typical unit is illustrated in Figure 8-1. Add 10 grams of ferrous ammonium sulfate and 10 ml of 10% borax solution. Distill as arsenic trichloride, passing a current of nitrogen or carbon dioxide. During the distillation add 15 ml of hydrochloric acid, 60 ml of 3:2 hydrochloric acid, and finally 50 ml of 3:2 hydrochloric acid. Add 1 gram of potassium chlorate to the distillate and let stand with cooling for 60 minutes. Evaporate to dryness. Take up the residue in 20 ml of water and add 5 ml of 4% solution of ammonium molybdate in 1:18 sulfuric acid. Add 2 ml of 15% tartaric acid solution and 1.5 mg of ascorbic acid. Dilute to 50 ml and heat at 100° for 10 minutes. Cool and read at 700 nm against a reagent blank.

Selenium.[34] Dissolve 1 gram in 5 ml of nitric acid and 2 ml of sulfuric acid. Evaporate to dryness, take up the residue in 25 ml of hydrochloric acid, and precipitate the selenium by mixing with 25 ml of saturated aqueous solution of sulfur dioxide. Add 2 ml of 0.15% solution of hydrazine sulfate and set aside for 3 hours. Filter and wash well on the filter. Make up to a known volume. Transfer a 100-ml aliquot and 100 ml of hydrochloric acid to a distilling unit (e.g., Figure 8-1). Add 1 gram of hydrazine sulfate and 0.5 gram of potassium bromide. Distill in a stream of carbon dioxide, absorbing the distillate in ice water.

[32]G. A. Butenko and E. M. Rodisnova, *Tr. Nauch.-Tekh. Obshch. Chernoi Met., Ukr. Resp. Pravl.* **4**, 119–125 (1956); N. S. Tkachenko and V. I. Sakunov, *ibid.*, **4**, 125–137 (1956).
[33]P. Povondra and Z. Sulcek, *Collect. Czech. Chem. Commun.* **24**, 2393–2397 (1959).
[34]G. G. Lystova and G. S. Dolgorukova, *Zavod. Lab.* **28**, 1319–1321 (1962).

Mix the distillate with 10 ml of 0.1% solution of ammonium molybdate in 1:5 sulfuric acid and 1 ml of 0.15% solution of hydrazine sulfate. Heat at 100° for 15 minutes, dilute to 100 ml and read at 830 nm.

Germanium Dioxide.[35] Dissolve 1 gram of sample in 20 ml of 1% solution of potassium hydroxide. Evaporate to dryness and take up the residue in 12 ml of aqua regia. Extract the germanium with 5 ml and 5 ml of carbon tetrachloride, discarding these extracts. Evaporate to dryness, take up in 5 ml of 1:5 sulfuric acid, and again evaporate to dryness. Complete as described for stainless steel from "Take up in 40 ml of ...."

Antimony Sulfide.[36] As the mixed molybdate reagent, add 30 ml of 1% solution of ammonium molybdate in 1:6 sulfuric acid to about 200 ml of water. Add 3 ml of 0.15% hydrazine sulfate solution and dilute to 300 ml. Prepare fresh daily.

As sample containing up to 0.05% of arsenic use 0.5 gram; for 0.05–0.125%, use 0.2 gram; for 0.125–1% use 0.1 gram. Dissolve in 10 ml of nitric acid and 5 ml of sulfuric acid and evaporate to sulfur trioxide fumes. Fume strongly for 7 minutes, cool, and add exactly 5 ml of water.

The distillation apparatus is shown in Figure 8-6. Add 20 ml of water to the graduate, and ice and water to the 400-ml beaker. Pour the solution of the sample into the distillation flask, setting the beaker aside. Add 0.5 gram of hydrazine sulfate and 2 ml of hydrobromic acid. Add 5 ml and 5 ml of hydrochloric acid to the retained beaker and from there to the distilling flask. Add no water. Regulate the flow of helium, argon, nitrogen, or carbon dioxide to deliver 5 bubbles per second. Adjust the Bunsen flame to about 0.75 inch high. Heat the contents of the flask to 93°, which will require about 4 minutes. Now brush the flame back and

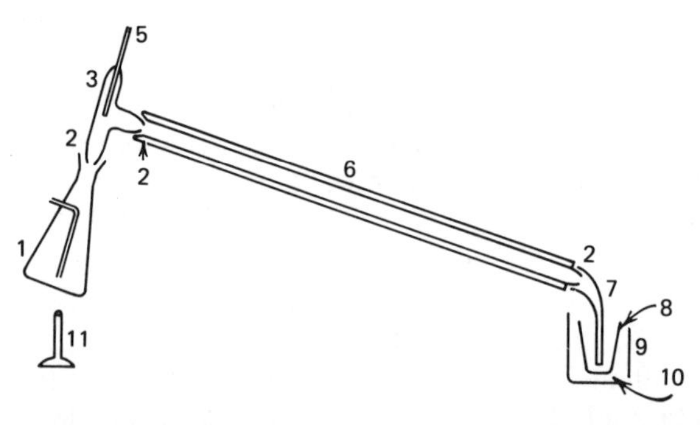

Figure 8–6 Apparatus for distillation of arsenic as trichloride: 1, distillation flask made from 250-ml Erlenmeyer flask by attaching inlet tube 7 mm wide; 2, 24/40 ground glass joint; 3, connecting tube (Scientific Glass Apparatus Co., J-225, Catalog J-52); 4, 10/30 ground glass joint; 5, thermometer, 0–110° (Scientific Glass Apparatus Co., J-2300, Catalog J-52); 6, condenser, 20 inches long; 7, adapter, vertical part is 4 inches long, made from tubing 7 mm wide; 8, 30-ml pharmaceutical graduate; 6, 400-ml beaker; 10, ice and water; 11, Bunsen burner.

[35]L. S. Vasilevskaya, L. S. Vlasova, and G. P. Chibisova, *Nauch. Tr. Nauchn.-Issled-proekt. Inst. Redkomet. Prom.* **1971** (33), 182–183.
[36]George Norwitz, Joseph Cohen, and Martin E. Everett, *Anal. Chem.* **32**, 1132–1135 (1960).

forth across the bottom of the flask to maintain a temperature of 92–95° for 5 minutes for up to 0.125% of arsenic (7 minutes if above that level).

The temperature must not go above 95° or some antimony may distill. Rinse the adapter into the graduate. For less than 0.125% arsenic, use the distillate for development; for larger amounts, dilute to 50 ml and take an aliquot containing 0.15–0.25 mg of arsenic.

To the portion to be developed, add 15 ml of nitric acid and evaporate to dryness with gentle heat, running a blank in parallel. Add 50 ml of the mixed molybdate reagent to dissolve the residue. Heat at 100° for 15 minutes and cool. Dilute to 50 ml and read at 660 nm, setting the blank at 100% transmittance.

Tellurium.[37] Dissolve 2 grams of sample in 20 ml of 1:1 hydrochloric acid and 5 ml of nitric acid. Add 30 ml of perchloric acid and heat to white fumes. Transfer to a silica still with 10 ml of 1:1 hydrobromic-hydrochloric acid and distill in a stream of carbon dioxide. After 5-minute intervals, add more 10-ml portions of the mixed acid until 10 ml of distillate has been accumulated in 50 ml of ice water. Add to the receiver 5 ml of 20% potassium iodide solution and 5 ml of 10% sodium thiosulfate solution. Set aside for 15 minutes and filter. Add 2 ml more of the potassium iodide and sodium thiosulfate solutions and cool. Shake with 30 ml of 0.1% diethylammonium diethyldithiocarbamate solution in chloroform. Evaporate the chloroform extract to dryness. Add 2 ml of nitric acid to the residue and again take to dryness. Take up in 5 ml of water. Add 1 ml of 0.15% hydrazine sulfate solution and 5 ml of 1% solution of ammonium molybdate in 1:5 sulfuric acid. Dilute to 25 ml and heat at 100° for 15 minutes. Cool and read at 830 nm.

The solution from which the arsenic has been extracted is appropriate for determination of tin by phenylfluorone.

Urine.[38] Add excess of 7% uranyl acetate solution to a 25-ml sample to precipitate phosphate and arsenate or arsenite. Test a drop with a crystal of potassium ferrocyanide to ensure excess of reagent. Stir occasionally for 15 minutes, then centrifuge. Dissolve the precipitate in 10 ml of hydrochloric acid. Transfer the solution to a conical flask with 10 ml of hydrochloric acid and 25 ml of water. Add 2 ml of 15% solution of potassium iodide and 0.5 ml of 40% stannous chloride solution. After 5 minutes add 1 ml of octanol or isobutanol and 3 grams of granular zinc. Immediately close with a tube having in its lower and broader end a pad of cotton impregnated with saturated lead acetate solution. The upper constricted part of the tube bears a circle of filter paper that has been impregnated with a 5% solution of mercuric chloride in ethanol and dried. Treat the paper with 4 ml of 0.005 *N* iodine solution at 100° until the paper is colorless. Add 0.5 ml of 1% ammonium molybdate solution and 0.5 ml of 0.15% hydrazine sulfate solution. Heat at 100° for 10 minutes, adjust to a known volume, and read at 850 nm against water. The precipitation of arsenic by uranyl acetate is better than 80% efficient.

Antimony.[39] Dissolve 1 gram of sample by heating with 10 ml of sulfuric acid.

[37] Yoshihiro Ishihara and Hideo Komuro, *Jap. Anal.* **12**, 380–385 (1963).
[38] Dusan Jovanovic, Emilija Krstic-Bogdanovic, and Ljubomir Stojadinovic, *Acta Pharm. Jugosl.* **27**, 1–6 (1971).
[39] Boguslaw Nagorski, *Chem. Anal.* (Warsaw) **20**, 225–227 (1975).

Add 30 ml of hydrochloric acid and 0.1 ml of 1% potassium iodide solution. Extract arsenic tri-iodide with 10 ml and 10 ml of benzene. Wash the organic phase with 10 ml of 9 *N* hydrochloric acid. Back-extract the arsenic with 10 ml and 5 ml of water. Add 2 ml of nitric acid and evaporate to dryness. Take up in water and add 25 ml of 5% solution of ammonium molybdate. Add 5 ml of 1% solution of hydrazine sulfate and heat at 100° for 10 minutes. Extract with 5 ml and 5 ml of isobutyl alcohol. Read at 800 nm.

Nickel.[40] Dissolve a 4-gram sample in 50 ml of nitric acid. Add 20 ml of sulfuric acid and evaporate to sulfur trioxide fumes. Add 0.5 gram of hydrazine sulfate and heat for 5 minutes. Cool, take up in 40 ml of water, and dilute to 50 ml. To a 10-ml aliquot add 30 ml of 0.1 *M* potassium iodide in hydrochloric acid. Extract the arsenic with 20 ml of carbon tetrachloride. Wash the extract with 10 ml of 9 *N* hydrochloric acid. Reextract the arsenic with 15 ml and 15 ml of water. Add 0.03 *N* potassium permanganate to the aqueous extract to give a definite pink. Discharge the color with 1.5% solution of hydrazine sulfate. As a molybdate-hydrazine reagent, mix 50 ml of 1% solution of ammonium molybdate tetrahydrate with 5 ml of 1.5% solution of hydrazine sulfate and dilute to 100 ml. Add 4 ml of this reagent to the aqueous solution. Dilute to 50 ml, and read at 690 nm.

Without Distillation

In this group of methods interferences may be masked or are more commonly avoided by extraction of the arsenic as a complex. Occasionally the arsenic is precipitated with a collector.

To segregate arsenic from gallium and indium in 1 : 1 hydrochloric acid, make the solution 0.5 *M* with potassium iodide and extract the arsenic as the bromide with benzene.[41a] Then extract with water.

For analysis of rhenium, form arsenomolybdic acid and phosphomolybdic acid.[41b] Extract the arsenomolybdic acid with 1 : 1 : 2 isobutyl alcohol–ethyl acetate–chloroform, then phosphomolybdic acid with 1 : 3 isobutyl alcohol–chloroform. In each case, remove unreacted molybdate by filtration through sodium sulfate.

Strip the arsenic or phosphorus from the extract with 5% potassium hydroxide solution and determine indirectly as the molybdate by zinc dithiol at 675 nm or by nitrosulfophenol M and hydroxylamine at pH 2.5–4 and 650 nm. The technic will determine 0.06–0.12% of arsenic or 0.07–0.12% of phosphorus.

Procedure. Make 1–10 ml of germanium-free sample solution containing 0.88–1.3 mg of trivalent arsenic 10 *N* with hydrochloric acid. Extract the arsenic trichloride

[40] L. S. Studeskaya, M. N. Kruglova, G. S. Dolgorukova, and V. A. Verbitskaya, *Tr. Vses. Nauchn.-Issled. Inst. Stand. Obrazstov Spektr. Etalonov* **2**, 29–32 (1965).

[41a] L. I. Ganaga, L. M. Bukhtseeva, and L. R. Starobinets, *Vest. Akad. Navuk. BSSR, Ser. Khim. Navuk.* **1969** (5), 49–53.

[41b] T. M. Malyutina, S. B. Savvin, V. A. Orlova, V. A. Mireeva, and T. A. Kirillova, *Zh. Anal. Khim.* **29**, 925–928 (1974).

with 2 ml, 2 ml, and 2 ml of chloroform. Extract the arsenic from the combined chloroform layers with 2 ml and 2 ml of 1:20 ammonium hydroxide. Neutralize with 1:1 hydrochloric acid and add 3 ml of 0.01 *N* hydrochloric acid. Oxidize the arsenic to the pentavalent form by adding a saturated solution of iodine to produce a pale yellow. Remove excess iodine with 2–3 drops of 30% ascorbic acid solution. Add 10 ml of a fresh mixture of 10 ml of 1% ammonium molybdate in 1:6 sulfuric acid, 1 ml of 0.15% hydrazine sulfate solution, and 89 ml of water. Heat at 100° for 20 minutes, cool, and read at 830 nm.

For greater sensitivity, extract the blue color with 2 ml, 2 ml, and 2 ml of 1:3 isopentyl alcohol–ether and read the combined extracts at 700 nm.

Alternatively, the arsenomolybdate blue is developed in amyl alcohol.[42] As the color reagent, dissolve 10 grams of ammonium molybdate tetrahydrate in 100 ml of hot water; cool, add 100 ml of sulfuric acid, and dilute to 600 ml. As reducing agent, dissolve 10 grams of stannous chloride dihydrate in 25 ml of hydrochloric acid and store under toluene. Just before use, add 1 ml to 25 ml of hydrochloric acid and add 125 ml of water.

To 10 ml of sample solution, add 0.4 ml of molybdate reagent, 5 ml of amyl alcohol, and 1 ml of the stannous chloride reagent. Shake vigorously and set aside for 10 minutes. Centrifuge the organic layer and read at 730 nm against a reagent blank. This will show 0.05–2 μg of arsenic in the extract.

Phosphate and Arsenate.[43a] Dissolve 1 gram of a substance in high purity, such as aluminum nitrate, in 8 ml of water and add 1 ml of 9 *N* hydrochloric acid. Add 1 ml of 5% ammonium molybdate solution and set aside for 10 minutes. Shake for 2 minutes with 3 ml of 3:1 benzene-butanol. For phosphorus, add to the organic layer 1 ml of butanol and 0.06 ml of 0.3% solution of stannous chloride in glycerol. Read at 740 nm against 1:1 benzene-butanol.

For arsenic, add 2 ml of butanol to the residual aqueous layer and shake for 2 minutes. Mix the resulting organic phase with 2 ml of benzene and 0.06 ml of 3% solution of stannous chloride in glycerol. Read at 740 nm against 1:1 benzene-butanol.

For phosphorus and arsenic in molybdenum use hydrazine sulfate as reducing agent.[43b] Read both at 825 ml. Remove arsenic, determine phosphorus similarly, and determine arsenic by difference.

Lead. As reagent, dissolve 1 gram of ammonium molybdate in 10 ml of water. Dilute to 100 ml with 1:4 sulfuric acid. Mix a 5-ml aliquot with 0.5 ml of 0.25% solution of hydrazine sulfate and dilute to 50 ml.

Dissolve 5 grams of lead containing 0.002–0.01 mg of arsenic in 12 ml of 1:1 nitric acid.[44] Evaporate to dryness and take up in 10 ml of water and 2 ml of 1:1 nitric acid with heating. Bring to a boil and add 20 ml of 1:11 sulfuric acid

[42]R. Castagnou and J. Jacq, *Bull. Soc. Pharm. Bordeaux* **99**, 77–87 (1960); R. Castagnou and M. Sylvestre, *ibid.* **107**, 27–30 (1968).

[43a]I. G. Shafran, M. V. Pavlova, and O. N. Khrupkina, *Tr. Vses. Nauchn.-Issled. Inst. Khim. Reakt. Osobo Chist. Khim. Veshchestv.* **1973** (35), 180–186; cf. P. Pakalns, *Anal. Chim. Acta* **47**, 225–236 (1968).

[43b]Michiko Namiki and Richinosuke Hirokawa, *Jap. Anal.* **25**, 715–717 (1976).

[44]H. Goszczynska and M. Kowalczyk, *Chem. Anal.* (Warsaw) **12**, 1261–1266 (1967).

dropwise. Cool, add 10 ml of ethanol, and dilute to 50 ml with 1:11 sulfuric acid. To the separated supernatant layer, add 2 ml of a solution containing 15% of potassium iodide and 2.5% of ascorbic acid. Add 2 ml of 10% solution of mercaptoacetic acid and set aside for 15 minutes. Extract the arsenic with 5 ml, 5 ml, and 5 ml of 1% solution of diethylammonium diethyldithiocarbamate in chloroform. Evaporate the combined extracts to dryness. Moisten the residue with 1 ml of nitric acid and evaporate to dryness. Carry out this step twice more. Add 8 ml of the molybdate reagent and heat at 100° for 15 minutes. Cool, dilute to 10 ml, and read at 840 nm against a reagent blank.

Lead or Zinc Concentrate.[45] Dissolve 1 gram of sample in 15 ml of nitric acid. Add 10 ml of 1:1 sulfuric acid and evaporate to sulfur trioxide fumes. Add 3 ml of water and 0.5 gram of hydrazine sulfate. Evaporate to fumes and continue heating for 5 minutes. Take up in 30 ml of water. If the sample contained less than 0.005% of arsenic, filter and concentrate to 5 ml. Otherwise dilute to 100 ml and take an aliquot containing 5–45 μg of arsenic. Add to the sample solution 3 volumes of 1.66% solution of potassium iodide in hydrochloric acid. Extract the arsenic by shaking with 20 ml of carbon tetrachloride for 2 minutes. Wash the extract with 10 ml of 9 *N* hydrochloric acid. Extract the arsenic from the carbon tetrachloride layer by shaking with 15 ml and 15 ml of water. Make the combined aqueous extracts pink with 0.1% potassium permanganate solution. After 5 minutes decolorize by dropwise addition of 0.15% hydrazine sulfate solution.

Prepare a mixture of 50 ml of 1% ammonium molybdate in 1:5 sulfuric acid and 5 ml of 0.15% hydrazine sulfate solution and dilute to 100 ml. Immediately add 4 ml of this reagent to the sample and boil for 5 minutes. Cool, dilute to 50 ml, and read at 830 nm.

Lead-Arsenic Alloy.[46] Dissolve a 4-gram sample by heating with 30 ml of 1:3 nitric acid and 10 ml of 30% hydrogen peroxide. Then add 10 ml of 1:1 sulfuric acid, boil briefly, cool, and dilute to 1 liter. After the precipitate of lead sulfate has settled, make a 10-ml aliquot alkaline with 12% sodium hydroxide solution. Boil to decompose hydrogen peroxide, neutralize with 1:10 sulfuric acid, and dilute to 500 ml. Add 2 ml of 1:1 nitric acid to a 50-ml aliquot and neutralize with 12% sodium hydroxide solution. Add 10 ml of 1:10 sulfuric acid, 4 ml of 2% ammonium molybdate solution, and 15 ml of 2% solution of hydrazine sulfate. Dilute to 100 ml and heat at 100° for 15 minutes. Read at 750 nm within an hour. A typical sample will contain 2–10% arsenic.

Silver Alloys.[47] Dissolve 1 gram of sample in 5 ml of 1:1 nitric acid and dilute to 50 ml. Add 2 ml of ferric nitrate solution containing 1 mg of iron per ml of 1:50 nitric acid. Heat nearly to boiling and precipitate with excess of ammonium hydroxide. Filter and wash with 1:10 ammonium hydroxide followed by water. Dissolve in 10 ml of hydrochloric acid and add 0.02 gram of potassium iodide.

[45]S. M. Milaev and K. P. Voroshnina, *Zavod. Lab.* **29**, 410–412 (1963); cf. Takao Watabiki and Takashi Miura, *Nat. Tech. Rep.* **2**, 188–191 (1956); cf. L. I. Maksai and A. A. Silaeva, *Sb. Tr. Vses. Nauch.-Issled. Gonomet. Inst. Tsvet. Met.* **1962** (7), 345–347.
[46]Takashi Miura and Masao Ozeki, *Jap. Anal.* **7**, 718–720 (1958).
[47]Z. Skorko-Trybula and J. Chwastowska, *Chem. Anal.* (Warsaw) **8**, 859–864 (1963).

Extract the iron and other interferences with 10 ml, 10 ml, and 10 ml of benzene. Wash the combined extracts with 10 ml, 10 ml, and 10 ml of water. Discard the benzene and evaporate the aqueous acid solution and washings to dryness. Add 10 ml of a reagent containing 0.1 gram of ammonium molybdate tetrahydrate and 7.5 mg of hydrazine sulfate per 100 ml. Heat at 100° for 15 minutes, cool, and read at 700 nm against a reagent blank.

Bismuth.[48] Dissolve 1 gram of sample in 5 ml of nitric acid and evaporate to a syrup. Add 4 ml of sulfuric acid and evaporate to sulfur trioxide fumes. Add 2 ml of water and 0.3 gram of hydrazine sulfate and evaporate to a syrup. Add 4 ml of water and 13 ml of 2% solution of potassium iodide in hydrochloric acid. Shake for 2 minutes with 5 ml of carbon tetrachloride. Wash the organic layer with 5 ml and 5 ml of 9 *N* hydrochloric acid. Extract the arsenic from the organic layer by shaking for 2 minutes with 5 ml of water. Add 5 drops of nitric acid to the extract and evaporate to dryness. Take up the residue in 5 ml of water and add 0.3 ml of a 1% solution of ammonium molybdate tetrahydrate in 1:5 sulfuric acid. Add 0.1 ml of 0.15% solution of hydrazine sulfate. Heat at 100° for 10 minutes and cool. Extract the color by shaking with 0.5 ml of pentyl alcohol and read at 700 nm.

Copper for Arsenic and Phosphorus. Dissolve 10 grams of sample in 80 ml of 1:1 nitric acid and dilute to 200 ml. Adjust the acidity of a 20-ml aliquot to *N*. Add 0.5 ml of sodium hypobromite solution and let stand for 10 minutes. Stir in 2 ml of 10% ammonium molybdate solution and let stand for 15 minutes. Shake for 1 minute with 10 ml of 1:3 butanol-chloroform to extract the phosphate. Repeat until the last two extracts show no color. Shake the combined organic layers for 30 seconds with 10 ml of 1% stannous chloride solution in 0.4 *N* hydrochloric acid. After allowing to stand for a time, read the aqueous layer at 725 nm as phosphate.

Adjust the solution from which phosphate has been extracted to 1.8 *N* with nitric acid. Extract the arsenate with 20 ml of 1:1 butanol–ethyl acetate. Add 15 ml of 1% solution of stannous chloride in 0.4 *N* hydrochloric acid to the organic layer and shake for 30 seconds. Discard the aqueous layer and repeat the extraction. Read the arsenomolybdate in the organic layer at 725 nm.

Copper and Copper-Base Alloys.[49a] Add 5 ml of hydrochloric acid to a 1-gram sample. Immerse the beaker in cold water and add 3 ml of 30% hydrogen peroxide. When the initial reaction subsides, add 7 ml more of the hydrogen peroxide. Warm to decompose excess peroxide and evaporate to dryness. Dissolve in 50 ml of hydrochloric acid, adjust the temperature to 20°, and add 3 ml of 50% hypophosphorous acid solution. After 5 minutes add 10 ml of hydrochloric acid. Shake with 25 ml of chloroform for 1 minute and discard the aqueous layer. Wash the chloroform extract with 10 ml of hydrochloric acid for 30 seconds.

Extract the arsenic by shaking the chloroform layer with 20 ml of water for 1 minute; discard the chloroform. Add 5 drops of 0.1 *N* iodine, 5 ml of 1% ammonium molybdate solution, and 2 ml of freshly prepared 0.15% solution of

[48] S. M. Milaev and T. V. Lyashenko, *Sb. Nauch. Tr. Vses. Nauchn.-Issled. Gornomet. Inst. Tsvet. Met.* **1965** (9), 34–36.

[49a] I. R. Scholes and W. R. Waterman, *Analyst* **88**, 374–379 (1963); cf. P. Niviere and M. Winterheimer, *Chim. Anal.* (Paris) **47**, 448–457 (1965).

hydrazine sulfate. Heat at 100° for 10 minutes, cool to 20°, dilute to 50 ml, and read at 840 nm.

As a variant adjust 30 ml of an ammoniacal solution containing 25–200 μg of pentavalent arsenic and 0.25–0.5 mg of cupric ion to pH 11.5.[49b] Shake for 10 minutes with 0.35 gram of silica gel to adsorb interfering ions. Decant and wash the gel with 5 ml of 1:5 ammonium hydroxide. Neutralize the decantate and washings with 2 *N* hydrochloric acid and dilute to 50 ml. Mix a 2.5-ml aliquot with 2.5 ml of 5% solution of ammonium molybdate and set aside for 15 minutes. Add 2.5 ml of perchloric acid and extract residual silica with 10 ml of iso-octyl alcohol. Add 1 ml of 1% solution of hydrazine sulfate and set aside for 1 hour. Read at 690 nm.

Brass.[50] Dissolve 0.5 gram of sample containing 0.02–0.06% arsenic in 15 ml of 1:1 sulfuric acid and 3 ml of 30% hydrogen peroxide. After the initial reaction subsides, add an additional 2 ml of the hydrogen peroxide and heat to drive off the excess peroxide. Cool and add 6 ml of 25% solution of sulfosalicylic acid to mask copper. After 3 minutes add 20 ml of sulfuric acid and dilute to 50 ml.

To a 20-ml aliquot, add 3.5 ml of 40% potassium bromide solution. Shake with 25 ml of carbon tetrachloride for 2 minutes to extract the arsenious bromide. Reextract the arsenic from the organic layer with 15 ml and 10 ml of water. Neutralize the combined aqueous layers with 10% sodium hydroxide solution. Add a drop of 1:11 sulfuric acid, 2 drops of 0.05 *N* iodine solution, and 5 ml of a fresh 0.8% solution of ammonium molybdate in 1:11 sulfuric acid. Add 0.5 ml of 0.8% solution of hydrazine sulfate and dilute to 50 ml. Heat at 100° for 10 minutes, cool, and read at 840 nm. As a blank, use a solution of copper sulfate and zinc sulfate, approximating the amounts in the sample solution.

Copper, Bronze, or Antimony.[51] Dissolve a 2-gram sample in 6 ml of nitric acid and evaporate almost to dryness. Add 25 ml of hydrochloric acid and 0.5 gram of sodium sulfite. Dilute with 1:1 hydrochloric acid to about 10 *N* and extract the arsenic trichloride with 3 ml, 3 ml, and 3 ml of chloroform. Combine the extracts and reextract the arsenic with 3 ml, 3 ml, and 3 ml of 1:20 ammonium hydroxide. Acidify with 2–3 ml of 0.1 *N* hydrochloric acid. Add a saturated solution of iodine until a pale yellow is obtained, and decolorize by dropwise addition of 30% solution of ascorbic acid.

Prepare a fresh reagent by diluting 10 ml of 10% ammonium molybdate solution and 1 ml of 0.15% hydrazine sulfate solution to 100 ml. Add not more than 1 ml for each 3 μg of arsenic expected to be present. Heat at 100° for 20 minutes, dilute to a known volume, and read at 700 nm. If the color is light, concentrate by extraction with 1:3 isoamyl alcohol–ethyl ether.

Hydrochloric Acid. Treat 25 ml of sample with 0.5 gram of sodium sulfite. Proceed as above from "Dilute with 1:1 hydrochloric acid...."

[49b] S. Singh, *J. Radioanalyt. Chem.* **33**, 237–242 (1976).

[50] Jarmila Spoustova, *Hutn. Listy* **25**, 129–131 (1970); cf. A. Gomez Coedo and M. T. Dorado, *Rev. Metal.* **9**, 382–384 (1973).

[51] L. B. Kristaleva and P. V. Krietalev, *Tr. Kom. Anal. Khim., Akad. Nauk SSSR* **14**, 279–280 (1963); cf. Shizuya Maekawa, Yoshio Yoneyama, and Eiichi Fujimori, *Jap. Anal.* **11**, 497–500 (1962).

Copper, Lead, and Zinc Alloys.[52] As reagent, mix 100 ml of 1% solution of ammonium molybdate in 1 : 6 sulfuric acid with 10 ml of 0.15% hydrazine sulfate solution and dilute to 1 liter. Prepare the mixture just before use.

Dissolve a 1-gram sample expected to contain 0.008–0.06 mg of arsenic in 20 ml of nitric acid. For copper and copper alloys, add 1 or 2 drops of bromine. Evaporate the bulk of the acid, add 10 ml of 1 : 1 sulfuric acid, and evaporate to sulfur trioxide fumes. Cool and add 20 ml of 1 : 10 hydrochloric acid. Warm but do not boil, and add 10 ml of 20% solution of hydrazine hydrochloride. Filter on a pad of paper pulp and wash with 5% ammonium chloride solution. Dilute the filtrate to about 50 ml. Add 40 ml of hydrochloric acid and 10 ml of 40% stannous chloride solution. Add 10 mg of a catalyst of 1 : 9 mercurous chloride–sodium chloride. Heat at 100° for 45 minutes, cool to 30–35°, add paper pulp, and filter. Wash the precipitate of arsenic five times with 1 : 1 hydrochloric acid and five times with 5% ammonium chloride solution. To the precipitate and pulp, add 50 ml of water and 1 gram of sodium peroxide. Boil for 10 minutes, cool, filter, and dilute the filtrate to 100 ml. Mix a 5-ml aliquot with 5 ml of nitric acid and evaporate to dryness. Add 2 ml of water and take to dryness. Bake the residue at 130° for 30 minutes; cool. Add 40 ml of reagent solution and heat at 100° for 15 minutes. Cool, dilute to 100 ml with reagent solution, and read at 830 nm.

Antimony.[53] Dissolve 0.2 gram in 3 ml of nitric acid and 7 ml of hydrochloric acid. Add 20 ml of 1 : 1 sulfuric acid and evaporate to fumes of sulfur trioxide. Add 0.5 gram of hydrazine sulfate and continue to heat for 10 minutes. Cool and take up in 10 ml of water. Add 75 ml of hydrochloric acid containing 2% of potassium iodide. Shake for 2 minutes with 20 ml of carbon tetrachloride to extract the arsenious iodide. As a wash solution, shake the carbon tetrachloride layer with 10 ml of 9 *N* hydrochloric acid for 30 seconds.

Extract the arsenic from the chloroform by shaking for 2 minutes with 15 ml and 15 ml of water. Add phenolphthalein indicator to the combined aqueous extracts, just neutralize by dropwise addition of 20% sodium hydroxide solution, and discharge the color of the indicator with a few drops of 1 : 5 sulfuric acid. Add 0.1% potassium permanganate solution dropwise to produce a pink color. Add 4 ml of 1% solution of ammonium molybdate. Discharge the permanganate color by adding 0.5% stannous chloride solution, and add 1 ml in excess. Dilute to 100 ml and set aside for 30 minutes. Read at 700 nm against a reagent blank. Hydrazine sulfate is an alternative to stannous chloride.

Chromium. Dissolve 1 gram of sample in 5 ml of 2 : 1 sulfuric acid. Add 0.5 gram of hydrazine sulfate and heat for 10 minutes. Cool and dilute to 100 ml. Take a 25-ml aliquot and complete as for antimony from "Add 75 ml of hydrochloric acid...."

Gallium.[54] Dissolve a sample expected to contain 0.5–5 μg of arsenic in excess of 1 : 6 sulfuric-hydrochloric acid. Complete as for antimony from "Add 20 ml of 1 : 1 sulfuric acid...."

[52]N. L. Babenko, *Tr. Altai. Gornomet. Nauch.-Issled. Inst.* **14**, 129–130 (1963).
[53]M. N. Kruglova, *Tr. Vses. Nauchn.-Issled. Inst. Stand. Obraztov.* **3**, 59–62 (1967).
[54]V. G. Goryushina, E. V. Romanova, and L. S. Razumova, *Zh. Anal. Khim.* **28**, 601–603 (1973).

Films of Arsenic and Antimony.[55] Dissolve the sample of film containing up to 0.05 mg each of arsenic and antimony in 0.75 ml of sulfuric acid and 1 drop of nitric acid. Let stand overnight at 100°, to allow the nitric acid to evaporate. Dilute to 5 ml. To a 1-ml aliquot, add 0.25 ml of 0.1% solution of potassium permanganate, 5 ml of 0.11% solution of ammonium molybdate in 0.37 *N* sulfuric acid, and 0.5 ml of 0.15% solution of hydrazine sulfate. Heat at 100° for 15 minutes, cool, and read at 830 nm.

Germanium.[56] To 1 gram of sample in a quartz beaker, add 20 ml of water and 40 ml of 30% hydrogen peroxide. Heat gently until the germanium is dissolved; then evaporate to dryness. Dissolve the residue in 25 ml of water and transfer to a distilling flask. Pass a stream of carbon dioxide preheated to 108° and distill until the white turbidity disappears. Transfer the residual solution to a quartz beaker. Add 2 ml of 1:4 sulfuric acid and 5 ml of nitric acid. Evaporate to 0.5 ml. Add 1 ml of hydrochloric acid and 15 ml of water and warm to 50°. Add 3 ml of 50% potassium iodide solution and 0.1 gram of sodium thiosulfate, and let stand for 15 minutes. Add 5 ml of 10% sodium acetate solution that has been adjusted to pH 5 with ammonium hydroxide. Add 3 ml of 2% solution of sodium diethyldithiocarbamate and extract with 5 ml and 5 ml of carbon tetrachloride. To the combined extracts, add 2 ml of nitric acid and 2 ml of 1:5 sulfuric acid. Evaporate to sulfur trioxide fumes and take up in 10 ml of water. Warm the solution and add 0.01 *N* potassium permanganate until pink. Add 2 ml of 5% solution of ammonium molybdate and 2 ml of a 2% solution of hydrazine sulfate. Dilute to 20 ml and heat at 100° for 15 minutes. Cool and read at 850 nm.

Germanium Dioxide.[57a] Dissolve 3 grams of sample in 10 ml of water with 15 grams of oxalic acid. Ignore traces of insoluble matter and do not heat excessively or too much oxalic acid may decompose. Finally add 1 ml of 30% hydrogen peroxide and heat for a couple of minutes to complete solution. Add 40 ml of 1:1 hydrochloric acid, and boil gently for 2 minutes to decompose most of the remaining hydrogen peroxide. Filter if necessary. Dilute to 100 ml and adjust to 50°. Add 2 ml of 20% potassium iodide solution and 2 ml of 5% sodium metabisulfite solution. Let stand for 15 minutes to reduce the arsenic. Cool to about 35°, a temperature at which oxalic acid will not crystallize from the sample. It often crystallizes from the blank, and if this happens, cool to room temperature and filter through glass wool without washing.

Add 15 ml of 0.1% solution of diethylammonium diethyldithiocarbamate solution in chloroform and shake for 1 minute. Drain the extract through a glass-wool plug. Add 0.5 ml of perchloric acid and 1 ml of nitric acid to the organic extract and boil gently. After the chloroform is expelled, heat at 175° until nitric acid is driven off and perchloric acid fumes are evolved copiously. Cool and add 2 ml of a mixture of 60 ml of hydrochloric acid, 40 ml of water, and 1 ml of nitric acid. Evaporate to fumes of perchloric acid. Then flame down to a volume of 0.2 ml. Use 2 ml of water to wash down the sides of the container. Add 1 drop of 0.02%

[55] L. Tomlinson, *Rep. U.K. At. Energ. Auth.* **AERE-M 1886**, 7 pp (1967).
[56] Hidehiro Goto and Yachiyo Kakita, *Sci. Rep. Res. Inst. Tohoku Univ.* Ser. A, **8**, 243–251 (1956).
[57a] C. L. Luke and Mary E. Campbell, *Anal. Chem.* **25**, 1588–1593 (1953).

solution of methyl red as the sodium salt and neutralize by dropwise addition of ammonium hydroxide. Bring back the red color by dropwise addition of 2:3 hydrochloric acid and add 0.5 ml in excess. Add 0.2 ml of 0.03% solution of potassium bromate, warm, and swirl to decolorize. Cool, and add 0.5 ml of 1% ammonium molybdate solution. Add 0.5 ml of 1.6% solution of hydrazine sulfate and dilute to 10 ml. Heat at 100° for 5 minutes and cool. Read promptly at 840 nm against water and subtract a blank.

Silicon.[57b] Transfer 0.5 gram of 150-mesh sample to a platinum dish and add 5 ml of 2% sodium hydroxide solution. Warm until violent reaction ceases and boil until dissolved, adding water if necessary. When reaction ceases, add 1 ml of 30% hydrogen peroxide and boil until excess peroxide is decomposed, Cool, dilute to 50 ml, and make just acid to Congo red paper with hydrochloric acid. Filter through paper to remove traces of insolubles and wash. Add 25 ml of hydrochloric acid and heat to 50°. Complete as for germanium dioxide from "Add 2 ml of 20% potassium iodide solution...."

Iron, Steel, and Ores.[58] Digest a 0.25-gram sample with 15 ml of 1:1 nitric acid. While still boiling, add 100 ml of hot water and 7 ml of 5% manganous sulfate solution. Add dropwise with stirring 2 ml of 3% potassium permanganate solution. Filter at once, and wash the precipitate of manganese dioxide carrying the arsenic with hot water. Dissolve the precipitate with 3 ml of 1:11 sulfuric acid, adding sufficient 3% hydrogen peroxide to complete the reaction. Boil to decompose excess hydrogen peroxide. Cool and neutralize with 12% sodium hydroxide solution. Add 4 ml of 1:11 sulfuric acid and 8 ml of 1% ammonium molybdate solution. It should now be 0.2–0.3 *N* with sulfuric acid. Dilute to 50 ml, heat at 100° for 10 minutes, and cool. Extract with 5-ml portions of amyl alcohol as long as color is extracted. Discard the extracts and add sufficient 1:11 sulfuric acid to build the final acidity to 0.4 *N*. Add 10 ml of 1% hydrazine sulfate and dilute to 100 ml. Heat at 100° for 10 minutes, cool, and read at 700 nm.

Steel.[59] Dissolve 0.25 gram of sample containing 0.01–0.06% arsenic in 3 ml of nitric acid and 6 ml of hydrochloric acid. Evaporate to dryness and take up the residue in 35 ml of hydrochloric acid. Add 10 ml of hydrochloric acid, 5 ml of 25% solution of stannous chloride in hydrochloric acid, and 3 ml of 30% solution of potassium iodide. Extract the arsenious iodide with 20 ml and 10 ml of chloroform. Extract the arsenic from the combined chloroform phases with 20 ml of water. To the aqueous extract, add successively, mixing after each addition, 5 drops of 1% iodine in 2% potassium iodide solution, 5 ml of 1% ammonium molybdate solution

[57b] *Ibid.*

[58] Arihiro Tominaga and Shiro Watanabe, *Jap. Anal.* **5**, 495–499 (1956); cf. Katsu Tanaka, *ibid.* **9**, 700–704 (1960).

[59] A. G. Fogg, D. R. Marriott, and D. Thorburn Burns, *Analyst* **97**, 657–662 (1972); cf. K. N. Ershova, *Tr. Vses. Nauch.-Issled. Inst. Zheleznodorozhn. Transp.* **1956** (116), 54–61; M. Jean, *Anal. Chim. Acta* **14**, 172–182 (1956); U. Bohnstedt and R. Budenz, *Z. Anal. Chem.* **159**, 95–102 (1957); S. W. Craven et al., *J. Iron Steel Inst.* (London) **188**, 331–337 (1958); W. R. Nall, *Analyst* **96**, 398–402 (1971); Elizabeth Migeon and Jean Migeon, *Chim. Anal.* (Paris) **43**, 276–279 (1961); Shizuya Maekawa, Yoshio Yoneyama, and Eiichi Fujimori, *Jap. Anal.* **11**, 493–500 (1962).

in 1 : 6 sulfuric acid, and 2 ml of 0.15% hydrazine sulfate solution. Heat at 100° for 10 minutes, cool, dilute to 50 ml, and read at 840 nm. For greater sensitivity, concentrate by extraction of the arsenomolybdate complex with a 1 : 4 mixture of hexanol and 3-methylbutanol.

Alternatively,[60] dissolve 0.4 gram of sample in 5 ml of hydrochloric acid and 2 ml of nitric acid. Add 15 ml of 2 : 1 sulfuric acid and evaporate to sulfur trioxide fumes. Cool and dilute to 90 ml. Add 20% titanous chloride solution until the ferric ion has been reduced, and add 1 ml in excess. Add 10 ml of 35% potassium iodide solution and set aside for 15 minutes. Extract the free iodine with 50 ml of ether and discard. Add 2 ml of 1% solution of α-mercapto-*N*,2-naphthylacetamide (thionalide) and extract the arsenic complex with 40 ml of ether. Extract the arsenic from the complex in ether with a mixture of 5 ml of 1 : 18 sulfuric acid and 3 ml of 2% sodium hypobromite solution. Add 1 ml of 5% solution of ammonium molybdate in 1 : 6 sulfuric acid and 0.5 ml of saturated solution of hydrazine sulfate. Dilute to 25 ml, heat at 40° for 30 minutes, and read at 650 nm.

Uranium, Uranium Compounds, and Thorium Oxide for Phosphorus and Arsenic.[61] The technic provides for first determining phosphorus, then arsenic. Dissolve a 0.5-gram sample in a mixture of nitric, boric, and hydrofluoric acids. Evaporate to dryness and take up in 20 ml of 1 : 15 nitric acid. Add 2 ml of 10% solution of ammonium molybdate and let stand for 15 minutes. Extract the molybdophosphoric acid by shaking with 5 ml and 5 ml of 3 : 1 chloroform-butanol. Develop the phosphomolybdenum blue by shaking the extract for 30 seconds with 10 ml of 1% solution of stannous chloride in 1 : 30 hydrochloric acid. Then read phosphorus in the aqueous phase at 700 nm.

After extraction of the phosphomolybdic acid, add to the aqueous solution 10 ml of 1 : 2 nitric acid. Extract the arsenomolybdic acid with 20 ml of 1 : 1 butanol–ethyl acetate. Wash this extract three times with 1 : 30 hydrochloric acid. Dilute the organic extract back to 20 ml with the same solvent and add 1 ml of 50% solution of stannous chloride in 1 : 1 hydrochloric acid. Read the arsenomolybdenum blue in the organic phase at 735 nm.

In analysis of uranium, arsenic as the arsenomolybdate can be extracted at 2–50 ppm by carbon tetrachloride from perchloric acid solution containing hydrochloric and hydriodic acids.[62] This separates it from phosphorus and silica.

For determination of arsenic in hydrolyzed uranium hexafluoride, an electrolytic cell is used for the evolution of arsine, which is oxidized to arsenate by absorption in a solution containing sodium bicarbonate and iodine. The excess iodine is removed with thiosulfate and arsenomolybdenum blue is developed with molybdate and stannous chloride.[63] The results are stated to be more accurate than by evolution of arsine by zinc and hydrochloric acid.

Arsenic can be developed as molybdenum blue with 2% ascorbic acid solution in 3.85 *M* perchloric acid containing 3% of molybdenum oxide or in 5 *M* sulfuric acid

[60] Shozo Nakaya, *Jap. Anal.* **12**, 483–486 (1963); cf. Shizuya Maekawa, Yoshio Yoneyama, and Eiichi Fujimori, *ibid.* **11**, 493–497 (1962).

[61] Tadao Hattori and Toshiaki Kuroda, *Furukawa Denko Jiho* **28**, 84–91 (1962).

[62] Daido Ishii and Tsugio Takeuchi, *Jap. Anal.* **11**, 118–119 (1962).

[63] D. Rogers and A. E. Heron, *Analyst* **71**, 414–417 (1946); J. A. Ryan, J. R. Sanderson, and T. Mason, *U.K. At. Energ. Auth., Ind. Group Hdq.* **SCS-R-308**, 10 pp (1959).

containing 3% of molybdenum oxide at a level of up to 0.01 mg of pentavalent arsenic per ml.[64] If phosphorus is present, the sum of phosphorus and arsenic must be determined, the arsenic reduced by sodium thiosulfate to the trivalent form (which does not react for determination of the phosphorus), and arsenic taken by difference. When heating at 100° for 10 minutes, up to 25-fold excess of silica and germanium not greater than the arsenic are tolerated. Excess of ascorbic acid reduces any ferric ion present.

Selenium.[65] As a mixed reagent, dilute 10 ml of 1% solution of ammonium molybdate in 1:5 sulfuric acid and 1 ml of 0.15% solution of hydrazine sulfate to 100 ml.

Dissolve a 2-gram sample in nitric acid and evaporate to dryness. Take up the residue in hydrochloric acid and precipitate the selenium by slowly adding an equal volume of saturated solution of sulfur dioxide. Stir well and let stand for at least 3 hours. Filter, and wash well with cold water. Pass nitrogen through the filtrate until the odor of sulfur dioxide can no longer be detected. Dilute to a known volume and take an aliquot containing 0.002–0.04 mg of arsenic. Add excess nitric acid and evaporate to sulfur trioxide fumes. Wash down the sides of the container and again take to sulfur trioxide fumes.

Take up the residue in 2 ml of water and neutralize to Congo red paper with ammonium hydroxide. Add 5 ml of the mixed reagent and heat at 100° for 30 minutes. Cool, dilute to a known volume, and read at 808 nm against a reagent blank.

Tellurium.[66] Dissolve 0.1 gram in 5 ml of 1:1 nitric acid and 2 ml of sulfuric acid. Evaporate to sulfur trioxide fumes, cool, and dilute to 250 ml. Raise a 5-ml aliquot to pH 7 by adding ammonium hydroxide. Add 1 ml of 1% solution of ammonium molybdate in 1:5 sulfuric acid and 0.4 ml of 0.15% solution of hydrazine sulfate. Heat at 50° for 30 minutes, cool, dilute to 25 ml, and read at 840 nm.

Ores.[67] Dissolve 1 gram of sample in 10 ml of 1:1 nitric acid. Boil to expel oxides of nitrogen, add 10 ml of water, and filter. Evaporate the filtrate to dryness and add 2 ml of sulfuric acid. Evaporate to dryness and take up in 20 ml of hydrochloric acid. Add 2 ml of 10% solution of phenylhydrazine sulfate and extract with 20 ml of carbon tetrachloride. Extract the arsenic from the organic layer with four successive 50-ml portions of water.

To a 50-ml aliquot, add excess of bromine water and boil off the excess bromine. Add 25 ml of 2.5% solution of ammonium molybdate in 1:2.5 sulfuric acid and 10 ml of 15% hydrazine sulfate solution. Dilute to 100 ml and heat at 100° for 10 minutes. Read at 700 nm.

Alternatively,[68] with less than 0.01% arsenic, dissolve a 2-gram sample in 20 ml of hydrochloric acid with the aid of 10 ml of 30% hydrogen peroxide and evaporate

[64]Louis Duval, *Chim. Anal.* (Paris) **51**, 415–424 (1969).
[65]James F. Reed, *Anal. Chem.* **30**, 1122–1124 (1958); E. Ebner, *Z. Anal. Chem.* **206**, 106–112 (1964).
[66]E. S. Beskova, G. I. Zhuravlev, and V. I. Dorofeeva, *Zavod. Lab.* **36**, 541 (1970).
[67]V. Fejfar and F. Pechar, *Chem. Prum.* **13**, 80–82 (1963).
[68]H. Pohl, *Erzmetall* **24**, 491–493 (1971).

to 3 ml. Add 20 ml of hydrochloric acid, 5 ml of hydrobromic acid, and 10 ml of perchloric acid. Extract arsenious bromide by shaking for 1 minute with 25 ml of benzene. Wash the benzene layer by shaking with 10 ml of hydrochloric acid. Extract the arsenic from the benzene layer by shaking with 25 ml of water. Add 10 ml of 1% ammonium molybdate solution in 1:5 sulfuric acid and 10 ml of 1% ascorbic acid solution. Boil for 2 minutes, cool, dilute to 100 ml, and read at 578 nm.

Pyrite.[69] For arsenic at 0.01–0.1% take 2 grams of sample, for 0.1–1% take 0.2 gram. Digest a finely ground sample with 10 ml of nitric acid and 20 ml of hydrochloric acid until all black particles disappear. Add 20 ml of 1:2 sulfuric acid and evaporate to white fumes. Cool and add 50 ml of water and 15 ml of 1:2 sulfuric acid. Boil until salts are dissolved and add 3 ml of 5% potassium iodide solution while still hot. Cool and add 5% sodium sulfite solution until the free iodine is decolorized. Dilute to 250 ml and filter.

Shake a 25-ml aliquot for 1 minute with 2 ml of 5% solution of potassium xanthate and 20 ml of carbon tetrachloride. Further extract the aqueous layer with 1 ml of the xanthate solution and 10 ml of carbon tetrachloride. Shake the combined carbon tetrachloride extracts with 25 ml of bromine water for 1 minute. Further extract the organic layer with 25 ml of water. Filter the combined aqueous extracts and dilute to 70 ml. Add 2 drops of 1.5% solution of phenol and mix well. Then add 5 ml of 1% solution of ammonium molybdate in 1:5 sulfuric acid and mix. Add 5 ml of a 0.15% solution of hydrazine sulfate and mix. Heat at 95° for 15 minutes, cool, and dilute to 100 ml. Read at 840 nm. The method tolerates 2% copper, 4% zinc and antimony.

Pyrite Slag.[70] Dissolve a 0.25-gram sample with 16 ml of nitric acid and 8 ml of hydrochloric acid. Dilute to 50 ml and filter. Dilute the filtrate to 250 ml. Take an aliquot containing up to 0.015 mg of arsenic and add ammonium hydroxide until ferric hydroxide begins to precipitate. Add 4% of the volume as hydrochloric acid to bring the solution to 0.5 *N* with that acid. Add 0.1 ml of 0.5% solution of ammonium molybdate in 1:5 sulfuric acid and let stand for 15 minutes. Extract the molybdoarsenate complex with 15 ml and 5 ml of 2:1 butanol–ethyl ether. Wash the combined extracts with 10 ml of 1:10 hydrochloric acid. To the solvent solution of the complex, add 0.1 ml of 0.1% solution of stannous chloride in 1:5 hydrochloric acid. Add a couple of grams of anhydrous sodium sulfate and dilute to 25 ml with the mixed solvent. Let stand for 15 minutes and read at 840 nm.

Chromic Oxide.[71] Dissolve a 0.5-gram sample in 5 ml of 1:1 hydrochloric acid. Add 1 ml of 0.1% solution of ferric sulfate. Dilute to 100 ml and add 0.1% potassium permanganate solution to a faint pink color. Neutralize with ammonium hydroxide and add 5 ml excess to precipitate ferric hydroxide and basic iron arsenate. Filter and wash with 1:100 ammonium hydroxide. Dissolve the precipi-

[69] G. Alterescu, *Rev. Chim.* (Bucharest) **16** (4), 221–222 (1965).
[70] T. Nikolov, M. Iosifova, and N. Lyakov, *Krimiya Ind.* **41**, 468–469.
[71] V. I. Kurbatova, L. N. Makogonova, G. N. Emasheva, and L. S. Fokina, *Tr. Vses. Nauchn.-Issled. Inst. Stand. Obraztsov.* **4**, 95–100 (1968).

tate from the paper with 5 ml of hot 1:1 hydrochloric acid. Add 2 ml of 0.15% solution of hydrazine sulfate and 1 ml of 40% potassium iodide solution. Dilute to 15 ml with hydrochloric acid and extract with 5 ml and 5 ml of carbon tetrachloride. Add 2 ml of nitric acid to the organic extract and evaporate to dryness. Take up in 2 ml of nitric acid and dilute to 40 ml. Add 5 ml of 1% ammonium molybdate solution and 1 ml of 40% stannous chloride solution inhydrochloric acid; dilute to 50 ml. Heat at 100° for 15 minutes, cool, and read at 840 nm. If the absorption is not as large as desired, extract with 5 ml and 5 ml of amyl alcohol and read the combined extracts.

Acid Copper-Plating Baths.[72] As reagent, mix 25 ml of 1% ammonium molybdate tetrahydrate solution in 1:5 sulfuric acid with 2.5 ml of 0.15% hydrazine sulfate solution and dilute to 50 ml.

To 100 ml of sample containing 1–6 mg of arsenic, if an equivalent amount of iron is not already present, add 0.1 gram of ferric sulfate. Heat nearly to boiling and add ammonium hydroxide until precipitation of iron is complete and the copper is all present as the ammonia complex. Filter while hot and wash well on the filter with hot 1:20 ammonium hydroxide. Discard the filtrate. Dissolve the precipitate from the paper with 25 ml of 1:10 sulfuric acid. Add ammonium hydroxide to the solution until precipitation is complete, heat nearly to boiling, and filter on the original paper. Wash well with hot 1:20 ammonium hydroxide and discard the filtrate. Dissolve the precipitate from the paper with 20 ml of 1:4 sulfuric acid. Add 20% sodium hydroxide solution to raise the solution to pH 3–7, and dilute to 100 ml.

To a 6-ml aliquot, add 0.01 *N* potassium permanganate dropwise to a pink coloration. Discharge the color with a few drops of 0.15% hydrazine sulfate solution. Add 40 ml of water, followed by 4 ml of the reagent. Dilute to 50 ml and heat at 100° for 20 minutes. Cool, adjust to 50 ml, and read at 840 nm against water.

Sulfuric Acid.[73a] Dilute a sample expected to contain 0.005 mg of arsenic to 10 ml. Add 0.16% potassium permanganate solution until the color remains pink. Add 2 ml of 3% ammonium molybdate solution and 1 ml of 1% ascorbic acid solution. Dilute to 25 ml and heat at 100° for 10 minutes. Cool and read at 820 nm.

Phosphoric Acid and Phosphates.[73b] Dissolve 0.5 gram in 5 ml of water and add 30 ml of hydrochloric acid. Add 3 ml of 2% solution of potassium iodide containing 1% of iodine. Extract the arsenate with 20 ml and 10 ml of chloroform. Back-extract the arsenate with 20 ml of water. Filter, add 0.03 *N* potassium permanganate dropwise to give a pink color, and set aside for 5 minutes. Decolorize with 0.15% solution of hydrazine sulfate and add 2 ml excess. Add 5 ml of 0.1% solution of ammonium molybdate in 6 *N* sulfuric acid. Heat at 100° for 15 minutes, cool, dilute to 100 ml, and read at 840 nm.

[72] T. A. Zot'eva, *Zavod. Lab.* **39**, 950–951 (1973).
[73a] S. Nakao, *Ryusan* **14**, 131–142 (1961).
[73b] E. D. Lapkina and G. B. Beisembaeva, *Zavod. Lab.* **42**, 783 (1976).

Seawater.[74] As a reagent, mix 50 ml of 1:5 sulfuric acid, 15 ml of 4.8% solution of ammonium molybdate tetrahydrate, 5 ml of 0.274% solution of potassium antimony tartrate, and 30 ml of 1.76% solution of ascorbic acid. Dilute to 125 ml.

As sample, use 1 liter of filtered seawater expected to contain about 2 μg of arsenic. Add 4 ml of 5% ascorbic acid solution and heat to 100° to render the arsenic trivalent. Let cool for 10 minutes and add 2 ml more of 5% ascorbic acid solution. At room temperature add 10 ml of 1:5 sulfuric acid and stir in 7 ml of 2% solution in acetone of thionalide, which is α-mercapto-*N*,2-naphthylacetamide. Stir for 5 minutes, set aside for 10 minutes, and boil for 10 minutes to evaporate the acetone. Cool with stirring and set aside overnight. Filter on paper with suction and wash with water. Heat the paper and precipitate with 7.5 ml of nitric acid until the solution is pale yellow and evaporate the nitric acid, completing it in a stream of carbon dioxide to avoid charring. Dissolve the light yellow residue in 1 ml of *N* sulfuric acid and add 2 ml of the freshly prepared reagent. Dilute to 10 ml, and after 30 minutes read at 866 nm against a blank. To prepare the blank, use seawater from which the arsenic has been coprecipitated with ferric hydroxide.

Marine Plants. Wet-ash samples with nitric acid and evaporate to dryness. Take up in 100 ml of 0.1 *N* sulfuric acid and proceed as for seawater from "Add 4 ml of 5% ascorbic acid...."

Wood.[75] This procedure is designed for determination of arsenic in wood treated with arsenic oxide as a preservative. As a reagent, mix 10 ml of 1% solution of ammonium molybdate in 1:6 sulfuric acid with 10 ml of 0.15% solution of hydrazine sulfate and dilute to 100 ml.

Slice the wood in sections not more than 300 nm thick. Heat a section of appropriate size to contain not more than 0.08 mg of arsenic with 20 ml of the reagent until the blue color develops and continue for 15 minutes longer. Cool, dilute to 25 ml, and read at 840 nm.

For analysis of preserved softwoods, convert to sawdust and dry.[76] Extract with 50 ml of 1:6 sulfuric acid and 10 ml of 30% hydrogen peroxide at 75° for 20 minutes. Mix with 150 ml of water, and allow to cool and settle. Filter and develop an aliquot as above.

Hair and Nails.[77] Digest the sample with 10 ml of sulfuric acid, 20 ml of fuming nitric acid, and 5 ml of perchloric acid until clear. Cool and neutralize with 40% sodium hydroxide solution. Add 4 ml of 1:9 sulfuric acid and 2 ml of 4% ammonium molybdate solution. Dilute to 50 ml and extract with 5 ml of amyl alcohol. Discard the extract and add 2 ml of 6% ascorbic acid solution to the aqueous layer. Heat at 100° for exactly 10 minutes, cool, and extract with 5-ml portions of 1:1 butanol–ethyl acetate. Dilute the combined extracts to 25 ml and read at 700 nm.

---

[74] J. E. Portmann and J. P. Riley, *Anal. Chim. Acta* **31**, 509–519 (1964).
[75] A. I. Williams, *Analyst* **93**, 611–617 (1968).
[76] A. I. Williams, *ibid.* **95**, 670–674 (1970).
[77] Bela Rengei, *Deut. Z. Ges. Gerichtl. Med.* **47**, 600–613 (1958).

Miscellaneous

Antimony can be extracted as the pentavalent chloride with 2-chloroethyl ether and discarded before extracting the arsenic as the tribromide.[78]

Diethylammonium diethyldithiocarbamate in chloroform is an appropriate reagent for extraction of arsenic from interfering ions before determination as arsenomolybdenum blue, as typified by the procedure given for arsenic in lead.[79]

Arsenic volatilized as arsine is absorbed on mercuric bromide paper and read by reflectance with 90% recovery of 0.1–0.5 μg.[80]

For arsenic in hydrofluoric acid, distill 400 ml to leave a residue of less than 1 ml. Take up in 10 ml of water and add 0.5% lanthanum nitrate solution to precipitate the residual fluoride. Separate by centrifuging. Mix a 5-ml aliquot with 10 ml of hydrochloric acid and extract the arsenic with benzene for determination as the arsenomolybdenum blue.

Arsenic in organic compounds requires combustion by an oxygen flask method.[81] Then arsenic is determined as arsenomolybdenum blue.

For determination of arsenic in zinc sulfide phosphors, 25 ml of hydrochloric acid is added in 5-ml increments in an atmosphere of argon.[82] The evolved gases are absorbed in 180 ml of bromine water. After evaporation to 30 ml, the solution is diluted to 100 ml and an aliquot is developed as arsenomolybdate.

When the sample solution contains germanium, silica, phosphorus, and arsenic, the germanium is first extracted as the phenylfluorone complex with isoamyl alcohol.[83] Then treatment of the aqueous phase with perchloric acid polymerizes the soluble silica. Then form the arsenomolybdate and phosphomolybdate in 0.7 *M* hydrochloric acid and extract the phosphomolybdate from 6 ml of solution with 10 ml of isobutyl acetate.[84] Then add to the aqueous phase 1 ml of 60% perchloric acid and 1 ml of 1% hydrazine sulfate solution to form the arsenomolybdate. Read at 690 nm after 1 hour.

Many reducing agents have been used in production of molybdenum blue, whether for determination of arsenic, phosphate, silicate, or germanium. An addition to the list is *N*,N′-dimethyl-2-pyrazoline-3,4-dicarboxamide.[85]

Although hydroxides of iron, lanthanum, and magnesium are used as collectors in determining 0.01–0.1 ppm of arsenic, ferric hydroxide is preferable because it precipitates over the widest pH range and yields the precipitate most easily handled.[86] One mg of arsenic can be coprecipitated with 10–20 mg of iron as ferric hydroxide from 200–500 ml of solution with little of other metals such as copper and lead. No more than minor amounts of phosphorus, tin, bismuth, and antimony are permissible. The precipitate is then dissolved in 15 ml of 1 : 1 hydrochloric acid and reduced with titanous chloride solution. Then after addition of 5 *M* potassium iodide solution to 0.5 *M*, hydrochloric acid is added to more than 6 *N*. Arsenic

[78] Zygmunt Marczenko, Miroslaw Mojski, and Hubert Skibe, *Chem. Anal.* (Warsaw) **17**, 881–889 (1972).
[79] H. Furrer, *Mitt. Lebens. Hyg.* (Bern) **52**, 286–298 (1961); *ibid.* **54**, 291–297 (1963).
[80] N. C. Maronowski, Robert E. Snyder, and Ralph O. Clark, *Anal. Chem.* **29**, 353–357 (1957); June M. O. Damon and M. G. Mellon, *ibid.* **30**, 1849–1855 (1958).
[81] R. Belcher, A. M. G. Macdonald, S. E. Pfang, and T. S. West, *J. Chem. Soc.* **1965**, 2044–2048.
[82] L. A. Gromov, V. D. Lukonenko, and V. A. Trofimov, *Zh. Prikl. Khim., Leningr.* **43**, 1438–1444 (1970).
[83] F. A. Sorrentino and J. Paul, *Microchem. J.* **15**, 446–451 (1970).
[84] J. Paul, *Mikrochim. Ichnoanal. Acta* **1965**, 830–835.
[85] Yu. G. Zhukovskii, *Zh. Anal. Khim.* **19**, 1361–1365 (1964).
[86] Zygmunt Marczenko and Miroslaw Mojski, *Chem. Anal.* (Warsaw) **14**, 495–504 (1969).

iodide is extracted with benzene, transferred to water, and developed as arsenomolybdenum blue.

For arsenic in biological material, wet-ash a 25-gram sample containing 0.002–0.08 mg of arsenic with nitric acid, sulfuric acid, and ammonium sulfate.[87] Reduce with hydrazine sulfate, generate the arsine, and burn it to form the arsenic mirror of the Marsh test. Dissolve the mirror in 3 ml of 0.16 *N* sodium hypobromite. Acidify with 5 ml of 1:18 sulfuric acid and add 1 ml of 5% ammonium molybdate solution in 1:6 sulfuric acid and 1 ml of 2% solution of hydrazine sulfate. Read at 830 nm against a reagent blank after 2 hours $\pm$ 10 minutes. As an alternative, absorb arsine in 1 ml of 2% sodium hydroxide and 3 ml of half-saturated bromine water.[88]

For analysis of water, arsenic can be quantitatively extracted with chloroform from 4 *N* sulfuric acid containing 0.2 *M* potassium iodide by adding 0.1% ethanolic thioanalide.[89] Alternatively it is collected by coprecipitation with ferric hydroxide.[90] After solution of the precipitate in nitric acid, arsenic is converted to the xanthate, extracted with carbon tetrachloride, and converted to arsenomolybdenum blue.

Arsenic and antimony can be recovered by electrolysis from a hydrochloric acid solution of the residue from a biological sample.[91] Arsenic in the deposit is then converted to the arsenomolybdenum blue.

Arsenomolybdic acid is reducible only when the acidity is below 0.7–0.8 *N*.[92] Hydrazine chloride is a better reducing agent than is the sulfate.

For analysis of arsenic in bismuth, the latter is extracted from the sample solution as the iodo complex with cyclohexanone, then discarded.[93] Thereafter the arsenic is determined as arsenomolybdenum blue without prior distillation as arsine.

In analysis of copper, zinc, and lead, the sample is dissolved and converted to a solution in hydrochloric acid.[94] Addition of a solution of sodium molybdate forms the yellow complex, which is extracted with 1:2 ether-butanol. After treatment of the extract with stannous chloride, arsenomolybdenum blue is formed. In determination of 0.02 mg of arsenic, there is no interference by mg amounts of silica, germanium, iron, manganese, and nickel.

Acetone stabilizes the system and increases the color intensity of arsenomolybdenum blue.[95]

For arsenic in zinc containing less than 0.5% arsenic, dissolve a sample in acid.[96] Oxidize with permanganate and add ammonium molybdate. Extract the molybdoarsenate complex with isobutyl alcohol. Wash the extract with dilute sulfuric acid, reduce to molybdenum blue with stannous chloride, and read at 645 nm.

[87]A. Dyfverman and R. Bonnichsen, *Anal. Chim. Acta* **23**, 491–500 (1960).

[88]Morris B. Jacobs and Jack Nagler, *Ind. Eng. Chem., Anal. Ed.* **14** (5), 442–444 (1942).

[89]Shozo Nakaya, *Jap. Anal.* **12**, 241–247 (1963).

[90]Ken Sugawara, Motoharu Tanaka, and Satoru Kanamori, *Bull. Chem. Soc. Jap.* **29**, 670–673 (1956); cf. David L. Johnson and Michael E. Q. Pilson, *Anal. Chim. Acta* **58**, 289–299 (1972).

[91]P. S. Kislitsyn and L. O. Druzhinin, *Tr. Inst., Khim., Akad. Nauk Kirg. SSR* **7**, 79–92 (1956).

[92]I. M. Litinskii and T. D. Pozdnyakova, *Nek. Vop. Farm., Sb. Nauch. Tr. Vyssh. Farm. Ucheb. Zaved. Ukr. SSR* **1956**, 58–62.

[93]E. Jackwerth, *Z. Anal. Chem.* **211**, 254–265 (1965).

[94]R. A. Karanov and A. N. Karolev, *Zh. Anal. Khim.* **20**, 639–640 (1965).

[95]R. A. Chalmers and A. G. Sinclair, *Anal. Chim. Acta* **33**, 384–390 (1965).

[96]B. Luft and E. Rill, *Neue Huette* **18**, 372–374 (1973).

For analysis of rhenium, form arsenomolybdic acid and phosphomolybdic acid.[97a] Extract the arsenomolybdic acid with 1 : 1 : 2 isobutyl alcohol–ethyl acetate–chloroform, then phosphomolybdic acid with 1 : 3 isobutyl alcohol–chloroform. In each case, remove unreacted molybdate by filtration through sodium sulfate.

Strip the arsenic or phosphorus from the extract with 5% potassium hydroxide solution and determine indirectly as the molybdate by zinc dithiol at 675 nm or by nitrosulfophenol M and hydroxylamine at pH 2.5–4 and 650 nm. The technic will determine 0.06–0.12% arsenic or 0.07–0.12% phosphorus.

For phosphorus and arsenic in a solution containing large amounts of tungsten and rhenium add aluminum ion and hydrogen peroxide. Make the solution alkaline to coprecipitate the phosphate and arsenate with aluminum hydroxide.[97b] Filter and dissolve in a minimal amount of 1 : 1 nitric acid.[97c] Add 1 ml of 10% solution of sodium molybdate and 1 ml of 15% solution of sodium chloride. Adjust to pH 1.5 and extract phosphorus with 3 ml, 3ml, and 3 ml of 2 : 3 isobutyl alcohol-chloroform. Read phosphorus at 372 nm. Increase the acidity of the aqueous layer to 10 *N* nitric acid and extract the arsenic with 2.5 ml of 1 : 1 ethyl acetate and *n*-butanol. Further, add 5 ml of chloroform and continue to shake for 1 minute. Make another similar extraction of arsenic. Read the combined extracts at 312 nm.

For arsenic in potatoes, ash with magnesium nitrate.[98] The solution of the ash in acid treated with acid-molybdate leads to extraction of phosphorus as molybdophosphoric acid with a mixture of butanol and chloroform. After adjustment of the acidity, arsenomolybdenum blue is developed.

BOUGAULT REAGENT

This nephelometric method is read photometrically.[99] Nitrate must be absent. For determination of arsenic in copper, the calibration curve must be prepared from arsenic-free copper plus known amounts of arsenic.

Procedure. *Copper Alloys.* As a reagent, dissolve 20 grams of monosodium phosphate in 20 ml of water and 200 ml of hydrochloric acid. Filter, let stand for a week, and refilter.

Dissolve 0.5 gram of sample in 25 ml of 1 : 2.5 sulfuric acid and 4 ml of nitric acid. Boil off oxides of nitrogen and add 2 ml of 1% potassium permanganate solution. Boil until clear, add 2 ml of saturated solution of sodium persulfate, and again boil until clear. Evaporate to copious fumes and continue to fume for 5 minutes. Cool, and dilute to 100 ml. Mix a 10-ml aliquot with 10 ml of reagent and heat at 100° for 30 minutes. Cool, adjust to 20 ml with 9 *N* hydrochloric acid, and read at 612 nm.

[97a]T. M. Malyutina, S. B. Savvin, V. A. Orlova, V. A. Mineeva, and T. A. Kirillova, *Zh. Anal. Khim.* **29**, 925–928 (1974).
[97b]E. I. Shelikhina, M. A. Chernysheva and V. P. Antonovich, *Zavod. Lab.* **42**, 1057–1058 (1976).
[97c]Toshiyasa Kiba and Mitsuru Ura, *J. Chem. Soc. Jap., Pure Chem. Sect.* **76**, 520–524 (1955).
[98]Donald J. Lisk, *J. Agr. Food Chem.* **8**, 121–123 (1960).
[99]M. C. Steele and L. J. England, *Analyst* **82**, 595–597 (1957).

BUTYLRHODAMINE B

Butylrhodamine B is the butyl ester of rhodamine B (basic violet 10); it is combined with molybdate and arsenic, presumably as a ternary complex.[100] This is extracted for reading. Silicate and phosphate must be absent. The result can also be read by fluorescence.

Procedure. *Brine.* Heat 35 ml of sample containing 10 grams of sodium chloride to 40° and add 5 ml of 1:3 sulfuric acid. Add 1.5 ml of 20% potassium iodide solution and 1 ml of 5% ascorbic acid solution to reduce arsenic to the trivalent state. Set aside for 15 minutes, then extract the arsenic with 5 ml and 5 ml of 1% solution of diethyldithiocarbamic acid in chloroform. Wash the combined extracts with 2.5 ml and 2.5 ml of 1:35 sulfuric acid. Add 5 ml of water to the chloroform layer and evaporate the latter. Then add 2 ml of nitric acid to the aqueous residue and evaporate to dryness. To the residue add 1.5 ml of 1:3 sulfuric acid, 1.3 ml of 0.01 *N* ammonium molybdate, and 0.3 ml of 0.02% solution of butylrhodamine B. After 10 minutes extract with 7 ml of ether by shaking for 1 minute. Wash the ether with 5 ml and 5 ml of water to remove excess of butylrhodamine B. Dilute the ether layer to 10 ml and read.

CURCUMIN

The reaction product of arsenic and curcumin follows Beer's law up to 10 ppm.[101] Phosphate does not interfere if less than 20 mg is present, but nitrate, chloride, sulfate, and acetate do.

Procedure. To a sample containing less than 0.05 mg of pentavalent arsenic in ethanol, add 1 ml of 0.2% solution of curcumin in ethanol, 1 ml of 5% ethanolic oxalic acid, and 0.1 ml of 1:11 hydrochloric acid. Heat at $70 \pm 5°$ for 3.5 hours. Dissolve the residue in 10 ml of butanol and read at 545 nm.

COLLOIDAL ARSENIC

The colloidal dispersion of arsenic in an acid medium can be read at 420 or 915 nm.[102] In an alkaline medium it is read at 765 nm.[103] Beer's law is followed for up

[100] A. K. Babko, Z. I. Chalaya, and V. F. Mikitchenko, *Zavod Lab.* **32**, 270–273 (1966).
[101] Kazuo Hiiro and Takashi Tanaka, *J. Chem. Soc. Jap., Pure Chem. Sect.* **83**, 1258–1262 (1962).
[102] M. Cyrankowska, *Chem. Anal.* (Warsaw) **8**, 679–684 (1963).
[103] L. Vignoli and B. Cristau, *Bull. Soc. Pharm. Marseille* **5**, 207–212 (1956).

to 80 μg of arsenic per ml. Interference is by elements that give colored solutions or are reduced by hypophosphite, such as silver, mercury, selenium, and tellurium. The development of the colloid can be sensitized by stannous chloride and the colloid stabilized with polyvidone.[104] This reagent is appropriate for alloys containing 8% antimony, 0.1% silver, or 3% tin, and about 0.5% arsenic.

Procedure. As ammonium citrate reagent, add 25 grams of citric acid to water, neutralize with ammonium hydroxide, and dilute to 100 ml. Dissolve 0.5 gram of sample in 10 ml of 1:3 nitric acid. Add 5 ml of 1:1 nitric acid and evaporate nearly to dryness. Dissolve the residue by boiling in 20 ml of 1:1 nitric acid. Add 50 ml of hydrochloric acid and dilute to 100 ml. To an aliquot containing less than 0.2 mg of arsenic, add 2 ml of 1% copper sulfate solution and 10 ml of freshly prepared 15% solution of sodium hypophosphite in 1:6 hydrochloric acid. Warm for 7 minutes, boil for 2 minutes, and dilute to 50 ml with 1:1 hydrochloric acid. Stir vigorously while cooling and read at once.

Ores in General.[105] Dissolve 0.5 gram of a sample containing 0.02–0.5% of arsenic in 5 ml of nitric acid and 5 ml of hydrochloric acid and evaporate to dryness. Moisten the residue with hydrochloric acid, add 10 ml of 1:1 sulfuric acid, and heat to sulfur trioxide fumes. Cool, add 2 ml of water, and again heat to sulfur trioxide fumes. Cool, take up in 50 ml of water, and add ammonium hydroxide until a slight odor is detectable. Filter on a No. 2 filter crucible and wash with hot 1:20 ammonium hydroxide. Dissolve the precipitated hydroxides with 2 ml of 1:1 hydrochloric acid, and rinse the crucible with 2 ml of 1:1 hydrochloric acid. Add 1 ml of 1% cupric sulfate solution in 1:1 hydrochloric acid, 0.5 ml of 1% gelatin solution, and 2.5 ml of a 40% solution of sodium hypophosphite in 1:1 hydrochloric acid. Dilute to 10 ml with 1:1 hydrochloric acid and heat at 100° for 10 minutes. Cool, adjust to volume, and read at 420 nm against a reagent blank. In the presence of molybdenum, cobalt, vanadium, or nickel, it may be necessary to vary the wavelength used.

Iron Ore.[106] Dissolve 1 gram of sample in 5 ml of hydrochloric acid and 2 ml of nitric acid. Add 2 ml of perchloric acid and take to fumes. Take up with 20 ml of water and add 40% stannous chloride solution in hydrochloric acid until the iron content is decolorized. Add 20 ml of 15% solution of sodium hypophosphite in 1:1 hydrochloric acid and filter. Add 5 ml of 10% sodium iodide solution and warm for 7 minutes. Boil for 2 minutes, cool with vigorous stirring, dilute to 50 ml with 1:1 hydrochloric acid, and read at 420 nm at once.

---

[104] B. Cristau, *Ann. Pharm. Fr.* **16**, 26–38 (1958).
[105] D. P. Shcherbov and K. I. Don, *Sovrem. Metody Anal. Met., Zb.* **1955**, 160–166.
[106] Takeo Morimoto and Takami Mizuno, *Fuji Seitetsu Giho* **8**, 207–213 (1959).

MOLYBDENUM THIOCYANATE

After development as arsenomolybdenum blue, the molybdenum so complexed can be separated and read as molybdenum thiocyanate. Arsenic present in natural water as arsenite and arsenate, along with phosphate, is determined as molybdenum thiocyanate and read at 475 nm.[107] The phosphate as molybdophosphate is extracted with 2:3 butanol-chloroform. The arsenite is oxidized to arsenate with potassium iodate, and the arsenomolybdate is extracted with 3:7:3 butanol-chloroform-acetone. The molybdenum of the arsenomolybdate is extracted from the organic layers by sodium hydroxide solution and converted to thiocyanate. Tolerances, in mg/liter, are as follows: germanate, 2; silicate, 4; titanic ion, 0.02; vanadate, 0.02; and tungstate, 0.2.

Procedure. *Red Phosphorus.*[108] Boil 0.5 gram of sample with 5 ml of 1:1 nitric acid and evaporate to a syrup. Add 25 ml of hydrochloric acid and 0.5 gram of potassium iodide. Extract the arsenic tri-iodide with 5 ml and 5 ml of benzene. Reextract the arsenic from the combined organic layers with 5 ml and 5 ml of water. Add an excess of iodine solution to oxidize the arsenic to arsenic acid. Add 1 ml of hydrochloric acid to make the solution *M* with that acid and add 0.5 ml of 5% ammonium molybdate solution. Extract the blue molybdophosphate with amyl alcohol and discard. Then extract the molybdoarsenate with 2 ml of 1:1 butanol-ethyl acetate. Wash this extract with 1:10 hydrochloric acid saturated with butanol. Then extract the molybdenum with 0.5 ml of 1:2 ammonium hydroxide. To the ammoniacal extract, add 0.4 ml of hydrochloric acid to make the solution 3 *M* in that acid. Add 1.5 ml of 25% ammonium thiocyanate solution and a drop of 0.04% copper sulfate solution. After 10 minutes extract the molybdenum thiocyanate with 2 ml of 1:1 butanol–ethyl acetate. Read at 465 nm.

MOLYBDOARSENOVANADATE

This reagent permits determination of phosphorus in acid so strong that arsenic does not react, then determination of both phosphorus and arsenic at a lower acidity, giving arsenic by difference. Phosphorus and arsenic can be read up to 0.75% in the sample.

Procedures

Deoxidized Copper Alloy

ARSENIC, PHOSPHORUS ABSENT.[109] As a mixed color reagent, dissolve 3.6 grams of sodium vanadate in 800 ml of water, add 48 grams of sodium molybdate

[107] Ken Sugawara and Satoru Kanamori, *Bull. Chem. Soc. Jap.* **37**, 1358–1363 (1964).
[108] P. V. Kristalev, L. B. Kristaleva, and N. A. Shor, *Tr. Kom. Anal. Khim., Akad. Nauk SSSR* **16**, 19–23 (1968).
[109] H. C. Baghurst and V. J. Norman, *Anal. Chem.* **29**, 778–782 (1957).

dihydrate, and dilute to 1 liter. As neutralized copper nitrate solution, dissolve 15 grams of copper in the minimum possible amount of nitric acid and boil off the nitrous oxide fumes. Dilute to about 300 ml and neutralize with ammonium hydroxide to the first appearance of a permanent precipitate. Add 1:7 nitric acid until the precipitate just dissolves, and dilute to 500 ml.

Dissolve 1.5 gram of sample in 10 ml of nitric acid. Boil off nitric oxide fumes and expel most of the acid by evaporation to a small volume. The arsenic has been oxidized to pentavalency. Dilute to 30 ml and neutralize with 1:1 ammonium hydroxide to the initial permanent precipitate. Add 2 *N* nitric acid dropwise until the precipitate just dissolves. Dilute to 100 ml and divide into 50-ml portions, one being used as the sample blank.

To one portion add 10 ml of 2 *N* nitric acid followed by 25 ml of the mixed color reagent. After 10 minutes dilute to 100 ml. To the second flask add 10 ml of 2 *N* nitric acid and dilute to 100 ml. As a reagent blank, mix 10 ml of 2 *N* nitric acid with 25 ml of mixed color reagent and dilute to 100 ml. Read the absorption of each at 420 nm. Subtract both blanks from the sample reading. The temperature difference at the time of reading must not exceed 0.5°.

Arsenic and Phosphorus. Proceed to dissolve 1.5 grams of sample as for arsenic. Dissolving with nitric acid ensures that all the phosphorus is present as orthophosphate. After dilution to 100 ml, take three 20-ml aliquots. To one portion add 40 ml of 4 *N* nitric acid and 25 ml of mixed color reagent, then dilute to 100 ml. Set aside for 15 minutes. Read the phosphorus at 420 nm as present in 1.6 *N* acid. To the second portion add 10 ml of 2 *N* nitric acid and 25 ml of the mixed color reagent. Dilute to 100 ml. Place this and a reagent blank in a bath at 17–30° and read the sum of arsenic and phosphorus at 420 nm. Prepare a sample blank with the third portion by adding 10 ml of 2 *N* nitric acid and diluting to 100 ml.

Three calibration curves are required: (1) phosphorus in 1.6 *N* acid containing 10 ml of the neutralized copper nitrate solution; (2) phosphorus in 2 *N* nitric acid containing 10 ml of the neutralized copper nitrate solution; and (3) arsenic in 2 *N* nitric acid containing 10 ml of the neutralized copper nitrate solution. The added copper nitrate solution corrects for that in the sample.

Calculation

Absorption due to phosphorus = absorption at 1.6 *N* acid − absorption of blank

Absorption due to arsenic = absorption of solution in 2 *N* acid
− (absorption of reagent blank
+ absorption of sample blank
+ phosphorus correction)

QUERCETIN

Quercetin, which is 5,7,3′,4′-flavonol, develops with pentavalent arsenic in ethanol.[110] The absorption is decreased by water. Trivalent arsenic, phosphate, and borate are tolerated, but the sample must not be alkaline.

Procedure. Mix 50 ml of solution containing up to 20 ppm of pentavalent arsenic with 1 ml of 0.03% ethanolic quercetin. Evaporate to dryness at $70 \pm 5°$ over a period of 90 minutes. Take up the residue in 10 ml of ethanol and set aside for 30 minutes. Read at 398 nm against a reagent blank.

QUINOLINE MOLYBDATE

Arsenic can be precipitated by a quinoline molybdate reagent and later extracted by tetramethylenedithiocarbamate in chloroform.[111]

Procedure. As reagent, dissolve 250 grams of sodium molybdate in 500 ml of water and pour into 460 ml of hydrochloric acid. Add a drop of 30% hydrogen peroxide. Mix with a solution of 28 ml of quinoline in 600 ml of 1 : 1 hydrochloric acid. Heat to boiling, let stand for 24 hours, and filter.

To 50 ml of neutral sample solution containing less than 0.01 mg of arsenic, add 1 gram of potassium chlorate and 5 ml of hydrochloric acid. Heat to boiling, add 200 ml of 10% solution of tartaric acid, and 20 ml of the quinoline molybdate reagent. Mix well and filter. Wash the precipitate with 1 : 5 hydrochloric acid, then with water, until the washings are neutral.

Wet-ash the precipitate with 5 ml of nitric acid and 2 ml of sulfuric acid, then evaporate to sulfur trioxide fumes. Take up in 10 ml of water and add 1 ml of 10% solution of hydrazine sulfate. Add 1 ml of 0.05% solution of EDTA and 2 ml of 10% solution of ammonium citrate. Adjust to pH 4.5 with ammonium hydroxide. Extract with 5 ml and 4 ml of 0.25% solution of tetramethylenedithiocarbamate in chloroform. Dilute the combined extracts to 10 ml and read at 350 nm.

SILVER DIETHYLDITHIOCARBAMATE

The use of a 0.5% solution of silver diethyldithiocarbamate in pyridine as an absorbant for arsine gives a comparatively simple technic.[112] An alternative ab-

[110]Takashi Tanaka and Kazuo Hiiro, *Jap. Anal.* **11**, 1180–1184 (1962).

[111]S. Meyer and O. G. Koch, *Z. Anal. Chem.* **158**, 434–438 (1957); E. Kovacs, H. Guyer, and W. Lüscher, *ibid.* **208**, 321–328 (1965).

[112]Vladimir Vasak and Vaclav Sedivek, *Chem. Listy* **46**, 341–344 (1952); *Collect. Czech. Chem. Commun.* **18**, 64–72 (1953); cf. E. Jackwerth, *Arch. Pharm.* (Berlin) **295**, 779–795 (1962); British Standards Institution, *BS 4404*, 11 pp (1968); *BS 2690*, Pt. **19**, 15 pp (1972); Halina Bahr and Henryk Bahr, *Chem. Anal.* (Warsaw) **16**, 427–431 (1971).

sorbant is 2.5 grams of silver diethyldithiocarbamate and 1.65 grams of ephedrine per liter of chloroform.[113] Another alternative is silver diethyldithiocarbamate and hexamine in chloroform, for reading at 492 nm. Because of possible interference by hydrogen sulfide, the reagent is preceded by lead acetate as an absorber. It is desirable to have the lead acetate on cotton that has been dried to the touch.

In analysis of iron and steel, the concentration of iron in the solution from which arsine is evolved must be standardized.[114] Antimony up to 0.1% or copper up to 1% does not interfere. Copper does not interfere up to 0.25 mg per ml of test solution if sufficient stannous chloride is added before evolving the arsine.[115] Mercury, cobalt, and nickel interfere. At 7.3 mg of stannous chloride dihydrate per ml, there is no interference by 0.19 mg of iron per ml, 10 mg of tungsten per ml, or 0.4 mg of vanadate.

Either hydrochloric or sulfuric acid may be used with zinc for evolution of arsine, and the relatively simple apparatus shown in Figure 8-7 has been elaborated on.[116]

Although the peak absorption is at 522 nm, it may be more accurate to read at 540 nm if antimony is present, even up to 20 times the arsenic.[117] The color developed can be read up to 0.04 mg of arsenic per 100 ml of pyridine.[118] It is a highly successful reagent, but it requires control.[119] The reagent must contain more than 20 mg of silver diethyldithiocarbamate per 10 ml of pyridine. The red product as studied by thin-layer chromatography appears to be a colloidal form of silver.

Arsenic at 6.5–10 ppm in the solution of copper or a copper salt can be isolated by coprecipitation as magnesium arsenate on magnesium ammonium phosphate.[120]

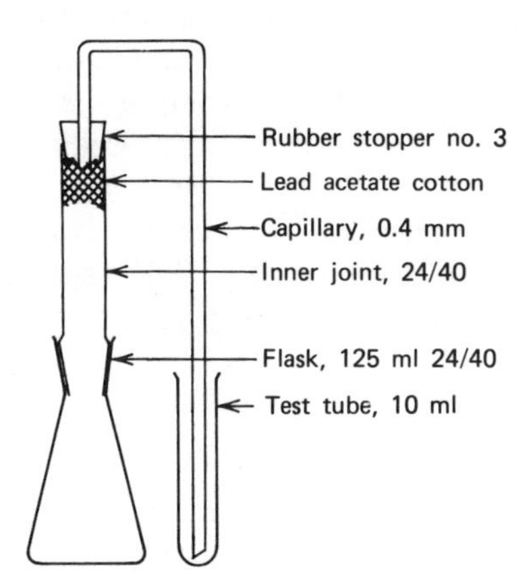

Figure 8–7 Apparatus for the evolution of arsine, followed by absorption by silver diethyldithiocarbamate.

[113] H. Bode and K. Hachmann, *Z. Anal. Chem.* **229**, 261–266 (1967); John F. Kopp, *Anal. Chem.* **45**, 1786–1787 (1973).
[114] O. P. Bhargava, J. F. Donovan, and W. G. Hines, *Anal. Chem.* **44**, 2902–2904 (1972); Z. Vecera and B. Bieber, *Giessereitechnik* **3**, 61–64 (1957).
[115] E. Gastinger, *Mikrochim. Acta* **1972**, 526–543.
[116] F. Martin and A. Floret, *Bull. Soc. Chim. Fr.* **1965**, 404–410.
[117] T. Teichman, C. J. Baker, A. Zdrojewski, and J. L. Monkman, *Mikrochim. Acta* **1969**, 185–192.
[118] Zdnek Vecera and Boleslav Bieber, *Slevarenstvi* **12**, 366–370 (1956); cf. L. M. Steckel and J. R. Hall, *USAEC* **Y-1406**, 18 pp (1962).
[119] H. Bode and K. Hachmann, *Z. Anal. Chem.* **241**, 18–30 (1968).
[120] J. Meyer, *ibid.* **210**, 84–89 (1965).

The precipitate and hydrochloric acid are transferred to the evolution flask for addition of zinc.

For determination of arsenic in silver and silver compounds, the silver is reduced and removed as an amalgam; arsenic is then volatilized as arsine for determination by silver diethyldithiocarbamate.[121]

In determination in samples containing antimony, the interference of evolution of stibine with the arsine is minimized by having potassium iodide and stannous chloride in the generator flask.[122a] Alternatively, heat another portion of sample solution with hydrochloric acid to 130° to expel arsenic chloride.[122b] Develop the residual antimony by the same technic as a correction.

For determination of arsenic on apples, homogenize and dry-ash a 50-gram portion with magnesium oxide, magnesium nitrate, and cellulose powder.[123] Dissolve the ash with 1:1 hydrochloric acid and filter. To the filtrate and washings at not less than 1:3 hydrochloric acid, add 10% potassium iodide solution and stannous chloride in hydrochloric acid. Refrigerate for 1 hour and develop as arsine by silver diethyldithiocarbamate in pyridine. Alternatively, for organoarsenic compounds on fruits, wash the surface with 2:1 ethyl acetate–acetone.[124] Add sulfuric acid and hydrogen peroxide. Evaporate to dryness, take up in water, and determine arsenic by the complex with silver diethyldithiocarbamate or by the Gutzeit method. The procedure is applicable to arsenic in poultry tissue dry-ashed with magnesium oxide and magnesium nitrate.[125a]

For arsenic in food-wrapping paper digest with sulfuric and nitric acids.[125b] Centrifuge or filter to remove detritus such as plastics. Develop by evolution as arsine and absorption in silver diethyldithiocarbamate.

For arsenic in fats and oils decompose the sample with sulfuric acid and hydrogen peroxide.[125c] Add ferric salt, precipitate by addition of ammonium hydroxide to pH 9, and filter. Dissolve in acid and transfer to an evolution flask. Add zinc, absorb the arsine evolved with silver diethyldithiocarbamate in pyridine, and read at 530 nm.

The method has been recommended for diagnostic purposes.[126] A collaborative study has recommended the Gutzeit method over this reagent for beer.[127] The silver diethyldithiocarbamate method as applied to iron and steel is superior to the arsenomolybdate determination.[128]

[121]J. Meyer, *ibid.* **231**, 241–252 (1962).

[122a]I. Hoffman and A. D. Gordon, *J. Assoc. Off. Agr. Chem.* **47**, 629–630 (1964); Adam Hulanicki and Stanislaw Glab, *Chem. Anal.* (Warsaw) **15**, 1089–1096 (1970).

[122b]I. Lauermann, *Zenthbl. Pharm. Pharmakother, u. Lab.*-Diagnostic **115**, 1151–1155 (1976).

[123]M. Nagy, *Deut. Lebensm. Rundsch.* **70**, 173–175 (1974).

[124]H. Berg and Heinz Sperlich, *Mittbl. GDCh-Fachgr. Lebensm. Gerichtl. Chem.* **28**, 298–304 (1974).

[125a]Joseph L. Morrison and Glenn M. George, *J. Assoc. Off. Anal. Chem.* **52**, 930–932 (1969).

[125b]Michele Scrima and Giuseppe Polidori, *Boll. Chim. Unione Ital. Lab. Prov.* **2**, 58–64 (1976).

[125c]Haruo Hirayama, Zenya Shimoda, Chiaki Noguchi, Akiharu Kobayashi, Takeshi Shige, Takeomi Takahashi, Kenji Tsuji, Satoshi Nakasato, M. Masahiko, Tateo Murui, Yukio Yamakawa, Taro Yamashita, Michio Yamashita, and Jiro Yoshida, *Yukagaku* **25**, 275–280 (1976).

[126]I. Lauermann, G. Matthey, and S. Rackow, *Zentbl. Pharm. Pharmakother. Lab. Diagn.* **112**, 561–564 (1973).

[127]Institute of Brewing Analysis Committee, *J. Inst. Brew.* **77**, 365–368 (1971).

[128]T. E. Clayton, W. W. Foster, and T. S. Harrison, *Lab. Pract.* **24**, 333–335, 348 (1975).

Procedures

Iron and Steel. Take a sample according to the following table.

Arsenic (%)	Sample (g)	Add Pure Iron (g)
0.003–0.05	0.100	Nil
0.05–0.10	0.050	0.050
0.10–0.15	0.030	0.070

To the sample and the added pure iron, add if necessary 5 ml of hydrochloric acid and 5 ml of nitric acid, followed by 2 ml of 1:2 sulfuric acid. Simmer until dissolution is complete, and evaporate to sulfur trioxide fumes, but avoid heating above 200°. Take up the residue in 20 ml of 1:3 hydrochloric acid. Cool and add 2 ml of freshly prepared 15% solution of potassium iodide. Add a 50% solution of stannous chloride in hydrochloric acid dropwise until the solution is colorless. Let stand for 15 minutes to ensure that all the arsenic is trivalent.

Place a plug of cotton saturated with lead acetate in the apparatus of Figure 8-7. Transfer the solution to the flask and dilute to 40 ml. Add 5 ml of 0.5% solution of silver diethyldithiocarbamate *in pyridine* to the absorber tube.

Quickly add 3 grams of granulated zinc to the flask. The angle of the end of the tube in the absorber promotes liberation of small bubbles for efficient absorption. Continue the reaction for 1 hour, swirling the flask from time to time, and read at 540 nm against a portion of the absorbing solution as a blank.

Tungsten Compounds.[129] Dilute 5 ml of sample solution containing about 0.5 gram of tungsten oxide to 22.5 ml. Add 3 grams of oxalic acid to complex tungstic acid and heat gently to dissolve. Cool and add 5 ml of hydrochloric acid dropwise. Add 0.5 ml of 40% solution of potassium iodide and 0.5 ml of 40% stannous chloride solution in hydrochloric acid. Mix well and set aside for 15 minutes. Complete as for iron and steel from "Place a plug of cotton..." but read at 560 nm after 12 hours.

Silicon.[130a] Dissolve 1 gram of sample in 20 ml of 20% potassium hydroxide solution. Neutralize and make to 3 *N* with hydrochloric acid. The authors call for evolution as arsine and absorption on mercuric bromide paper, but it should be equally applicable by silver diethyldithiocarbamate as described for iron and steel starting at "Cool and add 2 ml of freshly...."

Germanium. Dissolve 2 grams of sample in 5 ml of 6% oxalic acid solution containing 0.1 ml of 30% hydrogen peroxide. To the hot solution, add 0.5 ml of 5% manganous sulfate solution and 4 ml of 3% potassium permanganate solution.

[129] V. Blechta, *Chem. Prum.* **14**, 373–374 (1964).
[130a] A. A. Tumanov, A. N. Sidorenko, and F. S. Taradenkova, *Zavod. Lab.* **30**, 652–654 (1964).

Arsenic is coprecipitated with the manganese dioxide. Centrifuge after 3 hours and wash the precipitate with 1:5 hydrochloric acid to free it of germanium. Dissolve the precipitate in 20 ml of 1:5 hydrochloric acid by adding sufficient hydrazine hydrochloride solution to reduce and solubilize the precipitate. Complete as for iron and steel, starting at "Cool and add 2 ml of freshly...."

Germanium Dioxide.[130b] Dissolve 5 grams of sample by heating with 20 grams of oxalic acid and 150 ml of water. Cool to 60–65° and add 2 ml of 10% potassium iodide solution and 1 ml of 5% metabisulfite solution. Set aside for 15 minutes for the reduction of arsenic to trivalency. Add 5 ml of 0.25% solution of diethylammonium diethyldithiocarbamate in chloroform and 5 ml of chloroform. Shake for 40 seconds. Extract the aqueous solution with 5 ml and 5 ml of chloroform, and add to the first extract.

Shake the combined chloroform extracts with 10 ml of 5% oxalic acid solution to scavage any entrained germanium. Add 2 ml of 3:1 nitric-perchloric acids; evaporate the chloroform, and heat to fumes of perchloric acid. Let cool, add 2 ml of 1:1 hydrochloric acid containing 1% of nitric acid, and again take to perchloric acid fumes. This removes the last traces of germanium. Add 2 ml of sulfuric acid, 1 ml of perchloric acid, and 1 ml of nitric acid. Heat to sulfur trioxide fumes. During the fuming off of the perchloric acid, the solution turns yellow; when it has been completely removed, the solution is nearly colorless. Continue to heat until the sulfuric acid is condensing on the sides of the container. Let cool, and add 2 ml of 1:1 sulfuric acid and 12 ml of water. Transfer to the flask of the apparatus for evolution of arsine shown in Figure 8-7. Add 1 ml of 10% potassium iodide solution and 1 ml of 1% solution of stannous chloride in 1:2 hydrochloric acid. Set aside for 15 minutes. Complete as for iron and steel from "Quickly add 3 grams of granulated...."

Alternatively,[131] transfer 2 grams of sample to a Kjeldahl flask marked at 10 ml. Add 50 ml of hydrochloric acid and pass chlorine at 2 bubbles per second. After 5 minutes, heat the solution gently to avoid frothing. When the solution becomes clear, heat more strongly, continuing the passage of chlorine until evaporated to 10 ml. While still heating and passing chlorine, add 25 ml of hydrochloric acid and again reduce to 10 ml. The germanium will have been largely volatilized without loss of arsenic or antimony. Dilute to 24 ml with hydrochloric acid. Add 8 ml of water, 1 ml of saturated solution of hydroxylamine hydrochloride, 2 ml of 15% solution of potassium iodide, and 2 ml of 40% solution of stannous chloride in hydrochloric acid. Cool for 15 minutes. Complete as for iron and steel from "Place a plug of cotton saturated...."

Paint Flakes.[132] Heat 50 mg of flakes such as those scraped from toys with 50 mg of sodium carbonate at 450° for 30 minutes and cool. Add 1 ml of hydrochloric acid and 1 ml of nitric acid. Evaporate at 100°; then take up the residue in 25 ml of boiling 1:100 hydrochloric acid. Dilute a 5-ml aliquot to 25 ml with 1:3 hydrochloric acid and add 2 ml of 15% solution of potassium iodide. Complete as for iron and steel from "Add a 50% solution of stannous...."

[130b] E. W. Fowler, *Analyst* **88**, 380–386 (1963).
[131] R. Rezac and J. Ditz, *Zavod Lab.* **29**, 1176–1178 (1963).
[132] T. Hodson and D. W. Lord, *J. Assoc. Pub. Anal.* **9**, 60–63 (1971).

Sulfur.[133] Heat 10 grams of pulverized sample in a covered polypropylene beaker with 50 grams of 50% sodium hydroxide solution. Cool somewhat, add 10 ml of water and 5 ml of 30% hydrogen peroxide, and heat to boiling. Repeat until the sample is disintegrated. Make acid with 1:35 sulfuric acid and boil with more hydrogen peroxide until solution is complete. Evaporate nearly to dryness. Complete as for iron and steel from "Take up the residue in...."

Alternatively,[134] sulfur is treated with bromine and carbon tetrachloride, then oxidized with nitric acid. Thereafter it is reduced to arsine by zinc and absorbed by silver diethyldithiocarbamate.

Sulfuric Acid.[135] Mix 10 grams with 5 ml of 1:1 nitric acid and 5 ml of 5% solution of potassium sulfate. Heat almost to dryness, cool, and take up in 10 ml of 1:1 sulfuric acid. Dilute to 30 ml. Add 2 ml of 15% potassium iodide solution and 0.5 ml of 40% solution of stannous chloride in hydrochloric acid. Set aside for 15 minutes. Complete as for iron and steel from "Place a plug of cotton...."

Hydrochloric Acid.[136] To 50 grams of sample in a Teflon crucible, add a few crystals of potassium chlorate, causing liberation of chlorine vapor. Evaporate carefully to about 35 ml, adding more chlorate if the sample loses its yellow-green color. Complete as for iron and steel from "Cool and add 2 ml of freshly prepared...."

Nitric Acid. To 50 grams of sample in a Teflon crucible, add 1 ml of sulfuric acid and evaporate to sulfur trioxide fumes. Add 5 ml of water and 2 ml of saturated solution of ammonium oxalate. Again evaporate to sulfur trioxide fumes. Add 5 ml of water, then 30 ml of 1:1 hydrochloric acid. Complete as for iron and steel from "Cool and add 2 ml of freshly prepared...."

Hydrofluoric Acid.[137a] Mix a 10-gram sample in a polyethylene bottle with 15 ml of water, 10 ml of 1% stannous chloride solution in 1:2 hydrochloric acid, and 5 ml of 15% solution of potassium iodide. Complete as for iron and steel from "Let stand for 15 minutes to...."

Hexafluorosilicic Acid, Tetrafluoroboric Acid, and Alkali Fluorides. Proceed as above, but dissolve the sample in 30 ml of water and 20 ml of the stannous chloride solution.

Trivalent Arsenic.[137b] Add the sample solution to a generating flask plus 5 ml of *M*-citrate buffer for pH 4. Pass a stream of nitrogen to sweep out the arsine as generated and add 0.5 ml of 1% solution of sodium borohydride in 0.1 *N* sodium hydroxide at the rate of 20 drops a minute. Absorb the arsine in 4 ml of 5% solution of silver diethyldithiocarbamate in pyridine. Read at 530 nm.

[133] E. Heinerth, *Rev. Port. Quim.* **9**, 202–203 (1967).
[134] F. Fisher and Leschhorn, *Chemikerzeitung-Chem. Appar.* **91**, 416–419 (1967).
[135] Tadashi Kaneshige, Muneharu Takizawa, and Hiroshi Nagai, *Jap. Anal.* **13**, 780–788 (1964).
[136] J. Ditz, J. Dvorak, and J. Marecek, *Chem. Prum.* **15**, 677–678 (1965).
[137a] J. Meyer, *Z. Anal. Chem.* **229**, 409–413 (1966).
[137b] J. Aggett and A. C. Aspell, *Analyst* **101**, 912–913 (1976).

Water.[138] No interference is to be expected in surface waters. If necessary, add sulfuric acid and concentrate expected to contain 0.1–20 μg of arsenic in a 20-ml sample. Transfer to the flask of Figure 8-7, add 30 ml of 1 : 8 sulfuric acid, and adjust to 50 ml. Add 2 ml of 15% potassium iodide solution and 8 drops of a 40% solution of stannous chloride in hydrochloric acid. Mix well and set aside for 15 minutes. Complete as for iron and steel from "Place a plug of cotton,...." modifying procedure to use 6 grams of zinc and 3 ml of absorbant, and to read at 560 nm.

Alternatively,[139] to 250 ml of sample add 2 ml of a 7% solution of ferric nitrate monohydrate in 1 : 9 nitric acid. Add ammonium hydroxide to pH 8, filter, and wash with warm water. Dissolve in 2 ml of warm 1 : 3 sulfuric acid and transfer to the flask of the reactor, (Figure 8-7). Add 2 ml of 15% potassium iodide solution, and if the solution indicates the presence of antimony by turning yellow, extract with 10 ml of benzene. Add 1 ml of 40% solution of stannous chloride in hydrochloric acid, incubate at 50° for 30 minutes, and cool to 0°. Charge the reactor with 3 ml of 0.5% silver diethyldithiocarbamate solution. Add 6 grams of granulated zinc and close the reactor. After 1 hour read at 540 nm.

Urine.[140] Wet-ash 30 ml with 3 ml of sulfuric acid and 2 ml of nitric acid, adding increments of 30% hydrogen peroxide if necessary to decolorize. Evaporate to sulfur trioxide fumes and dilute to about 20 ml. Add 2 ml of 40% solution of potassium iodide. Set aside for 15 minutes. Complete as for iron and steel from "Place a plug of cotton...." Some workers recommend letting the urine stand 12 hours with the acid before ashing.[141]

Blood. Follow the technic for urine, but the sample need not stand with the acid before ashing.

Wines and Vinegars.[142] Digest 100 ml of sample with 5 ml of 1 : 1 sulfuric acid and 5 ml of nitric acid. Boil until the solution is colorless and oxides of nitrogen have been driven off. Cool, dilute to 100 ml, and take a 25-ml aliquot. Transfer to the evolution flask of Figure 8-7. Add 1 ml of 40% solution of stannous chloride in hydrochloric acid, 2 ml of 15% potassium iodide solution, and 2 ml of 2% copper sulfate solution; set aside for 15 minutes. Change the receiver with 5 ml of 0.5% silver nitrate in pyridine. Add 5 grams of 20-30 mesh zinc and let operate as long as the color darkens. Read at 538 nm.

[138] W. Fresenius and W. Schneider, *Z. Anal. Chem.* **203**, 417–422 (1964); cf. D. G. Ballinger, R. Lishka, and M. E. Gates, Jr., *J. Am. Water Works Assoc.* **54**, 1424–1428 (1962); G. Stratton and H. C. Whitehead, *ibid.* 861–864; cf. L. A. Khristianova, N. I. Udal'tsova, and S. S. Soldatova, *Gig. Sanit.* **1975**, 70–72.

[139] R. Thiel and G. Carpentier, *Bull. Cent. Rech. Pau* **4**, 243–246 (1970); cf. Hans Senften, *Mitt. Geb. Lebensmittelunters. Hyg.* **64**, 152–170 (1973); F. Tiemann and J. Maassen, *Z. Wass.-u. Abwass.-Forsch.* **9**, (2), 42–47 (1976).

[140] T. C. Roddy, Jr., and Stuart M. Wallace, *Am. J. Clin. Pathol.* **36**, 323–325 (1961); cf. Leo A. Dal Cortivo, Michael Cefola, and Charles J. Umberger, *Anal. Biochem.* **1**, 491–497 (1960).

[141] M. del Pilar, P. M. de Neuman, and A. Singerman, *Rev. Asoc. Bioquim. Argent.* **31**, 10–17 (1966).

[142] M. D. Garrido, M. L. Gil, and C. Llaguno, *An. Bromatol.* **28**, 167–175 (1974).

Food.[143] Mix 20 grams of sample with 3 grams of magnesium oxide and 10 ml of cellulose powder. Slurry with water and oven-dry. Prechar under an infrared lamp until smoking ceases, and cool. Cover with 3 grams of magnesium nitrate hexahydrate and place in a cool muffle furnace. Raise the temperature to 550° and ash for 2 hours. Cool, moisten with water, and dissolve in 50 ml of 1 : 1 hydrochloric acid. Transfer to the evolution flask of Figure 8-7 with the aid of 10 ml and 10 ml of 1 : 1 hydrochloric acid. Dilute to 75 ml. Add 5 ml of 0.5% solution of silver diethyldithiocarbamate to the receiving tube. Add 2 ml of 15% potassium iodide solution and 1 ml of 40% solution of stannous chloride dihydrate in hydrochloric acid to the evolution flask. Add 6 grams of 20-30 mesh zinc to the evolution flask, close the assembly, and let operate until evolution of hydrogen ceases. Read at 540 nm against a reagent blank within 4 hours.

Alternatively,[144] wet-ash a sample containing up to 0.005 mg of arsenic. Take up in water, dilute to 200 ml, and take a 50-ml aliquot. Add 7.5 ml of sulfuric acid and cool to 0°. Add 1 ml of 15% potassium iodide solution, 5 drops of 20% stannous chloride solution in hydrochloric acid, and 1 ml of isopropanol. Complete as for iron and steel from "Let stand for 15 minutes to...."

After wet-ashing of foodstuffs and destruction of nitrogenous residuals with hydrogen peroxide, the addition of hydrazine sulfate and hydrochloric acid permits distillation of arsenious chloride in a stream of nitrogen.[145] The arsenic chloride distillate is then an appropriate sample for reduction to arsine with zinc and its absorption by silver diethyldithiocarbamate in pyridine.

Organic Samples.[146] Digest a 5-mg sample with 2 ml of sulfuric acid and 2 ml of nitric acid, adding 30% hydrogen peroxide from time to time until the solution is colorless. Evaporate to sulfur trioxide fumes and let cool. Complete as for iron and steel from "Take up the residue in 20 ml...."

Naphthas.[147] A combination of chromatography and photometry determines arsenic in parts per billion. A 10 × 1 cm chromatographic column required has a male 14 × 35 $ joint at the bottom and a conical 500-ml reservoir with a 28/14 $ socket joint at the top. The detachable tip has a 14/35 $ joint and a No. 2 Teflon stopcock. Other equipment required is shown in Figure 8-8. A 50-ml heating mantle rated at 50 volts is required to reflux sulfuric acid. Prepare the adsorbant by mixing 20 ml of sulfuric acid with 50 grams of 60 to 200-mesh silica gel until free from lumps. Dry the cotton saturated with lead acetate solution by pressing and by vacuum-drying until dry to the touch.

As reagent, slowly add 0.1 *M* silver nitrate below 8° with vigorous stirring to an equal volume of 0.1 *M* sodium diethyldithiocarbamate trihydrate, equally chilled. Filter and wash three times with water chilled to below 8°. Dry under vacuum.

[143] Harvey K. Hundley and Joseph C. Underwood, *J. Assoc. Off. Anal. Chem.* **53**, 1176–1178 (1970).
[144] J. Szymczak and A. Zechalko, *Rocz. Zakl. Hig.* (Warsaw) **14**, 239–244 (1963).
[145] E. Kröller, *Lebensm. Rundsch.* **61**, 115–117 (1965).
[146] Hironobu Kashiwagi, Yoshiko Tukamoto, and Masami Kan, *Ann. Rep. Takedi Res. Lab.* **22**, 69–75 (1962).
[147] George W. Powers, Jr., Ronald L. Martin, Frank J. Piehl, and J. Marcus Griffin, *Anal. Chem.* **31**, 1589–1593 (1959).

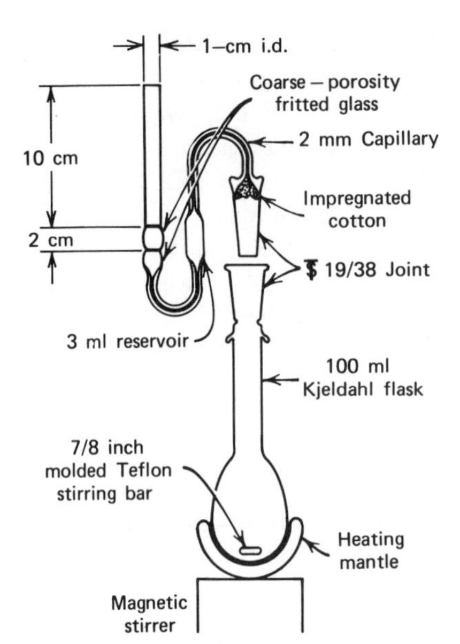

Figure 8–8 Digestion flask and arsine absorber.

Percolate pyridine over activated alumina, then over silica gel, and dissolve 1 gram of the dry reagent in 200 ml of the pyridine.

Place a wad of cotton in the stopcock section of the chromatographic column and fill the barrel with *n*-hexane. As a slurry in *n*-hexane, add sufficient silica gel to fill the stopcock section. Similarly, add enough adsorbant to form a column 10 cm deep. Drain excess hexane from the column and tap from time to time, if necessary to release air bubbles.

Transfer the desired volume of sample to the column and allow it to flow through under gravity or apply pressure if necessary. Rinse the sample bottle and the walls of the reservoir with 10 ml and 10 ml of hexane. Rinse the sample bottle with 12 ml of sulfuric acid, and transfer it to the reservoir. Rotate the column to rinse the entire surface of the reservoir with the acid wash. Remove the stopcock section with its contents and quickly insert the base of the column into the neck of the 100-ml Kjeldahl flask, which will later become part of Figure 8-8. When all the adsorbant and acid have passed into the Kjeldahl flask, rinse the bottle and the column with an additional 10 ml of sulfuric acid. Finally, rinse the column into the flask with 3 ml of perchloric acid and 5 ml of nitric acid.

Stir the contents of flask rapidly for 15 minutes without heating. Then reduce to the minimum possible rotation of the stirring bar and heat gradually to boil off the hydrocarbons. Apply more heat and increase the stirring but remove from the mantle and cool if the reaction becomes too vigorous. When vigorous reaction subsides, apply more heat, and if the mixture darkens, add more nitric acid. When reaction ceases, envelop the flask with asbestos fabric and reflux the sulfuric acid to within 2 inches of the top for 15 minutes. If a yellow to orange color persists, complete the oxidation with a few drops of perchloric acid. Cool and add 10 ml of water, then 10 ml of saturated solution of ammonium oxalate. Again heat, boiling off the water, and reflux the sulfuric acid for 15 minutes.

Cool and add 50 ml of water. Cool again and add 20 ml of isopropanol. Chill in ice for 5 minutes and add 2 ml of 15% solution of potassium iodide. Again chill in ice for 5 minutes. Stir with 12 drops of 40% solution of stannous chloride in hydrochloric acid. Let stand in an ice bath for 15 minutes.

To prepare the equipment, insert a small wad of impregnated cotton in the arsine absorber. Use water-soluble lubricant on the joint. Add 3 ml of reagent to the absorber and draw most of it through the fritted disk into the 3-ml reservoir.

Stir the contents of the flask, add 6 ± 1 grams of 20-mesh zinc, and immediately connect to the absorber. Stir slowly for 30 minutes and rapidly for 45 minutes. Draw the liquid in the absorber back and forth through the fritted-glass disk until the color is uniform. Read at 540 nm against a blank, which must not represent more than 0.2 mg of arsine.

Since a substantial amount of arsenic will be deposited on the walls of the sample container, it is essential that each be taken in a container of appropriate size and the arsenic be recovered from it.

Arsenic in naphthas is also determined in the parts per billion range by combustion in an oxyhydrogen burner.[148] The products of combustion are absorbed in sodium hydroxide solution, subsequently converted to arsine, and determined by the reagent above.

SODIUM DIETHYLDITHIOCARBAMATE

The precipitate of trivalent arsenic with this reagent is extracted at pH 6 by carbon tetrachloride, chloroform, benzene, toluene, and xylene.[149] The method will determine 5 μg.

Procedure. Mix a sample solution containing up to 1 mg of arsenic with 10 ml of a buffer solution for pH 5 and 2 ml of 2% solution of sodium diethyldithiocarbamate. Shake with 5 ml of carbon tetrachloride and read the extract at 340 nm against a blank.

SILVER DITHIZONATE

When the complex of trivalent arsenic with diethyldithiocarbamate contacts silver dithizonate in acid solution, the silver is displaced and the free dithizone read.[150a] Many elements interfere by forming diethyldithiocarbamate complexes that either absorb at 600 nm or exchange with silver dithizonate. Preliminary extraction at pH 8.5–9 in the presence of tartrate with 8-hydroxyquinoline in chloroform removes many interferences. Antimony is removed at pH 1.5 by extraction with cupferron in chloroform.

[148] D. Kendall Albert and Lawrence Granatelli, *ibid.* 1593–1597.

[149] Emiko Sudo, *Sci. Rep. Res. Inst., Tohoku Univ.* **6**, 142–146 (1954).

[150a] F. Sebesta and J. Stary, *Collect. Czech. Chem. Commun.* **33**, 3895–3898 (1968); cf. J. Ruzicka and J. Stary, *Talenta* **14**, 909–920 (1967).

Procedure. As the reagent, shake 20 ml of 0.25 m*M* dithizone in chloroform with 20 ml of 1:35 sulfuric acid containing 1 ml of 0.1 *N* silver nitrate solution. Filter the chloroform phase and store in the dark.

To 10 ml of neutral sample containing up to 0.02 mg of arsenic, add 1 ml of 0.77% ammonium acetate solution and 1 mg of sodium diethyldithiocarbamate. Shake for 2 minutes and add 5 ml of 0.1 *N* sulfuric acid to reduce to about pH 2. After 2 minutes, shake for 2 minutes with 10 ml of chloroform. Then mix 4 ml of the chloroform layer with 1 ml of the silver dithizonate reagent. Read the free dithizone at 620 nm after 3 minutes.

SILVER AND FERRIC IONS

When arsine is passed into a solution containing silver ion and ferric ion, the arsine is oxidized to arsenite and ferric ion is reduced to ferrous. The latter can then be determined as an indirect measure of the arsenic. For the technic given here, ferrozine, which is the disodium salt of 3-(2-pyridyl)-5, 6-bis(4-phenylsulfonic acid), is used. Silver ion must exceed 0.025 *M* for arsine absorption to be complete. Added ferric ion in the reagent facilitates oxidation of arsine and development of color. The absorbing solution must be below pH 4. Chloride ion in the developing solution must exceed the silver ion in the absorbing solution. Interference comes from metals that inhibit arsine, such as cobalt, mercury, nickel, platinum, silver, and palladium. Stannous chloride inhibits evolution of stibine.

Procedure. As the arsine evolution apparatus, connect one end of a 4-mm i.d. glass tube at a 90° angle to a 24/40 vacuum adapter. Bend the glass tubing at a 90° angle at a distance of 10 cm from the adapter.[150b] Place the adapter on a 125-ml 24/40 ground glass flask and adjust the tubing to reach the bottom of a 10-ml conical centrifuge tube used as the absorber.

As the absorbing solution, mix equal volumes of 0.1 *M* silver nitrate and 0.0025 *M* ferric ammonium sulfate in 0.1 *M* perchloric acid. As a buffer solution dissolve 28.6 ml of acetic acid with 68.04 grams of sodium acetate trihydrate in water and dilute to 1 liter. As a developing solution, mix 15 ml of 0.5 *M* sodium chloride and 15 ml of the buffer solution with 20 ml of 0.0078 *M* ferrozine and dilute to 50 ml.

To the generating flask for arsine, add a sample containing up to 2 μg of arsenic, 25 ml of 1:1 hydrochloric acid, and 75 ml of water. Next add 2 ml of 15% potassium iodide solution, then add 0.4 ml of 40% stannous chloride solution in hydrochloric acid. Cool in an ice bath for 15 minutes. Add 5 grams of 20-mesh granular zinc and attach at once to the arsine evolution apparatus. Use 2 ml of absorbing solution in the absorber. After evolution of arsine for 45 minutes, add 2 ml of developing solution to the absorber and dilute to 5 ml. Mix, centrifuge, set aside for 5 minutes, and read at 560 nm.

[150b]Gerald J. Kellen and Bruno Jaselskis, *Anal. Chem.* **48**, 1538–1539 (1976).

If arsenic is oxidized by excess of ferricyanide, the complex with **bi-*o*-anisidine** can be read at 470 nm.[151]

Trivalent arsenic at 7.5–37.5 μg is determined in 0.1 *N* sulfuric acid by oxidation with **ceric ion** catalyzed by octavalent osmium.[152] The reaction mixture is activated at 260 nm to read the fluorescence of the resulting cerous ion at 350 nm. There is interference by nitrate, ferric ion, EDTA, citrate, and tartrate.

The molybdoarsenate as well as the molybdosilicate complexes combine with basic dyes,[153] of which **crystal violet** (basic violet 3) or **malachite green** (basic green 4) are typical. Extract the molybdoarsenate complex with 4 : 1 cyclohexanone-toluene before treatment with the dye solution for reading at 582 nm. The reading is sensitive to 50 μg/ml.

When the arsinofluoride is reacted with the **ferrous 1 : 10-phenanthroline** complex known as **feroin,** the product can be extracted with butyronitrile.[154] At 0.55–5.5 m*M* this is read at 505 nm. The pH at 2.2–10.6 does not affect the result. More than 0.2 m*M* of fluoride ion interferes.

Arsenic is extracted as the **iodide** along with tin by cyclohexane from 4 *M* sulfuric acid–*M* sodium iodide.[155a] Arsenic is then read at 282 nm and tin at 364 nm. The 4 *M* sulfuric acid can be replaced by 5 *M* hydrochloric acid for extraction of arsenic.

For arsenic in water concentrate in a flash evaporator and filter if necessary. Make 3 *N* with sulfuric acid. Add 0.2 gram of potassium iodate to an aliquot. Extract the liberated iodine with carbon tetrachloride and read at 520 nm.[155b] As little as 2 μg of arsenic per liter can be determined.

After extraction as arsenious iodide with carbon tetrachloride, a mixture of 8 ml of extract, 1.5 ml of m*M* **8-mercaptoquinoline** in carbon tetrachloride and 0.5 ml of acetone is read at 380 nm.[156]

For determination of arsenic, mix 3 ml of sample containing arsenic acid with 0.3 ml of 0.4% ethanolic **morin,** which is 2′,3,4′,5,7-pentahydroxyflavone.[157] Add 1 ml of 15% ethanolic oxalic acid and 0.6 ml of water. Evaporate to dryness at 70–90° and hold at that temperature for 25 minutes. Cool, dissolve in 10 ml of ethanol, and read at 440 nm. Beer's law applies up to 1.2 ppm of arsenic. Alkali-metal salts decrease the absorption.

Evaporate a sample solution containing arsenic acid to dryness at 70±5° with 0.5 ml of 0.1% methanolic **rutin,** which is quercetin-3-rutinoside, 1 ml of 10% ethanolic oxalic acid, and 1 ml of water.[158] Heat the residue at 90±5° for 15 minutes. Dissolve in 10 ml of ethanol and read at 450 nm. Borate gives a similar color, and phosphate diminishes the intensity of the color.

The intensity of color developed by octavalent **osmium,** thiourea, and trivalent

[151] J. Artigas, F. Buscarons, and C. Rodrigues-Roda, *An. Real Soc. Espan. Fis. Quim.* (Madrid) **50B**, 373–376 (1960).
[152] G. F. Kirkbright, T. S. West, and C. Woodward, *Anal. Chim. Acta* **36**, 298–303 (1966).
[153] A. K. Babko and E. M. Ivashkovich, *Zh. Anal. Khim.* **27**, 120–127 (1972).
[154] V. S. Archer and F. G. Doolittle, *Talenta* **14**, 921–924 (1967).
[155a] Katu Tanaka and Nobuyuki Takagi, *Anal. Chim. Acta* **48**, 357–366 (1969).
[155b] Shingara S. Sandhu, *Analyst* **101**, 856–859 (1976).
[156] Vera Stara and Jiri Stary, *Talenta* **17**, 341–345 (1970); Vera Stara, *ibid.* **18**, 228–230 (1971).
[157] Kazuo Hiiro, Takashi Tanaka, and Shizuko Watanabe, *Jap. Anal.* **12**, 918–922 (1963).
[158] Takashi Tanaka and Kazuo Hiiro, *ibid.* **12**, 914–918 (1963).

arsenic is proportional to the arsenic.[159] Specifically, the reaction is in 1:9 sulfuric acid with osmium: thiourea, 1:7 for determination of 0.08–1 μg of arsenic per ml of the solution as developed.

As a kinetic method, arsenic serves as an auxiliary catalytic agent in the osmium catalysis of the redox reaction of bromate and iodide.[160] To 25 ml of sample containing up to 25 μg of trivalent arsenic, add as a mixture 5 ml of *M* acetate buffer for pH 4.8–5, 2 ml of 4.15% solution of potassium iodide, and 1 ml of 1% starch indicator solution. Add 5 ml of 0.0001 *M* solution of **osmium tetroxide** and read against water at 580 nm, 4 ± 0.2 minutes after mixing.

To 5 ml of sample solution containing up to 0.16 mg of arsenic, add 1 ml of 1:1 sulfuric acid and zinc granules pretreated with copper sulfate.[161] Pass the arsine and hydrogen developed for 1 hour through a mixture of 2 parts of 0.1 *M* silver nitrate, 2 parts of 0.1 *M* ***p*-sulfamoylbenzoic acid,** and 1 part of 4% sodium hydroxide solution. Read the brown solution at 420 nm.

To 1 ml of sample solution containing 0.3–1.5 μg of trivalent arsenic add 1 ml of 0.1 m*M* potassium iodate, 1 ml of 0.1 m*M* **Rhodamine B,** and 1 ml of 1:1 hydrochloric acid.[162] In the oxidation of arsenic in hydrochloric acid by iodate, there is concomitant formation of chloriodate ion that complexes with rhodamine B. Shake with 5 ml of benzene, activate the extract at 550 nm, and read the fluorescence at 590 nm. Beer's law applies to 0.06–0.3 μg of arsenic per ml of extract.

When arsenic in a solution acidified with sulfuric acid is precipitated by hydrogen sulfide in the presence of poly(vinylpyrrolidinone), the yellow **sulfide** is read at 400 nm.[163]

Arsenic is determined as the 1:1 complex of molybdoarsenic acid and **tetraphenylphosphonium chloride** stabilized with polyvinyl alcohol.[164] Treat a sample solution containing 8 micromoles of arsenic with 3 ml of 10 *N* nitric acid and 4 ml of 0.5 *M* ammonium molybdate. Dilute to 14 ml, incubate at 25° for 10 minutes, and add 5 ml of 5% polyvinyl alcohol. Add 1 ml of 0.02 *M* tetraphenylphosphonium chloride, and dilute to 20 ml. Read at 400 nm against a reagent blank within 30 minutes. Ferric, nickel, titanic, and silicate ions should be absent. If phosphate is present, it must be corrected for by a parallel determination of molybdophosphate at 420 nm. In the absence of the polyvinyl alcohol, as a stabilizer, the complex can be extracted into 1,2-dichloroethane as a 3:1 complex of tetraphenylphosphonium chloride with arsenic.

[159]P. P. Naidu and G. G. Rao, *Microchem. J.* **18**, 422–427 (1973).
[160]T. Tarumoto and H. Freiser, *Anal. Chem.* **47**, 180–182 (1975).
[161]Gb. Cinhandu and M. Rocsin, *Z. Anal. Chem.* **172**, 268–274 (1950); G. Ciuhandu, A. Roscovanu, and M. Cutui, *Chim. Anal.* (Paris) **50**, 489–493 (1968).
[162]Daijiro Yamomoto and Ken-Ichiro Kisu, *Jap. Anal.* **23**, 638–644 (1974).
[163]B. Cristau, *Ninth Congr. Soc. Pharm. Fr., Clermont-Ferrand* **1957**, 237–239.
[164]Yukio Nagaosa and Tatsuo Yonekubo, *Jap. Anal.* **22**, 850–856 (1973).

CHAPTER NINE

ANTIMONY

The predominant methods for the determination of antimony are by brilliant green, as the iodoantimosiate, by methyl violet or crystal violet, and by rhodamine B. Thus as with several other elements, the complexes with dyestuffs become more and more important.

The determination of arsenic as arsine by absorption in silver diethyldithiocarbamate is described in detail elsewhere. In the presence of potassium iodide, stannous chloride, hydrochloric acid, and metallic zinc, stibine will be evolved with the arsine.[1] Read at 525 nm, this measures the sum of arsenic and antimony. This is then corrected by subtraction of the arsenic determined, for example, by distillation as arsenic chloride.[2] The technic is applicable to a biological digestate.

For separation of arsenic, antimony, and tin by distillation as the halides, see Arsenic. For antimony in bismuth-antimony alloys by flame photometry, see Bismuth.

ANTIPYRINYLBIS[4-(BENZYLMETHYLAMINO)PHENYL]METHANOL

This reagent complexes with pentavalent antimony. Extracted into benzene, it conforms to Beer's law for 0.04–1 μg of antimony per ml.[3] There is interference by microgram amounts of thallic and auric ion and by mg amounts of ferric and stannic ion.

Procedure. *Tellurium.* Dissolve 0.5 gram of sample in 2 ml of nitric acid and dilute with water to precipitate some tellurous acid and coprecipitate antimony.

[1]Adam Hulanicki and Stanislaw Glab, *Chem. Anal.* (Warsaw) **15**, 1089–1096 (1970).

[2]Leo A. Dal Cortivo, Michael Cefola, and Charles J. Umberger, *Anal. Biochem.* **1**, 491–497 (1960).

[3]A. I. Busev, V. G. Tiptsova, E. S. Bogdanova, and A. M. Andreichuk, *Zh. Anal. Khim.* **20**, 812–814 (1965).

Filter and dissolve in 3 ml of 1 : 1 hydrochloric acid. Add 1 ml of 10% sodium nitrite. After 3 minutes add 1 ml of saturated solution of urea. Dilute to 20 ml and add 2 ml of 0.05% solution of the reagent in 1 : 10 hydrochloric acid. Extract with 5 ml and 4 ml of benzene. Dilute the combined extracts to 10 ml and read at 590 nm.

ANTIPYRINYL(4-DIMETHYLAMINOPHENYL)-(4-DIMETHYLAMINO-3-NITROPHENYL)METHANOL

This reagent complexes with pentavalent antimony.[4]

Procedure. *Copper.* Dissolve 0.5 gram of sample in 5 ml of 1 : 1 nitric acid and heat to expel oxides of nitrogen. Add 1.5 ml of sulfuric acid and heat to sulfur trioxide fumes. Cool, rinse down the walls of the beaker, and again evaporate to sulfur trioxide fumes. Take up the residue in 8 ml of 3 : 1 hydrochloric acid and add 5 drops of 2.3% solution of stannous chloride dihydrate in 1 : 1 hydrochloric acid. After 5 minutes add 3 ml of 0.05 *M* ceric sulfate, to oxidize the antimony to the pentavalent chloride. Add 3 drops of 1% hydroxylamine hydrochloride to reduce residual ceric ion. Dilute to *M* in hydrochloric acid and add 2 ml of 0.05% solution of the reagent in *M* hydrochloric acid. Extract with 5 ml of benzene and read the extract at 585 nm.

BRILLIANT GREEN

Pentavalent antimony as the chloride forms a 1 : 1 orange complex with brilliant green (basic green 1) which is extractable from 0.25–1.5 *N* hydrochloric acid for reading at 0.15–15 μg/ml.[5] The reagent is not soluble in benzene, toluene, or xylene. Gold and thallium interfere seriously. There is also some interference by chromate, bromide, thiocyanate, perchlorate, chlorate, and molybdate. Tenfold amounts of tin, mercury, cadmium, and iron form a green complex with the reagent which is stable for less than 15 minutes.[6] Iron can be masked by sodium fluoride.

The standard curve must have been prepared from the batch used to develop the sample.[7] Purification for more consistent results is by filtering a 0.33% solution, precipitating as the perchlorate at pH 1–2, and recrystallizing the perchlorate of the dyestuff from 20% ethanol.[8]

[4] A. I. Busev, E. S. Bogdonova, and V. G. Tiptsova, *ibid.*, **20**, 585–590 (1965).

[5] R. W. Burke and Oscar Menis, *Anal. Chem.* **38**, 1719–1722 (1966); A. A. Nakharova, E. S. Tovstopyat, and V. Ya. Eremenko, *Gidrokhim. Mater.* **51**, 196–202 (1969); cf. Chin-hung Ts'ao, Jun-sheng Chu, Chih-ming Yang, and T'eng-huang T'ang, *Yao Hsüeh Hsüeh Pao* **5**, 327–332 (1957); L. A. Soldatova, Z. G. Kilina, and G. A. Kataev, *Zh. Anal. Khim.* **19**, 1267–1269 (1964).

[6] L. B. Kristaleva, *Zavod. Lab.* **25**, 1294–1295 (1959).

[7] A. G. Fogg, J. Jillings, D. R. Marriott, and D. Thorburn Burns, *Analyst* **94**, 768–773 (1969).

[8] G. O. Kerr and G. R. E. C. Gregory, *ibid.* 1036–1037.

Antimony is extracted from solution in 1 : 5 hydrochloric acid containing 20% of magnesium chloride hexahydrate by 20% tributyl phosphate in toluene.[9] The complex with brilliant green is then formed in the organic phase.

For determination in ferromolybdenum or chromium, it is not necessary to separate the antimony, but for determination in copper solutions it is coprecipitated with ferric hydroxide.[10]

Bismuth and antimony are extracted from a solution of iron, molybdenum, copper, and nickel alloys by 2-ethylhexyl pyrophosphate in hexane. Thereafter the two elements are separated by extractions, and the antimony is determined by brilliant green. For details see Bismuth by Thiourea.

Procedure. Mix 10 ml of sample solution containing less than 0.02 mg of antimony in 1 : 1 hydrochloric acid with 0.5 ml of 0.1 *M* ceric sulfate in 1 : 18 sulfuric acid.[11] After 1 minute reduce the excess of ceric ion by dropwise addition of 1% solution of hydroxylamine hydrochloride. At once dilute to 25 ml and add 0.5 ml of 0.5% ethanolic solution of brilliant green. Extract with 10 ml of toluene by shaking for 1 minute and filter the extract. Add another 0.5 ml of the ethanolic reagent and shake with 10 ml of toluene. Dilute the combined extracts to 25 ml and read at 640 nm.

Lead or Zinc.[12] If the sample solution contains chlorides, evaporate repeatedly with sulfuric acid, finally taking up the solids in 10 ml of water. Add 0.5 ml of 10% ascorbic acid solution or 2 ml of 10% hydroxylamine sulfate solution to reduce thallic ion, preventing its extraction with antimony. Make 0.5 *N* with nitric acid or, if lead is absent, sulfuric acid may be used. Dilute to 20 ml and extract with 20 ml of 0.5 *N*-bis(2-ethylhexyl)hydrogen phosphate in benzene or toluene. Reextract the antimony from the organic phase with 20 ml of 9 *N* hydrochloric acid. Dilute the aqueous phase to 6 *N* with 10 ml of water. Oxidize the antimony to pentavelency with 1 ml of 10% sodium nitrite solution. After 5 minutes add 1 ml of saturated urea solution followed by 1 ml of 0.2% ethanolic solution of brilliant green. Extract with 10 ml of benzene and read the extract at 650 nm.

Zinc Sulfate.[13] To 20 ml of zinc sulfate solution thought to contain about 5 μg of antimony, add 5 ml of 1 : 3 sulfuric acid, 6.5 ml of hydrochloric acid, and 1.5 ml of 5% solution of sodium sulfite. Heat at 100° for 3 minutes and add 7 ml of hydrochloric acid. Add a 4% solution of ceric sulfate in *N* sulfuric acid dropwise until the yellow color persists and 7 drops in addition. After 3 minutes add 1 ml of 1% solution of hydroxylamine hydrochloride and mix until colorless. Mix with 2 ml

[9] A. A. Yadav and S. M. Khopkar, *Bull. Chem. Soc. Jap.* **44**, 693–696 (1971).

[10] V. I. Kurbatova, L. N. Makogonova, and V. K. Zharikova, *Tr. Vses. Nauchn.-Issled. Inst. Stand. Obraztsov. Spektr. Etalonov* **4**, 108–112 (1968).

[11] A. G. Fogg, C. Burgess, and D. Thorburn Burns, *Analyst* **98**, 347–350 (1973); cf. L. Narushkyavichyus and R. Kazlauskas, *Nauch. Tr. Vyssh. Ucheb. Zaved. Lit. SSR, Khim. Khim. Tekhnol.* **1968** (9), 47–53; cf. L. B. Kristaleva, *Tr. Tomsk. Univ.* **154**, 271–274 (1962).

[12] T. F. Rodina, I. S. Levin, M. N. Chepik, and E. G. Leont'eva, *Zavod. Lab.* **39**, 19–20 (1973).

[13] M. Kowalczyk, K. Ogiolda, and T. Pukas, *Rudy Met. Niezelaz.* **4**, 67–69 (1958).

of 0.2% solution of brilliant green and let stand for 3 minutes. Extract with 20 ml of benzene and read at 625 nm.

Titanium Dioxide.[14] Dissolve 0.8 gram of sample in 30 ml of sulfuric acid and 8 grams of ammonium sulfate. Cool and dilute to about 90 ml. When cold, dilute to 100 ml and take a 5-ml aliquot. Add 3 grams of sodium chloride and 3.3 ml of hydrochloric acid and dilute to about 15 ml. Keeping the temperature below 25°, add 5 ml of 0.5% solution of sodium nitrite. After 2 minutes for oxidation add 25 ml of 10% sodium hexametaphosphate solution and 5 ml of 0.05% solution of brilliant green. Extract with 5 ml of toluene and read the extract at 640 nm against a reagent blank.

Alternatively,[15] fuse 0.1 gram of sample with 4 grams of ammonium bisulfate for 4 minutes. Boil the product in 15 ml of 1 : 1 hydrochloric acid for 2 minutes. When cool complete as above from "Keeping the temperature below 25°,...."

Pharmaceutical Products. Digest 0.4 gram of a sample, which may contain titanium dioxide, with 2 ml of sulfuric acid acid and 2 ml of nitric acid, adding more nitric acid if necessary to prevent charring. Heat until evolution of brown fumes ceases, add 1 ml of water, and heat to sulfur trioxide fumes. Again add 1 ml of water and heat to sulfur trioxide fumes. Add 4 grams of ammonium bisulfate and fuse for 4 minutes. Complete as for titanium dioxide from "Boil the product in 15 ml of...."

Paint Flakes.[16] This technic is appropriate for such samples as scrapings from toys. Mix 50 mg of sample with 50 mg of sodium carbonate and heat at 450° for 30 minutes. Cool and add 1 ml of hydrochloric acid and 1 ml of nitric acid. Evaporate to dryness at 100°. Take up the residue by boiling with 20 ml of 1 : 100 hydrochloric acid, filter if necessary, cool, and dilute to 25 ml with that acid. Complete as for titanium dioxide from "Keeping the temperature below 25°,...."

Water. Evaporate 500 ml of sample with 50 ml of hydrochloric acid to dryness. Take up the residue in 30 ml of 1 : 1 hydrochloric acid. Add 2 drops of 25% solution of stannous chloride in hydrochloric acid and 1 ml of 10% sodium nitrite solution. After 2 minutes add 2 ml of 50% urea solution and 1 ml of 0.2% solution of brilliant green. Extract with 15 ml of toluene and dilute the extract to 20 ml. After 5 minutes read at 656 nm against toluene.

Soil.[17] As reagent made fresh daily, mix in 5 : 1 proportions a 10% solution of sodium hexametaphosphate and a solution containing 0.5 mg of brilliant green per ml.

Mix 1 gram of sieved sample with 1 gram of ammonium chloride in a test tube. Heat until the ammonium chloride has sublimed on the side of the tube. When cool, add 5 ml of 1 : 1 hydrochloric acid and heat at 100° until the sublimate had

[14] D. J. B. Galliford and J. T. Yardley, *Analyst* **88**, 653–654 (1963).
[15] R. J. M. Ratcliffe and S. G. E. Stevens, *Mfg. Chem.* **35** (9), 83–88 (1964).
[16] T. Hodson and D. W. Lord, *J. Assoc. Pub. Anal.* **9**, 60–63 (1971).
[17] R. E. Stanton and Alison J. McDonald, *Bull. Inst. Mining Met.* No. **667**, 517–522 (1962).

dissolved. Mix well, cool, and let settle. Take a 2-ml aliquot and add 1 ml of 1 : 1 hydrochloric acid. Add 0.1 ml of 5% sodium nitrite solution. Dilute to 10 ml with the reagent and add 1 ml of toluene. Shake for 30 seconds and read the toluene layer at 625 nm.

Alternatively,[18] fuse a sample expected to contain 0.4–4 μg of antimony with twice its weight of potassium bisulfate. Leach the melt with 8 ml of hydrochloric acid and filter. Wash with water and dilute to 15 ml. Add 0.5 ml of 5% solution of sodium nitrite and mix. Add 25 ml of 10% solution of sodium hexametaphosphate and 5 ml of 0.05% solution of brilliant green. Extract by shaking for 30 seconds with 5 ml of toluene. Filter the organic layer and read at 640 nm against toluene.

Tissue.[19] Grind 0.5 gram of sample in a glass mortar with 2 ml of water and 8 ml of 1 : 1 hydrochloric acid. Filter and take a 5-ml aliquot. Add 0.5 ml of 10% stannous chloride solution and heat at 100° for 5 minutes. Antimony is split from protein as the trivalent ion. Cool and oxidize to pentavalent antimony with 0.5 ml of 7% sodium nitrite solution. Destroy excess nitrite with 0.5 ml of 2% solution of potassium permanganate and 0.5 ml of 10% solution of glycerine in hydrochloric acid. Add 0.3 ml of 1% solution of brilliant green in 25% ethanol. Extract with 0.5 ml of toluene. Remove water and hydrochloric acid from the organic extract with a few crystals of disodium phosphate and read at 625 nm.

CATECHOL VIOLET

Pentavalent antimony forms a 1 : 1 complex with catechol violet in 0.02–0.15 *N* hydrochloric acid.[20] There is interference by twentyfold amounts of molybdenum or tungsten or by 100-fold amounts of copper. The reagent is sensitive to about 0.5 μg in ore by the technic below. There is interference by molybdenum, lanthanum, manganese, tin, germanium, niobium, iron, and gallium.

Procedure. Dilute the sample solution to 5 ml with 0.6 *M* hydrochloric acid.[21] Add 2.5 ml of 0.41 m*M* catechol violet and dilute to 25 ml. Set aside for 3 minutes and read at 560 nm against a reagent blank.

Ores. Treat 1 gram of sample with 10 ml of hydrofluoric acid. Add 5 ml of 1 : 1 sulfuric acid and evaporate to sulfur trioxide fumes. Take up the residue in 1 : 9 sulfuric acid and dilute to 100 ml with that acid. To a 5-ml aliquot, add 0.6% potassium permanganate to a pink coloration and decolorize with a solution of a ferrous salt. Add 5 ml of 0.05 *M* EDTA solution to mask most interfering ions; then add 2 ml of 0.04% solution of catechol violet. Add ammonium hydroxide to

[18]R. E. Stanton and Alison J. McDonald, *Analyst* **87**, 299–301 (1962).
[19]Yu. M. Fuzailov, *Nauch. Tr. Semark. Gos. Med. Inst.* **12**, 47–51 (1956).
[20]T. T. Bykhovtseva and I. A. Tserkovnitskaya, *Zavod. Lab.* **30**, 943 (1964).
[21]L. Naruskevicius, R. Kazlauskas, J. Skadauskas, and L. Vaskite, *Nauch. Tr. Vyssh. Ucheb. Zaved. Lit. SSR, Khim. Khim. Tekhnol.* **1973** (15), 61–66.

raise to pH 1 and dilute to 50 ml with 0.1 *N* hydrochloric acid. After 20 minutes read at 580 nm.

IODOANTIMONIATE

The complex of trivalent antimony with iodide ion is read at various wavelengths.[22] In 1:4 sulfuric acid, 425 nm is appropriate for reading against potassium iodide in 1:10 sulfuric acid. The extinction at 330 nm is greater but more subject to interference.[23] Bismuth interferes.

As developed from a solution of glass in hydrofluoric acid and read at 425 nm, no major component of glass interferes.[24] Lead can be separated as sulfate and chloride before development of the iodobismuthate.[25]

For antimony in sebacate-base lubricants, decompose the organic matter by heating with nitric and sulfuric acids and evaporate to sulfur trioxide fumes.[26] Dilute with water, filter, reduce with sodium hypophosphite, and add potassium iodide solution for reading as the iodoantimoniate.

For determination of antimony in an electrolyte for deposition of a lead-tin alloy concentrate with sulfuric and nitric acids, reduce the antimony with ascorbic acid, complex the copper with thiourea, and develop as the iodoantimoniate.[27]

By using sodium hypophosphite as the reducing agent for antimony, μg amounts can be determined in the presence of 200 ppm of gold.[28]

Procedure.[29] Mix a sample solution containing 0.2–1.2 mg of antimony with 25 ml of a solution containing 1% of ascorbic acid and 14% of potassium iodide. Dilute to 50 ml adjusting the acidity to 2.2–3.6 *N*. Read at 425 nm.

Copper.[30] Dissolve 3 grams in 20 ml of 1:4 nitric acid and evaporate to about 3 ml. Dilute to 25 ml and add ammonium hydroxide until a precipitate is formed. Redissolve by adding 1:4 nitric acid. Add 5 ml of 10% manganous sulfate solution followed by 2 ml of 5% potassium permanganate solution. Boil for 5 minutes and filter the manganese dioxide, which is a carrier for antimony. Filter, and wash with hot water. To the filtrate add another 5-ml portion of 10% manganous sulfate solution and 2 ml of 5% potassium permanganate solution. Boil for 5 minutes, filter, wash with hot water, and combine with the previous precipitate. Dissolve the precipitates in 5 ml of 1:1 hydrochloric acid by dropwise addition of 30% hydrogen peroxide. Add 5 ml of 1:4 sulfuric acid and evaporate to dryness. Take

[22] R. A. Washington, *Analyst* **90**, 502–503 (1965).
[23] F. P. Scaringelli and S. Blasof, *J. Assoc. Off. Ag. Chem.* **45**, 759–761 (1962).
[24] Horst Voelker and Guenter Schwarts, *Silikattechnik* **22**, 44–45 (1971).
[25] K. Lenhardt, *Z. Erzbergb. Metallhütt.* **19**, 139–141 (1966).
[26] George Norwitz and Michael Galen, *Anal. Chim. Acta* **61**, 413–420 (1972).
[27] A. F. Oparina and E. S. Bruile, *Zh. Prikl. Khim., Leningr.* **45**, 1368–1369 (1972).
[28] B. Yarash, *Chemist-Analyst* **56**, 28 (1967).
[29] E. V. Lapitskaya, F. P. Gorbenko, and E. I. Vel'shtein, *Zavod. Lab.* **34**, 1446–1447 (1968).
[30] A. G. Kon'kov, *Tr. Staling. S.-Kh. Inst.* **8**, 59–64 (1960).

up the residue in 5 ml of 1:4 sulfuric acid and add 5 ml of a solution of ammonium iodide containing 1 mg of iodide per ml. Add 10 ml of 1.5% solution of methyl violet in 1:19 hydrochloric acid. Filter the precipitate of oxidation products of methyl violet and of antimony. Wash with a solution containing 5 ml of sulfuric acid and 2 ml of 1.5% solution of methyl violet per 100 ml. Ash the paper and contents at 600°. Dissolve the ash in 10 ml of 1:4 sulfuric acid, filter, and dilute with 1:4 sulfuric acid to a known volume. Mix a 5-ml aliquot with 5 ml of a solution containing 12% of ammonium iodide and 1% of ascorbic acid. Dilute to 20 ml with 1:4 sulfuric acid and read at 330 nm.

As an alternative technic copper is precipitated as cupric sulfide in an alkaline medium.[31] Then the sulfides of antimony and tin are precipitated from the acidified filtrate as sulfides and dissolved in hydrochloric acid. Antimony is determined as the iodobismuthate and tin as molybdenum blue. The antimony in solutions of copper alloys can also be collected by precipitation with ferric hydroxide as the collector.[32]

Lead.[33] For 0.01–0.05% of antimony use a 1-gram sample, for 0.05–0.1%, use 0.5 gram. Dissolve in 50 ml of hydrochloric acid and 10 ml of a 1% solution of bromine in hydrochloric acid by gentle warming. The sample is desirably in the form of thin foil. When solution is complete, cool and oxidize with 1 ml of 0.2% solution of ceric sulfate in 0.1 *N* sulfuric acid. Dilute to 120 ml with hydrochloric acid and mix well. Add 40 ml of isopropyl ether and extract the antimony by shaking for 1 minute. Separate the organic layer. Extract the aqueous solution by again adding 1 ml of the ceric sulfate solution and 40 ml of isopropyl ether.

As a wash solution, mix 60 ml of hydrochloric acid with 28 ml of water. Add 5 ml of this wash solution to the combined extracts of antimony in isopropyl ether, shake, and discard the washings. Add the organic layer to 20 ml of 1:20 sulfuric acid and heat at 100°. After the organic solvent has evaporated, add 6.6 ml of 1:1 sulfuric acid and cool. Add 25 ml of a solution containing 11% of potassium iodide and 2% of ascorbic acid. Filter and dilute to 50 ml. After 5 minutes read at 425 nm against water.

Antimony-Arsenic Alloy Films.[34] Dissolve a film containing about 0.05 mg each of antimony and arsenic in 0.75 ml of 1:3 sulfuric acid and 0.05 ml of nitric acid. Keep at 100° overnight to volatilize the nitric acid. Dilute to 5 ml. Add to a 1-ml aliquot 1.73 ml of 1:3 sulfuric acid and 2 ml of a 15% solution of potassium iodide in 4% ascorbic acid solution. Dilute to 5 ml and read at 330 nm.

Grey Cast Iron. As a reagent, dissolve 50 grams of thiourea in 60 ml of warm water and add 175 grams of potassium iodide in 100 ml of water. Dilute to 1 liter.

Dissolve 0.4 gram of sample in 20 ml of 1:2 nitric acid and 5 ml of hydrofluoric acid in platinum.[35] Add 15 ml of 1:1 sulfuric acid and heat to sulfur trioxide

[31]Zygmunt Marczenko, *Chem. Anal.* (Warsaw) **2**, 160–167 (1957).
[32]M. L. Foglino, *Met. Ital.* **52** (6), 355–357 (1960).
[33]J. Bassett and J. C. H. Jones, *Analyst* **91**, 176–179 (1966).
[34]L. Tomlinson, *Rep. U.K. AEA AERE-M*, 1886, 7 pp. (1967).
[35]I. Janousek, *Hutn. Listy* **19**, 276–278 (1964); A. A. Amsheeva, *Izv. Vyssh. Ucheb. Zaved., Khim. Khim. Tekhnol.* **13**, 339–341 (1970); cf. A. I. Lazarev and V I. Lazareva, *Zavod. Lab.* **25**, 557 (1959).

fumes. Dissolve in 40 ml of water and add 5 ml of 2.5% solution of potassium permanganate. Heat to the boiling point and add a 10% solution of hydrogen peroxide dropwise until the precipitate is dissolved. Dilute to 100 ml and filter. Mix a 10-ml aliquot with 8 ml of 1:2.5 sulfuric acid and 20 ml of reagent. Dilute to 50 ml and read at 435 nm.

Alloy Steel.[36] Dissolve 4 grams of sample in 100 ml of 2:3 sulfuric acid, adding 30% hydrogen peroxide dropwise. Boil off excess hydrogen peroxide, add 0.1 gram of activated carbon, and filter. Dilute to a known volume and take two 50-ml aliquots. Add 10 ml of fresh 25% solution of ascorbic acid to each. To one add 5 ml of a solution containing 60% of potassium iodide and 5% of ascorbic acid. Dilute each to 100 ml and set aside for 1 hour. Read the developed sample against the blank at 436 nm. Not more than 0.2% of copper may be present.

Unalloyed Steel.[37] Dissolve 0.5 gram of sample in 5 ml of 3:2 nitric acid and add 5 ml of 1:1 sulfuric acid. Add 10% formic acid solution until the nitric acid has been destroyed and dilute to 50 ml. Add 0.5 gram of activated carbon and filter. To 10 ml of filtrate add 1 ml of 25% solution of ascorbic acid. Then add 1 ml of 50% potassium iodide. If cuprous iodide precipitates, filter it off. Dilute the filtrate to a known volume and read at 435 nm.

Gold-Antimony Alloys.[38] This technic is based on an alloy containing not less than 0.2% antimony. Transfer a sample not larger than 25 mg, containing 0.05–0.38 mg of antimony. Add 1 ml of 50% aqua regia and heat at 100° until solution is complete. Add 5 ml of water and 50 mg of sulfamic acid. Boil gently for 2 minutes, cool, and add 10 ml of 1:1 sulfuric acid. Dilute to 25 ml.

Mix a 10-ml aliquot with 8 ml of water. Add 5 ml of a fresh reagent containing 25 grams of potassium iodide and 2.5 grams of ascorbic acid in 41 ml of water. Dilute to 25 ml and read after 10 minutes but in less than 2 hours at 426 nm. Deduct a blank. Deduct a correction of 0.00035% of antimony for each 1% of gold present.

White Metal. For this technic, the tin content must be known and the antimony must not exceed 1%. Transfer a sample not larger than 50 mg, containing 0.4–3 mg of antimony and not more than 40 mg of tin or 10 mg of copper. Disintegrate by heating with 1 gram of potassium bisulfate and 5 ml of sulfuric acid. Cool, add 10 ml of water, and boil gently for 3 minutes. Cool and add 8 ml of 1:1 hydrochloric acid. Add 10 ml of 1:1 sulfuric acid and dilute to 100 ml.

After allowing any lead sulfate to settle, take a 5-ml aliquot. Add 6 ml of 1:3 sulfuric acid and 7 ml of water. Proceed as for gold-antimony alloys from "Add 5 ml of a fresh reagent...." In this case deduct a tin correction of 0.00653% of antimony for each 1% of tin present.

Thallium Solutions.[39] The sample solution contains more than 0.5 mg of

[36]Dagmar Blazjak-Ditges and Hans Klingeleers, *Z. Anal. Chem.* **248**, 18–20 (1969).
[37]H. Ploum, *Arch. Eisenhüttenw.* **33**, 601–604 (1962).
[38]A. Dym, *Analyst* **88**, 232–236 (1963).
[39]W. J. Maurice and R. L. M. van Lingen, *Anal. Chim. Acta* **28**, 91–92 (1963).

thallium and 0.05–0.6 mg of antimony. Add sulfuric acid in an amount sufficient to give 3 *N* when the solution is later diluted to 100 ml. Add 20 ml of a solution containing 56% of potassium iodide and 10% of ascorbic acid and dilute to 100 ml. After allowing to stand for 30 minutes or more, separate the precipitate of thallous iodide by centrifuging. Read at 425 nm against a blank.

Ores.[40] Dissolve a 2-gram sample in 20 ml of nitric acid and 10 ml of hydrochloric acid. Evaporate to a small volume and add 10 ml of 1 : 1 sulfuric acid. Evaporate to sulfur trioxide fumes. Add 2 ml of water and again evaporate to sulfur trioxide fumes. Take up in 5 ml of water and 30 ml of hydrochloric acid. Without filtration of insoluble matter, add 1 gram of cuprous chloride and 0.5 gram of potassium bromide. Extract the arsenic with 15 ml, 15 ml, and 15 ml of benzene, setting this aside for a separate determination or discarding it.

If sufficient iron is not present in the extracted solution, add it as ferric chloride. Make alkaline with ammonium hydroxide to coprecipitate the antimony with ferric hydroxide. Filter and wash with 1 : 20 ammonium hydroxide. Dissolve the precipitate from the filter with 5 ml of 1 : 3 sulfuric acid. Add 1 gram of potassium iodide and 0.5 gram of ascorbic acid. Extract with 15 ml, 15 ml, and 15 ml of benzene. Wash the extract with 30 ml of 1 : 3 sulfuric acid. Reextract the antimony by shaking the benzene extract with 50 ml of water. To this aqueous extract, add 15 ml of sulfuric acid, 20 ml of 10% potassium iodide solution, and 2 ml of 10% thiourea solution. Dilute to 100 ml and read at 430 nm.

Stibine.[41] This technic was developed for determination of stibine in air. Prepare an absorbant solution containing 10 grams of iodine, 80 grams of potassium iodide, and 80 ml of sulfuric acid per liter.

Pass the gas through 50 ml of the absorbant for a period expected to yield an amount of the order of 0.1 mg of stibine. Transfer to a stoppered bottle and add 1 gram of sodium hypophosphite. Set aside for 20 minutes to reduce the excess iodine. Read at 425 nm against water.

MALACHITE GREEN

The complex of malachite green (basic green 4) with antimony is extracted from 0.25–2 *N* hydrochloric acid.[42] Most elements do not interfere. The color is stable for 15 minutes. More than 5 μg of thallium interferes. For antimony as a transesterification or polymerization catalyst at 1–100 ppm in polyethylene terephthalate, use 0.1 gram of sample with this reagent.[43]

[40] K. Minisyan, *Prom. Arm.* **1966** (11), 40–42.
[41] R. Holland, *Analyst* **87**, 385–387 (1962).
[42] L. Narushkyavichyus and R. Kazlauskas, *Nauch. Tr. Vyssh. Ucheb. Zaved. Lit. SSR Khim. Khim. Tekhnol.*, **1968** (9), 47–53; cf. M. A. Popov and L. S. Kiparisova, *Uch. Zap. Tsent. Nauchn.-Issled. Inst. Olovyan. Prom.* **1965** (1) 55–60.
[43] Heinz Zimmermann, Horst Hoyme, and Annemarie Tryonadt, *Faserforsch TektTech.* **21**, 33–36 (1970).

Procedure. To 1 ml of sample containing 0.7–7 μg of antimony add 1 ml of 1 : 1 hydrochloric acid and heat for 5 minutes. Add 0.1 ml of 10% stannous chloride solution, 3 ml of water, and 0.5 ml of 5% sodium nitrite solution. After 1 minute add 1 ml of saturated urea solution, 15 ml of water, and 0.5 ml of 0.5% solution of malachite green in 25% ethanol. Extract with 25 ml of toluene and read at 610 nm.

Air.[44] Set up two absorbers in series, each containing 5 ml of 6% mercuric chloride solution in 1 : 1 hydrochloric acid. Pass 10 liters of air at about 500 ml/minute. Oxidize the antimony with 0.5 ml of 5% sodium nitrite solution. After 1 minute remove the excess nitrite with 10 ml of 10% disodium phosphate solution, which will also serve to mask iron and chromium. Shake until colorless, and if necessary add more phosphate to take care of excess iron. At once mix with 5 ml of fresh 0.05% solution of malachite green. Shake for 2 minutes with 5 ml of toluene and read the organic phase at 610 nm.

Blood.[45] Heat 1 ml of sample with 1 ml of nitric acid over a small flame until its color changes to amber. Add 0.5 ml of sulfuric acid and evaporate to sulfur trioxide fumes. The solution should be clear and colorless. Add 1 ml of 30% sodium chloride solution and 0.5 ml of 7% sodium nitrite solution. After 1 minute add 1 ml of 33% urea solution. Extract with 10 ml of water. Add 0.5 ml of 0.2% solution of malachite green and shake for 5 minutes. Read at 607 nm against a blank.

Biological Material.[46] Ash 100 grams of sample containing more than 1 mg of antimony with sulfuric acid and adjust to 63–83% of sulfuric acid in 50 ml. Mix a 3-ml aliquot with 2 ml of 1 : 4 sulfuric acid, 3 ml of 5 *N* hydrochloric acid, 0.1 ml of 5 *N* nitric acid, 0.75 ml of 1% solution of malachite green, and 0.2 gram of sodium sulfite. Extract with 5 ml of toluene. If the aqueous layer is no longer orange, add malachite green solution until the color is restored and again extract with 5 ml of toluene. Adjust the extract to 10 ml and read at 610 nm.

6-METHOXY-3-METHYL-2-[4-(*N*-METHYLANILINO)PHENYLAZO] BENZOTHIAZOLIUM CHLORIDE

This reagent forms a 1 : 1 complex with pentavalent antimony chloride in 1–4 *N* hydrochloric acid, 1–6 *N* sulfuric acid, or phosphoric acid. The complex is extractable with amyl acetate, and in this solvent it conforms to Beer's law for 0.1–5 μg of antimony per ml.[47] There is interference by thallic, auric, gallium, iodide, or thiocyanate ions.

Procedure. Adjust an appropriate sample to 1 : 1 hydrochloric acid or 1 : 3 sulfuric

[44] L. Palalau and E. Mihailescu, *Rev. Chim.* (Bucharest) **18**, 562–564 (1967).
[45] King-hung Tsao, Tun-chieng Loo, and Teng-han Tang, *Yao Hsüeh Hsüch Pao* **4**, 347–352 (1956).
[46] A. N. Krylova, *Sud.-Med. Ekspert., Min. Zdravookhr. SSSR* **2** (3), 31–36 (1959).
[47] P. P. Kish and Yu. K. Onishchenko, *Zavod. Lab.* **36**, 520–523 (1970).

acid. Add sodium chloride to make the solution more than 2 *M* in chloride ion. Add 5 ml of saturated chlorine water to oxidize the antimony, and after 2 minutes add 2 ml of saturated solution of urea to destroy the excess chlorine. Dilute to 3 *N* with hydrochloric acid or to 4 *N* with sulfuric acid. Add 1 ml of 0.05% solution of the reagent and extract with 6 ml of amyl acetate. Read the extract at 634 nm.

METHYLENE BLUE

The 1:1 complex of pentavalent antimony chloride with methylene blue is extracted with various solvents for reading.[48] Extraction from solution in 0.3–1 *N* sulfuric acid is as a basic chloride.[49] There is interference by gallium, thallic, auric, iodide, or thiocyanate ions, and by more than 100-fold amounts of nitrate ion. For steel analysis by this reagent, the antimony is isolated by distillation as the chloride.[50]

Procedure. To the sample solution containing up to 0.03 mg of antimony, add 1 gram of sodium chloride, 3 ml of 1:1 sulfuric acid, and water to make 5 ml. Oxidize the antimony with 0.2 ml of 10% solution of sodium nitrite. After 1 minute decompose excess oxidizing agent with 0.5 ml of saturated solution of urea. Add 0.5 ml of 0.1% methylene blue solution and 6 ml of 1:1 benzene-dichloroethane, or 5:1 benzene-nitrobenzene or chloroform. Shake for 1 minute and read the extract at either 660 or 630 nm against a blank.

METHYLENE GREEN B

Methylene green B (basic green 5) complexes with antimony at pH 0.3–0.45 for extraction by chloroform. Mercuric and thallic ions interfere.

Procedure. *Cadmium.*[51] Dissolve 0.5 gram of sample in 10 ml of hydrochloric acid. Evaporate to less than 3 ml and add 3 ml of water. Add 2 ml of 10% sodium nitrite solution and set aside for 5 minutes. Add 3 ml of 50% solution of urea to decompose excess sodium nitrite. Dilute to 50 ml and take a 10-ml aliquot. Add 9 *N* hydrochloric acid to pH 0.3–0.45. Mix with 0.25 ml of 0.04% solution of methylene green. Shake for 2 minutes with 10 ml of chloroform and read the extract at 655 nm.

---

[48] P. P. Kish and Yu. K. Onishchenko, *Zh. Anal. Khim.* **23**, 1651–1657 (1968).
[49] V. M. Tarayan and Zh. M. Arstamyan, *Army. Khim. Zh.* **26**, 124–128 (1973).
[50] W. Chetkowski, *Pr. Inst. Hutn.* **18**, 109–112 (1966).
[51] V. M. Tarayan and Zh. M. Arstamyan, *Army. Khim. Zh.* **27**, 557–563 (1974).

METHYLFLUORONE

At pH 2, trivalent antimony forms a 2:3 complex with methylfluorone which is 2,3,7-trihydroxy-9-methyl-6-fluorone.[52] The maximum absorption at 530 nm conforms to Beer's law for 0.24 μg/ml. The colloidal complex is stabilized with gelatin or polyvinyl alcohol and can be extracted with methyl isobutyl ketone, ethyl acetate, butyl acetate, or 4-methyl-2-pentanol.

Procedure. *Iron and Steel.* Dissolve 1 gram of sample in 30 ml of 1:2 nitric acid. Add 80 ml of water and 5 ml of 4% solution of manganous sulfate pentahydrate. Dropwise, add 2 ml of saturated solution of potassium permanganate. Boil, filter the manganese dioxide carrier of the antimony, and wash thoroughly with hot water. Dissolve the precipitate in 8 ml of 1:1 sulfuric acid and 2 ml of 10% hydroxylamine hydrochloride solution. To that solution add 8 ml of 1:1 sulfuric acid, 2 ml of phosphoric acid, and 5 ml of 20% potassium iodide solution. Extract the antimony iodide with 15 ml and 10 ml of benzene. Mix the organic phase with 10 ml of water and add 5% ascorbic acid solution in 1-ml portions until the solution is colorless. Discard the organic phase and add 2% sodium hydroxide solution to raise the aqueous phase to pH 2. Add 5 ml of hydrochloric acid–potassium chloride buffer solution for pH 2. Add 1 ml of a 1% solution of polyvinyl alcohol and 1 ml of 0.05% solution of methylfluorone in 50% ethanol. Dilute to 40 ml, adjust to pH 2, and dilute to 50 ml. Read at 495 nm against a blank prepared with pure iron powder.

METHYL VIOLET AND CRYSTAL VIOLET

The methods given here include crystal violet, which is pure hexamethylpararosaniline chloride, and methyl violet, which is mostly the pentamethyl derivative. Their complexes with antimony absorb in the same range. Increase in concentration of hydrochloric acid greatly increases the accuracy. Gold, mercury, and thallium react in the same way as antimony.[53] The complex is relatively stable in both aqueous and benzene solutions. Trichloroethylene is also a suitable solvent for the complex.[54]

Methyl violet is appropriate for the determination of 0.0001–0.1% of antimony in lead.[55] In solution less than 1:1 in hydrochloric acid or less than 1:35 in sulfuric acid, the antimony-cupferron complex is extracted with chloroform. Gold is not extracted if the solution is in hydrochloric acid. If more than 0.5 mg of iron is present, oxidize the antimony to pentavalency with permanganate. Add citrate and cupferron and extract the ferric cupferronate with chloroform. Then reduce the antimony to trivalency with thiourea and extract as the cupferron complex. The

[52] S. Meyer and O. G. Koch, *Z. Anal. Chem.* **179**, 175–186 (1961); O. G. Koch, *ibid.* **265**, 29–30 (1973).
[53] King-hung Tsao, Yun-chieng Loo, and Teng-han Tang, *Yao Hsiieh Hsiieh Pao* **4**, 107–116 (1956).
[54] K. Studlar and I. Janousek, *Chem. Commun.* **25**, 1965–1969 (1960); cf. I. A. Blyum, S. T. Solv'yan, and G. N. Shebalkova, *Zavod. Lab.* **27**, 950–956 (1961).
[55] M. Cyrankowska and J. Downarowicz, *Chem. Anal.* (Warsaw) **10**, 67–76 (1965).

technic is appropriate for antimony in tin and copper.[56] The reagent is also applicable to solder and to antimony in rhenium and molybdenum.[57]

Antimony in a solution of ferrotungsten is precipitated with thionalide, which is α-mercapto-*N*, 2-naphthylacetamide, and determined by methyl violet.[58] As another technic for analysis of ferrotungsten, antimony, tin, and arsenic are precipitated as the sulfides from a dilute sulfuric acid solution containing tartaric acid.[59] Then in a solution of the sulfides the antimony and tin are coprecipitated by formation of manganese dioxide from manganese and permanganate ions, redissolved, and the antimony determined with methyl violet.

Methyl violet is used for determining antimony in chromium and chromium-nickel alloys.[60] The optimum acidity for complex formation is 3–3.5 *N* hydrochloric acid. Nitric acid interferes. The method does not usually require separation of chromium and other components of the alloy.

For analysis of antimony in zinc and cadmium by methyl violet, interference by auric, thallic, and mercuric ions is avoided by development in 1:9 hydrochloric acid.[61]

Methyl violet will normally determine antimony in a solution of ammonium molybdate without interference.[62] When large amounts of molybdate are present, as in the analysis of molybdenum metal, a flocculent precipitate of molybdic acid and methyl violet absorbs the antimony complex. It is not then extractable with benzene or toluene. Citric acid is the best complexing agent for the molybdic acid.

Coprecipitation of 25–100 μg of antimony with metastannic acid successfully separates it from a solution of a copper alloy for determination with methyl violet.[63] Residual copper reduces the color developed. Results with methyl violet in water analysis have been reported to be unsatisfactory.[64]

Ethyl violet gives a greater extinction coefficient than methyl violet and has a maximum absorption in the same range—590 versus 595 nm.[65]

For antimony in lead and lead oxide by methyl violet, see Thallium by Crystal Violet.

Procedures

Cast Iron.[66] Dissolve 1 gram of sample in a distilling flask with 10 ml of nitric

[56]E. I. Ishutchenko and V. M. Eliseeva, *Zavod. Lab.* **21**, 791 (1955).

[57]D. I. Ryabchikov and A. I. Lazarev, *Renii, Tr. Vses. Soveshch. Probl. Reniya, Akad. Nauk SSSR, Inst. Met.* **1958**, 267–274.

[58]V. I. Kurbatova, L. Ya. Kalashnikova, G. N. Emasheva, I. N. Nikulina, and V. V. Feofanova, *Tr. Vses. Nauchn.-Issled. Inst. Standn. Obraztsov. Spektr. Etalonov*, **5**, 111–118 (1969).

[59]Hidehiro Goto, Yachiyo Kakita, and Masahiko Sase, *Nippon Kinzoku Gakkaishi* **21**, 385–387 (1957).

[60]L. S. Nadezhina, *Tr. Leningr., Politekh. Inst. M. I. Kalinina* **201**, 120–126 (1959); E. I. Nikitina, *Tr. Kom. Anal. Khim., Akad. Nauk SSSR, Inst. Geokhim. Anal. Khim.* **12**, 311–313 (1960).

[61]L. I. Maksai and A. A. Silaeva, *Sb. Tr. Vses. Nauch.-Issled. Gornmet. Inst. Tsvet. Met.* **1962** (7), 343–344.

[62]A. I. Lazarev and V. I. Lazariva, *Zavod. Lab.* **25**, 405–406 (1959).

[63]Yu. V. Morochevskii and D. G. Barbanel, *Uch. Zap. Leningr. Gosut. Univ. A. A. Zhdanova* No. **211**, *Ser. Khim. Nauk* (15), 62–75 (1957).

[64]A. A. Nakharova, E. S. Tovatopyat, and V. Ya. Eremenko, *Girdokhim. Mater.* **51**, 196–202 (1969).

[65]Tsutomu Matsuo, Tasuo Kuroyanagi, and Shinsaku Maguro, *Jap. Anal.* **12**, 515–520 (1963).

[66]Z. Vecema and B. Bieber, *Slevarenstvi* **11**, 272–274 (1963).

acid, 5 ml of sulfuric acid, 5 ml of phosphoric acid, and 50 ml of water. Add 35 ml of sulfuric acid and evaporate to sulfur trioxide fumes. Cool and add 20 ml of water, 2 grams of hydrazine sulfate, and 20 ml of hydrochloric acid. Heat to 155–165° and distill at that temperature, adding 100 ml of hydrochloric acid over a period of 60–90 minutes. Dilute the distillate of antimonous chloride to 200 ml.

To a 10-ml aliquot, add 3 drops of 0.5% ferric chloride solution. Decolorize by dropwise addition of 20% stannous chloride solution. After 2 minutes oxidize by adding 1 ml of 20% solution of sodium nitrite. After 4 minutes add 8 ml of ice water and 4 ml of ice-cold 10% urea solution. Dilute with 60 ml of ice water and add 1 ml of 0.5% solution of methyl violet. Shake for 2 minutes with 25 ml of toluene. Filter the toluene through a cotton plug and read at 610 nm.

Iron and Steel Products.[67] Dissolve 0.5 gram of sample by heating with 10 ml of hydrochloric acid and 5 ml of nitric acid. Dilute to 15 ml with hydrochloric acid and add 25% solution of stannous chloride in hydrochloric acid until the solution is dark green. The antimony has now been reduced to trivalency. Chill in ice and oxidize the iron to the ferric form by dropwise addition of a saturated solution of sodium nitrite until the color is orange. After 1 minute add 60 ml of water and cool to 5–6°. Add 8 drops of 30% hydrogen peroxide to reoxidize the antimony to pentavalency. One minute later add a saturated solution of hydroxylamine hydrochloride slowly until effervescence ceases. Add 5 ml excess to ensure absence of hydrogen peroxide. One minute later add 20 ml of 0.025% crystal violet solution. Extract with 10 ml, 5 ml, and 5 ml of toluene. Dilute the extracts to 25 ml with toluene and read at 615 nm against a reagent blank. The antimony standard for preparation of the calibration graph must contain an equivalent amount of iron.

Copper Alloy.[68] Dissolve 5 grams of sample by heating in a distilling flask with 40 ml of 1 : 1 nitric acid. Cool and add 25 ml of 1 : 1 sulfuric acid. Evaporate to sulfur trioxide fumes. Dissolve the residue in 25 ml of water. Add 1 gram of potassium bromide in 10 ml of water. Add 40 ml of sulfuric acid and 20 ml of hydrochloric acid. Heat in a glycerol bath and pass a stream of carbon dioxide through the flask and an attached condenser. Heat, and when the temperature reaches 160–170°, begin slow addition from a dropping funnel of a mixture of 10 ml of 10% potassium bromide solution and 20 ml of hydrochloric acid. The antimony will be distilled in 30–40 minutes. Dilute the distillate to 200 ml and complete as for cast iron from "To a 10-ml aliquot, add…."

Nickel.[69] Dissolve a 5-gram sample in 45 ml of 2 : 1 sulfuric acid and 10 ml of 1 : 1 hydrochloric acid. Add 1 ml of 10% sodium nitrite solution to oxidize to pentavalent antimony. After 5 minutes add 1 ml of 50% urea solution to decompose excess nitrite. Dilute to 150 ml and add 1 ml of 2% crystal violet solution. Extract with 20 ml of benzene and read the extract at 610 nm.

[67] A. Alderisio and F. Gauzzi, *Ric. Sci. Riv.* **36**, 1025–1028 (1966).

[68] Yu. V. Morachevskii and D. G. Barbanel, *Uch. Zap. Leningr. Gos. Univ. A. A. Zhdanova* No. **211**, *Ser. Khim. Nauk* (15), 76–82 (1957).

[69] L. S. Studeskaya, M. N. Kruglova, G. S. Dolgorukova, and V. A. Nerbitskaya, *Tr. Vses. Nauchn.-Issled. Inst. Stand. Obraztsov. Spektr. Etalonov* **7**, 29–32 (1965).

Converter Copper.[70] Dissolve 1 gram of sample in 10 ml of 1 : 3 nitric acid. Add 10 ml of 1 : 1 sulfuric acid and evaporate to sulfur trioxide fumes. Add 5 ml of water and again evaporate to sulfur trioxide fumes. Take up the residue in 50 ml of 1 : 1 hydrochloric acid, dilute to 100 ml, and filter. Mix a 5-ml aliquot of filtrate with 5 ml of hydrochloric acid and decolorize the ferric ion by dropwise addition of 10% solution of stannous chloride in 1 : 1 hydrochloric acid. Add 1 ml of 10% sodium nitrite solution and, after 5 minutes, 1 ml of saturated solution of urea. Dilute to 75 ml and add 0.5 ml of 0.2% solution of methyl violet. Extract with 30 ml of benzene or toluene and read at 530 nm.

Bismuth.[71] Dissolve 1 gram in 3 ml of sulfuric acid and heat to sulfur trioxide fumes. Dissolve the residue in 3 : 1 hydrochloric acid and dilute to 100 ml with that acid. Take a 5-ml aliquot and add 0.5 ml of 10% citric acid solution to mask bismuth. Add 2 drops of 10% stannous chloride solution and set aside for 1 minute. Add 1 ml of 10% sodium nitrite solution and set aside for 5 minutes. Destroy the residual nitrite with 0.5 ml of saturated solution of urea. When the evolution of gases is nearly complete, dilute to 80 ml and add 0.5 ml of 0.1% solution of crystal violet. Shake for 1 minute with 20 ml of toluene, dilute the extract to 25 ml, and after 5 minutes read at 590 nm.

Thallium.[72] Dissolve 1 gram of sample by heating with 5 ml of nitric acid and 10 ml of hydrochloric acid. Add 5 ml of 1 : 1 sulfuric acid and evaporate to sulfur trioxide fumes. Add 5 ml of water and again evaporate to sulfur trioxide fumes. Take up the residue in 20 ml of 1 : 1 hydrochloric acid, filter, and dilute to 100 ml with the same acid. Add 10 ml of 5% zinc chloride solution and 5 grams of tartaric acid and dilute to 200 ml.

Prepare a chromatographic column of 0.5–1 mm SBS cation exchange resin and free from iron by washing with 1 : 5 hydrochloric acid. Pass the solution through the column, followed by 5% tartaric acid solution. Elute thallium from the column with 50 ml of 1 : 1 hydrochloric acid. Then elute antimony with water and dilute the eluate to a known volume. Mix a 10-ml aliquot of eluate with 10 ml of hydrochloric acid. Decolorize by dropwise addition of 20% stannous chloride in 1 : 1 hydrochloric acid. Then oxidize with 1 ml of 10% solution of sodium nitrite. After 4 minutes dilute 1 : 1 with water and add 0.25 ml of saturated solution of urea. Dilute with water to a 1 : 9 ratio of hydrochloric acid to water. Add 0.5 ml of 2% solution of methyl violet. Extract with 10 ml of toluene and read the extract at 590 nm.

Ferroniobium.[73] To 0.5 gram of sample add 1 ml of 10% solution of stannous chloride and 2.5 ml of hydrofluoric acid. Add nitric acid dropwise until solution is complete. Cool, and add 0.4 gram of boric acid to mask fluoride ion. Oxidize the

[70] L. N. Krasil'nikova and K. N. Dolgorukova, *Sb. Nauch. Tr. Vses. Nauchn.-Issled. Gornomet. Inst. Tsvet. Met.* **1965** (9), 22–25.

[71] T. V. Lyashenko, *Met. Khim. Prom. Kaz., Nauch.-Tekhn. Sb.* **1962** [6(22)], 67–68.

[72] A. P. Kilimov, V. A. Polishchuk, and N. R. Gasparyants, *Sb. Nauch.-Tekh. Tr. Nauch.-Issled. Inst. Met. Chelyab. Sov. Nar. Khoz.* **1960** (1), 168–172.

[73] N. I. Shishkina, Ya. G. Sakharova, and L. N. Freidenzon, *Tr. Ural'. Nauchn.-Issled. Inst. Chern. Met.* **9**, 55–63 (1970).

antimony with 1 ml of 10% sodium nitrite. After 1 minute add 1 ml of 25% solution of urea, followed by 1 ml of 0.1% solution of crystal violet. Shake for 1 minute each with 30 ml and 15 ml of benzene. Filter the extracts, dilute to 50 ml, and read at 610 nm.

Germanium.[74] To 1 gram of finely powdered sample in a quartz beaker add 20 ml of water and 40 ml of 30% hydrogen peroxide. Heat gently until solution is complete and evaporate to dryness. If arsenic is present, distill at 108° as arsenious chloride. For details see Chapter 8. To the residue, after distillation of arsenic, add 2 ml of 1 : 5 sulfuric acid and 3 ml of nitric acid. Evaporate at 100° to about 0.5 ml. If arsenic was not distilled, use the residue on evaporation. Take up in 3 ml of 1 : 1 hydrochloric acid. Add 5 ml of 0.1% solution of ceric sulfate and let stand for 5 minutes. Reduce the excess of ceric ion by dropwise addition of 10% hydrazine sulfate solution. Add 1 ml of 0.2% solution of methyl violet and 5 ml of 5% solution of sodium citrate. Dilute to 25 ml and shake for 1 minute with 5 ml of amyl acetate. Read the organic layer at 600 nm.

Germanium Dioxide.[75] Transfer 2 grams of sample to a Kjeldahl flask marked at 10 ml. Add 50 ml of hydrochloric acid and pass chlorine at 2 bubbles per second. After 5 minutes heat the flask gently to avoid frothing. When the solution becomes clear, heat more strongly, continuing the passage of chlorine until evaporated to 5 ml. Cool, and remove dissolved chlorine by passing nitrogen through the solution for 10 minutes. Transfer to a 20-ml graduated tube with hydrochloric acid and dilute to volume with that acid.

Mix a 10-ml aliquot with 3 drops of 10% solution of stannous chloride in hydrochloric acid and 1 ml of 10% sodium nitrite solution. After 5 minutes add 10 ml of water, 1.5 ml of saturated solution of urea, and 75 ml of water. Add 0.6 ml of 0.2% solution of crystal violet with vigorous stirring during the addition. Extract by shaking with 10 ml of toluene for 3 minutes and read the organic phase at 530 nm.

Cadmium Oxide.[76] Dissolve 1 gram of sample in 8 ml of 1 : 1 hydrochloric acid and dilute to 25 ml. Mix a 10-ml aliquot with 1 ml of 20% sodium nitrite solution and heat at 100° for 10 minutes. Add 1 ml of saturated solution of urea to destroy excess nitrite, and dilute to 100 ml. Add 0.5 ml of 0.2% solution of crystal violet and extract with 25 ml of toluene. Read at 530 nm.

Zinc Electrolytes.[77] The technic is also applicable to solutions of zinc alloys. Mix the sample solution with 20 ml of hydrochloric acid to give not less than 2.5×10^{-5} *M* antimony. Add 1 ml of 10% sodium nitrite solution and let stand for 5 minutes. Add an equal volume of water, then 0.5 ml of saturated solution of urea. Add 0.6 ml of 0.2% solution of methyl violet and dilute to 50 ml. Extract with 25 ml of benzene and read the extract at 610 nm.

[74] Hidehiro Goto and Yachiyo Kakita, *Sci. Rep. Res. Inst. Tohoku Univ. Ser. A*, **8**, 243–251 (1956).
[75] R. Rezac and J. Ditz, *Zavod. Lab.* **29**, 1176–1178 (1963).
[76] S. A. Lomonosov, V. N. Podchainova, and V. E. Rybina, *ibid.* **31**, 420 (1965).
[77] P. N. Kovalenko and N. P. Moricheva, *Izv. Vyssh. Ucheb. Zaved., Khim. Khim. Tekhnol.* **2** (3), 322–327 (1959).

Blood.[78a] Digest 4 ml of sample with 1 ml of perchloric acid, 3 ml of nitric acid, and 3 ml of sulfuric acid. If necessary to avoid charring, add more nitric acid dropwise during the course of the digestion. When the perchloric acid is completely decomposed, chill in ice and add 5 ml of 10% sodium chloride solution. Mix well and extract the antimonic chloride with 5 ml of amyl acetate that has previously been washed with 5 ml of 1 : 1 hydrochloric acid and 5 ml of water. Shake the amyl acetate extract with 0.5 ml of 0.1% solution of crystal violet and 5 ml of 0.3 *N* hydrochloric acid. Centrifuge and read at 590 nm.

Urine. Digest 5 ml of sample with 0.5 ml of perchloric acid, 0.5 ml of nitric acid, and 3 ml of sulfuric acid. Proceed as for blood from "When the perchloric acid is...." In prewashing the amyl acetate, use 5 ml of 0.1% solution of crystal violet along with the hydrochloric acid.

Organic Compounds.[78b] Ignite 2 mg of an organic antimony compound with magnesium under a 40-mm layer of magnesium to form a compound of 3 atoms of magnesium with 2 atoms of antimony. Prepare an absorption solution by dissolving 5 grams of sodium nitrite in 250 ml of 1 : 1 hydrochloric acid and prepare an absorption cylinder containing 10 ml of the absorption solution. Decompose the melt with 10 ml of water followed by 40% sulfuric acid. Facilitate the absorption of the evolved stibine by passing an inert gas. As absorption proceeds add three successive 5-ml portions of the absorption solution to the absorber. In each case flush out the absorber with 5 ml of water. Add to the combined absorption solutions 80 ml of water, 10 ml of saturated solution of urea, and 1 ml of 2% solution of crystal violet. Shake for 1 minute each with 30 ml, 20 ml, and 15 ml of toluene. Dilute the extracts to 100 ml with ethanol and read at 610 nm.

Stibine in Air.[79] Mix 10 ml of 1 : 1 hydrochloric acid with 0.5 ml of 30% hydrogen peroxide in an absorber. Draw air through at up to 50 liters/hour until about 5 μg of stibine has been collected. To the contents of the absorber, add 5 drops of 10% solution of stannous chloride in 1 : 1 hydrochloric acid. Wait 1 minute and add 0.5 ml of 30% hydrogen peroxide. This sequence is essential to avoid formation of nonreactive antimony compounds. Let stand for 3 minutes and add 37 ml of 0.2% solution of crystal violet. Shake for 1 minute with 10 ml and 10 ml of toluene. Dilute the combined extracts to 25 ml with toluene and read at 590 nm against a reagent blank.

MOLYBDATE

Antimony can be read by the blue developed from molybdate near pH 1.4.[80a] Presence of a large amount of sodium sulfate minimizes the blank and improves

[78a]Yü-ch'ing Lü Leu, *Yao Hsüeh Hsüeh* **7**, 171–174 (1959).
[78b]Josef Jenik, *Chem. Listy* **51**, 1312–1315 (1957).
[79]G. Kh. Sorokin and S. A. Lomonosov, *Zavod. Lab.* **40**, 23–25 (1974).
[80a]R. M. Matulis and J. C. Guyon, *Anal. Chem.* **37**, 1391–1395 (1965); J. C. Guyon and R. M. Matulis, *Chemist-Analyst* **56**, 22–23 (1967).

reproducibility. The range of applicability is 0.02–1.6 ppm. There is serious interference by bismuth, chlorate, iodate, dichromate, silicate, tantalum, and phosphate. Because of interferences, the antimony must commonly be distilled as the chloride. That separation from arsenic as well as distillation of tin is described in Chapter 8

In an acid solution containing antimony, phosphate, and molybdate in the presence of ascorbic acid a molybdophosphoricantimonic acid is formed.[80b] This complex is extracted with butyl acetate for reading at 651 nm.

Procedure. As a reagent, mix 120 ml of 5% solution of sodium molybdate dihydrate, 400 ml of *M* sodium sulfate, and 40 ml of 1 : 3 sulfuric acid. Dilute to 1 liter and let stand 24 hours before use.

Dilute the sample containing up to 0.16 mg of antimony to about 40 ml. Add 1 drop of phenolphthalein indicator solution, then 2% sodium hydroxide solution to the first permanent pink color. Add 1 ml of 5% ascorbic acid solution. Start a timer, and simultaneously add 50 ml of the molybdate reagent. Dilute to 100 ml and after 30 minutes read at 725 nm against water.

MORIN

Morin, which is 2′,3,4′,5,7-pentahydroxyflavone, complexes with antimony and also with tin.[81] The two elements are separated before determination by fluorescence. The tin can also be read by absorption. Extractions as iodides and reextractions avoid interferences.

Procedure. *Silicate Minerals.* Disintegrate a 0.5-gram sample in platinum with 5 ml of sulfuric acid and 2 ml of hydrofluoric acid. Evaporate to sulfur trioxide fumes and take up in 15 ml of water. Filter, and wash on the filter with water. Ash the residue and fuse with 1 gram of sodium carbonate and 1 gram of borax. Take up the melt with 10 ml of 1 : 10 sulfuric acid and filter if necessary. Dilute the combined aqueous acid filtrates to 100 ml with 1 : 10 sulfuric acid.

Mix an aliquot containing 0.5–20 μg of antimony with 30 ml of 1 : 1 sulfuric acid. Add 4 ml of 1% solution of potassium iodide and dilute to 50 ml. Extract with 10 ml of benzene. Reextract the antimony from the benzene layer with 10 ml of 0.2 *N* hydrochloric acid. Add 1 ml of 0.05% solution of morin and read by fluorescence.

Mix an aliquot containing 0.05–5 μg of tin for later reading by fluorescence, or up to 0.05 mg if to be read absorbtimetrically, with 50 ml of 1 : 1 sulfuric acid and 4 ml of 1% potassium iodide solution. Dilute to 50 ml and extract the antimony with 10 ml of benzene. Discard this extract and add 4 ml of 10% potassium iodide solution to the aqueous layer. Now extract the tin with 10 ml of benzene. Reextract the tin from the benzene layer with 10 ml of 0.05 *N* hydrochloric acid and add 1 ml

[80b] J. E. Going and J. Thompson, *Microchem. J.* **21**, 98–105 (1976).
[81] D. P. Shcherbov, I. N. Istaf'eva, and R. N. Plotnikova, *Zavod. Lab.* **39**, 546 (1973).

of 0.05% morin solution. Read fluorimetrically or measure the absorption at 420 nm.

PHENYLFLUORONE

The complex of phenylfluorone with trivalent antimony is read as a stabilized colloid.[82] The extinction is constant in 0.2–0.5 *N* sulfuric acid and less than 0.2 *N* hydrochloric acid does not interfere. The optimum condition for reading is in 28% ethanol containing 8 ppm of phenylfluorone at pH 1.[83]

For analysis of animal feed, the sample is dissolved in sulfuric acid.[84] After adding tartaric acid, the antimony is extracted with cupferron in chloroform. Antimony is then separated from coextracted elements by radial chromatography on paper developed with butanol saturated with *N* hydrochloric acid. The dried paper, sprayed with 0.05% solution of phenylfluorone in methanol, gives a red antimony zone. This is extracted, and the antimony is reduced from pentavalency to trivalency and read at 495 nm.

Procedures

Gallium or Indium.[85] Dissolve a 1-gram sample in 6 ml of nitric acid and evaporate to dryness. Carefully add 2 ml of 85% formic acid solution; when the vigorous reaction abates, add 2 ml more. Evaporate to dryness. Dissolve the dry residue in 20 ml of 1 : 7 sulfuric acid and heat to 50–60°. Add 1 ml of a 3% solution of titanous sulfate in 1 : 5 sulfuric acid and let stand for 30 minutes.

To prepare fresh diethyldithiocarbamic acid solution, mix 50 ml of 1% solution of the sodium salt with 2 ml of 1 : 1 hydrochloric acid and extract with 50 ml of chloroform. Extract the sample solution with 10 ml and 5 ml of this reagent. To the combined chloroform extracts, add 2.5 ml of nitric acid and 0.5 ml of sulfuric acid. Evaporate to sulfur trioxide fumes. Take up in 2 ml of water, add 0.2 ml of 85% formic acid solution, and again evaporate to fumes. Take up this residue in 15 ml of water and add 5 ml of sulfuric acid. Add 0.5 gram of tartaric acid and 0.1 gram of ascorbic acid. Cool and add 1 ml of 10% potassium iodide solution and 0.6 ml of 10% pyridine solution. After allowing to stand for 15 minutes, extract with 5 ml, 3 ml, and 3 ml of ether. Combine the ether extracts and extract the antimony with 1.5 ml and 1.5 ml of 1.5 *N* sulfuric acid. Mix the combined acid extracts with 0.3 ml of 1% gelatin solution and 0.3 ml of 0.05% solution of phenylfluorone. After 30 minutes dilute to 5 ml and read at 530 nm.

Thallium. Proceed as for gallium until the residue is obtained on evaporation after treatment with formic acid. Dissolve the residue in 2 ml of sulfuric acid and

[82] T. V. Gurkina and K. M. Konovalova, *Tr. Kaz. Nauch.-Issled. Inst. Mineral. Syr'ya* **1961** (1), 249–254.
[83] S. Ganapathy Iyer, B. V. Kadam, and C. Venkateswarin, *Indian J. Chem.* **11**, 385–387 (1973).
[84] Gerda Boenig and H. Heigener, *Landw. Forsch.* **26**, 84–88 (1973).
[85] E. A. Biryuk, *Zavod. Lab.* **30**, 651–652 (1964).

dilute to 30 ml. Add 1 ml of 3% titanous sulfate solution in 1:5 sulfuric acid and 10 ml of hydrochloric acid. Filter the precipitated thallium chloride and wash it with 5 ml of water. Add 5 ml of sulfuric acid and 27 ml of hydrochloric acid to the filtrate and dilute to 80 ml. Complete as for gallium from "To prepare fresh diethyldithiocarbamic...."

Iron Cathodes.[86] The sample is a mass from alkaline accumulators. Treat 0.5 gram of sample containing less than 3% of antimony with 40 ml of 1:1 hydrochloric acid. When evolution of hydrogen ceases, heat almost to boiling. Oxidize metallic antimony by adding 3 ml of 5% solution of ferric chloride hexahydrate in 0.1 *N* hydrochloric acid. When black specks of antimony have been decomposed, dilute with an equal volume of water and filter. Cool and dilute to 250 ml. Reduce the iron in a 1-ml aliquot with 1 ml of 1% ascorbic acid solution. Add 1 ml of 1% solution of gum acacia and dilute to 9 ml with 0.5 *N* sulfuric acid. Add 1 ml of a solution of 25 mg of phenylfluorone and 0.5 ml of 1:1 hydrochloric acid made up to 100 ml with ethanol. After 10 minutes read at 530 nm against water. By solution in hydrochloric acid, there is no evolution of stibine.

Ores. Mix 1 gram of ore with 6 grams of 4:3 sodium carbonate–sulfur in a porcelain crucible. Place in a muffle furnace at 150° and raise the temperature to 300° over a 20-minute period. Hold at that temperature for 30 minutes, then raise to 400–450° for 30 minutes. Cool, and heat to 100° with 20 ml of water. Add 5 ml of 20% sodium sulfide solution and again heat to 100°. Filter, and wash the precipitate well with 1% sodium sulfide solution. Add to the filtrate and washings 8 ml of 1:1 sulfuric acid and 8 ml of 2:1 sulfuric acid–30% hydrogen peroxide. Evaporate to about 8 ml and add 3 ml of nitric acid. Evaporate to sulfur trioxide fumes. Wash down the sides of the beaker and again evaporate to sulfur trioxide fumes. Repeat this step. Add 8 ml of 1:1 sulfuric acid and dilute to 50 ml.

To a 10-ml aliquot add 5 ml of 1% thiourea solution and let stand for 5 minutes. Add 1 ml of 1% gelatin solution and 2.5 ml of 0.03% phenylfluorone solution. Dilute to 25 ml and read at 540 nm against a blank. Simple ores that will dissolve directly in acid can bypass the fusion step.

4-(2-PYRIDYLAZO)RESORCINOL

Trivalent antimony forms a 1:2 complex with this reagent, known as PAR, in 60% acetone at pH 1.2–1.9.[87] Beer's law holds for 0.006–0.1 mg of antimony in 25 ml. Extraction of the complex with benzene both concentrates the material and avoids many interferences.

Procedure. *Minerals.* Mix 0.25 gram of sample with 1.5 gram of a 4:3 mixture of sodium carbonate and sulfur. Place in a porcelain crucible and top with 0.3

[86] I. S. Mustafin and A. N. Ivanova, *Izv. Vyssh. Ucheb. Zaved. Khim. Khim. Teknol.* **5**, 504–505 (1962).
[87] Sh. T. Talipov, R. Kh. Dzhiyanbaeva, and A. V. Abdisheva, *Zavod. Lab.* **37**, 387–389 (1971).

gram of the mixture. Heat to 150° over a period of 30 minutes; then raise to 420° for 30 minutes. Cool, extract by boiling with 15 ml of water, and filter. Wash the residue with 1% solution of sodium sulfide. To the filtrate add 8 ml of a 2:1 mixture of sulfuric acid and 30% hydrogen peroxide to oxidize the polysulfides. Evaporate to about 8 ml and add 8 ml of nitric acid. Evaporate to a moist residue, which will contain a bead of sulfur. Heat the salts with 5 ml of sulfuric acid and then add 10 ml of hot water. Filter and wash the residue with a solution containing 1% of thiourea and 1% of potassium iodide. Dilute to 100 ml with 1:5 hydrochloric acid.

For direct reading, add hydrochloric acid to an aliquot to adjust to pH 1.5. Then add 15 ml of acetone and 1 ml of 1% solution of the reagent. Dilute to 25 ml and read at 540 nm.

For extraction, treat an aliquot with 2.5 ml of 20% potassium chloride solution and adjust to pH 1 with hydrochloric acid. Add 10 ml of acetone and 1 ml of 0.1% solution of the reagent. Dilute to 25 ml, extract with 5 ml of benzene, and read the extract at 540 nm.

PYROGALLOL RED

A 1:1 complex of antimonous ion with pyrogallol red conforms to Beer's law at 0.5–3 μg/ml if read at 520 nm or at 0.6–2.6 μg if read at 420 nm.[88] There is interference by titanium, zirconium, selenium, iron, aluminum, gallium, indium, bismuth, cerium, manganese, copper, and mercury. Tin or lead in an amount exceeding the antimony also interferes.

Procedure. Dilute 0.1–2 ml of a solution about 0.4 m*M* in antimony to 3 ml with 1:20 hydrochloric acid. Add 2.5 ml of 3% sodium hydroxide solution and 5 ml of an acetate buffer solution for pH 4.5. Add 2 ml of 0.41 m*M* pyrogallol red and dilute to 25 ml. After 3 minutes read against a reagent blank, selecting the wavelength as above.

2-(2-QUINOLYLAZO)-*p*-CRESOL

A chloroform solution of this reagent extracts trivalent antimony iodides or bromides.[89] The extracted color is stable for 8 hours. There are few interferences. The behavior of **6-(2-quinolylazo)-3,4-xylenol** is similar.[90]

[88] L. Naruskevicius, R. Kazlauskas, J. Skadauskas, and N. Karitonaite, *Nauch. Tr. Vyssh. Ucheb. Zaved. Lit. SSR, Khim. Khim. Tekhnol.* **1972** (14), 95–100.
[89] K. Rakhmatullaev, M. A. Rakhmatullaeva, Kh. R. Rakhimov, and Sh. T. Talipov, *Uzb. Khim. Zh.* **1970** (3), 23–26.
[90] K. Rakhmatullaev, M. A. Rakhmatulaeva, Kh. R. Rakhimov, and Sh. T. Talipov, *Zh. Anal. Khim.* **25**, 1132–1134 (1970).

Procedure. The sample solution containing 0.005–0.16 mg of antimony should be in 0.7–2.5 *N* sulfuric acid. Make it 0.1–0.35 *N* with potassium iodide and dilute to 6 ml. Shake for 3 minutes with 6 ml of a chloroform solution of the reagent containing 0.8 mg/ml and read the extract at 630 nm. Sodium bromide may replace potassium iodide.

1-(2-QUINOLYLAZO)-2-NAPHTHOL

This reagent forms a 1 : 1 complex with antimonious iodide in chloroform.[91] Ferric and cupric ion interfere.

Procedure. The sample solution should contain 0.005–0.17 mg of trivalent antimony. To prevent interference by ferric or cupric ion, add 0.12 gram of ascorbic acid and 1 ml of 2.6% potassium iodide solution. Filter the precipitated iodides. Dilute to 6 ml with 2.5 *N* sulfuric acid and add 6 ml of 0.035% solution of the reagent in chloroform. Shake for 4 minutes, filter the extract, and read at 625 nm.

RHODAMINE B

The complex of pentavalent antimony with rhodamine B (basic violet 10) is appropriate for extraction and reading.[92] Antimony in bismuth can be determined by rhodamine B without preliminary separation.[93a] For determination in steel, it is extracted from a solution in hydrochloric acid with isopropyl ether for determination by rhodamine B.[93b]

In analysis of silver alloys, the tin and antimony are separated by coprecipitation with manganese dioxide.[94] The solution of the precipitate in nitric and sulfuric acid is used for determination of antimony by rhodamine B.

For analysis of lead, the antimony can be coprecipitated with either manganese dioxide or bismuth oxynitrate.[95] Pentavalent antimony chloride is 95% extracted with ether from 1–8 *N* hydrochloric acid. In more than 9 *N* hydrochloric acid it is sorbed on Dowex 1 and eluted with 0.5 *N* sodium hydroxide for determination by rhodamine B. For analysis of copper- or lead-base alloys, manganese dioxide is an appropriate collector.[96] From lead and lead alloys in acetic acid, antimony penta-

[91] K. Rakhmatullaev, M. A. Rakmatullaeva, Sh. T. Talipov, and A. Ataev, *ibid.* **27**, 1793–1796 (1972).
[92] T. L. C. de Souza and J. D. Kerbyson, *Anal. Chem.* **40**, 1146–1148 (1968); cf. R. E. Van Aman, F. D. Hollibaugh, and J. H. Kanzelmeyer, *ibid.* **31**, 1783–1785 (1953); Masao Tanaka and Muyoshi Kawahara, *Jap. Anal.* **10**, 185–187 (1961); Hiromu Imai, *ibid.* **11**, 806–811 (1962).
[93a] S. I. Sinyakova and Ch. Ya. Krol, *Tr. Kom. Anal. Khim. Akad. Nauk SSSR, Inst. Geokhim. Anal. Khim.* **12**, 206–216 (1960).
[93b] L. Kidman and C. B. Waite, *Metallurgia* **66**, 143–146 (1962).
[94] J. Chwastowska and Z. Skorko-Trybula, *Chem. Anal.* (Warsaw) **9**, 123–130 (1964).
[95] Niro Matsuura and Masuo Kojima, *Jap. Anal.* **6**, 155–164 (1957).
[96] Masao Tanaka and Miyoshi Kawahara, *ibid.* **10**, 185–187 (1961).

chloride can be isolated by extraction with isopropyl ether for determination by rhodamine B.[97] If thallium is present, it must be separately determined and a correction applied.

For antimony in zone-refined lead, dissolve the sample in sulfuric acid and potassium bisulfate by addition of dilute hydrochloric acid, sulfuric acid, silicon carbide, sulfurous acid, and water.[98] Oxidize with ceric sulfate, add rhodamine B, extract with benzene, and read at 565 nm.

In oxidation of antimony with ceric ion and removal of the excess with hydroxylamine hydrochloride, the extraction of color with rhodamine B varies with time.[99] Reproducibility is better if the antimony in 1:1 hydrochloric acid is oxidized with 0.5 ml of 6.9% sodium nitrite solution and the excess is not removed. A 1:2 mixture of 2-butoxyethanol and benzene is a preferred extractant for the rhodamine B complex.

For isolation of antimony, a solution of platinum-rhodium alloy in aqua regia is adjusted to 7.7 *N* hydrochloric acid.[100] The antimony is extracted with isopropyl ether and read with rhodamine B.

As little as 0.1 μg of antimony per liter of seawater is determinable as follows.[101] Coprecipitate the antimony with manganese dioxide formed by reaction of potassium permangate and ethanol at pH 4–5. The solution of this in 1:6 sulfuric acid by reaction with hydrogen peroxide is made 0.01 *M* with potassium iodide. Antimony iodide is extracted with methyl isobutyl ketone, reextracted into 0.4 *N* hydrochloric acid, and converted to the rhodamine B complex.

After preparation of a sample of titanium dioxide as shown elsewhere for determination by brilliant green, addition of hydrochloric acid, sodium chloride, potassium chlorate, and phenol is followed by rhodamine S.[102] The isopropyl ether extract is read at 532 nm.

For analysis of rocks, the antimony can be liberated as stibine and absorbed in mercuric chloride solution before ceric oxidation and conversion to the rhodamine B chloroantimoniate.[103]

For determination of antimony in organic antimony compounds, extract it from 1:1 hydrochloric acid solution with isopropyl ether.[104]

Procedure.[105] As a reagent, dissolve 0.1 gram of rhodamine B in 25 ml of water and filter. Add 8 ml of 0.7% solution of sodium nitrite and dilute to 100 ml.

Mix a sample solution containing less than 15 μg of antimony in 1:1 hydrochloric acid with 2 ml of the reagent. Set aside for 30 minutes. Extract by shaking for 1 minute with 10 ml of 1:3 carbon tetrachloride–monochlorobenzene. Read within 2 hours at 565 nm against the solvent mixture and subtract a blank. Beer's law is followed for up to 1.5 μg of antimony per ml.

[97]British Standards Institution, *BS 3908*, Pt. **13**, 8 pp (1970).
[98]J. A. Corbett, *Metallurgia* **65**, 43–47 (1962).
[99]Hiromu Iwai, *Jap. Anal.* **11**, 806–811 (1962).
[100]Vladimir Widtmann, *Hutn. Listy* **25**, 733–735 (1970).
[101]J. E. Portmann and J. P. Riley, *Anal. Chim. Acta* **35**, 35–41 (1966).
[102]W. Z. Jablonski and C. A. Watson, *Analyst* **95**, 131–137 (1970).
[103]Marian M. Schnepfe, *Talenta* **20**, 175–184 (1973).
[104]Yayoe Kinoshita, *Yakugaku Zasshi* **78**, 315–318 (1958).
[105]R. Greenhalgh and J. P. Riley, *Anal. Chim. Acta* **27**, 305–306 (1962).

Arsenic. The distillation unit consists of a 250-ml flask, an entry socket for a supply reservoir containing 100 ml of hydrochloric acid, an exit socket for delivery of vapor to a water-cooled condenser, and a thermometer.

Transfer 1 gram of 100-mesh sample to a covered beaker. Carefully pour a mixture of 9 ml of hydrochloric acid and 3 ml of nitric acid down the side of the beaker. Initially the reaction may be quite violent. Warm until solution is complete, and evaporate to 2–5 ml to drive off chlorine and nitric acid. Add 10 ml of hydrochloric acid and transfer to the distillation flask containing 0.5 gram of hydrazine hydrochloride. Rinse in with 100 ml of hydrochloric acid. Connect, and distill at a few ml/minute, with the vapor temperature not exceeding 108°, until about 30 ml has been collected. Then begin addition of hydrochloric acid to maintain the volume in the distilling flask and to avoid precipitation of arsenic. After the 100 ml of hydrochloric acid has been added, continue distillation at a temperature not exceeding 108° until the volume in the flask is reduced to less than 25 ml. Discard the distillate of arsenic chloride and let the flask and contents cool.

Dilute the solution from the flask to 25 ml and take an aliquot containing 5–10 μg of antimony. By adding water or hydrochloric acid, adjust to 25 ml at acidity of 8 *N*. The distillation residue will be about 6 *N*. Let cool if necessary. Add 3.3% ceric sulfate solution in 1:18 sulfuric acid dropwise until the solution becomes yellow, usually 3 drops, to oxidize the antimony. Destroy excess ceric sulfate by dropwise addition of 0.1% hydroxylamine hydrochloride solution, usually 0.5 ml, plus 0.5 ml excess. Add 4 ml of 0.2% solution of rhodamine B at once, and shake for a minute with 25 ml of benzene. Swirl to separate the phases and discard the aqueous layer. Filter the benzene extract and read within 5 minutes at 565 nm against a blank. Adherence to a fixed time schedule is important for each step.

Lead.[106] Dissolve 1 gram in 10 ml of 1:3 nitric acid and boil off oxides of nitrogen. If the sample contains less than 15 μg of antimony, skip the following paragraph.

If more than 15 μg of antimony is present, add 50 ml of water and 1 ml of 1% potassium permanganate solution. Add a couple of crystals of silicon carbide and heat to 100°. Add 1 ml of 0.5% solution of manganese nitrate and boil for 2 minutes. Add another ml of 0.5% manganese nitrate solution and again boil for 2 minutes. Filter on paper at 60–70° and wash out lead salts. Discard the filtrate, and return the paper and precipitate to the original container. Add 4 ml of sulfuric acid, 10 ml of nitric acid, and 0.5 ml of perchloric acid. Heat until the paper is destroyed and the solution darkens by formation of permanganic ion. Take to copious fumes over a flame to drive off all perchloric acid and destroy most of the color. Cool. Dilute to 100 ml with 1:9 hydrochloric acid.

To the original sample solution containing up to 15 μg of antimony or an aliquot from the preceding paragraph containing no more than 15 μg of antimony, add 2 ml of nitric acid and 2 boiling chips. Bring the total volume of sulfuric acid up to 4 ml and evaporate hydrochloric acid if present. Then heat over a flame to copious fumes, to drive off all nitric acid. Cool somewhat and add 0.1 gram of hydrazine sulfate. Heat over a flame to copious fumes and a volume of 2.5 ml. Wash down the sides of the container with 5 ml of 1% phosphoric acid and transfer to a flask.

[106] C. L. Luke, *Anal. Chem.* **31**, 1680–1682 (1959).

Repeat the transfer with 5 ml of 1 : 1 hydrochloric acid and 5 ml of 1% phosphoric acid. Cool to 25 ± 0.5° and add 1 ml of 0.2 ml ceric sulfate in 1 : 40 sulfuric acid. Add 2 ml of 0.2% rhodamine B solution and 1 ml of 33% butyl Cellosolve solution. Shake for 1 minute with 15 ml of benzene and read the filtered benzene extract at 565 nm against benzene.

Copper.[107] Dissolve a 0.5-gram sample in 10 ml of 2 : 1 hydrochloric acid and 5 ml of 30% hydrogen peroxide. At 60–70° add 7 ml more of the hydrochloric acid. When solution is complete, adjust to 1 : 1 hydrochloric acid by adding 6 ml of water. Complete as shown later for rocks and soil from "When cooled below 25°, add 3 ml...."

Germanium.[108] Dissolve 2 grams of 150-mesh sample in 2 ml of perchloric acid, 8 ml of nitric acid, and 42 ml of hydrochloric acid, and evaporate to under 10 ml. When the solution begins to become cloudy and before perchloric acid fumes, add 15 ml of hydrochloric acid. Boil and agitate to dissolve any germanium oxide on the wall of the vessel. Heat to copious fumes of perchloric acid to remove any remaining nitric acid. Add 15 ml of hydrochloric acid and again heat to perchloric acid fumes to remove the last traces of germanium. Reduce to about 0.5 ml over a free flame. Cool, and add 2 ml of sulfuric acid and about 30 mg of powered sulfur. Evaporate to about 1 ml. Cool, and heat with 10 ml of water to dissolve. Dilute to 20 ml and add 2 ml of 1% cupferron solution. Shake for 30 seconds with 10 ml and 5 ml of chloroform, and filter the extracts through glass wool. Add 1 ml of nitric acid to the extract and expel the chloroform. Add 2 ml of sulfuric acid and 0.25 ml of perchloric acid and evaporate to 1 ml to destroy organic matter. Take up with 20 ml of water. Add 2 ml of 1% cupferron solution and extract with chloroform as before, but discard the chloroform extract, retaining the aqueous solution of antimony. Add 1 ml of nitric acid and evaporate to sulfur trioxide fumes. Add 1 ml of sulfuric acid and 0.25 ml of perchloric acid and evaporate to about 1 ml.

As a selenium solution, dissolve 0.5 gram of powered selenium in 5 ml of nitric acid. Add 10 ml of sulfuric acid and evaporate to sulfur dioxide fumes. Add 10 ml of water and again evaporate to sulfur trioxide fumes. Dilute to 100 ml with water.

Mix the 1 ml of sample solution with 1 ml of the selenium solution, add 5 ml of 6% sulfurous acid solution, and heat nearly to 100°. After 5 minutes filter through paper and wash well. Evaporate to sulfur trioxide fumes, add 3 ml of sulfuric acid and 0.25 ml of perchloric acid, and evaporate to 2.5 ml. From this point on there must be no delay. Cool quickly, and wash down the sides of the container with 5 ml of 1 : 1 hydrochloric acid. Add 1 ml of 0.2% ceric sulfate solution in 1 : 40 sulfuric acid. Mix, add 5 ml of 0.1% solution of rhodamine B, and mix. Add 1 ml of a 1 : 2 dilution of butyl Cellosolve with water and extract with 15 ml of benzene. Filter the benzene layer and read at 565 nm.

Germanium Dioxide. Dissolve 3 grams of sample in 2 ml of perchloric acid, 8 ml of nitric acid, and 42 ml of hydrochloric acid. When in solution, heat to copious fumes of perchloric acid to remove most of the nitric acid. All the germanium

[107] Charles M. Dozinel, *Z. Anal. Chem.* **157**, 401–405 (1957).
[108] C. L. Luke and Mary E. Campbell, *Anal. Chem.* **25**, 1588–1593 (1953).

should have been distilled, but in case of doubt add 5 ml of hydrochloric acid and again evaporate to copious fumes of perchloric acid. Complete as for germanium from "Reduce to about 0.5 ml over a free flame."

Zinc.[109] Dissolve a sample containing 5–20 μg of antimony in a minimum amount of 10:1 hydrochloric-nitric acid and add 1 ml of sulfuric acid. Evaporate to sulfur trioxide fumes and take up in 17 ml of hydrochloric acid. Shake for 30 seconds with 15 ml of isopropyl ether. Add 7 ml of water, cool at 25° for 10 minutes, and again shake for 30 seconds. Discard the aqueous layer and shake the organic layer for 30 seconds with 20 ml of 0.01% solution of rhodamine B in 0.5 *N* hydrochloric acid. Centrifuge the organic layer to eliminate water droplets and read at 550 nm against a water blank.

If an interfering amount of thallium is present, a correction can be applied from a separate determination. Preferably the sample solution is made *N* with hydrobromic acid and the thallium is separated by extraction with isopropyl ether. Thereafter, evaporate the aqueous layer to small volume at 100°, take up in 17 ml of hydrochloric acid, and proceed as above.

Zinc Oxide. Dissolve a sample containing 5–20 μg of antimony in 17 ml of hydrochloric acid. Add 2 mg of ceric sulfate and complete as for zinc from "Shake for 30 seconds with...."

Tellurium.[110] Dissolve 2 grams of sample in a distilling flask by heating with 20 ml of 1:1 hydrochloric acid, 5 ml of nitric acid, and 30 ml of sulfuric acid. When solution is complete, evaporate to sulfur trioxide fumes. Cool, and add 15 ml of water. Connect with a condenser and pass a slow current of carbon dioxide through the solution. Provide a dropping funnel with a mixture of equal parts of 1:1 hydrochloric acid and hydrobromic acid. Provide 50 ml of ice water in the receiver. Heat, and when the distillation temperature reaches 160–170°, add the acid mixture from the dripping funnel at about 2 ml/minute. Distillation should be complete in less than an hour after the distillation temperature has been reached.

Warm the distillate to 50°. Add 5 ml of 20% potassium iodide solution and 5 ml of 10% sodium metabisulfite solution. Cool, and extract with 10 ml of 1% solution of diammonium diethyldithiocarbamate in chloroform. Evaporate the chloroform extracts of antimony with 2 ml of nitric acid, 2 ml of perchloric acid, and 2 ml of sulfuric acid. Evaporate to sulfur trioxide fumes and add 10 ml of sulfuric acid. Cool, and add 3 ml of water and 15 ml of hydrochloric acid. Extract with 25 ml of isopropyl ether. Wash the extract with 15 ml of 1:2 hydrochloric acid; then shake with 5 ml of 0.02% solution of rhodamine B in 1:35 sulfuric acid for 30 seconds. Read the organic layer at 562 nm.

Bronze.[111] Dissolve 1 gram in 10 ml of sulfuric acid and 5 ml of water. Cool and mix with 50 ml of water. Cool and dilute to 100 ml with 1:5 sulfuric acid. Mix a 10-ml aliquot with 30 ml of 1:5 sulfuric acid, 10 ml of 10% thiourea solution, and

[109] R. E. Van Aman, F. D. Hollibaugh, and J. H. Kanzelmeyer, *ibid.* **31**, 1783–1785 (1959).
[110] Yoshihiro Ishikara, Moritaka Koga, and Hideo Komuro, *Jap. Anal.* **11**, 566–570 (1962).
[111] R. S. Volodarskaya, *Zavod. Lab.* **25**, 143–144 (1959).

20 ml of 10% potassium iodide solution. Dilute to 100 ml with 1 : 5 sulfuric acid and read at 550 nm.

Zinc Electrolytes.[112] Mix 5 ml of sample containing about 0.5 gram of zinc and 2.5 μg of antimony with 2.5 ml of 1 : 1 sulfuric acid, 2.5 ml of nitric acid, and 3 mg of ceric sulfate. Heat to sulfur trioxide fumes. Add water and fume again. Repeat that step. Cool, and add 10 ml of hydrochloric acid and 5 mg of ceric sulfate. When solution is complete, heat to 70°; then cool, and add 7 ml of hydrochloric acid. Shake for 1 minute with 15 ml of isopropyl ether. Add 7 ml of water, mix, and incubate in water at 25° for 10 minutes. Shake for 1 minute; then discard the aqueous layer. Shake the organic layer for 1 minute with 20 ml of 0.01% solution of rhodamine B in 0.5 *N* hydrochloric acid. Read the organic layer at 550 nm against water.

Catalyst Material.[113] An alumina-based catalyst is assumed to contain vanadium and titanium. Dissolve 0.5 gram of sample in 5 ml of 1 : 1 sulfuric acid and evaporate to sulfur trioxide fumes. Cool, and add 15 ml of 1 : 3 hydrochloric acid. Filter, and wash the residue with 5 ml of 1 : 3 hydrochloric acid followed by 5 ml of water. Add 5 ml of 4% solution of acetamide to the filtrate and set aside for 15 minutes. Filter on sintered glass and wash. Dissolve the precipitate from the filter with 20 ml of hot sulfuric acid and heat to sulfur trioxide fumes. Cool, dilute to 100 ml, and dilute a 5-ml aliquot to 50 ml. If necessary, aliquot further so that the sample solution contains 10–50 μg of antimony. Add 0.5 ml of 3% solution of ceric sulfate in *N* sulfuric acid. Set aside for 5 minutes; then add 0.5 ml of freshly prepared 1% solution of hydroxylamine hydrochloride. Set aside for 2 minutes, and add 5 ml of 0.02% solution of rhodamine B in *N* hydrochloric acid. Dilute to be 2 *N* in acid and extract with 10 ml and 10 ml of benzene. Dilute the extracts to 50 ml and read at 565 nm.

Talc.[114] Heat 3 grams of sample with 25 ml of 1 : 1 hydrochloric acid at 100° for 30 minutes, stirring from time to time. Add 1 ml of 1% solution of sodium sulfite heptahydrate and mix. Filter. Wash the filter with 5 ml of 1 : 1 hydrochloric acid and 5 ml, 5 ml, and 5 ml of hot water. When cool add to the filtrate 3 ml of 3.3% solution of ceric sulfate in 0.5 *M* sulfuric acid. After 5 minutes add 1 ml of 1% solution of hydroxylamine hydrochloride. After a minute add 70 ml of water, cool below 25°, and shake for 1 minute with 5 ml of isopropyl ether. Wash the organic phase with 2 ml of 1% solution of hydroxylamine hydrochloride in 1 : 11 hydrochloric acid. Shake the washed organic phase for 15 seconds with 2 ml of 0.02% solution of rhodamine B in 1 : 11 hydrochloric acid. Read the organic phase at 550 nm against a reagent blank.

Rocks and Soil.[115] Take 0.2 gram of soil or of 80-mesh rock and mix with 1.5 grams of recently fused and crushed sodium bisulfate in a culture tube. Fuse until

[112] Z. Hasek, *Hutn. Listy* **17**, 733–735 (1962).
[113] Halina Barr and Cecylia Jedras, *Chem. Anal.* (Warsaw) **19**, 895–897 (1974).
[114] H. D. Spitz and A. J. Goudie, *J. Pharm. Sci.* **56**, 1641–1643 (1967).
[115] F. N. Ward and H. W. Lakin, *Anal. Chem.* **26**, 1168–1173 (1954).

the organic matter is substantially destroyed and the tube is filled with white fumes. In cooling, rotate to give a thin film around the inner wall. Digest the contents with 6 ml of 1 : 1 hydrochloric acid by gentle heating, but do not boil. Add 1 ml of 1% solution of sodium sulfite and 3 ml of 1 : 1 hydrochloric acid, mix and filter. Rinse the residue in the tube and wash that on the paper with 3 ml and 3 ml of hot 1 : 1 hydrochloric acid. Finally wash with 2 ml of hot water. When cooled below 25°, add 3 ml of 3.3% solution of anhydrous ceric sulfate in *N* sulfuric acid. Add 0.5 ml of 1% hydroxylamine hydrochloride solution and let stand for 1 minute to destroy excess ceric ion. Add 45 ml of water and ensure that the temperature is below 25°. Shake for 30 seconds with 5 ml of isopropyl ether. Allow to stand for 5 minutes, and drain off all but about 0.5 ml of the aqueous phase. Add 2 ml of 1 : 11 hydrochloric acid and shake for a couple of seconds. Drain the aqueous phase. Add 2 ml of 0.02% solution of rhodamine B in 1 : 11 hydrochloric acid and shake for 10 seconds. Read the organic layer at 550 nm. For field use, comparison may be with the isopropyl ether extract corresponding to 0.5–4 μg of antimony.

Soil.[116] Ash 2 grams of sample at 500° and fuse with 10 grams of potassium pyrosulfate. Extract the melt with 30 ml of 1 : 1 hydrochloric acid. Oxidize by addition of 3 ml of 3.3% solution of ceric sulfate in 0.5 *M* sulfuric acid. Add 0.5 ml of 1% solution of hydroxylamine hydrochloride and extract with 5 ml and 5 ml of isopropyl ether. Shake the extract with 2 ml of 0.02% solution of rhodamine B in 1 : 11 hydrochloric acid and read the extract at 574 nm.

Water.[117] Evaporate 500 ml with 20 ml of hydrochloric acid to dryness. Take up the residue in 30 ml of 1 : 1 hydrochloric acid. Add 2 drops of 25% solution of stannous chloride in hydrochloric acid and 1 ml of 10% sodium nitrite solution. Add 2 ml of 50% urea solution followed by 1 ml of 0.2% solution of rhodamine B. Extract with 20 ml of toluene and read at 530 nm.

Biological Material.[118] The sample consists of blood, urine, or tissue expected to contain about 3μg of antimony. Digest with 3 ml of sulfuric acid and 2 ml of nitric acid, adding more nitric acid if necessary to get a clear digestate. Add 3 ml of water and evaporate to sulfur trioxide fumes. Add 3 ml of 1% sodium sulfite solution and again take down to sulfur trioxide fumes. The antimony is now trivalent. Chill at 0° and add 4 ml of 1 : 1 hydrochloric acid. Add 2% solution of ceric sulfate dropwise until oxidation is complete, about 0.2 ml. Dilute this solution of pentavalent antimony to 24 ml and extract with 5 ml of amyl acetate. Shake the organic extract with 9 ml of 1 : 4 hydrochloric acid and 1 ml of 2% solution of rhodamine B, adding more dye solution if needed. The aqueous layer should be yellowish orange. Separate the organic layer and read at 555 nm against amyl acetate.

---

[116] F. Popea and C. Jurascu, *Acad. Rep. Pop. Rom., Stud. Cercet. Chim.* **10**, 211–218 (1962).
[117] A. A. Nakharova, E. S. Tovstopyat, and V. Ya. Eremenko, *Gidrokhim. Mater.* **51**, 196–202 (1969).
[118] Mai-ling Shen, Hui-min Chang, and Huang-sheng Ting, *Sheng Li Hsüeh Pao* **21**, 127–132 (1957); Chen-kuan Ho, Cheng-chia Kung, Shi-ying Wang, and Liang Li, *ibid.* 167–173.

SAFRANINES

In 3.5 *N* hydrochloric acid pentavalent antimony chloride forms a 1:1 complex with safranine (basic red 2).[119] The subclasses of safranine as O and T appear to be interchangeable as reagents. The 1:1 complex has maxima at 280 and 518 nm, the lower one being somewhat more sensitive. Interference is rare when the complex is extracted from 2 *N* hydrochloric acid. Extracted with 7:3 benzene-acetone, it shows a maximum fluorescence at 580 nm over the range of 0.015–1 μg of antimony per ml.[120]

Procedure. Mix 2–3 ml of sample solution containing 0.001–0.05 mg of antimony with 0.5 ml of 10% stannous chloride solution and 7 ml of 10 *N* hydrochloric acid. Oxidize the antimony with 1 ml of 10% sodium nitrite solution. After 5 minutes destroy excess nitrite with 1 ml of saturated urea solution and add 2 ml of 7.5% solution of potassium chloride. When evolution of gas ceases add 1 ml of 0.1% solution of safranine T and dilute to 20 ml. Extract with 10 ml and 10 ml of benzene and read at 510 nm.

Steel.[121] Dissolve 1 gram of sample in 9 ml of hydrochloric acid and 1 ml of nitric acid. Boil off oxides of nitrogen and dilute to 25 ml with hydrochloric acid. Mix a 2-ml aliquot with 8 ml of 1:1 hydrochloric acid and cool below 5°. Add a 10% solution of stannous chloride in 1:1 hydrochloric acid until the yellow of ferric ion has been discharged. Oxidize the antimony with 1 ml of 10% sodium nitrite solution, and after 1 minute add 2 ml of saturated solution of urea. Add 40 ml of 40% potassium chloride solution and shake for 2 minutes, by which time effervescence should have ceased. Add 5 ml of 0.1% solution of safranine O in 0.1 *N* hydrochloric acid and extract with 10 ml and 10 ml of benzene. Filter the extracts, dilute to 25 ml, and read at 527 nm.

THIOCAPROLACTAM

The 2:1 complexes of thiocaprolactam with antimony trihalides are extracted for reading.[122] The iodide is read at 1–18 μg/ml, the bromide at 1–25 μg/ml. There is interference by trivalent bismuth, cupric, lead, ferric, and silver ions, and by EDTA.

Procedure. To the sample containing trivalent antimony, add 1 ml of 1:35 sulfuric acid and 3 ml of 5% solution of potassium iodide. Extract with 3 ml, 3 ml,

[119]A. T. Pilipenko and Nguyen Mong Shinh, *Ukr. Khim. Zh.* **34**, 1286–1291 (1968).
[120]M. A. Matveets, D. P. Shcherbov, and S. D. Akhmetova, *Zh. Anal. Khim.* **29**, 740–742 (1974).
[121]C. Burgess, A. G. Fogg, and D. Thorburn Burns, *Analyst* **98**, 605–609 (1973); cf. A. T. Pilipenko and Nguyen Mong Shinh, *Zavod. Lab.* **33**, 1074–1075 (1967).
[122]Halina Sikorska-Tomicka, *Chem. Anal.* (Warsaw) **15**, 795–808 (1970).

and 3 ml of 3% solution of thiocaprolactam in chloroform. Dilute the combined extracts to 10 ml and read at 400 nm against a blank. The iodide can be replaced by bromide.

THIOUREA

Antimony complexes with thiourea for reading.[123] The extinction is at a maximum in 1.3 *N* sulfuric acid. Beer's law applies to 0.005–0.2 mg of antimony per ml. The absorption is constant for at least a week. Addition of acetone in development is desirable.

Procedures

Bronze. Dissolve 0.1 gram in 20 ml of 1 : 1 sulfuric acid and heat to sulfur trioxide fumes. Cool and dilute with 20 ml of water. Add 60 ml of 10% thiourea solution and dilute to 100 ml with 1 : 5 sulfuric acid. Read within 5 minutes at 428 nm.

Zinc-Lead Ore.[124] As an iodide reagent, dissolve 350 grams of potassium iodide and 25 grams of ascorbic acid with 400 ml of hot water; dilute to 1 liter and filter.

Boil 1 gram of 200-mesh sample containing 0.02–0.6% of antimony with 10 ml of sulfuric acid for 15 minutes. Cool, and add 100 ml of water. Add 5 ml of 10% tartaric acid solution and boil. Filter, wash the filter with hot water, and dilute to 250 ml. To a 50-ml aliquot, add 15 ml of 1 : 1 sulfuric acid and 20 ml of iodide reagent. Add 5 ml of 10% solution of thiourea. Dilute to 100 ml and read at 430 nm.

3,4′,7-TRIHYDROXYFLAVONE

The complex of this reagent with trivalent antimony is read by fluorescence in perchloric acid solution.[125] Zirconium and hafnium are masked by orthophosphate. Interference by ferric ion absorbing the fluorescence due to antimony is avoided by reduction to ferrous ion by hydroxylamine, then complexing with sulfamic acid. Many elements produce turbidity, which can be filtered off. Titanium, germanium, tungsten, molybdenum, niobium, tantalum, and bismuth interfere by forming colored complexes with the reagent, as indicated by the appearance of a yellow color on addition of the reagent. Interference from all but bismuth is avoided by preextraction of the antimony with methyl isobutyl ketone. Tin, aluminum,

[123] R. S. Volodarskaya, *Zavod. Lab.* **25**, 143–144 (1959); USSR Patent *114,286* (1958); cf. E. Popper, Ileana Olteanu, H. Popescu, and Gh. Sucia, *Rev. Chim.* (Bucharest) **7**, 367–369 (1956).
[124] J. Jankovsky, *Hutn. Listy* **20**, 205–206 (1965).
[125] T. D. Filer, *Anal. Chem.* **43**, 725–729 (1971).

gallium, and indium interfere by forming fluorescent complexes with the reagent. Oxidation with nitric acid is necessary if iodides, bromides, or chlorides are present; otherwise some antimony halide is volatilized. Fluoride would interfere but is volatilized in a preliminary step. Phosphate in excess of that used as a masking agent is avoided by extraction of the antimony because it interferes, probably as pyrophosphate.

Procedure. As an acid mixture, dissolve 43 grams of disodium phosphate heptahydrate, 5 grams of sulfamic acid, and 86 ml of perchloric acid in water, and dilute to 500 ml. As iodide solution, dissolve 0.75 gram of sodium iodide in about 250 ml of water, add 139 ml of sulfuric acid, cool, and dilute to 500 ml. This solution must be used on the day it is made.

Take a sample solution containing up to 5 μg of antimony. If interfering elements are present, extract antimony from them as follows. To the solution, add 1 ml of a 17% solution of sodium bisulfate and 2 drops of sulfuric acid. Evaporate to about 2 ml. Add 3 drops of 25% hydroxyamine sulfate solution and evaporate to about 0.5 ml. Add 25 ml of 1:6 sulfuric acid containing 0.15% of sodium iodide. Shake for 3 minutes with 25 ml of methyl isobutyl ketone and discard the aqueous phase. Wash the organic phase with 25 ml of the acid-iodide solution.

Extract the antimony from the organic phase by shaking for 3 minutes with 10 ml, 10 ml, and 10 ml of 0.4 *N* hydrochloric acid. Combine these extracts.

If the sample is free from halides, the addition of nitric acid in the next step can be omitted. To the sample, add 1 ml of a 17% solution of monosodium phosphate, 5 drops of nitric acid, and 2 drops of sulfuric acid. Heat to fumes of sulfur trioxide and cool. Add 1 ml of water and 5 drops of nitric acid and again heat to sulfur trioxide fumes. Repeat this step, and heat until all the sulfuric acid has been volatilized.

To the residue of sodium hydrogen sulfate, add 2 ml of water and 3 drops of 25% solution of hydroxylamine sulfate. Boil until reduced to about 0.5 ml. Add 10 ml of the acid mixture and 2 drops of 25% solution of hydroxylamine sulfate. Boil and cool. Add 1 ml of 0.02% ethanolic solution of 3,4′,7-trihydroxyflavone solution and dilute to 25 ml. Store at 25° for 30 minutes and read the fluorescence by activation at 422 nm and reading at 475 nm. Subtract a blank.

VICTORIA BLUE 4R

In 5–9 *N* sulfuric acid, pentavalent antimony chloride forms a 1:2 complex with Victoria blue 4R (basic blue 8).[126] This is extractable with 5:1 benzene–isobutyl benzoate, 5:1 benzene–amyl acetate, or 10:1 benzene-acetophenone. The extract conforms to Beer's law for 0.16–4 μg of antimony per ml. There is interference by gold, gallium, thallic ion, thiocyanate, and iodide.

[126] P. P. Kish and Yu. K. Onishchenko, *Zh. Vses. Khim. Obshch.* **14**, 355–357 (1969).

Procedure. *Germanium Dioxide.* Dissolve a 5-gram sample in 20 ml of hydrochloric acid. Heat at 100° to a volume of about 1 ml, which will distill the germanium as the tetrachloride. Add 3 ml of water, 2 grams of sodium chloride; and 4.5 ml of 1 : 1 sulfuric acid. Dilute to 10 ml and oxidize the antimony to the pentavalent form with 0.2 ml of 10% sodium nitrite solution. Destroy excess nitrite with 1 ml of saturated solution of urea. Add 1 ml of 0.1% solution of Victoria blue and 6 ml of 5 : 1 benzene–amyl acetate. Shake for 1 minute and centrifuge. Read the organic layer at 608 nm or 590 nm against a blank.

Trivalent antimony as the iodide complexes with **acridine.**[127] The complex is isolated, then decomposed by nitric acid, and the iodine absorbed in carbon tetrachloride is read at 530 nm. Arsenic below 0.01 *M* does not interfere.

The 1 : 1 complex of antimonic chloride and **basic blue K** conforms to Beer's law for 0.02–0–14 μg of antimony per ml and has a maximum absorption at 638 nm.[128]

Trivalent antimony in 6 *N* hydrobromic acid is read as as the red fluorescence of the **bromo complex** at − 196°.[129] Activation at 360 nm and reading at 586 nm determine 0.01–0.25 ppm of antimony. Ferric ion at fifty-fold concentration and tetravalent tellurium at the twentyfold level interfere. Pentavalent antimony has only about one-tenth the fluorescence of the trivalent form.

Bromopyrogallol red complexes with trivalent antimony at pH 6.6–6.8.[130] Masking agents in the buffer solution used are disodium phosphate, monopotassium phosphate, EDTA, potassium cyanide, and sodium fluoride. To avoid interference by copper, add excess of 0.65% solution of potassium cyanide before adding the buffer solution. The final solution should be developed with 0.7 m*M* bromopyrogallol red in 50% ethanol, which should be 0.035 m*M* in the final solution. Read at 560 nm, Beer's law is followed up to 0.1 mg per 100 ml of final solution.

Butylrhodamine B is a preferred example of the ion association complexes formed between pentavalent antimony and chloride solutions of xanthene dyes.[131] They are extractable with benzene and its homologues or mixtures of carbon tetrachloride with other organic solvents. An optimum vehicle is 6–9 *N* hydrochloric acid, decreasing with increased basicity of the dye. Readings are around 520 nm.

Pentavalent antimony forms a 1 : 1 complex in methanol with **chloranilic acid.**[132] In 2 *N* hydrochloric acid, up to 0.17 mg of antimony can be read at 530 nm. Tin also complexes with the reagent.

Trivalent antimony as the iodide complexes with **1,1-diantipyrinyl-ethane** or

[127]C. Dragulescu, *Acad. Rep. Pop. Rom., Baza Cercet. Stiint. Timisoara, Stud. Cercet. Stiint., Ser. Stiin. Chim.* **6** (3–4), 43–46 (1959).

[128]L. Naruskevicius, R. Kazlauskas, and J. Skadauskas, *Zh. Anal. Khim.* **26**, 813–816 (1971).

[129]G. F. Kirkbright, C. G. Saw, J. V. Thompson, and T. S. West, *Talenta* **16**, 1081–1084 (1969).

[130]D. H. Christopher and T. S. West, *ibid.* **13**, 507–513 (1966).

[131]P. P. Kish and Yu. K. Onishchenko, *Zh. Anal. Khim.* **26**, 514–520 (1971).

[132]Chozo Yoshimura, Hayao Noguchi, Tadaomi Inoue, and Hiroshi Hara, *Jap. Anal.* **15**, 918–924 (1966).

1,1-diantipyrinyl-butane and, when extracted with chloroform, is read around 400 nm.[133a] A corresponding complex is formed with **αα-diantipyrinyltoluene.**

For antimony in seawater, treat a volume containing more than 0.3 μg of pentavalent antimony with 75 ml of hydrochloric acid per liter.[133b] Extract with 10 ml of 0.01 *M* **diantipyrinylmethane** in chloroform per liter of sample. Shake the extract with 10 ml of 10% ascorbic acid solution and 8 ml of 10% potassium iodide solution in 0.5 *N* hydrochloric acid for 1 minute. Read the organic phase at 340 nm. Beer's law is obeyed for 0.3–8 μg of antimony per liter.

For photometric determination by **2-(3,5-dibromo-2-pyridylazo)-5-diethylaminophenol,** the solution should contain 0.12–4.2 μg of trivalent antimony per ml.[134] To the sample solution, add 1 ml of 20% potassium chloride solution, 5 ml of 1 : 5 hydrochloric acid, 8 ml of acetone, and water to dilute to 18 ml. Add 2 ml of m*M* reagent in ethanol and read at 610 nm against a reagent blank. The complex developed from 1–25 μg of antimony with 5 ml of 1 : 5 hydrochloric acid and 2 ml of m*M* ethanolic reagent can be extracted with 10 ml of chloroform. Iron can be masked with ascorbic acid and copper with thiourea. Aluminum, lead, and stannic ion interfere.

Pentavalent antimony in acid solution forms a 1 : 2 complex in the presence of excess of **3-(4-diethylaminophenylazo)-1,4-dimethyl-1,2,4-triazolium chloride.**[135] When extracted with 5 : 1 benzene-nitrobenzene, the maximum absorption is at 552 nm, and Beer's law is obeyed at 0.1–5 μg/ml. There is interference by iodide, thallic, auric, and thiocyanate ions. Extracted from 5 *M* sulfuric acid with chlorobenzene, it is read at 540 nm.

Dimethylthionine, known as azure A, forms a 1 : 1 complex in 0.9–2.6 *N* hydrochloric acid extractable with 1 : 2 dichloroethane-trichloroethylene for reading at 640 nm.[136] Similarly **trimethylthionine,** known as azure B, forms the 1 : 1 complex in 1.8–3.6 *N* hydrochloric acid for extraction by 2 : 1-dichloroethane-trichloroethylene for reading at 655 nm. There is interference for both by stannous, auric, thallic, and mercuric ions, and for azure B the aqueous phase should be saturated with sodium phosphate if ferric ion is present.

5-Ethylamino-2-(2-pyridylazo)-*p*-cresol and its 3-bromo- and 3,5-dibromo-analogues form 1 : 1 complexes with antimony when extracted into benzene.[137] Addition of thiourea prevents interference by copper. Beer's law holds for 0.2–12 μg of antimony per ml in 0.01 *M* potassium iodide in 1 : 6 sulfuric acid. Under the same conditions,[138] **α-antipyrinyl-4-dimethylamino-α-(4-dimethylaminophenyl)-3-nitrobenzyl alcohol** covers the range of 0.1–4 μg of antimony per ml and **rezarson,** which

[133a] A. I. Busev and E. S. Bogdanova, *Zh. Anal. Khim.* **19**, 1346–1354 (1964).

[133b] Yu. A. Afanas'ev, A. I. Ryabinin, L. T. Azhipa and A. S. Romanov, *Zh. Anal. Khim.* **30**, 1830–1832 (1975).

[134] S. I. Gusev and L. V. Poplevina, *Zh. Anal. Khim.* **23**, 541–546 (1968); S. I. Gusev, L. V. Poplevina, and G. G. Shalamova, *Uch. Zap. Perm. Gos. Univ.* **1968** (178), 214–218.

[135] P. P. Kish and Yu. E. Onishchenko, *Zh. Anal. Khim.* **25**, 112–118, 500–504 (1970).

[136] V. M. Tarayan, E. N. Ovsepyan, and M. G. Ekimyan, *Uch. Zap. Erevan. Gos. Univ., Estestv. Nauk* **1972** [1(119)], 73–78.

[137] S. I. Gusev, L. V. Poplevina, and A. S. Pesis, *Zh. Anal. Khim.* **22**, 731–735 (1967).

[138] A. I. Fomina, N. A. Agrinskaya, and V. I. Petrashen', *Tr. Novocherk. Politekh. Inst.* **220**, 113–120 (1969).

is **5-chloro-3-(2,4-dihydroxyphenylazo)-2-hydrocybenzenearsonic acid,** at 0.5–5 μg/ml.

Chlorantimoniate ion is extracted with benzene from 2.5 *N* hydrochloric acid as a complex with crystal violet.[139] Then on shaking the benzene extract with **ethylrhodamine B** in 8–14 *N* sulfuric acid, the crystal violet is replaced with ethylrhodamine B. Then the ethylrhodamine B chlorantimoniate is read by fluorescence at 590–595 nm. Beer's law is followed for 0.0033–0.33 μg of antimony in the final extract. The optimum conditions for determination of antimony with ethylrhodamine B are 7 *N* sulfuric acid, 20 *N* bromide ion, and 0.2 mg of reagent per ml.[140] Extraction with benzene from 8 *N* sulfuric acid containing 20 mg of bromide per ml and 0.4 mg of reagent provides preconcentration.

Pentavalent antimony as the hexofluoroantimonate is determined photometrically with **ferroin**, which is the reaction product of ferrous sulfate and 1,10-phenanthroline. The complex is read at 506 nm.[141]

When pentavalent antimony reacts with **fluotitanic acid** and hydrogen peroxide, the more stable fluoantimonic acid is formed and titanic ions are released.[142] In the presence of hydrogen peroxide, the titanium ions become pertitanic acid with an absorption proportional to the antimony content. The solution should be in 2.5 *N* hydrochloric acid.

For determination of antimony with **fuchsin** (basic violet 14), first oxidize the antimony ion with ceric sulfate.[143] Reduce the excess ceric ion with hydrazine sulfate. Add a solution of fuchsin and adjust with sodium citrate solution to pH 1–1.2. Extract the antimony-fuchsin complex with amyl acetate and read at 555 nm. Copper, chromic ion, chromate, vanadate, molybdenum, iodine, tungsten, and stannous ion interfere. The complex can also be extracted with dichloroethane and read at 540 nm.[144]

Gallein, which is 4′,5′-dihydroxyfluorescein, complexes with trivalent antimony at pH 3–4.[145] With a color-stabilizing agent, it conforms to Beer's law at 0.1–0.7 μg/ml. The solution tolerates 3 μg of ferric and stannous ions. To the sample, add 20 ml of 0.005% solution of gallein, 10 drops of 0.5% solution of sulfonated castor oil, and 40 ml of water. Mix well and add 20 ml of water. Adjust to pH 4 with a buffer solution, *N* in sodium acetate and sulfuric acid. Dilute to 100 ml and read at 525 nm.

Antimony as a halide or thiocyanate is extracted at as low as 0.03 ppm from acids with **4-(6-methoxy-3-methylbenzothiazolin-2-ylazo-*N*-methyldiphenylamine** in amyl acetate and read at 634 nm.[146]

***N*-Methylanabasine-(α′-azo-6-)-*m*-aminophenol,** which is 5-amino-2-[3-(methyl-

[139] I. A. Blyum, F. P. Kalupina, and T. I. Tsenskaya, *Zh. Anal. Khim.* **29**, 1572–1576 (1974).
[140] I. A. Bochkareva and I. A. Blyum, *ibid.* **30**, 874–882 (1975).
[141] V. S. Archer and R. B. Twelves, *Talenta* **15**, 47–54 (1968).
[142] Hisao Fukamauchi, Ryuko Enohara, Mitsuko Uehara, Satoko Terui, and Toshiko Tokimoto, *Jap. Anal.* **8**, 353–356 (1959).
[143] Hidehiro Goto and Yachiyo Kakita, *Sci. Rep. Res. Inst. Tohoku Univ. Ser. A* **10**, 103–109 (1958); *J. Chem. Soc. Jap.* **78**, 1521–1524 (1957).
[144] Zh. M. Arstamyan, *Melodoi Nauch. Rebotnik. Estestv. Nauk* **1972** [2(16)], 65–71.
[145] Sha-wei Pang and Ming-lein Lu, *Hua Hsiieh Hsiieh Pao* **23**, 17–23 (1957).
[146] P. P. Kish, I. I. Zimomrya, I. I. Pogoida, Yu. K. Onishchenko, and G. M. Vitenko, *Tr. Khim. Khim. Tekhnol.* (Gor'kii) **1973** [4(35)], 71–73.

2-piperidyl)-2-pyridylazo]phenol, forms a 1 : 1 complex with trivalent antimony.[147] The maximum absorption is at 570 nm at pH 1.4–1.7. Beer's law is followed for 0.25–1.2 μg/ml. There is little interference.

Trivalent antimony complexes with ***N*-methylanabasine-α′-azodiethylaminophenol,** which is diethylamino-2-[3-(1-methyl-2-piperidyl)-2-piperidylazo]phenol, in 60% acetone in the presence of potassium chloride.[148] The pH should be 1.2–1.5 with hydrochloric acid. The reagent absorbs at 430 nm, the complex at 600 nm. Beer's law is followed for 0.2–3 μg/ml. Zinc, iron, copper, and tin are masked with tartaric acid, ascorbic acid, thiourea, and potassium iodide, respectively. Many other ions are tolerated at tenfold and more.

Trivalent antimony forms a 1 : 1 complex with **4-(2-*N*-methylanabasine-α′-azo)resorcinol,** which is 4[3-(1-methyl-2-piperidyl)-2-pyridylazo]resorcinol.[149] The optimum conditions are 60% acetone as solvent that is 0.1 *N* with hydrochloric acid and contains 0.8% of potassium chloride. Beer's law is followed at 540 nm for 0.2–4 μg of antimony per ml. Nickel, mercuric, cupric, and ferric ions must be masked.

Pentavalent antimony forms a 1 : 1 complex with **methyl green** (basic green 5) in hydrochloric acid at pH 0.35–0.75.[150] Extracted into 1 : 1 benzene–ethyl acetate, the maximum absorption is at 640 nm and it is stable for 24 hours. Beer's law holds for 0.06–4 μg per ml of extract. The optimum concentration of methyl green is 0.16 m*M*.[151] A suitable extraction medium is also 5 : 1 benzene-nitrobenzene. The sensitivity of this reagent is not as good as that of brilliant green or crystal violet.

Nile blue (basic blue 12) forms a 1 : 1 complex with pentavalent antimony.[152] Extracted with 3 : 5 benzene–ethyl acetate, it has a maximum absorption at 630 nm. Beer's law is followed for 0.05–3 μg of antimony per ml. There is interference by auric, thallic, mercuric, stannous, and lead ions.

Nitron, which is 1,4-diphenyl-3,5-endo-analino-4,5-dihydro-1,2,4-triazole, forms a ternary complex with antimony and iodide ion which is extractable with dichloroethane.[153]

At pH 2 the complex of **4-phenylazoresorcinol** in sulfuric or hydrochloric acid complexes with antimony for reading at 490 nm.[154a] The complex can be extracted with chloroform. Bismuth, aluminum, tin, and arsenic interfere.

Antimony forms a 1 : 2 complex with **1-phenyl-3-(2-pyridyl)thiourea** in ethanol at pH 4.2–9.6.[154b] Read at 410 nm it conforms to Beer's law up to 1.7 ppm of antimony.

Trivalent antimony as the bromide or iodide forms a 1 : 1 complex with **1-(2-pyridylazo)-2-naphthol** read in chloroform at 585–595 nm.[155] To a sample solution

[147] Sh. T. Talipov, R. Kh. Dzhiyanbaeva, and A. V. Abdisheva, *Uzb. Khim. Zh.* **1971** (3), 9–10.
[148] A. V. Abdisheva, *Nauch. Tr. Tashk. Gos. Univ.* **1970** (379), 177–181.
[149] Sh. T. Talipov, R. Kh. Dzhiyanbaeva, and A. V. Abdisheva, *Zh. Anal. Khim.* **27**, 1550–1553 (1972).
[150] Zh. M. Arstamyan and V. M. Tarayan, *Arm. Khim. Zh.* **25**, 117–122 (1972).
[151] P. P. Kish, Yu. K. Onishchenko, and I. I. Pogoida, *Zh. Anal. Khim.* **28**, 1746–1750 (1973).
[152] Zh. M. Arstamyan, *Melodoi Nauch. Rabotnik. Estestv. Nauk* **1972** [2(16)], 65–71.
[153] L. A. Mineeva and A. S. Babenko, *Ukr. Khim. Zh.* **38**, 808–812 (1972).
[154a] R. Kazlauskas, L. Naruskevicius, and J. Shkadauskas, *J. Nauch. Tr. Ucheb. Zaved. Lit. SSR, Khim. Khim. Tekhnol.* **1970** (12), 15–19.
[154b] S. P. Mathur and M. R. Bhandari, *Revta Latinoam. Quim.* **6**, 160 (1975).
[155] K. Rakhmatullaev, M. A. Rakhmatullaeva, Sh. T. Talipov, and A. Mamatov, *Zavod. Lab.* **37**, 1027–1029 (1971); cf. Li-shu Ho, Chih-sheng Shih, and Chin-lun Hu, *Chem. Bull.* (Peking) **1965** (3), 56–58.

containing 0.005–0.140 mg of antimony, add 1.6 ml of 20% solution of sodium bromide in 1 : 15 sulfuric acid. Dilute to 6 ml with 1 : 5 sulfuric acid and extract with 6 ml of 0.03% solution of the reagent in chloroform. An alternative is to add 0.6 ml of 33% solution of potassium iodide and dilute to 6 ml with 1 : 12 sulfuric acid before extraction. The bromide method determines trivalent antimony in the presence of the pentavalent form; the iodide technic determines total antimony.

Rhodamine 6Zh (basic red 1) performs much like rhodamine B, forming a 1 : 1 complex extractable into benzene for reading at 540 nm.[156] Thallium interferes.

Antimony in smelter zinc can be determined photometrically by **sodium pyrrolidine dithiocarbamate.**[157]

When antimony is reduced to stibine by zinc, the latter reduces **silver sulfamidobenzoic acid** in sodium hydroxide solution to form colloidal silver for reading at 420 nm.[158]

Trivalent antimony forms a 1 : 1 complex in 0.1 *N* hydrochloric acid with **3,4′,5,7-tetrahydroxyflavone** for reading at 420 nm.[159] Both the position and intensity of the maximum are pH dependent. Beer's law is followed for 1.9–7.8 ppm. There is serious interference by zirconium, titanium, tungsten, molybdenum, niobium, tantalum, ferric ion, fluoride, thiosulfate, ascorbic acid, and EDTA.

For 0.02–0.3 mg of antimony in iron and steel, dissolve and remove most of the iron as the sulfide or by extraction.[160] Then after adjusting to pH 8.6 and adding masking agents, extract with **tetramethylenedithiocarbamate** in chloroform and read at 350 or 380 nm.

4-(2-Thiazolylazo)catechol complexes at pH 5–5.5 with trivalent antimony in hydrochloric acid for reading at 480–530 nm.[161]

[156]L. Naruskevicius, R. Kazlauskas, J. Skadauskas, and D. Virbalite, *Nauch. Tr. Vyssh. Ucheb. Zaved. Lit. SSR, Khim. Khim. Tekhnol.* **13**, 59–63 (1971).

[157]E. Kovacs and H. Guyer, *Z. Anal. Chem.* **186**, 267–288 (1962).

[158]Gh. Ciuhandu and M. Rocsin, *ibid.* **174**, 118–121 (1960); Gh. Ciuhandu, *Rev. Chim.* (Bucharest) **11**, 530–532 (1960).

[159]B. S. Garg, K. C. Trikha, and R. P. Singh, *Talenta* **16**, 462–464 (1969).

[160]E. Kovacs and H. Guyer, *Z. Anal. Chem.* **208**, 255–262 (1965).

[161]V. Purmalis, J. Putnins, V. Barkane, and E. Gudriniece, *Izv. Akad. Nauk Latv. SSR, Ser. Khim.* **1972** (2), 234–235.

CHAPTER TEN

TIN

Besides expansion of the use of phenylfluorone, numerous reagents for tin have been added in recent years and the additional reagents have included various other fluorone derivatives. Methods on the one hand determine traces and on the other analyze ores and alloys by dilution and use of reagents of low sensitivity.

For determination of stannic ion, catechol violet, hematein, and quercetin are preferable to phenylfluorone, quercetin, 4-(2-pyridylazo)-2-naphthol, and xylenol orange.[1] The bases of evaluation were (1) the magnitude of the shift in absorption between the reagent and its stannic complex, (2) the difference in molecular extinction between the reagent and the complex, and (3) the pH ranges within which the complex remains stable.

Stannous ion at the level of 0.02 mg/ml is separated from more than 160-fold concentrations of alkalies, alkaline earths, rare earths, zirconium, titanium, gallium, aluminum, manganous ion, chromic ion, arsenic ion, and ferrous ion by extraction with 0.5 *N* bis-(2-ethylhexyl)phosphorodithioate in heptane from *N*–2*N* hydrochloric acid.[2] The stannous ion is stabilized with mercaptoacetic acid and is accompanied in the organic phase of the extraction by bismuth, trivalent arsenic and antimony, copper, mercury, hexavalent molybdenum, nickel, cadmium, zinc, and cobalt. Then stannous ion is stripped from the organic phase with 8 *N* hydrochloric acid.

For separation of stannic ion in 6–6.5 *N* hydrochloric acid, the organic solvent removes many of the ions, leaving in the aqueous phase stannic ion, titanium, zirconium, gallium, aluminum, and part of the ferric ion, cobalt, nickel, and molybdenum. Then addition of mercaptoacetic acid reduces the stannic ion and ascorbic acid reduces the ferric ion. Then another extraction of the aqueous phase with the organic solvent extracts the stannous ion and isolates it as described

[1] A. K. Babko and N. N. Karnaukhova, *Zh. Anal. Khim.* **22**, 868–875 (1967).
[2] I. S. Levin and V. A. Tarasova, *Zh. Anal. Khim.* **28**, 1341–1345 (1973).

earlier. The technic was developed for later oxidation of the stannous ion to stannic ion and determination by catechol violet.

Flame photometry with a nitrous oxide–acetylene flame will determine 0.5 ppm of tin for a signal equal to twice the root mean square of the background at 284 nm, with an optimum height of 10–12 nm for the red zone of the flame.[3] Addition of 1000 ppm of potassium chloride enhances the signal by suppression of ionization in the flame.

For separation of arsenic, antimony, and tin by distillation as the halides, see Arsenic. For tin and antimony in silicate minerals by morin, see Antimony. Tin can be read fluorimetrically or photometrically.

ANTIPYRINYL-[4-(BENZYLMETHYLAMINO)PHENYL]-4-DIMETHYLAMINOPHENYLMETHANOL

This reagent complexes with stannic ion in 0.7–0.8 *N* hydrochloric acid in the presence of ammonium thiocyanate.[4] The complex obeys Beer's law for 0.02–0.4 μg of tin per ml. Extraction of the iodine complex with diantipyrinylmethane into benzene-chloroform separates from aluminum, iron, manganese, zinc, nickel, and cobalt. The absorption is stable for 1 hour.

Procedure. *Aluminum Alloys.* Dissolve 0.2 gram of sample in 10 ml of 20% sodium hydroxide solution. Add 50 ml of water, filter, and wash the residue. Dilute to 100 ml and take a 20-ml aliquot. Adjust to pH 1.5–2 with 1:1 sulfuric acid and add 0.2 gram of thiourea to mask copper. Add 0.1 gram of ascorbic acid, 2 grams of diantipyrinyl methane, 0.8 gram of potassium iodide, and 15 ml of 1:1 benzene-chloroform. Shake for 10 minutes and set aside for 1 minute. There will be three phases. Separate the lower layer of the organic phase. Shake the aqueous phase with 5 ml of 1:1 benzene-chloroform, let separate, and withdraw the lower organic phase. Shake the combined organic phases so withdrawn for 1 minute with 20 ml of 1:20 ammonium hydroxide to reextract the tin. Wash this ammoniacal extract with 5 ml, 5 ml, and 5 ml of chloroform and discard these washings. Filter. Add 15 ml of nitric acid to the aqueous phase and evaporate to dryness to eliminate iodide. Add 5 ml and 5 ml of hydrochloric acid to the residue, taking to dryness after each addition to eliminate nitrates. Take up in water and dilute to 50 ml.

Adjust a 20-ml aliquot to pH 2 with hydrochloric acid and add 1.6 ml in excess. Dilute to 18 ml and add 0.05 gram of thiourea, 0.01 gram of ascorbic acid, 0.5 ml of 0.5% gelatin solution, and 3 ml of 6×10^{-4} *M* solution of the captioned reagent. Incubate at 20° for 15 minutes and add 1.2 ml of 50% solution of ammonium thiocyanate. Dilute to 25 ml and read at 610 nm against a reagent blank in 0.8 *N* hydrochloric acid.

---

[3] E. E. Pickett and S. R. Koirtyohann, *Spectrochim. Acta B*, **24** (6), 325–333 (1969).

[4] V. P. Zhivopistsev, E. A. Selezneva, Z. I. Bragina, and A. P. Lipchina, *Uc. Zap. Perm. Gos. Univ.* **1966** (141), 213–221.

BUTYLRHODAMINE S

Stannous chloride complexes with butylrhodamine S in an acidic medium containing cupferron.[5] The complex is extractable with benzene. The method will determine down to 2 ppm of tin.

Procedure. *Minerals and Ores.* Heat a 0.5-gram sample with a minimal amount of hydrofluoric acid until all silica has been volatilized. Add 2 ml of sulfuric acid and heat to dryness. Add 2 grams of a 2 : 1 mixture of sodium carbonate and borax and fuse at around 900°. Dissolve the melt in 20 ml of 1 : 1 hydrochloric acid with heating. Without cooling, add an excess of ammonium hydroxide, filter, and wash with hot 1 : 20 ammonium hydroxide. The borate would interfere later, which dictates careful washing of the precipitate of metal hydroxides. Dissolve the precipitate in 20 ml of 1 : 1 hydrochloric acid. Add 10 ml of 15% solution of EDTA and 0.5 ml of a 0.1% solution of beryllium chloride. Precipitate with excess of ammonium hydroxide. The EDTA masks gallium, antimony, cuprous ion, mercury, platinum, indium, thallium, and part of the tellurium. Let stand for 2 hours, filter, and wash with cold 1 : 20 ammonium hydroxide.

If the tellurium is less than 10 times the tin and the antimony less than 500 times the tin, this paragraph is applicable. Dissolve the precipitate in 10 ml of 5.75 *M* sulfuric acid. If the tin content exceeds 15 μg, dilute with 11.5 *N* sulfuric acid and take a 10-ml aliquot. Add 1 ml of 0.1% solution of butylrhodamine S. Add 0.3 ml of 4% cupferron solution and shake with 10 ml of benzene. Add 0.4 ml of 8% solution of titanous chloride and shake again. Abstract 7 ml of the benzene extract, stabilize with 3 ml of acetone, and read at 560 nm.

If the tellurium or antimony exceeds the limits cited in the previous paragraph, dissolve the precipitate in 10 ml of 8.5 *N* sulfuric acid and dilute it or an aliquot containing less than 15 μg of tin to 20 ml with 8.5 *N* sulfuric acid. Add 1 ml of 4% solution of cupferron and extract the tin with 5 ml of butyl acetate. Wash the extract with 5 ml of 5 *N* sulfuric acid, then add it to a mixture of 10 ml of 11.5 *N* sulfuric acid, 1 ml of 0.1% solution of butylrhodamine S, and 10 ml of benzene. Add 0.3 ml of 4% cupferron solution and shake. Add 0.4 ml of 8% solution of titanous chloride and shake. Dilute 10.5 ml of the extract to 15 ml with acetone and read at 560 nm.

CACOTHELINE

A complex of stannous ion is formed with cacotheline, which is a nitration product of brucine.[6] After fusion of cassiterite with sodium carbonate and sulfur, the solution with added potassium hydroxide and potassium cyanide is developed with cacotheline or dimethylglyoxime.[7]

[5] T. I. Shumova and A. I. Blyum, *Zavod. Lab.* **34**, 659–662 (1968).
[6] G. Fritz and H. Scheer, *Z. Anorg. Chem.* **331**, 151–153 (1964).
[7] René de Peyronnet, *Min. Fr. Outre-Mer, Cent. Geol., Notes Trav.* (1), 36–41 (1957).

For determination of tin in ferrotungsten, the sulfides of tin, antimony, and arsenic are precipitated from dilute sulfuric acid solution containing tartaric acid.[8] The sulfides are dissolved, then antimony and tin are coprecipitated with manganese dioxide. Tin is determined in the solution of this precipitate by cacotheline.

For iron and steel, the tin is coprecipitated with manganese dioxide and redissolved with hydrochloric acid and hydrogen peroxide.[9] To avoid interference by vanadium, evaporate the solution in hydrochloric acid to 18 ml. Add 2 ml of 0.5% solution of hydroxylamine hydrochloride and 0.16 gram of aluminum with the solution blanketed with carbon dioxide. Add 0.5 ml of 0.2% cacothelite solution and read after 10 minutes. Less than 0.3 mg of vanadium is tolerated.

Procedure. *Organotin Compounds.* Oxidize a sample containing 6–26 mg of tin in a Parr bomb. Details of a technic are given under determination of tin with Dithiol. Acidify the solution with 20 ml of hydrochloric acid and dilute to 100 ml. Dilute an aliquot expected to contain 1–2 mg of tin to 50 ml with 1 : 1 hydrochloric acid. Add 0.5 ml of aluminum powder and boil in a stream of carbon dioxide until solution is complete. Cool, add 5 ml of 0.25% solution of cacotheline, and dilute to 100 ml with 1 : 10 hydrochloric acid. Set aside for 5 minutes and read.

CATECHOL VIOLET

The complex of stannic ion with catechol violet obeys Beer's law for 0.1–0.6 μg per ml.[10] An initial 1 : 1 tin-dye complex absorbing at 555 nm is changed in 5–10 minutes in the presence of gelatin to a 2 : 1 complex absorbing at 655 nm. The latter is stable for 1 hour. Ferric ion is reduced with ascorbic acid to avoid interference.[11] There is interference by molybdate, tungstate, tetravalent titanium, and zirconium, and by trivalent antimony, gallium, indium, and bismuth. By extracting stannic iodide with benzene and reextracting into dilute sulfuric acid, most interferences are avoided. Both ferric ion and hexavalent chromium can be reduced with titanous ion. Tungstate can be masked with phosphate. EDTA masks molybdenum, zirconium, phosphate, and vanadate.[12]

The reaction of stannic ion with catechol violet is sensitized by addition of cetyltrimethylammonium bromide for reading at 662 nm.[13] An acetate buffer solution for pH 3.5 is desirable.[14a]

[8] Hidehiro Goto, Yachiyo Kakita, and Masahiko Sase, *Nippon Kinzoku Gakkaishi* **21**, 385–387 (1957).
[9] Shiro Morita and Naoya Inoyama, *Suiyokaishi* **14**, 47–52 (1959).
[10] P. Ya. Yakovlev and G. P. Razumova, *Zavod. Lab.* **31**, 130–138 (1965); cf. I. L. Ruzinova and V. Ya. Fedosova, *Izv. Sib. Otdel. Akad. Nauk SSSR, Ser. Khim. Nauk* **1963** (3), 56–60.
[11] M. Malat, *Z. Anal. Chem.* **187**, 404–409 (1962).
[12] Katu Tanaka, *Jap. Anal.* **13**, 725–729 (1964).
[13] R. M. Dagnall, T. S. West, and P. Young, *Analyst* **92**, 27–30 (1967); Homer B. Corbin, *Anal. Chem.* **45**, 534–537 (1973).
[14a] Katu Tanaka and Katutosi Yamayosi, *Jap. Anal.* **13**, 540–544 (1964); cf. Masakazu Deguchi, Yoshitake Sumida, and Mikio Yashiki, *J. Hyg. Chem.* (Japan) **20**, 233–235 (1974).

A 1:2:1 tin–catechol violet–diphenylguanidine complex is extractable with butanol or isobutyl alcohol at pH 5.[14b] At 580 nm the extract obeys Beer's law for 0.025–2.5 μg of tin per ml.

For tin in organometallic compounds, decompose a 10-mg sample by fusion with sodium peroxide and develop the solution with catechol violet.[14c]

Procedures

Steel. Dissolve 0.5 gram of sample in 25 ml of 1:4 sulfuric acid and 10 ml of perchloric acid. Evaporate to incipient fumes and take up in 60 ml of water. Add 8 ml of hydrochloric acid and 10 ml of 2% solution of thioacetamide. Boil and after 10 minutes add 10 ml more of the thioacetamide solution. Test on a spot plate with ammonium thiocyanate solution to be sure that all ferric ion has been reduced. Filter the precipitate of sulfur and sulfides of copper, molybdenum, and so on. Wash it and discard. Add ammonium hydroxide to the filtrate and washings to adjust to pH 2. Add hydrochloric acid to make the solution about 0.1 N with that acid. Add 0.5 ml of 1% solution of copper nitrate. Coprecipitate the copper and tin by heating to about 90° and adding 5 ml of 2% solution of thioacetamide. Maintain that temperature, and after 20 minutes add 5 ml more of the thioacetamide solution. After another 30 minutes at 90° set aside for 12 hours. Filter the precipitate and wash it with hot water. Heat the paper and precipitate with 50 ml of nitric acid and 10 ml of sulfuric acid. When organic matter has been destroyed, evaporate to incipient fumes. Wash down the sides of the vessel with water and evaporate to fumes.

If molybdenum is absent, skip this paragraph. Take up the residue in about 60 ml of water and add 5 ml of 1% solution of potash alum. Add ammonium hydroxide to coprecipitate aluminum and tin. Filter and wash with hot water. If the sample contained more than 5% molybdenum, dissolve this precipitate in 10 ml of hot 1:4 sulfuric acid and reprecipitate with ammonium hydroxide. Finally dissolve the precipitate in 10 ml of hot 1:4 sulfuric acid and evaporate to fumes.

Take up the sulfuric acid residue of tin together with copper and in some cases with aluminum in 10 ml of water. Cool, and add successively the following: 1 ml of 1% ascorbic acid solution, 2 drops of 0.01% solution of metanil yellow in 50% ethanol, 1:99 ammonium hydroxide to a yellow-pink color, 1 ml of 0.5% solution of gelatin, 1.5 ml of mM catechol violet. After 20 minutes dilute to 50 ml. After a further 20 minutes read at 655 nm against water.

Bronze.[15] Dissolve a sample containing 0.25–0.1 mg of tin in 3 : 1 hydrochloric-nitric acid using 8 ml/gram. Evaporate to dryness. Add 5 ml of hydrochloric acid and again evaporate to dryness. Take up the residue in 10 ml of 1:3 hydrochloric acid and pass it through a column of the anionic resin EDE-10P in chloride form. Tin is adsorbed, but nickel, iron, copper, aluminum, and manganese pass through. Wash the column with 1:3 hydrochloric acid. Elute the

[14b]N. L. Shestidesyatnaya, L. I. Kotelyanskaya, and M. I. Yanik, *Zh. Anal. Khim.* **31**, 67–71 (1976).
[14c]R. N. Fotsepkina, E. A. Bondarevskaya, and L. N. Kudryashova, *Zh. Anal. Khim.* **32**, 166–169 (1977).
[15]N. N. Karnaukhova, *Zavod. Lab.* **36**, 1047–1048 (1970).

tin by passing 0.5 *N* hydrochloric acid at 1 ml/minute. Evaporate the eluate to a small volume and oxidize antimony with 2 ml of 30% hydrogen peroxide. Boil off the excess hydrogen peroxide and neutralize with ammonium hydroxide. Add 3 ml of 2 m*M* catechol violet, 2 ml of 0.5% gelatin solution, 10 ml of an acetate buffer solution for pH 3, and a few drops of saturated thiourea solution. Dilute to 50 ml with the buffer solution, set aside for 20 minutes, and read at 610 nm.

Sea Water.[16] Dilute a sample containing about 0.5 μg of tin to 500 ml and make it 2 *N* with hydrochloric acid. Adsorb on a 110×11 mm column of 100-200 mesh Dowex 1-X8 resin in chloride form. Elute the tin with 100 ml of 1:7 nitric acid. Add 1.5 ml of 1:17 sulfuric acid and evaporate to dryness. Take up in 20 ml of 0.4 *N* hydrochloric acid and pass through a 60×10 mm column of the same resin, preconditioned with 0.4 *N* hydrochloric acid. Then elute the tin with 50 ml of 1:7 nitric acid. Add 0.5 ml of 1:17 sulfuric acid to the eluate and evaporate to sulfur trioxide fumes. Take up in 5 ml of water and add 1 ml of 0.3 m*M* catechol violet and 1 ml of 0.3 m*M* cetrimide trihydrate which is cetyltrimethylammonium bromide. Dilute to 10 ml and after 15 minutes read at 662 nm. Ion exchange has separated tin from titanium, zirconium, and antimony.

Organotin Compounds.[17] This technic is designed to separate an organotin compound from foodstuffs. As a buffer solution, dissolve 40 grams of sodium acetate trihydrate and 8.3 grams of sodium hydroxide in water and dilute to 250 ml. Add 1 ml to a mixture of 0.3 ml of *N* sulfuric acid, 0.9 ml of water, 0.4 ml of 1:2.2 hydrochloric acid, and 0.4 ml of 0.05% solution of catechol violet. If this does not give pH 3.8, adjust the buffer solution by minimal additions of solutions of sodium hydroxide or sodium acetate.

VINEGAR. Extract a sample continuously with petroleum ether for 2 hours at as rapid a rate as possible. A design for such an extractor is presented in Figure 10-1. Concentrate to about 1 ml. Apply the concentrated extract to the baseline of chromatographic paper, cut as in Figure 10-2, using a capillary pipet. Speed evaporation by blowing with air on the underside of the paper until dry. Wash the extraction flask with 0.2 ml and 0.2 ml of a 2% solution of formic acid in ether, transfer these sequentially to the baseline, and air-dry until free from the odor of formic acid. Develop the chromatogram with chloroform to the front line. Dry in air and carry out a second development. When air-dried, spray with a 5% solution of disodium 4-(2-pyridylazo)resorcinol to identify the pink area containing the organotin compound.

Cut out the area and lightly char by warming at 100° with 0.5 ml of 75% sulfuric acid. Add two drops of 30% hydrogen peroxide, and after it reacts continue such additions until a clear solution is obtained. Evaporate to dryness; then ignite at 300°. Cool, moisten the residue with 0.3 ml of *N* sulfuric acid, and evaporate at 100°. Cool, and add sequentially 0.9 ml of water, 0.4 ml of 1:2.2 hydrochloric acid, 0.4 ml of 0.05% solution of catechol violet, and 1 ml of the buffer solution to give pH 3.8. Set aside for 3 hours.

[16]Yukio Kodama and Hiroyuki Tsubota, *Jap. Anal.* **20**, 1554–1560 (1971).
[17]L. H. Adcock and Miss W. G. Hope, *Analyst* **95**, 868–874 (1970).

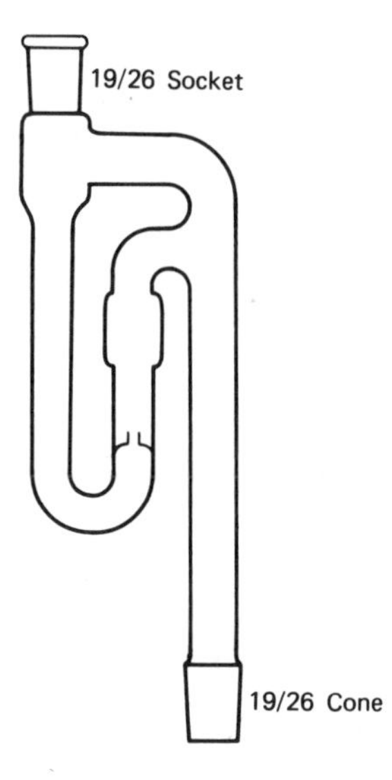

Figure 10–1 Apparatus for continuous extraction, approximately one-quarter actual size.

Disperse 0.1 gram of asbestos and 0.15 gram of cellulose powder in 20% solution of sodium acetate trihydrate adjusted to pH 3.8 with hydrochloric acid. Pour into a piece of 4 mm i.d. glass tubing not less than 150 mm long, lightly plugged with glass wool for a distance of 5 mm. Transfer the entire developed sample to this column. Wash in with 1.5 ml of the 20% sodium acetate solution at pH 3.8. Reject the drainings. Add 1.5 ml of 0.5% solution of a surface-active agent such as Tergitol NPX to elute the tin–catechol violet complex and read at a wavelength depending on the surface-active agent used. With Tergitol NPX it is 570 nm.

ORANGE DRINK. Extract and develop the chromatogram as for vinegar. Because the orange oil in the cut-out area reacts vigorously, carry out the digestion in a micro-Kjeldahl flask. Transfer the clear solution to a 10-ml beaker and wash in with 0.2 ml and 0.2 ml of *N* sulfuric acid. Then proceed as for vinegar from "Evaporate to dryness; then ignite at 300°."

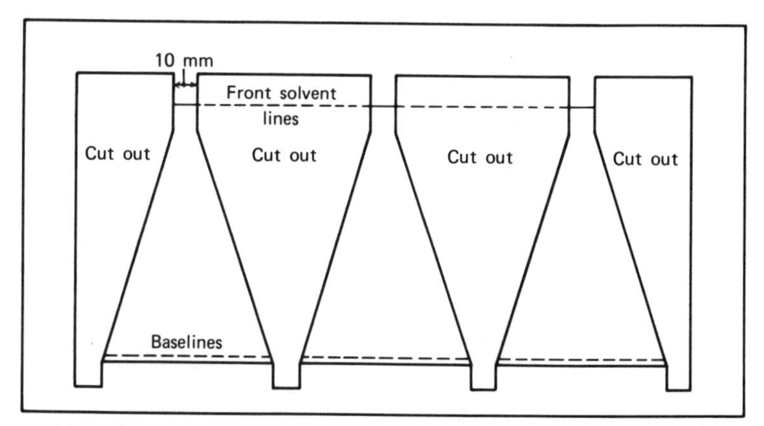

Figure 10–2 Chromatographic paper cut to shape, approximately one-quarter actual size.

SUNFLOWERSEED OIL. Activate 0.05–0.20 mm silica gel at 150° for 2 hours. Transfer 0.15 gram in petroleum ether to form a column in a 4 mm i.d. tube 150 mm long over a plug of cotton. Dilute 5 ml of sample with 75 ml of petroleum ether. Pass this through the column and wash with 1 ml of petroleum ether followed by three 1-ml portions of ether. Discard all effluents. Elute with four successive 1-ml portions of 2% solution of formic acid in ether. Transfer the eluate to the baseline of chromatographic paper cut as in Figure 10-2. Continue as for vinegar from "Wash the extraction flask with 0.2 ml...."

5,7-DIBROMO-8-HYDROXYQUINOLINE

This reagent, also known as 5,7-dibromo-8-quinolinol or simply as dibromooxine, complexes with stannic ion at pH 1.[18a] The extraction technic tolerates 1 mg of common ions.

Procedure. *Stannic Oxide.* This technic determines tin and stannous oxide in the dioxide. Heat 1 gram of sample with 0.2 gram of potassium chlorate and 20 ml of 1:5 hydrochloric acid at 75° for 20 minutes under a reflux condenser. Cool, dilute to 100 ml, and filter. To 50 ml of the filtrate add 10 ml of 10% hydroxylamine hydrochloride solution and 40 ml of 10% thiourea solution. Adjust to pH 1 with ammonium hydroxide and add 7.5 ml of 0.2% solution of the reagent in acetone. After 5 minutes extract with 30 ml and 30 ml of isobutanol, shaking each time for 2 minutes. Wash the combined extracts with 20 ml and 20 ml of dilute hydrochloric acid adjusted to exactly pH 1. Dilute the organic layer to 100 ml with isobutanol and read at 410 nm against a blank.

4,5-DIHYDROXYFLUORESCEIN

The complex of tin with this reagent is preferably read by the differential method. The color is stable and follows Beer's law up to 2 μg per ml.

For analysis of steel dissolve 1 gram of sample in dilute sulfuric acid and filter. Add a solution of cupric ion and coprecipitate copper and tin as the sulfides by adding thioacetamide. Filter, dissolve in 10 ml of hydrochloric acid, add hydrogen peroxide, and develop an appropriate aliquot.

Procedure.[18b] As a buffer solution dissolve 2.5 grams of sodium hydroxide in 40 ml of water. Dissolve 10.6 grams of monochloracetic acid with 2 grams of hydroxylamine hydrochloride in 40 ml of water. Mix and dilute to 100 ml.

Adjust a sample solution containing 10–20 μg of tin to *N* with hydrochloric acid.

[18a] Erick Ruf, *Z. Anal. Chem.* **162**, 9–17 (1958).
[18b] Srinivasan Ambujavalli and Nott Premavathi, *Anal. Chem.* **48**, 2152–2154 (1976).

Add sodium acetate to adjust to pH 2.5. Add 3.5 ml of the prepared buffer solution, 0.5 ml of 0.5% gelatin solution, and 3 ml of a freshly prepared and filtered 0.1% ethanolic 4,5-dihydroxyfluorescein. Set aside at 30° for 15 minutes and read at 520 nm against a similarly developed standard of 10 μg of tin.

DITHIOL

The reagent given the trivial name of dithiol is toluene-3,4-dithiol, which is 1-methyl-3,4-dimercaptobenzene. Decomposition in a bomb is appropriate for determination of tin in organic compounds.[19] The colloidal complex requires a dispersing agent for the stannous dithiolate, for which Santomerse 30X and sodium lauryl sulfate have been found to be satisfactory.[20] The dispersant should be added before the reagent.[21]

Procedures

Organotin Compounds. As 0.3% dithiol reagent, dissolve 0.15 gram of dithiol and 8 drops of thioglycollic acid in 50 ml of 2% sodium hydroxide solution. If not clear, filter through paper. This is stable for 3 days only if stored in a refrigerator.

Place 1 gram of sugar in a nickel fusion cup of a Parr bomb. Weigh a sample to contain 25–40 mg of tin into a gelatin capsule. Place on the sugar and add a level scoop of sodium peroxide. Seal the cup, mix the contents, and combust. When cool, remove the cover and wash into a beaker containing the fusion. Cover with warm water, and when the reaction has subsided, heat at 100° for 90 seconds. After removing the fusion cup and washing well, add 35 ml of hydrochloric acid in several portions. When this has reacted, add 60 ml more of hydrochloric acid. Dilute to 500 ml.

To a 10-ml aliquot, add 3.2 ml of sulfuric acid and 2 drops of perchloric acid. Evaporate to light fumes by heating without boiling. Then heat more vigorously until the perchloric acid effervesces and fume for 1 minute. Cool, wash down the sides of the beaker with 10 ml of water, and add 5 drops of hydrochloric acid. Heat to boiling, cool, and add 5 drops of thioglycollic acid. Dilute to 40 ml and add 2 ml of 2% sodium lauryl sulfate solution. Add 1 ml of 0.3% dithiol reagent, cool if necessary, and dilute to 50 ml. Read at 530 nm.

Silicates.[22] As the dithiol reagent, dissolve 0.1 gram in 50 ml of 1% sodium hydroxide solution and add 0.5 ml of thioglycollic acid as a stabilizer. Store under refrigeration and renew after a week.

To 1 gram of sample in platinum, add 5 ml of 1:1 sulfuric acid, 1 ml of nitric

[19]Marie Farnsworth and Joseph Pekola, *Anal. Chem.* **31**, 410–414 (1959); Wallace W. Sanderson and Arthur M. Hanson, *Sewage Ind. Wastes* **29**, 422–427 (1957); cf. J. J. Hefferren, *J. Pharm. Sci.* **52**, 1190–1196 (1963).

[20]Masuo Kojima, Yujiro Nomura, and Atsuo Suzuki, *Jap. Anal.* **6**, 34–35 (1957).

[21]T. C. J. Ovenston and C. Kenyon, *Analyst* **80**, 566–567 (1955).

[22]Hiroshi Onishi and E. B. Sandell, *Anal. Chim. Acta* **14**, 153–161 (1956); cf. Giulio De Nittis, *Rass. Chim.* **25**, 390–395 (1973).

acid, and 10 ml of hydrofluoric acid. Heat to sulfur trioxide fumes, cool, add 2 ml of 1:3 sulfuric acid, and again heat to sulfur trioxide fumes.

If minerals not decomposed by hydrofluoric acid (e.g., cassiterite) are absent, skip this paragraph. Add 30 ml of 1:3 hydrochloric acid to the fumed residue and keep close to boiling until nearly complete solution has occurred. Filter and wash with 1:5 hydrochloric acid. The filtrate and washings go to the distilling flask of the next paragraph. Ignite the paper in a nickel crucible and fuse the ash with 0.5 gram of sodium hydroxide, keeping the fusion at a dull red for 30 minutes. Dissolve the melt in 8 ml of water, transfer to a beaker, and add 10 ml of 1:1 hydrochloric acid. Warm for a few minutes, filter, and wash with 1:5 hydrochloric acid. The filtrate and washings go into the distilling flask. If the tin content of the sample is high, repeat the ignition, fusion, and solution of the residue.

Using 25 ml of 1:1 sulfuric acid, transfer the fumed residue to a 200-ml round-bottom flask having a thermometer well and gas inlet tube, carrying by standard taper joints a separatory funnel and an outlet tube to a condenser. If the fusion of the preceding paragraph was necessary, add the 25 ml of 1:1 sulfuric acid to the accumulated filtrate and washings. Add 5 grams of monopotassium phosphate, 1 gram of hydrazine sulfate, and 10 ml of 1:1 hydrochloric acid. Dip the end of the condenser into 30 ml of water. Pass a slow stream of carbon dioxide and boil gently. When the temperature reaches 160°, introduce 1:1 hydrochloric acid at a rate of 1 drop every 4 seconds while boiling at 155–165°. When 20 ml of hydrochloric acid has been so added, remove and discard the distillate of arsenic. Dip the end of the condenser in 15 ml of water. Add a mixture of 15 ml of 1:1 hydrochloric acid and 7 ml of hydrobromic acid drop by drop, while distilling at 145–160° in a stream of carbon dioxide. This distillation should require 15–20 minutes. and the distillate contains the tin.

Add 0.5 ml of 1:2 sulfuric acid and 5 ml of nitric acid and heat at 100°. After vigorous reaction ceases, evaporate to fumes and continue to fume for 5 minutes. Dilute to about 7 ml. Add a drop of thioglycollic acid and 0.5 ml of 5% solution of a surfactant, preferably sodium lauryl sulfate. Dilute to 10 ml and heat at $50 \pm 5°$ for 5 ± 1 minutes. Read at 530 nm promptly.

Iron and Steel. Dissolve 0.5 gram of sample containing up to 100 ppm of tin in 7 ml of 1:1 sulfuric acid and 10 ml of 1:1 nitric acid. Evaporate to sulfur trioxide fumes. Complete as for silicates from "Using 25 ml of 1:1 sulfuric acid...."

Sulfides. Treat 1 gram of sample containing 0.5–50 ppm of tin with 10 ml of fuming nitric acid. Warm gently, and after vigorous reaction ceases, evaporate to about 1 ml. Add 7 ml of 1:1 sulfuric acid and evaporate to sulfur trioxide fumes. Complete as for silicates from "Using 25 ml of 1:1 sulfuric acid...."

Organic Samples.[23] As the reagent prepared immediately before use, dissolve 0.2 gram of zinc dithiol in 1% sodium hydroxide solution containing a few drops of

[23]Metallic Impurities in Organic Matter Subcommittee of the Analytical Methods Committee of the Society for Analytical Chemistry, *Analyst* **93**, 414–416 (1968).

ethanol. Add 1 ml of thioglycollic acid and dilute to 100 ml with 1% sodium hydroxide solution.

Wet-ash by any appropriate method according to the nature of the sample. Evaporate to sulfur trioxide fumes and dilute to contain no more than 4 ml of sulfuric acid per 100 ml. Mix 10 ml of sample solution containing 0.03–0.15 mg of tin with 5 ml of 0.02% solution of dithizone in carbon tetrachloride. Shake and discard the dithizone layer. Extract with successive 5-ml portions of dithizone until the green color is no longer modified. Wash the aqueous solution with 5 ml and 5 ml of carbon tetrachloride.

Add 1 : 4 sulfuric acid to the aqueous layer so that it contains a total of 0.7–1 ml. Add 1 ml of 1% solution of sodium lauryl sulfate, followed by 1 ml of the zinc dithiol reagent. Dilute to 20 ml and heat in boiling water for 1 minute. Let cool to room temperature and read at 535 nm against a reagent blank.

Foods.[24] As a standard tin solution, dissolve 0.2 gram in 125 ml of hydrochloric acid and add 250 ml of water. Add 250 ml of 3 : 7 sulfuric acid, and dilute to 1 liter. This contains 200 ppm of tin. For a solution containing 4 ppm of tin, dilute 20 ml of the 200-ppm solution with 500 ml of 3 : 7 sulfuric acid and dilute to 1 liter with water.

As 0.3% dithiol reagent dissolve 4 drops of melted dithiol, about 0.075 gram, and 4 drops of thioglycollic acid in 10 ml of 5% sodium hydroxide solution and add 15 ml of water.

Homogenize a coarsely ground sample. Mix 100 grams with 125 ml of nitric acid, 25 ml of sulfuric acid, and a few glass beads. Mix occasionally for 20 minutes, then boil down to sulfur trioxide fumes. Cool somewhat, and complete the destruction of the organic matter with additions portionwise of about 10 ml of nitric acid and 2 ml of perchloric acid. Finally again take down to sulfur trioxide fumes, cool and add 75 ml of water. Again evaporate to sulfur trioxide fumes.

For isolation of tin by distillation, use the equipment illustrated in Figure 10-3. Add 25 ml of 1 : 1 sulfuric acid and 50 ml of water to the beaker, and set in an ice-water bath. Add 15 ml of sulfuric acid and 35 ml of water to the flask and mix. Lubricate the joint of the distilling head with sulfuric acid. Add 45 ml of 1 : 2 hydrochloric acid–hydrobromic acid to the flask, attach the head, and mix. Place on a preheated hot plate set at high heat with the outlet tube dipping 0.5–1 inch below the surface of the water in the beaker. Distill until sulfur trioxide fumes are visible, and continue for an additional 4–5 minutes so that sulfuric acid refluxes on the wall above the liquid. Lower the receiver below the outlet tube and continue with distillation for 1 minute.

Add 25 ml of nitric acid and 3 drops of perchloric acid to the distillate, insert a stirring rod with the end roughened to promote smooth boiling, and mix. Warm at medium heat until reaction starts. Heat to boiling and boil down to sulfur trioxide fumes at a moderate rate. Fume at high heat to remove perchloric acid and water.

[24]Homer B. Corbin, *J. Assoc. Off. Anal. Chem.* **53**, 140–146 (1970); cf. D. Dickinson and R. Holt, *Analyst* **79**, 104–106 (1954); T. W. Raven, *ibid.* **87**, 827–828 (1962); R. De Giacomi, *ibid.* **65**, 216–218 (1940); J. H. Shelton and J. M. T. Gill, *J. Assoc. Pub. Anal.* **2** (4), 98–100 (1964); Metallic Impurities in Organic Matter Subcommittee of the Analytical Methods Committee of the Society for Analytical Chemistry, *Analyst* **93**, 414–416 (1968).

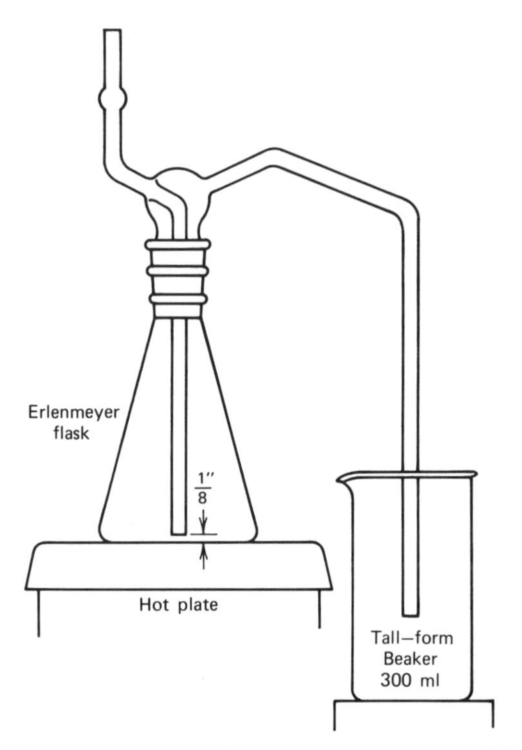

Figure 10–3 Distillation apparatus for determination of tin.

Cool, wash down with 10–15 ml of water, then remove the rod and evaporate below boiling to fumes. Fume off water; cover and fume strongly for 1 minute at high heat. The solution should be clear and water-white. The sample may be held at this point.

As an alternative, separate tin by extraction. Add 50 ml of water to the sample and cool. Transfer the solution to a separatory funnel. Rinse the walls of the flask with 25 ml of 20% potassium iodide solution, transfer to the funnel, and swirl to mix. Rinse the flask and beads with 50 ml of *n*-hexane and transfer to the funnel. Extract for 90 seconds and let the layers separate. Drain the aqueous layer into the original flask. Wash this hexane with 10 ml of a mixture of 3 volumes of 1:2 sulfuric acid with 1 volume of 20% potassium iodide solution. If the wash solution is more deeply colored than the reagent blank, repeat the washing with a second 10-ml portion.

Separate and decant the hexane from any insoluble matter. If the original digestate had more than a trace of insoluble matter present, such as considerable calcium, or if addition of potassium iodide solution caused precipitation, combine the solution and washings in the separatory funnel and extract a second time with 50 ml of *n*-hexane. Wash once or twice as in the first extraction, and combine with the first extract. If only one extraction is made, rinse the funnel with 5 ml of hexane, and combine the rinse with the extract.

Add 5 ml of nitric acid and let the hexane evaporate without heat. Add 4 ml of sulfuric acid and 3–4 drops of perchloric acid. Cover and heat until nitric acid refluxes from the cover glass. Remove the cover and heat the solution to fumes. Fume off perchloric acid. Then fume the sulfuric acid for 1 minute with the beaker covered. Cool, rinse down with 10 ml of water, and add 3 drops of perchloric acid. Evaporate below boiling and fume off water and perchloric acid. Cover with a flat watch glass and heat briefly to reflux sulfuric acid. The sample may be held at this point.

Rinse down the walls of the beaker from either of the foregoing separation procedures with 15 ml of water and cool in a water bath. Add 1 ml of standard tin solution containing 4 ppm of tin, 3 drops of thioglycollic acid, and 2 ml of 2% sodium lauryl sulfate solution. Swirl to mix. Dilute to 50 ml and read the absorption at 530 nm. Deduct the reading obtained from a blank solution containing 1 ml of tin standard plus other reagents.

If it is necessary to remove arsenic exceeding 1.5 ppm, add 0.2 gram of hydrazine sulfate to the sulfuric acid digest of the sample. Heat strongly for 2 minutes, cool, and add 0.2 gram of hydrazine sulfate, 25 ml of water, and 50 ml of hydrochloric acid. Evaporate to sulfur trioxide fumes and swirl the flask in a flame for 2 minutes.

If it is necessary to remove antimony exceeding 2 ppm, only the distillation technic is applicable. Dilute the sulfuric acid digest of the sample with 150 ml of water and add 10 ml of hydrochloric acid. Cool in ice water. Add 1% potassium permanganate solution dropwise to excess and transfer to a separatory funnel. Add 4 ml of 9% cupferron solution and mix. Extract with 20 ml, 10 ml, and 10 ml of chloroform, collecting the extracts under 25 ml of water in the original digestion-distillation flask. Add 25 ml of nitric acid to the flask and boil off the chloroform. Add 25 ml of sulfuric acid and take to fumes. Complete the digestion with nitric and perchloric acids, and remove nitric acid by boiling down with water as in the original digestion of the sample. Distill tin and proceed with the determination.

FERROUS DIMETHYLGLYOXIME

Tin forms a red color with ferrous ion and dimethylglyoxime in acid solution.

Procedure. *Zirconium Alloys.*[25] As reagent, mix a solution of 1.2 gram of ferrous sulfate heptahydrate, 40 ml of *N* hydrochloric acid, 100 ml of 0.1 *M* dimethylglyoxime in 0.2 *N* sodium hydroxide, and 200 ml of *N* sodium hydroxide. Add 155 ml of *N* hydrochloric acid. Adjust to pH 5 and dilute to 1 liter.

Dissolve 0.1 gram of sample in 6 ml of 1:10 hydrofluoric acid and 1 ml of sulfuric acid. Evaporate to dryness, take up in 3 ml of water, and again evaporate to dryness. Dissolve the residue with 1 ml of *N* hydrochloric acid, 0.4 gram of tartaric acid, and 10 ml of water. Neutralize to phenolphthalein with 4% sodium hydroxide solution. Add 20 ml of *N* hydrochloric acid and dilute to 100 ml. Mix a 5-ml aliquot containing 0.025–02 mg of tin with 25 ml of the reagent. Let stand for

[25]S. V. Elinson and V. T. Tsvetkova, *Zavod. Lab.* **37**, 662–664 (1971).

5 minutes and add 12 ml of N hydrochloric acid. Let stand for 30 minutes and dilute to 50 ml. Read at 460 or 530 nm.

FLAVONOL

The fluorescence in 0.1–0.5 N sulfuric acid of the 1:1 complex of tetravalent tin with flavonol, which is 3-hydroxylflavone, is sensitive to 0.1 μg/ml.[26] The complex and the reagent are almost insoluble in water. Dimethylformamide is a preferred solvent. Zirconium, fluoride, phosphate, and molybdenum interfere. The acid concentration prevents precipitation of tin. More than 0.33 N sulfuric acid greatly decreases the fluorescence. Chloride decreases the fluorescence. For reading in water, the pH should be 0.05–0.1 N as compared with 0.12 N in diluted dimethylformamide, and precipitation occurs after about 10 minutes. In solution in 60% methyl Cellosolve the color developed can be read at 400 nm and pH 3 for 0.5–3 μg/ml.[27]

Procedure. To an aqueous sample containing 0.1–0.7 mg of tin, add 7.5 ml of dimethylformamide. Add sufficient sulfuric acid to make the total in the sample equivalent to 1 ml of 1:11 acid. Add 2 ml of 0.05% flavonol solution in ethanol and dilute to 25 ml with water. After 15 minutes activate at 405 nm and read at 346 nm.

GALLEIN

Gallein, which is 4,5-dihydroxyfluorescein, structurally 3′,4′,5′,6′-tetrahydroxyspiro[isobenzofuran-1(3H),9′(9H)-xanthen]-3-one, complexes with tin.[28] Extraction separates from many interfering ions. Fluoride ion can be masked with zirconium ion. In analysis of iron, steel, and nonferrous alloys, tartaric acid prevents interference by antimony. Titanium, tungsten and zirconium, iron, and vanadium are masked by ascorbic acid.[29] Interference by molybdenum and titanium is prevented by hydrogen peroxide. An appropriate stabilizer is cetylpyridinium bromide, which prevents precipitation below pH 2.[30] Unless masked, ferric ion, bismuth, aluminum, thallium, and zirconium increase the absorption, whereas antimony decreases it. In 0.15 N hydrochloric acid the maximum absorption is at 575 nm, at 0.05 N it is at 500 nm.

For a light alloy or a ferroalloy, dissolve in an appropriate acid with hydrogen peroxide if needed.[31] Boil off free hydrogen peroxide if present. Add masking

[26] Charles F. Coyle and Charles E. White, *Anal. Chem.* **29**, 1486–1488 (1957).
[27] Yoshinaga Oka and Reiko Tanaka, *J. Chem. Soc. Jap.* **81**, 1846–1849 (1960).
[28] Hanspeter Heegn, *Freiberg. Forschungsh. A* **445**, 47–61 (1969).
[29] Masami Murano and Shigeru Miyazaki, *Jap. Anal.* **13**, 994–1000 (1964); **15**, 657–661 (1966).
[30] Itsuo Mori, *ibid.* **19**, 455–458 (1970).
[31] V. A. Popov, E. I. Rudenko, and T. A. Pal'chun, *Zavod. Lab.* **41**, 515–520 (1975).

agents and ethanolic gallein. Adjust to pH 1.4, diluting to any extent necessary. Extract with toluene, which will contain the tin complex as a precipitate. Filter, dissolve the precipitate in ethanol, and read at 495 nm. Ascorbic acid and tartaric acid mask iron, vanadium, arsenic, antimony, bismuth, and tungsten. Oxalic acid masks zirconium. Thiourea masks copper; hydrogen peroxide masks titanium and molybdenum.

For determination of stannic tin, gallein and 2′,7′-dihydroxyfluorescein have been recommended at pH 1.9 for extraction into cyclohexanone.[32a] For gallein, readings are up to 1.5 μg of tin per ml; for the fluorescein derivative, up to 6 μg of tin per 20 ml. Ascorbic acid masks interference by iron, manganese, vanadium, chromium, or cerium. Germanium cannot be masked.

Procedure. Neutralize the sample solution to phenolphthalein with sodium hydroxide solution or sulfuric acid. Add 2 ml of 1:9 sulfuric acid and 1 ml of 0.1% gallein solution. Dilute to 25 ml and set aside for 1 hour. Shake for 30 seconds with 10 ml of cyclohexanone. Centrifuge the organic layer and read at 495 nm.

Steel.[32b] As a buffer solution for pH 2.5, dissolve 2.6 grams of sodium hydroxide in 40 ml of water. Mix with 10.6 grams of chloroacetic acid and 2 grams of hydroxylamine hydrochloride in 40 ml of water. Dilute to 100 ml.

Dissolve a 1-gram sample in 10 ml of 1:3 sulfuric acid and filter. Add 1 ml of 1% copper sulfate solution and dilute to 250 ml. Heat to boiling. Slowly add 10 ml of 2.5% solution of thioacetamide and continue to boil for 1 minute. Filter the sulfide precipitate and wash with a dilute solution of thioacetamide. Dissolve in 10 ml of hydrochloric acid, adding 1 ml of 30% hydrogen peroxide. Boil, filter, and dilute to 100 ml. Adjust an aliquot containing 10–15 μg of tin to pH 2.5 with sodium acetate solution. Add 3.5 ml of the buffer solution, 3 ml of 10% ammonium iodide solution to mask copper, 0.5 ml of 0.5% solution of gelatin, and 3 ml of freshly filtered 0.1% solution of gallein. Dilute to 25 ml, set aside for 15 minutes, and read at 520 nm against a developed solution containing 10 μg of tin in a final solution of 25 ml.

Zinc, Magnesium, and Aluminum Alloys. Dissolve a 1-gram sample in 10 ml of hydrochloric acid and 1 ml of 30% hydrogen peroxide. Slowly add sodium hydroxide solution until a precipitate just forms and redissolve with a couple of drops of 1:1 sulfuric acid. If not already present, add 1 ml of 1% solution of copper sulfate in 1:3 sulfuric acid. Dilute to 250 ml. Complete as for steel from "Heat to boiling."

Antimony. Dissolve a 1-gram sample in 20 ml of aqua regia. Add 10 ml of 1:1 sulfuric acid and evaporate to sulfur trioxide fumes. Take up in 20 ml of 1:1 hydrochloric acid and precipitate antimony with 1.5 grams of powdered iron. Filter and reserve the filtrate. Dissolve the precipitate with 20 ml of hydrochloric acid containing 0.1 gram of potassium chlorate. Reprecipitate with iron as before and

[32a]Gerhard Ackermann and Hanspeter Heegn, *Talenta* **21**, 431–438 (1974).
[32b]Srinivasan Ambujavalli and Nott Premavathi, *Anal. Chem.* **48**, 2152–2154 (1976).

filter. Combine the filtrates, add 10 ml of nitric acid to oxidize ferrous ion, and heat to boiling. Dilute to 300 ml and add 15 grams of EDTA to mask iron. Add 25 mg of beryllium sulfate. Make alkaline with ammonium hydroxide and filter the tin and beryllium hydroxides. Wash with 2% ammonium nitrate until free from EDTA and iron. Dissolve in 10 ml of N hydrochloric acid and proceed as for steel from "Heat to boiling."

Copper. Dissolve 1 gram of sample in 15 ml of aqua regia. Add 25 mg of beryllium sulfate. Add excess ammonium hydroxide until copper is redissolved as the cuprammonium complex. Filter the tin and beryllium hydroxides and wash free of copper with 2% ammonium nitrate solution. Dissolve in 10 ml of N hydrochloric acid and proceed as for steel from "Heat to boiling."

Lead and Antimony-Lead. Dissolve 1 gram of sample in 25 ml of 1 : 4 nitric acid and 10 grams of citric acid. Add 12 ml of ammonium hydroxide, 10 ml of 0.5% ammonium chloride solution, and 2 grams of EDTA. Adjust to pH 5.5 with ammonium hydroxide or citric acid. Prepare a column of silica gel and wash it successively with hydrochloric acid, water, and a 0.5 M citrate buffer solution for pH 5.5. Pass the solution through the column; tin is retained but lead and antimony pass through. Wash the column with the citrate buffer solution for pH 5.5. Elute the tin with 10 ml of 1 : 1 hydrochloric acid, wash with water, and dilute to 50 ml. Mix a 5-ml aliquot with 20 ml of a buffer solution of chloroacetic acid, sodium chloroacetate, and hydroxylamine hydrochloride for pH 2.4–2.5. Add 5 ml of 0.005% solution of gallein in ethanol and dilute to 50 ml. Read at 520 nm.

Soil.[33] Mix 1 gram of sample with 1 gram of ammonium iodide, and heat in a test tube until sublimation of ammonium iodide to the cool top of the tube ceases. Boil the residue with 10 ml of N hydrochloric acid, filter, wash the filter with hot N hydrochloric acid, and dilute to 50 ml with the same acid. Add 10% solution of hydrazine hydrate to reduce the iodine in an aliquot of the solution, and make up the acidity to N in hydrochloric acid. Add 10 ml of a buffer solution containing 4.5% monochloroacetic acid and 7.5% sodium monochloroacetate and dilute to 100 ml. Add 0.1 ml of 0.05% ethanolic solution of gallein and extract with 10 ml of pentanol for 30 seconds. Read at 500 nm.

Black Currant Pulp.[34] Wet-ash the sample with sulfuric and nitric acids and expel nitric acid by repeated evaporation with water. Take up in water, add 1 gram of ammonium iodide, and dilute to 50 ml. Complete as for soil from "Add 10% solution of hydrazine hydrate...."

---

[33]R. E. Stanton and A. J. McDonald, *Trans. Inst. Min. Met.* (London) **71**, 27–29 (1961); A. J. McDonald and R. E. Stanton, *Analyst* **87**, 600–602 (1962); cf. A. Purushottam and M. R. Nayar, *Curr. Sci.* **38**, 565–566 (1969).

[34]R. E. Stanton, *Food Technol.* (Aust.) **22**, 236–237 (1970).

HEMATEIN

This reagent is the oxidation product of hematoxylin; it has the structure of 6a, 7-dihydro-3, 4, 6a, 10-tetrahydroxybenz[b]-indeno[1, 2-d]pyran-9(6H)-one and is not to be confused with hematin. Its complex with tin can be stabilized with gum arabic for photometric reading.[35] By taking a large sample and using a collector, the method will determine 0.0001% of tin. It forms 1:2 complexes with both stannous and stannic ions.[36] The color takes 40 minutes to develop at 10°, 20 minutes at 30°, and 10 minutes at 50°. If developed at a temperature not exceeding 30°, it is stable for 24 hours.

By coprecipitating tin and beryllium hydroxide with ammonium hydroxide, pentavalent antimony is not coprecipitated. By warming with EDTA at pH 3–4, the reaction with titanium and chromium is completed, preventing them from coprecipitating. If a large amount of chromium is present, a precaution is to reprecipitate the beryllium in the presence of hydrogen peroxide.

Procedures

Iron and Steel.[37] As the hematein reagent, reflux 1.5 gram of hematoxylin in 100 ml of ethanol with 10 ml of 5% hydrogen peroxide for 10 minutes and dilute to 500 ml.

Dissolve 0.5 gram of sample containing less than 0.2 mg of tin in 20 ml of 1:1 hydrochloric acid and 3 ml of 30% hydrogen peroxide. Add 2 ml of 5% solution of beryllium sulfate and 35 ml of 10% solution of EDTA. Dilute to 150 ml and heat to 80°. Add ammonium hydroxide until precipitation is complete, and 10 ml in excess. Boil for 2 minutes and filter. Wash the beryllium hydroxide coprecipitated with tin with 1:5 ammonium hydroxide. Dissolve the precipitate in 6 ml of 1:1 hydrochloric acid, dilute to 150 ml, precipitate as before, filter, and wash. Dissolve the precipitate in 5 ml of 1:1 hydrochloric acid, cool, and neutralize to *p*-nitrophenol with ammonium hydroxide. Dissolve the precipitate of beryllium hydroxide by adding *N* hydrochloric acid and add 4 ml in excess. Filter, and dilute to 70 ml. Add 5 ml of 10% solution of potassium acid phthalate and 5 ml of the oxidized hematoxylin reagent. Heat for 10 minutes at 100°, cool, and dilute to 100 ml. Read at 575 nm.

Pig Iron and Steel.[38] Dissolve 4 grams of sample in 40 ml of 1:2 hydrochloric acid with heating. Complete by adding more hydrochloric acid if necessary and by dropwise addition of 6 ml of 30% hydrogen peroxide. Evaporate to about 20 ml, cool, and dilute to 100 ml with hydrochloric acid. Shake a 20-ml aliquot with 20 ml of 2.5 *M* tributyl phosphate in benzene for 1 minute. Wash the organic phase, which contains most of the iron, with 5 ml and 5 ml of hydrochloric acid. To the

[35]Masuo Kojima, *Jap. Anal.* **6**, 139–146 (1957).
[36]E. Asmus, H. J. Altmann, and E. Thomasz, *Z. Anal. Chem.* **216**, 3–13 (1966).
[37]Shizuya Maekawa, Yoshio Yoneyama, and Eiichi Fujimori, *Jap. Anal.* **10**, 1335–1340 (1961).
[38]H. Specker and G. Graffmann, *Z. Anal. Chem.* **228**, 401–405 (1967).

aqueous phase plus washings add 10 ml of 5 *N* sulfuric acid and evaporate to sulfur trioxide fumes. Take up in 20 ml of water and add 20 ml of 4 *M* sodium perchlorate in 4 *M* sodium iodide solution. Add 10 ml of perchloric acid. Reduce the liberated iodine to no more than a faint yellow by adding solid sodium sulfite. Extract the stannic iodide by shaking with 25 ml of benzene. Add to the benzene or an aliquot, 8 ml of 1 : 1 hydrochloric acid and evaporate the benzene. Add 2 ml of 2% ascorbic acid solution to the tin solution and neutralize to Tropaeolin 00 by adding saturated sodium acetate solution. Add 5 ml of 10% polyvinyl alcohol solution. Add 2 ml of 0.05% solution of hematein in dioxane and dilute to 100 ml. Let stand for 60 minutes and read at 550 nm.

Lead. Dissolve 15 grams of sample in excess 1 : 1 nitric acid and evaporate to dryness. Take up the residue in 200 ml of water and if turbid, clarify with a small amount of nitric acid. Add 2 ml of a 10% solution of manganese nitrate and heat to boiling. Add 1 ml of 3% potassium permanganate solution and boil for 5 minutes. Separate the precipitate by centrifuging. Repeat the treatment with permanganate and manganese nitrate twice. Combine the three precipitates of manganese dioxide and dissolve in 30 ml of 1 : 2 hydrochloric acid. Pass the solution through a column of Dowex 1 in chloride form. Wash with 30 ml of 2 : 1 hydrochloric acid. Elute the tin with 30 ml of 1 : 15 nitric acid. Add 20 ml of hydrochloric acid and 2 drops of 1 : 9 sulfuric acid to the eluate and evaporate to dryness. Take up the residue in 3 ml of water and add 1 ml of 1% gum arabic solution. Add 1 ml of 0.5% solution of hematein. After allowing to stand for 50 minutes, dilute to 10 ml with 1 : 20 sulfuric acid and read at 550 nm.

Lead Alloys.[39] Dissolve a sample containing 0.02–0.1 mg of tin and not more than 0.2 gram of antimony in 10 ml of 1 : 4 nitric acid per gram of sample. Cool and add 25 ml of 1 : 4 hydrochloric acid. Cool below 5° with continous shaking. Filter the precipitate of lead chloride and wash with 5 ml, 5 ml, and 5 ml of ice-cold 1 : 1 hydrochloric acid. Add 1 ml of 1 : 1 sulfuric acid to the combined filtrates and evaporate to sulfur trioxide fumes. Take up the residue in 10 ml of water and 3 ml of perchloric acid. Dilute to 20 ml and add 20 grams of sodium iodide. Extract the tin with 10 ml and 5 ml of benzene. Wash the combined extracts by shaking for 1 minute with 10 ml of 5 *M* sodium iodide, 3 ml of perchloric acid, and 7 ml of water. Extract the washed benzene phase with 5 ml of hydrochloric acid; then heat the extract slowly, to evaporate benzene and drive off free iodine. Cool, and add 2 ml of 2% ascorbic acid solution. Add 2 drops of a saturated solution of Tropaeolin 00 followed by saturated sodium acetate solution until the red changes to yellow. Add 5 ml of 10% solution of polyvinyl alcohol and 2 ml of 0.05% methanolic solution of hematein. Dilute to 100 ml and read after 60 minutes at 590 nm. This method will determine 2 ppm of tin.

Copper Alloy.[40] As reagent, dissolve 0.5 gram of hematoxylin in 100 ml of ethanol. Add 4 ml of 3% hydrogen peroxide and 0.25 ml of 1 : 3 hydrochloric acid. Heat at 100° for 15 minutes and dilute to 500 ml. This reagent is stable for a week.

[39] R. Shirodker and E. Schibilla, *Z. Anal. Chem.* **248**, 173–176 (1969).
[40] Reizi Tanaka, *Jap. Anal.* **10**, 336–341 (1961).

Dissolve 0.5 gram of sample in 35 ml of 1 : 1 nitric acid and dilute to 200 ml. Add 5 ml of 10% solution of manganese nitrate and heat to boiling. With vigorous agitation, slowly add 5 ml of 1% solution of potassium permanganate. Boil for a few minutes and filter the manganese dioxide with the coprecipitated tin. Dissolve the precipitate in 20 ml of 1:3 hydrochloric acid by dropwise addition of 30% hydrogen peroxide. Boil off excess peroxide, evaporate to less than 10 ml, and dilute to 20 ml. Add 10% sodium hydroxide solution to precipitate the hydroxides. Add a few drops of 3% hydrogen peroxide, dissolve the hydroxides with 1:3 hydrochloric acid, and add 2 ml excess. Add 5 ml of 0.25% gelatin solution and 30 ml of the prepared reagent. Incubate at 30° for 20 minutes, cool, and read at 570 nm.

Aluminum Alloy. Dissolve 1 gram of sample in 15 ml of 20% sodium hydroxide solution. Add 35 ml of 1 : 1 nitric acid and dilute to 200 ml. Proceed as for copper alloy from "Add 5 ml of 10% solution...."

Zinc. Dissolve 15 grams of sample in excess 1 : 1 nitric acid and evaporate to dryness. Take up in 200 ml of water and 2 ml of nitric acid. Add 2 ml of 10% solution of manganese nitrate and heat to boiling. Add 1 ml of 3% potassium permanganate solution and boil for 5 minutes. Separate the precipitate by centrifuging. Repeat the treatment with permanganate and manganese nitrate. Dissolve the combined precipitates in hydrochloric acid and evaporate to dryness. Take up the residue in 3 ml of water, 2 drops of 1:9 sulfuric acid, 1 ml of 1% solution of gum arabic, and 1 ml of 0.5% solution of hematein. After allowing to stand for 50 minutes, dilute to 10 ml with 1:20 sulfuric acid and read at 550 nm.

Pharmaceutical Preparations.[41] If organic matter is present, dry-ash. Grind the residue with 1.5 grams of sodium carbonate and 0.5 gram of sodium cyanide. Fuse for 30 minutes, cool, and dissolve in 1 : 11 hydrochloric acid. Dilute to a known volume and take an aliquot containing 0.05–7 mg of tin. Add 5 ml of sulfuric acid and evaporate to sulfur trioxide fumes.[42] Dilute to 100 ml. Mix 5 ml of hematein reagent as described for analysis of copper alloy with a 25 ml aliquot, and dilute to 50 ml. After 1 hour for development, read at 515 nm and subtract a blank.

4-HYDROXY-3-NITROBENZENEARSONIC ACID

The complex of this reagent with stannic ion has serious interference only by titanium, zirconium, and hafnium.[43] The complex can also be read turbidimetrically.[44]

[41]R. Bontemps, *Arch. Pharm. Chem.* **68**, 207–211 (1961).
[42]Harry Teicher and Louis Gordon, *Anal. Chem.* **25**, 1182–1185 (1953).
[43]K. Vasadi, *Magy. Kem. Lapja* **23**, 344–346 (1968).
[44]H. J. G. Challis and J. T. Jones, *Anal. Chim. Acta* **21**, 58–67 (1959).

Procedure. *Lead-Antimony Alloys.* Add to 1 gram of sample containing 0.001–0.03% of tin, 2 grams of citric acid, 3 ml of lactic acid, and 10 ml of 1:2 nitric acid. Weigh the vessel. Heat to 70–80° and when dissolved, add 8 ml of 1:8 nitric acid. Cool, and add water until the original weight is increased by 9 grams. Add 6 ml of 2% solution of the captioned reagent in methanol. Mix, let stand for 3 hours, and read at 424 nm against a reagent blank.

8-HYDROXYQUINOLINE

Stannic ion and molybdenum ion complex with this reagent, also known as 8-quinolinol and by the trivial name of oxine.[45] Other elements forming colored complexes with oxine include copper, nickel, iron, and aluminum. By extraction at substantially under pH 1, separation is made from most interfering elements. Then by control of halide concentration, molybdenum is extracted, followed by tin, using a high concentration of oxine. Interference by tungsten, niobium, tantalum, and tin in the determination of molybdenum is avoided by addition of fluoride.

The complex of tin with oxine as extracted at pH 2.7–5.6 with chloroform can be read by the fluorescence in ultraviolet light.[46] Aluminum, indium, zinc, iron, copper, and germanium form fluorescent complexes under the same conditions.

For determination in iron and steel with this reagent, it is appropriate to coprecipitate with manganese dioxide.[47]

Procedures

Uranium Oxide. MOLYBDENUM AND TIN. Dissolve 5 grams of sample in 10 ml of 1:1 sulfuric acid and dilute to 250 ml. Take an aliquot, which will contain less than 0.4 mg of molybdenum and less than 0.5 mg of tin. Dilute to 50 ml and add 25 ml of 4% solution of oxine in pH 0.85 sulfuric acid. Adjust to pH 0.85 ± 0.1 at 25° with ammonium hydroxide. Transfer to a separatory funnel with the aid of not more than 15 ml of pH 0.85 sulfuric acid.

For molybdenum, shake for 2 minutes with 20 ml of chloroform and drain this extract of molybdenum into a separatory funnel containing 50 ml of 4% solution of ammonium chloride in pH 0.85 hydrochloric acid. Shake for 2 minutes. Read the value for molybdenum at 385 nm against chloroform.

For tin, wash the aqueous phase from which molybdenum has been extracted, with 10 ml of chloroform and discard the washings. Add 5 ml of a 20% solution of ammonium chloride in pH 0.85 hydrochloric acid. Shake for 2 minutes with 20 ml of chloroform to extract the tin. Drain this extract into a separatory funnel containing 50 ml of 4% solution of ammonium chloride in pH 0.85 hydrochloric

[45] A. R. Eberle and M. W. Lerner, *Anal. Chem.* **34**, 627–632 (1962).
[46] L. B. Ginzburg and E. P. Shkrobot, *Zavod. Lab.* **23**, 527–533 (1957).
[47] Shigeo Wakamatsu, *Jap. Anal.* **9**, 858–861 (1960).

acid. Shake for 2 minutes and read the chloroform layer at 385 nm against chloroform.

Steels. Dissolve 2 grams of sample in 10 ml of 1:1 sulfuric acid. If necessary, add 30% hydrogen peroxide and boil off the excess. Filter and dilute to 250 ml. Complete as for uranium oxide from "Take an aliquot...."

Copper- and Zinc-Base Alloys. Dissolve 2 grams of sample in 10 ml of nitric acid and 1 ml of perchloric acid. Add 5 ml of sulfuric acid and take to sulfur trioxide fumes. Cool and take up in 25 ml of water. Cool, dilute to 250 ml, and let stand for any lead sulfate to settle. Complete as for uranium oxide from "Take an aliquot...."

Zirconium. Dissolve 2 grams of sample in 25 ml of 1:10 hydrofluoric acid and 5 ml of sulfuric acid, and evaporate to dryness. Add 5 ml of water and evaporate to dryness. Repeat this step. Take up in water, filter if necessary, and dilute to 250 ml. Complete as for uranium oxide from "Take an aliquot...."

Thorium Oxide. Reflux 4 grams of sample with 20 ml of perchloric acid and 0.05 gram of sodium fluoride. After solution is complete, add 50 ml of water, 0.5 gram of boric acid, and 5 ml of 20% solution of ammonium chloride in pH 0.85 hydrochloric acid. Adjust to pH 0.85 with 1:1 ammonium hydroxide. Add 25 ml of 4% solution of oxine in pH 0.85 sulfuric acid. Extract tin and molybdenum by shaking with 20 ml and 20 ml of chloroform. Evaporate the chloroform extracts to dryness and fuse the residue with 3 grams of potassium persulfate. Take up the residue in 50 ml of pH 0.85 sulfuric acid and add 25 ml of 4% solution of oxine in pH 0.85 sulfuric acid. Complete as for uranium oxide from "Adjust to pH 0.85 ± 0.01...."

Oxide Inclusions in Copper-Tin Alloys.[48] Reflux 5 grams of sample with 20 ml of bromine in 200 ml of methanol at 60–65° for 2 hours. Filter the residue of stannic oxide and wash with methanol, then with water. Ignite in a nickel crucible, and fuse the residue with 2 grams of sodium peroxide. Dissolve the melt in 1:1 hydrochloric acid, and dilute to 100 ml with that acid.

Adjust a 10-ml aliquot to pH 0.85 with ammonium hydroxide. Complete as for tin in uranium oxide from "Add 5 ml of a 20% solution...."

Mixed Antimony and Tin Oxides.[49a] Dissolve 0.2 gram of sample expected to contain 65–99% tin in 5 ml of 1:1 sulfuric acid containing 2 grams of ammonium sulfate. Dilute to 250 ml and dilute a 25-ml aliquot to 250 ml. Take an aliquot containing 0.1–0.5 mg of tin and add 40 ml of 1:260 sulfuric acid, 5 ml of 20% ammonium chloride solution, and 25 ml of 4% solution of oxine in 1:260 sulfuric acid. Extract with 20 ml and 20 ml of chloroform. Dilute the combined extracts to 50 ml and read at 385 nm.

---

[48]Iwao Tsukahara, Toshimi Yamamoto, and Takashi Tonomura, *ibid.* **18**, 1229–1236 (1969).
[49a]E. V. Lapitskaya, F. P. Gorbenco, and E. I. Vel'shtein, *Zavod. Lab.* **34**, 1446–1447 (1968).

o-HYDROXYQUINOLPHTHALEIN

This reagent forms a 1:2 complex with stannic ion which can be read photometrically or fluorimetrically.

Procedures [49b]

By Photometry. Mix a sample solution containing not more than 16 micrograms of stannic ion with 0.5 ml of 1% gelatin solution and 1 ml of m*M* *o*-hydroxyquinolphthalein. Adjust to pH 0.8–1.8 with 3% sulfuric acid, dilute to 10 ml, and read at 515 nm against a reagent blank.

By Fluorometry. Mix a sample solution containing up to 0.55 μg of stannic ion with 0.5 ml of 0.5% solution of gelatin and 0.5 ml of m*M* reagent. Adjust to pH 2–2.8 with 3% sulfuric acid and dilute to 10 ml. Read the fluorescence at 520 nm against the reagent. The fluorescence is increased slightly by zirconium, antimony, bismuth, ferric ion, and molybdate.

LUMOGALLION

Lumogallion, which is 5-chloro-3-(2,4-dihydroxyphenylazo)-2-hydroxybenzenesulfonic acid, forms a 1:1 complex with stannic ion.[50] The optimum medium for forming the complex is 0.1 *N* nitric acid. At pH 5, extraction is possible only by salting out with sodium nitrite. Oxalic acid, EDTA, and fluoride interfere, and to a lesser extent, so do tartaric and citric acids. Thiourea does not interfere and can be used to mask lead, bismuth, and copper.

Procedure. *Indium.* Dissolve 0.2 gram of sample in 8 ml of 1:14 nitric acid and a few drops of hydrochloric acid. Boil off oxides of nitrogen and cool. Add 2 ml of 10% thiourea solution, 6 ml of 4% sodium hydroxide solution, and 1.2 ml of m*M* lumogallion. Add 1:3 nitric acid until the red changes to orange. Dilute to 25 ml with 0.1 *N* nitric acid. Let stand for 10 minutes, extract with 2 ml of butanol, and read at 510 nm.

MORIN

The yellow complex of morin with stannic ion can be read photometrically or by fluorescence in ultraviolet light. The fluorescence is most intense in 0.04–0.06 *N* hydrochloric acid.[51] For photometric reading, ferric ion, antimony, molybdenum,

[49b] Itsuo Mori, Yoshikazu Fujita, and Takehisa Enoki, *Jap. Anal.* **25**, 388–392 (1976).

[50] P. V. Marchenko and N. V. Obolonchik, *Zh. Anal. Khim.* **22**, 725–730 (1967); *Tr. Kom. Anal. Khim.* **16**, 41–46 (1968).

[51] L. B. Ginzburg and E. P. Shkrobot, *Zavod. Lab.* **23**, 527–533 (1957).

tungsten, aluminum, and fluoride must be absent or masked. At pH 2 it is read at 415 nm for 0.04–3.6 μg/ml.[52a]

For lead alloys extract the tin as the iodide with benzene from the sample solution.[52b] Back-extract the tin with 0.05 *N* hydrochloric acid. Develop the tin with a mixed reagent of morin and antipyrine in chloroform and ethanol. Read the fluorescence at 436 nm for more than 5 μg per 10 ml of the hydrochloric acid extract.

Procedure. *Foodstuffs.*[53] Digest 10 grams of sample with 20 ml of 1:1 sulfuric acid and 5 ml of nitric acid. According to the sample, it may be necessary to add from time to time more nitric acid, some perchloric acid, or 30% hydrogen peroxide. When the digestate is colorless, evaporate to copious sulfur dioxide fumes. Cool and dilute to 50 ml. Neutralize a 10-ml aliquot with 40% sodium hydroxide solution; then render just acid with 1:3 hydrochloric acid. Add 2 ml of 0.2% ethanolic solution of morin and dilute to 50 ml with ethanol. Read at 420 nm.

6-NITRO-2-NAPHTHYLAMINE-8-SULFONIC ACID

The fluorescence of stannous ion with the ammonium salt of the reagent above is both sensitive and accurate.[54] Uranate, vanadate, ferrous, and bivalent titanium ions fluoresce strongly. The curve is linear up to 2 μg of tin per ml. An atmosphere of carbon dioxide is essential to prevent stannous ion from going to stannic.

For this purpose, the apparatus in Figure 10-4 was employed. It consisted of a 150 ml Erlenmeyer flask *D* with a B_{19} Quickfit neck into which fitted a small funnel *E*, with a side arm to allow carbon dioxide to be passed into the flask. The carbon dioxide was generated from dry ice in a thermos flask *B*. The gas pressure was adjusted by a pressure regulator valve *A*, consisting of a glass cylinder with a constricted neck, which was filled with water to about 5 cm from the top. A piece of glass tubing reaching almost to the bottom of the cylinder was connected to the thermos flask by means of a well-fitting rubber stopper. The carbon dioxide from the thermos flask passed through an empty bubble counter *C*, which was connected to the side arm of the funnel.

Procedure. *Copper-Base Alloy.* To prepare the reagent, dissolve 33 grams of 2-aminonaphthalene-8-sulfonic acid in 450 ml of sulfuric acid by stirring it in as small portions. Add 12.9 grams of urea to prevent formation of nitrous acid during the nitration. Chill to −20° with ice and salt externally, and by adding dry ice to the solution. Add a mixture of 11.6 ml of nitric acid and 18 ml of sulfuric acid from a dropping funnel over a period of 2 hours while the solution is kept at −20° and

[52a]K. P. Stolyarov, V. G. Pogodaeva, and N. E. Kuz'minova, *Vest. Leningr. Gos. Univ., Ser. Fiz. Khim.* **1968** [2 (10)], 133–136.

[52b]A. T. Pilipenko, S. L. Lisichenok and A. I. Volkova, *Ukr. Khim. Zh.* **42**, 976–981 (1976).

[53]Bogdan Fitak and Teresa Zaklika, *Rocz. Panstw. Zakl. Hig.* **24**, 627–633 (1973).

[54]J. R. A. Anderson and S. Lenzer Lowy, *Anal. Chim. Acta* **15**, 246–253 (1956).

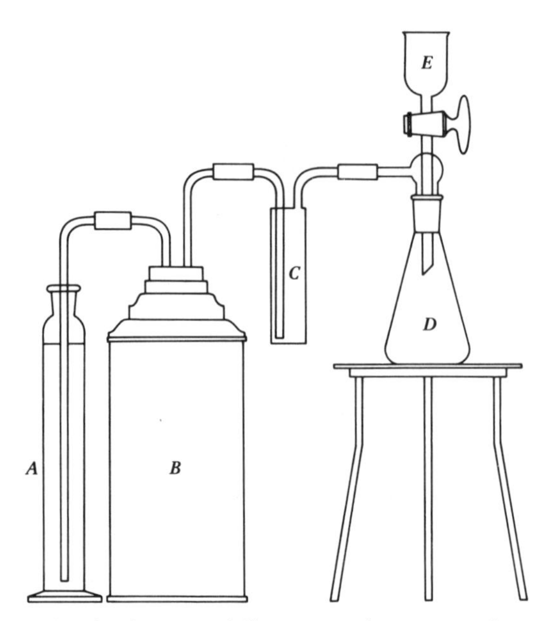

Figure 10-4 Apparatus for development of fluorescence by stannous ion and 6-nitro-2-naphthylamine-8-sulfonic acid.

slowly agitated mechanically. After another hour at −20° pour onto 1 kg of crushed ice. After 1 hour, filter the pale yellow precipitate and wash free from sulfuric acid with ice water. Mix with ammonium hydroxide until just alkaline and cool in the refrigerator. Decolorize with activated carbon and recrystallize from water.

Dissolve a sample containing about 10 mg of tin in 5 ml of nitric acid. Evaporate almost to dryness but do not bake. Add 30 ml of 1 : 5 nitric acid and incubate near 100° for 1 hour. Centrifuge at 90° and pour off the supernatant liquid. Wash the metastannic acid with 5 ml, 5 ml, and 5 ml of hot 1% ammonium nitrate solution. Add 1 gram of tartaric acid and 10 ml of 5 *N* hydrochloric acid. Insert into *D* of the equipment shown in Figure 10-4. Pass in carbon dioxide and heat gently until the solution clears. Cool, and add 1 ml of saturated mercuric chloride solution and 1 ml of 30% solution of hypophosphorous acid. Warm until the precipitate coagulates. Add 30 ml of 5 *N* ammonium hydroxide and cool. Add 10 ml of a 1% solution of the reagent and let stand for 2 hours to develop the maximum fluorescence. Dilute to 1 liter. Mix a 10-ml aliquot with 10 ml of 5 *N* ammonium hydroxide and dilute to 100 ml. Read the fluorescence, activating at 365 nm.

3-NITRO-4-HYDROXYBENZENEARSONIC ACID

The turbidimetric product of reaction of this reagent with tin can be read photometrically—for example, with a blue filter. Typically it is applied to copper containing 0.001–0.005% tin.[55]

[55]C. M. Dozinel and H. Gill, *Chemist-Analyst* **45**, 105, 109 (1956).

Procedure. *Copper.* Dissolve 1 gram of the reagent in 15 ml of methanol, dilute to 50 ml with water, and filter. Heat 1 gram of sample with 10 ml of hydrochloric acid and cautiously add 10 ml of 30% hydrogen peroxide. Boil off the excess of hydrogen peroxide, add 1 ml of sulfuric acid, and evaporate to sulfur trioxide fumes. Take up in 10 ml of warm water, add an additional 20 ml of water, and cool. Add 10 ml of the reagent solution and dilute to 50 ml. Let stand for 2 hours and read against water.

6-NITRO-2-NAPHTHYLAMINE-8-SULFONATE

This reagent as the ammonium salt gives fluorescence in the presence of stannous ion.[56] There is interference by uranyl, vanadyl, ferrous, and titanous ions. The fluorescence is linear up to 0.2 mg of tin per 100 ml.

Procedure. *Copper-Base Alloys.* Dissolve a sample expected to contain about 10 mg of tin in 5 ml of nitric acid in a centrifuge tube. Evaporate almost to dryness; to go to dryness adversely affects the solubility of metastannic acid. Add 5 ml of nitric acid and 25 ml of hot water. Maintain at 100° for 1 hour. Centrifuge at above 90° for 10 minutes. Decant the supernatant layer and wash the residue with 5 ml, 5 ml, and 5 ml of hot 1% ammonium nitrate solution. Add 1 gram of tartaric acid and 10 ml of 5 *N* hydrochloric acid to the metastannic acid. Pass carbon dioxide into the tube to blanket the solution, and heat gently until solution is complete. Add 1 ml of 1% mercuric chloride solution and 1 ml of 30% hypophosphorous acid solution. Warm until the black precipitate coagulates. Add 30 ml of 1:2 ammonium hydroxide, cool, and add 10 ml of a 1% solution of the captioned reagent. Let stand for 2 hours to develop the maximum fluorescence and dilute to 1 liter. Dilute a 10-ml aliquot and 10 ml of 1:2 ammonium hydroxide to 100 ml and read the fluorescence.

p-NITROPHENYLFLUORONE

This reagent is applicable to ores containing 0.001–0.002% tin.[57]

Procedures

Low-Silica Tin Ores. Fuse 1 gram of sample with 3 grams of sodium oxide in a nickel crucible. Take up the melt in water, adding 20 ml of 1:2 sulfuric acid and 1 ml of 30% hydrogen peroxide. When dissolved, add 1 ml of a solution containing 0.05 mg of iron and dilute to 130 ml. While mixing, add ammonium hydroxide

[56] J. R. A. Anderson and S. Lenzer Lowy, *Anal. Chim. Acta* **15**, 246–253 (1956).
[57] V. A. Nazarenko and N. V. Lebedeva, *Zavod. Lab.* **28**, 268–271 (1962).

until a precipitate forms, and add 2 ml excess. Filter. Dissolve the precipitate in 6 ml of 1:1 sulfuric acid and wash the filter well with water. Add 1 ml of 30% hydrogen peroxide to the solution and washings. Reprecipitate and redissolve from the filter as before. Boil for 10 minutes, then add 5 ml of 5% ascorbic acid solution and 3 ml of 10% mercaptoacetic acid solution. Dilute to 80 ml, heat at 100° for 30 minutes, and cool. Extract the tin with four 10-ml portions of 0.02% solution of diethyldithiocarbamic acid in chloroform. As a wash solution, mix 6 ml of 1:1 sulfuric acid, 5 ml of 5% ascorbic acid solution, 3 ml of 10% mercaptoacetic acid solution, and 68 ml of water. Wash the combined organic extracts with 10 ml, 10 ml, and 10 ml of this wash solution. Reextract the tin with 10 ml and 10 ml of a solution containing 5 ml of 2% potassium permanganate solution, 1.4 ml of 1:1 sulfuric acid, and 13.6 ml of water. To the combined aqueous extracts, add 4 ml of 5% ascorbic acid solution, 0.2 ml of 30% hydrogen peroxide, 8 ml of ethanol, 2 ml of 1% solution of gelatin, and 1.5 ml of 0.05% solution of *p*-nitrophenylfluorone, mixing after each addition. Dilute to 50 ml, let stand for 1 hour, and read at 530 nm.

High-Silica Tin Ores. Treat a 1-gram sample in platinum with 3 ml of 1:1 sulfuric acid and 5 ml of hydrofluoric acid. Heat to sulfur trioxide fumes and cool. Add 3 ml of water and again heat to sulfur trioxide fumes. Take up in 5 ml of water, filter, and ignite the paper and residue in a nickel crucible. Fuse the residue with 1 gram of sodium oxide, take up the melt in 5 ml of water, and add to the filtrate. Dilute to 100 ml. Add 1 ml of 30% hydrogen peroxide and 1 ml of a solution containing 0.05 mg of iron. Complete as for low silica samples from "While mixing, add ammonium hydroxide...."

PHENYLFLUORONE

Phenylfluorone, which is 2,6,7-trihydroxy-9-phenylisoxanthene-3-one as applied to tin in copper- or lead-base alloys, has no interferences.[58] The product is a 1:20 colloidal complex and requires the presence of sulfuric acid for its formation. Antimony, germanium, and iron interfere.[59] Addition of tartaric acid improves the reproducibility. The absorption is markedly affected by the time of standing and increases with temperature.

For determination in a 6–8 *N* hydrochloric acid solution of lead, the tin is extracted with a solution of Amberlite LA-2 in xylene.[60] The tin is then reextracted into 0.5 *N* nitric acid and evaporated; organic matter is decomposed, and tin is determined with phenylfluorone. There is no interference in this technic by up to 0.5% antimony in the sample. Tin in lead and cable-sheathing alloy is isolated by coprecipitation, extracted as the carbamate in chloroform, and determined with phenylfluorone.[61]

[58]C. L. Luke, *Anal. Chem.* **31**, 1803–1804 (1959).
[59]Masayoshi Ishibashi and Tsunenobu Shigematsu, *Jap. Anal.* **7**, 473–477 (1958); cf. J. D. Smith, *Analyst* **95**, 347–350 (1970).
[60]A. Hofer and B. Landl, *Z. Anal. Chem.* **244**, 103–105 (1969).
[61]A. Schottak and H. Schweiger, *Z. Erzberg. Metalluhütt.* **19** (4), 180–185 (1966).

For determination of tin in the solution of a silver alloy, coprecipitate tin and antimony with manganese dioxide.[62a] Dissolve the precipitate with nitric and sulfuric acids. Extract the antimony with sodium diethyldithiocarbamate solution; then determine tin with phenylfluorone.

By extraction with ethyl acetate from sulfuric acid solution in the presence of ammonium thiocyanate, tin is separated from antimony, bismuth, germanium, and titanium.[62b] Molybdenum interferes. EDTA permits extraction of molybdenum in the presence of ammonium thiocyanate while suppressing the extraction of tin. The ethyl acetate extract is destroyed in the presence of sulfuric acid, and the tin is determined thereafter with phenylfluorone.

For determination in salt solution or in seawater, stannic ion is coprecipitated with about 1 mg of iron as ferric hydroxide at pH 6–8.[63] Molybdenum is not precipitated. The solution of the precipitate in hydrochloric acid is passed through a column of Dowex 1-X8 resin. Elution with 0.5 *N* hydrochloric acid removes up to 1 mg of germanium, titanium, zirconium, antimony, selenium, and iron. Elution with 30 ml of 4% solution of sodium hydroxide recovers the tin with any mercury or bismuth that was present. When this eluate is made 4–7 *N* with hydrochloric acid and 0.5 *N* with sulfuric acid, phenylfluorone is added and the tin complex is extracted with methyl isobutyl ketone for reading at 515 nm.

For tin in canned foods, decompose 0.5 gram of sample with sulfuric acid and hydrogen peroxide.[64] Add perchloric acid to complete digestion and dilute. Extract tin as the cupferron complex with chloroform. Evaporate and oxidize the organic matter with sulfuric acid and hydrogen peroxide. Evaporate excess hydrogen peroxide, take up with water, and add hydroxylamine sulfate. Dilute with sulfurous acid solution and develop with phenylfluorone.

Biological material such as feeds and plants is dry-ashed, then dissolved by hydrochloric and hydrofluoric acids; tin is collected by precipitation of ferric hydroxide and the precipitate is dissolved in hydrochloric acid.[65] After addition of thiocyanate and ascorbic acid, an ethyl acetate extract is evaporated to dryness. The residue in sulfuric acid oxidized with hydrogen peroxide is developed with phenylfluorone using sodium lauryl sulfate as stabilizer.

Procedures

Separation by Oxine.[66] Adjust a chloride-free sample containing up to 15 μg of tin to about 40 ml at pH 0.85 ± 0.05 by addition of sulfuric acid. Add 25 ml of 4% solution of oxine in pH 0.85 sulfuric acid. Shake for 2 minutes with 15 ml of chloroform and discard this extract, which contains molybdenum, the only element interfering with tin in this determination. Add 5 ml of 20% solution of ammonium chloride in pH 0.85 sulfuric acid. Extract the tin with 15 ml and 15 ml of

[62a] J. Chwastowska and Z. Skorko-Trybula, *Chem. Anal.* (Warsaw) **9**, 123–130 (1964).

[62b] Masao Tanaka and Humiki Morikawa, *Jap. Anal.* **10**, 396–400 (1961).

[63] Kazuo Shimizu and Noboru Ogata, *ibid.* **12**, 526–531 (1963).

[64] M. Glathe, *Chem. Mikro-biol., Technol. Lebensm.* **3**, 125–128 (1974).

[65] W. Oelschläger, *Z. Anal. Chem.* **174**, 241–254 (1960).

[66] Jacques Kagan and T. J. Mabry, *Anal. Chem.* **37**, 288–289 (1965); E. N. Pollock and L. P. Zopotti, *ibid.* 290–291.

chloroform. Add the chloroform extracts to 15 ml of 1:9 sulfuric acid and evaporate to sulfur trioxide fumes. If necessary to remove residual organic matter, add a few drops of nitric acid and 30% hydrogen peroxide. Take up in 15 ml of water and transfer with no more than 10 ml of a buffer solution containing 2.25% of sodium acetate trihydrate and 1.2% of acetic acid. Add 1 ml of 1% gelatin solution. Add 10 ml of 0.01% solution of phenylfluorone in 1:99 hydrochloric acid. Dilute to 50 ml and read at 510 nm in 30 ± 5 minutes against water.

Separation by Acetylacetone. Adjust a sample solution containing up to 15 μg of tin to contain 1 ml of 1:3 sulfuric acid and dilute to 50 ml. Add 0.5 ml of 30% hydrogen peroxide. Extract by 2 minutes of shaking with 25 ml, 25 ml, and 25 ml of 1:1 acetylacetone-chloroform. Discard the extracts. Add 15 ml of 1:9 sulfuric acid to the aqueous layer and evaporate to sulfur trioxide fumes. Complete as for oxine separation from "If necessary to remove residual...."

Separation by Ion Exchange.[67] Pretreat a strong base, ion exchange resin with 1:1 hydrochloric acid, then with 24% sodium hydroxide solution. Finally convert to the acetate form with 5% acetic acid. Prepare a 55×8 mm column of the resin.

The technic described here was developed for tin on synthetic hydroxyapatite and on tooth surfaces, but it should be modifiable for broader application. Dissolve a sample in 4 ml of 1:1 hydrochloric acid and 1 ml of 3% hydrogen peroxide. Dilute to 10 ml and take an aliquot containing 4–40 μg of tin. Add to the column adjusted to flow at 1.4 ml/minute. Wash the column with 18 ml of 1:1 hydrochloric acid to remove phosphate and fluoride. Elute the tin with 12 ml of 24% sodium hydroxide solution. Chill the eluate with ice and add 7.2 ml of hydrochloric acid. Dilute to 50 ml with 1:11 hydrochloric acid. Take a 10-ml aliquot and adjust to pH 3.5 ± 0.1 with an acetate buffer solution. Dilute to 20 ml and add 5 ml of phenylfluorone solution containing 25 μg/ml. Let stand for 10 minutes and read at 530 nm.

Extraction with N-Phenylbenzohydroxamic Acid. STEEL.[68] Dissolve 0.5 gram of sample in 10 ml of 1:1 perchloric acid and 2 ml of 1:1 hydrochloric acid. Cool and filter. Ash the residue, treat with hydrofluoric acid and perchloric acid, and add to the filtrate. Make the filtrate to 15 ml, which should be 6 *M* with perchloric acid. Add 4 ml of 30% hydrogen peroxide to prevent interference by titanium. Set aside for 10 minutes; then add 1 ml of 1% solution of phenylfluorone in 50% acetic acid. Extract the zinc complex with 10 ml and 10 ml of chloroform. Filter and back-extract with 10 ml of 8 *N* hydrochloric acid, followed by 5 ml of water. Add 1 ml of sulfuric acid and evaporate to sulfur trioxide fumes. Take up the residue with 3.8 ml of perchloric acid, 15 ml of 6 *M* sodium perchlorate, 5 ml of 44.8% solution of sodium iodide, and 5.4 ml of water. Extract the tin complex with 5 ml of benzene. Wash the extract with a solution 1.5 *M* with perchloric acid, 0.5 *M* with sodium iodide, and 3 *M* with sodium perchlorate. Back-extract the tin with 5 ml of 0.5 *N* sulfuric acid. Add to 30 ml of phthalate–hydrochloric acid buffer solution for

[67] R. L. Speirs, *J. Dent. Res.* **41**, 909 (1962).
[68] S. Mareva, N. Iordanev, and S. Kadieva, *Zavod. Lab.* **41**, 660–661 (1975).

pH 2.6. Add 1 ml of 1% gum acacia solution and 10 ml of 0.01% methanolic phenylfluorone. Dilute to 50 ml and read at 510 nm against water.

Cast Iron.[69] Disintegrate a 0.5-gram sample with 20 ml of 1:3 nitric acid. Filter to separate the graphite, and wash. Add 5 ml of 5% solution of manganese sulfate and dilute to about 200 ml with 1.2 *N* nitric acid. Boil, slowly add 2.5 ml of 4% potassium permanganate solution, and continue to boil for 5 minutes. Let stand for 20 minutes, filter, and wash with 1:60 sulfuric acid. Dissolve in 10 ml of 2:5 sulfuric acid by adding as much 30% hydrogen peroxide as may be required. Add 20 ml of 5% solution of beryllium sulfate and 20 ml of 0.1 *M* EDTA. Dilute to about 200 ml and add a drop of phenol red indicator. Warm to 70–80°, neutralize with 1:9 ammonium hydroxide, and add 2 ml excess. Filter after 30 minutes while still hot. Wash the precipitate with 50 ml of 1% ammonium sulfate solution. Dissolve in 11 ml of hot 2:5 sulfuric acid, add 3 ml of 30% hydrogen peroxide, and evaporate to sulfur trioxide fumes. Add 3 ml of water and again evaporate to sulfur trioxide fumes. Take up the residue in water and dilute to 250 ml. To a 25-ml aliquot add 30 ml of a buffer solution containing 5 grams of potassium acid phthalate and 18 ml of hydrochloric acid per liter, 2 ml of 0.5% gelatin solution, and 5 ml of 0.4% phenylfluorone solution. Dilute to 100 ml and after 30 minutes read at 490 nm.

Steel.[70] Dissolve a sample containing 1–15 μg of tin in 4 ml of hydrochloric acid and 1 ml of nitric acid. Add 2 ml of sulfuric acid, followed by sufficient formic acid to reduce residual nitric acid. Evaporate to about 3 ml, thereby driving off most of the hydrochloric acid. Add 10 ml of 1:3 sulfuric acid, and heat to dissolve all salts. Cool, and add 32 ml of 1:3 sulfuric acid. Add 5 ml of 5 *M* sodium iodide. Add sulfurous acid in small portions until most of the iodine has been reduced, let stand a couple of minutes, and add further 0.25-ml portions of sulfurous acid until all iodine has disappeared. Extract stannous iodide with 10 ml of benzene. Wash down the inside of the mouth of the separatory funnel with 5 ml of a fresh mixture of 10 ml of 1:3 sulfuric acid and 1 ml of 5 *M* sodium iodide solution. Again shake for 10 seconds and discard the aqueous layer. Wash the benzene extract with another 5 ml of the sulfuric acid–sodium iodide solution. Transfer the benzene extract with the aid of 0.5 ml of nitric acid, followed by 4 ml of benzene. Evaporate the benzene and iodine at low heat with an air jet. Add 4 drops of perchloric acid with 4 drops of sulfuric acid and heat to destroy any residual organic matter. Evaporate to dryness and take up in 1 ml of 1:3 hydrochloric acid. Add 20 ml of water, 1 ml of 1% solution of ammonium oxalate, 0.5 ml of 30% hydrogen peroxide, and 1 gram of sodium chloride. Add 5 ml of 0.01% solution of phenylfluorone in methanol containing 0.2 ml of hydrochloric acid per 100 ml. Let stand for 10 minutes, and extract with 10 ml of methyl isobutyl ketone. Filter the organic layer and read at 530 nm.

Alternatively,[71] dissolve 1 gram of sample in 75 ml of hot 1:4 sulfuric acid,

[69] A. A. Amsheev, *ibid.* **34**, 789–790 (1968).
[70] C. L. Luke, *Anal. Chim. Acta* **37**, 97–101 (1967); **39**, 404–406 (1967).
[71] G. Picasso and A. Pizzimenti, *Met. Ital.* **55** (8), 361–364 (1963).

adding a few drops of 30% hydrogen peroxide, and dilute to 100 ml. Make additions to a 10-ml aliquot as follows: 10 ml of 5% ascorbic acid solution, 10 ml of a sodium acetate–acetic acid buffer solution for pH 5, 5 ml of 1% gum acacia solution, and 10 ml of 0.01% solution of phenylfluorone in methanol containing 0.2 ml of hydrochloric acid per 100 ml. Dilute to 50 ml, let stand for 10 minutes, and read at 510 nm against a blank containing the same amount of iron. There is no interference by less than 3% manganese, copper, chromium, or titanium, but more than 0.1% molybdenum interferes.

Ferrous Metals.[72] Dissolve 1 gram of sample with 10 ml of 1:4 sulfuric acid and 5 ml of 1:1 nitric acid. Evaporate to sulfur trioxide fumes and take up in 15 ml of water. Filter and wash. Add 1 ml of sulfuric acid and dilute to about 25 ml. Add 7.5 grams of sodium iodide and 20 ml of perchloric acid. Dilute to 50 ml and extract with 10 ml and 5 ml of benzene. Extract the tin from the combined benzene extracts with 5 ml and 5 ml of 1:72 sulfuric acid. Add 5 ml of 5% ascorbic acid solution, 10 ml of acetate buffer solution for pH 5, 1 ml of 0.5% gelatin solution, and 10 ml of 0.01% solution of phenylfluorone in methanol containing 0.2 ml of hydrochloric acid per 100 ml. Dilute to 50 ml and after 30 minutes read at 508 nm.

Ferrotungsten.[73] Fuse 0.5 gram of powdered sample in an iron crucible with 2 grams of sodium carbonate and 9 grams of sodium peroxide. Dissolve the melt in 150 ml of water, add 5 ml of ethanol, and boil with stirring for 5 minutes. Dilute to 250 ml and filter, discarding the first 25 ml of filtrate. Take the next 25 ml as an aliquot and add 5 ml of 10% tartaric acid solution. Add a drop of phenolphthalein indicator and neutralize with 10 *N* sulfuric acid. Add 0.5 ml excess of the acid and boil. Cool, neutralize with 10% sodium hydroxide solution, and make just acid with a drop of 10 *N* sulfuric acid. Add 10 ml of acetate buffer solution for pH 5.5 and 5 ml of 1% solution of sodium diethyldithiocarbamate. After 5 minutes extract for 1 minute with 25 ml of chloroform. Add 2 ml of the carbamate solution and extract with 10 ml of chloroform. Add to the combined chloroform layers 5 ml of 10 *N* sulfuric acid, 8 ml of water, and 2 ml of 2% potassium permanganate solution. Shake for 1 minute, add 2 ml of 3% hydrogen peroxide, and shake until colorless. Add 5 ml of 10% solution of tartaric acid, shake, and discard the chloroform layer. Filter and wash the filter with 5 ml and 5 ml of water. Add to the filtrate and washings 5 ml of 0.5% gelatin solution and 10 ml of 0.03% solution of phenylfluorone. Dilute to a known volume and set aside for 20 minutes. Read at 510 nm against a reagent blank.

Ferromolybdenum.[74] As a buffer solution, dissolve 20 grams of sodium acetate in 70 ml of hot water and filter. Add 48 ml of 70% acetic acid and dilute to 200 ml. As color reagent, dissolve 75 mg of phenylfluorone in 100 ml of ethanol.

Add 10 mg of antimony to a 1-gram sample and decompose with 50 ml of 1:3 nitric acid. Add 20 ml of water and some filter pulp. Let stand for 1 hour and filter

[72]A. M. Leblond and R. Boulin, *Chim. Anal.* (Paris) **50** (4), 171–177 (1968); cf. A. M. Dymov, I. G. Ivanov, and T. I. Romantseva, *Zh. Anal. Khim.* **26**, 2360–2363 (1971).
[73]J. Jankovsky, *Hutn. Listy* **18**, 276–278 (1963).
[74]E. V. Silaeva and V. I. Kurbatova, *Zavod. Lab.* **27**, 1462–1464 (1961).

Wash thoroughly with a 1% solution of ammonium nitrate in 1:99 nitric acid. Return the residue to the original beaker, add 10 ml of sulfuric acid and 20 ml of nitric acid, and heat to sulfur trioxide fumes. Take up in 1:9 sulfuric acid and dilute to 100 ml with that acid. To a 10-ml aliquot, add 5% potassium permanganate solution dropwise to a distinct pink color. Add an equal amount of 5% sodium nitrate solution. Neutralize to phenolphthalein with ammonium hydroxide and add 10 ml of 1:9 sulfuric acid. Cool and add 2 ml of 3% hydrogen peroxide. Add 10 ml of buffer solution, 3 ml of the color reagent, and 1 ml of 0.5% gelatin solution. After 5 minutes dilute to 50 ml with 1:9 sulfuric acid. Let stand for 40 minutes and read at 500 nm.

Copper-Base Alloys. Dissolve a 0.2-gram sample in 2 ml of nitric acid. Add 0.5 ml of perchloric acid and 17 ml of sulfuric acid. Heat to expel volatile acids and redissolve any lead sulfate. Fume vigorously, reducing the volume to about 15 ml. Cool, add 15 ml of water, and heat to boiling. Cool again, and dilute to about 98 ml. Finally, dilute to 100 ml and allow about 10 minutes for lead sulfate to settle if present.

Dilute a 5-ml aliquot to 100 ml with 15% sulfuric acid. To 10 ml of this, add 10 ml of a buffer solution for pH 5. Add 1 ml of 1% gum arabic solution. Add 10 ml of 0.01% solution of phenylfluorone in 1:99 hydrochloric acid–methanol and dilute to 49 ml. Cool to 25° for 8 minutes, dilute to 50 ml, and read at 510 nm against water 10 ± 0.5 minutes after adding reagent.

Lead-Base Alloys. Dissolve a 0.2-gram sample in 2 ml of 5:1 perchloric-nitric acid. Add 17 ml of sulfuric acid. Complete as for copper base alloys from "Heat to expel volatile acids...."

Heat-Resistant and Copper-Base Alloys.[75] Dissolve 2 grams of sample in aqua regia and evaporate to a small volume. Add 5 ml of hydrochloric acid and evaporate to a small volume. Repeat this step three times more. Take up in 15 ml of 3 *N* hydrochloric acid and filter. Prepare a 1.2×20 cm column of an anionic ion exchange resin in chloride form and wash well with 3 *N* hydrochloric acid. Pass the filtrate through this at 0.5 ml/minute. Wash with 60 ml of 3 *N* hydrochloric acid at 1 ml/minute to remove aluminum, copper, iron, nickel, and chromium. Then elute the tin with 150 ml of 0.5 *N* hydrochloric acid at 1 ml/minute. Evaporate the eluate nearly to dryness, add 10 ml of hydrochloric acid, and dilute to 100 ml.

Oxidize a 10-ml aliquot with 2 drops of 2.5% potassium permanganate solution. Stir for 3 minutes and add a drop of 5% sodium nitrite solution followed by 0.5 ml of 10% urea solution. Neutralize to phenolphthalein with ammonium hydroxide. Acidify with 1:9 sulfuric acid and add 10 ml excess. Cool and add 2 ml of 3% hydrogen peroxide and 10 ml of acetate buffer solution for pH 5. Add 3 ml of a phenylfluorone solution containing 75 mg of reagent per 250 ml of ethanol, to which 1 ml of 1:1 sulfuric acid was added. Add 1 ml of 0.5% gelatin solution and let stand for 5 minutes. Dilute to 50 ml with 1:9 hydrochloric acid and let stand for 60 minutes. Read at 530 nm.

[75]V. I. Kurbatova and V. V. Stepin, *Tr. Vses. Nauchn.-Issled. Inst. Stand. Obraztsov. Spektr. Etalonov* **1**, 14–18 (1964).

Zircaloy.[76] Dissolve 0.1 gram oi sample in sulfuric and hydrofluoric acids, adding nitric acid if necessary. Evaporate to sulfur trioxide fumes and adjust to 100 ml in 15% sulfuric acid. Dilute a 2-ml aliquot to 10 ml with 15% sulfuric acid. Add sequentially, 3 ml of sodium fluoride solution containing 1 mg of fluoride per ml, 10 ml of saturated solution of boric acid, 10 ml of sodium acetate–acetic acid solution for pH 5, 1 ml of 0.2% gelatin solution, and 10 ml of 0.01% ethanolic phenylfluorone. Set aside for 1 hour and read at 510 nm.

Lead and Antimony-Lead Alloys.[77] As a buffer solution, dissolve 900 grams of sodium acetate trihydrate or 540 grams of the anhydrous salt in about 700 ml of water. Filter, add 480 ml of acetic acid, and dilute to 2 liters. As the color reagent, dissolve 0.05 gram of phenylfluorone in 50 ml of methanol and 1 ml of hydrochloric acid. Dilute to 500 ml with methanol.

Dissolve 1 gram of sample in 2.5 ml of perchloric acid and 0.5 ml of nitric acid. Add 50 ml of water and 1 ml of 1% solution of potassium permanganate. Boil gently and add 1 ml of 0.4% manganese nitrate solution. Boil for 1 or 2 minutes to coagulate the precipitate. Filter, wash well, and discard the filtrate. Return the paper and precipitate to the original container. Add 10 ml of sulfuric acid and 10 ml of nitric acid. Digest until the organic matter is destroyed. Add 1 ml of perchloric acid and evaporate to 7 ml to drive off nitric and perchloric acids. Cool somewhat and add about 0.3 gram of hydrazine sulfate. Heat carefully until foaming subsides. Evaporate to 5 ml to destroy all hydrazine sulfate and expel sulfur dioxide. Cool, add 5 ml of water, and heat to boiling. Digest for a couple of minutes to complete solution of iron or chromium sulfates. Boil once more and cool. Add 30 ml of water. Filter, wash with 10 ml of water, and discard the paper and precipitate. Extract with 25 ml of 1% solution of diethyldithiocarbamate in chloroform for 30 seconds and discard the chloroform layer. Wash the aqueous layer with 5 ml and 5 ml of chloroform.

Heat the aqueous layer to 50° and add 2 ml of 10% mercaptoacetic acid solution, followed by 1 ml of a solution containing 2.5% of ascorbic acid in 15% potassium iodide solution. Let stand for 10 minutes to reduce the tin. Cool to room temperature and add 10 ml of 1% solution of diethyldithiocarbamate in chloroform. Shake for 30 seconds and separate the chloroform extract. Further extract with 5 ml of chloroform, combine the extracts, and discard the aqueous layer.

Add 2 ml of sulfuric acid, 1 ml of nitric acid, and 0.5 ml of perchloric acid. Evaporate to sulfur trioxide fumes. Cool, add 0.25 ml of perchloric acid, and heat until all acids are expelled. Cool, add 5 ml of 1:4 sulfuric acid, and heat just to sulfur trioxide fumes. Cool, and add successively 9 ml of water, 1 ml of 3% hydrogen peroxide, 10 ml of the buffer solution, 1 ml of 1% solution of gum arabic, and 10 ml of color reagent. After 5 minutes add 16 ml of 1:9 hydrochloric acid and dilute to 50 ml with that acid. Read at 510 nm at once against water.

Nickel.[78] Dissolve 1 gram of sample containing 0.001–0.005% tin in 15 ml of

[76]D. Rajkovic, *Z. Anal. Chem.* **263**, 334 (1973).
[77]C. L. Luke, *Anal. Chem.* **28**, 1276–1279 (1956); cf. Roy L. Bennett and Hilton A. Smith, *ibid.* **31**, 1441–1442 (1959).
[78]A. Hofer, *Z. Anal. Chem.* **240**, 229–232 (1968).

1 : 1 nitric acid. Evaporate to dryness and ignite to decompose to nickel oxide. Dissolve in 50 ml of 1 : 1 hydrochloric acid. Extract the tin with 25 ml and 25 ml of 10% Amberlite LA-2 solution in xylene. Wash the combined extracts with 10 ml, 10 ml, and 10 ml of 1 : 1 hydrochloric acid. Reextract the tin from the xylene extract with 30 ml, 30 ml, and 30 ml of 0.5 *N* nitric acid. Wash the combined aqueous extracts with 25 ml of xylene. Add 0.5 ml of 1 : 1 sulfuric acid to the aqueous phase and evaporate to dryness. Add 1 ml of sulfuric acid and 1 ml of 30% hydrogen peroxide and evaporate to sulfur trioxide fumes. Warm the residue with 5 ml of 1 : 18 sulfuric acid, cool, and add ammonium hydroxide to pH 5. Add 1 ml of 1% gum acacia solution, 1 ml of 10% tartaric acid solution, 1.6 ml of 1 : 1 sulfuric acid, and 10 ml of a solution containing 10 mg of phenylfluorone and 0.2 ml of hydrochloric acid per 100 ml of methanol. Dilute to 50 ml, let stand for 30 minutes, and read at 530 nm against a reagent blank. Under these conditions, nickel and lead are not extracted.

Tin is also determined in nickel by dissolving in nitric acid, precipitation by anthraquinone arsonic acid, filtration, ignition, solution in hydrochloric acid, and development with phenylfluorone.[79]

Zinc.[80] Dissolve a 1-gram sample in 10 ml of 1 : 1 nitric acid and 0.5 ml of 50% tartaric acid solution. Evaporate to a syrup and take up in 10 ml of 0.5 *N* nitric acid. Add 0.25 ml of 1% solution of potassium permanganate and neutralize with ammonium hydroxide until turbidity appears. Add 10 ml of 0.5 *N* nitric acid, 4 ml of 2% cupferron solution, and 5 ml of chloroform. Shake for 1 minute and separate the chloroform extract. Repeat the addition of cupferron solution and extraction with chloroform twice more. To the combined chloroform extracts, add 5 ml of nitric acid and evaporate to disappearance of the chloroform. Add 5 ml of 1 : 1 sulfuric acid and evaporate to 0.5 ml. If the solution is not clear and colorless, heat with a few crystals of ammonium nitrate. Cool, add 5 ml of 5% tartaric acid solution, and boil for 2 minutes. Cool and neutralize to phenolphthalein with ammonium hydroxide. Add 1.6 ml of 1 : 1 sulfuric acid, 2.5 ml of 0.5% gelatin solution, and 2.5 ml of 0.03% solution of phenylfluorone in ethanol containing 2 ml of sulfuric acid per 100 ml. Dilute to 25 ml and let stand for 20 minutes. Read at 500 nm.

In analysis of zinc, the zinc can be extracted from the solution as the cupferronate. For analysis of tin ores and ore wastes, silica is removed by hydrofluoric acid and the residue is fused with sodium peroxide. Then tin is precipitated with ammonium hydroxide in the presence of EDTA. The sulfuric acid solution of this precipitate is suitable for determination by phenylfluorone.

Antimony.[81] Dissolve a 0.5-gram sample in 20 ml of sulfuric acid and evaporate nearly to dryness. Take up in 30 ml of 25% tartaric acid solution and add 1.5 ml of 1 : 1 sulfuric acid. Add 10 ml of 50% solution of ammonium thiocyanate and dilute to about 70 ml. Separate from antimony, bismuth, germanium, and tin by extrac-

[79] P. Ya. Yakovlev, G. P. Razumova, and M. S. Dymova, *Sb. Tr. Tsent. Nauch.-Issled. Inst. Chem, Met.* **1962** (24) 168–171.
[80] Eugeniusz Raczka and Henryk Suchy, *Rudy Met. Niezaz.* **6**, 274–277 (1961).
[81] Masao Tanaka and Humiki Morikawa, *Jap. Anal.* **10**, 396–400 (1961).

tion with 20 ml of ethyl acetate. Wash this solution of stannic thiocyanate with 30 ml of *M* ammonium thiocyanate in 0.5 *N* sulfuric acid. Add 5 ml of 1 : 1 sulfuric acid and 5 ml of nitric acid to the ethyl acetate layer and evaporate to sulfur trioxide fumes. Take up the residue in 5 ml of 1 : 1 hydrochloric acid and dilute to 50 ml. To an aliquot, add 5 ml of 10% citric acid solution neutralized with 1 : 2 ammonium hydroxide and 5 ml of 0.5% solution of polyvinyl alcohol. Dilute to 40 ml and add 5 ml of a 0.01% solution of phenylfluorone in ethanol to which has been added 2% by volume of 5 *N* hydrochloric acid. Dilute to 49 ml and set aside for 20 minutes. Mask traces of bismuth and antimony by adding 1 ml of 30% solution of EDTA and read at 500 nm.

Lead Antimonate.[82] As reagent, dissolve 0.03 gram of phenylfluorone in 100 ml of ethanol containing 1 ml of 1 : 1 sulfuric acid.

Dissolve 1 gram of sample with 15 ml of 1 : 1 nitric acid and 1 ml of 50% solution of tartaric acid. Dilute to 50 ml and take an aliquot containing less than 0.03 mg of tin. Add 0.2 ml of 30% hydrogen peroxide and evaporate to dryness at 100°. Take up the residue by boiling for 1 minute with 10 ml of 0.5 *N* nitric acid and cool. Dilute to about 30 ml with 0.5 *N* nitric acid and add 0.5 ml of 0.3% potassium permanganate solution. Add 3 ml of 1% solution of cupferron and extract the tin complex with 5 ml of chloroform. Repeat the addition of cupferron and the chloroform extraction twice more. Wash the combined extracts with 5 ml, 5 ml, and 5 ml of 0.2 *N* nitric acid. Evaporate the chloroform extract to about 2 ml, add 5 ml of nitric acid, and complete the removal of chloroform. Add 5 ml of 1 : 1 sulfuric acid and evaporate to sulfur trioxide fumes. Add 2 drops of 30% hydrogen peroxide to destroy residual organic matter, and evaporate to a residue of about 0.3 ml. Add 5 ml of 5% tartaric acid solution with warming and cool. Neutralize to Congo red with ammonium hydroxide. Add 1.6 ml of 1 : 1 sulfuric acid and oxidize any traces of antimony with a couple of drops of bromine water. After 5 minutes decolorize the solution by dropwise addition of 1% solution of phenol. Add 2.5 ml of 5% solution of gelatin and 5 ml of the phenylfluorone reagent. Dilute to 25 ml and let stand for 20 minutes. Read at 500 nm and deduct a reagent blank.

Silicates.[83] Transfer a 1-gram sample containing not more than 3 μg of tin to a 500-ml polyethylene bottle. Add 20 ml of 1 : 1 hydrochloric acid and 5 ml of hydrofluoric acid. If the sample contains organic matter, add 5 ml of 50% hydrogen peroxide. Close securely with a vented screw cap and heat at 100° overnight.

Wash 100-200 mesh Amberlite 1R-400 resin to remove fines and prepare a 1 × 10 cm bed. Condition with 20 ml of 1 : 1 hydrochloric acid.

If hydrogen peroxide was used in solution of the sample, add 5 ml of water; otherwise, add 10 ml of water to the cooled sample solution and pass it through the column at 1 ml/minute. Wash the bottle with 10 ml and 10 ml of 1 : 1 hydrochloric acid and pass through the column. Wash the column successively with 25 ml of 1 : 1 hydrochloric acid, 25 ml of 20% citric acid solution, and 25 ml of water. Elute the tin with 45 ml of 1 : 17 sulfuric acid and add 9.7 ml of sulfuric acid to bring the eluate to 4.5 *M* sulfuric acid.

[82]S. D. Gur'ev and N. F. Saraeva, *Sb. Nauch. Tr. Gos. Nauch.-Issled. Inst. Tsvet. Met.* **1961** (18), 48–52.
[83]J. David Smith, *Anal. Chem. Acta* **57**, 371–378 (1971).

Add 10 ml of 5 *M* potassium iodide and extract stannic iodide by shaking for 2 minutes with 10 ml of toluene. Discard the aqueous layer and wash the toluene layer with 5 ml of a 1:5 mixture of 5 *M* potassium iodide and 1:3 sulfuric acid. Extract the tin from the toluene layer by shaking for 1 minute with 5 ml of 4% sodium hydroxide solution. Then shake the toluene layer for 30 seconds with 2 ml of 0.4% sodium hydroxide solution. Adjust the combined extracts to pH 1 ± 0.05, using a pH meter, by dropwise addition of 1:3 sulfuric acid. Decolorize liberated iodine with a drop of 5% ascorbic acid solution.

As the phenylfluorone reagent, just before use dissolve 4 mg in 10 ml of ethanol containing 5 drops of 1:3 sulfuric acid and dilute to 100 ml with ethanol. Add 5 ml of this reagent and dilute to a known volume with water. Let stand for 2 hours and read at 525 nm.

Limestone or Ferromanganese. The sample should not contain more than 3 μg of tin. Shake it overnight at room temperature with 20 ml of 1:1 hydrochloric acid. Centrifuge, and pass the decantate through the column above. Wash the residue with 10 ml and 10 ml of 1:1 hydrochloric acid and pass through the column. Complete as for silicates from "Wash the column successively with..."

Canned Fruit and Fruit Juice.[84] Digest the sample with sulfuric and nitric acids until colorless and evaporate to copious sulfur trioxide fumes. Take up in 10 ml of water and dilute to 100 ml with 1:4 sulfuric acid. Mix a 2-ml aliquot with 1 ml of 0.5% ascorbic acid solution and 10 ml of 5% sodium acetate solution. Dilute to about 30 ml with water and adjust to pH 2 with 1:4 sulfuric acid. Add 1.5 ml of 0.25% solution of gum acacia and 5 ml of 0.02% phenylfluorone solution. Dilute to 50 ml with water and set aside for 1 hour. Remove unchanged phenylfluorone by shaking with 25 ml of chloroform. Read the extracted aqueous layer at 480 nm against a reagent blank.

For tin in mashed vegetables, digest 10 grams in the cold with 15 ml of hydrochloric acid for 1 hour.[85] Dilute to 100 ml, filter, and determine tin with phenylfluorone.

1,10-PHENANTHROLINE—FERROUS COMPLEX

The complex of tin with the captioned reagent can be extracted with nitrobenzene.[86] Mercuric and titanic ion and hexavalent tungsten and molybdenum interfere. Beer's law applies to 4–20 m*M* extracts.

Procedure. *Iron and Steel.* Prepare the sample as described for cast iron by phenylfluorone through "Wash the precipitate with 50 ml of 1% ammonium sulfate

[84]Saburo Kanno and Hidemaro Ogura, *J. Food. Hyg. Soc. Jap.* **5**, 116–119 (1964); cf. Yukio Nakamura and Sachiko Kamiwada, *ibid.*, **14**, 352–356 (1973).
[85]R. Biston, F. Verstraeten, and P. Nangniot, *Anal. Chim. Acta* **59**, 453–460 (1972).
[86]Kuzuo Hiiro, Takashi Tanaka, Toyozo Shirai, and Yuroku Yamomoto, *Jap. Anal.* **18**, 563–569 (1969).

solution." Dissolve the precipitate in 5 ml of 1 : 1 hydrochloric acid, add 1 ml of 1 : 3 sulfuric acid, and evaporate nearly to dryness. Take up the residue in 1 ml of 1 : 11 hydrochloric acid and add 10 ml of 1 : 8 sulfuric acid. Add 2.5 ml of 1.8% oxalic acid solution and 8 ml of 2 m*M* reagent. Dilute to 25 ml and extract with 10 ml of nitrobenzene by shaking for 2 minutes. After 30 minutes separate and dry the organic extract with sodium sulfate. Read at 516 nm.

PHOSPHOMOLYBDATE BLUE

Since stannous chloride will serve to develop the phosphomolybdate blue, it follows that under appropriate conditions the blue will be proportional to the stannous ion.[87] Optimum conditions were defined as 3.6 *N* sulfuric acid, 0.4% monopotassium phosphate, 0.4% ammonium molybdate, and 0–15 ppm of tin with reduction to stannous ion by aluminum of high purity. Isolation of tin by coprecipitation with manganese dioxide is appropriate. The method is suitable for 0.1–0.5 mg of tin.

Procedure. Mix a sample containing 0.1–0.7 mg of tin with 10 ml of 1 : 1 sulfuric acid. Add 5 ml of hydrochloric acid and dilute to 40 ml. Add about two-thirds of a 0.2-gram portion of metallic aluminum. Filter off any precipitated antimony and copper and add the balance of the aluminum. Boil for 3 minutes, and cool in a stream of nitrogen or carbon dioxide to avoid air oxidation of the stannous ion. Add 2 ml of 10% solution of monopotassium phosphate and 2 ml of 10% ammonium molybdate solution. Dilute to 50 ml and read at 800 nm.

Water. Add 50 ml of nitric acid and 20 ml of 20% manganese sulfate solution to a liter sample. Boil gently while slowly adding 10 ml of 3% potassium permanganate solution. Boil for 5 minutes, let settle, and decant the supernatant layer. Dissolve the precipitate in 50 ml of 1 : 3 nitric acid and excess of 3% hydrogen peroxide. Boil off any excess of hydrogen peroxide and add 2 ml of 3% potassium permanganate solution to reprecipitate manganese dioxide. Filter and wash with water. Dissolve the precipitate with 10 ml of 1 : 1 sulfuric acid and 3% hydrogen peroxide. Evaporate to sulfur trioxide fumes. Take up with 40 ml of water and complete by the method above from "Add about two-thirds...."

Copper Alloy. Dissolve a sample containing 0.2–0.7 mg of tin in 50 ml of nitric acid. Boil off oxides of nitrogen, add 20 ml of 20% manganese sulfate solution, and dilute to 1 liter. Complete as for water from "Boil gently while slowly adding...."

Steel. Dissolve a sample expected to contain 0.1–0.7 mg of tin in 100 ml of 1 : 1 nitric acid. Boil off oxides of nitrogen, filter off the silica, and wash. Add 20 ml of 20% manganese sulfate solution and dilute to 1 liter. Complete as for water from "Boil gently while slowly adding...."

---

[87] Hiroshi Namiki, *ibid.* **10**, 895–903 (1961).

4-(2-PYRIDYLAZO)RESORCINOL

This reagent forms a 1 : 1 water-soluble complex with tin at pH 5.5–6.1.[88] Extraction as the iodocomplex avoids interferences. It compares favorably with phenylfluorone, although it is less sensitive, because pH control is not as important and a protective colloid is not required.

Procedure. *Zinc Concentrates or Titanium.* Dissolve 1 gram of sample in 25 ml of 1 : 1 sulfuric acid, oxidize with 2 ml of nitric acid, and evaporate to sulfur trioxide fumes. Take up in water, add 25 ml of 1 : 1 sulfuric acid, and dilute to 100 ml. Without excessive loss of sulfur trioxide fumes, this should approximate 9 *N*. Mix a 25-ml aliquot with 2.5 ml of 5 *M* potassium iodide and extract tin iodide with 10 ml of toluene. Wash the toluene extract with 10 ml and 10 ml of 1 : 3 sulfuric acid containing 1 ml of 5 *M* potassium iodide. Reextract the tin with 10 ml of 1 : 25 hydrochloric acid. Add 5 ml of 0.01% solution of the reagent and dilute to about 30 ml. Add 5 ml of acetate buffer solution for pH 5.7 and sufficient ammonium hydroxide to raise the pH to that level. Dilute to 50 ml, let stand for 15 minutes, and read at 515 nm against a reagent blank.

9-(3-PYRIDYL)FLUORONE

At pH 1–2 stannous ion forms a 1 : 1 complex with 9-(3-pyridyl)fluorone absorbing at 540–550 nm. Stannic ion similarly forms a 1 : 2 complex.[89] In concentrated solutions of hydrogen peroxide, the peroxide is first destroyed with ferrous ion.[90] The ferric ion is then reduced with ascorbic acid, and the tin is complexed with the 9-(3-pyridyl)fluorone. With a mercury lamp and an Hg 546 filter, the complex is read at 0.01–2 μg of tin per ml. Nitrate and phosphate do not interfere.

Procedures

Organotin Compounds.[91] This technic is applicable to such samples as tin tetraethyl and tin tetraphenyl. For more volatile compounds, decomposition in a Parr bomb is suggested as described for determination by Dithiol.

Heat 0.1 gram of sample with 5 ml of sulfuric acid. Cool to about 60° and add 5 ml of 30% hydrogen peroxide. Heat until clear and colorless. Add 35 ml of hydrochloric acid and dilute to 1 liter. Take an appropriate aliquot and dilute with 4 volumes of 1 : 20 hydrochloric acid. Mix 5 ml of this dilution with 10 ml of a buffer solution containing 5 parts of 0.2 *M* potassium chloride and 4 parts of 0.2 *N* hydrochloric acid. Add 10 ml of 5% solution of polyvinyl alcohol (Polyviol M 05/140) and 10 ml of 0.5 m*M* 9-(3-pyridyl)fluorone in 80% methanol containing 10

[88] K. Kasiura and K. Olesiak, *Chem. Anal.* (Warsaw) **14**, 139–144 (1969).
[89] E. Asmus and J. Kraetsch. *Z. Anal. Chem.* **223**, 401–410 (1966).
[90] E. Asmus and J. Jahny, *ibid.* **255**, 186–190 (1971).
[91] E. Asmus, B. Kropp and R.-M. Moczko, *ibid.* **256**, 276–278 (1971).

ml of hydrochloric acid per liter. Dilute to 50 ml, let stand for 45 minutes, and read with an Hg 546 filter.

Steel.[92] Dissolve a 1-gram sample free from molybdenum and titanium in 10 ml of 1:1 hydrochloric acid. Dilute to 250 ml. Complete as for organotin compounds from "Mix 5 ml of this dilution...."

Molybdenum Steel.[93] Dissolve a sample containing 3–60 μg of tin in 10 ml of 1:1 sulfuric acid with the assistance of 30% hydrogen peroxide. Boil off the excess of hydrogen peroxide and filter. To the filtrate and washings, approximating 20 ml, add 5 ml of 10% ascorbic acid to reduce the iron. Add 7 grams of potassium iodide and extract the stannous iodide with 20 ml of benzene. Wash the organic phase with 1:1 sulfuric acid. Add the organic phase to 10 ml of hydrochloric acid, 3 ml of water, and 1 ml of 10% ascorbic acid solution. Evaporate the benzene at 70° under reduced pressure. To the aqueous tin solution, add 10 ml of a buffer solution 2 *M* in sodium acetate and 2 *M* in dichloroacetic acid. Add 5 ml of 5% solution of polyvinyl alcohol (Polyviol M 05/140) and 10 ml of 0.05 *M* 9-(3-pyridyl)fluorone in methanol. Dilute to 50 ml with water and set aside for 30 minutes. Read with an Hg 546 filter. The molybdenum does not interfere with determining down to 3 ppm of tin in a 1-gram sample.

PYROCATECHOL VIOLET

At pH 2.5 stannic ion forms a 1:2 complex with pyrocatechol violet which is read at 555 nm.[94] There is interference by zirconium, titanium, bismuth, iron, antimony, gallium, and molybdenum. To avoid interference, adjust the sample solution to 0.5 *N* hydrochloric acid in 50% acetic acid.[95] Add 1% of *N*-phenylbenzohydroxamic acid and extract with chloroform. Back-extract the tin from the organic phase with 8 *N* hydrochloric acid and evaporate the extract to dryness. Take up in 0.1 *N* hydrochloric acid and develop.

Procedure. Take a sample containing 5–40 μg of tin with less than 4 milliequivalents of acids and salts. Adjust the volume to 5 ml and add 2 ml of 0.05% solution of pyrocatechol violet. Add 5 ml of 0.1 *N* hydrochloric acid saturated with potassium acid phthalate as a buffer solution for pH 2.5. Adjust to pH 2.5±0.2 by suitable addition of ammonium hydroxide if necessary. Dilute to 25 ml, let stand for 15 minutes, and read at 555 nm against a reagent blank.

Foods.[96] Dry-ash an appropriate sample and fuse the ash with sodium

[92]E. Asmus and U. Kossmann, *ibid.* **245**, 137–139 (1969).
[93]E. Asmus and H. Weinert, *ibid.* **249**, 179–181 (1970).
[94]W. J. Ross and J. C. White, *Anal. Chem.* **33**, 421–424 (1961); cf. V. P. Sagakova and A. I. Lyubivaya, *Tr. Ukr. Nauch.-Issled. Inst. Konserv. Prom.* **1** (2), 118–123 (1959).
[95]M. Kosva, S. Mareva, and N. Iordanov, *Anal. Chim. Acta* **75**, 464–467 (1975).
[96]V. P. Sagakova and A. I. Lyubivaya, USSR Patent *134,462* (1960); *Konserv. Ovoshch. Prom.* **15** (9), 37–40 (1960).

carbonate. Take up the fusion in 1:5 hydrochloric acid and neutralize to pH 3.5 with ammonium hydroxide. Dilute to 0.5–40 μg of tin per ml. Mix with a 1-ml aliquot 2 drops of 10% ascorbic acid solution to reduce the iron content, 0.5 ml of 0.1% ethanolic solution of pyrocatechol violet, and 1 ml of 13.6% solution of sodium acetate trihydrate. Let stand for 5 hours and read at 619 nm.

QUERCETIN

Quercetin, which is 3,3′,4′,5,7-pentahydroxyfluorone, forms a 1:complex with stannic ion for photometric reading.[97] Substantial amounts of antimony, bismuth, and phosphorus interfere. The complex can be extracted with 1:1 isoamyl alcohol–ethyl ether from aqueous solution.[98] Iron is masked with thiourea.

For determination of 0.1–0.001% in steel, distill the tin as the bromide and complex with quercetin.[99] An alternative is to adsorb the chloro complex on an anion exchange resin and elute with dilute nitric acid.[100] For analysis of ferrotitanium, pass the solution through a column of AN-31 anion active resin and determine the tin in the eluate with quercetin.[101] An alternative is precipitation with thionalide and an alternative development with hematein.

For analysis of zinc alloys, the sample is dissolved in hydrochloric acid and oxidized with hydrogen peroxide.[102] Then on treatment with thiourea, ascorbic acid, and quercetin, the tin-quercetin complex is extracted with methyl isobutyl ketone for reading at 435 nm.

For magnesium and magnesium alloys, dissolve the sample in 17% sulfuric acid, add thiourea and ascorbic acid, and form the complex with quercetin.[103] Thin-layer chromatography on silica gel, with or without diatomaceous earth, is appropriate for isolation of organotin compounds for development with quercetin.[104]

Procedures

Brass and Bronze. Decompose 1 gram of sample with 5 ml of 1:1 hydrochloric acid by adding 1 ml of 30% hydrogen perioxide from time to time. Boil off the excess of hydrogen peroxide and dilute to 50 ml. Neutralize an aliquot with 1:3 ammonium hydroxide until turbidity begins to appear. Clear with a drop of 1:3 hydrochloric acid and add 5 ml in excess. Add sufficient 10% thiourea solution to mask both copper and ferric ion. Add 5 ml of 0.2% ethanolic solution of quercetin. Add 25 ml of ethanol, dilute to 50 ml, and read at 435 nm.

[97] Karel Liska, *Chem. Listy* **49**, 1656–1660 (1955).
[98] T. N. Nazarchuk, *Zh. Anal. Khim.* **14**, 696–699 (1959).
[99] W. Chetkowski, *Pr. Inst. Hutn.* **18** (2), 109–112 (1966).
[100] Ivan Janousek and Karel Studlar, *Hutn. Listy* **15**, 889–893 (1960).
[101] V. I. Ponosov, L. Ya. Kalashnikova, V. I. Kurbatova, and V. V. Stepin, *Tr. Vses. Nauchn.-Issled. Inst. Stand. Obraztsov. Spektr. Etalnov* **6**, 79–87 (1970).
[102] J. L. Jimenez Seco and Aurora Gomez Coedo, *Rev. Met.* **5**, 161–177 (1969); British Standards Institution, *BS 3630*, Pt. **2**, 9 pp (1972).
[103] British Standards Institution, *BS 3907*, Pt. **13**, 8 pp (1972).
[104] H. Wisczorek, *Deut. Lebensm. Rundsch.* **65**, 74–78 (1969).

"Tin-Free" Bronze.[105] Dissolve 1 gram in 30 ml of 1:1 nitric acid and dilute to 100 ml. Neutralize an aliquot with ammonium hydroxide to the first formation of cupric hydroxide, clear the solution by dropwise addition of 1:4 hydrochloric acid, and add 5 ml of that acid in excess. Add 10 ml of 10% solution of thiourea, 3 ml of 10% antipyrine solution, and 5 ml of ethanolic mM quercetin. Adjust to pH 1.5–2, extract with 10 ml of chloroform, and read at 442 nm.

Lead, Bismuth, or Indium.[106] Dissolve 10 grams of a sample of high purity in 60 ml of 1:3 nitric acid and dilute to 400 ml. Add 10 ml of a solution containing 20 mg of beryllium and 60 ml of 25% solution of EDTA. Raise to pH 9.5 with ammonium hydroxide. Heat to 80°, let cool over a 2-hour period, and filter. Wash the precipitate with 1% solution of EDTA adjusted to pH 9.5 with ammonium hydroxide, then with ammonium nitrate solution at pH 9.5. Dissolve the precipiate in 10 ml of hot 1:4 sulfuric acid. Add 3 ml of 30% hydrogen peroxide and evaporate to sulfur trioxide fumes. Add 15 ml of water and again evaporate to sulfur trioxide fumes. Take up with 5 ml of water, add 15 ml of 1:1 hydrochloric acid, and dilute to 25 ml. Mix with 20 ml of 5% solution of thiourea and 5 ml of 2% solution of ascorbic acid. As a color reagent, dissolve 50 mg of quercetin in 60 ml of ethanol, add 2.5 ml of hydrochloric acid, and dilute to 100 ml with ethanol. Add 20 ml of the color reagent to the sample and set aside for 15 minutes. Extract with 15 ml of methyl isobutyl metone, wash the extract with 25 ml of 1:20 sulfuric acid, and read at 440 nm.

Alternatively, dissolve 10 grams of sample in 75 ml of 10% solution of sodium citrate in 1:3 nitric acid. Add 250 ml of 7.2% solution of EDTA and adjust to pH 5.5 with ammonium hydroxide. Wash a silica gel column with 0.5 M sodium citrate adjusted to pH 5.5 with hydrochloric acid. Pass the sample solution through this column at 4 ml/minute and wash the column with 200 ml of the sodium citrate solution. Elute the tin with 50 ml of 1:1 hydrochloric acid followed by 130 ml of water. Add 10 ml of 1:4 sulfuric acid to the eluate and evaporate to 10 ml. Add 7 ml of 1:1 hydrochloric acid and dilute to 25 ml. Complete as for the previous technic from "Mix with 20 ml of 5% solution...."

Ores and Concentrates.[107] Heat a sample containing 0.002–0.1 gram of tin with 20 ml of hydrofluoric acid and 10 ml of 1:1 sulfuric acid in platinum. Evaporate to sulfur trioxide fumes, add 2 ml of water, and again heat to sulfur trioxide fumes. Take up in 100 ml of water, heat, and precipitate with excess ammonium hydroxide. Filter, and wash with 5% solution of ammonium chloride. Ignite at 500°; then fuse with 5 grams of sodium hydroxide and 0.5 gram of sodium peroxide. Extract the melt with hot water, cool, and neutralize to Congo red with 1:1 sulfuric acid. Add 0.2 ml excess of acid and filter. Add 2 ml of 10% solution of manganese sulfate, followed by 1 ml of 4% potassium permanganate solution. Boil for 10 minutes, filter, and wash the precipitate with hot 1:100.sulfuric acid. Digest the

[105] N. L. Olenovich, G. I. Savenka, and T. D. Andraeva, *Ukr. Khim. Zh.* **31**, 882–883 (1975).
[106] E. Wunderlich and G. Bosse, *Erzmetallurgie* **24**, 537–542 (1971).
[107] V. L. Verand, Nauch. *Tr. Irkutsk. Nauch.-Issled. Inst. Redk. Met.* **1963** (11), 56–60.

precipitate and paper with 10 ml of 1:1 hydrochloric acid and filter. Wash the residue with hot water and dilute to 25 ml. Neutralize a 5-ml aliquot with ammonium hydroxide and acidify with 1 ml of 5 *N* sulfuric acid. Reduce ferric ion with 2 ml of saturated solution of urea and let stand for 20 minutes. Add 2 ml of a solution containing 1 mg of quercetin per ml of ethanol. Add 10 ml of ethanol and dilute to 25 ml with water. Let stand for 15 minutes and read at 420 or 450 nm.

Canned Food.[108] Wet-ash a sample expected to contain up to 1 mg of tin in silica with 5 ml of sulfuric acid and 5 ml of nitric acid. Addition of 30% hydrogen peroxide solution may prove advantageous for getting a colorless digestate, in which case the excess must be removed by boiling.[109] Cool, and dilute to 50 ml. To a 2-ml aliquot add 0.2 ml of 0.1% solution of 2,4-dinitrophenol in 50% ethanol. Add 10% solution of sodium carbonate dropwise to the first appearance of yellow. Discharge the color by dropwise addition of 2.5 *N* hydrochloric acid and add 5 ml excess of the acid. To mask ferric ion, add 3 ml of saturated solution of thiourea. Add 5 ml of 0.2% ethanolic solution of quercetin and 25 ml of ethanol. Dilute to 50 ml with water. Let stand for 30 minutes and read at 437 nm against a reagent blank.

High salt, acid, and phosphate content may reduce the accuracy of determination of less then 10 ppm of tin.[110]

Alternatively,[111a] digest a sample of juice from fruit or vegetables expected to contain 0.25–125 μg of tin with 4 ml of sulfuric acid and 5 ml of nitric acid. Add 30% hydrogen peroxide as necessary to produce a colorless digestate, and evaporate to sulfur trioxide fumes. Add 5 ml of water, evaporate to sulfur trioxide fumes, add 5 ml more of water, and again evaporate to sulfur trioxide fumes. Dilute to 5 ml with sulfuric acid, then to 50 ml with water. Mix a 5-ml aliquot with 5 ml of 1:9 sulfuric acid and dilute to 15 ml with water. Add 10 ml of 10% thiourea solution, followed by 5 ml of 0.2% ethanolic solution of quercetin. Dilute to 50 ml with ethanol and set aside for 20 minutes. Read at 435 nm.

Another technic is to convert the solution to stannic bromide and complex with brilliant green in hydrochloric and hydrobromic acids.[111b] Extract the weakly colored brilliant green complex with ethyl acetate. Add excess of quercetin solution in ethyl acetate and shake with 10% sodium chloride to convert to the more intensely colored quercetin complex. Read at 440 nm. Chromic ion must not exceed the stannic ion.

Another technic is to wet-ash foodstuff, extract from 1:3 sulfuric acid as the iodocomplex into toluene, back-extract with sodium hydroxide solution, acidify with 9 *N* sulfuric acid, add ascorbic acid, thiourea, and ethanol, and finally develop with quercetin.[112]

---

[108]R. S. Kirk and W. D. Pocklington, *Analyst* **94**, 71–74 (1969); cf. Yu Lyaskovskaya and T. Krasil'nikova, *Myas. Ind. SSSR* **1961** (4), 44–45.

[109]W. Eyrich, *Deut. Lebensm. Rundsch.* **68**, 280–282 (1972).

[110]H. Woidich, *Z. Lebensm.-Unters. Forsch.* **151**, 114–120 (1973).

[111a]M. Karvanek, D. Miler, *Prum. Potravin* **16**, 369–371 (1965).

[111b]K. Manolov, V. Stamatova, and A. Matschev, *Mikrochim. Acta* **II**, 343–347 (1976).

[112]A. Engberg, *Analyst* **98**, 137–145 (1973).

RHODAMINE 6Zh

In 3 N hydrochloric acid this dyestuff, which is basic red 1, forms a 1:1 complex with chlorstannate ion which is extractable with benzene. Beer's law is followed for 0.2–1.4 μg of zinc per ml. There is interference by selenium, tellurium, gold, palladium, or indium, as well as by more than twentyfold amounts of stibnic or nitrate ion.

Procedure. *Brass.*[113] Dissolve 30 mg in 10 ml of 1:1 nitric acid and 0.2 ml of hydrochloric acid. Evaporate oxides of nitrogen. Add 5 grams of ammonium nitrate and boil with 50 ml of water. Adjust to pH 6 with 1:1 ammonium hydroxide and heat at 100° for 30 minutes. Filter and wash with 1:5 ammonium hydroxide. Dissolve the precipitate with 3.5 N hydrochloric acid and dilute to 250 ml. Mix a 0.25-ml aliquot, 1 ml of mM rhodamine 6Zh, and 3.75 ml of 3.5 N hydrochloric acid. Extract the zinc complex with 5 ml of benzene. Add 2 ml of isobutyl alcohol and read at 530 nm against a reagent blank.

2-(SALICYLIDENEAMINO)BENZENE-THIOL

This reagent will determine 0.004–0.055% of tin in iron and steel.[114] Beer's law is followed for up to 70 μg of tin. There is interference by more than 6% chromium, more than 20 μg of niobium, or more than 30 μg of tantalum. To avoid interference by the latter two, stannous iodide is extracted into benzene and the tin is reextracted with 1:70 sulfuric acid. Chromium can be volatilized as chromyl chloride.

Procedures

Iron and Steel. Dissolve 0.5 gram of sample with 20 ml of hydrochloric acid and 5 ml of 30% hydrogen peroxide. Boil to decompose excess peroxide, cool, and dilute to 50 ml. Mix a 10-ml aliquot with 5 ml of 10% solution of ascorbic acid to mask the iron. Neutralize to 2,4-dinitrophenol with 20% sodium hydroxide solution. Add 2 ml of 20% lactic acid solution and 2 ml of 1% sodium thiosulfate solution. Add 5 ml of a solution containing 0.1% of the captioned reagent and 1% of ascorbic acid in ethanol. Set aside for 5 minutes; then shake with 10 ml of benzene for 1 minute. Read the extract at 415 nm.

Tin Ores.[115] Fuse 0.25 gram of sample with 2.5 grams of sodium peroxide and dissolve in 100 ml of 1:1 hydrochloric acid. Dilute to 500 ml and neutralize an aliquot containing less than 0.05 mg of tin to 2,4-dinitrophenol. Dilute to about 20 ml and add 2 ml of 20% lactic acid solution. Dissolve any precipitated hydroxides

[113] N. L. Shestidesyatna, N. M. Milyaeva, and L. I. Kotelyankaya, *Zavod. Lab.* **41**, 653–655 (1975).
[114] Shizuya Maekawa and Kiyotoshi Kato, *Jap. Anal.* **20**, 474–479 (1971).
[115] G. R. E. C. Gregory and P. G. Jeffery, *Analyst* **92**, 283–289 (1967).

by adding 1:1 hydrochloric acid dropwise. Add 1 ml of 1% solution of sodium thiosulfate and 5 ml of 1% ascorbic acid solution. Add 5 ml of 1% solution of the captioned reagent and set aside for 5 minutes. Read at 415 nm.

SILICOMOLYBDATE BLUE

This member of the molybdenum blue family is produced by reduction with metallic aluminum in the presence of molybdate ion, silicate ion, and stannous ion.[116] Iron and antimony produce a similar color. As developed for determination of organic tin compounds on fabric, the color tends toward green. Beer's law is followed for up to 1.5 mg in 200 ml of solution. At a higher tin concentration, a different blue complex is formed. The acidity must approximate 2 *N* hydrochloric acid. Although the method given digests the sample fabric, it is also feasible to extract the organotin compound and digest it.

Procedure. *Fabric.* The weight of sample should be such that diluted to 250 ml, an aliquot will contain 0.5–1.25 mg of tin. Cut the sample into pieces and place in a Kjeldahl flask. Carefully add 15 ml of a 1:1 mixture of fuming sulfuric acid and fuming nitric acid. Warm gently, and after vigorous reaction has ceased, add nitric acid in small amounts until the solution is clear. Evaporate to sulfur trioxide fumes. After cooling add 5 ml of water, and again heat to sulfur trioxide fumes to drive off the last of the nitric acid. Dilute to 250 ml.

To an aliquot containing 0.5–1.25 mg of tin, add 14 ml of hydrochloric acid and dilute to 100 ml. Add an aluminum pellet weighing 0.03–0.04 gram and warm so that when the pellet is completely dissolved, boiling is reached. Boil for 5 minutes and add 20 ml of a freshly mixed reagent containing 2 parts of 2.96% solution of sodium molybdate dihydrate in 1:50 sulfuric acid and 1 part of 2.54% solution of sodium metasilicate pentahydrate. Cool to 20° and dilute to 200 ml. After 15 minutes read at 680 nm against a blank.

SILICOMOLYBDOTUNGSTATE

This familiar member of the molybdenum blue family is also applicable to tin as determined by reduction of 10-molybdo-2-tungstosilicic acid by stannous ion.[117] After distillation to separate from interferences, the extinction developed is read for 0.04–0.12 mg of tin.

Procedure. *Brass.* Dissolve 0.4 gram of sample in 5 ml of 1:1 nitric acid and 2 ml of sulfuric acid and evaporate to sulfur trioxide fumes. Transfer to a distillation

[116] A. B. Russell, *Analyst* **84**, 712–716 (1959).
[117] E. F. Tkach, *Uch. Zap. Kishinev. Univ.* **1963** (68), 61–63.

flask with 5 ml of water and 10 ml of sulfuric acid. Add 0.2 ml of hydrochloric acid and heat to 200°. Distill at 200–220° for 1 hour, while adding 10 ml of hydrobromic acid dropwise. Collect the distilled stannic bromide in a series of 5-ml vessels. Dilute the distillate to 50 ml. Mix a 2-ml aliquot with 1 ml of hydrochloric acid, 3 ml of bismuth amalgam, and a few crystals of sodium bicarbonate. Agitate at 100° for 10 minutes and remove the amalgam. Add 4 ml of water and 1 ml of 2% solution of 10-molybdo-2-tungstosilicic acid. Dilute to 10 ml and read after 5 minutes.

3,4′,7-TRIHYDROXYFLAVONE

Stannic ion reacts with this reagent in an inert atmosphere to give fluorescence.[118] There is interference by antimony, zirconium, hafnium, aluminum, gallium, tungsten, molybdenum, niobium, and tantalum, and by halides. The method will detect 0.007 μg of tin and is accurate to 10% for more than 0.02 μg.

Procedure. As a buffer solution, mix 110 ml of sulfuric acid in 400 ml of water with 120 grams of sodium hydroxide in 400 ml of water and cool. Reagent grade sodium sulfate is not an acceptable substitute. Add 20 grams of sulfamic acid and dilute to 1 liter. The organic reagent is 5 mg of 3,4′,7-trihydroxyflavons per 100 ml of ethanol.

If interfering ions are present, add to the sample solution 1 ml of 17% solution of sodium bisulfate and 0.5 ml of sulfuric acid. Evaporate until all fuming has ceased. Add 2 ml of water and evaporate to about 0.5 ml. Take up in 25 ml of 1:3 sulfuric acid. Add 2.5 ml of freshly prepared 5 *M* potassium iodide solution. Extract by shaking for 2 minutes with 10 ml of toluene and discard the aqueous phase. The extract will contain some free iodine. Shake for 5 seconds with 15 ml of 1:3 sulfuric acid. Add 3 ml of 5 *N* potassium iodide, shake for 1 minute, and discard the aqueous phase. Add 5 ml of water to the toluene phase, then add 20% sodium hydroxide solution dropwise until the toluene layer is colorless. This usually requires 6–8 drops; then add 2 drops in excess. Shake for 30 seconds and separate the aqueous tin solution. Wash the toluene layer with 3 ml and 3 ml of 0.4% solution of sodium hydroxide and add to the previous extract. Add 1 ml of 17% solution of sodium bisulfate, 0.5 ml of sulfuric acid, and 5 drops of nitric acid. Cool and add 1 ml of water and 5 drops more of nitric acid. Evaporate to sulfur trioxide fumes. Repeat the addition of water and nitric acid and evaporation. This time carry the heating to a pyrosulfate fusion. In these evaporations and the fusion, nitric acid can be omitted if halides are absent.

Take up the residue in 2 ml of water and add 3 drops of 25% solution of hydroxylamine sulfate. Evaporate to about 0.5 ml and add 10 ml of the sodium sulfate–sulfamic acid buffer solution. Add 1 ml of the organic reagent and dilute to

[118] T. D. Filer, *Anal. Chem.* **43**, 1753–1757 (1971).

25 ml. After 20 minutes for equilibration, activate at 377 nm and read at 473 nm. Because of the high sensitivity, every step must be rigidly controlled.

XYLENOL ORANGE

Stannic ion forms a 2:1 complex in 0.025 *N* sulfuric acid with xylenol orange.[119] This is read at 536 nm, where absorption of the reagent is negligible. Beer's law is followed for 0.2–3 μg/ml. The reagent forms interfering complexes with antimony, niobium, gallium, vanadium, ferric ion, zirconium, and indium. Of these, indium is masked with sodium chloride, bismuth with potassium iodide, and ferric ion by reduction with ascorbic acid.

Procedure. *Indium-Tin Alloy.* Dissolve 5 mg of sample in 5 ml of 1:2 sulfuric acid and dilute to 50 ml. Adjust a 1-ml aliquot to about pH 2 with 0.4% sodium hydroxide solution. Add 2 ml of 10% solution of sodium chloride in 0.02 *N* sulfuric acid and a few crystals of ascorbic acid. Add 0.2 ml of 0.08% solution of xylenol orange in 0.01 *N* sulfuric acid containing a little mercuric iodide. Dilute to 25 ml with 0.02 *N* sulfuric acid and let stand for 20 minutes. Read at 536 nm.

Tin forms with **alizarin blue** a violet complex having a maximum absorption at 580 nm.[120] The complex is extractable with a mixture of cyclohexanone and ethyl acetate. Beer's law applies up to 10 μg/ml. Molybdenum, vanadium, iron, and copper interfere. Antimony and bismuth are masked with tartaric acid and EDTA.

When stannous ion is oxidized by excess ferricyanide at pH 1–2.8, there is a red color developed by **bi-*o*-anisidine.**[121] At 450 nm this conforms to Beer's law at 0.6–60 mg/liter. Other reducing agents must be absent.

A 1:1 complex is formed in a mixture of 0.17 m*M* **2,2′-biquinoxalinyl** in 1:1 hydrochloric acid and 6.2–42 m*M* stannous ion in 1:1 hydrochloric acid.[122] Read at 675 nm 25–50 minutes after mixing. There is interference from the following, in μg/ml: copper, 0.06; nickel, cobalt, or oxalate, 0.5; selenate, 1; ferric ion, 4.

Dissolve 0.5 gram of rock with phosphoric, nitric, and hydrofluoric acids.[123] Evaporate to a syrup and take up in 9 *N* sulfuric acid. Add 3 *M* sodium iodide and extract stannic iodide with toluene. Back-extract the stannic ion into 20% sodium hydroxide solution. Add 3,5-dinitrocatechol with which stannic ion forms a 1:3

[119] V. N. Danilova, *Zavod. Lab.* **29**, 407–409 (1963).

[120] Kichinosuke Hirokawa, *Rep. Res. Inst., Tohoku Univ., Ser. A* **13**, 426–432 (1961).

[121] J. Artigas, F. Buscarons, and C. Rodriguez-Roda, *An. Real Soc. Espan. Fis. y. quim.* (Madrid) **568**, 369–372 (1960).

[122] Ryszard Baranowski, Krystyna Grabowska, Zbigniew Gregorowicz, and Irena Baranowska, *Chem. Anal.* (Warsaw) **19**, 997–1001 (1974).

[123] V. A. Nazarenko, L. I. Vinarova, and N. V. Lebedeva, *Zh. Anal. Khim.* **30**, 617–619 (1975).

complex. Add **brilliant green** (basic green 1) to form an ion association complex, extract with carbon tetrachloride, and read at 630 nm. Beer's law is followed for 0.02–0.2 μg of tin per ml in the extract. Tenfold amounts of germanium, antimony, and molybdenum are tolerated with 5 μg of tin.

For determining tin in brass by brilliant green, dissolve 0.2 gram in 4 ml of hydrochloric acid and 1 ml of nitric acid.[124] Evaporate to dryness at 100°, add 1 ml of hydrochloric acid, and again evaporate to dryness. Take up the residue in 10 ml of 0.1 *M* phosphoric acid and take a 1-ml aliquot. Add 1 ml of 23.4% solution of sodium chloride and 1 ml of m*M* brilliant green. Dilute to 5 ml with 0.1 *M* phosphoric acid and extract with 5 ml of benzene. Read at 630 nm.

A ternary complex of stannous ion, **cationic pink**, which is 2- [4-(2-chloroethylmethylamino)styryl]-1,3,3-trimethyl-3*H*-indolium chloride, and chloride ion is extracted at pH 2–3 with benzene for reading at 540 nm.[125] Beer's law is followed for 0.2–6 μg of stannous ion per ml.

Stannic ion forms a green 1 : 1 complex with **chloranilic acid** having a maximum absorption at 510 nm.[126] Beer's law is followed for up to 0.08 mg of tin in 50 ml of 0.6 *N* hydrochloric acid.

Stannic ion in 1 : 1 hydrochloric acid can be read as the **chloro complex** at 233 nm up to 10 ppm of tin.[127] There is interference by trivalent and pentavalent arsenic, and by antimony, ferric ion, titanic ion, and pentavalent vanadium. Stannic chloride can be reduced with titanous ion in the presence of cupferron.[128] Then the solution is adjusted to pH 1 and the complex with crystal violet is extracted with 4-heptanone for reading at 595 nm. The complex fades in daylight.

Crystal violet (basic violet 3) forms a 1 : 1 complex with stannous ion in hydrochloric acid.[129] As extracted with benzene, it is read at 605 nm. Interfering ions are auric, mercuric, gallic, chromate, rhenate, perchlorate, and ferric, when at the same concentration as the stannous ion.

A 1 : 2 complex is formed by stannous chloride with **5,7-dichloro-8-hydroxyquinoline.**[130] For development, adjust 25 ml of solution more than 0.2 *N* with hydrochloric acid to pH 1 with ammonium hydroxide. Add 5 ml of 0.1% solution of the reagent in acetone. Let stand for 5 minutes and extract with 10 ml of carbon tetrachloride. Read at 405 nm for 1–15 μg of tin per ml of extract. Ferric ion, antimonic ion, gallium, and copper interfere.

Tin in iron and steel can be determined as the **diethyldithiocarbamate** complex.[131] The bulk of the iron is coprecipitated with manganese dioxide from a nitric acid solution. The tin is complexed by adding sodium diethyldithiocarbamate and the balance of iron and manganese extracted with petroleum ether. The tin complex is then extracted with benzene and read at 330 nm.

[124] A. I. Busev, N. L. Shestidesyatnaya, and G. G. Zimomrya, *Zh. Anal. Khim.* **26**, 1517–1520 (1971).
[125] A. I. Busev and N. L. Shestidesyatna, *Zh. Anal. Khim.* **29**, 905–909 (1974).
[126] Chozo Yoshimura, Hayao Noguchi, Tadaomi Inous, and Hiroshi Hara, *Jap. Anal.* **15**, 918–924 (1966).
[127] Masayoshi Ishibashi, Yoruko Yamamota, and Yasushi Inoue, *Bull. Inst. Chem. Res., Kyoto Univ.* **37**, 38–47 (1959).
[128] Lucien Ducret and Henri Maurel, *Anal. Chim. Acta* **21**, 74–79 (1959).
[129] G. Ackermann and J. Koethe, *Chem. Anal.* (Warsaw) **17**, 445–458 (1972).
[130] Tsutomu Matsuo and Kazuko Funayama, *J. Chem. Soc. Jap., Pure Chem. Sect.* **87**, 433–437 (1966).
[131] Hidehiro Goto and Yachiyo Kakita, *Sci. Rep. Res. Inst., Tohoku Univ.* **9A**, 253–261 (1957).

A 1 : 1 : 1 complex is formed in 30% ethanol by stannic dihydroxide ion in 0.2 *M* antipyrine containing a 10 *M* excess of **4-(2,4-dihydroxyphenylazo)-3-hydroxy-naphthalene-2-sulfonic acid.**[132] At pH 3 the complex is extracted with chloroform and read at 560 nm. Beer's law is followed for up to 25 μM stannic ion.

Stannic ion forms a dibasic complex acid with 3 molecules of **3,5-dinitrocatechol** around pH 2.[133] This in turn forms ion-association complexes with basic dyes such as brilliant green (basic green 1), methyl violet (basic violet 1), Nile blue A (basic blue 12), and methylene blue (basic blue 9). The complexes are extractable with chloroform, carbon tetrachloride, or benzene, for reading around 630 nm for 0.02–0.7 μg/ml. The stannic ion must first be isolated from many interfering ions.

The 1 : 1 complex of stannic ion with **dithizone** can be formed and read in the presence of a large excess of reagent in isopentyl alcohol.[134] The reagent has maximum absorption at 440 and 600 nm, a minimum at 510 nm. By contrast, the maximum for the complex is at 530 nm. Determination of dioctyl tin as a plastics stabilizer is applicable to 0.75 ppm by dithizone.[135]

The reaction of stannic ion with a solution in *N* hydrochloric acid of 1:4:1 sodium **fluotitanate**–sodium fluoride–hydrogen peroxide, a reagent that is 0.1 *M* in titanium, liberates titanic ion equivalent to the stannic ion.[136] The liberated titanic ion is read at 390 nm.

Mix 5 ml of the sample solution with 5 ml of 0.01 *M* **hexacyanochromate** and dilute to 50 ml with a buffer solution for pH 3.[137a] Expose to ultraviolet radiation for 2 hours and read 11.8–160 ppm of stannous ion at 460 nm. The method is also applicable to beryllium, mercuric ion, tetravalent vanadium, chromic ion, and molybdate.

Stannous ion in excess of 10 μg/ml is extracted from 2.5–6 *N* hydrochloric acid with 4–20 m*M* **hexa-hydrazepine-1-carbodithioate** in 3 : 1 chloroform-isoamyl alcohol for reading at 253 nm.[137b]

Stannous ion forms a 1:2 complex with **2-(2-hydroxyphenyl)benzothiazoline.**[138] Extracted into toluene around pH 2, the tin at 0.021–1.17 m*M* is read at 415 nm.

Tin is extracted from 5 *N* sulfuric acid containing *M* sodium **iodide** as iodide by cyclohexanone and read at 364 nm.[139] Stannic ion at 0.0001–0.001 *M* in 1.2 *N* hydrochloric acid containing 4 *M* potassium iodide is extractable with ethyl iodide from a solution in which free iodine has been removed by a small excess of sodium thiosulfate solution.[140] Lead and copper are precipitated as the iodides. Read at 440 nm, antimony at less than 0.01 *M* does not interfere; read at 410 nm, a correction must be applied for antimony greater than 0.001 *M*.

A ternary complex is formed by stannous ion, thiocyanate, and **malachite green** in 2 *N* sulfuric acid.[141] The maximum absorption at 610 nm is stable for 2 hours for

[132]N. L. Olenevich and G. I. Savenko, *Zh. Anal. Khim.* **30**, 1365–1368 (1975).
[133]V. A. Nazarenko, N. V. Lebedeva, and L. I. Vinarova, *Zh. Anal. Khim.* **28**, 1100–1103 (1973).
[134]M. Vancea and M. Volusniuc, *Stud. Cercet. Chim., Cluj* **13**, 203–211 (1962).
[135]J. Koch, *Dent. Lebensm. Rundsch.* **70**, 209–211 (1974).
[136]Hisao Fukamauchi, Ryuko Enohara, and Mitsuko Uehara, *Jap. Anal.* **8**, 315–317 (1959).
[137a]Wahid U. Malik and Krishna Dev Sharma, *Indian J. Chem.* **13**, 1232–1233 (1975).
[137b]A. I. Busev, A. P. Tereshchenko, and V. P. Naidina, *Zh. Anal. Khim.* **31**, 1159–1162 (1976).
[138]E. Uhlemann and V. Pohl, *Anal. Chim. Acta* **65**, 319–328 (1973).
[139]Katu Tanaka and Nobuyaki Takagi, *Anal. Chim. Acta* **48**, 357–366 (1969).
[140]Armine D. Paul and J. A. Gibson, Jr., *Anal. Chem.* **36**, 2321–2324 (1964).
[141]Chih-sheng Shih and Li-shu Ho, *Chem. Bull. Peking* **1965** (4), 240–243.

reading up to 0.6 μg/ml. Separation from tungsten, molydenum, zinc, and indium is necessary—by distillation as stannous iodide, for example.

The 1 : 1 complex of malachite green (basic green 4) in hydrochloric acid with chlorstannite ion is extracted with benzene and read at 628 nm.[142] Beer's law is followed for 1–4 μg of tin per ml. There is interference by auric, mercuric, gallic, chromate, rhenic, perchlorate, and ferric ions at the same concentration as the tin.

For tin in zinc and zinc alloys, dissolve 1 gram of sample in 10 ml of 2 : 1 hydrochloric acid with the aid of 30% hydrogen peroxide.[143] Add 3 ml of sulfuric acid and evaporate to sulfur trioxide fumes. Take up in 20 ml of water and filter. As reagent, dissolve 1 gram of **3-nitro-4-hydroxybenzenearsonic acid** in 15 ml of methanol, dilute to 50 ml, and filter. Add 10 ml of reagent, dilute to 50 ml, and set aside for 2 hours. Read at 436 nm against a blank.

When tin and antimony have been isolated by precipitation with thioanilide, the tin is read as a 1 : 2 compound with ***p*-nitrophenylfluorone**.[144] The complex is stabilized with gelatin in 20% ethanol. The color develops to a maximum in 1 hour in 0.1–0.3 *N* acid. It follows Beer's law at 0.05–5 μg/ml. For isolation of tin from niobium, vanadium, and silicon, chloroform extraction as the diethyldithiocarbamate from strongly acid solution is appropriate.

The 2 : 1 complex of stannous ion with **nitroso-R salt** has a maximum absorption at 370 nm.[145] The maximum sensitivity is with the sample solution at pH 2–2.6 and a reagent 1.2–2.5 *N* with hydrochloric acid.

Phenylfluorone and antipyrine form a ternary complex with tin for extraction with chloroform and reading at 500 nm.[146] Beer's law applies up to 2.5 μg of tin per ml. **Salicylfluorone** reacts similarly. With **4-dimethylaminophenylfluorone**, Beer's law applies up to 17.5 μg/ml.

For determination of stannous ion, mix 2.4 ml of sample solution in 0.1 *N* hydrochloric acid with 10 ml of saturated solution of **picrolonic acid**.[147] Filter the stannous picrolonate and read the excess picrolonic acid. Calcium and copper interfere.

For tin in bronze or antimony, make the sample solution to *M* with potassium iodide.[148] Reduce free iodine with 5% solution of sodium thiosulfate. If the sample is bronze, mask copper with 5 ml of 10% thiourea solution. Extract with 10 ml of 5% solution of **propylfluorone** in chloroform. Add to the separated organic phase 2 ml of ethanolic m*M* antipyrine and 2 ml more of 5% solution of propylfluorone in chloroform to form the 1 : 1 : 2 : 1 complex of stannic ion–propylfluorene–antipyrine–iodide ion: read at 500 nm. The optimum pH for formation of this complex is 2.4–2.6. Beer's law is followed for 0.54–2.02 μg of tin in the extract.

Bromstannic ion forms a 1 : 2 complex with **1-(2-pyridylazo)-2-naphthol**, extractable with 0.025–0.5% solution of the reagent in benzene from a solution 3 *M* in

[142] G. Ackermann and J. Koethe, *Chem. Anal.* (Warsaw) **17**, 445–458 (1972).
[143] H. Pohl, *Metall* **12**, 103–105 (1958).
[144] V. I. Kurbatova, L. Ya. Kalashnikova, G. N. Emasheva, I. N. Nikulina, and V. V. Feofanova, *Tr. Vses. Nauchn.-Issled. Inst. Stand., Obraztsov. Spektr. Etalonov* **5**, 111–118 (1969); N. V. Lebedeva and N. A. Nazarenko, *Tr. Kom. Anal. Khim., Akad. Nauk SSSR, Inst. Geokhim. Anal. Khim.* **11**, 287–298 (1960).
[145] A. Lopez Roman and M. Lachica Garrido, *An. Quim.* **67**, 751–765 (1971).
[146] A. Sh. Shakhabudinov and O. A. Tatev, *Zh. Anal. Khim.* **27**, 2382–2385 (1972).
[147] A. Aleksandrov and P. Vasileva-Aleksandrova, *Zh. Anal. Khim.* **18**, 905–906 (1963).
[148] N. L. Olenovich and G. I. Savenko, *Zavod. Lab.* **41**, 658–659 (1975).

sodium bromide and 0.1–2.5 *N* in sulfuric acid containing 20% of ethanol.[149] A similar complex is formed with iodostannic ion in 1.5 *M* potassium iodide solution. Beer's law is followed for 0.2–9 μg of tin per ml of extract. **6-(2-Quinolylazo)-3,4-xylenol** and **6-(2-pyridylazo)-3,4-xylenol** form similar complexes. Copper and bismuth are masked with thiourea, and there is interference by fluoride, antimony, indium, and palladium.

Tin in smelter zinc can be determined photometrically with sodium **pyrrolidinedithiocarbamate.**[150]

Stannic ion buffered by ammonium acetate at pH 3–3.3 reacts with **quinalizarin** for reading at 550 nm.[151] For iron and steel analysis, tin is collected by manganese dioxide and dissolved in hydrochloric acid with the aid of hydrogen peroxide. Beer's law applies up to 0.2 mg.

The 1 : 1 complex of tin with **rezarson**, which is 5-chloro-3-(2,4-dihydroxyphenylazo)-2-hydroxybenzenearsonic acid, requires 90% ethanol to prevent precipitation.[152] Conditions are *N* acetic acid and maximum absorption at 500 nm.

When a solution of **rhodamine B** (basic violet 10) in 1 : 5 hydrochloric acid is shaken with ethyl acetate, some red color is transmitted to the organic layer.[153] The presence of stannic ion increases this, so that read at 572.5 nm it conforms to Beer's law for up to 0.7 mg of tin for 20 ml. Zinc, mercuric ion, ferric ion, tri- and pentavalent antimony, and bismuth interfere.

When stannic ion in hydrobromic acid is extracted as the rhodamine B complex with benzene, toluene, or xylene, and activated at 555–560 nm, it fluoresces at 575–580 nm. Beer's law is followed up to 2 μg of tin per ml of extract.[154] Indium, thallic ion, gold, hexavalent chromium, and tungsten give a similar fluorescence.

A complex of stannic ion with **salicylfluorone** is formed at an optimum pH of 2 and read at 490 nm.[155]

The 1 : 2 complex of tin with **stilbazo**, which is diammonium 4,4′-bis-(3,4-dihydroxyphenylazo)stilbene-2,2′disulfonate, is stable for 40 minutes in an acetate buffer solution for pH 5.[156] The color intensity is markedly dependent on pH. Beer's law applies up to 0.85 ppm. There is pronounced interference by ferric ion, pentavalent vanadium, hexavalent chromium, tungsten, and molybdenum.

The complex of tin with 1,10-phenanthroline adds **tetrabromo-, tetraiodo-, or dichlorotetraiodofluorescein**, extractable at pH 9 with chloroform.[157] The maximum absorption is at 540 nm, or in 1 : 1 chloroform-acetone, fluorescence is at 580 nm. Fluorescence will determine 0.005 μg/ml, which is one order of magnitude more sensitive than spectrophotometry.

[149]K. Rakhmatullaev and Kh. Tashmamatov, *Zh. Anal. Khim.* **29**, 2402–2407 (1974).
[150]E. Kovacs and H. Guyer, *Z. Anal. Chem.* **186**, 367–288 (1962).
[151]Shigeo Wakamatsu, *Nippon Kinzoku Gakkaishi* **21**, 450–453 (1957).
[152]A. M. Lukin, N. A. Kaslina, V. I. Fadeeva, and G. S. Petrova, *Vest. Mosk. Gos. Univ., Ser. Khim.* **1972**, 247–249.
[153]R. T. Arnesen and A. R. Selmer-Olsen, *Anal. Chim. Acta* **33**, 335–338 (1965).
[154]Yashaharu Nishikawa, Keizo Hiraki, Takeshi Naganuma, and Shozo Niina, *Jap. Anal.* **19**, 1224–1229 (1970).
[155]O. A. Tataev, G. N. Bagdasarov, S. A. Akmedov, Kh. A. Mirzaeva, Kh. G. Buganov, and E. A. Yarysheva, *Izv. Sev-Kavkaz. Nauch. Tsent. Vyssh. Shk., Ser. Estestv. Nauk*, **1973** (2), 30–32.
[156]Masayoshi Ishibashi, Yuroku Yamamoto, and Riichi Todoroki, *Jap. Anal.* **10**, 1272–1275 (1961); cf. A. K. Babko and T. N. Nazarchuk, *Zh. Anal. Khim.* **14**, 174–180 (1959).
[157]D. N. Lisitsyna and D. P. Shchzrbov, *Zh. Anal. Khim.* **28**, 1203–1205 (1973).

3,4′,5,6-Tetrahydroxyflavone complexes with stannic ion in 50% ethanol, 0.1 *N* with hydrochloric acid, for reading 0.7–2.5 ppm at 470 nm.[158] There is interference by oxalate, fluoride, iodate, zirconic, titanic, antimonous, and antimonic ions. The reagent is precipitated by hexavalent molybdenum and tungsten as well as by pentavalent niobium and tantalum. Zirconium and antimony can be masked by tartrate, citrate, or thiosulfate.

Tin as the complex with **tetramethylenedithiocarbamate** is extractable with chloroform from a solution buffered with ammonium citrate at pH 4.5.[159] Accompanying metals can be separated at pH 8.6 for reading the tin complex at 400 or 440 nm.

4-(2-Thiazolylazo)catechol in aqueous ethanol complexes with stannic ion at pH 5–5.5 for reading in the 480–530 nm range.[160]

For analysis of tantalum, oxidize the sample and fuse with potassium hydroxide.[161] Because of many interferences, the tin must be separated, usually by chloroform extraction of the diethyldithiocarbamate, from strongly acid solution.[162] Then convert to solution in 0.2 *N* hydrochloric acid. Add 20% of ethanol and develop the 1:2 complex of tin with **2,6,7-trihydroxy-9-*p*-nitrophenyl-fluorone**. Stabilize with gelatin, set aside for 1 hour, and read at 530 nm. Beer's law is followed for 0.05–5 μg of tin per ml.

Tin in iron and steel can be determined by solvent extraction as the **tris-(1,10-phenanthrolinate)ferrous-trioxalato-stannate ion.**[163] Coprecipitate the tin from acid solution with beryllium hydroxide. Dissolve the precipitate in 5 ml of 1:1 hydrochloric acid, add 1 ml of 1:1 sulfuric acid, and evaporate almost to dryness. Take up in 1 ml of *N* hydrochloric acid and add 10 ml of 1:18 sulfuric acid. Add 2.5 ml of 0.2 *M* oxalic acid and 8 ml of 2 m*M*-tris-(1,10-phenanthrolinate)ferrous sulfate. Dilute to 25 ml and shake with 10 ml of nitrobenzene for 2 minutes. Set aside for 30 minutes, separate the organic layer, and dry it with sodium sulfate. Read at 516 nm. Beer's law is followed for 4–20 μg of tin. There is interference by silver, mercuric, tungstate, molybdate, and titanic ions. The technic is also applicable to the ash of canned juices.[164]

[158]B. S. Garg and R. P. Singh, *Microchem. J.* **18**, 509–519 (1973).
[159]E. Kovacs and H. Guyer, *Z. Anal. Chem.* **208**, 255–262 (1965).
[160]V. Purmalis, J. Putnins, V. Barkane, and E. Gudriniece, *Izv. Akad. Nauk Latv. SSR, Ser. Khim.* **1972** (2), 234–235.
[161]V. A. Nazarenko, E. A. Biryuk, M. B. Shustova, G. G. ShiFareva, S. Ya. Vinkovetskaya, and G. V. Flyantikova, *Zavod. Lab.* **32**, 267–269 (1966).
[162]N. V. Lebedova and V. A. Nazarenko, *Tr. Kom. Anal. Khim. Akad. Nauk SSSR* **11**, 287–298 (1960).
[163]Kazuo Hiiro, Takashi Tanaka, Toyozo Shirai, and Yuroku Yamamoto, *Jap. Anal.* **18**, 563–569 (1969).
[164]Yuroku Yamamoto, Taro Kobayashi, Kazuo Hiiro, and Takashi Tanaka, *J. Hyg. Chem.* **16** (3), 114–118 (1970).

CHAPTER ELEVEN

INDIUM

Reagents for indium include as classes hydroxyanthraquinones, triphenylmethane derivatives, and azo dyes, as well as stilbazo and 3,4-dihydroxyazobenzene.[1] These reagents are also applicable to gallium, and the latter two to aluminum. As that implies, there is a close parallel between the reactions of indium and gallium. The most satisfactory reagents are methylthymol blue, xylenol orange, eriochrome cyanine R 1-(2-pyridylazo-2-naphthol, 4-(2-pyridylazo)resorcinol, and quercetin.[2]

For separation of indium compounds in metallurgical dusts, indium sulfate is extracted with water, then indium sulfide with 3% solution of bromine in methanol, and finally the oxide is extracted by boiling with 1:3 hydrochloric acid.[3] Indium and gallium in 1:1 hydrochloric acid are readily separated, the gallium being extracted with butyl acetate, leaving indium in the aqueous phase.[4a] Addition of bromine water prevents interference by arsenic.

Extraction with molten naphthalene of complexes with 8-hydroxyquinoline is applicable at varying pH levels to indium, zinc, magnesium, cadmium, bismuth, cobalt, nickel, molybdenum, uranium, aluminum, iron, palladium, and copper.[4b]

BROMOPYROGALLOL RED

This reagent forms a 1:1 complex with indium.[4c] The maximum absorption is at 570 nm, but the difference of the complex from the reagent is greater at 610–620

[1]M. Z. Vampol'skii, *Uch. Zap., Kursk. Gos. Pedagog. Inst.* **4**, 116–127, 128–141 (1957).

[2]A. K. Babko and P. P. Kish, *Dopov. Akad. Nauk Ukr. RSR* **1961**, 1323–1326; *Zh. Anal. Khim.* **17**, 693–699 (1962); cf. A. I. Volkova, T. E. Get'man, and T. I. Kukibaev, *Ukr. Khim. Zh.* **35**, 844–850 (1969).

[3]L. N. Terzeman, *Zavod. Lab.* **38**, 1439–1440 (1962).

[4a]L. I. Ganaga, L. M. Bukhtseeva, and L. R. Stanobinets, *Vest. Akad. Navuk. BSSR, Ser. Khim. Navuk.* **1969** (5), 49–53.

[4b]T. Fujinaga, Masatada Sataka, and Masaki Shimizu, *Jap. Anal.* **25**, 313–318 (1976).

[4c]Sh. T. Talipov, Kh. S. Abdullaeva, and G. P. Gor'kovaya, *Uzb. Khim. Zh.* **1962** (5), 16–19.

nm. A maximum absorption can be obtained by ph 3.4–3.6 is the phthalate buffer solution. Color is developed immediately and is stable for 12 hours. Beer's law applies for 0.15–16 μg/ml. Separation of indium from possible interference by extraction as the bromide with butyl acetate is possible.

A technic calling for extraction of the 1:2 complex at pH 9 recovers only about 60% by a single extraction with benzyl alcohol but is reproducible. The sensitivity is superior to oxine but inferior to morin and quercetin. Beer's law then applies for 2.5–20 μg of indium in 30 ml of extract.

Procedure. *Zinc-Base Alloys.*[4d] Dissolve a 4-gram sample in hydrobromic acid and evaporate to dryness. Take up the residue in 5.5 *N* hydrobromic acid and dilute to 50 ml with that acid. Extract the indium with 25 ml of isopropyl ether and back-extract with 10 ml and 10 ml of water. Evaporate the aqueous solution to dryness. Add nitric and perchloric acid and evaporate to dryness to destroy organic matter. Take up the residue in 6 ml of a buffer solution containing 0.13 *M* ammonium chloride and 0.8 *M* ammonium hydroxide. Add 1 ml of 2% ascorbic acid solution and 3 ml of 5% potassium cyanide solution. Adjust to pH 9 and add 10 ml of 0.01% solution of pyrogallol red in 2% ammonium acetate solution. Dilute to 30 ml and extract the 1:2 indium-dye complex with 20 ml of benzyl alcohol. Read at 540 nm.

CATECHOL VIOLET

Indium forms a 1:2 complex with catechol violet, stable in the presence of gelatin.[5] The optimum pH is 5.3, the maximum absorption at 600 nm. Beer's law is followed for 0.1–0.7 μg/ml. The uncomplexed dye should be in about fiftyfold excess; therefore it calls for a correction.[6] There is interference by stannic, aluminum, bismuth, ferric, cupric, and lead ions.

When cetylpyridinium chloride is present, the 1:2:2 complex of indium, quaternary molecule, and catechol violet is extracted with butanol. Beer's law is followed for 0.5–3.2 μg of indium in the extract. The absorption is decreased by EDTA, tartrate, oxalate, phosphate, tetravalent zirconium, tin, or thorium, and by chromic ion. Positive errors are caused by cobalt, lead, copper, aluminum, gallium, manganous ion, and ferric ion.

Procedure.[7] To a sample solution containing 0.25–2.5 mg of indium, add 1 ml of 0.001 *M* catechol violet per 0.5 mg of indium. Add 0.5 ml of a buffer solution containing 86.7 ml of pyridine and 13.3 ml of acetic acid for pH 6. Dilute to 50 ml and let stand for 20 minutes. Read at 600 nm.

[4d] S. G. Jadhav, P. Murugaiyan, and C. Venkateswarlu, *Anal. Chim. Acta* **82**, 391–399 (1976).
[5] M. Malat and M. Hrachovcova, *Collect. Czech. Chem. Commun.* **29**, 1503–1505 (1964).
[6] C. Dragulescu, T. Simonescu, and N. Vilceanu, *Acad. Rep. Pop. Rom. Baza. Cercet. Stiint. Timisoara, Stud. Cercet. Stiint. Chim.* **9** (1–2) (1962).
[7] R. Staroscik and J. Terpilowski, *Chem. Anal.* (Warsaw) **7**, 803–808 (1962).

By Extraction.[8] Mix 1 ml of 0.5 m*M* catechol violet with 10 ml of hexamine buffer solution for pH 6.3 and 1 ml of *M* potassium chloride. Add a sample solution containing 5–32 μg of indium and dilute to 20 ml. Shake for 1 minute with 10 ml of 0.4% solution of cetylpyridinium chloride in butanol and read the extract at 600 nm.

Tin and Cadmium Alloys.[9] Dissolve 0.1 gram of sample in 5 ml of sulfuric acid. Cool, add 5 ml of 30% hydrogen peroxide, and dilute to 50 ml. Mix a 1-ml aliquot, 1 ml of m*M* solution of catechol violet, and 0.2 ml of 4% sodium hydroxide solution. Dilute to 4 ml with an acetate buffer solution for pH 5.5–6 and read at 580 nm.

CHROME AZUROL S

Chrome azurol S (mordant blue 29) absorbs at 490 nm at pH 4. Its indium complex absorbs at 530 nm at pH 3.5–5.[10] Beer's law is followed for 0.45–13.5 ppm. Even below 1 ppm, beryllium, scandium, yttrium, lanthanum, rare earths, thorium, titanic and zirconic ions, hafnium, hexavalent molybdenum, tungsten, ferrous and ferric ions, aluminum, gallium, borate, carbonate, oxalate, and tartrate interfere. Larger amounts of many other ions interfere.[11] The indium complex conforms to Beer's law for 0.2–50 μg/ml. Addition of cetyltrimethylammonium bromide increases the sensitivity.[12]

When developed with chrome azurol S in 50% ethanol, the system is buffered at 5.5 with acetate–hydrochloric acid, and indium is read at 590 nm.[13] The controlled pH eliminates interference by cobalt. Ferric ion can be masked with ascorbic acid.

Procedure. Add at least a fourfold excess of 0.1 *M* dye and adjust to pH 4. Dilute to 25 ml and let stand for 15 minutes. Read at 530 nm against a reagent blank.

5,7-DIBROMO-8-HYDROXYQUINOLINE

This reagent, also called 5,7-dibromo-8-quinolinol, is referred to as dibromooxine.[14] Neutral salts inhibit the rate of color development. There is interference by stannous, antimonous, ferric, aluminum, and cupric ions. Aluminum can be masked with sulfosalicylic acid. Beer's law is followed up to 0.1 mg of indium in 5 ml of chloroform extract. Both the blank and absorption are less than for the

[8] Tsutomu Ishito and Koichi Tonosaki, *Jap. Anal.* **20**, 689–693 (1971).
[9] G. N. Trochinskaya and P. O. Knizhko, *Ukr. Khim. Zh.* **36**, 950–953 (1970).
[10] Satendra P. Sangal, *Chemist-Analyst* **56**, 101 (1967).
[11] R. Staroscik and P. Ladogorski, *Chem. Anal.* (Warsaw) **9**, 97–102 (1964).
[12] B. Evtimova and D. Nonova, *Anal. Chim. Acta* **67**, 107–112 (1973).
[13] Prodrome B. Issopoulos and Andre Galinos, *Analysis* **2**, 672–674 (1973–1974).
[14] J. E. Johnson, M. C. Lavine, and A. J. Rosenberg, *Anal. Chem.* **30**, 2055–2056 (1958).

indium-oxine complex, which has a maximum at 395 nm. A large excess of reagent is essential. At pH 2 the complex is extracted with ethyl acetate and read at 413 nm.[15]

For analysis of semiconductors, sulfides are dissolved in 1:1 hydrochloric acid and selenides in sulfuric acid containing bromine.[16] Indium is extracted with 5,7-dibromo-8-hydroxyquinoline.

Procedure. To a sample containing up to 0.1 mg of indium in a 15-ml graduated glass-stoppered centrifuge tube, add 2.5 ml of 4% potassium acid phthalate. Adjust to pH 3.5–4.5 and dilute to 10 ml. Add 5 ml of 0.1% solution of 5,7-dibromo-8-quinolinol in chloroform. Shake at intervals over 5 minutes. Centrifuge and read the extract at 415 nm against chloroform.

Zinc and Lead Ores.[17] Dissolve 1 gram of sample in 10 ml of hydrochloric acid and 3 ml of nitric acid. Dilute to 50 ml, filter, and evaporate to dryness. Take up the residue in 20 ml of water and add 8 ml of 1:4 sulfuric acid. Filter off precipitated sulfates and evaporate to sulfur trioxide fumes. Add 2 ml of water and again take to sulfur trioxide fumes. Take up with 20 ml of water and make alkaline with 1:60 ammonium hydroxide containing 2.7% of ammonium chloride. Let the precipitated hydroxides settle and wash well by decantation. Dissolve the precipitate in a minimum amount of 1:1 nitric acid, dilute to 50 ml, and precipitate with 1:20 ammonium hydroxide. Wash by decantation. Again dissolve and reprecipitate. Dissolve the precipitate in 5 ml of nitric acid. Add 10% ascorbic acid solution dropwise to decolorize the ferric ion. Add 10 ml of 50% potassium iodide solution and extract the indium iodide with 5 ml and 5 ml of ether. Reextract the indium from the ether extract with 10 ml of water. Decolorize the aqueous extract by dropwise addition of 10% ascorbic acid solution. Add 10 ml of phthalate buffer solution for pH 3.5 and 2 ml of 10% sulfosalicylic acid solution. Dilute to 25 ml. Extract with 4 ml and 4 ml of 0.1% solution of dibromooxine in chloroform and read at 420 nm.

Silicate Solutions.[18] The sample may contain 500 parts of aluminum and 100 parts of iron. As a cyanide reagent, mix 253 ml of 1:7 ammonium hydroxide and 377 ml of 1:7 nitric acid; cool, and add 200 ml of 10% potassium cyanide solution. Dilute to 1 liter. As a phthalate buffer solution for pH 3, mix 10 ml of nitric acid and 10 ml of ammonium hydroxide. Add 1 gram of potassium phthalate and dilute to 1 liter.

Mix 10 ml of sample solution containing 1–5 mg of indium per ml with 10 ml of the cyanide reagent. Add 1 ml of 1% solution of hydroxylamine and dilute to 25 ml. Extract by shaking for 2 minutes with 5 ml, 5 ml, and 5 ml of 0.01% dithizone solution in chloroform. Wash the aqueous layer with 2 ml of chloroform. Shake the combined chloroform extracts for 1 minute with 20 ml of 1:99 nitric acid and wash the chloroform with 5 ml of water. Wash the nitric acid extract and washings with

[15] Nguyen Shi Zuong and F. G. Zharovskii, *Ukr. Khim. Zh.* **36**, 1273–1278 (1970).
[16] R. P. Pantaler, N. B. Lebed', and L. N. Semenova, *Tr. Kom. Anal. Khim.* **16**, 24–29 (1968).
[17] Z. Gregorowicz and M. Marczak, *Chem. Anal.* (Warsaw) **14**, 159–164 (1969).
[18] J. Minczewski, U. Stolarczyk, and Z. Marczenko, *Chem. Anal.* (Warsaw) **6**, 57–61 (1961).

2 ml of chloroform. Titrate the nitric acid solution to the methyl orange end point with 1:30 ammonium hydroxide. Add 25 ml of the phthalate buffer solution and dilute to 60 ml. Shake for 1 minute with 10 ml and 10 ml of either 1% oxine or 1% 5,7-dibromooxine solution in chloroform. Filter the extract and dilute to 25 ml with the extracting solution. Read at 450 or 420 nm, respectively.

Ores. Roast 1 gram of powdered sample at red heat for 30 minutes. Cool; add 0.5 ml of perchloric acid and 15 ml of hydrofluoric acid. Heat slowly to evolution of perchloric acid fumes. Add 0.5 ml of perchloric acid and 5 ml of hydrofluoric acid and heat to perchloric acid fumes. Repeat this step three times. Add 2 ml of perchloric acid and evaporate to perchloric acid fumes. Repeat this step twice more. Take up with 25 ml of water and heat for 1 hour to complete solution. Add 18 ml of 10% solution of sulfosalicylic acid neutralized to pH 9 with ammonium hydroxide and 1 ml of 10% hydroxylamine solution. Neutralize to pH 9 with ammonium hydroxide. Add 10 ml of the cyanide reagent as described for silicate solutions. Extract with four successive 10-ml portions of 0.01% solution of dithizone in chloroform. Complete as for silicates from "Shake the combined chloroform extracts...."

Cadmium Sulfide Crystals.[19] Dissolve 1 gram of sample in nitric acid and evaporate to dryness. Dissolve the residue in water and dilute to 50 ml. Mix an aliquot with 10 ml of a buffer solution consisting of equal volumes of 0.2 *N* acetic acid and 0.2 *N* ammonium hydroxide for pH 3.4. Dilute to 25 ml and add 5 ml of 0.1 *M* oxine or dibromooxine. Extract with 10 ml and 10 ml of chloroform. Dilute to 25 ml with chloroform and read at 451 or 415 nm, respectively. Cadmium is not extracted if the pH is below 5.5.

5,7-DICHLORO-8-HYDROXYQUINOLINE

This dichlorooxine complexes with indium for extraction by dichlorobenzene.[20] There is interference by the following ions: aluminum, bismuth, cobaltous, cupric, ferrous, ferric, gallic, nickel, antimonous, titanous, and zinc.

Procedure. *Tin.* Prepare a chromatographic column 1 cm in diameter of 3 grams of 50-100 mesh Dowex 1-X8 in chloride form. As an effluent solution, dissolve 1% of hydroxylamine hydrochloride in 1:11 hydrochloric acid.

Dissolve 1 gram of sample containing less than 5% indium in 5 ml of hydrochloric acid and evaporate nearly to dryness. Dilute to 100 ml with the effluent solution. Cover the column contents with the effluent solution and transfer an aliquot of the sample solution containing less than 5 mg of indium and less than 120 mg of metals to the column. Pass through the column at 5–10 ml/minute, following with effluent solution until about 70 ml of eluate has been collected. Add

[19] O. P. Kulik and I. B. Mizetskaya, *Zavod. Lab.* **31**, 150–151 (1965).
[20] B. A. Raby and C. V. Banks, *Anal. Chim. Acta* **29**, 532–538 (1963).

10 ml of nitric acid to the eluate. Evaporate nearly to dryness to oxidize the hydroxylamine and any stannous ion not adsorbed. Take up in water and dilute so that a 25-ml aliquot contains less than 1.5 mg of indium. Add 10 ml of a buffer solution containing 143 ml of acetic acid and 61.2 gram of sodium acetate trihydrate per liter for pH 4. Add 15 ml of water and shake for 5 minutes with 50 ml of 0.1% solution of dichlorooxine in 1,2-dichlorobenzene. Read the extract at 415 nm against a reagent blank. When preparing the calibration curve, appropriate amounts of tin should be present with the indium.

DITHIZONE

After appropriate separation of indium from interferences, it is extracted with dithizone in organic solvent. Lead and cadmium interfere. More than 0.2 *M* zinc decreases the extraction. Hydroxylamine reduces the iron, minimizes its interference, and improves the extraction of indium. Arsenic, antimony, tin, and thallium interfere. If large amounts of iron are present, they are removed by extraction with isopropyl ether from 7 *M* hydrochloric acid. Indium dithizonate obeys Beer's law for 0.25–10 μg/ml in chloroform.[21] At pH 8.3–9.6 in a solution containing ammonium citrate and potassium cyanide, 0.02% solution of dithizone in chloroform extracts indium, lead, bismuth, thallium, and stannous ion. Bismuth in large amounts can be preextracted at pH 3.5–4. Indium, lead, and thallium are back-extracted with 1 : 100 nitric acid. Then lead and bismuth are complexed at pH 5–6 with sodium thiosulfate. Since thallium is not extracted at that pH, the indium is individually extracted with 0.002% solution of dithizone in carbon tetrachloride.

Procedure. *Zinc or Zinc Oxide.*[22] As a stock dithizone solution, dissolve 25 mg in 500 ml of chloroform. As the working 0.002% solution, dilute 100 ml to 250 ml with chloroform. Refrigerate. To prepare a buffer solution, titrate 150 ml of 10% hydroxylamine hydrochloride solution to pH 9 with about 270 ml of 1 : 14 ammonium hydroxide. Add 40 ml of 5% potassium cyanide solution and dilute to 500 ml. Extract with 10 ml of 0.002% dithizone in chloroform and discard the extract.

Weigh 1 ± 0.1 gram of sample expected to contain up to 0.1 mg of indium and cover with water. Gradually add 10 ml of hydrochloric acid. Then add 3 ml of perchloric acid and evaporate to fumes of perchloric acid so strong that the sample will completely solidify on cooling. Dissolve in 50 ml of 6 *N* hydrobromic acid. Shake for 2 minutes with 25 ml of isopropyl ether and discard the aqueous layer. Wash the organic phase with 10 ml of 6 *N* hydrobromic acid.

Shake the organic phase with 25 ml of water to reextract the indium, and discard the isopropyl ether. Mix an aliquot of the aqueous phase containing less than 0.02 mg of indium with 50 ml of the buffer solution. Let stand for 1 hour to complete

[21] V. T. Athavale, T. P. Ramachandran, M. M. Tillu, and G. M. Vaidya, *ibid.* **22**, 56–60 (1960).
[22] T. A. Collins, Jr., and J. H. Kanzelmeyer, *Anal. Chem.* **33**, 245–247 (1961).

reduction of iron. Shake for 2 minutes with 20 ml of 0.002% solution of dithizone in chloroform. Read the chloroform extract at 510 nm against a blank.

GLYOXAL BIS(2-HYDROXYANIL)

This reagent is suitable for determination of indium as well as scandium in the presence of many other cations.[23] It also applies to solutions of yttrium, samarium, and europium isolated from other cations. A 100-fold amount of zinc, cadmium, or lanthanides is tolerated.

Procedure. Adjust a sample not exceeding 5 ml, containing 0.002–0.25 mg of indium, to pH 2.2–3.2. Add 2.5 ml of 16% solution of ammonium nitrate and 15 ml of 0.02 *M* reagent in ethanol. Heat at 80° for 15 minutes, cool, and dilute to 25 ml with ethanol. Read at 580 nm.

8-HYDROXYQUINOLINE

Indium complexes with 8-hydroxyquinoline, also known as 8-quinolinol and oxine. Readings are photometric or by fluoresence.[24] The reagent is applicable to ores containing sulfates and to lead and zinc sulfides.[25] The indium-oxine complex is extracted with methyl isobutyl ketone at pH 4–12 and is read at 451.1 nm.[26] Interference by iron, copper, and thallium is avoided by extracting their oxinates at appropriate pH levels. Iron can be reduced with ascorbic acid and copper can be masked by thiourea, but bismuth, zirconium, and thallium must be absent.

For analysis of gold, dissolve 10 grams of sample in aqua regia.[27] Dilute and extract chloraurate with isopropyl ether. Then determine indium with 8-hydroxyquinoline. Aluminum also reacts with that reagent.

Indium oxinate can be selectively extracted at pH 5.5 from the corresponding complexes of gallium and beryllium.[28] 8-hydroxy-2-methylquinoline, which is 2-methyloxine has also been used, but in absolute ethanol.[29a]

The yellowish-green fluorescence of indium oxinate in chloroform is read at 526 nm.[29b] The aluminum complex fluoresces at 510 nm, that of gallium at 526 nm. The lifetime of the fluorescence of the chelates varies.

[23]P. Bocek and M. Vrchlabsky, *Chem. Prum.* **16**, 625–627 (1966); M. Vrchlabsky and P. Bocek, *Spis. Prir. Fak. Univ. Brne, E* **33** (2), 63–70 (1967).

[24]L. B. Ginsburg, *Izv. Akad. Nauk Kaz. SSR, Ser. Khim.* **1957** (1), 94–98.

[25]V. F. Abramova, A. G. Nagaeva, M. Usubakumov, and S. V. Bleshinskii, *Izv. Akad. Nauk Kirgiz. SSR, Ser. Estestv. Tekh. Nauk* **2** (11), 67–77 (1960); P. K. Agasyan, *Nauch. Dok. Vyssh. Shk., Khim. Khim. Tekhnol.* **1958** (2), 308–313.

[26]Hidehiro Goto and Emiko Sudo, *Jap. Anal.* **10**, 456–462 (1961).

[27]Krzysztof Kasiura, *Chem. Anal.* (Warsaw) **20**, 809–816 (1975).

[28]Masoyoshi Ishibashi, Tsunenobu Shigmatsu, and Yashuharu Nishikawa, *J. Chem. Soc. Jap.* **78**, 1143–1146 (1957).

[29a]W. E. Ohnesorge, *Anal. Chem.* **35**, 1137–1142 (1963).

[29b]Yashaharu Nishikawa, Keizo Hiraka, Kiyotoshi Morishige, Koichi Takahashi, Tsunenobu Shigematsu, and Taro Nogami, *Jap. Anal.* **25**, 459–463 (1975).

Procedure.[30] Adjust a sample solution containing 0.04–0.36 mg of indium to 4 *N* with sulfuric acid. Add 3.3 grams of potassium iodide or 2.4 grams of potassium bromide and dilute to 20 ml with 1:35 sulfuric acid. If potassium iodide is present, add a few crystals of sodium thiosulfate and extract with 10 ml of a 0.3% solution of antipyrinylmethane or 1,1-diantipyrinylbutane in dichloroethane. If potassium bromide is present, use 10 ml of 0.3% diantipyrinylmethane in dichloroethane. Separate the extract, wash the aqueous layer with 3 ml of dichloroethane, and add the washings to the extract. Discard the aqueous phase. If potassium iodide was used, wash the organic phase with 10 ml of 16.6% solution of potassium iodide in 1:35 sulfuric acid. Correspondingly, if potassium bromide was used, wash with 10 ml of 12% solution of potassium bromide in 1:35 sulfuric acid. Add to the organic phase 6 ml of 0.5% solution of oxine in dichloroethane and shake with 5 ml of acetate buffer solution for pH 4.5. Separate the aqueous layer, wash it with 3 ml of dichloroethane, and add the washings to the organic extract. Dilute to 25 ml with dichloroethane and read at 395 nm.

Germanium.[31] As a buffer solution for pH 3.5, add successively 40 ml of nitric acid, 40 ml of ammonium hydroxide, and 4 grams of potassium acid phthalate to 2 liters of water, and dilute to 4 liters. Neutralize to pH 3.5 with ammonium hydroxide.

Treat 2 grams of sample with 2 ml of perchloric acid, 8 ml of nitric acid, and 42 ml of hydrochloric acid. Heat to reduce the volume to under 10 ml. When the solution starts to get cloudy, add 15 ml of hydrochloric acid and boil to dissolve the precipitated germanium oxide. Heat to copious fumes of perchloric acid to remove most of the nitric acid. Add 15 ml of hydrochloric acid and heat to perchloric acid fumes to expel the last of the germanium. Reduce to about 0.2 ml over a free flame. Wash down the sides of the container with 10 ml of water and add 2 ml of 10% ammonium citrate solution and a drop of *m*-cresol purple indicator solution. Neutralize by dropwise addition of ammonium hydroxide until the indicator turns yellow. Add 5 ml of 1:9 perchloric acid. Extract with 10 ml and 2 ml of 0.1% solution of dithizone in chloroform. Discard the chloroform layers, which contain bismuth. Wash the aqueous layer with 5 ml and 2 ml of chloroform. If the solution shows a pronounced pink because of the presence of chromium, add a couple of drops more of *m*-cresol purple indicator. Neutralize by dropwise addition of ammonium hydroxide, first to yellow, then to purple. Add 5 ml of 10% sodium cyanide solution, followed by 10 ml of 0.01% solution of dithizone in chloroform. Shake for 1 minute and drain the chloroform into another separatory funnel. Wash down the sides with 2 ml of chloroform and join this with the other extract. Repeat the extraction and washing step, adding the chloroform to the first dithizone extract. Wash with 50 ml of water containing 1 drop of ammonium hydroxide per 500 ml and drain the chloroform layer. Wash down the sides of the separatory funnel with 2 ml of chloroform and add to the chloroform extract. Add 0.5 ml of nitric acid and evaporate the chloroform. Add 1 ml of perchloric acid to the

[30] V. K. Akimov, A. I. Busev, and K. A. Zhgenti, *Zh. Anal. Khim.* **27**, 1941–1944 (1972); cf. Wei-ch'un Teng, *Khim., Fiz.-Khim. Spektr. Metody Issled. Rud Redk. Rasseyan. Elem. Min., Geol. Okhr. Nedr SSSR* **1961**, 47–56.

[31] C. L. Luke and Mary E. Campbell, *Anal. Chem.* **25**, 1588–1593 (1953); **28**, 1340–1342 (1956).

residual liquid and heat to copious fumes of perchloric acid. When oxidation is complete, evaporate to about 0.2 ml. Dilute to 20 ml and add a drop of *m*-cresol purple indicator solution and a piece of Congo red paper. Neutralize by dropwise addition of ammonium hydroxide until the solution starts to turn yellow. Then continue with 1:9 ammonium hydroxide until the edges of the Congo red paper just start to turn red. Transfer to a separatory funnel, using 25 ml of pH 3.5 buffer solution to promote the transfer. Extract with 20 ml of 0.1% solution of oxine in chloroform. Filter the extract through paper and read at 451 nm.

Germanium Dioxide. Dissolve 3 grams of sample in 2 ml of perchloric acid, 8 ml of nitric acid, and 42 ml of hydrochloric acid. When in solution, heat to copious fumes of perchloric acid to remove most of the nitric acid. All the germanium should have been distilled, but in case of doubt add 5 ml of hydrochloric acid and again evaporate to copious fumes of perchloric acid. Complete as for germanium from "Reduce to about 0.2 ml over a free flame."

Ores.[32] This technic provides separation from iron, tungsten, molybdenum, thallous ion, antimony, gallium, tin, ferrous sulfide, copper, lead, and zinc.

Digest 1 gram of sample with 20 ml of aqua regia. Evaporate to a small volume and add methyl orange indicator. Add 1:5 ammonium hydroxide until the pink color disappears, boil for a couple of minutes, and let stand for coagulation. Filter, and wash the precipitate four times with hot 2% solution of ammonium chloride. Dissolve the precipitate with 1:1 hydrochloric acid and evaporate nearly to dryness. Add 5 ml of hydrobromic acid and evaporate nearly to dryness. Add 2 ml of hydrobromic acid and 1 ml of bromine water and evaporate to dryness. Take up the residue in 10 ml of 1:11 hydrochloric acid and add 10% sodium thiosulfate solution dropwise to decolorize. Add a few drops in addition but a total of less than 5 ml. Filter and wash well with 1:11 hydrochloric acid. Shake with 25 ml of ether presaturated with 1:11 hydrochloric acid. Withdraw the aqueous layer and again extract with the presaturated ether. Saturate 1:11 hydrochloric acid with ether, add a crystal of sodium thiosulfate, and wash the combined ether extracts with 3 ml and 3 ml of this prepared wash solution. Extract these acid washings with 10 ml of ether saturated with 1:11 hydrochloric acid and add to the combined ether extracts. Extract indium from the ether with 20 ml and 20 ml of 1:1 hydrochloric acid containing a couple of drops of 30% hydrogen peroxide. Evaporate this acid extract to a known volume and take an aliquot. Add a drop of methyl orange indicator and 1:5 ammonium hydroxide until just alkaline. Add 0.75 ml of 1:11 hydrochloric acid, 1 ml of 1% thiourea solution, 1 ml of 10% solution of hydroxylamine hydrochloride, and 1 ml of 1.36% solution of sodium acetate trihydrate to buffer to pH 3.6. Shake with 3 ml of 15% oxine solution in carbon tetrachloride. Read the extract at 450 nm against a reagent blank.

Sulfide Ores.[33] The method is appropriate for ores containing iron, copper, lead, and zinc with indium down to 0.005%. Heat 1 gram of sample with 15 ml of hydrochloric acid until evolution of hydrogen sulfide ceases. Add 10 ml of nitric

[32]M. V. Kanyukova, *Min. Tsetn. Met. SSSR* **1957** (19), 1–13.
[33]V. K. Akimov, K. A. Zhgenti, and A. I. Busev, *Zavod. Lab.* **39**, 948 (1973).

acid to complete disintegration and evaporate nearly to dryness. Add 10 ml of 1 : 1 sulfuric acid and evaporate to dryness. Take up the residue in 15 ml of 1 : 35 sulfuric acid, filter, and wash with that acid. Add 10% ascorbic acid solution dropwise to reduce ferric ion and 2 ml of 5% thiourea solution to mask copper. Make the sample solution *M* with potassium iodide by adding 5 grams. Extract the indium by shaking with 10 ml of 0.3% solution of diantipyridinylbutane in chloroform. Separate the aqueous phase, wash with 3 ml of chloroform, and add the washings to the organic extract. Wash the combined chloroform extracts with 10 ml of 1 : 35 sulfuric acid containing 16.6% of potassium iodide. Add 6 ml of 0.5% solution of oxine in chloroform and shake with 5 ml of an acetate buffer solution for pH 4.5. Wash this aqueous phase with 3 ml of chloroform and add it to the main solution. Dilute the chloroform solution to 25 ml with chloroform and read at 451 nm for 0.04–0.36 mg of indium.

Silicate Solutions and Silicate Ores. See 5,7-dibromo-8-hydroxyquinoline.

Cadmium Sulfide Crystals. See 5,7-dibromo-8-hydroxyquinoline.

METHYLTHYMOL BLUE

Although methylthymol blue forms 1 : 1 and 1 : 2 complexes with indium, the latter is more appropriate for photometric determination.[34] Optimum conditions are an acetate-buffered medium for about pH 4, tenfold excess of reagent, and reading at 600 nm. Beer's law is followed for 0.2–4 μg of indium per ml. There is interference by aluminum, bismuth, nickel, and ferric ion and to some extent by other cations, EDTA, nitrilotriacetic acid, phosphate, citrate, and oxalate.[35] The 1 : 1 complex at pH 2–4 is read at 600 nm.[36]

Procedure. *Fuming Oxide.* The typical sample is defined as containing 11–14% lead, 60–65% zinc, 0.2% arsenic, 0.0173% indium, and about 0.1% chloride, 0.1% carbon, 0.02% fluoride, and 0.3% iron. The technic given below is also applicable with xylenol orange.

Dissolve 1 gram of sample in 8 ml of hydrochloric acid and 22 ml of nitric acid and evaporate to dryness. Add 10 ml of hydrochloric acid and evaporate to dryness. Repeat that step twice more. Dissolve in 50 ml of water with 0.3 ml of hydrochloric acid, heat to boiling, and set aside for 2 hours. Filter the lead chloride and add 5 ml of 30% ammonium chloride solution to the filtrate. Heat to boiling, precipitate the hydroxides with ammonium hydroxide, and wash with 3% ammonium chloride solution. Zinc, copper, and cadmium are separated in the filtrate. Dissolve the precipitate with 50 ml of 1 : 5 hydrochloric acid and reprecipitate with ammonium hydroxide. Wash the precipitate with 3% ammonium chloride

[34] B. P. Karadakov, P. N. Kovalenko, and K. N. Bagdasarov, *Zh. Anal. Khim.* **24**, 682–687 (1969).
[35] Koichi Tonosaki and Tsutomu Ishito, *Jap. Anal.* **18**, 1096–1100 (1969).
[36] V. N. Tikhonov, *Zh. Anal. Khim.* **21**, 1172–1178 (1966).

solution. Dissolve the precipitate with 10 ml of 5*N* hydrobromic acid and add a few drops of 10% titanous chloride to form a pale violet color that persists for about 4 minutes. This avoids coextraction of iron. Shake for 1 minute with 30 ml of butyl acetate to extract the indium. Wash the extract with 5 ml and 5 ml of 5 *N* hydrobromic acid. Extract the indium from the butyl acetate by shaking for 1 minute with 10 ml, 10 ml, and 10 ml of 1 : 1 hydrochloric acid containing 0.2 ml of 30% hydrogen peroxide. Evaporate the combined extracts to dryness and take up in 25 ml of water. To a 5-ml aliquot, add 10 ml of acetate buffer solution for pH 4 and 1 ml of 0.01 *M* methylthymol blue. Dilute to 50 ml and read at 510 nm against a reagent blank.

PHENOSAFRANINE

The complex of indium with phenosafranine, which is 3,7-diamino-5-phenylphenazinium chloride, is unstable in aqueous solution but is stable for more than 24 hours after extraction with an organic solvent.[37] The method will determine 0.15 μg of indium per ml. An equal amount of ferric or gold ion is tolerated.

Procedure. Mix a sample containing 1–50 μg of indium with 3 ml of hydrobromic acid and 2 ml of 0.4% solution of phenosafranine. Dilute to 10 ml and shake for 2 minutes with 20 ml of 2 : 1 benzene–methyl ethyl ketone. Separate the phases and rinse the container with 3 ml of the solvent. Dilute the solvent phase to 25 ml and read at 530 nm against a reagent blank.

PHENYLFLUORONE

Indium forms a 1 : 2 complex with phenylfluorone.[38] The useful concentration is 0.2–1.4 μg of indium per ml. Small amounts of lead and thallium do not interfere. Conditions for **2-hydroxyphenylfluorone** and **2,4-disulfophenylfluorone** are similar.[39]

Procedure. *Lead Concentrates.* Dissolve 1 gram of sample in 10 ml of 1 : 1 nitric acid and precipitate lead with 5 ml of 1 : 1 sulfuric acid. Filter and evaporate to dryness. Take up with 20 ml of water and add 2 ml of 2% solution of sulfosalicylic acid and 1 ml of 5% potassium cyanide solution. Extract indium, lead, and thallium with 10 ml of 0.001% solution of dithizone in carbon tetrachloride. Extract the indium from the organic layer with 10 ml of 1 : 100 nitric acid. Add 1 ml of 0.5% solution of gelatin, 1 ml of a pyridine–acetic acid buffer solution for pH 5.5, and 5

[37]G. Popa and C. Patroescu, *Rev. Chim.* (Bucharest) **18** (4), 244–245 (1967); **21**, 770–774 (1970).
[38]U. Stolarczykowa and J. Mincewski, *Chem. Anal.* (Warsaw) **9**, 151–160 (1964); U. Stolarczykowa, *ibid.* 161–166.
[39]E. A. Biryuk, V. A. Nazzrenko, and R. V. Ravitskaya, *Zh. Anal. Khim.* **24**, 1337–1340 (1969).

ml of 0.005% ethanolic solution of phenylfluorone. Dilute to 25 ml with 40% ethanol and read at 540 nm.

1-(2-PYRIDYLAZO)-2-NAPHTHOL

At pH 3 in ammonium acetate solution, indium forms a 1 : 1 complex with this reagent, often abbreviated as PAN, which is stable up to pH 6.[40] The complex is extractable with butanol or isoamyl alcohol. For determination, mask cadmium with potassium iodide.

With an acetate buffer solution at pH 5.3–6.7, a 1 : 2 complex is formed.[41] This is extractable with chloroform and read at 560 nm for less than 5 μg/ml. Ferric ion can be read at 775 nm, and iron and indium can be calculated by simultaneous equations.

Procedure.[42] As the reagent, dissolve 40 mg of 1-(2-pyridylazo)-2-naphthol in a mixture of 150 ml of dimethylformamide and 50 ml of 10% solution of ammonium acetate.

If gallium is present, extract up to 0.025 mg from 0.045 mg of indium in 7 *N* hydrochloric acid with 10 ml of isopropyl ether. Evaporate to dryness the aqueous phase, which contains other metals along with the indium. Take up the residue in 5 ml of 0.01 *N* hydrochloric acid, add 1 ml of the prepared reagent, and dilute to 10 ml. Set aside for 10 minutes and read at 545 nm against a reagent blank.

4-(2-PYRIDYLAZO)RESORCINOL

This reagent, abbreviated as PAR, absorbs at 390 nm at pH 4. The 1 : 2 complexes of indium or gallium with PAR show maxima at 510 nm; the complex with thallium exhibits a maximum at 520 nm.[43] The optimum pH range for indium and thallium is 3.4–4.5, that for gallium is 3.7–4.5. At least fivefold *M* concentration of the reagent is required. Aside from interference of gallium and thallium with the determination of indium, there is interference by thorium, zirconium, hexavalent uranium, pentavalent vanadium, and niobium, tantalum, ferric ion, nickel, cobalt, palladium, copper, zinc, cadmium, and aluminum. The complex conforms to Beer's law for 0.2–24 μg/ml.[44] The maximum extinction of the complex is not affected by

[40] A. I. Busev and V. M. Ivanov, *Izv. Vyssh. Ucheb. Zaved., Khim. Khim. Tekhnol.* **5**, 202–209 (1962).
[41] Shozo Shibate, *Anal. Chim. Acta* **23**, 434–438 (1960).
[42] K. L. Cheng and B. L. Goydish, *ibid.* **34**, 154–164 (1966).
[43] C. D. Dwivedi, K. N. Munshi, and A. K. Dey, *Chemist-Analyst* **55**, 13 (1966); cf. A. I. Busev and V. M. Ivanov, *Izv. Vyssh. Ucheb. Zaved., Khim. Khim. Tekhnol.* **5** (2), 202–209 (1962); M. Hnilickova, *Collect. Czech. Chem. Commun.* **29**, 1424–1431 (1964).
[44] P. P. Kish and S. T. Orlovskii, *Zh. Anal. Khim.* **17**, 1057–1062 (1962); cf. Kazuyoshi Hagawara and Isao Muraki, *Jap. Anal.* **10**, 1022–1024 (1961).

the presence of methanol, ethanol, propanol, butanol, isobutyl alcohol, isopentyl alcohol, methyl ethyl ketone, acetone, or dioxane.[45] A large excess of bromide ion has little effect.

Procedure. To a solution containing 1–2 ppm of indium, add at least 5 times the molar amount of PAR. Adjust to pH 4 ± 0.2 and let stand for 30 minutes. Read at 520 nm against water.

2-(2-PYRIDYL)BENZIMIDAZOLE

This reagent is appropriate for reading fluorescence due to indium down to the nanogram range, in the absence of interferences.[46] The reagent also fluoresces with gallium and zinc. The fluorescence is quenched by cobaltous, ceric, copper, ferric, mercuric, nickel, platinic, palladic, silver, molybdate, and sulfide ions. Cadmium enhances the fluorescence. Interference is avoided by reducing cerium and iron to cerous and ferrous ions.

Procedure. To 10 ml of sample solution containing 0.11–10 μg of indium, add 1 ml of 50% ammonium acetate solution as buffer and 1 ml of 25% sodium benzoate solution. Shake for 2 minutes with 3 ml and 3 ml of ethyl acetate. Combine the extracts, add 2 ml of water containing 2 drops of 0.1 *N* hydrochloric acid, and evaporate the solvent. To the residual aqueous phase containing the indium, add 0.5 ml of pH 5.2 buffer solution and 0.5 ml of 10^{-4} *M* reagent. Let stand for 30 minutes and irradiate at 335 nm. Read the fluorescence at 411 nm.

RHODAMINE B

The complex of rhodamine B (basic violet 10) with indium is appropriate for photometric reading for up to 0.025 mg of indium per ml.[47] For extraction from 4.5 *N* sulfuric acid–2 *M* potassium bromide, appropriate solvents are 6:3:1 benzene–isopropyl ether–acetylacetone and 5:3:2 benzene–isopropyl ether–methyl isobutyl ketone.[48]

Indium is separated from gallium on a column of 4% of trioctylamine on polytetrafluoroethylene by elution with 9 *N* hydrochloric acid.[49] The gallium is then eluted with 0.05 *N* hydrochloric acid.

For fluorescent reading, the complex is extracted with benzene from 3–6 *N*

[45] V. I. Ivanov, E. P. Tsintsevich, and A. N. Gorokhova, *Vest. Mosk. Univ., Ser. Khim.* **1964** (3), 69–72.

[46] L. S. Bark and A. Rixon, *Anal. Chim. Acta* **45**, 425–432 (1969).

[47] J. Jakovsky, *Hutn. Listy* **21**, 274–276 (1966).

[48] A. Garcic and L. Sommer, *Collect. Czech. Chem. Commun.* **35**, 1047–1065 (1970).

[49] I. P. Alimarin, T. A. Bol'shova, V. A. Lukashenkova, and N. V. Arslanova, *Zh. Anal. Khim.* **29**, 1558–1562 (1974).

sulfuric acid–2 *N* with potassium bromide and containing 5% of acetone.[50] The acetone affects the solubility of the complex but does not affect the solubility of rhodamine B.

Procedures

Zinc Concentrates. Dissolve 1 gram of sample in 15 ml of hydrochloric acid. Add 5 ml of nitric acid and drive off the brown fumes. Add 2 ml of 1:1 sulfuric acid and evaporate to dryness. Take up the residue in 10 ml of 7:3 hydrobromic acid, cool, and filter. Wash the filter with 5 ml, 5 ml, and 5 ml of 7:3 hydrobromic acid. Shake for 1 minute with 25 ml of isopropyl ether. Discard the aqueous layer and remove iron from the organic layer by shaking for 1 minute with 10 ml, 5 ml, and 3 ml of 3:2 hydrobromic acid–15% titanous sulfate solution. Reextract indium from the organic phase with 10 ml and 10 ml of water. Add 15 ml of 1:1 sulfuric acid to the combined aqueous extracts and boil for 5 minutes to drive off isopropyl ether. Dilute to 50 ml. To a 10-ml aliquot, add a few crystals of ascorbic acid and 3 grams of solid potassium bromide. When dissolved, add 1 ml of 0.25% solution of rhodamine B and 10 ml of 9:1 benzene-acetone. Shake for 1 minute and read the extract at 540 nm against a blank.

Cassiterite.[51] Grind 0.2 gram of sample in an agate mortar with 5 grams of 1:1 sodium carbonate–sodium tetraborate. Place in a platinum crucible and cover with a layer of the carbonate-borax mixture. Heat slowly until fusion begins and hold at that temperature for 15 minutes. Leach the fused mass with 50 ml of water and filter. Add 60 mg of iron as ferric chloride, dilute to 60 ml, and heat to boiling. Collect the coagulated precipitate on a coarse filter washed with 5% sodium hydroxide solution. Wash the precipitate three times with 5% sodium hydroxide solution, then twice with hot water. Rinse the crucible with 5 ml of hot 1:1 hydrochloric acid and use this to dissolve the precipitate from the paper. Evaporate to dryness. To convert to bromides, add 4 ml of hydrobromic acid and 4 ml of bromine water and evaporate to dryness. Repeat that step. Take up the residue in 12 ml of 5 *N* hydrobromic acid and add sufficient sodium thiosulfate solution to reduce the ferric ion. Shake for 2 minutes with 12 ml, 12 ml, and 12 ml of ether presaturated with 5 *N* hydrobromic acid. Wash the combined extracts with 2 ml and 2 ml of 5 *N* hydrobromic acid. Extract these washings with 2 ml of ether and add it to the main ether extract. Reextract the indium from the ether phase with 15 ml and 15 ml of 1:1 hydrochloric acid, adding 2 drops of 30% hydrogen peroxide to each. Evaporate this extract to dryness, add 1 ml of 5 *N* hydrobromic acid and 1 ml of bromine water, and again take to dryness. Dissolve the residue in 2 ml of 2 *N* hydrobromic acid. Add 2 drops of 10% ascorbic acid solution and 2 drops of 0.2% solution of rhodamine B. Dilute to 10 ml with the 2 *N* acid. Add 3 ml of benzene and 1 ml of ether and shake. Read the extract at 540 nm or, if less than 2 μg of indium is present, read the extract by fluorescence.

---

[50] A. K. Babko and Z. I. Chalya, *ibid.* **18**, 570–574 (1963).

[51] A. A. Rozbianskaya, *Tr. Inst. Mineral., Geokhim. Kristallokhim. Redk. Elem. Akad. Nauk SSSR* **1961** (6), 138–141.

RHODAMINE 3B

The complex of rhodamine 3B (basic violet 11) with indium is read by fluorescence.[52] The value is linear for 0.02–0.6 μg per ml of extract.

Procedure. Dilute 0.5 ml of sample in 1-6 *N* sulfuric acid to 7 ml with 24% solution of potassium bromide. Add 2 ml of 2:1 sulfuric acid and 1 ml of a rhodamine 3B solution containing 1 mg/ml. Shake with 5 ml of benzene. Activate the extract at 565 nm and read the fluorescence at 588 nm against a blank.

RHODAMINE 6G

The complex of indium with rhodamine 6G (basic red 1) can be extracted with benzene from 0.3 *M* hydrobromic acid–10 *N* sulfuric acid in the presence of 800–2500-fold amounts of gallium.[53] The optimum concentration of reagent is about 0.1 m*M*. The complex is read fluorimetrically or photometrically against a reagent blank.

Procedures

Indium Telluride.[54] Dissolve 1 gram of sample in 10 ml of 1:1 hydrochloric acid, adding nitric acid dropwise to promote solution. Evaporate nearly to dryness and take up in 5 ml of 1:1 hydrobromic acid. Evaporate substantially to dryness and take up in 2 *N* hydrobromic acid, diluting to a known volume with that acid to contain about 0.5 μg of indium per ml. Mix 2 ml with 0.5 ml of 0.1% solution of rhodamine 6G and 5 ml of 1:1 sulfuric acid. Extract with 5 ml of benzene and read the fluorescence of the extract.

Tin-Containing Samples.[55] As an extractant, treat 150 ml of 6-methyl-heptanol or -octanol in 150 ml of octane with 75 grams of phosphorus pentoxide at 65° for 2 hours. When cool, add 500 ml of octane. Wash with 125 ml of water, then with 125-ml portions of 1:35 sulfuric acid until the washings show no phosphate ion.

Fuse 1 gram of sample with 7 grams of sodium peroxide and 0.1 gram of magnesium oxide by slowly raising the temperature to 700°; hold at 700° for 8 minutes. Dissolve the melt in 60 ml of water and filter. Wash the residue with 2% ammoniacal ammonium sulfate and discard the filtrate and washings. Dissolve the metastannic acid in 30 ml of 1:8 sulfuric acid. Add 10% ascorbic acid solution,

[52] Ya. Glovadskii, A. P. Golovina, L. V. Levshin, and Yu. A. Mittsel', *Zh. Anal. Khim.* **19**, 693–696 (1964).
[53] I. P. Alimarin, A. P. Golovina, N. B. Zorov, and E. P. Tsintsevich, *Izv. Akad. Nauk SSSR, Ser. Khim.* **1968**, 2678–2682.
[54] T. B. Vesene, *Zavod. Lab.* **35**, 32–33 (1969).
[55] I. S. Levin and T. G. Azarenko, *ibid.* **28**, 1313–1316 (1962); cf. Gerhard Roebisch, *Z. Chem.* **13**, 64–65 (1973).

usually 1–2 ml, dropwise, until a test with thiocyanate on filter paper shows a pale pink.

Shake the solution of sample with 30 ml of the organic extractant. If an emulsion forms, add 10 ml of 2% ammonium fluoride solution. Discard the aqueous phase and wash the organic phase with 10 ml and 10 ml of 1:35 sulfuric acid. Extract indium from the organic phase with 2 ml, 2 ml, and 2 ml of 5:1 hydrobromic acid. Dilute the combined extracts to 25 ml. Mix an aliquot containing less than 10 μg of indium with 1 ml of 1% solution of rhodamine 6G, 10 ml of 1:1 sulfuric acid, 1 ml of 10% ascorbic acid solution, 10 ml of benzene, and a few drips of titanous sulfate solution. Shake for 1 minute and read the organic phase at 625 nm.

RHODAMINE 6Zh

The complex of indium with this dyestuff (basic red 1) is read fluorescently.[56]

Procedure. *Gallium.* Prepare a 150×4 mm column packed to a depth of 100 mm with powdered polytetrafluoroethylene suspended in tributyl phosphate. Dissolve 1 gram of gallium, suspected to contain about 4 μg of indium, in 0.8 *M* hydrobromic acid and dilute to 25 ml with that acid. Saturate the sample with tributyl phosphate and apply to the column at the rate of 1 drop every 2 seconds. Then elute with 30 ml of 0.8 *M* hydrobromic acid. Collect the last 2 ml of eluate. Add 0.5 ml of 0.6 m*M* rhodamine 6Zh, 0.5 ml of 7.4 *M* hydrobromic acid, 3.5 ml of 10 *M* sulfuric acid, 1 ml of water, and 5 ml of benzene. Mix, and read the fluorescence in the organic layer immediately.

RHODIUM ZV

Indium, after isolation from interfering ions, complexes with rhodium ZV for reading by fluorescence.[57]

Procedure. *Ores.* Treat 0.1 gram of sample in the cold with 10 ml of nitric acid and 5 ml of hydrochloric acid. Then heat at 100° for 2 hours and evaporate nearly to dryness. Add 5 ml of hydrobromic acid and evaporate to dryness, but do not bake. Again add 5 ml of hydrobromic acid and take to dryness. Take up the residue in 5 ml of 5 *N* hydrobromic acid and decant the aqueous layer. Add 5 ml of 5 *N* hydrobromic acid to the residue and ash. Take up with 5 ml of 5 *N* hydrobromic acid and add to the decantate already obtained. Extract with 10 ml of butyl acetate. Wash the organic extract with 5 ml and 5 ml of 5 *N* hydrobromic acid. Reextract the indium from the organic layer with 10 ml and 10 ml of 2:1

[56] I. M. Ivanova and N. B. Zorov, *Vest. Mosk. Gos. Univ., Ser. Khim.* **15**, 475–477 (1974).
[57] I. A. Blyum and T. K. Dushina, *Zavod. Lab.* **25**, 137–139 (1959).

hydrochloric acid. Add to the aqueous extract 2.5 mg of ferric chloride hexahydrate as a collector, 5 ml of 1 : 1 sulfuric acid, and 1 ml of 30% hydrogen peroxide. Evaporate to sulfur trioxide fumes, and if residual organic matter darkens, add more hydrogen peroxide with water and again take to sulfur trioxide fumes. Take up the residue with 40 ml of water and make definitely alkaline with ammonium hydroxide. Filter and wash with 1:400 ammonium hydroxide. Dissolve the precipitate with 5 ml of 2.5 *N* hydrobromic acid and wash the paper with 5 ml of that acid. Add 0.2 gram of reduced iron to the filtrate and set aside for 2 hours. Filter and add 2 drops of 0.1% solution of rhodamine ZV. Extract the complex with benzene and read the fluorescence of the extract.

SULFONAZO

Sulfonazo which is bis[3-(8-amino-3,6-disulfo-1-hydroxy-2-naphthylazo)-4-hydroxyphenyl]sulfone or its dimethyl or dibromo derivatives, forms 1 : 1 complexes with indium and gallium at pH 3.6–5.[58]

Procedure. Dissolve 0.5 gram of sample in 5 ml of 1 : 1 hydrochloric acid and 2 ml of nitric acid. Evaporate to dryness and take up in 10 ml of 1 : 3 hydrochloric acid. Add 2 grams of cadmium shavings, and let stand for 15 minutes to reduce iron and precipitate heavy metals. Filter, and add to the filtrate 10 ml of 0.05% titanous chloride solution in hydrochloric acid. Extract with 20 ml and 20 ml of ether saturated with 1 : 1 hydrochloric acid. Wash the combined ether extracts with 3 ml and 3 ml of 1 : 1 hydrochloric acid saturated with ether. Add 0.5 ml of 10% solution of sodium chloride to the ether extract and evaporate. Take up in 5 ml of acetate buffer solution for pH 3.6. Add 0.5 ml of 5.15×10^{-6} *M* solution of suljfonazo or its dimethyl or dibromo derivative. Dilute to 10 ml with the buffer solution and read at 594 nm.

TRIHYDROXYFLUORONES

The complex of indium with trihydroxyfluorones is extracted with ether for reading.[59] Salicylfluorone, which is 2,6,7-trihydroxy-9-(2-hydroxyphenyl)-3*H*-xanthene-3-one, or diphenyl fluorone, are appropriate reagents.

Procedure. *Cassiterite.* Fuse 1 gram of sample with 6 grams of sodium peroxide

[58]E. P. Shkrobot, *Zh. Anal. Khim.* **17**, 311–317 (1962); V. G. Brudz, D. A. Drapkina, K. A. Smirnova, V. I. Titov, I. E. Pokrowskaya, E. P. Osiko, N. I. Droshina, and V. I. Maslinovska, USSR Patent *119,709* (1959).

[59]V. A. Nazarenko and R. V. Ravitskaya, *Zavod. Lab.* **31**, 1301–1303 (1965); R. V. Ravitskaya, *ibid.* **33**, 565 (1967).

and 0.1 gram of magnesium oxide in a corundum crucible. Alternatively, add 0.5 gram to 3 grams of fused sodium tetraborate in platinum, cover with 3 grams of sodium carbonate, and heat at 900° to a homogeneous melt. In either case, extract the melt with 100 ml of water and add about 0.15 gram of ferric chloride or sulfate. Filter the precipitated hydroxide and wash well. Dissolve in 25 ml of 1 : 1 hydrochloric acid and dilute to 200 ml. Add 2 ml of 30% hydrogen peroxide and a few drops of 0.1% thymol blue solution. Make alkaline with 20% potassium hydroxide solution, filter, and wash. Dissolve in 25 ml of 1 : 1 hydrochloric acid, add 20 ml of 30% hydrogen peroxide, dilute to 200 ml, and precipitate with an excess of ammonium hydroxide. Filter, wash with 1% solution of ammonium chloride in 1 : 100 ammonium hydroxide, and dissolve in 50 ml of 1 : 18 sulfuric acid. Add 5 ml of 50% potassium iodide solution and decolorize by dropwise addition of sodium thiosulfate solution. Extract the indium triiodide with 5 ml and 5 ml of ether. Reextract the indium from the combined ether extracts with 10 ml of water.

To this extract or an aliquot containing no more than 40 μg of indium, add 0.5 ml of 1% solution of ammonium fluoride, 0.5 ml of 2% ascorbic acid solution, 1 ml of 0.25% solution of 1,10-phenanthroline, and 2 ml of 1% gelatin solution. As color reagent, add 2 ml of 0.05% solution of either salicylfluorone or diphenylfluorone in 99 : 1 ethanol–*N* hydrochloric acid. Add 20 ml of acetate buffer solution for pH 4.6. After 45 minutes, extract with 5 ml and 5 ml of ether. Read the combined extracts at 535 or 530 nm, respectively.

Indium forms a 1 : 2 complex with **alizarin red S** (mordant red 3) which is sodium alizarin-3-sulfonate.[60] The color is constant at pH 3.8–4.5, has a maximum absorption at 530 nm, and obeys Beer's law for 0.23–27 ppm.[61] Temperature does not affect the color. Indium also forms an insoluble ternary complex with the 1,3-diphenylguanidinium cation, which is extractable with butyl acetate in the presence of acetate buffer solution for pH 5.3–5.9. This conforms to Beer's law for 0.3–2.8 ppm and is read at 525 nm. There is interference by EDTA and other cations that complex with the reagent. The complex is also extractable with chloroform.[62]

Buffered at pH 6, ***o*-arsonophenylazochromotropic acid**, which has the trivial name of **neothoron,** forms a 3 : 2 complex with indium for reading at 580 nm.[63] Beer's law is followed for 2–10 μg/ml. Zinc, lead, and copper interfere.

Arsenazo, which is *o*-(1,8-dihydroxy-3,6-disulfo-2-naphthyl)benzenearsonic acid, absorbs at 510 nm; its complex with indium absorbs at 550 nm.[64] Treat a sample containing 0.8–50 μg indium per ml with 0.005 mole of reagent. Adjust to pH 4.5–5.2 and add a few drops of 50% solution of hexamethylenetetramine. Read after 30 minutes at 590 nm, where the difference between the absorption by the

[60]Kailash N. Munshi, Krishna K. Sakena, and Arun K. Dey, *J. Prakt. Chem.* **26** (3–4), 113–119 (1964).
[61]Makoto Otomo and Koichi Tonosaki, *Talenta* **18**, 438–441 (1971).
[62]E. A. Biryuk and V. A. Nazarenko, *Zh. Anal. Khim.* **30**, 1720–1723 (1975).
[63]Teiichi Matsumae, *Jap. Anal.* **8**, 97–99 (1959).
[64]V. I. Kuznetsov and I. S. Levin, *Izv. Sib. Otdel. Akad. Nauk SSSR* **1958** (7), 131–132; Arun P. Joshi and Kailash N. Munshi, *Microchem. J.* **18**, 277–287 (1973).

reagent and by the complex is greatest. Beer's law is followed for 0.61–9.18 ppm. There is interference by cupric, ferrous, ferric, thallic, ceric, and uranyl ions. Gallium forms a similar complex.

To the sample solution, add 4 ml of hydrobromic acid and 4 ml of 0.36% solution of **astrazone blue 5 Gl** (basic blue 5).[65] Dilute to 10 ml, extract the 1 : 1 indium-bromoindic ion with 10 ml of benzene, and read at 620 nm. Beer's law is obeyed for 0.5–5 μg of indium per ml of extract.

With **astrazone pink FG** (basic red 13), mix 0.5–1.5 ml of sample in *M* hydrobromic acid with 1.75 ml of hydrobromic acid and add 2 ml of m*M* reagent.[66] Dilute to 10 ml and extract with 20 ml of 2 : 1 benzene–methyl ethyl ketone. Read the organic layer at 548 nm within the range of 0.12–1.7 μg/ml.

Azastron blue 5 (basic blue 45) forms a 1 : 1 complex with indium and also with gallium.[67] The complex with indium is extracted from 2.7–3.4 *N* hydrobromic acid by benzene–methyl ethyl ketone for reading at 640 nm. Beer's law is followed for 0.5–10 μg of indium per ml. There is interference by ferric, antimonous, and pentavalent vanadium ions.

Brilliant cresyl blue, which is 7-diethylamino-3-imino-8-methylphenoxazinium chloride, complexes with indium.[68] To a solution containing 1–8 μg of indium per ml, add 3 ml of hydrobromic acid and 4 ml of 0.4% solution of the reagent. Dilute to 10 ml and extract with 20 ml and 5 ml of 1 : 1 ether-benzene. Read the extract at 640 nm. The reagent absorbs at 525 and 600 nm. Alternative extractants are 4 : 1 benzene-acetone and 4 : 1 benzene–methyl ethyl ketone. Ferric and gold ions must not exceed the concentration of indium.

The complex of **brilliant green** with indium is determined photometrically.[69] To 5 ml of sample solution, add 5 ml of 5 *M* hydrobromic acid, 1 ml of 0.5% solution of brilliant green, and 0.5 ml of acetone. The absorption is a straight line for 0.02–0.16 mg of indium per ml.

Indium complexes with **1-(5-bromo-2-pyridylazo)-5-ethylamino-*p*-cresol.**[70] To a sample solution containing 3.4–37 mg of indium, add 1 ml of ethanol and 3 ml of 0.15 m*M* solution of the reagent. Dilute to 9 ml with an acetate buffer solution for pH 3.5 and read at 550 nm. The 2-(3,5-dibromo)-compound reacts similarly.

The 3 : 1 complex of **4-(5-bromo-2-pyridylazo)resorcinol** with indium, when read at 520 nm, conforms to Beer's law for 0.2–1 mg/ml.[71]

Butylrhodamine B as a complex with indium is extracted from 3.6–4 *N* hydrobromic acid by 7 : 3 benzene–isopropyl ether for reading at 561 nm.[72] Interference by thallic and ferric ions is avoided by reduction with ascorbic acid. Stannous or stannic ions interfere.

Indium and gallium in 0.1 *N* nitric acid complex with ***N N*-bis(carboxymethyl)-anthranilic acid.**[73] Indium is read for up to 60 μg/ml at 289.5 nm, which is the

[65]Ana Serbanescu, *Rev. Chim.* (Bucharest) **26**, 68–71 (1975).

[66]G. Popa, C. Patroescu, and G. M. Costache, *ibid.* **23** (10); *Chim. Anal.* (Paris) **2** (3), 219–221.

[67]A. Serbanescu, P. Constantinescu, C. Constantinescu, and G. Banateanu, *Bul. Inst. Petrol., Gaze Geol.* **19**, 227–234 (1972).

[68]G. Popa and C. Patroescu, *Rev. Chim.* (Bucharest) **18**, 300–302 (1967); **21**, 770–774 (1970).

[69]C. Liteanu and E. Cordos, *Bull. Inst. Politeh., Iasi* **7** (3–4), (1961).

[70]S. I. Gusev and E. M. Nikolaeva, *Zh. Anal. Khim.* **21**, 1183–1190 (1966).

[71]A. I. Busev, V. M. Ivanov, and N. S. Khlybova, *ibid.* **22**, 547–551 (1967).

[72]A. Garcic and L. Sommer, *Collect. Czech. Chem. Commun.* **35**, 1047–1065 (1970).

[73]Septimia Policec, T. Simonescu, and C. Dragulescu, *Rev. Roum. Chim.* **17**, 231–237 (1972).

isobestic point of the gallium system. Similarly, gallium is read for up to 40 μg/ml at 260 and 316.5 nm, the isobestic points of the indium system. Zinc, cadmium, lead, and cupric ions interfere.[74]

Indium forms a 1:2 complex at pH 5 with **1-(4-carboxy-2-thiazolylazo)-2-naphthol**, which is 2-(2-hydroxy-1-naphthylazo)thiazole-4-carboxylic acid.[75] At pH 3 the ratio is 1:1. The absorption of the 1:2 complex is read at 570 nm. The reagent can determine down to 15 ng of indium per ml. Zinc interferes if greater than the indium.

Chlorosulfophenol R, which is 4-(5-chloro-2-hydroxy-3-sulfophenylazo)-3-hydroxynaphthalene-2,7-disulfonic acid, forms a 1:1 complex with indium having an absorption maximum at 510 nm.[76] It will determine 3–6 μg of indium per ml.

The violet complex of indium with **chrome azurol S** (mordant blue 29) is read at 555 nm.[77] Beer's law applies for 0.4–10 ppm of indium at pH 5.45–5.9. In 50 ml there should be 2 ml of 0.257% solution of the reagent and less than 10 ml of 8% hexamine solution as a buffer. There is interference by copper, ferric ion, beryllium, gallium, titanium, zirconium, and hexavalent molybdenum.

Indium precipitates from 2.4 *N* hydrochloric acid as a **cobaltamine** complex, $In(NH_3)_6CoCl_6$.[78] The precipitate dissolves in 1.5 *N* hydrochloric acid, and sodium sulfide decomposes the complex. Without separation, the solution is dried and treated with nitric acid, and the cobalt is determined by nitroso-R salt as an indirect measure of indium in the complex.

Alternatively,[79] heat the sample solution in 1:2 hydrochloric acid almost to boiling. Add 0.05 *M* cobalt ammonium chloride until the solution becomes orange, and allow to cool. Filter through glass and wash with 1:5 hydrochloric acid. Dissolve the precipitate with hot water, dilute to a known volume, and read at 420 nm. The technic is applicable to 1–50 mg.

The complex of indium in a solution containing 0.01% of **crystal violet** in 1.5 *N* sulfuric acid, and 0.4 *N* iodic acid is readily extracted with benzene.[80] The benzene extract has peaks at 550 and 610 nm and conforms to Beer's law up to 1 mg/ml. Interfering ions are cadmium, thallous, stannous, stannic, antimonous, bismuth, cupric, lead, tungstate, molybdate, and chloride. The corresponding **methyl violet** is also used.[81]

Indium is determined indirectly by its displacement of copper from **cupric diethylenetriaminepentaacetate**.[82] The complex of copper with another reagent is then read. To a neutral or slightly acid sample, add 5 ml of 2 m*M* cupric diethylenetriaminepentaacetate, 1 ml of 1% ascorbic acid solution, and 2.5 ml of 40% solution of sodium acetate. Dilute to 50 ml and mix well with 5 ml of 1% solution of bathocuproine in methanol. Extract by shaking for 1 minute with 5 ml of hexanol. Read the extract at 470 nm. The method is also applicable to gallium, lanthanum, and zirconium.

[74]C. Dragulescu, S. Policec, and T. Simonescu, *Talenta* **13**, 1451–1457 (1966).
[75]S. N. Drozdova, A. P. Momsenko, and M. Z. Yampol'skii, *Zh. Anal. Khim.* **26**, 291–296 (1971).
[76]V. D. Salikhova, Yu. M. Dedkov, and M. Z. Yampol'skii, *ibid.* **24**, 368–373 (1969).
[77]Yoshizo Horiuchi and Hiroshi Nishida, *Jap. Anal.* **16**, 1146–1152 (1967).
[78]Toru Nozaki, *J. Chem. Soc. Jap.* **77**, 1751–1752 (1956).
[79]Tomitaro Ishimori and Kaoru Ueno, *Jap. Anal.* **5**, 329–331 (1956).
[80]Sheng-chieh Hsü and Cheng-chi Wang, *H's Hsüeh T'ung Pao* **1959** (13), 430.
[81]N. S. Poluektov and N. K. Kiseleva, *Zh. Anal. Khim.* **13**, 555–561 (1958).
[82]E. Jackworth and G. Graffmann, *Z. Anal. Chem.* **257**, 265–268 (1971).

Indium will rapidly displace copper from the **cupric-EDTA** complex at pH 4.3.[83a] Thus the decrease in absorption by the cupric-EDTA complex at 740 nm measures the indium content.

In 30% ethanol at pH 3–6 indium forms a 1 : 3 complex with **6,7-dihydroxy-2,4-diphenylbenzopyrilium chloride**. The optimum reagent concentration is 40 μM.[83b] Stabilization with 0.04% of gelatin is necessary. As read at 550 nm Beer's law is followed for 0.06–0.9 μg of indium per ml.

A 1 : 2 : 4 complex is formed of indium with **3-(4-diethylaminophenylazo)-1,4-dimethyl-1,2,4-triazolium chloride** and halide ion.[84] This is extractable into benzene from 2 *M* potassium iodide at pH 2 with iodide as the halide. From 0.5 *M* potassium bromide in 5 *N* sulfuric acid, the solvent is 2 : 1 benzene-nitrobenzene and the halide is the bromide. For the chloride derivative, extract with nitrobenzene from *M* sodium chloride in 10 *N* sulfuric acid. Beer's law is followed for 0.1–4 μg of indium per ml of extract. Absorption maxima fall at 542–556 nm. Thallic ion can be masked with thiosulfate.

3,4-Dihydroxyazobenzene-4′-sulfonic acid forms a 1 : 1 complex with indium suitable for photometric reading at pH 6.[85]

The complex of indium with **2-(2,4-dihydroxyphenylazo)-4-phenyl-5-benzoylthiazole**, which is 4-(5-benzoyl-4-phenylthiazol-2-ylazo)resorcinol, in 36% ethanol at pH 4.28, is read at 520 nm.[86] The reagent also forms colored complexes with gallium, thorium, beryllium, and cerium. The color develops in 20 minutes, is stable for 24 hours, and obeys Beer's law for 0.1–1.2 μg/ml.

When 1 drop of 1% **diphenylamine** solution and 1 drop of **potassium ferricyanide** are added to 25 ml of 1 : 20 sulfuric acid, the indium content can be read.[87] There is interference by copper, nickel, chromium, iron, and thallium.

For indium in germanium, the latter is first volatilized. Then after reducing in hydrobromic acid solution with titanous chloride, the indium is extracted with ether to separate from iron and determined by **diphenylcarbazone**.[88]

The optimum conditions for determination of indium with **ethylrhodamine B** are 10 *N* sulfuric acid, 20 *N* bromide ion, and 0.2 mg of reagent per ml.[89] Extraction with benzene from 8 *N* sulfuric acid containing 20 mg of bromide per ml and 0.4 mg of reagent provides preconcentration.

In air-hydrogen or oxyhydrogen flames, the strongest lines of indium by **flame photometry** are at 451.1 and 410.2 nm, followed by 325.6 and 303.9 nm.[90] In an oxyhydrogen flame, the 451.1-nm line provides a limit of detection of 0.01 ppm in aqueous solution with a Beckman flame photometer and a multiplier phototube with accuracy up to 100 ppm. Hydrochloric acid increases the background and decreases the emission. An acetylene-air flame is appropriate for 100–1000 mg of

[83a]Hiroshi Onishi and Hitoshi Nagai, *Jap. Anal.* **13**, 429–433 (1969).

[83b]N. L. Olenovich, A. A. Bazilevich, V. A. Nazarenko, and O. D. Dira, *Zh. Anal. Khim.* **29**, 2287–2290 (1974).

[84]P. P. Kish and I. I. Pogoida, *Zh. Anal. Khim.* **28**, 1923–1932 (1973).

[85]M. Z. Yampol'skii, *Uch. Zap. Kursk. Gos. Pedagog. Inst.* **1957** (4), 128–141.

[86]V. S. Korol'kova, J. Putnins, E. Gudriniece, and E. I. Bruk, *Izv. Akad. Nauk Latv. SSR, Ser. Khim.* **1970**, 444–448.

[87]E. N. Deichman and I. V. Tananaev, *Zh. Anal. Khim.* **13**, 196–200 (1958).

[88]V. A. Nazarenko, E. A. Biryuk, and R. V. Ravitskaya, *ibid.* **13**, 445–448 (1958).

[89]I. A. Bochkareva and I. A. Blyum, *ibid.* **30**, 874–882 (1975).

[90]P. T. Gilbert, Jr., *Spectrochim. Acta* **12**, 397–400 (1958).

indium per liter.[91] For indium at 451.1 nm, the background is read at 448 or 454 nm.[92] By use of 80% acetone containing 0.5 *N* hydrochloric acid, the flame emission is increased about fiftyfold. With an ether extract in an oxyhydrogen flame, the limit of detection of indium is 0.06 ppm.[93]

For silver solutions in nitric acid, the silver is reduced by formic acid and removed as an amalgam.[94] Then indium is determined by flame photometry at 417.2 and 451.1 nm. The oxinates of indium, titanium, gallium, nickel, and cobalt are extractable with methyl isobutyl ketone, amyl acetate, ethyl acetate, or chloroform for reading in an oxyhydrogen flame.[95]

Indium chloride in the air-acetylene flame is read at 451.1 nm at such concentrations as 1 mg/ml. The sensitivity in 50% isopropanol is increased.[96] Zinc interferes, but indium as the bromide can be separated from it by extraction with isopropyl ether.

The detection limit, defined as the concentration in ppm for signals equal to twice the mean square of the background, is 0.002 at 451.13 nm at a height of 3 mm in the red zone of the flame.[97] This detection limit is one order of magnitude better than for atomic absorption.

When a solution of a magnesium alloy in 1:2 hydrochloric acid is sprayed into an acetylene-air flame, the intensity of 20 mg of indium per 100 ml is reduced by 8% by 5 grams of magnesium.[98] Reduction in intensity is also caused by neodymium, praseodymium, lanthanum, yttrium, and lead; some increase is due to thallium, erbium, and lutecium. The method will determine 0.5–1% indium to ±0.02%.

Gallein, which is 3′,4′,5′,6′-tetrahydroxyfluoran, forms a 1:1 complex with indium having a maximum absorption at 534 nm around pH 3.5–4.[99] This conforms to Beer's law for 4–50 μg in 25 ml. Aluminum is masked with fluoride, copper with thiosulfate. There is interference by lead, stannous and stannic ions, ferrous and ferric ions, antimonous ion, chromate, bichromate, ferrocyanide, ferricyanide, iodate, EDTA, and citric and tartaric acids. At pH 4.5, reading at 610 nm with sixfold excess of reagent, the limit of detection is 0.3 μg/ml.[100a]

To a solution containing 4–50 μg of indium in 0.5 *M* perchloric acid add 3 ml of acetate buffer solution for pH 4–4.5 and dilute to 10 ml.[100b] Shake for 3 minutes with 5 ml of m*M* gallein in butanol. Dilute the organic layer to 10 ml with ethanol and read at 530 nm against a blank.

The 1:1 complex of **gallion** with indium will detect 0.3 μg/ml. The desirable pH

[91] J. Malinowski, D. Dancewicz, and S. Szymczak, *Pol. Acad. Sci. Inst. Nuclear Res.* Rep. No. **113/VIII**, 7 pp (1959).

[92] Helmut Bode and Horst Fabian, *Z. Anal. Chem.* **170**, 387–399 (1959).

[93] H. Brandenberger and H. Bader, *Helv. Chim. Acta* **47**, 353–358 (1964).

[94] J. Meyer, *Z. Anal. Chem.* **231**, 241–252 (1967).

[95] Emiko Sudo and Hidehiro Goto, *Sci. Rep. Res. Inst., Tohoku Univ. Ser. A* **14**, 220–230 (1962).

[96] I. Perman, *Xe Congr. Groupe Av. Methodes Anal. Spectrog. Met.* (Paris) **1957**, 167–183.

[97] E. E. Pickett and S. R. Koirtyohann, *Spectrochim. Acta, B* **24**, 325–333 (1969).

[98] N. A. Kanaev, *Zavod. Lab.* **32**, 168–169 (1966).

[99] S. T. Orlovskii and P. P. Kish, *Ukr. Khim. Zh.* **27**, 687–692 (1961).

[100a] Arun P. Joshi and Kailash Munshi, *Mikrochim. Acta* **1971**, 526–530.

[100b] Jerzy Minczewski, Zofia Trybula, and Malgorzata Krzyzanowska, *Chem. Anal.* (Warsaw) **21**, 311–320 (1976).

is 4.5, with sixfold excess of reagent for reading at 610 nm. There is interference by cupric ion, iron, gallium, and ceric ion.

Gallocyanine (mordant blue 10) forms a 2:1 complex with indium which at pH 4.94 conforms to Beer's law at 496 nm for 5–40 μg of indium per ml.[101a]

A 1:2:1 complex of indium–**glycinecresol red**–diphenylguanidine in an acetate–ammonium hydroxide buffer solution is extracted by alcohols for reading at 530–542 nm.[101b] Gallium and aluminum perform similarly.

Add 1 ml of 1% solution of **8-hydroxyquinaldine** in *N* hydrochloric acid to the solution containing indium.[102] Add 2 ml of 20% ammonium acetate solution, dilute to 50 ml, and adjust to pH 7.5 with ammonium hydroxide. Extract with 10 ml, 10 ml, and 10 ml of benzene. Dilute the extracts to 50 ml, filter, and read the fluorescence.

For indium by **3-hydroxy-4-(2-thiazolylazo)naphthalene-2,7-disulfonic acid**, see Gallium.

Indium is determined by 3,4′,5,7-tetrahydroxyflavone, known as **kaempferol**.[103a] To a sample solution containing 6–28 μg of indium, add 1 ml of 0.3% solution of the reagent in ethanol. Raise the pH to 4.5 and dilute to 10 ml containing 50% of ethanol. Read at 430 nm for up to 2.8 ppm. Gallium gives a similar reaction. Several cations, as well as fluoride, thiosulfate, iodate, EDTA, and nitrilotriacetate, also interfere.

For reading by the yellow-green fluorescence in 40% ethanol add **kaempferol**, adjust to pH 5.9–6.3 with a hexamine buffer, activate at 366 nm, and read at 555 nm.[103b] Indium tolerates an equal amount of nickel, molybdenum, zirconium, phosphate, or urea. Beer's law is followed for 0.03–16.6 μ9 of indium in 10 ml with sensitivity to 5.7 μg per ml.

Lumogallion, which is 5-chloro-3-(2,4-dihydroxyphenylazo)-2-hydroxybenzenesulfonic acid, complexes with indium at pH 4.1–5 for reading at 500 nm.[104] The method is applicable to 2-5 μg/ml. Gallium and aluminum interfere, and ferric ion must be reduced.

In hydrobromic acid, indium complexes with **malachite green** (basic green 4) to give a maximum absorption at 635 nm.[105] Increase of the concentration of the dyestuff increases the absorption. The appropriate range for indium is 0.01–0.065 mg per 25 ml. More than 5 μg of ferrous or ferric ion interferes. From a sample solution in 3 *N* sulfuric acid containing 2.5 *N* potassium iodide and malachite green, the complex extracted with carbon tetrachloride is 1:1:4 indium–malachite green–iodide ion.[106] This is read at 633 nm.

[101a] M. Z. Yampol'skii, A. E. Okun, and L. N. Orlova, *Uch. Zap. Kursk. Gos. Pedagog. Inst.* **11**, 134–142 (1958).

[101b] M. L. Akhmedli, E. L. Glushchenko, and F. T. Aslanova, *Azerb. Khim. Zh.* **1974**, 126–130.

[102] Masayoshi Ishibashi, Tsunenobu Shigematsu, and Yashuharu Nishikawa, *J. Chem. Soc. Jap.* **77**, 1479–1482 (1956).

[103a] B. S. Garg and R. P. Singh, *Talenta* **18**, 761–766 (1971).

[103b] Z. T. Maksimycheva, V. Ya. Artemova, and Sh. T. Talipov, *Manuscript No. 664–74 deposited at Vsesoyuznyi Institut Nauchnoi i Tekhnicheskoi Informatsii, Moscow* 1974.

[104] V. D. Salikhov and M. Z. Vampol'skii, *Zh. Anal. Khim.* **22**, 998–1003 (1967), A. T. Pilipenko, A. I. Zhebentyaev, and A. I. Volkova, *Ukr. Khim Zh.* **42**, 998–1000 (1976).

[105] Tsutomu Matsuo, Shunsuke Funada, Hiroyuki Koide, and Mikio Suzuki, *Jap. Anal.* **13**, 763–767 (1964).

[106] P. P. Kish and I. I. Pogoida, *Zh. Anal. Khim.* **29**, 52–57 (1974).

The 1:3 complex of indium and **8-mercaptoquinoline** can be extracted with toluene at pH 4–12.5 for reading of 0.5–60 μg.[107] In alkaline solution, potassium cyanide masks copper, silver, gold, iron, cobalt, nickel, palladium, and platinum. There is interference by manganese, zinc, mercury, gallium, thallium, tin, lead, antimony, and bismuth.

Indium as a halide or thiocyanate is extracted at as low as 2 ppm from aluminum or 10 ppm from gallium with **4-(6-methoxy-3-methylbenzothiazolin-2-ylazo-*N*-methyldiphenylamine** in benzene and read at 634 nm.[108]

***N*-Methylanabasine-α′-azo-diethylaminophenol**, which is diethylamino-2-[3-(1-methyl-2-piperidyl)-2-pyridylazo]phenol, complexes with indium at an optimum of pH 5.[109] It also complexes with nickel, cobalt, copper, bismuth, vanadium, and gallium in acid solution. Beer's law is followed for 1–20 μg in 25 ml. Fluoride masks iron; tartaric acid masks aluminum, molybdenum, tungsten, and chromium.

The reagent absorbs at 440 nm; its 1:1 indium complex absorbs around 550–590 nm.[110] Complexes are also formed with nickel, cobalt, pentavalent vanadium, copper, and bismuth.

At pH 4–6 indium forms a 1:1 complex with ***N*-methylanabasine-α′-azo-3,4-dimethylphenol**, which is 2-[3-(1-methyl-2-piperidyl)-2-pyridylazo]-3,4-xylenol.[111] The maximum absorptions for the reagent and for the complex are at 415 and 570 nm, respectively. Beer's law is followed for 0.2–4.4 μg/ml. Nickel and cobalt interfere seriously. Masking is necessary for many cations.

At pH 4-5 indium forms a 1:1 complex with ***N*-methylanabasine-α-azo-β-naphthol**, which is 1-[3-(1-methyl-2-piperidyl)-2-pyridylazo]-2-naphthol, read at 560 nm.[112] The complex is insoluble in water but soluble in alcohols, acetone, and chloroform. Beer's law is followed for 0.2–6.4 μg/ml. An equal amount of lead, gallium, bismuth, stannous ion, or molybdenum is tolerated.

Indium as well as gallium forms stable complexes with **4-(6-methyl-2-pyridylazo)resorcinol.**[113] For indium, the maximum absorption is at 420 nm; the optimum pH is 6.5. Beer's law is followed for 5–45 μg of indium in 25 ml containing 2 ml of 0.1% solution of the reagent in ethanol and 15 ml of acetate buffer. There is interference by ferric, cobalt, nickel, zinc, and cadmium ions.

The 1:2 complex of indium with **2-[3-(1-methyl-2-piperidyl)-2-pyridylazo]-1-naphthol** has a maximum absorption at 570 nm, as contrasted with that of the reagent at 460 nm.[114] The optimum pH is 5–6, and Beer's law applies with twentyfold excess of reagent at 0.4–2.5 μg of indium per ml. Gallium can be masked with citric or tartaric acid. The color is stable for 45 minutes.

[107] J. Bankovskis, J. Cirule, and A. Ievins, *ibid.* **16**, 562–572 (1961).

[108] P. P. Kish, I. I. Zimomrya, I. I. Pogoida, Yu. K. Onishchenko, and G. M. Vitenko, *Tr. Khim. Khim. Tekhnol.* (Gor'kii) **1973** [4(35)], 71–73.

[109] Sh. T. Talipov, R. Kh. Dzhiyanbaeva, and A. E. Martirosov, *Izv. Akad. Nauk Kaz. SSR, Ser. Khim.* **1968** (6), 1–4.

[110] A. E. Martirosov, Sh. T. Talipov, and R. Kh. Dzhiyanbaeva, *Tr. Tashk. Gos. Univ.* **1968** (323), 73.

[111] K. Rakhmatullaev, M. A. Rakhmatulaeva, Sh. T. Talipov, and Kh. Tashmatov, *Dokl. Akad. Nauk Uzb. SSR* **1970** (12), 26–27.

[112] L. V. Chaprasova, Sh. T. Talipov, R. Kh. Dzhiyanbaeva, and N. F. Bagdasarova, *Izv. Akad. Nauk Kaz. SSR, Ser. Khim. Nauk* **1971** (1), 5–9; Ch. Ibraimov, Sh. T. Talipov, L. A. Grilov, and E. A. Likhonina, *Zh. Anal. Khim.* **29**, 2318–2323 (1974).

[113] N. Kulmuratov, N. B. Babaev, and Sh. T. Talipov, *Dokl. Akad. Nauk Uzb. SSR* **1969** (4), 29–30.

[114] V. D. Podgornova, Kh. S. Abdullaeva, and Sh. T. Talipov, *Uzb. Khim. Zh.* **11** (5), 25–29 (1967).

Indium shows a maximum absorption at 410–415 nm with **morin**, which is 2′,3,4′,5,7-pentahydroxyflavone.[115] The optimum pH is 3.6 and may be obtained by acetate or phthalate buffer solution. Chlorides decrease the value. The maximum absorption is at a 1:1 ratio of metal to reagent. The maximum absorption of morin at 355 nm is independent of pH over the range of 1.2–5.

A 1:1:4:2 complex is formed at pH 4.3–4.7 as indium-morin-antipyrine-perchlorate ion.[116] This is extractable with chloroform for reading at 430 nm or determination by fluorescence at 520 nm. Beer's law is followed for 0.02–2.5 μg/ml. Fluorescence will detect 3 ng of indium per ml. Ferric ion is masked with thiourea.

Nile blue A (basic blue 12) forms a 1:1 complex with indium.[117] To a sample containing 0.025–0.2 mg of indium, add 3 ml of hydrobromic acid, 5 ml of 0.2% solution of Nile blue A, and water to make 10 ml. Shake for 20 minutes with 20 ml of 1:1 benzene-ether. Wash the aqueous phase with 5 ml of the solvent. Read the combined extracts at 610 or 640 nm against a reagent blank. The absorption is stable for 12 hours.

A 1:1 complex is formed in faintly acid solution between **4-(2-pyridiylazo)-1-naphthol** and indium.[118] The aqueous phase buffered at pH 5.2 is extracted with chloroform for reading at 540 nm. Beer's law is followed for 1–6 μg of indium per ml. Masking with thiourea and cyanide prevent interference by copper, cobalt, iron, and zinc.

The 2:3 complex of indium with **pyrocatechol violet** at pH 5.7–6 absorbs at 619 nm.[119] Mix a sample containing 0.25–2.5 mg of indium with 1 ml of a solution of pyrocatechol violet at 0.3864 mg/ml for each 0.5 mg of indium. Add 0.5 ml of a buffer containing 87% pyridine and 13% acetic acid. Dilute to 50 ml and read after 20 minutes. Stannous, bismuth, ferric, cupric, and lead ions interfere.

A bromoindate of **rhodamine C** is extracted with benzene from 2.5 *N* hydrobromic acid containing acetone.[120] There is interference by auric ion, molybdenum, tungsten, thallic ion, ferric ion, telluric ion, and selenium.

Because of its greater availability, **quercetin**, which is 3,3′,4′,5,7-pentahydroxyflavone, can replace morin as a reagent of similar structure.[121] Beer's law applies to 0.01–0.1 mg of indium in 55% ethanol. Aluminum, fluoride, oxalate, and tartrate interfere.

Rhodamine 4Zh, which is 9-(2-ethoxycarbonylphenyl)-6-ethylamino)xanthene-3-ylidene diethylammonium chloride, is more sensitive as a reagent for indium or gallium than rhodamine 3ZhO, which is the 6-amino-7-methyl analogue. The complex is extracted as the bromoindate with benzene for fluorescent reading.[122] The solution for extraction should be 8–9.5 *N* sulfuric acid and 0.2–0.5 *M*

[115] A. I. Busev and E. P. Shkrobot, *Vest. Mosk. Univ., Ser. Mat., Mekh., Astron., Fiz. Khim.* **14** (4), 199–207 (1959).

[116] N. L. Olenovich, L. I. Koval'chuk, and E. P. Lozitskaya, *Zh. Anal. Khim.* **29**, 47–51 (1974).

[117] G. Popa and C. Patroescu, *Rev. Roum. Chim.* **12**, 447–451 (1967); *Rev. Chim.* (Bucharest) **21**, 770–774 (1970).

[118] S. I. Gusev and E. M. Nikolaeva, *Zh. Anal. Khim.* **21**, 166–171 (1966).

[119] Rudolf Staroscik and Janusz Terpilowski, *Chem. Anal.* (Warsaw) **7**, 803–808 (1962).

[120] N. S. Poluektov, L. I. Kononenko, and R. S. Lauer, *Zh. Anal. Khim.* **13**, 396–401 (1958).

[121] I. P. Alimarin, A. P. Golovina, and V. G. Torgov, *Zavod. Lab.* **26**, 709–711 (1960).

[122] A. P. Golovina, Z. M. Khvatkova, N. B. Zorov, and I. P. Alimarin, *Vest. Mosk. Gos. Univ., Ser. Khim.* **13**, 551–555 (1972).

hydrobromic acid. Rhodamine G[123] and rhodamine 6Zh[124] are also applicable.

A solution is passed through a column of resin AV-17 in its carbonate form, and the column is washed with water. Cadmium, zinc, aluminum, and gallium are desorbed with 1:4 ammonium hydroxide; then indium with 3 N ammonium carbonate, lead with 4% sodium hydroxide solution, and iron with 1:11 hydrochloric acid. The indium fraction is developed with **rhodamine G** (basic red 8).[125]

In a solution buffered at pH 6–6.5, the complex of indium with **salicylaldehyde isonicotinoylhydrazone** is read at 380 nm, that of gallium at 390 nm.[126]

Solochrome cyanine R (mordant blue 3) is a sensitive reagent for 0.8–2.5 ppm of indium at pH 3.5.[127]

Solochrome dark blue (mordant black 17) in sixfold excess complexes with indium during the course of an hour.[128] The reagent absorbs at 550 nm. The maximum absorption for indium is at 570 nm, but reading is preferably at 590 nm, where the reagent absorbs only slightly. Gallium is also read with this reagent. The appropriate range is 0.73–4.8 ppm of indium, and the complex is stable for 10 hours. There is serious interference by oxalate, nitrite, ferrous, ferric, cerous, chromic, and thallic ions.

For determination of indium with **stilbazo**, the ammonium salt of 4,4′-bis(3,4-dihydroxyphenylazo)-2,2′-stilbenedisulfonic acid, the optimum pH is 4.9 as buffered with hexamine.[129] An appropriate solution contains 3 ml of 1:25 hydrochloric acid and 4 ml of 0.05% solution of stilbazo in 50 ml of 8% hexamine solution. Reading is at 520 nm from 30 to 50 minutes after mixing.

Sulfonazo, which is di[3-(8-amino-1-hydroxy-3,6-disulfo-2-naphthylazo)-4-hydroxyphenyl]sulfone, and its dimethyl and dibromo derivatives, form 1:1 complexes with indium as well as gallium.[130] The optimum pH is 5, for reading at 600 nm. Interference by zinc, copper, and aluminum is avoided by adding metallic cadmium.

For determination of indium by **superchrome garnet Y** (mordant red 5), mix a solution containing 5–30 μg of indium in 15–20 ml with 2 ml of 0.01% solution of the dyestuff, adjust to pH 5, dilute to 25 ml, heat at 50° for 10 minutes, and read at 565 nm.[131a] The reagent is also used for determination of aluminum, gallium, and scandium and has numerous interferences.

Indium forms a 1:2:2 complex with **4,5,6,7-tetrachloro-3′,4′,5′,6′-tetrahydroxyfluoran** and cetylpyridinium chloride at pH 4.2–5.2.[131b] Mix the sample solution with 2 ml of 0.01 M cetylpyridinium chloride, 3 ml of acetic acid-sodium acetate buffer solution for pH 4.4, and 2 ml of mM methanolic reagent. Dilute to 10 ml

[123] A. I. Blyum, S. T. Solv'yan, and G. N. Shebalkova, *Zavod. Lab.* **27**, 950–956 (1961).

[124] D. I. Eristavi, V. D. Eristavi, and G. Sh. Kutateladze, *Tr. Gruz. Politekh. Inst.* **1969** [3(131)], 36–46.

[125] D. I. Eristavi, V. I. Eristavi, and G. Sh. Kutateladze, *Zh. Anal. Khim.* **26**, 2234–2235 (1971); English translation *ibid.* 1999–2000.

[126] G. S. Vasilikiotis, T. A. Kouimtzis, and V. C. Vasiliades, *Microchem. J.* **20**, 173–179 (1975).

[127] Arun P. Joshi and Kailash N. Munshi, *ibid.* **12**, 447–453 (1967).

[128] Arun P. Joshi and Kailash N. Munshi, *J. Prakt. Chem.* **38**, 305–310 (1968).

[129] Toshio Ozawa, *Jap. Anal.* **19**, 1389–1393 (1970); cf. M. Z. Yampol'skii, *Tr. Kom. Anal. Khim., Akad. Nauk SSSR, Inst. Geokhim. Anal. Khim.* **8**, 141–151 (1958).

[130] E. P. Shkrobot, *Zh. Anal. Khim.* **17**, 311–317 (1962).

[131a] Keizo Hiraki, *Bull. Chem. Soc. Jap.* **45**, 1395–1399 (1972).

[131b] Itsuo Mori, Yoshikazu Fujita, Masako Ida, and Takehisa Enoki, *Jap. Anal.* **25**, 239–242 (1976).

and set aside for 30 minutes. Read at 620 nm against a reagent blank. Beer's law is followed up to 4.5 μg of indium per ml in the developed solution.

Indium complexes with **4-(2-thiazolylazo)resorcinol** for reading at 530 nm.[132] Make 10 ml of sample solution acid to pentamethoxyl red, which is below pH 3.5; add 10 ml of 0.047% ethanolic solution of the reagent, dilute with water to 50 ml, and read after 15 minutes.

Indium is determined with **thiothenoxyltrifluoroacetone**, which is 1, 1, 1-trifluoro-4-mercapto-4-(2-thienyl)-but-3-en-2-one.[133] Extract 25 ml of sample containing about 0.04 mg of indium at pH 4.5–5 with 10 ml of m*M* reagent in carbon tetrachloride by shaking for 10 minutes. Read at 480 nm. The extract can be read for 1–20 μg/ml and is stable for 72 hours. There is interference by cupric, cadmium, cobalt, nickel, tellurite, phosphate, oxalate, and EDTA ions.

To a solution containing 12.6–125 μg of indium ion, add 4 ml of m*M* **4-(2-thiazolylazo)-3-hydroxynaphthalene-2, 7-disulfonic acid.**[134] Adjust to pH 3–4, dilute to 25 ml, and read at 570–580 nm. Oxalate interferes seriously. Interference by gallium can be prevented by a tenfold amount of tartrate without affecting the indium complex.

Indium can be read by the complex with Thoron 1 at pH 3.5 and 523 nm over the range of 0.015–0.14 mg per 25 ml.[135] There is interference by 0.25 mg of bismuth, 0.025 mg of lead, and 10 mg of thallium. Add the reagent in 0.1% solution to the sample in sulfuric or perchloric acid at pH 2, add 10% sodium hydroxide solution until the solution becomes red, and adjust to pH 3.5 with 1 : 10 sulfuric acid.

Other dyes applicable as complexes of indium are **toluidine blue** (basic blue 17) and **pyronine G.**[136]

A 1 : 1 complex of indium with **trihydroxyfluorone** is formed at an optimum pH of 4 and read at 530 nm.[137]

Victoria blue B (basic blue 26) forms a 1 : 1 complex as the bromide with indium.[138] In 1 : 5 sulfuric acid containing 2 gram-ions of bromide per liter, an aliquot complexed with 1.5 ml of 0.1% solution of Victoria blue B can be extracted with an equal volume of 10 : 1 benzene-acetone. Titanous chloride minimizes the effects of mercuric, selenic, antimonic, thallic, stannic, and ferric ions.

Victoria pure blue BO (basic blue 7) in *M*–2 *M* hydrobromic acid forms a 1 : 1 complex with the tetrabromoindate ion.[139] Extracted into benzene, it is read at 625 nm, in contrast to 610–615 nm for the dye. Beer's law is followed for 0.4–16 μg of indium per ml. Thallium and gold interfere.

The 1 : 1 complex of indium with **xylenol orange** has a maximum absorption at

[132]M. Langova-Hnilickova and L. Sommer, *Talenta* **16**, 681–690 (1969).

[133]K. R. Solanke and S. M. Khopkar, *Anal. Chim. Acta* **66**, 307–310 (1973).

[134]A. I. Busev, T. N. Zholondkovskaya, L. S. Krysina, and N. A. Golubkova, *Zh. Anal. Khim.* **27**, 2165–2169 (1972).

[135]H. A. Mottola, *Talenta* **11**, 715–717 (1964).

[136]G. Popa and C. Patroescu, *Rev. Chim.* (Bucharest) **21**, 770–774 (1970).

[137]O. A. Tataev, G. N. Bagdasarov, S. A. Akmedov, Kh. A. Mirzaeva, Kh. G. Buganov, and E. A. Yarysheva, *Izv. Sev-Kavkaz. Nauch. Tsent. Vyssh. Shk., Ser. Estestv. Nauk* **1973** (2), 30–32.

[138]A. I. Ivankova, *Tr. Kaz. Nauch.-Issled. Inst. Mineral. Syr'ya* **1962** (7), 214–219.

[139]Cecelia Constantinescu and Lucia Antonescu, *Rev. Roum. Chim.* **20**, 985–991 (1975).

560 nm and pH 3.5.[140] Beer's law is followed for 0.2–3 μg/ml. There is interference by bismuth, stannous ion, tetravalent zirconium, hexavalent uranium, pyrophosphate, bichromate, oxalate, sulfosalicylic acid, and EDTA. Ammonium fluoride will mask 0.3 mg of aluminum, ascorbic acid 0.5 mg of iron, and thiosulfate 0.6 mg of copper. Less than 1 mg of copper, nickel, and cobalt is masked by 1,10-phenanthroline at 1 ml of 1% solution per 25 ml.[141]

The determination of indium with methylthymol blue, as given in detail elsewhere, is also applicable to xylenol orange. This reagent is applied to determination of indium in thin films of indium-tellurium.[142]

The complex is 1:2:2 indium–xylenol orange–trioctylamine when extracted from a 0.5 m*M* solution of the dye in sulfuric or nitric acid at pH 2.8–3.8 by a 10 m*M* solution of trioctylamine in chloroform.[143] Then the maximum absorption is at 540 nm and follows Beer's law for 1–10 μg of indium per ml. Up to 13 μg of gallium per ml does not interfere.

When fourfold excess of xylenol orange is added to 0.7–8 ppm of indium or 0.5–2 ppm gallium at pH 4 ± 0.2, set aside for 30 minutes. Then read indium or gallium at 560 nm.[144] From a solution containing gallium and indium at pH 2.5–3, the gallium is removed by extraction with butyl acetate and the indium is then developed with xylenol orange.[145] Indium can be read with xylenol orange at pH 3–6 in the presence of cadmium and bismuth.[146a]

Adjust a sample solution containing 10–150 μg of indium to pH 3–3.6 with ammonium hydroxide and dilute to 10 ml. Extract the indium with 10 ml of 2 *M* 2-bromobutyric acid in chloroform.[146b] Mix the organic layer with 10 ml of saturated ethanolic xylenol orange and add 5 ml of ethanol. Read the ternary complex at 570 nm. Gallium performs similarly for reading at 560 nm.

[140] S. T. Orlovskii and P. P. Kish, *Ukr. Khim. Zh.* **29**, 209–213 (1963).
[141] Nasumi Ishiwatara, Hitoshi Nagai, and Yukio Toita, *Jap. Anal.* **12**, 603–608 (1963).
[142] T. B. Vesiene, *Zh. Anal. Khim.* **24**, 1890–1891 (1969).
[143] I. V. Pyatnitskii and S. G. Pinaeva, *ibid.* **28**, 671–677 (1973).
[144] Chandra D. Dwivedi and Arun K. Dey, *Mikrochim. Acta* **4**, 708–711 (1968).
[145] L. I. Ganaga, L. M. Bukhtseeva, and L. R. Starobinets, *Vest. Akad. Navuk. BSSR, Ser. Khim. Navuk.* **1969** (5), 49–53.
[146a] K. N. Bagdasarov, P. N. Kovalenko, and M. A. Shemyakina, *Zh. Anal. Khim.* **23**, 515–520 (1968).
[146b] I. V. Pyatnitskii, L. L. Kolomits, and I. L. Sadovskaya, *ibid.* **30**, 2131–2137 (1975).

CHAPTER TWELVE
GALLIUM

In general, gallium complexes with hydroxyanthraquinones, triphenylmethane derivatives, and azo dyes.[1] Known reagents for fluorimetric determination include carmine, morin, morin plus cupferron, 8-hydroxyquinoline (oxine), dibromoquinolinol, pontachrome blue, rhodamine B, rhodamine S, rhodamine 6G, *o*-salicylideneaminophenol, solochrome red, solochrome black, and (sulfo)(hydroxy)naphthoylazoresorcinol.

Photometric reagents in the presence of iron, aluminum, and indium include gallion, stilbazo, catechol violet, arsenazo 1, eriochrome cyanine (mordant blue 3), xylenol orange, methylthymol blue, diphenylcarbazone, phenylfluorone, and rhodamine B (basic violet 10).[2] They are applicable to 7–56 μg of gallium in 25 ml of solution. The most sensitive are stilbazo and xylenol orange; the most accurate is arsenazo 1. The most selective at pH 2.8 are gallion and stilbazo.

The decomposition of rocks, refractory silicates, and minerals is rapid and complete by fusion in platinum with 6 parts of lithium fluoride and 4 parts of boric acid, but sodium fluoride is even more efficient than lithium fluoride and will also decompose chromite.[3]

For isolation of gallium, adjust 10 ml of solution to pH 5–7 containing 0.1 *M* sodium perchlorate.[4] Add an excess of 6% cupferron solution and extract with 10 ml of chloroform. To recover the gallium, evaporate the chloroform and digest the residue with sulfuric acid and hydrogen peroxide.

Tributyl phosphate is appropriate for extraction of gallium from chloride solutions.[5] If reagents need to be purified from gallium present as an impurity, dissolve

[1]M. Z. Yampol'skii, *Uch. Zap., Kursk. Gost. Pedagog. Inst.* **4**, 116–141 (1957); cf. A. K. Babko, A. I. Volkova, and T. E. Get'man, *Ukr. Khim. Zh.* **35**, 69–76 (1969).

[2]M. K. Akhmedli and E. L. Glushchenko, *Zh. Anal. Khim.* **19**, 556–560 (1964); M. K. Akhmedi, E. A. Bashirov, and E. L. Glushchenko, *Uch. Zap. Azerb. Gos. Univ., Ser. Khim. Nauk.* **1965** (4), 9–15.

[3]V. S. Biskupsky, *Anal. Chim. Acta* **33**, 333–334 (1965); *Chemist-Analyst* **56**, 49–51 (1967).

[4]K. G. Vadasdi, *Anal. Chim. Acta* **44**, 471–472 (1969).

[5]I. P. Alimarin, N. I. Ershova, and T. A. Bol'shova, *Vest. Mosk. Gos. Univ., Ser. Khim.* **1967** (4), 51–54.

in 1:1 hydrochloric acid, add titanous chloride in the same acid, and extract with 9:1 benzene-ether.[6] Gallium can also be extracted from 8.5 *N* hydrochloric acid with isopropyl ether.

ACRIDINE ORANGE NO

The 1:1 complex of acridine orange NO (solvent orange 15) with chlorogallate is extracted with dichloroethane from 10–12 *N* sulfuric acid containing *N* sodium chloride for reading photometrically or fluorescently.[7] Virtually no dye is extracted. The latter technic can detect 0.01 ppm in a 5-gram sample.[8] There is interference by gold, antimony, or mercury, and only limited amounts of scandium, yttrium, or copper are tolerated.

Procedure. *Aluminum Silicates or Aluminum Salts.* For minerals, fuse 1 gram of sample with sodium carbonate and dissolve in 50 ml of 1:1 hydrochloric acid. Dissolve salts in 50 ml of 1:1 hydrochloric acid. Extract the gallium with 50 ml of butyl acetate. Reextract from the organic solvent with 50 ml of water and concentrate to about 10 ml. Add 1.75 ml of 0.01% solution of acridine orange, 1.75 ml of hydrochloric acid, and 1.25 ml of 15% titanous chloride in hydrochloric acid. Dilute to 25 ml with 1:1 hydrochloric acid. Dilute a 2.5-ml aliquot to 5 ml with 1:1 hydrochloric acid, fortifying if necessary with a known amount of gallium chloride solution. Shake for 1 minute with 6 ml of dichloroethane. Activate the extract at 500 nm and read the fluorescence at 525 nm or read the absorption at 505 nm.

ASTRAZONE BLUE B

The complex of gallium with astrazone blue B (basic blue 5) in 1:1 hydrochloric acid is extracted for reading at up to 70 ppm in the extract.[9] The method will determine 5 μg of gallium in the presence of 5 mg of antimony or molybdenum.

Procedure. Mix 4 ml of sample in 1:1 hydrochloric acid with 0.5 ml of 20% solution of titanous chloride in 1:1 hydrochloric acid and 0.5 ml of 2% solution of astrazone blue B in 1:1 hydrochloric acid. Shake for 5 minutes with 5 ml of 4:1 chlorobenzene–carbon tetrachloride. Set aside for 10 minutes, filter the organic layer, and read at 620 nm.

---

[6]D. P. Shcherbov and N. V. Kagarlitskaya, *Zavod. Lab.* **28**, 30–33 (1962).

[7]V. M. Tarayan, E. N. Ovsepyan, and A. N. Pogosyan, *Arm. Khim. Zh.* **23**, 410–413 (1970); **24**, 865–870 (1971).

[8]L. A. Grigoryan, A. N. Pogasyan, and V. M. Tarayan, *ibid.* **25**, 931–935 (1972).

[9]C. Dragulescu and P. Costinescu, *Stud. Cercet. Chim.* **13**, 1217–1222 (1965); cf. A. Serbanescu, P. Costinescu, C. Constantinescu, and G. Banateanu, *Bul. Inst. Petrol., Gaze Geol.* **19**, 227–234 (1972).

4-(4-BIPHENYLAZO)CATECHOL

At pH 4–6 gallium forms a 1 : 3 complex with this reagent.[10] The maximum absorptions of the reagent and of the complex are at 380 and 530 nm, respectively. Beer's law is followed for 0.2–20 μg of gallium per ml. Color develops at once and is stable for several hours.

Procedure. *Minerals.* Decompose 1 gram of sample containing 0.0002–0.004% of gallium in 5 ml of 1 : 1 hydrochloric-nitric acid at 100°. If the mineral is a silicate, evaporate to dryness. Add 5 ml of hydrofluoric acid and 1 ml of 1 : 1 sulfuric acid. In that case take to dryness and take up in 5 ml of 1 : 1 hydrochloric acid.

Add 5 ml of phosphoric acid and 50 ml of 1 : 1 hydrochloric acid. Extract the gallium with 10 ml, 10 ml, and 10 ml of chloroform. Wash residual ferric ion from the combined extracts with 5 ml and 5 ml of 1 : 1 hydrochloric acid. Extract gallium from the chloroform with 7 ml, 7 ml, and 7 ml of water and dilute the aqueous extracts to 25 ml. To an aliquot containing 0.2–20 μg of gallium, add 1.5 ml of 0.1% solution of the captioned reagent in ethanol, 2 ml of ethanol, and 1 ml of 0.5% gelatin solution. Dilute to 25 ml with ethanol and read at 530 nm.

BRILLIANT GREEN

The complex of gallium with brilliant green (basic green 1) can be extracted from 1 : 1 hydrochloric acid with benzene at a partition coefficient of 0.56.[11] As applied to semiconductor materials, the presence of antimony trichloride and of aluminum chloride improves the extraction.

At 635 nm the absorption conforms to Beer's law for 0.1–1.5 μg/ml.[12] Ferric and thallic ions behave similarly, but reduction by titanous chloride removes their interference.

Procedure. To two 2-ml portions of sample in 1 : 1 hydrochloric acid containing less than 10 μg of gallium per ml, add 0.2 ml and 0.4 ml of a standard gallium chloride solution in 1 : 1 hydrochloric acid containing 10 μg of gallium per ml. To these and a third 2-ml portion with no added gallium, add 5% titanous chloride in 1 : 1 hydrochloric acid to a pale violet color. After 10 minutes add 0.5 ml of 0.5% solution of brilliant green in 1 : 1 hydrochloric acid and dilute each to 3 ml with 1 : 1 hydrochloric acid. Shake each for 2 minutes with 3 ml of benzene and read each at 640 nm against benzene.

[10] Sh. T. Talipov, Kh. S. Abdullaeva, and N. A. Romanova, *Ubz. Khim. Zh.* **1964** (3), 16–20.

[11] V. K. Kuznetsova and N. A. Tananaev, *Nauch. Dokl. Vyssh. Shk., Khim. Khim. Tekhnol.* **1959** (2), 289–292; V. K. Kuznetsova, *Tr. Ural. Politekh. Inst.* **1967** (163), 14–19; *Tr. Kom. Anal. Khim.* **16**, 33–40 (1968); cf. V. Armeanu and P. Costinescu, *An. Stiint. Univ. "Al. I. Cuza," Iasi, Sect. 1*, **6**, 943–952 (1960).

[12] Kazuyoshi Hagiwara, Masonori Nakane, Yasuaki Osumi, Eiichi Ishii, and Yoshizo Miyake, *Jap. Anal.* **10**, 1374–1378 (1961).

Alternatively,[13] dissolve a 0.5-gram sample in 10 ml of hydrochloric acid and 1 ml of nitric acid. Evaporate to dryness, add 2 ml of hydrochloric acid, and again evaporate to dryness. Take up with 12 ml of 1 : 1 hydrochloric acid and add 0.5 ml of 10% titanous chloride solution in 1 : 1 hydrochloric acid. Extract the gallium as the chloride with 5 ml and 5 ml of ether. Add 1 ml of 2% sodium chloride solution to the ether extracts and evaporate to dryness. Take up the residue in 3 ml of 1 : 5 hydrochloric acid. Add 0.1 ml of 10% titanous chloride solution. Add 0.5 ml of 0.5% solution of brilliant green. Extract with 10 ml of benzene and read the benzene layer at 635 nm against a blank.

BUTYLRHODAMINE B

Butylrhodamine B is more sensitive than rhodamine B.[14]

Procedure. *Minerals.* Fuse 0.5 gram of sample in a corundum crucible with 1 gram of sodium peroxide and 0.5 gram of sodium hydroxide at 800°. Leach the melt with water, acidify with hydrochloric acid, and evaporate to dryness at 100°. Take up in 15 ml of 1 : 1 hydrochloric acid and shake for 2 minutes each with 5 ml, 5 ml, and 5 ml of 0.1% titanous chloride solution in amyl acetate. Combine the extracts and wash with 5 ml, 5 ml, and 5 ml of 1 : 1 hydrochloric acid. Extract the gallium by shaking with 10 ml of water. Evaporate the aqueous extract to dryness at 100°. Take up the residue with 5 ml of 1 : 1 hydrochloric acid, washing it in with 2 ml of 1 : 1 hydrochloric acid. Add 2 ml of 20% titanous chloride solution in 1 : 1 hydrochloric acid and shake. Add 1 ml of 0.1% solution of butylrhodamine B in 1 : 1 hydrochloric acid. Shake for 2 minutes with 10 ml of toluene, filter the extract through cotton, discarding the first portion, and read at 565 nm.

CATECHOL VIOLET

Catechol violet will determine 0.35–5.68 μg of gallium in 20 ml of sample in the presence of cetylpyridinium bromide.[15] Diphenyl guanidine is an alternative additive.[16] There is interference by aluminum, indium, iron, scandium, bismuth, titanous ion, pentavalent vanadium, mercuric ion, tin, pentavalent zirconium, thorium, and hexavalent tungsten, an equal amount of copper, or a twofold amount of molybdenum.

Gallium also forms a 1 : 2 : 1 complex at pH 5–10 with catechol violet and quinine.[17] The maximum absorption at 680 nm follows Beer's law for 2–20 m*M* of gallium. Gelatin stabilizes the complex.

[13] *ibid.*
[14] L. M. Skerbkova, *Zh. Anal. Khim.* **16**, 422–425 (1961).
[15] Tsutomo Ishito, *Jap. Anal.* **21**, 752–755 (1972).
[16] M. K. Akhmedi, E. L. Glushchenko, and Z. L. Gazanova, *Zh. Anal. Khim.* **26**, 1947–1952 (1971).
[17] M. I. Shtokalo and L. V. Livadko, *ibid.* **28**, 484–488 (1973).

A 1:2:1 complex is formed at pH 3–6 by gallium, catechol violet, and phenanthroline for reading at 600 nm.[18] Add gelatin as a stabilizer to the gallium and dye before adding phenanthroline. Beer's law is observed at 2–20 μg of gallium per ml.

Procedures

With Cetylpyridinium Bromide. To 20 ml of sample, add 1 ml of 0.5 m*M* solution of catechol violet, 10 ml of hexamine buffer solution for pH 5.8, and 1 ml of 4% butanolic solution of cetyl pyridinium bromide. Add 9 ml of butanol and shake for 3 minutes. Let stand for 10 minutes and read the extract at 600 nm.

With Diphenylguanidine. To a sample solution containing 1–8 μg of gallium in 0.1 *N* hydrochloric acid, add 1 ml of 0.04% solution of catechol violet, 0.5 ml of 2% solution of diphenylguanidine, and 10 ml of 2.9% solution of sodium chloride. Dilute to 25 ml with a hexamine buffer solution for pH 5. Extract with 10 ml of 1:1 chloroform-butanol and read at 600 nm against a reagent blank.

Gallium Arsenide.[19] Dissolve a 0.25-gram sample in 3 ml of hydrochloric acid and 0.3 ml of nitric acid at 100°. Add 2 ml of hydrochloric acid and 1 ml of 10% sodium chloride solution to prevent loss of gallium by hydrolysis. Dilute to 200 ml. Mix a 5-ml aliquot with 4 ml of 0.16% solution of catechol violet. Neutralize with 1:15 ammonium hydroxide and add 40 ml of acetate buffer solution for pH 5.8. Dilute to 100 ml and read at 580 nm against a comparable solution of gallium.

C-CYANO-*NN'*DI-(2-HYDROXYPHENYL)FORMAZAN

This reagent, also designated as *NN'*-bis(2-hydroxyphenyl)-*C*-cyanoformazan, forms complexes with gallium absorbing at 634 nm at pH 2–5 and at 630 nm at pH 5.[20] The reagent tolerates up to 0.4 mg per 20 ml of zinc, lead, chromium, or cadmium, and 0.04 mg of indium, germanium, copper, or nickel.

Procedure. Adjust a sample solution containing not more than 0.4 μg of gallium to pH 3 with *M* sodium acetate and *M* hydrochloric acid. Dilute to 20 ml with an acetate buffer solution for pH 3. Add 1 ml of 0.0025 *M* reagent in ethanol and incubate at 60° for 2 hours. Cool, and extract with 5 ml of benzene for 1 minute to remove copper, nickel, and excess reagent. Read the aqueous layer at 634 nm.

---

[18] M. I. Shtokalo, N. V. Ovchinnikova, N. A. Daeva, and V. V. Khrushch, *Zh. Neorg. Khim.* **16**, 2104–2107 (1971).
[19] M. K. Akhmedli and L. I. Ivanova, *Azerb. Khim. Zh.* **1969**, 106–109.
[20] N. L. Vasil'eva and M. I. Ermokova, *Zh. Anal. Khim.* **18**, 43–51 (1963).

4-(5-CHLORO-2-HYDROXYPHENYLAZO)RESORCINOL

The complex of this reagent with gallium is read by fluorescence at levels down to 23 ng of gallium in 5 ml.[21]

Procedures

Silicates.[22] Treat 0.05 gram of sample in platinum with 3 ml of hydrofluoric acid and 1 ml of 1 : 1 sulfuric acid. Evaporate to dryness and take up in 10 ml of 1 : 1 hydrochloric acid. Add 10% solution of titanous chloride in 1 : 1 hydrochloric acid until a pale violet color is obtained. Cool and extract with 8 ml, 3 ml, and 3 ml of ether. Wash the combined extracts with 3 ml and 3 ml of 1 : 1 hydrochloric acid saturated with ether. Add 1 ml of water to the ether phase and evaporate to dryness. Take up the residue in 10 ml of 0.01 *N* hydrochloric acid. Mix a 2-ml aliquot with 0.5 ml of ethanol, 0.1 ml of 0.8 m*M* reagent, and 2 ml of 0.2 *N* acetic acid. Dilute to 5 ml and shake for 15 minutes with 0.5 ml of hexanol. Read the fluorescence. The technic may be varied for silicon and for sulfide ores.

Seawater. Adjust a 250-ml sample to pH 3 with acetic acid and add 0.5 ml of 0.5 m*M* reagent in ethanol. After 15 minutes extract with 15 ml and 10 ml of hexanol by shaking for 1 minute with each. Combine the extracts and reextract the gallium with 10 ml of 1 : 5 hydrochloric acid. Add 1 ml of 0.25% solution of sodium chloride and evaporate the aqueous extract to dryness at 100°. Ignite the residue at 600–700° for 10 minutes, cool, and dissolve in 6 ml of 1 : 1 hydrochloric acid. Extract with 6 ml and 6 ml of ether presaturated with 1 : 1 hydrochloric acid. Evaporate the ether extracts to dryness and take up the residue in 1.5 ml of 0.01 *N* hydrochloric acid. Add 0.5 ml of ethanol, previously distilled over potassium hydroxide, 0.1 ml of 0.5 m*M* reagent in ethanol, and 2 ml of 0.2 *N* acetic acid. Dilute with water to 5 ml and after 20 minutes extract with 0.4 ml of hexanol. Read the fluorescence of this extract.

CRYSTAL VIOLET

The 1 : 1 complex of crystal violet (basic violet 3) or its close relatives such as methyl violet with gallium in 1 : 1 hydrochloric acid is extracted for reading at 0.3–5 μg/ml.[23]

Procedure. Mix 1 ml of sample solution in 1 : 1 hydrochloric acid containing about 0.01 mg of gallium with sufficient 10% solution of titanous chloride in 1 : 1

[21] Liam Ngog Thu, *ibid.* **22**, 636–637 (1967).
[22] Liam Ngog Thu, R. M. Dranitskaya, and V. A. Nazarenko, *Ukr. Khim. Zh.* **34**, 186–189 (1968).
[23] V. K. Kuznetsova, *Zh. Anal. Khim.* **18**, 1326–1331 (1963); cf. V. Armeanu and P. Costinescu, *Stud. Cert. Chim.* **13**, 1223–1229 (1965).

hydrochloric acid to reduce ferric ion to the noninterfering ferrous form. Add 0.5 ml of hydrochloric acid and dilute to 3 ml with 1 : 1 hydrochloric acid. Shake with 3 ml of chloroform and 1 ml of acetone for 5 minutes. Separate the organic extract and wash it with 1.5 ml and 1.5 ml of 6 : 17 *N* hydrochloric acid–acetone containing a couple of drops of the titanous chloride solution. This washing removes the last of copper, aluminum, and indium. Dilute the washed extract to 6 ml with chloroform. Shake it with 2 ml of a solution in 1 : 1 hydrochloric acid containing 2% of crystal violet with 0.01% of diphenylamine as a stabilizer. Filter the organic layer and read at 587 nm against a blank.

Bauxite.[24] Decompose 0.25 gram of sample with 2 ml of 1 : 1 sulfuric acid and 5 ml of hydrofluoric acid. Evaporate to sulfur trioxide fumes and fuse with 2 ml of potassium pyrosulfate. Digest the melt for 30 minutes with 10 ml of hot 1 : 1 hydrochloric acid and filter. Wash the residue with water. Add 0.5 gram of cadmium powder and shake until the solution becomes colorless, indicating complete reduction of iron—about 20 minutes. Add 5 ml of hydrochloric acid and dilute to 25 ml with 1 : 1 hydrochloric acid. Add a 20-ml aliquot of the sample solution to 0.5 ml of 20% titanous chloride in 1 : 1 hydrochloric acid. Shake for 2 minutes with 20 ml of ether, then for 1 minute with 5 ml of ether and 0.2 ml of the titanous chloride solution. Wash the combined ether extracts with 5 ml of 1 : 1 hydrochloric acid saturated with ether. Add 0.5 ml of 10% sodium chloride solution to the ether and evaporate to dryness at 100°. Take up the residue with 3 ml of 1 : 1 hydrochloric acid. Add 2 ml of 2% solution of crystal violet in 1 : 1 hydrochloric acid containing 0.005% of diphenylamine. Shake for 5 minutes with 5 ml of 6 : 1 chloroform-acetylacetone. After 10 minutes filter the organic layer through glass wool and read at 589 nm.

CUPRIC DIETHYLENETRIAMINEPENTAACETATE

In the presence of bathocuproine, gallium displaces copper from the reagent and it is read as the bathocuproine complex.[25]

Procedure. To a neutral or slightly acid sample, add 5 ml of 2 m*M* solution of the captioned reagent, 1 ml of 1% ascorbic acid solution, and 2.5 ml of 40% solution of sodium acetate. Dilute to 50 ml and add 0.5 ml of 1% methanolic solution of bathocuproine. Shake for 1 minute with 5 ml of hexanol and read the extract at 470 nm.

---

[24] C. Dragulescu and P. Costinescu, *Stud. Ceret. Chim.* **14**, 67–65 (1965); *Rev. Roum. Chim.* **10**, 67–75 (1965).

[25] E. Jackwerth and G. Graffmann, *Z. Anal. Chem.* **257**, 265–268 (1971).

GALLION

Gallion, which is 5-amino-3-(3-chloro-2-hydroxy-5-nitrophenylazo)-4-hydroxy-naphthalene-2,7-disulfonic acid, complexes with gallium to give a complex that is stable for 40 hours. This is extracted with solvents such as ether, amyl acetate, or isoamyl alcohol.[26] Ferric ion is reduced with hydroxylamine and copper with thiosulfate to avoid interference.

Gallion also forms a 1 : 1 : 1 complex with gallium and 8-hydroxyquinolin.[27] This is extractable at pH 3 into alcohols for reading at 640 nm.

Procedure. *Semiconductor Alloys.*[28] Dissolve 0.02 gram of sample in 4 ml of aqua regia. Evaporate to dryness at 100°. Add 2 ml and 2 ml of hydrochloric acid, evaporating to dryness after each addition to volatilize tin and germanium. Take up the residue with 10 ml of 1 : 1 hydrochloric acid. If the alloy contains gold, extract it with dithizone in carbon tetrachloride. Extract the gallium with four 15-ml portions of ether saturated with 1 : 1 hydrochloric acid. Evaporate the ether extracts, take up the residue with 1 ml of *N* hydrochloric acid, and dilute to 50 ml.

Mix a 2-ml aliquot with 3 ml of 0.01 *N* hydrochloric acid; add 1 ml of 0.01% solution of gallion. Dilute to 25 ml with an acid phthalate buffer solution for pH 3.5. Read at 610 nm against a blank.

GLYCINECRESOL RED

The intensely red complex of this reagent with gallium is applicable for 0.2–3 μg/ml.[29] The maximum absorption for glycinecresol red is at 413 nm, for the complex at 536 nm. Aluminum five-fold, scandium ten-fold, zirconium 25-fold, and uranium fifty-fold are permissible. Iron is reduced with ascorbic acid and copper by thiosulfate.

Procedure. *Zinc and Zinc Oxide.* Dissolve 0.2 gram of sample in 10 ml of 1 : 1 hydrochloric acid, add 1 ml of 2% sodium chloride solution, and evaporate to dryness. Dissolve the residue in 5 ml of water, add 2 ml of 0.1% solution of glycinecresol red, and dilute to 25 ml with a buffer solution for pH 4. Let stand for 15 minutes and read at 536 nm.

---

[26] G. G. Karanovich, L. A. Ionova, and B. L. Podol'skaya, *Zh. Anal. Khim.* **13**, 439–444 (1958).

[27] M. K. Akhmedii, E. L. Glushchenko, and A. K. Kyazimova, *Azerb. Khim. Zh.* **1971** (3), 102–107; cf. A. M. Lukin and G. B. Zavarikhina, *Zh. Anal. Khim.* **13**, 66–71 (1958); D. P. Shcherbov and A. I. Ivankova, *Zavod. Lab.* **24**, 667–674 (1958).

[28] V. S. Sotnikov, A. M. Kononova, L. I. Kononova, and M. E. Yaskevich, *Zavod. Lab.* **33**, 10–12 (1967).

[29] P. P. Kish and Yu. K. Onishchenko, *Zh. Vses. Khim. Obshch.* **10**, 447–478 (1965).

GLYCINETHYMOL BLUE

Glycinethymol blue forms a 2 : 1 complex with gallium at pH 4.75–5.4.[30] The absorption of the reagent at that pH is negligible for 560–570 nm. Heating or standing is necessary to obtain the maximum absorption. There is serious interference by ferric, cupric, and stannous ions; somewhat less by thallic, titanous, hexavalent molybdenum, aluminum, arsenous, fluoride, citrate, tartrate, and oxalate ions. Beer's law applies to 0.12–2.8 μg/ml.

Procedure. To the sample solution, add 1.5 ml of 0.1% solution of the reagent and 5 ml of a buffer solution for pH 4.75. Dilute to 25 ml and heat at 90° for 15 minutes. Read against water at 582 nm.

3-HYDROXY-4-(2,4-DIHYDROXYPHENYLAZO)NAPHTHALENE-1-SULFONIC ACID

At pH 3–6 the 1 : 1 and 2 : 1 complexes of the captioned reagent with gallium have maxima at 540–546 nm, as compared with 484 nm for the reagent.[31] Extraction with isopropyl ether separates from all interfering ions except ferric, auric, and thallic. After reduction with stannous chloride, they are not extracted.

Procedure. Adjust a sample solution containing 1–10 μg of gallium to 7.5 *N* with hydrochloric acid. Add an excess of 15% stannous chloride solution in 7.5 *N* hydrochloric acid. Let stand 5 minutes for complete reduction, then extract with 10 ml and 10 ml of isopropyl ether. Wash the combined extracts with 10 ml of 7.5 *N* hydrochloric acid containing 0.1 ml of the stannous chloride solution, then with 10 ml and 10 ml of 7.5 *N* hydrochloric acid. Evaporate the organic phase to dryness at 100° and take up the residue in 5 ml of water plus a drop of 1 : 10 hydrochloric acid. Dilute to 25 ml. Mix an aliquot with 5 ml of 0.3603% solution of the reagent buffered at pH 5.8. Let stand for 30 minutes and read at 560 nm against a blank.

8-HYDROXYQUINOLINE

For the use of this reagent, popularly known as oxine, the sample can be adjusted to 1 : 1 hydrochloric acid and the gallium extracted into isopropyl ether.[32] Then the

[30] M. K. Akhmedli, L. I. Ivanova, and A. A. Kafarova, *Azerb. Khim. Zh.* **1972** (2), 112–116; K. C. Srivastava and Usha Mehndiratta, *J. Clin. Chem. Soc.* (Taipei) **21**, 53–57 (1974).

[31] Teh-Liang Chang and John H. Yoe, *Anal. Chim. Acta* **29**, 344–349 (1963); V. A. Nazarenko, Liam-Ngog-Thu, and R. M. Dranitskaya, *Zh. Anal. Khim.* **22**, 346–352 (1967).

[32] R. Keil, *Z. Anal. Chem.* **249**, 172–174 (1970); cf. L. B. Ginzburg, *Izv. Akad. Nauk. Kaz. SSR, Ser. Khim* **1957** (1), 94–98; Yashaharu Nishikawa, Keizo Hiraka, Kiyotoshi Morishige, Koichi Takahashi, Tsunenobu Shigematsu, and Taro Nogami, *Jap. Anal.* **25**, 459–463 (1975).

gallium is reextracted with water before adding the masking agents and oxine for chloroform extraction of the complex.

Gallium oxinate gives a fluorescence peak at 492 nm when extracted into chloroform, from a solution at pH 3.9–5.5, and activated at 365 nm.[33] This is proportional to 1–30 μg per 50 ml of extract. Copper and citric acid interfere seriously, thallium and indium only slightly, at pH 3.9. At that pH gallium can be selectively extracted from indium and beryllium. The fluorescence is stable for 2 hours.

Procedure. To an acid sample solution, add 0.5 gram of ascorbic acid, 0.5 gram of citric acid, and 0.5 gram of potassium thiocyanate. Adjust to pH 2.5 ± 0.1 with 1:7.5 ammonium hydroxide. Add 2.5 ml of 5% solution of oxine in acetic acid at pH 2.5. Extract with 10 ml of chloroform, filter the extract, and read at 393 nm.

Silicate Minerals.[34] Treat a 0.2-gram sample with 10 ml of hydrochloric acid and 1 ml of sulfuric acid. Evaporate to dryness and fuse the residue with 2 grams of potassium bisulfate. Take up the melt in 10 ml of 1:1 hydrochloric acid. Extract with 10 ml and 10 ml of ether, which removes gallium, iron, and several other elements. Evaporate the ether extracts to dryness and take up the residue in 5 ml of 1:5 hydrochloric acid. Add 1 gram of hydroxylamine hydrochloride and 1 ml of 1% solution of oxine in *N* acetic acid. Adjust to pH 3.9 with 5 ml of 20% ammonium acetate solution. Extract with 10 ml, 10 ml, and 10 ml of chloroform, dilute the combined extracts to 50 ml, and read by fluorescence at 492 nm.

Nonferrous Ores.[35] Decompose a 0.5-gram sample in platinum with 3 ml of nitric acid, 2 ml of sulfuric acid, and 5 ml of hydrofluoric acid. Evaporate to dryness, add 5 ml of water, and again take to dryness. Heat with 2 ml of 30% hydrogen peroxide and 1 ml of sulfuric acid as long as sulfur trioxide can be driven off. Dissolve in 10 ml of 1:3 hydrochloric acid and heat to 40°. Add 1 gram of metallic cadmium and mix for 10 minutes. Add more cadmium, which must not darken if displacement is complete. In some cases as long as 12 hours may be required to complete the reaction. Filter through cotton wet with 1:3 hydrochloric acid. Wash the residue with 3 ml and 3 ml of 1:3 hydrochloric acid. Add to the filtrate 8 ml of hydrochloric acid and 0.5 ml of 10% titanous chloride solution in 1:1 hydrochloric acid. After 1 minute for reduction of the iron, shake with 25 ml and 25 ml of ether saturated with 1:1 hydrochloric acid. Wash the combined ether extracts with 3 ml and 3 ml of 1:1 hydrochloric acid saturated with ether. Add the ether extract to 0.5 ml of 10% sodium chloride solution and evaporate to dryness. Take up the residue in 3 ml of 0.2 *N* hydrochloric acid. Add 2 ml of water, 1 ml of 1:4 ammonium hydroxide, 1 ml of 5% thiourea solution, and 6 ml of 0.2 *M*

[33]Masayoshi Ishibash, Tsunenobu Shigmatsu, and Yashuharu Nishikawa, *J. Chem. Soc. Jap.* **78**, 1039–1046 (1957); *Bull. Inst. Chem. Res., Kyoto Univ.* **37**, 191–197 (1959); cf. M. F. Landi, *Met. Ital.* **52** (6) 366–371 (1960).

[34]Yauharu Nishikawa, *J. Chem. Soc. Jap.* **79**, 236–238 (1958); cf. S. R. Desai and K. K. Sudhalatha, *Indian J. Appl. Chem.* **30**, 116–120 (1967).

[35]I. F. Goryunova, *Sb. Nauch. Tru. Irkut. Nauch.-Issledl. Inst. Red. Met.* **1959** (8), 33–37; cf. S. M. Milaev, *Sb. Tr. Vses. Nauch.-Issledl. Gorno Met. Inst. Tsve. Met.* **1959** (5), 41–45.

potassium acid phthalate solution. Dilute to 18 ml and let stand for 30 minutes. Add 2 ml of 0.1% solution of oxine and 3 ml of chloroform. Let stand for 30 minutes and again shake for 1 minute. Read by fluorescence.

Germanium.[36] Dissolve 2 grams of finely ground sample in 2 ml of perchloric acid, 8 ml of nitric acid, and 42 ml of hydrochloric acid. Boil vigorously until reduced to less than 10 ml. When the solution starts to get cloudy but before perchloric acid fumes appear, add 15 ml of hydrochloric acid. Heat to boiling, taking care to dissolve germanium from the walls of the container, and reduce to perchloric acid fumes, thus removing most of the remaining nitric acid. Cool slightly, add 15 ml of hydrochloric acid, and heat to perchloric acid fumes, thus driving off any remaining germanium. Then evaporate to dryness over a flame to expel all perchloric acid. Dissolve in 25 ml of 1 : 1 hydrochloric acid. Acid-wash 200 ml of ether with 30 ml, 30 ml, and 30 ml of 1 : 1 hydrochloric acid containing 50 mg of sodium sulfite. Extract the sample by vigorous shaking with 25 ml of the acid-washed ether, venting the pressure from time to time. Discard the aqueous layer and wash the ether with 2 ml and 2 ml of 1 : 1 hydrochloric acid. Extract gallium from the ether by shaking with 15 ml and 10 ml of water. Heat the aqueous extracts until the odor of ether disappears and cool. Add a drop of 0.1% *m*-cresol purple indicator solution in dilute sodium hydroxide. Add ammonium hydroxide until the indicator turns yellow and at once add 5 ml of 4% sodium cyanide solution. After mixing, shake with 20 ml of 1% oxine solution in chloroform. Filter the extract through a dry paper and read at 400 nm against chloroform.

Germanium Dioxide. To 3 grams of sample add 2 ml of perchloric acid, 8 ml of nitric acid, and 42 ml of hydrochloric acid. Heat to boiling and, making sure that all the sample is dissolved, evaporate to copious perchloric acid fumes. All the germanium should have been volatilized, but in case of doubt add 5 ml of hydrochloric acid and again take to fumes. Complete as for germanium from "Dissolve in 25 ml of 1 : 1 hydrochloric acid."

LUMOGALLION

Lumogallion, which is 5-chloro-3-(2,4-dihydroxyphenylazo)-2-hydroxybenzenesulfonic acid, is used fluorimetrically for determining down to 0.1 ppm of gallium.[37] Up to 1 μg of iron, 5 μg of copper, or 10 μg of aluminum does not interfere with reading 0.05–0.5 μg of gallium in 10 ml. There is interference by nickel, cobalt, hexavalent chromium, stannic ion, titanium, pentavalent vanadium, and scandium.[38] Beer's law is followed for 0.5–5 μg of gallium in 25 ml of solution containing 0.5–0.6 ml of 0.01% lumogallion solution. Reaction is complete in 20 minutes at 80°. The complex is extractable with amyl alcohol.

[36] C. L. Luke and Mary E. Campbell, *Anal. Chem.* **25**, 1588–1593; 1340–1342 (1956).

[37] N. B. Lebed and R. P. Pantaler, *Zh. Anal. Khim.* **20**, 59–61 (1965).

[38] Yashuharu Nishikawa, Keizo Hiraki, Kiyotoshi Morishiga, and Tsunenobu Shigematsu, *Jap. Anal* **16**, 692–697 (1967).

Addition of polyoxyethylene glycol monolauryl ether to gallium and lumogallion increases the intensity of fluorescence-emission and with excitation at 490 nm changes the peak wavelength from 570 to 553 nm.[39]

Around pH 1.5 the complex is 1 : 1, and reading at 490 nm has been suggested.[40] The reagent does not absorb above 520 nm.[41] For analysis of semiconductors, sulfides are dissolved in 1 : 1 hydrochloric acid and selenides in sulfuric acid containing bromide.[42a]

A variant, 5-chloro-3-(5-hexyl-2,4-dihydroxyphenylazo)-2-hydroxybenzene-sulfonic acid, is known as hexyl-lumogallion.[42b] The 1 : 1 chelate at pH 2.7–4 is extracted into methyl isobutyl ketone, activated at 493 nm and read at 580 nm. The fluorescence in the extract is 25-fold that of the aqueous solution. In the determination of less than 1 μg/ml the addition of 5% of polyoxyethyleneglycol monolauryl ether further increases the fluorescence 10-fold.

Procedures

Cadmium Sulfide. Dissolve 0.5 gram of sample in 5 ml of hydrochloric acid and evaporate to dryness in a silica crucible. Take up the residue in 10 ml of potassium acid phthalate buffer solution for pH 2.4. Add 0.2 ml of 0.01% solution of the reagent and set aside for 30 minutes. Read the fluorescence at 570 nm against a blank.

Cadmium Chloride or Sulfate. Dissolve 0.5 gram in 10 ml of the buffer solution and proceed as for cadmium sulfide.

Biological Material.[43] Boil 1 gram of sample with 0.5 ml of sulfuric acid and 0.5 ml of nitric acid, adding more nitric acid if necessary to prevent charring. When brown fumes have nearly ceased, add 0.25 ml of perchloric acid and heat until colorless and sulfur trioxide fumes are evolved. Take up in 5 ml of 1 : 1 hydrochloric acid, centrifuging and washing with that acid if a precipitate is present. Dilute to 10 ml with 1 : 1 hydrochloric acid and pass through a 2.5 $\times$ 70 mm column of 50-100 mesh Dowex 1-X8 in sodium form. Wash the column with 20 ml of 1 : 1 hydrochloric acid. Elute the gallium with 10 ml of 1 : 11 hydrochloric acid. Add to the eluate 0.1 ml of 1% ascorbic acid solution and 0.1 ml of 0.25% sodium thiosulfate solution. Adjust to pH 2.25 $\pm$ 0.05 with 16% sodium carbonate solution, taking care not to let the solution become alkaline. Add 0.8 ml of 0.01% lumogallion solution in 0.06 *N* hydrochloric acid. Allow 1 hour for complex formation, and extract with 3 ml of isoamyl alcohol. Activate the extract at 490 nm and read the fluorescence at 570 nm. This technic will detect 15 ng of gallium per gram of sample.

---

[39] Kenyu Kina and Nobuhiko Ishibashi, *Microchem. J.* **19**, 26–31 (1974).
[40] V. D. Salikov and M. Z. Yampol'skii, *Zh. Anal. Khim.* **20**, 1299–1305 (1965).
[41] A. K. Babko, A. I. Volkova, and T. E. Get'man, *Ukr. Khim. Zh.* **35**, 190–194 (1969).
[42a] R. P. Pantaler, N. B. Lebed', and L. N. Semenova, *Tr. Kom. Anal. Khim.* **16**, 24–29 (1968).
[42b] Kenyu Kina, Katsuhiko Shiraishi, and Nobuhiko Ishibashi, *Jap. Anal.* **25**, 501–505 (1976).
[43] Ruth A. Zweidinger, Lois Barnett, and Colin G. Pitt, *Anal. Chem.* **45**, 1563–1564 (1973).

MALACHITE GREEN

Photometric reading of the complex of malachite green with gallium is appropriate for analysis of bauxite and many other ores.[44]

For isolation of gallium from solutions of industrial products, *N*-bis-(2-ethylhexyl)dihydrogen pyrophosphate in heptane extracts it, along with several other elements from *N*–3 *N* sulfuric, nitric, perchloric, or hydrobromic acid. Prereduction of ferric ion with ascorbic acid avoids extraction of iron.[45] Then washing the organic phase with 3–4 *N* hydrobromic acid or 2–4 *N* sulfuric acid in the presence of sodium chloride reextracts titanium, zirconium, hafnium, and scandium. Then gallium is extracted from the heptane solution with 7% oxalic acid solution, treated with 0.1% malachite green solution in 1 : 3 hydrochloric acid; the complex is extracted with benzene and the reading taken at 610 nm. For analysis of clays, interfering elements must be separated by cation exchange.[46]

Procedure. *Ores.*[47] Decompose a 1-gram sample with 3 ml of nitric acid and 7 ml of hydrochloric acid. Add 2 ml of 1 : 1 sulfuric acid and evaporate to white fumes. Add 3 ml of water and again evaporate to white fumes. Take up in 10 ml of 1 : 1 hydrochloric acid and filter. If the residue appears to warrant it, ash the filter in platinum, add 2 ml of 1 : 1 sulfuric acid and 5 ml of hydrofluoric acid, and evaporate to white fumes. Take up with 5 ml of 1 : 1 hydrochloric acid and add to the main filtrate and washings. Add 5 ml of 1 : 1 sulfuric acid, dilute to 200 ml, and heat. Pass in hydrogen sulfide for 10 minutes, add 200 ml more of water, and continue the hydrogen sulfide for 10 minutes. Filter and evaporate the filtrate to white fumes. Take up the residue with 5 ml of 1 : 1 sulfuric acid and 5 ml of 1 : 1 hydrochloric acid. Dilute to 50 ml with 1 : 1 hydrochloric acid.

Take a 5-ml aliquot and mix with 0.5 ml of hydrochloric acid and 0.3 ml of 15% titanous chloride solution in 1 : 1 hydrochloric acid. Let stand for 5 minutes and add 1 ml of 2% solution of malachite green in hydrochloric acid. Shake for 2 minutes with 5 ml of benzene and read the extract at 640 nm. The method will determine 0.001–0.0001% of gallium in the 1-gram sample in the presence of large amounts of antimony, tin, copper, and lead.

BIS-(*p*-METHYLBENZYLAMINOPHENYL)ANTIPYRYLCARBINOL

From the synthesis of four antipyrine dyes, the reagent above was selected for analysis of ores, bauxite, and clays.[48] To avoid interference, thallic, antimonic, and ferric ions must be reduced with a titanous reagent.

[44]Joseph Jankovski, *Talenta* **2**, 29–37 (1959).
[45]I. S. Levin and N. A. Balakireva, *Zh. Anal. Khim.* **22**, 1975–1981 (1967).
[46]Maria Kubonova and Kvetuse Poljakova, *Hutn. Listy* **24**, 309–311 (1969).
[47]P. Klein and V. Skrivanek, *Chem. Prum.* **13**, 250–251 (1959).
[48]A. I. Busev, L. M. Skrebkova, and V. P. Zhivopistsev, *Zh. Anal. Khim.* **17**, 685–692 (1962).

Procedure. *Ores.* To prepare the reagent, heat 10.5 grams of 4,4′-bis(methylamino)benzophenone, 14.1 gram of antipyrine, 5 ml of toluene, and 5 ml of phosphoryl chloride at 100° for 3 hours. Then heat at 115° for 4 hours. Cool to 60–70° and treat with 200 ml of 1:3 hydrochloric acid. Filter and wash. Remove ketones by washing with 10-ml portions of benzene until the latter, shaken with 1:5 hydrochloric acid, turns faintly yellow. Chill to below 5° and extract the excess antipyrine by washing with 10% potassium hydroxide solution.

To 0.5 gram of sample in platinum, add 0.25 ml of water, 3 ml of hydrofluoric acid, and 5 ml of sulfuric acid. Evaporate to sulfur trioxide fumes, add 3 grams of potassium persulfate, and fuse. Take up with 25 ml of 1:1 hydrochloric acid and add 10% solution of titanous chloride in 1:1 hydrochloric acid until the solution becomes violet. Let stand a few minutes and extract the gallium with 5 ml and 5 ml of butyl acetate. Wash the combined extracts with 3 ml and 3 ml of 1:1 hydrochloric acid. Reextract the gallium from the organic layer with 15 ml of water, add 0.1 gram of sodium hydroxide, and evaporate to dryness at 100°. Take up the residue with 5 ml of 1:1 hydrochloric acid and add 10% titanous chloride in 1:1 hydrochloric acid until violet. Add 2 ml of 0.6% solution of the captioned reagent in 1:1 hydrochloric acid and shake for 10 minutes with 10 ml of toluene. Filter the extract and read at 585 nm.

METHYLENE BLUE

Methylene blue, which is *NNN′N′*-tetramethylthionine, complexes with gallium, as do also the dimethyl and trimethyl derivatives.[49] The optimum pH is 5.5–6.5; the best extractant is 3:2 dichloroethane-trichloroethane. Beer's law applies up to 5 μg per ml of extract for reading at 655 nm. The complex can also be extracted from a solution of gallium chloride in aqueous sulfuric acid with benzene-1,2-dichloroethane for reading around 600 nm.[50]

Procedure. Dilute a sample containing 1–10 μg of gallium in 1:1 hydrochloric acid to 12.5 ml with that acid.[51] Add 1.5 ml of 0.25 *M* ascorbic acid in 1:1 hydrochloric acid to reduce the iron, heat to 100°, and cool. Add 2 ml of 0.5% solution of methylene blue in 1:1 hydrochloric acid, and extract by shaking for 5 minutes with 5 ml and 5 ml of 3:1 chloroform-acetone. Read at 600 nm against the solvent mixture.

[49]V. M. Tarayan, E. N. Ovsepyan, and A. N. Pogosyan, *Arm. Khim. Zh.* **23**, 996–1003 (1970).
[50]P. P. Kish and A. M. Bukovich, *Zh. Anal. Khim.* **24**, 1653–1660 (1969).
[51]I. K. Guzeinov and E. T. Allakhverdieva, *Azerb. Khim. Zh.* **1968** (5), 99–103; I. L. Bagbanly, I. K. Guseinov, and E. T. Allakhverdieva, *Uch. Zap. Azerb. Gos. Univ., Ser. Khim. Nauk* **1969** (2), 47–52.

METHYL GREEN

A 1 : 1 complex of methyl green with gallium in 1 : 1 hydrochloric acid is extractable for photometric reading.[52] Beer's law holds for 0.05–5.5 μg per ml of extract. Addition of titanous chloride avoids extraction of interfering amounts of the dye. Thallic ion must not exceed the gallium and yttrium must not be more than twice as much.

Procedure. *Silicates.* Fuse 0.5 gram of sample with 2.5 grams of sodium carbonate and dissolve with 1 : 1 hydrochloric acid. Evaporate to dryness, take up in 20 ml of 1 : 1 hydrochloric acid, and filter off the silica. Wash with 1 : 1 hydrochloric acid and dilute to 25 ml with that acid. Extract the gallium from a 20-ml aliquot with 10 ml and 10 ml of butyl acetate. Wash the combined extracts with 5 ml of 1 : 1 hydrochloric acid. Extract the gallium with 10 ml, 10 ml, and 10 ml of water and evaporate these extracts nearly to dryness. Dilute to 25 ml with 1 : 1 hydrochloric acid and add 0.5 ml of 15% titanous chloride in 1 : 1 hydrochloric acid. Add 4 ml of 1% solution of methyl green in 1 : 5 hydrochloric acid and extract with 5 ml and 5 ml of 1 : 3 dichloroethane-trichloroethylene. Read at 640 nm.

METHYL-8-HYDROXYQUINOLINE

This reagent, methyl oxine, gives fluorescence with the gallium complex at 0.01–0.0001%.[53]

Procedure. *Germanium.* Dissolve 1 gram of sample by warming with 20 ml of 20% sodium hydroxide solution and 10 ml of 30% hydrogen peroxide. Cool, acidify with hydrochloric acid, and dilute to 100 ml. Dilute a 10-ml aliquot to 40 ml. Add 1 ml of 1% 2-methyloxine and 5 ml of 20% solution of ammonium acetate. Adjust the pH to 3.9 with 1 : 11 hydrochloric acid. Extract with 10 ml, 10 ml, and 10 ml of chloroform. Dilute the extracts to 50 ml with chloroform, dry with sodium sulfate, and read the fluorescence at 495 nm.

METHYLTHYMOL BLUE

Gallium forms 1 : 1 and 1 : 2 complexes with this reagent. By use of a large excess of reagent, the latter complex is formed.[54] Beer's law is followed for 7–84 μg of

[52] V. M. Tarayan, E. N. Ovsepyan, and A. N. Pogosyan, *Zavod. Lab.* **36**, 656–658 (1970).
[53] Tsunenobu Shigematsu, *Jap. Anal.* **7**, 787–788 (1958); cf. Koichi Mukai, *ibid.* **9**, 631–633 (1960).
[54] K. N. Bagdasarov and S. A. Popova, *Khim. Ind.* **45**, 108–110 (1973); Koichi Tonosaki and Koyko Sakai, *Jap. Anal.* **14**, 495–500 (1965; cf. V. A. Nazarenko and E. M. Nevshaya, *Zh. Anal. Khim.* **24**, 839–843 (1969). V. N. Tikhonov, *ibid.* **21**, 1172–1178 (1966).

gallium in 25 ml for reading at 565 nm. Buffered with acetate at 4.5–5.5, a 1:2 complex is formed with a maximum absorption at 515 nm over the range 7–70 μg of gallium in 25 ml.

Procedure. Mix 5 ml of sample solution containing 3–37 μg of gallium with 5 ml of 3.1 m*M* methylthymol blue.[55] Add 15 ml of 1:16 acetic acid and boil for 3 minutes. Cool, dilute to 50 ml with 1:16 acetic acid, and read at 510 nm.

Zinc Alloy. Dissolve 1 gram in 1:1 hydrochloric acid and dilute to 100 ml with that acid. To a 5-ml aliquot, add 10% titanous chloride solution dropwise to a pale violet, which persists for several minutes. This avoids coextraction of iron. Extract by shaking for 1 minute with 5 ml and 5 ml of ether presaturated with 1:1 hydrochloric acid. Wash the combined extracts with 3 ml and 3 ml of 1:1 hydrochloric acid presaturated with ether. Extract the gallium from the ether by shaking with 25 ml of water. Mix 5 ml of the aqueous extract with 10 ml of acetate buffer solution for pH 4. Add 1 ml of 0.01% solution of methylthymol blue, dilute to 50 ml, and heat to 80°. Cool and read at 510 nm.

Aluminum.[56] Dissolve 0.3 gram in 5 ml of 1:1 hydrochloric acid. To mask aluminum, iron, titanium, thallium, and vanadium, add 1.25 grams of ammonium fluoride and 1 ml of 20% triethanolamine solution. Add 1 ml of 0.05% solution of methylthymol blue and then add 20% sodium hydroxide solution until a pale lilac-pink develops. Add 10 ml of a buffer solution for pH 1.85–2.4 and heat at 100° for 9 minutes. Filter, dilute the filtrate to 25 ml with the buffer solution, and read at 570 nm.

1-(2-PYRIDYLAZO)-2-NAPHTHOL

For analysis of aluminum in hydrochloric acid solution, the iron is reduced with titanous chloride and the gallium extracted with isopropyl ether.[57] The residue on evaporation of the extract is dissolved in hydrochloric acid and buffered to pH 4.3 with acetate. On shaking with the captioned reagent in chloroform, the organic layer is read at 550 nm.

Procedure. Adjust a 10-ml sample containing up to 0.025 mg of gallium to 7 *N* hydrochloric acid and extract the gallium with 10 ml of isopropyl ether.[58]

[55]N. D. Lukomskaya, T. V. Mal'kova, and K. B. Yatsimirskii, *Izv. Vyssh. Ucheb. Zaved., Khim. Khim. Tekhnol.* **10**, 994–996 (1967).
[56]M. K. Akhmedli, E. A. Bashirov, and E. L. Glushchenko, *Uc. Zap. Azerb. Gos. Univ., Ser. Khim. Nauk* **1969** (2), 42–46.
[57]Toshio Suzuki, *Jap. Anal.* **12**, 655–656 (1963); cf. L. Sommer and Z. Chromy, *Scripta* **3**; *Chemia* (2) 77–90 (1973).
[58]K. L. Cheng and B. L. Goydish, *Anal. Chim. Acta* **34**, 154–164 (1966); cf. Shozo Shibata, *ibid.* **25**, 348–359 (1961).

Evaporate the extract to dryness at under 200° and take up the residue in 5 ml of water saturated with isopropyl ether. Add 1 ml of a solution containing 40 mg of the reagent in 150 ml of dimethylformamide and 50 ml of 10% solution of ammonium acetate. Dilute to 10 ml with water and set aside for 10 minutes. Read at 545 nm against a reagent blank.

Germanium.[59] This technic is appropriate for alloys containing 0.01–1% of gallium. Add 20 ml of 5% sodium hydroxide solution to 0.3 gram of finely ground sample. Dissolve by adding 4 ml of 30% hydrogen peroxide, add 4 ml of 1:4 sulfuric acid, and dilute to 50 ml. To a 4-ml aliquot, add 7 ml of a buffer solution containing 8.2% of sodium acetate and 5% of acetic acid. Add 1 ml of 0.1% solution of the reagent and dilute to 25 ml. After 8 minutes read at 540 nm against a blank.

Gold. This technic is appropriate for samples containing 1–5% of gallium. Dissolve a 25-mg sample in 2.5 ml of hydrochloric acid and 0.5 ml of nitric acid at 100°. Dilute to 250 ml and treat as for germanium from "To a 4-ml aliquot add...."

High Purity Aluminum. See the alternative method for determination by rhodamine B.

4-(2-PYRIDYLAZO)RESORCINOL

This reagent absorbs at 390 nm, as compared with the 2:1 gallium complex at 510 nm.[60] Development is instantaneous, and the absorption is stable for 24 hours. The optimum pH is 3.7–4.5. Aside from indium and thallium, there is interference by thorium, zirconium, hexavalent uranium, pentavalent vanadium and niobium, tantalum, ferric ion, cobalt, nickel, palladium, cupric ion, zinc, cadmium, aluminum, and nitrate ion. Copper, cadmium, nickel, zinc, and iron can be masked by cyanide. Beer's law is followed for 1–15 μg per 25 ml.

In 0.1 N hydrochloric acid, 0.05 mg of gallium in 5 ml of solution can be extracted by shaking for 5 minutes with 5 ml of tributyl phosphate.[61] Then the gallium can be reextracted from the organic solvent with an aqueous solution of the captioned reagent for reading. From 1:1 hydrochloric acid butyl acetate, extracts gallium, leaving indium behind.[62] The gallium is then reextracted with water. The extinction maximum of the complex is not affected by the presence of methanol, ethanol, propanol, butanol, isobutyl alcohol, isopentyl alcohol, methyl ethyl ketone, acetone, or dioxan.[63a]

[59] I. Dobes and M. Salamon, *Chem. Listy* **60**, 68–71 (1966).

[60] C. D. Dwivedi, K. N. Munshi, and A. K. Dey, *Chemist-Analyst* **55**, 13 (1966); cf. Kazuyoshi Hagiwara, Masanori Nakane, Yasuski Osumi, Elichi Ishii, and Yoshizo Miyake, *Jap. Anal.* **10**, 1397–1382 (1961); M. Hnilickova and L. Sommer, *Z. Anal. Chem.* **193**, 171–178 (1963); E. P. Tsintsevich, V. M. Ivanov, and V. A. Tsapel', *Vest. Mosk. Univ., Ser. Khim.* **1963** (5), 54–56.

[61] I. P. Alimarin, N. I. Ershova, and T. A. Bol'shova, *Vest. Mosk. Univ. Ser. Khim,* **1967** (4), 51–54.

[62] L. I. Ganaga, L. M. Bukhtseeva, and L. R. Starobinets, *Vest. Akad. Navuk. BSSR, Ser. Khim. Navuk.* **1969** (5), 49–53.

[63a] V. I. Ivanov, E. P. Tsintsevich, and A. N. Gorokhova, *Ves. Mosk. Univ., Ser. Khim* **1964** (3), 69–72.

At pH 4.5 a complex is extracted with chloroform of gallium with PAR, antipyrine, and thiocyanate ion.[63b] Aluminum, indium, and thallic ion interfere.

Procedure. Mix 10 ml of sample solution in 7 *N* hydrochloric acid containing 2–8 μg of gallium with 1 ml of 17% solution of titanous chloride in 1 : 1 hydrochloric acid.[64a] Extract with 10 ml of 1 : 1 isopropyl ether–butyl acetate. Wash the organic phase with a mixture of 10 ml of 7 *N* hydrochloric acid and 1 ml of the titanous chloride solution, then with 5 ml and 5 ml of 7 *N* hydrochloric acid. To 5 ml of the organic extract add 1 ml of a methanolic reagent containing 0.05% of 4-(2-pyridylazo)resorcinol and 0.5% of sodium acetate trihydrate. Add 0.2 ml of pyridine and dilute to 10 ml with a solution containing 1% of sodium acetate trihydrate and 0.3% of sodium diethyldithiocarbamate in methanol. Let stand for 10 minutes and read at 510 nm.

Alternatively, take a sample solution containing 1–10 μg of gallium.[64b] If ferric ion or pentavalent vanadium is present, mask with ascorbic acid. Add 1 ml of 0.02 *M* tetraphenyl-arsonium or -phosphonium chloride and adjust to 3 *N* with hydrochloric acid. If copper is present mask with cyanide. Extract for 2 minutes with 5 ml of 1,2-dichlorobenzene. Mix the organic layer with 2 ml of m*M* 4-(2-pyridylazo)resorcinol, 0.5 ml of 0.02 *M* tetraphenyl-arsonium or -phosphonium chloride, and 2.5 ml of acetate buffer solution for pH 6. Shake for 5 minutes, centrifuge, and read at 510 nm. Beer's law is followed for 0.2–1.5 ppm of gallium. Stannous ion equal to the gallium interferes, mercuric ion and zinc in ten-fold quantities. Tungstate and large amounts of oxalate or tartrate interfere.

Tungsten Solutions.[65] The sample is an alkaline solution containing 5–60 μg of gallium and up to 100 mg of tungsten as tungstate. Add 30% hydrogen peroxide so that there are 2–4 moles per mole of tungsten. Add 3 ml of 10% solution of potassium cyanide, then add acetic acid until the pH reaches 6.5 ± 0.1. Mix 0.3 ml of fresh 0.5% solution of the captioned reagent with 2 ml of water and add to the sample. Dilute to 50 ml, let stand for 5 minutes, and read at 510 nm.

2-(2-PYRIDYL)BENZIMIDAZOLE

The complex of gallium with this reagent is read by fluorescence.[66] Ceric and ferric ions can be reduced to cerous and ferrous ions to avoid interference. Cadmium enhances the fluorescence; cobalt, copper, nickel, tetravalent platinum, palladic, silver, molybdate, and sulfide ions quench the fluorescence.

Procedure. To a 10-ml sample containing 0.07–0.7 μg of gallium per ml, add 1 ml

[63b] V. A. Nazarenko, E. A. Biryuk, and R. V. Ravitskaya, *Zh. Anal. Khim.* **30**, 1724–1727 (1975).
[64a] Kenji Bansho and Yoshimi Umezaki, *Bull. Chem. Soc. Jap.* **40**, 326–329 (1967).
[64b] M. Siroki and M. J. Herak, *Anal. Chim. Acta* **87**, 193–199 (1976).
[65] Karoly G. Vadasdi, *Chem. Anal.* (Warsaw) **14**, 733–739 (1969).
[66] L. S. Bark and A. Rixon, *Anal. Chem. Acta* **45**, 425–432 (1969).

of 50% ammonium acetate solution and 1 ml of 25% sodium benzoate solution. Extract the gallium and indium with 3 ml and 3 ml of ethyl acetate. Add 2 ml of water and 2 drops of 0.1 *N* hydrochloric acid to the combined organic extracts. Evaporate until the ethyl acetate is removed. Add 0.5 ml of buffer solution for pH 4.4 and 0.5 ml of 0.001 *M* solution of the reagent; set aside for 30 minutes. Irradiate at 347 nm and read the fluorescence at 413 nm.

PYROGALLOL RED

At pH 3 pyrogallol red and gallium form a 1:1 complex that is read at 525 nm.[67] There is interference by iron, aluminum, indium, tin, cupric ion, hexavalent molybdenum, thorium, zirconium, rare earths, phosphate, EDTA, and nitriloacetic acid.

Procedure. To a sample solution containing 0.007–0.119 mg of gallium add a fourfold excess of m*M* pyrogallol red.[68] Adjust to pH 4.5±0.1 with an acetate buffer solution and dilute to 25 ml. Keep at 15° for 30 minutes and read at 530 nm.

PYRROLIDINE-1-CARBODITHIOATE

The complex of this reagent with gallium is extractable with chloroform for reading at 2–25 ppm.[69]

Procedure. Adjust 20 ml of sample solution containing 0.02–0.25 mg of gallium to pH 2.5 with a phosphate buffer. Add 2 ml of 0.1% solution of the captioned reagent in 0.1 *N* ammonium hydroxide. After 2 minutes extract with 10 ml of chloroform and read the extract at 308 nm.

RHODAMINE B

Rhodamine B (basic violet 10) is applicable to samples containing less than 2 μg of gallium per gram.[70] Extraction with isopropyl ether from 7 *N* hydrochloric acid is appropriate.[71] However fluorescence is more sensitive by a factor of 40.[72] For

[67] L. I. Ivanova, E. A. Bashirov, and M. K. Akhmedli, *Uch. Zap. Azerb. Gos. Univ., Ser. Khim. Nauk* **1970** (3) 33–41.
[68] K. C. Srivastava, *Chim. Anal.* (Paris) **53**, 525–528 (1971).
[69] W. Likussar, C. Sagan, and D. F. Boltz, *Mikrochim. Acta* **1970**, 683–687.
[70] A. Szucs and O. M. Klug, *Magy. Kem. Lapja* **21**, 328–331 (1966).
[71] G. N. Lypka and A. Chow, *Anal. Chim. Acta* **60**, 65–70 (1972.
[72] M. A. Matveets and D. P. Shcherbov, *Zh. Pril. Spektr.* **2** (2), 111–114 (1965).

fluorescence, the optimum concentration of hydrogen ion is 13.5 *N*. The chloride ion should not be less than *N*, but the use of sulfuric acid and sodium chloride has no advantage over the use of hydrochloric acid.[73]

As applied to by-products of the zinc industry, reduce ferric and thallic ion in 1:1 hydrochloric acid with titanous chloride in 1:1 hydrochloric acid.[74] If molybdenum is present remove it with benzoin α-oxime. Extract gallium chloride with ether, evaporate and take up in 1:1 hydrochloric acid. Add a few drops of titanous chloride solution with the rhodamine B solution, and extract with 3:1 benzene-ether for fluorimetric determination.

As applied to plutonium of high purity, the sample is dissolved in hydrochloric acid and the rhodamine B complex extracted with 3:1 chlorobenzene–carbon tetrachloride without separation of the plutonium and other elements commonly found to be present.[75] At 0.5–100 ppm it is read at 562 nm.

Indium is separated from gallium on a column of 4% of trioctylamine on polytetrafluoroethylene by elution with 9 *N* hydrochloric acid.[76] The gallium is then eluted with 0.05 *N* hydrochloric acid and developed with rhodamine B.

For fluorescent reading, the optimum conditions are 0.08% of rhodamine B in the aqueous phase, which is 6 *N* with hydrochloric acid, or 4 *N* with hydrochloric acid plus 5 *N* with sulfuric acid.[77] The best extraction medium for the colored complex is either benzene-acetone or 9:1 benzene-ether. Isolation of gallium chloride from interference is by butyl acetate from 1:1 hydrochloric acid or a mixture of hydrochloric and sulfuric acids such as that cited earlier.

For analysis of gold, dissolve 10 grams of sample in aqua regia.[78] Dilute and extract chloraurate with isopropyl ether. Then determine gallium with rhodamine B.

Procedure. To 10 ml of sample solution containing 1–10 μg of gallium in 7 *N* hydrochloric acid, add 2 ml of 20% solution of titanous chloride in hydrochloric acid.[79] After 5 minutes extract with 10 ml and 5 ml of isopropyl ether. Wash the combined ether extracts with a mixture of 1 ml of 7 *N* hydrochloric acid and 0.5 ml of the titanous chloride solution. Repeat the washing, then wash with 1 ml of 7 *N* hydrochloric acid. Add 0.3 ml of 10% sodium chloride solution and evaporate the ether. Take up in 5 ml of 1:1 hydrochloric acid. Add 0.4 ml of 0.5% rhodamine B solution in 1:1 hydrochloric acid. Shake with 10 ml of 3:1 chloroform–carbon tetrachloride for 1 minute and read the extract at 565 nm against a blank.

Alternative extraction media are benzene, 1:3 ether-benzene and *o*-dichlorobenzene.[80] Ascorbic acid may replace titanous chloride.

[73] D. P. Shcherbov and N. V. Kagarlitskaya, *Tr. Kaz. Nauch.-Issled. Inst. Mineral. Syr'ya* **1961** (5), 255–259.

[74] V. Armeanu and P. Continescu, *Bul. Inst. Politeh., Iasi* **8**, 123–128 (1962).

[75] R. D. Gardner, C. T. Apel. and W. J. Ashley, *USAEC* **Rep. LA-3248** (1965).

[76] I. P. Alimarin, T. A. Bol'shova, V. A. Lukashenkova, and N. V. Arslanova. *Zh. Anal. Khim.* **29**, 1558–1562 (1974).

[77] Yu. N. Knipovich and V. M. Krasikova, *Tr. Vses, Nauch.-Issled. Geol. Inst.* **117**, 105–110 (1964).

[78] Krzysztof Kasiura, *Chem. Anal.* (Warsaw) **20**, 809–816 (1975).

[79] Hiroshi Onishi and E. B. Sandell, *Anal. Chim. Acta* **13**, 159–164 (1955); F. Culkin and J. P. Riley, *Analyst* **83**, 208–212 (1958); V. S. Saltykova and E. A. Fabrikova, *Zh. Anal. Khim.* **13**, 63–65 (1958).

[80] W. Rutkowski and M. Basinska, *Chem. Anal.* (Warsaw) **13**, 641–648 (1968).

Alternatively,[81] for absorption, if the sample contains carbon, ash 0.5 gram at 600°. Fuse the ash or 0.5 gram of mineral in an iron crucible with 3 grams of sodium hydroxide at 600°. Take up the melt with 40 ml of hot water. Add 0.2 ml of 30% hydrogen peroxide and boil for a couple of minutes. Dilute to 100 ml.

Neutralize a 10-ml aliquot with 1:1 hydrochloric acid, add 2 ml in excess, and evaporate to dryness. Take up the residue in 10 ml of 1:1 hydrochloric acid. To a 2-ml aliquot add 0.5 ml of 7.5% solution of titanous chloride in 1:1 hydrochloric acid to reduce ferric, thallic, antimonic, or auric ion. After 2 minutes for reaction, add 0.5 ml of 0.2% solution of rhodamine B in 1:1 hydrochloric acid. Shake for 3 minutes with 3 ml of 3:1 benzene-ether. Set aside for 30 minutes and read the organic layer at 564 nm against the solvent.

Separation by 2-Thenoyltrifluoroacetone.[82] Adjust 25 ml of sample solution containing up to 0.2 mg of gallium to pH 4.5–6. Extract with 10 ml of 0.03 *M* 2-thenoyltrifluoroacetone in xylene. Then shake the organic phase with 10 ml of 6 *N* hydrochloric acid containing 0.5% of rhodamine B. Read at 565 nm. This separates from indium, thallium, and aluminum. Citrate, ascorbate, and EDTA must be absent.

Tungsten.[83] For 1–10 ppm of gallium take a 2-gram sample, for 0.1–1 ppm, use 5 grams. Dissolve in 20 ml of 30% hydrogen peroxide, more if necessary for the larger sample. Add 15 ml of 20% sodium hydroxide solution and bring to a boil to complete conversion to sodium tungstate. Cool and dilute to 50 ml. Add to 50 ml of 8 *N* hydrochloric acid and set aside for 2 hours. Filter the precipitated tungstic acid. Take 50 ml of filtrate and add 3 ml of 30% hydrogen peroxide. Add 10 ml of 9% solution of aluminum chloride hexahydrate and sufficient ammonium hydroxide to raise to pH 7. Filter and dissolve in 10 ml of 8 *N* hydrochloric acid. Add 1 ml of 10% titanous chloride solution in 8 *N* hydrochloric acid and 0.4 ml of 0.5% solution of rhodamine B in 8 *N* hydrochloric acid. Shake with 10 ml of 3:1 chlorobenzene–carbon tetrachloride for 10 minutes. Adjust the organic layer to a known volume with ethanol and read at 562 nm.

Tungsten and Tungstates. Fuse 0.5 gram of sample with 5 grams of sodium hydroxide for 10 minutes at 1000°. Extract the cooled melt with 20 ml of 5% tartaric acid solution, filter, and dilute to 25 ml. To an aliquot containing less than 2 μg of gallium, add 0.5 gram of tartaric acid and sufficient hydrochloric acid to make the solution 1:1. If iron is present, add 1 ml of 10% titanous chloride in hydrochloric acid and set aside for 5 minutes. Add 0.4 ml of 0.5% solution of rhodamine B in 1:1 hydrochloric acid. Shake for 2 minutes with 4 ml of 3:1 benzene-ether and read the extract at 564 nm.

Aluminum.[84] Dissolve 0.5 gram of sample containing 0.001–0.02% gallium in 10 ml of 1:1 hydrochloric acid containing 0.3 mg of nickel as nickel chloride. Add 1

[81] V. V. Chekalin, *Nauch. Tr. Irkut. Nauch.-Issled, Inst. Redk. Met.* **1961** (10), 93–96.
[82] P. V. Dhond and S. M. Khopkar, *Talenta* **23**, 51–53 (1976).
[83] Piroska Buxbaum and Karoly G. Vadasdi, *Chem. Anal.* (Warsaw) **14**, 429–436 (1964).
[84] Shigeo Inowe, Sango Oseki, and Hideo Unno, *Jap. Anal.* **24**, 137–139 (1975); cf V. K. Kuznetsova and N. A. Tananaev, *Nauch. Dokl. Vyssh. Shk., Met.* **1958** (4), 258–260.

ml of 30% hydrogen peroxide and boil off the excess. Cool and dilute to 100 ml. Mix a 10-ml aliquot with 10 ml of hydrochloric acid and 1 ml of 20% titanous chloride solution. Set aside for 10 minutes, then add 1 ml of 0.5% rhodamine B solution. Extract for 1 minute with 25 ml of 1:6 methyl isobutyl ketone–benzene. Dry the organic layer and read at 563 nm.

High Purity Aluminum.[85] Dissolve a 5-gram sample in 10 ml of hydrochloric acid and evaporate to dryness. Take up the residue in 50 ml of 7 *N* hydrochloric acid. Extract by shaking for 3 minutes with 10 ml, 10 ml, and 10 ml of methyl isobutyl ketone. Combine the extracts and reextract the iron and gallium with four 10-ml portions of water. Evaporate the aqueous extracts to incipient crystallization. Add 4 ml of 1:1 hydrochloric acid and 0.5 ml of 20% titanous chloride solution in 1:1 hydrochloric acid. Set aside for 5 minutes. Add 0.4 ml of 0.5% solution of rhodamine B in 1:1 hydrochloric acid. Shake for 2 minutes with 1.5 ml of benzene and 0.5 ml of ether. Read the extract at 564 nm.

Alternatively, after evaporation to incipient crystallization add 1 ml of 1:11 hydrochloric acid and 2 ml of 10% tartaric acid solution. Adjust to pH 5–6 with ammonium hydroxide. Add 1 ml of 0.5% ethanolic solution of 1-(2-pyridylazo)-2-naphthol. Dilute to 10 ml with ethanol and read at 550 nm.

Zinc.[86] Decompose 1 gram of sample with 20 ml of 1:1 hydrochloric acid, adding 0.5 ml of 2% solution of sodium chloride. Add 0.1 ml of 5% titanous chloride solution in 1:1 hydrochloric acid. Shake for 3 minutes with 20 ml of butyl acetate and discard the aqueous layer. Wash the organic extract with 3 ml and 3 ml of 1:1 hydrochloric acid. Extract the gallium with 15 ml and 10 ml of water. Add 0.5 ml of 20% sodium hydroxide solution to the combined aqueous extracts and evaporate to dryness at 100°. Add 5 ml of 1:1 hydrochloric acid to the residue. Add 0.1 ml of 5% titanous chloride in 1:1 hydrochloric acid, 2 ml of benzene, 0.5 ml of butyl acetate, and 0.6 ml of 0.5% solution of rhodamine B in 1:1 hydrochloric acid. Shake for 2 minutes, filter the extract, and read it fluorimetrically.

Indium.[87] Dissolve a 0.5-gram sample in 2 ml of hydrochloric acid and 6 ml of nitric acid and evaporate to dryness. Add 3 ml of hydrochloric acid and evaporate to dryness. Take up in 20 ml of 1:1 hydrochloric acid and extract the indium with 10 ml and 10 ml of ether. Wash the combined ether extracts with 5 ml of 1:1 hydrochloric acid to remove entrained indium. Extract the gallium from the ether with 5 ml of water. Add 5 ml of hydrochloric acid and 1 ml of 10% titanous chloride in 1:1 hydrochloric acid and set aside for 5 minutes. Add 1 ml of 0.001% solution of rhodamine B in 1:1 hydrochloric acid. Extract with 10 ml of 13:2 benzene–methyl isobutyl ketone and read at 550 nm.

Alternatively,[88a] for molybdenum, iron, and gallium, dissolve 10 grams of sample in 25 ml of hydrochloric acid and 0.1 ml of nitric acid. Concentrate until a precipitate appears. Add 25 ml of 7 *N* hydrochloric acid and extract the gallium,

[85] A. I. Szücs and O. N. Klug, *Chem. Anal.* (Warsaw) **12**, 939–947 (1967); cf. G. Parissakis and P. B. Issopoulos, *Mikrochim. Ichnoanal. Acta* **1965**, 28–32.
[86] G. I. Kuchmistaya, *Zavod. Lab.* **27**, 377–379 (1961).
[87] Koichi Nishimura and Teruo Imai, *Jap. Anal.* **13**, 518–524 (1964).
[88a] K. Kasiura, *Chem. Anal.* (Warsaw) **13**, 849–855 (1968).

gold, iron, molybdenum, and thallium with 10 ml, 10 ml, and 10 ml of isoamyl acetate by shaking for 1 minute each. Reextract gallium, iron, and molybdenum with 5 ml and 5 ml of 1 : 3 hydrochloric acid. Evaporate almost to dryness, dilute to 25 ml with water, and acidify to pH 1.5 with 1 : 1 hydrochloric acid. Pass through a column of Amberlite IR-120 in hydrogen form. Elute the molybdenum with 1.5 *N* hydrochloric acid as a sample for reading as the thiocyanate at 467 nm. Then elute the iron and gallium with 1 : 2 hydrochloric acid. Add about 50 mg of lanthanum as chloride, dilute to 100 ml, and precipitate with ammonium hydroxide. Filter the precipitate as a sample to be dissolved in 25 ml of 1 : 4 hydrochloric acid for determination of iron by thiocyanate. Acidify the ammoniacal filtrate containing the gallium with 1 : 1 hydrochloric acid and evaporate to dryness. Take up with 5 ml of water and complete as for the previous method from "Add 5 ml of hydrochloric acid and...."

Steel.[88b] Dissolve 0.2 gram of sample in dilute hydrochloric acid. If insoluble gallium trioxide is present, add 2 ml of perchloric acid. Evaporate to fumes. Add 40 ml of 7 *N* hydrochloric acid and 10 ml of 15% titanous chloride solution. Extract gallium with 40 ml of isopropyl ether. Wash the extract with 5 ml and 5 ml of 7 *N* hydrochloric acid containing 1 ml of 15% titanous chloride solution. Wash with 5 ml of 7 *N* hydrochloric acid. Add 0.3 ml of 10% sodium chloride solution to the washed extract and evaporate to dryness. Dissolve the residue in 10 ml of 1 : 1 hydrochloric acid. Add 0.8 ml of 0.5% solution of rhodamine B in 1 : 1 hydrochloric acid and 0.2 ml of 10% solution of hydroxylamine hydrochloride. Extract with 20 ml of benzene and read at 570 nm.

Bauxite.[89] Fuse 1 gram of sample with 4 grams of sodium carbonate for 10 minutes at 1000°. Take up with 10 ml of water, neutralize with 1 : 1 hydrochloric acid, and evaporate to dryness. Take up the residue in 10 ml of 1 : 1 hydrochloric acid. Add 1 ml of 15% titanous chloride solution in 1 : 1 hydrochloric acid to reduce the iron and 1 ml of 0.5% solution of rhodamine B in 1 : 1 hydrochloric acid. Extract with 3-ml portions of 1 : 3 benzene-ether until the extracts are colorless. Read the combined extracts at 565 nm. A delay of 30 minutes before separation of the organic layer and retention of it for 15 minutes after separation before reading has been recommended.[90]

Silicate Ores. Treat 1 gram of sample with 2 ml of sulfuric acid and 5 ml of hydrofluoric acid and evaporate to dryness. Fuse the residue with 4 grams of sodium carbonate for 10 minutes at 1000°. Complete as for bauxite from "Take up with 10 ml of water, neutralize...."

Sulfide Ores. Treat 1 gram of sample with 5 ml of hydrochloric acid and 0.1 ml of 30% hydrogen peroxide and evaporate to dryness. Complete as for bauxite from "Take up the residue in 10 ml of 1 : 1...."

[88b]Karl-Heinz Sauer and Martin Nitsche, *Arch. Eisenhütt.* **47**, 153–156 (1976).
[89]V. Gluck and S. Ioan, *Rev. Chim.* (Bucharest) **13**, 551–552 (1962).
[90]Z. Soljic and V. Marjanovic, *Chim. Anal.* (Paris) **51** (3), 121–124 (1969).

Minerals.[91] Treat a 0.2-gram sample with 5 ml of nitric acid and evaporate to a syrup. Add 1 ml of 1 : 1 sulfuric acid and 5 ml of hydrofluoric acid and evaporate to dryness. Add 2 grams of sodium pyrosulfate and fuse at 500° for 10 minutes. Dissolve the melt in 25 ml of 6 *N* hydrochloric acid.

Take a 5-ml aliquot and add 15% solution of titanous chloride in 1 : 1 hydrochloric acid until the color is violet. Add 2 ml in excess and dilute to about 12 ml with 1 : 1 hydrochloric acid. Let stand for 10 minutes and then add 1 ml of 1% solution of rhodamine B in 1 : 1 hydrochloric acid. Shake for 1 minute with 5 ml of 1 : 9 ether-benzene. Separate the organic layer. As a wash solution mix 50 ml of 15% titanous chloride in 1 : 1 hydrochloric acid, 20 ml of hydrochloric acid, and 6 ml of 1% solution of rhodamine B in 1 : 1 hydrochloric acid. To the organic layer, add 2 ml of this wash solution and shake for 1 minute. Separate the organic layer and after 10 minutes read at 564 nm against water.

Silicates.[92] Nephthalines are typical of those to which this technic is applicable. Mix 0.2 gram of finely ground sample with 1 gram of ammonium fluoride in platinum. Heat at 300–400° until decomposition is complete, normally 5 minutes. Take up the contents in 1 : 1 hydrochloric acid and dilute to 25 ml with that acid. Take a 2-ml aliquot, add 0.2 ml of 10% titanous chloride in 1 : 1 hydrochloric acid to reduce the iron, and dilute to 3 ml with 1 : 1 hydrochloric acid. Let stand for 5 minutes and add 0.4 ml of 0.5% solution of rhodamine B in 1 : 1 hydrochloric acid. Add 1 ml of acetone and shake with 5 ml of benzene for 2 minutes. Let stand for 10 minutes and read the organic layer at 564 nm.

RHODAMINE S

The complex of rhodamine S with gallium as extracted with organic solvent is read by fluorescence.[93] It is more selective fluorescently than rhodamine 6G[94] (basic red 1). It is also read photometrically.

Procedures

Silicates and Bauxites. Treat a 0.25-gram sample with 3 ml of 1 : 1 sulfuric acid and 5 ml of hydrofluoric acid. Evaporate to dryness, add 2 grams of potassium pyrosulfate, and fuse to a clear melt. Dissolve in 1 : 1 hydrochloric acid and dilute to 25 ml with that acid. Mix a 2-ml aliquot with 0.2 ml of 15% titanous chloride solution in 1 : 1 hydrochloric acid and heat to boiling. Let cool and again heat to

[91] V. A. Oshman and Yu. N. Tishchenko, *Tr. Ural. Nauch.-Issled. Proekt. Inst. Medn. Prom.* **1963**, 423–428; cf. W. Lemke, *Z. Angew. Geol.* **11**, 552–558 (1965).
[92] V. K. Kuznetsova and N. A. Tananaev, *Iz. Vyssh. Ucheb. Zaved., Khim. Khim. Tekhnol.* **2**, 840–842 (1959).
[93] P. I. Vasil'ev, R. L. Podval'naya, and M. A. Voronkova, *Mineral. Syr'ya, Mosk., Sb.* **1960** (1), 302–306
[94] D. P. Shcherbov, I. T. Solov'yan, A. I. Ivankova, and A. V. Drobachenko, *Tr. Kaz. Nauch.-Issledl. Inst. Mineral. Syr'ya* **1**, 188–195 (1959); D. P. Shcherbov and I. T. Solov'yan, *ibid.* 196–199.

boiling. Let cool and add 0.5 ml of 15% titanous chloride solution in 1 : 1 hydrochloric acid, 0.4 ml of 0.5% solution of rhodamine S in 1 : 1 hydrochloric acid, and 3 ml of 1 : 6.6 ether-benzene. Shake well and read the fluorescence of the extract.

Sulfide Ores. Add 15 ml of 1 : 1 hydrochloric acid to a 0.5-gram sample, gradually bring to 100° and boil for 10 minutes. Add 5 ml of nitric acid and boil until evolution of gas ceases. Evaporate to dryness, add 3 ml of hydrochloric acid, and again evaporate to dryness. Repeat the last step twice. Complete as for silicates from "Dissolve in 1 : 1 hydrochloric acid and . . . ."

Coal. Ash at 400° and treat as for silicates and bauxites.

Polymetallic Ores and Concentrates.[95] To 1 gram of sample in platinum add 5 ml of hydrofluoric acid, 5 ml of nitric acid, and 5 ml of sulfuric acid. Evaporate at 100°, then heat to voluminous fumes of sulfur trioxide. Add 5 ml of water and again take to fumes. Take up in 10 ml of 1 : 3 hydrochloric acid and add 2 grams of fine cadmium wire. After 30 minutes filter through cotton and wash with 1 : 3 hydrochloric acid. Add 9 ml of hydrochloric acid and reduce the iron with 1 ml of 15% titanous chloride solution in 1 : 1 hydrochloric acid. Extract with 25 ml and 20 ml of ether saturated with 1 : 1 hydrochloric acid. Wash the combined ether extracts with 5 ml of 1 : 1 hydrochloric acid saturated with ether. Add 0.5 ml of 10% sodium chloride solution to the ether extracts and evaporate to dryness at 100°. Take up the residue in 3 ml of 1 : 1 hydrochloric acid. Add 0.5 ml of 15% titanous chloride in 1 : 1 hydrochloric acid and 0.5 ml of hydrochloric acid. Set aside for 3 minutes. Add 0.5 ml of 0.5% solution of rhodamine S. Shake for 3 minutes with 10 ml of 3 : 1 benzene-ether. Read the extract at 564 nm against the solvent.

SULFONAPHTHOLRESORCINOL

The complex of this reagent with gallium is read by fluorescence for analysis of semiconductor materials.[96]

Procedures

Silicon. Dissolve 1 gram of sample in 3 ml of nitric acid and 10 ml of hydrofluoric acid and evaporate to dryness. Add 2 ml of 1 : 2 sulfuric acid and 40 mg of sodium sulfate, evaporate to dryness, and heat at 700° for 30 minutes. Moisten the residue with 2 ml of 1 : 1 sulfuric acid and 1 ml of hydrofluoric acid

[95] V. S. Saltkova and E. A. Fabrikova, *Zh. Anal. Khim.* **13**, 63–65 (1958); S. M. Milaev and L. I. Maksai, *Sb. Nauch. Tr. Vsses. Nauch.-Issled. Gornomet. Inst. Tsve. Me.* **1960** (6), 459–471.

[96] V. A. Nazarenko and S. Ya. Vinkovetskaya, and R. V. Ravitskaya, *Zh. Anal. Khim.* **13**, 327–331 (1958); V. A. Nazarenko, S. Ya. Vinkovetskaya, and R. V. Ravitskaya, *Ukr. Khim. Zh.* **28**, 726–728 (1962).

and take to dryness. Dissolve in 10 ml of 1:1 hydrochloric acid and add 1 ml of 10% titanous chloride in 1:1 hydrochloric acid to reduce the iron. Extract the gallium with 5 ml and 5 ml of ether. Wash the extract with 5 ml of 1:1 hydrochloric acid and evaporate the extract along with 1 ml of 2% sodium chloride solution. Dissolve the residue in 0.2 ml of 0.2 *N* hydrochloric acid and add 1 ml of water. Adjust to pH 3 with 0.1 ml of monochloroacetic acid buffer solution. Add 0.2 ml of 10% hydroxylamine sulfate solution, 2 ml of ethanol, and 0.2 ml of 0.1% solution of the reagent in ethanol. Read by fluorescence.

Silicon Tetrachloride. Mix 1 ml with 2 ml of water and add 15 ml of hydrofluoric acid. Evaporate to dryness and complete as for silicon from "Add 2 ml of 1:2 sulfuric acid. . . ."

Silica. Dissolve 1 gram in 15 ml of hydrofluoric acid and evaporate to dryness. Complete as for silicon from "Add 2 ml of 1:2 sulfuric acid. . . ."

Zinc. Dissolve 1 gram of sample in 5 ml of 1:1 hydrochloric acid and evaporate to dryness. Take up the residue in 5 ml of water, add 2 ml of 25% solution of ammonium carbonate, and add 1:1 ammonium hydroxide to pH 9. Pass through a column of EDE-10P anion exchange resin in carbonate form. Wash the column with 25% ammonium carbonate solution, followed by water. Elute the gallium with 5 ml of 2.5 *N* hydrochloric acid, followed by 5 ml of water. Evaporate the eluate to dryness and complete as for silicon from "Dissolve in 10 ml of 1:1 hydrochloric acid and add. . . ."

TETRALINE GREEN

The complex of gallium with tetraline green, which is 4- [4-dimethylamine-α-(2-naphthyl)benzylidene] cyclohexa-2,5-diene-1-ylidene(dimethyl)ammonium chloride, follows Beer's law for 0.1–2 μg per ml of extract.[97]

Procedure. *Bauxite.* Treat 0.25 gram of sample with 3 ml of 1:1 sulfuric acid and 5 ml of hydrofluoric acid. Evaporate to dryness, add 2 grams of potassium pyrosulfate, and fuse to a clear melt. Dissolve in 1:1 hydrochloric acid, add 0.25 gram of powdered cadmium, and dilute to 25 ml with 1:1 hydrochloric acid. To a 4-ml aliquot add 0.5 ml of 15% titanous chloride in 1:1 hydrochloric acid and 0.5 ml of 0.02% solution of tetraline green. Let stand for 2 hours and extract with 5 ml of benzene. Read the extract at 606 nm.

---

[97] V. Armeanu, Polixenia Costinescu, and Georgeta Camelia Calin, *Chim. Anal.* (Paris) **2**, 18–21 (1972).

XYLENOL ORANGE

Whereas gallium forms 1:1 and 1:2 complexes with this reagent, by sufficient excess of xylenol orange, the latter is formed.[98] Thus when a fourfold excess of xylenol orange is added to 0.5–2 ppm of gallium or 0.7–8 ppm of indium at pH 4 ± 0.2 and set aside for 30 minutes, indium or gallium is read at 560 nm.[99] Beer's law applies up to 3 ppm of gallium. No bismuth or EDTA is tolerated. Fluoride partially masks aluminum, thorium, and zirconium. Reducing with ascorbic acid or sulfite avoids interference by ferric, palladic, or thallic ions.[100]

The technic described elsewhere for determination of gallium with methylthymol blue is also applicable with xylenol orange with the exception that 10 ml of citrate buffer solution for pH 2 replaces 10 ml of acetate buffer solution for pH 4.

Gallium also forms a ternary complex with xylenol orange and 8-hydroxyquinoline.[101] The optimum pH is 3.3–3.4, and after extraction with butanol it is read at 560 nm. Beer's law applies up to 1.4 μg per ml of extract. Aluminum stabilizes the complex. Interference by ferric, cupric, and vanadic ions is masked by mercaptoacetic acid and EDTA. Aluminum can be masked with fluoride, ferric ion can be reduced with ascorbic acid, and copper can be eliminated with thiourea.[102]

Gallium forms a ternary complex with xylenol orange and 2-bromobutyric acid according to the technic for forming a similar complex with indium. For details see Indium.

Procedure. To a sample solution containing 6–33 mM of gallium add tenfold excess of 0.001 M xylenol orange solution to form the 1:2 complex. Adjust to pH 2 and heat at 80° for 10 minutes. Read at 540 nm.

Acid chrome dark-green S forms a 1:1 complex with gallium.[103] The optimum pH is 2 for reading at 620 nm. A tenfold excess of reagent is necessary, an appropriate range is 1–2.5 μg of gallium per ml.

Gallium forms a 1:2 chelate with **alizarin red S** (mordant red 3), which is 3,4-dihydroxyanthraquinone-2-sulfonic acid.[104] The complex is stable at pH 3–5, follows Beer's law for 0.55–14 ppm, and is read at 490 nm. The complex fluoresces at 521 nm, as contrasted with 480 nm for the dye.[105] There is interference by oxalate, thioglycollate, thiourea, EDTA, and the following ions: ferric, scandium, titanous, molybdate, and vanadate.

[98] R. Doichiva, S. Popova, and E. Mitropolitskaya, *Talenta* **13**, 1345–1351 (1966); cf. P. P. Kish and M. I. Golovei, *Zh. Anal. Khim.* **20**, 794–799 (1965).
[99] Chandra D. Dwivedi and Arun K. Dey, *Mikrochim. Acta* **4**, 708–711 (1968).
[100] Makoto Otomo, *Bull. Chem. Soc. Jap.* **38**, 624–629 (1965).
[101] E. L. Glushchenko, M. K. Akhmedli, and A. K. Kyazimova, *Zh. Anal. Khim.* **26**, 75–79 (1971).
[102] L. M. Khokhlov, *Uch. Zap. Kursk. Gos. Pedagog. Inst.* **57**, 128–134 (1969).
[103] T. M. Devyatova and M. Z. Yampol'skii, *Zh. Anal. Khim.* **23**, 1468–1475 (1968).
[104] Chandra D. Dwivedi, Kailash N. Munshi, and Arun K. Dey, *Microchem. J.* **9**, 218–226 (1965).
[105] M. K. Akhmedli, D. A. Efendiev, and F. I. Ruvinova, *Uc. Zap. Azerb. Gos. Univ., Ser. Khim. Nauk* **1973** (4), 10–15.

As a ternary complex, 1,3-diphenylguanadinium ion becomes the third component.[106] The adduct without the gallium absorbs at 435 nm, the ternary complex at 525 nm. The optimum pH is 4.5–4.7. The complex is extractable with 1:4 butanol-benzene.

The complex of gallium with **aluminon**, which is the ammonium salt of aurintricarboxylic acid is read at 525 nm in an acetate buffer solution for pH 3.8.[107] Separation from aluminum, iron, and beryllium is necessary. The sensitivity increases with the age of the reagent. The concentrations of both buffer and aluminon affect the absorption, which does not obey Beer's law. The predominantly 1:1 complex contains some 3:2 form. The absorption increases with temperature but reverts on cooling.

Alumocreson, which is the triammonium salt of trimethylaurinetricarboxylic acid, forms a 1:1 complex with basic gallium at pH 3.5–4 for reading at 505 nm.[108] Beer's law is followed for 0.12–2 μg of gallium per ml. Copper is masked with thiosulfate, ferric ion with ascorbic acid. Titanium interferes. This reagent is more selective than xylenol orange, gallion, or rhodamine B.

Gallium at 0.2–3 μg per ml of final solution is read at pH 4.3–4.7 adjusted with sodium acetate–acetic acid buffer solution as the complex with the sodium salt of **6-amino-5-(2-hydroxy-4-nitrophenylazo)naphthalene-1-sulfonic acid**.[109] The maximum absorption is at 595 nm with 0.02% of dye. Similarly, the complexes with **4-amino-3-(2-hydroxy-3, 5-dinitrophenylazo)-1-sulfonic acid** and **3-(2-amino-1-naphthylazo)-4-hydroxy-5-nitrobenzene sulfonic acid** are read at 575 nm. Colored complexes are also formed by these reagents with cobalt, nickel, iron, copper, palladium, and chromium; therefore gallium must be isolated from them.

The complex of gallium with **antipyrinylbis-(4-dimethylaminophenyl)methanol**, when extracted into toluene or benzene, is read at 585 nm.[110] It will determine 0.0012–0.0065% of gallium in copper-zinc ore, bauxite, or soil.

In the presence of a sixfold excess of **arsenazo 1** both gallium and indium form 1:1 complexes.[111] That for gallium at pH 3–5.5 has a maximum absorption at 550 nm. The color develops immediately and is stable for 18 hours. Beer's law applies for 0.37–5.57 ppm.

The 1:1 complex of **basic blue K** with gallium chloride is at a maximum in 1:11 sulfuric acid, provided the chloride content is standardized by addition of sodium chloride or potassium chloride, or in 1:3 hydrochloric acid.[112] The complex is not extractable with benzene or toluene alone, but it is by 5:3 benzene-acetone. It is read at 620 nm for 0.35–8 μg per ml of extract. Large amounts of zinc adversely affect the stability of the color, which is normally stable for days.

The complex of **brilliant alizarin blue G** (mordant blue 31) with gallium is read by its red-yellow fluorescence in an acid or basic medium.[113] Aluminum and scandium react similarly.

[106]Makoto Otomo, Shuchi Masuda, and Koichi Tonosaki, *J. Chem. Soc. Jap., Pure Chem. Sect.* **92**, 739–740 (1971).

[107]Emile Rinck and Pierre Feschotte, *Bull. Soc. Chim. Fr.* **1957**, 230–234.

[108]N. F. Lisenko, G. S. Petrova, and N. B. Etingen, *Zh. Anal. Khim.* **29**, 1729–1733 (1974).

[109]W. Likussar and M. Czermak, *Mikrochim. Acta* **1974**, 475–482.

[110]A. I. Busev, L. M. Skrebkova, and V. P. Zhivopistsev, *Zh. Anal. Khim.* **17**, 685–692 (1962).

[111]Arun P. Joshi and Kailash N. Munshi, *Microchem. J.* **18**, 277–287 (1973).

[112]I. A. Shmeleva, *Uch. Zap. Kursk. Gos. Pedagog-Inst.* **1969** (57), 160–165.

[113]Keizo Hiraki, *Bull. Chem. Soc. Jap.* **45**, 789–793 (1972).

The complex of gallium with **Capri blue** in 1 : 1 hydrochloric acid is extracted with benzene for reading at 655 nm.[114] Beer's law is followed for 1–15 μg per 25 ml of extract. Ferrous or ferric ion cannot be tolerated, but up to 0.5 mg of other cations and up to 1 mg of anions do not interfere.

***NN*-Bis(carboxymethyl)anthranilic acid**, as a 1 : 1 complex with gallium, is read at 278 nm in 0.001 *M* reagent solution and pH 4.[115] Beer's law applies up to 55 μg of gallium per ml. There is interference by silver, zinc, cadmium, lead, and cupric ions. Indium and gallium in 0.1 *N* nitric acid complex with *NN*-bis(carboxymethyl)anthranilic acid.[116] Indium is read for up to 60 μg/ml at 289.5 nm, which is the isobestic point of the gallium system. Similarly, gallium is read for up to 40 μg/ml at 260 and 316.5 nm, the isobestic points of the indium system.

Gallium forms a positively charged fluorescent 1 : 1 complex with **1-(5-chloro-2-hydroxyphenylazo)-2-naphthol** at pH 3–4 and a neutral nonfluorescent 1 : 2 complex at pH 6–7. The respective absorption maxima are at 538 and 570 nm.[117a]

In the presence of indium both gallium and aluminum are determined with **chlorsulfophenol S**, which is 3,6-bis(5-chloro-2-hydroxy-3-sulfophenylazo) chromotropic acid.[117b] Dissolve 0.1 gram of indium expected to contain no more than 0.15% of gallium in not more than 2 ml of hydrochloric acid and dilute to 100 ml. Mix a 5-ml aliquot with 10 ml of 2 *M* acetate buffer for pH 4.5 and 2 ml of 0.1% solution of chlorsulfophenol S. Heat for 10 minutes at 70°, dilute to 25 ml, and read at 667 nm. Beer's law is applicable to 0.3–1.5 μg of gallium per ml. The complex can be extracted into a m*M* solution of benzylhexadecyldimethyl ammonium chloride in organic solvent.

Chrome azurol S (mordant blue 29), also called alberon, forms a ternary complex with cetyltrimethylammonium chloride, also called cetrimonium chloride, and gallium for reading at 612 nm.[118] Beer's law is not obeyed. The desirable pH is 4.4–6.8 and the desirable range 0.6–6 m*M*. There is interference by aluminum, vanadium, copper, iron, gold, thorium, uranium, titanium, zirconium, and tin. Hydrosulfite will mask copper and ascorbic acid reduce ferric ion.[119]

A desirable ratio is 1 part of the dyestuff to 2 parts of the cation active surfactant, the presence of the latter increasing the sensitivity of the dyestuff.[120]

Chrome azurol S[121] is used without the surfactant for reading at 547 nm for 0.1–1.6 ppm of gallium. A 50-ml solution should be buffered to pH 4.2–4.4 and should contain 0.257% of chrome azurol S and 5 ml of 0.25 *M* sodium acetate. Then beryllium, aluminum, titanium, zirconium, stannic ion, and hexavalent molybdenum interfere.

With **chromotrope 2R** (acid red 29), add a fiftyfold excess of the dye to a solution containing 1.7–7.3 ppm of gallium.[122] Adjust to pH 3.4 and dilute to 25 ml. Set aside for 30 minutes and read at 580 nm. Aluminum, ferric, and thoric ions can be

[114]Tsutomo Matsuo, Shunsake Funada, and Mikio suzuki, *Bull. Chem. Soc. Jap.* **38**, 326–328 (1965).
[115]C. Dragulescu, C. Policec, and T. Simonescu, *Talenta* **13**, 1451–1457 (1966).
[116]Septimia Policec, T. Simonescu, and C. Dragulescu, *Rev. Roum. Chim.* **17**, 231–237 (1972).
[117a]V. A. Nazarenko, Liam Ngog Thu, and R. M. Dranitskaya, *Zh. Anal. Khim.* **22**, 518–524 (1967).
[117b]N. A. Ivanov and N. G. Todorov, Dolk. *Bolg. Akad. Nauk* **29**, 1775–1778 (1976).
[118]Yoshio Shijo and Tsugio Takeuchi, *Jap. Anal.* **20**, 137–141 (1971).
[119]L. M. Khokhlov, *Uc. Zap. Kafedry Obshch. Khim., Kursk. Gos. Pedagog, Inst.* **35**, 138–148 (1967).
[120]B. Evtimova and D. Nonova, *Anal. Chim. Acta* **67**, 107–112 (1973).
[121]Yoshizo Horiuchi and Hiroshi Nishida, *Jap. Anal.* **16**, 1146–1152 (1967).
[122]S. C. Dhupar, K. C. Srivastava, and Semir K. Banerji, *J. Indian Chem. Soc.* **49**, 935–938 (1972).

masked with fluoride. Pentavalent vanadium can be reduced to tetravalent with ascorbic acid. There remains interference by ceric, stannous, magnesium, hexavalent tungsten, and borate ions, and several organic acids.

The 1 : 1 complex of **chromoxane blue** (mordant blue 42) with gallium has an optimum pH of 3.0 and of absorption at 560 nm.[123] Aluminum, ferric ion, scandium, indium, and cupric ion interfere by forming colored complexes. Ascorbic acid will reduce the ferric ion and sodium hydrosulfite the cupric ion. The reagent will determine 2–3 μg of indium per ml in the presence of 1 mg of zinc or lead.

Gallium is determined indirectly by its displacement of copper from **cupric diethylenetriaminepentaacetate.**[124] The complex of copper with another reagent is read. To a neutral or slightly acid sample, add 5 ml of 2 m*M* cupric diethylenetriaminepentaacetate, 1 ml of 1% ascorbic acid solution, and 2.5 ml of 40% solution of sodium acetate. Dilute to 50 ml and mix well with 5 ml of 1% methanolic solution of bathocuproine. Extract by shaking for 1 minute with 5 ml of hexanol. Read the extract at 470 nm. The method is also applicable to indium, lanthanum, and zirconium.

The complexes of gallium with **5,7-dichloro-8-hydroxyquinoline** and **5,7-dibromo-8-hydroxyquinoline** are extracted with chloroform for reading by fluorescence at 450–650 nm at about 0.2 μg/ml.[125] The gallium must be isolated by ether extraction to avoid interference by copper, zinc, aluminum, and tungstate. This was applied to metallic aluminum. Gallium in silicate ores can be determined with 5,7-dibromo-8-hydroxyquinoline, dibromooxine.[126]

In 25% ethanol at pH 3.4, m*M* **6,7-dihydroxy-2,4-diphenyl-1-benzopyrylium** chloride forms a 3 : 1 complex with gallium which is read at 510 nm.[127a] Gelatin must be present as a stabilizer. Beer's law is followed for 0.03–0.6 μg of gallium per ml. The complex is extractable with chloroform at pH 2.5–3 from m*M* reagent.[127b] Read at 545 nm, the extract obeys Beer's law for 0.07–0.7 μg of gallium per ml. To separate from interference adjust the sample solution to 1 : 1 hydrochloric acid and complex gallium with αα′-diantipyrinyltoluene. Extract with chloroform and back-extract with water.

For determination of gallium by **3,4-dihydroxyazobenzene-2′-carboxylic acid,** which is 2-(3,4-dihydroxyphenylazo)benzoic acid, a 1 : 1 complex is read at 520 nm against a reagent blank.[128] Buffer the sample containing 0.1–2 μg of gallium to pH 3.05, add 5 ml of 1.4 m*M* reagent in ethanol and dilute with water to 25 ml.

The complex of gallium with **2-(2,4-dihydroxyphenylazo)-4-phenyl-5-benzoylthiazole**, which is 4-(5-benzoyl-4-phenylthiazol-2-ylazo)resorcinol, in 36%

[123] L. M. Khokhlov, *Uch. Zap. Kafedry Obshch. Khim., Kursk. Gos. Pedagog, Inst.* **35**, 138–148 (1968); *Uch. Zap. Kursk. Gos. Pedagog. Inst.* **57**, 124–127 (1970).

[124] E. Jackworth and G. Graffmann, *Z. Anal. Chem.* **257**, 265–268 (1971).

[125] Yasuharu Nishikawa, *J. Chem. Soc. Jap.* **79**, 631–637 (1958).

[125] Yasuharu Nishikawa, *J. Chem. Soc. Jap.* **79**, 631–637 (1958).

[126] Heinrich M. Koester, *Ber. Deut. Keram. Ges.* **46**, 247–253 (1969); cf. Nguyen Shi Zuong and F. G. Zharovskii, *Ukr. Khim. Zh.* **36**, 1273–1278 (1970).

[127a] A. A. Brazilevich, N. L. Olenovich, and V. A. Nazarenko, *Zh. Anal. Khim.* **28**, 2047–2050 (1973).

[127b] N. L. Olenovich, A. A. Bazilevich, A. A. Nazarenko, and N. P. Bidenko, *Zh. Anal. Khim.* **31**, 2120–2124 (1976).

[128] E. R. Oskotskaya and M. Z. Yampol'skii, *Zh. Anal. Khim.* **23**, 1307–1314 (1968); cf. E. R. Oskotskaya, *Uc. Zap. Kafedry Obshch. Khim., Kursk. Gos. Pedagog, Inst.* **35**, 121–127 (1967).

ethanol at pH 4.28 is read at 520 nm.[129] The reagent also forms colored complexes with indium, thorium, beryllium, and cerium. The color develops in 20 minutes, is stable for 24 hours, and obeys Beer's law for 0.1–1.2 μg/ml.

At pH 5–6 **dioxythiazo** forms a 3 : 1 complex with gallium, readily extracted with butanol.[130] The maximum absorption for the reagent is at 460 nm, that for the complex at 570 nm. Color develops immediately and is constant for 6 hours.

The reaction of gallium with dithizone in carbon tetrachloride is relatively insensitive.[131] With dithizone in acetone, it is multiplied 1000-fold and will detect 10 μg in 250 *μM* solution in the absence of masking by citrate or tartrate.

The 1 : 1 complex of **eriochrome azurol B** with gallium obeys Beer's law for 0.1–5 μg/ml when read at 533 nm.[132]

Gallium forms a positively charged, fluorescent 1 : 1 complex with **eriochrome blue-black R** (mordant black 17) at pH 3–4 and a neutral nonfluorescent 1 : 2 complex at pH 6–7.[133] The respective maxima are at 570 and 590 nm.

The 1 : 2 complex of gallium with **eriochrome cyanine R** (mordant blue 3) at pH 3.3 has a maximum absorption at 550 nm after standing for 1 hour.[134a] Aluminum, ferric ion, scandium, and cupric ion interfere. The presence of organic solvent causes more rapid color development and permits greater acidity.[134b] To a sample solution containing 5 μg of gallium add 2.5 ml of acetate buffer solution for pH 3, 5 ml of ethanol, and 1 ml of 0.1% solution of eriochrome cyanine R. Dilute to 10 ml, set aside for 5 minutes, and read at 550 nm against a reagent blank.

For reading in the presence of indium, mix 10 ml of sample solution containing 0.008–0.03 mg of gallium with 10 ml of 0.0053% solution of the reagent.[135] After allowing to stand for 2 hours, read at 540 nm. In addition to the elements cited, tin and zirconium interfere.

Gallium forms a 1 : 2 complex with **eriochrome red B** at pH 3–3.9 for maximum fluorescence.[136] The complex is developed by heating at 70–80° for 10 minutes and is stable for 2 hours. Interference by ferric ion, vanadic acid, thallic ion, dichromate, and permanganate is eliminated by addition of hydroxylamine. Aluminum interferes strongly.

For **flame photometry**, gallium is read at 403.3 nm with a nitrous oxide–acetylene flame.[137] The detection limit is 0.01 ppm at a height of 3 mm in the red zone. Spectral interferences are eliminated at the less sensitive line at 417.21 nm. Addition of 1000 ppm of potassium chloride suppresses ionization in the flame, thereby enhancing the signal.

[129]V. S. Korol'kova, J. Putnins, E. Gudriniece, and E. I. Bruk, *Izv. Akad. Nauk Latv. SSR, Ser. Khim.* **1970**, 444–448.

[130]V. S. Korol'kova, Ya. K. Putnin', and E. Yu. Gudrinietze, *Latv. PSR Zinat. Akad. Vest., Ser. Khim.* **1968**, 423–427.

[131]G. Iwantscheff and C. Jorrens, *Anal. Chim. Acta* **38**, 470–473 (1967).

[132]M. Z. Yampol'skii, A. E. Okun, and L. N. Orlova, *Uch. Zap., Kursk. Gosudarst. Pedagog. Inst.* **11**, 134–142 (1958).

[133]V. A. Nazerenko, Liam-Ngog-Thu, and R. M. Dranitskaya, *Zh. Anal. Khim.* **22**, 518–524 (1967).

[134a]L. M. Khokhlov, *Uch. Zap. Kafedry Obshch. Khim., Kursk. Gos. Pedagog, Inst.* **35**, 128–137 (1967); cf. Vikash C. Garg, Suresh C. Shrivastava, and Arun K. Dey, *Mikrochim. Acta* **1969**, 668–672.

[134b]R. Borisova and N. Ivanov, *C. R. Acad. Bulg. Sci.* **29**, 1313–1316 (1976).

[135]P. U. Sakellaridis and B. S. Roufogalis, *Chim. Chron.* **29**, 113–117 (1964).

[136]Yasuharu Nishikawa, Jap. Anal. **7**, 549–553 (1958).

[137]E. E. Pickett and S. R. Koirtyohann, *Spectrochim. Acta B* **24**, 325–333 (1969).

Extracted as the thiocyanate with ether and sprayed into an oxyhydrogen flame in that solvent, the limit of detection is 0.4 ppm.[138] When a silver solution in dilute nitric acid is reduced by formic acid and the silver is removed as an amalgam, the gallium can be obtained by flame photometry.[139] The acetylene-air flame is also appropriate.[140]

By the use of 80% acetone containing 0.5 *N* hydrochloric acid and reading at 417.2 nm with background reading at 414 or 420 nm, the emission is increased about 50 times.[141] Extraction of gallium, indium, and thallium with ether from 5.5 *N* hydrochloric acid eliminates all interferences.

Extraction of the 8-hydroxyquinolinates of gallium, titanium, and indium with methyl isobutyl ketone, amyl acetate, ethyl acetate, or chloroform gives a suitable medium for use with an oxyhydrogen flame.[142a]

A 1:2:2 complex of gallium–**glycinecresol red**–diphenylguanidine in an acetate–ammonium hydroxide buffer solution is extracted by alcohols for reading at 530–532 nm.[142b] Indium and aluminum perform similarly.

Gallium and **hematin** form 1:1 and 1:2 complexes dependent on pH and reagent concentration.[143] By reading at the isobestic point, the extinction is independent of the position of the equilibrium, and 38 ng of gallium per ml can be detected.

Gallium complexes with **4-(2-hydroxyphenyl)resorcinol.**[144] At pH 3–6 the 1:1 and 1:2 complexes have maxima at 484 and 500 nm, as compared with the reagent at 428 nm. Around pH 3 **4-(2-hydroxyphenylazo)resorcinol** shows a maximum fluorescence at 575 nm when activated at 485 nm.[145]

At pH 3, gallium forms a stable 1:1 complex with ***N,N'*-bis(2-hydroxy-5-sulfophenyl)-*s*-cyanoformazen.**[146] At 634 nm it conforms to Beer's law for 0.05–2.5 μg/ml. There is interference by copper, nickel, cobalt, titanium, vanadium, and iron.

With **3-hydroxy-4-(2-thiazolylazo)naphthalene-2,7-disulfonic acid**, both indium and gallium are read.[147] To a solution containing 10–45 μg of gallium and 35–100 μg of indium, add 4 ml of m*M* reagent, adjust to pH 3.5, dilute to 25 ml, and read the sum of gallium and indium at 530 nm against water. To a similar sample, add 1 ml of 0.23% sodium tartrate solution to mask the gallium, adjust the pH, add the reagent, and read as before. The result is indium. Obtain gallium by difference.

The 1:1 complex of **kaempferol,** which is 3,4'-5,7-tetrahydroxyflavone, with gallium is read at 425 nm in 50% ethanol.[148]It follows Beer's law for up to 1.44

[138]H. Brandenberger and H. Bader, *Helv. Chim. Acta* **47**, 353–358 (1964).

[139]J. Meyer, *Z. Anal. Chem.* **231**, 241–252 (1967).

[140]J. Malinowski, D. Dancewicz, and S. Szymeczak, *Pol. Acad. Sci. Inst. Neuclear Res. Rep.* No. **113/VIII,** 7 pp (1959).

[141]Helmut Bode and Horst Fabian, *Z. Anal. Chem.* **170**, 387–399.

[142a]Emiko Sudo and Hidehiro Goto, *Sci. Rep. Res. Inst., Tohoku Univ. Ser. A* **14**, 220–230 (1962).

[142b]M. L. Akhmedli, E. L. Glushchenko, and F. T. Aslanova, *Azerb. Khim. Zh.* **1974**, 126–130.

[143]G. Graffmann, E. Jackwerth, J. Lohmar, and W. Ufer, *Anal. Chem.* **246**, 12–18 (1969).

[144]V. A. Nazarenko, Liam Ngog Thu, and R. M. Dranitskaya, *Zh. Anal. Khim.* **22**, 346–352 (1967).

[145]Keizo Hiraki, *Bull. Chem. Soc. Jap.* **46**, 2438–2443 (1973).

[146]N. L. Vasileva, M. I. Ermakova, and I. Ya. Postovskii, *Zh. Vses. Khim. Obshch. D. I. Mendeleeva,* **5**, 110 (1960); M. I. Ermokova, N. L. Vasil'eva, and I. Ya. Postovskii, *Zh. Anal. Khim.* **16** (1), 8–13 (1961).

[147]T. N. Zholondkovskaya, L. S. Krysina, and N. A. Golubkova, *Dokl. Mosk. S.-Kh. Akad. K. A. Timiryaz.* **1972** (176), 289–292; *Izv. Timiryazev. S.-Kh. Akad.* **1973** (4), 175–181.

[148]B. S. Garg and R. P. Singh, *Talenta* **18**, 761–766 (1971).

ppm. Indium gives a similar complex, having a maximum at 430 nm. Many cations, fluoride, thiosulfate, iodate, EDTA, and nitrilotriacetic acid interfere.

For determination of gallium by fluorescence with kaempferol in 25% ethanol in a sodium acetate–hydrochloric acid buffer solution for pH 3–4, activate at 355–365 nm and read at 485–495 nm.[149] A maximum is reached in 1 hour, then remains constant for several hours. The sensitivity is 2 ng in 10 ml of solution. Beer's law is followed for 0.002–0.024 μg/ml. An equal amount of bismuth, chromium, manganese, magnesium, barium, strontium, calcium, ferric ion, or pentavalent antimony is tolerated.

Magneson IREA, which is 5-chloro-2-hydroxy-3-(2-hydroxy-1-naphthylazo)benzenesulfonic acid, forms a 1 : 1 complex with gallium.[150] It is extracted into butanol from pH 2.6–4.2 for reading at 574 nm. Beer's law is followed for 0.25–2.5 μg/ml. Copper can be masked with mercaptoacetic acid; iron with ascorbic acid; and zirconium, aluminum, thorium, and scandium with fluoride. There is also interference by stannous, titanous, mercuric, lead, vanadic, bismuth, terbium, hexavalent molybdenum and tungsten, chromic, and beryllium ions.

For determination of gallium by **meldola blue** (basic blue 6), dissolve 0.2 gram of sample in 1 : 1 hydrochloric acid and evaporate to a moist residue.[151] Take up in 10 ml of 1 : 1 hydrochloric acid, heat to boiling, and filter. Wash the residue a few times with 1 : 20 hydrochloric acid, add 1 gram of ascorbic acid, and dilute to 50 ml. To a 5-ml aliquot add 1 ml of water, 2 ml of 0.2% solution of meldola blue, and 12 ml of hydrochloric acid. Extract with 10 ml and 10 ml of 5 : 1 benzene-acetone. Filter the extracts, dilute to 25 ml with the mixed solvent, and read at 580 nm against a reagent blank.

The 1 : 1 complex of gallium and **8-mercaptoquinoline** is extracted with methyl isobutyl ketone, activated at 395 nm, and read by fluorescence at 505 nm.[152] As applied to analysis of aluminum, the gallium can be extracted from 7 *N* hydrochloric acid with isopropyl ether to avoid interference due to zinc, cadmium, copper, nickel, and cobalt. Beer's law is followed for 0.1–3 μg of gallium per ml.

***N*-Methyl-anabasine-α′-diethylaminophenol** is also known as MAAF II.[153] The optimum for the complex with gallium is 5. It also forms colored complexes with nickel, cobalt, copper, vanadium, bismuth, and indium. The complex conforms to Beer's law for 1–20 μg of gallium in 25 ml. Iron can be masked with fluoride; aluminum, molybdenum, tungsten, and chromium with sodium tartrate.

Gallium as a dihydroxy ion forms a 1 : 1 complex with 10 *M* **4-methyl-2-(2-hydroxy-1-naphthylazo)thiazole** in an ethanolic acetate buffer solution for pH 4.[154] Extracted with chloroform, the complex is read at 578 nm. Beer's law applies for 0.14–5.5 μg of gallium per ml. There is interference by copper, bismuth, iron, lead, zinc, cadmium, phosphate, citrate, oxalate, tartrate, and EDTA. For determination of gallium in aluminum in 1 : 1 hydrochloric acid, add titanous chloride to facilitate

[149] Z. T. Maksimycheva, Sh. T. Talipov, Z. P. Pakudina, and A. S. Sadykov, *Izv. Vyssh. Uch. Zaved., Khim. Tekhnol.* **17**, 348–351 (1974).

[150] M. K. Akhmedli, E. L. Glushchenko, and A. K. Dyazimova, *Azerb. Khim. Zh.* **1972** (1), 138–142.

[151] A. T. Pilipenko and Nguyen Dyk Tu, *Ukr. Khim. Zh.* **35**, 200–202 (1969).

[152] Kunihiro Watanabe and Kyozo Kawagaki, *Bull. Chem. Soc. Jap.* **48**, 1812–1815 (1975).

[153] Sh. T. Talipov, R. Kh. Dzhiyanbaeva, and A. E. Martirosov, *Izv. Akad. Nauk Kaz. SSR, Ser. Khim.* **1968** (6), 1–4.

[154] A. T. Pilipenko and E. A. Shevchenko, *Zh. Anal. Khim.* **30**, 1101–1105 (1975).

separation from iron, zinc, and copper. Extract gallium with isopropyl ether and back-extract with 0.1 *N* nitric acid.

The 2:1 complex of **2-[3-(1-methyl-2-piperidyl)-2-pyridylazo]-1-naphthol** has a maximum absorption at 570 nm, as compared with 460 nm for the reagent.[155] A pH of 4–5 with eight to tenfold excess of reagent is desirable. Beer's law applies for 0.4–4 μg/ml. As a typical solution, mix a sample containing 0.01–0.1 mg of gallium, add sequentially 10 ml of ethanol, 2 ml of 0.1% solution of the reagent, and 5 ml of buffer solution for pH 4.79. Dilute to 25 ml and let stand for 15 minutes before reading against a blank. Tartrate and citrate interfere. Thallic ion is reduced with bisulfite. Copper is masked with thiourea; aluminum and bismuth with fluoride and oxalate.

4-(6-Methyl-2-pyridylazo)resorcinol complexes at pH 5.5 for reading 5–40 μg of gallium at 530 nm.[156] The final solution should contain 2 ml of 0.1% solution of the reagent in ethanol and 15 ml of acetate buffer solution in a total volume of 25 ml. Indium reacts similarly, and there is conflict with ferric, cobalt, nickel, zinc, and cadmium ions.

For forming a 1:1:2 complex gallium is treated with **methylthymol blue** and digested at 70° for 10 minutes.[157] Cool and add 8-hydroxyquinoline in acetic acid. Adjust to pH 2.9–3.3 with an acetate buffer solution and extract with 10 ml of 1:1 isoamyl alcohol–chloroform. Read in the range of 585–600 nm.

Morin, which is 2′,3,4′,5,7-pentahydroxyflavone, forms a yellow-green complex with gallium to give both strong absorption and strong fluorescence.[158] The reagent absorbs at 355 nm for pH 1.2–5. The gallium complex absorbs at 415–420 nm with pH 3.4 and an acetate buffer solution. The maximum fluorescence is at pH 3.6. Indium reacts similarly and cannot be masked with citric, tartaric, oxalic, or sulfosalicylic acid because the complexes of both gallium and indium with the masking agents are more stable than with morin.

Neothoron, which is 2-(*o*-arsenophenylazo)chromotropic acid, forms a 1:1 stable red-purple complex with gallium in acid solution.[159] To a sample containing up to 0.1 mg of gallium, add 2 ml of 0.1% solution of neothoron and 5 ml of a buffer solution for pH 3.5. Dilute to 25 ml, let stand for 5 minutes, and read at 570 nm against a blank. There is interference by aluminum, beryllium, uranium, and copper.

Gallium forms a 1:2 complex with **4-phenylazocatechol** at pH 5–6 for reading at 450 nm.[160] Similarly, that with **4-(3,4-dihydroxyphenylazo)benzenesulfonic acid** is read at 460 nm.

Gallium forms a 1:1 complex with **9,9′-*p*-phenylenedifluorone** at pH 6 and a 1:3 complex at pH 9.5.[161] In the presence of gelatin and boric acid, the latter is read at 535 nm and follows Beer's law up to 10 m*M*. There is interference by aluminum,

[155] V. S. Podgornova, Kh. S. Abdullaeva, and Sh. T. Talipov, *Uzeb. Khim. Zh.* **1967** (6), 9–11.
[156] N. Kulmuratov, N. B. Babaev, and Sh. T. Talipov, *Dokl. Akad. Nauk Uzeb. SSR* **1969** (4), 29–30.
[157] M. K. Akhmedii, E. L. Glushchenko, and Z. L. Gasanova, *Uch. Zap. Azerb. Gos. Univ., Ser. Khim. Nauk,* **1972** (2), 14–19.
[158] A. I. Busev and E. P. Shkrobot, *Vest. Mosk. Univ., Ser. Mat., Mekh., Astronl, Fiz. Khim.* **14** (4), 199–207 (1959).
[159] Yoshio Ishiguro, Shozo Shibata, and Teiichi Matsumae, *Nagoya Kogyo Gijutsu Shikenosho Kokoku* **10**, 809–812 (1961).
[160] V. A. Nazarenko and S. Ya. Vinkovetskaya, *Zh. Anal. Khim.* **22**, 181–186 (1967).
[161] A. V. Fedin and S. I. Kravchuk, *ibid.* **29**, 1734–1740 (1974).

zinc, indium, bismuth, nickel, copper, ferric ion, and EDTA. Gallium can be separated from interference by extraction from 7 *N* hydrochloric acid with ether.

Picramine RG, which is 4-hydroxy-3-(2-hydroxy-3,5-dinitrophenylazo)naphthalene-2,7-disulfonic acid, forms a 1:1 complex with gallium.[162] The maximum absorption of the reagent is at 495 nm, that of the complex at 560 nm.

The 1:1 complex of gallium with **(pyridylazo)-3,4-xylenol** is extracted by chloroform from a medium buffered at pH 3–4.5.[163] The maximum extinction at 570 nm is stable for 12 hours. The reagent has a maximum at 430 nm. Sixfold excess of reagent is necessary. Pentavalent vanadium forms a similar complex with a maximum at 630 nm.

Quercetin, which is 3,3′,4′,5,7-pentahydroxyflavone, forms a 1:1 complex with gallium, best read at 428 nm for 0.08–2.28 μg/ml.[164] It is buffered at pH 4.5 in 50% methanol. A period of 15 minutes is required for color development at more than 0.28 μg/ml. The reagent is also applicable to indium. Aluminum, fluoride, oxalate, citrate, and tartrate ions interfere.

Gallium forms a fluorescent complex with **quercetin 3′-glucoside,** a reagent extracted from cotton with hot methanol, in a solution buffered with acetate at pH 2.8–3.6.[165] Activation is at 415–425 nm and reading at 550 nm. Beer's law is followed for 0.14–4.2 μg of gallium per 25 ml. Aluminum, titanium, and zinc interfere, and only tenfold indium is tolerated.

After suitable isolation from interfering ions, gallium in steel is developed with **quinalizarin.**[166]

Resorcinthiazo, which is 4-(4,5,6,7-tetrahydrobenzothiazol-2-ylazo)resorcinol, forms a 2:1:1 complex with 1 mole of gallium dihydroxyl ion and 1 mole of water.[167] Mix 2 ml of sample solution containing 0.2–2 μg of gallium with 5 ml of 0.5 m*M* reagent in ethanol and 8 ml of acetate buffer solution for pH 5.9. Dilute to 25 ml with ethanol and read at 530 nm. Beer's law applies to 0.008–0.8 μg of gallium per ml.

The 1:1 complex of gallium with **rezarson,** which is 5-chloro-3-(2,4-dihydroxyphenylazo)-2-hydroxybenzearsonic acid, requires 50% ethanol to prevent precipitation.[168] Conditions are pH 1–3 and maximum absorption at 500 nm.

The purple solution of **rhodamine C** absorbs at 555 nm and gives an intense orange-yellow fluorescence.[169] The solution of the complex with 0.1–1.1 μg of gallium per ml is read by fluorescence. Only ytterbium, gold, antimony, and tellurium interfere. Extraction with benzene in the presence of titanous ion separates gallium from most interference.

Gallium complexes with orange-red rhodamine 6G (basic red 1) to give an

[162] V. D. Salikhov, L. M. Khokhlov, and Yu. M. Dedkov, *ibid.* **26**, 69–74 (1971).

[163] K. Rakhmatullaev, M. A. Rakhmatulaeva, and N. Shoazizov, *Uzb. Khim. Zh.* **1970** (5), 25–28.

[164] I. P. Alimarin, A. P. Golovina, and V. G. Torgov, *Zavod. Lab.* **26**, 709–711 (1960); Gr. Popa, D. Negoiu, C. Luca, and Gh. Baiulescu, *Rev. Chim. Acad. Rep. Pop. Roum.* **6**, 87–94 (1961).

[165] Sh. T. Talipov, Z. T. Maksimycheva, Z. P. Pakudina, and A. S. Sakykov, *Izv. Vyssh. Ucheb. Zaved., Khim. Khim. Tekhnol.* **16**, 1154–1156 (1973).

[166] Oleg Engel and Stanislav Kral, *Hutn. Listy* **25**, 432–435 (1970).

[167] J. Kalina, J. Putnins, and E. Gudriniece, *Izv. Akad. Nauk Latv. SSR, Ser. Khim* **1972** (2), 145–149.

[168] A. M. Lukin, N. A. Kaslina, V. I. Fadeeva, and G. S. Petrova, *Vest. Mosk. Gos. Univ., Ser. Khim.* **1972**, 247–249.

[169] D. P. Shcherbov and A. I. Ivankova, *Zavod. Lab.* **24**, 667–674 (1958).

intense yellow complex.[170] Extraction with benzene from a solution in 12 *N* sulfuric acid–*N* hydrochloric acid separates from a 2500-fold amount of indium.

Rhodamine 4Zh is extracted as the chlorogallate complex by benzene for reading by fluorescence.[171] The solution should be in 1:1 hydrochloric acid.

In a solution buffered at pH 6–6.5 the complex of gallium with **salicylaldehyde isonicotinoylhydrazone** is read at 390 nm, that of indium at 380 nm.[172]

At pH 3–4 gallium complexes with **4-salicylideaminoantipyrine.**[173] In 5 ml of solution, 5 ng is determinable by fluorescence at 505 nm.

The chelate of gallium with **2-salicylideneaminophenol** at pH 4 in an acetate buffer solution is read by fluorescence.[174] The appropriate range is 0.1–10m*M*. It is also read photometrically.[175]

The complex of gallium with **solochrome cyanine R** (mordant blue 3) is appropriate for reading.[176] Indium and thallium also form violet complexes.

At pH 2–9 gallium forms a 1:2 complex with the sodium salt of **solochrome dark blue** (mordant black 17).[177] Sodium chloride should be present and the dyestuff about 0.05 m*M* at pH 3. The maximum absorption after 1 hour is at 580 nm and is stable for 36 hours. The reagent absorbs at 550 nm and reading gallium at 590 nm results in smaller blank. An appropriate range is 0.44–2.2 ppm of gallium. There is serious interference by oxalate, nitrite, ferrous, ferric, cerous, chromic, and thallic ions.

For determination of gallium with **superchrome garnet Y** (mordant red 5) by fluorescence, add 0.8 ml of 0.01% solution of the reagent for up to 5 μg of gallium or 1.7 ml for larger amounts.[178] Add 2 ml of 20% ammonium acetate solution and adjust to pH 3. Dilute to 25 ml, activate at 485 nm, and read at 565 nm. There is interference by aluminum, cobalt, cupric ion, ferric ion, zirconium, pentavalent vanadium, and hexavalent molybdenum and chromium.

One can determine 1–5 μg of gallium with **stilbazo** in the presence of 50–70 μg of indium.[179] The intensity of absorption is proportional to gallium content up to 2.3 μg/ml at pH 6.1–6.5. Mix the sample with 2 ml of ammoniacal acetate buffer solution for pH 6.5 and 0.5 ml of 0.001% solution of stilbazo and dilute to 3 ml. Read at 530 nm against water.

The complex of gallium with **sulfonaphtholazoaminocresol,** which is 4-(4-ethylamino-6-hydroxy-*m*-tolylazo)-3-hydroxynaphthalene-1-sulfonic acid, is suitable for

[170] I. P. Alimarin, A. P. Golovina, N. B. Zorov, and E. P. Tsintsevich, *Izv. Akad. Nauk SSSR, Ser. Khim.* **1968**, 2678–2682; cf. note 169.

[171] A. P. Golvina, Z. M. Khvatokova, N. B. Zorov, and I. P. Alimarin, *Vest. Mosk. Gos. Univ., Ser. Khim.* **13**, 551–555 (1972).

[172] G. S. Vasilikoiotis, T. A. Kouimtzis, and V. C. Vasiliades, *Microchem. J.* **20**, 173–179 (1975).

[173] Sh. T. Talipov, A. T. Tashkodzhaev, L. E. Zel'tser, and Kh. Khikmatov, *Izv. Vyssh. Ucheb. Zaved., Khim. Khim. Tekhnol.* **15**, 1109–1110 (1972).

[174] R. M. Dagnall, R. Smith, and T. S. West, *Chem. Ind.* (London) **1965** (39), 1499–1500.

[175] A. A. Tumanov and V. S. Efimychev, *Zh. Anal. Khim.* **22**, 700–705 (1967).

[176] Arun P. Joshi and Kailash N. Munshi, *Microchem. J.* **12**, 447–453 (1967).

[177] Arun P. Joshi and Kailash N. Munshi, *J. Prakt. Chem.* **38**, 305–310 (1968); *Chim. Anal.* (Paris) **51** (4), 182–186 (1969).

[178] Keizo Hiraki, *Bull. Chem. Soc. Jap.* **45**, 1395–1399 (1972).

[179] M. Z. Yampol'skii, *Uch. Zap., Kursk. Gos. Pedagog. Inst.* **1957** (4), 128–141; **1958** (7), 67–82; *Tr. Kom. Anal. Khim., Akad. Nauk SSSR, Inst. Geokhim. Anal. Khim.* **8**, 141–151 (1958); cf. E. R. Oskotskaya, *Uch. Zap. Kafedry Obshch. Khim., Kursk. Gos. Pedagog. Inst.* **35**, 121–127 (1967).

photometric reading.[180] In samples of aluminum or zinc, the gallium is isolated by extraction from 1 : 1 hydrochloric acid with butyl acetate.

The complex of gallium with **sulfonaphtholazoresorcinol,** which is 1-(2,4-dihydroxyphenylazo)-2-naphthol-4-sulfonic acid, is read by fluorescence.[181] for the determination, mix 1 ml of sample in 0.04 *N* hydrochloric acid, 0.2 ml of monochloroacetic acid buffer solution for pH 3, and 2 ml of ethanol.

At pH 2.9–5 in an acetate medium, gallium forms a 1 : 1 complex with **sulfonazo,** which is di-[3-(8-amino-4-hydroxy-3,6-disulfo-2-naphthylazo)-4-hydroxyphenyl] sulfone.[182] The maximum extinction at 610 nm conforms to Beer's law up to 0.2 μg of gallium per ml. There is interference by magnesium, calcium, lead, titanium, thallium, aluminum, iron, and indium.

The optimum pH is 3.6, and heating aids in development of the absorption.[183] Interference by zinc, copper, and aluminum can be avoided by displacement of the ions with metallic cadmium. The dimethyl and dibromo derivatives are also satisfactory reagents. For more details, see Indium, with which sulfonazo and its dimethyl and dibromo derivatives also form complexes. All are read at 594 nm.

Gallium forms a 1 : 1 complex with **sulfonitrazo E,** which is 8-hydroxy-5-nitro-3-sulfophenylazo)naphthalene-1,6-disulfonic acid, at pH 2.4 for reading at 560 nm.[184] Beer's law is followed for 0.2–1.6 μg of gallium per ml. Equal amounts of nickel or indium and threefold lead or lanthanum are tolerated.

To a solution containing 5–62.5 μg of gallium ion, add 4 ml of m*M* **4-(2-thiazolylazo)-3-hydroxynaphthalene-2,7-disulfonic acid.**[185] Adjust to pH 3–4, dilute to 25 ml and read at 570–580 nm. Oxalate interferes seriously. Interference by thallic ion can be prevented by a tenfold amount of tartrate. Indium interferes.

For gallium by 4-(2-thiazolylazo)resorcinol, mix 10 ml of sample solution, neutral to pentamethoxyl red (which defines it as under pH 3.5), with 10 ml of 0.047% ethanolic solution of the reagent and 2.5 ml of *M* pyridine buffer for pH 5–5.2.[186] Dilute to 50 ml, set aside for 15 minutes, and read at 525 nm. Indium reacts similarly, and there are many interferences.

Tiron, which is disodium 4,5-dihydroxy-*m*-benzenedisulfonate, forms a 1 : 1 complex with gallium at pH 2.5–2.9 which is suitable for reading at 310 nm for 7–35 μg/ml.[187] Mix the sample solution with 0.33% solution of the reagent and read after 24 hours.

Gallium forms a 1 : 1 complex with **tolulyene blue,** which is 4-[(4,6-diamino-*m*-tolylimino)cyclohexa-2,5-diene-1-ylidene]dimethylammonium chloride.[188] This is extractable into benzene-acetone. The absorption maxima of the aqueous reagent and of the extract of the complex are at 525 and 540 nm, respectively. At pH 1.05–1.5, extraction of the reagent is negligible. The extracted complex is stable for

[180] S. I. Gusev and L. G. Dazhina, *Zh. Anal. Khim.* **24**, 1506–1511 (1969).
[181] V. A. Nazarenko and S. Ya. Vinkovetskaya, *ibid.* **13**, 327–331 (1958).
[182] M. N. Gordeeva and A. M. Ryndina, *Vest. Leningr. Gos. Univ., Ser. Fiz. Khim.* **1968** [1(4)], 132–136.
[183] E. P. Shkrobot, *Zh. Anal. Khim.* **17**, 311–317 (1962).
[184] P. D. Tyutyunnikova, Yu. M. Dedkov, and V. D. Salikhov, *ibid.* **30**, 508–513 (1975).
[185] A. I. Busev, T. N. Zholondkovskaya, L. S. Krysina, and N. A. Golubkova, *ibid.* **27**, 2165–2169 (1972).
[186] M. Langova-Hnilickova and L. Sommer, *Talenta* **16**, 681–690 (1969).
[187] Wladyslaw Reksc, *Chem. Anal.* (Warsaw) **14**, 795–802 (1969).
[188] I. L. Bagbanly, N. Kh. Rustamov, and E. G. Allakhverdieva, *Uch. Zap. Azerb. Gos. Univ., Ser. Khim. Nauk* **1971**, 27–31.

24 hours. To 1.5 ml of neutral sample containing 1–30 μg of gallium, add 2.5 of 42% lithium chloride solution, 0.45 ml of 1 : 11 hydrochloric acid, and 0.5 ml of 0.05% solution of toluylene blue. Extract with 5 ml of benzene-acetone and read the extract at 540 nm. Beryllium, gold, and iron interfere.

A 1 : 1 complex of gallium with **trihydroxyfluorone** is formed at an optimum pH of 4 and read at 530 nm.[189]

Triphenyltetrazolium chloride in 1 : 1 hydrochloric acid gives a white precipitate with gallium that is soluble in water and extractable with benzene for reading at 278 nm.[190] Beer's law applies for 0.1–1 mg/ml.

The complex of gallium with Victoria blue B (basic blue 26) is read in the range 600–660 nm after extraction.[191] To 6 ml of sample in 1 : 3 sulfuric or hydrochloric acid containing 0.5–20 μg of gallium, add 1 ml of 0.1% solution of the dye and extract with 6 ml of 10 : 1 benzene-acetone. Titanous chloride should be present as a reducing agent.

Victoria pure blue BO (basic blue 7) complexes with the tetrachlorogallate ion in 1 : 1 hydrochloric acid containing 0.5% of the dye.[192] The extract with 8 : 1 benzene–methyl ethyl ketone is read at 630 nm. Addition of titanous chloride suppresses interference by thallic, indium, antimonic, ferric, vanadic, and auric ions.

[189] O. A. Tataev, G. N. Bagdasarov, S. A. Akmedov, Kh. A. Mirazeva, Kh. G. Buganov, and E. A. Yarysheva, *Izv. Sev-Kavkaz. Nauch. Tsent. Vyssh. Shk., Ser. Estestv. Nauk* **1973** (2), 30–32.
[190] Tadeusz Pukas, *Chem. Anal.* (Warsaw) **5**, 513 (1960).
[191] A. I. Ivankova, *Tr. Kazk. Nauch.-Issled. Inst. Miner. Syr'ya* **1961** (5), 245–248.
[192] Polixenia Costinescu, Cecelia Constantinescu, and G. Banateanu, *Rev. Chim.* (Bucharest) **24**, 740–742 (1973).

CHAPTER THIRTEEN
SCANDIUM

Butyric acid in isobutyl alcohol containing sulfosalicylic acid will extract scandium ion at pH 4 from most common elements and from the rare earths.[1] This is appropriate for analysis of silicate ores.

For determination of scandium in cassiterite, reduce the sample in a stream of dry hydrogen at 600° and dissolve in hydrochloric acid.[2a] Distill the tin. Extract titanium and iron from 1:4 hydrochloric acid with chloroform as cupferron complexes. Destroy organic matter by heating with nitric acid, then convert to chlorides by evaporations with hydrochloric acid. Concentrate the scandium by two precipitations from tartaric acid solution 0.5–1*M* with ammonia. Separate scandium and yttrium by ascending chromatography on paper. Prepare the solvent by shaking 1:1:1 acetone-ether-butanol or isobutyl alcohol with 50 ml and 50 ml of saturated ammonium nitrate solution and add 5 ml of nitric acid. Wash the paper with 1:50 hydrochloric acid, then with water; saturate with 20% ammonium nitrate solution and dry. Treat the chromatogram with 0.1% solution of arsenazo 1, cut out the violet zone, and ash. Complete photometrically with alizarin red S or arsenazo III. The reduction of cassiterite can be replaced by fusion with 1:1 sodium peroxide–potassium hydroxide or sodium hydroxide, or by fusion with 1:1 sodium carbonate–borax.

In 7 *N* nitric acid zirconium is extracted from scandium by 0.5 m*M* 4-benzoyl-3-methyl-1-phenylpyrazolin-5-one in 0.01 *M* phosphorodiamidate in benzene leaving the scandium for determination.[2b]

For determination of scandium by flame photometry, see Chapter 50, Rare Earths.

[1] L. . Galkins and S. A. Strel'tsova, *Zh. Anal. Khim.* **25**, 889–893 (1970).
[2a] M. P. Belopol'skii and N. P. Popov, *Tr. Vses. Nauch.-Issled. Geol. Inst.* **117**, 53–62 (1964).
[2b] V. I. Fadaeva, V. S. Putuline, and I. P. Alimarin, *Zh. Anal. Khim.* **29**, 1918–1923 (1974).

ALIZARIN RED S

After isolation from interferences, scandium is complexed with alizarin red S (mordant red 3), which is the sodium salt of alizarin sulfonic acid.[3] Copper, fluoride, and phosphate interfere. Scandium in coal ash can be isolated by paper chromatography and determined with this reagent.[4]

In a weakly acidic medium, scandium is precipitated as a 1:3 complex by *N*-benzoylphenylhydroxylamine.[5] Thorium is similarly precipitated. The scandium precipitate can be extracted with butanol, amyl acetate, benzene, chloroform, and isoamyl alcohol for determination by alizarin red S.

Procedures

Zirconium and Thorium Absent.[6] As an ammonium acetate buffer solution, dissolve 100 grams in 300 ml of water, adjust to pH 3.5 with hydrochloric acid, and dilute to 500 ml.

Dilute a sample containing 0.01–0.12 mg of scandium in hydrochloric acid to 25 ml with hydrochloric acid. Add 0.5 ml of 30% hydrogen peroxide and 25 ml of tributyl phosphate. Shake well and discard the aqueous phase. Wash the organic phase with 25 ml, 25 ml, and 25 ml of hydrochloric acid. Add 70 ml of water and extract the scandium by shaking for 30 seconds. Extract residual tributyl phosphate with 25 ml of ether. Add 5 ml of 0.5% yttrium chloride in 1:19 hydrochloric acid to the aqueous layer. Add 25 ml of 40% solution of ammonium tartrate in 8% ammonium hydroxide. Add ammonium hydroxide with mixing until the solution is alkaline and a few ml in excess. Heat nearly to boiling to expel dissolved ether, stir for a few minutes and filter insoluble tartrates on sintered glass. Wash the precipitate with 5 ml of 40% ammonium tartrate solution in 8% ammonium hydroxide diluted with 20 ml of water. Dissolve the tartrates with 50 ml of 1:4 hydrochloric acid, collecting in the beaker used for their precipitation. Reprecipitate the tartrates, filter, wash, and dissolve in 1:4 hydrochloric acid. Repeat the step once more, ending with a solution of the tartrates in 50 ml of 1:5 hydrochloric acid. Add 2 ml of 0.1% solution of alizarin red S and follow with ammonium hydroxide until the solution just becomes red, in effect titrating to that end point. Add 20 ml of 20% ammonium acetate buffer solution at pH 3.5 and dilute to 100 ml. Read at 520 nm against a reagent blank.

Zirconium and Thorium Present. Dilute a sample containing 0.01–0.12 mg of scandium in sulfuric acid, free of nitrate or fluoride ions to 100 ml with water or acid so that 10% of acid is present. Add 10 ml of 6% cupferron solution and extract with 25 ml, 25 ml, and 25 ml of chloroform. Add 5 ml of 6% cupferron solution and extract the residual cupferrates and excess cupferron with four 25-ml portions of chloroform. Dilute the aqueous solution to about 300 ml and boil off residual

[3]Allan W. Ashbrook, *Analyst* **88**, 113–116 (1963).
[4]M. P. Belopol'skii, K. K. Gumbar, and N. P. Popov, *Zavod. Lab.* **28**, 921–922 (1962); B. Ya. Kaplan and I. V. O.'shevskaya, *ibid.* **29**, 26–27 (1963).
[5]I. P. Alimarin and Yun-Syan Tsze, *ibid.* **25**, 1435–1437 (1959).
[6]A. R. Eberle and M. W. Lerner, *Anal. Chem.* **27**, 1551–1554 (1955).

chloroform. Make alkaline with ammonium hydroxide, if no precipitation occurs, acidify with nitric acid and add about 200 mg of aluminum nitrate. Precipitate with ammonium hydroxide and add filter-paper pulp. Filter on sintered glass. Dissolve the hydroxides with 100 ml of warm 1:4 nitric acid. Add 75 ml of a saturated solution of potassium iodate in 1:4 nitric acid and stir frequently for 15 minutes. Filter the iodates on sintered glass with 25 ml of the potassium iodate precipitating solution. Dilute the filtrate and washings, which contain the scandium, to 300 ml; make alkaline with ammonium hydroxide and heat to 80°. Add paper pulp and filter. Dissolve the hydroxides from the filter with 25 ml of hydrochloric acid. Wash the funnel with 25 ml of hydrochloric acid and hold this in reserve. Extract the acid solution with 25 ml of tributyl phosphate for 30 seconds. Discard the aqueous phase and wash the organic phase with the reserved hydrochloric acid wash, then with 25 ml and 25 ml of hydrochloric acid. Complete as for zirconium and thorium absent from "Add 70 ml of water and extract the scandium by ...."

Uranium Compounds. Fuse a 0.5-gram sample in platinum with 5 grams of sodium peroxide at 400–450° for 1 hour. Dissolve in water, acidify with 1:1 nitric acid, boil, cool, filter, and dilute to 50 ml. To an aliquot containing up to 0.2 mg of scandium, add 10 ml of 20% solution of tartaric acid and adjust to pH 5 with ammonium hydroxide. Cool, add 2 ml of 30% hydrogen peroxide, and adjust to pH 5–5.5. Extract with 20 ml of 0.2 *M* 2-thenoyltrifluoroacetone in xylene. Wash the organic layer with 10 ml portions of 5% tartaric acid solution adjusted to pH 5 with ammonium hydroxide until the washings are colorless. Then wash with 10 ml and 10 ml of sodium nitrite solution. Extract the scandium from the organic solvent with 10 ml of dilute nitric acid adjusted to pH 1. Add to the extract 5 ml of 0.05% solution of alizarin red S. Follow with 1:10 ammonium hydroxide until the color just changes from yellow to red. Add 5 ml of 10% sodium acetate–acetic acid buffer solution for pH 3.5 and dilute to 100 ml. Read at 530 nm against a blank.

Coal Ash. Decompose 0.5 gram of sample with 5 ml of hydrofluoric acid and 2 ml of 1:1 sulfuric acid and evaporate to dryness. Add 3 ml of water, evaporate to dryness, and repeat this step. Take up in 5 ml of water and filter. Ash the filter, fuse the residue with 2 grams of sodium pyrosulfate, dissolve in water, and add to the main solution. Unless already present, add about 20 mg of ferric chloride. Neutralize with 10% sodium hydroxide solution and add 1 ml excess per 10 ml of sample solution. Filter, wash, and dissolve in 10 ml of 1:1 hydrochloric acid. Add 25 ml of 40% tartaric acid solution and about 25 mg of yttrium oxide as the chloride to serve as a collector. Add phenol red indicator solution, dilute to 100 ml, and precipitate with ammonium hydroxide. Boil for 3 minutes, add 5 ml of 1:3 ammonium hydroxide, and let stand overnight. Filter the basic tartrates of scandium and yttrium and wash with 4% tartaric acid solution neutralized to ammonium tartrate. Ignite at 600°. Dissolve in 2 ml of 1:1 hydrochloric acid and evaporate to dryness at 100°. Take up the residue in 0.25 ml of 1:2 hydrochloric acid.

Pretreat 26×2 cm chromatographic paper with 10% ammonium nitrate and dry. Apply 0.1 ml of the sample solution containing 0.01–0.1 μg of scandium oxide as a 0.5-cm wide streak 2 cm from the end and dry under an infrared lamp. Mix 100 ml of isopentyl alcohol, 100 ml of 1:5 hydrochloric acid, 100 ml of 30% ammonium thiocyanate solution, 25 ml of benzene, and 25 ml of acetone. Transfer 50 ml of the upper organic layer to a 17.5×15×30 cm jar. Suspend the chromographic paper; together with a blank and a control having a basic tartrate precipitate containing 0.15 mg of scandium oxide as a solution of the basic tartrate. Develop for 20 hours; at this time the scandium zone should have moved 15–18 cm. Dry in air, spray with 0.5% arsenazo 1 solution, and hold in ammonia vapor to locate the scandium zone. Cut out the scandium zones and place in 25-ml tubes with 0.5 ml of 1:1 hydrochloric acid, 10 ml of water, and 1 ml of 20% hydroxylamine hydrochloride solution, the latter to reduce to ferric ion. Heat the tubes at 100° for 5 minutes and cool. Add 0.5 ml of 0.1% alizarin red S solution to each, sufficient 1:9 ammonium hydroxide to give a color change, and 2.5 ml of 3 *M* acetate buffer solution for pH 3.5. Dilute to 25 ml and read at 530 nm.

ANTHRARUFIN-2,6-DISULFONIC ACID

When passed through a column of Dowex 1-X8 for reading by this reagent, scandium chloride is weakly adsorbed and comes through later than many unadsorbed ions.[7] Other weakly adsorbed ions such as chromic, titanous, thallous, andtetravalent vanadium can be separated by extraction with thenoyltrifluoroacetone in cyclohexane. To obtain a low reagent blank, a low concentration of reagent is essential. Fresh reagent solution is required daily. Fluorides and phosphates interfere. Pyrex containers rather than polyethylene should be used. A buffer decreases the sensitivity. Beer's law is followed up to 2 ppm of scandium.

Procedure. Adjust 10 ml of sample solution containing 0.2–20 μg of scandium to pH 2. Shake for 1 minute with 10 ml of a 2.2% solution of thenoyltrifluoroacetone in cyclohexane. Discard the aqueous layer and wash the organic extract with 10 ml and 10 ml of 0.01 *N* hydrochloric acid. Extract the scandium by shaking for 1 minute with 1 ml of *N* hydrochloric acid and transfer to a beaker. Wash down the inside surface of the separatory funnel with 5 ml of *N* hydrochloric acid and add to the beaker. Again extract by shaking with 1 ml of *N* hydrochloric acid. Evaporate the combined extracts to dryness and dissolve the residue in 1 ml of hydrochloric acid.

Prepare a 23×0.5 cm column of 200-400 mesh Dowex 1-X8 resin and wash with hydrochloric acid. When it drops to the surface of the resin, add the 1 ml of sample in hydrochloric acid containing 0.2–20 μg of scandium. Maintain a hydrostatic head of up to 1 ml of hydrochloric acid until scandium begins to come through as shown by a calibration curve. Collect the scandium effluent and evaporate to dryness. Add 5 ml of 0.01 *N* hydrochloric acid and incubate at 105° for an hour. Add 1 ml of a solution of 11 mg of color reagent in 25 ml of water. Adjust to pH 4–7 with 0.4% solution of sodium hydroxide, the reagent serving as the pH indicator. Dilute to 10 ml and adjust to pH 3.9–5.1 with a glass electrode. Read at 495 nm.

---

[7] John C. MacDonald and John H. Yos, *Anal. Chim. Acta* **28**, 264–270 (1963).

ARSENAZO I

Arsenazo I, which is 3-(2-arsenophenylazo)-4,5-dihydroxy-2,7-naphthalenedisulfonic acid, as the disodium salt forms a 1 : 1 complex with scandium.[8] Separation from rare earths and many other ions is obtained by extraction with a solution of 2-thenoyltrifluoroacetone in xylene. The scandium is back-extracted with 1 : 5 hydrochloric acid. Aluminum, copper, and ferric ion are separated by extraction with oxine in chloroform. Uranyl chloride is separated by Dowex 1-X8, an anion exchange resin that also adsorbs thorium nitrate.

Procedure. *Silicates and Ash.* For preparation of samples, see Propylfluorone. Carry out the separation through "Take up in 2 ml of 8 *N* hydrochloric acid and evaporate at 100°." Take up in 23 ml of 0.1 *N* hydrochloric acid. Add 0.5 ml of 5% ascorbic acid solution, 1 ml of 0.25% solution of 1 : 10-phenanthroline, and 1 ml of 0.1% solution of arsenazo I. Dilute to 50 ml with 0.05 *M* borate buffer solution for pH 7.9. Let stand for 20 minutes and read at 570 nm.

ARSENAZO III

Scandium is determined by arsenazo III, which is 3,6-di(2-arsonophenylazo)-4,5-dihydroxynaphthalene-2,7-disulfonic acid.[9] Extraction at pH 1.8 avoids interferences. The reagent also determines thorium, uranium, protactinium, neptunium, zirconium, and hafnium.[10]

Procedures

Iron and Steel. Dissolve 0.2 gram of sample in 10 ml of 1 : 1 hydrochloric acid and dilute to 100 ml. Add 10 ml of 1% solution of ascorbic acid and 1 ml of 13.6% solution of sodium acetate trihydrate. Adjust to pH 1.8 and add 5 ml of 0.1 *M* potassium chloride–hydrochloric acid buffer solution for pH 1.8. Extract by shaking for 3 minutes with 15 ml of 0.5 *M*-thenoyltrifluoroacetone in benzene. Separate the organic extract and reextract with 15 ml of benzene. Then add 3 ml of nitric acid and 3 ml of perchloric acid to the combined extracts and heat to white fumes. Take up the residue in 10 ml of water and add 10 ml of 1% solution of ascorbic acid and 1 ml of 13.6% solution of sodium acetate trihydrate. Adjust to pH 1.8 and add 15 ml of 0.1 *M* potassium chloride–hydrochloric acid buffer solution for pH 1.8. Add 7 ml of 0.05% solution of arsenazo III and dilute to 50 ml. Set aside for 5 minutes, then read at 675 nm.

[8] Hiroshi Onishi and C. V. Banks, *ibid.* **29**, 240–248 (1963); V. A. Nazarenko and E. A. Biryuk, *Ukr. Khim. Zh.* **29**, 198–204 (1963); Tsuneo Shimizu, *Anal. Chim. Acta.* **37**, 75–80 (1967).
[9] Ohiko Kammori, Isamu Taguchi, and Kenzi Yoshikawa, *Jap. Anal.* **15**, 458–466 (1966).
[10] S. B. Savin, *Talenta* **11**, 1–6 (1964); cf. S. B.Savvin and A. A. Muk, *Bull. Inst. Nucllar Sci. "Boris Kidrich"* (Belgrade) **12**, 97–107 (1961).

Ores.[11] Fuse 2 grams of sample with 10 grams of sodium hydroxide and 0.5 gram of sodium peroxide in an iron crucible. Extract the melt with hot water, filter, and wash the residue with hot 0.4% sodium hydroxide solution. Treat the filtrate with 5 ml of 14% calcium chloride solution, 2 grams of oxalic acid, and 3 ml of 30% hydrogen peroxide. Adjust to pH 2, heat to boiling, and maintain at 100° for several hours. Filter and dissolve in 20 ml of 1:3 hydrochloric acid. Repeat the precipitation with 5 ml of 14% calcium chloride solution, 2 grams of oxalic acid, and 3 ml of 30% hydrogen peroxide. Filter and dissolve the precipitate in 25 ml of 1:2 hydrochloric acid. Extract with 20 ml of 0.01 *M* dibutyl phosphate in petroleum ether. Wash the organic extract with 20 ml and 20 ml of 1:2 hydrochloric acid. Evaporate the organic extract to dryness and heat the residue at 100° with 5 ml of hydrobromic acid and 5 ml of acetic acid for 3 hours. Evaporate to dryness and take up in 3 ml of hydrochloric acid. Boil, add 10 ml of water, and evaporate to about 10 ml. Add another 10 ml of water and evaporate to about 2 ml. Neutralize with 2% sodium hydroxide solution, add 1.5 ml of 0.1 *N* hydrochloric acid diluted to 50 ml. To a 10-ml aliquot, add 1 ml of 0.1% solution of arsenazo III and dilute to 25 ml with 3 m*M* hydrochloric acid. Read at 670 nm as a measure of scandium plus thorium.

To another 10-ml aliquot, add 10 ml of hydrochloric acid, 3 ml of 10% solution of oxalic acid, and 2 ml of 0.06% solution of arsenazo III. Dilute to 25 ml and read thorium at 660 nm. Determine scandium by difference.

CATECHOL VIOLET

An anionic 1:1 complex is formed by catechol violet with scandium.[12]

Procedure. Add a drop of methyl red indicator solution to the sample solution. Add 10% hexamine solution until the indicator turns to orange. Add 1 ml of 0.075% solution of the dyestuff and 5 ml of acetate buffer solution for pH 5–5.7. Dilute to 25 ml with water and set aside for 20 minutes. Read against a blank at 585 nm.

CHLORPHOSPHONAZO III

Chlorphosphonazo III is 3,6-bis-(4-chloro-2-phosphonophenylazo)-4,5-dihydroxynaphthalene-2,7-disulfonic acid.[13] This reacts with scandium to immediately form a blue complex. Titanium, zirconium, iron, and aluminum interfere but are masked by tartaric acid. Fluoride, oxalate, and EDTA must be absent. The reaction is sensitive to 0.1–0.2 μg of scandium per ml. The complex forms an ionic combination with such organic cations as diphenylguanidinium chloride, tetraphenyl-

[11] E. Upor, Mrs. J. Szalay, and K. Klesch, *Magy. Kem. Foly.* **74**, 438–440 (1968).
[12] S. P. Onosova and G. K. Kuntsevich, *Zh. Anal. Khim.* **20**, 802–804 (1965).
[13] I. P. Alimarin and V. I. Fadeeva, *Vest. Mosk. Univ.* **1963** (4), 67–69.

arsonium chloride, or dioctylamine. The maximum extraction into butanol is at pH 2, with or without the associated cation.[14]

Beer's law is followed for 1–1.2 μg/ml.[15] The complexes with thorium, zirconium, and titanium also absorb at the same wavelength, 690 nm.

Procedure. *Wolframite.* Fuse 1 gram of sample with 10 grams of sodium carbonate and dissolve in hot water. Decant from the residue through a filter and wash twice with 1% sodium carbonate solution. Combine the residue from the filter with that from decantation. Evaporate to dryness and heat with 35 ml of sulfuric acid. Cool and add 50 ml of water. Reduce the manganese with 1 ml of 30% formaldehyde and evaporate the excess formaldehyde from the colorless solution. Dilute to 250 ml, filter, and discard the residue. Precipitate hydroxides with 1:4 ammonium hydroxide and filter. Dissolve the precipitate with 10 ml of 1:1 hydrochloric acid and wash the filter with 1:10 hydrochloric acid. Evaporate almost to dryness and take up the residue in 30 ml of 0.1 *N* hydrochloric acid. Add 30 grams of ammonium thiocyanate and a few crystals of ascorbic acid. Extract with 50 ml of ether. Add 4 ml of 1:11 hydrochloric acid to the aqueous layer and extract with 50 ml of ether. Repeat the procedure described in last sentence. To the 150 ml of ether extracts, add 3 ml of 1:1 hydrochloric acid and evaporate to dryness. Take up in 25 ml of 1:11 hydrochloric acid and dilute to 100 ml. Mix a 1-ml aliquot with 5 ml of buffer solution for pH 2.5. Add 1 ml of 0.05% solution of chlorphosphonazo III, dilute to 25 ml, and read at 690 nm against a reagent blank.

CHLORSULFOPHENOL R

The complex of scandium with chlorsulfophenol R is extracted with butanol for reading.[16]

Procedure. *Rare Earth Oxides.* As a guanidine reagent, dissolve 130 grams of *N*-diphenylguanidine in 500 ml of 1:11 hydrochloric acid, cool, filter, and adjust to pH 2.5.

Dissolve 20 mg of sample containing less than 0.01% of scandium in 2.5 ml of 1:11 hydrochloric acid. Add 5 ml of 0.1% solution of thoron 1 reagent and 2 ml of 0.05% solution of the captioned reagent. Dilute to about 35 ml and add 3 ml of the guanidine reagent. Extract with 25 ml of 1:1 chlororoform-butanol. Repeat the addition of thoron 1 and guanidine reagent and extract with 25 ml of mixed solvent. Repeat the last sentence. Discard the 75 ml of extracts and add 1 ml of 0.25% ascorbic acid to the aqueous sample. After 5 minutes add 5 ml of 0.05% solution of the captioned reagent, adjust to pH 2.5–2.9 and dilute to 50 ml. Add 3 ml of guanidine reagent and set aside for 2 minutes. Extract with 25 ml of butanol.

[14]V. I. Fadeeva and O. I. Kuchinskaya, *Vest. Mosk. Gos. Univ., Ser. Khim.* **1967** (1), 67–69.
[15]V. I. Fedeeva and I. P. Alimarin, *Zh. Anal. Khim.* **17**, 1020–1023 (1962).
[16]D. I. Ryabchikov, S. B. Savin, and Yu. M. Dedkov, *Zavod. Lab.* **31**, 154–155 (1965).

Wash the butanol layer with 25 ml of water containing 1.5 ml of guanidine reagent and repeat that washing. Read the butanol layer at 535 nm.

3-CYANO-1,5-BIS(2-HYDROXY-5-SULFOPHENYL)FORMAZAN

This reagent, sometimes known as formazan 1, forms a 1 : 1 complex with scandium.[17]

Procedure. *Rare Earth Metal Oxides.* Dissolve a 0.5-gram sample in 1 : 1 hydrochloric acid and evaporate to incipient dryness. Take up in 100 ml of water. Take an aliquot containing 0.01–0.05 mg of scandium and neutralize to the orange color of methyl orange. Add 10 ml of acetate buffer solution for pH 3.8 and 2 ml of 2 m*M* reagent. Dilute to 25 ml and read at 590 nm against a reagent blank.

1,5-DIANTIPYRINYL-3-CYANOFORMAZAN

There is serious interference with the complex of the reagent above by aluminum, bismuth, ferric ion, thallium, zirconium, hafnium, yttrium, uranium, lanthanum, vanadium, chromic ion, EDTA, oxalate, citrate, tartrate, sulfate, phosphate, and fluoride.[18] In most cases isolation by ion exchange appears to be indicated.

Procedure. To a sample solution containing 0.05–0.5 micromole of scandium, add 1 ml of 14% solution of hexamethylenetetramine and 1 ml of 0.8% solution of potassium cyanide. Add 0.5 *M* perchloric acid to adjust to pH 4.5 ± 0.3 and add 5 ml of 0.2 m*M* reagent in ethanol. Dilute to 25 ml with water, let stand for 10 minutes, and read at 552 nm against a reagent blank.

ERIOCHROME BRILLIANT VIOLET B

Eriochrome brilliant violet B (mordant violet 28) forms a 2 : 1 complex with scandium.[19] Because of interference by rare earth metals, aluminum, beryllium, cadmium, chromic ion, cupric ion, thallium, ferric ion, lead, and yttrium, it is necessary to isolate the scandium by columm chromatography. The complex is bleached by oxalate, citrate, fluoride, and EDTA.

Procedure. Adjust the sample solution to 0.1 *M* in ammonium sulfate and 0.025

[17]S. P. Onosova, G. K. Kuntsevich, and D. C. Lisenko, *Zh. Anal. Khim.* **22**, 1469–1474 (1967.
[18]B. W. Budesinsky and J. Svec, *Microchem. J.* **16**, 253–258 (1971).
[19]Katsuya Uesugi, *Anal. Chim. Acta* **49**, 597–602.

M in sulfuric acid. Apply to a 9×210 mm column of 100-200 mesh Dowex 1-X8 previously treated with the ammonium sulfate solution at a flow rate of 0.5–1.5 ml/minute. Elute interfering metals with 60 ml of the acid ammonium sulfate solution. Then elute the scandium with 50 ml of 1 : 11 hydrochloric acid. Since removal of ferric ion is not complete, add 3 ml of 1% solution of ascorbic acid to the scandium eluate. Then to an aliquot of the eluate containing 1–15 μg of scandium, add 2 ml of 0.1% ethanolic solution of the dye. Add 10 ml of 0.1 M sodium acetate–acetic acid buffer solution for pH 6. Dilute to 25 ml and set aside for 20 minutes, then read at 562 nm.

GLYOXAL BIS(*o*-HYDROXYANIL

Scandium in largely organic solvent forms a 1 : 2 complex at pH 4.5–5.5 in the presence of at least 150-fold excess of glyoxal bis(*o*-hydroxyanil), which is 2,2′-(ethanedixylidenedinitrilo)diphenol.[20] Beer's law is followed for 0.4–6.2 μg/ml. Only erbium gives the same reaction with ammonium nitrate present. There is interference by uranyl, cupric, cerous, ceric, titanic, and thallic ions.

At pH 4.8–5.9 in a medium containing acetate and perchlorate ion, a reddish-violet 1 : 1 complex is formed.[21] The complex is extracted by shaking with butanol for 2 hours, has a maximum absorption at 560 nm, and at pH 5 obeys Beer's law up to 3.8 ppm of scandium. The complex is stable for 42 hours.

Procedure. Mix the sample solution containing 10–15 μg of scandium with 2.5 ml of 8% ammonium nitrate solution and dilute to 5 ml. Add 10 ml of ethanol and a large excess of 0.1 M solution of the reagent in dimethylformamide. Heat at 80° for 5 minutes, cool, and dilute to 25 ml with ethanol. Read at 560 nm.

HYDRAZO III

Hydrazo III, which is the 2-hydroxy-1-naphthylmethylenehydrazide of 4-methoxybenzoic acid, forms a fluorescent complex in 50% acetone with scandium at pH 2–3.5 with a maximum at 505 nm.[22] With fiftyfold excess of reagent, it conforms to Beer's law at 0.01–3 μg of scandium in 50 ml. Ascorbic acid or another reducing agent is necessary if ferric ion is present. There is interference by 0.1 mg of titanium, 0.15 mg of thorium, niobium, or tantalum, or 0.2 mg of vanadium.

Procedure. *Rare Earth Oxides.* This technic is applicable to as little as 0.03 ppm of scandium in the sample. Dissolve 1 gram in 5 ml of hydrochloric acid and dilute

[20] A. Okac and M. Vrchlabsky, *Z. Anal. Chem.* **195**, 338–342 (1963); M. Vrchlabsky and P. Bocek, *Spis. Prir. Fak. Univ. Brne, E* **33** (2), 63–70 (1967).
[21] B. S. Garg, Nisha Nagpal, and R. P. Singh, *Curr. Sci.* **44**, 156–157 (1975).
[22] A. V. Dolgorev, N. N. Pavlova, and V. A. Ershova, *Zavod. Lab.* **39**, 658–660 (1973).

to 50 ml. Neutralize a 10-ml aliquot to Congo red with 1 : 1 ammonium hydroxide. Add 10 ml of an acetate buffer solution for pH 2.22, 0.5 ml of 5% ascorbic acid solution, 15 ml of acetone, and 10 ml of hydrazo III solution in acetone containing 5 mg of the reagent adjusted to pH 2.5. Dilute to 50 ml with acetone adjusted to pH 2.5 and set aside for 10 minutes. Read the fluorescence at 505 nm.

PROPYLFLUORONE

The complex of scandium with propylfluorone can be determined after separation from interferences as below, inserting a special step if thorium or zirconium is present.[23]

Procedures

Silicates and Ash. Decompose 0.5 gram of sample in platinum with 20 ml of hydrofluoric acid and 10 ml of sulfuric acid. Evaporate to about 2 ml and take up in 5 ml of water.

If the sample contains more than 2–3 mg of thorium or zirconium, add 10 ml of nitric acid and 15 ml of 10% potassium iodate solution. Heat at 100° for 15 minutes and let stand at room temperature for 50 minutes. Filter and wash with 0.8% solution of potassium iodate. In the absence of those amounts of thorium or zirconium, skip this paragraph.

Add 5 ml of 30% hydrogen peroxide, neutralize with 20% potassium hydroxide solution, and add 10 ml excess. Filter the precipitate and wash. Dissolve in 30 ml of 7 *N* hydrochloric acid. Saturate 30 ml of ether with 1 : 1 hydrochloric acid and use it to extract the sample. Evaporate the extracted aqueous layer to 10 ml and dilute to 50 ml. Add 5 ml of a 0.5% suspension of yttrium oxide and 25 ml of 40% tartaric acid solution. Neutralize with 1 : 3 ammonium hydroxide. Boil and set aside for a day in an atmosphere of ammonia vapor. Filter and wash with 10% ammonium tartrate solution. Dissolve in 4 ml of 1 : 3 hydrochloric acid and wash the paper with 11 ml of water. Add 0.1 gram of ascorbic acid and 13 grams of ammonium thiocyanate to the filtrate. Extract with 30 ml and 20 ml of ether. Wash the combined ether extracts with a solution of 26 grams of ammonium thiocyanate and 0.1 gram of ascorbic acid in 30 ml of 0.5 *N* hydrochloric acid. Extract the scandium from the ether with 50 ml of water, evaporate the extract in quartz to dryness, and calcine at 700°. Take up the residue in 0.5 ml of nitric acid and 1.5 ml of hydrochloric acid and evaporate to dryness.

Take up in 2 ml of 8 *N* hydrochloric acid and evaporate at 100°. Take up in 10 ml of 1 : 23 hydrochloric acid and dilute to 100 ml. Take a 25-ml aliquot. Add a drop of thymol blue indicator solution and add 20% sodium acetate solution to a neutral reaction. Add 0.5 ml of 2% solution of ascorbic acid, 1 ml of 0.25% solution of *o*-phenanthroline, 1 ml of 1% gelatin solution, 2.5 ml of 20% solution of sodium

[23] V. A. Nazarenko and E. A. Biryuk, *Zavod. Lab.* **28**, 401–406 (1962).

acetate, and 2 ml of 0.05% solution of phenylfluorone in 0.6 *N* ethanolic hydrochloric acid. Read at 530 nm.

If only 2–10 μg of scandium is present, follow the procedure above to the evaporation to dryness with nitric and hydrochloric acids. Take up the residue in 2 ml of 0.1 *N* hydrochloric acid and add 0.8 ml of 40% ammonium tartrate solution. Neutralize with 1 : 4 ammonium hydroxide and filter. Wash the precipitate with 1 ml of 10% ammonium sulfate solution. Dissolve in 3.5 ml of 0.5 *N* hydrochloric acid. Complete as above from "Add a drop of thymol blue...."

Wolframite. Fuse 0.5 gram of sample with 5 grams of sodium hydroxide in a nickel crucible. Dissolve in 10 ml of water and neutralize with 1 : 1 sulfuric acid. Evaporate to about 5 ml. Complete as for silicates from "If the sample contains...."

Cassiterite. Fuse 0.2 gram of sample with 3 grams of sodium peroxide in an iron crucible. Dissolve in 10 ml of water and neutralize with 1 : 1 sulfuric acid. Evaporate to about 5 ml and complete as for silicates from "If the sample contains...."

QUERCETIN

The 1 : 1 complex with quercetin, which is 3,3′,4′,5,7-pentahydroxyflavone, is applicable to reading 0.1–3 ppm of scandium.[24] Because of interference by nickel, ferric, tri- and hexavalent chromium, hexavalent uranium, thorium, phosphate, sulfate, citrate, tartrate, and oxalate, the scandium usually must be isolated by ion exchange. Ascorbic acid or hydroxylamine by reducing ferric ion will avoid that interference. At pH 5–6.5 two fluorescent complexes can be read at 435 and 470 nm.[25]

Procedure. Evaporate the sample solution containing 0.018–0.075 mg of scandium to dryness and take up in 10 ml of hydrochloric acid. Pass through a 1 × 15 cm column of Diaion SA-100 a strongly basic anion exchange resin that is in chloride form because of pretreatment with hydrochloric acid. Elute lanthanum and aluminum ion with hydrochloric acid. Then elute thorium and scandium with 10 ml of 1 : 1 hydrochloric acid. Evaporate this eluate to dryness and take up in 10 ml of 8 *N* hydrochloric acid. Pass this through a 1 × 15 cm column of Diaion resin in nitrate form. Elute with 10 ml of 8 *N* nitric acid. Evaporate this eluate to dryness and take up the residue in 5 ml of hot water containing 3 drops of 0.1 *N* perchloric acid. To this solution add 10 ml of ethanol and sufficient acetate buffer solution to adjust to pH 4.4. Add 2 ml of 0.1% ethanolic solution of quercetin and dilute to 25 ml with water. Let stand for 5 minutes and read at 435 nm against a reagent blank.

[24] H. Hamaguchi, R. Kuroda, R. Sugisita, N. Onuma, and T. Shimizu, *Anal. Chim. Acta* **28**, 61–67 (1963).
[25] V. A. Nazarenko and V. P. Antonovich, *Zh. Anal. Khim.* **22**, 1812–1817 (1967).

To separate a few mg of scandium from 100 mg of lanthanum, pass the solution through a 1×5 cm column of Diaion SKI in hydrogen form. Elute the scandium with 15 ml of 2% oxalic acid solution. The lanthanum remains adsorbed.

SALICYLALDEHYDE SEMICARBAZONE

This reagent, the product of condensation of salicylaldehyde with semicarbazide hydrochloride, forms a 1 : 1 fluorescent complex with scandium.[26] Activated at 370 nm, it shows a maximum fluorescence at 455 nm. The technic is applicable to 0.002–0.02 ppm of scandium. The fluorescence develops in a few minutes and is constant for more than 24 hours. Extraction of cupferrates removes interfering ions such as iron and vanadium, and subsequent extraction with organic phosphate and back-extraction separate from other interfering ions such as aluminum and yttrium. The method appears to be as sensitive as with morin and substantially more sensitive than methods by absorption.

Procedure. As a buffer solution, dissolve 100 grams of hexamine in 800 ml of water, adjust to pH 6 with hydrochloric acid, and dilute to 1 liter.

Mix an aliquot of sample solution containing 4.5–22.5 μg of scandium with 2.5 ml of hydrochloric acid and 5 ml of 1% cupferron solution. Dilute to 25 ml, extract with 25 ml and 25 ml of chloroform, and discard the extracts. Evaporate the sample and washings of the container to about 5 ml. Add to 25 ml of tri-*n*-butyl orthophosphate and 25 ml of hydrochloric acid. Shake for 1 minute and discard the aqueous layer. Wash the organic layer with 25 ml and 25 ml of hydrochloric acid, discarding the washings. Extract the scandium with 25 ml of water and 1 ml of ammonium hydroxide, noting that this aqueous extract is still acid. Wash the aqueous extract with 25 ml and 25 ml of chloroform to remove residual organic phosphate. Add 10 ml of ammonium hydroxide to the aqueous extract and evaporate to about 20 ml. Cool, dissolve crystallized ammonium chloride with water, and add 2 ml of the buffer solution for pH 6. The pH must now be pH 6 ± 0.05, preferably 6 ± 0.02. Add 25 ml of a solution of 0.1788 gram of the reagent per liter of ethanol and dilute to 100 ml with water. Set aside for 30 minutes, activate at 370 nm, and read the fluorescence at 455 nm.

STILBAZOCHROME

Stilbazochrome, which is 4,4′-bis-(1,8-dihydroxy-3,6-disulfo-2-naphthylazo)stilbene-2,2′-disulfonic acid, is suitable for reading of scandium in 0.025 *M* acetic acid.[27] There is interference by thorium, niobium, and tantalum.

[26] I. M. Korenman and V. S. Efimychev, *ibid.* **17**, 425–428 (1962); G. F. Kirkbright, T. S. West, and C. Woodward, "Proceedings of the SAC Conference, Nottingham 1965," pp 474–480; *Analyst* **91**, 23–26 (1966).

[27] N. M. Alykov and A. I. Cherkesov, *Zh. Anal. Khim.* **20**, 870–871 (1965).

Procedure. Decompose 0.1 gram of sample in platinum with 3 ml of hydrofluoric acid and 5 ml of 1:4 sulfuric acid. Evaporate to sulfur trioxide fumes, add 3 ml of water, and again evaporate to sulfur trioxide fumes. Repeat that addition and evaporation twice more. Take up with 3 ml of hydrochloric acid and dilute to 50 ml with water. Neutralize to Congo red with 5% sodium hydroxide solution and add 5 ml in excess. Heat at 100° to coagulate the precipitate. Filter and wash the precipitate with 1% sodium chloride solution. Dissolve the precipitate with 10 ml of hydrochloric acid and evaporate to dryness. Dissolve in 25 ml of water and add 0.5 ml of 0.1 *M* yttrium salt as a coprecipitant. Neutralize to a faint turbidity with ammonium hydroxide and redissolve with 1:1 hydrochloric acid. Add 20 ml of 40% solution of ammonium tartrate. Neutralize to neutral red with ammonium hydroxide and add 5 ml in excess. Heat to coagulate and set aside for 12 hours. Filter and ignite. Take up in 3 ml of 1:1 hydrochloric acid and evaporate to dryness. Take up in 1 ml of 1:1 hydrochloric acid and dilute to 100 ml. Mix a 15-ml aliquot, 20 ml of 0.1 *N* acetic acid, and 5 ml of 0.1% solution of the reagent. Dilute to 100 ml and read.

XYLENOL ORANGE

Scandium forms a violet 1:1 complex at pH 1.2–5.5 and a red 1:2 complex at pH 3.5–6.0.[28] Both ranges have been recommended, with a preponderance of investigators preferring the 1:1 complex. Beer's law is followed for up to 2 μg/ml. At pH 1.5–5 the color develops in 10 minutes and is stable for 2 days. There is interference by zirconium, thorium, indium, bismuth, and gallium. Addition of ascorbic acid to reduce ferric and ceric ion avoids their interference. There is also interference by sulfate, phosphate, fluoride, oxalate, and nitrilotriacetic acid.[29] The optimum conditions for reading scandium are pH 2.8 and 560 nm.[30]

A ternary complex is formed by scandium, xylenol orange, and ethyltridodecylammonium bromide.[31] This is formed by extraction of the binary complex at pH 6–6.7 with the quaternary compound in benzene or toluene. The complex is read at 520 nm and obeys Beer's law up to 1.4 μg of scandium per ml. Many ions are masked by triethylenetetramine and 1,10-phenanthroline, but there is serious interference by vanadium, aluminum, beryllium, thorium, zirconium, uranium, bismuth, lanthanum, nitrite, perchlorate, citrate, and EDTA.

Procedure. To a sample containing 4–90 μg of scandium in dilute hydrochloric acid, add 1:20 ammonium hydroxide until it changes Congo red paper from blue to violet.[32] Add 2 ml of acetate buffer solution for pH 2.6 and 3.5 ml of 0.03%

[28] V. P. Antonovich and V. A. Nazarenko, *ibid.* **23**, 1143–1151 (1968); R. S. Volodarskaya and G. N. Derevyanko, *USSR* Patent *143,785* (1962).
[29] Mei-ling Gong, *A*cta Chim. Sin. **29**, 223–226 (1963).
[30] B. Budesinsky, *Collect. Czech. Chem. Commun.* **28**, 1858–1866 (1963).
[31] Y. Shijo, *J. Chem. Soc. Jap.* **1974**, 889–895.
[32] O. V. Kon'kova, *Zh. Anal. Khim.* **19**, 68–73 (1964).

solution of xylenol orange. Dilute to 25 ml and read at 556 nm. If ferric ion is present, add ascorbic acid to reduce it to ferrous.

Separation of Scandium and Thorium.[33] Adjust a sample solution containing 0.01–0.1 mg of thorium and 3–15 μg of scandium to pH 3.2–3.4. Extract the thorium by shaking with 5 ml of 0.1% solution of 5,7-dichloro-8-hydroxyquinoline in chloroform. This leaves scandium in the aqueous phase. To determine thorium, add 5 ml of 0.1% solution of xylenol orange in ethanol to the organic extract and read at 540 nm. Adjust the aqueous solution of scandium to pH 5–5.5 with hexamine. Extract the scandium with 5 ml of 0.1% solution of 5,7-dichloro-8-hydroxyquinoline in chloroform. Add 5 ml of 0.1% ethanolic solution of xylenol orange and read at 530 nm. Ferric ion extracted with the thorium causes a positive error.

Copper Alloys.[34] The sample may contain nickel, titanium, and zirconium. Dissolve 1 gram of sample in 10 ml of 1 : 1 hydrochloric acid, add 10 ml of sulfuric acid, and evaporate to sulfur trioxide fumes. Take up in 25 ml of water, add an aluminum chloride solution containing about 5 mg of aluminum as a collector, and dilute to about 100 ml. Prepare a solution of cupferron by mixing 10 ml of 1.5 *N* sulfuric acid and 5 ml of 6% aqueous solution of cupferron and extracting it with 10 ml of chloroform. Extract the titanium, zirconium, iron, and part of the copper with 5 ml and 5 ml of the cupferron solution so prepared. Then extract the aqueous phase with 10 ml, 10 ml, and 10 ml of chloroform. Discard all the chloroform extracts. Warm the aqueous phase until it no longer has an odor of chloroform and add about 1 mg of iron as ferric chloride. Make alkaline with ammonium hydroxide to precipitate the scandium, aluminum, and iron. Filter the precipitated hydroxides and wash well with 1 : 100 ammonium hydroxide. Dissolve in 25 ml of 1 : 1 hydrochloric acid and dilute to 50 ml. Take an aliquot containing 0.02–0.05 mg of scandium. Add 0.1 ml of 10% solution of ascorbic acid, adjust to pH 1.5, add 1 ml of 0.1% solution of xylenol orange, and read after 20 minutes at 555 nm.

Copper-Scandium Alloys.[35] Dissolve 1 gram of sample in 5 ml of 1 : 1 nitric acid and evaporate to dryness. Add 3 ml of hydrochloric acid, evaporate to dryness, and repeat this step to eliminate nitrates. Dissolve in 1 : 1 hydrochloric acid and dilute to 50 ml with that acid. Take an aliquot containing no more than 0.2 mg of copper. Apply it to a 20×200 mm column of Dowex 1-X8 previously washed with 1 : 1 hydrochloric acid. Pass at a rate of 15–20 ml/hour and wash with 75 ml of 1 : 1 hydrochloric acid. Evaporate the eluate and washings to dryness, finishing at 100°. To bake would give low results. Take up in 0.01 *M* perchloric acid, add 2 ml of 0.05% solution of xylenol orange in 0.01 *M* perchloric acid, and dilute to 25 ml with the same acid. Read at 553 nm. It may be read at once and is stable for 24 hours.

Aluminum Solutions.[36] Mix an acid solution containing about 0.05 mg of

[33] Che-ming Nee and Shu-chuan Liang, *Acta Chim. Sin.* **30**, 296–300 (1964).
[34] I. S. Postnikova, *Zavod. Lab.* **36**, 542 (1970).
[35] S. S. Berman, G. R. Duval, and D. S. Russell, *Anal. Chem.* **35**, 1392–1394 (1963).
[36] Maksymilian Kranz, Wojciech Duczmal, and Krystyna Langowska, *Chem. Anal.* (Warsaw) **16**, 399–406 (1971).

scandium with 10 ml of a buffer solution for pH 4 and 20 ml of isopentyl alcohol. Add either 1.25 gram of a 1 : 50 mixture of *N*-(2-hydroxy-1-naphthyl)methylidene-*o*-arsanilic acid or 2 ml of 0.05% solution of 2-(2-hydroxy-1-naphthylazo)benzenearsonic acid. Shake for 30 seconds and separate the organic phase. Wash with 5 ml of water. Extract the scandium from the organic phase with 15 ml of 1:30 hydrochloric acid. Wash the organic phase with 5 ml of water and combine with the aqueous extract. Neutralize by adding 0.5 ml of ammonium hydroxide and add 2 ml of 0.04% solution of xylenol orange in chloroacetate buffer solution for pH 2.6. Dilute to 25 ml and set aside for 10 minutes. Read at 555 nm against water.

Magnesium.[37] Dissolve 1 gram of sample containing 0.001–0.005% of scandium or 0.1 gram containing 0.01–0.05% of scandium in 10 ml of 1:1 hydrochloric acid. Add 5 ml of 2% solution of ascorbic acid, followed by 50% sodium acetate solution until Congo red paper turns to lilac. Add 5 ml of a mixture of 263 ml of 0.2 *N* hydrochloric acid and 500 ml of 0.2 *N* potassium chloride as a buffer. Add 5 ml of 0.1% solution of xylenol orange and dilute to 100 ml. Set aside for 20 minutes and read at 555 nm. If zirconium is present, it must be separated by benzenearsonic acid.

Coal Ash.[38] Prepare the sample as for determination by alizarin red S through "Ignite at 600°. Dissolve in 2 ml of 1:1 hydrochloric acid, and evaporate to dryness." Take up the residue in 5 ml of 0.01 *N* hydrochloric acid. Add 20 ml of *N* acetic acid, 2 ml of 10% hydroxylamine hydrochloride solution, and 2.5 ml of 0.2% solution of xylenol orange in 80% ethanol. Dilute to 50 ml, set aside for 15 minutes, and read at 560 nm.

Aluminosilicates. Treat as for coal ash.

Blood.[39] Dry 5 grams of sample at 180° for 2 hours and calcine at 250° for 2 hours, 300° for 2 hours, 400° for 1 hour, 500° for 1 hour, and 700° for 2 hours. Heat the residue with 5 ml of hydrochloric acid for 2 hours, then evaporate to 3 ml, and dilute to 10 ml. Add 4 ml of 5% solution of ascorbic acid and adjust to pH 2.35 by adding *N* sodium acetate. Add 2 ml of 0.2% xylenol orange solution and dilute to 50 ml with a buffer solution for pH 2.35. Set aside for 1 hour and read at 550 nm against a reagent blank.

Scandium forms a 1:2 complex with **3-acetyl-1,5-bis-(3,5,6-trichloro-2-hydroxyphenyl)formazan** having a maximum absorption at 675 nm.[40]

The 1:1 complex of scandium with **acid monochrome Bordeaux S** (mordant red 30) has a maximum absorption at 500 nm, as contrasted with 450 nm for the

[37] R. S. Volodarskaya and G. N. Derevyanko, *Zavod. Lab.* **29**, 148–149 (1963).
[38] M. P. Belopol'skii and N. P. Popov, *ibid.* **30**, 1441–1442 (1964).
[39] N. P. Tszyu, R. I. Bocharova, and A. A. Men'kov, *Lab. Delo* **1969** (5), 275–278.
[40] V. M. Dziomoko, V. M. Ostrovskaya, and O. V. Kon'kova, *Zh. Anal. Khim.* **25**, 267–271 (1970).

reagent.[41] The color is stable for 24 hours and conforms to Beer's law for 0.05–3 μg/ml. For determination, mix 3 ml of sample, 1 ml of acetate buffer solution for pH 5, and 0.6 ml of 0.1% solution of the dye. Set aside for 5 minutes before reading. There are many interferences, although there are fewer if the pH is reduced to 2.

Scandium is determined by **arsenazo M**, which is 3-(2-arsonophenylazo)-6-(sulfophenylazo)chromotropic acid, in the presence of yttrium by reading at pH 2.7 and 670 nm.[42]

For reading scandium with **azonol A**, which is 3-(4-antipyrinylazo)pentane-2,4-dione, mix a sample solution containing 0.1–0.5 micromole of scandium with 3 ml of 0.05 *M* potassium cyanide solution.[43] Adjust to pH 4–7 with 0.5 *M* perchloric acid or 2% sodium hydroxide solution and add 5 ml of 0.5 *M* perchloric acid. Add 5 ml of 0.5 m*M* azonol A, dilute to 25 ml, and set aside for 10 minutes. Read at 610 nm. Bismuth, thallium, hafnium, and ferric ion interfere.

Azophosphon, which is 2-(2,4-dihydroxyphenylazo)benzenephosphonic acid, forms a 1:1 complex with scandium down to 0.04 μg/ml.[44] The optimum pH is 0.75–3.5. Addition of 70% of ethanol or acetone increases the stability of the complex, which has a maximum at 415 nm, as compared with 385 nm for the reagent. The method tolerates 2.5-fold amounts of molybdenum, zirconium, or beryllium and 0.5-fold thorium. Reading is preferably at 470 nm, at which wavelength the reagent shows little absorption.

Scandium is read fluorescently with **brilliant alizarin blue G** (mordant blue 31).[45] The dye also combines with aluminum, gallium, yttrium, lanthanum, lutecium, indium, and magnesium.

Bromocresol green as a 1:1 complex with scandium at pH 4–4.6 has a maximum absorption at 520 nm that follows Beer's law for 1–33 μg/ml.[46] Correspondingly, **bromocresol purple** at pH 4–4.2 is read at 470 nm for 1–35 μg/ml and **bromophenol blue** at pH 4–4.2 is read at 480 nm for 0.5–28 μg/ml.

The complex of scandium with **bromopyrogallol red** is read at pH 0.1 and 610 nm.[47] Many common ions interfere. As applied to igneous rocks and ocean sediments, the sample is decomposed with hydrofluoric and perchloric acids. The residue on evaporation is dissolved in 1:1 hydrochloric acid and in acid ammonium sulfate solution separated by cation and anion exchange chromatography. The scandium is then coprecipitated with ferric hydroxide, dissolved in 1:1 hydrochloric acid, and isolated from the iron by ion exchange for development with bromopyrogallol red.

The 1:1 complex of scandium with ***NN*-bis(carboxymethyl)anthranilic acid** at pH 4 is read at 285–290 nm.[48] Yttrium and lanthanum react similarly.

The 1:1 complex of **chromal blue G** (mordant blue 55) with scandium buffered

[41] I. M. Korenman and N. V. Zaglyadimova, *Tr. Khim. Khim. Tekhnol.* (Gor'kii) **1965** [1(12], 99–104; [3(14)], 191–194.

[42] R. I. Bocharova, A. F. Kusheleva, and A. A. Men'kov, *Zh. Anal. Khim.* **26**, 1505–1510 (1971).

[43] B. Budesinsky and J. Svecova, *Anal. Chim. Acta* **49**, 231–240 (1970).

[44] A. M. Lukin, N. N. Vysokova, and N. A. Bolotina, *Tr. Vses. Nauchn.-Issled. Inst. Khim. Reakt. Osobo Chist. Khim. Veshchestv.* **1971** (33), 162–169.

[45] Keizo Hariki, *Bull. Chem. Soc. Jap.* **45**, 789–793 (1972).

[46] Yu. G. Eremin and V. S. Katochinka, *Zh. Anal. Khim.* **25**, 68–71 (1970).

[47] Tsuneo Shimizu, *Talenta* **14**, 473-9 (1967); *Bull. Chem. Soc. Jap.* **42**, 1561–1569 (1969).

[48] C. Dragalescu, S. Policec, and T. Simonescu, *Talenta* **13**, 1543–1548 (1966).

with acetate at pH 6 has a maximum absorption at 590 nm.[49] Beer's law is followed up to 1.2 ppm. There is interference by cupric, cobalt, beryllium, ferric, chromic, and yttrium ions, and by the rare earths.

Chrome azurol S (mordant blue 29) forms a 2:1 complex with scandium for reading at 4–27 μg per 10 ml.[50] Adjust to pH 5.6 with *M* hexamine–*M* perchloric acid buffer solution, add 2 ml of m*M* chrome azurol S, and dilute to 10 ml. Set aside for 90 minutes and read at 550 nm. There is interference by fluoride, phosphate, oxalate, tartrate, citrate, EDTA, aluminum, beryllium, hexavalent uranium or tungsten, iron, thorium, cupric ion, yttrium, and zirconium.

The addition of zephiramine increases the sensitivity fivefold.[51] At 625 nm Beer's law is followed up to 8 μg in 25 ml. To a neutral solution, add 0.5 ml of 0.2 *N* hydrochloric acid, 0.3 ml of 0.3% solution of the dyestuff, 1 ml of 7% solution of hexamine, and 2 ml of 0.5% solution of zephiramine solution at pH 5.5. Iron can be masked with ascorbic acid and copper with 1:10 phenanthroline. Aluminum, gallium, indium, phosphate, and fluoride ions interfere.

A 1:2:1 complex of scandium, chrome azurol S, and 4,4′-diantipyrinylmethane has an absorption maximum at 620 nm.[52] Beer's law is followed for 0.12–1.76 μg of scandium per ml. Optimum conditions are 0.224 m*M* dye, 2.24 m*M* added reagent, and ammonium acetate buffer solution for pH 7.2.

The 1:1 chelate of **chromotrope 2R** (acid red 29) has an optimum pH of 4.5 for reading at 560 nm. The dye has a maximum absorption at 530 nm.[53] Yttrium also forms a chelate absorbing at the same wavelength.

Scandium forms a ternary complex with **diantipyrinylmethane** and **3-(4-sulfophenylazo)chromotropic acid**.[54] Buffered at pH 7.8–8, it conforms to Beer's law for 0.1–3 μg of scandium at 580 nm. Zinc, copper, titanic ion, nickel, EDTA, and rare earths interfere.

Scandium forms a 1:1 complex with **1,5-dihydroxyanthraquinone-2,6-disulfonate**.[55] Yttrium forms a similar complex.

Heat 0.1 gram of mineral for 3 hours with 30 ml of sulfuric acid, then evaporate to dryness.[56] Take up with 4 ml of hydrochloric acid and dilute to 75 ml. Filter. Add 70 mg of calcium, adjust to pH 2, and coprecipitate scandium and calcium as oxalates. Separate centrifugally. Add perchloric acid and evaporate to dryness. Dissolve the residue in 5 ml of 1:1 hydrochloric acid and adjust to pH 1.6. Extract scandium with 10 ml of 0.2 *M* 2-thenoyltrifluoroacetone in benzene. Back-extract the scandium with 10 ml of *N* hydrochloric acid. Adjust to pH 6 and add **2,4-dihydroxybenzaldehyde semicarbazone**. Activate at 360 nm and read the fluorescence at 425 nm.

More than sixfold excess of **2-(2,4-dihydroxyphenylazo)pyridine** as reagent is

[49]Katsuya Uesugi, *Bull. Chem. Soc. Jap.* **42**, 2051–2054 (1969).

[50]Ryoei Ishida and Norio Hasagawa, *ibid.* **40**, 1153–1158 (1967);Surendra N. Sinha, Satendra P. Sangal, and Arun K. Dey, *Mikrochim. Acta* **1968**, 899–902.

[51]Yoshizo Horiuchi and Horoshi Nishida, *Jap. Anal.* **17**, 1486–1491 (1968).

[52]L. A. Alinovskaya and L. I. Ganaga, *Zh. Anal. Khim.* **28**, 661–665 (1973).

[53]Shrikant B. Dabhade and Satendra P. Sangal, *Microchem. J.* **14**, 190–198 (1969).

[54]L. I. Ganago and L. A. Alinovskaya, *ibid.* **27**, 261–265 (1972).

[55]J. C. MacDonald, *Thesis, Univ. Virginia* **1962**, 112 pp; *Dis. Abstr.* **23**, 2678 (1963).

[56]Kiyotoshi Morisige, Sakae Sasaki, Keigo Hiraki, and Yasharu Nishikawa, *Jap. Anal.* **24**, 321–324 (1975).

required to give the maximum absorption with scandium of the 1 : 1 complex.[57] At the maximum of 515 nm there is little absorption by the reagent. By buffering with 20% ammonium acetate solution, an optimum pH of 6–7.3 is obtained. Beer's law applies to 0.05–2 μg/ml. Interferences are by zinc, cobalt, nickel, cadmium, copper, thallium, indium, gallium, zirconium, thorium, uranium, iron, and aluminum.

Scandium is read as a 1 : 1 complex at 420 nm with **7-(4,8-disulfo-2-naphthylazo-8-hydroxyquinoline-5-sulfonic acid** at 0.2–2.8 μg/ml.[58]

Scandium in an acid medium above pH 5 forms a 1 : 2 complex with **eriochrome azurol G** (mordant blue 47) with a maximum at 610 nm.[59] Beer's law is followed for 0.1–1 ppm, and the complex is stable for 1 hour. There is interference by beryllium, cupric ion, aluminum, ferric ion, yttrium, thorium, and rare earths. Oxalate, fluoride, citrate, and EDTA bleach the complex.

The red 1 : 2 chelate of scandium with **eriochrome cyanine R** (mordant blue 3) is read at 535 nm.[60] Yttrium and lanthanum form similar complexes. To a sample solution containing 0.5–2.5 ppm of scandium, add a fivefold excess of the dye and adjust to pH 5. Read after not less than 30 minutes or more than 90 minutes. There is interference by beryllium, aluminum, gallium, indium, ferric ion, chromic ion, niobium, and palladous ion. At pH 7–8 there are even more interferences.

Scandium complexes at pH 5–7 with **5-ethylamino-2-(2-pyridylazo)-*p*-cresol** for reading at 540 nm.[61] The useful range is 0.135–2.25 μg/ml. Similarly at pH 6–7, **5-ethylamino-2-(2-thiazolylazo)-*p*-cresol** at 0.09–1.8 μg of scandium per ml is read at 550 nm. There is interference by zinc, cobalt, nickel, lead, ferric ion, indium, gallium, thorium, and zirconium.

To the sample solution add 5 ml of 0.01 *M* **2-ethyl-5-hydroxy-3-methylchromone** in methanol, 2.5 ml of methanol, and 5 ml of 0.25 *M* boric acid–sodium hydroxide buffer solution for pH 11.5–12.[62] Dilute to 25 ml and set aside for 30 minutes. Extract the scandium complex with 10 ml of benzene, dry the extract with sodium sulfate, and activate at 405 nm for reading fluorescence above 430 nm against fluorescein sodium at 1 μg/ml as a standard. The complex is stable for 30 minutes, and Beer's law is followed for up to 4 μg of scandium per ml.

For **flame photometry**, scandium is read at 402.36 nm in the inner cone of an acetylene-air flame.[63]

Glycinecresol red[64] complexes with scandium for reading at 520–530 nm in an acetate-buffered medium of pH 5–6. The zirconium complex absorbs at 520 nm at pH 4.5. There is interference by zinc, ferric ion, aluminum, cobalt, zirconium, cupric ion, pentavalent vanadium, hexavalent molybdenum, gallium, titanic ion,

[57] A. I. Busev and Fang Chang, *Vest. Mosk. Univ., Ser. 11, Khim* **15** (6), 46–51 (1960).

[58] A. I. Busev and G. E. Lunina, *Zh. Anal. Khim.* **20**, 1069–1072 (1965).

[59] Katsuya Uesugi, *Bull. Chem. Soc. Jap.* **42**, 2398–2401 (1969).

[60] Taitiro Fujinaga, Toru Kuwamoto, and Kazuo Kuwabara, *Jap. Anal.* **12**, 399–400 (1963); Taitiro Fujinaga, Toru Kuwomoto, Sigeo Tsurubo, and Kazuo Kuwabara, *ibid.* **13**, 127–132 (1964); Kailash N. Munshi, Suresh C. Shrivastava, and Arun K. Dey, *J. Indian Chem. Soc.* **45**, 817–820 (1968).

[61] S. I. Gusev and L. M. Shurova, *Uch. Zap. Perm. Gos. Univ.* **1973** (289), 271–277.

[62] Motoshi Nakamura and Akira Murata, *Jap. Anal.* **22**, 1474–1480 (1973).

[63] L. A. Ovchar and N. S. Poluektov, *Zh. Anal. Khim.* **22**, 45–49 (1967).

[64] O. A. Tataev, E. T. Beschetnova, E. A. Yarysheva, and V. K. Guseinov, *ibid.* **24**, 255–257 (1969); M. K. Akhmedli, D. G. Gambarov, and R. F. Fati-Zade, *Azerb. Khim. Zh.* **1973**, (5-6), 101–108.

fluoride, EDTA, and tartaric acid. Nickel and erbium must not exceed scandium and europium is only tolerated at 2-fold.

Scandium forms a chloroform-soluble complex with **8-hydroxyquinoline** (oxine) and some derivatives.[65] Extracted at pH 8–9, the complex with oxine gives a maximum absorption at 370 nm. Correspondingly, **5,7-dichlorooxine** extracted at pH 8.5–9.5 has a maximum at 397 nm and **5,7-dibromooxine** extracted at pH 8–9 has a maximum at 402 nm. With dichlorooxime, Beer's law is followed for 0.04–0.6 μg of scandium per ml. To separate from aluminum, gallium, yttrium, lanthanum, and lutecium, extract with 2-thenoyltrifluoroacetone in benzene at pH 1.5. Then reextract into 1 : 11 hydrochloric acid.

The chloroform extracts are also read by fluorescence.[66] That of the complex with oxine has a maximum at 510 nm. Other fluorescent maxima of derivatives are as follows: 5,7-dichloro-, 526 nm; 5,7-dibromo-, 520 nm; 5,7-diiodo-, 515 nm; and 2-methyl-, 505 nm. That with the dichloro derivative is most reproducible. Treat 20 ml of sample containing less than 0.01 mg of scandium with 0.25 ml of 0.2% reagent, add 5 ml of 20% ammonium chloride solution, and adjust to pH 9.5. Extract by shaking for 3 minutes with 25 ml of chloroform.

Mix a sample solution containing 10–90 μg of scandium with 5 ml of a buffer solution for pH 4.1 consisting of 0.8 *M* acetic acid and 0.2 *M* sodium acetate.[67] Add 1 ml of 0.2% solution of **indoferron**, which is 5-[bis(carboxymethyl)aminomethyl]-2,6-dibromo-3′-methylindophenol. Dilute to 25 ml and read at 600 nm. Beer's law is followed for 0.6–3.2 ppm of scandium as developed.

The 1 : 1 complex of **kaempferol**, which is 3,4′,5,7-tetrahydroxyflavone, with scandium at pH 3 in 50% ethanol is read at 415 nm for 0.4–1.8 ppm.[68] There is serious interference by hexavalent uranium, thorium, molybdenum, ferric ion, titanium, zirconium, beryllium, aluminum, nickel, cobalt, vanadium, phosphate, sulfate, oxalate, tartrate, and citrate. Severalfold amounts of yttrium and the rare earths are tolerated.

With **lumogallion**, which is 5-chloro-3-(2,4-dihydroxyphenylazo)-2-hydroxybenzenesulfonic acid, at below pH 6, scandium forms a 1 : 1 complex, but at higher pH it is 1 : 2.[69] The 1 : 1 complex at pH 2 is read at 530 nm against a reagent blank. Beer's law is followed for 0.1–0.8 μg of scandium per ml. Although 0.5-fold amounts of thorium, ferric ion, and cupric ion are tolerated, larger amounts of ferric ion are reduced with ascorbic acid and larger amounts of cupric ion masked with thiourea.

Lumomagneson, which is 5-(5-chloro-2-hydroxy-3-sulfophenylazo)barbituric acid, in threefold excess forms a 1 : 1 complex with scandium.[70] Reading is at pH 4.5–4.7 and 500 nm, where the difference between the absorption of the complex and of the reagent is at a maximum. It is applied for 0.18–1.26 μg of scandium per ml.

[65] Yasuharu Nishikawa, Keizo Hiraki, Shiro Goda, and Tsunenobu Shigematsu, *J. Chem. Soc. Jap. Pure Chem. Sect.* **83**, 1264–1267 (1962).

[66] Yashuharu Nishikawa, Keizo Hiraki, and Tsunenobo Shigematsu, *ibid.* **90**, 483–486 (1969).

[67] Tsuneo Shimizu and Kazuko Ogami, *Talenta* **16**, 1527–1533 (1969).

[68] B. S. Garg, K. C. Trikha, and R. P. Singh, *Anal. Chim. Acta* **42**, 343–346 (1968).

[69] V. P. Antonovich and V. A. Nazarenko, *Zh. Anal. Khim.* **24**, 676–681 (1969); M. K. Akhmedli and D. G. Gambarov, *ibid.* **22**, 276–278 (1967).

[70] M. K. Akhmedli and D. G. Gambarov, *Uch. Zap. Azerb. Gos. Univ., Ser. Khim Nauk* **1965** (4), 23–29.

Ascorbic acid masks iron and thiourea copper. There is interference by aluminum, ferric ion, zirconium, bismuth, cupric ion, lead, mercurous ion, oxalate, tartrate, EDTA, and rare earths.

Magneson IREA, which is 5-chloro-2-hydroxy-3-(2-hydroxy-1-naphthylazo)benzenesulfonic acid, as a 1 : 1 complex with scandium has been recommended as preferable to eriochrome black T, eriochrome blue-black R, lumogallion, or lumomagneson.[71] It is extractable with butanol or cyclohexanol, but the extraction agent affects the reading. Therefore isoamyl acetate with 88% extraction at pH 8.4 is preferable. Reading is at 515 nm, and in the aqueous phase Beer's law is followed for 0.2–2 μg/ml.

The 1 : 1 complex of **methylthymol blue** with scandium has a maximum absorption at 570 nm and is read at 0.09–1.26 μg/ml.[72] For the determination add 2 ml of m*M* reagent to the sample solution and dilute to 25 ml with a buffer solution for pH 2. Phosphate and fluoride interfere. Ferric ion is reduced with ascorbic acid, aluminum masked with sulfosalicylic acid. Thorium, zirconium, and hafnium are masked with a 0.1 *M* mixture of citric and tartaric acids.

Buffered at pH 3.1, a 1 : 2 complex is formed which has a maximum absorption at 582 nm and follows Beer's law for up to 1 μg of scandium per ml.[73]

The red-violet complex of scandium with **methylxylenol blue** has a maximum absorption at 585 nm and pH 2.3–2.7.[74] Beer's law is followed for 0.4–1.6 μg of scandium per ml. Ferric ion is masked by ascorbic acid. There is interference by indium, bismuth, thorium, fluoride, and oxalate.

In the presence of anions of a strong acid such as perchloric, scandium forms a ternary complex with **morin** and **phenazone**.[75] At pH 3.3–3.4 the complex is extracted by chloroform for reading down to 0.1 μg/ml at 425 nm or down to 0.01 μg/ml fluorimetrically. There is interference by aluminum, gallium, indium, titanium, thorium, zirconium, and lutecium.

The chelate of scandium with **murexide**, which is 5,5′-nitrilodibarbituric acid ammonium compound and also known as ammonium purpurate, is read at pH 6 and 480 nm for 2–5 ppm.[76] The complexes with yttrium and lanthanum are read at the same wavelength. Most common ions do not interfere.

With scandium at pH 4.1–6.3, the complex of **naphthyl azoxine S**, which is 8-hydroxy-7-(6-sulfo-2-naphthylazo)quinoline-5-sulfonic acid, has a maximum absorption at 420 nm.[77] In the presence of 5 ml of 0.1 *M* acetate buffer for pH 4.5 and 0.8 ml of 0.5% solution of the reagent, Beer's law is followed up to 0.08 mg of scandium in 25 ml.

The 1 : 1 complex of scandium with **nitrobromarsenazo**, which is 4-(3-bromo-4,5-dihydroxyphenylazobenzenesulfonic acid, at pH 4.5–5.5, is read at 580 nm.[78]

Mix the sample solution with fourfold excess of 0.001 *M* ***p*-nitrophenyla-**

[71] I. P. Alimarin, A. K. Pigaga, and I. M. Gibalo, *Vest. mosk. Gos. Univ., Ser. Khim.* **1969** (5), 94–95.
[72] M. K. Akhmedli and D. G. Gambarov, *Zh. Anal. Khim.* **22**, 1460–1467 (1968).
[73] A. E. Okun' and L. D. Fomenko, *Uch. Zap. Kafedry Obshch. Khim., Kursk. Gos. Pedagog. Inst.* **25**, 149–153 (1967).
[74] J. Ueda, *J. Chem. Soc. Jap.* **1973**, 1467–1473.
[75] V. A. Nazarenko and V. P. Antonovich, *Zh. Anal. Khim.* **22**, 1812–1817 (1967); **24**, 358–361 (1969).
[76] S. P. Sangal, *Microchem. J.* **11**, 508–512 (1966).
[77] Tsuneo Shimizu, *Jap. Anal.* **16**, 233–238 (1967).
[78] A. I. Busev, G. E. Lunina, and N. N. Basargin, *Zh. Anal. Khim.* **21**, 1414–1419 (1966).

zochromotropic acid (acid red 176) and adjust to pH 5.5.[79] Dilute to 25 ml, set aside for 30 minutes, and read at 580 nm. Beer's law is obeyed for 0.2–3 ppm of scandium. Yttrium forms a similar complex. There is interference by copper, lanthanum, thallium, zirconium, hafnium, and thorium.

The 1:2 complex of scandium with **pontachrome azure blue B** (mordant blue 1) at pH 6.5 has a maximum absorption at 610 nm and follows Beer's law for 2–25 μg per 25 ml.[80] A colored complex is also formed with beryllium, copper, aluminum, iron, chromium, bismuth, titanium, and the rare earths.

Pontacyl violet 4B5N (acid violet 3) forms a 2:1 complex with scandium at pH 6.4–7.2 which at 630 nm conforms to Beer's law up to 1.2 ppm in 25 ml of solution containing 5 ml of acetate buffer solution for pH 6.8 and 2.5 ml of 0.1% solution of the dye.[81] There is interference by aluminum, ferric ion, cupric ion, nickel, chromic ion, beryllium, and phosphate.

In the presence of acetylacetone at pH 3–7, a red 1:1 complex of scandium with **9-propyl-2,3,7-trihydroxy-6-fluorone** is formed, at an optimum at pH 5.6.[82] This absorbs at 510 nm, the reagent at 480 nm. Color develops in 5 minutes and lasts for several hours. It is applicable for 0.04–2 μg of scandium per ml.

For determination with **4-(2-pyridylazo)resorcinol**, mix a faintly acid sample solution containing 0.1–2.4 μg of scandium per ml with 10 ml of 0.0228% solution of the reagent.[83] Adjust to pH 3–5 with ammonium hydroxide or perchloric acid and the total ionic strength to 0.4 with 10.6% solution of sodium carbonate. Dilute to 50 ml and set aside for 10 minutes. Read at 530 nm against a reagent blank. Because of many interferences, the scandium should have been isolated.

A 1:1 complex of scandium with **β-resorcylaldehyde acetylhydrazone** in 40% ethanol activated at 406 nm in pH 6 acetate buffer solution can be read fluorescently for 1–18 μg.[84] Nickel, cobalt, chromium, and ferric ion decrease this fluorescence. Zinc, aluminum, and gallium as well as scandium show blue fluorescence when activated at 365 nm.

β-Resorcylaldehyde semicarbazone complexes with scandium at pH 6.[85] When activated at 360 nm, the fluorescence at 425 nm is read for 2–400 ng of scandium per ml.

In analysis of coal, the scandium in hydrochloric acid is complexed with **rhodamine B** (basic violet 10), extracted with benzene and read at 555–560 nm.[86] Tartaric acid masks iron and titanium. Germanium and gallium are not coextracted. Europium, ytterbium, gadolinium, and samarium interfere.

A 1:2:1 complex of scandium, cinchophen (which is 2-phenylcinchoninic acid), and rhodamine B is formed.[87a] Mix up to 2 ml of faintly acid sample solution containing up to 0.045 mg of scandium with 0.4 ml of 2.5% solution of potassium 2-phenylcinchoninate, 0.2 ml of 0.5% rhodamine B solution, 0.5 ml of acetate

[79] Satendra P. Sangal, *Microchem. J.* **8**, 313–316 (1964).

[80] Tsunenobu Shigematsu, Katsuya Uesugi, and Masayuki Tabushi, *Jap. Anal.* **12**, 267–270 (1963).

[81] Tsunenobu Shigematsu and Katsuya Uesugi, *ibid.* **16**, 467–469 (1967).

[82] E. A. Biryuk and V. A. Nazarenko, *Zh. Anal. Khim.* **14**, 298–302 (1959).

[83] L. Sommer and M. Hnilickova, *Anal. Chim. Acta* **27**, 241–247 (1962).

[84] Z. Urner, *Collect. Czech. Chem. Commun.* **33**, 1078–1090 (1968).

[85] Kiyotoshi Morisige, *Anal. Chim. Acta* **73**, 245–254 (1974).

[86] N. A. Suvorovskaya and I. A. Shmarinova, *Tr. Inst. Obogashch. Tverd. Goryuch. Iskop. Mosk. Ugol. Prom SSSR* **1**, 127–131 (1972).

[87a] S. V. Bel'tyukova and N. S. Poluektov, *Ukr. Khim. Zh.* **37**, 1277–1279 (1971); also see note 87b.

buffer solution for pH 4.2, and water to make 5 ml. Extract with 5 ml of benzene and read at 556 nm. There is interference by rare earths, thorium, aluminum, gallium, zirconium, hafnium, and titanium. If one works below pH 4, the method will tolerate 0.5 mg of yttrium or lanthanum. Addition of 0.2 ml of 80% mercaptoacetic acid before the extraction masks up to 1 mg of iron, bismuth, or indium.

Another ternary complex is formed by scandium and rhodamine B with salicylic acid.[87b] This is extracted with benzene from a medium buffered at pH 4 with acetate solution for reading at 552 nm. Both it and the ternary complex with cinchophen are read by fluorescence at 580 nm. Activation is by a mercury lamp. The reagent will determine down to 0.04% of scandium in mixtures with rare earths.

A complex of **rutin**, which is quercetin-3-rutinoside, with scandium at pH 6, absorbs at 430–440 nm and is determinable at 5–30 m*M*.[88a]

Scandium forms a 1 : 1 complex in a medium buffered at pH 2 by potassium chloride–hydrochloric acid solution with **sulfonitrazo R** which is 3-hydroxy-4-(2-hydroxy-5-nitro-3-sulfophenylazo)-naphthalene-2,7-disulfonic acid.[88b] This is extracted with butanol from a medium containing 1% of diphenylguanadine and read at 540 nm. Beer's law is obeyed for 0.04–0.8 μg of scandium per ml of extract. For rock analysis, dissolve 1 gram in hydrofluoric acid–nitric acid and evaporate to dryness. Take up in water and extract the scandium with a solution of cupferron in chloroform. Developed with sulfonitrazo R, this will determine 1 ppm.

For determination with **superchrome garnet Y** (mordant red 5), take a sample solution containing 5–40 μg of scandium.[89] Add 2 ml of 20% ammonium acetate solution and 0.8 ml of 0.01% solution of the dye. Adjust to pH 5.5, activate at 485 nm, and read by fluorescence at 565 nm. There are numerous interferences.

Scandium forms a violet to violet-blue complex with **sulfonazo**, which is di[3-(8-amino-1-hydroxy-3,6-disulfo-2-naphthylazo)-4-hydroxyphenyl]sulfone.[90] At pH 4–5.5 in acetate or hexamine buffer solution the complex is stable for several hours. At 610–620 nm it follows Beer's law up to 4 μg/ml. Vanadium, cobalt, and gallium interfere. **Stilbazochrome**, which is 4,4′-bis-(1,8-dihydroxy-3,6-disulfo-2-naphthylazo)stilbene-2,2′-disulfonic acid, is used similarly.

The complex of **4-(2-thioazolylazo)resorcinol** with scandium has maximum absorption at 540 nm at pH 8.1 and is applicable to 0.12–1.6 ppm.[91] Because of numerous interferences, the scandium must usually be isolated by ion exchange.

Scandium is read as a complex with **thoron II**, which is 3,3′-di-(2-hydroxy-3,6-disulfo-1-naphthylazo)biphenyl-4,4′-diarsonic acid, in hydrochloric acid at pH 4.[92] The reagent also complexes with yttrium and lanthanum.

The colorless complex of scandium with **Tiron**, which is the disodium salt of

[87b]L. I. Kononenko, S. V. Bel'tyukova, V. N. Drobyazko, and N. S. Poluektov, *Zh. Anal. Khim.* **30**, 1716–1719 (1975).

[88a]V. A. Nazarenko and V. P. Antonovich, *Zh. Anal. Khim.* **22**, 1812–1817 (1967).

[88b]T. K. Bakhmotova, Yu. M. Dedkov, and V. A. Ershova, *Zh. Anal. Khim.* **31**, 292–297 (1976).

[89]Keizo Hiraki, *Bull. Chem. Soc. Jap.* **45**, 1395–1399 (1972).

[90]V. G. Brudz and D. A. Drapkina, *Sb. Statei, Vses. Nauchn.-Issled. Inst. Khim. Reak. Osobo Chist. Khim. Veshchestv.* **1961** (24), 60–72; V. G. Brudz', V. I. Titov, E. P. Osiko, D. A. Drapkina, and K. A. Smirnova, *Zh. Anal. Khim.* **17**, 568–573 (1962); A. I. Cherkesov and N. M. Alykov, *ibid.* **19**, 1067–1072 (1964).

[91]Tsuneo Shimizu and E. Momo, *Anal. Chim. Acta* **52**, 146–149 (1970).

[92]N. N. Sentyurina, *Zh. Anal. Khim.* **17**, 442–446 (1962).

4,5-dihydroxy-*m*-benzenedisulfonic acid, absorbs strongly at 310 nm.[93] There is interference by aluminum, titanium, iron, cerium, molybdenum, thorium, and zirconium, as well as many organic ions. Therefore isolation of the scandium by column chromatography on Dowex 50-X8 is often required. Elution with *M* ammonium thiocyanate–0.5 *M* hydrochloric acid separates from rare earths and thorium. Iron and aluminum will leave the column with that elution medium before scandium. To the eluate containing 0.003–0.25 mg of scandium, add 3 ml of 0.1% solution of Tiron and 20 ml of acetate buffer solution for pH 6. Dilute to 50 ml and read.

A 1 : 1 complex of scandium with **trihydroxyfluorone** is formed at an optimum pH of 6 and read at 530 nm.[94]

[93]Hiroshi Hamaguchi, Noaki Onuma, Rokuro Kuroda, and Ryuichiro Sugisata, *Talenta* **9**, 563–571 (1962).

[94]O. A. Tataev, G. N. Bagdasarov, S. A. Akmadov, Kh. A. Mirzaeva, Kh. G. Buganov, and E. A. Yaryshena, *Izv. Sev-Kavkaz. Nauch. Tsent. Vyssh. Shk. Ser. Estestv. Nauk,* **1973** (2), 30–32.

CHAPTER FOURTEEN
GERMANIUM

The methods of isolation of germanium from most interferences are distillation or extraction as the chloride from 7.5 *N* hydrochloric acid. Germanium chloride can be extracted from 7.5 *N* hydrochloric acid by methyl isobutyl ketone with 90% efficiency.[1] In an acetate buffer medium, complexing of germanic acid with EDTA takes so long that either may be present without interfering with the other.[2]

The addition of hydrogen peroxide accelerates the decomposition of sulfide ores with sulfuric and nitric acids and prevents loss of germanium.[3]

In strongly acidic solution germanium reacts with the vicinal carbony-hydroxy group of quercetin, quercetinsulfonic acid, morin, flavonal, 3-hydroxy-4′-methoxyflavone, and kojic acid.[4] In a neutral solution, quercetin, quercetinsulfonic acid, and rutin react by their vicinal diphenol group, but the others cited do not. Thus the complex of germanium with quercetinsulfonic acid has a maximum absorption at 400 nm in neutral solution but at 450 nm in acid solution. Substituted anthraquinones are applicable both photometrically and fluorimetrically.[5]

For complexing with germanium, an organic compound must have two OH groups in the ortho position.[6] Fluorone and coumarin derivatives react better than derivatives of naphthoquinone at low pH. This has led to synthesis of such derivatives as 2,6,7-trihydroxy-9-phenylfluorone and the methyl analogue. In com-

[1]Paschoel Senice, *Sel. Chim.* (16), 63–73 (1957).
[2]N. Konopik, *Z. Anal. Chem.* **225**, 416 (1967).
[3]N. N. Proshenkova, A. I. Salova, and G. A. Agarkova, *Izv. Vyssh. Ucheb. Zaved., Khim. Khim. Technol.* **8**, 1029–1030 (1965).
[4]Takuji Kanno, *Sci. Rep. Res. Inst., Tohuku Univ., Ser. A* **11**, 141–151 (1959).
[5]I. M. Korenman, N. V. Kurina, and E. A. Emelin, *Tr. Khim. Khim. Tekhnol.* **1** (1), 134–137 (1958).
[6]Kenjiro Kimura, Kazuo Asito, and Masako Asada, *Bull. Chem. Soc. Jap.* **29**, 635–640 (1956).

paring photometric reagents for germanium, stilbazo and hematoxylin are preferable to alizarin red S, pyrogallol red, bromopyrogallol red, and quinalizarin.[7]

A column of 100-200 mesh Dowex 1-X8 resin in chloride form adsorbs tri- and penta-valent arsenic and many other ions from a 9:1 acetic acid–9 *N* hydrochloric acid medium, but no germanium.[8]

Citric acid will mask up to thirtyfold stannic ion and 100-fold titanous ion in the presence of germanium.[9]

At 10–30 μg/liter in 2 *M* zinc sulfate, 0.5 gram of ferric chloride is added and the hydroxides are precipitated with ammonium hydroxide.[10] The precipitate is dissolved in 1:1 hydrochloric acid and the germanium chloride distilled into 10% sodium hydroxide solution.

For analysis of tar, the most satisfactory method is to decompose with sulfuric and nitric acids, evaporate to sulfur trioxide fumes, dilute, neutralize, add hydrochloric acid, reduce with stannous chloride, and distill.[11] Excess of stannous chloride does not interfere. Simple distillation is unsatisfactory because of bumping by heat and acid. On drying and ignition, some germanium oxide is volatilized. Nevertheless ashing at 550° followed by distillation from 6 *N* hydrochloric acid has been recommended.[12]

Photometric methods and fluorimetric methods for determining germanium have been reviewed.[13]

Tetravalent germanium serves as a catalyst for oxidation of 40 m*M* iodide by 20 m*M* ammonium molybdate.[14] In sulfuric acid the pH should be 0.9 unless silicon is present, in which case it should be 0.8. In hydrochloric acid the pH should be 0.7. If arsenic is present the pH should be 1 with either acid. The reaction rate versus concentration of germanium is linear for 20 micromoles to m*M* germanium. The sensitivity is 0.1 ppm in sulfuric acid and 0.03 ppm in hydrochloric acid.

For isolation by distillation, the sample must be free from nitric acid and other oxidizing agents.[15] To 25 ml of neutral sample in a distillation apparatus, add 25 ml of hydrochloric acid. Heat to boiling over a period of 5 minutes, then distill at 2 ml/minute, collecting 20 ml of distillate in ice water.

ALIZARIN

At pH 4.7 a 1:2:2 complex is formed by germanium, diphenylguanidine, and alizarin, which is 1,2-dihydroxyanthraquinone.[16] When extracted with 1:4 chloroform-acetone it has a maximum absorption at 475 nm. Beer's law is followed for 0.1–5 μg of germanium in 8 ml of solution.

[7]R. A. Alieva, M. K. Akhmedi, and A. M. Ayubova, *Azerb. Khim. Zh.* **1969** (6), 129–134.
[8]J. Krokisch and F. Feik, *Sep. Sci.* **2** (1), 1–9 (1967).
[9]R. M. Dranitskaya, A. I. Gavril'chenko, and L. A. Okhitina, *Zh. Anal. Khim.* **25**, 1740–1743 (1970).
[10]A. Fernandez Segura, A. A. Garmendia, and E. L. Pella, *An. Asoc. Quim. Arqent.* **45**, 126–135 (1957).
[11]L. Lechner and Z. Ferenczy, *Nehezvegyip. Kut. Int. Kozl.* **2**, 353–356 (1963).
[12]K. P. Medvedev, L. M. Khar'kina, V. M. Petropol'skaya, and K. A. Nikitina, *Zavod. Lab.* **29**, 805 (1963).
[13]D. P. Shcherbov and R. N. Plotnikova, *Z. Anal. Khim* **27** 740–747 (1972).
[14]I. I. Alekseeva and I. I. Nemzer, *ibid.* **24**, 1939–1400 (1969).
[15]H. J. Cluley, *Analyst* **76**, 523–530 (1951).
[16]D. I. Zul'fugarly, I. K. Guseinov, and Kh. N. Kulieva, *Azerb. Khim. Zh.* **1972** (2), 173–179.

Procedure. *Ores, Industrial Concentrates, and Slags.*[17] Decompose 1 gram of sample with 2 ml of nitric acid, 2 ml of hydrofluoric acid, and 2 ml of phosphoric acid. Evaporate to a syrup and take up in 25 ml of water. Add 75 ml of hydrochloric acid and extract with 12.5 ml and 12.5 ml of chloroform. Wash the combined chloroform extracts with 10 ml, 10 ml, and 10 ml of 2% solution of hydroxylamine hydrochloride in 9 *N* hydrochloric acid. To the chloroform layer add 3 ml of m*M* alizarin and add ammonium hydroxide until the solution is crimson. Add 20 ml of ammonium acetate buffer solution for pH 7 and 5 ml of 0.1 *M* diphenylguanidine in 0.1 *N* hydrochloric acid. Shake, separate the chloroform layer, dilute to 25 ml with chloroform, and read at 500 nm.

ALIZARIN RED S

Alizarin red S (mordant red 3), which is 2,4-dihydroxyanthraquinone sulfonate, forms a 1:2 complex with germanium with the optimum pH at 6.[18] The complex obeys Beer's law for 0.5–2 μg/ml and is stable for 48 hours. Ascorbic acid, sulfosalicylic acid, thiosulfate, thiocyanate, thiourea, fluoride, and borate must be absent, since they mask germanium. EDTA masks copper, ferric ion, zinc, nickel, or lead. Aluminum interferes. Normally the germanium will have been isolated by extraction as the chloride with carbon tetrachloride, then reextracted into water.

Procedure. Follow that given for alizarin, but use an ammonium acetate buffer solution for pH 6 and read at 480 nm.

ASTRAZONE PINK FG

The red-violet complex of astrazone pink FG (basic red 13) with germanium and molybdic acid is insoluble in water and most solvents but dissolves in 1.5–2 *N* hydrochloric acid.[19] Beer's law holds for 0.05–0.88 μg of germanium per ml.

Procedure. To a sample solution containing 2.5–44 μg of germanium, add successively 1 ml of 0.5% ammonium molybdate in 0.5 *N* sulfuric acid, 2.5 ml of 1% gelatin solution, 8 ml of 0.5% solution of the dyestuff, and 7 ml of hydrochloric acid. Dilute to 50 ml and read at 520 nm.

[17] V. A. Nazarenko, G. V. Flyantikova, and T. N. Selyutina, *Z. Anal. Khim.* **27**, 2369–2376 (1972).
[18] M. K. Akhmedli, D. G. Gasanov, and R. A. Alieva, *Uch. Zap. Azerb. Univ., Ser. Khim. Nauk* **1964** (1), 75–81; *Azerb. Khim. Zh.* **1965** (4), 92–95; cf. M. Finkelsteinaite, J. Garenite, and R. Piletskaite, *Nauchn. Trudy Vyssh. Zaved. Lit. SSR, Khim. Khim. Tekhnol.* **11**, 33–37 (1970).
[19] G. Popa and I Paralescu, *Rev. Chim.* (Bucharest) **21**, 43–45 (1970).

2,2′-BIPYRIDYL

Germanium as the catecholate complexes with 2,2′-bipyridyl and the complex is extractable with chloroform.[20] The extract conforms to Beer's law for 4–36 μg of germanium per ml of extract.

The nitrocatecholate of germanium complexes with ferrous 2,2′-bipyridyl.[21] This is extractable at pH 3–4.5 with 1,2-dichloroethane and then conforms to Beer's law for 3–13 m*M*. EDTA at 150-fold excess to the germanium masks ferric ion, aluminum, cadmium, lead, zinc, magnesium, and calcium.

Procedures

Ores As the Catecholate. To a sample of the solution containing 0.1–0.9 mg of germanium, add 5 ml of 10% solution of catechol, 0.1 gram of 2,2′-bipyridyl, and 12 ml of a chloroacetate buffer solution for pH 2.4–2.6. Dilute to 25 ml and extract with 6 ml, 6 ml, and 6 ml of chloroform. Filter the combined extracts, dilute to 25 ml with chloroform, and read.

As the Nitrocatecholate. To the sample solution, add 2 ml of 3 m*M* 4-nitrocatechol, 2 ml of 10% solution of hydroxylamine hydrochloride, 2 ml of chloroacetate buffer solution for pH 4, and 2 ml of 3.7 m*M* ferrous 2,2′-bipyridyl. Dilute to 25 ml and shake with 10 ml of 1,2-dichloroethane for 4 minutes. Dry the organic layer with anhydrous sodium sulfate and read at 526 nm.

BRILLIANT GREEN

The complex of molybdogermanate with brilliant green (basic green 1) can be extracted with butyl acetate for reading at 620 nm. A ternary complex of germanium, 3,5-dinitrocatechol, and brilliant green can be extracted with carbon tetrachloride for reading at 625 nm. This conforms to Beer's law for 0.5–7 μg of germanium per ml.

Germanium is precipitated as a ternary complex with brilliant green and 2,3,4- or 3,4,5-trihydroxybenzoic acid.[22] The solution of this complex in acetone is read at 425 nm and conforms to Beer's law from 25 ng/ml to 1.4 μg/ml. There is interference by arsenic, nickel, niobium, hexavalent chromium, tellurium, tungsten, and molybdenum.

Procedures

Lead-Tin-Germanium Glasses.[23] Fuse 0.1 gram of sample at 1000° in platinum

[20]Sh. T. Talipov, R. Kh. Dzhiyanbaeva, and V. S. Anishkina, *Uzb. Khim. Zh.* **1962** (5), 25–28.

[21]Shizehiko Hayashi, Keiya Kotsugi, Ryuichiro Hayashi, Mikio Hara, and Koji Hirakawa, *Jap. Anal.* **20**, 1152–1157 (1971).

[22]A. I. Busev, N. E. Dzotsenidze, and V. K. Akimov, *Zh. Anal. Khim.* **24**, 556–560 (1969).

[23]L. I. Ganago, I. A. Prostak, and I. A. Afonskaya, *Izv. Akad. Nauk Beloruss. SSR, Ser. Khim. Nauk*, **1971** (3), 118–120.

with 0.8 gram of sodium carbonate. Leach with 15 ml of water and neutralize to methyl red with hydrochloric acid. Boil off the carbon dioxide. Add 15 ml of hydrochloric acid, heat to dissolve, and dilute to 100 ml.

Mix an aliquot containing 0.25-3 μg of germanium, 1.5 ml of 1:10 nitric acid, and 1 ml of 0.02 *M* ammonium molybdate. Dilute to 10 ml and set aside for 5 minutes. Mask excess molybdate with 2 ml of 0.45% oxalic acid solution and add 0.7 ml of nitric acid. Add 1.5 ml of 0.274 m*M* solution of brilliant green. Extract with 5 ml of butyl acetate and wash the organic layer with 5 ml and 5 ml of 1:30 nitric acid. Combine the extracts, add 5 ml of acetone, and read at 620 nm.

Arsenic.[24] Dissolve 2 grams of sample in 15 ml of nitric acid and 1 ml of sulfuric acid. Concentrate to 3 ml. Add 10 ml of water and 30 ml of 9 *N* hydrochloric acid. Then extract for 2 minutes with 10 ml and 10 ml of carbon tetrachloride. Wash the combined extracts with 10 ml of 9 *N* hydrochloric acid, then back-extract the germanium with 10 ml of water. Complete as for slags, concentrates, and ash [see below] from "Add a drop of 0.1% solution...."

Slags, Concentrates, and Ash.[25] To 1 gram of sample in platinum, add 5 ml each of nitric, sulfuric, and hydrofluoric acid. Evaporate at 100°, then ignite to dryness. If sulfur or carbon is present, add 2.5 ml of 40% solution of magnesium nitrate hexahydrate and ignite at 850°. If necessary, repeat this step. Dissolve the residue in 60 ml of 9 *N* hydrochloric acid. Extract with 10 ml and 10 ml of carbon tetrachloride. Wash the combined extracts with 10 ml and 10 ml of 9 *N* hydrochloric acid. Then back-extract germanium with 10 ml of water. Add a drop of 0.1% solution of *p*-nitrophenol and add 0.4% sodium hydroxide solution to a yellow end point. Add 1 ml of 1:5 hydrochloric acid, 1.5 ml of 0.02% solution of 3,5-dinitrocatechol in ethanol, and 6 ml of 0.05% solution of brilliant green. Shake and separate the organic phase. Extract the aqueous phase with 10 ml of carbon tetrachloride and add to the organic phase. Add 5 ml of ethanol, dilute to 50 ml with carbon tetrachloride, and read at 625 nm.

BROMOPYROGALLOL RED

This reagent as a 3:1 complex with germanium conforms to Beer's law for 0.2–15 μg/ml.[26] The selectivity and sensitivity are better than with gallein or catechol violet.

The binary complex also forms a ternary complex with diphenylguanidine which can be extracted with the higher alcohols.[27]

Procedures

Cuprous Materials. Dissolve 1 gram of sample in 2 ml each of nitric, phos-

[24] Jadwiga Chwastowska and Elzbieta Grzegrzolka, *Chem. Anal.* (Warsaw) **23**, 1065–1070 (1975).
[25] V. A. Nazarenko, N. V. Lebedeva, and L. I. Vinarova, *Zh. Anal. Khim.* **27**, 128–133 (1972).
[26] G. Popa and I. Paralescu, *Talenta* **15**, 272–274 (1968).
[27] V. A. Nazarenko and N. I. Makrinich, *Zh. Anal. Khim*, **24**, 1694–1698 (1969).

phoric, and hydrofluoric acids. Evaporate to dryness, add 2 ml of water, and again take to dryness. Dissolve the residue in 2 ml of water and add 10 ml of hydrochloric acid. Add 0.5 ml of 10% hydroxylamine hydrochloride solution and extract the germanium with 10 ml and 5 ml of carbon tetrachloride. Combine the organic layers and reextract the germanium with 10 ml of water. Adjust to pH 2–3 with 0.1 *N* hydrochloric acid. Add 5 ml of 0.5% gelatin solution and 15 ml of 0.025% solution of bromopyrogallol red in 50% ethanol. Dilute to 50 ml and set aside for 15 minutes. Read at 550 nm against a blank.

Poly(ethylene) Terephthalate.[28] Add 5 ml of 4% sodium hydroxide solution to a 1-gram sample, evaporate to dryness, ash, and heat at 650° for 1 hour. Dissolve the residue in 20 ml of water and mix with 50 ml of hydrochloric acid. Extract the germanium chloride with 10 ml, 10 ml, and 10 ml of carbon tetrachloride. Reextract the germanium from the combined organic layers with 10 ml and 10 ml of water. Dilute the aqueous extracts to 25 ml and take a 10-ml aliquot. Add 15 ml of sodium citrate buffer solution for pH 2.6, 5 ml of 0.5% solution of gelatin, and 15 ml of 0.02% solution of bromopyrogallol red in ethanol. Dilute to 50 ml, set aside for 15 minutes, and read at 570 nm.

CATECHOL VIOLET

In aqueous solution at pH 4.8, germanium forms a complex with catechol violet that is purple in the absence of gelatin and has a maximum absorption at 555 nm.[29] In the presence of gelatin at pH 3.4–3.6 the complex is green and has a maximum absorption at 650 nm. Beer's law is applicable at 0.15–0.5 m*M* in the absence of gelatin but only at 0.3–0.5 m*M* in its presence. There is interference by pentavalent vanadium, hexavalent molybdenum, gallium, thallic ion, antimony, stannous ion, and ferric ion. The last can be masked with tartrate.

When a gaseous stream containing the chlorides of germanium, arsenic, antimony, tin, and selenium is passed through kerosene, the germanium chloride is absorbed.[30] Dilute an aliquot of sample containing 8–75 μg of germanium to 12.5 ml with kerosene. Add 12.5 ml of 0.032% solution of catechol violet in butanol, wait for 30 seconds, and read at 610 nm against 1 : 1 kerosene-butanol.

A ternary complex is formed by germanium, catechol violet, and cetyltrimethylammonium bromide in 0.5 *N* hydrochloric acid.[31] It will read 0.1–1 ppm of germanium at 655 nm. Interferences require separation of the germanium by the usual extraction technic.

The color develops instantly in a strongly acid medium but starts to fade.[32] It

[28]Horst Hoyme, Antoaneta Seganowa, and Heinz Zimmermann, *Faserforsch. TextTech.* **22**, 419–421 (1971).
[29]V. A. Nazarenko and L. I. Vinarova, *Zh. Anal. Khim.* **18**, 1217–1221 (1963); A. K. Babko and G. I. Gridchina, *Tr. Kom. Anal. Khim.* **17**, 67–72 (1969).
[30]N. A. Agrinskaya, V. A. Golosnitskaya, and E. V. Kovalenko, *Zavod. Lab.* **33**, 923–924 (1967).
[31]C. L. Leong, *Talenta* **18**, 845–848 (1971).
[32]N. A. Agrinskaya, V. A. Golosnitskaya, and E. V. Kovalenko, *Tr. Novocherk. Politekh. Inst.* **220** (20), 1–6 (1969).

can be stabilized for 6 hours in hydrochloric acid by adding gelatin and warming to 30°. The optimum media are 3.2–6 N sulfuric acid or 0.9–2.5 N hydrochloric acid. The reagent must be twelvefold. Beer's law holds for 0.5–10 μg of germanium per ml read at 610–615 nm. Only half as much hexavalent molybdenum or trivalent antimony as the germanium content is tolerated. Results agree with those obtained with phenylfluorone.

Procedure. Adjust the sample solution to 25 ml in 1 : 1 hydrochloric acid and extract with 5 ml, 5 ml, and 5 ml of 30% solution of tributyl phosphate in kerosene.[33] Evaporate the kerosene from the combined extracts at 100°. Take up the residue with 30 ml of 2-methoxyethanol and 5 ml of hydrochloric acid. Pass this solution through a column of Dowex 1-X8 ion exchange resin. Elute with 25 ml of 6 : 1 : 3 2-methoxyethanol–hydrochloric acid–tributyl phosphate. Mix a 10-ml aliquot with 5 ml of a saturated solution of hexamine and 5 ml of 0.03% solution of catechol violet. Read at 555 nm.

Coal and Coke.[34] Mix 1 gram of ash with 5 ml of phosphoric acid, 10 ml of hydrofluoric acid, and 5 ml of nitric acid. Evaporate to a syrup free from hydrofluoric acid. If decomposition appears to be incomplete, add 5 ml more of hydrofluoric acid and repeat the evaporation. Take up in 15 ml of water and add 20 ml of hydrochloric acid. Dilute to 50 ml and take an aliquot to yield about 10 μg of germanium. Extract with 10 ml and 6 ml of carbon tetrachloride. To the combined extracts, add 5 ml of propanol containing 0.5 ml of 1 : 1 hydrochloric acid per 500 ml. Add 2 ml of 0.1% solution of catechol violet in propanol. Dilute to 25 ml with carbon tetrachloride and read at 555 nm against a reagent blank.

7,8-DIHYDROXY-2,4-DIMETHYL-1-BENZOPYRYLIUM CHLORIDE

For determination of germanium with this reagent, the error is reduced by a factor of 2 or 3 by differential photometry.[35]

Procedure. *Industrial Concentrates.* For 10% germanium, use 0.1 gram of sample varying to 1 gram if about 0.01% is present. Heat with 5 ml of nitric acid, then add 8 ml of phosphoric acid and complete the dissolution with 5 ml of hydrofluoric acid. Evaporate almost to dryness and dilute to 10 ml with water. To an aliquot containing 0.1–0.7 mg of germanium, add 3 volumes of hydrochloric acid. Extract the germanium chloride with 15 ml and 10 ml of carbon tetrachloride. Wash the combined extracts with 10 ml of 9 N hydrochloric acid. Reextract the germanium with 8 ml and 7 ml of water and filter the combined aqueous extracts. Neutralize to phenolphthalein with 0.8% sodium hydroxide solution. Add 1.5 ml of 0.1 N

[33] W. Koch and J. Korkish, *Mikrochim. Acta* **1973**, 101–112.
[34] R. M. Saginashvii and V. I. Petrashen', *Zavod. Lab.* **32**, 661–663 (1966).
[35] V. F. Vozisova and V. N. Podchainova, *Tr. Ural. Politekh. Inst.* **1967** (163), 5–9.

hydrochloric acid and 10 ml of 0.1% solution of the reagent. Dilute to 50 ml and read against a standard containing 0.09 μg of germanium in the 50-ml volume.

DIMETHYLAMINOPHENYLFLUORONE

This reagent, which is 2,6,7-trihydroxy-9-dimethylaminophenylfluorone, forms a 3 : 1 complex with germanium in *N* hydrochloric acid.[36] The maximum absorptions of the reagent and of the complex are at 460–475 and 500 nm, respectively. Niobium, tantalum, tungsten, and gallium interfere. The colloidal complex is more stable in ethanol than in aqueous media. For maximum sensitivity, a sixtyfold excess of reagent is essential.[37] Conditions of addition of reagents and times of standing must be standardized. The extinction is temperature dependent.

Procedure. *Zinc.* Adjust the solution of sample in hydrochloric acid to 9 *N*. Extract the germanium with 10 ml and 10 ml of carbon tetrachloride. Combine the organic extracts and reextract the germanium with 20 ml of water. Dilute the aqueous extract to 25 ml and take a 5-ml aliquot. Add 2.5 ml of hydrochloric acid and 5 ml of 0.03% solution of the reagent in ethanol. Dilute to 25 ml, set aside for 10 minutes, and read at 470 nm against a blank.

GERMANOMOLYBDATE BLUE

The familiar molybdenum blue reaction is applicable to germanium but requires standardization of detail.[38] Variability in results by different laboratories is attributed to differences in the time interval between the addition of the molybdate and addition of the reducing agent.[39] Arsenic, phosphorus, silicon, and tungsten interfere seriously.

Addition of acetone stabilizes the β-form of the heteropoly acid, which would otherwise change with time to the less intensely colored α-form.[40] It also increases the rate of formation of the heteropoly acid.

For analysis of biological material, digest a sample with sulfuric and nitric acid, distill as germanium chloride, hydrolyze to the dioxide, complex with ammonium molybdate, and reduce with ferrous ammonium sulfate to the blue complex which is read at 880 nm.[41] The heteropolymolybdate of germanium is also produced by

[36] Chun Khvan Kim, En Khe Tsoi, and Yan Sep Kim, *J. Anal. Chem. Korea* **1**, 6–17 (1962); cf. N. F. Kazarinova and N. L. Vasil'eva, *Zh. Anal. Khim.* **13**, 677–681 (1958).
[37] A. Campe and J. Hoste, *Talenta* **8**, 453–460 (1961).
[38] Z. F. Shakhova, R. K. Motorkina, and N. N. Mal'tseva, *Zh. Anal. Khim.* **12**, 95–99 (1957); Teru Yuasa, *J. Chem. Soc. Jap.* **80**, 1201–1202 (1959); *Tokyo Kogyo Shikensho Hokoku* **54**, 109–115 (1959); A. Halasz and E. Pungor, *Talenta* **18**, 557–575 (1971); A. Halasz, E. Pungor, and K. Polyak, *ibid.* 577–586.
[39] Elwood R. Shaw, *U.S. Dept. Comm., Off. Tech. Serv., PB Rep.* **127,584**, 12 pp (1957).
[40] R. A. Chalmers and A. G. Sinclair, *Anal. Chim. Acta* **33**, 384–390 (1965).
[41] George Rosenfeld, *Anal. Biochem.* **1**, 469–477 (1960).

reducing with $N N'$-dimethyl-2-pyrazoline-3,4-dicarboxamide, to give absorption with a maximum at 810 nm.[42]

If the sample solution for methyl isobutyl ketone extraction contains phosphoric acid, precipitate it with zirconium oxychloride. Avoid interference by arsenate by reducing it to arsenite. Mask aluminum with fluoride ion. Then mix 1 ml of slightly acid sample solution containing 5–20 μg of germanium with 0.1 ml of sulfuric acid, 0.1 ml of saturated solution of cupferron, and 0.5 ml of methyl isobutyl ketone.[43] Shake. Break any emulsion with a drop of saturated sodium perchlorate solution. Extract with 0.7 ml and 0.7 ml of methyl isobutyl ketone. Evaporate the combined extracts to dryness, add 1 ml of nitric acid, and ash. Dissolve in acid for development as germanomolybdate blue.

Procedure. As ammonium molybdate solution, dissolve 5 grams in warm water, cool, add 2.8 ml of sulfuric acid, and dilute to 100 ml.[44] As reductant, dissolve 5 grams of ferrous ammonium sulfate hexahydrate in 250 ml of water containing 1 ml of 1:8 sulfuric acid. Just before using, mix 5 ml of this 2% solution with 20 ml of 1:8 sulfuric acid to give a 0.4% solution.

Mix a neutral sample containing 0.1–0.3 mg of germanium with 2 ml of 1:8 sulfuric acid and dilute to 36 ml. Add 8 ml of the ammonium molybdate reagent and mix. Add 50 ml of the 0.4% reductant, 1 ± 0.2 minute later, and dilute to 100 ml. Set aside for 30 minutes and read at 825 nm.

Alternatively,[45] mix a sample solution containing less than 0.1 mg of germanium with 10 ml of 0.13% solution of ammonium molybdate tetrahydrate. Add 2.75 ml of 1:11 hydrochloric acid and dilute to 25 ml. At once add 1 ml of 0.3% ascorbic acid solution, mix, and heat at 100° for 25 minutes. Cool in cold water, let stand for 6 minutes, and read at 825 nm.

As an alternative form of reagent, dissolve 8.15 grams of ammonium molybdate in 60 ml of water.[46] Mix 25 ml of this solution with 12.5 ml of hydrochloric acid and dilute to 50 ml. Reduce by shaking vigorously with 10 ml of mercury for 5 minutes and filter to give a reddish-brown color. Mix 30 ml of the initial unreduced molybdate solution, 56 ml of sulfuric acid, and 40 ml of the reduced molybdate solution. Dilute to 200 ml. This contains hexavalent and pentavalent molybdenum in a 3:2 ratio.

By Metal Reduction.[47] Mix a neutral sample solution containing up to 0.25 mg of germanium with 15 ml of acetic acid and dilute to 40 ml. Add 1 ml of 10% solution of ammonium molybdate. After 1 minute and within 10 minutes, add 1 ml of a reducing solution containing 15% of sodium thiosulfate, 1% of sodium sulfite heptahydrate, and 0.5% of metol, which is methylaminophenol sulfate. Set aside for 30 minutes, dilute to 50 ml, and read at 827 nm.

[42]Yu. G. Zhukovskii, *Zh. Anal. Khim.* **19**, 1361–1365 (1964).
[43]I. Paschoal Senise and Lilia Sant'Agostino, *Mikrochim. Acta* **1956**, 1445–1455.
[44]Elwood R. Shaw and James F. Corwin, *Anal. Chem.* **30**, 1314–1516 (1958).
[45]Joseph K. Samuels III and D. F. Boltz, *Microchem. J.* **15**, 638–641 (1970).
[46]F. Lucena-Conde and L. Prat, *15th International Congress of Pure and Applied Chemistry*, Lisbon (1959).
[47]Z. Rezac and L. Ruzickova, *Collect. Czech. Chem. Commun.* **25**, 2242–2245 (1960).

By 2-Ethylhexanol Extraction.[48] Mix 2 ml of sample with 2 ml of 1:5 hydrochloric acid and 2 ml of 5% solution of ammonium molybdate. Set aside for 15 minutes and extract the phosphorus by shaking for 1 minute with 10 ml of isobutyl acetate. Adjust the aqueous phase to pH 0.4 by adding 1 ml of perchloric acid, and shake for 30 seconds with 10 ml of 2-ethylhexanol to extract the germanium. Reextract the germanium by shaking for 30 seconds with 6 ml of water. To the aqueous extract, add 1 ml of perchloric acid and 1 ml of a mixed solution containing 0.2% of 4-amino-3-hydroxynaphthalene-1-sulfonic acid, 2.4% of sodium sulfite heptahydrate, and 12% of sodium metabisulfite. Set aside for 1 hour and read at 690 nm.

Calcium and Rare Earth Germanates and Silicogermanates.[49] Add 3 ml of hydrofluoric acid and 1 ml of 1:1 sulfuric acid to 0.05 gram of sample. Evaporate to dryness. Fuse the residue with 0.5 gram of sodium carbonate and 0.5 gram of borax. Dissolve the melt in water adding 7 ml of hydrochloric acid and dilute to 500 ml. Dilute an appropriate aliquot to 70 ml. Add 10 ml of acetone as a stabilizer, 2 ml of 1:2 hydrochloric acid, and 5 ml of 5% solution of ammonium molybdate. Let stand for 15 minutes; then add 5 ml of reducing solution containing 1% of ascorbic acid and 5% of citric acid. Dilute to 100 ml and set aside for 75 minutes. Read at 690 nm.

MOLYBDOGERMANIC ACID

This complex acid can be read directly in the ultraviolet region, extracted for ultraviolet reading or decomposed and read as the molybdate.[50] To avoid instability, a large excess of molybdate must be present at the time the yellow acid is formed. The complex is 12:1 molybdenum-germanium. For the maximum absorption, the ratio of molybdenum to germanium in the solution must be at least 36:1, to provide the essential excess of molybdate. Color is fully developed at pH 1.5 in 30 minutes.

Procedures

Direct Ultraviolet Method. To a sample containing up to 0.125 mg of germanium, add 10 ml of 0.13% ammonium molybdate solution. Add 1 ml of *N* hydrochloric acid and dilute to 25 ml. The pH should approximate 1.5. Let stand for 30 minutes and read at 315 nm against a blank.

Ultraviolet Method by Extraction. Mix a neutral sample containing not more than 0.1 mg of germanium with 10 ml of 0.13% ammonium molybdate solution and 1 ml of *N* hydrochloric acid. Set aside for 30 minutes. Add 10 ml of 1:4

[48] J. Paul, *Anal. Chim. Acta* **35**, 200–205 (1966).
[49] M. M. Piryutko and T. G. Kostyreva, *Zavod. Lab.* **36**, 276 (1970).
[50] Robert Jakubiec and D. F. Boltz, *Anal. Chem.* **41**, 78–81 (1969); cf. R. E. Kitson and M. G. Mellon, *Ind. Eng. Chem., Anal. Ed.* **16**, 128–130 (1944).

pentanol-ether and 40 ml of 1.4 *N* hydrochloric acid saturated with ether. Shake for 1 minute and discard the aqueous phase. Wash with 30 ml of 1.4 *N* hydrochloric acid saturated with ether. Dilute the organic phase to 25 ml with mixed solvent and read at 295 nm against the mixed solvent.

Indirect Ultraviolet Method. As a buffer solution for pH 9.3, dissolve 53.5 grams of ammonium chloride in water, add 70 ml of ammonium hydroxide, and dilute to 1 liter.

Mix a sample solution containing not more than 25 μg of germanium with 1 ml of *N* hydrochloric acid and 5 ml of 0.13% solution of ammonium molybdate. Set aside for 30 minutes. Add 10 ml of 1:4 pentanol-ether and 40 ml of 1.4 *N* hydrochloric acid saturated with ether. Shake for 1 minute and discard the aqueous phase. Wash the organic layer with 30 ml of 1.4 *N* hydrochloric acid saturated with ether. Shake the organic phase with 10 ml and 10 ml of the buffer solution at pH 9.3 for 30 seconds. Dilute the combined buffer extracts to 25 ml with the buffer solution and read at 226 nm against the buffer solution. Correct for a blank.

Copper-Nickel-Germanium Alloy.[51a] Dissolve 0.1 gram of sample in 3 ml of nitric acid and 3 ml of hydrofluoric acid and dilute to 25 ml. Mix a 5-ml aliquot with 35 ml of hydrochloric acid and extract the germanium by shaking for 3 minutes with 25 ml of chloroform. Separate the extract, add 10 ml of glacial acetic acid, and extract the germanium by shaking for 3 minutes with 20 ml of water. To the aqueous layer add 2.5 ml of 8% solution of ammonium molybdate in acetic acid. Dilute to 100 ml and read at 420 nm.

Organogermanium Compounds.[51b] Mix 5 mg of sample with 4 grams of sucrose and burn in oxygen in a 500-ml oxygen flask coated inside with a polychlorofluoroethylene. Absorb the combustion products in 10 ml of 0.2 *N* sodium hydroxide. Set aside for 10 minutes, then mix with 5 ml of 10% solution of ammonium heptamolybdate tetrahydrate, 3 ml of 2 *N* sulfuric acid, and 12 ml of acetone. Dilute to 50 ml, set aside for 15 minutes and read the β-molybdogermanic acid at 430 nm.

OXYHYDROQUINONE PINK

This reagent is 2-(2,6,7-trihydroxy-3-oxoxanthen-9-yl)benzene sulfonic acid.[52] It obeys Beer's law for 0.09–1.44 μg of germanium per ml. The method tolerates an equal amount of lead and sevenfold hexavalent chromium.

Procedure. *Coal Ash.* Dissolve 1 gram of sample with 1 ml of nitric acid, 1 ml of 1:1 sulfuric acid, and 5 ml of hydrofluoric acid. Evaporate the hydrofluoric acid at 100°; then heat until sulfur trioxide is no longer given off. Dissolve the residue in 2

[51a]Masayoshi Huseya, *Jap. Anal.* **12**, 555–559 (1963).
[51b]Mary M. Masson, *Mikrochim. Acta* **I**, 385–390 (1976).
[52]M. Finkelsteinate and J. Burskiene, *Nauch. Tr. Vyssh. Ucheb. Zaved. Lit. SSR, Khim, Khim. Tekhnol.* **13**, 47–52 (1971).

ml of water. Add 5 ml of water, neutralize with 5.6% potassium hydroxide solution, filter, and wash. Add 2 ml of 5% EDTA solution and 2 drops of 30% hydrogen peroxide. Adjust to pH 4 with 5.6% potassium hydroxide solution and add 4 ml of 0.25 m*M* reagent. Dilute to 50 ml and read at 540 nm.

An alternative technic is to dissolve the sample with nitric, phosphoric, and hydrofluoric acids, make the solution 9 *N* with hydrochloric acid, extract with carbon tetrachloride, reextract with water, and determine as above.

PHENYLFLUORONE

Phenylfluorone, which is 2,6,7-trihydroxy-9-phenyl-3*H*-xanthene-3-one, complexes with germanium. Because of low sensitivity, concentration in an organic solvent is desirable. Excess reagent in the solvent will give a high background color.[53] Gum arabic is a suitable colloid, EDTA will mask several elements, and Beer's law is obeyed for 0.1–0.6 ppm of germanium. Molybdenum, bismuth, antimony, tin and iron interfere, but interference is avoided by distillation or by extraction of germanium tetrachloride from 9 *N* hydrochloric acid. The reaction product of germanium and phenylfluorone is water insoluble and this property can be used as a method for concentation. Acetone or dimethylformamide may be used to dissolve the precipitate.[54]

The 2-nitro-, 4-nitro-, 2,4-dinitro-, and disulfo- derivatives of phenylfluorone are more sensitive than is the parent.[55]

The method is applicable for a sample of 0.1–0.2 gram down to 0.005% germanium and to a 1-gram sample for lesser amounts.[56] The reagent is generally applicable to the products of controlled combustion of organic materials,[57a] to coal ash, and to ores. As applied to sulfite ores by acid decomposition results are low due to germanium being adsorbed on silica during acid digestion.[57b]

In a solution containing arsenic, germanium, phosphorus, and silicon, the germanium can first be developed with phenylfluorone and read at 525 nm. Extraction of the germanium complex with isoamyl alcohol then yields a sample for the other ingredients.[58]

Before extraction of the germanium from 9 *N* hydrochloric acid with carbon tetrachloride, decompose sulfide ores with nitric and phosphoric acids, silicates, and coal ash with hydrofluoric acid and 1 : 1 sulfuric acid, iron ores with 1 : 1 hydrochloric acid.[59]

For determination of germanium in organogermanium compounds, burn to the

[53] J. D. Burton and J. P. Riley, *Mikrochim. Acta* **1959**, 586–591; cf. M. V. Kanyukova, *Tr. Vses. Magadan. Nauch.-Issled. Inst. Min. Tsvet. Met. SSSR* **1957** (19), 1–13.

[54] Toshio Suzuki and Takeshi Sotobayashi, *Jap. Anal.* **12**, 376–380 (1963).

[55] V. A. Nazarenko and N. V. Lebedeva, *Zavod. Lab.* **25**, 899–903 (1959).

[56] L. B. Ginzburg, S. D. Gur'ev, and A. P. Shibarenkova, *Sb. Nauch. Tr. Gos. Nauch.-Issled. Inst. Tsvet. Met.* **1955** (10), 378–386.

[57a] V. A. Nazarenko, N. V. Levedeva, and R. N. Ravitskaya, *Zavod. Lab.* **24**, 9–13 (1958).

[57b] Hiroshi Nishida, *Jap. Anal.* **5**, 17–20 (1956).

[58] F. A. Sorrentino and J. Paul, *Microchem. J.* **15**, 446–551 (1970).

[59] Ya. E. Yudovich, *Mater. Geol. Polezn. Iskop. Yakutsk. SSSR* **1961** (7), 124–127.

oxide by the Schöniger method and absorb in dilute sodium hydroxide solution.[60] Then read by phenylfluorone or rezarson. Halogen may be present in the sample. The reagent is applicable to solutions of technical germanium oxide.[61]

A large addition of phosphoric acid will mask substantial amounts of iron in production solutions.[62a]

Superconducting films are sputtered films approximating three moles of niobium with one mole of germanium.[62b] Dissolve the film by heating with a mixture of sulfuric, nitric, and hydrofluoric acids. Heat to copious fumes, take up with water, and reheat to copious fumes. Make 9 *N* with hydrochloric acid and extract the germanium with carbon tetrachloride. Back-extract the germanium from the organic layer with water. Add hydrochloric acid and 1% gelatin solution to the aqueous layer. Add phenylfluorone reagent, set aside for 1 hour, and read at 504 nm.

Procedure. As a buffer solution for pH 5, dissolve 225 grams of sodium acetate trihydrate in 170 ml of water and filter. Add 125 ml of acetic acid and dilute to 500 ml. As the reagent, dissolve 0.05 gram of phenylfluorone in 50 ml of methanol. Add 1 ml of hydrochloric acid and dilute to 500 ml with methanol.

Extract a sample solution containing up to 1.5 μg of germanium in 15 ml of 9*N* hydrochloric acid.[63] Shake for 1 minute with 5 ml of carbon tetrachloride. Wash the extract for 1 minute with 5 ml of buffer solution for pH 5 and 5 ml of water. Add 0.75 ml of 1 : 1 sulfuric acid and 5 ml of reagent to the extract. Mix and put aside for 5 minutes. Add 10 ml of 1 : 9 hydrochloric acid and shake for 15 seconds. Slowly drain off the carbon tetrachloride layer, retaining a precipitate at the interface. Decant the aqueous layer. Add 5 ml of 80% acetone and shake to dissolve the precipitate. Read at 479 nm.

Alternatively,[64] as a reagent, dissolve 0.15 gram of phenylfluorone in ethanol, add 20.8 ml of hydrochloric acid, and dilute to 250 ml with ethanol.

Adjust a sample containing 0.2–10 μg of germanium to 9 *N* hydrochloric acid in a volume of 20 ml. Shake for 2 minutes with 10 ml of carbon tetrachloride to extract the germanium chloride. To a 5-ml aliquot of the extract, add 3 ml of the reagent and dilute to 10 ml with ethanol. Let stand for 5 minutes and read at 507 nm against a blank.

Extraction with Mesityl Oxide.[65] There is strong interference with this technic by silver, stannous ion, platinum, rhenium, mercuric ion, ceric ion, tungstate, and molybdate. As the color reagent, dissolve 0.1 gram of phenylfluorone in 50 ml of methanol and 1 ml of hydrochloric acid and dilute to 500 ml with methanol.

[60] I. V. Dudova and G. F. Dikaya, *Zh. Anal. Khim.* **23**, 784–786 (1968).
[61] E. Krouzek and V. Patrovsky, *Chem. Prum.* **11**, 24–26 (1961).
[62a] N. V. Lebedeva, *Zavod. Lab.* **30**, 1331 (1964).
[62b] J. B. Bodkin and D. A. Rogowski, *Analyst* **102**, 110–113 (1977).
[63] C. L. Luke, *Chemist-Analyst* **54**, 109–110 (1965); Andres Huesca Moreno and Oscar Hector Saggese, *Publ. Inst. Invest. Microquim., Univ. Nacl. Litoral* (Rosario, Arg.) **21**, 35–8 (1954).
[64] Fumio Yamauchi and Akira Murata, *Jap. Anal.* **9**, 959–961 (1960).
[65] P. V. Dhond and S. M. Khopkar, *Anal. Chim. Acta* **59**, 161–164 (1972).

Adjust 25 ml of sample containing 3–90 μg of germanium to 5 *N* with hydrochloric acid and to contain 8.5% of lithium chloride. Shake with 10 ml of mesityl oxide for 2 minutes. Extract the germanium from the organic phase with 10 ml and 10 ml of water. To the combined aqueous extracts, add 5 ml of 1% gum acacia solution and 5 ml of hydrochloric acid. Add 5 ml of phenylfluorone reagent and dilute to 50 ml. Let stand 30 minutes and read at 510 nm.

Extraction with Benzyl Alcohol.[66] Adjust a sample containing 1.25–12.5 μg of germanium to 25 ml in 0.5 *N* hydrochloric acid. Add 5 ml of 0.03% solution of phenylfluorone and set aside for 5 minutes. Extract with 10 ml, 5 ml, and 5 ml of benzyl alcohol. Dilute the combined extracts to 25 ml with ethanol and read at 505 nm against a blank.

Separation by Distillation. Transfer 25 ml of sample solution containing up to 1 mg of germanium to a distilling flask connected by a ground-glass joint to a water-cooled condenser. Add 25 ml of hydrochloric acid, or less if the sample already contains that acid, making the final solution 6 *N* with hydrochloric acid. Distill at about 2 ml/minute, collecting 25 ml of distillate in an ice-cooled receiver. Dilute to 100 ml.

Adjust an aliquot containing up to 20 μg of germanium to contain the equivalent of 2 ml of hydrochloric acid. Add 1 ml of 1% gelatin solution and 1 ml of 0.08% ethanolic phenylfluorone. Dilute to 25 ml and set aside for 30 minutes. Read at 530 nm.

Zirconium. As the color reagent, dissolve 0.05 gram of phenylfluorone in methanol, add 1 ml of hydrochloric acid, and dilute to 500 ml with methanol. As a buffer solution for pH 5, dissolve 225 grams of sodium acetate trihydrate in water, add 120 ml of acetic acid, and dilute to 500 ml.

Fuse a sample expected to contain 0.04 mg of germanium with 5 times its weight of potassium bisulfate. Take up in hot water and dilute to a known volume. To an aliquot add 2 ml of 1 : 1 sulfuric acid and twice its volume of hydrochloric acid. Extract with 20 ml and 5 ml of carbon tetrachloride. Wash the combined extracts with 2 ml of 9.8 *N* hydrochloric acid. Extract the germanium from the organic phase with 10 ml of water and filter. Add 1.5 ml of 1 : 1 sulfuric acid, 10 ml of the buffer solution for pH 5, 3 ml of 1% solution of gum acacia, and 10 ml of phenylfluorone reagent. Let stand for 5 minutes, dilute to 50 ml with 1 : 9 hydrochloric acid, and read at 510 nm.

Solutions.[67] To 2 liters of sample containing 0.1–10 μg of germanium, add 4 ml of 20% ferric chloride solution. Add 20% sodium hydroxide solution to pH 9. Filter the precipitate and shake with 40 ml of 4 *M* octanoic acid in petroleum ether. Wash the extract of germanium and iron with 15 ml, 15 ml, and 15 ml of 10 *N* hydrochloric acid, to extract the iron and any titanium, tin, or antimony present. Back-extract the germanium with 6 ml and 6 ml of water. Add 2 ml of hydrochloric acid and 1 ml of 1% gelatin solution. Add 1.5 ml of 0.05% solution of phenylfluo-

[66] A. Hillebrant and J. Hoste, *ibid.* **18**, 569–574 (1958).
[67] A. M. Andrianov and V. E. Polaydan, *Zavod. Lab.* **40**, 1064 (1974).

rone in ethanolic 0.03 *N* hydrochloric acid and dilute to 25 ml. Set aside for 30 minutes and read at 530 nm.

Minerals.[68] Treat 0.2 gram of sample with 2 ml of nitric acid, 2 ml of hydrofluoric acid, and 2 ml of phosphoric acid. Evaporate to a syrup and add 20 ml of 9 *N* hydrochloric acid. Shake for 2 minutes with 10 ml of carbon tetrachloride. Wash the organic extract with 5 ml of 9 *N* hydrochloric acid, to remove interfering phosphoric acid. To 5 ml of the organic extract, add 2 ml of ethanol and 3 ml of 0.03% ethanolic phenylfluorone. Read at 530 nm.

Coal and Lignite.[69] Mix 1 gram of 80-100 mesh sample with 10 ml of sulfuric acid and 3 ml of nitric acid. Heat on a sand bath to fumes. Add 2 ml portions of nitric acid and continue to heat until decomposition is complete. Heat until nitric acid is driven off, cool, dissolve in water, and filter. Evaporate to about 25 ml. Ash the filter in platinum and treat with 1 ml of 1:3 sulfuric acid, 0.2 ml of nitric acid, and 1 ml of hydrofluoric acid. Add this to the previous solution and add 100 ml of hydrochloric acid. Distill, collecting 20 ml of distillate for less than 30 μg of germanium, taking 30 ml for larger amounts. Add 5 ml of 0.4% gum arabic solution to the distillate, then 5 ml of 0.03% solution of ethanolic phenylfluorone. Adjust the hydrochloric acid concentration to 1.2–2 *N* and dilute to 50 ml. Let stand overnight and read at 510 nm.

Coal.[70] Mix 2 grams of sample with 3 grams of anhydrous sodium sulfate and fuse at 900° for 1 hour. Leach the melt and filter the solution of sodium thiogermanate. Add 2 ml of 30% hydrogen peroxide and boil down to 10 ml. Add 3 drops of hydrochloric acid and cool. Prepare a 50×8 mm column of a 7:3 mixture of strongly acid cation exchange resin Amberlite IR-120 in hydrogen form, to adsorb ferric ion, and weakly basic anion exchange resin AN-2F in chloride form, to adsorb tin, antimony, arsenic, and molybdenum. Pass the sample solution through the column at 1 ml/minute and wash with water. Mix the percolate and washings with 2 ml of hydrochloric acid and 1 ml of 1% gelatin solution. Add 1.5 ml of a reagent containing 50 mg of phenylfluorone in 100 ml of ethanol and 0.5 ml of 1:1 hydrochloric acid. Dilute to 25 ml, set aside for 30 minutes, and read at 500 nm.

Alternatively,[71] mix 1 gram of sample with 30 ml of phosphoric acid and 25 grams of potassium dichromate. Heat to 140° and while heating, add 40 ml of sulfuric acid dropwise. When decomposition is complete, dissolve with a minimum volume of water. Add 2 ml of sulfurous acid and a fourfold volume of hydrochloric acid. Distill the germanium and develop by the technic given for coal and lignite. Coal can also be decomposed by boiling with sulfuric acid, copper sulfate, and potassium sulfate.[72]

As yet another ashing procedure for coal and coke, mix 1 gram of sample with

[68]N. I. Shuvalova, *Tr. Vses. Nauch.-Issled. Geol. Inst.* **117**, 73–74 (1964).
[69]Bunseki Nishida, *Jap. Anal.* **5**, 389–392 (1956).
[70]G. E. Salikova and N. N. Sevryukov, *Zavod. Lab.* **36**, 25–27 (1970).
[71]Mitsuru Ura, *J. Chem. Soc. Jap.* **78**, 316–320 (1957).
[72]N. N. Nurminskii, *Tr. Irkutsk. Polytekh. Inst.* **1970** (44), 148–155.

0.5 gram of lime and 6 ml of saturated solution of calcium nitrate.[73] Evaporate to dryness and burn off most of the carbon at 400–450°. Raise the temperature to about 800° until all carbon is gone. Add 5 ml of nitric acid and evaporate to dryness. Add 5 ml of hydrofluoric acid and again evaporate to dryness. Add 5 ml of hydrofluoric acid and 10 ml of phosphoric acid and evaporate to a syrup. Distill, and develop the germanium.

Coal Ash.[74] Moisten 1 gram of sample with water and add 5 ml of nitric acid, 5 ml of hydrofluoric acid, and 5 ml of phosphoric acid. When decomposition is complete, wash down the walls of the vessel and evaporate to a transparent syrup. Dilute to 50 ml and filter. To an aliquot containing up to 0.08 mg of germanium, add sufficient phosphoric acid to make a total of 7 ml. Add 4 ml of hydrochloric acid, 2 ml of 1% gelatin solution, and 3 ml of 0.05% ethanolic solution of phenylfluorone. Set aside for 40 minutes, dilute to 50 ml, and read at 530 nm against a reagent blank.

Sulfide Ores.[75] This technic is designed for molybdenites, enargite, sphalerite, chalcopyrite, bornite, or pyrite. Mix 1 gram of sample with 0.5 gram of calcium oxide. Add 8 ml of saturated solution of potassium nitrate and evaporate to dryness at 100°. Place in a furnace, gradually heat to 600°, and retain at that temperature for 1 hour. Cool, and add 6 ml of phosphoric acid. Heat on a water bath; then take to a syrupy state on a sand bath. Add 25 ml of water and 75 ml of hydrochloric acid. Complete as for zirconium samples from "Extract with 20 ml and 5 ml...."

Silicates.[76] As a reagent, dissolve 0.3 gram of phenylfluorone in 850 ml of ethanol, add 50 ml of 1:6 sulfuric acid, and dilute to 1 liter with ethanol.

Treat a 0.4-gram sample in platinum with 0.5 ml of 1:4 sulfuric acid, 0.5 ml of 1:4 nitric acid, and 4 ml of hydrofluoric acid. Evaporate to dryness, add 2 ml of hydrofluoric acid and evaporate to dryness. Dissolve in 2 ml of water, add 40 ml of 1:1 hydrochloric acid, and distill the germanium as shown earlier in this chapter. Dilute to a known volume and take an appropriate aliquot. Add 1 ml of sulfurous acid, 5 ml of phenylfluorone reagent, 2 ml of 0.5% gelatin solution, and sufficient 1:1 hydrochloric acid to make the concentration 1.3 *N*. Set aside for 30 minutes and read at 497 nm.

As an alternative for oxide and silicate ores, treat 1 gram of sample with 5 ml of phosphoric acid, 10 ml of hydrofluoric acid, and 5 ml of nitric acid. Evaporate to a syrup free from hydrofluoric acid. Proceed as above from "Dissolve in 2 ml of water."

Phosphorites.[77] Coprecipitation with ferric hydroxide provides separation from phosphorus and silicon. Fuse 0.5 gram of sample with 4 grams of sodium hydroxide

[73] V. A. Nazarenko, N. V. Lebedeva, and R. V. Ravitskaya, *Zavod. Lab.* **24**, 9–13 (1958).

[74] N. V. Lebedeva and R. A. Lyakh, *ibid.* **32**, 1333–1335 (1966); cf. M. A. Menkovskii and A. N. Aleksandrova, *ibid.* **25** 161 (1959); K. Pavlic, *Kem. Ind.* (Zagreb) **11**, 703–706 (1962).

[75] D. A. Dakhtrikyan, *Dokl. Akad. Nauk Arm. SSR* **28**, 213–216 (1959).

[76] V. Tyman and J. Vrbsky, *Rudy* **12** (6), 208–210.

[77] N. V. Lebedeva and L. I. Vinarova, *Zavod. Lab.* **39**, 798, 1056 (1973).

in an iron crucible. Dissolve in hot water, neutralize with 1:1 hydrochloric acid, and filter if necessary. Add 0.5 ml of 5% solution of ferric chloride and add ammonium hydroxide until precipitation is complete. Filter. Acidify the filtrate with 1:1 hydrochloric acid, add 0.5 ml of 5% ferric chloride solution, and precipitate with excess of ammonium hydroxide. Filter. Combine the precipitates in platinum and ignite for 10 minutes at 600°. Cool, moisten with water, add 5 ml of hydrofluoric acid, and evaporate to dryness. Add 5 ml of nitric acid and again evaporate to dryness. Add 5 ml each of hydrofluoric, nitric, and phosphoric acid. Evaporate to a syrup. Take up in 5 ml of water and add 15 ml of hydrochloric acid. Extract the germanium from this 9 *N* hydrochloric acid solution with 10 ml and 10 ml of carbon tetrachloride. Wash the combined extracts with 10 ml of 2% solution of hydroxylamine hydrochloride in *N* hydrochloric acid, then wash with 10 ml of 9 *N* hydrochloric acid. Extract the germanium with 10 ml, 10 ml, and 10 ml of water. To the combined extracts, add 5 ml of 0.02% solution of phenylfluorone in ethanol. Dilute to 50 ml and read at 530 nm.

Ores. Refer to Quinalizarin as reagent for preparation of the sample through "Evaporate to dryness but do not bake." Add 5 ml of *N* sulfuric acid to the residue and evaporate to dryness. Take up in 10 ml of *N* sulfuric acid by warming. Add 1 ml of 0.5% gelatin solution and 5 ml of 0.05% ethanolic phenylfluorone. Dilute to 25 ml with *N* sulfuric acid, let stand for 30 minutes, and read at 510 nm.

Lead-Zinc Ores.[78] Mix 1 gram of sample with 10 grams of Eschka mixture, cover with 3 grams of Eschka mixture, and heat at 800° for 2 hours. Moisten with 10 ml of water and add 1:1 sulfuric acid until reaction ceases. Add 10 ml of 1:1 sulfuric acid and dilute to 200 ml, heating if necessary to dissolve sulfates. If permanganate color is shown, add 30% hydrogen peroxide dropwise to decolorize. Add about 50 mg of iron as ferrous sulfate and heat to about 70°. Coprecipitate germanium and iron as hydroxides by addition of ammonium hydroxide. Filter the hydroxides and insoluble residue and wash with hot 1% solution of ammonium sulfate in 1:20 ammonium hydroxide.

Dissolve the hydroxides from the filter with 10 ml of 1:1 hydrochloric acid. Ash the paper and treat the ash with 5 ml of hydrofluoric acid and 1 ml of 1:1 sulfuric acid. Evaporate to sulfur trioxide fumes, add 2 grams of potassium persulfate, and fuse. Dissolve the fusion with 10 ml of 1:1 hydrochloric acid and add to the previous filtrate. Add 10 ml of hydrochloric acid and extract germanium chloride with 10 ml and 10 ml of carbon tetrachloride. Wash the combined organic extracts with 5 ml of 9 *N* hydrochloric acid. Reextract germanium with 10 ml and 5 ml of water. Filter the aqueous extract and dilute to 25 ml. To an appropriate aliquot add 1.5 ml of 1:1 hydrochloric acid and 10 ml of a buffer solution for pH 5. Add 5 ml of 0.5% gelatin solution and 10 ml of 0.01% methanolic phenylfluorone containing 0.2 ml of hydrochloric acid per 100 ml. Let stand for 10 minutes, dilute to 50 ml with 1:9 hydrochloric acid, and read at 510 nm.

Lead, Zinc, and Copper Ores.[79] Fuse a 0.5-gram sample with 3 grams of sodium

[78]T. L. Belopol'skaya, F. Ya. Saprykin, and I. O. Baranova, *Tr. Vses. Nauch. Issled. Geol. Inst.* **117**, 75–77 (1964).
[79]S. M. Milaev, *Sb. Tr. Vses. Nauch.-Issled. Gornomet. Inst. Tsvet. Met.* **1959** (5), 28–40.

peroxide in an iron crucible. Dissolve in 25 ml of 1:1 hydrochloric acid, filter, and wash. Precipitate as hydroxides with excess ammonium hydroxide, filter, and wash with 1:20 ammonium hydroxide. Dissolve the precipitate with 10 ml of 1:1 hydrochloric acid and add 10 ml of hydrochloric acid. Extract with 10 ml and 10 ml of carbon tetrachloride and wash the combined extracts with 10 ml of 9 *N* hydrochloric acid. Extract the germanium from the organic layer with 10 ml and 5 ml of water. To the combined extracts, add 2.5 ml of hydrochloric acid, 2.5 ml of 1% gelatin solution, and 2 ml of 0.05% solution of phenylfluorone. Dilute to 25 ml, set aside for 30 minutes, and read at 500 nm.

Manganese Ore.[80] Agitate a 5-gram sample in 30 ml of 1:4 sulfuric acid, with addition of increments of 30% hydrogen peroxide until no dark particles remain and the solution is colorless. Evaporate the solution and any colorless residue to dryness, add 10 ml of hydrofluoric acid, and again evaporate to dryness. Take up in 100 ml of 1:1 hydrochloric acid and add 0.5 gram of potassium permanganate. Distill at about 2 ml/minute until 45 ml has been collected in ice water. Add 10% sodium sulfite solution dropwise to reduce any free chlorine and dilute to 50 ml. The acidity at this point should be at least 4*N*. Mix a 10-ml aliquot with 30 ml of water, 2 ml of 1% gelatin solution, and 3 ml of 0.05% solution of phenylfluorone. Dilute to 50 ml, set aside for 30 minutes, and read at 510 nm.

Alternatively, after evaporation to dryness with hydrofluoric acid, take up in 40 ml of 9 *N* hydrochloric acid and extract germanium chloride with 20 ml and 20 ml of carbon tetrachloride. Reextract the germanium with 10 ml and 10 ml of water and adjust with hydrochloric acid to 4 *N* and 50 ml. Then proceed as before from "Mix a 10-ml aliquot...."

Quartz-Sulfide Ores.[81] Moisten 1 gram of sample in platinum with water, add 5 ml of nitric acid, and evaporate to dryness at 100°. Repeat this step twice more. Add to the dry residue 10 ml of 1:9 sulfuric acid and 15 ml of hydrofluoric acid. Heat almost to dryness, and if decomposition is not complete, add 5 ml more of hydrofluoric acid. Then heat to sulfur trioxide fumes. Wash down the sides of the dish with water and again evaporate to sulfur trioxide fumes. Repeat this step twice more. Dissolve the salts with 15 ml of water, dilute to 50 ml, and filter. Mix a 10-ml aliquot with 3.5 ml of hydrochloric acid and 3.5 ml of phosphoric acid. Add 1 ml of 1% gelatin solution and 2 ml of 0.05% solution of phenylfluorone. Dilute to 25 ml with 1:50 hydrochloric acid and set aside for 1 hour. Read at 530 nm against a blank.

Flue Gases.[82] Draw a 500-liter sample at 10–15 liters/minute through fine filter paper, which should collect germanium dioxide, germanates, and germanosilicates. Follow this with two absorbers, each containing 10 ml of 0.5% solution of sodium hydroxide, to collect germanium tetrachloride and fine particles passing the filters. If the filter paper blackens, ignite it. Treat the ash with 2 ml of nitric acid, 2 ml of

[80]E. G. Davitashvili and S. G. Kurashvili, *Soobshch. Akad. Nauk Gruz. SSR* **29**, 143–149 (1962).
[81]S. A. Dekhtrikyan, *Izv. Akad. Nauk Arm. SSR, Nauka Zemled.* **19** (3), 97–99 (1966).
[82]N. V. Lebedeva, L. I. Vinarova, and M. F. Grigor'a, *Zavod. Lab.* **32**, 1208–1209 (1966).

hydrofluoric acid, and 2 ml of phosphoric acid. Evaporate to sulfur trioxide fumes and take up in 10 ml of water.

Boil the contents of the absorbers with the filter paper or the solution of ash of the filter paper for 10 minutes, and filter. To the filtrate, add 2 ml of phosphoric acid, 4 ml of hydrochloric acid, 2 ml of 1% solution of gelatin, and 3 ml of 0.05% ethanolic phenylfluorone. Dilute to 50 ml and set aside for 30 minutes. Read at 497 nm.

Quartz-Topaz Greisens.[83] Fuse 0.5 gram of sample with 0.5 gram of sodium hydroxide and 2 grams of sodium peroxide in a corundum crucible, steadily increasing the temperature from 400 to 650°. Take up the melt with 25 ml of water, and if not already present, add several mg of iron as ferric chloride. Neutralize with 1:1 hydrochloric acid until solution is complete. Make alkaline with ammonium hydroxide, filter the precipitate, and wash with 1:100 ammonium hydroxide. Dissolve the precipitate with 100 ml of 9 *N* hydrochloric acid. Shake with 20 ml, 20 ml, and 20 ml of carbon tetrachloride for 5 minutes each. Wash the combined extracts with 10 ml and 10 ml of 9 *N* hydrochloric acid. Extract with 6 ml, 6 ml, and 6 ml of water, shaking for 2 minutes each. To the combined extracts, add 2 ml of hydrochloric acid, 1 ml of 1% gelatin solution, and 1.5 ml of 0.05% solution of phenylfluorone. Dilute to 25 ml, set aside for 30 minutes, and read at 560 nm.

Solutions Containing Hydrogen Peroxide and Hydrogen Fluoride.[84] As reagent, dissolve 0.2 gram of phenylfluorone in 700 ml of ethanol, add 160 ml of sulfuric acid, dilute to 1 liter with ethanol, and filter.

To 5 ml of sample solution containing 1–10 μg of germanium, add 1 gram of boric acid to mask fluoride ion. Add 2 ml of 1:1 sulfuric acid, and titrate to a pink color with 6% potassium permanganate solution. Add 0.05 gram of solid hydroxylamine sulfate and 30 ml of hydrochloric acid. Extract germanous chloride with 10 ml and 10 ml of carbon tetrachloride. Combine the organic extracts and reextract the germanium from them with 10 ml of water. Filter the aqueous phase and take 2 ml as an aliquot. Add 1 ml of 0.5% solution of gum acacia and 3 ml of reagent. Dilute to 10 ml and set aside for 40 minutes. Read at 505 nm against a blank.

Plant Material.[85] As the reagent, dissolve 0.2 gram of phenylfluorone in 500 ml of methanol or ethanol, add 160 ml of hydrochloric acid, and dilute to 1 liter with the alcohol.

Mix 0.3 gram of sample with 0.5 gram of calcium oxide in a porcelain dish and cover with 1.5 grams of the oxide. Heat in a muffle furnace to 480° to drive off volatile matter, then heat at 800° for 30 minutes. Cool, and take up in 30 ml of 9 *N* hydrochloric acid. Extract with 20 ml and 20 ml of carbon tetrachloride and dilute the extracts to 50 ml. To a 10-ml aliquot, add 2 ml of 0.5% gelatin solution and 6 ml of the reagent. Dilute to 20 ml and read at 510 nm.

[83] I. I. Tychinskaya, *Mater. 3-ei Nauch.-Tekh. Konf. Molodykh Uch.* **1957**, *Novosib., Sib. Otdel. Akad. Nauk SSSR Sb.* 53–57 (1960).

[84] E. Pungor and A. Halasz, *Magy. Kem. Foly.* **73**, 451–454 (1967).

[85] H. Kick and H. Arent, *Z. Pflanzenernähr. Düng. Bodenk.* **81**, 153–157 (1958).

Organometallic Compounds.[86] Decompose 15 mg of sample with sodium peroxide in a bomb and dissolve in 10 ml of water. To an aliquot, add 4 ml of hydrochloric acid, 2 ml of 1% gelatin solution, and 5 ml of 0.05% ethanolic phenylfluorone. Dilute to 50 ml, set aside for 30 minutes, and read at 500 nm.

QUINALIZARIN

Ores containing germanium dissolve in hot concentrated phosphoric acid. Chlorides must be absent, to avoid volatilization of germanium chloride.[87] Then the germanium is extracted from hydrochloric acid solution with carbon tetrachloride and reextracted as trioxalatogermanic acid. Then the oxalates are destroyed, and the germanium is dissolved in ammonium oxalate–oxalic acid buffer solution for pH 5 and developed with quinalizarin, or alternatively, with phenylfluorone. Numerous ores dissolve readily, but with silicates the decomposition must not be prolonged or a hard insoluble crust will form on the bottom of the beaker, invalidating the determination. Properly handled, the silica is gelatinous and easily dissolved. For refractory silicates, decompose with 4 ml of 1 : 1 sulfuric acid, 5 ml of nitric acid, and 5 ml of hydrofluoric acid before dissolving with phosphoric acid.

A chloroform-soluble ternary complex is formed at pH 4 by germanium, quinalizarin, and diphenylguanidine.[88] At 500 nm it obeys Beer's law for 0.1–2 μg of germanium per ml.

Procedure. *Ores.*[89] As a buffer solution, dissolve 5 grams of oxalic acid and 5 grams of ammonium oxalate in water, dilute to 1 liter, and adjust the pH to 5.

Mix 0.5 gram of sample with 5 ml of nitric acid and 5 ml of phosphoric acid. Heat gently until the nitric acid is evaporated and then at full heat of a hot plate. Remove from heat as soon as the sample is completely decomposed. Cool and dissolve the viscous, semiglassy mass in 25 ml of hydrochloric acid. Shake for 2 minutes with 15 ml and 15 ml of carbon tetrachloride. Extract the combined extracts for 2 minutes with 10 ml and 10 ml of the buffer solution. To the combined oxalate extracts, add 20 ml of nitric acid. Evaporate to dryness but do not bake. Dissolve the residue in 5 ml of the oxalate–oxalic acid buffer solution by warming. Cool and add 1 ml of 0.5% gelatin solution and 1 ml of 0.1% methanolic quinalizarin acetate. Let stand for 30 minutes, then dilute to 25 ml and read at 500 nm. Ores can also be developed by phenylfluorone, as indicated under that topic.

---

[86] A. P. Terent'ev, E. A. Bondarevskaya, R. N. Potsepkina, and O. D. Kuleshova, *Zh. Anal. Khim.* **27**, 812–813 (1972); cf. S. I. Obtemperanskaya and I. V. Dudova, *Vest. Mosk. Gos. Univ., Ser. Khim.* **11**, 461–465 (1970).

[87] E. H. Strickland, *Analyst* **80**, 548–550 (1955); cf. M. Finkelsteinate, J. Garjonite, and V. Skominaite, *Nauch. Tr. Vyssh. Ucheb. Zaved. Lit. SSR, Khim. Khim. Tekhnol.* **11**, 69–74 (1970).

[88] G. V. Flyantikova, V. A. Nazarenko, and I. G. Kostenko, *Zh. Anal. Khim.* **30**, 814–817 (1975).

[89] C. K. N. Nair and J. Gupta, *J. Sci. Ind. Res. India* **10B**, 300 (1951); **11B**, 274–276 (1952).

REZARSON

Rezarson, which is 5-chloro-3-(2,4-dihydroxyphenylazo)-2-hydroxybenzenearsonic acid, determines a minimum of 0.3 μg of germanium as a 1:1 complex in 5 ml of solution.[90] The reagent absorbs at 440 nm, the complex at 495 nm. The complex also has a maximum absorption at 360 nm and fluorescence at 600 nm. The presence of 100-fold amounts of aluminum or 1000-fold amounts of indium or zinc markedly enhances the fluorescence. Beer's law applies to 0.13–15 μg of germanium in 4 ml. As applied to silicate and sulfide minerals and to coal, it will determine down to 0.01 ppm of germanium by fluorescence.[91]

Procedures

Absorption.[92] To 2.5 ml of sample solution containing 0.3–15 μg of germanium, add 1.5 ml of a 0.02% solution of rezarson in 50% ethanol. Set aside for 30 minutes and read at 500 nm.

Coal Ash.[93] FLUORESCENCE. Mix 0.1 gram of sample with 1 ml of 1% sodium carbonate solution in platinum and take to dryness. Add 5 ml of hydrofluoric acid and again evaporate to dryness. Take up the residue in 2.8 ml of water. Add 0.2 ml of hydrochloric acid, 1 ml of phosphoric acid, and 0.8 ml of 0.005% solution of rezarson. Let stand for 15 minutes and read the fluorescence at 600 nm.

Arylgermanium Compounds or Tetraalkyl Germanes.[94] Ignite 3–5 mg of the compound containing halogen–germanium bonds in a Schöniger flask. Use a silica-thread sample holder; a platinum holder causes erroneous results. Absorb the combustion products in 2 ml of 0.2 *N* potassium hydroxide solution and neutralize. Complete as for coal ash from "Add 0.2 ml of hydrochloric acid, ..." The color can also be developed as germanomolybdate blue.

2,3,7-TRIHYDROXY-9-PHENYL-6-FLUORONE

The complex of this reagent with germanium is suitable for determination of 0.0001%.[95]

Procedure. *Zinc Ores.* Fuse a 0.4-gram sample with 1.6 gram of sodium hydrox-

[90] A. M. Lukin, K. A. Smirnova, and G. B. Zavarikhina, *Tr. Vses. Nauchn.-Issled. Inst. Khim. Reakt.* **1966** (29), 282–289.
[91] D. P. Shcherbov, R. N. Plotnikova, and I. N. Astaf'eva, *Zavod. Lab.* **36**, 528–530 (1970).
[92] A. M. Lukin, O. A. Efremenko, and B. L. Podol'skaya, *Zh. Anal. Khim.* **21**, 970–975 (1966).
[93] A. M. Lukin, G. V. Serebryakova, E. A. Bozhevol'nov, and G. B. Zavarikhina, *Tr. Vses. Nauchn.-Issled. Inst. Khim. Reakt.* **1967** (30), 161–166.
[94] T. M. Shanina, N. E. Gel'man, and T. V. Bychkova, *Zh. Anal. Khim.* **28**, 2424–2427 (1973).
[95] Zbigniew Gregorowicz, *Chem. Anal.* (Warsaw) **4**, 829–835 (1959).

ide and 0.4 gram of sodium peroxide. Take up with 10 ml of water and neutralize with 1 : 1 sulfuric acid. Add hydrochloric acid until the solution is 2 *N*; then add an equal volume of hydrochloric acid. Extract with 20 ml and 20 ml of carbon tetrachloride. Reextract the germanium from the combined organic extracts with 12.5 ml of 1 : 5 hydrochloric acid. Add 0.5 ml of 1% solution of gum arabic and 7.5 ml of 0.03% solution of the reagent in ethanol. Dilute to 25 ml, set aside for 30 minutes and read.

At pH 7, germanium and **aesculetin**,[96] which is 6,7–dihydroxycoumarin, form a 1 : 4 complex that is read at 380 nm. Absorption decreases with time. At pH 10 the ratio is 1 : 3, and as read at 315 nm, the complex remains stable.

Mix a sample solution containing 12–150 μg of germanium with 5 ml of 0.5% solution of gum acacia, 6 ml of 2 m*M* **alizarin complexan**, and 10 ml of phosphate buffer solution for pH 8.[97] Dilute to 25 ml and set aside for 15 minutes. Read the 1 : 4 germanium-ligand complex at 450 nm. In an ammonia–ammonium acetate medium at pH 7, the complex is 1 : 3; it is read at 445 nm and follows Beer's law for 0.2–4 μg of germanium dioxide per ml.[98]

A 1 : 3 : 5 complex is formed at pH 5–6 by germanium–alizarin complexan–rhodamine 6G (basic red 1).[99] This is separated by flotation with 1 : 4 carbon tetrachloride–chloroform, filtered, washed with 0.1 *M* sodium chloride solution, and dissolved in 1 : 1 ethanol–0.1 *N* sodium hydroxide. The complex breaks down in the solution. It is read by absorption at 520 nm or by fluorescence at 543 nm. Beer's law holds by absorption for 0.02–1 μg of germanium dioxide with sixty-fold *M* complexan and 100-fold *M* rhodamine. The fluorescence is linear for 2–100 ng of germanium dioxide per ml. There is interference by chloride, bromide, nitrate, iodide, and perchlorate ions.

Mix the sample solution with 10 ml of ethanolic m*M* **anthrapurpurin**, which is 1,2,7-trihydroxyanthraquinone. Add 0.1 ml of 0.1 *N* ammonium hydroxide and dilute to 25 ml with ethanol.[100] The final solution should contain 0.5–4 ppm of germanium and 10% water. Set aside for 30 minutes and read the 1 : 4 metal-reagent complex at 470 nm.

By complexing with germanium, the color of **anthrazo**, which is 1-(4-dimethylaminophenylazo)anthraquinone hydrochloride, is decreased.[101] Beer's law is followed at 0.2–1.44 μg of germanium per ml at 530 nm as read against the anthrazo reagent. Rigid standardization is required. There is interference by zirconium, titanium, antimony, iron, tin, zinc, and bismuth.

Germanium forms a blue 1 : 1 complex with **3-cyano-1,5-bis(2-hydroxy-5-**

[96]M. Finkelšhteinaite, G. Streckite, and Zh. V. Penina, *Nauch. Tr. Vyssh. Ucheb. Zaved. Lit. SSR, Khim. Khim. Tekhnol.* **1973** (15), 53–56.

[97]M. Roman Ceba and A. Fernandez-Gutierrez, *Quim. Anal.* **29**, 281–287 (1975).

[98]V. A. Nazarenko, G. V. Flyantikova, and L. I. Korolenko, *Zh. Anal. Khim.* **30**, 1354–1368 (1975).

[99]G. V. Flyantikova and L. I. Korolanko, *ibid.* 1349–1353.

[100]M. Roman Ceba and A. Fernandez-Gutierrez, *Quim. Anal.* **29**, 203–208 (1975).

[101]M. Finkelšhteinaite, J. Garenite, and V. Skominaite, *Nauchn. Tr. Vyssh. Ucheb. Zaved. Lit. SSR, Khim. Khim. Tekhnol.* **11**, 19–23 (1970).

sulfophenyl)formazan.[102] At pH 6 the maximum absorption is at 650 nm. Beer's law is obeyed for 6–60 μg/ml.

A 1:2 colorless complex of germanium with **dihydroxycumarone**, which is 6,7-dihydroxy-2*H*,3*H*-benzofuran-one, at pH 3.5–6, obeys Beer's law up to 4 μg/ml.[103] Mix a solution containing 1–100 μg of germanium with 1 ml of 0.15% solution of the reagent and 5 ml of acetate buffer solution for pH 5. Dilute to 25 ml, set aside for 30 minutes, and read at 330 nm against a blank.

A 1:3 complex is formed by germanium with **dioxythiazo**, which is 4-(5-benzoyl-4-phenylthiazol-2-ylazo)catechol.[104] Buffered at pH 4–7 with acetate, the maximum absorption is at 540 nm. The complex can be extracted with butanol and then is stable for 24 hours.

Germanium is read by **flame photometry** in the nitrous oxide–acetylene flame at 265.12 nm with a 10–12 mm height in the red zone.[105] The detection limit is 1 ppm, and there is interference from aluminum, vanadium, molybdenum, tungsten, cobalt, chromium, and potassium.

In 2.8–6.4 *M* sulfuric acid, **gossypol** complexes with germanium.[106] The complex is extractable with chloroform. The maximum absorption is at 545 nm, that of the reagent at 435 nm. Beer's law is followed for 2–120 μg of germanium in 25 ml.

A 1:2 complex is formed when a solution containing 0.001–0.1 mg of germanium is treated with 5 ml of 1% gelatin solution and 5 ml of 0.001 *M* **hematein.**[107] After 1 hour add 5 ml of 0.5 *M* potassium acid phthalate, dilute to a known volume, and read at 580 nm. The maximum sensitivity is at pH 4. There is interference by tin, antimony, and bismuth.

The optimum pH for forming the complex of **hematoxylin** with germanium is 5.6.[108] The maximum extinction is at 570 nm, and Beer's law is followed for 0.5–2 μg/ml. The complex is stable for 30 minutes. Ascorbic acid, sulfosalicylic acid, thiosulfate, thiocyanate, thiourea, fluoride, and borate interfere because they react with germanium. Aluminum interferes. EDTA masks copper, ferric ion, zinc, nickel, or lead.

Germanium combines with **8-hydroxyquinoline** in chloroform as a 1:2 complex and also as a 1:1 ion.[109] The result can be read by fluorescence.

When extracted from 6 *M* sulfuric acid–0.5 *M* sodium iodide with cyclohexane, germanium **iodide** has a maximum absorption at 360 nm.[110]

At pH 7–8 the complex of germanium with **lumogallion**, which is 5-chloro-3[(2,4-dihydroxyphenyl)azo]2-hydroxybenzenesulfonic acid, reaches a maximum

[102]V. F. Vozisova and V. N. Podchainova, *Zh. Anal. Khim.* **19**, 640–642 (1964).
[103]V. A. Nazarenko and E. N. Poluektova, *Zh. Anal. Khim.* **22**, 895–899 (1967).
[104]V. S. Korol'kova, Ya. K. Putnin', and E. Yu. Gudrinietse, *Latv. PSR Zinat. Akad. Vest., Khim. Ser.* **1967**, 529–534.
[105]E. E. Pickett and S. R. Koirtyohann, *Spectrochim. Acta, B* **24**, 325–333 (1969); R. M. Dagnall, G. F. Kirkbright, T. S. West, and R. Wood, *Analyst* **95**, 425–430 (1970).
[106]Sh. T. Talipov, R. Kh. Dzhiyanbaeva, A. Inoyatov, L. V. Chaprasova, and M. Ziyaeva, *Zh. Anal. Khim.* **25**, 1420–1422 (1970).
[107]J. J. R. Frausto de Silva, J. Goncalves Calado, and M. LeGrande de Moura, *Rev. Port. Quim.* **5** (2), 65–71 (1963).
[108]M. K. Akhmedli, D. G. Gasanov, and R. A. Alieva, *Uch. Zap. Azerb. Univ., Ser. Khim. Nauk* **1964** (1), 75–81.
[109]L. B. Ginzburg, *Inv. Akad. Nauk. Kaz. SSR, Ser. Khim.* **1957** (1), 94–98; N. P. Rudenko and L. V. Kovtun, *Tr. Komi. Anal. Khim., Akad. Nauk SSSR* **14**, 209–217 (1963).
[110]Katu Tanaka and Nobuyuki Takagi, *Anal. Chim. Acta* **48**, 357–366 (1969).

extinction in 15 minutes and is stable for 30 minutes thereafter.[111] The maximum absorption is at 490 nm. The complex can be extracted into cyclohexane for reading, or the unreacted lumogallion can be extracted with butanol. The absorption obeys Beer's law for 0.8–13 μg of germanium per ml. Interfering ions are copper, titanium, zinc, gallium, ferrous, stannous and zirconium.

Germanium forms a 1:4 complex with **methylaesculetin** which is 6,7-dihydroxy-4-methylcoumarin.[112a] The complex is read at pH 6.5.

To determine 0.5–2 μg of germanium add to the sample solution 5 ml of glycine buffer solution for pH 3, 2 ml of 13.7 mM gallic acid, and 1 ml of 1.37 mM **methyl violet**.[112b] Separate the insoluble ternary complex by flotation on 5 ml of toluene. Wash the precipitate with 5 ml and 5 ml of 5 mM sulfuric acid. Then add 5 ml of ethanol to the toluene layer and shake to dissolve the complex. Read at 590 nm. Pyrogallol or catechol may replace the gallic acid.

Purpurin, which is 1,2,4-trihydroxyanthraquinone, complexes with germanium in 40% ethanol at pH 5.[113a] It must be read at 490 nm between 4 and 15 minutes after mixing. Beer's law is followed for 3.4–15 μg/ml. There is interference by iron, copper, titanium, and sulfide.

A 1:3:3 complex is formed by germanium–**pyrogallol red**–diphenylguanidine at pH 4 for reading at 540 nm for 0.7–5 μg of germanium per ml.[113b] Aluminum, ferric ion, and calcium interfere.

The absorption maxima of three dyes with molybdogermanic acid in 3.75–4.25 N hydrochloric acid are as follows:[114] **pyronine G** (color index 45005), 544 nm; **rhodamine S** (basic red 11), 545 nm; **rhodamine 6G** (basic red 1), 546 nm. For **rhodamine B** (basic violet 10) in 0.75–1.5 N hydrochloric acid, it is 576 nm.

For **quercetin-6′-sulfonic acid**, dissolve 2 grams of quercetin in 20 grams of sulfuric acid.[115] After 24 hours pour into ice water, extract with ethyl acetate, then with ether. Neutralize the aqueous layer to pH 3.5 with 50% sodium hydroxide solution to give the reagent having absorption maxima at 261 and 365 nm. The 1:3 germanium-ligand complex in neutral solution has a maximum at 400 nm. In 2.8 N hydrochloric acid the maximum is at 450 nm. Maximum color is developed at pH 5.6–6.6. Beer's law is followed up to 1.06 μg of germanium per ml. Added ammonium chloride increases the absorption.

Germanium forms a 1:3 complex with **rutin**, which is 3,3′,4′,5,7-pentahydroxyflavone-3-rutinoside, that is read at 400 nm.[116] There is interference by titanium, zirconium, vanadium, molybdenum, and iron.

[111] M. Finkel'shteinaite and Ya. Gar'enite, *Nauch. Tr. Vyssh. Vchels, Zaved. Lit. SSR, Khim Khim. Tekhnol.* **10**, 69–73 (1969).

[112a] M. Finkel'shteinaite and G. Streckite, *ibid.* **1972** (14), 63–67.

[112b] I. A. Prortak and L. I. Ganago, *Izv. Vyssh. Ucheb. Zaved., Khim Khim. Tekhnol.* **17**, 559–561 (1974).

[113a] L. Finkel'shteinaite and Ya. Gar'enite, *Zh. Vses. Khim. Obshch.* **13**, 593–594 (1968); M. Finkel'shteinaite, J. Garjonite, and V. Skominaite, *Nauch. Tr. Vyssh. Ucheb. Zaved. Lit. SSR, Khim. Khim. Tekhnol.* **11**, 69–74 (1970).

[113b] R. A. Alieva, *Azerb. Khim. Zh.* **1974**, 93–96.

[114] G. Popa and I. Paralescu, *Talenta* **16**, 315–321 (1969).

[115] Takuji Kanno, *Sci. Rep. Res. Insts., Tohoku Univ., Ser. A* **10**, 207–211 (158); *J. Chem. Soc. Jap.* **79**, 306 (1958).

[116] M. Finkel'shteinaite, J. Burskiene, and Ts. Brener, *Nauch. Tr. Vyssh. Ucheb. Zaved. Lit. SSR, Khim Khim. Tekhnol.* **13**, 41–45 (1971).

Germanium at 0.05–1.2 ppm in 0.5 *N* hydrochlroic acid is read at 510–520 nm as the complex with **2,6,7-trihydroxy-9-(4-dimethylaminophenyl)fluorone.**[117] There is possible interference by stannic, antimony, zirconium, ceric, ferric, and permanganate ions.

A 1 : 1 : 2 : 1 complex is formed by germanium, **trihydroxyflavone**, antipyrine, and the anion of a strong acid.[118] This is extractable with chloroform. The optimum pH is 0.8–2; the optimum wavelength is 510–515 nm. Beer's law holds up to 3 μg of germanium per ml of extract. An appropriate anion is bromide.

Ternary complexes are formed between germanium, molybdate, and such **triphenylmethane dyes** as crystal violet (basic violet 3), methyl violet (basic violet 1), brilliant green (basic green 1), and malachite green (basic green 4).[119] All absorb at 600 nm and obey Beer's law for 0.03–0.8 μg/ml.

[117]Kenjiro Kimura and Masako Adada, *Bull. Chem. Soc. Jap.* **29**, 812–815 (1956).
[118]V. A. Nazarenko, N. I. Makrinich, and M. B. Shustova, *Zh. Anal. Khim.* **25**, 1595–1602 (1970).
[119]L. I. Ganago and I. A. Prostak, *ibid.* **26**, 104–110 (1971).

CHAPTER FIFTEEN
ALUMINUM

Aluminum is one of the elements whose ubiquitous presence may cause large blanks or require purification of reagents. It is no surprise to find that many of the methods consist of lake formation with dyestuffs and that in some cases a stabilizer or dispersing agent is desirable.

For determination of down to 0.1% aluminum in plain carbon steel and alloy steel without separating the iron, eriochrome cyanine R is preferable to chrome azurol S or aluminon.[1]

As applied to the ash of viscose pulp, similar accuracy is obtainable with aluminon and eriochrome cyanine R, but aluminon is slightly more precise.[2]

At pH 3–6 in 40–60% ethanol, various hydroxyflavones form soluble aluminum complexes that fluoresce yellow-green with an intensity proportional to the aluminum.[3] Morin, datiscetin, datiscin, and quercetin as 0.01% solution range in sensitivity from 0.005 μg/ml for morin to 0.04 μg/ml for quercetin.

In combustion of 0.5 gram of polyethylene in an oxygen bomb, the small amount of nitrogen present is oxidized so that the metallic ions are present as nitrates in the aqueous phase.[4] Taken up in 5 ml of 5 *N* hydrochloric acid and diluted to 50 ml, stilbazo is recommended for determination of aluminum.

As an indirect method, aluminum is precipitated from acetic acid solution as the phosphate.[5] The precipitate is filtered, washed with 5% ammonium nitrate solution, and dissolved in 1 : 2 hydrochloric acid; the phosphorus is determined as molybdenum blue.

[1] I. P. Kharlamov, G. V. Eremina, and T. A. Borcheva, *Zavod. Lab.* **38**, 8–12 (1972).
[2] Jan Gelo, *Przegl. Pap.* **27** (3), 77–81 (1971).
[3] A. P. Golovina, I. P. Alimarin, D. I. Kuznetsov, and A. D. Filyugina, *Zhur. Anal. Khim.* **21**, 163–165 (1966).
[4] Shizuo Fujiwara and Hisatake Narasaki, *Jap. Anal.* **10**, 1268–1272 (1961).
[5] A. M. Dymov and V. V. Koreneva, *Izv. Vyssh. Ucheb. Zaved, Chern Met.* **1961** (3), 192–196.

For analysis of semiconductors, sulfides are dissolved in 1:1 hydrochloric acid and selenides in sulfuric acid containing bromine.[6] If the sample was cadmium sulfide, determine aluminum fluorimetrically with 2-salicylidenaminophenol. If the sample was zinc sulfide, use aluminon.

Iron, copper, manganese, and aluminum can be determined in the same solution in that sequence.[7] Iron is extracted with chloroform as the 1,10-phenanthroline–ferrous complex and read at 490 nm. Addition of sodium diethyldithiocarbamate forms the copper complex, extracted into butyl acetate and read at 440 nm. The aqueous phase is made alkaline with ammonium hydroxide extracted with chloroform, and the chloroform is discarded. Addition of sodium diethyldithiocarbamate forms the manganese complex in the aqueous phase, extracted with isobutyl acetate for reading at 345 nm. Finally the aluminum is extracted by carbon tetrachloride as the hydroxyquinolate for reading at 400 nm. The iron may be up to 0.02 mg/ml, copper or manganese 0.03 mg/ml, and aluminum 0.08 mg/ml.

The technic for nonmetallic inclusions in sheet steel[8] assumes deoxidation with ferrocerium. The finished solution provides aliquots for determination of silica, aluminum, iron, manganese, chromium, and cerium.

Provide anodic dissolution of the sample in an electrolyte containing 3% of ferrous sulfate, 1% of sodium chloride, and 0.2% of sodium potassium tartrate. Filter the residue. Dissolve silicates from it with 50 ml of a solution of 3% sodium carbonate decahydrate and 5% of sodium citrate. Filter and wash. Use 5 ml of 5% ammonium citrate to remove the precipitate from the paper. Add 5 ml of a solution containing 30% of aluminum copper chloride, 5% of aluminum citrate and 0.75% of ferrous sulfate. Heat for 1 hour at 80° and filter. Wash the insoluble residue with hot water, 1:4 ammonium hydroxide, and 5% solution of sodium citrate. Heat the residue with 10 ml of 30% iodine solution for 1 hour, filter, and wash with 5% potassium iodide solution followed by 5% ammonium citrate solution. Treat with 60 ml of 1:2.5 hydrochloric acid. Filter this solution containing cerium from the sulfide phase and iron and manganese from the oxide phase. Wash the insoluble oxide phase of the nonmetallic inclusions. Wash with hot water, ignite, and weigh as total oxide inclusions.

Fuse with 0.5 gram each of sodium carbonate and potassium carbonate. Dissolve the melt with 40 ml of 3% sulfuric acid. Filter, ash, and fuse with 1 gram of potassium pyrosulfate. Dissolve this melt and add to the main solution. Use aliquots for determination of the elements listed earlier.

ACID CHROME PURE BLUE B

This dyestuff (mordant blue 1) at pH 5.4 has a golden yellow color but complexes with aluminum to violet. Interference by copper is masked by thiosulfate and that by ferric ion by ascorbic acid. Sulfate ion greatly reduces the color developed.[9]

[6] R. P. Pantalar, N. B. Lebed', and L. N. Semenova, *Tru. Kom. Anal. Khim.* **16**, 24–29 (1968).
[7] J. Paul and Sharad M. Paul, *Microchem. J.* **19**, 204–209 (1972).
[8] E. V. Dashkevich, *Zavod. Lab.* **34**, 938–939 (1968).
[9] R. B. Golubtsova, USSR Patent *128,200* (1960).

Procedure. *Sulfur.*[10] Shake 1 gram of finely ground sample with 10 ml of 0.1 *N* hydrochloric acid for 2 hours. Centrifuge. Mix a 2-ml aliquot of the extract, 4 ml of buffer solution for pH 5.4, 0.5 ml of ethanol, and 0.15 ml of 0.05% solution of the dye. Set aside for 15 minutes and read.

ALIZARIN RED S

This reagent, which is the sodium salt of 1,2-dihydroxyanthraquinone sulfonic acid, complexes with aluminum. The commercial dyestuff contains large amounts of sodium and potassium chlorides and sulfates. By purification, the upper limit for Beer's law is extended by over 100%. Pass the aqueous solution of the dyestuff through a column of water-saturated Sephadex G-10. The salts pass through first. Concentrate the colored eluate, adjust to pH 4 with 4% sodium hydroxide, and obtain the sodium salt of the dyestuff by evaporation. The aluminum complex can be read at either 490 or 392 nm.[11]

Alizarin red S is suitable for reading the aluminum content of soil extracts prepared with 0.75% solution of potassium chloride.[12] The reagent is added in ammoniacal solution after acidification of the sample solution with an ammonium acetate–acetic acid buffer solution. Possible interference by manganese, iron, and calcium must be provided for.

For aluminum in lithium salts, separate magnesium and aluminum from the lithium by chromatographing on a cation exchange resin such as Wofatit KPS 200.[13] Elute with 1:2 hydrochloric acid. Remove interfering ions as oxinates and read aluminum with alizarin red S.

For analysis of aluminum in phosphate rock, the interfering iron is preextracted as the cupferron complex.[14] This reagent is also appropriate for determination of aluminum in the ash from coal and coke.[15]

Procedures

Iron and Steel.[16] Dissolve 0.5 gram of sample in 20 ml of 1:5 perchloric acid and electrolyze with a mercury cathode until all iron, chromium, and molybdenum are removed. Filter and dilute to 100 ml. Adjust a 20-ml aliquot containing up to 0.07 mg of aluminum to about 0.5 *N* in perchloric acid. Add 2 ml of 1:1 hydrochloric acid and 1 ml of 2% cupferron solution. Set aside for 5 minutes. Extract with 20-ml portions of chloroform, adding 1 ml of cupferron solution each

[10] I. M. Korenman, F. R. Sheyanova, and N. I. Osipova, *Tr. Khim. Khim. Tekhnol.* (Gor'kii) **1962**, 395–401.
[11] W. Oelschläger, *Z. Anal. Chem.* **154**, 321–329 (1957).
[12] W. Schoenberg and E. Schoenfeld, *Albracht-Thare-Arch.* **4**, 37–43 (1960).
[13] J. Bosholm, *J. Prakt. Chem.* **29**, 65–71 (1965).
[14] William A. Jackson, *J. Agr. Food. Chem.* **7**, 628–630 (1969).
[15] F. H. Gibson and W. H. Ode, *Rep. Invest. U.S. Bur. Min.* **No. 6036**, 23 pp. (1962).
[16] J. A. Corbett and B. D. Guerin, *Analyst* **91**, 490–498 (1966).

time until the extract is colorless. Evaporate the aqueous layer to about 5 ml and add 1 ml of 1:1 nitric acid. Heat to fumes of perchloric acid. Take up with 10 ml of water and render faintly alkaline to phenolphthalein by adding 8% sodium hydroxide solution. Then neutralize with 0.2 *N* hydrochloric acid and add 1 ml in excess. Add 1 ml of 1.4% calcium chloride solution, 10 ml of an acetate buffer solution for pH 4.75, and 5 ml of 0.14% solution of alizarin red S. Dilute to 100 ml and read at 490 nm against a reagent blank.

Silicate Minerals.[17] To 0.5 gram of sample add 5 ml of hydrofluoric acid and 1 ml of 1:1 sulfuric acid. Heat to sulfur trioxide fumes, add 3 ml of water, and again evaporate to sulfur trioxide fumes. Take up in water and dilute to 50 ml. To an appropriate aliquot according to the specific mineral, add sequentially 2 ml of 1.56% calcium chloride solution, 1 ml of 10% solution of hydroxylamine hydrochloride, 1 ml of 0.75% solution of potassium ferricyanide, and 2 ml of 4% mercaptoacetic acid solution. Let stand for 5 minutes and add 10 ml of acetate buffer solution for pH 4.8. Let stand 10 minutes more and add 10 ml of 0.05% solution of alizarin red S. Dilute to 100 ml, let stand for 1 hour, and read at 490 nm against a reagent blank.

Leaves.[18] Dry a sample and dry-ash 1 gram or wet-ash 2 grams. Ignite the residue at 400° for 1 hour. Digest with 8 ml of 1:4 nitric acid, filter, wash, and dilute to 100 ml. To a 25-ml aliquot, add 2 ml of 0.14 *M* calcium chloride, 1 ml of 10% solution of hydroxylamine hydrochloride, 1 ml of 0.75% solution of potassium ferricyanide, and 2 ml of 4% solution of mercaptoacetic acid. After 5 minutes add 10 ml of acetate buffer solution for pH 5.5. Then after 10 minutes add 10 ml of 0.5% solution of alizarin red S and dilute to 100 ml. Let stand for 20 minutes and read at 490 nm. The original study was of *Hevea brasiliensis*. The method is applicable to the AutoAnalyzer.[19]

ANTHRAZOCHROME

Aluminum forms a red-violet 1:2 complex at pH 4.8–5.2 with anthrazochrome, which is 3-(2-carboxyphenylazo)chromotropic acid.[20] The color develops within 3 minutes and is stable for more than 24 hours. The absorption maxima of the reagent and of the complex are at 530 and 555 nm, respectively. Beer's law is followed for 0.04–0.4 μg of aluminum per ml. The tolerance of some elements in ratio to aluminum is as follows: hexavalent chromium, 30; pentavalent vanadium, 7; indium, 3; gallium, 2; titanium, 2; ferric ion, 1.5; hexavalent tungsten or molybdenum, 1.

[17]H. G. C. King and G. Pruden, *Ibid*, **93**, 601–605 (1968); G. Pruden and K. G. C. King, *Clay Miner*. **8** (1), 1–13 (1969); cf. Leonard Shapiro and W. W. Brannock, *U.S. Geol. Surv., Bull*. No. **1144A**, 56 pp (1962); F. Hegemann and H. Thomann, *Ber. Deut. Keram. Ges*. **37**, 127–35 (1960); I. Roelandts and J. C. Duchesne, *Annls. Soc. Geol. Belg*. **91**, 159–164 (1968).
[18]Lai-Aim Lancaster and Mum Kong Lum, *J. Sci. Food. Agr*. **24**, 349–355 (1973).
[19]Lai Aim Lancaster and Rajadurai Balasubramaniam, *ibid*. **25**, 381–386 (1974).
[20]A. A. Kafarova, M. K. Ahmedli, and N. N. Basargin, *Uch. Zap. Azerb. Gos. Univ., Ser. Khim. Nauk*. **1969** (1), 23–30.

Procedure. *Steel.*[21] Dissolve 0.2 gram of sample in 10 ml of hydrochloric acid and 0.5 ml of nitric acid. Evaporate to 3 ml, dilute to 25 ml, filter, and wash. Ash the paper and residue in platinum, and treat the ash with 5 ml of hydrofluoric acid and 1 ml of 1 : 1 sulfuric acid. Evaporate to dryness, take up in 5 ml of 1 : 5 hydrochloric acid, and add to the main filtrate. Dilute to 200 ml and take a 10-ml aliquot. Add 2 drops of methyl orange indicator and neutralize with 1 : 20 ammonium hydroxide to a strong orange color. Add 0.2 ml of 1% ascorbic acid solution and 5 ml of 0.025% solution of anthrazochrome. Dilute to 25 ml with 1 : 2 5% solution of hexamine–0.1 *N* hydrochloric acid, a buffer solution for pH 4.8. Read at 555 nm against a reagent blank.

ARSENAZO I

This reagent, which is disodium-3-(2-arsenophenylazo)-4,5-dihydroxy-2,7-naphthalenedisulfonate, obeys Beer's law as a complex with 0.1–0.8 μg of aluminum per ml.[22] Copper and iron must be masked. The complex is read at 580 nm in a medium at pH 6.[23]

A ternary complex is formed by aluminum ion, 0.5 m*M* arsenazo I, and 0.1 *M* diphenylguanidine, which is extractable at pH 6–8 with pentanol.[24]

Procedures

Aerosols. Filter a known volume of air through paper at 10 liters/minute. Ash the paper in porcelain and ignite at 600°. Fuse the residue with 0.5 gram of potassium pyrosulfate by slowly raising the temperature from 200 to 600°. Dissolve in water and dilute to 50 ml. To a 5-ml aliquot, add 0.2 ml of 1 : 5 hydrochloric acid, 0.5 ml of 0.5% ascorbic acid solution to mask iron, and 1 ml of 5% thiourea solution to mask copper. Add 1 ml of 0.05% solution of arsenazo I and 0.5 ml of 25% hexamine solution. Dilute to 10 ml, set aside for 15 minutes, and read at 536 nm.

Nonmetallic Inclusion Residues from Steel.[25] Fuse 7 mg of sample with 0.5 gram of 5 : 2 boric acid–sodium carbonate for 3 hours at 900°. Dissolve the melt in 30 ml of 1 : 5 hydrochloric acid and dilute to 50 ml. Complete as for aerosols from "To a 5-ml aliquot. . . ."

[21] M. K. Akhmedli, N. N. Basargin, and A. A. Kafarova, *Zavod. Lab.* **38**, 263–265 (1972).
[22] A. F. Kosternaya and N. A. Zavorovskaya, *Sb. Nauch. Rab. Inet. Okhr. Tr. VTsSPS* **1962** (1), 82–87.
[23] L. V. Bogova, *Tr., Vses. Inst. Nauchn.-Issled. Proekt. Rab. Ogneuporn. Prom.* **1968** (40), 131–143.
[24] V. N. Tolmachev, L. A. Kvichko, and V. D. Konkin, *Zh. Anal. Khim.* **22**, 11–14 (1967).
[25] K. N. Ersheva and G. M. Orlova, *Zavd. Lab.* **34**, 276–277 (1968).

ARSENAZO II

The complex with arsenazo II, which is 3,3′-bis(1,8-dihydroxy-3,6-disulfo-2,7-naphthylazo)biphenyl-4,4′-diarsonic acid, is sensitive to 0.04 μg of aluminum per ml.[26]

Procedure. *Tungstic Oxide.* Fuse 1 gram of sample with 1 gram of sodium carbonate. Dissolve in 10 ml of 1:1 hydrochloric acid and precipitate the tungstic acid with excess benzo [f] quinoline. Filter and make the filtrate strongly alkaline with 20% sodium hydroxide solution. Filter the precipitated benzo [f] quinoline. Add 2 drops of phenolphthalein indicator solution and neutralize with 1:10 hydrochloric acid. Add 1 drop of 10% ascorbic acid solution, 1 ml of 0.05% solution of arsenazo II, and 0.2 ml of 25% solution of hexamine. Dilute to 50 ml and read.

AURINETRICARBOXYLIC ACID

The ammonium salt of this reagent is known as aluminon. The color with aluminum requires time to develop, an alternative being a brief heating period. The optimum pH is 4.8–5 for reading 0.01–0.03 mg of aluminum oxide per 100 ml.[27] At pH 4 and 525 nm the minimum detectable amount of aluminum in 50 ml is 1 μg.[28] At pH 3.2–5.4 there is interference by the following ions: cupric, ferrous, chromic, beryllium, uranyl, thorium, zirconium, cerous, and ceric.[29]

A reagent containing gelatin gives the best sensitivity, highest stability of color, and best adherence to Beer's law.[30]

When ferric ion is reduced with thioglycollic acid it decolorizes aluminon but does not adversely affect the aluminum-aluminon lake.[31] Another mask for ferric ion is 0.02 *M* EDTA. And 0.02 *M* zinc chloride masks iron for reading aluminum with aluminon.[32]

The sum of aluminum and iron can be determined by aluminon around pH 4.5 and the iron subtracted as determined by thiocyanate.[33] An alternative is to read at 530 and 570 nm and determine aluminum and iron from an empirical diagram.[34]

Interfering ions in a solution of steel can be eliminated by chromatographing on an anion exchange resin such as Amberlite IRA-400-OH.[35] Electrolysis at a mercury cathode is also applicable.[36]

[26] V. G. Shcherbakev and Z. K. Stagenda, *Sb. Vses. Nauch.-Issled. Inst. Tverdy. Splavev* **1964** (5), 528–288.

[27] L. V. Bogova, *Tr. Vses Inst. Nauch.-Issled. Proakt. Rab. Ogneuporn. Prem.* **1968** (40), 131–43; cf. P. Chichile, *J. Assoc. Off. Agr. Chem.* **47**, 1019–1027 (1964).

[28] Kenneth E. Shull, *J. Am. Water Works Assoc.* **52**, 779–785 (1960).

[29] Anil K. Mukhenji and Arun K. Doy, *Proc. Nat. Acad. Sci., India* **A26**, Pt. 2, 138–153 (1957).

[30] V. N. Tikhonov and N. A. Gordeeva, *Zavod. Lab.* **39**, 408–409 (1973).

[31] Helmut Lilie and Brunhilde Sturzebecher, *Chem. Tech.* (Berlin) **8**, 672–675 (1956).

[32] Fumikazu Kawamura and Hireshi Namiki, *Bull. Fac. Eng., Yokohama Nat. Univ.* **8**, 261–265 (1959).

[33] L. A. Molot and L. M. Kul'berg, *Uch. Zap. Saratv. Univ.* **42**, 79–83 (1955).

[34] Hiroshi Nishida, *Jap. Anal.* **12**, 56–57 (1963).

[35] M. L. Foglino and G. P. Spagliardi, *Met. Ital.* **50**, 372–374 (1958).

[36] A. Devoti and A. Sommariva, *Met. Ital.* **50**, 355–366 (1958).

For analysis of silver, dissolve 10 grams of sample in nitric acid.[37] Add lanthanum nitrate solution and coprecipitate gold, lead, bismuth, iron, and aluminum along with the lanthanum by adding ammonium hydroxide. Filter and wash. Dissolve the precipitated hydroxides in 10 ml of 1 : 1 hydrochloric acid and dilute to 50 ml. Then determine aluminum in an aliquot by aluminon.

For analysis of aluminum in zirconium alloys by aluminon, the solution is twice treated with cupferron and extracted with chloroform, followed by a final chloroform extraction to complete the removal of zirconium and cupferron.[38]

For analysis of tin and lead alloys, the sample is dissolved in sulfuric acid.[39] By heating at 200–220° and adding hydrobromic acid, tin, antimony, and arsenic are volatilized. Lead is separated as lead sulfate, other interfering metals by electrolysis with the mercury cathode. Thereafter conditions must be rigidly standardized to get reproducible color development with aluminon.

For aluminum in iron-chromium-nickel alloys, dissolve sample in hydrochloric and nitric acids and neutralize excess acid with ammonium hydroxide.[40] Add ethanol, and complex iron and chromium with potassium acid phthalate in ethanol. Add ascorbic acid, adjust to pH 4.4, and read with aluminon at 536 nm. Suitably modified, this technic can determine aluminum all the way from 0.01 to 1.8%. Titanium interferes seriously but can be removed with cupferron.

For aluminum in beryllium bronze containing 0.05–0.24% aluminum and 1.7–2.5% beryllium, remove most of the copper electrolytically.[41] Precipitate residual copper, iron, and some other metals as sulfides. Precipitate aluminum with 8-hydroxyquinoline. This may be extracted for reading or filtered and developed with aluminon.

For determination of aluminum in chromite, sinter 0.2 gram of sample with 3 grams of 5 : 5 : 10 : 2 sodium carbonate–potassium carbonate–magnesium oxide–borax for 25 minutes.[42a] Dissolve in 90 ml of hot 1 : 5 sulfuric acid and dilute to 250 ml. Adjust to pH 4.2–4.4, reduce ferric ion with ascorbic acid, and read with aluminon at 525 nm.

For aluminum in an alkali silicate solution, add sulfuric acid followed by hydrofluoric acid and evaporate to dryness to volatilize the silicon as the tetrafluoride.[42b] Dissolve the residue in water and dilute to 200 ml. Add ammonium hydroxide to a 20-ml aliquot until it is only slightly acid. Add 2 ml of 10% solution of potassium thiocyanate and extract the ferric thiocyanate formed with amyl alcohol or isoamyl alcohol. Add 20 ml of acetate buffer solution for pH 5.5 and 2 ml of 0.1% solution of aluminon. Heat to 100° for 3 hours and cool. Read 3–21 μg per ml at 597 nm against water.

For as little as 0.05 ppm of aluminum in potassium chloride solution, coprecipitate with ferric hydroxide, dissolve in hydrochloric acid, reduce the iron with ascorbic acid, and read the aluminum with aluminon.[43]

For free aluminum in lanthanum aluminide, dissolve it with 3% sodium hy-

[37] Z. Marczenke and K. Kasiura, *Chem. Anal.* (Warsaw) **9**, 87–95 (1964).

[38] S. V. Elinson, L. I. Pobedina, and N. A. Mirzoyan, *Zh. Anal. Khim.* **15**, 334–338 (1960).

[39] Kazuo Ota, *Jap. Anal.* **7**, 162–166 (1958).

[40] G. Signorelli and A. Alderisie, *Ann. Chim.* (Rome) **56**, 143–150 (1966).

[41] E. Ya. Khatkevich, L. A. Kir'yanova, and N. V. Stashkova, *Tr. Vses. Nauchn.-Issled. Inst. Standn. Obraztsov. Spaktr. Etalonov* **5**, 122–126 (1969).

[42a] Z. L. Nazarenko, *Sb. Nauch. Tr. Ukr. Nauchn.-Issled. Inst. Ogneuporn.* **1965** [8(55)], 307–314.

[42b] M. S. Movsesyan and Sh. K. Manukhyan, *Armyan. Khim. Zh.* **29**, 481–487 (1976).

[43] A. Glasner and S. Skurnik, *Isr. J. Chem.* **3** (4), 143–150 (1965).

droxide solution, filter from the aluminide, and determine with aluminon.[44] Treat titanium aluminide similarly but use 1% sodium hydroxide solution or 1% ferric sulfate solution. In the latter case, reduce ferric ion in the filtrate with ascorbic acid.

When aluminum and tellurium in 6 *N* hydrochloric acid are passed through a column of the anion exchange resin EDE-10p in chloride form, the tellurium is completely adsorbed and the aluminum is in the eluate.[45]

For aluminum in a solution of thorium, adsorb the thorium by passage through a strong base ion exchange resin such as Merck III, a quaternary ammonium crosslinked polystyrene.[46] Wash the resin with water acidified to pH 2.1 ± 0.1 and determine aluminum in the eluate with aluminon.

For determination of aluminum in plutonium-aluminum alloys, dissolve sample material in perchloric acid.[47] Dilute to *M* acidity and electrolyze with a mercury cathode to remove iron. Precipitate plutonium as the insoluble iodate, centrifuge, and determine aluminum by aluminon.

For aluminum in steel, first remove the bulk of the iron with a mercury cathode or by extraction with butyl acetate from a chloride solution. Oxidize chromic ion to chromate.[48] Add titanic ion and precipitate metallic hydroxides with ammonium hydroxide. Filter, ignite, and fuse with sodium carbonate. Extract aluminum as sodium aluminate. Precipitate vanadate from the extract with cupferron and beryllium with oxine. Finally read aluminum with aluminon.

For analysis of organoaluminum materials such as diethyl(triphenylsiloxy)aluminum by aluminon, fuse the sample with sodium peroxide in an oxygen-filled bomb.[49]

The reagent is applied to alunite,[50] to cement,[51] to the raw cement slurry,[52] and to Martin slag.[53]

Procedure. As indicator, add a drop of 0.25% solution of 2,6-dinitrophenol to the sample solution and adjust to pH 1.8 with *N* hydrochloric acid or sodium hydroxide, at which value the solution is colorless.[54] Mask iron by adding 2 ml of 1% mercaptoacetic acid solution and heating at 90° for 30 minutes. Cool and dilute to 30 ml. Add 10 ml of 0.035% solution of aluminon in 5% sodium acetate solution adjusted to pH 4.2. Dilute to 50 ml and check that the pH is 3.7–4. Set aside for 2 hours and read at 530 nm. The yellow indicator does not interfere.

[44] V. P. Kopylova and T. N. Nazarchuk, *Zh. Anal. Khim.* **20**, 892–893 (1965).

[45] N. P. Strel'nikova and V. N. Pavlova, *Zavod. Lab.* **26**, 425–426 (1960).

[46] V. T. Athavale and A. R. Subramanian, *J. Sci. Ind. Res.* (India) **19B**, 431–432 (1960).

[47] Maynard E. Smith, *USAEC* **LA-1953**, 25 pp (1959).

[48] E. de la Torre Gonzalez, V. Torner Carilla, and R. Suarez Acosta, *Inst. Hierro Acero* **17**, 192–198 (1964).

[49] A. P. Teren'ev, E. A. Bondarevskaya, N. A. Gradskova, and E. D. Kropotova, *Zh. Anal. Khim.* **22**, 454–456 (1967).

[50] L. M. Kul'berg, N. V. Chugreeva, and L. A. Molet, *Tsement* **18** (6), 21–23 (1952); M. K. Akhmedli, E. A. Bashirov, and I. S. Lozovskaya, *Uch. Zap. Azerb. Gos. Univ. S. M. Kirova* **1955** (2), 27–31.

[51] N. M. Neshchadimova and I. V. Bogdanova, *Tsement* **24** (3), 18–23 (1958).

[52] K. G. Kondratova, E. A. Sazhina, and E. E. Gurevich, *ibid.* **25** (6), 26–27 (1959).

[53] V. I. Zhuravskaya, *Zavod. Lab.* **16**, 1302–1304 (1950).

[54] G. S. R. Krishna Murti, V. A. K. Sarma, and P. Rengasamy, *Indian J. Technol.* **12**, 270–271 (1974).

Titanium and Titanium Alloys.[55] As an acetate buffer solution, mix 950 ml of ammonium hydroxide and 860 ml of glacial acetic acid. Cool, adjust the pH to give a value of 5.25–5.35 when the solution is diluted 1:20, and dilute to 2 liters. As 0.033% aluminon reagent, dissolve 0.7 gram in 400 ml of water, add 140 ml of a 10% methanolic benzoic acid and dilute to 700 ml. Add 700 ml of the buffer solution and 700 ml of 1% gelatin solution. Stabilize by storing 3 days before use. As cupferron, wash solution just before use; mix 15 ml of 5% cupferron solution, 50 ml of hydrochloric acid, and 435 ml of water. Keep the wash solution ice cold.

For aluminum 0.1–1%, dissolve 1 gram of sample by heating with 50 ml of 1:1 sulfuric acid. Add 30% hydrogen peroxide to oxidize the titanium, boil off the excess, and dilute to 500 ml. Transfer an aliquot containing 0.02–0.03 mg of aluminum, add 6 ml of hydrochloric acid, and dilute to 50 ml. Chill in ice water and add 6 ml of 5% cupferron solution slowly with stirring to produce a flocculent precipitate. Filter, and wash with the cupferron wash solution until the filtrate totals about 150 ml. Add 10 ml of nitric acid and boil down to 40 ml. Add 10 ml of nitric acid and 10 ml of perchloric acid. Evaporate to perchloric acid fumes, wash down the sides of the container with water, and fume until free from organic matter.

If chromium is present, bring up the volume to 8–10 ml with perchloric acid. Boil vigorously, adding hydrochloric acid dropwise until the chromium has been volatilized as chromyl chloride. Allow time between drops for reoxidation of the residual chromium. If chromium is absent, omit this paragraph.

Wash down the sides of the container and evaporate to 3–4 ml. Add 15 ml of water and 1 ml of 1% thioglycollic acid solution. Let stand for 5 minutes. Add 1 drop of 0.1% metacresol purple in dilute sodium hydroxide solution. Add 1:1 ammonium hydroxide to a fading red but not to yellow, and dilute to 25 ml. Add 15 ml of 0.033% aluminon reagent and develop by heating at 100° for 30 minutes. Cool, dilute to 100 ml, and read at 540 nm against water.

For aluminum 1–10%, dissolve 0.5 gram of sample in 25 ml of 1:1 sulfuric acid. Add 30% hydrogen peroxide to oxidize the titanium, and boil off the excess. Cool and dilute to 500 ml. Take an aliquot containing about 0.12 mg of aluminum, add 6 ml of hydrochloric acid, and dilute to 50 ml. Continue as for aluminum 0.1–1% from "Chill in ice water and..." but use 0.66% aluminon reagent and subtract a blank.

Ferrotitanium.[56] Heat 0.2 gram of sample with 25 ml of sulfuric acid. While heating, add 5 ml of nitric acid and 10 ml of hydrochloric acid. Continue heating for 15 minutes to disintegrate carbides and decompose excess nitric acid. Cool, dissolve precipitated salts by adding 50 ml of water, and dilute to 500 ml. To a 3-ml aliquot add 1 ml of 10% ascorbic acid solution and 50 ml of a buffer solution containing 1.36% of sodium acetate and 0.75% of acetic acid. Add 4 ml of 0.1% solution of aluminon, set aside for 15 minutes, and dilute to 100 ml. Read at 540 nm.

Ferrovanadium.[57] Disintegrate 0.5 gram of sample with 15 ml of 1:1 nitric acid

[55] Dilip K. Banerjee, *Anal. Chem.* **29**, 55–60 (1957); cf. M. Codell and G. Norwitz, *ibid.* **25**, 1437 (1953).
[56] G. L. Povolotskaya, *Met. Khim. Prom. Kaz., Nauchn.-Tekh. Sb.* **1961** (6), 66–68.
[57] J. Musil, *Hutn. Listy* **23**, 649–650 (1968).

and filter. Ignite the filter in platinum and fuse the residue with 1 gram of potassium bisulfate. Dissolve the melt in 1 ml of 1 : 1 sulfuric acid, add to the previous filtrate, and dilute to 500 ml. To a 5-ml aliquot, add 2 ml of 15% hydrogen peroxide and 3 grams of the resin Wofatit L-150. Add 10 ml of water, set aside for 10 minutes, and filter. Wash the filter with 25 ml of water. Boil the filtrate for 3 minutes, cool, and add 2 ml of 2.5% ascorbic acid solution. Add 15 ml of 0.25% solution of aluminon in an acetate buffer solution containing 0.05% of benzoic acid and 0.25% of gelatin adjusted to pH 4.5. Heat at 100° for 5 minutes, cool, and dilute to 100 ml. Let stand for 15 minutes and read at 520 nm against a reagent blank.

Steel, Alloys, and Metals.[58] The first technic is applicable to iron, nickel, manganese, molybdenum, steel, and alloys containing less than 20% copper or tungsten, less than 10% niobium, less than 6% titanium, less than 3% cobalt, and less than 0.2% vanadium.

Dissolve 0.1 gram of sample in 5 ml of hydrochloric acid and 1 ml of nitric acid. Evaporate almost to dryness and dissolve in 2 ml of hydrochloric acid. Boil with 50 ml of water and filter the insoluble residue. If necessary, as when niobium or tungsten is present, add paper pulp before filtering and dilute further to 100 ml.

To each of two 5-ml aliquots, add 3 ml of acetate buffer solution for pH 4.7 and 30 ml of water. One is a blank. To the other aliquot add 2 ml of 1% solution of aluminon. Let stand for 10 minutes and dilute to 100 ml with pH 4.7 acetate buffer solution. Read the sample at 540 nm against the blank.

This second technic is applicable to steel and alloys containing 0.2–5% vanadium and to nickel, molybdenum, and manganese. Dissolve 0.1 gram of sample in 5 ml of hydrochloric acid and 1 ml of nitric acid. Evaporate to dryness, add 5 ml of hydrochloric acid, and again evaporate to dryness. Repeat the evaporation with hydrochloric acid twice more. Take up the residue in 10 ml of hydrochloric acid, and if the sample contained less than 5% iron, add 2 ml of a 2.5% solution of ferric chloride hexahydrate. Add 40 ml of water, followed by 4% potassium permanganate, to a permanent pink color. Add paper pulp and coprecipitate the iron and vanadium with 20 ml of 6% cupferron solution. Let stand for 30 minutes and filter. Wash the filter thoroughly with water. Add 50 ml of nitric acid to the filtrate and evaporate to about 4 ml. Add 1 ml of sulfuric acid and 5 ml of nitric acid and evaporate to sulfur trioxide fumes. Add 5 ml more of nitric acid, heat to complete oxidation of organic matter, and ignite to drive off sulfur trioxide. Heat with 5 ml of hydrochloric acid, evaporate nearly to dryness, and repeat that step. Dilute to 200 ml and complete by the final paragraph under the first technic.

This third technic is applicable to steel and alloys containing more than 3% cobalt. Dissolve 0.1 gram of sample in 5 ml of hydrochloric acid and 1 ml of nitric acid. Evaporate almost to dryness and dissolve in 50 ml of buffer solution for pH 6. Pass the solution at 1 ml/minute through a column of silica gel and wash the column at 2 ml/minute with 200 ml of the buffer solution. Elute aluminum from the column with 70 ml of 0.2 *N* hydrochloric acid heated to 50°. Follow this with

[58] A. A. Federov and G. P. Sokolova, *Sb. Tr. Tsent. Nauch.-Issled. Inst. Chern. Met.* **1963** (31), 162–169; cf. N. I. Shishkina, *Byull. Nauch.-Tekh. Inf. Ural. Nauch.-Issled. Inst. Chem. Met.* **1957** (3), 173–182; Theo Kurt Wilmer, *Arch. Eisenhüttenw.* **29**, 159–164 (1958); M. Herrmann and H. Weber, *Z. Anal. Chem.* **267**, 13–16 (1973).

water until the percolate is neutral. Evaporate the percolate to about 50 ml and dilute to 200 ml. Complete by the final paragraph under the first technic.

Carbonaceous and Low Alloy Steel.[59a] Dissolve 1 gram of sample in 50 ml of 1:5 hydrochloric acid. Filter the insoluble carbides and wash with water. Ignite the filter and fuse the ash with 1 gram of potassium pyrosulfate. Dissolve the melt in water, add to the original filtrate, and dilute to 100 ml. To a 10-ml aliquot, add 2.5 ml of 5% solution of ascorbic acid and neutralize with 1:2 ammonium hydroxide. Add 2 drops of 1:4 hydrochloric acid, 30 ml of water, and 30 ml of a 1:1 mixture of 0.1 *N* acetic acid–0.1 *N* sodium acetate as a buffer solution. Add 1 ml of 0.1% solution of aluminon and dilute to 100 ml with the buffer solution. Let stand for 20 minutes and read at 540 nm against water.

Alloy Steel.[59b] Dissolve 0.1 gram of sample containing more than 0.01% of aluminum in 10 ml of aqua regia, add 10 ml of 1:6 sulfuric acid, and evaporate to sulfur trioxide fumes. Take up in 50 ml of water and boil. Add 5 ml of butanol and 20 ml of 20% sodium hydroxide solution. Cool and dilute to 200 ml. Adjust an aliquot to pH 2.4 with 1:6 sulfuric acid. Add 10 ml of acetate buffer solution for pH 5 and 8 ml of 0.033% solution of aluminon to which benzoic acid and ammonium acetate have been added. Add 3 ml of 0.5% solution of gelatin and dilute to 50 ml with acetate buffer solution for pH 5. Set aside for 20 minutes and read at 520 nm.

Steel. ALUMINUM AND TITANIUM.[60] Dissolve 0.2 gram of sample in 25 ml of 1:1 hydrochloric acid, dilute to 100 ml, and filter. Ash the filter, fuse the residue with 1 gram of potassium bisulfate, dissolve in water, and add to the previous filtrate. Dilute to about 400 ml and heat to about 80°. Add sequentially 12 ml of glacial acetic acid, 50 ml of 30% sodium thiosulfate solution, 20 ml of 10% diammonium phosphate solution, a few grams of paper pulp, and 20 ml of 20% ammonium acetate solution. Boil to coagulate the sulfur that was liberated. Filter the precipitate, which contains the aluminum and titanium as phosphates. Wash with 5% ammonium nitrate solution. Ash the paper and residue, then fuse with 1.5 grams of potassium bisulfate. Dissolve the melt with 10 ml of 1:9 sulfuric acid. Add 1 ml of phosphoric acid and 4 ml of 30% hydrogen peroxide. Read titanium at 390 nm.

Pass the solution through a column of KU-2 cation exchange resin and wash the column with 100 ml of 13.2% solution of ammonium sulfate in 0.5 *N* sulfuric acid. Elute the aluminum with 50 ml of 1:3 hydrochloric acid and concentrate the eluate to about 30 ml. Add 1 ml of 1:9 hydrochloric acid and 1 ml of 0.2% solution of aluminon. Dilute to 50 ml and read at 540 nm.

Nonmetallic Inclusions in Carbon Steel.[61] Decompose the sample by electrolysis

[59a] A. A. Fedorov and G. P. Sokolova, *Sb. Tr. Tsent. Nauch.-Issled. Inst. Chern. Metall.* **1962** (24), 128–129; cf. M. L. Tsap, *Vest. S.-Kh. Nauk* **1963** (2), 122–128.

[59b] M. F. Mal'tsev, E. N. Pashchenko, N. P. Volkova, V. P. Novak and V. E. Tubol'tseva, *Zavod. Lab.* **43**, 31–32 (1977).

[60] A. M. Dymov, L. Z. Kozel', and R. D. Iskandaryan, *Izv. Vyssh. Ucheb. Zaved., Chern. Met.* **1969** (9), 184–186.

[61] Yu. A. Klyachko, M. M. Shapiro, and E. F. Yakovleva, *Sb. Tr. Tsent. Nauch.-Issled, Inst. Chern. Met.* **1962** (24), 75–81.

to give a sludge of the inclusions. Add 100 ml of 8% ammonium citrate solution and 100 ml of 20% cuprammonium chloride solution to the filtered and washed sludge. Warm below boiling for 1 hour, then add 200 ml of water and set aside overnight. Siphon off the supernatant liquid and centrifuge the precipitate. Wash it with water, 1:20 ammonium hydroxide, and water. To complete destruction of carbides, add 25 ml of electrolyte as used for the electrolysis, 25 ml of 8% solution of ammonium citrate, and 25 ml of 20% cuprammonium chloride solution. Warm as before, let stand, centrifuge, and wash. If more than 0.5% of carbon is present in the original sample, this step may have to be repeated a third time to complete decomposition of carbides. Finally disperse the washed precipitate in 20 ml of 10% ammonium citrate solution. Filter, wash, and ignite at 800–1000° to give the total weight of nonmetallic inclusions. Treat with 5 ml of hydrofluoric acid and 1 ml of 1:1 sulfuric acid to volatilize the silica. Take to dryness and fuse with 0.2 gram of potassium bisulfate. Dissolve in water and dilute to 100 ml. Complete as for steel, alloys, and metals from "To each of two 5-ml aliquots...."

Aluminum Oxide in Aluminum.[62] The train for isolation of the sample consists sequentially of sulfuric acid, a silica tube of platinized asbestos at 450°, a trap of solid carbon dioxide and acetone, and a molybdenum-glass tube heated by an external coil. Place a boat containing 0.5 gram of granular aluminum in the molybdenum glass tube. Pass hydrogen at 5 ml/minute and hydrogen chloride at 50 ml/minute through the train for 30 minutes. Heat the furnace and boat to 300° and continue to pass the gases for 4 hours. Let the furnace cool, then turn off the hydrogen chloride. Treat the contents of the boat with 1:500 hydrochloric acid, and filter. Ignite the paper and aluminum oxides, as well as other oxides at 600–700° in a silica dish. Fuse with 1 gram of potassium pyrophosphate, let cool, and dissolve in 5 ml of water. Complete as for the next to last paragraph of titanium and titanium alloys, starting at "Add 15 ml of water and 1 ml...."

Aluminum Oxide in Aluminum Bronze.[63] As the reagent, dissolve 0.5 gram of aluminon, 25 grams of sodium acetate, and 1 gram of benzoic acid in 40 ml of glacial acetic acid.

Reflux a 5-gram sample gently with 200 ml of methanol and 10 ml of bromine. Add to the yellow solution 5 ml of bromine and 50 ml of methanol and heat again for 1 hour. Then add 5 ml of bromine and 50 ml of methanol and heat again for 1 hour. Cool, filter on a double paper, and wash the residue, first with methanol, then with 0.1 *N* hydrochloric acid. Ash the paper, ignite to 1100°, and fuse the residue with 0.25 gram of potassium bisulfate. Dissolve this fusion of the alumina in water and dilute to 100 ml. To a 10-ml aliquot add 15 ml of 0.05% aluminon solution and heat at 100° for 5 minutes. Cool, dilute to 100 ml, and read at 525 nm.

Boron Carbide.[64] Fuse the sample with sodium carbonate and potassium nitrate. Dissolve the melt in hydrochloric acid and dilute to 250 ml. Neutralize a

[62] A. A. Federov and F. V. Liukova, *ibid.* **1962** (24), 172–178.
[63] R. Dufek and L. Kopa, *Hutn. Listy* **14**, 620–622 (1959); cf. Jozefa Wolna and Jerzy Studencki, *Rudy Met. Niezel.* **14**, 207–210 (1969).
[64] M. V. Kharitonova, *Tr. Vses. Nauchn.-Issled. Inst. Abraz. Shlifovaniya* **1974** (15), 66–68.

50-ml aliquot with ammonium hydroxide and heat to 70°. Suppress the effect of boron by adding 5 ml of 10% solution of hexamine. Maintain at 70° for 15 minutes to coagulate the precipitate. Filter, wash well with 5% solution of hexamine, and dissolve the precipitate in 15 ml of 1 : 3 hydrochloric acid. Dilute to 50 ml. Dilute a 10-ml aliquot to 25 ml with 1 : 9 hydrochloric acid, add 10 ml of 0.05% solution of aluminon in 10% ammonium acetate solution, and read at 540 nm.

Iron Ores, Agglomerates, and Slags.[65] This technic is designed for samples containing 0.9–10% aluminum oxide. Treat 0.1 gram with 15 ml of hydrochloric acid and filter. Ash the filter, and fuse the residue with 3 grams of sodium carbonate. Dissolve the cooled melt in the filtrate from treatment of the sample. Evaporate to dryness and treat with 5 ml of hydrofluoric acid and 1 ml of 1 : 1 sulfuric acid. Evaporate to sulfur trioxide fumes. Take up in water and dilute to 250 ml.

Dilute a 5-ml aliquot to 10 ml, add 2 drops of phenolphthalein indicator solution, and neutralize with 1 : 20 ammonium hydroxide. Decolorize with *N* hydrochloric acid and add 2 ml in excess. Reduce ferric ion with 1 ml of 1.5% ascorbic acid solution. Add 25 ml of a 0.02% solution of aluminon in an acetate buffer for pH 4.7. Dilute to 100 ml, set aside for 1 hour, and read at 550 nm. The sample may contain 10% of chromic or manganese oxide or 5% of titanium dioxide without interference.

Manganese Ore.[66] As a buffered reagent, dissolve 500 grams of ammonium acetate in 1 liter of hot water. Add 80 ml of glacial acetic acid. Add 1 gram of aluminon in 50 ml of water containing a couple of drops of 1 : 20 ammonium hydroxide. Add 2 grams of benzoic acid in 20 ml of methanol, filter, and dilute to 2 liters. Add 1 liter of 1% gelatin solution and store in the dark for 3 days before using.

Dissolve 0.5 gram of sample by heating with 20 ml of 1 : 1 hydrochloric acid, adding small portions of 30% hydrogen peroxide until disintegrated. Dilute to 50 ml and filter. Ash the paper and the residue. Fuse with 2 grams of potassium pyrosulfate, dissolve in water, and add to the previous filtrate. Dilute to 500 ml and take a 50-ml aliquot, which should show no more than 0.2 *N* acidity. Add 2 ml of 0.5% ferric chloride solution and precipitate with excess of ammonium hydroxide. Filter, and wash with 2% ammonium chloride solution in 1 : 100 ammonium hydroxide. Dissolve the precipitate with 10 ml of 1 : 50 hydrochloric acid and dilute to 100 ml with 1 : 100 hydrochloric acid. Take an aliquot containing 0.01–0.1 mg of aluminum and dilute to 20 ml with 1 : 100 hydrochloric acid. Add 3 drops of 0.1% solution of pentamethoxy red in 70% ethanol and 3 ml of 5% solution of phenylarsonic acid. Decolorize the solution by dropwise addition of 1 : 20 ammonium hydroxide and add 3 ml of 0.2 *N* hydrochloric acid. Boil for about 4 minutes and keep hot for 1 hour, maintaining the volume by addition of hot water. Filter the

[65] O. P. Lebedev and N. I. Vlasova, *Sb. Nauch. Tr. Nauch.-Issled. Gornorud. Inst. Ukr. SSR* 7, 293–298 (1963).

[66] L. D. Dolaberidze, Yu. H. Politova, L. T. Gveslesiani, and A. G. Dzhaliashvili, *Zavod. Lab.* **30** 1439–1441 (1964).

titanium precipitate, wash with hot water, and discard. Add 3 ml of fresh 5% ascorbic acid solution and maintain at 100° for 4 minutes. Add 15 ml of the buffered solution of aluminon and maintain at 100° for 3 minutes. Cool quickly, dilute to 100 ml, and read at 540 nm.

Fused Zinc-Plating Bath.[67] Dissolve 1 gram of sample containing 0.05–0.15% of aluminum in 10 ml of 1:1 hydrochloric acid and evaporate to dryness. Dissolve in 50 ml of water containing 0.25 ml of 1:1 hydrochloric acid. Filter and dilute to 200 ml. To a 5-ml aliquot add 10 ml of 10% solution of ammonium acetate, 2 ml of 1:9 hydrochloric acid, and 2 ml of 0.2% solution of aluminon. Dilute to 50 ml and set aside for 20 minutes. Read at 550 nm against a blank.

Rock Salt. ALUMINUM, TITANIUM, IRON, AND MAGNESIUM.[68] Dissolve a 50-gram sample in 200 ml of water and 10 ml of 1:6 hydrochloric acid. Heat to 70° and add an excess of 5% solution of oxine in acetic acid. Neutralize with ammonium hydroxide to thymolphthalein, pH 9.5–10.5. Hold at 60–70° for 20 minutes and let stand for 2 hours. Filter, and wash the precipitate of aluminum, titanium, iron, and magnesium with 1:20 ammonium hydroxide. Transfer the precipitate and paper to 8 ml of sulfuric acid and heat at 100° to vigorous fuming and blackening of the residue. Cover and heat, adding nitric and perchloric acids intermittently until oxidation is complete and evolution of nitrogen oxides ceases. Evaporate to sulfur trioxide fumes and take up in 50 ml of 1:50 perchloric acid. Heat nearly to boiling and filter silica and calcium sulfate, washing with 1:100 perchloric acid. Dilute to 400 ml.

For aluminum, take a 20-ml aliquot and add 2 ml of 2% solution of ascorbic acid. Let stand for 5 minutes and add ammonium hydroxide to pH 2. Add 5 ml of 10% solution of polyvinyl alcohol in 10% ethanol, 10 ml of a 10% solution of ammonium acetate, 10 ml of 0.2% solution of aluminon, and water to make 50 ml. Let stand for 15 minutes and read at 536 nm.

For titanium, take a 20-ml aliquot. Add 5 ml of 2% ascorbic acid solution and 2 ml of a solution containing 2% of sodium chromotropate and 0.1% of ascorbic acid. Neutralize to pH 2–3 by dropwise addition of ammonium hydroxide. Dilute to about 35 ml and heat to 70°. Add 10 ml of a buffer solution containing 57 ml of formic acid and 28 grams of sodium hydroxide per liter. Cool and dilute to 50 ml. Set aside for 15 minutes and read at 650 nm.

For iron, take a 20-ml aliquot and add 2 ml of 2% ascorbic acid solution. Set aside for 5 minutes, then add 1 ml of 20% ammonium citrate solution and neutralize to about pH 4 with ammonium hydroxide. Dilute to 25 ml and add 5 ml of a 0.25% solution of 1,10-phenanthroline. Add 5 ml of a 25% solution of ammonium acetate, dilute to 50 ml, set aside for 5 minutes, and read at 396 nm.

For magnesium, take a 20-ml aliquot and add 5 ml of 0.04% solution of titan yellow. Add 5 ml of 1% solution of polyvinyl alcohol in 10% ethanol. Dilute to 25 ml and add 4% sodium hydroxide solution dropwise until the color changes, and add 2 ml in excess. Dilute to 50 ml, set aside for 10 minutes, and read at 550 nm against water.

[67] R. Sirmanov, *Mashinostroenie* **11** (5), 29–30 (1962).
[68] Z. Marczenko and A. Stepfen, *Chem. Anal.* (Warsaw) **5**, 247–259 (1960).

Liming Materials.[69] As a reagent, prepare as separate solutions: 0.5% aluminon, 5% filtered gum acacia, and 25% ammonium acetate. Add 56 ml of hydrochloric acid to the ammonium acetate solution, adjust to pH 4.5, and add the other solutions.

Grind the sample to pass 100-mesh. To 0.5 gram of limestone or 0.2 gram of silicate in a nickel crucible, add 0.3 gram of potassium nitrate and 1.5 gram of sodium hydroxide. Heat to a dull redness for 5 minutes and swirl to cool the melt on the sides. Disintegrate with 50 ml of water. Add 15 ml of 1 : 1 perchloric acid and dilute to 100 ml.

Dilute an aliquot containing less than 0.08 mg of aluminum to 20 ml. Add 2 ml of 1% solution of thioglycollic acid, 0.5 ml of a dispersion of 0.03 gram of Dow-Corning Antifoam A in 100 ml of water, and 10 ml of the aluminon reagent. Heat at 100° for 20 minutes, let cool, and dilute to 100 ml. Read at 525 nm against a reagent blank.

Soil Extracts.[70] Take an aliquot of an extract in 0.01 *M* calcium chloride containing 2.5–25 μg of aluminum. Add 2 ml of 1 : 1 nitric acid, evaporate to dryness, and heat in a muffle furnace at 400° for 15 minutes. Take up the residue in 2 ml of 1 : 1 hydrochloric acid and evaporate to dryness at 100°. Again heat in a muffle furnace at 100°, this time for 1 hour. Take up in 1 ml of 1 : 1 hydrochloric acid, set aside for 30 minutes, dilute to 15 ml, and centrifuge. Take an aliquot of the clear layer and add 0.1 *N* hydrochloric acid and 0.1 *M* calcium chloride such that the test solution contains 1 millimole of each. Add 1 ml of 5% hydroxylamine hydrochloride solution and 15 ml of 0.1% aluminon solution containing 1 ml of 10% methanolic benzoic acid per 100 ml. Heat at 100° for 10 minutes, cool, and read at 530 nm. The technic for plants that follows is also applicable to soil extracts.

Plants.[71] Ash 0.2 gram of sample overnight at 450°. Moisten the ash, add 1 ml of 1 : 1 hydrochloric acid, and evaporate to dryness. Take up the residue in 10 ml of 0.05 *N* hydrochloric acid and let insoluble matter settle for 1 hour. Dilute an aliquot containing up to 50 μg of aluminum to 10 ml and add 1 ml of 0.5% solution of ascorbic acid and 2 ml of *N* hydrochloric acid. Heat at 100° for 30 minutes and cool. Dilute to 35 ml, add 10 ml of a reagent containing 0.035% of aluminon, 12% by volume of acetic acid, and 2.4% of sodium hydroxide. Dilute to 50 ml and set aside for 1 hour. Read at 530 nm against a reagent blank.

Chromic–Phosphoric Acid Solutions.[72] This technic is designed for analysis of corrosion products stripped by 2% chromic–5% phosphoric acid solution.

To prepare a buffered aluminon reagent, dissolve 136 grams of sodium acetate trihydrate, 57 grams of acetic acid, and 0.160 gram of aluminon in water. Dilute to 1 liter and store for 3 days before using. As mercaptoacetic acid reagent, dilute 5

[69] P. Chicilo, *J. Assoc. Off. Agr. Chem.* **47**, 620–626 (1964).

[70] C. R. Frink and M. Peech, *Soil. Sci.* **93**, 317–324 (1962); cf. Pa Ho Hsu, *ibid.* **96**, 230–238 (1963); S. A. Harris, *J. Sci. Food Agr.* **14**, 259–263 (1963); Chang Wang and F. A. Wood, *Can. J. Soil Sci.* **53**, 237–239 (1973).

[71] T. C. Z. Jayman and S. Sivasubramaniam, *Analyst* **99**, 296–301 (1974).

[72] Cornelius Groot, R. M. Peekema, and V. H. Troutner, *Anal. Chem.* **28**, 1571–1576 (1956).

ml to 100 ml, titrate to pH 4.7 with about 10 ml of 1 : 1 ammonium hydroxide, and dilute to 250 ml.

Dilute an aliquot of the stripping solution containing 2–40 mg of aluminum with at least 6 volumes of water and neutralize to pH 5–5.5 with ammonium hydroxide. Let stand overnight, filter, and wash free of dichromate color. Dissolve with four 5-ml portions of hot 1 : 1 hydrochloric acid, then wash with water to a volume of 100 ml.

To an aliquot containing 10–40 μg of aluminum add 25 ml of buffered aluminon reagent and 2 ml of 2% mercaptoacetic acid reagent at pH 4.7. Dilute to 50 ml. At 30 ± 0.5 minutes after adding the aluminon reagent, read at 530 nm.

Alternatively, prepare an anion exchange column as shown in Figure 15-1. Use 50-100 mesh Dowex 1-X8 in chloride form. Pass an aliquot containing 0.1–40 mg of aluminum through the column. Follow with 10 ml, 10 ml, and 10 ml of wash water, and dilute to 100 ml. Complete an aliquot as above.

Beer.[73] As a reagent, dissolve in 250 ml of water sequentially 50 grams of hydroxylamine hydrochloride, 10 grams of sodium mercaptoacetate, 3 ml of hydrochloric acid, and 135 grams of ammonium acetate. Add 1.5 gram of aluminon in 100 ml of water and refrigerate in dark glass.

To 10 ml of degassed beer, add 50 ml of water and 2 ml of reagent. Set aside for 30 minutes and read at 525 nm. The calibration curve is not rectilinear but is reproducible.

Leather.[74] As a buffer, mix 47 ml of ammonium hydroxide with water and 43 ml of acetic acid. Dilute to nearly 1 liter, titrate to pH 5.3–5.4, and dilute to volume. As aluminon reagent, dissolve 1 gram in 100 ml of 1% gelatin solution.

Digest 0.5 grams of sample with 5 ml of perchloric acid and 5 ml of 1 : 1 nitric

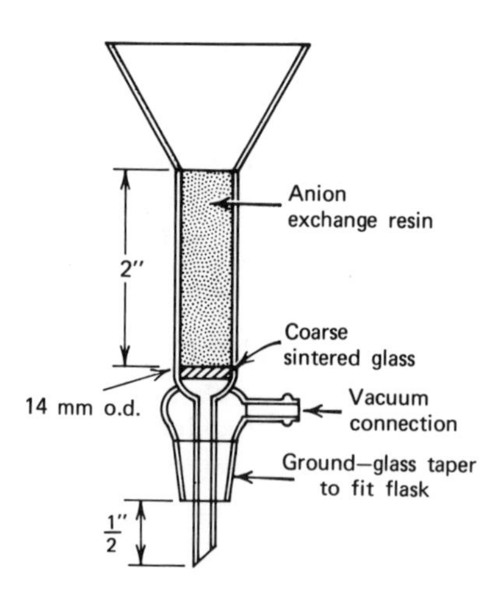

Figure 15-1 Anion exchange column.

[73] H. O. Etian and M. A. Rovella, *Proc. Am. Soc. Brew. Chem.* **1966**, 177–179.

[74] Ferdinand Langmaier, Drohoslav Kokes, and Zdenka Slovacekova, *Kozarstvi* **8**, 166–168, 198–199 (1958).

acid. After 15 minutes boil for 3 hours. Add 20 ml of water and dilute to 100 ml. To a 5-ml aliquot, add 3 ml of the buffer solution and 5 ml of aluminon reagent. Dilute to 50 ml and digest at 80° for 15 minutes. Cool and read at 535 nm.

CALCICHROME

The complex of calcichrome with aluminum is stable for 24 hours. It has maxima at 317 and 620 nm when read against the reagent. The maximum in the visible spectrum is more convenient, but that in the ultraviolet range is more sensitive. It follows Beer's law up to 1 ppm of aluminum. There is interference by cobalt, chromic, cupric, ferric, mercuric, magnesium, titanic, pentavalent vanadium, zirconium, oxalate, and thiosulfate ions.

Procedure. Mix the sample solution with 5 ml of 0.2 m*M* solution of calcichrome and 5 ml of sodium acetate–acetic acid buffer solution for pH 6.[75] Dilute to 25 ml and set aside for 30 minutes. Read at 620 nm against a reagent blank.

CATECHOL VIOLET

The 2:1 complex of catechol violet with aluminum conforms to Beer's law for 0.2–0.8 ppm of aluminum oxide.[76] As applied to beer, there is interference only by iron, which causes high results. If the aluminum exceeds 1.5 ppm, the sample must be diluted, preferably with beer of low aluminum content. As applied to alloys, zirconium and titanic ions must be absent, ferric ion masked with mercaptoacetic or ascorbic acid, and cupric ion masked with thiosulfate.[77] The reaction is instantaneous, and the color is stable for hours.

A 1:2:5 ternary complex is formed by aluminum, catechol violet, and hexadecyltrimethylammonium bromide.[78a] Then there is interference by beryllium, chromium, rare earths, vanadate, zirconium, and tartrate. A ternary complex is also formed with 1,10-phenanthroline.

The binary complex is not extractable with organic solvents, but as a blue-green ternary complex with benzyldimethylphenylammonium chloride it is extractable with butanol or isobutanol.[78b] Optimum conditions for extraction of 0.08 m*M* aluminum are pH 6.5–7.5, fivefold excess of catechol violet, and 1000-fold excess of the quaternary. Beer's law is followed for 0.05–0.5 μ9/ml. There is interference by manganous ion, zinc, cupric ion, cobalt, nickel, ferric ion, chromic ion, and gallium.

At pH 6.3–7.9 a 1:3:1 complex formed in 0.01 m*M* aluminum–0.05 m*M*

[75] Hajime Ishii and Hisahiko Einaga, *Bull. Chem. Soc. Jap.* **39**, 1721–1724 (1966).
[76] I. Stone, C. S. Gantz, and L. T. Saletan, *Proc. Am. Soc. Brew. Chem.* **1963**, 149–157.
[77] I. S. Mustafin, L. A. Molot, and A. S. Arkhangel'skaya, *Zh. Anal. Khim.* **22**, 1808–1811 (1967).
[78a] J. E. Chester, R. M. Dagnall, and T. S. West, *Talenta* **17**, 13–19 (1970).
[78b] M. M. Tananaiko and O. P. Vdovenko, *Ukr. Khim. Zh.* **42**, 752–755 (1976).

catechol violet–0.1 *M* diphenylguanidine is extractable with isobutyl alcohol for reading at 590 nm.[79a] The minimum is 0.03 mg of aluminum per ml. There is interference by nickel, cobalt, ferric ion, cupric ion, zinc, gallium, chromic ion, and manganous ion.

Aluminum at pH 7.4–10 forms a ternary complex in the presence of fivefold catechol violet and eightyfold hexadecylpyridinium ion.[79b] Extracted with butanol, this is read at 540 nm and obeys Beer's law for 3–30 mg of aluminum per ml. There is interference by antimonic, tetra- and pentavalent vanadium, ferric, titanic, and zirconium ions.

For aluminum in fused silica, heat 5 grams of sample nearly to dryness with water, hydrofluoric acid, and nitric acid.[80a] Take up with water and develop with catechol violet.

For aluminum in anode sludge, decompose the sample with perchloric acid and hydrofluoric acid.[80b] Filter. Ignite the residue, fuse with sodium carbonate and sodium peroxide, dissolve the melt, and add to the prior filtrate. Add mercapto-acetic acid and potassium cyanide solutions as masking agents. Extract aluminum as the complex with *N*-phenylbenzohydroxamic acid in benzene. Back-extract the aluminum with 2 *N* hydrochloric acid and develop with catechol violet.

For aluminum in zirconium metal dissolve by heating with sulfuric acid and ammonium sulfate. Take up in water, dilute to 2 *N* acid and boil to depolymerize the zirconium. Add cupferron and extract the zirconium complex with chloroform. Evaporate the aqueous layer to sulfur trioxide fumes adding hydrogen peroxide if necessary to decompose organic matter. Dilute with water and develop with catechol violet.

Procedure. As a buffer solution for pH 5, mix 77 ml of pyridine with 63 ml of acetic acid.[81] Take an aliquot of sample containing 0.3–1.5 mg of aluminum. Add 10 ml of buffer solution and 2 ml of a reagent containing 0.3864 gram of catechol violet per 100 ml. Dilute to 100 ml and read at 615 nm against a reagent blank. An acetate buffer for pH 6 can be substituted.[82]

As the Ternary Complex with a Quaternary. To 25 ml of sample solution, add 2 ml of 50% ammonium acetate solution, 2 ml of 2% solution of 1,10-phenanthroline, 1 ml of 10% hydroxylamine hydrochloride solution, and 5 ml of 10% ammonium benzoate solution. Adjust to pH 7.5–9 and extract the aluminum with 10 ml of 5% solution of benzoic acid in ethyl acetate. Reextract the aluminum from the organic layer with 10 ml and 10 ml of *N* hydrochloric acid. Mix the combined aqueous extracts with 25 ml of a reagent containing 0.00773% of catechol violet and 0.0729% of hexadecyltrimethylammonium bromide. Add 5 ml of 5% ascorbic acid solution and follow with ammonium hydroxide until the solution is green. Add 5

[79a] M. M. Tananaiko, O. P. Vdovenko, and V. M. Zatsarevnyi, *Zh. Anal. Khim.* **29**, 1724–1728 (1974).
[79b] M. M. Tananaiko and O. P. Vdovenko, *Zh. Anal. Khim.* **30**, 1095–1100 (1975).
[80a] K. F. Sagawara and Yao-sin Su, *Anal. Chim. Acta* **80**, 143–151 (1975).
[80b] E. Komarkova, *Rep. Natn. Inst. Metall.* No. **1808**, 9 pp. (1976).
[81] Anthony Anton, *Anal. Chem.* **32**, 725–726 (1960).
[82] Katu Tanaka and Katutosi Yamayosi, *Jap. Anal.* **13**, 540–544 (1964).

ml of an ammoniacal buffer solution for pH 10.2. Set aside for 20 minutes and add 5 ml of 0.1 *M* EDTA. Dilute to 100 ml and read at 670 nm.

As the Ternary Complex with 1,10-Phenanthroline. To 35 ml of sample solution, add 1 ml of 0.1% solution of 1,10-phenanthroline in 10% hydroxylamine hydrochloride, 2 ml of 0.0375% solution of catechol violet, and 10 ml of 30% ammoniacal hexamine buffer solution.[83] Set aside for 10 minutes and read ar 585 nm. Fluoride should not exceed 1 mg/liter. The method will determine 0.05–0.3 mg of aluminum per liter.

Magnesium Alloys or Silicates.[84] To 2 ml of sample solution containing 1–56 μg of aluminum, add 0.25 ml of 0.5% solution of ascorbic acid to mask iron or chromium. Add 3 ml of 4% boric acid to prevent interference by calcium or magnesium and 10 ml of succinate-borate buffer solution for pH 5. Add 5 ml of 0.03% catechol violet solution, dilute to 50 ml, and read at 582 nm.

Silicate Minerals.[85] Fuse a 5-mg sample with 1 gram of sodium hydroxide in a gold crucible. Dissolve in water and add hydrochloric acid to pH 6. To a blank in parallel, add amounts of iron and titanium equivalent to these in the sample. Add to each, 4.5 ml of 0.12 *N* perchloric acid and 5 ml of 4% solution of mercaptoacetic acid. Set aside for 10 minutes; then add 10 ml of 0.03% solution of catechol violet. Let stand for 10 minutes; then dilute to 80 ml. Add 10 ml of 50% ammonium acetate solution, dilute to 100 ml, and set aside for 1 hour. Read at 580 nm.

As an alternative,[86] to 20 ml of slightly acid sample solution containing not more than 0.04 mg of aluminum, add successively 2 ml of 10% hydroxylamine hydrochloride solution, 2 ml of 0.15% solution of 1,10-phenanthroline, 2 ml of 0.15% solution of catechol violet, and 5 ml of ammonium acetate–acetic acid buffer solution for pH 6.2. Adjust to pH 6.1–6.2 if necessary, add 50 ml more of buffer solution, and dilute to 100 ml. Let stand for 2 hours, read at 580 nm, and subtract a reagent blank.

If interfering amounts of iron, titanium, thorium, zirconium, and vanadium are present, add an equal volume of hydrochloric acid to make the solution at least 6 *N*. Add 5 ml of 8% solution of cupferron. Extract the interfering elements with 10 ml and 10 ml of chloroform. Then proceed as above, but omit the addition of 1,10-phenanthroline, which was to mask iron.

Lanthanum Aluminate–Calcium Titanate.[87] Fuse 0.1 gram of sample with 1 gram of sodium hydroxide and dissolve in 15 ml of water. Neutralize with 1 : 1 hydrochloric acid, heat with 25 ml of hydrochloric acid, to give a clear solution, and dilute to 100 ml. Shake a 5-ml aliquot with 30 ml of 2% solution of

[83] W. K. Dougan and A. L. Wilson, *Analyst* **99**, 413–430 (1974); cf. Torstein Dale and Arne Hendriksen, *Vatten* **31**, 91–93 (1975).

[84] V. F. Mal'tsev, N. P. Volkova, E. N. Pashchenko, and P. P. Dubinskaya, *Ukr. Khim. Zh.* **40**, 539–541 (1974).

[85] Robert Meyrowitz, *U.S. Geol. Survey Prof. Pap.* No. **700-D**, D225–D229 (1970).

[86] A. D. Wilson and G. A. Sergeant, *Analyst* **88**, 109–112 (1963).

[87] L. A. Molot, A. S. Arkhangel'skaya, M. I. Trusova, and L. S. Mustafin, *Zavod. Lab.* **34**, 408 (1968).

N-phenylbenzohydroxamic acid in chloroform. Repeat the extraction with 15 ml and 15 ml of the reagent in chloroform to complete removal of the titanium complex. Heat the aqueous phase to evaporate residual chloroform and dilute to 50 ml. Mix a 10-ml aliquot with 2 ml of 0.1% solution of catechol violet and 20 ml of a buffer solution for pH 6–6.3. Heat to 60–70°, cool, dilute to 50 ml, and read at 580 nm.

Beer.[88] Take two 10-ml portions of the sample and two 10-ml portions of fourfold dilution. Add to each, 25 ml of 0.2 *M* disodium phosphate. To one undiluted and one diluted sample, add 2 ml of fresh 0.02% solution of catechol violet. Dilute each to 50 ml and set aside for 30 minutes. Read the treated samples at 645 nm against the untreated blanks.

CHLORCYANFORMAZAN

This reagent, which is the disodium salt of 1,5-bis(5-chloro-2-hydroxy-3-sulfophenyl)-3-cyanoformazan, forms a blue 1 : 1 complex with aluminum around pH 4.

Procedure. *Titanium Dioxide.*[89a] As a buffer solution, mix 859 ml of 0.2 *N* acetic acid with 141 ml of 0.2 *N* ammonium hydroxide. Fuse 0.1 gram of sample with 2 grams of sodium carbonate. Dissolve in 50 ml of water, filter, and dilute to 250 ml. Add 2 drops of 2,4-dinitrophenol indicator solution to a 25-ml aliquot and neutralize with 1 : 10 hydrochloric acid. Add 5 ml of 0.1% solution of the reagent and 100 ml of buffer solution. Heat to 100°. Cool, dilute to 250 ml with the buffer solution, and read at 670 nm against a reagent blank.

CHLORSULFOPHENOL S

The captioned reagent is 3,6-bis(5-chloro-2-hydroxy-3-sulfophenolazo)chromotropic acid. A 5–6-fold concentration of the complex is effected by extraction with a m*M* solution of benzylhexadecyldimethyl ammonium chloride in organic solvent.[89b]

Procedure. *Steel.* Dissolve 0.5 gram of sample in 40 ml of 1 : 1 hydrochloric acid and add sufficient nitric acid to oxidize the ferrous iron. Boil for 10 minutes and add 60 ml of 20% sodium hydroxide solution. Dilute to 250 ml and filter. To an aliquot of the filtrate containing 5–40 mg of aluminum, usually 5 or 10 ml, add 1 : 1

[88] R. Parsons, *Proc. Am. Soc. Brew. Chem.* **1964**, 152–154.
[89a] V. A. Malevannyi, M. I. Ermakova, and Yu. L. Lel'chuk, *Zavod. Lab.* **35**, 414–417 (1969).
[89b] N. A. Ivanov and N. G. Todorov, Dolk. *Bolg. Akad. Nauk* **29**, 1775–1778 (1976).

sulfuric acid until neutral to phenolphthalein. Add 25 ml of acetate buffer solution for pH 4 and 3 ml of 0.05% solution of chlorsulfophenol S. Dilute to 50 ml and heat at 60° for 10 minutes. Cool, and read at 660 nm against a blank prepared from aluminum-free steel.

CHROMAZOL KS

This reagent is the trisodium salt of 2-hydroxy-3-(1,8-dihydroxy-3,6-disulfo-2-naphthylazo)-4-sulfobenzoic acid. At pH 3.9 the aluminum complex is stable and obeys Beer's law for 0.08–1.1 μg of aluminum per ml. In all cases the reagent solution shoud be added to a test solution at a pH below 1.5.

Procedures

Ferrotitanium.[90] Dissolve 50 mg of sample containing 1–10% of aluminum in 4:1:2 mixture of 1:1 sulfuric-nitric-hydrochloric acid. Evaporate to sulfur trioxide fumes and take up with 25 ml of 1:1 hydrochloric acid. Filter and dilute to 500 ml. Mix a 2-ml aliquot with 1 ml of 2% solution of ascorbic acid, 1 ml of 1% solution of tartaric acid, 5 ml of 0.25% solution of chromazol KH, and 25 ml of water. Add 5% hexamine solution to adjust to pH 3.9 and dilute to 50 ml. Set aside for 20 minutes and read at 600 nm against a reagent blank.

Ferrotungsten. Fuse 0.2 gram of sample with 2.5 grams of sodium peroxide and dissolve in water. Acidify with excess of hydrochloric acid and evaporate to dryness. Take up in 25 ml of 1:1 hydrochloric acid, filter, and dilute to 250 ml. Mix a 2-ml aliquot with 1 ml of 2% solution of ascorbic acid and 5 ml of 0.25% solution of chromazol KH. Add 5% hexamine solution to adjust to pH 3.9 and dilute to 50 ml. Set aside for 20 minutes and read at 600 nm against a reagent blank.

Steel. Dissolve 1 gram of sample containing 0.1–1% aluminum in 15 ml of 1:1 hydrochloric acid and 0.25 ml of nitric acid. Dilute to 100 ml. Mix a 1-ml aliquot with 1 ml of 2% ascorbic acid solution, 1 ml of 1% anthranilic acid solution, and 5 ml of 0.25% solution of chromazol KS. Add 5% hexamine solution to adjust to pH 3.9 and dilute to 50 ml. Set aside for 20 minutes and read at 600 nm against a low carbon steel.

CHROME AZUROL S

Aluminum forms a 1:1 complex with this reagent at pH 2.9 and a 1:2 complex at pH 6.4. In determining aluminum with chrome azurol S (mordant blue 29),

[90] N. N. Basargin, P. Ya. Yakoviev, and O. N. Morozova, *ibid.* **40**, 1322–1325 (1974).

masking avoids interference by cations other than by beryllium, tin, titanium, thorium, and zirconium.[91] Beer's law is followed up to 1.2 μg of aluminum per ml. Up to 0.5% of chromium, copper, and vanadium can be tolerated. Tartrate, EDTA, oxalate, citrate, phosphate, and fluoride interfere.[92] Reports about the time required for development of color and the duration of stability vary quite widely. Beer's law is followed up to 0.1 μg of aluminum per ml as developed.

The result of the determination is affected by the order of mixing the reagents.[93] The following is recommended: 5 ml of 0.1 *N* hydrochloric acid, adjustment to pH 5.7–5.8 with 5 ml of 2 *N* sodium acetate, and addition of 2 ml of 0.1% solution of the dye.

Although both ascorbic acid and mercaptoacetic acid mask ferric ion by reducing it, they also decrease the extinction and color stability of the complex of aluminum with chrome azurol S.[94] Ascorbic acid will mask about 6 times as much iron as mercaptoacetic acid, but when it is used the reading should be taken within 10 minutes after development of the color.

For analysis of magnesium and magnesium alloys, dissolve in sulfuric acid, mask iron with ascorbic acid and copper with sodium thiosulfate, buffer, and determine with chrome azurol S.[95] The reagent is also applicable to cement analysis,[96] in which case interference by calcium must be avoided.[97]

When ferric ion has been reduced to the ferrous state, the reading of aluminum will vary unless the same amount is present in the standard, a problem avoided by having a large excess present. Mercaptoacetic acid is a common reducing agent for the iron and also masks copper. Because it decreases the absorption of the aluminum chelate slightly, the amount of mercaptoacetic acid must be rigorously standardized. As an alternative to reducing it, a large amount of iron can be removed by extraction with methyl isobutyl ketone.

Addition of cetrimide, which is cetyltrimethylammonium bromide, to form an orange ternary complex at pH 5.6, leads to absorption at 615 nm as compared with 430 nm for the reagent.[98a] A typical composition in 50 ml is less than 4 μg of aluminum, 10 ml of 0.05% chrome azurol S containing 0.02% of the quaternary compound, and 5 ml of 4 *M* acetate buffer solution for pH 5.8–6. Gallium, titanium, and thorium interfere with the ternary complex. With up to 0.01 mg of aluminum, 5 ml of 0.05% dye solution, and 5 ml of 0.1% cetrimide solution, the full color develops in 15 minutes. Beer's law is followed up to 0.2 μg of aluminum per ml. With ascorbic acid present, iron, copper, and nickel are masked with 1,10-phenanthroline.

As another application of a surfactant, mix 10 ml of sample 0.0002–0.002 m*M* in aluminum with 1 ml of 0.18% solution of chrome azurol S, 2 ml of 0.09% solution

[91] P. Pakalns. *Anal. Chim. Acta* **32**, 57–63 (1965).

[92] V. P. Novak, A. P. Marynov, and V. F. Mal'tsev, *Zh. Anal. Khim.* **28**, 657–660 (1973), cf. V. B. Sokolovich, Yu. L. Lel'chuk, and B. N. Besprozvanykh, *Izv. Tomsk. Politekh. Inst.* **128**, 112–116 (1964).

[93] V. N. Tikhonov, *Nauch. Tr. Tul'sk. Gos. Pedagog. Inst.* **1968** (1), 19–27.

[94] V. N. Tikhonov, *Zh. Anal. Khim.* **26**, 65–68 (1971).

[95] British Standards Institution, *BS 3907*, Pt. **12**, 8 pp (1971).

[96] A. I. Voinovitch, R. Barbas, G. Cohort, G. Koelbel, G. Legrand, and J. Louvrier, *Chim. Anal.* (Paris) **50**, 334–349 (1968).

[97] M. E. Ribiero Coelho, *Tecnica* (Lisbon) **29**, 533–536 (1967).

[98a] Yoshio Shijo and Tsugio Takeuchi, *Jap. Anal.* **17**, 61–65 (1968); V. N. Tikhonov and L. V. Yarkova, *Zavod. Lab.* **41**, 1180–1183 (1975).

of Dispergator 80, and 10 ml of 4 *M* sodium acetate.[98b] Dilute to 25 ml, set aside for 20 minutes and read at 645 nm against a blank. Titanic ion interferes and only small amounts of manganese, lead, vanadium, and tungsten are tolerated.

For determination of aluminum in tin, the latter is volatilized as stannic chloride.[99] Up to 20% aluminum in titanium and magnesium alloys can be read with this reagent without more than ±0.5% error.[100a] A determination of aluminum in solder alloys by chrome azurol S is included under Copper as occurring before the determination of the latter by biscyclohexanone oxalyldihydrazone, also known as cuprizone. For aluminum in solder alloys by chrome azurol S, see Copper by Cuprizone.

Procedure. Add to the sample solution 8 ml of 0.18% solution of chrome azurol S and 40 ml of *M* ammonium acetate–acetic acid buffer solution for pH 6.[100b] Dilute to 100 ml and read at the isobestic point, 568 nm.

Iron and Steel.[101] Dissolve 0.5 gram of sample in 20 ml of 1:1 hydrochloric acid and 1 ml of nitric acid. Evaporate to dryness, dissolve the residue in 20 ml of 1:1 hydrochloric acid, and filter. Ash the filter and treat the residue with 5 ml of hydrofluoric acid and 1 ml of 1:1 sulfuric acid. Evaporate to sulfur trioxide fumes, then fuse with 1 gram of potassium pyrosulfate. Dissolve in water and add to the main filtrate. Evaporate to about 10 ml and add 14 ml of hydrochloric acid. Extract with 20 ml and 20 ml of methyl isobutyl ketone and discard these extracts. Evaporate to dryness. Add 4 ml of hydrochloric acid and 1 ml of nitric acid. Evaporate to dryness, take up the residue in 5 ml of 1:1 hydrochloric acid, and dilute to 100 ml. To a 10-ml aliquot add 1 ml of 10% mercaptoacetic acid solution, then add 2 ml of 0.1% solution of chrome azurol S. Adjust to pH 5.6–6.8 with 0.2 *M* sodium acetate–acetic acid, and dilute to 100 ml. After 20 minutes read at 550 nm and subtract a blank.

Cast Iron with Zephiramine.[102] Dissolve 0.2 gram of sample in 1 ml of hydrochloric acid and 5 ml of water. Add 0.5 ml of nitric acid, boil off oxides of nitrogen, and filter. Wash the insoluble matter with 1:100 nitric acid and dilute to 100 ml. Take an aliquot expected to contain up to 5 μg of aluminum. Mask the iron with 1 ml of 8% mercaptoacetic acid solution. Add 1 ml of 0.2% solution of zephiramine and 4 ml of 7% solution of hexamine. Adjust to pH 4.9 with 2% sodium hydroxide solution. Dilute to 25 ml, set aside for 30 minutes, and read at 620 nm against a blank.

[98b] Gerhard Roebisch, *Chemia Analit.* **21**, 1061–1068 (1976).
[99] M. I. Shvaiger and E. I. Rudenko, *Zavod. Lab.* **26**, 939 (1960).
[100a] V. N. Tikhonov, *Zh. Anal. Khim.* **19**, 1204–1209 (1964).
[100b] G. Roebisch, *Anal. Chim. Acta* **82**, 207–211 (1976).
[101] Minoru Hosoya, Yachiyo Kakita, and Hidehiro Goto, *Sci. Rep. Res. Inst., Tohoku Univ., Ser. A* **13**, 206–216 (1961); cf. H. Brockmann and H. Keller, *Arch. Eisenhüttenw.* **35**, 367–369 (1964); V. A. Verbitskaya, V. V. Stepin, I. A. Onorina, and L. S. Studenskaya, *Tr. Vses. Nauchn.-Issled. Inst. Stand. Obraztsov. Spektr. Etalonov* **2**, 52–53 (1965); L. Buck, *Chim. Anal.* (Paris) **47**, 10–16 (1965); Detlev Thierig, *Arch. Eisenhüttenw.* **41**, 895–897 (1970); V. I. Putkh, G. I. Chadova, A. S. Aksel'rod, and K. N. Mamina, *Zavod. Lab.* **43**, 30–31 (1977).
[102] Hiroshi Nashida, *Jap. Anal.* **22**, 971–975 (1973).

Titanium-Bearing Steel.[103] Dissolve 1 gram of sample with 50 ml of 1:20 sulfuric acid. Filter, wash the residue with hot water, and reserve it for separate analysis. Add 10 ml of nitric acid to the filtrate and washings to oxidize the iron. Then add 20% sodium hydroxide solution until only faintly acid. Pour into 20 ml of 20% sodium hydroxide solution to precipitate the iron and titanium. Cool, dilute to 200 ml, and filter. Add 5 drops of 10% mercaptoacetic acid and 2 drops of methyl orange indicator solution to a 10-ml aliquot. Titrate with 1:20 sulfuric acid. To another 10-ml aliquot add 5 drops of 10% mercaptoacetic acid, the volume of 1:20 sulfuric acid required for the titration, and 0.2 ml in addition. Add 0.3 ml of 0.3% solution of chrome azurol S and dilute to 50 ml with an acetate buffer solution for pH 5.8 –6. Let stand for 20 minutes and read at 545 nm.

Ash the paper and fuse the insoluble residue with 2 grams of potassium pyrosulfate. Dissolve in 20 ml of 1:1 nitric acid to oxidize the iron. Complete as above from "Then add 20% sodium hydroxide..."

Uranium Alloys.[104] Dissolve 0.1 gram of sample in 10 ml of 1:1 hydrochloric acid, add 1 ml of nitric acid, and heat until the color is golden yellow. Add 5 ml of hydrochloric acid and evaporate nearly to dryness. Repeat that step twice more. Dissolve in hydrochloric acid and dilute with water so that an aliquot will contain 5–20 μg of aluminum and 0.06 ml of hydrochloric acid. Add 0.5 mg of ammonium tartrate to mask zirconium, thorium, niobium, and molybdenum. Add 2 ml of 10% solution of hydroxylamine hydrochloride, which will mask 3 mg of uranium. If iron is present, add 2 ml of 1% solution of ascorbic acid. Add 5 ml of 0.05% solution of chrome azurol S and dilute to 30 ml. Add 5 ml of 25% solution of sodium acetate and dilute to 50 ml. Set aside for 30 minutes and read at 545 nm.

Ore, Sinter, or Slag.[105a] As a standard solution, dissolve 1 gram of iron in 20 ml of hydrochloric acid and 0.2 ml of nitric acid. Evaporate to approximately 2 ml and add 15 ml of 1:1 hydrochloric acid. Dilute to 250 ml.

Mix 0.25 gram of sample for levels below 1% of alumina or 0.1 gram for up to 5% of alumina with 1.5 gram of sodium peroxide in a nickel crucible. Fuse at a dull red until no undissolved specks are visible. Cool, and dissolve with 10 ml of water. Wash the crucible with 40 ml of water, add 15 ml of 1:1 hydrochloric acid, and warm to a clear solutuon. Dilute to 250 ml for samples containing up to 2.5% of alumina or to 500 ml for those containing up to 5%.

If fluoride is present, treat a 5-ml aliquot with 2 ml of perchloric acid and heat through vigorous fuming to the absence of fumes. Dissolve the residue in 5 ml of standard iron solution and cool. In the absence of fluoride take a 5-ml aliquot and mix with 5 ml of standard iron solution.

In either case, now add 10 ml of 1.6% ascorbic acid solution and let stand for exactly 10 minutes. Add 30 ml of 7.5% solution of sodium acetate trihydrate followed by 5 ml of a 0.04% solution of chrome azurol S in 50% ethanol. Dilute to 100 ml. The color develops in 5 minutes and is stable for 20 minutes. Read at 545 nm after 10 minutes and subtract a blank.

[103] V. D. Konkin and L. A. Kvichko, *Zavod. Lab.* **37**, 538–539 (1971).
[104] L. S. Mal'tseva and L. B. Kubareva, *ibid.* **35**, 1299–1301 (1969).
[105a] Om. P. Bhargava and W. Grant Hines, *Anal. Chem.* **40**, 413–415 (1968).

Water

IONIC ALUMINUM.[105b] Mix 50 ml of sample, 5 ml of 0.1 *N* hydrochloric acid, 1 ml of 10% solution of hydroxylamine hydrochloride, 0.6 ml of 0.1% solution of chrome azurol S, and 5 ml of 2 *N* sodium acetate. Dilute to 100 ml, set aside for 15 minutes, and read at 545 nm.

TOTAL ALUMINUM. Evaporate 50 ml to dryness and ignite to destroy organic matter. Take up the residue in 5 ml of 0.1 *N* hydrochloric acid and complete as above from mixing with "1 ml of 10% solution of...." Alternatively, boil 50 ml of sample for 15 minutes with 0.5 gram of ammonium persulfate. Then complete from mixing with "5 ml of 0.1 *N* hydrochloric...."

CHROMOXANE VIOLET R

Chromoxane violet R (mordant violet 16) can be used for reading the 1 : 1 complex with aluminum at 500 nm, any iron present having been reduced with ascorbic acid.[106] The 1 : 1 complex with iron can be determined by difference in another sample at 530 nm. The desirable pH is around 4.75. Although 3 hours is necessary for full development of the aluminum complex at room temperature, this time requirement can be bypassed by heating. The method will determine in 10 ml of solution less than 1 μg of iron in the presence of more than 24 μg of aluminum and less than 0.5 μg of aluminum in the presence of 40 μg of iron.

Procedure. *Aluminum and Iron.*[107] Add 1 ml of nitric acid to the sample solution to oxidize the iron, and take two 3-ml aliquots. To one, add 2 drops of 1% solution of ascorbic acid. Dilute each to 8.5 ml with ammoniacal acetate buffer solution for pH 4.75. To each, add 1.5 ml of 0.01% solution of the dye. Heat to boiling, cool, and read the aluminum at 530 nm in the solution in which the iron was masked. Read the total iron and aluminum in the unmasked sample at 530 nm. Determine iron by difference.

CYANOFORMAZAN 2

Aluminum complexes with cyanoformazan 2, which is 3-cyano-1,5-bis(2-hydroxy-5-sulfophenyl)formazan.[108]

[105b] N. F. Titkova, *Nauch. Dokl. Vyssh. Shk., Biol. Nauk.* **1968** (9), 126–130.
[106] I. S. Mustafin and L. F. Lisenko, *Zh. Anal. Khim.* **17**, 1052–1056 (1962).
[107] N. F. Lisenko, I. S. Mustafin, and L. A. Molot, *Izv. Vyssh. Ucheb. Zaved. Khim. Khim. Tekhnol.* **5**, 712–716 (1962).
[108] V. A. Malevannyi, M. I. Ermakova, and Yu. L. Lel'chuk, *Khim. Volonkna* **1969**, (1) 72–73.

Procedure. *Titanium Dioxide.* As a buffer solution for pH 4, mix 141 ml of 0.2 *N* ammonium hydroxide with 859 ml of 0.2 *N* acetic acid. Fuse 0.5 gram of sample in platinum with 5 grams of sodium carbonate at 950°. Extract the melt with 100 ml of hot water, filter, and wash the residue with 30 ml, 30 ml, and 30 ml of hot water. Dilute the filtrate to 200 ml. To a 100-ml aliquot, add 4 drops of saturated solution of 2,4-dinitrophenol. Discharge the color of the indicator with 1 : 1 hydrochloric acid; then bring the color back to pale yellow with 1 : 3 ammonium hydroxide. Add 1 ml of 5% ascorbic acid solution and 5 ml of 0.1% solution of the color reagent. Add 50 ml of the buffer solution and heat to 100°. Cool, dilute to 200 ml, and read.

ERIOCHROME BLUE BLACK B

With this reagent, which is mordant black 3, aluminum is read by fluorescence up to 0.4 μg/ml.

Procedures

Antimony.[109] Decompose 1 gram of sample with 2 ml of nitric acid, add 2 ml of hydrobromic acid, and evaporate to dryness. Repeat the addition of 2 ml of hydrobromic acid and evaporation to dryness 3 more times to volatilize the antimony. Add 1 ml of hydrochloric acid and evaporate to dryness. Take up the residue in 4 ml of 10 *N* hydrochloric acid and transfer to a test tube. Add 1 ml of melted paraffin and 2 ml of pentyl acetate. Place in hot water, and when the contents are liquid, stopper and shake for 1 minute. Let cool until the paraffin is solid. It will have extracted interfering amounts of iron and antimony. Pour off the aqueous layer, evaporate it to dryness, and ignite. Dissolve the residue in 0.1 ml of 0.25 *N* hydrochloric acid and dilute to 1.6 ml. Add 0.4 ml of 5% potassium iodide solution to mask any residual antimony, and set aside for 1 hour. Add 0.2 ml of a buffer solution containing 1 : 1 *M* sodium acetate–0.1 *M* acetic acid. Add 0.1 ml of 0.25% ethanolic 1, 10-phenanthreline, 0.1 ml of 0.002% ethanolic eriochrome blue black B, and 1 ml of ethanol. Read the fluorescence.

Titanium.[110] Add 4 ml of hydrofluoric acid, 1 ml of nitric acid, and 1 ml of 1 : 1 sulfuric acid to a 0.1-gram sample. Evaporate to sulfur trioxide fumes, dissolve in 0.5 *N* sulfuric acid, and dilute to 100 ml with that acid. To remove titanium extract with 15 ml, 15 ml, and 15 ml of 6% solution of cupferron in chloroform. Then extract with 15 ml, 15 ml, and 15 ml of chloroform. Evaporate the extracted aqueous solution to dryness. Add 0.25 ml of 1 : 1 hydrochloric acid and evaporate to dryness. Dissolve in 0.1 ml of 0.25 *N* hydrochloric acid and filter, washing through the paper with 1 ml of water. Add 0.2 ml of buffer solution which is 1 : 1

[109] V. A. Nazarenko, M. B. Shustova, R. V. Ravitskaya, and M. P. Nikonova, *Zavod. Lab.* **28**, 537–539 (1962).

[110] V. A. Nazarenko, M. B. Shustova, G. G. Shitareva, G. Ya. Yagnyatinskaya, and R. V. Ravitskaya, *ibid.* **28**, 645–648 (1962).

M-sodium acetate–0.1 *M* acetic acid, 0.1 ml of 0.25% ethanolic 1,10-phenanthroline, 0.1 ml of 0.02% solution of eriochrome blue black B, and 1 ml of ethanol. Read the fluorescence.

ERIOCHROME CYANINE R

The 1:1 violet complex of aluminum with erochrome cyanine R (mordant blue 3) can be read in a solution from which interfering ions have been removed. The use of this reagent is hampered by the variable composition of the dyestuff from different suppliers.[111a]

The presence of organic solvent causes more rapid color development and permits greater acidity.[111b] To a sample solution containing 5 μg of aluminum, add 2.5 ml of acetate buffer solution for pH 3, 5 ml of ethanol, and 1 ml of 0.1% solution of eriochrome cyanine R. Dilute to 10 ml, set aside for 5 minutes, and read at 575 nm against a reagent blank.

A ternary complex is formed by addition of a quaternary ammonium compound, and as would be predicted, this raises the maximum absorption at pH 5.2–6.3 from 535 to 587 nm. The absorption is linear for 0.01–0.08 μg of aluminum per ml. There is interference by pentavalent vanadium and tantalum, hexavalent uranium, zirconium, copper, ferric ion, chromium, gallium, bismuth, titanium, platinic ion, germanium, beryllium, and thorium. Copper and ferrous ion can be masked with 1,10-phenanthroline. The color is stable for at least 5 hours. Titanium, vanadium, and zirconium can be removed with cupferron.

The reagent is used for iron and steel analysis,[112] but removal of iron by solvent extraction has been called unsatisfactory.[113] After precipitation of interfering ions with sodium hydroxide solution or their removal on a mercury cathode, the desirable pH is 5.8–6.2 for reading at 530 nm.[114] Complexing the iron with mercaptoacetic acid is unsatisfactory if titanium or vanadium is present.[115] If chromium is present, an analogous amount must be used in the standard in preparing calibration curves.

After mercury cathode separation of the bulk of the iron, sodium hydroxide precipitation of small residual amounts has been recommended.[116] An alternative to complete after removal of most of the iron is treatment with hydrogen sulfide.[117] The interfering ions can also largely be precipitated with cationic zinc. The interfering ions can also be separated on an ion exchange resin.[118]

[111a] A. J. Hegedus, *Mikrochim. Ichnoanal. Acta* **1963**, 831–850.

[111b] R. Borisova and N. Ivanov, *C. R. Acad. Bulg. Sci.* **29**, 1313–1316 (1976).

[112] P. H. Scholes and D. V. Smith, *J. Iron Steel Inst.* **200**, 729–734 (1962); cf. procedure with solochrome cyanine R, *Analyst* **83**, 615 (1958); and that with chrome azurol S, Hiroshi Nishida, *Jap. Anal.* **22**, 971–975 (1973).

[113] P. B. Dunnill, J. A. Greenwood, and P. H. Scholes, *BISRA Open Rep.* **MG/D/603/68**, 9 pp (1968).

[114] Alois Bocek and Anna Morafkova, *Hutn. Listy* **24**, 307–309 (1969).

[115] Helmut Lilie and Hartmut Rosin, *Z. Anal. Chem.* **160**, 261–267 (1958).

[116] Sigmar Spauszus, *Neue Huette* **6**, 653–659 (1961).

[117] M. Kurjakovic-Bogunovic and R. Pleplic, *Kem. Ind. (Zagrab)* **11**, 700–703 (1962).

[118] Chen-chiang Liao and Ch'ing-hsien Chang, *Un Han Ta Hsüeh, Tzu Jan K'o Hsüeh Hsüeh Pao* **1959** (1), 24–37.

For determination in the presence of iron, a recommended form of reagent is as follows:[119] dissolve 1 gram of erochrome cyanine R and 10 grams of hydroxylamine hydrochloride in 900 ml of water, adjust to pH 5.5 with ammonium hydroxide, and dilute to 1 liter.

Aluminum in gold is determined photometrically after isolation of iron.[120] Dissolve a 1-gram sample in 6 ml of hydrochloric acid and 2 ml of nitric acid. Evaporate to dryness and take up the residue in 10 ml of 1 : 5 hydrochloric acid. Extract the gold with 10 ml and 10 ml of isopentyl acetate and discard. Evaporate the solution to about 1 ml and dilute to 10 ml with 1 : 1 hydrochloric acid. Extract the iron with 10 ml and 10 ml of isopentyl acetate as a sample for determination with bathophenanthroline. After extraction of silver and copper as dithizonates, concentrate the aqueous solution. Add a zirconium carrier and adjust to pH 4.5. Precipitate hydroxides, dissolve, and determine with eriochrome cyanine R.

For aluminum in 15% ferrosilicon, dissolve sample material in hydrofluoric and nitric acids. Add sulfuric acid and evaporate to dryness. Fuse with potassium bisulfate and dissolve in water.[121] Then mask iron in an aliquot with ascorbic acid and adjust the pH with acetate buffer solution.

As applied to copper-zinc alloys, normal amounts of zinc, lead, nickel, tin, and manganese do not interfere.[122] Iron is masked with ascorbic acid and copper with thiosulfate.

In analysis of a copper alloy in nitric acid, the copper, lead, and zinc are removed by electrolysis.[123] Neutralize the alkaline solution from which zinc has been deposited to pH 4.8 with hydrochloric acid and buffer with sodium acetate. Complex iron and the remaining interfering elements with mercaptoacetic acid.

When the reagent is used for water analysis it should have been adjusted to about pH 2.9 with acetic acid and readings made at 20–25° between 5 and 15 minutes after adding the dye.[124] Iron and manganese are masked with ascorbic acid, complex phosphates are hydrolyzed by boiling, and fluoride must be absent.

Eriochrome cyanine is applied to cement analysis,[125] to silicates[126] as fused in a nickel crucible, to analysis of calcium and magnesium carbonates,[127] and to analysis of New Caledonia laterites.[128] Samples with this reagent are automated for analysis of slags from iron and steel manufacture, using ascorbic acid to complex the iron.[129]

[119] M. Barrachina Gomez, L. Gasco Sanchez, and R. Fernandez Cellini, *An. Real Soc. Espan. Fis. Quim.* (Madrid) **56B**, 861–868 (1960).

[120] G. Ackermann and J. Köthe, *Z. Anal. Chem.* **231**, 252–261 (1967); Zygmunt Marczenko, Krzysztof Kasiura, and Maria Krasiejko, *Chem. Anal.* (Warsaw) **14**, 1277–1287 (1969).

[121] Zdenek Marek and Lubomir Kabrt, *Hutn. Listy* **16**, 743–745 (1961).

[122] V. F. Mal'tsev, L. P. Luk'yanenko, and D. M. Kukui, *Zavod. Lab.* **27**, 807–808 (1961).

[123] Helmut Lilie, *Chem. Tech.* (Berlin) **9**, 421 (1957).

[124] K. E. Shull and G. R. Guthan, *J. Am. Water Works Assoc.* **59**, 1456–1468 (1967).

[125] Maria Wallraf, *Zement, Kalk, Gips* **9**, 186–194 (1956); J. A. Fifield and R. G. Blezard, *Chem. Ind.* (London) **1969** (37), 1286–1291.

[126] A. Sass, *Inf. Chim.* **1972**, 227–235.

[127] Adam Hulanicki, Malgorzata Galus, Jedral Wojciech, Regina Karwowska, and Marek Trojanowski, *Chem. Anal.* (Warsaw) **16**, 1011–1019 (1971).

[128] G. Valence and S. Marques, *Chim. Anal.* (Paris) **49**, 275–284 (1967).

[129] P. H. Scholes and C. Thulbourne, *Analyst* **88**, 702–712 (1963).

For the most accurate results, the solution of the dye used as reagent contains nitric acid and urea.[130]

Procedure. Adjust a sample containing 0.2–0.9 ppm of aluminum to pH 4.5.[131] Add 2 ml of m*M* eriochrome cyanine R, dilute to 25 ml, and read at 540 nm.

As the Ternary Complex.[132] To 25 ml of neutral sample containing 0.5–4 μg of aluminum, add 2 ml of 0.5 m*M* acetic acid, 1 ml of 0.05% solution of cetyltrimethylammonium chloride, 10 ml of 0.02% solution of eriochrome cyanine R, and 5 ml of 40% solution of ammonium acetate. Dilute to 50 ml and read at 587 nm against a reagent blank.

Steel

TOTAL ALUMINUM.[133] As eriochrome cyanine R reagent, dissolve 0.35 gram of sample in 2 ml of 1:2 nitric acid and dilute to 75 ml. Add 0.25 gram of urea and dilute to 1 liter. As an acetate buffer solution, dissolve 320 grams of ammonium acetate in water and dilute to 1 liter. Adjust so that 5 ml used as a buffer in the procedure will produce a pH of 5.9 in the final solution.

If the aluminum content exceeds 1%, large cuttings are desirable. If finely divided, both the acid-soluble and acid-insoluble figures are larger. Dissolve 1 gram of sample by heating with 25 ml of 1:1 hydrochloric acid and 15 ml of 1:4 nitric acid. When dissolved, add 0.25 gram of ammonium persulfate, boil, and filter. Wash the precipitate with 1:100 hydrochloric acid followed by water. Ash the paper in platinum and fuse the residue with 1 gram of potassium bisulfate. Extract the fusion with water, filter, and combine the filtrate with the previous one. Evaporate to about 75 ml. Add 10 ml of a solution containing 220 grams of zinc sulfate heptahydrate and 1 ml of sulfuric acid per liter. Heat to boiling, and pour as a thin continuous stream over 15 grams of sodium hydroxide in a polystyrene beaker suitably controlled with respect to spattering by a polystyrene plate. When reaction ceases, dilute to about 150 ml and cool. Dilute to 250 ml or, if more than 1.2% of aluminum is present, to 500 ml. Filter at once through a fluted filter, discarding the first 50 ml of filtrate. Take an aliquot containing not more than 30 μg of aluminum and add a drop of 1% *p*-nitrophenol indicator solution. Decolorize by dropwise addition of 1:3 hydrochloric acid and add 3 drops in excess. Cool below 20° and add 5 ml of eriochrome cyanine R reagent. Let stand for 2 minutes and add 5 ml of acetate buffer solution. Let stand 2 minutes longer, dilute to 50 ml, and read at 535 nm against a reagent blank.

[130] V. N. Tikhonov and M. M. Shashkina, *Izv. Vyssh. Ucheb. Zaved., Khim. Khim. Tekhnol.* **16**, 1117–1118 (1973).
[131] Vikash C. Garg, Suresh C. Shrivastava, and Arun K. Dey, *Mikrochim. Acta* **1969**, 668–672; cf. S. Henry and P. Haniset, *Ind. Chim. Belg.* **27**, 24–27 (1962).
[132] Yoshio Shijo and Tsugio Takeuchi, *Jap. Anal.* **17**, 323–327 (1968).
[133] H. Studlar and E. Eichler, *Chemist-Analyst* **51**, 68–69 (1962); cf. Uno T. Hill, *Anal. Chem.* **31**, 429–431 (1959); G. Picasso, *Met. Ital.* **53**, 260–273, 276 (1961); A. Neuberger, *Stahl Eisen* **85**, 1446–1451 (1965); *BISRA, Open Rep.* **MD/D/561/67** (1967); B. Zagorchev, R. Doicheva, L. Dodova, M. Koeva, and N. Ruseva, *Khim. Ind.* (Zagreb) **92**, 303–306 (1970).

ACID-SOLUBLE ALUMINUM. Carry through the procedure for total aluminum, omitting the portion from "Ash the paper in platinum..." to "and combine the filtrate with the previous one." In other words, discard the insoluble residue.

As an alternative technic for acid-soluble aluminum, dissolve 0.5 gram of sample by heating with 25 ml of 1:11 sulfuric acid and 25 ml of 1:4 nitric acid.[134] Boil off the oxides of nitrogen and oxidize by dropwise addition of saturated potassium permanganate solution plus 2 drops in excess. Reduce the excess permanganate by dropwise addition of either 10% sodium nitrite solution or a saturated solution of sulfur dioxide. Cool and dilute to 250 ml.

Dilute 40 ml of 1:1 sulfuric acid to 250 ml to serve as a blank. Take 2 ml of sample and 2 ml of blank. Add to each 2 ml of 0.1% solution of eriochrome cyanine R. Complete as for alloy steels [below] from "Let stand for a few minutes or...."

ACID-INSOLUBLE ALUMINUM. Carry through the procedure for total aluminum with these following modifications. Discard the first, main filtrate. Proceed to ash the paper in platinum, and fuse the residue with 1 gram of potassium bisulfate. Extract the fusion with water, and filter. To this filtrate add 25 ml of 1:1 hydrochloric acid, 15 ml of 1:4 nitric acid, and 0.25 gram of ammonium persulfate. Heat to boiling, adjust to 75 ml, and proceed from "Add 10 ml of a solution containing...."

Alloy Steels.[135] As a buffer solution, dissolve 320 grams of ammonium acetate in water, add 25 grams of sodium sulfite, and dilute to 1 liter. Adjust to pH 7.6 with ammonium hydroxide or acetic acid. As a working ammonium acetate–polycyclic ketoamine buffer solution, add 0.7 gram of polycyclic ketoamine to 100 ml of the ammonium acetate buffer solution.

Add 5 ml of 30% hydrogen peroxide to 0.4 gram of sample. Cautiously add hydrochloric acid to dissolve the sample. Then add 10 ml of nitric acid and 15 ml of perchloric acid. Evaporate to fumes of perchloric acid and remove chromium by volatilizing with hydrochloric acid. Cool in ice water and dilute to 50 ml. Add sufficient cold 6% cupferron solution to precipitate the vanadium, titanium, and zirconium. Let stand for a few minutes and dilute to 100 ml. Filter a portion through paper, discarding the first 20 ml. Mix a 1-ml aliquot with 2 ml of 0.1% solution of eriochrome cyanine R. Let stand for a few minutes or warm briefly at 66–100°. Add 2 ml of 2% mercaptoacetic acid solution and 2 ml of working ammonium acetate–polycyclic ketoamine buffer solution. Allow 2 minutes for reaction and dilute to 25 ml. Heat at 100° for 30 seconds and let stand at room temperature for about 7 minutes. Cool and read at 595 nm against a blank.

Spelters. Dissolve 0.2 gram of sample with 25 ml of 1:11 sulfuric acid and 25 ml of 1:4 nitric acid by heating. Boil off the oxides of nitrogen and dilute to 250 ml. Dilute 40 ml of 1:11 sulfuric acid to 250 ml to serve as a blank. Take 2 ml of sample and 2 ml of blank. To each add 2 ml of 0.1% eriochrome cyanine R solution. Complete as for alloy steels from "Let stand for a few minutes...."

[134] Uno T. Hill, *Anal. Chem.* **38**, 654–656 (1966); cf. Helmut Lilie, *Chem. Tech.* (Berlin) **9**, 364 (1957); K.-D. Wille, *Z. Anal. Chem.* **250**, 23–26 (1970).

[135] Uno T. Hill, *Anal. Chem.* **38**, 654–656 (1966).

Iron Ores. Add 3 drops of 1 : 3 sulfuric acid and 2 ml of hydrofluoric acid to 0.1 gram of sample in platinum. Evaporate to dryness and ignite briefly. Add 2 grams of potassium bisulfate and fuse. Cool, and dissolve in 40 ml of 1 : 11 sulfuric acid and 40 ml of 1 : 4 nitric acid. Dilute to 250 ml. Proceed as for alloy steels from "Mix a 1-ml aliquot with 2 ml. . . ."

High Alloy Steel.[136] Dissolve 0.1 gram of sample in 50 ml of 1 : 5 sulfuric acid. Add nitric acid dropwise to decompose the sample, and boil for 20 minutes. Add 150 ml of water and 25 ml of 25% ammonium persulfate solution to oxidize manganese to permanganate. Boil for 20 minutes. Cool, add 4 grams of ammonium chloride, and precipitate hydroxides with excess ammonium hydroxide. This separates from chromium. Wash the precipitate, then dissolve it in 20 ml of hot 1 : 1 hydrochloric acid and dilute to 100 ml. To a 5-ml aliquot, add 8 ml of 0.5% solution of mercaptoacetic acid to reduce the iron. Add 8% sodium hydroxide solution until the color becomes brown-violet. Add 4 ml of *N* hydrochloric acid and 4 ml of a buffer solution containing 30% of sodium acetate and 0.6% of acetic acid. Add 5 ml of 0.1% solution of eriochrome cyanine R and heat at 100° for 5 minutes. Cool, dilute to 100 ml, and let stand for 20 minutes. Read at 535 nm against a blank.

Tin Foil.[137] Wash a 1-gram sample of 0.1–0.2 mm thickness with hydrochloric acid and water. Dissolve in 2 ml of hydrochloric acid and 2.5 ml of hydrobromic acid by adding 2 ml of bromine dropwise. Evaporate nearly to dryness, add 1 ml of hydrochloric acid, and take to dryness. Add 0.5 ml of hydrochloric acid and again evaporate to dryness. Dissolve the residue in 6 drops of 1 : 4 hydrochloric acid and reduce the ferric ion with 0.5 ml of fresh 5% solution of ascorbic acid. Add 0.5 ml of 0.075% solution of eriochrome cyanine R and dilute to 25 ml with a buffer solution for pH 8.3 consisting of 540 ml of 2 *N* ammonium hydroxide and 460 ml of 2 *N* acetic acid. Set aside for 7 minutes, read at 535 nm against water, and subtract a blank.

Brass or Bronze.[138] Dissolve a 1-gram sample in 10 ml of nitric acid and dilute to 100 ml. Add 5 ml of 20% sodium hydroxide solution to a 10-ml aliquot, and boil for a couple of minutes to eliminate interference by magnesium, iron, copper, and arsenic. Cool, and dilute to 100 ml. Take a 5-ml aliquot of the clear supernatant layer and add to 20 ml of 0.005% solution of eriochrome cyanine R. Add hydrochloric acid dropwise until the blue color is changed to red. Read at 550 nm.

Manganin.[139] Dissolve 1 gram of this alloy in 1 : 1 nitric acid. Remove the copper by electrolysis and dilute the remaining solution to 100 ml. To an aliquot, add a drop of methyl orange indicator solution. Neutralize with ammonium hydroxide; then add *N* hydrochloric acid dropwise until the color changes. Add 10 ml of acetate buffer solution for pH 5.9, 10 ml of 0.1 *N* sodium thiosulfate, 0.1 ml of 5% ascorbic acid solution, and 2.5 ml of 0.1% solution of eriochrome cyanine R. Dilute to 50 ml, set aside for 2 minutes, and read against water at 536 nm.

[136] T. V. Sinitsyna, *Tr. Vses. Nauchn.-Issled. Konstr.-Tekhnol. Inst. Podshipnik. Prom.* **1968** [2(54)], 60–66.
[137] Yu. L. Lel'chuk, V. B. Sokolovich, and O. A. Drelina, *Izv. Tomsk. Politekh. Inst.* **128**, 101–105 (1964).
[138] M. C. Steele and L. J. England, *Anal. Chim. Acta* **16**, 148–149 (1957).
[139] I. V. Aronina, *Izv. Akad. Nauk Mold. SSR, Ser. Biol. Khim. Nauk* **1967** (10), 28–31.

Alkaline Earth Titanates.[140] Dissolve 50 mg in 10 ml of 1:1 hydrochloric acid and dilute to 100 ml. Add 25 ml of a saturated solution of cupferron, extract with 25 ml and 25 ml of 3:1 amyl alcohol–benzene, and discard the extracts. Wash the aqueous phase with 20 ml and 20 ml of chloroform and adjust to pH 5–6. Add 20 ml of 10% solution of sodium acetate and 5 ml of 0.1% solution of eriochrome cyanine R. Dilute to 100 ml, set aside for 1 hour, and read at 530 nm against water.

Silicate Rocks.[141] To a 0.5-gram sample in platinum, add 15 ml of hydrofluoric acid and 0.5 ml of 1:1 sulfuric acid. Evaporate to sulfur trioxide fumes. Add 5 ml of hydrofluoric acid and evaporate to dryness. Fuse with 2 grams of potassium pyrophosphate and cool. Dissolve in 20 ml of 1:1 hydrochloric acid and dilute to 250 ml. To a 5-ml aliquot, add 5 ml of 5% ascorbic acid solution and 20 ml of acetic acid–sodium acetate buffer solution for pH 5.5. Dilute to 200 ml and add 5 ml of 0.5% solution of eriochrome cyanine R. Dilute to 250 ml and set aside for 1 hour. Read at 535 nm.

Iron Ore. This technic, with suitable modification, is applicable to steels and to copper-aluminum alloys. Dissolve 0.1 gram of sample in 3 ml of 1:1 nitric acid and dilute to 10 ml with nitric acid. Prepare a 10×150 mm column of 60-80 mesh Zerolite FF in chloride form, pretreated with hydrochloric acid. Pass the sample through the column. Elute the aluminum with 10 ml of 9 *N* hydrochloric and wash through with 20 ml more of the acid. Dilute the eluate to 100 ml. Take an aliquot containing 0.01–0.1 mg of aluminum and add a few drops of 0.1% *p*-nitrophenol. Add 1:1 ammonium hydroxide until the indicator is yellow, then 1:9 hydrochloric acid until it becomes colorless. Add 5 ml of 0.02% solution of eriochrome cyanine R and heat at 70° for 5 minutes. Cool, dilute to 100 ml, and read at 535 nm.

Refractory Products.[142] Fuse 0.05 gram of sample with 1 gram of 1:3 borax–sodium carbonate. Dissolve in 100 ml of 1:3 hydrochloric acid and dilute to 250 ml. Dilute a 10-ml aliquot to 30 ml and add 5 ml of 1% solution of ascorbic acid. Make alkaline to Congo red with 1:5 ammonium hydroxide. Add 3 ml of 0.1% solution of erichrome cyanine R and 10 ml of acetate buffer solution for pH 4.8–5. Heat at 100° for 3 minutes, cool, dilute to 100 ml, and read at 540 nm.

Metal Oxides.[143] This procedure is applicable to solutions of ferric oxide, nickel oxide, lithium carbonate, magnesium hydroxide, and manganese carbonate.

Dilute a sample solution containing 5–20 μg of aluminum to 100 ml and, if not already present, add 1 ml of 2 *N* hydrochloric acid containing 2 mg of iron. Add 1 drop of 0.1% solution of methyl red in 0.004 *N* sodium hydroxide, 1 gram of sodium chloride and, dropwise, 20% solution of sodium acetate until the color changes to orange. Boil the solution for 2 minutes and filter. Wash well with hot 1% ammonium chloride solution. Dissolve the precipitate from the filter with 2 ml of hot 1:1 hydrochloric acid. Add 1 ml of 2% ascorbic acid solution, then 10% sodium hydroxide solution until the pH is raised to 2–2.5. Let stand for 5 minutes, and add

[140] Natasa Hlasivcova and Josef Novak, *Silikaty* **13**, 157–162 (1969).
[141] Chen-chiang Liao, *Hua Hauch Pao* **25**, 152–159 (1959).
[142] L. V. Bogova, *Ogneupory* **1969** (6), 52–56.
[143] E. Grzegrzolka and C. Rozycki, *Chem. Anal.* (Warsaw) **12**, 1319–1323 (1967).

2 ml of 0.1% solution of eriochrome cyanine R. Add 5 ml of 50% solution of sodium acetate and adjust to pH 6 with 20% acetic acid. Let stand for 5 minutes, dilute to 50 ml, and read at 535 nm.

Iron and Manganese Ores.[144] Dissolve 1 gram of finely powdered sample in 10 ml of hydrochloric acid. Add 3 ml of nitric acid and evaporate to dryness. Add 5 ml of hydrochloric acid and again evaporate to dryness. Take up in 5 ml of hydrochloric acid, add 50 ml of hot water, and filter. Wash well with hot 1:9 hydrochloric acid and hot water. Ignite the filter in platinum. Add 1 ml of 1:1 sulfuric acid and 5 ml of hydrofluoric acid and evaporate to sulfur trioxide fumes. Add 2 grams of potassium pyrosulfate and fuse. Dissolve the melt in 5 ml of 1:5 hydrochloric acid and add to the original filtrate. Filter if barium sulfate is present, and dilute to 250 ml. This solution is also suitable for determination of copper, iron, manganese, titanium, and phosphorus.

To a 10-ml aliquot, add 10 ml of 1% solution of ascorbic acid and 1 drop of 1% solution of *p*-nitrophenol. Add 10% sodium hydroxide solution until the indicator becomes yellow. Then add 1:3 hydrochloric acid until colorless and add 2 drops in excess. Add 10 ml of 0.035% solution of eriochrome cyanine R and dilute to 50 ml. Set aside for 10 minutes and read at 595 nm against a reagent blank.

Silicon Tetrachloride.[145] ALUMINUM AND IRON. Chill 25 ml of water containing 4 drops of sulfuric acid to 0°. Add 10 ml of sample and 8 ml of hydrofluoric acid. Evaporate the silicon tetrafluoride and take to white fumes. Add 5 ml of water and again take down to white fumes. Dissolve in 5 ml of water and add 0.5 ml of 1% solution of ascorbic acid, 2 ml of 0.5% solution of 4,7-diphenyl-1,10-phenanthroline, and 3 drops of 50% solution of ammonium acetate. Adjust to pH 4.2±0.5 and extract with 5 ml and 2 ml of chloroform. Dilute the extracts to 10 ml with ethanol and read the iron at 533 nm against a blank. Adjust the aqueous solution containing less than 10 mg of aluminum to pH 1–2. Evaporate to 5 ml and dilute to 15 ml. Add 1 ml of 1% solution of ascorbic acid and sufficient ammonium hydroxide to obtain pH 2. Let stand for 5 minutes and add 2 ml of 0.1% solution of eriochrome cyanine R adjusted to pH 2.5. Add 2 ml of 50% ammonium acetate solution and with ammonium hydroxide raise to pH 6.1–6.2. Dilute to 25 ml and read aluminum at 535 nm against a blank.

FERRON

Ferron, which is 8-hydroxy-7-iodoquinoline-5-sulfonic acid, reacts with both iron and aluminum, permitting determination of the latter by difference. There is interference by copper, nickel, zinc, fluoride ion, pyrophosphate ion, oxalate, citrate, and tartrate.[146] The presence of cetyltrimethylammonium chloride improves

[144] J. Jankovsky and E. Pavlikova, *Sb. Pr. Ustavu Pro Vyzk. Rud* **IV**, 229–241 (1962); cf. Ko-jen Hu and Chich-hung Ch'an, *Wu Han Ta Hsüeh, Tzu Jan K'o Hsüeh Pao* **5**, 83–88 (1959).
[145] Zygmunt Marczenko, Krzysztof Kasiura, and Miroslaw Mojski, *Chem. Anal.* (Warsaw) **16**, 203–210 (1971).
[146] Katsumi Goto, Hiroki Tamura, Mitsuko Onodera, and Masaichi Nagayama, *Talenta* **21**, 183–189 (1974).

the stability of the complex. Iron is corrected by reading at 385 and 455 nm. To correct for zinc, copper, cadmium, and manganese, read the extinction; then determine aluminum by difference after masking it with sodium fluoride.

For determination of aluminum in glass, this reagent is applied by the difference technic.[147] Large amounts of titanium require correction. In application of the reagent by "chemical differential photometry," a known amount of aluminum is masked with diaminocyclohexane-$NNN'N'$-tetraacetic acid, and the residual amount is read at 366 nm.[148]

This reagent can be applied to the determination of 0.15–10% ferric oxide in phosphate rock and fertilizers, by the Auto Analyzer.[149]

Procedure. To a neutral sample solution containing not more than 50 μg of aluminum, add 1 ml of hydrochloric acid and 1 ml of nitric acid.[150] Heat to oxidize the iron, and cool. Add 5 ml of 10% ammonium acetate solution and 2 ml of 2% solution of ferron. Dilute to 25 ml and read the sum of iron and aluminum at 370 nm. Read the iron alone at 600 nm and obtain aluminum by difference.

Soil Extract.[151] Digest 0.5 ml of sample with 2 ml of hydrochloric acid and 5 ml of nitric acid. Evaporate to dryness. Dissolve in 1 ml of 1 : 1 hydrochloric acid and 1 ml of 1 : 9 nitric acid. Dilute to 15 ml and add 5 ml of 10% solution of ammonium acetate. Add 2 ml of a 0.2% solution of ferron and read at 470 nm and 390 nm. The value at 390 nm is the sum of aluminum and iron. Subtract the iron as read at 470 nm to obtain the value for aluminum. Each reading requires correction for interference of the other element. A formula for such calculation appears under determination of aluminum in alkalies by oxine.

FLAME PHOTOMETRY

Aluminum can be read by either the atomic band or the oxide band at 5–40 μg/ml. Aspirating in 4-methyl-2-pentanone rather than water increeases the emissivity 100-fold, the sensitivity in aqueous solution often being inadequate. Only iron and calcium interfere by the extraction technics given here. Iron is conveniently removed by electrolysis with the mercury electrode.

In the nitrous oxide–acetylene flame aluminum is read at a height of 4 mm in the red zone at 396.15 nm.[152]

When a silicate is dissolved in hydrofluoric and perchloric acids, a medium of hydrochloric acid, methanol, ammonium chloride, acetic acid, and ammonium acetate enhances the emission at 484 nm.[153] Calcium, magnesium, and potassium

[147] O. Loodin, *Glastek. Tidskr.* **13**, 43–45 (1958).
[148] K. Potzl, *Z. Anal. Chem.* **223**, 10–16 (1966).
[149] Jack L. Hoyt and Donald E. Jordan, *J. Assoc. Off. Anal. Chem.* **52**, 1121–1126 (1969).
[150] N. I. Belyaeva, *Pochvovedenie* **1966** (2), 106–108; F. J. Langmyhr and A. R. Storm, *Acta Chem. Scand.* **15**, 1461–1466 (1961).
[151] L. C. Blackemore, *N. Z. J. Agr. Res.* **11**, 515–520 (1968).
[152] E. E. Pickett and S. R. Koirtyohann, *Spectrochim. Acta*, B, **24**, 325–333 (1969).
[153] F. Hegemann and O. Osterried, *Ber. Deut. Keram. Ges.* **40**, 424–427 (1963).

interfere slightly. Addition of oxine intensifies the emission of aluminum in an oxyacetylene flame to detect 3μg of aluminum oxide per ml.[154]

When read at 484.7 nm, the temperature of combustion does not affect the emission up to 15 mg of aluminum oxide per 100 ml.[155] Added cyclic compounds increase the sensitivity more than linear compounds. Phenol and 1,2-dimethylhydrazine were most effective of 30 compounds studied in preventing interference by calcium, titanium, and iron. The oxyacetylene flame gives greater sensitivity than the oxyhydrogen flame.

Addition of fluoride ion increases the emission by aluminum.[156] Butanol has a similar effect. With both present, 1 mg of aluminum oxide per 100 ml is detectable. Sulfate and phosphate must be absent.

For ores and slags fuse with sodium peroxide, dissolve in water, and acidify with hydrochloric acid.[157] Adjust to 0.02–0.5 ppm of aluminum chloride and to about 0.6% of sodium ion. Read in a nitrous oxide–acetylene flame at 396.5 nm. Large amounts of calcium should not be present.

Procedures

Magnesium-Base Alloys.[158] Dissolve a sample in a minimum amount of 1:35 sulfuric acid and dilute to a known volume. Adjust an aliquot containing 0.1–0.5 mg of aluminum to pH 2.5–4.5 with 7.7% ammonium acetate solution and dilute to 30 ml. Add 3 ml of 1.6% cupferron solution followed by 10 ml of 4-methyl-2-pentanone. Shake for 2 minutes after any precipitate formed by the cupferron has disappeared. Aspirate the organic phase into an oxyacetylene or oxyhydrogen flame. Read the emission at 396.2 nm with the background of 395 nm or at 484 nm with the background of 482 nm. The first is the atomic line, the second the oxide band.

An alternative calls for adjustment with ammonium acetate to pH 5.5–6. Then extract with 10 ml of 1.1% solution of 2-thenoyltrifluoroacetone in 4-methyl-2-pentanone by shaking for 5 minutes. If large amounts of alkalies and alkaline earths are present, backwash the organic layer with 30 ml of 0.1 *N* nitric acid before aspirating. This technic gives slightly higher emission values.

Zinc-Base Alloys. Dissolve a sample in a minimum amount of 1:15 nitric acid or 1:35 sulfuric acid. Dilute to a known volume and take an aliquot containing 0.1–1 mg of aluminum. Adjust to pH 6–6.5 with a 15% solution of ammonium acetate. Add excess of 5% solution of sodium diethyldithiocarbamate. Each ml of that solution will complex 13 mg of zinc or manganese. Shake with 20 ml of chloroform for 2 minutes and discard the extract. Extract with further 10-ml portions of chloroform until no further color is extracted; usually four such extractions are required. Adjust to pH 5.8–6 with 1:16 acetic acid and extract with 10 ml of 1.1% solution of 2-thenoyltrifluoroacetone in 4-methyl-2-pentanone by

[154] Jeannine Debras-Guedon and Igor A. Vomovitch, *Compt. Rend.* **249**, 242–244 (1959).
[155] I. A. Voinovich, G. Legrand, G. Hameau, L. Katz, and J. Louvrier, *Zh. Anal. Khim.* **22**, 682–688 (1967).
[156] K. Konopicky and W. Schmidt, *Z. Anal. Chem.* **174**, 262–268 (1960).
[157] R. J. Guest and D. R. MacPherson, *Anal. Chim. Acta* **78**, 299–306 (1975).
[158] H. C. Eshelman and John A. Dean, *Anal. Chem.* **31**, 183–187 (1959).

shaking for 5 minutes. Backwash the organic layer with 30 ml of 0.1 *N* nitric acid. Complete as for magnesium-base alloys from "Aspirate the organic phase. . . ."

Steels

ACID-SOLUBLE ALUMINUM. Dissolve a sample containing 0.1–0.5 mg of aluminum in the minimum amount of 5 *N* perchloric acid. Filter, and wash with hot 1% perchloric acid. Electrolyze with a mercury cathode until spot tests show less than 1 mg of iron remaining. Evaporate to 20 ml and add 10 ml of 7.7% ammonium acetate solution. Adjust to pH 2.5–4.5 with 1 : 15 ammonium hydroxide and add 5 ml of 1.6% cupferron solution followed by 10 ml of 4-methyl-2pentanone. Complete as for magnesium-base alloys from "Shake for 2 minutes after. . . ."

Alternatively,[159] dissolve 1 gram of sample in 20 ml of 1 : 1 hydrochloric acid. Add 0.5 ml of nitric acid and filter. Evaporate the filtrate almost to dryness and take up with 50 ml of 7 *N* hydrochloric acid. Extract the iron with 50 ml of isobutyl ketone and discard the extract. Evaporate almost to dryness, dilute to 50 ml, and read in the oxyacetylene flame at 396.15 nm for up to 20 ppm.

ACID-INSOLUBLE ALUMINUM. Ignite the filter from determination of acid-soluble aluminum. Fuse the residue with 2 ml of potassium pyrosulfate and dissolve the melt in 10 ml of water. Add 5 grams of ammonium acetate and 5 ml of a 2% solution of oxine in pH 4.8–5.2 acetic acid. Extract aluminum with 10 ml of methyl isobutyl ketone. Dilute the organic layer to 50 ml with methanol and read as above.

Bronzes. Dissolve a sample containing 0.1–0.5 mg of aluminum in 10 ml of 2 *N* hydrochloric acid and 5 ml of 30% hydrogen peroxide. Add 10 ml of 5 *N* perchloric acid and evaporate to copious fumes of that acid. Dilute to 10 ml and complete as for acid-soluble aluminum in steels from "Filter, and wash with hot 1% perchloric acid. . . ."

Silicates and Glasses. Moisten a sample containing 0.2–1 mg of aluminum in platinum with water. Add 5 ml of hydrofluoric acid and 0.5 ml of sulfuric acid. Heat to fumes of sulfur trioxide and add 2 ml of hydrofluoric acid. Evaporate to dryness and ignite as long as sulfur trioxide is evolved. Take up in 20 ml of 1 : 9 hydrochloric acid. Add 5 ml of 1.6% cupferron solution and shake for 2 minutes with 25 ml of 4-methyl-2-pentanone to extract heavy metals. To the aqueous phase, add 10 ml of 7.7% ammonium acetate solution and adjust to pH 2.5–4.5 with 1 : 15 ammonium hydroxide. Add 5 ml of 1.6% solution of cupferron followed by 10 ml of 4-methyl-2-pentanone. Proceed as for magnesium-base alloys from "Shake for 2 minutes after. . . ."

Refractory Products.[160] ALUMINUM AND IRON. Adjust a sample solution containing more than 5 mg of aluminum oxide and more than 10 mg of ferric oxide to pH 4–5. Extract with 1 : 1 acetylacetone-chloroform and dilute to 100 ml with that

[159] Hidehiro Goto, Yachiyo Kakita, and Michiko Namiki, *Jap. Anal.* **19**, 1211–1214 (1970).
[160] W. Schmidt, K. Konopicky, and J. Kostyra, *Z. Anal. Chem.* **206**, 174–185 (1964).

solvent mixture. Spray into an oxyhydrogen flame and read aluminum at 484 nm and iron at 386 nm.

HEMATOXYLINE

This reagent forms a 3 : 1 complex with aluminum which must be developed by heating at 100° for 10 minutes, or it must be allowed to stand.[161] It should be read within 6 hours. A twentyfold excess of reagent is required. There is interference by iron, tetravalent titanium, cupric ion, chromic ion, mangenous ion, and calcium. Complex-forming agents such as fluoride, oxalate, citrate, or EDTA may be present.

Aluminum is determined by hematoxylin in the aluminum black deposits on electrodes.[162] A solution of the sample is buffered with potassium hydroxide–ammonium sulfate solution containing ammonium uranate as a carrier. Ferric hydroxide is precipitated but aluminum and chromium are not. Then aluminum is developed in an aliquot of the filtrate. Aluminum in uranyl nitrate is also determined with hematoxylin.[163]

Procedure. *Soil Extract.*[164] To a sample at pH 2–5 containing 1–5 μg of aluminum add 1 ml of 2.5% solution of potassium cyanide. Set aside for 5 minutes, then add 5 ml of acetate buffer solution for pH 8.2. Add 1 ml of 0.1% ethanolic hematoxylin. Dilute to 25 ml and 1 hour later read at 620 nm against a blank.

2-HYDROXY-1-NAPHTHALDEHYDE-BENZOHYDRAZONE

The fluorescence of this reagent with aluminum is linear at 0.1–1 μg/ml. The hydrazones of acetohydrazide, phenylacetohydrazide, and isonazid also fluoresce with aluminum.

Procedure. Mix 4 ml of acid sample with 1 ml of 0.5 *M* acetate buffer solution for pH 4.6.[165] Add 6 ml of 2 : 1 methanol-dimethylformamide and 2 ml of a solution of 4 mg of the reagent per 100 ml of 50% ethanol. Incubate at 40° for 30 minutes and dilute to 20 ml with 50% methanol. Activate at 395 nm and read the fluorescence at 475 nm.

---

[161] V. A. Malevannyi and Yu. L. Lel'chuk, *Izv. Tomsk. Politskh. Inst.* **1971** (174), 101–106; cf. Luigi Guerreschi and Rachele Romiat, *Ric. Sci.* **29**, 2178–2185 (1959).

[162] R. Guerreschi and R. Romita, *Ric. Sci., R. C., A.* **2** (6), 603–616 (1962).

[163] J. Nowicka-Jankowska, A. Golkowska, I. Pietrzak, and W. Zmijewska, *Pol. Akad. Nauk., Inst. Baden Jadh.* No. **45**, 42 pp (1958).

[164] R. C. Dalal, *Plant Soil* **36**, 223–231 (1972).

[165] Toyozo Uno and Hirokazu Taniguchi, *Jap. Anal.* **20**, 1123–1128 (1971).

2-HYDROXY-3-NAPTHOIC ACID

The complex of this reagent with aluminum or beryllium gives a fluorescence at 460 nm which permits reading aluminum at 2 ppb and beryllium at 0.2 ppb.[166]

Procedure. As a buffer solution for pH 5.8, add 40 ml of glacial acetic acid to 800 ml of water. Adjust to pH 5.8 with approximately 40 ml of ammonium hydroxide and dilute to 1 liter. As an alternative, use a buffer solution for pH 5.8 of sodium acetate and acetic acid.

To a sample solution containing 0.2–2.5 μg of aluminum, add 5 ml of the buffer solution for pH 5.8. Add 10 ml of a reagent containing 0.2102 gram of 2-hydroxy-3-naphthoic acid per liter and dilute to 100 ml. After 1 hour activate at 370 nm and read the fluorescence at 460 nm.

8-HYDROXYQUINOLINE

This reagent is also called 8-quinolinol and has the trivial name of oxine. The aluminum complex is read either photometrically or fluorimetrically.[167] The latter technic will detect 0.02 ppm. The complex is extractable with chloroform. But there is an element of instability with aluminum oxinate, even in the dark.[168] A decrease of absorption of 2–4% in 1 hour depends on the grade of chloroform used and is accelerated by light and oxygen.

By use of a carbonate medium and masking agents, *N*-benzoyl-*N*-phenylhydroxylamine is selective for separation of aluminum. Beer's law is followed for the oxine complex in 10 ml of chloroform up to 50 μg. Fluoride, citrate, and EDTA interfere. Aluminum can be separated from many interferences by solution as sodium aluminate. At pH 9 oxine combines not only with aluminum but also with antimony, bismuth, cadmium, cerium, cobalt, copper, lead, mercury, iron, nickel, tin, titanium, uranium, and zinc. An ammoniacal buffer for pH 9 eliminates interference by zirconium, hexavalent molybdenum, pentavalent vanadium, beryllium, magnesium, manganese, and rare earths. Plutonium is separated from aluminum on an anion exchange resin as the anionic chloride complex; aluminum does not form such a complex. Cyanide eliminates interference by nickel, copper, and cobalt.

In analysis of industrial water, the amounts tolerated fluorimetrically and photometrically were, respectively, in ppm: iron 5, 3; copper, 1, 0.5; chromium, 10, 10; zinc, 5, 5; manganous ion, 0.3, 0.3; boron, 5, 5; fluoride, 20, 20; polyphosphate or orthophosphate, 30, 30; chromate, 60, 60; others, 100 or more.

As applied to less than 0.1% aluminum in uranium and substances containing uranium, an ammonium hydroxide–ammonium carbonate buffer solution for pH

[166] Gordon F. Kirkbright, T. S. West, and Colin Woodward, *Anal. Chem.* **37**, 137–143 (1965).
[167] W. T. Rees, *Analyst* **87**, 202–206 (1962).
[168] Robert H. Linwell and Fredrick Raab, *Anal. Chem.* **33**, 154–155 (1961).

9.5–10 effectively maintains the uranium in solution.[169] Interference by up to 0.1 mg of cobalt, nickel, zinc, copper, lead, and iron is prevented by extraction with a chloroform solution of 8-hydroxyquinaldine. Manganese is extracted as the diethyldithiocarbamate with chloroform. As an alternative technic iron is reduced with thioglycollic acid.[170] Zirconium, titanium, and vanadium are extracted from perchloric acid solution by *N*-benzoyl-*N*-phenylhydroxylamine in chloroform. EDTA and cyanide complex other interfering ions before extracting the oxine complex from alkaline carbonate solution.

For analysis of aluminum chloride–vanadyl chloride solution, heat to 100° and adjust to pH 6–7 with 1 : 1 ammonium hydroxide.[171] Filter and wash the aluminum hydroxide; the vanadium is in the filtrate. Dissolve in 1 : 20 sulfuric acid and dilute to 100 ml.

For determination of aluminum in metallic nickel, the sample is dissolved in nitric acid. The solution is buffered to pH 9 with ammonium acetate. Nickel, iron, copper, and zinc are masked with either potassium cyanide or 2, 3-dimercaptopropanesulfonate. Then aluminum oxinate is extracted with chloroform and read at 390 nm against chloroform.[172] Magnesium can be masked by EDTA.

Aluminum in deodorants and antiperspirants is extracted with dilute sulfuric acid and determined as the oxinate in chloroform by fluorometry.[173]

The procedure by extracting the complex of aluminum with 8-hydroxyquinoline by chloroform has been successfully applied to analysis of magnesium, tungsten, molybdenum, niobium, titanium, nickel, iron, and zinc.[174]

Although aluminum and iron form oxinates at pH 4.5–10.7 in the presence of hydrogen peroxide, titanium does not.[175] Therefore under these conditions the aluminum and iron oxinates are extracted with chloroform. Then aluminum is read at 390 nm and the iron at 470 nm. The same applies to separation of aluminum and iron from vanadium.

For analysis of selenium, treat with nitric and sulfuric acids and volatilize the selenium.[176] Eliminate silica with hydrofluoric acid and dissolve in a sodium acetate–acetic acid buffer solution. Extract interfering ions with a solution of diethyldithiocarbamate in chloroform. Then extract aluminum by oxine in chloroform.

In analysis of metallic hromium, the aluminum is volatilized as aluminum chloride in a current of chlorine at 700° for 30–40 minutes and condensed on the cool part of the tube.[177] Aluminum oxide is not affected. The aluminum chloride is

[169] Allan W. Ashbrook and G. M. Ritcoy, *Can. J. Chem.* **39**, 1109–1112 (1961); cf. Kenji Motojima and Kimie Isawa, *Nippon Genshiryoku Gakkaishi* **2**, 253–259 (1960); Hiroshi Hashitani, *J. At. Energ. Soc., Jap.* **4**, 287–293 (1962).
[170] C. O. Granger, *UKAEA Develop. Eng. Group* **DEG Rep. 219(C)**, 1–17 (1960).
[171] N. N. Ruban, K. A. Vanogradova, S. M. Isaeva, and Yu. A. Avetisyan, *Tr. Inst. Met. Obogashch., Alma-Ata* **12**, 120–124 (1965).
[172] M. N. Gorlova and L. I. Chichkova, *Tekhnol. Lagk. Splavov Nauchn.-Tekh. Yull. VILS* **1970** (6), 97–99; cf. British Standards Institution, *BS 3727*; Pt. **1**, 7 pp (1966).
[173] C. Hozdic, *J. Assoc. Off. Anal. Chem.* **49**, 1187–1189 (966).
[174] M. N. Gorlova, *Zavod. Lab.* **40**, 27–28 (1974).
[175] Hiroshi Hashitani and Kenji Motojima, *Jap. Anal.* **7**, 478–482 (1958).
[176] Masuo Miyamoto, *ibid.* **10**, 98–102 (1961).
[177] A. A. Tumanov and V. G. Patukhova, *Zavod. Lab.* **35**, 654–655 (1969).

washed out with 0.1 *N* hydrochloric acid, converted to the oxinate, and extracted with carbon tetrachloride. Silicon is volatilized as the tetrachloride, absorbed in 1% sodium carbonate solution, and determined as molybdenum blue. The metallic chromium is dissolved in sulfuric acid, evaporated to dryness, and ignited to the oxide at 1000°. Then 50 mg of the chromium oxide is heated at 900° for 40 minutes in a current of chlorine to volatilize the chromium and leave the aluminum oxide. This residue fused with potassium pyrosulfate is determined with oxine.

After removal of the bulk of base materials from an alloy solution with the mercury cathode, adjust to pH 9.8 and add 8-hydroxyquinaldine.[178] Extract with chloroform to separate cadmium, copper, iron, manganese, cerium, nickel, lead, antimony, tin, zinc, and titanium. Add hydrogen peroxide to prevent interference by niobium, thorium, and uranium.

For samples containing less than 0.1% aluminum, oxidize the solution in hydrochloric acid with nitric acid. Dilute with hot water, partially neutralize with sodium hydroxide solution, and slowly run through a capillary into the middle of a solution of 60 grams of sodium hydroxide in 100 ml of water being stirred with a silver stirrer.[179] Filter, acidify an aliquot of the filtrate with hydrochloric acid and develop with oxine.

As a general procedure for ferrous and nonferrous metals and alloys, dissolve a sample containing less than 20 μg of aluminum in hydrochloric acid and oxidize with nitric acid.[180] Extract the ferric chloride with ether. Eliminate other interfering metals by chloroform extraction of their complexes with diethyldithiocarbamate. Adjust the aqueous solution to pH 8.3 to avoid interference by chromium, and extract aluminum as the oxinate.

For separation of aluminum from less than half its weight of iron, add a 1% solution of oxine in acetic acid to the solution in nitric acid.[181] Adjust to pH 5 with sodium acetate solution and extract aluminum oxinate with butyl acetate. Read the organic layer at 395 nm.

For analysis of gold, dissolve 10 grams of sample in aqua regia.[182] Dilute and extract chloraurate with isopropyl ether. Then determine aluminum with oxine.

In analysis of alloy steel by the alkaline separation method, addition of borax prevents less of aluminum by adsorption on the precipitated ferric hydroxide.[183] The method is applicable to 0.005–1.5% aluminum.

The oxine complex is applied to determination of aluminum in solutions of glass.[184a] For aluminum as the 8-hydroxyquinolate, after determining iron, copper, and manganese sequentially, see Copper by Diethyldithiocarbamate.

The yellowish-green fluorescence of aluminum oxinate is at a maximum at 510 nm.[184b] Indium oxinate and gallium oxinate have maxima at 526 and 528 nm respectively.

[178] Robert J. Hynok and Lewis J. Wrangell, *Anal. Chem.* **28**, 1520–1527 (1956).
[179] Sigmar Spauszus, *Neue Huette* **6**, 653–659 (1961).
[180] D. Filipov and I. Nachev, *C. R. Acad. Bulg. Sci.* **22**, 687–690 (1969).
[181] F. G. Zharovskii and V. L. Ryzhenko, *Zh. Anal. Khim.* **22**, 1142–1145 (1967).
[182] Krzysztof Kasiura, *Chem. Anal.* (Warsaw) **20**, 809–816 (1975).
[183] Dagmar Blazejak-Ditges, *Z. Anal. Chem.* **246**, 241–243 (1969).
[184a] Wilhelm Geilmann and Günther Tölg, *Glastech. Ber.* **51**, 260–268 (1958).
[184b] Yashaharu Nishikawa, Keizo Hiraka, Kiyotoshi Morishige, Koichi Takahashi, Tsunenobu Shigematsu, and Taro Nogami, *Jap. Anal.* **25**, 459–463 (1975).

Procedure. As a buffer solution dissolve 5 grams of sodium metaphosphate and 200 grams of ammonium carbonate in 900 ml of water. Extract with 10 ml, 10 ml, and 10 ml of 2% solution of oxine in chloroform and wash with 10 ml and 10 ml of chloroform. Discard the extracts, dilute to 1 liter, and store in polyethylene.

Mix an acidified sample containing 5–30 μg of aluminum with 1 ml of thioglycollic acid.[185] Add the metaphosphate-carbonate buffer solution until the sample solution is alkaline. Then add about 3 ml in excess. Add 1 ml of 5% solution of 1.2% potassium cyanide and 1 ml of hydrogen peroxide. Dilute to about 30 ml and add 1 ml of 2% solution of *N*-phenylbenzihydroxamic acid in ethanol or acetone. After allowing to stand for 5 minutes, extract the aluminum complex by shaking for 1 minute with 15 ml of benzene. Discard the aqueous phase and wash the organic phase with 15 ml of water. Then extract the aluminum by shaking the benzene phase for 1 minute with 15 ml of 0.2 *N* hydrochloric acid. Dilute the aqueous extract to about 30 ml and add 1 ml of 2% solution of oxine. Add 20% solution of ammonium carbonate until the solution is alkaline and shake for 1 minute with 10 ml of chloroform. Read the organic phase at 390 nm.

Nickel-, Iron-, and Copper-Base Alloys.[186] As a buffer solution for pH 9, dissolve 207 grams of ammonium chloride in water, add 266 ml of ammonium hydroxide, and dilute to 2 liters.

Select a sample expected to contain 0.01–0.1 mg of aluminum according to the following table.

Aluminum (%)	Sample (grams)	Aliquot from 250-ml Flask	μg per Aliquot
5.0	0.10	5	100
1.0	0.25	10	100
0.5	0.25	20	100
0.2	0.5	25	100
0.05	1.0	25	50
0.02	1.0	50	40
0.005	2.0	50	20
0.0005	10.0	50	10

Dissolve in 20 ml of hydrochloric acid and 5 ml of nitric acid. Boil off oxides of nitrogen. If chromium is present, add 5 ml of perchloric acid and evaporate to fumes to oxidize the chromium to the hexavalent state. Dilute to 40 ml. Unless more than 0.1 gram of iron is already present, add about 0.4 gram of iron as ferric nitrate. Dissolve 40 grams of sodium hydroxide, 6 grams of boric acid, and 10 grams of sodium cyanide in 100 ml of water in a Teflon container or in platinum. If prepared shortly before use, it will be at 70±10°; otherwise heat to that tempera-

[185] Robert Villarreal, John R. Krsul, and Spence A. Barker, *Anal. Chem.* **41**, 1420–1422 (1969).
[186] Keith E. Burke, *ibid.* **38**, 1608–1611 (1966).

ture. Pour the sample into the alkaline solution with stirring. Cool and dilute to 250 ml. Filter, discarding the first 15 ml, and take an aliquot according to the table. Add hydrochloric acid to lower to pH 9.5–10. Add 10 ml of the buffer solution for pH 9. Dilute to 100 ml and add 10 ml of 1% solution of oxine in chloroform. Shake for 1 minute and draw off the chloroform layer. Wash the aqueous layer with 5 ml and 5 ml of chloroform, adding the washings to the previous extract. Dilute the chloroform extract to 50 ml with acetone and read at 390 nm, subtracting a blank.

Heat-Resistant Alloys.[187a] Dissolve 0.1 gram of sample by warming with 15 ml of hydrochloric acid. Near the end, add a few drops of nitric acid to dissolve carbides. Add 15 ml of perchloric acid and evaporate to fumes to oxidize chromium. Take up in 50 ml of hot water, filter if necessary, wash with 1:9 hydrochloric acid, and dilute to 200 ml. To an aliquot add 10 ml of 10% perchloric acid followed by 1 ml of 6% cupferron solution. Shake for 2 minutes with 20 ml and 20 ml of chloroform, discarding the extracts. Add 2 ml of 10% ammonium tartrate solution, then 10 ml of a saturated solution of sodium sulfite. Let stand for 3 minutes and add 10 ml of 10% sodium cyanide solution. Using pH paper, adjust the pH 9 with ammonium hydroxide or hydrochloric acid. Stabilize by adding 5 ml of 50% butyl Cellosolve. Shake for 3 minutes each with 20 ml, 10 ml, and 10 ml of 2% solution of oxine in chloroform. Dilute the combined extracts to 100 ml with chloroform and read against a blank at either 380 or 425 nm.

High-Temperature Alloys.[187b] Dissolve 0.1 gram of sample in 10 ml of 1:1:1 nitric acid-hydrochloric acid-water. Add 10 ml of 1:1 sulfuric acid and evaporate to sulfur trioxide fumes. Take up in water and dilute to 25 ml. Mix an aliquot containing 10–100 μg of aluminum with 5 ml of 5% oxalic solution and 5 ml of 2.5% solution of oxine. Adjust to pH 5.7. Extract with 10 ml of chloroform, then with 10 ml, 10 ml, and 10 ml of 1% solution of oxine in chloroform. This extracts the oxinates of iron, nickel, and other interfering metals. Wash the aqueous phase with 3 ml of 1% solution of oxine in chloroform. Add 1 ml of *M* potassium cyanide, adjust to pH 9.5, and extract aluminum oxinate with 20 ml of chloroform. Wash manganese from this extract with 50 ml of 1% solution of hydroxylamine hydrochloride. Read at 390 nm.

If the sample contains substantial amounts of niobium or tantalum precipitate them with cupferron before extracting aluminum oxinate.

Plutonium-Aluminum Alloys.[188] This technic is designed for operation in a glovebox hood. Transfer 0.5 gram of sample to a tube, cover with a minimal amount of water, and add 5 ml of hydrochloric acid. Heat to complete solution, cool, and dilute to 50 ml. Transfer an aliquot containing 0.1–0.5 mg of aluminum to a tube and evaporate to less than 0.5 ml. Add 3 ml of nitric acid and evaporate to about 0.2 ml. Add 5 ml of hydrochloric acid until evolution of gas occurs.

The recommended column vacuum apparatus is illustrated in Figure 15-2. As the

[187a]Harvey Zibulsky, M. F. Slowinski, and J. A. White, *ibid.* **31**, 280–281 (1959).
[187b]Hiroshi Hashitani and Takeo Atachi, *Jap. Anal.* **24**, 49–52 (1975).
[188]H. B. Evans and Hiroshi Hashitani, *ibid.* **36**, 2032–2034 (1964); cf. I. G. Jones and G. Phillips, *AERE* (Gt. Britain) **R-2879**, 11 pp (1960).

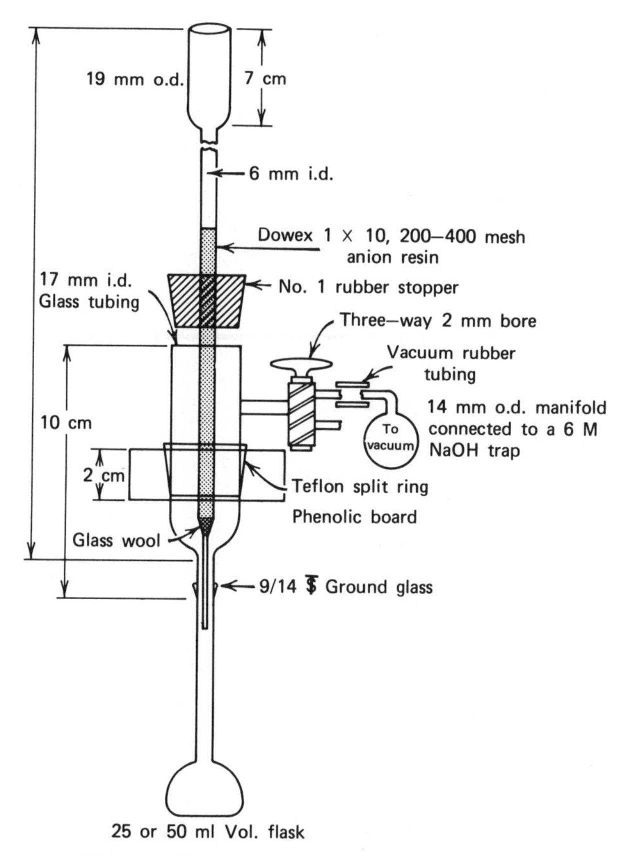

Figure 15-2 Side view of column vacuum apparatus.

resin, use Bio-Rad AG1 × 10, 200-400 mesh, chloride form. Fit the column with a small glass-wool plug and fill with water to the midpoint of the reservoir. Pour in an aqueous slurry of the resin to a height of 7 inches. Wash alternately with 10 ml of 8 *N* hydrochloric acid, 10 ml of water, and 10 ml of 8 *N* hydrochloric acid. Control the rate of passage of the sample through the column by adjustment of the plug.[189]

Add the cooled solution to the resin column. Make hydrochloric acid 0.1 *N* with nitric acid. Use 5 ml, 5 ml, and 5 ml of this to wash out the sample container and elute the aluminum at 40 ± 5 drops per minute. Wash the column with 10 ml of 1 : 1 hydrochloric acid. Evaporate the eluate and washings almost to dryness. To dispose of the plutonium, eluate it to waste with 0.5 *N* hydrochloric acid.

Dilute to about 10 ml and heat to 70°. Precipitate the aluminum with 1 ml of 2% solution of oxine, 1 ml of 6.5% potassium cyanide solution, and 3 ml of 15.4% solution of ammonium acetate. Digest at 70° for 15 minutes and cool. Filter with the apparatus shown in Figure 15-3. Wash the precipitate with small portions of water while evacuating. Turn the cock and dissolve the precipitate with hydrochlo-

[189] Harold B. Evans, Carol A. Bloomquist, and John P. Hughes, *Anal. Chem.* **34**, 1692–1695 (1962).

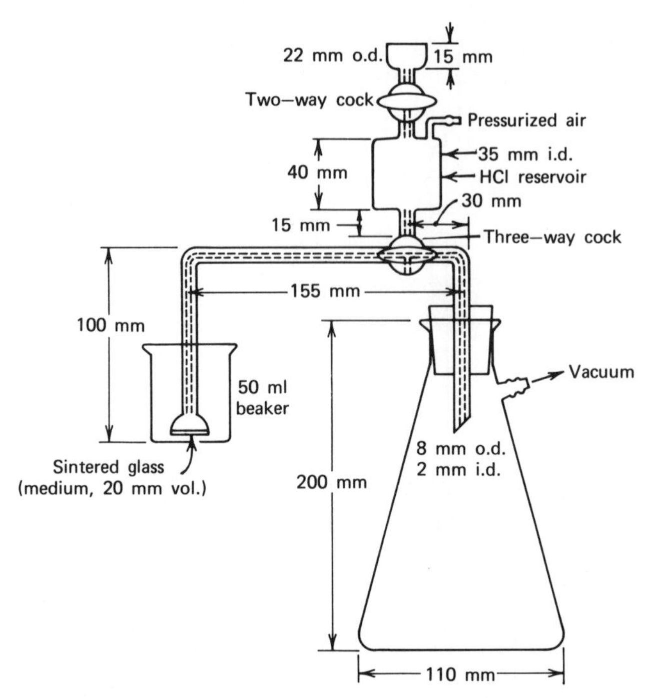

Figure 15-3 Apparatus for filtering aluminum oxinate.

ric acid, which is transferred to the beaker from the reservoir, with the aid of pressurized air. Dissolve the adhering precipitate on the walls of the beaker and on the filter stick with hydrochloric acid from a dropping bottle. Do not use more than 15 ml of hydrochloric acid.

Dilute to 100 ml and read at 360 nm against water.

Cast Iron and Steel[190]

ACID-SOLUBLE ALUMINUM. Dissolve a 5.5-gram sample containing 0.002–0.012% of aluminum with 50 ml of hydrochloric acid and 15 ml of nitric acid. Filter and reserve the residue for acid-insoluble aluminum. Extract the filtrate with 150 ml of isoamyl acetate and discard the extract. Add 10 ml of perchloric acid and take to fumes for 3 minutes. Take up by boiling with 25 ml of water. Transfer to a mercury cathode cell and electrolyze for 30 minutes at 4 amperes. Add 10% sodium hydroxide solution to pH 11, boil, cool, and dilute to 55 ml. Filter the manganese precipitate and take 50 ml of filtrate. Adjust to pH 4 with hydrochloric acid and add 2 ml of 2% solution of oxine. Extract the aluminum complex with 20 ml and 15 ml of chloroform. Dry with sodium sulfate and read the fluorescence against quinine sulfate as a standard.

ACID-INSOLUBLE ALUMINUM. Ash the retained insoluble matter. Treat with 5 ml

[190]H. Green, M. Laband, and M. A. Bryan, *Brit. Cast Iron Res. Assoc., J. Res. Develop.* **12**, 749–753 (1964).

of hydrofluoric acid and 1 ml of 1:1 sulfuric acid. Evaporate to dryness and ignite. Fuse with 2 grams of sodium carbonate and dissolve the fusion in 10 ml of 1:5 perchloric acid. Proceed as for acid soluble aluminum from "Add 10% sodium hydroxide solution. . . ."

As little as 0.001% aluminum in 1 gram of cast iron can be determined by oxine when the iron is extracted with butyl acetate and the other interfering elements as diethyldithiocarbamates.[191] Aluminum is then extracted as the cupferrate, which is ashed for extraction of the aluminum as the oxinate by chloroform.

Plain Carbon Steel.[192] Dissolve a sample containing 0.067–0.27 mg of aluminum, usually 0.1–2 gram, in 10 ml of hydrochloric acid, adding sufficient nitric acid to oxidize the iron to the ferric state. Filter through a small pulp pad, washing with 20 ml of hydrochloric acid. Extract the bulk of the iron with 50 ml and 15 ml of amyl acetate.

Transfer the pad and residue to a platinum crucible and ash. Add 1 ml of hydrofluoric acid and 0.5 ml of sulfuric acid, and heat to dryness. Fuse the residue with 2 grams of potassium bisulfate. Take up in 10 ml of water and add to the reserved filtrate. Dilute to 50 ml.

Take an aliquot of the composite sample, usually 5 ml, and dilute to 25 ml. Add 5 ml of 10% solution of hydroxylamine hydrochloride solution to reduce any residual ferric ion. Add 5 ml of 0.2% solution of 1,10-phenanthroline solution, 5 ml of ammonium hydroxide, and 1 ml of 10% potassium cyanide solution. Adjust to pH 5.5 with acetic acid.

Add 10 ml of 0.5% solution of oxine in chloroform and shake for 2 minutes. Shake the chloroform layer with 20 ml of buffer solution for pH 9.2 and 2 ml of 2% sodium sulfide solution for 1 minute and withdraw the chloroform layer. Add a further 10 ml of 0.5% oxine solution to the aqueous sample solution and shake for 1 minute. Then shake with the buffer solution and sodium sulfide. Combine the chloroform extracts, dry by filtering through paper, and dilute to 25 ml with chloroform. Read at 385 nm against a reagent blank.

As an alternative technic, pass the sample in 9:1 acetic acid–hydrochloric acid through a column of Dowex 50W-X8 resin.[193] Aluminum is retained. Wash the iron through with 5% solution of ammonium thiocyanate. Then elute the aluminum with 1:3 hydrochloric acid and read at 405 nm as the oxine complex in benzene.

Zircaloy-2.[194] Dissolve 1 gram of sample with 10 ml of sulfuric acid, 10 ml of 50% ammonium sulfate solution, and 2 ml of nitric acid. Dilute to 100 ml. Take an aliquot containing less than 0.4 gram of zirconium, 40 μg of aluminum, and 50 μg of manganese. Adjust to 1–2 *N* in sulfuric acid content and add at under 10° 15 ml of 8% cupferron solution for each 0.1 gram of zirconium. Shake for 2 minutes with

[191] R. C. Reeney, *ibid.* **7**, 436–441 (1958).

[192] R. M. Dagnall, T. S. West, and P. Young, *Analyst* **90**, 13–18 (1965); cf. J. Bernal Nievas and M. Herrero Lancina, *Inst. Hierre Acere* **17**, 517–522 (1964).

[193] Rolf Oberhauser, *Materialpruefung*, **15**, 85–89 (1973); cf. Haruno Okochi, *Jap. Anal.* **20**, 1381–1386 (1971).

[194] Hiroshi Hashitani, Kazuo Katsuyama, Chiaki Sugawa, and Kenji Motojima, *Jap. Anal.* **16**, 596–601 (1967).

a volume of chloroform equal to the volume of cupferron solution added. Discard the extract, add 5 ml of the cupferron solution, and further extract with 10 ml and 10 ml of chloroform. Evaporate the aqueous layer to almost 5 ml and add 2 ml each of nitric acid and hydrochloric acid. Evaporate to sulfur trioxide fumes and dilute to 50 ml. Add 2 ml of 4% solution of 8-hydroxy-2-methylquinoline in 10% acetic acid. Add 15% ammonium acetate solution and ammonium hydroxide to adjust to pH 5 ± 0.2. Heat at 60° for 5 minutes. Extract the chromium by shaking with 10 ml, 10 ml, and 10 ml of chloroform and discard. Acidify the aqueous layer to pH 2 and heat to drive off residual chloroform. Add 3 ml of 1% solution of oxine in 2.5% acetic acid and raise to pH 5 ± 0.2 with 1:3 ammonium hydroxide. Add 1 ml of 6.5% solution of potassium cyanide and shake for 1 minute with 10 ml of chloroform. Dry the organic extract with sodium sulfate and read at 390 nm.

In analysis of zirconium and zircaloy, the bulk of the zirconium can also be separated from aluminum as fluozirconate by ion exchange.[195]

Chromium.[196] Heat 3 grams of sample with 40 ml of perchloric acid until white fumes are evolved. Take up in 80 ml of water and add 5 ml of 0.4% solution of chromic chloride in 1:5 hydrochloric acid as a collector. Add ammonium hydroxide until alkaline, boil for 15 minutes, and centrifuge. Dissolve the precipitate in 20 ml of 1:2 perchloric acid and evaporate to white fumes. Add 5 ml of hydrochloric acid and again evaporate to white fumes. Add 5 ml of water and 2 ml of 20% sodium acetate solution. Adjust to pH 5 with ammonium hydroxide, then add 1 ml of 5% solution of sodium diethyldithiocarbamate. Extract with 10 ml and 10 ml of chloroform and discard the extracts of iron and residual chromium. Add 1 ml of hydrochloric acid and 1 drop of 30% hydrogen peroxide and heat to 100°. Add 2 ml of 10% ammonium citrate solution and adjust to pH 5 with ammonium hydroxide. Add 3 ml of 1% solution of oxine in 3% acetic acid and 1 ml of 30% hydrogen peroxide. Raise to pH 9 by addition of ammonium hydroxide and dilute to 35 ml. Extract with 10 ml of chloroform and read the extract at 390 nm.

Tungsten.[197] Fuse 79.3 mg of pulverized metallic tungsten with 0.1 gram of sodium potassium carbonate and 0.1 gram of sodium tetraborate in platinum. Dissolve the melt in 2 ml of water and agitate with 0.5 ml of 0.5% solution of oxine in chloroform. Acidify to pH 4.6 with 50% acetic acid and add 2 ml of chloroform. Shake for 30 seconds, adjust to pH 9 with 1:3 ammonium hydroxide, and shake for 1 minute. Separate the chloroform layer and repeat the extraction. Dry the combined chloroform layers with sodium sulfate and read at 390 nm.

High Purity Copper.[198] This procedure will determine 0.1 ppm of aluminum in a 1-gram sample. Dissolve 1 gram in 6 ml of nitric acid and 2 ml of hydrochloric acid. Evaporate substantially to dryness and take up in 20 ml of 7 *N* hydrochloric acid. Extract interfering ions with 10 ml and 10 ml of methyl isobutyl ketone. Raise

[195]R. M. Hurd, G. W. Goward, M. D. Hartman, M. A. McCracken, and B. B. Wilson, *USAEC* **WAPD-C (AR)-143**, 7 pp (1957).

[196]Akina Kawass, *Jap. Anal.* **11**, 844–850 (1962); cf. A. A. Tumanov and N. S. Maslennikova, *Zavod. Lab.* **35**, 1439–1440 (1969).

[197]A. Danneil, *Tech.-Wiss. Abhandl. Osram-Ges* **7**, 350–356 (1958).

[198]B. Krässner and W. Angermann, *Z. Chem.* **7**, 438 (1967).

the aqueous phase to pH 5.6–5.9 by addition of 1 : 1 ammonium hydroxide. Make the solution *M* with thiosulfate to mask copper. Extract the aluminum with four 5-ml portions of 0.5% solution of oxine in chloroform. Wash the combined extracts with 10 ml and 10 ml of a solution containing 2% of EDTA and 2% of potassium cyanide. Read at 400 nm.

Uranium.[199] Dissolve 0.2 gram of sample at 100° in a few drops of hydrochloric and nitric acids and evaporate to crystals. Dissolve in 20 ml of 1 : 1 hydrochloric acid and shake with 10 ml of ether. Return the aqueous layer to the original beaker and add the ether to a beaker containing 2 ml of water and a few crystals of potassium thiocyanate to indicate the presence of iron. Repeat the extraction of the aqueous layer until all the iron is removed. Combine the ether layers for determination of iron if desired.

Dilute the aqueous layer to 100 ml, take a 20-ml aliquot, and evaporate to dryness. Add 10 ml of 8% sodium carbonate solution and boil for 10 minutes to mask uranium. Cool, add 1 ml of 2% solution of oxine in *N* acetic acid, and shake for 5 minutes with 5 ml and 5 ml of chloroform. Filter the chloroform through paper and read the fluorescence against quinine sulfate as a standard.[200]

Uranium Oxide. Dissolve sample material in a few drops of nitric and hydrochloric acids at 100° and evaporate to crystals. Take up in 15 ml of 1 : 1 hydrochloric acid and filter. Ash the filter and fuse the residue with 1 gram of sodium carbonate. Dissolve the melt in 5 ml of 1 : 1 hydrochloric acid and add to the previous filtrate. Complete as for uranium from "and shake with 10 ml of ether."

Magnesium.[201] Quartz equipment is recommended throughout. Add 0.2 gram of sample to 5 ml of water and dissolve by dropwise addition of 1 : 1 hydrochloric acid. Evaporate to a syrup. Add 10 ml of 1 : 1 hydrochloric acid and shake for 5 minutes with 10 ml of ether. After separation for 10 minutes, return the aqueous phase to the original beaker and add the ether to another beaker containing a few crystals of potassium thiocyanate and 1 ml of water. Repeat extractions of the aqueous layer until the ether extract gives no reaction for iron. Combine the ether extracts for determination of iron if desired. Evaporate the aqueous layer to a syrup, take up in water, and dilute to 200 ml. To a 10-ml aliquot, add 1 ml of 8% solution of sodium bicarbonate and 1 ml of 2% solution of oxine in *N* acetic acid. Shake for 5 minutes with 5 ml and 5 ml of chloroform. Adjust the combined chloroform extracts to 10 ml and read the fluorescence.

Bismuth.[202] To 12 ml of neutral sample solution, add 0.5 ml of nitric acid. Electrolyze for 1 hour at 2 amperes. Evaporate the electrolyte and washings to dryness and take up in 10 ml of 1 : 1 hydrochloric acid. Complete as for magnesium from "and shake for 5 minutes with 10 ml of ether."

[199] A. F. Fioletova, *Zh. Anal. Khim.* **17**, 302–304 (1962).
[200] Edward Coon, Joseph E. Petley, Warren H. McMullen, and Stephen E. Wiberley, *Anal. Chem.* **25**, 308–610 (1953).
[201] A. F. Fioletova, *Zh. Anal. Khim.* **14**, 739–740 (1959).
[202] A. F. Fioletova, *ibid.* **17**, 520–521 (1962).

Magnesium and Magnesium-Base Alloys.[203] Dissolve a 0.5-gram sample in 10 ml of hydrochloric acid and dilute to 50 ml. To a 5-ml aliquot, add 5 ml of acetate buffer solution for pH 5 and 2 drops of 0.1% Congo red indicator solution. Neutralize to a rose color with 1:1 ammonium hydroxide. Shake for 3 minutes with 2 ml of 1% solution of oxine in chloroform. Read the extract at 390 nm against chloroform.

Addition of hydroxylamine hydrochloride and 1,10 phenanthroline as masking agents has been recommended.[204] The sum of iron and aluminum can also be read at 390 nm and the iron as read at 470 nm subtracted.[205]

Tellurium.[206] Dissolve 0.5 gram of sample in 6 ml of hydrochloric acid and 2 ml of nitric acid. Add 1 ml of 0.1% solution of ammonium tartrate, then sufficient ammonium hydroxide to give a clear solution. Add 20 ml of 2 *M* ammonium chloride buffer solution for pH 10 and 5 ml of 0.1% solution of cyclohexane-*N,N,N′,N′*-tetraacetic acid. The latter masks up to 15 μg of iron, lead, copper, bismuth, and magnesium. Add 0.5 ml of 1% solution of potassium cyanide. Add 3 ml of 1% solution of oxine in 3% acetic acid and shake for 1 minute with 10 ml of carbon tetrachloride. Dry the organic layer with sodium sulfate and read at 395 nm.

Silicon Carbide.[207] ALUMINUM AND IRON. Fuse 0.5 gram of sample with 5 grams of sodium hydroxide and 5 grams of potassium nitrate. Neutralize with hydrochloric acid and evaporate to dryness. Add 5 ml of hydrofluoric acid and 1 ml of 1:1 sulfuric acid. Again evaporate to dryness, take up in water, and dilute to 200 ml. Mix a 10-ml aliquot with 5 ml of 0.6% solution of oxine in 2% acetic acid. Add 5 ml of 15% solution of ammonium acetate and raise to pH 7 with ammonium hydroxide. Dilute to 100 ml and shake for 5 minutes with 10 ml of chloroform. Read aluminum at 390 nm and iron at 470 nm and correct each for the interference of the other element. A formula for such calculation appears under determination of aluminum in alkalies by oxine.

Metals and Salts.[208] Extract 20 ml of sample solution in 7 *N* hydrochloric acid with 10 ml of methyl isobutyl ketone and discard the extract. Add 10 ml of 10% sodium acetate solution and adjust to pH 5.5. Extract the aluminum by shaking for 5 minutes with 10 ml of a 2% solution of oxine in chloroform. Wash the organic layer with 10 ml of a solution containing 2% of potassium cyanide and 2% of EDTA adjusted to pH 10. Read at 400 nm against chloroform.

Mixed Aluminum and Titanium Chlorides.[209] Dissolve 0.1 gram of sample in 10 ml of *N* sulfuric acid and dilute to 25 ml with that acid. Mix a 5-ml aliquot with 10

[203] M. N. Gorlova and L. I. Chichkova, *Tekhnol. Legk. Splavov. Nauchn.-Tekh. Byull. VILSa* **1972** (3), 125–126.

[204] *UKAEA* **Rep. PG 348 (S)**, 4 pp (1962).

[205] Kenji Motojima, Hiroshi Hashitani, and Kazuo Katsuyama, *Jap. Anal.* **9**, 517–520 (1960).

[206] Teru Yuasa, *ibid.* **11**, 1269–1273 (1962).

[207] Hiroshi Hirata and Masanao Arai, *ibid.* **16**, 820–825 (1967).

[208] Walter Angermann, Bernd Kaessner, Gerd Lorenz, and Manfred Geissler, *Chem. Anal.* (Warsaw) **16**, 261–270 (1971).

[209] Bohumil Pokorny and Ivan Cadersky, *Chim. Prum.* **20**, 434–436 (1970).

ml of N sulfuric acid, 2 ml of 30% hydrogen peroxide, and 2 ml of 2% solution of oxine in methanol. Add 1:5 ammonium hydroxide to raise to pH 8.5 ± 0.2. Extract with 10 ml, 10 ml, and 10 ml of chloroform. Dilute the combined extracts to 50 ml, dry a portion with sodium sulfate, and read at 389 nm. There is no interference by twentyfold excess of titanium.

Thorium Compounds.[210] ALUMINUM, IRON, AND CHROMIUM. Dissolve 1 gram of sample in 20 ml of 6 N nitric acid. Extract the thorium with 30 ml and 30 ml of 1:1 tributylphosphate-chloroform. Add 5 ml of acetic acid and 3 ml of 2% solution of oxine. Adjust to pH 2 with 1:1 ammonium hydroxide, extract ferric oxinate with 10 ml of chloroform, and read at 465 nm. Adjust to pH 5.1 with 1:1 ammonium hydroxide and incubate at 75° for 40 minutes. Cool and dilute to 100 ml. Extract the oxinates of chromium, nickel, copper, and aluminum with 10 ml of chloroform. Wash the copper and nickel from the organic layer with 50 ml of 2:1 30% ammonium carbonate solution–13% potassium cyanide solution. Read aluminum in the washed chloroform extract at 385 nm and chromium at 450 nm.

Sodium Silicate Solution.[211] Evaporate 1 ml of sample in platinum. Add 5 ml of hydrofluoric acid and 1 ml of 1:1 sulfuric acid. Evaporate to sulfur trioxide fumes. Take up in water and dilute to 50 ml. Adjust a 10-ml aliquot to pH 5. If the sample contains iron, reduce with 0.5 ml of 10% hydroxylamine hydrochloride solution, and mask with 0.5 ml of 2% solution of 2,2′-bipyridyl. Shake for 3 minutes with 5 ml of 0.1% solution of oxine in carbon tetrachloride. Read at 390 nm.

Alkalies.[212] ALUMINUM AND IRON. Dissolve the sample of sodium or potassium hydroxide or carbonate in water, neutralize with hydrochloric acid, make 0.1 N with that acid, and dilute to a known volume. Take an aliquot containing not more than 150 μg of ferric oxide and not more than 50 μg of aluminum oxide. Dilute to about 35 ml and add 3 ml of 1% oxine solution in ethanol. Add 3 ml of 15% ammonium acetate solution and adjust to pH 5–5.5 with 1:6 ammonium hydroxide. Shake for 1 minute with 10 ml of chloroform and dry the organic layer with sodium sulfate. Read at 470 and 390 nm.

$$Fe_2O_3 = \frac{A_{470}}{a_{470}^{Fe}}$$

$$Al_2O_3 = \frac{(a_{470}^{Fe}A_{390}) - a_{390}^{Fe}A_{470}}{a_{390}^{Al}a_{470}^{Fe}}$$

A_{390} and A_{470} = observed optical densities

a_{390}^{Fe} and a_{470}^{Fe} and a_{390}^{Al} = optical density of 1 μg of iron and aluminum at the specified wavelengths

[210] Kwok-yuan Chan, Pu Sun, and Ton-pol Tam, *Chemistry* (Taipei) **1968**, 1–6.
[211] A. B. Kiss and E. D. Walko, *Magy. Kem. Lapja.* **23**, 720–723 (1968).
[212] Hiroshi Awaya, Senpachi Miyoshi, and Kenji Motojima, *Jap. Anal.* **6**, 503–507 (1957).

Silicate Minerals.[213] Decompose 0.5 gram of sample in platinum with 5 ml of hydrofluoric acid and 1 ml of 1:1 sulfuric acid. Evaporate to sulfur trioxide fumes and fuse with 2 grams of potassium pyrosulfate. Take up with 25 ml of 1:10 sulfuric acid and dilute to 100 ml. To a 10-ml aliquot add 0.5 ml of 6% solution of cupferron. Extract with 5-ml portions of chloroform until the aqueous layer is colorless. Add 1.5 ml of ammonium hydroxide, 0.2 gram of sodium sulfite, and 20 ml of acetate buffer for pH 5.4. Extract traces of iron and titanium with 5 ml of chloroform and discard. Extract the aqueous solution with 5 ml, 5 ml, and 5 ml of 1% solution of oxine in chloroform. Dry the combined extracts with sodium sulfate, dilute to 50 ml with chloroform, and read at 386 nm.

Industrial Water[214]

ALUMINUM. As a buffer solution, mix 238 ml of ammonium hydroxide with 500 ml of water, add 111 ml of acetic acid and dilute to 1 liter.

Mix an appropriate volume of sample with 5 ml of hydrochloric acid and evaporate to about 1 ml. For water untreated and uncycled (e.g., from wells and rivers), a 100-ml sample is appropriate. For treated and cycled waters, such as cooling waters that contain chromate ion, 25 ml is adequate. Add 5 ml of nitric acid and evaporate to dryness at 100° but do not bake. Dissolve the residue with 2 ml of 0.1 *N* hydrochloric acid and add 80 ml of water. Add 4 ml of a solution of 2.2105 grams of sodium fluoride per liter. Transfer to a shaking jar containing 20 ml of moist, 20-50 mesh beads of IRC-50 ion exchange resin. Add 0.5 ml of the buffer solution and shake at 12 strokes per 5 seconds for 20 ± 3 minutes. Decant the sample, add 2 ml of 2% solution of oxine in 6% acetic acid, and let stand for 3 minutes. Extract by shaking for 1 minute each with 25 ml and 10 ml of chloroform. Discard these extracts. Now add 10 ml of the buffer solution and 2 ml of the oxine reagent. After 2 minutes extract the aluminum by shaking for 1 minute each with 25 ml and 10 ml of chloroform. Filter these combined extracts on paper through sodium sulfate and dilute to 50 ml with chloroform. Read photometrically at 380 nm or fluorimetrically with a 360-nm primary filter and a 415-nm secondary filter against chloroform.

SOLUBLE ALUMINUM. Repeat as above with a clear, filtered sample.

Alternatively,[215] mix 30 ml of sample containing less than 4 ppm of aluminum with 0.5 ml of 1:10 hydrochloric acid and 1 ml of 10% hydroxylamine hydrochloride solution. Warm and cool. Mask iron with 3 ml of 0.12% solution of 1,10-phenanthroline. Add 0.5 ml of 40% sodium acetate solution, then 3 ml of 1% solution of oxine, followed by 2 ml more of 40% sodium acetate solution. Extract with 10 ml of chloroform, wash the extract with water, and dry with sodium sulfate. Read at 410 nm.

[213] Frantisek Vlacil and Vladimir Zatka, *Chem. Prum.* **11**, 139 (1961); cf. Christina C. Miller and Robert A. Chalmers, *Analyst* **78**, 686–694 (1953); Robert A. Chalmers and Mohammed Abdul Basit, *ibid.* **93**, 629–632 (1968).

[214] Charles A. Noll and Louis J. Stefanelli, *Anal. Chem.* **35**, 1914–1916 (1963).

[215] Katsumi Goto, *Chem. Ind.* (London) **1957**, 329.

Seawater.[216] ALUMINUM AND IRON. To 1 liter of sample, add 5 ml of hydrochloric acid and 13 mg of beryllium as beryllium sulfate. Heat to boiling and filter. To a 250-ml aliquot of the filtrate, add 3 ml of 15% ammonium acetate solution and 3 ml of 1% solution of oxine in 3% acetic acid. Add ammonium hydroxide to adjust to pH 5–5.5 and dilute to 400 ml. Extract the oxinates of iron, aluminum, copper, and nickel with 10 ml of chloroform. Shake the organic phase with 10 ml of 0.65% solution of potassium cyanide to wash out the oxinates of copper and nickel. Read the organic phase at 390 nm for aluminum. Read at 470 nm as the sum of aluminum and iron, and determine iron by difference applying a suitable correction for the absorption of aluminum oxinate at 470 nm. A formula is given under determination in alkalies.

Tartrate Solutions.[217] The solution is assumed to contain iron, chromium, and aluminum. Extract the iron from a 50-ml sample with 10 ml and 10 ml of 1:1 acetylacetone-chloroform. Adjust the aqueous layer to pH 9, extract with 10 ml of 1% oxine in chloroform, and read at 390 nm. Chromium does not interfere.

Wine.[218] ALUMINUM AND IRON. Mix 0.5 ml of sample with 10 ml of water and 1 ml of 0.1% solution of oxine. Add saturated sodium acetate solution until adjusted to pH 5.4–5.7. Extract with 10 ml of chloroform and read the sum of iron and aluminum at 390 nm. Read the iron at 470 nm and get aluminum by difference.

Polyethylene.[219] Add 2 grams of powder or pellets to 20 ml of sulfuric acid and heat until the polymer is completely charred. Slowly add 20 ml of nitric acid dropwise to oxidize all the carbon, using more nitric acid if necessary. Add 5 ml of perchloric acid dropwise to the amber solution and heat to a volume of about 5 ml, adding more sulfuric acid if necessary to ensure volatilization of the other acids. Take up with 50 ml of water, heating to dissolve, and dilute to 100 ml.

As an alternative, the sample may be dry-ashed. Mix 5 grams of pellets and 2 grams of 3:1 potassium sulfate–potassium nitrate in a covered platinum crucible. Heat with a flame until the polymer is melted. Heat further until the polymer ignites and burns with a flame not more than 3 inches high. As the next step, burn the carbon from the sides and cover of the crucible. When cooled, add 1 ml of sulfuric acid and heat as long as oxides of nitrogen are evolved. Continue to heat to quiescent boiling and cool. Add 10 ml of 1:9 sulfuric acid, heat to dissolve, and dilute to 100 ml.

Mix 50 ml of the solution with 1 ml of 10% thioglycollic acid solution and a few drops of phenolphthalein indicator solution. Neutralize with ammonium hydroxide and add 10 ml excess. Add 1 ml of 5% sodium cyanide solution, 5 ml of 2% solution of oxine, and 10 ml of trichloroethylene. Shake for 1 minute and dry the organic layer by filtering on paper. Read at 390 nm against a reagent blank. The optimum range is 5–30 μg of aluminum in the 10 ml of extract.

[216]Hiroshi Hashitani and Katsumi Yamamoto, *J. Chem. Soc. Jap.* **80**, 727–731 (1959); Kenji Motojima and Hiroshi Hastitani, *Jap. Anal.* **6**, 642–646 (1957).
[217]E. Uhlemann, G. Meissner, E. Butter, and H. Weinelt, *J. Prakt. Chem.* **25**, 272–278 (1964).
[218]Yu. S. Lyalikov, B. V. Lipis, and L. G. Madan, *Sadovod. Vinograd. Vinod. Mold.* **1961** (8), 45–46.
[219]William T. Bolleter, *Anal. Chem.* **31**, 210–203 (1959).

Polyethylene can also be wet-ashed with 50% hydrogen peroxide.[220]

Polypropylene.[221] In a silver Kjeldahl flask with 15 ml of sulfuric acid, heat 1 gram of sample until black and viscous. Cool, add 5 ml of 30% hydrogen peroxide, and heat again. Add more hydrogen peroxide and continue heating until the solution is clear and colorless. Boil off the excess of hydrogen peroxide and dilute to 100 ml.

Dilute a 20-ml aliquot containing 0.02–0.1 mg of aluminum to 50 ml. Add 1 ml of 50% tartaric acid solution, 1 ml of 50% ammonium acetate solution, and 5 ml of a solution of 2.5 grams of 8-hydroxyquinaldine in 5 ml of acetic acid and 80 ml of water.[222] Add ammonium hydroxide to pH 8–9 and extract with 15 ml of chloroform. Repeat twice adding 2.5 ml of the hydroxyquinaldine solution and extracting with 8 ml of chloroform. Discard these extracts.

Add 1.5 ml of 2.5% solution of oxine in 5% acetic acid to the aqueous layer. Extract with 7 ml of chloroform. Repeat the addition of oxine and chloroform extraction. Dilute the chloroform extracts to 25 ml and read at 395 nm.

Silicone Polymers.[223] ALUMINUM AND IRON. To 0.2 gram of solid or a corresponding amount of liquid in a gelatin capsule, add 10 ml of sulfuric acid. Heat gently to dissolve and more strongly to char, swirling to avoid deposition of silica on the sides. Cool, add a drop of nitric acid, and heat to oxidize residual carbon. Repeat if necessary, but avoid excess of nitric acid. Cool, filter, and dilute the filtrate to 250 ml.

Ash the paper in platinum and ignite. Add 2 ml of perchloric acid and 10 ml of hydrofluoric acid. Heat to 100°, then evaporate to perchloric acid fumes. Add 1 ml of perchloric acid and repeat the evaporation. Then heat under infrared lamp not quite to dryness. Add 2 ml of perchloric acid, cool and dilute to 100 ml. This is a second fraction for separate development.

For aluminum, dilute an aliquot of the sample containing 1–15 μg of aluminum to 30 ml. Add 1 ml of 3% hydrogen peroxide, 1 ml of a saturated solution of sodium sulfite, 1 ml of 15% solution of sodium cyanide, and 2 ml of 1% solution of oxine in 3% acetic acid. Adjust to pH 9 with N ammonium hydroxide or acetic acid and shake for 2 minutes with 20 ml of benzene. Allow 2 hours for separation, and withdraw the aqueous phase. Filter the organic phase through paper to dry it and read at 390 nm against a reagent blank.[224]

For iron, take a 10-ml aliquot. Add 10 ml of 4% sodium hydroxide solution, 2 ml of , 0% hydroxylamine hydrochloride solution, 5 ml of 0.1% solution of 2,2′-bipyridyl, and 10 ml of 2 N sodium acetate. The pH should be 3.5–9. Dilute to 50 ml and let stand for 15 minutes. Read at 517 nm against a reagent blank.

Plant Tissue.[225] Wet-ash 1 gram of dried sample with 3 ml of nitric acid, 1 ml of

[220]W. Mischa, E. Schroeder, and K. Thinius, *Plaste Kautsch.* **16** (1), 23–25 (1969).
[221]K. Novak and V. Mika, *Chem. Prum.* **13**, 260–262 (1963).
[222]Kenji Motojima and Hiroshi Hashitani, *Bull. Chem. Soc. Jap.* **29**, 458–460 (1956).
[223]Shizuo Fujiwara and Hisatake Narasaki, *Anal. Chem.* **36**, 206–207 (1964).
[224]Cf. Nobuhiko Iritani and Hitoshi Aoyama, *Shizuoka Yukka Daiguku Kaigaku 5-Shunen Kinen Rombunshi* **1958**, 141–143.
[225]C. R. Frink and D. E. Peaslee, *Analyst* **93**, 469–474 (1968).

hydrochloric acid, and 1 ml of sulfuric acid. If necessary to get a colorless digest, add 30% hydrogen peroxide dropwise. Evaporate to sulfur trioxide fumes, take up in 25 ml of water, and filter. Wash the residual silica with 1:1 hydrochloric acid and dilute to 100 ml. To an aliquot containing 2–20 μg of aluminum, add 1 ml of *N* acetic acid and 2 drops of thymol blue indicator solution. Neutralize with 1:1 ammonium hydroxide. Add 5 ml of ammonium acetate buffer solution for pH 6.2, dilute to 50 ml, and add 2 ml of 3% solution of sodium diethyldithiocarbamate. Extract with 5 ml and 5 ml of chloroform and discard the extracts. Extract the aluminum with 5 ml of 2% solution of oxine in chloroform and set aside for 15 minutes. Read at 385 nm against a reagent blank.

Alternatively,[226] dry-ash a 0.6-gram sample such as grass or straw and heat at 450° for 2 hours. Dissolve the ash in 6 ml of hot 1:1 hydrochloric acid, dilute to 10 ml, and set aside overnight. Filter, and dilute the filtrate to 50 ml.

For iron, mix a 2-ml aliquot with 1 ml of 10% hydroxylamine hydrochloride solution and 4 ml of a 0.1 *M* ammonium hydroxide–5 *M* acetic acid buffer for pH 2.8. After 5 minutes add 4 ml of 0.1% solution of 1,10-phenanthroline. Set aside for 20 minutes and add 1 ml of perchloric acid. Extract the iron complex with 10 ml of chloroform and read at 490 nm against a blank.

For aluminum, add 10 ml of 1:1 ammonium hydroxide to the aqueous phase and set aside for 5 minutes. Add 4 ml of 10% EDTA solution and set aside for 5 minutes more. Add 2 ml of 5% ethanolic oxine and set aside for 1 hour. Extract the oxine complex with 20 ml of carbon tetrachloride, filter, and read at 400 nm against carbon tetrachloride.

Alternatively,[227] take an aliquot of extract of aluminum leachable by 7.5% potassium chloride solution to contain not over 20 μg of aluminum. Add sequentially 50 ml of water, 2 ml of 2% solution of oxine, and 2 ml of an ammonium acetate–ammonium hydroxide buffer solution for pH 9. Shake for 2 minutes with 10 ml and 10 ml of chloroform. Dilute the combined extracts to a known volume and set aside for 30 minutes. Read fluorimetrically.

Yet another technic is to digest with perchloric acid,[228] reduce ferric ion to the ferrous state, complex with 4,7-diphenyl-1,10-phenanthroline at pH 3–4, extract with 9:1 chloroform–isoamyl alcohol, adjust to pH 4.7–5.1, and extract aluminum as the oxine complex to read fluorescently.

Soil Extracts. The medium is 7.5% potassium chloride solution. Follow the second technic given for iron and aluminum in plant tissue, but instead of adding 1 ml of perchloric acid, which would precipitate potassium perchlorate, add 4 ml of 10% trichloroacetic acid solution.

Leaves.[229] ALUMINUM AND IRON. Digest a sample expected to contain 8–32 μg of aluminum and 20–80 μg of iron with 5 ml of nitric acid, 2 ml of perchloric acid, and 2 ml of 1:1 sulfuric acid. After a clear digestate has been obtained, evaporate to incipient sulfur trioxide fumes. Add 25 ml of water and evaporate to about 5 ml. Add 10 ml of water and 1 ml of 0.6% solution of oxine in 2% acetic acid. Neutralize

[226] J. Paul, *Mikrochim. Acta* **1966**, 1075–1087.
[227] M. G. Cook, *Proc. Soil Sci. Soc. Am.* **32**, 292–293 (1968).
[228] E. J. Rubins and G. R. Hagstrom, *J. Agr. Food Chem.* **7**, 722–724 (1959).
[229] K. R. Middleton, *Analyst* **89**, 421–427 (1964).

with 1:2 ammonium hydroxide. Add 5 ml of acetate buffer solution for pH 2.85 and shake with 10 ml of 0.3% solution of oxine in chloroform. Read the organic layer at 470 nm against a reagent blank for determination of the iron content. Add 4.5 ml of 1:2 ammonium hydroxide to the aqueous layer and extract with 10 ml of 0.3% solution of oxine in chloroform. Read the organic layer at 385 nm against a reagent blank as a measure of the aluminum content. The method was developed for analysis of leaves of the rubber tree, *Hevea brasiliensis*.

JUGLONE

Juglone, which is 5-hydroxy-1,4-naphthaquinone, forms a red-purple complex with aluminum appropriate for reading 0.1–0.26 mg in 25 ml.

Procedure. *Aluminum Sulfate or Potash Alum.*[230] Dissolve 0.2 gram in 10 ml of hot water and precipitate sulfates by adding a solution of 0.25 gram of barium chloride. Filter, wash the barium sulfate, and dilute the filtrate and washings to 500 ml. To a 2.5-ml aliquot, add 2 ml of 0.04% sodium bicarbonate solution and 1 ml of 0.2% solution of juglone in ethanol. Dilute to 10 ml with ethanol and read.

LUMOGALLION

Lumogallion, which is 5-chloro-3-(2,4-dihydroxyphenylazo)-2-hydroxybenzenesulfonic acid, as a complex with aluminum is read by fluorescence.[231] Addition of Antarox CO 890, a nonionic surfactant, shifts the maximum from 558 to 548 nm, increases the fluorescence about sixfold, thus permits determination of 0.5 ng of aluminum per ml. Dodecybenzenesulfonate, an anionic surfactant, has no effect. Benzyltetradecylammonium chloride, a cationic surfactant, first increases the fluorescent intensity, then causes the complex to dissociate. The optimum pH is 5 with ammonium acetate buffer solution, and the calibration is a straight line for 0.1–2 μg of aluminum in 25 ml of solution containing 0.5 ml of 0.01% solution of lumogallion.[232] At 80° reaction is complete within 20 minutes. Amyl alcohol will extract the complex. The presence of iron, nickel, copper, cobalt, hexavalent chromium, pentavalent vanadium, stannic ion, or scandium causes low results.

Procedures

Seawater.[233] To a sample containing 0.003–2 μg of aluminum, add 0.5 ml of

[230] L. N. Aizenberg, A. I. Suprunenko, and R. S. Aizenberg, *Tr. Kishinev. S.-Kh. Inst.* **43**, 193–201 (1966).
[231] Nobuhiko Ishibashi and Kenyu Kina, *Anal. Lett.* **5**, 637–641 (1972).
[232] Yasuharu Nishikawa, Keizo Hiraki, Kiyotoshi Morishige, and Tsunenobu Shigematsu, *Jap. Anal.* **16**, 692–697 (1967).
[233] Yasuharu Nishikawa, Keizo Hiraki, Kiyotoshi Morishige, Akira Tsuchiyama, and Tsunenobu Shigematsu, *ibid.* **17**, 1092–1097 (1968).

0.01% solution of lumogallion. Add 5 ml of 20% ammonium acetate solution and adjust to pH 5 with 1:1 hydrochloric acid. Dilute to 25 ml and heat at 80° for 20 minutes. Activate at 365 nm and read the fluorescence at 558 nm.

Natural Water.[234] Add 2 ml of 4% solution of hydroxylamine hydrochloride to a 15-ml sample and set aside for 15 minutes to reduce ferric ion. Add 1 ml of 1% solution of 1,10-phenanthroline to mask the iron, 0.5 ml of 0.01% solution of lumogallion, and 2 ml of 2% ammonium acetate solution, to buffer to pH 5. Dilute to 25 ml, heat at 80° for 20 minutes, and cool. Activate at 485 nm and read the fluorescence at 576 nm. Use rhodamine B at 0.168 μg/ml as a reference.

METHYLTHYMOL BLUE

Aluminum forms two violet 1:1 complexes with methylthymol blue, as well as a red 1:2 complex; the same is true of gallium.[235] Both types of complex can coexist. With this reagent, mercaptoacetic acid used in conjunction with EDTA is more effective than ascorbic acid for masking interfering ions.[236] In the 50 ml of final solution, 2 ml of 10% solution of mercaptoacetic acetic acid, 2.5 ml of 0.1% solution of methylthymol blue, and 3 ml of 0.025 *M* EDTA will mask 5 mg of iron in reading 0.01–0.03 mg of aluminum at either 590 or 540 nm.

Procedure. Mix 5 ml of sample containing 1.7–13.5 μg of aluminum with 5 ml of 3.1 m*M* methylthymol blue and 15 ml of *M* acetic acid.[237] Boil for 3 minutes, cool, dilute to 50 ml with *M* acetic acid, and read at 590 nm against a blank.

Titanium.[238] Dissolve 0.5 gram of sample in 30 ml of 1:1 hydrochloric acid. Add a further 20 ml of 1:1 hydrochloric acid and dilute to 250 ml. To a 50-ml aliquot, add 5 ml of nitric acid to oxidize titanous chloride. Add 20 ml of 6% cupferron solution and shake until the precipitate coagulates. Extract with successive 20-ml portions of chloroform until a colorless extract is obtained, and discard the extracts. Add 5 ml of nitric acid and 3 ml of perchloric acid and evaporate to dryness. Dissolve the residue in 20 ml of 1:1 hydrochloric acid, adjust to pH 2 with ammonium hydroxide, and dilute to 100 ml. Mix a 20-ml aliquot with 2 ml of 2% ascorbic acid solution and 1 ml of 0.1% methylthymol solution. Add 5 ml of a buffer solution consisting of 17.9 ml of *N* hydrochloric acid adjusted to pH 3 by addition of 10% glycine solution and diluted to 100 ml. Heat at 100° for 3 minutes, cool, and mask manganese, zinc, copper, and nickel by adding 3 ml of 0.025 *M* EDTA. Dilute to 50 ml, set aside for 30 minutes, and read at 590 nm.

[234] Tsunenobu Nishikawa, Keizo Hiraki, and Noriko Nagano, *ibid.* **19**, 551–554 (1970); cf. D. J. Hydes and P. S. Liss, *Analyst* **101**, 922–931 (1976).

[235] V. A. Nazarenko and E. M. Nevskaya, *Zh. Anal. Khim.* **24**, 839–843 (1969).

[236] V. N. Tikhonov, *Izv. Vyssh. Ucheb. Zaved., Khim. Khim. Tekhnol.* **15**, 1789–1791 (1972).

[237] N. D. Lukomskaya, T. V. Mal'keva, and K. B. Yatsimirskii, *Izv. Vyssh. Ucheb. Zaved., Khim. Khim. Tekhnel.* **10**, 994–996 (1967); N. D. Lukomskaya, *ibid.* **16**, 1157–1158 (1973).

[238] V. N. Tikhenev and M. Ya. Grankina, *Zavod. Lab.* **32**, 278–279 (1966).

Bismuth.[239] Dissolve 0.5 gram of sample in a minimal amount of 1:2 nitric acid and add 2 ml in excess. Dilute to 100 ml, and add 3 ml of 40% formaldehyde solution. Partially neutralize with 10% potassium hydroxide solution and boil for 2 minutes to precipitate the bismuth. Filter, and wash with hot water. Dilute the filtrate to 200 ml and take a 20-ml aliquot. Neutralize to phenolphthalein with acetic acid, add 10 ml of a buffer solution for pH 3, and heat at 100° for 10 minutes. Add 30 ml of water and 1 ml of 0.1% methylthymol blue solution. When cool, mask zinc, cadmium, lead, and residual bismuth with 1 ml of 10% EDTA solution and dilute to 100 ml. Read at 590 nm against a reagent blank. The addition of potassium hydroxide solution to cause reduction of bismuth by formaldehyde requires careful control.

MORIN

The fluorescence of the complex with morin, which is 2′,3,4′,5,7-pentahydroxyflavone, at pH 3 will determine aluminum in the parts per billion range. As applied to water, tolerances of ions, in ppb, are as follows: chromic, 30; cupric, 20; fluoride, 5; ferric, 100; magnesium, 200; ammonium, 500; phosphate, 3.

Procedure. *Water.*[240] Add 1 ml of 1:1 acetic acid to 99 ml of sample. For high quality water, this gives a pH of 3. Dilute 5 ml of morin in ethanol to 100 ml with the acidified sample. After 20 minutes, activate at 440 nm and read the fluorescence at 525 nm. The waiting period can be avoided by heating the mixture with morin at 100° for at least 30 seconds and cooling before reading.

PHENOXYDINAPHTHOFUCHSONEDICARBOXYLIC ACID

This reagent, which is mordant green 31, forms a 3:2 complex with aluminum at pH 5–8.[241] The technic reads 0.3–3 μg aluminum in 25 ml. The sensitivity is 0.016 as compared with hematoxylin 0.1, oxine 0.4, alizarin red S 0.05, aluminon 0.05, and stilbazo 0.04.

Procedure. *Cesium Iodide.* Dissolve 0.5 gram of sample in 10 ml of m*M* hydrochloric acid. Add 0.2 ml of 10% ascorbic acid solution to mask ferric ion and 0.2 ml of 10% sodium thiosulfate solution to mask copper. Add 1 ml of 0.5 m*M* reagent and 5 ml of ammoniacal acetate buffer solution for pH 6.15 to suppress interference by cadmium, nickel, cobalt, and magnesium. Dilute to 25 ml, set aside for 30 minutes, and read at 640 nm against a reagent blank.

---

[239] A. V. Vinegradov and M. P. Filippova, *ibid.* **35**, 1165–1166 (1969).
[240] Fritz Will III, *Anal. Chem.* **33**, 1360–1362 (1961).
[241] L. P. Adamovich, A. P. Mirnaya, and A. K. Khukhryanskaya, *Zavod. Lab.* **35**, 481–784 (1969).

PONTACHROME BLUE BLACK R

The fluorescence of the complex of pontachrome blue black R (mordant black 17) with aluminum is read fluroescently down to 0.002 ppm.[242] There is interference by iron, gallium, cobalt, vanadate, copper, titanium, and nickel. These are negligible in natural water except iron, and up to 25-fold amounts of it are masked by 4,7-diphenyl-1,10-phenanthroline. A maximum of 0.5 mg of dye is used in 50 ml of solution and attains a maximum fluorescence in 65 minutes. The complex is readily extracted with amyl alcohol at pH 4.8–5.4.[243] The maximum fluorescence is at 593 nm.

The reagent has been applied to differentiation of forms of aluminum in seawater as (1) that in true solution, (2) that so finely divided as to pass through a 0.45 μm Millipore filter, and (3) that retained on the filter.[244]

Procedure. Add 5 ml of 10% ammonium acetate solution to the sample solution and dilute to 40 ml. Use 0.1% pontachrome blue black R in ethanol.[245] For 0.2–1 μg of aluminum, add 1 ml; for 1–12 μg, add 1.5 ml; and for 12–18 μg, add 2 ml. Adjust the solution to pH 4.8 and dilute to 50 ml. Heat at 100° for 10 minutes, cool, activate at 535 nm, and read the fluorescence at 600 nm.

Organic Matter.[246] Heat a sample containing up to 10 μg of aluminum in platinum to destroy organic matter. Dissolve the residue in 2 ml of 1:4 sulfuric acid and add a few ml of water. Adjust to pH 4.5 with about 2.5 ml of 15% ammonium acetate solution. Add 5 ml of 0.01% solution of pontachrome blue black and dilute to 20 ml. Heat at 100° for 15 minutes. Cool, activate at 535 nm, and read the fluorescence at 595 nm.

Zinc Sulfide Fluorescent Pigment.[247] Heat 0.1 gram of sample with 1 ml of sulfuric acid in platinum to fumes of sulfur trioxide. Take up with 10 ml of water and electrolyze to remove interfering ions. Add 0.75 ml of 0.1% solution of the dye in ethanol containing 0.25 gram of ammonium acetate. Adjust to pH 5 and dilute to 25 ml. Let stand for 2 hours in polyethylene. Activate at 535 nm and read the fluorescence at 595 nm.

PYROCATECHOL VIOLET

The 3:2 complex of pyrocatechol violet with aluminum at pH 5 absorbs at a maximum of 615 nm. The maximum for the reagent is at 450 nm, and it has only a

[242]D. E. Donaldson, *U.S. Geol. Survey Prof. Pap.* No. **550-d, 258-61** (1966); cf. J. F. Possidoni de Albinati, *An. Asoc. Quim. Argent.* **51**, 96–122 (1963).
[243]Masayoshi Ishibashi, Tsunenobu Shigematsu, and Vasuharu Nishikawa, *J. Chem. Soc. Jap.* **81**, 259–262 (1960).
[244]W. Sackett and G. Arrhenius, *Geochim. Cosmechim. Acta* **26**, 955–968 (1962).
[245]Masayoshi Ishibashi, Tsunenobu Shigematsu, and Yasuharu Nishikawa, *Jap. Anal.* **6**, 568–571 (1957).
[246]A. Puech. G. Kister, and J. Chanal, *Zentbl. Pharm. Pharmakether. Lab.-Diagn.* **111**, 7–18 (1972).
[247]A. Danneil, *Tech.-Wiss. Abhandl. Osram-Ges.* **7**, 350–356 (1958).

minimal absorption at 615 nm. It is suitable for reading 3–15 ppm of aluminum at pH 5, where zinc and magnesium do not interfere; above pH 7 they complex. The reagent is 3,3′,4′-trihydroxyfuchsone-2″-sulfonic acid. The absorption of different lots of the reagent varies substantially. At pH 6.1 the complex is formed by 1.5-fold excess of reagent, but use of a threefold excess enhances the stability of the complex. Because of nonselectivity of the reagent, aluminum must be separated from interfering ions. Common masking agents such as tartrate, citrate, and EDTA inhibit formation of the aluminum complex.

Procedure. To a sample solution containing 0.3–1.5 mg of aluminum, add 10 ml of a buffer solution prepared by mixing 77 ml of pyridine with 63 ml of acetic acid.[248] Add 2 ml of 0.3864% solution of pyrocatechol violet and dilute to 100 ml. Read at 615 nm against a reagent blank.

Molybdenum and Tungsten.[249] Heat 0.5 gram of powdered sample in platinum at 600–640° for 1 hour to convert to oxide. Add 3 grams of sodium carbonate and fuse. Add 25 ml of water to the sample and crucible, and disintegrate with 5 ml of 1:1 sulfuric acid. At this stage the tungsten compound is insoluble. Remove the crucible and dilute to 50 ml with the wash water. Add successively 5 ml of 30% hydrogen peroxide, 5 ml of 10% acetylacetone in 20% ethanol, and 10 ml of 50% ammonium acetate solution. If tungsten is present, adjust to approximately pH 8 to dissolve insoluble compounds, and acidify with a few drops of 1:1 sulfuric acid. Adjust to pH 6.5±0.1 with ammonium hydroxide and dilute to 100 ml. Extract with 10 ml, 5 ml, and 5 ml of chloroform, shaking for 2 minutes each time. Shake the combined chloroform extracts for 3 minutes with 5 ml of hydrochloric acid and discard the chloroform phase. Dilute the acid extract of aluminum to 50 ml and add 3 ml of a fresh solution containing 1% of ammonium pyrrolidinedithiocarbamate and 3% of cupferron. After 5 minutes shake for 1 minute with 10 ml of chloroform; discard the chloroform phase containing copper and iron. Add 2 ml of carbamate-cupferron solution, extract with 5 ml of chloroform, and wash the aqueous phase with 5 ml and 5 ml of chloroform. Add 5 ml of 1:1 sulfuric acid to the aqueous phase, warm to drive off chloroform, and evaporate to 25 ml. Add 3 ml of perchloric acid, 3 ml of hydrochloric acid, and 1 ml of nitric acid. Boil to destroy organic matter and evaporate to approximately 1.5 ml. If a yellow or brown color persists, add more oxidizing acids and repeat the treatment to end with a colorless solution in sulfuric acid. Now evaporate to sulfur trioxide fumes, but not to dryness. Dissolve in water and dilute to 25 ml.

To an aliquot containing up to 40 μg of aluminum, add 2 ml of 5% ascorbic acid solution, 4 ml of 0.08% solution of pyrocatechol violet, and 1 ml of 50% solution of ammonium acetate. Adjust the pH to 6.1±0.03 with dilute ammonium hydroxide. Add 5 ml of buffer solution consisting of 10% ammonium acetate adjusted to pH 6.1 with about 0.3% of acetic acid. Dilute to 100 ml, read at 578 nm against water, and subtract a blank.

Iron, Ferrovanadium, and Steel. Mix 0.2–2 grams of sample containing not more

[248]Anthony Anton, *Anal. Chem.* **32**, 725–726 (1960).
[249]Elsie M. Donaldson, *Talenta* **18**, 905–915 (1971).

than 2 mg of aluminum with 30 ml of water and 15 ml of 1:1 sulfuric acid. Heat until reaction ceases. If the sample contains more than 0.5% silicon, evaporate to sulfur trioxide fumes to dehydrate the silica and take up in 50 ml of water. Filter, wash well, and put the filtrate aside. Ash the paper and residue in platinum. To the residue add 1 ml of 1:1 sulfuric acid, 2 ml of hydrofluoric acid, and 1 ml of nitric acid. Warm to dissolve the residue, evaporate to dryness, and ignite at 500°. Fuse the residue with 2 grams of potassium pyrosulfate. Dissolve in 50 ml of water and 3 ml of 1:1 sulfuric acid. Combine with the original filtrate and dilute to 200 ml. Transfer a 50-ml aliquot to a mercury cathode cell, dilute to 100 ml, and electrolyze at 10 amperes for 1 hour. Filter the electrolyte, add 5 ml of hydrochloric acid, and evaporate to about 1.5 ml. Take up with 50 ml of water. Some flocculent silica will not interfere. Proceed as for molybdenum and tungsten from "Add successively 5 ml of 30% hydrogen peroxide,...."

Copper-Base Alloys. For aluminum greater than 0.1%, to a sample containing not more than 2 mg of aluminum, add 40 ml of 1:3 sulfuric acid, 2 ml of nitric acid, and 2 ml of hydrofluoric acid. Heat until dissolved, and evaporate to sulfur trioxide fumes. Cool, wash down with 5 ml of water, and again evaporate to sulfur trioxide fumes. Take up in 50 ml of water and dilute to 200 ml. Proceed as for iron, ferrovanadium, and steel from "Transfer a 50-ml aliquot to a mercury cathode cell,...."

For aluminum less than 0.1%, decompose 1 gram of sample as above, but use 14 ml of 1:3 sulfuric acid instead of 40 ml.

2-QUINALIZARINSULFONIC ACID

Aluminum forms a 1:1 complex with this reagent, which is the sodium salt of 1,4-dihydroxy-2-anthraquinonesulfonic acid. It gives a violet color with aluminum in absolute methanol. Beer's law is followed up to 1.7 ppm. Many interfering ions are removed by the mercury cathode. Those not so removed, and their tolerance limits in ppm for an error not exceeding 7%, are as follows: beryllium, 0.2; scandium, 0.2; thorium, 0.5; titanium, 0.2; yttrium, 0.2; zirconium, 0.2; phosphate, 1. Even 0.5 ppm of fluoride causes a negative error of 28 ppm. The reagent is sensitive to 1 part of aluminum in 50 million parts of ether or ethanol. At low aluminum concentrations no color develops in aqueous solution. The complex is also read fluorimetrically.

Procedures

Steel or Bronze.[250] Take 10 ml of a sample solution from which interfering ions have been removed by the mercury cathode. The sample solution should contain 0.2–1.5 ml of sulfuric acid and 0.03–0.07 mg of aluminum. Evaporate nearly to dryness in silica, then drive off sulfur trioxide fumes to dryness. Add 0.5 ml of hydrochloric acid and 0.25 ml of nitric acid to the cooled residue. Wash down the

[250] E. Guy Owens II and John H. Yoe, *Anal. Chem.* **31**, 384–387 (1959).

sides with water to a volume approximating 5 ml. Again evaporate to about a drop, and take up in 5 ml of absolute methanol. Add 10 ml of 0.032% solution of the reagent in methanol and dilute to 50 ml with absolute methanol. The pH should be 0.3–0.5; if necessary, adjust with hydrochloric acid. Let stand for 1 hour and read at 560 nm against a reagent blank.

By Fluorescence.[251] Take a sample that when diluted to 50 ml, will give 11–54 ppb of aluminum. Add 2 ml of 0.1 m*M* reagent and 2 ml of an acetic acid–sodium acetate buffer solution for pH 4.76. Dilute to 50 ml and set aside for 1 hour. Then activate at 500 nm and read the fluorescence at 558 nm.

2–SALICYLIDENEAMINOPHENOL

This reagent permits determination of 2 ng of aluminum by fluorescence of its 2 : 1 complex.[252] Interference by antimony, bismuth, and hexavalent uranium is avoided by extraction of their complex with diethyldithiocarbamate. If present beyond tenfold excess; tungsten causes yellow color and low results. Pentavalent vanadium and hexavalent cerium can be reduced with ascorbic acid before the extraction. Zirconium causes low results but can be masked by a large excess of mandelic acid after the extraction. Platinum can be masked with cyanide. Chromic ion causes low results, and scandium forms a fluorescent complex. Fluoride, tartrate, citrate, and oxalate interfere. EDTA reduces the fluorescence. Copper and iron must not exceed 0.02 μg/ml. The intensity of the fluorescence is proportional to the aluminum content at pH 5.6–6.2 up to 0.1 μg/ml.[253]

At pH 4–7 phthalate and acetate ions are bonded to the aluminum and decrease the color and fluorescent intensity.[254] A 0.2 *M* hexamine buffer solution had no such effect.

Procedure. As a buffer solution for pH 5.6, dissolve 50 grams of ammonium acetate in 400 ml of water and add 10 ml of acetic acid.

Mix 10 ml of sample solution containing up to 2.7 μg of aluminum with 10 ml of the buffer solution for pH 5.6.[255] Add 30 ml of ethyl acetate and 10 ml of a 0.2% solution of sodium diethyldithiocarbamate. Shake for 30 seconds and discard the extract. Add 5 ml of 0.1% solution of the reagent to the aqueous phase, dilute to 100 ml, and set aside for 20 minutes. Activate at 410 nm and read the fluorescence at 520 nm.

Cadmium Salts.[256] Dissolve 0.5 gram of sample containing not less than 1 ppm

[251] F. Capitan Garcia, M. Roman Ceba, and A. Guiraum, *An. Quim.* **70**, 508–514 (1974).

[252] R. M. Dagnall, R. Smith, and T. S. West, *Chem. Ind.* (London) **1965** (34), 1499–1500; R. Smith and A. E. Lawson, *Spectrovision* **1971** (26), 11–12.

[253] E. A. Bozhevol'nov and V. N. Yanishevskaya, *Zh. Vsesz. Khim. Obshch. D. I. Mendelaeva* **5**, 356–357 (1960).

[254] A. K. Babko and S. L. Lisichenok, *Ukr. Khim. Zh.* **35**, 98–100 (1969).

[255] R. M. Dagnall, R. Smith, and T. W. West, *Talenta* **13**, 609–617 (1966).

[256] N. B. Lebed' and R. P. Pantaler, *Zavod. Lab.* **31**, 163–164 (1965).

of aluminum and not more than 500 ppm of iron or copper in 10 ml of acetate buffer solution for pH 6.2–6.4. The buffer should have been purified by lumogallion IREA and by EDE-10 anionite. Add 0.3 ml of 0.01% solution of 2-salicylideneaminophenol and set aside for 30 minutes. Read the fluorescence at 530 nm.

Sodium Iodide or Ferrous Ammonium Sulfate.[257] Adjust a solution of the sample to pH 6. Add sufficient reagent in acetone containing 42% of sodium perchlorate to make the sample 0.2 m*M* with the reagent. Dilute to 5 ml and set aside for 50 minutes. Extract with 5 ml of 1 : 1 chloroform–tributyl phosphate by shaking for 10 minutes. Activate at 410 nm and read the fluorescence of the extract at 520 nm.

Lead Salts.[258] As a solvent, mix 7.5 ml of 2 *M* acetic acid with 92.5 ml of 2 *M* sodium acetate. This also serves as a buffer for pH 6 and to complex lead ion to prevent interference.

Dissolve 1 gram of lead acetate or nitrate in 7–8 ml of the solvent. Add 0.3 ml of 0.1% solution of the reagent and dilute to 10 ml with the solvent. Set aside for 50 minutes, activate at 410 nm, and read the fluorescence at 520 nm.

Lead Sulfide. Decompose 1 gram of sample by heating with 8 ml of nitric acid. Evaporate to dryness, and calcine to burn off the free sulfur. Dissolve the lead sulfate in the buffer solution and proceed as above.

SALINAZID

Salinazid, which is 1-isonicotinoyl-2-salicylidenehydrazine, forms a yellow 1 : 1 complex with aluminum at pH 3.5–5.5[259] which is stable in light for at least 12 hours. Beer's law is followed for 0.5–3.5 μg of aluminum per ml. There is interference by cupric ion, nickel, ferric ion, lead, hexavalent chromium, molybdenum, thorium, pentavalent vanadium, tungstate, and phosphate.

Procedure. To the sample solution, add 2 ml of m*M* salinazid in ethanol and 5 ml of 12.5% solution of sodium perchlorate. Dilute to 20 ml, adjust to pH 5 with 0.01 *M* perchloric acid or sodium carbonate, and dilute to 25 ml. Set aside for 40 minutes and read at 375 nm against a reagent blank.

SOLOCHROME CYANINE R

The complex with aluminum of this reagent, which is mordant blue 3, requires a blank derived from the sample solution.[260] If iron is present, complex it in the

[257] L. A. Demina, O. M. Petrukhin, Yu. A. Zolotov, and G. V. Serebryakova, *Zh. Anal. Khim.* **27**, 1731–1735 (1972).
[258] V. V. Klimov, O. S. Didkovskaya, and V. N. Kozachenko, *Zavod. Lab.* **28**, 652–654 (1962).
[259] G. S. Vasilikiotis and J. A. Tossidis, *Microchem. J.* **14**, 380–384. (1969).
[260] A. G. Knight, *Proc. Soc. Water Treat. Exam.* **9**, Pt. 1, 72–80 (1960); H. Bennett, R. P. Eardley, W. G. Hawley, and I. Thwaites, *Trans. Brit. Ceram. Soc.* **61**, 636–666 (1962).

sample with thioglycollic acid. The EDTA added to the sample blank will complex iron as well as aluminum.

Procedures

Iron and Steel.[261] As a concentrated buffer solution, dissolve 275 grams of ammonium acetate and 110 grams of sodium acetate trihydrate in 1 liter of water and add 10 ml of acetic acid. As a dilute buffer solution, mix 1 volume of the concentrate with 5 volumes of water and adjust to pH 6.1 with acetic acid or sodium hydroxide solution.

Dissolve 1 gram of sample in 15 ml of 3:20 sulfuric acid. If the sample contains more than 1% silicon, evaporate to sulfur trioxide fumes and redissolve with 15 ml of water. In either case filter on a pad of filter-paper pulp, wash, and reserve the filtrate. Ignite the pad in platinum, moisten the residue with 4 drops of 1:4 sulfuric acid, and add 2 ml of hydrofluoric acid. Evaporate under an infrared lamp, continuing until evolution of fumes ceases, and heat at 800°. Fuse the residue with 0.5 gram of sodium bisulfate and dissolve in 10 ml of water. Combine with the previous filtrate.

If the sample contains more than 0.5% chromium, neutralize to the first appearance of a permanent precipitate with saturated solution of sodium carbonate. Redissolve by dropwise addition of 3:20 sulfuric acid and add 1 ml in excess.

Transfer to a beaker containing 20 ml of mercury and dilute to 60 ml. Electrolyze at 2 amperes for 1 hour, and make a ferricyanide spot test for completeness of removal of iron. When iron has been completely removed, wash down the sides of the beaker and the cover glass and electrolyze 15 minutes longer or, if chromium is present, until the green color has disappeared. Filter the electrolyte and evaporate to about 5 ml. Pour slowly into 10 ml of 40% sodium hydroxide solution. Cautiously add 5 ml of 3% hydrogen peroxide; then boil gently for 10 minutes. Add some paper pulp and set aside for 5 minutes. Filter and wash with warm water. Add 2 drops of phenolphthalein indicator solution and neutralize with 1:1 hydrochloric acid. Add 0.5 ml of excess acid, cool, and dilute to 100 ml.

For samples containing up to 0.035% aluminum, take a 20-ml aliquot (for others take a 10-ml aliquot), and add 10 ml of water. Add 5 ml of 3% hydrogen peroxide, neutralize with 8% sodium hydroxide solution, and add 1 drop in excess. Immediately neutralize with 0.2 N hydrochloric acid and add 1 ml in excess. Add 5 ml of 0.1% solution of solochrome cyanine R and 50 ml of dilute buffer solution. Dilute to 100 ml and set aside for 30 minutes. Read at 535 nm against a reagent blank.

An alternative is to remove iron, nickel, chromium, copper, and molybdenum on a mercury cathode, then to extract residual iron, vanadium, and titanium as cupferrates with chloroform.[262]

Water. Neutralize two 25-ml aliquots with 0.02 N acetic acid. To one, add 1 ml

[261] P. H. Scholes and D. Valerie Smith, *Analyst* **83**, 615–623 (1958); *J. Iron Steel Inst.* (London) **195**, 190–195 (1960).

[262] D. Blair, K. Power, D. L. Griffiths, and J. H. Wood, *Talenta* **7**, 80–89 (1960).

of 0.01 M EDTA to complex the aluminum and serve as a blank. To each, add 10 ml of a buffer solution containing 13.6% of sodium acetate trihydrate and 4% of N acetic acid. Then add 5 ml of 0.01% solution of solochrome cyanine R to each and dilute to 50 ml. After 10 minutes read at 605 nm.

STILBAZO

Aluminum at pH 3.4 in an acetate buffer solution has a maximum at 508 nm with stilbazo (mordant blue 29), which is 4,4′-bis-(3,4-dihydroxyphenylazo)stilbene-2,2′-disulfonic acid.[263] The color develops in 10 minutes and is stable for 45 minutes. The 1 : 1 complex conforms to Beer's law for 0.1–0.8 ppm of aluminum.[264] Low concentrations of oxalate, tartrate, citrate, fluoride, nitrilotriacetic acid, and EDTA bleach the color of the complex. Chromic ion causes fading, but the color is intensified by traces of cupric, molybdate, tungstate, titanous, titanic, tetra- and pentavalent vanadium, ferrous, and ferric ions. The maximum of the reagent itself is in the ultraviolet range.

The color of the aluminum lake is most intense at pH 5.2–5.8.[265] The same color is produced by titanium, beryllium, iron, copper, and indium. The application is broadly to raw materials and fluxes of lead production.

The reagent is not applicable in the presence of a high concentration of electrolytes because of salting out.[266] A pH above 5 or a concentration of stilbazo above 0.2 mg/ml causes opalescence.[267] Prolonged standing may produce increased intensity of color.

Because of wider separation of maximum absorption of reagent and aluminum complex, stilbazo is more suitable for visual use than aluminon.[268] The sensitivity of stilbazo is twice as great. In the cold, the color with stilbazo develops in 5 minutes; with aluminon it requires 18 hours.

As applied to analysis of solutions derived from metallic silver, interfering ions such as silver and iron must first be removed by electrolysis at a mercury cathode. With added zephiramine, a ternary complex is formed.[269] As read at 540 nm this has a maximum absorption at pH 5–5.2 in an acetate buffer solution at 0.02–0.12 μg/ml. Typically, 50 ml should contain 10 ml of 0.1 M acetate buffer for pH 5, 1 ml of 0.08% stilbazo, and 1.5 ml of 0.5% solution of zephiramine. Standing for 20 minutes is necessary before reading.

Procedure. Add 5 ml of nitric acid to 25 ml of sample solution and evaporate to

[263] O. L. Kabanova and M. A. Danyushchenkova, *Zh. Anal. Khim.* **18**, 780–781 (1963); V. B. Sokolovich, Yu. L. Lel'chuk, and B. N. Besprozvanykh, *Isv. Tomsk. Politekh. Inst.* **128**, 112–116 (1964).
[264] C. U. Wetlesen and S. H. Omang, *Anal. Chim. Acta* **24**, 294–297 (1961).
[265] L. N. Krasil'nikova and L. I. Maksai, *Sb. Tr. Vse. Nauch.-Issledl. Inst. Tsvet. Met.* **1956**, 159–164; **1958** (3), 253–257.
[266] N. A. Vlasov and E. A. Morgan, *Izv. Fiz.-Khim. Nauch.-Issled. Inst. Irkutsk. Univ.* **6**, 200–203 (1964).
[267] M. F. Landi and L. Braicovich, *Met. Ital.* **54**, 389–393 (1962).
[268] N. A. Aginskaya and V. I. Petrashen, *Tr. Novocherk. Politekh. Inst.* **72**, 13–22 (1958).
[269] Toshio Ozawa, *Jap. Anal.* **18**, 745–749 (1969).

about 8 ml.[270] Add 5 ml of sulfuric acid and evaporate to sulfur trioxide fumes. Take up in water and dilute to 50 ml. Dilute a 10-ml aliquot of this solution to 100 ml. To 1 ml of this dilution containing not more than 5 μg of aluminum, add 10% sodium hydroxide to pH 5.5–5.6. Dilute to 2 ml and add 0.5 ml of 0.02% solution of stilbazo. Add 0.15 ml of 5% ascorbic acid solution and dilute to 5 ml with an acetate buffer solution for pH 5.4. Set aside for 5 minutes and read at 508 nm.

The technic as described is for a sample containing aluminum, beryllium, and iron as separated by ion exchange as oxalate complexes, the oxalate ion then having been destroyed.

Steel.[271] Dissolve 1 gram of sample in 10 ml of 1:1 hydrochloric acid, oxidize to ferric ion with 3 ml of nitric acid and evaporate to dryness. Redissolve with 10 ml of 1:1 hydrochloric acid, filter, and wash the residue. Burn off the paper and fuse the residue with 1 gram of sodium carbonate. Dissolve the melt in 10 ml of 1:1 hydrochloric acid, dilute to 100 ml, heat to 100°, and make alkaline with ammonium hydroxide. Filter, wash, and discard the filtrate. Dissolve the hydroxides from the filter and add to the earlier filtrate and washings. Evaporate to dryness. Take up the residue in hydrochloric acid with not more than 2 ml of water and dilute to 25 ml with hydrochloric acid. Pass a 5-ml aliquot through a 1.3×15 cm column of Dowex 1-X8 resin in chloride form, and elute the aluminum with 25 ml of hydrochloric acid. This will retain at least 10 mg of titanic ion, but much chromium will interfere. Evaporate nearly to dryness and take up in 10 ml of water. Add 10 ml of 0.02% solution of stilbazo. Add 2.4% solution of sodium acetate trihydrate dropwise to pH 4.8 ± 0.05. Dilute to 50 ml and read at 500 nm against a reagent blank.

Cast Iron.[272] Dissolve 0.05 gram of sample in 6 ml of 1:1 hydrochloric acid and evaporate just to dryness. Dissolve the residue in 0.25 ml of 1:1 hydrochloric acid and filter. Add 1 ml of 10% solution of ascorbic acid and 10 ml of a buffer for pH 5.65, which is 85.5% 0.2 *N* sodium acetate and 14.5% 0.2 *N* acetic acid. Add 3 ml of 0.8% solution of stilbazo, dilute to a known volume, and read at 508 nm.

Ferrovanadium or Ferrotitanium.[273] Dissolve 0.5 gram of sample in 15 ml of 2:1 hydrochloric acid and oxidize to the ferric state with 5 ml of nitric acid. Filter, ignite the residue, and fuse with 1 gram of sodium pyrosulfate. Take up in water and add to the previous filtrate. Dilute to 100 ml and take a 10-ml aliquot. Add 5 ml of 1:1 hydrochloric acid and 10 ml of 9% cupferron solution. Extract interfering vanadium, titanium, zirconium, and niobium with 20 ml, 10 ml, and 5 ml of chloroform. Add 2 ml of nitric acid and 5 ml of perchloric acid to the aqueous layer and heat to white fumes. Take up with 20 ml of water and add 2 drops of *p*-nitrophenol indicator solution. Neutralize with 1:1 ammonium hydroxide, then acidify with 2 ml of 1:1 hydrochloric acid and dilute to 100 ml. Take an aliquot

[270] N. A. Suvorovskaya, M. M. Voskresenskaya, and T. A. Mel'nikova, *Nauch. Soobshch. Inst. Gorn. Dela, Akad. Nauk SSSR* **10**, 148–152 (1961).
[271] C. U. Wetlesen, *Anal. Chim. Acta* **26**, 191–194 (1962).
[272] N. A. Agrinskaya and V. I. Petrashen, *Tr. Novocherk. Politekh. Inst.* **31**, 63–71 (1955); N. A. Agrinskaya, *Zavod. Lab.* **23**, 279–280 (1957).
[273] Shizuya Maskawa and Kiyotoshi Kato, *Jap. Anal.* **17**, 70–75 (1957).

containing less than 50 μg of aluminum and add 1 ml of 5% ascorbic acid solution, 10 ml of acetate buffer solution for pH 5.6, and 2.5 ml of 0.08% solution of stilbazo. Dilute to a known volume, let stand for 5 minutes, and read at 495 nm.

Ferrozirconium or Ferroniobium. Dissolve a 0.5-gram sample with 5 ml of hydrofluoric acid and 2 ml of nitric acid. Add 10 ml of 1:1 sulfuric acid and heat to sulfur trioxide fumes. Take up in 25 ml of water and complete as for ferrovanadium from "Filter, ignite the residue,... ."

Copper- and Lead-Smelting Slags.[274] As a buffer solution for pH 5.4, mix 900 ml of 5.44% solution of sodium acetate with 100 ml of 2.4% solution of acetic acid.

Fuse a 0.5-gram sample in an iron crucible with 3 grams of sodium hydroxide and 2 grams of sodium peroxide. Extract the melt with 50 ml of hot water and add 10 ml of hydrochloric acid. When solution of ferric hydroxide is complete, dilute to 500 ml. Dilute a 10-ml aliquot to 200 ml. To a 10-ml aliquot of this dilution, add 5 ml of 2.2% ascorbic acid solution. Set aside for 3 minutes, then add 4.5 ml of 0.8% sodium hydroxide solution to neutralize the ascorbic acid. Add 30 ml of the buffer solution for pH 5.4 and 15 ml of 0.02% solution of stilbazo. Dilute to 100 ml and read at 490 nm.

Phosphorus Furnace Slags.[275] Mix 0.1 gram of sample with 3 ml of hydrochloric acid, 4 ml of 1:1 nitric acid, and 4 ml of 1:9 sulfuric acid. Boil for 5 minutes and evaporate to dryness. Take up in 20 ml of 0.25 *N* hydrochloric acid, dilute to 250 ml, and filter. Dilute 5 ml of filtrate to 40 ml and add 0.2 ml of 5% ascorbic acid solution. Add 0.2 ml of 25% solution of sodium acetate and 2 ml of 0.05% solution of stilbazo. Dilute to 100 ml with acetate buffer solution for pH 5.4, set aside for 10 minutes, and read at 508 nm.

Alkaline Solutions.[276] To 10 ml of sample containing about 10 mg of sodium hydroxide, add sufficient acetic acid to give a ratio of acetate ion to acetic acid of 8:1. This should be about pH 5.4. If iron is present, add 2 ml of 0.5% solution of ascorbic acid. Add 8 ml of 0.05% solution of stilbazo, dilute to 50 ml, set aside for 10 minutes, and read at 413 nm.

Wine or Wort.[277] Wet-ash 20 ml of sample by boiling to a small volume with 3 ml of sulfuric acid and 1 ml of nitric acid, adding more nitric acid if necessary to decolorize. Neutralize with *N* ammonium hydroxide, make just acid with 0.1 *N* sulfuric acid, and add 5 ml of excess acid. Dilute to 200 ml. Mix a 10-ml aliquot with 0.15 ml of 5% ascorbic acid solution and add 10 ml of water. Add 5 ml of 0.01% solution of stilbazo and dilute to 50 ml with an acetate buffer solution for pH 5.4. Set aside for 10 minutes and read at 508 nm.

[274] S. Yu. Fainberg and A. A. Blyakhman, *Anal. Rud Tvet. Met. Prod. Pererab.* **1956** (12), 119–129.

[275] B. B. Evzlinea, *Soobshchen. Nauch.-Issledl. Rabot. Novoi Tekh. Nauch. Inst. Udobr. Insektofungits.* **1958** (8), 87–90.

[276] E. A. Morgen, N. A. Vlasov, and E. A. Anisimova, *Nauch. Tr. Irkutsk. Nauch.-Issled. Inst. Redk. Met.* **1963** (11), 67–73.

[277] E. V. Zobov and M. S. Shchelkunova, *Tr. Mold. Nauch.-Issled. Inst. Pishch. Prom.* **1**, 137–140 (1961).

Vaccines.[278] Mix 1 ml with 0.5 ml of 1.5 *N* hydrochloric acid and set aside for 20 minutes so that aluminum oxide or aluminum phosphate will dissolve. Dilute to 50 ml, take a 5-ml aliquot, and dilute it to 50 ml. To a 5-ml aliquot add 10 ml of *M*-acetate buffer for pH 5.65 and 5 ml of 0.06% solution of stilbazo. Dilute to 25 ml and after 30 minutes read at 530 nm against a reagent blank.

STILBAZOCHROME

Stilbazochrome, which is 4,4′-bis(1,8-dihydroxy-3,6-disulfo-2-naphthylazo)stilbene-2,2′-disulfonic acid, forms a 1:1 complex with aluminum at pH 4.8–6.[279]

Procedures

Brass. Dissolve 0.15 gram of sample in 10 ml of 1:1 nitric acid. Filter any metastannic acid, add 2 ml of sulfuric acid, and heat to sulfur trioxide fumes. Take up in 50 ml of water, neutralize with 10% sodium hydroxide solution, and dilute to 250 ml. Dilute 10 ml to 100 ml and take a 10-ml aliquot of that dilution. Add 25 ml of a buffer solution for pH 5, 1 ml of 15% sodium thiosulfate solution, and 1.1 ml of m*M* stilbazochrome. Dilute to 50 ml and after 20 minutes read at 665 nm.

Chalk. Add 1:1 hydrochloric acid dropwise to a 1-gram sample until reaction is complete. Add 50 ml of water, neutralize to pH 4–5 with 10% sodium hydroxide solution, and dilute to 100 ml. Mix a 1-ml aliquot with 25 ml of a buffer solution for pH 5 and add 1.1 ml of m*M* stilbazochrome. Mask iron with 1 ml of 10% mercaptoacetic acetic acid and dilute to 50 ml. Set aside for 20 minutes, then read at 665 nm.

5-SULFO-4′-DIETHYLAMINO-2′,2-DIHYDROXYAZOBENZENE

This reagent was specifically designed for determination of aluminum as a 1:1 complex in the presence of beryllium.[280] The latter causes absorption with aluminon, eriochrome cyanine R, alizarin red S, and chrome azurol S. Beryllium cannot be effectively masked for those reagents. Difficulties arise with oxine. Large amounts of beryllium depress the absorption of aluminum with the captioned reagent. Ferrocyanide masks iron and cobalt. Many interferences are extractable with triisooctylamine. Others are masked with EDTA. The bright pink color follows Beer's law up to 0.6 μg/ml.

Procedure. For synthesis of the reagent, suspend 7 grams of 2-aminophenol-4-

[278]Takeichi Sakaguchi, Tomoko Yoshida, and Sachiko Hara, *Jap. Anal.* **19**, 196–200 (1970).
[279]A. I. Cherkesov, B. I. Kazakov, and V. I. Shchepko, *Zavod. Lab.* **34**, 786 (1968).
[280]T. M. Florence, *Anal. Chem.* **37**, 704–707 (1965).

sulfonic acid in 40 ml of water and dissolve by dropwise addition of 40% sodium hydroxide solution. Cool below 5°, and add 9 ml of 10 *N* hydrochloric acid. Over a period of 15 minutes, diazotize by adding 2.8 gram of sodium nitrite in 10 ml of water, keeping the temperature below 5°. Keep in an ice bath for 30 minutes; then destroy free nitrous acid by addition of 0.5 gram of urea. Dissolve 5.3 grams of *m*-ethylaminophenol in 5 ml of ethanol. Stir this into the solution of the diazonium salt solution and let stand at room temperature for 5 hours. Filter the azo dye on a Büchner funnel, wash well with water, and dry in a desiccator over silica gel.

Evaporate a sample almost to dryness, take up in 5 ml of 8 *N* hydrochloric acid, and add 5 ml of 8 *N* hydrochloric acid. Add 10 ml of a 5% solution of triisooctylamine that has been shaken with 8 *N* hydrochloric acid, and extract for 1 minute. Wash the organic extract with 5 ml of 8 *N* hydrochloric acid, add the washings to the original aqueous layer, and discard the organic layer. Add 1 ml of nitric acid and 5 ml of perchloric acid. Evaporate almost to dryness and dilute to 100 ml.

To an aliquot containing less than 15 μg of aluminum and less than 10 mg of beryllium, add 2 ml of 0.1% potassium ferrocyanide solution, 2.5 ml of *M* sodium acetate–acetic acid buffer for pH 4.7, and 3 ml of 0.03% solution of the captioned reagent. Dilute to about 20 ml and incubate at 40–55° for 15 minutes. Cool and add 2 ml of 2% solution of EDTA. Dilute to 25 ml, set aside for 15 minutes, and read at 540 nm. If necessary apply a correction for the beryllium content of the aliquot read.

3-(4-SULFOPHENYLAZO)CHROMOTROPIC ACID

The complex of this reagent with aluminum is read at 582 nm, where the difference in absorption by the reagent and by the complex is greatest.[281]

Procedure. *Copper Alloys.* Dissolve 0.1 gram of sample in 5 ml of nitric acid and evaporate to dryness. Take up the residue in 1.5 ml of 1:4 sulfuric acid and dilute to 100 ml. To a 5-ml aliquot add 1 ml of 0.1 *M* sodium thiosulfate and set aside for 15 minutes. Add 5 ml of m*M* reagent and dilute to 25 ml with an acetate buffer solution for pH 6. Read at 582 nm against a reagent blank.

XYLENOL ORANGE

In acetate buffer solution xylenol orange forms two red complexes with aluminum which are in equilibrium with each other. The following interfering ions must be absent: sulfate, phosphate, tartrate, citrate, fluoride, nitrilotriacetate, EDTA, bismuth, ferric, nickel, yttrium, tetravalent titanium, thorium, vanadium, and zirconium.

Both ferrous and ferric ions form interfering colors that absorb strongly at 550

[281] L. I. Kotelyanskaya and P. P. Kish, *Zavod. Lab.* **36**, 523–524 (1970).

nm. Interference is avoided by reducing ferric ion to ferrous with ascorbic acid, or preferably mercaptoacetic acid, and adding EDTA.[282] As another version, iron is read at 550 nm where development of aluminum is insignificant or can be corrected for. Then after heating and standing, the sum of aluminum and iron is read.

Mercaptoacetic acid masks ferric ion, bismuth, cupric ion, and lead.[283a] The reagent will determine down to 0.004 ppm of aluminum in water, acids, and alkalies.

For masking interfering ions in determination of aluminum by xylenol orange, nitrilotriacetic acid reduces the absorption by 6–8% within an hour.[283b] The reduction is 3.5–5% by 1,2-diaminocyclohexanetetra-acetic acid and by diethylenetriaminepenta-acetic acid. The latter two are more effective in masking than EDTA.

By reading at the isobestic point, 536 nm, rather than at 555 nm, the maximum of the 1 : 1 complex, or 480 nm the maximum of the 1 : 2 complex, the effects of pH and the amount of xylenol orange are minimized.[284] Under optimum conditions of 0.1 mM xylenol orange, mM EDTA, and acetate buffer solution for pH 3–3.5, Beer's law is followed up to 0.4 μg of aluminum per ml.

By an automated technic, the rate of color development of aluminum with xylenol orange is linear up to 2.7 μg of aluminum per ml.[285]

Procedures

Aluminum.[286] Mix a sample solution containing 5–25 μg of aluminum with 10 ml of buffer solution for pH 3.4 and 2 ml of 0.001 M xylenol orange. Heat at 100° for a few minutes, cool, and dilute to 25 ml. Incubate at 25° for 15 minutes and read at 536 nm, the isobestic point, against a reagent blank.

Aluminum and Iron. To the sample solution containing 1–30 μg of iron and 1–10 μg of aluminum, add buffer solution for pH 2.6 and 2 ml of 0.0002 M xylenol orange. Dilute to 10 ml with the buffer solution and read immediately at 540 nm. Heat in a sealed container at 100° for 15 minutes, cool, let stand for 30 minutes, and read again at 540 nm. The difference between the two readings measures the aluminum.[287a] The first reading measures the iron but each is subject to a correction.

Ferrosilicon.[287b] For 0.8–1.2% of aluminum in ferrosilicon alloys, fuse 0.25

[282] V. N. Tikhonov and L. F. Petrova, *Izv. Vyssh. Uncheb. Zaved., Khim. Khim. Tekhnol.* **16**, 798–800 (1973).

[283a] J. Dvorak and E. Nyvltova, *Mikrochim. Acta* **1966**, 1082–1093.

[283b] V. N. Tikhonov and T. V. Andreeva, *Izv. Vyssh. Ucheb. Zaved., Khim. Khim. Tekhnol.* **19**, 1615–1616 (1976).

[284] V. N. Tikhonov and L. F. Petrova, *Zh. Anal. Khim.* **28**, 1413–1415 (1973).

[285] A. Dodson and V. J. Jennings, *Talenta* **19**, 801–803 (1972).

[286] Makoto Otomo, *Bull. Chem. Soc. Jap.* **36**, 809–813 (1963).

[287a] L. A. Molot, I. S. Mustafin, and R. F. Zagrebina, *Izv. Vyssh. Unchab. Zaved., Khim. Khim. Tekhnol.* **9**, 873–595 (1966).

[287b] V. N. Tikhonov, T. V. Andreeva, and O. V. Sidorova, *Zh. Anal. Khim.* **31**, 608–609 (1976).

gram of sample with 8 grams of sodium peroxide in an iron crucible at 800° for 1 minute. Cool, take up in 200 ml of water and filter. Dilute a 20-ml aliquot to 100 ml. Adjust a 10-ml aliquot of this dilution to pH 3 with hydrochloric acid. Add 1 ml of 1% solution of ascorbic acid, 4 ml of 0.1% solution of xylenol orange, and 5 ml of an acetate buffer solution for pH 3. Heat at 100° for 3 minutes, cool, and add 3 ml of 25 m*M* disodium EDTA. Dilute to 50 ml and read at 500 nm for 0.1–0.8 μg of aluminum per ml.

Copper Alloys.[288] Dissolve 0.1 gram of sample in 2 ml of nitric acid, add 0.5 ml of sulfuric acid, and evaporate to sulfur trioxide fumes. Take up in water and dilute to 250 ml. Dilute a 20-ml aliquot to 200 ml. Adjust a 10-ml aliquot of this dilution to pH 2 with 2 *M* sodium acetate or 0.1 *N* hydrochloric acid. Add successively 1 ml of 1% ascorbic acid solution to reduce ferric ion, 5 ml of 5% solution of thiourea to mask copper, 5 ml of 0.1% solution of 1,10-phenanthroline to mask nickel, 5 ml of 0.1% solution of xylenol orange, and 5 ml of acetate buffer solution for pH 3. Heat at 100° for 3 minutes, cool, and add 3 ml of 0.025 *M* EDTA solution. Dilute to 50 ml, let stand for 10 minutes, and read at 536 nm against a reagent blank.

Uranium.[289] Dissolve 0.125 gram of sample in 2 ml of nitric acid and evaporate to dryness *in vacuo*. Dissolve the resulting uranyl nitrate in 10 ml of water. Add 10 ml of 0.0002 *M* xylenol orange and dilute to 25 ml. Set aside for 1 hour and read at 555 nm against a blank containing an equivalent amount of uranium. Pentavalent vanadium and niobium interfere. Fluoride ion will mask iron, thorium, zirconium, bismuth, scandium, yttrium, indium, titanium, and rare earths.

Titanium Dioxide.[290] As a buffer solution for pH 3.2–3.3, dissolve 75.85 grams of glycine and 50.5 grams of sodium chloride in water and dilute to 1 liter. Then mix 90 ml of this solution with 10 ml of *N* hydrochloric acid and dilute to 1 liter.

Fuse 0.1 gram of sample at 700° with 3 grams of potassium hydroxide. Extract with boiling water, filter, and wash with hot water. Dilute to 250 ml. To an aliquot, add a drop of 0.01% solution of tropaeolin 00. Add hydrochloric acid dropwise to develop a red color of the indicator. Add 50 ml of the buffer solution and 10 ml of 0.1% xylenol orange solution. Heat to 100°, cool, and dilute to 250 ml with the buffer solution. Read at 555 nm.

Natural Iron-Oxide Pigments.[291] Fuse 1 gram of sample with 3 grams of potassium hydroxide at 700° for 1 hour. Dissolve in 25 ml of hot water, filter, and wash very thoroughly. Dilute to 250 ml. For an aluminum content under 2%, take a 50-ml aliquot, for 2–15% take 10 ml, and for over 15% take 2 ml. Add a drop of 0.01% solution of tropaeolin 00 solution followed by hydrochloric acid dropwise to a red color. Neutralize excess acid with 5% potassium hydroxide solution and add 50 ml of buffer solution for pH 3.3. Add 10 ml of 0.1% solution of xylenol orange and heat to boiling. Cool, dilute to 200 ml, and read at 555 nm.

[288] V. N. Tikhonov and L. F. Petrova, *Zavod. Lab.* **39**, 672–673 (1973).
[289] B. Budesinsky, *Zh. Anal. Khim.* **18**, 1071–1074 (1963).
[290] V. A. Malevannyi and Yu. L. Lelchuk, *Izv. Tomsk. Politekh. Inst.* **1971** (174), 95–100.
[291] V. A. Malevannyl and V. A. Shumina, *Lakokras. Mat. Primen.* **1967** (4), 75–76.

Nephaline-Apatite Ores.[292] Decompose 0.2 gram of sample with 10 ml of hydrofluoric acid and 5 ml of 1 : 1 sulfuric acid. Evaporate to sulfur trioxide fumes, add 5 ml of water, and again take to sulfur trioxide fumes to remove the last traces of fluoride ion. Dissolve in 10 ml of 1 : 1 hydrochloric acid, neutralize with 20% sodium hydroxide solution, and add 20 ml in excess. Heat to 100°, cool, and dilute to 200 ml. Filter, and take an aliquot containing 0.13–0.3 mg of aluminum. Add 5 ml of m*M* xylenol orange and 20 ml of acetate buffer solution for pH 3.5. Heat to 100° for 3 minutes, cool, and dilute to 50 ml with the buffer solution. Let stand for 30 minutes and read at 555 nm.

Refractory Products.[293] Fuse 0.05 gram of sample with 1 gram of 1:3 borax–sodium carbonate. Dissolve in 100 ml of 1:3 hydrochloric acid and dilute to 250 ml. Dilute a 10-ml aliquot to 30 ml and add 5 ml of 1% solution of ascorbic acid. Make alkaline to Congo red with 1:5 ammonium hydroxide. Add 3 ml of 0.1% solution of xylenol orange and 10 ml of a glycine buffer solution for pH 3. Heat at 100° for 3 minutes, cool, dilute to 100 ml, and read at 540 nm.

Soil Extracts.[294a] As an extraction reagent, dissolve 24.9 grams of ammonium oxalate monohydrate and 12.6 grams of oxalic acid dihydrate in water and dilute to 1 liter.

Crush the sample to pass a 2-mm sieve. Shake 2 grams overnight with 100 ml of the extraction reagent and centrifuge. Treat a 25-ml aliquot with excess 30% hydrogen peroxide until organic matter is destroyed. Add 10 ml of 1:7 sulfuric acid and evaporate to sulfur trioxide fumes. Take up in 50 ml of water, adjust to pH 2, and dilute to 100 ml. Take an aliquot containing up to 0.06 mg of aluminum and add 25 ml of a sodium acetate–hydrochloric acid buffer solution for pH 3.8. Add 10 ml of 0.15% solution of xylenol orange and heat at 40° for 90 minutes. Cool, add 5 ml of 0.05 *M* EDTA, and dilute to 100 ml. Let stand for 1 hour and read at 550 nm.

A binary complex is formed by **acid chrome dark blue** with aluminum ion at pH 4.5 for reading at 580 nm.[294b] Beer's law is followed for 3.1–154 μM aluminum. There is interference by fluoride, phosphate, oxalate, ascorbate, tartrate, citrate, hydroxylamine, magnesium, zinc, manganous ion, copper, iron, nickel, chromium, hexavalent molybdenum, titanic ion, and gallium. EDTA decomposes the complex. Beryllium forms a similar but noninterfering complex at a different pH absorbing at a different wavelength.

Either aluminum or beryllium forms a 1 : 1 : 2 complex of the metal with the dye and diphenylguanidine. At pH 6 the aluminum complex can be extracted with butanol to read at 600 nm. Up to 50-fold amounts of beryllium do not interfere with extraction and reading of 35 μg of aluminum. Beer's law is followed for

[292] V. Ya. Artem'eva, *Zavod. Lab.* **33**, 426 (1967).
[293] L. V. Bogova, *Ogneupory* **1969** (6), 52–56.
[294a] D. T. Pritchard, *Analyst* **92**, 103–106 (1967).
[294b] L. G. Anisimova and E. T. Beschetnova, *Zh. Anal. Khim.* **31**, 1302–1305 (1976).

0.44–21 mM aluminum. A 4-fold molar excess of the dye and 320-fold molar excess of diphenylguanidine is required. In addition to ions interfering with the binary complex there is interference by silver and thallous ions.

Alizarin fluorine blue, also called alizarin complexan, can be used to determine aluminum in the presence of iron and titanium.[295] Adjust 20 ml of a sample solution containing 1–12 ppm of aluminum and up to 5 ppm of iron to pH 2. Add 20 ml of 0.5 mM color reagent, 20 ml of 1,4-dioxan, and 10 ml of formate buffer solution for pH 4.1–4.3. Heat at 70° for 1 hour, cool, and dilute to 100 ml. Read against a sample in which 10 ml of 0.02 M sodium fluoride is added before the dioxan to mask the aluminum.

The complex of aluminum with **alizarin red S** (mordant red 3) fluoresces at 521 nm, as contrasted with 480 nm for the dye.[296] The maximum fluorescence is at pH 5. The complex is stable for 80 minutes. The reading will determine 2–19 ng of aluminum in 25 ml of solution. There is interference by ferric ion, scandium, titanous ion, indium, gallium, chromic ion, and fluoride, as well as by thiourea, EDTA, molybdate, and vanadate.

Alumocreson is triammonium trimethylaurintricarboxylate.[297a] As a 1:1 complex with aluminum ion, it is read in an ammonium acetate buffered solution at pH 4.2–4.4 and 500 nm. Beer's law is followed up to 1.6 μg of aluminum per ml.

At pH 4.8–6.5 aluminum produces an intense blue fluorescence with **3-amino-5-sulfosalicylic acid.**[297b] At pH 5.8 and ionic strength of 1.5 it will determine 0.005 μg of aluminum per ml at 460 nm. Beryllium acts similarly. The fluorescence is quenched by uranyl, ferric, oxalate, fluoride, tartrate, and EDTA ions.

For determining up to 2 ppm of aluminum, treat 25 ml of a neutralized solution with 15 ml of 0.061% methanolic **anthrapurpurin**, which is 1,2,7-trihydroxyanthraquinone.[298] Dilute with methanol to 50 ml, set aside for 2.5 hours, and read at 500 nm. The complex is stable for at least 3 hours.

Optimum conditions for forming the 1:2 complex of aluminum with **apigenin**, which is 4′,5,7-trihydroxyflavone, in 50% methanol, are 0.2 mM ligand at pH 5.1.[299] This is read at 388 nm. The pH must be carefully controlled. The complex is stable for 24 hours.

The 1:1 violet complex of 10 ml of 0.182 mM aluminum with 19 ml of 74.5 μM **arsenazo III** with 5 ml of phthalate or acetate buffer solution for pH 3 is read at 550 or 583 nm.[300] Ferric ion must be removed. There is serious interference by ferrous, silver, titanous, fluoride, oxalate, tartrate, and citrate ions. Beer's law is followed for 0.01–0.60 μg of aluminum per ml of final solution.

Aluminum and **bromopyrogallol red** form a 1:1 complex at pH 3.8 with a maximum absorption at 525 nm.[301] A corresponding 1:2 complex is formed at pH 8 with a maximum absorption at 620 nm. At pH 3–4.5 a ternary complex formed

[295] Folke Ingman, *Talenta* **20**, 999–1007 (1973).
[296] M. K. Akhmedli, D. A. Efendiev, and F. I. Ruvinova, *Uch. Zap. Azerb. Gos. Univ., Ser. Khim. Nauk* **1973** (4), 10–15.
[297a] T. I. Romantseva and S. S. Gosteva, *Zh. Anal. Khim.* **30**, 2342–2345 (1975).
[297b] N. M. Alykov and A. V. Brunin, *Izv. Vyssh. Ucheb. Zaved., Khim. Khim. Tekhnol.* **1974**, 1254–1255.
[298] F. Capitan-Garcia and M. Roman, *Rev. Univ. Ind. Santander* **9** (3–4), 17–24 (1967).
[299] Shuhji Abe, Akio Hakita, and Toshio Hoshino, *Jap. Anal.* **24**, 666–668 (1975).
[300] V. Mikhailova, *Acta Chim. Hung.* **76**, 221–228 (1973).
[301] E. I. Bashirov and A. M. Ayubova, *Azerb. Khim. Zh.* **1971** (4), 134–138.

by aluminum ion, bromopyrogallol red, and triethylhexyldecylammonium iodide is read at 620 nm.[302] Beer's law is obeyed for 0.04–0.6 μg of aluminum per ml. Iron can be masked with ascorbic acid, but titanium, zirconium, ferric ion, tungsten, and molybdenum interfere.

Brilliant alizarin blue G (mordant blue 31) complexes with aluminum to give a red-yellow fluorescence.[303] A similar fluorescence is given by gallium, scandium, yttrium, lanthanum, lutetium, indium, and magnesium.

For determination by **calmagite**, which is 3-hydroxy-4-(6-hydroxy-*m*-tolylazo-naphthalene-1-sulfonic acid, mix up to 5 ml of sample containing 1.1–5.4 μg of aluminum with 5 ml of borate buffer for pH 8.6 and 4 ml of 0.4 m*M* calmagite.[304] Let stand for 30 minutes and shake for 2 minutes with 10 ml of 0.025 *M* Aliquot 336 in chloroform. Filter the organic layer through glass wool and read at 570 nm. Preliminary separation of aluminum from an alkaline medium with oxine is desirable, using cyanide and EDTA as masking agents, particularly against the transition elements.

Catechol-4-sulfonate forms a 1:1 complex with aluminum at pH 4.5–6, a 2:1 complex at pH 6.5–8.5, and a 3:1 complex above pH 9.[305a] The 2:1 complex has maxima at 255 and 298 nm. For the latter mix 1 ml of 0.01 *M* reagent and 5 ml of 0.05 *M* Tris buffer for pH 7, and add a sample containing less than 1 μg of aluminum. Dilute to 25 ml and read at 298 nm.

Compounds having an azo group adjacent to an hydroxy group as typified by **3-(4-chloro-2-hydroxy-3-sulfonphenylazo)chromotropic acid** form 1:1 complexes with aluminum.[305b] They are read by spectrophotometry around 500 nm or by fluorescence around 620 nm. Buffered at pH 6 with an acetate–ammonium hydroxide buffer solution, aluminum obeys Beer's law at 0.023–0.190 μg/ml. Copper, ferrous ion, chromic ion, and boron interfere.

Aluminum can be read fluorimetrically as the complex formed with **6-(5-chloro-2-hydroxy-3-sulfophenylazo)-5-hydroxy-1-naphthalenesulfonic acid** (mordant blue 9).[306] As compared with pontachrome blue-black R, it gives a much weaker fluorescence spectrum, a response proportional to concentration over a much wider range, is more stable in solution, and is equally sensitive down to 5×10^{-4} μg of aluminum per ml. Many ions interfere.

At pH 5–6 aluminum forms a 1:2 red-violet complex with **dioxythiazo.**[307] The color develops immediately, is readily extractable into butanol, and is stable for 5 hours. At pH 5.5 the absorption maxima of the reagent and of the complex are at 460 and 560 nm, respectively.

Although the maximum extinction of **eriochrome garnet L** (mordant red 5) is at 400 nm and the maximum extinction of the aluminum lake is at 450 nm, the

[302] A. S. Arkhangel'skaya and L. A. Molot, Manuscript No. 1743-74 deposited at Vsesoyuznyi Institut Nauchnoi i Tekhnicheskoi Informatsii, Moscow, 1974.

[303] Keizo Hiraki, *Bull. Chem. Soc. Jap.* **45**, 789–793 (1972).

[304] C. Woodward and H. Freiser, *Talenta* **15**, 321–325 (1968).

[305a] Takao Yotsuyanagi, Yasuo Kudo, and Kazuo Aomura, *Jap. Anal.* **18**, 619–626 (1969).

[305b] R. K. Chernova, L. M. Kudryavtseva, I. K. Petrova, and K. I. Gur'ev, *Zh. Anal. Khim.* **31**, 37–43 (1976).

[306] J. F. Possidoni de Albinati, *An. Asoc. Quim. Argent.* **53** (2), 61–73 (1965).

[307] V. S. Korol'kova, Ya. K. Putnin', and E. Yu. Gudrinietse, *Altv. PSR Zinat. Akad. Vest., Ser. Khim.* **19GD** (1969) (6).

maximum difference between the reagent and the lake is at 480 nm.[308] Optimum conditions for 100 ml are as follows: 5–85 μg of aluminum, pH 4.6, 5 ml of *M* acetic acid–ammonium acetate buffer, 5 ml of 0.02% solution of the dye in 50% ethanol, 2 ml of 10% mercaptoacetic acid solution. Heat at 65° for 15 minutes, cool, set aside for 1 hour, and read at 480 nm.

Titanic ion is formed instantly in *N* hydrochloric acid by **fluotitanate ion** and aluminum ion.[309a] The reading at 420 nm is then proportional to the aluminum for molar ratios of aluminum to titanium of 1 : 10 to 1 : 1. The reagent is 2.08 grams of sodium fluotitanate, 1.6 grams of sodium fluoride, and 1.1 ml of 30% hydrochloric acid in 100 ml of *N* hydrochloric acid.

A 1 : 1 : 1 complex of aluminum–**glycinecresol red**–diphenylguanidine in an acetate–ammonium hydroxide buffer solution is extracted by alcohols for reading.[309b] Gallium and indium perform similarly.

Chelates of aluminum with **4-hydroxy-3-(2-hydroxy-4-methoxyphenylazo)benzenesulfonic acid** and **4-(2-hydroxyphenyl)resorcinol** are read fluorescently.[310] The maximum of the former is at pH 2.7–5, activated at 495 nm and read at 570 nm. For the latter the desirable pH is 5, activation at 480 nm, reading at 568 nm. The respective ranges are 0.05–2 μg and 0.05–1 μg of aluminum per 25 ml.

In a buffer solution for pH 9.8, aluminum is read at 660 nm with **3-hydroxy-4-(6-hydroxy-*m*-tolylazo)naphthalene-1-sulfonic acid.**[311]

Aluminum forms a complex with **3-hydroxy-2-naphthoic acid** at pH 5.5. Activate at 375 nm and read the fluorescence at 470–500 nm.[312a]

Mix a sample solution containing up to 3.2 μg of aluminum with 3 ml of a phosphate buffer solution for pH 8.8 and 1 ml of 0.5 m*M* methanolic ***o*-hydroxyquinolphthalein.**[312b] Set aside at not over 25° for 20 minutes and read at 530 nm against a reagent blank.

To 5 ml of sample solution in very dilute hydrochloric acid solution, add 0.5 ml of 0.1 m*M* **kaempferol**, which is 3,4′,5,7-tetrahydroxyfluorone, 0.5 ml of 55% ethanol, and 1 ml of a buffer solution for pH 4–5.[313] Dilute to 10 ml, set aside for 10 minutes, and read the fluorescence at 482 nm. Lanthanum at twofold is tolerated.

Aluminum 8-quinolinate reduces **molybdovanadophosphoric**, **molybdovanadosilicic**, or **molybdotungstosilicic acid** to a molybdenum blue.[314] To determine aluminum, it is precipitated as the oxinate, filtered, dissolved in 20 ml of 1 : 1 ethanol–2 *N* hydrochloric acid and diluted to 100 ml. A 1-ml aliquot with 4 ml of 8% sodium hydroxide solution and 1 ml of 1% solution of the triheteropolyacid is diluted to 6 ml and read. Beer's law is followed for 2–18 μg of aluminum per ml.

[308] M. F. Landi, L. Braicovich, and A. Battaglia, *Met. Ital.* **55**, 355–360 (1963).

[309a] Hisao Fukamauchi, Ryuko Ideno, Mitsuko Yanagida, and Chiyo Inotsume, *Jap. Anal.* **10**, 1029–1030 (1961).

[309b] M. L. Akhmedli, E. L. Glushchenko, and F. T. Aslanova, *Azerb. Khim. Zh.* **1974**, 126–130.

[310] Kaizo Hiraki, *Bull. Chem. Soc. Jap.* **46**, 2438–2443 (1973).

[311] Yobou Bokra and C. Luca, *Chim. Anal., Bucur.* **I** (1), 44–50 (1971).

[312a] Krzysztof Kasiura, *Chem. Anal. (Warsaw)* **20**, 389–395 (1975).

[312b] Yoshikazu Fujita, Itsuo Mori, and Takehisa Enoki, *Jap. Anal.* **24**, 253–256 (1975).

[313] Z. T. Maksimycheva, Sh. T. Talipov, and V. Ya. Artemova, *Nauch. Tr. Tashkt. Gos. Univ.* **1973** (435), 39–43.

[314] A. I. Kokorin and V. M. Ropot, *Uch. Zap. Kishinev. Univ.* **56**, 105–109 (1960).

A 1 : 1 complex of aluminum with **nitrosulfophenol M**, which is 3-(2-hydroxy-5-nitro-3-sulfophenylazo)-6-(3-sulfophenylazo)chromotropic acid, is read in the 600–640 nm region, as are also palladium, niobium, plutonium, and gallium.[315a]

The 3 : 1 complex of aluminum with **phenylenebis(fluorone)**, which is 9, 9′-*p*-phenylenebis-(2, 6, 7-trihydroxy-3*H*-xanthen-3-one, when stabilized with gelatin, is read at 505 nm.[315b] Excess reagent, which absorbs at 505 nm, is complexed with boric acid. Copper, ferric ion, stannous ion, tannic ion, and nickel interfere seriously.

Pontachrome violet SW (mordant violet 5) complexes with aluminum in an acetate–acetic acid buffer solution for pH 5 to reach a maximum fluorescence in 1 hour.[316] Beer's law applies for 0.2–2 μg/ml.

Aluminum can be read as the 1 : 3 complex with sodium **purpurin sulfonate.**[317] Add 25 ml of m*M* reagent to the sample solution containing 25–150 μg of aluminum. Adjust to pH 5.5 with sodium acetate–acetic acid buffer solution and dilute to 50 ml. Set aside for 45 minutes, then read at 580 nm. The complex is stable for 2.5 hours.

Aluminum forms a 1 : 3 complex with the sodium salt of **3-(3-pyridylazo)chromotropic acid** at pH 5 with an optimum range of 0.5–2.5 ppm of aluminum for reading at 580 nm.[318] A 1 : 1 complex of aluminum with **3-(2-carboxy-3-pyridylazo)chromotropic acid** at pH 5 has an optimum range of 0.2–2 ppm of aluminum and is also read at 580 nm.

The complex of aluminum and **1-(2-pyridylazo)-2-naphthol** in ethanol fluoresces at 590 nm when activated at 350 or 545 nm.[319] This will detect 10^{-6} gram of aluminum per ml. The fluorescence is enhanced by m*M* nitric, hydrochloric or hydrobromic acids, and as much as tenfold in the presence of the maximum possible percentage of ethanol.[320]

At pH 4.8–5.2 aluminum complexes with **pyrogallol red** to give a maximum at 525 nm up to 0.5 ppm.[321] The color is stable for 2 hours. There is serious interference by gallium, indium, fluoride, ferric ion, tetravalent zirconium, hexavalent tungsten, and tartrate.

Quercimeritrin, which is 3, 3′, 4′, 5, 7-pentahydroxyflavone 4-β-D-glucopyranoside, forms a fluorescent 1 : 1 complex with aluminum in 50% ethanol buffered with acetate to pH 4.8–5.15.[322] The fluorescence develops in 40 minutes and is then constant for 4 hours. Activation at 420–430 nm gives fluorescence at 545–555 nm. Beer's law is followed for 1–20 ng of aluminum per ml. An equal amount of titanium, molybdenum, bismuth, vanadium, or zirconium is tolerated.

A complex of aluminum and **quercetin** in methanol is read at 425 nm.[323] For glass analysis, fuse with sodium carbonate in platinum and add to the aqueous

[315a] S. B. Savvin, R. F. Propistsova, and L. A. Okhanova, *Zh. Anal. Khim.* **24**, 1134–1142 (1969).

[315b] A. V. Fedin and G. P. Vakar', *ibid* **30**, 2125–2130 (1975).

[316] J. Bognar and M. P. Szabo, *Mikrochim. Acta* **1969**, 221–224.

[317] F. Capitan Garcia, M. Roman Ceba, and E. Alvarez-Manzaneda, *Bol. Soc. Quim. Peru* **39**, 125–140 (1973).

[318] A. K. Majumdar and A. B. Chatterjee, *Mikrochim. Acta* **1967**, 663–669.

[319] J. G. Surak, M. F. Herman, and D. T. Haworth, *Anal. Chem.* **37**, 428–429 (1965).

[320] P. R. Haddad, P. W. Alexander, and L. E. Smythe, *Talenta* **21**, 123–130 (1974).

[321] Takashi Tanaka, Yoshinori Nakagawa, and Shieru Handa, *Jap. Anal.* **10**, 1148–1149 (1961).

[322] Z. T. Maksimycheva, Sh. T. Talipov, A. P. Pakudina, and A. S. Sadykov, *Dokl. Akad. Nauk Uzb. SSR* **1973** (2), 36–37.

[323] N. V. Chernaya and V. G. Matyashov, *Ukr. Khim. Zh.* **40**, 80–82 (1974).

extract ammonium thiocyanate, sodium thiosulfate and quercetin in methanol. Dilute with methanol. In 40% methanol a 1 : 1 : 1 complex is aluminum ion–quercetin–antipyrine.[324] At pH 5.5–6 5 this is extracted with dichloroethane and read at 419 nm. Beer's law is obeyed for 0.03–1.2 μg of aluminum per ml using tenfold *M* quercetin and 1250-fold antipyrine. Ferric ion can be masked with thiosulfate, but zinc, vanadium, molybdate, titanate, and tin interfere.

Aluminum can be read as the 2 : 3 lake with **quinalizarin.**[325] The reagent is added in alkaline solution, gelatin stabilizes the lake, and it is formed at pH 4.9.

At pH 5.1–5.5 the yellow 1 : 1 complex of aluminum with **resorcylal-*m*-aminophenol**, which is 4-(4-hydroxyphenyliminomethyl)resorcinol, fluoresces at 495–505 nm.[326] Development is complete within 40 minutes after mixing.

In 40% acetone, aluminum forms a 1 : 1 complex with **2-(resorcylidene-amino)phenol** at pH 1–5.4.[327] The fluorescence at 544–554 nm conforms to Beer's law for 0.01–20 μg of aluminum in 25 ml of solution. There is interference by cupric, ferrous, and nickel ions. Only twofold amounts of zinc, beryllium, and mercuric ion are tolerated.

For fluorimetric reading aluminum is complexed with **salicylaldehyde.**[328] The maximum excitation is at 397 nm, maximum emission at 486 nm. The graph is linear for 10–100 μ*M* aluminum. Beryllium forms a fluorescent complex, and several metals quench the fluorescence. Therefore aluminum should be separated by paper chromatography with butanol saturated with 2 *N* hydrochloric acid, the fluorescent spot developed with 0.5% salicylaldehyde in ethanol as the spray, and the spot extracted with 50% ethanol.

The complex of aluminum with **salicylaldehyde formylhydrazone** is read fluorimetrically at 0.7–22 μg per 25 ml.[329] Iron, nickel, zinc, and chromium are masked with thioglycollic acid, copper by thiosulfate.

2-Salicylideneaminobenzenearsonic acid forms a 1 : 1 complex with aluminum at pH 4–5 which has a maximum fluorescence at 545 nm.[330] The reaction is sensitive to 0.1 μg of aluminum per ml of solution.

For activation at 445 nm and fluorimetric reading at 520 nm, ***N*-salicylidene-2-amino-hydroxyfluorene** is effective for aluminum.[331] To the sample, add 1 ml of m*M* reagent in ethanol and 1 ml of acetate buffer solution for pH 5.2 and dilute to 10 ml. Ethanol at 10% is necessary to maintain the reagent in solution. Beer's law is followed for 0.001–0.04 micromole of aluminum. Iron must not exceed the aluminum, and the reagent should be 4–16 times the aluminum.

Sendachrome AL which is α-methylformaurin-3, 3′-dicarboxylic acid, has a maximum absorption at 360 nm. The maximum for its aluminum complex is at 510

[324] N. V. Chernaya and V. G. Matyashov, *Zh. Anal. Khim.* **30**, 698–702 (1975).

[325] F. Burriel and S. Bolle Taccheo, *Anal. Chim. Acta* **14**, 553–557 (1956).

[326] P. A. Shakirova, A. T. Tashkodzhaev, G. S. Andrushko, and Sh. T. Talipov, Manuscript No. 1715-74 deposited at Vsesoyuznyi Institut Nauchnoi i Tekhnicheskol Informatsii, Moscow, 1974.

[327] T. A. Shakirova, Sh. T. Talipov, G. S. Andrushko, and A. T. Tashkhodzhaev, *Nauch. Tr. Tashkt. Gos. Univ.* **1973** (435), 68–71.

[328] Lina Ben-Dor and E. Jungreis, *Isr. J. Chem.* **8**, 951–954 (1970).

[329] Z. Holzbecker and P. Pulkrab, *Collect. Czech. Chem. Commun.* **27**, 1142–1149 (1963).

[330] R. Kh. Dzhiyanbaeva, A. T. Tashkhodzhaev, L. E. Zel'tser, and Kh. Khikmatov, *Tr. Tashkt. Gos. Univ.* **1972** (419), 84–88.

[331] C. E. White, H. C. E. Mc Farlane, J. Fogt, and R. Fuchs, *Anal. Chem.* **39**, 367–369 (1967).

nm.[332] With not more than 25 μg of iron present, heat 10 ml of sample, 2 ml of acetate buffer solution for pH 3.8, and 1 ml of 2.5% solution of hydroxylamine hydrochloride for 3 minutes at 50°. Add 2 ml of 0.05% solution of the reagent and dilute to 15 ml. Heat at 50° for 5 minutes, cool, and read. The absorption is decreased by calcium, magnesium, zinc, fluoride, and phosphate ions, increased by chromic and cupric ions. The curve is almost linear for 0.3–2 μg of aluminum per ml.

Solochrome azurine BS (mordant blue 1) in a sodium acetate–acetic acid buffer solution for pH 5.5 obeys Beer's law up to 5 ppm of aluminum with threefold excess of dye.[333] The maximum absorption is at 560 nm. Many ions interfere.

To a sample solution containing 1–15 μg of aluminum at pH 4–5, add 2 ml of 0.1% solution of **SPADNS**, which is 3-(4-sulfophenylazo)chromotropic acid.[334] Add 2 ml of fresh 1% starch solution, dilute to 25 ml, set aside for 1 hour, and read at 590 nm.

A 1:2 complex is formed at pH 6 by **stilbazogall I**, which is 4,4′-bis(2,3,4-trihydroxyphenylazo)stilbene-2,2′-disulfonic acid; it has a maximum absorption at 490 nm.[335]

Aluminum can be read with a tenfold excess of **sulfosalicylic acid**, but to form the complex in the presence of 0.06–0.3 m*M* phosphate, this must be increased to thirtyfold.[336]

To 20 ml of sample solution containing 0.1–4 μg of aluminum, add 0.7 ml of 0.01% solution of **superchrome garnet Y** (mordant red 5) and 2 ml of 20% ammonium acetate solution. Adjust to pH 5, dilute to 25 ml, heat at 50° for 10 minutes, and cool. Activate at 485 nm for reading the fluorescence at 565 nm.[337] There is interference by cobalt, cupric ion, ferric ion, zirconium, pentavalent vanadium, and hexavalent chromium.

Tiron forms a 2:1 complex with aluminum having absorption maxima at 258 and 310 nm.[338] At 310 nm and 2 ml of 0.01 *M* tiron, Beer's law applies up to 0.05 mg of aluminum in 25 ml, provided 5 ml of *M* hexamine buffer for pH 5.4 is present. A 3:1 complex is formed above pH 8.

A 1:1 complex of aluminum with **trihydroxyfluorone** is formed at an optimum pH of 6 and read at 530 nm.[339]

An automated method depends on aluminum displacing iron from its complex with **2,4,6-tripyridyl-1,3,5-triazine.**[340] The decrease in extinction is read at 593 nm.

[332] Yoshijiro Arikawa and Takio Kato, *Tech. Rep. Tohoku Univ.* **25**, 55–63 (1960); Tsugio Takeuchi and Masami Suzuki, *Jap. Anal.* **10**, 58–61 (1961).
[333] C. L. Sharma, S. N. Tandon, Dilbagh Rai, and A. K. Sabharwal, *Indian J. Chem.* **10**, 744–745 (1972).
[334] G. Banerjee, *Z. Anal. Chem.* **271**, 284 (1974).
[335] S. V. Elinson, Yu. V. Pushinov, and V. T. Tsvetkova, *Zh. Anal. Khim.* **26**, 718–721 (1971).
[336] M. P. Strukova, V. N. Kotova, and L. M. Gulyaeva, *ibid.* **24**, 565–569 (1969).
[337] Keizo Hiraki, *Bull. Chem. Soc. Jap.* **45**, 1395–1399 (1972).
[338] Takao Yotsuyanagi, Katsumi Goto, and Nagayama Masaichi, *J. Chem. Soc. Jap. Pure Chem. Sect.* **88**, 1282–1287 (1967).
[339] O. A. Tateev, G. N. Bagdasarov, S. A. Akemedov, Kh. A. Mirzaeva, Kh. G. Buganov, and E. A. Yarsheva, *Izv. Sev-Kavkaz. Nauch. Tsent. Vyssh. Shk. Ser. Estestv. Nauk*, **1973** (2), 30–32.
[340] K. B. Wrightman and R. F. Mc Cadden, *Ann. N.Y. Acad. Sci.* **130**, 827–834 (1965).

CHAPTER SIXTEEN
BERYLLIUM

Although morin would appear to be the most widely used reagent for beryllium in spite of its known difficulties, beryllon III and beryllon IV have been described as the most suitable.[1] Their structures appear elsewhere in this chapter. Dyes containing an anthraquinone or naphthoquinone group, a diazo group, or the triarylmethane or salicylic acid group, are suitable for complexing with beryllium.[2]

Beryllium is read by absorption with quinalizarine, 8-hydroxyquinaldine, acetylacetone, *p*-nitrobenzeneazo-orcinol, 2-phenoxyquinizarin-3,4′-disulfonic acid, and others. By fluorescence it is read by morin, 8-hydroxyquinaldine, and quinizarin (1,4-dihydroxyanthraquinone), and others. Fluorescence with morin is often stated to be the most sensitive method known for beryllium. Acetylacetone and beryllon are recommended for routine use.[3]

Beryllium oxide in beryllium metal is isolated by dissolving the metal with methanol and bromine.[4] A similar technic is described elsewhere in detail for aluminum oxide in aluminum metal. This has also been applied to beryllium oxide in copper-beryllium alloys.[5]

ACETYLACETONE

The acetylacetone complex of beryllium is extracted with chloroform for reading in the ultraviolet region. If the sample is high in aluminum, it can interfere to give high results.

[1] V. I. Kuznetsov, L. I. Bolshakova, and Ming-o Fang, *Zh. Anal. Khim.* **18**, 160–165 (1963).
[2] L. P. Adamovich, *Uch. Zap. Khar'k. Gost. Univ. A. M. Gor'kogo* **95**, *Tr. Khim. Fak. Nauch.-Issledl. Inst. Khim.* No. **18**, 143–151 (1957).
[3] T. J. Hayes, *UKAEA, Prod. Group, P. G. Rep.* **171**, 15–24 (1960).
[4] A. R. Eberle and M. W. Lerner, *Metallurgia* **59**, 49–52 (1959).
[5] Shih-chang Chou, F. Sh. Muratov, and A. V. Novoselova, *Zavod. Lab.* **25**, 1292–1293 (1959).

For analysis of manganese-zinc alloys, add EDTA as a masking agent at pH 7.2–9.4 and an inorganic sodium salt as a salting-out agent.[6] Then extract the acetylacetone complex with chloroform and read at 295 nm within 5 minutes.

Procedures

Iron and Steel.[7] Dissolve 0.25 gram of sample in 10 ml of perchloric acid. If the chromium content exceeds 3%, volatilize the bulk of it as chromyl chloride by intermittent additions of small amounts of sodium chloride while heating strongly. Dilute to 250 ml and take a 10-ml aliquot. Add 5 ml of 20% citric acid solution and 10 ml of 5% solution of EDTA as masking agents. Add 1 ml of 5% solution of acetylacetone and 10 ml of 20% sodium chloride solution. Reduce to pH 7–7.5 by adding 4% sodium hydroxide solution. Extract the complex with 10 ml of chloroform. Remove excess acetylacetone from the organic layer by washing with 50 ml and 50 ml of 0.4% solution of sodium hydroxide. Read the organic layer at 295 nm against chloroform.

Aluminum Alloys.[8] Dissolve 0.5 gram of sample in 10 ml of 1 : 1 hydrochloric acid and dilute to 100 ml. To a 10-ml aliquot expected to contain 0.2–4 μg of beryllium, add 2 ml of 10% solution of EDTA and 2.5 grams of sodium chloride. Dilute to 30 ml and add ammonium hydroxide solution to raise to pH 7–8. Add 0.8 ml of 5% acetylacetone solution and again adjust to pH 7–8. Shake with 20 ml of chloroform. Wash the organic layer with 50 ml and 50 ml of 0.4% sodium hydroxide solution to remove excess acetylacetone. Read at 295 nm against a blank.

Magnesium-Aluminum Alloys.[9] Dissolve 0.5 gram of sample in 10 ml of 1 : 1 hydrochloric acid and dilute to 250 ml. To 50-ml aliquot, add 2.5 ml of 10% solution of EDTA. Adjust to pH 3.5 with 20% sodium hydroxide solution. Prepare a 6 × 1 cm column of 50-100 mesh Dowex 50-X8 resin in sodium form. Wash the column with 100 ml of 2% solution of sodium chloride 0.0004 *N* with sodium hydroxide and containing 1.02% of potassium acid phthalate. Then pass the solution through the column and wash it with 25 ml of a buffer solution for pH 3.5–4. Elute the beryllium and magnesium from the column with 40 ml of *N* nitric acid. Evaporate the eluate to dryness and take it up in 20 ml of water. Add 20 ml of 10% EDTA solution and 1 ml of 5% solution of acetylacetone. Raise to pH 7–8 with 10% sodium hydroxide solution and extract with 10 ml of chloroform. Wash the organic layer with 50 ml and 50 ml of 0.4% sodium hydroxide solution to remove excess acetylacetone. Read at 295 nm.

Sediments.[10] Pulverize the sample and dry for several hours at 90°. To 2 grams in platinum add 6 ml of hydrofluoric acid followed by 4 ml of sulfuric acid and heat to sulfur trioxide fumes. Cool, add 2 ml of hydrofluoric acid, and 1 ml of

[6] Kazuo Ota, *Nippon Kink. Gakh.* **28**, 338 (1964).
[7] Shizuya Maekawa and Yoshio Yoneyama, *Jap. Anal.* **10**, 732–736 (1961).
[8] Tsunenobu Shigematsu and Masayuki Tabushi, *J. Chem. Soc. Jap.* **80**, 162–165 (1959).
[9] Emiko Sudo and Haruno Ogawa, *Jap. Anal.* **13**, 406–411 (1964).
[10] John R. Merrill, Masatake Honda, and James R. Arnold, *Anal. Chem.* **32**, 1420–1426 (1960).

sulfuric acid, and heat to dryness. Fuse with 8 grams of potassium bisulfate for 3 minutes. Dissolve the melt with 0.2 ml of hydrochloric acid and 20 ml of water. Centrifuge to separate sediment, and wash the precipitate. Partially neutralize the combined decantates with 20% sodium hydroxide solution to pH 1.

Just before use, dissolve 2 ml of acetylacetone in 30 ml of 21% solution of tetrasodium EDTA. Add 20 ml to the sample and add hydrochloric acid if necessary to bring to pH 5–6. No precipitation should occur. Set aside for 15 minutes. Extract the approximately 90 ml of solution with 15 ml and 15 ml of carbon tetrachloride. Add 8 ml of hydrochloric acid and 4 ml of nitric acid to the combined extracts, evaporate slowly almost to dryness without charring, and take up in water. Add 30 ml of 10% solution of disodium EDTA and adjust to pH 7 by addition of 10% sodium hydroxide solution. Add 5 ml of 5% solution of acetylacetone and set aside for 10 minutes. Extract the beryllium with 15 ml of carbon tetrachloride. Add 8 ml of hydrochloric acid and 4 ml of nitric acid and evaporate as before. Take up in water, add 20 ml of 10% solution of disodium EDTA, and 5 ml of 5% acetylacetone solution. Adjust to pH 7 with 10% sodium hydroxide solution. Extract with 15 ml of carbon tetrachloride, add acids, and evaporate as before. Add 5 ml of water and boil until colorless. Add 1 ml of 10% disodium EDTA solution, adjust to pH 7 with 10% sodium hydroxide solution, and add 0.5 ml of 5% solution of acetylacetone. Recheck the pH as 7 and set aside for 10 minutes. Extract with 10 ml of chloroform. Wash the organic layer with 20 ml and 20 ml of 0.4% sodium hydroxide solution to remove excess acetylacetone, shaking each for 90 seconds. Filter the chloroform layer, dilute to 10 ml, and read at 295 nm.

Biological Samples.[11] Ash an appropriate sample in platinum at less than 600°. Add 3 ml of hydrofluoric acid and 1 ml of 1 : 1 sulfuric acid and evaporate to sulfur trioxide fumes. Take up in 10 ml of 1 : 9 hydrochloric acid and add 1 ml of 5% EDTA solution. Add about 20 mg of iron as ferric chloride and precipitate the hydroxides around pH 11 with ammonium hydroxide. Filter, wash, and dissolve in 10 ml of 1 : 5 hydrochloric acid. Add 1 ml of 5% EDTA solution and 2 ml of saturated sodium chloride solution. Add 0.8 ml of 5% solution of acetylacetone and raise to pH 7 with ammonium hydroxide. Extract with 20 ml of chloroform and wash the organic layer with 50 ml and 50 ml of 0.4% sodium hydroxide solution. Read at 295 nm.

Water.[12] Add 10 ml of 10% ferric chloride solution to a 2-liter sample. Precipitate hydroxides, filter, wash, and ash. Dissolve in 10 ml of 1 : 5 hydrochloric acid. Proceed as for biological samples from "Add 1 ml of 5% EDTA solution. . . ."

ALKALINE PHOSPHATASE

The inhibiting effect on hydrolysis by alkaline phosphatase is a measure of beryllium content. The reading is a ratio of net extinction from the inhibited

[11] Tsunenobu Shigematsu, Masayuki Tabushi, and Fumio Isojima, *Jap. Anal.* **11**, 752–756 (1962).
[12] Cf. T. G. Kornienko and A. I. Samchuk, *Ukr. Khim. Zh.* **38**, 914–916 (1972).

reaction to net extinction from an uninhibited reaction as against beryllium concentration. With *p*-nitrophenylphosphate as the substrate, the released *p*-nitrophenol is read. With umbelliferone phosphate as the substrate, the high fluorescence of the released umbelliferone is read.[13]

Procedure. Mix 3 ml of 5m*M* sodium salt of *p*-nitrophenylphosphate, 1 ml of Tris buffer solution for pH 9.8, 1 ml of sample solution containing 18–19 ng of beryllium, and 1 ml of 0.01 *M* sodium diethyldithiocarbamate.[14] Incubate at 25° for 10 minutes. Add 1 ml of 0.011% aqueous solution of calf intestinal alkaline phosphatase of nominal activity of 0.54 m*M*/mg per minute. After exactly 3 minutes at 25° stop the activity of the enzyme with 1 ml of 2% solution of sodium hydroxide. Read the absorption due to nitrophenol at 410 nm against a reagent blank from which the enzyme was omitted.

ARSENAZO I

The complex of this reagent with beryllium conforms to Beer's law up to 0.5 ppm of beryllium in the final solution.[15] The reagent is also applied to the residue of beryllium oxide left from solution of copper-beryllium alloys with bromine and absolute methanol.[16] The product of fusion with potassium pyrosulfate dissolved in 0.3 *N* hydrochloric acid is developed and read at 580 nm.

A ternary complex is formed by beryllium ion, 0.5 m*M* arsenazo I, and 0.1 *M* diphenylguanidine, which is extractable at pH 6–8 with pentanol.[17]

Procedures

Aluminum Alloy.[18] Dissolve 1 gram of sample containing 0.01–1% beryllium in 25 ml of 1:1 hydrochloric acid. Dilute to 100 ml. Neutralize a 10-ml aliquot with 10% sodium hydroxide solution. Add 10 ml of 0.1% solution of arsenazo I. Add 10% sodium hydroxide solution to raise to pH 12.5 and dilute to 100 ml. Read at 580 nm.

Magnesium Alloy. Proceed as for aluminum alloy, but add 30 ml of 10% EDTA solution to mask the magnesium.

Copper Alloy. Dissolve 1 gram of sample in 25 ml of 1:1 nitric acid and proceed as for aluminum alloy, but mask copper with 25 ml of 10% potassium cyanide solution.

[13] G. G. Guilbault, H. H. Sadar, and M. Zimmer, *Anal. Chim. Acta* **44**, 361–367 (1969).
[14] A. Townshend and A. Vaughan, *Talenta* **16**, 929–937 (1969).
[15] D. I. Eristavi, V. D. Eristavi, and Sh. A. Kekeliya, *Zh. Anal. Khim.* **26**, 1430–1431 (1971).
[16] Iwao Tsukahara, *Jap. Anal.* **19**, 1502–1507 (1970).
[17] V. N. Tolmachev, L. A. Kvichko, and V. D. Konkin, *Zh. Anal. Khim.* **22**, 11–14 (1967).
[18] Tadao Hattori, Iwao Tsukahara, and Toshimi Yamamoto, *Jap. Anal.* **15**, 41–46 (1966).

Bronzes. Dissolve 0.5 gram of a sample containing beryllium, nickel, copper, aluminum, and iron in 10 ml of 1 : 1 nitric acid, filter if necessary, and dilute to 500 ml. Dilute a 5-ml aliquot to 25 ml and adjust to pH 2.5. Pass the solution at 1 ml/minute through a column of AV-17 resin in carbonate form. Wash the column with 15 ml of water. Elute nickel, copper, and aluminum with 250 ml of 1 : 4 ammonium hydroxide. Then elute beryllium with either 70 ml of 4% sodium hydroxide solution or 450 ml of 3 *N* ammonium carbonate. In either case, take an aliquot equivalent to 0.05 gram of sample, 10% of the eluate. Neutralize, and add 1 ml of 10% EDTA solution and 10 ml of 1% solution of arsenazo I. Raise to pH 12.5 with 10% sodium hydroxide solution, dilute to 100 ml, and read at 580 nm.

AURINETRICARBOXYLIC ACID

The ammonium salt of the captioned reagent is known by the trivial name of aluminon. As a 1 : 1 complex with beryllium, it follows Beer's law up to 0.8 μg/ml. If iron is present, a correction is necessary. The sensitivity is better with a sodium acetate buffer than with an ammonium acetate buffer.Absorption is increased by increase of aluminon and by improved dispersion of the lake due to gelatin. Absorption is decreased by increased concentration of EDTA, ammonium ions, or alkali ions. Interference by calcium, iron, aluminum, and copper is masked by EDTA but 30, 20, 60, and 60 μg, respectively, do not interfere with 0.6 μg of beryllium per ml.[19] A large excess of EDTA must be avoided because it reacts with aluminon. An appropriate pH is 5. The curve of absorption is not linear but is reproducible.[20] At least fiftyfold excess of aluminon is required.[21]

In the analysis of zinc alloys, interfering amounts of copper, iron, and nickel are removed by chloroform extraction of complexes with sodium diethyldithiocarbamate and 8-hydroxyquinoline.[22] Then EDTA is added and beryllium determined with aluminon at 4–10 ppm.

For the analysis of solutions of mixed carbonates of calcium, strontium, and barium, the major ions do not interfere and aluminum is masked with EDTA.[23]

Procedure. As a reagent, dissolve 0.5 gram of aluminon, 272 grams of sodium acetate trihydrate, and 27 ml of acetic acid in water and dilute to 1 liter.

To a sample solution containing 4–32 μg of beryllium, add 5 ml of 2.5% solution of EDTA, 10 ml of buffered aluminon reagent, and 5 ml of 0.5% solution of gelatin.[24] Boil gently for 5 minutes, cool, and dilute to 50 ml, checking that the pH is 5.1–5.5. Read at 540 nm.

[19] E. A. Morgen and E. A. Anisimova, *Nauch. Tr. Irkutsk. Nauch.-Issled. Inst. Redk. Met.*, **1963** (11), 61–66.

[20] L. S. Serdyuk and G. P. Federova, *Ukr. Khim. Zh.* **24**, 384–387 (1958).

[21] Anil K. Mukherji and Arun K. Dey, *Chim. Anal.* (Paris) **40**, 200–303 (1958).

[22] I. Nachev and D. Filipov, *C. R. Acad. Bulg. Sci.* **20**, 201–204 (1967).

[23] L. A. Molot and N. S. Frumina, *Uch. Zap. Sarat. Univ.* **75**, 90–95 (1962).

[24] Yohe Tsuchiya, *Jap. Anal.* **9**, 934–939 (1960); cf. George E. Kosel and W. F. Neuman, *Anal. Chem.* **22**, 936–939 (1950).

Isolation with Mesityl Oxide.[25] Take an aliquot containing about 20 μg of beryllium. Make it 0.5 *N* with hydrochloric acid and 5 *M* with potassium thiocyanate at a volume of 25 ml. Extract for 3 minutes with 10 ml of mesityl oxide. Strip beryllium from the organic phase by shaking with 10 ml of water. To this aqueous extract add 2 ml of 20% sodium acetate solution and adjust to pH 6.6 with 0.4% sodium hydroxide solution or 0.1 *M* acetic acid. Add 3 ml of 0.1% solution of aluminon and dilute to 25 ml. Set aside for 20 minutes and read at 530 nm.

Iron and Steel.[26] Dissolve 2 grams of sample with 10 ml of hydrochloric acid, 10 ml of nitric acid, and 30 ml of perchloric acid. Heat to white fumes and dilute to 100 ml. For less than 0.05% of beryllium in the sample, take a 20-ml aliquot, for 0.05–0.5% take 2 ml, and for 0.5–1.5% take 1 ml. Add to 30 ml of 25% sodium hydroxide solution and 5 ml of 3% hydrogen peroxide. To separate titanium and iron, boil for 3 minutes, dilute to 100 ml, and filter. Neutralize a 50-ml aliquot of the filtrate to phenolphthalein with 1:1 hydrochloric acid and add 0.25 ml in excess. Dilute to 75 ml and mask aluminum with 3 ml of 2.5% EDTA solution. Add 15 ml of a solution containing 0.2% of aluminon, 50% of ammonium acetate, and 8% of acetic acid. Add 1 ml of 5% solution of gum acacia and heat at 100° for 4 minutes. Cool and read at 530 nm.

Copper Alloys.[27] As a buffer, prepare a 3:1 mixture of 0.1 *N* acetic acid and 0.1 *N* ammonium hydroxide. As the color reagent, dissolve 1.25 gram of EDTA and 0.1055 gram of aluminon in 100 ml of the buffer solution.

Dissolve 0.1 gram of sample containing 0.3–3% of beryllium in 10 ml of 1:5 nitric acid and dilute to 100 ml. Take a 10-ml aliquot and add ammonium hydroxide until a faint turbidity appears. Add a few drops of nitric acid to clarify this and dilute to 100 ml. Mix an aliquot of this diluted sample with an equal volume of the color reagent. Dilute to 250 ml with the buffer solution and heat at 100° for 5 minutes. Read at 540 nm against a blank prepared from beryllium-free copper.

Bronze.[28] As a buffer for pH 4 mix 1 volume of 0.1 *N* ammonium hydroxide with 3 volumes of 0.1 *N* acetic acid. As color reagent, dissolve 1.25 gram of EDTA and 0.1182 gram of aluminon in 100 ml of the buffer solution.

Dissolve 0.1 gram of sample in 5 ml of 1:5 nitric acid and dilute to 100 ml. Neutralize a 10-ml aliquot with ammonium hydroxide, acidify with a drop of nitric acid, and dilute to 100 ml. Mix a 25-ml aliquot of this dilution with 25 ml of the color reagent and dilute to 250 ml with the buffer solution. Heat at 100° for 5 minutes or set aside for 30 minutes and read at 520 nm.

Niobium Alloys.[29] As the aluminon reagent, dissolve 1 gram in 50 ml of water.

[25] P. V. Dhond and S. M. Khopkar, *Anal. Chem.* **45**, 1937–1938 (1973).
[26] Kimio Mukaewaki, *Jap. Anal.* **11**, 388–393 (1962).
[27] L. P. Adamovich and B. V. Yutsis, *Uch. Zap. Khar'k. Gos. Univ.* **1963**, 133; *Tr. Khim. Fak. Nauch.-Issled. Inst. Khim.* **19**, 135–139.
[28] L. P. Adamovich and B. V. Yutsis, *Zavod. Lab.* **28**, 920–921 (1962).
[29] B. S. Tsyvina and M. B. Ogareva, *ibid.* **28**, 917–919 (1962).

Add 2 grams of benzoic acid in 25 ml of ethanol and 100 ml of acetate buffer solution for pH 5.2. Filter and dilute to 250 ml.

Ignite 0.1 gram of sample at 700–800°. Then fuse with 2 grams of potassium pyrosulfate. Dissolve the melt in 10 ml of 10% tartaric acid solution and add 25 ml of water. Add 1:3 ammonium hydroxide until the pH is raised to 6 and filter. Dilute to 50 ml and take a 10-ml aliquot. Add 10 ml of an acetate buffer solution for pH 5.2, 2 ml of 5% EDTA solution, and 2 ml of the aluminon reagent. Dilute to 100 ml, set aside for 20 minutes, and read at 506 nm.

Airborne Dust.[30] As reagent, dissolve 8 grams of benzoic acid in 400 ml of methanol. Add 200 ml of water and 10 ml of acetic acid. Adjust to pH 4.8 with 20% solution of sodium hydroxide. Dissolve 1 gram of aluminon in the solution and dilute to 2 liters.

As a concentrated masking buffer solution, dissolve 100 grams of triethanolamine and 40 grams of EDTA in 500 ml of water. Add 20 grams of potassium cyanide in 100 ml of water. Adjust to pH 4.8 with acetic acid, which will require about 80 ml, and dilute to 2 liters. For use as a dilute masking solution, dilute 233 ml of this concentrate to 2 liters.

Collect the dust by filtration through paper, which should retain particles down to at least 0.5 μm. Ash the paper at 700°, add 1 ml of *N* sulfuric acid, and heat to dryness. Dissolve in 9 ml of dilute masking solution and set aside for 30 minutes. Add 1 ml of aluminon reagent. If the color indicates less than 10 μg of beryllium, set aside for 60 minutes and read at 530 nm against a reagent blank. If more than that amount of beryllium appears to be present, add 9 ml of concentrated masking solution and 9 ml of aluminon reagent. Dilute to 100 ml and after 60 minutes read at 530 nm against a reagent blank.

BERYLLON II

Beryllon II, the product of coupling chromotropic acid with diazo H-acid, is 4, 5-dihydroxy-3-(8-hydroxy-3, 6-disulfo-1-naphthylazo) naphthalene-2, 7-disulfonic acid. It gives a light blue color in sodium hydroxide solution at pH 12–13.2.[31] There is interference by calcium, magnesium, nickel, copper, cobalt, manganese, molybdenum, and chromic ion, all of which are masked by disodium EDTA. Ferric ion must be reduced with ascorbic acid.

Amounts of magnesium, copper, iron, manganese, titanium, and zinc commonly present in aluminum alloys do not interfere.[32] For analysis of zirconium alloys, adjust the solution to *N* sulfuric acid and extract the zirconium twice with

[30] J. P. McCloskey, *Microchem. J.* **12**, 32–45 (1967); cf. R. H. A. Crawley, *Anal. Chim. Acta* **22**, 413–420 (1960).

[31] G. G. Karanovich, *Zh. Anal. Khim.* **11**, 400–404 (1956); cf. A. M. Lukin and G. B. Zavarikhina, *ibid.* **11**, 393–399 (1956); M. S. Bykhovskaya, *Gig. Tr. Prof. Zabol.* **1** (6), 49–53 (1957); L. P. Adamovich and A. P. Mirnaya, *Zh. Anal. Khim.* **18**, 292–297 (1963); T. A. Belyavskaya and I. F. Kolosova, *ibid.* **19**, 1162–1163 (1964).

[32] L. M. Budanova and N. A. Zhukova, *Zavod. Lab.* **25**, 411–413 (1950).

cupferron in chloroform followed by once with chloroform.[33] Then read beryllium with beryllon II.

As applied to coal, the sample is destroyed by wet-ashing.[34] At pH 6 iron, aluminum, and some other interfering elements are masked with EDTA. The acetylacetone complex is extracted with chloroform or carbon tetrachloride and the extract evaporated with nitric acid. Develop at pH 12.05 ± 0.05 with beryllon II and read at 625 nm.

Procedure. To 25 ml of sample solution containing beryllium, aluminum, and iron, add 5 ml of nitric acid and evaporate to 8–10 ml.[35] Add 5 ml of sulfuric acid and evaporate to sulfur trioxide fumes. Take up in water and dilute to 50 ml. Dilute a 10-ml aliquot to 100 ml. Titrate a 10-ml aliquot to a methyl orange end point with 10% sodium hydroxide solution, and discard. To another 10-ml aliquot, add the same amount of 10% sodium hydroxide solution. Dilute to 35 ml and add 0.5 ml of 10% EDTA solution, which has been adjusted to pH 12.5 with 10% sodium hydroxide solution using tropaeolin O as indicator, add 2 ml of 0.05% beryllon II solution, adjust to pH 12.05 ± 0.05, and read at 625 nm.

Copper Alloy.[36] Take an aliquot of the sample solution containing 5–25 μg of beryllium oxide. Add 3 ml of a solution 0.1 *M* with triethanolamine and 0.4 *M* with EDTA. Add 5 ml of 0.05% solution of beryllon II and 4% solution of sodium hydroxide until the color changes from reddish violet to blue plus 2 drops in excess. Dilute to 50 ml with 0.04 *M* EDTA solution adjusted to pH 12.2. The final pH must be 11.9–12.3. Read at 625 nm against a blank consisting of 3 ml of the solution 0.1 *M* with triethanolamine and 0.4 *M* with EDTA, 3 drops of 5% ferric sulfate solution, 4% sodium hydroxide solution to discharge the color, plus 5 drops in excess, and 5 ml of 0.05% solution of beryllon II diluted to 50 ml with 0.04 *M* EDTA solution.

Bronze.[37] Dissolve 0.25 gram of sample in 5 ml of nitric acid and evaporate almost to dryness, adding 1 ml of sulfuric acid if necessary to complete dissolution. Evaporate to dryness. Dissolve the residue in 5 ml of hydrochloric acid and evaporate to about 2 ml. Add 10 ml of saturated solution of sodium chloride and 6 ml of 5% solution of EDTA. Boil for a few minutes, cool, and adjust to pH 9–10 by adding ammonium hydroxide. Add 15 ml of butyric or valeric acid and heat to 100°. Cool, and shake for 1 minute with 15 ml and 15 ml of chloroform. Copper, manganese, cobalt, titanium, chromium, lead, and zirconium are left in the aqueous phase. Wash the combined extracts with 20 ml and 20 ml of saturated sodium chloride solution containing 0.2 ml of 5% EDTA solution. Reextract the beryllium with 30 ml of warm nitric acid followed by 10 ml of nitric acid mixed with 3 ml of water. Evaporate the combined extracts to under 5 ml and dilute to 50 ml.

[33] S. V. Elinson, L. I. Pobedina, and N. A. Mirzoyan, *Zh. Anal. Khim.* **15**, 334–338 (1960).
[34] Roy F. Abernathy and Elizabeth A. Hattman, *Rep. Invest. U.S. Bur. Mines*, **RI 7452**, 8 pp (1970).
[35] N. A. Suvorovskaya, M. M. Voskresenskaya, and T. A. Melnikova, *Nauch. Soobshch. Inst. Gorn. Dela Akad. Nauk SSSR* **10**, 148–152 (1961).
[36] L. A. Bychkov and A. N. Nevzorov, *Zavod. Lab.* **38**, 927–928 (1972).
[37] A. L. Markman and L. L. Galkina, *Uzb. Khim. Zh.* **1962** (4), 5–7.

Neutralize a 10-ml aliquot to litmus with 5% sodium hydroxide solution, add 10 ml more of the sodium hydroxide solution and dilute to 50 ml. Mix a 10-ml aliquot with 2 ml of 0.01% beryllon II solution, dilute to 50 ml, and read at 600 nm against a blank.

Ore.[38] Decompose 0.5 gram of sample by fusing to a transparent melt with 2.5 grams of potassium bifluoride. Cool, add 10 ml of sulfuric acid, and evaporate to dryness. Fuse in a muffle furnace and cool. Dissolve in 10 ml of hot 1:2 hydrochloric acid and dilute to 100 ml. Add 10 ml of 10% EDTA solution to a 10-ml aliquot. Add 0.25 ml of acetylacetone and adjust to pH 6–8 with ammonium hydroxide, using phenol red as indicator. Extract with 10 ml and 10 ml of carbon tetrachloride, adjusting to pH 6–8 again after the first extraction. Wash the combined extracts with 20 ml of 0.1% solution of EDTA adjusted to pH 7. Add 3 ml of nitric acid and 3 ml of perchloric acid. Drive off carbon tetrachloride, then evaporate almost to dryness. Take up in 10 ml of water, add tropaeolin O indicator solution and raise to pH 12.5. Add 5 ml of 0.08% solution of beryllon and dilute to 50 ml. Read at 600 nm against a blank.

Water.[39] If the sample contains calcium bicarbonate, adjust 1 liter to pH 2 with 1:1 hydrochloric acid and boil for 10 minutes to drive off carbon dioxide, thus avoiding precipitation of calcium carbonate at the next step. If the sample does not contain 5 mg of ferric ion and 2 mg of aluminum ion, add them as sulfate solutions. At about 70° add ammonium hydroxide to pH 8–8.5, to coprecipitate beryllium with ferric and aluminum hydroxides. Filter, and wash with 1:100 ammonium hydroxide. Dissolve the hydroxides from the filter with 7 ml of 0.1 *N* sulfuric acid and dilute to 25 ml. Pass this solution at 1 ml/minute through a 20 × 1.6 cm column of Anionite AV-17 in fluoride form. Wash the column with 25 ml of water. Elute the beryllium with 180 ml of 3% sodium fluoride solution at 5 ml per minute. Add 10 ml of sulfuric acid and evaporate this eluate to sulfur trioxide fumes to drive off fluoride. Take up the residue in 10 ml of 1:9 sulfuric acid and dilute to 100 ml. Adjust a 10-ml aliquot to pH 12–13 with 10% sodium hydroxide solution. Add 2 ml of 0.01% solution of beryllon II and dilute to 50 ml. After 15 minutes read at 600 nm against a blank.

Alternatively,[40] neutralize a sample containing 1–3 μg of beryllium, usually 100–500 ml, to methyl orange. Add 1 ml of 2.5% solution of EDTA and 0.25 ml of saturated solution of acetylacetone. Add 3 drops of 0.1% solution of phenol red, followed by ammonium hydroxide until the solution turns red, pH 8.4. Extract the beryllium complex by shaking for 2 minutes with 10 ml of carbon tetrachloride. Add 0.1 ml of the acetylacetone solution to the aqueous layer, bring back the color of the indicator with more ammonium hydroxide and again extract for 2 minutes with 10 ml of carbon tetrachloride. Reextract the beryllium from the combined carbon tetrachloride extracts by shaking for 2 minutes with 8 ml of water and 6 ml

[38] L. G. Kamantseva and I. A. Stolyarova, *Tr. Vses. Nauch.-Issled. Geol. Inst.* **117**, 41–44 (1964); cf. N. A. Suvorovskaya, M. M. Voskresenskaya, and T. A. Mel'nikova, *Nauch. Soobshchen. Inst. Gorn. Dela Akad. Nauk SSSR* **6**, 63–66 (1960).

[39] D. I. Eristavi, F. I. Brouchek, and V. D. Eristavi, *Bul. Inst. Politsh. Iasi* **13**, 201–206 (1967).

[40] O. V. Yanter and E. A. Orlova, *Uch. Zap. Tsent. Nauchn.-Issled. Inst. Olovyan. Prom.* **1969** (2), 89–93.

of hydrochloric acid. Dilute the aqueous acid extract to 35 ml and add 3 drops of tropaeolin O indicator solution. Add 10% sodium hydroxide solution until the pH is raised to 12.5. Add 1 ml of borate buffer solution for pH 13 and 2 ml of 0.08% solution of beryllon II. Dilute to 50 ml and read at 600 nm.

Sewage.[41] This technic is also applicable to water containing 2–16 ppm of beryllium. Add 1 ml of 1 : 1 hydrochloric acid to 1 liter, evaporate to under 50 ml, and dilute to that volume. Take an aliquot containing 2–200 μg of beryllium. Add 5 ml of 1 : 1 hydrochloric acid, 2 ml of 0.2% titanous chloride solution, 1 drop of nitric acid, 10 ml of 15% solution of EDTA, 10 ml of 10% solution of diammonium phosphate, and 2 drops of methyl orange indicator solution. Make alkaline with 10% sodium hydroxide solution. Add 2 drops of hydrochloric acid and 10 ml of hot 1 : 4 hydrochloric acid. Add 3 ml of 3% hydrogen peroxide and discharge the yellow color with 20% sodium hydroxide solution. Add 12.5 ml of 0.5 *N* sodium hydroxide and dilute to 50 ml. Dilute an aliquot containing 0.5–4 μg of beryllium to 9 ml with 0.125 *N* sodium hydroxide. Add 1 ml of 0.02 *N* beryllon II, set aside for 20 minutes, and read at 600 nm.

Urine.[42] Digest 100 ml of sample with 20 ml of sulfuric acid, 25 ml of nitric acid, and 2 ml of perchloric acid overnight. Add 100 ml of nitric acid and evaporate to sulfur trioxide fumes. If not colorless, add more nitric and perchloric acids and again evaporate to sulfur trioxide fumes. When colorless, remove the last traces of nitric acid with 3 drops of formaldehyde solution and 7 ml of water, adding 30% hydrogen peroxide dropwide until colorless to diphenylamine. Add 15 ml of hot water, and dilute to 38 ml. Let settle, and centrifuge a portion to clarify. To 5 ml, add 0.5 ml of potash alum solution containing 1 mg of aluminum per ml. Add 1 ml of saturated solution of disodium phosphate and 3 drops of 0.1% solution of bromocresol green. Neutralize with 40% solution of sodium hydroxide. Acidify until a slight cloudiness appears and make slightly alkaline with 4% sodium hydroxide solution. Let stand for 30 minutes, centrifuge, and decant. Wash the precipitate with 3 ml and 3 ml of 0.9% sodium chloride solution. Dissolve the precipitate in 0.2 ml of 10% sodium hydroxide solution. To the sample and a blank, add 0.5 ml of 5% solution of EDTA. Dilute to 4.5 ml, add 0.5 ml of 0.02% solution of beryllon II, and centrifuge. Read at 600 nm against the blank.

Blood. Treat 10 ml of blood with 5 ml of sulfuric acid, 8 ml of nitric acid, and 2 ml of perchloric acid. Decompose as for urine. After removal of nitric acid, add 1 ml of 5% solution of potassium thiocyanate. If potassium ferrocyanide is not precipitated, add 5% ferric chloride solution dropwise. Set aside overnight and centrifuge. Complete as for urine from "To 5 ml, add 0.5 ml of potash alum...."

Tissue.[43] Digest 100 grams of sample with 50 ml of sulfuric acid and 50 ml of nitric acid until colorless, adding more nitric acid if required. Boil off oxides of nitrogen and add ammonium hydroxide to bring the digestate to pH 1. Evaporate

[41] Yu. N. Dunaeva, *Tr. Sverdl. S.-Kh. Inst.* **11**, 509–512 (1964).
[42] P. A. Rozenberg, *Lab. Delo* **9** (4), 11–15 (1963).
[43] A. F. Rubtsov, *Sb. Tr. Sudebnmed. Sudebnkhim., Perm.* **1961**, 236–240.

to about 50 ml and precipitate iron with 7 ml of 6% solution of cupferron. Extract the cupferron complex with 10 ml of carbon tetrachloride. If the organic layer is not colorless, extract with a further 5 ml of carbon tetrachloride. Raise the aqueous layer to pH 8 with ammonium hydroxide and add 15 ml of 0.05 *N* tetrasodium EDTA. Add 5 ml of 15% solution of acetylacetone and again adjust to pH 8. Extract the acetylacetone complex with 8 ml of carbon tetrachloride. Twice add 5 ml of acetylacetone and 0.1 ml of ammonium hydroxide and extract with 8 ml of carbon tetrachloride. Combine the carbon tetrachloride extracts and reextract the beryllium from them with 15 ml of hydrochloric acid and 20 ml of water. Evaporate the extract to dryness and add 0.25 ml of sulfuric acid. Add 2 ml of 30% hydrogen peroxide and heat to destroy the organic matter. Take up the peroxide-free residue in 10 ml of water, neutralize with ammonium hydroxide, and dilute to 25 ml. Mix a 3-ml aliquot with 0.3 ml of 5% ascorbic acid solution, 0.2 ml of 10% sodium hydroxide solution, and 0.5 ml of 0.02% solution of beryllon II. Dilute to 5 ml, set aside for 5 minutes, and read at 600 nm.

Alternatively, digest 0.1 gram of dried tissue with 1 ml of sulfuric acid, 1 ml of nitric acid, and 0.5 ml of perchloric acid and complete as for urine from "Add 100 ml of nitric acid...."

BERYLLON III

Beryllon III, which is 4-hydroxy-5-(4-diethylamino-2-hydroxyphenylazo)naphthalene-2,7-disulfonic acid, determines up to 15 μg of beryllium in 50 ml.[44] There is interference by titanium and by large amounts of manganese and zirconium.

Procedure.[45] To less than 5 ml of sample solution containing less than 10 μg of beryllium, add 2.5 ml of a solution containing 10% of EDTA and 6% of triethanolamine. Add 10% sodium hydroxide as a predetermined amount to neutralize the solution, plus 2 ml in excess. After 5 minutes dilute to 35 ml and again set aside for 5 minutes. Add 10 ml of 0.02% solution of beryllon III and dilute to 50 ml. If ferric ion, manganese, chromic ion, thorium, or zirconium is present read within 10 minutes.

Zinc Alloy. Dissolve 0.5 gram of sample with 2 ml of sulfuric acid, 3 ml of nitric acid, and 5 ml of 1:1 hydrochloric acid. Dilute to 100 ml and take a 10-ml aliquot. Add 4 ml of 10% EDTA solution and adjust to pH 10–11 with 8% sodium hydroxide solution. Add 1 ml of 10% potassium cyanide solution and set aside for 10 minutes. Add 3 ml of 0.1% solution of beryllon III and dilute to 50 ml. After 10 minutes read at 525 nm.

---

[44] Shozo Shibata, Kazuo Goto, Takehiko Amano, and Yoshihiro Miyazaki, *Jap. Anal.* **18**, 604–607 (1969); cf. V. I. Kuznetsov, L. I. Bolshakova, and M. Y. Fang, *Zh. Anal. Khim.* **18**, 160–165 (1963).
[45] P. Pakalns and W. W. Flynn, *Analyst* **90**, 300–303 (1965).

BERYLLON IV

Beryllon IV, which is 2-(6-dicarboxymethylamino-1-hydroxy-3-sulfo-2-naphthylazo)benzenearsonic acid, is preferred to beryllon III because aluminum can be masked with EDTA.[46] It is applied to aluminum alloys containing 0.01–0.9% beryllium.

Procedure. *Aluminum Alloy.*[47] Dissolve 1 gram of sample in 40 ml of 1:1 hydrochloric acid, adding 5–6 drops of nitric acid during the process, and dilute to 250 ml. Mix a 5-ml aliquot with 5 ml of 10% solution of EDTA, 5 ml of 0.1% solution of beryllon IV, and 30 ml of 25% solution of hexamine. Dilute to 100 ml and read at 530 nm against a blank prepared from pure aluminum.

CHLOROPHOSPHONAZO R

This reagent is 4-[(4-chloro-2-phosphonophenyl)azo]-3-hydroxynaphthalene-2,7-disulfonic acid. The yellow color with beryllium does not require masking for iron, provided the amount does not exceed 10%.

Procedure. *Ores.*[48] Fuse 0.5 gram of sample with 2 grams of potassium acid fluoride to a transparent melt. Cool, add 10 ml of sulfuric acid, and evaporate to dryness. Fuse the residue, dissolve in 1:9 hydrochloric acid, and dilute to 100 ml with that acid. Take two aliquots containing 10–50 μg of beryllium oxide and add 5 ml of 0.05 *M* EDTA to each. Then add 5 ml of 25% triethanolamine solution and 2 drops of 0.1% ethanolic thymolphthalein to each. Add 20% sodium hydroxide until the indicator is blue and just discharge the color by dropwise addition of 1:1 hydrochloric acid. To develop one portion, add 2 ml of 0.1% solution of chlorphosphonazo R and dilute to 50 ml with a borate buffer solution for pH 9.4. To the other portion as a blank, add 1 ml of 10% acetylacetone solution; add 10% sodium hydroxide solution until it turns blue, and just discharge the color by dropwise addition of 1:1 hydrochloric acid. Read the developed solution at 540 nm against the blank.

CHROME AZUROL S

Chrome azurol S (mordant blue 29) has the trivial name of **alberon** and is structurally 2, 6-dichloro-3-sulfophenylbis(3-carboxy-4-hydroxy-5-methylphenyl)-methane. It forms a blue-violet 1:1 complex with beryllium.[49] Disodium EDTA

[46] V. I. Kuznetsov, L. I. Bolshakova, and Ming-o Fang, *Zh. Anal. Khim.* **18**, 160–165 (1963).
[47] L. M. Budanova and S. N. Pinaeva, *Zavod. Lab.* **32**, 401–402 (1966).
[48] V. F. Luk'yanov, A. M. Lukin, E. M. Knyazeva, and I. D. Kalinea, *Zh. Anal. Khim.* **18**, 562–566 (1963).
[49] L. P. Adamovich, O. V. Morgul-Meshkova, and B. V. Yutsis, *ibid.* **17**, 678–684 (1962).

serves to mask many ions. The desirable pH is 4.5 for reading at 525 nm in a pyridine–hydrochloric acid buffer solution. The color develops in 15 minutes and is stable for at least 5 hours. It is used for reading 0.2–10 μg of beryllium in 50 ml. The color is bleached by sodium chloride, ammonium chloride, ammonium sulfate, sodium carbonate, EDTA, fluoride, and most seriously by sodium tartrate. Interference by iron or uranium can be masked with hydroxylamine hydrochloride. There is also interference by thorium, copper, zirconium, and aluminum. At pH 4.6 in the presence of EDTA, only chromic and fluoride ions interfere seriously.[50]

Varied conditions are applicable for nanogram amounts.[51] (1) At pH 4.9 the beryllium complexes with the tetrabasic acid hydrate of the reagent in acetate buffer solution or *M* pyridine–*M* nitric acid for reading at 568 nm. (2) At pH 6.7 ± 0.2 in the presence of *M* Tris buffer, beryllium is read with the reagent at 540 nm. (3) At pH 6.5 ± 0.4 in the presence of polyvinyl alcohol and *M* hexamine buffer, beryllium is read with this reagent at 615 nm. (4) At pH 10 ± 0.2 the tetrabasic form of the reagent in the presence of 2.5×10^{-2} *M* cyclohexanediaminetetraacetic acid and 3×10^{-2} *M* calcium nitrate is read with beryllium at 500 nm.

When the ternary complex is formed of beryllium, chrome azurol S, and cetrimonium chloride, there is interference by aluminum, vanadium, copper, iron, gold, thorium, uranium, titanium, zirconium, and tin.[52] The desirable pH is 5.3–5.6 at a concentration of 0.6-6 m*M*. Beer's law is not followed. The blue ternary complex of beryllium with chrome azurol S and the hexadecyltrimethylammonium salt above is extractable with butanol at pH 6.6–7 for reading at 596 nm.[53] The absorption of the extract is constant for hours and obeys Beer's law up to 7 m*M* beryllium.

Procedure. Take an aliquot of sample solution containing 1–80 μg of beryllium.[54] Titrate to a methyl orange end point with 10% sodium hydroxide solution and discard. To a similar aliquot, add 1 ml of 1% solution of ascorbic acid, 2 ml of 10% solution of EDTA, 5 ml of a buffer solution containing 23.8% sodium acetate trihydrate and 10.5% acetic acid, and the amount of 10% sodium hydroxide solution as predetermined. Dilute to 20 ml, add 2 ml of 0.165% solution of the dyestuff, and dilute to 25 ml. Read at 569 nm against a reagent blank.

With Cation-Active Agent.[55a] To 10 ml of sample solution in 0.02 *N* hydrochloric acid containing 0.1–3 μg of beryllium, add 1 ml of 0.5% EDTA solution, 0.6 ml of 0.25% solution of chrome azurol S, 2 ml of 0.5% solution of zephiran, which is benzyldimethyltetradecylammonium chloride, and 7% hexamine buffer solution to give pH 5.1. Dilute to 25 ml, set aside for 15 minutes, and read at 610 nm.

[50] P. Pakalns, *Australiz At. Energ. Comm.* **Rep AAEC/TM-175** 13 pp (1963).
[51] L. Sommer and V. Kuban, *Anal. Chim. Acta* **44**, 333–344 (1969).
[52] Yoshio Shijo and Tsugio Takeuchi, *Jap. Anal.* **20**, 137–141 (1971).
[53] R. Ishida and K. Tonosaki, *J. Chem. Soc. Jap.* **1974**, 1077–1082.
[54] P. Pakalns, *Anal. Chim. Acta* **31**, 575–583 (1964).
[55a] Yoshizo Horiuchi and Hiroshi Nishida, *Jap. Anal.* **18**, 180–184 (1969).

With Nonionic Surfactant.[55b] Mix a sample of solution containing up to 2 μg of beryllium with 1 ml of 1% solution of disodium EDTA and 5 ml of 5 *M* sodium perchlorate. Adjust to pH 4.5 with *M* sodium acetate and evaporate to 10 ml. Add 1 ml of 0.25% solution of chrome azurol S and 2 ml of 0.5% solution of polyoxyethylene-dodecylamine. Dilute to 25 ml, set aside for 15 minutes, and read at 605 nm.

Cast Iron.[55c] Dissolve 0.25 gram by heating with 15 ml of 1:1 hydrochloric acid, adding nitric acid to complete solution. Cool, filter the silica, and wash. Add 17.5 ml of 10% disodium EDTA solution and 10 ml of 0.25% solution of stannic chloride. Boil for 5 minutes and dilute to 170 ml. Heat to 100° and adjust to pH 9.6–10 with ammonium hydroxide. Set aside for 20 minutes, filter, and wash with 1:50 ammonium hydroxide. Dissolve the precipitate from the filter with 20 ml of 1:1 hydrochloric acid, add 5 ml of hydrochloric acid, and dilute to 250 ml.

Mix a 5-ml aliquot, 2 ml of 0.025 *M* disodium EDTA, 5 ml of acetate buffer solution for pH 4.6, and 3 ml of 0.165% solution of chrome azurol S. Dilute to 50 ml and read at 582 nm.

Bronze.[56] Dissolve 1 gram of sample in 10 ml of 1:1 nitric acid and evaporate to dryness at 100°. Add 3 ml of 1:1 hydrochloric acid and again evaporate to dryness. Take up in 1:100 hydrochloric acid and dilute to 200 ml with that acid. To a 0.5-ml aliquot, add 8 ml of a buffer solution for pH 4.4–4.8, 0.5 ml of 5% solution of Trilon B, which is disodium EDTA, and 1 ml of 1% solution of chrome azurol S. After 10 minutes read at 570 nm.

Beryl. Mix 0.1 gram of finely divided sample with 3 grams of potassium bifluoride in platinum. Calcine at 500°, then fuse at 800 ml. Heat at 100° with 10 ml of 1:1 sulfuric acid; then dry and heat at 500° to a transparent mass that no longer solidifies. Dissolve in 1:100 hydrochloric acid and dilute to 200 ml with the same acid. Filter if necessary. To a 0.5-ml aliquot add 8 ml of buffer solution for pH 4.4–4.8. If a turbidity appears, add 0.5 ml of 1% solution of gum arabic. Add 0.5 ml of 5% solution of disodium EDTA and 1 ml of 1% solution of chrome azurol S. After 10 minutes read at 570 nm.

Alternatively, fuse 0.5 ml of a 1:9 mixture of beryl and granite at 900°. Add 25 ml of 1:1 hydrochloric acid and evaporate to dryness at 100°. Dissolve in 0.01 *N* hydrochloric acid, dilute to 200 ml with that acid, filter off the silica, and proceed as for bronze from "To a 0.5-ml aliquot,...."

Uranium Sulfate.[57a] Prepare a solution at pH 1.5 and pass through a column of IRA-400 resin, which will retain the uranium. Add 0.5 ml of perchloric acid to the eluate and evaporate to fumes. Take up in water and dilute to a known volume. Adjust an aliquot containing up to 10 μg of beryllium to pH 6±0.1. Add 3 ml of

[55b] Hiroshi Nishida, Taeko Nishida, and Hiroshi Ohtomo, *Bull. Chem. Soc. Jap.* **49**, 571–572 (1976).
[55c] A. A. Amsheeva, *Visn. Kharkiv. Politekh. Inst. Tekhnol. Neorg. Regovin* (6), 63–66 (1974).
[56] I. S. Mustafin and L. O. Natveev, *Zavod. Lab.* **24**, 259–262 (1958).
[57a] Louis Silverman and Mary E. Shideler, *Anal. Chem.* **31**, 152–155 (1959); *USAEC* **NAA-SR-2686**, 21 pp. (1958).

2% hydroxylamine hydrochloride solution and 2 ml of a buffer solution for pH 6 consisting of 35 ml of hydrochloric acid in 215 ml of pyridine. Add 1 ml of 0.05% solution of chrome azurol S containing 0.2% of gum arabic. The pH of the sample should be 6 at this point. Dilute to 50 ml and set aside for 15 minutes. Read at 575 nm against a reagent blank.

Air.[57b] Collect the sample by filtration through filter paper under pressure. Leach the paper with 20 ml of 0.1 *N* sulfuric acid and wash with 10 ml of water. Add 1 ml of 1% solution of chrome azurol S, 10 ml of 1% solution of gum acacia, and 2 ml of 1% solution of disodium EDTA. Adjust to pH 2. Add 1 ml of 1% solution of cetylpyridinium bromide dropwise with constant stirring. Add 3 ml of 5% solution of hexamine and adjust to pH 5. Dilute to 50 ml and read at 605 nm. The sensitivity is similar to that with morin by fluorometry.

ERIOCHROME BRILLIANT VIOLET B

A purified form of this dyestuff (mordant violet 28) forms a 2:1 complex with beryllium.[58] Over the pH range of 5.8–7, Beer's law is followed for 0.02–0.15 ppm of beryllium with maximum absorption at 560 nm. There is serious interference by aluminum, cupric ion, ferric ion, chromic ion, thorium, hexavalent uranium, yttrium, scandium, rare earths, fluoride, citrate, and oxalate.

Procedure. To a neutral solution containing 0.5–4 μg of beryllium, add 2 ml of 0.01% solution of the dyestuff in ethanol and 10 ml of sodium acetate–acetic acid buffer solution for pH 6.2. Dilute to 25 ml, set aside for 20 minutes, and read at 560 nm against a reagent blank.

ERIOCHROME CYANINE R

Beryllium forms a red complex with eriochrome cyanine R (mordant blue 3) which is read at pH 9.8 and 512 nm. Interfering ions can be masked with cyanide and disodium EDTA for many samples. Iron should be in the ferrous state. The method is applicable to 10 μg of beryllium in 25 ml of final solution. The addition of tartrate extends the applicability of Beer's law and increases the color intensity. The reagent is applicable to monitoring the concentration of beryllium in air.[59]

When benzyldimethyltetradecylammonium chloride is added, it forms 1:2 complex with beryllium–eriochrome cyanine R with the maximum absorption at 595 nm and pH 6.7–7.2.[60]

[57b] R. M. Sathe, *Analyst* **102**, 137–139 (1977); cf. Gerhard Ackermann and Harald Koenig, *Z. Chemie* (Leipzig) **16**, 284–286 (1976).
[58] Katsuya Uesugi, *Anal. Chim. Acta* **49**, 89–95 (1970).
[59] Genevieve Chuiton, *Rapp. CEA* **N-1446**, 22 pp (1971).
[60] Hitoshi Kohara, Nobuhiko Ishibashi, and Kazumi Fukamachi, *Jap. Anal.* **17**, 1400–1406 (1968).

Procedure. Mix 1 ml of sample solution containing up to 10 μg of beryllium with 10 ml of 5% sodium potassium tartrate solution and 10 ml of 0.05 *M* borax solution.[61a] Let stand for 1 hour and add 1 ml of 0.5% solution of eriochrome cyanine R. Read at 527 nm after 3 hours.

The presence of organic solvent causes more rapid color development and permits greater acidity.[61b] To a sample solution containing up to 5 μg of beryllium, add 2.5 ml of acetate buffer solution for pH 6.8, 5 ml of ethanol, and 1 ml of 0.1% solution of eriochrome cyanine R. Dilute to 10 ml, set aside for 5 minutes, and read at 560 nm against a reagent blank.

Tributyl Phosphate Extraction as Thiocyanate.[62] Dilute a sample solution containing up to 12 μg of beryllium to 25 ml, making it 0.5 *N* with hydrochloric acid and 5 *M* with ammonium thiocyanate. Shake with 10 ml of 50% solution of tributyl phosphate in toluene for 5 minutes. Back-extract the beryllium from the organic phase with 15 ml and 15 ml of *N* sodium hydroxide. Add 3 ml of hydrochloric acid to the extracts, then 5 ml of 0.1% solution of eriochrome cyanine R, and 2 ml of 2.5% solution of EDTA. Adjust to pH 9.5 with 0.1 *N* sodium hydroxide or 0.1 *N* hydrochloric acid and dilute to 50 ml. Read at 510 nm against a reagent blank.

Aluminum.[63a] Dissolve 1 gram of sample in 20 ml of 1 : 1 hydrochloric acid and dilute to 100 ml. To a 1-ml aliquot, add 5 ml of 5% disodium EDTA solution, 5 ml of 5% sodium acetate solution, and 5 ml of 0.09% solution of eriochrome cyanine R. Raise to pH 9.7–9.8 by titration with 10% sodium hydroxide solution. The color changes through red and yellow finally to dark purple. Then add 5 drops of 10% hydrochloric acid, dilute to 50 ml, and read at 512 nm against a blank prepared with beryllium-free aluminum.

Steel. Process as for aluminum but adjust the pH with 10% potassium hydroxide solution and use beryllium-free steel to prepare the blank.

Copper. Dissolve 1 gram of sample in 20 ml of 1 : 1 nitric acid, boil off the oxides of nitrogen, and dilute to 100 ml. To a 1-ml aliquot, add 5 ml of 5% disodium EDTA solution, 5 ml of 5% sodium acetate solution, and 5 ml of 0.09% solution of eriochrome cyanine R. Add sufficient 10% potassium cyanide solution to discharge the blue of copper complexes, usually 2 ml. Complete as for aluminum from "Raise to pH 9.7–9.8..." but produce the blank with beryllium-free copper.

Bronze.[63b] Dissolve 5 mg of sample in 0.5 ml of nitric acid and evaporate to dryness. Take up in 2 ml of 1 : 1 hydrochloric acid and again evaporate to dryness. Dissolve in 2 ml of 1 : 1 hydrochloric acid and dilute to 100 ml. To a 10-ml aliquot add 2 ml of 0.05 *M* EDTA, 5 ml of 50% solution of ammonium acetate, and 2 ml of

[61a]Shunji Umemoto, *Bull. Chem. Soc. Jap.* **29**, 845–852 (1976).
[61b]R. Borisova and N. Ivanov, *C. R. Acad. Bulg. Sci.* **29**, 1313–1316 (1976).
[62]S. Kalyanaraman and S. M. Khopkar, *Anal. Chem.* **47**, 2041–2043 (1975).
[63a]Uno T. Hill, *ibid.* **28**, 1419 (1956); **30**, 521–524 (1958).
[63b]Krzystzof Kasiura, *Chim. Anal.* (Paris) **16**, 407–411 (1971).

0.1% solution of eriochrome cyanine R. Raise to pH 9.7 by dropwise addition of 1:1 ammonium hydroxide and dilute to 25 ml. Read at 525 nm.

Titanium. Dissolve 1 gram of sample in 25 ml of hydrochloric acid and dilute to 100 ml. To a 1-ml aliquot, add 4 drops of 30% hydrogen peroxide and proceed as for aluminum from "add 5 ml of disodium EDTA solution..." but substitute potassium hydroxide for sodium hydroxide and prepare the blank with beryllium-free titanium.

Oxides. Digest a 1-gram sample with 25 ml of hydrochloric acid, take to dryness, and bake to dehydrate silica. Add 10 ml of hydrochloric acid and warm; add 25 ml of water and boil. Add paper pulp and filter, washing with hot 1:10 hydrochloric acid and with hot water. Ash the filter in platinum, treat with 5 ml of hydrofluoric acid and 1 ml of 1:1 sulfuric acid, and evaporate to sulfur trioxide fumes. Fuse the residue with 1 gram of potassium bisulfate and dissolve in the earlier filtrate. Dilute to 100 ml. Complete as for aluminum from "To a 1-ml aliquot, add..." using an appropriate blank.

FLAME PHOTOMETRY

The presence of beryllium ion reduces absorption by the flame of strontium.

Procedure. *Brasses.*[64] Dissolve a sample containing 1–5 mg of beryllium in hot 1:5 nitric acid and boil off oxides of nitrogen. Add 50 ml of 5% EDTA solution and neutralize with 1:5 ammonium hydroxide. Filter the precipitate of beryllium hydroxide, wash with water, and dissolve in a minimal volume of 1:1 hydrochloric acid. Dilute to 25 ml with strontium chloride solution containing 750 mg of strontium per liter. Read at 461 nm against a blank. The decrease in absorption by the strontium chloride solution is a measure of the beryllium.

FAST SULFON BLACK F

The effect of beryllium in reducing the absorption by this dye is a direct method of determination.[65] The complex formed is 1:1. The maximum absorption develops in 90 minutes; letting it stand for 3 hours is a precaution, and it is stable for 24 hours. Beer's law is followed up to 7.2 μg of beryllium in 100 ml. With cyanide and EDTA as masking agents, the only interferences are chromic and stannous ions. They can be avoided by oxidation to a higher valence. Reducing agents such as ascorbic acid, sulfide, and hydroxylamine must be absent.

[64] J. Malinowski and D. Dancewicz, *Chem. Anal.* (Warsaw) **6**, 177–182 (1961).

[65] A. M. Cabrera and T. S. West, *Anal. Chem.* 35, 311–313 (1963); cf. T. S. West, *Ind. Chem.* **39**, 379–381 (1963).

Procedure. As a buffer for pH 11–11.2, mix 0.1 *M* sodium hydroxide solution with an equal volume of a solution 0.1 *M* with both glycine and sodium chloride.

To a sample solution containing up to 7.2 μg of beryllium, add 3 ml of 0.065% solution of potassium cyanide, 2 ml of 0.372% solution of the disodium salt of EDTA dihydrate, and 10 ml of 0.00155% solution of fast sulfon black F. Raise to pH 11–11.2 either with 10 ml of the buffer solution or 0.5–1 ml of ammonium hydroxide as indicated necessary by a glass electrode. Set the solution aside for 3 hours, use that containing beryllium as zero, and read the absorption by the blank at 630 nm. It may be necessary in some samples to increase the concentrations of masking agents.

2-HYDROXY-3-NAPHTHOIC ACID

This reagent permits reading beryllium as a 1 : 1 complex down to 0.2 ppb by fluorescence.[66] The intensity decreases less than 10% in the dark. Aluminum forms a similar complex but can be largely masked with EDTA, provided a noninterfering cation is present to combine with the excess EDTA. Aside from aluminum, interfering ions are bismuth, ceric, chromic, ferrous, ferric, scandium, stannic, titanic, and thorium. The complex has also been reported as a 2 : 1 ratio.[67]

Procedure. As a buffer solution for pH 7.5, dilute 50 ml of ammonium hydroxide to 800 ml. Add acetic acid, about 42 ml, to adjust to pH 7.5 and dilute to 1 liter. As Ca/CDTA solution, mix 25 ml of 0.1 *M* calcium chloride solution with 6 ml of 0.1 *M* solution of *trans*-1,2-diaminocyclohexane-*N*,*N*,*N*′,*N*′-tetraacetate.

To a sample solution containing 0.018–0.18 μg of beryllium, add 5 ml of Ca/CDTA reagent and raise to pH 8 with 1 : 10 ammonium hydroxide. Add 5 ml of buffer solution for pH 7.5 and 5 ml of 0.02102% solution of the captioned reagent. Dilute to 100 ml and set aside for 30 minutes. Activate at 380 nm and read the fluorescence within 30 minutes at 460 nm, as compared with quinine sulfate at 350 and 450 nm, expressing it as a ratio to the quinine sulfate at 450 nm.

Bronze.[68] Dissolve 0.02 gram of sample in 10 ml of 1 : 1 nitric acid and evaporate to dryness. Add 10 ml of water and again evaporate to dryness. Dissolve the residue in water and dilute to 100 ml. Adjust an aliquot containing 1.5–2.2 μg of beryllium to pH 3–4 by dropwise addition of 2 *N* sodium acetate solution. Add 0.5 ml of 0.0001 *M* reagent and 2 ml of 0.0001 *M* disodium EDTA. Dilute to 10 ml, activate at 380 nm, and read the fluorescence at 460 nm.

[66] Gordon F. Kirkbright, T. S. West, and Colin Woodward, *Anal. Chem.* **37**, 137–143 (1965).
[67] A. I. Cherkesov and T. S. Zhigalkine, *Tr. Astrak. Tekh. Inst. Rybn. Prom. Khoz.* **1962** (8), 25–49.
[68] A. I. Cherkesov and T. S. Zhigalkina, *Zavod. Lab.* **27**, 658–659 (1961).

8-HYDROXYQUINALDINE

The complex of 8-hydroxyquinaldine with beryllium is read directly or extracted with chloroform. For beryllium in Magnox, dissolve a 1-gram sample and precipitate the iron and beryllium as the hydroxides.[69] Dissolve the hydroxides in hydrochloric acid, add oxine, and extract the aluminum with chloroform. Complete with the captioned reagent. When EDTA is used as a mask, it lowers the absorption of the beryllium complex. With gallium, indium, and beryllium present, they are sequentially extracted with chloroform as the quinaldine complexes at pH 3.9, 5.5, and 8.1, respectively, for fluorimetric reading.[70]

Procedures

Beryllium-Aluminum Alloy.[71] Dissolve 1 gram of sample with 25 ml of 1:1 hydrochloric acid and 5 ml of 3% hydrogen peroxide. Boil off the excess peroxide and dilute to 500 ml. Dilute a 10-ml aliquot to 500 ml. To a 10-ml aliquot of this dilution, add 4 ml of 0.1 *M* EDTA. Add 5 ml of 10% solution of ammonium chloride and 3 ml of 1% solution of 8-hydroxyquinaldine. Add 1:3 ammonium hydroxide to raise to pH 8 ± 0.2 and dilute to 50 ml. Let stand for 30 minutes, then shake for 1 minute with 10 ml of chloroform. Read the extract at 380 nm against a blank.

Beryllium-Copper Alloy. Proceed as for the aluminum alloy, but as masking agent substitute 3 ml of 0.65% potassium cyanide solution for the EDTA solution.

Airborne Dust.[72] Collect the suspended dust from 2–5 cubic meters of air by filtration through paper. Wet-ash the paper and contents with 10 ml of nitric acid and 3 ml of sulfuric acid. Add 0.25 ml of perchloric acid and evaporate to white fumes. Add 2 grams of sodium sulfate and expel all fumes. Take up in 30 ml of 1:10 hydrochloric acid, filter, and dilute to 100 ml. Take an aliquot containing 5–20 μg of beryllium. Add 5 ml of 6% solution of EDTA, 5 ml of 10% solution of ammonium chloride, and 3 ml of 1% solution of 8-hydroxyquinaldine. Add 1:3 ammonium hydroxide to pH 8.2 ± 0.2 and dilute to 50 ml. Shake for 1 minute with 10 ml of chloroform and read the extract at 380 nm against a blank.

---

[69]Kenji Motojima, Hiroshi Hashitani, and Kazuo Katsuyama, *Jap. Anal.* **9**, 628–629 (1960); cf. Kenji Motojima, *Proc. 2nd UN Int. Conf. Peaceful Uses At. Energy, Geneva* **28**, 667–675 (1958); Rolf Keil, *Mikrochim. Acta* **1973**, 919–932.
[70]Masayoshi Ishibashi, Tsunenobu Shigematsu, and Yashuharu Nishikawa, *J. Chem. Soc. Jap.* **78**, 1139–1142 (1957).
[71]Katsuzo Kida, Mitsunobu Abe, Susumu Nishigaki, and Kazuo Kobayashi, *Jap. Anal.* **9**, 1031–1035 (1960).
[72]Katsuzo Kida, Mitsunobu Abe, Susumu Nishigaki, and Takeshi Kasaka, *ibid.* **10**, 663–664 (1961).

N-METHYLANABASINE-(α'-AZO-4)-*m*-AMINOPHENOL

This reagent is 3-amino-4-[3-(1-methyl-2-piperidyl)-2-pyridylazo]phenol. It can be used for reading beryllium, provided a tenfold amount of aluminum is not present.

Procedure. *Aluminum Alloys.*[73] Dissolve a 1-gram sample in 100 ml of 1:1 hydrochloric acid and evaporate to dryness. Moisten the residue with 1 ml of hydrochloric acid and again evaporate to dryness. Take up in 10 ml of 1:1 hydrochloric acid and filter off the silica. Wash with hot water and dilute the filtrate to 100 ml. Take a 10-ml aliquot and adjust the pH to 4–5 with ammonium hydroxide. Add 20 ml of 0.05 *M* EDTA. Add 5 ml of 15% solution of acetylacetone and 2 drops of ammonium hydroxide. Extract by shaking for 6 minutes with 8 ml of carbon tetrachloride. Repeat the extraction twice, starting at "Add 5 ml of 15% solution...." Combine the carbon tetrachloride extracts and reextract the beryllium with 20 ml of water and 15 ml of hydrochloric acid. Evaporate this extract almost to dryness and dilute to 20 ml. Add 2 ml of 0.01% solution of the reagent, dilute to 25 ml, and read the complex.

METHYLTHYMOL BLUE

A 1:1 complex of beryllium with methylthymol blue follows Beer's law for 0.04–0.36 ppm at pH 5–5.5.[74] The pH is maintained by a perchloric acid–hexamine buffer solution. The full color develops at 80° in 20 minutes. EDTA masks common bivalent ions, but interference by aluminum, bismuth, thorium, and zirconium cannot be masked. There is interference by oxalate, citrate, tartrate, fluoride, high concentrations of EDTA, more than 2 m*M* nitrilotriacetic acid, and more than fivefold amounts of aluminum, ferric ion, and titanic ion.

Procedure. *Silicate Minerals.*[75] Dissolve 0.5 gram of sample by heating with 5 ml of hydrofluoric acid and 1 ml of 1:1 sulfuric acid. Evaporate to dryness and fuse with 1 gram of sodium pyrosulfate. Dissolve the fusion in 10 ml of 1:1 hydrochloric acid and dilute to 100 ml. To a 10-ml aliquot add 0.8 ml of 5% acetylacetone solution and adjust to pH 7 with ammonium hydroxide. Extract beryllium as the acetylacetone complex with 20 ml of carbon tetrachloride. Wash excess acetylacetone from the organic layer with 50 ml of 0.4% sodium hydroxide solution. Extract interferences from the organic layer with 20 ml of 0.1% EDTA solution. Extract the beryllium from the organic layer with 20 ml of 0.5 *N* hydrochloric acid and heat the aqueous extract to 100° to clarify it. Cool, add 1:1 ammonium hydroxide to pH 3–4, and dilute to 50 ml. To an aliquot containing not more than 25 μg of

[73]R. Kh. Dzhiyanbaeva, N. T. Turukhanova, and Sh. T. Talipov, *Tr. Tashk. Gos. Univ.* **1967** (288), 54–57.

[74]K. C. Srivastava and Samir K. Banerjee, *J. Indian Chem. Soc.* **47**, 225–230 (1970).

[75]A. T. Pilipenko and Yu. D. Balyaeva, *Ukr. Khim. Zh.* **37**, 193–195 (1971).

beryllium, add 1.5 ml of 0.0625 *M* EDTA, 1 ml of 0.1 *M* calcium nitrate, 20 ml of acetone, 5 ml of 0.8 m*M* methylthymol blue in 50% acetone, and 20 ml of hexamine–perchloric acid buffer solution for pH 5–5.5. Dilute to 50 ml and read at 510 nm against a reagent blank.

MORIN

Morin, which is 2′,3,4′,5,7-pentahydroxyflavone, forms a 1:1 monomer, a 1:1 dimer, and a 1:2 complex with beryllium.[76] The fluorescence is linear for 0.001 to 0.4 μg of beryllium in 25 ml with an appropriate concentration of morin. In strongly alkaline solution morin is nearly specific for beryllium. In acid solution there is significant interference by thorium, lutetium, yttrium, ceric ion, chromic ion, zirconium, lanthanum, scandium, and lithium. Use of diethylenetriamine–pentaacetic acid masks scandium, yttrium, and lanthanum. Calcium and zinc can be masked by EDTA and cyanide. The presence of aluminum, ethanol, or methanol increases the fluorescence of the beryllium-morin complex.

Beryllium in mineral water can be isolated by adsorption on silica gel from neutral, slightly acidic, or slightly alkaline solution.[77] The beryllium is then eluted with hydrochloric acid for masking of interfering ions with EDTA or tartrate and reading fluorimetrically with morin.

Procedure.[78] As a buffer solution, dissolve 15 grams of recrystallized diethylenetriamine pentaacetic acid (DTPA) in 200 ml of water, add 75 ml of redistilled piperidine, and cool. Add 20 grams of anhydrous sodium sulfite in 150 ml of water and dilute to 500 ml.

As a complexing solution, dissolve 60 grams of sodium hydroxide and 320 grams of anhydrous sodium perchlorate in 250 ml of water. Filter through a double glass-fiber filter in a Büchner funnel. Dissolve 13 grams of DTPA and 10 ml of 20% triethanolamine solution in 50 ml of water and 20 ml of the sodium hydroxide–sodium perchlorate solution. Add the balance of the alkaline reagent and dilute to 500 ml. Acidify a sample and test for oxidizing capacity. If this is found to be present, correct by small additions of sodium sulfite.

The pretreatment of various samples that follows yields the beryllium in concentrated perchloric acid solution.

To 0.5 ml of sample in perchloric acid solution free from interfering ions, add 1 ml of 4.9% solution of aluminum sulfate octadecahydrate in 1:100 perchloric acid. Add 3 ml of the complexing solution and 3 drops of 0.01% solution of quinine in 1% perchloric acid. Neutralize by dropwise addition of perchloric acid until a brilliant blue fluorescence is given under ultraviolet light. Add 1 drop of excess perchloric acid. Then add *N* sodium hydroxide until the fluorescence is ex-

[76] Mary H. Fletcher, *Anal. Chem.* **37**, 550–557 (1965).
[77] Z. Sulcek, J. Dolozal, and J. Michal, *Collect. Czech. Chem. Commun.* **26**, 246–254 (1961).
[78] Claude W. Sills and Conrad P. Willis, *Anal. Chem.* **31**, 398–608 (1959); Claude W. Sills, Conrad P. Willis, and J. Kenneth Flygare, Jr., *ibid.* **33**, 1671–1684 (1961).

tinguished. Add 5 ml of the buffer solution and 1 ml of 0.0075% solution of morin in 40% ethanol. Dilute to 25 ml and incubate for 20 minutes at a standardized temperature. Read the fluorescence with a fluorometer at 360 nm.

Alternatively, using a tungsten filament bulb, read the fluorescence with a Corning filter no. 5860 between the light source and the sample. This passes wavelengths centering at 360 nm and cuts off at 390 nm. Between the sample and the detector unit, use a Corning filter no. 4015, which has a peak transmittance at 540 nm and transmits little below 460 nm.

Extraction. If the concentration of beryllium in the prepared sample is insufficient for fluorescent reading, concentrate as follows.

Cool the solution thoroughly in cold running water to reduce heating of the solution during subsequent neutralization. Add 10 ml of 10% disodium EDTA solution to complex heavy metals while the solution is still acidic. Otherwise, metal hydroxides or phosphates will precipitate when the pH is raised and will not redissolve on addition of EDTA. Add 2 drops of 0.1% phenol red indicator solution and partly neutralize with ammonium hydroxide. Cool to dissipate the heat produced if considerable acid is present.

Add 0.25 ml of acetylacetone and shake the solution vigorously for a few seconds. Add ammonium hydroxide dropwise until the indicator color just changes to the red alkaline form. If a turbidity is produced, reacidify, add more EDTA, and repeat the addition of ammonium hydroxide. Extract with 10 ml and 10 ml of chloroform, shaking vigorously for 2 minutes each time. Add an additional 3 to 4 drops of acetylacetone before shaking the second time.

As wash solution add 1 ml of sulfuric acid and 2 ml of 10% disodium EDTA solution to 100 ml of water. Add 2 drops of 0.1% phenol red indicator solution followed by ammonium hydroxide to change the color to red.

Shake the combined extracts for 1 minute with 20 ml of wash solution. Draw off the chloroform extract into 3 ml of perchloric and 3 ml of nitric acid. Extract the wash solution with 10 ml of chloroform and 2 drops of acetylacetone and add to the main extract. Evaporate the chloroform. After the chloroform has evaporated and the rather vigorous reaction between nitric acid and acetylacetone has subsided, boil until most of the perchloric acid has been expelled and the bottom of the beaker is just slightly wet with 0.5 ml of the acid.

Proceed as under the general procedure from "To 0.5 ml of sample in perchloric acid...."

Uranium. Mix 1 gram of metallic uranium, 10 ml of water, 5 ml of hydrochloric acid, and 0.7 ml of 30% hydrogen peroxide and heat. A vigorous reaction soon takes place, and the metal dissolves rapidly to a clear yellow solution. If insufficient hydrogen peroxide is present to oxidize the uranium completely, a dark color or precipitate forms that clears immediately on addition of a few more drops of 30% hydrogen peroxide. After the reaction has subsided, add 3 ml of sulfuric acid and evaporate to sulfur trioxide fumes. Add 5 grams of anhydrous sodium sulfate and heat until a fusion is obtained and the sample has dissolved completely to give a clear yellow solution. Evaporate as much sulfuric acid as possible to minimize heating during the subsequent neutralization. Cool the melt, add 35 ml of water, and warm until the cake has dissolved completely. Cool the solution in running water. Add 25 ml of 10% disodium EDTA solution, 6 drops of 0.1% phenolsulfonephthalein solution, and ammonium hydroxide dropwise until the deep yellow changes to a distinct orange-red that does not deepen with 1 or 2 more drops. Allow sufficient time between drops for the local precipitate of ammonium diuranate to redissolve and stop the addition before a permanent precipitate is formed. A slight precipitate of ammonium diuranate will redissolve in the next step.

Add 2 grams of sodium dithionite. Let the solution stand for a few minutes until the dark color that is formed at first has changed to dark green and is perfectly clear. Sodium dithionite causes a pronounced decrease in pH and destroys the indicator. To readjust the pH after reduction of uranium is complete, add 5 drops of ammonium hydroxide alternately with a few drops of phenolsulfonaphthalein indicator solution and observe the color of the indicator as it enters the solution before being reduced by the dithionite. Add 5 drops of ammonium hydroxide after the alkaline color of the indicator has been produced.

Add 10 ml of chloroform and 0.5 ml of acetylacetone and shake for 2 minutes. If a yellow color appears in the chloroform layer, add more dithionate and repeat the shaking. Draw off the chloroform extract, add an additional 5 drops of acetylacetone, and repeat the extraction with another 10 ml of chloroform. Add 50 ml of the wash solution to the combined extracts. Add 2 drops of 30% hydrogen peroxide and shake for 15 seconds. Add 2 drops of phenolsulfonaphthalein indicator solution and ammonium hydroxide until the indicator color changes to red. Shake for 1 minute and draw off the chloroform. Add 2 drops of acetylacetone, shake the aqueous solution with 5 ml of chloroform, and add it to the main extract. Add 3 ml of perchloric acid and 3 ml of nitric acid and warm. When the chloroform has evaporated and the reaction with acetylacetone is completed, evaporate to about 0.5 ml of perchloric acid.

Complete according to general procedure from "To 0.5 ml of sample in perchloric...."

Zirconium. Mix 0.5 gram of zirconium metal, 5 grams of anhydrous sodium sulfate, and 10 ml of sulfuric acid. Heat until the metallic zirconium has nearly completely dissolved. A white precipitate of zirconium sulfate will generally be present. Heat until excess sulfuric acid has been driven off and a clear pyrosulfate fusion is obtained. Any remaining particles of metal and the white precipitate will dissolve easily and completely in the fusion. Cool the melt, add 1 ml of sulfuric acid, and dissolve the cake in 30 ml of 10% disodium EDTA solution. Add 30 ml of water, 4 drops of 0.1% phenolsulfonaphthalein indicator solution, and then ammonium hydroxide, dropwise until the indicator changes to its red alkaline color. Allow to stand until both the EDTA and the remainder of the melt have dissolved completely. Cool the solution thoroughly and add perchloric acid until the indicator reverts to its yellow acidic form. Add 5 drops of excess acid and 0.5 ml of acetylacetone.

Add ammonium hydroxide until the full red alkaline color of the indicator is produced, followed by 5 drops of excess ammonium hydroxide. Extract with 10 ml and 10 ml of chloroform, shaking vigorously for 2 minutes each time. Add an additional 3–4 drops of acetylacetone before shaking the second time.

Shake the combined extracts with 20 ml of a wash solution consisting of 1%

sulfuric acid and 0.2% disodium EDTA adjusted to yellow to phenol red with ammonium hydroxide. Add the organic layer to 3 ml of perchloric acid and 3 ml of nitric acid. Extract the wash solution with 10 ml of chloroform and 2 ml of acetylacetone, adding this to the main chloroform extract. Evaporate the chloroform. After the reaction of acetylacetone with nitric acid has subsided evaporate to 0.5 ml.

Take up with 50 ml of water, add 5 ml of 10% solution of disodium EDTA, 0.5 ml of acetylacetone, and 4 drops of phenolsulfonaphthalein indicator solution. Repeat the extraction from "Add ammonium hydroxide until the full red..." to again end with evaporation to 0.5 ml. This step ensures complete removal of zirconium.

Complete according to the general procedure from "To 0.5 ml of sample in perchloric...."

Thorium. Mix 1 gram of thorium metal powder, 7 grams of anhydrous sodium sulfate, and 3 ml of sulfuric acid. To prevent spattering, heat slowly at first until all moisture has been removed. Then heat over a blast lamp until excess sulfuric acid has been driven off and a clear pyrosulfate fusion is obtained. The metal does not dissolve significantly until the actual high temperature fusion is obtained, and the solution may solidify before dissolution is complete if the fusion is heated too strongly.

When fusion is complete, cool and add 25 ml of 0.25 *M* disodium diethylenetriamine tetraacetic acid that contains 98.3 grams of the recrystallized acid and 20 grams of sodium hydroxide per liter. Add 50 ml of water and boil gently until solution is complete. Partial neutralization with a few drops of ammonium hydroxide will facilitate dissolving the cake, but the neutralization must not be carried too far because the solution must be definitely acidic before transferring, to prevent serious losses of beryllium. Cool. Add 0.5 ml of acetylacetone and 4 drops of 0.1% phenolsulfonaphthalein indicator solution. Treat as described for zirconium from "Add ammonium hydroxide until the full red...."

Aluminum. Mix 0.1 gram of aluminum chips with 10 ml of 1:1 hydrochloric acid, and heat gently until the vigorous reaction has subsided. When the metal has dissolved, add 3 ml of sulfuric acid down the side of the vessel to prevent spattering, followed by 4 grams of anhydrous sodium sulfate. Evaporate to fumes of sulfur trioxide. Aluminum sulfate will precipitate in the concentrated acid in a finely crystalline form that does not stick to the bottom of the container or cause bumping during the subsequent fusion. Heat the fuming solution at a high temperature until a clear salt is obtained. Roll the solution around the sides of the container while heating the sides continuously to dissolve the small quantity of aluminum salts that will have spattered.

Cool, and add 30 ml of 10% disodium EDTA solution and 40 ml of water. Swirl the solution for a few seconds until the melt has become detached from the sides. Add 4 drops of 0.1% phenolsulfonaphthalein indicator solution and ammonium hydroxide until the indicator changes from yellow to red. Without waiting for complete dissolution of the cake, heat the solution to boiling. As the cake dissolves, add ammonium hydroxide dropwise as required to keep the indicator light orange, with only the first trace of the red alkaline form showing in the yellow solution.

When the cake has dissolved completely, add 2 drops of 1:1 sulfuric acid to discharge the orange color and boil the solution gently for about 5 minutes at about pH 6. Do not permit reduction in volume by more than about 15 ml. Cool, add 10 drops of acetylacetone, and shake vigorously for 10 seconds. Add ammonium hydroxide dropwise just until the first red color of the indicator is produced; then shake vigorously with 10 ml chloroform for 2 minutes. Repeat the extraction with 3 additional drops of acetylacetone and 10 ml of chloroform. Complete as under zirconium from "Shake the combined extracts with 20 ml of...."

Aluminum Oxide. Treat a 0.2-gram sample as described for aluminum, adding it to the 3 ml of sulfuric acid and 4 grams of anhydrous sodium sulfate.

Copper Alloys. Mix 0.5 gram of sample and 10 ml of hydrobromic acid. Add 1 gram of solid sodium bromate in small portions and evaporate nearly to dryness. The solution will spatter severely near the end of the evaporation, and the beaker should be removed from the hot plate when highly colored droplets begin to appear on the cover glass. Add an additional 5 ml of hydrobromic acid and about 0.25 gram of sodium bromate and repeat the evaporation. Add 10 ml of perchloric acid and boil until the refluxing acid has cleaned the sides of the container. The deep purplish crystals of sodium bromocuprate are insoluble in perchloric acid but are transposed on boiling to the light green cupric perchlorate, which will dissolve in the concentrated acid. A light yellow precipitate of silver bromide will be present with alloys containing silver but will be metathesized and solubilized on further boiling with perchloric acid.

If tin, antimony, and arsenic are absent, the bromide volatilization can be eliminated and the copper alloy dissolved directly in perchloric acid. Add enough more perchloric acid to make a total of 5 ml and 30 ml of water. Evaporate to 10 ml to hydrolyze any pyrophosphate that may have been formed. Dilute to 50 ml and take an aliquot containing less than 1 mg of copper. Dilute to 0.5 ml with perchloric acid and develop by the general procedure from "To 0.5 ml of sample in perchloric...."

With beryllium bronzes, the beryllium concentration generally is so large that further dilutions must be made to bring the concentration within the range accommodated by the present procedure.

Stainless Steel. Dissolve 0.5 gram of sample by boiling with 10 ml of perchloric acid to give the red color of hexavalent chromium. Add 0.5 gram of sodium chloride and heat to volatilize chromium as chromyl chloride. Heat to fumes to reoxidize residual chromium. Repeat this step twice more, ending at the point when salts begin to separate. Boil for a few minutes and filter any silica that separates. Cool and add 40 ml of 10% solution of disodium EDTA. Add 10 drops of acetylacetone and follow with ammonium hydroxide until the ferric-EDTA complex becomes red. Decolorize by adding sodium dithionate in small portions, avoiding a substantial excess, which can reduce acetylacetone. Add 2 ml of 1:1 sulfuric acid, a drop of 30% hydrogen peroxide, 50 ml of the wash solution described under extraction, and ammonium hydroxide to raise the pH to the phenol red end point. Proceed with Extraction from "Add 0.25 ml of acetylacetone and shake the solution vigorously...."

Monazite Sand. Moisten 0.5 gram of monazite sand or similar material in platinum with water. Cautiously add 1 ml of hydrofluoric acid dropwise. Slowly evaporate to dryness. Spread 3 grams of anhydrous potassium fluoride over the residue and fuse. If necessary, add up to 1 gram of sodium pyrosulfate to promote fusion. Add 5 ml of sulfuric acid and heat cautiously to volatilize hydrofluoric acid and give a pyrosulfate fusion. Addition of sodium sulfate and more sulfuric acid may be required to give a smooth fusion. Volatilize excess sulfuric acid from the fusion. Add 40 ml of 1:3 hydrochloric acid and heat the solution with occasional swirling until the cake is detached completely from the dish. Remove the platinum dish and rinse it with about 20 ml of water. Add two boiling chips and boil the solution vigorously until the volume has been reduced to 10 to 15 ml. If the evaporation is carried further, pyrophosphate will be re-formed and recovery of beryllium will be seriously incomplete. A finely divided crystalline precipitate of the insoluble double sulfates will be present. Troublesome bumping may occur if too much water is added initially. Add 25 ml of water, cool to room temperature, and filter through a double filter paper with a small amount of paper pulp. Wash the precipitate with water before permitting the cake to suck dry. Add 25 ml of 0.25 *M* disodium diethylenetriamine–pentaacetic acid and 4 drops of 0.1% phenolsulfonaphthalein indicator solution to the filtrate. Neutralize most of the excess acid with ammonium hydroxide. Cool the solution and add 0.5 ml of acetylacetone. Add ammonium hydroxide until the indicator changes to its red alkaline form and 5 drops in excess. Complete as for zirconium from "Extract with 10 ml and 10 ml of chloroform...."

Airborne Dusts. Because of the extreme sensitivity of morin, an adequate sample for determination of 2 μg of beryllium per cubic meter is 20 liters of air, larger samples being appropriately aliquoted. Collect by paper filtration. Heat the paper with 3 ml of sulfuric acid and 10 ml of nitric acid. When most of the nitric acid has been driven off and the paper is carbonized, add more nitric acid dropwise until the solution can be taken to sulfur trioxide fumes at a yellow to red color. Then add perchloric acid dropwise until all organic matter is destroyed. Add 3 grams of anhydrous sodium sulfate and heat with a flame until all free sulfuric acid has been driven off and a fusion results. Boil the cake with water, filter if necessary, and dilute to 50 ml. Take an aliquot equivalent to 20 liters of air. It is not necessary to hydrolyze complex phosphates or mask aluminum. Evaporate with 0.5 ml of perchloric acid to fumes and cool. Complete by the general procedure from "To 0.5 ml of sample in perchloric acid solution...."

Minerals.[79] As a complexing solution, dissolve 1 mg of potassium metabisulfite, 2.5 grams of ascorbic acid, 2.5 grams of citric acid, and 5 grams of disodium EDTA in water and dilute to 100 ml. If necessary to complete solution of the EDTA, add 10% sodium hydroxide solution dropwise. As a borate buffer solution, dissolve 28.6 grams of boric acid and 96 grams of sodium hydroxide in water and dilute to 1 liter.

[79]D. P. Shcherbov and T. N. Plotnikova, *Zavod. Lab.* **27**, 1058–1062 (1961); cf. J. M. Riley, *U.S. Bur. Mines, Rep. Invest.* **5282**, 9 pp (1956).

Treat 0.1 gram of sample in platinum with 1 ml of sulfuric acid, 1 ml of nitric acid, and 5 ml of hydrofluoric acid. Evaporate to dryness. Fuse for 5 minutes at 1000° with 0.5 gram of sodium tetraborate and 1.5 grams of sodium carbonate. Take up with 10 ml of 1 : 1 sulfuric acid, add 40 ml of water, and filter. Dilute to 100 ml and take a 10-ml aliquot. Add 2 ml of complexing solution and make just alkaline to Congo red paper with 5% sodium hydroxide solution. Add 1 ml of borate buffer solution and 1 ml of 0.02% ethanolic morin. Dilute to 10 ml, set aside for 5 minutes, and read the fluorescence at 360 nm.

As alternatives,[80] the beryllium is extracted with acetylacetone or coprecipitated with titanium phosphate. In either case the solution thereafter is treated with the complexing solution.

For determination in minerals of low beryllium content or in the presence of little iron, separate the beryllium by the titanium phosphate method.[81] Details of the technic appear under Biological Samples. Titanium sulfate prepared from potassium fluotitonate is preferred.

For analysis of silicate minerals, after volatilizing silica, precipitate hydroxides in the presence of mercaptoacetic acid.[82] Filter, dry, and ash. Fuse with sodium hydroxide and extract with water. Filter, neutralize with hydrochloric acid, and dilute to a known volume.

To an aliquot, add either 2.5 ml of saturated solution of tetrasodium pyrophosphate or 1 ml of 10% solution of EDTA. Add 1 ml of 5% solution of potassium cyanide and 1 ml of 1% solution of stannous chloride in *N* sodium hydroxide. The aluminum content must be adjusted to correspond with that in the standards of the calibration graph. Add 10% sodium hydroxide solution to raise to pH 11.5. Add 0.2 ml of 0.05% solution of morin in acetone and let stand for 2 minutes. Activate at 375 nm and read the fluorescence at 400 nm. Fluorescein at 2.5 μg/ml is an appropriate comparison standard.

Beryl and Bertrandite.[83] Fuse a 0.5-gram sample in platinum with 3 grams of 3 : 1 sodium carbonate–sodium tetraborate. Dissolve, warming if necessary, with 50 ml of 3 : 7 hydrochloric acid containing a few drops of ethanol, and dilute to 1 liter. Mix a 1-ml aliquot with 2.5 ml of neutralized 0.5% solution of EDTA and 15 ml of water. Add 1 ml of 4% sodium hydroxide solution and mix. Add 1 ml of 0.005% solution of morin in 10% ethanol. Dilute to 25 ml and read the fluorescence with secondary filters peaking at 520 nm and primary filters with a maximum transmittance at 430 nm.

Complex Silicates. ISOLATION AS A BUTYRATE.[84] Fuse 0.2 gram of sample in platinum with 3 grams of potassium bifluoride to a clear melt. Cool, add 10 ml of 1 : 1 sulfuric acid, heat until fuming ceases, and fuse at 500–600°. Dissolve in 10 ml

[80] R. N. Plotnikova, R. P. Ashaeva, and D. P. Shcherbov, *Zavod. Lab.* **32**, 1063–1064 (1966).
[81] N. A. Suvorovskaya, M. M. Voskresenskaya, and T. A. Mel'nikova, *Nauch. Soobsheh. Inst. Gorn. Dala A. A. Skochinskii* **13**, 30–32 (1962); cf. J. Minczewski and W. Rutkowski, *Chem. Anal.* (Warsaw) **7**, 1107–1118 (1962).
[82] W. Rutkowski, *ibid.* **8**, 389–394 (1963).
[83] Irving May and F. S. Grimaldi, *Anal. Chem.* **33**, 1251–1253 (1961).
[84] L. L. Galkina, *Byull. Nauchn.-Tekh. Inf. Min. Geol. SSSR, Ser. Izuch. Veshchestv. Sostava Mine. Syr'ya Tekhnol. Obogasch. Rud* **1967** (3), 19–23; cf. L. L. Galkina and A. L. Markman, *Izv. Vyssh. Ucheb. Zaved. Khim. Khim. Tekhnol.* **6**, 735–738 (1963).

of 1 : 1 hydrochloric acid and evaporate to about 3 ml. Add 15 ml of saturated solution of sodium chloride and boil for 5 minutes. Dilute to 25 ml with saturated solution of sodium chloride and filter. Take 20 ml of filtrate as an aliquot and add 5 ml of saturated solution of EDTA. Raise to pH 9–10 by adding 1 : 4 ammonium hydroxide. Add 10 ml of butyric acid and shake for 1 minute to extract beryllium butyrate. Wash the organic layer with 10 ml and 10 ml of saturated sodium chloride solution containing 2 drops of a saturated solution of EDTA and 2 drops of 30% hydrogen peroxide. Extract the beryllium from the organic layer by shaking for 1 minute with 5 ml of chloroform and 5 ml of nitric acid. Evaporate the aqueous acid layer almost to dryness, add 3 ml of 1 : 1 sulfuric acid and evaporate to dryness. Dissolve the residue in water and dilute to 25 ml. Make a 5-ml aliquot alkaline to Congo red with 10% sodium hydroxide solution. For pH 13, add 1 ml of a buffer solution containing 2.86% boric acid and 9.6% sodium hydroxide. Add 1 ml of 0.02% solution of morin and dilute to 10 ml. Read the fluorescence.

Bone. Boil a 1-gram sample of bone ash with 5 ml of perchloric acid. After the sample has dissolved, evaporate until most of the perchloric acid has been driven off and the solution has a thick syrupy consistency, but not to the point at which the solution begins to solidify. During ignition beryllium will probably have been converted to a refractory oxide or phosphate, much more readily soluble in the strong phosphoric acid above than in boiling perchloric acid. Add 3 ml of perchloric acid and 30 ml of water and boil vigorously to hydrolyze the pyrophosphates that will have been produced. Evaporate until the first indications of perchloric acid fumes are observed, but not until strong fumes appear. Cool, and add 40 ml of 10% disodium EDTA solution to complex heavy metals while the solution is still acidic. Otherwise, metal hydroxides or phosphates will precipitate when the pH is raised and will not redissolve on addition of EDTA. Add 2 drops of 0.1% phenol red indicator solution and partly neutralize with ammonium hydroxide. Add ammonium hydroxide dropwise until the indicator color just changes to the red alkaline form. If a turbidity is produced, reacidify, add more EDTA, and repeat the addition of ammonium hydroxide. Extract with 10 ml and 10 ml of chloroform, shaking vigorously for 2 minutes each time. Add an additional 3 to 4 drops of acetylacetone before shaking the second time.

As wash solution, add 1 ml of sulfuric acid and 2 ml of 10% disodium EDTA solution to 100 ml of water. Add 2 drops of 0.1% phenol red indicator solution followed by ammonium hydroxide to change the yellow color to red.

Shake the combined extracts for 1 minute with 20 ml of wash solution. Draw off the chloroform extract into 3 ml of perchloric and 3 ml of nitric acid. Extract the wash solution with 10 ml of chloroform and 2 drops of acetylacetone and add to the main extract. Evaporate the chloroform. After the chloroform has evaporated and the rather vigorous reaction between nitric acid and acetylacetone has subsided, boil until most of the perchloric acid has been expelled and the bottom of the beaker is just slightly wet with acid, estimated at 0.5 ml. Complete by the general procedure from "To 0.5 ml of sample in perchloric. . . ."

Natural Water.[85] To 30 ml of sample containing up to 1.2 μg of beryllium, add

[85] E. A. Morgan, N. A. Vlasov, and A. Z. Serykh. *Zh. Prikl. Khim., Leningr.* **43**, 2744–2745 (1970).

10 ml of 10% EDTA solution and 1 ml of 5% solution of acetylacetone. Adjust to pH 6–8 and extract the beryllium complex with 10 ml and 10 ml of chloroform, adjusting the pH between extractions with 5% sodium hydroxide solution. Wash the combined extracts with 10 ml of 0.1% solution of EDTA. Add the organic layer to 2 ml of nitric acid and 1 ml of perchloric acid. Evaporate almost to dryness and take up in 10 ml of water. Add 2.5 ml of a solution containing 2.86% of boric acid and 9.6% of sodium hydroxide, which is a buffer for pH 13. Add 0.5 ml of 0.02% ethanolic morin. Dilute to 25 ml, set aside for 20 minutes, and read the fluorescence.

Biological Samples.[86] Dissolve 15 grams of disodium EDTA in ammonium hydroxide and dilute to 100 ml. As titanium solution, either heat 2.7 grams of potassium fluotitanate in 100 ml of 1 : 1 sulfuric acid until hydrofluoric acid is completely evaporated and dilute to 100 ml, or dissolve 0.83 gram of titanium dioxide and 8.8 grams of potassium pyrosulfate in 1 : 3 sulfuric acid and dilute to 100 ml with that acid.

Digest 100 ml of urine, 10 ml of blood, or 10 grams of wet tissue with 10 ml of sulfuric acid, 10 ml of nitric acid, and 5 ml of perchloric acid until clear and colorless. Dilute somewhat and add 5 ml of 10% disodium phosphate solution, 10 ml of 15% disodium EDTA solution, 1 ml of titanium solution, and 4 drops of methyl orange indicator solution. Add ammonium hydroxide until the indicator becomes orange and 10 ml of 20% sodium acetate solution. Filter the manganese and titanium phosphates. Wash with 2% ammonium sulfate solution. Dissolve the precipitate in 5 ml of 2% sodium hydroxide solution, add 5 ml of water, and separate the sodium beryllate solution by centrifuging. Dilute to a known volume and take a 1-ml aliquot. Add 4 ml of 1% sodium hydroxide solution, 0.5 ml of 5% solution of disodium EDTA, and 0.1 ml of 0.02% ethanolic morin. Dilute to 10 ml and read the fluorescence at 360 nm.

Urine.[87] To 500 ml of sample, add 10 ml of saturated solution of disodium EDTA. Add 10% sodium hydroxide solution to raise to pH 8. Add a few drops of 10% solution of tannin and 20 ml of 1% solution of methylene blue. Stir in 7 ml more of tannin solution, let settle, and filter. Wash the precipitate with 0.5% solution of ammonium chloride adjusted to pH 8. Dry the paper, ash, and ignite at 500°. Add 2 ml of perchloric acid and 0.1 ml of hydrofluoric acid. Heat almost to dryness, take up with 3 ml of 1 : 6 hydrochloric acid. If necessary, aliquot so that 0.01–0.1 μg of beryllium is being developed. Add 2.5 ml of 5% solution of tetrasodium EDTA and 1 ml of 3% solution of triethanolamine to mask heavy metals. Raise to pH 11.6–11.9 by addition of 10% sodium hydroxide solution. Cool in ice for 10 minutes and add 0.2 ml of 0.005% solution of morin in 10% ethanol. Dilute to 10 ml and store in ice for 10 minutes. Transfer 5 ml to a fluorometer, let it warm to $15 \pm 0.1°$, and read the fluorescence. The low temperature prevents fading at these low concentrations of beryllium.

---

[86] M. S. Bykhovskaya, *Gig. T. Prof. Zabol.* **5** (11), 55–57 (1961).
[87] S. R. Desai and Kum. K. Sudhalatha, *Talenta* **14**, 1346–1349 (1967).

NAPHTHOCHROME GREEN G

Naphthochrome green G (mordant green 31) is phenoxydinaphthofuchsonedicarboxylic acid.[88] It forms 1:1 complexes with beryllium at pH 5–8 and 11.5–13. The reagent will determine 0.5 ppm of beryllium in lithium fluoride of special purity.

Procedures

Lithium Fluoride. Add 2 ml of perchloric acid to a 0.2-gram sample in platinum. Heat over a period of 20–30 minutes, ending with evolution of white fumes. Take up in water, adjust to pH 5.5–6, and add 1 ml of 3% EDTA solution also at that pH. Add 2.5 ml of 1.5 m*M* naphthochrome green G and 5 ml of a buffer solution for pH 7.2–7.4. Dilute to 25 ml and read at 640 nm against a reagent blank.

Tissue.[89] As the dye reagent, which will be obtained as the sodium salt, dissolve 20 grams of naphthochrome green G in 300 ml of boiling water. Cool, and add 1:1 hydrochloric acid until acid to Congo red. Filter the precipitated dyestuff, wash, and dry at 70°. For use, dissolve 0.15 gram of the dye in ethanol, add 1 ml of *N* hydrochloric acid, and dilute to 100 ml with ethanol.

To a sample containing up to 2 μg of beryllium, add 2 ml of sulfuric acid and 5 ml of nitric acid. Digest with further additions of nitric acid as necessary until colorless or faintly yellow. Evaporate to sulfur trioxide fumes. Cool, add 2 ml of water, and evaporate to sulfur troxide fumes. Repeat the last step. Take up in 10 ml of water and add 1 ml of 10% potassium ferrocyanide solution. If the pale blue solution does not coagulate, add a drop of 5% solution of ferric chloride. Dilute to 15 ml, let stand overnight, and centrifuge. Transfer 5 ml of the colorless or pale yellow solution to a 15-ml centrifuge tube. Add 0.5 ml of 2.63% solution of potash alum, 1 ml of saturated solution of disodium phosphate, and 3 drops of 0.1% solution of bromocresol green. Add 40% sodium hydroxide solution dropwise until the indicator turns blue, showing that the solution is alkaline. Revert to yellow with *N* sulfuric acid, then to blue with 0.4% sodium hydroxide solution. After allowing to stand for 30 minutes, centrifuge for 5 minutes. Decant the supernatant layer and drain. Suspend the precipitate in 0.5 ml of 1% sodium chloride solution, add 5 ml more of 1% sodium chloride solution, and centrifuge for 3 minutes. Decant the supernatant layer and wash the precipitate again with 1% sodium chloride solution. Decant and drain. Add to the precipitate 0.08 ml of *N* sodium hydroxide and 2 ml of 10% trisodium phosphate solution. Add 0.2 ml of the dye reagent and dilute to 5 ml with 10% trisodium phosphate solution. Incubate at 30° for 20 minutes and read the decrease in absorption at 650 nm as against a reagent blank.

As an alternative,[90] develop with 1 ml of 0.2% solution of the dye, 1 ml of 1%

[88] L. P. Adamovich, A. P Mirnaya, and A. K. Khukhrynaskaya, *Zh. Anal. Khim.* **24**, 1816–1822 (1969).
[89] W. N. Aldridge and H. F. Liddell, *Analyst* **73**, 607–613 (1948).
[90] Taitiro Fujinaga, Tooru Kuwamoto, Kazuo Kuwabara, and Shozo Ikezawa, *Jap. Anal.* **13**, 1213–1218 (1964).

gelatin, and M acetate buffer to dilute to 25 ml. Read the decrease in absorption at 650 nm.

4-(*p*-NITROPHENYLAZO)ORCINOL

This dyestuff in buffered alkaline solution is nearly specific for beryllium by formation of a red lake.[91] The commonly associated ions are masked by EDTA. The calibration curve deviates slightly from the linear. The reagent has been applied to airborne dust.[92]

Procedure. *Titanium Alloys.*[93a] As a buffer solution, dissolve 116 grams of citric acid, 58.7 grams of anhydrous sodium borate, and 216 grams of sodium hydroxide. Dilute to 1 liter. As color reagent, dissolve 0.15 gram of 4-(*p*-nitrophenylazo)-orcinol in 500 ml of 0.4% solution of sodium hydroxide by stirring for 5 hours and filtering through asbestos.

Dissolve a 5-gram sample in 25 ml of 1 : 1 hydrochloric acid or 1 : 4 sulfuric acid and dilute to 100 ml. Take an aliquot containing 0.05–0.6 mg of beryllium. This would usually be 10 ml and must not contain more than 5 mg of magnesium, 20 mg of calcium, 10 mg of iron, 35 mg of aluminum, or 100 mg of titanium. Dilute to 15 ml and add 15 ml of 3% hydrogen peroxide. Add 5 ml of 12% solution of tetrasodium EDTA and adjust to pH 5.5 with 8% sodium hydroxide solution. Let stand for 5 minutes, add 10 ml of the buffer solution, and again let stand for 5 minutes. Add 10 ml of the color reagent and dilute to 100 ml. After 10 minutes read at 515 nm.

PERTITANIC ACID

Titanic ion and hydrogen peroxide react with beryllium in 2 N hydrochloric acid. Ferric ion up to 0.5 mM is tolerated. Aluminum up to 0.2 M is masked with EDTA. On addition of fluoride ion to the pertitanic complex, hydrogen peroxide is displaced to form a fluotitanate. Conversely, when a metal with a strong affinity for fluoride is added to a mixture of pertitanate and fluotitanate, fluoride is displaced and the yellowish-orange peroxo complex can be read.[93b]

Procedure. As a reagent, dissolve 2.08 grams of sodium titanate and 1.68 grams of sodium fluoride in 2 N hydrochloric acid. Add 1.1 ml of 30% hydrogen peroxide and dilute to 100 ml with 2 N hydrochloric acid.

Mix 4 ml of sample solution containing more than 0.25 mg of beryllium with 3

[91] A. S. Komorovskii and N. S. Poluekov, *Mikrochemie* **14**, 315–317 (1934).
[92] R. A. Hiser, H. M. Donaldson, and C. W. Schwenzfeier, *Am. Ind. Hyg. Assoc. J.* **24**, 280–285 (1961).
[93a] L. C. Covington and M. J. Miles, *Anal. Chem.* **28**, 1728–1730 (1956).
[93b] Chiyo Matsubara and Kiyoko Takamura, *Anal. Chim. Acta* **77**, 255–262 (1975).

ml of 0.2 *M* disodium EDTA.[93c] Neutralize with hydrochloric acid and add 4 ml of hydrochloric acid. Add 5 ml of reagent, dilute to 25 ml, and read at 420 nm.

As a reference solution, mix 10 ml of sample solution with 8 ml of 0.2 *M* EDTA and adjust to pH 7 with sodium hydroxide solution. Shake with 15 ml of 0.4% solution of benzoylacetone in chloroform for 10 minutes. Raise to pH 8–9 and extract with 15 ml of 0.4% solution of benzoylacetone in chloroform. Wash with 10 ml of chloroform. Neutralize the aqueous phase with hydrochloric acid, add 4 ml of hydrochloric acid, and dilute to 25 ml. Take two 1-ml aliquots and add 1-ml and 2-ml portions of 0.05 *M* beryllium chloride in 2 *N* hydrochloric acid. To each, add 4 ml of hydrochloric acid and 5 ml of reagent. Dilute each to 25 ml and read the second solution at 420 nm versus the first solution. Calculate from the absorption of the sample.

2–PHENOXYQUINIZARIN–3, 4′–DISULFONIC ACID

The stable violet complex of this reagent with beryllium is read at 550 nm.[94] Many interferences are masked by calcium EDTA. Those still interfering seriously are chromic ion, magnesium, zirconium, thorium, fluoride, and phosphate.

Procedures

Beryl. As the sequestering agent, dissolve 18.6 grams of disodium EDTA and 12 grams of calcium nitrate tetrahydrate in 350 ml of water. Adjust to pH 4.3 by adding about 6 ml of hydrochloric acid and dilute to 500 ml.

Fuse 0.1 gram of sample in platinum with 0.4 gram of sodium carbonate for 30 minutes. Dissolve the melt in 5 ml of 1 : 1 hydrochloric acid. Add 10 ml of hydrochloric acid and evaporate to dryness. Bake at 110–120° for 30 minutes. Moisten the residue with 5 ml of hydrochloric acid and add 70 ml of water. Bring to a boil, filter, and dilute to 1 liter.

To an aliquot containing 5–35 μg of beryllium, add 10 ml of the calcium EDTA sequestering agent, 10 ml of 0.1% solution of the dipotassium salt of the captioned reagent, and 5 ml of 4 *M* ammonium acetate adjusted to pH 6 with acetic acid. Dilute to 50 ml, adjust to pH 6 ± 0.1 if necessary, and let stand for 1 hour. Read at 550 nm against a reagent blank.

Aluminum. Dissolve 0.1 gram of sample with 6 ml of 1 : 1 hydrochloric acid and add 0.25 ml of 30% hydrogen peroxide. Add 30 ml of 5% solution of disodium EDTA and raise to pH 3–3.5 with 10% sodium hydroxide solution. Dilute to 100 ml and complete as for beryl from "To an aliquot containing 5–35. . . ."

Alternatively, transfer a portion of the sample solution containing about 1 mg of beryllium in 60–80 ml to a 25×0.8 mm column of Amberlite IR-120 cation exchange resin in sodium form, passing it at 1 ml/minute. Wash the column with 10 ml of 0.5% EDTA solution adjusted to pH 3.5 followed by 10 ml of water. Elute

[93c] Chiyo Matsubara, *Jap. Anal.* **23**, 878–883 (1974).
[94] E. Guy Owens II and John H. Yoe, *ibid.* **32**, 1345–1349 (1960).

the beryllium with 150 ml of 1:3 hydrochloric acid, collecting the first 100 ml. Complete as for beryl from "To an aliquot containing 5–35. . . ."

QUINALIZARIN

For reading of beryllium with quinalizarin, bivalent interferences are masked with EDTA. The color can be stabilized with sodium sulfide.[95]

Procedure. *Beryl Ores.*[96] Fuse 0.2 gram of sample with 3 grams of potassium bifluoride. Add 5 ml of sulfuric acid to the melt and heat to copious sulfur trioxide fumes. Cool, dissolve in 10 ml of 1:10 hydrochloric acid, and dilute to 100 ml. To a 10-ml aliquot, add 2 ml of 0.5% solution of EDTA, adjust to pH 6–7 with ammonium hydroxide, and dilute to 100 ml. Prepare a column of 20 grams of 60–100 mesh Amberlite IRC-50 resin in hydrogen form. Prewash with 250 ml of 1:5 ammonium hydroxide and with 200 ml of 0.1 *M* EDTA. Pass the sample through the column and wash it with 100 ml of 0.1 *M* EDTA followed by 200 ml of water. Elute the beryllium with 200 ml of 1:5 hydrochloric acid and dilute the eluate to 250 ml. Evaporate an aliquot containing 0.2–0.5 mg of beryllium oxide almost to dryness and take up in 25 ml of water. Add sequentially 10 ml of 20% sodium hydroxide solution, 2.5 ml of 0.1 *M* EDTA, 2 ml of 10% solution of hydroxylamine hydrochloride, and 5 ml of 0.05% solution of quinalizarine in 50% pyridine. Dilute to 50 ml and read at 650 nm against a reagent blank.

RHODAMINE B

This dyestuff forms a ternary complex with benzoic acid and beryllium. Beer's law is followed for up to 25 μg of beryllium extracted into 5 ml of toluene. EDTA masks copper, mercury, thallium, indium, scandium, tin, silver, hafnium, zirconium, thallium, bismuth, and antimony. Tartrate masks ferric ion and aluminum. When palmitic acid replaces benzoic acid with lesser sensitivity, the only interfering ions are scandium, hexavalent uranium, copper, mercurous, indium, and zinc.

Procedure. To a solution containing 5–25 μg of beryllium, add 1 ml of 2.5% benzoic acid solution, 2 ml of 1% rhodamine B solution, and 0.25 ml of 40% solution of hexamine.[97] Dilute to 5 ml and shake with 5 ml of toluene for 2 minutes. Let stand for 15 minutes, and read the organic layer at 550 nm. For less

[95] P. I. Vasil'ev, *Sb. Nauch.-Tekh. Inf. Min. Geol. Okhr. Nedr.* **1955** (1), 131.

[96] V. M. Karve, *Indian J. Chem.* **3**, 537–538 (1965); cf. M. V. Kanyukova, *Tr. Vses. Magadan. Nauch.-Issled. Inst. Min. Tset. Met. SSSR* **1957** 1–13.

[97] N. S. Poluektov, S. B. Meshkova, S. V. Bel'tyukova, and E. I. Tselik, *Zh. Anal. Khim.* **27**, 1721–1725 (1972).

than 5 μg of beryllium, reduce the amounts of solution of benzoic acid and rhodamine B to 0.5 ml.

RUFIGALLOL

The crimson complex of rufigallol is read for 2–10 μg of beryllium per ml. Extraction of interfering ions is as oxine derivatives in chloroform.

Procedure.[98] Adjust 0.5 ml of sample containing 0.02–0.07 mg of beryllium to pH 7. Add 1.5 ml of water, 2 ml of dimethylsulfoxide, and 1 ml of 0.07% solution of rufigallol. Set aside for 5 minutes and take a 1-ml aliquot. Add 4 ml of 60% solution of dimethylsulfoxide and read at 540 nm.

SULFOSALICYLIC ACID

Beryllium forms a 1:2 complex with sulfosalicylic acid for reading in the ultraviolet.[99] Aluminum is masked with EDTA. Ferric ion and copper produce serious interference, as do also nitrate and phosphate ions. The complex forms immediately and is stable for at least 2 months.

Procedures

Aluminum. Dissolve a 0.2-gram sample in a minimal amount of sulfuric, perchloric, or hydrochloric acid and dilute to 100 ml. Add 10 ml of 0.01 *M* salicylic acid and 50 ml of 0.5 *M* sodium or ammonium EDTA at pH 8–9. Add ammonium hydroxide to raise to pH 9.2–10.8, preferably 10. Dilute to 200 ml and read at 317 nm.

Magnesium Alloys.[100] Dissolve 1 gram in 25 ml of 1:1 hydrochloric acid and dilute to 100 ml. To a 10-ml aliquot add 2 ml of 10% hydroxylamine hydrochloride solution to reduce the iron, 20 ml of 0.5 *M* EDTA to mask magnesium, and 10 ml of 0.01 *M* sulfosalicylic acid. Raise to pH 10–11 by adding ammonium hydroxide, dilute to 100 ml, and read at 320 nm.

THORON

Thoron, which is also called thorin, is 1-(*o*-arsenophenylazo)-2-naphthol-3,6-disulfonic acid. It forms a 1:1 complex with beryllium, read around pH 12.[101] The

[98] M. A. Azim and A. A. Ayaz, *Mikrochim. Acta* **1969**, 153–159.
[99] H. V. Meek and Charles V. Banks, *Anal. Chem.* **22**, 1512–1516 (1950).
[100] N. I. Bugaeva and L. K. Mironenko, *Zavod. Lab.* **30**, 419 (1964).
[101] L. P. Adamovich and B. V. Yutsis, *Ukr. Khim. Zhur.* **22**, 523–526 (1956); cf. L. P. Adamovich and A. P. Mirnaya, *Sovrem. Matody Anal. Met., Sb.* **1955**, 172–175.

complex has also been reported to be 2:1, thoron to beryllium.[102] There is interference by calcium, magnesium, cobalt, cupric ion, ferrous and ferric ions, manganese, nickel, thorium, and zirconium, but they are masked by EDTA. Chromic ion also interferes and is not so masked.

Procedure. Mix 15 ml of sample solution containing up to 0.05 mg of beryllium with 5 ml of 50% ammonium citrate solution and 10 ml of 5% EDTA as masking agents.[103] Add 20% sodium hydroxide solution to raise to pH 10. Add 5 ml of 0.1% solution of thoron and dilute to 50 ml. Let stand for 30 minutes and read at 522 nm.

Steel.[104] Dissolve 1 gram of sample by heating with 25 ml of 1:1 hydrochloric acid. Add 3 ml of 1:1 nitric acid and boil to decompose carbides and precipitate yellow tungstic oxide. Evaporate to dryness, add 5 ml of hydrochloric acid, and again evaporate to dryness. Take up in 10 ml of 1:1 hydrochloric acid, filter the tungstic oxide and silica, and wash with 10 ml of 1:1 hydrochloric acid. Extract the bulk of the iron with 20 ml and 20 ml of ether. Heat to expel the dissolved ether, add 10 ml of sulfuric acid, and evaporate to sulfur trioxide fumes. Take up in water and dilute to 200 ml. Add 1 ml of 0.1 *N* nitric acid, heat to boiling, and add several small portions of ammonium persulfate to oxidize chromium and vanadium to the corresponding acids. Cool, and make alkaline with 1:1 ammonium hydroxide to precipitate beryllium, aluminum, and residual iron. Chromium, vanadium, molybdenum, copper, and nickel remain in solution. Filter and wash with 1% ammonium sulfate solution. Dissolve the precipitate in 20 ml of 1:3 hydrochloric acid and make alkaline with 4% sodium hydroxide solution to precipitate iron and nickel. Filter, wash, and dilute to 100 ml. To a 5-ml aliquot, add 4 ml of a sodium borate buffer solution for pH 12.4 and 1 ml of 0.0426% solution of thoron. Read at 522 nm against a reagent blank.

Beryllium Alloys.[105] To a sample solution containing up to 0.1 mg of beryllium as chloride, nitrate, or sulfate, add 5 ml of 10% EDTA solution. Add ammonium hydroxide to raise to pH 4–5. Add 1 ml of 15% solution of acetylacetone and 2 drops of ammonium hydroxide, which should further raise the pH to 7–8. Extract with 5 ml of carbon tetrachloride by shaking for 7 minutes. Carry out three similar extractions, in each case adding 5 ml of carbon tetrachloride and 1 ml of 15% solution of acetylacetone. Extract the beryllium from the combined carbon tetrachloride extracts by shaking for 7 minutes with 7 ml of 5 *N* hydrochloric acid. Add 2 ml of perchloric acid and evaporate to fumes. Take up in 10 ml of water and add 10 ml of 0.1% solution of thoron. Add 20 ml of a sodium hydroxide–borate buffer solution for pH 12 and dilute to 100 ml. Read at 470 nm against a reagent blank.

[102]Hisanhiko Einaga and Hajime Ishii, *Anal. Chim. Acta* **54**, 113–120 (1971).
[103]R. Keil, *Z. Anal. Chem.* **262**, 273–275 (1972).
[104]L. P. Adamovich and B. V. Yutsis, *Ukr. Khim. Zh.* **23**, 784–787 (1957).
[105]V. T. Athavale, C. S. Padmanabha Iyer, M. M. Tillu, and G. M. Vaidya, *Anal. Chim. Acta* **24**, 263–269 (1961).

Beryllium Bronze.[106a] Prepare 0.078% solution of cupric sulfate pentahydrate, add 5 drops of indigo carmine indicator solution, and titrate to pH 12.4 with 4% sodium hydroxide solution. Dissolve 0.2 gram of a bronze containing less than 3% of beryllium in 15 ml of 1 : 1 nitric acid and dilute to 100 ml. Dilute a 10-ml aliquot to 100 ml. Mix 5 ml with 5 ml of 0.5% solution of disodium EDTA and 5 drops of indigo carmine indicator solution. Titrate with 4% sodium hydroxide solution until the color matches that of the 0.078% solution of cupric sulfate with the indicator at pH 12.4. Discard this. Take a fresh 5-ml aliquot, add the amount of 4% sodium hydroxide solution required for pH 12.4, and dilute to 10 ml with 0.5% solution of disodium EDTA. Mix a 3-ml aliquot of this, 3 ml of 0.00426% solution of thoron, and 4 ml of sodium borate buffer solution at pH 12.4. Read at 522 nm against a reagent blank.

A binary complex is formed by **acid chrome dark blue** with beryllium ion at pH 6.5 for reading at 620 nm.[106b] Beer's law is followed for 1.6–80 μM beryllium. There is interference by fluoride, phosphate, oxalate, ascorbate, tartrate, citrate, hydroxylamine, magnesium, zinc, manganous ion, copper, iron, nickel, chromium, hexavalent molybdenum, titanic ion, and gallium. Aluminum forms a similar but noninterfering complex at a different pH absorbing at a different wavelength.

Either beryllium or aluminum forms a 1 : 1 : 2 complex of the metal with the dye and diphenylguanidine. At pH 6 the beryllium complex can be extracted with butanol to read at 635 nm. Up to 50-fold amounts of beryllium do not interfere with extraction and reading of 35 μg of aluminum. Beer's law is followed for 0.39–10 mM beryllium. A 4-fold molar excess of the dye and 120-fold molar excess of diphenylguanidine is required. In addition to ions interfering with the binary complex there is interference by silver and thallous ions.

Acid chrome yellow N, which is 5-(4-sulfophenylazo)salicylic acid, forms a 1 : 1 complex with beryllium around pH 12.[107] For analysis of bronze, remove copper electrolytically. Mask aluminum and ferric ions with EDTA, and form the complex in sodium hydroxide–glycine buffer solution at pH 11.7 for reading at 375 nm.

Alizarin red S (mordant red 3) forms a 1 : 1 complex with beryllium at pH 5.4–5.6 having a maximum absorption at 180 nm.[108a] A twentyfold excess of reagent is required. Beer's law is followed for 0.2–4.7 ppm of beryllium. There is interference by lead, zinc, copper, cadmium, cobalt, nickel, vanadium, molybdenum, and ceric ion.

Buffered at pH 7 the 1 : 3 complex of beryllium with **3-amino-5-sulfosalicylic acid** has a maximum absorption at 370 nm and a maximum fluorescence at 465 nm.[108b] The fluorescence is read at 5–100 ng/ml. Uranyl, phosphate, and citrate ions quench the fluorescence. Ferric, aluminum, or zinc ions are masked with disodium EDTA. Other metals are tolerated for 10-fold or more.

In weakly acid solution **arsenazo III** forms a 1 : 1 complex with beryllium having

[106a] L. P. Adamovich and B. V. Yutsis, *Ukr. Khim. Zh.* **22**, 805–808 (1956).
[106b] L. G. Anisimova and E. T. Beschetnova, *Zh. Anal. Khim.* **31**, 1302–1305 (1976).
[107] L. P. Adamovich and Va Van Nyan, *Zh. Anal. Khim.* **23**, 994–1001 (1968).
[108a] Pradip K. Govil and Samir K. Banarji, *J. Inst. Chem. India* **44**, 128–130 (1972).
[108b] N. M. Alykov and A. I. Cherkesov, *Zh. Anal. Khim.* **31**, 1104–1108 (1976).

a maximum absorption at 580 nm.[109] Optimum conditions are pH 6–6.5, 1.5- to 2-fold excess of arsenazo III, and up to 0.4 μg of beryllium per ml. EDTA, tartaric acid, and triethanolamine can be added as masking agents.

Beryllium complexes with the disodium salt of **5,5′-[4,4′-biphenylenebisazo]bis (salicylic acid).**[110] At the optimum pH of 8 Beer's law is followed for 0.01–0.45 μg of beryllium per ml for reading at 450 nm. EDTA masks many interferences.

For reading as a complex with beryllium, **4-(3-carboxy-4-hydroxyphenylazo)-3-methyl-1-phenylpyrazol-4-one** has an optimum pH of 8.4.[111] With a final volume of 10 ml, 1–1.5 ml of the reagent in ethanol, 1 ml of 0.5% gelatin, and 0.6 ml of 1% EDTA solution, 0.044–0.89 μg of beryllium can be read at 530 nm or 0.022–0.356 μg/ml at 505 nm. Masking with EDTA is required if copper, cobalt, nickel, or iron is present.

The 1:1 complex of beryllium with **4-[bis(carboxymethyl)aminomethyl]-3-hydroxy-2-naphthoic acid** shows maximum fluorescence at pH 6.8 when activated at 360 nm and read at 450 nm.[112a] The emission conforms to Beer's law for 0.09–1.8 μg/ml. There are substantially no interferences.

At pH 8 the complex of beryllium with the sodium salt of **3-[4′-(3-carboxy-4-hydroxyphenylazo)-3,3′-dimethoxybiphenyl-4-ylazo]chromotropic acid** has a maximum absorption at 580 nm.[112b] The maximum absorption of the reagent is at 520 nm. Color develops in 1 hour at room temperature or in 5 minutes at 60°. Beer's law is followed for 0.015–0.45 μg of beryllium per ml. There is interference by 3-fold aluminum, 5-fold copper, 8-fold iron, or 10-fold cobalt. On addition of EDTA a ternary complex is formed read at 0.01–0.3 μg/ml. Then the interfering concentrations become 20-fold aluminum, 60-fold copper, 50-fold iron, 47-fold cobalt.

Beryllium can be read as a complex with **chrome blue K** at pH 9–10 and 600 nm.[113] Many ions are masked with disodium EDTA.

Beryllium forms a 1:2 complex with **chromal blue G** (mordant blue 55) at pH 6 and a 1:1 complex below pH 4.5.[114a] The maximum absorption is at pH 5.8–6.4 as read at 610 nm. Beer's law is followed up to 0.3 ppm of beryllium. The color is stable for 20 minutes at up to 40°. There is serious interference by cupric ion, aluminum, ferric ion, nickel, scandium, yttrium, rare earths, oxalate, fluoride, citrate, phosphate, and EDTA. The 1:2 complex adds cetrimonium chloride from a 3 m*M* solution in an acetate buffer solution at pH 5.5[114b] This is read at 626 nm for 0.012–0.12 ppm of beryllium. Fluoride, citrate, and tartrate must be absent and seriously interfering cations masked with EDTA.

Chromotrope 2C, which is disodium(*o*-carboxyphenylazo)chromotropate forms a reddish-violet 1:1 complex with beryllium at pH 6.[115] At 0.1–0.4 ppm this is read

[109] Sh. T. Talipov, L. A. Khadeeva, and R. Popova, *Izv. Vyssh. Ucheb. Zaved., Khim. Khim. Tekhnol.* **14**, 343–345 (1971).
[110] G. A. Baiulescu and Viorica Nistreanu, *Rev. Chim.* (Bucharest) **25**, 322–324 (1974).
[111] G. Popa, G. Baiulescu, N. Barbulescu, and V. A. Ilie, *Acad. Rep. Pop. Rom. Stud. Cercet. Chim.* **11**, 291–296 (1963).
[112a] B. Budesinsky and T. S. West, *Anal. Chim. Acta* **42**, 455–465 (1968).
[112b] Viorica Nistrianu, *Rev. Chim.* (Bucharest) **27**, 709–12, 983–985 (1976).
[113] L. P. Kalinichenko, *Tr. XX Godn. Nauch. Sess. Sverdlk. Med. Inst., Sb.* **22**, 52–53 (1957).
[114a] Katsuya Uesugi, *Bull. Chem. Soc. Jap.* **42**, 2998–3000 (1969).
[114b] Katsuya Uesugi and Mitsuo Miyawaki, *Microchem J.* **21**, 438–444 (1976).
[115] A. K. Majumjar and C. P. Savariar, *Z. Anal. Chem.* **176**, 170–174 (1960).

at 590 nm. **Chromotrope 2R** (acid red 29) is disodium phenylazochromotropate.[116a] The maximum absorption of the 1:1 complex with beryllium is at 533 nm and pH 6–8.8. Beer's law holds for 0.045–0.45 μg of beryllium per ml.

Beryllium forms a 1:2 complex with **9,10-dihydro-1-hydroxy-9,10-dioxo-2-anthroic acid.**[116b] Mix a sample solution containing 0.4–1.2 ppm of beryllium with 7.5 ml of 99% ethanol, 5 ml of 1.85 m*M* reagent in 99% ethanol, and 0.5 ml of 0.2 *N* sodium hydroxide. Dilute to 25 ml and read at 470–480 nm. Several cations, magnesium for example, interfere.

1,4-Dihydroxyanthraquinone-2,6-disulfonic acid has a maximum absorption at 470 nm and as a beryllium complex at 530 nm.[117] As applied to analysis of bronze, a veronal buffer for pH 7 is used and copper is masked with thiosulfate.

5,5′-(3,3′-Dimethyl-4,4′-biphenylenebisazo)bis(salicylic)acid forms a 2:1 complex with beryllium at pH 6.5 which is read at 430 nm for 0.015–0.5 μg of beryllium per ml.[118] EDTA masks aluminum, iron, cobalt, copper, and nickel. The dimethoxy analogue at pH 9.5 is read at 470 nm.

For determination of beryllium, mix 2.5 ml of 20 m*M* **2-ethyl-5-hydroxy-3-methylchromone** or **5-hydroxy-7-methoxy-2-methylchromone** in methanol with 2.5 ml of 0.3 *M* borax buffer for pH 7.5 and dilute to 25 ml.[119] After 1 hour extract with 10 ml of carbon tetrachloride, dry over sodium sulfate, activate at 405 nm, and read at 483 nm. Beer's law is followed for 0.01–0.25 μg of beryllium. The complexes with aluminum, scandium, titanium, zirconium, antimony, bismuth, gallium, and yttrium are similarly extracted.

The **fluotitanic acid–hydrogen peroxide** complex formed in dilute sulfuric acid by beryllium, titanic, and fluoride ions in the presence of hydrogen peroxide is read at 420 nm.[120] At 1.5–6 mg of beryllium in 20 ml, the maximum absorption is proportional to beryllium content with beryllium-titanium at 1:1–1:4.

Gossypin, a glycoside of the flavanol gossypetin, forms a 2:1 complex with beryllium which is read at 450 nm and pH 7 in the presence of EDTA.[121]

As reagent, mix 2.5 ml of 10 m*M* potassium **hexacyanochromate** with either 2.5 ml of 10 m*M* *p*-anisidine or 5 ml of 10 m*M* *p*-phenylenediamine hydrochloride.[122] Add 1 ml of sample solution containing beryllium up to 10 m*M*. Dilute to 50 ml with a buffer solution for pH 3. For reading, dilute an aliquot to 10 ml with the buffer solution and read at 395 nm. There is interference by mercuric, chromic, vanadyl, and hexavalent molybdenum ions.

Alternatively, mix 5 ml of the sample solution with 5 ml of 0.01 *M* hexacyanochromate and dilute to 50 ml with a buffer solution for pH 3. Expose to ultraviolet radiation for 2 hours and read 0.9–8.1 ppm of beryllium ion at 395 nm. The method is also applicable to mercuric ion, stannous ion, tetravalent vanadium, chromic ion, and molybdate.

[116a]Gr. Popa, D. Negoiu, and Claudia Vasilescu, *Acad. Rep. Pop. Rom., Stud. Cercet. Chim.* **9**, 629–639 (1961).

[116b]F. Capitan Garcia, F. Salinas, and L. M. Franquelo, *An. Quim.* **72**, 529–533 (1976).

[117]L. O. Matveev and I. S. Mustafin, *Tr. Kom. Anal., Khim., Akad. Nauk SSSR, Inst. Geokhim. Anal. Khim.* **11**, 217–222 (1960).

[118]George E. Baiulescu and Viorica Nistreanu, *Rev. Chim.* (Bucharest) **26**, 764–767 (1975).

[119]Takushi Ito and Akira Murata, *Jap. Anal.* **20**, 335–340, 1422–1427 (1971).

[120]Hisao Fukamauchi, Ryuko Ideno, Mitsuko Yanagida, and Chiyo Inotsume, *ibid.* **11**, 121–122 (1962).

[121]J. C. Bose, K. Srinwasulu, and Bh. S. V. Raghava Rao, *Anal. Chem.* **162**, 93–95 (1958).

[122]W. U. Malik and K. D. Sharma, *ibid.* **276**, 379 (1975); *Indian J. Chem.* **13**, 1232–1233 (1975).

Mix 10 ml of a sample solution containing 30–130 ppb of beryllium, 12 ml of ethanol, and 1 ml of 0.925 m*M* ethanolic **1-hydroxy-2-carboxyanthraquinone**, which is 9,10-dihydro-1-hydroxy-9,10-dioxo-2-anthroic acid.[123] Add 0.5 ml of 0.2 *N* sodium hydroxide and dilute to 25 ml. Set aside for 30 minutes, activate at 470 nm, and read the fluorescence at 580 nm. Zinc, aluminum, and yttrium interfere.

Beryllium forms a 1:1 complex with **3-hydroxy-2-naphthoic acid** at pH 6.5–7.5.[124] For beryllium in solutions of copper ores, extract with acetylacetone at pH 6–7. For beryllium in phosphoric acid pickling solution, separate at pH 1.5 on a column of Amberlite IR-120. Activate at 375 nm and read the fluorescence at 470–500 nm. Most of the interferences can be masked with EDTA.

4-Hydroxy-3-(salicylideneamino)benzenesulfonic acid complexes with beryllium at pH 9.7.[125] The fluorescence at 430 nm when activated at 343 nm is read for 2–400 ng of beryllium per ml.

At pH 1.8–3 beryllium forms a 1:2 complex with ***N*-methylanabasine-(α′-azo-6)-*m*-aminophenol**, which is 5-amino-2-[3-(1-methyl-2-piperidyl)-2-pyridylazo]phenol.[126] The complex obeys Beer's law for 0.1–0.8 μg of beryllium per ml. To the sample, add 4 ml of 0.2% solution of the reagent in ethanol and 10 ml of a potassium chloride–hydrochloric acid buffer solution for pH 2.2. Dilute to 25 ml and read at 530 nm against a blank. Fluoride interferes, but many cations are masked by EDTA, of which 150-fold is tolerated.

A beryllium complex with **methyl cyanide** can be extracted for reading 0.0005–0.2% of beryllium in aluminum and its alloys.[127]

Naphthazarin as a complex with beryllium develops color in 15 minutes for reading at 580 nm and is stable for 30 minutes.[128] An acetate buffer solution for pH 5.6 is added. Iron, aluminum, cobalt, nickel, and zinc are masked with tartaric acid; copper is masked with rhodamine ion and thiosulfate. Acetone can improve the stability of the color to 18 hours.

Neothoron, which is *o*-arsonophenylazochromotropic acid, complexes with beryllium at pH 6.[129] Color develops within 5 minutes and at 0.15–1 ppm is read at 570 nm. Another complex at pH 11.3 is read at 580 nm for up to 0.15 ppm.

For determination of beryllium, add to 5 ml of sample solution containing 2–6.5 μg of beryllium, 2 ml of 5% disodium EDTA solution, and an equivalent amount of magnesium sulfate in 1 ml of solution.[130] Add 20 mg of **2-oxocyclopentanecarboxanilide** dissolved in a minimal amount of 2% sodium hydroxide solution. Neutralize with 1:7 nitric acid and dilute to 4 ml. Warm and add 1:7 ammonium hydroxide to above pH 6.8. Extract with 5 ml and 5 ml of methyl isobutyl ketone and read the combined extracts at 332 nm.

The precipitate of beryllium with **2-phenylacylpyridine** when extracted with chloroform is read at 380 nm.[131] Interferences by copper and aluminum in analysis

[123] F. Capitan Garcia, F. Salinas, and L. M. Franquelo, *Anal. Lett.* **8**, 753–761 (1975).
[124] Krzysztof Kasiura, *Chem. Anal.* (Warsaw) **20**, 389–395 (1975).
[125] Kiyotoshi Morisige, *Anal. Chim. Acta* **73**, 245–254 (1974).
[126] R. Kh. Dzhiyanbaeva, Sh. T. Talipov, and N. T. Turakhanova, *Nauch. Tr. Tash. Univ.* **1967** (284), 36–41.
[127] G. Matelli and V. Vicentini, *Alluminio* **32**, 377–381 (1963).
[128] G. G. Karanovich, *Tr. Vses. Nauch.-Issled. Inst. Khim. Reakt.* **1956** (21), 43–47.
[129] Shozo Shibata, Fukuo Takeuchi, and Teiichi Matsumae, *Bull. Chem. Soc. Jap.* **31**, 888–889 (1958).
[130] N. K. Chaudhuri and J. Das, *Anal. Chim. Acta* **57**, 193–199 (1971).
[131] Erhard Uhlemann and P. Fritzsche, *Z. Anorg. Chem.* **327**, 79–83 (1964).

of magnesium alloys is avoided by addition of EDTA. In other applications cobalt, nickel, iron, magnesium, and fluoride are masked with cyanide and acetic acid after reduction of iron with sulfite.[132]

A blue complex of beryllium with **plasmocorinth B** (mordant blue 13) at pH 6.5 is read at 610 nm and obeys Beer's law for 0.02–0.3 ppm of beryllium.[133] Add the sample to 10 ml of sodium acetate–acetic acid buffer solution and 2 ml of 0.1% solution of the reagent. Dilute to 25 ml. The color is stable for less than 30 minutes.

Beryllium at 0.05–0.3 ppm is read with **pontachrome azure blue B** at 520 nm and pH 5.8.[134] Interferences are precipitated with oxine.

Pontacyl violet 4 BSN (acid violet 3) complexes with beryllium at pH 6.6–7.2 to give a maximum absorption at 620 nm.[135] With 2.5 ml of 0.1% solution of reagent, 5 ml of 0.1 *M* sodium acetate buffer for pH 6.8, and 2.5 ml of acetone diluted to 25 ml, Beer's law is followed for a content up to 0.3 ppm. There is interference by copper, chromium, aluminum, scandium, hexavalent uranium, iron, and phosphate.

The sodium salt of **3-(2-pyridylazo)chromotropic acid** complexes with beryllium at pH 6.5 to give a maximum absorption at 580 nm, obeying Beer's law for 0.25–2 ppm.[136a] There is interference by zinc, cadmium, copper, lead, mercury, nickel, cobalt, iron, and thorium, but not by aluminum. The sodium salt of **3-(3-pyridylazo)chromotropic acid** similarly complexes with beryllium at pH 6.5 to give a maximum absorption at 590 nm. This obeys Beer's law for 0.05–0.8 ppm. In this case there is interference by copper, nickel, cobalt, aluminum, oxalate, and tartrate. With the sodium salt of **3-(2-carboxy-3-pyridylazo)chromotropic acid**, the maximum complex with beryllium is also at pH 6.5 and the maximum absorption at 590 nm over the range of 0.1–2 ppm. Interference is only by iron, aluminum, oxalate, and tartrate.

At pH 12.7–12.9 in 50% ethanol, the blue fluorescence of beryllium with **2-2-(pyridyl)phenol**, also known as *o*-pyridinophenol, is activated at 365 nm.[136b] The limit of detection is 1.6 ng/ml, and Beer's law holds up to 0.54 μg/ml. Disodium EDTA masks many interferences. Germanium interferes slightly. There is serious interference by ferric, zirconium, chromic, and fluoride ions.

Treat a solution of beryllium with 1 ml of 0.5 m*M* **quinizarinsulfonic acid** in ethanol and 2 ml of 0.1 *N* acetic acid; dilute with water to 6 ml, then dilute to 50 ml with ethanol.[137] The concentration of beryllium should now be 1–7 ppb. Set aside for 1 hour, activate at 475 nm, and read the fluorescence at 475 nm.

2-Salicylideneaminobenzenearsonic acid will determine beryllium by luminescence down to 0.9 ng/ml.[138] A 100-fold excess of aluminum, copper, or zinc is tolerated.

When beryllium is complexed with **SPADNS** in borate buffer solution for pH 6.3, it is read at 2–10 mg/ml at 580 nm.[139] The complex forms within 15 minutes. There is interference by iron, aluminum, and copper.

[132] Erhard Uhlemann, Joachim Hoppe, and Karl Friedrich Ratzsch, *Z. Chem.* **9**, 315–316 (1969).
[133] Katsuya Uesugi and Yukiteru Katsube, *Bull. Chem. Soc. Jap.* **39**, 194 (1966).
[134] Yukiteru Katsube, Katsuya Uesugi, and John H. Yoe, *ibid.* **34**, 72–76 (1961).
[135] Tsuninobu Shigematsu, Masayuki Tabushi, and Katsuya Uesugi, *Jap. Anal.* **15**, 1369–1373 (1966).
[136a] A. K. Majumdar and A. B. Chatterjee, *Z. Anal. Chem.* **202**, 323–331 (1964).
[136b] L. Kabrt and Z. Holzbecher, *Collect. Czech. Chem. Commun.* **41**, 540–547 (1976).
[137] A. Guiraun and J. L. Vilchez, *Quim. Anal.* **29**, 265–271 (1975).
[138] Sh. T. Talipov, A. T. Tashkhodzhaev, L. E. Zel'tser, and Kh. Khikmatov, *Nauch. Tr. Tash. Univ.* **1972** (419), 89–93.
[139] Kuang-pi Pai and Shu-shen Luan, *Chem. Bull.* (Peking) **1965** (5), 49–51.

Beryllium complexes with **tetracycline** to give greatly increased fluorescence in the pH range 6.2–7.5.[140] The sensitivity is in the nanogram range.

At pH 9 the fluorescence is greatest with 1.5–2 ml of 0.02% solution of tetracycline hydrochloride and more than 1.5 ml of 0.02% methanolic solution of 5,5-diethyl-2-thiobarbituric acid in 50 ml of solution containing 5 ml of 0.05 *M* borate buffer. A 1 : 1 : 1 complex is formed. After aging for 10 minutes, activation at 406 nm gives the maximum fluorescence at 506 nm. In that volume Beer's law is followed for 5–15 μg of beryllium. There is interference by ferric ion, magnesium, calcium, barium, and aluminum.

The 1 : 1 complex of beryllium in 0.01–0.1 m*M* solution of **thymol blue** shows a maximum absorption against the reagent blank at pH 5 and 500 nm.[141a] The solution must be heated at 95° for 15 minutes to develop the full color, which is then stable for 6 hours. Beer's law is followed for up to 9 μg of beryllium per ml.

To an acidic sample solution containing 12–55 μg of beryllium, add an excess of a solution of **uramildiacetic acid**, which is (hexahydro-2,4,6-trioxo-pyrimidin-5-yl)iminodiacetic acid.[141b] Add 5 ml of acetate buffer solution for pH 4.2. Dilute to 25 ml and read the stable 1 : 1 complex at 267 nm. Similar reactions occur with cobalt, titanic ion, hexavalent uranium, cadmium, cupric ion, nickel, chromic, and ferric ions.

Beryllium forms a 1 : 1 complex with **xylenol orange** below pH 5 and a 1 : 2 complex at higher pH.[142] To the sample solution containing up to 6 μg of beryllium, add 10 ml of perchloric acid–hexamine buffer solution for pH 5.8 and 2.5 ml of 0.001% solution of xylenol orange. Dilute to 25 ml and read at 495 nm against a reagent blank. The color of the complex is reduced by phosphate, oxalate, citrate, and fluoride. Masking with EDTA avoids interference by cations other than cobalt, aluminum, bismuth, thorium, and zirconium.

[140]T. V. Alykova, A. I. Cherkesov, and N. M. Alykov, *Izv. Vyssh. Ucheb. Zaved., Khim. Khim. Tekhnol.* **15**, 1107–1109 (1972).

[141a]K. C. Srivastava and Samir K. Banerji, *Analysis* **1**, 132–134 (1972).

[141b]M. Jose Presas-Barrosa, F. Bermejo-Martinez, and J. A. Rodriguez-Vazquez, *Anal. Chim. Acta* **88**, 395–398 (1977).

[142]Makoto Otomo, *Bull. Chem. Soc. Jap.* **38**, 730–734 (1965).

CHAPTER SEVENTEEN
CHROMIUM

The most widely used photometric reagent for chromium is *s*-diphenylcarbazide, also known as 1,5-diphenylcarbohydrazide.

Chromium in bauxite or red mud is oxidized by fusion with sodium peroxide and sodium hydroxide.[1] Aluminum is dissolved in 1 : 3 25% sulfuric acid–nitric acid or in 15% sodium hydroxide solution. In either case chromium is oxidized to the chromate with hydrogen peroxide. Concentration of the chromate and separation of interfering ions may be by (1) extraction of interferences as cupferrates by chloroform from an acid medium with ferric ion as a carrier, (2) extraction of interferences at pH 3.4–4.2 in the presence of ferric ion as 8-hydroxyquinolates in chloroform, or (3) separation from 3 *M* hydrochloric acid on Varion AD ion exchange resin in chloride form and elution with 0.005 *M* hydrochloric acid. The chromate is determined by *o*-dianisidine, *N*,*N*-dimethyl-*p*-phenylenediamine, or 3,3′-dimethylnaphthylidine.

In 20 m*M* acridine solution at pH less than 4, the quenching effect of chromate ion on the fluorescence will determine less than 1 μg/ml.[2] Halides have the same effect.

For chromium as the chromate read along with copper, nickel, and cobalt developed with biuret and solved by simultaneous equations, see Copper.

For a solution of nonmetallic inclusions in sheet steel for determination of chromium, see Aluminum. For chromium in thorium compounds by oxine, see Aluminum.

For chromium by 2-mercaptobenzoic acid, see separation by that reagent for determination by 1,5-diphenylcarbazide.

[1] O. N. Klug and A. I. Metlenko, *Acta Chim. Hung.* **49**, 123–130 (1966).
[2] K. P. Stolyarov, N. N. Grigor'ev, and G. A. Khomenok, *Vest. Leningr. Gos. Univ.* **1972** (22), *Fiz. Khim.* (4), 120–126.

ACETYLACETONE

Shaking a chromic solution at a low pH with 1 : 1 acetylacetone-chloroform easily extracts aluminum, ferric ion, titanic ion, hexavalent molybdenum, and vanadium in the tri-, tetra-, and pentavalent forms without extracting chromium. At a pH of 7 and an elevated temperature, the chelate of chromium and acetylacetone is formed.

Procedures

Ferrous Alloys[3]

DIRECT DETERMINATION. Dissolve a 1-gram sample with 50 ml of 1 : 1 hydrochloric acid and 20 ml of 1 : 1 sulfuric acid. Evaporate to about 25 ml and oxidize with 2.5 ml of nitric acid. Evaporate until salts start to separate, and cool. Add 50 ml of water and heat to 100° to dissolve. Filter, and wash the residue with 1 : 50 sulfuric acid. Most of the silica, tungsten, niobium, and tantalum remain on the paper. Evaporate to about 25 ml and cool. Add 20 ml of ammonium hydroxide from a buret with mixing, such that ferric hydroxide does not precipitate. Cool in ice and add more ammonium hydroxide to raise the pH to 2. Extract with 25 ml of acetylacetone and discard the extract. Readjust the aqueous layer to pH 2. Add 25 ml of chloroform, shake for 2 minutes and discard the extract. Extract the aqueous layer with 50 ml of 1 : 1 acetylacetone-chloroform and discard this extract. Neutralize the aqueous layer to litmus with ammonium hydroxide and add 2 ml in excess. Add 30 ml of acetylacetone and a magnetic stirring bar. Reflux with stirring for 30 minutes to form chromic acetylacetonate. Extract the aqueous layer with 25 ml of chloroform, add this to the chromic acetylacetonate solution, and discard the aqueous layer. Backwash the organic extract four times with 50-ml portions of 1 : 5 sulfuric acid. Extract the combined backwashings with chloroform and add this to the organic extract. Add 2 grams of sodium sulfate and dilute the organic extract to 50 ml with acetylacetone. Read at 560 nm.

INDIRECT DETERMINATION BY DIPHENYLCARBAZIDE. If the chromium in the original sample was less than 2 mg, the absorption of the acetylacetonate is too low for accurate reading and this indirect method becomes necessary. Transfer an aliquot of the organic extracts above, to contain not more than 0.3 mg of chromium, and boil off the solvent at 100°. Add 5 ml of nitric acid to the dry residue and evaporate at 100°. Add 5 ml of sulfuric acid and evaporate to strong fumes of sulfur trioxide. Add nitric acid dropwise to the fuming mixture to destroy the last traces of organic matter. Cool and boil with 100 ml of water. Add 0.4 ml of 0.3% solution of potassium permanganate and continue to boil for 10 minutes. If the pink color disappears, add additional drops of the permanganate solution. Add 6 ml of 1 : 9 hydrochloric acid to destroy excess permanganate and boil for 5 minutes after the color disappears. Cool, dilute to about 230 ml, and add 5 ml of 0.25% solution of diphenylcarbazide in ethanol. Dilute to 250 ml, allow 20 minutes for color development, and read at 525 nm.

---

[3] James P. McKaveney and Harry Freiser, *Anal. Chem.* **30**, 1965–1968 (1958).

ALUMINUM GLYCINATE

Chromate complexes with aluminum glycinate, known as Alamine 336-S, for extraction into chloroform. The method is applicable to alloys or the product of fusion of silicates containing 0.18–12.7% chromium.

Procedure. *Alloys.*[4a] Dissolve 0.5 gram of sample in 10 ml of hydrochloric acid and 3 ml of nitric acid. Evaporate nearly to dryness at 100°, add 1 ml of 1:1 sulfuric acid, and evaporate to sulfur trioxide fumes. Cool, and dilute to 100 ml. Boil a 10-ml aliquot with 0.5 gram of potassium iodate for 5 minutes. Cool and add 10 ml of 2 *N* sulfuric acid. Add 3 ml of saturated solution of sodium chloride and dilute to 50 ml. Extract with 5 ml of 5% solution of Alamine 336-S in chloroform. Filter, and after 5 minutes read at 450 nm.

2-AMINOPHENOL

Hexavalent chromium at 0.5–10 ng/ml catalyzes the oxidation of 2-aminophenol by hydrogen peroxide at pH 5.6–5.8. The optimum concentration of 2-aminophenol is 0.6–1 m*M*, that of hydrogen peroxide 25–40 m*M*. The absorption of the oxidation products increases linearly for 5–8 minutes. The catalytic reaction is accelerated to a greater extent than the noncatalytic reaction by *M* chloride, nitrate, or sulfate ion.

Procedure. *Silica.*[4b] Evaporate 0.2 gram of sample in platinum to dryness with 2 ml of hydrofluoric acid. Dissolve the residue in 2 ml of 4% sodium hydroxide solution. Heat with 0.5 ml of 3% hydrogen peroxide to oxidize to chromate. Neutralize with *M* nitric acid and dilute to 5 ml. To three 1-ml portions add 0, 0.2 ml, and 0.4 ml, respectively, of a solution containing 0.1 μg of chromium per ml. Also prepare a reagent blank. To each add 1 ml of saturated solution of potassium chloride, 1 ml of 0.1 *M* EDTA, 0.1 ml of 3% hydrogen peroxide, and 0.1 ml of *M* hexamine. Dilute each to 4.6 ml and add 0.4 ml of 10 m*M* 2-aminophenol in 0.04 *M* nitric acid. Read at 430 nm periodically over 3–4 minutes.

AMMONIUM PYRROLIDINE-1-CARBODITHIOATE

The complex of this reagent with chromate tolerates 2.5-fold molybdate or nickel and 5-fold vanadate or silver. There is interference by cupric, ferrous, ferric, cobalt, mercuric, lead, and stannic ions.

[4a] Jiri Adam and Rudolf Pribil, *Hutn. Listy* **25**, 580–581 (1970).
[4b] S. U. Kreingol'd, A. N. Vasnev, and J. V. Serebryakova, *Zavod. Lab.* **40**, 6–8 (1974).

Procedure.[4c] Mix a sample solution containing not more than 35 μg of chromium with 3 ml of hydrochloric acid and 1 ml of 0.5% solution of the reagent. Dilute to 20 ml and heat at 100° for 2 minutes. Cool, dilute to 25 ml, and read at 265 nm against a reagent blank.

ANTIPYRINYL-BIS-*p*-(4-METHYLBENZYLAMINO)PHENYL METHANOL

The complex of this reagent with chromic ion is read in 0.5–1.5 *N* hydrochloric acid.[5] Beer's law is followed for 2–12 μg of chromium per ml in the extract. There is interference by tin, antimony, bismuth, mercury, and lead.

Procedure. *Steel.* Dissolve 0.2 gram of sample in 15 ml of 2 *N* sulfuric acid and 0.2 ml of nitric acid. Boil off oxides of nitrogen and add 50 ml of water. Add 0.5 gram of lead dioxide and boil for 10 minutes. While hot, neutralize with 20% sodium hydroxide solution and add 4 ml in excess. Filter and wash the precipitate. Dilute the filtrate and washings to 100 ml. Neutralize a 25-ml aliquot with 1 : 1 hydrochloric acid and add 4 ml in excess. Add 5 ml of 0.02% solution of the reagent in *N* hydrochloric acid and shake for 3 minutes with 25 ml of benzene. Centrifuge the organic layer and read with a green filter.

2,2′-BIPYRIDINEPEROXYCHROMATE

The complex of peroxychromate ion with 2,2′-bipyridine is conveniently formed and read in ethyl acetate. The readings conform to Beer's law up to 10 ppm of chromium. The peroxychromate ion is formed at 10° or below in aqueous solution and even then is somewhat unstable. The solution in ethyl acetate is stable for at least 30 minutes. The complex with 2,2′-bipyridine is then stable at room temperature. The following ions must be absent: antimony, arsenous, ferrocyanide, iodide, ferrous, nitrite, permanganate, sulfite, thiocyanate, thiosulfate, and stannous.

Procedure. Mix an aliquot of less than 20 ml of neutral sample containing 0.05–0.25 mg of chromium as chromate with 1 ml of 2 *N* sulfuric acid and dilute to 20 ml.[6] Add 20 ml of ethyl acetate and cool at 10° for 30 minutes. Add 1 ml of 3% solution of hydrogen peroxide, also precooled to 10°, and shake for 30 seconds. Discard the aqueous layer. Add 10 ml of 0.01% solution of 2,2′-bipyridine precooled to 10° and shake for 30 seconds. Dilute the ethyl acetate layer to 25 ml and read at 308 nm against a reagent blank.

[4c] Yoshikazu Yamamoto, Takumi Murata, and Shunzo Ueda, *Jap. Anal.* **25**, 851–854 (1976).
[5] V. Kh. Aitova, *Uch. Zap. Perm. Univ.* **25**, 117–119 (1963).
[6] Gordan A. Parker and D. F. Boltz, *Anal. Chem.* **40**, 420–422 (1968).

CADMIUM IODIDE—STARCH

Chromate reacts with a cadmium iodide–starch reagent to give the blue starch-iodide complex. A desirable pH is 0.7–0.9.

Procedure. Dissolve 1.1 gram of cadmium iodide in 34 ml of water and boil gently for 15 minutes to drive off any free iodine, maintaining the volume constant.[7] Boil 0.26 gram of soluble starch in 46 ml of water for about 5 minutes, which should give a clear solution. Add the starch solution to the gently boiling cadmium iodide solution and boil for 5 minutes. Filter through glass fiber or paper, cool, and dilute to 100 ml.

Dilute a sample containing 0.5–5 μg of chromium in 2.2 ml of 1.63 *N* sulfuric acid to 9 ml. Add 1 ml of reagent, set aside for 20 minutes, and read at 610 nm.

CATALYSIS OF THE CERIC–TELLUROUS ACID REACTION

The catalytic effect on oxidation of tellurous acid to telluric acid is a method of determination of the concentration of chromic ion. Allowing 1 hour by the technic below measures 0.1–1 ppm of chromium, 3 hours measures 0.01–0.1 ppm, and 5 hours measures 0.001–0.014 ppm. There is interference by iodates, manganous, or ferrous ion, and any compounds that reduce ceric or chromic ion.

Procedure. For determination of 0.002–1 ppm of chromic ion, mix 1 ml of 0.01 *M* ceric sulfate in *N* sulfuric acid, 2 ml of 0.2 *M* potassium tellurite, and 2 ml of 0.5 *N* sulfuric acid.[8] Heat at 70° in a thermostat and add 1 ml of sample. After a measured period such as 1, 3, or 5 hours, according to the chromium concentration, stop the reaction by cooling and adding 1 ml of sulfuric acid. The decrease in absorption is a measure of the chromic ion concentration.

Water. To 500 ml of sample containing as little as 75 ng of chromic ion, add 0.5 gram of oxine in acetic acid and adjust to pH 3 with ammonium hydroxide or sulfuric acid. Reflux for 30 minutes and cool. Extract the chromic complex with 10 ml, 10 ml, and 10 ml of chloroform. The manganese complex is not extracted. Wash the combined chloroform extracts by shaking for 10 minutes with 20 ml of *M* ammonium sulfate adjusted to pH 1 with sulfuric acid. Evaporate the chloroform layer to dryness and dissolve the residue with 2 ml of perchloric acid. Add 2 ml of 5 *N* sulfuric acid containing 0.1 mg of mercuric ion and evaporate to sulfur trioxide fumes. Take up in water and dilute to 5 ml. Use a 1-ml aliquot for the general procedure.

---

[7] Geoffrey Halliwell, *ibid.* **32**, 1041–1042 (1960).
[8] Taitiro Fujinaga and Takejiro Takamatsu, *J. Chem. Soc. Jap., Pure Chem. Sect.* **91**, 1159–1169 (1970).

CATECHOL VIOLET

The complex of chromium as chromate is read with catechol violet. Large amounts of aluminum and titanium can be masked with fluoride. Vanadium and iron interfere.

Procedures

Chromic Ion and Chromate.[9] To a sample solution containing 15–85 μg of chromium in either form, add 5 ml of 10% sodium perchlorate solution, 10 ml of 5% sodium acetate solution, and 20 ml of 0.001 M catechol violet. Adjust to pH 5.4 with perchloric acid; absorption is high at this pH and small changes in pH do not affect it. Heat at 80–90° for 20 ± 1 minutes and cool quickly. Check the pH and dilute to 50 ml. Read at 590 nm against a reagent blank.

Anodic Powders.[10] The sample referred to is the product of electrolytic dissolution of titanium alloys. Dissolve 0.01 gram of sample in 15 ml of 1 : 2 sulfuric acid to which a few drops of nitric acid are added. Evaporate to sulfur trioxide fumes, add 3 ml of water, and evaporate to sulfur trioxide fumes. Repeat that step once more. Take up in water and dilute to 50 ml. Treat a 5-ml aliquot with 2 ml of 10% sodium perchlorate solution, 5 ml of 5% sodium acetate solution, and 1 drop of 0.1% solution of catechol violet. Adjust to pH 5–6 with 10% solution of sodium hydroxide and add 2 ml of 0.1% solution of catechol violet. Heat at 90° for 20 minutes. Then boil, cool, and dilute to 25 ml. Read at 605 nm against a reagent blank.

CHROMATE

The absorption of chromate ion or in acid solution as bichromate ion is one of the methods applicable to samples containing relatively high percentages of chromium.

Chromic ion in a solution acidified with nitric acid is oxidized to bichromate by silver oxide, boiled to destroy the excess of silver oxide, and read at 349 nm.[11]

By what amounts to an electrical short-circuit, chromic ion in alkaline solution is converted to chromate for reading as such.[12] Chromate ion in 0.1 N potassium hydroxide solution can be read at 373 nm without interference by up to 8 mM hydrogen peroxide.[13]

In reading chromium as chromate at 565 nm and manganese as permanganate at 440 nm, 0.7 N sulfuric acid and 0.6 N phosphoric acid are appropriate.[14]

[9] G. De Angelis and E. Chiacchierini, *Ric. Sci., Riv.* **36**, 53–55 (1966).
[10] R. D. Golubtsova and A. D. Yaroshenko, *Zavod. Lab.* **36**, 147–148 (1970).
[11] F. Lucena Conde, J. Hernandez Mendez, and L. Gonzalez Lopez, *Inf. Quim. Anal. Pura Apl. Ind.* **24**, 44–48 (1970).
[12] A. Lipcinsky and I. Kuleff, *Z. Anal. Chem.* **191**, 260–267 (1962).
[13] Farhataziz and G. A. Mirza, *Analyst* **90**, 509–510 (1965).
[14] Takuji Kanno, *Nippon Kinzoku Gakkaishi* **18**, 625–629 (1954).

For determination of chromium in tin, a solution is oxidized to chromate and extracted with methyl isobutyl ketone; ultimately the chromate is read at 375 nm.[15]

For determination of chromium in a solution of aluminum or aluminum alloys, oxidize with ammonium persulfate in the presence of silver nitrate.[16] Destroy the manganate ion by addition of hydrochloric acid and read the chromate ion at 565 nm.

For chromium in leather, dissolve the ash in nitric acid, add perchloric acid, and read at 395 nm.[17] Iron does not interfere.

Procedure. Dissolve a sample containing up to 400 mg of chromium in 1:1 hydrochloric acid, adding nitric and perchloric acid as may be needed according to the nature of the sample.[18] Dilute to 100 ml and take a 10-ml aliquot. Add 35 ml of perchloric acid and boil down to 30 ml. Cool and dilute to 250 ml. Read at 400 nm.

Steel.[19] Heat 0.1 gram of sample with perchloric acid until fuming ceases. Add 50 ml of water, filter, and dilute to 100 ml. Mix a 25-ml aliquot with 4 ml of phosphoric acid and dilute to 50 ml. Read at 440 nm.

Vanadium Steel.[20] Dissolve a 1-gram sample with 20 ml of 1:5 sulfuric acid and 4 ml of 1:1 phosphoric acid. After reaction ceases, add 2 ml of nitric acid and boil off oxides of nitrogen. Dilute to 50 ml and add 1 ml of 1.7% silver nitrate solution and 5 ml of 20% potassium persulfate solution. Boil until the purple color of permanganate is present and destroy that color with 5 ml of 1:1 hydrochloric acid. Evaporate to about 30 ml and neutralize with 24% sodium hydroxide solution; about 30 ml will be required. Filter, wash the precipitate with 10 ml and 10 ml of water, and dilute to 100 ml.

Mix a 5-ml aliquot with 5 ml of 1:1 hydrochloric acid. Shake for 30 seconds with 25 ml of methyl isobutyl ketone and read the organic layer at 444 nm as a measure of chromate ion.

Add 0.5 gram of sodium metabisulfite to the aqueous layer and boil for 10 minutes. Cool and dilute to 10 ml. Read vanadium at 760 nm.

Copper-Chromium Alloys.[21] Dissolve a 0.5-gram sample in 10 ml of nitric acid. Add 12 ml of 1:1 sulfuric acid and 2 ml of phosphoric acid. Heat to a clear solution evolving sulfur trioxide fumes. Add 20 ml of perchloric acid and heat for 75 seconds. Cool at once and dilute to 200 ml. Read at 430 nm.

Chromite.[22] Mix 0.2 gram of sample with 3 grams of 5:5:10:2 sodium

[15] V. G. Pogodaeva and K. P. Stolyarov, *Uch. Zap. Leningr. Gos. Univ. A. A. Zhdanova* No. **297**, *Ser. Khim. Nauk* No. **19**, 170–174 (1960).
[16] E. Szezesny, *Aluminum* **33**, 263 (1957).
[17] C. H. Perrin and P. A. Ferguson, *J. Am. Leather Chem. Assoc.* **65**, 424–425 (1970).
[18] Kichinosuke Hirokawa, *Sci. Rep. Res. Inst., Tohoku Univ. Ser. A* **13**, 419–425 (1961).
[19] V. M. Bhuchar and V. P. Kukreja, *Met. Metal Form.* **40**, 91–92 (1973); cf. Toru Nozaki and Kaoru Ueno, *Jap. Anal.* **8**, 185–189 (1959).
[20] S. A. Katz, W. M. McNabb and J. F. Hazel, *Anal. Chim. Acta* **25**, 193–199 (1961).
[21] H. Wiedmann, *Metall.* **12**, 1000 (1958).
[22] Z. L. Nazarenko, *Sb. Nauch. Tr. Ukr. Nauch.-Issled. Inst. Ogneupor.* **1965** [8(55)], 307–314.

carbonate–potassium carbonate–magnesium oxide–sodium tetraborate. Sinter at around 1100° for 20 minutes, cool, and dissolve in 90 ml of hot 1:5 sulfuric acid. Dilute to 250 ml and read at 430 nm. An alternative is to form the complex by heating an aliquot with EDTA solution as described elsewhere.

Basic Slag.[23] Fuse 2 grams of sample with 20 ml of sodium peroxide at 700–800°. Extract the melt with water and dilute to 100 ml. Dilute a 10-ml aliquot to 100 ml and read at 380 nm.

Silicate Minerals.[24] Fuse a 0.5-gram sample with 2 grams of sodium carbonate and 2 grams of sodium nitrate. Leach with 20 ml of hot water, filter, and wash the residue with hot 1% sodium carbonate solution. Dilute to 50 ml and read at 430 nm.

Match-Head Composition.[25] As a coagulating agent, dissolve 12.5 grams of potassium aluminum sulfate dodecahydrate and 25 grams of ferric ammonium sulfate dodecahydrate with 3 ml of sulfuric acid in water and dilute to 1 liter.

Boil a 2-gram sample with 120 ml of water for 15 minutes. Add 0.5 ml of ammonium hydroxide and 100 ml of the coagulating agent. Filter, and wash the residue with hot water. Dilute the cooled filtrate to 250 ml. Take a 25-ml aliquot, add 2 ml of ammonium hydroxide and 0.5 gram of ammonium persulfate, and heat until there is no further change of color. Add 10 ml of 1:3 sulfuric acid, dilute to 50 ml, and read at 565 nm.

Biological Samples.[26] Oxidize 100 grams of sample with sulfuric, nitric, and perchloric acids to complete destruction of organic matter. Evaporate to sulfur trioxide fumes, take up in water, and dilute to 100 ml. Add 0.2 gram of sodium sulfite and heat to 100°. Cool and raise to pH 6 with 40% sodium hydroxide solution. Add 20 ml of 25% solution of pyridine and heat to coagulate the precipitate. Filter and wash with hot 1% solution of ammonium chloride. Dissolve the precipitate with 25 ml of hydrochloric acid containing a crystal of potassium chlorate and evaporate nearly to dryness. Cool. Add 3 ml of hydrochloric acid saturated with ether and 0.2 ml of water. Extract with 15 ml and 15 ml of ether and discard the extracts.

Dilute the aqueous layer to 50 ml and add 25 ml of 0.1 *M* EDTA. Add 2 drops of phenolphthalein indicator solution, heat to 100°, and neutralize with ammonium hydroxide. Boil for 4 minutes, cool, and dilute to 100 ml. Mix a 10-ml aliquot with 10 ml of acetate buffer solution for pH 5 and read at 570 nm.

Feeding Stuff.[27] As a digestion mixture, add 10 grams of sodium molybdate dihydrate in 150 ml of water to 150 ml of sulfuric acid and 200 ml of perchloric acid.

To 0.5 g of sample containing 0.3–4% of chromic oxide, add 5 ml of digestion

[23] Shigeo Wakamatsu, *Tetsu to Hagane* **46**, 492–497 (1960).
[24] G. Csajaghy, *Magy. Kem. Lapja*. **22**, 333–334 (1967).
[25] Zygmunt Marczenko and Zofia Skorko-Trybulowa, *Chem. Anal.* (Warsaw) **5**, 71–77 (1960).
[26] V. D. Yablochkin, *Sudebnomed. Ekspert.* **6** (3), 45–49 (1963).
[27] H. Petry and W. Rapp, *Z. Tierphysiol. Tierernaehr. Futtermittelk.* **27**, 181–189 (1971).

mixture. Heat until the color changes to orange-red, indicating oxidation to chromate, usually 5 minutes. To the cooled solution, add 2 ml of perchloric acid and heat for a further 5 minutes. Cool, dilute to 15 ml, neutralize with 32% solution of sodium hydroxide, and dilute to 25 ml. Centrifuge to separate the precipitate of silicates. Wash it with 5 ml and 5 ml of 32% sodium hydroxide solution. Dilute the clear greenish-yellow liquid with 0.4% sodium hydroxide solution to contain 2–10 mg of chromic oxide per liter. Read at 370 nm against water.

Feed or Feces.[28] Heat 0.4 gram of dried sample with 1.5 ml of sulfuric acid, 1.5 ml of perchloric acid, and 2 ml of nitric acid until organic matter is destroyed. At that point the solution will be orange with bichromate. Take up in water and centrifuge to remove silica. Dilute to a known volume and read at 430 nm.

Alternatively,[29] ash in a nickel crucible. Then fuse the ash with potassium nitrate, sodium carbonate, and sodium hydroxide. Take up in water, dilute to a known volume, and read the chromate ion at 400 nm.

CHROMIC ION

Chromium can be read with identical results as the chromic ion of chromic sulfate in water or in 20% sulfuric acid, with maxima at 410 and 580 nm.[30] Increased concentration of sulfuric acid displaces the maximum and somewhat increases the optical density by varying the composition of aquocomplexes and formation of sulfate complexes. Between 400 and 700 nm, copper and manganese ions do not interfere. Cobalt, nickel, and iron absorb to some degree. Beer's law is observed for 2, 3, and 4 *M* chromic sulfate in sulfuric acid at 570 nm.

Bivalent chromium does not absorb at 504 nm.[31] Therefore that wavelength can be used for chromic ion in both the green and violet forms in 0.25 *M* sulfuric acid.

To determine chromic ion in a chromate solution, boil to convert to the green form, adjust to pH 1, and read.[32] A typical solution contains 0.001–4 grams of chromic oxide, 100–200 grams of chromic acid, and 50–200 grams of sodium sulfate per liter.

Procedures

Steel. Dissolve a 0.5-gram sample in 30 ml of 4 *M* sulfuric acid and add sufficient nitric acid to oxidize the ferrous ion. Boil off brown fumes and dilute to

[28] F. H. Cheong and F. J. Salt, *Lab. Pract.* **17**, 199–200 (1968); cf. Jadwige Czarnocki, I. R. Sibbald, and E. V. Evans, *Can. J. Agr. Sci.* **41**, 167–179 (1961).

[29] Germain J. Brisson, *Can. J. Agr. Sci.* **36**, 210–212 (1956); cf. C. J. K. Mink, R. H. G. C. Schefman-Van Neer, and L. Habets, *Clin. Chim. Acta* **24**, 183–185 (1969).

[30] V. N. Tolmachev and L. S. Prikhod'ko, *Iz. Vyssh. Ucheb. Zaved., Khim. Khim. Tekhnol.* **3**, 985–987 (1960); cf. N. A. Kornaev, A. I. Levin, and N. L. Kotovskaya, *Zavod. Lab.* **28**, 547–548 (1962).

[31] Maksymilian Kranz and Wojciech Duczmal, *Chem. Anal.* (Warsaw) **18**, 413–417 (1973).

[32] N. A. Karnaev, A. I. Levin, N. L. Kotovskaya, and A. A. Proskurnikov, *Tr. Ural. Politekh. Inst.* **1966** (148), 130–134.

50 ml. Read at 580 nm. For high carbon steels, destroy the carbides by heating to sulfur trioxide fumes for not more than 3.5 minutes. Then take up in water and filter before diluting to 50 ml.

Ferrochromium.[33] Dissolve 0.1 gram of sample by heating gently with 5 ml of phosphoric acid and 10 ml of sulfuric acid. After fuming slightly, cool, dilute to 50 ml, and read at 610 nm.

For a more sophisticated technic, fuse the sample with sodium peroxide.[34] Dissolve the melt in dilute sulfuric acid and boil with ethanol. The chromate is reduced to chromic ion and the acetaldehyde is boiled off. Add excess of ammonium hydroxide to precipitate hydroxides; filter and wash. Dissolve the precipitated hydroxides in 1:1 sulfuric acid. Add ammonium persulfate and boil. Separate ferric ion and chromic ion by passage through a column of KU-2 ion exchange resin.

1,2-DIAMINOCYCLOHEXANETETRAACETIC ACID

This close relative of EDTA forms a 1:1 complex with chromium on heating. Beer's law is followed for 0.2–8 mg of chromium per 100 ml.

Adjust a solution of chromic ion to pH 4.5 and separately adjust 0.05 *M* reagent. Mix and boil for 10 minutes.[35] Cool, and adjust to pH 3.7–4.5 with sulfuric acid. Extract with 5 ml, and 5 ml of 5% solution of methyltrioctylammonium chloride in chloroform. Back-extract the chromium complex with 5 ml of *M* potassium nitrate and read at 540 nm. Beer's law is followed up to 130 μg of chromium per ml. Nitrate and phosphate ions interfere seriously, and aluminum must be absent. An equal amount of chloride ion is tolerated.

Procedures

Chromium Trioxide.[36a] To a 0.1-gram sample in 25 ml of water, add 1 ml of 1% zinc chloride solution followed by 24% solution of sodium hydroxide until the orange color changes to yellow. Centrifuge the hydrous oxides of chromium, iron, and zinc, and wash twice with hot water adjusted to pH 10 with sodium hydroxide solution. Dissolve the precipitated oxides with 10 ml of 10% perchloric acid, reprecipitate, and wash as before.

Dissolve the precipitated oxides in 10 ml of 10% perchloric acid and add 4 ml of 4% solution of the captioned reagent. Add 4% sodium hydroxide solution to pH 3 and boil for 20 minutes. Cool and dilute to 100 ml. Read this as a diluted aliquot at 540 nm against water.

[33] Shuichiro Mizoguchi, *Jap. Anal.* **5**, 452–455 (1956).
[34] B. Zagorchev, L. Bozadzhieva, and E. Mitropolitska, *C. R. Acad. Bulg. Sci.* **15**, 483–486 (1962).
[35] Jiri Adam and R. Pribl, *Talenta* **21**, 1205–1207 (1974); G. M. Kinhikar and S. S. Dara, *ibid.* 1208–1210.
[36a] D. L. Fuhrman and G. W. Latimer, *ibid.* **14**, 1199–1203 (1967).

Dichromate Solution.[36b] Mix a sample solution containing 0.2–2.2 mg of dichromate with 3 ml of 0.1 *M* reagent, 3 ml of acetate-acetic acid buffer solution for pH 5, and 2 ml of 0.0036% solution of toluidine blue. Heat and stir at 60° for 3 minutes and expose to radiation from a 1000-W white-light lamp for 4 minutes. Cool, add 0.2 ml of 30% hydrogen peroxide, and dilute to 25 ml. Read the complex of the reagent with chromic ion at 390 nm or 540 nm for 6–86 μg of bichromate per ml. Vanadate and cobalt interfere.

---

o-DIANISIDINE

Chromate oxidizes *o*-dianisidine, which is 3,3-dimethoxybenzidine, for reading at 440 nm. Vanadate reacts similarly. Between pH 1.2 and 2.7 the color of chromate with *o*-dianisidine develops in 5 minutes and is constant for 20 minutes.[37] Antimony, tin, and lead lower the sensitivity. Iron, gold, cerium, osmium, and vanadium cause a similar coloration, although that by iron can be masked with fluoride ion or phosphate.

The oxidation of *o*-dianisidine by hydrogen peroxide is catalyzed by chromate ion, leading to a dynamic method.[38] For analysis of gallium arsenide, pass the sample in 1 : 1 hydrochloric acid through a column of tributyl phosphate on polytetrafluoroethylene to remove the gallium. Then adjust the percolate to pH 1.5 with sodium hydroxide solution and pass through a column of KU-2 ion exchange resin in hydrogen form. Elute the chromate in impure form with 3 *N* hydrochloric acid. Adjust to pH 6–7 and extract interferences with a solution of dithizone in carbon tetrachloride. The aqueous phase can then be evaluated by the catalytic effect of its chromic ion. The technic will determine 0.018 ppm of chromium in a 0.1-gram sample.

In a study of samples recovered from atmospheric air on polyvinylchloride filters, the method determined 1–25 μg of chromium as chromate with a procedure involving dilution that would permit determination of much smaller amounts. Major interferences in the reaction are lead, cupric ion, chromic ion, ferric ion, and vanadate, which are eliminated in the procedure given.

Procedures

Chromate or Vanadate.[39] To a sample solution containing 5 ppm of hexavalent chromium or 15 ppm of pentavalent vanadium if ferric ion is present, add 5 ml of phosphoric acid. Then add 1 ml of 1% solution of *o*-dianisidine in 1% acetic acid

[36b] F. Sierra Jimenez, C. Sanchez-Pedreno, T. Perez Ruiz, and C. Martinez Lozano, *An. Quim.* **71**, 382–385 (1975).
[37] F. Buscarons and J. Artigas, *Anal. Chim. Acta* **16**, 452–454 (1957).
[38] I. F. Dolmanova, G. F. Zolotova, T. N. Shekhovtsova, and V. M. Peshkova, *Zh. Anal. Khim.* **25**, 3136–3138 (1970); I. F. Dolmanova, T. A. Bol'shova, T. N. Shekhovtsova, and V. M. Peshkova, *ibid.* **27**, 1848–1851 (1972).
[39] M. Ariel and J. Manka, *Anal. Chim. Acta* **25**, 248–256 (1961).

and dilute to 50 ml with 10 *N* sulfuric acid. Set aside for 30 minutes and read vanadium or chromium at 440 nm.

Chromate and Vanadate. For a mixture, determine the sum of the two at 440 nm as above. Then acidify an equal aliquot with a few drops of 0.5 *N* sulfuric acid to about pH 2.5. Pass through a 5×300 mm cation exchange column of 40–60 mesh Amberlite IR-120 in hydrogen form at 3 ml/minute and wash with 20 ml of water. To the effluent and washings add 1 ml of 1% solution of *o*-dianisidine in 1% acetic acid and dilute to 50 ml with 10 *N* sulfuric acid. Set aside for 3 minutes and read chromium at 440 nm. Determine vanadium by difference.

Catalytic Oxidation in an Industrial Atmosphere.[40] Fit a 0.5×12 cm column with a Teflon stopcock. Add washed Dowex 50W-X8 to a depth of 1.5 cm and apply 5 ml of *M* sodium chloride. Wash until chloride-free.

Filter the atmosphere with polyvinyl chloride filter paper. Treat with 15 ml of water in a sealed system by ultrasonics for 20 minutes and dilute to 50 ml. Pass 1 ml through the column in 20 seconds and discard. Pass 5 ml of sample solution. Mix a 1-ml aliquot with 4 ml of ethanol, 1 ml of 5 *M* hydrogen peroxide, and 1 ml of 0.879% solution of *o*-dianisidine in absolute ethanol. Dilute to 10 ml and read periodically at 440 nm for 15 minutes.

DIANTIPYRINYLMETHANE

The complex of this reagent with chlorochromate ion is extracted with chloroform for reading, most fully in 0.3 *N* hydrochloric acid. A tenfold excess of the reagent is needed and must allow for complexing with other ions such as iron or titanium. Beer's law is followed up to 40 μg of chromium per ml.

Procedure. *Cast Iron.*[41] Dissolve a 0.2-gram sample in 10 ml of 1 : 1 sulfuric acid, adding 2 ml of nitric acid to decompose carbides. Evaporate to sulfur trioxide fumes, add 3 ml of water, and again evaporate to sulfur trioxide fumes. Add 2 ml of water and 0.2 gram of lead dioxide. Heat until a maximum pink color develops, cool, and dilute to 50 ml. Adjust a 10-ml aliquot to 0.3 *N* with hydrochloric acid and add 2 grams of diantipyrinylmethane. Dilute to 50 ml and shake for 3 minutes with 10 ml of chloroform. Filter the organic phase and read at 355 or 365 nm.

1,5-DIPHENYLCARBAZIDE

This reagent is *s*-diphenylcarbazide, also known as diphenylcarbohydrazide. It complexes with chromate ion and is widely used. A large excess of diphenylcarba-

[40] B. M. Kneebone and H. Freiser, *Anal. Chem.* **47**, 595–598 (1975).

[41] A. A. Minin and L. L. Milyutina, *Uch. Zap. Perm. Gos. Univ.* **1966** (141), 241–246.

zide is essential, desirably 100 : 1.[42] The literature is variable as to reported stability of the reaction product, sometimes calling for reading within 5 minutes, in other cases stating that the color is stable for 20 minutes. Beer's law is followed for 1–10 μg of chromium in 6 ml of developed solution. Extraction of hexavalent chromium from *N* hydrochloric acid separates from most interfering ions. This reagent is also applied to the ash of the acetylacetone chelate if too low for direct reading. For details, see Ferrous Alloys by Acetylacetone.

Chromic ion can be oxidized to chromate with xenon difluoride for determination by diphenylcarbazide.[43]

For oxide inclusions in copper alloys, decompose with bromine in methanol.[44] This does not affect chromic oxide. Filter the inclusions, fuse with potassium pyrosulfate, and dissolve in water. Develop with diphenylcarbazide. Decomposition with 1 : 2 nitric acid dissolves only chromium present as a solid solution in the copper, leaving both chromic oxide and metallic chromium unaffected. Fused as above, this gives the sum of the two. Metallic chromium is then determined by difference.

In determination of chromium in steel by diphenylcarbazide, amounts of vanadium and molybdenum under 5% do not interfere.[45] In reading as the bichromate, tungsten, vanadium, and molybdenum cause high values. In the presence of large amounts of iron and nickel, the chromium is oxidized to chromate with ammonium hexanitrocerate.[46] In *N* hydrochloric acid cooled below 10°, the chromate is extracted with methyl isobutyl ketone. Then reextraction with water gives the solution for reaction with diphenylcarbazide.

To determine chromium in a solution containing iron and a three- to fourfold excess of tartrate the separation depends on the different rates at which the tartrate complexes undergo conversion into metal acetoacetonates.[47] The iron is extracted with 1 : 1 acetylacetone-chloroform. Then the chromium solution is treated with sodium peroxide and sodium hydroxide, acidified, and the chromium developed with diphenylcarbazide.

For a mixture containing 3–10 μg each of chromate and chromic ions, adjust 45 ml to 0.03 *N* sulfuric acid.[48] Add cupferron, and extract the chromate complex with chloroform. Then for chromic ion add 5 ml of 1 : 5 sulfuric acid and 5 ml of nitric acid. Heat for 5 minutes, then neutralize to pH 7 with 20% sodium hydroxide solution. Add 2 ml of 0.1% potassium permanganate solution and 7 ml of 2 *N* sulfuric acid. Boil for 5 minutes, cool, and add 10 ml of 20% urea solution. Decolorize by dropwise addition of 10% sodium nitrite solution. Dilute to 70 ml and add 1 ml of 1% diphenylcarbazide solution. After 2 minutes add 15 grams of sodium chloride and 10 ml of benzyl alcohol. Shake, dry the organic phase with sodium sulfate, and read at 552 nm against a reagent blank. In another sample reduce chromate to chromic ion with ascorbic acid and determine total chromium with diphenylcarbazide. Get chromate by difference.

[42] G. Boda, *Acad. Rep. Pop. Rom., Fil. Cluj Stud. Cercet., Stiint. Ser. I*, **6**, 217–227 (1955).
[43] Anna Erdey and Kornelia Kozmutza, *Acta Chim. Hung.* **76**, 179–182 (1973).
[44] Iwao Tsukahara, *Jap. Anal.* **19**, 1496–1501 (1970).
[45] A. Devoti and A. Sommariva, *Met. Ital.* **50**, 355–366 (1958).
[46] P. D. Blundy, *Analyst* **83**, 555–558 (1958).
[47] E. Uhlemann, G. Meissner, E. Butter, and H. Weinelt, *J. Prakt. Chem.* **25**, 272–278 (1964).
[48] Kazuko Fujii, Takashi Kusuyama, and Kazuo Konishi, *Jap. Anal.* **24**, 332–336 (1975).

Alternatively, for a mixture of chromic ion and chromate ion, first determine the latter with diphenylcarbazide in acetone.[49] Then oxidize the chromic ion to chromate with ammonium persulfate catalyzed with silver nitrate, and redetermine.

For chromium in beryllium metal, dissolve in a controlled excess of 1 : 11 sulfuric acid and oxidize to chromate with potassium permanganate solution.[50] Destroy the excess permanganate with sodium azide solution, add diphenylcarbazide solution, and read at 540 nm. The final acidity must not exceed 0.6 *N*.

Chromium in copper can be determined as the chromate with diphenylcarbazide.[51] The same solution can be used for iron as the thiocyanate and manganese as permanganate.

For determination of chromium in molybdenum oxide, fuse with sodium carbonate and potassium nitrate.[52] Extract molybdenum from a solution of the melt with acetylacetone in chloroform at pH 2. Then develop with diphenylcarbazide. Chromium is also determined in the aluminum black of an electrode deposit with diphenylcarbazide.[53]

For determination of chromium in yttrium vanadate phosphors reduce the chromium to chromic ion with hydrogen peroxide in *N*–3.9 *N* sulfuric acid.[54] Extract vanadium with chloroform as the cupferrate. Then the chromium is oxidized to chromate with potassium permanganate and the excess permanganate reduced with manganous ion. Then read 60–250 ppm of chromium in the original sample with diphenylcarbazide.

For analysis of biphenyl, digest 1 gram with 1 ml of sulfuric acid, 5 ml of nitric acid, and 1 ml of 5% sodium chloride solution.[55] Evaporate to sulfur trioxide fumes, take up in water, and after adjustment of the acidity, develop chromate with diphenylcarbazide.

For analysis of preserved softwoods it is not necessary to ash. Convert to sawdust and dry.[56] Extract with 50 ml of 1 : 6 sulfuric acid and 10 ml of 30% hydrogen peroxide at 75° for 20 minutes. Mix with 150 ml of water and let cool and settle. Filter and determine chromium with 1,5-diphenylcarbazide.

For analysis of water, add saturated sodium chloride to prevent extraction of molybdenum. Extract chromate with Aliquat S 336, which is methyl tricaprylyl ammonium chloride in chloroform.[57] Add 5% solution of diphenylcarbazide in chloroform and read at 550 nm. To prevent interference by molybdenum, add saturated solution of sodium chloride to the sample before extraction. More than 0.1% of potassium nitrate interferes.

For chromium in water, oxidize to the chromate with sodium hypobromite or potassium permanganate.[58] Perchloric acid as the oxidizer gives low results.

For water high in alkaline earths, add 1 ml of *N* sodium hydroxide to a sample

[49] I. V. Kurtova and E. I. Mingulina, *Tr. Mosk. Energ. Inst.* **1974** (184), 22–24.

[50] J. Walkden, *UKAERE* **AM18**, 3 pp (1959).

[51] V. F. Mal'tev and V. Ya. Sych, *Byull. Nauch.-Tekh. Inf. Ukr. Nauch.-Issled. Trubn. Inst.* **1959** (6–7) (189–194).

[52] Elzbieta Wieteska and Urszula Stolarczyk, *Chem. Anal.* (Warsaw) **15**, 183–189 (1970).

[53] L. Guerreschi and R. Romita, *Ric. Sci. R. C., A*, **2**, 603–616 (1962).

[54] Masayoshi Ezawa, *Jap. Anal.* **19**, 1026–1031 (1970).

[55] D. Monnier, W. Haerdi, and E. Martin, *Helv. Chim. Acta* **46**, 1042–1046 (163).

[56] A. I. Williams, *Analyst* **95**, 670–674 (1970).

[57] Jiri Jirovac and Jiri Adam, *Hutn. Listy* **29**, 739–740 (1974).

[58] Katsuya Uesugi, *J. Waterworks Sewerage Assoc., Jap.* **1962** (329), 61–64.

containing 5–100 μg of chromium.[59] Heat to 90°, add 30% hydrogen peroxide, and set aside for 40 minutes. Boil off the excess of hydrogen peroxide and add 2 ml of *M* trichloroacetic acid. Dilute to 100 ml and set aside for 10 minutes. Read at 545 nm against a blank.

For analysis of metal-finishing effluents, fume the sample with nitric and sulfuric acids to destroy organic matter.[60] Then extract as the cupferrate and determine chromium with diphenylcarbazide.

For up to 50 ng of chromium in 1 ml of human serum, decompose with 3 : 1 : 1 nitric acid–perchloric acid–sulfuric acid.[61] Dilute and extract the chromium with methyl isobutyl ketone. Read with diphenylcarbazide at 540 nm, but use a special cell for less than 3 ng.

For chromium in urban air, filter at 75 cubic meters/hour through a glass-fiber filter. Extract the filter with benzene and discard the extract.[62] Dry the filter and grind it. Extract a sample of the filter with hot 2 *N* nitric acid and develop the extract with diphenylcarbazide.

For preparation of a sample of dialdehyde starch for determination of chromium by diphenylcarbazide, see Lead.

Procedures

Separation by 2-Mercaptobenzoic Acid.[63] To a slightly acid sample containing 1–10 μg of chromium, add sufficient 0.1 *M* EDTA to mask interfering ions. Add 1 ml of 2% solution of 2-mercaptobenzoic acid in pyridine and adjust to pH 6.5–8.5 with pyridine or sulfuric acid. Dilute to 10 ml and set aside for 5 minutes. Then extract with 8 ml of chloroform. With relatively large amounts of chromium, 6.5–110 μg, this can be read at 365 nm. But for the lesser amount specified above, evaporate the chloroform extract. Add 2 ml of 8% sodium hydroxide solution and 5 drops of 30% hydrogen peroxide to the residue. Heat at 120° for 5 minutes and cool. Add 1.5 ml of 4 *N* sulfuric acid, then 1 ml of 0.5% solution of 1,5-diphenylcarbazide. Dilute to 10 ml and read at 540 nm.

Extraction with Tributyl Phosphate in Xylene.[64] Take a sample containing about 10 μg of chromium as chromate. Add sufficient ammonium chloride and hydrochloric acid to make the solution *M* in each when diluted to 25 ml, and dilute to that volume. Shake for 10 minutes with 10 ml of 30% solution of tributyl phosphate in xylene. To the organic phase, add 0.2 ml of sulfuric acid and 1 ml of 0.25% solution of diphenylcarbazide in acetone. Shake and dry over sodium sulfate. Dilute to 10 ml with xylene and read at 590 nm against a reagent blank. Platinum, ferric ion, and hexavalent molybdenum interfere with this technic.

[59] A. Bilikova and V. Bilik, *Chem. Zvesti* **22**, 873–878 (1968).
[60] Earl J. Serfass, Ralph F. Muraca, and Donald G. Gardner, *Plating* **42**, 64–68 (1955); cf. R. Weiner, *Electroplat. Metal Finish.* **17** (2), 50–51 (1964).
[61] J. Agterdenbos, L. van Broekhaven, B. A. H. G. Juette, and J. Schuring, *Talenta* **19**, 341–345 (1972).
[62] Halina Wyszynska, Artur Strusinski, and Maria Borkowska, *Rocz. Panstw. Zakl. Hig.* **21**, 483–491 (1970).
[63] M. M. L. Khosla and S. P. Rao, *Z. Anal. Chem.* **269**, 29 (1974).
[64] A. A. Yadav and S. M. Khopkar, *Indian J. Chem.* **8**, 290–292 (1970).

Isolation by Mesityl Oxide.[65] This extractant is 4-methyl-3-en-2-one. To a sample solution containing about 15 μg of chromium as the chromate, add sufficient hydrochloric acid to adjust to *N* and sufficient potassium chloride to make it 2.5 *N* in a volume of 10 ml. Dilute to 10 ml and shake for 1 minute with 10 ml of mesityl oxide. Separate the organic phase and back-extract with 10 ml of 0.1 *N* ammonium hydroxide, followed by 10 ml of water. Combine the aqueous extracts and neutralize to Congo red paper with 1 : 1 sulfuric acid. Cool, add 2.5 ml of 1 : 9 sulfuric acid, and 0.5 ml of 1% methanolic diphenylcarbazide. Dilute to 50 ml, set aside for 1 minute, and read at 540 nm.

Isolation by Trioctylamine.[66] In the absence of other metals, extract a 10-ml sample, containing about 0.1 mg of chromium as chromate, in 0.1–2 *M* sulfuric acid with 10 ml of 0.05 *M* trioctylamine in benzene. Centrifuge. Reextract a 5-ml aliquot of the organic phase with 5 ml and 5 ml of 0.05 *N* sodium hydroxide solution. Mix the combined alkaline extracts with 3 ml of 2.5 *N* sulfuric acid and 4 ml of 0.15% solution of diphenylcarbazide in 50% acetone. Dilute to 50 ml and read at 540 nm.

For 0.05–0.1 mg of chromium in 10 ml of 0.1 *M* sulfuric acid containing 10–30 mg each of iron, cobalt, nickel, copper, and mercury or in a medium of 2 *M* sulfuric acid also containing 10–30 mg of vanadium, add 2 ml of 1% solution of ammonium ceric nitrate in 0.5 *M* sulfuric acid. Heat at 100° for 30 minutes to oxidize chromic ion to chromate, and cool. Add a few drops of 2% solution of sodium azide to reduce excess ceric ion. Extract with 10 ml of 0.05 *M* trioctylamine in benzene. Proceed as above from "Centrifuge."

Oxidation by Silver Oxide.[67] Adjust a 20-ml sample containing 0.01–0.1 mg of chromium to *N* with nitric acid and to 0.5 *N* with sulfuric acid. Add about 20 mg of silver oxide in small portions with stirring until a brown color appears. Continue to stir for 15 minutes, then heat to 85° to destroy excess silver oxide. Cool, and add 1 ml of fresh 0.5% solution of diphenylcarbazide in acetone. Dilute to 50 ml and set aside for 15 minutes. Read at 540 nm against a reagent blank.

Metals in General.[68] Boil a sample containing 0.03–0.5 mg of chromium gently with 40 ml of 1 : 3 nitric acid until decomposition is complete and filter. Add 2 ml of phosphoric acid and adjust to 40 ml. Add 1 ml of 1.7% solution of silver nitrate followed by addition of 2 grams of potassium persulfate in several portions. Heat at 100° for 15 minutes, cool, and dilute to 250 ml. Dilute a 10-ml aliquot to about 18 ml and add 2 ml of phosphoric acid. Add 10 ml of 4-methyl-2-pentanone and chill in an ice bath for 5 minutes. Add 2 ml of 10 *N* hydrochloric acid and shake for 1 minute. Discard the aqueous layer and if the iron in the aliquot taken exceeds 8 mg, wash the organic layer with 20 ml of *N* hydrochloric acid. To 5 ml of the organic extract add 1 ml of 0.25% solution of diphenylcarbazide in 4-methyl-2-pen-

[65] V. M. Shinde and S. M. Khopkar, *Z. Anal. Chem.* **249**, 239–241 (1970).
[66] C. Deptula and K. Moszynska, *Chem. Anal.* (Warsaw) **13**, 211–216 (1968).
[67] J. Appelbaum and J. Marshall, *Anal. Chim. Acta* **35**, 409–410 (1966).
[68] John A. Dean and Mary Lee Beverly, *Anal. Chem.* **30**, 977–979 (1958).

tanone previously saturated with *N* hydrochloric acid. Let stand for 15 minutes and read at 540 nm within 10 minutes against a reagent blank.

STEEL.[69] Dissolve a 0.1-gram sample in 5 ml of 1:1 sulfuric acid, add 0.5 ml of 1:1 nitric acid, and boil off oxides of nitrogen. Dilute to 25 ml, filter, and wash the residue. Add 5 ml of 0.25% silver nitrate solution and 4 ml of 10% ammonium persulfate solution. Boil to oxidize chromium to chromate and manganese to manganate. Add a little paper pulp to reduce the manganate; cool and filter. Decolorize the iron with a few drops of 2% sodium fluoride solution and suppress interference by manganese and molybdenum with a few drops of 1% solution of oxalic acid. Add 0.5 ml of 0.25% ethanolic diphenylcarbazide. If vanadium is present in a ratio to chromium not greater than 1:2, color due to vanadium will fade within 15 minutes. Read at 545 nm.

As a variant, dissolve the sample of steel in nitric and hydrochloric acid without heating.[70] Then extract iron with isobutyl acetate. Chromic ion is oxidized to chromate by boiling with perchloric acid, and oxidation is completed with potassium permanganate. The chromate is read at 540 nm with diphenylcarbazide.

HIGH PURITY IRON.[71] Dissolve 0.1 gram of sample in 10 ml of hydrochloric acid and 3 ml of nitric acid. Evaporate almost to dryness and take up in 25 ml of 5:3 hydrochloric acid. Extract the iron by shaking for 1 minute with 25 ml of 2:1 methyl isobutyl ketone–pentyl acetate. Add 5 ml of nitric acid to the aqueous layer and evaporate to dryness. Dissolve in 10 ml of 1:24 sulfuric acid and boil for 3 minutes with 1 ml of 0.1% permanganate solution. Mix with 10 ml of 20% solution of urea and decolorize by dropwise addition of 10% sodium nitrite solution. Dilute to 25 ml and mix with 3 ml of 0.2% ethanolic diphenylcarbazide. Add 8 ml of 1:1 sulfuric acid and 5 grams of anhydrous sodium sulfate. Shake for 1 minute with 10 ml of isopentyl alcohol, filter the organic layer through dry paper, and read at 543 nm.

PIG IRON AND IRON CASTINGS.[72] Dissolve 0.8 gram of sample in 10 ml of 1:4 nitric acid and 10 ml of perchloric acid. Evaporate to perchloric acid fumes, cool, add 50 ml of water, and heat to 100°. If the chromium is less than 0.1% add 1 ml of 30% hydrogen peroxide and heat for 3 minutes more. Filter, wash, and dilute to 200 ml. Take a 50-ml aliquot and add 10 ml of 0.3% potassium permanganate solution. Boil for 3 minutes. Then add 30 ml of 6% sodium hydroxide solution and a few drops of 30% hydrogen peroxide. Boil for 5 minutes, cool, and dilute to 100 ml. Filter. To 25 ml of filtrate, add 15 ml of a reagent containing 0.1 gram of diphenylcarbazide and 10 ml of ethanol in 150 ml of 1:2 sulfuric acid. Read within 1 minute at 540 nm.

Large amounts of ferric ion render the calibration curve nonlinear.[73] Phosphoric

[69] A. Dutillieux and R. Duhayon, *Chim. Tech.* 5 (4), 14–16 (1958); cf. W. B. Sobers, *Foundry* **88**, 110–113 (1960); Ecaterina Szocs and Stefan Szocs, *Rev. Chim.* (Bucharest) **27**, 713–715 (1976).
[70] P. H. Scholes and D. V. Smith, *Metallurgia* (Manch) **67**, 153–157 (1963).
[71] Ohiko Kammori and Akihiro Ono, *Jap. Anal.* **14**, 1137–1140 (1965).
[72] Laszlo Egri, *Ontode* **9**, 194–196 (1958); cf. Ohiko Kammori, *J. Jap. Chem. Soc.* **20**, 255–258 (1956); Theo. Kurt Willmer, *Arch. Eisenhüttenw.* **29**, 159–164 (1958).
[73] R. K. Korabel'nik, *Iz. Yvssh. Ucheb. Zaved., Khim. Khim. Tekhnol.* **1960** (3), 189–192.

acid slightly reduces the effect of ferric ion but decreases the intensity of color and makes the readings more erratic.

FERROMANGANESE.[74] Dissolve 0.2 gram of sample in 5 ml of nitric acid, 6 ml of 1 : 1 sulfuric acid, and 40 ml of water. Add 2 ml of phosphoric acid, 5 ml of 0.5% solution of silver nitrate, and 2 ml of 10% solution of ammonium persulfate. Boil. Then when fully oxidized, add 20 ml of 10% urea solution. Add 10% solution of sodium nitrite dropwise until the permanganate has been decolorized. Dilute to 100 ml and take a 25-ml aliquot. Add 3 ml of 0.3% ethanolic diphenylcarbazide and 10 ml of 1 : 1 sulfuric acid. Dilute to 50 ml and read at 540 nm.

ALUMINUM AND ALUMINUM ALLOYS.[75] This technic is designed for samples containing 0.002–1.5% chromium. Prepare mixed acid containing 200 ml of hydrochloric acid, 200 ml of nitric acid, and 120 ml of sulfuric acid per liter.

Dissolve a 1-gram sample in 50 ml of the mixed acid. Evaporate to sulfur trioxide fumes and continue to fume for 10 minutes. For chromium up to 0.05% in the sample, dilute to 100 ml and take a 20-ml aliquot. For 0.03–0.3% chromium in the sample, dilute to 250 ml, take a 10-ml aliquot, and add to it 8 ml of 2% solution of aluminum. For 0.2–1.5% chromium in the sample, dilute to 1 liter, take a 10-ml aliquot, and add 9.5 ml of 2% solution of aluminum.

Add to the aliquot 2 ml of 0.04 *N* ammonium nitrate–ceric nitrate in 2 *N* sulfuric acid. Dilute to 30 ml and heat at 100° for 25 minutes. Cool to below 10° and dilute to 45 ml. Add 4.5 ml of hydrochloric acid and extract for 1 minute with 25 ml and 25 ml of methyl isobutyl ketone. Wash the combined extracts with 25 ml of 0.5 *N* hydrochloric acid. Reextract the chromium with 25 ml, 25 ml, and 25 ml of water. To the aqueous phase add 2.5 ml of 8 *N* sulfuric acid and 2 ml of 1% solution of diphenylcarbazide in acetone. Dilute to 100 ml and read at 545 nm.

Alternatively,[76] digest 1 gram of sample with 20 ml of 20% sodium hydroxide solution. Add 20 ml of 5 : 1 : 5 sulfuric acid–nitric acid–water. Boil to a clear solution and dilute to 50 ml. Mix a 5-ml aliquot with 10 ml of water and 5 drops of saturated solution of potassium permanganate. Boil for 2 minutes, then destroy excess permanganate by dropwise addition of hydrochloric acid. Cool, dilute to 60 ml, and add 5 ml of fresh 0.1% solution of diphenylcarbazide. Dilute to 100 ml and read at 540 nm within 15 minutes against a blank.

ANTIMONY.[77] Dissolve 1 gram of sample with 2 ml of nitric acid. Add 2 ml of hydrobromic acid and evaporate to dryness. Repeat the steps of the last sentence three times more. Take up the residue in 5 ml of hydrobromic acid, transfer to a platinum crucible, and evaporate to dryness. Add 0.5 ml of 2% solution of sodium carbonate and a crystal of potassium chlorate. Evaporate to dryness and fuse. Dissolve the melt in 1.5 ml of water and filter. Add to the filtrate 0.3 ml of 1.5 *N*

[74] Takuya Imai and Shinko Nagumo, *Kogyo Kagaku Zasshi* **61**, 53–55 (1958).

[75] G. Matelli and V. Vicentini, *Allum. Nuova Met.* **39**, 645–650 (1970); cf. British Standards Institution, *BS 1728*, pt. **16**, 7 pp (1968).

[76] M. C. Steele and L. J. England, *Metallurgia* **59**, 153–156 (1959); cf. C. M. Dozinel, *Metall.* **20**, 242–245 (1966).

[77] V. A. Nazarenko, M. B. Shustova, R. V. Ravitskaya, and M. P. Nikonova, *Zavod. Lab.* **28**, 537–539 (1962).

sulfuric acid and 0.3 ml of 0.25% solution of diphenylcarbazide in acetone. Let stand for 5 minutes and read at 545 nm.

TIN.[78] Dissolve 0.25 gram of sample containing 0.0004–0.01% chromium in 3 ml of hydrochloric acid and 0.5 ml of nitric acid. Evaporate to about 1 ml, add 4.5 ml of water, and adjust to pH 6–7 with 40% solution of sodium hydroxide. Add an additional 2 ml of 40% sodium hydroxide solution to dissolve the precipitate, add 1 ml of 10% ammonium persulfate solution, and boil for 5 minutes. Cool, neutralize with 1:1 sulfuric acid, and add 1 ml more to dissolve the precipitate. Boil for 5 minutes to destroy excess of persulfate, and cool. Add 25 ml of 30% solution of sodium chloride and mask the iron with 1 ml of phosphoric acid. Add 3 ml of 0.5% solution of diphenylcarbazide in acetone and dilute to 50 ml. If the chromium is 0.001–0.01%, read at 536 nm against water. For lesser amounts, extract with 10 ml and 10 ml of isoamyl alcohol, dilute the combined extracts to 25 ml with isoamyl alcohol, filter, and read at 536 nm against isoamyl alcohol.

NICKEL.[79] Dissolve 0.5 gram of sample in 10 ml of 1:1 nitric acid, adding a drop of hydrofluoric acid if tungsten is absent. Boil off oxides of nitrogen, cool, and add 45 ml of water. Filter the tungstic acid. Reduce the chromium with 3 drops of 6% sulfurous acid. Shake for 30 seconds with 35 ml and 10 ml of 0.1% solution of dithizone in chloroform, discarding these extracts of nickel. Add 0.5 ml of 1:1 sulfuric acid and 10 ml of perchloric acid to the aqueous phase. Evaporate until copious fumes of perchloric acid begin to be evolved. Add 50 ml of water, a drop of 1% potassium permanganate solution, and a drop of 1% manganese sulfate solution. Boil gently for 3 minutes, cool to about 70°, and add 25 ml of ammonium hydroxide. Filter, wash well, and dilute to 100 ml. Transfer a 25-ml aliquot, add 50 ml of water, and neutralize with 1:1 sulfuric acid until a piece of suspended Congo red paper just turns blue. Cool, add 5 ml of 1:9 sulfuric acid, and 1 ml of freshly prepared 1% methanolic diphenylcarbazide. Dilute to 100 ml, let stand a minute for color development, and read at once at 540 nm.

TITANIUM.[80] Add 4 ml of hydrofluoric acid, 1 ml of nitric acid, and 1 ml of 1:1 sulfuric acid to a 0.1-gram sample. Evaporate to sulfur trioxide fumes, dissolve in 0.5 *N* sulfuric acid, and dilute to 100 ml with that acid. To remove titanium, extract with 15 ml, 15 ml, and 15 ml of 6% solution of cupferron in chloroform. Then extract with 15 ml, 15 ml, and 15 ml of chloroform. Evaporate the extracted aqueous solution to dryness. Add 0.25 ml of hydrochloric acid and evaporate to dryness. Add 0.5 ml of 2% sodium carbonate solution and 5 drops of saturated solution of potassium chlorate. Evaporate to dryness and ignite. Take up in 1.5 ml of water, add 0.3 ml of 1.5 *N* sulfuric acid, and drive off the carbon dioxide. Add 0.3 ml of 0.25% solution of diphenylcarbazide in acetone, dilute to 5 ml, and read at 540 nm.

[78] A. T. Pilipenko, A. I. Voronina, and B. I. Nabivanets, *ibid.* **36**, 273 (1970).
[79] C. L. Luke, *Anal. Chem.* **30**, 359–361 (1958); cf. Thomas L. Allen, *ibid.* 447–450; British Standards Institution, *BS 3727*, pt. **4**, 7 pp (1964).
[80] V. A. Nazerenko, M. B. Shustova, G. G. Shitareva, G. Ya. Yagnyatinskaya, and R. V. Ravitskaya, *Zavod. Lab.* **28**, 645–648 (1962); cf. George Norwitz and Maurice Codell, *Metallurgia* **57**, 261–270 (1958).

HIGH PURITY TANTALUM.[81] Add 10 ml of hydrofluoric acid and 2 ml of nitric acid to a 1-gram sample in platinum. Evaporate to 2 ml. Add 20 ml of 20% tartaric acid solution and 20 ml of 4% boric acid solution. Neutralize with ammonium hydroxide and add 2 ml in excess. Add 30 ml of acetylacetone and reflux for 1 hour with stirring. Cool, and separate the phases. Extract the aqueous phase with 5 ml of chloroform, add the extract to the acetylacetone phase, and discard the aqueous phase. Wash the organic layer with 50 ml, 50 ml, and 50 ml of 1:5 sulfuric acid. Extract the combined washings with 10 ml of chloroform and add the extract to the organic phase. Evaporate the organic phase to dryness. Add 1 ml of sulfuric acid and a few drops of 30% hydrogen peroxide. Heat to destruction of the organic matter and evaporate sulfur trioxide fumes to under 0.5 ml. Add 10 ml of perchloric acid and heat to strong fumes. Cool, and dilute to 50 ml. Add 5 drops of 0.3% solution of potassium permanganate and a drop of 1% manganous sulfate solution. Boil for 3 minutes, cool to 70°, and add 25 ml of ammonium hydroxide. Cool to room temperature, filter, and dilute to 100 ml. Neutralize a 50-ml aliquot to Congo red with 1:1 sulfuric acid. Add 5 ml of 1:9 sulfuric acid and 1 ml of 1% methanolic diphenylcarbazide. Dilute to 100 ml and after 1 minute read at 540 nm against a blank.

As an alternative method of preparation of tantalum, fuse an oxidized sample with potassium pyrosulfate and dissolve in water.[82]

NICKEL-VANADIUM PELLETS.[83] Dissolve a 0.1-gram sample in 5 ml of hydrochloric acid and 1.5 ml of nitric acid. Add 0.2 ml of 1:1 sulfuric acid and evaporate to dryness. Dissolve in 5 ml of water and cool. Add 2 drops of methyl orange indicator solution and neutralize to pH 3.7 with 0.1 *N* sodium carbonate solution or 0.1 *N* sulfuric acid. Dilute to 100 ml and take a 10-ml aliquot. Add 1.3 ml of 5.25% ethanolic 8-hydroxyquinoline. When the precipitation of vanadium appears to be complete, shake with 20 ml of chloroform for 1 minute. Then extract with 4 successive 5-ml portions of chloroform. Filter the aqueous layer and add 2.5 ml of 2 *N* sulfuric acid. Add 1 ml of 1.5% silver nitrate solution and 0.25 gram of ammonium persulfate. Boil for 10 minutes, filter, and dilute to 20 ml. Add 2 ml of 2.5% solution of diphenylcarbazide in 50% acetone and dilute to 25 ml. Let stand for 1 hour and read at 540 nm.

PERMANENT MAGNET ALLOYS.[84] Large amounts of vanadium are assumed to be present and phosphate absent. Dissolve a sample containing 3–7 mg of chromium in 10 ml of 1:1 nitric acid and 2 ml of 1:1 sulfuric acid. Evaporate to sulfur trioxide fumes, cool, and dilute to 100 ml. Mix a 1-ml aliquot with 3 ml of 2:23 sulfuric acid and a solution containing 2 mg of ferric ion. Dilute to 20 ml and boil, adding 0.3% potassium permanganate until the pink color persists. Cool, and add 5 ml of 20% solution of urea. Decolorize by dropwise addition of 0.3% solution of sodium nitrite. Add 5 ml of 0.4% ethanolic diphenylcarbazide, dilute to 50 ml, and after 5 minutes read at 540 nm.

[81] A. Hofer and R. Heidinger, *Z. Anal. Chem.* **233**, 415–418 (1968).
[82] V. A. Nazarenko, E. A. Biryuk, M. B. Shustova, G. G. ShiFareva, S. Ya. Vinkovetskaya, and G. V. Flyantikova, *Zavod. Lab.* **32**, 267–269 (1966).
[83] J. T. Mc Aloren and G. F. Reynolds, *Metallurgia* **57**, 52–56 (1958).
[84] Shigetoshi Ito, *Jap. Anal.* **14**, 15–20 (1965).

THIN FILMS OF NICKEL-CHROMIUM ALLOY.[85] Dissolve a sample containing 5–20 μg of chromium in 2 ml of 1 : 1 sulfuric acid and dilute to 25 ml. Mix a 5-ml aliquot with 0.8 ml of 4% sodium hydroxide solution. Add 5 ml of *N* sulfuric acid and 0.5 ml of 0.3% potassium permanganate solution. Heat at 75° for 20 minutes, then destroy excess permanganate by dropwise addition of 5% sodium azide solution. Add 1 ml of 0.125% solution of diphenylcarbazide in acetone and dilute to 25 ml. After allowing to stand for 40 minutes, read at 546 nm.

Cyanide Copper-Plating Solutions.[86] Mix a 10-ml sample with 10 ml of sulfuric acid and 3 ml of nitric acid. Evaporate to sulfur trioxide fumes, cool, and dilute to 250 ml. To a 10-ml aliquot, which will contain 6–26 mg of copper, add 2 drops of perchloric acid, 1 ml of 1 : 5 sulfuric acid, and 1 ml of nitric acid. Evaporate to initial crystallization. Add 7 ml of 7.5% silver nitrate solution, 0.5 gram of ammonium persulfate, and a drop of 3% potassium permanganate solution. Heat to boiling and reduce the permanganate by dropwise addition of 1 : 5 hydrochloric acid. Add 1 ml of 0.5% solution of sodium fluoride to mask the iron. Add 1 ml of 0.25% ethanolic diphenylcarbazide, dilute to 25 ml, filter, and read at 525 nm against a blank.

Minerals.[87] Fuse a 0.5-gram sample with 4 grams of sodium peroxide in a corundum crucible. Dissolve the melt in water and boil to decompose the peroxide. Filter, and make 0.2 *N* with sulfuric acid. Pass through a column of cation exchange resin to remove ferric ion. Add 1 ml of 1% methanolic diphenylcarbazide. Dilute to 100 ml and read at 540 nm.

An alternative to fusion is the classical decomposition with hydrofluoric and sulfuric acids.[88]

Calcium Fluoride.[89] The sample is of a mold lining. Fuse a 1-gram sample in platinum with 2.5 grams each of sodium carbonate and potassium carbonate and 0.5 gram of potassium nitrate for 3–4 minutes. Cool rapidly, add water, and let stand overnight at 100°. Leach the soft mass repeatedly with hot water, and filter. Neutralize the filtrate with 1 : 1 nitric acid. At this point the volume should be about 80 ml. Add more water if necessary, and maintain that volume through the next three steps. Add 5 grams of boric acid and boil for 10 minutes. Add 3 ml of 1 : 1 nitric acid and 5 ml of 5% solution of potassium bromate. Boil for 15 minutes. Add 5 ml of 5% sodium chloride solution and boil for 25 minutes more. Cool and dilute nearly to 99 ml. Add 1 ml of 0.1% ethanolic diphenylcarbazide, dilute to 100 ml, and read at 545 nm. Ammonium persulfate can replace potassium bromate as the oxidizing agent.

Ruby.[90] Fuse 3 pellets of potassium hydroxide in a shallow zirconium crucible

[85] E. Dominguez, G. San Miguel, and C. S. Martin, *Electron. Fis. Apl.* **15**, 84–86 (1972).
[86] R. P. Cope, Jr., *Plating* **52**, 774–776 (1965).
[87] Friedrich Fröhlich, *Z. Anal. Chem.* **170**, 383–387 (1959); cf. A. Funke and H. J. Laukner, *ibid.* **249**, 26–30 (1970).
[88] Heinrich M. Koester, *Ber. Deut. Keram. Ges.* **46**, 247–253 (1969).
[89] G. Robertson, *UKAEA Ind., Group Hdq.* **SCS-R-172**, 9 pp (1959).
[90] E. M. Dodson, *Anal. Chem.* **34**, 966–971 (1962); cf. R. C. Chirnside, H. J. Cluley, R. J. Powell, and P. M. C. Proffitt, *Analyst* **88**, 851–863 (1963).

and allow it to cool. Dust 0.02 gram of finely powdered sample on the surface and add a few more pellets of potassium hydroxide. Heat for 150 minutes at 400–450° until fusion is complete. Boil with 10 ml of water, repeat that extraction, and boil for 1 minute with 10 ml of 10% sulfuric acid. Combine these extracts. Fuse 3 pellets of potassium hydroxide in the crucible, dissolve in 10 ml of water, and add to the extracts. Make just acid with 1 : 1 sulfuric acid and if there is cloudiness, remove by centrifuging. Dilute to 50 ml.

Take an aliquot to contain not more than 20 μg of chromium. Adjust to pH 1 with 1 : 20 sulfuric acid. Add 1 ml of 10% silver nitrate solution and 5 ml of 10% ammonium persulfate solution and boil for 15 minutes. Alternatively, heat to 100° and make distinctly red with 3% potassium permanganate solution. Boil for 5 minutes, adding more permanganate solution if the color bleaches. Decolorize with 5% solution of sodium azide.

Dilute to 15 ml and adjust to pH 1. Add 5 ml of fresh 0.25% solution of diphenylcarbazide and dilute to 25 ml. Read at 540 nm within 10 minutes against a reagent blank.

Alternatively,[91] fuse 0.05 gram of sample in platinum with 1 gram of 3 : 2 sodium carbonate–sodium tetraborate decahydrate, finally raising the temperature to 1200° for 30 minutes. Dissolve in 50 ml of hot water and filter through a filter paper prewashed with 5% sodium carbonate solution. Wash the insoluble matter, which contains the iron and titanium, with 1% sodium carbonate solution. Neutralize the filtrate with 1 : 2 sulfuric acid, and add 5 ml in excess. Add 10 ml of nitric acid and evaporate to sulfur trioxide fumes. Add 5 ml of water and again evaporate to sulfur trioxide fumes. Take up with 30 ml of water, add 1 ml of 1% methanolic diphenylcarbazide, and dilute to 50 ml. After 5 minutes read at 540 nm.

Chromia-Alumina Catalysts[92]

HEXAVALENT CHROMIUM. Extract 5 mg by warming with 100 ml of water for 10 minutes and filter. To an aliquot, add 1 ml of 1 : 5 sulfuric acid and 1 ml of 0.5% diphenylcarbazide solution. Dilute to 25 ml and after a couple of minutes read at 540 nm.

TOTAL CHROMIUM. Fuse 5 mg of sample with 1 gram of sodium carbonate and 0.2 gram of sodium sulfate. Dissolve in water, filter if necessary, and dilute to 100 ml. Proceed as above from "To an aliquot, ..."

Uranium Oxide and Uranyl Nitrate.[93] Dissolve 25 grams of uranium oxide in 100 ml of 1 : 1 nitric acid and filter. Ignite the residue, fuse with 1 gram of potassium bisulfate, dissolve in water, and add to the main solution. As sample, take an aliquot of this or of a solution of uranyl nitrate equivalent to 1 gram of uranium.

[91] V. G. Sil'nichenko and M. M. Gritsenko, *Zavod. Lab.* **31**, 657–658 (1965).
[92] V. A. Klimova and F. B. Sherman, *Izv. Akad. Nauk SSSR, Ser. Khim.* **1966** (11), 2035–2037.
[93] *UKAEA* **Rep. PG 380 (S)**, 6 pp (1962); cf. O. A. Vita and L. R. Mullins, Jr., *USAEC* **GAT-T-843**, 11 pp (1960); J. Nowicke-Jankowska, A. Golkowska, I. Pietrzak, and W. Zmijewska, *Pol. Akad. Nauk, Inst. Badan Jadronych.* No. **45**, 42 pp (1958).

Evaporate to dryness and dissolve in 1.5 ml of 1:6 sulfuric acid. Add 1 ml of 1% silver nitrate solution, followed by 5 ml of 10% ammonium persulfate solution, and boil for 15 minutes. Cool and dilute to 15 ml. Adjust to pH 1 and add 5 ml of 0.2% solution of diphenylcarbazide. Dilute to 25 ml and read promptly at 540 nm against a reagent blank. The developed solution may contain up to 160 ppm of chromium.

Titanomagnetite Concentrate.[94] Fuse a 0.4-gram sample in an iron crucible with 7 grams of sodium peroxide. Boil the melt with 100 ml of water, cool, dilute to 250 ml, and filter. Neutralize a 10-ml aliquot of the filtrate with *N* sulfuric acid to the methyl orange end point. Add 3 ml of the acid in excess and 1 ml of 1% methanolic diphenylcarbazide. Dilute to 50 ml and after 5 minutes read at 540 nm.

Ilmenite or Rutile.[95] Fuse 1 gram of sample ground to finer than 200 mesh with 15 grams of anhydrous sodium bisulfate. Dissolve the melt in 50 ml of 2 *N* sulfuric acid. Add 30 ml of water, cool, filter, and wash the residue with *N* sulfuric acid. Ash the paper and residue and treat it with 3 ml of hydrofluoric acid and 1 ml of 1:1 sulfuric acid. Evaporate to sulfur trioxide fumes and fuse the residue with 0.5 gram of anhydrous sodium bisulfate. Dissolve the melt in 5 ml of 1:9 sulfuric acid and add to the main solution. Dilute to 200 ml and take an aliquot containing up to 50 μg of chromium. Add 1 ml of 1:1 sulfuric acid to avoid precipitation of basic titanium sulfate, dilute to 15 ml, and add dropwise 1 ml of 0.16% potassium permanganate solution. Boil gently for 5 minutes and cool to below 5°. Add 5 ml of 1:1 hydrochloric acid and dilute to 25 ml.

As oxidized phosphoric acid, heat 1:2 phosphoric acid to 60°, add 0.16% solution of potassium permanganate to a faint pink color, and heat further until decolorized. Add 0.5 ml of this oxidized phosphoric acid to mask iron. Equilibrate methyl isobutyl ketone with *N* hydrochloric acid. Cool the sample to 5° and shake for 40 seconds with 25 ml of the methyl isobutyl ketone. Wash the extract with 5 ml of *N* hydrochloric acid. Reextract the chromium from the organic layer by shaking at 70° with 40 ml of water for 30 seconds. Wash the organic layer with an additional 5 ml of water. Add 3 ml of oxidized phosphoric acid to the aqueous extracts. Cool below 10° and add 2 ml of 0.25% solution of diphenylcarbazide in acetone, adjust to 50 ml, and set aside for 10 minutes. Read at 546 nm.

Residues from Anodic Dissolution of Chromium-Nickel Alloys.[96] Dissolve 0.1 gram of sample in 10 ml of 3:1 hydrochloric-nitric acid. Filter and add 3 ml of sulfuric acid. Evaporate to sulfur trioxide fumes, add 3 ml of water, and again evaporate to sulfur trioxide fumes. Repeat this step twice more, and the last time fume for 3 minutes. Take up in water and dilute to 100 ml.

Boil a 1-ml aliquot with 0.3 gram of ammonium persulfate and 0.5 ml of 1% solution of silver nitrate until the excess of persulfate is decomposed. Cool, and make just alkaline to Congo red with sodium carbonate. If iron is present, add 1 ml

[94] V. I. Zolotavin and N. D. Fedorova, *Tr. Vses. Nauchn.-Issled. Inst. Stand. Obraztsov. Spektr. Etalonov* **2**, 92–96 (1965).
[95] E. S. Pilkington and P. R. Smith, *Anal. Chim. Acta* **39**, 321–328 (1967).
[96] R. B. Golbtsova and S. M. Savvateeva, *Zavod. Lab.* **32**, 150–151 (1966).

of 1% ascorbic acid solution. Add 1.5 ml of 1 : 1 sulfuric acid and 1 ml of 0.1% solution of diphenylcarbazide in 10% ethanol. Dilute to 25 ml and read at 550 nm.

Ash of Solid Fuels.[97] Fuse a 1-gram sample with 6 grams of sodium carbonate. Extract the melt with 50 ml of hot water, and reduce manganese with a few drops of ethanol. Filter, and wash well with hot 1% solution of sodium carbonate. Dilute the filtrate and washings to 100 ml.

Titrate a 25-ml aliquot to the methyl orange end point with 1 : 8 sulfuric acid, record the volume required, and discard. To another 25-ml aliquot add the volume just noted of 1 : 8 sulfuric acid, 2 ml of 0.25% solution of diphenylcarbazide in acetone, and 3 ml of 1 : 8 sulfuric acid. Dilute to 100 ml and read at 550 nm.

Soil[98]

TOTAL CHROMIUM. Calcine a 1-gram sample at 500° for 2 hours. Fuse with 3 grams of potassium carbonate. Dissolve in 25 ml of water and 0.5 ml of 1 : 2 hydrochloric acid. Evaporate to dryness, add 5 ml of 1 : 2 hydrochloric acid, and again evaporate to dryness. Dry at 120°. Take up in 50 ml of water and 3 ml of 1 : 6 sulfuric acid. Filter and wash the insoluble residue with hot water. Neutralize the filtrate and washings to litmus with *N* sulfuric acid and dilute to 100 ml. To a 25-ml aliquot, add 2 ml of 1 : 6 sulfuric acid and 1 ml of 0.25% solution of diphenylcarbazide. Read at 540 nm.

LABILE CHROMIUM. Shake a 10-gram sample of dry soil for 30 minutes with 50 ml of ammonium acetate buffer solution for pH 4.8. Filter, and dilute the filtrate to 50 ml. Evaporate a 25-ml aliquot to dryness and ignite at 500°. Dissolve the residue in 10 ml of perchloric acid and add 25 ml of water. Add 1 ml of phosphoric acid and heat at the boiling point until the solution becomes yellow. Heat 3 minutes longer, cool, and add 25 ml of water. Boil for 5 minutes to drive off chlorine. Cool, add 2 ml of 0.25% solution of diphenylcarbazide, dilute to 100 ml, and read at 540 nm.

Seawater.[99] Acidify a 2-liter sample with hydrochloric acid, add 1 gram of ferric chloride, and adjust to pH 7.5 with a borate buffer solution. Allow to stand overnight, and filter. Dissolve the precipitate in 5 ml of hydrochloric acid. Pass the solution through a column of De-Acidite FF ion exchange resin and wash through with hydrochloric acid. Add 2 ml of 30% hydrogen peroxide to the solution and washings, and evaporate to dryness. Take up in 2 ml of 1 : 9 sulfuric acid, add 2 ml of 1% solution of ceric ammonium nitrate, and boil. Destroy excess ceric ion by dropwise addition of a 1% solution of sodium azide. Add 1 ml of 1% solution of diphenylcarbazide, dilute to 25 ml, and after 15 minutes read at 540 nm. As a blank, use seawater stripped of chromium by precipitation.

[97] N. P. Fedorovskaya and L. V. Miesserova, *Tr. Inst. Goryuch. Iskop., Min. Ugol'n. Prom. SSSR* **23** (3), 3–7.

[98] O. K. Dobrolyubskii and G. M. Viktorova, *Pochvovdenie* **1969** (5), 126–31.

[99] L. Chuecas and J. P. Riley, *Anal. Chim. Acta* **35**, 240–246 (1966).

A parallel technic is applicable to acid industrial effluents.[100]

Air.[101a] MANGANESE AND CHROMIUM. Pass the sample through a filter. Extract the filter by heating for 15 minutes with 2 *N* nitric acid. The contents of manganese and chromium should each fall in the range of 1–30 μg. Filter, evaporate to dryness, and ignite the residue at 500° for 30 minutes. Dissolve in 7 ml of 0.5 *N* sulfuric acid, filter, and evaporate to sulfur trioxide fumes. Take up in 8 ml of water. Add 1 ml of phosphoric acid, 0.2 ml of 1% silver nitrate solution, and 0.5 gram of ammonium persulfate. Warm just below 100° for 15 minutes and dilute to 10 ml. Read manganese at 533 nm.

Heat to about 95° and decolorize the permanganate by dropwise addition of 2% solution of sodium azide. Cool and add 2 ml of 0.25% ethanolic diphenylcarbazide. Dilute to 25 ml and read chromium at 540 nm.

Alternatively,[101b] draw 100 liters of air at 2 liters/minute through a 37-mm polyvinylchloride filter of 5 μ*M* pore size. Wash the filter with 1:5 sulfuric acid and filter the washings. Add 0.5 ml of 0.25% solution of diphenylcarbazide in 50% acetone and dilute to 15 ml with 1:5 sulfuric acid. Set aside for 10 minutes and read at 540 nm. The filters may be stored for 2 weeks before washing and development.

Chromate Solution.[102] To a 10-ml sample containing up to 5 μg of chromic ion, add in a centrifuge tube 1 ml of 1% aluminum nitrate solution. Make alkaline with ammonium hydroxide to precipitate the aluminum hydroxide as a carrier for chromic hydroxide. Centrifuge, and wash the hydroxides with 1 ml of water. Dissolve the precipitate in 0.5 ml of 1:1 sulfuric acid, add 10 ml of water, reprecipitate the hydroxides as before, and wash. Dissolve in 0.5 ml of 1:1 sulfuric acid and dilute to 25 ml. Take two 10-ml aliquots, reserving one to be a reagent blank. To one aliquot add 1.5 ml of bromine water and decolorize to a faint yellow by dropwise addition of 40% sodium hydroxide solution, finally adding 5 drops in excess. Heat at around 95° for 15 minutes; then add 1:5 sulfuric acid until the color of bromine reappears. Add 2 drops in excess of the 1:5 sulfuric acid and boil off the bromine. Cool, and add 4 drops of 1:5 sulfuric acid and 1 drop of phenol. To the oxidized sample and reserved blank, add 1 ml of 0.3% solution of diphenylcarbazide in 50% acetone. Read at 540 nm.

Beer.[103] Evaporate 100 ml of degassed sample to dryness. Place in an oven at 200°, gradually raise to 500°, and ash overnight. Cool, moisten the ash with 3 ml of nitric acid, reevaporate, and ash for 2 hours at 500°. Repeat the steps described in the preceding sentence until no carbon is left. Take up in 2 ml of 1:9 sulfuric acid, filter, and wash. Evaporate the filtrate and washings to a small volume. Add 30%

[100]M. Zuse, *Galvanotechnik*, **57**, 500–504 (1966); cf. British Standards Institution, **BS 2690**, Pt **12**, 19 pp (1972).

[101a]Artur Strusinski, *Rocz. Panstw. Zakl. Hig.* **21**, 13–19 (1970).

[101b]Martin T. Abell and John R. Carlberg, *Am. Ind. Hyg. Assn. J.* **35**, 229–233 (1974).

[102]R. W. Cline, R. E. Simmons, and W. R. Rossmassler, *Anal. Chem.* **30**, 1117–1118 (1958).

[103]A. R. Deschreider and R. Meaux, *Rev. Ferment.* **17**, 73–76 (1962); *Brass. Malt.* **12** (11), 387–389 (1962).

solution of sodium hydroxide dropwise until a slight opalescence appears. Add 5 ml of 5% sodium hypobromite solution and heat at 100° for 1 hour. Add 1:9 sulfuric acid to adjust to pH 1.3–1.7. Remove excess bromine by adding 5 ml of 1.2% solution of phenol. Add 3 ml of 0.25% solution of diphenylcarbazide and dilute to 25 ml. Read at 544 nm.

Food.[104] Ash a 10-gram sample at 540° and dissolve the ash in 25 ml of 0.1 *N* sulfuric acid. Add 3% potassium permanganate solution to a definite pink color, then decolorize with 5% solution of sodium azide and dilute to 50 ml with 0.1 *N* sulfuric acid. Take a 20-ml aliquot and add 1.5 ml of 4 *M* monosodium phosphate. Add 1 ml of 0.5% solution of diphenylcarbazide in acetone and dilute to 25 ml with 0.1 *N* sulfuric acid. Let stand for 15 minutes and read at 540 nm.

Wine and Must.[105] Add 10 ml of nitric acid to a 100-ml sample and ash. Take up in water and dilute to 50 ml. To a 10-ml aliquot, add 0.5 ml of phosphoric acid, 0.2 gram of ammonium persulfate, and 50 μg of silver nitrate. Boil for 30 seconds and chill with ice water. Add 2 ml of hydrochloric acid and extract the chromium with 10 ml and 10 ml of methyl isobutyl ketone. Wash the combined organic extracts with 2 ml and 2 ml of ice-cold hydrochloric acid. Reextract the chromate with 2 ml and 2 ml of 1:7 ammonium hydroxide. Evaporate this extract to half volume and adjust to pH 4–5. Shake with 5 ml of 1% solution of 8-hydroxyquinoline in chloroform to remove traces of iron and vanadium. Extract the aqueous phase with 1 ml of chloroform to remove residual 8-hydroxyquinoline. Evaporate the aqueous phase to dryness, add 0.5 ml of hydrochloric acid and 0.5 ml of nitric acid, and again evaporate to dryness. Add 0.1 ml of fresh 1% solution of sodium peroxide and 0.1 ml of water. Heat in an oven at 120° for 30 minutes. Add 0.1 ml of a freshly prepared 0.4% solution of diphenylcarbazide in acetic acid mixed 1:1 with phosphoric acid. Dilute to 5 ml and read at 550 nm against a reagent blank.

Tissue.[106] Digest a 1-gram sample with 1 ml of sulfuric acid, 2 ml of perchloric acid, and 2 ml of nitric acid. If necessary, add more nitric acid dropwise to avoid charring. Evaporate to sulfur trioxide fumes, boil with 10 ml of water, and cool. Add a drop of thymol blue indicator solution and neutralize to a peach color with 40% sodium hydroxide solution. Add 0.5 ml of 0.3% potassium permanganate solution and heat to 80°. If the pink color fades, add 0.3 ml more of permanganate solution. After 20 minutes decolorize the permanganate by dropwise addition of 10% solution of sodium azide. Add 2 ml of 60% solution of monosodium phosphate and 1 ml of 0.1% solution of diphenylcarbazide. Read at 540 nm against water.

Blood. Digest a 5-ml sample with 1 ml of sulfuric acid, 2 ml of perchloric acid, and 0.5 ml of nitric acid. When the digest darkens, remove from the heat and add 0.2 ml of nitric acid. Allow the reaction to subside, and reheat, repeating as necessary until clear. Complete as for tissue from "Evaporate to sulfur trioxide fumes, ..."

[104]Teresa Zawadzka, *Rocz. Panstw. Zakl. Hig.* **24**, 315–323 (1973).
[105]K. Beyermann and H. Eschnauer, *Z. Lebensmittelunter.* **118**, 308–311 (1962).
[106]C. Y. Gooderson and F. J. Salt, *Lab. Pract.* **17**, 921 (1968); cf. S. Yarbro and H. A. Flaschka, *Microchem. J.* **21**, 415–423 (1976).

Diets or Feces.[107] Ash a sample containing 0.5–3 mg of chromic oxide in a nickel crucible. Fuse the ash with 2 grams of sodium peroxide, cool, and dissolve in water. Dilute to 250 ml, let stand to settle, and take a 1-ml aliquot of the clear supernatant layer. Add 1.5 ml of 1:5 sulfuric acid and 0.5 ml of 0.5% ethanolic diphenylcarbazide. Dilute to 10 ml and read at 530 nm against a reagent blank within 12 minutes.

Leather.[108] Treat a 0.5-gram sample with 5 ml of perchloric acid and 5 ml of 1:1 nitric acid. After 15 minutes heat to boiling for 2.5 hours, at which time it should be clear. Add 20 ml of water, cool, and dilute to 100 ml. To a 3-ml aliquot, add 10 ml of 1:9 hydrochloric acid and 5 ml of 1% solution of diphenylcarbazide in acetone. Dilute to 50 ml, let stand for 10 minutes, and read at 540 nm.

Preservative-Treated Wood.[109] The samples are in sections 0.3 mm thick. Heat a weighed sample at 120° for 5 minutes with 2 ml of 8% sodium hydroxide solution and a drop of 30% hydrogen peroxide. Cool, neutralize, and make 0.3 N with sulfuric acid. Add 2 ml of 1% solution of diphenylcarbazide in acetone, and dilute to 100 ml. Read at 540 nm.

Organochromium Complexes in Drilling Fluids.[110] Evaporate a 50-ml sample containing less than 5 mg of dry solids and less than 52 μg of chromium to dryness at 100°. Add 1 ml of nitric acid and 1 ml of 5% solution of potassium chlorate. Again evaporate to dryness, at which point all organic matter should have been destroyed. Dissolve in 5 ml of water and add 2 ml of 1:6 sulfuric acid. Add 2 drops of 2.5% silver nitrate solution and boil. Then add 0.1 gram of ammonium persulfate and simmer for 30 minutes. Add 0.4 ml of 5% solution of sodium chloride and filter. Add 1 ml of freshly prepared 1% methanolic diphenylcarbazide, dilute to 50 ml, and read at 540 nm after 5 minutes.

EDTA

The chromic complex with EDTA can be read in the presence of ferrous ion, but the formation of the complex requires heating. The bicarbonate ion catalyzes the formation of the complex.[111] There must be at least 2 moles of EDTA to 1 of chromium.[112] At pH 4.5 the absorption maxima are at 390 and 450 nm. Beer's law is followed for 5–150 μg of chromium per ml.

For determination of chromium in silicate rocks by EDTA, see Copper.

[107] M. Yoshida, *Nogyo Gijutsu Kenkyusho Hokoku Ser. G* **19**, 127–131 (1960).

[108] Ferdinand Langmair, Drohoslav Kikes, and Zdenka Slovacekova, *Kozarstvi* **8**, 166–168, 198–199 (1958).

[109] A. I. Williams, *Analyst* **93**, 611–617 (1968).

[110] E. A. Kalinovskaya and L. S. Sil'vestrova, *Zavod. Lab.* **34**, 30–32 (1968).

[111] V. Kameswara Rao, D. S. Sundar, and M. N. Sastri, *Chemist-Analyst* **54**, 86 (1965).

[112] S. Arribas Jimeno, R. Moro Garcia, M. L. Alvarez Bartolome, and C. Garcia Bao, *Inf. Quim. Anal. Pura Apl. Ind.* **27**, 201–208 (1973).

Procedure. To a nearly neutral solution containing 0.03–0.19 millimoles of chromic chloride, add 10 ml of 0.3 *M* EDTA, which will buffer the solution to pH 4.5.[113a] Add a pinch of zinc dust to reduce the chromium to bivalency. Shake, set aside for 5 minutes, filter, and dilute to 50 ml. Read at 545 nm against water.

In Dimethyl Sulfoxide.[113b] To 5 ml of a sample solution containing 25–175 μg of chromium as chromate per ml in dimethyl sulfoxide, add an equal volume of 0.5 *M* EDTA in the same solvent. Heat to 80° for 10 minutes, cool, and adjust to 10 ml. Read the violet 2:3 metal-ligand complex at 555 nm.

Cobalt and Nickel Absent.[114] Transfer 2.5 ml of sample containing 0.15–4 mg of chromic ion to a centrifuge tube. If ferrous ion is present, add bromine water to oxidize it and boil off the excess bromine. Add 1 ml of 0.2 *M* EDTA and heat at 90° for 30 minutes. Cool and add ammonium hydroxide until the purple color of the solution changes to blue. Add 1 ml of buffer solution for pH 10 and 1 ml of 0.6 *M* calcium chloride. Again heat for 30 minutes, this time at 95°. Cool, centrifuge, and decant. Wash the precipitate with 1 ml and 1 ml of 2% ammonium chloride solution made slightly ammoniacal. Dilute the solution and washings to 50 ml with a buffer solution for pH 10 and read at 590 nm. Alternatively dilute to volume with a buffer solution for pH 4 and read at 545 nm.

Cobalt or Nickel Present. Transfer 2.5 ml of solution containing 0.7–4 mg of chromic ion to a centrifuge tube. Neutralize with ammonium hydroxide and if a precipitate is obtained, dissolve it with a drop of hydrochloric acid. Add 3 ml of 0.2 *M* EDTA and heat at 90° for 30 minutes. Add a few drops of ammonium hydroxide to change the purple color to blue, 1 ml of a buffer solution for pH 10, 1.5 ml of an 8% solution of thioacetamide, and 200 mg of calcium chloride dihydrate. Heat for 30 minutes at 85°. Cool to 70°, centrifuge, and remove the supernatant layer. Wash with 2 ml and 2 ml of water. Add the solution and washings to 20 ml of buffer solution for pH 10, dilute to 100 ml, and read at 590 nm.

Steel.[115] Dissolve a 1-gram sample in 30 ml of 1:2 hydrochloric acid and 5 ml of nitric acid. When reaction ceases, add 20 ml of nitric acid and evaporate to about 20 ml. Add 20 ml of 1:1 sulfuric acid and evaporate to sulfur trioxide fumes. Cool, then boil with 100 ml of water to dissolve salts. Filter, wash with hot 1:99 sulfuric acid, and dilute to 250 ml. Tungsten is in the residue on the paper.

Add a 5-ml aliquot to 25 ml of water and heat to boiling. Add 20% sodium hydroxide solution until a permanent precipitate is obtained. Clarify this with a few drops of 1:1 sulfuric acid and acidify with 6 ml of 1:4 sulfuric acid. Add 0.15 gram of ascorbic acid and heat to 100°. Add 0.9 gram of disodium EDTA dihydrate in 10 ml of hot water if more than 8% of chromium is present, 0.5 gram

[113a]H. M. N. H. Irving and W. R. Tomlinson, *Chemist-Analyst* **55**, 14–15 (1966).
[113b]J. Areses, F. Bermejo Martinez, A. Concheiro, and J. A. Rodriguez Vazquez, *Quim. Anal.* **29**, 212–215 (1975).
[114]G. den Boef, W. J. de Jong, G. C. Krijn, and H. Poppe, *Anal. Chim. Acta* **23**, 557–564 (1960).
[115]Walter D. Nordling, *Chemist-Analyst* **49**, 78–80 (1960).

for a lesser chromium content. Dilute to 80 ml with hot water and heat at 100° for 30 minutes. Add 10 ml of 3% sulfuric acid solution, and cool. Dilute to 100 ml with 3% sulfuric acid solution, and read at 538 nm.

As applied to steel in the presence of ascorbic acid and mannitol, this reagent is particularly useful for martensitic and structural steels.[116]

Aluminum. Dissolve a 2.5-gram sample in 50 ml of 30% fluoboric acid solution. After reaction ceases, boil for 1 minute. Filter hot and wash with hot water. The residue on the paper contains the copper. Add 0.2 gram of ascorbic acid and 6 ml of 1:4 sulfuric acid. Heat to 100° and add 0.5 gram of disodium EDTA dihydrate in hot water. Dilute with hot water to about 80 ml and heat at 100° for 30 minutes. Add 10 ml of 1:50 sulfuric acid, cool, and dilute to 100 ml with 1:50 sulfuric acid. Read at 538 nm.

Nichrome.[117] Dissolve 0.2 gram of sample in 4 ml of nitric acid and 8 ml of hydrochloric acid. Adjust the solution to 6 *N* with hydrochloric acid. Extract the ferric ion with 10 ml, 10 ml, and 10 ml of ether. Dilute the solution to 250 ml and take a 10-ml aliquot. Neutralize to Congo red with ammonium hydroxide and add 2 ml of *N* acetic acid. Add 5 ml of *N* EDTA and boil for 5 minutes. Cool, dilute to 100 ml, and read at 538 nm.

Copper-Chromium Alloys.[118] Decompose 0.1 gram of sample with 40 ml of acid mixture comprising 1200 ml of 1:1 nitric acid, 300 ml of 1:1 sulfuric acid, and 100 ml of phosphoric acid. Heat to sulfur trioxide fumes. Add 20 ml of water and mask the copper with 20 grams of thiourea. Mix and add 100 ml of 20% sodium acetate–6% ammonium nitrate as a buffer solution for pH 7.5. Add 20 ml of 10% solution of EDTA. Heat to boiling for 1 minute and dilute to 200 ml. Filter and read at 550 nm.

Chrome Ore.[119] Fuse a 0.5-gram sample with 7 grams of sodium carbonate and 2.5 grams of boric acid by gradually raising the temperature to 950°. Dissolve the cooled melt in 85 ml of water and 25 ml of 1:1 sulfuric acid. Add 1:3 ammonium hydroxide until a slight precipitate appears, and dissolve it by dropwise addition of 1:1 sulfuric acid. Dissolve 50 ml of Amberlite LA-2 resin in 450 ml of chloroform. Extract the sample solution with 25 ml and 10 ml of the resin solution. Add 10 ml of 1:9 sulfuric acid to the aqueous solution and extract with 10 ml of the resin solution followed by 20 ml and 20 ml of chloroform. Extract the combined organic extracts, now totaling about 85 ml, with 50 ml and 50 ml of 5.6% potassium hydroxide solution. Wash the combined alkaline extracts with 10 ml and 10 ml of chloroform. Neutralize with hydrochloric acid and add 6 drops in excess. Heat to 100° to remove residual chloroform, cool, and dilute to 250 ml. To a 20-ml aliquot add 1 ml of hydrochloric acid and 10 ml of 5% solution of sodium sulfite. Dilute to 50 ml, boil for 5 minutes, and cool. Add 10 ml of 5% solution of EDTA followed

[116] M. Jean, *Analysis* **1** (5), 358–361 (1972).
[117] V. N. Polyanskii, *Sb. Tr. Mosk. Vech. Met. Inst.* **1957** (2), 255–256.
[118] W. Nielsch and G. Böltz, *Metall.* **10**, 916–920 (1956).
[119] H. Bennett and K. Marshall, *Analyst* **88**, 877–881 (1963); cf. Yu. I. Usatenko and E. A. Klimkovich, *Tr. Kom. Anal. Khim., Akad. Nauk SSSR, Inst. Geokhim. Anal. Khim.* **8**, 169–177 (1958).

by 1:3 ammonium hydroxide until a precipitate just forms. Dissolve the precipitate by adding 1:1 acetic acid, dilute to 200 ml, and boil. Cool, dilute to 250 ml, and read at 550 nm.

Chrome Brick or Chrome Slag.[120] Treat 0.5 gram of sample with 5 ml of hydrofluoric acid and 1 ml of 1:1 sulfuric acid. Evaporate to sulfur trioxide fumes and fuse the residue with 2 grams of potassium bisulfate. Take up in 100 ml of 1:4 hydrochloric acid and add 5 ml of 2% EDTA solution to mask ferric, aluminum, chromic, and manganous ions. Boil for 2 minutes, cool, and make alkaline with ammonium hydroxide. Add 5% solution of oxine in 5% acetic acid in slight excess at 60–70° to precipitate ferric and aluminum oxinates. Filter. Adjust the filtrate to pH 8 and dilute to 200 ml. Extract a 50-ml aliquot with 30 ml and 20 ml of carbon tetrachloride to remove oxine. Adjust to pH 3.5, boil off residual carbon tetrachloride, and dilute to 80 ml. Add 10 ml of 2% EDTA solution and heat to 100° for 5 minutes. Dilute to 100 ml and read chromium at 538 nm.

Chrome-Alum Solution.[121] Dilute the sample solution to 50 ml and neutralize any free acid to Congo red with 1:5 ammonium hydroxide. Add 3 ml of 0.5 *N* acetic acid and 5 ml of 0.5 *N* EDTA solution. Boil for 5 minutes, cool, and dilute to 100 ml. Read at 538 nm against a corresponding sample to which no EDTA was added. There is no interference by large amounts of sodium sulfate or ammonium sulfate.

Silicate Rocks.[122] Transfer 0.5 gram of sample to platinum and mix with 2 ml of sulfuric acid and 10 ml of hydrofluoric acid. Evaporate to fumes and take up with 5 ml of water. Add 5 ml of hydrochloric acid and extract the iron with 5 ml of pentyl acetate. If the sample contains more than 3% iron, repeat this extraction. Evaporate the aqueous phase to sulfur trioxide fumes and nearly neutralize with 10% sodium hydroxide solution. Add 15 ml of 0.2 *M* EDTA. Adjust to pH 4 and boil for 5 minutes. Cool, dilute to 25 ml, and read at 545 nm.

FLAME PHOTOMETRY

Chromium is oxidized to the hexavalent state by potassium persulfate and extracted with 4-methyl-2-pentanone from solution in *N* hydrochloric acid. The extract aspirated into an oxyacetylene flame gives fiftyfold emission as compared to the aqueous solution and avoids interferences other than iron. Even when at more than 0.5 mg/ml in the aqueous layer iron shows some extraction, it does not interfere with the chromium line at 359.4 nm.

The strongest spectral line for chromium is at 425 nm.[123] Hydrochloric acid and nitric acid do not affect the emission, but sulfuric acid and phosphoric acid decrease the intensity. Alkali metals increase the intensity, alkaline earths decrease

[120] Yoshide Endo and Hajime Takagi, *Jap. Anal.* **9**, 624–627 (1960).
[121] V. I. Zabiyako and G. N. Sharapova, *Tr. Ural'. Nauchn.-Issled. Khim. Inst.* **1964** (11), 14–15.
[122] K. P. Stolyarov and F. B. Agrest, *Zh. Anal. Khim.* **19**, 457–466 (1964).
[123] Shigero Ikada, *Sci. Rep. Res. Inst. Tohoku Univ., Ser. A* **8**, 449–456 (1956).

it. High concentrations of aluminum, cobalt, and copper decrease the emission. For alloy analysis, 4–30% of chromium can be determined by flame photometry.[124]

Procedures

Alloys.[125a] Boil a sample containing 0.3–0.8 mg of chromium gently with 30 ml of water and 8 ml of sulfuric acid until decomposition is complete or reaction ceases. Add 5 ml of nitric acid in 1-ml portions. If much carbonaceous residue remains, add 5 ml more of nitric acid and evaporate to copious sulfur trioxide fumes to ensure complete decomposition of carbides. Dilute to 80 ml and boil to complete solution of salts. Cool and add 2 ml of 1.7% solution of silver nitrate. Add 2 grams of potassium persulfate. Swirl to dissolve, and heat at 100° for 10 minutes. Cool and dilute to 100 ml. Dilute a 10-ml aliquot to 18 ml. Add 2 ml of 10 *N* hydrochloric acid and 10 ml of 4-methyl-2-pentanone. Chill in an ice bath for 5 minutes and shake for 1 minute. Aspirate the organic phase and measure under one or more of the following conditions:

Line Emission (nm)	Background (nm)
357.9	358.6
359.4	360.0
425.4	424.4

Minerals. Transfer a sample containing 0.3–1 mg of chromic oxide to a platinum dish. Add 5 ml of hydrofluoric acid and 1 ml of 1 : 1 sulfuric acid. Heat to sulfur trioxide fumes, add 2 ml of hydrofluoric acid, and evaporate to dryness. Ignite to drive off the free sulfuric acid and fuse for 10 minutes with 2 grams of sodium carbonate. Dissolve in 70 ml of 1 : 35 sulfuric acid and add 2 ml of 1.7% solution of silver nitrate. Complete as for alloys from "Add 2 grams of potassium persulfate."

Feces.[125b] Ash a 50-gram sample expected to contain 100 mg of chromium. Dissolve the ash in perchloric acid, dilute appropriately, and read in an oxyacetylene flame at 425 nm. Accuracy is comparable to chemical methods.

Alternatively,[126] digest a sample with nitric acid until all organic matter is destroyed. Add a few drops of perchloric acid to oxidize chromic ion to bichromate. Cool, and dilute with water to a known volume. Dilute an appropriate aliquot with *N* hydrochloric acid and extract with methyl isobutyl ketone previously equilibrated with *N* hydrochloric acid. Remove the organic layer and use for flame photometry at 425.4 or 427.5 nm.

[124] C. A. Waggoner, *Can. J. Spectrosc.* **10**, 51–57 (1965).
[125a] H. Alden Bryan and John A. Dean, *Anal. Chem.* **29**, 1289–1292 (1957).
[125b] J. Anderson and I. Weinbren, *Clin. Chim. Acta* **6**, 648–651 (1961).
[126] J. R. Daly and H. B. Anstall, *ibid.* **9**, 576–580 (1964).

8-HYDROXYQUINALDINE

Chromic ion complexes with 8-hydroxyquinaldine and the complex is extractable with chloroform. Acetic acid masks uranium. Interference due to iron, copper, nickel, molybdenum, vanadium, and titanium are removed by backwashing. A correction factor can replace the complete removal of iron. Cobalt, gallium, indium, and zirconium interfere.

Substantially, the technic for determination by 8-hydroxyquinaldine is also applicable by 8-hydroxyquinoline, the chloroform extract being read at 425 nm.[127]

Procedure. *Uranium.*[128] As a color reagent, dissolve 2 grams of steam-distilled 8-hydroxyquinaldine in 5 ml of glacial acetic acid and dilute to 100 ml. As a wash solution, mix 200 ml of 23% solution of ammonium carbonate, 100 ml of 6.5% solution of potassium cyanide, and 200 ml of 16.4% solution of hydroxylamine sulfate. Dilute to 1 liter. This should be at pH 8–9. Saturate by shaking with chloroform and filter through wet paper before use.

Dissolve a 1-gram sample in 10 ml of 1 : 1 hydrochloric acid. After decomposition add 5 ml of 0.33 *M* potassium chlorate and continue to heat for a few minutes to drive off excess chlorine. Add 5 ml of glacial acetic acid and 3 ml of the color reagent. Add ammonium hydroxide to raise to pH 5.3–6, for which methyl red paper will serve. Heat to about 70° for 10 minutes to form the complex. Dilute to about 100 ml and shake for 1 minute with 10 ml of chloroform. Add the organic layer to 50 ml of the wash solution and shake for 5–8 minutes to remove interfering ions. Dry the organic layer with sodium sulfate and read at 410 and 580 nm.

Then calculate as

$$C_{\mathrm{Cr}}=\frac{(A_{410}-A_{580})\times a_{410}^{\mathrm{Fe}}/a_{580}^{\mathrm{Fe}}}{a_{410}^{\mathrm{Cr}}}$$

The values of a are absorbances per microgram of the element; typical values are as follows:

$$a_{410}^{\mathrm{Cr}}=0.0152$$

$$a_{410}^{\mathrm{Fe}}=0.00577$$

$$a_{580}^{\mathrm{Fe}}=0.00820$$

The use of this correction eliminates the necessity for complete removal of iron.

---

[127] J. P. Tandon and R. C. Mehrotra, *Z. Anal. Chem.* **176**, 87–90 (1960).
[128] Henji Motojima and Hiroshi Hashitani, *Anal. Chem.* **33**, 239–242 (1961).

ISONICOTINIC ACID HYDRAZIDE AND 2,3,5-TRIPHENYLTETRAZOLIUM CHLORIDE

The captioned reagents in equimolecular proportions combine with chromic ion as well as with thallic and ferric ions.

Procedure. To a sample solution containing 5–15 mg of chromic ion, add 5 ml of 1:1 nitric acid and 0.5 gram of silver oxide.[129] After 10 minutes heat to destroy excess silver oxide and add 3 ml of 10% solution of potassium chloride to precipitate silver chloride. Add 10% potassium hydroxide solution to raise to pH 5.5, and filter. Add 2 ml of a reagent containing 0.01% each of isonicotinyl hydrazide and 2,3,5-triphenyltetrazolium chloride. Heat at 60° for 3 minutes. Cool, shake with 5 ml of *n*-amyl alcohol, and centrifuge. Dilute the organic layer to 100 ml with ethanol and read at 480 or 308 nm.

METHYL VIOLET

Methyl violet forms a complex as 2 moles combined with a bichromate ion. The deep blue complex is extractable with benzene and will detect 0.03 μg of chromium per ml. Crystal violet performs similarly. Titanium and antimony are removed by coprecipitation with ferric hydroxide.

Procedure. Dissolve 1 gram of sample in 10 ml of hydrochloric acid and 5 ml of nitric acid.[130a] Evaporate to about 4 ml, add 5 ml of 1:1 sulfuric acid, and heat to sulfur trioxide fumes. Take up in 500 ml of water, add 0.25 ml of nitric acid, and boil off any oxides of nitrogen. Add 10 ml of a solution containing 0.5% of cobalt sulfate heptahydrate and 1.5% of nickel sulfate heptahydrate. Bring to a boil, slowly add 15 ml of 5% ammonium persulfate solution, and continue to boil until the excess persulfate is decomposed. Cool and add 0.05 gram of ferric chloride. Precipitate by dropwise addition of a slight excess of ammonium hydroxide. Filter, wash with 1:50 ammonium hydroxide, and dilute to 100 ml. Add a 10-ml aliquot to 20 ml of water and add hydrochloric acid to make it 0.016 *N*. Add 0.75 ml of 0.01% solution of methyl violet and extract with 10 ml and 10 ml of benzene. Dilute the combined benzene extracts to 25 ml and read at 455 nm.

MOLYBDOSILICIC ACID

Both chromic and gallic ions decrease the color of molybdosilicic acid in proportion to their concentration. A twofold excess of iron is tolerated; 25-fold for most other cations.

[129] M. H. Hashmi, Abdur Rashid, Hamid Ahmed, A. A. Ayaz, and Farooq Azam, *ibid.* **37**, 1027–1029 (1965).
[130a] E. I. Savichev, E. I. Iskhakova, and L. F. Flyazhnikova, *Zavod. Lab.* **28**, 412 (1962) ; W. U. Malik, R. Bembi, P. P. Bhargava, and R. Singh, *Z. Anal. Chem.* **282**, 140 (1976).

Procedure. *Ore.*[130b] Fuse 0.2 gram of sample with 1.5 grams of 1 : 1 sodium carbonate–potassium carbonate. Dissolve the melt in 5 ml of hot 1 : 1 hydrochloric acid and evaporate to dryness. Dissolve with 0.5 ml of hydrochloric acid, add 10 ml of hot water, and filter off the silicic acid. Dilute the filtrate to 100 ml. To an aliquot containing 0.02–2 mg of chromic ion, add 4.6 ml of 0.054 *M* ammonium molybdate and dilute to 40 ml. Add 1 ml of *N* sulfuric acid, which should give pH 2, and 0.7 ml of 0.02 *M* disodium silicate. Dilute to 50 ml and read at 360 nm against water.

PYRIDINE-2, 6-DICARBOXYLIC ACID

This reagent develops a color with chromic ion on heating. As applied to analysis of steel, the resulting bis(2, 6-pyridinedicarboxylato)chromate is read at 550 nm for 5–250 μg of chromic ion per ml.[131] Corrections are required for cobalt, nickel, and ferric ion.

Procedure. To a weakly acid solution containing 0.3–7.5 mg of chromic ion, add sodium sulfate to give a twofold excess of sulfate ion over the total anions in the solution and dilute to 25 ml.[132] To a 10-ml aliquot, add 1 ml of a buffer solution for pH 3.9 containing 5.4% of sodium acetate and 10% of acetic acid. Add 0.3 *M* pyridine-2, 6-dicarboxylic acid to give a tenfold excess over the chromic ion present and a twofold excess over other cations. Heat at 100° for 5 minutes, cool, dilute to 25 ml, and read at 550 nm. Compare with a similar solution unheated as a blank.

4-(2-PYRIDYLAZO)RESORCINOL

Chromic ion complexes with 4-(2-pyridylazo)resorcinol in boiling acetate buffer solution at pH 5 to give a 1 : 3 complex.[133] This forms an ion association compound with benzyldimethyltetradecylammonium cations in chloroform for reading at 540 nm. Beer's law is followed for up to 1 μg of chromium per ml in water and up to 0.9 μg/ml in chloroform. If EDTA is added after forming the complex in the aqueous phase, only iron, cobalt, and nickel interfere seriously. They can be preextracted as cupferron complexes.

[130b] Z. F. Shakhova and E. N. Dorokhova, *Vest. Mosk. Gos. Univ., Ser. khim.* **19** (5), 77–80 (1964).
[131] Heinrich Hartkamp, *Z. Anal. Chem.* **187**, 16–29 (1962).
[132] G. den Boef and B. C. Poeder, *ibid.* **199**, 348–352 (1963).
[133] Takao Yotsuyanigi, Yasuo Takeda, Ryuji Yamashita, and Kazuo Aomura, *Anal. Chim. Acta* **67**, 297–306 (1973); cf. S. G. Nagarkar and M. C. Eshwar, *Indian J. Technol.* **13**, 377–378 (1975).

Procedures

Steel.[134a] CHROMIUM AND COBALT. This technic is applicable to an alloy steel containing 22.7% nickel, 0.2% molybdenum, and 1.1% vanadium, in addition to 3.85% cobalt and 2.64% chromium.

Dissolve a 1-gram sample in 15 ml of 1:4 sulfuric acid. Add 1 ml of nitric acid and boil to destroy carbides. Boil off oxides of nitrogen and dilute to 100 ml. To a 15-ml aliquot, add 10 ml of 10% ammonium fluoride to mask iron. Add 5 ml of 0.1% solution of 4-(2-pyridylazo)resorcinol and adjust to pH 5. Dilute to 50 ml with 2 *N* sulfuric acid and read cobalt at 540 nm.

Treat another portion similarly, but after adjustment to pH 5, boil for 5 minutes before dilution with 2 *N* sulfuric acid. After dilution, read the sum of chromium and cobalt at 540 nm, giving chromium by difference.

Effluents[134b]

CHROMIC ION, CHROMATE ION, AND A VARIETY OF CATIONS. Evaporate a sample containing 2–8 μg of chromium to 10 ml and adjust to pH 3 with hydrochloric acid. Add 2 ml of acetate buffer solution for pH 5 and 2 ml of 0.1% reagent solution. Boil for 5 minutes and add 2 ml more of buffer solution. Cool, dilute to 50 ml, and read at 540 nm. Prepare the blank in the same way as the sample without boiling.

COMPLEX CHROMATES PRESENT. Adjust 12 ml of 0.2% hydrazine hydrochloride solution to pH 6 with ammonium hydroxide and add 2 ml of acetate buffer solution for pH 7. Add 2 ml of 0.1% solution of the reagent followed by a sample containing 2–8 μg of chromium concentrated to 10 ml. Set aside for 30 minutes, dilute to 50 ml, and read at 540 nm. As the blank prepare the same as the sample but do not add the reagent until 10 minutes before final dilution and reading. Potassium dichromate is an appropriate standard for both cases above.

4-(2-THIAZOLYLAZO)RESORCINOL

The red complex of chromic ion with this reagent is stable for 3 days. Beer's law is follows for 0.06–1.1 μg of chromium per ml. Many interferences are masked by EDTA. Ferric ion can be extracted by mesityl oxide, cobalt ion by 1-(2-pyridylazo)-2-naphthol in chloroform.

Procedure. *Alloy Steel.*[134c] Dissolve 0.5 gram in 20 ml of 1:4 sulfuric acid and dilute to 100 ml. Acidify an aliquot containing not over 27 μg of chromic ion to

[134a] S. A. Akhmedov, O. A. Tatsev, and R. R. Abdullaev, *Zavod. Lab.* **37**, 756–758 (1971).
[134b] M. P. Volkova, B. P. Kolesnikov, and V. G. Kossykh. *Zavod. Lab.* **43**, 268–269 (1977).
[134c] B. Subrahmanyam and M. C. Eshwar, *Microchim. Acta* **II**, 579–584 (1976).

3.5 N with hydrochloric acid and extract ferric ion with 10 ml of mesityl oxide. Evaporate the aqueous layer substantially to dryness and take up in water. Adjust to pH 5 by adding 10 ml of 0.5 M acetate buffer. Add 2 ml of a 0.1% solution of the reagent in 2-methylpropan-2-ol. Heat at 100° for 45 minutes. Cool and add 2 ml of 2-methylpropan-2-ol and 5 ml of 5% solution of EDTA. Dilute to 25 ml and read at 525 nm against a reagent blank.

TRIOCTYLAMINE

Trioctylamine, commercially known as Alamine 336-S, extracts chromate from solution in 0.2–0.4 N sulfuric acid.

Procedure. To a sample solution containing 0.2–1 mg of chromium as chromate, add 1 ml of saturated sodium sulfate solution, 10 ml of saturated sodium chloride solution, and 10 ml of 2 N sulfuric acid.[135] Dilute to 50 ml and shake for 2 minutes with 5 ml of 5% solution of trioctylamine in chloroform. Filter the organic layer and read at 450 nm against water.

TRIOCTYLMETHYLAMMONIUM CHLORIDE

This reagent is sold commercially as Aliquat 336-S. When chromate is extracted with the reagent in chloroform, added sodium sulfate promotes phase separation. Vanadate, uranate, and molybdate are extractable, but added sodium chloride partially or completely blocks this. Nitrates should not exceed 0.2%. Beer's law is followed for 2–20 μg of chromium as chromate per ml of extract.

Procedure. *Alloys.*[136] Dissolve 1 gram of sample in a minimal amount of aqua regia and evaporate nearly to dryness. Add 2 ml of 1 : 1 sulfuric acid and evaporate to sulfur trioxide fumes. Cool, add water, and regardless of residue, dilute to 250 ml. To a 10-ml aliquot, add 0.2 gram of potassium periodate and boil for 5 minutes. Cool, and add 10 ml of 0.2 N sulfuric acid with 3 ml of saturated solution of sodium chloride. Dilute to 50 ml and extract with 5 ml of 5% solution of the reagent in chloroform for 2 minutes. Read the organic layer at 450 nm.

XYLENOL ORANGE

A 1 : 1 complex is formed by boiling a solution of chromic ion with xylenol orange at pH 3–6 for 5 minutes.[137] It is not decomposed by N nitric acid, which

[135]Jiri Adam and Ruldolf Pribil, *Talenta* **18**, 91–95 (1971).
[136]Jiri Adam and Rudolf Pribil, *ibid.* **18**, 91–95 (1971).
[137]Koichi Tonosaki, Makoto Otomo, and Koichi Tanaka, *Jap. Anal.* **15**, 683–686 (1966).

decomposes complexes of cobalt, nickel, copper, vanadium, molybdenum, titanium, tungsten, and rare earths immediately, and those of aluminum and ferric ions within 10 minutes. Nitrilotriacetic acid and EDTA interfere.

Procedure. Add 0.5 ml of a chloroacetic acid–sodium acetate buffer solution for pH 3 to the sample and dilute to about 8 ml.[138] Add 4 ml of m*M* xylenol orange and heat at 100° for 20 minutes. Cool, dilute to 25 ml, and read at 530 nm.

Steel and Alloys.[139] For 0.05–0.5% chromium, take 0.2–0.5 gram of sample; for 0.5–2%, take 0.1 gram. Dissolve in 10 ml of hydrochloric acid, add 3 ml of nitric acid to destroy carbides, and boil off oxides of nitrogen. Add 2 ml of 1% solution of ascorbic acid, 5 ml of 0.5% solution of xylenol orange, and 20 ml of water. Adjust to pH 4.5 and boil for 5 minutes. Cool, and dilute to 50 ml with 1 : 7 nitric acid. Let stand for 10 minutes and read at 550 nm.

Chromium is read at 490–500 nm after adding 0.5% ethanolic **alizarin** to a neutral solution containing 0.03–0.19 mg of chromic oxide per ml.[140]

A 1 : 1 complex is formed at pH 3–8 between hexavalent chromium and **alizarin red S** (mordant red 3).[141] Beer's law is followed by 0.2–10.4 ppm of chromium at pH 3–6 if read at 525 nm.

The 2 : 3 complex of chromium with the ***o*-aminoanalide of benzenesulfonic acid** is effective for reading chromium in a potash solution.[142]

***o*-Aminophenyldithiocarbamic acid** is suitable for determining 0.1–1 μg of chromium per ml.[143] Read at 480 nm a solution containing in 100 ml, 50 ml of 0.05% solution of the reagent, 35 ml of water, 5 ml of *N* hydrochloric acid, and 10 ml of chromate solution containing 10–100 μg of chromium.

Chromate ion oxidizes **8-aminoquinoline** at an optimum pH of 2.4–2.8.[144] Adjust 50 ml of sample solution containing 0.01–0.1 mg of chromium to pH 2.4 with perchloric acid or 10% sodium hydroxide solution. Add 5 ml of 0.002 *M* solution of 8-aminoquinoline and heat at 100° for 40 minutes. Cool, adjust to pH 1.8, and extract with 5 ml of 1 : 1 chloroform–benzyl alcohol. Read at 550 nm. Vanadate, permanganate, and ferric ion give a similar reaction.

Chromic ion forms an **azide**, which conforms to Beer's law for 4–320 ppm.[145] The desirable pH is 4.2–5.2. The interference of cupric, ferric, and uranyl ions is reduced by addition of EDTA.

[138] K. L. Cheng, *Talenta* **14**, 875–877 (1967).
[139] O. A. Tataev and R. R. Abdulaev, *Zavod. Lab.* **36**, 1173–1175 (1970).
[140] G. C. Bhattacharya, *J. Sci. Ind. Research* (India) **20B**, 351–352 (1961).
[141] Satendra P. Sangal, *Chim. Anal.* (Paris) **46**, 492–494 (1964); cf. Samir K. Banerji and Arun K. Dey, *J. Indian Chem. Soc.* **38**, 121–126 (1961).
[142] I. U. Martynchenko and L. R. Karabanova, *Uch. Zap. Khar'-k. Univ.* No. **110**, *Tr. Khim. Fak. Nauchn.-Issled. Inst. Khim. Khar'k. Gos. Univ.* No. **17**, 177–182 (1961).
[143] E. Gagliardi and W. Haas, *Z. Anal. Chem.* **147**, 321–326 (1955).
[144] Katsumi Yamomoto, Kohei Ametani, and Kiyohisa Amagai, *Jap. Anal.* **16**, 229–233 (1967).
[145] F. G. Sherif, W. M. Oraby, and Hussein Sadek, *J. Inorg. Nucl. Chem.* **24**, 1373–1379 (1962).

Arsenazo III forms a 1 : 1 complex with chromic ion at pH 2.5 on boiling for 10 minutes for reading at 580 nm.[146] Hydroxylamine masks ferric, chromic, permanganate, molybdenum, and tungsten ions.

Similarly, boiled **thoron** forms a 2 : 1 complex with chromic ion for reading at 500 nm. There is interference by zirconium, gallium, nitrite, and EDTA. Ferric ion must be masked.

Chromium forms a 1 : 3 complex with **azorubine** (acid red 14).[147] Treat 5 ml of neutral sample solution containing up to 3 ppm of chromium as chromate with 0.5 ml of 0.35% solution of the dyestuff and 1 ml of *N* sulfuric acid. Heat at 100° for 20 minutes and cool. Extract excess reagent with 5 ml of isoamyl alcohol and read the aqueous layer at 620 nm. Interferences include ferric and vanadate ions.

The 1 : 3 complex of chromate with **carmoasine**, which is sodium 2-(sulfo-1-naphthylazo)-1-naphthol-4-sulfonic acid, is developed in sulfuric acid by heating at 70° for 15 minutes.[148] The final solution should be 0.1 *N* with sulfuric acid. At least 0.12% of the color reagent must be present. Even traces of ferric ion cause low results.

After oxidation to the bichromate ion and separation on Dowex 50-X8, adjust to pH 2–4 and read the complex with **chromotropic acid** at 469 nm.[149] Beer's law is followed for 1.7–34 μg of chromium in 50 ml.

Oxidation by chromate of **3,3′-diaminobenzidine** in acid solution is applicable in the absence of permanganate, cerate, vanadate, and other oxidizing agents, including ferric ion in excess of 1.2 mg/ml.[150] To a sample solution containing 0.01–0.1 mg of chromium as chromate, add 1 ml of phosphoric acid and dilute to 20 ml. Add 1 ml of 0.01 *M* reagent, dilute to 25 ml, and set aside for 5 minutes. Read at 375 or 460 nm.

Chromic ion forms a stable, water-soluble 1 : 1 complex with **1,2-diaminocyclohexanetetraacetic acid** on boiling for 5 minutes at pH 4.65.[151] The complex has maxima at 395 and 540 nm. At the latter wavelength, Beer's law is followed for 5–150 μg/ml. Under the same conditions, cupric ion forms a complex absorbing at 720 nm. Cobalt and nickel also form colored complexes with the reagent.

Chromate ion can be measured by its oxidizing of **4,4-diaminodiphenylamine**.[152] To 10 ml of sodium acetate buffer solution for pH 3.8 containing 0.025–1 μg of chromium per ml as chromate, add 0.2 ml of 0.2% solution of the color reagent. Other oxidizing agents interfere.

To a slightly acid solution containing less than 12.5 mg of chromic ion, add 10 ml of 0.1 *M* **1,2-diaminopropane-*N N′ N′ N′*-tetraacetic acid.**[153a] Boil, cool, dilute to

[146]O. A. Tataev and V. K. Guseinov, *Z. Anal. Khim.* **30**, 935–938 (1975).

[147]F. Bosch Serrat, *Inf. Quim. Anal. Pura Apl. Ind.* **25**, 17–35 (1971).

[148]N. M. Kravtsova, *Tr. Kom. Anal. Khim. Akad. SSSR, Inst. Geokhim. Anal. Khim.* **8**, 161–168 (1958); N. M. Kravtsova and V. I. Petrashen, *Tr. Novocherk. Politekh. Inst.* **72**, 3–12 (1958).

[149]G. Popa, E. Radulescu-Grizore, and M. Basamac, *Anal. Univ. Bucur. Ser. Stiint. Nat. Chim.* **13**, 185–191 (1964).

[150]K. L. Cheng and B. L. Goydish, *Chemist-Analyst* **52**, 73–76 (1963).

[151]A. R. Selmer-Olsen, *Anal. Chim. Acta* **26**, 482–486 (1962); cf. M. R. Verma, V. M. Bhuchar, K. C. Agrawal, and R. K. Sharma, *Mikrochim. Acta* **1959**, 766–769; H. Khalifa, J. E. Roberts, and M. M. Khater, *Z. Anal. Chem.* **188**, 428–434 (1962); M. R. Verma, K. C. Agrawal, and V. K. Amar, *Indian J. Chem.* **5**, 79–80 (1967).

[152]Noburu Hara, *Bull. Nat. Inst. Ind. Health* (Japan) No. **1**, 33–41 (1958).

[153a]S. Vicente-Perez, L. Hernandez, and J. Rosas, *Quim. Anal.* **28** 283–288 (1974).

25 ml, and read the 1 : 1 complex at 540 nm. Cupric, nickel, and cobalt ions also complex with this reagent.

To 5 ml of a solution of 44–265 μg of chromium as chromate per ml in dimethyl sulfoxide, add an equal volume of 0.5 *M* **diethylenetriamine** ***NNN′N″N″*-penta-acetic acid** in the same solvent.[153b] Heat to 80° for 10 minutes, cool, and adjust to 10 ml. Read the violet 1 : 1 metal-ligand complex at 555 nm.

At pH 2–5 above 70°, chromic ion forms a 1 : 1 complex with **eriochrome cyanine R** (mordant blue 3). The complex does not form at room temperature but remains stable on cooling. Beer's law holds for 0.04–2 μg of chromium per ml. There is serious interference by phosphate, fluoride, nitrite, carbonate, oxalate, tartrate, citrate, and EDTA. If the solution is made *N* with nitric acid, complexes of aluminum and zirconium are destroyed within 10 minutes, and the complexes of all other metals except chromium are destroyed at once.

For details of determination of chromium by **1-(4-[(5-ethyl-1,3,4-thiadiazol-2-yl)sulfamoyl]phenyl)-3-methyl-5-(1-methylbenzimidazol-2-yl)formazan**, see Copper by that reagent.

The complex of chromic ion with potassium **ferrocyanide** above pH 2 forms in about 100 minutes at room temperature or 15 minutes at 50°.[154] Beer's law is followed up to 40 μg of chromium per ml at 415 nm, but acetate interferes seriously.

Alternatively,[155] mix 5 ml of sample solution containing 3–50 ppm with 1 ml of m*M* potassium ferrocyanide. Add 4 ml of acetate-hydrochloride buffer solution for pH 4. Set aside in the dark at 30° for 24 hours, or expose for 6 hours to radiation at 360 nm. Read at 235 nm. Iron, aluminum, and copper interfere. Other interferences removed by precipitation with ferrocyanide are manganous ion, zinc, beryllium, bismuth, and silver.

By oxidation of ferrocyanide in the presence of ferrous ion, the blue color as read at 715 nm is linear for 0.4–50 ppm of bichromate.[156] Gelatin and EDTA stabilize the prussian blue dispersion.

The 1 : 1 complex of chromate with **glycinethymol blue** is read at 570 nm.[157] Develop at pH 6.5–7 in a perchloric acid–hexamine buffer solution. Beer's law is followed at 1–5 μg of chromium per ml.

As reagent, mix 2.5 ml of 10 m*M* potassium **hexacyanochromate** with either 2.5 ml of 10 m*M* *p*-anisidine or 5 ml of 10 m*M* *p*-phenylenediamine hydrochloride.[158] Add 1 ml of sample solution containing chromic ion up to 10 m*M*. Dilute to 50 ml with a buffer solution for pH 3. Dilute an aliquot to 10 ml with the buffer solution and read at 405 nm. There is interference by beryllium, mercuric ion, vanadyl ion, and molybdate.

Alternatively, mix 5 ml of the sample solution with 5 ml of 0.01 *M* hexacyanochromate and dilute to 50 ml with a buffer solution for pH 3. Expose to

[153b] J. Areses, F. Bermejo Martinez, A. Concheiro, and J. A. Rodriguez Vazquez, *Quim. Anal.* **29**, 212–215 (1975).
[154] Yoshio Matsumoto and Michiko Shirai, *Bull. Chem. Soc. Jap.* **39**, 55–57 (1966).
[155] W. U. Malik and Ramesh Bembi, *Mikrochim. Acta* **1**, 681–684 (1975).
[156] S. A. Rahim, H. Abdulahad, and N. E. Milad, *J. Indian Chem. Soc.* **52**, 853–854 (1975).
[157] K. C. Srivastava and M. Malathi, *J. Clin. Chem. Soc.* (Taipei) **22**, 145–149 (1975).
[158] W. U. Malik and K. D. Sharma, *Z. Anal. Chem.* **276**, 379 (1975); *Indian J. Chem.* **13**, 1232–1233 (1975).

ultraviolet radiation for 2 hours and read 5.2–46.8 ppm of chromic ion at 405 nm. The method is also applicable to beryllium, stannous ion, tetravalent vanadium, and molybdate.

Chromate can be measured by its catalytic effect on oxidation of **indigo carmine** by hydrogen peroxide.[159] A typical composition contains 8.9 m*M* indigo carmine, 0.167 *N* hydrochloric acid, and 0.51 *M* hydrogen peroxide. The rate of reaction is plotted as a function of the chromate concentration as measured at 597 nm. As a more sensitive variant, 1.95 m*M* 2,2′-bipyridyl is included. Cupric ion interferes.

The complex of chromic ion with **isocinchomeronic acid**, which is pyridine-2,5-dicarboxylic acid, is read at 300 nm in acetate buffer solution for pH 5.4.[160]

Mercaptoacetic acid, also known as thioglycollic acid, reduces chromate to chromic ion at pH 7–10 and 100° to form a green complex.[161] This is read at 632 nm against a blank. Beer's law is followed for 0.02–0.15 mg of chromium per ml. Vanadium and uranium must be absent and iron interferes. A complex can also be read at pH 2 and 350 nm.[162] At that pH complexes are also formed with uranium, tungsten, and nickel.

8-Mercaptoquinoline reduces hexavalent chromium to chromic ion at pH 3.5–7.[163] Then the reagent forms a 1:3 metal-ligand complex extractable with chloroform at pH 3–5.5 for reading at 150 nm. This follows Beer's law for 0.1–6 μg of chromic ion per ml.

Methylthymol blue serves as a reagent according to the technic given for xylenol orange except that the maximum absorption is at 560 nm.

At pH 2–2.5 in a tartaric or oxalic acid solution, a 2:3 complex is formed by hexavalent chromium and **1-naphthylamine**.[164] This is read at 345 and 520 nm. Beer's law is followed for 0.18–9 μg of chromium per ml. For steel dissolve the sample in 1:5 sulfuric acid, add silver ion as a catalyst, and oxidize chromic ion to chromate with persulfate at pH 1–2. After decomposition of excess persulfate add hydrochloric acid to decompose permanganate ion. Precipitate iron with ammonium hydroxide, filter, and boil off the excess ammonia. Mix an aliquot with 1 ml of 10 μ*M* 1-naphthylamine in m*M* tartaric acid. Set aside for 20 minutes, then extract the complex with 5 ml of isoamyl alcohol. Read at 345 nm.

The technic of determination of chromic ion with EDTA as described elsewhere is also applicable as complexing agents to **nitrilotriacetic acid, 1,2-bis(2-aminoethoxy)ethane-*NNN′N′*-tetraacetic acid**, ***N*-hydroxyethylenediamine-*NN′N′*-triacetic acid**, **diethylamine-*NNN′N″N″*-pentaacetic acid**, and **1,2-diaminocyclohexane-*NNN′N′*-tetraacetic acid.**

Although the brown product of oxidation of **1-phenylthiosemicarbazide** by chromate is unstable in mineral acid solution, it is stable in acetic acid over 3 *N*.[165a] The product is extracted into chloroform and conforms to Beer's law for 6–200 μg of chromium per 100 ml. At least 2 *M* excess of reagent must be present. Mask ferric

[159]E. Jasinskiene and E. Bilidiene, *Z. Anal. Khim.* **22**, 741–745 (1967).
[160]G. De Angelis, E. Chiacchierini, and G. D'Ascenzo, *Gass. Chim. Ital.* **96**, 39–59 (1966).
[161]E. Jacobsen and W. Lund, *Anal. Chim. Acta* **36**, 135–137 (1966).
[162]V. M. Bhuchar, *Nature* **191**, 489–490 (1961).
[163]Yu. I. Usatenko, V. I. Suprunovich, and V. V. Velichko, *Zh. Anal. Khim.* **29**, 807–810 (1974).
[164]S. V. Vartanyan and V. M. Tarayan, *Armyan. Khim. Zh.* **29**, 303–307 (1976); cf. V. N. Polyanskii, *Sb. Tr. Mosk. Vech. Met. Inst.* **1957**, (2), 253–254.
[165a]Sumio Komatsu and Koji Takahashi, *J. Chem. Soc. Jap., Pure Chem. Sect.* **83**, 879–882 (1962).

ion with fluoride, mercuric ion with chloride ion, and cupric ion with EDTA; extract vanadium with 8-hydroxyquinoline. Permanganate gives a similar color but can be reduced.

Dissolve 0.5 gram of stainless steel in 10 ml of 1:4 sulfuric acid and dilute to 20 ml.[165b] Mix a 4-ml aliquot with an equal volume of 3.5 *N* hydrochloric acid and extract the iron with 10 ml of mesityl oxide. Evaporate the solution to dryness, take up in water, and dilute to 100 ml. Adjust an aliquot containing up to 20 μg of chromium to pH 3.5. Add 5 ml of a buffer solution for pH 3.5 consisting of 0.2 *M* sodium acetate–0.2 *M* acetic acid. Mask nickel by adding 1 ml of 0.5% solution of dimethylglyoxime. Add 1 ml of 1% methanolic **1-(2-pyridylazo)-2-naphthol** and dilute to 15 ml. Heat at 100° for 35 minutes, cool, and dilute to 25 ml. Extract chromium with 10 ml of butanol, dry with sodium sulfate, and read at 555 nm against a reagent blank.

Add sodium carbonate and sodium peroxide to the sample solution and boil for 30 minutes.[166] Filter, and wash with 0.1% sodium hydroxide solution. The filtrate contains chromium, molybdenum, vanadium, and tungsten. To an aliquot, add 3 drops of 0.01% *p*-nitrophenol solution and neutralize with 1:9 sulfuric acid. With continuous agitation, add 5 ml of 50% ammonium acetate solution, 5 ml of saturated solution of ammonium sulfite, then 5 ml of 20% **pyrocatechol** solution. Dilute to 50 ml and set aside for 40 minutes. Read chromium at 585 nm against a reagent blank.

Sunchromine blue black R (mordant black 17), which is 3-hydroxy-4-(2-hydroxy-1-naphthylazo)naphthalene-1-sulfonic acid, is read at 630 nm with chromic ion.[167] To a sample solution containing 5–75 μg of chromic ion, add 5 ml of 0.1% solution of the dyestuff and 10 ml of borate buffer for pH 10. Dilute to 50 ml and read against a reagent blank. Beer's law is followed up to 1.2 ppm of chromic ion. Interfering ions are aluminum, beryllium, ferric, scandium, calcium, cobalt, zinc, and titanium.

Chromic ion complexes with **3-thianaphthenoyltrifluoroacetone**, which is 1-(benzo[b]thiophen-3-yl)-4,4,4-trifluorobutene-1,3-dione.[168] To a sample solution containing up to 1 mg of chromic ion, add 10 ml of 7.5% solution of potassium chloride and adjust to pH 4–4.5. Add 2 ml of 5% solution of the monohydrate of the reagent in ethanol. Heat at 100° for 30 minutes, cool, extract with xylene, and read at 460 nm.

The complex formed by chromic ion with **thiocyanate** is read at 574 nm in either water or acetone.[169] There is no interference by excess thiocyanate ion or quaternary compounds. An appropriate pH range is 2.6–6.1. A typical composition contains 0.05–0.5 mg of chromic sulfate, 0.15 gram of potassium thiocyanate, and 2–3 drops of sulfuric acid in 500 ml. The color is developed by heating at 80° for 1 hour.

Chromate is developed with ***o*-tolidine** much as in the reaction with chlorine.[170] To 2 ml of sample solution containing 2–50 μg of bichromate ion or 6–50 μg of

[165b] B. Subrahmanyam and M. C. Eshwar, *Bull. Chem. Soc. Jap.* **49**, 347–348 (1976).

[166] C. G. Nestler and M. Nobis, *Z. Anal. Chem.* **167**, 81–90 (1959).

[167] Katsuya Vesugi and Yukiteru Katsube, *Bull. Chem. Soc. Jap.* **38**, 2010–2011 (1965).

[168] J. R. Johnston and W. J. Holland, *Mikrochim. Acta* **1972**, 321–325.

[169] P. Spacu, G. Gheorghiu, and El. Antonescu, *Z. Anal. Chem.* **174**, 81–87 (1960).

[170] I. Florea, *Rev. Roum. Chim.* **18**, 1993–1999 (1973).

chromate ion, add 1.5 ml of 0.5% solution of *o*-tolidine in 20% acetic acid. Dilute to 25 ml with 1:2 sulfuric acid. After 20 minutes read at 441 nm.

Triazinylstilbexon, which is the hexasodium salt of 4,4′-bis{4,6-bis[(carboxymethyl)amino]1,3,5-triazin-2-ylamino}stilbene-2,2′-disulfonic acid, forms a 1:1 complex with chromic ion at pH 2.5–3.5.[171] The fluorescence of the reagent is quenched by formation of the complex, thus permits determination of 4 ng of chromium per ml. Reading is at 450 nm.

Chromium at pH 1–2 can be extracted as **triphenylselenium bichromate** with dichloromethane.[172a] It is then read at 362 or 445 nm.

At pH 3.5–4.8 **uramildiacetic acid** forms a green complex with chromic ion.[172b] At 420 nm this will read for 0.5–1.5 mg of chromium. There is interference by ferric, arsenate, and tungstate ions.

Chromate is read at 570 nm with **variamine blue.**[173] To 25 ml of neutral or faintly acid solution containing 0.2–4 μg of chromium as chromate, add 2 ml of 0.5% solution of variamine blue and dilute to 50 ml with a buffer for pH 3. Vanadate will give the same color after 10 minutes. When both are present, the sum of chromate and vanadate is read, both are reduced with ferrous sulfate, and the vanadate is reoxidized with permanganate. Under those conditions the chromium is not reoxidized and is determined by difference. Both oxidizing and reducing agents interfere.

[171]V. Ya. Temkina, E. A. Bozhevol'nov, N. M. Dyatlova, S. U. Kreingol'd, G. F. Yaroshenko, V. N. Antonov, and R. P. Lastovskii, *Zh. Anal. Khim.* **22**, 1830–1835 (1967).

[172a]M. Ziegler and K.-D. Pohl, *Z. Anal. Chem.* **204**, 413–420 (1964).

[172b]A. Alvarez Devesa, C. Baluja Santos, and Laura Ces Viqueira, *Quim. Anal.* **30**, 267–270 (1976).

[173]Laszlo Erdey and Ferenc Szabadvary, *Magy. Kem. Foly.* **63**, 153–158 (1957).

CHAPTER EIGHTEEN
IRON

Of the numerous reagents for iron, the technic with thiocyanate is probably the simplest, development with *o*-phenanthroline the best, and that with bathophenanthroline the most sensitive of those widely used. One author has suggested that more than 100 photometric methods are known, but that is conservative; more than 200 are included here. Many are evolved from a structure $-N={}^1C-{}^1C=N-$, have high molar absorptivities, and are from the work of G. Frederick Smith and co-workers.

Extraction of ferric ion from 5.5–7 *N* hydrochloric acid by an equal volume of 4-methyl-2-pentanone is better than 99.9% effective.[1] Then recovery by reextraction with water is quantitative. Methyl isobutyl ketone is suitable for extraction of microgram amounts of ferric ion from large amounts of copper.[2] It is then readily stripped by aqueous hydroxylamine hydrochloride.

Ferric ion in approximately 1.3 m*M* solution is quantitatively extracted with *M* hexanoic acid in chloroform at pH 2.5–5.[3] Correspondingly, it is extracted with *M* 2-bromohexanoic acid at pH 2–4. These solvents will also extract the ferric ion at pH 10–12 in the presence of 0.5 *M* 2-aminopyridine, -butylamine, -quinoline, or antipyrine, or 0.5 *M*-2,2′-bipyridyl or 1,10-phenanthroline.

Iron and titanium in a mole ratio of 10:1 to 1:10 in 0.6 *N* sulfuric acid are passed through a bed of SBS ion exchange resin in hydrogen form.[4] Iron is retained and subsequently eluted with 0.2 *N* sulfuric acid.

For determining ferric ion with polyhydric phenols, tiron, protocatechuic acid, sulfosalicylic acid, kojic acid, and 4-aminosalicylic acid have been recommended.[5]

The complex of iron with 1,10-phenanthroline adds tetrabromo-, tetraiodo-, or

[1] Hermann Specker and Wilhelm Doll, *Z. Anal. Chem.* **152**, 178–185 (1956).
[2] D. Ader, *Isr. J. Chem.* **1**, 13–14 (1963).
[3] I. V. Pyatnitskii, V. V. Sukhan, and V. A. Frankovskii, *Zh. Anal. Khim.* **28**, 1696–1704 (1973).
[4] Z. P. Suranova, O. Ya. Grabchuk, and L. M. Kutovaya, *Zh. Prikl. Khim., Leningr.* **44**, 1566–1568 (1971).
[5] L. Sommer, *Acta Chim. Acad. Sci. Hung.* **33**, 23–30 (1962).

dichlorotetraiodofluorescein, extractable at pH 9 with chloroform.[6] The maximum absorption is at 540 nm or in 1 : 1 chloroform-acetone fluorescence at 580 nm. Fluorescence will determine 0.05–10 μg in 10 ml, one order of magnitude more sensitive than spectrophotometry.

As a method of isolation of iron from soil, mix 5 grams of 70-mesh sample with 5 grams of Zeo-Karb 215 cation exchange resin and 200 ml of water.[7] Shake for 2 hours at $25 \pm 1°$. Separate the resin from the soil on a 70-mesh sieve. Elute iron as well as copper, manganese, and zinc with 2 *N* sulfuric acid.

For determination of iron in ferrochrome and chromite, fuse with sodium peroxide.[8] Dissolve the melt in 1 : 1 sulfuric acid, add ethanol, and heat. The chromate is reduced to chromic ion and acetaldehyde is evolved. Dilute the solution and add a slight excess of ammonium hydroxide. Filter the hydroxides, wash, and dissolve in 1 : 1 sulfuric acid. Add ammonium persulfate solution and heat to convert to ferric ion. Pass through a cation exchange column of KU-2 to separate ferric and chromic ions.

Ferric ion at 0.16–3.5 ppm can be extracted from 2 grams of niobium pentoxide and tantalum pentoxide with pentyl acetate.[9a] The solution must be 5 *M* with hydrofluoric acid and 8 *M* with hydrochloric acid containing hydrogen peroxide to mask niobium and tantalum. The ferric ion can then be developed with thiocyanate or chrome azurol S.

In Chapter 12, Gallium, the iron is separated from an indium sample through chromatography on Amberlite IR-120. Precipitated from the eluate with lanthanum as a carrier, the solution in 1 : 4 hydrochloric acid is appropriate for reading iron as the thiocyanate at 500 nm.

For separation of iron, manganese, and copper by column chromatography, see Copper. For a solution of nonmetallic inclusions in sheet steel for determination of iron, see Aluminum. The oxalate complex of iron can be read at 370 nm; see Copper. For iron in water by chromaxane violet R in the presence of aluminum, see Aluminum. For iron in electrolytic nickel by sodium diethyldithiocarbamate, see Copper. For iron by thiocyanate and pyridine along with copper, nickel, and cobalt, see Copper.

Methods for iron include complexing with formaldoxime.[9b] For iron in aluminum by 8-mercaptoquinoline, see Copper by Di-8-quinolyl-disulfide. For iron by 1-nitroso-2-naphthol along with copper, nickel, and manganese in highly pure alkali–metal halide crystals, see Copper by Lead Diethyldithiocarbamate. For isolation of iron from niobium with diethyldithiocarbamate followed by column chromatography, see Zinc. For cobalt, copper, and iron as thiocyanates by simultaneous equations, see Cobalt.

ACETYLACETONE

Acetylacetone is pentane-2,4-dione. At pH 6–8 ferric acetylacetonate can be extracted with butyl acetate.[10] The chelate shows maxima at 273, 353, and 438 nm.

[6] D. N. Lisitsyna and D. P. Shcherbev, *Zh. Anal. Khim.* **28**, 203–205 (1973).
[7] D. K. Acquaye, A. B. Ankomah, and I. Kando, *J. Sci. Food Agr.* **23**, 1035–1044 (1972).
[8] B. Zagorchev, L. Bozadzhieva, and E. Mitropolitska, *C. R. Acad. Bulg. Sci.* **15**, 483–486 (1962).
[9a] T. M. Malyutina, V. A. Orlova, and B. Ya. Spivakov, *Zh. Anal. Khim.* **29**, 790–796 (1974).
[9b] Z. Marczenko, *Anal. Chim. Acta* **31**, 224–232 (1964).
[10] Masayoshi Tabushi, *Bull. Inst. Chem. Res., Kyoto Univ.* **37**, 245–251 (1959).

The last is commonly used. The color is stable and conforms to Beer's law for 0.5–10 ppm. There is interference by titanium, chromium, or bismuth, and by large amounts or uranium or copper.

To a solution containing iron and chromium, add fourfold excess of sodium tartrate and extract the iron with 1 : 1 acetylacetone-chloroform.[11] Dilute the extract with chloroform, dry with sodium sulfate, and read at 440 nm.

Both iron and uranium complexes with acetylacetone are extracted with butyl acetate at pH 7.[12] Then iron is read at 440 nm, uranium at 365 or 375 nm. The sample may contain 20–150 μg of iron, 50–1000 μg of uranium.

For analysis of silicate rocks, decompose with hydrofluoric acid and sulfuric acid.[13] Evaporate the excess hydrofluoric acid and develop with acetylacetone.

Procedures

Extraction with Propylene Carbonate.[14] Oxidize 60 ml of a sample solution containing 1.5–6 mM of iron with 1 ml of chlorine water or 1 ml of 3% hydrogen peroxide. Add 1 ml of acetylacetone and adjust to pH 4–9 by addition of a phosphate buffer solution for pH 6.84. Add 10 ml of saturated sodium chloride solution and saturate the solution with propylene carbonate. Extract with 7 ml, 5 ml, and 3 ml of propylene carbonate. Dilute the combined extracts to 25 ml with propylene carbonate and read at 440 nm.

By Ion Exchange.[15] Prepare the acetylacetonate by boiling zinc dust or magnesium turnings with acetylacetone in ethanol. Filter the acetylacetonate and recrystallize from ethanol Use as a freshly prepared 0.3% solution in dioxan or a saturated solution in chloroform.

Mix 3 ml of the solution of acetylacetonate in dioxan with an aqueous sample solution containing 0.1 mg of ferric ion. Dilute to 7 ml with dioxan and read at 440 nm. Chromium interferes.

Alternatively, shake 10 ml of the chloroform solution with an aqueous sample solution containing up to 0.18 mg of ferric ion. Read the chloroform phase at 440 nm.

Metallic Iron in Reduced Iron Ores.[16] Heat 0.5 mg of sample with 10 ml of acetylacetone and 3 grams of acetylacetone-2-thenoyltrifluoroacetone at 70° for 90 minutes. Cool and dilute to 50 ml with carbon tetrachloride. Filter and read at 470 nm.

---

[11] E. Uhlemann, G. Meissner, E. Butter, and H. Weinelt, *J. Prakt. Chem.* **25**, 272–278 (1964).

[12] Masayoshi Ishibashi, Tsunenobu Shigematsu, and Masayuki Tabushi, *J. Jap. Chem. Soc.* **80**, 1018–1021 (1959).

[13] B. Martinet, *Chim. Anal.* (Paris) **44**, 64–66 (1962).

[14] B. G. Stephens, James C. Loftin, William C. Looney, and Kenneth A. Williams, *Analyst* **96**, 230–234 (1971).

[15] M. M. Aly, *Anal. Chim. Acta* **58**, 467–469 (1972).

[16] Ohiko Kammori and Isamu Taguchi, *Jap. Anal.* **15**, 1223–1227 (1966).

ANTHRANILIC ACID

Ferric anthranilate can be extracted with 1-pentanol in the presence of ferrous ion. Then the residual ferrous ion can be oxidized to ferric ion and similarly determined. It is feasible for the ferric ion to be determined in 65-fold ferrous ion and for the ferrous ion to be determined in fifteenfold ferric ion. The total iron must not exceed 425 ppm.

Procedures

Ferrous Ion.[17] As color reagent, dissolve 5 grams of anthranilic acid in 400 ml of water. Add 37 ml of 4% sodium hydroxide solution. Add *N* sulfuric acid until the pH is reduced to 4.5 and dilute to 500 ml. As an organic mixture, combine 130 ml of amyl acetate, 100 ml of acetophenone, and 19 ml of 1-pentanol.

Degas the sample solution with oxygen-free nitrogen. Add 25 ml of degassed color reagent to the sample. Add 10 ml of degassed 1-pentanol, shake for 2 minutes, and centrifuge. Read the organic phase at 475 nm against a blank.

Total Iron. Add 2 ml of 30% hydrogen peroxide to the aqueous phase and enough 0.5 *N* nitric acid to adjust to pH 4.5. After 15 minutes for oxidation of the iron, saturate it with approximately 0.01 gram of anthranilic acid. Add 10 ml of the organic mixture, shake for 2 minutes, and centrifuge. Read the organic phase at 465 nm against a blank.

ASCORBIC ACID

This reagent, widely used for reduction of ferric to ferrous ion for complexing with other reagents, also complexes with ferrous ion for reading in the near-ultraviolet region.

Procedure. As the reagent solution, dissolve 0.1 gram of ascorbic acid in water. Add 10 ml of *N* ammonium hydroxide and 10 ml of 5.3% ammonium chloride solution. Dilute to 100 ml.

Boil a sample solution containing 0.3–1.5 mg of iron with a few drops of 30% hydrogen peroxide and cool.[18] Add hydrochloric acid so that the solution is 1 : 1 and dilute to 100 ml at that acid concentration. Extract a 10-ml aliquot with 10 ml, 5 ml, and 5 ml of ether. Wash the combined ether extracts with 10 ml of 1 : 1 hydrochloric acid. Evaporate the ether extract with 1 ml of 1 : 1 hydrochloric acid. Add 10 ml of ascorbic acid reagent and dilute to 25 ml. Read against a reagent blank.

---

[17] Donald L. Dinsel and Thomas R. Sweet, *Anal. Chem.* **35**, 2077–2081 (1963).
[18] K. P. Stolyarov and R. Gerbach, *Vest. Leningr. Univ., Ser. Fiz. Khim.* **1963** [2(10)], 116–121.

ASTRAZONE PINK FG

This reagent conforms to Beer's law for 0.14–2.24 μg of iron per ml at 530 nm. Gold, gallium, and thallic ion interfere.

Procedure. *Mineral Water.*[19] To 500 ml of sample add 5 ml of hydrochloric acid. After effervescence ceases, take a 100-ml aliquot. Add 0.2 ml of 30% hydrogen peroxide and boil for 3 minutes to destroy excess peroxide. Cool and adjust to 100 ml. Take a 0.25-ml aliquot. Add 2.5 ml of hydrochloric acid and 5 ml of 0.14% solution of astrazone pink FG (basic red 13). Dilute to 10 ml and extract with 10 ml of 2 : 1 benzene–methyl ethyl ketone. Read at 530 nm.

AZIDE

Ferric azide is read as a 1 : 1 complex at pH 4.5–5.5 and is stable for hours. There is interference at various levels by copper, nickel, cobalt, chromium, bismuth, tin, and titanium. Color is bleached somewhat by tartrate, citrate, phosphate, and fluoride.

Procedure. Take a sample solution containing 0.05–4 mg of iron.[20] If ferrous ion is present, add 2 ml of 1 : 1 nitric acid, heat, cool, and neutralize with ammonium hydroxide. Add 30 ml of 0.1 *N* hydrochloric acid and 5 ml of 10% solution of sodium azide. Dilute to 100 ml and read at 460 nm.

2,4-BDTPS

This trivial name designates tetraammonium-2, 4-bis-5, 6-bis(4-phenylsulfonic acid)-1,2,4-triazin-3-yl pyridine. It is the sulfonated form of the most sensitive reagent known for ferrous ion. There is significant interference by chromic ion, cobalt, nickel, cyanide, and pyrophosphate. The reagent determines cupric ion, but there is interference by cuprous ion. The color is stable for days. Heating at 40° for 15 minutes is required to develop the color for iron.

Procedures

Blood Serum.[21] IRON. As extractant-reductant, prepare 0.5% solution of hydroxylamine hydrochloride in 0.2 *N* hydrochloric acid. As acetate buffer solution, make 0.3 *M* ammonium acetate to 2.5 *N* with ammonium hydroxide so that added to a

[19] C. Patroescu, Anca Tasca, and Maria Cojocaru, *Rev. Chim.* (Bucharest) **25**, 328–330 (1974).
[20] Kazimierz Kapitanczyk, Zbigniew Kurzawa, and Zygmunt Pryminski, *Chem. Anal.* (Warsaw) **5**, 413–417 (1960); cf. Ernest K. Dukes and Richard M. Wallace, *Anal. Chem.* **33**, 242–244 (1961).
[21] George L. Traister and Alfred A. Schilt, *Anal. Chem.* **48**, 1216–1220 (1976).

treated serum sample, it will give pH 5–6. As tartrate buffer solution, make any necessary adjustments to a *M* solution of disodium tartrate so that added to a treated serum sample it will give pH 2.5–3.5.

Mix 0.5 ml of blood serum with 0.5 ml of extractant-reductant and set aside for 10 minutes. As protein precipitant, add 0.5 ml of 12% solution of trichloroacetic acid and centrifuge. Mix 1 ml of the clear supernatant layer with 0.3 ml of 0.48% solution of 2,4-BDTPS and 0.2 ml of tartrate buffer solution. Check that the pH is 2.5–3.5 and incubate at 40° for 15 minutes. Cool, read at 563 nm, and correct for a blank.

Copper and Iron. Reduce and deproteinize as for iron. Add 0.3 ml of 0.48% solution of 2,4-BDTPS and 0.2 ml of the acetate buffer solution. Set aside for 10 minutes and read at 563 and 460 nm. Subtract a blank. Calculate copper and iron by simultaneous equations.

Water. COPPER AND IRON. Acidify a sample containing 16–160 μg of copper and/or 4–40 μg of iron with 0.15 ml of 1 : 1 hydrochloric acid. Heat nearly to boiling and cool. Add 1 ml of 0.48% solution of 2,4-BDTPS and 1 ml of 10% hydroxylamine hydrochloride solution. Mix and add 1 ml of 10 *M* ammonium acetate to adjust to pH 5–7. Dilute to 25 ml and read at 563 and 460 nm. Subtract a blank and solve by simultaneous equations.

Alternatively, read at 463 nm. Add 30 mg of sodium cyanide, set aside for 15 minutes, and read again. The final reading measures iron, the decrease measures copper.

Beer. IRON. Evaporate 25 ml of degassed sample to near dryness. Cool, and add 5 ml of 1 : 1 nitric acid–perchloric acid. Warm to drive off oxides of nitrogen, then heat to fumes of perchloric acid. Cool. Add 5 ml of water, 1 ml of 0.48% solution of 2,4-BDTPS, 2 ml of 10% solution of hydroxylamine hydrochloride, and 3 ml of *M* disodium tartrate. Adjust to pH 2.5–3.1 with about 0.5 ml of ammonium hydroxide and develop the color by heating at 40–60° for 15 minutes. Dilute to 25 ml and read at 563 nm against a blank.

2,2′-BIPYRIDYL

This reagent is also known as 2,2′-bipyridine, 2,2′-dipyridyl, 2,2′-dipyridine, and α,α′-dipyridyl. Only ferrous ion and cuprous ion form complexes with 2,2′-bipyridyl, which absorb strongly in the visible region. At maximum absorption ferrous ion forms a 1 : 3 complex. The reagent should be present to at least 250-fold.

The reducing agent, if needed, and the color-forming reagent should be added before pH adjustment; otherwise the result will be low.[22] By calculation, iron can be read at 530 nm in the presence of copper at 685 nm.[23] For total iron, hydroxylamine is an appropriate reducing agent.[24]

[22] Nellie F. Davis, Clyde E. Osborne, Jr., and Harold A. Nash, *ibid.* **30**, 2035 (1958).
[23] Giordano Trabanelli, *Atti Accad. Sci. Ferrara* **35**, 133–142 (1957–1958).
[24] British Standards Institution, *BS 4258*, Pt. **2**, 8 pp (1968).

2,2′-Bipyridyl is appropriate for reading iron at 520 nm in such glass raw materials as red lead, zinc oxide, cadmium sulfide, nickel oxide, cobalt oxide, arsenous oxide, and antimony trioxide.[25] For reading ferrous ion with 2,2′-bipyridyl in the presence of ferric ion, filtration with methyl red in *M* sulfuric acid effectively screens out the absorption by the ferric ion.[26]

Ferrous ion can be completely extracted at pH 6–7 by *M* hexanoic acid in chloroform containing 0.1 *M* 2,2′-bipyridyl for reading at 520 nm.[27] This is applicable to separation from large amounts of copper masked with EDTA.

A trace of iron in pure uranyl nitrate can be determined with this reagent.[28] For analysis of biphenyl, digest 1 gram with 1 ml of sulfuric acid, 5 ml of nitric acid, and 1 ml of 5% sodium chloride solution.[29] Evaporate to sulfur trioxide fumes, take up in water, and after adjustment of the acidity, develop iron with 2,2′-bipyridyl.

For iron in blood serum make 2 *N* with hydrochloric acid and precipitate proteins with trichloroacetic acid.[30] Centrifuge, reduce iron with hydroxylamine hydrochloride, and develop. For total iron in blood, digest a hemolyzed sample with a 1:1 mixture of perchloric and sulfuric acids.[31] Dilute and neutralize to phenolphthalein with 35% potassium hydroxide solution. Add 0.5 *M* acetate buffer for pH 5.5 to bring to that level. Treat with hydroquinone and 2,2′-bipyridyl and let stand overnight. Read at 520 nm. The complex can be extracted with chloroform[32] for reading at 520 nm. Urine is wet-ashed for similar determination.[33]

For analysis of multivitamin preparations, wet-ash with 3:1 nitric-hydrochloric acids and evaporate to sulfur trioxide fumes with sulfuric acid.[34] For analysis of gelatin, dry-ash with nitric acid at 1 ml per gram of sample to avoid chlorides.[35] Then reduce to ferrous ion with hydroxylamine sulfate. A ternary complex is formed with nitroprusside.

For iron in silicone polymers by 2,2′-bipyridyl, see Aluminum by 8-Hydroxyquinoline.

Procedures

High Purity Antimony.[36] As a salt mixture, dissolve 7.5 grams of sodium chloride, 15 grams of anhydrous sodium sulfate, 30 grams of tartaric acid, and 10 grams of sodium hydroxide in 150 ml of water.

Dissolve 3 grams of finely ground sample in 20 ml of 1:1 hydrochloric acid and 4 ml of 1:1 nitric acid. Evaporate to dryness at under 100° and take up the residue

[25] A. Maresova, *Sklar Keram.* **12** [7(Add.)], 17–20 (1962).
[26] F. Santamaria, *Acta Cient. Venez.* **15** (5), 166–171 (1964).
[27] I. V. Pyatnitskii, V. V. Sukhan, and E. N. Gritsenko, *Zh. Anal. Khim.* **25**, 1949–1953 (1970).
[28] J. Nowicka-Jankowska, A. Godkowska, I. Pietrzak, and W. Zmijewska, *Pol. Akad. Nauk., Inst. Badan Jadrowych.* No. **45**, 42 pp (1958).
[29] D. Monnier, W. Haerdi, and E. Martin, *Helv. Chim. Acta* **46**, 1042–1046 (1963).
[30] I. Dezso and T. Fulop, *Mikrochim. Acta* **1961**, 154–157.
[31] Junjiro Okamoto, *Igaku Seibutsugaku* **40**, 115–118 (1956).
[32] W. N. M. Ramsay, *Clin. Chim. Acta* **2**, 214–220 (1957).
[33] E. S. N. Littlejohn and D. N. Raine, *ibid.* **14**, 793–796 (1966).
[34] Lilo E. Guerello, *Proanalisis* **3**, 29–33 (1970).
[35] Jaroslav Benes, *Chem. Prum.* **8**, 84–85 (1958).
[36] B. M. Lipshits, G. K. Smirnova, and F. S. Kulikov, *Zavod. Lab.* **27**, 1199–1200 (1961).

in 50 ml of the salt mixture. Adjust to pH 4–5 and boil for 5 minutes. Add 45 ml of 0.01% solution of 2,2′-bipyridyl and let stand for 18 hours. Shake for 3 minutes with 70 ml of cresol. Filter the aqueous layer, and if it shows that iron is still present, extract with 20 ml of cresol. Adjust the cresol extract or extracts to 100 ml and read at 522 nm.

High Purity Tin.[37] As a mixed reagent, add 2 ml of 2% solution of 2,2′-bipyridyl in 0.01 *N* hydrochloric acid to 500 ml of a solution containing 7.5 grams of sodium chloride, 30 grams of tartaric acid, 15 grams of sodium sulfite, and 10 grams of sodium hydroxide.

Dissolve 1 gram of sample as 0.1–0.2 mm foil in 5 ml of hydrochloric acid and 2 ml of nitric acid. Evaporate to dryness at 90°, thereby removing most of the tin. Dissolve the residue in 10 ml of the mixed reagent and add 2 ml of 2% solution of 2,2′-bipyridyl in 0.01 *N* hydrochloric acid. Dilute to 25 ml and set aside for 1 hour. Read at 508 nm against water and subtract a blank.

Alternatively, dissolve 1 gram of sample in 2.5 ml of hydrochloric acid and 2.5 ml of hydrobromic acid. Cool, and add 1.5 ml of bromine dropwise. Evaporate almost to dryness, add 0.5 ml of hydrochloric acid, and take to dryness. Add 0.5 ml of hydrochloric acid and again evaporate to dryness. Take up in 0.5 ml of hydrochloric acid and add 10 ml of the mixed reagent. Boil for 3 minutes, add 2 ml of 2% solution of 2,2′-bipyridyl in 0.01 *N* hydrochloric acid, and set aside for 1 hour. Dilute to 25 ml and read at 508 nm.

Copper Alloys.[38] Dissolve 1 gram of sample in 5 ml of 1 : 1 nitric acid and dilute to 100 ml. To a 10-ml aliquot add 10 ml of 0.5 *M* benzoic acid in chloroform that is 25 m*M* with 2,2′-bipyridyl. Adjust to pH 5–6 and add 4 ml of 0.2 *M* EDTA. Shake, separate the organic phase, and read at 522 nm.

Alumina.[39] Fuse 2 grams of sample with 10 grams of sodium carbonate and 4 grams of sodium tetraborate. Dissolve in 200 ml of 1 : 5 hydrochloric acid, evaporate to dryness, and bake at 130°. Moisten with hydrochloric acid, add 200 ml of hot water, and digest at 100° until dissolution is complete. Filter the silica and dilute the filtrate to 500 ml. Mix a 20-ml aliquot with 20 ml of 0.1% solution of 2,2′-bipyridyl and set aside for 1 hour. Read at 522 nm.

Mineral Water[40]

FERROUS IRON. Mix 100 ml of sample with 20 ml of 0.1% solution of 2,2′-bipyridyl and read at 522 nm.

TOTAL IRON. Acidify 1 ml with 5 ml of 1 : 1 hydrochloric acid and heat at 100° for 30 minutes. Raise to pH 2–3 by addition of ammonium hydroxide and add 0.4

[37] Yu. L. Lel'chuk, P. V. Kristalev, L. L. Skripova, and L. B. Kristeleva, *Izv. Tomsk. Politekh. Inst.* **128**, 96–100 (1964); cf. P. V. Kristalev and L. B. Kristaleva, *Sb. Nauch. Tr. Perm. Politekhn. Inst.* **1963** (14), 65–67.

[38] I. V. Pyatnitskii and T. A. Slobodenyuk, *Ukr. Khim. Zh.* **40**, 1333–1335 (1974).

[39] E. Krejzova, J. Kruml, and L. Plocek, *Sklar. Keram.* **9**, 244 (1959).

[40] W. Fresenius and W. Schneider, *Z. Anal. Chem.* **209**, 340–341 (1965).

gram of ascorbic acid. Add 10% sodium acetate solution to raise to pH 5–6 and 20 ml of 0.1% solution of 2,2′-bipyridyl. Read at 522 nm.

Phosphate Rock.[41] Add 1 gram of sample to 10 ml of hydrofluoric acid in platinum and let stand overnight. Add 10 ml of perchloric acid and evaporate to dryness. Add a further 10 ml of perchloric acid with 30 ml of water, heat to dissolve the solids, and evaporate to dryness. Repeat the steps described in the last sentence twice more to ensure removal of fluorides. Take up in 10 ml of 5 *N* hydrochloric acid and dilute to 100 ml.

To a 10-ml aliquot containing not more than 1 mg of ferric oxide, add 50 mg of ascorbic acid and 10 ml of 0.15% solution of 2,2′-bipyridyl. Set aside for 10 minutes. Add 5 ml of 9.5% solution of chloroacetic acid and raise to pH 3 by dropwise addition of 1 : 3 ammonium hydroxide. Dilute to 100 ml and read at 525 nm. Subtract a blank.

Condensed Phosphates.[42] Dissolve a sample containing 0.05–2 mg of ferric oxide in water and acidify with hydrochloric acid. Then add an equal volume of hydrochloric acid and heat at 100° for 30 minutes. Centrifuge to remove silica and other insolubles and cool. Add 5 ml of 20% solution of 2,2′-bipyridyl, 5 ml of 20% solution of sodium sulfite adjusted to pH 4, and 5 ml of 2% ethanolic hydroquinone. Raise to pH 4–5 with ammonium hydroxide and add 10 ml of a 4% solution of 2,2′-bipyridyl in 1 : 40 hydrochloric acid. Heat at 100° for 3 hours, cool, and dilute to 100 ml. Read at 522 nm against a blank.

Hydrofluoric Acid.[43] As a reagent, dissolve 7.5 grams of sodium chloride, 30 grams of tartaric acid, 15 grams of sodium sulfite, and 10 grams of sodium hydroxide in 500 ml of water.

Evaporate 2 ml of sample to dryness. Add 0.5 ml of 1 : 1 hydrochloric acid and 10 ml of the reagent. Heat to 100° and add 0.5 ml of 2% solution of 2,2′-bipyridyl in 1 : 50 hydrochloric acid. Dilute to 25 ml, set aside for 15 minutes, and read at 508 nm against water.

Antimony Trioxide or Diammonium Hydrogen Phosphate. Dissolve a 0.5-gram sample in 0.5 ml of 1 : 1 hydrochloric acid and add 10 ml of the reagent. Complete as above from "Heat to 100° and. . . ."

Hydrochloric Acid.[44] Dilute the sample to 8 *N* and pass it at 2–3 ml/minute through a column of 0.1–0.25 mm anionite AN-2F in chloride form. Wash the column with 5 ml, 5 ml, and 5 ml of 8 *N* hydrochloric acid. Desorb the iron from the column with 0.1 *N* hydrochloric acid at 2–3 ml/minute until a thin, dark ring remains in the lower part of the resin column. Add phenolphthalein indicator and ammonium hydroxide until the percolate is just alkaline. Decolorize by dropwise addition of 1 : 1 hydrochloric acid. Add 5 ml of acetate buffer solution for pH 3.4

[41] A. D. Wilson, *Analyst* **88**, 18–25 (1963).
[42] K. Gassner, *Z. Anal. Chem.* **153**, 81–83 (1956).
[43] Y. L. Lel'chuk, L. L. Skripova, and B. N. Besprozvannykh, *Izv. Tomsk. Politekh. Inst.* **163**, 127–129 (1970).
[44] Yu. L. Lel'chuk and R. D. Glukhovskaya, *Tr. Kom. Anal. Khim.* **16**, 15–18 (1968).

and 1 ml of 10% solution of hydroxylamine hydrochloride. Heat to 55° and add 0.5 ml of 2% solution of 2,2′-bipyridyl in 1:50 hydrochloric acid. Let stand for 20 minutes, dilute to a known volume, and read at 508 nm against water.

Calcium Carbonate, Calcium Phosphate, and Ammonium Phosphate. Dissolve the sample in 8 *N* hydrochloric acid and proceed as for hydrochloric acid.

Beer.[45] Heat 25 ml of sample with 3 ml of 4 *N* sulfuric acid at 100° under conditions to minimize evaporation. Cool, and add 2 ml of 10% solution of hydroxylamine hydrochloride. Set aside for 1 minute, then add 5 ml of 0.1% solution of 2,2′-bipyridyl in 0.1 *N* hydrochloric acid and 20 ml of acetate buffer solution for pH 4.38. Adjust to pH 4.3 with 8% solution of sodium hydroxide. Add a drop of Teepol 610 and shake for 5 minutes with 20 ml of 1,2-dichloroethane. Filter the organic phase and read at 528 nm.

Drug Solutions or Extracts[46]

FERROUS ION. Dilute a sample containing 0.3–0.35 mg of total iron in 0.1 *N* hydrochloric acid to 50 ml. Add 15 ml of acetate buffer solution for pH 4.5 and 6 ml of 0.1% solution of 2,2′-bipyridyl. Dilute to 100 ml and read at 525 nm against a reagent blank.

TOTAL IRON. To the sample as above, add 10 mg of ascorbic acid to reduce the ferric ion to the ferrous state. Then develop as above.

Sugar-Beet Juice.[47] As a preliminary purification, acidify the sample with hydrochloric acid and add 3% potassium permanganate solution in excess. Filter, add potassium iodide, and decolorize with sodium thiosulfate.

To 50 ml of clarified sample, add 1 ml of 10% solution of hydroxylamine hydrochloride and 1 ml of 0.1% solution of 2,2′-bipyridyl. Add 25 ml of 20% ammonium acetate solution and read at 603 nm.

Iron-Binding Capacity of Serum.[48] To 1 ml of serum or heparinized but not oxalated plasma, add 2 ml of ferric chloride solution containing 5 μg of iron per ml in 0.005 *N* hydrochloric acid. Set aside for 5 minutes and add 100 mg of finely ground magnesium carbonate. Shake frequently for 1 hour and then centrifuge. To 2 ml of the clear supernatant layer, add 0.5 ml of 0.2 *M* sodium sulfite and 0.5 ml of 0.2% 2,2′-bipyridyl solution in 6% acetic acid. Read at 520 nm and correct for the added iron.

Horse Serum.[49] As a reagent, dissolve 6.6045 grams of sodium acetate in 9 ml of acetic acid, add 0.5 gram of 2,2′-bipyridyl, and dilute to 100 ml.

[45] Institute of Brewing Analysis Committee, *J. Inst. Brew.* **76**, 299–301 (1970).
[46] Christos Zachariades and William C. McGavock, *J. Pharm. Sci.* **60**, 918–919 (1971); cf. Daniel J. Sullivan, *J. Assn. Off. Anal. Chem.* **59**, 1156–1161 (1976).
[47] M. Karvanek and R. Bretschneider, *Listy Cukr.* **78**, 226–230 (1962).
[48] W. N. M. Ramsey, *Clin. Chim. Acta* **2**, 221–226 (1957).
[49] Ibrehim Ruzdic and Vladimir Gregorovic, *Mikrochim. Acta* **1959**, 294–298.

Dilute 2 ml of sample serum with 6 ml of water and add 2 ml of 20% trichloroacetic acid solution. Heat at 90° for 10 minutes, cool, and filter. To 5 ml of filtrate, add 0.3 ml of 10% sodium hydroxide solution, 2 ml of reagent, and 0.2 ml of 2% solution of hydroquinone. Dilute to 15 ml and read at 520 nm against a blank.

Nonheme Iron in Tissue.[50] Homogenize intestinal tissue; add 1:1 25% trichloro-acetic acid solution–4% tetrasodium pyrophosphate solution. Heat at 100° for 15 minutes to extract iron bound to transferrin or ferritin without appreciable breakdown of hemoglobin. Centrifuge. Separate the supernatant layer, and either pass through a 0.45 nm bacterial filter or supercentrifuge. Repeat the extraction twice.

To a 3-ml aliquot of the extract, add 1 ml of 2% solution of sodium ascorbate, 1 ml of 0.4% solution of 2,2′-bipyridyl in 0.005 *N* hydrochloric acid, and 2 ml of 2 *M* acetate buffer solution for pH 4.78. Set aside for 30 minutes and read at 520 nm.

Ethanolamine.[51] To 10 ml of the commercial product, which will contain di- and triethanolamine as well as iron, add 5 ml of sulfuric acid slowly with cooling. Add 20 ml of saturated solution of potassium persulfate, boil for 20 minutes, cool, and dilute to 1 liter. To an aliquot, add 25% sodium acetate solution to adjust to pH 3.5. Add 4 ml of 10% solution of hydroxylamine hydrochloride and 4 ml of 1% ethanolic 2,2′-bipyridyl. After 1 hour dilute to 100 ml and read at 522 nm.

Polypropylene.[52] Heat 1 gram of sample with 15 ml of sulfuric acid until black and viscous. Cool, add 5 ml of 30% hydrogen peroxide, and heat again. Add more hydrogen peroxide and continue heating until the solution is clear and colorless. Boil off the excess of hydrogen peroxide and dilute to 100 ml.

To a 25-ml aliquot containing not more than 0.4 mg of iron, add 1 ml of 10% solution of hydroxylamine hydrochloride and sufficient ammonium hydroxide to adjust to pH 4. Add 2 ml of 0.2% solution of 2,2′-bipyridyl and dilute to 100 ml for reading at 523 nm.

Vegetable Matter.[53] Treat a 2-gram sample with 5 ml of 30% hydrogen peroxide and 2 ml of perchloric acid. Evaporate to dryness and complete by igniting at 500°. Cool, add 2 ml of water, and dissolve in 5 ml of hydrochloric acid. Add 5 ml of sulfuric acid and evaporate to sulfur trioxide fumes. Cool, and dilute to 50 ml. To a 5-ml aliquot, add 1 ml of 0.1% solution of 2-(4-hydroxyphenyl)glycine in 0.4 *N* sulfuric acid and 1 ml of 0.5% solution of 2,2′-bipyridyl in 10% acetic acid. Adjust to pH 4 with 4% sodium hydroxide solution and dilute to 25 ml. Read at 490 nm.

Vulcanization Accelerators.[54] Typical samples are 1,3-diphenylguanidine and *N*-cyclohexylbenzothiazole-2-sulfenamide. Dry-ash a 2-gram sample and ignite at 550°. Dissolve the ash in 1 ml of hydrochloric acid and 1 ml of nitric acid. Dilute

[50] A. L. Foy, H. L. Williams, S. Cortell, and M. E. Conrad, *Anal. Biochem.* **18**, 559–563 (1967).
[51] E. Levin and V. Cioromela, *Rev. Chim.* (Bucharest) **18**, 240–241 (1967).
[52] K. Novak and V. Mika, *Chem. Prum.* **13**, 360–362 (1963).
[53] Anna Krauze and Danuta Domska, *Chem. Anal.* (Warsaw) **14**, 679–681 (1969).
[54] M. Uchman and M. Stackowski, *ibid.* **13**, 655–658 (1968).

to 50 ml and take a 5-ml aliquot. Add 1 ml of 10% solution of hydroxylamine hydrochloride and raise to pH 4 with ammonium hydroxide. Add 2 ml of 0.2% solution of 2,2′-bipyridyl, dilute to 25 ml, and read at 522 nm.

Ternary Complex with Nitroprusside.[55] To a sample solution containing 2–14 μg of ferrous ion, add 1 ml of 0.1 m*M* ascorbic acid, 5 ml of m*M* 2,2′-bipyridyl, and 5 ml of m*M* nitroprusside. Dilute to 25 ml and adjust to pH 3–4. Shake for 2 minutes with 10 ml of nitromethane. Separate the organic phase and add 1 ml of ethanol. Adjust to 10 ml with nitromethane and read at 520 nm against a blank.

2,2′-BIPYRIMIDINE

Like 2,2′-bipyridyl, the ferrous complex of this reagent has absorption in the visible region only with ferrous and cuprous ions. At maximum absorption the metal-ligand complex is 1:3, analogous to 2,2′-bipyridyl. The performance is also closely related to that of the widely used 1,10-phenanthroline.

Procedure. *Borosilicate Glass.*[56] Take a 200-mesh sample containing 0.2–2 mg of iron. Treat in platinum with 15 ml of 1:1 hydrofluoric acid and evaporate to dryness. Add 10 ml of hydrofluoric acid and again evaporate to dryness. Add 1 ml of sulfuric acid and 5 ml of hydrofluoric acid and evaporate to sulfur trioxide fumes. Add 10 ml of hydrochloric acid and evaporate to 5 ml to complete removal of arsenic. Boil with water and if necessary filter off zirconites and titanates. Dilute to 100 ml.

To a 25-ml aliquot, add 4 ml of a buffer consisting of equal parts of 0.8 *M* sodium acetate and acetic acid. Add 6 ml of 10% hydroxylamine hydrochloride solution. Adjust to pH 4.5±0.5 with dilute hydrochloric acid or ammonium hydroxide. Add 10 ml of 2% solution of 2,2′-bipyrimidine and dilute to 50 ml. Read at 490 nm against water and subtract a reagent blank.

BRILLIANT CRESYL BLUE

This reagent conforms to Beer's law for 0.25–5 μg of ferric ion per ml. Gold, gallium, and thallic ion interfere.

Procedure. *Mineral Water.*[57] To a 500-ml sample, add 5 ml of hydrochloric acid. After effervescence ceases, take a 100-ml aliquot and add 2.5 mg of iron. Add 0.2 ml of 30% hydrogen peroxide and boil for 3 minutes to destroy excess peroxide.

[55] Franciszek Buhl and Kyrstyna Kania, *ibid.* **20**, 1055–1063 (1975).
[56] D. D. Bly and M. G. Mellon, *Anal. Chem.* **35**, 1386–1392 (1963).
[57] C. Patroescu, Anca Tasca, and Maria Cojocaru, *Rev. Chim.* (Bucharest) **25**, 328–330 (1974); Constantin Patroescu and Marieta Cojocaru, *ibid.* **27**, 444–445 (1976).

Cool and adjust to 100 ml. Take a 1-ml aliquot. Add 4 ml of hydrochloric acid and 4 ml of 0.4% solution of brilliant cresyl blue (color index 51010). Dilute to 10 ml and extract with 10 ml of 2:1 benzene–methyl ethyl ketone. Read at 630 nm and correct for the added iron.

CHROME AZUROL S

A 1:1 complex of ferric ion with chrome azurol S (mordant blue 29) in fourfold excess is read above pH 4.5 at 570 nm.[58] Beer's law is followed for 0.11–8 ppm of iron. There is interference by aluminum, gallium, stannous ion, palladium, copper, beryllium, phosphate, thiosulfate, citrate, oxalate, borate, EDTA, nitrilotriacetic acid, and hydrogen peroxide.

With large excess of chrome azurol S and cetyltrimethyl–ammonium chloride, ferric ion forms a 1:6:6 chelate at pH 3.1–3.8.[59] This has a maximum absorption at 630 nm and follows Beer's law for 2–16 μg of iron in 50 ml containing 10 ml of 0.05% solution of the dyestuff, 2 ml of 0.05% solution of the quaternary, and 5 ml of 4 *M* ammonium acetate–4 *M* acetic acid as buffer.

When zephiramine, which is benzyldimethyltetradecylammonium chloride, is the third component, the maximum absorption is at 640 nm.[60] The sensitivity is improved by a factor of 2.5, interference by copper and titanium disappears, but aluminum interferes seriously. The calibration graph is linear for 0.014–0.36 ppm of iron in 25 ml of acetate buffer solution for pH 4.3 containing 2 ml of 0.25% solution of chrome azurol S and 2 ml of 0.5% solution of zephiramine.

Procedure. Dilute a faintly acid solution containing 3.4–70 μg of ferric ion to 30 ml.[61] Add 0.6 ml of 0.275% solution of chrome azurol S and adjust to pH 5.3 with 0.2 *M* sodium acetate. Dilute to 50 ml and read at 575 nm against a reagent blank.

Lead.[62] Dissolve 5 grams of sample in 1:3 nitric acid and evaporate to dryness. Take up in 25 ml of water, add 1:1 sulfuric acid in moderate excess, and dilute to 100 ml. Set aside for lead sulfate to settle. Take a 20-ml aliquot and adjust with ammonium hydroxide to pH 2–3. Add 3 ml of 0.5% solution of cetrimonium chloride, 10 ml of 0.05% solution of chrome azurol S, and 5 ml of acetate buffer solution for pH 3.5. Adjust to pH 3.5 with acetic acid, dilute to 50 ml, and read at 630 nm against a reagent blank.

Copper. Dissolve 5 grams of sample in nitric acid and evaporate to dryness. Dissolve in 50 ml of 1:1 hydrochloric acid and extract the iron with 20 ml of 1:1

[58] S. P. Sangal, *Chim. Anal.* (Paris) **49**, 361–362 (1967).
[59] Yoshio Shijo and Tsugio Takeuchi, *Jap. Anal.* **17**, 1519–1523 (1968); cf. Hiroshi Nishida, *ibid.* **20**, 410–415 (1971).
[60] Yoshizo Horiuchi and Hiroshi Nishida, *ibid.* **17**, 756–764 (1968); cf. Hiroshi Nishida, *ibid.* **20**, 410–415 (1970).
[61] Yoshizo Horiuchi and Hiroshi Nishida, *ibid.* **16**, 769–775 (1967).
[62] Yasuchi Nakamura, Hiroshi Nagai, Daishiro Kubota, and Syunji Himeno, *ibid.* **22**, 1156–1162 (1973).

methyl isobutyl ketone–isopentyl acetate. Back-extract the iron with 10 ml of 1 : 1 hydrochloric acid and dilute to 100 ml. Complete as for lead from "Take a 20-ml aliquot. . . ."

Indium. Dissolve 5 grams of sample in 30 ml of 1 : 1 hydrochloric acid and 2 ml of 3% hydrogen peroxide. Evaporate to dryness and dissolve the residue in 40 ml of 1 : 1 hydrochloric acid. Extract the iron with 30 ml of ether and wash with 10 ml and 10 ml of 1 : 1 hydrochloric acid. Back-extract the iron with 10 ml and 10 ml of water and dilute to 100 ml. Complete as for lead from "Take a 20-ml aliquot. . . ."

Selenium. Dissolve 10 grams of sample in nitric acid and evaporate to less than 5 ml. Add 50 ml of 1 : 1 hydrochloric acid and extract the iron with 20 ml of 1 : 1 methyl isobutyl ketone–isopentyl acetate. Wash the extract with 10 ml and 10 ml of 1 : 1 hydrochloric acid. Back-extract the iron with 10 ml and 10 ml of water. Add 1 ml of sulfuric acid and evaporate to dryness. Take up with 0.5 ml of 1 : 1 hydrochloric acid and dilute to 100 ml. Complete as for lead from "Take a 20-ml aliquot. . . ."

Chlorosilanes.[63] Heat a 0.1-gram sample gently to volatilize the main volume. Add 1 ml of 1 : 1 sulfuric acid and 2 ml of hydrofluoric acid. Heat to sulfur trioxide fumes, take up in water, and dilute to 25 ml. Neutralize a 5-ml aliquot with 0.4% sodium hydroxide solution and dilute to 15 ml. Add 1 ml of *N* hydrochloric acid and 1 ml of 0.1% solution of chrome azurol S. Add 25 ml of a buffer solution for pH 6, dilute to 50 ml, and set aside for 15 minutes. Read at 546 nm.

CUPFERRON

Iron can be extracted with amyl acetate at pH 1–3 as the cupferrate. Titanium and vanadium interfere.

Procedures

Nonferrous Metals.[64] This technic is applicable to such samples as lead, zinc, nickel, and manganese. Dissolve a 2-gram sample with a suitable volatile acid. Add 2 ml of perchloric acid and heat to white fumes. Take up in 10 ml of water and add 10 ml of 5% solution of EDTA. Add 10 ml of 50% solution of sodium acetate and raise to pH 3 with ammonium hydroxide. Dilute to 50 ml. Add 5 ml of 5% solution of cupferron and shake for 1 minute with 10 ml of amyl acetate. Read the organic layer at 420 nm.

Tin. Dissolve 2 grams in 10 ml of 1 : 1 hydrochloric acid. Add 3 ml of bromine and evaporate nearly to dryness, driving off the bromine and tin in the process.

[63] E. A. Bondarevskaya, E. D. Kropotova, and N. A. Gradskova, *Zavod. Lab.* **37**, 539–540 (1971).
[64] Shigeo Wakamatsu, *Jap. Anal.* **8**, 298–302 (1959).

Add 2 ml of perchloric acid and heat to dense white fumes. Complete as above from "Take up in 10 ml of water and. . . ."

Rapeseed.[65] Ash 5 grams of sample at 500°. Extract the residue with 5 ml and 5 ml of hydrochloric acid. Dilute to 50 ml and extract the copper from a 10-ml aliquot by shaking for 2 minutes with 10 ml of 0.1% lead diethyldithiocarbamate in chloroform. Discard the extract. Add 2 ml of 1% solution of cupferron to the aqueous phase. Extract ferric cupferrate with 10 ml of chloroform and read at 420 nm. The technic can be applied to fat samples.[66]

Plant Material.[67] Ash a 10-gram sample at 450°. Extract the ash with 10 ml and 10 ml of 1 : 3 hydrochloric acid and filter the extracts. Proceed as above from "Add 2 ml of 1% solution of. . . ."

2-DIAMINOPHENOL HYDROCHLORIDE AND BROMATE

The oxidation of 2-diaminophenol hydrochloride by bromate is catalyzed by ferric ion to a degree dependent on pH. This kinetic method of estimation of ferric ion reaches an optimum at pH 2.5. With bromate at 0.02 *M*, the rate of oxidation is negligible in the absence of ferric ion. The slope of the plot of extinction with time over the range called for in the procedure is a straight line. The oxidation is also catalyzed by vanadate, chromate, molybdate, and cupric ions, which must therefore be absent.

Procedure. Mix a sample solution containing 19–130 ng of iron with 5 ml of 0.1 *M* potassium bromate and 5 ml of freshly prepared 0.1% solution of 2,4-diaminophenol hydrochloride.[68] Add 5 ml of a potassium citrate–sulfuric acid buffer solution for pH 2.5 and dilute to 25 ml. Read periodically at 510 nm and plot the extinction against time.

DIANTIPYRYLMETHANE

At pH 2–3.5, this reagent forms a red complex with ferric ion having a maximum absorption at 450 nm.[69] With 4 ml of 1% solution of the reagent in ethanol and 5 ml of *M* acetate buffer for pH 3 in a total volume of 15 ml, Beer's law is followed up to 10 ppm of iron. There is interference by molybdenum, tungsten, lead, perchlorate, chromate, thiosulfate, oxalate, phosphate, and titanic ions.

[65] Milan Karvanek, *Prum. Potravin.* **15**, 282–283 (1964).
[66] Milan Karvanek, *Sb. Vys. Sk. Chem.-Technol., Odd. Fak. Potrav. Technol.* **5**, Pt. 1, 203–212 (1961).
[67] Milan Karvanek, *ibid.* **23**, 13–21 (1969).
[68] A. Seteu, A. I. Crisan, and Zsuzsanna Kiss, *Stud. Univ. Babes-Bolyai, Ser. Chem.* **14** (2), 57–59 (1969).
[69] Hajime Ishii, *Jap. Anal.* **6**, 460–463 (1967).

Procedures

Bronze.[70] Dissolve 0.2 gram of sample in 15 ml of nitric acid and boil off oxides of nitrogen. Add 20 ml of hydrochloric acid and evaporate to a sirup. Dilute to 100 ml and take a 5-ml aliquot. Add 15 ml of 0.5% solution of diantipyrylmethane. Add 25% solution of sodium acetate until the solution turns Congo red paper to red; then add hydrochloric acid dropwise until the paper turns to lilac. Dilute to 100 ml and after 25 minutes read at 450 nm.

Magnesium Alloys. Dissolve a 1-gram sample in 20 ml of hydrochloric acid and a few drops of nitric acid. Boil off oxides of nitrogen, cool, and dilute to 50 ml. To a 25-ml aliquot add 10 ml of 0.5% solution of diantipyrylmethane. Complete as for bronze from "Add 25% solution of..." but make the final dilution 50 ml.

5,7-DIBROMO-8-HYDROXYQUINOLINE

This reagent is also known as 5,7-dibromo-8-hydroxyquinol or, less exactly, as dibromooxine. For the determination, ferric ion is isolated by paper chromatography. There is interference by the following ions: cerous, thoric, hexavalent uranium, vanadic, chromic, hexavalent molybdenum, cobalt, nickel, cadmium, mercurous, mercuric, and stannous, and to a lesser degree, magnesium, manganous, zinc, and aluminum.

Procedure. Mix 1 gram of 1-phenyl-1,3-butanedione, 100 ml of 1-butanol, and 100 ml of 15% solution of trichloroacetic acid.[71] Allow the phases to separate. In a unit for descending paper chromatography, place the upper layer of solvent in the trough and the lower layer on the tank bottom. Use carboxycellulose paper.

Follow this technic for each of four blank zones and for each of four loaded zones. Shred the zone, warm with 14 ml of ethanol, and filter through sintered glass. Repeat the extraction with 7 ml and 7 ml of ethanol and rinse with 4 ml and 4 ml of ethanol. Add 4 ml of 0.1 *N* nitric acid and 6 ml of 0.04% solution of the reagent in ethanol. Dilute to 50 ml with ethanol and set aside for 10 minutes. Read at 465 nm using a blank zone as reference.

DI(*o*-HYDROXYPHENYL)ACETIC ACID

The ferric complex of this reagent reaches a maximum absorption over a wide pH range and is so stable that a large excess of reagent is not necessary. Beer's law is followed up to 14 ppm of iron, but a range of 3.2–8 ppm has minimal error. Cobalt and chromium interfere even at 3 : 1, but most cations do not, even at 7 : 1.

[70] L. Ya. Polyak, *Zavod. Lab.* **27**, 388–389 (1961); USSR Patent *120,677* (1959).
[71] I. I. M. Elbeih and M. A. Abou-Elnagh, *Chemist-Analyst* **55**, 43–45 (1966); **56**, 99–100 (1967).

Procedure. *Aluminum Alloy.*[72] As reagent, suspend 1.8 grams of solid di(*o*-hydroxyphenyl)-acetic acid in 50 ml of water and add 24% sodium hydroxide solution dropwise, checking with a pH meter. The solid slowly dissolves with stirring, and solution is almost complete at pH 8.5–9. Dilute to 100 ml and filter. As *M* acetate buffer, dissolve 60 ml of glacial acetic acid in 200 ml of water, add 24% sodium hydroxide solution until the pH reaches 5, and dilute to 1 liter.

Cover 0.1 gram of sample with 1 ml of water and dissolve by dropwise addition of hydrochloric acid. Complete the reaction by warming, add 0.5 ml of nitric acid, and boil the solution gently for several minutes. Dilute to 10 ml and add 24% sodium hydroxide solution dropwise until a slight permanent precipitate is obtained. Add 2 ml of reagent solution and 5 ml of acetate buffer for pH 5. Dilute to 25 ml. The solution is slightly cloudy at this stage. Filter through a slow paper, discard the first 5 ml, and read at 470 nm.

2,4-DIHYDROXYPROPIOPHENONE OXIME

This colorless reagent forms a stable purple 1 : 1 complex with ferric ion at pH 2–3 with an optimum at 2.6–2.8. Color develops immediately and is stable for 10 hours. Beer' s law is followed for 1–40 ppm of iron. No significant interference was found.

Procedure. To 2 ml of sample approximately 0.01 *M* in ferric ion, add 20 ml of 0.01 *M* solution of the reagent in 40% ethanol and 5 ml of *M* sodium perchlorate.[73] Dilute to 50 ml and read at 510 nm.

DIMETHYLGLYOXIME

The 1 : 1 complex of ferric ion with this reagent is formed in both acid and alkaline solutions.[74]

Procedures

Aluminum Alloy.[75] Digest 0.2 gram of sample with 8 ml of 20% sodium hydroxide solution. When reaction ceases, dilute to 50 ml and boil for 5 minutes. Filter, and wash the residue with 2% solution of sodium hydroxide. Discard this solution. Add 10 ml of 1 : 1 hydrochloric acid and 1 ml of 30% hydrogen peroxide to the residue. Heat to decompose hydrogen peroxide, cool, and dilute to 100 ml. To a 10-ml aliquot, add 2 ml of 0.02 *N* solution of vanadyl sulfate, $VOSO_4$, and a

[72] A. L. Underwood, *Anal. Chem.* **30**, 44–47 (1958).
[73] Mrinalini H. Gandhi and Mahendra N. Desai, *ibid.* **39**, 1643–1644 (1967).
[74] Yu. V. Morachevskii, L. I. Lebedeva, and Z. G. Golubstrova, *Zh. Anal. Khim.* **15**, 472–475 (1960).
[75] Hiroshi Iinuma and Takayoshi Yoshimori, *Jap. Anal.* **5**, 149–152 (1956).

few ml of zinc amalgam. Shake for a few minutes and separate the aqueous layer. Add 4 ml of 10% hydroxylamine hydrochloride solution, 4 ml of 40% solution of sodium potassium tartrate, 4 ml of 1% ethanolic dimethylglyoxime, and 10 ml of ammonium hydroxide. Dilute to 100 ml and set aside for 20 minutes. Read at 530 nm.

Copper Alloy. Digest 0.5 gram of sample with 20 ml of 1 : 1 hydrochloric acid. Add several 0.5-ml portions of 30% hydrogen peroxide until the sample is dissolved. Add 10 ml of 1 : 1 sulfuric acid and heat to sulfur trioxide fumes. Cool. Take up with 70 ml of water. Remove the copper by electrolysis. Dilute to 100 ml and complete as for aluminum alloy from "To a 10-ml aliquot, add 2 ml. . . ."

Copper and Copper Alloys.[76] Dissolve 1 gram of sample in 10 ml of 1 : 1 nitric acid. Add 2 ml of potassium aluminum sulfate solution containing 5 mg of aluminum per ml and dilute to 200 ml. Add ammonium hydroxide in excess until copper is complexed and iron and aluminum hydroxides precipitated. Filter and wash with 1% ammonium nitrate solution in 1 : 100 ammonium hydroxide until free from copper. Dissolve the precipitate in 10 ml of hot 1 : 1 hydrochloric acid and wash the paper. To the solution, add sequentially 1 ml of copper sulfate solution containing 0.25 mg of copper, 2 ml of 10% hydroxylamine hydrochloride solution, and 1 ml of 1% ethanolic dimethylglyoxime. Add 10 ml of ammonium hydroxide, dilute to 50 ml, set aside for 10 minutes, and read at 532 nm against a reagent blank.

DIMETHYL-*p*-PHENYLENEDIAMINE

Ferric ion oxidizes this reagent to give a red proportional to the amount of ferric ion. The color develops immediately and after 6 minutes starts to decrease slowly, the decrease being 1% at 10 minutes. Ceric and permanganate ion give about half as much color as the same concentration of ferric ion.

Procedure. To 1 ml of sample at pH 2 containing 1–20 μg of iron, add 3 ml of water and 1 ml of 0.1% solution of dimethyl-*p*-phenylenediamine in water or ethanol.[77] Read within 10 minutes at 515 nm.

Biological Samples. Digest the sample with 5 ml of sulfuric acid and 1 ml of nitric acid, adding 30% hydrogen peroxide to the extent required to obtain a colorless solution. Evaporate to dryness. Dissolve the residue in 2 ml of 2 *N* sulfuric acid by heating for 15 minutes at 100°. Add 2 ml of a solution of 4 grams of sodium acetate in 10 ml of water and 1 ml of 0.1% solution of the reagent in ethanol. Read at 515 nm within 30 minutes.

---

[76] Tetsuro Murakami, *ibid.* **6**, 172–173 (1957).

[77] Giovanni Ceriotti and Luigi Spandrio, *Anal. Chem.* **33**, 579–580 (1961); *Clin. Chim. Acta* **6**, 233–236 (1961).

DINITROSORESORCINOL

The complex of ferrous ion and dinitrosoresorcinol has a maximum absorption at 640–690 nm.[78] Increased reagent concentration lowers the wavelength of maximum absorption. Beer's law is followed for 0.1–10 ppm of iron. There are maxima in both the acid and alkaline ranges, with the absorption higher on the acid side. Desirable pH values are 4.4 and 10. Hydroxylamine or sulfite will serve as the reducing agent. Ferric ion also complexes with this reagent, but at a lower absorption.

Procedure. To 25 ml of sample containing 0.2–20 ppm of iron, add 1 ml of 10% hydroxylamine hydrochloride solution followed by 3.75 ml of a 0.186% solution of 2,4-dinitrosoresorcinol in a 0.106% solution of sodium carbonate.[79] Add 0.5 ml of 0.53% solution of sodium carbonate and dilute to 50 ml. Read at 630 nm against a reagent blank.

4,7-DIPHENYL-1,10-PHENANTHROLINE

This reagent, commonly called bathophenanthroline, is one of the ferroin family and is the most sensitive of the widely used members. It is appropriate for reading ferrous ion in the presence of ferric ion.[80] The 1 : 3 complex is water insoluble, but in 10% ethanol at pH 2–6, it conforms to Beer's law for 5–100 ppb of iron. The color is very stable. There is interference by copper, cobalt, cyanide, and citrate. It is commonly extracted for reading. Such a liquid-liquid extraction eliminates interference by chromium and vanadium and avoids colloids or precipitates from hydrolysis of tantalum or titanium during development. Sodium hydrosulfite reduces iron in the presence of large amounts of complexing agents such as citrate or tartrate necessary to keep metals in solution at the pH for extraction. The reagent does not absorb at 533 nm, a common wavelength for reading. For extraction, pH 4–6 is desirable. If manganese, cobalt, or nickel is a major component, a prior separation of ferrous cupferrate by chloroform is required. With lesser amounts of these metals, a large excess of reagent will avoid interference. Unlike many other reagents for iron, perchloric acid is tolerated.

For determination of ferrous ion in the presence of up to twentyfold excess of ferric ion, mask the latter with sodium pyrophosphate.[81] Form the batho phenanthroline complex at pH 4.2–4.7 and extract with butanol. Separation of nickel, chromium, and other minor components of steel is not necessary.[82]

For water analysis, the solution is desirably 10% ethanol, less than 0.1 ppm of ferrous ion, read at 533 nm, pH 2–6, and 5 *M* excess of bathophenanthroline.[83] When determining ferrous ion in water, if the ratio of ferrous to ferric ion is 1 : 4,

[78] S. E. Zayan, R. M. Issa, and Jacqueline Maghrabi, *Microchem. J.* **18**, 662–669 (1973).
[79] Hassan El Khadem and Saad Eldin Zayan, *Anal. Chem.* **34**, 1382–1384 (1962).
[80] Fumito Nakashima and Kaoru Sakai, *Jap. Anal.* **10**, 89–98 (1961).
[81] Takayuki Mizuno, *Talenta* **19**, 369–372 (1972).
[82] A. Moauro, *Com. Naz. Energ. Nucl.* **RT/CHI(67) 16**, 9 pp (1967).
[83] Fumito Nakajima and Kaoru Sakai, *Jap. Anal.* **10**, 89–93 (1961).

there is appreciable increase in ferrous ion within 8 hours.[84] The reading should be taken at pH 5 within 10 minutes after adding the reagent.

For iron in fused silica, heat 5 grams of sample nearly to dryness with water, hydrofluoric acid, nitric acid, and sulfuric acid.[85] Take up with water and develop as the bathophenanthroline complex in hexanol.

For analysis of very high purity cobalt, extract 1–15 μg of iron from 8 *N* hydrochloric acid solution containing potassium chlorate.[86] Use 1 : 1 methyl isobutyl ketone–amyl acetate and develop with bathophenanthroline.

For up to 40 ppm of iron in molybdates, complex the molybdate with tartrate.[87] Then reduce the iron with hydroxylamine hydrochloride and extract the bathophenanthroline complex in the presence of perchlorate, using chloroform as solvent. Read at 522 nm.

For solutions of uranium-bearing materials, eliminate interference of copper by extraction of the iron from 6.5 *N* hydrochloric acid into ethyl acetate.[88] Then back-extract the iron with water, reduce with hydroxylamine hydrochloride, complex with bathophenanthroline, and read at 535 nm.

For analysis of high purity gallium arsenide, dissolve in hydrochloric and nitric acid, reduce with hydroxylamine hydrochloride, add citric and ascorbic acid, and form the bathophenanthroline complex.[89] Extract with chloroform at pH 4–5.5 and read at 533 nm.

For iron in butter fat, extract with nitric acid at 80° and cool.[90] Remove the fatty layer with petroleum ether and add perchloric acid. Digest until colorless. Reduce with hydroxylamine hydrochloride, add bathanthroline, extract with hexanol, and read at 533 nm.

For biological material in general, wet-ash the sample, evaporate excess acid, reduce with mercaptoacetic acid, neutralize with sodium acetate solution, develop with bathophenanthroline, and extract with isoamyl alcohol to read at 535 nm.[91]

Digest 4 grams of black olive pulp with nitric, sulfuric, and perchloric acids, and dilute to a known volume.[92] Mix an aliquot with 2 ml of 10% hydroxylamine hydrochloride solution and 5 ml of 0.1% ethanolic bathophenanthroline. Make alkaline with ammonium hydroxide, extract with carbon tetrachloride, and read the extract at 538 nm.

Bathophenanthroline can be sulfonated with chlorosulfonic acid to give water solubility, obviating the need for extraction.[93a]

For iron in tantalum pentoxide by bathophenanthroline, see determination by Thiocyanate. For iron in silicon tetrachloride by bathophenanthroline, see Aluminum by Eriochrome Cyanine R. For copper by neocuproine and iron by bathophenanthroline in serum, see Copper.

[84] M. M. Ghosh, J. T. O'Connor, and R. S. Engelbrecht, *J. Am. Water Works Assoc.* **59**, 897–905 (1967).
[85] K. F. Sagawara and Yao-sin Su, *Anal. Chim. Acta* **80**, 143–151 (1975).
[86] G. Uny, C. Mathien, J. P. Tardief, and Tran Van Danh, *ibid.* **53**, 109–116 (1971).
[87] D. J. B. Galliford and E. J. Newman, *Analyst* **87**, 68–70 (1962).
[88] R. J. Guest and F. P. Roloson, *Can. Dep. Mines Tech. Surv., Mines Branch, Radioact. Div. Top. Rep.* **TR-137/57**, 20 pp (1956).
[89] M. Knizek and A. Galik, *Z. Anal. Chem.* **213**, 254–259 (1965).
[90] Hermann Timmen and Jutta Bluethgen, *Z. Lebensm. Unters. Forsch.* **153**, 283–288 (1973).
[91] M. Van de Bogart and H. Beinert, *Anal. Biochem.* **20**, 325–334 (1967).
[92] M. A. Albi and A. Garrido Fernandez, *Grasas Aceit.* **26**, 133–135 (1975).
[93a] James W. Landers and Bennie Zak, *Tech. Bull. Reg. Med. Technol.* **28**, 98–100 (1958); *Am. J. Clin. Pathol.* **29**, 590–592 (1958).

Procedure. To 100 ml of sample solution, add 4 ml of 15% solution of ascorbic acid and 0.1 ml of hydrochloric acid.[93b] Store at near 100° for 1 hour and cool. Add 5 ml of 10% solution of sodium acetate and 4 ml of mM reagent. Mix well and set aside for 30 minutes. Extract with 10 ml of isoamyl alcohol and set aside for 8 hours. Read at 533 nm.

Isolation by Isopropyl Ether.[94] For determination with 1,10-phenanthroline or bathophenanthroline, iron must be isolated from phosphates or zirconium.

Take a sample containing 0.03–0.25 mg of iron if determination is to be with 1,10-phenanthroline or 0.01–0.06 mg if with bathophenanthroline. Dissolve in 18 ml of 8 N hydrochloric acid, add 1 drop of N potassium permanganate, and mix. Extract the iron by shaking for 1 minute each with 25 ml and 25 ml of isopropyl ether, keeping the extracts separate. Reextract the iron from the second isopropyl ether extract with 25 ml of water. Use this aqueous extract to reextract the iron from the first isopropyl ether extract. Develop with either of the reagents above.

Extraction with Propylene Carbonate.[95] To a sample solution containing 0.025–0.5 micromole of iron, add 3 ml of 10% solution of hydroxylamine hydrochloride and 5 ml of 0.001 M bathophenanthroline in 50% ethanol. Add 5 ml of 10% solution of sodium acetate as a buffer. Saturate the solution with propylene carbonate and add sufficient to give a 3-ml layer in excess. Shake for 1 minute, separate the organic layer, dilute to 10 ml with ethanol, and read at 533 nm.

Copper.[96] Take a solution of a copper sample containing up to 0.25 gram of copper. For copper salts, dissolve 1 gram in water. In either case dilute to 100 ml adding in the process sufficient sulfuric acid to give a final concentration of 0.2–0.3 N. Add 15 ml of 7% solution of ammonium thiocyanate and 15 ml of 6% solution of hydroxylamine hydrochloride. Heat at 100° for 45 minutes, and cool. Set aside for 10 minutes and filter through a medium-porosity sintered-glass crucible. To an aliquot of the filtrate containing not more than 40 μg of iron, add 5 ml of freshly prepared 0.1% ascorbic acid solution and 5 ml of 10% solution of sodium acetate trihydrate. If necessary, adjust to pH 4–6 with ammonium hydroxide. Add 10 ml of a solution of 0.0332 gram of bathophenanthroline per 100 ml of ethanol and set aside for 5 minutes. Extract with 5 ml and 5 ml of chloroform, dilute the filtered combined extracts to 25 ml with ethanol, and read at 533 nm against a reagent blank.

Gold.[97] Dissolve a 1-gram sample in 3 ml of nitric acid and 7 ml of hydrochloric acid and 2 ml of nitric acid. Evaporate to dryness and take up the residue in 10 ml of 1 : 5 hydrochloric acid. Extract the gold with 10 ml and 10 ml of isopentyl acetate and discard. Add 5 ml of 10% solution of hydroxylamine hydrochloride and

[93b] R. D. Perry and C. L. San Clemente, *Analyst* **102**, 114–119 (1977).
[94] F. H. Lohmann, D. F. Kuemmel, and E. M. Sallee, *Anal. Chem.* **31**, 1739–1740 (1959).
[95] B. G. Stephens and H. A. Suddeth, *ibid.* **39**, 1478 (1967).
[96] R. P. Hair and E. J. Newman, *Analyst* **89**, 42–48 (1964).
[97] G. Ackermann and J. Köthe, *Z. Anal. Chem.* **231**, 252–261 (1967); Zygmunt Marczenko, Krzysztof Kasiura, and Maria Krasiejko, *Chem. Anal.* (Warsaw) **14**, 1277–1278 (1969); cf. Krzysztof Kasiura, *ibid.* **20**, 809–816 (1975); Masuo Miyamoto, *Jap. Anal.* **9**, 748–753 (1960).

boil for 20 minutes. Cool. Add 5 ml of 2% solution of bathophenanthroline in 50% ethanol and 5 ml of a saturated solution of ammonium citrate. Extract with 5 ml and 5 ml of 1 : 1 isoamyl alcohol–isopropanol. Dilute the extracts to 25 ml with the mixed solvent and read at 530 nm.

Bismuth.[98] Dissolve 1 gram of sample in 10 ml of 1 : 1 hydrochloric acid and 1 ml of nitric acid. Evaporate almost to dryness, add 1 ml of hydrochloric acid, and dilute to 5 ml. Add 0.2 ml of 10% stannous chloride solution, boil, and cool. Add 10 ml of ethanol and 2 ml of 2 *N* hydrochloric acid. Add 4 ml of 0.2% solution of bathophenanthroline in 70% ethanol, 20 ml of 4% EDTA solution, and 10 ml of 50% solution of sodium citrate. Set aside for 5 minutes and extract with 10 ml of hexyl alcohol. Separate the organic layer and add 0.2 ml of 2 *N* hydrochloric acid. Dilute to 25 ml with ethanol and read at 533 nm against a blank.

High Purity Niobium, Tantalum, Molybdenum, and Tungsten.[99] Add 5 ml of water and 2 ml of hydrofluoric acid to a powdered 0.5-gram sample in a Teflon beaker. Add nitric acid dropwise until solution is complete. If the sample is molybdenum, add 2 ml of hydrochloric acid and heat until the solution is pale yellow. Add 3 ml of formic acid, and warm until nitric acid is destroyed. Evaporate to 5 ml and add 10 ml of water. Add 10 ml of 25% ammonium tartrate solution and warm below boiling until clear. Add 5 ml of 5% boric acid solution and set aside for 20 minutes. Add 2 ml of 1 : 1 sulfuric acid, 10 ml of 10% ascorbic acid solution, and 5 ml of 10% solution of hydroxylamine hydrochloride. Mix and set aside for 10 minutes. Raise to pH 5.5 with ammonium hydroxide and dilute to 100 ml.

To a 20-ml aliquot and to a blank, add 5 ml of water. Add 2 ml of 5% solution of thiourea. Five minutes later add 4 ml of 0.001-*M* bathophenanthroline, mix, and set aside for 15 minutes. Extract with 10 ml of pentanol and read the extract against the blank at 536 nm.

High Purity Chromium.[100] Dissolve 1 gram in excess of 1 : 1 hydrochloric acid and evaporate to a syrup. Take up in 10 ml of water and add 5 ml of 10% solution of sodium acetate. Adjust to pH 3 with 1 : 10 ammonium hydroxide. Add 2 ml of 10% solution of hydroxylamine hydrochloride and 1 ml of 0.01 *M* solution of the color reagent in ethanol. Extract with 4 ml and 4 ml of isoamyl alcohol, dilute the combined extracts to 10 ml with isoamyl alcohol, and read at 535 nm against that solvent.

High Purity Plutonium.[101] Dissolve 0.25 gram of sample in 2.5 ml of 1 : 1 hydrochloric acid. Add 2 ml of 10% solution of hydroxylamine hydrochloride and 5 ml of 0.001 *M* bathophenanthroline in 50% ethanol. Add 10% sodium acetate solution to bring to pH 4–5, followed by 5 ml of 5% EDTA solution. Extract with 5 ml and 5 ml of chloroform and read the combined extracts at 535 nm.

Thallium.[102] Treat a sample solution containing 2 grams of thallium as nitrate

[98] E. Booth and T. W. Evett, *Analyst* **83**, 80–82 (1958).
[99] E. M. Penner and W. R. Inman, *Talenta* **9**, 1027–1036 (1962).
[100] Tadashi Yanagihara, Nobuhisa Matano, and Akira Kawase, *Jap. Anal.* **10**, 414–417 (1961).
[101] *UKAEA*, **PG 729 (W)**, 6 pp (1966).
[102] E. Jackwerth, J. Lohmar, and G. Schwark, *Z. Anal. Chem.* **260**, 101–106 (1972).

or sulfate with 1 ml of 0.01 *M* EDTA and dilute to 55 ml as the nitrate or to 70 ml as the sulfate. Add 15 ml of 15% solution of sodium iodide, shake for 3 minutes, and centrifuge. Mix an aliquot of the clear upper layer with 2 ml of 1 : 1 sulfuric acid and evaporate to dryness. Take up in 1 ml of hydrochloric acid, dilute to 20 ml, and neutralize with 1 : 3 ammonium hydroxide. Dilute to 40 ml and add 50 ml of acetate buffer solution for pH 4.5. Add 2 ml of 1% sodium diethyldithiocarbamate solution and extract with 5 ml, 5 ml, and 5 ml of chloroform. Discard the extract and add 5 ml of 10% solution of hydroxylamine hydrochloride. Add 5 ml of 2% solution of bathophenanthroline in 50% ethanol. Extract with 10 ml of pentanol and read at 536 nm.

Vanadium.[103] Dissolve 1 gram of sample in 10 ml of hydrochloric acid and 3 ml of nitric acid. Add 10 ml of 1 : 1 sulfuric acid and evaporate to light fumes of sulfur trioxide. Cool, and add 10 ml of water and 5 ml of sulfurous acid. Again evaporate to light fumes of sulfur trioxide. Cool, and dilute to 100 ml. Dilute an aliquot containing 0.005–0.12 mg of iron to 30 ml. Add an excess of 10% sodium citrate solution and adjust to pH 4–6 with ammonium hydroxide. Add 20 ml of freshly prepared 10% sodium hydrosulfite solution. Let stand for exactly 15 minutes and add 10 ml of 0.1% solution of bathophenanthroline in ethanol or methanol. Extract with 15 ml and 10 ml of chloroform. Dilute the combined extracts to 50 ml with ethanol and read at 533 nm against a reagent blank.

Vanadium Pentoxide. Dissolve a 1-gram sample in 20 ml of hydrochloric acid by warming. Add 5 ml of sulfurous acid and evaporate most of the hydrochloric acid. Complete as for vanadium from "Cool, and dilute to 100 ml."

Ammonium Metavanadate. Heat 1 gram of sample with 30 ml of 1 : 2 hydrochloric acid, stirring occasionally until completely dissolved, and evaporate most of the hydrochloric acid. Add 5 ml of sulfurous acid and heat to reduce the vanadium. Complete as for vanadium from "Cool, and dilute to 100 ml."

Chromium. Dissolve 1 gram of metal by warming with 30 ml of 1 : 2 hydrochloric acid and evaporate most of the hydrochloric acid. Complete as for vanadium from "Cool, and dilute to 100 ml."

Niobium and Tantalum. Transfer 0.5 gram of metal to a platinum crucible and add 5 ml of hydrofluoric acid. Add 0.7 ml of nitric acid dropwise and evaporate nearly to dryness. Cool. Add 1 ml of hydrofluoric acid and 5 ml of water. Add 25 ml of 10% sodium citrate solution and complete as for vanadium from "Add 20 ml of freshly prepared. . . ."

Titanium Carbide. Transfer 0.5 gram of sample to a platinum dish. Add 5 ml of hydrofluoric acid, 1 ml of nitric acid, and 10 ml of sulfuric acid. Heat to sulfuric acid fumes until the solution is colorless and, if necessary, add more nitric acid. After evaporating off any extra nitric acid added, complete as for vanadium from "Cool, and dilute to 100 ml."

[103] A. R. Gahler, R. M. Hamner, and R. C. Shubert, *Anal. Chem.* **33**, 1937–1941 (1961).

Titanium or Titanium-Aluminum-Vanadium Alloy. Transfer 0.5 gram of sample to a platinum dish. Add 20 ml of 1 : 3 hydrochloric acid and 0.7 ml of hydrofluoric acid. Heat until dissolved, and add 15 ml of 30% sodium citrate solution. Add 25 ml of 4% boric acid solution and dilute to 100 ml. Complete as for vanadium from "Dilute an aliquot containing 0.005–0.12 mg. . . ."

Tungsten.[104] Dissolve 1 gram of tungsten containing up to 0.012% iron in platinum with 2 ml of hydrofluoric acid and 1 ml of nitric acid. Add 20 ml of 10% citric acid solution and evaporate to dryness at 100°. Add 10 ml of 10% sodium hydroxide solution and warm to dissolve tungstic oxide. Check that the solution is alkaline and, if necessary, add a few drops more of sodium hydroxide solution. Add 20 ml of 10% citric acid solution and check that the pH is 4–7. Dilute to 100 ml. Complete as for vanadium from "Dilute an aliquot containing 0.005–0.12 mg. . . ."

Uranium Carbide. Transfer a 0.5-gram sample to a weighing bottle; if particle size is very small, use a dry box flushed with argon. Transfer the weight from the bottle to a beaker. Cautiously add 15 ml of water, then 15 ml of nitric acid dropwise. Warm until action ceases, add 10 ml of 1 : 1 sulfuric acid, and evaporate to sulfur trioxide fumes. Cool, add 25 ml of water, and dilute to 250 ml. Complete as for vanadium from "Dilute an aliquot containing 0.005–0.12 mg. . . ."

Uranium Oxide. Dissolve 1 gram in 5 ml of nitric acid and 10 ml of 1 : 1 sulfuric acid. Evaporate to light fumes of sulfur trioxide. Complete as for vanadium from "Cool, and add 10 ml of water and 5 ml of sulfurous acid."

Corrosion Products.[105] Transfer a 10-mg sample containing not more than 3 mg of ferrous iron to a 100-ml flask that has been deaerated with carbon dioxide, and continue to pass carbon dioxide through a tube extending to the bottom of the flask while preparing the sample. Add 10 ml of 1 : 1 hydrochloric acid and while agitated by the carbon dioxide, heat below boiling to dissolve. Add water up to the base of the neck of the flask and cool. Remove the tube and dilute to volume. Store under carbon dioxide.

Mix 5 ml of 10% ammonium dihydrogen phosphate solution with 2 ml of acetic acid. Add an aliquot of the sample solution, which should contain no more than 30 μg of ferrous ion, and if less than 10 ml, make up to that volume with water. Neutralize to Congo red paper by dropwise addition of ammonium hydroxide or, if the aliquot is very small, with hydrochloric acid. The pH approximates 2.2. Pipet 10 ml of isoamyl acetate to form a separate supernatant layer. With a minimum of agitation, add 1 ml of 0.01 *M* bathophenanthroline that has been deaerated with carbon dioxide. Shake at a moderate rate for 4 minutes. After 1 hour, drain and discard the colorless lower layer and a few drops of the organic layer. Filter the organic layer through filter paper and read at 530 nm against isoamyl acetate.

[104] R. H. A. Crawley and M. L. Aspinal, *Anal. Chim. Acta* **13**, 376–378 (1955).
[105] Lewis J. Clark, *Anal. Chem.* **34**, 348–352 (1962); cf. E. N. Pollock and A. N. Miguel, *ibid.* **39**, 272 (1967).

Zircon.[106] Heat 50 mg of powdered sample with 0.5 gram of carbonate-borate flux in platinum at 150–200°. Raise gradually to 925° and hold there for 30 minutes. When cool, disintegrate with 5 ml of water, then dissolve with 12.4 ml of 1 : 1 hydrochloric acid and dilute to 25 ml. To a 5-ml aliquot add 2 ml of 10% hydroxylamine hydrochloride solution, 5 ml of 0.0015 *M* bathophenanthroline, and 5 ml of 1.25% solution of tartaric acid adjusted to pH 4.8–5 with ammonium hydroxide. Dilute to 25 ml and read at 535 nm against water.

Ruby.[107] Sinter 1 gram of finely divided sample with 2 grams of sodium carbonate and 2 grams of sodium tetraborate. Dissolve the melt in 10 ml of 1 : 1 hydrochloric acid and dilute to 25 ml. To a 5-ml aliquot, add 1 ml of 10% hydroxylamine hydrochloride solution and 1 ml of 2% ethanolic bathophenanthroline. Dilute to 25 ml and read at 535 nm against a reagent blank.

Soil.[108] Add 0.1 gram of 100-mesh sample to 75 ml of 1 : 2 sulfuric acid and boil for 1 minute. Then add 10 ml of hydrofluoric acid and add to 100 ml of 1 : 10 sulfuric acid. Add 25 ml of saturated solution of boric acid and filter. Wash the filter with 1 : 50 sulfuric acid and after cooling, dilute to 500 ml. Mix a 5-ml aliquot with 5 ml of 0.01 *M* bathophenanthroline in 50% ethanol and 5 ml of saturated solution of ammonium citrate. If necessary, reduce ferric ion with 5 ml of 20% hydroxylamine hydrochloride solution. Extract with 5 ml and 5 ml of nitrobenzene. Dilute the combined extracts to 25 ml with ethanol and read at 538 nm.

Reagent-Grade EDTA.[109] Decompose 1 gram of sample with 6 ml of nitric acid. Evaporate the clear solution to about 1 ml and dilute to 20 ml. Add 2 ml of 10% solution of hydroxylamine hydrochloride, followed by 2 ml of 0.03% ethanolic bathophenanthroline. Adjust to pH 4.5 with a 20% solution of ammonium acetate. Extract with 10 ml of 1 : 1 isopropyl alcohol–isopropyl ether and read the extract at 530 nm.

Rainwater.[110] To 500 ml of sample, add 1 ml of 10% solution of hydroxylamine hydrochloride, 1 ml of 0.005 *M* bathophenanthroline in 50% ethanol, and a couple of drops of 12% sodium perchlorate solution. Extract the complex, which includes perchlorate ion, with 10 ml of nitrobenzene and read at 522 nm.

In analysis of water with this reagent, erratic results have been traced to variations in the concentration of added hydrochloric acid and exposure of the acidified samples to sunlight.[111] Storage of the sample in the dark after acidification for at least 10 minutes is recommended.

Reactive Iron in Boiler Feed Water.[112] Collect the sample in a vessel containing

[106] Frank Cuttitta and Jesse J. Warr, *U.S. Geol. Surv., Prof. Papers* No. **424-C**, 383–384 (1961).
[107] R. C. Chirnside, H. J. Cluley, R. J. Powell, and P. M. C. Proffett, *Analyst* **88**, 851–863 (1963).
[108] J. L. Walker and G. D. Sherman, *Soil Sci.* **93**, 325–328 (1962).
[109] Keishi Nakahara and Taeko Danzuka, *Jap. Anal.* **13**, 20–22 (1964).
[110] F. P. Gorbenko, L. Ya. Zakora, I. A. Shevchuk, and F. P. Lapshin, *Gidrokhim. Mater.* **40**, 194–197 (1965).
[111] J. McMahon, *Limnol. Oceanogr.* **12**, 437–442 (1967).
[112] A. L. Wilson, *Analyst* **89**, 389–401 (1964).

sufficient 1 : 1 hydrochloric acid to make the sample 0.1 *N*. To a 200-ml portion of sample, add 2 ml of 10% solution of hydroxylamine hydrochloride. Add 2 drops of metacresol purple indicator, then add ammonium hydroxide until the indicator turns yellow. Add 5 ml of acetate buffer solution for pH 4. Extract with 4 ml of 0.1% bathophenanthroline solution in isopentyl alcohol, then with 10 ml and 5 ml of isopentyl alcohol. Dilute the combined extracts to 25 ml with ethanol and read at 534 nm. Correct for a reagent blank.

Total Iron in Boiler Feed Water.[113] Collect the sample in a polyethylene bottle containing 4 ml of mercaptoacetic acid. Heat at 80° for 1 hour and cool. Take a 200-ml aliquot and complete as above from "Add 5 ml of acetate buffer solution. . . ."

Jet Fuels.[114] Shake 250 ml of sample in a 16-ounce screw-cap bottle with 5 ml of 1 : 3 hydrochloric acid for 30 seconds. Heat at 80° for 90 minutes, shaking for 30 seconds at 20-minute intervals. Remove from the oven and shake horizontally for 10 minutes. Add 25 ml of water and shake for 5 minutes. Let cool to room temperature and take a 20-ml aliquot of the aqueous layer. Add 10 ml of 0.87 *N* sodium hydroxide, 2 ml of 10% hydroxylamine hydrochloride solution, 4 ml of 10% solution of sodium acetate, and 4 ml of 0.001 *M* bathophenanthroline, shaking briefly after each addition. Shake for 2 minutes with 6 ml of isoamyl alcohol. After 5 minutes discard the aqueous layer and dilute the organic layer to 10 ml with ethanol. Read at 533 nm against 3 : 2 isoamyl alcohol–ethanol and subtract a reagent blank. The result as reported includes both particulate and soluble iron in the sample.

Plant Tissue.[115] Digest 1 gram of a sample such as tobacco with 4 ml of sulfuric acid and 2.5 ml of nitric acid. After all solid matter has disintegrated and oxides of nitrogen have boiled off, dilute to 25 ml. Take an aliquot containing 0.5–10 μg of iron. Add 1 ml of 10% solution of hydroxylamine hydrochloride, 0.4 ml of 0.005 *M*-bathophenanthroline, and 10 ml of 10% sodium acetate solution. If not turbid, add 1-ml aliquots of the 10% sodium acetate solution until turbidity appears on mixing. Extract with 4 ml of hexanol. Mix 2 ml of the extract with 1 ml of ethanol and read at 536 nm.

Mitochondria and Submitochondrial Fractions[116]

TOTAL IRON. Mix 0.1 ml of sample, 0.1 ml of 5% mercaptoacetic acid solution, and 0.2 ml of acetic acid. Shake vigorously, add 0.28 ml of a saturated solution of sodium acetate, and dilute to 1 ml. The pH should be 4–5. Add 1 ml of 0.083% solution of bathophenanthroline in isopentyl alcohol and read the organic layer at

[113] J. A. Tetlow and A. L. Wilson, *ibid.* 442–452.

[114] Frank R. Short, William G. Scribner, and H. Clyde Eyster, *Chem. Eng. News.* **44** (39), 60 (1966); Frank R. Short, H. Clyde Eyster, and William G. Scribner, *Anal. Chem.* **39**, 251–253 (1967).

[115] D. E. Quinsland and D. C. Jones, *Talenta* **16**, 281–283 (1969); cf. E. J. Rubins and G. R. Hagstrom, *J. Agr. Food Chem.* **7**, 722–724 (1959).

[116] K. A. Doeg and D. M. Ziegler, *Arch. Biochem. Biophys.* **97**, 37–40 (1962).

535 nm against isopentyl alcohol. Correct for a blank without the sample and for one without the bathophenanthroline.

NONHEME IRON. Mix 0.1 ml of sample with 0.1 ml of freshly prepared 0.2% sodium metabisulfite solution and 0.7 ml of ethanol in a glass-stoppered tube. Add 0.05 ml of 0.2% ethanolic bathophenanthroline and 0.05 ml of *M* sodium acetate buffer for pH 4.6. Incubate at 38° for 5 minutes, centrifuge, and read at 535 nm against water, using 2 blanks as above.

Serum[117]

IRON. Mix 1 ml of serum with 0.5 ml of 1 : 1 hydrochloric acid and set aside for 10 minutes. Add 1 ml of 20% solution of trichloroacetic acid, set aside for 10 minutes, and centrifuge. To 1 ml of the clear upper layer, add for pH 4.5 1 ml of a buffer solution containing 4.1% of anhydrous sodium acetate and 6% of acetic acid. Add 1 drop of 10% ascorbic acid solution and set aside for 30 minutes. Read at 535 nm against water as E_1. Add 1 drop of 0.5% solution of bathophenanthroline and read at once against a reagent blank as E_2. The standardized value is E_2 minus E_1.

IRON-BINDING CAPACITY.[118] Mix 1 ml of serum and 1 ml of a reagent containing 4.9 mg of ferric chloride hexahydrate per 100 ml of 0.005 *N* hydrochloric acid. Shake for 10 minutes, then add 100 mg of basic magnesium carbonate and shake to precipitate excess iron. Develop as above from "...add for pH 4.5 1 ml of a buffer solution...."

IRON BY EXTRACTION.[119a] Mix 1 ml of sample, 0.5 ml of 0.4 *N* hydrochloric acid, and 1 drop of 80% mercaptoacetic acid solution. Set aside for 15 minutes, add 0.5 ml of 30% solution of trichloroacetic acid, and centrifuge. To 1 ml of the supernatant liquid, add 0.25 ml of 50% solution of potassium acetate and 0.5 ml of 0.025% solution of bathophenanthroline in isopropanol. After 10 minutes for color development, extract with 1.5 ml of 1 : 4 ethanol-chloroform. Read at 533 nm against a blank. Teepol 710 is helpful in separation of ferric ion from transferrin.[119b]

Urine.[120] Reflux 10 ml of sample with 5 ml of nitric acid and 0.75 ml of sulfuric acid at 250° until colorless. Add 5 ml of 30% hydrogen peroxide and when reaction

[117] J. Bouda, *Clin. Chim. Acta* **21**, 159–160 (1968); **23**, 511–512 (1969); cf. R. J. Henry, Ch. Sobel, and N. Chiamori, *ibid.* **3**, 523–530 (1958); R. N. Beale, J. O. Bostrom, and R. F. Taylor, *J. Clin. Pathol.* **14**, 488–495 (1961); D. Webster, *ibid.* **13**, 246–248 (1960); A. Castaldo, A. Vitale, and M. Cangiano, *Diagn. Napoli* **21**, 125–129 (1970); Robert E. Megraw, Ann M. Hritz, Arthur L. Babson, and James J. Carroll, *Clin. Biochem.* **6**, 266–272 (1973).

[118] Arne Koine, *Z. Ges. Inn. Med.* **25**, 125–128 (1970).

[119a] D. T. Forman, *Am. J. Clin. Pathol.* **42**, 103–108 (1964); *Tech. Bull. Reg. Med. Technol.* **34** (6), 93–98 (1964); cf. G. Ceriotti and L. Spandrio, *Biochim. Appl.* **11**, 104–110 (1964); V. T. Positano and L. L. Wiesel, *J. Lab. Clin. Med.* **52**, 912–914 (1958).

[119b] G. Schmidt, *Deut. Gesundhlitw.* **22**, 1686–1688 (1967).

[120] Marvin J. Seven and Ralph E. Peterson, *Anal. Chem.* **30**, 2016–2018 (1958); cf. M. Barry, *J. Clin. Pathol.* **21**, 166–168 (1968).

ceases, follow it with 2 ml of 30% hydrogen peroxide. After heating for 1 hour, add 0.2% solution of potassium permanganate until a pink coloration persists. Add 3 ml of 20% solution of hydroxylamine hydrochloride, 15 ml of a saturated solution of sodium acetate, and 4 ml of 0.0025 *M* bathophenanthroline in isoamyl alcohol. Shake well, and read the organic layer at 533 nm against a reagent blank.

Wine[121]

TOTAL IRON. Mix 3 ml of sample, 2 ml of 10% hydroxylamine hydrochloride, and 5 ml of acetate buffer solution for pH 4. Heat at 100° for 5 minutes and let cool. Add 2 ml of ethanol and 4 ml of water. Add 1 ml of 0.33% ethanolic bathophenanthroline. Extract with 10 ml, 5 ml, and 5 ml of isoamyl alcohol. Dilute the combined extracts to 25 ml with isoamyl alcohol, dry by filtering through sodium sulfate, and read at 535 nm against a blank.

FERROUS IRON. To 3 ml of sample, add 1 ml of the color reagent and 2 ml of ethanol. Complete as for total iron from "Extract with 10 ml, 5 ml, and 5 ml. . . ."

Culture Media.[122] To a sample expected to yield 2–3 μg of iron, add 0.5 ml of sulfuric acid and 2 ml of nitric acid. Heat until charring begins, and let cool. Add 1 ml of nitric acid and 1 ml of perchloric acid. Reheat until clear, adding more nitric acid if further charring occurs. Heat to sulfur trioxide fumes. Cool and dilute to 5 ml. Neutralize to phenolphthalein with ammonium hydroxide and discharge the pink color with 2 *N* hydrochloric acid. Add 1 ml of 10% solution of hydroxylamine hydrochloride, 1 ml of 0.1% solution of bathophenanthroline in 70% ethanol, and 1 ml of 40% solution of sodium acetate. Boil for 20 seconds to decompose ferric pyrophosphate. Add 2 ml of ethanol and extract with 15 ml of hexanol. Dilute the extract to 10 ml with ethanol and read at 533 nm.

4,7-DIPHENYL-1,10-PHENANTHROLINEDISULFONATE

Because of the water insolubility of the ferrous complex of bathophenanthroline, the reagent has quite logically been sulfonated to give a water-soluble complex. Results by this reagent on digestates of low iron diets agree with results of atomic absorption.[123] In the presence of mercaptoacetic acid, the red complex has a color intensity more than double that of conventional reagents.[124] The reagent can be applied to automatic analysis for serum iron.[125] For iron in silicon tetrachloride by 4,7-diphenyl-1,10-phenanthroline, see Aluminum by Eriochrome Cyanine R.

[121]O. Colagrande and A. Del Re, *Ind. Agrar.* **7**, 206–209 (1969).
[122]P. A. Seamer, *Nature* **184**, Suppl. No. 9, 636–637 (1959).
[123]Muriel I. Davies, K. Bush, and I. Motzok, *J. Assoc. Off. Anal. Chem.* **55**, 1206–1210 (1972).
[124]P. Trinder, *J. Clin. Pathol.* **9**, 170–172 (1956).
[125]H. P. T. Werkman, J. M. F. Trijbels, P. J. J. Van Munster, E. D. A. M. Schretlen, and Coby Moerkerk, *Clin. Chim. Acta* **31**, 395–401 (1971).

Procedures

Chromium Trioxide.[126] To an aqueous solution of sample, add 1 ml of 1% solution of zinc chloride. Follow with a 24% solution of sodium hydroxide until the color changes from orange to yellow, and centrifuge to separate the hydrous oxides of chromium, iron, and zinc. Wash twice with hot water adjusted to pH 10 with sodium hydroxide solution. Dissolve the precipitated oxides in 5 ml of 1:4 hydrochloric acid, add 10 ml of ethanol, and heat to reduce chromate to chromic ion. Add sequentially 3 ml of a 45% solution of sodium acetate trihydrate, 10 ml of 20% solution of hydroxylamine hydrochloride, 5 ml more of the sodium acetate solution, and 4 ml of 0.4% solution of the captioned reagent. Dilute to 100 ml, set aside for 15 minutes, and read at 533 nm against a reagent blank.

Plant Material.[127] As a reagent, stir 0.4 gram of bathophenanthroline with 4 ml of fuming sulfuric acid containing 20% of sulfur trioxide. After 30 minutes pour into 400 ml of water. Neutralize to pH 4–5 with ammonium hydroxide and dilute to 1 liter. Just before use, mix 3 volumes of this dilution with 4 volumes of 33% solution of sodium acetate trihydrate and 1 volume of 2.5% solution of hydroxylamine hydrochloride.

Dry 0.4-0.7 mesh material at 105° for 3 hours. Digest 0.5 gram with 1 ml of perchloric acid, 6 ml of nitric acid, and 1 ml of sulfuric acid. Heat for 15 minutes after the solution becomes colorless, and cool. Add 15 ml of water, boil for 10 minutes, filter, and dilute to 100 ml. Treat a 20-ml aliquot containing up to 0.03 mg of iron with 16 ml of the prepared reagent and dilute to 50 ml. Read at 536 nm and correct for a reagent blank. The method can be adapted to the AutoAnalyzer.

Spinach[128]

FREE IRON. Homogenize 100 grams of washed sample with 130 ml of water. Stir 25 ml of homogenate for 45 minutes with 25 ml of *N* hydrochloric acid. To deproteinize a 10-ml aliquot, add 5 ml of 1.23 *M* trichloroacetic acid, heat at 90° for 10 minutes, and centrifuge. To a 4-ml aliquot of the supernatant layer, add 0.2 ml of 0.56 *M* sulfonated 4,7-diphenyl-1,10-phenanthroline in acetate buffer solution for pH 4.6 containing 0.1 *N* sodium metabisulfite and 0.016 *M* (methylamino)phenol. Read at 546 nm.

PROTEIN-BOUND IRON. Homogenize 100 grams of washed sample with 130 ml of Tris buffer solution for pH 8 at 20°. Filter. Precipitate the protein from a 5-ml aliquot by dilution to 100 ml with acetone. Filter the precipitate, resuspend in 25 ml of the Tris buffer solution for pH 8, and centrifuge. Dialyze the supernatant layer at 4° for 12 hours against 0.025 *N* Tris buffer. Determine iron in 4 ml of the dialyzate as above from "...add 0.2 ml of 0.56 *M*...."

[126] D. L. Fuhrman and G. W. Latimer, Jr., *Talenta* **14**, 1199–1203 (1967).
[127] C. Quarmby and H. M. Grimshaw, *Analyst* **92**, 305–310 (1967).
[128] A. Zanobini and A. Firenzuoli, *Boll. Soc. Ital. Biol. Sper.* **50**, 892–894 (1974).

Cereal Flour.[129] As a buffer solution, dissolve 300 grams of sodium acetate and 5 grams of magnesium sulfate in a liter of water; add 50 ml of 4% sodium hydroxide solution and filter. As the color reagent, take 50 ml of the buffer solution; add 0.5 gram of ascorbic acid and 0.5 ml of 1% solution of 4,7-diphenyl-1,10-phenanthrolinedisulfonate.

Dissolve 5 grams of sample by warming with 20 ml of 30% hydrogen peroxide and 5 ml of sulfuric acid. Cool, and dilute to 25 ml. Warm a 0.5-ml aliquot with more hydrogen peroxide until colorless. Add 3 ml of the color reagent, allow 1 minute for development, and read at 550 nm against a blank.

Hemoglobin Iron.[130] Dilute to 2.5 ml 0.1 ml of the red-cell hemolyzate prepared from heparinized whole blood. Digest a 0.1-ml aliquot with 2 ml of a 3% solution of calcium carbonate in nitric acid. As a standard, mix 4 ml of an acid 0.1% solution of iron with 4 ml of 0.4381% solution of monopotassium phosphate and dilute to 100 ml. In parallel, treat 0.1 ml of this standard in the same way as the hemolyzate.

As a color reagent, mix 5.6 grams of sodium acetate, 60 mg of sulfonated 4,7-diphenyl-1,10-phenanthroline, 3 ml of acetic acid, and 1 ml of mercaptoacetic acid. Dilute to 500 ml. To sample and standard, add 2 ml of the color reagent and read at 535 nm against a blank.

Serum[131]

IRON. As a buffer solution for pH 4.5, mix 100 ml of 60% solution of sodium acetate with 40 ml of acetic acid. In each of two tubes take 0.5 ml of serum. Add 0.5 ml of 1% solution of ascorbic acid in the buffer solution for pH 4.5.[132] Mercaptoacetic acid is an alternative reducing agent. To the sample tube, add 0.1 ml of 0.1% solution of 4,7-diphenyl-1,10-phenanthrolinedisulfonate, and to the serum blank add 0.2 ml of water. Prepare a reagent blank. Incubate at 37° for 40 minutes or at 45° for 20 minutes. Cool and read the test sample against the serum blank at 535 nm. Subtract the reagent blank.

For iron content of serum, it can be deproteinized with trichloroacetic acid and ascorbic acid.[133]

IRON-BINDING CAPACITY. Mix 1 ml of serum with 2 ml of a solution containing 0.5 mg of ferric ion per ml. Set aside for 5 minutes. Add 0.5 gram of powdered magnesium carbonate and shake for 30 seconds. Mix occasionally for 30 minutes, then centrifuge. Treat two 0.5-ml portions of the clear layer as described for serum.

Alternatively, to measure iron-binding capacity in serum, add an excess of

[129]K. Lauber and H. Aebi, *Mitt. Geb. Lebensm.-Unters. u Hyg.* **57**, 363–366 (1966).

[130]Eugene S. Baginski, Piero P. Foa, S. M. Suchocka, and Bennie Zak, *Microchem. J.* **14**, 293–297 (1969).

[131]J. F. Goodwin, B. Murphy, and M. Guillemette, *Clin. Chem.* **12**, 47–57 (1966); cf. R. Askavold and O. D. Veblar, *Scand. J. Clin. Lab. Invest.* **20**, 122–128 (1967); C. V. Nelson, *Am. J. Med. Technol.* **30**, 71–80 (1964); D'a Kok and F. Wild, *J. Clin. Pathol.* **13**, 241–245 (1960); H. G. Eisener, *Z. Klin. Chem. Klin. Biochem.* **13**, 21–24 (1975).

[132]J. Hoeflmayr and R. Fried, *Med. Klin.* **61** (46), 1820–1823 (1966); J. F. Goodwin and B. Murphy, *Clin. Chem.* **12**, 58–69 (1966).

[133]Ruth Watkins, Lawrence M. Wiener, and Bennie Zak, *Microchem. J.* **16**, 14–23 (1971).

iron–nitrilotriacetic acid chelate at pH 8.4–8.7.[134] Develop the excess iron chelate with the captioned reagent and read at 535 nm.

For sutomation, the bound iron is released by lowering the pH to 4.[135] This is reduced to the ferrous state with mercaptoacetic acid, and dialyzed into acetate buffer solution at pH 4.65, after which the captioned reagent is added. For unsaturated iron-binding capacity, incubate the serum for 4 minutes at pH 8.5 with a solution containing ascorbic acid and 2 μg of ferric ion per ml. Then add the captioned reagent and more ascorbic acid simultaneously to the diluted serum. After 3 minutes for development, read at 534 nm.

DI-2-PYRIDYL KETOXIME

The complex of ferrous ion with this reagent follows Beer's law up to 4 ppm and is stable for 24 hours. Read at 534 nm, the reagent shows no absorption. There is interference by magnesium, calcium, strontium, barium, and zinc. That of cerium is masked by EDTA, that of thorium by fluoride.

For iron in potassium hydroxide, dissolve sample material in water and reduce with sodium hydrosulfite.[136] Add 1% ethanolic di-2-pyridyl ketoxime and heat at 100° for 20 minutes. Extract with isoamyl alcohol, dry the extract with sodium sulfate, and read at 578 nm against a blank prepared with iron-free potassium hydroxide.

Procedure. To a sample solution containing 6–60 μg of iron, add 1 ml of 10% hydroxylamine hydrochloride solution.[137] Then as a masking agent, add 2 ml of 40% solution of sodium citrate. Add 2 ml of 1% ethanolic di-2-pyridyl ketoxime. Raise to pH 10.5–13.5 with 20% potassium hydroxide solution and heat to 90° for 30 minutes. Cool, dilute to 25 ml, and read at 534 nm against a reagent blank.

DIPYRONE

The complex of ferric ion with dipyrone conforms with Beer's law up to 4 μg/ml.

Procedure. *Serum.*[138] Mix 1 ml of sample with 2 ml of 10% trichloroacetic acid and centrifuge. To 1 ml of the supernatant layer, add 0.3 ml of fresh 20% solution of dipyrone. After 30 seconds add 3 ml of 30% hydrogen peroxide. Set aside for 20 minutes, then read at 440 nm.

---

[134] Junji Morikawa, Kazuei Takase, and Ryuzaburo Osawa, *Jap. Anal.* **22**, 275–280 (1973).
[135] Howard S. Friedman and Chandler S. Cheek, *Clin. Chim. Acta* **31**, 315–327 (1971).
[136] M. De Pooter, W. Holland, and J. Bozic, *ibid.* **II**, 443–448 (1975).
[137] W. J. Holland, J. Bozic, and J. T. Gerard, *Anal. Chim. Acta* **43**, 417–422 (1968).
[138] Fayez K. Guirgis and Yehia A. Habib, *Analyst* **35**, 614–618 (1970).

EDTA

When EDTA forms a 1 : 1 complex with iron, it not only serves as the color-forming reagent when hydrogen peroxide is added, it prevents interference of many other cations. The addition of hydrogen peroxide slightly increases the absorption by the copper complex. Cobalt must be oxidized to the trivalent form to avoid interference. Since the chromic complex slowly decomposes to chromate, the reading must be taken promptly if chromium is present. Though usually read at 520 nm, it is also read in the ultraviolet region.[139]

Procedure. To the slightly acid sample solution containing 0.2–5 mg of ferric ion, add at least twice as much EDTA as is needed to complex the iron.[140] If cobalt is present, add 0.5 gram of sodium nitrite plus 1 ml of acetic acid and heat at 100° for 2 hours. If only chromium must be provided for, heat at 100° for 30 minutes without the added reagents. Dilute to 25 ml and take two 10-ml aliquots. To each, add 10 ml of buffer solution for pH 10 containing 7 grams of ammonium chloride, and 57 ml of 25% ammonium hydroxide per 100 ml. To one portion, add 2 ml of 30% hydrogen peroxide. Dilute each to 25 ml, and read the one to which hydrogen peroxide was added at 515 nm against the other as a blank.

When diethylenetriamine–N,N,N',N'',N''-pentaacetic acid replaces EDTA in this method, the amount of chromic ion tolerated increases but interference by cobalt also increases.[141] Reading is at 520 nm.

Extraction as the Caproic Acid Derivative.[142] To a sample dissolved in an appropriate concentration of nitric acid, add 5 grams of sodium nitrite and dilute to 80 ml. Raise to pH 6.5 with ammonium hydroxide. Extract with 20 ml, 10 ml, and 10 ml of 1.5% solution of caproic acid in chloroform. Evaporate chloroform from the combined extracts and heat to drive off caproic acid, which boils just under 200°. Dissolve the residue in 5 ml of 1 : 1 hydrochloric acid and add 10% EDTA solution to exceed double that required to complex the iron. Take up the general procedure at "Dilute to 25 ml and take two 10-ml aliquots."

Lead.[143] Dissolve 2 grams of sample in 10 ml of 1 : 1 nitric acid and boil off oxides of nitrogen. If the iron content is low, add a known amount of standard ferric chloride solution and dilute to 25 ml. To a 10-ml aliquot add 5 ml of *N* EDTA and raise the pH above 10 with ammonium hydroxide. Add 2 ml of 3% hydrogen peroxide, dilute to 25 ml, and read at 520 nm against a parallel sample to which hydrogen peroxide was not added.

[139] Yasumitsu Uzumasa and Masakichi Nishimura, *Bull. Chem. Soc. Jap.* **28**, 88–89 (1955).

[140] B. C. Poeder, G. den Boef, and C. E. M. Franswa, *Anal. Chim. Acta* **27**, 339–344 (1962); H. J. Cluley and E. J. Newman, *Analyst* **88**, 3–17 (1963); cf. G. C. Krijn, C. J. J. Rouws, and G. den Boef, *Anal. Chim. Acta* **23**, 186–190 (1960).

[141] G. den Boef and M. F. Riemersma, *Anal. Chim. Acta* **31**, 185–186 (1964); cf. F. Bermejo Martinez, A. Blas Perez, and J. A. Rodriguez Campos, *Inf. Quim. Anal. Pura Apl. Ind.* **19**, 86–96, 106 (1965).

[142] W. Pietsch, *Anal. Chim. Acta* **53**, 287–294 (1971).

[143] A. K. Babko and N. V. Loriya, *Zavod. Lab.* **34**, 1305–1606 (1968).

Bismuth. Dissolve a 3.75-gram sample in 10 ml of 1 : 1 nitric acid and boil off oxides of nitrogen. If the iron content, is low add a known amount of standard ferric sulfate solution and dilute to 25 ml. To a 5-ml aliquot, add 5 ml of *M* EDTA and proceed as above.

Cadmium. Dissolve 5 grams of sample in 10 ml of 1 : 1 nitric acid and boil off oxides of nitrogen. If the iron content is low, add a known amount of standard ferric chloride solution and dilute to 25 ml. To a 5-ml aliquot, add 10 ml of *M* EDTA and proceed as above.

Samples Containing Copper, Vanadium, and Titanium.[144] This technic was developed for analysis of the material obtained in the process of removal of vanadium from titanium tetrachloride with copper powder.

Dissolve 1 gram of sample in 5 ml of nitric acid and 20 ml of 1 : 1 sulfuric acid. Evaporate to sulfur trioxide fumes, rinse down the walls of the vessel with 5 ml of water, and again evaporate to sulfur trioxide fumes. Cool, dilute to 100 ml, and filter. Ash the paper. Fuse the residue with 2 grams of sodium carbonate and 1 gram of sodium tetraborate, then dissolve in 30 ml of 1 : 3 sulfuric acid. Add to the main solution and dilute to 250 ml.

To a 10-ml aliquot add 3 ml of 0.2 *M* EDTA and 2 ml of 5% hydrogen peroxide. Add ammonium hydroxide until pH 9 is reached, add 5 ml of an ammoniacal buffer solution for pH 10 and 5 ml of ammonium hydroxide. Dilute to 50 ml and read at 540 nm against a blank.

Aluminum-Bronze.[145] Dissolve 2 grams of sample in 60 ml of 1 : 1 : 3 sulfuric acid–nitric acid–water and expel nitrous fumes. Add 10 ml of hydrochloric acid and evaporate to sulfur trioxide fumes. Cool, dissolve in hot water, and set aside for 1 hour. Filter and wash free from copper with 1 : 20 sulfuric acid. Dilute the filtrate to 300 ml and deposit copper by electrolysis. Dilute the residual electrolyte to 500 ml.

Take two 10-ml aliquots and add to each successively 1 ml of 10% tartaric acid solution, 10 ml of 0.05 *M* EDTA, and 10 ml of ammonium hydroxide. To one add 4 ml of 10% hydrogen peroxide. Dilute each to 100 ml and read the solution to which hydrogen peroxide was added at 520 nm against the other as a blank.

Clay.[146] Fuse 1 gram of sample with 5 grams of sodium carbonate and dissolve the fusion in a minimal amount of water. Add 20 ml of hydrochloric acid and evaporate to dryness. Heat to about 160° to dehydrate silica without decomposing sulfates. Take up in 25 ml of water and simmer for 30 minutes. Filter the digested solution to remove silica and dilute to 100 ml. Mix a 5-ml aliquot with 10 ml of 0.1 *M* EDTA solution. Add 3 ml of ammonium hydroxide and 4 ml of 30% hydrogen peroxide. Dilute to 25 ml and read at 520 nm.

[144] V. N. Tikhonov, *ibid.* **32**, 1053–1055 (1966).
[145] M. Freegarde and Mrs. B. Allen, *Analyst* **85**, 731–735 (1960).
[146] Peter F. Lott and K. L. Cheng, *Anal. Chem.* **29**, 1777–1778 (1957).

Electroplating Solution.[147] To a sample containing 0.5–5 mg of iron, add 10% solution of EDTA in slight excess. Add 5 ml of ammonium hydroxide and 1 ml of 30% hydrogen peroxide. Cool and dilute to 50 ml. Read at 520 nm against a second aliquot from which the step of adding hydrogen peroxide was omitted.

Wine.[148] To 25 ml add 5 ml of 30% hydrogen peroxide. Evaporate to 3 ml by boiling gently. If the solution is yellow, add 10 ml of water and reconcentrate. The solution should be colorless but may be slightly turbid. Add 10 ml of 0.02 *M* disodium EDTA and 3 ml of ammonium hydroxide. Dilute to 50 ml and read at 520 nm.

ETHYL ACETOACETATE

This reagent is sensitive to 0.4 μg of iron per ml. Addition of thiocyanate masks copper.

Procedure. *Cathodic Nickel.*[149] Dissolve 1 gram of sample in 25 ml of 3:2 nitric acid. Evaporate to a small volume and cool. Adjust to pH 5 by addition of 25% sodium acetate solution. Add 30 ml of 0.2 *N* buffer consisting of 1:9 acetic acid–sodium acetate and dilute to 100 ml. Add 2 ml of ethyl acetoacetate. Extract by shaking for 5 minutes with 10 ml, 5 ml, and 5 ml of chloroform. Dilute the combined extracts to 25 ml with chloroform and read at 450 nm.

ETHYL 4,6-DIHYDROXY-5-NITROSONICOTINATE

This reagent forms a blue 4:1 complex with ferrous ion. The color develops in full within 10 minutes and is stable for days. The reagent must be present in a mole ratio approximating 6:1. A large excess of the yellow reagent causes a visual green. The optimum pH range is 2.2–4.1. The tolerance for foreign ions, in ppm, is as follows: cobaltous, 1; cupric, 2; molybdate, 2; nickel, 3; chromic, 16; vanadate, 20.

Procedure. To 15 ml of sample solution containing 2.5 ppm of iron, add excess 1% hydroquinone solution buffered to pH 3.5 with 9:1 *M* acetic acid–*M* sodium acetate to reduce to ferrous ion.[150] If necessary, preadjust the sample solution to pH 2.2–4.1. Add 2 ml of 0.1% solution of the reagent and dilute to 25 ml. Set aside for 15 minutes and read at 653 nm against a reagent blank.

---

[147] K. L. Cheng and B. L. Joydish, *Chemist-Analyst* **51**, 45–48 (1962).
[148] P. Armandola and A. Mastrototaro, *Ind. Agrar.* **5**, 307–309 (1967).
[149] S. E. Kreimer, A. V. Stogova, and A. S. Lomekhov, *Zavod. Lab.* **26**, 1104–1106 (1960).
[150] Curtis W. McDonald and John H. Bedenbaugh, *Anal. Chem.* **39**, 1476–1477 (1967).

ETHYLENEDIAMINEBIS(*o*-HYDROXYPHENYLACETIC ACID)

This reagent is *N,N*-ethylenebis[2-(2-hydroxyphenyl)glycine] and is structurally closely related to EDTA. The 1 : 1 chelate with ferric ion is very stable. There is serious interference by cobalt and chromium; less serious by uranyl, thoric, cupric, nickel, and zinc ions.

The chelate can be extracted by methyltrioctylammonium chloride solution in chloroform.[151] As read at 480 nm, it will determine 1 μg of ferric ion per ml. Addition of borate masks fluoride ion. Chromate and vanadate must be removed before the reagent is added.

As applied to quick-frozen and freeze-dried plant samples, grind to 80-mesh. Homogenize with 0.2 *M* sodium acetate buffer solution for pH 4.6 and tetrachloroethylene. Centrifuge, separate the aqueous layer, and extract the residual material with more buffer solution. Filter the combined aqueous extracts, evaporate at 35° *in vacuo* to a small volume, dilute, and develop the chelate for reading at 480 nm.[152]

Procedure. *Aluminum Alloy.*[153] As solution of the color reagent, suspend 1.8 grams of sample in 50 ml of water. Add 24% sodium hydroxide solution while checking with a pH meter. At pH 8.5–9 solution is almost complete. Dilute to 100 ml and filter. As a buffer solution, dissolve 60 ml of acetic acid in water, add 24% sodium hydroxide solution to pH 5, and dilute to 1 liter.

Cover 0.1 gram of sample with 1 ml of water and decompose by dropwise addition of hydrochloric acid, warming to complete the reaction. Add 0.5 ml of nitric acid and boil gently. Dilute to 12 ml and add 24% sodium hydroxide solution until a slight permanent precipitate appears. Add 2 ml of color reagent and 5 ml of buffer solution for pH 5. Dilute to 25 ml, and filter to remove a slight cloudiness. Read at 470 nm against a blank.

FERRIC ION

In direct reading of ferric ion at 345 nm, the chloride ion can be 2–9 *N*.[154] In analysis of carbonate rocks in a solution 3.5 *N* in chloride ion, interference is avoided by calculation from the difference between readings at 345 and 385 nm. The method is applicable to 40 ppm to 45% of iron. Ferric ion is also read as the sulfate.

When read in 5.45 *N* hydrochloric acid for 0.03 m*M* to 1.6 *M*, the wavelength selected for reading varies with the concentration of ferric ion, the most sensitive

[151] Jira Adam, *Hutn. Listy* **27** (4), 287–288 (1972); R. Pribil and Jira Adam, *Collect. Czech. Chem. Commun.* **37**, 1277–1283 (1972).
[152] T. W. McCreary and R. H. Maier, *Anal. Biochem.* **13**, 165–168 (1965).
[153] A. L. Underwood, *Anal. Chem.* **30**, 44–47 (1958); cf. Gordon V. Johnson and Ralph A. Young, *ibid.* **40**, 354–357 (1968).
[154] P. Avinur, D. H. Yaalon, and I. Barzily, *Isr. J. Chem.* **4** (4), 129–134 (1966).

being 365 nm, the least sensitive 616 nm.[155] With a spectrophotometer, the error is ±4%; with a filter photometer, ±10%.

Extract 5 ml of a solution 0.1–1 m*M* in ferric ion in 7.75 *N* hydrochloric acid with 5 ml of 1:1 benzene–butyl acetate for reading at 362 nm.[156]

For iron and molybdenum in titanium alloys, extract as the chlorides from the sample solution with methyl isobutyl ketone.[157] Evaporate the solvent, ash, and dissolve the residue in 1:2 hydrochloric acid. Read the ferric ion at 340 nm and molybdenum at 270 nm, but correct each by simultaneous equations.

Iron in such materials as phenol, maleic anhydride, and arylsulfonic acid is read in solution in ethanol and hydrochloric acid at 360 nm.[158a] A phosphoric acid–ethanol solution of the sample serves as a blank to correct for organic background.

For iron in sulfuric acid dilute a sample with reagent grade sulfuric acid to contain 20–200 μg of iron in 25 ml.[158b] Read at 280 nm. Only sulfur dioxide interferes.

Procedures

High Purity Cobalt.[159] Dissolve 3 grams of sample in 10 ml of 1:1 nitric acid and 5 ml of hydrofluoric acid. Evaporate to dryness, add 5 ml of hydrochloric acid, and again evaporate to dryness. Repeat the addition of hydrochloric acid and evaporation to dryness twice more. Take up in hydrochloric acid and dilute to 25 ml with that acid. Extract the iron with 5 ml and 5 ml of 2-chloroethyl ether. Reextract the iron from the combined extracts with 10 ml of 0.6 *N* hydrochloric acid. Read the extract at 340 nm.

High Purity Copper. Dissolve 5 grams of sample in 15 ml of hydrochloric acid, adding a few drops of nitric acid. Evaporate to a small volume, then dilute to 50 ml with hydrochloric acid. Extract the iron with 7 ml and 7 ml of 2-chloroethyl ether. Wash the combined extracts free from copper with 10 ml of hydrochloric acid. Reextract the iron from the solvent with 5 ml of 0.6 *N* hydrochloric acid and read at 340 nm.

Chromium.[160] Dissolve 0.2 gram of finely powdered sample in 15 ml of 1:8 sulfuric acid and 10 ml of 10% solution of sodium persulfate. Extract with 7 ml, 7 ml, and 7 ml of amyl alcohol and dilute the combined extracts to 25 ml with amyl alcohol. After 10 minutes read at 500 nm against amyl alcohol.

Nickel Alloys Containing Vanadium.[161] Dissolve 0.2 gram of sample in 15 ml of

[155] R. H. H. Wolf and M. Orhanovic, *Z. Anal. Chem.* **216**, 405–408 (1966).
[156] C. Dragelescu and R. Pomoje, *Rev. Roum. Chim.* **13**, 1585–1600 (1968).
[157] Oleg Engel, *Hutn. Listy* **28**, 816–818 (1973).
[158a] R. G. White and R. E. Seeber, *Appl. Spectrosc.* **18** (5), 158–159 (1964).
[158b] B. W. Budesinsky, *Analyst* **102**, 211–213 (1977).
[159] P. I. Artyukhin, E. N. Gil'bert, V. A. Pronin, and V. M. Moralev, *Zavod. Lab.* **33**, 926–927 (1967).
[160] J. Minczewski and J. Chwastowska, *Chem. Anal.* (Warsaw) **6**, 715–723 (1961).
[161] I. P. Kharlamov and E. I. Dodin, *Zavod. Lab.* **33**, 147–149 (1967).

hydrochloric acid and 5 ml of nitric acid. Add 5 ml of sulfuric acid and evaporate to sulfur trioxide fumes. Add 5 ml of water and again evaporate to sulfur trioxide fumes. Take up in water and add 5 ml of 2% solution of citric acid. This reduces pentavalent vanadium to the tetravalent form, which does not interfere. Let cool, dilute to 100 ml, and filter. Read at 320 nm against water.

Ferrochrome, Brass, or Duralumin.[162] Dissolve a 1-gram sample in 5 ml of hydrochloric acid and 2 ml of nitric acid. Add 10 ml of 1:1 sulfuric acid and evaporate to sulfur trioxide fumes. Add 5 ml of water, evaporate to sulfur trioxide fumes, and repeat that step to complete removal of nitric acid. Take up in 50 ml of water, filter, dilute to 100 ml, and read at 340 nm. This will tolerate 2 mg of molybdenum, 3 mg of titanium, or 0.1 mg of vanadium. Nitrate must be absent.

Phosphating Baths.[163] Mix 10 ml of sample, which may contain ferrous acid phosphate, manganese nitrate, and sodium nitrate, with 10 ml of 1:1 sulfuric acid. Heat to copious sulfur trioxide fumes, add 5 ml of water, and again heat to sulfur trioxide fumes to ensure the absence of nitrate. Dilute to 100 ml and read at 320 nm.

FERRIC 1,10-PHENANTHROLATE

When the red ferrous complex of 1,10-phenanthroline is oxidized with ceric ion, a greenish-blue ferric complex is formed. Only large amounts of phosphate interfere. The low intensity of the color permits the use for determining large concentrations of iron.

Procedure. Decompose a 1-gram sample by solution in acids or by fusion according to the type. Dilute the sample solution to 500 ml and take an appropriate aliquot according to the following table.[164]

Expected Ferric Oxide Concentration (%)	Recommended Sample Size (gram)
0.1–6	0.2–0.4
6–12	0.2
12–24	0.1
24–48	0.05

As ceric sulfate solution, dissolve 40 grams of ceric bisulfate in 50 ml of 1:1 sulfuric acid and 350 ml of water by heating. Cool, add 500 ml of 1:1 sulfuric acid, and dilute to 1 liter.

[162] I. P. Kharlamov, P. Ya. Yakovlev, T. A. Borcheva, and M. I. Lykova, *ibid.* **32**, 512–514 (1966).
[163] E. V. Tokmakova and I. P. Kharlamov, *ibid.* **35**, 666–667 (1969).
[164] H. L. Watts, *Anal. Chem.* **36**, 364–366 (1964).

If the sample solution contains nitric or perchloric acid, add 5 ml of 1 : 1 sulfuric acid and evaporate to sulfur trioxide fumes. Take up in 50 ml of water. Add 10 ml of a solution containing 2.5% of stannous chloride dihydrate and 10% of sodium gluconate in 1 : 3 hydrochloric acid. Mix and dilute to 100 ml. Add 20 ml of 0.1% solution of 1, 10-phenanthroline in 20% ethanol and 5 ml of 50% solution of sodium acetate trihydrate. Adjust to pH 3–4 with 10% sodium hydroxide solution. Add 50 ml of ceric sulfate solution and dilute to 500 ml. Read at 585 nm against water.

FERROCYANIDE

The blue color of ferric ion with potassium ferrocyanide, prussian blue, conforms to Beer's law up to 3 μg of iron per ml.

Procedure. *Thorium Ore.*[165] Treat 1 gram of sample with 10 ml of nitric acid, 20 ml of 1 : 3 sulfuric acid, 2 ml of perchloric acid, and 5 ml of hydrofluoric acid. Heat almost to dryness. Take up with 25 ml of 1 : 10 nitric acid, filter, and evaporate to dryness. Heat the residue with 1 ml of nitric acid, then dissolve in 25 ml of 5% phosphoric acid and dilute to 250 ml. Mix a 0.1-ml aliquot with 1 ml of 1% potassium ferrocyanide solution and dilute to 25 ml with 5% phosphoric acid. Set aside for 15 minutes and read at 610 nm.

FERRON

The reagent having the trivial name of ferron is 7-iodo-8-hydroxyquinoline-5-sulfonic acid.[166] The ferric complex forms an insoluble salt with tributylammonium acetate which can be extracted with amyl alcohol. Nickel and copper require masking with cyanide. Citrate is necessary to mask titanium, antimony, tin, molybdenum, and tungsten. Bismuth must be absent. Chromate must be reduced to chromic ion. Oxalic, hydrofluoric, and pyrophosphoric acids must be absent.

For iron in soil extracts by this reagent, see Aluminum. The determination with this reagent has been automated.[167a]

Procedure. Adjust 25 ml of sample containing up to 0.15 mg of iron to pH 2.5–3.[167b] Add 5 ml of a buffer solution containing 5% of sodium acetate in 50%

[165] F. I. Nagi, Azhar Ali, and Jamshed Anwar, *Rep. Paki. Atom. Energ. Comm.* **AEMC/Chem-134**, 17 pp (1975).

[166] Peter Mandt, *Keram. Z.* **11**, 632–637 (1959); cf. Max Ziegler, Oskar Glemser, and Norbert Petri, *Mikrochim. Acta* **1957**, 215.

[167a] Jack L. Hoyt and Donald E. Jordan, *J. Assoc. Off. Anal. Chem.* **52**, 1121–1126 (1969).

[167b] Max Ziegler, Oskar Glemser, and Norbert Petri, *Z. Anal. Chem.* **154**, 170–182 (1957); *Angew. Chem.* **69**, 174–177 (1957).

acetic acid, 4 ml of 0.1% solution of ferron, 1 ml of tributylammonium acetate, and 5 ml of isoamyl alcohol. Shake, separate the organic layer, and dilute it to 10 ml with isoamyl alcohol. Read at 610 nm against the solvent.

Mineral Water.[168a] Add 1 gram of ferron to 500 ml of sample. If there is iron in suspension, shake occasionally for a week, during which period the precipitate should dissolve. To a suitable aliquot, add excess of 30% hydrogen peroxide and adjust to pH 4–8. Read at 600 nm.

Wine.[168b] Ash 10 ml of sample with 3 ml of nitric acid and 2 ml of 30% hydrogen peroxide. Dissolve the ash in 10 ml of 1:3 hydrochloric acid, filter, and dilute to 25 ml. Evaporate a 5-ml aliquot to dryness and dissolve in 9 ml of 0.05 *N* hydrochloric acid. Add 1 ml of 1% tartaric acid, set aside for 10 minutes, and read at 650 nm.

FERROZINE

This reagent is the disodium salt of 3-(2-pyridyl)-5,6-bis(4-sulfophenyl)-1,2,4-triazine.[169] It forms a complex with ferrous ion. After reduction with ascorbic acid, the reagent can be applied in acetate buffer solution for pH 4.65 for reading at 578 nm by continuous flow.[170] Cobaltous, cuprous, cyanide, nitrite, and oxalate ions interfere. Neocuproine will mask copper for reading at 562 nm.[171]

For copper and iron in serum, deproteinize with hydrochloric and trichloroacetic acids.[172] Centrifuge. For copper, warm at 37° for 10 minutes with a solution of bis(piperidinothiocarbonyl)disulfide in acetic acid and read at 420 nm. Then add thiourea and mercaptoacetic acid. Warm at 37° for 10 minutes to reduce cupric and ferric ions and to mask the copper. Add ferrozine and ammonium chloride, warm at 37°, and read iron at 560 nm.

For iron and copper by ferrozine, also see Copper. Solution is by simultaneous equations.

For analysis of serum by the AutoAnalyzer, the sample is mixed with an acid diluent containing mercaptoacetic acid, hydrochloric acid, and polysorbate 20.[173a] This is diffused into a solution of ferrozine, hydrochloric acid, sodium chloride, and polysorbate 20. Mix the diffusate with acetate buffer solution for pH 4.7 and read at 560 nm. For iron-binding capacity, saturate the transferrin with ferric ammonium citrate and adsorb the excess reagent on magnesium carbonate.

[168a]J. Pastor, *Bull. Soc. Pharm. Marseilles* **4** (13), 63–70 (1955).
[168b]I. G. Mokhnachev and L. G. Serdyuk, *Vinodel. Vinograd. SSSR* **24** (1), 22–26 (1964).
[169]Lawrence L. Stookey, *Anal. Chem.* **42**, 779–781 (1970); Ching-hong Chen and Joyce Lewin, *J. Sci. Food Agr.* **23**, 1355–1357 (1973).
[170]H. Hirsch, *Z. Klin. Chem. Klin. Biochem.* **11**, 465–470 (1973).
[171]Paul Carter, *Anal. Biochem.* **40**, 450–458 (1971).
[172]Hugh Y. Yee and Jesse F. Goodwin, *Clin. Chem.* **20**, 188–191 (1974).
[173a]Hugh Y. Yee and Anastasia Zin, *ibid.* **17**, 950–953 (1971).

Procedures

Beer.[173b] As reagent, dissolve 0.257 gram of ferrozine in 50 ml of a buffer solution for pH 4.3 containing 7.5% of ammonium acetate and 15% of acetic acid. To 40 ml of decarbonated beer, add 2 ml of the reagent solution and 25 mg of ascorbic acid. Mix, dilute to 50 ml, and read at 560 nm against a blank. It obeys Beer's law up to 0.1 mg of iron and avoids heating.

Hemoglobin.[174] Mix 1 ml of 5.4% solution of sodium hypochlorite at pH 11.4 and 1 ml of 5% solution of polyoxyethylene lauryl ether. Add 0.02 ml of oxalated blood and set aside for 5 minutes. Add 3 ml of a solution containing 8.2% of sodium acetate, 6.06% of acetic acid, and 3% of ascorbic acid. Add 1 ml of solution containing 15 mg of ferrozine. Set aside for 20 minutes and read at 562 nm.

Serum[175a]

IRON. As a color reagent, dissolve 250 mg of ferrozine, 10 grams of ascorbic acid, and 30 grams of Triton X-100 in 1 liter of glycine–hydrochloric acid buffer solution for pH 3.1. Mix 300 μl of sample with 800 μl of color reagent and set aside for 45 minutes. Read at 560 nm.

Alternatively, to 1 ml of serum add 1 drop of 20% ascorbic acid solution and set aside for 5 minutes.[175b] Then shake for 30 seconds with 1 ml of 40% solution of trichloroacetic acid in carbon tetrachloride. As a blank use 1 ml of 8% solution of poly(vinylpyrrolidone). Centrifuge, and develop ferric ion in the aqueous layer with ferrozine.

UNSATURATED IRON-BINDING CAPACITY. To a Tris buffer solution for pH 7.95, add Triton X-100 and ascorbic acid. Mix 300 μl of serum with 800 μl of the Tris buffer solution and 150 μl of a solution 200 mM with iron. Set aside for 10 minutes, then add 150 μl of 0.135% solution of ferrozine. Set aside for 45 minutes and read at 560 nm.

FLAME PHOTOMETRY

This technic is appropriate at 386 or 372 nm. For the latter wavelength, iron is extracted from 1 : 1 hydrochloric acid solution with acid-pretreated methyl isobutyl ketone and determined in an oxyhydrogen flame.[176] In alloy analysis, flame photometry is appropriate for 0.3–2% iron.[177]

[173b] H. G. Ulloa, A. M. Canales, and B. D. Zapata, *Proc. Am. Soc. Brew. Chem.* **33**, 167–170 (1975).
[174] A. Manastenski, R. Watkins, E. S. Baginski, and B. Zak, *Z. Klin. Chem. Klin. Biochem.* **11**, 335–338 (1973).
[175a] Risto Ruuto, *Clin. Chim. Acta* **61**, 229–232 (1975); cf. Eugene W. Rice and Henry E. Fenner, *ibid.* **53**, 391–393 (1967); P. Reinouts van Haga, *Mikrochim. Acta* **II**, 543–547 (1976).
[175b] M. C. Goren, *Rev. Asoc. Bioquim. Argent.* **40**, 162–171 (1975).
[176] Hidehiro Goto and Emiko Sudo, *Jap. Anal.* **9**, 213–215 (1960).
[177] C. A. Waggoner, *Can. J. Spectrosc.* **10** (3), 51–57 (1965).

For extraction of ferrous ion from mM solutions containing perchlorate ion buffered to pH 7 with 0.1 M ammonium acetate, nitrobenzene is the most effective solvent, giving 100% recovery.[178a] The extract is then suitable for aspiration in flame photometry at 386 nm.

In determination of iron, manganese, and copper in plant material, the sample is dry-ashed.[178b] The iron is precipitated as the oxinate from a solution of the ash. That in turn is ashed and dissolved in hydrochloric acid, thus reducing the background in the flame. Oil-well scale is dissolved directly in hydrochloric acid for flame photometry.[179]

For column chromatographic separation of copper, manganese, and iron for flame photometry, see Copper. For separation of iron from cobalt mattes and concentrates for flame photometry, see Copper. For iron in refractory products by flame photometry, see Aluminum. For iron by flame photometry after separation from copper and manganese by photometry, see Copper.

Procedures

Thorium Oxide.[180] Dissolve 2 grams of sample by heating with 5 ml of nitric acid, 15 ml of perchloric acid, and 2 drops of hydrofluoric acid. Dilute to 100 ml and take an aliquot containing 20–80 μg of iron. Add 5 ml of hydrochloric acid and dilute to 10 ml. Add 10 ml of 4-methyl-2-pentanone and extract by repeated inversions for 2 minutes to avoid emulsion formation. Adjust to 10 or 25 ml, and read iron by flame photometry at 372 nm and the background at 368 nm. In the presence of high concentrations of diverse ions, wash the organic phase with 10 ml of 5 N hydrochloric acid that has been preequilibrated with 4-methyl-2-pentanone before aspirating to the flame.

Uranyl Sulfate and Copper Sulfate Solutions. Take an aliquot of sample containing 20–80 μg of iron and complete as above from "Add 5 ml of hydrochloric acid...."

Alloys. Dissolve a 2-gram sample with 20 ml of 1 : 1 hydrochloric acid and 0.5 ml of nitric acid. If there is a substantial amount of tin present, volatilize it as stannic bromide. Dilute to 50 ml with N hydrochloric acid. Take an aliquot containing 20–80 μg of iron and complete as above from "Add 5 ml of hydrochloric acid...."

Crude Tin.[181] For 0.05–3.5% iron, dissolve 0.7 gram of sample in hydrochloric acid, adding a few drops of 30% hydrogen peroxide. Add 3.5 ml of butanol to double the sensitivity, and dilute to 50 ml with water. Read with an air-acetylene flame at 386.5 nm. Beer's law is followed for 10–300 μg of iron per ml.

[178a]Alfred A. Schilt, Rose L. Abraham, and John E. Martin, *Anal. Chem.* **45**, 1808–1811 (1973).
[178b]A. D. Berneking and W. G. Schrenk, *J. Agr. Food Chem.* **5**, 742–745 (1957); cf. H. Seiler, *Mitt. Geb. Lebensm.-Unters. Hyg.* **63**, 180–187 (1972).
[179]G. L. Gates and W. H. Caraway, *Rep. Invest. U.S. Bur. Mines* No. **6602**, 10 pp (1965).
[180]Oscar Menis and T. C. Rains, *Anal. Chem.* **32**, 1837–1841 (1960).
[181]I. G. Yudelevich, L. V. Zelentsova, and L. P. Chabovskii, *Uch. Zap. Tsent. Nauchn.-Issled. Inst. Olovyan. Prom.* **1970** (3), 76–79.

Refractory Products.[182] ALUMINUM AND IRON. Adjust a sample solution containing more than 5 mg of aluminum oxide and more than 10 mg of ferric oxide to pH 4–5. Extract with 1:1 acetylacetone-chloroform and dilute to 100 ml with that solvent mixture. Spray into an oxyhydrogen flame, and read aluminum at 484 nm and iron at 386 nm.

2-FLUOROBENZOIC ACID

Ferric ion forms a 1:3 complex with 2-fluorobenzoic acid, following Beer's law for 0–20 ppm of iron at 525 nm. The reagent precipitates with silver, bismuth, mercuric, niobium, lead, silicate, stannic, tantalum, and thoric ions. It forms a colorless complex with aluminum, arsenate, fluoride, and phosphate ions, and forms a yellow complex with titanic, uranyl, and vanadate ions. The color is not affected by ferrous ion. The maximum absorption develops at pH 3–3.5 and is stable for 24 hours.

Procedure. To a solution free from interfering ions, containing 0.1–1 mg of iron, add 20 ml of saturated solution of 2-fluorobenzoic acid.[183] Add hydrochloric acid dropwise until decolorized; then add 5 ml of 7.5% ammonium formate solution to restore the color. Dilute to 50 ml and read at 525 nm against water.

FORMALDOXIME

In aqueous acid solution formaldoxime forms a violet ferric complex.

Procedures

Water.[184] To 25 ml of sample acidified with hydrochloric acid, add 1 ml of *M* formaldoxime and 10 ml of ammonium hydroxide. Dilute to 50 ml and read. The value is the sum of iron and manganese. To another 25 ml of the acidified sample, add 0.1 gram of hydroxylamine hydrochloride and 0.1 gram of potassium cyanide. Complete as above from "add 1 ml of *M* formaldoxime...." The result is manganese, and subtracted from the previous figure, it gives the iron content.

Potatoes. Dry the sample at 120° and ash at 500°. Moisten the ash with nitric acid and add 25 ml of 1:1 nitric acid. Heat for 20 minutes, then dilute to 250 ml. Proceed as for water.

[182] W. Schmidt, K. Konopicky, and J. Kostyra, *Z. Anal. Chem.* **206**, 174–185 (1964).
[183] E. B. Buchanan, Jr., and Walter Wagner, *Anal. Chem.* **29**, 754–756 (1957).
[184] Z. Marczenko and J. Minczewski, *Chem. Anal.* (Warsaw) **5**, 903–915 (1960); cf. Meisetsu Kajiwara, *Jap. Anal.* **13**, 529–532 (1964).

Kaolin. Moisten a 0.5-gram sample in platinum. Add 10 ml of perchloric acid and 15 ml of hydrofluoric acid. Heat to white fumes, add 5 ml of hydrofluoric acid, and again heat to white fumes. Add 5 ml of water and evaporate to white fumes to remove hydrofluoric acid. Dilute to 250 ml and complete as for water above.

Brass. Dissolve 1 gram of sample in 15 ml of 1 : 1 nitric acid. Add 10 ml of water and boil off oxides of nitrogen. Cool; make alkaline with ammonium hydroxide. Filter, and wash with 1 : 50 ammonium hydroxide. Acidify the filtrate with hydrochloric acid and dilute to 100 ml. Complete as for water.

GLYCINECRESOL RED

Ferrous or ferric ions form a 1 : 3 complex with this reagent at pH 2.2–2.9. This obeys Beer's law up to 1 μg of iron per ml. There is serious interference by bismuth, palladium, zirconium, EDTA, fluoride, oxalate, and citrate. The color is stable for 3 hours.

Procedure. To a solution containing 2.8–40 μg of iron, add 0.7 ml of 5% solution of sodium hydrosulfite, 8 ml of hexamine–potassium nitrate–nitric acid buffer solution for pH 5.2 and 3.5 ml of 0.001 *M* reagent.[185] Incubate at 60° for 15 minutes, cool, dilute to 25 ml, and read at 515 nm against a reagent blank.

Aluminum Alloys.[186] Dissolve 0.1 gram of sample in 5 ml of 1 : 1 hydrochloric acid and 1 ml of nitric acid. Dilute to 25 ml, filter, and dilute to 100 ml. Mix a 5-ml aliquot with 9 ml of hydrochloric acid and dilute to 20 ml. Extract the ferric ion with 10 ml of methyl isobutyl ketone. Back-extract the iron with 10 ml of water. Add 2 ml of 0.4 *N* nitric acid and boil off residual organic solvent. Cool and add 4 ml of m*M* glycinecresol red. Adjust to pH 2.3 with 0.5 *N* sodium hydroxide and dilute to 50 ml. Set aside for 10 minutes and read at 520 nm against a reagent blank.

HEMATOXYLIN

The color of ferric ion with hematoxylin follows Beer's law for up to 5 μg of iron per ml. For reading up to 0.3 μg/ml in the presence of aluminum, an empirical correction curve is required.[187] For general application, aluminum and fluoride interfere seriously. There is also interference by cobalt, nickel, thorium, and mercury.

For analysis of the aluminum black found as electrode deposits in production of

[185] Takayoshi Sakai and Ikuko Sato, *J. Chem. Soc. Jap., Pure Chem. Sect.* **87**, 372–375 (1966).
[186] Toshio Azawa, Tadao Okutani, and Satori Utsumi, *Jap. Anal.* **23**, 284–288 (1974).
[187] Luigi Guerreschi and Rachele Romita, *Ric. Sci.* **29**, 2178–2185 (1959).

aluminum, dissolve the sample and add ammonium uranate as a carrier.[188] Buffer the solution with potassium hydroxide–ammonium sulfate. Ferric hydroxide is precipitated and aluminum and chromium are not. Dissolve the precipitate of ferric hydroxide and develop with hematoxylin.

Procedures

Serum[189]

TOTAL IRON. Mix 0.5 ml of sample with 1 ml of water and 1 ml of 10% trichloroacetic acid solution. Digest at 90° for 10 minutes. Mix well, cool, and centrifuge. To 1.5 ml of clear centrifugate, add 0.3 ml of a buffer solution for pH 7.74 containing 3.32 grams of ammonium acetate and 7.6 ml of 4% sodium hydroxide solution per 100 ml. Add 0.6 ml of 0.1% solution of hematoxylin, set aside for 10 minutes, and read at 700 nm.

IRON-BINDING CAPACITY. Mix 0.5 ml of serum with 1 ml of ferric chloride solution containing 1 mg of iron per 100 ml. Set aside for 5 minutes and add 100 mg of magnesium carbonate. Shake occasionally for 30 minutes and centrifuge. Determine the excess iron as for total iron from "To 1.5 ml of clear centrifugate,...."

HEXAMETHYLPHOSPHORAMIDE

This reagent forms a yellow complex with ferric ion at pH 1. At 363 nm when extracted with dichloromethane, this conforms to Beer's law for 0.2–4 μg of iron per ml of extract.[190] The reagent will determine down to 10 ppm of iron in lead, nickel, cobalt, aluminum, and light alloys without interference.

Procedure. Adjust the sample solution to pH 1.1 with sodium acetate solution. Add 10 grams of sodium chloride to an aliquot and dilute to 50 ml with 0.11 *N* hydrochloric acid. Add 10 ml of 10% solution of hexamethylphosphoramide and extract with 10 ml of dichloromethane. Add 5 ml more of reagent and extract with 5 ml and 5 ml of dichloromethane. Filter the combined extracts, dilute to 25 ml, and set aside until 1 hour after starting the extractions. Read at 313 or 363 nm against a blank.

---

[188] W. Guerreschi and R. Romita, *Ric. Sci. R. C., A*. **2**, 603–616 (1962).
[189] D. Micak-Devic, *Clin. Chim. Acta* **24**, 293–298 (1969); cf. S. Chandra, R. K. Sahgal, and D. Prakash, *Lab. Pract*. **24**, 347 (1975).
[190] M. Ziegler, B. Bitterling, and H. Winkler, *Z. Anal. Chem*. **228**, 15–17 (1967).

8-HYDROXYQUINOLINE

This reagent is also known as 8-quinolinol and familiarly as oxine. By extraction with chloroform at pH 2.5–3.3, ferric oxinate is separated from moderate amounts of chromium, manganese, and nickel and read at 470 or 580 nm. Aluminum and ferric oxinates are extracted with chloroform at pH 4.5–10.7 and read at 390 and 470 nm.[191]

Ferric ion and nickelous ion are read as the oxinates at 5–100 μg of each.[192] Extract the oxinates at pH 5.5–8 and read at 370 and 470 nm. Calculate iron from the value at 370 nm. In calculating nickel from the value at 470 nm, correct for the absorption of iron at that wavelength.

Ferric oxinate and the oxinates of cobalt, zinc, scandium, and the rare earth metals are quantitatively coprecipitated with organic carriers such as 1-naphthol and 2-phenylphenol.[193] The coprecipitate of ferric oxinate at pH 4.5 is dissolved in benzene for reading at 470 nm for iron up to 10 ppm.[194] The coprecipitate with 2-phenylphenol is filtered on a sintered-glass crucible, dissolved in methyl isobutyl ketone, and read at 465 nm. Titanic ion, vanadate, and molybdate interfere seriously.

For iron in uranium metal, dissolve a 0.5-gram sample in hydrochloric and perchloric acids.[195a] Add oxine and acetic acid. Adjust to pH 5.3–5.7 and extract with chloroform. Wash the extract with a solution of potassium cyanide and ammonium carbonate to remove oxinates of uranium, copper, and nickel. Then read aluminum at 390 nm and iron at 470 nm.

Extraction with molten naphthalene of complexes with 8-hydroxyquinoline is applicable at varying pH levels to iron, zinc, magnesium, cadmium, bismuth, cobalt, nickel, molybdenum, uranium, aluminum, copper, palladium, and indium.[195b]

For iron in silicon carbide by 8-hydroxyquinoline, see Aluminum. For iron and aluminum in seawater by 8-hydroxyquinoline, see Aluminum. For iron in alkalies as read as the oxinate along with aluminum, see Aluminum. For iron in thorium compounds by oxine, see Aluminum by Oxine. For iron in wine, see Aluminum by 8-Hydroxyquinoline. For iron in leaves by 8-hydroxyquinoline, see Aluminum by that reagent.

Procedures

Vanadium Present.[196a] Take two 10-ml portions of sample which may contain

[191] Hiroshi Hashitani and Kenji Motojima, *Jap. Anal.* **7**, 478–482 (1958); Nobuhiko Iritani and Hitoshi Aoyama, *Shizuoka Yukka Daigaku Kaigaku 5-Shunen Kinen Rombunshu* **1958**, 141–143.
[192] P. B. Issopoulos, *Hem. Hron., A*, **33**, 75–78 (1968).
[193] Masakazu Matsui, Megumu Munakata, and Tsunenobu Shigematsu, *Bull. Inst. Chem. Res. Kyoto Univ.* **45**, 273–281 (1967).
[194] Tsunenobu Shigematsu, Masakazu Matsui, Megumu Munakata, and Takayuki Sumida, *Bull. Chem. Soc. Jap.* **41**, 609–613 (1968).
[195a] Kenji Motojima and Kimie Izawa, *Nippon Genshiryoku Gakkaishi* **2**, 253–259 (1960).
[195b] T. Fujinaga, Masatada Satake, and Masaaki Shimizu, *Jap. Anal.* **25**, 313–318 (1976).
[196a] Michiko Kodama and Takshiko Tominaga, *Chem. Lett.* **1976**, 789–791.

up to 10 ppm of ferric ion and pentavalent vanadium. To each add 5 ml of 1% solution of oxine as precipitating agent. Dissolve the precipitates with 4 ml of 0.625 *M* sodium dodecyl sulfate. Adjust one sample to pH 8, dilute to 50 ml, and read iron at 580 nm.

Adjust the other sample to pH 4.6 and dilute to 50 ml. Read vanadium at 560 nm correcting for the absorption of the ferric ion.

Tellurium.[196b] Dissolve 1 gram of sample in 7 ml of hydrochloric acid and 3 ml of nitric acid. Neutralize with ammonium hydroxide and add 5 ml of ammonium chloride buffer solution for pH 10. Add 2 ml of 0.01 *M* EDTA, 5 ml of 1% potassium cyanide solution, and 0.5 ml of 2.5% solution of oxine in 5% acetic acid. Shake with 10 ml of chloroform for 10 minutes and read the organic layer at 470 nm.

Blood.[197] Digest 1 ml of oxalated whole blood with 5 ml of sulfuric acid and 1.5 ml of perchloric acid until colorless. Cool and dilute to 100 ml. Adjust a 20-ml aliquot to pH 2.5 with ammonium hydroxide. Extract with four successive 5-ml portions of freshly prepared 0.01 *M* oxine in chloroform. Dilute the combined extracts to 50 ml with chloroform and read at 468 nm.

ISONICOTINIC ACID HYDRAZIDE AND 2,3,5-TRIPHENYLTETRAZOLIUM CHLORIDE

Equimolar amounts of these reagents give a yellow color with ferric ion. The method will determine 1 μg of iron per ml. Iodate, bromate, thallic ion, and chromic ion interfere.

Procedures

Ferric Solutions.[198] Take a sample containing 16–160 μg of ferric ion. Adjust to pH 2.03 with 1 : 10 nitric acid and dilute to 5 ml. Add 2 ml of a solution containing 1% of each of the captioned reagents and heat for 3 minutes at 60°. Cool, shake with 5 ml of *n*-amyl alcohol, and centrifuge. Dilute the organic layer to 100 ml with ethanol and read with a green filter. Up to tenfold ferrous ion does not interfere.

Ferrous Solutions. To a sample solution containing 16–160 μg of iron, part of which may be ferric ion, add an excess of bromine water. Evaporate to substantially less than 5 ml and cool. Complete as for ferric solutions from "Adjust to pH 2.03 with...." Subtract the value for ferric ion as determined in the presence of ferrous ion.

[196b] Teru Yuasa, *Jap. Anal.* **10**, 1292–1293 (1961).
[197] William K. Easley, *J. Tenn. Acad. Sci.* **33**, 192–194 (1958).
[198] M. H. Hashmi, Abdur Rashid, Hamid Ahmad, A. A. Ayaz, and Faroog Azam, *Anal. Chem.* **27**, 1027–1029 (1965).

MERCAPTOACETIC ACID

This reagent is also known as thioglycollic acid. Iron at 0.001–0.1% in zinc is determined by reaction with mercaptoacetic acid after separation of the zinc from the solution.[199] It is applied to the product of wet-ashing of drugs.[200]

For analysis of uranium, dissolve in hydrochloric acid and hydrogen peroxide; then boil off the excess peroxide with added sodium sulfite.[201] Various other forms are converted to solutions. Add sulfamic acid, ammonium citrate, and mercaptoacetic acid. Then add ammonium hydroxide until the color is deep orange and read at 540 nm.

For iron in uranium mono- and dicarbides containing more than 50 ppm of iron, ignite 5 grams of sample to the oxide.[202] Dissolve in nitric acid and filter. Ash the residue, fuse with potassium bisulfate, extract with nitric acid, and add to the previous filtrate. Destroy nitrites with sulfamic acid. Add mercaptoacetic acid and ammonium citrate. Make alkaline with ammonium hydroxide and add ammonium carbonate to destroy uranium mercaptoacetate. Read at 605 nm.

For acid-soluble iron in chemical stoneware, extract with boiling 1 : 1 hydrochloric acid and develop with mercaptoacetic acid.[203]

Iron in tin and in tin melts for tinning sheet steel is read as the mercaptoacetate.[204] As applied to silicates of the ceramic industry after decomposition with sulfuric and hydrofluoric acids or by fusion with sodium carbonate, any aluminum is held in solution by addition of tartaric acid. In ammoniacal solution at pH 9.2–9.5, the reagent reduces any ferric ion.[205]

Procedures

Zirconium Alloys.[206] Dissolve 1 gram of sample in 10 ml of nitric acid and 1 ml of hydrofluoric acid. Add 2 ml of sulfuric acid and evaporate to sulfur trioxide fumes. Take up in water and dilute to 100 ml. To a 25-ml aliquot, add 2 ml of 15% mercaptoacetic acid solution and 5 ml of 1% solution of citric acid. Make alkaline with ammonium hydroxide and read at 604 nm.

Uranium Tetrafluoride.[207] Dissolve 5 grams of sample in 1 : 4 nitric acid containing 3% boric acid. Evaporate to dryness and dissolve in water. To a portion containing not more than 1 gram of uranium, add sulfamic acid to destroy nitrous acid. Add 5 ml of 1% solution of ammonium citrate and 2 ml of 15% mercaptoacetic acid. Add ammonium hydroxide to produce a deep orange color. Destroy uranium mercaptoacetate with ammonium carbonate. Add more ammonium hydroxide to a maximum color intensity and read at 605 nm.

[199] W. Kemula, A. Hulanicki, and S. Rubel, *Przem. Chem.* **11**, 99–102 (1955).
[200] J. Hollos and I. Horvath-Gabai, *Pharmazie* **20**, 207–210 (1965).
[201] *UKAEA, Rep.* **PG276 (S)**, 9 pp (1973).
[202] *UKAEA Rep.* **PG516 (S)**, 34 pp (1964).
[203] British Standards Institution, *BS 784* 9 pp (1973).
[204] Heinrich Ploum, *Arch. Eisenhuttenw.* **29**, 169–172 (1958).
[205] Aurelio Gisondi, *Ceramica* (Milan) **15** (6), 47–51 (1960).
[206] C. F. Bush, *UKAEA Rep.* **AERE-AM 91**, 5 pp (1962).
[207] *UKAEA Rep.* **PG 614 (S)**, 7 pp (1965).

Magnox MN70.[208] Dissolve 5 grams of sample in 10 ml of 1:1 hydrochloric acid and dilute to 100 ml. To a 10-ml aliquot, add 1 ml of 15% mercaptoacetic acid solution, 10 ml of 15% solution of ammonium chloride, and ammonium hydroxide until the solution turns red. Add 1 ml more of ammonium hydroxide and dilute to 50 ml. Read at 605 nm.

Benzenesulfonic Acid.[209] Neutralize 50 ml of sample solution with ammonium hydroxide. Add 2 ml of 10% solution of ammonium mercaptoacetate and 10 ml of 1:3 ammonium hydroxide. Dilute to 100 ml and read at 530 nm.

7-NITROSO-8-QUINOLINOL-5-SODIUM SULFONATE

The 1:3 complex of ferrous ion with this reagent forms a dark green color immediately. Readings are linear at 0.17–2.38 μg of ferrous ion per ml. Ferric ion also forms a chelate absorbing near that of ferrous ion but this can be masked with citrate.

Procedure. *Tablets Containing Ferrous Sulfate.*[210a] Dissolve a powdered sample equivalent to about 0.3 gram of ferrous sulfate and dilute to 100 ml. Mix 4 ml of citric acid-phosphate buffer solution for pH 5 and 2 ml of 0.0025 *M* reagent. Add 2 ml of sample solution and dilute to 10 ml. Read at 710 nm against a blank.

NALIDIXIC ACID

This reagent is 1-ethyl-1,4-dihydro-7-methyl-4-oxo-1,8-naphthyridine-3-carboxylic acid. Its yellow 3:1 complex with ferric ion will determine 1 part of iron in the presence of 2.5 parts of aluminum and 25 parts of chromium. Beer's law is followed for 0.43–17 μg of iron per ml. The color is formed immediately.

Procedure. As a reagent, dissolve 0.1 gram of nalidixic acid in the stochiometric amount of sodium hydroxide in solution and dilute to 100 ml.[210b] To a sample containing 4–170 μg of iron, add 5 ml of the reagent, dilute to 10 ml, and read at 410 nm.

Ferromanganese. Dissolve a 0.5-gram sample in nitric acid and evaporate to dryness. Add 2 ml of hydrochloric acid, evaporate to dryness, and repeat that step. Take up with hot water and filter off the silica. Adjust the filtrate to pH 3–4 and

[208] *UKAEA Rep*. **PG 437 (S)**, 4 pp (1963).
[209] T. D. Prishletsova, N. V. Prokhorova, and V. N. Pavlov, *Zavod. Lab*. **35**, 558–559 (1969).
[210a] M. A. Eldawy and S. R. Elshabouri, *Anal. Chem*. **47**, 1944–1946 (1975).
[210b] I. Dick and N. Murgu, *Rev. Chim*. (Bucharest) **15**, 757–758 (1964); **16**, 516–517 (1965).

dilute to 100 ml. To a 0.5-ml aliquot add 5 ml of 0.1% solution of the sodium salt of the reagent. Dilute to 10 ml and read at 410 nm against water.

NITROSO-R SALT

This reagent is 3-hydroxy-4-nitroso-2,7-naphthalene disulfonic acid. Ferrous ion forms a 1:3 complex with nitroso-R salt in the presence of oxalate at pH 8–11.[211] The maximum absorption is at 710 nm. Optimum conditions are pH 10, 100-fold excess of reagent, and heating for 15–20 minutes. Beer's law is followed for 2.4–29.3 μg of ferrous ion per 50 ml. In analysis of oxalate solutions such as etching baths the ferric ion is masked with tartrate.

With sufficient excess of reagent present ferric ion is reduced to ferrous. The optical density is at a maximum at pH 4.5–6. With 100-fold excess of reagent nickel does not affect the reading for iron, even at 80:1. Chromium interferes at 6 times the concentration of iron. Molybdate interferes at more than 12 times the iron concentration. There is some interference by tungstate and vanadate ions.

The complex of ferrous ion with nitroso-R salt is more sensitive than those with sulfosalicylic acid, 1,10-phenanthroline, 8-hydroxy-7-iodoquinoline-5-sulfonic acid, and mercaptoacetic acid, provided the pH is adjusted precisely before and after adding the reagent.[212] Many metallic ions react with nitroso-R salt, but added EDTA destroys the complexes with all metals other than iron.[213]

After clarification of urine, tomato juice, and syrups with barium chloride and sodium sulfate in the presence of hydroxylamine, the iron is complexed in sodium acetate buffer solution for pH 8.5–10 with nitroso-R salt. Copper, cobalt, and nickel do not interfere.[214]

The reagent is appropriate for reading ferrous ion in ammonium acetate–acetic acid extracts of soil at pH 5.3 and 710 nm.[215] The reagent does not absorb at that wavelength.

Procedures

Iron-Nickel Alloys.[216] Heat a 1-gram sample for 90 minutes with 5 ml of 1:1 hydrochloric acid and 1 ml of 1:1 nitric acid. Add 5 ml of 1:1 sulfuric acid and heat to copious sulfur trioxide fumes. Dissolve the residue in 25 ml of water, filter, wash well with hot water, and dilute to 250 ml. To a 5-ml aliquot add 5 ml of water, 10 ml of acetate buffer solution for pH 6, and 10 ml of 0.006 *M* nitroso-R salt. Heat at 50° for 5 minutes, cool, and read at 720 nm.

[211] V. F. Barkovskii and V. G. Solonenko, *Zh. Anal. Khim.* **25**, 128–131 (1970).
[212] R. Quast, *Acta Chem. Scand.* **21**, 873–878 (1967).
[213] British Standards Institution, *BS 3727*, Pt. **7**, 7 pp (1964).
[214] O. Mitoseru and I. Herinean, *Ind. Aliment., Bucur.* **24**, 143–145 (1973).
[215] O. Makitie, *Ann. Agr. Fenn.* **7**, 117–122 (1968).
[216] V. N. Tolmachev and E. V. Bashkinskii, *Izv. Vyssh. Ucheb. Zaved., Khim. Khim. Tekhnol.* **3**, 815–818 (1960).

Nuclear-Grade Graphite.[217] Ash 5 grams of sample containing 2–75 ppm of iron in platinum at 800°. Dissolve the residue in 10 ml of 1:1 hydrochloric acid and dilute to 100 ml. Take a 10-ml aliquot and add 2 ml of 10% hydroxylamine hydrochloride solution. Add sufficient 15% ammonium acetate solution to buffer to pH 6–7. Add 5 ml of 1% solution of nitroso-R salt and dilute to 25 ml. Incubate at 50° for 5 minutes, cool, and read at 600 or 715 nm.

For analysis of urania-graphite fuel, ash, dissolve, and coprecipitate the iron with manganese dioxide formed by manganous ion and hydrogen peroxide in ammoniacal carbonate solution. Interfering copper and chromium as well as the uranium are thus left in solution. Dissolve the manganese dioxide in acidified hydroxylamine for development.

Paper.[218] Cut the sample into small pieces, moisten with 25 ml of 10% sodium carbonate solution, and evaporate to dryness in platinum. Ash. Dissolve the ash with 6 ml of 1:1 nitric acid and complete as for Iron-Nickel Alloys from "Add 5 ml of 1:1 sulfuric. . . ."

Serum Iron.[219] As a buffer solution, make up *M* sodium acetate in 0.15 *M* acetic acid. As reagent, mix 10 ml of 1% solution of nitroso-R salt with 100 ml of the buffer solution and dilute to 200 ml.

To 0.5 ml of serum as a blank, add 1.25 ml of the buffer solution and 1.25 ml of water. As the test add 2.5 ml of the reagent to 0.5 ml of serum. Incubate at 50° for 30 minutes, cool, and read at 720 nm against the blank.

---

o-NITROSORESORCINOL

This reagent is used to determine ferric acetylacetonate used as a catalyst in polymerization of epoxy-crosslinked butadiene, acrylic acid, and acrylonitrile terpolymer. For determination, the ferric compound is reduced to the ferrous form. A 1:1 chloroform-methanol solvent is required. The desirable pH is 5–6.2. The color develops in 20 minutes and is constant for 24 hours.

Procedure. *Ferric Acetylacetonate in Uncured Terpolymer.*[220] As a buffer solution, dissolve 38.6 grams of ammonium acetate in 200 ml of methanol and 688 ml of acetic acid; dilute to 1 liter with methanol.

Dissolve a sample expected to contain 4–12 mg of ferric acetylacetonate in chloroform and dilute to 200 ml with chloroform. To a 25-ml aliquot add with sequential mixing 25 ml of methanol, 5 ml of 4% methanolic hydroxylamine hydrochloride, 5 ml of 0.113% methanolic *o*-nitrosoresorcinol, 5 ml of buffer

[217]Masatoshi Miyamoto, *Bull. Chem. Soc. Jap.* **34**, 1435–1440 (1961).
[218]Tsutomu Matsuo, Shigeru Kematsu, and Koichi Tozawa, *Jap. Anal.* **11**, 1194–1196 (1962).
[219]A. T. Ness and H. C. Dickerson, *Clin. Chim. Acta* **12**, 579–588 (1965); cf. Masao Noda, *Seikagaku* **29**, 199–204 (1957); J. Pre, P. Giraudet, and P. Cornillot, *Clin. Chim. Acta* **22**, 429–432 (1968); Robert G. Martinek, *ibid.* **43**, 73–80 (1973).
[220]Robert B. Lew, *Anal. Chem.* **40**, 438–440 (1968).

solution, and 25 ml of chloroform. Dilute to 200 ml with methanol. Set aside for 30 minutes and read at 695 nm against a reagent blank. Read standards of 0.5, 1, 1.5, and 2 mg of ferric acetylacetonate at the same time.

---

o-NITROSORESORCINOLMONOMETHYL ETHER

This reagent complexes with ferrous ion for extraction and reading.

Procedure. *Serum.*[221] Acidify a 2-ml sample with 0.5 ml of 1 : 1 hydrochloric acid and set aside for 10 minutes. Add 1 ml of water and 0.5 ml of 20% solution of trichloroacetic acid. Centrifuge and take a 3-ml aliquot of the clear upper layer. Add 0.5 ml of 10% hydroxylamine hydrochloride and set aside for 15 minutes. Neutralize to *p*-nitrophenol with 1 : 1 ammonium hydroxide. Add 2 ml of acetic acid–sodium acetate buffer solution for pH 4.5 and 0.5 ml of 5% sodium thiosulfate solution to mask copper. Set aside for 5 minutes, then add 1 ml of saturated solution of *o*-nitrosoresorcinolmonomethyl ether. Dilute to 10 ml and after 15 minutes extract with 2 ml, 2 ml, and 2 ml of carbon tetrachloride. Centrifuge to clarify the aqueous layer, and read at 700 nm.

PENTAMETHYLENEDITHIOCARBAMATE

This reagent is appropriate for determination of ferric ion in the absence of interference by cobalt, cupric, or manganic ion. Beer's law is followed for 0.01–0.3 mg of iron per ml. Reducing substances and those containing Si–H and Si–Si linkages interfere.

Procedure. *Resin.*[222] Dissolve 2 grams of sample in 10 ml of benzene. Add 1 ml of pyridine and 2 ml of 1.5% solution of the reagent in benzene. Set aside for 1 hour and read at 445 nm against a reagent blank.

1,10-PHENANTHROLINE

Though not one of the oldest reagents for photometric determination of iron, this ferroin chromogen has become one of the most widely used. That has led to various derivatives, of which bathophenanthroline, 4,7-diphenyl-1,10-phenanthroline, is an important example. The ferrous complex follows Beer's law for 0.2–4 μg/ml.

[221] Shunji Umemoto and Yasuhisa Yamomoto, *Bull. Chem. Soc. Jap.* **31**, 1–3 (1958).
[222] P. Dostal, J. Cermak, and J. Kartons, *Collect. Czech. Chem. Commun.* **33**, 1539–1548 (1968).

If the pH adjustment is made before adding the color reagent, results may be low.[223] This has been attributed to partial oxidation of ferrous ion to the ferric state. For purification of reagents to lower the blank, extraction of the ferrous–1,10-phenanthroline–perchlorate complex is appropriate.[224] Iron can be extracted with chloroform as the 1,10-phenanthroline–ferrous complex in the presence of copper, manganese, and aluminum and read at 490 nm.[225]

In determination of iron with 1,10-phenanthroline, citric acid will mask aluminum, molybdenum, antimony, tin, thorium, titanium, uranium, tungsten, and zirconium.[226] EDTA will mask bismuth, cadmium, chromium, and zinc. Mercaptoacetic acid masks copper and tartaric acid masks tantalum. In determination of ferrous ion by 1,10-phenanthroline, more than 10 ppm of platinum interferes by forming a sparingly soluble complex.[227]

As reducing agents, the usual hydroxylamine salts, ascorbic acid, sulfite, and hydroquinone are applicable. Ammonium thioglycollate[228] and mercaptoacetic acid are also applicable. Also in 0.1–1 *N* sulfuric acid, ferric ion is reduced by hypophosphite in the cold or by heating with phosphite using palladous chloride as the catalyst.[229] The reducing agent must be present in excess of 100-fold.

A convenient method of concentration of ferric 1,10-phenanthrolate is by adsorption from an aqueous buffered solution on Amberlite XAD-2. This is then eluted with methanol containing hydroxylamine hydrochloride.

The stability of the EDTA chelate of ferrous ion is such that it will pass through a cation exchange column without being absorbed.[230] Thereafter by heating, 1,10-phenanthroline displaces the EDTA.

Sulfonation of 1,10-phenanthroline gives the 5-sulfonic acid and 3-sulfonic acid.[231] The spectrophotometric characteristics of their red ferrous compounds do not differ significantly from those of the parent. They do differ in that the complexes with perchlorates are soluble.

The 1:2:2 complex of ferrous ion, 1,10-phenanthroline, and iodide ion is extractable with such solvents as chloroform and nitrobenzene.[232] EDTA masks the interfering elements.

A ternary complex of ferric ion, 1,10-phenanthroline, and a sulfonaphthalein dye, of which bromophenol blue is a typical example, is extractable at pH 8.7–8.9 with 4:1 chloroform–isoamyl alcohol stabilized with ethanol.[233] Then, reading at 610 nm will determine 0.06 ppm of iron in a salt.

[223] Nellie F. Davis, Clyde E. Osborne, Jr., and Harold A. Nash, *Anal. Chem.* **30**, 2035 (1958); Keisuke Tachibana, *Jap. Anal.* **10**, 61–62 (1961).

[224] M. Knizek and J. Provaznik, *Chemist-Analyst* **54**, 6 (1965).

[225] J. Paul and Sharad M. Paul, *Microchem. J.* **19**, 204–209 (1972).

[226] S. S. Yamamura and J. H. Sikes, *Anal. Chem.* **38**, 793–795 (1966); cf. Saichiro Onuki, Kunihiko Watanuki, and Yukichi Yoshino, *Jap. Anal.* **13**, 23–27 (1964); G. S. R. Krishna Murti, A. V. Moharir, and V. A. H. Sarma, *Microchem. J.* **15**, 585–589 (1970).

[227] S. Das Gupta, S. Kumar, and B. C. Sinha, *Trans. Indian Ceram. Soc.* **24**, 66–70 (1965).

[228] Frantisek Vydra and Rudolf Pribit, *Z. Anal. Chem.* **180**, 295–300 (1962).

[229] G. Somidevamma and G. Gopalo Rao, *ibid.* **187**, 183–187 (1962).

[230] R. Fernandez Cellini and F. Ruiz Sanchez, *USAEC* **JEN-11**, 17 pp (1959).

[231] David E. Blair and Harvey Diehl, *Anal. Chem.* **33**, 867–870 (1961).

[232] Frantisek Vydra and Rudolf Pribil, *Talenta* **3**, 72–80 (1959).

[233] S. O. Kobyákova, V. M. Savostina, and N. L. Dobychina, *Zh. Anal. Khim.* **25**, 1348–1352 (1970).

A ternary complex of ferrous ion–thiosalicylic acid–1,10-phenanthroline can be extracted with chloroform at pH 5–6 for reading at 510 nm.[234] Similar complexes are formed with cobalt and nickel. Likewise, a ternary complex of ferrous ion–1,10-phenanthroline and methyl orange is read at 420 nm.[235] A ternary complex is also formed with nitroprusside.

For analysis of lead and lead alloys, the lead is precipitated as the sulfide.[236] The ferric ion is then reduced with hydroxylamine hydrochloride and the ferrous ion developed with 1,10-phenanthroline.

For iron in aluminum fluoride at 200 ppm or more as ferric oxide, fuse the sample at 550° with 3:1 sodium carbonate–boric acid, then raise to 750°.[237] Dissolve in water acidified with hydrochloric or nitric acid, reduce with hydroxylamine hydrochloride, adjust to pH 3.5–4.2, add 1,10-phenanthroline solution, and read at 510 nm.

In analysis of nickel ferrite and hematite, the ferrous ion is extracted with 1:1 nitrobenzene-chloroform as a ferrous–1,10-phenanthroline–perchlorate complex from an acetate-citrate buffer medium at pH 4–5.[238] Red light or amber glassware avoids photoreduction of the complex.

For iron in niobium or niobium compounds, fuse with potassium hydroxide or potassium bisulfate, dissolve with 0.5 *N* sulfuric acid, and develop with 1,10-phenanthroline.[239]

Dissolve uranium-molybdenum alloys in hydrochloric and nitric acids.[240] Fuse niobium-molybdenum alloys with potassium bisulfate and dissolve in hydrochloric acid. Tartrate suppresses interference by uranium or niobium. After reducing to the ferrous state to develop iron with 1,10-phenanthroline, heat to 70° for uranium alloys and 80° for niobium alloys.

For iron in metallic copper the ferrous–1,10-phenanthrolate is formed.[241] Then addition of cyanide ion converts it to dicyanobis(1,10-phenanthroline)iron, which can be extracted with an organic solvent. The same reaction holds for ferrous-bathophenanthroline.

For iron in uranium mono- and dicarbides, ignite 5 grams of sample to oxide.[242] Dissolve in nitric acid and filter. Ash the residue, fuse with potassium bisulfate, extract with nitric acid, and add to the previous filtrate. Evaporate an aliquot containing about 10 μg of iron and not more than 1 gram of uranium to dryness and dissolve in hydrochloric acid. Add ammonium tartrate solution and adjust to pH 6–7. Reduce the iron with hydroxylamine hydrochloride and add 1,10-phenanthroline and potassium iodide. Extract the 1:2:2 complex with chloroform and read at 510 nm.

[234] I. V. Pyatnitskii and G. N. Trochinskaya, *ibid.* **28**, 704–708 (1973).
[235] Adam Hulanicki, Malgorzata Galus, Wojciech Jedral, Regina Karwowska, and Marek Trojanowski, *Chem. Anal.* (Warsaw) **16**, 1011–1019 (1971).
[236] British Standards Institution, *BS 3908* Pt. **15**, 7 pp (1972); cf. J. A. Corbett, *Metallurgia*, **65**, 43–47 (1962).
[237] British Standards Institution, *BS 4993*, Pt. **2**, 8 pp (1974).
[238] J. Novak, A. Funke, and P. Kleinert, *Z. Anal. Chem.* **237**, 339–347 (1968).
[239] H. Klimczyk, Z. Kubas, K. Pytel, and L. Suski, *Hutn. Katowice* **35**, 462–465 (1968).
[240] R. Dickinson and J. R. Sanderson, *UKAEA, Ind. Group* **IGO R/S-17**, 20 pp (1956).
[241] Harvey Diehl and E. B. Buchanan, Jr., *Talenta* **1**, 76–79 (1958).
[242] *UKAEA Rep.* **PG 516 (S)**, 34 pp (1964).

For low levels of iron in uranium, dissolve in nitric acid and evaporate to dryness.[243] Dissolve in hydrochloric acid, add ammonium tartrate, and adjust to pH 6–7. Add hydroxylamine hydrochloride, 1,10-phenanthroline, and potassium iodide. Extract the complex with chloroform and read at 510 nm.

For iron in the presence of uranium nitrate, add 10 ml of buffer solution for pH 4.5–5 containing 1.3% of acetic acid and 2.7% of sodium acetate.[244] Reduce the iron with 0.1 gram of hydroquinone and add 10 ml of 0.2% 1,10-phenanthroline. Dilute to 25 ml and read at 510 nm. Up to 12:1 uranium does not interfere.

For iron in yttrium or yttrium oxide, prepare a solution in 1:1 hydrochloric acid.[245] Extract ferric ion with a solution of tri-*n*-octylphosphine oxide in cyclohexane. Back-extract the iron with 6 *N* sulfuric acid. Mix with an equal volume of hydrochloric acid and again extract with the reagent in cyclohexane. Reduce and develop a portion of the extract with 1,10-phenanthroline.

For analysis of bismuth, dissolve a 5-gram sample and extract the bismuth as an iodo complex with cyclohexanone.[246] Then determine iron with 1,10-phenanthroline.

In analysis of chromium, the solution in hydrochloric acid is aerated to oxidize the chromic ion.[247] Then solution is completed by adding potassium chlorate. An anion exchange resin separates chromium from iron, copper, zinc, and cobalt. Elution with 4 *N* hydrochloric acid elutes cobalt; then 0.5 *N* hydrochloric acid elutes iron. The latter is then reduced and read with 1,10-phenanthroline.

For analysis of a multicomponent electrolytic bath containing nickel, cobalt, and iron, first oxidize with hydrogen peroxide.[248] Precipitate nickel as hexaaminenickel perchlorate and filter. Destroy unconsumed perchlorate and proceed to develop iron with 1,10-phenanthroline.

For iron in solutions of purified cadmium or zinc, coprecipitate it with manganese dioxide in ammoniacal solution.[249] Dissolve in 0.1 *N* sulfuric acid containing hydroxylamine and adjust to pH 6–7, adding 1,10-phenanthroline. EDTA will mask manganese, zinc, and cadmium to permit determination of 0.03 ppm of iron.

For determination of iron in aluminum alloys with 1,10-phenanthroline or sulfosalicylic acid, the standard should contain a corresponding amount of silicon unless the standard addition technic is used.[250]

In determination of iron in soil or plant materials, masking of aluminum with fluoride is ineffective.[251] Precipitate hydroxides with ammonium hydroxide, centrifuge to separate, and extract the precipitate with sodium hydroxide solution. Then dissolve the residual ferric hydroxide in hydrochloric acid. Dilute an aliquot to 15

[243] *UKAEA Rep.* **PG 345 (S)**, 6 pp (1962).

[244] R. Pipan, I. Eger, and N. Bojan, *Stud. Cercet. Chim.* **13**, 873–877 (1964); *Rev. Roum. Chim.* **9**, 829–833 (1964).

[245] J. O. Hibbits, W. F. Davis, and M. R. Menke, *Talenta* **6**, 28–29 (1960).

[246] E. Jackwerth, *Z. Anal. Chem.* **211**, 254–265 (1965).

[247] Tadashi Yanagihara, Nobuhishi Matano, and Akira Kawase, *Jap. Anal.* **9**, 439–443 (1960).

[248] Andrzej Jacklewicz, *Chem. Anal.* (Warsaw) **19**, 207–210 (1974).

[249] V. G. Tiptsova and O. I. Kopnina, *Izv. Vyssh. Ucheb. Zaved., Khim. Khim. Tekhnol.* **11**, 364–365 (1968).

[250] Viktor Kemula, Stanislaw Rubel, and Wanda Stefanska, *Chem. Anal.* (Warsaw) **15**, 361–370 (1970).

[251] T. C. Z. Jayman, S. Sivasubramanian, and M. A. Wijedasa, *Analyst* **100**, 716–720 (1975).

ml, add 5 ml of sodium acetate buffer solution for pH 4, 2 ml of freshly prepared 1% solution of hydroquinone, and 1 ml of 0.5% solution of 1,10-phenanthroline. Dilute to 25 ml and read at 490 nm.

For iron in food, heat with 15 ml of nitric acid and 45 ml of saturated solution of ammonium nitrate for each gram of dry weight.[252] When colorless, evaporate to dryness; then heat for 10 minutes with 20 ml of hydrochloric acid. Develop with thiocyanate or 1,10-phenanthroline.

In combustion of 0.5 gram of polyethylene in an oxygen bomb, the small amount of nitrogen present is oxidized so that the metallic ions are present as nitrates in the aqueous phase.[253] Taken up in 5 ml of 5 *N* hydrochloric acid and diluted to 50 ml, 1,10-phenanthroline is recommended for determination of iron.

The broad application of the reagent for iron determination includes samples of arsenic,[254] alumina,[255] uranium,[256] magnesium and magnesium alloys,[257] dolomite,[258] iron- and steel-making slags by the AutoAnalyzer,[259] cement raw materials by the AutoAnalyzer,[260] oxalic acid and alkali-metal oxalates,[261] brass,[262] high silica and aluminosilicate materials,[263] silicate, carbonate, and phosphate rocks,[264] chromite or chrome ore,[265] ash of coal and coke,[266] nonferrous metals in general,[267] titanium dioxide,[268] electrolytic chromium,[269] biological samples,[270] and wheat flour and bread.[271]

For iron in rock salt by 1,10-phenanthroline, see Aluminum by Aurintricarboxylic Acid. For a sequential determination of iron as the 1,10-phenanthroline complex followed by copper, manganese, and aluminum, see Copper by Diethyldithiocarbamate. For separation of iron from manganese ores, ferromanganese, and manganese metal by column chromatography for determination by 1,10-phenanthroline, see Lead. For iron in fatty acids by 1,10-phenanthroline, see Copper by Lead Diethyldithiocarbamate. For iron in plant tissue by 1,10-phenanthroline, see Aluminum by 8-Hydroxyquinoline. For preparation of the

[252] Ma. T. Valdehita and A. Carballido, *Ana. Bromatol.* (Madrid) **191**, 437–453 (1954).
[253] Shizuo Fujiwara and Hisatake Narasaki, *Jap. Anal.* **10**, 1268–1272 (1961).
[254] L. B. Kristaleva, *Zavod. Lab.* **25**, 1294–1295 (1959).
[255] R. J. Julietti, *J. Brit. Ceram. Soc.* **5**, 47–58 (1968).
[256] Daido Ishii, *Jap. Anal.* **9**, 693–697 (1960).
[257] British Standards Institution, *BS 3907*, Pt. **2** (1966).
[258] A. K. Kutarkina and R. G. Syrova, *Tr. Ural'. Nauchn.-Issled. Khim. Inst.* **1964** (11), 19–22.
[259] P. H. Scholes and C. Thulbourne, *Analyst* **88**, 702–712 (1963).
[260] J. A. Fifield and R. G. Blezard, *Chem. Ind.* (London) **1969** (37), 1286–1291.
[261] M. Knizek and M. Musilova, *Chemist-Analyst* **55**, 108–109 (1966).
[262] J. Le Polles, *Bull. Soc. Pharm. Bordeaux* **101** (3), 189–193 (1962).
[263] H. Bennett, R. P. Eardley, W. G. Hawley, and I. Thwaites, *Trans. Brit. Ceram. Soc.* **61**, 636–666 (1962).
[264] Leonard Shapiro and W. W. Brannock, *U.S. Geol. Surv., Bull.* **1144A**, 56 pp (1962); H. Bennett, R. P. Eardley, W. G. Hawley, and I. Thwaites, *Trans. Brit. Ceram. Soc.* **61**, 636–666 (1962); A. Sass, *Inf. Chim.* **1972**, 227–235; K. Langer and P. Baumann, *Z. Anal. Chem.* **277**, 359–368 (1975).
[265] Joseph Dinnin, *U.S. Geol. Survey Bull.* **1084-B**, 31–68 (1959).
[266] F. H. Gibson and W. H. Ode, *Rep. Invest. U.S. Bur. Mines* **6036**, 23 pp (1962).
[267] Tadao Hattori and Toshiaki Kuroha, *Jap. Anal.* **11**, 727–730 (1962).
[268] A. A. Tumanov and A. N. Sidorenko, *Tr. Khim. Khim. Tekhnol.* (Gor'kii) **1962**, 378–383.
[269] A. A. Fedorov, F. A. Ozerskaya, R. D. Malinina, Z. M. Sokolova, and F. V. Linkova, *Sb. Tr. Tsentl. Nauch.-Issled. Inst. Chern. Met.* **1960** (19), 7–21.
[270] P. Punnett, R. V. Iyer, and B. W. Ellinwood, *Anal. Biochem.* **7**, 328–334 (1964).
[271] C. P. Czerniejowski, C. W. Shank, W. G. Bechtel, and W. B. Bradley, *Cereal Chem.* **41**, 65–72 (1964).

sample of niobium and sequential separation of nickel, cobalt, copper, iron, and zinc by column chromatography, see Copper by 1-(2-Pyridylazo)-2-naphthol. For extraction as complexes of ferric ion, 1,10-phenanthroline, and thiocyanate, see Thiocyanate.

The red ferrous 1,10-phenanthrolate can be oxidized with cerate to a blue ferric complex. This comforms to Beer's law up to 50 μg of iron per ml. Since the intensity of color is much less than that of the ferrous complex, it is more appropriate for determining substantial amounts of iron.

Procedures

Column Extraction.[272] Use a 7.2 mm i.d.$\times$15 cm column with a Teflon plug in the stopcock, a 25-ml reservoir, and a glass-wool plug to support the contents. Screen Fluoropak through 70-mesh and drench 200–400 ml with 2-octanol. Remove excess solvent on a Büchner funnel and draw air through for 10 minutes. Add 10 ml of 7 *N* hydrochloric acid to the reservoir and transfer a portion of the treated Fluoropak. Slurry, and drain the acid to give a bed of Fluoropak 5 inches high, gently tamped for uniformity.

Add a sample of solution containing 5–100 μg of iron to the reservoir. Adjust the hydrochloric acid concentration to 7 ± 1 *N*. If the volume will exceed 10 ml, transfer portionwise. Pass through the column at 3–5 ml/minute. Wash with 25 ml of 7 *N* hydrochloric acid added in 5-ml portions and drain to the surface of the bed. Strip the 2-octanone–ferric chloride with 18 ml of 2-propanol into 0.5 ml of 0.1 *N* hydroquinone. Add 2 ml of 0.03 *M* 1,10-phenanthroline and mix. Add 3 ml of pyridine, cool, and dilute to 25 ml with 2-propanol. After 15 minutes read at 507 nm against a reagent blank.

Isolation with Amberlite XAD-2.[273] Pack a 30-cm column, 11 mm i.d., with a slurry of 100-150 mesh resin in methanol to a depth of 5–6 cm.

To a sample solution containing up to 50 μg of iron, add 5 ml of 10% solution of hydroxylamine hydrochloride, 3 ml of 0.25% solution of 1,10-phenanthroline, and 5 ml of 10% solution of sodium acetate. Pass through the column and wash with 4 ml of water. Elute the red complex with a 1:19 mixture of 5% hydroxylamine hydrochloride solution and methanol. Dilute to 10 ml and read at 510 nm.

Extraction with Propylene Carbonate.[274] To a sample solution containing 0.05–1 micromole of iron, add 3 ml of 10% solution of hydroxylamine hydrochloride, 5 ml of 0.005 *M* 1,10-phenanthroline in 50% ethanol, and 5 ml of *M* sodium perchlorate. Add 5 ml of 10% solution of sodium acetate as a buffer. Saturate the solution with propylene carbonate and add sufficient to give a 3-ml layer in excess. Shake for 1 minute, separate the organic layer, dilute to 10 ml with ethanol, and read at 510 nm.

[272]Marven A. Wade and Stanley S. Yamamura, *Anal. Chem.* **36**, 1861–1862 (1964).
[273]Raymond B. Willis and Darrel Sangster, *ibid.* **48**, 59–62 (1976).
[274]B. G. Stephens and H. A. Suddeth, *ibid.* **39**, 1478–1480 (1967).

Extraction with Trioctylamine.[275] ALLOYS. Dissolve a 0.5-gram sample in 20 ml of hydrochloric acid and dilute to 100 ml. Extract the iron from a 10-ml aliquot with 10 ml of a 5% solution of trioctylamine in carbon tetrachloride by shaking for 10 minutes. The aqueous solution is appropriate for determination of cobalt. Reextract the ferric ion from the organic layer by shaking with 25 ml of 1% sodium chloride solution. Add 10 ml of acetic acid–sodium acetate buffer solution for pH 4, then 2 ml of 1% solution of hydroxylamine hydrochloride and 5 ml of 0.2% solution of 1,10-phenanthroline. Dilute to 50 ml, set aside for 30 minutes, and read at 510 nm.

Ferrous Ion in the Presence of Ferric.[276] By masking ferric ion with fluoride ion below pH 2.5 to prevent air oxidation, 7 ppm of ferrous ion can be determined in the presence of 2500 ppm of ferric ion. To 15 ml of sample in polyethylene, add sequentially 1 ml of 1:4 sulfuric acid, 2 ml of 2 *M* ammonium fluoride, 2 ml of 1% solution of 1,10-phenanthroline, and 3 ml of 3 *M* hexamethylenetetramine buffer. Dilute to 25 ml and read at 510 nm against a reagent blank after 30 minutes and within 1 hour.

As Ferric 1,10-Phenanthrolate.[277] Dissolve 5 grams of stannous chloride dihydrate in 100 ml of 1:1 hydrochloric acid. Dissolve 20 grams of sodium gluconate in 100 ml of water. Mix.

To a sample solution containing up to 25 mg of iron, add 10 ml of the stannous chloride–sodium gluconate reagent and dilute to more than 100 ml. Add 20 ml of 1% solution of 1,10-phenanthroline in 20% ethanol and 5 ml of 50% solution of sodium acetate. Adjust to pH 3–4 with sodium hydroxide solution. Add 50 ml of 4% ceric bisulfate solution in 1:2 sulfuric acid and dilute to 500 ml. Read at 585 nm.

As a Ternary Complex with Azide.[278] Adjust a sample solution containing up to 3 μg of iron per ml to pH 3.5–6 with sodium acetate–hydrochloric acid buffer solution. Add 1 ml of 0.01 *M* 1,10-phenanthroline and 2 ml of *M* sodium azide. Dilute to 25 ml. The sodium azide reduces ferric ion to the ferrous form. Extract with 25 ml of nitrobenzene and set the organic layer aside for 30 minutes. Read at 510–515 nm.

If cupric and hexavalent tungsten ions interfere, they can be masked with EDTA. Then ferric ion should be reduced with hydroxylamine hydrochloride at pH 7–8 before EDTA is added.

As a Ternary Complex with Nitroprusside.[279] To a sample solution containing 1–10 μg of ferrous ion, add 1 ml of 0.1 m*M* ascorbic acid, 5 ml of m*M* 1,10-phenanthroline, and 5 ml of m*M* nitroprusside. Dilute to 25 ml and adjust to

[275] B. E. McClellan and V. M. Benson, *ibid.* **36**, 1985–1987 (1964).
[276] Hiroki Tamura, Katsumi Goto, Takao Yotsuyanagi, and Masaichi Nagayama, *Talenta* **21**, 314–318 (1974).
[277] H. L. Watts, *Anal. Chem.* **36**, 364–366 (1964).
[278] V. Pandu Rango Rao and P. V. R. Bhaskara Sarma, *Mikrochim. Acta* **1970**, 783–786.
[279] Franciszek Buhl and Krystyna Kania, *Chem. Anal.* (Warsaw) **20**, 1055–1063 (1975).

pH 3–4. Shake for 2 minutes with 8 ml of nitromethane. Separate the organic phase, and add 1 ml of ethanol. Adjust to 10 ml with nitromethane and read at 510 ml against a blank.

Citrate-EDTA Masking.[280] To a sample solution containing 2–100 μg of iron, add 0.5 ml of hydrochloric acid. Add 1 ml of 2 *M* hydroxylamine hydrochloride, 5 ml of 0.5 *M* diammonium citrate–0.25 *M* disodium EDTA, and 2 ml of 0.05 *M* ethanolic 1, 10-phenanthroline. Add 3 ml of pyridine and dilute to 25 ml. If the sample contains a high concentration of organic complexing agents, heat to 60° for 15 minutes; otherwise omit this step. Set aside for 25 minutes. Read at 507 nm against a blank.

Lead.[281] Dissolve a 0.2-gram sample in 10 ml of hot 1:3 nitric acid. Add 5 ml of perchloric acid and evaporate to 2 ml. Add 40 ml of water. To eliminate interference by more noble metals such as copper, add 0.5 gram of finely granulated lead containing less than 0.001% iron. Boil gently for 15 minutes, cool, and filter. Add 10 ml of a buffer solution containing 272 grams of sodium acetate monohydrate and 240 ml of acetic acid per liter. Add 2 ml of 1% solution of hydroxylamine hydrochloride and 5 ml of 0.2% solution of 1, 10-phenanthroline. Dilute to 100 ml and read at 490 nm against a blank.

Zone-Refined Aluminum.[282a] As color reagent, mix 30 ml of 0.2% solution of 1, 10-phenanthroline, 30 ml of 5% solution of hydroxylamine hydrochloride, and 30 ml of 4 *M* sodium acetate. Dilute to 100 ml.

Dissolve a 5-gram sample by adding in small portions 125 ml of 1:1 hydrochloric acid containing 5 drops of 5% mercuric chloride solution. Add a few drops of nitric acid and evaporate until crystals begin to appear. Add 10 ml of hydrochloric acid and dilute to 125 ml. Extract for 5 minutes each with 15 ml and 10 ml of methyl isobutyl ketone. Shake the combined extracts with 10 ml of the color reagent. Wash the organic layer with 5 ml and 5 ml of water. Dilute the color reagent and washings to 25 ml and read at 570 nm against a reagent blank.

Aluminum-Base Hardener Alloys.[282b] Dissolve 5 grams in 125 ml of 1:1 hydrochloric acid and a few drops of nitric acid. Evaporate until the crystals begin to appear at around 10 ml. Cool, add 31 ml of hydrochloric acid, and dilute to 50 ml. Shake for 5 minutes each with 15 ml and 10 ml of 2:1 methyl isobutyl ketone–benzene. Shake the combined extracts with 10 ml of the color reagent, dilute to 25 ml, and read at 510 nm against a blank. For iron in aluminum of high purity, 1, 10-phenanthroline is preferable to sulfosalicylic acid or mercaptoacetic acid.[283]

Aluminum-Vanadium Alloys.[284] Dissolve 0.5 gram of sample in 5 ml of nitric

[280]Stanley S. Yamamura and John H. Sikes, *Anal. Chem.* **38**, 793–795 (1966).
[281]Haydee Armandola de Alderuccio, *Rev. Obras Sanit. Nac.* (Buenos Aires) **22** (174), 20–24 (1958).
[282a]H. Jackson and D. S. Phillips, *Analyst* **87**, 712–723 (1962); cf. G. Matelli and E. Attini, *Alluminio* **27**, 119–121 (1958); G. Matelli and V. Vicentini, *Allum. Nuova Met.* **38**, 559–566 (1969).
[282b]Cf. British Standards Institution, *BS 1728*, Pt. **8**, 6 pp (1957).
[283]T. B. Gorodentseva, G. S. Dolgorukova, K. F. Vorozhbitskaya, V. A. Vorbitskaya, L. S. Studenskaya, and V. S. Shvarev, *Tr. Vses. Nauchn.-Issled. Inst. Stand. Obraztsov. Spektr. Etalonov*, **2**, 34–39 (1965).
[284]Noboru Yamamoto, Shigeo Soejima, and Masao Harada, *Jap. Anal.* **19**, 1356–1360 (1970).

acid and 15 ml of 1:1 sulfuric acid. Evaporate to sulfur trioxide fumes, take up in water, and dilute to 250 ml. Dilute an aliquot containing less than 150 μg of iron and less than 50 mg of vanadium to 30 ml. Add 20 ml of 10% hydroxylamine hydrochloride solution to mask the vanadium. Heat at 100° for 10 minutes and cool. Add 3 ml of 10% solution of sodium acetate and raise to pH 4–4.5 with 10% sodium hydroxide solution. Then lower to pH 2 with 2 *N* sulfuric acid and dilute to 70 ml. Add 20 ml of 0.2% solution of 1,10-phenanthroline, dilute to 100 ml, and read at 500 nm within 30 minutes.

Beryllium.[285] Take a sample containing up to 1 gram of beryllium. Because of the intensity of reaction, add 1:1 hydrochloric acid dropwise until reaction subsides. Add more 1:1 hydrochloric acid to a total of 20 ml and evaporate to 5 ml. Filter if an insoluble residue remains. In that case, ash the residue, fuse with 1 gram of potassium bisulfate, and take up in 10 ml of 1:5 sulfuric acid. Add dropwise 0.3 ml of hydrofluoric acid, 1 ml of hydrochloric acid, and 1 ml of nitric acid. Heat to fumes of sulfur trioxide, take up in water, and add to the filtrate.

Add 1 gram of citric acid, adjust to 1:1 hydrochloric acid, and add 4 drops of bromine water. Shake for 2 minutes with 25 ml of 0.1 *M* tri-*n*-octylphosphine oxide in chloroform. Wash the organic layer with 30 ml of 1:1 hydrochloric acid. Extract the iron from the organic layer with 30 ml and 10 ml of 1:5 sulfuric acid, shaking for 2 minutes with each. To the combined extracts, add 40 ml of hydrochloric acid. Shake for 2 minutes with 10 ml of 0.01 *M* tri-*n*-octylphosphine oxide in chloroform. Centrifuge the organic phase. To 3 ml of the clear extract, add 15 ml of isopropanol, 1 ml of 3% isopropanolic hydroquinone, and 2 ml of 3% isopropanolic 1,10-phenanthroline. Dilute to 25 ml with isopropanol, let stand for 2 hours, and read at 510 nm against a blank.

Beryllium Oxide. Dissolve 1 gram of sample in 30 ml of 1:1 sulfuric acid and evaporate to about 15 ml. Proceed as for beryllium from "Add 1 gram of citric acid, adjust. . . ."

Tin and Lead Alloys.[286] COPPER AND ZINC ABSENT. Digest 1 gram of sample containing 0.02–0.06% iron with 10 ml of sulfuric acid in a 100-ml flask. Insert a dropping funnel with an air-intake side arm. Heat to 200° while intermittently bubbling air through it. Carefully drop hydrobromic acid into the solution to evaporate tin, antimony, and arsenic. Cool, transfer to a beaker, and add 2 ml of nitric acid. Heat to sulfur trioxide fumes, cool, and dilute to 50 ml. Filter lead sulfate and wash with 3:100 sulfuric acid. Evaporate to about 5 ml and dissolve the salts with water. Cool, add a couple of drops of *p*-nitrophenol indicator solution, and neutralize with ammonium hydroxide. Add 2 ml of 1:1 sulfuric acid to give a clear solution. Add 25 ml of acetic acid–sodium acetate buffer solution for pH 4, 3 ml of 1% solution of hydroxylamine hydrochloride, and 5 ml of 0.2% solution of 1,10-phenanthroline. Dilute to 100 ml and maintain above 20° for 20 minutes. Read at 510 nm.

[285] J. O. Hibbits, W. F. Davis, and M. R. Menke, *Talenta* **4**, 61–66 (1960); cf. J. Walkden, *UKAERE* **AM6**, 3 pp (1959).
[286] Kazuo Ota, *Jap. Anal.* **5**, 3–7 (1956).

Copper or Zinc Present. Proceed as above to "Evaporate to about 5 ml. . . ." Add 7 ml of 2% solution of potassium aluminum sulfate and heat. Add 1 : 1 ammonium hydroxide until aluminum hydroxide starts to precipitate. Add 10 ml more of 1 : 1 ammonium hydroxide, boil for 5 minutes, and filter. Wash with warm 1 : 20 ammonium hydroxide. Dissolve the precipitate in 50 ml of 1 : 5 sulfuric acid and precipitate as before. This time after filtering and washing, dissolve in 50 ml of 1 : 5 sulfuric acid and evaporate to about 5 ml. Add enough water to dissolve the salts, and neutralize with 1 : 1 ammonium hydroxide until aluminum hydroxide begins to precipitate. Proceed as above from "Add 2 ml of 1 : 1 sulfuric acid to give a clear solution."

Zinc, Zinc Alloys, and Alloys Containing Zinc.[287] Dissolve 1 gram of sample in an appropriate acid and add 10 ml of 50% ammonium acetate solution. Add 25 ml of *M* EDTA, 1 ml of 1% ascorbic acid solution, and 5 ml of 0.2% solution of 1,10-phenanthroline. Adjust to pH 5 ± 0.5. To overcome the effect of EDTA in retarding the color development, heat to boiling and cool. Dilute to 100 ml and read at 508 nm.

Refined Zinc.[288] Dissolve a 1.5-gram sample in hydrochloric acid, evaporate to dryness, and take up in water. Add 25 ml of *M*-EDTA and 5 ml of 10% hydroxylamine hydrochloride solution. Adjust to pH 7.5, add 2 ml of 0.1 *M* 1,10-phenanthroline, and dilute to 50 ml. After 10 minutes read at 510 nm.

Refined Cadmium. Use 2.5 grams of sample; treat as for zinc but delay reading for 25 minutes after dilution.

Titanium Alloys.[289] For 0.2–1% iron, take a 0.5-gram sample; for 0.05–0.5%, a 1-gram sample. Dissolve in 15 ml of 1 : 1 hydrochloric acid and 15 ml of sulfuric acid, adding a drop of 30% hydrogen peroxide to promote action. Add nitric acid dropwise to discharge the blue color, plus 5 drops in excess. Add 1 ml of hydrochloric acid and evaporate to sulfur trioxide fumes. Take up with 35 ml of water and add 60 ml of hydrochloric acid and 2 drops of nitric acid. Cool, adding 25 ml of 1 : 1 hydrochloric acid and 2 ml of perchloric acid. Shake with 75 ml of methyl isobutyl ketone for 3 minutes and discard the aqueous phase. Wash the organic phase with 25 ml and 25 ml of 1 : 1 hydrochloric acid. Extract the iron with 25 ml and 25 ml of water. Add 10 ml of perchloric acid and concentrate to 50 ml. Add 10 ml of isopropanol and dilute to 100 ml. Mix a 10-ml aliquot with 15 ml of 1% solution of hydroxylamine hydrochloride, 15 ml of 0.5% solution of 1,10-phenanthroline solution in 0.01 *N* hydrochloric acid, and 25 ml of a buffer solution containing 170 grams of sodium acetate trihydrate and 850 ml of acetic acid per liter. Adjust to pH 3.5 with 1 : 1 hydrochloric acid or 1 : 1 ammonium hydroxide, dilute to 100 ml, and set aside for 30 minutes. Read at 510 nm.

[287]Tadao Hattori and Toshiaki Kuroha, *ibid.* **11**, 727–730 (1962).
[288]F. Vydra and V. Markova, *Z. Anal. Chem.* **192**, 347–350 (1963).
[289]H. C. Naumann, *Metall.* **27**, 247–249 (1973); cf. George Norwitz and Maurice Codell, *Anal. Chim. Acta* **11**, 350–358 (1954); *Metallurgia* **57**, 261–270 (1958).

Titanium-Iron Alloys.[290] As a buffered reagent dissolve 2.5 grams of 1,10-phenanthroline in 50 ml of water, add 40 grams of tartaric acid and 500 grams of sodium acetate, and make up to 1 liter.

The alloy may contain up to 10% of iron. Dissolve 0.5 gram of sample in 50 ml of 1:3 sulfuric acid, adding sufficient nitric acid to destroy the titanium color. Evaporate to sulfur trioxide fumes and take up in 50 ml of water. Boil for 1 minute, cool, and dilute to 250 ml.

Mix a 5-ml aliquot with 10 ml of 10% hydroxylamine hydrochloride solution. Set aside for 10 minutes, then add 10 ml of buffered reagent. Dilute to 100 ml, set aside for 30 minutes, and read at 510 nm against a solution from which the addition of the reagent was omitted.

Titanium and Magnesium Products.[291] Heat a 0.5-gram sample with 30 ml of 1:1 hydrochloric acid, filter, and wash the residue with hot water. Ignite the residue in platinum and fuse with 3 grams of sodium carbonate and 1.5 grams of sodium tetraborate. Extract the melt with 30 ml of 1:1 hydrochloric acid, add to the previous filtrate, and dilute to 500 ml.

To a 5-ml aliquot add 10 ml of 5% solution of sodium tartrate, 10 ml of fresh 2% hydroxylamine hydrochloride solution, 10 ml of 25% solution of sodium acetate, 10 ml of 0.2% solution of 1,10-phenanthroline, and water to make 100 ml. After 30 minutes read against a solution containing either 0.5 or 1 mg of iron.

Tungsten and Tungsten Alloys.[292] Warm a 2-gram sample containing less than 0.09% iron in platinum with 5 ml of hydrofluoric acid and 4 ml of nitric acid. If necessary, add more of the acids to dissolve, giving a final volume of 4–7 ml. Add 30 ml of water, then 50 ml of 20% ammonium tartrate solution, followed by 90 ml of 4.5% sodium borate solution. Dilute to 220 ml, cool, and dilute to 250 ml.

Use a 50-ml aliquot. Add 1 ml of 10% solution of hydroxylamine hydrochloride and 5 ml of 0.2% solution of 1,10-phenanthroline. Dilute to about 90 ml and heat at 60–70° for 30 minutes. Dilute to 100 ml and read at 510 against a reagent blank.

Tungsten-Iron, Tungsten-Nickel, Tungsten-Copper, and Tungsten-Cobalt Alloys. Dissolve a 1-gram sample as above and dilute to 500 ml. Take an aliquot containing 0.15–0.3 mg of iron and proceed as above from "Add 1 ml of 10% solution of. . . ." More than 0.8 mg of nickel, 1.6 mg of copper, or 0.8 mg of cobalt in the aliquot will cause low results.

Bismuth.[293] Dissolve 2 grams of sample containing 5–50 ppm of iron in 5 ml of nitric acid and evaporate to dryness. Add 3 ml of hydrochloric acid, evaporate to dryness, and repeat this step twice more. Redissolve in hydrochloric acid by dropwise addition. Add 1 ml of 5% solution of hydroxylamine hydrochloride, 10 ml of 3.72% solution of disodium EDTA, 30 ml of 50% sodium citrate solution, and 2

[290] G. S. Grigorovskaya, *Tekhnol. Legk. Splavov. Nauchn.-Tekh. Byull. VILS* **1970** (6), 99–100.
[291] V. N. Tikhonov and N. D. Rychkova, *Zh. Anal. Khim.* **18**, 1131–1133 (1963).
[292] George Norwitz and Herman Gordon, *Anal. Chem.* **37**, 417–419 (1965).
[293] D. G. Holmes, *Analyst* **82**, 523–529 (1957).

ml of 0.25% solution of 1, 10-phenanthroline. Dilute to 100 ml, set aside for 1 hour, and read at 510 nm.

Indium.[294] Dissolve a 0.5-gram sample in 1 ml of nitric acid and 2 ml of 7 *N* hydrochloric acid. Evaporate to dryness and take up in 5 ml of 7 *N* hydrochloric acid. Extract the iron with 5 ml and 5 ml of isopropyl ether. Wash the extract with 5 ml and 5 ml of 7 *N* hydrochloric acid. Evaporate the extract to dryness and calcine. Take up in 0.5 ml of *N* hydrochloric acid. Add 3 ml of phthalate buffer solution for pH 3, 2 ml of 10% hydroxylamine hydrochloride solution, 2 ml of 0.5% solution of 1, 10-phenanthroline, and 0.1 ml of 2.5 *M* sodium perchlorate or lithium perchlorate. Extract with 0.5 ml of nitrobenzene and read at 510 nm.

Plutonium.[295] As hydroxylamine reagent, dissolve 108 grams of hydroxylamine hydrochloride in water, add 60 ml of acetic acid, and dilute to 5 liters. As 1, 10-phenanthroline reagent, dissolve 2 grams in 25 ml of ethanol. Add to 80 grams of sodium acetate trihydrate in water and dilute to 1 liter. As a buffer solution, mix 10 ml of 28.5% solution of sodium acetate trihydrate with 1 ml of *M* sodium perchlorate.

This procedure through the extraction must be carried out in a glovebox. Dissolve 0.1 gram of sample in 1 ml of 1:1 hydrochloric acid. Add 5 ml of hydroxylamine reagent and set aside for 30 minutes. Add 5 ml of 1, 10-phenanthroline reagent and 10 ml of the sodium acetate–perchlorate buffer solution. The pH should be 3.5–4.5. Shake for 30 seconds with 25 ml of nitrobenzene. Remove a portion of the organic layer, dry with sodium sulfate, and read at 515 nm against a reagent blank.

Antimony Sulfide.[296] Dissolve 0.5 gram of sample in 20 ml of hydrochloric acid by boiling for 5 minutes to drive off the hydrogen sulfide. Cool, add 20 ml of 10% tartaric acid solution, and dilute to 100 ml. For up to 0.7% iron, take a 20-ml aliquot; for 0.7–1.4% iron, a 10-ml aliquot. Add 30 ml of water and 10 ml of 5% solution of hydroxylamine hydrochloride. Set aside for 15 minutes. Add 10 ml of 0.2% solution of 1, 10-phenanthroline and 20 ml of 50% solution of sodium acetate trihydrate. Dilute to 200 ml and set aside for 30 minutes. Read at 500 nm.

Electrocorundum[297]

METALLIC IRON. Reflux a 1-gram sample ground to less than 0.1 mm with 80 ml of methanol and 4 ml of bromine for 1 hour, agitating from time to time. Cool and filter, washing the insoluble matter with methanol until colorless, then with water. Acidify the filtrate and washings with 5 ml of 1:1 sulfuric acid, and boil off the

[294] V. A. Nazarenko and G. V. Flyantikova, *Zavod. Lab.* **27**, 1339–1341 (1961); cf. Koichi Nishimura and Teruo Imai, *Jap. Anal.* **13**, 713–717 (1964).
[295] C. E. Plock and C. E. Caldwell, *Anal. Chem.* **39**, 1472–1473 (1967); **40**, 331 (1968); cf. T. K. Marshall, J. W. Dahlby, and G. K. Waterbury, *Rep. USAEC* **LA-3781**, 13 pp (1967).
[296] George Norwitz, Joseph Cohen, and Martin E. Everett, *Anal. Chem.* **32**, 1132–1135 (1960).
[297] I. Ya. Rivlin, E. A. Totsman, and I. V. Gromozova, *Tr. Vses. Nauchn.-Issled. Inst. Abrazivov Shlifovianiya* **1967** (4), 44–56; cf. A. M. Leblond, R. Wendling, and J. M. Bourdieu, *Chim. Anal.* (Paris) **50**, 431–438 (1968).

bromine. Precipitate hydroxides from the hot solution with ammonium hydroxide. Filter and wash the precipitate, first with methanol, then with 2% solution of ammonium nitrate. Dissolve the precipitate from the filter with 10 ml of hot 1 : 1 hydrochloric acid, wash with water, dilute to 250 ml, and take a 50-ml aliquot. Add 5 ml of 1 : 1 hydrochloric acid and 2 ml of 10% hydroxylamine hydrochloride solution. Agitate for 10 minutes. Add 5 ml of 0.25% 1,10-phenanthroline reagent. Add 50% ammonium acetate solution until neutral to Congo red. Let stand for 30 minutes, dilute to 100 ml, and read at 510 nm.

SURFACE IRON. To 0.5 gram of the washed and dried insoluble matter from removal of metallic iron, add in a platinum dish 40 ml of 1 : 1 sulfuric acid and 10 ml of hydrofluoric acid. Boil for 1 hour in an atmosphere of carbon dioxide and cool while still passing carbon dioxide. Add 20 ml of saturated solution of boric acid. Determine the surface iron as for metallic iron, starting at "Add 5 ml of 1 : 1 hydrochloric acid and 2 ml. . . ."

INTERNAL IRON. Mix 0.2 gram of the washed and dried insoluble matter from determination of metallic iron with 1.2 grams of sodium fluoborate. Cover with more sodium fluoborate and blanket the contents of the crucible with carbon dioxide. Heat to red heat, fuse for 2 minutes, and cool for 20 minutes in a stream of carbon dioxide. Heat 140 ml of water with 8 grams of boric acid and 20 ml of 1 : 1 sulfuric acid to boiling. Saturate with carbon dioxide and add the crucible. Boil for 30 minutes in a stream of carbon dioxide and cool. Dilute the solution and crucible washings to 250 ml. For ferrous ion take a 50-ml aliquot and complete as under metallic iron from "Add 5 ml of 0.25% 1,1-phenanthroline reagent." Obtain total iron in a 50-ml aliquot as under metallic iron from "Add 5 ml of 1 : 1 hydrochloric acid. . . ."Obtain ferric iron by difference.

Ferrous Iron in Ferric Oxide.[298] Remove air from 0.1 gram of sample by passing argon through it for 5 minutes. Add 10 ml of hydrochloric acid and boil for 2 minutes. Add 10 ml more of hydrochloric acid and cool. Add 30 ml of water, then 10 ml of 1 : 1 ammonium hydroxide, and dilute to 100 ml.

As a buffer solution, dissolve 300 grams of acetic acid and 205 grams of sodium hydroxide in 1500 ml of water. Add 200 grams of trisodium citrate dihydrate. Adjust to pH 5 and dilute to 2000 ml.

Mix 10 ml of chloroform, 1 ml of *M* sodium perchlorate, 5 ml of 0.25% solution of 1,10 phenanthroline in 1 : 1000 hydrochloric acid, and 20 ml of the buffer solution for pH 5. In red light add an aliquot of the sample solution and shake for 1 minute. Further exract with 7 ml, 7ml, and 7ml of chloroform. Dilute the combined extracts to 50 ml with chloroform and read at 530 nm.

Ferrous Iron in Silicate Rocks.[299] To a 1-gram sample in platinum, add 1 ml of 1 : 1 sulfuric acid and 0.6 ml of hydrofluoric acid. Stir with a platinum spatula. Cover the crucible with a funnel and pass carbon dioxide through to prevent air oxidation of ferrous ion.

[298] J. Novak, *Chem. Zvesti* **20**, 545–549 (1966).
[299] Keisuke Tachibana, *Mem. Fac. Sci. Kyushu Univ., Ser. C* **4**, 239–246 (1961).

Mix 150 ml of water, 2 ml of acetate buffer solution for pH 4.6, and 6 ml of 0.2% solution of 1, 10-phenanthroline. Add the crucible containing the decomposed sample and stir. Raise to pH 4.6 by dropwise addition of 1 : 1 ammonium hydroxide. Filter, wash, and dilute to 250 ml. Read at 510 nm 20 minutes after adjusting the pH.

Ferrocenes.[300] Add 10 ml of hydrochloric acid to 0.2 gram of sample. After 10 minutes add 10 ml of nitric acid. After a few minutes boil gently. After 30 minutes add 10 ml more of nitric acid and boil for 1 hour. Cool, add 5 ml of 30% hydrogen peroxide, and boil for 15 minutes. Cool, add 100 ml of water, filter if necessary, and dilute to 500 ml. Titrate a 10-ml aliquot with 16% sodium hydroxide. Add that amount of alkali to another 10-ml aliquot and dilute to 100 ml. To a 10-ml aliquot of this dilution add 2 ml of 2.5% solution of hydroquinone in 2% sulfuric acid, 2 ml of 20% solution of sodium sulfite, and 1 ml of 0.5% ethanolic 1, 10-phenanthroline. Set aside for 15 minutes, then add 10 ml of 20% solution of ammonium acetate. Dilute to 200 ml and read at 510 nm.

Vanadium Pentoxide.[301] Dissolve 0.25 gram of sample in 20 ml of 2.5% ammonium hydroxide. Cool and add 2 ml of 0.25% 1, 10-phenanthroline solution. Dilute to 25 ml, set aside for 10 minutes, and read at 510 nm.

Tantalum Oxide.[302] Fuse a 1-gram sample with 3 grams of sodium hydroxide in a gold crucible for 10 minutes. Dissolve in 10 ml of water in a polyethylene beaker and acidify with 20 ml of 1 : 1 hydrochloric acid. Add 1 ml of aluminum chloride solution containing about 10 mg of aluminum. Make alkaline with ammonium hydroxide and heat at 100° for 20 minutes. Filter the precipitate of iron, aluminum, and tantalum and wash with 2% ammonium chloride solution. Wash back into the original polyethylene beaker with hot water. Add 2 ml of hydrofluoric acid and heat at 100° for 20 minutes. Add 20 ml of 25% tartaric acid solution, 20 ml of 5% boric acid solution, 10 ml of 10% hydroxylamine hydrochloride solution, and 13 ml of 1 : 1 ammonium hydroxide. Adjust to pH 5.5–6.5 and add 3 ml of 0.2% solution of 1, 10-phenanthroline. Heat at 100° for 30 minutes and cool. Dilute to 100 ml and read at 517 nm against a blank.

Rhenium.[303] Dissolve 0.02 gram of powdered sample in 1 ml of 1 : 1 nitric acid at 20°. Evaporate to a syrup at 100° and dissolve in 2 ml of water acidified with 0.1 ml of 1 : 1 nitric acid. Add 0.5 ml of 10% hydroxylamine hydrochloride solution, 1 ml of 50% solution of sodium acetate, 0.8 ml of 8% solution of thiourea, and 1 ml of 0.5% solution of 1, 10-phenanthroline in *N* acetic acid. Dilute to 10 ml and read at 510 nm.

Lead Molybdate.[304] As EDTA-tartrate solution, dissolve 140 grams of sodium

[300] J. Decombe and J.-P. Ravoux, *Bull. Soc. Chim. Fr.* **1964**, 1405–1406.

[301] S. Ya. Vinkovetskaya and T. I. Levitskaya, *Zavod. Lab.* **34**, 278 (1968).

[302] Masao Yoshida and Noboru Kitamura, *Jap. Anal.* **11**, 744–747 (1962).

[303] S. Jedrzejewski and H. Jedrzejewska, *Chem. Anal.* (Warsaw) **12**, 1283–1289 (1967).

[304] F. P. Gorbenko, L. Ya. Enel'eva, and V. S. Smirnaya, *Tr. Vses. Nauchn.-Issled. Inst. Khim. Reakt.* **1966** (29), 80–83.

potassium tartrate in 250 ml of water. Add 75 grams of EDTA in 400 ml of water and dilute to 1 liter. Add 10 ml of 10% solution of hydroxylamine hydrochloride and 20 ml of 0.5% solution of 1,10-phenanthroline. Set aside for 12 hours; then add 3 ml of 12% solution of sodium perchlorate. Extract with successive 25-ml portions of dichloroethane until the extracts are colorless. Add 5 ml of 10% solution of hydroxylamine hydrochloride, 5 ml of 0.5% solution of 1,10-phenanthroline, and 2 ml of 12% solution of sodium perchlorate to the aqueous solution. Again extract with successive 25-ml portions of chloroform until the extracts are colorless to give the purified reagent.

Dissolve a 3-gram sample in 100 ml of the EDTA-tartrate solution. Add 2 ml of 10% solution of hydroxylamine hydrochloride and 2 ml of 0.5% solution of 1,10-phenanthroline. Set aside for 1 hour; then add 1 ml of 12% solution of sodium perchlorate. Shake for 1 minute with 25 ml of dichloroethane, filter the extract, and read at 510 nm against the solvent.

Calcium Borate.[305] Dissolve 1 gram of sample in 50 ml of boiling 1:9 hydrochloric acid, dilute to 250 ml, and filter. Neutralize a 5-ml aliquot of filtrate to Congo red with 1:1 ammonium hydroxide. Add 2 drops of 1:1 hydrochloric acid, 3 ml of 10% hydroxylamine hydrochloride solution, and 20% solution of sodium acetate trihydrate until the solution turns red. Add 3 ml of 0.2% solution of 1,10-phenanthroline, dilute to 50 ml, and read at 510 nm.

Carbonate Minerals, Oxidized Manganese Compounds Present.[306] To prepare Reinhardt solution, dissolve 200 grams of manganese sulfate tetrahydrate in a liter of water, and add 400 ml of phosphoric acid and 1600 ml of 1:3 sulfuric acid.

Heat 0.1 gram of sample with 20 ml of 0.2 N oxalic acid, 2 ml of Reinhardt solution, and 5 ml of 1:4 sulfuric acid at 100° for 10 minutes. Add 20 ml of 1:4 sulfuric acid, 5 ml of phosphoric acid, and 2 ml of hydrofluoric acid. Heat until dissolved, add 1 gram of boric acid, and dilute to 200 ml. Mix a 2-ml aliquot, 4 ml of 20% ammonium acetate solution, 2 ml of 10% sodium citrate solution, and 2 ml of 0.2% solution of 1,10-phenanthroline in 0.1 N hydrochloric acid. Dilute to 200 ml and read at 508 nm against a reagent blank.

Chrome-Tanning Agents.[307] Dissolve a 2.5-gram sample in water and dilute to 250 ml. Filter, and discard the first 10 ml. Mix a 10-ml aliquot with 2 drops of 1:1 hydrochloric acid, 2 ml of 10% solution of hydroxylamine hydrochloride, 4 ml of 2.5% solution of sodium citrate, and 4 ml of 1.2% solution of EDTA. Neutralize with 1:1 hydrochloric acid to Congo red paper. Add 3 drops of 1:1 hydrochloric acid, 5 ml of 0.2% solution of 1,10-phenanthroline, and 3 ml of 20% sodium acetate solution. Dilute to 100 ml, set aside for 30 minutes, and read against a sample blank at 510 nm.

Chromic Chloride Hexahydrate. Dissolve 2.5 grams of sample in 80 ml of water

[305] A. K. Kutarkina, A. A. Moshkina, and M. D. Chetverkina, *Tr. ural.' Nauchn.-Issled. Khim. Inst.* **1964** (11), 78–80.

[306] T. L. Radovskaya and K. G. Borovik, *Zavod. Lab.* **40**, 141–142 (1974).

[307] I. M. Novoselova, L. I. Vasserman, A. E. Aleshechkina, and V. M. Masalovich, *Tr. Ural'. Nauchn.-Issled. Khim. Inst.* **1973** (27), 239–242.

and 4 ml of 1:1 hydrochloric acid. Boil for 5 minutes, cool, and dilute to 250 ml. Proceed as above from "Mix a 10-ml aliquot...."

Chromium Phosphate. Dissolve a 0.5-gram sample in 10 ml of 1:1 hydrochloric acid by boiling for 10 minutes. Cool and dilute to 100 ml. Proceed as above from "Mix a 10-ml aliquot...."

EDTA Solution.[308] To a sample containing the equivalent of 3 grams of the free acid, add 30 ml of hot water and adjust to pH 7.5. Cool to 50° and add 5 ml of 10% solution of hydroxylamine hydrochloride adjusted to pH 7.5. Add 2 ml of 2% solution of 1,10-phenanthroline slightly acidified with 1:4 sulfuric acid. Cool and dilute to 50 ml. After 15 minutes read at 500 nm against a blank.

Water[309]

FERROUS AND FERRIC ION. As a reagent, mix 5 volumes of 25 m*M* 1,10-phenanthroline hydrochloride, 5 volumes of 0.5 *M* glycine adjusted to pH 2.9 with 0.5 *N* hydrochloric acid, and 2 volumes of 0.1 *M* sodium nitrilotriacetate.

To 50 ml of sample containing up to 25 μg of iron, add 10 ml of freshly mixed reagent and 1 ml of *M* sodium perchlorate. Set aside for 2 minutes. Extract the ferrous–1,10-phenanthroline–perchlorate complex by shaking with 10 ml and 10 ml of nitrobenzene. Read the extract as ferrous ion at 575 nm against a reagent blank.

Add 5 ml of 1:1 hydrochloric acid and 1 ml of 0.25 *M* ascorbic acid to the residual aqueous phase. Set aside for 5 minutes. Meanwhile titrate an aliquot with 2.5 *N* sodium hydroxide, then exactly neutralize the balance. Complete as above from "add 10 ml of freshly mixed..." reading as ferric ion.

BOILER FEED WATER.[310] For this purpose the adduct formed with 1,10-phenanthroline in the presence of a large excess of an anionic surfactant is extracted. Mix 200 ml of sample, which may contain as little as 2 μg of iron, with 4 ml of 7 *N* hydrochloric acid. Heat at 80° for 30 minutes, cool, and add 4 ml of 10% solution of hydroxylamine hydrochloride. Add 10 ml of 0.12% solution of 1,10-phenanthroline hydrochloride and sufficient ammonium hydroxide to raise to pH 5.8. Add 5 ml of *M* acetate buffer for pH 4.3 and then 1:1 hydrochloric acid to reduce to pH 2.4. Add 3 ml of 1% solution of sodium dioctylsulfosuccinate. Shake for 2 minutes each with 10 ml, 7 ml, and 7 ml of chloroform. Dilute the combined extracts to 25 ml with ethanol and read at 515 nm.

Readings for ferrous ion in water should be made within 20 minutes after adding the reagents.[311]

Petroleum Oils.[312] Transfer 25 grams of sample to a platinum crucible and add

[308]F. Vydra and R. Pribil, *Chemist-Analyst* **51**, 76–77 (1962).

[309]Hubert Fadrus and Josef Maly, *Anal. Chim. Acta* **77**, 315–316 (1975); *Analyst* **100**, 549–554 (1975).

[310]Kin'ya Sono, Hiroto Watanabe, Yoshimichi Mitsukami, and Tatsuo Nakashima, *Jap. Anal.* **14**, 213–218 (1965).

[311]A. R. Ghosh and I. I. Radhakrishnan, *J. Proc. Inst. Chem. India* **39**, 168–172 (1967); cf. N. J. Nicholson, *Proc. Soc. Water Treat. Exam.* **15**, 157–158 (1966).

[312]E. J. Agazzi, D. C. Burtner, D. J. Crittenden, and D. R. Patterson, *Anal. Chem.* **35**, 332–335 (1963).

a weight of sulfur equivalent to 10% of that of the oil. This addition prevents loss of metals. Heat gently with a burner until the contents ignite and burn readily. Remove the burner and allow the oil to burn. When the flame extinguishes, reignite and insert in the air bath (Figure 18-1). Since external heat is required, the crucible can be heated with a burner, but this calls for considerable operator attention. Setting the crucible on a hot plate is only moderately successful because the heat conducted up the crucible is insufficient to maintain burning. Thus the top of a hot plate was removed and covered with a shallow box with holes in which the crucibles were inserted and held just above the heating coils. Some oils form a crust of carbon over the surface during the burning process which may entrap enough gas to cause the partially burned material to expand over the top and sides of the crucible. To prevent this, heat the sides of the crucible projecting above the transite top of the air bath until the carbon shrinks enough to allow the combustible gases to escape. Continue the burning process until only carbon remains. Place the crucible in a muffle furnace at 550±50°, leaving the door slightly ajar, until no carbon remains. Cool, and dissolve the ash with 0.5 ml of 1:1 sulfuric acid and a couple of drops of hydrofluoric acid. Dilute to 5 ml. This sample solution is also appropriate for determination of nickel and vanadium.

To an aliquot containing 0.02–60 μg of iron, add sequentially 2 drops of 1:1 hydrochloric acid, 0.2 ml of 10% hydroxylamine hydrochloride solution, and 0.2 ml of a buffer solution containing 5 grams of diammonium citrate and 10 grams of ammonium acetate in 85 ml of water. Add 1 ml of 0.2% 1,10-phenanthroline solution and a small piece of Congo red paper. Add 1:4 ammonium hydroxide dropwise until the indicator paper turns from blue to red, and add 1 drop in excess. Dilute to 5 ml, set aside for 10 minutes, and read at 510 nm.

Red Phosphorus.[313] To a 0.5-gram sample, add 10 ml of water and 35 ml of nitric acid. Evaporate at a high heat to the sudden appearance of red nitrogen oxide fumes at a volume of about 1 ml. Remove from the heat at once. Foaming will start just before that reaction, and the sample should be removed before foaming ceases. If heated to white fumes, discard and start over, since the glass is attacked and high results will be obtained. Let cool, add 75 ml of water, and evaporate to 35 ml to hydrolyze pyrophosphate. Cool, and add 10 ml of 5%

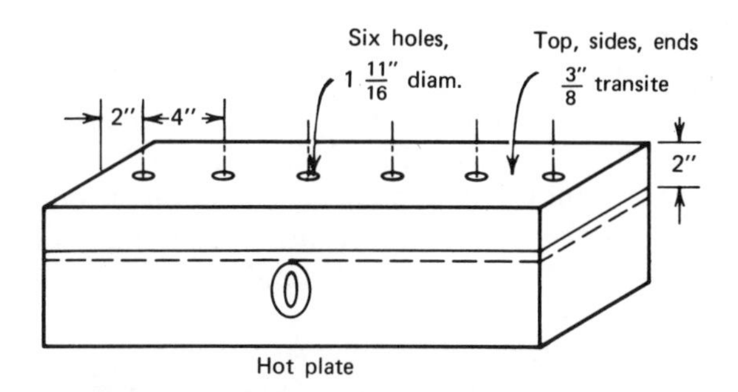

Figure 18-1 Hot plate modified for burning oils.

[313]George Norwitz, Joseph Cohen, and Martin E. Everett, *ibid.* **36**, 142–144 (1964).

solution of hydroxylamine hydrochloride. Let stand for 15 minutes and add 10 ml of 0.2% solution of 1,10-phenanthroline. Add 15 ml of 20% sodium citrate solution to buffer to pH 5.5 and dilute to 100 ml. Set aside for 1 hour and read at 510 nm against a reagent blank.

Citrus Fruit Juice.[314] Digest 50 grams of fresh or reconstituted juice with 5 ml of nitric acid at 50° for 30 minutes. Add 1 ml of sulfuric acid, 6 ml of nitric acid, and 1 ml of perchloric acid. Heat to white fumes at 200°, adding more acid if necessary to decolorize. Take up in 1:1 hydrochloric acid and dilute to 100 ml with that acid.

Neutralize a 20-ml aliquot with 40% sodium hydroxide solution to pH 6 for the bulk of the reaction, finally adjusting with 2 *M* sodium acetate. Add 1 ml of 1% solution of hydroquinone in a 0.1 *M* buffer of 65:35 acetic acid–sodium acetate. Add 5 ml of 0.5% solution of 1,10-phenanthroline and store in the dark for 15 minutes. Dilute to 50 ml and read at 485 nm.

Fermentation Broths.[315] Digest a sample containing 0.02–0.15 mg of iron for 2 hours with 1 ml of perchloric acid and 1 ml of sulfuric acid. Cool, add 1 ml of fresh 20% solution of sodium hydrosulfite, and boil for 1 hour. Cool, slowly add 2 ml of water, then add a drop of 1% ethanolic *p*-nitrophenol. Add 1:2 ammonium hydroxide until the color changes, then acidify to pH 2 with 1:2 sulfuric acid. Cool, add 1 ml of 10% solution of hydroxylamine hydrochloride, and set aside for 15 minutes. Add 5 ml of 0.25% solution of 1,10-phenanthroline, followed by neutral 25% solution of sodium citrate, until the pH is 3.5–4. Dilute to 50 ml, warm at 35° for 30 minutes, cool, and read at 510 nm.

High Purity Acids.[316] As the color reagent, dissolve 10 grams of hydroxylamine hydrochloride and 0.1 gram of 1,10-phenanthroline in 200 ml of water and add 100 ml of acetate buffer solution for pH 5.

Evaporate hydrofluoric acid in a polytetrafluoroethylene beaker; use quartz for hydrochloric, sulfuric, or acetic acid. Dissolve the residue in 0.03 ml of hydrochloric acid and 2 ml of water. Adjust to pH 3 with 1:9 ammonium hydroxide. Add 0.03 ml of 12% acetic acid and 1 ml of the color reagent. Heat at 100° for 10 minutes, cool, dilute to 5 ml, and read at 500–520 nm.

Wine.[317] Boil 20 ml with 20 ml of 30% hydrogen peroxide until concentrated to less than 5 ml. Cool, and add ammonium hydroxide until the solution is opalescent and has an odor of ammonia; about 3–4 ml will be required. Cool and dilute to 100 ml with *N* hydrochloric acid. To a 20-ml aliquot, add 2 ml of 2.5% solution of hydroquinone in 1:100 sulfuric acid, 2 ml of 20% solution of sodium sulfite, and 1 ml of 0.5% ethanolic 1,10-phenanthroline. Set aside for 15 minutes, then add 10 ml of 20% solution of ammonium acetate. Dilute to 50 ml and read at 500 nm.

[314]Guiseppe Safina, *Conserv. Deriv. Agrm.* (Palermo) **9**, 169–170 (1960).

[315]L. Carta de Angeli and F. Dentice di Accadia, *R.C. Ist Suppl. Sanit.* **25**, 966–971 (1962).

[316]I. G. Shafren, I. F. Vzorova, A. V. Patlakh, and L. K. Fidlon, *Tr. Vses. Nauchn.-Issled. Inst. Khim. Reakt.* **1969** (31), 229–234.

[317]L. Laporta, *Boll. Lab. Chim. Prov.* **15**, 239–248 (1964).

Dietetic Milk Products.[318] Dilute a sample containing 0.1–0.5 mg of iron to 100 ml and heat to 100°. Add 50 ml of 20% trichloroacetic acid solution and 25 ml of 2:1 hydrochloric acid with continuous mixing during the addition and the subsequent 20 minutes. Filter and develop as for wine from "To a 20-ml aliquot, add...."

Serum.[319] Agitate 6 ml of serum with 3 ml of 1:1 hydrochloric acid and set aside for 10 minutes. Add 3 ml of 20% solution of trichloroacetic acid and again set aside for 10 minutes. Filter and transfer 5 ml to a calibrated stoppered tube. Add a drop of 1% solution of dinitrophenol in isopropanol. Add ammonium hydroxide to the yellow end point. Cool, and decolorize by dropwise addition of 0.5 N hydrochloric acid. Add a drop of 2% solution of hydroquinone in isopropanol and 2 drops of 2% solution of 1,10-phenanthroline hydrochloride in isopropanol. Set aside for 1 hour, dilute to 10 ml, and read at 500 nm. Serum is also wet-ashed for more conventional analysis.[320]

Saponifiable Oils.[321] Heat 20 grams of sample in a silica dish with 2 ml of perchloric acid, 2 ml of nitric acid, and 1 ml of hydrochloric acid. Keep the temperature below 150° for 30 minutes, then raise it to 250°, and finally ash at 800°. Dissolve the ash in 0.5 ml of nitric acid and take to dryness in platinum. Add 0.5 ml of 10% sodium carbonate solution, evaporate, and fuse. Dissolve in water and dilute to 50 ml. Complete as for wine from "To a 20-ml aliquot, add...."

Alternatively,[322] transfer 1 gram of sample as a 50% solution in petroleum ether to a screw-capped 40-ml test tube. Add 5 ml of petroleum ether and 1 ml of 10% solution of hydroxylamine hydrochloride. Add 5 ml of a buffer solution containing 8.3 grams of anhydrous sodium acetate and 1 ml of acetic acid per 100 ml. Add 2 ml of 0.1% solution of 1,10-phenanthroline and mix for 2 minutes. Read the aqueous layer at 510 nm.

Black Olives.[323] Pulp 100 grams of sample with 100 ml of water. Add 5 ml of a saturated ethanolic magnesium nitrate and ash below 550°. Dissolve in 2 ml of 1:1 hydrochloric acid and dilute to 50 ml. Complete as for wine from "To a 20-ml aliquot, add...."

Cumene and Cumene Hydroperoxide.[324] To 2 grams of sample, add 5 ml of 10% sodium hydroxide solution and digest to destroy the hydroperoxide. Neutralize with 1:1 sulfuric acid and evaporate to a small volume. Add 5 ml of 1:1 sulfuric acid, evaporate to dryness, and ignite at 500°. Cool, take up in 5 ml of 1:10

[318] P. Boersma, *Neth. Milk Dairy J.* **17**, 289–297 (1963); cf. Libuse Dedicova and Jiri Dolezalek, *Prum. Potravin.* **26**, 218–222 (1975).
[319] S. M. Idu, Lucia Cociumian, and S. Ciorapciu, *Acad. Rep. Popl R., Inst. Biochim., Stud. Cercet. Biochim.* **3** (1), 33–38 (1960); cf. Johannes Lorenz, Roentgenerox. Laboratoriumsprax. **12**, L62–L64 (1959); E. A. Efimova, *Lab. Delo* **9** (4), 19–21 (1963).
[320] Bennie Zak and N. Ressler, *Clin. Chem.* **4**, 40–48 (1958).
[321] Tsugio Takeuchi and Tamotsu Tanaka, *J. Chem. Soc. Jap. Ind. Chem. Sect.* **64**, 305–307 (1961).
[322] T. P. Labuza and M. Karel, *J. Food Sci.* **32**, 572–575 (1967).
[323] A. Garrido Fernandez, M. A. A.bi, and M. J. Fernandez Diez, *Graces. Aceit.* **24**, 287–292 (1973).
[324] Adam Hulanicki and Urszula Kozicka, *Chem. Anal.* (Warsaw) **18**, 723–726 (1973).

hydrochloric acid, dilute to 20 ml, and adjust the pH to 3–4. Add 1 ml of 0.25% 1,10-phenanthroline solution. Dilute to 25 ml and read at 512 nm.

Organic Samples.[325] Digest a sample containing 5–25 μg of iron with 0.1 ml of 95:5 mixture of nitric and sulfuric acids in a sealed tube at 300±20°. To the reaction product add 10 ml of water, 1 ml of 1% solution of hydroquinone, 5 ml of 2 *N* 1:9 mixture of acetic acid and sodium acetate, and 5 ml of 0.25% solution of 1,10-phenanthroline. Dilute to 25 ml, let stand for 5 hours, and read at 509 nm.

Animal and Poultry Feed.[326] Dry 5 grams of a premix or 50 grams of a complete feed for 2 hours at 130°. Stir for 5 minutes with 50 ml of carbon tetrachloride. Centrifuge, and decant the clear upper layer. Add another 50 ml of carbon tetrachloride, agitate, and filter. After evaporation of the solvent, boil the solid residue with 50 ml of 1:1 hydrochloric acid for 15 minutes. Add 0.5 gram of activated carbon, dilute to 100 ml, and filter. Complete as for wine from "To a 20-ml aliquot, add...."

Wood Pulp.[327] Ash 10 grams of pulp. Add 5 ml of 1:1 hydrochloric acid and evaporate to dryness at 100°. Again add 5 ml of 1:1 hydrochloric acid and evaporate to dryness at 100°. Take up in water, and heat at 100° for 5 minutes. Dilute to 25 ml, treating any residue with 2.5 ml of 1:1 hydrochloric acid. Add successively 1 ml of 2% solution of hydroxylamine hydrochloride, 1 ml of 1% solution of 1,10-phenanthroline hydrochloride, and 15 ml of 4 *M* sodium acetate. Dilute to 50 ml and filter or centrifuge if necessary. Read at 510 nm against a blank.

Delustered Acrylic Fibers.[328] Transfer a 5-gram sample to an 8-ounce narrow-mouth bottle. As a blank, prepare a similar amount. Add 60 ml of dimethylformamide to each and shake for 1 hour. Add 5 ml of 8% solution of hydroxylamine hydrochloride in dimethylformamide and shake for 15 minutes. Add 5 ml of 0.4% solution of 1,10-phenanthroline in dimethylformamide to the sample and 5 ml of dimethylformamide to the blank. Shake for 30 minutes. Measure the absorption by reflection at 510 and 610 nm, using a magnesium oxide reference. Subtract the absorption at 610 nm from that at 510 nm. The solutions as read are essentially opaque.

PHENOTHIAZINE DERIVATIVES

The derivatives cited in the procedure form colored oxidation products with ferric ion in acid solution. There is interference by oxalate, phosphate, fluoride, bichromate, vanadate, ceric ion, ferrous ion, and nitric acid.

[325] T. R. F. W. Fennell and J. R. Webb, *Talenta* **9**, 96–97, 795–797 (1962).
[326] A. Amati and A. Minguzzi, *Ind. Agr.* **8**, 249–257 (1970).
[327] Scandanavian Pulp, Paper and Board Testing Committee, **SCAN-C13:62**, *Norsk Skogind.* **16**, 395–396 (1962); cf. British Standards Institution, **BS 4897**, 11 pp (1973).
[328] M. E. Gibson, Jr., D. A. Hoes, J. T. Chesnutt, and R. H. Heidner, *Anal. Chem.* **32**, 639–642 (1960).

Procedure. *Cement.*[329] Heat a 0.2-gram sample with 25 ml of 1:4 hydrochloric acid. Filter, and wash the filter with hot water. Add 1 ml of 3% hydrogen peroxide to the filtrate and washings. Heat to 100°, cool, and dilute to 100 ml. Mix a 10-ml aliquot with 1 ml of 1:9 hydrochloric acid, add 4 ml of 2% solution of chlorpromazine hydrochloride, and dilute to 25 ml. Read at 525 nm.

Alternatively, use promazine hydrochloride and read at 510 nm, or use thioridazine and read at 640 nm.

N-PHENYLBENZOHYDROXAMIC ACID

The complex of ferric ion with this reagent is extractable with carbon tetrachloride, chloroform, pentyl acetate, or methyl isobutyl ketone, with the absorption greatest in carbon tetrachloride.[330] Extraction is quantitative at pH 1.8–6. Addition of hydrogen peroxide and tartrate prevents interference by tin, vanadium, molybdenum, tungsten, or titanium. Ferrous ion can also be extracted with this reagent, but the absorption is only about one-third as great.

Procedure. To a sample in dilute nitric acid solution containing less than 50 μg of iron, add sodium acetate solution to adjust to pH 2.5–3. Add 5 ml of 30% hydrogen peroxide and 1 gram of tartaric acid. Extract by shaking for 2 minutes with 10 ml of 0.2% solution of the reagent in carbon tetrachloride. Read at 440 nm.

Silicates.[331] Heat 0.5 gram of sample, 5 ml of 1:4 sulfuric acid, and 5 ml of hydrofluoric acid. Evaporate to sulfur trioxide fumes. Take up in water, add 5 ml of 1:1 hydrochloric acid, and dilute to 200 ml. Adjust a 5-ml aliquot to pH 3 with ammonium hydroxide and add 1 ml of *M* chloroacetic acid buffer solution for pH 3. Dilute to 10 ml and add 5 ml of 0.1% solution of the reagent. Extract with 5 ml of benzene for 3 minutes, dry the extract with sodium sulfate, and read at 440 nm.

3-(4-PHENYL-2-PYRIDYL)-5,6-DIPHENYL-1,2,4-TRIAZINE

This reagent is another ferroin chromogen. The 1:3 ferrous complex follows Beer's law at pH 3–8 and is soluble in ethanol, isoamyl alcohol, benzene, and nitrobenzene. Only cuprous and cobaltous ions form colored extractable complexes, and these interfere only if the concentration is considerably greater than that of iron. The cuprous complex can be significantly reduced in intensity by washing the extract with an ammoniacal buffer solution. Cyanide must be absent; excess of reagent allows for complexing noninterfering cations.

[329]Mikolaj Tarasiewicz, *Chem. Anal.* (Warsaw) **16**, 1179–1187 (1971).
[330]Fumio Aoki and Hideo Tomioka, *Jap. Anal.* **13**, 1024–1029 (1964).
[331]Hajime Ishii and Hishahiko Einaga, *ibid.* **17**, 1296–1302 (1968).

Procedure. To a sample containing 1–15 μg of iron, add 1 ml of 10% hydroxylamine hydrochloride solution, 2 ml of 0.005% solution of the captioned reagent in slightly acidified ethanol, 5 ml of a buffer solution for pH 4–5, molar with sodium acetate and molar with acetic acid, and 2 ml of 50% solution of sodium perchlorate.[332] Extract with 9 ml of isoamyl alcohol and discard the lower layer. If copper is present, wash the organic layer with a buffer solution for pH 8.5–9.2, which is molar with ammonium chloride and molar with ammonium hydroxide. If the washings are not alkaline, repeat that step. Dilute the organic layer to 10 ml with ethanol, and read at 561 nm against a blank within 30 minutes after washing with the ammoniacal buffer solution.

Seawater. Use a 100-ml sample by the general procedure.

Beer. To 10 ml of degassed sample, add 5 ml of perchloric acid and 5 ml of nitric acid. Heat gently until evolution of oxides of nitrogen ceases; then heat more strongly until reaction ceases and copious fumes of perchloric acid are evolved for 2–3 minutes. Cool, add 25 ml of water, and boil briefly to drive off chlorine. Add 2 ml of 10% solution of hydroxylamine hydrochloride, 2 ml of 0.005 *M* solution of the captioned reagent in slightly acidified ethanol, 2 ml of buffer solution for pH 4–5, molar in sodium acetate and acetic acid, and about 4 ml of ammonium hydroxide to adjust to pH 4–7. Complete as above from "Extract with 9 ml of isoamyl alcohol...."

Skim Milk. To 5 ml of sample, add 5 ml of nitric acid and 5 ml of perchloric acid. Heat below boiling. The stages of wet ashing are (1) evolution of oxides of nitrogen, (2) a quiescent stage of evaporation of water, (3) a vigorous oxidation stage for about 10 minutes, and (4) a quiescent stage with copious evolution of perchloric acid fumes. Continue the last stage for at least 10 minutes. Continue as for beer from "Cool, add 25 ml of water,...."

PHENYL-2-PYRIDYL KETOXIME

Ferrous ion forms a bluish-purple 1:2 complex with phenyl-2-pyridyl ketoxime in aqueous ethylene amine solutions. This is extractable with a mixture of ethanol and isoamyl alcohol. The complex has maxima at 588, 509, and 405 nm. The absorption conforms to Beer's law for 0.01–0.2 mg of iron in 25 ml of solution and after development by heating is stable for at least 24 hours.[333] The complex differs from the 1:3 complex formed in aqueous alkaline solution.

For iron and copper in ferrite by phenyl-2-pyridyl ketoxime, see Copper.

[332] Alfred A. Schilt and William C. Hoyle, *Anal. Chem.* **39**, 114–117 (1967); cf. Alfred A. Schilt and Paul J. Taylor, *ibid.* **42**, 220–224 (1970).
[333] H. J. Cluley and E. J. Newman, *Analyst* **88**, 3–17 (1963); cf. Fred Trusell and Harvey Diehl, *Anal. Chem.* **31**, 1978–1980 (1959).

Procedures

Ethylene Amines.[334] Dilute to 25 ml a 10-ml sample containing not more than 20 ppm of ferrous ion. Add 2 ml of 10% solution of sodium hydrosulfite and set aside for 10 minutes to reduce ferric ion. Add 5 ml of 0.2% solution of phenyl-2-pyridyl ketoxime and heat at around 90° for 10 minutes. Cool, and extract with 15 ml of 15 : 2 isoamyl alcohol–ethanol. Filter the extract, wash the paper with isoamyl alcohol, and dilute to 25 ml with that solvent. Read at 588 nm against a reagent blank.

Groundwater[335]

FERROUS ION. To 100 ml of sample, add 2 ml of 20% solution of tartaric acid, 1 ml of 1% solution of the color reagent in ethanol, and 5 ml of 10% sodium carbonate solution. Let stand for 1 minute and extract with 10 ml, 10 ml, and 10 ml of chloroform. Dilute the filtered extracts to 50 ml with chloroform and read at 570 nm.

TOTAL IRON. Repeat the determination, adding as an additional reagent 2 ml of 10% ascorbic acid solution.

PONTACHROME BLUE BLACK R

The decrease in fluorescence of the complex of aluminum with pontachrome blue black R (mordant black 17) caused by ferric ion is a measure of 0.02–0.2 μg of iron per ml.[336] Thus the sensitivity is three times that of bathophenanthroline. The determination of 1 μg of iron tolerates 10 μg of chromic or zinc ion, 3 μg of manganese, nickel, or aluminum ion, or 1 μg of titanic or zirconium ion. There is slight interference by cobaltic, cupric, and vanadic ion. Fluoride and phosphate interfere seriously.

Procedure. As a reagent, dissolve 30 mg of potash alum and add 12.5 grams of ammonium acetate. Add 1 : 1 sulfuric acid until the solution is clear. Add 37.5 ml of 0.1% ethanolic pontachrome blue black R and dilute to 100 ml.

Dilute a sample containing 0.5–10 μg of iron to 15 ml. Add 2 ml of the reagent and adjust to pH 4.8 ± 0.1. Heat nearly to boiling, cool, and dilute to 25 ml. Activate with an ultraviolet lamp and read the fluorescence at 595 nm.

[334] Robert Chernin and E. R. Simonsen, Anal. Chem. **36**, 1093–1095 (1964).
[335] T. Kempf, *Z. Anal. Chem.* **231**, 200–203 (1967).
[336] J. Block and E. Morgan, *Anal. Chem.* **34**, 1647–1650 (1962).

POTASSIUM CYANIDE—POTASSIUM FERROCYANIDE

The solubility of ferric ferrocyanide in potassium cyanide solution leads to the use of this mixed reagent. Increase of potassium cyanide lessens the absorption. Increase of pH above that selected decreases the absorption. Beer's law is followed for 5–50 μg of ferric ion in 10 ml. Pyrophosphate must not exceed tenfold. Chromate, dichromate, thiosulfate, sulfide, permanganate, cadmium, and zinc interfere. The sensitivity approximates one-tenth of that of thiocyanate. The absorption remains constant for 5 hours.

Procedure. As a color reagent, mix equal parts of 5% solution of potassium cyanide, 5% solution of potassium ferrocyanide in 2% solution of sodium carbonate, and acetic acid.[337] As a buffer solution for pH 1.42, add 200 ml of *M* sodium acetate to 240 ml of *N* hydrochloric acid and dilute to 1 liter.

Take a sample containing 5–50 μg of iron in 1% sulfuric acid. If nitrite is present, decompose by heating for 15 minutes. If thallous ion is present, oxidize it by adding 0.2 ml of nitric acid and heating. Add 0.2 ml of 5% potassium cyanide, and if a precipitate appears, indicating the presence of barium, strontium, or mercurous ion, remove it by centrifuging. If thoric or yttrium ions are present, add sodium fluoride solution containing 2 mg of fluoride ion. If zirconium is present, add sodium tartrate solution to give 2 mg of tartrate. Add 0.2 ml of the color reagent and 5 ml of buffer solution. Dilute to 10 ml and read at 710 nm.

PROLINETHYMOL BLUE

Ferrous ion complexes with prolinethymol blue to give a purple blue color. The optimum pH is 6.6–7.5 buffered with hexamine and hydrochloric acid. The rate of formation of the complex is increased by lowering pH and temperature. Beer's law is followed for 0.2–5 μg of ferrous ion in 25 ml. There is interference by citrate, borate, oxalate, phosphate, ferrocyanide, ferricyanide, and strong oxidizing or reducing agents.

Procedure. *Soil.*[338] Stir a 20-gram sample of dry soil with 50 ml of water for 3 minutes, store at 20° for 24 hours, and filter. To 10 ml of filtrate, add 0.4 ml of *N* hydrochloric acid, 3 drops of 20% solution of sodium metabisulfite, 0.2 ml of 0.1% solution of prolinethymol blue, and 5 ml of *M* hexamine. After 30 minutes add 0.6 ml of *N* hydrochloric acid, dilute to 25 ml, and read at 545 nm.

---

[337] Mohsin Qureshi and K. G. Varshney, *ibid.* **39**, 1064–1035 (1967).
[338] M. Koch and I. Kochova, *Sb. Vys. Sk. Zemed. Praze, Fac. Agron.* **16**, 167–172 (1968).

2,6-PYRIDINEDIAMIDOXIME

This reagent is particularly adapted to determination of ferrous ion in sodium, potassium, or lithium hydroxide. It is slightly soluble in water but readily soluble in acid. By the technic below, copper, nickel, and cobalt precipitate. It conforms to Beer's law for 0.4–7 ppm of ferrous ion. When developed by heating, the color is stable for 6 hours.

Procedure. *Strong Alkalies.*[339] Dissolve a sample in water to give a 6–8 *M* solution. To a 25-ml portion of this sample solution, add 5 ml of 0.1% solution of the reagent, 5 ml of a freshly prepared 5% solution of sodium dithionate, and 5 ml of 4% sodium hydroxide solution. Heat at 100° for 10 minutes, cool, dilute to 50 ml, and read at 523 nm against a blank.

1-(2-PYRIDYLAZO)-2-NAPHTHOL

This reagent, which has the trivial name of PAN, forms extractable complexes at pH 5 with the following ions: vanadyl, ferric, cobaltic, nickel, palladium, cupric, zinc, gallium, indium, and thallic. Shaking with EDTA solution at pH 5 leaves only the first five ions in the organic layer. Vanadium can be masked with hydrogen peroxide. Then the ferric complex is read at 764 nm, cobalt at 640 nm, and nickel at 565 nm. Applying corrections for the iron and nickel complexes at 640 nm and the iron and cobalt complexes at 565 nm permits reporting iron up to 80 μg, cobalt up to 80 μg, and nickel up to 25 μg in 25 ml of solution, but palladium interferes.

After standing for 5 minutes at pH 4–8, the ferric complex is extractable with benzene for reading at 775 nm.[340] For simultaneous reading of ferric and indium ions at 775 and 560 nm as developed with PAN, see Indium.

The complex of ferric ion with naphthalene can be extracted with molten naphthalene, a technic that has been described elsewhere for extraction of nickel.

Procedures

Molybdenum and Tungsten

ORES.[341] Fuse a 1-gram sample in porcelain with 5 grams of potassium pyrosulfate. Leach with water, adding not more than 10 ml of ammonium hydroxide. Adjust to pH 1 by dropwise addition of 1 : 1 sulfuric acid. Add 0.5 gram of ascorbic acid, 5 ml of an ammoniacal buffer solution for pH 10 containing 7% ammonium chloride and 57% of ammonium hydroxide, and 2 ml of 0.025% methanolic PAN.

[339]Milan W. Wehking, Ronald T. Pflaum, and E. Scott Tucker III, *Anal. Chem.* **38**, 1950–1951 (1966).
[340]S. Shibata, *Anal. Chim. Acta* **25**, 348–359 (1961).
[341]R. Püschel, E. Lassner, and K. Katzenbruber, *Z. Anal. Chem.* **223**, 414–426 (1966); *Chemist-Analyst* **56**, 63–65 (1967).

Heat to 60° and cool. Extract with 5 ml of chloroform, and further extract with 3-ml portions until the extract is colorless. Mix the extracts with 5 ml of ethanol and dilute to 25 ml with chloroform. Read at 630 nm against a reagent blank.

POWDERS. Heat 1 gram of sample in porcelain at 500°, then flux with 5 grams of potassium pyrosulfate. Proceed as above from "Leach with water,...."

COMPACT METALS. Mix 1 gram of sample with 5 ml of hydrofluoric acid in platinum; while warming, add nitric acid dropwise until dissolution is complete. Evaporate to dryness and heat at 400–500° to complete expulsion of hydrofluoric acid. Fuse with 5 grams of potassium pyrosulfate. Complete as above from "Leach with water...."

High-Purity Yttrium Oxide.[342] Dissolve 0.2 gram of sample containing more than 1 μg of iron in 5 ml of 1:1 hydrochloric acid and evaporate nearly to dryness. Add 10 ml of water, 4 ml of 0.2% solution of the reagent in methanol, and 10 ml of acetate buffer solution for pH 4. Set aside for 15 minutes, then extract with 10 ml of chloroform. Read the organic layer at 770 nm for up to 3 ppm of iron.

Beer.[343] IRON, COBALT, AND NICKEL. Evaporate a 250-ml sample and ash. Take up the residue in 1:1 hydrochloric acid and dilute to 100 ml with that acid. To a 25-ml aliquot, add 2 ml of 0.25% ethanolic PAN. Neutralize with ammonium hydroxide and add 5 ml of acetate buffer solution for pH 5. Heat to 80° and cool. Extract the turbid solution with 20 ml of chloroform. Wash the extract with 10 ml and 10 ml of 0.1 *M* EDTA. Add 4 ml of ethanol to the washed extract and dilute to 25 ml with chloroform. Read at 764, 640, and 565 nm. Solve by simultaneous equations.

4-(2-PYRIDYLAZO)RESORCINOL

This reagent is familiarly known as PAR. A 1:3 complex is formed by ferrous ion at 1 ml of 0.05% reagent solution per 50 ml in ammoniacal buffer solution for pH 8.8–10.3.[344] Beer's law is followed up to 40 μg in 50 ml; the molecular extinction coefficient is 5 times that of the 1,10-phenanthroline complex. Most interfering ions except cobalt and nickel are masked by adding ascorbic acid and thiourea to the sample solution and EDTA and potassium cyanide after formation of the complex.

The presence of nitrilotriacetate causes the absorption to be constant over the range pH 8–9.3 and also masks some interfering ions.[345] Cyanide masks cobalt and nickel. Interference that cannot be masked by the reagents mentioned is caused by copper, gallium, titanium, zirconium, hafnium, pentavalent vanadium, tantalum,

[342] Shozo Shibata, Kazuo Goto, and Ryozo Nakashima, *Anal. Chim. Acta* **46**, 146–148 (1969).
[343] T. Vondenhof and H. Beindorf, *Mschr. Brau.* **21**, 156–157 (1968).
[344] Takao Yotsuyanagi, Katsumo Goto, and Masaichi Nagayama, *Jap. Anal.* **18**, 184–188 (1969).
[345] Tsugio Takeuchi and Yoshio Hijo, *Jap. Anal.* **14**, 903–934 (1965).

bismuth, chromium, hexavalent uranium, stannic ion, and gold. At 536 nm Beer's law is followed up to 40 μg of iron per 50 ml of solution containing 1 ml of 0.1 *M* nitrilotriacetate, 5 ml of 0.05 *M* sodium tetraborate, and 1 ml of 0.1% solution of PAR at pH 9.

In the presence of benzyldimethyltetradecylammonium chloride, a ternary complex is extracted with chloroform for reading at 522 nm. The appropriate buffer for pH 10 is monosodium borate–sodium hydroxide.

For iron in serum, precipitate protein with trichloroacetic acid.[346] Reduce iron in the supernatant layer with ascorbic acid. Adjust to pH 10 with ammonium hydroxide, develop with PAR, and read at 500 nm. For iron-binding capacity, treat the sample with ferric ammonium citrate solution, ion exchange resin, and barbitone buffer solution. Determine the residual iron as above.

Procedure. Dilute a sample solution containing up to 20 μg of iron to 50 ml and add 10 mg of sodium ascorbate to reduce ferric ion.[347] Add 3 ml of 0.1% solution of PAR and adjust to pH 10 with 2 ml of 0.05 *M* borax and 2 ml of 0.4% solution of sodium hydroxide. After 2 minutes add 1.5 ml of 0.05 *M* EDTA and boil for 30 minutes. Cool, dilute to 50 ml, and read at 500 nm.

5-PYRIDYLBENZODIAZEPIN-2-ONE

This reagent is 7-bromo-1-(3-dimethylaminopropyl)-1, 2-dihydro-5-(2-pyridyl)-2*H*-1,4-benzodiazepin-2-one dihydrochloride. The use for serum iron and iron-binding capacity has been adapted to automatic operation.[348]

Procedures

Serum Iron.[349] Mix 2 ml of serum with 1 ml of 0.02% solution of ascorbic acid in 1 : 5 hydrochloric acid solution. Mix intermittently for 10 minutes, then add 3 ml of 10% trichloroacetic acid solution. Centrifuge, and take 4 ml of the clear supernatant layer. Add 2 ml of 1% solution of the captioned reagent in 25% ammonium acetate solution. Set aside for 5 minutes and read at 580 nm against a reagent blank.

Iron-Binding Capacity. Mix 1 ml of serum with 3 ml of a solution containing 3 μg of iron per ml. Mix intermittently for 10 minutes, then add 160 mg of magnesium carbonate for absorption. Mix at intervals for 10 minutes; centrifuge,

[346]Daniel C. Paschal, Richard J. Carter, Nancy Selfridge, and Deborah Thomas, *Anal. Lett.* **8**, 741–751 (1975).
[347]Takao Yotsuyanagi, Ryuji Yamashita, and Kazuo Aomura, *Anal. Chem.* **44**, 1091–1096 (1972); *Jap. Anal.* **19**, 981–982 (1970); cf. Ryuji Yamashita, Takao Yotsuyanagi, and Kazuo Aomura, *ibid.* **20**, 1282–1288 (1971); D. Ch. Nonova and B. E. Evtimova, *C. R. Acad. Bulg. Sci.* **26**, 791–794 (1973).
[348]Bernard Klein, Norman Kleinman, and Ronald L. Searcy, *Clin. Chem.* **16**, 495–499 (1970).
[349]Bernard Klein, Lois B. Lucas, and Ronald L. Searcy, *Clin. Chim. Acta* **26**, 517–523 (1969).

and take 2 ml of the supernatant layer. Complete as above from "Add 2 ml of 1% solution...."

3-(2-PYRIDYL)-5,6-DIPHENYL-1,2,4-TRIAZINE

This ferroin reagent forms a ternary complex with ferrous ion and thiocyanate which is appropriate for extraction with chloroform from acid solutions. The applicability of this reagent in strong acids differentiates it from other ferroins. The method is appropriate for 1–20 μg of iron and follows Beer's law over that range. With large samples it is adaptable to 2 μg/liter. The following ions are tolerated at the indicated ppm: chromium, 10; cobalt, 2; copper, 50; manganese, 50; nickel, 2; oxalate, 50; vanadium, 2. All others are tolerated at 100 ppm.

For determination of iron and copper by complexing with 3-(2-pyridyl)-5,6-diphenyl-1,2,4-triazine, see Copper.

Procedure. *Acids.*[350] For acids, the concentration in the sample for extraction should be as follows: sulfuric, 1–4 *M*; perchloric, 1–4 *M*; hydrochloric, 1–3 *M*; nitric, 1–2 *M*; acetic, *M* to beyond 4 *M*; phosphoric, *M* to beyond 4 *M*. Take an appropriate sample for ultimate dilution to 100 ml. Dilute to 80 ml and cool if necessary. Add 10 ml of a solution containing 10% of ascorbic acid and 41% of sodium thiocyanate and dilute to 100 ml. Mix an aliquot containing 2–20 μg of iron with 10 ml of 0.25% solution of the color reagent in chloroform. Shake for 40 seconds and read the organic layer at 555 nm against chloroform. With iron preextracted from the reagents, a blank is unnecessary.

2-PYRIDYL-2′-HYDROXYMETHANE SULFONIC ACID

The 2:1 complex of this reagent with ferrous ion is read at pH 10 buffered to within a half-unit. Beer's law is followed up to 4 ppm. Copper and cobalt interfere mildly.

Procedure. As color reagent, dissolve 0.2 gram of the captioned compound in 1 ml of 1:3 ammonium hydroxide and dilute to 100 ml. As buffer solution for pH 10, dissolve 70 grams of ammonium chloride in water, add 570 ml of ammonium hydroxide, and dilute to 1 liter.

Mix a sample solution containing 2.5–100 μg of iron with 5 ml of 5% solution of hydroxylamine hydrochloride, 5 ml of color reagent, and 5 ml of buffer solution for pH 10. Dilute to 25 ml and read at 525 nm against a reagent blank.

---

[350] C. D. Chriswell and A. A. Schilt, *Anal. Chem.* **46**, 992–996 (1974); cf. A. Schilt and Paul J. Taylor, *ibid.* **42**, 220–224 (1970).

PYRIDIL β-MONOXIME

Pyridil β-monoxime complexes with ferrous ion. Beryllium, titanic, and tetravalent tellurium ions must be absent. Trivalent and hexavalent chromium, cuprous and cupric ions, and nickel can be masked.

Procedure. As reagent solution, dissolve 5 grams of pyridil β-monoxime in 200 ml of dimethyl sulfoxide. Add 100 ml of ethanol and dilute to 500 ml with 2:1 pyridine-ethanol.

To a sample solution containing 8–96 μg of iron, add 1 ml of 10% solution of hydroxylamine hydrochloride and adjust to pH 6.1.[351a] Add 5 ml of 2:1 pyridine-ethanol and 5 ml of the reagent solution. Dilute to 40 ml and heat at 100° for 30 minutes. Cool, and extract with 10 ml, 7 ml, and 7 ml of chloroform. Dilute the combined extracts to 25 ml with chloroform and read at 558 nm.

PYROPHOSPHATE

At pH 7–8 ferric pyrophosphate shows a maximum absorption at 270–285 nm. The pyrophosphate ion exhibits no absorption. The technic is appropriate for determining 0.15–14 μg of iron.

Procedures

Phosphoric Acid.[351b] Dilute 3 grams of sample to 50 ml and take a 2-ml aliquot. Add 5 ml of 10% tetrasodium pyrophosphate solution and titrate to pH 7.5–8 with 0.1 *N* ammonium hydroxide. To another 2-ml aliquot and 5 ml of 10% tetrasodium pyrophosphate solution, add the amount of 0.1 *N* ammonium hydroxide called for by the titration. Add 25 ml of an acetate buffer solution for pH 7.5–8 and 10 ml of 12% solution of sodium perchlorate. Dilute to 100 ml and read at 285 nm against a reagent blank.

Pyrophosphoric Acid. Approximately neutralize an appropriate sample with ammonium hydroxide and dilute to 50 ml. Titrate a 2-ml aliquot with 0.1 *N* ammonium hydroxide. To another 2-ml aliquot, add the amount of ammonium hydroxide called for by the titration. Complete as above from "Add 25 ml of an acetate buffer...."

Acid Sodium Pyrophosphate. Render an appropriate sample slightly acid with acetic acid and dilute to 50 ml. Complete as for pyrophosphoric acid from "Titrate a 2-ml aliquot...."

[351a]H. R. Notenboom, W. J. Holland, and D. Soules, *Mikrochim. Acta* **1973**, 187–192.
[351b]T. N. Vladimirskaya, M. L. Chepelevetskii, and A. A. Levshina, *Zavod. Lab.* **34**, 1287–1288 (1968).

RHODAMINE B

Rhodamine B (basic violet 10) forms a 1 : 1 complex with ferric ion.[352] To the ferric solution, add rhodamine B and 2-butoxyethanol and adjust to 5 *N* with hydrochloric acid. Extract with 10 : 1 benzene–methyl isobutyl ketone and read the extract at 550 or 562 nm against a blank. Beer's law is followed up to 1.4 μg of iron per ml of extract.

Procedure. *Petroleum Industry Water.*[353] Evaporate 10 ml of filtered sample to dryness and ignite to destroy organic matter. Dissolve in 5 ml of 1 : 1 hydrochloric acid. Oxidize the iron with 5 drops of 0.2% solution of ammonium persulfate. Add 1 ml of 0.5% solution of rhodamine B in 1 : 1 hydrochloric acid, and extract with 5 ml of benzene. Read at 550 or 562 nm against a blank.

SALICYLIC ACID

At pH 0.7–2.5 the complex of ferric ion with salicylic acid has a maximum absorption at 530 nm. With rise in pH above those levels, the maximum shifts to lower levels. As applied to plant materials, it will determine 10–1000 ppm of iron.[354]

Procedures

Ferric Ion, Ferrous Ion Present.[355] Adjust the sample solution to pH 2.8 with an amount of chloroacetic acid–sodium hydroxide buffer solution to give a final chloroacetate concentration of 0.5 *M*. Alternatively, adjust to pH 3.5 with *M* acetic acid–sodium acetate buffer solution to give a final acetate concentration of 0.05 *M*. Add ethanolic salicylic acid to give a final concentration of 0.036 *M*. Dilute to the predetermined volume and set aside for 2 hours. At pH 2.8 read at 515 nm or at pH 3.5 read at 490 nm. At pH 2.8 or 3.5, up to 10 ppm of ferric ion can be read in the presence of 100 ppm of ferrous ion. Ferrous ion interferes at higher pH levels.

Tungsten Alloy.[356a] IRON AND TITANIUM. Dissolve 0.2 gram of sample in aqua regia. Add 0.2 ml of sulfuric acid and evaporate to a moist residue. Add 20 ml of 1 : 1 hydrochloric acid and boil for 5 minutes. Add 50 ml of water, cool, and filter tungstic acid. Wash with 1 : 1 hydrochloric acid, then with 2% ammonium nitrate solution, and dilute to 250 ml. Mix a 10-ml aliquot, 10 ml of acetate buffer solution

[352] Hiromu Imai and Tadahiko Yamada, *Bull. Chem. Soc. Jap.* **42**, 232–240 (1969).
[353] I. L. Bagbanly and T. R. Mirzoeva, *Azerb. Neft. Khoz.* **1967** (3), 43–44.
[354] Baldev Kapoor and C. Dakshinamurti, *J. Sci. Ind. Res.* (India) **18B**, 89–90 (1959).
[355] Kin'ya Ogawa and Nobuko Tobe, *Bull. Chem. Soc. Jap.* **39**, 227–232 (1966).
[356a] A. A. Bulatova, V. B. Aleskovskii, and M. I. Bulatov, *Izv. Vyssh. Ucheb. Azved., Khim. Khim. Tekhnol.* **17**, 345–347 (1974).

for pH 4, 2 ml of saturated solution of salicylic acid, and sufficient *N* sodium hydroxide to adjust to pH 4. Dilute to 100 ml and read iron at 571 nm. Then read iron and titanium at 360 nm and subtract a correction for iron to give the titanium content.

SALICYLOHYDROXAMIC ACID

The complex of this reagent obeys Beer's law for 0.01–10 ppm of ferric ion. It tolerates 10 ppm of fluoride ion.

Procedure. *China Clay.*[356b] Fuse 1 gram of powdered sample with 5 grams of potassium bisulfate in borosilicate. Leach the melt with dilute hydrochloric acid. Filter, wash, and dilute to 50 ml. Mix a 10-ml aliquot with 5 ml of sodium acetate–acetic acid buffer solution for pH 5. Extract with 10 ml of 0.01 *M* salicylohydroxamic acid in isoamyl alcohol. Dry the organic layer with sodium sulfate and read at 450 nm.

SULFIDE

When iron is converted to ferrous sulfide, the blue-green colloidal solution will pass a 0.22-nm Millipore filter and can be read at 0.04–1 ppm in water or 5–50 ppm in a solution of a sample of aluminum.

Procedures

Water.[357] Add 2 ml of 0.5% solution of hydrazine sulfate to a 75-ml sample. Neutralize with 0.5% solution of sodium hydroxide and add 5 ml of 0.3 *M* sodium sulfide. Dilute to 100 ml, incubate at 30° for 45 minutes, and read at 430 nm against water.

Aluminum. Dissolve 1 gram of sample in 10 ml of 20% sodium hydroxide solution. Dilute to 50 ml and add 2 ml of 0.5% solution of hydrazine sulfate. Add 10 ml of 0.3 *M* sodium sulfide solution and dilute to 100 ml. Incubate at 30° for 60 minutes and read at 430 nm against water. The solution used in preparation of the calibration curve should contain the amount of aluminum in the sample solution.

[356b]Chanchal Kumar Pal and Amiya Kumar Chakraburtty, *Technology* (Sindri) **11**, 392–394 (1974).
[357]Tsutomo Fukasawa, Masaaki Iwatsuki, and Hiraku Asakawa, *Jap. Anal.* **20**, 444–450 (1971).

SULFOSALICYLIC ACID

At pH 1–2.5 the 1 : 1 complex of ferric ion and sulfosalicylic acid has a maximum absorption at 510 nm. With an increase in pH, the maximum is shifted toward shorter wavelengths. Three complexes are formed.[358] The absorption is very temperature sensitive. In a phosphate-free medium a tenfold excess of sulfosalicylic acid is sufficient, but in the presence of 60–300 m*M* of phosphate ion, at least fiftyfold excess of reagent is required.[359] At pH 4.6–5.5 the red 2 : 1 sulfosalicylate: ferric ion complex has a maximum at 468 nm.[360]

For analysis of organometallic compounds, decompose 10 mg by fusion with sodium peroxide in a bomb.[361] Dissolve in 15 ml of water and add 17 ml of 10 *N* sulfuric acid. Dilute to 100 ml and develop iron in an aliquot with sulfosalicylic acid.

For iron in vegetable oil, extract a 130-gram sample in a liquid-liquid extractor with azeotropic (21%) hydrochloric acid for 12 hours.[362] Neutralize the acid extract to Congo red with ammonium hydroxide. Develop iron in an aliquot with sulfosalicylic acid.

For iron in aluminum chloride or aluminum nitrate, reading as the thiocyanate at 490 nm is as satisfactory as reading the sulfosalicylate at 436 or 465 nm.[363] For iron in soil, adjust the sample to pH 1.3 with tropaeolin OO as indicator and develop with sulfosalicylic acid.[364]

Other applications of the reagent are to the less than 10% iron in catalysts used in the manufacture of acrylonitrile,[365] to aluminosilicates,[366] and to cement.[367]

For iron in rhodium plating electrolyte as separated by column chromatography for determination by sulfosalicylic acid, see Copper. For separation of iron from manganese ores, ferromanganese, and manganese metal by column chromatography for determination by sulfosalicylic acid, see Lead.

Procedures

Ferric Ion, Ferrous Ion Present.[368] Adjust the sample to pH 3.4 or 4.5. Use *M* acetic acid–sodium acetate buffer solution to give a 0.5 *M* acetate concentration when later diluted to 100 ml. Add sulfosalicylic acid solution to give a final concentration of 0.016 *M*. Dilute to 100 ml and read within 3 hours for pH 3.4 at 480 nm, or for pH 4.5 at 470 nm. This will read 2–10 ppm of ferric ion in the presence of not more than 100 ppm of ferrous ion.

[358] M. I. Vakhtel' and M. M. Chebotareva, *Zavod. Lab.* **34**, 160 (1968).
[359] M. P. Strukova, V. N. Kotova, and L. M. Gulyaeva, *Zh. Anal. Khim.* **24**, 565–569 (1969).
[360] S. I. Smyshyaev, *Izv. Vyssh. Ucheb. Zaved., Khim. Khim. Tekhnol.* **12**, 384–387 (1969).
[361] N. A. Gradskova, E. A. Bondarevskaya, and A. P. Terent'ev, *Zh. Anal. Khim.* **28**, 1846–1848 (1973).
[362] A. D. Popov, B. S. Ivanova, and N. V. Yanishlieva, *Nahrung* **13**, 39–42 (1969).
[363] F. S. From, N. P. Plovinkina, and S. V. Sheshulina, *Tr. Khim. Khim. Tekhnol.* **2**, 574–578 (1959).
[364] Yu. I. Dobritskaya, *Pochvovedenie* **1962** (9), 88–96.
[365] A. S. Kuznetsova, *Zavod. Lab.* **34**, 410 (1968).
[366] J. Debras-Guedon, *Ind. Ceram.* **1963**, 345–352.
[367] I. A. Voinovitch, R. Barabas, G. Cohort, G. Koelbel, G. LeGrand, and J. Louvrier, *Chim. Anal.* (Paris) **50**, 334–349 (1968); Maria Wallraf, *Zement-Kalk-Gips* **9**, 186–194 (1956).
[368] Kin'ya Ogawa and Nobuko Tobe, *Bull. Chem. Soc. Jap.* **39**, 223–227 (1966).

Extraction with Mono- and Di-2-ethylhexyl Phosphates.[369] To prepare the extractant, stir 60 grams of phosphorus pentoxide with 50 ml of 2-ethylhexanol and 50 ml of octane at no higher than 80°. Add 50 ml of 9 *N* hydrochloric acid and stir at 80–85° for 3 hours. Cool and wash the organic phase with 50 ml of 0.2 *M* potassium iodide in 10 *N* hydrochloric acid, then with 50 ml of 5 *N* hydrochloric acid. Dilute the organic phase with octane to give a *N* solution.

Shake 10 ml of sample solution containing not more than 1 μg of ferric oxide in *N*–2 *N* hydrochloric or sulfuric acid with 10 ml of the extractant. Wash the extract with 10 ml and 10 ml of 1:6 hydrochloric acid. Extract the iron from the organic phase with 6 ml of freshly prepared 0.2 *M* potassium iodide in 7 *N* hydrochloric acid by shaking for 10 minutes. Add 2 ml of 25% solution of sulfosalicylic acid to the extract and cool. Add 7 ml of 12 *N* ammonium hydroxide and dilute to 25 ml. Set aside for 20 minutes, then read at 413 nm.

Aluminum.[370] Dissolve 2 grams of sample in 100 ml of 5% sodium hydroxide solution. Add 10 ml of 10% solution of sodium sulfide, boil, and filter. Dissolve the precipitate in 20 ml of 1:1 nitric acid and dilute to 100 ml. To a 5-ml aliquot, add sequentially 20 ml of 12% solution of sulfosalicylic acid and 10 ml of 25% solution of sodium metabisulfite. Dilute to 100 ml and read at 430 nm.

Copper Alloys.[371] For less than 4% iron, dissolve a 1-gram sample in 10 ml of nitric acid. Boil off oxides of nitrogen and dilute to 50 ml with hot water. Add 2 grams of ammonium sulfate, followed by 10 ml of saturated sodium bisulfite solution to precipitate lead, copper, and tin. Add 20 ml of 20% ammonium thiocyanate solution, dilute to 200 ml, and filter. To 25 ml of filtrate add 10 ml of 8% sulfosalicylic acid solution and 10 ml of ammonium hydroxide. Dilute to 100 ml and read at 415 nm.

Red Brass and Bronze.[372] Dissolve 1 gram of sample in 25 ml of 1:1 nitric acid and 5 ml of 10% sodium chloride solution. Dilute to 100 ml and take a 5-ml aliquot. Add 20 ml of 10% sulfosalicylic acid solution, 20 ml of 40% ammonium acetate solution, and 10 ml of 20% sodium thiosulfate solution. Dilute to 100 ml and read at 415 nm.

Ferromolybdenum.[373] IRON AND MOLYBDENUM. Dissolve 0.1 gram of sample in 30 ml of 1:2 nitric acid and evaporate to dryness. Add 50 ml of water, boil for 2 hours, and evaporate to dryness. Take up the residue in 10 ml of water and add 5 ml of 30% solution of sulfosalicylic acid. Add ammonium hydroxide dropwise until the initial precipitate just dissolves. Adjust to pH 2 with hydrochloric acid. Pass the solution at 1–2 ml/minute through a column, 1 cm in diameter, of Vionit CS-3 in hydrogen form. Wash the column with 15 ml of water and complete the elution of

[369] Yu. B. Kletenik and A. I. Bykhovskaya, *Zavod. Lab.* **29**, 1306–1307 (1963).

[370] A. Bradvarov and M. Dimitrova, *Mashinostroene* (Sofia) **9** (9), 32–33 (1960); cf. C. M. Dozinel, *Metall* **20**, 242–245 (1966).

[371] M. Hoffman, *Chem. Anal.* (Warsaw) **9**, 495–500 (1964).

[372] H. Wiedmann, *Metall* **12**, 1005–1007 (1958); cf. I. I. Kalinichenko and N. S. Zakharova, *Metody Anal. Chern. Tsvet. Met.* **1953**, 75–78.

[373] G. Popa, V. Dumitrescu, and P. Toma, *Rev. Chim.* (Bucharest) **25**, 239–240 (1974).

molybdenum with 35 ml of 0.75 *N* hydrochloric acid. Adjust the the eluate to pH 5–6 and add 1 ml of 2% solution of 4-[2-(2-hydroxy-2-isopropylamino)ethyl] catechol. Dilute to 100 ml and read molybdenum at 400 nm.

Elute iron from the column with 75 ml of 1:2 hydrochloric acid. Adjust to pH 5.5. Add 10 ml of 10% sulfosalicylic acid solution and 10 ml of 40% ammonium acetate solution. Dilute to 100 ml with 20% sodium thiosulfate solution and read at 415 nm.

Silver Chloride.[374] Dissolve 1 gram of sample in 15 ml of 1:3 ammonium hydroxide and dilute to 50 ml. Add 5 ml of 20% solution of calcium chloride and 5 ml of 20% solution of ammonium carbonate in 1:9 ammonium hydroxide. After 30 minutes filter through a No. 3 sintered-glass funnel. Wash with 1:20 ammonium hydroxide until silver-free. Dissolve the precipitate in 5 ml of 1:4 hydrochloric acid and add 2 ml of 10% solution of sulfosalicylic acid. Neutralize with 1:1 ammonium hydroxide and add 5 ml in excess. Dilute to 25 ml and read at 430 nm.

Iron Tungstide in Iron Alloys.[375] Dissolve the sample by electrolysis for 2 hours at 0.025 ampere/cm^2 in an electrolyte containing 300 grams of potassium chloride, 50 grams of triammonium citrate, and 200 ml of hydrochloric acid per liter. Filter the anodic precipitate and wash with 0.3% ammonium sulfate in 3% citric acid solution to remove ferrous ion. Then wash thoroughly with 10% sodium hydroxide solution and finally with water. Ignite at under 800° and fuse with potassium persulfate. Dissolve the melt in 50 ml of water and add 10% sodium hydroxide to precipitate ferric hydroxide and excess to form sodium tungstate. Dilute to 200 ml and filter. The filtrate is appropriate for reduction by titanous chloride and determination of tungsten by thiocyanate.

Wash the ferric hydroxide precipitate with 1:20 ammonium hydroxide, dissolve in 25 ml of hot 1:4 hydrochloric acid, and dilute to 100 ml. To a 10-ml aliquot add 2 ml of 10% solution of sulfosalicylic acid. Neutralize with 1:1 ammonium hydroxide and add 5 ml in excess. Dilute to 25 ml and read at 430 nm.

Iron-Niobium-Carbon Alloys. Dissolve sample material by electrolysis as for iron tungstide. Separate the precipitate by centrifuging and wash free of ferrous ion. Transfer to a platinum dish, add 30 ml of hydrofluoric acid, and evaporate almost to dryness. Add 30 ml of hydrofluoric acid and again take almost to dryness. Dilute to 100 ml and filter the niobium carbide and nitride on a platinum funnel, collecting the filtrate containing the iron niobide in a platinum dish. Wash the contents of the funnel with hot water. Evaporate the filtrate to a small volume, add 5 ml of sulfuric acid, add a few drops of nitric acid, and evaporate to sulfur trioxide fumes. Add 5 ml of water and again evaporate to sulfur trioxide fumes.

Take up in 50 ml of water, adding 5 grams of anhydrous sodium sulfite and some paper pulp. Boil for 5 minutes to hydrolyze the niobium. Digest at 60° for 30 minutes, filter, and wash with 2% hydrochloric acid. Dilute the filtrate to 100 ml. To a 10-ml aliquot, add 2 ml of 10% solution of sulfosalicylic acid. Neutralize with

[374] I. G. Shafran, L. Ya. Mazo, and T. N. Karskaya, *Sb. Statei, Vses. Nauchn.-Issled. Inst. Khim. Reakt. Osobo Chist. Khim. Veshchestv.* **1961** (24), 255–257.
[375] Yu. A. Klyachko and E. F. Yakovleva, *Sb. Tr. Tsent. Nauch-Issled. Chern. Met.* **1962** (24), 30–38.

1:1 ammonium hydroxide and add 5 ml in excess. Dilute to 25 ml and read at 430 nm.

Manganese Alloys.[376] Heat a 0.06-gram sample with 3 ml of 1:1 nitric acid. Using a platinum dish as cathode and a platinum plate as anode, electrolyze at 2.5 volts and 0.5 ampere. The deposit on the anode will be lead dioxide and manganese dioxide. Copper will be plated on the dish. Dissolve the deposit from the anode with 10 ml of 1:1 nitric acid and a few drops of 0.45% solution of oxalic acid. Combine this with the residual solution from the dish and dilute to 100 ml. To a 10-ml aliquot, add 2 ml of 10% solution of sulfosalicylic acid. Neutralize with 1:1 ammonium hydroxide and add 5 ml in excess. Dilute to 25 ml and read at 430 nm.

Chromite.[377] Fuse 0.2 gram of sample with 3 grams of 5:5:10:2 sodium carbonate–potassium carbonate–magnesium oxide–sodium tetraborate. Dissolve in 90 ml of hot 1:5 sulfuric acid and dilute to 250 ml.

Dilute an aliquot containing 0.1–0.8 mg of iron to 50 ml. Add 5 ml of 10% solution of sulfosalicylic acid, 5 ml of 10% solution of hydroxylamine hydrochloride, and 0.5 ml of 5% solution of potassium cyanide. Add ammonium hydroxide until the solution turns to yellow and 1 ml in excess. Dilute to 100 ml and read at 430 nm against another sample similarly treated but with the reagent replaced by 5 ml of water.

Sulfuric Acid.[378] Dilute a 2.5-ml sample to 500 ml. To a 50-ml aliquot, add 1 ml of hydrochloric acid and 1 ml of 3% hydrogen peroxide. Boil for 10 minutes and dilute to 100 ml. To a 50-ml aliquot, add 5 ml of 30% solution of sulfosalicylic acid. Add 1:9 ammonium hydroxide to pH 8.5–11, which will be shown by a yellow color. Read at 430 nm against the solution previously boiled and diluted to 100 ml made up as a blank without the color reagent.

Nitric Acid.[379] Partially neutralize a 10-gram sample with 1:3 ammonium hydroxide. When approaching neutrality, add a couple of drops of 0.1% ethanolic hexamethoxytriphenylcarbinol as indicator, and complete the neutralization. Dilute to 20 ml and add 2 ml of 10% solution of sulfosalicylic acid in 1:9 hydrochloric acid. Set aside for 5 minutes, add 2.5 ml of ammonium hydroxide, adjust to 25 ml, and read at 468 nm.

Phosphoric Acid and Phosphorites.[380] Dissolve a 2.5-gram sample in 70 ml of 1:1 hydrochloric acid, dilute to 250 ml, and filter. To an aliquot of the filtrate containing 0.2–1.3 mg of ferric oxide, add 25 ml of 20% solution of sulfosalicylic acid. Neutralize with 1:3 ammonium hydroxide to a yellow color and add 0.5 ml in excess. Dilute to 100 ml and read at 468 nm. By interposing a 1-mm cell filled with

[376] I. V. Aronina, *Izv. Akad. Nauk Mold. SSR* **1967** (9), 25–29.
[377] Z. L. Nazarenko, *Sb. Nauch. Tr. Ukr. Nauchn.-Issled. Inst. Ogneuporn.* **1965** [8(55)], 307–314.
[378] T. M. Cheremichkina, M. S. Finker, and G. N. Bezdvernyi, *Koks Khim.* **1967** (12), 43–49.
[379] I. G. Shafran, L. Ya. Mazo, and T. N. Karskaya, *Sb. Statei, Vses. Nauchn.-Issled. Inst. Khim. Reakt. Osobo Chis. Khim. Veshchestv.* **1961** (24), 189–191.
[380] E. P. Panteleeva and I. N. Krupina, *Zavod. Lab.* **38**, 1323 (1972).

0.1 N potassium permanganate, the upper limit of reading of 0.6 mg of ferric oxide per 100 ml is raised to 1.3 mg per 100 ml.

Metallic Iron in Iron Oxide, Ferrosilicon, and Silicide Alloys.[381] Reflux 100 mg of ferric oxide or ferrosilicon or 200 mg of an alloy containing silicides at 60° for 1 hour. The medium is 20 ml of methanol containing 0.5 gram of sodium salicylate and 0.5 gram of mercuric chloride. Cool, dilute with water, and filter. To an appropriate aliquot of the filtrate, add 1 drop of 1 : 1 hydrochloric acid and 3 ml of 5% solution of sulfosalicylic acid. Neutralize with ammonium hydroxide and add 5 ml of an acetate buffer solution for pH 5.2. Dilute to 25 ml and read at 450 nm.

Chlorobenzene, m-Chloroaniline, p-Chloroaniline.[382] Dissolve a 2-gram sample in 7 ml of isopropanol. Add 5 ml of 10% isopropanolic sulfosalicylic acid. Add ammonium hydroxide dropwise until the solution turns yellow and then add 5 ml in excess. Heat until fine bubbles cease and read at 430 nm.

Dyes.[383] Dissolve 1 gram of sample in 25 ml of hydrochloric acid by heating at 80°. Cool to room temperature, add 20 ml more of hydrochloric acid, and dilute to 50 ml with water. Let stand for 30 minutes and filter through a double paper. To a 10-ml aliquot of the filtrate, add 10 ml of 20% solution of sulfosalicylic acid. Add ammonium hydroxide until the color of the solution ceases to change, dilute to 100 ml, and read at 468 nm against a sample blank.

Soap.[384] Reflux 100 grams of grated sample with 100 ml of water and 12 ml of sulfuric acid for 1 hour. Cool and filter. Wash the fatty acids with four 20-ml portions of 5% sulfuric acid and dilute the filtrate and washings to 200 ml. Neutralize a 10-ml aliquot to Congo red with ammonium hydroxide and add 5 ml of 10% sulfosalicylic acid solution. Add 2 ml of 1 : 5 hydrochloric acid, then 5 ml of 2 : 3 ammonium hydroxide. Dilute to 25 ml and read at 425 nm against a blank.

THIOCYANATE

The color of ferric thiocyanate in this classical method suffers from fading on exposure to light. Addition of methyl ethyl ketone and acetone avoids that defect and increases the sensitivity of the method. Various extractable ternary complexes are formed. Details of those with 1,10-phenanthroline and thiocyanate, with thiocyanate and tributylammonium ion, or with 2-furyldiantipyrinylmethane follow in some detail. One is formed with 0.5 M bis(2-ethylhexyl)hydrogen phosphate in cyclohexane,[385] another with dibenzyl sulfoxide in dichloromethane.[386a]

[381] A. A. Tumanov and L. P. Zolotova, *ibid.* **36**, 276–277 (1970); cf. N. S. Tkachenko, P. I. Davidenko, and A. V. Dobrzkanskii, *ibid.* **29**, 536–538 (1963).
[382] E. A. Arkhangel'skaya and I. P. Krasnova, *Tr. Khim. Khim. Tekhnol.* (Gor'kii) **1966** [2(16)], 298–300.
[383] L. M. Golomb and V. V. Karpov, *Khim. Volokna* **1960** (6), 55–56.
[384] A. Popov and B. Ivanova, *Zavod. Lab.* **35**, 161–162 (1969).
[385] E. Cerrai and G. Ghersini, *Analyst* **91**, 662–664 (1966).
[386a] Max Ziegler and Georges Stephan, *Mikrochim. Acta* **1970**, 1270–1276.

A solution of diantipyrinylmethane in chloroform or dichloroethane extracts red ferric thiocyanate from a dilute hydrochloric acid solution. Extraction as the ferric thiocyanate permits determination as that complex when phosphate, sulfate, oxalate, and acetate would adversely affect the sensitivity of the determination.[386b]

In extraction of ferric ion from chloride solution with bis(2-ethylhexyl)phosphate and development by adding potassium thiocyanate to the organic layer, interference occurs with aluminum, indium, beryllium, thorium, tungsten, titanium, zirconium, hafnium, the rare earths, acetate, fluoride, oxalate, and large amounts of chloride.[387] Extraction may also be from 0.1 *N* nitrate, sulfate, or perchlorate. By a 2 : 1 ratio of aqueous phase to organic phase, the method will determine down to 0.05 μg/ml in the aqueous phase.

Ferric thiocyanate is extractable by ethyl acetate[388] or by tributyl phosphate.[389] For iron in bismuth in the presence of antimony, manganese, and tellurium, extract from the solution of the sample as ferric thiocyanate.[390] Iron in lead and in lead cable-sheathing alloys is also read as the ferric thiocyanate complex.[391] Iron in cement and cement raw materials is determined as the thiocyanate.[392]

Traces of iron, copper, zinc, and lead in a solution of gelatin are adsorbed on a column of cation exchange resin Dowex 50W and eluted with 1 : 3 hydrochloric acid.[393] Iron is extracted from the acid medium as the thiocyanate by methyl isobutyl ketone.

For iron in urban air, filter at 75 cubic meters/hour through a glass-fiber filter.[394] Extract the filter with benzene, discard the extract, dry the filter, and grind it. Extract a sample of the filter with 1 : 1 hydrochloric acid–7 *N* nitric acid and determine as ferric thiocyanate.

For iron in high purity molybdenum trioxide, dissolve 1 gram of sample in water by addition of ammonium hydroxide.[395] Neutralize to Congo red with 1 : 1 hydrochloric acid and add 3 grams of ammonium chloride. Add 20 ml of 0.45% solution of potash alum and precipitate iron and aluminum hydroxides with ammonium hydroxide. Filter, and wash the precipitate with 1 : 20 ammonium hydroxide. Dissolve the precipitate in 5 ml of 1 : 1 hydrochloric acid and determine as ferric thiocyanate.

For iron in 99.999% pure tellurium, a preliminary concentration is necessary.[396] Dissolve in acid, add cupferron, and extract ferric cupferrate with chloroform. Evaporate the chloroform and decompose the residue with nitric, sulfuric, and perchloric acids. Take up the residue in 8 *N* hydrochloric acid and extract the iron with isopropyl ether. Evaporate the organic layer and dissolve the residue in a 30%

[386b] E. A. Selezneva and V. P. Zhivopistaev, *Uch. Zap. Perm. Gos. Univ.* **1974** (324), 201–206.
[387] E. Cerrai and G. Ghersini, *Proceedings of the SAC Conference, Nottingham* **1965**, 462–473.
[388] Yoshaburo Okura, *Jap. Anal.* **12**, 279–283 (1963).
[389] E. Jackwerth and E.-L. Schneider, *Z. Anal. Chem.* **207**, 188–192 (1965).
[390] S. I. Sinyakova and Ch. Ya. Krol, *Tr. Kom. Anal. Khim., Akad. Nauk SSSR, Inst. Geokhim. Anal. Khim.* **12**, 206–216 (1960).
[391] A. Schottak and H. Schweiger, *Z. Erzbergh. Metallhüttw.* **19** (4), 180–185 (1966).
[392] N. M. Neshchadimova and I. V. Bogdanova, *Tsement* **24** (3), 18–23 (1958).
[393] J. Saulnier, *Sci. Ind. Photogr.* **35**, 1–7 (1964).
[394] Halina Wyszynska, Artur Strusinski, and Maria Borkowska, *Rocz. Panstw. Zakl. Hig.* **21**, 483–491 (1970); cf. Artur Strusinski and Halina Wyszynska, *ibid.* **22**, 649–656 (1971).
[395] V. G. Shcherbakov, *Sb. Tr. Vses. Nauch.-Issled. Inst. Tverd. Splavov* **1964** (5), 303–306.
[396] Jan Dobrowlski and Tadeusz Wilczewski, *Chem. Anal.* (Warsaw) **15**, 839–844 (1970).

solution of lithium chloride in 1 : 1 hydrochloric acid. Extract the iron with methyl isobutyl ketone. Add a 30% solution of ammonium thiocyanate to the organic layer and read at 48 nm.

For analysis of high purity titanium slag, fuse the sample with potassium hydroxide and boric acid.[397] Dissolve in hot dilute sulfuric acid and read as the thiocyanate.

Deposits in an internal combustion engine are from fuel and organometallic additives in the lubricant.[398] Determine iron by thiocyanate.

For multivitamin and mineral preparations, digest a sample containing 20–30 μg of iron with 2 ml of hydrochloric acid, 0.1 ml of nitric acid, and 5 ml of water for 30 minutes at 100°. Cool, add 5 ml of 20% potassium thiocyanate solution, dilute to 25 ml, and read at 480 nm.[399a]

For iron in textiles, ash in platinum.[399b] Then treat with hydrofluoric acid and sulfuric acid as silica interferes with determination as ferric thiocyanate.

For iron in urine, extract as ferric ion from a strongly acidified sample into methyl isobutyl ketone.[400a] Reextract with water and add potassium thiocyanate to the aqueous solution. Extract ferric thiocyanate with methyl isobutyl ketone and read at 500 nm.

For analysis of Sabouraud culture medium to a sample containing 1–10 ppm of iron, add 0.2 ml of 30% hydrogen peroxide and make 0.4 *N* with nitric acid.[400b] Add 1 ml of 10% solution of ammonium thiocyanate and dilute to 10 ml. Extract the red complex with ethyl acetate and read at 500 nm.

A ternary complex is formed by ferric ion with thiocyanate and benzyldimethyl-octadecylammonium perchlorate which can be extracted with dichloroethane.[401a] Optimum conditions are 0.5 *M* potassium thiocyanate, pH 2 with hydrochloric acid, and 0.02 *M* quaternary. Beer's law is followed for 3–0.9 m*M* iron. The color is stable for 20 hours.

A stable complex of ferric thiocyanate is extracted from an acid solution by 50 m*M* octyl α-analinobenzylphosphonate in chloroform and read at 450 nm.[401b] There is interference by hexavalent uranium and molybdenum, and by oxalate.

For iron by thiocyanate in the presence of copper, nickel, and cobalt, see Copper. For iron in indium by thiocyanate see Gallium.

Procedures

Stabilization with Methyl Ethyl Ketone and Acetone.[402] Take a sample of 1–20 ml

[397] Tsuyoshi Nakayama, *Jap. Anal.* **9**, 119–122 (1960).
[398] T. Hammerich and H. Gondermann, *Erdöl Kohle Erdgas, Petrochem.* **16**, 303–308 (1963).
[399a] H. C. Chiang, *Formosan Sci.* **14**, 239–242 (1960).
[399b] K. Schliefer and A. E. Jabali, *Melliand Textilber. Int.* **57**, 776 (1976).
[400a] P. Tavenier, C. E. Raaijmakers, H. G. van Eijk, and P. Leijnse, *Chlin. Chim. Acta* **32**, 63–66 (1971).
[400b] H. M. Sammour, S. A. El-Kinawy, F. A. Aly, and F. M. Abdel-Gawad, *J. Drug Res. Egypt* **7**, 179–189 (1976).
[401a] A. S. Babenko and E. A. Alferov, Manuscript No. 6365-73 deposited at Vsesoyuznyi Institut Nauchnoi i Teknhicheskoi Informatsii, Moscow, **1973**.
[401b] B. Tamhina, B. Jagodic, and M. J. Herak, *Z. Anal. Chem.* **282**, 220 (1976).
[402] Paul Baily, *Anal. Chem.* **29**, 1534–1536 (1957).

of solution containing 10–200 ppm of iron. Dilute nearly to 30 ml and adjust to approximately pH 1.5, usually by addition of 2 ml of 2 *N* hydrochloric acid. Add 20 ml of acetone and 40 ml of methyl ethyl ketone and mix. Add 5 ml of 40% solution of potassium thiocyanate and dilute to 100 ml. Read at 490 nm.

As Ferric–1,10-Phenanthroline–Thiocyanate Complex.[403] Mix a sample containing up to 120 μg of iron with 15 ml of 0.1 *N* hydrochloric acid, 0.5 ml of 0.01 *M* 1,10-phenanthroline, and 1 ml of 10% potassium thiocyanate solution. Dilute to 25 ml and shake for 1 minute with 25 ml of methyl isobutyl ketone. Dry the organic phase with sodium sulfate and read at 525 nm. Iron must have been separated from cobaltic, cupric, nickel, vanadate, and tungstate ions.

As Ferric-Tributylammonium-Thiocyanate Complex.[404] To 20 ml of sample containing up to 20 mg of iron, made acid with hydrochloric or sulfuric acid, add 2 ml of a buffer solution for pH 3–4, 4 ml of 50% solution of potassium thiocyanate, 1 ml of tributylammonium acetate, and a few mg of thiourea. Extract with 4 ml and 1 ml of isoamyl acetate and read at 480 nm.

As Ferric–2–Furyldiantipyrinylmethane–Thiocyanate Complex.[405] NATURAL WATERS. The sensitivity is manyfold that of just thiocyanate. Add 1 ml of hydrochloric acid to 20 ml of sample. Oxidize any ferrous ion with a few drops of potassium permanganate solution. If phosphate, oxalate, or acetate is present in substantial amount, add 3 ml more of hydrochloric acid. Add 5 ml of 20% solution of ammonium thiocyanate and immediately add 5 ml of a saturated solution of 2-furyldiantipyrinylmethane in 0.05 *N* hydrochloric acid. After 3–4 minutes filter on paper and wash with 2% solution of ammonium thiocyanate in *N* hydrochloric acid. Wash with a saturated solution of the 2-furyl compound in 0.05 *N* hydrochloric acid to remove colored compounds of chromium, nickel, and so on. Dissolve the precipitate in acetone, add 0.5 ml of 20% ammonium thiocyanate solution, dilute to 10 ml with acetone and read at 490 nm.

As a Complex with a Quaternary Perchlorate.[406] THIN FILMS. A typical sample would be a magnetic iron film on beryllium bronze wire. Dissolve a sample containing at least 2 μg of iron in 5 ml of 1 : 1 nitric acid, boil off oxides of nitrogen, and dilute to 20 ml. Add a solution containing 2 mg of aluminum or lanthanum as a carrier and precipitate the hydroxides at 60° with ammonium hydroxide. Dissolve the precipitate with 5 ml of *N* hydrochloric acid and dilute to 25 ml. Mix with 5 ml of 20% potassium thiocyanate solution and extract with 10 ml of 0.02 *M* benzyldimethyloctadecylammonium perchlorate in dichloroethane. Filter the extract and read at 480 nm against a blank. If copper is absent in the sample, the precipitation step is not necessary.

[403] V. Pandu Ranga Rou, K. Venugopalo Rao, and P. V. R. Bhaskara Sarma, *Talenta* **16**, 277–280 (1969); cf. C. Liteanu, I. A. Crisan, and M. Marian, *Stud. Univ. Babes-Bolyai, Ser. Chem.* **13**, 39–42 (1968).
[404] Max Ziegler, O. Glemser, and N. Petri, *Angew. Chem.* **69**, 174–177 (1957).
[405] V. P. Zhivopistsev and V. S. Minina, *Uch. Zap. Molotov. Gos. Univ. A. M. Gor'kogo* **8** (3), *Mat., Fiz. Khim.* 37–41 (1954).
[406] E. A. Alferov, I. P. Danilov, and G. G. Chernenko, *Zavod. Lab.* **42**, 19 (1976).

Silver Alloys.[407] Dissolve 1 gram of sample in 5 ml of 1 : 1 nitric acid. Add 10 ml of water, then heat, and add 1 : 1 ammonium hydroxide dropwise until the solution turns blue. Set aside for 20 minutes, filter, and wash with 1 : 20 ammonium hydroxide. Dissolve the precipitate with 20 ml of hot 1 : 1 hydrochloric acid. To a 5-ml aliquot containing 1–2 μg of iron, add 5 ml of 10% potassium thiocyanate solution. Extract with 5 ml, 5 ml, and 5 ml of isopentyl alcohol. Adjust the extracts to 20 ml and read at 500 nm.

Uranium–Fission Element Alloys.[408] Transfer a solution of the sample containing 2–30 μg of iron to a centrifuge tube containing an excess of ammonium hydroxide. Centrifuge and discard the supernatant liquid. Dissolve the precipitate in 10 ml of 7.5 *N* hydrochloric acid and 0.5 ml of 30% hydrogen peroxide. Add 10 ml of *n*-butyl acetate and shake for 1 minute. Wash the organic phase with 10 ml and 10 ml of 7.5 *N* hydrochloric acid. Add 10 ml of methyl isobutyl ketone to dilute the organic phase, and shake with 20 ml of 20% ammonium thiocyanate solution for 1 minute. Set aside for 15 minutes and read the organic phase at 500 nm.

Pure and Electrolytic Copper.[409] Dissolve a 1-gram sample in 5 ml of 1 : 1 hydrochloric acid adding 30% hydrogen peroxide as necessary to complete solution. Boil off excess hydrogen peroxide and dilute to 50 ml. Add 4 grams of pure cadmium metal and heat until the copper has been displaced to give a colorless solution. Filter, cool, and add 2 drops of 30% hydrogen peroxide. Add 25 ml of 12% potassium thiocyanate solution, dilute to 100 ml, and read at 500 nm against water.

Copper Alloys.[410] Dissolve a 1-gram sample in 15 ml of 1 : 1 nitric acid and evaporate to 5 ml. Add 100 ml of water, boil, set aside for 30 minutes, and filter. Dilute the filtrate to 100 ml and take a 5-ml aliquot. Add 2 ml of *M* tartaric acid, 20 ml of *M* octanoic acid in chloroform, and 4 ml of 1 : 4 ammonium hydroxide. Shake for 1 minute and discard the organic layer. Neutralize with 1 : 1 hydrochloric acid and add 5 ml in excess. Add 10 ml of 10% potassium thiocyanate solution and dilute to 25 ml. Read at 500 nm.

High Purity Aluminum.[411] Dissolve a 2-gram sample in 10 ml of 1 : 1 hydrochloric acid and slowly evaporate to dryness. Take up the residue in 50 ml of 7 *N* hydrochloric acid and shake for 2 minutes each with 10 ml, 10 ml, and 10 ml of methyl isobutyl ketone. Combine the extracts and reextract the iron with four successive 10-ml portions of water. Evaporate the combined aqueous extracts to

[407] Z. Skorko-Trybula and J. Chwastowska, *Chem. Anal.* (Warsaw) **8**, 859–864 (1963); cf. Z. Marczenko and K. Kasiura, *ibid.* **9**, 87–95 (1964).
[408] J. J. McCown and D. E. Kudera, *Anal. Chem.* **34**, 870–871 (1962).
[409] C. M. Dozinel and H. Gill, *Metall* **10**, 1042–1044 (1956).
[410] V. V. Sukhan and S. F. Skachkova, *Zavod. Lab.* **36**, 1029–1031 (1970); cf. V. F. Mal'tsev and V. Ya. Sych, *Byull. Nauch.-Tekh. Inf. Ukr. Nauch.-Issled. Trwbn. Inst.* **1959** (6–7), 189–194; Ivor Ilmet, *Chemist-Analyst* **54**, 71–72 (1965).
[411] A. I. Szücs and O. N. Klug, *Chem. Anal.* (Warsaw) **12**, 939–947 (1967); cf. M. I. Abramov, *Uch. Zap. Azerb. Univ., Ser. Khim. Nauk.* **1963** (3), 31–38.

incipient crystallization. Add 2 ml of 0.5 *N* hydrochloric acid, 3 ml of acetone, and 2 ml of 10% solution of potassium thiocyanate. Read at 480 nm.

For aluminum and aluminum alloys such as Zieral and Osmagal, dissolve in sodium hydroxide and hydrogen peroxide.[412] Acidify with nitric acid and boil off oxides of nitrogen. Develop iron with potassium thiocyanate as above.

Ultrapure Tin.[413] Melt the sample and allow it to drip into a large porcelain basin from a height such that each individual drop forms a thin sheet. Dissolve a 10-gram sample in 20 ml of hydrochloric acid and cool. Add 15 ml of bromine in 25 ml of hydrobromic acid. If necessary, warm gently to complete solution of the tin. Heat, covered, on a hot plate until nearly dry. Remove the cover but do not bake. Add 5 ml of 1 : 4 hydrochloric acid and evaporate to 2.5 ml. Cool; add 0.5 ml of 10% ammonium thiocyanate solution and 0.5 ml of 1% ammonium persulfate solution. Dilute to 10 ml. Read at 603 nm and subtract a blank.

Molybdenum.[414] Dissolve a 0.05-gram sample in 3 ml of hydrochloric acid, 0.5 ml of nitric acid, and 0.25 ml of perchloric acid by gentle warming. Then heat until the perchloric acid has been removed and molybdenum precipitates. Cool, add 5 ml of hydrochloric acid, and evaporate to 1 ml. Add 1 ml of water. Add 25 ml of 30% lithium chloride solution and 1 gram of ammonium tartrate. Heat to 100° to dissolve the tartrate and cool. Extract by shaking for 30 seconds with 10 ml of methyl isobutyl ketone. Mix an aliquot of the organic layer with 0.5 gram of ammonium thiocyanate, filter, and read at 490 nm against water.

Tungsten. Dissolve 0.05 gram of sample in platinum with 0.5 ml of hydrofluoric acid, 2 drops of nitric acid, and 2 drops of perchloric acid. Evaporate to dryness. Dissolve the tungstic acid in 0.5 ml of 10% sodium hydroxide solution by warming. Add 5 ml of hydrochloric acid and heat until only a small, flocculent, white precipitate remains. Evaporate to 1 ml. Add 1 ml of water and 1 gram of ammonium tartrate. Heat until most of the tungstic oxide has dissolved. Complete as for molybdenum from "Add 25 ml of 30% lithium. . . ."

Vanadium. Dissolve 0.1 gram of sample by gently warming with 5 ml of 1 : 1 nitric acid, and evaporate to moist dryness. Add 5 ml of hydrochloric acid and evaporate to moist dryness to drive off nitric acid. Repeat the last step. Add 25 ml of 7 *N* hydrochloric acid and heat to 100° to dissolve salts. When the solution is blue, cool it, and extract by shaking with 10 ml of methyl isobutyl ketone. Discard the aqueous layer and add 25 ml of 30% lithium chloride solution containing 1 gram of ammonium tartrate. Shake for 30 seconds to remove vanadium from the organic layer. Mix an aliquot of the organic layer with 0.5 gram of ammonium thiocyanate, filter, and read at 490 nm against water.

Tantalum.[415] Dissolve 0.05 gram of sample by gentle warming in platinum in 1

[412] M. Bednara, *Z. Erzbergb. Metallhüttenw.* **5**, 149–152 (1952).
[413] G. Bradshaw and J. Rands, *Analyst* **85**, 76–78 (1960).
[414] C. L. Luke, *Anal. Chim. Acta* **36**, 122–129 (1966).
[415] Cf. V. A. Nazarenko, E. A. Biryuk, M. B. Shustova, G. G. ShiFareva, S. Ya. Vinkovetskaya, and G. V. Flyantikova, *Zavod. Lab.* **32**, 267–269 (1966).

ml of hydrofluoric acid and 3 drops of nitric acid. Evaporate to dryness and dissolve in 2 drops of hydrofluoric acid. Add 25 ml of 30% lithium chloride solution and 5 ml of 5% boric acid solution. Complete as for molybdenum from "Extract by shaking for 30 seconds...." Results can be expected to be about 10% low, corrected by the calibration graph.

Gallium.[416] Dissolve a 0.5-gram sample in 1 ml of nitric acid and 2 ml of 7 *N* hydrochloric acid. Evaporate to dryness and take up in 5 ml of 5 *N* hydrochloric acid. Shake with 2 ml of 1% solution of cupferron in chloroform and follow with 2 ml of chloroform. Wash the combined iron cupferrate extracts with 3 ml and 3 ml of 5 *N* hydrochloric acid. Evaporate the chloroform extract with 0.1 ml of sulfuric acid and 0.5 ml of 30% hydrogen peroxide. Heat until iron cupferrate is completely decomposed, adding additional peroxide if required. Take to dryness and dissolve the residue in 0.25 ml of *N* hydrochloric acid. Add 2 ml of water and 1 ml of 25% solution of potassium thiocyanate. Extract with 0.4 ml of isoamyl alcohol and read the organic phase at 500 nm.

Indium.[417] Dissolve 10 grams of sample in 25 ml of hydrochloric acid, adding a couple of drops of nitric acid. Concentrate until a precipitate appears, and add 25 ml of 7 *N* hydrochloric acid. Extract by shaking for 1 minute each with 10 ml, 10 ml, and 10 ml of isoamyl acetate. Reextract iron, gallium, and molybdenum from the pooled isoamyl acetate solutions with 5 ml and 5 ml of 1:3 hydrochloric acid. Evaporate almost to dryness. Take up in 25 ml of water and adjust to pH 1.5 with hydrochloric acid. Pass through a column of Amberlite IR-120 in hydrogen form. Elute the molybdenum with 1.5 *N* hydrochloric acid. Then elute the iron and gallium with 1:2 hydrochloric acid. Evaporate almost to dryness, take up in 25 ml of water, and add a few mg of lanthanum ion. Dilute to 200 ml and precipitate with ammonium hydroxide. Let stand overnight, filter, and wash with 1:100 ammonium hydroxide. Dissolve in 1:5 hydrochloric acid and dilute to 50 ml with that acid. To a 5-ml aliquot, add 0.5 ml of 10% ammonium thiocyanate solution and 0.5 ml of 1% solution of ammonium persulfate. Dilute to 10 ml and read at 500 nm.

Titanium.[418] Dissolve 0.2 gram of sample in 5 ml of hydrofluoric acid and 2 ml of nitric acid. Add 4 drops of sulfuric acid and evaporate to sulfur trioxide fumes. Take up in 50 ml of water and add 5 ml of 30% hydrogen peroxide. Add 5 ml of 1% solution of oxine in *N* acetic acid. Raise to pH 9–10 by addition of ammonium hydroxide and extract with 10 ml, 10 ml, and 10 ml of chloroform. Wash the combined extracts with 1:20 ammonium hydroxide containing 1% of hydrogen peroxide. Evaporate the washed chloroform extract to dryness and ignite. Add 4 drops of 1:1 hydrochloric acid and dissolve with 0.5 ml of 1.5 *N* sulfuric acid. Add 1 ml of water and 0.5 ml of 5% ammonium thiocyanate solution. Extract with 1 ml of isopentyl alcohol and read the extract at 500 nm.

[416] V. A. Nazarenko and G. V. Flyantikova, *ibid.* **27**, 1339–1341 (1961).
[417] K. Kasiura, *Chem. Anal.* (Warsaw) **13**, 849–855 (1968).
[418] V. A. Nazarenko, M. B. Shustova, G. G. Shitareva, G. Ya. Yagnyatinskaya, and R. V. Ravitskaya, *Zavod. Lab.* **28**, 645–648 (1962); cf. Shikao Hashimoto and Toshio Sawada, *Nippon Kinzoku Gakkaishi* **18**, 417–420 (1954); Hidehiro Goto and Shuro Takeyama, *Sci. Rep. Res. Inst., Tohoku Univ.* **6**, 424–430 (1954).

Copper Alloys.[419] Dissolve a sample in 1:1 nitric acid and dilute to a known volume. To a 10-ml aliquot add 10 ml of 0.5 *M* benzoic acid in chloroform. Adjust to pH 3–3.2 and extract. Add to the extract 2.5 ml of 20% sodium thiocyanate solution and 0.1 ml of 1:1 nitric acid. Dilute to a known volume with methanol and read at 490 nm.

Nickel and Nickel Salts.[420] Adjust the sample solution to approximate neutrality and make it a 20% solution of ammonium thiocyanate. Extract with 15 ml of tributylphosphate. Wash the organic layer with 20 ml and 20 ml of 0.5 *N* hydrochloric acid. Dilute to 25 ml with the solvent and read at 503 nm. The end product is a complex of ferric thiocyanate with 3 moles of the solvent. When analyzing cathode nickel, make the solution strongly acid with hydrochloric acid, oxidize with ammonium persulfate, and add potassium thiocyanate. Extract the complex with 1:2 isobutyl alcohol–butanol by shaking for 2 minutes, and read at 500 nm for 10–50 μg of iron.[421]

Nickel Alloys.[422] Dissolve 0.1 gram of sample in 3 ml of nitric acid and 7 ml of 1:1 hydrochloric acid. Dilute to 100 ml and take a 5-ml aliquot. Add 20 ml of 1:4 nitric acid and 25 ml of 4% solution of ammonium thiocyanate. Read at 500 nm. If the alloy contains more than 15% chromium, it must be included in the calibration curve.

Nichrome.[423] Dissolve 0.2 gram of sample in 5 ml of nitric acid and 10 ml of hydrochloric acid, add 1 ml of 1:1 sulfuric acid, and evaporate to sulfur trioxide fumes. Take up in 10 ml of 1:1 hydrochloric acid. Extract the iron with 10 ml, 5 ml, and 5 ml of ether. Add 5 ml of water to the combined extracts and evaporate the ether. Add 20 ml of 1:4 nitric acid and 20 ml of 5% solution of ammonium thiocyanate. Dilute to 50 ml and read at 500 nm.

Tin-Base Babbitt Metal.[424] Dissolve 1 gram of sample containing 0.02–0.1% iron in 5 ml of hydrochloric acid and 2 ml of nitric acid. Dilute to 100 ml. Mix a 5-ml aliquot with 5 ml of hydrochloric acid. Reduce the iron to the ferrous state with 1 ml of 5% solution of sodium thiosulfate. Blanket the solution with carbon dioxide and extract the tin with 10 ml and 5 ml of methyl isobutyl ketone. Add 2 ml of 30% hydrogen peroxide to the aqueous layer and boil to decompose thiosulfate and excess peroxide. Cool, and extract the ferric ion with 25 ml of methyl isobutyl ketone. Shake the organic layer with 5 ml of 20% solution of ammonium thiocyanate and read at 490 nm.

Tantalum Pentoxide.[425] Dissolve 0.1 gram of sample in platinum by heating for

[419] I. V. Pyatnitskii and T. A. Slobodenyuk, *Ukr. Khim. Zh.* **40**, 1333–1335 (1974).
[420] E. Jackwerth, *Z. Anal. Chem.* **206**, 335–344 (1964).
[421] H. Gielczewska and M. Kleczynska, *Pr. Inst. Mech. Precyz.* **15**, 41–46 (1967).
[422] V. F. Mal'tsev, *Byull. Nauch.-Tekh. Inf., Ukr. Nauch.-Issled. Trubn. Inst.* **1959** (6–7), 186–189; cf. Wiktor Kemula, Krystyna Brajter, Stefania Cieslik, and Hanna Lipinska-Kostrowicka, *Chem. Anal.* (Warsaw) **5**, 229–234 (1960).
[423] V. N. Polyanskii, *Sb. Tr. Mosk. Vech. Met. Inst.* **1957** (2), 255–256.
[424] Nobuo Tajima and Moriji Kurobe, *Jap. Anal.* **9**, 399–402 (1960).
[425] I. G. Shafran and I. F. Vzorova, *Tr. Vses. Nauchn.-Issled. Inst. Khim. Reakt.* **1967** (30), 233–237.

15 minutes with 3 ml of hydrofluoric acid and a drop of nitric acid while covered with a larger platinum dish of cold water as a condenser. Remove the upper dish, add 2 ml of sulfuric acid, and evaporate to sulfur trioxide fumes. Add a drop of hydrofluoric acid, then mask fluoride with 20 ml of 1.5% boric acid solution. Add 2 ml of 30% ammonium thiocyanate solution and extract with 10 ml of isoamyl alcohol. Read the organic layer at 500 nm.

As an alternative, after evaporation to sulfur trioxide fumes take up in 5 ml of water. Add 3 ml of 10% sodium acetate solution and 0.5 ml of 10% hydroxylamine hydrochloride solution. After 2 minutes shake for 1 minute with 1 ml of 0.08% bathophenanthroline solution in isoamyl alcohol. Read the organic layer at 533 nm.

Germanium Tetrachloride.[426] Shake 10 ml of sample for 5 minutes with 1 ml of 9 *N* hydrochloric acid. Mix the extract with 5 ml of 1% solution of potassium thiocyanate and 10 ml of water. Extract with 1 ml of 1:2 tributylphosphate-benzene. Read at 500 nm.

Titanium Tetrachloride.[427] Mix 10 ml of sample, 25 ml of 7 *N* hydrochloric acid, and 5 ml of 30% hydrogen peroxide. Cool, and add 10 ml of hydrochloric acid. Extract with 10 ml of amyl acetate and wash the organic layer with 5 ml of 8 *N* hydrochloric acid. Shake with 5 ml of methyl isobutyl ketone and 5 ml of 10% ammonium thiocyanate solution. Discard the aqueous layer. Add 1 ml of ethanol to the organic layer and read at 490 nm.

Vanadium Tetrachloride. Boil 10 ml of sample, 0.5 ml of nitric acid, and 25 ml of 7 *N* hydrochloric acid for 2–3 minutes; cool. Add 10 ml of 50% citric acid solution and 10 ml of hydrochloric acid. Proceed as above from "Extract with 10 ml of. . . ."

Zirconium Tetrachloride and Hafnium Tetrachloride. Boil 5 grams of sample with 20 ml of hydrochloric acid and 1 ml of 30% hydrogen peroxide until dissolution is complete. Add 2 grams of citric acid to the hot solution and cool. Proceed as above from "Extract with 10 ml of. . . ."

Tantalum Chlorides and Niobium Chlorides. Dissolve 10 grams of sample in 35 ml of hydrochloric acid, heating if necessary. Add 5 ml of 30% hydrogen peroxide and dilute to 50 ml. Set aside until evolution of bubbles ceases, then proceed as above from "Extract with 10 ml of. . . ."

Titanic Chloride.[428] Add 10 ml of sample in small portions, while stirring, to 25 ml of 7 *N* hydrochloric acid. Add 0.5 ml of 1:1 nitric acid and warm for a couple of minutes to oxidize any ferrous ion. Cool, add 6 ml of hydrochloric acid, and extract the ferric ion with 10 ml of amyl acetate. Wash the organic layer with 5 ml of hydrochloric acid to remove titanium. Reextract the ferric ion with 9 ml of

[426] L. N. Filatova and I. F. Golovaneva, *ibid.* **1965** (27), 215–218.
[427] T. M. Malyutina and V. A. Orlova, *Tr. Nauchn.-Issled. Proekt. Inst. Redkomet. Prom.* **1973** (47), 66–76.
[428] T. M. Malyutina and V. A. Orlova, *Zavod. Lab.* **34**, 277–278 (1968).

water. Add 3 ml of acetone and 1 ml of 30% ammonium thiocyanate solution to the extract and read at 500 nm against water.

Titanium Dioxide.[429] Heat a 0.5-gram sample with 8 ml of sulfuric acid and 5 grams of ammonium sulfate. Cool, and dilute to 100 ml. Add 5 ml of 10% potassium thiocyanate solution and set aside for 12 minutes. Shake for 2 minutes with 10 ml of 1:3 tributylphosphate-benzene and read the organic layer at 517 nm.

Orthovanadates of Rare Earth Metals.[430] Moisten 1 gram of sample with 2 ml of water and dissolve by adding 4 ml of 8 *N* hydrochloric acid and 4 ml of 10% hydroxylamine hydrochloride solution. Evaporate to about 5 ml, cool, add 4 ml of 8 *N* hydrochloric acid and dilute to 25 ml. Take four 5-ml aliquots. Add 2 mg of ferric ion to one and 4 mg to another. To each and to the two sample solutions add 3 ml of water and 3 ml of 20% potassium cyanide solution. After 2 minutes shake each with 10 ml of 35% solution of tributylphosphate in chloroform. Wash each extract with a solution of 2 grams of potassium thiocyanate and 1 ml of 8 *N* hydrochloric acid in 100 ml of water. Read at 518 nm.

Copper, Cobalt, Cadmium, and Zinc Salts.[431] Dissolve a sample containing about 1 gram of iron in water. Pretreat a column of Lewatit M-11 with ammonium hydroxide. Pass the sample solution through the column. Wash the column with 100 ml of 10% solution of ammonium nitrate in ammonium hydroxide, then with 100 ml of water. Elute the iron with 100 ml of *N* sulfuric acid or, in the case of copper salts, with 100 ml of 0.5 *N* hydrochloric acid. Dilute the eluate to 500 ml. To a 10-ml aliquot, add 1 ml of 5% ammonium persulfate solution and 5 ml of 20% ammonium thiocyanate solution. Dilute to 50 ml and read at 488 nm against a reagent blank.

Soils and Clays.[432] FREE IRON. For pH 4.75, mix 0.5 gram of soil in a 15-ml centrifuge tube with a buffer solution that is 0.15 *M* with sodium citrate and 0.05 *M* with citric acid. Stopper and shake in a water bath at 50° for 30 minutes, which should be at least 10 minutes after the sample appears to be completely bleached. Filter and dilute to 25 ml. To a 10-ml aliquot add 0.1 ml of hydrochloric acid and 2 ml of 10% solution of ammonium thiocyanate. Add 3 drops of 30% hydrogen peroxide, dilute to 25 ml, and read at 500 nm.

Alunite Minerals.[433] Take an aliquot of sample solution containing about 0.4 mg of iron. Add excess of 10% ammonium thiocyanate solution and extract with 10 ml and 10 ml of 1:4 butanol-ether. Dilute to 50 ml with isoamyl alcohol and read at 533 nm.

Iron and Manganese Ores.[434] Dissolve 1 gram of finely powdered sample in 10

[429] Kazuyoshi Onishi, *Jap. Anal.* **12**, 534–539 (1963).
[430] N. S. Poluektov, R. S. Lauer, and S. F. Ognichenko, *Zavod. Lab.* **37**, 1050–1051 (1971).
[431] W. Kemula, K. Brajter, and H. Kostrowicka, *Chem. Anal.* (Warsaw) **6**, 463–468 (1961).
[432] D. E. Coffin, *J. Soil Sci.* **43**, 7–17 (1963).
[433] M. K. Akhmedli and E. A. Bashirov, *Uch. Zap., Azerb. Gos. Univ. S. M. Kirova* **1955** (7), 25–28.
[434] J. Jankovsky and E. Pavlikova, *Sb. Pr. Ustavu. Pro Vyzk. Rud* **IV**, 229–241 (1962).

ml of hydrochloric acid. Add 3 ml of nitric acid and evaporate to dryness. Add 5 ml of hydrochloric acid and again evaporate to dryness. Take up in 5 ml of hydrochloric acid, add 50 ml of hot water, and filter. Wash well with hot 1:9 hydrochloric acid and hot water. Ignite the filter in platinum. Add 1 ml of 1:1 sulfuric acid and 5 ml of hydrofluoric acid and evaporate to sulfur trioxide fumes. Add 2 grams of potassium pyrosulfate and fuse. Dissolve the melt in 5 ml of 1:5 hydrochloric acid, and add to the original filtrate. Filter if barium sulfate is present and dilute to 250 ml.

Mix a 10-ml aliquot with 10 ml of 1:1 hydrochloric acid. Dilute to 80 ml and add 1 ml of 10% solution of potassium persulfate. Add 10 ml of 10% solution of ammonium thiocyanate, dilute to 100 ml, and read at 500 nm.

Thorium Ore.[435] Fuse 0.5 gram of sample in platinum with 4 grams of sodium carbonate for 1 hour at 900°. Leach the melt with 25 ml of 1:10 nitric acid and filter. Evaporate to dryness, add 1 ml of nitric acid, and again evaporate. Take up with 25 ml of 1:10 nitric acid and dilute to 250 ml. Mix a 0.1-ml aliquot with 10 ml of 7.5 *M* potassium thiocyanate and dilute to 25 ml. Read at 480 nm.

Iron Carbonyl in Water Gas.[436] Pass the gas at 1 cubic foot/hour through a sintered bubbler containing *M* iodinemonochloride in acetic acid. Transfer to a beaker and wash the bubbler with 10 ml of 5% sulfuric acid. Dissolve sulfur deposited in the bubbler with nitric acid and add to the beaker. Evaporate to dryness, add a few drops of nitric acid, and ash over a flame. Dissolve the residue in 10 ml of 1:1 hydrochloric acid. Add 50 ml of water, 1 ml of 10% ammonium persulfate solution, and 10 ml of 10% solution of ammonium thiocyanate. Dilute to 100 ml and read at 500 nm. Because the solution tends to be turbid, comparison with standards may be preferable to photometry.

Nickel Plating Baths.[437] To 5 ml of sample, add 1 ml of 1:3 sulfuric acid and 5 drops of 3% hydrogen peroxide. Heat to 100° and add 6 ml of ammonium hydroxide saturated with ammonium chloride. Filter and wash the precipitate with hot 10% ammonium chloride solution, then with hot water. Dissolve the precipitate in 5 ml of hot 1:20 sulfuric acid and dilute to 10 ml. Add 1 ml of 1:3 sulfuric acid, 2 drops of 3% solution of ammonium persulfate, and 4 drops of a saturated solution of ammonium thiocyanate. Dilute to 25 ml and read at 500 nm.

m-Chlorophenylisocyanate.[438] Shake 2 grams of sample with 5 ml of saturated bromine water. Filter the upper layer. Add to the filtrate 2 drops of hydrochloric acid, 5 ml of 40% solution of ammonium thiocyanate, and 10 ml of water. Extract with 10 ml of isoamyl alcohol and read the organic layer at 500 nm.

Isopropyl Chloroformate. Reflux 2 grams of sample with 5 ml of 2% solution of sodium hydroxide and cool. Complete as above from "Filter the upper layer."

[435] F. I. Nagi, Azhar Ali, and Jamshed Anwar, *Rep. Pak. Atom. Energ. Comm.* **AEMC/Chem-134** 17 pp (1975).
[436] A. B. Densham, P. A. A. Beale, and R. Palmer, *J. Appl. Chem.* **13**, 576–580 (1963).
[437] G. V. Loshkareva, *Tr. Sverdl. Gorn. Inst. V. V. Vakhrusheva* **1960** (36), 95–99.
[438] E. A. Arkhangel'skaya and I. P. Krasnova, *Tr. Khim. Khim. Tekhnol.* (*Gor'kii*) **1966** [2(16)], 298–300.

Milk.[439] Mix 50 grams of sample with 10 ml of hydrochloric acid. Heat for 20 minutes at 70°, mixing frequently, and filter. Add 1 ml of 10% solution of ammonium persulfate and 10 ml of 10% solution of ammonium thiocyanate. Dilute to 100 ml and read at 500 nm.

Cream and Butter. Add 25 ml of water to a 25-gram sample at 40°, mix well, and add 10 ml of hydrochloric acid. Complete as for milk from "Heat for 20 minutes at 70°,... ."

Edible Oils.[440] Reflux 10 grams of sample for 45 minutes with 10 ml of 1:6 nitric acid, using an air condenser. Cool, shake with 10 ml of petroleum ether, and discard the extract. Filter the aqueous layer and dilute to 25 ml. To an appropriate aliquot, add 1 ml of 10% solution of ammonium persulfate and 10 ml of 10% solution of ammonium thiocyanate. Dilute to 25 ml and read at 500 nm.

For lesser amounts of iron, dissolve 50 grams of sample in 150 ml of petroleum ether. Shake at 120 cycles/minute with 100 ml of 1:2.5 nitric acid for 1 hour. Filter the aqueous layer, evaporate to dryness, and calcine. Take up in 5 ml of 1:2.5 nitric acid and determine as above from "To an appropriate aliquot,... ."

Dyes.[441] Burn 0.25 gram of sample in a dish, then heat in a muffle furnace for a couple of hours to oxidize iron completely. Add 5 ml of hydrochloric acid and evaporate to dryness. Repeat that step twice more. Take up in 5 ml of hydrochloric acid, evaporate to a small volume, and take up in 50 ml of water. Filter, and wash until the washings no longer give a chloride test. Add 2 ml of 1:1 nitric acid and 25 ml of 10% ammonium thiocyanate solution. Dilute to 200 ml and read at 488 nm.

Pyrethrum Extract.[442] Digest not more than 0.1 gram of sample containing 5–30 μg of iron by heating with 2 ml of a 25% solution of potassium sulfate in sulfuric acid. Cautiously add small amounts of nitric acid until the clear solution is colorless or light yellow. Heat until the sulfuric acid begins to distill. Cool, add 5 ml of water, and again heat until sulfuric acid begins to distill. If not nearly colorless, add more nitric acid and evaporate it with water as before. Cool, and add 10 ml of *N* nitric acid. Mix a 5-ml aliquot with 1 ml of 20% potassium thiocyanate solution. Read at 480 nm within the next 15 minutes.

Cerebrospinal Fluid.[443] NONHEME IRON. Immediately after withdrawal of sample, mix 1 ml with 0.1 ml of 1:11 hydrochloric acid and 0.1 ml of 50% trichloroacetic acid solution. Boil for 15 minutes, cool, and centrifuge to separate proteins. To the supernatant layer, add 0.1 ml of 0.54% solution of potassium persulfate and 0.2 ml of 50% potassium thiocyanate solution. Extract with 0.4 ml of methyl isobutyl

[439] Celina Barska, W. Bednarczyk, and Maria Luczak, *Int. Dairy Congr., Proc. 15th Congr., London* **3**, 1754–1760 (1959).
[440] C. G. Macarovici, V. Farcasan, G. Schmidt, V. Bota, M. Macarovici, A. Dorutiu, I. Pirvu, and E. Tesler, *Stud. Univ. Babes-Bolyai, Cluj, Chim.* **1961** (2), 103–108; cf. Toshihisa Maruta and Tomoyuki Mukoyama, *Jap. Anal.* **18**, 1312–1316 (1969).
[441] V. K. Ponomarev and T. B. Filicheva, *Khim. Volokna* **1960** (3), 60–62.
[442] R. A. G. Marshall, *Analyst* **96**, 675–678 (1971).
[443] B. G. Bleijenberg, H. G. van Eijk, and B. Leijnse, *Clin. Chim. Acta* **31**, 277–281 (1971).

ketone and read at 500 nm. The determination of plasma iron is only slightly influenced by hemoglobin iron.[444]

Feces.[445] Mix 10 grams of homogenized sample with 35 ml of 1:2 hydrochloric acid. Boil for 5 minutes, cool, filter, and wash the paper until 50 ml is collected. Dilute an aliquot containing 10–50 μg of iron to 5 ml. Add 1 ml of hydrochloric acid and mix. Add 0.2 ml of 0.25% solution of potassium persulfate and mix. Add 2 ml of 20% solution of potassium thiocyanate and shake for 2 minutes with 10 ml of isobutyl alcohol. After 10 minutes read the organic layer at 485 nm against a reagent blank.

Liver.[446] Heat 0.1 gram of dry tissue with 0.2 ml of perchloric acid until clear and colorless. Dilute to 3 ml. Mix a 1-ml aliquot with 0.1 ml of 0.1 *N* hydrochloric acid, 0.1 ml of 0.01 *M* potassium persulfate, and 0.5 ml of 5 *M*-potassium thiocyanate. Shake with 0.4 ml of methyl isobutyl ketone and centrifuge. Withdraw 0.3 ml of the organic layer, add 0.3 ml more of methyl isobutyl ketone to the sample, shake, and centrifuge. Combine 0.3 ml of the organic layer with that previously withdrawn and read at 500 nm.

TIRON

Tiron, which is the disodium salt of 4,5-dihydroxy-*m*-benzenedisulfonic acid, is appropriate for determining iron in such salts as sodium iodide and calcium and cadmium tungstates. The red color is produced at pH 10.5–11.5 but fades after 10–15 minutes. A purity of 50% for the reagent is satisfactory. Up to 0.25% thallium does not interfere.

Ferric ion forms a 1:3 complex with tiron for reading at 500 nm.[447a] At pH 5–6 in the presence of diphenylguanidinium acetate, it is extracted with 1:1 chloroform–isoamyl alcohol. At pH 2 a preliminary extraction of complexes of tiron with hexavalent molybdenum, tetra- and pentavalent vanadium, and a small amount of titanic ion avoids interference. Large amounts of titanium interfere. Hexavalent uranium is masked with fluoride and cupric ion by thiourea. In the determination of iron and titanium with tiron, beryllium effectively masks fluoride.[447b]

For iron in sphalerite, dissolve the sample in aqua regia and evaporate to dryness.[448] Take up in hydrochloric acid. Dilute, add an ammonium acetate–acetic acid buffer solution, and develop with tiron. The reagent is applicable to the residue from decomposing a sample of glass with hydrofluoric acid and perchloric acid.[449]

[444] E. Scala, A. Castaldo, and G. Ruiz, *Boll. Soc. Ital. Biol. Sper.* **36**, 1048–1052 (1960).

[445] H. J. Ybema, B. Leijnse, and W. F. Wiltink, *Clin. Chim. Acta* **11**, 178–180 (1965).

[446] H. G. van Eijk, W. F. Wiltink, Gre Bos, and J. P. Goosens, *ibid.* **50**, 275–280 (1974).

[447a] A. I. Busev, G. P. Rudzit, and I. A. Tsurika, *Zh. Anal. Khim.* **25**, 2151–2154 (1970).

[447b] Allen F. M. Barton and Stephen R. McConnel, *Anal. Chem.* **48**, 363–364 (1976).

[448] Leonard Shapiro and Martha S. Toulmin, *U.S. Geol. Survey, Prof. Paper* No. **424-B**, B-328–B-329 (1961).

[449] W. Geilmann, E. Guenon, and G. Tolg, *Glastech. Ber.* **35**, 138–145 (1962).

Procedures

Sodium Iodide.[450] Add 10 ml of nitric acid to a 10-gram sample and evaporate to dryness. Add 3.5 ml of sulfuric acid and again take to dryness. Dissolve the residue in 20 ml of water and neutralize to pH 9 with 20% sodium hydroxide solution. Add 5 ml of 1% solution of tiron and dilute to 100 ml with a phosphate buffer solution for pH 10.5. Set aside for 5 minutes and read against a blank.

Brass.[451] Dissolve 0.5 gram of fine drillings in 20 ml of hydrochloric acid and 2 ml of 30% hydrogen peroxide. Boil to decompose excess peroxide, cool, and dilute to 200 ml. Add ammonium hydroxide to a 10-ml aliquot until the solution is a clear blue; then decolorize with hydrochloric acid. Add 20 ml of a buffer solution for pH 4.7, 2 ml of 5% potassium cyanide solution, and 1 ml of 1% solution of tiron. Dilute to 50 ml and read at 500 nm against a blank.

Aluminum Oxide Whiskers.[452] Fuse 5 mg of dried sample with 150 mg of potassium bisulfate, heating slowly to 250°, then strongly to a clear melt. Dissolve the melt in 1.5 ml of 1 : 1 hydrochloric acid and dehydrate. Repeat the dehydration. Take up in 25 ml of 1 : 5 hydrochloric acid, filter, and dilute to 250 ml with 1 : 100 hydrochloric acid. To a 25-ml aliquot, add 10 ml of 1 : 1 *M* sodium acetate–*M* acetic acid. Add a drop of 30% hydrogen peroxide and 5 ml of 4% solution of tiron. Dilute to 50 ml and read at 565 nm. For very low values, compare with standards.

2,4,6-TRI(2-PYRIDYL)-1,3,5-TRIAZINE

This reagent forms an intense violet color with ferrous ion. In the presence of perchlorate, the complex is extractable with nitrobenzene. The reagent is applied with an AutoAnalyzer to ferrous ion in deuterium oxide and after addition of hydroxylamine to determination of total iron.[453]

For ferrous iron in water at less than 2.4 mg/liter, add a buffer solution for a pH above 3.3 and the triazine reagent.[454] Read at 595 nm against a blank. For total iron, add hydroxylamine hydrochloride solution with the buffer solution.

The reagent is satisfactory for determination of iron in limestone, blast-furnace slag, and cement-kiln dust.[455] For simultaneous determination of iron by bis-2,4,6-tri(2-pyridyl)-1,3,5-triazine and of copper by 2,9-dimethyl-1,10-phenanthroline by extraction with propylene carbonate, see Copper.

For multivitamin products, prepare a solution containing 1 μg of iron per ml and

[450] A. M. Bulgakova, A. B. Blank, A. K. Khurkryanskii, and G. S. Plotnikova, *Stsintill. Stsintill. Materily 2-go [Vtorogo] Koord. Soveshch.* **1957**, 281–290; A. M. Bulgakova and A. K. Khukhryanskii, *Sb. Statei, Vses. Nauchn.-Issled. Inst. Khim. React. Osobo Chist. Khim. Veshchestv.* **1961** (24), 183–188.
[451] B. Lehky, *Hutn. Listy* **15**, 554–555 (1960).
[452] F. W. Vahldisk, C. T. Lynch, and L. B. Robinson, *Anal. Chem.* **34**, 1667–1668 (1962).
[453] R. D. Britt, Jr., *Tech. Equ. Assain.* **1966** (238), 29–33.
[454] C. Hammerton, *Proc. Soc. Water Treat. Exam.* **16**, 293–295 (1967).
[455] P. Chichilo, *J. Assoc. Off. Agr. Chem.* **47**, 1019–1027 (1964).

develop with the captioned reagent at pH 4 in the presence of ascorbic acid.[456] Read at 600 nm. This technic has been automated.

For analysis of brandy, determination without ashing is not accurate, but the results are satisfactory after wet oxidation with sodium hypochlorite and hydrogen peroxide.[457]

For iron in blood serum filtrates by this reagent, see Copper by Bathocuproinedisulfonate. This reagent, with addition of desferrioxamine, has been applied to the AutoAnalyzer.[458]

Procedures

Extraction with Propylene Carbonate.[459] This solvent is 4-methyl-1,3-dioxolan-2-one. Mix a sample solution containing 0.025–0.5 micromole of iron with 3 ml of 10% solution of hydroxylamine hydrochloride and 5 ml of 0.001 *M* color reagent. Add 5 ml of 2 *M* sodium acetate–acetic acid as a buffer. Saturate the solution with propylene carbonate and add sufficient to give a 3-ml layer in excess. Shake for 1 minute, separate the organic layer, dilute to 10 ml with ethanol, and read at 593 nm.

Burnt Refractories.[460] As a 2 *M* buffer, dissolve 164 grams of sodium acetate and 115 ml of acetic acid in water and dilute to 1 liter.

In a silver crucible, mix a sample containing 3–5 mg of ferric oxide with 1 gram of sodium carbonate and 1 gram of sodium borate decahydrate. Heat gently until moisture is vaporized, then heat further until completely decomposed. Solidify on the sides of the crucible, cool, and dissolve in 15 ml of 1:2 hydrochloric acid, adding more hydrochloric acid if necessary. A residue of silver chloride and silica remains. Dilute to 250 ml and filter or centrifuge a portion. To a 5-ml aliquot add 2 ml of 10% solution of hydroxylamine hydrochloride, 5 ml of 0.001 *M* color reagent, and 10 ml of the acetate buffer solution. Dilute to 50 ml and read at 593 nm, correcting for a blank.

Low Iron Silicate. Fuse a 3-gram sample with 5 grams of sodium carbonate and 5 grams of sodium borate decahydrate in a 50-ml silver crucible until a clear melt is obtained. Cool on the sides of the crucible, and dissolve in 300 ml of 1:2 hydrochloric acid. Cool, and dilute to 1 liter. Add 5 ml of hydrochloric acid to a 25-ml aliquot and heat for hours to precipitate silica. Cool and dilute to 250 ml. Filter silica and silver chloride from part of the solution. To a 15-ml aliquot, add 2 ml of 10% solution of hydroxylamine hydrochloride, 5 ml of 0.001 *M* color reagent, and 10 ml of the acetate buffer solution. Dilute to 50 ml and read at 593 nm correcting for a blank.

High Iron Silicate. Fuse 0.22 gram of sample with 1 gram of sodium carbonate

[456] W. F. Beyer and K. G. Zipple, *J. Pharm. Sci.* **57**, 653–657 (1968).
[457] M. K. Meredith, Sidney Baldwin, and A. A. Andreasen, *J. Assoc. Off. Anal. Chem.* **53**, 12–16 (1970).
[458] J. R. Evans and A. M. M. Shepherd, *Clin. Chim. Acta* **60**, 401–404 (1975).
[459] B. G. Stephens and H. A. Suddeth, *Anal. Chem.* **39**, 1478 (1967).
[460] Peter F. Collins, Harvey Diehl, and G. Frederick Smith, *ibid.* **31**, 1862–1867 (1959).

and 1 gram of sodium borate decahydrate in a silver crucible. Decomposition should require about 15 minutes. Cool, dissolve in 30 ml of 1:2 hydrochloric acid, and dilute to 500 ml. Dilute a 50-ml aliquot to 500 ml. Centrifuge a portion to remove silica and silver chloride. To a 15-ml aliquot, add 2 ml of 10% hydroxylamine hydrochloride solution, 5 ml of 0.001 *M* color reagent, and 10 ml of the acetate buffer solution. Dilute to 50 ml and read at 593 nm against a blank.

Glass and Glass Sand. Transfer a sample containing 0.5–1 mg of ferric oxide to a platinum crucible. If the sample is glass, add 2 ml of water and 4 ml of hydrofluoric acid; for glass sand, omit the water. After reaction subsides, add 1 ml of perchloric acid and evaporate to dryness without boiling. Cool, add 2 ml of hydrofluoric acid, and again take to dryness. Treat with 20 ml of hydrochloric acid and 50 ml of water.

If solution is complete, dilute to 250 ml. Mix a 25-ml aliquot with 2 ml of 10% hydroxylamine hydrochloride solution and 5 ml of 0.001 *M* color reagent. Add ammonium hydroxide dropwise until, on mixing, the color of the iron derivative remains violet. Add 10 ml of the acetate buffer solution, dilute to 50 ml, and read at 593 nm. Correct for a blank.

If the fusion did not dissolve completely, filter the insoluble residue. Wash with 1:100 hydrochloric acid, then with water. Ash the filter in a silver crucible and fuse the ash with 1 gram of sodium carbonate and 1 gram of sodium borate decahydrate to a clear melt. Cool and dissolve in 15 ml of 1:2 hydrochloric acid. Add, including the silica residue and silver chloride, to the previous filtrate and washings, and dilute to 250 ml. Centrifuge or filter a portion and treat as if solution had been complete from "Mix a 25-ml aliquot with... ."

Limestone. To a sample containing 0.5–1 mg of ferric oxide, add 30 ml of 1:2 hydrochloric acid and heat gently. After reaction ceases, filter and complete as for glass or glass sand where there was a residue from fusion. Pick up at "Wash with 1:100 hydrochloric acid... ."

Liming Materials.[461] Grind the sample to pass 100-mesh. To 0.5 gram of limestone or 0.2 gram of silicate in a nickel crucible, add 0.3 gram of potassium nitrate and 1.5 gram of sodium hydroxide. Heat to a dull redness for 5 minutes, and swirl to cool the melt on the sides. Cool, and disintegrate with 50 ml of water. Add 15 ml of 1:1 perchloric acid and dilute to 100 ml. To an aliquot containing less than 100 μg of iron add 3 ml of 10% solution of hydroxylamine hydrochloride and 10 ml of 0.05% solution of the captioned reagent. Add ammonium hydroxide dropwise until the color remains constant. Add 10 ml of a buffer solution containing 16.4% of anhydrous sodium acetate and 11.5% of acetic acid. Dilute to 100 ml and read at 593 nm.

Boiler Water.[462] Add 3 ml of 1:3 hydrochloric acid to 80 ml of sample and heat at 80° for 20 minutes. Add 1 ml of 10% solution of hydroxylamine hydrochloride, 2 ml of 1:4 ammonium hydroxide, 2 ml of 30% sodium acetate solution, and 5 ml of 0.001 *M* reagent. Dilute to 100 ml and read at 595 nm.

[461] P. Chichilo, *J. Assoc. Off. Anal. Chem.* **47**, 620–626 (1964).
[462] Fumito Nakashima and Kaoru Sakai, *Jap. Anal.* **11**, 73–77 (1962).

Serum.[463] Mix a 1-ml sample with 2 ml of 3% ascorbic acid solution and 0.1 ml of 5% sodium acetate solution. Add 0.1 ml of 0.75% solution of the captioned reagent. Incubate at 37° for 10 minutes, cool in water for 5 minutes, and read at 595 nm.

For iron-binding capacity, mix 1 ml of serum with 2 ml of a solution containing 5 μg of iron per ml. After 5 minutes add 170 mg of magnesium carbonate and shake for 20 minutes. Centrifuge, and carry 1 ml of the supernatant layer through the technic for serum.

XYLENOL ORANGE

In the presence of an excess of xylenol orange in 0.084 *M* perchloric acid, ferric ion forms a 1:1 complex with a maximum absorption at 545 nm.[464] In 0.25 *M* acetate buffer solution for pH 6, a 2:1 complex is formed with a maximum at 500 nm. Each will determine down to 3 μg of iron in 25 ml, and each follows Beer's law.

Although both ferric and uranyl ions complex with xylenol orange in faintly acid solution, up to 160 μg of iron can be read at 575 nm in the presence of 500 mg of uranium.[465]

For analysis of thin magnetic films used in computers, dissolve the sample in a few drops of nitric acid and dilute to 10 ml with 0.05 *N* perchloric acid.[466] Add an aliquot containing 5–50 μg of iron to xylenol orange solution, dilute with 0.05 *N* perchloric acid, and read at 550 nm.

For iron and aluminum by xylenol orange by their differential rate of development, see Aluminum.

Procedure. To a sample containing 5–45 μg of iron, add 5 ml of 0.001 *M* xylenol orange and 1 ml of 5% solution of sodium hydrosulfite.[467] Buffer at pH 5.5–6 with a 10% solution of hexamine and perchloric acid. Dilute to a known volume for reading at 500 nm. There are many interferences.

Liquid from Flotation Pulps.[468] As a buffer solution for pH 3, dissolve 27.2 grams of sodium acetate in water, add 194 ml of *N* hydrochloric acid, and dilute to 1 liter. To a solution containing 0.01–0.15 mg of iron, add 20 ml of the buffer

[463] G. Piccardi, M. Nyssen and J. Dorche, *Clin. Chim. Acta* **40**, 214–228 (1972); cf. D. S. Fischer and D. C. Price, *Clin. Chem.* **10**, 21–31 (1964); M. London and J. H. Marymont, Jr., *Clin. Chim. Acta* **12**, 227–229 (1965); Tokuro Chida, Tokiko Osaka, and Kenji Kojima, *ibid.* **22**, 271–275 (1968); John A. O'Malley, Anne Hassan, Judith Shiley, and Henry Traynor, *Clin. Chem.* **16**, 92–96 (1970); Egon Rosner and Anna Molnar, *Kiserl. Orvostud.* **23**, 220–224 (1971).

[464] Makoto Otomo, *Jap. Anal.* **14**, 677–682 (1965); cf. B. Budesinsky, *Z. Anal. Chem.* **188**, 266–272 (1962).

[465] B. Budesinsky, *ibid.* **188**, 266–272 (1962).

[466] K. L. Cheng and B. L. Goydish, *Microchem. J.* **7**, 166–178 (1963).

[467] Makoto Otomo, *Jap. Anal.* **14**, 45–52 (1965).

[468] E. A. Morgen and N. A. Vlasov, *Izv. Vyssh. Ucheb. Zaved., Khim. Khim. Tekhnol.* **10**, 1090–1093 (1967).

solution containing 1.5 mM xylenol orange. Add 2 ml of 3% hydrogen peroxide and dilute to 50 ml. Heat at 100° for 5 minutes, cool, and read at 500 nm.

Ferric ion forms a 1 : 3 complex with **acetothioacetanalide**, which is 4-analino-4-thioxobutan-2-one, at pH 7.1–8.3 as adjusted with ammonium hydroxide.[469] This is extracted with chloroform for reading at 500 nm. Beer's law is obeyed for 4–35 μg/ml in the extract. Manganous ion is masked with ammonium bifluoride.

Acetylacetone oxime, which is isonitrosoacetylacetone, forms a 2 : 1 complex with ferrous ion which in a borate buffer solution for pH 7 is stable for 10–30 minutes.[470] It is read at 590 nm for 0.05–0.45 mg of ferrous ion per ml. There is interference by sulfide, cyanide, ferricyanide, and ferrocyanide, but not by any metals. **Acetophenone oxime** acts similarly at pH 8, and the complex is read at 650 nm.[471] The complexes are preferably extracted for reading: that of acetyl oxime with butanol, that of acetophenone with chloroform. Masking agents are thiosulfate for cobalt, chloride ion for mercury, and thiourea for ruthenium. For the acetophenone complex, nickel can be masked with tartrate.

The 1 : 3 orange-red complex of ferrous ion with **3-acetyl-4-hydroxycoumarin** in 50% ethanol obeys Beer's law for 1.5–5.3 ppm of iron at 400 nm.[472] Interferences, in ppm, are as follows: tartrate, 200; oxalate, 10; nickel, 5.8; lead, 115; manganous ion, 15; beryllium, 4.5; magnesium, 125.

The ferric complex of ***N*-acetyl-*N*-phenylhydroxylamine** at pH 2.8 is read at 470 nm.[473] The color is stable for 2 hours. Fluoride and oxalate interfere. Beer's law holds for 1–12 μg of iron per ml.

The complex of **2-acetylpyridine oxime** with ferrous ion at pH 10 is read at 525 nm.[474] The color is stable for several hours. Beer's law has an optimum at 1–4 ppm but holds for 0.05–10 ppm. There is serious interference by copper, cobalt, nickel, uranium, vanadate, and chromate.

The complex of ferric ion with **alizarin black SN** (mordant black 25) can be extracted into butanol and read at 600 nm.[475] Addition of C_1–C_3 alcohols increases the absorption. The reagent forms similar complexes with copper, uranium, bismuth, thorium, molybdenum, and vanadium.

Alizarin cyanine RC (mordant blue 3) forms complexes with ferric ion read at 680 nm.[476] At pH 4 the range is 0.27–4.86 μg of iron per ml; at pH 5 it is 0.135–4.05 μg/ml, and at pH 9 it is 0.135–2.7 μg/ml.

When aluminum is read at 530 nm with **aluminon**, the iron can be read at 570 nm, and 5–45 μg can be determined by an empirical diagram.[477a]

[469] Krishna De and Jyotirmoy Das, *J. Indian Chem. Soc.* **52**, 1026–1028 (1975).
[470] Andrzej Mrozowski and Tadeusz Lipiec, *Acta Pol. Pharm.* **28**, 291–295 (1971).
[471] U. B. Talwar and B. C. Haldar, *J. Indian Chem. Soc.* **49**, 785–792 (1972).
[472] A. N. Bhat and B. D. Jain, *Talenta* **5**, 271–275 (1960).
[473] H. K. L. Gupta and N. C. Sogani, *J. Indian Chem. Soc.* **37**, 769–772 (1960).
[474] Keshav Kumar Tripathi and D. Banerjea, *Sci. Culture* (Calcutta) **23**, 611–612 (1958).
[475] E. Gagliardi and M. Khadem-Awal, *Mikrochim. Acta* **1969**, 882–887.
[476] M. Finkel'steinate, V. Budraitene, and A. Guzauskaite, *Nauch. Tr. Vyssh. Ucheb. Zaved. Lit. SSR, Khim. Khim. Tekhnol.* **1973** (15), 45–48.
[477a] Hiroshi Nishida, *Jap. Anal.* **12**, 56–57 (1963).

Adjust less than 100 ml of sample solution containing not more than 100 μg of iron to pH 3–4 with 2 ml of *M* acetate buffer. Add 10 ml of 0.01 *M* **aluminum-cupferron** in chloroform.[477b] Shake for 3 minutes and read the organic layer at 400 nm against a reagent blank. Beer's law is followed for 10–100 μg of iron entering the chloroform layer by ion exchange. To avoid interference by vanadium, add 1 ml of 3% hydrogen peroxide.

Ferric ion can be extracted from 7–8 *N* hydrobromic acid into a 1% solution of **Amberlite LA-1** in xylene. The organic layer is read at 475 nm.[478] The color is unchanged for a week. Copper is coextracted but can be removed by washing the organic layer with 7 *N* hydrobromic acid. Beer's law is followed for 1–9 ppm of iron in the extract.

Aminoacetonediacetic acid, which is 1-amino-propon-2-one-*N*,*N*-diacetic acid, forms a 2 : 1 complex with ferric ion in the presence of hydroxylamine hydrochloride and hydrogen peroxide.[479] At pH 5.5 and 370 nm Beer's law is followed for 2–20 μg of iron per ml. Recommended conditions are 2 ml of 0.03 *M* reagent, 2 ml of *M* hydroxylamine hydrochloride, 1 ml of 30% hydrogen peroxide, and 5 ml of *M* acetate buffer for pH 5.5 per 25 ml of final solution.

For reading a solution containing 0.06–0.6 mg of ferric ion, add 5 ml of acetate buffer solution for pH 5.5 and 5 ml of m*M* **2-aminoacetophenone-*N*,*N*-diacetic acid**, which is β-(hydroxyiminophenethyl)imino-diacetic acid.[480] Dilute to 25 ml and read at 420 nm. The reagent is less sensitive but more specific than a phenanthroline type. Bivalent copper interferes, but twofold amounts of cobalt or chromic ion are tolerated. There is interference by fluoride, phosphate, and vanadate ions.

Ferric ion complexes with **bis(2-aminoethyl)ether of ethylene glycol-*N*,*N*,*N*′,*N*′-tetracetic acid**, sometimes known as EGTA.[481] At pH 5 and 430 nm it follows Beer's law for 3–100 μg/ml. There is interference by nickel, manganese, cobalt, fluoride, cyanide, and large amounts of phosphate.

For ferrous ion as a 1 : 3 complex, add 5 ml of 4 m*M* **2-amino-5-nitrosopyrimidine-4,6-diol** and 4 ml of hydrochloric acid–sodium acetate buffer solution for pH 5 to the sample solution.[482] Read at 640 nm for up to 2.7 μg of iron per ml. There is serious interference by oxalate, EDTA, citrate, borate, cupric ion, vanadium, ferric ion, and cobalt.

To a solution containing 0.05–0.75 mg of iron, add 10 ml of 0.2 *M* hydrochloric acid–0.2 *M* potassium chloride as a buffer for pH 1.2–1.6. Add 5 ml of 0.3% solution of ***o*-aminophenol** in 0.1 *N* hydrochloric acid and dilute to 25 ml. Set aside for 20 minutes and read at 470 nm.[483] There is interference by hydrogen peroxide, chromic ion, pentavalent vanadium, ceric ion, and bromate.

8-Aminoquinoline at pH 1.25–1.5 is oxidized by ferric ion at 100°.[484] The stable, reddish-violet oxidation product is extracted with 1 : 1 benzyl alcohol–chloroform at pH 1.6–2 and read at 550 nm. Beer's law is followed for 1.5–150 μg of iron in 50 ml. Phosphate must be totally absent.

[477b] Yoshimi Sasaki, *ibid.* **25**, 103–107 (1976).
[478] Toshio Suzuki and Takeshi Sotobayashi, *ibid.* **13**, 866–871 (1964).
[479] Hiroko Takama, Takeshi Ando, and Keihei Ueno, *ibid.* **13**, 346–350 (1964).
[480] Takashi Ando and Keihei Ueno, *Bull. Chem. Soc. Jap.* **39**, 2400–2405 (1966).
[481] F. Bernij Martinez and M. Paz Castro, *Inf. Quim. Anal.* (Madrid) **13** (1), 1–7 (1950).
[482] S. K. Kundra, Mohan Katyal, and S. P. Singh, *Curr. Sci.* **44** (5), 81–82.
[483] A. Suteu, T. Hodisan, and Fiametta Kormos, *Rev. Roum. Chim.* **14**, 1613–1616 (1969).
[484] V. K. Gustin and T. R. Sweet, *Anal. Chem.* **35**, 1395–1397 (1963).

To 10 ml of a solution containing 1–15 μg of ferric ion, add 2 ml of 2% solution of sodium tetraborate as a buffer solution for pH 9–10 and 5 ml of 10% solution of sodium ***p*-aminosalicylate**. Heat at 60° for 15 minutes, cool, and read at 420 nm.

In the presence of 2,2′-bipyridyl, **aniline** is oxidized by ferric ion, which increases the absorption due to the bipyridyl.[485] Thus to a 10-ml sample containing not more than 20 μg of iron, add 0.5 ml of 10% solution of aniline in 1 : 4 hydrochloric acid and adjust to pH 4.7 by adding 15% sodium acetate solution, using bromophenol blue as indicator. Add 0.2 ml of 0.75% ethanolic 2,2′-bipyridyl. Heat at 100° for 5 minutes, cool, dilute to 15 ml, and read at 500 nm. Subtract a reagent blank.

For total iron, adjust the sample solution to pH 4.5, treat with **anthranilohydroxamic acid**, which is 2-aminobenzohydroxamic acid,[486] extract with isobutanol, and read at 450 nm.

Ferrous ion as well as thorium ion is extracted with amyl alcohol as a ternary complex with **arsenazo I** and **diphenylguanadine** at pH 4–8.[487]

The 1 : 2 complex of ferrous ion with **α,α′-azinodi(2-picoline)** at pH 4–7 in the presence of hydroxylamine hydrochloride forms a stable red color with a maximum absorption at 480 nm.[488] Beer's law is followed for 0.1–2 ppm of iron in 10 ml of acetate buffer solution at pH 4.2 containing 1 ml of 0.01 *M* reagent in ethanol and 1 ml of 0.1 *M* hydroxylamine hydrochloride. The absorption becomes stable after 20 minutes.

Ferric benzilate can be precipitated by adding a slight excess of 5% solution of sodium **benzilate.**[489] The filtered and washed precipitate is dissolved in ethanol and read at 375 nm over a range of 0.05–0.112 μg of ferric ion per ml. There is interference by the following ions: ferrous, cupric, titanous, gallium, indium, thoric, lanthanum, tetravalent zirconium, pentavalent vanadium, phosphate, cyanide, and thiocyanate.

To a weakly acidic sample containing less than 20 μg of ferrous ion, add 5 ml of methanolic 0.025% **benzil α-dioxime** and 2 ml of 5% solution of 4-picoline.[490] Adjust to pH 9, add 3 ml of methanol, and dilute to 20 ml. Extract the complex with 10 ml of chloroform and read at 559 nm.

For iron in cereal grains calcine at 600°. Dissolve the ash in 5 ml of hydrochloric acid and 3 ml of nitric acid.[491] Evaporate to dryness, dissolve in 50 ml of warm 1 : 9 hydrochloric acid, and filter. Wash the filter with warm water and dilute the filtrate to 100 ml. Mix a 5-ml aliquot with 5 ml of 0.5% solution of **benzohydroxamic acid** in 20% methanol. Read at 480 nm.

When a sample solution containing 25–200 μg of iron is treated with 2 ml of 10% sodium tartrate, it masks most metal ions.[492] Addition of 2 ml of 2% solution of **1-(benzo[*b*]thiophen-3-yl)-4,4,4-trifluorobutane-1,3-dione** in ethanol is followed by adjustment of pH to 1.5–2.5. Color of a 3 : 1 complex with iron is developed by heating near boiling for 25 minutes. On cooling, extract with 10 ml, 5 ml, and 5 ml

[485] Gyula Almassy and Maria Kavai, *Magy. Kem. Foly.* **61**, 246–248 (1955).

[486] R. L. Dutta, *J. Indian Chem. Soc.* **37**, 167–170 (1960).

[487] L. A. Kvichko, V. N. Tolmachev, and V. D. Konkin, *Ukr. Khim. Zh.* **36**, 494–496 (1970).

[488] Atsushi Sugii, Motoko Dan, Yoko Inoue, and Hiromi Nakamura, *Jap. Anal.* **14**, 1133–1137 (1965).

[489] S. S. Gupta and D. Mukerjee, *Clin. Anal.* **49**, 37 (1967).

[490] Hisahiko Einaga, *Mikrochim. Acta* **1** (1), 67–73 (1976).

[491] Yahya Abdoh, Mohammed H. Khorgami, and Mohammed T. Kowsar-Nechan, *Lebensm.-Wiss. Technol.* **5** (6), 219–220 (1972).

[492] J. Gerard, W. J. Holland, A. E. Veel, and J. Bozic, *Mikrochim. Acta* **1969**, 724–730.

of chloroform and dilute the combined extracts to 25 ml with chloroform. Read at 516 nm against a reagent blank. Beer's law is followed for 1–14 ppm of iron with an optimum of 3.1–6.9 ppm. There is serious interference by EDTA, citrate, oxalate, phosphate, tin, and vanadium.

For reading 10–40 ppm of ferric ion, it is complexed in 60% ethanol at pH 3 with **benzoylacetanilide**, which is 1-analino-3-phenylpropane-1,3-dione.[493a] There is interference by fluoride, thiocyanate, oxalic acid, and other organic acids.

Shake 10 ml of 7 *N* nitric acid sample solution with 10 ml of 0.1 *M* **4-benzoyl-3-methyl-1-phenyl-3-pyrazolin-5-one** in benzene for 5 minutes.[493b] Dilute the organic phase to 25 ml with benzene and read at 460 nm. As an alternative the solvent of the sample may be *N* hydrochloric acid or 7.2 *N* sulfuric acid. Beer's law is followed for 50–250 μg of iron in the sample. Chromic ion interferes.

The condensation product of **biacetyl** and **hydrazine** formed *in situ* gives a stable red complex with ferrous ion.[494] To the sample solution containing 25–175 mg of ferrous ion, add 3 ml of 1% solution of biacetyl in ethanol, 6 ml of 0.5% solution of hydrazine, and 1 gram of sodium bicarbonate to raise to pH 9.4. Heat at 60° for 5 minutes, then add 20% acetic acid to lower to pH 4–7. Continue at 60° for 10 minutes, cool, and read at 490 nm. Sodium perchlorate may be added, the complex extracted with nitrobenzene, and the extinction read at 498 nm.

For 1–20 μg of ferric ion per ml in 0.25 *M* perchloric acid, mix 2 ml of sample solution, 0.5 ml of 25% ascorbic acid solution, and 5 ml of 0.04 *M* **biacetyl monoxime thiosemicarbazone.**[495] Dilute to 10 ml and set aside for 30 minutes. Read at 507 nm. Interfering elements are bismuth, cobalt, nickel, copper, silver, and mercury.

The ferric complex of ***N,N*-bis(carboxymethyl)anthranilic acid** is formed at pH 1.4–1.6 and read at 370 nm.[496] Copper, nickel, and cobalt interfere.

To a weakly acid sample containing up to 0.4 mg of ferric ion, add 5 ml of *M* acetic acid–*M* sodium acetate as a buffer.[497] Add 20 ml of ethanol and 5 ml of 0.4% solution in ethanol of **bithionol**, which is 2,2′-thiobis(4,6-dichlorophenol). Dilute to 50 ml and extract with 8 ml, 4 ml, and 3 ml of chloroform. Dilute the combined extracts to 50 ml with ethanol and read at 484 nm. Oxalate, citrate, and phosphate interfere.

Ferrous ion forms 1 : 3 complexes with **7-bromo-1,3-dihydro-5-(2-pyridyl)-1,4-benzodiazepin-2-one** and the dihydrochloride of its 1-(3-dimethylaminopropyl) analogue at pH 5–7.[498] Each has maximum absorption at 580 nm and follows Beer's law for 2.5–30 μg of iron per 100 ml of developed solution. The complex is extractable from ammoniacal solution with dichloromethane and is then linear for 5–30 μg of iron per 25 ml.

[493a] K. P. Srivastava and A. D. Taneja, *Microchem. J.* **17**, 540–545 (1972).

[493b] G. Nageswara and H. C. Arora, *Microchem. J.* **21**, 1–4 (1976).

[494] I. Nunez de Castro, E. Graciani, J. Gasch, and F. Pino Perez, *Inf. Quim. Anal. Pura Apl. Ind.* **20**, 128–142 (1966); E. Constante Graciani and J. M. Olias Jiminez, *An. Quim.* **67**, 615–622 (1971).

[495] M. Valentova and L. Sucha, *Collect. Czech. Chem. Commun.* **38**, 1497–1501 (1973); cf. D. Perez Bendito and F. Pino Perez, *Inf. Quim. Anal. Pura Apl. Ind.* **22**, 177–192, 201 (1968).

[496] Donald L. Dinsel and Thomas R. Sweet, *Anal. Chem.* **33**, 1078–1080 (1961).

[497] A. G. Fogg, A. Gray, and D. Thorburn Burns, *Anal. Chim. Acta* **45**, 196–198 (1969); **47**, 151–153 (1969).

[498] J. D. Sabatino, O. W. Weber, G. R. Padmanabhan, and B. Z. Senkowshi, *Anal. Chem.* **41**, 905–909 (1969).

A ternary complex is formed by **4-(2-bromo-4,5-dihydroxyphenylazo)benzenesulfonate**, cetylpyridinium chloride, and ferric ion in hexamine–hydrochloric acid buffer solution for pH 5.5–6.5.[499] After setting aside for 30 minutes, extract with chloroform and read at 565 nm. Beer's law is followed for up to 0.7 μg of ferric ion per ml. Masking avoids interference by cupric, zirconium, molybdate, and lead ions. There is serious interference by gallium, indium, titanium, uranate, oxalate, and EDTA.

7-Bromo-1-(3-dimethylaminopropyl)-1,2-dihydro-5-(2-pyridyl-2*H*-1,4-benzodiazepin-2-one dihydrochloride forms a brilliant violet-blue with ferrous ion. To prepare the reagent, dissolve 3 grams of ascorbic acid and 0.33 gram of the compound above in 80 ml of *M* acetate buffer for pH 4.8, dilute to 100 ml with the buffer solution, and add 0.05 ml of 0.02% solution of Sterox SE.[500] For determination of hemoglobin, mix 0.02 ml of whole blood with 2 ml of 2.5% sodium hypochlorite solution and set aside for 3 minutes. Mix with 4 ml of prepared reagent, allow 5 minutes for color development, and read at 580 nm against a reagent blank.

Iron in serum and iron-binding capacity are similarly determined.[501]

The 1:1 ferric complex of **5-bromo-8-hydroxyquinoline-7-sulfonate** has a maximum absorption at 585 nm and is stable at pH 2.5–4.[502] There is interference by vanadium, cobalt, manganese, cadmium, zinc, nickel, palladium, and uranium. It follows Beer's law for 5–50 ppm of iron with an optimum range of 10–40 ppm.

To a sample solution containing up to 20 μg of ferrous ion, add 6 ml of 0.1 m*M* **bromophenol red**, 1 ml of m*M* 1,10-phenanthroline, and a crystal of ascorbic acid.[503] Adjust to pH 9, dilute to 10 ml, and extract with 10 ml of nitromethane. Adjust the extract to 10 ml and read at 420 nm.

Ferrous ion forms 1:3 complexes with **7-bromo-5-(2-pyridyl)-2*H*-1,4-benzodiazepin-2-one** and with **7-bromo-1,3-dihydro-1-(3-dimethylaminopropyl)-5-(2-pyridyl)-2*H*-1,4-benzodiazepin-2-one dihydrochloride**. The optimum pH is 5–7, and each has a maximum absorption at 580 nm.[504] Beer's law is followed for 2.5–30 μg of iron per 100 ml. The complexes can be extracted from ammoniacal solution with dichloromethane and the extract follows Beer's law for 20–120 μg of iron per 100 ml of extract.

Ferric ion complexes with **2-bromo-3-thiophenecarbohydroxamic acid** at pH 2–9.[505] At pH 2 the maximum absorption is at 500 nm, shifting with increase of pH to shorter wavelengths. Above pH 4, a protective colloid is required to avoid turbidity. The reagent also forms colored, water-soluble complexes with vanadium, hexavalent molybdenum, and tungsten, and with uranyl ion.

Calcichrome forms a 1:1 ferric complex at pH 4 in an acetate buffer solution with a maximum absorption at 310 nm.[506] It forms a 2:1 ferric complex at pH 6 in

[499] Toshinobu Wakamatsu and Makoto Otomo, *Anal. Chim. Acta* **79**, 322–325 (1975).

[500] B. Klein, B. K. Weber, L. B. Lucas, J. A. Foreman, and R. L. Searcy, *Clin. Chim. Acta* **26**, 77–84 (1969).

[501] A. Bernegger, H. Keller, and R. Wenger, *Z. Klin. Chem. Klin. Biochem.* **10** (8), 359–362 (1972).

[502] K. B. Ba.achandran and Samir K. Banerji, *Mikrochim. Acta* **1968**, 1138–1142; *Chem. Age India* **19**, 455–457 (1968).

[503] Francizek Buhl and Barbara Mikula, *Chem. Anal.* (Warsaw) **19**, 1225–1229 (1974).

[504] J. D. Sabatino, O. W. Weber, G. R. Padmanabhan, and B. Z. Senkowski, *Anal. Chem.* **41**, 905–909 (1969).

[505] J. Minczewski and Z. Skorko-Trybula, *Chem. Anal.* (Warsaw) **5**, 163–165 (1960).

[506] Hajime Ishii and Hisahiko Einaga, *J. Chem. Soc. Jap., Pure Chem. Sect.* **87**, 410–447 (1966).

glycine buffer solution with a maximum at 540 nm. In 0.04 m*M* calcichrome Beer's law is followed for 0.2–1.5 and 0.4–3 ppm of iron, respectively. At 555 nm the difference in absorption between the complexes and the reagent is greatest. The reagent also complexes with cupric, aluminum, titanic, vanadate, and zirconium ions. Mercury and nickel interfere at pH 6 but not at pH 4. Chromic ion interferes at pH 4. Oxalate decreases the absorption at both pH levels.

Calcichrome also forms a 2:1 complex with ferrous ion in neutral solution.[507a] The extinction reaches maxima at 310 and 545 nm in 20 minutes. At 555 nm Beer's law is followed up to 1.2 ppm of iron in acetate buffer solution for pH 6.2 with 0.04 m*M* calcichrome. Positive errors are caused by copper, aluminum, titanium, vanadium, zirconium, mercury, nickel, and cobalt; a negative error is due to chromium.

At pH 5.6–7.3 ferric ion complexes with the sodium salt of ***N*-(*o*-carboxybenzoyl)-*N*-phenylhydroxylamine** which is *N*-hydroxy-*N*-phenylphthalamate for reading 2.6–10 ppm at 480 nm.[507b] There is interference by vanadium, platinum, palladium, molybdenum, tungsten, fluoride, EDTA, and oxalate.

A ternary complex of ferric ion–***N*,*N*-bis(carboxymethyl)anthranilic acid**–hydrogen peroxide is formed at an optimum molar ratio of 1:2:100.[508] The maximum absorption is around 500 nm. Manganese interferes.

1-*o*-(Carboxyphenyl)-3-hydroxy-3-methyltriazen, which is 2-(3-hydroxy-3-methyltriazeno)benzoic acid, forms a 1:2 complex with ferric ion at pH 4–9.4 having an absorption maximum at 570 nm and following Beer's law for 8.9–35.8 ppm of iron.[509] At pH 1.5–2 the complex is 1:1, has a maximum absorption at 660 nm, and conforms to Beer's law for 3.9–11.2 ppm of iron.

A 1:2:4 complex of ferrous ion with **catechol and aniline** is extracted with chloroform or benzene from a solution buffered with acetate to 5.1–5.6.[510] The maximum absorption is at 540 nm and Beer's law is followed for 2–70 μg of ferrous ion per ml.

As a **catechol violet** reagent, mix 0.4 ml of 0.001 *M* solution with 5 ml of *M* ammonium acetate, 0.5 ml of 0.0486 *M* acetic acid, and 1 ml of 0.92 *M* hydrogen peroxide.[511] Add to the sample solution, dilute to 50 ml, and set aside for 12 minutes before reading at 610 nm. There is interference by the following ions: cupric, molybdate, aluminum, magnesium, antimony, EDTA, and phosphate.

There is a 1:1:1 complex of catechol violet, ferric ion, and diphenylguanidine.[512] Buffered at pH 5.9 with ammonium hydroxide–ammonium acetate, the concentration of diphenylguanidine is desirably 7 m*M*. Chloroform extracts the complex for reading down to 0.02 μg of iron per ml. Beer's law is followed up to 0.6 μg of iron per ml.

Ferric ion forms varied complexes with **2-carbethoxy-5-hydroxy-1-(*p*-tolyl)-4-pyridone**, which is ethyl-1,4-dihydro-5-hydroxy-4-oxo-1-(*p*-tolyl)picolinate.[513] To 1

[507a] Hajime Ishii and Hisahiko Einaga, *Jap. Anal.* **15**, 577–581 (1966).

[507b] S. P. Mathur, C. S. Bhandari, and N. C. Sogani, *Revta Latinoam. Quim.* **7**, 63–64 (1976).

[508] C. Dragulescu and M. Pirlea, *Zh. Anal. Khim.* **23**, 224–226 (1968).

[509] A. K. Majumdar, B. C. Bhattacharyya, and B. C. Roy, *Anal. Chim. Acta* **67**, 307–315 (1973).

[510] T. D. Ali-Zade, G. A. Gamid-Zade, and E. M. Ganieva, *Uch. Zap. Azerb. Gos. Univ., Ser. Khim. Nauk.* **1972** (4), 30–35.

[511] J. Birmantas and E. Jasinskiene, *Zh. Anal. Khim.* **20**, 811–813 (1965).

[512] R. Jurevicius and C. Valiukevicius, *Nauch. Tr. Vyssh. Ucheb. Zaved. Lit. SSR, Khim. Khim. Tekhnol.* **1972** (14), 53–56.

[513] M. J. Harek, M. Janko, and B. Tamhina, *Mikrochim. Acta* **1973**, 783–795.

ml of neutral or slightly acid sample solution containing 0.01–0.1 mg of ferric ion, add 2 ml of citrate buffer solution for pH 3 and 5 ml of 5 mM reagent in 50% ethanol. Dilute to 10 ml and read at 443 nm.

Alternatively, shake a sample solution containing 0.01–0.1 mg of ferric ion in 0.5–1 M hydrochloric acid with 5 ml and 3 ml of 0.01 M reagent in chloroform. Dilute the combined extracts to 10 ml with chloroform and read at 432 nm. Interferences are phosphate, cyanide, oxalate, tartrate, zirconium, titanic ion, and pentavalent vanadium.

Iron in water is measured by its catalytic effect on the oxidation in acetic acid of **4,4′-bis[bis(carboxymethyl)amino]stibine-2,2′-disulfonic acid** by hydrogen peroxide.[514] The result is read fluorometrically with a sensitivity of 1 ng/ml.

Neutralize 50 ml of a solution containing up to 1.5 mg of iron in acid solution.[515] Use 2 drops of 0.2% *p*-nitrophenol indicator and 8% sodium hydroxide solution, adding 2 drops of the alkali in excess. Add 20 ml of 2% solution of **catechol** and set aside for 15 minutes. Add 10 ml of 0.067 M phosphate buffer for pH 5.3 and dilute to 100 ml. Let stand for 1 hour and read at 580 nm. Beer's law is obeyed up to 15 ppm. Vanadium must be absent.

Although of less importance than in the past, the reading of ferric ion as the **chloride** persists.[516a] A solution of diantipyrinylmethane in chloroform or dichloroethane extracts yellow ferric chloride from a dilute hydrochloric acid solution.[516b] This separates iron from titanium, aluminum, copper, and cobalt. When iron-cobalt catalyst for synthesis of hydrocarbons from water gas is dissolved in hydrochloric acid, the two elements are read as chlorides with $\pm 10\%$ accuracy.[517] For analysis of high purity bismuth, dissolve in nitric acid, evaporate to dryness, convert to the chloride by evaporating with hydrochloric acid, and read at 390 nm.[518] Beer's law is followed up to 20 μg/ml. For iron and copper in zinc sulfate solution read as the chlorides and determined by simultaneous equations, see Copper.

2-Chloro-10-(3-dimethylaminopropyl)phenothiazine hydrochloride at a concentration greater than 0.2% gives an absorption at 512 nm conforming to Beer's law for 1–8 μg of ferric ion per ml.[519a] The color is stable for 15 minutes. EDTA interferes.

At pH 2.2–3.7 a blue complex is formed by ferric ion and ***N*-4-chlorophenyl-*N*-hydroxy-*N*′-phenylbenzamidine** in 60% ethanol for reading at 590 nm.[519b] This is stable at 15–40° for 2 hours. Beer's law is followed for 1–14 ppm of iron. There is serious interference by cupric ion, aluminum, molybdate, tungstate, fluoride, and citrate.

A 1 : 1 chelate of ferric ion with **chromotrope 2R** at pH 1.8–3.5 is read at 530 nm for 2.6–8.9 ppm of iron.[520]

[514]A. A. Abraztsov and V. G. Bocharova, *Tr. Voronezh. Gos. Univ.* **82**, 182–184 (1971).

[515]A. N. Smith, *Analyst* **84**, 516 (1959).

[516a]Masayoshi Ishibashi, Tsunenobu Shigematsu, Yuuroku Yamanoto, Masayuki Tabushi, and Toyokichi Kitagawa, *Bull. Chem. Soc. Jap.* **30**, 433–437 (1957); cf. W. Davison and E. Regg, *Analyst* **101**, 634–638 (1976).

[516b]E. A. Selezneva and V. P. Zhivopistaev, *Uch. Zap. Perm. Gos. Univ.* **1974** (324), 201–206.

[517]Yu. L. Polyakin, *Tr. Grozn. Neft. Inst.* **1955** (20), 24–32.

[518]J. H. High and P. J. Placito, *Analyst* **83**, 522–525 (1958).

[519a]Masao Maruyama, Tomonori Miki, and Shogo Ueda, *Bull. Chem. Soc. Jap.* **31**, 998–999 (1958).

[519b]K. K. Deb and R. K. Mishra, *Curr. Sci.* **45**, 134–135 (1976).

[520]A. Prakash and S. P. Mushran, *Chim. Anal.* (Paris) **49**, 473–476 (1967).

At a pH below 6, ferric ion and **chromotropic acid** form a 1 : 1 complex having a maximum absorption at 600 nm.[521] Above pH 7 the complex is 1 : 2 and has a maximum absorption at 660 nm. Ferric ion forms a yellow-green 1 : 1 complex with **3,6-dichlorochromotropic acid** at pH 2 with a maximum absorption at 760 nm; at pH 5 it forms a 1 : 1 complex with a maximum at 660 nm. All the foregoing complexes conform to Beer's law for 0.04–0.4 m*M* ferric ion.

The sum of aluminum and ferric ion with **chromoxane violet R** (mordant violet 16) is read at 530 nm, which is the maximum for ferric ion.[522a] Then ferric ion is reduced to ferrous ion in another portion of the sample and read to give a correction for the aluminum, since ferrous ion does not give a color with this reagent. For iron in water by chromoxane violet R, see Aluminum in water by that reagent.

A stable 1 : 3 orange-red complex is formed by ferric ion with **cyclopiroxolamine**, which is the ethanolamine adduct of 6-cyclohexyl-1-hydroxy-4-methyl-2-pyridone.[522b] This is extracted with chloroform and the dried extract read at 415 nm against a reagent blank. When formed at pH 1 only aluminum and stannic ion interfere. At higher pH levels there are numerous interferences.

The complex of ferric ion with **croconic acid**, which is 4,5-dihydroxycyclopent-4-ene-1,2,3-trione, is read at 500 nm for 0.2–3 mg of iron in 50 ml.[523]

Crystal violet complexes with a bivalent ferric hexachloride, $(FeCl_6)^{3-}$, ion for reading. Adjust a 25-ml sample containing 0.5–4 mg of ferric ion to pH 1–5.[524] Add 25 ml of 8 *N* hydrochloric acid and extract with 3 ml, 3 ml, and 3 ml of 0.05% solution of crystal violet in chloroform. Dilute the extracts to 10 ml with methanol and read at 590 nm against a reagent blank.

To a sample solution, alkaline to litmus, add 3 ml of 10% solution of pyridine, 0.5 ml of 1% solution of **1,2-cyclohexanedione dioxime**, and 5 ml of 20% solution of sodium potassium tartrate.[525] Dilute to 20 ml and extract with 5 ml and 5 ml of benzene. Read the combined extracts at 520 nm.

Cyclohexane-1,2-dione dithiosemicarbazone is a reagent for ferrous and ferric ions as well as for cobalt and nickel.[526] For ferrous ion, mix the sample with 5 ml of 0.25% solution of the reagent in dimethylformamide, 10 ml of sodium acetate–acetic acid buffer solution for pH 5.5, and 10 ml of ethanol. Dilute to 50 ml and read at 450 or 500 nm. For ferric ion, treat the sample with 2 ml of 0.25% solution of the reagent in dimethylformamide, 10 ml of ethanol, and 10 ml of phthalate–hydrochloric acid buffer solution to adjust to pH 2.5–4.2. Dilute to 50 ml and read at 400 nm. Beer's law is followed for 1.5–9 ppm of ferrous ion and for 1–3.2 ppm of ferric ion.

For determination of iron with **2,5-diacetylpyridine dioxime**, it is reduced by

[521] V. A. Nazarenko and E. A. Biryuk, *Zh. Anal. Khim.* **24**, 44–47 (1969); cf. Lumir Sommer, *Chem. Listy* **52**, 1485–1500 (1958).

[522a] I. S. Mustafin and N. F. Lisenko, *Zh. Anal. Khim.* **17**, 1052–1056 (1962).

[522b] Takeo Kuriki, Tsuyoshi Tsujiyama, Noriko Suzuki, and Nobuo Suzuki, *Jap. Anal.* **24**, 112–115 (1975).

[523] L. P. Adamovich and M. S. Kravchenko, *ibid.* **26**, 545–547 (1971).

[524] W. Likussar, O. Wawschinek, and W. Beyer, *Anal. Chim. Acta* **40**, 538–539 (1968).

[525] V. M. Peshkova and V. M. Bochkova, *Metody Anal. Redk. Tsvet. Met. Sb.* **1956**, 15–23.

[526] J. A. Munoz Leyva, J. M. Cano Pavon, and F. Pino Perez, *Quim. Anal.* **28**, 90–98 (1974).

hydroxylamine. The ferrous complex is formed at pH 12.5 and read at 490 nm.[527] Beer's law is followed at 0.1–5 μg/ml.

When ferric ion is complexed with **di(2-aminoethyl)ether-*NNN'N'*-tetraacetic acid**, it is read at 420 nm for 20–280 μg of iron per ml at pH 2–10.[528]

At 260 nm, 0.2–3 ppm of ferric ion can be read with **1,2-diaminocyclohexane-*NNN'N'*-tetraacetic acid, diethylenetriamine-*NNN'N''N''*-pentaacetic acid, *N*-(i-hydroxyethyl)-ethylenediamine-*NN'N'*-triacetate**, and **1,2-bis(2-amino-ethoxy)ethane-*NNN'N'*-tetraacetic acid.**[529] There is interference by copper, hexavalent uranium, pentavalent vanadium, bismuth, lead, manganese, mercury, and cadmium.

The purple complex of ferric ion with the first of the reagents above can be read as a ternary complex with hydrogen peroxide at 530 nm.[530] To prevent interference, add excess reagent to 0.5–80 μg of iron per ml and adjust to pH 10.8–12 with 20% sodium hydroxide solution.

Dianisidine, which is 3,3'-dimethoxybenzidine, applied at pH 2.5–2.6, is allowed to develop for 15 minutes and read at 450 nm.[531] It is applied to the solution of the ash of blood.

To a sample at pH 2.5–3 containing less than 20 μg of ferric ion, add 0.5 ml of 5% solution of **dibenzoylmethane** in acetone and heat at 70° for 15 minutes.[532] Extract with 20 ml of butyl acetate and read the extract at 410 nm. Beer's law holds up to 2 ppm of ferric ion. There is serious interference by cupric, molybdate, titanic, and EDTA ions.

Alternatively,[533] extract 10 ml of sample in 1:1 hydrochloric acid containing 10–30 μg of ferric ion by shaking for 1 minute with 3 ml of methyl isobutyl ketone and dehydrate the extract with sodium sulfate. Mix a 2-ml aliquot of the extract with 2 ml of 0.5% solution of dibenzoylmethane in acetone and 0.5 ml of pyridine. Dilute to 10 ml with methanol and read at 410 nm against a reagent blank. Beer's law is followed up to 4 mg of iron per ml.

Ferric ion is read as a blue-green complex with **5,7-dibromo-8-hydroxyquinoline** in 2:25 0.1 *N* nitric acid–ethanol.[534] Separation from interfering ions is by descending paper chromatography on carboxymethylcellulose, using as solvent 100 ml of butanol and 100 ml of 15% solution of trichloroacetic acid containing 1 gram of 1-phenylbutane-1,3-dione. Extract the spots with warm ethanol. Then add 4 ml of 0.1 *N* nitric acid plus 6 ml of 0.04% solution of the color reagent. After dilution to 50 ml with ethanol and setting aside for 10 minutes, read at 465 nm. Beer's law is followed for 0.4–4 μg of iron per ml.

A 1:1 complex of ferric ion with **3,6-dichlorochromotropic acid** at pH 2 has a maximum absorption at 760 nm.[535] Beer's law applies for 1–28 μg of iron per ml.

[527]E. Gagliardi and P. Presinger, *Mikrochim. Ichnoanal. Acta* **1965**, 1047–1052.
[528]F. Bernejo-Martinez and A. G. Blas-Perez, *Inf. Quim. Anal.* **16** (5), 129–132 (1962).
[529]Masako Idemori, *J. Chem. Soc. Jap., Pure Chem. Sect.* **85**, 331–335 (1964).
[530]F. Bermejo Martinez and A. Blas Perez, *Microchem. J., Symp. Ser.* **2**, 333–342 (1962).
[531]Const. Vassiliades and George Manoussakis, *Bull. Soc. Chim. Fr.* **1961**, 582–583.
[532]Tsunenobo Shigematsu, Masayuki Tabushi, and Tsunehiko Tarumoto, *Bull. Inst. Chem. Res., Kyoto Univ.* **40**, 388–399 (1962).
[533]Yoshimi Umezaki, *Bull. Chem. Soc. Jap.* **37**, 70–73 (1964).
[534]I. I. M. Elbeih and M. A. Abou-Elnaga, *Chemist-Analyst* **56**, 99–100 (1967).
[535]N. N. Basargin and Hz. I. Nemtseva, *Zh. Anal. Khim.* **20**, 966–975 (1965).

Interference by titanic ion, hexavalent molybdenum or tungsten, and pentavalent niobium necessitates the removal of these metals.

For iron in electrolytic nickel by sodium **diethyldithiocarbamate** after extraction of the copper complex, see Copper.

To a sample solution, add 7.5 ml of 0.1 *M* **2′,4′-dihydroxyacetophenone** in 40% ethanol and adjust to pH 2.95.[536] Dilute to 50 ml and set aside for 30 minutes. Read at 470 nm against a sample blank. Beer's law is followed for 1–56 ppm of iron in the final solution. Phosphate, fluoride, citrate, and oxalate interfere.

For ferric ion add 2 ml of 1% ethanolic **2,3-dihydroxybenzoic acid** to the ferric salt in the corresponding acid.[537] Dilute to 25 ml and read at 600 nm against water. Beer's law is followed for 5–27 μg of iron per ml. Large amounts of iodide, thiocyanate, phosphate, or fluoride interfere.

A 1 : 1 complex of **2′,4′-dihydroxybutyrophenone** with ferric ion is read at pH 2.9–3 and 470 nm.[538] Beer's law is followed for 1–56 ppm of iron. Development takes 30 minutes. There is interference by phosphate, fluoride, citrate, oxalate, and vanadate.

For ferric ion by **2,4′-dihydroxybutyrophenone oxime** at 0.002–0.05 *M* in 40% ethanol, adjust to pH 2.7 and maintain the ionic strength at 0.1 *M* sodium perchlorate.[539] The ethanol in the final solution is maintained at 24% for reading at 510 nm. Beer's law is followed for 1–56 ppm of ferric ion. There is interference by phosphate, fluoride, citrate, and oxalate.

2′,4′-Dihydroxychalcone forms a complex with ferric ion at pH 2.8–3.2 which has a maximum absorption at 440 nm and is stable for 20 hours.[540a] There is interference by zirconium, ceric, and vanadate ions. Beer's law does not apply.

To a sample solution containing 25–175 μg of ferrous ion add 1 ml of 1% solution of hydroquinone, 5 ml of sodium acetate-acetic acid buffer solution for pH 3.5, and 4 ml of 0.1% ethanolic **2,4-dihydroxy-3-nitrosopyridine.**[540b] Dilute to 50 ml, set aside for 15 minutes, and read at 655 nm.

The complex of ferrous ion with **4,7-dihydroxy-1,10-phenanthroline** in alkaline solution is read at 515 nm for 0.6–50 μg of iron in 25 ml of solution.[541]

The blue 1 : 1 complex of ferric ion with **2,2-dihydroxypyridine** in *N* strong acid follows Beer's law up to 0.8 m*M*.[542]

A 1 : 1 complex of ferric ion with **2′,4′-dihydroxyvalerophenone oxime** is formed at pH 2.6–2.8 using fifteenfold excess of reagent and read at 510 nm.[543] Beer's law is followed for 1–56 ppm of ferric ion, preferably 9–29 ppm. Phosphate, fluoride, citrate, and oxalate interfere.

When 0.2–6 ml of ferric nitrate solution containing 58 μg of iron per ml is

[536]M. H. Gandhi and Mahendra N. Desai, *Anal. Chim. Acta* **43**, 338–340 (1968).

[537]M. L. Finkel'shteinaite, *Zh. Anal. Khim.* **19**, 516–517 (1964).

[538]M. H. Gandhi and Mahendra N. Desai, *Analyst* **93**, 528–531 (1968).

[539]M. H. Gandhi and M. N. Desai, *Indian J. Appl. Chem.* **32**, 360–364 (1969).

[540a]K. Syamasunder, *Proc. Indian Acad. Sci., A*, **59**, 241–250 (1964).

[540b]S. Musumeci, E. Rizzarelli, I. Fragala, S. Smartano, and R. P. Bonomo, *Z. Anal. Chem.* **282**, 221 (1976).

[541]E. Gagliari and P. Höhn, *Mikrochim. Ichnoanal. Acta* **1964**, 1036–1042.

[542]J. A. Thomson and G. F. Atkinam, *Anal. Chim. Acta* **49**, 531–535 (1970).

[543]Jai Singh, S. P. Gupta, and O. P. Malik, *Indian J. Chem.* **13**, 1217–1220 (1975); Jai Singh and S. P. Gupta, *Curr. Sci.* **44**, 612–614 (1975).

treated with 1 ml of methanolic **3,5-diiodosalicylic acid** and diluted with methanol to 10 ml, the absorption at 558 nm conforms to Beer's law.[544]

The complex of ferric ion with **3,3-dimethoxylbenzidine** at pH 2.5–3 is let stand for 40 minutes and read at 450 nm.[545] Ferrous ion does not react. There is interference by an equal concentration of gold and nitrite and by a twofold concentration of silver, palladous ion, ruthenium, and platinic ion. Tartrates, chromates, dichromates, and iodides interfere.

For determination of iron in aluminum, dissolve 0.2 gram of sample in 20 ml of hydrochloric acid and evaporate nearly to dryness.[546] Take up the residue in 5 ml of hot water and filter if necessary. Add 3 ml of 10% solution of hydroxylamine hydrochloride and 5 ml of 4 m*M* ***N,N'*-dimethylamide**, which is *N,N'*-dimethylpyridine-2,4-di(thioamide). Dilute to 25 ml with 8% solution of sodium hydroxide and read at 588 nm. Copper must be masked with thiourea.

Mix a sample solution containing less than 5 μg of iron with 0.5 ml of 1% solution of hydroxylamine hydrochloride and 5 ml of 0.2% solution of **5-dimethylamino-2-nitrosophenol** in 0.1 *N* hydrochloric acid.[547] Add buffer solution and adjust to pH 7–9.5. Dilute to 50 ml and read at 750 nm. The complex is extractable with zephiramine in chloroform. Nickel and copper interfere but can be masked with EDTA.

At pH 4–7, **3,3'-bi(5,6-dimethyl-1,2,4-triazinyl)** forms an intensely orange ferrous complex having maximum absorption at 408 and 493 nm.[548a] Adjust a 5-ml sample containing 0.4–7 ppm of ferrous ion to pH 4, add 2 ml of 10% hydroxylamine hydrochloride, 5 ml of acetate buffer solution, and at least a fortyfold excess of the reagent. There is interference by copper, nickel, ruthenium, chromium, and thiosulfate.

At pH 3–6 resorcinol and nitrite react to form **2,4-dinitrosoresorcinol** in the solution. This immediately forms a 3:1 monovalent ferrous complex ion.[548b] This can be extracted as a ternary complex with tetrabutylammonium ion or tributylammonium ion in chloroform at pH 5–7.5 for reading at 690 nm.

The complex of ferrous ion with **diphenylthiovioluric acid** at 0.2–1.2 ppm is read at 660 nm.[549]

In methanol or ethanol at pH 5–6, ferrous ion can be read as the salt of **dipicolinic acid**, which is pyridine-2,6-dicarboxylic acid, for less than 17 mg of iron per 100 ml.[550] Chromium, cobalt, and nickel must be corrected for. Copper, manganese, cerium, and vanadium must be masked. The optimum pH is 4–8 with reading at 468 nm.

A deep-yellow, water-soluble 2:1 complex is formed by **1,2-di-2-pyridylethanediol** with ferric ion at pH 6.5–7.[551a] The maximum absorption is at 360 nm,

[544]I. C. Ciurea, G. Popa, and C. Lazar, *Anal. Univ. Bucur., Ser. Stiint. Nat. Chim.* **14**, 113–120 (1965).
[545]Const. Vassiliades and G. Manoussakis, *Bull. Soc. Chim. Fr.* **1960**, 390–393.
[546]E. Gagliardi and A. Ban, *Mikrochim. Acta* **1973**, 763–769.
[547]Kyoji Toei, Shoji Motomizu, and Takashi Kozenaga, *Analyst* **100**, 629–636 (1975).
[548a]R. E. Jensen and R. T. Pflaum, *Anal. Chim. Acta* **32**, 235–244 (1965).
[548b]V. M. Peshkova, T. V. Polenova, and Yu. A. Barbalat, *Zh. Anal. Khim.* **32**, 471–477 (1977).
[549]Lal C. Malik and R. P. Singh, *J. Indian Chem. Soc.* **33**, 335–338 (1956).
[550]Heinrich Hartkamp, *Z. Anal. Chem.* **190**, 66–80 (1962); cf. Ichiro Morimoto, Takako Hayashi, and Hideo Oida, *J. Chem. Soc. Jap.* **82**, 203–205 (1961).
[551a]R. D. Gupta and B. D. Jain, *Indian J. Appl. Chem.* **29** (2–3), 64–68 (1966).

and there is conformity to Beer's law for 0.84–7.5 ppm of iron. Silver interferes seriously. The complex is stable for 32 hours.

To 40 ml of sample solution containing 40–150μg of iron add 2 ml of 10% solution of ascorbic acid.[551b] Set aside for 1 minute and add 2.5 ml of 0.01 *M* ethanolic **2,2′-dipyridyl-2-pyridylhydrazone**. Adjust to pH 1.6–3.4 with dilute solution of sodium hydroxide or perchloric acid and dilute to 50 ml. Read at 538 nm against a reagent blank. Cobalt interferes seriously.

Dipyridylglyoxal dithiosemicarbazone forms a 1:1 complex with ferrous ion at a pH less than 2.5.[552] The maximum absorption is at 550 nm. Over the range of pH 5–10 it forms a 2:1 complex which is extractable with chloroform for reading at 630 nm. The presence of ascorbic acid is necessary to avoid oxidation to ferric ion in both cases. Beer's law is followed for the 1:1 complex at 2–8 ppm and for the 2:1 complex at 2–9 ppm.

To 50 ml of sample solution, add 5 ml of 8 *N* hydrochloric acid, 1 ml of 10% solution of hydroxylamine hydrochloride, 2 ml of ethylenediamine, and 5 ml of 0.012% solution of **di(2-pyridyl)glyoxime**.[553] Dilute to 100 ml, set aside for 30 minutes, and read at 595 nm. At pH 7.5 it conforms to Beer's law up to 4 ppm of iron. There is interference by chromous ion, zirconium, thorium, cobalt, copper, molybdate, and tungstate.

For iron by 8, 8-diquinolyldisulfide, see Copper by that reagent.

Ferric ion at pH 3.6–5 forms a stable 1:2 complex with **di(salicylaldehydo)ethylenediamine**, which is α,α'-ethylenedinitrilobis(*o*-cresol).[554] This is read at 500 nm. There is interference by cupric, oxalate, tartrate, citrate, chromate, bichromate, arsenite, arsenate, phosphate, and EDTA ions. In 3 *N* ammonium hydroxide, the complex is 2:3 ferric ion–chelate, which is read at 460 nm. There is interference by cupric, zinc, cerous, manganous, cobalt, nickel, acetate, tartrate, chromate, bichromate, arsenate, phosphate, and EDTA ions. Each obeys Beer's law up to 80 μg of ferric ion per ml.

The 1:2 complex of ferrous ion with **dithizone** is extractable with carbon tetrachloride at pH 6.5–9.5 or with chloroform at pH 6.9–9.8.[555a] The maximum absorption is at 560 nm. Because of slow decomposition, it is not used for analysis for iron. Ferric ion does not react.

By condensation of an EDTA ester with hydrazine hydrate in ethanol **EDTA tetrahydrazide** is formed.[555b] Purify by recrystallization from 60% ethanol. At pH 4.5 a 1:1 chelate is formed with ferric ion, read at 530 nm for 1–20 μg of iron per ml. At pH 11 the reagent forms a 2:3 chelate with ferric ion, read at 450 nm for 0.5–12 μg per ml. The optimum ranges are 3–18 and 2–14 μg respectively. Interference by 10-fold zinc, manganous ion, uranyl ion, nickel, pentavalent

[551b]H. Alexaki-Tzivanidou, *Anal. Chim. Acta* **75**, 231–234 (1975).

[552]J. L. Bahamonde, D. Perez Benedito, and F. Pino Perez, *Talenta* **20**, 694–696 (1973).

[553]Sakuro Wakimoto, *Jap. Anal.* **10**, 968–971 (1961).

[554]Danuta Dabrowska, Janina Lukasiak, and Stanislaw Ostrowski, *Acta Pol. Pharm.* **32**, 615–622 (1975); Danuta Dabrowska, Krzysztof Karpow, Janina Lukasiak, and Stanislaw Ostrowski, *Acta Pol. Pharm.* **33**, 229–233 (1976).

[555a]Zygmunt Marczenko and Miroslaw Mojski, *Chim. Anal.* (Paris) **53**, 529–534 (1971).

[555b]F. Bermejo Martinez, J. M. Grana-Molares and J. A. Rodriguez-Vasquez, *Microchem. J.* **21**, 261–266 (1976).

vanadium, copper, and cobalt as well as by tungstate, acetate, oxalate, and nitrite must be eliminated.

Adjust a sample solution containing 0.2–2 μg of ferric ion in 2 *N* hydrochloric acid to pH 6 with an acetate buffer solution.[556] Make the solution 0.5 m*M* with **eriochrome cyanine R** (mordant blue 3). Extract as a ternary complex with 0.8 m*M* tridodecylammonium bromide in xylene and read at 613 nm. There is serious interference by lead, mercuric, stannous, chromic, beryllium, thorium, uranyl, nitrate, iodide, thiocyanate, and perchlorate ions, as well as by such chelating agents as EDTA.

Ethyl di(1-sodiotetrazol-5-ylazo)acetate complexes with both ferrous and ferric ions for reading at 520 nm.[557] The optimum pH is 2–3 and Beer's law applies up to 2.4 μg of iron per ml. Nickel and cobalt interfere.

In weakly acid solution ferric ion forms a 1:2 complex with **disulfodiantipyrinylmethane**, which is bis[2,3-dimethyl-1-(*p*-sulfophenyl)pyrazol-5-onyl] methane.[558] Mix 3 ml of 10% solution of the potassium salt of the reagent with a sample containing 0.5–3 μg of iron per ml in 0.05–0.1 *N* hydrochloric acid. Dilute to 25 ml with 0.05 *N* hydrochloric acid, set aside for 15 minutes, and read at 460 nm.

Esculetin, which is 6,7-dihydroxycoumarin, forms a green complex with ferric ion at pH 4 which is read at 365 nm.[559] At pH 10 it forms a red complex, read at 440 nm. **Methylexculetin** at pH 3.5 and 10 forms similar complexes. The sensitivity for either is 2 ng/ml in an alkaline medium or 5 ng/ml for the methyl derivative at pH 10.

To a sample solution containing ferrous ion, add ***N,N′*-ethylenebis(4-methoxy-1,2-benzoquinone 1-oxime-2-imine)**.[560a] Adjust to pH 1.3–8 with acetic acid. Set aside for 10 minutes; then make alkaline with ammonium hydroxide and read at 688 nm. If large amounts of copper, cobalt, or nickel are present, they must be preextracted.

From a ferric solution in 0.75 *N* nitric acid containing 2 *M* thiocyanate ion, extract the complex with *M* **bis(2-ethylhexyl)methylphosphonate** in benzene.[560b] Read at 490 nm. The color of the extract is stable for 24 hours.

To a ferric chloride solution add 5 ml of 0.5% ethanolic potassium **ethylxanthate**.[561] Dilute to 25 ml with ethanol and read within 5 minutes at 428 or 465 nm. Beer's law is followed for 3–27 μg of iron per ml.

Iron is read as the buffered **ferric acetate** complex at 3–1200 ppm.[562] To a sample containing 0.05–30 mg of iron in dilute mineral acid solution other than phosphoric, add 10 ml of 2 *M* buffer solution of sodium acetate–acetic acid. Dilute to 25 ml and read at 337.5 or 465 nm. There is interference at the lower wavelength by

[556] Yoshio Shijo, *Bull. Chem. Soc. Jap.* **48**, 2793–2796 (1975).
[557] N. S. Frumina, N. N. Goryunova, and I. S. Mustafin, *Zh. Anal. Khim.* **24**, 1049–1052 (1969).
[558] V. P. Zhivopistsev and V. V. Parkacheva, *Uch. Zap. Perm. Gos. Univ.* **1964** (111), 141–145.
[559] G. Streekite and Zh. V. Panina, *Nauch. Tr. Vyssh. Ucheb. Zaved. Lit. SSR, Khim. Khim. Tekhnol.* **1973** (15), 49–52; cf. B. D. Jain and H. B. Singh, *J. Indian Chem. Soc.* **40**, 263–264 (1963).
[560a] Michio Mashima, *J. Jap. Chem. Soc.* **80**, 1260–1263 (1959).
[560b] V. V. Rublev and T. I. Martyshova, *Zh. Anal. Khim.* **31**, 402–404 (1976).
[561] C. Isvoranu-Panuit, D. Negoiu, and M. Marica, *Anal. Univ. Bucur., Ser. Stiint. Nat. Chim.* **13**, 113–126 (1964).
[562] D. D. Perrin, *Anal. Chem.* **31**, 1181–1182 (1959).

ceric, chromic, and bichromate ions as well as to a lesser degree by fluoride. At the higher wavelength, which is only 4% as sensitive, only chromic and fluoride ions interfere seriously.

For soil analysis, the iron in the sample is reduced to ferrous ion with hydroxylamine and developed with potassium **ferricyanide**.[563] After 90 minutes the Turnbull's blue is read at 700 nm.

A complex of ferric ion with **feroin α-oxime** is read at 530 nm and follows Beer's law up to 4 μg of iron per ml.[564a]

Ferric ion forms a stable 1:2 complex with **ferricon**, which is disodium-3-hydroxynaphthalene-2,7-disulfonate.[565a] Excess reagent is added to the sample at pH 2.5–2.8, and the color is read at 580 nm. Beer's law is followed for 0.4–20.2 ppm of iron at 5–95°. There is interference by fluoride, tetraborate, phosphate, oxalate, tartrate, and citrate.

Ferrocene, which is dicyclopentadienyliron, reduces ferric ion as well as those of molybdenum and rhenium in organic solvent. The oxidation product is read. For 0.1–1 mg of ferric ion add sequentially to the sample solution 5 ml of 20% sulfuric acid, 23 ml of acetone, and 2.5 ml of 10 m*M* ferrocene in acetone.[565b] Dilute to 50 ml and read at 615 nm. Molybdate and copper interfere.

For a sample of a tablet or capsule boil with 0.005% solution of ascorbic acid or in the case of ferrous fumarate use 0.1 *N* hydrochloric acid.[565c] Mix 1 ml of sample solution containing about 65 μg of ferrous ion with 2 ml of phosphate buffer for pH 5.3 containing 0.012% of ascorbic acid and dilute to 25 ml. Add 1 ml of 0.5% solution of **ferrozine** which is the sodium salt of [3-(2-pyridyl)-5,6-bis(4-sulfophenyl)]-1,2,4-triazine. Dilute to 50 ml, mix for 10 minutes, and read at 562 nm against a reagent blank. Cuprous and cobalt ions interfere.

When ferric ion in *N* nitric acid is reacted with 1:4:1 sodium **fluotitanate–sodium fluoride–hydrogen peroxide**, the absorption at 420 nm is proportional to 6–30 μg of iron per 25 ml.[566] The ferric ion displaces an equivalent amount of titanium, which is read as pertitanic acid.

If one adds 5 ml of 1% solution of the Schiff base derived from **3-formylsalicylic acid** and **ethylenediamine** to a ferric ion solution, adjusts to pH 7.8, and dilutes to 25 ml, the maximum absorption of the complex is at 480 nm.[567] Beer's law is followed up to 16 ppm of iron. The color is stable for 24 hours. A large excess of vanadate ion interferes.

To a sample solution containing ferric ion to give a final concentration of 15–112 ppm, add 8 ml of 0.05 *M* sodium **furfuryliminodiacetate** and 10 ml of hydrochloric acid–potassium chloride buffer solution for pH 2.[568] Dilute to 25 ml and read at 400 nm. There is interference by bismuth, palladous, and titanic ions.

[563]M. L. Tsap, *Vest. S.-Kh. Nauk* **1963** (2), 122–128.

[564]V. Armeanu and E. Diamandescu, *Rev. Chim.* (Bucharest) **19**, 226–228 (1968).

[565a]S. N. Sinha and A. K. Dey, *Chim. Anal.* (Paris) **45** (5), 224–236 (1963).

[565b]V. T. Solomatin, G. V. Kozina, and S. P. Rzhavichev, *Zh. Anal. Khim.* **32**, 302–307 (1977).

[565c]Richard Juneau, *J. Pharm. Sci.* **66**, 140–141 (1977).

[566]Hisao Fukamauchi, Ryuko Enohara, Mitsuko Uehara, and Chiyo Inotsume, *Jap. Anal.* **10**, 363–365 (1961).

[567]Sailendra Nath Poddar and Kamalendu Dey, *J. Indian Chem. Soc.* **43**, 359–362 (1966).

[568]E. J. Alonzo and M. C. Valencia, *Quim. Anal.* **29**, 292–295 (1975).

When ferric ion is reacted with **gallic acid**, the complex is stabilized with sodium sulfite at pH 6.8–7.42 for reading at 500 nm.[569] Alternatively, it is read at 630 nm and pH 3.25–3.50. At 0.22–105 μg of iron, Beer's law is followed at either wavelength. Phosphate interferes strongly.

Ferric ion forms a 1:1 complex with **gentisic acid**, which is 2,5-dihydroxybenzoic acid, having a maximum absorption at 582 nm and pH 2.5–3.5.[570] Since the initial blue gradually changes to yellow, reading is taken 90 seconds after adding the reagent. Beer's law is obeyed for 0.13–120 μg of iron per ml. Oxalate, acetate, and phosphate interfere.

Ferric ion can be measured by its catalytic effect on oxidation of **H-acid**, which is 4-amino-5-hydroxynaphthalene-2,7-disulfonic acid, by hydrogen peroxide.[571a] It is applied to analysis of salts. Dissolve 1 gram of sample in a minimum volume of water, add 0.1 ml of *N* hydrochloric acid, boil for 3 minutes, and dilute to 10 ml. To 1-ml portions in three tubes, add 0, 0.1 ml, and 0.2 ml of a solution of 1 μg of ferric ion per ml in 0.17 *N* acetic acid. Prepare a blank set in which the sample is omitted. To each of the tubes add 0.5 ml of 0.3% solution of hydrogen peroxide and dilute to 4.5 ml with 0.17 *M* acetic acid. Add 0.5 ml of 10 m*M* *H*-acid and mix with one of the solutions. Similarly treat the others at 1-minute intervals. Read each at 508 nm after 15 minutes. The method is sensitive to 5 ng of iron in the final solution.

In a solution of an acetate buffer at pH 5.5–6 ferric ion forms a bluish complex with **hematoxylin**.[571b] As read at 530 nm it obeys Beer's law up to 40 μg/ml. Aluminum and ferric ion interfere seriously.

Mix a sample solution containing 10–100 μg of iron with 1.5 ml of glycine buffer solution for pH 8.7 and 5 ml of 0.01 *M* **hesperatin** in methanol.[572] Dilute to 10 ml with methanol and read at 480 nm. There is interference by the following ions: aluminum, cupric, pentavalent vanadium, hexavalent uranium, beryllium, phosphate, carbonate, tetraborate, fluoride, oxalate, and tartrate.

Adjust 10 ml of sample solution containing 10–100 μg of ferric ion to pH 1.5–2.5 with *N* perchloric acid or *N* sodium hydroxide.[573] Extract the 1:3 metal-ligand complex with five portions of 10 ml each of 0.02 *M* **hexafluoroacetylacetone** in carbon tetrachloride. Adjust the combined extracts to 50 ml and read at 455 nm against carbon tetrachloride. Molybdate and tungstate interfere.

At pH 4–6 ferrous ion complexes with **hydrazinium hydrazinecarbothioate** for reading at 570 nm.[574] With a 10 *M* excess of reagent, Beer's law is followed for 10–300 μg of iron per ml.

Hydrazinophthalazine and **dihydrazinophthalazine** complex with ferric ion, preferably at pH 11.2.[575] The former is read at 535 nm. There is interference by all

[569] M. Finkel'shteinate and G. Matsaite, *Nauch. Tr. Vyssh. Ucheb. Zaved. Lit. SSR, Khim. Khim. Tekhnol.* **10**, 75–79 (1969).

[570] V. Paleckite, G. Stretskite, and M. Finkelshteinaite, *ibid.* **12**, 21–26 (1970).

[571a] S. U. Kreingol'd, E. A. Bozhevol'nov, and V. N. Antonov, *Zavod. Lab.* **34**, 260–263 (1968).

[571b] Indu Prakash, Suresh Chandra, and H. C. Saxena, *Curr. Sci.* **45**, 674–675 (1976).

[572] Anna Korkuc, *Chem. Anal.* (Warsaw) **15**, 441–444 (1970).

[573] I. Sarghie and S. Fisel, *Rev. Chim.* (Bucharest) **25**, 744–745 (1974).

[574] V. M. Byr'ko, A. I. Busev, T. I. Tikhonova, V. N. Baibakova, and L. I. Shepel', *Zh. Anal. Khim.* **30**, 1885–1891 (1975).

[575] R. Ruggieri, *Anal. Chim. Acta* **16**, 242–245 (1957).

oxidizing agents, oxalate, phosphate, and the following ions: cobalt, nickel, chromic, molybdate, tungstate, and manganous.

Ferric ion can be read as the complex with **hydrazoic acid** in nitric acid.[576] The concentration of nitric acid is a third factor that enters into reading ferric azide at 460 nm.

Ferrous ion complexes with the **hydrazonium salt** of **dithiocarbazic acid** for reading at 570 nm.[577] Readings are of 0.1–2 μg/ml. There is interference by zinc, aluminum, tungsten, vanadium, nickel, chromium, and uranyl ions.

The Schiff base of ***o*-hydroxyacetophenone** and **ethylenediamine** complexes with ferric ion for reading at 490 nm and pH 3.7–7.1.[578] The complex tolerates 40 ppm of phosphate and 10 ppm of citrate.

Ferric ion complexes with ***o*-hydroxybutyrophenone oxime** in 60% ethanol at pH 2.9 to have a maximum absorption at 510 nm.[579a] The optimum working range is 5–20 ppm. There is interference by citrate, oxalate, phosphate, and bifluoride.

Mix 5 ml of sample solution with 10 ml of 2 *M* ammonium thiocyanate and 5 ml of 50 m*M* **N′-hydroxyethylethylenediamine-*NNN*′-triacetate** in 0.001 X*N* sodium hydroxide.[579b] Adjust to pH 3.5 with 2 *N* sodium hydroxide or perchloric acid and read at 425 nm. Beer's law is followed for 1.6–8 μg of ferric ion per ml.

Ferric ion forms a 1:3 complex with **3-hydroxyflavone.**[580] To a 10-ml sample containing up to 40 m*M* of iron, add 1 ml of 0.025% solution of 3-hydroxyflavone in dimethyl sulfoxide and 2 ml of 0.5 *M* acetate buffer for pH 3.5–4. Shake with 10 ml of benzene for 10 minutes. Read the extract at 407 nm or, preferably, at 470 nm.

At pH 3.5–4.5 ***N*-(α-hydroxyiminobenzyl)benzenesulfonamide** forms a red-violet complex with ferric ion having a maximum absorption at 540 nm.[581] The reagent has almost no absorption at that wavelength. Beer's law is followed for 0.2–60 μg of iron per ml.

The 1:3 complex of ferric ion with **2′-hydroxy-5′-methylacetophenone** at pH 1.5–2.2 has an absorption maximum at 400 nm.[582] Beer's law is followed for 5–40 ppm.

Ferric ion forms a 1:1 complex with **2′-hydroxy-5′-methylacetophenone thiosemicarbazone** at pH 2.2 for reading at 645 nm.[583] Beer's law is followed for 0.5–8 μg of ferric ion per ml. There is interference by vanadate and cupric ion.

Ferric ion complexes with **2-hydroxymethyl-6-(2-hydroxymethyl-5-hydroxy-4-pyrone-6-yl)pyrano-[3,2-b]pyran-4,8-dione** in a buffer solution that is *M* with ammonium acetate and 0.35 *M* with acetic acid for reading at 408 nm.[584] Beer's law

[576]Ernest K. Dukes and Richard M. Wallace, *Anal. Chem.* **33**, 242–244 (1961).
[577]A. Musil and W. Haas, *Mikrochim. Acta* **1959**, 481–487.
[578]S. N. Poddar, M. M. Ray, and K. Dey, *Sci. Cult.* (Calcutta) **29**, 309–310 (1963).
[579a]M. H. Gandhi and M. N. Desai, *Chim. Anal.* (Paris) **50** (4), 167–170 (1968).
[579b]K. Yamomoto and Kousaburo Ohashi, *Anal. Chim. Acta* **88**, 141–146 (1977).
[580]Hitoshi Kohara and Hakozaki Ishibashi, *Jap. Anal.* **16**, 470–475 (1967).
[581]E. A. Abrazhanova, G. A. Butenko, A. S. Grzhegorzhevskii, and T. Ya. Kalenchenko, *Zh. Anal. Khim.* **23**, 1408–1410 (1968).
[582]K. Adinarayana Reddy, *Curr. Sci.* **40**, 372 (1971).
[583]M. C. Patel, J. R. Shah, and R. P. Patel, *J. Indian Chem. Soc.* **50**, 560–561 (1973).
[584]Ray F. Wilson and R. C. Daniels, *Z. Anal. Chem.* **187**, 100–104 (1962).

applies up to 220 micromoles of ferric ion. The following ions must be absent: ruthenium, platinum, osmium, rhodium, uranium, dihydrogen phosphate, and oxalate.

For ferric ion by **3-hydroxy-2-methyl-1-phenylpyridine-4-one**, adjust 8 ml of sample solution containing 0.01–0.1 mg of iron to pH 2–2.5 with sodium hydroxide solution or hydrochloric acid.[585] Add 2 ml of glycine buffer solution for pH 2.2. Shake for 2 minutes with 5 ml of a 0.01 *M* solution of the reagent in chloroform. Extract with 3 ml of chloroform, dilute the combined extracts to 10 ml with chloroform, and read at either 420 or 470 nm. Interfering ions are as follows: hexavalent uranium, titanic, vanadic, molybdate, tantalum, and cupric.

At pH 8.5–9.5 ferric ion forms a weak brown-red 1:3 complex with **2′-hydroxy-5′-methylpropiophenone oxime** that can be extracted into chloroform for reading at 500 nm.[586] This will determine up to 45 ppm of ferric ion. The sample solution should contain 20% of ethanol to prevent extraction of excess reagent by chloroform. There is interference by phosphate, fluoride, tetraborate, molybdate, tungstate, EDTA, manganous, chromic, and aluminum ions. At pH 2.1 in 75% ethanol, ferric ion forms a weak blue, extractable 1:1 complex absorbing at 580 nm.

Ferric ion at pH 1–6 complexes with **3-hydroxy-2-methylpyran-4-one** for reading at either 546 or 492 nm.[587] There is interference by nickel, bismuth, antimony, hexavalent chromium, ceric ion, and complex-forming anions.

To a ferric ion solution at pH 2–5 containing less than 4 mg of iron, add at least 20% excess of a 1% solution of **1-hydroxy-4-methylpyridine-2-thione** in 0.4% sodium hydroxide solution.[588] Heat at 100° for 10 minutes and cool. Extract with 10 ml of chloroform and read at 577 nm.

For ferric ion by **3-hydroxy-2-methylpyrone**, add 1 ml of *N* sulfuric acid and 2 drops of 3% hydrogen peroxide.[589a] Set aside for 10 minutes. Then add 5 ml of 0.2% solution of the color reagent, allow 12 minutes for color development, dilute to 50 ml, and read at 496 nm. Beer's law is followed for 0.2–2 μg of iron per ml.

Adjust 8 ml of sample solution containing 10–100 μg of ferric ion to pH 2–2.5. Add 2 ml of glycine buffer solution for pH 2.2 and extract with 5 ml and 3 ml of 5 m*M* **3-hydroxy-2-methyl-1-(4-tolyl)pyridin-4-one** in chloroform.[589b] Dilute the extracts to 10 ml and read at 418 or 465 nm. There is interference by hexavalent uranium, titanic, vanadate, molybdate, tantalate, cyanide, and oxalate ions.

Absorption maxima for ferric complexes are as follows: **3-hydroxy-2-naphthoic acid**, 590 nm; **3-hydroxy-1-nitro-2-naphthoic acid**, 430 nm; and **4-bromo-3-hydroxy-2-naphthoic acid**, 570 nm.[590] They are applicable at 0.1–0.7 μg of iron per ml and remain constant for 6 hours.

The absorption maxima of 1:1 complexes of ferric ion with sodium salts of

[585] B. Tamhina and M. J. Herak, *Croat. Chem. Acta* **48**, 603–610 (1973).
[586] Shri Prakash, Yag Dutt, and Rajendra Pal Singh, *Microchem. J.* **18**, 412–421 (1973).
[587] Klaus Lorentz and Barbara Flatter, *Mikrochim. Acta* **1969**, 1023–1026.
[588] M. Edressi and A. Massoumi, *Microchem. J.* **16**, 353–358 (1971).
[589a] H. E. Jungnickel and W. Klinger, *Pharmazie* **17**, 221–222 (1962).
[589b] M. J. Herak and B. Tamhina, *Croat. Chem. Acta* **46**, 237–241 (1974).
[590] V. Armeanu, P. Costinescu, and C. G. Calin, *Rev. Chim.* (Bucharest) **19**, 174–177 (1968).

phenolic acids are as follows: **1-hydroxy-2-naphthoic acid**, 580 nm; **sulfo-*p*-cresotic acid**, 560 nm; and **acetyl-*p*-cresotic acid**, 540 nm.[591]

The blue complex of ferric ion with **2-hydroxy-3-naphthoic acid** has a maximum absorption at 590 nm and pH 2.9–3.1.[592]

At pH 3.5 ferric ion forms a 1 : 1 complex with **3-hydroxy-2-naphthol** that above pH 8 becomes a 1 : 2 complex. Each follows Beer's law for 2–10 μg of iron per ml. Salicylic acid or hydrofluoric acid destroys the complex.[593]

5-(2-Hydroxy-1-naphthylazo)pyrazole-4-carboxylic acid complexes with ferric ion in an acid medium.[594] Below pH 4 there are no interferences. Oxidize the sample with nitric acid. Mix an aliquot with a buffer solution for pH 3.9 and add a 0.05% solution of the reagent in dioxan. Dilute with water and dioxan to give a 1 : 1 ratio of the solvents. Let stand for 20 minutes and read at 625 nm. Beer's law is followed for 1.2–6 μg of ferric ion per ml.

Ferric iron at pH 3 complexes with **8-hydroxy-7-nitroquinoline-5-sulfonic acid** for reading at 550 nm against a reagent blank.[595] Beer's law applies for 4 μM to 0.64 mM iron.

5-Hydroxy-4-nitrosobenzo-2, 1, 3-thiadiazole forms a 2 : 1 complex with ferrous ion in 50% ethanol at pH 5.6 which can be read over the range of 570–660 nm.[596]

Ferrous ion in cesium iodide is determined after reduction with ascorbic acid and complexing with **4-hydroxy-3-nitrosonaphthalene-1-sulfonic acid.**[597] Reading at 690 nm permits determination down to 0.01 ppm.

To a sample containing ferrous ion, add 4 ml of 0.2 M acetate buffer for pH 5 and 2 ml of 0.2% solution of disodium **8-hydroxy-7-nitrosoquinoline-5-sulfonic acid.**[598a] Read at 704 nm against a blank. To reduce interference mix 5 ml of sample solution 50 μM to 0.5 mM in ferric ion with 10 ml of sodium acetate-acetic acid buffer solution for pH 5.[598b] Add 5 ml of saturated solution of the reagent. After 1 minute add hydrochloric acid dropwise to give pH 0. Dilute to 25 ml and read immediately at 705 nm. Equimolar amounts of platinic ion, trivalent ruthenium and osmium, palladous ion, silver, and cobalt are tolerated.

The ferric chelate of **6-hydroxy-1, 7-phenanthroline** is read in 40% propanol with greater sensitivity than that with 1, 10-phenanthroline.[599] A borate buffer solution for pH 7–9 is appropriate. Beer's law is followed for 1.2–8 ppm of iron.

The complex of ferric ion with **3-hydroxy-1-phenyl-3-methyltriazene** at pH 3.1–4.5 has a maximum absorption at 625 nm.[600]

[591] R. C. Aggarwal, S. P. Agarwal, and T. N. Srivastava, *Z. Anorg. Allg. Chem.* **304**, 337–343 (1960).

[592] A. K. Majumdar and C. P. Savariar, *Anal. Chim. Acta* **21**, 47–52 (1959).

[593] A. I. Cherkesov and T. S. Zhigalkina, *Tr. Astrakh. Tekh. Inst. Rybn. Prom. Khoz* **1962** (8), 25–49.

[594] Boleslaw Janik and Tadeusz Gancarczyk, *Acta Pol. Pharm.* **30**, 303–306 (1973).

[595] S. K. Patel, K. P. Soni, and I. M. Bhatt, *Lab. Pract.* **21**, 337–339 (1972).

[596] S. A. D'yachenko, S. O. Gerasimova, I. A. Belen'kaya, and V. G. Pesin, *Zh. Anal. Khim.* **29**, 877–880 (1974).

[597] A. B. Blank, V. G. Chepurnaya, L. P. Eksperiandova, and V. Ya. Vakulenko, *ibid.* **29**, 1705–1709 (1974).

[598a] M. M. Aly, *Z. Anal. Chem.* **270**, 32 (1974).

[598b] Shahid Abbas Abbasi, B. G. Bhat, and R. S. Singh, *ibid.* **282**, 222 (1976); *Indian J. Chem. A*; **14**, 215–216 (1976).

[599] Joan M. Duswalt, *Dis. Abstr.* **23**, 813 (1962); Joan M. Duswalt and M. G. Mellon, *Anal. Chem.* **33**, 1782–1786 (1961).

[600] H. K. L. Gupta and N. C. Sogani, *J. Indian Chem. Soc.* **36**, 87–91 (1959).

The complex of ferric ion with **3-hydroxy-3-phenyl-1-(*p*-sulfophenyl)triazene** at pH 2–4.3 has absorption maxima at 520 and 650 nm.[601] At least fifteenfold reagent is required. Beer's law is followed for 1–20 ppm of iron.

Ferric ion and **2-(3-hydroxy-3-phenyltriazino)benzoic acid** form a pink 1:2 complex in ammoniacal tartrate at pH 6–10.5.[602] The complex is extracted into chloroform for reading at 500 nm. Beer's law is followed at 1–12 ppm of iron.

2-(3-Hydroxy-3-phenyltriazeno)-5-sulfobenzoic acid forms a 2:1 complex with ferric ion at pH 3.3–4.5 with a maximum absorption at 410 nm.[603] Beer's law is followed for 1–12 ppm. There is interference by oxalate, EDTA, tartrate, citrate, phosphate, fluoride, cupric ion, palladous ion, hexavalent molybdenum, pentavalent vanadium, and thiosulfate.

The 1:2 complex of **3-(2-hydroxyphenyl)-1-*H*-1,2,4-triazole** with ferric ion is formed at pH 3.5–5 for reading at 530 nm.[604] Beer's law is followed for 0.5–10 ppm of iron in the presence of more than 0.5 ml of 1% ethanolic reagent in 0.1 *M* acetate buffer solution for pH 4. Copper and mercury interfere.

At pH 1–4 in sulfuric acid, ferric ion oxidizes ***N,N'*-bis(2-hydroxypropyl)-*o*-phenylenediamine** to a red phenazine derivative having a maximum absorption at 525 nm.[605a] The color is stable for 3 hours. Beer's law is followed for 0.5–15 μg of ferric ion per ml. Oxidizing agents and tenfold amounts of cupric or nickel ion interfere.

Iron and titanium are extracted from acid solution with **8-hydroxyquinaldine** and read at 380 and 580 nm.[605b]

When a sample is mixed with 3 ml of 2% solution of **8-hydroxyquinoline-7-sulfonic acid** as the sodium salt, buffered to pH 4 and diluted to 25 ml, it should be set aside for 10 minutes.[606] Then as read at 600 nm, it follows Beer's law for 1–20 ppm of iron. There is interference by vanadium, cobalt, uranium, tungstate, molybdate, and citrate.

The complex with **8-hydroxyquinoline-5-sulfonic acid** in the presence of zephiramine is extracted with chloroform at pH 4.5–8.5.[607a] There are absorption maxima at 465 and 595 nm. Beer's law holds for the latter wavelength at 0.14–1.26 μg of iron per ml of the organic layer.

Mix a sample solution containing up to 3.3 μg of iron with 3 ml of a phosphate buffer solution for pH 8.8 and 1 ml of 0.5 m*M* methanolic ***o*-hydroxyquinolphthalein.**[607b] Dilute to 10 ml and heat at 60° for 30 minutes and cool. If aluminum is present read at 620 nm or otherwise at 605 nm against water.

Ferric ion complexes with **1-hydroxyxanthone.**[608] The complex is read at 425 nm, pH 1.8–2.3. Beer's law is followed for 1.1–11 ppm of iron. There is interference by the following: sulfate, oxalate, tartrate, citrate, fluoride, phosphate, zirconium, titanium, beryllium, copper, lead, and aluminum.

[601] H. K. L. Gupta and N. C. Sogani, *ibid.* **37**, 97–100 (1960).
[602] A. K. Majumdar and S. C. Saha, *Anal. Chim. Acta* **44**, 85–93 (1969).
[603] A. K. Majumdar and D. Chakraborti, *ibid.* **23**, 127–134 (1971).
[604] Atsushi Sugii and Hideo Sumie, *Jap. Anal.* **12**, 364–368 (1963).
[605a] S. Ostrowski and B. Kasterka, *Chem. Anal.* (Warsaw) **10**, 43–48 (1965).
[605b] Kenji Motojima and Hiroshi Hashitani, *Bull. Chem. Soc. Jap.* **20**, 458–460 (1956).
[606] K. Balachandran and Samir K. Banerji, *Microchem. J.* **13**, 599–603 (1968).
[607a] Tominito Kambara, Shiyo Matsumae, and Kiyoshi Hasebe, *Jap. Anal.* **19**, 462–466 (1970).
[607b] Yoshikazu Fujita, Itsuo Mori, and Takehisa Enoki, *Jap. Anal.* **24**, 253–256 (1975).
[608] Brahm Dev and B. D. Jain, *J. Indian Chem. Soc.* **39** (4), 247–250 (1962).

A 1 : 2 complex of ferric ion and **iodoferron**, which is 5′-[bis(carboxymethyl)aminomethyl]-2, 6-dibromo-3′-methylindophenol, has a maximum absorption at 590 nm and pH 3.9–4.1.[609] Beer's law is followed up to 2 μg of iron per ml.

A 1 : 2 : 3 complex is formed by ferric ion, 1, 3-diphenylguanadine, and **indoferron** which is extracted at pH 3 by chloroform.[610] As read at 635 nm, it conforms to Beer's law for 0.15–1 μg of ferric ion per ml. Added sodium chloride promotes the extraction. There is interference by titanic ion, thorium, tetravalent vanadium, EDTA, and nitrilotriacetate. These interferences with the ternary complex are much less serious than those with the binary complex.

Ferric ion forms a 1 : 3 complex with **maltol**, which is 3-hydroxy-2-methylpyran-4-one, in the presence of at least fiftyfold concentration of the reagent and pH above 3.4.[611] Under other conditions two other complexes are formed. Desirable conditions are 10 m*M* reagent, acetate or pyridine nitrate buffer solution for pH 4.7–4.8, and reading at 415 nm. Beer's law is followed for 0.5–18 ppm of iron. Interferences are vanadate, molybdate, uranate, phthalate, oxalate, phosphate, EDTA, tartrate, and citrate.

A stable yellow complex of ferric ion with **mannitol** is formed at pH 1–3 with a maximum absorption at 410 nm.[612]

To a sample solution containing less than 30 μg of iron, add 5 ml of 0.01 *M* **2-mercaptopyridine-3-ol** in methanol.[613] Add potassium acid phthalate solution to buffer to pH 5 and dilute to 10 ml. Read at 600 nm.

For iron and manganese by **8-mercaptoquinoline**, see Copper by 8, 8′-Diquinolyldisulfide.

Iron is developed by **mercaptosuccinic acid**, which is thiomalic acid, at pH 8–10.5 for reading at 358–372 nm.[614]

Ferric ion is extracted as an ion association complex with **4-(6-methoxy-3-methylbenzothiazol-2-ylazo)-*N*-methyldiphenylamine.**[615] Make a sample, which may be as large as 150 ml, 6 *N* in sulfuric acid and 3 *M* in sodium chloride. Add 1 ml of m*M* reagent and extract with 6 ml of 5 : 1 benzene-cyclohexane. Read the organic phase at 640 nm. Alternatively, make the sample solution 18 *N* with sulfuric acid and 2 *M* with sodium chloride; then extract with 6 ml of 5 : 1 benzene-nitrobenzene. Beer's law applies for 0.1–2.8 μg of iron per ml of extract. A twentyfold amount of thallic ion, stannic ion, gallium, pentavalent antimony, auric, thiocyanate, or iodide ion is tolerated.

Ferric ion forms a 1 : 3 complex with **4-methoxybenzothiohydroxamic acid** at pH 2–9 which has a maximum absorption at 375 nm.[616]

Allowing 12 minutes for color development, ferric ion is read with **2-methyl-3-hydroxypyrone** at 496 nm.[617] Beer's law is followed for 0.2–2 μg of iron per ml.

Ferric ion in the presence of chloride ion complexes with **6-methoxy-3-**

[609] Takyoshi Sakai, *ibid.* **43**, 3171–3175 (1970).

[610] Yoshinobu Wakamatsu and Makoto Otomo, *Jap. Anal.* **19**, 537–541 (1970).

[611] A. Stefanovic, J. Havel, and L. Sommer, *Collect. Czech. Chem. Commun.* **33**, 4198–4214 (1968).

[612] I. V. Pyatnitskii and A. Kh. Klibus, *Ukr. Khim. Zh.* **29**, 480–489 (1963).

[613] Mohan Katyal, Veena Kushwaha, and R. P. Singh, *Analyst* **98**, 659–662 (1973).

[614] F. Bermejo Martinez and M. del Carmen Meijon Mourino, *Inf. Quim. Anal.* **18** (1), 7–11, 20 (1964).

[615] L. I. Kotelyanskaya, P. P. Kish, and I. I. Pogoida, *Zh. Anal. Khim.* **27**, 1128–1133 (1972).

[616] Z. Skorko-Trybula, *Chem. Anal.* (Warsaw) **12**, 815–825 (1967); Z. Skorko-Trybula and B. Debska, *ibid.* **13**, 557–573 (1968).

[617] H. E. Jungnickel and W. Klinger, *Pharmazie* **17**, 221–222 (1962).

methyl-2-[4-(*N*-methylanalino)phenylazo]benzothiazolium chloride.[618] To a sample solution, add 1 gram of sodium chloride, 2 ml of 1 : 1 sulfuric acid, and 1 ml of m*M* reagent, and dilute to 6 ml. Extract with 6 ml of 5 : 1 benzene-cyclohexanone and read the organic layer at 640 nm against a reagent blank. Beer's law holds for 0.1–2.8 μg of iron per ml.

Ferrous ion complexes with **4′-(4-methoxyphenyl)-2,2′ : 6′,6″-terpyridyl.**[619] To a volume of sample solution containing 0.5–10 μg of iron, add 0.1 volume of 0.03 *M* ascorbic acid, 1 volume of 3.5 *M* sodium acetate, and 2 volumes of 0.5 m*M* reagent in methanol or ethanol. Dilute to a known volume with 50% ethanol, set aside for 15 minutes, and read at 570 nm. Beer's law applies for 0.05–0.1 μg of iron in the developed solution. The complex can be concentrated by extraction with chloroform, butanol, or pentanol.

The reagent fluoresces, but the ferric complex does not. By activation at 376 nm and reading at 468 nm, as little as 0.01 μg of iron per ml can be read by difference.[620] The sulfonate can also be used.

Ferrous ion in 10% ethanol at pH 2.7–6.5 forms a 1 : 2 complex with ***N*-methylanabasine-α′-azo-*p*-cresol**, which is 2-[5-(1-methyl-2-piperidyl)-2-pyridylazo]-*p*-cresol.[621] At 510 nm it obeys Beer's law for 0.4–2.5 μg of iron per ml. It is necessary to mask zinc, mercury, lead, thorium, molybdenum, tungsten, and copper. There is interference by EDTA, oxalate, tartrate, and thiosulfate. Ascorbic acid reduces ferric ion to ferrous.

Ferric ion catalyzes the photooxidation of **methyl orange.**[622] Add methyl orange to sulfuric or perchloric acid at pH 1.9–2.1. Read at 510 nm and expose for 10 minutes to radiation with a wavelength less than 350 nm. Read again at 510 nm. The loss in absorption measures ferric ion in the absence of copper ion.

The ferrous complexes of the 5-**methyl-**, 5,6-**dimethyl-**, 5-**nitro-**, and 5-**chloro-derivatives** of **1,10-phenanthroline** all absorb in the range of 508–520 nm.[623] The methyl derivatives are extractable with chloroform or higher alcohols and obey Beer's law in the extracts.

For iron and copper by **methyl-2-pyridyl ketoxime**, see Copper.

Methylthymol blue in excess forms a 2 : 1 complex with ferric ion having a maximum absorption at 515 nm in approximately neutral solution.[624] A technic for iron in liquid from flotation pulps by xylenol orange is applicable with methylthymol blue under identical conditions.

Adjusted with hexamine–perchloric acid solution, the red complex of ferrous ion at pH 5.8 has a maximum absorption at 510 nm; but buffered to pH 2.5, a green complex has a maximum at 600 nm. Both complexes are stable for 10 hours and follow Beer's law up to 40 μg of iron per ml.[625] There is interference by oxalate,

[618] L. I. Kotelyanskaya, P. P. Kish, and E. P. Gavrilets, *Zh. Vses. Khim. Obshch.* **17**, 230–231 (1972).
[619] P. Stanchev, *Chem. Anal.* (Warsaw) **16**, 243–249 (1971).
[620] R. Schmidt, W. Weis, V. Klingmüller, and H. Staudinger, *Z. Klin. Chem. Klin. Biochem.* **5**, 304–309 (1967).
[621] E. L. Krukovskaya, Sh. T. Talipov, and I. R. Kaiger, *Zh. Anal. Khim.* **30**, 131–135 (1975).
[622] I. P. Kharlamov, E. I. Dodin, and A. J. Mantsevich, *ibid.* **22**, 371–375 (1967); I. P. Kharlamov and A. D. Mantsevich, *Tr. Tsent. Nauchn.-Issled. Inst. Tekhnol. Mashinostr.* **1972** (110), 47–51.
[623] E. Gagliardi and E. Wolf, *Mikrochim. Ichnoanal. Acta* **1964**, 700–711.
[624] B. Karadakov, D. Kancheva, and P. Nenova, *Talenta* **15**, 525–534 (1968), cf. V. N. Tikhonov, *Zh. Anal. Khim.* **21**, 1172–1178 (1966).
[625] K. C. Srivastava and Samir K. Banerji, *Microchem. J.* **13**, 621–629 (1968); cf. Koichi Tonosaki, *Bull. Chem. Soc. Jap.* **39**, 425–428 (1966).

nitrilotriacetate, EDTA, aluminum, gallium, indium, thallium, bismuth, ferric ion, thorium, vanadium, and zirconium.

Developed in solution at pH 3–5.5 in 10 ml of acetate buffer solution and 1 ml of 0.04 *M* **morin** in dimethylformamide, the brown ferric complex can be extracted with 10 ml of isoamyl alcohol and read at 500–600 nm.[626] Beer's law is followed for 0.14–33.6 μg of iron at 400 nm.[627] **Quercetin** reacts quite similarly at 434 nm.

After separation by descending paper chromatography with 2 grams of benzoylacetone in 200 ml of 1 : 1 butanol–20% trichloroacetic acid, the ferric ion is extracted with warm ethanol and read as its morin complex at pH 5 and 433 nm.[628]

When ferrous ion is used as the reducing agent in producing the blue color of **molybdotungstophosphoric acid**, the maximum absorption is at 725 nm.[629] Beer's law is followed for 0.02–1 mg of ferrous ion in 25 ml of final solution at pH 2–4. Ferric ion can be masked with fluoride; but stannous, antimonous, and arsenous ions, as well as many other reducing agents, must be absent.

The reaction of morpholine and carbon bisulfide produces **morpholinium morpholine-4-carbodithioate.**[630] In its application, hydrolyze serum at room temperature with hydrochloric acid for 20 minutes. Add trichloroacetic acid solution to deproteinize, and filter. Add the reagent, and read iron at 510 nm. Copper can be read in the same sample at 438 nm. The complex can be extracted with chloroform at pH 2–6.

For determination with **morpholinium 3-oxapentamethylenedithiocarbamate**, ferrous and ferric ions are separated by thin-layer chromatography on a 0.25-mm layer of cellulose powder with 3 : 1 ethanol or butanol–6 *N* hydrochloric acid as the mobile phase. The spots at R_F values of 0.15–0.55 for ferrous ion and 0.6–1 for ferric ion are extracted with 0.1 *N* hydrochloric acid, and a 1% solution of the reagent is added for 3–30 μg of iron. Then adjust to pH 1.5 with sulfuric acid and extract with chloroform for reading at 510 nm.

In 50 ml of hydrochloric or perchloric acid solution at pH 1, containing 40% of ethanol, 0.12–0.7 mg of ferric ion complexes with **nalidixic acid**, which is 1-ethyl-1, 4-dihydro-7-methyl-4-oxo-1, 8-naphthyridine-3-carboxylic acid.[631] This is read at 425 nm. **Noralidixic acid**, which is the 4-hydroxy-7-methyl derivative, performs similarly.

Add 0.1 gram of solid **naphthalene-2, 3-diol** to a solution containing 10–50 μg of iron.[632] Add a buffer solution to adjust to pH 4.5–4.7 and read at 500–510 nm.

Although ferric ion forms three complexes with **nicotinohydroxamic acid**, the yellow complex at pH 5.25 is the one suitable for analytical purposes.[633] It is read at 430 nm for 1–10 mg of iron per ml.

The complex of **1-nitroso-2-naphthol** with ferrous ion at pH 4.3–10 can be extracted with a mixture of ethyl acetate and butyl acetate for reading at 660 nm.[634]

[626]Hitoshi Kohara, Kiniki Ueno, and Nobohiko Ishibashi, *Jap. Anal.* **15**, 1252–1257 (1966).

[627]V. Paleckite and M. Finkel'shteinaite, *Zh. Anal. Khim.* **24**, 1550–1553 (1969).

[628]I. I. M. Elbeih and M. A. Abou-Elnaga, *Chemist-Analyst* **55**, 43–45 (1966).

[629]A. Suteu and C. Maniu, *Rev. Roum. Chim.* **13**, 1201–1205 (1968).

[630]W. Beyer, *Clin. Chim. Acta* **38**, 119–126 (1972); W. Beyer and W. Likusser, *Mikrochim. Acta* **1971**, 610–614.

[631]E. Ruzicka, J. Lasovsky, and P. Brazdil, *Chem. Zvesti* **29**, 517–520 (1975).

[632]V. Patrovsky, *Collect. Czech. Chem. Commun.* **35**, 1599–1604 (1970).

[633]Z. Skorko-Trybula and J. Minczewski, *Chem. Anal.* (Warsaw) **6**, 523–529 (1961).

[634]Masao Kawahata, Heiichi Mochizuki, and Takeshi Misaki, *Jap. Anal.* **10**, 15–18 (1961).

Extraction with isoamyl alcohol and reading at either 410 or 700 nm has also been recommended.[635] The complex of **2-nitroso-1-naphthol** is unstable in ethyl acetate–butyl acetate but is extracted with methyl isobutyl ketone at pH 4.9–10 for reading at 670 nm.

On a 100×5 mm polytetrafluoroethylene column with dibutylphosphate as the stationary phase, gallium, cobalt, copper, and ferric ion in *N* ammonium thiocyanate as the mobile phase are adsorbed.[636] Thus a 0.6-gram sample of aluminum is dissolved in hydrochloric acid, evaporated to dryness, and taken up in *N* ammonium thiocyanate. This is passed through the column at 0.5–1 ml/minute and washed with 50 ml of *N* ammonium thiocyanate. Copper and iron are then eluted with 0.5 *N* hydrochloric acid, and the iron is determined with 1-nitroso-2-naphthol.

The 1:3 complex of ferrous ion with **2-nitroso-1-naphthol-4-sulfonate** at pH 2–8 has a maximum absorption at 710 nm.[637] Reduction of ferric ion is with hydroxylamine hydrochloride.

The complex of ferric iron with **oxalate** ion is read at 255–280 nm.[638] To a solution containing not more than 0.18 mg of ferric ion and 0.5–1.4 ml of 0.1 *N* perchloric acid, add 10 ml of 0.02 *M* ammonium oxalate and dilute to 25 ml. Read after 30 minutes. Cupric and vanadate ions interfere.

Ferrous ion complexes with alcoholic **oximidobenzotetronic acid** at pH 2.5–10 in the presence of hydroxylamine hydrochloride for reading at 625 nm.[639a] The complex is soluble in chloroform as well as in 50% ethanol and follows Beer's law for 0.54–5.4 ppm of iron. Cobalt, nickel, ceric, and tetravalent zirconium ions also give colored complexes that interfere if there is more than 50 times the ferrous ion concentration.

Adjust the sample solution containing 13.5–135 μg of ferric ion to pH 2.2–5 with a potassium acid phthalate-hydrochloric acid buffer solution.[639b] Extract with 10 ml of 0.01 *M* **5-oxo-3-phenylisooxazoline-4-carboxylic acid ethyl ester** in 4-methylpentan-2-ol and read the organic phase at 490 nm. Beer's law is followed up to 15 ppm of ferric ion. Reducing agents interfere.

Mix 1 ml of 0.06–0.33 m*M* ferric sample solution with 5 ml of 0.2 *M* acetate buffer for pH 5 and 1 ml of 0.2 *M* **5-oxopyrrolidine-2-carbohydroxamic acid.**[640] Dilute to 10 ml and read at 430 nm. There is interference by hexavalent uranium, pentavalent vanadium, bivalent molybdenum, cupric, oxalate, and citrate ions.

In **perchloric acid** the absorption maxima are as follows: iron, 240 nm; lead, 208 nm; and bismuth, 222 nm.[641] All three can be read in the same sample solution.

[635]A. B. Blank and A. M. Bulgakova, *Zh. Anal. Khim.* **15**, 605–609 (1960); A. B. Blank, I. I. Fedorova, and L. E. Tete, *Ibid.* **24**, 1367–1375 (1969).

[636]I. P. Alimarin, N. I. Ershova, and T. A. Bol'shova, *Vest. Mosk. Gos. Univ., Ser. Khim.* **1969** (6), 79–84.

[637]V. N. Tolmachev, G. N. Podol'naya, and L. N. Serpukhova, *Zh. Neorg. Khim.* **2**, 2073–2077 (1957); F. H. Pollard, J. F. W. McOmie, A. J. Banister, and G. Nickless, *Analyst* **82**, 780–800 (1957); V. N. Tolmachev and L. M. Serpukhova, *Tr. Komi. Anal. Khim., Akad. Nauk SSSR, Inst. Geokhim. Anal. Khim.* **8**, 115–124 (1958).

[638]Toru Nozaki and Hirondo Kurihara, *J. Chem. Soc. Jap.* **82**, 710–712 (1961); cf. V. M. Bhuchar and V. P. Kukreja, *Indian J. Chem.* **5**, 562–565 (1967).

[639a]A. N. Bhat and B. D. Jain, *Anal. Chim. Acta* **25**, 343–347 (1961).

[639b]F. Corigliano and S. Di Pasquale, *Talenta* **23**, 545–546 (1976).

[640]M. Pecar, N. Kujundzic, H. Mezaric, and I. Jelcic, *Microchem. J.* **20**, 401–408 (1975).

[641]Masayoshi Ishibashi, Yuroku Yamomoto, and Kazuo Hiiro, *Bull. Inst. Chem. Res., Kyoto Univ.* **36**, 24–29 (1958).

Although the maximum absorption of iron in perchloric acid solution is at 240 nm, reading at 260 nm is more favorable with respect to copper, lead, and bismuth.[642]

As one technic, precipitate ferric ion with ammonium hydroxide in the presence of ammonium chloride.[643] Dissolve the precipitate in perchloric acid to give a solution at 0.05–0.1 *M* acid. Dilute to give 1.4–10 μg of iron per ml and read at 223 nm. This is the isobestic point for ferric ion and $FeOH^{2+}$.

Iron is read at 240 nm in the digestates of biological samples as the perchlorate in *N* perchloric acid.[644] There is a slight interference by sodium, potassium, calcium, magnesium, and copper.

In analysis of lead alloys, the complex of iron with ***o*-phenanthrol** at pH 4.5 is read.[645]

A kinetic method for iron depends on the catalytic effect of ferrous or ferric ion on oxidation of ***p*-phenetidine** with potassium periodate in the presence of 2,2′-bipyridyl.[646] Optimum conditions are 0.1 m*M* *p*-phenetidine, 0.45 m*M* 2,2′-bipyridyl, 2 m*M* potassium periodate, and acetate buffer solution for pH 4.5 at 25°. The reaction rate is measured at 536 nm. The increase in concentration of the reaction products is a straight line for 0.5–5 ng of iron per ml. Manganese in amounts equivalent to the iron interferes.

Mix 50 ml of serum, which need not be deproteinized, successively with 2.5 ml of 0.2 *M* hydrogen peroxide, 1 ml of 0.75 m*M* 1,10-phenanthroline, and 4.95 ml of phthalate buffer solution for pH 2.8.[647] Place in a bath at 40°. After 5 minutes add 1.5 ml of 0.05 *M* ***p*-phenetidine hydrochloride**. After 30 minutes at 40°, read the *N*-(ethoxyphenyl)-1,4-benzoquinoneimine with a green filter. The first-order oxidation by hydrogen peroxide is catalyzed by ferric ion, and 1,10-phenanthroline serves as an activator. For preparation of standards, the sequence is as follows: ferric chloride, hydrogen peroxide, buffer solution, heat to 40°, add 1,10-phenanthroline, and finally the *p*-phenetidine hydrochloride.

Ferric ion is extracted at pH 2.8 by *M* **phenylacetic acid** in chloroform and read at 340 nm against chloroform.[648] For less than 1 μg/ml in water, extract a 50-ml sample with 10 ml of the reagent solution. Beer's law is followed up to 18 μg of iron per ml of the extract. If lead or silver is high in the sample, substitute decanoic acid as solvent and extract at pH 2.9.

For simultaneous extraction of iron and copper by ***N*-phenylbenzohydroxamic acid**, see Copper.

For determination of ferric ion with ***N*-phenylcinnamohydroxamic acid** in the presence of pentavalent vanadium and hexavalent uranium, make the solution 4 *N* with hydrochloric acid and extract the vanadium with a 0.1% solution of the reagent in chloroform.[649] Then adjust the aqueous layer to pH 1 with ammonium

[642] Robert Bastian, Richard Weberling, and Frank Palilla, *Anal. Chem.* **28**, 459–462 (1956).

[643] M. T. Escot, N. Parry, D. Chatonier, and J. Dauphin, *Trav. Soc. Pharm. Montpellier* **33**, 295–298 (1973); cf. Masayuki Tabushi, and Toyokichi Kitagawa, *Bull. Chem. Soc. Jap.* **29**, 57–60 (1956).

[644] I. D. P. Wooton, *Biochem. J.* **68**, 197–199 (1958).

[645] Haydee Armandola de Alderuccio, *Rev. Obras Sanit. Nac.* (Buenos Aires) **22**, 20–24 (1958).

[646] I. F. Dolmanova, V. I. Rychkova, and V. M. Peshkova, *Zh. Anal. Khim.* **28**, 1763–1767 (1973).

[647] A. Aleksiev, P. Bonchev, and D. Raikova, *Mikrochim. Acta* **1974**, 751–758; cf. Hartwig Shurig, Helmut Mueller, and P. R. Bonchev, *Z. Chemie (Leipzig)* **16**, 190–191 (1976).

[648] R. W. Cattrall and M. J. Walsh, *Microchem. J.* **19**, 123–129 (1974).

[649] F. G. Zharovskii and R. I. Sukhomlin, *Zh. Anal. Khim.* **21**, 59–64 (1966).

hydroxide and add a 0.5% solution of the reagent in ethanol. Extract the iron complex with isopentyl alcohol and read at 453 nm.

The 2 : 1 complex formed by **phenylfluorone** with ferric ion in 0.001 *N* sodium hydroxide has a maximum absorption at 580 nm.[650] At 600 nm the absorption of the reagent is less. Stabilization by gelatin is necessary. Other complexes are formed at 1 : 1 and 2 : 3.[651] Beer's law applies up to 8 m*M* iron. Isolation of ferric ion by extraction from 1 : 1 hydrochloric acid is essential to avoid interferences.

The bluish-green 1 : 3 complex of ferrous ion with **picolinaldehyde thiosemicarbazone** in ammoniacal solution can be extracted with chloroform and is read at 610 nm.[652] Beer's law applies for 1–7 ppm of iron.

1-Picolinoyl-3-thiosemicarbazide forms a yellow 2 : 1 complex with ferric ion at pH 3–4.[653] Beer's law is followed for 0.5–5 ppm of iron at 405 nm. The color is stable for 90 minutes. There is no interference by 0.2 ppm of cobalt, mercurous ion, or mercuric ion; 1 ppm of copper, cadmium, or nickel; 2 ppm of ferrous ion; 5 ppm of silver or oxalate.

Ferric ion can be read with **picramic acid** at 428 nm in 1.2 *N* hydrochloric acid.[654] Picramates absorbing at the same wavelength are formed by vanadium, gold, cerium, and vanadium.

The complex of 0.5–3 ppm of iron with **pontachrome azure blue B** at pH 3.5 can be read at 610 nm after removal of interferences by ion exchange.[655]

Mix 2 ml of sample solution containing 0.08–1.75 mg of iron, 4 ml of 0.05 *M* **pyrazine-2,3-dicarboxylic acid** in 2 *M* ammonium acetate, and 5 ml of 10% hydroxylamine hydrochloride solution that has been adjusted to pH 5.5.[656] Dilute to 50 ml and read at 475 nm against a blank. The color is stable for 48 hours. The concentration above conforms to Beer's law. There is interference by oxalate, molybdate, cupric, cadmium, cobalt, chromic, and aluminum ions.

A complex of **pyridazine-3,6-diol** with ferric ion in nitric acid at pH 3–3.5 is read at 480 nm.[657] Beer's law is followed up to 10 μg of iron per ml. The ions that interfere most are zirconium, silver, trivalent ruthenium, and stannic.

Incubate a solution containing 25–275 mg of ferrous ion for 10 minutes at 60° with 16 ml of a saturated solution of **pyridil** in absolute ethanol and 16 ml of 0.25% solution of hydrazine as the sulfate with 1 gram of sodium perchlorate added.[658] Adjust to pH 5–6 with acetic acid–sodium acetate buffer solution and heat at 60° for 30 minutes. Let cool for 1 hour and read at 488 nm. Alternatively, heat the ferrous solution with 1% ethanolic **di(2-pyridil)glyoxime** at pH 4.5–7.8 for 15 minutes at 60°. Let cool for 1 hour and read at 488 nm. As a variant on the latter technic, add 0.3 gram of sodium chlorate to the ferrous solution at pH 4.5–7 and extract with 0.05% solution of di(2-pyridil)glyoxime in nitrobenzene. Dry the extract and read at 486 nm.

[650] J. Minczewski and U. Stolarczykowa, *Chem. Anal.* (Warsaw) **9**, 1135–1138; **11**, 531–541 (1966).
[651] U. Stolarczyk, *ibid.* **12**, 1113–1118 (1967).
[652] J. M. Cano Pavon, D. Perez Bendito, and F. Pino Perez, *An. Quim.* **67**, 299–307 (1971).
[653] Atsushi Sugii, Motoko Dan, and Hiroko Okazawa, *Jap. Analyst* **13**, 51–54 (1964).
[654] G. Popa, I. Paralescu, and D. Mircea, *Z. Anal. Chem.* **184**, 353–355 (1961).
[655] Yuhiteru Katsube, Katsuya Uesugi, and John H. Yoe, *Bull. Chem. Soc. Jap.* **34**, 72–76 (1961).
[656] G. S. Sanyal and S. Mookherjea, *Z. Anal. Chem.* **276**, 71 (1975).
[657] M. H. Hashmi, Tehseen Quereshi, and A. I. Ajmal, *Microchem. J.* **16**, 626–632 (1971).
[658] E. Graciani Constante, *An. Quim.* **67**, 607–614 (1971).

Ferrous ion complexes with **pyridine-2-aldoxime** in dilute ethanol at pH 5.5–7.4 for reading at 510 nm.[659] The complex is stable for 5 hours and conforms to Beer's law for 0.52–12.9 ppm of iron. Above pH 10 the complex has a maximum absorption at 520 nm and follows Beer's law for 0.26–7.74 ppm. At concentrations around 50 ppm there is interference by nickel, thorium, vanadium, tungsten, and zirconium.

For reading as the 1:3 complex, take a sample containing up to 0.7 mg. Add 1 ml of 10% hydroxylamine hydrochloride solution, 5 ml of ammoniacal solution of pyridine-2-aldoxime, 5 ml of 10% solution of sodium citrate, and 5 ml of ammonium hydroxide. Warm to cause complex formation. If fluoride or phosphate is present, heat at 100° for 10 minutes. Cool, dilute to 100 ml, and read at 525 nm. If vanadium, chromium, cobalt, nickel, manganese, or copper is present, extract the iron with methyl isopropyl ketone before color development.[660]

Ferrous ion forms a 1:2 complex with **pyridine-2,6-dicarboxylic acid** at pH 4.3–7.9.[661] The complex is stable only in aqueous solution in the presence of hydroxylamine hydrochloride. The maximum absorption is at 485 nm, and there is conformity to Beer's law for 1–20 ppm of iron. As a suitable technic, the ferrous salt is extracted from hydrochloric acid solution with methyl isobutyl ketone.[662] Then it is back-extracted with hydrochloric acid solution of hydroxylamine hydrochloride and developed with the reagent.

The 1:2 chelate of ferrous ion with **pyridine-2,4,6-tricarboxylic acid** is stable at pH 5.2–6 in the presence of 0.4% hydroxylamine hydrochloride.[663a] The maximum absorption is at 520 nm, and Beer's law holds up to 20 μg of iron per ml. Copper, nickel, zinc, and manganese interfere.

For iron along with copper by **3-(2-pyridyl)-5,6-diphenyl-1,2,4-triazine**, see Copper.

Ferrous ion forms a 1:3 complex with **1,2-bis(α-2-pyridylbenzylidene)hydrazine**, which is the azine of phenyl-2-pyridyl ketone.[663b] To the sample solution, add 10 ml of acetic acid–sodium acetate buffer solution for pH 4.7, 2 ml of 2% ascorbic acid solution, 5 ml of 0.5 *M* potassium nitrate, and 10 ml of 0.1% ethanolic reagent. Dilute to 50 ml, which should contain about 4 ppm of ferrous ion, and read at 630 nm. Alternatively, use a borate buffer solution for pH 7, heat at 60° for 1 hour before diluting to volume, and read at 480 nm.

For determination by extraction, add to the sample solution 1 ml of 2% ascorbic acid solution and 3 grams of sodium perchlorate. Then add 10 ml of 0.1% solution of the reagent in methyl isobutyl ketone and 10 ml of the acetic acid–sodium acetate buffer solution for pH 4.7. Shake for 2 minutes, dry the organic phase with sodium sulfate, and read at 650 or 690 nm.

A ferrous complex with **2-pyridyl-2′-hydroxymethane sulfonic acid** has a maximum absorption at 520 nm and conforms to Beer's law up to 4 ppm of iron.[664] To a

[659] A. N. Bhat and B. J. Jain, *J. Sci. Ind. Res.* (India) **21B**, 576–577 (1962).

[660] Heinrich Hartkamp, *Z. Anal. Chem.* **170**, 399–407 (1959).

[661] Ichiro Morimoto and Susumu Tanaka, *Anal. Chem.* **35**, 141–145 (1963); H. Poppe and G. den Boef, *Z. Anal. Chem.* **228**, 244–257 (1967).

[662] Ichiro Morimoto and Susumu Tanaka, *Jap. Anal.* **11**, 861–863 (1962); cf. Ichiro Morimoto and Katsushi Furata, *ibid.* **10**, 1294–1296 (1961).

[663a] Ichiro Morimoto and Susumu Tanaka, *J. Chem. Soc. Jap., Pure Chem. Sect.* **83**, 357 (1962); *Anal. Chem.* **35**, 141–144 (1963).

[663b] M. D. Luque de Castro, M. Valcarcel, and F. Pino Perez, *An. Quim.* **72**, 382–391 (1976).

[664] E. E. H. Pitt and V. M. Stanway, *ibid.* **41**, 981–983 (1969).

sample containing 2.5–100 μg of iron, add 5 ml of 5% solution of hydroxylamine hydrochloride, 5 ml of 0.2% solution of the color reagent in faintly ammoniacal solution, and 5 ml of a buffer solution containing 7% of ammonium chloride and 57% of ammonium hydroxide. Dilute to 25 ml and read after 1 hour. The maximum shifts somewhat with pH and time.

To 5 ml of sample solution containing 0.01–0.2 mg of iron, add excess hydroxylamine hydrochloride solution and warm to reduce the iron.[665] Add 1 gram of sodium potassium tartrate as a masking agent and 10 ml of freshly prepared 0.5% solution of **2-pyridyl ketoxime**. Adjust to pH 10.5 with a solution of sodium or potassium hydroxide. Dilute to 25 ml and read at 525 nm against water.

Dilute a sample solution containing 20 ml of 1 : 30 2 *M* acetic acid–2 *M* sodium acetate buffer, 5 ml of 10% sodium sulfite solution, and 4 ml of 30% **pyrogallol** solution to 50 ml.[666a] Read at 610 nm for up to 5 μg of ferric ion per ml in the pH 6.5 solution above. Addition of 1% EDTA solution reduces the absorption.

Add ascorbic acid and sodium perchlorate to a ferrous solution.[666b] Adjust with an acetate buffer solution to pH 2–4.2. Shake for 15 minutes with a 50-fold excess of 3.75 *M* **8-(2-pyridylmethyleneamino)quinoline** in nitrobenzene. Read the organic layer at 663 nm. Beer's law is followed up to 0.12 m*M* ferrous ion in the extract.

At pH 4.8–5.3 a 1 : 2 : 3 complex of ferric ion-**pyrogallol-aniline** is extractable with carbon tetrachloride and amyl alcohol for reading at 500 nm.[666c] This is stable for 4 hours and will determine 2–60 μg of iron per ml of extract.

The complex of ferric ion with **pyrogallolcarboxylic acid**, which is 2,3,4-trihydroxybenzoic acid, is read at 410 nm at pH 1.75–3.3.[667] A 75-fold amount of color reagent is desirable. Beer's law applies to 0.22–82 μg of iron per ml of final solution. There is serious interference by cobalt, ferrous ion, cupric ion, phosphate, and thiosulfate.

To a 200-ml sample of seawater containing up to 10 ppm of iron, add 20 ml of 2.5% sodium fluoride solution, 4 ml of 0.05% solution of **quinaldehyde-2-quinolylhydrazone**, and 20 ml of 0.2 *M* disodium phosphate.[668a] After 20 minutes extract with 20 ml and 5 ml of benzene. Adjust the combined extracts to 25 ml and read at 555 nm within 1 hour against a blank.

Mix 3.6 ml of sample solution containing 12–120 μg of ferric ion in 1 : 99 hydrochloric acid with 8 ml of 10 m*M* methanolic **quinaldic acid *N*-oxide**.[668b] Add 2.6 ml of *M* acetic acid–*M* ammonium acetate as a buffer for pH 4.6. Dilute to 20 ml and extract with 20 ml of chloroform. Read the extract at 380 nm or for somewhat higher concentrations of iron at 395 nm.

Quinisatin oxime forms a 3 : 1 complex with ferrous ion.[669] The reagent slowly reduces ferric ion. To a sample, add 5 ml of 0.0015 *M* reagent in ethanol containing 3% of dimethylformamide and 3 ml of 0.1 *M* potassium acid phthalate–sodium hydroxide buffer for pH 5. Set aside for 20 minutes, dilute to 25 ml, and read at 660 nm against a reagent blank.

[665] D. Banerjea and K. K. Tripathi, *ibid.* **32**, 1196–1200 (1960).
[666a] Kunika Sugiwara, Koichi Tanino, and Jisuki Seki, *Jap. Anal.* **22**, 1559–1568 (1973).
[666b] Makoto Otomo and Kazunobu Kodama, *Bull. Chem. Soc. Japan* **48**, 906–910 (1975).
[666c] G. A. Gamid-Zade, T. D. Ali-Zade, and E. M. Ganieva, *Azerb. Khim. Zh.* **1975**, (4), 102–105.
[667] M. L. Finkel'shteinate and G. I. Matsaite, *Zh. Vses. Khim. Obshch.* **12**, 598 (1967).
[668a] J. Abraham, M. Winpe, and D. E. Ryan, *Anal. Chim. Acta* **48**, 431–432 (1969).
[668b] A. Hom Chaudhuri and F. Umland, *Z. Anal. Chem.* **281**, 361–364 (1976).
[669] G. H. Ayres and M. K. Roach, *ibid.* **26**, 332–339 (1962).

Ferrous ion is precipitated in 0.8% solution of **quinoline-8-carboxylic acid** at pH 3–6.[670] Then extracted with 1 : 9 pyridine-chloroform, this is read at 530 nm. Beer's law is followed for 6–46 μg of iron per ml of extract.

Ferric ion forms a 2 : 1 chelate with **2R acid**, which is 6-amino-4-hydroxy-naphthalene-2,7-disulfonic acid, having a maximum absorption at pH 1.2–2.5 and 470 nm.[671]

To a sample containing ferric ion, add a fiftyfold excess of ethanolic **resacetophenone oxime** and adjust with 0.01 *N* sodium hydroxide to pH 6–8.3.[672] Make up the volume of ethanol to 10% and read at 580 nm. Beer's law is followed for 1–14 ppm of ferric ion. There is interference by phosphate, carbonate, cobalt, nickel, and palladium.

Ferric ion is read with **resorcinol** at pH 2.9 and 450 nm over the range $0.25–1.42 \times 10^{-3}$ *M*.[673] Ferrous ion does not react.

A 1 : 1 : 3 complex is formed by ferrous ion, **rhodamine B** and **4-chloro-2-nitrosophenol**.[674] At pH 4.3–5.3 this is readily extracted into benzene, has a maximum absorption at 558 nm, and obeys Beer's law up to 10 m*M* iron. The color is stable for a week. For total iron, add hydroxylamine hydrochloride. The method is suitable for water analysis. For 2–20 μg of iron per liter in seawater, add 20 ml of acetate buffer solution for pH 4.7 to 80 ml of sample. Extract with 10 ml of 2 m*M* 4-chloro-2-nitrosophenol and 10 m*M* rhodamine B in toluene and read at 558 nm.

Rufigallic acid forms a complex with ferric ion read at 320 nm for 0.27–4.86 μg of iron per ml.[675]

Below pH 6.5 **rutin** in aqueous methanol forms a 1 : 1 complex with ferric ion.[676] Using a glycine buffer for pH 3.5, it conforms to Beer's law for 0.5–5 μg of ferric ion per ml. Above pH 6.5 a twofold excess of rutin forms a 2 : 1 red-brown complex with ferrous ion which is read at 600 nm.[677a] The complex conforms to Beer's law for 1–10 μg of iron per ml and is stable for 6 hours. Above pH 7 rutin also forms a 3 : 1 complex with ferric ion. There is interference by the following ions: magnesium, cupric, chromic, beryllium, uranyl, aluminum, vanadate, phosphate, oxalate, carbonate, iodide, fluoride, tetraborate, and tartrate.

Add excess of 0.01 *M* solution of **salicylaldehyde** in acetone to a ferric solution in at least 60% acetone to maintain solubility.[677b] Adjust to pH 3.5–5, extract the complex with carbon tetrachloride, and read at 510 nm. Beer's law is followed for 0.72–3.6 μg/ml.

Salicylaldehyde-ethylenediamine complexes with ferric ion.[678a] At pH 5 the complex is extracted with chloroform and read at 495 nm. There is interference by cupric ion, cyanide, fluoride, and EDTA.

[670] J. M. Zehner and T. R. Sweet, *ibid.* **35**, 135–137 (1966).

[671] Roshan L. Seth and Arun K. Dey, *Z. Anal. Chem.* **194**, 271–276 (1963).

[672] G. S. Chowdary and N. Appalaraju, *Curr. Sci.* **44**, 343–344 (1975).

[673] G. S. Rao, *Anal. Chim. Acta* **21**, 564–568 (1959).

[674] Takashi Korenaga, Shoji Motomizu, and Kyoji Toei, *ibid.* **65**, 335–346 (1973); *Talenta* **21**, 645–649 (1974).

[675] M. Finkel'shteinate, V. Budraitene, and A. Guzauckaite, *Nauch. Tr. Vyssh. Ucheb. Zaved. Lit. SSR, Khim. Khim. Tekhnol.* **1973** (15), 45–48.

[676] Anna Korkuc, *Acta Pol. Pharm.* **31**, 47–52 (1974).

[677a] Anna Korkuc, *ibid.* **32**, 193–198 (1975).

[677b] H. L. Ray, B. S. Garg, and R. P. Singh, *J. Chin. Chem. Soc., Taipai* **23**, 47–51 (1976).

[678a] Nobuichi Oi, *Toyama Daigaku Kogakubu Kiyo* **9**, 1–4 (1958).

Salicylaldehyde thiosemicarbazone in a solution buffered at pH 10 with ammonium hydroxide and ammonium chloride forms a 1 : 2 complex with ferrous or ferric ion for reading at 590 nm.[678b] Beer's law is followed for 1–28 ppm.

When a solution containing 2–40 μg of ferric ion is complexed with 0.5 ml of 0.5% solution of **salicylic acid–formaldehyde** polymer in 4% sodium hydroxide solution, and the pH is adjusted to 5.5, the maximum absorption is at 500 nm.[679] There is interference by tartrate, oxalate, phosphate, aluminum, chromous ion, ceric ion, zirconium, thorium, pentavalent vanadium, cupric ion, nickel, titanic ion, and ferrous ion.

When 5 ml of a 0.4% solution of the complex of **bis-salicylidene**, **ethylene diamine**, and **hypophosphorous acid** reacts with ferric ion at up to 15 μg/ml, it is read at 495 nm.[680] The pH range is 3–8, and reading is taken after 30 minutes.

1,3-Bis(salicylideneamino)propan-2-ol forms a reddish–yellow 1 : 1 complex with ferric ion at pH 6.7–10.2 having a maximum absorption at 480 nm.[681] Beer's law is followed up to 7.2 ppm of iron. Color develops instantaneously and is stable for 3 weeks. Fluoride and phosphate interfere.

Solochrome azurine BS (mordant blue 1) forms a 3 : 1 complex with ferric ion in an acetate buffer solution for pH 3.5–4.8.[682] Read at 600 nm, Beer's law is followed for iron at 1–35 ppm.

When **5-sulfo-β-resorcylic acid**, which is 2,4-dihydroxy-5-sulfobenzoic acid, is added as excess of 0.01 *M* solution to ferric ion, the latter is read at 370 nm within 30 minutes.[683] Beer's law is followed for 50–150 ppm of iron. Interfering ions are sulfate, thiocyanate, chromate, nickel, and cobalt.

Mix a sample containing ferric ion with 10 ml of 7 : 3 0.2 *M* acetic acid–*M* sodium acetate as a buffer solution for pH 4.1–4.4.[684] Add 1 ml of m*M* **tannic acid**, dilute to 20 ml, and read at 550 nm.

Serum iron and total iron-binding capacity can be determined on clear sera with **Teepol 610** but not on lipaemic sera.[685]

Terosite sulfonate, which is (4,4′,4″-triphenyl-2,2′ : 6′,2″-terpyridyl)sulfonate, complexes with ferrous ion for reading at 583 nm.[686] For iron in hemoglobin, the technic is identical with that given elsewhere for ferrozine. For serum iron, deproteinize with 20% trichloroacetic acid in 1% mercaptoacetic acid solution and centrifuge.[687] Add the color reagent to an aliquot of the supernatant layer and read at 583 nm. The stable color is formed at once.

Mix 2 ml of sample solution containing up to 50 μg of ferrous ion with 1 ml of 20% ascorbic acid solution, 5 ml of 0.025 *M* **1,2,3,4-tetrahydro-6-hydroxy-5-**

[678b] D. Perez Bendito and F. Pino Perez, *Mikrochim Acta* **I**, 613–622 (1976).
[679] P. Umapathy, A. M. Hundekar, and D. N. Sen, *Indian J. Appl. Chem.* **31**, 185–188 (1968).
[680] W. J. P. Neish and Linda Key, *Experimentia* **25**, 788–789 (1969).
[681] Sailandra Nath Poddar and Kamalendu Dey, *Indian J. Appl. Chem.* **28** (2), 49–52 (1965).
[682] C. L. Sharma, S. N. Tandon, and Surinder Kumar, *Z. Anal. Chem.* **255**, 368 (1971).
[683] G. C. Shivahare, S. Mathur, and M. K. Mathur, *ibid.* **261**, 126 (1972).
[684] Manzoor Ahmad Chaudhry, Misbahul-Haq Javed, and Riaz-ur-Rehman, *Pak. J. Sci. Ind. Res.* **16**, 172–175 (1973).
[685] H. J. Colenbrander and C. L. J. Vink. *Clin. Chim. Acta* **28**, 175–184 (1970).
[686] A. Manasterski, R. Watkins, E. S. Baginski, and B. Zak, *Z. Klin. Chem. Klin. Biochem.* **11**, 335–338 (1973).
[687] Bennie Zak, Eugene S. Baginski, Emmanuel Epstein, and Lawrence M. Weiner, *Clin. Chim. Acta* **29**, 77–82 (1970); *Clin. Toxicol.* **4**, 621–629 (1971).

nitrosopyrimidine-3,5-dione, and 2 ml of 2 *M* ammonium chloride–ammonium hydroxide buffer for pH 9.2.[688a] Dilute to 10 ml and read at 630 nm against a reagent blank. Beer's law holds for up to 5 μg/ml as developed. Citrate, tartrate, or sulfite interferes.

To a sample solution containing 33.5–558.5 μg of ferric ion add 2.5 ml of 4 m*M* solution of **sorbohydroxamic acid** and adjust to pH 1.8 with 0.1 *M* perchloric acid.[688b] Dilute to 50 ml and read at 550 nm. There is interference in ppm by 20 phosphate, 1 fluoride, 0.2 oxalate, 0.4 lead, and 1.3 cupric ion.

Adjust 5 ml of sample solution containing 0.3–5.6 ppm of iron to pH 4 and add 2 ml of 10% solution of hydroxylamine hydrochloride.[689] Add 5 ml of a buffer containing 2 *M* acetic acid–2 *M* sodium acetate, then 10 ml of 0.001 *M* **2,3,5,6-tetra-2-pyridylpyrazine**. Dilute to 50 ml and read at 575 nm.[690] The maximum absorption at 575 nm is shifted to 590 nm by strong acid. The complex follows Beer's law for 0.28–5.6 ppm of iron.[691] The perchlorate salt of the complex is extractable with chloroform or nitrobenzene. There is interference by cobaltous, cupric, nickel, and cyanide ions. Chromic and fluoride ions are tolerated in limited amounts. Bismuth and stannous ions precipitate.

For 2–100 μg of ferric ion in 10 ml of 2 *N* nitric acid, 9 *M* with ammonium nitrate, extract by shaking for 10 minutes at 45° with 10 ml of 15% solution of **thenoyltrifluoroacetone** in xylene.[692] Read the organic phase at 510 nm against a reagent blank. Beer's law is followed for 0.2–10 μg of ferric ion per ml. The sample solution may be that of uranium, cobalt, aluminum, or nickel. Cerous and ceric ions interfere seriously. The presence of phosphate, tartrate, and citrate decreases the efficiency of extraction.[693]

Pyridine acts synergistically for determination of ferrous ion.[694] To a sample solution containing 7–100 μg of iron, add 2 ml of 10% solution of hydroxylamine hydrochloride. Adjust to pH 5 with *M* sodium acetate–0.5 *M* acetic acid. Extract with 10 ml and 10 ml of 1 : 1 0.2 *M* pyridine–0.02 *M* 2-thenoyltrifluoroacetone in benzene. Dilute the extracts to 25 ml and read at 580 nm.

Adjust a sample solution containing up to 120 μg of ferric ion to pH 1 with hydrochloric acid.[695] Extract for 20 minutes with 10 ml of benzene 0.1 *M* with thenoyltrifluoroacetone and m*M* with Capriquat, which is methyltrioctylammonium chloride. Dry the extract with sodium sulfate and read at 500 nm. In this case mg amounts of sulfate, nitrate, vanadate, pyrovanadate, cupric ion, molybdate, and tungstate interfere.

To a sample solution, add 10 ml of 0.4% ethanolic **2-thiobarbituric acid** and 5 ml of potassium acid phthalate buffer solution for pH 3.[696] Dilute to 25 ml and read at 675 nm. This will read 4–40 μg of iron per ml. There is interference by bismuth,

[688a]Masaomi Tsuchiya and Hirooki Sasaki, *Jap. Anal.* **24**, 691–694 (1975).

[688b]Chuen-Ling Liu and Peng-Joung Sun, *J. Chin. Chem. Soc.* (Taipai) **23**, 219–227 (1976).

[689]R. T. Pflaum, C. J. Smith, Jr., E. B. Buchanan, Jr., and R. E. Jensen, *Anal. Chim. Acta* **31**, 341–347 (1964).

[690]R. E. Jensen, J. A. Carlson, and M. L. Grant, *ibid.* **44**, 123–128 (1969).

[691]C. J. Smith, Jr., *Dis. Abstr.* **22**, 2566 (1962).

[692]C. Testa, *Anal. Chim. Acta* **25**, 525–532 (1961); cf. S. M. Khopkar and A. K. De, *ibid.* **22**, 223–228 (1960).

[693]Hiroshi Kawamoto and Hideo Akaiwa, *Chem. Lett.* **1973**, 259–260.

[694]H. Akaiwa, H. Kawamoto, and M. Hara, *Anal. Chim. Acta* **43**, 297–301 (1968).

[695]Hiroshi Kawamoto and Hideo Akaiwa, *Jap. Anal.* **23**, 495–500 (1974).

[696]H. Sikorska-Tomicka, *Z. Anal. Chem.* **234**, 414–417 (1968).

cupric ion, stannous ion, mercurous and mercuric ions, silver, cobalt, chromic ion, acetate, and oxalate.

Ferrous ion is extractable at pH 6.5 with **thiodibenzoylmethane**, which is 1,3-diphenyl-3-thioxopropan-1-one, in hexane containing some benzene.[697] Read at 480 nm. Beer's law is obeyed for 0.3–15 ppm of iron. There is serious interference by cupric, cobalt, nickel, zinc, platinic, silver, cadmium, and EDTA ions.

The complex with **thione**, which is **2-mercaptopyridine *N*-oxide**, is formed by either ferrous or ferric ion.[698] To prepare the reagent, dissolve 10 grams of thiurone, which is *S*-2-pyridylthiouronium bromide-*N*-oxide, in 40 ml of 4% sodium hydroxide solution. Boil for 2 minutes to complete hydrolysis, and acidify. Filter, wash, and dry the precipitate. Prepare a 0.25% solution in chloroform. Shake 25 ml of sample solution in *M* sulfuric acid containing up to 0.4 mg of iron with 25 ml of reagent solution. Filter the organic layer and read at 550 nm against a reagent blank. Cupric and nitrite ions interfere seriously.

The blue complex of ferrous ion with **thiopicolinamide** at pH 10–12 has an absorption maximum at 615 nm and conforms to Beer's law up to 5 μg of iron per ml.[699]

To 25 ml of sample solution made acid with hydrochloric or sulfuric acid, add 2 ml of 5% sodium acetate–50% acetic acid as a buffer solution. Add 4 ml of 50% solution of ammonium thiocyanate, 0.5 ml of freshly prepared tributylamine acetate, and a few mg of thiourea.[700] Extract **tributylammoniumhexathiocyantoferrate** with 4 ml of amyl acetate and read at 480 nm against that solvent. Fluoride can be masked with aluminum chloride. Nickel, cobalt, copper, and cadmium interfere but can be preextracted. Precipitation of tin and titanium is prevented by adding tartrate. The yellow color formed by bismuth, molybdate, and tungstate is also extracted.

Tris(*o*-hydroxyphenyl)phosphine oxide complexes with ferric ion in 0.3–2.5 *N* nitric acid to give a violet-red color absorbing at 530–550 nm.[701] It is stable for 16 hours. Ceric ion interferes, and the absorption is reduced by acetate, borate, sulfate, chloride, biphosphate, and fluoride.

Ferrous ion complexes with **2,3,5-triphenyltetrazolium chloride** at pH 8–10.5 in the presence of sodium arsenite and citric acid.[702] The complex in dimethylformamide is read at 480 nm for 50–160 μg of ferrous ion per ml. A twofold concentration of ferric ion is tolerated, but cuprous and cupric ions interfere seriously. When ferrous ion is extracted with propylene carbonate as the violet chelate with **bis-2,4,6-tri(2-pyridyl)-1,3,5-triazine**, it is read at 458 nm. For details, see Copper by Neocuproine.

At pH 3.5–7 ferric ion forms a 1:3 complex with **tropolone-5-sulfonic acid**, which is 4-hydroxy-5-oxocyclohepta-1,3,6-triene-1-sulfonic acid.[703] It has peaks at 332, 418, 550, and 595 nm. Beer's law is followed up to 0.07 μg of iron per ml.

Unithiol, which is sodium-2,3-dimercaptopropane-1-sulfonate,[704] complexes with

[697] R. R. Mulye and S. M. Khopkar, *ibid.* **272**, 283 (1974).
[698] J. A. W. Dalziel and M. Thompson, *Analyst* **91**, 98–101 (1966).
[699] O. Wawschinek and K. Felice, *Mikrochim. Ichnoanal. Acta* **1964**, 694–699.
[700] Max Ziegler, Oskar Glemser, and Norbert Petri, *Z. Anal. Chem.* **154**, 81–98 (1957).
[701] M. J. Holdoway and J. L. Willans, *Anal. Chim. Acta* **18**, 376–380 (1954).
[702] M. H. Hashmi, L. Ahmad, A. Subhan, and A. A. Ayaz, *Microchem. J.* **17**, 403–409 (1972).
[703] Yoshinaga Oka, Misao Umehara, and Tetsuo Nozoe, *J. Chem. Soc. Jap., Pure Chem. Sect.* **83**, 703–708 (1962).
[704] Kh. K. Ospanov, N. E. Makletsova, and N. I. Tember, *Zh. Anal. Khim.* **22**, 444–445 (1967).

ferric ion. For analysis of reagent grade salts, dissolve 5 grams of sample in water, add 5 ml of 1 : 3 ammonium hydroxide, and 4 ml of 0.1 *M* unithiol. Dilute to 50 ml and read with a blue filter. Beer's law is followed at 0.05–100 μg of iron per ml.

The effect of ferric ion in catalysis of the oxidation of **variamine blue B** (azoic diazo component 35) by hydrogen peroxide is read as a kinetic method at 530 nm.[705]

A mixture of 1% solution of **violuric acid** with an equal volume of *M* hydroxylamine gives a dark blue 2 : 1 complex with ferrous ion in a borate buffer solution at pH 8.9.[706] It is stable for 30 minutes. There is interference by copper, cobalt, zinc, cadmium, silver, mercury, and cyanide ions.

[705]S. U. Kreingol'd and L. I. Sosenkova, *ibid.* **26**, 332–337 (1971).
[706]Pavel Cerny, *Chem. Listy* **51**, 735–738 (1957).

CHAPTER NINETEEN
NICKEL

Although trivalent compounds of nickel are known, the bivalent form is so much more common that it is rarely referred to as the nickelous ion. An outstanding reagent for nickel is dimethylglyoxime, partly because it has multiple purposes. The complex of bivalent nickel ion and dimethylglyoxime is extractable with chloroform and can be read. But the greater sensitivity often leads to oxidation of nickel for determination by dimethylglyoxime or other reagents.

An acetate buffer solution at pH 5–8 is an appropriate medium for extraction of nickel dimethylglyoxime with chloroform.[1] Either aniline or pyridine is superior to chloroform for extraction of the complex because the solubility is about 4 times as great.[2] The absorption shows a plateau at 415–450 nm in aniline and at 430–460 nm in pyridine. More commonly that is a step in separation of bivalent nickel from interfering ions. Then the nickel is extractable from the complex with 0.5 *N* hydrochloric acid. Then the nickel can be oxidized to the tetravalent form, which is again complexed with dimethylglyoxime for reading or the nickel ion is developed with another reagent.

The various dioximes, dimethylglyoxime, furil-α-dioxime, nioxime, heptoxime, and benzil-α-dioxime, form nickel complexes extractable from tartaric acid solution.[3] Only that with benzil-α-dioxime is extractable from citric acid solution. Other oxime reagents are oxamidoxime, 1,2-cycloheptanedione dioxime, and 1,2-cyclohexanedione dioxime.

The most sensitive reagents are xylenol orange, 4-(2-pyridylazo)resorcinol, and salicylfluorone.[4] Other significant reagents are methylthymol blue, glycine thymol blue, and glycine cresol red.

[1] Hiroshi Kitagowa and Norio Shibata, *Jap. Anal.* **7**, 284–287 (1958).
[2] Federico Minutilli and Piero Ruggieri, *Rass. Chim.* **14**, 97–99 (1962).
[3] V. M. Peshkova, V. M. Savostina, E. K. Astakhova, and N. A. Minaeva, *Tr. Kom. Anal. Khim.* **15**, 104–110 (1965); cf. Gheorghe Marcu, Csaba Varhelyi, and Laszlo Raduly, *Studia Univ. Babes-Bolyai, Ser. Chem.* **20**, 15–19 (1975).
[4] O. A. Tataev, S. A. Akhmedov, and B. A. Magomedova, *Zh. Anal. Khim.* **25**, 1229–1231 (1970).

For concentration of copper and nickel by coprecipitation with ferric hydroxide, see Copper. For simultaneous reading of nickel, cobalt, and copper extracted as the dithiocarbamate, see Copper. For nickel as developed with biuret along with copper and cobalt, and chromium as the chromate, solved by simultaneous equations, see Copper. For nickel by thiocyanate and pyridine along with copper, cobalt, and iron, see Copper. For determination of nickel along with copper by disodium ethyl bis(5-tetrazolylazo)acetate trihydrate, see Copper. The nickel oxalate complex can be read at 390 nm; see Copper. For nickel in beer by cuprethol, see Copper by that reagent. For isolation of nickel from niobium with diethyldithiocarbamate followed by column chromatography, see Zinc.

ACETOPHENONE OXIME

This reagent is also called isonitrosacetophenone. Nickel forms with it a 1:2 complex that as the chloroform extract conforms to Beer's law for 0.1–3 μg of nickel per ml. Various masking agents are used, or the reagent may be an intermediate in the transfer of nickel to hydrochloric acid solution for development with dimethylglyoxime.

Procedure. Add 1 ml of 1% solution of acetophenone oxime in 0.4% sodium hydroxide solution to a sample containing 1–30 μg of nickel.[5] Adjust to pH 7.6 and extract with 5 ml and 5 ml of chloroform. Adjust the combined extracts to 10 ml and read at 340 nm.

ACETOTHIOACETANILIDE

This reagent is 4-analino-4-thioxobutan-2-one. It forms a 1:2 metal-ligand complex above pH 6.9 as adjusted with triethanolamine. Extracted with chloroform, the complex follows Beer's law for up to 150 μg of nickel per ml. Iron is masked with either triethanolamine or sulfosalicylic acid.

Procedure. *Steel.*[6] Dissolve 0.5 gram of sample in 1:1 hydrochloric acid–nitric acid. Add 1 ml of sulfuric acid and evaporate to sulfur trioxide fumes. Take up in water, filter, and dilute to 25 ml. Mix a 10-ml aliquot with 5 ml of 20% solution of sodium potassium tartrate and 1 ml of 5% ethanolic acetothioacetanilide. Adjust to pH 7 with 20% solution of triethanolamine. Extract by shaking for 5 minutes with 10 ml of chloroform. Add 1 gram of ammonium bifluoride in water and shake. Dry the chloroform phase with sodium sulfate and read at 496 nm.

---

[5] U. B. Talwar and B. C. Haldar, *Anal. Chim. Acta* **51**, 53–59 (1970).
[6] Krishna De and Jyotirmoy Das, *J. Indian Chem. Soc.* **52**, 1026–1028 (1975).

2-AMINO-1-CYCLOPENTENE-1-DITHIOCARBOXYLIC ACID

This reagent is visually sensitive to 20 ppb of nickel and more sensitive than dimethylglyoxime or diethyldithiocarbamate. The color conforms to Beer's law for 0.5–5 ppm. There is interference by ferrous, ferric, manganous, and zinc ions.

Procedure. To 100 ml of sample solution containing 50–500 μg of nickel ion, add 10 ml of a buffer solution consisting of 40% 0.1 *M* disodium citrate and 60% 0.1 *N* hydrochloric acid to give pH 3.[7] Add 0.2 ml of 0.64% ethanolic reagent. Shake for 30 seconds each with 50 ml and 50 ml of chloroform. Adjust the combined extracts to 100 ml and read immediately at 530 nm.

α-BENZILDIOXIME

This reagent, which is diphenylglyoxime, willl determine 5×10^{-7}% of nickel in the presence of cobalt and copper.

Procedure. To a sample containing 0.01–0.1 mg of nickel, add 5 ml of 0.5% solution of polyvinyl alcohol, 6 ml of 0.02% ethanolic diphenylglyoxime, 8 ml of a saturated aqueous solution of iodine, and 0.25 ml of 2 *N* ammonium hydroxide.[8] After shaking for 2 minutes, dilute to a known volume and set aside for 5 minutes. Read at 420 nm.

Indium.[9] Dissolve 1 gram of sample in 10 ml of nitric acid and evaporate to dryness. Dissolve in 20 ml of 20% citric acid. Add 10 ml of 20% solution of sodium potassium tartrate followed by 10% sodium hydroxide solution to raise to pH 8–11. Add 2 ml of 0.02% ethanolic α-benzildioxime. Add sufficient ethanol to make the solution 20% in that solvent. Shake for 2 minutes with 5 ml of chloroform. Wash the chloroform layer with 5 ml and 5 ml of 4% sodium hydroxide solution. Read at 775 nm.

BENZOIN-α-OXIME

The nickel complex of this reagent is extracted at pH 8.2–9.7 with chloroform. In the presence of more than 0.5 mg of reagent per 20 μg of nickel, Beer's law applies up to 5 μg of nickel per ml of chloroform. Copper, cobalt, and manganous ions cause positive errors; aluminum and chromous ion cause negative errors. Tartrate masks lead, manganese, cadmium, and zinc.

[7] Masataka Yokoyama and Tatsuo Takeshima, *Anal. Chem.* **40**, 1344–1345 (1968).
[8] Zbigniew Gregorowicz, *Acta Chim. Acad. Sci. Hung.* **18**, 79–84 (1959).
[9] V. M. Peshkova, V. M. Bochkova, and E. K. Astakhova, *Zh. Anal. Khim.* **16**, 596–598 (1961).

Procedure. *High Purity Beryllium Oxide.*[10] Heat a 0.5-gram sample with 1 ml of 1:4 sulfuric acid and 10 ml of hydrofluoric acid. Evaporate to dryness and take up the residue in 10 ml of 0.8 *M* hydrofluoric acid containing 0.06 *M* hydrochloric acid. Pass through a 25×0.9 cm column of a strongly acidic cation exchange resin. Wash the column with 50 ml of the hydrofluoric-hydrochloric acid. Elute the nickel with 25 ml of 1:1 hydrochloric acid and evaporate the eluate to dryness. Take up in 10 ml of hydrochloric acid and pass through a 30×1 cm column of strongly basic anion exchange resin. Wash the column with 25 ml of hydrochloric acid and evaporate the percolate and washings to dryness. Take up the residue in 5 ml of *N* hydrochloric acid and raise to pH 3–4 by addition of 1:1 ammonium hydroxide. Add 0.5 ml of 0.5% ethanolic benzoin-α-oxime, 2 ml of 0.05% sodium tetraborate solution to raise to pH 9, and water to 20 ml. Shake for 10 minutes with 10 ml of chloroform and read the extract at 330 nm.

4-*tert*-BUTYL-1,2-CYCLOHEXANEDIONEDIOXIME

The 1:2 complex of nickel with this reagent is extractable with xylene for reading at 3–10 μg of nickel per ml of solvent. The color is stable for 24 hours. Citrate slows complex formation. Ceric and chromic ions are effectively masked at pH 5 with ammonium acetate and tiron, copper at pH 7 with thiocyanate, ferric ion at pH 7 with fluoride, rare earths at pH 5 with tartaric acid, thoric ion at pH 5 with tiron, stannic ion reduced with hydroxylamine hydrochloride at pH 11 with tiron, tetravalent titanium and pentavalent vanadium at pH 8 with tartaric acid, and tetravalent zirconium at pH 11 with tiron. Nickel is separated from cobalt in hydrochloric acid by anion exchange. Cyanide must be absent.

Procedure. To a sample solution containing 8.5–170 μg of nickel, preferably 30–100 μg, add 3 ml of saturated solution of the captioned reagent for every 17 μg of nickel present.[11] Adjust to approximately pH 7 and add sufficient 10 *M* sodium acetate to make the final solution *M* with acetate. Extract with 10 ml of xylene and read at 386 nm against a reagent blank.

CHROME AZUROL S

Although the complex of nickel with chrome azurol S (mordant blue 29) has a maximum absorption at 550 nm, the addition of hydroxydodecyltrimethylammonium bromide shifts it to 639 nm.[12] At pH 10.5–11.3. Beer's law is followed for 0.01–0.2 μg of nickel per ml.

[10] Hisahiko Einaga and Hajime Ishii, *Jap. Anal.* **18**, 439–446 (1969).
[11] Mary M. Barling and Charles V. Banks, *Anal. Chem.* **36**, 2359–2360 (1964).
[12] Yoshio Shijo and Tsugio Takeuchi, *Jap. Anal.* **17**, 1192–1197 (1968).

Procedure. To 20 ml of sample containing 0.5–1 μg of nickel, add 2.5 ml of 20% solution of pyridine, 5 ml of 0.02% solution of chrome azurol S, 5 ml of 0.05 *M* borate buffer for pH 10.8, and 10 ml of 0.09% solution of hydroxydodecyltrimethylammonium bromide. Set aside for 10 minutes, dilute to 50 ml, and read at 639 nm.

CYANIDE

Nickel can be read as the stable tetracyanonickelate in the ultraviolet region. Beer's law is followed for 1–100 ppm of nickel. There is interference by cobalt, copper, iron, platinum, rhodium, and manganese. Nickel can be separated from cobalt, copper, and iron by anion exchange resin, eluting the nickel with hydrochloric acid. Manganese interference can be removed by alkaline oxidation of manganous ion with hydrogen peroxide to manganese dioxide, leaving the manganese dioxide in suspension in forming the complex ion. Interference by iron is removed by filtration of the hydroxide. Nitrate interference is relatively mild.

Procedure. For solutions containing nitric acid, add sulfuric acid and evaporate to sulfur trioxide fumes.[13] Take up in water. To an aliquot containing not more than 1 mg of nickel, add 10 ml of 2% solution of potassium cyanide in 1:9 ammonium hydroxide and dilute to 100 ml. Read at 267 nm against a reagent blank. There is a maximum of about 40% of that intensity at 284 nm.

Hydrocarbons. Shake a sample solution containing not more than 100 mg of nickel, diluted with *n*-hexane if necessary, with 25 ml and 25 ml of 2% solution of potassium cyanide in 1:9 ammonium hydroxide. Dilute an aliquot of the combined extracts containing no more than 1 mg of nickel to 100 ml and read at 267 nm against a reagent blank.

1,2-CYCLOHEXANEDIONE DIOXIME

This oxime for complexing with nickel has the trivial name of nioxime. For analysis of plant material, ash 20 grams of sample at 450°.[14] Leach with 10 ml and 10 ml of 1:3 hydrochloric acid and dilute the combined extracts to 100 ml. In an aliquot, determine successively copper with lead diethyldithiocarbamate, iron with cupferron, and nickel with nioxime.

The red 1:6 nickel-reagent complex is stable for at least 4 hours at 17–40°. Beer's law is obeyed for a 0.01–0.15 m*M* solution of nickel. Ferric ion can be

[13] M. W. Scoggins, *ibid.* **42**, 301–302 (1970); cf. T. M. Florence, *Anal. Chim. Acta* **19**, 548–551 (1958).
[14] Milan Karvanek, *Sb. Vys. Sk. Chem.-Tekhnol. Praze, E*, **23**, 13–21 (1969).

masked with tartrate. Titanium produces small positive errors. Excess nioxime is essential in the presence of manganese, copper, and cobalt.

Procedure. Mix a sample solution containing 30–400 μg of nickel, 20 ml of 20% sodium hydroxide solution, 6 mg of nioxime, and a solution of 10 mg of potassium ferricyanide.[15a] Dilute to 50 ml, set aside for 15 minutes, and read at 460 nm.

Aluminum-Bronze.[16] Dissolve 2 grams of sample in 60 ml of 1 : 1 : 3 sulfuric acid–nitric acid–water and expel nitrous fumes. Add 10 ml of hydrochloric acid and evaporate to sulfur trioxide fumes. Cool, dissolve in hot water, and set aside for 1 hour. Filter, and wash the residue free from copper with 1 : 20 sulfuric acid. Dilute the residual electrolyte to 500 ml.

Mix a 10-ml aliquot of the electrolyte with 10 ml of a buffer solution containing 50 grams of citric acid and 70 ml of ammonium hydroxide per 100 ml. Add 5 ml of 10% solution of gum arabic and dilute to about 90 ml. Add 1 ml of Teepol and 4 ml of 0.8% solution of nioxime in 10% ethanol. Dilute to 100 ml and read at 550 nm against a blank containing 2 mg of iron and 1.6 mg of nickel without color reagent.

Foodstuffs.[17a] Ash a 5-gram sample at 500°. Dissolve the ash with 25 ml of 1 : 3 hydrochloric acid and filter. To the filtrate, add 2 ml of 5% thiourea solution and 2 ml of 5% catecholdisulfonic acid solution. Titrate with 0.1 *N* ammonium hydroxide to the blue of bromocresol green or, if more than 10 μg of iron is present, titrate without indicator to a deep red. Add 1 ml of 0.8% solution of nioxime and set aside for 10 minutes. Extract the nickel complex with 10 ml of benzene. Wash the organic layer, then reextract the nickel with 10 ml of 0.2 *N* hydrochloric acid. To the acid extract, add 1 ml of 10% potassium bromide solution, 3 ml of 0.2 *N* sodium hypochlorite solution, and 6 ml of 5 *N* ammonium hydroxide. After mixing, add 4 ml of 0.8% solution of nioxime and dilute to 25 ml. Set aside for 20 minutes and read at 430 nm.

1,2,3-CYCLOHEXANETRIONE TRIOXIME

The nickel complex of this reagent, which has the trivial name of nicon, has maximum absorption at 560 and 430 nm. The reagent has negligible absorption at 560 nm. The optimum range for reading is at 6–22.5 ppm of nickel. The pH may be 3–6. Stabilization with gelatin is necessary to avoid precipitation. Development is essentially instantaneous. Copper, iron, and cobalt interfere but can be separated by ion exchange resin. In the presence of less than 2.5 ppm of cobalt, adjustment to pH 5–6 minimizes interference by this metal.

[15a] C. Gonzalez Perez, L. Polo Diez, and A. Sanchez Perez, *Anal. Chim. Acta* **87**, 233–237 (1976).
[16] M. Freegarde and Mrs. B. Allen, *Analyst* **85**, 731–735 (1960).
[17a] Milan Karvanek, *Sb. Vys. Sk. Chem.-Technol. Praze, Potravim. Technol.* **8** (1), 13–30 (1964); *Prum. Potravin.* **15**, 282–283 (1964).

Procedure. To a sample solution containing 1–25 μg of nickel, add 3 ml of 1% gelatin solution and 0.75 ml of a solution of 0.291 gram of nicon per 100 ml in ethanol.[17b] Dilute to 10 ml and read at 560 nm against a reagent blank.

DIETHYLDITHIOCARBAMATE

Iron and copper interfere with determination of nickel with this reagent but can be separated by ion exchange. Lead interferes but can be adsorbed on calcium carbonate. The complex is extracted by isoamyl alcohol for reading.

For nickel in serum, freeze-dry, then digest with nitric acid and hydrogen peroxide.[18] Add dimethylglyoxime and extract the nickel complex with chloroform. Wash the organic phase with dilute ammonium hydroxide. Reextract the nickel with 0.5 *N* hydrochloric acid. Raise to pH 9.5 and determine nickel with sodium diethyldithiocarbamate by extraction of the complex with amyl alcohol. Read at 325 nm and apply an Allen correction from readings at 295 and 355 nm.

For analysis of metallic uranium, read the absorption at 328 nm and correct the absorption due to nickel for interference by cobalt.[19]

Procedures

Cobalt.[20] Dissolve 2 grams of sample in 50 ml of 0.1 *N* hydrochloric acid. Add 16 grams of ammonium thiocyanate. Extract by shaking for 30 seconds with 80 ml of methyl isobutyl ketone and 20 ml of ethanol. Then extract with 50 ml of methyl isobutyl ketone. Extract the combined organic phases with 25 ml and 25 ml of 12.5% solution of ammonium thiocyanate in 0.25 *N* hydrochloric acid. Evaporate the combined aqueous layers to dryness and add 20 ml of nitric acid dropwise. When effervescence ceases, add 30 ml more of nitric acid and evaporate to dryness. Take up the residue in 40 ml of 0.5 *N* hydrochloric acid. Add 2 ml of 20% citric acid and raise to pH 9.5 by addition of ammonium hydroxide. Add 5 ml of 0.2% solution of sodium diethyldithiocarbamate. Extract the nickel complex with 10 ml of isoamyl alcohol and read at 325 nm against a blank.

Blood.[21] As the ion exchange column, add Dowex-1 resin as a slurry to give a bed of 28 cm $\times$ 0.2 cm^2 in a buret. Similarly prepare a bed 8 cm $\times$ 0.2 cm^2 of pure calcium carbonate in a standard, 1-mm bore, straight stopcock.

Digest 10 ml of sample with 5 ml of nitric acid and 0.5 ml of 30% hydrogen peroxide at 100° overnight. Then heat to 330° until reaction ceases—about 15 minutes. Cool, add 0.5 ml of nitric acid, heat at 330° to dryness, and for 15 minutes

[17b] W. Joe Frierson and Nina Marable, *Anal. Chem.* **34**, 210–212 (1962).
[18] F. W. Sunderman, Jr., *Clin. Chem.* **13**, 115–125 (1967).
[19] A. Borges Pimenta, *Rev. Port. Quim.* (Lisbon) **6** (2), 63–66 (1964).
[20] Tran Van Danh, J. P. Tardif, and J. P. Spitz, *Anal. Chim. Acta* **42**, 341–343 (1968).
[21] Maxwell L. Cluett and John H. Yoe, *Anal. Chem.* **29**, 1265–1269 (1957).

thereafter. Repeat once or more if necessary until the residue is white. Add 0.5 ml of 8 *N* hydrochloric acid and 0.5 ml of 30% hydrogen peroxide. Heat at 330° to dryness and repeat the step until no oxides of nitrogen are given off. Dissolve the residue in 5 ml of 8 *N* hydrochloric acid.

Transfer the solution to a polyethylene beaker placed in the evaporating dish of the apparatus illustrated in Figure 19-1. Evaporate to about 1 ml or until precipitation just begins. Prewash the ion exchange column with 60 ml of 0.005 *N* hydrochloric acid and acidify it with 14 ml of 8 *N* hydrochloric acid. Collect the last 5 ml to use as a blank. Transfer the sample to the column, using three 0.5-ml rinses to complete the transfer. Continue the elution with 8 *N* hydrochloric acid at a flow rate of 0.25 ml/minute, discarding the "free volume" and collecting the next 5 ml.

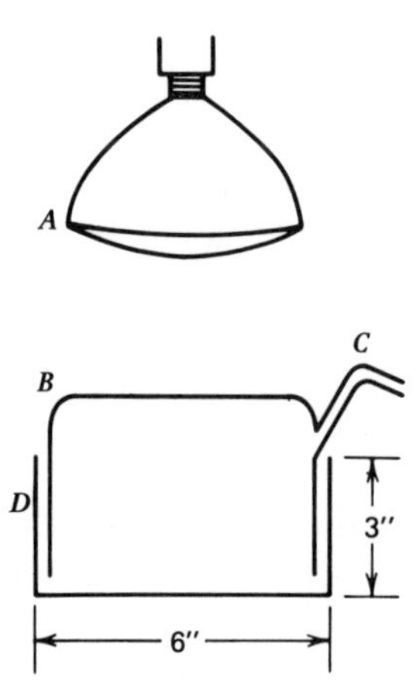

Figure 19-1 Evaporation unit.

Evaporate the sample and blank to about 0.1 ml and dilute to 1 ml. Add 3 drops of 20% citric acid solution and a piece of red litmus paper to each. Neutralize by dropwise addition of ammonium hydroxide. Transfer sample and blank each to a calcium carbonate column with three 0.2-ml rinses of water. Elute with 2 ml of water. To each effluent add 0.5 ml of 20% citric acid solution and a piece of red litmus paper. Neutralize with ammonium hydroxide and raise to pH 9.5. Add 1 ml of 0.2% solution of sodium diethyldithiocarbamate and dilute to 10 ml. Extract the nickel complex and the blank by shaking for 2 minutes with 3 ml of isoamyl alcohol. Read at 325 nm against the blank.

DIETHYLENETRIAMINE

This reagent complexes with nickel to give a maximum absorption at 860 nm. In the vicinity of 460 nm, both trivalent cobalt and nickel absorb as complexes; thus cobalt is determined by applying a correction based on the nickel as determined at 860 nm. In the procedure given here, cobalt is oxidized by air but an alternative oxidizing agent is ammonium persulfate.[22] Copper, chromium, and manganese form interfering colored complexes with the reagent. Iron or tin is precipitated and filtered.

Procedure. *Nickel and Cobalt.*[23] Dissolve the sample in nitric acid of appropriate dilution. Add 20% sodium hydroxide solution dropwise until the first precipitate of hydroxides appears. Clarify the solution by adding 1 or 2 drops of nitric acid. Add 2 ml of diethylenetriamine for each 10 ml of solution. Oxidize the cobalt by bubbling air through the solution for 15 minutes. If a precipitate forms, filter and wash it. Dilute to a volume at which nickel and cobalt each will be 0.01–0.06 gram-ion/liter. Read the nickel content at 850 nm. Read cobalt and nickel at 460 nm and, subtracting a correction for nickel, read the cobalt content.

5-DIMETHYLAMINO-2-(2-THIAZOLYLAZO)PHENOL

In ammoniacal solution nickel forms a stable 1 : 2 complex with this reagent. Preliminary separation of nickel is necessary. The complex can be read at 520 or 575 nm. When read at 575 nm, there is no absorption by excess reagent.

Procedure. *Indium.*[24] Dissolve a 1-gram sample in 10 ml of 1 : 1 hydrochloric acid. Evaporate to dryness and take up in 10 ml of water. Add 5 ml of 20% ammonium citrate solution, 0.1 gram of hydroxylamine hydrochloride, 2 ml of 1% ethanolic dimethylglyoxime, and 3 drops of ammonium hydroxide. Extract the nickel complex with 10 ml and 10 ml of chloroform. Reextract nickel from the organic phase with 10 ml and 5 ml of 0.5 *N* hydrochloric acid. Add 1 ml of nitric acid and 1 ml of 1 : 1 sulfuric acid to the acid extract and evaporate to dryness. Take up in 10 ml of water and add 1 ml of 1% sodium potassium tartrate solution. Add 1 ml of ammonium hydroxide and extract with 5 ml of 0.4 m*M* captioned reagent in chloroform. Dilute to 50 ml with chloroform and read at 575 nm.

DIMETHYLGLYOXIME

Nickel forms 1 : 1 and 1 : 2 complexes with dimethylglyoxime, and the latter is more stable.[25] The established technics call for oxidation to tetravalent nickel for determination with this reagent. Bromine water, iodine, or persulfate is commonly

[22] Harold L. Howling and Edward S. Shanley, *Chem. Can.* **12** (1), 48–49 (1960).
[23] R. D. Wheatly and S. O. Colgate, *Anal. Chem.* **28**, 1897–1898 (1956).
[24] K. Kasiura and Z. Sytniewska, *Chem. Anal.* (Warsaw) **13**, 177–183 (1968).
[25] Kenichiro Yamasaki and Chuya Matsumoto, *J. Chem. Soc. Jap.* **78**, 833–836 (1957); Zbigniew Gregorowicz, Stanislaw Grochowski, and Jerzy Kubala, *Chem. Anal.* (Warsaw) **2**, 322–325 (1957); Zbigniew Gregorowicz, *Z. Anal. Chem.* **168**, 241–246 (1959); Jose O. Nieto, *Bol. Inf. Petrol.* (Buenos Aires), No. **318**, 641–643 (1959); V. G. Hill and A. C. Ellington, *Econ. Geol.* **56**, 982–984 (1961).

used for the purpose. This complex is at least 3 times as sensitive as that with bivalent nickel. The presence of excess oxidizing agent is desirable but not absolutely essential. Addition of a protective colloid such as starch, gelatin, or gum arabic broadens the range that conforms to Beer's law.[26]

In the presence of cobalt, cyanide may be added and the cobalt converted by oxidation with hydrogen peroxide to the cobalticyanide.[27] Thereafter, in the presence of formaldehyde, the nickel glyoxinate is extracted with chloroform.

To convert nickel to the most stable complex, oxidize the 1 : 1 complex of nickel and dimethylglyoxime with 0.1 *M* iodine in 0.05 *M* sodium hydroxide at pH 12.5.[28] Extract the resulting complex of nickel with 1.5 *M* diphenylguanidinium chloride as a ternary complex in 1 : 1 chloroform–isopentyl alcohol. Read at 490 nm.

For nickel in sodium iodide, dissolve in an ammonium citrate solution at pH 8.5.[29] Wash out free iodine with chloroform. Add sodium diethyldithiocarbamate solution and extract the carbamates of manganese, cobalt, and nickel with chloroform. Evaporate the chloroform layer and destroy the organic residue with nitric and sulfuric acids. Dissolve in water and determine nickel with dimethylglyoxime.

For nickel in blister and refined copper, dissolve in hydrochloric and nitric acids, deposit the copper electrolytically, destroy oxides of nitrogen with urea, and determine with dimethylglyoxime.[30] When nickel-dimethylglyoxime complex is extracted with carbon tetrachloride from products of copper production, the copper can be washed out of the extract with 1 : 50 ammonium hydroxide.[31] For analysis of copper ore or refined products, EDTA can be used to mask low copper content; but for a high copper content, the bulk must be removed by electrolysis.[32]

For nickel in uranium mono- and dicarbides, ignite 5 grams of sample to oxides.[33] Dissolve in nitric acid and filter. Ash the residue, fuse with potassium bisulfate, extract with nitric acid, and add to the filtrate. To an aliquot containing about 1 gram of uranium, add hydroxylamine hydrochloride and sodium citrate. Make alkaline with ammonium hydroxide, add dimethylglyoxime, and extract with chloroform. Wash the organic phase with water and extract the nickel from it with hydrochloric acid. To the acid extract, immediately add iodine in aqueous potassium iodide and dimethylglyoxime solution. Make alkaline with sodium hydroxide solution and read at 540 nm.

For nickel in triuranium octaoxide and yellow cake, mix tetrahydrofuran with 5% of 12 *N* nitric acid.[34] As eluting solvent, make this 0.1 *M* with trioctylphosphine oxide. Dissolve 1 gram of triuranium octaoxide or 0.2 gram of yellow cake in nitric acid and evaporate to dryness. Dissolve in the eluting solvent and apply to a column of Dowex 50-X8 in hydrogen form. Wash the uranium out of the column

[26] M. Z. Yampol'skii, *Uch. Zap., Kursk. Gos. Pedagog. Inst.* **1958** (7), 73–82.

[27] Walter Nielsch, *Metallk.* **50**, 234–236 (1959).

[28] Yu. A. Zolotov and G. E. Vlasova, *Zh. Anal. Khim.* **28**, 1540–1543 (1973).

[29] A. M. Bulgakova, A. B. Blank, A. K. Khurkryanskii, and G. S. Plotnikova, *Stsintill. Stsintill. Materily 2-go (Vtorogo) Koord. Soveshch.* **1957**, 281–290.

[30] Silvio Barabas and W. Charles Cooper, *Metallurgia* **56**, 101–105 (1957).

[31] L. N. Krasil'nikova, *Sb. Tr. Vses. Nauch.-Issled. Inst. Tsvet. Met.* **1956** (1), 165–168.

[32] Ting-hui Chu, Tu-wen Chang, Yen-ts'ai Li, Ch'u-ch'ih T'ang, Tsun-che Ch'en, Chao-pi Sung, Shêng-hsiang Ch'en, Te-hsiang T'ang, Chung-ch'ing Liao, Ying-ming Li, and Ta-hung Ch'en. *Wu Han Ta Hsüeh*, and *Tsu Jan K'a Hsüeh Hsüeh Pao* **5**, 99–104 (1959).

[33] *UKAEA Rep.* **PG-516 (S)**, 34 pp (1964).

[34] J. Korkisch and H. Huebner, *Mikrochim. Acta* **1975**, II (2), 219–226.

with eluting solvent, then wash the column with a 5% solution of 12 N nitric acid in tetrahydrofuran. Elute with 1:1 hydrochloric acid. The eluate contains nickel, copper, zinc, cadmium, and iron. Isolate nickel from the interfering ions, and determine by dimethylglyoxime.

For nickel in pig iron, the preferred reagent is dimethylglyoxime at a nickel content of the order of 0.03%.[35] The ferrous ion is oxidized to ferric and masked with tartrate.[36] Copper is masked with thiosulfate. The complex is extracted at pH 5.2–5.3 with chloroform and read at 380 nm.

For nickel in beryllium and beryllium oxide, take a sample solution containing about 0.1 mg of nickel.[37] Add 100 mg of cadmium ion and adjust the medium to contain chloride, sulfate, and tartrate at pH 8.5. Precipitate the nickel and some interfering elements with benzotriazole. Filter, and dissolve in nitric and perchloric acids. Separate the nickel by ion exchange for developing with dimethylglyoxime and reading at 465 nm.

For nickel in thin magnetic films used in computers, dissolve the sample in a few drops of nitric acid and dilute to 10 ml with 0.05 N perchloric acid.[38] Oxidize an aliquot containing 5–50 μg with ammonium persulfate using osmium tetroxide as a catalyst, develop with dimethylglyoxime, and read at 445 or 536 nm.

For separation of iron and copper from nickel in copper and copper ores, dissolve 1 gram of sample in nitric acid.[39] If hydrochloric acid is added, also add 1:1 sulfuric acid. Evaporate to a syrup, take up in 25 ml of water, and adjust with 50% sodium acetate solution to pH 5.6. Extract the iron and copper with 10 ml and 5 ml of a 20% solution of C_7–C_9 fatty acids in chloroform. Scavenge the last of the copper by extraction with 10 ml of 1% solution of lead diethyldithiocarbamate in chloroform. Develop the nickel in the aqueous phase with dimethylglyoxime.

For nickel carbonyl in water gas, bubble 0.1–2 cubic feet through 10 ml of M-iodine monochloride and transfer with 10 ml of 5% sulfuric acid. Dissolve deposited sulfur with nitric acid and add to the solution. Evaporate to dryness and ignite.[40] Develop the residue in 1:1 hydrochloric acid with dimethylglyoxime. Alternatively, absorb the nickel carbonyl in acidified solution of chloramine B.[41]

For analysis of biphenyl, digest 1 gram of sample with 1 ml of sulfuric acid, 5 ml of nitric acid, and 1 ml of 5% sodium chloride solution.[42] Evaporate to sulfur trioxide fumes, take up in water, oxidize with bromine water, and develop with dimethylglyoxime.

For nickel in fats and oils, ash, extract with hydrochloric acid, determine copper by lead diethyldithiocarbamate in chloroform, extract iron as the cupferrate with chloroform; then determine nickel with dimethylglyoxime.[43]

[35] Theo Kurt Willmer, *Arch. Eisenhüttenw.* **29**, 159–164 (1958).

[36] Zdenek Hasek, *Hutn. Listy* **16**, 281–283 (1961).

[37] J. O. Hibbets and S. Kallmann, *Talenta* **10**, 181–188 (1963); cf. J. Walkden, *UKAERE* **AM22**, 3 pp (1959).

[38] K. L. Cheng and B. L. Goydish, *Microchem. J.* **7**, 166–178 (1963).

[39] S. E. Kreimer, N. V. Tuzhilina, and A. S. Lomekhov, *Zh. Anal. Khim.* **18**, 1080–1082 (1963).

[40] A. B. Densham, P. A. A. Beale, and R. Palmer, *J. Appl. Chem.* **13**, 576–580 (1963); cf. V. S. Fikhtengol'ts and N. P. Kozlova, *Zavod. Lab.* **23**, 917 (1957); Hiroshi Suzuki and Kazuhiko Oishi, *Jap. Anal.* **12**, 1011–1017 (1963).

[41] A. A. Belyakov, *Zavod. Lab.* **26**, 158–159 (1960).

[42] D. Monnier, W. Haerdi, and E. Martin, *Helv. Chim. Acta* **46**, 1042–1046 (1963).

[43] M. Karvanek, *Sb. Vys. Sk. Chem.-Tekhnol., Odd. Fak. Potravin.-Tekhnol.* **5**, Pt. 1, 203–212 (1961).

For nickel in grease, extract the bulk of the fatty material with benzene.[44] Oxidize the residue with nitric, sulfuric, and perchloric acids. Take up in water and adjust to pH 8.8–9.4 with 1 : 1 ammonium hydroxide. Add excess of 30% hydrogen peroxide to oxidize ferrous ion. At once add dimethylglyoxime in ethanol and set aside for 15 minutes. Extract with chloroform, dilute to under 0.8 ppm of nickel in that solution, and read at 329 nm.

For analysis of cement for cobalt and nickel, disintegrate the sample with 9 *N* hydrochloric acid and a few drops of nitric acid.[45] Centrifuge to remove silica. Pass through a column of Dowex 1. Cobalt is adsorbed but nickel is not. Determine the latter with dimethylglyoxime at 540 nm. Elute the cobalt with 4 *N* hydrochloric acid and read with nitroso-R salt at 520 nm.

For copper, nickel, cobalt, zinc, and cadmium as impurities in sodium chloride, complex with EDTA and extract from the solution with chloroform.[46] Evaporate to dryness, add sulfuric and nitric acids, and heat to sulfur trioxide fumes. Determine nickel by dimethylglyoxime.

Applications of determination of nickel by dimethylglyoxime include various residues developed in the petroleum industry,[47] or in crude oil or its ash,[48] nickel in high alloy iron and steel,[49] traces of nickel in uranyl nitrate,[50] nickel in vanadium pentoxide with the vanadium masked with citrate,[51] silicates after solution in hydrofluoric and perchloric acids,[52] in a buffered solution of lead oxidized with bromine water,[53] silver in silver alloys after electrolytic deposition of silver and copper,[54] in cobalt-arsenic ores,[55] in semiconductors,[56] in hard and soft lead, and in antimony,[57] cadmium, or zinc,[58] as well as tantalum,[59] high purity niobium, molybdenum, and tungsten,[60] magnesium alloys,[61] cesium, potassium, and sodium hydroxides,[62] optical glass,[63] capping material,[64] and laterites.[65]

For nickel in rhodium-plating electrolyte as isolated by column chromatography

[44] H. Ssekaalo, *Anal. Chim. Acta* **62**, 220–222 (1972).
[45] F. J. Valle Fuentes and F. Burriel-Marti, *Rev. Cien. Apl.* **22**, 15–20 (1968).
[46] Cezary Rozychi and Janusz Rogozinski, *Chem. Anal.* (Warsaw) **20**, 107–111 (1975).
[47] Z. Gregorowicz, *Nafta* **13**, 39–41 (1957).
[48] A. Buzas, *Magy. Kem. Lapja.* **23**, 716–719 (1968).
[49] W. B. Sobers, *Foundry* **88**, 110–113 (1960).
[50] J. Nowicka-Jankowska, A. Golkowska, I. Pietrzak, and W. Zmijewska, *Pol. Akad. Nauk., Inst. Badan Jadrowych.* No. **45**, 42 pp (1958).
[51] S. Ya. Vinkovetskaya, *Nauch. Tr. Nauchn.-Issled. Proekt. Inst. Redkomet. Prom.* **1972**, 202–204.
[52] Heinrich M. Koester, *Ber. Deut. Keram. Ges.* **46**, 247–253 (1969).
[53] British Standards Institution, *BS 3908*, Pt. **5**, 7 pp (1968).
[54] J. Chwastowska and Z. Skorko-Trybula, *Chem. Anal.* (Warsaw) **9**, 123–130 (1964).
[55] Yu. N. Lushchik, *Zavod. Lab.* **25**, 801–802 (1959).
[56] R. P. Pantaler, N. B. Lebed', and L. N. Semenova, *Tr. Kom. Anal. Khim.* **16**, 24–29 (1968).
[57] Maria Cyrankowska, *Chem. Anal.* (Warsaw) **15**, 209–213 (1970).
[58] V. P. Tiptsova and O. I. Kopnina, *Zh. Anal. Khim.* **22**, 1108–1110 (1967).
[59] R. D. Gardner, C. H. Ward, and W. H. Ashley, *USAEC Rep.* **LA-3152**, 10 pp (1964); V. A. Nazarenko, E. A. Biryuk, M. B. Shustova, G. G. ShiFareva, S. Ya. Vinkovetskaya, and G. V. Flyantikova, *Zavod. Lab.* **32**, 267–269 (1966).
[60] E. M. Penner and W. R. Inman, *Talenta* **10**, 997–1003 (1963).
[61] British Standards Institution, *BS 3907*, Pt. **7**, 7 pp (1969).
[62] M. M. Godneva and R. D. Vodyannikova, *Zh. Anal. Khim.* **20**, 831–835 (1965).
[63] A. I. Krupkin, *Tr. Kom. Anal. Khim., Akad. Nauk SSSR, Inst. Geokhim. Anal. Klim.* **8**, 204–209 (1958).
[64] British Standards Institution, *BS 3338*, Pt. **20**, 7 pp (1970).
[65] François Ruf, *Rep. Malgache, Rapp. Ann. Serv. Geol.* **1961**, 103–105.

for determination by dimethylglyoxime, see Copper. For nickel by dimethylglyoxime along with copper, iron, and manganese in highly pure alkali–metal halide crystals, see Copper by Lead Diethyldithiocarbamate. For nickel in tellurium by dimethylglyoxime after separation and determination of cobalt by 1-nitroso-2-naphthol, see Cobalt.

Procedures

Extraction with Naphthalene.[66] Dilute a sample solution containing 3–25 μg of nickel to 30 ml and add 0.5 ml of 0.5% ethanolic dimethylglyoxime. Add 2 ml of *M* ammonium hydroxide–ammonium chloride buffer solution for pH 5.5–9.6. Heat to 90°, add 5 ml of naphthalene, and shake. Cool, filter, and wash the crystals with water. Dissolve the naphthalene in 10 ml of chloroform and read at 375 nm.

Hydrazine Added.[67] Prepare a reagent that is 6 *N* with potassium hydroxide and 0.5 *M* with hydrazine. Mix 10 ml of sample solution, 10 ml of reagent, and 10 ml of nitric acid. Evaporate until evolution of brown fumes ceases. Add 1 ml of nitric acid and dilute to 50 ml. Add 5 ml of 5% tartaric acid solution and 4 ml of saturated bromine water. Add 1 : 1 ammonium hydroxide until the solution is decolorized, plus 1 ml in excess. Add 1 ml of 1% ethanolic dimethylglyoxime, dilute to 100 ml, and set aside for 5 minutes. Read at 540 nm.

Aluminum.[68] Digest 1 gram of sample with 20 ml of 20% sodium hydroxide solution. Add 20 ml of 1 : 1 sulfuric acid and 2 ml of nitric acid. Boil to a clear solution and dilute to 50 ml. To a 5-ml aliquot, add 2 ml of a solution of 50 grams of citric acid and 50 ml of ammonium hydroxide per 100 ml. Add 75 ml of water and 2 ml of a saturated solution of iodine in 10% potassium iodate solution. Set aside for 15 minutes and add 15 ml of 1% solution of dimethylglyoxime in ammonium hydroxide. Dilute to 250 ml, set aside for 30 minutes, and read against a blank at 430 or 530 nm.

Aluminum Alloys.[69] Dissolve 2 grams of sample in 100 ml of 5% sodium hydroxide solution. Add 10 ml of 10% solution of sodium sulfide, boil, and filter. Dissolve the precipitate in 20 ml of 1 : 1 nitric acid and dilute to 100 ml. To a 10-ml aliquot add 1 ml of 1 : 1 sulfuric acid and 10 ml of 12% solution of sodium sulfide. Heat to boiling, cool, and dilute to 100 ml. Filter and boil a 10-ml aliquot until free from hydrogen sulfide. Cool, and add sequentially 10 ml of 20% tartaric acid solution, 20 ml of 6% sodium hydroxide solution, 10 ml of 3% solution of potassium persulfate, and 10 ml of 1% solution of dimethylglyoxime in 6% sodium hydroxide solution. Dilute to 100 ml and read at 540 nm.

[66] Taitiro Fujinaga, Masatada Satake, and Tatsuo Yonekubo, *Jap. Anal.* **20**, 1255–1259 (1971).
[67] E. I. Mingulina, N. G. Ryzhova, and E. P. Klochek, *Tr. Mosk. Energ. Inst.* **1972** (112), 102–107.
[68] M. C. Steele and L. J. England, *Metallurgia* **59**, 153–156 (1959).
[69] A. Bradvarov and M. Dimitrova, *Mashinostroene* (Sofia) **9** (9), 32–33 (1960); cf. British Standards Institution, *BS 1728*, Pt. **15**, 8 pp (1966).

Tungsten and Tungsten Alloys.[70] For preparation of the sample solution, see Copper by Neocuproine. To a 50-ml aliquot, add 5 ml of bromine water. Add ammonium hydroxide until the bromine color disappears, and add 5 ml of ammonium hydroxide in excess. Within 2 minutes, add 5 ml of 1% ethanolic dimethylglyoxime. Dilute to 100 ml and within 15 ± 2 minutes read at 540 nm against a reagent blank.

Titanium.[71] To 1 gram of sample in platinum add 10 ml of hydrofluoric acid followed by 2 ml of nitric acid. Evaporate to 5 ml and heat with 10 ml of 10% sodium citrate solution. Cool and add 80 ml of water, 10 ml of 1:1 ammonium hydroxide, and 5 ml of 1% ethanolic dimethylglyoxime. Set aside for 15 minutes. Extract with 5 ml, 5 ml, and 5 ml of chloroform. Wash the combined extracts with 1 ml and 1 ml of *N* hydrochloric acid. Evaporate the acid extract to dryness. Add 0.1 ml of nitric acid and 0.3 ml of hydrochloric acid. Again evaporate to dryness and dissolve in 4 ml of water. Add 0.1 ml of ammonium hydroxide, 0.1 ml of 1% ethanolic dimethylglyoxime, and 0.1 ml of fresh bromine water. Dilute to 5 ml and read at 540 nm.

Titanium Alloy.[72] Dissolve a 5-gram sample in 100 ml of 1:1 sulfuric acid and 2 ml of hydrofluoric acid. After vigorous action ceases, warm to complete solution. Add 30% hydrogen peroxide until the yellow titanium complex is obtained, then evaporate to sulfur trioxide fumes. Cool and take up in water. Cool again and dilute to 500 ml.

Complete as for tungsten and tungsten alloys from "To a 50-ml aliquot,... ."

Chromium.[73] Dissolve 1 gram of sample in 10 ml of 1:4 sulfuric acid and dilute to 50 ml. Complete as for aluminum from "To a 5-ml aliquot,... ."

Chromium Oxide Pellets.[74] Fuse 1 gram of sample with 5 grams of sodium carbonate in platinum, extract with 20 ml of *N* sulfuric acid, and dilute the filtrate to 100 ml. Neutralize a 10-ml aliquot to methyl orange with *N* sodium carbonate. Add 1 ml of sulfuric acid and 0.5 ml of 10% hydrogen peroxide. Boil for 10 minutes, cool, add 5 ml of 6% cupferron solution, and dilute to 20 ml. Extract for 1 minute each with 20 ml, 5 ml, 5 ml, and 5 ml of chloroform. Evaporate the combined extracts to dryness in platinum and fuse with 0.5 gram of sodium carbonate. Extract with 5 ml of *N* sulfuric acid and dilute to 50 ml. Complete as for aluminum from "To a 5-ml aliquot,... ."

Nichrome.[75] Dissolve 0.2 gram of sample in 1 ml of nitric acid and 2 ml of hydrochloric acid. Dilute to 10 ml with 1:1 hydrochloric acid and extract the iron

[70] George Norwitz and Herman Gordon, *Anal. Chem.* **37**, 417–419 (1965); cf. T. E. Green, *ibid.* **37**, 1595–1596 (1965); G. I. Postogvard, *Zh. Anal. Khim.* **25**, 950–952 (1970).

[71] V. A. Nazarenko, M. B. Shustova, G. G. Shitareva, G. Ya. Yagnyatinskaya, and R. V. Ravitskaya, *Zavod. Lab.* **28**, 645–648 (1962).

[72] George Norwitz and Maurice Codell, *Metallurgia* **57**, 261–270 (1958).

[73] A. A. Fedorov, F. A. Ozerskaya, R. D. Malinina, Z. M. Sokolova, and F. V. Linkova, *Sb. Tr. Tsent. Nauch.-Issled. Inst. Chern. Met.* **1960** (19), 7–21.

[74] J. T. McAloren and G. F. Reynolds, *Metallurgia* **57**, 52–56 (1958).

[75] V. N. Polyanskii, *Sb. Tr. Mosk. Vech. Met. Inst.* **1957** (2), 255–256.

with 10 ml and 10 ml of ether. Dilute to 250 ml and take a 10-ml aliquot. Neutralize with ammonium hydroxide and add 5 ml of 10% solution of potassium tartrate. Add 2 ml of 0.1% ethanolic dimethylglyoxime. Then add 1 : 1 ammonium hydroxide to give a faint ammoniacal odor. Extract with 15 ml, 15 ml, and 15 ml of chloroform. Dilute the extracts to 50 ml and read at 375 nm.

Crude Copper.[76] NICKEL AND ZINC. Dissolve 5 grams of sample in 30 ml of 1 : 1 nitric acid and dilute to 250 ml. Adjust a 10-ml aliquot to pH 3 with ammonium hydroxide and pass it through a 30 cm × 14 mm column of 60-80 mesh Amberlite IR-120 resin in sodium form at 3 ml/minute. Wash the column with water. Elute the copper with 200 ml of 0.1 *M* sodium thiosulfate and again wash the column with water. Elute nickel and zinc with 150 ml of 2 *M* sodium chloride solution. Pass this eluate through a 20 cm × 9 mm column of Amberlite IRA-400 in chloride form. Dilute the percolate to 250 ml and take a 10-ml aliquot. Complete as for nichrome from "Neutralize with ammonium hydroxide...."

Wash the column with water and elute the zinc with 150 ml of *M* sulfuric acid for development with dithizone.

Nickel Oxide in Copper-Nickel Alloy.[77] Heat 5 grams of sample, 15 ml of bromine, and 150 ml of methanol at 35° for 20 hours. Filter the residue, wash with methanol, and ash. Fuse with 5 grams of potassium pyrosulfate, dissolve the melt in water, and dilute to 100 ml. Mix an aliquot containing 5–100 μg of nickel with 10 ml of 30% solution of ammonium citrate. Adjust to pH 9 with ammonium hydroxide and add 1 ml of 1% ethanolic dimethylglyoxime. Extract with 10 ml of chloroform and read at 375 nm.

As an alternative to disintegration with bromine, dissolve 5 grams of sample with 50 ml of 1 : 1 nitric acid at under 35°. Filter the residue, wash, and ash. Complete as above from "Fuse with 5 grams of...."

Red Brass and Bronzes.[78] As a buffer solution, add ammonium hydroxide to 100 grams of ammonium acetate in 250 ml of water to raise to pH 9.2. Add 200 grams of sodium thiosulfate and dilute to 500 ml.

Dissolve 1 gram of sample in 25 ml of 1 : 1 nitric acid and 5 ml of 10% sodium chloride solution. Cool, add 0.16 gram of lead nitrate, and dilute to 100 ml. To a 5-ml aliquot add 4 ml of 1 : 8 ammonium hydroxide, then 5 ml of 10% sodium tartrate solution. Add 7 ml of the buffer solution, then 5 ml of 20% solution of hydroxylamine hydrochloride, and 2 ml of 0.5% ethanolic dimethylglyoxime. Extract the nickel complex with 25 ml of chloroform, dry with sodium sulfate, and read at 375 nm.

Copper Alloys.[79] Dissolve 0.1 gram of sample in 5 ml of 1 : 3 nitric acid. Drive off nitrogen oxides, cool, and dilute to 250 ml. To a 10-ml aliquot add 2 ml of 10%

[76] Tetsuo Katsura, *Jap. Anal.* **10**, 1207–1210 (1961); cf. W. Nielsch, *Z. Anal. Chem.* **150**, 114–118 (1956); C. M. Dozinel, *Chim. Anal.* (Paris) **44**, 436–438 (1962).

[77] Iwao Tsukahara, *Jap. Anal.* **20**, 596–601 (1971).

[78] H. Wiedmann, *Metall* **12**, 1005–1007 (1958); British Standards Institution, *BS 3630*, Pt. **16** (1976).

[79] I. I. Kalinichenko and A. A. Kuyaseva, *Iz. Vyssh. Ucheb. Zaved., Khim. Khim. Tekhnol.* **3**, 418–421 (1960).

solution of sodium potassium tartrate, 5 ml of 10% solution of ammonium persulfate, 5 ml of 4% solution of sodium hydroxide, and 5 ml of 1% solution of dimethylglyoxime in 4% sodium hydroxide solution. After 5 minutes add 5 ml of 0.1 *N* tetrasodium EDTA and 5 ml of 5% solution of ammonium chloride. Dilute to 100 ml and read at 375 nm against a blank in which 4% sodium hydroxide solution has replaced the dimethylglyoxime reagent. The later addition of EDTA destroys any interfering copper complex without affecting that derived from oxidized nickel ion.

Tin-Containing Copper Alloys.[80] Dissolve a 0.1-gram sample in 10 ml of hydrochloric acid and 2 ml of 30% hydrogen peroxide. Evaporate to dryness and take up in 10 ml of 8 *N* hydrochloric acid. Pass through a column of anionite EDE-10P in chloride form to adsorb copper, tin, iron, lead, and zinc. Percolate the column with 100 ml of 8 *N* hydrochloric acid to elute the nickel. Add 5 ml of nitric acid and 5 ml of 1 : 1 sulfuric acid to the percolate. Evaporate to sulfur trioxide fumes, take up in water, and dilute to 50 ml. To a 10-ml aliquot, add 10 ml of 2% solution of tartaric acid, 40 ml of 5% sodium hydroxide solution, 10 ml of 3% solution of ammonium persulfate, and 10 ml of 1% solution of dimethylglyoxime in 1% sodium hydroxide solution. Dilute to 100 ml, set aside for 10 minutes, and read at 540 nm.

Manganese Alloys.[81] Heat 0.06 gram of sample with 3 ml of 1 : 1 nitric acid. Using a platinum dish as cathode and a platinum plate as anode electrolyze at 2.5 volts and 0.5 ampere. The deposit on the anode will be lead dioxide and manganese dioxide. Copper will be plated on the dish. Dissolve the deposit from the anode with 10 ml of 1 : 1 nitric acid and a few drops of 0.45% solution of oxalic acid. Combine this with the residual solution from the dish and dilute to 100 ml. Complete as for tin-containing copper alloys from "To a 10-ml aliquot, add. . . ."

Uranium.[82] Dissolve 2 grams of sample in 25 ml of 1 : 1 hydrochloric acid. Add 12.5 ml of 0.33 *M* potassium chlorate to oxidize uranium to the hexavalent state. Add 10 ml of 30% ammonium citrate solution and 1 ml of 1% ethanolic dimethylglyoxime. Raise to pH 8.8–10 by adding ammonium hydroxide, and dilute to 50 ml. Extract the complex with 10 ml of chloroform and read at 375 nm. Ammonium carbonate and ammonium citrate effectively mask uranium, thorium, and iron.[83]

For greater sensitivity, dissolve 1 gram of sample in 5 ml of hydrochloric acid and 2 ml of 30% hydrogen peroxide.[84] Evaporate to dryness and dissolve in 30 ml of water. Add 2 ml of bromine water; 5 ml of 20% solution of ammonium citrate, 5 ml of 1 : 1 ammonium hydroxide, and 1 ml of 1% ethanolic dimethylglyoxime. Dilute to 50 ml and read at 562 nm.

[80] M. N. Kruglova, L. S. Studenskaya, and V. V. Stepin, *Tr. Vses. Nauchn.-Issled. Inst. Stand. Obraztsov.* **4**, 146–149 (1968).

[81] I. V. Aronina, *Izv. Akad. Nauk Mold. SSR* **1967** (9), 25–29.

[82] Kenji Motojima, Hiroshi Hashitani, and Kazuo Katsuyama, *Nippon Genshicyoku Gakkaishi* **3**, 89–92 (1961); cf. Masami Suzuki, *Jap. Anal.* **8**, 432–435 (1959); Masami Suzuki and Tsugio Takeuchi, *ibid.* **9**, 708–709 (1960).

[83] V. T. Atharale, L. M. Mahajan, N. R. Trakoor, and M. S. Varde, *Anal. Chim. Acta* **21**, 491–492 (1959).

[84] Yoshio Morimoto, Takashi Ashizawa, and Sadao Araya, *Jap. Anal.* **10**, 1387–1391 (1961); cf. A. B. Crowther and H. T. Neville, *UKAEA Ind. Group Hd.* **SCS-R-45**, 11 pp (1959).

Uranium Tetrafluoride.[85] Dissolve 5 grams of sample in 1:4 nitric acid containing 3% of boric acid, evaporate to dryness, and take up in water. To an aliquot equivalent to 1 gram of uranium, add 5 ml of 20% ammonium citrate solution, 2 ml of saturated bromine water, 2.5 ml of ammonium hydroxide, and 1 ml of 1% ethanolic dimethylglyoxime. Dilute to a known volume and read at 562 nm.

High Purity Thorium Compounds.[86] Dissolve 10 grams of sample in 20 ml of 6 *N* nitric acid. Extract the thorium with 30 ml and 30 ml of 1:1 tributylphosphate-chloroform. Evaporate the aqueous solution to dryness and dissolve in 25 ml of water. To a 10-ml aliquot, add 10 ml of 30% citric acid solution, 1 ml of 1% ethanolic dimethylglyoxime, and 100 ml of water. Raise to pH 9–10 with 6 *N* ammonium hydroxide. Extract the nickel complex with 10 ml of chloroform, dry the extract with sodium sulfate, and read at 375 nm. This will determine 1 ppm of nickel in the sample.

Selenium.[87] Dissolve a 0.5-gram sample in 10 ml of hydrochloric acid and 2 ml of nitric acid. Evaporate to dryness and take up in 2 ml of hydrochloric acid. Add 5 ml of water and 30 ml of 20% solution of sodium citrate. Neutralize to phenolphthalein with ammonium hydroxide and add 2 ml of 1% diethyldithiocarbamate solution. Extract with 25 ml and 5 ml of chloroform. Reextract the combined organic layers with 10 ml, 10 ml, 5 ml, and 5 ml of 1:1 nitric acid. Add 5 ml of hydrochloric acid and evaporate to dryness. Take up in 25 ml of water. Complete as for high purity thorium compounds from "To a 10-ml aliquot, ..."

Steel.[88] As a sample, take 0.2 gram of steel containing 5–10% nickel, or 0.1 gram if the nickel content exceeds 12%. Dissolve in 20 ml of 1:1 hydrochloric acid. Add 0.25 ml of nitric acid and heat to destroy the carbides. Cool and dilute to 200 ml. To a 20-ml aliquot, add 4 ml of 0.2% solution of chromic nitrate, 12 ml of 0.1% solution of ferric nitrate, 5 ml of 20% solution of sodium potassium tartrate, 10 ml of 10% solution of ammonium persulfate, 15 ml of 10% sodium hydroxide solution, and 5 ml of 1% solution of dimethylglyoxime in 4% sodium hydroxide solution. Let stand for 1 hour, dilute to 100 ml, and read at 540 nm.

Up to 4% of nickel is determined in steel by dimethylglyoxime and iodine using the AutoAnalyzer.[89]

Copper-Nickel Concentrates.[90] Dissolve 0.25 gram of sample containing 5–70% of nickel with 8 ml of phosphoric acid and 10 ml of perchloric acid. Evaporate to fumes, boil with 50 ml of water, and dilute to 250 ml. To a 5-ml aliquot add sequentially 2.5 ml of 20% solution of sodium potassium tartrate tetrahydrate, 30 ml of 5% solution of sodium hydroxide, 15 ml of 1% solution of dimethylglyoxime in 5% sodium hydroxide solution, and 4.5 ml of 5% solution of ammonium

[85] *UKAEA Rep.* **PG 614 (S)**, 7 pp (1965).
[86] Kwok-yuen Chan and Ton-pol Tam, *Chemistry* (Taipei) **1969** (3), 55–61.
[87] N. P. Strel'nikova, G. G. Lystsova, and G. S. Dolgorukova, *Zavod. Lab.* **28**, 1319–1321 (1962).
[88] V. F. Barkovskii and I. N. Vtorygina, *ibid.* **28**, 275276 (1962); cf. Ohiko Kammori, *Nippon Kinzoku Gahkaishi* **20**, 255–258 (1956); A. Sommariva and A. Devoti, *Met. Ital.* **52**, 341–354 (1960); M. Simek, *Hutn. Listy* **19**, 663–665 (1964).
[89] K. Braithwaite and J. D. Hobson, *Metallurgia* **81**, 205–208 (1970).
[90] N. I. Shebarshina and E. P. Shkrobot, *Zavod. Lab.* **40**, 625–627 (1974).

persulfate. Dilute to 200 ml and after 15 minutes read at 470 nm against a solution differing from the sample solution by about 0.5 mg of nickel per 200 ml.

Nickel and Cobalt Concentrates.[91] Treat 0.1 gram of ignited concentrate with 0.5 gram of ammonium fluoride, 5 ml of hydrochloric acid, and 5 ml of nitric acid. Evaporate to a couple of ml, add 10 ml of 1:1 sulfuric acid, and evaporate to sulfur trioxide fumes. Cool, add 3 ml of water, and again take to sulfur trioxide fumes. Boil with 40 ml of water, cool, filter, and dilute to 250 ml. Dilute a 25-ml aliquot to 250 ml. To a 25-ml aliquot of this dilution, add 10 ml of 20% solution of sodium potassium tartrate, 10 ml of 5% solution of sodium hydroxide, 10 ml of 3% solution of ammonium persulfate, and 10 ml of 1% solution of dimethylglyoxime in 5% sodium hydroxide solution. Dilute to 100 ml and read at 500–550 nm.

Ores in General.[92] Digest 0.5 gram of sample containing 0.005–0.2% nickel with 5 ml of hydrochloric acid and 2 ml of nitric acid. Add 10 ml of 1:1 sulfuric acid and evaporate to sulfur trioxide fumes. Cool, take up in 50 ml of water, and filter. Dilute to 250 ml and take a 25-ml aliquot. Add 2 ml of 50% citric acid solution, neutralize with ammonium hydroxide, and add 1 ml in excess. Add 2 ml of 1% ethanolic dimethylglyoxime. Extract with 5 ml, 5 ml, and 3 ml of chloroform. Strip the nickel from the combined chloroform extracts by shaking for 30 seconds with 10 ml and 5 ml of 0.5 *N* hydrochloric acid. Dilute the acid extracts to 80 ml, and add 10 ml of 25% ammonium chloride solution and 2 ml of bromine water. Let the mixture stand for 1 minute, neutralize with ammonium hydroxide, and add 2 ml in excess. Cool, and add 2 ml of 1% solution of dimethylglyoxime in 1% sodium hydroxide solution. Dilute to 100 ml, set aside for 5 minutes, and read at 430 nm within 30 minutes.

COPPER ORE.[93] This technic is applicable up to a copper:nickel ratio of 1000:1. Dissolve 0.25 gram of sample in 18 ml of hydrochloric acid, add 5 ml of nitric acid, and evaporate to dryness. Add 6 ml of 1:1 sulfuric acid and heat to sulfur trioxide fumes. If manganese is present, neutralize with ammonium hydroxide, add 0.5 gram of ammonium persulfate, and heat for 20 minutes. Dilute to 100 ml and filter. To a 10-ml aliquot, add sequentially 5 ml of 10% solution of sodium potassium tartrate, 5 ml of 5% solution of ammonium persulfate, 5 ml of 1% ethanolic dimethylglyoxime, and 10 ml of 10% solution of sodium hydroxide. Three minutes later add 5 ml of 10% solution of EDTA. Dilute to 50 ml and read at 500 nm.

IRON ORE.[94] Dissolve a 0.2-gram sample in 15 ml of hydrochloric acid and 5 ml of nitric acid. Add 20 ml of 1:1 sulfuric acid and evaporate to sulfur trioxide fumes. Add 5 ml of water and again evaporate to sulfur trioxide fumes. Take up in 25 ml of 1:4 hydrochloric acid and filter. Mask iron and bismuth in the filtrate with 5 ml of 20% ammonium tartrate solution. Render the solution alkaline with ammonium hydroxide and add 2 ml of 1% ethanolic dimethylglyoxime. Extract the

[91] L. B. Ginzburg and E. P. Shkrobot, *Sb. Nauch. Tr. Gos. Nauch.-Issled. Inst. Tsvet. Met.* (*Anal. Rud Tsevt. Met. i Prod. Pererabot.*) **1956** (12), 52–56.
[92] Kiyoshi Isano, *Jap. Anal.* **6**, 557–561 (1957).
[93] Siao-wei Kuo, *Hua Hsueh Hsueh Pao* **24**, 252–255 (1958).
[94] D. P. Shcherbov and D. N. Perminova, *Zavod. Lab.* **33**, 921–923 (1967).

nickel complex with 6 ml and 4 ml of either benzene or chloroform. Wash the combined extracts with 10 ml of 1:50 ammonium hydroxide to prevent interference by cobalt or copper. Read at 350 nm against the solvent. Manganese interferes.

Exchangeable Nickel in Soil.[95] Mix 10 grams of soil with 100 ml of *M* ammonium acetate adjusted to pH 7, and store for 24 hours. Filter. To a 20-ml aliquot, add 10 ml of 10% citric acid solution and 2 ml of saturated bromine water. Add ammonium hydroxide to raise to above pH 8. Dilute to 50 ml with 1% ethanolic dimethylglyoxime. Read at 445 nm.

Marine Sediments.[96] Digest 1 gram of -120 mesh sample with 4 ml of perchloric acid and 15 ml of hydrofluoric acid in platinum overnight at 100°. Evaporate gently until no fumes are visible, then evaporate almost to dryness under an infrared lamp. Add 2 ml of perchloric acid and repeat the evaporation. Add 1 ml of perchloric acid and 15 ml of water. Warm to dissolve, dilute to 100 ml, and take a 25-ml aliquot. Complete as for ores in general from "Add 2 ml of 50% citric acid solution, ..."

Zinc Electrolyte.[97] Mix 10 ml of sample, 10 ml of 23.2% solution of sodium citrate, 16 ml of ammonium hydroxide of d. 0.934, 4 ml of 1% ethanolic dimethylglyoxime, and 2 drops of 20% solution of hydroxylamine hydrochloride. Set aside for 10 minutes, then extract with 6 ml, 3 ml, and 3 ml of chloroform. Extract the nickel from the combined chloroform extracts with 10 ml, 5 ml, and 5 ml of 0.5 *N* hydrochloric acid. Wash the combined acid extracts with 5 ml and 5 ml of chloroform. Mix with 5 ml of hydrochloric acid and 5 ml of saturated bromine water. Add 2 ml of 1% ethanolic dimethylglyoxime and dilute to 100 ml. Set aside for 10 minutes and read at 590 nm.

Seawater.[98] Filter the sample through an HA Millipore filter. To 750 ml, add 25 ml of 20% sodium citrate solution and 5 ml of 1% ethanolic dimethylglyoxime. Add ammonium hydroxide dropwise, 4–5 drops, to raise to pH 9–10. Extract the nickel-dimethylglyoxime with 10 ml, 5 ml, 5 ml, and 5 ml of chloroform. Extract the nickel from the combined chloroform extracts with 5 ml, 5 ml, and 5 ml of *N* hydrochloric acid. Add 1 ml of saturated bromine water and set aside for 15 minutes. Adjust to pH 10.4 ± 0.05 and add 1 ml of 1% ethanolic dimethylglyoxime. Dilute to 50 ml and read at 442 nm against a reagent blank. The length of time between development and reading should be standardized.

Synthetic Detergents.[99] Decompose a 5-gram sample with 20 ml of hydrochloric acid and 5 ml of nitric acid. Evaporate to dryness and ignite at red heat. Cool. Take up in 5 ml of hydrochloric acid and 1 ml of nitric acid. Dilute to 100 ml. Add 5 ml of 10% solution of citric acid to a 25-ml aliquot. Add ammonium hydroxide to

[95] Naoharu Mizuno and Kenjiro Hayashi, *Jap. Anal.* **16**, 38–41 (1967).
[96] R. Chester and M. J. Hughes, *Trans. Inst. Min. Met., B*, **77**, B37–B41 (1968).
[97] Marian Kowalczyk and Konrad Ogiolda, *Rudy Met. Niezelaz.* **6**, 118–119 (1961).
[98] Edward Kentner, D. Bruce Armitage, and Harry Zeitlin, *Anal. Chim. Acta* **45**, 343–346 (1969).
[99] S. Baselgia, *Dermatologica* **135**, 497–501 (1967).

raise to pH 7.5. Add 2 ml of 1% ethanolic dimethylglyoxime plus a further 2 ml for each 10 mg of copper or 4 mg of cobalt present. Extract with 5 ml, 5 ml, and 5 ml of chloroform. Wash copper and cobalt from the combined extracts with 5 ml of 0.5 *N* ammonium hydroxide. Scavenge any residual nickel in the aqueous phase with 3 ml of chloroform. Extract nickel from the combined organic phases with 5 ml and 5 ml of 0.5 *N* hydrochloric acid. Add 1 ml of bromine water to the combined acid extracts and follow with 2 ml of ammonium hydroxide. Add 1 ml of 1% ethanolic dimethylglyoxime and dilute to 25 ml. Set aside for 15 minutes, then read at 445 nm.

Wheat Bran, Feed, or Flour.[100] Treat a 2-gram sample with excess of nitric and sulfuric acid, then with fuming nitric acid. Heat until no more nitric acid is given off. Add 5 ml of 1 : 1 sulfuric acid and heat to sulfur trioxide fumes. Let cool, add 5 ml of water, and again heat to sulfur trioxide fumes. Add water and evaporate to about 20 ml. Cool somewhat, add 10 ml of 20% solution of sodium tartrate, and heat to 80°. Add 4 ml of 1% ethanolic dimethylglyoxime, raise to pH 8–9 with ammonium hydroxide, and set aside overnight. Filter on sintered glass, treat with 1 ml of bromine water, and wash with hot water, collecting the filtrate in the original beaker. Repeat the treatment of the filter with bromine water and washing. Boil off the bromine from the combined filtrates and washings, adding hot water to maintain the volume. Cool. Add sequentially 1 ml of ethanol, 3 ml of 20% solution of sodium tartrate, 5 ml of saturated solution of potassium persulfate, 5 ml of ammonium hydroxide, and 2 ml of 1% ethanolic dimethylglyoxime. Dilute to 50 ml and read at 465 nm.

Petroleum Oils.[101] For the detailed combustion to give a solution of the sample, see Iron by 1, 10-Phenanthroline.

To an aliquot of sample solution containing up to 60 μg of nickel in not more than 2.5 ml, add 0.4 ml of 10% solution of diammonium citrate and 0.4 ml of 20% solution of sodium potassium tartrate. Add a piece of Congo red paper and neutralize with 8% sodium hydroxide solution until the paper becomes a full red. Add 0.5 ml more of the 8% sodium hydroxide solution and 0.5 ml of saturated bromine water. At once add 0.2 ml of 0.1% ethanolic dimethylglyoxime in ethanol. After 30 seconds add 0.2 ml of 10% EDTA solution and dilute to 5 ml. Read against water at 470 nm.

Hydrogenated Fat.[102] Ash 2 grams of sample and ignite to redness. Add 0.2 ml of hydrochloric acid and 0.2 ml of nitric acid. Evaporate to dryness. Moisten with 3 drops of 1 : 4 hydrochloric acid and dissolve in 5 ml of hot water. Add 2 ml of 20% solution of sodium potassium tartrate and 1 ml of 1% ethanolic iodine. Add 1 : 9 ammonium hydroxide dropwise until the color disappears. Add 1 ml of 1% ethanolic dimethylglyoxime. Dilute to 50 ml and read at 540 nm after 10 minutes.

Margarine.[103] Melt a 25-gram sample with a few ml of kaolin and dry at 120°.

[100] I. Hoffman, *Analyst* **87**, 650–652 (1962).
[101] A. J. Agazzi, D. C. Burtner, D. J. Crittenden, and D. R. Patterson, *Anal. Chem.* **35**, 332–335 (1963).
[102] N. Kh. Kameneva and A. G. Koblyanskii, *Tr. Krasnoyarpk. Inst. Pichchevoi Prom.* **1955** (11), 39–42.
[103] A. Rudnicki and H. Niewiadomski, *Chem. Anal.* (Warsaw) **13**, 755–762 (1968).

Ignite to burn off the fat and calcine the residue at 600° for 1 hour. Cool, add 2 ml of 30% hydrogen peroxide, and evaporate to dryness. Add 2 ml of hydrochloric acid and again evaporate to dryness. Dissolve the residue in 2 ml of 0.1 *N* hydrochloric acid and wash the crucible with 1 ml of that acid. Add 0.5 ml of 10% citric acid solution, 0.5 ml of 3% solution of ammonium persulfate, 0.5 ml of 1% solution of dimethylglyoxime in 5% sodium hydroxide solution, and 0.5 ml of 10% solution of sodium hydroxide. Set aside for 10 minutes, centrifuge, and read at 460 nm.

Rapeseed Oil.[104] Dissolve a 2-gram sample in 15 ml of benzene. Extract with 10 ml of 1:3 hydrochloric acid at room temperature or, if the sample shows visible nickel dust, at 70°. After shaking for 5 minutes, filter the aqueous phase. To a 2-ml aliquot, add 3 drops of a saturated solution of sodium potassium tartrate, 3 ml of 10% sodium hydroxide solution, 1 ml of 3% solution of ammonium persulfate, and 1 ml of 1% solution of dimethylglyoxime in 5% sodium hydroxide solution. Dilute to 10 ml and read at 540 nm.

Biological Materials.[105] Ash a 2-gram sample and dissolve the ash in 5 ml of 1:4 hydrochloric acid. Filter and dilute to 25 ml. Dilute a 5-ml aliquot with 8 ml of water, and extract with successive 5-ml portions of 0.01% solution of dithizone in carbon tetrachloride until no further extraction of copper dithizonate occurs. Add ammonium hydroxide until the solution is alkaline. Extract the nickel with 5 ml and 5 ml of 0.1% solution of dimethylglyoxime in chloroform. Reextract the nickel from the organic layer with 5 ml and 5 ml of 0.5 *N* hydrochloric acid. To the combined acid extracts, add 0.5 ml of 1% ethanolic dimethylglyoxime. Add 1 ml of 1% ethanolic iodine and discharge the color of the iodine by dropwise addition of 28% solution of sodium hydroxide, adding 3 drops in excess. Filter, dilute to 25 ml, set aside for 20 minutes, and read at 540 nm.

Urine.[106] Evaporate 100 ml to dryness with 10 ml of nitric acid and 5 ml of 30% hydrogen peroxide. Take up the residue in 8 ml of water, add 1 ml of hydrochloric acid, and adjust to pH 4 with 4% sodium hydroxide solution. Add 0.5 ml of 0.5% solution of sodium diethyldithiocarbamate, 0.4 ml of 0.4 *N* cadmium sulfate, and 0.3 ml of 0.4 *N* sodium sulfide. Set aside for 20 minutes, then centrifuge until clear. Decant the clear layer, and dissolve the precipitate in 4 drops of hydrochloric acid and 1 ml of nitric acid. Evaporate to dryness and dissolve in 1 ml of water. Add 1 ml of 20% solution of sodium potassium tartrate and adjust to pH 4. Add about 4 mg of ammonium persulfate and 2 ml of 1:3 ammonium hydroxide. Centrifuge until clear and add 1 ml of saturated solution of dimethylglyoxime in 1:7 ammonium hydroxide. Read at 540 nm against water.

---

[104] Jadwiga Witwicka, *Tluszcze Srodki Piorace* **2**, 157–163 (1957); cf. M. Karvanek, *Prum. Potravin.* **15**, 282–283 (1964).
[105] S. D. Taktakishvili, *Lab. Delo* **10**, 153–156 (1964).
[106] L. E. Gorn and A. D. Miller, *ibid.* **1966**, 163–164.

DIPHENYLCARBAZONE

Nickel after extraction as the dimethylglyoxime complex can be determined by its catalysis of the oxidation of diphenylcarbazone by hydrogen peroxide.[107] The change of absorption at 420 nm is measured. Cobalt interferes. By increase of sample size to 0.5 gram, 0.1 ppm of nickel can be determined.

Procedure. *Tantalum or Niobium Pentoxide.*[108] Dissolve a 0.1-gram sample by heating in Teflon with 5 ml of hydrofluoric acid and 0.5 ml of nitric acid. Add 20 ml of 83:210:210 sulfuric acid–hydrofluoric acid–water. Extract with 10 ml, 10 ml, and 10 ml of methyl isobutyl ketone and discard the extract. Add 5 ml of 5% boric acid solution and 1 ml of 25% tartaric acid solution to the aqueous layer. Raise to pH 7 by addition of 2 *N* ammonium hydroxide. Add 10 ml of borate buffer solution for pH 9.2 and 5 ml of 1% ethanolic dimethylglyoxime. Set aside for 10 minutes, then extract the nickel with 2.5 ml and 2.5 ml of chloroform. Wash the combined extracts with 10 ml of 1:10 ammonium hydroxide, then evaporate almost to dryness. Add 5 ml of nitric acid and 0.5 ml of sulfuric acid. Destroy organic matter by evaporating to dryness. Take up the residue in 0.5 ml of water and adjust to pH 7 with 8% sodium hydroxide solution. Dilute to 10 ml with borate buffer solution for pH 8.9.

Mix the buffered sample at 25° with 1 ml of 0.9 *M* hydrogen peroxide and 1 ml of 0.4 m*M* diphenylcarbazone. One minute after mixing, monitor the extinction for 10 minutes. The record is a graph of log-extinction versus time.

DITHIOOXALATE

For nickel as metallized plastic films, dissolve sample material with a few drops of nitric acid under an infrared lamp and evaporate to dryness.[109] Dissolve in 1 ml of *N* sulfuric acid. If less than 50 μg of nickel is present, add 8 ml of water and 1 ml of 0.05% solution of potassium dithiooxalate. Read at once at 500 nm. If more than 50 μg of nickel is present, develop an aliquot. For catholyte and washings from electrolytic cells, evaporate to dryness under an infrared lamp and proceed as above.

Procedure. *Steel.*[110] Dissolve 3 grams of sample in 50 ml of warm 1:1 hydrochloric acid. If copper is present, add aluminum granules to displace it. Filter, and wash the residue with 1:2 hydrochloric acid. Add 30% hydrogen peroxide to

[107] K. B. Yatsimirskii, E. M. Emel'yanov, V. K. Pavlova, and Ya. S. Savichenko, *Okeanologiya* **10**, 1111–1116 (1970); V. M. Peshkova, E. K. Astakhova, I. F. Dolmanova, and V. M. Savostina, *Acta Chim. Hung.* **53**, 121–125 (1967); I. F. Dolmanova, G. A. Zolatova, T. P. Kabanova, and V. M. Peshkova, *Vest. Mosk. Gos. Univ., Ser. Khim.* **1968** (6), 96–100.

[108] I. F. Delmanova, G. A. Zolotova, R. D. Voronina, and V. M. Peshkova, *Zavod. Lab.* **39**, 386–387 (1973).

[109] Y. Ujihira and J. C. Roy, *Can. J. Chem.* **46**, 1233–1236 (1968).

[110] A. T. Pilipenko and N. N. Maslei, *Ukr. Khim. Zh.* **34**, 174–177 (1968).

oxidize iron to ferric ion and boil to decompose excess hydrogen peroxide. Dilute to 200 ml. To an 8.3-ml aliquot containing 7.5–125 μg of nickel, add 6.5 ml of 45% ammonium fluoride solution and adjust to pH 4.5. Add 4 ml of 0.01 *M* potassium dithiooxalate and dilute to 25 ml. Centrifuge if necessary to separate ferric fluoride, and read at once at 490 nm.

DITHIZONE

The optimum conditions for extraction of nickel as the dithizonate in carbon tetrachloride are pH 8.3–8.8 and a citrate buffer solution.[111] When cobalt is also present, the nickel is read at 670 nm and cobalt at 605 nm. A ternary nickel–dithizone–1,10–phenanthroline complex forms much more rapidly than the nickel-dithizone complex and is more rapidly extracted with chloroform.[112]

Procedures

Gold.[113] Dissolve a 1-gram sample in 6 ml of hydrochloric acid and 2 ml of nitric acid. Evaporate to dryness, and take up the residue in 10 ml of 1:5 hydrochloric acid. Extract the gold with 10 ml and 10 ml of isopentyl acetate and discard. Evaporate the solution to about 1 ml and dilute to 10 ml with 1:1 hydrochloric acid. Extract the iron with 10 ml and 10 ml of isopentyl acetate as a sample for determination with bathophenanthroline.

Add 1 ml of 1:1 sulfuric acid to the aqueous layer and evaporate to sulfur trioxide fumes. Take up the residue in water and dilute to 10 ml. Shake for 1 minute with 10 ml of benzene containing 0.1 mg of dithizone. Wash the extract with 10 ml and 10 ml of 1:100 ammonium hydroxide to remove excess dithizone and read at 520 nm.

As a Ternary Complex.[114] Adjust 10 ml of sample solution containing 1–25 μg of nickel to pH 6 with ammonium hydroxide, phthalate buffer solution, or acetate buffer solution. Shake with 10 ml of chloroform 70 μ*M* with dithizone and 30 μ*M* with 1,10-phenanthroline. Wash excess dithizone from the organic layer with 10 ml of 0.4% solution of sodium hydroxide and read at 520 nm.

EDTA

The complex of nickel with EDTA at pH 4.55–6.82 in 20% sodium acetate solution is read at 580–750 nm.[115] High concentrations of ammonium ion interfere. There is

[111] Takashi Ashizawa, *Jap. Anal.* **10**, 350–354 (1961).
[112] B. S. Freiser and H. Freiser, *Talenta* **17**, 540–543 (1970).
[113] G. Ackermann and J. Köthe, *Z. Anal. Chem.* **231**, 252–261 (1967); Zygmunt Marczenko, Krzysztof Kasiura, and Maria Krasiejko, *Chem. Anal.* (Warsaw) **14**, 1277–1287 (1969).
[114] K. S. Math, K. S. Bhatki, and H. Freiser, *Talenta* **16**, 412–414 (1969).
[115] Walter Nielsch and Gerhard Böltz, *Anal. Chim. Acta* **11**, 367–375 (1954); Loron D. Brake, Wallace M. McNabb, and J. Fred Hazel, *ibid.* **19**, 39–42 (1958).

interference by cerium, bismuth, silver, barium, cobalt, copper, magnesium, and vanadium. Nitroso-R salt serves to mask cobalt.

Procedures

Silicate Rocks.[116] Transfer 0.5 gram of sample to platinum. Mix with 2 ml of sulfuric acid and 10 ml of hydrofluoric acid. Evaporate to fumes and take up with 5 ml of water. Add 5 ml of hydrochloric acid and extract the iron with 5 ml of pentyl acetate. If the sample contains more than 3% iron, repeat this extraction. Evaporate the aqueous phase to fumes and nearly neutralize with 10% sodium hydroxide solution. Add 15 ml of 0.2 *M* EDTA. Adjust to pH 4 and boil for 5 minutes. Cool, dilute to 25 ml, and read nickel for 0.1–4.5 mg at 380 nm or 590 nm.

Hydrosulfuration Catalysts.[117] Dissolve 4 grams of such a mixed oxide catalyst on an aluminum oxide support by refluxing with 100 ml of 1 : 1 hydrochloric acid. Add 25 ml more of 1 : 1 hydrochloric acid and dilute to 200 ml. Mix a 5-ml aliquot with 40 ml of 0.2 *M* disodium EDTA and 20 ml of 3 *M* sodium acetate. Adjust to pH 5, dilute to 100 ml, and read at 590 nm.

Electroplating Solution.[118] To an aliquot containing 0.5–5 millimoles of nickel, add a slight excess of 0.04 *M* EDTA and 5 ml of 50% solution of sodium acetate trihydrate as a buffer for pH 3.5. Dilute to 50 ml and read at 590 nm against water.

FLAME PHOTOMETRY

Nickel can be isolated from most other elements by extraction from a mannitol solution adjusted to pH 9.6 with ammonium hydroxide. The solvent is salicylaldoxime in 4-methylpentan-2-one. Aspirated directly into the flame, it is about 11 times as sensitive as an aqueous sample.

Available wavelengths are as follows:

Line Emission (nm)	Background (nm)
341.5	340.0
352.4	353.5
361.9	363.0

When the complex of nickel with oxine is extracted with ethyl acetate at pH 5.5–8.8 and developed by flame photometry at 352.4 nm, the sensitivity is about 30 times as great as when the aqueous solution is used.[119] Interference by copper,

[116] K. P. Stolyarov and F. B. Agrest, *Zh. Anal. Khim.* **19**, 457–466 (1964).

[117] T. Fernandez, J. M. Rocha, N. Rufino, and A. Garcia-Luis, *An. Quim.* **69**, 981–989 (1973).

[118] K. L. Cheng and B. L. Goydish, *Chemist-Analyst* **51**, 45–48 (1962).

[119] Hidehiro Goto and Emiko Sudo, *Jap. Anal.* **10**, 463–467 (1961); Emiko Sudo and Hidehiro Goto, *Sci. Rep. Res. Inst., Tohoku Univ. Ser. A* **14**, 220–230 (1962).

gallium, indium, iron, and manganese is avoided by preextraction of their oxine complexes above pH 10.

For application of flame photometry in alloys containing 5–35% of nickel, interference effects do not necessitate preliminary separation.[120] For nickel in aluminum alloy along with copper and manganese by flame photometry, see Copper for separation as diethyldithiocarbamates in chloroform. For nickel by flame photometry in the presence of copper and manganese, see Copper.

Procedures

Steel.[121] Dissolve a sample in a minimum amount of 1:2 nitric acid. Filter, wash the residue with hot 1% nitric acid, and dilute to a known volume. To an aliquot containing 20–100 μg of nickel, add 30 ml of 15% solution of mannitol. Add 10 ml of a 1% solution of salicylaldoxime in 4-methyl-pentan-2-one. Adjust to pH 9.6$\pm$0.2 with 1:4 ammonium hydroxide. Shake for 3 minutes and use the organic phase for flame photometry. The following instrument settings are appropriate.

Selector switch, position	0.1
Sensitivity control, ERA % adjust	
For aqueous aerosols	70
For organic aerosols	50
Phototube resistor (megohms)	22
Phototube (RCA 1P28) volts/dynode	60
Slit (mm)	0.030
Half-bandwidth (nm) (at 352.4 nm)	0.40
Acetylene flow (cubic feet/hour)	
For aqueous aerosols	3.3
For organic aerosols	2.9
Oxygen flow (cubic feet/hour)	5.7
Oxygen pressure (psi)	10.0

Brass. Dissolve a sample in a minimum amount of 1:2 nitric acid, heating to coagulate stannic oxide. Filter, and wash with 1% nitric acid. Dilute to a suitable volume with 2% nitric acid. Electrolyze with platinum electrodes at 3 amperes until the solution just becomes colorless, thus avoiding deposition of nickel. A small amount of undeposited copper will not interfere. Dilute to a known volume and proceed as above from "To an aliquot containing 20–100 μg. . . ."

FORMALDEHYDE OXIME

Nickel combines with this reagent, formaldoxime, to give a 1:6 complex.

[120] C. A. Waggoner, *Can. J. Spectrosc.* **10** (3), 51–57 (1965).
[121] Howard C. Eshelman and John A. Dean, *Anal. Chem.* **33**, 1339–1342 (1961).

Procedures

Aluminum.[122] Dissolve 5 grams of sample in 40 ml of hot 20% sodium hydroxide solution and dilute to 150 ml. Add 2 ml of 10% sodium sulfide solution, and filter. Dissolve the precipitate in the minimum possible amount of hot 1:4 nitric acid and concentrate to 20 ml. Add 5 ml of 20% ammonium citrate solution, 0.1 gram of hydroxylamine hydrochloride, and 2 ml of 1% ethanolic dimethylglyoxime. Neutralize to phenolphthalein with ammonium hydroxide. Extract for 1 minute each with 10 ml and 5 ml of chloroform. Extract the combined chloroform layers for 1 minute each with 10 ml and 5 ml of 0.1 *N* hydrochloric acid. To the combined acid extracts, add 50 ml of *M* formaldoxime followed immediately by 5 ml of 4% sodium hydroxide solution. Dilute to 50 ml and read after 15 minutes.

Aluminum Silicate. Moisten 1 gram in platinum with 1 ml of perchloric acid and 10 ml of hydrofluoric acid. Evaporate to white fumes. Add 10 ml of hydrofluoric acid and again evaporate to white fumes. Add 5 ml of water and 10 ml of 1:20 perchloric acid. Heat to a clear solution and cool. Add 10 ml of 20% ammonium citrate solution and proceed as above from "... 0.1 gram of hydroxylamine hydrochloride, and... ."

α-FURILDIOXIME

The yellow complex of this reagent with nickel is insoluble in water but extractable with chloroform. For analysis of a trace of nickel in indium or aluminum, form the α-furildioxime complex and extract it with chloroform at pH 8.7–9.3 in the presence of citrate or tartrate ion to prevent precipitation of nickel hydroxides.[123] Cobalt, iron, and copper show almost no absorption under those conditions.[124]

For nickel in molybdenum oxide, fuse with potassium pyrosulfate and dissolve in water.[125] Coprecipitate the metals with magnesium at pH 11–11.3. Dissolve the precipitate in dilute sulfuric acid and extract copper at pH 0.9 with dithizone in carbon tetrachloride. Then develop nickel with α-furildioxime.

After fusion of chromite and chrome ore in a zirconium crucible with sodium peroxide, the nickel in the aqueous extract is determined with α-furildioxime.[126] Silicate rocks are analyzed with this reagent.[127] For separation of nickel as the α-furildioxime from other metals, see "Isolation of Cadmium, Zinc, Cobalt, and Nickel," in Chapter 6, Cadmium.

Procedure. To 25 ml of sample solution containing more than 0.5 μg of nickel in 0.1 *N* hydrochloric acid, add 5 ml of 0.2% solution of ethylenediamine and set aside

[122] Z. Marczenko and Z. Kasiura, *Chem. Anal.* (Warsaw) **6**, 353–364 (1961); Z. Marczenko, *Anal. Chim. Acta* **31**, 224–232 (1964).
[123] V. M. Peshkova, V. M. Bochkova, and L. I. Lazareva, *Zh. Anal. Khim.* **15**, 610–613 (1960).
[124] A. R. Jafar, F. N. Masondi, and I. M. Ali Shafouh, *J. Indian Chem. Soc.* **52**, 635–638 (1975).
[125] Elzbieta Wieteska and Urszula Stolarczyk, *Chem. Anal.* (Warsaw) **15**, 183–189 (1970).
[126] Joseph Dinnin, *U.S. Geol. Surv., Bull.* **1084-B**, 31–68 (1959).
[127] D. E. Bodart, *Chem. Geol.* **6**, 133–142 (1970).

for 5 minutes.[128] Add 10 ml of a solution containing 2.3% of sodium hydroxide, 40% sodium acetate trihydrate, 2.5% tartaric acid, and 3.5% sodium tetraborate decahydrate. Add 5 ml of 50% solution of sodium thiosulfate pentahydrate. Oxidize ferrous ion with a drop of 30% hydrogen peroxide. Add 1 ml of 0.5% solution of the reagent in ethanol. Extract with 8 ml of chloroform, filter the extract through cotton, and read at 435 nm.

Sodium Hydroxide and Sodium Hypochlorite Solutions.[129] Neutralize a sample containing about 5 μg of nickel with hydrochloric acid and evaporate to dryness. Take up in water and dilute to 50 ml. To a 20-ml aliquot, add 10 ml of 10% solution of sodium potassium tartrate made alkaline to litmus with ammonium hydroxide. Neutralize to litmus with ammonium hydroxide. Add 10 ml of 0.1% α-furildioxime in chloroform and shake. Then extract with 10 ml and 10 ml of chloroform. Combine the extracts plus 1 ml of aqueous phase from the last extraction. Extract with 50 ml of 0.1% solution of sodium potassium tartrate in 1:50 ammonium hydroxide to reduce interference by copper. Read the chloroform layer at 435 nm against a blank.

Boiler Feed Water.[130] Add hydrochloric acid to make 200 ml of sample 0.1 *N*. Heat to boiling to dissolve any suspended matter. Cool, and add 0.2 ml of 30% hydrogen peroxide. Add 5 ml of 2% solution of sodium potassium tartrate and make alkaline to phenolphthalein. Decolorize by dropwise addition of 1:10 hydrochloric acid. Add 1 ml of 0.5% ethanolic α-furildioxime. Render alkaline with ammonium hydroxide and extract with 10 ml and 5 ml of chloroform. Read at 435 nm against chloroform.

8-HYDROXYQUINOLINE

In *N* hydrochloric acid in acetone, a complex of nickel with 8-hydroxyquinoline, also called quinolinol and oxine, absorbs at 365 nm but not at 700 nm. The complex of cobalt with oxine absorbs at 700 and 365 nm. Therefore the concentration of cobalt can be read directly and that of nickel can be determined by difference. Beer's law holds below 20 μg of metal per ml. Iron and chromium form complexes in the same range and must be absent.

Ferric oxinate and nickel oxinate are read in chloroform in the range of 5–100 μg of each.[131a] Extract the oxinates at pH 5.5–8 and read at 370 and 470 nm. Calculate iron from the value at 370 nm. In calculating nickel from the value at 470 nm, correct for the absorption of iron at that wavelength.

Extraction with molten naphthalene of complexes with 8-hydroxyquinoline is applicable at varying pH levels to nickel, zinc, magnesium, cadmium, bismuth, cobalt, copper, molybdenum, uranium, aluminum, iron, palladium, and indium.[131b]

[128] D. E. Bodart, *Z. Anal. Chem.* **247**, 32–36 (1969).
[129] F. Mains and R. E. Ragett, *Chemist-Analyst* **50**, 4–6 (1961).
[130] A. L. Wilson, *Analyst* **93**, 83–92 (1968); cf. British Standards Institution, *BS 2690*, Pt. **12** 19 pp (1972).
[131a] P. B. Issopoulos, *Hem. Hron., A*, **33**, 75–78 (1968).
[131b] T. Fujinaga, Masatada Satake, and Masaaki Shimizu, *Jap. Anal.* **25**, 313–318 (1976).

Procedure. *Nickel and Cobalt.*[132] As N hydrochloric acid in acetone, dilute 133.3 ml of 7.8 N hydrochloric acid to 1 liter with acetone.

Mix a sample solution containing 0.03–0.5 mg of nickel and cobalt with 5 ml of borax–sodium hydroxide buffer solution for pH 9.5 in a 50-ml centrifuge tube. Add excess of 4% ethanolic oxine to give complete precipitation of the complexes. Digest at 60° for 30 minutes, and centrifuge. Remove excess reagent by washing the precipitate with the buffer solution for pH 9.5, then with water. Dry the precipitate at 110°. Dissolve in N hydrochloric acid in acetone and dilute to 25 ml with that solvent. Read at 365 and 700 nm against the solvent. Calculate cobalt from the value at 700 nm. Calculate nickel from the value at 365 nm as corrected for the cobalt content.

4-ISOPROPYL-1,2-CYCLOHEXANEDIONEDIOXIME

The complex of this reagent conforms to Beer's law for 1–12 μg of nickel per ml in xylene. Toluene has also been used as the extractant.[133] Interferences from ferric ion, cupric ion, and cobalt can be masked with fluoride, sulfide, and hydrogen peroxide plus cyanide, respectively.

Procedure. Adjust a sample solution containing 10–150 μg of nickel to pH 7 and add sufficient 10 M ammonium acetate to make the solution M with acetate ion.[134] Add 2 ml of saturated solution of the captioned reagent, which will be about 0.004 M, and if the volume of sample exceeds 100 ml, add 2 ml for each additional 100 ml. After 30 minutes shake with 10 ml of xylene for 2 minutes. Read the organic layer against a blank.

8-MERCAPTOQUINOLINE

This reagent, also known as thiooxine, complexes with both nickel and cobalt, but the latter can be corrected. Beer's law is followed for up to 10 μg of nickel in the final organic extract.

Procedure. *Steel.*[135] Dissolve 0.25 gram of sample in aqua regia and dilute to 250 ml in 1 : 1 hydrochloric acid. Mix a 10-ml aliquot with 10 ml of 1 : 1 hydrochloric acid and extract the iron with 30 ml and 20 ml of methyl isobutyl ketone. Wash the organic layer with 10 ml of 1 : 1 hydrochloric acid and discard the organic layer.

[132] A. J. Mukhedkar and N. V. Deshpande, *Anal. Chem.* **35**, 47–48 (1963).
[133] P. D. Blundy and M. P. Simpson, *UKAERE* **CE/M-215** (1958).
[134] B. L. McDowell, A. S. Meyer, Jr., R. E. Feathers, Jr., and J. C. White, *Anal. Chem.* **51**, 931–934 (1959).
[135] Kunihiro Watanabe and Kyozo Kawagaki, *Jap. Anal.* **23**, 510–514 (1974).

Evaporate the aqueous phase to a small volume to remove most of the hydrochloric acid. Add 5 ml of 5% solution of ascorbic acid and dilute to 50 ml. Adjust to pH 4, and mask copper with 5 grams of sodium thiosulfate. Add 0.6 ml of 0.2% methanolic thiooxine. Extract the nickel complex by shaking for 2 minutes with 10 ml of either chloroform or methyl isobutyl ketone. Read the chloroform extract at 540 nm or the ketone extract at 550 nm.

If cobalt is present, extract the nickel with 30 ml of 4 : 5 hydrochloric acid and read again. Apply that as a correction to the previous reading of nickel plus cobalt in the organic layer.

N-METHYLANABASINE-α′-AZO-2-NAPHTHOL-6-SULFONIC ACID

This reagent, which is 6-hydroxy-5-[3-(1-methyl-2-piperidyl)-2-pyridylazo]naphthalene-2-sulfonic acid, forms a 2 : 1 complex with nickel. This can be read in aqueous solution or extracted with chloroform. The absorption maxima are at 480 nm for the reagent, and at 530 and 570 nm for the complex. The color increases with pH for 2–5.2, then is constant to pH 14. Interference by copper can be masked with thiourea; that of iron, cadmium, manganous ion, and lead by tartaric acid. Aluminum and titanium can be masked with triethanolamine. Oxalate, zinc, and palladous ion interfere.

Procedure. Dissolve 1 gram of sample in 15 ml of 1 : 1 hydrochloric acid.[136] Add 10 ml of nitric acid and evaporate to dryness. Dissolve the residue in 10 ml of 1 : 1 hydrochloric acid, add 70 ml of water, and filter. After washing the residue with hot water, evaporate the filtrate to half-volume. Make it alkaline with ammonium hydroxide, filter, and wash with hot water. Evaporate the filtrate to 50 ml. To an aliquot containing 5–40 μg of nickel, add 1 : 1 hydrochloric acid to give pH 7. Add 2 ml of 0.2% solution of the color reagent in ethanol, 10 ml of acetate buffer solution for pH 5.2, and sufficient EDTA to mask any cobalt present. Dilute to 25 ml and read at 560 nm.

For extraction, as a buffer solution mix 50 ml of *N* sodium acetate solution with 10 ml of *N* hydrochloric acid and dilute to 250 ml.[137] To the sample solution, add 10 ml of buffer solution and 2 ml of 0.2% solution of the captioned reagent in ethanol. Dilute to 25 ml and extract with 10 ml, 5 ml, and 5 ml of chloroform. Adjust the combined organic layers to 20 ml and read at 530 or 570 nm.

NICKEL ION

Nickel ion in a plating bath permits reading nickel sulfate at 350–400 grams/liter.[138] Changes in pH or variation in the quantity of boric acid or sodium chloride

[136] R. Kh. Dzhiyanbaeva, Sh. T. Talipov, U. Mansurkhozhdaev, and N. Smaglyuk, *Tr. Tashk. Gos. Univ.* **1968** (323), 21–27.
[137] L. V. Chaprasova, Sh. T. Talipov, and R. Kh. Dzhiyanbaeva, *Nauch. Tr. Tashk. Univ.* **1967** (288), 5–53.
[138] M. M. Menkina, L. A. Fridman, and L. E. Granchel, *Mashinostr. Belorus.* (Minsk) *Ab.* **1956** (1), 17–118; cf. G. Kopczyk, *Galvano-Tecknick* **53**, 25–26 (1962).

do not affect the result. Interference by copper is avoided by reduction to cuprous ion. In 3 *M* sulfuric acid, the maximum absorption of nickel ion is at 390 nm and conforms to Beer's law for 0.04–0.2 *M*.[139] For further information about reading nickel, copper, and cobalt ions in sulfuric acid solution, see Copper.

In analysis of alloys, cupric ion is read at 820 nm, nickel ion at 395 nm, and cobalt ion at 515 nm in *N* sulfuric acid.[140] Simultaneous equations are then applied. Chloride ion interferes. Ferric ion amounting to less than 5% of the nickel can be masked with phosphoric acid in reading the latter. Higher concentrations of iron are extracted with isopropyl ether from 8 *N* hydrochloric acid before driving off the chloride by heating with sulfuric acid.

Procedures

Nickel and Cobalt.[141] Dissolve a sample containing about 0.5 gram of nickel and 0.01–0.02 gram of cobalt in 20 ml of 30% sulfuric acid and dilute to 100 ml. To a 20-ml aliquot, add 10 ml of 5% solution solution of sodium metabisulfite in 1:3 ammonium hydroxide. Heat to 70° for 30 minutes, cool, and dilute to 50 ml. Filter and read nickel at 585 nm and cobalt at 440 nm.

Nickel Silver.[142] Dissolve 1 gram in 10 ml of 1:1 nitric acid. Add 3 ml of perchloric acid and heat to strong fumes. Take up in 1:3 hydrochloric acid, and add to 4 grams of sodium hypophosphite. Dilute to 50 ml with 1:3 hydrochloric acid and read at 650 nm.

Fluoborate Alloy-Plating Solutions.[143] NICKEL AND COBALT ION. Add 10 ml of 5% hydrazine sulfate solution to a 25-ml sample solution and heat to 60°. Let cool and adjust to pH 3.3 with *N* sulfuric acid or a saturated solution of sodium tetraborate. Dilute to 50 ml and read nickel at 512 nm and cobalt at 394 nm. Apply a correction for the absorption due to iron.

NITROXAMINOAZO

At pH 5–6 this reagent forms a 1:2 complex with nickel which is read at 585 nm for 0.1–1.5 μg of nickel per ml. Iron up to 4 mg/ml can be masked with tartrate and ascorbic acid; cobalt up to 0.4 mg/ml by phosphate and tartrate; molybdenum or vanadium up to 0.4 mg/ml by phosphate; copper by thiourea. Oxalate and EDTA must be absent. The reagent is more selective than dimethylglyoxime.

Procedure. *Steel.*[144] Mix an aliquot of sample solution containing 5–75 μg of nickel with 2 ml of 30% solution of potassium pyrophosphate and 1 ml of 5%

[139] N. P. Komar' and L. S. Palagina, *Zh. Anal. Khim.* **23**, 75–79 (1968).
[140] D. V. Jayawant and T. K. S. Murthy, *Indian J. Technol.* **9**, 396–400 (1971).
[141] J. Truchly and T. Sramco, *Chem. Zvesti* **19**, 767–773 (1965).
[142] Charles Goldberg, *Chemist-Analyst* **39**, 56–57 (1950).
[143] T. Keily and R. C. Woodford, *Analyst* **97**, 872–876 (1972).
[144] Yu. M. Dedkov and G. V. Kochatova, *Zavod. Lab.* **40**, 1325–1327 (1974).

solution of ascorbic acid. Set aside for 5 minutes and add 1 ml of 5% solution of thiourea. Again wait 5 minutes; then add 10 ml of 10% solution of sodium tartrate and 5 ml of 20% trisodium phosphate solution. Adjust to pH 5–6 by adding 15 ml of acetate buffer solution for pH 5. Heat at 60° for 5 minutes, then add 2 ml of 0.2% solution of nitroxaminoazo. Dilute to 50 ml with the buffer solution, set aside for 5 minutes, and read at 585 nm against a reagent blank.

OXAMIDOXIME

This reagent precipitates a 1:1 complex of nickel at pH 8–9.5 but forms a water-soluble cobalt complex under those conditions. The system obeys Beer's law for 1–50 ppm of cobalt. A solution of the nickel-amidoxime at pH 3–5 has a maximum absorption at 233 nm and obeys Beer's law for 1–30 ppm of nickel. Iron and copper react with the reagent but are removed along with zinc by an anion exchange resin. Gross amounts of aluminum, chromic ion, lead, and manganous ion interfere with determination of nickel. Pentavalent vanadium interferes with the determination of cobalt.

Procedure. *Nickel and Cobalt.*[145] Evaporate the sample to 1 ml and take up in 15 ml of hydrochloric acid. Add to a 20 cm × 0.75 cm^2 column of 50-100 mesh Dowex 1-X8 resin. Elute with 20 ml of 1:2 hydrochloric acid and adjust the volume to 50 ml at pH 6. Add 10 ml of 0.1 *M* oxamidoxime for each mg of the sum of cobalt and nickel ion. Dissolve sodium acetate in the sample solution to raise to pH 8–9.5. Filter the precipitate of nickel oxamidoxime, which may also contain hydrous oxides of other heavy metal ions, on a medium-porosity sintered-glass filter. Wash with 5 ml and 5 ml of 0.1 *M* sodium acetate solution. Adjust the combined filtrate and washings to a known volume with 0.1 *M* sodium acetate and read the cobalt at 350 nm.

For determination of nickel, dissolve the precipitate with 10 ml of 1:5 hydrochloric acid and dilute to 100 ml with water. Read at 233 nm.

PICOLINALDEHYDE 2-QUINOLYLHYDRAZONE

This reagent is applicable in chloride solutions at pH 3–4. Masking agents are required in the presence of cadmium, cobalt, copper, iron, mercury, and zinc.

Procedure. *Seawater.*[146] As the color reagent, dissolve 1% in water by adding hydrochloric acid to solubilize. Then adjust to pH 3–4 with 1:10 ammonium hydroxide.

[45] George A. Pearse, Jr., and Ronalt T. Pflaum, *Anal. Chem.* **32**, 213–215 (1960).
[46] B. K. Afghan and D. E. Ryan, *Anal. Chim. Acta* **41**, 167–170 (1968).

To 250 ml of sample, add 10 ml of 10% solution of sodium citrate and 20 ml of an ammonium chloride–ammonium hydroxide buffer solution for pH 10. Add 2.5 ml of 5% solution of potassium cyanide and 10 ml of 20% solution of ascorbic acid. Heat to 70–80°, cool, and add 2.5 ml of 5% solution of formaldehyde. Adjust to pH 5 with 1:1 hydrochloric acid. Add 2.5 ml of the prepared color reagent and 5 ml of mercaptoacetic acid. Raise to pH 10 by adding ammonium hydroxide. Set aside for 30 minutes, then extract with 25 ml and 25 ml of benzene. Adjust the volume of the extracts to 50 ml and read at 515 nm.

PYRIDINE AND AZIDE

The ternary complex of nickel, pyridine, and azide is extractable with chloroform. Similar complexes are formed with iron, cobalt, copper, palladium, gold, and uranium. The technic is applicable to copper and palladium as well as to nickel.

Procedure. To a sample solution containing 50–200 μg of nickel, add 1 ml of pyridine and 2 ml of *M* sodium azide.[147] Dilute to 10 ml and adjust to pH 8.5–11. Shake with 10 ml of chloroform for 30 seconds and read the extract at 320 nm.

1-(2-PYRIDYLAZO)-2-NAPHTHOL

The 1:2 complex of nickel with this reagent, often abbreviated as PAN, is extractable with chloroform. It is applicable to determination of 0.05–1 ppm of nickel at pH 5–9.[148] The complex extracted with chloroform is read at 575 nm for up to 1.5 μg of nickel per ml. Copper and zinc can be masked with cyanide.

For determining nickel and manganese, extract both as the PAN complex at pH 9.5 with chloroform. Read both at 575 nm. Then extract manganese from the organic layer with a buffer solution for pH 5 and read nickel at 575 nm.

For analysis of high purity calcium and magnesium carbonates, dissolve the sample in hydrochloric acid and extract the iron with isopropyl ether.[149] Then extract nickel and cobalt at pH 5 with chloroform as the complex with the captioned reagent. Shake the extract with EDTA solution to minimize interference. Read nickel at 565 nm, correcting for residual iron.

Cobalt can be extracted with trioctylamine from 8 *N* hydrochloric acid, leaving the nickel content unaffected.[150] Thereafter raise the pH to 5–6 with an acetate

[147] R. G. Clem and E. H. Huffman, *Anal. Chem.* **38**, 926–928 (1966).

[148] Shozo Shibata, Yoshinobu Niimi, and Teiichi Matsumae, *Nagoya Kogyo Gijutsu Shikensho Hokaku* **11**, 275–278 (1962); *Rep. Gov. Ind. Res. Inst., Nagoya* **11**, 275–278 (1962).

[149] Adam Hulanicki, Malgorzata Galus, Wojciech Jedral, Regina Karwowski, and Marek Trojanowski, *Chem. Anal.* (Warsaw) **16**, 1011–1119 (1971).

[150] Tsurumatsu Dono, Genkichi Nakagawa, and Hiroko Wada, *J. Jap. Chem. Soc.* **82**, 540–544 (1961); *Jap. Anal.* **11**, 654–656 (1962).

buffer solution, mask residual cobalt with pyrophosphate and potassium periodate, and develop nickel with PAN.

For nickel, cobalt, and iron by 1-(2-pyridylazo)-2-naphthol, see Iron. For nickel in beer by this reagent, see Iron.

A structural isomer[151] **4-(2-pyridylazo)-1-naphthol** has also been studied in its relations to nickel, cupric, and zinc ions.

Procedures

Cobalt Present. Take a 5-ml sample containing up to 20 μg of nickel and 0.1–0.5 mg of cobalt.[152] Add 5 ml of 25% solution of tetrasodium pyrophosphate and adjust to pH 8–9. Add 3 ml of 1% solution of potassium periodate and oxidize the cobalt by heating at 100° for 5 minutes. Reduce to pH 5–6 and heat at 90° for 5 minutes to decompose excess periodate. Add 0.3 ml of 0.2% solution of PAN, warm for 5 minutes, and cool. Extract the nickel complex with 5 ml and 5 ml of chloroform. If interfering amounts of copper are present, wash the combined chloroform extracts with 5 ml of 25% solution of sodium thiosulfate or with 5 ml of 0.01 *M* EDTA. Read at 570 nm.

High Purity Molybdenum.[153] Dissolve 1 gram of sample in 5 ml of hydrochloric acid and 2 ml of nitric acid. Evaporate nearly to dryness to drive off excess acid. Add 10 ml of water, make just alkaline with 1 : 1 ammonium hydroxide, and dilute to 100 ml. Mix a 20-ml aliquot with 20 ml of water. Mask iron and copper by dissolving 0.1 gram of tiron in the sample. Add 3 ml of buffer solution for pH 10 containing 7% ammonium chloride and 57% ammonium hydroxide. Add 5 ml of 0.1% solution of the color reagent in methanol or ethanol. Heat to 60° and cool. Extract with 10 ml, 5 ml, and 5 ml of chloroform. Combine the extracts, add 4 ml of methanol or ethanol, and adjust to 25 ml with chloroform. Read at 570 nm against a reagent blank.

High Purity Tungsten. Dissolve 1 gram of sample in 5 ml of hydrofluoric acid and 2 ml of nitric acid. Proceed as for molybdenum from "Evaporate nearly to dryness...."

4-(2-PYRIDYLAZO)RESORCINOL

The complex of nickel with 4-(2-pyridylazo)resorcinol in a neutral or faintly basic medium is extracted with chloroform in the presence of benzyldimethylcyclohexanetetraacetic acid, EDTA, and a buffer solution and to read at 505 nm.[154] The appropriate buffer solution is monosodium borate at pH 9.3.

The 1 : 3 complex of nickel with this reagent at pH 8 has a maximum absorption

[151] D. Betteridge, Paula K. Todd, Quintus Fernando, and Henry Freiser, *Anal. Chem.* **35**, 729–733 (1963).

[152] Genkichi Nakagawa and Hiroko Wada, *Jap. Anal.* **10**, 1008–1012 (1961).

[153] R. Puschel and E. Lassner, *Mikrochim. Ichnoanal. Acta* **1965** (1), 17–20.

[154] Takao Yotsuyanagi, Ryuji Yamashita, and Kazuo Aomura, *Jap. Anal.* **19**, 981–982 (1970).

at 495 nm.[155] Beer's law is followed at 0.024–0.82 μg of nickel per ml. For nickel in the presence of cobalt, EDTA serves as a masking reagent and bismuth as an unmasking reagent.

For analysis of mineral water, oxidize with hydrogen peroxide and adjust to pH 4.5 with 2 *M* sodium acetate.[156a] Pass through a column of Dowex A-1 resin in ammonium form. Elute nickel and other cations with 2 *N* hydrochloric acid. Evaporate the eluate to dryness and take up in hydrochloric acid. Pass through a column of Dowex 1-X10 in chloride form, which will retain many interfering cations. Mask others with EDTA, develop 0.5–25 μg of nickel with 4-(2-pyridylazo)resorcinol, and read the complex at 496 nm.

For analysis of saponifiable oil decompose the sample with sulfuric acid and evaporate to dryness.[156b] Take up in water, adjust to pH 1, and add a solution of cupferron. Extract ferric ion, pentavalent vanadium, hexavalent molybdenum, titanium, zirconium, and antimonous ion with chloroform and discard the extract. Neutralize to pH 9 and add a buffer solution for pH 8.5–9.5. Add benzyldimethyltetradecylammonium chloride and EDTA. Extract the ternary complex with chloroform and read 0.02–1.2 μg of nickel per ml in the extract at 500 nm.

Procedure. To a sample containing 12–25 μg of nickel, add 5 ml of a buffer solution for pH 9.3 and dilute to 30 ml.[157] Add 1 ml of 0.5 *M* sodium citrate, 5 ml of 0.05 *M* sodium tetraborate, and 1 ml of 1% solution of 4-(2-pyridylazo)resorcinol. Set aside for 10 minutes. To mask copper, zinc, cadmium, mercury, lead, aluminum, chromium, beryllium, zirconium, gallium, indium, silver, and the rare earth metals, add 5 ml of 0.05 *M* EDTA. To mask ferric ion, add 4 grams of potassium fluoride. To mask cobalt, add 4 ml of 1% solution of potassium iodate with 5 ml of 20% solution of tetrasodium pyrophosphate and heat at 90° for 5 minutes. Dilute to 50 ml and read at 494 nm.

By Extraction.[158] To the sample, add 2 ml of acetate buffer solution for pH 4 and 1 ml of 0.05% solution of the sodium salt of the reagent. Extract with 8 ml and 8 ml of ethyl acetate. Add 5 ml of water to the combined extracts and evaporate the ethyl acetate at 100°. To the resulting aqueous solution, add 2 ml of borate buffer solution for pH 9.7, dilute to 25 ml, and read at 496 nm. Ferric ion must have been preextracted with ether.

Cobalt Present.[159] Adjust the sample solution to pH 8 with 0.8% solution of sodium hydroxide. Add 0.5 ml of 30% hydrogen peroxide and heat at 50° for 20 minutes. Adjust 0.586 m*M* bismuth solution to pH 2 with 0.2 *N* nitric acid and 0.8% sodium hydroxide solution. Add 2 ml of the bismuth solution and heat at 70°

[155] D. Nonova and B. Evtimova, *Anal. Chim. Acta* **49**, 103–108 (1970).
[156a] V. Nevoral and A. Okac, *Cslka Farm.* **17**, 478–482 (1968).
[156b] Takao Yotsuyanagi, Ryuji Yamashita, Hitoshi Hoshino, Kazuo Aomura, Hideo Sato, and Nobosuke Masuda, *Anal. Chim. Acta* **82**, 431–434 (1976).
[157] Yoshio Shijo and Tsugio Takeuchi, *Jap. Anal.* **14**, 511–515 (1965).
[158] D. Nonova and N. Likhareva, *C. R. Acad. Bulg. Sci.* **27**, 815–818 (1974).
[159] Katsumi Yamamoto, Housaburo Ohashi, and Ichiro Hirako, *Bull. Chem. Soc. Jap.* **44**, 2254–2255 (1971).

for 40 minutes. Destroy excess hydrogen peroxide by adding 1 ml of 0.5 *M* sodium sulfite. Chill in ice water. Add 1 ml of 0.047 *M* sodium tetraborate and 1 ml of 2 m*M* 4-(2-pyridylazo)resorcinol. Set aside for 10 minutes, then add 2 ml of 0.022 *M* EDTA to convert any bismuth complex with the reagent to bismuth-EDTA. Read the nickel complex at 496 nm. Manganous and cupric ions interfere.

QUINOXALINE-2,3-DITHIOL

This reagent in ammoniacal solution is suitable for reading 0.03–3 ppm of nickel. Because of the high absorption of the reagent at 520 nm, an unnecessarily large excess must be avoided and the amount added must be rigidly standardized. Citrate will mask ferric, aluminum, and manganous ions. The following ions interfere: silver, cupric, palladous, cadmium, mercuric, stannous, lead, bismuth, and platinic.

Because the captioned reagent is unstable, it is convenient to use as the reagent *S*-(3-mercaptoquinoxalin-2-yl)thiouronium chloride, which is stable indefinitely as a solid and for several days as a solution in 80% ethanol. It decomposes in aqueous solution to form the captioned reagent.

Procedure. To a neutral sample solution containing 2–25 μg of nickel, add 10 ml of ammonium hydroxide and 3 ml of 0.39% solution of the color reagent in ethanol, prepared the same day.[160] Dilute to 25 ml and set aside for at least 30 minutes. Read at 520 nm against a blank.

Nickel and Cobalt.[161] Prepare a buffer solution containing 70 grams of ammonium chloride, 6 grams of zinc sulfate heptahydrate, and 570 ml of ammonium hydroxide. Dilute to 1 liter.

Take a sample solution containing 12.5–125 μg of nickel and 7.5–75 μg of cobalt. Add 1 ml of the buffer solution and 5 ml of 0.1% solution of the thiouronium chloride reagent in 80% ethanol. Dilute to 25 ml and heat at 60° for about 15 minutes. Cool and dilute to 50 ml. Read at 472 and 520 nm.

Calculate from equations developed fo the particular spectrophotometer used, the following being typical:

$$\mathrm{Ni} \times 10^5 = 6.74E_{520} - 3.43E_{472}$$

$$\mathrm{Co} \times 10^5 = 3.95E_{472} - 2.23E_{520}$$

Alternatively,[162] to a sample containing 10–35 μg of cobalt and 22–58 μg of nickel, add 1 ml of 0.1% solution of formic acid in 80% dimethylformamide (DMF). Dilute to about 20 ml with 80% DMF. Add 1 ml of 0.1% solution of the

[160] D. A. Skoog, Ming-gon Lai, and Arthur Furst, *Anal. Chem.* **30**, 365–368 (1958).

[161] J. A. W. Dalziel and A. K. Slawinski, *Talenta* **15**, 367–372 (1968).

[162] Gilbert H. Ayres and Robert R. Annancl, *Anal. Chem.* **35**, 33–36 (1963); cf. R. W. Burke and J. H. Yoe, *ibid.* **34**, 1378–1382 (1962).

captioned reagent in 80% DMF. Set aside for 5 minutes and dilute to 25 ml with 80% DMF. Read nickel at 650 nm and cobalt at 505 nm.

Seawater.[163] Collect a 1-liter sample in a plastic container and add 50 ml of 5% sodium carbonate solution. Let settle for a minimum of 7 days and filter through an HA Millipore filter. Wash twice with water, then dissolve in 15 ml of 1:1 hydrochloric acid. Add 15 ml of 10% sodium citrate solution and gently evaporate to about 25 ml. Adjust to pH 8 and add 2 ml of 1% ethanolic dimethylglyoxime. Cool, and shake for 2 minutes with 3 ml and 3 ml of chloroform. Wash the combined chloroform extracts with 5 ml and 5 ml of 1:49 ammonium hydroxide. Extract the combined washings with 3 ml of chloroform. Extract nickel from the combined chloroform extracts with 5 ml and 5 ml of 1:1 hydrochloric acid. To the acid extracts, add 10 ml of ammonium hydroxide; then add 3 ± 0.1 ml of 0.39% solution of quinoxaline-2,3-dithiol in ammonium hydroxide. Dilute to 50 ml and set aside for 30 minutes. Read at 520 nm against a reagent blank. Prepare the calibration curve from spiked seawater.

1-(2-THIAZOLYLAZO)-2-NAPHTHOL

This reagent will determine 0.03–40 ppm of nickel. The 1:2 nickel complex can be extracted at pH 4–10 with chloroform. Cobalt is masked as the cobaltic ammine complex. Copper, zinc, manganese, and iron are removed from 0.01 *M* diethyldithiocarbamate solution by back-extraction with 0.01 *N* hydrochloric acid. An alternative, if EDTA, cyanide, citrate, phosphate, cupric ion, zinc, cadmium, cobalt, or manganous ion is present, is to extract as the dimethylglyoxime complex with chloroform and back-extract the nickel with 0.5 *N* hydrochloric acid.[164] Beer's law is followed for up to 1.2 μg of nickel per ml, and the color is stable for hours.

As a general technic, add 25% solution of tetrasodium pyrosulfate and 0.5% solution of potassium periodate to a sample solution containing 0.1–2 μg of nickel per ml.[165] Warm the mixture, and adjust to pH 9–10 with 2 *M* ammonium hydroxide–ammonium chloride buffer solution. Warm, and add 0.2% solution of the reagent in methanol. Extract the complex with chloroform and read at 595 nm.

Procedure. *High Purity Chromium.*[166] Dissolve 2 grams of sample in perchloric acid, evaporate almost to dryness, and take up in 50 ml of water. Add 10 ml of 10% solution of ammonium citrate and raise to pH 9 with ammonium hydroxide. Add 2 ml of 1% ethanolic dimethylglyoxime. Extract with 10 ml, 5 ml, and 5 ml of chloroform. Wash the combined extracts with 10 ml of 1:50 ammonium hydroxide. Reextract the nickel with 10 ml and 5 ml of 0.5 *N* hydrochloric acid. Add 1 ml of nitric acid and 0.2 ml of perchloric acid and heat to white fumes. Take up in 10 ml of 0.4 *M* sodium acetate. Mask iron and zinc with 1 ml of 10% solution of

[163] W. Forster and H. Zeitlin, *ibid.* **38**, 649–650 (1966).
[164] Krzysztof Kasiura, *Chem. Anal.* (Warsaw) **14**, 375–380 (1969).
[165] Hiroko Wada and Genkichi Nakagawa, *Anal. Lett.* **1**, 687–695 (1968).
[166] Akira Kawase, *Jap. Anal.* **13**, 609–614 (1964).

tetrasodium pyrophosphate decahydrate. Adjust to pH 5.5 and mask copper with 1 ml of 25% solution of sodium thiosulfate pentahydrate. Add 1 ml of 0.0255% ethanolic 1-(2-thiazolylazo)-2-napthol. Set aside for 10 minutes, then extract with 10 ml of chloroform. Read at 596 nm. If cobalt is present, it must be removed by a preliminary extraction with dithizone in chloroform.

THIOSEMICARBAZONE OF BENZOTHIAZOL-2-YL PHENYL KETONE

This reagent gives a visible absorption with nickel, copper, cobalt, cadmium, and zinc.[167] By reading nickel at pH 2.5 and 446 nm, cadmium at pH 4 and 418 nm, and zinc at pH 6 and 420 nm, the three can be determined by simultaneous equations. Silver, palladium, iron, and lead interfere.

Procedure. Mix 2 ml of sample with 0.5 ml of an ammoniacal buffer solution for pH 8. Add 1 ml of 0.1% ethanolic reagent and dilute to 5 ml with ethanol. Set aside for 30 minutes. Read nickel at 446 nm.

XYLENOL ORANGE

The complex of nickel with xylenol orange follows Beer's law at pH 6.1–6.3 for 3–40 μg per 25 ml.[168] Reading is at 585 nm, and a 1.5-fold excess of xylenol orange is necessary.[169] There is interference by cobalt, copper, lead, manganous ion, calcium, iron, hydroxylamine, EDTA, tartaric and ascorbic acids, and fluoride ion.

A 1 : 2 : 4 complex having extreme sensitivity is formed by nickel, xylenol orange, and cetylpyridinium bromide. The rare earths form similar complexes. Even without the cation active agent, the sensitivity with a 5-cm cell can be 5 ng/ml.[170]

Procedure. Mix 0.37 ml of 10^{-3} *M* xylenol orange in 20% methanol, 2.5 ml of 0.1% cetylpyridinium bromide solution, and 2.5 ml of tris(hydroxymethyl)aminomethane-borate buffer solution for pH 7.5.[171] Add a neutral sample solution containing up to 8 μg of nickel ion and dilute to 25 ml. Set aside for 1 hour and read at 610 nm against a reagent blank.

---

[167] Toyozo Uno and Sumiyuki Akihama, *ibid.* **10**, 941–945 (1961).
[168] Makoto Otomo, *ibid.* **14**, 45–52 (1965).
[169] E. A. Bashirov, A. M. Ayubova, T. E. Abdullaeva, and F. T. Aslanova, *Uch. Zap. Azerb. Gos. Univ., Ser. Khim. Nauk* **1971** (2), 28–32.
[170] M. I. Bulatov, *Zh. Anal. Khim.* **24**, 1053–1057 (1969).
[171] W. J. de Wet and G. B. Behrens, *Anal. Chem.* **40**, 200–202 (1968).

Nickel ion can be measured by its catalytic effect on decomposition of 0.01 *M* permanganate ion in 10% sodium hydroxide solution containing 0.25% **acetodiphosphonic acid**, which is 1-hydroxyethylidene-1, 1-diphosphonic acid.[172] Measurement is of the time for the absorption to reach 0.300. Silver, copper, cobalt, and iron interfere, and preliminary extraction of these metals is necessary. Trace amounts of nickel require concentrating. The method will determine 0.1–0.7 ppm of nickel, but preferably more than 0.25 ppm. Oxalate and phosphate retard the reaction.

As a ternary complex for determination of nickel, mix 1–10 ml of 0.1 m*M* sample, 2 ml of succinate buffer solution for pH 4.5, 10 ml of 0.1 m*M* lanthanum ion, and 20 ml of 0.1 m*M* **alizarin complexon.**[173] Dilute to 200 ml, set aside for 2 hours, and read at 550 nm against a reagent blank.

A 2:1 complex of **6-amino-4-hydroxy-2-mercapto-5-nitrosopyrimidine** has a maximum absorption at pH 9.4.[174a] It does not strictly conform to Beer's law, but it will determine 0.3–2 μg of nickel per ml. It is stable for 3 hours.

Reading a complex of nickel with **4-amino-6-hydroxy-2-methyl-5-nitrosopyrimidine** is interfered with by stannic ion, tetravalent vanadium, cupric, ferrous, and ferric ions.[174b]

To 2 ml of sample solution containing up to 50 μg of nickel, add 5 ml of ammonium hydroxide–ammonium chloride buffer solution for pH 11 and dilute to 10 ml.[175a] Add 8 ml of dimethylformamide and set aside for 10 minutes. Add 2 ml of 0.01 *M* **3-aminoquinoxaline-2-thiol** and dilute to 20 ml with 50% dimethylformamide. Set aside for 5 minutes and read at 495 nm against a reagent blank. There is minor interference by ferric, cupric, and palladous ions. Cobalt forms a similar complex.

At pH 3.5–6 a crimson complex of nickel and **7-(4-antipyrinylazo)-8-hydroxyquinoline** is read at 485 nm for 0.04–4 μg/ml of nickel.[175b]

Atidan, which is 1-(5-amino-1,3,4-thiadiazol-2-ylazo)2-naphthol, forms a 2:1 complex with nickel at pH 10.8 which can be extracted with pentanol and read at 570 nm for 0.06–1.3 μg of nickel per ml. Cobalt, copper, cyanide, and EDTA interfere.[176]

Mix 2 ml of sample with 2 ml of 0.05% solution of **benzothiazol-2-yl-hydrazone of benzothiazole-2-aldehyde** in dioxan, dilute to 5 ml with dioxan, and read nickel at 474 nm and pH 5.42.[177] The same derivative of **naphtho-[2, 1-]thiazole-2-aldehyde** is read at 468 nm and pH 5.42 within 10 minutes after mixing. Beer's law is followed for 0.1–2.9 μg of nickel. The corresponding reactions are given by copper and cobalt. Zinc and palladium interfere. With **benzothiazol-2-yl phenyl ketoxime**, mix 2 ml of sample, 0.5 ml of ammoniacal buffer solution for pH 9, 1 ml of 0.2% ethanolic reagent, and 1.5 ml of dioxan. Read at 400 nm for nickel. Beer's law is

[172]D. Mealor and A. Townshend, *Anal. Chim. Acta* **39**, 235–244 (1967).
[173]M. A. Leonard and F. I. Nagi, *Talenta* **16**, 1104–1108 (1969).
[174a]S. Przeszlakowski and A. Waksmundzki, *Chem. Anal.* (Warsaw) **9**, 919–924 (1964).
[174b]Mohan Katyal, P. Jain, W. Kushwaha, and R. P. Singh, *Indian J. Chem., A* **14**, 811–812 (1976).
[175a]Kazuo Ohira, Yoshinori Kidani, and Hsashi Koike, *Jap. Anal.* **23**, 658–664 (1974).
[175b]Sh. T. Talipov, N. G. Smaglyuk, and A. Khodzhaev, *Dokl. Akad. Nauk Uzbek. SSR* **1974**, (3), 37–38.
[176]Eugenia Domagalina and Stanislaw Zareba, *Chem. Anal.* (Warsaw) **16**, 883–890 (1971).
[177]Toyozo Uno and Sumiyuki Akihama, *Jap. Anal.* **10**, 822–832 (1961).

followed for 0.1–10 μg. Copper can be read at 362 nm and cobalt at 358 nm, and all three can be solved by simultaneous equations. Palladium interferes with all three, cadmium with copper and nickel, and silver with cobalt.

The 1:2 complex of nickel with **2-(2-benzothiazolylazo)-1-naphthol** is extractable with chloroform above pH 6.[178] The maximum absorption is at 634 nm. A copper complex absorbs at 610 nm, a cobalt complex at 620 nm.

Mix 10 ml of sample with 8 ml of an appropriate buffer solution to maintain at a pH above 6.2 and add 10 ml of 0.1 *M* **benzoyltrifluoroacetone** in isoamyl alcohol.[179] Extract for 10 minutes, dilute the organic layer to 25 ml, and read nickel at 380 nm. There is interference by ferric, cupric, and cobalt ions. Beer's law applies to the extract at 0.1–1 m*M*.

The 1:2 complex of nickel with **biacetyl monothiosemicarbazone** in a weakly alkaline potassium or sodium phosphate buffer solution is read at 400 nm.[180] Added as 0.05% solution in ethanol, the reagent gives a maximum absorption at 340 nm with pH 11.4 and a boric acid–sodium hydroxide buffer solution. Beer's law is followed at 0.5–2.5 ppm. **Picolinaldehyde semicarbazone** similarly is read with nickel at 350 nm, pH 9.5 with ammonium hydroxide–ammonium chloride buffer solution, and 0.2–1.2 ppm. Both reagents also complex with copper and cobalt.

At more than 10 μg per ml, nickel is read with **2,2′-bipyridyl** at 400 nm.[181] Iron absorbs at 530 nm with this reagent and copper at 685 nm. All three can be read with simultaneous equations.

The complex of nickel with ***NN′*-bis(2-aminobenzylidene)ethylenediamine** is extracted with ether, chloroform, or benzene and read at 486 nm.[182] At pH 5–5.5 the cobaltous complex can be oxidized by air to the water-soluble cobaltic complex. After extraction of the nickel complex, oxidize, adjust to pH 7–9, and read the cobalt complex at 430 nm.

Extract a nickel solution buffered to pH 6–8 with an equal volume of 0.02 *M* ***NN′*-bis[1-(2-aminophenyl)ethylidene]ethylenediamine** in benzene.[183] For 0.5–25 μg of nickel per ml of extract, read at 496 nm.

The complex of nickel with **2,2′-bis(carboxymethyl[-mercapto]ethyl) ether**, which is 3-(oxapentamethylenedithio)diacetic acid, follows Beer's law at 380 and 600 nm for 0.82–7.34 m*M*.[184]

Nickel can be converted to a boron chelate such as **bis-diphenylborylbis-dimethylglyoximotonickel.**[185] Photometric determination of the boron content of the chelate permits indirect determination of nanogram amounts of nickel. The technic also has application to determination of ferrous, cobaltous, cupric, and palladous ions.

[178]Akira Kawase, *ibid.* **17**, 56–60 (1968).

[179]G. N. Rao and J. S. Thakur, *Z. Anal. Chem.* **271**, 286 (1974).

[180]F. Sanchez Burgos, P. Martinez Martinez, and F. Pino Perez, *Inf. Quim. Anal. Pura Apl. Ind.* **23**, 17–27 (1969); M. P. Martinez Martinez, D. Perez Benedito, and F. Pino Perez, *ibid.* **21**, 31–37 (1967); *An. Quim.* **69**, 747–756 (1973).

[181]Giordano Trabanelli, *Atti Accad. Sci. Ferrara* **35**, 133–142 (1957–1958).

[182]H. Berge and H. Mennenga, *Z. Anal. Chem.* **213**, 346–349 (1965); cf. Michio Mashima, *Jap. Anal.* **9**, 269–271 (1960).

[183]E. Uhlemann and W. Wischnewski, *Anal. Chim. Acta* **42**, 247–252 (1968).

[184]E. Cassassas, J. J. Arias, and A. Mederos, *An. Quim.* **69**, 1121–1131 (1973).

[185]F. Umland and D. Thierig, *Z. Anal. Chem.* **197**, 151–160 (1963).

Bis(hydroxyethyl)dithiocarbamate forms complexes with maxima for nickel at 390 nm, for copper at 440 nm, and for cobalt at 390 nm.[186] If disodium EDTA precedes the reagent, cobalt does not react but nickel does. Then copper and nickel can be solved by simultaneous equations. Sodium citrate interferes in determination of nickel.

Nickel is extracted with toluene from solutions of uranium, thorium, copper, iron, and chromium as the complex with **bis-1,4-methylcyclohexane-1,2-dione dioxime** and read at 365 nm.[187] Copper is masked with mercaptoacetic acid. If iron is present, tartaric acid prevents formation of ferric mercaptoacetate. Tartrate also prevents hydrolysis of thorium.

The 1 : 1 complex of nickel with **2,3-bis(salicylideneamino)benzofuran** can be extracted with chloroform from an ammoniacal solution above pH 7.7.[188] To 30 ml of sample solution, add 1 ml of 0.15% solution of the color reagent in dioxan and 1 ml of *M* ammoniacal buffer solution for pH 10–10.5. Extract with 10 ml of chloroform and read at 508 or 535 nm. There is interference by cobalt, cupric, chromic, ferric, manganous, zinc, tartrate, and citrate ions.

To a sample solution, add 10 ml of 0.5% solution of ***N,N'*-bis(*m*-sulfobenzyl) dithiooxamide** and 10 ml of a borate buffer solution or an ammonium chloride–ammonium hydroxide buffer solution for pH 10.[189] Dilute to 50 ml, set aside for at least 90 minutes, and read the 1 : 1 complex at 500 nm.

For nickel, cobalt, and chromium in steel by **biuret** applying simultaneous equations, see Copper.

Calcichrome at pH under 12 forms a 1 : 1 complex with nickel having a maximum absorption around 550 nm.[190] At about pH 13, the reagent forms a 2 : 1 complex with a maximum around 585 nm. For 25 ml of solution containing 5 ml of 0.2 m*M* calcichrome buffered at pH 11.5 with 0.1 *M* glycine–sodium hydroxide, the graph is rectilinear at 615 nm down to 8.5 ng of nickel per ml or at 626 nm down to 4 ng of nickel per ml. Color development takes 30 minutes at pH 11.5 and 45 minutes at pH 13. There is serious interference by calcium, magnesium, strontium, cobalt, manganese, copper, aluminum, ferric ion, titanium, pentavalent vanadium, chromium, and phosphate.

Nickel forms a 1 : 1 complex with **2-carboxy-1-hydroxyanthraquinone.**[191] Mix 2.5 ml of ethanol, 10 ml of ethanolic 0.925 m*M* reagent, and 1 ml of 2 *N* ammonium hydroxide, and test solution to give 2.2–6.6 ppm of nickel when diluted to 25 ml. Read at 495 nm. There is interference by more than 2 ppm of lead, manganous, zinc, cobalt, chloride, arsenate, or tetraborate ions.

For analysis of 0.4–6 μm films of nickel and chromium, dissolve 2–10 cm^2 from the glass or silica base with 1–5 ml of nitric acid in a silica crucible.[192] Add a few drops of sulfuric acid and volatilize the nitric acid. Dilute to 100 ml. To an aliquot, add 10% by volume of buffer solution for pH 10 and a solution of **chlorindazon DS,**

[186] M. Pinkas, *Bull. Soc. Pharm. Lille* **1960** (3), 93–97; P. Balistre and M. Pinkas, *Chim. Anal.* (Paris), **43**, 433–438 (1961).

[187] P. D. Blundy and M. P. Simpson, *Analyst* **83**, 558–561 (1958).

[188] Hajime Ishii and Hisahiko Einaga, *Jap. Anal.* **19**, 1351–1356 (1970).

[189] A. A. Janssens, G. L. van de Capelle, and M. A. Herman, *Anal. Chim. Acta* **31**, 425–430 (1964).

[190] Hajime Ishii and Hisahiko Einaga, *Jap. Anal.* **16**, 322–327 (1967).

[191] F. Salinas and L. M. Franquelo, *Quim. Anal.* **29**, 319–322 (1975).

[192] Dieter Molch, Hartmut Koenig, and Eberhard Than, *Z. Chem.* **14**, 408–410 (1974).

which is 4-(6-chlorindazol-3-ylazo)-3-hydroxynaphthalene-2, 7-disulfonic acid. Dilute to 50 ml, set aside for 30 minutes, and read the violet nickel complex at 575 nm. Beer's law holds up to 1.5 μg of nickel per ml.

Chrome azurol S complexes with nickel and cobalt at pH 9.7 to give a maximum absorption at 567 nm.[193] The graph is rectilinear for 0.06–1 ppm of nickel and for 0.08–2.4 ppm of cobalt in the presence of 3 ml of 0.25% solution of the color reagent and 3 ml of 0.05 *M* borate buffer per 25 ml. There is interference by aluminum, beryllium, ferric, chromic, gallium, stannic, zirconium, and sulfate ions, as well as by hydrogen peroxide, phosphate, tartrate, oxalate, and EDTA.

To a sample solution containing up to 9 μg of nickel per ml, add 7 ml of 0.05% solution of **chromotrope 2B** (acid red 176) and 1 ml of 1% solution of ammonium hydroxide. Dilute to 25 ml and read at 574 nm.[194] Beer's law is followed for 1–9 μg of nickel per ml. Cobalt reacts similarly.

The 1:2 chelate of nickel with **chromotrope 2R** (acid red 29) at pH 7–10 is read at 530 nm for 0.5–4.4 ppm of nickel.[195]

Nickel forms a complex with **cuprizone**, also known as bicyclohexanone oxalyldihydrazone at pH 9.5 of an ammonium chloride–ammonium hydroxide buffer solution having a maximum absorption at 330 nm.[196a] Cobalt forms a similar complex with a maximum at 292 nm. Ferric, manganous, chromic, and lead ions precipitate the reagent at that pH.

To a sample solution containing 1–5 μg of nickel, add 5 ml of 0.02% solution of **cycloheptane-1, 2-dione dioxime**, dilute to 15 ml, and adjust to pH 5.4–12.75 with sodium acetate solution or with ammonium hydroxide.[196b] Set aside for 1 hour, then extract the nickel complex by shaking for 5 minutes each with 5 ml, 5 ml, and 5 ml of chloroform. Dilute the combined extracts to 25 ml with chloroform and read at 377 nm against a reagent blank. If necessary to avoid interference by copper, wash the organic extract with 6 *N* ammonium hydroxide.

For application to cadmium carbonate, dissolve the sample in hydrochloric acid, adjust to pH 4, and extract the dioxime complex with chloroform.[197] Wash the organic layer with 4% sodium hydroxide solution and water before reading at 377 nm. Similarly, dissolve zirconyl nitrate in hydrofluoric acid for determination of nickel.

For determination of nickel, add to the sample solution 3 ml of 0.25% solution of **cyclohexane-1, 2-dione dithiosemicarbazone** in dimethylformamide and 10 ml of ethanol.[198] Add 1:10 hydrochloric acid to adjust to pH 2.4–3.4, and dilute to 50 ml. Read at 400 nm.

A complex of nickel with **1, 2-diaminopropane-*NNN'N'*-tetraacetic acid** is read as a 1:1 complex at 380 nm.[199] Chromic, cupric, and cobalt ions also complex with this reagent.

A ternary complex of nickel with **diantipyrinylmethane** and **dithiooxalate** is

[193]Yoshigo Horiuchi and Hiroshi Nishida, *Jap. Anal.* **16**, 576–580 (1967).
[194]D. Negoiu, A. Kriza, and M. Negoiu, *Anal. Univ. Bucur., Ser. Stiint. Nat. Chim.* **13**, 149–156 (1964).
[195]O. Prakash and S. P. Mushran, *Chim. Anal.* (Paris) **49**, 473–476 (1967).
[196a]S. H. Omang and A. R. Selmer-Olsen, *Anal. Chim. Acta* **27**, 335–338 (1962).
[196b]V. M. Peshkova and N. G. Ignat'eva, *Zh. Anal. Khim.* **17**, 1086–1090 (1962).
[197]V. M. Savostina, S. O. Kobyakova, and V. M. Peshkova, *ibid.* **23**, 938–940 (1968).
[198]J. A. Munoz Leyva, J. M. Cano Pavon, and F. Pino Perez, *Quim. Anal.* **28**, 90–98 (1974).
[199]S. Vicente-Perez, L. Hernandez, and J. Rosas, *ibid.* **28**, 283–288 (1974).

extracted by chloroform.[200] The maximum absorption of the complex is at 505 nm, the same wavelength as the nickel dithiooxalate complex, but the sensitivity is much greater. Iron, molybdenum, cobalt, copper, and vanadium interfere. The iron can be masked with ammonium fluoride.

To a sample containing 0.45–1.6 mg of nickel, add 5 ml of 3% solution of **2-(diethylamino)ethanethiol hydrochloride** and raise to pH 10 with 0.4% solution of sodium hydroxide.[201] Add a boric acid–potassium chloride–sodium hydroxide buffer solution for pH 10 and dilute to 25 ml. Read the 1:2 complex of nickel to reagent immediately at 520 nm.

At pH 4.5–5.5, nickel forms a 1:2 complex with **diethylamino[3-(1-methyl-2-piperidyl)-2-piperidylazo]phenol.**[202] The maximum absorption is at 560 nm. Beer's law is followed for 1–20 μg of nickel per 25 ml.

A complex of nickel with **2′,4′-dihydroxyvalerophenone oxime** is extracted with chloroform and read at 580 nm.[203] Beer's law is obeyed for 38–70 ppm. The cobalt complex is insoluble in water.

In ethanol at pH 3±0.5, the complex of nickel with **2,3-dimercaptoquinoxaline** is read at 565 nm.[204]

To a sample solution containing 0.3–3 mg of nickel, add 3 ml of 3% solution of **2-(dimethylamino)ethanethiol hydrochloride** for each 0.125 mg of nickel present.[205a] Raise to pH 10 with 0.4% solution of sodium hydroxide and add 5 ml of boric acid–potassium chloride–sodium hydroxide buffer solution for pH 10. Dilute to 25 ml and read at 510 nm. Interfering cations are palladous, pentavalent vanadium, calcium, ferrous, ferric, cobaltous, and uranium.

Mix a sample solution containing less than 30 μg of nickel with 0.5 ml of 0.5 *M* sodium potassium tartrate, 2 ml of 0.1 *M* potassium pyrophosphate, 2 ml of ammoniacal buffer solution for pH 8, 10 mg of *N*-(dithiocarboxy)glycine, and 3 ml of 0.1% solution of **5-dimethylamino-2-(2-thiazolylazo)phenol** in a 20% solution of Triton X-100.[205b] Dilute to 25 ml, set aside for 10 minutes, and read at 560 nm against a reagent blank for 3–30 μg of nickel. Interferences are ferric ion, pentavalent vanadium, zinc, manganese, cadmium, copper, and cobalt.

At pH 5.2 **di-2-pyridylglyoxal dithiosemicarbazone** complexes with nickel.[206] This is extractable with chloroform, in contrast to the corresponding cobalt complex. Each determines 1 ppm in the presence of 5 ppm of the other. To the sample solution, add 15 ml of 0.1% solution of the reagent in ethanol and 20 ml of acetic acid–sodium acetate buffer solution for pH 5.2. After 30 minutes extract with four 5-ml portions of chloroform. Dilute the extracts to 25 ml with chloroform and read nickel at 410 nm.

At pH 9, nickel forms a blue-violet 1:1 complex with **dithiooxamide**, also known as **rubianic acid.**[207] Above pH 11 with 25-fold excess of reagent in 50% methanol, a

[200] A. T. Pilipenko, N. N. Maslei, and E. T. Skorokhod, *Zh. Anal. Khim.* **23**, 227–230 (1968).
[201] F. Bermejo Martinez and J. F. Vila Brion, *Inf. Quim. Anal. Pura Apl. Ind.* **26**, 109–115 (1972).
[202] A. E. Martirosov, Sh. T. Talipov, and R. Kh. Dzhiyanbaeva, *Uzb. Khim. Zh.* **1968** (1), 23–25.
[203] Jai Singh, S. P. Gupta, and O. Pl Malik, *Indian J. Chem.* **13**, 1217–1220 (1975).
[204] R. W. Burke, *Diss. Abstr.* **23**, 3614 (1963).
[205a] F. Bermejo Martinez and J. Vila Brion, *Acta Cient.* Compostelana **1970** 107–118.
[205b] Hideki Ishii and Hiroto Watanabe, *Jap. Anal.* **26**, 86–91 (1977).
[206] J. L. Bahamonde, D. Perez Benedito, and F. Pino Perez, *Analyst* **99**, 355–359 (1974).
[207] Katarzyna Lesz and Joanna Borowiecka, *Chem. Anal.* (Warsaw) **20**, 631–634 (1975).

yellow 2:1 metal-ligand complex is obtained. This is read at 460 nm, after 20 minutes for development. Beer's law is followed for 1–10 μg of nickel per ml. There is interference by heavy metal cations, cyanide, ferrocyanide, ferricyanide, sulfide, sulfite, thiosulfate, chromate, bichromate, phosphate, and carbonate.

To a sample solution containing 0.05–0.4 mg of nickel, add 2 drops of phenolphthalein indicator solution and 1 ml of 5% solution of gelatin.[208] Make the solution slightly alkaline with 1:10 ammonium hydroxide, then just acid with 0.2 *N* hydrochloric acid. Add 1 ml of 5% sodium acetate solution as a buffer and 1 ml of 0.2% solution of the reagent. Dilute to 50 ml, let stand for 30 minutes, and read at 570 nm. Copper and ferric ion interfere.

To a sample solution containing 10–50 μg of nickel, add 5 ml of 10% ammonium acetate solution and dilute to 20 ml.[209] Add 1 ml of 0.5% solution of **dithiosalicylic acid**. Extract the complex with 10 ml of ethyl acetate and read the extract at 550 nm.

Nickel can be read at 510 nm by disodium **ethyl bis(5-tetrazolylazo)acetate**.[210] This is applicable along with copper at 600 nm by dicyclohexanone oxalyldihydrazone, also known as cuprizone.

Ethyl-di(1-sodiotetrazol-5-ylazo)acetate complexes with nickel at pH 6–9 for reading at 490 nm.[211] The sensitivity is to 0.01 μg of nickel per ml, and Beer's law is followed up to 2 μg/ml. For this determination, iron and aluminum are masked with fluoride, and copper with EDTA.

4-Ethyl-1-phenylthiosemicarbazide forms a 2:1 complex with nickel in an alkaline medium.[212] It is insoluble in water but extractable with organic solvents. In ethyl acetate it is read at 510 nm. Beer's law is followed for 0.73–6.6 μg of nickel per ml. There is interference by cyanide, citrate, oxalate, tartrate, EDTA, and cupric ion. A tenfold excess of cobalt is tolerated.

The optimum pH for a complex of nickel with potassium **ethylxanthate** is 9.[213] To a sample solution containing 50 μg of nickel per ml, add 8 ml of 0.05% solution of the color reagent and dilute to 25 ml with ethanol. Read at 420 or 470 nm within 15 minutes. Beer's law is followed for 5–50 μg of nickel per ml.

For extraction, add 1 ml of 0.5% solution of potassium ethylxanthate to a sample solution at pH 4–9 and set aside for 5 minutes.[214] Shake for 10 minutes with 10 ml of chloroform. Dilute the separated extract to 25 ml with chloroform and read the 1:2 metal-reagent complex at 415 nm. Beer's law is obeyed for 3–16 μg of nickel per ml.

Down to 0.3 ppm of nickel can be read with **fast sulfon black F** (acid black 32).[215] There is interference by cupric, beryllium, cobaltous, manganous, ferric, or chromic ions.

[208] Antal Janosi and Laszlo Farady, *Veszpremi Vegyip. Egyelem Tud. Ulesszakanak Elodasai* **1957** (2), 10–18; J. Xavier and Priyadaranjan Ray, *J. Indian Chem. Soc.* **35**, 432–444 (1958).

[209] J. Weyers and T. Gancarczyk, *Z. Anal. Chem.* **235**, 418–421 (1968).

[210] M. Freegarde and Miss B. Jones, *Analyst* **84**, 716–719 (1959).

[211] N. S. Frumina, N. N. Goryunova, and I. S. Mustafin, *Zh. Anal. Khim.* **22**, 1523–1526 (1967).

[212] Boleslaw Janik and Dorota Holiat, *Chem. Anal.* (Warsaw) **14**, 357–361 (1969).

[213] C. Isvoranu-Panait, D. Negoiu, and M. Crutescu, *Anal. Univ. Bucur., Ser. Stiint. Nat. Chim.* **13**, 127–140 (1964).

[214] P. K. Paria and S. K. Majundar, *Z. Anal. Chem.* **278**, 364 (1976).

[215] F. Burriel-Marti, A. Cabrera-Martin, and T. S. West, *An. Real Soc. Esp. Fis. Quim., B,* **61**, 839–842 (1965).

To a sample solution containing 12–60 mg of nickel, add 1 ml of *M* **gluconic acid** and raise to pH 12.6 with a sodium hydroxide–sodium tetraborate buffer solution.[216] Dilute to 25 ml and read at 395 nm. For analysis of Monel metal, which is about 60% nickel, precipitate copper as the thiocyanate and mask iron with tartaric acid.

To read as the nickel **hydrosol**, dilute a solution containing 1–10 mg of nickel as the sulfate and 450 mg of gelatin to 95 ml.[217] Add 5 ml of 10% potassium hydroxide solution, set aside for 1 hour, and read.

The yellowish, water-insoluble 1 : 2 complex of nickel with **2′-hydroxyacetophenone hydrazone** at pH 10–10.5 is stable for 24 hours.[218] The complex is soluble in 75% acetone, has a maximum absorption at 425 nm, and obeys Beer's law up to 11.6 ppm of nickel. Cadmium and manganous ions are masked by citrate. There is serious interference by EDTA, thiocyanate, fluoride, palladous ion, cobaltic ion, bismuth, and cupric ion.

Nickel in ammoniacal leach liquors from copper–nickel sulfide concentrates forms a 1 : 2 complex with LIX-64N which is largely 2- and 3-**hydroxybenzophenone oximes.**[219] This is extractable at pH 8.5–9 with carbon tetrachloride, chloroform, cyclohexane, or benzene. The green extract is read at 590 nm. Beer's law is obeyed for 0.02–0.5 mg of nickel per ml of extract. There is interference by various metals, including uranium, aluminum, iron, and thorium, as well as by oxalate and EDTA. Copper and cobalt are tolerated up to 5% of the nickel; larger amounts of copper are masked with thiosulfate.

Mix a sample solution containing 3–40 μg of nickel with 15 ml of an acetate buffer solution for pH 3 and 1 ml of 0.1% solution of the zinc salt of **4-hydroxydithiobenzoic acid** in ethanol.[220] Extract with 10 ml of isoamyl alcohol. Wash with 10 ml, 10 ml, and 10 ml of 0.4% solution of sodium hydroxide in 2% sodium sulfate solution. Then extract with 10 ml of *N* hydrochloric acid and read the extract at 530 nm. Copper can be masked with thiourea.

1-Hydroxydithio-2-naphthoic acid forms a 2 : 1 complex with nickel in a solution containing at least 70% of ethanol.[221] Buffered at pH 4.75–9.9, the maximum absorption is at 510–520 nm and conforms to Beer's law for 0.2–3 μg of nickel per ml. There is interference by silver, lead, mercuric, cupric, cadmium, ferric, cobalt, and chromic ions.

The complex of nickel with the zinc salt of **bis(2-hydroxyethyl)dithiocarbamate** is read at 390 nm.[222] Add 1 ml of 20% citric acid solution to the sample solution. Adjust to pH 5.5–9.5 with *N* hydrochloric acid or 1 : 3 ammonium hydroxide. Add 1 ml of 0.5% solution of the reagent, and dilute to 100 ml. Read against a blank within 25 minutes.

The complex of nickel with **hydroxyiminoacetylacetone** as extracted with chloroform at pH 7.6–8.2 is read at 340 nm.[223]

[216]F. Bermejo Martinez and G. Branas Miguez, *Acta Cient. Compostelana* **7**, 195–202 (1970).
[217]Kaimierz Kapitanczyk and Stanislaw Kiciak, *Chem. Anal.* (Warsaw) **4**, 762 (1959).
[218]Hem Lata Ray, B. S. Garg, and R. P. Singh, *Curr. Sci.* **42**, 852–853 (1973).
[219]K. S. Koppiker and A. B. Chakraborty, *Indian J. Technol.* **13**, 529–531 (1975).
[220]G. Rudzitis, S. Pastare, and E. Jansons, *Zh. Anal. Khim.* **25**, 2407–2413 (1970).
[221]B. Janik and H. Gawron, *Mikrochim. Acta* **1967**, 843–846.
[222]Hitoshi Yoshida, Masahiro Yamamoto, and Seiichiro Hikime, *Jap. Anal.* **11**, 197–201 (1962).
[223]U. B. Talwar and B. C. Haldar, *Indian J. Chem.* **9**, 593–596 (1971).

Mix 2 ml of 5×10^{-4} *M* methanolic **2-(2-hydroxy-5-methoxyphenylazo)-4-methylthiazole** with a sample solution containing less than 10 μg of nickel at pH 7 and extract with 10 ml of isoamyl alcohol.[224] Read the 1:2 metal-ligand complex at 620 nm.

Nickel, copper, and palladium form 1:1 complexes with **2-(3-hydroxy-3-phenyltriazeno)benzoic acid.**[225] That of nickel is read at 410 nm for 0.25–2 ppm at pH 6.8–8.3.

3-Hydroxypicolinaldehyde, which is α,α′-azinodi(3-hydroxy-2-picoline), complexes in dimethylformamide solution with nickel and cobalt at pH 4.5 in acetate buffer solution for pH 4.5.[226] Nickel is read at 480 nm and cobalt at 540 nm against a reagent blank. EDTA interferes seriously.

A ternary complex is formed by 2.57–25.7 μg of nickel with 10 ml of m*M* **8-hydroxyquinoline-5-sulfonate** and 3 ml of 5 m*M* zephiramine at pH 7.8 as established with 0.05 *M* phosphate buffer.[227] Dilute to 50 ml, extract with 10 ml of chloroform, and read at 400 nm.

When an acidic sample at pH 6–8 is passed through a column of powdered **iron**, the nickel liberates an equivalent of ferrous ion.[228] After oxidation to ferric ion, it is determined at 510 nm by 1,10-phenanthroline or at 458 nm with salicylhydroxamic acid. The same reaction is given by cobalt, lead, cupric, silver, or mercuric ions.

Nickel, zinc, cobalt, and copper in a solution containing ammonium carbonate or sulfate are extracted with 0.5% solution of **Kelex 100**, which is a β-alkaline-8-hydroxyquinoline in Solvesso 10 containing 10% of isodecyl alcohol.[229] The metals are back-extracted with 15% sulfuric acid.

Meditan, which is 1-(5-mercapto-1,3,4-thiadiazol-2-ylazo)-2-naphthol, forms a 1:2 complex with 0.06–0.94 μg of nickel per ml.[230] The complex is stable for 72 hours. It is read at 638 nm and pH 6.3–10.6. There is interference by cobalt, ferric, and chromic ions.

Addition of **4-(mercaptoacetamido)benzenesulfonate** to a nickel solution produces a complex that is stable for 3 hours.[231] Adjust to pH 6–7.5 with a tartrate–hydrochloric acid buffer solution and read at 390 nm. Beer's law is followed for 4–15 ppm of nickel. There is interference by cobalt, iron, palladium, and molybdate.

A stable complex of nickel with **mercaptoacetic acid** is formed at pH 7–9 for reading at 522 nm over the range $2\text{–}20\times10^{-4}$ *M*.[232] Chromium, uranium, and tungsten form somewhat similar complexes.

From 0.4 to 16 ppm of nickel is determined by extraction as the **2-mercaptobenzothiazole** complex at pH 7.7–8.1.[233] The maximum absorption is at 750 nm. Iron, cobalt, copper, and bismuth are removed by an anion exchange column. Lead is

[224]Tadashi Yanagihara, Nobuhisa Matano, and Akira Kawase, *Jap. Anal.* **8**, 14–17 (1959).
[225]A. K. Majumdar and D. Chakraborti, *Z. Anal. Chem.* **257**, 33–36 (1971).
[226]A. Garcia de Torres, M. Valcarcel, and F. Pino Perez, *Anal. Chim. Acta* **79**, 257–263 (1975).
[227]Tomishito Kambara, Masao Sugawara, and Kiyoshi Hasebe, *Jap. Anal.* **19**, 1239–1244 (1970).
[228]Eugeniusz Kloczko, *Chem. Anal.* (Warsaw) **16**, 1291–1298 (1971).
[229]G. M. Ritcey and B. H. Lucak, *C.I.M. Bull., Canada* **68** (754), 105–113 (1975).
[230]Eugenia Domagalina and Stanislaw Zarbeda, *Chem. Anal.* (Warsaw) **15**, 1227–1231 (1970).
[231]J. P. C. Jaimni, M. R. Bhandari, and N. C. Sogani, *J. Proc. Inst. Chem. India* **37** (4), 162–166 (1965).
[232]V. M. Bluchar, *Nature* **191**, 489–490 (1961).
[233]E. G. Walliczek, *Talenta* **11**, 573–579 (1964).

precipitated by hydrogen sulfide at pH 1. Fluoridation of alkali metals is necessary to prevent them from complexing with the reagent.

β-Mercaptohydrocinnamic acid forms a 2:1 complex with nickel and a 3:1 complex with cobalt.[234] Either can be read up to 100 ppm. At pH 8–9.5 nickel is read at 520 nm, cobalt at 560 nm. With **β-mercaptocinnamic acid** at pH 11.5–13, complexes of the same ratios are formed. At 10–100 ppm nickel is read at 495 nm, cobalt at 635 nm. The times for complete reaction are 3 and 15 minutes, respectively. Thereafter they are stable for 15 hours. The absorption is lowered by zinc, iron, and lead.

To a sample of nickel solution, add **β-mercaptohydrocinnamic acid isopentyl ester**, extract with chloroform, and read 0.4–7 μg of nickel per ml of extract at 455 nm.[235a]

Above pH 6.5 nickel forms a 1:2 complex with **4-methoxy-2-(2-thiazolylazo) phenol** with more than 15-fold concentration of the ligand for reading at 625 nm.[235b]

At pH 4–8.5 nickel forms a 1:2 complex with ***N*-methylanabasine-α′-azo-cresol**, which is 2-[5-(1-methyl-2-piperidyl)-2-pyridylazo]-*p*-cresol.[236] The complex absorbs at 585 nm. Beer's law is obeyed up to 2 μg of nickel per ml. Cobalt, vanadium, EDTA, and citrate interfere. The reagent is more sensitive than dimethylglyoxime, and less sensitive than dithizone or 1-(2-pyridylazo)-2-naphthol.

***N*-Methylanabasine-α′-azodiethylaminophenol**, which is diethylamino-2-[3-(1-methyl-2-piperidyl)-2-pyridylazo]phenol, forms a 2:1 complex with nickel as well as with cobalt, read in the region of 550–590 nm.[237] It also forms 1:1 complexes with vanadium, copper, indium, and bismuth. The reagent absorbs at 440 nm.[238] It is also known by the trivial name MAAF II. Iron can be masked with fluoride and aluminum; molybdenum, tungsten, and chromium with 1% solution of sodium tartrate.

The more complex **4-(*n*-methyl-2-pyridylazo)resorcinol** derivatives of nickel function, like 4-(2-pyridylazo)resorcinol, and offer little advantage.[239] Similar derivatives are formed with cobalt, copper, zinc, and uranyl ion.

Methylthiourea sulfate complexes with nickel for maximum absorption at 495–510 nm and pH 9.14–9.51.[240] Beer's law is followed for 1–7 μg/ml. A fortyfold excess of reagent is required, and development takes 1 hour. Cobalt reacts similarly.

Nickel can be developed as **molybdenum blue**.[241] Nickel complexes with molybdophosphoric acid at pH 4.3. After masking excess of the acid with sodium citrate, the nickel complex is reduced with stannous chloride and the blue developed is read at 695 nm. Temperature and time between addition of the sodium citrate and

[234] Hisashi Tanana, Yukio Sugiura, and Akira Yokoyama, *Jap. Anal.* **17**, 129–134 (1968).
[235a] A. I. Busev and D. Kh. Vin', *Zh. Anal. Khim.* **21**, 1082–1088, 1311–1318 (1966).
[235b] V. Kuban, L. Sommer, and J. Havel, *Colln. Czech. Chem. Commun.* **40**, 604–633 (1975).
[236] E. L. Krukovskaya, Sh. T. Talipov, and A. D. Volodina, *Izv. Akad. Nauk Kaz. SSR, Ser. Khim.* **19GD** (1974).
[237] A. E. Martirosov, Sh. T. Talipov, and R. Kh. Dzhiyanbaeva, *Tr. Tashk. Gos. Univ.* **1968** (323), 73.
[238] Sh. T. Talipov, R. Kh. Dzhiyanbaeva, and A. E. Martirosov, *Izv. Akad. Nauk Kaz. SSR, Ser. Khim.* **1968** (6), 1–4.
[239] W. J. Geary and F. Bottomly, *Talenta* **14**, 537–542 (1967).
[240] S. K. Siddhanta and S. N. Banerjee, *J. Indian Chem. Soc.* **35**, 547–552 (1958).
[241] R. L. Heller and J. C. Guyon, *Talenta* **17**, 865–871 (1970).

annous chloride must be controlled. The method will detect 0.1 ppm of nickel n, and Beer's law is followed up to 5 ppm.

Nickel forms a water-insoluble 1:3 complex with **3-nitroso-4-hydroxy-5:6-ben-coumarin**, which is 1-hydroxy-2-nitrosonaphtho[2,1-b]pyran-3-one.[242] The com-ex is soluble in aqueous acetone for reading at 395 nm and pH 4.9–10.2. Beer's w is best applied at 0.29–1.3 ppm of nickel.

Nickel can be separated from numerous ions by thin-layer chromatography on owdered cellulose by development with 1:3 ratio 1:1 hydrochloric acid–ace-ne.[243] The deposit of 2–10 μg of nickel is extracted with 5 ml of water and treated ith 2 ml of **morpholinium 3-oxapentamethylenedithiocarbamate**. The precipitated omplex is extracted with 4.5 ml and 4.5 ml of chloroform and read at 395 nm. uoride masks titanium and aluminum, but manganese in large excess interferes.

The maximum absorption of nickel as the **myristate** or **palmitate** in pyridine is ad at 650 nm.[244]

Reading **nickel ammonium sulfate** in the ultraviolet region at 215 nm is 100 times ore sensitive than reading it at 582 nm.[245a]

For nickel in tellurium by **1-nitroso-2-naphthol** after removal of cobalt, see obalt.

Mix 5 ml of sample solution containing 0.25–3.75 μg of nickel with 5 ml of etate buffer solution for pH 5.35 and 5 ml of 0.18 m*M* **oxyquinolinethiazo**, which 4,5,6,7-tetrahydro-2-(8-hydroxyquinolin-5-ylazo)benzothiazole.[245b] Dilute to 25 l with ethanol and read. Indium and cobalt interfere.

A mixed complex of nickel with **1,10-phenanthroline** and **thiosalicylate** is ex-acted with chloroform at pH 5–6 and read at 365 nm.[246] The complex can be termined along with cobalt and iron.

The complex of nickel with **1,10-phenanthroline** adds tetrabromo-, tetraiodo-, or chlorotetraiodofluorescein, extractable at pH 9 with chloroform.[247] The maxi-um absorption is at 540 nm or in 1:1 chloroform-acetone fluorescence at 580 nm. uorescence will determine 0.05–10 μg in 10 ml, one order of magnitude more nsitive than spectrophotometry.

For nickel, adjust to pH 9.3 with a buffer solution of ammonium hydroxide and nmonium chloride. Add 0.1% solution of **phthalaldehyde bisthiosemicarbazone** in methylformamide and read at 450 nm.[248a] The complex can be extracted with loroform at pH 8.8 with the same type of buffer solution. Beer's law is followed to 4 ppm of nickel. Cobalt complexes similarly for reading at 385 or 410 nm.

For analysis of an industrial catalyst containing nickel and cobalt reflux in aqua gia for 3 hours and dilute.[248b] Nickel and cobalt form a 1:2 complex with **colinaldehyde 4-phenyl-3-thiosemicarbazone** at pH 8–9 in an acetate medium. This read at 390 nm in dimethylformamide. Convert to a hydrochloric acid medium at I 1 and read cobalt at 430 nm. Determine nickel by difference. Beer's law is

Nitin Kohli and Rajendra Pal Singh, *Curr. Sci.* **42**, 142–143 (1973).
E. Gagliardi and G. Pokorny, *Mikrochim. Acta* **1966**, 577–586.
Wahid U. Malik and Rizwanul Hague, *Z. Anal. Chem.* **189**, 179–182 (1962).
aN. M. Paunovic and M. M. Paunovic, *Bull. Sci., Cons. Acad. RPF Yougosl.* **5**, 99–100 (1960).
bJ. Kalina, J. Putnins, and E. Gudriniece, *Izv. Akad. Nauk. Latv. SSR, Ser. Khim.* **1974**, (2), 194–198.
I. V. Pyatnitskii and G. N. Trochinskaya, *Zh. Anal. Khim.* **28**, 704–708 (1973).
D. N. Lisitsyna and D. P. Shcherbev, *ibid.* **28**, 1203–1205 (1973).
aM. Rueda Rueda and J. Munez Leiva, *Quim. Anal.* **29**, 122–128 (1975).
bJ. L. Gomez Ariza and J. M. Cano Pavon, *Analyt. Lett.* **9**, 677–686 (1976).

applicable for 0.8–1.5 ppm of nickel. Alternatively extract the nickel complex wi chloroform and read 0.8–2 ppm at 400 nm. In determining 1 ppm of nickel there interference by more than 1 ppm of silver, lead, iron, cupric ion, cobalt, zin pentavalent vanadium, mercury, cadmium, bismuth, tin, auric ion, or pallado ion.

The absorption of the nickel complex with **picolinaldehyde thiosemicarbazone** constant for pH 7–11 and of the copper complex for pH 8.9–10.7.[249] Beer's law followed for 5 ppm of nickel or 7 ppm of copper. Interference by the reagent minimal at 415 nm, but the sensitivity is better at 360–400 nm. For nickel b difference, read copper and nickel at pH 9.6 and subtract copper as read at pH 2.

The complex of nickel with **picolinaldehyde 2-quinolylhydrazone** in the presen of mercaptoacetic acid at pH 3–4 is extractable with chloroform.[250a] Read at 4 nm, this will determine 0.1–1 ppm of nickel.

The nickel chelate of **2-picolylamine**, which is 2-aminomethylpyridine, at p 7–8.5 is read as a 1:3 metal-ligand complex at 535 nm.[250b] Beer's law is obeyed f 1.8–14.4 m*M* nickel.

Adjust a sample solution containing up to 150 μg of nickel to pH 8.1 with buffer solution. Add 15 ml of m*M* sodium **purpurinsulfonate** and dilute to 50 ml.[2 Set aside for 30 minutes and read at 520 nm. Cobalt interferes.

The complex of nickel with **pyridine** is read at 370–380 nm or at 620–640 nm.[2 Cobalt does not interfere.

Nickel forms a pale yellow 1:2 chelate with **pyridine-2,6-dialdoxime** havi maximum absorption at 330 nm and a pH of 3–6.7.[253] The reagent forms 1 chelates with cobalt and manganese and a 1:1 chelate with copper.

Nickel is read as a 1:2 complex with **pyridine-2,6-dicarboxylic acid** at p 2.5–10.5.[254] Chromic and ferric ions reduce the accuracy.

The complex of **3-(2-pyridylazo)chromotropic acid** at pH 7.5 with nickel has maximum absorption at 570 nm, that of cobalt at 640 nm.[255] In both cases cyani ion and EDTA must be absent. Beer's law is followed for 0.125–0.8 ppm of nick preferably not more than 0.25 ppm, and for 0.16–1.2 ppm of cobalt. In addition ions forming colored complexes with the reagent, the following ions must removed or masked: barium, strontium, bismuth, antimony, lead, and mercuric.

The sampling and subsequent separation of nickel by column chromatograp from cobalt, copper, iron, and zinc in niobium samples appears under determin tion of copper in niobium by **1-(2-pyridylazo)-2-naphthol.**[256]

4-(2-pyridylazo)thymol complexes with nickel at pH 4.5–7.5.[257] To the samp add 5 ml of acetate buffer solution for pH 5 and 1.5 ml of 0.2% ethanolic reage

[249] J. M. Cano Pavon, J. Vazguez Allen, D. Perez Benedito, and F. Pino Perez, *Inf. Quim. Anal. Pura A Ind.* **25** (5), 149–158 (1971); J. M. Cano Pavon and D. Perez Benedito, *ibid.* **27**, 20–30 (1972).

[250a] S. P. Singhal and D. E. Ryan, *Anal. Chim. Acta* **37**, 91–96 (1967).

[250b] Tong-ming Hseu, Yen-shiang Tsai, and Chi-wang Cheng, *J. Chin. Chem. Soc.* (Taipai) **22**, 299– (1975).

[251] M. Roman Ceba and E. Alvarez-Manzaneda, *Quim. Anal.* **29**, 253–260 (1975).

[252] L. V. Nazarova, *Uch. Zap., Kishinev. Gos. Univ.* **56**, 35–37 (1960).

[253] Saswati P. Bag, Quintus Fernando, and Henry Freiser, *Anal. Chem.* **35**, 719–722 (1963).

[254] G. Den Boef and H. Poppa, *Talenta* **15**, 1058–1060 (1968).

[255] A. K. Majumdar and A. B. Chatterjee, *ibid.* **13**, 821–828 (1966).

[256] O. Grossmann and H. Grosse-Ruyken, *Z. Anal. Chem.* **233**, 14–20 (1968).

[257] T. M. Mirzakasimov, K. Z. Rakhmatullaev, and Sh. T. Talipov, *Uzb. Khim. Zh.* **1968** (5), 29–33.

Heat at 100° for 5 minutes, cool, and add 2 ml of acetone. Dilute to 25 ml and extract with 6 ml and 4 ml of chloroform. Filter the combined extracts, adjust to 10 ml, and read at 610 nm. Beer's law is followed for 0.25–3.25 μg of nickel per ml. Sodium tartrate will mask zinc, cadmium, pentavalent vanadium, and hydrolyzed ions. Thiourea masks copper.

The complex with **4-(2-quinolylazo)thymol** is developed analogously at pH 5.5–7.5. Use 5 ml of acetate buffer solution for pH 6. Read at 640 nm for 0.2–2.5 μg of nickel per ml.

Similarly, ***N*-methylanabasine-α′-azothymol** is developed at pH 6.5–8.5. Use 5 ml of acetate buffer solution for pH 7. Read at 620 nm for 0.25–3.25 μg of nickel per ml.

The yellow complex of nickel with **pyrrole-2-aldehyde-ethylenediimine**, which is 1,2-bis(pyrrol-2-ylmethyleneimino)ethane, at pH 11.4 when extracted with chloroform, has a maximum absorption at 395 nm. Beer's law applies up to 4 μg/ml. There is interference by manganese, aluminum, copper, iron, zirconium, tin, chromium, and cobalt.

Nickel and **quinoxaline-2,3-dithiol** form maxima at 650 and 598 nm at an optimum of 0.9–2.3 ppm.[258] Color develops rapidly at room temperature in 80% dimethylformamide. There is interference by cobalt, palladium, platinum, copper, manganese, iridium, and silver.

Nickel at 6–1500 ppm can be determined as the 1:2 complex by extraction at pH 8–11 with **resacetophenone** in cyclohexanone.[259] The maximum absorption is at 595 nm.

Nickel forms a 1:2 complex with **salicylaldehyde isonicotinoylhydrazone**.[260] This is almost insoluble in water, but it obeys Beer's law in 1:1 1,4-dioxan–water for 0.5–3 ppm of nickel at 390–400 nm. This reagent also complexes with cadmium, manganese, cobalt, and zinc.

For nickel in bronze, dissolve 0.1 gram of sample in nitric acid, add sulfuric acid, and evaporate to dryness.[261] Take up in water, remove copper by electrolysis, and dilute to 200 ml. Take a 5-ml aliquot, add 0.2 gram of tartaric acid, adjust to pH 3–5, and extract nickel by shaking for 2 minutes with 0.025% solution of **salicylaldoxime** in chloroform. Read at 365 nm. The extraction is also quantitative at pH 8.5–9.4 as maintained by an ammonium citrate buffer solution.[262] Zinc, cobalt, and magnesium interfere. Ferric ion can be masked with pyrophosphate.

The 1:1 complex of nickel with **2-(salicylideneamino)benzenethiol** has a maximum absorption at 424 nm in neutral or basic solution.[263] Extracted into benzene the maximum is at 429 nm. Shake a 5-ml sample containing less than 40 μg of nickel with 1 ml of *M* ammoniacal buffer for pH 9 and 10 ml of 0.01% solution of the reagent in benzene for 10 minutes. Alternatively, buffer a sample solution containing up to 130 μg of nickel to pH 10, add 7.5 ml of 0.01% ethanolic reagent, and dilute to 25 ml. There is interference by cobalt, copper, ferric ion, lead, zinc,

[258] R. R. Annans, *Dis. Abstr.* **24**, 1813 (1963).
[259] K. S. Bhatki, A. T. Rane, and M. B. Kabadi, *Bull. Chem. Soc. Jap.* **36**, 1689–1693 (1963).
[260] G. S. Vasilikiotis and T. A. Kouimtzis, *Microchem. J.* **18**, 85–94 (1973).
[261] K. N. Bagdasarov, O. D. Kashparova, and M. S. Chernov'yants, *Zavod. Lab.* **37**, 1046–1047 (1971).
[262] Yoshikazu Yamamoto, Kazumasa Ueda, and Shunzo Ueda, *J. Chem. Soc. Jap., Pure Chem. Sect.* **89**, 288–291 (1968).
[263] Hajime Ishii and Hisahiko Einaga, *ibid.* **90**, 175–180 (1969).

and citrate with both technics. Cadmium and pentavalent vanadium also interfere with the reading in aqueous solution. Chromic and tetravalent vanadium ions can be masked with M sodium acetate.

To a sample solution containing less than 30 μg of nickel, add 1 ml of ammonium chloride–ammonium hydroxide buffer solution for pH 9 and dilute to 30 ml.[264] Shake for 10 minutes with 0.01% solution of **2-salicylideneaminophenol** in chloroform, dry the organic phase with sodium sulfate, and read at 418 nm. There is interference by cobaltous, cuprous, cupric, chromous, lead, manganous, and citrate ions.

The complex of nickel with **solochrome red ERS** in a solution buffered at pH 5.8 with sodium acetate is read at 510 nm.[265] Sodium fluoride masks aluminum, titanium, and zirconium. There is interference by ferrous, cupric, and cobalt ions, EDTA, and some organic acids.

For determination of 1.2–12 μg of nickel in 10 ml of sample, add 5 ml of 0.2 mM **4-(5-sulfo-2-thiazolylazo)-2-nitroresorcinol**, which is 2-(2, 4-dihydroxy-3-nitrophenylazo)thiazole-5-sulfonic acid.[266] Dilute to 25 ml with acetate buffer solution for pH 6. Read at 510 nm against an undeveloped portion of the sample.

To a sample solution containing 0.18–1.6 mg of nickel, add 10 ml of buffer solution for pH 5.5–8.[267] Mix with 10 ml of 0.15 M **2-thenoyltrifluoroacetone** in acetone. Then shake for 5 minutes with a mixture of 5 ml of acetone and 5 ml of benzene. Dilute the organic layer to 25 ml with benzene and read at 420 nm against a reagent blank. Copper is masked with thiosulfate ion at pH 6. Cobalt is preextracted as its thiocyanate complex with amyl alcohol–ether. Manganous, chromic, and hexavalent uranium ions must be substantially absent.

A mixture of equal volumes of 0.0016 M **2-thenoyltrifluoroacetone** in benzene and 0.7 M **isoquinoline** in benzene extracts nickel from a phosphate solution buffered to pH 7 as a 1:2:2 complex.[268] This is read at 390 nm. Cupric, cobalt, manganese, and zinc ions must be absent.

Dissolve 0.05 gram of uranium-based alloy in aqua regia and evaporate to dryness.[269a] Take up with 3 ml of hydrochloric acid and 10 ml of water. Dilute to 100 ml and take a 5-ml aliquot. Add 0.4 ml of 10% sodium potassium tartrate solution and dilute to 10 ml with 2 N ammonium hydroxide. Add 1 ml of 0.1 mM **5-(2-thiazolylazo)-*p*-cresol** in chloroform and 9 ml of chloroform. Shake, and read the extract at 610 nm against a reagent blank. Beer's law is followed for 0.1–2 μg of nickel per ml. Iron is masked with fluoride. Cobalt interferes.

At pH 8–10 nickel forms a 1:2 complex with **4-(2-thiazolylazo)-3-hydroxynaphthalene-2,7-disulfonic acid** in the presence of fourfold excess of the reagent.[269b] This is read at 596 nm or less desirably at 550 nm for 0.1–1.7 μg/ml. Copper and zinc interfere. Ferric ion and aluminum are masked with tartrate. Fivefold cobalt and manganous ion and tenfold chromium are tolerated. The sensitivity is six times that of dimethylglyoxime.

[264]Hajime Ishii and Hisakiko Einaga, *Bull. Chem. Soc. Jap.* **42**, 1558–1561 (1969).

[265]G. E. Janauer and J. Korkisch, *Z. Anal. Chem.* **177**, 407–412 (1960).

[266]L. P. Adamovich, A. L. Gershuns, O. O. Oliinik, and L. I. Degtyar'ova, *Visn. Kharkiv. Univ.* **1972** (84), *Khim.* (3), 66–68.

[267]Anil K. De and M. Syedur Rahaman, *Anal. Chim. Acta* **27**, 591–594 (1962).

[268]Hideo Akaiwa, Hiroshi Kawamoto, and Minoru Hara, *Jap. Anal.* **17**, 183–187 (1968).

[269a]S. I. Gusev, M. V. Zhvakina, I. A. Kozhevnikova, and L. S. Mal'tseva, *Zavod. Lab.* **42**, 19–20 (1976).

[269b]A. I. Busev, T. N. Zholondkovskaya, L. S. Krysina, and O. A. Barinova, *Zh. Anal. Khim.* **29**, 1758–1761 (1974).

The 1:2 insoluble complex of **1-(2-thiazolylazo)-2-naphthol** with nickel in an acetate-borate buffer solution for pH 7 can be extracted with chloroform and read at 595 nm.[270] The complex can be dispersed at pH 4.7–10 with Triton X-100, a nonionic surfactant.[271a] To a sample solution containing less than 110 μg of nickel, add 2 ml of 0.5 *M* sodium potassium tartrate, 1 ml of 0.5 *M* potassium pyrophosphate, 0.95 ml of *N* ammonium hydroxide at pH 9.2, 20 mg of *N*-(dithiocarboxy)glycine, and 2 ml of a reagent consisting of 0.1 gram of 1-(2-thiazolylazo)-2-naphthol in 80 ml of water containing 20 grams of Triton X-100. Set aside for 10 minutes and adjust to pH 6.7 with *M* ammonium citrate. Dilute to 50 ml and read at 595 nm against a reagent blank. Beer's law is obeyed for 11–110 μg of nickel. The *N*-(dithiocarboxy)glycine, tartrate, and pyrophosphate mask zinc, cobalt, copper, bismuth, and vanadium.

At pH 8.8–9.4 nickel forms a red complex with **thiobenzoylacetone**, which is 3-mercapto-1-phenyl-but-2-en-1-one.[271b] Extracted with benzene it is read at 500 nm for 0.5–10 μg/ml. There is serious interference by lead, copper, zirconium, mercury, palladium, cobalt, cyanide, and EDTA.

The complex of nickel with **thiodibenzoylmethane**, which is 1,3-diphenyl-3-thiopropan-1-one, at pH 8.5 is extracted with cyclohexane and read at 430 nm.[272] It is applicable to 0.1–1 μ*M*. Salicylaldoxime masks cobalt. Palladium interferes, as does tenfold copper, tin, and bismuth.

Adjust a sample solution containing 0.1–1 mg of nickel to pH 10–11 with ammonium hydroxide.[273] Shake with 10 ml, 10 ml, 5 ml, and 5 ml of 0.2 *M* **2,2′-thiodiethanethiol** in chloroform. Extract with 10 ml and 5 ml of chloroform. Dilute the combined extracts to 50 ml with chloroform and read at 520 nm against chloroform. Cobalt and palladium form interfering colored complexes. Copper suppresses color development.

Take a sample acid with hydrochloric or nitric acid containing 5 μg to 3 mg of nickel.[274] Add 1 ml of 0.5% solution of **1-thioglycerol**, which is 3-mercaptopropane-1,2-diol, for each 250 μg of nickel in the sample. Adjust to pH 8.6 with 0.04% solution of sodium hydroxide and add 5 ml of a borate buffer solution for that pH. Dilute to 25 ml and read at once at 330, 410, or 540 nm. The complex is 1:2 nickel-reagent, is stable for only a short time, and is unaffected by change of pH 8–10. To avoid interference, the nickel should have been isolated by a preliminary extraction as the dimethylglyoxime complex.

Take a sample containing more than 0.05 mg of nickel; it must be free of ammonium salts.[275a] Add 10 ml of phosphate buffer solution for pH 8, 1 ml of 0.1 *M* mixed **thionaphthenic acids** (mean molecular weight 227) in carbon tetrachloride, and 4 ml of carbon tetrachloride. Shake for 1 minute, filter the extract, and read at 470 nm against 0.02 *M* reagent in carbon tetrachloride. Beer's law holds for 0.005–0.1 mg of nickel per ml. Cobalt forms a similar complex absorbing at 465 nm.

[270]Akira Kawase, *Jap. Anal.* **12**, 810–817 (1963).

[271a]Hiroto Watanabe and Hideki Matsunaga, *ibid.* **25**, 35–39 (1976); Jun'ichiro Miura, Hideki Ishii, and Hiroto Watanabe, *Jap. Anal.* **25**, 808–809 (1976); Hiroto Watanabe and Jun'ichiro Miura, *ibid.* 667–670.

[271b]M. V. R. Murti and S. M. Khopkar, *Indian J. Chem. A* **14**, 455–456 (1971).

[272]E. Uhlemann and B. Schuknecht, *Anal. Chim. Acta* **63**, 236–240 (1973).

[273]J. Segall, M. Ariel, and L. M. Shorr, *Analyst* **88**, 314–317 (1963).

[274]F. Bermejo Martinez and E. Mendez Domeneck, *Inf. Quim. Anal. Pura Apl. Ind.* **22** (3), 85–99 (1968).

[275a]A. A. Alekperova, R. A. Alekperov, and Yu. A. Zolotov, *Zh. Anal. Khim.* **25**, 2283–2286 (1970).

For determination of nickel as a ternary 1:2:2 nickel-pyridine-**thiosalicylic acid** complex a single extraction with chloroform at pH 5.5–7 gives 85–90% yield.[275b]

The complex of nickel with the **thiosemicarbazone** of **benzothiazol-2-yl phenyl ketone** is read at 446 nm.[276] Mix 2 ml of sample solution, 0.5 ml of ammoniacal buffer solution for pH 8, 1 ml of 0.1% ethanolic color reagent, and 1.5 ml of ethanol. Read after 30 minutes. Beer's law is followed at 0.2–7.5 μg of nickel. The reagent also reacts with copper, cobalt, cadmium, and zinc. By reading at pH 2.5, 4, and 6, nickel, cadmium, and zinc can be determined by simultaneous equations.

Add an excess of solution of **thiovioluric acid**, which is 5-hydroxyimino-2-thiobarbituric acid, to the sample solution and adjust to pH 8.05–9.1 with sodium hydroxide solution.[277] Dilute, and read an aliquot against a reagent blank. Beer's law is followed up to 3 ppm of nickel. There is interference by tartrate, citrate, thiourea, phosphate, pyrophosphate, EDTA, nitrilotriacetic acid, cobalt, iron, copper, palladium, rhodium, platinum, and ruthenium. Also see Cobalt in Steel by this reagent.

In a solution containing nickel and cobalt, oxidize cobaltous ion to cobaltic with potassium persulfate.[278] Then form the cobaltic and nickelous complexes of **triethylenediamine** at pH 11.3. Read nickel at 875 nm, cobalt at 463 nm. The concentration of nickel must not exceed 10^{-2} *M*, and for accuracy, the cobalt-nickel ratio should not exceed 10:1.

As a reagent, dissolve 5 grams of **triethylenetetramine** disulfate in a minimal amount of 20% sodium hydroxide solution and dilute to 100 ml. To a sample solution containing 0.25–18 mg of nickel, add 3 ml of the reagent.[279] Adjust to pH 8.8 by adding *N* sodium hydroxide. Add 10 ml of sodium hydroxide–boric acid buffer solution for pH 8.8 and dilute to 25 ml. Read the 1:1 complex at 355 nm. If the copper content is less than 40 μg/ml, it can be ignored.

Nickel buffered at pH 4.6 is complexed and extracted by **1,1,1-trifluoro-4-mercaptopent-3-en-2-one** in hexane.[280] Read at 251 nm, the extract conforms to Beer's law up to 4 ppm of nickel. Interference by cupric ion is avoided by precipitating it with hydrogen sulfide. Seriously interfering ions are mercurous, mercuric, and cobaltous.

A 1:1 complex of nickel with **trihydroxyfluorone** is formed at an optimum pH of 9.5 and read at 545 nm.[281]

Triphenylmethylarsonium bis(dithiooxalato)nickelate at pH 4–5 is extracted with 55:45 acetophenone-chloroform and read at 505 nm.[282] Cobalt, copper, and ferric ion are preextracted as their thiocyanates at pH 3–4.

A complex of nickel ion is formed at pH 9–11 with **zinc tetramethylenediselenocarbamate**, which is zinc pyrrolidine-1-carbodiselenate.[283] The complex is extractable with chloroform. Copper, cobalt, and bismuth form similar complexes.

[275b] I. V. Pyatnitski and G. N. Trochinskaya, *Ukr. Khim. Zh.* **41**, 262–268 (1975).
[276] Toyozo Uno and Sumiyuki Akihama, *Jap. Anal.* **10**, 941–945 (1961).
[277] R. S. Chawla and R. P. Singh, *J. Indian Chem. Soc.* **52**, 169 (1975).
[278] C. Caullet and B. Villecroze, *Chim. Anal.* (Paris) **46**, 308–312 (1964).
[279] F. Bermejo-Martinez and J. A. Rodriquez-Vasguez, *Acta Cient. Compostelana* **7**, 135–146 (1970); *Rev. Univ. Ind. Santander* **1970** (3), 33–39.
[280] R. S. Barratt, R. Belcher, W. I. Stephen, and P. C. Uden, *Anal. Chim. Acta* **58**, 107–114 (1972).
[281] O. A. Tataev, G. N. Bagdasarov, S. A. Akmadev, Kh. A. Mirzaeva, Kh. G. Buganov, and E. A. Yarysheva, *Izv. Sev-Kavkaz. Nauch. Tsent. Vyssh. Shk., Ser. Estestv. Nauk* **1973** (2), 30–32.
[282] A. J. Cameron and N. A. Gibson, *Anal. Chim. Acta* **24**, 360–364 (1961).
[283] Janusz Terpilowski and Pawel Ladogorski, *Farm. Pol.* **30**, 1011–1015 (1974).

CHAPTER TWENTY
COBALT

Although cobalt has bivalent and trivalent forms, the former is so much more common that the term "cobalt" usually means the cobaltous ion. Reagents that have developed into major methods for cobalt are nitroso-R salt and 2-nitroso-1-naphthol. For cobalt in cement, see Nickel by Dimethylglyoxime. For cobalt ion in fluoborate alloy-plating solutions, see Nickel Ion. For cobalt in beer by cuprethol, see Copper by that reagent. For cobalt in mollusks there is loss in dry-ashing, even at 110°, but none in wet-ashing with hydrogen peroxide.[1]

For cobalt as read as the biuret complex along with copper and nickel, and chromium as the chromate, solved by simultaneous equations, see Copper. For cobalt by 8-hydroxyquinoline, see Nickel by that reagent. For cobalt by diethylene triamine, see Nickel with that reagent. For cobalt by 3-(2-pyridylazo)chromotropic acid, see Nickel. For cobalt by oxamidoxime, see Nickel by that reagent. For cobalt by *N,N'*-bis(2-aminobenzylidene)ethylenediamine, see Nickel. For the complex of cobalt with cuprizone, also known as dicyclohexanone oxalyldihydrazone, see Nickel. For cobalt by oxine blue, see Copper.

In determination of cobalt gravimetrically as cobalt ammonium phosphate, the residual cobalt in the filtrate is determined by nitroso-R salt at 545 nm.[2a]

Extraction with molten naphthalene of complexes with 8-hydroxyquinoline is applicable at varying pH levels to cobalt, zinc, magnesium, cadmium, bismuth, copper, nickel, molybdenum, uranium, aluminum, iron, palladium, and indium.[2b]

ALIZARIN RED S

The measurement of catalytic oxidation of alizarin red S (mordant red 3) by hydrogen peroxide will determine cobalt at 0.05 ng/ml. Rather than mask impuri-

[1] P. Etrohal, S. Lulic, and O. Jelisaveic, *Analyst* **94**, 678–680 (1969).
[2a] A. G. Foster and W. J. Williams, *Anal. Chim. Acta* **22**, 538–546 (1960).
[2b] T. Fujinaga, Masatada Satake, and Masaaki Shimizu, *Jap. Anal.* **25**, 313–318 (1976).

ties, the cobalt can be isolated by ion exchange resin. The reading is of the decrease in absorption as the dyestuff is destroyed.

Procedure. Mix 2 ml of 0.01% solution of the dyestuff, 2 ml of a buffer solution consisting of 1% of boric acid in 2 N sodium hydroxide, and 5 ml of the sample solution containing 0.2–20 ng of cobalt.[3] Read at once. Incubate at a controlled temperature such as 50° for 2 ng of cobalt. After exactly 10 minutes, add 1 ml of 1% EDTA solution and read at 510 nm.

Water.[4] To the sample solution, add 5 ml of 0.2 M disodium phosphate and titrate to pH 11 with 8% sodium hydroxide solution. Add 1 ml of 1.5×10^{-3} M alizarin red S and dilute to 20 ml. Add 1 ml of 4.41 M hydrogen peroxide and dilute to 25 ml. Record the reading at 15-second intervals for 3 minutes.

By Ion Exchange. Prepare a 3×40 mm column of 100-200 mesh Amberlite CG400 ion exchange resin in chloride form. Adjust the cobalt sample to 8 M with hydrochloric acid, pass it through the column, and discard the eluate. Elute the cobalt with 50 ml of water and evaporate this eluate to 15 ml. Treat as above in full from "To the sample solution,… ."

4–ANALINO–1–PHENYL–1,2,4–TRIAZOLIUM CHLORIDE

This reagent determines 1–6 μg of cobalt in the aqueous phase by concentration by extraction. More than 0.1 mM of copper interferes. Nitrate must not exceed mM.

Procedure. To 5 ml of faintly acid solution containing up to 40 μg of cobalt, add 5 ml of 30% solution of ammonium thiocyanate and 20 ml of 0.01 M reagent.[5] If ferric ion is present and does not exceed 0.5 mg/ml, add 2 grams of sodium fluoride. Dilute to 38 ml and set aside for 5 minutes. Dilute to 50 ml and shake for 3 minutes with 10 ml of 1,2-dichloroethane. Read the organic layer at 626 nm and correct for a blank.

Chromium-Nickel Alloy.[6] Dissolve 1 gram of sample in 10 ml of hydrochloric acid and 3 ml of nitric acid and dilute to 100 ml. To a 5-ml aliquot containing up to 0.2 mg of cobalt, add 2.5 ml of 30% solution of ammonium thiocyanate and 0.1 ml of 10% ammonium fluoride solution. Add saturated sodium bicarbonate solution dropwise to give a pH of 3. Add a few drops of 10% solution of ammonium fluoride, followed by 10 ml of 0.01 M reagent at a pH of 4. Dilute to 20 ml and extract with 5 ml of 1,2-dichloroethane. Read at 625 nm.

[3] V. I. Vershinin, V. T. Chuiko, and B. E. Reznik, *Zh. Anal. Khim.* **26**, 1710–1718 (1971).
[4] G. E. Batley, *Talenta* **18**, 1225–1232 (1971).
[5] Claudio Calzolari and Luciano Favretto, *Analyst* **93**, 494–497 (1968).
[6] Claudio Calzolari, Luciano Favretto, L. Favretto Gabrielli, and G. Pertoldi Marletta, *Rass. Chim.* **21**, 135–141 (1969).

Cobaltiferrous Rock. Dissolve 1 gram of sample with 5 ml of hydrofluoric acid and 2 ml of 1 : 1 sulfuric acid. Evaporate to sulfur trioxide fumes and dilute to 100 ml. Complete as above from "To a 5-ml aliquot. . . ."

4-(5-BROMO-2-THIAZOLYLAZO)RESORCINOL AND 4-(5-BROMO-2-PYRIDYLAZO)RESORCINOL

Cobalt forms 1 : 2 complexes with the captioned compounds.[7] With the thiazolylazo compound the appropriate pH is 6.5–8; with the pyridylazo compound, 5.5–8. The maximum absorption for either reagent is at 510 nm. Beer's law is followed for either one at 0.25–2 μg per ml.

Procedures

Thiazolyl Compound.[8] To a sample containing 2–12 μg of cobalt, add 2 ml of dimethylformamide, 2 ml of triethanolamine buffer solution for pH 7.5–8, 1 ml of *M* thiourea, and 1 ml of 1.25 m*M* 4-(5-bromo-2-thiazolylazo)resorcinol in dimethylformamide. Set aside for 1 hour. Add 1 ml of 0.01 *M* EDTA, dilute to 25 ml, and read at 570 or 580 nm against water.

Pyridyl Compound. Follow the technic above with 1–10 μg of cobalt and 4-(5-bromo-2-pyridylazo)resorcinol. Read at 530, 550, or 570 nm against a reagent blank.

Water. Adjust a sample containing 0.2–1 μg of cobalt to pH 5–6 with 4% sodium hydroxide solution. Add 1 ml of dimethylformamide and 2 ml of triethanolamine buffer solution for pH 7.5–8. Add 0.5 ml of 0.25 m*M* pyridyl compound in dimethylformamide and set aside for 10 minutes. Add 0.5 ml of 0.01 *M* EDTA and read against a reagent blank.

Copper Alloys and Salts. Remove the copper by electrolysis in a solution containing 2 ml of sulfuric acid and 1 ml of nitric acid per 250 ml. Neutralize with 4% sodium hydroxide solution and develop with the general technic for either the thiazolyl compound or the pyridyl compound.

Nickel and Nickel Salts. Dissolve metallic nickel in 1 : 1 nitric acid, evaporate nearly to dryness, and take up in water. Dissolve nickel salts in water and acidify with nitric acid. To the solution, add 1.5 ml of 27% solution of ammonium chloride, 0.15 gram of aluminum nitrate, and 2 ml of 30% hydrogen peroxide. Dilute to 40 ml, heat nearly to boiling, and neutralize to methyl red with 1 : 6 ammonium hydroxide. After 15 minutes at 60° let stand for 45 minutes at room temperature. Filter and wash with 2% ammonium chloride solution neutralized to methyl red. Dissolve in 10 ml of 1 : 5 hydrochloric acid and 1 ml of 30% hydrogen peroxide.

[7] A. I. Busev, Zh. I. Nemtseva, and V. M. Ivanov, *Zh. Anal. Khim.* **24**, 1376–1380 (1969).
[8] V. M. Ivanov, A. I. Busev, Zh. I. Nemtseva, and L. I. Smirnova, *Zavod. Lab.* **35**, 1042–1044 (1969).

Evaporate to dryness and dissolve in 2 ml of *M* tartaric acid. Neutralize with 4% sodium hydroxide solution and develop all or an aliquot by the general technics above.

COBALT ION

The extinction of cobalt ion in hydrochloric acid at 628 nm is markedly affected by the concentration of the hydrochloric acid.[9] In 7:3 acetone–8 *N* hydrochloric acid, variations in hydrochloric acid concentration are far less significant. This permits reading of cobalt down to 0.5% in aluminum-copper alloys.

Cobalt can be read at 625 nm in 10 *N* hydrochloric acid in the presence of copper, provided the latter is no more than a few times the cobalt concentration.[10] Thus the method is applicable to many cobalt-copper alloys but not to copper alloys in general. Cobalt ion in hydrochloric acid when made up to 90% by volume of dimethylformamide is read at 675 nm, presumably of a cobalt-chloride complex formed with the solvent.[11]

Cobalt down to 0.005% in plain and alloy steel is isolated in strong hydrochloric acid by column chromatography and amyl acetate extraction.[12] Cobalt is then read as cobalt ion. When iron-cobalt catalyst for synthesis of hydrocarbons from water gas is dissolved in hydrochloric acid, the two elements are read as chlorides with ±10% accuracy.[13]

Cobalt sulfate at 0.1–0.3 *M* in 3 *M* sulfuric acid has a maximum absorption of the cobaltous ion at 510 nm.[14] Beer's law is followed for 0.0025–0.25 gram-ion of cobaltous ion per liter.

When cobalt is converted to cobaltous perchlorate by evaporation with perchloric acid, the reading of the solution is at 511 nm for 14–26 grams of cobaltous ion per liter.[15] Chromic ion must be converted to chromate to avoid interference. A small negative error is introduced by nickelous, ferrous, cupric, and manganous ions.

In analysis of alloys, cupric ion is read at 820 nm, nickel ion at 395 nm, and cobalt ion at 515 nm in 0.5 *M* sulfuric acid.[16] Simultaneous equations are then applied. Chloride ion interferes. Ferric ion amounting to less than 5% of the nickel can be masked with phosphoric acid in reading the latter. Higher concentrations of iron are extracted with isopropyl ether from 8 *M* hydrochloric acid solution before driving off the chloride by heating with sulfuric acid.

For cobalt ion read in nickel solution, see Nickel as Nickel Ion.

Procedures

Copper Alloys.[17] Dissolve 0.2 gram of sample in a minimal amount of nitric acid and evaporate almost to dryness. Boil with 50 ml of water to dissolve salts,

[9]C. R. Walker and O. A. Vita, *Anal. Chim. Acta* **47**, 9–11 (1969).
[10]G. Lindley, *R and D* **1963** (25), 23–26.
[11]Leroy Pike and John H. Yoe, *Talenta* **12**, 657–664 (1965).
[12]A. E. Sherwood, *Metallurgia* **64**, 47–50 (1961).
[13]Yu. L. Polyakin, *Tr. Grozn. Neft. Inst.* **1955** (20), 24–32.
[14]N. F. Komar' and L. S. Palagrina, *Zh. Anal. Khim.* **23**, 75–79 (1968).
[15]A. Pall, G. Svehla, and L. Erdey, *Talenta* **11**, 1383–1390 (1964).
[16]D. V. Jayawant and T. K. S. Murthy, *Indian J. Technol.* **9**, 396–400 (1971).
[17]G. Lindley, *Metallurgia* **62**, 45–49 (1960).

digest for 30 minutes while hot, and filter off metastannic acid. Electrolyze to remove copper, and evaporate to dryness. Add 0.5 ml of hydrochloric acid and again evaporate to dryness. Take up in hydrochloric acid, dilute to 20 ml with hydrochloric acid, and read at 625 nm against hydrochloric acid.

Cobalt Naphthenate.[18] Boil 0.2 gram of sample with 35 ml of 1 : 3 sulfuric acid for 10 minutes. Cool, and extract with 50-ml portions of carbon tetrachloride until there is no further color removal. Boil the aqueous layer to evaporate retained carbon tetrachloride. Cool, add 10 ml of 1 : 3 phosphoric acid, and dilute to 100 ml. Read cobalt ion at 517 nm.

Products of the Oxo Process.[19] Shake 5 grams of light-colored sample or 0.5 gram of darker sample for 2 minutes with 20 ml of 8 *N* hydrochloric acid. The aqueous layer should become light colored. Shake the aqueous layer for 2 minutes with 20 ml and 20 ml of isopropyl ether to separate iron. Read the cobalt in the aqueous layer at 590 nm.

CYANATE

When potassium cyanate and acetone are added to a solution of a cobalt salt, the reading conforms to Beer's law up to 0.725 mg of cobalt ion per ml.[20] If cadmium, zinc, or bismuth is present it is precipitated and removed by centrifuging. In the presence of iron and chromium, add ammonium fluoride and centrifuge. Lead and copper interfere.

Procedure. Neutralize a sample solution contains up to 30 mg of cobalt and 300 mg of nickel with 20% sodium hydroxide solution.[21] Add 40 ml of 20% potassium cyanate solution and 15 ml of 15% solution of ammonium acetate. Pass this solution at pH 6.5–8 through a column of Dowex 1-X8 resin and wash the nickel complex through with 50 ml of water. The nickel complex will not be adsorbed below pH 9. Transfer the resin with water to 10 ml of 1 : 4 hydrochloric acid and stir for 5 minutes. Filter and wash with 1 : 4 hydrochloric acid. Neutralize the filtrate and washings with 20% sodium carbonate solution and evaporate to 30 ml. Add 10 ml of 15% ammonium acetate solution and 40 ml of 20% potassium cyanate solution. Dilute to 100 ml and read at 620 nm.

1,2,3-CYCLOHEXANETRIONE TRIOXIME

The complex of this reagent with cobalt shows no maximum absorption but decreases steadily from 285 to 575 nm. Copper, nickel, and ferric ion interfere above 0.5 ppm.

[18]Kohai Shinoda, Taro Noguchi, and Shigeyuki Yamada, *Jap. Anal.* **6**, 435–438 (1957); cf. Vladimir Zvonar, *Chem. Prum.* **7**, 512–514 (1957).
[19]M. Singliar and G. Reckova, *Chem. Prum.* **16**, 535–537 (1966).
[20]Florin Modreanu, *Acad. Rep. Pop. Rom., Fil. Iasi, Stud. Cercet. Stiint. Ser. I*, **6**, 273–290 (1955).
[21]Max Ziegler and Walter Rittner, *Z. Anal. Chem.* **165**, 197–200 (1959).

Procedure. To a sample containing 10–40 μg of cobalt, add 0.8 ml of 0.1473% ethanolic reagent and adjust to pH 3–4.[22] Dilute to 10 ml and read at 400 nm against a reagent blank.

Beer.[23] Evaporate 50 ml of degassed sample with 1 ml of 1 : 1 sulfuric acid and ash at 650°. Moisten the carbon-free ash with 1 ml of 1 : 1 hydrochloric acid and evaporate to dryness. Take up in 10 ml of water, heat at 90° for 5 minutes, and cool. Add 1.6 ml of 8.6×10^{-3} *M* reagent in ethanol. Adjust to pH 3.5 with 0.1 *N* hydrochloric acid or 0.4% sodium hydroxide solution. Filter, and wash the residue with water. Dilute to 20 ml and read at 400 nm. Ion exchange separation is required only if iron exceeds 0.4 ppm.

DIETHYLDITHIOCARBAMATE

The optimum pH range for extraction of the complex of cobalt with this reagent is 4–10. Under appropriate conditions, small amounts of cobalt can be isolated from large amounts of other metallic ions. When the cobalt complex is extracted with butyl acetate, Beer's law applies to 20–200 μg of cobalt per 25 ml. The color is stable for 30 hours. For simultaneous reading of cobalt, nickel, and copper extracted as the diethyldithiocarbamate, see Copper. For isolation of cobalt from niobium with diethyldithiocarbamate followed by column chromatography, see Zinc. For cobalt in soil by diethyldithiocarbamate after removal of copper and nickel by nitroso-R salt, see Copper by Diethyldithiocarbamate.

Procedure. Dissolve 1 gram of sample in 15 ml of hydrochloric acid and 3 ml of nitric acid.[24] Dilute to 40 ml, filter, and evaporate to dryness. Add 1 ml of hydrochloric acid and heat to drive off oxides of nitrogen. Take up with 10 ml of 1 : 1 hydrochloric acid and dilute to 50 ml.

Large Amounts of Nickel and Manganese Present. Shake a 20-ml aliquot of sample solution with 4 ml of 1% solution of tetrasodium EDTA for 2 minutes. Add 16 ml of a buffer solution for pH 10 containing 1 : 1 0.1 *M* sodium carbonate–0.1 *M* hydrochloric acid. Add 4 ml of 0.5% solution of sodium diethyldithiocarbamate and 1 ml of acetic acid. Shake for 2 minutes with 10 ml of carbon tetrachloride. Then add 2 ml of 0.5% solution of mercuric acetate, shake for 4 minutes, and separate the organic layer. Extract the aqueous layer with another 10 ml of carbon tetrachloride and filter the combined extracts. Dilute the filtrate to 25 ml with carbon tetrachloride and read at 365 nm.

Large Amounts of Iron Present. Mix a 20-ml aliquot of sample solution with 4 ml of 1% pyrocatechol solution. Add ammonium hydroxide dropwise until the

[22]W. Joe Frierson, Nancy Patterson, Harriet Harril, and Nina Marable, *Anal. Chem.* **33**, 1096–1098 (1961).
[23]E. Segal and A. F. Lautenbach, *Proc. Am. Soc. Brew. Chem.* **1964**, 49–54.
[24]K. P. Stolyarov, *Zh. Anal. Khim.* **16**, 452–457 (1961).

solution turns red. Add 15 ml of the buffer solution for pH 10 and 4 ml of 0.5% solution of sodium diethyldithiocarbamate. Shake for 2 minutes with 10 ml of carbon tetrachloride. Shake the extract with 15 ml of buffer solution for pH 10 and 2 ml of 0.5% solution of mercuric acetate for 2 minutes. Again shake the organic layer with 15 ml of buffer solution for pH 10 and 2 ml of 0.5% solution of mercuric acetate. Dilute the organic layer to 25 ml and read at 365 nm.

Large Amounts of Copper Present. Shake a 20-ml aliquot of sample solution with 2 ml of 1% solution of sodium EDTA, 17 ml of buffer solution for pH 10, and 2 ml of 0.5% solution of sodium diethyldithiocarbamate. Extract with 10-ml portions of carbon tetrachloride as long as the extracts are colored, and discard the extracts. Add 2 ml of 2% solution of sodium diethyldithiocarbamate and 1 ml of acetic acid, which should give a pH of 4. Extract with 10 ml of carbon tetrachloride. Continue as for large amounts of iron present from "Again shake the organic layer with 15 ml. . . ."

Large Amounts of Zinc Present. Mix a 20-ml aliquot with 19 ml of a buffer solution for pH 4. Add 4 ml of 0.5% solution of sodium diethyldithiocarbamate and shake for 2 minutes with 10 ml of carbon tetrachloride. Add 2 ml of 0.5% mercuric acetate solution and shake for another 2 minutes. Filter the organic layer. Extract the aqueous layer with another 10 ml of carbon tetrachloride, shake with mercuric acetate solution, and filter. Dilute the combined filtrates to 25 ml with carbon tetrachloride and read at 365 nm.

Molybdenum-Cobalt-Aluminum Catalyst.[25a] Dilute a solution of the catalyst to 25 ml and adjust to pH 5 with hydrochloric acid or ammonium hydroxide. Add 5 ml of 2% solution of tiron to mask trivalent metals. Add 5 ml of 5% solution of sodium diethyldithiocarbamate and 20 ml of butyl acetate. Shake for 30 seconds. This extracts cobalt, copper, nickel, and manganese. Wash down the walls with ethanol and discard the aqueous phase. Wash the organic phase with successive portions of 0.05% mercuric chloride solution until one remains clear. Mercuric ion displaces the positive ions other than cobalt. Dilute the organic phase to 25 ml with butyl acetate and read at 410 nm.

---

p-DINITROPHENYLHYDRAZONE OF DIACETYLMONOXIME

After chromatographic isolation of cobalt this reagent in the presence of ethylenediamine complexes with the element. The color development is complete after 3 hours at 90°. Then excess reagent is extracted with ether before reading. Beer's law is followed at 0.05–0.6 ppm of cobalt. The sensitivity is three times that of nitroso-R salt.

Procedure.[25b] *Biological Material.* As acid digestion agent mix 1 volume of

[25a] S. Popescu and M. Papagheorge, *Rev. Chim.* (Bucharest) **19**, 735–756 (1968).
[25b] R. M. Pearson and H. J. Selm, *Anal. Chem.* **49**, 580–582 (1977).

sulfuric acid and 1 volume of perchloric acid with 10 volumes of nitric acid. As color reagent[25c] dissolve 1 gram of diacetylmonoxime in 10 ml of water. Add to 1.5 gram of *p*-nitrophenylhydrazine in 100 ml of hot water and keep warm for 30 minutes. Cool, filter, and recrystallize from ethanol.

The ion exchange column is shown in Figure 20-1. Prepare it with 50–100 mesh Dowex 1-X8 resin. If the column has already been used elute copper, molybdenum, and iron with 75 ml of *N* hydrochloric acid. Then elute zinc with 175 ml of 0.005 *N* hydrochloric acid. Treat with 50 ml of 9 *N* hydrochloric acid before use.

Treat a dried and finely ground sample containing 1–5 μg of cobalt with 50 ml of the acid digestion agent and simmer until decomposition gives a colorless solution. Evaporate to dryness and cool. Take up with 50 ml of 1:20 hydrochloric acid and simmer for 15 minutes. Filter the silicious residue and wash with 1:20 hydrochloric

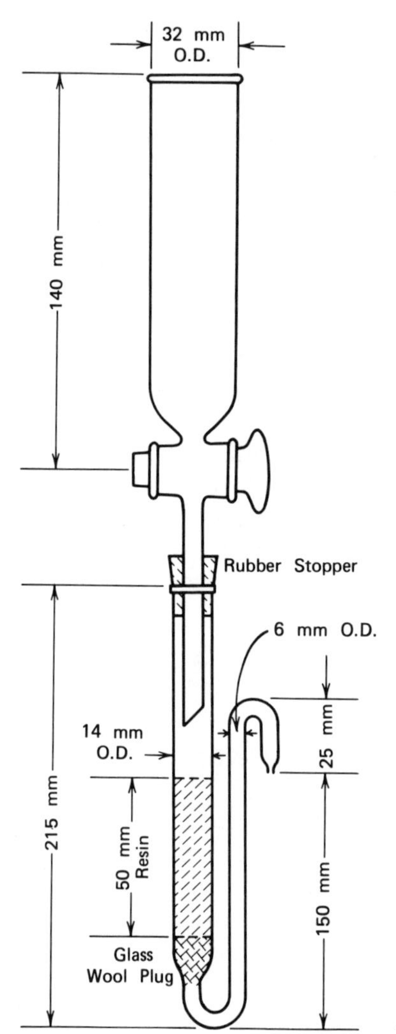

Figure 20–1 Ion exchange column. From *Anal. Chem.* **29**, 443 (1957).

[25c]F. Feigl and D. Goldstein, *Analyst* **81**, 709–710 (1956).

acid. Evaporate the filtrate to dryness. Take up with 15 ml of 9 *N* hydrochloric acid by warming for 10 minutes. Centrifuge insoluble salts such as calcium sulfate.

Decant the sample solution into the column and adjust the flow rate to 1 ml per minute, a rate to be standardized throughout the use of the column. Wash the beaker which contained the sample with 3 ml, 3 ml, and 3 ml of 9 *N* hydrochloric acid adding these washings to the centrifuge tube. Stir the residue and centrifuge. Decant these washings into the column. When the sample solution has entered the resin wash the column with 40 ml of 6 *N* hydrochloric acid to remove manganese and foreign cations. Then elute the cobalt with 50 ml of 4 *N* hydrochloric acid and evaporate the eluate to dryness. Take up the residue by warming with 5 ml of water. Add 0.8 ml of 10% solution of ethylenediamine and 1.2 ml of the color reagent. Mix, warm at 90° for 3 hours, and cool. Dilute to 10 ml and extract excess reagent with 25 ml and 25 ml of ether. Read at 520 nm.

DI-2-PYRIDYL KETOXIME

This reagent determines cobalt in the absence of EDTA and cyanide. Two ml of 5% tetrasodium pyrophosphate solution will mask up to 5 mg of zinc, beryllium, lead, arsenous, antimonic, or ferric ion and up to 1 mg of manganous or rhodium ion. Two ml of 5% sodium thiosulfate solution will mask 5 mg of silver, gold, or mercuric ion. One ml of 10% thiourea masks 5 mg of trivalent iridium ion. In alkaline solution, ammonium ion interferes.

Procedures

Without Extraction.[26] To a sample that contains 18–40 μg of cobalt and may contain a 150-fold content of nickel, add 3 ml of 40% sodium citrate solution and 5 ml of 1% ethanolic di-2-pyridyl ketoxime. Adjust to pH 8.5 and set aside for 5 minutes. Then add 3 ml of hydrochloric acid to decompose the nickel chelate. Dilute to 25 ml and read at 388 nm.

By Extraction.[27] Adjust a sample solution containing 6–60 μg of cobalt to pH 2. Extract palladium and copper with 10 ml of 1 : 1 acetylacetone-chloroform. Add 3 ml of 50% sodium citrate solution and a drop of phenolphthalein indicator solution. Adjust to a distinct pink with 10% potassium hydroxide solution. Dilute to 20 ml and add 3 ml of 1% ethanolic di-2-pyridyl ketoxime. Extract with 8 ml and 8 ml of chloroform and dilute the extracts to 25 ml with chloroform. To remove nickel, wash the chloroform extract with 0.7 ml of hydrochloric acid. This decomposes the nickel complex without affecting the cobalt chelate. Read cobalt at 388 nm.

---

[26] W. J. Holland, M. DePooter, and J. Bozic, *Microchim. Acta* **1974**, 99–103.
[27] W. J. Holland and J. Bozic, *Talenta* **15**, 843–847 (1968).

DITHIZONE

In a typical case, cobalt in a concentration greater than 0.1 ppm is extracted with dithizone in carbon tetrachloride from a solution buffered at pH 8.5.[28] When cobalt and nickel dithizonates have been adsorbed on a column of alumina, both are eluted with chloroform as a method of separation from copper and zinc.[29]

In analysis of gold, after extraction of the gold with isopentyl acetate and of iron with the same solvent, cobalt is determined with dithizone.[30] For separation of copper and cobalt on a column of silicic acid and determination of each by dithizone, see Copper.

Procedure. *Uranium.*[31] Dissolve 1 gram of sample in 10 ml of hydrochloric acid and 2 ml of nitric acid. Evaporate to dryness and take up the residue in 40 ml of water. Add 5 ml of 20% ammonium citrate solution and adjust the pH to 8.5 with ammonium hydroxide. Extract cobalt along with several other metals with 10 ml of 10^{-3} *M* dithizone in carbon tetrachloride. Wash the organic layer with 10 ml of 1:2 hydrochloric acid. Pass the dithizone solution through a 7×50 mm column of alumina on which dithizone complexes are adsorbed. Elute with 5 ml of chloroform isolating the violet cobalt complex fraction. Dilute to 2 ml and read at 550 nm.

EDTA

In analysis of nickel alloys, iron and chromium are separated as hydroxides by pyridine. Cobaltic ion is read with EDTA without interference by copper and nickel.[32] For determination of cobalt in silicate rocks by EDTA, see Copper.

Procedure. Treat a solution of 0.3–0.5 mg of cobalt ion in 10 ml of 0.1 *N* hydrochloric acid with 2 ml of 0.1 *M* disodium EDTA and 5 ml of 3% hydrogen peroxide.[33] Boil for 1 minute, cool, and dilute to 45 ml. Adjust to pH 1.5–2.5 and dilute to 50 ml. Set aside for 10 minutes and read at 520 nm.

Steel.[34] Dissolve 1 gram of sample by heating with 18 ml of 1:5 sulfuric acid and 1.5 ml of phosphoric acid. Slowly add 8 ml of 1:1 nitric acid to dissolve any black residue. Boil off nitrous fumes, cool, and dilute to 100 ml. Mix a 10-ml aliquot with 10 ml of 0.1 *M* EDTA and add 5 grams of ammonium fluoride. Heat until the yellow color of the ferric complex disappears. Chromic ion forms a green

[28]Yoshio Morimoto, Takashi Ashizawa, and Teruko Takenouchi, *Jap. Anal.* **11**, 167–170 (1962); cf. Takashi Ashizawa, *ibid.* **10**, 683–688 (1961).
[29]Takashi Ashizawa, *ibid.* **10**, 443–448 (1961).
[30]G. Ackermann and J. Köthe, *Z. Anal. Chem.* **231**, 252–261 (1967); Zygmunt Marczenko, Krzysztof Kasiura, and Maria Krasiejko, *Chem. Anal.* (Warsaw) **14**, 1277–1287 (1969).
[31]Yoshio Morimoto, Takashi Ashizawa, and Teruko Takenouchi, *Jap. Anal.* **10**, 167–170 (1961).
[32]N. M. Silverstone and B. B. Bach, *Metallurgia* **63**, 205–207 (1961).
[33]Luigi Giuffre and Franco Maitan Capizzi, *Ann. Chim.* (Rome) **51**, 558–562 (1961).
[34]R. Moro Garcia and M. L. Alvarez Bartolome, *Inf. Quim. Anal. Pura Apl. Ind.* **24**, 15–22 (1970).

precipitate. Cool, add 2 ml of 10% solution of sodium periodate, and immediately reduce manganic ion with 4 ml of 10% hydroxylamine hydrochloride. Set aside for 10 minutes and read the cobaltic-EDTA complex at 540 nm.

Silicate Rocks.[35] Transfer 0.5 gram of sample to platinum and mix with 2 ml of sulfuric acid and 10 ml of hydrofluoric acid. Evaporate to fumes and take up with 5 ml of water. Add 5 ml of hydrochloric acid and extract the iron with 5 ml of pentyl acetate. If the sample contains more than 3% iron, repeat this extraction. Evaporate the aqueous phase to fumes and nearly neutralize with 10% sodium hydroxide solution. Add 15 ml of 0.2 *M* EDTA. If titanium or vanadium is to be determined, add 5 drops of 30% hydrogen peroxide. Adjust to pH 4 and boil for 5 minutes. Cool, dilute to 25 ml, and read cobalt at 465 nm for 0.05–4.25 μg.

ETHYLENEDIAMINE

A complex of cobaltic ion with ethylenediamine is isolated from interferences by chromatography on silica at pH 8. Many interfering ions are masked by EDTA, which does not interfere with adsorption of the cobalt complex. After elution, the color development is by the reduction-oxidation reaction converting ferric ion to the ferrous complex with 1,10-phenanthroline at pH 3.5.

Procedure. *Nickel.*[36] As a buffer solution, dissolve 31 grams of boric acid in water, add ammonium hydroxide to adjust to pH 9, and dilute to 500 ml.

Dissolve a 0.2-gram sample containing 0.02–0.1% cobalt in 5 ml of 1:1 nitric acid and evaporate to dryness. Add 2 ml of water, again evaporate, and take up in 5 ml of water. Add 1 ml of ethylenediamine and 3.7 ml of *M* EDTA. Add 5 ml of the buffer solution for pH 9 and dilute to 25 ml. Heat at 40–50° until the color changes from violet to blue.

Pack a 12×50 mm column with 0.15–0.20 mm chromatographic-grade silica such as VUPL, Prague. This has a pore size of 85 Å, a surface area of 568 m^2/gram. Wash the column with 5 ml of the buffer solution diluted to 25 ml. Pass the sample solution through the column at about 5 ml/minute. Follow with a mixture of 5 ml of the buffer solution for pH 9, 0.5 ml of ethylenediamine, 3 ml of 0.1 *M* EDTA, and 16.5 ml of water. Elute the cobalt with 25 ml of 1:4 hydrochloric acid. Boil the eluate for 2 minutes, and cool. Add 2 ml of 0.01 *M* ferric chloride and 1 ml of 0.1 *M* 1,10-phenanthroline. Raise to pH 3.5 by adding ammonium hydroxide. Heat to about 50° and cool. Dilute to 50 ml and set aside for 10 minutes. Read at 500 nm against a reagent blank.

FERROZINE

This reagent is the disodium salt of 3-(2-pyridyl)-5,6-bis(4-sulfophenyl)-1,2,4-triazine. It complexes with cobalt, copper, and iron. Nickel must be absent.

[35] K. P. Stolyarov and F. B. Agrest, *Zh. Anal. Khim.* **19**, 457–466 (1964).
[36] F. Vydra and V. Markova, *Chemist-Analyst* **54**, 69 (1965).

Procedure. Mix 5 ml of sample solution, 5 ml of 0.01 *M* ferrozine, and 5 ml of a phosphate buffer solution for pH 6.5.[37] Dilute to 25 ml and read at 500 nm against a reagent blank.

FLAME PHOTOMETRY

When extracted as cobalt quinolinate with methyl isobutyl ketone, ethyl acetate, amyl acetate, or chloroform, the solution is injected into the oxyhydrogen flame for reading at 352.7 nm.[38] The sensitivity is 5–30 times that of injection of an aqueous solution.

For flame photometry, extract 0.1–0.4 mg of cobalt as its 8-hydroxyquinoline complex at pH 7.3–8.2 by pentyl acetate.[39] When read by flame photometry at 352.7 nm, the sensitivity approximates 10 times that of the aqueous solution. To avoid interference, the corresponding complexes of copper, gallium, indium, iron, and manganese are preextracted with 8-hydroxyquinoline at a pH above 10 in the presence of citrate and tartrate. Nickel interferes.

For separation from iron to determine cobalt in cobalt mattes and concentrates by flame photometry, see Copper.

FORMIC ACID HYDRAZIDE

This reagent forms colored complexes with cobalt, copper, and gold. There is interference by silver, mercurous, mercuric, ferrous, stannous, and stannic ions and to a lesser degree by lead, zinc, and bismuth.

Procedure. To a neutral sample solution containing 0.3–1.5 mg of cobalt, add 0.1 ml of 0.25 *N* sulfuric acid and dilute to 8 ml.[40] Add 2 ml of 10% solution of formic acid hydrazide and read at 480 nm. The final pH should be 3–4.5.

3-HYDROXY-4-NITROSONAPHTHALENE-2,7-DISULFONATE

The complex of cobalt with this reagent can be separated from interfering ions by adsorption on alumina.

Procedure. *Paper or Wood Pulp.*[41] Ash 100 grams of sample in platinum at

[37] S. R. Kundra and R. P. Singh, *Anal. Chem.* **46**, 1605–1606 (1974).
[38] Emiko Sudo and Hidehiro Goto, *Sci. Rep. Res. Inst., Tohoku Univ. Ser. A* **14**, 220–238 (1962).
[39] Hidehiro Goto and Emiko Sudo, *Jap. Anal.* **10**, 463–467 (1961).
[40] M. H. Hashmi, Abdur Rashid, Mohammad Umar, and Farooq Azam, *Anal. Chem.* **38**, 439–441 (1966).
[41] O. Lagerstrom and H. Haglund, *Svensk. Papp-Tidn.* **67**, 4–5 (1964).

$575 \pm 25°$ for 4 hours. Dissolve the ash in 10 ml of hydrochloric acid and evaporate to dryness at 100°. Add 50 ml of water and 5 ml of perchloric acid. Boil for 5 minutes, then precipitate sulfate with 2 ml of saturated barium perchlorate solution. Heat to boiling, filter, and dilute to 100 ml. Dilute an aliquot containing 10–50 μg of cobalt to 50 ml and add 5 ml of 1% solution of the reagent. Adjust to pH 5 with 4 *M* ammonium acetate and heat to boiling. For each ml of ammonium acetate used, add 0.5 ml of perchloric acid. Pass the solution through a 4-cm column of alumina. Wash the column with ten 10-ml portions of 0.75 *N* nitric acid at 70°. Elute the red cobalt complex with *N* sulfuric acid until 50 ml is collected and read it immediately at 410 nm against a blank.

MERCAPTOACETIC ACID

The complex of cobalt with mercaptoacetic acid, also known as thioglycollic acid, is read at 358 nm. Interference by copper, nickel, iron, chromium, molybdenum, uranium, vanadium, and some other ions can be bypassed by first separating the cobalt as the dithizonate. Beer's law is followed up to 10 μg of cobalt per ml, preferably more than 0.25 μg/ml. The maximum color is developed at pH 4.8–5.25. Excess reagent lessens the absorption.

Procedure. To a sample containing 0.1–100 μg of cobalt, add 2 ml of *M* mercaptoacetic acid.[42] Adjust to pH 5.2 with 4% solution of sodium hydroxide. Dilute to 10 ml and heat at 100° for 20 minutes. Cool, adjust to 10 ml, and read at 358 nm against water.

Steel. Dissolve a sample containing 10–30 mg of cobalt in 15 ml of hydrochloric acid and 5 ml of nitric acid. If necessary in the case of alloys, boil with 20 ml of hydrochloric acid and dissolve by dropwise addition of fuming nitric acid. Evaporate to dryness. Take up with 5 ml of hydrochloric acid and evaporate to dryness. Repeat that step to remove the last traces of nitric acid. Dissolve in the minimum possible amount of hydrochloric acid. Add 50 ml of water and boil for a few minutes. Cool and filter. Wash on the filter with 1 : 50 hydrochloric acid, then with water. Dilute to 250 ml and take an aliquot containing not more than 1 mg of cobalt.

Dilute the aliquot to 10 ml and add 5 ml of 10% sodium citrate solution. Add ammonium hydroxide to reach pH 9. Shake for 2 minutes with 5 ml of 0.01% solution of dithizone in carbon tetrachloride. Repeat with 5-ml portions until the extract is no longer red. Wash the combined extracts of cobalt dithizonate with 10 ml of water. Evaporate the carbon tetrachloride extracts and ignite at red heat. Cool, dissolve with 1 ml of hydrochloric acid, and dilute to 100 ml. Develop a 10-ml aliquot by the general procedure.

---

[42] V. D. Anand, G. S. Deshmukh, and C. M. Panday, *Anal. Chem.* **33**, 1933–1937 (1961).

N-MERCAPTOACETYL-p-TOLUIDINE

Cobalt forms a 1:3 complex with this reagent. Beer's law is followed for 0.2–5 μg of cobalt per ml of the solvent extract. Cyanide, oxalate, molybdate, and palladium interfere.

Procedure. *Steel.*[43] Dissolve 0.1 gram of sample by heating with 10 ml of 1:1 hydrochloric acid and 5 ml of nitric acid. Evaporate to about 2 ml, take up with water, and dilute to 25 ml. Dilute a 10-ml aliquot to 100 ml. To a 5-ml aliquot of this dilution, add 1 gram of ammonium tartrate, 5 ml of a borate buffer solution for pH 7.09, and 2 ml of 1% solution of the captioned reagent in 45% ethanol. Shake for 2 minutes with 10 ml of 2:1 chloroform–isopentyl alcohol. After 20 minutes read the organic phase at 490 nm against a blank.

3-METHOXY-2-NITROSOPHENOL

The cobalt complex of this reagent is ether extractable.

Procedure. *Beer.*[44] Mix 25 ml of degassed beer with 0.5 ml of 30% hydrogen peroxide. Add 5 ml of a saturated solution of the reagent and 0.5 ml of 1:4 ammonium hydroxide. Set aside for 10 minutes, then extract with 45 ml of ether. Wash the extract with 10 ml of 10% sodium carbonate solution. Dilute to 50 ml with ether and read at 380 nm.

2-(3-METHYL-5-PROPYL-2-PYRROLYL)PHENOL

The 2:1 complex of this reagent with cobalt at pH 6–10 has a maximum absorption at 560 nm. In 1:2 dioxan-water the maximum is shifted to 440 nm. Ferric ion, nickel, and copper must be masked with EDTA. An appropriate range is 2–10 μg of cobalt in 25 ml of developed solution.

Procedure. *Nickel Salts.*[45] Mix 10 ml of sample solution with 5 ml of 0.1 *N* hydrochloric acid and heat to 80°. Add 5 ml of 3% solution of dimethylglyoxime followed by dropwise addition of ammonium hydroxide until the odor of ammonia is perceptible. Filter, and wash the precipitate with 50 ml of hot water. Evaporate the filtrate and washings to 10 ml. Destroy the dimethylglyoxime with 5 ml of nitric acid. Add sufficient 10% sodium acetate solution to give a pH of 7–7.2. Add 0.5 ml

[43] A. I. Busev and A. Naku, *Zh. Anal. Khim.* **19**, 475–479 (1964).
[44] F. V. Harold and E. Szobolotsky, *J. Inst. Brew.* **69**, 253–258 (1963).
[45] A. I. Cherepakhin, *Zh. Anal. Khim.* **21**, 502–504 (1966).

of mM reagent in 1 : 2 dioxan-water and 2 ml of saturated solution of EDTA. Dilute to 25 ml with 20% ammonium citrate solution and read at 560 nm.

MOLYBDOPHOSPHORIC ACID

In acetic acid solution at pH 4.5–5, cobaltous ion reduces molybdophosphoric acid to molybdenum blue. Beer's law is followed for 0.5–3 μg of cobalt in 50 ml.

Procedures

Nickel.[46] Dissolve 3 grams of sample in 30 ml of hydrochloric acid and 15 ml of 1 : 1 nitric acid. Add 5 ml of sulfuric acid and evaporate to about 20 ml. Cool, and dilute to 100 ml. Mix a 5-ml aliquot with 10 ml of acetate buffer solution for pH 4.8. Add 12 ml of saturated solution of EDTA and 2 ml of 5% solution of molybdophosphoric acid. Store at 35° for 10 minutes, cool, and dilute to 50 ml. Read at 656 nm against a sample from which the color reagent was omitted.

Steel. Dissolve a 0.5-gram sample in 5 ml of hydrochloric acid and 5 ml of 1 : 1 nitric acid. Add 40 ml of water and 5 ml of sulfuric acid. Boil off oxides of nitrogen, cool, and dilute to 100 ml. Proceed as for nickel from "Mix a 5-ml aliquot. . . ."

NITRILOTRIACETIC ACID

This close relative of EDTA as a cobaltic-NTA complex is oxidized by sodium perborate to give a purple color.[47] It is desirable to separate potentially interfering ions such as nickel, iron, and chromium by column chromatography. Phosphoric acid masks ferric ion.

Procedure. *Nickel Alloy or Steel.*[48] For a sample containing 0.005–0.03% cobalt, take 2 grams, for 0.02–1% take 1 gram. Dissolve in 15 ml of hydrochloric acid and 5 ml of nitric acid by warming. Evaporate nearly to dryness but do not bake. Dissolve in 10 ml of 1 : 1 hydrochloric acid, dilute to 25 ml, and add some filter pulp. Boil for a few minutes, filter while hot, and wash with 0.1 N hydrochloric acid. Add 5 ml of 30% hydrogen peroxide and evaporate nearly to dryness. Take up in 20 ml of 9 N hydrochloric acid. Prepare an anion exchange column of 100-200 mesh Dowex 1-X8 in a 125-ml Jones reductor tube and wash it with 9 N hydrochloric acid. If a 2-gram sample was used, split it between two such columns.

[46] S. P. Klochkovskii and V. D. Chistota, *Zavod. Lab.* **36**, 911–912 (1970).
[47] K. L. Cheng, *Anal. Chem.* **30**, 1035–1039 (1958).
[48] K. L. Cheng and Francis J. Warmuth, *Chemist-Analyst* **48**, 96–97 (1959).

Use portions of 9 N hydrochloric acid to complete transfer to the column, and use portions of that acid to wash down the column above the bed.

Elute the column with 125 ml of 9 N hydrochloric acid at 1–1.5 ml/minute. This eluate contains nickel, chromium, manganese, and other metals. Before this elution is complete, similarly apply 125 ml of 1 : 2 hydrochloric acid. This elutes the cobalt. Evaporate the second eluate to about 5 ml, cool, and dilute to 25 ml. To a 10-ml aliquot, add 1 ml of 0.2 N solution of nitrilotriacetic acid and adjust to pH 8–9 with sodium carbonate. Add about 0.1 gram of sodium perborate and set aside for 10 minutes. Add 2 ml of 4 M phosphoric acid and dilute to 25 ml. Read at 560 nm against a reagent blank.

---

o-NITROSOCRESOL

This reagent complexes with cobalt for reading. Extraction with petroleum ether separates the complex from all interferences except ferric and palladous ions; that by ferric ion is avoided by reduction to ferrous ion. In analysis of plant material, a dithizone extraction separates copper, cobalt, and zinc from possibly interfering ions.

Procedure. *Plant Material.*[49] For *o*-nitrosocresol reagent, dissolve 8.4 grams of anhydrous cupric chloride and 8.4 grams of hydroxylamine hydrochloride in 900 ml of water. Add 8 ml of *m*-cresol and stir vigorously, while slowly adding 24 ml of 30% hydrogen peroxide. Stir for 2 hours. Add 25 ml of hydrochloric acid and extract with four successive 150-ml portions of petroleum ether. Wash the combined extracts with 100 ml and 100 ml of 0.1 N hydrochloric acid, then with 100 ml and 100 ml of water. Shake the organic layer with 100-ml portions of 1% cupric acetate solution until the aqueous extracts are no longer blood red, only a light purple. Discard the organic phase. Add 25 ml of hydrochloric acid to the combined aqueous extracts and extract the reagent with 500 ml and 500 ml of petroleum ether. Wash this solution of the reagent with 200 ml and 200 ml of 0.1 N hydrochloric acid, followed by several 200 ml portions of water. Store the *o*-nitrosocresol in petroleum ether at 4°.

As a sodium borate buffer solution for pH 7.8, dissolve 20 grams of boric acid in 1 liter of water and add 50 ml of N sodium hydroxide. As a sodium borate buffer solution for pH 9.1, mix 1 liter of the sodium borate buffer solution for pH 7.8 with 120 ml of N sodium hydroxide.[50] Adjust the pH if necessary.

For sodium *o*-nitrosocresol reagent, extract 100 ml of the reagent solution in petroleum ether with 50 ml and 50 ml of the borate buffer solution for pH 9.1.

Ash 6 grams of oven-dried sample at 500° overnight in platinum. Cool, moisten the ash with water, and add 5 ml of perchloric acid dropwise. Add 5 ml of

[49] Richard L. Gregory, Clayton J. Morris, and Gordon H. Ellis, *J. Assoc. Off. Agr. Chem.* **34**, 710–716 (1951); cf. C. P. Czerniejewski, C. W. Shank, W. G. Bechtel, and W. B. Bradley, *Cereal Chem.* **41**, 65–72 (1964).

[50] Richard L. Gregory, Clayton J. Morris, and Gordon H. Ellis, *J. Assoc. Off. Agr. Chem.* **34**, 710–716 (1951).

hydrofluoric acid, evaporate at 100°, and then heat to fumes. Heat at 600° for 1 hour. Cool and add 5 ml of 1 : 1 hydrochloric acid and 10 ml of water. Warm to dissolve and dilute to 50 ml.

To a 25-ml aliquot, add 5 ml of 40% ammonium citrate solution. Add a drop of phenolphthalein indicator solution and neutralize with 1 : 1 ammonium hydroxide. If a precipitate forms, add more ammonium citrate solution. Shake for 5 minutes with 10 ml of 0.05% dithizone solution in carbon tetrachloride. Repeat the extraction with 5-ml portions of dithizone solution as long as color is extractable. Add 2 ml of perchloric acid to the combined extracts and evaporate just to dryness. Take up with 5 ml of 0.01 *N* hydrochloric acid. Add 5 ml of sodium borate buffer solution for pH 7.8 and 2 ml of freshly prepared sodium *o*-nitrosocresol reagent. Shake for 10 minutes with 5 ml of petroleum ether. To remove excess reagent, shake the petroleum ether extract for 1 minute with 5 ml of 1% solution of cupric acetate. Wash the organic layer with 5 ml of water. Reduce the iron in the organic layer by shaking for 1 minute with a buffer solution for pH 5.1 containing 2% of hydroxylamine hydrochloride and 1.9% of anhydrous sodium acetate. Read at 360 nm.

1-NITROSO-2-NAPHTHOL

This reagent forms a 3 : 1 complex with cobalt. At pH 2.5–7.2 extraction of the complex by benzene is complete.[51] Washing the extract with hydrochloric acid removes many interferences. Washing with 4% sodium hydroxide solution removes excess reagent.

For analysis of silver alloys, the copper and silver are removed by electrodeposition to give a solution for determination of cobalt by 1-nitroso-2-naphthol.[52] At pH 4.5–8 cobalt is extracted with chloroform as the 1 : 2 complex with 4-methoxybenzothiohydroxamic acid for subsequent determination with 1-nitroso-2-naphthol.[53]

Dissolve 0.5 gram of high temperature alloy or up to 3 grams of uranium in acid.[54] Adjust an aliquot containing up to 40 μg of cobalt to pH 3–4, adding citrate ion and 1-nitroso-2-naphthol. Extract the complex from a chloride medium with benzene. Wash with hydrochloric acid, water, *N* sodium hydroxide, and *N* sodium hydroxide containing 0.5% of hydroxylamine hydrochloride. This removes interfering metals and excess reagent. Read at 410 nm.

For copper, nickel, cobalt, zinc, and cadmium as impurities in sodium chloride, complex with EDTA and extract cobalt from the solution with chloroform.[55] Evaporate to dryness, add sulfuric and nitric acids, and heat to sulfur trioxide fumes. Determine cobalt by 1-nitroso-2-naphthol.

Procedure. Mix 2 ml of a solution containing cobalt, nickel, and iron with 0.2 ml

[51] Masao Kawahata, Heiichi Mochizuki, and Takeshi Misaki, *Jap. Anal.* **9**, 1023–1027 (1960).
[52] J. Chwastowska and Z. Skorko-Trybula, *Chem. Anal.* (Warsaw) **9**, 123–130 (1964).
[53] Z. Skorko-Trybula and B. Debska, *Chem. Anal.* (Warsaw) **13**, 557–573 (1968).
[54] Hiroshi Hashitani, Hideyo Yoshida, and Takeo Adachi, *Jap. Anal.* **24**, 452–454 (1975).
[55] Cezary Rozychi and Janusz Rogozinski, *Chem. Anal.* (Warsaw) **20**, 107–111 (1975).

of 30% hydrogen peroxide and 3 drops of 5% solution of sodium hydroxide.[56] Warm. Add 2 ml of acetic acid and 5 ml of 1% solution of 1-nitroso-2-naphthol in acetic acid. Dilute to 25 ml and extract with 5 ml and 5 ml of benzene. Destroy the nickel and iron complexes by washing with 10 ml of 1 : 1 hydrochloric acid. Wash out excess of the reagent with 15 ml of 8% sodium hydroxide solution. Wash with 20 ml and 20 ml of water and read at 416 or 436 nm.

Tellurium

COBALT.[57] Dissolve 3 grams of sample in 20 ml of 1 : 1 nitric acid. Adjust to pH 4.6 ± 0.4 with 1 : 1 ammonium hydroxide. Add 10 ml of 3% hydrogen peroxide and set aside for 5 minutes. Add 2 ml of 1% solution of the color reagent in acetic acid, adjust to pH 4–5, and set aside for 30 minutes. Extract with 10 ml and 5 ml of toluene. Dilute the combined extracts to 25 ml with toluene and read at 417 nm against toluene.

NICKEL. Wash the aqueous phase with 10 ml of chloroform and discard the washings. Add 5 ml of 20% hydroxylamine hydrochloride solution and 5 ml of 0.5% solution of dimethylglyoxime in ammonium hydroxide. Adjust to pH 10 with 1 : 1 ammonium hydroxide and extract the nickel complex with 10 ml of chloroform. Separate the organic layer and wash with 5 ml of 0.4% sodium hydroxide solution. Then wash with 5 ml of water and read at 263 nm.

Calcium Barium Titanate.[58] A cobalt additive is present in the sample. Heat 0.1 gram of sample for 10 minutes with 75 ml of hydrochloric acid. Reduce titanium with 2.5 grams of hydroxylamine hydrochloride and close the flask with a trap containing 5% sodium carbonate solution. Boil until the sample is dissolved, 30 minutes or more, and dilute to 100 ml. Add 0.2 gram of tartaric acid to a 25-ml aliquot. Add 4% sodium hydroxide solution to raise to pH 3. Add 5 ml of 0.1% solution of 1-nitroso-2-naphthol and set aside for 12 hours. Shake for 5 minutes each with 4 successive 10-ml portions of carbon tetrachloride. Wash the combined extracts successively with 20 ml and 20 ml of hydrochloric acid, 40 ml of water, 20 ml of 20% sodium hydroxide solution, and 40 ml of water. Filter, dilute to 50 ml with carbon tetrachloride, and read at 417 nm.

Seawater.[59] Prepare a basic wash solution consisting of 50 ml of 4% sodium hydroxide solution, 10 ml of 20% solution of sodium citrate, and 1 ml of 30% hydrogen peroxide diluted to 100 ml.

To 750 ml of filtered sample, add 25 ml of 20% solution of sodium citrate, 1 ml of 30% hydrogen peroxide, and 1 ml of 1% ethanolic 1-nitroso-2-naphthol. Set aside for 10 minutes. Extract with 8 ml, 4 ml, 4 ml, and 4 ml of chloroform. Remove the excess of reagent from the combined extracts by shaking for 1 minute each with 5

[56] V. M. Peshkova and V. M. Bochkova, *Tr. Kom. Anal. Khim. Akad. Nauk SSSR, Inst. Geokhim. Anal. Khim.* **8**, 125–134 (1958); C. Rozycki, *Chem. Anal.* (Warsaw) **12**, 131–136 (1967).

[57] Franciszek Kozera and Barbara Kasterka, *Chem. Anal.* (Warsaw) **16**, 925–932 (1971).

[58] F. P. Gorbenko, E. V. Lapitskaya, and Yu. K. Tselinskii, *Tr. Vses. Nauchn.-Issled. Inst. Khim. Reakt.* **1966** (24), 99–109.

[59] Edward Kentner and Harry Zeitlin, *Anal. Chim. Acta* **49**, 587–590 (1970).

ml, 5 ml, and 5 ml of the basic wash solution. Extract copper and nickel from the organic phase with 5 ml of 2 *N* hydrochloric acid. Extract any residual reagent from the organic phase with 5 ml of basic wash solution. Again wash the organic phase with 5 ml of 2 *N* hydrochloric acid. Dilute the organic phase to 50 ml with chloroform and read at 410 nm against a reagent blank.

Blood.[60] Evaporate 10 ml of sample and ignite at 500–550°. Add 0.5 ml of sulfuric acid and again ignite at 500–550°. Add 2 ml of 1:4 hydrochloric acid and evaporate to a few drops. Add 2 ml of 1% sulfuric acid and heat to sulfur trioxide fumes. Warm with 5 ml of water. Add 5 ml of 20% potassium citrate solution and heat to 60°. Add 5 ml of 0.5% solution of dithiooxamide. Raise to pH 7.2 with ammonium hydroxide and add 2.5 ml of 0.04% solution of 1-nitroso-2-naphthol. Extract with 5 ml of 1:1 carbon tetrachloride–toluene. The cobalt-dithiooxamide complex forms a thin green layer at the interface. Transfer the organic layer to separatory funnel and wash out excess reagent by shaking with 10 ml of 5% potassium hydroxide solution. Adjust the organic layer to 5 ml with toluene and read at 420 nm.

Plant Material.[61] Ash 50 grams of air-dried sample containing 5–20 μg of cobalt at 500°. Add 5 ml of 30% hydrogen peroxide to the cooled ash and heat slowly to 500°. Wet with water, dissolve in 15 ml of 1:3 hydrochloric acid, and dilute to 100 ml. Add 1 ml of nitric acid and boil. Add 10 ml of a hot 25% solution of sodium citrate in 0.5 *N* sodium hydroxide solution. Cool, add 30 ml of 50% sodium acetate solution, and mix well. Add 1 ml of 1% solution of 1-nitroso-2-naphthol in acetic acid. Shake for 30 minutes, add 25 ml of toluene, and set aside for 20 minutes. Shake for 1 minute and separate the organic layer. Wash it with 20 ml and 20 ml of 8% sodium hydroxide solution. Read at 550 nm.

2-NITROSO-1-NAPHTHOL

This reagent complexes rapidly with cobalt at pH 4–9. The 3:1 complex is extractable with isoamyl acetate, carbon tetrachloride, chloroform, and toluene. Citrate prevents precipitation of hydroxides such as those of iron, aluminum, and calcium at pH 8.5. Beer's law is followed for 0.05–6 ppm of cobalt in the final solution.[62] Excess reagent must be removed. Manganese and platinum interfere. Sodium citrate complexes calcium. Iron is oxidized by hydrogen peroxide solution.

Cobalt in tantalum and plutonium solutions is reacted with 2-nitroso-1-naphthol and extracted with chloroform at pH 4 in the presence of citrate.[63] Excess reagent is washed from the extract before reading at 530 nm.

When a fertilizer sample is dissolved with 5:1:2 nitric-sulfuric-perchloric acids, centrifuge the diluted sample solution.[64] Adjust to pH 8.5 and treat with 2-

[60] E. D. Zhukovskaya and L. I. Idel'son, *Lab. Delo* **9** (4), 16–22 (1963).
[61] A. Krause, *Chem. Anal.* (Warsaw) **6**, 711–714 (1961).
[62] H. Schuller, *Mikrochim. Acta* **1**, 107–121 (1959).
[63] T. K. Marshall, J. W. Dahlby, and G. R. Waterbury, *USAEC Rep.* **LA-3124**, 16 pp (1964).
[64] J. F. Hodgson and V. A. Lazar, *J. Assoc. Off. Agr. Chem.* **48**, 412–415 (1965).

nitroso-1-naphthol and sodium thiosulfate. Extract the complex with isopentyl acetate and read at 530 nm. Chromium may cause low results.

The reagent is applied to determination of cobalt in soil, plants, and biological samples by extraction of the 2-nitroso-1-naphthol complex at pH 7.2–7.4 with toluene.[65]

For analysis of sodium metal, react with water, neutralize with hydrochloric acid, and extract the complex of cobalt with 2-nitroso-1-naphthol with carbon tetrachloride.[66] Read at 535 nm. In dilute solutions, chloride ion must be present. Ammonium ion interferes.

This reagent is applicable to determination of cobalt in nickel for use in electronic tubes and valves[67] and to analysis of ferrites[68] and laterites.[69] For separation of cobalt as the 2-nitroso-1-naphthol complex from other metals, see "Isolation of Cadmium, Zinc, Cobalt, and Nickel" in Chapter 6, Cadmium.

Procedure. Buffer 2 ml of a solution containing cobalt, nickel, and iron with 5 ml of saturated solution of sodium acetate, which should give a pH of 5.4.[70] Add 3 ml of 0.5% ethanolic 2-nitroso-1-naphthol. Extract with 5 ml and 5 ml of benzene. Destroy the nickel and iron complexes by washing with 10 ml of 1 : 5 hydrochloric acid, followed by 10 ml of water. Wash out excess reagent with 5 ml and 5 ml of 8% sodium hydroxide solution. Wash with 20 ml and 20 ml of water and read at 360 nm. This reagent is preferable to 1-nitroso-2-naphthol.

Iron and Steel.[71] Dissolve 0.5 gram of sample by warming with 15 ml of hydrochloric acid and 5 ml of nitric acid. Add 2 ml of hydrofluoric acid to help dissolve high silicon samples and to remove silica. Evaporate to a small volume. If the sample is high in chromium, add 5 ml of perchloric acid and partially volatilize the chromium by evaporating nearly to dryness. Add 10 ml of hydrochloric acid an warm to promote solution and reduce residual chromium. Dilute to 100 ml. After insoluble material has settled take an appropriate aliquot according to the following table.

Cobalt (%)	Solution Aliquot (ml)	Amyl Acetate Volume (ml)
0.25–0.6	5	100
0.04–0.3	5	50
0.006–0.06	10	50
0.001–0.008	25	25

[65] L. F. Gorin, A. A. Nemodruk, and V. V. Stasyuchenko, *Izv. Vyssh. Ucheb. Zaved., Khim. Khim. Tekhnol.* **6**, 385–389 (1963).

[66] Louis Silverman and Rachel L. Seitz, *Anal. Chim. Acta* **20**, 340–343 (1959).

[67] British Standards Institution, *BS 3727*, Pt. **4**, 7 pp (1964).

[68] A. Funke and H. J. Laukner, *Z. Anal. Chem.* **249**, 26–30 (1970).

[69] G. Valence and S. Marques, *Chim. Anal.* (Paris) **49**, 275–284 (1967).

[70] V. M. Peshkova and V. M. Bochkova, *Tr. Kom. Anal. Khim., Akad. Nauk SSSR, Inst. Geokhim. Anal. Khim.* **8**, 125–134 (1958); V. M. Peshkova, E. K. Astakhova, I. F. Dolmanova, and V. M. Savostina, *Acta Chim. Hung.* **53**, 121–125 (1967); cf. Walter Nielsch, *Mikrochim. Acta* **1959**, 725–734.

[71] Manuel Needleman, *Anal. Chem.* **38**, 915–917 (1966); cf. R. C. Rooney, *Metallurgia* **58**, 205–208 (1958); **62**, 175–180 (1960); *UKAEA Rep.* **PG 599 (S)**, 8 pp (1963); Tran Van Danh, J. Spitz, and C. Mathien, *Anal. Chim. Acta* **36**, 204–209 (1966).

Dilute to 50 ml and add 25 ml of 2 *M* sodium acetate. Add *M* sodium fluoride until the color of iron has disappeared. Add 10 ml of 2.5% solution of 2-nitroso-1-naphthol in acetone, mix, and let stand for 2 minutes. Add 20 ml of hydrochloric acid and shake for 30 seconds with the appropriate volume of amyl acetate. Discard the aqueous layer and wash the organic layer successively with 10-ml portions of water, hydrochloric acid, water, 4% sodium hydroxide solution, water, hydrochloric acid, water, 4% sodium hydroxide solution, and two portions of water. If an emulsion forms at any stage, break it by addition of a few grams of potassium chloride. Dilute to volume with amyl acetate and dry with sodium sulfate. Filter and read at 362 nm against a reagent blank.

Tungsten Alloys. Decompose 0.5 gram of sample with 5 ml of sulfuric acid, 5 ml of nitric acid, and 1 ml of phosphoric acid. Add 50 ml of water to complete solution, cool, and dilute to 100 ml. Take an aliquot. Complete as above from "Dilute to 50 ml and ... ."

In an analysis of iron and its alloys, the chromium may first be oxidized to chromate and extracted with a solution of methyltrioctylammonium chloride in chloroform. Then iron, copper, lead, and uranium are extracted from slightly alkaline or slightly acid solution by phenylacetic acid in chloroform.[72] Zinc and cadmium are partially extracted. Cobalt, nickel, and manganese remain in the aqueous phase for determination of the cobalt.

Soils and Rocks.[73] As the color reagent, dissolve 0.04 gram of 2-nitroso-1-naphthol in 0.4 ml of 4% sodium hydroxide solution and dilute to 100 ml.

Take a 10-ml aliquot of the sample solution equivalent to 0.2 gram of sample; it may contain up to 10 μg of cobalt. Add sequentially 0.5 ml of saturated bromine water, 10 ml of 20% ammonium citrate solution, 1 ml of 10% sodium thiosulfate solution, and a drop of 1% phenolphthalein indicator solution in ethanol. Add ammonium hydroxide solution dropwise to a distinct pink. Add 2 ml of the color reagent and 5 ml of isoamyl acetate. Shake for 1 minute and set aside for 1 hour. Discard the aqueous layer. Add 5 ml of *N* hydrochloric acid to the organic layer, shake for 30 seconds, and let stand for 2 minutes. Repeat this three times, then discard the aqueous layer. This removes copper and helps later steps in removal of nickel. Add 5 ml of 4% sodium hydroxide solution, shake for 30 seconds, and let stand for 2 minutes. Repeat the shaking and separation. Discard the aqueous layer. Repeat the previous step from "Add 5 ml of 4% sodium... ." Finally shake for 30 seconds with 5 ml of *N* hydrochloric acid, let stand for 2 minutes, and shake again. These steps remove excess reagent as well as the indicator and the last of the nickel. Discard the washings, filter the organic layer through a pledget of cotton, and read at 530 nm against isoamyl acetate.

Nonferrous Ores.[74] To the solution of the ore, add 20 ml of 2 *M* sodium acetate, 1 ml of hydrochloric acid, and 10 ml of 0.5% methanolic 2-nitroso-1-naphthol. Mix for 30 seconds, set aside for 5 minutes, and extract with 20 ml of benzene. Wash the benzene extract successively with 10 ml of water, 10 ml of 1:5 hydrochloric

[72] Jiri Adam and Rudolf Pribil, *Hutn. Listy* **27**, 137–138 (1972).

[73] Lewis J. Clark and J. H. Axley, *Anal. Chem.* **27**, 2000–2003 (1955); Lewis J. Clark, *ibid.* **30**, 1153–1156 (1958); cf. D. E. Bodart, *Chem. Geol.* **6**, 133–142 (1970).

[74] Yu. L. Lel'chuk, L. L. Skripova, and P. V. Kristalev, *Izv. Sibir. Otdel. Akad. Nauk SSSR* **1960** (11), 63–70.

acid, 10 ml of water, 10 ml of 4% sodium hydroxide solution, and 10 ml of water. Filter, dilute to 50 ml with benzene, and read at 530 nm. If the ore contains iron, remove it from the solution of the sample by use of cupferron or mask it by chelation.[75]

Nitric Acid.[76] Evaporate 10 grams of sample almost to dryness in silica. Add 10 ml of phosphate buffer solution for pH 6.1 and 2 ml of 0.02% solution of 2-nitroso-1-naphthol to the residue. Heat almost to boiling, cool, and shake for 10 minutes with 5 ml of chloroform. Wash the extract by shaking for 20 minutes each with, successively, 5 ml of 16% sodium hydroxide solution, 5 ml of 1:2 hydrochloric acid, and 5 ml of 16% sodium hydroxide solution. Read at 307 nm against a blank.

Sodium Hydroxide. Dissolve 10 grams of sample in 25 ml of water, neutralize with nitric acid, and add 3 ml in excess. Evaporate almost to dryness and take up in 20 ml of water. Proceed as for nitric acid from "Add 10 ml of phosphate buffer solution for pH 6.1. . . ."

Alkali Solutions.[77] This technic is designed for solutions used in depolymerization of cellulose. Neutralize 50 ml with 1:1 hydrochloric acid and adjust to pH 6.5 with a sodium tetraborate–ammonium hydroxide buffer solution. Add 5 ml of 0.02% solution of 2-nitroso-1-naphthol. Shake for 3 minutes with 20 ml of chloroform. Wash the organic phase with 10 ml of 4% sodium hydroxide solution containing a drop of 10% potassium cyanide solution, then with 10 ml of 10% sodium chloride solution. Read at 530 nm.

Seawater.[78] Add 20 ml of 1:1 hydrochloric acid to 1 liter of sample and boil for 15 minutes. Cool. Mask iron, aluminum, and calcium with 5 ml of 20% ammonium citrate solution. Adjust to pH 7.35–7.4 with ammonium hydroxide. Add 2 ml of 0.04% solution of 2-nitroso-1-naphthol and shake for 4 minutes. Then add 10 ml of 0.04% solution of the reagent in toluene and shake for 5 minutes. Let stand over night, and separate the organic layer. Wash successively with 6 ml of 1:2 hydrochloric acid, 30 ml of water without shaking, 6 ml of 10% sodium hydroxide solution, then 40 ml and 20 ml of water without shaking. Filter, read at 530 nm against toluene, and correct for a reagent blank.

Peas.[79] Incinerate 150 grams of sample at 540°. Dissolve the ash in 15 ml of hydrochloric acid and evaporate to dryness at 100°. Dissolve the residue in 25 ml of water, filter, and wash. Add 2 drops of methyl orange indicator solution and neutralize with 10% sodium hydroxide solution. Add 5 ml of 30% sodium acetate solution and boil. Filter. Wash the precipitate sequentially with 3 ml of 20% solution of sodium thiosulfate, 5 ml of 5% monosodium phosphate solution, and finally with water. Add 2 ml of 0.5% solution of 2-nitroso-1-naphthol to the filtrate

[75]Yu. L. Lel'chuk and L. L. Skripova, *Izv. Tomsk. Politekhn. Inst.* **III**, 55–58 (1961).
[76]V. M. Savostina and V. M. Peshkova, *Tr. Kom. Anal. Khim., Akad. Nauk SSSR* **14**, 298–302 (1963).
[77]Mrs. L. Buzas, *Magy. Kem. Lapja* **26**, 593–594 (1971).
[78]L. I. Rozhanskaya, *Tr. Sevastop. Biol. Sta* **15**, 503–511 (1964).
[79]E. Szyszko and B. Choznicka, *Rocz. Panstw. Zakl. Hig.* (Warsaw) **12**, 125–130 (1961).

and set it aside in the dark for 30 minutes. Shake for 10 minutes with 5 ml of carbon tetrachloride. Add 10 ml of water and 2 ml of 10% sodium hydroxide and shake again. Reject the bulk of the aqueous phase, add 10 ml of 2% hydrochloric acid to the balance and shake. Filter the organic layer and read at 455 nm.

Fish.[80] Ash 200 grams of sample at 550° and dissolve the ash in 20 ml of hydrochloric acid. Evaporate to dryness, take up in 25 ml of water, and filter. Wash the filter with hot water, add 2 drops of methyl orange indicator solution, and neutralize with 10% sodium hydroxide solution. Add 5 ml of 30% sodium acetate solution and boil. Filter. Wash the precipitate successively with 3 ml of 20% sodium thiosulfate solution, 5 ml of 5% monosodium phosphate solution, and water. Add 2 ml of 0.5% solution of 2-nitroso-1-naphthol and set aside in the dark for 30 minutes. Shake for 10 minutes with 5 ml of carbon tetrachloride. Add 10 ml of water and 2 ml of 10% sodium hydroxide solution and shake again. Reject most of the aqueous phase, add 10 ml of 2% hydrochloric acid to the balance, and shake. Filter the organic layer and read at 455 nm.

Plant Material.[81] Oxidize 10 grams of dried sample with 5 ml of nitric acid and 10 ml of 1:1 sulfuric acid. According to the nature of the sample, it may be necessary to add 30% hydrogen peroxide or perchloric acid. When colorless, evaporate to sulfur trioxide fumes. Take up with 80 ml of water, filter, and wash the filter. Add 5 ml of 50% solution of sodium citrate and raise to pH 3.4 with 40% sodium hydroxide solution. Add 2 ml of 30% hydrogen peroxide and 4 ml of 1% solution of 2-nitroso-1-naphthol in acetic acid. Extract the cobalt complex with 20 ml and 10 ml of chloroform. If more than 25 mg of copper is present, wash the combined extracts with 20 ml of 2 *N* sulfuric acid. Evaporate the chloroform. To the residue, add 0.3 ml of sulfuric acid, 3 ml of nitric acid, and 1 ml of perchloric acid. Heat to sulfur trioxide fumes and take up with 10 ml of water. Add 0.5 ml of 50% solution of sodium citrate and adjust to pH 3–4. Add 2 ml of 1% solution of 2-nitroso-1-naphthol in acetic acid and set aside for 45 minutes. Extract the cobalt complex with 5 ml of chloroform. Wash excess reagent from the chloroform with 3 ml of 8% sodium hydroxide solution. Read at 367 nm.

NITROSO-R SALT

The red complex of cobalt with nitroso-R salt, which is the disodium salt of 3-hydroxy-4-nitroso-2,7-naphthalene disulfonic acid, formed at pH 5–7 is stable in strongly acid solution.[82] The absorption of the reagent below 500 nm interferes with reading the cobalt complex at its maximum absorption at 415 nm; therefore unreacted excess reagent must be decomposed for reading at the maximum. A satisfactory alternative is to read at 500–600 nm, where excess reagent does not interfere. Warming is necessary to avoid slow development of color. Iron is satisfactorily masked with tetrasodium pyrophosphate. The colored complexes with

[80]B. Choznicka and E. Szyszko, *ibid.*; **15**, 23–26 (1964).
[81]H. Ssekaalo, *Anal. Chim. Acta* **51**, 503–508 (1970).
[82]Yoshinaga Oka and Saburo Ayusawa, *Nippon Kinsoku Gakkaishi* **19**, 68–69 (1957).

chromium, nickel, and copper are readily decomposed with sulfuric acid. The reagent is unstable to heat and light, and this property provides a mechanism for destruction of excess reagent. The reagent is as accurate as thiocyanate, and results are more reproducible.

Cobalt is separated chromatographically with nitroso-R salt from 10,000-fold nickel, 2000-fold iron, 750-fold chromium, and 500-fold copper.[83] Pass the solution, containing 350–1000-fold excess of nitroso-R salt relative to the cobalt, and up to 2 *N* with hydrochloric acid, through a 15×1 cm column of Amberlite IRA-400 resin. Elute nickel with 0.3 *N* hydrochloric acid, ferric ion with 0.5 *N* phosphoric acid, chromic ion with 0.5 *N* hydrochloric acid, and copper with 2 *N* hydrochloric acid. Then elute the cobalt with 5 *N* perchloric acid for determination with nitroso-R salt.

For determination of cobalt in uranium oxide (known as yellow cake), uranium minerals, and geological samples, add perchloric acid and hydrofluoric acid to remove silica, add 1:1 hydrochloric acid, and evaporate to dryness.[84] Take up in 1:1 hydrochloric acid and add tetrahydrofuran and 2-methoxyethanol. Treat a column of Dowex 1-X8 resin in chloride form with 1:5:4 hydrochloric acid–tetrahydrofuran–2-methoxyethanol. Pass the sample through the column and wash with the 1:5:4 solvent mixture to remove iron, vanadium, molybdenum, alkalies, and alkaline earths. Elute the cobalt with 1:1 hydrochloric acid. Evaporate an aliquot of the eluate containing less than 10 μg of cobalt to dryness. Add perchloric acid and again evaporate to dryness. Dissolve the residue in 1:1 hydrochloric acid and develop cobalt with nitroso-R salt.

For cobalt in chromium, dissolve in hydrochloric acid, pass air through the solution to form chromic ion, and add potassium chlorate to complete the oxidation.[85] Pass through an anion exchange resin, then elute cobalt with 1:2 hydrochloric acid for determination by nitroso-R salt.

For analysis of magnesium, the cobalt from the solution of a 5-gram sample can be coprecipitated at pH 8.6–8.8 with 5 mg of aluminum for determination with nitroso-R salt.[86]

For cobalt in tungsten and tungsten alloys, dissolve in hydrofluoric and nitric acids.[87] Adjust to pH 6 with an ammonium tartrate–sodium borate buffer solution and develop with nitroso-R salt.

For analysis of reactor grade zirconium, dissolve 0.1 gram of sample in hydrochloric and nitric acids with addition of a few drops of hydrofluoric acid.[88] Evaporate nearly to dryness with 0.1 gram of aluminum chloride and take up in 9 *N* hydrochloric acid. Pass through a column of Dowex 1-X8 resin, then elute the cobalt with 25 ml of 1:2 hydrochloric acid. Adjust the eluate to pH 8.2 with sodium hydroxide solution, and add citrate. Extract the cobalt with 5 ml and 5 ml of 0.05% dithizone in ether. Evaporate to dryness, add 1 ml of nitric acid, and take to dryness. Add 5 ml of perchloric acid and evaporate to dryness. Add 2 ml of 0.1 *M* sodium chloride solution, 1 ml of hydrochloric acid, and 2 ml of water.

[83]Krystna Brajter, *Chem. Anal.* (Warsaw) **18**, 125–135 (1973).
[84]J. Korkish and D. Dimitriadis, *Microchim. Acta* **1974**, 549–560.
[85]Tadashi Yanagihara, Nobuhisa Matano, and Akira Kawase, *Jap. Anal.* **9**, 439–443 (1960).
[86]Shizo Hirano and Atsushi Mizuike, *ibid.* **9**, 623–624 (1960).
[87]George Norwitz and H. Gordon, *Anal. Chem.* **37**, 417–419 (1965).
[88]J. Vogel, D. Monnier, and W. Haerdi, *Anal. Chim. Acta* **24**, 55–60 (1961).

Evaporate to dryness and dissolve in 2 ml of water. Develop with nitroso-R salt and read at 520 nm against a blank.

For cobalt in zirconium and Zircaloy, dissolve in hydrofluoric acid, buffer, and extract with chloroform as the nitrosonaphtholate complex.[89] Evaporate the extract and destroy the organic matter. Dissolve in hydrochloric acid, buffer, and add nitroso-R salt. Decompose complexes of other heavy metals with hydrochloric acid and read at 510 nm.

In analysis of vanadium pentoxide, cobalt is separated by extraction as the 1-nitroso-2-naphthol complex at pH 3–4, the vanadium being masked with citrate.[90] Cobalt is then determined with nitroso-R salt.

In analysis of stainless steel, separate tungsten as the tungstate and reduce vanadium to the tetravalent form. Extract the iron from 8 *N* hydrochloric acid solution with 2:1 methyl isobutyl ketone–pentyl acetate.[91] Then adsorb 0.001–0.2% of cobalt in the aqueous phase on Amberlite IRA-400, an anion exchange resin, to separate it from chromium and nickel. Elute the cobalt with 2.5 *N* hydrochloric acid, adjust to pH 7.8–8 with citrate buffer solution, add 0.2% of nitroso-R salt, and read at 420 nm.

For cobalt in 18/8 steels, separate iron from the solution by ether extraction.[92] Separate the cobalt on Dowex 1-X8 and elute with 1:2 hydrochloric acid. Extract with dithizone in ether, convert to the nitroso-R salt complex, and read at 520 nm.

For analysis of copper electrolyte, convert iron to its ferric complex with ammonium bifluoride.[93] Pass through a column of 60-80 mesh Amberlite IR-120 in sodium form. Copper, cobalt, zinc, and manganese are absorbed. Elute copper with 0.1 *M* sodium thiosulfate. Then elute cobalt, zinc, and manganese with 2.5 *M* sodium chloride. Determine cobalt in the eluate with nitroso-R salt.

For analysis of iron, steel, ores, and slags dissolved in acids, precipitate basic ferric acetate by boiling with sodium acetate solution and filter.[94] To an aliquot of the filtrate, add nitroso-R salt. Set aside for 5 minutes, or 15 minutes if the nickel content is high, add nitric acid, and again set aside for 5 minutes. Dilute to volume and read at 546 nm against a reagent blank. Oxidation with bromine will eliminate interference by chromic, ferrous, stannous, or tetravalent vanadium ions.

For analysis of ferrochrome, the solution is passed through AV-17 anion exchange resin in chloride form.[95] Nickel, chromium, and manganese are eluted with 9 *N* hydrochloric acid. Elution with 4 *N* hydrochloric acid then recovers the cobalt with a noninterfering fraction of the iron.

In analysis of samples high in iron, boiling with sodium acetate will form the basic ferric acetate, after which development is achieved by boiling with nitroso-R salt, addition of nitric acid, and boiling.[96] For analysis of solutions of samples high in iron, such as ferric ammonium alum, precipitate iron as the phosphate.[97] Extract

[89]P. Tate, N. R. Wilson, G. W. Goward, and E. W. Beiter, *USAEC* **WAPD-CTA(GLA)-326**, 5 pp (1957).
[90]S. Ya. Vinkovetskaya, *Nauch. Tr. Nauchn.-Issled. Proekt. Inst. Redkomet. Prom.* **1972**, 202–204 (1972).
[91]Shizo Hirano, Atsushi Mizuike, Yoshio Iida, and Nobuhiko Kokubu, *Jap. Anal.* **10**, 326–330 (1961).
[92]D. Monnier, W. Haerdi, and J. Vogel, *Anal. Chim. Acta* **23**, 577–584 (1960).
[93]Tetsuo Katsura, *Jap. Anal.* **10**, 1320–1330 (1961).
[94]K. H. Koch, K. Ohls, E. Sebastini, and G. Riemer, *Z. Anal. Chem.* **249**, 307–312 (1970).
[95]E. V. Novikova, V. I. Kurbatova, and V. V. Stepin, *Tr. Vses. Nauchn.-Issled. Inst. Stand. Obraztsov. Spaktr. Etalonov* **6**, 94–97 (1970).
[96]D. P. Shcherbov and D. N. Perminova, *Tr. Kaz. Nauchn.-Issled. Inst. Miner. Syr'a* **1960** (3), 333–337.
[97]I. G. Shafran and L. Ya. Mazo, *Sb. Statei, Vses. Nauchn.-Issled. Inst. Khim. Reakt. Osobo Chist. Khim. Veshchestv.* **1961** (24), 258–264.

the cobalt from the filtrate as the 1-nitroso-2-naphthol complex with carbon tetrachloride and determine by nitroso-R salt.

For analysis of seaweed, ash the sample and add 1 : 1 hydrochloric acid to the ash.[98] Evaporate to dryness and take up in 50 ml of 2 *N* nitric acid. Filter, treat the insoluble residue with hydrofluoric acid, dissolve, and add to the main sample. Extract the cobalt with 1-nitroso-2-naphthol in chloroform and determine with nitroso-R salt.

For cobalt in sodium iodide, dissolve in ammonium citrate solution at pH 8.5.[99] Wash out free iodine with chloroform. Add sodium diethyldithiocarbamate solution, and extract the complexes of manganese, cobalt, and nickel with chloroform. Evaporate the chloroform layer and destroy the residue with nitric and sulfuric acids. Dissolve in water and determine cobalt with nitroso-R salt.

For analysis of a multicomponent electrolytic bath containing nickel, cobalt, and iron, first oxidize with hydrogen peroxide.[100] Precipitate nickel as hexaamminenickel perchlorate and filter. Then develop cobalt with nitroso-R salt.

Nitroso-R salt is applicable to concentrates rich in copper, nickel, and iron and to analysis of titanium, zirconium, and their alloys.[101] For analysis of manganese ores, ferromanganese, and manganese metal, separate cobalt by column chromatography before determination by nitroso-R salt.[102a]

Procedures

Extraction by Trioctylamine.[102b] As solvent, prepare a 10% solution of trioctylamine in carbon tetrachloride and wash it with an equal volume of 1 : 1 hydrochloric acid, then with an equal volume of 0.1 *N* hydrochloric acid.

Mix 25 ml of nitrate-free sample solution with 50 ml of hydrochloric acid. Extract the cobalt with 10 ml and 10 ml of the prepared trioctylamine solvent. Reextract the cobalt from the combined solvent solutions with 10 ml and 10 ml of 0.1 *N* hydrochloric acid and dilute to 100 ml. Boil a 25-ml aliquot with 5 ml of 50% solution of sodium acetate for 2 minutes. Add 10 ml of 0.1% solution of nitroso-R salt and continue to boil for 2 minutes. Add 10 ml of nitric acid and continue to boil for 2 minutes. Cool, dilute to 100 ml, and read at 500 nm.

Separation of Cobalt and Iron.[103] ALLOYS. Dissolve a 0.5-gram sample in hydrochloric acid and dilute to 100 ml. Adjust a 10-ml aliquot to 2 *N* with hydrochloric acid. Extract the iron by shaking for 10 minutes with 10 ml of 5% solution of trioctylamine in carbon tetrachloride. Dilute the aqueous phase with hydrochloric acid to give 25 ml of a 7.7 *N* solution. Extract the cobalt by shaking for 10 minutes with 25 ml of 5% solution of octylamine in carbon tetrachloride. Strip the cobalt

[98] Toshio Yamamoto, Tetsuo Fujita, and Masayoshi Ishibashi, *J. Chem. Soc. Jap., Pure Chem. Sect.* **86**, 49–53 (1965).
[99] A. M. Bulgokova, A. B. Blank, A. K. Khurkryanskii, and G. S. Plotnikova, *Stsintill. Stsintill. Materily 2-go (Vtorogo) Koordi. Soveshch.* **1957**, 281–290.
[100] Andrzej Jaklewicz, *Chem. Anal.* (Warsaw) **19**, 207–210 (1974).
[101] L. D. Dolaberidze and A. G. Dzhaliashvili, *Tr. Kavkaz. Inst. Miner. Syr'ya* **1971** [9(11)], 389–397.
[102a] D. F. Wood and R. T. Clark, *Talenta* **2**, 1–11 (1959).
[102b] E. Sh. Ioffe and A. D. Karaseva, *Zavod. Lab.* **33**, 1502 (1967).
[103] B. E. McClellan and V. M. Benson, *Anal. Chem.* **36**, 1985–1987 (1964).

from the organic layer with 25 ml of *N* hydrochloric acid and dilute the extract to 100 ml. Complete as for extraction by trioctylamine from "Boil a 25-ml aliquot. . . ."

Separation by Trioctylmethylammonium Chloride.[104] This reagent is known commercially as Aliquat. To the sample solution containing up to 20 μg of cobalt, add 5 ml of 0.0188% solution of nitroso-R salt and 0.5 ml of 0.5 *M* sodium acetate. Set aside for 15 minutes and add 2 ml of 0.1 *N* hydrochloric acid. Dilute to 25 ml and shake for 2 minutes with 5 ml of 5% solution of the quaternary in chloroform. Filter the organic layer and read it at 500 nm. Iron, copper, or nickel does not interfere.

Hydrazine Added.[105] Prepare a reagent that is 6 *N* with potassium hydroxide and 0.5 *M* with hydrazine. Mix 10 ml of sample solution, 10 ml of reagent, and 10 ml of nitric acid. Evaporate until evolution of brown fumes ceases. Take up with 20 ml of water. Add 5 ml of 0.1% solution of nitroso-R salt and 5 ml of 50% solution of sodium acetate. Boil for 1 minute, cool, and dilute to 50 ml with 50% solution of sodium acetate. Read at 500 nm.

Solutions.[106] Adjust a 25-ml sample to pH 4.5 by adding 5 ml of acetate–acetic acid buffer. Add 0.2 ml of 30% hydrogen peroxide and heat to 100° to destroy the excess. Add 10 ml of 40% sodium acetate solution and 10 ml of 0.1% solution of nitroso-R salt. Boil for 2 minutes, add 1 ml of nitric acid, continue to boil for 1 minute, and cool. Dilute to 50 ml and read at 500 nm.

Samples High in Iron.[107] More than 100 μg of iron per ml interferes with determination of 0.4 μg of cobalt per ml with nitroso-R salt. The iron can be extracted with ether from 1 : 1 hydrochloric acid solution or precipitated as follows. A typical sample would be tailings from magnetic separation.

Moisten 1 gram of sample with 2 ml of water, add 10 ml of hydrochloric acid, and boil for 10 minutes. Add 15 ml of hydrochloric acid and 5 ml of nitric acid and evaporate to a couple of ml. Add 10 ml of 1 : 1 sulfuric acid and evaporate to less than 1 ml. Cool, boil with 50 ml of water to dissolve soluble matter, and cool. Add a thick creamy suspension of zinc oxide in small portions until precipitation of iron is complete, dilute to 100 ml, and filter. To a 10-ml aliquot, add 0.5 ml of 1 : 4 sulfuric acid and 5 ml of 50% solution of sodium acetate. Boil for 2 minutes, add 10 ml of 0.1% solution of nitroso-R salt, and boil for 1 minute. Cool, dilute to 50 ml, and read at 520 nm. This technic is also applicable to samples high in copper if the copper and iron are coprecipitated.

An alternative is to precipitate with cupferron, extract with chloroform, and calcine with nitric acid and hydrogen peroxide. Then dissolve in dilute nitric and hydrochloric acids for development with nitroso-R salt.[108]

[104] Jiri Adam and Rudolf Pribil, *Talenta* **18**, 733–737 (1971).
[105] E. I. Mingulina, N. G. Ryzhova, and E. Klochek, *Tr. Mosk. Energ. Inst.* **1972** (112), 102–107.
[106] Cf. W. Haerdi, J. Vogel, D. Monnier, and P. E. Wenger, *Helv. Chim. Acta* **42**, 2334–2342 (1959).
[107] L. N. Krasil'nikova and K. N. Dolgorukova, *Sb. Nauch. Tr. Vses. Nauchn.-Issled. Gornomet. Inst. Tsvet. Met.* **1965** (9), 30–33; cf. Shigeo Wakamatsu, *Jap. Anal.* **8**, 830–832 (1959); Theo Kurt Willmer, *Arch. Eisenhüttenw.* **29**, 159–164 (1958).
[108] L. T. Ikramov, *Tr. Tash. Farm. Inst.* **2**, 341–344 (1960).

Alloy Steel.[109] Dissolve 1 gram of sample containing 0.001–0.005% cobalt in 10 ml of hydrochloric acid. Add 1 ml of nitric acid and evaporate nearly to dryness. Dissolve in 6 ml of 8 *N* hydrochloric acid, filter, and dilute to 50 ml with 8 *N* hydrochloric acid. Apply a 15-ml aliquot to a 28×1 cm column of AV17-X8 resin in chloride form. Wash nickel, chromium, titanium, manganese, and most of the niobium through the column with 100 ml of 8 *N* hydrochloric acid at 8 drops/minute. Elute the cobalt with 100 ml of 4.8 *N* hydrochloric acid at 0.6 ml/minute. Molybdenum can then be eluted with 400 ml of 2 *N* hydrochloric acid and iron by 200 ml of 0.5 *N* hydrochloric acid.

Evaporate the cobalt eluate to dryness and take up in 25 ml of an acid mixture containing 25 ml of hydrochloric acid and 5 ml of nitric acid per liter. Add 2.5 ml of 0.2% solution of nitroso-R salt and 10 ml of 50% sodium acetate solution. Boil for 1 minute, add 5 ml of nitric acid, and again boil for 1 minute. Cool, dilute to 50 ml, and read at 440 or 490 nm against a reagent blank.

Nickel.[110] Prepare a 9×1.2 cm column of 100-200 mesh Amberlite IRA-400 in chloride form. Dissolve a 10-gram sample containing 0.01–0.2 μg of cobalt in hydrochloric acid, make the solution 9 *N* with hydrochloric acid, and pass through the column. Only the cobalt is adsorbed. Wash out the nickel with 70 ml of 9 *N* hydrochloric acid. Elute the cobalt with 30 ml of 0.1 *N* hydrochloric acid and complete as for alloy steel from "Evaporate the cobalt eluate. . . ."

High Purity Nickel.[111] Dissolve 0.5 gram of sample by heating with 5 ml of 1:1 nitric acid and boil off brown fumes. Add 15 ml of water and 5 ml of 10% citric acid solution. Heat just to boiling. Remove from the heat and add 2 ml of 3% hydrogen peroxide. Mix and add 20 ml of ammonium hydroxide. Set aside for 30 seconds and add 10 ml of 1:2 perchloric acid. Cool to under 20° in cold water and filter with suction on a Büchner funnel. Wash the precipitate on the filter with the washings of the reaction vessel. Discard the precipitate and evaporate the filtrate and washings to 25 ml. Add 3 grams of sodium hydroxide and 2 drops of hydrofluoric acid. Boil until the odor of ammonia is gone, and concentrate to 15 ml. Add 10 ml of water and neutralize with nitric acid to a reddish purple of Congo red paper. Add 5 ml of sodium acetate solution as buffer and 2 ml of 1% solution of nitroso-R salt. Heat just below boiling for a minute, add 5 ml of 1:2 nitric acid, heat just to boiling, and cool to 25° in water. Dilute to 50 ml and read immediately at 515 nm. If during reading the solution darkens or clouds up, let stand for 4 minutes, filter, and read.

Cobalt and Nickel Concentrates.[112] Treat 0.1 gram of ignited concentrate with 0.5 gram of ammonium fluoride, 5 ml of hydrochloric acid, and 5 ml of nitric acid.

[109]V. P. Novak, A. P. Marynov, and V. F. Mal'tsev, *Zavod. Lab.* **37**, 412–413 (1971); cf. V. P. Novak, S. S. Bedovik, V. I. Bogovina, and V. F. Mal'tsev, *ibid.* 912–913; Shizo Hirano, Atsushi Mizuike, Yoshio Iida, and Nobuhiko Kokubun, *Jap. Anal.* **10**, 326–330 (1961).

[110]Yoshio Iida, Atsushi Mizuike, and Shizo Hirano, *J. Chem. Soc. Jap., Ind. Chem. Sect.* **67**, 2042–2045 (1964); cf. Atsushi Mizuike, Yoshio Iida, and Shizo Hirano, *ibid.* **61**, 1459–1460 (1958); S. E. Kreimer, N. V. Tuzhilina, V. A. Golovina, and R. A. Tyabina, *Zavod. Lab.* **24**, 262–264 (1958).

[111]C. L. Luke, *Anal. Chem.* **32**, 836–837 (1960).

[112]L. B. Ginzburg and E. P. Shkrobot, *Sb. Nauch. Tr. Gos. Nauch.-Issled. Inst. Tsvet. Met.; Anal. Rud Tsevt. Met. Prod. Pererab.*) **1956** (12), 52–69.

Evaporate to 2 ml. Add 10 ml of 1 : 1 sulfuric acid and evaporate to sulfur trioxide fumes. Add water and dilute to 250 ml. Dilute a 25-ml aliquot to 250 ml. Dilute a 25-ml aliquot of this dilution to 50 ml and add 5% sodium hydroxide solution dropwise until it becomes cloudy. Clarify with 1 : 1 sulfuric acid, add 5 ml of 50% sodium acetate solution, and boil for 5 minutes. Add 10 ml of 0.2% solution of nitroso-R salt and boil for 1 minute. Add 5 ml of nitric acid, boil for 1 minute, and dilute to 100 ml. Read at 500 nm.

Copper Ores and Concentrates[113]

1.4–14% COPPER. To 0.5 gram of sample, add 16 ml of hydrochloric acid and 4 ml of nitric acid. Evaporate to dryness, add 5 ml of sulfuric acid, and heat to sulfur trioxide fumes. If the solution is colorless, evaporate almost to dryness; if dark, add 5 ml of 3 : 1 nitric acid–sulfuric acid, and evaporate to sulfur trioxide fumes, repeating until the solution is colorless. Add 20 ml of water, heat for 10 minutes, and cool. Filter, and wash the residue with a minimal amount of water. Concentrate to 20 ml and neutralize with ammonium hydroxide. Acidify with 2 drops of 1 : 2 sulfuric acid, add 5 ml of 50% sodium acetate solution, and boil for 3 minutes. Add 10 ml of 0.5% solution of nitroso-R salt and boil for 2 minutes. Add 5 ml of 1 : 1 nitric acid and boil for 1 minute. Dilute to 100 ml and read at 500 nm.

UNDER 1.4% COPPER. Heat 2 grams of sample with 3 ml of hydrochloric acid. After a few minutes add 3 ml of nitric acid and evaporate to dryness. Repeat the addition of nitric acid and evaporation to dryness several times. Add 10 ml of hydrochloric acid and evaporate to dryness. Take up in 5 ml of hydrochloric acid and 70 ml of water. Boil until solubles are dissolved, filter, and wash the filter. Concentrate to 80 ml, and add 50% sodium hydroxide solution until the solution becomes turbid. Acidify with acetic acid, add 50 ml of 50% potassium nitrite solution, and set aside for 10 minutes. Filter, and wash the cobaltinitrite precipitate with 2% potassium nitrite solution until the washings are colorless. Dissolve the precipitate with hot 1 : 1 hydrochloric acid, filter, and wash the filter with hot water. Evaporate the filtrate to dryness and heat the residue to dissolution with 0.5 ml of hydrochloric acid and 40 ml of water. If turbid, filter and wash with water. Dilute to 100 ml. To a 25-ml aliquot add 5 ml of 50% sodium acetate solution and boil for 2 minutes. Add 10 ml of 0.2% solution of nitroso-R salt and boil for 2 minutes. Add 5 ml of 1 : 1 nitric acid and boil for 1 minute. Dilute to 50 ml and read at 500 nm.

Bismuth.[114] Grind the sample to a coarse powder, then wash with hot 1 : 1 hydrochloric acid followed by water. Dissolve 1 gram in 3 ml of nitric acid and evaporate the oxides of nitrogen. Cool, and add 20 ml of 50% solution of sodium tartrate. Add 3 drops of phenolphthalein indicator solution and neutralize by dropwise addition of 40% sodium hydroxide solution plus 1 ml in excess. Add 1 ml of 1% solution of 1-nitroso-2-naphthol in acetic acid. Add 3 ml of 2 *N* nitric acid and heat at 100° for 2 hours. Cool, and extract with 5 ml, 5 ml, and 5 ml of chloroform. Wash the combined chloroform extracts with 5 ml of 0.1 *N* hydrochloric acid by shaking for 2 minutes. Transfer the chloroform extracts to a quartz

[113] K. Kunz and E. Duczyminska, *Rudy Met. Niezelaz.* **8**, 302–305 (1963).
[114] V. A. Nazarenko and G. G. Shitareva, *Zavod. Lab.* **24**, 932–934 (1958).

crucible, add 2 ml of nitric acid, and evaporate the chloroform by heating. Add 0.1 ml of perchloric acid and evaporate to dryness at 100°. Finish on a sand bath and ignite at 400–500°. Cool, add 1 ml of hydrochloric acid, and evaporate at 100°. Wet the residue with 0.2 ml of 1:2 nitric acid. Take up with 2.5 ml of water and add 0.2 ml of 0.1% solution of nitroso-R salt. Add 0.5 gram of sodium acetate and heat at 100° for 2 minutes. Add 0.5 ml of nitric acid and heat at 100° for 1 minute. Cool, let stand for 10 minutes, and read at 500 nm.

Uranium Metal.[115] Dissolve 1 gram of sample in 3 ml of nitric acid and evaporate to dryness. Take up in 10 ml of water and add 10 ml of 50% solution of ammonium citrate. Neutralize with ammonium hydroxide and add 1 ml in excess. Dilute to 25 ml and add 2 ml of 1% solution of sodium diethyldithiocarbamate. Shake for 3 minutes each with 5 ml, 5 ml, and 5 ml of chloroform. Wash the combined chloroform extracts with 20 ml of water. Evaporate the chloroform and add to the residue 3 drops each of nitric acid and sulfuric acid. Destroy organic matter by heating to sulfur trioxide fumes. Take up in 5 ml of water and adjust to pH 6 with saturated solution of sodium acetate. Add 0.5 ml of 0.1% solution of nitroso-R salt and heat at 100° for 10 minutes. Add 3 ml of 1:1 hydrochloric acid and 3 drops of 30% hydrogen peroxide. Destroy excess reagent by heating at 100° for 30 minutes. Cool, dilute to 15 ml, and read at 420 nm.

Magnetic Ceramics.[116] Fuse 1 mg of sample with 10 mg of ammonium persulfate. Dissolve in 4 ml of 1:5 hydrochloric acid and adjust to pH 1–1.5. Mask iron with 0.5 ml of 10% oxalic acid solution. Add 0.2 ml of 1% solution of nitroso-R salt and 0.2 ml of 20% solution of diphenylguanidinium chloride as a coprecipitating agent. Add 40% ammonium acetate solution to adjust to pH 5–7. Boil with 5 drops of 1:1 nitric acid to destroy complexes of chromium, copper, and nickel. Cool, and add 0.2 ml of 5% solution of 2-naphthol and 0.2 ml of 5% solution of phenolphthalein in acetone. As a wash solution, dissolve 0.5 gram of oxalic acid, 0.4 gram of nitroso-R salt, 0.8 gram of diphenylguanidinium chloride, 0.2 gram of 2-naphthol, and 0.2 gram of phenolphthalein in 100 ml of water. Filter the precipitate and wash with the wash solution. Dissolve the washed precipitate in 5 ml of ethanol and read at 500 nm.

Magnetic Film.[117] The product referred to is normally on a glass substrate. As a buffer solution for pH 8, dissolve 6.2 grams of boric acid and 35.6 grams of disodium phosphate dihydrate in a liter of 2% sodium hydroxide solution.

With ethanol, wash the substrate, carrying about 100 μg of film, and dissolve the film in 2–3 drops of hot 1:1 nitric acid. Heat to drive off oxides of nitrogen, and dilute to 10 ml. Evaporate a 1-ml aliquot to dryness and dissolve the residue in 0.5 ml of 4.2% solution of citric acid monohydrate and 0.6 ml of the buffer solution for pH 8. Add 0.5 ml of 0.1% solution of nitroso-R salt and boil for 1 minute. Add 0.25 ml of nitric acid and continue to boil for 1 minute. Cool, dilute to 5 ml, and read at 500 nm against a reagent blank.

[115]Masami Suzuki and Tsugio Takeuchi, *Jap. Anal.* **9**, 179–181 (1960).
[116]V. V. Gorshkov, L. I. Mekhryuscheva, and L. A. Smakhtin, *Zavod. Lab.* **37**, 396–398 (1971).
[117]A. S. Babenko and T. T. Volodchenko, *ibid.* **33**, 1059–1061 (1967).

Aluminum-Cobalt Wire.[118] As a buffer solution, dissolve 36 grams of disodium phosphate dihydrate and 6.2 grams of boric acid in 500 ml of 4% sodium hydroxide solution and dilute to 1 liter.

Dissolve 46 mg of sample in 3 ml of 1:1 hydrochloric acid, add 1 ml of 10% phosphoric acid solution, and adjust to pH 3.5. Add 5 ml of 1% solution of PAN in 1:1 acetic acid. Let stand for 1 hour, then extract with 8 ml, 8 ml, and 8 ml of chloroform. To the combined extracts, add 1 ml of nitric acid, 0.75 ml of perchloric acid, and 50 mg of sodium nitrate, the latter to minimize sorption of cobalt on glass. Heat until the solvent is gone and the organic matter destroyed to give a clear solution. Dilute to 25 ml and take an aliquot expected to contain 2 μg of cobalt. Add 1 ml of 0.2 *M* citric acid to prevent precipitation of metal hydroxides, then 1.2 ml of the buffer solution. Adjust to pH 7.6 ± 0.4 with 4% sodium hydroxide solution. Add 0.2 ml of 0.5% solution of nitroso-R salt and set aside for 30 minutes. Add 1 ml of nitric acid and dilute to 10 ml. Heat at 100° for 10 ± 0.1 minute, cool, and read at 500 nm.

Selenium.[119a] Dissolve 0.5 gram of sample in 10 ml of hydrochloric acid and 2 ml of nitric acid. Evaporate to dryness, take up in 2 ml of hydrochloric acid, and add 5 ml of water. Add 30 ml of 20% solution of sodium citrate, and neutralize to phenolphthalein. Add 2 ml of 1% solution of sodium diethyldithiocarbamate and set aside for 10 minutes. Extract the cobalt with 25 ml and 5 ml of chloroform. Reextract the cobalt from the chloroform with 10 ml, 10 ml, 5 ml, and 5 ml of 1:1 nitric acid. Evaporate to dryness and take up in 10 ml of water. Adjust to pH 6 with 10% sodium acetate solution. Add 0.2 ml of 0.1% solution of nitroso-R salt and heat at 100° for 2 minutes. Add 0.5 ml of nitric acid and destroy excess reagent by heating at 100° for 2 minutes. Cool, set aside for 10 minutes, and read at 500 nm.

Ores, Tailings, and Concentrates.[119b] Decompose the sample by fusion with sodium carbonate and zinc oxide. Dissolve the fusion in hydrochloric acid and dilute as necessary. Neutralize an aliquot with ammonium hydroxide, add 5 ml of acetic acid, and boil. Add 5 ml of 0.2% solution of nitroso-R salt. Bring to a boil again, add 5 ml of 1:1 nitric acid, and boil again. Cool, dilute to 50 ml, and read at 520 nm.

Industrial Dust.[120] Draw a known volume of air through a polyvinylchloride filter to collect the sample. Boil the filter with 5 ml of sulfuric acid and 2 grams of ammonium sulfate until a clear solution results. Cool, add 20 ml of water, 0.5 gram of ammonium persulfate, and 20 ml of 20% sodium hydroxide solution. Boil for 25 minutes and filter. Wash the precipitate with 2% sodium hydroxide solution.

TUNGSTEN. Dilute the filtrate and washings to 100 ml with 2% sodium hydroxide solution. Mix a 5-ml aliquot with 2 ml of 50% ammonium thiocyanate solution, 25 ml of 1:1 hydrochloric acid, and 2 ml of 10% titanous chloride solution. Dilute to 50 ml and set aside for 10 minutes. Read at 400 nm.

[118] R. S. Ondrejcin, *Anal. Chem.* **36**, 937–938 (1964).
[119a] N. P. Strel'nikova, G. G. Lystsova, and G. S. Dolgorukova, *Zavod. Lab.* **28**, 1319–1321 (1962).
[119b] M. I. Kazartseva and T. K. Butorina, *Zavod. Lab.* **42**, 1275 (1976).
[120] T. M. Urusova, *Gig. Sanit.* **1969** (4), 72–74.

COBALT AND TITANIUM. Dissolve the washed precipitate above from the filter with 10 ml of hot 1:3 sulfuric acid and dilute to 100 ml. Add 0.2 ml of 0.1% solution of nitroso-R salt to a 10-ml aliquot, along with 5 ml of 20% sodium acetate solution. Heat at 100° for 2 minutes, add 0.5 ml of nitric acid, and continue to heat for 2 minutes. Cool, dilute to 25 ml, and after 10 minutes read cobalt at 500 nm.

Marine Sediments.[121] For preparation of this sample, see Lead by Dithizone. Develop cobalt in an aliquot by the general procedure.

Carbonate Soil.[122] Mix 20 grams of air-dried sample with 100 ml of 10% solution of potassium nitrate. Add nitric acid to decompose the carbonate and give pH 3. Shake for 1 hour, filter, and evaporate the filtrate to 100 ml. If the solution is colored, boil with 0.5 ml of 30% hydrogen peroxide, cool and dilute to 100 ml. Evaporate a 20-ml aliquot to about 10 ml and add 5 drops of nitric acid. Add *N* potassium permanganate solution until a faint pink persists. Add 0.3 gram of sodium acetate and 1.5 ml of 1% nitroso-R salt solution. Boil for 20 seconds and cool. Add 2.5 ml of phosphoric acid and 0.5 ml of nitric acid, dilute to 20 ml, and read at 500 nm.

For preparation of a sample solution of noncarbonate soil suitable for determination of cobalt by nitroso-R salt, see Copper by Diethyldithiocarbamate. For cobalt in soil by nitroso-R salt after separation from copper and zinc, see Chapter 5, Copper, under determination by diethyldithiocarbamate.

Water.[123] Acidify 1 liter of sample with hydrochloric acid and filter. Add 5 ml of 1% ascorbic acid solution and 5 ml of 2% solution of potassium thiocyanate. Add 5 ml of sorption solution that is 5:4:1 tetrahydrofuran–2-methoxyethanol–hydrochloric acid. Set aside for 6 hours. Pass through a column of Dowex 1-X8, which adsorbs the cobalt-thiocyanate. Wash the column with sorption solution, then elute the cobalt with 1:1 hydrochloric acid. Evaporate to dryness, add 2 ml of perchloric acid, and again evaporate to dryness. Dissolve the residue in 10 ml of 0.2 *M* citric acid, add 10 ml of boric acid–disodium phosphate buffer solution for pH 8, and 1 ml of 0.2% nitroso-R salt solution. Boil for 2 minutes, add 1 ml of nitric acid, and continue to boil for 2 minutes. Cool, dilute to 100 ml, and read at 420 nm.

For cobalt in sea water, make alkaline with sodium carbonate solution and let stand for at least 7 days.[124] Filter the precipitate, which contains the cobalt, and dissolve in dilute hydrochloric acid. Buffer the solution with sodium citrate and raise to pH 8.5 with ammonium hydroxide. Extract the cobalt with dithizone in carbon tetrachloride. Evaporate the extract and destroy the organic matter with nitric acid. Take up the residue in water, buffer by addition of phosphate and borate, and adjust to pH 7.5 with sodium hydroxide solution. Add nitroso-R salt solution and set aside for 30 minutes. Add nitric acid and heat to destroy excess nitroso-R salt. Read at 425 nm. The method will determine 0.5 ppb.

[121] R. Chester and M. J. Hughes, *Trans. Inst. Min. Met., B*, **77** (1), B37–B41 (1968).
[122] A. N. Gyul'akhmedov, *Izv. Akad. Nauk Azerb. SSR, Ser. Biol. Med. Nauk* **1961** (7), 57–63.
[123] J. Korkisch and D. Dimitriadis, *Talenta* **20**, 1287–1293 (1973).
[124] W. Forster and H. Zeitlin, *Anal. Chim. Acta* **34**, 211–224 (1966).

Uranium Liquors.[125] These may be precipitation liquors or mother liquors. Take as sample a solution containing nitric acid with not more than 2 grams of uranium and 0.3 mg of cobalt. Evaporate nearly to dryness. Mix with 50 ml of 25% solution of sodium citrate and 10 ml of acetic acid. Add 25 ml of 5% solution of potassium ethyl xanthate. Extract the cobalt complex with six successive 10-ml portions of chloroform. Combine the extracts and filter. Evaporate substantially to dryness, add 4 ml of nitric acid, and heat to dryness to destroy organic matter. Add 4 ml more of nitric acid and again heat to dryness. Dissolve the residue in 2 ml of hydrochloric acid, add 40 ml of water, and 10 ml of a saturated solution of sodium acetate. Add 3 ml of 0.4% solution of nitroso-R salt and heat to boiling. Add 5 ml of nitric acid and boil for 2 minutes. Cool, dilute to 100 ml, and read at 500 nm against a blank.

Copper-Refining Electrolyte.[126] Add 5 ml of nitric acid to 25 ml of sample containing 1–5 mg of cobalt. Heat, cool, and dilute to 250 ml. To a 20-ml aliquot, add ammonium hydroxide until a precipitate appears. Clarify by addition of less than 20 ml of 1% ammonium bifluoride solution and dilute to 50 ml. Pass through a 240×12.5 mm column of 60-80 mesh Amberlite IR-120 resin in sodium form. Elute copper, cadmium, and lead with 60 ml of 0.1 *M* sodium thiosulfate. Then elute zinc, manganese, and cobalt with 120 ml of 15% sodium chloride solution. Complete by the procedure for solutions. Zinc and manganese do not interfere.

Manganese Electrolytes.[127] Adjust a 50-ml sample to pH 4.5 with acetate buffer and acetic acid solution. Add 6 drops of acetic acid and 0.5 ml of 15% hydrogen peroxide. Boil for 15 minutes to decompose excess hydrogen peroxide. Add 15 ml of 40% solution of sodium acetate and 20 ml of 0.3% solution of nitroso-R salt. Boil for 1 minute, add 10 ml of hydrochloric acid, and boil for another minute. Cool, dilute to 200 ml, and read at 530 nm against a blank.

Beer.[128] Boil and cool 50 ml of degassed sample. Adjust to pH 5–5.5 with 4 *M* sodium acetate. Add 2 ml of 1% solution of nitroso-R salt, boil for 2 minutes, and cool rapidly. Pass through a 5×15 mm alumina column and wash with water. Elute coloring matter and excess nitroso-R salt with *N* nitric acid at 80°. When the eluate comes through clear wash with cold water. Elute the cobalt with 2 *N* sulfuric acid, dilute to 25 ml, and read at 500 nm.

Rubber.[129] Ash 1 gram of sample. Dissolve in 10 ml of hydrochloric acid and evaporate to dryness at 100°. Add 5 ml of nitric acid and again evaporate to dryness. Take up in 15 ml of hot 1 : 1 hydrochloric acid, add 20 ml of water, and boil to dissolve. Filter and dilute to 500 ml. Neutralize a 25-ml aliquot with ammonium hydroxide and add 5 ml of 50% solution of sodium acetate. Boil for 2 minutes, add 10 ml of 0.1% solution of nitroso-R salt, and boil for a couple of minutes to develop the maximum color. Add 5 ml of 1 : 1 nitric acid and boil for 2 minutes. Cool, dilute to 100 ml, and read at 500 nm.

[125]C. J. Riley and R. M. Lees, *UKAEA, Ind. Group Hdq.* **SCS-R-94**, 10 pp (1959).
[126]Tesuo Katsura, *Jap. Anal.* **10**, 1320–1323 (1961).
[127]Z. Hasek, *Hutn. Listy* **22**, 416–417 (1967).
[128]R. Parsons, *Proc. Am. Soc. Brew. Chem.* **1964**, 152–54.
[129]L. L. Bogina and I. P. Martyukhina, *Kauch. Rezina* **18** (11), 58–59 (1959).

Animal and Poultry Feed.[130] Dry 5 grams of a premix or 50 grams of a complete feed for 2 hours at 130°. Stir for 5 minutes with 50 ml of carbon tetrachloride. Centrifuge and decant the clear layer of solvent. Add another 50 ml of carbon tetrachloride, agitate, and filter. After evaporation of the solvent, boil the solid residue with 50 ml of 1:1 hydrochloric acid for 15 minutes. Add 0.5 gram of activated carbon, dilute to 100 ml, and filter. Develop an aliquot by the procedure for solutions.

Biological Material.[131] Warm a moist sample containing up to 5 μg of cobalt with 25 ml of nitric acid until reaction subsides. Cool and filter fat on sintered glass. Add 2 ml of sulfuric acid and boil gently with further additions of nitric acid until the solution is light brown. Add perchloric acid and nitric acid alternately, dropwise, until the digest is colorless. Heat to sulfur trioxide fumes for 5 minutes. Take up in 50 ml of water and boil. Cool and filter. Add 10 ml of 10% sodium citrate solution adjusted as a buffer solution to pH 4 with 20% sodium hydroxide solution. Heat to boiling and add 3 ml of 0.52% ethanolic 1-nitroso-2-naphthol. Boil gently for 3 minutes, and while boiling, add 2 ml of 1:1 sulfuric acid. Cool, and extract with 5 ml, 5 ml, and 5 ml of chloroform. Evaporate the solvent from the extracts and oxidize the residue with a few drops of sulfuric acid and perchloric acids. Boil for 1 minute with 5 ml of water, make alkaline with ammonium hydroxide, and boil off excess ammonia. While warm, add 1 ml of the citrate buffer solution for pH 4 and 1 ml of 0.2% solution of nitroso-R salt. Mask ferrous ion with a few drops of phosphoric acid. Set aside for 5 minutes, add 1 ml of nitric acid, and boil for 1 minute. Decolorize excess reagent with 0.25 ml of saturated bromine water and boil off the excess bromine. Dilute to 10 ml and read at 480 nm against a reagent blank.

1,10-PHENANTHROLINE

Cobaltous ion reacts to reduce ferric ion and in the presence of 1,10-phenanthroline, the red ferrous complex is formed.[132] A large excess of 1,10-phenanthroline is essential to avoid interferences by other metals and to expedite the reduction to form the color. The complex can be adsorbed on silica for separation from interfering ions.

The complex of cobalt with 1,10-phenanthroline adds tetrabromo-, tetraiodo-, or dichlorotetraiodofluorescein, extractable at pH 9 with chloroform.[133] The maximum absorption is at 540 nm or, in 1:1 chloroform-acetone, fluorescence at 580 nm. Fluorescence will determine 0.05–10 μg in 10 ml, which is one order of magnitude more sensitive than spectrophotometry.

Procedure. To an acid sample solution containing up to 20 mg of cobalt, add 3[illegible] ml of 0.1% solution of 1,10-phenanthroline and 5 ml of *M* acetate buffer for p[illegible]

[130] A. Amati and A. Minguzzi, *Ind. Agr.* **8**, 249–257 (1970).
[131] D. W. Dewey and H. R. Marston, *Anal. Chim. Acta* **57**, 45–49 (1971).
[132] F. Vydra and R. Pribil, *Collect. Czech. Chem. Commun.* **26**, 3081–3085 (1961).
[133] D. N. Lisitsyna and D. P. Shcherbov, *Zh. Anal. Khim.* **28**, 1203–1205 (1973).

4.6.[134] Add ammonium hydroxide to pH 5–6 and heat at 80° for 5 minutes. Cool at once, dilute to 50 ml, and read at 420 nm. Beer's law is followed for 20–400 ppm of cobalt. There is interference by ferrous, cupric, and nickel ions.

Alternatively, add to the cobalt sample solution 2 ml of 0.1% solution of 1,10-phenanthroline, 1 ml of 0.1 *M* EDTA, and 2 ml of the buffer solution for pH 4.6. Adjust to pH 4–5 with ammonium hydroxide and add 1 ml of 0.1 *M* potassium iodide. Extract with 10 ml of ethylene chloride and read at 420 nm. Beer's law is followed for 2–40 ppm of cobalt in the extract. Nickel must be absent but ferrous ion is masked by the EDTA.

Isolation on Silica.[135] To the sample containing 5–100 μg of cobalt and not in excess of 25 mg of iron, add sufficient EDTA solution to complex the iron plus at least 50% excess. Adjust to pH 4.5, add 1 ml of 0.1 *M* 1,10-phenanthroline, and dilute to 25 ml. Prepare a 16×50 mm column of silica, pore size 85 Å, particle size 0.15–0.20 mm, activated at 120°. Pass the solution through the column at about 2 ml/minute. Wash the column with 25 ml of 0.1 *M* buffer solution for pH 4.5, then with 10 ml of water. Elute the cobalt phenanthrolate with 15 ml of *M* ammonium formate at pH 9.5, containing 1 ml of 0.1 *M* 1,10-phenanthroline and 15 ml of methanol per 100 ml. Wash the eluate through with 10 ml of water. Adjust the eluate to pH 3 with 1 : 1 hydrochloric acid. Add 1 ml of 0.1 *M* 1,10-phenanthroline and 2 ml of 0.001 *M* ferric chloride and dilute to 50 ml. Set aside for 15 minutes and read at 510 nm against a reagent blank.

1-(2-PYRIDYLAZO)-2-NAPHTHOL

This reagent has the trivial name PAN. It forms with trivalent cobalt a chloroform-extractable complex that has maxima at 590 and 640 nm. There are fewer interferences at 640 nm. Copper interferes seriously at 1 : 1 and nickel at 10 : 1. Amounts of zinc, cadmium, and ferric ion less than 1 mg/ml are readily masked with citrate.

For analysis of high purity nickel, the cobalt complex with PAN is stable under strongly acid conditions where most other PAN complexes are destroyed.[136] At room temperature the cobalt ion is displaced rapidly from the EDTA complex, whereas the nickel ion is displaced very slowly. Dissolve the sample in nitric acid, adjust to pH 10 with ammonium hydroxide, and add EDTA to chelate the metallic ions, avoiding excess. Mask ferric ion with phosphate ion; cupric and palladous ion with thiourea. Reduce to pH 2.5 and add bismuth nitrate to displace cobalt. Add ethanol to further slow down displacement of nickel, and form the complex of cobalt with PAN at pH 2. Extract the complex with chloroform. Add a few drops of hydrochloric acid to the extract, dilute with ethanol, and read at 625 nm.

For analysis of high purity calcium and magnesium carbonates, dissolve the sample in hydrochloric acid and extract the iron with isopropyl ether.[137] Then

[134] Keisuke Tachibana, *Mem. Fac. Sci., Kyushu Univ. Ser. C* **4**, 229–238 (1961).
[135] Frantisek Vydra, *Talenta* **11**, 433–439 (1964).
[136] H. Flashka and R. M. Speights, *Microchem. J.* **14**, 490–499 (1969).
[137] Adam Hulanicki, Malgorzata Galus, Wojciech Jedral, Regina Karwonski, and Marek Trojanowski, *Chem. Anal.* (Warsaw) **16**, 1011–1019 (1971).

extract cobalt and nickel at pH 5 with chloroform as the PAN complexes. Shake the extract with EDTA solution to minimize interference. Read at 565, 628, or 764 nm, correcting for residual iron.

The sampling and subsequent separation of cobalt by column chromatography from copper, iron, and zinc in niobium samples appears under determination of copper in niobium by PAN. For cobalt, nickel, and iron by PAN, see Iron. For cobalt in the presence of iron and nickel by PAN, see Iron. For cobalt in beer by PAN, see Iron. The reagent is appropriate for determination of cobalt in nickel- or iron-base alloys.[138]

Procedures

With Surfactant.[139] Mix a sample solution containing 0.5–3.2 μg of cobalt with 5 ml of 0.1 *M* ammonium oxalate. This masks ferric, bismuth, stannic, and aluminum ions. Add 2 ml of a surfactant solution containing 20% of Triton X-100 and 2.5% of sodium dodecylbenzenesulfonate. Then add 2 ml of 0.1% solution of PAN, 5 ml of acetate buffer solution for pH 5, and 0.05 gram of ammonium persulfate. Set aside for 3 minutes, then add 5 ml of 0.05 *M* EDTA as a further masking agent. Dilute to 50 ml and read at 620 nm against a reagent blank. Ferrous ion must be absent.

Instead of adding EDTA, interferences also may be avoided by forming the cobaltic complex of the reagent below pH 0.5 or by adding 1:1 hydrochloric acid after the color complex is formed.

Molybdenum.[140] Dissolve 1 gram of sample in the minimum amount of nitric acid, dilute to 200 ml, and take a 20-ml aliquot. Add 0.5 ml of saturated solution of sodium pyrophosphate, 1 ml of 10% solution of thiourea, 10 ml of 0.05% ethanolic PAN, and 1 ml of 10% hydrogen peroxide. Adjust to pH 4.5–5 by adding solid ammonium acetate. Add 30 ml of water, and set aside for 5 minutes. Extract with successive 5-ml portions of chloroform until the last extract is colorless, and combine the extracts.

If interfering ions are present, add 3 ml of buffer solution for pH 5 containing 27% of sodium acetate trihydrate and 6% of acetic acid. Add 5 ml of 0.1 *M* EDTA and shake for 5 minutes. Drain the chloroform layer, wash the aqueous layer with 5 ml of chloroform, and add this to the chloroform extract. Add 5 ml of ethanol and dilute to 50 ml with chloroform. Read at 630 nm against chloroform.

Titanium Carbide and Titanium Boride. This technic is designed for ball-milled samples and also applies to other hard materials. Fuse a sample containing 0.1–1 mg of cobalt with 10 grams of potassium persulfate in a porcelain crucible. Dissolve the melt in 50 ml of 1:1 sulfuric acid and 10 ml of 10% hydrogen peroxide. Dilute to 200 ml. Do not let the solution stand at this stage, since hydrolysis will precipitate hydrated titanium oxide with adsorbed cobalt. Mix

[138] S. A. Akhmedov and O. A. Tataev, *Tr. Novocherk. Politekh. Inst.* **266**, 13–19 (1972).
[139] Hiroto Watanabe, *Talenta* **21**, 295–302 (1974).
[140] R. Puschel, E. Lassner, and A. Iblaszewicz, *Chemist-Analyst* **55**, 40–42 (1966).

20-ml aliquot with 10 ml of 10% oxalic acid solution and complete as above for molybdenum from "Add 0.5 ml of saturated solution...."

Nickel Alloys and Steel.[141] Dissolve a 0.1-gram sample in 20 ml of hydrochloric acid with dropwise addition of nitric acid as an oxidizing agent. Dilute to 100 ml. Mix a 1-ml aliquot with 5 ml of 0.2% solution of PAN in acetone. Set aside for 15 minutes, then add 20 ml of hydrochloric acid, and extract with 5 ml of chloroform. Read at 640 nm against water.

Thorium Oxide.[142] Dissolve a 3-gram sample or an equivalent amount of slurry by warming with 20 ml of 1:1 nitric acid and a few drops of hydrofluoric acid. Evaporate to dryness. Add 10 ml of hydrochloric acid and again evaporate to dryness. Repeat the last step.

Prepare an anion exchange column 50×10 mm using 50-100 mesh Dowex 1-X10. Condition by washing with 50 ml of 10 *N* hydrochloric acid. Dissolve the residue of the sample in 10 ml of 10 *N* hydrochloric acid and heat just to boiling. Pass the hot solution through the column at not more than 2 ml/minute. Wash the column with 20 ml of hot 10 *N* hydrochloric acid and discard the effluent and washings. Elute cobalt from the column with 30 ml of 1:2 hydrochloric acid and evaporate the eluate to dryness. Dissolve the residue in 15 ml of water containing 5 drops of hydrochloric acid. Add 2 ml of 10% citric acid solution, 2 ml of 50% solution of ammonium acetate, and 1 drop of saturated solution of potassium periodate. The pH of the solution should be approximately 4.5. Add 0.5 ml of 0.1% ethanolic PAN. Extract by shaking for 3 minutes with 5 ml of chloroform, and allow 15 minutes for phase separation. Read at 640 nm against a reagent blank.

4-(2-PYRIDYLAZO)RESORCINOL

This reagent has the trivial name of PAR. The 1:2 complex of cobalt with PAR is read at 2.5–20 μg of cobalt in 50 ml. It is applicable to steel containing up to 10% cobalt. The usual alloying elements do not interfere. In other solutions at pH 6.5–8.2, up to 1000-fold amounts of EDTA may be present, masking iron, copper, nickel, manganese, and zinc for reading at 510 nm.[143]

The complex of cobalt with PAR can be extracted with chloroform from a neutral or weakly alkaline solution in the presence of zephiramine, which is benzyldimethyltetradecylammonium chloride,[144] EDTA, and an appropriate buffer. Ferrous, nickel, and vanadium ions act similarly. The maximum absorption is at 420 nm and the appropriate buffer solution for pH 8.3 consists of monosodium borate and monopotassium phosphate.

Buffered at pH 6.8, copper and bismuth can be masked with EDTA and iron, mercury, vanadium, and tantalum with potassium cyanide. With such masking agents present, the amount of reagent should be increased.

[141] N. S. Enshova, V. V. Orlov, V. M. Ivanov, and A. I. Busev, *Zavod. Lab.* **41**, 913–914 (1975).
[142] Gerald Goldstein, D. L. Manning, and Oscar Menis, *Anal. Chem.* **31**, 192–194 (1959).
[143] A. D. Gololobov and I. A. Vakhrameeva, *Pochvovedenie* **1965** (2), 81–88.
[144] Takao Yotsuyanagi, Ryuji Yamashita, and Kazuo Aomura, *Jap. Anal.* **19**, 981–982 (1970).

At pH 3 in the presence of EDTA, there is no interference by iron, nickel, manganese, aluminum, or chromium, and none by small amounts of copper, zinc, bismuth, or vanadium. The reagent is appropriate for determination of cobalt in nickel- or iron-base alloys.[145]

Procedure. Boil a sample solution containing not more than 5 μg of cobalt with 0.5 ml of 15% hydrogen peroxide.[146] Cool, and add 0.5 ml of 0.1% solution of the reagent in 0.04% sodium hydroxide solution. Add 2 ml of sodium tetraborate–phosphate buffer solution for pH 8.2 and 1.5 ml of 0.05 *M* EDTA. Boil for 30 minutes, cool, add 1.5 ml of 0.05 *M* zephiramine, and dilute to 50 ml. Shake with 10 ml of chloroform for 10 minutes and read the organic layer at 520 nm.

Steel.[147] Dissolve 0.05 gram of sample in 5 ml of nitric acid and 1 ml of hydrochloric acid. Evaporate nearly to dryness, add 20 ml of water, and heat at 100° to dissolve salts. Filter and dilute to 100 ml. Dilute a 10-ml aliquot to 100 ml and take a 2.5 ml aliquot. Add 5 ml of 20% ammonium citrate solution, and adjust to pH 7–8 with 10% sodium acetate solution. Add 5 ml of 0.025% solution of PAR and 1 ml of saturated solution of EDTA. Heat at 80° for 30 minutes and cool. Dilute to 50 ml with 10% sodium acetate solution and read at 500 nm.

Iron and Steel with Added Zephiramine.[148] Dissolve a 0.1-gram sample with 5 ml of hydrochloric acid and 1 ml of nitric acid. Evaporate almost to dryness and take up in 20 ml of 5:3 hydrochloric acid. Unless the sample contains more than 0.01% cobalt, remove most of the iron by extraction with 20 ml of methyl isobutyl ketone. For higher cobalt contents, this extraction is unnecessary. Add 1 ml of nitric acid and 0.5 ml of perchloric acid to the aqueous layer, and evaporate nearly to dryness. Add 2 ml of 0.5 *N* sulfuric acid and heat to sulfur trioxide fumes. Take up with water and dilute to 45 ml. Add 2 ml of 0.5 *M* sodium citrate. Add 5.3 ml of 0.125 *M* sodium tetraborate and check that the pH is 7–7.5. Set aside for 3 minutes. Add 0.5 ml of 0.1% solution of PAR in 0.01 *N* sodium hydroxide and 1.5 ml of 0.05 *M* zephiramine. Set aside for 2 minutes. Add 2 ml of 0.05 *M* 1, 2-diaminocyclohexanetetraacetate and 2 ml of 0.05 *M* EDTA. Boil for 5 minutes. This prevents interference by up to 50 μg of nickel and up to 5 mg of iron. Cool and shake for 2 minutes with 10 ml of chloroform. Read at 530 nm.

Alloy Steel.[149] Dissolve 0.1 gram of sample in 10 ml of 1:4 sulfuric acid. Add 2 ml of nitric acid, and heat to destroy carbides. Boil off oxides of nitrogen and dilute to 100 ml. Mix a 15-ml aliquot with 5 ml of 0.1% solution of PAR and neutralize to pH 5. Then dilute to 50 ml with 2 *N* sulfuric acid. This acidity destroys complexes with iron, copper, nickel, titanium, and vanadium. Read cobalt at 540 nm.

[145] S. A. Akhmedov and O. A. Tataev, *Tr. Novocherk. Politekh. Inst.* **266**, 13–19 (1972).
[146] Ryuji Yamashita, Takao Yotsuyanagi, and Kazuo Aomura, *Jap. Anal.* **20**, 1282–1288 (1971); *Anal. Chem.* **44**, 1091–1093 (1972); cf. F. H. Pollard, P. Hanson, and W. J. Geary, *Anal. Chim. Acta* **20**, 26–31 (1950); Yoshio Shijo and Tsugio Takeuchi, *Jap. Anal.* **13**, 536–540 (1964).
[147] A. I. Busev and V. M. Ivanov, *Zh. Anal. Khim.* **18**, 208–215 (1963).
[148] Haruno Okochi, *Jap. Anal.* **21**, 51–56 (1972).
[149] S. A. Akhmedov, O. A. Tataev, and R. R. Abdullaev, *Zavod. Lab.* **37**, 756–758 (1971).

Iron Ore.[150] Decompose 0.5 gram of sample with 6 ml of hydrochloric acid and 2 ml of nitric acid. Add 2 ml of hydrofluoric acid and 2 ml of 1:1 sulfuric acid. Evaporate to sulfur trioxide fumes. Take up with 5 ml of hydrochloric acid and 30 ml of water. Dilute to 100 ml and filter. Add 5 ml of 0.1 *M* EDTA to a 10-ml aliquot, adjust to pH 3 with an ammoniacal acetate solution, and add 10 ml of acetate buffer solution for pH 3. Add 4 ml of 0.1% solution of PAR and dilute to 50 ml with the acetate buffer solution for pH 3. Set aside for 1 hour and read at 536 nm against a reagent blank.

PYRROLIDINEDITHIOCARBAMATE

The complexes of cobalt, cadmium, bismuth, and molybdenum with this reagent are extractable from acid solutions with such solvents as chloroform and methyl isobutyl ketone. Their absorption maxima are in the ultraviolet region.

Procedure. As a buffer solution, dissolve 22.1 grams of sodium acetate trihydrate in water, add 5 ml of acetic acid, and dilute to 1 liter.

Mix a sample solution containing 0.05–0.15 mg of cobaltous ion with 10 ml of the acetate buffer solution.[151] Add 5 ml of 0.2% solution of the captioned reagent, mix, and set aside for 5 minutes. Extract with 25 ml and 15 ml of chloroform. Dilute the combined extracts to 50 ml with chloroform. Read at 324 nm against a reagent blank.

QUINOLINAZO R

This reagent is 3-hydroxy-4-(8-quinolylazo)naphthalene-2,7 disulfonic acid. It complexes with cobalt ion, preferably at pH 1–1.8, and obeys Beer's law for 0.02–2 μg of cobalt per ml.

Procedure. *Steel.*[152] Dissolve 0.1 gram of sample in 7 ml of hydrochloric acid and 3 ml of nitric acid. Evaporate to dryness, add 1 ml of 1:4 sulfuric acid, and again evaporate to dryness. Take up in water and dilute to 100 ml. Mix a 5-ml aliquot with 0.5 ml of 5% ascorbic acid solution to reduce ferric and vanadium ions. Add 0.5 ml of 10% thiourea solution to mask copper. Add 4 ml of 0.2 *N* sulfuric acid and 2 ml of 0.2% solution of quinolinazo R. Let stand for 3 minutes; then destroy the interfering nickel complex by adding 9 ml of 1:1 sulfuric acid. Dilute to 25 ml and read at 570 nm.

---

[150] M. I. Zaboeva, G. N. Zus', and M. P. Dorofeeva, *ibid.* **35**, 1158–1159 (1969); cf. T. L. Radovskaya, *Zavod. Lab.* **42**, 398–399 (1976).
[151] Melvyn B. Kalt and D. F. Boltz, *Anal. Chem.* **40**, 1086–1091 (1968).
[152] N. N. Basargin, A. V. Kadomtseva, and V. I. Petrashen', *Zavod. Lab.* **35**, 16–17 (1969); N. N. Basargin and A. V. Kadomtseva, *Tr. Novocherk. Politekh. Inst.* **220**, 32–36 (1969).

QUINOXALINE-2, 3-DITHIOL

The complex of cobalt with quinoxaline-2,3-dithiol is read photometrically for 0.4–1.4 ppm of cobalt.[153] Nickel also reacts, and the reagent can determine both if the nickel-cobalt ratio is between 5:1 and 1:3. Color develops rapidly in 80% dimethylformamide. There is interference by copper, manganese, mercury, palladium, platinum, and silver. In 80–95% ethanol at pH 3 ± 0.5, the peak absorption for cobalt is at 510 nm, that of nickel at 656 nm. For determination in aqueous solution, mix the cobalt solution with 10 ml of 10 *N* sulfuric acid. Add a tenfold excess of reagent as compared with the cobalt content, and dilute to 25 ml with 10 *N* sulfuric acid.[154]

For reading cobalt along with nickel by quinoxaline-2,3-dithiol, see Nickel by that reagent.

Procedure. *Beer.*[155] Mix 10 ml of 1% solution of 2-methoxyethanol in acetic acid with 0.4 ml of 1:1 hydrochloric acid and 5 ml of a solution of 0.5 mg of quinoxaline-2,3-dithiol per ml of 2% ascorbic acid solution. Add 10 ml of degassed beer and read at 540 nm against water.

SODIUM-*p*-(MERCAPTOACETAMIDO)BENZENESULFONATE

This reagent forms a cobaltous complex at pH 6.5–7.5 having a maximum absorption at 475 nm and one at pH 8.5–11 with a maximum at 390 nm. Beer's law is followed for 0.3–2 ppm of cobalt at either wavelength. The color developed is stable in terms of both time and temperature. Nickel interferes. Cobalt can be separated from it by precipitation as potassium cobaltinitrite and dissolved in 2 *N* sulfuric acid for development. To avoid interference by ferrous and ferric ions, dilute the prepared sample to about 40 ml. Add 1.5 ml of 1:9 hydrochloric acid to give a pH of 2–4 before diluting to 50 ml, and read within 15 minutes.

Procedure. As a buffer solution for pH 7–7.2, dissolve 68 grams of potassium dihydrogen phosphate and 15.73 grams of sodium hydroxide in water and dilute to 1 liter.[156]

To a sample solution containing 15–100 μg of cobalt, add 2 ml of 10% solution of sodium potassium tartrate, 5 ml of buffer solution for pH 7–7.2, and 5 ml of 1% solution of the reagent. If necessary, adjust to pH 7–7.2 with 0.1% solution of sodium hydroxide. Dilute to 50 ml, set aside for 30 minutes, and read at 475 nm against water.

[153] R. R. Annand, *Dist. Abstr.* **24**, 1813 (1963); cf. R. W. Burke and John H. Yoe, *Anal. Chem.* **34**, 1378–1382 (1962).
[154] L. I. Chernomorchenko, T. V. Chuiko, and A. G. Akhmetshin, *Zh. Anal. Khim.* **27**, 2262–2265 (1972).
[155] I. Stone, *Proc. Am. Soc. Brew. Chem.* **1965**, 151–156.
[156] H. K. L. Gupta and N. C. Sogani, *Anal. Chem.* **31**, 918–920 (1959).

Alternatively, before dilution to 50 ml, heat at 100° for about 6 minutes and cool. Then dilute to 50 ml and read at 475 nm without waiting.

SULFARSAZAN

Sulfarsazan, which is 5-nitro-2-[3-(4-*p*-sulfoazophenyl)-1-triazeno]benzenearsonic acid, forms a soluble 1:1 complex in slightly alkaline solution. Beer's law is followed for 1–20 μg of cobalt per 100 ml. There is interference by lead, zinc, cadmium, copper, nickel, manganese, silver, mercury, uranium, iron, aluminum, titanium, tin, zirconium, vanadium, and rare earths. This is bypassed by separation of the cobalt as the cobaltinitrite before development with sulfarsazan. Ammonium salts in excess of 0.5 gram/liter noticeably reduce the color intensity.

Procedure. *Ores or Slags.*[157] Decompose 2 grams of sample with 10 ml of hydrochloric acid; add 2 ml of nitric acid, and evaporate to dryness. Add 5 ml of hydrochloric acid and again take to dryness. Take up in 50 ml of water and filter. Neutralize with 20% potassium hydroxide solution and add 15 ml of 80% acetic acid. Add 50 ml of 60% potassium nitrite solution and set aside for 2 hours. Filter the precipitate of cobaltinitrite and wash with 2% solution of potassium nitrite, slightly acidified with acetic acid. Dissolve the precipitate with 10 ml of hot 1:1 hydrochloric acid and evaporate to 1 ml. Dilute to 50 ml. To an aliquot, add 1 ml of a buffer solution containing 5.4% of ammonium chloride and 35% of 25% ammonia solution. Add 2 ml of 0.05% solution of sulfarsazan in 0.05 *M* sodium tetraborate, dilute to 100 ml, and read at 515 nm.

THIOCYANATE

The blue complex of cobaltous thiocyanate is extractable with 1:5 isoamyl alcohol–ethyl acetate at pH 4.2, showing maxima at 320 and 620 nm.[158] At 320 nm Beer's law is followed for 1–10 μg/ml, at 620 nm for 10–50 μg/ml.

Acetylacetone extracts cupric, beryllium, manganous, ferric, tetravalent zirconium, and tetravalent vanadium ions from aqueous solution at pH 4 before thiocyanate ion is added.

Various organic solvents intensify the blue color of cobalt thiocyanate.[159] In at least 20% concentration by volume, the effect is enhanced by *t*-butyl alcohol to 1.09 times, by acetone 1.26 times, and by sulfolan, which is tetrahydrothiophen 1,1-dioxide, 2.8 times. The blue color when sulfolan is added can be extracted into nitrobenzene.

[157] M. A. Yagodnitsyn, *Zavod. Lab.* **35**, 788 (1969); cf. E. A. Bashirov and A. M. Ayubova, *Uch. Zap. Azerb. Gos. Univ., Ser. Khim. Nauk*, **1971** (4), 28–30.
[158] Shigero Ikeda, *Sci. Rep. Res. Inst., Tohoku Univ. A*, **6**, 417–423 (1954).
[159] H. Flaschka and R. Barnes, *Anal. Chim. Acta* **63**, 489–490 (1973).

Tricaprylmethylammonium chloride, which is Aliquat 336, as the thiocyanate in benzene extracts cobaltous thiocyanate at pH 8.[160] Ferric ion is masked with citrate. Washing with 0.7 *M* sodium citrate, 0.2 *M* sodium thiocyanate, and 0.9 *M* sodium thiosulfate removes interfering cupric and nickelous ions.

Zephiramine, which is benzyldimethyltetradecylammonium chloride, forms a blue precipitate with cobalt thiocyanate.[161] This is soluble in chloroform for reading at 625 nm. Extraction may be at pH 2–7, reading at 1.2–12 ppm of cobalt. Typically, extract by shaking 50 ml of solution containing 4 ml of *M* potassium thiocyanate and 10 ml of 0.01 *M* quaternary compound with 10 ml of chloroform for 10 minutes. Only iron in excess of 0.1 ppm and more than 0.8 ppm of copper interfere.

Cobalt-diantipyrinylmethane-thiocyanate is formed in 0.01–1 *N* hydrochloric acid or sulfuric acid in the presence of 3000-fold excess of thiocyanate ion and 100-fold excess of the added reagent.[162] The maximum absorption is at 620 nm. The complex is extracted with chloroform for reading more than 0.05% cobalt. For lesser amounts, it should be converted to the nitroso-R complex. In chloroform extraction, nickel does not interfere. Cupric and ferric ions are masked with ascorbic acid.

The methyl-, *o*-hydroxyphenyl-, and *p*-hydroxyphenyl- derivatives of diantipyrylmethane can be similarly used.[163]

Trioctylammonium thiocyanate in benzene effectively extracts cobalt thiocyanate from a solution at pH 8 containing citrate to mask iron.[164] Wash the extract with a solution of sodium thiocyanate, sodium citrate, and sodium thiosulfate to remove cupric and nickel ions. Then read the organic layer at 321 nm.

Tributyl phosphate extracts cobalt thiocyanate from aqueous solution.[165] Complexes are formed in aqueous solution between cobalt thiocyanate and acetone, the picolines which are 2-, 3-, or 4- methylpyridine, or triethanolamine.

The 2:1 complex between triphenyltetrazolium ion and cobalt thiocyanate is extracted by chloroform for reading at 620 nm.[166] The same reaction is given by several other cations. That with copper, mercury, molybdenum, and iron can be masked with thiosulfate and ammonium fluoride, that with zinc by excess thiocyanate. Tungsten is precipitated by acidifying. Vanadium and nickel interfere.

For analysis of soil for cobalt, decompose the sample with aqua regia and perchloric acid.[167] The cobalt is isolated by extraction with dithizone in chloroform, then determined as the thiocyanate.

For cobalt in multivitamin preparations, ammonium fluoride is added and the cobalt thiocyanate extracted with amyl alcohol.[168] Cobalt in polyethylene

[160] A. M. Wilson and O. K. McFarland, *Anal. Chem.* **35**, 302–307 (1963).

[161] Hiroshi Matsuo, Shokichi Chaki, and Shigeki Hara, *Jap. Anal.* **14**, 935–938 (1965).

[162] I. Adamiec, *Chem. Anal.* (Warsaw) **14**, 115–123 (1969); cf. N. A. Ugol'nikov and V. S. Kirsa, *Tr. Tomsk. Gos. Univ. V. V. Kuibysheva, Ser. Khim.* 145, *5-ya Nauch. Konf.* **1954**, 63–66.

[163] E. V. Sokolova, A. S. Pesis, and N. I. Panova, *Zh. Anal. Khim.* **12**, 489–494 (1957).

[164] A. M. Wilson and O. K. McFarland, *USAEC Rep.* **TID-17745**, 22 pp (1962).

[165] M. R. Verma and P. K. Gupta, *Z. Anal. Chem.* **196**, 187–190 (1963); E. Jackwerth and E.-L. Schneider, *ibid.* **207**, 188–192 (1965).

[166] A. V. Aleksandrov, P. Vasileva-Aleksandrova, and E. Kovacheva, *Mikrochim. Acta* **1967**, 579–584.

[167] A. Duca and D. Stanescu, *Acad. Rep. Pop. Rom., Fil. Cluj, Stud. Cercet. Chim.* **8** (1–2), 75–83 (1957).

[168] Lilo E. Guerello, *Proanalisis* **3** (6), 29–33 (1970).

terephthalate is determined as the complex with thiocyanate and tributyl ammonium ion.[169]

For cobalt by thiocyanate and pyridine along with copper, nickel, and iron, see Copper.

Procedures

With Tri-n-Butylamine Acetate.[170] As a buffer solution for pH 3.7, dissolve 50 grams of sodium acetate in water, add 160 ml of acetic acid, and dilute to 1 liter. As tributylamine acetate reagent, dissolve colorless *n*-tributylamine in water with acetic acid and adjust to pH 3–4.

To 20 ml of sample solution at pH 2, add sequentially 10 ml of buffer solution for pH 3.7, 1 ml of 3% solution of potassium fluoride, 3 ml of 25% solution of potassium thiocyanate, and 0.5 ml of *n*-tributylamine acetate solution. Shake with 8 ml of amyl alcohol and set aside for 20 minutes. Filter the organic phase, dilute to 10 ml with amyl acetate, and read at 321 nm. Large amounts of iron may require an increase in potassium fluoride. The method has been applied to steels, alloys of nickel, chromium and zinc, glasses, zinc oxide, and iron pyrites.

With Acetylacetone.[171] To a sample solution containing 50–500 μg of cobalt, add 5 ml of 20% solution of stannous chloride in 1:1.5 hydrochloric acid. Then add a freshly prepared mixture of 25 ml of acetone and 2.5 ml of 50% solution of ammonium thiocyanate. Dilute to 50 ml and read at 625 nm.

Steel.[172] Dissolve 1 gram of sample in 10 ml of hydrochloric acid and 3 ml of nitric acid. Evaporate nearly to dryness. Take up in 50 ml of water and add 1:5 ammonium hydroxide to reach pH 4. Dilute to 100 ml.

Extract a 10-ml aliquot with successive portions of acetylacetone until the extract remains colorless. Discard the organic phases and adjust the aqueous phase to 10 ml. To a 5-ml aliquot, add 5 ml of 1.1 *M* sodium thiocyanate solution presaturated with acetylacetone. Add 10 ml of acetylacetone saturated with water. Shake for 1 minute and read the organic phase at 625 nm against a reagent blank.

Lead Alloys.[173] Dissolve a sample in nitric acid and precipitate the lead with sodium chloride. Filter, and dilute to a known volume. To a 10-ml aliquot containing 2–200 μg of cobalt, add 2 ml of 50% solution of ammonium thiocyanate, 2 ml of 0.2% solution of triphenyltetrazolium chloride, 0.5 gram of ascorbic acid, 0.5 gram of sodium thiosulfate, and 5 ml of chloroform. Shake for 2 minutes and read the extract at 620 nm.

[169] Heinz Zimmermann, Horst Hoyme, and Annemarie Tryonadt, *Faserforsch. TextTech.* **21**, 33–36 (1970).

[170] Max Ziegler, O. Glemser, and E. Preisler, *Angew. Chem.* **68**, 436–437 (1956); *Z. Anal. Chem.* **158**, 358–360 (1958).

[171] R. E. Kitson, *Anal. Chem.* **22**, 664–667 (1950).

[172] W. B. Brown and J. F. Steinbach, *ibid.* **31**, 1805–1806 (1959).

[173] A. Aleksandrov, *Mikrochim. Acta* **1972**, 664–668.

Nickel.[174] COBALT AND IRON. Dissolve 0.1 gram of sample containing more than 0.01% cobalt in 10 ml of 1 : 1 nitric acid. Evaporate to dryness and take up in 0.5 ml of 10 *N* hydrochloric acid. Add 4.5 ml of water and 5 ml of 8 *M* ammonium thiocyanate. Extract with 10 ml of methyl isobutyl ketone. Read cobalt at 630 nm and iron at 520 nm. Alternatively, add stannous chloride solution before extraction and iron will not be extracted.

Beryllium and Beryllium Oxide.[175] Dissolve a sample containing about 0.1 mg of cobalt in 5 ml of hydrochloric acid and 10 ml of sulfuric acid. Add 5 ml of 50% citric acid solution, dilute to 100 ml, and neutralize to pH 4 with 50% sodium hydroxide solution. Extract with successive 20-ml portions of acetylacetone until color is no longer extracted, and discard these extracts. Add to the aqueous layer 30 ml of 50% sodium thiocyanate solution and 25 ml of acetylacetone. Shake, separate the organic layer, and adjust to 25 ml. Read at 625 ml against a blank.

Soil and Sediments.[176] As a buffered reagent, dissolve 170 grams of sodium acetate, 40 grams of tetrasodium pyrophosphate, and 5 grams of hydroxylamine hydrochloride with 140 ml of 1 : 1 hydrochloric acid in water and dilute to 1 liter. For use mix 9 : 1 daily with 36% potassium thiocyanate solution.

Fuse 0.25 gram of sample with 1 gram of potassium bisulfate for 2 minutes after frothing ceases. Digest with 5 ml of 0.5 *N* hydrochloric acid until disintegrated, and dilute to 10 ml. Mix a 2-ml aliquot with 10 ml of the buffered reagent. Extract with 0.5 ml of 10% butylamine in amyl alcohol and read the extract at 625 nm.

Biological Material.[177] This technic was developed for samples from animals that have inhaled aerosols containing the cobalt-EDTA anion.

Ash the sample at 500° and take up the ash in 1 ml of 1 : 1 hydrochloric acid. Transfer with water to a centrifuge tube and evaporate to dryness at 85°. Add 1.5 ml of 2 *N* hydrochloric acid and 0.5 ml of 30% hydrogen peroxide and heat to fumes. Repeat this step twice. Add 1 ml of 1 : 1 hydrochloric acid and 5 ml of saturated solution of sodium acetate, which should give a pH of 5.2–5.4. If the iron content exceeds 25% of that present as cobalt, mask with sodium fluoride at this point. Add 5 ml of 50% solution of ammonium thiocyanate. Extract by shaking for 2 minutes with 5 ml and 2 ml of isopentyl acetate. Extract further if residual color is present. Centrifuge the combined organic layers, adjust to a known volume, and read at 623 nm.

2–THIOVIOLURIC ACID

This reagent forms a 1 : 3 cobalt complex when buffered with borate at pH 8.8. There is interference by cupric, ferrous, ferric and nickel ions, and by EDTA and nitrilotriacetic acid.[178]

[174]Hidehiro Goto, Yachiyo Kakita, and Michiko Namiki, *J. Jap. Inst. Metals* (Sendai) **25**, 181–184 (1961).

[175]J. O. Hibbits, A. F. Rosenberg, and R. T. Williams, *Talenta* **5**, 250–253 (1960).

[176]R. E. Stanton and A. J. McDonald, *Trans. Inst. Min. Met.* **71**, 511–516 (1962); *Bull. Inst. Mining Met.* No. **667**, 511–516 (1962).

[177]M. Hoebel and R. Roessle, *Z. Anal. Chem.* **244**, 317 (1969).

[178]R. S. Chawla and R. P. Singh, *Mikrochim. Acta* **1970**, 332–336.

Procedure. Add 5 ml of sample solution containing up to 1.3 ppm of cobalt to 75 ml of 0.2% solution of 2-thiovioluric acid. Adjust to pH 7.1–9.3, dilute to 100 ml, and read at 425 nm against a reagent blank.

Steel [179]

NICKEL PRESENT. Dissolve 0.05 gram of sample in 6 ml of hydrochloric acid and 2 ml of nitric acid. Evaporate to dryness and take up in 25 ml of water. Filter and dilute to 100 ml. Dilute a 10-ml aliquot to 50 ml and take a 10-ml aliquot. Add 10 ml of borate buffer solution for pH 8.8 and 2 ml of 10 m*M* sodium thioviolurate. Heat at 80° for 2 minutes, and cool. Add 2 ml of 10% solution of hydroxylamine hydrochloride, 2.5 ml of 20% solution of ammonium citrate, and 1 ml of saturated solution of EDTA. Dilute to 25 ml and read at 413 nm against a reagent blank.

UP TO 100-FOLD NICKEL PRESENT. To a portion of a sample solution above containing 2–10 μg of cobalt, add 5 ml of borate buffer solution for pH 8.8 and 4 ml of 10 m*M* sodium thioviolurate. Set aside for 2 minutes; then add 2.5 ml of saturated solution of EDTA. Set aside for 3 minutes; then dilute to 20 ml. Read at 413 nm against a reagent blank.

TIRON

The oxidation of Tiron, the disodium salt of 4,5-dihydroxy-*m*-benzenedisulfonic acid, by hydrogen peroxide as catalyzed by cobalt, permits determination of 1–12 ng of cobalt by reading after 4 minutes, 0.1–1.2 ng by reading after 1 hour, or less than 0.12 ng by reading after 24 hours.

In reading by absorption at 340 nm, Beer's law is obeyed up to 10 ppm. There is interference by nickel, ferrous, ferric, chromic, cupric, and titanic ions. An optimum range to read is 4–8 ppm.

Procedures

By Catalytic Oxidation. [180] MOLYBDENUM. Dissolve 0.1 gram of molybdenum or 0.15 gram of molybdenum oxide in 2 ml of hydrofluoric acid and 1 ml of nitric acid in a Teflon crucible, and evaporate to dryness. Add 0.5 ml and 0.5 ml of nitric acid and evaporate to dryness after each addition to remove hydrofluoric acid. Dissolve in 2 ml of 5 *N* hydrochloric acid, 1 ml of 30% hydrogen peroxide, and 7 ml of water. Pass through a 12 × 120 mm column filled to a depth of 50 mm with a strongly basic ion exchange resin. Wash the column with 10 ml of the solvent and evaporate the eluate in a Teflon crucible. Dissolve the residue in 5 drops of 9 *N* hydrochloric acid and pass it through a 1.8 × 80 mm microcolumn filled to a depth of 40 mm with the anion exchange resin. Rinse the crucible with 3 drops of 9 *N*

[179] L. V. Ershova, *Zavod. Lab.* **38**, 905–907 (1972).
[180] R. Kucharkowski and H. G. Döge, *Z. Anal. Chem.* **238**, 241–251 (1968); R. Kucharkowski, *ibid.* **249**, 22–25 (1971).

hydrochloric acid and add to the column that now retains nickel and cobalt. Eluate nickel with 0.5 ml of 9 *N* hydrochloric acid, then, cobalt with 0.5 ml of 1:2 hydrochloric acid. Evaporate the first 10 drops of the cobalt eluate and dissolve in 10 ml of water. Add 5 ml of boric acid–sodium hydroxide buffer solution for pH 10.4 and 5 ml of 1.78×10^{-3} *M* Tiron. Set aside for 30 minutes at 25°. Add 5 ml of 0.5 *M* hydrogen peroxide and record the absorption at 336 nm after 4 minutes. If the absorption is less than 0.25, read after 1 hour; if less than 0.2, read after 24 hours. Apply a blank.

By Absorption.[181] Mix a test solution containing up to 0.4 mg of cobalt and 10 ml of 3.4 m*M* Tiron. Adjust to pH 9.6 with dilute ammonium hydroxide and dilute to 50 ml with an ammonium hydroxide–ammonium chloride buffer solution for that pH. Read at 340 nm.

2,4,6-TRIS(2′-PYRIDYL-*s*-1,3,5-TRIAZINE)

Cobalt forms a 2:1 complex with this reagent in aqueous ethanol at pH 8.5 having absorption maxima at 485 and 404 nm. An appropriate range is 4–20 ppm of cobalt. Extraction with tri-*n*-butyl phosphate separates cobalt as the chloride from nickel and iron. The reagent may require recrystallization from aqueous ethanol; otherwise the peak at 404 nm may not appear. There is interference at 485 nm by ruthenium, osmium, iridium, and platinum. At 404 nm iron, nickel, copper, rhodium, and palladium also interfere.

Procedure. To a sample of not more than 2 ml of solution containing 0.05–0.2 mg of cobalt, add 4 ml of 0.01 *M* solution of the captioned reagent in ethanol.[182] Add 1 ml of borax–sodium hydroxide buffer solution for pH 10 and 2.6 ml of ethanol. Allow 45 minutes for color development; then dilute to 10 ml. This should contain 65–67% of ethanol at pH 8.5 ± 0.1. Read at 485 or 404 nm against a blank.

XYLENOL ORANGE

Cobalt forms a 1:1 complex with xylenol orange which has a maximum at 580 nm.[183] The optimum pH is 5.7–5.9. Beer's law is followed for 3–35 μg per 25 ml in the presence of 2–4 ml of 0.001 *M* reagent. There is interference by nickel, lead, manganous, cupric, magnesium, calcium, aluminum, and iron ions, as well as by thiourea and oxalates.

[181] C. K. Bhaskare and S. K. Deshmukh, *ibid.* **277**, 127 (1975).
[182] Mirjchan J. Jammohamed and Gilbert H. Ayres, *Anal. Chem.* **44**, 2263–2268 (1972).
[183] Makoto Otomo, *Jap. Anal.* **14**, 45–52 (1965).

Procedure. *Minerals.*[184] Dissolve 0.3 gram of sample in 8 ml of hydrochloric acid and 2 ml of nitric acid. If silica is present, volatilize with hydrofluoric acid. Dilute to 100 ml. Precipitate cobalt in an appropriate aliquot with hydrogen sulfide and filter. Dissolve the precipitate in hydrochloric acid with addition of hydrogen peroxide and dilute to 50 ml.

Neutralize a 10-ml aliquot with ammonium hydroxide. Add 3 ml of 0.5 m*M* xylenol orange and dilute to 25 ml with an acetate buffer solution for pH 6. Read at 580 nm.

Cobalt ion forms a 1:3 complex with **acetothioacetanalide**, which is 4-analino-4-thioxobutan-2-one, at above pH 8.4 as adjusted with ammonium hydroxide.[185] This is extracted with chloroform for reading at 625 nm. Beer's law is obeyed for 20–200 μg/ml in the extract. Iron is masked with triethanolamine or sulfosalicylic acid.

Mix 1 ml of sample solution containing 0.25–6 μg of cobaltous ion with 1 ml of 0.0238% solution of **acid alizarin black SN** (mordant black 25).[186] Add 5 ml of dimethylformamide and 1 ml of *N* hydrochloric acid. Dilute to 10 ml with dimethylformamide and read at 623 nm against a reagent blank. Cupric, ferric, and vanadium ions in an amount equal to the cobalt interfere, as well as aluminum and molybdenum in tenfold excess.

To 0.75 ml of sample add 0.75 ml of 0.2% solution of **acid monochrome green S** (mordant green 14) and dilute to 5 ml with 23.4% solution of sodium chloride in borate buffer solution for pH 10.[187] Extract the complex with 5 ml of butanol and read at 625 nm for 0.2–2 μg of cobalt per ml. EDTA and nickel interfere.

For cobalt in suspended matter in seawater, filter 2.5 liters of sample through a membrane filter having pore size 0.7 μm.[188] Wet-ash the filter, dissolve the ash, and separate the cobalt as the 2-nitroso-1-naphthol complex. Determine the cobalt by its catalysis of the oxidation of **alizarin** by hydrogen peroxide at pH 12.4.

Cobalt can be extracted from 9 *N* hydrochloric acid solution by a 10% solution of **Amberlite LA-2** in chloride form in xylene.[189a] It is read as the chloro complex in the organic phase at 665 nm. The extract follows Beer's law up to 0.2 mg of cobalt per ml.

The complex of cobalt with **5-(8-amino-1-hydroxy-3,6-disulfo-2-naphthylazo-1,2,4-triazole-3-carboxylic acid** is read at 620 nm for 0.2–1 μg/ml.[189b]

[184]E. A. Bashirov and A. M. Ayubova, *Uch. Zap. Azerb. Gos. Univ., Ser. Khim. Nauk* **1971** (3), 24–29.
[185]Krishna De and Jyotirmoy Das, *J. Indian Chem. Soc.* **52**, 1026–1028 (1975).
[186]D. Monnier, A. Marcantonatos, and M. Marcantonatos, *Helv. Chim. Acta* **49** (special issue), 57–65 (1966).
[187]V. V. Bagreev and Yu. A. Zolotov, *Talenta* **15**, 988–991 (1968).
[188]K. B. Yatsimirskii, E. M. Emel'yanov, V. K. Pavlova, and Ya. S. Savichenko, *Okeanologiya* **10**, 1111–1116 (1970).
[189a]Yoshio Iida, *Tech. Rep. Seikei Univ.* **1**, 49 (1964).
[189b]M. Israilov, M. Yu. Yusupov, D. N. Pachadzhanov, and Kh. Dadazhonov, *Manuscript No. 1360-74 deposited at Vsesoyuznyi Institut Nauchnoi i. Teknicheskoi Informatsii, Moscow* 10 pp. (1974).

6-Amino-4-hydroxy-2-mercapto-5-nitrosopyrimidine reacts slowly with cobalt at pH 8–9.8 to form a 3 : 1 complex.[190] Treat a sample solution containing 6–80 μg of cobalt with 25 ml of 0.0005 *M* solution of the reagent. Add 10 ml of ammoniacal buffer solution for pH 9.2 and dilute to 100 ml. Set aside for 1 hour, then read at 400 nm. There is interference by ferric, chromic, uranyl, and cyanide ions. Addition of 10 ml of 0.01 *M* EDTA 30 minutes before reading prevents interference by the colored complexes of nickel and copper with the reagent.

Cobalt catalyzes oxidation of the disodium salt of **5-{4-[amino-1-hydroxy-7-(4-nitrophenylazo)-3,6-disulfo-2-naphthylazo]phenylazo}-2-hydroxybenzoic acid** by hydrogen peroxide.[191] To 2 ml of 0.01% solution of the reagent, add 5 ml of 4% sodium hydroxide solution and 5 ml of hydrogen peroxide solution containing 9.6–18.4 mg/ml. Add the sample solution and dilute to 25 ml. Mix for 1 minute and follow the absorption at 660 nm for 20 minutes. Beer's law is followed for 32.6–136 ng of cobalt per ml. There is interference by more than 1000-fold amounts of ferrous, ferric, and octavalent osmium ions.

The 1 : 1 complex of cobalt with **5-aminopyrazole-4-carboxylic acid** has a maximum absorption at 550–560 nm in sodium hydroxide solution and at 530–540 nm in ammonium hydroxide solution.[192] Beer's law is followed for 4–20 μg of cobalt in 25 ml of solution containing 1 ml of 0.1 *N* reagent and 1 ml of 4% sodium hydroxide solution heated at 70° for 10 minutes. There is interference by phosphate, cyanide, citrate, tartrate, and tenfold excess of nickel.

To 2 ml of sample solution containing up to 30 μg of cobalt, add 5 ml of ammonium hydroxide–ammonium chloride buffer solution for pH 9.6 and dilute to 10 ml.[193] Add 8 ml of dimethylformamide and set aside for 10 minutes. Add 2 ml of 0.01 *M* **3-aminoquinoxaline-2-thiol** and dilute to 20 ml with 50% dimethylformamide. Set aside for 5 minutes and read at 428 nm against a reagent blank. There is minor interference by ferric, cupric, and palladous ions. Nickel forms a similar complex.

To a sample solution containing 0.003–0.3 mg of cobalt, add sequentially 5 ml of 4 *M* sodium ***p*-aminosalicylate**, 10 ml of 16% sodium hydroxide solution, and a few grains of sodium peroxide.[194] Dilute to 25 ml and read the brown color. Copper interferes.

At pH 4, **7-antipyrinylazo-8-hydroxyquinoline** forms a 1 : 1 complex with cobalt having a maximum absorption at 490 nm.[195] Beer's law is followed for 0.4–4 μg of cobalt per ml.

Cobalt is quantitatively precipitated by **antipyrinyldithioformic acid** in ammoniacal solution.[196] Mix 50 ml of sample solution containing up to 0.05 mg of cobalt with 5 ml of ammonium hydroxide and heat to 100°. Add a fifteenfold excess of 2% solution of the reagent in ammonium hydroxide. Set aside at 100° for an hour, cool, filter, and wash. Dissolve the precipitate in 5 ml of pyridine, dilute to 50 ml,

[190] A. Waksmundzki and S. Przeszlakowski, *Chem. Anal.* (Warsaw) **9**, 69–76 (1964).

[191] D. Costache, *Rev. Roum. Chim.* **16**, 565–568 (1971).

[192] Boleslaw Janik and Tadeusz Gancarczyk, *Chem. Anal.* (Warsaw) **14**, 293–297 (1969).

[193] Kazuo Ohira, Yoshinori Kidani, and Hisashi Koike, *Jap. Anal.* **23**, 658–664 (1974).

[194] Radu Ralea and Neculai Iorga, *An. Stiint. Univ. "Al. I. Cuza" Iasi, Sect. I* [N.S.] **2**, 211–226 (1956).

[195] Sh. T. Talipov, N. G. Smaglyuk, and A. Khodzhaev, *Manuscript 6204-73 deposited at Vsesoyuznyi Institut Nauchnoi i Tekhnicheskoi Informatsii, Moscow* (1975).

[196] B. Janik, B. Sawicki, and J. Weyers, *Mikrochim. Ichnoanal. Acta* **1965**, 810–813; B. Janik and B. Sawicki, *ibid.*, 386–388.

and read at 610 nm. Beer's law is followed for 0.2–1 μg of cobalt per ml. There is interference by nickel, silver, gold, and copper.

For cobalt by **atidan** in the presence of nickel, see Nickel.

To a solution containing 6–30 μg of cobalt, add 3 ml of 0.05 *M* EDTA.[197] Adjust to pH 4–7 with 0.5 *M* perchloric acid or 2% solution of sodium hydroxide. Mix with 5 ml of 0.5 *M* perchloric acid. Add 5 ml of 0.5 m*M* **azonol A**, which is 3-(4-antipyrinylazo)pentane-2,4-dione; dilute to 10 ml, and set aside for 10 minutes. Read at 610 nm. There is interference by oxalate, cyanide, pentavalent vanadium, palladous ion, cupric ion, scandium, zinc, and ferrous ion.

Above pH 13.5, cobalt forms a purple complex with **benzamidoxime.**[198] To the sample solution, add 2 ml of 0.2% solution of the reagent and 2 ml of 28% solution of potassium hydroxide. Dilute to 50 ml and read at 575 nm for up to 24 ppm of cobalt in the developed solution. The reaction is complete within 3 minutes, and the color is stable for 1 hour. There is interference by magnesium, ferric, manganous, cyanide, thiocyanate, and EDTA ions.

The 1:2 complex of cobalt at pH 1–9.25 with **benzenesulfonylthiocarbamate** is read at 640 nm.[199] Beer's law is followed for 30–140 ppm of cobalt.

Mix 1 ml of sample solution with 15 ml of 1% ethanolic **benzil mono(2-pyridyl) hydrazone** and adjust to pH 7.5 with 4% sodium hydroxide solution.[200] Set aside for 30 minutes, then add 6 ml of 8 *N* hydrochloric acid. This destroys other complexes without affecting that with cobaltic ion. Adjust to 25 ml and read at 535 nm. Beer's law is followed for 6.1–61 μ*M* cobalt. Citrate, tartrate, chromium, tungsten, and uranium interfere.

The 1:2 complex of cobalt with **2-(2-benzothiazolylazo-1-naphthol)** is extractable with chloroform above pH 6.[201] The maximum absorption is at 620 nm. A copper complex absorbs at 610 nm, a nickel complex at 634 nm.

Mix 2 ml of sample with 2 ml of 0.05% solution of **benzothiazol-2-yl-hydrazone of benzothiazole-2-aldehyde** in dioxan, dilute to 5 ml with dioxan, and set aside for at least 10 minutes.[202] Read cobalt at 475 nm and pH 8. The same derivative of naphtho[2,1-d]thiazole-2-aldehyde is read at 462 nm and pH 8. Beer's law is followed for 0.1–5.2 μg of cobalt per ml. The corresponding reactions are given by copper and nickel; zinc and palladium interfere. With **benzothiazol-2-yl phenyl ketoxime**, mix 2 ml of sample, 0.5 ml of ammoniacal buffer solution for pH 9, 1 ml of 0.2% ethanolic reagent, and 1.5 ml of dioxan. Read at 358 nm for cobalt. Beer's law is followed for 0.1–10 μg. Copper can be read at 362 nm and nickel at 400 nm, with all three solved by simultaneous equations. Palladium interferes with all three, cadmium with copper and nickel, silver with cobalt.

Cobalt complexes at pH 8 with **benzothiazol-2-yl phenyl ketone** in 50% ethanol.[203] Mix 2 ml of sample solution, 0.5 ml of buffer solution, 1 ml. of 0.1% ethanolic reagent, and 1.5 ml of ethanol. Set aside for 30 minutes and read at 466 nm. Beer's law is followed for 0.2–15 μg of cobalt. Similar complexes are formed with copper,

[197] B. Budesinsky and J. Svecova, *Anal. Chim. Acta* **49**, 231–240 (1970).
[198] Nobusuke Masuda and Meisetsu Kajiwara, *Jap. Anal.* **18**, 1466–1471 (1969).
[199] S. N. Tandon, P. K. Srivastava, and S. R. Joshi, *Z. Anal. Chem.* **253**, 36 (1971).
[200] Ronald T. Pflaum and E. Scott Tucker III, *Anal. Chem.* **43**, 458–459 (1971).
[201] Akira Kawase, *Jap. Anal.* **17**, 56–60 (1968).
[202] Toyozo Uno and Sumiyuki Akihama, *ibid.* **10**, 822–832 (1961).
[203] Toyozo Uno and Sumiyuki Akihama, *ibid.* **10**, 941–945 (1961).

nickel, cadmium, and zinc. There is interference by silver, palladium, iron, and lead.

Cobalt forms a soluble red 1:2 complex with **1-benzoyl-4-phenylthiosemicarbazide** in ethanol at pH 8.5–10.[204] The color is stable for 4 hours. Beer's law applies for 1.3–17 ppm of cobalt. For analysis for vitamin B_{12}, which is cyanocobalamin, add 0.25 ml of sulfuric acid to 1 ml of sample and boil for 1 minute. Add 0.3 ml of 30% hydrogen peroxide, and boil as long as gas is evolved. Cool, make slightly alkaline with 1% hydrazine hydrate solution, and dilute to a known volume. Take an aliquot equivalent to 0.15–1.5 μg of cobalt and add 0.5 ml of 0.1% solution of the reagent and 1 ml of 30% sodium hydroxide solution. Dilute to 10 ml with 50% ethanol and read at 400 nm.

Mix 10 ml of sample with 8 ml of an appropriate buffer solution to maintain a pH above 6 and add 10 ml of 0.1 *M* **benzoyltrifluoroacetone** in isoamyl alcohol.[205] Shake for 10 minutes, dilute the organic layer to 25 ml, and read cobalt at 390 nm. There is interference by copper, nickel, and ferric ion.

When 10 ml of cobalt sample solution in 0.01 *M* acetic acid at pH 5.5–6 is shaken for 10 minutes with 0.001 *M* **benzoyltrifluoroacetone** and 0.01 *M* **trioctylphosphine oxide** in cyclohexane, the complex in the extract is 1:2:1.[206a] The maximum absorption is at 367 nm, but it is better read at 380 or 390 nm, where the absorption is rectilinear up to 20 ppm of cobalt. Nickel, manganese, ferric ion, copper, citrate, and EDTA interfere.

For cobalt at 30 ppm in cadmium, dissolve the sample in 1:1 hydrochloric acid, adjust to pH 9 with ammonium hydroxide, and dilute to 100 ml. Extract an aliquot with 1 ml of 0.02% solution of **1-benzyl-benzamidazole-2-aldoxime** in chloroform and 4 ml of chloroform.[206b] Read the 1:2 metal-ligand complex at 343 nm against chloroform.

A ternary complex is formed by cobalt ion with thiocyanate and **benzyldimethyloctadecylammonium perchlorate**, which can be extracted with dichloroethane.[207] Optimum conditions are 0.5 *M* potassium thiocyanate, pH 2 with hydrochloric acid, and 0.02 *M* quaternary. Beer's law is followed for 50 μ*M*–2 m*M* cobalt. The color is stable for 20 hours.

Biacetylmonoxime-*p*-nitrophenylhydrazone has the trivial name of **cobaltone-I.**[208] To the sample solution, add 1.5 ml of ammonium hydroxide and 2 ml of freshly prepared 0.1% ethanolic reagent. Add 20 ml of ethanol and 1 ml of *N* ammonium acetate to destroy the color developed by the reagent with ammonium hydroxide. Dilute to 50 ml and read at 520 nm. Beer's law is followed for 0.5–5 ppm of cobalt, but excess reagent must be extracted with ether and the solution kept cold and in the dark because both air exposure and light destroy the complex. Iron, chromium, vanadium, and manganese are precipitated and filtered.

[204]R. Craciuneanu and E. Florean, *Rev. Roum. Chim.* **13**, 105–108 (1968); *Pharmazie* **24**, 462–464 (1969).
[205]G. N. Rao and J. S. Thakur, *Z. Anal. Chem.* **271**, 286 (1974).
[206a]Tsunenobu Shigematsu and Takaharu Honjyo, *Jap. Anal.* **18**, 68–71 (1969).
[206b]K. N. Bagdasarov, M. S. Chernov'yants, T. M. Chernoivanova, and E. B. Tsupak, *Zavod. Lab.* **42**, 143–144 (1976).
[207]A. S. Babenko and E. A. Alferov, *Manuscript No. 6365-73 deposited at Vsesoyuznyi Institut Nauchnoi i Tekhnicheskoi Informatsii, Moscow*, **1973**.
[208]Hsi-hua Hu, Yu-cheng Leu, Hsueh-tze Soong, and Ping-tsing Chu, *Hua Hsüeh Hsüeh Pao* **24**, 255–257 (1958); G. S. Deshmukh, V. D. Anand, and C. M. Pandev, *Z. Anal. Chem.* **182**, 170–177 (1961).

For the complex of cobalt with **biacetylmonoxime semicarbazone**, the optimum buffer solution contains ammonium chloride–ammonium hydroxide for pH 10.[209] The optimim absorption is at 330 nm and Beer's law is followed for 1–8 ppm of cobalt. For **piconaldehyde semicarbazone**, the same buffer for pH 10 gives a maximum absorption at 345 nm for 0.5–2.2 ppm. Cobalt forms a 2 : 3 complex with **biacetyl thiosemicarbazone oxime** in ammoniacal solution and a 1 : 1 complex in acetic acid.[210] In either medium reading is at 350 nm and Beer's law is followed for 1–7 ppm of cobaltous ion. Nickel interferes. For chromium, nickel, and cobalt in steel by **biuret** applying simultaneous equations, see Copper.

The 1 : 1 complex of cobalt with **2,2′-bis(carboxymethyl)mercaptoethyl ether**, which is 3-oxapentamethylenedithiodiacetic acid, follows Beer's law for 2.1–17.1 m*M* at 490 nm.[211]

In approximately neutral solution, cobalt is read at 10–40 μg as a complex with potassium **bis(2-hydroxyethyl)dithiocarbamate**.[212] EDTA for masking purposes does not interfere. Copper and nickel also complex with this reagent. Adjust a sample containing 50–500 μg of cobalt with hydrochloric acid to pH 2.5–3. Add 3 ml of 5% solution of the reagent and 10 ml of 5% solution of disodium EDTA. Set aside for 5 minutes; then make alkaline with 1 : 2 ammonium hydroxide. Add 3 drops of 10% solution of sodium cyanide and dilute to 100 ml. Read at 420 nm.

Mix with a sample solution 3 ml of m*M* ***N,N′*-bis(*m*-sulfobenzyl)dithiooxamide** and 20 ml of sodium hydroxide–borate buffer solution for pH 10.[213] Dilute to 50 ml, set aside for 60 minutes, and read at 400 nm against a reagent blank. Beer's law is followed for 0.3–5 μg of cobalt per ml. There is serious interference by the following ions: cadmium, chromic, copper, mercuric, manganous, nickel, lead, palladous, trivalent rhodium, and zinc.

Three ***N*-(bromophenyl)** derivatives of **2-thiocarbamoyldimedone** chelate at pH 7–7.4 with cobalt.[214] After standing for 30 minutes, they are extracted with benzene for reading at 405 nm. Beer's law is followed for 10–20 μg of cobalt per ml, in contrast to a similar absorption for the copper complex at 0.1–6 μg/ml. There is interference by iron, nickel, manganese, silver, and gold.

Cobalt at 0.8–3.9 ng/ml catalyzes the oxidation of **bromopyrogallol red** by hydrogen peroxide.[215] Readings are at 533 nm and pH 11.5–13.4. More than 1000 : 1 amounts of nickel, ferrous, and pentavalent vanadium ions interfere.

Calcichrome forms a 2 : 1 complex with cobaltous ion above pH 11 and a 1 : 1 complex below pH 10.[216] The 2 : 1 complex has a maximum absorption at 520 nm, and at pH 12 it follows Beer's law at 595 nm down to 4 ng of cobalt per ml in a solution containing 5 ml of 0.2 m*M* calcichrome per 25 ml. The 1 : 1 complex has a maximum absorption at 550 nm, and at pH 9.9 and 580 nm it follows Beer's law down to 2 ng of cobalt per ml.

[209] M. P. Martinez Martinez, D. Perez Bendito, and F. Pino Perez, *An. Quim.* **69**, 747–756 (1973).
[210] D. Perez Bendito and F. Pino Perez, *Inf. Quim. Anal. Pura Apl. Ind.* **22**, 1–12 (1968).
[211] E. Cassassas, J. J. Arias, and A. Mederos, *An. Quim.* **69**, 1121–1131 (1973).
[212] M. Pinkas, *Bull. Soc. Pharm. Lille* **1960** (3), 93–97; L. Nebbia and V. Bellotti, *Chim. Ind.* (Milan) **46**, 956–957 (1964).
[213] A. A. Janssens and M. A. Herman, *Bull. Soc. Chim. Belg.* **79**, 161–166 (1970).
[214] R. Mocanu and S. Fisel, *Anal. Stiint. Univ. "Al. I. Cuza"* **17**, 9–11 (1971).
[215] D. Costache and G. Popa, *Rev. Roum. Chim.* **15**, 1349–1353 (1970).
[216] Hajime Ishii and Hisahiko Einaga, *Jap. Anal.* **16**, 328–333 (1967).

To a sample solution containing up to 50 μg of cobalt, add 5 ml of 0.2 *M* boric acid–potassium chloride buffer solution for pH 9.2.[217] If necessary, adjust to pH 9.2 with 0.2 *M* sodium carbonate. Dilute to 25 ml and shake for 5 minutes with 10 ml of 25 m*M* **5-chloro-8-hydroxy-7-iodoquinoline** in chloroform. Read the 1 : 2 metal-ligand complex at 430 nm against a reagent blank. Beer's law is followed up to 5 ppm.

The 1 : 2 complex of cobalt with **chlorindazon DS**, which is 4-(6-chloroindazol-3-ylazo)-3-hydroxynaphthalene-2,7-disulfonic acid, at pH 10 with an ammoniacal buffer solution has a maximum absorption at 620 nm.[218] A less satisfactory peak is at 415 nm.

2-(5-Chloro-5-pyridylazo)-5-diethylaminophenol forms a 2 : 1 complex with cobalt ion at pH 5–9.[219] This complex is stable at up to 4 *N* sulfuric acid, but it complexes with other metals decomposed at a lower pH. The cobalt complex is extracted with chloroform and shows a maximum absorption at 580 nm. Beer's law is followed for 0.04–1 μg of cobalt per ml.

Adjust a mixture of cobalt ion and ethanolic **4-(5-chloro-2-pyridylazo)-*m*-phenylenediamine** to pH 5 with an acetate buffer solution and add 10 ml of 1 : 1 hydrochloric acid.[220] Read the absorption of the 1 : 2 cobalt-ligand complex at 588 nm and of the reagent at 425 nm. The decrease in absorption by the reagent at 425 nm due to complex formation is then added to the increase at 588 nm due to formation of the complex. The calibration curve is linear for 1–4 ppb of cobalt. This is dual-wavelength spectrophotometry. There is interference by ferric ion, hexavalent tungsten, and chromium.

To a sample solution containing about 30 μg of cobalt, add 1 ml of ethanolic **3-(5-chloro-2-pyridylazo)pyridine-2,6-diamine.**[221] Adjust to pH 5 with 0.2 *M* acetic acid–0.2 *M* sodium acetate. Add 5 ml of 1 : 1 hydrochloric acid and dilute to 25 ml. Read at 620 nm. Ferric ion and large amounts of copper interfere.

The complex of cobalt in the presence of an excess of **chrome azurol S** (mordant blue 29) shows a maximum absorption at 567 nm.[222] At pH 10 with a borate buffer solution it is suitable for reading 0.08–2 ppm of cobalt. A similar color is given by rare earths, nickel, zinc, lead, and manganese. In the presence of hydroxydodecyltrimethylammonium bromide and pyridine, the extinction is stable at pH 10.6–11.5 and is read at 654 nm.[223] Beer's law applies for 0.02–0.32 μg of cobalt per ml. Fluoride ion masks antimony, bismuth, and the rare earth and alkaline earth ions. There is interference by tenfold excess of the following ions: zinc, manganese, zirconium, magnesium, copper, ferric, nickel, chromic, tetravalent uranium, titanium, platinic, and beryllium. To 25 ml of sample containing 0.5–16 μg of cobalt, add 3 ml of 20% pyridine solution, 4 ml of 0.02% solution of chrome azurol S, 5 ml of 0.1% solution of the quaternary reagent, and 5 ml of 0.05 *M* borate buffer solution for pH 10.9. Dilute to 50 ml. Set aside for 5 minutes before reading at 654 nm. For cobalt by chrome azurol S, also see nickel.

[217] Naoichi Ohta, Shin-ichi Tama, Hisao Nunokawa, and Minoru Tarai, *ibid.* **23**, 931–935 (1974).
[218] Dieter Molch, Hartmut Koenig, and Eberhard Than, *Z. Chem.* **15**, 361–362 (1975).
[219] S. I. Gusev and L. G. Dazhina, *Zh. Anal. Khim.* **29**, 810–813 (1974).
[220] Shozo Shibata, Masamichi Furukawa, and Tadashi Honkawa, *Anal. Chim. Acta* **78**, 487–491 (1975); cf. Shozo Shibata, Masamichi Furukawa, Yoshio Ishiguro, and Shozo Sasaki, *ibid.* **55**, 231–237 (1971).
[221] Shozo Shibata, Masamichi Furukawa, and Kazuo Goto, *Talenta* **20**, 424–430 (1973).
[222] Hiroshi Nishida, *Jap. Anal.* **19**, 34–39 (1970).
[223] Yoshio Shijo, Tsugio Takeuchi, and Shogo Yoshizawa, *ibid.* **18**, 204–207 (1969).

To a sample solution containing up to 9 μg of cobalt per ml, add 7 ml of 0.05% solution of **chromotrope 2B** (acid red 176) and 1 ml of 1% solution of ammonium hydroxide. Dilute to 25 ml and read at 574 nm.[224] Beer's law is followed for 1–9 μg of nickel per ml. Nickel reacts similarly.

The 1:2 chelate of cobalt with **chromotrope 2R** (acid red 29) at pH 7–10 is read at 530 nm for 0.9–4.7 ppm of cobalt.[225]

For cobalt determination, add 10 ml of 0.25% solution of **cyclohexane-1,2-dione dithiosemicarbazone** in dimethylformamide.[226] Adjust to pH 5.5 by addition of 10 ml of acetic acid–sodium acetate buffer solution and 10 ml of ethanol. Dilute to 50 ml and read at either 450 or 500 nm. Beer's law is followed for 2.2–8.7 ppm of cobalt. Nickel and ferrous or ferric ion react.

Cobaltic ion complexes with **1,2-diaminocyclohexanetetraacetic acid**, known as complexon IV.[227] Beer's law is followed for 1–100 μg of cobalt per ml. To determine cobaltous ion, add to the sample solution 10 ml of 0.096 *M* reagent as the sodium salt, 5 ml of 30% hydrogen peroxide, and 10 ml of sodium acetate–acetic acid buffer solution for pH 4.65. Boil for 10 minutes, cool, and dilute to 100 ml. Read at 545 nm. Chromic ion interferes seriously.

The 1:1 complex of cobalt with **1,2-diaminopropane-NNN′N′-tetraacetic acid** is read at 490 nm.[228] Chromic, nickel, and cupric ions also complex with this reagent.

Decompose 1 gram of silicate with hydrofluoric and sulfuric acids, treat with hydrofluoric and nitric acids, and evaporate to sulfur trioxide fumes.[229] For analysis of iron meteorites, dissolve 50 mg of drillings in aqua regia and evaporate nearly to dryness. In either case take up with water and dilute to 250 ml. To an aliquot of not more than 25 ml containing not more than 25 μg of cobalt, add 5 ml of 20% solution of ammonium fluoride, 2 ml of 2 m*M* **4-(3,5-dibromo-4-methyl-2-pyridylazo)-*m*-phenylenediamine**, and 5 ml of 4 *M* ammonium acetate. Warm for 10 minutes, cool, add 10 ml of hydrochloric acid, and dilute to 50 ml. Read at 590 nm against a reagent blank.

The chelate of **3′,5′-dichloro-2′-hydroxyacetophenone oxime** is formed in 10 minutes at pH 7–8 by mixing a solution containing 0.01 *M* reagent and a solution containing up to 50 μg of cobalt.[230a] This complex is extracted with chloroform for reading at 400 or 420 nm for 1.7–10 ppm.

Mix 2 ml of sample solution with 2 ml of acetate buffer solution for pH less than 6 or 2 ml of phosphate buffer solution for pH above 6.[230b] Add 2 ml of m*M* rhodamine 6G and dilute to 10 ml. Extract with 10 ml of m*M* **5,7-dichloro-8-hydroxyquinoline** in benzene. Dilute to 25 ml with benzene and read the ternary complex at 542 nm. At pH 6.3 as much as 40 m*M* cobalt is so extracted.

Mix a sample solution containing up to 10 μg of cobalt with 1 ml of 0.1% ethanolic **5-(3,5-dichloro-2-pyridylazo)toluene-2,4-diamine**. If other metals may in-

[224]D. Negoiu, A. Kriza, and M. Negoiu, *Anal. Univ. Bucur., Ser. Stiint. Nat. Chim.* **13**, 157–164 (1964).
[225]O. Prokash and S. P. Mushran, *Chim. Anal.* (Paris) **49**, 473–476 (1967).
[226]J. A. Munoz Leyva, J. M. Cano Pavon, and F. Pino Perez, *Quim. Anal.* **28**, 90–98 (1974).
[227]E. Jacobsen and A. R. Selmer-Olsen, *Anal. Chim. Acta* **25**, 476–481 (1961).
[228]S. Vicente-Perez, L. Hernandez, and J. Rosas, *Quim. Anal.* **28**, 283–288 (1974).
[229]E. Kiss, *Anal. Chim. Acta* **77**, 320–323 (1975).
[230a]Keemti Lal and S. P. Gupta, *Indian J. Chem.* **13**, 973–974 (1975).
[230b]Jerzy Minczewski, Jadwiga Chwastowska and Elwina Lachowicz, *Chem. Anal.* (Warsaw) **21**, 373–381 (1976).

terfere, add 5 ml of a disodium phosphate–citric acid buffer solution for pH 3–4. Alternatively, adjust to pH 3–5 with 5 ml of acetate buffer solution. Add 5 ml of either 1:1 hydrochloric acid or 1:1 sulfuric acid. Dilute to 25 ml and read at 590 nm against a reagent blank. Beer's law applies at 0.01–0.04 ppm of cobalt. Palladium, cyanide, and EDTA interfere.

For cobalt in seawater, add 10 ml of 40% solution of sodium citrate dihydrate to a 2-liter sample.[231] Add 20 ml of 0.2% solution of **2-diethylamino-5-nitrosophenol** in 0.01 *N* hydrochloric acid. Set aside for 30 minutes, then add 10 ml of 10% solution of EDTA. Add 20 ml of 1,2-dichloroethane and shake mechanically for 10 minutes. Wash the organic phase successively with 5 ml, 5 ml, and 5 ml of 1:2 hydrochloric acid, 5 ml of 5.6% solution of potassium hydroxide, and 5 ml of 1:2 hydrochloric acid. Filter, and read at 462 nm.

For steel analysis, dissolve 50 mg in 1:3 nitric acid–hydrochloric acid and evaporate nearly to dryness.[232a] Take up in water, and dilute to 100 ml. To an aliquot, add 2.5 ml of 20% ammonium nitrate solution, 2 ml of 1:9 ammonium hydroxide, and 2 ml of 0.03% ethanolic **5-diethylamino-2-(2-pyridylazo)-phenol**. Add 1 ml of saturated solution of EDTA and heat at 80° for 30 minutes. Dilute to 25 ml with acetate buffer solution for pH 8 and read at 582 nm.

At pH 6.5 ethanolic **2,6-dihydroxy-3-(2-thiazolylazo)benzoic acid** forms a red complex with cobalt.[232b] At 560 nm it follows Beer's law for 0.26–1.30 ppm of cobalt. There are many interferences.

Mix a sample solution containing less than 22 μg of cobalt with 5 ml of 0.75 m*M* zinc **dimercaptomaleonitrile** and 5 ml of 0.1 *M* hexamine-hydrochloric acid buffer solution for pH 5.1.[232c] Dilute to 25 ml and read the 1:3 complex of cobalt and ligand at 288 nm within 45 minutes against a reagent blank. If nickel is present modify this to 312 nm. More than 5 μg of iron, nickel, or copper interfere seriously.

A 1:3 complex of cobalt with **5-dimethylamino-2-nitrosophenol** has a maximum extinction at pH 5–7 in a citric acid buffer solution.[233] The complex and excess reagent are extracted with 1, 2-dichloroethane, benzyl alcohol, or nitrobenzene. Iron, nickel, and cupric ion are largely masked by the citrate. Residual amounts and excess reagent are washed out with 1:2 hydrochloric acid. Beer's law applies for 2–14 μg/ml. Reading is at 456 nm.

To a sample solution containing 0.15–1 μg of cobalt per ml, add two volumes of 5 m*M* solution of **diethyl-3,4-dioxoadipate**, also known as kethiazone, in 1:1 ethanol-cyclohexanone.[234] Set aside for 3 hours and read at 386 nm against a blank.

Cobalt complexes with **dimedone**, which is 5,5-dimethylcyclohexane-1,3-dione.[235] The maximum absorption at 400 nm is stable for 1 hour at pH 9–9.5. Interference by copper and nickel is avoided by acidifying to decompose their complexes and extracting that of cobalt with methyl isobutyl ketone.

[231]Shoji Motomizu, *Anal. Chim. Acta* **64**, 217–224 (1973).
[232a]S. I. Gusev and N. N. Kiryukhina, *Zh. Anal. Khim.* **24**, 210–215 (1969).
[232b]F. Garcia Montelongo and J. A. Lopez Cancio, *An. Quim.* **73**, 248–253 (1977).
[232c]Sadanobu Inoue, Takao Yotsuyanagi, Mitsuo Sasaki, and Kazuo Aomura, *Jap. Anal.* **25**, 534–539 (1976).
[233]Shoji Motomizu, *J. Chem. Soc. Jap., Pure Chem. Sect.* **92**, 726–730 (1971); *Anal. Chim. Acta* **56**, 415–426 (1971); *Talenta* **21**, 654–660 (1974).
[234]Wojciech Gorski and Sylwia Podlasin-Paradowska, *Pol. Pharm.* **29**, 31–34 (1972).
[235]R. Belcher, S. A. Ghonaim, and A. Townshend, *Talenta* **21**, 191–198 (1974).

Dissolve 0.2 gram of iron or steel with 4 ml of hydrochloric acid and 1 ml of nitric acid.[236] Evaporate to under 2 ml and dilute to 100 ml. Take a 5-ml aliquot. Mask nickel and iron with 2 ml of 2 *M* citrate buffer for pH 5.3. Add 1 ml of 5 m*M* **5-dimethylamino-2-nitrosophenol** in 0.01 *N* hydrochloric acid. Set aside for 5 minutes, then shake for 30 seconds with 5 ml of 1, 2-dichloroethane. Wash the organic layer with 5 ml of 1:2 hydrochloric acid, filter through paper, and read at 456 nm.

Mix 1 ml of sample solution containing 1.5–35 μg of cobalt ion with 1 ml of 0.1 *M* **2,4-dimethylbenzamidooxime** in 50% ethanol.[237] Add 1 ml of 8% solution of sodium hydroxide in 50% ethanol and dilute to 10 ml. Set aside for 30 minutes and read at 578 nm against water.

To 7 ml of acid sample solution containing 2–20 μg of cobalt, add 0.8 gram of potassium iodide.[238] Adjust to pH 4.5–5 with 10% sodium bicarbonate solution. Add 1 ml of an acetate buffer solution for pH 6 and 1 ml of 0.1% ethanolic **dimethylglyoxime**. Read the complex, which is the bis(dimethylglyoximato)diiododocobaltate of bivalent cobalt. By treatment with dimethylglyoxime followed by sodium azide, a maximum absorption for 1–75 ppm of cobalt is developed in 3 hours and read at 540 nm.[239]

To a cobalt sample solution, add 1 ml of ethanol, 0.5 ml of 0.05 *M* ethanolic benzidine, and 0.5 ml of 0.1 *M* ethanolic dimethylglyoxime to form a 1:1:2 complex.[240a] Adjust to pH 6–7. Extract the solution or an aliquot with 10 ml of benzyl alcohol. Read the organic phase. Beer's law holds up to 20 μg of cobalt per ml. An alternative to benzidine is *o*-phenylenediamine as the ternary complex at pH 4–5.

At pH 3–6 resorcinol and nitrite react to form **2,4-dinitrosoresorcinol** in the solution and immediately forms a 3:1 monovalent complex ion with trivalent cobalt.[240b] This can be extracted as a ternary complex with tetrabutylammonium ion or tributylammonium ion in chloroform at pH 5–7.5 for reading at 420 nm.

For steel analysis, dissolve 0.8 gram of sample in hydrochloric and nitric acids or in hydrochloric acid and hydrogen peroxide and concentrate to 25 ml.[241] Add 10 ml of 50% potassium thiocyanate solution and discharge the color with 40% ammonium fluoride solution. Add 30 ml of acetone and adjust the pH to 7 ± 0.2 with 1:2 ammonium hydroxide. Add 2 ml of 50% sodium thiosulfate solution and 25 ml of 1.2% solution of **diphenyliodonium chloride**. Extract with 10 ml, 10 ml, and 10 ml of chloroform, filter the combined extracts, and dilute to 50 ml with chloroform. Read at 625 nm. As given, this technic is applicable to 0.05–0.17% cobalt in the sample.

di-2-Pyridylglyoxal dithiosemicarbazone complexes with nickel at pH 5.2.[242] This is extractable with chloroform, in contrast to the corresponding cobalt complex.

[236] Shoji Motomizu, *Jap. Anal.* **20**, 1507–1512 (1971); **22**, 695–699 (1973); Kyoji Toei and Shoji Motomizu, *Analyst* **101**, 497–511 (1976).
[237] K. Manolov and N. Motekov, *Mikrochim. Acta* **1974**, 231–234.
[238] K. Burger and I. Ruff, *Acta Chim. Hung.* **45**, 77–86 (1965).
[239] Kuladaranjan Ray and B. Das Sarma, *Sci. Cult.* (Calcutta) **23**, 617 (1958).
[240a] I. V. Pyatnitskii, P. B. Mikhel'son, and L. T. Moshkovskaya, *Ukr. Khim. Zh.* **41**, 739–742 (1975).
[240b] V. M. Peshkova, T. V. Polenova, and Yu. A. Barbalat, *Zh. Anal. Khim.* **32**, 471–477 (1977).
[241] Arnold J. Fogg, C. T. Higgins, and D. Thorburn Burns, *Mikrochim. Acta* **1969**, 546–549.
[242] J. L. Bahamonde, D. Perez Benedito, and F. Pino Perez, *Analyst* **99**, 355–359 (1974).

Each determines 1 ppm in the presence of 5 ppm of the other. To the sample solution, add 15 ml of 0.1% ethanolic reagent and 20 ml of acetic acid–sodium acetate buffer solution for pH 5.2. After 30 minutes dilute to 50 ml. Extract nickel if present, and read cobalt at 410 nm.

di-(2-Pyridyl)ketone-2-pyridylhydrazone complexes instantly with cobalt ion at pH 2–11.[243] Adjust 5 ml of sample containing 5–50 μg of cobalt to pH 3–8 with ammonium hydroxide or perchloric acid. Add 2 ml of 10 m*M* reagent in ethanol and 5 ml of perchloric acid. Dilute to 25 ml. If palladous ion is present, either heat at 80° for 20 minutes or add 5 ml more of perchloric acid and set aside for 20 minutes. Oxidize ferrous ion to the ferric form. If cupric, ferric, or nickel ion is present in more than 100-fold excess over the cobalt, increase the amount of reagent. Read at 500 nm, a wavelength at which the reagent does not absorb. Beer's law is followed for 0.15–2 ppm of cobalt.

To a sample solution containing 3–20 μg of cobalt add 0.8 ml of 0.1% solution of **di-2-quinolyl oxime** in ethanol.[244] Adjust to pH 10 with 10% potassium hydroxide solution. Extract with benzene and read the dried extract at 365 nm for 0.04–1.2 ppm of cobalt.

Dissolve 0.1 gram of steel in 5 ml of phosphoric acid and 5 ml of sulfuric acid, adding 0.5 ml of nitric acid dropwise to destroy carbides.[245] Boil off oxides of nitrogen and dilute to 100 ml. Treat an aliquot containing 0.01–0.1 mg of cobalt with 5 ml of acetate buffer solution for pH 5.5–6 and 5 ml of 1% ascorbic acid solution. Add 1 ml of 0.01 *M* **(dithiocarboxy)iminodiacetic acid** as the tripotassium salt. Set aside for 3 minutes, dilute to 50 ml with 1:5 hydrochloric acid, and read the 1:2 complex at 315 nm against a portion of the undeveloped sample. Beer's law is followed at that concentration. The cobalt complex is stable up to 4 *N* hydrochloric acid, but the complexes with nickel and ferrous ions break down at under 2 *N* hydrochloric acid.

To the sample of solution, add 20 ml of m*M* ethanolic **β-dithionaphtholic acid**, which is 1-hydroxy-naphthalene-2-carbodithiolic acid, and 2 ml of 2 *N* sodium hydroxide.[246] Dilute to 100 ml and set aside for 20 minutes. Read at 535 nm against a reagent blank. Beer's law is followed for 0.02–1.6 μg of cobalt per ml. There is interference by an equal amount of nickel, copper, stannous ion, or bismuth, and by tenfold amounts of lead, mercuric ion, silver, cadmium, ferric ion, or aluminum.

To 25 ml of sample containing cobalt in 10:1 ethanol-pyridine, add 1 ml of 0.1% ethanolic **dithiooxamide**, also known as **rubeanic acid**.[247] Read at pH 6.5–7.2 and 400 nm. The absorption is stable for 2 hours. Chloride masks nickel, thiourea masks copper and silver, fluoride masks iron. Even small amounts of bismuth, chromium, cadmium, zinc, lead, mercury, tin, platinum, palladium, and gold interfere.

Cobalt complexes with **eriochrome black A** at pH 10, giving a proportional reduction in absorption at 620 nm.[248] EDTA masks nickel, zinc, lead, and calcium.

[243]G. S. Vasilikiotis, T. Kouimtzis, C. Apostolopoulou, and A. Voulgaropoulos, *Anal. Chim. Acta* **70**, 319–326 (1974).

[244]S. Stupavsky and W. J. Holland, *Mikrochim. Acta* **1971**, 559–564.

[245]F. M. Tulyupa, Yu. I. Usatenko, and V. A. Pavlichenko, *Zavod. Lab.* **34**, 14–16 (1968).

[246]H. Gorniak and B. Janik, *Z. Anal. Chem.* **273**, 127 (1975).

[247]J. Xavier and Pryadaranjan Ray, *J. Indian Chem. Soc.* **35**, 432–444 (1958); cf. R. Bondivenne, G. Beau, and R. Y. Mauvernay, *Chim. Anal.* (Paris) **44**, 114–121 (1962).

[248]Abdel-Aziz M. Amin, A. A. Abd El Raheem, and Farouk A. Osman, *Z. Anal. Chem.* **167**, 8–16 (1959).

There is interference by cadmium, copper, barium, strontium, manganese, and magnesium.

The complexes of cobalt and zinc with **eriochrome blue black R** both absorb at 625 nm.[249] Addition of EDTA decolorizes the zinc complex, permitting reading of cobalt and of zinc by difference; Beer's law is followed for 0.05–0.6 ppm of each. To a sample containing up to 60 μg of cobalt, add 2 ml of 0.1 *M* ammonium chloride that has been titrated to pH 9.9. Add sufficient 0.025% solution of the dye to give an absorption of about 0.6 when diluted. Add 1 drop of 0.04 *M* EDTA, dilute to 100 ml, and read at 625 nm.

The pink complex of cobalt chloride with **ethanol** can be read at 553 and 665 nm up to 50 μg/ml,[250] the blue complex with **acetone** at 665 nm up to 77 μg/ml.

Cobalt forms a 1:2 complex with **5-ethylamino-2-(2-pyridylazo)-*p*-cresol**.[251] Dissolve 50 mg of steel in 1:3 nitric acid–hydrochloric acid, evaporate nearly to dryness, and dilute to a known volume. To a 10-ml aliquot containing 0.05–0.5 μg of cobalt per ml, add 2.5 ml of 20% ammonium citrate solution, 2 ml of 1:5 ammonium hydroxide, 0.5 ml of 0.05% solution of the reagent, and 1 ml of saturated solution of EDTA. Heat at 80° for 30 minutes, cool, and dilute to 25 ml with buffer solution for pH 8. Read at 530 nm. The technic is also applicable with the 5-bromo- and 3,5-dibromo-2-pyridylazo- analogues of the reagent.

Ethyl-di(1-sodiotetrazol-5-ylazo)acetate complexes with cobalt at pH 2–3 for reading at 610 nm.[252] The sensitivity is to 0.06 μg of cobalt per ml, and Beer's law is followed up to 4 μg/ml. For the determination, iron and aluminum are masked with fluoride and copper with EDTA.

The complex of cobalt with ***NN'*-ethylenebis(4-methoxy-1,2-benzoquinone)-1-oxime-2-imine** is soluble in ethanol, acetone, and chloroform.[253a] Take a sample solution containing 1–18 μg of cobalt. Add EDTA to mask iron, copper, nickel, and palladium. Adjust the pH to 1–3.5 with sulfuric acid or acetic acid. Add an aqueous solution of the reagent and extract the complex with chloroform. Remove palladium and decomposition products of the reagent by shaking with 2 *N* ammonium hydroxide, then with 2 *N* sulfuric acid. Read at 381 nm.

Buffered at pH 10 with ammonium acetate, cobalt forms a 1:1 complex with **ethylenediaminebis(acetylacetone)** which is 5,8-diaza-4,9-dimethyldodeca-4,8-diene-2,11-dione.[253b] Read at 345 nm it follows Beer's law for 1.2–31 μg of cobalt per ml. Ferric and cupric ions interfere.

The chelate of cobaltic ion with **ethylene glycol bis(2-aminoethyl ether)-*N,N,N',N'*-tetraacetic acid** is read at 10–250 μg/ml.[254] Major interferences are ferric, manganous, cupric, nickelous, and chromic ions. Below pH 6.5 oxidation to cobaltic ion is incomplete; above pH 9.5 the complex formed is unstable.

A green 1:2 complex of cobalt with **ethylxanthate** is extracted with carbon tetrachloride at pH 6.5–9.[255] A peak at 356 nm can be read for up to 0.025 m*M* cobalt.

[249]D. W. Rogers, *Anal. Chem.* **34**, 1657–1659 (1962).
[250]M. K. Akhmedli and E. A. Bashirova, *Tr. Azerb. Gos. Univ. S. M. Kirova, Ser. Khim.* **1959**, 11–20.
[251]S. I. Gusev, N. N. Kiryukhina, and Z. A. Bitovt, *Zh. Anal. Khim.* **23**, 889–893 (1968).
[252]N. r. Frumina, N. N. Goryunova, and I. S. Mustafin, *ibid.* **22**, 1523–1526 (1967).
[253a]Michio Mashima, *J. Chem. Soc. Jap.* **80**, 1260–1265 (1959).
[253b]D. G. Gambarov, S. B. Bilalov, and A. K. Babaev, *Zh. Anal. Khim.* **31**, 1731–1733 (1976).
[254]Francisco Bermejo Martinez and M. Paz Castro, *Inf. Quim. Anal.* (Madrid) **5**, 129–134 (1959).
[255]Kenjiro Hayashi, Yoshiaki Sasaki, and Kazuko Nagano, *Jap. Anal.* **20**, 727–731 (1971).

For determination of cobalt as a 1:3 complex, take a sample solution containing 0.5–17.5 μg. Add 2 ml of 0.045% solution of **fast grey RA** (mordant black 15) and 5 ml of dioxan. Dilute to 25 ml, heat at 100° for 1 minute, and cool at once. Read at 554 nm against a reagent blank. Interference by ferric, vanadate, zirconium, cupric, gallium, and iodide ions is serious; that by cadmium, scandium, bromide, tungstate, and tartrate ions is less serious.

Cobalt forms a 3:4 complex with **fast navy 2R** (mordant blue 9), which has a maximum absorption at 550 nm and is read at an inflexion at 620 nm.[256] Beer's law applies to 0.2–5 μg of cobalt in 25 ml. The color is stable for 3 hours. EDTA masks calcium, magnesium, manganese, lead, or zinc. Lead is masked by tartaric acid, manganese by ascorbic acid. Masking nickel with EDTA requires 15 minutes at pH 8.5.

Fast sulphon black F (acid black 32) permits reading of cobalt in ammoniacal solution down to 0.5 ppm.[257] Cupric, beryllium, nickel, and manganous ions enhance the color; ferric and chromic ions diminish it.

For cobalt in steel, dissolve 0.1 gram of sample in an acid mixture containing 14% of sulfuric acid and 18% of phosphoric acid by volume.[258] Add nitric acid and evaporate to sulfur trioxide fumes. Take up in water. Add 60 ml of ammonium citrate solution containing 20% of citric acid and 5% of 25% ammonium hydroxide. Dilute to 100 ml. To 50 ml, add 0.2 ml of 25% potassium **ferrocyanide**. Read against the other part of the sample as a blank.

For determination with **α-furilmonoxime**, the sample solution may contain up to 0.1 mg of cobalt if nickel and copper are present, up to 0.2 mg if they are absent.[259a] Adjust to pH 5–6 with 5% solution of sodium hydroxide. Mask iron and aluminum with 20 ml of a saturated solution of sodium fluoride. Add 5 ml of a 10% solution of α-furilmonoxime in pyridine and set aside for 10 minutes. Add 15 ml of 1:1 hydrochloric acid and extract the cobalt complex with 12 ml and 12 ml of chloroform. Add 1 ml of pyridine to the combined extracts to clarify, adjust to 25 ml, and read the orange complex at 405–410 nm. Pyridine, α-, and β-picoline accelerate the formation of the complex.[259b] Benzene is an alternative extraction medium. By making the solution strongly acid before extraction, nickel, copper, and manganese do not interfere but platinum and palladium still interfere.

To 10 ml of sample containing 50–250 μg of cobalt add 1 ml of *M* **gluconic acid** to form the 1:1 complex.[260] Add 4% sodium hydroxide to raise to pH 10.8–11.3. Add 5 ml of a buffer solution containing 50 ml of 0.1 *M* sodium carbonate and 3 ml of 0.1 *N* hydrochloric acid per 100 ml. Dilute to 25 ml, set aside for 30 minutes, and read at 330 nm against water. There is serious interference by iron, copper, uranium, nickel, and cerium.

Cobalt reacts with **glycerol** in a strongly alkaline medium to form a stable blue compound.[261] Nickel and iron do not interfere. The cobalt content must not exceed 0.3 mg/ml; 7 μg can be detected.

[256] M. M. Dokhana, *Z. Anal. Chem.* **189**, 389–396 (1962).

[257] F. Burriel-Marti, A. Cabrera Martin, and T. S. West, *An. Real. Soc. Esp. Fis. Quim., B*, **61**, 703–706 (1965).

[258] V. F. Mal'tsev, *Proizvod. Trub. Khark., Sb.* **1960** (3), 154–158.

[259a] Jaroslav Martinek and Vaclav Hovorka, *Chem. Listy* **50**, 1450–1455 (1956).

[259b] F. Vlacil and A. Jehlickova, *Colln. Czech. Chem. Commun.* **40**, 539–545 (1975); *ibid.* **41**, 3749–3753 (1976).

[260] F. Bermejo Martinez and G. Branas Miguez, *Z. Anal. Chem.* **253**, 353–354 (1971).

[261] G. D. Nessonova and E. K. Pogosyants, *Nauch.-Issled. Tr. Mosk. Tekstil. Inst.* **18**, 114–120 (1956); G. D. Nessonova, E. K. Pogosyants, and M. O. Lishevskaya, *Zavod. Lab.* **25**, 786–789 (1959).

The complex of cobalt with **glyoxime** develops in 3 minutes for a maximum absorption at 286.2 nm.[262] Beer's law is followed up to 4 ppm of cobalt in 50 ml of solution containing 2 ml of 0.02 *M* glyoxime and *M* ammonium chloride buffer for pH 9–9.5. Interference by cupric, nickel, ferric, and aluminum ions is masked by adding EDTA after forming the chelate.

Cobalt catalyzes the oxidation of **hematoxylin** by hydrogen peroxide in a solution buffered with 0.5% solution of boric acid in 0.24 *N* sodium hydroxide.[263] The reaction is followed at 533 nm. The method will determine 15.5–77.5 pg of cobalt per ml. If more than 1000 times the cobalt, ferrous, ferric, palladous, and pentavalent vanadium ions interfere.

To a solution containing 30–300 μg of cobalt, add 5 ml of 0.5 *M* acetate buffer solution for pH 4.6, 5 ml of *M* potassium thiocyanate, and 2 ml of 1% **hyamine 1622**, which is benzethonium chloride.[264] Dilute to 25 ml, extract with 10 ml of benzene, and read the extract at 624 nm.

To 3 ml of sample solution, add 5 ml of ethanol, 1 ml of 1% solution of the hydrazinium salt of **hydrazinedithiocarboxylic acid**, and read at 400 nm.[265] Large amounts of nickel, ferric, and chromic ions interfere.

For cobalt in seawater, extract 2.5 liters of sample with a 1% solution of 8-hydroxyquinoline in chloroform.[266] Distill the solvent, treat the residue with hydrochloric acid, and ignite. Dissolve in hydrochloric acid, adjust to pH 10–11 with ammonium hydroxide, and add **1-hydrazinophthaleine**. Heat at 60° for 10 minutes, cool, and read at 450 nm.

Above pH 9 cobalt forms a water-insoluble complex with **2-hydroxy-2′-(hydroxy-1-naphthylazo)-5-methylazobenzene** which has the trivial name of **azo-azoxy BN**.[267] The optimum condition is to extract with the reagent in carbon tetrachloride from a 0.8–1.4 *M* solution of cobalt in sodium hydroxide solution and read at 600–650 nm. The reagent does not absorb in that region. Beer's law is followed up to 1.6 μg of cobalt if read within 40 minutes of extraction. A 50% excess of reagent is necessary. Most masking agents do not interfere.

The condensation product of **2-hydroxy-1-naphthaldehyde** and **4-chlorobenzyl dithiocarbamate** in 50% dimethylformamide, when activated at 470 nm, shows an intense fluorescence at 520 nm.[268] Added cobalt ion complexes with the reagent to reduce the fluorescence, permitting determination of 0.1–1.1 ppm. Cupric ion and cadmium react similarly and must be absent. Silver, nickel, and zinc also interfere.

Mix 2 ml of 5×10^{-4} *M* solution of **2-(2-hydroxy-5-methoxyphenylazo)-4-methylthiazole** in methanol with less than 10 μg of cobalt at pH 8 and extract with 10 ml of isoamyl alcohol.[269] Read the 1:2 metal-ligand complex at 607 nm.

For cobalt in ferrocobalt and C-cobalt tablets in powder form, take a sample containing 7.5–20 mg of cobalt.[270] Extract with 1:3 nitric acid, dilute, and centrifuge, washing the residue with water. Make an aliquot alkaline with *M* ammonium hydroxide and add 0.05% solution of **5-(2-hydroxy-1-naphthylazo)pyra-**

[262] Nobusuke Masuda and Meisetsu Kajiwara, *Jap. Anal.* **17**, 1353–1358 (1968).
[263] D. Costache, *Rev. Roum. Chim.* **15**, 1181–1185 (1970).
[264] N. Gundersen and E. Jacobsen, *Anal. Chim. Acta* **42**, 330–333 (1968).
[265] W. Haas, *Mikrochim. Ichnoanal. Acta* **1963**, 274–278.
[266] V. S. Kublanovskii and E. A. Mazurenko, *Tr. Odessk. Gedrometeoral. Inst.* **1961** (27), 39–43.
[267] Elzbieta Wieteska and Marian Kamela, *Chem. Anal.* (Warsaw) **17**, 85–91 (1972).
[268] Imre Kasa and Gabor Bajnoczy, *Period. Polytech. Chem. Eng.* **18**, 289–294 (1974).
[269] Tadashi Yanagihara, Nobuhisa Matano, and Akira Kawase, *Jap. Anal.* **8**, 14–17 (1959).
[270] Boleslaw Janik and Tadeusz Gancarczyk, *Chem. Anal.* (Warsaw) **16**, 413–417 (1971); *Acta Pol. Pharm.* **31**, 61–64 (1974).

zole-4-carboxylic acid. Acidify with 1:1 hydrochloric acid and extract the 2:1 cobalt complex with pentanol. Read the extract at 620 nm. The maximum absorption above pH 9.3 in water is at 580 nm. Beer's law is followed for 0.8–6.4 μg of cobalt per ml of final solution. Acidification before extraction decomposes complexes of ferric, cupric, and nickel ions. There is interference by EDTA, cyanide, tartrate, citrate, and oxalate.

At pH 6 cobalt forms a 1:3 complex with **4-hydroxy-5-nitrosobenzo-2,1,3-thiadiazole**.[271] This is extracted with chloroform for reading at 470 nm. Beer's law is obeyed for 0.1–10 μg of cobalt per ml. Iron and palladium interfere. The analogous 5-hydroxy-4-nitroso compound also complexes at pH 6 for extraction with chloroform and reading at 410, 490, or 510 nm.

A solution containing 9–120 μg of cobalt and 0.05% of **8-hydroxy-7-nitrosoquinoline-5-sulfonic acid** is read at 525 nm and pH 5.[272a] Iron, manganese, and chloroform interfere. To reduce interference mix 5 ml of sample solution 50 μ*M* to 0.5 m*M* in cobalt ion with 10 ml of sodium acetate-acetic acid buffer solution for pH 5.[272b] Add 5 ml of saturated solution of the reagent. After 1 minute add hydrochloric acid dropwise to give pH 0. Dilute to 25 ml and read immediately at 705 nm. Equimolar amounts of platinic ion, trivalent ruthenium and osmium, palladous ion, silver, and cobalt are tolerated.

For cobalt in the presence of iron, develop with **3-hydroxy-3-phenyl-1-(3-pyridyl)triazane**, extract with chloroform, and read at 347 nm.[273] Mask iron with sodium tartrate in dilute sodium hydroxide solution.

To a sample containing 4–40 μg of cobalt, add 4 ml of acetate buffer solution for pH 4.5 and 5 ml of 0.02% solution of **3-hydroxypicolinaldehyde** in dimethylformamide.[274] Heat at 50° for 15 minutes, then dilute to 25 ml and read at 545 nm against water. For cobalt and nickel with 3-hydroxypicolinaldehyde, see Nickel.

When a sample at pH 6–8 is passed through a column of powdered **iron**, the cobalt ion liberates an equivalent of ferrous ion.[275] After oxidation to ferric ion, it is determined at 510 nm by 1,10-phenanthroline or at 458 nm with salicylhydroxamic acid. The same reaction is given by nickel, lead, cupric, silver, or mercuric ions.

For cobalt by **8-hydroxyquinoline** in acetone, see Nickel.

The complex of cobalt in 0.025% solution of **isonitrosodimedon** at pH 4–7 is extracted with isoamyl alcohol for reading at 374 nm.[276] There is interference by ferric, cupric, chromous, and nickelous ions.

For analysis of cobalt in steel, dissolve with perchloric phosphoric, and nitric acids.[277] If necessary, eliminate chromium by volatilizing as the oxychloride, iron by extraction with isopropyl ether, and nickel by ion exchange. To an appropriate

[271] S. A. D'yachenko, S. O. Gerasimova, I. A. Belen'kaya, and V. G. Pesin, *Zh. Anal. Khim.* **29**, 877–880 (1974).

[272a] I. M. Issa and M. M. Aly, *Z. Anal. Chem.* **266**, 127 (1973).

[272b] Shahid Abbas Abbasi, B. G. Bhat, and R. S. Singh, *ibid.* **282**, 222 (1976); cf. M. A. Eldawy, A. S. Tawfik, and S. R. Elshabouri, *J. Pharm. Sci.* **65**, 664–666 (1976).

[273] E. D. Korotkaya, A. Kh. Klibus, V. E. Pochinok, and T. S. Shul'gach, *Izv. Vyssh. Ucheb. Zaved. Khim. Khim. Tekhnol.* **18**, 1695–1699 (1975).

[274] A. Garcia de Torres, M. Valcarcel, and F. Pino-Perez, *Anal. Chim. Acta* **68**, 466–469 (1974).

[275] Eugeniusz Kloczko, *Chem. Anal.* (Warsaw) **16**, 1291–1298 (1971).

[276] W. van den Bossche and J. Hoste, *Anal. Chim. Acta* **18**, 564–568 (1958).

[277] R. Bonlin, A. M. Leblond, and M. Jean, *ibid.* **56**, 45–54 (1971).

aliquot, then add 10 ml of 0.2% solution of **isonitrosomalonylguanidine**, which is 5-hydroxyimino-2,6-diiminopyrimidin-4-one, in dilute perchloric acid. Add 10 ml of 50% ammonium acetate solution, 20 ml of 10% solution of diammonium phosphate, and 10 ml of hypochlorous acid, $d = 1.67$. Dilute to 100 ml and read at 370 nm against a blank.

Copper, nickel, zinc, and cobalt in a solution containing ammonium carbonate or sulfate are extracted with 0.5% solution of **Kelex 100**, which is a β-alkalene-8-hydroxyquinoline, in Solvesso 10 containing 10% of isodecyl alcohol.[278] They are back-extracted with 15% sulfuric acid.

Cobalt can be measured by its catalytic effect on oxidation by hydrogen peroxide of **lucigenin**, which is 10,10′-dimethyl-9,9′-biacridinium dinitrate.[279] Measurement of the luminescence evaluates 1–60 μg of cobalt in 5 ml. Silver and cupric ion also catalyze the reaction and therefore interfere.

Adjust a sample of cobalt solution to pH 9 in *M* sodium thiocyanate containing a fourfold *M* excess of **malachite green.**[280] Extract the 1:3:1 complex of cobalt-thiocyanate-dyestuff with 7:1 carbon tetrachloride–cyclohexanone and read at 630 nm. Beer's law is followed for 0.017–1 μg of cobalt per ml of extract. Copper, zinc, bismuth, and gold must be masked.

Dilute a sample solution to 25 ml to contain 0.25–4.5 μg of cobalt per ml. Add 0.5 ml of 5% **malic acid** solution, 0.5 ml of 5% hydrogen peroxide, and 2 ml of 4% sodium hydroxide solution.[281] Heat at 100° for 10 seconds, cool, and dilute to 50 ml. Read at 370 nm. To avoid interference by positive ions, preextract the cobalt with dithizone from a citrate medium at pH 9.

Meditan, which is 1-(5-mercapto-1:3,4-thiadiazol-2-ylazo)-2-naphthol, forms a 1:2 complex with 0.12–1.65 μg of cobalt per ml.[282] The complex is stable for only 3 hours; is read at 680 nm and pH 6.3–10.6. In determination of cobalt, there is interference by ferric, chromic, cupric, ferrous, and nickel ions.

The 1:3 complex of cobalt with **2-mercaptobenzoic acid** at pH 5.2–7.2 is read at 570 nm.[283] Addition of ascorbic acid accelerates color development and improves sensitivity. Readings are for 2.9–29 μg of cobalt in 10 ml of solution. Copper, silver, and cadmium are masked with thiourea. To avoid interference by bismuth, iron, manganese, molybdenum, nickel, vanadium, and tungsten, add EDTA. Extract the cobalt as an ammoniacal-pyridine complex with chloroform, evaporate to dryness, dissolve with ammonium hydroxide, and develop.

At pH 7.5–11 the blue complex of cobaltous ion with **β-mercaptohydrocinnamic acid** is rapidly oxidized by air to a yellow-brown cobaltic complex.[284] The maximum absorption at 380 nm follows Beer's law for 0.5–12 μg of cobalt per ml at pH 8.5–9.5. EDTA masks nickel. For more details of determination of cobalt by β-mercaptohydrocinnamic acid and by β-mercaptocinnamic acid, see Nickel.

To a sample containing 5.75–112 μg of cobalt, add 2 ml of 0.5% solution of

[278] G. M. Ritcey and B. H. Lucas, *C.I.M. Bull. Canada* **68** (754), 105–113 (1975).
[279] J. Bognar and L. Sipos, *Mikrochim. Ichnoanal. Acta* **1963**, 442–455 (1963).
[280] L. I. Kotelyanskaya and P. P. Kish, *Zh. Anal. Khim.* **28**, 1999–2004 (1973).
[281] F. Bermejo-Martinez and E. Mendez-Domenech, *Inf. Quim. Anal. Pura Apl. Ind.* **21** (5), 161–170 (1967).
[282] Eugenia Domagalina and Stanislaw Zarbeda, *Chem. Anal.* (Warsaw) **15**, 1227–1231 (1970).
[283] M. M. L. Khosla and S. P. Rao, *Mirochem. J.* **17**, 388–395 (1972).
[284] A. I. Busev and D. Kh. Vin′, *Zh. Anal. Khim.* **20**, 1347–1352 (1965).

mercaptopropane-1,2-diol.[285] Adjust to pH 10 and add 5 ml of boric acid–potassium chloride–sodium hydroxide buffer solution for pH 10. Dilute to 25 ml and read immediately at 450 nm.

Cobalt forms a 1:3 complex with **3-mercaptopropionic acid** in ammoniacal solution at pH 7–10.[286] As read at 370 nm, this conforms to Beer's law for 3–35 μg of cobalt in 20 ml of solution containing 1 ml of 0.01 *M* reagent. Development of the color requires 20 minutes. It is stable for 3 hours thereafter. Copper can be masked with 2 ml of 5% potassium cyanide solution, but iron and nickel interfere.

For cobalt in rock, fuse 3 grams of sample with 6 grams of sodium carbonate in platinum.[287] Dissolve the melt in 20 ml of hydrochloric acid and evaporate to dryness. Take up the residue in 20 ml of 1:3 hydrochloric acid, filter, and dilute to 50 ml. Neutralize with 0.4% sodium hydroxide solution. Add 2 ml of 10% sodium acetate solution and 2 ml of 10% solution of sodium potassium tartrate hexahydrate. Add 1 ml of 1.5% ethanolic **3-mercapto-*p*-propionophenetdide** and a drop of phenolphthalein indicator solution. Raise to pH 8.2–9 with 0.4% sodium hydroxide solution and extract with 10 ml of chloroform. Set aside for 30 minutes and read at 485 nm.

For soil, calcine 4 grams of sample at 450°. Treat with 25 ml of perchloric acid and evaporate to dryness. Take up in 30 ml of *N* hydrochloric acid, boil for 30 minutes, and filter. Complete as above from "Neutralize with 0.4%. . . ."

Above pH 9, cobalt can be read as a complex with **4-methoxybenzothiohydroxamic acid.**[288] Iron or manganese performs similarly.

For analysis of water and wastewater, a complex with **3-methoxy-2-nitrosophenol** is read at 410 nm.[289] Interference by nickel and copper can be prevented, that of iron reduced, and within limits, chromic and chromate ions can be tolerated.

***N*-Methylanabasine-α'-azo-*p*-cresol** forms a violet complex with cobaltous ion and a green complex with cobaltic ion; the latter is more useful.[290] To a sample solution containing up to 60 μg of cobalt in hydrochloric acid, add sequentially 1 ml of 2% potassium nitrite solution, 5 ml of 0.45 *M* potassium citrate to mask copper, 0.5 ml of *N* nitric acid, 3 ml of m*M* reagent in ethanol, and 1 ml of 2% urea solution. Heat the solution, then mask iron, bismuth, and vanadium with 5 ml of 0.1 *M* EDTA. Evaporate to 10 ml, cool, add 10 ml of *M* acetate–hydrochloric acid buffer for pH 0.65–0.7, and dilute to 25 ml. Read at 630 nm.

***N*-Methylanabasine-α'-azodiaminopyridine**, which is 3-[(1-methyl-2-piperidyl)-2-pyridylazo]pyridine-2,6-diamine, forms a 3:1 complex with cobalt.[291] In an acid medium the absorption maxima are at 570 and 620 nm; in an alkaline medium the maximum is at 590 nm. The maximum absorption by the reagent is at 440–490 nm. Beer's law is followed for 1–30 μg of cobalt in 25 ml. Cobalt can be determined in the presence of iron, nickel, and copper without masking agents.

[285] F. Bermejo Martinez and J. F. Vila Brion, *Acta Cient. Compostelana* **7**, 153–163 (1970).
[286] Shigeo Hara, *Jap. Anal.* **14**, 42–45 (1965).
[287] A. Nacu, Didina Nacu, and Doina Bilba, *Anal. Stiint. Univ. "Al. I. Cuza," 1C*, **16**, 123–127 (1970).
[288] Z. Skorko-Trybula, *Chem. Anal.* (Warsaw) **12**, 815–825 (1967).
[289] K. Benford, M. Gilbert, and S. H. Jenkins, *Water Res.* **1**, 695–715 (1967).
[290] E. L. Krukovskaya, Sh. T. Talipov, L. M. Zhabitskaya, and T. Lobanova, *Zh. Anal. Khim.* **27**, 2427–2431 (1972).
[291] V. S. Podgornova, S. N. Kosolapova, and Sh. T. Talipov, *ibid.* **24**, 945–949 (1969).

N-Methylanabasine-α′-azodiethylaminophenol, which is diethylamino-2-[3-(1-methyl-2-piperidyl)-2-pyridylazo]phenol, has a trivial name of **MAAF II**.[292] It forms with cobalt as well as with nickel a 2:1 complex that is read at 585 nm. The reagent has a maximum absorption at 440 nm. It also forms 1:1 complexes with vanadium, copper, indium, and bismuth. The reagent absorbs at 440 nm. Beer's law is followed for 0.04–0.6 μg of cobalt per ml. Iron can be masked with fluoride; aluminum, molybdenum, tungsten, and chromium with 1% sodium tartrate solution. Nickel and copper interfere seriously.

To a sample solution, add 5 ml of ethanol and 1 ml of 0.05% solution of **N-methylanabasine-α′-azoheptylresorcinol**, which is 2-[3-(1-methyl-2-piperidyl)-2-pyridylazo]heptylresorcinol.[293] Add 0.5 ml of 0.5% gelatin solution and 5 ml of buffer solution for pH 4.5. Dilute to 25 ml and read at 520–530 nm against a reagent blank. Interference by twofold excess of iron is prevented by adding 0.5 ml of 2% solution of ascorbic acid. Fluoride or EDTA does not destroy the complex.

N-Methylanabasine-α′-azo-2-naphthol, which is 1-[3-(1-methyl-2-piperidyl)-2-pyridylazo]-2-naphthol, forms a green 1:1 complex with cobalt in 0.1–6 *N* hydrochloric acid.[294] It follows Beer's law for 0.04–2 μg of cobalt per ml. To a sample solution containing 5–50 μg of cobalt, add 2 ml of 0.2% reagent solution in ethanol. Dilute to 25 ml with 1:5 hydrochloric acid and read at 580 nm. Interference arises with fivefold ferric ion, fourfold cadmium, threefold zinc, 2.5-fold stannous ion, twofold hexavalent tungsten or uranium, and an equal amount of manganous ion.

The 1-naphthol equivalent of the reagent above forms a 4:1 complex with cobalt.[295] To a sample containing 5–50 μg of cobalt, add 3 ml of 0.1% solution of the reagent and 10 ml of a buffer solution for pH 5.35. Extract with 5 ml and 5 ml of 2:1 chloroform–amyl alcohol. Read the filtered extracts at 610 or 650 nm.

N-Methylanabasine-α′-azo-2-naphthol-6-sulfonic acid, which is 6-hydroxy-5-[3-(1-methyl-2-piperidyl)-2-pyridylazo]naphthalene-2-sulfonic acid, forms a 2:1 complex with cobalt.[296] The optimum pH is 1; the maximum absorption is at 580 nm. Beer's law is followed for 5–50 μg of cobalt per ml. *N*-Methylanabasine-α′-azo-1-naphthol-5-sulfonic acid is an isomer of the preceding reagent and is used similarly.[297]

Cobalt complexes with **methylglyoxime** at pH 8.5 for reading at 260.5 nm.[298] Cupric, nickel, and ferric ions also complex. Beer's law is obeyed for up to 7 ppm of cobalt in 50 ml of solution containing 2 ml of *M* ammonium chloride at pH 8.5. Addition of EDTA masks nickel; but iron, copper, thiocyanate, and cyanide interfere.

[292] A. E. Martirosov, Sh. T. Talipov, and R. Kh. Dzhiyanbaeva, *Tr. Tashk. Gos. Univ.* **1968** (323), 73; Sh. T. Talipov, R. Kh. Dzhiyanbaeva, and A. E. Martirosov, *Izv. Akad. Nauk Kaz. SSR, Ser. Khim.* **1968** (6), 1–4.

[293] I. P. Shesterova, R. Kh. Dzhiyanbaeva, Sh. T. Talipov, and N. F. Kostylev, *Uzb. Khim. Zh.* **1967** (6), 19–20.

[294] L. V. Chaprasova, Sh. T. Talipov, and R. Kh. Dzhiyanbaeva, *Nauch. Tr. Tashkt. Univ.* **1967** (284), 42–48.

[295] V. S. Podgornova and I. P. Shesterova, *Tr. Tashkt. Gos. Univ.* **1967** (288), 116–120.

[296] R. Kh. Dzhiyanbaeva, Sh. T. Talipov, U. Mansurkhozhdaev, and N. Smaglyuk, *ibid.* **1968** (323), 21–27.

[297] Sh. T. Sharipova, R. Kh. Dzhiyanbaeva, and Sh. T. Talipov, *ibid.* 39–45.

[298] Nobusuke Masuda and Meisetsu Kajiwara, *Jap. Anal.* **19**, 1613–1619 (1970).

The more complex **4-(*n*-methyl-2-pyridylazo)resorcinol** derivatives of cobalt function like 4-(2-pyridylazo)resorcinol and offer little advantage.[299] Similar derivatives are formed with nickel, copper, zinc, and uranyl ions.

A complex of cobalt with ***s*-methylthiourea sulfate** at pH 8.75–9.70 is read at 495 nm for 0.5–7 μg/ml.[300] Preaging for 2 hours with 180-fold concentration of the reagent is required to develop the color.

Cobalt **myristate** and **palmitate** are read in pyridine at 550 nm.[301] Copper and nickel soaps are similarly determined. The solubility in pyridine is limited.

Mix 10 ml of neutral sample solution containing 5–90 μg of cobalt with 1 ml of 50 m*M* ethanolic **1-naphthamidoxime** and 1 ml of 0.4 *M* potassium hydroxide.[302] Set aside for 10 minutes, then extract with 5 ml of isobutyl alcohol. Dry the extract with sodium sulfate and read at 581 nm. Suitably modified, the technic will detect 1 ppm of cobalt. Oxidizing agents, nickel, chromic ion, and manganous ion must be absent. Cupric ion can be masked with tartrate, ferric ion with fluoride.

At pH 10.5–11 in 40% ethanol, cobalt is read in the presence of excess **nicotinamideoxime** as a 1:3 complex at 580 nm.[303] Beer's law is followed for 2–16 ppm of cobalt. There are many interfering cations.

Cobalt complexes with **nioxime**, which is 1,2-cyclohexanedione dioxime.[304] The complex is extracted with 3:1 chloroform–amyl alcohol for reading at 470 nm. To avoid interference by iron and nickel, adjust the aqueous phase to pH 1 and add a saturated solution of iodine in potassium iodide solution before extracting the complex. Then after separating the complex, wash the iodine from the organic phase with sodium thiosulfate solution.

Adjust a sample solution containing up to 15 μg of cobalt to pH 2.2 with dilute hydrochloric acid and add 10 ml of 85 μ*M* **6-nitroquinoxaline-2,3-dithiol.**[305] Extract with 10 ml of methyl isobutyl ketone and read the extract at 530 nm against a reagent blank. The color is stable for 30 minutes. Beer's law is obeyed up to 3.5 ppm of cobalt. There is serious interference by nickel, cadmium, platinic ion, cupric ion, palladic ion, and EDTA.

At pH 7 cobalt forms a stable yellow complex with **3-nitrosopyridine-2,6-diol** with maximum absorption at 411 nm.[306]

Cobalt in steel at 0.1–24% is read at 560, 580, or 600 nm as the complex formed with **sodium 1-nitroso-2-naphthol-3,6-disulfonate.**[307]

To a sample solution containing 0.005–0.114 mg of cobalt, add 10 ml of a solution of 0.5 mg/ml of **2-nitroso-1-naphthol-4-sulfonic acid** neutralized with sodium hydroxide, and dilute to 40 ml.[308] Color develops at 100° in 15 minutes or

[299] W. J. Geary and F. Bottomly, *Talenta* **14**, 537–542 (1967).
[300] S. K. Siddhanta and S. N. Banerjee, *J. Indian Chem. Soc.* **35**, 547–552 (1958).
[301] Wahid U. Malik and Rizwanul Hague, *Z. Anal. Chem.* **189**, 179–182 (1962).
[302] N. Motekov and K. Manolov, *ibid.* **272**, 48 (1974).
[303] K. K. Tripathi and D. Banerjee, *ibid.* **168**, 407–410 (1959).
[304] A. R. Jafar, *J. Indian Chem. Soc.* **48**, 857–861 (1971).
[305] C. K. Bhaskare and U. D. Jagadale, *Z. Anal. Chem.* **278**, 127 (1976).
[306] Curtis W. McDonald, Thornton Rhodes, and John H. Bedenbaugh, *Mikrochim. Acta* **1972**, 298–302.
[307] A. Sommariva and A. Devoti, *Met. Ital.* **52** (6), 341–354 (1960).
[308] V. N. Telmachev and L. N. Serpukhova, *Tr. Kom. Anal. Khim., Akad. Nauk SSSR, Inst. Geokhim. Anal. Khim.* **8**, 115–124 (1958); Sumio Komatsu, *J. Chem. Soc. Jap.* **82**, 1064–1067 (1961).

in 50 minutes at room temperature. Add 20 ml of 1:2 nitric acid or 10 ml of hydrochloric acid and set aside for 15 minutes. Extract with 10 ml of amyl alcohol and read at 532 nm. Copper, nickel, palladium, and ferric ion interfere.

For cobalt, boil 0.2 gram of a product of the metallurgy of copper with 0.5 gram of ammonium fluoride and 5 ml of phosphoric acid.[309] Add 5 ml of nitric acid and 3 ml of sulfuric acid. Evaporate to white fumes, cool, add 30 ml of water, and boil. Add 0.5 gram of aluminum foil, boil until colorless, filter, and dilute to 100 ml. Add 1 ml of ammonium hydroxide to a 25-ml aliquot and dissolve the precipitated aluminum hydroxide with 1:1 sulfuric acid. Add 5 ml of 50% sodium acetate solution and 10 ml of 10% citric acid solution. Boil for 2 minutes and add 10 ml of 0.2% solution of **nitroso red**. Boil for 1 minute, add 3 ml of nitric acid, boil for 30 seconds, and cool. Dilute to 50 ml and read at 530 nm.

The 1:1 complex of cobalt with **orotic acid** at pH 8 is stable for 4 hours.[310] To 5 ml of sample solution containing up to 50 ppm of cobalt, add 10 ml of 0.08% orotic acid solution and read at 315 nm.

Above pH 10 cobalt complexes rapidly with **oximidobenzotetronic acid**, which is 4-hydroxy-3-nitrosocoumarin.[311] At 485 nm this conforms to Beer's law up to 4.7 ppm of cobalt. The complex is stable for more than a week. Ferrous and ferric ions interfere. At pH 3.5–8.5 the complex at 0.2–3 ppm is extractable with a solution of the reagent in benzene from the solution containing iron. Then cobalt can be read in the organic phase at 430 nm. There is serious interference in the extraction technic by EDTA, cyanide, and tetravalent cerium and vanadium.

The complex of cobalt precipitated by **3-oxapentamethylenedithiocarbamate** from 0.2 *N* hydrochloric acid can be extracted with chloroform for reading at 645 nm.[312] Ferric, cupric, and nickel ions must be removed.

For cobalt by **oxine blue** along with copper and zinc, see Copper by Oxine Blue.

The yellow-orange complex of cobaltic ion with **phenanthraquinone monoxime** in ethanol at pH 4.4–8 is extracted with chloroform for reading at 470 nm.[313] Heating is desirable to accelerate the reaction. Ferrous, copper, and nickel ions also complex with this reagent. The reagent is stable to heat and light. Beer's law is followed up to 3 ppm of cobalt. EDTA and nitrilotriacetic acid interfere.

A mixed complex of cobalt with **1,10-phenanthroline** and **thiosalicylate** is extracted with chloroform at pH 5–6 and read at 450 nm.[314a] The complex can be determined along with nickel and iron.

On addition of a slight excess of ***N*-(1-phenylazoethylidene)hydroxylamine** in 1:1 ammonium hydroxide and adjustment to pH 8 with ammonium hydroxide a dark brown 1:3 complex of cobaltic ion with the reagent is precipitated.[314b] This can be extracted with benzene and read at 490, 390, or 315 nm.

To a sample solution containing up to 2% cobalt, add 15 ml of ammonium

[309] Mou-sen Chang and Chen-tren Huang, *Wu Han Ta Hsueh, Tzu Jan K'o Hsueh Hsueh Pao* **1959**, 105–111.

[310] F. Capitan Garcia and A. Arrebola, *Inf. Quim. Anal. Pura Apl. Ind.* **21**, 1–8 (1967).

[311] A. N. Bhat and B. D. Jain, *J. Indian Chem. Soc.* **38**, 779–783 (1961); *Talenta* **16**, 1431–1433 (1969).

[312] W. Beyer and R. D. Ott, *Mikrochim. Ichnoanal. Acta* **1965**, 1130–1135.

[313] K. C. Trikha, Mohan Katyal, and R. P. Singh, *Talenta* **14**, 977–980 (1967).

[314a] I. V. Pyatnitskii and G. N. Trochinskaya, *Zh. Anal. Khim.* **28**, 704–708 (1973).

[314b] K. C. Kalia and Anil Kumar, *Indian J. Chem., Sect. A* **14**, 545–546 (1976).

hydroxide and 5 ml of 0.5% solution of **1-phenylthiosemicarbazide** in acetone.[315] Dilute to 50 ml with water or acetone and read.

For cobalt, adjust a sample to pH 9.3 with a buffer solution of ammonium hydroxide and ammonium chloride. Add 0.1% solution of **phthalaldehyde bisthiosemicarbazone** in dimethylformamide and read at 385 or 410 nm.[316a]

The complex of cobalt ion with **picolinaldehyde 4-phenyl-3-thiosemicarbazone** is stable in strongly acid media.[316b] At pH 1 it is read at 430 nm; at pH 4.7 at 390 nm. Beer's law is followed for 0.2–2 ppm of cobalt. At pH 1 it tolerates 1000 ppm of iron. There is interference by 1 ppm of auric, cupric, pentavanadium, silver, palladous, or platinic ion; 2 ppm of stannous ion; 5 ppm of mercury; 25 ppm of chromic or nickel ion. The complex can be extracted with chloroform.

A ternary complex of cobalt, **picolinaldehyde 2-pyridylhydrazone**, and eosin at pH 5–6 is extracted with 7:3 chloroform-acetone.[316c] Read at 547 nm, this follows Beer's law for 0.04–0.4 ppm of cobalt. Activated at 530 nm, it is read by fluorescence at 558 nm for 0.02–0.2 ppm. There is interference by cupric ion, nickel, iron, palladium, and mercury.

Picolinaldehyde 2-quinolylhydrazone forms colored complexes with several metals, but only that with cobalt is stable at pH 3.[317] At 510 nm this will read 0.2–2 ppm of cobalt.

The complex of cobalt with **picolinaldehyde thiosemicarbazone** in acetic acid–sodium acetate buffer solution absorbs at 356 nm, but because of absorption of the reagent, it is preferably read at 410 or 425 nm.[318] There is serious interference by ferrous, nickel, and cupric ions, and to a lesser degree by silver, cadmium, or zinc.

At pH 2–3 cobalt is precipitated as a 1:2 cherry-red complex with **picraminazo-4-cyclohexylresorcinol.**[319] This is extractable with chloroform or 9:1 chloroform–isoamyl alcohol for reading at 555 nm. Beer's law is followed for 0.5–5 μg of cobalt per ml. An equal amount of ferrous ion is tolerated. Iron, copper, palladium, and vanadium can be masked. Evaporate a sodium acetate extract of soil to dryness and take up in water. Add 1.5 ml of 0.05% solution of the reagent, 4 ml of 10% solution of sodium potassium tartrate, and 1.5 ml of triethanolamine. Dilute to 20 ml with 2 *N* hydrochloric acid, extract with 9:1 chloroform–isoamyl alcohol, and read at 555 nm.

In 2 *N* sulfuric acid **picraminazodiaminopyridine**, which is 2-(2,6-diamino-3-pyridylazo)-4,6-dinitrophenol, forms a 2:1 complex with cobalt having a maximum absorption at 540 nm.[320] The absorption is stable for 20 minutes and obeys Beer's law for 1–20 μg of cobalt per 25 ml of solution.

Picraminazo-*m*-phenylenediamine, which is 2-(2,4-diaminophenylazo)-4,6-dinitrophenol, forms a complex with cobalt that is soluble in 25% solution of dimethyl-

[315]N. V. Koshkin and N. M. Shreiner, *ibid.* **18**, 757–760 (1963).
[316a]M. Kueda Rueda and J. Munez Leiva, *Quim. Anal.* **29**, 122–128 (1975).
[316b]J. L. Gomez Ariza, J. M. Cano Pavon, and F. Pino Perez, *Talenta* **23**, 460–462 (1976).
[316c]P. R. Haddad, P. W. Alexander, and L. E. Smythe, *Talenta* **23**, 275–281 (1976).
[317]S. P. Singal and D. E. Ryan, *Anal. Chim. Acta* **37**, 91–96 (1967).
[318]J. M. Cano Pavon, D. Perez Benedito, and F. Pino Perez, *An. Real Soc. Esp. Fis. Quim. B* **65**, 667–672 (1969).
[319]L. V. Chaprasova, A. T. Tashkhodzhaev, Sh. T. Talipov, and N. Urmanov, *Zavod. Lab.* **40**, 1327–1328 (1974).
[320]S. N. Kosolapova, V. S. Podgornova, and Sh. T. Talipov, *Zh. Anal. Khim.* **24**, 880–883 (1969).

formamide.[321] At pH 9 the maximum absorption of the complex is at 610 nm, that of the reagent at 500 nm. To the sample solution containing 5–35 μg of cobalt, add 3 ml of 0.63 m*M* reagent in dimethylformamide and 3 ml of dimethylformamide. Dilute to 25 ml with 2 *N* hydrochloric acid and read at 610 nm.

For cobaltic ion in silicone resins, dissolve 2 grams of sample in 10 ml of benzene. Add 1 ml of pyridine and 2 ml of 1.5% solution of **piperidinium pentamethylenedithiocarbamate** in benzene.[322] Dilute to 25 ml with benzene and read 0.2–0.7 mg of cobalt per ml after 60 minutes at 445 nm against a reagent blank. Cupric, ferric, and manganic ions interfere. Reducing substances and those containing Si–H and Si–Si linkages also interfere.

Polypropylene green BM, which is α-[(methoxy-2-benzothiazolyl)amino]-*p*-cresol, is suitable for reading trivalent cobalt in acetic acid at 726 nm.[323] Organic peroxides such as *t*-butyl hydroperoxide must be absent. For cobaltous ion, develop with **ferricyanide** at 510 nm without interference by cobaltic ion. Organic peroxides may be present.

The complex of cobalt with **pontachrome blue black R** (mordant black 17) has an absorption maximum at 280 nm.[324] A sensitivity of 1 ng of cobalt per ml can be reached by measuring the inhibition of fluorescence of aluminum and pontachrome blue black R. Extraction of the aluminum complex will even improve that.

Adjust a sample solution containing up to 125 μg of cobalt to pH 5.35 with a buffer solution. Add 20 ml of m*M* sodium **purpurinsulfonate** and dilute to 50 ml.[325] Set aside for 30 minutes and read at 590 nm. Nickel interferes.

At pH 8–11 **pyrazinecarboxamide** forms a 3 : 1 complex with cobalt for reading at 380 nm.[326] Beer's law is obeyed for 1–10 μg of cobalt per ml. There is interference by silver, copper, iron, lead, bismuth, and mercuric ion, as well as by large amounts of tartrate, citrate, and EDTA.

The chelate of cobalt with **pyridine-2-aldoxime** at a pH above 4 is read at 380 nm.[327] Beer's law is followed up to about 7 μg of cobalt per ml. For higher concentrations, less sensitive readings are made at 400 or 420 nm.

Cobalt forms a 1 : 2 chelate with **pyridine-2,6-dialdoxime** having a maximum absorption at 330 nm over the pH range of 3–6.7.[328] Similar complexes are formed with cupric, manganous, nickel, and zinc ions.

2,6-Pyridinedicarboxylic acid in nitric or sulfuric acid when heated with cobalt ion and potassium bromate forms a 1 : 1 cobaltic compound absorbing at 514 nm.[329]

To a sample solution containing cobalt, add 5 ml of a solution containing **pyridine-2,4,6-tricarboxylic acid** at 0.25 mg/ml and 5 ml of 1 : 2 hydrochloric acid.[330] Dilute to 25 ml, set aside for 15 minutes, and read at 277 nm. Beer's law

[321] Z. B. Nasyrova, S. N. Kosolapova, and N. Kh. Maksudov, *Uzb. Khim. Zh.* **1973** (4), 26–29.
[322] P. Dostal, J. Cermak, and J. Kartous, *Collect. Czech. Chem. Commun.* **33**, 1539–1548 (1968).
[323] I. I. Chuev and M. K. Shchennikova, *Tr. Khim. Khim. Tekhnol.* (Gor'kii) **1968** [1 (19)], 146–150.
[324] J. F. Possidoni de Albinati, *An. Asoc. Quim. Argent.* **55**, 61–74 (1967).
[325] M. Roman Ceba and E. Alvarez-Manzaneda, *Quim. Anal.* **29**, 253–260 (1975).
[326] Halina Sikorska-Tomicka, *Chem. Anal.* (Warsaw) **20**, 1025–1030 (1975).
[327] E. Gagliardi and P. Presinger, *Mikrochim. Ichnoanal. Acta* **1965**, 791–797.
[328] Saswati P. Bag, Quintus Fernando, and Henry Freiser, *Anal. Chem.* **35**, 719–722 (1963).
[329] Heinrich Hartkamp, *Z. Anal. Chem.* **182**, 259–269 (1961).
[330] M. Franz, *J. Pharm. Belg.* **24**, 357–364 (1969).

applies for 10–200 μg of cobalt. An equal concentration of ferric or copper ion interferes, as does tenfold concentration of silver, lead, and barium.

Cobalt forms a 1:2 complex with **1-(2-pyridylazo)phenanthren-2-ol** having a maximum absorption at 570 nm for pH 1.15–10.5.[331] At least fivefold excess of reagent is required. The complex dissolves in 70% methanol or 3:10:7 1,4-dioxan–methanol–water and is stable for 20 hours. Beer's law applies up to 1.61 ppm of cobalt, but the complex is preferably read at 0.49–0.98 ppm of cobalt. There is interference by stannic, phosphate, tartrate, fluoride, mercuric, iron, gallium, molybdate, vanadate, and EDTA ions.

For cobalt by **3-(2-pyridylazo)chromotropic acid**, see Nickel.

In determination of cobalt with **4-(2-pyridylazo)-*m*-phenylenediamine** and such analogues as the 5-chloro- and 5-bromo-derivatives in 2.4 *N* hydrochloric acid, the maximum absorption is at 565–590 nm.[332a] Only iron, which can be masked with fluoride, and hexavalent chromium, which can be masked with hydroxylamine, interfere. In *N* hydrochloric acid a 1:2 metal-ligand complex is formed, read at 555 nm for 0.4–1.2 μg of cobalt per ml.[332b] Then the tolerance for vanadium, zirconium, mercuric ion, and beryllium is only tenfold.

To a sample solution containing up to 40 μg of cobalt, add at least a twofold concentration of **3-(2-pyridylazo)pyridine-2,6-diamine.**[333] Mix for 1 minute and adjust to 25 ml, 3 *N* in sulfuric acid. Read the 1:2 complex at 582 or 610 nm.

To a sample solution containing 9–90 μg of cobalt, add 5 ml of 1% ethanolic **2,2′-pyridil monoxime** and adjust to pH 5.[334] Set aside for 5 minutes; then extract with 5 ml, 5 ml, and 5 ml of chloroform. Dilute the combined extracts to 25 ml and read the 1:3 cobalt-ligand complex at 408 nm. A final concentration of 0.7–2.8 ppm of cobalt is appropriate. Aside from species that can be masked, there is interference by strontium, chromic, trivalent ruthenium, and EDTA ions, and by large amounts of cyanide ions.

Mix a sample solution containing 7–60 μg of cobalt with 5 ml of 40% solution of sodium citrate and 25 ml of water. Add 2 drops of phenolphthalein indicator solution and neutralize with potassium hydroxide solution or with hydrochloric acid. Add 3 ml of 1% ethanolic **2-pyridyl-2-thienyl-β-ketoxime.**[335] After 10 minutes add 10 ml of hydrochloric acid. Let stand for 10 minutes and extract with 10 ml and 10 ml of chloroform. Dilute the extracts to 25 ml with chloroform and read at 412 nm. The extract obeys Beer's law for 0.15–2.7 ppm of cobalt. There is interference by the following ions: permanganate, chromate, EDTA, fluoride, palladous, platinous, and platinic.

To a sample solution containing 0.08–7 μg of cobalt ion, add 5 ml of a borate buffer and 5 ml of 0.04% solution of **5,8-quinolinedione dioxime** in isoamyl alcohol.[336] Extract the complex and read at 418–422 nm. The reagent has other complexes with maxima as follows: nickel, 370 and 445–450; palladium, 403 nm;

[331] A. K. Rishi, B. S. Garg, and Rajendra Pal Singh, *Indian J. Chem.* **10**, 1037–1038 (1972).

[332a] E. Kiss, *Anal. Chim. Acta* **66**, 385–396 (1973).

[332b] N. Kh. Maksudov, Z. B. Nasyrova, and S. N. Kosolapova, *Zh. Anal. Khim.* **29**, 2063–2065 (1974).

[333] Sh. T. Talipov, V. S. Podgornova, and S. N. Kosolapova, *Zh. Anal. Khim.* **24**, 409–413 (1969).

[334] S. Stupavsky and W. J. Holland, *Mikrochim. Acta* **1970**, 115–118.

[335] H. R. Notenboom, W. J. Holland, and R. G. Billinghurst, *ibid.* **1973**, 467–473.

[336] A. P. Golovina, I. P. Alimarin, and D. I. Kuznetsov, *Vest. Mosk. Univ., Ser. Mat., Mekh., Astron., Fiz. Khim.* **12** (5), 187–191 (1957).

titanium, 313 nm; zirconium, 316–318 nm; thorium, 325 nm; beryllium and aluminum, 310 nm; ferric ion, 330 nm. The iron can be complexed with fluoride.

The 1:2 complex of cobaltous ion with **1-(2-quinolylazo)phenanthren-2-ol** at pH 8 has a maximum absorption at 550 nm.[337a] It follows Beer's law up to 2.2 ppm of cobalt, preferably being read at 0.3–1.6 ppm. There is interference by cyanide, EDTA, palladium, aluminum, mercury, and transition metals.

Add excess of 0.01 *M* solution of **salicylaldehyde** in acetone to a cobalt solution in at least 75% acetone to maintain solubility.[337b] Adjust to pH 6.2–7 and heat at 100° for 30 minutes. Replace evaporated acetone, extract the complex with carbon tetrachloride, and read at 450 nm. Beer's law is followed for 0.58–2.32 μg/ml.

The 1:2 complex of cobalt with **salicylaldehyde isonicotinoylhydrazone** is read in 50% 1,4-dioxan at 390–400 nm.[338] Beer's law is followed for 0.5–3 ppm of cobalt. Similar complexes are formed with manganous, nickel, zinc, and cobalt ions.

Cobaltous ion complexes with **succinimide** in ammoniacal solution.[339] At pH 10–10.2 this is appropriate for reading 1–200 ppm of cobalt at 565 nm. A fortyfold excess of reagent is required. There is serious interference by the following ions: chromous, uranyl, vanadyl, ceric, lanthanum, thoric, stannic, carbonate, sulfide, arsenate, and phosphate. Masking is applicable to ferrous ion by sodium citrate and hydrogen peroxide, mercuric ion by thiourea, cupric ion by thiosulfate, and antimonous ion by tartaric acid.

The 1:2 complex of cobalt with **2-(3′-sulfobenzoyl)pyridine-2-pyridylhydrazone,** which is α-2-pyridyl-α-(2-pyridylhydrazono)toluene-*m*-sulfonic acid, is water soluble and has a sensitivity approaching that of dithizone, 1-(2-pyridylazo)-2-naphthol, or 4-(2-pyridylazo)resorcinol.[340] The spectrum of the ligand does not overlap that of the complex.

Dissolve 0.3 gram of sample with hydrochloric acid, adding nitric acid if necessary, and dilute to 100 ml. Mix a 1-ml aliquot with 0.5 ml of 5% ascorbic acid solution. Dilute to 25 ml with 0.1 *M* solution of **thiosemicarbazide** and set aside for 5 minutes.[341] Read for 2.5–50 μg of cobalt per ml at 428 or 533 nm. Developed at pH 9.2–9.6 with more than twelvefold excess of reagent, color appears in 30 minutes and is read at 420 nm.[342] Sodium potassium tartrate prevents precipitation. Nickel, cupric, auric, palladous, and platinic ions interfere.

To a solution containing less than 4 mg of cobalt in perchloric acid, add 6 ml of 0.1 *M* **sulfosalicylic acid.**[343] Raise to pH 9 with ammonium hydroxide and add 5 ml of phthalate buffer solution for that pH. Dilute to 100 ml and read at 325 nm against a reagent blank. Citrate and tartrate interfere.

Cobaltous ion is read by quenching of the fluorescence of the aluminum complex of **superchrome blue black extra** (mordant black 17).[344] Buffered at pH 4.8, the

[337a]R. N. Virmani, *J. Prakt. Chem.* **314**, 965–968 (1972).

[337b]H. L. Ray, B. S. Garg, and R. P. Singh, *J. Clin. Chem. Soc., Taipai* **23**, 47–51 (1976).

[338]G. S. Vasilikiotis and T. A. Kouimtzis, *Microchem. J.* **18**, 85–94 (1973).

[339]W. U. Malik and C. L. Sharma, *Z. Anal. Chem.* **244**, 317 (1969).

[340]J. E. Gonig and C. Sykora, *Anal. Chim. Acta* **70**, 127–132 (1974).

[341]M. K. Akhmedli and A. M. Sadykova, *Elmi Eserler Azerb. Univ. Kimja Elmleri Ser.* [*Uch. Zap. Azerb. Univ., Ser. Khim. Nauk*] **1963** (1), 23–26.

[342]Sumio Komatsu, *J. Chem. Soc. Jap.* **83**, 48–51 (1962).

[343]Oliver R. Hunt and G. A. Parker, *Mikrochim. Acta* **1974**, 59–63; cf. Ajit Kumar, *Indian J. Chem.* **10**, 306–307 (1972).

[344]S. B. Zamochnick and G. A. Rechnitz, *Z. Anal. Chem.* **199**, 424–429 (1963).

fluorescence is read at 405 nm with activation at 365 nm. The method is applicable to 0.5–10 μg of cobalt in 50 ml.

Cobalt can be determined by oxidizing the pink cobaltous **tartrate** complex to the green cobaltic form for reading.[345] Mix a sample containing 6–60 mg of cobaltous ion in water or dilute nitric acid with 3 grams of sodium potassium tartrate and 5 ml of 56% potassium hydroxide solution. Add 5 ml of 3% hydrogen peroxide and dilute to 100 ml. Set aside for 10 minutes and read at 640 nm. There is interference by magnesium, calcium, iron, cadmium, copper, chromium, manganese, thiocyanate, and oxalate.

For alloy analysis, dissolve a 0.1-gram sample in 0.5 ml of nitric acid and dilute to 200 ml. Mix a 2-ml aliquot with 5 ml of acetate buffer solution for pH 5.5. Add 5 ml of m*M* **tetra[ethyl-di-(1-sodiotetrazol-5-ylazo)]acetate**, and dilute to 25 ml with the buffer solution.[346] Read at 620 nm against water.

To a sample solution containing 76–300 μg of cobalt, add 5 ml of 6 *N* acetic acid. Oxidize cobaltous ion to cobaltic ion with 0.2 gram of sodium nitrate and stir for 2 minutes. Add 10 ml of 4% solution of citric acid and 0.5 gram of disodium phosphate dodecahydrate to give pH 3.2–3.4.[347a] Shake for 10 minutes with 10 ml of 0.5 *M* **2-thenoyltrifluoroacetone** in xylene. Read the filtered extract at 460 nm against a reagent blank. Beer's law is followed for 3–12 μg of cobalt per ml. Iron is masked with fluoride ion, manganese with tartaric acid. Copper and EDTA interfere.

A 1:2:2 cobalt-diketone-triphenylphosphine complex is more readily extracted with benzene or chloroform than the binary metal-diketone complex.[347b] The triphenylphosphine should be 0.1 *M* and the cobalt 10 μ*M*. The diketone component is 0.01 *M* 2-thenoyltrifluoroacetone, trifluoroacetylacetone, or hexafluoroacetylacetone.

In a neutral medium cobaltous ion forms a purple or blue 1:2 chelate with **1-(2-thiazolylazo)-2-naphthol**.[348] Cobaltic ion in the presence of 0.1% potassium periodate and 0.0001 *M* reagent at pH 5 is read at 618 nm as a green chelate. To 40 ml of sample containing 6–30 μg of cobalt add 2.5 ml of 0.1% solution of the reagent in acetone. Oxidize the cobalt to the trivalent form with 1 ml of saturated solution of potassium periodate. Adjust to pH 3–6 with an acetate buffer and extract with 20 ml of chloroform. Read the extract at 660–670 nm.[349]

To 5 ml of neutral or slightly alkaline solution containing 0.1–10 μg of cobalt as chloride or sulfate, add 0.5 ml of 0.001 *M* **4-(2-thiazolylazo)resorcinol** in 0.05 *N* ammonium hydroxide. Iodide, bromide, nitrate, perchlorate, and thiocyanate seriously inhibit the extraction.[350] Shake for 15 minutes with 5 ml of 0.01 *M* tetraheptylammonium chloride in benzene. Add 2 ml of 0.1 *M* EDTA and adjust to pH 4.5–5 with sodium acetate–acetic acid buffer solution. Shake for 2 hours longer. Read the organic layer at 570 nm. Beer's law is followed for 0.1–2 ppm of cobalt in

[345] H. Basinska and K. Polak, *Chem. Anal.* (Warsaw) **12**, 253–260 (1967).
[346] N. A. Batasheva, O. V. Sivanova, and N. N. Gritsienko, *Zavod. Lab.* **39**, 17–18 (1973).
[347a] Anil K. De and S. K. Majumdar, *Anal. Chim. Acta* **27**, 153–157 (1962); Anil K. De and M. S. Rahaman, *ibid.* **34**, 233–234 (1966).
[347b] V. G. Makarov, G. G. Goroshko, and O. M. Petrukhin, *Zh. Anal. Khim.* **31**, 460–464 (1976).
[348] Akira Kawase, *Jap. Anal.* **12**, 817–821, 904–910 (1963).
[349] N. S. Ershova, V. M. Ivanov, and A. I. Busev, *Zh. Anal. Khim.* **28**, 2220–2226 (1973).
[350] O. Navratil and R. W. Frei, *Anal. Chim. Acta* **52**, 221–227 (1970).

the extract. For 0.25–3 μg of cobalt without extraction at pH 7.5–8, read the 1:2 complex with 4-(2-thiazolylazo)resorcinol at 510 nm.[351a]

The complex of cobalt at pH 8.4–9.1 with **thiobenzoylacetone**, which is 3-mercapto-1-phenylbut-2-en-1-one, is extracted with benzene and read at 440 nm.[351b] Beer's law is obeyed for 0.2–4.58 μg of cobalt in the extract, which is stable for 6 days. There is serious interference by mercuric ion, cyanide, and EDTA. To avoid interference, silver and bismuth are masked with thiourea; thorium, aluminum, and beryllium are masked with fluoride. Preliminary separation by extraction is necessary to avoid interference by antimony, iron, lead, chromium, copper, and palladium. Since the complex with nickel absorbs at 500 nm, cobalt and nickel can be determined by simultaneous equations.

The complex of cobalt with **thiodibenzoylmethane**, which is 1,3-diphenyl-3-thiopropan-1-one, is extracted at pH 9.5 with cyclohexane.[352] After destroying excess reagent, it is read at 410 nm for 0.04–0.4 μ*M* cobalt. At pH 9 both cobalt and nickel are extracted. Then for reading cobalt, the nickel is back-extracted from the organic phase with an alkaline cyanide solution. Palladium or tenfold copper, tin, or bismuth interferes.

To a sample containing more than 0.02 mg of cobalt, add 5 ml of ammoniacal buffer solution for pH 10. Add 1 ml of 0.1 *M* mixed **thionaphthenic acids** (mean molecular weight 227) in carbon tetrachloride, and 4 ml of carbon tetrachloride.[353a] Shake for 1 minute. Wash the organic layer with 5 ml and 5 ml of 1 : 10 ammonium hydroxide to avoid interference by nickel. Read at 465 nm against 0.02 *M* reagent in carbon tetrachloride. Beer's law holds for 0.01–0.1 mg of cobalt per ml. Nickel forms a similar complex absorbing at 470 nm. Ferrous ion interferes.

For determination of cobalt as a ternary 1:2:2 cobalt-pyridine-**thiosalicylic acid** complex, a single extraction with chloroform at pH 4–5.5 gives 85–90% yield.[353b]

The **thiosemicarbazone of benzothiazol-2-yl phenyl ketone** gives a visible absorption with cobalt, copper, nickel, cadmium, and zinc.[354] Silver, palladium, iron, and lead interfere. Mix 2 ml of sample solution with 0.5 ml of an ammoniacal buffer solution for pH 8. Add 1 ml of 0.1% ethanolic solution of the reagent and dilute to 5 ml with ethanol. Set aside for 30 minutes. Read cobalt at 466 nm.

Mix a sample solution containing about 20 μg of cobalt with a buffer solution for pH 6.[355] Extract with a m*M* solution of **thiothenoyltrifluoroacetone**, which is 1,1,1-trifluoro-4-mercapto-4-(2-thienyl)but-3-en-2-one in carbon tetrachloride. Extraction may also be with the reagent in cyclohexane. In that case shake for 15 minutes and separate the organic layer. Wash it with *N* hydrochloric acid, followed by washing with a buffer solution for pH 9.[356a]

Read the 1:4 complex in the organic phase at 490 nm against a reagent blank. Beer's law is followed for 0.5–6 μg of cobalt per ml. Palladous, lead, auric, and nickel ions interfere seriously. Thorium, antimonous, and molybdate ions must be

[351a] A. I. Busev, V. M. Ivanov, and Zh. I. Nemtseva, *Zh. Anal. Khim.* **24**, 414–421 (1969).
[351b] M. V. R. Murti and S. M. Khopkar, *Talenta* **23**, 246–248 (1976).
[352] E. Uhlemann and B. Schucknecht, *Anal. Chim. Acta* **63**, 236–240 (1973).
[353a] A. A. Alekperova, R. A. Alekperov, and Yu. A. Zolotov, *Zh. Anal. Khim.* **25**, 2283–2286 (1970).
[353b] I. V. Pyatnitski and G. N. Trochinskaya, *Ukr. Khim. Zh.* **41**, 262–268 (1975).
[354] Toyozo Uno and Sumiyuki Akihama, *Jap. Anal.* **10**, 941–945 (1961).
[355] R. R. Mulye and S. M. Khopkar, *Mikrochim. Acta* **1973**, 55–60.
[356a] Takaharu Honjijo and Toshiyasu Kiba, *Bull. Chem. Soc. Jap.* **45**, 185–191 (1972).

removed by extraction or ion exchange. Aluminum, chromic, manganous, and zirconium ions can be masked.

For steel analysis dissolve 1 gram of sample with aqua regia, dilute, and filter.[356b] Add solid potassium fluoride to an aliquot containing 0.4–2.5 mg of cobalt. Centrifuge. Neutralize the solution with solid sodium bicarbonate and add 0.1 gram excess. Add 1 ml of 3% hydrogen peroxide. Filter. As a wash solution dissolve 2 grams of potassium fluoride in 10 ml of saturated solution of sodium bicarbonate and add 2 ml of 3% hydrogen peroxide. Wash the precipitate with 3 ml and 1 ml of this wash solution and add to the prior filtrate. Add 0.1 gram of sodium bicarbonate and dilute to 10 ml. Set aside for 20 minutes and read the resulting **tricarbonatocobaltate** at 640 nm.

Mix 2 ml of sample solution containing 25–200 μg of cobalt with 2 ml of 0.085 *M* ***p*-toluamidoxime** in ethanol and 2.5 ml of 4% solution of sodium hydroxide.[357] Dilute to 25 ml and read at 580 nm. Alternatively, take a sample containing 10–80 μg of cobalt and add 1 ml of 0.17 *M* solution of the color reagent in ethanol. Add 1 ml of 0.4% solution of sodium hydroxide and dilute to 10 ml. Extract for 1 minute with 10 ml of butanol, isobutyl alcohol, or heptanol. Read the extract at 580 nm.

When cobalt ion is oxidized with potassium persulfate, the **triethylenediamine** complexes of cobaltic and nickelous ions are formed at pH 11.3.[358] They are read at 875 and 463 nm.

To a sample solution containing 1.5–750 μg of cobalt, add 2 ml of 0.5% solution of **triethylenetetraamine** in 4% sodium hydroxide solution.[359] Adjust to pH 10 with 4% sodium hydroxide solution, and add a borate buffer solution. Dilute to a known volume and read at 520 nm. There is interference by cyanide, thiocyanate, nitrite, cupric, nickelous, and palladous ions. Filterable precipitates are given by ferric, aluminum, bismuth, chromic, and manganous ions.

The oxidation of **2,6,7-trihydroxy-9-phenylxanthen-3-one** by hydrogen peroxide is catalyzed by cobaltous ion.[360a] As a kinetic method, the basic solution is 3 ml of a buffer solution of 0.05% boric acid in 1% sodium hydroxide solution, 1 ml of reagent in ethanol, and 2 ml of 0.3% hydrogen peroxide with dilution to 25 ml. The added cobalt in a sample in ng leads to reading at 460 nm at 1-minute intervals for 15 minutes. Readings are for 0.08–0.64 ng of cobalt per ml of the solution.

To an acidic sample solution containing 10–157 μg of cobalt add an excess of a solution of **uramildiacetic acid**, which is (hexahydro-2,4,6-trioxo-pyrimidin-5-yl)iminodiacetic acid.[360b] Add 5 ml of acetate buffer for pH 5. Dilute to 25 ml and read the stable 1:1 complex at 266 nm. Similar reactions occur with beryllium, titanic ion, hexavalent uranium, cadmium, cupric ion, nickel, chromic, and ferric ions.

To a sample solution containing 6–20 μg of cobalt, add 4 ml of buffer solution for pH 8.6 and 10 ml of 0.01 *M* **violuric acid**, which is 5-hydroxyiminobarbituric

[356b] A. Sanz Medel, A. Cobo Guzman and J. A. Perez-Bustamante, *Analyst* **101**, 860–866 (1976).

[357] A. I. Busev, T. N. Zholondkovskaya, and G. N. Teplova, *Zh. Anal. Khim.* **26**, 1133–1138 (1971).

[358] C. Caullet and B. Villecroze, *Chim. Anal.* (Paris) **46**, 308–312 (1964).

[359] F. Bermejo-Martinez and J. A. Rodriguez-Vazguez, *Inf. Quim. Anal. Pura Apl. Ind.* **24**, 183–189 (1970); **26**, 55–62, 70 (1972).

[360a] G. Popa and D. Costache, *Rev. Roum. Chim.* **12**, 963–968 (1967).

[360b] M. Jose Presas-Barrosa, F. Bermejo-Martinez, and J. A. Rodriguez-Vazquez, *Anal. Chim. Acta* **88**, 395–398 (1977).

acid.[361] Heat at 100° for 1 minute and add 5 ml of a saturated solution of EDTA to the hot solution. Cool, dilute to 50 ml, and read at 360 nm against a reagent blank.

The complex of cobalt with **zinc bis(2-hydroxyethyl)dithiocarbamate** is read at 635 nm.[362] Add 1 ml of 20% citric acid solution to the sample solution, adjust to pH 5.5–9.5 with *N* hydrochloric acid or 1 : 3 ammonium hydroxide, add 1 ml of 0.5% solution of the reagent, and dilute to 100 ml. Read at 470 nm against a blank within 25 minutes. The complex is extractable with tributyl phosphate.

When an acid solution of cobalt is shaken with **zinc diethyldithiocarbamate** in benzene, Beer's law is followed for 1–15 μg of cobalt per 10 ml of chloroform.[363] Interference by many ions is avoided by washing the organic extract with potassium cyanide solution. Lead or bismuth may be displaced by copper, then the copper is washed out. Interference by mercury, palladium, and platinum is not avoided in that way.

[361] L. V. Ershova and V. V. Noskov, *Zh. Anal. Khim.* **26**, 2406–2409 (1971).
[362] Hitoshi Yoshida, Masahiro Yamamoto, and Seiichiro Hikime, *Jap. Anal.* **11**, 197–201 (1962).
[363] Kenji Motojima and Nori Tamura, *Anal. Chim. Acta* **45**, 327–332 (1969).